THE CIA WORLD FACTBOOK 2012

CENTRAL INTELLIGENCE AGENCY

Skyhorse Publishing

Skyhorse Publishing books may be purchased in bulk at special discounts for sales promotion, corporate gifts, fund-raising, or educational purposes. Special editions can also be created to specifications. For details, contact the Special Sales Department, Skyhorse Publishing, 307 West 36th Street, 11th Floor, New York, NY 10018 or info@skyhorsepublishing.com.

www.skyhorsepublishing.com

ISBN: 978-1-61608-332-8

10 9 8 7 6 5 4 3 2 1

In general, information available as of June 2011 was used in preparation of this edition.

Printed in the United States of America

CONTENTS

INTRODUCTION

The World Factbook is prepared by the Central Intelligence Agency for the use of US Government officials, and the style, format, coverage, and content are designed to meet their specific requirements. Information is provided by Antarctic Information Program (National Science Foundation), Armed Forces Medical Intelligence Center (Department of Defense), Bureau of the Census (Department of Commerce), Bureau of Labor Statistics (Department of Labor), Central Intelligence Agency, Council of Managers of National Antarctic Programs, Defense Intelligence Agency (Department of Defense), Department of Energy, Department of State, Fish and Wildlife Service (Department of the Interior), Maritime Administration (Department of Transportation), National Geospatial-Intelligence Agency (Department of Defense), Naval Facilities Engineering Command (Department of Defense), Office of Insular Affairs (Department of the Interior), Office of Naval Intelligence (Department of Defense), US Board on Geographic Names (Department of the Interior), US Transportation Command (Department of Defense), Oil & Gas Journal, and other public and private sources.

The *Factbook* is in the public domain. Accordingly, it may be copied freely without permission of the Central Intelligence Agency (CIA). The official seal of the CIA, however, may **NOT** be copied without permission as required by the CIA Act of 1949 (50 U.S.C. section 403m). Misuse of the official seal of the CIA could result in civil and criminal penalties.

Comments and queries are welcome and may be addressed to:

Central Intelligence Agency
Attn.: Office of Public Affairs
Washington, DC 20505
Hours: Monday-Friday 8:00 AM-4:30 PM Eastern Standard Time
Telephone: [1] (703) 482-0623
FAX: [1] (703) 482-17

A BRIEF HISTORY OF BASIC INTELLIGENCE AND *THE WORLD FACTBOOK*

The Intelligence Cycle is the process by which information is acquired, converted into intelligence, and made available to policymakers. ***Information*** *is raw data from any source, data that may be fragmentary, contradictory, unreliable, ambiguous, deceptive, or wrong.* ***Intelligence*** *is information that has been collected, integrated, evaluated, analyzed, and interpreted.* ***Finished intelligence*** *is the final product of the Intelligence Cycle ready to be delivered to the policymaker.*

The three types of finished intelligence are: basic, current, and estimative. Basic intelligence provides the fundamental and factual reference material on a country or issue. Current intelligence reports on new developments. Estimative intelligence judges probable outcomes. The three are mutually supportive: basic intelligence is the foundation on which the other two are constructed; current intelligence continually updates the inventory of knowledge; and estimative intelligence revises overall interpretations of country and issue prospects for guidance of basic and current intelligence. *The World Factbook, The President's Daily Brief*, and the *National Intelligence Estimates* are examples of the three types of finished intelligence.

The United States has carried on foreign intelligence activities since the days of George Washington but only since World War II have they been coordinated on a government-wide basis. Three programs have highlighted the development of coordinated basic intelligence since that time: (1) *the Joint Army Navy Intelligence Studies* (JANIS), (2) *the National Intelligence Survey* (NIS), and (3) *The World Factbook* .

During World War II, intelligence consumers realized that the production of basic intelligence by different components of the US Government resulted in a great duplication of effort and conflicting information. The Japanese attack on Pearl Harbor in 1941 brought home to leaders in Congress and the executive branch the need for integrating departmental reports to national policymakers. Detailed and coordinated information was needed not only on such major powers as Germany and Japan, but also on places of little previous interest. In the Pacific Theater, for example, the Navy and Marines had to launch amphibious operations against many islands about which information was unconfirmed or nonexistent. Intelligence authorities resolved that the United States should never again be caught unprepared.

In 1943, Gen. George B. Strong (G-2), Adm. H. C. Train (Office of Naval Intelligence–ONI), and Gen. William J. Donovan (Director of the Office of Strategic Services–OSS) decided that a joint effort should be initiated. A steering committee was appointed on 27 April 1943 that recommended the formation of a Joint Intelligence Study Publishing Board to assemble, edit, coordinate, and publish the *Joint Army Navy Intelligence Studies* (JANIS). JANIS was the first interdepartmental basic intelligence program to fulfill the needs of the US Government for an authoritative and coordinated appraisal of strategic basic intelligence. Between April 1943 and July 1947, the board published 34 JANIS studies. JANIS performed well in the war effort, and numerous letters of commendation were received, including a statement from Adm. Forrest Sherman, Chief of Staff, Pacific Ocean Areas, which said, "JANIS has become the indispensable reference work for the shore-based planners."

The need for more comprehensive basic intelligence in the postwar world was well expressed in 1946 by George S. Pettee, a noted author on national security. He wrote in *The Future of American Secret Intelligence* (Infantry Journal Press, 1946, page 46) that world leadership in peace requires even more elaborate intelligence than in war. "The conduct of peace involves all countries, all human activities–not just the enemy and his war production."

The Central Intelligence Agency was established on 26 July 1947 and officially began operating on 18 September 1947. Effective 1 October 1947, the Director of Central Intelligence assumed operational responsibility for JANIS. On 13 January 1948, the National Security Council issued Intelligence Directive (NSCID) No. 3, which authorized the *National Intelligence Survey* (NIS) program as a peacetime replacement for the wartime JANIS program. Before adequate NIS country sections could be produced, government agencies had to develop more comprehensive gazetteers and better maps. The US Board on Geographic Names (BGN) compiled the names; the Department of the Interior produced the gazetteers; and CIA produced the maps.

The Hoover Commission's Clark Committee, set up in 1954 to study the structure and administration of the CIA, reported to Congress in 1955 that: "The National Intelligence Survey is an invaluable publication which provides the essential elements of basic intelligence on all areas of the world. There will always be a continuing requirement for keeping the Survey up-to-date." The *Factbook* was created as an annual summary and update to the encyclopedic NIS studies. The first classified *Factbook* was published in August 1962, and the first unclassified version was published in June 1971. The NIS program was terminated in 1973 except for the *Factbook*, map, and gazetteer components. The 1975 *Factbook* was the first to be made available to the public with sales through the US Government Printing Office (GPO). The *Factbook*was first made available on the Internet in June 1997. The year 2010 marks the 63rd anniversary of the establishment of the Central Intelligence Agency and the 67th year of continuous basic intelligence support to the US Government by *The World Factbook* and its two predecessor programs.

The Evolution of The World Factbook

National Basic Intelligence Factbook produced semiannually until 1980. Country entries include sections on Land, Water, People, Government, Economy, Communications, and Defense Forces.

1981—Publication becomes an annual product and is renamed*The World Factbook*. A total of 165 nations are covered on 225 pages.

1983—Appendices (Conversion Factors, International Organizations) first introduced.

1984—Appendices expanded; now include: A. The United Nations, B. Selected United Nations Organizations, C. Selected International Organizations, D. Country Membership in Selected Organizations, E. Conversion Factors.

1987—A new Geography section replaces the former separate Land and Water sections. UN Organizations and Selected International Organizations appendices merged into a new International Organizations appendix. First multi-color-cover *Factbook*.

1988—More than 40 new geographic entities added to provide complete world coverage without overlap or omission. Among the new entities are Antarctica, oceans (Arctic, Atlantic, Indian, Pacific), and the World. The front-of-the-book explanatory introduction expanded and retitled to Notes, Definitions, and Abbreviations. Two new Appendices added: Weights and Measures (in place of Conversion Factors) and a Cross-Reference List of Geographic Names. *Factbook*size reaches 300 pages.

1989—Economy section completely revised and now includes an Overview briefly describing a country's economy. New entries added under People, Government, and Communications.

1990—The Government section revised and considerably expanded with new entries.

1991—A new International Organizations and Groups appendix added. *Factbook* size reaches 405 pages.

1992—Twenty new successor state entries replace those of the Soviet Union and Yugoslavia. New countries are respectively: Armenia, Azerbaijan, Belarus, Estonia, Georgia, Kazakhstan, Kyrgyzstan, Latvia, Lithuania, Moldova, Russia, Tajikistan, Turkmenistan, Ukraine, Uzbekistan; and Bosnia and Hercegovina, Croatia, Macedonia, Serbia and Montenegro, Slovenia. Number of nations in the *Factbook* rises to 188.

1993—Czechoslovakia's split necessitates new Czech Republic and Slovakia entries. New Eritrea entry added after it secedes from Ethiopia. Substantial enhancements made to Geography section.

1994—Two new appendices address Selected International Environmental Agreements. The gross domestic product (GDP) of most developing countries changed to a purchasing power parity (PPP) basis rather than an exchange rate basis. *Factbook* size up to 512 pages.

1995—The GDP of all countries now presented on a PPP basis. New appendix lists estimates of GDP on an exchange rate basis. Communications category split; Railroads, Highways, Inland waterways, Pipelines, Merchant marine, and Airports entries now make up a new Transportation category. *The World Factbook* is first produced on CD-ROM.

1996—Maps accompanying each entry now present more detail. Flags also introduced for nearly all entities. Various new entries appear under Geography and Communications. Factbook abbreviations consolidated into a new Appendix A. Two new appendices present a Cross-Reference List of Country Data Codes and a Cross-Reference List of Hydrogeographic Data Codes. Geographic coordinates added to Appendix H, Cross-Reference List of Geographic Names. *Factbook* size expands by 95 pages in one year to reach 652.

1997—*The World Factbook* introduced onto the Internet. A special printed edition prepared for the CIA's 50th anniversary. A schema or Guide to Country Profiles introduced. New color maps and flags now accompany each country profile. Category headings distinguished by shaded backgrounds. Number of categories expanded to nine with the addition of an Introduction (for only a few countries) and Transnational Issues (which includes Disputes-international and Illicit drugs).

1998—The Introduction category with two entries, Current issues and Historical perspective, expanded to more countries. Last year for the production of CD-ROM versions of the *Factbook*.

1999—Historical perspective and Current issues entries in the Introduction category combined into a new Background statement. Several new Economy entries introduced. A new physical map of the world added to the back-of-the-book reference maps.

2000—A new "country profile" added on the Southern Ocean. The Background statements dramatically expanded to over 200 countries and possessions. A number of new Communications entries added.

2001—Background entries completed for all 267 entities in the *Factbook*. Several new HIV/AIDS entries introduced under the People category. Revision begun on individual country maps to include elevation extremes and a partial geographic grid. Weights and Measures appendix deleted.

2002—New entry on Distribution of Family income—Gini index added. Revision of individual country maps continued (process still ongoing).

2003—In the Economy category, petroleum entries added for oil production, consumption, exports, imports, and proved reserves, as well as natural gas proved reserves.

2004—Bi-weekly updates launched on *The World Factbook*website. Additional petroleum entries included for natural gas production, consumption, exports, and imports. In the Transportation category, under Merchant marine, subfields added for foreign-owned vessels and those registered in other countries. Descriptions of the many forms of government mentioned in the Factbook incorporated into the Definitions and Notes.

2005_In the People category, a Major infectious diseases field added for countries deemed to pose a higher risk for travelers. In the Economy category, entries included for Current account balance, Investment, Public debt, and Reserves of foreign exchange and gold. The Transnational issues category expanded to include Refugees and internally displaced persons. Size of the printed *Factbook* reaches 702 pages.

2006—In the Economy category, national GDP figures now presented at Official Exchange Rates (OER) in addition to GDP at purchasing power parity (PPP). Entries in the Transportation section reordered; Highways changed to Roadways, and Ports and harbors to Ports and terminals.

2007—In the Government category, the Capital entry significantly expanded with up to four subfields, including new information having to do with time. The subfields consist of the name of the capital itself, its geographic coordinates, the time difference at the capital from coordinated universal time (UTC), and, if applicable, information on daylight saving time (DST). Where appropriate, a special note is added to highlight those countries with multiple time zones. A Trafficking in persons entry added to the Transnational issues category. A new appendix, Weights and Measures, (re)introduced to the online version of the *Factbook*.

2008—In the Geography category, two fields focus on the increasingly vital resource of water: Total renewable water resources and Freshwater withdrawal. In the Economy category, three fields added for: Stock of direct foreign investment—at home, Stock of direct foreign investment—abroad, and Market value of publicly traded shares. Concise descriptions of all major religions included in the Definitions and Notes. Responsibility for printing of *The World Factbook*turned over to the Government Printing Office.

2009—The online *Factbook* site completely redesigned with many new features. In the People category, two new fields provide information on education in terms of opportunity and resources: School Life Expectancy and Education expenditures. Additionally, the Urbanization entry expanded to include all countries. In the Economy category, five fields added: Central bank discount rate, Commercial bank prime lending rate, Stock of narrow money, Stock of broad money, and Stock of domestic credit.

2010—Weekly updates inaugurated on the *The World Factbook* website. The dissolution of the Netherlands Antilles results in two new listings: Curacao and Sint Maarten. In the Communications category, a Broadcast media field replaces the former Radio broadcast stations and TV broadcast stations entries. In the Geography section, under Natural hazards, a Volcanism subfield added for countries with historically active volcanoes. In the Government category, a new National anthems field introduced. Concise descriptions of all major Legal systems incorporated into the Definitions and Notes. In order to facilitate comparisons over time, dozens of the entries in the Economy category expanded to include two (and in some cases three) years' worth of data.

2011—In the People section, a Major cities—population field provides the number of persons living in a country's capital and up to four of its major cities.

NOTES AND DEFINITIONS

Abbreviations This information is included in **Appendix A: Abbreviations**, which includes all abbreviations and acronyms used in the *Factbook*, with their expansions.

Acronyms An acronym is an abbreviation coined from the initial letter of each successive word in a term or phrase. In general, an acronym made up solely from the first letter of the major words in the expanded form is rendered in all capital letters (NATO from North Atlantic Treaty Organization; an exception would be ASEAN for Association of Southeast Asian Nations). In general, an acronym made up of more than the first letter of the major words in the expanded form is rendered with only an initial capital letter (Comsat from Communications Satellite Corporation; an exception would be NAM from Nonaligned Movement). Hybrid forms are sometimes used to distinguish between initially identical terms (ICC for International Chamber of Commerce and ICCt for International Criminal Court).

Administrative divisions This entry generally gives the numbers, designatory terms, and first-order administrative divisions as approved by the US Board on Geographic Names (BGN). Changes that have been reported but not yet acted on by the BGN are noted.

Age structure This entry provides the distribution of the population according to age. Information is included by sex and age group (*0-14 years, 15-64 years, 65 years and over*). The age structure of a population affects a nation's key socioeconomic issues. Countries with young populations (high percentage under age 15) need to invest more in schools, while countries with older populations (high percentage ages 65 and over) need to invest more in the health sector. The age structure can also be used to help predict potential political issues. For example, the rapid growth of a young adult population unable to find employment can lead to unrest.

Agriculture—products This entry is an ordered listing of major crops and products starting with the most important.

Airports This entry gives the total number of airports or airfields recognizable from the air. The runway(s) may be paved (concrete or asphalt surfaces) or unpaved (grass, earth, sand, or gravel surfaces) and may include closed or abandoned installations. Airports or airfields that are no longer recognizable (overgrown, no facilities, etc.) are not included. Note that not all airports have accommodations for refueling, maintenance, or air traffic control.

Airports—with paved runways This entry gives the total number of airports with paved runways (concrete or asphalt surfaces) by length. For airports with more than one runway, only the longest runway is included according to the following five groups—(1) *over 3,047 m* (over 10,000 ft), (2)*2,438 to 3,047 m* (8,000 to 10,000 ft), (3) *1,524 to 2,437 m* (5,000 to 8,000 ft), (4) *914 to 1,523 m* (3,000 to 5,000 ft), and (5) *under 914 m* (under 3,000 ft). Only airports with usable runways are included in this listing. Not all airports have facilities for refueling, maintenance, or air traffic control. The type aircraft capable of operating from a runway of a given length is dependent upon a number of factors including elevation of the runway, runway gradient, average maximum daily temperature at the airport, engine types, flap settings, and take-off weight of the aircraft.

Airports—with unpaved runways This entry gives the total number of airports with unpaved runways (grass, dirt, sand, or gravel surfaces) by length. For airports with more than one runway, only the longest runway is included according to the following five groups—(1) *over 3,047 m* (over 10,000 ft), (2) *2,438 to 3,047 m* (8,000 to 10,000 ft), (3) *1,524 to 2,437 m* (5,000 to 8,000 ft), (4)*914 to 1,523 m* (3,000 to 5,000 ft), and (5) *under 914 m* (under 3,000 ft). Only airports with usable runways are included in this listing. Not all airports have facilities for refueling, maintenance, or air traffic control. The type aircraft capable of operating from a runway of a given length is dependent upon a number of factors including elevation of the runway, runway gradient, average maximum daily temperature at the airport, engine types, flap settings, and take-off weight of the aircraft.

Appendixes This section includes *Factbook*-related material by topic.

Area This entry includes three subfields. *Total area* is the sum of all land and water areas delimited by international boundaries and/or coastlines. *Land area* is the aggregate of all surfaces delimited by international boundaries and/or coastlines, excluding inland water bodies (lakes, reservoirs, rivers). *Water area* is the sum of the surfaces of all inland water bodies, such as lakes, reservoirs, or rivers, as delimited by international boundaries and/or coastlines.

Area—comparative This entry provides an area comparison based on total area equivalents. Most entities are compared with the entire US or one of the 50 states based on area measurements (1990 revised) provided by the US Bureau of the Census. The smaller entities are compared with Washington, DC (178 sq km, 69 sq mi) or The Mall in Washington, DC (0.59 sq km, 0.23 sq mi, 146 acres).

Background This entry usually highlights major historic events and current issues and may include a statement about one or two key future trends.

Birth rate This entry gives the average annual number of births during a year per 1,000 persons in the population at midyear; also known as crude birth rate. The birth rate is usually the dominant factor in determining the rate of population growth. It depends on both the level of fertility and the age structure of the population.

Broadcast media This entry provides information on the approximate number of public and private TV and radio stations in a country, as well as basic information on the availability of satellite and cable TV services.

Budget This entry includes *revenues*, *expenditures*, and capital expenditures. These figures are calculated on an exchange rate basis, i.e., not in purchasing power parity (PPP) terms.

Capital This entry gives the *name* of the seat of government, its *geographic coordinates*, the *time difference* relative to **Coordinated Universal Time** (UTC) and the time observed in Washington, DC, and, if applicable, information on *daylight saving time* (**DST**). Where appropriate, a special *note* has been added to highlight those countries that have multiple time zones.

Central bank discount rate This entry provides the annualized interest rate a country's central bank charges commercial, depository banks for loans to meet temporary shortages of funds.

Climate This entry includes a brief description of typical weather regimes throughout the year.

Coastline This entry gives the total length of the boundary between the land area (including islands) and the sea.

Commercial bank prime lending rate This entry provides a simple average of annualized interest rates commercial banks charge on new loans, denominated in the national currency, to their most credit-worthy customers.

Communications This category deals with the means of exchanging information and includes the telephone, radio, television, and Internet host entries.

Communications—note This entry includes miscellaneous communications information of significance not included elsewhere.

Constitution This entry includes the dates of adoption, revisions, and major amendments.

Coordinated Universal Time (UTC) UTC is the international atomic time scale that serves as the basis of timekeeping for most of the world. The hours, minutes, and seconds expressed by UTC represent the time of day at the Prime Meridian (0 longitude) located near Greenwich, England as reckoned from midnight. UTC is calculated by the Bureau International des Poids et Measures (BIPM) in Sevres, France. The BIPM averages data collected from more than 200 atomic time and frequency standards located at about 50 laboratories worldwide. UTC is the basis for all civil time with the Earth divided into time zones expressed as positive or negative differences from UTC. UTC is also referred to as "Zulu time." See the Standard Time Zones of the World map included with the **Reference Maps**.

Country data codes See **Data codes**.

Country map Most versions of the *Factbook* provide a country map in color. The maps were produced from the best information available at the time of preparation. Names and/or boundaries may have changed subsequently.

Country name This entry includes all forms of the country's name approved by the US Board on Geographic Names (Italy is used as an example): *conventional long form* (Italian Republic), *conventional short form* (Italy), *local long form* (Repubblica Italiana), *local short form* (Italia), *former*(Kingdom of Italy), as well as the *abbreviation*. Also see the **Terminology** note.

Crude oil See entry for oil.

Current account balance This entry records a country's net trade in goods and services, plus net earnings from rents, interest, profits, and dividends, and net transfer payments (such as pension funds and worker remittances) to and from the rest of the world during the period specified. These figures are calculated on an exchange rate basis, i.e., not in purchasing power parity (PPP) terms.

Data codes This information is presented in **Appendix D: Cross-Reference List of Country Data Codesand Appendix E: Cross-Reference List of Hydrographic Data Codes.**

Date of information In general, information available as of January in a given year is used in the preparation of the printed edition.

Daylight Saving Time (DST) This entry is included for those entities that have adopted a policy of adjusting the official local time forward, usually one hour, from Standard Time during summer months. Such policies are most common in mid-latitude regions.

Death rate This entry gives the average annual number of deaths during a year per 1,000 population at midyear; also known as crude death rate. The death rate, while only a rough indicator of the mortality situation in a country, accurately indicates the current mortality impact on population growth. This indicator is significantly affected by age distribution, and most countries will eventually show a rise in the overall death rate, in spite of continued decline in mortality at all ages, as declining fertility results in an aging population.

Debt—external This entry gives the total public and private debt owed to nonresidents repayable in internationally accepted currencies, goods, or services. These figures are calculated on an exchange rate basis, i.e., not in purchasing power parity (PPP) terms.

Dependency status This entry describes the formal relationship between a particular nonindependent entity and an independent state.

Dependent areas This entry contains an alphabetical listing of all nonindependent entities associated in some way with a particular independent state.

Diplomatic representation The US Government has diplomatic relations with 189 independent states, including 187 of the 192 UN members (excluded UN members are Bhutan, Cuba, Iran, North Korea, and the US itself). In addition, the US has diplomatic relations with 2 independent states that are not in the UN, the Holy See and Kosovo, as well as with the EU.

Diplomatic representation from the US This entry includes the *chief of mission*, *embassy* address, *mailing address*, *telephonenumber*, *FAX* number, *branch office* locations, *consulate general* locations, and *consulate*locations.

Diplomatic representation in the US This entry includes the *chief of mission*, *chancery*, *telephone*, *FAX*, *consulate general*locations, and *consulate* locations.

Disputes—international This entry includes a wide variety of situations that range from traditional bilateral boundary disputes to unilateral claims of one sort or another. Information regarding disputes over international terrestrial and maritime boundaries has been reviewed by the US Department of State. References to other situations involving borders or frontiers may also be included, such as resource disputes, geopolitical questions, or irredentist issues; however, inclusion does not necessarily constitute official acceptance or recognition by the US Government.

Distribution of family income—Gini index This index measures the degree of inequality in the distribution of family income in a country. The index is calculated from the Lorenz curve, in which cumulative family income is plotted against the number of families arranged from the poorest to the richest. The index is the ratio of (a) the area between a country's Lorenz curve and the 45 degree helping line to (b) the entire triangular area under the

45 degree line. The more nearly equal a country's income distribution, the closer its Lorenz curve to the 45 degree line and the lower its Gini index, e.g., a Scandinavian country with an index of 25. The more unequal a country's income distribution, the farther its Lorenz curve from the 45 degree line and the higher its Gini index, e.g., a Sub-Saharan country with an index of 50. If income were distributed with perfect equality, the Lorenz curve would coincide with the 45 degree line and the index would be zero; if income were distributed with perfect inequality, the Lorenz curve would coincide with the horizontal axis and the right vertical axis and the index would be 100.

Economy This category includes the entries dealing with the size, development, and management of productive resources, i.e., land, labor, and capital.

Economy—overview This entry briefly describes the type of economy, including the degree of market orientation, the level of economic development, the most important natural resources, and the unique areas of specialization. It also characterizes major economic events and policy changes in the most recent 12 months and may include a statement about one or two key future macroeconomic trends.

Education expenditures This entry provides the public expenditure on education as a percent of GDP.

Electricity—consumption This entry consists of total electricity generated annually plus imports and minus exports, expressed in kilowatt-hours. The discrepancy between the amount of electricity generated and/or imported and the amount consumed and/or exported is accounted for as loss in transmission and distribution.

Electricity—exports This entry is the total exported electricity in kilowatt-hours.

Electricity—imports This entry is the total imported electricity in kilowatt-hours.

Electricity—production This entry is the annual electricity generated expressed in kilowatt-hours. The discrepancy between the amount of electricity generated and/or imported and the amount consumed and/or exported is accounted for as loss in transmission and distribution.

Elevation extremes This entry includes both the highest point and the lowest point.

Entities Some of the independent states, dependencies, areas of special sovereignty, and governments included in this publication are not independent, and others are not officially recognized by the US Government. "Independent state" refers to a people politically organized into a sovereign state with a definite territory. "Dependencies" and "areas of special sovereignty" refer to a broad category of political entities that are associated in some way with an independent state. "Country" names used in the table of contents or for page headings are usually the short-form names as approved by the US Board on Geographic Names and may include independent states, dependencies, and areas of special sovereignty, or other geographic entities. There are a total of 266 separate geographic entities in *The World Factbook* that may be categorized as follows:

INDEPENDENT STATES

194 Afghanistan, Albania, Algeria, Andorra, Angola, Antigua and Barbuda, Argentina, Armenia, Australia, Austria, Azerbaijan, The Bahamas, Bahrain, Bangladesh, Barbados, Belarus, Belgium, Belize, Benin, Bhutan, Bolivia, Bosnia and Herzegovina, Botswana, Brazil, Brunei, Bulgaria, Burkina Faso, Burma, Burundi, Cambodia, Cameroon, Canada, Cape Verde, Central African Republic, Chad, Chile, China, Colombia, Comoros, Democratic Republic of the Congo, Republic of the Congo, Costa Rica, Cote d'Ivoire, Croatia, Cuba, Cyprus, Czech Republic, Denmark, Djibouti, Dominica, Dominican Republic, Ecuador, Egypt, El Salvador, Equatorial Guinea, Eritrea, Estonia, Ethiopia, Fiji, Finland, France, Gabon, The Gambia, Georgia, Germany, Ghana, Greece, Grenada, Guatemala, Guinea, Guinea-Bissau, Guyana, Haiti, Holy See, Honduras, Hungary, Iceland, India, Indonesia, Iran, Iraq, Ireland, Israel, Italy, Jamaica, Japan, Jordan, Kazakhstan, Kenya, Kiribati, North Korea, South Korea, Kosovo, Kuwait, Kyrgyzstan, Laos, Latvia, Lebanon, Lesotho, Liberia, Libya, Liechtenstein, Lithuania, Luxembourg, Macedonia, Madagascar, Malawi, Malaysia, Maldives, Mali, Malta, Marshall Islands, Mauritania, Mauritius, Mexico, Federated States of Micronesia, Moldova, Monaco, Mongolia, Montenegro, Morocco, Mozambique, Namibia, Nauru, Nepal, Netherlands, NZ, Nicaragua, Niger, Nigeria, Norway, Oman, Pakistan, Palau, Panama, Papua New Guinea, Paraguay, Peru, Philippines, Poland, Portugal, Qatar, Romania, Russia, Rwanda, Saint Kitts and Nevis, Saint Lucia, Saint Vincent and the Grenadines, Samoa, San Marino, Sao Tome and Principe, Saudi Arabia, Senegal, Serbia, Seychelles, Sierra Leone, Singapore, Slovakia, Slovenia, Solomon Islands, Somalia, South Africa, Spain, Sri Lanka, Sudan, Suriname, Swaziland, Sweden, Switzerland, Syria, Tajikistan, Tanzania, Thailand, Timor-Leste, Togo, Tonga, Trinidad and Tobago, Tunisia, Turkey, Turkmenistan, Tuvalu, Uganda, Ukraine, UAE, UK, US, Uruguay, Uzbekistan, Vanuatu, Venezuela, Vietnam, Yemen, Zambia, Zimbabwe

OTHER

2 Taiwan, European Union

DEPENDENCIES AND AREAS OF SPECIAL SOVEREIGNTY

6 Australia–Ashmore and Cartier Islands, Christmas Island, Cocos (Keeling) Islands, Coral Sea Islands, Heard Island and McDonald Islands, Norfolk Island
2 China–Hong Kong, Macau 2 Denmark–Faroe Islands, Greenland
8 France–Clipperton Island, French Polynesia, French Southern and Antarctic Lands, New Caledonia, Saint Barthelemy, Saint Martin, Saint Pierre and Miquelon, Wallis and Futuna 3 Netherlands–Aruba, Curacao, Sint Maarten 3 New Zealand–Cook Islands, Niue, Tokelau 3 Norway–Bouvet Island, Jan Mayen, Svalbard 17 UK–Akrotiri, Anguilla, Bermuda, British Indian Ocean Territory, British Virgin Islands, Cayman Islands, Dhekelia, Falkland Islands, Gibraltar, Guernsey, Jersey, Isle of Man, Montserrat, Pitcairn Islands, Saint Helena, South Georgia and the South Sandwich Islands, Turks and Caicos Islands
14 US–American Samoa, Baker Island*, Guam, Howland Island*, Jarvis Island*, Johnston Atoll*, Kingman Reef*, Midway Islands*, Navassa Island, Northern Mariana Islands, Palmyra Atoll*, Puerto Rico, Virgin Islands, Wake Island (* consolidated in United States Pacific Island Wildlife Refuges entry)

MISCELLANEOUS

6 Antarctica, Gaza Strip, Paracel Islands, Spratly Islands, West Bank, Western Sahara

OTHER ENTITIES

5 oceans–Arctic Ocean, Atlantic Ocean, Indian Ocean, Pacific Ocean, Southern Ocean 1 World **266 total**

Environment—current issues This entry lists the most pressing and important environmental problems. The following terms and abbreviations are used throughout the entry:

Acidification–the lowering of soil and water pH due to acid precipitation and deposition usually through precipitation; this process disrupts ecosystem nutrient flows and may kill freshwater fish and plants dependent on more neutral or alkaline conditions (see acid rain). *Acid rain*–characterized as containing harmful levels of sulfur dioxide or nitrogen oxide; acid rain is damaging and potentially deadly to the earth's fragile ecosystems; acidity is measured using the pH scale where 7 is neutral, values greater than 7 are considered alkaline, and values below 5.6 are considered acid precipitation; note–a pH of 2.4 (the acidity of vinegar) has been measured in rainfall in New England.

Aerosol—a collection of airborne particles dispersed in a gas, smoke, or fog.
Afforestation—converting a bare or agricultural space by planting trees and plants; reforestation involves replanting trees on areas that have been cut or destroyed by fire.
Asbestos—a naturally occurring soft fibrous mineral commonly used in fireproofing materials and considered to be highly carcinogenic in particulate form.
Biodiversity—also biological diversity; the relative number of species, diverse in form and function, at the genetic, organism, community, and ecosystem level; loss of biodiversity reduces an ecosystem's ability to recover from natural or man-induced disruption.
Bio-indicators—a plant or animal species whose presence, abundance, and health reveal the general condition of its habitat.
Biomass—the total weight or volume of living matter in a given area or volume.
Carbon cycle—the term used to describe the exchange of carbon (in various forms, e.g., as carbon dioxide) between the atmosphere, ocean, terrestrial biosphere, and geological deposits.
Catchments—assemblages used to capture and retain rainwater and runoff; an important water management technique in areas with limited freshwater resources, such as Gibraltar.
DDT (dichloro-diphenyl-trichloro-ethane)—a colorless, odorless insecticide that has toxic effects on most animals; the use of DDT was banned in the US in 1972.
Defoliants—chemicals which cause plants to lose their leaves artificially; often used in agricultural practices for weed control, and may have detrimental impacts on human and ecosystem health.
Deforestation—the destruction of vast areas of forest (e.g., unsustainable forestry practices, agricultural and range land clearing, and the over exploitation of wood products for use as fuel) without planting new growth.
Desertification—the spread of desert-like conditions in arid or semi-arid areas, due to overgrazing, loss of agriculturally productive soils, or climate change.
Dredging—the practice of deepening an existing waterway; also, a technique used for collecting bottom-dwelling marine organisms (e.g., shellfish) or harvesting coral, often causing significant destruction of reef and ocean-floor ecosystems.
Drift-net fishing—done with a net, miles in extent, that is generally anchored to a boat and left to float with the tide; often results in an over harvesting and waste of large populations of non-commercial marine species (by-catch) by its effect of "sweeping the ocean clean."
Ecosystems—ecological units comprised of complex communities of organisms and their specific environments.
Effluents—waste materials, such as smoke, sewage, or industrial waste which are released into the environment, subsequently polluting it.
Endangered species—a species that is threatened with extinction either by direct hunting or habitat destruction.
Freshwater—water with very low soluble mineral content; sources include lakes, streams, rivers, glaciers, and underground aquifers.
Greenhouse gas—a gas that "traps" infrared radiation in the lower atmosphere causing surface warming; water vapor, carbon dioxide, nitrous oxide, methane, hydrofluorocarbons, and ozone are the primary greenhouse gases in the Earth's atmosphere.
Groundwater—water sources found below the surface of the earth often in naturally occurring reservoirs in permeable rock strata; the source for wells and natural springs.
Highlands Water Project— a series of dams constructed jointly by Lesotho and South Africa to redirect Lesotho's abundant water supply into a rapidly growing area in South Africa; while it is the largest infrastructure project in southern Africa, it is also the most costly and controversial; objections to the project include claims that it forces people from their homes, submerges farmlands, and squanders economic resources.
Inuit Circumpolar Conference (ICC)—represents the roughly 150,000 Inuits of Alaska, Canada, Greenland, and Russia in international environmental issues; a General Assembly convenes every three years to determine the focus of the ICC; the most current concerns are long-range transport of pollutants, sustainable development, and climate change.
Metallurgical plants—industries which specialize in the science, technology, and processing of metals; these plants produce highly concentrated and toxic wastes which can contribute to pollution of ground water and air when not properly disposed.
Noxious substances—injurious, very harmful to living beings.
Overgrazing—the grazing of animals on plant material faster than it can naturally regrow leading to the permanent loss of plant cover, a common effect of too many animals grazing limited range land.
Ozone shield—a layer of the atmosphere composed of ozone gas (O3) that resides approximately 25 miles above the Earth's surface and absorbs solar ultraviolet radiation that can be harmful to living organisms.
Poaching—the illegal killing of animals or fish, a great concern with respect to endangered or threatened species.
Pollution—the contamination of a healthy environment by man-made waste.
Potable water—water that is drinkable, safe to be consumed.
Salination—the process through which fresh (drinkable) water becomes salt (undrinkable) water; hence, desalination is the reverse process; also involves the accumulation of salts in topsoil caused by evaporation of excessive irrigation water, a process that can eventually render soil incapable of supporting crops.
Siltation—occurs when water channels and reservoirs become clotted with silt and mud, a side effect of deforestation and soil erosion.
Slash-and-burn agriculture—a rotating cultivation technique in which trees are cut down and burned in order to clear land for temporary agriculture; the land is used until its productivity declines at which point a new plot is selected and the process repeats; this practice is sustainable while population levels are low and time is permitted for regrowth of natural vegetation; conversely, where these conditions do not exist, the practice can have disastrous consequences for the environment.
Soil degradation—damage to the land's productive capacity because of poor agricultural practices such as the excessive use of pesticides or fertilizers, soil compaction from heavy equipment, or erosion of topsoil, eventually resulting in reduced ability to produce agricultural products.
Soil erosion—the removal of soil by the action of water or wind, compounded by poor agricultural practices, deforestation, overgrazing, and desertification.
Ultraviolet (UV) radiation—a portion of the electromagnetic energy emitted by the sun and naturally filtered in the upper atmosphere by the ozone layer; UV radiation can be harmful to living organisms and has been linked to increasing rates of skin cancer in humans.
Waterborne diseases—those in which bacteria survive in, and are transmitted through, water; always a serious threat in areas with an untreated water supply.

Environment—international agreements This entry separates country participation in international environmental agreements into two levels - *party to* and *signed, but not ratified*. Agreements are listed in alphabetical order by the abbreviated form of the full name.

Environmental agreements This information is presented in **Appendix C: Selected International Environmental Agreements, which includes the name, abbreviation, date opened for signature, date entered into force, objective, and parties by category.**

Ethnic groups This entry provides an ordered listing of ethnic groups starting with the largest and normally includes the percent of total population.

Exchange rates This entry provides the official value of a country's monetary unit at a given date or over a given period of time, as expressed in units of local currency per US dollar and as determined by international market forces or official fiat. The International Organization for Standardization (ISO) 4217 alphabetic currency code for the national medium of exchange is presented in parenthesis.

Executive branch This entry includes several subfields. *Chief of state* includes the name and title of the titular leader of the country who represents the state at official and ceremonial functions but may not be involved with the day-to-day activities of the government. *Head of government*includes the name and title of the top administrative leader who is designated to manage the day-to-day activities of the government. For example, in the UK, the monarch is the chief of state, and the prime minister is the head of government. In the US, the president is both the chief of state and the head of government. *Cabinet* includes the official name for this body of high-ranking advisers and the method for selection of members. *Elections* includes the nature of election process or accession to power, date of the last election, and date of the next election. *Election results* includes the percent of vote for each candidate in the last election.

Exports This entry provides the total US dollar amount of merchandise exports on an f.o.b. (free on board) basis. These figures are calculated on an exchange rate basis, i.e., not in purchasing power parity (PPP) terms.

Exports—commodities This entry provides a listing of the highest-valued exported products; it sometimes includes the percent of total dollar value.

Exports—partners This entry provides a rank ordering of trading partners starting with the most important; it sometimes includes the percent of total dollar value.

Flag description This entry provides a written flag description produced from actual flags or the best information available at the time the entry was written. The flags of independent states are used by their dependencies unless there is an officially recognized local flag. Some disputed and other areas do not have flags.

Flag graphic Most versions of the *Factbook* include a color flag at the beginning of the country profile. The flag graphics were produced from actual flags or the best information available at the time of preparation. The flags of independent states are used by their dependencies unless there is an officially recognized local flag. Some disputed and other areas do not have flags.

Freshwater withdrawal (domestic/industrial/agricultural) This entry provides the annual quantity of water in cubic kilometers removed from available sources for use in any purpose. Water drawn-off is not necessarily entirely consumed and some portion may be returned for further use downstream. Domestic sector use refers to water supplied by public distribution systems. Note that some of this total may be used for small industrial and/or limited agricultural purposes. Industrial sector use is the quantity of water used by self-supplied industries not connected to a public distribution system. Agricultural sector use includes water used for irrigation and livestock watering, and does not account for agriculture directly dependent on rainfall. Included are figures for *total*annual water withdrawal and *per capita* water withdrawal.

GDP (official exchange rate) This entry gives the gross domestic product (GDP) or value of all final goods and services produced within a nation in a given year. A nation's GDP at official exchange rates (OER) is the home-currency-denominated annual GDP figure divided by the bilateral average US exchange rate with that country in that year. The measure is simple to compute and gives a precise measure of the value of output. Many economists prefer this measure when gauging the economic power an economy maintains vis-ß-vis its neighbors, judging that an exchange rate captures the purchasing power a nation enjoys in the international marketplace. Official exchange rates, however, can be artificially fixed and/or subject to manipulation–resulting in claims of the country having an under- or over-valued currency–and are not necessarily the equivalent of a market-determined exchange rate. Moreover, even if the official exchange rate is market-determined, market exchange rates are frequently established by a relatively small set of goods and services (the ones the country trades) and may not capture the value of the larger set of goods the country produces. Furthermore, OER-converted GDP is not well suited to comparing domestic GDP over time, since appreciation/depreciation from one year to the next will make the OER GDP value rise/fall regardless of whether home-currency-denominated GDP changed.

GDP (purchasing power parity) This entry gives the gross domestic product (GDP) or value of all final goods and services produced within a nation in a given year. A nation's GDP at purchasing power parity (PPP) exchange rates is the sum value of all goods and services produced in the country valued at prices prevailing in the United States. This is the measure most economists prefer when looking at per-capita welfare and when comparing living conditions or use of resources across countries. The measure is difficult to compute, as a US dollar value has to be assigned to all goods and services in the country regardless of whether these goods and services have a direct equivalent in the United States (for example, the value of an ox-cart or non-US military equipment); as a result, PPP estimates for some countries are based on a small and sometimes different set of goods and services. In addition, many countries do not formally participate in the World Bank's PPP project that calculates these measures, so the resulting GDP estimates for these countries may lack precision. For many developing countries, PPP-based GDP measures are multiples of the official exchange rate (OER) measure. The differences between the OER- and PPP-denominated GDP values for most of the wealthy industrialized countries are generally much smaller.

GDP—composition by sector This entry gives the percentage contribution of *agriculture*, *industry*, and *services* to total GDP. The distribution will total less than 100 percent if the data are incomplete.

GDP—per capita (PPP) This entry shows GDP on a purchasing power parity basis divided by population as of 1 July for the same year.

GDP—real growth rate This entry gives GDP growth on an annual basis adjusted for inflation and expressed as a percent.

GDP methodology In the **Economy** category, GDP dollar estimates for countries are reported both on an official exchange rate (OER) and a purchasing power parity (PPP) basis. Both measures contain information that is useful to the reader. The PPP method involves the use of standardized international dollar price weights, which are applied to the quantities of final goods and services produced in a given economy. The data derived from the PPP method probably provide the best available starting point for comparisons of economic strength and well-being between countries. In contrast, the currency exchange rate method involves a variety of international and domestic financial forces that may not capture the value of domestic output. Whereas PPP estimates for OECD countries are quite reliable, PPP estimates for developing countries are often rough approximations. In developing countries with weak currencies, the exchange rate estimate of GDP in dollars is typically one-fourth to one-half the PPP estimate. Most of the GDP estimates for developing countries are based on extrapolation of PPP numbers published by the UN International Comparison Program (UNICP) and by Professors Robert Summers and Alan Heston of the University of Pennsylvania and their colleagues. GDP derived using the OER method should be used for the purpose of calculating the share of items such as exports, imports, military expenditures, external debt, or the current account balance, because the dollar values presented in the *Factbook* for these items have been converted at official exchange rates, not at PPP. One should use the OER GDP figure to calculate the proportion of, say, Chinese defense expenditures in GDP, because that share will be the same as one calculated in local currency units. Comparison of OER GDP with PPP GDP may also indicate whether a currency is over- or under-valued. If OER GDP is smaller than PPP GDP, the official exchange rate may be undervalued, and vice versa. However, there is no strong historical evidence that market exchange rates move in the direction implied by the PPP rate, at least not in the short- or medium-term. Note: the numbers for GDP and other economic data should not be chained together from successive volumes of the*Factbook* because of changes in the US dollar measuring rod, revisions of data by statistical agencies, use of new or different sources of information, and changes in national statistical methods and practices.

Geographic coordinates This entry includes rounded latitude and longitude figures for the purpose of finding the approximate geographic center of an entity and is based on the locations provided in the Geographic Names Server (GNS), maintained by the National Geospatial-Intelligence Agency on behalf of the US Board on Geographic Names.

Geographic names This information is presented in **Appendix F: Cross Reference List of Geographic Names. It includes a listing of various alternate names, former names, local names, and regional names referenced to one or more related *Factbook* entries. Spellings are normally, but not always, those approved by the US Board on Geographic Names (BGN). Alternate names and additional information are included in parentheses.**

Geography This category includes the entries dealing with the natural environment and the effects of human activity.

Geography—note This entry includes miscellaneous geographic information of significance not included elsewhere.

Gini index See entry for **Distribution of family income–Gini index**

GNP Gross national product (GNP) is the value of all final goods and services produced within a nation in a given year, plus income earned by its citizens abroad, minus income earned by foreigners from domestic production. The *Factbook*, following current practice, uses GDP rather than GNP to measure national production. However, the user must realize that in certain countries net remittances from citizens working abroad may be important to national well-being.

Government This category includes the entries dealing with the system for the adoption and administration of public policy.

Government—note This entry includes miscellaneous government information of significance not included elsewhere.

Government type This entry gives the basic form of government. Definitions of the major governmental terms are as follows. (Note that for some countries more than one definition applies.):

Absolute monarchy–a form of government where the monarch rules unhindered, i.e., without any laws, constitution, or legally organized opposition.

Anarchy–a condition of lawlessness or political disorder brought about by the absence of governmental authority.

Authoritarian–a form of government in which state authority is imposed onto many aspects of citizens' lives.

Commonwealth–a nation, state, or other political entity founded on law and united by a compact of the people for the common good.

Communist–a system of government in which the state plans and controls the economy and a single–often authoritarian–party holds power; state controls are imposed with the elimination of private ownership of property or capital while claiming to make progress toward a higher social order in which all goods are equally shared by the people (i.e., a classless society).

Confederacy (Confederation)–a union by compact or treaty between states, provinces, or territories, that creates a central government with limited powers; the constituent entities retain supreme authority over all matters except those delegated to the central government.

Constitutional–a government by or operating under an authoritative document (constitution) that sets forth the system of fundamental laws and principles that determines the nature, functions, and limits of that government.

Constitutional democracy–a form of government in which the sovereign power of the people is spelled out in a governing constitution.

Constitutional monarchy–a system of government in which a monarch is guided by a constitution whereby his/her rights, duties, and responsibilities are spelled out in written law or by custom.

Democracy–a form of government in which the supreme power is retained by the people, but which is usually exercised indirectly through a system of representation and delegated authority periodically renewed.

Democratic republic–a state in which the supreme power rests in the body of citizens entitled to vote for officers and representatives responsible to them.

Dictatorship–a form of government in which a ruler or small clique wield absolute power (not restricted by a constitution or laws).

Ecclesiastical–a government administrated by a church.

Emirate–similar to a monarchy or sultanate, but a government in which the supreme power is in the hands of an emir (the ruler of a Muslim state); the emir may be an absolute overlord or a sovereign with constitutionally limited authority.

Federal (Federation)–a form of government in which sovereign power is formally divided–usually by means of a constitution–between a central authority and a number of constituent regions (states, colonies, or provinces) so that each region retains some management of its internal affairs; differs from a confederacy in that the central government exerts influence directly upon both individuals as well as upon the regional units.

Federal republic–a state in which the powers of the central government are restricted and in which the component parts (states, colonies, or provinces) retain a degree of self-government; ultimate sovereign power rests with the voters who chose their governmental representatives.

Islamic republic–a particular form of government adopted by some Muslim states; although such a state is, in theory, a theocracy, it remains a republic, but its laws are required to be compatible with the laws of Islam.

Maoism–the theory and practice of Marxism-Leninism developed in China by Mao Zedong (Mao Tse-tung), which states that a continuous revolution is necessary if the leaders of a communist state are to keep in touch with the people.

Marxism–the political, economic, and social principles espoused by 19th century economist Karl Marx; he viewed the struggle of workers as a progression of historical forces that would proceed from a class struggle of the proletariat (workers) exploited by capitalists (business owners), to a socialist "dictatorship of the proletariat," to, finally, a classless society–Communism.

Marxism-Leninism–an expanded form of communism developed by Lenin from doctrines of Karl Marx; Lenin saw imperialism as the final stage of capitalism and shifted the focus of workers' struggle from developed to underdeveloped countries.

Monarchy–a government in which the supreme power is lodged in the hands of a monarch who reigns over a state or territory, usually for life and by hereditary right; the monarch may be either a sole absolute ruler or a sovereign–such as a king, queen, or prince–with constitutionally limited authority.

Oligarchy–a government in which control is exercised by a small group of individuals whose authority generally is based on wealth or power.

Parliamentary democracy–a political system in which the legislature (parliament) selects the government–a prime minister, premier, or chancellor along with the cabinet ministers–according to party strength as expressed in elections; by this system, the government acquires a dual responsibility: to the people as well as to the parliament.

Parliamentary government (Cabinet-Parliamentary government)–a government in which members of an executive branch (the cabinet and its leader–a prime minister, premier, or chancellor) are nominated to their positions by a legislature or parliament, and are directly responsible to it; this type of government can be dissolved at will by the parliament (legislature) by means of a no confidence vote or the leader of the cabinet may dissolve the parliament if it can no longer function.

Parliamentary monarchy–a state headed by a monarch who is not actively involved in policy formation or implementation (i.e., the exercise of sovereign powers by a monarch in a ceremonial capacity); true governmental leadership is carried out by a cabinet and its head—a prime minister, premier, or chancellor—who are drawn from a legislature (parliament).

Presidential—a system of government where the executive branch exists separately from a legislature (to which it is generally not accountable).

Republic—a representative democracy in which the people's elected deputies (representatives), not the people themselves, vote on legislation.

Socialism—a government in which the means of planning, producing, and distributing goods is controlled by a central government that theoretically seeks a more just and equitable distribution of property and labor; in actuality, most socialist governments have ended up being no more than dictatorships over workers by a ruling elite.

Sultanate—similar to a monarchy, but a government in which the supreme power is in the hands of a sultan (the head of a Muslim state); the sultan may be an absolute ruler or a sovereign with constitutionally limited authority.

Theocracy—a form of government in which a Deity is recognized as the supreme civil ruler, but the Deity's laws are interpreted by ecclesiastical authorities (bishops, mullahs, etc.); a government subject to religious authority.

Totalitarian—a government that seeks to subordinate the individual to the state by controlling not only all political and economic matters, but also the attitudes, values, and beliefs of its population.

Greenwich Mean Time (GMT) The mean solar time at the Greenwich Meridian, Greenwich, England, with the hours and days, since 1925, reckoned from midnight. GMT is now a historical term having been replaced by UTC on 1 January 1972. See **Coordinated Universal Time**.

Gross domestic product See GDP

Gross national product See GNP

Gross world product See GWP

GWP This entry gives the gross world product (GWP) or aggregate value of all final goods and services produced worldwide in a given year.

Heliports This entry gives the total number of heliports with hard-surface runways, helipads, or landing areas that support routine sustained helicopter operations exclusively and have support facilities including one or more of the following facilities: lighting, fuel, passenger handling, or maintenance. It includes former airports used exclusively for helicopter operations but excludes heliports limited to day operations and natural clearings that could support helicopter landings and takeoffs.

HIV/AIDS—adult prevalence rate This entry gives an estimate of the percentage of adults (aged 15-49) living with HIV/AIDS. The adult prevalence rate is calculated by dividing the estimated number of adults living with HIV/AIDS at yearend by the total adult population at yearend.

HIV/AIDS—deaths This entry gives an estimate of the number of adults and children who died of AIDS during a given calendar year.

HIV/AIDS—people living with HIV/AIDS This entry gives an estimate of all people (adults and children) alive at yearend with HIV infection, whether or not they have developed symptoms of AIDS.

Household income or consumption by percentage share Data on household income or consumption come from household surveys, the results adjusted for household size. Nations use different standards and procedures in collecting and adjusting the data. Surveys based on income will normally show a more unequal distribution than surveys based on consumption. The quality of surveys is improving with time, yet caution is still necessary in making inter-country comparisons.

Hydrographic data codes See Data codes

Illicit drugs This entry gives information on the five categories of illicit drugs—narcotics, stimulants, depressants (sedatives), hallucinogens, and cannabis. These categories include many drugs legally produced and prescribed by doctors as well as those illegally produced and sold outside of medical channels.

Cannabis (*Cannabis sativa*) is the common hemp plant, which provides hallucinogens with some sedative properties, and includes marijuana (pot, Acapulco gold, grass, reefer), tetrahydrocannabinol (THC, Marinol), hashish (hash), and hashish oil (hash oil).

Coca (mostly *Erythroxylum coca*) is a bush with leaves that contain the stimulant used to make cocaine. Coca is not to be confused with cocoa, which comes from cacao seeds and is used in making chocolate, cocoa, and cocoa butter.

Cocaine is a stimulant derived from the leaves of the coca bush.

Depressants (sedatives) are drugs that reduce tension and anxiety and include chloral hydrate, barbiturates (Amytal, Nembutal, Seconal, phenobarbital), benzodiazepines (Librium, Valium), methaqualone (Quaalude), glutethimide (Doriden), and others (Equanil, Placidyl, Valmid).

Drugs are any chemical substances that effect a physical, mental, emotional, or behavioral change in an individual.

Drug abuse is the use of any licit or illicit chemical substance that results in physical, mental, emotional, or behavioral impairment in an individual.

Hallucinogens are drugs that affect sensation, thinking, self-awareness, and emotion. Hallucinogens include LSD (acid, microdot), mescaline and peyote (mexc, buttons, cactus), amphetamine variants (PMA, STP, DOB), phencyclidine (PCP, angel dust, hog), phencyclidine analogues (PCE, PCPy, TCP), and others (psilocybin, psilocyn).

Hashish is the resinous exudate of the cannabis or hemp plant (*Cannabis sativa*).

Heroin is a semisynthetic derivative of morphine.

Mandrax is a trade name for methaqualone, a pharmaceutical depressant.

Marijuana is the dried leaf of the cannabis or hemp plant (*Cannabis sativa*).

Methaqualone is a pharmaceutical depressant, referred to as mandrax in Southwest Asia and Africa.

Narcotics are drugs that relieve pain, often induce sleep, and refer to opium, opium derivatives, and synthetic substitutes. Natural narcotics include opium (paregoric, parepectolin), morphine (MS-Contin, Roxanol), codeine (Tylenol with codeine, Empirin with codeine, Robitussin AC), and thebaine. Semisynthetic narcotics include heroin (horse, smack), and hydromorphone (Dilaudid). Synthetic narcotics include meperidine or Pethidine (Demerol, Mepergan), methadone (Dolophine, Methadose), and others (Darvon, Lomotil).

Opium is the brown, gummy exudate of the incised, unripe seedpod of the opium poppy.

Opium poppy (*Papaver somniferum*) is the source for the natural and semisynthetic narcotics.

Poppy straw is the entire cut and dried opium poppy-plant material, other than the seeds. Opium is extracted from poppy straw in commercial operations that produce the drug for medical use.

Qat (kat, khat) is a stimulant from the buds or leaves of *Catha edulis* that is chewed or drunk as tea.

Quaaludes is the North American slang term for methaqualone, a pharmaceutical depressant.

Stimulants are drugs that relieve mild depression, increase energy and activity, and include cocaine (coke, snow, crack), amphetamines (Desoxyn, Dexedrine), ephedrine, ecstasy (clarity, essence, doctor, Adam), phenmetrazine (Preludin), methylphenidate (Ritalin), and others (Cylert, Sanorex, Tenuate).

Imports This entry provides the total US dollar amount of merchandise imports on a c.i.f. (cost, insurance, and freight) or f.o.b. (free on board) basis. These figures are calculated on an exchange rate basis, i.e., not in purchasing power parity (PPP) terms.

Imports—commodities This entry provides a listing of the highest-valued imported products; it sometimes includes the percent of total dollar value.

Imports—partners This entry provides a rank ordering of trading partners starting with the most important; it sometimes includes the percent of total dollar value.

Independence For most countries, this entry gives the date that sovereignty was achieved and from which nation, empire, or trusteeship. For the other countries, the date given may not represent "independence" in the strict sense, but rather some significant nationhood event such as the traditional founding date or the date of unification, federation, confederation, establishment, fundamental change in the form of government, or state succession. For a number of countries, the establishment of statehood was a lengthy evolutionary process occurring over decades or even centuries. In such cases, several significant dates are cited. Dependent areas include the notation "none" followed by the nature of their dependency status. Also see the **Terminology** note.

Industrial production growth rate This entry gives the annual percentage increase in industrial production (includes manufacturing, mining, and construction).

Industries This entry provides a rank ordering of industries starting with the largest by value of annual output.

Infant mortality rate This entry gives the number of deaths of infants under one year old in a given year per 1,000 live births in the same year; included is the total death rate, and deaths by sex, *maleand female*. This rate is often used as an indicator of the level of health in a country.

Inflation rate (consumer prices) This entry furnishes the annual percent change in consumer prices compared with the previous year's consumer prices.

International disputes see Disputes—international

International organization participation This entry lists in alphabetical order by abbreviation those international organizations in which the subject country is a member or participates in some other way.

International organizations This information is presented in **Appendix B: International Organizations and Groups which includes the name, abbreviation, date established, aim, and members by category.**

Internet country code This entry includes the two-letter codes maintained by the International Organization for Standardization (ISO) in the ISO 3166 Alpha-2 list and used by the Internet Assigned Numbers Authority (IANA) to establish country-coded top-level domains (ccTLDs).

Internet hosts This entry lists the number of Internet hosts available within a country. An Internet host is a computer connected directly to the Internet; normally an Internet Service Provider's (ISP) computer is a host. Internet users may use either a hardwired terminal, at an institution with a mainframe computer connected directly to the Internet, or may connect remotely by way of a modem via telephone line, cable, or satellite to the Internet Service Provider's host computer. The number of hosts is one indicator of the extent of Internet connectivity.

Internet users This entry gives the number of users within a country that access the Internet. Statistics vary from country to country and may include users who access the Internet at least several times a week to those who access it only once within a period of several months.

Introduction This category includes one entry, **Background**.

Investment (gross fixed) This entry records total business spending on fixed assets, such as factories, machinery, equipment, dwellings, and inventories of raw materials, which provide the basis for future production. It is measured gross of the depreciation of the assets, i.e., it includes investment that merely replaces worn-out or scrapped capital.

Irrigated land This entry gives the number of square kilometers of land area that is artificially supplied with water.

Judicial branch This entry contains the name(s) of the highest court(s) and a brief description of the selection process for members.

Labor force This entry contains the total labor force figure.

Labor force—by occupation This entry lists the percentage distribution of the labor force by occupation. The distribution will total less than 100 percent if the data are incomplete and may range from 99-101 percent due to rounding.

Land boundaries This entry contains the *total* length of all land boundaries and the individual lengths for each of the contiguous *border countries*. When available, official lengths published by national statistical agencies are used. Because surveying methods may differ, country border lengths reported by contiguous countries may differ.

Land use This entry contains the percentage shares of total land area for three different types of land use: *arable land* - land cultivated for crops like wheat, maize, and rice that are replanted after each harvest; *permanent crops* - land cultivated for crops like citrus, coffee, and rubber that are not replanted after each harvest; includes land under flowering shrubs, fruit trees, nut trees, and vines, but excludes land under trees grown for wood or timber; *other* - any land not arable or under permanent crops; includes permanent meadows and pastures, forests and woodlands, built-on areas, roads, barren land, etc.

Languages This entry provides a rank ordering of languages starting with the largest and sometimes includes the percent of total population speaking that language.

Legal system This entry provides the description of a country's legal system; it also includes information on acceptance of International Court of Justice (ICJ) jurisdiction. The legal systems of nearly all countries are generally modeled upon elements of five main types: civil law (including French law, the Napoleonic Code, Roman law, Roman-Dutch law, and Spanish law); common law (including United State law); customary law; mixed or pluralistic law; and religious law (including Islamic law). An additional type of legal system—international law, which governs the conduct of independent nations in their relationships with one another—is also addressed below. The following list describes these legal systems, the countries or world regions where these systems are enforced, and a brief statement on the origins and major features of each.

Civil Law—The most widespread type of legal system in the world, applied in various forms in approximately 150 countries. Also referred to as European continental law, the civil law system is derived mainly from the Roman *Corpus Juris Civilus*, (Body of Civil Law), a collection of laws and legal interpretations compiled under the East Roman (Byzantine) Emperor Justinian I between A.D. 528 and 565. The major feature of civil law systems is that the laws are organized into systematic written codes. In civil law the sources recognized as authoritative are principally legislation—especially codifications in constitutions or statutes enacted by governments—and secondarily, custom. The civil law systems in some countries are based on more than one code.

Common Law—A type of legal system, often synonymous with "English common law," which is the system of England and Wales in the UK, and is also in force in approximately 80 countries formerly part of or influenced by the former British Empire. English common law reflects Biblical influences as well as remnants of law systems imposed by early conquerors including the Romans, Anglo-Saxons, and Normans. Some legal scholars attribute the formation of the English common law system to King Henry II (r.1154-1189). Until the time of his reign, laws customary among England's various manorial and ecclesiastical (church) jurisdictions were administered locally. Henry II established the king's court and designated that laws were "common" to the entire English realm. The foundation of English common law is "legal precedent"—referred to as *stare decisis*, meaning "to stand by things decided." In the English common law system, court judges are bound in their decisions in large part by the rules and other doctrines developed—and supplemented over time—by the judges of earlier English courts.

Customary Law—A type of legal system that serves as the basis of, or has influenced, the present-day laws in approximately 40 countries—mostly in Africa, but some in the Pacific islands, Europe, and the Near East. Customary law is also referred to as "primitive law," "unwritten law," "indigenous law," and "folk law." There is no single history of customary law such as that found in Roman civil law, English common law, Islamic law, or the Napoleonic Civil Code. The earliest systems of law in human society were customary, and usually developed in small agrarian and hunter-gatherer communities. As the term implies, customary law is based upon the customs of a community. Common attributes of customary legal systems are that they are seldom written down, they embody an organized set of rules regulating social relations, and they are agreed upon by members of the community. Although such law systems include sanctions for law infractions, resolution tends to be reconciliatory rather than punitive. A number of African states practiced customary law many centuries prior to colonial influences. Following colonization, such laws were written down and incorporated to varying extents into the legal systems imposed by their colonial powers.

European Union Law— A sub-discipline of international law known as "supranational law" in which the rights of sovereign nations are limited in relation to one another. Also referred to as the Law of the European Union or Community Law, it is the unique and complex legal system that operates in tandem with the laws of the 27 member states of the European Union (EU). Similar to federal states, the EU legal system ensures compliance from the member states because of the Union's decentralized political nature. The European Court of Justice (ECJ), established in 1952 by the Treaty of Paris, has been largely responsible for the development of EU law. Fundamental principles of European Union law include:*subsidiarity* - the notion that issues be handled by the smallest, lowest, or least centralized competent authority; *proportionality* - the EU may only act to the extent needed to achieve its objectives; *conferral* - the EU is a union of member states, and all its authorities are voluntarily granted by its members; *legal certainty* - requires that legal rules be clear and precise; and *precautionary principle* - a moral and political principle stating that if an action or policy might cause severe or irreversible harm to the public or to the environment, in the absence of a scientific consensus that harm would not ensue, the burden of proof falls on those who would advocate taking the action.

French Law—A type of civil law that is the legal system of France. The French system also serves as the basis for, or is mixed with, other legal systems in approximately 50 countries, notably in North Africa, the Near East, and the French territories and dependencies. French law is primarily codified or systematic written civil law. Prior to the French Revolution (1789-1799), France had no single national legal system. Laws in the northern areas of present-day France were mostly local customs based on privileges and exemptions granted by kings and feudal lords, while in the southern areas Roman law predominated. The introduction of the Napoleonic Civil Code during the reign of Napoleon I in the first decade of the 19th century brought major reforms to the French legal system, many of which remain part of France's current legal structure, though all have been extensively amended or redrafted to address a modern nation. French law distinguishes between "public law" and "private law." Public law relates to government, the French Constitution, public administration, and criminal law. Private law covers issues between private citizens or corporations. The most recent changes to the French legal system—introduced in the 1980s—were the decentralization laws, which transferred authority from centrally appointed government representatives to locally elected representatives of the people.

International Law—The law of the international community, or the body of customary rules and treaty rules accepted as legally binding by states in their relations with each other. International law differs from other legal systems in that it primarily concerns sovereign political entities. There are three separate disciplines of international law: public international law, which governs the relationship between provinces and international entities and includes treaty law, law of the sea, international criminal law, and international humanitarian law; private international law, which addresses legal jurisdiction; and supranational law—a legal framework wherein countries are bound by regional agreements in which the laws of the member countries are held inapplicable when in conflict with supranational laws. At present the European Union is the only entity under a supranational legal system. The term "international law" was coined by Jeremy Bentham in 1780 in his*Principles of Morals and Legislation*, though laws governing relations between states have been recognized from very early times (many centuries B.C.). Modern international law developed alongside the emergence and growth of the European nation-states beginning in the early 16th century. Other factors that influenced the development of international law included the revival of legal studies, the growth of international trade, and the practice of exchanging emissaries and establishing legations. The sources of International law are set out in Article 38-1 of the Statute of the International Court of Justice within the UN Charter.

Islamic Law—The most widespread type of religious law, it is the legal system enforced in over 30 countries, particularly in the Near East, but also in Central and South Asia, Africa, and Indonesia. In many countries Islamic law operates in tandem with a civil law system. Islamic law is embodied in the sharia, an Arabic word meaning "the right path." Sharia covers all aspects of public and private life and organizes them into five categories: obligatory, recommended, permitted, disliked, and forbidden. The primary sources of sharia law are the Qur'an, believed by Muslims to be the word of God revealed to the Prophet Muhammad by the angel Gabriel, and the Sunnah, the teachings of the Prophet and his works. In addition to these two primary sources, traditional Sunni Muslims recognize the consensus of Muhammad's companions and Islamic jurists on certain issues, called ijmas, and various forms of reasoning, including analogy by legal scholars, referred to as qiyas. Shia Muslims reject ijmas and qiyas as sources of sharia law.

Mixed Law—Also referred to as pluralistic law, mixed law consists of elements of some or all of the other main types of legal systems—civil, common, customary, and religious. The mixed legal systems of a number of countries came about when colonial powers overlaid their own legal systems upon colonized regions but retained elements of the colonies' existing legal systems.

Napoleonic Civil Code—A type of civil law, referred to as the Civil Code or *Code Civil des Francais*, forms part of the legal system of France, and underpins the legal systems of Bolivia, Egypt, Lebanon, Poland, and the US state of Louisiana. The Civil Code was established under Napoleon I, enacted in 1804, and officially designated the *Code Napoleon*in 1807. This legal system combined the Teutonic civil law tradition of the northern provinces of France with the Roman law tradition of the southern and eastern regions of the country. The Civil Code bears similarities in its arrangement to the Roman *Body of Civil Law* (see Civil Law above). As enacted in 1804, the Code addressed personal status, property, and the acquisition of property. Codes added over the following six years included civil procedures, commercial law, criminal law and procedures, and a penal code.

Religious Law—A legal system which stems from the sacred texts of religious traditions and in most cases professes to cover all aspects of life as a seamless part of devotional obligations to a transcendent, imminent, or deep philosophical reality. Implied as the basis of religious law is the concept of unalterability, because the word of God cannot be amended or legislated against by judges or governments. However, a detailed legal system generally requires human elaboration. The main types of religious law are sharia in Islam, halakha in Judaism, and canon law in some Christian groups. Sharia is the most widespread religious legal system (see Islamic Law), and is the sole system of law for countries including Iran, the Maldives, and Saudi Arabia. No country is fully governed by halakha, but Jewish people may decide to settle disputes through Jewish courts and be bound by their rulings. Canon law is not a divine law as such because it is not found in revelation. It is viewed instead as human law inspired by the word of God and applying the demands of that revelation to the actual situation of the church. Canon law regulates the internal ordering of the Roman Catholic Church, the Eastern Orthodox Church, and the Anglican Communion.

Roman Law–A type of civil law developed in ancient Rome and practiced from the time of the city's founding (traditionally 753 B.C.) until the fall of the Western Empire in the 5th century A.D. Roman law remained the legal system of the Byzantine (Eastern Empire) until the fall of Constantinople in 1453. Preserved fragments of the first legal text, known as the Law of the Twelve Tables, dating from the 5th century B.C., contained specific provisions designed to change the prevailing customary law. Early Roman law was drawn from custom and statutes; later, during the time of the empire, emperors asserted their authority as the ultimate source of law. The basis for Roman laws was the idea that the exact form–not the intention–of words or of actions produced legal consequences. It was only in the late 6th century A.D. that a comprehensive Roman code of laws was published (see Civil Law above). Roman law served as the basis of law systems developed in a number of continental European countries.

Roman-Dutch Law–A type of civil law based on Roman law as applied in the Netherlands. Roman-Dutch law serves as the basis for legal systems in seven African countries, as well as Guyana, Indonesia, and Sri Lanka. This law system, which originated in the province of Holland and expanded throughout the Netherlands (to be replaced by the French Civil Code in 1809), was instituted in a number of sub-Saharan African countries during the Dutch colonial period. The Dutch jurist/philosopher Hugo Grotius was the first to attempt to reduce Roman-Dutch civil law into a system in his *Jurisprudence of Holland* (written 1619-20, commentary published 1621). The Dutch historian/lawyer Simon van Leeuwen coined the term "Roman-Dutch law" in 1652.

Spanish Law–A type of civil law, often referred to as the Spanish Civil Code, it is the present legal system of Spain and is the basis of legal systems in 12 countries mostly in Central and South America, but also in southwestern Europe, northern and western Africa, and southeastern Asia. The Spanish Civil Code reflects a complex mixture of customary, Roman, Napoleonic, local, and modern codified law. The laws of the Visigoth invaders of Spain in the 5th to 7th centuries had the earliest major influence on Spanish legal system development. The Christian Reconquest of Spain in the 11th through 15th centuries witnessed the development of customary law, which combined canon (religious) and Roman law. During several centuries of Hapsburg and Bourbon rule, systematic recompilations of the existing national legal system were attempted, but these often conflicted with local and regional customary civil laws. Legal system development for most of the 19th century concentrated on formulating a national civil law system, which was finally enacted in 1889 as the Spanish Civil Code. Several sections of the code have been revised, the most recent of which are the penal code in 1989 and the judiciary code in 2001. The Spanish Civil Code separates public and private law. Public law includes constitutional law, administrative law, criminal law, process law, financial and tax law, and international public law. Private law includes civil law, commercial law, labor law, and international private law.

United States Law–A type of common law, which is the basis of the legal system of the United States and that of its island possessions in the Caribbean and the Pacific. This legal system has several layers, more possibly than in most other countries, and is due in part to the division between federal and state law. The United States was founded not as one nation but as a union of 13 colonies, each claiming independence from the British Crown. The US Constitution, implemented in 1789, began shifting power away from the states and toward the federal government, though the states today retain substantial legal authority. US law draws its authority from four sources: *constitutional law, statutory law, administrative regulations*, and *case law*. Constitutional law is based on the US Constitution and serves as the supreme federal law. Taken together with those of the state constitutions, these documents outline the general structure of the federal and state governments and provide the rules and limits of power. US statutory law is legislation enacted by the US Congress and is codified in the United States Code. The 50 state legislatures have similar authority to enact state statutes. Administrative law is the authority delegated to federal and state executive agencies. Case law, also referred to as common law, covers areas where constitutional or statutory law is lacking. Case law is a collection of judicial decisions, customs, and general principles that began in England centuries ago, that were adopted in America at the time of the Revolution, and that continue to develop today.

Legislative branch This entry contains information on the structure (unicameral, bicameral, tricameral), formal name, number of seats, and term of office. *Elections* includes the nature of the election process or accession to power, date of the last election, and date of the next election.*Election results* includes the percent of vote and/or number of seats held by each party in the last election.

Life expectancy at birth This entry contains the average number of years to be lived by a group of people born in the same year, if mortality at each age remains constant in the future. The entry includes *total population* as well as the *male* and *female* components. Life expectancy at birth is also a measure of overall quality of life in a country and summarizes the mortality at all ages. It can also be thought of as indicating the potential return on investment in human capital and is necessary for the calculation of various actuarial measures.

Literacy This entry includes a *definition* of literacy and Census Bureau percentages for the *total population, males*, and *females*. There are no universal definitions and standards of literacy. Unless otherwise specified, all rates are based on the most common definition–the ability to read and write at a specified age. Detailing the standards that individual countries use to assess the ability to read and write is beyond the scope of the *Factbook*. Information on literacy, while not a perfect measure of educational results, is probably the most easily available and valid for international comparisons. Low levels of literacy, and education in general, can impede the economic development of a country in the current rapidly changing, technology-driven world.

Location This entry identifies the country's regional location, neighboring countries, and adjacent bodies of water.

Major cities—population This entry provides the population of the capital and up to four major cities defined as urban agglomerations with populations of at least 750,000 people. An *urban agglomeration* is defined as comprising the city or town proper and also the suburban fringe or thickly settled territory lying outside of, but adjacent to, the boundaries of the city. For smaller countries, lacking urban centers of 750,000 or more, only the population of the capital is presented.

Major infectious diseases This entry lists major infectious diseases likely to be encountered in countries where the risk of such diseases is assessed to be very high as compared to the United States. These infectious diseases represent risks to US government personnel traveling to the specified country for a period of less than three years. The **degree of risk** is assessed by considering the foreign nature of these infectious diseases, their severity, and the probability of being affected by the diseases present. The diseases listed do not necessarily represent the total disease burden experienced by the local population. The risk to an individual traveler varies considerably by the specific location, visit duration, type of activities, type of accommodations, time of year, and other factors. Consultation with a travel medicine physician is needed to evaluate individual risk and recommend appropriate preventive measures such as vaccines. Diseases are organized into the following six exposure categories shown in italics *and listed in typical descending order of risk*. Note: The sequence of exposure categories listed in individual country entries may vary according to local conditions.

food or waterborne diseases acquired through eating or drinking on the local economy:

Hepatitis A–viral disease that interferes with the functioning of the liver; spread through consumption of food or water contaminated with fecal matter, principally in areas of poor sanitation; victims exhibit fever, jaundice, and diarrhea; 15% of victims will experience prolonged symptoms over 6-9 months; vaccine available

Hepatitis E–water-borne viral disease that interferes with the functioning of the liver; most commonly spread through fecal contamination of drinking water; victims exhibit jaundice, fatigue, abdominal pain, and dark colored urine.

Typhoid fever–bacterial disease spread through contact with food or water contaminated by fecal matter or sewage; victims exhibit sustained high fevers; left untreated, mortality rates can reach 20%.

vectorborne diseases acquired through the bite of an infected arthropod:

Malaria–caused by single-cell parasitic protozoa *Plasmodium*; transmitted to humans via the bite of the female Anopheles mosquito; parasites multiply in the liver attacking red blood cells resulting in cycles of fever, chills, and sweats accompanied by anemia; death due to damage to vital organs and interruption of blood supply to the brain; endemic in 100, mostly tropical, countries with 90% of cases and the majority of 1.5-2.5 million estimated annual deaths occurring in sub-Saharan Africa.

Dengue fever–mosquito-borne (*Aedes aegypti*) viral disease associated with urban environments; manifests as sudden onset of fever and severe headache; occasionally produces shock and hemorrhage leading to death in 5% of cases.

Yellow fever–mosquito-borne viral disease; severity ranges from influenza-like symptoms to severe hepatitis and hemorrhagic fever; occurs only in tropical South America and sub-Saharan Africa, where most cases are reported; fatality rate is less than 20%.

Japanese Encephalitis–mosquito-borne (*Culex tritaeniorhynchus*) viral disease associated with rural areas in Asia; acute encephalitis can progress to paralysis, coma, and death; fatality rates 30%.

African Trypanosomiasis–caused by the parasitic protozoa *Trypanosoma*; transmitted to humans via the bite of bloodsucking Tsetse flies; infection leads to malaise and irregular fevers and, in advanced cases when the parasites invade the central nervous system, coma and death; endemic in 36 countries of sub-Saharan Africa; cattle and wild animals act as reservoir hosts for the parasites.

Cutaneous Leishmaniasis–caused by the parasitic protozoa *leishmania*; transmitted to humans via the bite of sandflies; results in skin lesions that may become chronic; endemic in 88 countries; 90% of cases occur in Iran, Afghanistan, Syria, Saudi Arabia, Brazil, and Peru; wild and domesticated animals as well as humans can act as reservoirs of infection.

Plague–bacterial disease transmitted by fleas normally associated with rats; person-to-person airborne transmission also possible; recent plague epidemics occurred in areas of Asia, Africa, and South America associated with rural areas or small towns and villages; manifests as fever, headache, and painfully swollen lymph nodes; disease progresses rapidly and without antibiotic treatment leads to pneumonic form with a death rate in excess of 50%.

Crimean-Congo hemorrhagic fever–tick-borne viral disease; infection may also result from exposure to infected animal blood or tissue; geographic distribution includes Africa, Asia, the Middle East, and Eastern Europe; sudden onset of fever, headache, and muscle aches followed by hemorrhaging in the bowels, urine, nose, and gums; mortality rate is approximately 30%.

Rift Valley fever–viral disease affecting domesticated animals and humans; transmission is by mosquito and other biting insects; infection may also occur through handling of infected meat or contact with blood; geographic distribution includes eastern and southern Africa where cattle and sheep are raised; symptoms are generally mild with fever and some liver abnormalities, but the disease may progress to hemorrhagic fever, encephalitis, or ocular disease; fatality rates are low at about 1% of cases.

Chikungunya–mosquito-borne (*Aedes aegypti*) viral disease associated with urban environments, similar to Dengue Fever; characterized by sudden onset of fever, rash, and severe joint pain usually lasting 3-7 days, some cases result in persistent arthritis.

water contact diseases acquired through swimming or wading in freshwater lakes, streams, and rivers:

Leptospirosis–bacterial disease that affects animals and humans; infection occurs through contact with water, food, or soil contaminated by animal urine; symptoms include high fever, severe headache, vomiting, jaundice, and diarrhea; untreated, the disease can result in kidney damage, liver failure, meningitis, or respiratory distress; fatality rates are low but left untreated recovery can take months.

Schistosomiasis–caused by parasitic trematode flatworm *Schistosoma*; fresh water snails act as intermediate host and release larval form of parasite that penetrates the skin of people exposed to contaminated water; worms mature and reproduce in the blood vessels, liver, kidneys, and intestines releasing eggs, which become trapped in tissues triggering an immune response; may manifest as either urinary or intestinal disease resulting in decreased work or learning capacity; mortality, while generally low, may occur in advanced cases usually due to bladder cancer; endemic in 74 developing countries with 80% of infected people living in sub-Saharan Africa; humans act as the reservoir for this parasite.

aerosolized dust or soil contact disease acquired through inhalation of aerosols contaminated with rodent urine:

Lassa fever–viral disease carried by rats of the genus *Mastomys*; endemic in portions of West Africa; infection occurs through direct contact with or consumption of food contaminated by rodent urine or fecal matter containing virus particles; fatality rate can reach 50% in epidemic outbreaks.

respiratory disease acquired through close contact with an infectious person:

Meningococcal meningitis–bacterial disease causing an inflammation of the lining of the brain and spinal cord; one of the most important bacterial pathogens is *Neisseria meningitidis* because of its potential to cause epidemics; symptoms include stiff neck, high fever, headaches, and vomiting; bacteria are transmitted from person to person by respiratory droplets and facilitated by close and prolonged contact resulting from crowded living conditions, often with a seasonal distribution; death occurs in 5-15% of cases, typically within 24-48 hours of onset of symptoms; highest burden of meningococcal disease occurs in the hyperendemic region of sub-Saharan Africa known as the "Meningitis Belt" which stretches from Senegal east to Ethiopia.

animal contact disease acquired through direct contact with local animals:

Rabies–viral disease of mammals usually transmitted through the bite of an infected animal, most commonly dogs; virus affects the central nervous system causing brain alteration and death; symptoms initially are non-specific fever and headache progressing to neurological symptoms; death occurs within days of the onset of symptoms.

Manpower available for military service This entry gives the number of males and females falling in the military age range for a country (defined as being ages 16-49) and assumes that every individual is fit to serve.

Manpower fit for military service This entry gives the number of males and females falling in the military age range for a country (defined as being ages 16-49) and who are not otherwise disqualified for health reasons; accounts for the health situation in the country and provides a more realistic estimate of the actual number fit to serve.

Manpower reaching militarily significant age annually This entry gives the number of males and females entering the military manpower pool (i.e., reaching age 16) in any given year and is a measure of the availability of military-age young adults.

Map references This entry includes the name of the *Factbook* reference map on which a country may be found. Note that boundary representations on these maps are not necessarily authoritative. The entry on **Geographic coordinates** may be helpful in finding some smaller countries.

Maritime claims This entry includes the following claims, the definitions of which are excerpted from the United Nations Convention on the Law of the Sea (UNCLOS), which alone contains the full and definitive descriptions:

territorial sea–the sovereignty of a coastal state extends beyond its land territory and internal waters to an adjacent belt of sea, described as the territorial sea in the UNCLOS (Part II); this sovereignty extends to the air space over the territorial sea as well as its underlying seabed and subsoil; every state has the right to establish the breadth of its territorial sea up to a limit not exceeding 12 nautical miles; the normal baseline for measuring the breadth of the territorial sea is the mean low-water line along the coast as marked on large-scale charts officially recognized by the coastal state; where the coasts of two states are opposite or adjacent to each other, neither state is entitled to extend its territorial sea beyond the median line, every point of which is equidistant from the nearest points on the baseline from which the territorial seas of both states are measured; the UNCLOS describes specific rules for archipelagic states.

contiguous zone–according to the UNCLOS (Article 33), this is a zone contiguous to a coastal state's territorial sea, over which it may exercise the control necessary to: prevent infringement of its customs, fiscal, immigration, or sanitary laws and regulations within its territory or territorial sea; punish infringement of the above laws and regulations committed within its territory or territorial sea; the contiguous zone may not extend beyond 24 nautical miles from the baselines from which the breadth of the territorial sea is measured (e.g., the US has claimed a 12-nautical mile contiguous zone in addition to its 12-nautical mile territorial sea); where the coasts of two states are opposite or adjacent to each other, neither state is entitled to extend its contiguous zone beyond the median line, every point of which is equidistant from the nearest points on the baseline from which the contiguous zone of both states are measured.

exclusive economic zone (EEZ)–the UNCLOS (Part V) defines the EEZ as a zone beyond and adjacent to the territorial sea in which a coastal state has: sovereign rights for the purpose of exploring and exploiting, conserving and managing the natural resources, whether living or non-living, of the waters superjacent to the seabed and of the seabed and its subsoil, and with regard to other activities for the economic exploitation and exploration of the zone, such as the production of energy from the water, currents, and winds; jurisdiction with regard to the establishment and use of artificial islands, installations, and structures; marine scientific research; the protection and preservation of the marine environment; the outer limit of the exclusive economic zone shall not exceed 200 nautical miles from the baselines from which the breadth of the territorial sea is measured.

continental shelf–the UNCLOS (Article 76) defines the continental shelf of a coastal state as comprising the seabed and subsoil of the submarine areas that extend beyond its territorial sea throughout the natural prolongation of its land territory to the outer edge of the continental margin, or to a distance of 200 nautical miles from the baselines from which the breadth of the territorial sea is measured where the outer edge of the continental margin does not extend up to that distance; the continental margin comprises the submerged prolongation of the landmass of the coastal state, and consists of the seabed and subsoil of the shelf, the slope and the rise; wherever the continental margin extends beyond 200 nautical miles from the baseline, coastal states may extend their claim to a distance not to exceed 350 nautical miles from the baseline or 100 nautical miles from the 2,500-meter isobath, which is a line connecting points of 2,500 meters in depth; it does not include the deep ocean floor with its oceanic ridges or the subsoil thereof.

exclusive fishing zone–while this term is not used in the UNCLOS, some states (e.g., the United Kingdom) have chosen not to claim an EEZ, but rather to claim jurisdiction over the living resources off their coast; in such cases, the term exclusive fishing zone is often used; the breadth of this zone is normally the same as the EEZ or 200 nautical miles.

Market value of publicly traded shares This entry gives the value of shares issued by publicly traded companies at a price determined in the national stock markets on the final day of the period indicated. It is simply the latest price per share multiplied by the total number of outstanding shares, cumulated over all companies listed on the particular exchange.

Median age This entry is the age that divides a population into two numerically equal groups; that is, half the people are younger than this age and half are older. It is a single index that summarizes the age distribution of a population. Currently, the median age ranges from a low of about 15 in Uganda and Gaza Strip to 40 or more in several European countries and Japan. See the entry for "Age structure" for the importance of a young versus an older age structure and, by implication, a low versus a higher median age.

Merchant marine Merchant marine may be defined as all ships engaged in the carriage of goods; or all commercial vessels (as opposed to all nonmilitary ships), which excludes tugs, fishing vessels, offshore oil rigs, etc. This entry contains information in four fields - *total*, ships *by type*, *foreign-owned*, and *registered in other countries*.

Total includes the number of ships (1,000 GRT or over), total DWT for those ships, and total GRT for those ships. DWT or dead weight tonnage is the total weight of cargo, plus bunkers, stores, etc., that a ship can carry when immersed to the appropriate load line. GRT or gross register tonnage is a figure obtained by measuring the entire sheltered volume of a ship available for cargo and passengers and converting it to tons on the basis of 100 cubic feet per ton; there is no stable relationship between GRT and DWT.

Ships *by type* includes a listing of barge carriers, bulk cargo ships, cargo ships, chemical tankers, combination bulk carriers, combination ore/oil carriers, container ships, liquefied gas tankers, livestock carriers, multifunctional large-load carriers, petroleum tankers, passenger ships, passenger/cargo ships, railcar carriers, refrigerated cargo ships, roll-on/roll-off cargo ships, short-sea passenger ships, specialized tankers, and vehicle carriers.

Foreign-owned are ships that fly the flag of one country but belong to owners in another.

Registered in other countries are ships that belong to owners in one country but fly the flag of another.

Military This category includes the entries dealing with a country's military structure, manpower, and expenditures.

Military—note This entry includes miscellaneous military information of significance not included elsewhere.

Military branches This entry lists the service branches subordinate to defense ministries or the equivalent (typically ground, naval, air, and marine forces).

Military expenditures This entry gives spending on defense programs for the most recent year available as a percent of gross domestic product (GDP); the GDP is calculated on an exchange rate basis, i.e., not in terms of purchasing power parity (PPP).

Military service age and obligation This entry gives the required ages for voluntary or conscript military service and the length of service obligation.

Money figures All money figures are expressed in contemporaneous US dollars unless otherwise indicated.

National anthem A generally patriotic musical composition–usually in the form of a song or hymn of praise–that evokes and eulogizes the history, traditions, or struggles of a nation or its people. National anthems can be officially recognized as a national song by a country's constitution or by an enacted law, or simply by tradition. Although most anthems contain lyrics, some do not.

National holiday This entry gives the primary national day of celebration–usually independence day.

Nationality This entry provides the identifying terms for citizens - *noun* and *adjective*.

Natural gas—consumption This entry is the total natural gas consumed in cubic meters (cu m). The discrepancy between the amount of natural gas produced and/or imported and the amount consumed and/or exported is due to the omission of stock changes and other complicating factors.

Natural gas—exports This entry is the total natural gas exported in cubic meters (cu m).

Natural gas—imports This entry is the total natural gas imported in cubic meters (cu m).

Natural gas—production This entry is the total natural gas produced in cubic meters (cu m). The discrepancy between the amount of natural gas produced and/or imported and the amount consumed and/or exported is due to the omission of stock changes and other complicating factors.

Natural gas—proved reserves This entry is the stock of proved reserves of natural gas in cubic meters (cu m). Proved reserves are those quantities of natural gas, which, by analysis of geological and engineering data, can be estimated with a high degree of confidence to be commercially recoverable from a given date forward, from known reservoirs and under current economic conditions.

Natural hazards This entry lists potential natural disasters. For countries where volcanic activity is common, a *volcanism* subfield highlights historically active volcanoes.

Natural resources This entry lists a country's mineral, petroleum, hydropower, and other resources of commercial importance, such as rare earth elements (REEs).

Net migration rate This entry includes the figure for the difference between the number of persons entering and leaving a country during the year per 1,000 persons (based on midyear population). An excess of persons entering the country is referred to as net immigration (e.g., 3.56 migrants/1,000 population); an excess of persons leaving the country as net emigration (e.g., -9.26 migrants/1,000 population). The net migration rate indicates the contribution of migration to the overall level of population change. The net migration rate does not distinguish between economic migrants, refugees, and other types of migrants nor does it distinguish between lawful migrants and undocumented migrants.

Oil—consumption This entry is the total oil consumed in barrels per day (bbl/day). The discrepancy between the amount of oil produced and/or imported and the amount consumed and/or exported is due to the omission of stock changes, refinery gains, and other complicating factors.

Oil—exports This entry is the total oil exported in barrels per day (bbl/day), including both crude oil and oil products.

Oil—imports This entry is the total oil imported in barrels per day (bbl/day), including both crude oil and oil products.

Oil—production This entry is the total oil produced in barrels per day (bbl/day). The discrepancy between the amount of oil produced and/or imported and the amount consumed and/or exported is due to the omission of stock changes, refinery gains, and other complicating factors.

Oil—proved reserves This entry is the stock of proved reserves of crude oil in barrels (bbl). Proved reserves are those quantities of petroleum which, by analysis of geological and engineering data, can be estimated with a high degree of confidence to be commercially recoverable from a given date forward, from known reservoirs and under current economic conditions.

People This category includes the entries dealing with the characteristics of the people and their society.

People—note This entry includes miscellaneous demographic information of significance not included elsewhere.

Personal Names—Capitalization The *Factbook* capitalizes the surname or family name of individuals for the convenience of our users who are faced with a world of different cultures and naming conventions. The need for capitalization, bold type, underlining, italics, or some other indicator of the individual's surname is apparent in the following examples: MAO Zedong, Fidel CASTRO Ruz, George W. BUSH, and TUNKU SALAHUDDIN Abdul Aziz Shah ibni Al-Marhum Sultan Hisammuddin Alam Shah. By knowing the surname, a short form without all capital letters can be used with confidence as in President Castro, Chairman Mao, President Bush, or Sultan Tunku Salahuddin. The same system of capitalization is extended to the names of leaders with surnames that are not commonly used such as Queen ELIZABETH II. For Vietnamese names, the given name is capitalized because officials are referred to by their given name rather than by their surname. For example, the president of Vietnam is Tran Duc LUONG. His surname is Tran, but he is referred to by his given name–President LUONG.

Personal Names—Spelling The romanization of personal names in the *Factbook* normally follows the same transliteration system used by the US Board on Geographic Names for spelling place names. At times, however, a foreign leader expressly indicates a preference for, or the media or official documents regularly use, a romanized spelling that differs from the transliteration derived from the US Government standard. In such cases, the *Factbook* uses the alternative spelling.

Personal Names—Titles The *Factbook* capitalizes any valid title (or short form of it) immediately preceding a person's name. A title standing alone is not capitalized. Examples: President PUTIN and President BUSH are chiefs of state. In Russia, the president is chief of state and the premier is the head of the government, while in the US, the president is both chief of state and head of government.

Petroleum See entries under **Oil**.

Petroleum products See entries under **Oil**.

Pipelines This entry gives the lengths and types of pipelines for transporting products like natural gas, crude oil, or petroleum products.

Piracy Piracy is defined by the 1982 United Nations Convention on the Law of the Sea as any illegal act of violence, detention, or depredation directed against a ship, aircraft, persons, or property in a place outside the jurisdiction of any State. Such criminal acts committed in the territorial waters of a littoral state are generally considered to be armed robbery against ships. Information on piracy may be found, where applicable, in the **Transportation–note**.

Political parties and leaders This entry includes a listing of significant political organizations and their leaders.

Political pressure groups and leaders This entry includes a listing of a country's political, social, labor, or religious organizations that are involved in politics, or that exert political pressure, but whose leaders do not stand for legislative election. International movements or organizations are generally not listed.

Population This entry gives an estimate from the US Bureau of the Census based on statistics from population censuses, vital statistics registration systems, or sample surveys pertaining to the recent past and on assumptions about future trends. The total population presents one overall measure of the potential impact of the country on the world and within its region. Note: Starting with the 1993 *Factbook*, demographic estimates for some countries (mostly African) have explicitly taken into account the effects of the growing impact of the HIV/AIDS epidemic. These countries are currently: The Bahamas, Benin, Botswana, Brazil, Burkina Faso, Burma, Burundi, Cambodia, Cameroon, Central African Republic, Democratic Republic of the Congo, Republic of the Congo, Cote d'Ivoire, Ethiopia, Gabon, Ghana, Guyana, Haiti, Honduras, Kenya, Lesotho, Malawi, Mozambique, Namibia, Nigeria, Rwanda, South Africa, Swaziland, Tanzania, Thailand, Togo, Uganda, Zambia, and Zimbabwe.

Population below poverty line National estimates of the percentage of the population falling below the poverty line are based on surveys of sub-groups, with the results weighted by the number of people in each group. Definitions of poverty vary considerably among nations. For example, rich nations generally employ more generous standards of poverty than poor nations.

Population growth rate The average annual percent change in the population, resulting from a surplus (or deficit) of births over deaths and the balance of migrants entering and leaving a country. The rate may be positive or negative. The growth rate is a factor in determining how great a burden would be imposed on a country by the changing needs of its people for infrastructure (e.g., schools, hospitals, housing, roads), resources (e.g., food, water, electricity), and jobs. Rapid population growth can be seen as threatening by neighboring countries.

Ports and terminals This entry lists major ports and terminals primarily on the basis of the amount of cargo tonnage shipped through the facilities on an annual basis. In some instances, the number of containers handled or ship visits were also considered.

Public debt This entry records the cumulative total of all government borrowings less repayments that are denominated in a country's home currency. Public debt should not be confused with external debt, which reflects the foreign currency liabilities of both the private and public sector and must be financed out of foreign exchange earnings.

Railways This entry states the *total* route length of the railway network and of its component parts by gauge, which is the measure of the distance between the inner sides of the load-bearing rails. The four typical types of gauges are: *broad*, *standard*, *narrow*, and *dual*. Other gauges are listed under *note*. Some 60% of the world's railways use the standard gauge of 1.4 m (4.7 ft). Gauges vary by country and sometimes within countries. The choice of gauge during initial construction was mainly in response to local conditions and the intent of the builder. Narrow-gauge railways were cheaper to build and could negotiate sharper curves, broad-gauge railways gave greater stability and permitted higher speeds. Standard-gauge railways were a compromise between narrow and broad gauges.

Rare earth elements Rare earth elements or REEs are 17 chemical elements that are critical in many of today's high-tech industries. They include lanthanum, cerium, praseodymium, neodymium, promethium, samarium, europium, gadolinium, terbium, dysprosium, holmium, erbium, thulium, ytterbium, lutetium, scandium, and yttrium. Typical applications for REEs include batteries in hybrid cars, fiber optic cables, flat panel displays, and permanent magnets, as well as some defense and medical products.

Reference maps This section includes world and regional maps.

Refugees and internally displaced persons This entry includes those persons residing in a country as *refugees* or internally displaced persons (*IDPs*). The definition of a refugee according to a United Nations Convention is "a person who is outside his/her country of nationality or habitual residence; has a well-founded fear of persecution because of his/her race, religion, nationality, membership in a particular social group or political opinion; and is unable or unwilling to avail himself/herself of the protection of that country, or to return there, for fear of persecution." The UN established the Office of the UN High Commissioner for Refugees (UNHCR) in 1950 to handle refugee matters worldwide. The UN Relief and Works Agency for Palestine Refugees in the Near East (UNRWA) has a different operational definition for a Palestinian refugee: "a person whose normal place of residence was Palestine during the period 1 June 1946 to 15 May 1948 and who lost both home and means of livelihood as a result of the 1948 conflict." However, UNHCR also assists some 400,000 Palestinian refugees not covered under the UNRWA definition. The term "internally displaced person" is not specifically covered in the UN Convention; it is used to describe people who have fled their homes for reasons similar to refugees, but who remain within their own national territory and are subject to the laws of that state.

Religions This entry is an ordered listing of religions by adherents starting with the largest group and sometimes includes the percent of total population. The core characteristics and beliefs of the world's major religions are described below.

Baha'i—Founded by Mirza Husayn-Ali (known as Baha'u'llah) in Iran in 1852, Baha'i faith emphasizes monotheism and believes in one eternal transcendent God. Its guiding focus is to encourage the unity of all peoples on the earth so that justice and peace may be achieved on earth. Baha'i revelation contends the prophets of major world religions reflect some truth or element of the divine, believes all were manifestations of God given to specific communities in specific times, and that Baha'u'llah is an additional prophet meant to call all humankind. Bahais are an open community, located worldwide, with the greatest concentration of believers in South Asia.

Buddhism—Religion or philosophy inspired by the 5th century B.C. teachings of Siddhartha Gautama (also known as Gautama Buddha "the enlightened one"). Buddhism focuses on the goal of spiritual enlightenment centered on an understanding of Gautama Buddha's Four Noble Truths on the nature of suffering, and on the Eightfold Path of spiritual and moral practice, to break the cycle of suffering of which we are a part. Buddhism ascribes to a karmic system of rebirth. Several schools and sects of Buddhism exist, differing often on the nature of the Buddha, the extent to which enlightenment can be achieved—for one or for all, and by whom—religious orders or laity.

Basic Groupings Theravada Buddhism: The oldest Buddhist school, Theravada is practiced mostly in Sri Lanka, Cambodia, Laos, Burma, and Thailand, with minority representation elsewhere in Asia and the West. Theravadans follow the Pali Canon of Buddha's teachings, and believe that one may escape the cycle of rebirth, worldly attachment, and suffering for oneself; this process may take one or several lifetimes.

Mahayana Buddhism, including subsets Zen and Tibetan Buddhism: Forms of Mahayana Buddhism are common in East Asia and Tibet, and parts of the West. Mahayanas have additional scriptures beyond the Pali Canon and believe the Buddha is eternal and still teaching. Unlike Theravada Buddhism, Mahayana schools maintain the Buddha-nature is present in all beings and all will ultimately achieve enlightenment.

Christianity—Descending from Judaism, Christianity's central belief maintains Jesus of Nazareth is the promised messiah of the Hebrew Scriptures, and that his life, death, and resurrection are salvific for the world. Christianity is one of the three monotheistic Abrahamic faiths, along with Islam and Judaism, which traces its spiritual lineage to Abraham of the Hebrew Scriptures. Its sacred texts include the Hebrew Bible and the New Testament (or the Christian Gospels).

Basic Groupings Catholicism (or Roman Catholicism): This is the oldest established western Christian church and the world's largest single religious body. It is supranational, and recognizes a hierarchical structure with the Pope, or Bishop of Rome, as its head, located at the Vatican. Catholics believe the Pope is the divinely ordered head of the Church from a direct spiritual legacy of Jesus' apostle Peter. Catholicism is comprised of 23 particular Churches, or Rites—one Western (Roman or Latin-Rite) and 22 Eastern. The Latin Rite is by far the largest, making up about 98% of Catholic membership. Eastern-Rite Churches, such as the Maronite Church and the Ukrainian Catholic Church, are in communion with Rome although they preserve their own worship traditions and their immediate hierarchy consists of clergy within their own rite. The Catholic Church has a comprehensive theological and moral doctrine specified for believers in its catechism, which makes it unique among most forms of Christianity. Mormonism (including the Church of Jesus Christ of Latter-Day Saints): Originating in 1830 in the United States under Joseph Smith, Mormonism is not characterized as a form of Protestant Christianity because it claims additional revealed Christian scriptures after the Hebrew Bible and New Testament. The Book of Mormon maintains there was an appearance of Jesus in the New World following the Christian account of his resurrection, and that the Americas are uniquely blessed continents. Mormonism believes earlier Christian traditions, such as the Roman Catholic, Orthodox, and Protestant reform faiths, are apostasies and that Joseph Smith's revelation of the Book of Mormon is a restoration of true Christianity. Mormons have a hierarchical religious leadership structure, and actively proselytize their faith; they are located primarily in the Americas and in a number of other Western countries. Orthodox Christianity: The oldest established eastern form of Christianity, the Holy Orthodox Church, has a ceremonial head in the Bishop of Constantinople (Istanbul), also known as a Patriarch, but its various regional forms (e.g., Greek Orthodox, Russian Orthodox, Serbian Orthodox, Ukrainian Orthodox) are autocephalous (independent of Constantinople's authority, and have their own Patriarchs). Orthodox churches are highly nationalist and ethnic. The Orthodox Christian faith shares many theological tenets with the Roman Catholic Church, but diverges on some key premises and does not recognize the governing authority of the Pope.

Protestant Christianity: Protestant Christianity originated in the 16th century as an attempt to reform Roman Catholicism's practices, dogma, and theology. It encompasses several forms or denominations which are extremely varied in structure, beliefs, relationship to state, clergy, and governance. Many protestant theologies emphasize the primary role of scripture in their faith, advocating individual interpretation of Christian texts without the mediation of a final religious authority such as the Roman Pope. The oldest Protestant Christianities include Lutheranism, Calvinism (Presbyterians), and Anglican Christianity (Episcopalians), which have established liturgies, governing structure, and formal clergy. Other variants on Protestant Christianity, including Pentecostal movements and independent churches, may lack one or more of these elements, and their leadership and beliefs are individualized and dynamic.

Hinduism—Originating in the Vedic civilization of India (second and first millennium B.C.), Hinduism is an extremely diverse set of beliefs and practices with no single founder or religious authority. Hinduism has many scriptures; the Vedas, the Upanishads, and the Bhagavad-Gita are among some of the most important. Hindus may worship one or many deities, usually with prayer rituals within their own home. The most common figures of devotion are the gods Vishnu, Shiva, and a mother goddess, Devi. Most Hindus believe the soul, or *atman*, is eternal, and goes through a cycle of birth, death, and rebirth (*samsara*) determined by one's positive or negative karma, or the consequences of one's actions. The goal of religious life is to learn to act so as to finally achieve liberation (*moksha*) of one's soul, escaping the rebirth cycle.

Islam—The third of the monotheistic Abrahamic faiths, Islam originated with the teachings of Muhammad in the 7th century. Muslims believe Muhammad is the final of all religious prophets (beginning with Abraham) and that the Qu'ran, which is the Islamic scripture, was revealed to him by God. Islam derives from the word submission, and obedience to God is a primary theme in this religion. In order to live an Islamic life, believers must follow the five pillars, or tenets, of Islam, which are the testimony of faith (*shahada*), daily prayer (*salah*), giving alms (*zakah*), fasting during Ramadan (*sawm*), and the pilgrimage to Mecca (*hajj*).

Basic Groupings The two primary branches of Islam are Sunni and Shia, which split from each other over a religio-political leadership dispute about the rightful successor to Muhammad. The Shia believe Muhammad's cousin and son-in-law, Ali, was the only divinely ordained Imam (religious leader), while the Sunni maintain the first three caliphs after Muhammad were also legitimate authorities. In modern Islam, Sunnis and Shia continue to have different views of acceptable schools of Islamic jurisprudence, and who is a proper Islamic religious authority. Islam also has an active mystical branch, Sufism, with various Sunni and Shia subsets. Sunni Islam accounts for over 75% of the world's Muslim population. It recognizes the Abu Bakr as the first caliph after Muhammad. Sunni has four schools of Islamic doctrine and law—Hanafi, Maliki, Shafi'i, and Hanbali—which uniquely interpret the *Hadith*, or recorded oral traditions of Muhammad. A Sunni Muslim may elect to follow any one of these schools, as all are considered equally valid. Shia Islam represents 10-20% of Muslims worldwide, and its distinguishing feature is its reverence for Ali as an infallible, divinely inspired leader, and as the first Imam of the Muslim community after Muhammad. A majority of Shia are known as "Twelvers," because they believe that the 11 familial successor imams after Muhammad culminate in a 12th Imam (al-Mahdi) who is hidden in the world and will reappear at its end to redeem the righteous.

Variants Ismaili faith: A sect of Shia Islam, its adherents are also known as "Seveners," because they believe that the rightful seventh Imam in Islamic leadership was Isma'il, the elder son of Imam Jafar al-Sadiq. Ismaili tradition awaits the return of the seventh Imam as the Mahdi, or Islamic messianic figure. Ismailis are located in various parts of the world, particularly South Asia and the Levant.

Alawi faith: Another Shia sect of Islam, the name reflects followers' devotion to the religious authority of Ali. Alawites are a closed, secretive religious group who assert they are Shia Muslims, although outside scholars speculate their beliefs may have a syncretic mix with other faiths originating in the Middle East. Alawis live mostly in Syria, Lebanon, and Turkey.

Druze faith: A highly secretive tradition and a closed community that derives from the Ismaili sect of Islam; its core beliefs are thought to emphasize a combination of Gnostic principles believing that the Fatimid caliph, al-Hakin, is the one who embodies the key aspects of goodness of the universe, which are, the intellect, the word, the soul, the preceder, and the follower. The Druze have a key presence in Syria, Lebanon, and Israel.

Jainism—Originating in India, Jain spiritual philosophy believes in an eternal human soul, the eternal universe, and a principle of "the own nature of things." It emphasizes compassion for all living things, seeks liberation of the human soul from reincarnation through enlightenment, and values personal responsibility due to the belief in the immediate consequences of one's behavior. Jain philosophy teaches non-violence and prescribes vegetarianism for monks and laity alike; its adherents are a highly influential religious minority in Indian society.

Judaism—One of the first known monotheistic religions, likely dating to between 2000-1500 B.C., Judaism is the native faith of the Jewish people, based upon the belief in a covenant of responsibility between a sole omnipotent creator God and Abraham, the patriarch of Judaism's Hebrew Bible, or *Tanakh*. Divine revelation of principles and prohibitions in the Hebrew Scriptures form the basis of Jewish law, or *halakhah*, which is a key component of the faith. While there are extensive traditions of Jewish halakhic and theological discourse, there is no final dogmatic authority in the tradition. Local communities have their own religious leadership. Modern Judaism has three basic categories of faith: Orthodox, Conservative, and Reform/Liberal. These differ in their views and observance of Jewish law, with the Orthodox representing the most traditional practice, and Reform/Liberal communities the most accommodating of individualized interpretations of Jewish identity and faith.

Shintoism—A native animist tradition of Japan, Shinto practice is based upon the premise that every being and object has its own spirit or *kami*. Shinto practitioners worship several particular *kamis*, including the *kamis* of nature, and families often have shrines to their ancestors' *kamis*. Shintoism has no fixed tradition of prayers or prescribed dogma, but is characterized by individual ritual. Respect for the *kamis* in nature is a key Shinto value. Prior to the end of World War II, Shinto was the state religion of Japan, and bolstered the cult of the Japanese emperor.

Sikhism—Founded by the Guru Nanak (born 1469), Sikhism believes in a non-anthropomorphic, supreme, eternal, creator God; centering one's devotion to God is seen as a means of escaping the cycle of rebirth. Sikhs follow the teachings of Nanak and nine subsequent gurus. Their scripture, the Guru Granth Sahib—also known as the Adi Granth—is considered the living Guru, or final authority of Sikh faith and theology. Sikhism emphasizes equality of humankind and disavows caste, class, or gender discrimination.

Taoism—Chinese philosophy or religion based upon Lao Tzu's Tao Te Ching, which centers on belief in the Tao, or the way, as the flow of the universe and the nature of things. Taoism encourages a principle of non-force, or wu-wei, as the means to live harmoniously with the Tao. Taoists believe the esoteric world is made up of a perfect harmonious balance and nature, while in the manifest world—particularly in the body—balance is distorted. The Three Jewels of the Tao-compassion, simplicity, and humility—serve as the basis for Taoist ethics.

Zoroastrianism—Originating from the teachings of Zoroaster in about the 9th or 10th century B.C., Zoroastrianism may be the oldest continuing creedal religion. Its key beliefs center on a transcendent creator God, Ahura Mazda, and the concept of free will. The key ethical tenets of Zoroastrianism expressed in its scripture, the Avesta, are based on a dualistic worldview where one may prevent chaos if one chooses to serve God and exercises good thoughts, good words, and good deeds. Zoroastrianism is generally a closed religion and members are almost always born to Zoroastrian parents. Prior to the spread of Islam, Zoroastrianism dominated greater Iran. Today, though a minority, Zoroastrians remain primarily in Iran, India, and Pakistan.

Reserves of foreign exchange and gold This entry gives the dollar value for the stock of all financial assets that are available to the central monetary authority for use in meeting a country's balance of payments needs as of the end-date of the period specified. This category includes not only foreign currency and gold, but also a country's holdings of Special Drawing Rights in the International Monetary Fund, and its reserve position in the Fund.

Roadways This entry gives the *total* length of the road network and includes the length of the *paved*and *unpaved* portions.

School life expectancy (primary to tertiary education) School life expectancy (SLE) is the total number of years of schooling (primary to tertiary) that a child can expect to receive, assuming that the probability of his or her being enrolled in school at any particular future age is equal to the current enrollment ratio at that age. Caution must be maintained when utilizing this indicator in international comparisons. For example, a year or grade completed in one country is not necessarily the same in terms of educational content or quality as a year or grade completed in another country. SLE represents the expected number of years of schooling that will be completed, including years spent repeating one or more grades.

Sex ratio This entry includes the number of males for each female in five age groups - *at birth*, *under 15 years*, *15-64 years*, *65 years and over*, and for the *total population*. Sex ratio at birth has recently emerged as an indicator of certain kinds of sex discrimination in some countries. For instance, high sex ratios at birth in some Asian countries are now attributed to sex-selective abortion and infanticide due to a strong preference for sons. This will affect future marriage patterns and fertility patterns. Eventually, it could cause unrest among young adult males who are unable to find partners.

Stock of broad money This entry covers all of "Narrow money," plus the total quantity of time and savings deposits, credit union deposits, institutional money market funds, short-term repurchase agreements between the central bank and commercial deposit banks, and other large liquid assets held by nonbank financial institutions, state and local governments, nonfinancial public enterprises, and the private sector of the economy. National currency units have been converted to US dollars at the closing exchange rate for the date of the information. Because of exchange rate movements, changes in money stocks measured in national currency units may vary significantly from those shown in US dollars, and caution is urged when making comparisons over time in US dollars. In addition to serving as a medium of exchange, broad money includes assets that are slightly less liquid than narrow money and the assets tend to function as a "store of value"–a means of holding wealth.

Stock of direct foreign investment—abroad This entry gives the cumulative US dollar value of all investments in foreign countries made directly by residents–primarily companies–of the home country, as of the end of the time period indicated. Direct investment excludes investment through purchase of shares.

Stock of direct foreign investment—at home This entry gives the cumulative US dollar value of all investments in the home country made directly by residents–primarily companies–of other countries as of the end of the time period indicated. Direct investment excludes investment through purchase of shares.

Stock of domestic credit This entry is the total quantity of credit, denominated in the domestic currency, provided by financial institutions to the central bank, state and local governments, public non-financial corporations, and the private sector. The national currency units have been converted to US dollars at the closing exchange rate on the date of the information.

Stock of narrow money This entry, also known as "M1," comprises the total quantity of currency in circulation (notes and coins) plus demand deposits denominated in the national currency held by nonbank financial institutions, state and local governments, nonfinancial public enterprises, and the private sector of the economy, measured at a specific point in time. National currency units have been converted to US dollars at the closing exchange rate for the date of the information. Because of exchange rate movements, changes in money stocks measured in national currency units may vary significantly from those shown in US dollars, and caution is urged when making comparisons over time in US dollars. Narrow money consists of more liquid assets than broad money and the assets generally function as a "medium of exchange" for an economy.

Suffrage This entry gives the age at enfranchisement and whether the right to vote is universal or restricted.

Telephone numbers All telephone numbers in *The World Factbook* consist of the country code in brackets, the city or area code (where required) in parentheses, and the local number. The one component that is not presented is the international access code, which varies from country to country. For example, an international direct dial telephone call placed from the US to Madrid, Spain, would be as follows: 011 [34] (1) 577-xxxx, where 011 is the international access code for station-to-station calls; 01 is for calls other than station-to-station calls, [34] is the country code for Spain, (1) is the city code for Madrid, 577 is the local exchange, and xxxx is the local telephone number. An international direct dial telephone call placed from another country to the US would be as follows: international access code + [1] (202) 939-xxxx, where [1] is the country code for the US, (202) is the area code for Washington, DC, 939 is the local exchange, and xxxx is the local telephone number.

Telephone system This entry includes a brief general assessment of the system with details on the domestic and international components. The following terms and abbreviations are used throughout the entry:

Arabsat–Arab Satellite Communications Organization (Riyadh, Saudi Arabia).

Autodin–Automatic Digital Network (US Department of Defense).

CB–citizen's band mobile radio communications.

Cellular telephone system–the telephones in this system are radio transceivers, with each instrument having its own private radio frequency and sufficient radiated power to reach the booster station in its area (cell), from which the telephone signal is fed to a telephone exchange.

Central American Microwave System–a trunk microwave radio relay system that links the countries of Central America and Mexico with each other.

Coaxial cable–a multichannel communication cable consisting of a central conducting wire, surrounded by and insulated from a cylindrical conducting shell; a large number of telephone channels can be made available within the insulated space by the use of a large number of carrier frequencies.

Comsat–Communications Satellite Corporation (US).

DSN–Defense Switched Network (formerly Automatic Voice Network or Autovon); basic general-purpose, switched voice network of the Defense Communications System (US Department of Defense).

Eutelsat–European Telecommunications Satellite Organization (Paris).

Fiber-optic cable–a multichannel communications cable using a thread of optical glass fibers as a transmission medium in which the signal (voice, video, etc.) is in the form of a coded pulse of light.

GSM–a global system for mobile (cellular) communications devised by the Groupe Special Mobile of the pan-European standardization organization, Conference Europeanne des Posts et Telecommunications (CEPT) in 1982.

HF–high frequency; any radio frequency in the 3,000- to 30,000-kHz range.

Inmarsat–International Maritime Satellite Organization (London); provider of global mobile satellite communications for commercial, distress, and safety applications at sea, in the air, and on land.
Intelsat–International Telecommunications Satellite Organization (Washington, DC).
Intersputnik–International Organization of Space Communications (Moscow); first established in the former Soviet Union and the East European countries, it is now marketing its services worldwide with earth stations in North America, Africa, and East Asia.
Landline–communication wire or cable of any sort that is installed on poles or buried in the ground.
Marecs–Maritime European Communications Satellite used in the Inmarsat system on lease from the European Space Agency.
Marisat–satellites of the Comsat Corporation that participate in the Inmarsat system.
Medarabtel–the Middle East Telecommunications Project of the International Telecommunications Union (ITU) providing a modern telecommunications network, primarily by microwave radio relay, linking Algeria, Djibouti, Egypt, Jordan, Libya, Morocco, Saudi Arabia, Somalia, Sudan, Syria, Tunisia, and Yemen; it was initially started in Morocco in 1970 by the Arab Telecommunications Union (ATU) and was known at that time as the Middle East Mediterranean Telecommunications Network.
Microwave radio relay–transmission of long distance telephone calls and television programs by highly directional radio microwaves that are received and sent on from one booster station to another on an optical path.
NMT–Nordic Mobile Telephone; an analog cellular telephone system that was developed jointly by the national telecommunications authorities of the Nordic countries (Denmark, Finland, Iceland, Norway, and Sweden).
Orbita–a Russian television service; also the trade name of a packet-switched digital telephone network.
Radiotelephone communications–the two-way transmission and reception of sounds by broadcast radio on authorized frequencies using telephone handsets.
PanAmSat–PanAmSat Corporation (Greenwich, CT).
SAFE–South African Far East Cable
Satellite communication system–a communication system consisting of two or more earth stations and at least one satellite that provide long distance transmission of voice, data, and television; the system usually serves as a trunk connection between telephone exchanges; if the earth stations are in the same country, it is a domestic system.
Satellite earth station–a communications facility with a microwave radio transmitting and receiving antenna and required receiving and transmitting equipment for communicating with satellites.
Satellite link–a radio connection between a satellite and an earth station permitting communication between them, either one-way (down link from satellite to earth station–television receive-only transmission) or two-way (telephone channels).
SHF–super high frequency; any radio frequency in the 3,000- to 30,000-MHz range.
Shortwave–radio frequencies (from 1.605 to 30 MHz) that fall above the commercial broadcast band and are used for communication over long distances.
Solidaridad–geosynchronous satellites in Mexico's system of international telecommunications in the Western Hemisphere.
Statsionar–Russia's geostationary system for satellite telecommunications.
Submarine cable–a cable designed for service under water.
TAT–Trans-Atlantic Telephone; any of a number of high-capacity submarine coaxial telephone cables linking Europe with North America.
Telefax–facsimile service between subscriber stations via the public switched telephone network or the international Datel network.
Telegraph–a telecommunications system designed for unmodulated electric impulse transmission.
Telex–a communication service involving teletypewriters connected by wire through automatic exchanges.
Tropospheric scatter–a form of microwave radio transmission in which the troposphere is used to scatter and reflect a fraction of the incident radio waves back to earth; powerful, highly directional antennas are used to transmit and receive the microwave signals; reliable over-the-horizon communications are realized for distances up to 600 miles in a single hop; additional hops can extend the range of this system for very long distances.
Trunk network–a network of switching centers, connected by multichannel trunk lines.
UHF–ultra high frequency; any radio frequency in the 300- to 3,000-MHz range.
VHF–very high frequency; any radio frequency in the 30- to 300-MHz range.

Telephones—main lines in use This entry gives the total number of main telephone lines in use.

Telephones—mobile cellular This entry gives the total number of mobile cellular telephone subscribers.

Terminology Due to the highly structured nature of the *Factbook* database, some collective generic terms have to be used. For example, the word **Country** in the **Country name** entry refers to a wide variety of dependencies, areas of special sovereignty, uninhabited islands, and other entities in addition to the traditional countries or independent states. **Military** is also used as an umbrella term for various civil defense, security, and defense activities in many entries. The **Independence** entry includes the usual colonial independence dates and former ruling states as well as other significant nationhood dates such as the traditional founding date or the date of unification, federation, confederation, establishment, or state succession that are not strictly independence dates. Dependent areas have the nature of their dependency status noted in this same entry.

Terrain This entry contains a brief description of the topography.

Time difference This entry is expressed in *The World Factbook* in two ways. First, it is stated as the difference in hours between the capital of an entity and **Coordinated Universal Time (UTC)** during Standard Time. Additionally, the difference in time between the capital of an entity and that observed in Washington, D.C. is also provided. Note that the time difference assumes both locations are simultaneously observing Standard Time or Daylight Saving Time.

Time zones Ten countries (Australia, Brazil, Canada, Indonesia, Kazakhstan, Mexico, New Zealand, Russia, Spain, and the United States) and the island of Greenland observe more than one official time depending on the number of designated time zones within their boundaries. An illustration of time zones throughout the world and within countries can be seen in the Standard Time Zones of the World map included in the **Reference Maps** section of *The World Factbook*.

Total fertility rate This entry gives a figure for the average number of children that would be born per woman if all women lived to the end of their childbearing years and bore children according to a given fertility rate at each age. The total fertility rate (TFR) is a more direct measure of the level of fertility than the crude birth rate, since it refers to births per woman. This indicator shows the potential for population change in the country. A rate of two children per woman is considered the replacement rate for a population, resulting in relative stability in terms of total numbers. Rates above two children indicate populations growing in size and whose median age is declining. Higher rates may also indicate difficulties for families, in some situations, to feed and educate their children and for women to enter the labor force. Rates below two children indicate populations decreasing in size and growing older. Global fertility rates are in general decline and this trend is most pronounced in industrialized countries, especially Western Europe, where populations are projected to decline dramatically over the next 50 years.

Total renewable water resources This entry provides the long-term average water availability for a country in cubic kilometers of precipitation, recharged ground water, and surface inflows from surrounding countries. The values have been adjusted to account for overlap resulting from surface flow recharge of groundwater sources. Total renewable water resources provides the water total available to a country but does not include water resource totals that have been reserved for upstream or downstream countries through international agreements. Note that these values are averages and do not accurately reflect the total available in any given year. Annual available resources can vary greatly due to short-term and long-term climatic and weather variations.

Trafficking in persons Trafficking in persons is modern-day slavery, involving victims who are forced, defrauded, or coerced into labor or sexual exploitation. The International Labor Organization (ILO), the UN agency charged with addressing labor standards, employment, and social protection issues, estimates that 12.3 million people worldwide are enslaved in forced labor, bonded labor, forced child labor, sexual servitude, and involuntary servitude at any given time. Human trafficking is a multidimensional threat, depriving people of their human rights and freedoms, risking global health, promoting social breakdown, inhibiting development by depriving countries of their human capital, and helping fuel the growth of organized crime. In 2000, the US Congress passed the Trafficking Victims Protection Act (TVPA), reauthorized in 2003 and 2005, which provides tools for the US to combat trafficking in persons, both domestically and abroad. One of the law's key components is the creation of the US Department of State's annual *Trafficking in Persons Report*, which assesses the government response (i.e., the *current situation*) in some 150 countries with a significant number of victims trafficked across their borders who are recruited, harbored, transported, provided, or obtained for forced labor or sexual exploitation. Countries in the annual report are rated in three tiers, based on government efforts to combat trafficking. The countries identified in this entry are those listed in the *2010 Trafficking in Persons Report* as *Tier 2 Watch List* or *Tier 3* based on the following *tier rating* definitions:

Tier 2 Watch List countries do not fully comply with the minimum standards for the elimination of trafficking but are making significant efforts to do so, and meet one of the following criteria:

1. they display high or significantly increasing number of victims,
2. they have failed to provide evidence of increasing efforts to combat trafficking in persons, or,
3. they have committed to take action over the next year.

Tier 3 countries neither satisfy the minimum standards for the elimination of trafficking nor demonstrate a significant effort to do so. Countries in this tier are subject to potential non-humanitarian and non-trade sanctions.

Transnational issues This category includes four entries - **Disputes—international, Refugees and internally displaced persons, Trafficking in persons**, and **Illicit drugs** - that deal with current issues going beyond national boundaries.

Transportation This category includes the entries dealing with the means for movement of people and goods.

Transportation—note This entry includes miscellaneous transportation information of significance not included elsewhere.

Unemployment rate This entry contains the percent of the labor force that is without jobs. Substantial underemployment might be noted.

Urbanization This entry provides two measures of the degree of urbanization of a population. The first,*urban population*, describes the percentage of the total population living in urban areas, as defined by the country. The second, *rate of urbanization*, describes the projected average rate of change of the size of the urban population over the given period of time. Additionally, the World entry includes a list of the *ten largest urban agglomerations*. An *urban agglomeration* is defined as comprising the city or town proper and also the suburban fringe or thickly settled territory lying outside of, but adjacent to, the boundaries of the city.

UTC (Coordinated Universal Time) See entry for Coordinated Universal Time.

Waterways This entry gives the total length of navigable rivers, canals, and other inland bodies of water.

Weights and Measures This information is presented in **Appendix G: Weights and Measures and includes mathematical notations (mathematical powers and names), metric interrelationships (prefix; symbol; length, weight, or capacity; area; volume), and standard conversion factors.**

Years All year references are for the calendar year (CY) unless indicated as fiscal year (FY). The calendar year is an accounting period of 12 months from 1 January to 31 December. The fiscal year is an accounting period of 12 months other than 1 January to 31 December.

NOTE: INFORMATION FOR THE US AND US DEPENDENCIES WAS COMPLIED FROM MATERIAL IN THE PUBLIC DOMAIN AND DOES NOT REPRESENT INTELLIGENCE COMMUNITY ESTIMATES.

GUIDE TO COUNTRY PROFILES

INTRODUCTION

Background

GEOGRAPHY

Location
Geographic coordinates
Map references
Area
total
land
water
Area—comparative
Land boundaries
total
border countries
Coastline
Maritime claims
territorial sea
contiguous zone
exclusive economic zone
continental shelf
exclusive fishing zone
Climate
Terrain
Elevation extremes
lowest point
highest point
Natural resources
Land use
arable land
permanent crops
other
Irrigated land
Total Renewable Water Resources
Freshwater withdrawal (domestic/industrial/agricultural)
total
per capita
Natural hazards
volcanism
Environment—current issues
Environment—international agreements
party to
signed, but not ratified
Geography—note

PEOPLE

Population
Age structure
0-14 years
15-64 years
65 years and over
Median Age
total
male
female
Population growth rate
Birth rate
Death rate
Net migration rate
Sex ratio
at birth
under 15 years
15-64 years
65 years and over
total population
Infant mortality rate
total
male
female
Life expectancy at birth
total population
male
female
Total fertility rate
HIV/AIDS— adult prevalence rate
HIV/AIDS—people living with HIV/AIDS
HIV/AIDS—deaths
Major infectious diseases
degree of risk
food or waterborne diseases
vectorborne diseases
water contact diseases
aerosolized dust or soil contact disease
respiratory disease
animal contact disease
Nationality
noun
adjective
Ethnic groups
Religions
Languages
Literacy
definition
total population
male
female
School life expectancy (primary to tertiary)
Education expenditures
People—note

GOVERNMENT

Country name
conventional long form
conventional short form
local long form
local short form
former
abbreviation
Dependency status
Government type
Capital
name
geographic coordinates
time difference
daylight saving time
Administrative divisions
Dependent areas
Independence
National holiday
Constitution
Legal system
Suffrage
Executive branch
chief of state
head of government
cabinet
elections
election results
Legislative branch
elections
election results
Judicial branch
Political parties and leaders
Political pressure groups and leaders
International organization participation
Diplomatic representation in the US
chief of mission
chancery
telephone
FAX
consulate(s) general
consulate(s)
Diplomatic representation from the US
chief of mission
embassy
mailing address
telephone
FAX
consulate(s) general
consulate(s)
branch office(s)
Flag description
Government—note

ECONOMY

Economy—overview
GDP (purchasing power parity)
GDP (official exchange rate)
GDP—real growth rate
GDP—per capita (PPP)
GDP—composition by sector
agriculture
industry
services

Labor force
Labor force—by occupation
agriculture
industry
services
Unemployment rate
Population below poverty line
Household income or consumption by percentage share
lowest 10%
highest 10%
Distribution of family income—Gini index
Investment (gross fixed)
Budget
revenues
expenditures
Public debt
Inflation rate (consumer prices)
Central bank discount rate
Commercial bank prime lending rate
Stock of money
Stock of quasi money
Stock of domestic credit
Market value of publicly traded shares
Agriculture—products
Industries
Industrial production growth rate
Electricity—production
Electricity—consumption
Electricity—exports
Electricity—imports
Oil—production
Oil—consumption
Oil—exports
Oil—imports
Oil—proved reserves
Natural gas—production
Natural gas—consumption
Natural gas—exports
Natural gas—imports
Natural gas—proved reserves
Current account balance
Exports
Exports—commodities
Exports—partners
Imports
Imports—commodities
Imports—partners
Reserves of foreign exchange and gold
Debt—external
Stock of direct foreign investment—at home
Stock of direct foreign investment—abroad
Exchange rates

COMMUNICATIONS

Telephones—main lines in use
Telephones—mobile cellular
Telephone system
general assessment
domestic
international
Broadcast media
Internet country code
Internet hosts
Internet users
Communications—note

TRANSPORTATION

Airports
Airports—with paved runways
total
over 3,047 m
2,438 to 3,047 m
1,524 to 2,437 m
914 to 1,530 m
under 914 m
Airports—with unpaved runways
total
over 3,047 m
2,438 to 3,047 m
1,524 to 2,437 m
914 to 1,530 m
under 914 m
Heliports
Pipelines
Railways
total
broad gauge
standard gauge
narrow gauge
dual gauge
Roadways
total
paved
unpaved
Waterways
Merchant marine
total
ships by type
foreign-owned
registered in other countries
Ports and terminals
Transportation—note

MILITARY

Military branches
Military service age and obligation
Manpower available for military service
males age 16-49
females age 16-49
Manpower fit for military service
males age 16-49
females age 16-49
Manpower reaching military age annually
males
females
Military expenditures—percent of GDP
Military—note

TRANSNATIONAL ISSUES

Disputes—international
Refugees and internally displaced persons
refugees
IDPs
Trafficking in persons
current situation
tier rating
Illicit drugs

AFGHANISTAN

INTRODUCTION

Background: Ahmad Shah DURRANI unified the Pashtun tribes and founded Afghanistan in 1747. The country served as a buffer between the British and Russian Empires until it won independence from notional British control in 1919. A brief experiment in democracy ended in a 1973 coup and a 1978 Communist counter-coup. The Soviet Union invaded in 1979 to support the tottering Afghan Communist regime, touching off a long and destructive war. The USSR withdrew in 1989 under relentless pressure by internationally supported anti-Communist mujahedin rebels. A series of subsequent civil wars saw Kabul finally fall in 1996 to the Taliban, a hardline Pakistani-sponsored movement that emerged in 1994 to end the country's civil war and anarchy. Following the 11 September 2001 terrorist attacks in New York City and Washington, D.C., a US, Allied, and anti-Taliban Northern Alliance military action toppled the Taliban for sheltering Osama BIN LADIN. The UN-sponsored Bonn Conference in 2001 established a process for political reconstruction that included the adoption of a new constitution, a presidential election in 2004, and National Assembly elections in 2005. In December 2004, Hamid KARZAI became the first democratically elected president of Afghanistan and the National Assembly was inaugurated the following December. KARZAI was re-elected in August 2009 for a second term. Despite gains toward building a stable central government, a resurgent Taliban and continuing provincial instability–particularly in the south and the east–remain serious challenges for the Afghan Government.

GEOGRAPHY

Location: Southern Asia, north and west of Pakistan, east of Iran
Geographic coordinates: 33 00 N, 65 00 E
Map references: Asia
Area: *total:* 652,230 sq km
country comparison to the world: 41
land: 652,230 sq km
water: 0 sq km
Area—comparative: slightly smaller than Texas
Land boundaries: *total:* 5,529 km
border countries: China 76 km, Iran 936 km, Pakistan 2,430 km, Tajikistan 1,206 km, Turkmenistan 744 km, Uzbekistan 137 km
Coastline: 0 km (landlocked)
Maritime claims: none (landlocked)
Climate: arid to semiarid; cold winters and hot summers
Terrain: mostly rugged mountains; plains in north and southwest
Elevation extremes: *lowest point:* Amu Darya 258 m
highest point: Noshak 7,485 m
Natural resources: natural gas, petroleum, coal, copper, chromite, talc, barites, sulfur, lead, zinc, iron ore, salt, precious and semi-precious stones
Land use: *arable land:* 12.13%
permanent crops: 0.21%
other: 87.66% (2005)
Irrigated land: 27,200 sq km (2003)
Total renewable water resources: 65 cu km (1997)
Freshwater withdrawal (domestic/industrial/agricultural): *total:* 23.26 cu km/yr (2%/0%/98%)
per capita: 779 cu m/yr (2000)
Natural hazards: damaging earthquakes occur in Hindu Kush mountains; flooding; droughts
Environment—current issues: limited natural freshwater resources; inadequate supplies of potable water; soil degradation; overgrazing; deforestation (much of the remaining forests are being cut down for fuel and building materials); desertification; air and water pollution
Environment—international agreements: *party to:* Biodiversity, Climate Change, Desertification, Endangered Species, Environmental Modification, Marine Dumping, Ozone Layer Protection
signed, but not ratified: Hazardous Wastes, Law of the Sea, Marine Life Conservation
Geography—note: landlocked; the Hindu Kush mountains that run northeast to southwest divide the northern provinces from the rest of the country; the highest peaks are in the northern Vakhan (Wakhan Corridor)

PEOPLE

Population: 29,835,392 (July 2011 est.)
country comparison to the world: 40
note: this is a significantly revised figure; the previous estimate of 33,609,937 was extrapolated from the last Afghan census held in 1979, which was never completed because of the Soviet invasion
Age structure: *0–14 years:* 42.3% (male 6,464,070/female 6,149,468)
15–64 years: 55.3% (male 8,460,486/female 8,031,968)
65 years and over: 2.4% (male 349,349/female 380,051) (2011 est.)
Median age: *total:* 18.2 years
male: 18.2 years
female: 18.2 years (2011 est.)
Population growth rate: 2.375% (2011 est.)
country comparison to the world: 33
Birth rate: 37.83 births/1,000 population (2011 est.)
country comparison to the world: 17
Death rate: 17.39 deaths/1,000 population (July 2011 est.)
country comparison to the world: 2
Net migration rate: 3.31 migrant(s)/1,000 population (2011 est.)
country comparison to the world: 27
Urbanization: *urban population:* 23% of total population (2010)
rate of urbanization: 4.7% annual rate of change (2010-15 est.)
Major cities—population: KABUL (capital) 3.573 million (2009)
Sex ratio: *at birth:* 1.05 male(s)/female
under 15 years: 1.05 male(s)/female
15–64 years: 1.05 male(s)/female
65 years and over: 0.92 male(s)/female
total population: 1.05 male(s)/female (2011 est.)
Infant mortality rate: *total:* 149.2 deaths/1,000 live births
country comparison to the world: 2
male: 152.75 deaths/1,000 live births
female: 145.47 deaths/1,000 live births (2011 est.)
Life expectancy at birth: *total population:* 45.02 years
country comparison to the world: 221
male: 44.79 years
female: 45.25 years (2011 est.)
Total fertility rate: 5.39 children born/woman (2011 est.)
country comparison to the world: 13
HIV/AIDS—adult prevalence rate: 0.01% (2001 est.)
country comparison to the world: 168
HIV/AIDS—people living with HIV/AIDS: NA
HIV/AIDS—deaths: NA
Major infectious diseases: *degree of risk:* high
food or waterborne diseases: bacterial and protozoal diarrhea, hepatitis A, and typhoid fever
vectorborne disease: malaria
animal contact disease: rabies
note: highly pathogenic H5N1 avian influenza has been identified in this country; it poses a negligible risk with extremely rare cases possible among US citizens who have close contact with birds (2009)
Drinking water source: *Improved:*
urban: 78% of population
rural: 39% of population
total: 48% of population
Unimproved: urban: 22% of population
rural: 61% of population
total: 52% of population (2008)
Sanitation facility access: *Improved:*
urban: 60% of population
rural: 30% of population
total: 37% of population
Unimproved: urban: 40% of population
rural: 70% of population
total: 63% of population (2008)
Nationality: *noun:* Afghan(s)
adjective: Afghan
Ethnic groups: Pashtun 42%, Tajik 27%, Hazara 9%, Uzbek 9%, Aimak 4%, Turkmen 3%, Baloch 2%, other 4%
Religions: Sunni Muslim 80%, Shia Muslim 19%, other 1%
Languages: Afghan Persian or Dari (official) 50%, Pashto (official) 35%, Turkic languages (primarily Uzbek and Turkmen)

11%, 30 minor languages (primarily Balochi and Pashai) 4%, much bilingualism
Literacy: *definition:* age 15 and over can read and write
total population: 28.1%
male: 43.1%
female: 12.6% (2000 est.)
School life expectancy (primary to tertiary education): *total:* 9 years
male: 11 years
female: 7 years (2009)
Education expenditures: NA

GOVERNMENT

Country name: *conventional long form:* Islamic Republic of Afghanistan
conventional short form: Afghanistan
local long form: Jomhuri-ye Eslami-ye Afghanestan
local short form: Afghanestan
former: Republic of Afghanistan
Government type: Islamic republic
Capital: *name:* Kabul
geographic coordinates: 34 31 N, 69 11 E
time difference: UTC+4.5 (9.5 hours ahead of Washington, DC during Standard Time)
Administrative divisions: 34 provinces (welayat, singular–welayat); Badakhshan, Badghis, Baghlan, Balkh, Bamyan, Daykundi, Farah, Faryab, Ghazni, Ghor, Helmand, Herat, Jowzjan, Kabul, Kandahar, Kapisa, Khost, Kunar, Kunduz, Laghman, Logar, Nangarhar, Nimroz, Nuristan, Paktika, Paktiya, Panjshir, Parwan, Samangan, Sar-e Pul, Takhar, Uruzgan, Wardak, Zabul
Independence: 19 August 1919 (from UK control over Afghan foreign affairs)
National holiday: Independence Day, 19 August (1919)
Constitution: constitution drafted 14 December 2003-4 January 2004; signed 16 January 2004; ratified 26 January 2004
Legal system: mixed legal system of civil, customary, and Islamic law
International law organization participation: has not submitted an ICJ jurisdiction declaration; accepts ICCt jurisdiction
Suffrage: 18 years of age; universal
Executive branch: *chief of state:* President of the Islamic Republic of Afghanistan Hamid KARZAI (since 7 December 2004); First Vice President Mohammad FAHIM Khan (since 19 November 2009); Second Vice President Abdul Karim KHALILI (since 7 December 2004); note–the president is both the chief of state and head of government
head of government: President of the Islamic Republic of Afghanistan Hamid KARZAI (since 7 December 2004); First Vice President Mohammad FAHIM Khan (since 19 November 2009); Second Vice President Abdul Karim KHALILI (since 7 December 2004)
cabinet: 25 ministers; note–ministers are appointed by the president and approved by the National Assembly
(For more information visit the World Leaders website)
elections: the president and two vice presidents elected by direct vote for a five-year term (eligible for a second term); if no candidate receives 50% or more of the vote in the first round of voting, the two candidates with the most votes will participate in a second round; election last held on 20 August 2009 (next to be held in 2014)
election results: Hamid KARZAI reelected president; percent of vote (first round)– Hamid KARZAI 49.67%, Abdullah ABDULLAH 30.59%, Ramazan BASHARDOST 10.46%, Ashraf GHANI 2.94%; other 6.34%; note–ABDULLAH conceded the election to KARZAI following the first round vote
Legislative branch: the bicameral National Assembly consists of the Meshrano Jirga or House of Elders (102 seats, one-third of members elected from provincial councils for four-year terms, one-third elected from local district councils for three-year terms, and one-third nominated by the president for five-year terms) and the Wolesi Jirga or House of People (no more than 250 seats); members directly elected for five-year terms
note: on rare occasions the government may convene a Loya Jirga (Grand Council) on issues of independence, national sovereignty, and territorial integrity; it can amend the provisions of the constitution and prosecute the president; it is made up of members of the National Assembly and chairpersons of the provincial and district councils
elections: last held on 18 September 2010 (next election expected in 2015)
election results: results by party–NA; note– ethnicity is the main factor influencing political alliances; compositon of Loya Jirga seats by ethnic groups–Pashtun 96, Hazara 61, Tajik 53, Uzbek 15, Aimak 8, Arab 8, Turkmen 3, Nuristani 2, Baloch 1, Pahhai 1, Turkic 1; women hold 68 seats
Judicial branch: the constitution establishes a nine-member Stera Mahkama or Supreme Court (its nine justices are appointed for 10-year terms by the president with approval of the Wolesi Jirga) and subordinate High Courts and Appeals Courts; there is also a minister of justice; a separate Afghan Independent Human Rights Commission established by the Bonn Agreement is charged with investigating human rights abuses and war crimes
Political parties and leaders: Afghanistan Peoples' Treaty Party [Sayyed Amir TAHSEEN]; Afghanistan's Islamic Mission Organization [Abdul Rasoul SAYYAF]; Afghanistan's Islamic Nation Party [Toran Noor Aqa Ahmad ZAI]; Afghanistan's National Islamic Party [Rohullah LOUDIN]; Afghanistan's Welfare Party [Meer Asef ZAEEFI]; Afghan Social Democratic Party [Anwarul Haq AHADI]; Afghan Society for the Call to the Koran and Sunna [Mawlawee Samiullah NAJEEBEE]; Comprehensive Movement of Democracy and Development of Afghanistan Party [Sher Mohammad BAZGAR]; Democratic Party of Afghanistan [Al-hajj Mohammad Tawos ARAB]; Democratic Party of Afghanistan [Abdul Kabir RANJBAR]; Elites People of Afghanistan Party [Abdul Hamid JAWAD]; Freedom and Democracy Movement of Afghanistan [Abdul Raqib Jawid KOHISTANEE]; Freedom Party of Afghanistan [Abdul MALEK]; Freedom Party of Afghanistan [Dr. Ghulam Farooq NEJRABEE]; Hizullah-e-Afghanistan [Qari Ahmad ALI]; Human Rights Protection and Development Party of Afghanistan [Baryalai NASRATI]; Islamic Justice Party of Afghanistan [Mohammad Kabir MARZBAN]; Islamic Movement of Afghanistan [Mohammad Ali JAWID]; Islamic Movement of Afghanistan Party [Mohammad Mukhtar MUFLEH]; Islamic Party of Afghanistan [Mohammad Khalid FAROOQI, Abdul Hadi ARGHANDIWAL]; Islamic Party of the Afghan Land [Mohammad Hassan FEROZKHEL]; Islamic People's Movement of Afghanistan [Al-haj Said Hussain ANWARY]; Islamic Society of Afghanistan [Ustad RABBANI]; Islamic Unity of the Nation of Afghanistan Party [Qurban Ali URFANI]; Islamic Unity Party of Afghanistan [Mohammad Karim KHALILI]; Islamic Unity Party of the People of Afghanistan [Haji Mohammad MOHAQQEQ]; Labor and Progress of Afghanistan Party [Zulfiqar OMID]; Muslim People of Afghanistan Party [Besmellah JOYAN]; Muslim Unity Movement Party of Afghanistan [Wazir Mohammad WAHDAT]; National and Islamic Sovereignty Movement Party of Afghanistan [Ahmad Shah AHMADZAI]; National Congress Party of Afghanistan [Abdul Latif PEDRAM]; National Country Party [Ghulam MOHAMMAD]; National Development Party of Afghanistan [Dr. Assef BAKTASH]; National Freedom Seekers Party [Abdul Hadi DABEER]; National Independence Party of Afghanistan [Taj Mohammad WARDAK]; National Islamic Fighters Party of Afghanistan [Amanat NINGARHAREE]; National Islamic Front of Afghanistan [Pir Sayed Ahmad GAILANEE]; National Islamic Moderation Party of Afghanistan [Qara Baik IZADYAR]; National Islamic Movement of Afghanistan [Sayed NOORULLAH]
National Islamic Unity Party of Afghanistan [Mohammad AKBAREE]; National Movement of Afghanistan [Ahmad Wali MASOOUD]; National Party of Afghanistan [Abdul Rashid ARYAN]; National Patch of Afghanistan Party [Sayed Kamal SADAT]; National Peace Islamic Party of Afghanistan [Shah Mohammood Popal ZAI]; National Peace & Islamic Party of the Tribes of Afghanistan [Abdul Qaher SHARIATEE]; National Peace & Unity Party of Afghanistan [Abdul Qader IMAMI]; National Prosperity and Islamic Party of Afghanistan [Mohammad Osman SALEKZADA]; National Prosperity Party [Mohammad Hassan JAHFAREE]; National Solidarity Movement of Afghanistan [Pir Sayed Eshaq GAILANEE]; National Solidarity Party of Afghanistan [Sayed Mansoor NADREEI]; National Sovereignty Party [Sayed Mustafa KAZEMI]; National Stability Party [Mohammad Same KHAROTI]; National Stance Party [Habibullah JANEBDAR]; National Tribal Unity Islamic Party of Afghanistan [Mohammad Shah KHOGYANI]; National Unity Movement [Sultan Mohammad GHAZI]; National Unity

Movement of Afghanistan [Mohammad Nadir AATASH]; National Unity Party of Afghanistan [Abdul Rashid JALILI]; New Afghanistan Party [Mohammad Yunis QANUNI]; Peace and National Welfare Activists Society [Shamsul al-Haq Noor SHAMS]; Peace Movement [Shahnawaz TANAI]; People's Aspirations Party of Afghanistan [Ilhaj Saraj-u-din ZAFAREE]; People's Freedom Seekers Party of Afghanistan [Feda Mohammad EHSAS]; People's Liberal Freedom Seekers Party of Afghanistan [Ajmal SUHAIL]; People's Message Party of Afghanistan [Noor Aqa WAINEE]; People's Movement of the National Unity of Afghanistan [Abdul Hakim NOORZAI]; People's Party of Afghanistan [Ahmad Shah ASAR]; People's Prosperity Party of Afghanistan [Ustad Mohammad ZAREEF]; People's Sovereignty Movement of Afghanistan [Hayatullah SUBHANEE]; People's Uprising Party of Afghanistan [Sayed Zahir Qayedam Al-BELADI]; People's Welfare Party of Afghanistan [Miagul WASIQ]; People's Welfare Party of Afghanistan [Mohammad Zubair PAIROZ]; Progressive Democratic Party of Afghanistan [Mohammad Wali ARYA]; Republican Party [Sebghatullah SANJAR]; Solidarity Party of Afghanistan [Abdul Khaleq NEMAT]; The Afghanistan's Mujahid Nation's Islamic Unity Movement [Saeedullah SAEED]; The People of Afghanistan's Democratic Movement [Mohammad Sharif NAZARI]; Tribes Solidarity Party of Afghanistan [Mohammad Zarif NASERI]; Understanding and Democracy Party of Afghanistan [Ahamad SHAHEEN]
United Afghanistan Party [Mohammad Wasil RAHIMEE]; United Islamic Party of Afghanistan [Wahidullah SABAWOON]; Young Afghanistan's Islamic Organization [Sayed Jawad HUSSINEE]; Youth Solidarity Party of Afghanistan [Mohammad Jamil KARZAI]; note–includes only political parties approved by the Ministry of Justice

Political pressure groups and leaders: *other:* religious groups; tribal leaders; ethnically based groups; Taliban

International organization participation: ADB, CICA, CP, ECO, FAO, G-77, IAEA, IBRD, ICAO, ICRM, IDA, IDB, IFAD, IFC, IFRCS, ILO, IMF, Interpol, IOC, IOM, IPU, ISO (correspondent), ITSO, ITU, MIGA, NAM, OIC, OPCW, OSCE (partner), SAARC, SACEP, UN, UNCTAD, UNESCO, UNIDO, UNWTO, UPU, WCO, WFTU, WHO, WIPO, WMO, WTO (observer)

Diplomatic representation in the US: *chief of mission:* Ambassador Eklil Ahmad HAKIMI
chancery: 2341 Wyoming Avenue NW, Washington, DC 20008
telephone: [1] (202) 483-6410
FAX: [1] (202) 483-6488
consulate(s) general: Los Angeles, New York

Diplomatic representation from the US: *chief of mission:* Ambassador Karl W. EIKENBERRY
embassy: The Great Masood Road, Kabul
mailing address: U.S. Embassy Kabul, APO, AE 09806
telephone: [93] 0700 108 001
FAX: [93] 0700 108 564

Flag description: three equal vertical bands of black (hoist side), red, and green, with the national emblem in white centered on the red band and slightly overlapping the other two bands; the center of the emblem features a mosque with pulpit and flags on either side, below the mosque are numerals for the solar year 1298 (1919 in the Gregorian calendar, the year of Afghan independence from the UK); this central image is circled by a border consisting of sheaves of wheat on the left and right, in the upper-center is an Arabic inscription of the Shahada (Muslim creed) below which are rays of the rising sun over the Takbir (Arabic expression meaning "God is great"), and at bottom center is a scroll bearing the name Afghanistan; black signifies the past, red is for the blood shed for independence, and green can represent either hope for the future, agricultural prosperity, or Islam
note: Afghanistan had more changes to its national flag in the 20th century than any other country; the colors black, red, and green appeared on most of them

National anthem: *name:* "Milli Surood" (National Anthem)
lyrics/music: Abdul Bari JAHANI/Babrak WASA
note: adopted 2006; the 2004 constitution of the post-Taliban government mandated that a new national anthem should be written containing the phrase "Allahu Akbar" (God is Great) and mentioning the names of Afghanistan's ethnic groups

ECONOMY

Economy—overview: Afghanistan's economy is recovering from decades of conflict. The economy has improved significantly since the fall of the Taliban regime in 2001 largely because of the infusion of international assistance, the recovery of the agricultural sector, and service sector growth. Despite the progress of the past few years, Afghanistan is extremely poor, landlocked, and highly dependent on foreign aid, agriculture, and trade with neighboring countries. Much of the population continues to suffer from shortages of housing, clean water, electricity, medical care, and jobs. Criminality, insecurity, weak governance, and the Afghan Government's inability to extend rule of law to all parts of the country pose challenges to future economic growth. Afghanistan's living standards are among the lowest in the world. While the international community remains committed to Afghanistan's development, pledging over $67 billion at four donors' conferences since 2002, the Government of Afghanistan will need to overcome a number of challenges, including low revenue collection, anemic job creation, high levels of corruption, weak government capacity, and poor public infrastructure.

GDP (purchasing power parity): $27.36 billion (2010 est.)
country comparison to the world: 111
$25.28 billion (2009 est.)
$20.92 billion (2008 est.)
note: data are in 2010 US dollars

GDP (official exchange rate): $15.61 billion (2010 est.)

GDP—real growth rate: 8.2% (2010 est.)
country comparison to the world: 17
20.9% (2009 est.)
3.6% (2008 est.)

GDP—per capita (PPP): $900 (2010 est.)
country comparison to the world: 217
$900 (2009 est.)
$800 (2008 est.)
note: data are in 2010 US dollars

GDP—composition by sector:
agriculture: 31%
industry: 26%
services: 43%
note: data exclude opium production (2008 est.)

Labor force: 15 million (2004 est.)
country comparison to the world: 40

Labor force—by occupation:
agriculture: 78.6%
industry: 5.7%
services: 15.7% (FY08/09 est.)

Unemployment rate: 35% (2008 est.)
country comparison to the world: 182
40% (2005 est.)

Population below poverty line: 36% (FY08/09)

Household income or consumption by percentage share: *lowest 10%:* NA%
highest 10%: NA%

Budget: *revenues:* $1 billion
expenditures: $3.3 billion
note: Afghanistan has also received $2.6 billion from the Reconstruction Trust Fund and $63 million from the Law and Order Trust Fund (FY09/10 est.)

Inflation rate (consumer prices): 13.3% (2009 est.)
country comparison to the world: 213
20.7% (2008 est.)

Commercial bank prime lending rate: 15% (31 December 2009 est.)
country comparison to the world: 44
14.92% (31 December 2008 est.)

Stock of narrow money: $3.943 billion (31 December 2009)
country comparison to the world: 102
$2.819 billion (31 December 2008)

Stock of broad money: $4.149 billion (31 December 2009)
country comparison to the world: 125
$2.915 billion (31 December 2008)

Stock of domestic credit: $363.6 million (31 December 2008 est.)
country comparison to the world: 167
$20.06 million (31 December 2007 est.)

Market value of publicly traded shares: $NA

Agriculture—products: opium, wheat, fruits, nuts; wool, mutton, sheepskins, lambskins

Industries: small-scale production of textiles, soap, furniture, shoes, fertilizer, apparel, food-products, non-alcoholic beverages, mineral water, cement; handwoven carpets; natural gas, coal, copper

Industrial production growth rate: NA%

Electricity—production: 285.5 million kWh (2009 est.)
country comparison to the world: 168

Electricity—consumption: 231.1 million kWh (2009 est.)
country comparison to the world: 176

Electricity—exports: 0 kWh (2008 est.)

Electricity—imports: 230 million kWh (2007 est.)
Oil—production: 0 bbl/day (2009 est.)
country comparison to the world: 147
Oil—consumption: 5,000 bbl/day (2009 est.)
country comparison to the world: 164
Oil—exports: 0 bbl/day (2007 est.)
country comparison to the world: 145
Oil—imports: 4,404 bbl/day (2007 est.)
country comparison to the world: 162
Oil—proved reserves: 0 bbl (1 January 2010 est.)
country comparison to the world: 102
Natural gas—production: 30 million cu m (2008 est.)
country comparison to the world: 83
Natural gas—consumption: 30 million cu m (2008 est.)
country comparison to the world: 108
Natural gas—exports: 0 cu m (2008 est.)
country comparison to the world: 49
Natural gas—imports: 0 cu m (2008 est.)
country comparison to the world: 141
Natural gas—proved reserves: 49.55 billion cu m (1 January 2010 est.)
country comparison to the world: 65
Current account balance: $-2.475 billion (2009 est.)
country comparison to the world: 160
$85 million (2008 est.)
Exports: $547 million (2009 est.)
country comparison to the world: 163
$603 million (2008 est.)
note: not including illicit exports or reexports
Exports—commodities: opium, fruits and nuts, handwoven carpets, wool, cotton, hides and pelts, precious and semi-precious gems
Exports—partners: US 26.47%, India 23.09%, Pakistan 17.36%, Tajikistan 12.51% (2009)
Imports: $5.3 billion (2008 est.)
country comparison to the world: 111
$4.5 billion (2007)
Imports—commodities: machinery and other capital goods, food, textiles, petroleum products
Imports—partners: Pakistan 26.78%, US 24.81%, India 5.15%, Germany 5.06%, Russia 4.04% (2009)
Debt—external: $2.7 billion (FY08/09)
country comparison to the world: 134
$8 billion (2004)
Exchange rates: afghanis (AFA) per US dollar–46.45 (2010), 50.23 (2009)

COMMUNICATIONS

Telephones—main lines in use: 129,300 (2009)
country comparison to the world: 140
Telephones—mobile cellular: 12 million (2009)
country comparison to the world: 57
Telephone system: *general assessment:* limited fixed-line telephone service; an increasing number of Afghans utilize mobile-cellular phone networks
domestic: aided by the presence of multiple providers, mobile-cellular telephone service continues to improve rapidly
international: country code–93; multiple VSAT's provide international and domestic voice and data connectivity (2009)
Broadcast media: state-owned broadcaster, Radio Television Afghanistan (RTA), operates a series of radio and television stations in Kabul and the provinces; an estimated 50 private radio stations, 8 TV networks, and about a dozen international broadcasters are available; more than 30 community-based radio stations broadcasting (2007)
Internet country code: .af
Internet hosts: 46 (2010)
country comparison to the world: 211
Internet users: 1 million (2009)
country comparison to the world: 100
Communications—note: Internet access is growing through Internet cafes as well as public "telekiosks" in Kabul (2005)

TRANSPORTATION

Airports: 53 (2010)
country comparison to the world: 90
Airports—with paved runways: *total:* 19
over 3,047 m: 4
2,438 to 3,047 m: 3
1,524 to 2,437 m: 8
914 to 1,523 m: 2
under 914 m: 2 (2010)
Airports—with unpaved runways: *total:* 34
2,438 to 3,047 m: 5
1,524 to 2,437 m: 14
914 to 1,523 m: 6
under 914 m: 9 (2010)
Heliports: 11 (2010)
Pipelines: gas 466 km (2010)
Roadways: *total:* 42,150 km
country comparison to the world: 86
paved: 12,350 km
unpaved: 29,800 km (2006)
Waterways: 1,200 km; (chiefly Amu Darya, which handles vessels up to 500 DWT) (2008)
country comparison to the world: 61
Ports and terminals: Kheyrabad, Shir Khan

MILITARY

Military branches: Afghan Armed Forces: Afghan National Army (ANA, includes Afghan Air Force (AAF)) (2011)
Military service age and obligation: 22 years of age; inductees are contracted into service for a 4-year term (2005)
Manpower available for military service: *males age 16–49:* 7,056,339
females age 16–49: 6,653,419 (2010 est.)
Manpower fit for military service: *males age 16–49:* 4,050,222
females age 16–49: 3,797,087 (2010 est.)
Manpower reaching militarily significant age annually: *male:* 392,116
female: 370,295 (2010 est.)
Military expenditures: 1.9% of GDP (2009)
country comparison to the world: 77

TRANSNATIONAL ISSUES

Disputes—international: Afghan, Coalition, and Pakistan military meet periodically to clarify the alignment of the boundary on the ground and on maps; Afghan and Iranian commissioners have discussed boundary monument densification and resurvey; Iran protests Afghanistan's restricting flow of dammed Helmand River tributaries during drought; Pakistan has sent troops across and built fences along some remote tribal areas of its treaty-defined Durand Line border with Afghanistan which serve as bases for foreign terrorists and other illegal activities; Russia remains concerned about the smuggling of poppy derivatives from Afghanistan through Central Asian countries
Refugees and internally displaced persons: *IDPs:* 132,246 (mostly Pashtuns and Kuchis displaced in south and west due to drought and instability) (2007)
Illicit drugs: world's largest producer of opium; poppy cultivation decreased 22% to 157,000 hectares in 2008 but remains at a historically high level; less favorable growing conditions in 2008 reduced potential opium production to 5,500 metric tons, down 31 percent from 2007; if the entire opium crop were processed, 648 metric tons of pure heroin potentially could be produced; the Taliban and other antigovernment groups participate in and profit from the opiate trade, which is a key source of revenue for the Taliban inside Afghanistan; widespread corruption and instability impede counterdrug efforts; most of the heroin consumed in Europe and Eurasia is derived from Afghan opium; vulnerable to drug money laundering through informal financial networks; regional source of hashish (2008)

AKROTIRI

(UK SOVEREIGN BASE AREA)

INTRODUCTION

Background: By terms of the 1960 Treaty of Establishment that created the independent Republic of Cyprus, the UK retained full sovereignty and jurisdiction over two areas of almost 254 square kilometers–Akrotiri and Dhekelia. The southernmost and smallest of these is the Akrotiri Sovereign Base Area, which is also referred to as the Western Sovereign Base Area.

GEOGRAPHY

Location: Eastern Mediterranean, peninsula on the southwest coast of Cyprus
Geographic coordinates: 34 37 N, 32 58 E
Map references: Europe

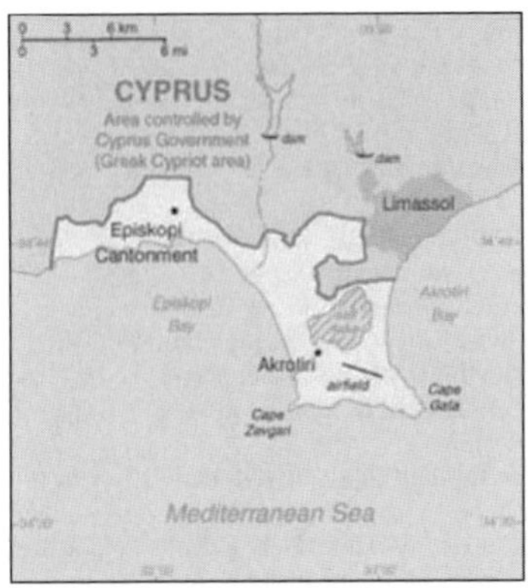

Area: *total:* 123 sq km
country comparison to the world: 222
note: includes a salt lake and wetlands
Area—comparative: about 0.7 times the size of Washington, DC
Land boundaries: *total:* 47.4 km
border countries: Cyprus 47.4 km
Coastline: 56.3 km
Climate: temperate; Mediterranean with hot, dry summers and cool winters
Environment—current issues: hunting around the salt lake; note–breeding place for loggerhead and green turtles; only remaining colony of griffon vultures is on the base
Geography—note: British extraterritorial rights also extended to several small off-post sites scattered across Cyprus; of the Sovereign Base Area (SBA) land, 60% is privately owned and farmed, 20% is owned by the Ministry of Defense, and 20% is SBA Crown land

PEOPLE

Population: approximately 15,700 live on the Sovereign Base Areas of Akrotiri and Dhekelia including 7,700 Cypriots, 3,600 Service and UK-based contract personnel, and 4,400 dependents
country comparison to the world: 218
Languages: English, Greek
School life expectancy (primary to tertiary education): NA
Education expenditures: NA

GOVERNMENT

Country name: *conventional long form:* none
conventional short form: Akrotiri
Dependency status: a special form of UK overseas territory; administered by an administrator who is also the Commander, British Forces Cyprus
Capital: *name:* Episkopi Cantonment (base administrative center for Akrotiri and Dhekelia)
geographic coordinates: 34 40 N, 32 51 E
time difference: UTC+2 (7 hours ahead of Washington, DC during Standard Time)
daylight saving time: +1hr, begins last Sunday in March; ends last Sunday in October
Constitution: Sovereign Base Areas of Akrotiri and Dhekelia Order in Council 1960, effective 16 August 1960, functions as a basic legal document
Legal system: the Sovereign Base Area Administration has its own court system to deal with civil and criminal matters; laws applicable to the Cypriot population are, as far as possible, the same as the laws of the Republic of Cyprus
Executive branch: *chief of state:* Queen ELIZABETH II (since 6 February 1952)
head of government: Administrator Air Vice Marshall Graham STACEY (since 4 November 2010); note–reports to the British Ministry of Defense; the Chief Officer is responsible for the day-to-day running of the civil government of the Sovereign Bases
elections: none; the monarchy is hereditary; the administrator appointed by the monarch
Judicial branch: The Court headed by a resident judge and senior judges from the UK as needed; note–the laws of the Sovereign Base Areas are kept as close as possible to the laws of the Republic of Cyprus
Diplomatic representation in the US: none (overseas territory of the UK)
Diplomatic representation from the US: none (overseas territory of the UK)
Flag description: the flag of the UK is used
National anthem: note: as a United Kingdom area of special sovereignty, "God Save the Queen" is official (see United Kingdom)

ECONOMY

Economy—overview: Economic activity is limited to providing services to the military and their families located in Akrotiri. All food and manufactured goods must be imported.
Exchange rates: note: uses the euro

COMMUNICATIONS

Broadcast media: British Forces Broadcast Service (BFBS) provides multi-channel satellite TV service as well as BFBS radio broadcasts to the Akrotiri Sovereign Base (2009)

MILITARY

Military—note: Akrotiri has a full RAF base, Headquarters for British Forces Cyprus, and Episkopi Support Unit

ALBANIA

INTRODUCTION

Background: Albania declared its independence from the Ottoman Empire in 1912, but was conquered by Italy in 1939. Communist partisans took over the country in 1944. Albania allied itself first with the USSR (until 1960), and then with China (to 1978). In the early 1990s, Albania ended 46 years of xenophobic Communist rule and established a multiparty democracy. The transition has proven challenging as successive governments have tried to deal with high unemployment, widespread corruption, a dilapidated physical infrastructure, powerful organized crime networks, and combative political opponents. Albania has made progress in its democratic development since first holding multiparty elections in 1991, but deficiencies remain. International observers judged elections to be largely free and fair since the restoration of political stability following the collapse of pyramid schemes in 1997; however, there have been claims of electoral fraud in every one of Albania's post-communist elections. The 2009 general elections resulted in no single party gaining a majority of the 140 seats in Parliament, and the Movement for Socialist Integration (LSI) and the Democratic Party (DP) combined to form a coalition government, the first such in Albania's history. The Socialist Party (SP) has, in effect, boycotted Parliament since it convened in September 2009 and has called for investigations into alleged electoral fraud in the June 2009 elections. Albania joined NATO in April 2009 and is a potential candidate for EU accession. Although Albania's economy continues to grow, the country is still one of the poorest in Europe, hampered by a large informal economy and an inadequate energy and transportation infrastructure.

GEOGRAPHY

Location: Southeastern Europe, bordering the Adriatic Sea and Ionian Sea, between Greece in the south and Montenegro and Kosovo to the north
Geographic coordinates: 41 00 N, 20 00 E
Map references: Europe
Area: *total:* 28,748 sq km
country comparison to the world: 144
land: 27,398 sq km
water: 1,350 sq km

Area—comparative: slightly smaller than Maryland
Land boundaries: *total:* 717 km
border countries: Greece 282 km, Macedonia 151 km, Montenegro 172 km, Kosovo 112 km
Coastline: 362 km
Maritime claims: territorial sea: 12 nm
continental shelf: 200 m depth or to the depth of exploitation
Climate: mild temperate; cool, cloudy, wet winters; hot, clear, dry summers; interior is cooler and wetter
Terrain: mostly mountains and hills; small plains along coast
Elevation extremes: *lowest point:* Adriatic Sea 0 m
highest point: Maja e Korabit (Golem Korab) 2,764 m
Natural resources: petroleum, natural gas, coal, bauxite, chromite, copper, iron ore, nickel, salt, timber, hydropower
Land use: *arable land:* 20.1%
permanent crops: 4.21%
other: 75.69% (2005)
Irrigated land: 3,530 sq km (2003)
Total renewable water resources: 41.7 cu km (2001)
Freshwater withdrawal (domestic/industrial/agricultural): *total:* 1.71 cu km/yr (27%/11%/62%)
per capita: 546 cu m/yr (2000)
Natural hazards: destructive earthquakes; tsunamis occur along southwestern coast; floods; drought
Environment—current issues: deforestation; soil erosion; water pollution from industrial and domestic effluents
Environment—international agreements: *party to:* Air Pollution, Biodiversity, Climate Change, Climate Change-Kyoto Protocol, Desertification, Endangered Species, Hazardous Wastes, Law of the Sea, Ozone Layer Protection, Wetlands
signed, but not ratified: none of the selected agreements
Geography—note: strategic location along Strait of Otranto (links Adriatic Sea to Ionian Sea and Mediterranean Sea)

PEOPLE

Population: 2,994,667 (July 2011 est.)
country comparison to the world: 136
Age structure: *0–14 years:* 21.4% (male 337,364/female 303,669)
15–64 years: 68.1% (male 996,666/female 1,043,472)
65 years and over: 10.5% (male 148,151/female 165,345) (2011 est.)
Median age: *total:* 30.4 years
male: 29.2 years
female: 31.6 years (2011 est.)
Population growth rate: 0.267% (2011 est.)
country comparison to the world: 171
Birth rate: 12.15 births/1,000 population (2011 est.)
country comparison to the world: 164
Death rate: 6.15 deaths/1,000 population (July 2011 est.)
country comparison to the world: 159
Net migration rate: -3.34 migrant(s)/1,000 population (2011 est.)
country comparison to the world: 181
Urbanization: *urban population:* 52% of total population (2010)
rate of urbanization: 2.3% annual rate of change (2010-15 est.)
Major cities—population: TIRANA (capital) 433,000 (2009)
Sex ratio: *at birth:* 1.118 male(s)/female
under 15 years: 1.1 male(s)/female
15–64 years: 1.05 male(s)/female
65 years and over: 0.87 male(s)/female
total population: 1.04 male(s)/female (2011 est.)
Infant mortality rate: *total:* 14.61 deaths/1,000 live births
country comparison to the world: 122
male: 16.23 deaths/1,000 live births
female: 12.79 deaths/1,000 live births (2011 est.)
Life expectancy at birth: *total population:* 77.41 years
country comparison to the world: 60
male: 74.82 years
female: 80.3 years (2011 est.)
Total fertility rate: 1.48 children born/woman (2011 est.)
country comparison to the world: 187
HIV/AIDS—adult prevalence rate: NA
HIV/AIDS—people living with HIV/AIDS: NA
HIV/AIDS—deaths: NA
Drinking water source: *Improved:*
urban: 96% of population
rural: 98% of population
total: 97% of population
Unimproved: urban: 4% of population
rural: 2% of population
total: 3% of population (2008)
Sanitation facility access: *Improved:*
urban: 98% of population
rural: 98% of population
total: 98% of population
Unimproved: urban: 2% of population
rural: 2% of population
total: 2% of population (2008)
Nationality: *noun:* Albanian(s)
adjective: Albanian
Ethnic groups: Albanian 95%, Greek 3%, other 2% (Vlach, Roma (Gypsy), Serb, Macedonian, Bulgarian) (1989 est.)
note: in 1989, other estimates of the Greek population ranged from 1% (official Albanian statistics) to 12% (from a Greek organization)
Religions: Muslim 70%, Albanian Orthodox 20%, Roman Catholic 10%
note: percentages are estimates; there are no available current statistics on religious affiliation; all mosques and churches were closed in 1967 and religious observances prohibited; in November 1990, Albania began allowing private religious practice
Languages: Albanian (official–derived from Tosk dialect), Greek, Vlach, Romani, Slavic dialects
Literacy: *definition:* age 9 and over can read and write
total population: 98.7%
male: 99.2%
female: 98.3% (2001 census)
School life expectancy (primary to tertiary education): *total:* 11 years
male: 11 years
female: 11 years (2004)
Education expenditures: NA

GOVERNMENT

Country name: *conventional long form:* Republic of Albania
conventional short form: Albania
local long form: Republika e Shqiperise
local short form: Shqiperia
former: People's Socialist Republic of Albania
Government type: parliamentary democracy
Capital: *name:* Tirana (Tirane)
geographic coordinates: 41 19 N, 19 49 E
time difference: UTC+1 (6 hours ahead of Washington, DC during Standard Time)
daylight saving time: +1hr, begins last Sunday in March; ends last Sunday in October
Administrative divisions: 12 counties (qarqe, singular–qark); Berat, Diber, Durres, Elbasan, Fier, Gjirokaster, Korce, Kukes, Lezhe, Shkoder, Tirane, Vlore
Independence: 28 November 1912 (from the Ottoman Empire)
National holiday: Independence Day, 28 November (1912) also known as Flag Day
Constitution: approved by parliament 21 October 1998; adopted by popular referendum 22 November 1998; promulgated 28 November 1998
Legal system: civil law system except in the northern rural areas where customary law known as the "Code of Leke" prevails
International law organization participation: has not submitted an ICJ jurisdiction declaration; accepts ICCt jurisdiction
Suffrage: 18 years of age; universal
Executive branch: *chief of state:* President of the Republic Bamir TOPI (since 24 July 2007)
head of government: Prime Minister Sali BERISHA (since 10 September 2005)
cabinet: Council of Ministers proposed by the prime minister, nominated by the president, and approved by parliament
(For more information visit the World Leaders website)
elections: president elected by three-fifths the Assembly for a five-year term (eligible for a second term); four election rounds held between 8 and 20 July 2007 (next election to be held in 2012); prime minister appointed by the president
election results: Bamir TOPI elected president; Assembly vote, fourth round (three-fifths majority, 84 votes, required): Bamir TOPI 85 votes, Neritan CEKA 5 votes
Legislative branch: unicameral Assembly or Kuvendi (140 deputies; 100 deputies elected directly in single member electoral zones with an approximate number of voters; 40 deputies elected from multi-name lists of parties or party coalitions according to their respective order; elected for a 4-year term)
elections: last held on 28 June 2009 (next to be held in 2013)
election results: percent of vote by party–NA; seats by party–PD 68, PS 65, LSI 4, other 3
Judicial branch: Constitutional Court consists of 9 members appointed by the president with the consent of the Assembly who serve 9-year terms (chairman is elected by the People's Assembly for a four-year term); the High Court members appointed by the president with the consent of the Assembly for a 9-year term; note–there are also courts of appeal and courts of first instance
Political parties and leaders: Democratic Party or PD [Sali BERISHA]; New

Democracy Party or PDR [Genc POLIO]; Party for Justice and Integration or PDI [Tahir MUCHEDINI]; Republican Party or PR [Fatmir MEDIU]; Social Democracy Party or PDS [Paskel MILO]; Social Democratic Party or PSD [Skender GJINUSHI]; Socialist Movement for Integration or LSI [Ilir META]; Socialist Party or PS [Edi RAMA]; Unity for Human Rights Party or PBDNJ [Vangjel DULE]

Political pressure groups and leaders: Citizens Advocacy Office [Kreshnik SPAHIU]; Confederation of Trade Unions of Albania or KSSH [Kastriot MUCO]; Front for Albanian National Unification or FBKSH [Gafur ADILI]; Mjaft Movement [Elton KACIDHJA]; Omonia [Ligorag KARAMELO]; Union of Independent Trade Unions of Albania or BSPSH [Gezim KALAJA]

International organization participation: BSEC, CE, CEI, EAPC, EBRD, FAO, IAEA, IBRD, ICAO, ICRM, IDA, IDB, IFAD, IFC, IFRCS, ILO, IMF, IMO, Interpol, IOC, IOM, IPU, ISO (correspondent), ITU, ITUC, MIGA, NATO, OIC, OIF, OPCW, OSCE, SECI, UN, UNCTAD, UNESCO, UNIDO, UNWTO, UPU, WCO, WFTU, WHO, WIPO, WMO, WTO

Diplomatic representation in the US:
chief of mission: Ambassador Gilbert GALANXHI
chancery: 2100 S Street NW, Washington, DC 20008
telephone: [1] (202) 223-4942
FAX: [1] (202) 628-7342
consulate(s) general: New York

Diplomatic representation from the US:
chief of mission: Ambassador Alexander ARVIZU
embassy: Rruga e Elbasanit, Labinoti #103, Tirana
mailing address: US Department of State, 9510 Tirana Place, Dulles, VA 20189-9510
telephone: [355] (4) 2247285
FAX: [355] (4) 2232222

Flag description: red with a black two-headed eagle in the center; the design is claimed to be that of 15th-century hero George Castriota SKANDERBEG, who led a successful uprising against the Turks that resulted in a short-lived independence for some Albanian regions (1443-1478); an unsubstantiated explanation for the eagle symbol is the tradition that Albanians see themselves as descendants of the eagle; they refer to themselves as "Shkypetars," which translates as "sons of the eagle"

National anthem: *name:* "Hymni i Flamurit" (Hymn to the Flag)
lyrics/music: Aleksander Stavre DRENOVA/ Ciprian PORUMBESCU
note: adopted 1912

ECONOMY

Economy—overview: Albania, a formerly closed, centrally-planned state, is making the difficult transition to a more modern open-market economy. Macroeconomic growth averaged around 6% between 2004-08, but declined to about 3% in 2009-10. Inflation is low and stable. The government has taken measures to curb violent crime, and recently adopted a fiscal reform package aimed at reducing the large gray economy and attracting foreign investment. Remittances, a significant catalyst for economic growth have declined from 12-15% of GDP to 9% of GDP in 2009, mostly from Albanians residing in Greece and Italy; this helps offset the towering trade deficit. The agricultural sector, which accounts for almost half of employment but only about one-fifth of GDP, is limited primarily to small family operations and subsistence farming because of lack of modern equipment, unclear property rights, and the prevalence of small, inefficient plots of land. Energy shortages because of a reliance on hydropower, and antiquated and inadequate infrastructure contribute to Albania's poor business environment and lack of success in attracting new foreign investment needed to expand the country's export base. FDI is among the lowest in the region, but the government has embarked on an ambitious program to improve the business climate through fiscal and legislative reforms. The completion of a new thermal power plant near Vlore has helped diversify generation capacity, and plans to upgrade transmission lines between Albania and Montenegro and Kosovo would help relieve the energy shortages. Also, with help from EU funds, the government is taking steps to improve the poor national road and rail network, a long-standing barrier to sustained economic growth.

GDP (purchasing power parity): $23.86 billion (2010 est.)
country comparison to the world: 115
$23.06 billion (2009 est.)
$22.32 billion (2008 est.)
note: data are in 2010 US dollars
Albania has an informal, and unreported, sector that may be as large as 50% of official GDP

GDP (official exchange rate): $11.77 billion (2010 est.)

GDP—real growth rate: 3.5% (2010 est.)
country comparison to the world: 110
3.3% (2009 est.)
7.7% (2008 est.)

GDP—per capita (PPP): $8,000 (2010 est.)
country comparison to the world: 124
$7,700 (2009 est.)
$7,500 (2008 est.)
note: data are in 2010 US dollars

GDP—composition by sector:
agriculture: 18.9%
industry: 23.5%
services: 57.6% (2010 est.)

Labor force: 1.06 million (2010 est.)
country comparison to the world: 141

Labor force—by occupation:
agriculture: 47.8%
industry: 23%
services: 29.2% (September 2010 est.)

Unemployment rate: 13.5% (September 2010 est.)
country comparison to the world: 139
13.8% (2009 est.)
note: these are official rates, but actual rates may exceed 30% due to preponderance of near-subsistence farming

Population below poverty line: 12.5% (2008 est.)

Household income or consumption by percentage share: *lowest 10%:* 3.2%
highest 10%: 25.9% (2005)

Distribution of family income—Gini index: 26.7 (2005)
country comparison to the world: 126

Investment (gross fixed): 27.7% of GDP (2010 est.)
country comparison to the world: 25

Budget:
revenues: $3.205 billion
expenditures: $3.571 billion (2010 est.)

Public debt: 59.3% of GDP (2010 est.)
country comparison to the world: 34
59.7% of GDP (2009 est.)

Inflation rate (consumer prices): 3.6% (2010 est.)
country comparison to the world: 102
3.7% (2009 est.)

Central bank discount rate: 5% (31 December 2010)
country comparison to the world: 77
5.25% (31 December 2009)

Commercial bank prime lending rate: 12.46% (31 December 2009 est.)
country comparison to the world: 73
11.75% (31 December 2008 est.)

Stock of narrow money: $2.566 billion (31 December 2010 est.)
country comparison to the world: 115
$2.982 billion (31 December 2009 est.)

Stock of broad money: $9.245 billion (31 December 2009)
country comparison to the world: 104
$9.136 billion (31 December 2008)

Stock of domestic credit: $7.701 billion (31 December 2010 est.)
country comparison to the world: 103
$8.231 billion (31 December 2009 est.)

Market value of publicly traded shares: $NA

Agriculture—products: wheat, corn, potatoes, vegetables, fruits, sugar beets, grapes; meat, dairy products

Industries: food processing, textiles and clothing; lumber, oil, cement, chemicals, mining, basic metals, hydropower

Industrial production growth rate: 3% (2010 est.)
country comparison to the world: 114

Electricity—production: 5.201 billion kWh (2009 est.)
country comparison to the world: 113

Electricity—consumption: 6.593 billion kWh
country comparison to the world: 102
note: 35% of electricity is lost in the system as a result of transmission inefficiencies and theft (2009 est.)

Electricity—exports: 0 kWh (2009 est.)

Electricity—imports: 1.884 billion kWh (2009 est.)

Oil—production: 5,400 bbl/day (2009 est.)
country comparison to the world: 94

Oil—consumption: 36,000 bbl/day (2009 est.)
country comparison to the world: 107

Oil—exports: 748.9 bbl/day (2005 est.)
country comparison to the world: 122

Oil—imports: 24,080 bbl/day (2007 est.)
country comparison to the world: 106

Oil—proved reserves: 199.1 million bbl (1 January 2010 est.)
country comparison to the world: 59

Natural gas—production: 30 million cu m (2008 est.)
country comparison to the world: 84

Natural gas—consumption: 30 million cu m (2008 est.)
country comparison to the world: 109
Natural gas—exports: 0 cu m (2010 est.)
country comparison to the world: 50
Natural gas—imports: 0 cu m (2010 est.)
country comparison to the world: 144
Natural gas—proved reserves: 849.5 million cu m (1 January 2010 est.)
country comparison to the world: 100
Current account balance: $-1.245 billion (2010 est.)
country comparison to the world: 141
$-1.845 billion (2009 est.)
Exports: $1.55 billion (2010 est.)
country comparison to the world: 137
$1.08 billion (2009 est.)
Exports—commodities: textiles and footwear; asphalt, metals and metallic ores, crude oil; vegetables, fruits, tobacco
Exports—partners: Italy 50.8%, Kosovo 6.2%, Turkey 5.9%, Greece 5.4%, China 5.5% (2010 est.)
Imports: $4.59 billion (2010 est.)
country comparison to the world: 123
$4.53 billion (2009 est.)
Imports—commodities: machinery and equipment, foodstuffs, textiles, chemicals
Imports—partners: Italy 28%, Greece 13%, China 6.3%, Turkey 5.6%, Germany 5.6% (2010 est.)
Reserves of foreign exchange and gold: $1.992 billion (31 December 2010 est.)
country comparison to the world: 103
$2.37 billion (31 December 2009 est.)
Debt—external: $2.81 billion (2009)
country comparison to the world: 132
$1.55 billion (2004)
Exchange rates: leke (ALL) per US dollar–104.08 (2010), 94.98 (2009), 79.546 (2008), 92.668 (2007), 98.384 (2006)

COMMUNICATIONS

Telephones—main lines in use: 363,000 (2009)
country comparison to the world: 108
Telephones—mobile cellular: 4.162 million (2009)
country comparison to the world: 102
Telephone system: *general assessment:* despite new investment in fixed lines teledensity remains low with roughly 10 fixed lines per 100 people; mobile-cellular telephone use is widespread and generally effective
domestic: offsetting the shortage of fixed line capacity, mobile-cellular phone service has been available since 1996; by 2009, four companies were providing mobile services and mobile teledensity exceeded 130 per 100 persons; Internet broadband services initiated in 2005 but growth has been slow; Internet cafes are popular in Tirana and have started to spread outside the capital
international: country code–355; submarine cable provides connectivity to Italy, Croatia, and Greece; the Trans-Balkan Line, a combination submarine cable and land fiber-optic system, provides additional connectivity to Bulgaria, Macedonia, and Turkey; international traffic carried by fiber-optic cable and, when necessary, by microwave radio relay from the Tirana exchange to Italy and Greece (2009)
Broadcast media: 3 public television networks, one of which transmits by satellite to Albanian-language communities in neighboring countries; more than 60 private television stations operating; many viewers can pick up Italian and Greek TV broadcasts via terrestrial reception; cable TV service is available; 2 public radio networks and roughly 25 private radio stations; several international broadcasters are available (2010)
Internet country code: .al
Internet hosts: 15,098 (2010)
country comparison to the world: 117
Internet users: 1.3 million (2009)
country comparison to the world: 91

TRANSPORTATION

Airports: 5 (2010)
country comparison to the world: 177
Airports—with paved runways: *total:* 4
2,438 to 3,047 m: 3
1,524 to 2,437 m: 1 (2010)
Airports—with unpaved runways: *total:* 1
914 to 1,523 m: 1 (2010)
Heliports: 1 (2010)
Pipelines: gas 339 km; oil 207 km (2010)
Railways: *total:* 339 km
country comparison to the world: 117
standard gauge: 339 km 1.435-m gauge (2010)
Roadways: *total:* 18,000 km
country comparison to the world: 116
paved: 7,020 km
unpaved: 10,980 km (2002)
Waterways: 41 km (on the Bojana River) (2010)
country comparison to the world: 104
Merchant marine: *total:* 25
country comparison to the world: 93
by type: bulk carrier 1, cargo 23, roll on/roll off 1
foreign-owned: 1 (Turkey 1)
registered in other countries: 4 (Antigua and Barbuda 1, Panama 3) (2010)
Ports and terminals: Durres, Sarande, Shengjin, Vlore

MILITARY

Military branches: Joint Force Command (includes Land, Naval, and Aviation Brigade Commands), Joint Support Command (includes Logistic Command), Training and Doctrine Command (2010)
Military service age and obligation: 19 years of age (2004)
Manpower available for military service: *males age 16–49:* 731,111
females age 16–49: 780,216 (2010 est.)
Manpower fit for military service: *males age 16–49:* 622,379
females age 16–49: 660,715 (2010 est.)
Manpower reaching militarily significant age annually: *male:* 31,986
female: 29,533 (2010 est.)
Military expenditures: 1.49% of GDP (2005 est.)
country comparison to the world: 102

TRANSNATIONAL ISSUES

Disputes—international: none
Illicit drugs: increasingly active transshipment point for Southwest Asian opiates, hashish, and cannabis transiting the Balkan route and–to a lesser extent–cocaine from South America destined for Western Europe; limited opium and expanding cannabis production; ethnic Albanian narcotrafficking organizations active and expanding in Europe; vulnerable to money laundering associated with regional trafficking in narcotics, arms, contraband, and illegal aliens

ALGERIA

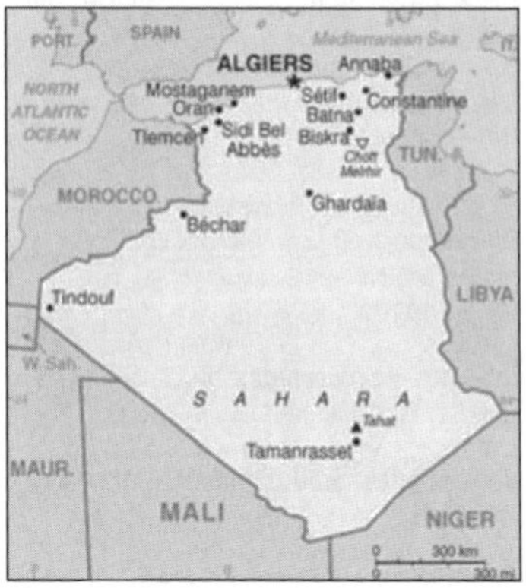

INTRODUCTION

Background: After more than a century of rule by France, Algerians fought through much of the 1950s to achieve independence in 1962. Algeria's primary political party, the National Liberation Front (FLN), was established in 1954 as part of the struggle for independence and has largely dominated politics since. The Government of Algeria in 1988 instituted a multi-party system in response to public unrest, but the surprising first round success of the Islamic Salvation Front (FIS) in the December 1991 balloting spurred the Algerian army to intervene and postpone the second round of elections to prevent what the secular elite feared would be an extremist-led government from assuming power. The army began a crackdown on the FIS that spurred FIS supporters to begin attacking government targets, and fighting escalated into an insurgency, which saw intense violence between 1992-98 resulting in over 100,000 deaths–many attributed to indiscriminate massacres of villagers by extremists. The government gained the upper hand by the late-1990s, and FIS's armed wing, the Islamic Salvation Army, disbanded in January 2000. Abdelaziz BOUTEFLIKA, with the backing of the military, won the presidency in 1999 in an election widely viewed as fraudulent. He was reelected to a second term in 2004, and overwhelmingly won a third term in 2009 after the government amended the constitution in 2008 to remove presidential term limits. Longstanding problems continue to face BOUTEFLIKA, including large-scale unemployment, a shortage of housing, unreliable electrical and water supplies, government inefficiencies and corruption, and the continuing activities of extremist militants. The Salafist Group for Preaching and Combat (GSPC) in 2006 merged with al-Qai'da to form al-Qai'da in the Lands of the Islamic Maghreb, which has launched an ongoing series of kidnappings and bombings targeting the Algerian Government and Western interests. The Arab uprising across the Near Eastern and North African region beginning in December 2010, coupled with a sudden rise in the cost of food staples, triggered a wave of protests across Algeria during early 2011. Organizers of many of these protests included opposition political parties, labor and trade unions, and human rights organizations. The government's initial response to protester demands included lifting the 19-year state of emergency laws and temporarily cutting taxes and duties on selected food items, but demonstrators continued to attempt to organize protests. An overwhelming Algerian police response managed to contain most demonstrations with some reports of violence. In mid-April 2011, President BOUTEFLIKA said he would seek to amend the nation's constitution to reinforce representative democracy, propose changes to the election laws, and submit the proposals for national referendum.

GEOGRAPHY

Location: Northern Africa, bordering the Mediterranean Sea, between Morocco and Tunisia
Geographic coordinates: 28 00 N, 3 00 E
Map references: Africa
Area: *total:* 2,381,741 sq km
country comparison to the world: 11
land: 2,381,741 sq km
water: 0 sq km
Area—comparative: slightly less than 3.5 times the size of Texas
Land boundaries: *total:* 6,343 km
border countries: Libya 982 km, Mali 1,376 km, Mauritania 463 km, Morocco 1,559 km, Niger 956 km, Tunisia 965 km, Western Sahara 42 km
Coastline: 998 km
Maritime claims: territorial sea: 12 nm
exclusive fishing zone: 32-52 nm
Climate: arid to semiarid; mild, wet winters with hot, dry summers along coast; drier with cold winters and hot summers on high plateau; sirocco is a hot, dust/sand-laden wind especially common in summer
Terrain: mostly high plateau and desert; some mountains; narrow, discontinuous coastal plain
Elevation extremes: *lowest point:* Chott Melrhir -40 m
highest point: Tahat 3,003 m
Natural resources: petroleum, natural gas, iron ore, phosphates, uranium, lead, zinc
Land use: *arable land:* 3.17%
permanent crops: 0.28%
other: 96.55% (2005)
Irrigated land: 5,690 sq km (2003)
Total renewable water resources: 14.3 cu km (1997)
Freshwater withdrawal (domestic/industrial/agricultural): *total:* 6.07 cu km/yr (22%/13%/65%)
per capita: 185 cu m/yr (2000)
Natural hazards: mountainous areas subject to severe earthquakes; mudslides and floods in rainy season
Environment—current issues: soil erosion from overgrazing and other poor farming practices; desertification; dumping of raw sewage, petroleum refining wastes, and other industrial effluents is leading to the pollution of rivers and coastal waters; Mediterranean Sea, in particular, becoming polluted from oil wastes, soil erosion, and fertilizer runoff; inadequate supplies of potable water
Environment—international agreements: *party to:* Biodiversity, Climate Change, Climate Change-Kyoto Protocol, Desertification, Endangered Species, Environmental Modification, Hazardous Wastes, Law of the Sea, Ozone Layer Protection, Ship Pollution, Wetlands
signed, but not ratified: none of the selected agreements
Geography—note: second-largest country in Africa (after Sudan)

PEOPLE

Population: 34,994,937 (July 2011 est.)
country comparison to the world: 35
Age structure: *0–14 years:* 24.2% (male 4,319,295/female 4,144,863)
15–64 years: 70.6% (male 12,455,378/female 12,242,604)
65 years and over: 5.2% (male 845,116/female 987,681) (2011 est.)
Median age: *total:* 27.6 years
male: 27.4 years
female: 27.8 years (2011 est.)
Population growth rate: 1.173% (2011 est.)
country comparison to the world: 99
Birth rate: 16.69 births/1,000 population (2011 est.)
country comparison to the world: 121
Death rate: 4.69 deaths/1,000 population (July 2011 est.)
country comparison to the world: 196
Net migration rate: -0.27 migrant(s)/1,000 population (2011 est.)
country comparison to the world: 126
Urbanization: *urban population:* 66% of total population (2010)
rate of urbanization: 2.3% annual rate of change (2010-15 est.)
Major cities—population: ALGIERS (capital) 2.74 million; Oran 770,000 (2009)
Sex ratio: *at birth:* 1.05 male(s)/female
under 15 years: 1.04 male(s)/female
15–64 years: 1.02 male(s)/female
65 years and over: 0.86 male(s)/female
total population: 1.01 male(s)/female (2011 est.)
Infant mortality rate: *total:* 25.81 deaths/1,000 live births
country comparison to the world: 80
male: 28.8 deaths/1,000 live births
female: 22.67 deaths/1,000 live births (2011 est.)
Life expectancy at birth: *total population:* 74.5 years
country comparison to the world: 98
male: 72.78 years
female: 76.31 years (2011 est.)
Total fertility rate: 1.75 children born/woman (2011 est.)
country comparison to the world: 162
HIV/AIDS—adult prevalence rate: 0.1%; 0.1% note–no country specific models provided (2009 est.)
country comparison to the world: 108
HIV/AIDS—people living with HIV/AIDS: 18,000 (2009 est.)
country comparison to the world: 80
HIV/AIDS—deaths: fewer than 1,000 (2009 est.)
country comparison to the world: 66
Drinking water source: *Improved:*
urban: 85% of population
rural: 79% of population
total: 83% of population
Unimproved:
urban: 15% of population
rural: 21% of population
total: 17% of population (2008)
Sanitation facility access: *Improved:*
urban: 98% of population
rural: 88% of population
total: 95% of population
Unimproved:
urban: 2% of population
rural: 12% of population
total: 5% of population (2008)
Nationality: *noun:* Algerian(s)
adjective: Algerian
Ethnic groups: Arab-Berber 99%, European less than 1%
note: although almost all Algerians are Berber in origin (not Arab), only a minority identify themselves as Berber, about 15% of the total population; these people live mostly in the mountainous region of Kabylie east of Algiers; the Berbers are also Muslim but identify with their Berber rather than Arab cultural heritage; Berbers have long agitated, sometimes violently, for autonomy; the government is unlikely to grant autonomy but has offered to begin sponsoring teaching Berber language in schools
Religions: Sunni Muslim (state religion) 99%, Christian and Jewish 1%
Languages: Arabic (official), French, Berber dialects
Literacy: *definition:* age 15 and over can read and write
total population: 69.9%
male: 79.6%
female: 60.1% (2002 est.)
School life expectancy (primary to tertiary education): *total:* 13 years
male: 13 years
female: 13 years (2005)
Education expenditures: 4.3% of GDP (2008)
country comparison to the world: 90

GOVERNMENT

Country name: *conventional long form:* People's Democratic Republic of Algeria
conventional short form: Algeria
local long form: Al Jumhuriyah al Jaza'iriyah ad Dimuqratiyah ash Sha'biyah
local short form: Al Jaza'ir

Government type: republic
Capital: *name:* Algiers
geographic coordinates: 36 45 N, 3 03 E
time difference: UTC+1 (6 hours ahead of Washington, DC during Standard Time)
Administrative divisions: 48 provinces (wilayat, singular—wilaya); Adrar, Ain Defla, Ain Temouchent, Alger, Annaba, Batna, Bechar, Bejaia, Biskra, Blida, Bordj Bou Arreridj, Bouira, Boumerdes, Chlef, Constantine, Djelfa, El Bayadh, El Oued, El Tarf, Ghardaia, Guelma, Illizi, Jijel, Khenchela, Laghouat, Mascara, Medea, Mila, Mostaganem, M'Sila, Naama, Oran, Ouargla, Oum el Bouaghi, Relizane, Saida, Setif, Sidi Bel Abbes, Skikda, Souk Ahras, Tamanghasset, Tebessa, Tiaret, Tindouf, Tipaza, Tissemsilt, Tizi Ouzou, Tlemcen
Independence: 5 July 1962 (from France)
National holiday: Revolution Day, 1 November (1954)
Constitution: 8 September 1963; revised 19 November 1976; effective 22 November 1976; revised 3 November 1988, 23 February 1989, 28 November 1996, 10 April 2002, and 12 November 2008
Legal system: mixed legal system of French civil law and Islamic law; judicial review of legislative acts in ad hoc Constitutional Council composed of various public officials including several Supreme Court justices
International law organization participation: has not submitted an ICJ jurisdiction declaration; non-party state to the ICCt
Suffrage: 18 years of age; universal
Executive branch: *chief of state:* President Abdelaziz BOUTEFLIKA (since 28 April 1999); note—the president is both the chief of state and head of government; a November 2008 constitutional amendment separated the position of head of government from that of the prime minister
head of government: President Abdelaziz BOUTEFLIKA (since 28 April 1999)
cabinet: Cabinet of Ministers appointed by the president
(For more information visit the World Leaders website)
elections: president elected by popular vote for a five-year term; note—a November 2008 constitutional amendment abolished presidential term limits; election last held on 9 April 2009 (next to be held in April 2014)
election results: Abdelaziz BOUTEFLIKA was reelected president for a third term; percent of vote—Abdelaziz BOUTEFLIKA 90.2%, Louisa HANOUNE 4.2%, Moussa TOUATI 2.3%, Djahid YOUNSI 1.4%, Ali Fawzi REBIANE less than 1%, Mohamed SAID less than 1%
Legislative branch: bicameral Parliament consists of the Council of the Nation (upper house; 144 seats; one-third of the members appointed by the president, two-thirds elected by indirect vote to serve six-year terms; the constitution requires half the Council to be renewed every three years) and the National People's Assembly (lower house; 389 seats; members elected by popular vote to serve five-year terms)
elections: Council of the Nation—last held on 29 December 2009 (next to be held in December 2012); National People's Assembly—last held on 17 May 2007 (next to be held in 2012)
election results: Council of the Nation—percent of vote by party—NA; seats by party—NA; National People's Assembly—percent of vote by party—NA; seats by party—FLN 136, RND 61, MSP 52, PT 26, RCD 19, FNA 13, other 49, independents 33;
Judicial branch: Supreme Court
Political parties and leaders: Ahd 54 [Ali Fauzi REBAINE]; Algerian National Front or FNA [Moussa TOUATI]; Movement of the Society of Peace or MSP [Boudjerra SOLTANI]; National Democratic Rally (Rassemblement National Democratique) or RND [Ahmed OUYAHIA]; National Liberation Front or FLN [Abdelaziz BELKHADEM, secretary general]; National Reform Movement or Islah [Ahmed ABDESLAM] (formerly MRN); Rally for Culture and Democracy or RCD [Said SADI]; Renaissance Movement or EnNahda Movement [Fatah RABEI]; Socialist Forces Front or FFS [Hocine Ait AHMED]; Workers Party or PT [Louisa HANOUNE]
note: a law banning political parties based on religion was enacted in March 1997
Political pressure groups and leaders: The Algerian Human Rights League or LADDH [Hocine ZEHOUANE]; SOS Disparus [Nacera DUTOUR]
International organization participation: ABEDA, AfDB, AFESD, AMF, AMU, AU, BIS, FAO, G-15, G-24, G-77, IAEA, IBRD, ICAO, ICC, ICRM, IDA, IDB, IFAD, IFC, IFRCS, IHO, ILO, IMF, IMO, IMSO, Interpol, IOC, IOM, IPU, ISO, ITSO, ITU, ITUC, LAS, MIGA, MONUSCO, NAM, OAPEC, OAS (observer), OIC, OPCW, OPEC, OSCE (partner), UN, UNCTAD, UNESCO, UNHCR, UNIDO, UNITAR, UNWTO, UPU, WCO, WHO, WIPO, WMO, WTO (observer)
Diplomatic representation in the US: *chief of mission:* Ambassador Abdallah BAALI
chancery: 2118 Kalorama Road NW, Washington, DC 20008
telephone: [1] (202) 265-2800
FAX: [1] (202) 667-2174
Diplomatic representation from the US: *chief of mission:* Ambassador David D. PEARCE
embassy: 05 Chemin Cheikh Bachir, El-Ibrahimi, El-Biar 16000 Algiers
mailing address: B. P. 408, Alger-Gare, 16030 Algiers
telephone: [213] 770-08-2000
FAX: [213] 21-60-7355
Flag description: two equal vertical bands of green (hoist side) and white; a red, five-pointed star within a red crescent centered over the two-color boundary; the colors represent Islam (green), purity and peace (white), and liberty (red); the crescent and star are also Islamic symbols, but the crescent is more closed than those of other Muslim countries because the Algerians believe the long crescent horns bring happiness
National anthem: *name:* "Kassaman" (We Pledge)
lyrics/music: Mufdi ZAKARIAH/Mohamed FAWZI
note: adopted 1962; ZAKARIAH wrote "Kassaman" as a poem while imprisoned in Algiers by French colonial forces

ECONOMY

Economy—overview: Algeria's economy remains dominated by the state, a legacy of the country's socialist post-independence development model. Gradual liberalization since the mid-1990s has opened up more of the economy, but in recent years Algeria has imposed new restrictions on foreign involvement in its economy and largely halted the privatization of state-owned industries. Hydrocarbons have long been the backbone of the economy, accounting for roughly 60% of budget revenues, 30% of GDP, and over 95% of export earnings. Algeria has the eighth-largest reserves of natural gas in the world and is the fourth-largest gas exporter. It ranks 16th in oil reserves. Thanks to strong hydrocarbon revenues, Algeria has a cushion of $150 billion in foreign currency reserves and a large hydrocarbon stabilization fund. In addition, Algeria's external debt is extremely low at about 1% of GDP. Algeria has struggled to develop industries outside of hydrocarbons in part because of high costs and an inert state bureaucracy. The government's efforts to diversify the economy by attracting foreign and domestic investment outside the energy sector have done little to reduce high poverty and youth unemployment rates. In 2010, Algeria began a five-year, $286 billion development program to update the country's infrastructure and provide jobs. The costly program will boost Algeria's economy in 2011 but worsen the country's budget deficit. Long-term economic challenges include diversification from hydrocarbons, relaxing state control of the economy, and providing adequate jobs for younger Algerians.
GDP (purchasing power parity): $251.1 billion (2010 est.)
country comparison to the world: 49
$243 billion (2009 est.)
$237.4 billion (2008 est.)
note: data are in 2010 US dollars
GDP (official exchange rate): $160.3 billion (2010 est.)
GDP—real growth rate: 3.3% (2010 est.)
country comparison to the world: 116
2.4% (2009 est.)
2.4% (2008 est.)
GDP—per capita (PPP): $7,300 (2010 est.)
country comparison to the world: 128
$7,100 (2009 est.)
$7,000 (2008 est.)
note: data are in 2010 US dollars
GDP—composition by sector:
agriculture: 8.3%
industry: 61.5%
services: 30.2% (2010 est.)
Labor force: 9.877 million (2010 est.)
country comparison to the world: 49
Labor force—by occupation:
agriculture: 14%
industry: 13.4%
construction and public works: 10%
trade: 14.6%
government: 32%
other: 16% (2003 est.)
Unemployment rate: 9.9% (2010 est.)

country comparison to the world: 109
10.2% (2009 est.)
Population below poverty line: 23% (2006 est.)
Household income or consumption by percentage share: *lowest 10%:* 2.8%
highest 10%: 26.8% (1995)
Distribution of family income—Gini index: 35.3 (1995)
country comparison to the world: 86
Investment (gross fixed): 27.5% of GDP (2010 est.)
country comparison to the world: 28
Budget: *revenues:* $66.48 billion
expenditures: $85.57 billion (2010 est.)
Public debt: 25.7% of GDP (2010 est.)
country comparison to the world: 97
20% of GDP (2009 est.)
Inflation rate (consumer prices): 5% (2010 est.)
country comparison to the world: 139
5.7% (2009 est.)
Central bank discount rate: 4% (31 December 2009)
country comparison to the world: 102
4% (31 December 2008)
Commercial bank prime lending rate: 8% (31 December 2009 est.)
country comparison to the world: 114
8% (31 December 2008 est.)
Stock of narrow money: $79.07 billion (31 December 2010 est.)
country comparison to the world: 36
$68.13 billion (31 December 2009 est.)
Stock of broad money: $109.7 billion (31 December 2010 est.)
country comparison to the world: 50
$98.82 billion (31 December 2009 est.)
Stock of domestic credit: $12.29 billion (31 December 2009 est.)
country comparison to the world: 90
$21.71 billion (31 December 2008 est.)
Market value of publicly traded shares: $NA
Agriculture—products: wheat, barley, oats, grapes, olives, citrus, fruits; sheep, cattle
Industries: petroleum, natural gas, light industries, mining, electrical, petrochemical, food processing
Industrial production growth rate: 4.8% (2010 est.)
country comparison to the world: 73
Electricity—production: 34.98 billion kWh (2007 est.)
country comparison to the world: 61
Electricity—consumption: 28.34 billion kWh (2007 est.)
country comparison to the world: 61
Electricity—exports: 273 million kWh (2007 est.)
Electricity—imports: 279 million kWh (2007 est.)
Oil—production: 2.125 million bbl/day (2009 est.)
country comparison to the world: 16
Oil—consumption: 325,000 bbl/day (2009 est.)
country comparison to the world: 39
Oil—exports: 1.891 million bbl/day (2007 est.)
country comparison to the world: 12
Oil—imports: 14,320 bbl/day (2007 est.)
country comparison to the world: 129
Oil—proved reserves: 13.42 billion bbl (1 January 2010 est.)
country comparison to the world: 16
Natural gas—production: 86.5 billion cu m (2008 est.)
country comparison to the world: 7
Natural gas—consumption: 26.83 billion cu m (2008 est.)
country comparison to the world: 28
Natural gas—exports: 59.67 billion cu m (2008 est.)
country comparison to the world: 4
Natural gas—imports: 0 cu m (2008 est.)
country comparison to the world: 142
Natural gas—proved reserves: 4.502 trillion cu m (1 January 2010 est.)
country comparison to the world: 10
Current account balance: $3.959 billion (2010 est.)
country comparison to the world: 34
$-4.185 billion (2009 est.)
Exports: $52.66 billion (2010 est.)
country comparison to the world: 50
$43.69 billion (2009 est.)
Exports—commodities: petroleum, natural gas, and petroleum products 97%
Exports—partners: US 23.2%, Italy 17.23%, Spain 10.83%, France 7.97%, Canada 7.65%, Netherlands 5.19%, Turkey 4.22% (2009)
Imports: $37.07 billion (2010 est.)
country comparison to the world: 52
$39.1 billion (2009 est.)
Imports—commodities: capital goods, foodstuffs, consumer goods
Imports—partners: France 19.7%, China 11.72%, Italy 10.19%, Spain 8.13%, Germany 5.77%, Turkey 5.05% (2009)
Reserves of foreign exchange and gold: $150.1 billion (31 December 2010 est.)
country comparison to the world: 13
$149.3 billion (31 December 2009 est.)
Debt—external: $4.138 billion (31 December 2010 est.)
country comparison to the world: 115
$5.413 billion (31 December 2009 est.)
Stock of direct foreign investment—at home: $19.34 billion (31 December 2010 est.)
country comparison to the world: 69
$17.34 billion (31 December 2009 est.)
Stock of direct foreign investment—abroad: $1.844 billion (31 December 2010 est.)
country comparison to the world: 68
$1.644 billion (31 December 2009 est.)
Exchange rates: Algerian dinars (DZD) per US dollar–76 (2010), 72.65 (2009), 63.25 (2008), 69.9 (2007), 72.647 (2006)

COMMUNICATIONS

Telephones—main lines in use: 2.576 million (2009)
country comparison to the world: 52
Telephones—mobile cellular: 32.73 million (2009)
country comparison to the world: 30
Telephone system: *general assessment:* privatization of Algeria's telecommunications sector began in 2000; three mobile cellular licenses have been issued and, in 2005, a consortium led by Egypt's Orascom Telecom won a 15-year license to build and operate a fixed-line network in Algeria; the license will allow Orascom to develop high-speed data and other specialized services and contribute to meeting the large unfulfilled demand for basic residential telephony; Internet broadband services began in 2003
domestic: a limited network of fixed lines with a teledensity of less than 10 telephones per 100 persons is offset by the rapid increase in mobile-cellular subscribership; in 2009, combined fixed-line and mobile-cellular teledensity was roughly 100 telephones per 100 persons
international: country code–213; landing point for the SEA-ME-WE-4 fiber-optic submarine cable system that provides links to Europe, the Middle East, and Asia; microwave radio relay to Italy, France, Spain, Morocco, and Tunisia; coaxial cable to Morocco and Tunisia; participant in Medarabtel; satellite earth stations–51 (Intelsat, Intersputnik, and Arabsat) (2009)
Broadcast media: state-run Radio-Television Algerienne operates the broadcast media and carries programming in Arabic, Berber dialects, and French; use of satellite dishes is widespread, providing easy access to European and Arab satellite stations; state-run radio operates several national networks and roughly 40 regional radio stations (2007)
Internet country code: .dz
Internet hosts: 572 (2010)
country comparison to the world: 176
Internet users: 4.7 million (2009)
country comparison to the world: 49

TRANSPORTATION

Airports: 143 (2010)
country comparison to the world: 39
Airports—with paved runways: *total:* 57
over 3,047 m: 12
2,438 to 3,047 m: 28
1,524 to 2,437 m: 11
914 to 1,523 m: 5
under 914 m: 1 (2010)
Airports—with unpaved runways:
total: 86
2,438 to 3,047 m: 3
1,524 to 2,437 m: 19
914 to 1,523 m: 41
under 914 m: 23 (2010)
Heliports: 2 (2010)
Pipelines: condensate 2,600 km; gas 16,360 km; liquid petroleum gas 3,447 km; oil 7,611 km; refined products 144 km (2010)
Railways: *total:* 3,973 km
country comparison to the world: 44
standard gauge: 2,888 km 1.435-m gauge (283 km electrified)
narrow gauge: 1,085 km 1.055-m gauge (2008)
Roadways: *total:* 108,302 km
country comparison to the world: 38
paved: 76,028 km (includes 645 km of expressways)
unpaved: 32,274 km (2004)
Merchant marine: *total:* 35
country comparison to the world: 81
by type: bulk carrier 6, cargo 8, chemical tanker 2, liquefied gas 9, passenger/cargo 3, petroleum tanker 4, roll on/roll off 3
foreign-owned: 12 (UK 12) (2010)
Ports and terminals: Algiers, Annaba, Arzew, Bejaia, Djendjene, Jijel, Mostaganem, Oran, Skikda

MILITARY

Military branches: People's National Army (Armee Nationale Populaire, ANP), Land Forces (Forces Terrestres, FT), Navy of the Republic of Algeria (Marine de la Republique Algerienne, MRA), Air Force (Al-Quwwat al-Jawwiya al-Jaza'eriya, QJJ), Territorial Air Defense Force (2009)
Military service age and obligation: 19-30 years of age for compulsory military service; conscript service obligation—18 months (6 months basic training, 12 months civil projects) (2006)
Manpower available for military service: *males age 16–49:* 10,273,129
females age 16–49: 10,114,552 (2010 est.)
Manpower fit for military service: *males age 16–49:* 8,622,897
females age 16–49: 8,626,222 (2010 est.)
Manpower reaching militarily significant age annually: *male:* 342,895
female: 330,098 (2010 est.)
Military expenditures: 3.3% of GDP (2006)
country comparison to the world: 37

TRANSNATIONAL ISSUES

Disputes—international: Algeria, and many other states, rejects Moroccan administration of Western Sahara; the Polisario Front, exiled in Algeria, represents the Sahrawi Arab Democratic Republic; Algeria's border with Morocco remains an irritant to bilateral relations, each nation accusing the other of harboring militants and arms smuggling; dormant disputes include Libyan claims of about 32,000 sq km still reflected on its maps of southeastern Algeria and the FLN's assertions of a claim to Chirac Pastures in southeastern Morocco
Refugees and internally displaced persons: refugees (country of origin): 90,000 (Western Saharan Sahrawi, mostly living in Algerian-sponsored camps in the southwestern Algerian town of Tindouf)
IDPs: undetermined (civil war during 1990s) (2007)
Trafficking in persons: current situation: Algeria is a transit country for men and women trafficked from sub-Saharan Africa to Europe for the purposes of commercial sexual exploitation and involuntary servitude; criminal networks of sub-Saharan nationals in southern Algeria facilitate transit by arranging transportation, forged documents, and promises of employment
tier rating: Tier 2 Watch List—Algeria is placed on the Tier 2 Watch List because it does not fully comply with the minimum standards for the elimination of trafficking, however, it is making significant efforts to do so; in January 2009, the government approved new legislation that criminalizes trafficking in persons for the purposes of labor and sexual exploitation representing an important step toward complying with international standards; despite these efforts, the government did not show overall progress in punishing trafficking crimes and protecting trafficking victims and continued to lack adequate measures to protect victims and prevent trafficking (2009)

AMERICAN SAMOA

(TERRITORY OF THE US)

INTRODUCTION

Background: Settled as early as 1000 B.C., Samoa was "discovered" by European explorers in the 18th century. International rivalries in the latter half of the 19th century were settled by an 1899 treaty in which Germany and the US divided the Samoan archipelago. The US formally occupied its portion–a smaller group of eastern islands with the excellent harbor of Pago Pago–the following year.

GEOGRAPHY

Location: Oceania, group of islands in the South Pacific Ocean, about half way between Hawaii and New Zealand
Geographic coordinates: 14 20 S, 170 00 W
Map references: Oceania
Area: *total:* 199 sq km
country comparison to the world: 214
land: 199 sq km
water: 0 sq km
note: includes Rose Island and Swains Island
Area—comparative: slightly larger than Washington, DC
Land boundaries: 0 km
Coastline: 116 km
Maritime claims: territorial sea: 12 nm
exclusive economic zone: 200 nm
Climate: tropical marine, moderated by southeast trade winds; annual rainfall averages about 3 m; rainy season (November to April), dry season (May to October); little seasonal temperature variation
Terrain: five volcanic islands with rugged peaks and limited coastal plains, two coral atolls (Rose Island, Swains Island)
Elevation extremes: *lowest point:* Pacific Ocean 0 m
highest point: Lata Mountain 964 m
Natural resources: pumice, pumicite
Land use: *arable land:* 10%
permanent crops: 15%
other: 75% (2005)
Irrigated land: NA
Natural hazards: typhoons common from December to March
volcanism: American Samoa experiences limited volcanic activity on the Ofu and Olosega Islands, neither has erupted since the 19th century
Environment—current issues: limited natural freshwater resources; the water division of the government has spent substantial funds in the past few years to improve water catchments and pipelines
Geography—note: Pago Pago has one of the best natural deepwater harbors in the South Pacific Ocean, sheltered by shape from rough seas and protected by peripheral mountains from high winds; strategic location in the South Pacific Ocean

PEOPLE

Population: 67,242 (July 2011 est.)
country comparison to the world: 201
Age structure: *0–14 years:* 31.9% (male 10,910/female 10,518)
15–64 years: 63.9% (male 21,764/female 21,228)
65 years and over: 4.2% (male 1,322/female 1,500) (2011 est.)
Median age: *total:* 23.7 years
male: 23.6 years
female: 23.9 years (2011 est.)
Population growth rate: 1.211% (2011 est.)
country comparison to the world: 96
Birth rate: 22.84 births/1,000 population (2011 est.)
country comparison to the world: 73
Death rate: 4.1 deaths/1,000 population (July 2011 est.)
country comparison to the world: 206
Net migration rate: -6.63 migrant(s)/1,000 population (2011 est.)
country comparison to the world: 201
Urbanization: *urban population:* 93% of total population (2010)
rate of urbanization: 1.8% annual rate of change (2010-15 est.)
Major cities—population: PAGO PAGO (capital) 60,000 (2009)
Sex ratio: *at birth:* 1.06 male(s)/female
under 15 years: 1.04 male(s)/female
15–64 years: 1.03 male(s)/female
65 years and over: 0.88 male(s)/female
total population: 1.02 male(s)/female (2011 est.)
Infant mortality rate: *total:* 9.66 deaths/1,000 live births
country comparison to the world: 151
male: 12.56 deaths/1,000 live births
female: 6.6 deaths/1,000 live births (2011 est.)
Life expectancy at birth: *total population:* 74.21 years
country comparison to the world: 105
male: 71.27 years
female: 77.32 years (2011 est.)
Total fertility rate: 3.16 children born/woman (2011 est.)

country comparison to the world: 55
HIV/AIDS—adult prevalence rate: NA
HIV/AIDS—people living with HIV/AIDS: NA
HIV/AIDS—deaths: NA
Nationality: *noun:* American Samoan(s) (US nationals)
adjective: American Samoan
Ethnic groups: native Pacific islander 91.6%, Asian 2.8%, white 1.1%, mixed 4.2%, other 0.3% (2000 census)
Religions: Christian Congregationalist 50%, Roman Catholic 20%, Protestant and other 30%
Languages: Samoan 90.6% (closely related to Hawaiian and other Polynesian languages), English 2.9%, Tongan 2.4%, other Pacific islander 2.1%, other 2%
note: most people are bilingual (2000 census)
Literacy: *definition:* age 15 and over can read and write
total population: 97%
male: 98%
female: 97% (1980 est.)
School life expectancy (primary to tertiary education): NA
Education expenditures: NA

GOVERNMENT

Country name: *conventional long form:* Territory of American Samoa
conventional short form: American Samoa
abbreviation: AS
Dependency status: unincorporated and unorganized territory of the US; administered by the Office of Insular Affairs, US Department of the Interior
Government type: NA
Capital: *name:* Pago Pago
geographic coordinates: 14 16 S, 170 42 W
time difference: UTC-11 (6 hours behind Washington, DC during Standard Time)
Administrative divisions: none (territory of the US); there are no first-order administrative divisions as defined by the US Government, but there are three districts and two islands* at the second order; Eastern, Manu'a, Rose Island*, Swains Island*, Western
Independence: none (territory of the US)
National holiday: Flag Day, 17 April (1900)
Constitution: ratified 2 June 1966; effective 1 July 1967
Legal system: mixed legal system of US common law and customary law
Suffrage: 18 years of age; universal
Executive branch: *chief of state:* President Barack H. OBAMA (since 20 January 2009); Vice President Joseph R. BIDEN (since 20 January 2009)
head of government: Governor Togiola TULAFONO (since 7 April 2003)
cabinet: Cabinet made up of 12 department directors
(For more information visit the World Leaders website)
elections: under the US Constitution, residents of unincorporated territories, such as American Samoa, do not vote in elections for US president and vice president; however, they may vote in Democratic and Republican presidential primary elections; governor and lieutenant governor elected on the same ticket by popular vote for four-year terms (eligible for a second term); election last held on 4 and 18 November 2008 (next to be held in November 2012)
election results: Togiola TULAFONO reelected governor; percent of vote—Togiola TULAFONO 56.5%, Afoa Moega LUTU 43.5%
Legislative branch: bicameral Fono or Legislative Assembly consists of the Senate (18 seats; members are elected from local chiefs to serve four-year terms)and the House of Representatives (21 seats; 20 members are elected by popular vote and 1 is an appointed, nonvoting delegate from Swains Island; members serve two-year terms)
elections: House of Representatives—last held on 2 November 2010 (next to be held in November 2012); Senate—last held on 4 November 2008 (next to be held in November 2012)
election results: House of Representatives—percent of vote by party—NA; seats by party—independents 20; Senate—percent of vote by party—NA; seats by party—independents 18
note: American Samoa elects one nonvoting representative to the US House of Representatives; election last held on 2 November 2010 (next to be held in November 2012); results—Eni F. H. FALEOMAVAEGA reelected as delegate
Judicial branch: High Court (chief justice and associate justices are appointed by the US Secretary of the Interior)
Political parties and leaders: Democratic Party [Oreta M. TOGAFAU]; Republican Party [Tautai A. F. FAALEVAO]
Political pressure groups and leaders: Population Pressure LAS (addresses the growing population pressures)
International organization participation: AOSIS, Interpol (subbureau), IOC, SPC, UPU
Diplomatic representation in the US: none (territory of the US)
Diplomatic representation from the US: none (territory of the US)
Flag description: blue, with a white triangle edged in red that is based on the fly side and extends to the hoist side; a brown and white American bald eagle flying toward the hoist side is carrying two traditional Samoan symbols of authority, a war club known as a "Fa'alaufa'i" (upper; left talon), and a coconut fiber fly whisk known as a "Fue" (lower; right talon); the combination of symbols broadly mimics that seen on the US Great Seal and reflects the relationship between the United States and American Samoa
National anthem: *name:* "Amerika Samoa" (American Samoa)
lyrics/music: Mariota Tiumalu TUIASOSOPO/Napoleon Andrew TUITELELEAPAGA
note: local anthem adopted 1950; as a territory of the United States, "The Star-Spangled Banner" is official (see United States)

ECONOMY

Economy—overview: American Samoa has a traditional Polynesian economy in which more than 90% of the land is communally owned. Economic activity is strongly linked to the US with which American Samoa conducts most of its commerce. Tuna fishing and tuna processing plants are the backbone of the private sector, with canned tuna the primary export. The two tuna canneries account for 80% of employment. In late September 2009, an earthquake and the resulting tsunami devastated American Samoa and nearby Samoa, disrupting transportation and power generation, and resulting in about 200 deaths. The US Federal Emergency Management Agency is overseeing a relief program of nearly $25 million. Transfers from the US Government add substantially to American Samoa's economic well being. Attempts by the government to develop a larger and broader economy are restrained by Samoa's remote location, its limited transportation, and its devastating hurricanes. Tourism is a promising developing sector.
GDP (purchasing power parity): $575.3 million (2007 est.)
country comparison to the world: 211
$510.1 million (2003 est.)
GDP (official exchange rate): $462.2 million (2005)
GDP—real growth rate: 3% (2003)
country comparison to the world: 125
GDP—per capita (PPP): $8,000 (2007 est.)
country comparison to the world: 123
$5,800 (2005 est.)
GDP—composition by sector:
agriculture: NA%
industry: NA%
services: NA%
Labor force: 17,630 (2005)
country comparison to the world: 210
Labor force—by occupation:
agriculture: 34%
industry: 33%
services: 33% (1990)
Unemployment rate: 29.8% (2005)
country comparison to the world: 175
Population below poverty line: NA%
Household income or consumption by percentage share: *lowest 10%:* NA%
highest 10%: NA%
Budget: *revenues:* $155.4 million (FY07)
expenditures: $183.6 million (FY07)
Inflation rate (consumer prices): NA%
Agriculture—products: bananas, coconuts, vegetables, taro, breadfruit, yams, copra, pineapples, papayas; dairy products, livestock
Industries: tuna canneries (largely supplied by foreign fishing vessels), handicrafts
Industrial production growth rate: NA%
Electricity—production: 185 million kWh (2007 est.)
country comparison to the world: 178
Electricity—consumption: 172.1 million kWh (2007 est.)
country comparison to the world: 181
Electricity—exports: 0 kWh (2008 est.)
Electricity—imports: 0 kWh (2008 est.)
Oil—production: 0 bbl/day (2009 est.)
country comparison to the world: 149
Oil—consumption: 4,000 bbl/day (2009 est.)
country comparison to the world: 172
Oil—exports: 0 bbl/day (2007 est.)
country comparison to the world: 147
Oil—imports: 4,140 bbl/day (2007 est.)

country comparison to the world: 166
Oil—proved reserves: 0 bbl (1 January 2010 est.)
country comparison to the world: 104
Natural gas—production: 0 cu m (2008 est.)
country comparison to the world: 96
Natural gas—consumption: 0 cu m (2008 est.)
country comparison to the world: 145
Natural gas—exports: 0 cu m (2008 est.)
country comparison to the world: 53
Natural gas—imports: 0 cu m (2008 est.)
country comparison to the world: 146
Natural gas—proved reserves: 0 cu m (1 January 2010 est.)
country comparison to the world: 107
Exports: $445.6 million (FY04 est.)
country comparison to the world: 170
Exports—commodities: canned tuna 93%
Imports: $308.8 million (FY04 est.)
country comparison to the world: 194
Imports—commodities: raw materials for canneries 56%, food, petroleum products, machinery and parts
Debt—external: $NA
Exchange rates: the US dollar is used

COMMUNICATIONS

Telephones—main lines in use: 10,400 (2009)
country comparison to the world: 201
Telephones—mobile cellular: 2,200 (2004)
country comparison to the world: 213
Telephone system: *general assessment:* NA
domestic: good telex, telegraph, facsimile, and cellular telephone services; domestic satellite system with 1 Comsat earth station
international: country code—1-684; satellite earth station—1 (Intelsat-Pacific Ocean)
Broadcast media: 3 television stations broadcasting; multi-channel pay-per-view television services are available; about a dozen radio stations, some of which are repeater stations (2009)
Internet country code: .as
Internet hosts: 1,676 (2010)
country comparison to the world: 157
Internet users: NA

TRANSPORTATION

Airports: 3 (2010)
country comparison to the world: 191
Airports—with paved runways:
total: 3
over 3,047 m: 1
914 to 1,523 m: 1
under 914 m: 1 (2010)
Roadways: *total:* 241 km (2008)
country comparison to the world: 205
Ports and terminals: Pago Pago

MILITARY

Manpower fit for military service: *males age 16–49:* 14,562
females age 16–49: 14,129 (2010 est.)
Manpower reaching militarily significant age annually: *male:* 775
female: 762 (2010 est.)
Military—note: defense is the responsibility of the US

TRANSNATIONAL ISSUES

Disputes—international: Tokelau included American Samoa's Swains Island (Olohega) in its 2006 draft independence constitution

ANDORRA

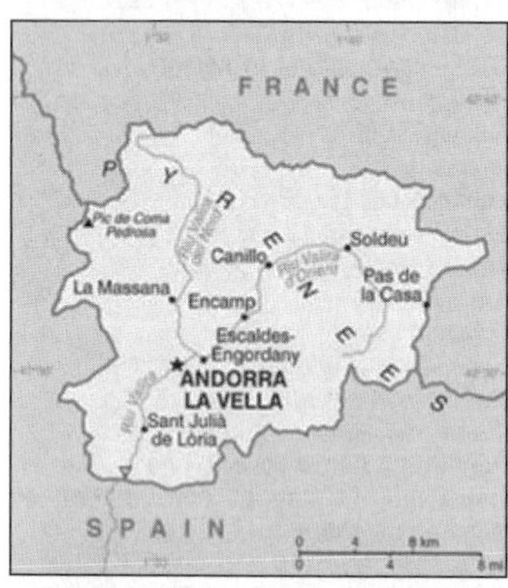

INTRODUCTION

Background: For 715 years, from 1278 to 1993, Andorrans lived under a unique co-principality, ruled by French and Spanish leaders (from 1607 onward, the French chief of state and the Spanish bishop of Seu d'Urgell). In 1993, this feudal system was modified with the titular heads of state retained, but the government transformed into a parliamentary democracy. For decades Andorra enjoyed its status as a small refuge of fiscal and banking freedom and benefitted from Spanish and French tourists attracted to the country's duty-free shopping. The situation has changed in recent years as Andorra started to tax foreign investment and other sectors. Tourism accounts for over 80% of Andorra's gross domestic product.

GEOGRAPHY

Location: Southwestern Europe, Pyrenees mountains, on the border between France and Spain
Geographic coordinates: 42 30 N, 1 30 E
Map references: Europe
Area: *total:* 468 sq km
country comparison to the world: 194
land: 468 sq km
water: 0 sq km
Area—comparative: 2.5 times the size of Washington, DC
Land boundaries: *total:* 120.3 km
border countries: France 56.6 km, Spain 63.7 km
Coastline: 0 km (landlocked)
Maritime claims: none (landlocked)
Climate: temperate; snowy, cold winters and warm, dry summers
Terrain: rugged mountains dissected by narrow valleys
Elevation extremes: *lowest point:* Riu Runer 840 m
highest point: Pic de Coma Pedrosa 2,946 m
Natural resources: hydropower, mineral water, timber, iron ore, lead
Land use: *arable land:* 2.13%
permanent crops: 0%
other: 97.87% (2005)
Irrigated land: NA
Natural hazards: avalanches
Environment—current issues: deforestation; overgrazing of mountain meadows contributes to soil erosion; air pollution; wastewater treatment and solid waste disposal
Environment—international agreements: *party to:* Biodiversity, Desertification, Hazardous Wastes, Ozone Layer Protection
signed, but not ratified: none of the selected agreements
Geography—note: landlocked; straddles a number of important crossroads in the Pyrenees

PEOPLE

Population: 84,825 (July 2011 est.)
country comparison to the world: 197
Age structure: *0–14 years:* 15.6% (male 6,799/female 6,440)
15–64 years: 71.4% (male 31,545/female 29,037)
65 years and over: 13% (male 5,502/female 5,502) (2011 est.)
Median age: *total:* 40.5 years
male: 40.8 years
female: 40.2 years (2011 est.)
Population growth rate: 0.33% (2011 est.)
country comparison to the world: 167
Birth rate: 9.66 births/1,000 population (2011 est.)
country comparison to the world: 197
Death rate: 6.35 deaths/1,000 population (July 2011 est.)
country comparison to the world: 153
Net migration rate: 0 migrant(s)/1,000 population (2011 est.)
country comparison to the world: 72
Urbanization: *urban population:* 88% of total population (2010)
rate of urbanization: 1.1% annual rate of change (2010-15 est.)
Major cities—population: ANDORRA LA VELLA (capital) 25,000 (2009)
Sex ratio: *at birth:* 1.066 male(s)/female
under 15 years: 1.06 male(s)/female
15–64 years: 1.09 male(s)/female
65 years and over: 0.99 male(s)/female
total population: 1.07 male(s)/female (2011 est.)
Infant mortality rate: *total:* 3.8 deaths/1,000 live births
country comparison to the world: 204
male: 3.76 deaths/1,000 live births
female: 3.84 deaths/1,000 live births (2011 est.)
Life expectancy at birth: *total population:* 82.43 years
country comparison to the world: 4
male: 80.35 years
female: 84.64 years (2011 est.)

Total fertility rate: 1.35 children born/woman (2011 est.)
country comparison to the world: 206
HIV/AIDS—adult prevalence rate: NA
HIV/AIDS—people living with HIV/AIDS: NA
HIV/AIDS—deaths: NA
Drinking water source: *Improved:*
urban: 100% of population
rural: 100% of population
total: 100% of population (2008)
Sanitation facility access: *Improved:*
urban: 100% of population
rural: 100% of population
total: 100% of population (2008)
Nationality: *noun:* Andorran(s)
adjective: Andorran
Ethnic groups: Spanish 43%, Andorran 33%, Portuguese 11%, French 7%, other 6% (1998)
Religions: Roman Catholic (predominant)
Languages: Catalan (official), French, Castilian, Portuguese
Literacy: *definition:* age 15 and over can read and write
total population: 100%
male: 100%
female: 100%
School life expectancy (primary to tertiary education): *total:* 12 years
male: 11 years
female: 12 years (2008)
Education expenditures: 3.2% of GDP (2008)
country comparison to the world: 123

GOVERNMENT

Country name:
conventional long form: Principality of Andorra
conventional short form: Andorra
local long form: Principat d'Andorra
local short form: Andorra
Government type: parliamentary democracy (since March 1993) that retains as its chiefs of state a coprincipality; the two princes are the president of France and bishop of Seu d'Urgell, Spain, who are represented in Andorra by the coprinces' representatives
Capital: *name:* Andorra la Vella
geographic coordinates: 42 30 N, 1 31 E
time difference: UTC+1 (6 hours ahead of Washington, DC during Standard Time)
daylight saving time: +1hr, begins last Sunday in March; ends last Sunday in October
Administrative divisions: 7 parishes (parroquies, singular–parroquia); Andorra la Vella, Canillo, Encamp, Escaldes-Engordany, La Massana, Ordino, Sant Julia de Loria
Independence: 1278 (formed under the joint sovereignty of the French Count of Foix and the Spanish Bishop of Seu d'Urgel)
National holiday: Our Lady of Meritxell Day, 8 September (1278)
Constitution: Andorra's first written constitution was drafted in 1991; approved by referendum 14 March 1993; effective 28 April 1993
Legal system: mixed legal system of civil and customary law with canon (religious) law influences
International law organization participation: has not submitted an ICJ jurisdiction declaration; accepts ICCt jurisdiction
Suffrage: 18 years of age; universal
Executive branch: *chief of state:* French Coprince Nicolas SARKOZY (since 16 May 2007); represented by Christian FREMONT (since September 2008) and Spanish Coprince Archbishop Joan-Enric VIVES i Sicilia (since 12 May 2003); represented by Nemesi MARQUES i Oste (since 30 July 2003)
head of government: Executive Council President (or Cap de Govern) Antoni MARTI PETIT (since 12 May 2011)
cabinet: Executive Council or Govern designated by the Executive Council president (For more information visit the World Leaders website)
elections: Executive Council president elected by the General Council (Andorran Parliament) and formally appointed by the coprinces for a four-year term; election last held on 3 April 2011 (next to be held in April 2015)
election results: Antoni MARTI PETIT to be elected Executive Council president; percent of General Council vote–NA; note–the leader of the party which wins a majority of seats in the General Council is usually elected president
Legislative branch: unicameral General Council of the Valleys or Consell General de las Valls (28-42 seats; members are elected by direct popular vote, 14 from a single national constituency and 14 to represent each of the seven parishes; to serve four-year terms)
elections: last held on 3 April 2011 (next to be held in April 2015)
election results: percent of vote by party–DA 55%, PS 35%, Andorra for Change 7%, Andorran Green 3%; seats by party–DA 20, PS 6, Lauredian Union 2
note: under usual circumstances, the next election would have been held in 2013, but because the General Council was unable to pass important laws such as the budget, it was dissolved and an election was held
Judicial branch: Tribunal of Judges or Tribunal de Batlles; Tribunal of the Courts or Tribunal de Corts; Supreme Court of Justice of Andorra or Tribunal Superior de Justicia d'Andorra; Supreme Council of Justice or Consell Superior de la Justicia (coprinces appoint the members); Constitutional Tribunal or Tribunal Constitucional (coprinces appoint the members)
note: all judges elected for a 6-year term renewable term
Political parties and leaders: there are four political parties at the national level: Andorra for Change or ApC [Eusebio NOMEN CALVET]; Democrats for Andorra or DA [Antoni MARTI PETIT], coalition including Liberal Party (PRA) and Reformist Coalition; Greens of Andorra [Isabel LOZANO MUNOZ]; Social Democratic Party or PS [Jaume BARTUMEU CASSANY]; note–there are also several smaller parties at the Parish level (one is Lauredian Union)
Political pressure groups and leaders: NA
International organization participation: CE, FAO, ICAO, ICRM, IFRCS, Interpol, IOC, IPU, ITU, OIF, OPCW, OSCE, UN, UNCTAD, UNESCO, Union Latina, UNWTO, WCO, WHO, WIPO, WTO (observer)
Diplomatic representation in the US: *chief of mission:* Ambassador Narcis CASAL de Fonsdeviela
chancery: 2 United Nations Plaza, 27th Floor, New York, NY 10017
telephone: [1] (212) 750-8064
FAX: [1] (212) 750-6630
Diplomatic representation from the US: the US does not have an embassy in Andorra; the US Ambassador to Spain is accredited to Andorra; US interests in Andorra are represented by the US Consulate General's office in Barcelona (Spain); *mailing address:* Paseo Reina Elisenda de Montcada, 23, 08034 Barcelona, Spain; *telephone:* [34] (93) 280-2227; *FAX:* [34] (93) 280-6175
Flag description: three vertical bands of blue (hoist side), yellow, and red, with the national coat of arms centered in the yellow band; the latter band is slightly wider than the other two so that the ratio of band widths is 8:9:8; the coat of arms features a quartered shield with the emblems of (starting in the upper left and proceeding clockwise): Urgell, Foix, Bearn, and Catalonia; the motto reads VIRTUS UNITA FORTIOR (Strength United is Stronger); the flag combines the blue and red French colors with the red and yellow of Spain to show Franco-Spanish protection
note: similar to the flags of Chad and Romania, which do not have a national coat of arms in the center, and the flag of Moldova, which does bear a national emblem
National anthem: *name:* "El Gran Carlemany" (The Great Charlemagne)
lyrics/music: Joan BENLLOCH i VIVO/Enric MARFANY BONS
note: adopted 1921; the anthem provides a brief history of Andorra in a first person narrative

ECONOMY

Economy—overview: Tourism, commerce, and finance are the mainstays of Andorra's tiny, well-to-do economy, accounting for more than three-quarters of GDP. An estimated 9 million tourists visit annually, attracted by Andorra's duty-free status for some products and by its summer and winter resorts. Andorra's comparative advantage eroded when the borders of neighboring France and Spain opened, providing broader availability of goods and lower tariffs. The banking sector also contributes substantially to the economy. Agricultural production is limited–only 2% of the land is arable–and most food has to be imported. The principal livestock activity is sheep raising. Manufacturing output and exports consist mainly of perfumes and cosmetic products, products of the printing industry, electrical machinery and equipment, clothing, tobacco products, and furniture. Andorra is a member of the EU Customs Union and is treated as an EU member for trade in manufactured goods (no tariffs) and as a non-EU member for agricultural products.

GDP (purchasing power parity): $3.3 billion (2009 est.)
country comparison to the world: 174
$4.22 billion (2008 est.)
$3.66 billion (2007 est.)
GDP (official exchange rate): $NA
GDP—real growth rate: 3.8% (2009 est.)
country comparison to the world: 104
2.6% (2008 est.)
2% (2007 est.)
GDP—per capita (PPP): $46,700 (2009 est.)
country comparison to the world: 12
$44,900 (2008 est.)
$42,500 (2007 est.)
GDP—composition by sector:
agriculture: NA%
industry: NA%
services: NA%
Labor force: 38,220 (2010)
country comparison to the world: 200
Labor force—by occupation:
agriculture: 0.4%
industry: 4.7%
services: 94.9% (2010)
Unemployment rate: 2.9% (2009)
country comparison to the world: 23
2.7% (2008)
Population below poverty line: NA (2008)
Household income or consumption by percentage share: *lowest 10%:* NA%
highest 10%: NA%
Budget: *revenues:* $872 million
expenditures: $868.4 million (2009)
Inflation rate (consumer prices): 1.6% (2010)
country comparison to the world: 41
0% (2009)
Agriculture—products: small quantities of rye, wheat, barley, oats, vegetables; sheep
Industries: tourism (particularly skiing), cattle raising, timber, banking, tobacco, furniture
Industrial production growth rate: NA%
Electricity—production: 101 million kWh (2009)
country comparison to the world: 190
Electricity—consumption: 598.7 million kWh (2009)
country comparison to the world: 156
Electricity—exports: 9 million kWh (2009)
Electricity—imports: 497.7 million kWh
note: most electricity supplied by Spain and France; Andorra generates a small amount of hydropower (2009)
Exports: $64 million (2009)
country comparison to the world: 200
$89.5 million (2008)
Exports—commodities: tobacco products, furniture
Imports: $1.474 billion (2009)
country comparison to the world: 164
$1.801 billion (2008)
Imports—commodities: consumer goods, food, electricity
Debt—external: $NA
Exchange rates: euros (EUR) per US dollar–0.755 (2010), 0.7179 (2009), 0.6827 (2008), 0.7306 (2007), 0.7964 (2006)

COMMUNICATIONS

Telephones—main lines in use: 37,900 (2009)
country comparison to the world: 172
Telephones—mobile cellular: 64,500 (2009)
country comparison to the world: 191
Telephone system: *general assessment:* modern automatic telephone system
domestic: modern system with microwave radio relay connections between exchanges
international: country code–376; landline circuits to France and Spain
Broadcast media: 1 public television station and 2 public radio stations; about 10 commercial radio stations operating; good reception of radio and TV broadcasts from stations in France and Spain; upgraded to terrestrial digital television broadcasting in 2007; roughly 25 international television channels available (2010)
Internet country code: .ad
Internet hosts: 26,773 (2010)
country comparison to the world: 100
Internet users: 67,100 (2009)
country comparison to the world: 170

TRANSPORTATION

Roadways: *total:* 320 km (2008)
country comparison to the world: 202

MILITARY

Military branches: no regular military forces, Police Service of Andorra (2010)
Manpower available for military service: *males age 16–49:* 22,390 (2010 est.)
Manpower fit for military service: *males age 16–49:* 17,977
females age 16–49: 17,069 (2010 est.)
Manpower reaching militarily significant age annually: *male:* 397
female: 347 (2010 est.)
Military—note: defense is the responsibility of France and Spain

TRANSNATIONAL ISSUES

Disputes—international: none

ANGOLA

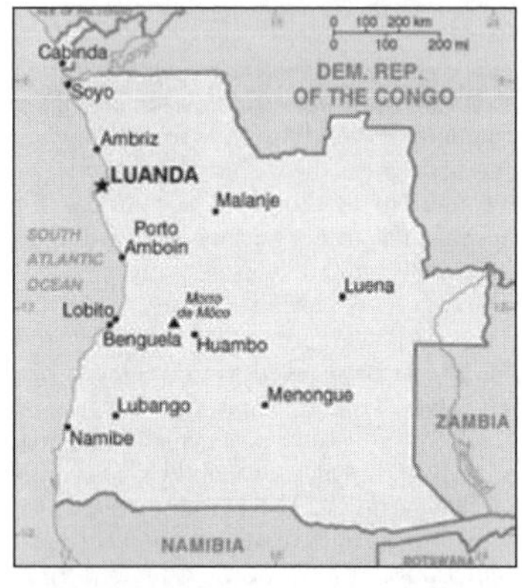

INTRODUCTION

Background: Angola is rebuilding its country after the end of a 27-year civil war in 2002. Fighting between the Popular Movement for the Liberation of Angola (MPLA), led by Jose Eduardo DOS SANTOS, and the National Union for the Total Independence of Angola (UNITA), led by Jonas SAVIMBI, followed independence from Portugal in 1975. Peace seemed imminent in 1992 when Angola held national elections, but fighting picked up again by 1996. Up to 1.5 million lives may have been lost–and 4 million people displaced–in the quarter century of fighting. SAVIMBI's death in 2002 ended UNITA's insurgency and strengthened the MPLA's hold on power. President DOS SANTOS held legislative elections in September 2008 and, despite promising to hold presidential elections in 2009, has since pushed through a new constitution that calls for elections in 2012.

GEOGRAPHY

Location: Southern Africa, bordering the South Atlantic Ocean, between Namibia and Democratic Republic of the Congo
Geographic coordinates: 12 30 S, 18 30 E
Map references: Africa
Area: *total:* 1,246,700 sq km
country comparison to the world: 23
land: 1,246,700 sq km
water: 0 sq km
Area—comparative: slightly less than twice the size of Texas
Land boundaries: *total:* 5,198 km
border countries: Democratic Republic of the Congo 2,511 km (of which 225 km is the boundary of discontiguous Cabinda Province), Republic of the Congo 201 km, Namibia 1,376 km, Zambia 1,110 km
Coastline: 1,600 km
Maritime claims: territorial sea: 12 nm
contiguous zone: 24 nm
exclusive economic zone: 200 nm
Climate: semiarid in south and along coast to Luanda; north has cool, dry season (May to October) and hot, rainy season (November to April)
Terrain: narrow coastal plain rises abruptly to vast interior plateau
Elevation extremes: *lowest point:* Atlantic Ocean 0 m
highest point: Morro de Moco 2,620 m
Natural resources: petroleum, diamonds, iron ore, phosphates, copper, feldspar, gold, bauxite, uranium
Land use: *arable land:* 2.65%
permanent crops: 0.23%
other: 97.12% (2005)
Irrigated land: 800 sq km (2003)
Total renewable water resources: 184 cu km (1987)
Freshwater withdrawal (domestic/industrial/agricultural): *total:* 0.35 cu km/yr (23%/17%/60%)
per capita: 22 cu m/yr (2000)
Natural hazards: locally heavy rainfall causes periodic flooding on the plateau

Environment—current issues: overuse of pastures and subsequent soil erosion attributable to population pressures; desertification; deforestation of tropical rain forest, in response to both international demand for tropical timber and to domestic use as fuel, resulting in loss of biodiversity; soil erosion contributing to water pollution and siltation of rivers and dams; inadequate supplies of potable water

Environment—international agreements: *party to:* Biodiversity, Climate Change, Climate Change-Kyoto Protocol, Desertification, Law of the Sea, Marine Dumping, Ozone Layer Protection, Ship Pollution
signed, but not ratified: none of the selected agreements

Geography—note: the province of Cabinda is an exclave, separated from the rest of the country by the Democratic Republic of the Congo

PEOPLE

Population: 13,338,541 (July 2011 est.)
country comparison to the world: 70

Age structure: *0–14 years:* 43.2% (male 2,910,981/female 2,856,527)
15–64 years: 54.1% (male 3,663,400/female 3,549,896)
65 years and over: 2.7% (male 157,778/female 199,959) (2011 est.)

Median age: *total:* 18.1 years
male: 18.1 years
female: 18.1 years (2011 est.)

Population growth rate: 2.034% (2011 est.)
country comparison to the world: 47

Birth rate: 42.91 births/1,000 population (2011 est.)
country comparison to the world: 7

Death rate: 23.4 deaths/1,000 population (July 2011 est.)
country comparison to the world: 1

Net migration rate: 0.82 migrant(s)/1,000 population (2011 est.)
country comparison to the world: 57

Urbanization: *urban population:* 59% of total population (2010)
rate of urbanization: 4% annual rate of change (2010-15 est.)

Major cities—population: LUANDA (capital) 4.511 million; Huambo 979,000 (2009)

Sex ratio: *at birth:* 1.05 male(s)/female
under 15 years: 1.02 male(s)/female
15–64 years: 1.03 male(s)/female
65 years and over: 0.79 male(s)/female
total population: 1.02 male(s)/female (2011 est.)

Infant mortality rate: *total:* 175.9 deaths/1,000 live births
country comparison to the world: 1
male: 187.86 deaths/1,000 live births
female: 163.34 deaths/1,000 live births (2011 est.)

Life expectancy at birth: *total population:* 38.76 years
country comparison to the world: 222
male: 37.74 years
female: 39.83 years (2011 est.)

Total fertility rate: 5.97 children born/woman (2011 est.)
country comparison to the world: 9

HIV/AIDS—adult prevalence rate: 2% (2009 est.)
country comparison to the world: 29

HIV/AIDS—people living with HIV/AIDS: 200,000 (2009 est.)
country comparison to the world: 28

HIV/AIDS—deaths: 11,000 (2009 est.)
country comparison to the world: 25

Major infectious diseases: *degree of risk:* very high
food or waterborne diseases: bacterial and protozoal diarrhea, hepatitis A, typhoid fever
vectorborne diseases: malaria, African trypanosomiasis (sleeping sickness)
water contact disease: schistosomiasis (2009)

Drinking water source: *Improved:*
urban: 60% of population
rural: 38% of population
total: 50% of population
Unimproved:
urban: 40% of population
rural: 62% of population
total: 50% of population (2008)

Sanitation facility access: *Improved:*
urban: 86% of population
rural: 18% of population
total: 57% of population
Unimproved:
urban: 14% of population
rural: 82% of population
total: 43% of population (2008)

Nationality: *noun:* Angolan(s)
adjective: Angolan

Ethnic groups: Ovimbundu 37%, Kimbundu 25%, Bakongo 13%, mestico (mixed European and native African) 2%, European 1%, other 22%

Religions: indigenous beliefs 47%, Roman Catholic 38%, Protestant 15% (1998 est.)

Languages: Portuguese (official), Bantu and other African languages

Literacy: *definition:* age 15 and over can read and write
total population: 67.4%
male: 82.9%
female: 54.2% (2001 est.)

School life expectancy (primary to tertiary education): *total:* 9 years (2006)

Education expenditures: 2.6% of GDP (2006)
country comparison to the world: 148

GOVERNMENT

Country name: *conventional long form:* Republic of Angola
conventional short form: Angola
local long form: Republica de Angola
local short form: Angola
former: People's Republic of Angola

Government type: republic; multiparty presidential regime

Capital: *name:* Luanda
geographic coordinates: 8 50 S, 13 14 E
time difference: UTC+1 (6 hours ahead of Washington, DC during Standard Time)

Administrative divisions: 18 provinces (provincias, singular—provincia); Bengo, Benguela, Bie, Cabinda, Cuando Cubango, Cuanza Norte, Cuanza Sul, Cunene, Huambo, Huila, Luanda, Lunda Norte, Lunda Sul, Malanje, Moxico, Namibe, Uige, Zaire

Independence: 11 November 1975 (from Portugal)

National holiday: Independence Day, 11 November (1975)

Constitution: adopted by National Assembly 5 February 2010

Legal system: civil legal system based on Portuguese civil law; no judicial review of legislative acts

International law organization participation: has not submitted an ICJ jurisdiction declaration; non-party state to the ICCt

Suffrage: 18 years of age; universal

Executive branch: *chief of state:* President Jose Eduardo DOS SANTOS (since 21 September 1979); Vice President Fernando da Piedade Dias DOS SANTOS (since 2 February 2010); note—the president is both chief of state and head of government
head of government: President Jose Eduardo DOS SANTOS (since 21 September 1979); Vice President Fernando da Piedade Dias DOS SANTOS (since 2 February 2010)
cabinet: Council of Ministers appointed by the president
(For more information visit the World Leaders website)
elections: president indirectly elected by National Assembly for a five-year term (eligible for a second consecutive or discontinuous term) under the 2010 constitution; President DOS SANTOS was selected by the party to take over after the death of former President Augustino NETO (1979) under a one-party system and stood for reelection in Angola's first multiparty elections on 29-30 September 1992 (next were to be held in September 2009 but were postponed)
election results: Jose Eduardo DOS SANTOS 49.6%, Jonas SAVIMBI 40.1%, making a run-off election necessary; the run-off was never held leaving DOS SANTOS in his current position as the president

Legislative branch: unicameral National Assembly or Assembleia Nacional (220 seats; members elected by proportional vote to serve four-year terms)
elections: last held on 5-6 September 2008 (next to be held in September 2012)
election results: percent of vote by party—MPLA 81.6%, UNITA 10.4%, PRS 3.2%, ND 1.2%, FNLA 1.1%, other 2.5%; seats by party—MPLA 191, UNITA 16, PRS 8, FNLA 3, ND 2

Judicial branch: Constitutional Court or Tribunal Constitucional; Supreme Court or Tribunal Supremo; Court of Auditions or Tribunal de Contas; Supreme Military Court or Supremo Tribunal Militar; judges for all courts appointed by the president

Political parties and leaders: National Front for the Liberation of Angola or FNLA [Ngola KABANGU]; National Union for the Total Independence of Angola or UNITA [Isaias SAMAKUVA] (largest opposition party); New Democracy Electoral Union or ND [Quintino de MOREIRA]; Popular Movement for the Liberation of Angola or MPLA [Jose Eduardo DOS SANTOS] (ruling party in power since 1975); Social Renewal Party or PRS [Eduardo KUANGANA]
note: nine other parties participated in the legislative election in September 2008 but won no seats

Political pressure groups and leaders: Front for the Liberation of the Enclave

of Cabinda or FLEC [N'zita Henriques TIAGO, Antonio Bento BEMBE]
note: FLEC's small-scale armed struggle for the independence of Cabinda Province persists despite the signing of a peace accord with the government in August 2006
International organization participation: ACP, AfDB, AU, CPLP, FAO, G-77, IAEA, IBRD, ICAO, ICRM, IDA, IFAD, IFC, IFRCS, ILO, IMF, IMO, Interpol, IOC, IOM, IPU, ISO (correspondent), ITSO, ITU, ITUC, MIGA, NAM, OAS (observer), OPEC, SADC, UN, UNCTAD, UNESCO, UNIDO, Union Latina, UNWTO, UPU, WCO, WFTU, WHO, WIPO, WMO, WTO
Diplomatic representation in the US: *chief of mission:* Ambassador Josefina Perpetua Pitra DIAKITE
chancery: 2108 16th Street NW, Washington, DC 20009
telephone: [1] (202) 785-1156
FAX: [1] (202) 785-1258
consulate(s) general: Houston, New York
Diplomatic representation from the US: *chief of mission:* Ambassador Dan MOZENA
embassy: number 32 Rua Houari Boumedienne (in the Miramar area of Luanda), Luanda
mailing address: international mail: Caixa Postal 6468, Luanda; pouch: US Embassy Luanda, US Department of State, 2550 Luanda Place, Washington, DC 20521-2550
telephone: [244] (222) 64-1000
FAX: [244] (222) 64-1232
Flag description: two equal horizontal bands of red (top) and black with a centered yellow emblem consisting of a five-pointed star within half a cogwheel crossed by a machete (in the style of a hammer and sickle); red represents liberty, black the African continent, the symbols characterize workers and peasants
National anthem: *name:* "Angola Avante" (Forward Angola)
lyrics/music: Manuel Rui Alves MONTEIRO/Rui Alberto Vieira Dias MINGAO
note: adopted 1975

ECONOMY

Economy—overview: Angola's high growth rate in recent years was driven by high international prices for its oil. Angola became a member of OPEC in late 2006 and in late 2007 was assigned a production quota of 1.9 million barrels a day (bbl/day), somewhat less than the 2-2.5 million bbl/day Angola's government had wanted. Oil production and its supporting activities contribute about 85% of GDP. Diamond exports contribute an additional 5%. Subsistence agriculture provides the main livelihood for most of the people, but half of the country's food is still imported. Increased oil production supported growth averaging more than 15% per year from 2004 to 2008. A postwar reconstruction boom and resettlement of displaced persons has led to high rates of growth in construction and agriculture as well. Much of the country's infrastructure is still damaged or undeveloped from the 27-year-long civil war. Land mines left from the war still mar the countryside, even though peace was established after the death of rebel leader Jonas SAVIMBI in February 2002. Since 2005, the government has used billions of dollars in credit lines from China, Brazil, Portugal, Germany, Spain, and the EU to rebuild Angola's public infrastructure. The global recession temporarily stalled economic growth. Lower prices for oil and diamonds during the global recession led to a contraction in GDP in 2009, and many construction projects stopped because Luanda accrued $9 billion in arrears to foreign construction companies when government revenue fell in 2008 and 2009. Angola abandoned its currency peg in 2009, and in November 2009 signed onto an IMF Stand-By Arrangement loan of $1.4 billion to rebuild international reserves. Although consumer inflation declined from 325% in 2000 to under 14% in 2010, Luanda has been unable to reduce inflation below 10%. The Angolan kwanza depreciated again in mid 2010, which, along with higher oil prices, should boost economic growth in all sectors. Corruption, especially in the extractive sectors, also is a major challenge.
GDP (purchasing power parity): $107.3 billion (2010 est.)
country comparison to the world: 68
$105.6 billion (2009 est.)
$103.1 billion (2008 est.)
note: data are in 2010 US dollars
GDP (official exchange rate): $85.31 billion (2010 est.)
GDP—real growth rate: 1.6% (2010 est.)
country comparison to the world: 159
2.4% (2009 est.)
13.8% (2008 est.)
GDP—per capita (PPP): $8,200 (2010 est.)
country comparison to the world: 121
$8,300 (2009 est.)
$8,200 (2008 est.)
note: data are in 2010 US dollars
GDP—composition by sector:
agriculture: 9.6%
industry: 65.8%
services: 24.6% (2008 est.)
Labor force: 7.977 million (2010 est.)
country comparison to the world: 57
Labor force—by occupation:
agriculture: 85%
industry and *services:* 15% (2003 est.)
Unemployment rate: NA
Population below poverty line: 40.5% (2006 est.)
Household income or consumption by percentage share: *lowest 10%:* 0.6%
highest 10%: 44.7% (2000)
Investment (gross fixed): 15.9% of GDP (2010 est.)
country comparison to the world: 127
Budget: *revenues:* $40.41 billion
expenditures: $37.38 billion (2010 est.)
Public debt: 20.3% of GDP (2010 est.)
country comparison to the world: 110
21.7% of GDP (2009 est.)
Inflation rate (consumer prices): 13.3% (2010 est.)
country comparison to the world: 214
13.7% (2009 est.)
Central bank discount rate: 30% (31 December 2009)
country comparison to the world: 12
19.57% (31 December 2008)
Commercial bank prime lending rate: 15.68% (31 December 2009 est.)
country comparison to the world: 62
12.53% (31 December 2008 est.)
Stock of narrow money: $8.74 billion (31 December 2010 est.)
country comparison to the world: 74
$9.792 billion (31 December 2009 est.)
Stock of broad money: $24.92 billion (31 December 2010 est.)
country comparison to the world: 77
$29.04 billion (31 December 2009 est.)
Stock of domestic credit: $17.52 billion (31 December 2010 est.)
country comparison to the world: 84
$22.06 billion (31 December 2009 est.)
Agriculture—products: bananas, sugarcane, coffee, sisal, corn, cotton, manioc (tapioca), tobacco, vegetables, plantains; livestock; forest products; fish
Industries: petroleum; diamonds, iron ore, phosphates, feldspar, bauxite, uranium, and gold; cement; basic metal products; fish processing; food processing, brewing, tobacco products, sugar; textiles; ship repair
Industrial production growth rate: 5% (2010 est.)
country comparison to the world: 69
Electricity—production: 3.722 billion kWh (2007 est.)
country comparison to the world: 121
Electricity—consumption: 3.173 billion kWh (2007 est.)
country comparison to the world: 124
Electricity—exports: 0 kWh (2008 est.)
Electricity—imports: 0 kWh (2008 est.)
Oil—production: 1.948 million bbl/day (2009 est.)
country comparison to the world: 17
Oil—consumption: 70,000 bbl/day (2009 est.)
country comparison to the world: 90
Oil—exports: 1.407 million bbl/day (2007 est.)
country comparison to the world: 16
Oil—imports: 28,090 bbl/day (2007 est.)
country comparison to the world: 101
Oil—proved reserves: 13.5 billion bbl (1 January 2010 est.)
country comparison to the world: 15
Natural gas—production: 680 million cu m (2008 est.)
country comparison to the world: 65
Natural gas—consumption: 680 million cu m (2008 est.)
country comparison to the world: 92
Natural gas—exports: 0 cu m (2008 est.)
country comparison to the world: 52
Natural gas—imports: 0 cu m (2008 est.)
country comparison to the world: 145
Natural gas—proved reserves: 271.8 billion cu m (1 January 2010 est.)
country comparison to the world: 42
Current account balance: $2.089 billion (2010 est.)
country comparison to the world: 42
$-1.668 billion (2009 est.)
Exports: $51.65 billion (2010 est.)
country comparison to the world: 52

$40.08 billion (2009 est.)
Exports—commodities: crude oil, diamonds, refined petroleum products, coffee, sisal, fish and fish products, timber, cotton
Exports—partners: China 35.65%, US 25.98%, France 8.83%, South Africa 4.13% (2009)
Imports: $18.1 billion (2010 est.)
country comparison to the world: 74
$15.74 billion (2009 est.)
Imports—commodities: machinery and electrical equipment, vehicles and spare parts; medicines, food, textiles, military goods
Imports—partners: Portugal 18.71%, China 17.39%, US 8.51%, Brazil 8.22%, South Korea 6.72%, France 4.51%, Italy 4.28%, South Africa 4.02% (2009)
Reserves of foreign exchange and gold: $16.89 billion (31 December 2010 est.)
country comparison to the world: 48
$13.64 billion (31 December 2009 est.)
Debt—external: $17.98 billion (31 December 2010 est.)
country comparison to the world: 74
$13.64 billion (31 December 2009 est.)
Stock of direct foreign investment—at home: $91.55 billion (31 December 2010 est.)
country comparison to the world: 35
$79.88 billion (31 December 2009 est.)
Stock of direct foreign investment—abroad: $4.883 billion (31 December 2010 est.)
country comparison to the world: 60
$3.933 billion (31 December 2009 est.)
Exchange rates: kwanza (AOA) per US dollar–92.08 (2010), 79.33 (2009), 75.023 (2008), 76.6 (2007), 80.4 (2006)

COMMUNICATIONS

Telephones—main lines in use: 303,200 (2009)
country comparison to the world: 113
Telephones—mobile cellular: 8.109 million (2009)
country comparison to the world: 73
Telephone system: *general assessment:* limited system; state-owned telecom had monopoly for fixed-lines until 2005; demand outstripped capacity, prices were high, and services poor; Telecom Namibia, through an Angolan company, became the first private licensed operator in Angola's fixed-line telephone network; by 2010, the number of fixed-line providers had expanded to 5; Angola Telecom established mobile-cellular service in Luanda in 1993 and the network has been extended to larger towns; a privately-owned, mobile-cellular service provider began operations in 2001
domestic: only about two fixed-lines per 100 persons; combined fixed-line and mobile-cellular teledensity about 65 telephones per 100 persons in 2009
international: country code–244; landing point for the SAT-3/WASC fiber-optic submarine cable that provides connectivity to Europe and Asia; satellite earth stations–29 (2009)
Broadcast media: state controls all broadcast media with nationwide reach; state-owned Televisao Popular de Angola (TPA) provides terrestrial TV service on 2 channels; a third TPA channel is available via cable and satellite; TV subscription services are available; state-owned Radio Nacional de Angola (RNA) broadcasts on 5 stations; about a half dozen private radio stations broadcast locally (2008)
Internet country code: .ao
Internet hosts: 3,717 (2010)
country comparison to the world: 142
Internet users: 606,700 (2009)
country comparison to the world: 114

TRANSPORTATION

Airports: 193 (2010)
country comparison to the world: 32
Airports—with paved runways: *total:* 31
over 3,047 m: 5
2,438 to 3,047 m: 9
1,524 to 2,437 m: 13
914 to 1,523 m: 4 (2010)
Airports—with unpaved runways:
total: 162
over 3,047 m: 2
2,438 to 3,047 m: 4
1,524 to 2,437 m: 31
914 to 1,523 m: 78
under 914 m: 47 (2010)
Pipelines: gas 2 km; oil 87 km (2010)
Railways: *total:* 2,764 km
country comparison to the world: 59
narrow gauge: 2,641 km 1.067-m gauge; 123 km 0.600-m gauge (2010)
Roadways: *total:* 51,429 km
country comparison to the world: 79
paved: 5,349 km
unpaved: 46,080 km (2001)
Waterways: 1,300 km (2010)
country comparison to the world: 54
Merchant marine: *total:* 7
country comparison to the world: 126
by type: cargo 1, passenger/cargo 2, petroleum tanker 3, roll on/roll off 1
foreign-owned: 1 (Spain 1)
registered in other countries: 15 (Bahamas 5, Liberia 1, Malta 7, former Netherlands Antilles 2) (2010)
Ports and terminals: Cabinda, Lobito, Luanda, Namibe

MILITARY

Military branches: Angolan Armed Forces (Forcas Armadas Angolanas, FAA): Army, Navy (Marinha de Guerra Angola, MGA), Angolan National Air Force (Forca Aerea Nacional Angolana, FANA) (2011)
Military service age and obligation: 20-45 years of age for compulsory and 18-45 years for voluntary military service; conscript service obligation–2 years; Angolan citizenship required; minimum age for women volunteers is 20; the Marinha de Guerra Angola (Navy) is entirely staffed with volunteers (2011)
Manpower available for military service:
males age 16–49: 3,062,438
females age 16–49: 2,964,262 (2010 est.)
Manpower fit for military service: *males age 16–49:* 1,546,781
females age 16–49: 1,492,308 (2010 est.)
Manpower reaching militarily significant age annually: *male:* 155,476
female: 152,054 (2010 est.)
Military expenditures: 3.6% of GDP (2009)
country comparison to the world: 32

TRANSNATIONAL ISSUES

Disputes—international: DROC accuses Angola of shifting monuments
Refugees and internally displaced persons: refugees (country of origin): 12,615 (Democratic Republic of Congo)
IDPs: 61,700 (27-year civil war ending in 2002; 4 million IDPs already have returned) (2007)
Illicit drugs: used as a transshipment point for cocaine destined for Western Europe and other African states, particularly South Africa

ANGUILLA

(OVERSEAS TERRITORY OF THE UK)

INTRODUCTION

Background: Colonized by English settlers from Saint Kitts in 1650, Anguilla was administered by Great Britain until the early 19th century, when the island–against the wishes of the inhabitants–was incorporated into a single British dependency along with Saint Kitts and Nevis. Several attempts at separation failed. In 1971, two years after a revolt, Anguilla was finally allowed to secede; this arrangement was formally recognized in 1980 with Anguilla becoming a separate British dependency.

GEOGRAPHY

Location: Caribbean, islands between the Caribbean Sea and North Atlantic Ocean, east of Puerto Rico
Geographic coordinates: 18 15 N, 63 10 W
Map references: Central America and the Caribbean
Area: *total:* 91 sq km
country comparison to the world: 225
land: 91 sq km
water: 0 sq km
Area—comparative: about one-half the size of Washington, DC
Land boundaries: 0 km
Coastline: 61 km
Maritime claims: territorial sea: 3 nm
exclusive fishing zone: 200 nm
Climate: tropical; moderated by northeast trade winds
Terrain: flat and low-lying island of coral and limestone
Elevation extremes: *lowest point:* Caribbean Sea 0 m
highest point: Crocus Hill 65 m
Natural resources: salt, fish, lobster
Land use: *arable land:* 0%
permanent crops: 0%

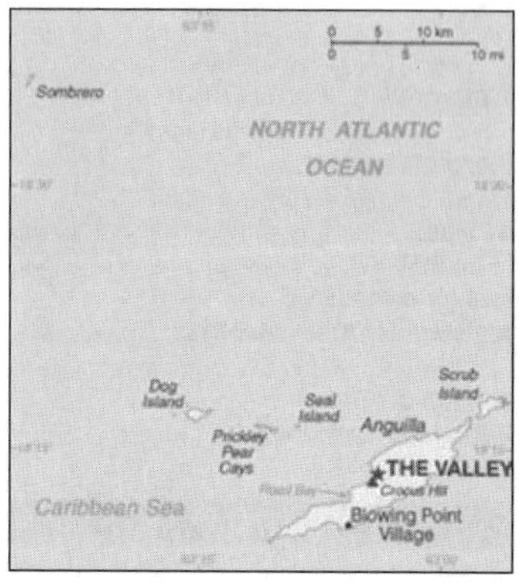

other: 100% (mostly rock with sparse scrub oak, few trees, some commercial salt ponds) (2005)
Irrigated land: NA
Natural hazards: frequent hurricanes and other tropical storms (July to October)
Environment—current issues: supplies of potable water sometimes cannot meet increasing demand largely because of poor distribution system
Geography—note: the most northerly of the Leeward Islands in the Lesser Antilles

PEOPLE

Population: 15,094 (July 2011 est.)
country comparison to the world: 221
Age structure: *0–14 years:* 24% (male 1,861/female 1,764)
15–64 years: 68.1% (male 4,855/female 5,427)
65 years and over: 7.9% (male 577/female 610) (2011 est.)
Median age: *total:* 33.3 years
male: 31.9 years
female: 34.7 years (2011 est.)
Population growth rate: 2.173% (2011 est.)
country comparison to the world: 39
Birth rate: 12.92 births/1,000 population (2011 est.)
country comparison to the world: 155
Death rate: 4.44 deaths/1,000 population (July 2011 est.)
country comparison to the world: 201
Net migration rate: 13.25 migrant(s)/1,000 population (2011 est.)
country comparison to the world: 7
Urbanization: *urban population:* 100% of total population (2010)
rate of urbanization: 1.7% annual rate of change (2010-15 est.)
Major cities—population: THE VALLEY (capital) 2,000 (2009)
Sex ratio: *at birth:* 1.031 male(s)/female
under 15 years: 1.05 male(s)/female
15–64 years: 0.9 male(s)/female
65 years and over: 0.93 male(s)/female
total population: 0.94 male(s)/female (2011 est.)
Infant mortality rate: *total:* 3.47 deaths/1,000 live births
country comparison to the world: 210
male: 3.91 deaths/1,000 live births
female: 3.01 deaths/1,000 live births (2011 est.)
Life expectancy at birth: *total population:* 80.87 years
country comparison to the world: 19
male: 78.32 years
female: 83.51 years (2011 est.)
Total fertility rate: 1.75 children born/woman (2011 est.)
country comparison to the world: 161
HIV/AIDS—adult prevalence rate: NA
HIV/AIDS—people living with HIV/AIDS: NA
HIV/AIDS—deaths: NA
Drinking water source: *Improved:*
urban: 60% of population
total: 60% of population
Unimproved: urban: 40% of population
total: 40% of population (2000)
Sanitation facility access:
Improved:
urban: 99% of population
total: 99% of population
Unimproved:
urban: 1% of population
total: 1% of population (2008)
Nationality: *noun:* Anguillan(s)
adjective: Anguillan
Ethnic groups: black (predominant) 90.1%, mixed, mulatto 4.6%, white 3.7%, other 1.5% (2001 census)
Religions: Anglican 29%, Methodist 23.9%, other Protestant 30.2%, Roman Catholic 5.7%, other Christian 1.7%, other 5.2%, none or unspecified 4.3% (2001 census)
Languages: English (official)
Literacy: *definition:* age 12 and over can read and write
total population: 95%
male: 95%
female: 95% (1984 est.)
School life expectancy (primary to tertiary education): *total:* 11 years
male: 11 years
female: 11 years (2008)
Education expenditures: 3.5% of GDP (2008)
country comparison to the world: 120

GOVERNMENT

Country name: *conventional long form:* none
conventional short form: Anguilla
Dependency status: overseas territory of the UK
Government type: NA
Capital: *name:* The Valley
geographic coordinates: 18 13 N, 63 03 W
time difference: UTC-4 (1 hour ahead of Washington, DC during Standard Time)
Administrative divisions: none (overseas territory of the UK)
Independence: none (overseas territory of the UK)
National holiday: Anguilla Day, 30 May (1967)
Constitution: Anguilla Constitutional Order 1 April 1982; amended 1990
Legal system: common law based on the English model
Suffrage: 18 years of age; universal
Executive branch: *chief of state:* Queen ELIZABETH II (since 6 February 1952); represented by Governor Alistair HARRISON (since 21 April 2009)
head of government: Chief Minister Hubert HUGHES (since 16 February 2010)
cabinet: Executive Council appointed by the governor from among the elected members of the House of Assembly
(For more information visit the World Leaders website)
elections: the monarchy is hereditary; governor appointed by the monarch; following legislative elections, the leader of the majority party or the leader of the majority coalition usually appointed chief minister by the governor
Legislative branch: unicameral House of Assembly (11 seats; 7 members elected by direct popular vote, 2 ex officio members, and 2 appointed; members serve five-year terms)
elections: last held on 15 February 2010 (next to be held in 2015)
election results: percent of vote by party–NA; seats by party–AUM 4, AUF 2, APP 1
Judicial branch: High Court (judge provided by Eastern Caribbean Supreme Court)
Political parties and leaders: Anguilla Progressive Party or APP [Roy ROGERS]; Anguilla Strategic Alternative or ANSA [Edison BAIRD]; Anguilla United Front or AUF [Osbourne FLEMING, Victor BANKS] (a coalition of the Anguilla Democratic Party or ADP and the Anguilla National Alliance or ANA); Anguilla United Movement or AUM [Hubert HUGHES]
Political pressure groups and leaders: NA
International organization participation: Caricom (associate), CDB, Interpol (subbureau), OECS, UPU
Diplomatic representation in the US: none (overseas territory of the UK)
Diplomatic representation from the US: none (overseas territory of the UK)
Flag description: blue, with the flag of the UK in the upper hoist-side quadrant and the Anguillan coat of arms centered in the outer half of the flag; the coat of arms depicts three orange dolphins in an interlocking circular design on a white background with a turquoise-blue field below; the white in the background represents peace; the blue base symbolizes the surrounding sea, as well as faith, youth, and hope; the three dolphins stand for endurance, unity, and strength
National anthem: *name:* "God Bless Anguilla"
lyrics/music: Alex RICHARDSON
note: local anthem adopted 1981; as a territory of the United Kingdom, "God Save the Queen" is official (see United Kingdom)

ECONOMY

Economy—overview: Anguilla has few natural resources, and the economy depends heavily on luxury tourism, offshore banking, lobster fishing, and remittances from emigrants. Increased activity in the tourism industry has spurred the growth of the construction sector contributing to economic growth. Anguillan officials have put substantial effort into developing the offshore financial sector, which is small but growing. In the medium term, prospects for the economy will depend largely on the tourism sector and, therefore, on revived income growth in the industrialized nations as well as on favorable weather conditions.

GDP (purchasing power parity): $175.4 million (2009 est.)
country comparison to the world: 216
$191.7 million (2008 est.)
$108.9 million (2004 est.)
GDP (official exchange rate): $175.4 million (2009 est.)
GDP—real growth rate: -8.5% (2009 est.)
country comparison to the world: 215
GDP—per capita (PPP): $12,200 (2008 est.)
country comparison to the world: 95
GDP—composition by sector: *agriculture:* 4%
industry: 18%
services: 78% (2002 est.)
Labor force: 6,049 (2001)
country comparison to the world: 218
Labor force—by occupation: agriculture/fishing/forestry/mining: 4%
manufacturing: 3%
construction: 18%
transportation and utilities: 10%
commerce: 36%
services: 29% (2000 est.)
Unemployment rate: 8% (2002)
country comparison to the world: 90
Population below poverty line: 23% (2002)
Household income or consumption by percentage share: *lowest 10%:* NA%
highest 10%: NA%
Budget: *revenues:* $22.8 million
expenditures: $22.5 million (2000 est.)
Inflation rate (consumer prices): 5.3% (2006 est.)
country comparison to the world: 146
Central bank discount rate: 6.5% (31 December 2009)
country comparison to the world: 59
6.5% (31 December 2008)
Commercial bank prime lending rate: 9.27% (31 December 2009 est.)
country comparison to the world: 94
9.51% (31 December 2008 est.)
Stock of narrow money: $19.03 million (31 December 2009)
country comparison to the world: 186
$19.57 million (31 December 2008)
Stock of broad money: $458.9 million (31 December 2009)
country comparison to the world: 172
$470.6 million (31 December 2008)
Stock of domestic credit: $529.6 million (31 December 2008 est.)
country comparison to the world: 163
$447.7 million (31 December 2007 est.)
Agriculture—products: small quantities of tobacco, vegetables; cattle raising
Industries: tourism, boat building, offshore financial services
Industrial production growth rate: NA%
Electricity—production: NA kWh
Current account balance: $-42.87 million (2003 est.)
country comparison to the world: 70
Exports: $119.5 million (2009 est.)
country comparison to the world: 191
Exports—commodities: lobster, fish, livestock, salt, concrete blocks, rum
Imports: $143 million (2006)
country comparison to the world: 205
Imports—commodities: fuels, foodstuffs, manufactures, chemicals, trucks, textiles
Debt—external: $8.8 million (1998)
country comparison to the world: 192
Exchange rates: East Caribbean dollars (XCD) per US dollar–2.7 (2010), 2.7 (2009), 2.7 (2005), 2.7 (2004), 2.7 (2003)

COMMUNICATIONS

Telephones—main lines in use: 6,300 (2009)
country comparison to the world: 209
Telephones—mobile cellular: 27,000 (2009)
country comparison to the world: 203
Telephone system: *general assessment:* modern internal telephone system
domestic: fixed-line teledensity is roughly 40 per 100 persons; mobile-cellular teledensity is roughly 180 per 100 persons
international: country code–1-264; landing point for the East Caribbean Fiber System (ECFS) submarine cable with links to 13 other islands in the eastern Caribbean extending from the British Virgin Islands to Trinidad; microwave radio relay to island of Saint Martin/Sint Maarten (2009)
Broadcast media: 1 private television station; multi-channel cable TV subscription services are available; about 10 radio stations, one of which is government-owned (2007)
Internet country code: .ai
Internet hosts: 271 (2010)
country comparison to the world: 186
Internet users: 3,700 (2009)
country comparison to the world: 207

TRANSPORTATION

Airports: 3 (2010)
country comparison to the world: 195
Airports—with paved runways: *total:* 1
1,524 to 2,437 m: 1 (2010)
Airports—with unpaved runways: *total:* 2
under 914 m: 2 (2010)
Roadways: *total:* 175 km
country comparison to the world: 209
paved: 82 km
unpaved: 93 km (2004)
Ports and terminals: Blowing Point, Road Bay

MILITARY

Manpower available for military service: *males age 16–49:* 3,641 (2010 est.)
Manpower fit for military service: *males age 16–49:* 3,009
females age 16–49: 3,397 (2010 est.)
Manpower reaching militarily significant age annually: *male:* 111
female: 113 (2010 est.)
Military—note: defense is the responsibility of the UK

TRANSNATIONAL ISSUES

Disputes—international: none
Illicit drugs: transshipment point for South American narcotics destined for the US and Europe

ANTARCTICA

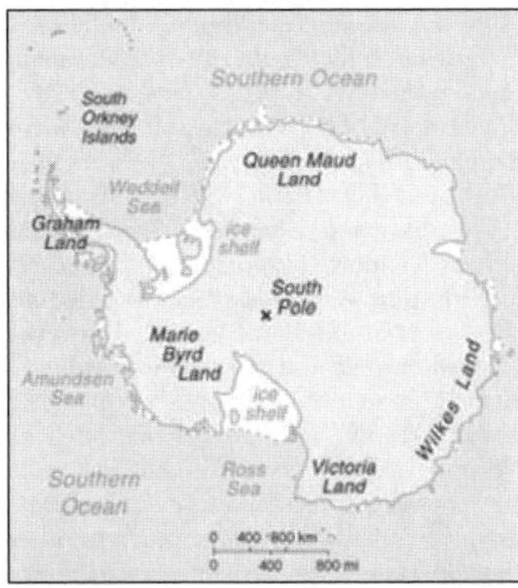

INTRODUCTION

Background: Speculation over the existence of a "southern land" was not confirmed until the early 1820s when British and American commercial operators and British and Russian national expeditions began exploring the Antarctic Peninsula region and other areas south of the Antarctic Circle. Not until 1840 was it established that Antarctica was indeed a continent and not just a group of islands or an area of ocean. Several exploration "firsts" were achieved in the early 20th century, but generally the area saw little human activity. Following World War II, however, there was an upsurge in scientific research on the continent. A number of countries have set up a range of year-round and seasonal stations, camps, and refuges to support scientific research in Antarctica. Seven have made territorial claims, but not all countries recognize these claims. In order to form a legal framework for the activities of nations on the continent, an Antarctic Treaty was negotiated that neither denies nor gives recognition to existing territorial claims; signed in 1959, it entered into force in 1961.

GEOGRAPHY

Location: continent mostly south of the Antarctic Circle
Geographic coordinates: 90 00 S, 0 00 E
Map references: Antarctic Region
Area: *total:* 14 million sq km
land: 14 million sq km (280,000 sq km ice-free, 13.72 million sq km ice-covered) (est.)
note: fifth-largest continent, following Asia, Africa, North America, and South America, but larger than Australia and the subcontinent of Europe
Area—comparative: slightly less than 1.5 times the size of the US
Land boundaries: 0 km
note: see entry on Disputes–international
Coastline: 17,968 km
Maritime claims: Australia, Chile, and Argentina claim Exclusive Economic Zone (EEZ) rights or similar over 200 nm extensions seaward from their continental claims, but like the claims themselves, these zones are not accepted by other countries;

21 of 28 Antarctic consultative nations have made no claims to Antarctic territory (although Russia and the US have reserved the right to do so) and do not recognize the claims of the other nations; also see the Disputes–international entry

Climate: severe low temperatures vary with latitude, elevation, and distance from the ocean; East Antarctica is colder than West Antarctica because of its higher elevation; Antarctic Peninsula has the most moderate climate; higher temperatures occur in January along the coast and average slightly below freezing

Terrain: about 98% thick continental ice sheet and 2% barren rock, with average elevations between 2,000 and 4,000 meters; mountain ranges up to nearly 5,000 meters; ice-free coastal areas include parts of southern Victoria Land, Wilkes Land, the Antarctic Peninsula area, and parts of Ross Island on McMurdo Sound; glaciers form ice shelves along about half of the coastline, and floating ice shelves constitute 11% of the area of the continent

Elevation extremes: *lowest point:* Bentley Subglacial Trench -2,540 m
highest point: Vinson Massif 4,897 m
note: the lowest known land point in Antarctica is hidden in the Bentley Subglacial Trench; at its surface is the deepest ice yet discovered and the world's lowest elevation not under seawater

Natural resources: iron ore, chromium, copper, gold, nickel, platinum and other minerals, and coal and hydrocarbons have been found in small uncommercial quantities; none presently exploited; krill, finfish, and crab have been taken by commercial fisheries

Land use: *arable land:* 0%
permanent crops: 0%
other: 100% (ice 98%, barren rock 2%) (2005)

Natural hazards: katabatic (gravity-driven) winds blow coastward from the high interior; frequent blizzards form near the foot of the plateau; cyclonic storms form over the ocean and move clockwise along the coast; volcanism on Deception Island and isolated areas of West Antarctica; other seismic activity rare and weak; large icebergs may calve from ice shelf

Environment—current issues: in 1998, NASA satellite data showed that the Antarctic ozone hole was the largest on record, covering 27 million square kilometers; researchers in 1997 found that increased ultraviolet light passing through the hole damages the DNA of icefish, an Antarctic fish lacking hemoglobin; ozone depletion earlier was shown to harm one-celled Antarctic marine plants; in 2002, significant areas of ice shelves disintegrated in response to regional warming

Geography—note: the coldest, windiest, highest (on average), and driest continent; during summer, more solar radiation reaches the surface at the South Pole than is received at the Equator in an equivalent period; mostly uninhabitable

PEOPLE

Population: no indigenous inhabitants, but there are both permanent and summer-only staffed research stations
note: 29 nations, all signatory to the Antarctic Treaty, operate through their National Antarctic Program a number of seasonal-only (summer) and year-round research stations on the continent and its nearby islands south of 60 degrees south latitude (the region covered by the Antarctic Treaty); the population doing and supporting science or engaged in the management and protection of the Antarctic region varies from approximately 4,400 in summer to 1,100 in winter; in addition, approximately 1,000 personnel, including ship's crew and scientists doing onboard research, are present in the waters of the treaty region; peak summer (December-February) population—4,490 total; Argentina 667, Australia 200, Australia and Romania jointly 13, Belgium 20, Brazil 40, Bulgaria 18, Chile 359, China 90, Czech Republic 20, Ecuador 26, Finland 20, France 125, France and Italy jointly 60, Germany 90, India 65, Italy 102, Japan 125, South Korea 70, NZ 85, Norway 44, Peru 28, Poland 40, Russia 429, South Africa 80, Spain 50, Sweden 20, Ukraine 24, UK 217, US 1,293, Uruguay 70 (2008-2009); winter (June-August) station population—1,106 total; Argentina 176, Australia 62, Brazil 12, Chile 114, China 29, France 26, France and Italy jointly 13, Germany 9, India 25, Japan 40, South Korea 18, NZ 10, Norway 7, Poland 12, Russia 148, South Africa 10, Ukraine 12, UK 37, US 337, Uruguay 9 (2009); research stations operated within the Antarctic Treaty area (south of 60 degrees south latitude) by National Antarctic Programs: year-round stations—40 total; Argentina 6, Australia 3, Brazil 1, Chile 6, China 2, France 1, France and Italy jointly 1, Germany 1, India 1, Japan 1, South Korea 1, NZ 1, Norway 1, Poland 1, Russia 5, South Africa 1, Ukraine 1, UK 2, US 3, Uruguay 1 (2009); a range of seasonal-only (summer) stations, camps, and refuges—Argentina, Australia, Belgium, Bulgaria, Brazil, Chile, China, Czech Republic, Ecuador, Finland, France, Germany, India, Italy, Japan, South Korea, New Zealand, Norway, Peru, Poland, Romania (with Australia), Russia, South Africa, Spain, Sweden, Ukraine, UK, US, and Uruguay (2008-2009); in addition, during the austral summer some nations have numerous occupied locations such as tent camps, summer-long temporary facilities, and mobile traverses in support of research (May 2009 est.)

GOVERNMENT

Country name: *conventional long form:* none
conventional short form: Antarctica

Government type: Antarctic Treaty Summary—the Antarctic region is governed by a system known as the Antarctic Treaty System; the system includes: 1. the Antarctic Treaty, signed on 1 December 1959 and entered into force on 23 June 1961, which establishes the legal framework for the management of Antarctica, 2. Recommendations and Measures adopted at meetings of Antarctic Treaty countries, 3. The Convention for the Conservation of Antarctic Seals (1972), 4. The Convention for the Conservation of Antarctic Marine Living Resources (1980), and 5. The Protocol on Environmental Protection to the Antarctic Treaty (1991); the 33rd Antarctic Treaty Consultative Meeting was held in Punta del Este, Uruguay in May 2010; at these periodic meetings, decisions are made by consensus (not by vote) of all consultative member nations; by April 2010, there were 48 treaty member nations: 28 consultative and 20 non-consultative; consultative (decision-making) members include the seven nations that claim portions of Antarctica as national territory (some claims overlap) and 21 non-claimant nations; the US and Russia have reserved the right to make claims; the US does not recognize the claims of others; Antarctica is administered through meetings of the consultative member nations; decisions from these meetings are carried out by these member nations (with respect to their own nationals and operations) in accordance with their own national laws; the years in parentheses indicate when a consultative member-nation acceded to the Treaty and when it was accepted as a consultative member, while no date indicates the country was an original 1959 treaty signatory; claimant nations are—Argentina, Australia, Chile, France, NZ, Norway, and the UK; nonclaimant consultative nations are—Belgium, Brazil (1975/1983), Bulgaria (1978/1998), China (1983/1985), Ecuador (1987/1990), Finland (1984/1989), Germany (1979/1981), India (1983/1983), Italy (1981/1987), Japan, South Korea (1986/1989), Netherlands (1967/1990), Peru (1981/1989), Poland (1961/1977), Russia, South Africa, Spain (1982/1988), Sweden (1984/1988), Ukraine (1992/2004), Uruguay (1980/1985), and the US; non-consultative members, with year of accession in parentheses, are-Austria (1987), Belarus (2006), Canada (1988), Colombia (1989), Cuba (1984), Czech Republic (1962/1993), Denmark (1965), Estonia (2001), Greece (1987), Guatemala (1991), Hungary (1984), North Korea (1987), Monaco (2008), Papua New Guinea (1981), Portugal (2010), Romania (1971), Slovakia (1962/1993), Switzerland (1990), Turkey (1996), and Venezuela (1999); note—Czechoslovakia acceded to the Treaty in 1962 and separated into the Czech Republic and Slovakia in 1993;
Article 1—area to be used for peaceful purposes only; military activity, such as weapons testing, is prohibited, but military personnel and equipment may be used for scientific research or any other peaceful purpose; Article 2—freedom of scientific investigation and cooperation shall continue; Article 3—free exchange of information and personnel, cooperation with the UN and other international agencies; Article 4—does not recognize,

dispute, or establish territorial claims and no new claims shall be asserted while the treaty is in force; Article 5–prohibits nuclear explosions or disposal of radioactive wastes; Article 6–includes under the treaty all land and ice shelves south of 60 degrees 00 minutes south and reserves high seas rights; Article 7–treaty-state observers have free access, including aerial observation, to any area and may inspect all stations, installations, and equipment; advance notice of all expeditions and of the introduction of military personnel must be given; Article 8–allows for jurisdiction over observers and scientists by their own states; Article 9–frequent consultative meetings take place among member nations; Article 10–treaty states will discourage activities by any country in Antarctica that are contrary to the treaty; Article 11–disputes to be settled peacefully by the parties concerned or, ultimately, by the ICJ; Articles 12, 13, 14–deal with upholding, interpreting, and amending the treaty among involved nations; other agreements–some 200 recommendations adopted at treaty consultative meetings and ratified by governments; a mineral resources agreement was signed in 1988 but remains unratified; the Protocol on Environmental Protection to the Antarctic Treaty was signed 4 October 1991 and entered into force 14 January 1998; this agreement provides for the protection of the Antarctic environment through six specific annexes: 1) environmental impact assessment, 2) conservation of Antarctic fauna and flora, 3) waste disposal and waste management, 4) prevention of marine pollution, 5) area protection and management and 6) liability arising from environmental emergencies; it prohibits all activities relating to mineral resources except scientific research; a permanent Antarctic Treaty Secretariat was established in 2004 in Buenos Aires, Argentina

Legal system: Antarctica is administered through annual meetings–known as Antarctic Treaty Consultative Meetings–which include consultative member nations, non-consultative member nations, observer organizations, and expert organizations; decisions from these meetings are carried out by these member nations (with respect to their own nationals and operations) in accordance with their own national laws; more generally, access to the Antarctic Treaty area, that is to all areas between 60 and 90 degrees south latitude, is subject to a number of relevant legal instruments and authorization procedures adopted by the states party to the Antarctic Treaty; note–US law, including certain criminal offenses by or against US nationals, such as murder, may apply extraterritorially; some US laws directly apply to Antarctica; for example, the Antarctic Conservation Act, 16 U.S.C. section 2401 et seq., provides civil and criminal penalties for the following activities unless authorized by regulation of statute: the taking of native mammals or birds; the introduction of nonindigenous plants and animals; entry into specially protected areas; the discharge or disposal of pollutants; and the importation into the US of certain items from Antarctica; violation of the Antarctic Conservation Act carries penalties of up to $10,000 in fines and one year in prison; the National Science Foundation and Department of Justice share enforcement responsibilities; Public Law 95-541, the US Antarctic Conservation Act of 1978, as amended in 1996, requires expeditions from the US to Antarctica to notify, in advance, the Office of Oceans, Room 5805, Department of State, Washington, DC 20520, which reports such plans to other nations as required by the Antarctic Treaty; for more information, contact Permit Office, Office of Polar Programs, National Science Foundation, Arlington, Virginia 22230; *telephone:* (703) 292-8030, or visit its website at www.nsf.gov

ECONOMY

Economy—overview: Scientific undertakings rather than commercial pursuits are the predominate human activity in Antarctica. Fishing off the coast and tourism, both based abroad, account for Antarctica's limited economic activity. Antarctic fisheries, targeting three main species–Patagonian and Antarctic toothfish (Dissostichus eleginoides and D. mawsoni), mackerel icefish (Champsocephalus gunnari), and krill (Euphausia superba)–reported landing 141,147 metric tons in 2008-09 (1 July–30 June). (Estimated fishing is from the area covered by the Convention on the Conservation of Antarctic Marine Living Resources (CCAMLR), which extends slightly beyond the Antarctic Treaty area.) Unregulated fishing, particularly of Patagonian toothfish (also known as Chilean sea bass), is a serious problem. The CCAMLR determines the recommended catch limits for marine species. A total of 37,858 tourists visited the Antarctic Treaty area in the 2008-09 Antarctic summer, down from the 46,265 visitors in 2007-2008 (estimates provided to the Antarctic Treaty by the International Association of Antarctica Tour Operators (IAATO); this does not include passengers on overflights). Nearly all of them were passengers on commercial (nongovernmental) ships and several yachts that make trips during the summer.

COMMUNICATIONS

Telephone system: *general assessment:* local systems at some research stations
domestic: commercial cellular networks operating in a small number of locations
international: country code–none allocated; via satellite (including mobile Inmarsat and Iridium systems) to and from all research stations, ships, aircraft, and most field parties (2007)
Internet country code: .aq
Internet hosts: 7,765 (2010)
country comparison to the world: 135

TRANSPORTATION

Airports: 26 (2010)
country comparison to the world: 126
Airports—with unpaved runways: *total:* 26
over 3,047 m: 5
2,438 to 3,047 m: 5
1,524 to 2,437 m: 1
914 to 1,523 m: 9
under 914 m: 6 (2010)
Heliports: 53
note: all year-round and seasonal stations operated by National Antarctic Programs stations have some kind of helicopter landing facilities, prepared (helipads) or unprepared (2010)
Ports and terminals: McMurdo Station; most coastal stations have sparse and intermittent offshore anchorages; a few stations have basic wharf facilities
Transportation—note: US coastal stations include McMurdo (77 51 S, 166 40 E) and Palmer (64 43 S, 64 03 W); government use only except by permit (see Permit Office under "Legal System"); all ships at port are subject to inspection in accordance with Article 7, Antarctic Treaty; relevant legal instruments and authorization procedures adopted by the states parties to the Antarctic Treaty regulating access to the Antarctic Treaty area to all areas between 60 and 90 degrees of latitude south have to be complied with (see "Legal System"); The Hydrographic Commission on Antarctica (HCA), a commission of the International Hydrographic Organization (IHO), is responsible for hydrographic surveying and nautical charting matters in Antarctic Treaty area; it coordinates and facilitates provision of accurate and appropriate charts and other aids to navigation in support of safety of navigation in region; membership of HCA is open to any IHO Member State whose government has acceded to the Antarctic Treaty and which contributes resources or data to IHO Chart coverage of the area

MILITARY

Military—note: the Antarctic Treaty prohibits any measures of a military nature, such as the establishment of military bases and fortifications, the carrying out of military maneuvers, or the testing of any type of weapon; it permits the use of military personnel or equipment for scientific research or for any other peaceful purposes

TRANSNATIONAL ISSUES

Disputes—international: the Antarctic Treaty freezes, and most states do not recognize, the land and maritime territorial claims made by Argentina, Australia, Chile, France, New Zealand, Norway, and the United Kingdom (some overlapping) for three-fourths of the continent; the US and Russia reserve the right to make claims; no formal claims have been made in the sector between 90 degrees west and 150 degrees west; the International Whaling Commission created a sanctuary around the entire continent to deter catches by countries claiming to conduct scientific whaling; Australia has established a similar preserve in the waters around its territorial claim

ANTIGUA AND BARBUDA

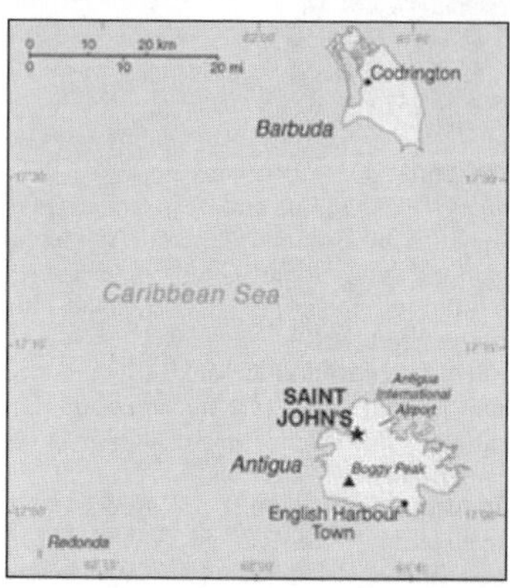

INTRODUCTION

Background: The Siboney were the first to inhabit the islands of Antigua and Barbuda in 2400 B.C., but Arawak Indians populated the islands when COLUMBUS landed on his second voyage in 1493. Early settlements by the Spanish and French were succeeded by the English who formed a colony in 1667. Slavery, established to run the sugar plantations on Antigua, was abolished in 1834. The islands became an independent state within the British Commonwealth of Nations in 1981.

GEOGRAPHY

Location: Caribbean, islands between the Caribbean Sea and the North Atlantic Ocean, east-southeast of Puerto Rico
Geographic coordinates: 17 03 N, 61 48 W
Map references: Central America and the Caribbean
Area: *total:* 442.6 sq km (Antigua 280 sq km; Barbuda 161 sq km)
country comparison to the world: 199
land: 442.6 sq km
water: 0 sq km
note: includes Redonda, 1.6 sq km
Area—comparative: 2.5 times the size of Washington, DC
Land boundaries: 0 km
Coastline: 153 km
Maritime claims: territorial sea: 12 nm
contiguous zone: 24 nm
exclusive economic zone: 200 nm
continental shelf: 200 nm or to the edge of the continental margin
Climate: tropical maritime; little seasonal temperature variation
Terrain: mostly low-lying limestone and coral islands, with some higher volcanic areas
Elevation extremes: *lowest point:* Caribbean Sea 0 m
highest point: Boggy Peak 402 m
Natural resources: NEGL; pleasant climate fosters tourism
Land use: *arable land:* 18.18%
permanent crops: 4.55%
other: 77.27% (2005)
Irrigated land: NA
Total renewable water resources: 0.1 cu km (2000)
Freshwater withdrawal (domestic/industrial/agricultural): *total:* 0.005 cu km/yr (60%/20%/20%)
per capita: 63 cu m/yr (1990)
Natural hazards: hurricanes and tropical storms (July to October); periodic droughts
Environment—current issues: water management–a major concern because of limited natural freshwater resources–is further hampered by the clearing of trees to increase crop production, causing rainfall to run off quickly
Environment—international agreements: *party to:* Biodiversity, Climate Change, Climate Change-Kyoto Protocol, Desertification, Endangered Species, Environmental Modification, Hazardous Wastes, Law of the Sea, Marine Dumping, Ozone Layer Protection, Ship Pollution, Wetlands, Whaling
signed, but not ratified: none of the selected agreements
Geography—note: Antigua has a deeply indented shoreline with many natural harbors and beaches; Barbuda has a large western harbor

PEOPLE

Population: 87,884 (July 2011 est.)
country comparison to the world: 196
Age structure: *0–14 years:* 25.8% (male 11,530/female 11,174)
15–64 years: 67.4% (male 27,599/female 31,592)
65 years and over: 6.8% (male 2,592/female 3,397) (2011 est.)
Median age: *total:* 30.3 years
male: 28.7 years
female: 31.7 years (2011 est.)
Population growth rate: 1.289% (2011 est.)
country comparison to the world: 91
Birth rate: 16.31 births/1,000 population (2011 est.)
country comparison to the world: 125
Death rate: 5.73 deaths/1,000 population (July 2011 est.)
country comparison to the world: 173
Net migration rate: 2.32 migrant(s)/1,000 population (2011 est.)
country comparison to the world: 37
Urbanization: *urban population:* 30% of total population (2010)
rate of urbanization: 1.4% annual rate of change (2010-15 est.)
Major cities—population: SAINT JOHN'S (capital) 27,000 (2009)
Sex ratio: *at birth:* 1.05 male(s)/female
under 15 years: 1.03 male(s)/female
15–64 years: 0.87 male(s)/female
65 years and over: 0.76 male(s)/female
total population: 0.9 male(s)/female (2011 est.)
Infant mortality rate: *total:* 14.63 deaths/1,000 live births
country comparison to the world: 121
male: 16.85 deaths/1,000 live births
female: 12.29 deaths/1,000 live births (2011 est.)
Life expectancy at birth: *total population:* 75.48 years
country comparison to the world: 85
male: 73.47 years
female: 77.59 years (2011 est.)
Total fertility rate: 2.05 children born/woman (2011 est.)
country comparison to the world: 124
HIV/AIDS—adult prevalence rate: NA
HIV/AIDS—people living with HIV/AIDS: NA
HIV/AIDS—deaths: NA
Drinking water source: *Improved:*
urban: 95% of population
rural: 89% of population
total: 91% of population
Unimproved: urban: 5% of population
rural: 11% of population
total: 9% of population (2000)
Sanitation facility access: *Improved:*
urban: 98% of population
rural: 94% of population
total: 95% of population
Unimproved:
urban: 2% of population
rural: 6% of population
total: 5% of population (2000)
Nationality: *noun:* Antiguan(s), Barbudan(s)
adjective: Antiguan, Barbudan
Ethnic groups: black 91%, mixed 4.4%, white 1.7%, other 2.9% (2001 census)
Religions: Anglican 25.7%, Seventh Day Adventist 12.3%, Pentecostal 10.6%, Moravian 10.5%, Roman Catholic 10.4%, Methodist 7.9%, Baptist 4.9%, Church of God 4.5%, other Christian 5.4%, other 2%, none or unspecified 5.8% (2001 census)
Languages: English (official), local dialects
Literacy: *definition:* age 15 and over has completed five or more years of schooling
total population: 85.8%
male: NA
female: NA (2003 est.)
School life expectancy (primary to tertiary education): *total:* 14 years
male: 14 years
female: 14 years (2009)
Education expenditures: 2.7% of GDP (2009)
country comparison to the world: 142

GOVERNMENT

Country name: *conventional long form:* none
conventional short form: Antigua and Barbuda
Government type: constitutional monarchy with a parliamentary system of government and a Commonwealth realm
Capital: *name:* Saint John's
geographic coordinates: 17 07 N, 61 51 W
time difference: UTC-4 (1 hour ahead of Washington, DC during Standard Time)
Administrative divisions: 6 parishes and 2 dependencies*; Barbuda*, Redonda*, Saint George, Saint John, Saint Mary, Saint Paul, Saint Peter, Saint Philip
Independence: 1 November 1981 (from the UK)
National holiday: Independence Day (National Day), 1 November (1981)
Constitution: 1 November 1981
Legal system: common law based on the English model
International law organization participation: has not submitted an ICJ jurisdiction declaration; accepts ICCt jurisdiction

Suffrage: 18 years of age; universal
Executive branch: *chief of state:* Queen ELIZABETH II (since 6 February 1952); represented by Governor General Louisse LAKE-TACK (since 17 July 2007)
head of government: Prime Minister Winston Baldwin SPENCER (since 24 March 2004)
cabinet: Council of Ministers appointed by the governor general on the advice of the prime minister
(For more information visit the World Leaders website)
elections: the monarchy is hereditary; governor general chosen by the monarch on the advice of the prime minister; following legislative elections, the leader of the majority party or the leader of the majority coalition usually appointed prime minister by the governor general
Legislative branch: bicameral Parliament consists of the Senate (17 seats; members appointed by the governor general) and the House of Representatives (17 seats; members are elected by proportional representation to serve five-year terms)
elections: House of Representatives–last held on 12 March 2009 (next to be held in 2014)
election results: percent of vote by party–UPP 50.9%, ALP 47.2%, BPM 1.1%; seats by party–UPP 9, ALP 7, BPM 1
Judicial branch: Eastern Caribbean Supreme Court consisting of a High Court of Justice and a Court of Appeal (based in Saint Lucia; two judges of the Supreme Court are residents of the islands and preside over the Court of Summary Jurisdiction); Magistrates' Courts; member of the Caribbean Court of Justice
Political parties and leaders: Antigua Labor Party or ALP [Lester Bryant BIRD]; Barbuda People's Movement or BPM [Thomas H. FRANK]; Barbuda People's Movement for Change [Arthur NIBBS]; Barbudans for a Better Barbuda [Ordrick SAMUEL]; United Progressive Party or UPP [Baldwin SPENCER] (a coalition of three parties–Antigua Caribbean Liberation Movement or ACLM, Progressive Labor Movement or PLM, United National Democratic Party or UNDP)
Political pressure groups and leaders: Antigua Trades and Labor Union or ATLU [William ROBINSON]; People's Democratic Movement or PDM [Hugh MARSHALL]
International organization participation: ACP, AOSIS, C, Caricom, CDB, FAO, G-77, IBRD, ICAO, ICRM, IDA, IFAD, IFC, IFRCS, ILO, IMF, IMO, IMSO, Interpol, IOC, ISO (subscriber), ITU, ITUC, MIGA, NAM, OAS, OECS, OPANAL, OPCW, PetroCaribe, UN, UNCTAD, UNESCO, UPU, WFTU, WHO, WIPO, WMO, WTO
Diplomatic representation in the US: *chief of mission:* Ambassador Deborah Mae LOVELL
chancery: 3216 New Mexico Avenue NW, Washington, DC 20016
telephone: [1] (202) 362-5122
FAX: [1] (202) 362-5225
consulate(s) general: Miami, New York
Diplomatic representation from the US: the US does not have an embassy in Antigua and Barbuda; the US Ambassador to Barbados is accredited to Antigua and Barbuda
Flag description: red, with an inverted isosceles triangle based on the top edge of the flag; the triangle contains three horizontal bands of black (top), light blue, and white, with a yellow rising sun in the black band; the sun symbolizes the dawn of a new era, black represents the African heritage of most of the population, blue is for hope, and red is for the dynamism of the people; the "V" stands for victory; the successive yellow, blue, and white coloring is also meant to evoke the country's tourist attractions of sun, sea, and sand
National anthem: *name:* "Fair Antigua, We Salute Thee"
lyrics/music: Novelle Hamilton RICHARDS/Walter Garnet Picart CHAMBERS
note: adopted 1967; as a Commonwealth country, in addition to the national anthem, "God Save the Queen" serves as the royal anthem (see United Kingdom)

ECONOMY

Economy—overview: Tourism continues to dominate Antigua and Barbuda's economy, accounting for nearly 60% of GDP and 40% of investment. The dual-island nation's agricultural production is focused on the domestic market and constrained by a limited water supply and a labor shortage stemming from the lure of higher wages in tourism and construction. Manufacturing comprises enclave-type assembly for export with major products being bedding, handicrafts, and electronic components. Prospects for economic growth in the medium term will continue to depend on tourist arrivals from the US, Canada, and Europe and potential damages from natural disasters. After taking office in 2004, the SPENCER government adopted an ambitious fiscal reform program, and was successful in reducing its public debt-to-GDP ratio from 120% to about 90% in 2008. However, the global financial crisis that began in 2008, has led to a significant increase in the national debt, which topped 130% at the end of 2010. The Antiguan economy experienced solid growth from 2003 to 2007, reaching over 12% in 2006 driven by a construction boom in hotels and housing associated with the Cricket World Cup, but growth dropped off in 2008 with the end of the boom. In 2009, Antigua's economy was severely hit by the global economic crisis, suffering from the collapse of its largest financial institution and a steep decline in tourism. This decline continued in 2010 as the country struggled with a yawning budget deficit.
GDP (purchasing power parity): $1.425 billion (2010 est.)
country comparison to the world: 194
$1.486 billion (2009 est.)
$1.631 billion (2008 est.)
note: data are in 2010 US dollars
GDP (official exchange rate): $1.105 billion (2010 est.)
GDP—real growth rate: -4.1% (2010 est.)
country comparison to the world: 211
-8.9% (2009 est.)
1.8% (2008 est.)
GDP—per capita (PPP): $16,400 (2010 est.)
country comparison to the world: 68
$17,400 (2009 est.)
$19,300 (2008 est.)
note: data are in 2010 US dollars
GDP—composition by sector: *agriculture:* 3.8%
industry: 22%
services: 74.3% (2002 est.)
Labor force: 30,000 (1991)
country comparison to the world: 204
Labor force—by occupation: *agriculture:* 7%
industry: 11%
services: 82% (1983)
Unemployment rate: 11% (2001 est.)
country comparison to the world: 121
Population below poverty line: NA%
Household income or consumption by percentage share: *lowest 10%:* NA%
highest 10%: NA%
Budget: *revenues:* $229.5 million
expenditures: $293.4 million (2009 est.)
Inflation rate (consumer prices): 1.5% (2007 est.)
country comparison to the world: 36
Central bank discount rate: 6.5% (31 December 2009)
country comparison to the world: 60
6.5% (31 December 2008)
Commercial bank prime lending rate: 10.07% (31 December 2009 est.)
country comparison to the world: 82
10.43% (31 December 2008 est.)
Stock of narrow money: $233.5 million (31 December 2009)
country comparison to the world: 170
$266.7 million (31 December 2008)
Stock of broad money: $1.186 billion (31 December 2009)
country comparison to the world: 159
$1.236 billion (31 December 2008)
Stock of domestic credit: $1.13 billion (31 December 2008 est.)
country comparison to the world: 147
$1.002 billion (31 December 2007 est.)
Agriculture—products: cotton, fruits, vegetables, bananas, coconuts, cucumbers, mangoes, sugarcane; livestock
Industries: tourism, construction, light manufacturing (clothing, alcohol, household appliances)
Industrial production growth rate: NA%
Electricity—production: 110 million kWh (2007 est.)
country comparison to the world: 188
Electricity—consumption: 102.3 million kWh (2007 est.)
country comparison to the world: 189
Electricity—exports: 0 kWh (2008 est.)
Electricity—imports: 0 kWh (2008 est.)
Oil—production: 0 bbl/day (2009 est.)
country comparison to the world: 146
Oil—consumption: 5,000 bbl/day (2009 est.)
country comparison to the world: 166
Oil—exports: 219 bbl/day (2007 est.)
country comparison to the world: 130
Oil—imports: 4,690 bbl/day (2007 est.)
country comparison to the world: 160
Oil—proved reserves: 0 bbl (1 January 2010 est.)
country comparison to the world: 101

Natural gas—production: 0 cu m (2008 est.)
country comparison to the world: 94
Natural gas—consumption: 0 cu m (2008 est.)
country comparison to the world: 144
Natural gas—exports: 0 cu m (2008 est.)
country comparison to the world: 48
Natural gas—imports: 0 cu m (2008 est.)
country comparison to the world: 140
Natural gas—proved reserves: 0 cu m (1 January 2010 est.)
country comparison to the world: 105
Current account balance: $-211 million (2007 est.)
country comparison to the world: 92
Exports: $84.3 million (2007 est.)
country comparison to the world: 197
Exports—commodities: petroleum products, bedding, handicrafts, electronic components, transport equipment, food and live animals
Imports: $522.8 million (2007 est.)
country comparison to the world: 189
Imports—commodities: food and live animals, machinery and transport equipment, manufactures, chemicals, oil
Debt—external: $359.8 million (June 2006)
country comparison to the world: 165
Exchange rates: East Caribbean dollars (XCD) per US dollar–2.7 (2010), 2.7 (2009), 2.7 (2005), 2.7 (2004), 2.7 (2003)

COMMUNICATIONS

Telephones—main lines in use: 37,400 (2009)
country comparison to the world: 173
Telephones—mobile cellular: 134,900 (2009)
country comparison to the world: 178
Telephone system: *general assessment:* good automatic telephone system
domestic: fixed-line teledensity roughly 40 per 100 persons; mobile-cellular teledensity is some 150 per 100 persons
international: country code–1-268; landing points for the East Caribbean Fiber System (ECFS) and the Global Caribbean Network (GCN) submarine cable systems with links to other islands in the eastern Caribbean extending from the British Virgin Islands to Trinidad; satellite earth stations–2; tropospheric scatter to Saba (Netherlands) and Guadeloupe (France) (2009)
Broadcast media: state-controlled Antigua and Barbuda Broadcasting Service (ABS) operates 1 TV station; multi-channel cable TV subscription services are available; 1 radio station operated by ABS; roughly 15 radio stations, some broadcasting on multiple frequencies (2007)
Internet country code: .ag
Internet hosts: 9,795 (2010)
country comparison to the world: 122
Internet users: 65,000 (2009)
country comparison to the world: 171

TRANSPORTATION

Airports: 3 (2010)
country comparison to the world: 190
Airports—with paved runways: *total:* 2
2,438 to 3,047 m: 1
under 914 m: 1 (2010)
Airports—with unpaved runways: *total:* 1
under 914 m: 1 (2010)
Roadways: *total:* 1,165 km
country comparison to the world: 181
paved: 384 km
unpaved: 781 km (2002)
Merchant marine: *total:* 1,219
country comparison to the world: 9
by type: barge carrier 1, bulk carrier 53, cargo 703, carrier 6, chemical tanker 4, container 412, liquefied gas 12, petroleum tanker 1, refrigerated cargo 9, roll on/roll off 16, vehicle carrier 2
foreign-owned: 1,186 (Albania 1, Colombia 1, Denmark 20, Estonia 20, Germany 1050, Greece 5, Iceland 9, Isle of Man 2, Latvia 16, Lithuania 4, Mexico 2, Netherlands 18, Norway 9, NZ 2, Poland 2, Russia 3, Slovenia 1, Sweden 1, Switzerland 7, Turkey 7, US 6)
note: this country allows large numbers of ships owned by foreign entities to be registered in its national shipping registry and to fly its flag; these ships operate under the laws of the flag state (2010)
Ports and terminals: Saint John's

MILITARY

Military branches: Royal Antigua and Barbuda Defense Force (includes Antigua and Barbuda Coast Guard) (2011)
Military service age and obligation: 18 years of age for voluntary military service; no conscription (2010)
Manpower available for military service: *males age 16–49:* 21,141
females age 16–49: 24,056 (2010 est.)
Manpower fit for military service: *males age 16–49:* 17,676
females age 16–49: 19,960 (2010 est.)
Manpower reaching militarily significant age annually: *male:* 806
female: 799 (2010 est.)
Military expenditures: 0.5% of GDP (2009)
country comparison to the world: 162

TRANSNATIONAL ISSUES

Disputes—international: none
Illicit drugs: considered a minor transshipment point for narcotics bound for the US and Europe; more significant as an offshore financial center

ARCTIC OCEAN

INTRODUCTION

Background: The Arctic Ocean is the smallest of the world's five oceans (after the Pacific Ocean, Atlantic Ocean, Indian Ocean, and the recently delimited Southern Ocean). The Northwest Passage (US and Canada) and Northern Sea Route (Norway and Russia) are two important seasonal waterways. In recent years the polar ice pack has thinned allowing for increased navigation and raising the possibility of future sovereignty and shipping disputes among countries bordering the Arctic Ocean.

GEOGRAPHY

Location: body of water between Europe, Asia, and North America, mostly north of the Arctic Circle
Geographic coordinates: 90 00 N, 0 00 E
Map references: Arctic
Area:
total: 14.056 million sq km
note: includes Baffin Bay, Barents Sea, Beaufort Sea, Chukchi Sea, East Siberian Sea, Greenland Sea, Hudson Bay, Hudson Strait, Kara Sea, Laptev Sea, Northwest Passage, and other tributary water bodies
Area—comparative: slightly less than 1.5 times the size of the US
Coastline: 45,389 km
Climate: polar climate characterized by persistent cold and relatively narrow annual temperature ranges; winters characterized by continuous darkness, cold and stable weather conditions, and clear skies; summers characterized by continuous daylight, damp and foggy weather, and weak cyclones with rain or snow
Terrain: central surface covered by a perennial drifting polar icepack that, on average, is about 3 meters thick, although pressure ridges may be three times that thickness; clockwise drift pattern in the Beaufort Gyral Stream, but nearly straight-line movement from the New Siberian Islands (Russia) to Denmark Strait (between Greenland and Iceland); the icepack is surrounded by open seas during the summer, but more than doubles in size during the winter and extends to the encircling landmasses; the ocean floor is about 50% continental shelf (highest percentage of any ocean) with the remainder a central basin interrupted by three submarine ridges (Alpha Cordillera, Nansen Cordillera, and Lomonosov Ridge)
Elevation extremes: *lowest point:* Fram Basin -4,665 m
highest point: sea level 0 m
Natural resources: sand and gravel aggregates, placer deposits, polymetallic nodules,

oil and gas fields, fish, marine mammals (seals and whales)
Natural hazards: ice islands occasionally break away from northern Ellesmere Island; icebergs calved from glaciers in western Greenland and extreme northeastern Canada; permafrost in islands; virtually ice locked from October to June; ships subject to superstructure icing from October to May
Environment—current issues: endangered marine species include walruses and whales; fragile ecosystem slow to change and slow to recover from disruptions or damage; thinning polar icepack
Geography—note: major chokepoint is the southern Chukchi Sea (northern access to the Pacific Ocean via the Bering Strait); strategic location between North America and Russia; shortest marine link between the extremes of eastern and western Russia; floating research stations operated by the US and Russia; maximum snow cover in March or April about 20 to 50 centimeters over the frozen ocean; snow cover lasts about 10 months

ECONOMY

Economy—overview: Economic activity is limited to the exploitation of natural resources, including petroleum, natural gas, fish, and seals.

TRANSPORTATION

Ports and terminals: Churchill (Canada), Murmansk (Russia), Prudhoe Bay (US)
Transportation—note: sparse network of air, ocean, river, and land routes; the Northwest Passage (North America) and Northern Sea Route (Eurasia) are important seasonal waterways

TRANSNATIONAL ISSUES

Disputes—international: Canada and the United States dispute how to divide the Beaufort Sea and the status of the Northwest Passage but continue to work cooperatively to survey the Arctic continental shelf; Denmark (Greenland) and Norway have made submissions to the Commission on the Limits of the Continental shelf (CLCS) and Russia is collecting additional data to augment its 2001 CLCS submission; record summer melting of sea ice in the Arctic has renewed interest in maritime shipping lanes and sea floor exploration; Norway and Russia signed a comprehensive maritime boundary agreement in 2010

ARGENTINA

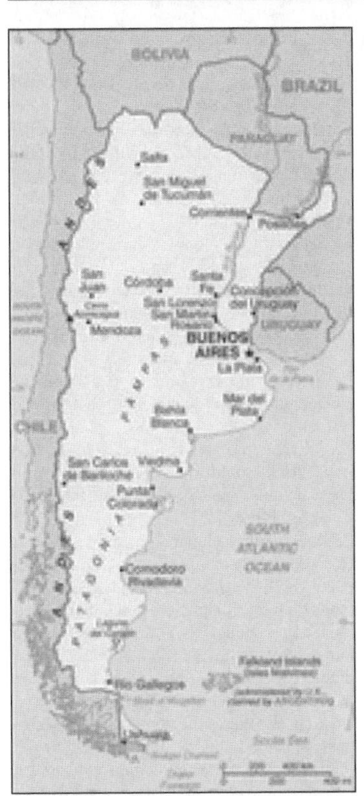

INTRODUCTION

Background: In 1816, the United Provinces of the Rio Plata declared their independence from Spain. After Bolivia, Paraguay, and Uruguay went their separate ways, the area that remained became Argentina. The country's population and culture were heavily shaped by immigrants from throughout Europe, but most particularly Italy and Spain, which provided the largest percentage of newcomers from 1860 to 1930. Up until about the mid-20th century, much of Argentina's history was dominated by periods of internal political conflict between Federalists and Unitarians and between civilian and military factions. After World War II, an era of Peronist populism and direct and indirect military interference in subsequent governments was followed by a military junta that took power in 1976. Democracy returned in 1983 after a failed bid to seize the Falkland (Malvinas) Islands by force, and has persisted despite numerous challenges, the most formidable of which was a severe economic crisis in 2001-02 that led to violent public protests and the successive resignations of several presidents.

GEOGRAPHY

Location: Southern South America, bordering the South Atlantic Ocean, between Chile and Uruguay
Geographic coordinates: 34 00 S, 64 00 W
Map references: South America
Area: *total:* 2,780,400 sq km
country comparison to the world: 8
land: 2,736,690 sq km
water: 43,710 sq km
Area—comparative: slightly less than three-tenths the size of the US
Land boundaries: *total:* 9,861 km
border countries: Bolivia 832 km, Brazil 1,261 km, Chile 5,308 km, Paraguay 1,880 km, Uruguay 580 km
Coastline: 4,989 km
Maritime claims: territorial sea: 12 nm
contiguous zone: 24 nm
exclusive economic zone: 200 nm
continental shelf: 200 nm or to the edge of the continental margin
Climate: mostly temperate; arid in southeast; subantarctic in southwest
Terrain: rich plains of the Pampas in northern half, flat to rolling plateau of Patagonia in south, rugged Andes along western border
Elevation extremes: *lowest point:* Laguna del Carbon -105 m (located between Puerto San Julian and Comandante Luis Piedra Buena in the province of Santa Cruz)
highest point: Cerro Aconcagua 6,960 m (located in the northwestern corner of the province of Mendoza)
Natural resources: fertile plains of the pampas, lead, zinc, tin, copper, iron ore, manganese, petroleum, uranium
Land use: *arable land:* 10.03%
permanent crops: 0.36%
other: 89.61% (2005)
Irrigated land: 15,500 sq km (2003)
Total renewable water resources: 814 cu km (2000)
Freshwater withdrawal (domestic/industrial/agricultural): *total:* 29.19 cu km/yr (17%/9%/74%)
per capita: 753 cu m/yr (2000)
Natural hazards: San Miguel de Tucuman and Mendoza areas in the Andes subject to earthquakes; pamperos are violent windstorms that can strike the pampas and northeast; heavy flooding in some areas
volcanism: Argentina experiences volcanic activity in the Andes Mountains along the Chilean border; Copahue (elev. 2,997 m) last erupted in 2000; other historically active volcanoes include Llullaillaco, Maipo, Planchon-Peteroa, San Jose, Tromen, Tupungatito, and Viedma
Environment—current issues: environmental problems (urban and rural) typical of an industrializing economy such as deforestation, soil degradation, desertification, air pollution, and water pollution
note: Argentina is a world leader in setting voluntary greenhouse gas targets
Environment—international agreements: *party to:* Antarctic-Environmental Protocol, Antarctic-Marine Living Resources, Antarctic Seals, Antarctic Treaty, Biodiversity, Climate Change, Climate Change-Kyoto Protocol, Desertification, Endangered Species, Environmental Modification, Hazardous Wastes, Law of the Sea, Marine Dumping, Ozone Layer Protection, Ship Pollution, Wetlands, Whaling
signed, but not ratified: Marine Life Conservation

Geography—note: second-largest country in South America (after Brazil); strategic location relative to sea lanes between the South Atlantic and the South Pacific Oceans (Strait of Magellan, Beagle Channel, Drake Passage); diverse geophysical landscapes range from tropical climates in the north to tundra in the far south; Cerro Aconcagua is the Western Hemisphere's tallest mountain, while Laguna del Carbon is the lowest point in the Western Hemisphere

PEOPLE

Population: 41,769,726 (July 2011 est.)
country comparison to the world: 32
Age structure: *0–14 years:* 25.4% (male 5,429,488/female 5,181,289)
15–64 years: 63.6% (male 13,253,468/female 13,301,530)
65 years and over: 11% (male 1,897,144/female 2,706,807) (2011 est.)
Median age: *total:* 30.5 years
male: 29.5 years
female: 31.6 years (2011 est.)
Population growth rate: 1.017% (2011 est.)
country comparison to the world: 116
Birth rate: 17.54 births/1,000 population (2011 est.)
country comparison to the world: 112
Death rate: 7.38 deaths/1,000 population (July 2011 est.)
country comparison to the world: 121
Net migration rate: 0 migrant(s)/1,000 population (2011 est.)
country comparison to the world: 114
Urbanization: *urban population:* 92% of total population (2010)
rate of urbanization: 1.1% annual rate of change (2010-15 est.)
Major cities—population: BUENOS AIRES (capital) 12.988 million; Cordoba 1.493 million; Rosario 1.231 million; Mendoza 917,000; San Miguel de Tucuman 831,000 (2009)
Sex ratio: *at birth:* 1.052 male(s)/female
under 15 years: 1.05 male(s)/female
15–64 years: 1 male(s)/female
65 years and over: 0.7 male(s)/female
total population: 0.97 male(s)/female (2011 est.)
Infant mortality rate:
total: 10.81 deaths/1,000 live births
country comparison to the world: 145
male: 12.08 deaths/1,000 live births
female: 9.48 deaths/1,000 live births (2011 est.)
Life expectancy at birth: *total population:* 76.95 years
country comparison to the world: 68
male: 73.71 years
female: 80.36 years (2011 est.)
Total fertility rate: 2.31 children born/woman (2011 est.)
country comparison to the world: 98
HIV/AIDS—adult prevalence rate: 0.5% (2009 est.)
country comparison to the world: 68
HIV/AIDS—people living with HIV/AIDS: 110,000 (2009 est.)
country comparison to the world: 40
HIV/AIDS—deaths: 2,900 (2009 est.)
country comparison to the world: 46
Major infectious diseases: *degree of risk:* intermediate
food or waterborne diseases: bacterial diarrhea, hepatitis A
water contact disease: leptospirosis (2009)
Drinking water source: *Improved:*
urban: 98% of population
rural: 80% of population
total: 97% of population
Unimproved:
urban: 2% of population
rural: 20% of population
total: 3% of population (2008)
Sanitation facility access: *Improved:*
urban: 91% of population
rural: 77% of population
total: 90% of population
Unimproved:
urban: 9% of population
rural: 23% of population
total: 10% of population (2008)
Nationality: *noun:* Argentine(s)
adjective: Argentine
Ethnic groups: white (mostly Spanish and Italian) 97%, mestizo (mixed white and Amerindian ancestry), Amerindian, or other non-white groups 3%
Religions: nominally Roman Catholic 92% (less than 20% practicing), Protestant 2%, Jewish 2%, other 4%
Languages: Spanish (official), Italian, English, German, French
Literacy: *definition:* age 15 and over can read and write
total population: 97.2%
male: 97.2%
female: 97.2% (2001 census)
School life expectancy (primary to tertiary education): *total:* 16 years
male: 15 years
female: 17 years (2007)
Education expenditures: 4.9% of GDP (2007)
country comparison to the world: 61

GOVERNMENT

Country name: *conventional long form:* Argentine Republic
conventional short form: Argentina
local long form: Republica Argentina
local short form: Argentina
Government type: republic
Capital: *name:* Buenos Aires
geographic coordinates: 34 36 S, 58 40 W
time difference: UTC-3 (2 hours ahead of Washington, DC during Standard Time)
daylight saving time: none scheduled for 2011
Administrative divisions: 23 provinces (provincias, singular—provincia) and 1 autonomous city* (distrito federal); Buenos Aires, Buenos Aires Capital Federal*, Catamarca, Chaco, Chubut, Cordoba, Corrientes, Entre Rios, Formosa, Jujuy, La Pampa, La Rioja, Mendoza, Misiones, Neuquen, Rio Negro, Salta, San Juan, San Luis, Santa Cruz, Santa Fe, Santiago del Estero, Tierra del Fuego—Antartida e Islas del Atlantico Sur (Tierra del Fuego), Tucuman
note: the US does not recognize any claims to Antarctica
Independence: 9 July 1816 (from Spain)
National holiday: Revolution Day, 25 May (1810)
Constitution: 1 May 1853; amended many times starting in 1860
Legal system: civil law system based on West European legal systems; note—efforts at civil code reform begun in the mid-1980s has stagnated
International law organization participation: has not submitted an ICJ jurisdiction declaration; accepts ICCt jurisdiction
Suffrage: 18-70 years of age; universal and compulsory
Executive branch: *chief of state:* President Cristina FERNANDEZ DE KIRCHNER (since 10 December 2007); Vice President Julio COBOS (since 10 December 2007); note—the president is both the chief of state and head of government
head of government: President Cristina FERNANDEZ DE KIRCHNER (since 10 December 2007); Vice President Julio COBOS (since 10 December 2007)
cabinet: Cabinet appointed by the president
(For more information visit the World Leaders website)
elections: president and vice president elected on the same ticket by popular vote for four-year terms (eligible for a second term); election last held on 28 October 2007 (next election to be held in 23 October 2011)
election results: Cristina FERNANDEZ DE KIRCHNER elected president; percent of vote—Cristina FERNANDEZ DE KIRCHNER 45%, Elisa CARRIO 23%, Roberto LAVAGNA 17%, Alberto Rodriguez SAA 8%, other 7%
Legislative branch: bicameral National Congress or Congreso Nacional consists of the Senate (72 seats; members are elected by direct vote; presently one-third of the members elected every two years to serve six-year terms) and the Chamber of Deputies (257 seats; members are elected by direct vote; one-half of the members elected every two years to serve four-year terms)
elections: Senate—last held on 28 June 2009 (next to be held in 2011); Chamber of Deputies—last held on 28 June 2009 (next to be held in 2011)
election results: Senate—percent of vote by bloc or party—NA; seats by bloc or party—FpV 8, ACyS 14, PJ disidente 2; Chamber of Deputies—percent of vote by bloc or party—NA; seats by bloc or party—FpV 45, ACyS 42, PRO 20, PJ disidente 12, other 8; note—as of 1 February 2011, the composition of the entire legislature is as follows: Senate—seats by bloc or party—FpV 32, UCR 16, PJ disidente 14, other 10; Chamber of Deputies—seats by bloc or party—FpV 87, ACyS 43, PRO 11, PJ disidente 28, CC 19, PS 6, other 63
Judicial branch: Supreme Court or Corte Suprema (the Supreme Court judges are appointed by the president with approval of the Senate)
note: the Supreme Court has seven judges; the Argentine Congress in 2006 passed a bill to gradually reduce the number of Supreme Court judges to five
Political parties and leaders: Civic and Social Accord or ACyS (a now-defunct center-left alliance that included the CC, UCR, and Socialist parties-created ahead of the 2009 legislative elections); Civic

Coalition or CC (a broad coalition loosely affiliated with Elisa CARRIO); Dissident Peronists or PJ Disidente (a sector of the Justicialist Party opposed to the Kirchners); Front for Victory or FpV (a broad coalition, including elements of the UCR and numerous provincial parties) [Cristina FERNANDEZ DE KIRCHNER]; Justicialist Party or PJ [Daniel SCIOLI]; Radical Civic Union or UCR [Ernesto SANZ]; Republican Proposal or PRO [Mauricio MACRI] (including Federal Recreate Movement or RECREAR [Esteban BULLRICH]; Socialist Party or PS [Ruben GIUSTINIANI]; Union For All [Patricia BULLRICH] (associated with the Civic Coalition); numerous provincial parties

Political pressure groups and leaders: Argentine Association of Pharmaceutical Labs (CILFA); Argentine Industrial Union (manufacturers' association); Argentine Rural Confederation or CRA (small to medium landowners' association); Argentine Rural Society (large landowners' association); Central of Argentine Workers or CTA (a radical union for employed and unemployed workers); General Confederation of Labor or CGT (Peronist-leaning umbrella labor organization); White and Blue CGT (dissident CGT labor confederation); Roman Catholic Church
other: business organizations; Peronist-dominated labor movement; Piquetero groups (popular protest organizations that can be either pro or anti-government); students

International organization participation: AfDB (nonregional member), Australia Group, BCIE, BIS, CAN (associate), FAO, FATF, G-15, G-20, G-24, G-77, IADB, IAEA, IBRD, ICAO, ICC, ICRM, IDA, IFAD, IFC, IFRCS, IHO, ILO, IMF, IMO, IMSO, Interpol, IOC, IOM, IPU, ISO, ITSO, ITU, ITUC, LAES, LAIA, Mercosur, MIGA, MINURSO, MINUSTAH, NAM (observer), NSG, OAS, OPANAL, OPCW, Paris Club (associate), PCA, RG, SICA (observer), UN, UNASUR, UNCTAD, UNESCO, UNFICYP, UNHCR, UNIDO, Union Latina (observer), UNTSO, UNWTO, UPU, WCO, WFTU, WHO, WIPO, WMO, WTO, ZC

Diplomatic representation in the US: *chief of mission:* Ambassador Alfredo Vicente CHIARADIA
chancery: 1600 New Hampshire Avenue NW, Washington, DC 20009
telephone: [1] (202) 238-6400
FAX: [1] (202) 332-3171
consulate(s) general: Atlanta, Chicago, Houston, Los Angeles, Miami, New York

Diplomatic representation from the US: *chief of mission:* Ambassador Vilma MARTINEZ
embassy: Avenida Colombia 4300, C1425GMN Buenos Aires
mailing address: international mail: use embassy street address; APO address: US Embassy Buenos Aires, Unit 4334, APO AA 34034
telephone: [54] (11) 5777-4533
FAX: [54] (11) 5777-4240

Flag description: three equal horizontal bands of light blue (top), white, and light blue; centered in the white band is a radiant yellow sun with a human face known as the Sun of May; the colors represent the clear skies and snow of the Andes; the sun symbol commemorates the appearance of the sun through cloudy skies on 25 May 1810 during the first mass demonstration in favor of independence; the sun features are those of Inti, the Inca god of the sun

National anthem: *name:* "Himno Nacional Argentino" (Argentine National Anthem)
lyrics/music: Vicente LOPEZ y PLANES/ Jose Blas PARERA
note: adopted 1813; Vicente LOPEZ was inspired to write the anthem after watching a play about the 1810 May Revolution against Spain

ECONOMY

Economy—overview: Argentina benefits from rich natural resources, a highly literate population, an export-oriented agricultural sector, and a diversified industrial base. Although one of the world's wealthiest countries 100 years ago, Argentina suffered during most of the 20th century from recurring economic crises, persistent fiscal and current account deficits, high inflation, mounting external debt, and capital flight. A severe depression, growing public and external indebtedness, and a bank run culminated in 2001 in the most serious economic, social, and political crisis in the country's turbulent history. Interim President Adolfo RODRIGUEZ SAA declared a default–the largest in history–on the government's foreign debt in December of that year, and abruptly resigned only a few days after taking office. His successor, Eduardo DUHALDE, announced an end to the peso's decade-long 1-to-1 peg to the US dollar in early 2002. The economy bottomed out that year, with real GDP 18% smaller than in 1998 and almost 60% of Argentines under the poverty line. Real GDP rebounded to grow by an average 8.5% annually over the subsequent six years, taking advantage of previously idled industrial capacity and labor, an audacious debt restructuring and reduced debt burden, excellent international financial conditions, and expansionary monetary and fiscal policies. Inflation also increased, however, during the administration of President Nestor KIRCHNER, which responded with price restraints on businesses, as well as export taxes and restraints, and beginning in early 2007, with understating inflation data. Cristina FERNANDEZ DE KIRCHNER succeeded her husband as President in late 2007, and the rapid economic growth of previous years began to slow sharply the following year as government policies held back exports and the world economy fell into recession. The economy has rebounded strongly from the 2009 recession, but the government's continued reliance on expansionary fiscal and monetary policies risks exacerbating already high inflation.

GDP (purchasing power parity): $596 billion (2010 est.)
country comparison to the world: 24
$554.5 billion (2009 est.)
$571.6 billion (2008 est.)
note: data are in 2010 US dollars

GDP (official exchange rate): $370.3 billion (2010 est.)

GDP—real growth rate: 7.5% (2010 est.)
country comparison to the world: 26
-3% (2009 est.)
5% (2008 est.)

GDP—per capita (PPP): $14,700 (2010 est.)
country comparison to the world: 78
$13,700 (2009 est.)
$14,100 (2008 est.)
note: data are in 2010 US dollars

GDP—composition by sector:
agriculture: 8.5%
industry: 31.6%
services: 59.8% (2010 est.)

Labor force: 16.62 million
country comparison to the world: 36
note: urban areas only (2010 est.)

Labor force—by occupation:
agriculture: 5%
industry: 23%
services: 72% (2009 est.)

Unemployment rate: 7.9% (2010 est.)
country comparison to the world: 84
8.7% (2009 est.)

Population below poverty line: 30%
note: data are based on private estimates (2010)

Household income or consumption by percentage share: *lowest 10%:* 1.7%
highest 10%: 29.5% (3rd Quarter, 2010)

Distribution of family income—Gini index: 41.4 3rd quarter, 2010
country comparison to the world: 53

Investment (gross fixed): 22% of GDP (2010 est.)
country comparison to the world: 68

Budget: *revenues:* $87.63 billion
expenditures: $86.85 billion (2010 est.)

Public debt: 50.3% of GDP (2010 est.)
country comparison to the world: 51
48.6% of GDP (2009 est.)
note: official data

Inflation rate (consumer prices): 22% (2010 est.)
country comparison to the world: 221
16% (2009 est.)
note: data are derived from private estimates

Central bank discount rate: NA%

Commercial bank prime lending rate: 11.3% (31 December 2010 est.)
country comparison to the world: 87
10% (31 December 2009 est.)

Stock of narrow money: $40.35 billion (31 December 2010)
country comparison to the world: 47
$32.22 billion (31 December 2009)

Stock of broad money: $112.9 billion (31 December 2010 est.)
country comparison to the world: 49
$85.18 billion (31 December 2009 est.)

Stock of domestic credit: $113.9 billion (31 December 2010 est.)
country comparison to the world: 47
$84.92 billion (31 December 2009 est.)

Market value of publicly traded shares: $48.93 billion (31 December 2009)
country comparison to the world: 50
$52.31 billion (31 December 2008)
$86.68 billion (31 December 2007)

Agriculture—products: sunflower seeds, lemons, soybeans, grapes, corn, tobacco, peanuts, tea, wheat; livestock

Industries: food processing, motor vehicles, consumer durables, textiles, chemicals and petrochemicals, printing, metallurgy, steel
Industrial production growth rate: 8.9%
country comparison to the world: 25
note: based on private estimates (2010 est.)
Electricity—production: 109.5 billion kWh (2007 est.)
country comparison to the world: 30
Electricity—consumption: 99.21 billion kWh (2007 est.)
country comparison to the world: 30
Electricity—exports: 2.628 billion kWh (2007 est.)
Electricity—imports: 10.28 billion kWh (2007 est.)
Oil—production: 796,300 bbl/day (2009 est.)
country comparison to the world: 26
Oil—consumption: 622,000 bbl/day (2009 est.)
country comparison to the world: 26
Oil—exports: 314,400 bbl/day (2007 est.)
country comparison to the world: 38
Oil—imports: 52,290 bbl/day (2007 est.)
country comparison to the world: 84
Oil—proved reserves: 2.386 billion bbl (1 January 2010 est.)
country comparison to the world: 34
Natural gas—production: 41.36 billion cu m (2009 est.)
country comparison to the world: 20
Natural gas—consumption: 43.14 billion cu m (2009 est.)
country comparison to the world: 20
Natural gas—exports: 890 million cu m (2008 est.)
country comparison to the world: 38
Natural gas—imports: 2.66 billion cu m (2009 est.)
country comparison to the world: 43
Natural gas—proved reserves: 398.4 billion cu m (1 January 2010 est.)
country comparison to the world: 35
Current account balance: $6.976 billion (2010 est.)
country comparison to the world: 29
$11.29 billion (2009 est.)
Exports: $68.5 billion (2010 est.)
country comparison to the world: 42
$55.67 billion (2009 est.)
Exports—commodities: soybeans and derivatives, petroleum and gas, vehicles, corn, wheat
Exports—partners: Brazil 18.78%, China 9.26%, Chile 7.11%, US 6.38% (2009)
Imports: $56.44 billion (2010 est.)
country comparison to the world: 45
$38.78 billion (2009 est.)
Imports—commodities: machinery, motor vehicles, petroleum and natural gas, organic chemicals, plastics
Imports—partners: Brazil 31.12%, US 13.69%, China 10.26%, Germany 4.69% (2009)
Reserves of foreign exchange and gold: $53.61 billion (31 December 2010 est.)
country comparison to the world: 23
$48.03 billion (31 December 2009 est.)
Debt—external: $160.9 billion (30 September 2010 est.)
country comparison to the world: 32
$147.1 billion (31 December 2009 est.)
Stock of direct foreign investment—at home: $86.8 billion (31 December 2010 est.)
country comparison to the world: 37
$80.1 billion (31 December 2009 est.)
Stock of direct foreign investment—abroad: $30.16 billion (31 December 2010 est.)
country comparison to the world: 38
$29.46 billion (31 December 2009 est.)
Exchange rates: Argentine pesos (ARS) per US dollar—3.8983 (2010), 3.7101 (2009), 3.1636 (2008), 3.1105 (2007), 3.0543 (2006)

COMMUNICATIONS

Telephones—main lines in use: 9.764 million (2009)
country comparison to the world: 22
Telephones—mobile cellular: 51.891 million (2009)
country comparison to the world: 22
Telephone system: *general assessment:* the "Telecommunications Liberalization Plan of 1998" opened the telecommunications market to competition and foreign investment encouraging the growth of modern telecommunications technology; fiber-optic cable trunk lines are being installed between all major cities; major networks are entirely digital and the availability of telephone service is improving
domestic: microwave radio relay, fiber-optic cable, and a domestic satellite system with 40 earth stations serve the trunk network; fixed-line teledensity is increasing gradually and mobile-cellular subscribership is increasing rapidly; broadband Internet services are gaining ground
international: country code—54; landing point for the Atlantis-2, UNISUR, South America-1, and South American Crossing/Latin American Nautilus submarine cable systems that provide links to Europe, Africa, South and Central America, and US; satellite earth stations—112; 2 international gateways near Buenos Aires (2009)
Broadcast media: government owns a TV station and a radio network; more than 2 dozen TV stations and hundreds of privately-owned radio stations; high rate of cable TV subscription usage (2007)
Internet country code: .ar
Internet hosts: 6.025 million (2010)
country comparison to the world: 16
Internet users: 13.694 million (2009)
country comparison to the world: 28

TRANSPORTATION

Airports: 1,141 (2010)
country comparison to the world: 6
Airports—with paved runways: *total:* 156
over 3,047 m: 4
2,438 to 3,047 m: 27
1,524 to 2,437 m: 65
914 to 1,523 m: 51
under 914 m: 9 (2010)
Airports—with unpaved runways:
total: 985
over 3,047 m: 1
2,438 to 3,047 m: 1
1,524 to 2,437 m: 43
914 to 1,523 m: 530
under 914 m: 410 (2010)
Heliports: 2 (2010)
Pipelines: gas 29,401 km; liquid petroleum gas 41 km; oil 6,166 km; refined products 3,631 km (2010)
Railways: *total:* 36,966 km
country comparison to the world: 8
broad gauge: 26,475 km 1.676-m gauge (94 km electrified)
standard gauge: 2,780 km 1.435-m gauge (42 km electrified)
narrow gauge: 7,711 km 1.000-m gauge (2010)
Roadways: *total:* 231,374 km
country comparison to the world: 22
paved: 69,412 km (includes 734 km of expressways)
unpaved: 161,962 km (2004)
Waterways: 11,000 km (2007)
country comparison to the world: 11
Merchant marine: *total:* 43
country comparison to the world: 74
by type: bulk carrier 3, cargo 7, chemical tanker 4, container 1, passenger/cargo 3, petroleum tanker 23, refrigerated cargo 2
foreign-owned: 12 (Brazil 1, Chile 6, Spain 3, UK 2)
registered in other countries: 17 (Liberia 3, Panama 7, Paraguay 5, Uruguay 2) (2010)
Ports and terminals: Arroyo Seco, Bahia Blanca, Buenos Aires, La Plata, Punta Colorada, Rosario, San Lorenzo-San Martin, Ushuaia

MILITARY

Military branches: Argentine Army (Ejercito Argentino), Navy of the Argentine Republic (Armada Republica; includes naval aviation and naval infantry), Argentine Air Force (Fuerza Aerea Argentina, FAA) (2011)
Military service age and obligation: 18-24 years of age for voluntary military service (18-21 requires parental consent); no conscription (2001)
Manpower available for military service:
males age 16–49: 10,038,967
females age 16–49: 9,959,134 (2010 est.)
Manpower fit for military service: *males age 16–49:* 8,458,362
females age 16–49: 8,414,460 (2010 est.)
Manpower reaching militarily significant age annually:
male: 339,503
female: 323,170 (2010 est.)
Military expenditures: 0.8% of GDP (2009)
country comparison to the world: 147
Military—note: the Argentine military is a well-organized force constrained by the country's prolonged economic hardship; the country has recently experienced a strong recovery, and the military is implementing a modernization plan aimed at making the ground forces lighter and more responsive (2008)

TRANSNATIONAL ISSUES

Disputes—international: Argentina continues to assert its claims to the UK-administered Falkland Islands (Islas Malvinas), South Georgia, and the South Sandwich Islands in its constitution, forcibly occupying the Falklands in 1982, but in 1995 agreed no longer to seek settlement by force; UK continues to reject Argentine requests for sovereignty talks; territorial claim in Antarctica partially overlaps UK and Chilean claims; uncontested dispute between Brazil and Uruguay over Braziliera/Brasiliera Island in the Quarai/Cuareim River leaves the tripoint with Argentina in question; in 2010, the ICJ ruled in favor of Uruguay's operation of two paper mills on the Uruguay River, which forms the border with Argentina; the two countries formed a joint pollution monitoring regime; the joint boundary commission, established by Chile and Argentina in 2001 has yet to map and demarcate the delimited boundary in the inhospitable Andean Southern Ice Field (Campo de Hielo Sur); contraband smuggling, human trafficking, and illegal narcotic trafficking are problems in the porous areas of the border with Bolivia

Illicit drugs: a transshipment country for cocaine headed for Europe, heroin headed for the US, and ephedrine and pseudoephedrine headed for Mexico; some money-laundering activity, especially in the Tri-Border Area; law enforcement corruption; a source for precursor chemicals; increasing domestic consumption of drugs in urban centers, especially cocaine base and synthetic drugs (2008)

ARMENIA

INTRODUCTION

Background: Armenia prides itself on being the first nation to formally adopt Christianity (early 4th century). Despite periods of autonomy, over the centuries Armenia came under the sway of various empires including the Roman, Byzantine, Arab, Persian, and Ottoman. During World War I in the western portion of Armenia, Ottoman Turkey instituted a policy of forced resettlement coupled with other harsh practices that resulted in an estimated 1 million Armenian deaths. The eastern area of Armenia was ceded by the Ottomans to Russia in 1828; this portion declared its independence in 1918, but was conquered by the Soviet Red Army in 1920. Armenian leaders remain preoccupied by the long conflict with Azerbaijan over Nagorno-Karabakh, a primarily Armenian-populated region, assigned to Soviet Azerbaijan in the 1920s by Moscow. Armenia and Azerbaijan began fighting over the area in 1988; the struggle escalated after both countries attained independence from the Soviet Union in 1991. By May 1994, when a cease-fire took hold, ethnic Armenian forces held not only Nagorno-Karabakh but also a significant portion of Azerbaijan proper. The economies of both sides have been hurt by their inability to make substantial progress toward a peaceful resolution. Turkey closed the common border with Armenia in 1994 because of the Armenian separatists' control of Nagorno-Karabakh and surrounding areas, further hampering Armenian economic growth. In 2009, senior Armenian leaders began pursuing rapprochement with Turkey, aiming to secure an opening of the border; this process is currently dormant.

GEOGRAPHY

Location: Southwestern Asia, between Turkey (to the west) and Azerbaijan
Geographic coordinates: 40 00 N, 45 00 E
Map references: Middle East
Area: *total:* 29,743 sq km
country comparison to the world: 142
land: 28,203 sq km
water: 1,540 sq km
Area—comparative: slightly smaller than Maryland
Land boundaries: *total:* 1,254 km
border countries: Azerbaijan-proper 566 km, Azerbaijan-Naxcivan exclave 221 km, Georgia 164 km, Iran 35 km, Turkey 268 km
Coastline: 0 km (landlocked)
Maritime claims: none (landlocked)
Climate: highland continental, hot summers, cold winters
Terrain: Armenian Highland with mountains; little forest land; fast flowing rivers; good soil in Aras River valley
Elevation extremes: *lowest point:* Debed River 400 m
highest point: Aragats Lerrnagagat' 4,090 m
Natural resources: small deposits of gold, copper, molybdenum, zinc, bauxite
Land use: *arable land:* 16.78%
permanent crops: 2.01%
other: 81.21% (2005)
Irrigated land: 2,860 sq km (2003)
Total renewable water resources: 10.5 cu km (1997)
Freshwater withdrawal (domestic/industrial/agricultural): *total:* 2.95 cu km/yr (30%/4%/66%)
per capita: 977 cu m/yr (2000)
Natural hazards: occasionally severe earthquakes; droughts
Environment—current issues: soil pollution from toxic chemicals such as DDT; the energy crisis of the 1990s led to deforestation when citizens scavenged for firewood; pollution of Hrazdan (Razdan) and Aras Rivers; the draining of Sevana Lich (Lake Sevan), a result of its use as a source for hydropower, threatens drinking water supplies; restart of Metsamor nuclear power plant in spite of its location in a seismically active zone
Environment—international agreements: *party to:* Air Pollution, Biodiversity, Climate Change, Climate Change-Kyoto Protocol, Desertification, Environmental Modification, Hazardous Wastes, Law of the Sea, Ozone Layer Protection, Wetlands
signed, but not ratified: Air Pollution-Persistent Organic Pollutants
Geography—note: landlocked in the Lesser Caucasus Mountains; Sevana Lich (Lake Sevan) is the largest lake in this mountain range

PEOPLE

Population: 2,967,975 (July 2011 est.)
country comparison to the world: 137
Age structure: *0–14 years:* 17.6% (male 279,304/female 242,621)
15–64 years: 72.4% (male 1,006,312/female 1,141,430)
65 years and over: 10.1% (male 112,947/female 185,361) (2011 est.)
Median age: *total:* 32.2 years
male: 29.5 years
female: 35 years (2011 est.)
Population growth rate: 0.063% (2011 est.)
country comparison to the world: 188
Birth rate: 12.85 births/1,000 population (2011 est.)
country comparison to the world: 156
Death rate: 8.46 deaths/1,000 population (July 2011 est.)
country comparison to the world: 86
Net migration rate: -3.76 migrant(s)/1,000 population (2011 est.)
country comparison to the world: 186
Urbanization: *urban population:* 64% of total population (2010)
rate of urbanization: 0.5% annual rate of change (2010-15 est.)
Major cities—population: YEREVAN (capital) 1.11 million (2009)
Sex ratio: *at birth:* 1.124 male(s)/female
under 15 years: 1.15 male(s)/female
15–64 years: 0.88 male(s)/female
65 years and over: 0.62 male(s)/female
total population: 0.89 male(s)/female (2011 est.)
Infant mortality rate: *total:* 18.85 deaths/1,000 live births
country comparison to the world: 101
male: 23.38 deaths/1,000 live births
female: 13.75 deaths/1,000 live births (2011 est.)
Life expectancy at birth: *total population:* 73.23 years
country comparison to the world: 119
male: 69.59 years
female: 77.31 years (2011 est.)
Total fertility rate: 1.37 children born/woman (2011 est.)
country comparison to the world: 205

HIV/AIDS—adult prevalence rate: 0.1% (2009 est.)
country comparison to the world: 115
HIV/AIDS—people living with HIV/AIDS: 1,900 (2009 est.)
country comparison to the world: 137
HIV/AIDS—deaths: fewer than 100 (2009 est.)
country comparison to the world: 119
Drinking water source: *Improved:*
urban: 98% of population
rural: 93% of population
total: 96% of population
Unimproved:
urban: 2% of population
rural: 7% of population
total: 4% of population (2008)
Sanitation facility access: *Improved:*
urban: 95% of population
rural: 80% of population
total: 90% of population
Unimproved:
urban: 5% of population
rural: 20% of population
total: 10% of population (2008)
Nationality: *noun:* Armenian(s)
adjective: Armenian
Ethnic groups: Armenian 97.9%, Yezidi (Kurd) 1.3%, Russian 0.5%, other 0.3% (2001 census)
Religions: Armenian Apostolic 94.7%, other Christian 4%, Yezidi (monotheist with elements of nature worship) 1.3%
Languages: Armenian (official) 97.7%, Yezidi 1%, Russian 0.9%, other 0.4% (2001 census)
Literacy: *definition:* age 15 and over can read and write
total population: 99.4%
male: 99.7%
female: 99.2% (2001 census)
School life expectancy (primary to tertiary education): *total:* 12 years
male: 12 years
female: 13 years (2009)
Education expenditures: 3% of GDP (2007)
country comparison to the world: 134

GOVERNMENT

Country name: *conventional long form:* Republic of Armenia
conventional short form: Armenia
local long form: Hayastani Hanrapetut'yun
local short form: Hayastan
former: Armenian Soviet Socialist Republic, Armenian Republic
Government type: republic
Capital: *name:* Yerevan
geographic coordinates: 40 10 N, 44 30 E
time difference: UTC+4 (9 hours ahead of Washington, DC during Standard Time)
daylight saving time: +1hr, begins last Sunday in March; ends last Sunday in October
Administrative divisions: 11 provinces (marzer, singular—marz); Aragatsotn, Ararat, Armavir, Geghark'unik', Kotayk', Lorri, Shirak, Syunik', Tavush, Vayots' Dzor, Yerevan
Independence: 21 September 1991 (from the Soviet Union)
National holiday: Independence Day, 21 September (1991)
Constitution: adopted by nationwide referendum 5 July 1995; amendments adopted through a nationwide referendum 27 November 2005
Legal system: civil law system
International law organization participation: has not submitted an ICJ jurisdiction declaration; non-party state to the ICCt
Suffrage: 18 years of age; universal
Executive branch: *chief of state:* President Serzh SARGSIAN (since 9 April 2008)
head of government: Prime Minister Tigran SARGSIAN (since 9 April 2008)
cabinet: Council of Ministers appointed by the prime minister
(For more information visit the World Leaders website)
elections: president elected by popular vote for a five-year term (eligible for a second term); election last held on 19 February 2008 (next to be held in February 2013); prime minister appointed by the president based on majority or plurality support in parliament; the prime minister and Council of Ministers must resign if the National Assembly refuses to accept their program
election results: Serzh SARGSIAN elected president; percent of vote—Serzh SARGSIAN 52.9%, Levon TER-PETROSSIAN 21.5%, Artur BAGHDASARIAN 16.7%, other 8.9%
Legislative branch: unicameral National Assembly (Parliament) or Azgayin Zhoghov (131 seats; members elected by popular vote, 90 members elected by party list and 41 by direct vote; to serve five-year terms)
elections: last held on 12 May 2007 (next to be held in the spring of 2012)
election results: percent of vote by party—HHK 33.9%, Prosperous Armenia 15.1%, ARF (Dashnak) 13.2%, Rule of Law 7.1%, Heritage Party 6%, other 24.7%; seats by party—HHK 64, Prosperous Armenia 18, ARF (Dashnak) 16, Rule of Law 9, Heritage Party 7, independent 17
Judicial branch: Constitutional Court; Court of Cassation (Appeals Court)
Political parties and leaders: Armenian National Congress or ANC (bloc of independent and opposition parties) [Levon TER-PETROSSIAN]; Armenian National Movement or ANM [Ararat ZURABIAN]; Armenian Revolutionary Federation ("Dashnak" Party) or ARF [Hrant MARKARIAN]; Heritage Party [Raffi HOVHANNISIAN]; People's Party of Armenia [Stepan DEMIRCHIAN]; Prosperous Armenia [Gagik TSARUKIAN]; Republican Party of Armenia or HHK [Serzh SARGSIAN]; Rule of Law Party (Orinats Yerkir) [Artur BAGHDASARIAN]
Political pressure groups and leaders: Aylentrank (Impeachment Alliance) [Nikol PASHINIAN]; Yerkrapah Union [Manvel GRIGORIAN]
International organization participation: ADB, BSEC, CE, CIS, CSTO, EAEC (observer), EAPC, EBRD, FAO, GCTU, IAEA, IBRD, ICAO, ICRM, IDA, IFAD, IFC, IFRCS, ILO, IMF, Interpol, IOC, IOM, IPU, ISO, ITSO, ITU, MIGA, NAM (observer), OAS (observer), OIF (associate member), OPCW, OSCE, PFP, UN, UNCTAD, UNESCO, UNIDO, UNWTO, UPU, WCO, WFTU, WHO, WIPO, WMO, WTO
Diplomatic representation in the US: *chief of mission:* Ambassador Tatoul MARKARIAN
chancery: 2225 R Street NW, Washington, DC 20008
telephone: [1] (202) 319-1976
FAX: [1] (202) 319-2982
consulate(s) general: Los Angeles
Diplomatic representation from the US: *chief of mission:* Ambassador Marie L. YOVANOVITCH
embassy: 1 American Ave., Yerevan 0082
mailing address: American Embassy Yerevan, US Department of State, 7020 Yerevan Place, Washington, DC 20521-7020
telephone: [374](10) 464-700
FAX: [374](10) 464-742
Flag description: three equal horizontal bands of red (top), blue, and orange; the color red recalls the blood shed for liberty, blue the Armenian skies as well as hope, and orange the land and the courage of the workers who farm it
National anthem: *name:* "Mer Hayrenik""(Our Fatherland)
lyrics/music: Mikael NALBANDIAN/ Barsegh KANACHYAN
note: adopted 1991; based on the anthem of the Democratic Republic of Armenia (1918-1922) but with different lyrics

ECONOMY

Economy—overview: After several years of double-digit economic growth, Armenia faced a severe economic recession with GDP declining more than 14% in 2009, despite large loans from multilateral institutions. Sharp declines in the construction sector and workers' remittances, particularly from Russia, were the main reasons for the downturn. The economy began to recover in 2010 with nearly 5% growth. Under the old Soviet central planning system, Armenia developed a modern industrial sector, supplying machine tools, textiles, and other manufactured goods to sister republics, in exchange for raw materials and energy. Armenia has since switched to small-scale agriculture and away from the large agroindustrial complexes of the Soviet era. Armenia has managed to reduce poverty, slash inflation, stabilize its currency, and privatize most small- and medium-sized enterprises. Since the breakup of the Soviet Union in 1991, Armenia had made progress in implementing some economic reforms, including privatization, price reforms, and prudent fiscal policies, but geographic isolation, a narrow export base, and pervasive monopolies in important business sectors have made Armenia particularly vulnerable to the sharp deterioration in the global economy and the economic downturn in Russia. The conflict with Azerbaijan over the ethnic Armenian-dominated region of Nagorno-Karabakh contributed to a severe economic decline in the early 1990s and Armenia's borders with Turkey remain closed. Armenia is particularly dependent on Russian commercial and governmental support and most key Armenian infrastructure is Russian-owned and/or managed, especially in the energy sector.

The electricity distribution system was privatized in 2002 and bought by Russia's RAO-UES in 2005. Construction of a pipeline to deliver natural gas from Iran to Armenia was completed in December 2008, and gas deliveries are slated to expand due to the April 2010 completion of the Yerevan Thermal Power Plant. Armenia has some mineral deposits (copper, gold, bauxite). Pig iron, unwrought copper, and other nonferrous metals are Armenia's highest valued exports. Armenia's severe trade imbalance has been offset somewhat by international aid, remittances from Armenians working abroad, and foreign direct investment. Armenia joined the WTO in January 2003. The government made some improvements in tax and customs administration in recent years, but anti-corruption measures have been ineffective and the current economic downturn has led to a sharp drop in tax revenue and forced the government to accept large loan packages from Russia, the IMF, and other international financial institutions. Armenia will need to pursue additional economic reforms in order to regain economic growth and improve economic competitiveness and employment opportunities, especially given its economic isolation from two of its nearest neighbors, Turkey and Azerbaijan.

GDP (purchasing power parity): $16.86 billion (2010 est.)
country comparison to the world: 133
$16.43 billion (2009 est.)
$19.14 billion (2008 est.)
note: data are in 2010 US dollars

GDP (official exchange rate): $9.389 billion (2010 est.)

GDP—real growth rate: 2.6% (2010 est.)
country comparison to the world: 133
-14.2% (2009 est.)
6.9% (2008 est.)

GDP—per capita (PPP): $5,700 (2010 est.)
country comparison to the world: 141
$5,500 (2009 est.)
$6,400 (2008 est.)
note: data are in 2010 US dollars

GDP—composition by sector:
agriculture: 22%
industry: 46.6%
services: 31.4% (2010 est.)

Labor force: 1.481 million (2007 est.)
country comparison to the world: 132

Labor force—by occupation:
agriculture: 46.2%
industry: 15.6%
services: 38.2% (2006 est.)

Unemployment rate: 7.1% (2007 est.)
country comparison to the world: 74

Population below poverty line: 26.5% (2006 est.)

Household income or consumption by percentage share: *lowest 10%:* 1.6%
highest 10%: 41.3% (2004)

Distribution of family income—Gini index: 37 (2006)
country comparison to the world: 76
44.4 (1996)

Investment (gross fixed): 33.3% of GDP (2010 est.)
country comparison to the world: 12

Budget:
revenues: $2.063 billion
expenditures: $2.607 billion (2010 est.)

Inflation rate (consumer prices): 6.9% (2010 est.)
country comparison to the world: 172
3.4% (2009 est.)

Central bank discount rate: NA% (31 December 2009)
country comparison to the world: 52
7.25% (2 December 2008)
note: this is the Refinancing Rate, the key monetary policy instrument of the Armenian National Bank

Commercial bank prime lending rate: 18.76% (31 December 2009 est.)
country comparison to the world: 29
17.05% (31 December 2008 est.)

Stock of narrow money: $1.131 billion (31 December 2010 est.)
country comparison to the world: 138
$1.071 billion (31 December 2009 est.)

Stock of broad money: $3.507 billion (31 December 2010 est.)
country comparison to the world: 130
$3.339 billion (31 December 2009 est.)

Stock of domestic credit: $1.821 billion (31 December 2010 est.)
country comparison to the world: 128
$1.733 billion (31 December 2009 est.)

Market value of publicly traded shares: $140.5 million (31 December 2009)
country comparison to the world: 115
$176 million (31 December 2008)
$105 million (31 December 2007)

Agriculture—products: fruit (especially grapes), vegetables; livestock

Industries: diamond-processing, metal-cutting machine tools, forging-pressing machines, electric motors, tires, knitted wear, hosiery, shoes, silk fabric, chemicals, trucks, instruments, microelectronics, jewelry manufacturing, software development, food processing, brandy

Industrial production growth rate: 8% (2010 est.)
country comparison to the world: 35

Electricity—production: 5.584 billion kWh (2007 est.)
country comparison to the world: 109

Electricity—consumption: 4.776 billion kWh (2007 est.)
country comparison to the world: 113

Electricity—exports: 451.3 million kWh; note–exports an unknown quantity to Georgia; includes exports to Nagorno-Karabakh region in Azerbaijan (2007 est.)

Electricity—imports: 418.7 million kWh; note–imports an unknown quantity from Iran (2007 est.)

Oil—production: 0 bbl/day (2009 est.)
country comparison to the world: 148

Oil—consumption: 49,000 bbl/day (2009 est.)
country comparison to the world: 96

Oil—exports: 0 bbl/day (2007 est.)
country comparison to the world: 146

Oil—imports: 45,200 bbl/day (2007 est.)
country comparison to the world: 91

Oil—proved reserves: 0 bbl (1 January 2010 est.)
country comparison to the world: 103

Natural gas—production: 0 cu m (2008 est.)
country comparison to the world: 95

Natural gas—consumption: 1.93 billion cu m (2008 est.)
country comparison to the world: 80

Natural gas—exports: 0 cu m (2008 est.)
country comparison to the world: 51

Natural gas—imports: 1.93 billion cu m (2008 est.)
country comparison to the world: 46

Natural gas—proved reserves: 0 cu m (1 January 2010 est.)
country comparison to the world: 106

Current account balance: $-1.138 billion (2010 est.)
country comparison to the world: 139
$-1.326 billion (2009 est.)

Exports: $846 million (2010 est.)
country comparison to the world: 158
$722.3 million (2009 est.)

Exports—commodities: pig iron, unwrought copper, nonferrous metals, diamonds, mineral products, foodstuffs, energy

Exports—partners: Germany 16.47%, Russia 15.45%, US 9.64%, Bulgaria 8.6%, Georgia 7.57%, Netherlands 7.48%, Belgium 6.71%, Canada 4.91% (2009)

Imports: $2.988 billion (2010 est.)
country comparison to the world: 141
$2.817 billion (2009 est.)

Imports—commodities: natural gas, petroleum, tobacco products, foodstuffs, diamonds

Imports—partners: Russia 24.02%, China 8.72%, Ukraine 6.15%, Turkey 5.39%, Germany 5.36%, Iran 4.07% (2009)

Reserves of foreign exchange and gold: $2.247 billion (31 December 2010 est.)
country comparison to the world: 98
$2.004 billion (31 December 2009 est.)

Debt—external: $5.227 billion (30 June 2010)
country comparison to the world: 104
$3.449 billion (31 December 2008)

Exchange rates: drams (AMD) per US dollar–374.29 (2010), 363.28 (2009), 303.93 (2008), 344.06 (2007), 414.69 (2006)

COMMUNICATIONS

Telephones—main lines in use: 630,000 (2009)
country comparison to the world: 93

Telephones—mobile cellular: 2.62 million (2009)
country comparison to the world: 120

Telephone system: *general assessment:* telecommunications investments have made major inroads in modernizing and upgrading the outdated telecommunications network inherited from the Soviet era; now 100% privately owned and undergoing modernization and expansion; mobile-cellular services monopoly terminated in late 2004 and a second provider began operations in mid-2005
domestic: reliable modern fixed-line and mobile-cellular services are available across Yerevan in major cities and towns; significant but ever-shrinking gaps remain in mobile-cellular coverage in rural areas
international: country code–374; Yerevan is connected to the Trans-Asia-Europe fiber-optic cable through Iran; additional international service is available by microwave radio relay and landline connections to the other countries of the Commonwealth of Independent States, through the Moscow international switch, and by satellite to the rest of the world; satellite earth stations–3 (2008)

Broadcast media: 2 public television networks operating alongside more than 40 privately-owned television stations that provide local to near nationwide coverage; major Russian broadcast stations are widely available; subscription cable TV services are available in most regions; Public Radio of Armenia is a national, state-run broadcast network that operates alongside about 20 privately-owned radio stations; several major international broadcasters are available (2008)
Internet country code: .am
Internet hosts: 65,279 (2010)
country comparison to the world: 83
Internet users: 208,200 (2009)
country comparison to the world: 138

TRANSPORTATION

Airports: 11 (2010)
country comparison to the world: 153
Airports—with paved runways: *total:* 10
over 3,047 m: 2
2,438 to 3,047 m: 2
1,524 to 2,437 m: 4
914 to 1,523 m: 2 (2010)
Airports—with unpaved runways: *total:* 1
914 to 1,523 m: 1 (2010)
Pipelines: gas 2,233 km (2010)
Railways: *total:* 869 km
country comparison to the world: 96
broad gauge: 869 km 1.520-m gauge (818 km electrified)
note: some lines are out of service (2010)
Roadways: *total:* 8,888 km
country comparison to the world: 139
paved: 7,079 km (includes 1,561 km of expressways)
unpaved: 1,809 km (2008)

MILITARY

Military branches: Armenian Armed Forces: Ground Forces, Air Force and Air Defense; "Nagorno-Karabakh Republic": Nagorno-Karabakh Self-Defense Force (NKSDF) (2011)
Military service age and obligation: 18-27 years of age for voluntary or compulsory military service; 2-year conscript service obligation (2010)
Manpower available for military service:
males age 16–49: 805,847
females age 16–49: 854,296 (2010 est.)
Manpower fit for military service: *males age 16–49:* 644,372
females age 16–49: 717,272 (2010 est.)
Manpower reaching militarily significant age annually: *male:* 23,470
female: 21,417 (2010 est.)
Military expenditures: 2.8% of GDP (2010)
country comparison to the world: 51

TRANSNATIONAL ISSUES

Disputes—international: the dispute over the break-away Nagorno-Karabakh region and the Armenian military occupation of surrounding lands in Azerbaijan remains the primary focus of regional instability; residents have evacuated the former Soviet-era small ethnic enclaves in Armenia and Azerbaijan; Turkish authorities have complained that blasting from quarries in Armenia might be damaging the medieval ruins of Ani, on the other side of the Arpacay valley; in 2009, Swiss mediators facilitated an accord reestablishing diplomatic ties between Armenia and Turkey, but neither side has ratified the agreement and the rapprochement effort has faltered; local border forces struggle to control the illegal transit of goods and people across the porous, undemarcated Armenian, Azerbaijani, and Georgian borders; ethnic Armenian groups in Javakheti region of Georgia seek greater autonomy from the Georgian Government
Refugees and internally displaced persons: refugees (country of origin): 113,295 (Azerbaijan)
IDPs: 8,400 (conflict with Azerbaijan over Nagorno-Karabakh, majority have returned home since 1994 ceasefire) (2007)
Illicit drugs: illicit cultivation of small amount of cannabis for domestic consumption; minor transit point for illicit drugs—mostly opium and hashish—moving from Southwest Asia to Russia and to a lesser extent the rest of Europe

ARUBA

(PART OF THE KINGDOM OF THE NETHERLANDS)

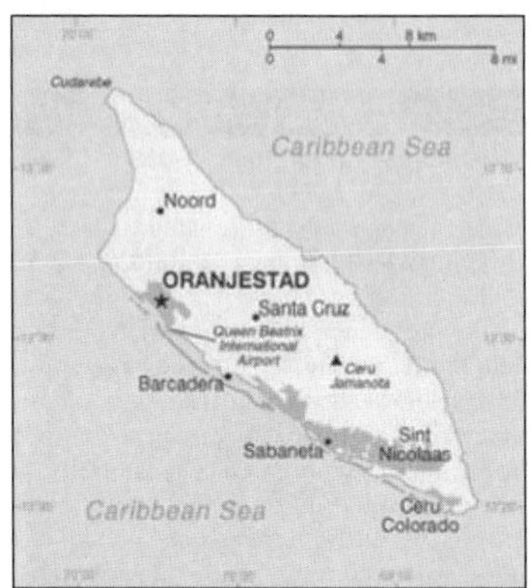

INTRODUCTION

Background: Discovered and claimed for Spain in 1499, Aruba was acquired by the Dutch in 1636. The island's economy has been dominated by three main industries. A 19th century gold rush was followed by prosperity brought on by the opening in 1924 of an oil refinery. The last decades of the 20th century saw a boom in the tourism industry. Aruba seceded from the Netherlands Antilles in 1986 and became a separate, autonomous member of the Kingdom of the Netherlands. Movement toward full independence was halted at Aruba's request in 1990.

GEOGRAPHY

Location: Caribbean, island in the Caribbean Sea, north of Venezuela
Geographic coordinates: 12 30 N, 69 58 W
Map references: Central America and the Caribbean
Area: *total:* 180 sq km
country comparison to the world: 216
land: 180 sq km
water: 0 sq km
Area—comparative: slightly larger than Washington, DC
Land boundaries: 0 km
Coastline: 68.5 km
Maritime claims: territorial sea: 12 nm
Climate: tropical marine; little seasonal temperature variation
Terrain: flat with a few hills; scant vegetation
Elevation extremes: *lowest point:* Caribbean Sea 0 m
highest point: Ceru Jamanota 188 m
Natural resources: NEGL; white sandy beaches
Land use: *arable land:* 10.53%
permanent crops: 0%
other: 89.47% (2005)
Irrigated land: 0.01 sq km (1998 est.)
Natural hazards: hurricanes; lies outside the Caribbean hurricane belt and is rarely threatened
Environment—current issues: NA
Geography—note: a flat, riverless island renowned for its white sand beaches; its tropical climate is moderated by constant trade winds from the Atlantic Ocean; the temperature is almost constant at about 27 degrees Celsius (81 degrees Fahrenheit)

PEOPLE

Population: 106,113 (July 2011 est.)
country comparison to the world: 190
note: estimate based on a revision of the base population, fertility, and mortality numbers, as well as a revision of 1985-99 migration estimates from outmigration to inmigration, which is assumed to continue into the future; the new results are consistent with the 2000 census
Age structure: *0–14 years:* 18.4% (male 9,847/female 9,729)
15–64 years: 70.3% (male 35,809/female 38,816)
65 years and over: 11.2% (male 4,698/female 7,214) (2011 est.)
Median age: *total:* 38.2 years
male: 36.4 years
female: 40 years (2011 est.)
Population growth rate: 1.436% (2011 est.)
country comparison to the world: 83
Birth rate: 12.78 births/1,000 population (2011 est.)
country comparison to the world: 157
Death rate: 7.84 deaths/1,000 population (July 2011 est.)
country comparison to the world: 111
Net migration rate: 9.42 migrant(s)/1,000 population (2011 est.)

country comparison to the world: 9
Urbanization: *urban population:* 47% of total population (2010)
rate of urbanization: 0.6% annual rate of change (2010-15 est.)
Major cities—population: ORANJESTAD (capital) 33,000 (2009)
Sex ratio: *at birth:* 1.021 male(s)/female
under 15 years: 1.01 male(s)/female
15–64 years: 0.92 male(s)/female
65 years and over: 0.66 male(s)/female
total population: 0.9 male(s)/female (2011 est.)
Infant mortality rate: *total:* 12.92 deaths/1,000 live births
country comparison to the world: 128
male: 17.07 deaths/1,000 live births
female: 8.68 deaths/1,000 live births (2011 est.)
Life expectancy at birth: *total population:* 75.72 years
country comparison to the world: 83
male: 72.68 years
female: 78.82 years (2011 est.)
Total fertility rate: 1.84 children born/woman (2011 est.)
country comparison to the world: 150
HIV/AIDS—adult prevalence rate: NA
HIV/AIDS—people living with HIV/AIDS: NA
HIV/AIDS—deaths: NA
Drinking water source: *Improved:*
urban: 100% of population
rural: 100% of population
total: 100% of population (2008)
Nationality: *noun:* Aruban(s)
adjective: Aruban; Dutch
Ethnic groups: mixed white/Caribbean Amerindian 80%, other 20%
Religions: Roman Catholic 80.8%, Evangelist 4.1%, Protestant 2.5%, Jehovah's Witnesses 1.5%, Methodist 1.2%, Jewish 0.2%, other 5.1%, none or unspecified 4.6%
Languages: Papiamento (a Spanish-Portuguese-Dutch-English dialect) 66.3%, Spanish 12.6%, English (widely spoken) 7.7%, Dutch (official) 5.8%, other 2.2%, unspecified or unknown 5.3% (2000 census)
Literacy:
definition: age 15 and over can read and write
total population: 97.3%
male: 97.5%
female: 97.1% (2000 census)
School life expectancy (primary to tertiary education): *total:* 13 years
male: 13 years
female: 13 years (2009)
Education expenditures: 5% of GDP (2008)
country comparison to the world: 58

GOVERNMENT

Country name: *conventional long form:* none
conventional short form: Aruba
Dependency status: constituent country of the Kingdom of the Netherlands; full autonomy in internal affairs obtained in 1986 upon separation from the Netherlands Antilles; Dutch Government responsible for defense and foreign affairs
Government type: parliamentary democracy
Capital: *name:* Oranjestad
geographic coordinates: 12 31 N, 70 02 W
time difference: UTC-4 (1 hour ahead of Washington, DC during Standard Time)
Administrative divisions: none (part of the Kingdom of the Netherlands)
Independence: none (part of the Kingdom of the Netherlands)
National holiday: Flag Day, 18 March (1976)
Constitution: 1 January 1986
Legal system: civil law system based on the Dutch civil code
Suffrage: 18 years of age; universal
Executive branch: *chief of state:* Queen BEATRIX of the Netherlands (since 30 April 1980); represented by Governor General Fredis REFUNJOL (since 11 May 2004)
head of government: Prime Minister Michiel "Mike" Godfried EMAN (since 30 October 2009)
cabinet: Council of Ministers elected by the Staten
(For more information visit the World Leaders website)
elections: the monarchy is hereditary; governor general appointed for a six-year term by the monarch; prime minister and deputy prime minister elected by the Staten for four-year terms; election last held in 2009 (next to be held by 2013)
election results: Mike EMAN elected prime minister; percent of legislative vote—NA
Legislative branch: unicameral Legislature or Staten (21 seats; members elected by direct popular vote to serve four-year terms)
elections: last held on 25 September 2009 (next to be held in 2013)
election results: percent of vote by party—AVP 48%, MEP 35.9%, PDR 5.7%; seats by party—AVP 12, MEP 8, PDR 1
Judicial branch: Common Court of Justice, Joint High Court of Justice (judges appointed by the monarch)
Political parties and leaders: Aliansa/Aruban Social Movement or MSA [Robert WEVER]; Aruban Liberal Organization or OLA [Glenbert CROES]; Aruban Patriotic Movement or MPA [Monica ARENDS-KOCK]; Aruban Patriotic Party or PPA [Benny NISBET]; Aruban People's Party or AVP [Mike EMAN]; People's Electoral Movement Party or MEP [Nelson O. ODUBER]; Real Democracy or PDR [Andin BIKKER]; RED [Rudy LAMPE]; Workers Political Platform or PTT [Gregorio WOLFF]
Political pressure groups and leaders: *other:* environmental groups
International organization participation: Caricom (observer), FATF, ILO, IMF, Interpol, IOC, ITUC, UNESCO (associate), UNWTO (associate), UPU
Diplomatic representation in the US: none (represented by the Kingdom of the Netherlands); note—Mr. Henry BAARH, Minister Plenipotentiary for Aruba at the Embassy of the Kingdom of the Netherlands
Diplomatic representation from the US: the US does not have an embassy in Aruba; the Consul General to Curacao is accredited to Aruba
Flag description: blue, with two narrow, horizontal, yellow stripes across the lower portion and a red, four-pointed star outlined in white in the upper hoist-side corner; the star represents Aruba and its red soil and white beaches, its four points the four major languages (Papiamento, Dutch, Spanish, English) as well as the four points of a compass, to indicate that its inhabitants come from all over the world; the blue symbolizes Caribbean waters and skies; the stripes represent the island's two main "industries": the flow of tourists to the sun-drenched beaches and the flow of minerals from the earth
National anthem: *name:* "Aruba Deshi Tera" (Aruba Precious Country)
lyrics/music: Juan Chabaya 'Padu' LAMPE/Rufo Inocencio WEVER
note: local anthem adopted 1986; as part of the Kingdom of the Netherlands, "Het Wilhelmus" is official (see Netherlands)

ECONOMY

Economy—overview: Tourism is the mainstay of the small open Aruban economy, together with offshore banking. Oil refining and storage ended in 2009. The rapid growth of the tourism sector over the last decade has resulted in a substantial expansion of other activities. Over 1.5 million tourists per year visit Aruba with 75% of those from the US. Construction continues to boom with hotel capacity five times the 1985 level. Tourist arrivals rebounded strongly following a dip after the 11 September 2001 attacks. The government has made cutting the budget and trade deficits a high priority.
GDP (purchasing power parity): $2.258 billion (2005 est.)
country comparison to the world: 182
$2.205 billion (2004 est.)
GDP (official exchange rate): $2.258 billion (2005 est.)
GDP—real growth rate: 2.4% (2005 est.)
country comparison to the world: 140
GDP—per capita (PPP): $21,800 (2004 est.)
country comparison to the world: 59
GDP—composition by sector: *agriculture:* 0.4%
industry: 33.3%
services: 66.3% (2002 est.)
Labor force: 41,500 (2004 est.)
country comparison to the world: 193
Labor force—by occupation: *agriculture:* NA%
industry: NA%
services: NA%
note: most employment is in wholesale and retail trade and repair, followed by hotels and restaurants
Unemployment rate: 6.9% (2005 est.)
country comparison to the world: 69
Population below poverty line: NA%
Household income or consumption by percentage share: *lowest 10%:* NA%
highest 10%: NA%
Budget: *revenues:* $507.9 million
expenditures: $577.9 million (2005 est.)
Public debt: 46.3% of GDP (2005)
country comparison to the world: 57
Inflation rate (consumer prices): 3.4% (2005)
country comparison to the world: 95

Central bank discount rate: 3% (31 December 2009)
country comparison to the world: 84
5% (31 December 2008)
Commercial bank prime lending rate: 10.77% (31 December 2009 est.)
country comparison to the world: 76
11.23% (31 December 2008 est.)
Stock of narrow money: $865 million (31 December 2009)
country comparison to the world: 142
$781 million (31 December 2008)
Stock of broad money: $1.771 billion (31 December 2009 est.)
country comparison to the world: 146
$1.671 billion (31 December 2008 est.)
Stock of domestic credit: $1.333 billion (31 December 2009)
country comparison to the world: 140
$1.321 billion (31 December 2008)
Agriculture—products: aloes; livestock; fish
Industries: tourism, transshipment facilities
Industrial production growth rate: NA%
Electricity—production: 850 million kWh (2007 est.)
country comparison to the world: 149
Electricity—consumption: 790.5 million kWh (2007 est.)
country comparison to the world: 148
Electricity—exports: 0 kWh (2008 est.)
Electricity—imports: 0 kWh (2008 est.)
Oil—production: 2,235 bbl/day (2009 est.)
country comparison to the world: 103
Oil—consumption: 8,000 bbl/day (2009 est.)
country comparison to the world: 154
Oil—exports: 231,100 bbl/day (2007 est.)
country comparison to the world: 51
Oil—imports: 236,400 bbl/day (2007 est.)
country comparison to the world: 39
Oil—proved reserves: 0 bbl (1 January 2010 est.)
country comparison to the world: 100
Natural gas—production: 1 cu m (2008 est.)
country comparison to the world: 93
Natural gas—consumption: 1 cu m (2008 est.)
country comparison to the world: 111
Natural gas—exports: 1 cu m (2008 est.)
country comparison to the world: 47
Natural gas—imports: 1 cu m (2008 est.)
country comparison to the world: 71
Natural gas—proved reserves: 0 cu m (1 January 2010 est.)
country comparison to the world: 104
Exports: $124 million (2006)
country comparison to the world: 190
note: includes oil reexports
Exports—commodities: live animals and animal products, art and collectibles, machinery and electrical equipment, transport equipment
Exports—partners: Panama 23.84%, former Netherlands Antilles 20.49%, Colombia 17.48%, Venezuela 12.61%, US 9.12%, Netherlands 7.5% (2009)
Imports: $1.054 billion (2006)
country comparison to the world: 170
Imports—commodities: machinery and electrical equipment, crude oil for refining and reexport, chemicals; foodstuffs
Imports—partners: US 49.51%, Netherlands 16.15%, UK 4.94% (2009)
Debt—external: $478.6 million (2005 est.)
country comparison to the world: 163
Exchange rates: Aruban guilders/florins (AWG) per US dollar–NA (2007), 1.79 (2006), 1.79 (2005), 1.79 (2004), 1.79 (2003)

COMMUNICATIONS

Telephones—main lines in use: 38,300 (2009)
country comparison to the world: 170
Telephones—mobile cellular: 128,000 (2009)
country comparison to the world: 180
Telephone system: *general assessment:* modern fully automatic telecommunications system
domestic: increased competition through privatization; 3 mobile-cellular service providers are now licensed
international: country code–297; landing site for the PAN-AM submarine telecommunications cable system that extends from the US Virgin Islands through Aruba to Venezuela, Colombia, Panama, and the west coast of South America; extensive interisland microwave radio relay links (2007)
Broadcast media: 2 commercial television stations; cable TV subscription service provides access to foreign channels; about 20 commercial radio stations broadcast (2007)
Internet country code: .aw
Internet hosts: 25,080 (2010)
country comparison to the world: 101
Internet users: 24,000 (2009)
country comparison to the world: 187

TRANSPORTATION

Airports: 1 (2010)
country comparison to the world: 210
Airports—with paved runways: *total:* 1
2,438 to 3,047 m: 1 (2010)
Ports and terminals: Barcadera, Oranjestad, Sint Nicolaas

MILITARY

Military branches: no regular military forces (2010)
Manpower available for military service: *males age 16–49:* 24,891
females age 16–49: 26,202 (2010 est.)
Manpower fit for military service: *males age 16–49:* 20,527
females age 16–49: 21,493 (2010 est.)
Manpower reaching militarily significant age annually: *male:* 767
female: 743 (2010 est.)
Military—note: defense is the responsibility of the Kingdom of the Netherlands

TRANSNATIONAL ISSUES

Disputes—international: none
Illicit drugs: transit point for US- and Europe-bound narcotics with some accompanying money-laundering activity; relatively high percentage of population consumes cocaine

ASHMORE AND CARTIER ISLANDS

(TERRITORY OF AUSTRALIA)

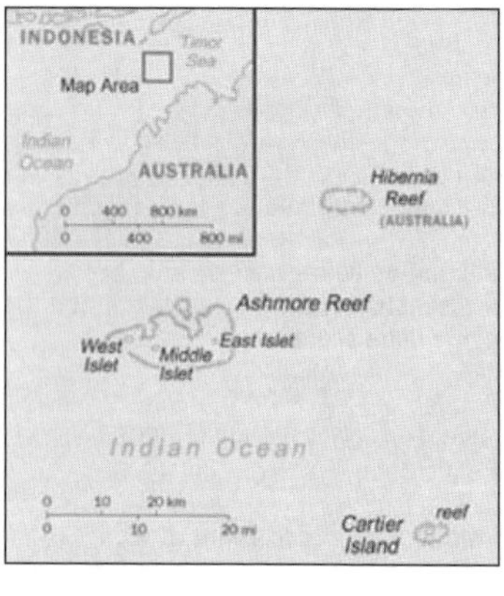

INTRODUCTION

Background: These uninhabited islands came under Australian authority in 1931; formal administration began two years later. Ashmore Reef supports a rich and diverse avian and marine habitat; in 1983, it became a National Nature Reserve. Cartier Island, a former bombing range, became a marine reserve in 2000.

GEOGRAPHY

Location: Southeastern Asia, islands in the Indian Ocean, midway between northwestern Australia and Timor island
Geographic coordinates: 12 14 S, 123 05 E
Map references: Oceania
Area: *total:* 5 sq km
country comparison to the world: 246
land: 5 sq km
water: 0 sq km
note: includes Ashmore Reef (West, Middle, and East Islets) and Cartier Island
Area—comparative: about eight times the size of The Mall in Washington, DC
Land boundaries: 0 km
Coastline: 74.1 km
Maritime claims: territorial sea: 12 nm
contiguous zone: 12 nm
exclusive fishing zone: 200 nm
continental shelf: 200 m depth or to the depth of exploitation
Climate: tropical
Terrain: low with sand and coral
Elevation extremes: *lowest point:* Indian Ocean 0 m

highest point: unnamed location 3 m
Natural resources: fish
Land use: *arable land:* 0%
permanent crops: 0%
other: 100% (all grass and sand) (2005)
Irrigated land: 0 sq km
Natural hazards: surrounded by shoals and reefs that can pose maritime hazards
Environment—current issues: illegal killing of protected wildlife by traditional Indonesian fisherman, as well as fishing by non-traditional Indonesian vessels, are ongoing problems
Geography—note: Ashmore Reef National Nature Reserve established in August 1983; Cartier Island Marine Reserve established in 2000

PEOPLE

Population: no indigenous inhabitants
note: Indonesian fishermen are allowed access to the lagoon and fresh water at Ashmore Reef's West Island; access to East and Middle Islands is by permit only

GOVERNMENT

Country name: *conventional long form:* Territory of Ashmore and Cartier Islands
conventional short form: Ashmore and Cartier Islands
Dependency status: territory of Australia; administered from Canberra by the Department of Regional Australia, Regional Development and Local Government
Legal system: the laws of the Commonwealth of Australia and the laws of the Northern Territory of Australia, where applicable, apply
Diplomatic representation in the US: none (territory of Australia)
Diplomatic representation from the US: none (territory of Australia)
Flag description: the flag of Australia is used

ECONOMY

Economy—overview: no economic activity

TRANSPORTATION

Ports and terminals: none; offshore anchorage only

MILITARY

Military—note: defense is the responsibility of Australia; periodic visits by the Royal Australian Navy and Royal Australian Air Force

TRANSNATIONAL ISSUES

Disputes—international: Australia has closed parts of the Ashmore and Cartier reserve to Indonesian traditional fishing; Indonesian groups challenge Australia's claim to Ashmore Reef

ATLANTIC OCEAN

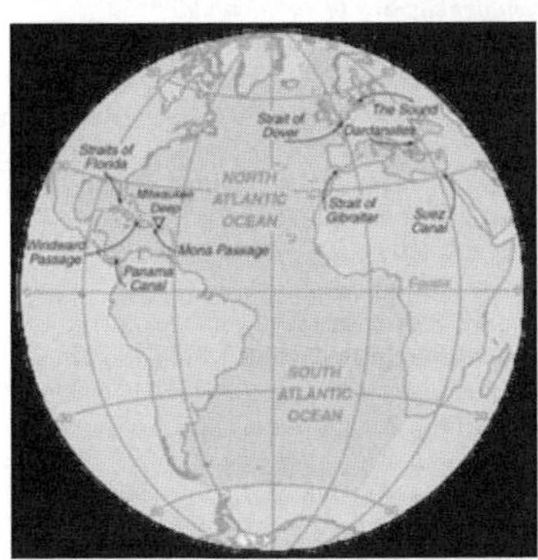

INTRODUCTION

Background: The Atlantic Ocean is the second largest of the world's five oceans (after the Pacific Ocean, but larger than the Indian Ocean, Southern Ocean, and Arctic Ocean). The Kiel Canal (Germany), Oresund (Denmark-Sweden), Bosporus (Turkey), Strait of Gibraltar (Morocco-Spain), and the Saint Lawrence Seaway (Canada-US) are important strategic access waterways. The decision by the International Hydrographic Organization in the spring of 2000 to delimit a fifth world ocean, the Southern Ocean, removed the portion of the Atlantic Ocean south of 60 degrees south latitude.

GEOGRAPHY

Location: body of water between Africa, Europe, the Arctic Ocean, the Americas, and the Southern Ocean
Geographic coordinates: 0 00 N, 25 00 W
Map references: Political Map of the World
Area: *total:* 76.762 million sq km
note: includes Baltic Sea, Black Sea, Caribbean Sea, Davis Strait, Denmark Strait, part of the Drake Passage, Gulf of Mexico, Labrador Sea, Mediterranean Sea, North Sea, Norwegian Sea, almost all of the Scotia Sea, and other tributary water bodies
Area—comparative: slightly less than 6.5 times the size of the US
Coastline: 111,866 km
Climate: tropical cyclones (hurricanes) develop off the coast of Africa near Cape Verde and move westward into the Caribbean Sea; hurricanes can occur from May to December but are most frequent from August to November
Terrain: surface usually covered with sea ice in Labrador Sea, Denmark Strait, and coastal portions of the Baltic Sea from October to June; clockwise warm-water gyre (broad, circular system of currents) in the northern Atlantic, counterclockwise warm-water gyre in the southern Atlantic; the ocean floor is dominated by the Mid-Atlantic Ridge, a rugged north-south centerline for the entire Atlantic basin
Elevation extremes: *lowest point:* Milwaukee Deep in the Puerto Rico Trench -8,605 m
highest point: sea level 0 m
Natural resources: oil and gas fields, fish, marine mammals (seals and whales), sand and gravel aggregates, placer deposits, polymetallic nodules, precious stones
Natural hazards: icebergs common in Davis Strait, Denmark Strait, and the northwestern Atlantic Ocean from February to August and have been spotted as far south as Bermuda and the Madeira Islands; ships subject to superstructure icing in extreme northern Atlantic from October to May; persistent fog can be a maritime hazard from May to September; hurricanes (May to December)
Environment—current issues: endangered marine species include the manatee, seals, sea lions, turtles, and whales; drift net fishing is hastening the decline of fish stocks and contributing to international disputes; municipal sludge pollution off eastern US, southern Brazil, and eastern Argentina; oil pollution in Caribbean Sea, Gulf of Mexico, Lake Maracaibo, Mediterranean Sea, and North Sea; industrial waste and municipal sewage pollution in Baltic Sea, North Sea, and Mediterranean Sea
Geography—note: major chokepoints include the Dardanelles, Strait of Gibraltar, access to the Panama and Suez Canals; strategic straits include the Strait of Dover, Straits of Florida, Mona Passage, The Sound (Oresund), and Windward Passage; the Equator divides the Atlantic Ocean into the North Atlantic Ocean and South Atlantic Ocean

ECONOMY

Economy—overview: The Atlantic Ocean provides some of the world's most heavily trafficked sea routes, between and within the Eastern and Western Hemispheres. Other economic activity includes the exploitation of natural resources, e.g., fishing, dredging of aragonite sands (The Bahamas), and production of crude oil and natural gas (Caribbean Sea, Gulf of Mexico, and North Sea).

TRANSPORTATION

Ports and terminals: Alexandria (Egypt), Algiers (Algeria), Antwerp (Belgium), Barcelona (Spain), Buenos Aires (Argentina), Casablanca (Morocco), Colon (Panama), Copenhagen (Denmark), Dakar (Senegal), Gdansk (Poland), Hamburg (Germany), Helsinki (Finland), Las Palmas (Canary Islands, Spain), Le Havre (France), Lisbon (Portugal), London (UK), Marseille (France), Montevideo (Uruguay), Montreal (Canada), Naples (Italy), New Orleans (US), New York (US), Oran (Algeria), Oslo (Norway), Peiraiefs or Piraeus (Greece), Rio de Janeiro (Brazil), Rotterdam (Netherlands), Saint Petersburg (Russia), Stockholm (Sweden)
Transportation—note: Kiel Canal and Saint Lawrence Seaway are two important waterways; significant domestic commercial and recreational use of Intracoastal

Waterway on central and south Atlantic seaboard and Gulf of Mexico coast of US; the International Maritime Bureau reports the territorial waters of littoral states and offshore Atlantic waters as high risk for piracy and armed robbery against ships, particularly in the Gulf of Guinea off West Africa, the east coast of Brazil, and the Caribbean Sea; numerous commercial vessels have been attacked and hijacked both at anchor and while underway; hijacked vessels are often disguised and cargoes stolen; crews have been robbed and stores or cargoes stolen

TRANSNATIONAL ISSUES

Disputes—international: some maritime disputes (see littoral states)

AUSTRALIA

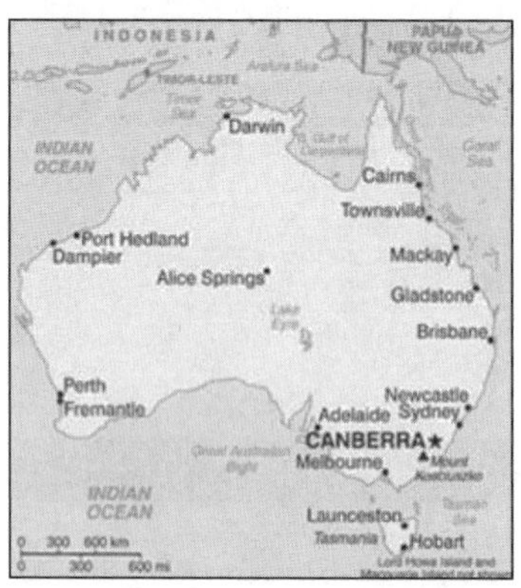

INTRODUCTION

Background: Aboriginal settlers arrived on the continent from Southeast Asia about 40,000 years before the first Europeans began exploration in the 17th century. No formal territorial claims were made until 1770, when Capt. James COOK took possession of the east coast in the name of Great Britain (all of Australia was claimed as British territory in 1829 with the creation of the colony of Western Australia). Six colonies were created in the late 18th and 19th centuries; they federated and became the Commonwealth of Australia in 1901. The new country took advantage of its natural resources to rapidly develop agricultural and manufacturing industries and to make a major contribution to the British effort in World Wars I and II. In recent decades, Australia has transformed itself into an internationally competitive, advanced market economy. It boasted one of the OECD's fastest growing economies during the 1990s, a performance due in large part to economic reforms adopted in the 1980s. Long-term concerns include ageing of the population, pressure on infrastructure, and environmental issues such as frequent droughts.

GEOGRAPHY

Location: Oceania, continent between the Indian Ocean and the South Pacific Ocean
Geographic coordinates: 27 00 S, 133 00 E
Map references: Oceania
Area:
total: 7,741,220 sq km
country comparison to the world: 6
land: 7,682,300 sq km
water: 58,920 sq km
note: includes Lord Howe Island and Macquarie Island
Area—comparative: slightly smaller than the US contiguous 48 states
Land boundaries: 0 km
Coastline: 25,760 km
Maritime claims: territorial sea: 12 nm
contiguous zone: 24 nm
exclusive economic zone: 200 nm
continental shelf: 200 nm or to the edge of the continental margin
Climate: generally arid to semiarid; temperate in south and east; tropical in north
Terrain: mostly low plateau with deserts; fertile plain in southeast
Elevation extremes: *lowest point:* Lake Eyre -15 m
highest point: Mount Kosciuszko 2,229 m
Natural resources: bauxite, coal, iron ore, copper, tin, gold, silver, uranium, nickel, tungsten, rare earth elements, mineral sands, lead, zinc, diamonds, natural gas, petroleum
note: Australia is the world's largest net exporter of coal accounting for 29% of global coal exports
Land use: *arable land:* 6.15% (includes about 27 million hectares of cultivated grassland)
permanent crops: 0.04%
other: 93.81% (2005)
Irrigated land: 25,450 sq km (2003)
Total renewable water resources: 398 cu km (1995)
Freshwater withdrawal (domestic/industrial/agricultural): *total:* 24.06 cu km/yr (15%/10%/75%)
per capita: 1,193 cu m/yr (2000)
Natural hazards: cyclones along the coast; severe droughts; forest fires
volcanism: volcanic activity occurs on the Heard and McDonald Islands
Environment—current issues: soil erosion from overgrazing, industrial development, urbanization, and poor farming practices; soil salinity rising due to the use of poor quality water; desertification; clearing for agricultural purposes threatens the natural habitat of many unique animal and plant species; the Great Barrier Reef off the northeast coast, the largest coral reef in the world, is threatened by increased shipping and its popularity as a tourist site; limited natural freshwater resources
Environment—international agreements: *party to:* Antarctic-Environmental Protocol, Antarctic-Marine Living Resources, Antarctic Seals, Antarctic Treaty, Biodiversity, Climate Change, Climate Change-Kyoto Protocol, Desertification, Endangered Species, Environmental Modification, Hazardous Wastes, Law of the Sea, Marine Dumping, Marine Life Conservation, Ozone Layer Protection, Ship Pollution, Tropical Timber 83, Tropical Timber 94, Wetlands, Whaling
signed, but not ratified: none of the selected agreements
Geography—note: world's smallest continent but sixth-largest country; population concentrated along the eastern and southeastern coasts; the invigorating sea breeze known as the "Fremantle Doctor" affects the city of Perth on the west coast and is one of the most consistent winds in the world

PEOPLE

Population: 21,766,711 (July 2011 est.)
country comparison to the world: 55
Age structure: *0–14 years:* 18.3% (male 2,040,848/female 1,937,544)
15–64 years: 67.7% (male 7,469,092/female 7,266,143)
65 years and over: 14% (male 1,398,576/female 1,654,508) (2011 est.)
Median age: *total:* 37.7 years
male: 37 years
female: 38.4 years (2011 est.)
Population growth rate: 1.148% (2011 est.)
country comparison to the world: 101
Birth rate: 12.33 births/1,000 population (2011 est.)
country comparison to the world: 159
Death rate: 6.88 deaths/1,000 population (July 2011 est.)
country comparison to the world: 142
Net migration rate: 6.03 migrant(s)/1,000 population (2011 est.)
country comparison to the world: 14
Urbanization: *urban population:* 89% of total population (2010)
rate of urbanization: 1.2% annual rate of change (2010-15 est.)
Major cities—population: Sydney 4.429 million; Melbourne 3.853 million; Brisbane 1.97 million; Perth 1.599 million; CANBERRA (capital) 384,000 (2009)
Sex ratio: *at birth:* 1.055 male(s)/female
under 15 years: 1.05 male(s)/female
15–64 years: 1.03 male(s)/female
65 years and over: 0.84 male(s)/female
total population: 1 male(s)/female (2011 est.)
Infant mortality rate: *total:* 4.61 deaths/1,000 live births
country comparison to the world: 190
male: 4.93 deaths/1,000 live births
female: 4.27 deaths/1,000 live births (2011 est.)
Life expectancy at birth: *total population:* 81.81 years
country comparison to the world: 9
male: 79.4 years
female: 84.35 years (2011 est.)
Total fertility rate: 1.78 children born/woman (2011 est.)
country comparison to the world: 157

HIV/AIDS—adult prevalence rate: 0.1% (2009 est.)
country comparison to the world: 116
HIV/AIDS—people living with HIV/AIDS: 20,000 (2009 est.)
country comparison to the world: 78
HIV/AIDS—deaths: fewer than 100 (2009 est.)
country comparison to the world: 120
Drinking water source: *Improved:*
urban: 100% of population
rural: 100% of population
total: 100% of population (2008)
Sanitation facility access: *Improved:*
urban: 100% of population
rural: 100% of population
total: 100% of population (2008)
Nationality: *noun:* Australian(s)
adjective: Australian
Ethnic groups: white 92%, Asian 7%, aboriginal and other 1%
Religions: Catholic 25.8%, Anglican 18.7%, Uniting Church 5.7%, Presbyterian and Reformed 3%, Eastern Orthodox 2.7%, other Christian 7.9%, Buddhist 2.1%, Muslim 1.7%, other 2.4%, unspecified 11.3%, none 18.7% (2006 Census)
Languages: English 78.5%, Chinese 2.5%, Italian 1.6%, Greek 1.3%, Arabic 1.2%, Vietnamese 1%, other 8.2%, unspecified 5.7% (2006 Census)
Literacy: *definition:* age 15 and over can read and write
total population: 99%
male: 99%
female: 99% (2003 est.)
School life expectancy (primary to tertiary education):
total: 21 years
male: 20 years
female: 21 years (2008)
Education expenditures: 4.5% of GDP (2007)
country comparison to the world: 81

GOVERNMENT

Country name: *conventional long form:* Commonwealth of Australia
conventional short form: Australia
Government type: federal parliamentary democracy and a Commonwealth realm
Capital: *name:* Canberra
geographic coordinates: 35 17 S, 149 13 E
time difference: UTC+10 (15 hours ahead of Washington, DC during Standard Time)
daylight saving time: +1hr, begins first Sunday in October; ends first Sunday in April
note: Australia is divided into three time zones
Administrative divisions: 6 states and 2 territories*; Australian Capital Territory*, New South Wales, Northern Territory*, Queensland, South Australia, Tasmania, Victoria, Western Australia
Dependent areas: Ashmore and Cartier Islands, Christmas Island, Cocos (Keeling) Islands, Coral Sea Islands, Heard Island and McDonald Islands, Macquarie Island, Norfolk Island
Independence: 1 January 1901 (from the federation of UK colonies)
National holiday: Australia Day, 26 January (1788); ANZAC Day (commemorates the anniversary of the landing of troops of the Australian and New Zealand Army Corps during World War I at Gallipoli, Turkey), 25 April (1915)
Constitution: 9 July 1900; effective 1 January 1901
Legal system: common law system based on the English model
International law organization participation: accepts compulsory ICJ jurisdiction with reservations; accepts ICCt jurisdiction
Suffrage: 18 years of age; universal and compulsory
Executive branch: *chief of state:* Queen of Australia ELIZABETH II (since 6 February 1952); represented by Governor General Quentin BRYCE (since 5 September 2008)
head of government: Prime Minister Julia Eileen GILLARD (since 24 June 2010); Deputy Prime Minister Wayne Maxwell SWAN (since 24 June 2010)
cabinet: prime minister nominates, from among members of Parliament, candidates who are subsequently sworn in by the governor general to serve as government ministers
(For more information visit the World Leaders website)
elections: the monarchy is hereditary; governor general appointed by the monarch on the recommendation of the prime minister; following legislative elections, the leader of the majority party or leader of a majority coalition is sworn in as prime minister by the governor general
Legislative branch: bicameral Federal Parliament consists of the Senate (76 seats; 12 members from each of the six states and 2 from each of the two mainland territories; one-half of state members are elected every three years by popular vote to serve six-year terms while all territory members are elected every three years) and the House of Representatives (150 seats; members elected by popular vote to serve terms of up to three-years; no state can have fewer than 5 representatives)
elections: Senate—last held on 21 August 2010; House of Representatives—last held on 21 August 2010 (the latest a simultaneous half-Senate and House of Representative elections can be held is 30 November 2013)
election results: Senate (effective 1 July 2011)—percent of vote by party—NA; seats by party—Liberal/National Party 34, Australian Labor Party 31, Greens 9, others 2; House of Representatives—percent of vote by party—Australian Labor Party 38.1%, Liberal Party 30.4%, Greens 11.5%, Liberal National Party of Queensland 9.3%, independents 6.6%, The Nationals 3.7%, Country Liberals 0.3%; seats by party—Australian Labor Party 72, Liberal Party 44, Liberal National Party of Queensland 21, The Nationals 7, Country Liberals 1, Greens 1, independents 4
Judicial branch: High Court (the chief justice and six other justices are appointed by the governor general acting on the advice of the government)
Political parties and leaders: Australian Greens [Bob BROWN]; Australian Labor Party [Julia GILLARD]; Family First Party [Steve FIELDING]; Liberal Party [Tony ABBOTT]; The Nationals [Warren TRUSS]
Political pressure groups and leaders: *other:* business groups; environmental groups; social groups; trade unions
International organization participation: ADB, ANZUS, APEC, ARF, ASEAN (dialogue partner), Australia Group, BIS, C, CP, EAS, EBRD, FAO, FATF, G-20, IAEA, IBRD, ICAO, ICC, ICRM, IDA, IEA, IFC, IFRCS, IHO, ILO, IMF, IMO, IMSO, Interpol, IOC, IOM, IPU, ISO, ITSO, ITU, ITUC, MIGA, NEA, NSG, OECD, OPCW, OSCE (partner), Paris Club, PCA, PIF, SAARC (observer), Sparteca, SPC, UN, UNCTAD, UNESCO, UNHCR, UNMIS, UNMIT, UNRWA, UNTSO, UNWTO, UPU, WCO, WFTU, WHO, WIPO, WMO, WTO, ZC
Diplomatic representation in the US: *chief of mission:* Ambassador Kim Christian BEAZLEY
chancery: 1601 Massachusetts Avenue NW, Washington, DC 20036
telephone: [1] (202) 797-3000
FAX: [1] (202) 797-3168
consulate(s) general: Atlanta, Chicago, Honolulu, Los Angeles, New York, San Francisco
Diplomatic representation from the US: *chief of mission:* Ambassador Jeffrey L. BLEICH
embassy: Moonah Place, Yarralumla, Canberra, Australian Capital Territory 2600
mailing address: APO AP 96549
telephone: [61] (02) 6214-5600
FAX: [61] (02) 6214-5970
consulate(s) general: Melbourne, Perth, Sydney
Flag description: blue with the flag of the UK in the upper hoist-side quadrant and a large seven-pointed star in the lower hoist-side quadrant known as the Commonwealth or Federation Star, representing the federation of the colonies of Australia in 1901; the star depicts one point for each of the six original states and one representing all of Australia's internal and external territories; on the fly half is a representation of the Southern Cross constellation in white with one small five-pointed star and four larger, seven-pointed stars
National anthem: *name:* "Advance Australia Fair"
lyrics/music: Peter Dodds McCORMICK
note: adopted 1984; although originally written in the late 19th century, the anthem did not become used for all official occasions until 1984; as a Commonwealth country, in addition to the national anthem, "God Save the Queen" is also played at Royal functions (see United Kingdom)

ECONOMY

Economy—overview: Australia's abundant and diverse natural resources attract high levels of foreign investment and include extensive reserves of coal, iron ore, copper, gold, natural gas, uranium, and renewable energy sources. A series of major investments, such as the US$40 billion Gorgon Liquid Natural Gas project, will significantly expand the resources sector. Australia also has a large services sector and is a significant exporter of natural resources, energy, and food. Key tenets of

Australia's trade policy include support for open trade and the successful culmination of the Doha Round of multilateral trade negotiations, particularly for agriculture and services. The Australian economy grew for 17 consecutive years before the global financial crisis. Subsequently, the Rudd government introduced a fiscal stimulus package worth over US$50 billion to offset the effect of the slowing world economy, while the Reserve Bank of Australia cut interest rates to historic lows. These policies–and continued demand for commodities, especially from China–helped the Australian economy rebound after just one quarter of negative growth. The economy grew by 1.2% during 2009–the best performance in the OECD–and by 3.3% in 2010. Unemployment, originally expected to reach 8-10%, peaked at 5.7% in late 2009 and fell to 5.1% in 2010. As a result of an improved economy, the budget deficit is expected to peak below 4.2% of GDP and the government could return to budget surpluses as early as 2015. Australia was one of the first advanced economies to raise interest rates, with seven rate hikes between October 2009 and November 2010. The GILLARD government is focused on raising Australia's economic productivity to ensure the sustainability of growth, and continues to manage the symbiotic, but sometimes tense, economic relationship with China. Australia is engaged in the Trans-Pacific Partnership talks and ongoing free trade agreement negotiations with China, Japan, and Korea.
GDP (purchasing power parity): $882.4 billion (2010 est.)
country comparison to the world: 18
$858.8 billion (2009 est.)
$847.5 billion (2008 est.)
note: data are in 2010 US dollars
GDP (official exchange rate): $1.236 trillion (2010 est.)
GDP—real growth rate: 2.7% (2010 est.)
country comparison to the world: 131
1.3% (2009 est.)
2.6% (2008 est.)
GDP—per capita (PPP): $41,000 (2010 est.)
country comparison to the world: 18
$40,400 (2009 est.)
$40,300 (2008 est.)
note: data are in 2010 US dollars
GDP—composition by sector:
agriculture: 4%
industry: 24.8%
services: 71.2% (2010 est.)
Labor force: 11.62 million (2010 est.)
country comparison to the world: 46
Labor force—by occupation:
agriculture: 3.6%
industry: 21.1%
services: 75% (2009 est.)
Unemployment rate: 5.1% (2010 est.)
country comparison to the world: 47
5.6% (2009 est.)
Population below poverty line: NA%
Household income or consumption by percentage share:
lowest 10%: 2%
highest 10%: 25.4% (1994)
Distribution of family income—Gini index: 30.5 (2006)
country comparison to the world: 110
35.2 (1994)
Investment (gross fixed): 27.4% of GDP (2010 est.)
country comparison to the world: 29
Budget: *revenues:* $396.1 billion
expenditures: $426.5 billion (2010 est.)
Public debt: 22.4% of GDP (2010 est.)
country comparison to the world: 108
22.1% of GDP (2009 est.)
Inflation rate (consumer prices): 2.9% (2010 est.)
country comparison to the world: 82
1.8% (2009 est.)
Central bank discount rate: 4% (31 March 2010)
country comparison to the world: 101
4.25% (3 December 2008)
note: this is the Reserve Bank of Australia's "cash rate target," or policy rate
Commercial bank prime lending rate: 6.02% (31 December 2009 est.)
country comparison to the world: 101
8.91% (31 December 2008 est.)
Stock of narrow money: $347.1 billion (31 December 2010 est.)
country comparison to the world: 14
$290.8 billion (31 December 2009 est.)
Stock of broad money: $1.134 trillion (31 December 2010 est.)
country comparison to the world: 15
$976.6 billion (31 December 2009 est.)
Stock of domestic credit: $1.731 trillion (31 December 2010 est.)
country comparison to the world: 13
$1.407 trillion (31 December 2009 est.)
Market value of publicly traded shares: $1.258 trillion (31 December 2009)
country comparison to the world: 13
$675.6 billion (31 December 2008)
$1.298 trillion (31 December 2007)
Agriculture—products: wheat, barley, sugarcane, fruits; cattle, sheep, poultry
Industries: mining, industrial and transportation equipment, food processing, chemicals, steel
Industrial production growth rate: 3% (2010 est.)
country comparison to the world: 106
Electricity—production: 239.9 billion kWh (2007 est.)
country comparison to the world: 17
Electricity—consumption: 222 billion kWh (2007 est.)
country comparison to the world: 15
Electricity—exports: 0 kWh (2008 est.)
Electricity—imports: 0 kWh (2008 est.)
Oil—production: 589,200 bbl/day (2009 est.)
country comparison to the world: 30
Oil—consumption: 946,300 bbl/day (2009 est.)
country comparison to the world: 19
Oil—exports: 311,900 bbl/day (2008 est.)
country comparison to the world: 39
Oil—imports: 716,700 bbl/day (2008 est.)
country comparison to the world: 19
Oil—proved reserves: 3.318 billion bbl (1 January 2010 est.)
country comparison to the world: 29
Natural gas—production: 42.33 billion cu m (2009 est.)
country comparison to the world: 19
Natural gas—consumption: 26.59 billion cu m (2009 est.)
country comparison to the world: 29
Natural gas—exports: 22.3 billion cu m (2009 est.)
country comparison to the world: 10
Natural gas—imports: 6.56 billion cu m (2009 est.)
country comparison to the world: 29
Natural gas—proved reserves: 3.115 trillion cu m (1 January 2010 est.)
country comparison to the world: 12
Current account balance: $-35.23 billion (2010 est.)
country comparison to the world: 183
$-41.33 billion (2009 est.)
Exports: $210.7 billion (2010 est.)
country comparison to the world: 21
$154.8 billion (2009 est.)
Exports—commodities: coal, iron ore, gold, meat, wool, alumina, wheat, machinery and transport equipment
Exports—partners: China 21.81%, Japan 19.19%, South Korea 7.88%, India 7.51%, US 4.95%, UK 4.37%, NZ 4.1% (2009)
Imports: $200.4 billion (2010 est.)
country comparison to the world: 21
$160.4 billion (2009 est.)
Imports—commodities: machinery and transport equipment, computers and office machines, telecommunication equipment and parts; crude oil and petroleum products
Imports—partners: China 17.94%, US 11.26%, Japan 8.36%, Thailand 5.81%, Singapore 5.54%, Germany 5.3% (2009)
Reserves of foreign exchange and gold: $38.62 billion (31 December 2010 est.)
country comparison to the world: 33
$41.74 billion (31 December 2009 est.)
Debt—external: $1.169 trillion (31 December 2010 est.)
country comparison to the world: 14
$1.094 trillion (31 December 2009 est.)
Stock of direct foreign investment—at home: $329.1 billion (31 December 2010 est.)
country comparison to the world: 14
$295.9 billion (31 December 2009 est.)
Stock of direct foreign investment—abroad: $245.9 billion (31 December 2010 est.)
country comparison to the world: 17
$221.1 billion (31 December 2009 est.)
Exchange rates: Australian dollars (AUD) per US dollar–1.0902 (2010), 1.2822 (2009), 1.2059 (2008), 1.2137 (2007), 1.3285 (2006)

COMMUNICATIONS

Telephones—main lines in use: 9.02 million (2009)
country comparison to the world: 24
Telephones—mobile cellular: 24.22 million (2009)
country comparison to the world: 37
Telephone system: *general assessment:* excellent domestic and international service
domestic: domestic satellite system; significant use of radiotelephone in areas of low population density; rapid growth of mobile telephones
international: country code–61; landing point for the SEA-ME-WE-3 optical telecommunications submarine cable with links to Asia, the Middle East, and Europe; the Southern Cross fiber optic submarine cable provides links to New Zealand

and the United States; satellite earth stations—19 (10 Intelsat—4 Indian Ocean and 6 Pacific Ocean, 2 Inmarsat—Indian and Pacific Ocean regions, 2 Globalstar, 5 other) (2007)
Broadcast media: the Australian Broadcasting Corporation (ABC) runs multiple national and local radio networks and TV stations, as well as Australia Network, a TV service that broadcasts throughout the Asia-Pacific region and is the main public broadcaster; Special Broadcasting Service (SBS), a second large public broadcaster, operates radio and TV networks broadcasting in multiple languages; several large national commercial TV networks, a large number of local commercial TV stations, and hundreds of commercial radio stations are accessible; cable and satellite systems are available (2008)
Internet country code: .au
Internet hosts: 13.361 million (2010)
country comparison to the world: 8
Internet users: 15.81 million (2009)
country comparison to the world: 25

TRANSPORTATION

Airports: 465 (2010)
country comparison to the world: 17
Airports—with paved runways: *total:* 326
over 3,047 m: 11
2,438 to 3,047 m: 13
1,524 to 2,437 m: 148
914 to 1,523 m: 140
under 914 m: 14 (2010)
Airports—with unpaved runways:
total: 139
1,524 to 2,437 m: 17
914 to 1,523 m: 110
under 914 m: 12 (2010)
Heliports: 1 (2010)
Pipelines: gas 27,900 km; liquid petroleum gas 240 km; oil 3,257 km; oil/gas/water 1 km (2010)
Railways: *total:* 38,445 km
country comparison to the world: 7
broad gauge: 3,355 km 1.600-m gauge
standard gauge: 21,674 km 1.435-m gauge (650 km electrified)
narrow gauge: 9,539 km 1.067-m gauge (2,067 km electrified); 3,877 km 1.000-m gauge (2010)
Roadways: *total:* 812,972 km
country comparison to the world: 9
paved: 341,448 km
unpaved: 471,524 km (2004)
Waterways: 2,000 km (mainly used for recreation on Murray and Murray-Darling river systems) (2006)
country comparison to the world: 44
Merchant marine: *total:* 45
country comparison to the world: 73
by type: bulk carrier 10, cargo 8, liquefied gas 4, passenger 6, passenger/cargo 6, petroleum tanker 6, roll on/roll off 5
foreign-owned: 20 (Canada 7, Germany 2, Netherlands 1, Norway 1, Singapore 2, UK 5, US 2)
registered in other countries: 29 (Dominica 1, Fiji 2, Liberia 2, Marshall Islands 1, Netherlands 1, NZ 1, Panama 5, Singapore 11, Tonga 1, UK 1, US 1, Vanuatu 2) (2010)
Ports and terminals: Brisbane, Cairns, Dampier, Darwin, Fremantle, Gladstone, Geelong, Hay Point, Hobart, Jervis Bay, Melbourne, Newcastle, Port Adelaide, Port Dalrymple, Port Hedland, Port Kembla, Port Lincoln, Port Walcott, Sydney

MILITARY

Military branches: Australian Defense Force (ADF): Australian Army, Royal Australian Navy, Royal Australian Air Force, Special Operations Command (2006)
Military service age and obligation: 17 years of age for voluntary military service (with parental consent); no conscription; women allowed to serve in Army combat units in non-combat support roles (2010)
Manpower available for military service:
males age 16–49: 5,316,464
females age 16–49: 5,116,722 (2010 est.)
Manpower fit for military service: *males age 16–49:* 4,411,958
females age 16–49: 4,239,985 (2010 est.)
Manpower reaching militarily significant age annually: *male:* 143,565
female: 135,800 (2010 est.)
Military expenditures: 3% of GDP (2009)
country comparison to the world: 44

TRANSNATIONAL ISSUES

Disputes—international: In 2007, Australia and Timor-Leste signed agreed to a 50-year development zone and revenue sharing arrangement and deferred a maritime boundary; Australia asserts land and maritime claims to Antarctica; Australia's 2004 submission to Commission on the Limits of the Continental Shelf (CLCS) extends its continental margins over 3.37 million square kilometers, expanding its seabed roughly 30 percent beyond its claimed exclusive economic zone; a 1997 treaty between Indonesia and Australia settled some parts of their maritime boundary but some outstanding issues, especially around Timor Leste, remain; Indonesian groups challenge Australia's claim to Ashmore Reef; Australia has closed parts of the Ashmore and Cartier reserve to Indonesian traditional fishing
Illicit drugs: Tasmania is one of the world's major suppliers of licit opiate products; government maintains strict controls over areas of opium poppy cultivation and output of poppy straw concentrate; major consumer of cocaine and amphetamines

AUSTRIA

INTRODUCTION

Background: Once the center of power for the large Austro-Hungarian Empire, Austria was reduced to a small republic after its defeat in World War I. Following annexation by Nazi Germany in 1938 and subsequent occupation by the victorious Allies in 1945, Austria's status remained unclear for a decade. A State Treaty signed in 1955 ended the occupation, recognized Austria's independence, and forbade unification with Germany. A constitutional law that same year declared the country's "perpetual neutrality" as a condition for Soviet military withdrawal. The Soviet Union's collapse in 1991 and Austria's entry into the European Union in 1995 have altered the meaning of this neutrality. A prosperous, democratic country, Austria entered the EU Economic and Monetary Union in 1999.

GEOGRAPHY

Location: Central Europe, north of Italy and Slovenia
Geographic coordinates: 47 20 N, 13 20 E
Map references: Europe
Area: *total:* 83,871 sq km
country comparison to the world: 113
land: 82,445 sq km
water: 1,426 sq km
Area—comparative: slightly smaller than Maine
Land boundaries: *total:* 2,562 km
border countries: Czech Republic 362 km, Germany 784 km, Hungary 366 km, Italy 430 km, Liechtenstein 35 km, Slovakia 91 km, Slovenia 330 km, Switzerland 164 km
Coastline: 0 km (landlocked)
Maritime claims: none (landlocked)
Climate: temperate; continental, cloudy; cold winters with frequent rain and some snow in lowlands and snow in mountains; moderate summers with occasional showers
Terrain: in the west and south mostly mountains (Alps); along the eastern and northern margins mostly flat or gently sloping
Elevation extremes: *lowest point:* Neusiedler See 115 m
highest point: Grossglockner 3,798 m
Natural resources: oil, coal, lignite, timber, iron ore, copper, zinc, antimony, magnesite, tungsten, graphite, salt, hydropower

Land use:
arable land: 16.59%
permanent crops: 0.85%
other: 82.56% (2005)
Irrigated land: 40 sq km (2003)
Total renewable water resources: 84 cu km (2005)
Freshwater withdrawal (domestic/industrial/agricultural): *total:* 3.67 cu km/yr (35%/64%/1%)
per capita: 448 cu m/yr (1999)
Natural hazards: landslides; avalanches; earthquakes
Environment—current issues: some forest degradation caused by air and soil pollution; soil pollution results from the use of agricultural chemicals; air pollution results from emissions by coal- and oil-fired power stations and industrial plants and from trucks transiting Austria between northern and southern Europe
Environment—international agreements: *party to:* Air Pollution, Air Pollution-Nitrogen Oxides, Air Pollution-Persistent Organic Pollutants, Air Pollution-Sulfur 85, Air Pollution-Sulphur 94, Air Pollution-Volatile Organic Compounds, Antarctic Treaty, Biodiversity, Climate Change, Climate Change-Kyoto Protocol, Desertification, Endangered Species, Environmental Modification, Hazardous Wastes, Law of the Sea, Ozone Layer Protection, Ship Pollution, Tropical Timber 83, Tropical Timber 94, Wetlands, Whaling
signed, but not ratified: none of the selected agreements
Geography—note: landlocked; strategic location at the crossroads of central Europe with many easily traversable Alpine passes and valleys; major river is the Danube; population is concentrated on eastern lowlands because of steep slopes, poor soils, and low temperatures elsewhere

PEOPLE

Population: 8,217,280 (July 2011 est.)
country comparison to the world: 92
Age structure: *0–14 years:* 14% (male 590,855/female 563,300)
15–64 years: 67.7% (male 2,793,725/female 2,769,840)
65 years and over: 18.2% (male 627,456/female 872,104) (2011 est.)
Median age: *total:* 43 years
male: 41.9 years
female: 44 years (2011 est.)
Population growth rate: 0.034% (2011 est.)
country comparison to the world: 190
Birth rate: 8.67 births/1,000 population (2011 est.)
country comparison to the world: 215
Death rate: 10.14 deaths/1,000 population (July 2011 est.)
country comparison to the world: 54
Net migration rate: 1.81 migrant(s)/1,000 population (2011 est.)
country comparison to the world: 42
Urbanization: *urban population:* 68% of total population (2010)
rate of urbanization: 0.6% annual rate of change (2010-15 est.)
Major cities—population: VIENNA (capital) 1.693 million (2009)
Sex ratio: *at birth:* 1.051 male(s)/female
under 15 years: 1.05 male(s)/female
15–64 years: 1.01 male(s)/female
65 years and over: 0.71 male(s)/female
total population: 0.95 male(s)/female (2011 est.)
Infant mortality rate:
total: 4.32 deaths/1,000 live births
country comparison to the world: 195
male: 5.23 deaths/1,000 live births
female: 3.35 deaths/1,000 live births (2011 est.)
Life expectancy at birth: *total population:* 79.78 years
country comparison to the world: 32
male: 76.87 years
female: 82.84 years (2011 est.)
Total fertility rate: 1.4 children born/woman (2011 est.)
country comparison to the world: 200
HIV/AIDS—adult prevalence rate: 0.3% (2009 est.)
country comparison to the world: 87
HIV/AIDS—people living with HIV/AIDS: 15,000 (2009 est.)
country comparison to the world: 85
HIV/AIDS—deaths: fewer than 100 (2009 est.)
country comparison to the world: 121
Drinking water source: *Improved:*
urban: 100% of population
rural: 100% of population
total: 100% of population (2008)
Sanitation facility access: *Improved:*
urban: 100% of population
rural: 100% of population
total: 100% of population (2008)
Nationality: *noun:* Austrian(s)
adjective: Austrian
Ethnic groups: Austrians 91.1%, former Yugoslavs 4% (includes Croatians, Slovenes, Serbs, and Bosniaks), Turks 1.6%, German 0.9%, other or unspecified 2.4% (2001 census)
Religions: Roman Catholic 73.6%, Protestant 4.7%, Muslim 4.2%, other 3.5%, unspecified 2%, none 12% (2001 census)
Languages: German (official nationwide) 88.6%, Turkish 2.3%, Serbian 2.2%, Croatian (official in Burgenland) 1.6%, other (includes Slovene, official in Carinthia, and Hungarian, official in Burgenland) 5.3% (2001 census)
Literacy: *definition:* age 15 and over can read and write
total population: 98%
male: NA
female: NA
School life expectancy (primary to tertiary education): *total:* 15 years
male: 15 years
female: 15 years (2008)
Education expenditures: 5.4% of GDP (2007)
country comparison to the world: 47

GOVERNMENT

Country name: *conventional long form:* Republic of Austria
conventional short form: Austria
local long form: Republik Oesterreich
local short form: Oesterreich
Government type: federal republic
Capital: *name:* Vienna
geographic coordinates: 48 12 N, 16 22 E
time difference: UTC+1 (6 hours ahead of Washington, DC during Standard Time)
daylight saving time: +1hr, begins last Sunday in March; ends last Sunday in October
Administrative divisions: 9 states (Bundeslaender, singular—Bundesland); Burgenland, Kaernten (Carinthia), Niederoesterreich (Lower Austria), Oberoesterreich (Upper Austria), Salzburg, Steiermark (Styria), Tirol (Tyrol), Vorarlberg, Wien (Vienna)
Independence: 12 November 1918 (republic proclaimed); notable earlier dates: 976 (Margravate of Austria established); 17 September 1156 (Duchy of Austria founded); 11 August 1804 (Austrian Empire proclaimed)
National holiday: National Day, 26 October (1955); note—commemorates the passage of the law on permanent neutrality
Constitution: 1920; revised 1929; reinstated 1 May 1945; note—during the period 1 May 1934-1 May 1945 there was a fascist (corporative) constitution in place
Legal system: civil law system; judicial review of legislative acts by the Constitutional Court
International law organization participation: accepts compulsory ICJ jurisdiction; accepts ICCt jurisdiction
Suffrage: 16 years of age; universal; note—reduced from 18 years of age in 2007
Executive branch: *chief of state:* President Heinz FISCHER (SPOe) (since 8 July 2004)
head of government: Chancellor Werner FAYMANN (SPOe) (since 2 December 2008); Vice Chancellor Michael SPINDELEGGER (OeVP) (since 21 April 2011)
cabinet: Council of Ministers chosen by the president on the advice of the chancellor
(For more information visit the World Leaders website)
elections: president elected for a six-year term (eligible for a second term) by direct popular vote and formally sworn into office before the Federal Assembly or Bundesversammlung; presidential election last held on 25 April 2010 (next to be held on 25 April 2016); chancellor formally chosen by the president but determined by the coalition parties forming a parliamentary majority; vice chancellor chosen by the president on the advice of the chancellor
election results: Heinz FISCHER reelected president; percent of vote—Heinz FISCHER 79.33%, Barbara ROSENKRANZ 15.24%, Rudolf GEHRING 5.43%
note: government coalition—SPOe and OeVP
Legislative branch: bicameral Federal Assembly or Bundesversammlung consists of Federal Council or Bundesrat (62 seats; delegates appointed by state parliaments with each state receiving 3 to 12 seats in proportion to its population; members serve five- or six-year terms) and the National Council or Nationalrat (183 seats; members elected by popular vote for a five-year term under a system of proportional representation with partially-open party lists)
elections: National Council—last held on 28 September 2008 (next to be held by September 2013)

election results: National Council–percent of vote by party–SPOe 29.3%, OeVP 26%, FPOe 17.5%, BZOe 10.7%, Greens 10.4%, other 6.1%; seats by party–SPOe 57, OeVP 51, FPOe 34, BZOe 21, Greens 20; note–seats by party since 2010–SPOe 57, OeVP 51, FPOe 39, BZOe 16, Greens 20
Judicial branch: Supreme Judicial Court or Oberster Gerichtshof; Administrative Court or Verwaltungsgerichtshof; Constitutional Court or Verfassungsgerichtshof
Political parties and leaders: Alliance for the Future of Austria or BZOe [Josef BUCHER]; Austrian People's Party or OeVP [Josef PROELL]; Freedom Party of Austria or FPOe [Heinz Christian STRACHE]; Social Democratic Party of Austria or SPOe [Werner FAYMANN]; The Greens [Eva GLAWISCHNIG]
Political pressure groups and leaders: Austrian Trade Union Federation or OeGB (nominally independent but primarily Social Democratic); Federal Economic Chamber; Labor Chamber or AK (Social Democratic-leaning think tank); OeVP-oriented Association of Austrian Industrialists or IV; Roman Catholic Church, including its chief lay organization, Catholic Action
other: three composite leagues of the Austrian People's Party or OeVP representing business, labor, farmers, and other nongovernment organizations in the areas of environment and human rights
International organization participation: ADB (nonregional member), AfDB (nonregional member), Australia Group, BIS, BSEC (observer), CE, CEI, CERN, EAPC, EBRD, EIB, EMU, ESA, EU, FAO, FATF, G-9, IADB, IAEA, IBRD, ICAO, ICC, ICRM, IDA, IEA, IFAD, IFC, IFRCS, ILO, IMF, IMO, Interpol, IOC, IOM, IPU, ISO, ITSO, ITU, ITUC, MIGA, MINURSO, NEA, NSG, OAS (observer), OECD, OIF (observer), OPCW, OSCE, Paris Club, PCA, PFP, Schengen Convention, SECI (observer), UN, UNCTAD, UNDOF, UNESCO, UNFICYP, UNHCR, UNTSO, UNWTO, UPU, WCO, WFTU, WHO, WIPO, WMO, WTO, ZC
Diplomatic representation in the US:
chief of mission: Ambassador Christian PROSL
chancery: 3524 International Court NW, Washington, DC 20008-3035
telephone: [1] (202) 895-6700
FAX: [1] (202) 895-6750
consulate(s) general: Chicago, Los Angeles, New York
Diplomatic representation from the US:
chief of mission: Ambassador William C. EACHO III
embassy: Boltzmanngasse 16, A-1090, Vienna
mailing address: use embassy street address
telephone: [43] (1) 31339-0
FAX: [43] (1) 3100682
Flag description: three equal horizontal bands of red (top), white, and red; the flag design is certainly one of the oldest–if not the oldest–national banners in the world; according to tradition, in 1191, following a fierce battle in the Third Crusade, Duke Leopold V of Austria's white tunic became completely blood-spattered; upon removal of his wide belt or sash, a white band was revealed; the red-white-red color combination was subsequently adopted as his banner
National anthem: *name:* "Bundeshymne" (Federal Hymn)
lyrics/music: Paula von PRERADOVIC/ Wolfgang Amadeus MOZART or Johann HOLZER (disputed)
note: adopted 1947; the anthem is also known as "Land der Berge, Land am Strome" (Land of the Mountains, Land on the River); Austria adopted a new national anthem after World War II to replace the former imperial anthem composed by Franz Josef HAYDN, which had been appropriated by Germany in 1922 and was now associated with the Nazi regime

ECONOMY

Economy—overview: Austria, with its well-developed market economy and high standard of living, is closely tied to other EU economies, especially Germany's. Its economy features a large service sector, a sound industrial sector, and a small, but highly developed agricultural sector. Following several years of solid foreign demand for Austrian exports and record employment growth, the international financial crisis and global economic downturn in 2008 led to a sharp but brief recession. Austrian GDP contracted 3.9% in 2009 but saw positive growth of about 2% in 2010. Unemployment has not risen as steeply in Austria as elsewhere in Europe, partly because its government has subsidized reduced working hour schemes to allow companies to retain employees. Stabilization measures, stimulus spending, and an income tax reform pushed the budget deficit to 3.5% of GDP in 2009 and 4.7% in 2010, from only about 1.3% in 2008. The international financial crisis caused difficulties for Austria's largest banks whose extensive operations in central, eastern, and southeastern Europe faced large losses. The government provided bank support–including in some instances, nationalization–to prevent insolvency and possible contagion. In the medium-term all large Austrian banks will need additional capital. Even after the global economic outlook improves, Austria will need to continue restructuring, emphasize knowledge-based sectors of the economy, and encourage greater labor flexibility and labor participation to offset growing unemployment and Austria's aging population and low fertility rate.
GDP (purchasing power parity): $332 billion (2010 est.)
country comparison to the world: 36
$325.6 billion (2009 est.)
$338.8 billion (2008 est.)
note: data are in 2010 US dollars
GDP (official exchange rate): $376.8 billion (2010 est.)
GDP—real growth rate: 2% (2010 est.)
country comparison to the world: 146
-3.9% (2009 est.)
2.2% (2008 est.)
GDP—per capita (PPP): $40,400 (2010 est.)
country comparison to the world: 19
$39,700 (2009 est.)
$41,300 (2008 est.)
note: data are in 2010 US dollars
GDP—composition by sector:
agriculture: 1.5%
industry: 29.4%
services: 69.1% (2010 est.)
Labor force: 3.7 million (2010 est.)
country comparison to the world: 94
Labor force—by occupation:
agriculture: 5.5%
industry: 27.5%
services: 67% (2009 est.)
Unemployment rate: 4.5% (2010 est.)
country comparison to the world: 43
4.8% (2009 est.)
Population below poverty line: 6% (2008)
Household income or consumption by percentage share: *lowest 10%:* 4%
highest 10%: 22% (2007)
Distribution of family income—Gini index: 26 (2007)
country comparison to the world: 129
31 (1995)
Investment (gross fixed): 21% of GDP (2010 est.)
country comparison to the world: 78
Budget: *revenues:* $172.1 billion
expenditures: $189.4 billion (2010 est.)
Public debt: 70.4% of GDP (2010 est.)
country comparison to the world: 24
67.5% of GDP (2009 est.)
Inflation rate (consumer prices): 1.9% (2010 est.)
country comparison to the world: 48
0.5% (2009 est.)
Commercial bank prime lending rate: 4.82% (30 November 2010 est.)
country comparison to the world: 146
5.03% (31 December 2009 est.)
Stock of narrow money: $173.4 billion (31 December 2010 est.)
country comparison to the world: 18
$175.6 billion (31 December 2009 est.)
note: see entry for the European Union for money supply for the entire euro area; the European Central Bank (ECB) controls monetary policy for the 17 members of the Economic and Monetary Union (EMU); individual members of the EMU do not control the quantity of money circulating within their own borders
Stock of broad money: $402.8 billion (31 December 2010 est.)
country comparison to the world: 23
$402.2 billion (31 December 2009 est.)
Stock of domestic credit: $659.2 billion (31 December 2009 est.)
country comparison to the world: 20
$606.2 billion (31 December 2008 est.)
Market value of publicly traded shares: $118 billion (31 December 2010)
country comparison to the world: 37
$107.2 billion (31 December 2009)
$72.3 billion (31 December 2008)
Agriculture—products: grains, potatoes, wine, fruit; dairy products, cattle, pigs, poultry; lumber
Industries: construction, machinery, vehicles and parts, food, metals, chemicals, lumber and wood processing, paper and paperboard, communications equipment, tourism
Industrial production growth rate: 7% (2010 est.)
country comparison to the world: 47

Electricity—production: 68.85 billion kWh (2009 est.)
country comparison to the world: 40
Electricity—consumption: 65.67 billion kWh (2009 est.)
country comparison to the world: 39
Electricity—exports: 18.76 billion kWh (2009 est.)
Electricity—imports: 19.54 billion kWh (2009 est.)
Oil—production: 21,880 bbl/day (2009 est.)
country comparison to the world: 73
Oil—consumption: 247,700 bbl/day (2009 est.)
country comparison to the world: 50
Oil—exports: 50,410 bbl/day (2009 est.)
country comparison to the world: 77
Oil—imports: 273,000 bbl/day (2009 est.)
country comparison to the world: 37
Oil—proved reserves: 89 million bbl (1 January 2010 est.)
country comparison to the world: 73
Natural gas—production: 1.58 billion cu m (2009)
country comparison to the world: 60
Natural gas—consumption: 8.13 billion cu m (2009)
country comparison to the world: 52
Natural gas—exports: 3.961 billion cu m (2009 est.)
country comparison to the world: 30
Natural gas—imports: 9.46 billion cu m (2009)
country comparison to the world: 23
Natural gas—proved reserves: 24.8 billion cu m (1 January 2010 est.)
country comparison to the world: 75
Current account balance: $9.9 billion (2010 est.)
country comparison to the world: 23
$11.1 billion (2009 est.)
Exports: $157.4 billion (2010 est.)
country comparison to the world: 29
$130.3 billion (2009 est.)
Exports—commodities: machinery and equipment, motor vehicles and parts, paper and paperboard, metal goods, chemicals, iron and steel, textiles, foodstuffs
Exports—partners: Germany 30.96%, Italy 8.17%, Switzerland 4.99%, US 3.99% (2009)
Imports: $156 billion (2010 est.)
country comparison to the world: 29
$135.6 billion (2009 est.)
Imports—commodities: machinery and equipment, motor vehicles, chemicals, metal goods, oil and oil products; foodstuffs
Imports—partners: Germany 45.07%, Switzerland 6.76%, Italy 6.66%, Netherlands 4.03% (2009)
Reserves of foreign exchange and gold: $21.89 billion (31 December 2010 est.)
country comparison to the world: 44
$18.05 billion (31 December 2009 est.)
Debt—external: $755 billion (30 June 2010)
country comparison to the world: 17
$864.2 billion (31 December 2008)
Stock of direct foreign investment—at home: $154 billion (30 September 2010 est.)
country comparison to the world: 24
$144.7 billion (31 December 2009 est.)
Stock of direct foreign investment—abroad: $154.1 billion (30 September 2010 est.)
country comparison to the world: 22
$146.3 billion (31 December 2009 est.)
Exchange rates: euros (EUR) per US dollar–0.755 (2010), 0.7198 (2009), 0.6827 (2008), 0.7345 (2007), 0.7964 (2006)

COMMUNICATIONS

Telephones—main lines in use: 3.253 million (2009)
country comparison to the world: 47
Telephones—mobile cellular: 11.773 million (2009)
country comparison to the world: 59
Telephone system: *general assessment:* highly developed and efficient
domestic: fixed-line subscribership has been in decline since the mid-1990s with mobile-cellular subscribership eclipsing it by the late 1990s; the fiber-optic net is very extensive; all telephone applications and Internet services are available
international: country code–43; satellite earth stations–15; in addition, there are about 600 VSATs (very small aperture terminals) (2007)
Broadcast media: Austria's public broadcaster, ORF, was the main broadcast source until commercial radio and television service was introduced in the 1990s; cable and satellite TV are available, including German TV stations (2008)
Internet country code: .at
Internet hosts: 3.266 million (2010)
country comparison to the world: 29
Internet users: 6.143 million (2009)
country comparison to the world: 43

TRANSPORTATION

Airports: 55 (2010)
country comparison to the world: 85
Airports—with paved runways: *total:* 25
over 3,047 m: 1
2,438 to 3,047 m: 5
1,524 to 2,437 m: 1
914 to 1,523 m: 4
under 914 m: 14 (2010)
Airports—with unpaved runways: *total:* 30
1,524 to 2,437 m: 1
914 to 1,523 m: 3
under 914 m: 26 (2010)
Heliports: 1 (2010)
Pipelines: gas 3,028 km; oil 663 km; refined products 157 km (2010)
Railways: *total:* 6,399 km
country comparison to the world: 29
standard gauge: 5,927 km 1.435-m gauge (3,853 km electrified)
narrow gauge: 384 km 1.000-m gauge (15 km electrified); 88 km 0.760-m gauge (10 km electrified) (2010)
Roadways: *total:* 107,262 km
country comparison to the world: 39
paved: 107,262 km (includes 1,696 km of expressways) (2006)
Waterways: 358 km (2011)
country comparison to the world: 90
Merchant marine: *total:* 2
country comparison to the world: 141
by type: cargo 2
registered in other countries: 4 (Cyprus 1, Malta 1, Saint Vincent and the Grenadines 2) (2010)
Ports and terminals: Enns, Krems, Linz, Vienna

MILITARY

Military branches: Land Forces (KdoLdSK), Air Forces (KdoLuSK)
Military service age and obligation: 18-35 years of age for compulsory military service; 16 years of age for male and female voluntary service; service obligation 6 months of training, followed by an 8-year reserve obligation; conscripts cannot be deployed in military operations outside Austria (2009)
Manpower available for military service: *males age 16–49:* 1,941,110
females age 16–49: 1,910,434 (2010 est.)
Manpower fit for military service: *males age 16–49:* 1,579,862
females age 16–49: 1,554,130 (2010 est.)
Manpower reaching militarily significant age annually: *male:* 48,108
female: 45,752 (2010 est.)
Military expenditures: 0.8% of GDP (2009)
country comparison to the world: 146

TRANSNATIONAL ISSUES

Disputes—international: none
Illicit drugs: transshipment point for Southwest Asian heroin and South American cocaine destined for Western Europe; increasing consumption of European-produced synthetic drugs

AZERBAIJAN

INTRODUCTION

Background: Azerbaijan–a nation with a majority-Turkic and majority-Muslim population–was briefly independent from 1918 to 1920; it regained its independence after the collapse of the Soviet Union in 1991. Despite a 1994 cease-fire, Azerbaijan has yet to resolve its conflict with Armenia over Nagorno-Karabakh, a primarily Armenian-populated region that Moscow recognized as part of Soviet Azerbaijan in the 1920s after Armenia and Azerbaijan disputed the status of the territory. Armenia and Azerbaijan began fighting over the area in 1988; the struggle escalated after both countries attained independence from the Soviet Union in 1991. By May 1994, when a cease-fire took hold, ethnic Armenian forces held not only Nagorno-Karabakh but also seven surrounding provinces in the territory of Azerbaijan. Corruption in the country is ubiquitous, and the government, which eliminated presidential term limits in a 2009 referendum, has been accused of authoritarianism. Although the poverty rate has been reduced in recent years due to revenue from oil production, the promise of widespread wealth resulting from the

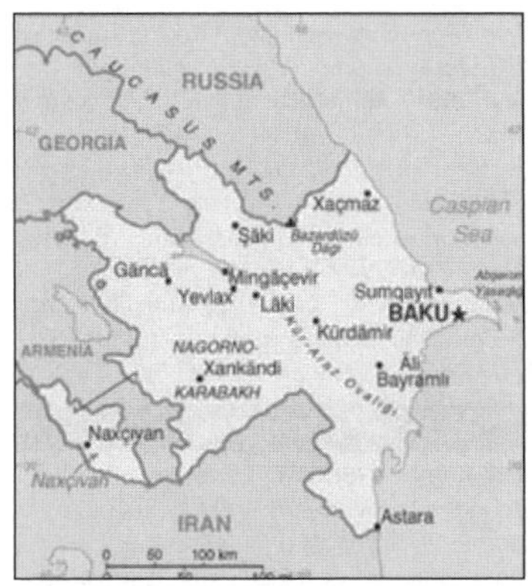

continued development of Azerbaijan's energy sector remains largely unfulfilled.

GEOGRAPHY

Location: Southwestern Asia, bordering the Caspian Sea, between Iran and Russia, with a small European portion north of the Caucasus range
Geographic coordinates: 40 30 N, 47 30 E
Map references: Middle East
Area: *total:* 86,600 sq km
country comparison to the world: 112
land: 82,629 sq km
water: 3,971 sq km
note: includes the exclave of Naxcivan Autonomous Republic and the Nagorno-Karabakh region; the region's autonomy was abolished by Azerbaijani Supreme Soviet on 26 November 1991
Area—comparative: slightly smaller than Maine
Land boundaries: *total:* 2,013 km
border countries: Armenia (with Azerbaijan-proper) 566 km, Armenia (with Azerbaijan-Naxcivan exclave) 221 km, Georgia 322 km, Iran (with Azerbaijan-proper) 432 km, Iran (with Azerbaijan-Naxcivan exclave) 179 km, Russia 284 km, Turkey 9 km
Coastline: 0 km (landlocked); note–Azerbaijan borders the Caspian Sea (713 km)
Maritime claims: none (landlocked)
Climate: dry, semiarid steppe
Terrain: large, flat Kur-Araz Ovaligi (Kura-Araks Lowland) (much of it below sea level) with Great Caucasus Mountains to the north, Qarabag Yaylasi (Karabakh Upland) in west; Baku lies on Abseron Yasaqligi (Apsheron Peninsula) that juts into Caspian Sea
Elevation extremes: *lowest point:* Caspian Sea -28 m
highest point: Bazarduzu Dagi 4,485 m
Natural resources: petroleum, natural gas, iron ore, nonferrous metals, bauxite
Land use: *arable land:* 20.62%
permanent crops: 2.61%
other: 76.77% (2005)
Irrigated land: 14,550 sq km (2003)
Total renewable water resources: 30.3 cu km (1997)
Freshwater withdrawal (domestic/industrial/agricultural): *total:* 17.25 cu km/yr (5%/28%/68%)
per capita: 2,051 cu m/yr (2000)
Natural hazards: droughts
Environment—current issues: local scientists consider the Abseron Yasaqligi (Apsheron Peninsula) (including Baku and Sumqayit) and the Caspian Sea to be the ecologically most devastated area in the world because of severe air, soil, and water pollution; soil pollution results from oil spills, from the use of DDT pesticide, and from toxic defoliants used in the production of cotton
Environment—international agreements: *party to:* Air Pollution, Biodiversity, Climate Change, Climate Change-Kyoto Protocol, Desertification, Endangered Species, Hazardous Wastes, Marine Dumping, Ozone Layer Protection, Ship Pollution, Wetlands
signed, but not ratified: none of the selected agreements
Geography—note: both the main area of the country and the Naxcivan exclave are landlocked

PEOPLE

Population: 8,372,373 (July 2011 est.)
country comparison to the world: 91
Age structure: *0–14 years:* 23.2% (male 1,029,931/female 912,639)
15–64 years: 70.3% (male 2,896,785/female 2,993,092)
65 years and over: 6.4% (male 195,853/female 344,073) (2011 est.)
Median age: *total:* 28.8 years
male: 27.2 years
female: 30.6 years (2011 est.)
Population growth rate: 0.846% (2011 est.)
country comparison to the world: 129
Birth rate: 17.85 births/1,000 population (2011 est.)
country comparison to the world: 108
Death rate: 8.25 deaths/1,000 population (July 2011 est.)
country comparison to the world: 92
Net migration rate: -1.14 migrant(s)/1,000 population (2011 est.)
country comparison to the world: 152
Urbanization: *urban population:* 52% of total population (2010)
rate of urbanization: 1.4% annual rate of change (2010-15 est.)
Major cities—population: BAKU (capital) 1.95 million (2009)
Sex ratio: *at birth:* 1.116 male(s)/female
under 15 years: 1.13 male(s)/female
15–64 years: 0.97 male(s)/female
65 years and over: 0.58 male(s)/female
total population: 0.97 male(s)/female (2011 est.)
Infant mortality rate: *total:* 51.08 deaths/1,000 live births
country comparison to the world: 46
male: 56.55 deaths/1,000 live births
female: 44.97 deaths/1,000 live births (2011 est.)
Life expectancy at birth: *total population:* 67.36 years
country comparison to the world: 157
male: 63.2 years
female: 72.01 years (2011 est.)
Total fertility rate: 2.02 children born/woman (2011 est.)
country comparison to the world: 128
HIV/AIDS—adult prevalence rate: 0.1% (2009 est.)
country comparison to the world: 114
HIV/AIDS—people living with HIV/AIDS: 3,600 (2009 est.)
country comparison to the world: 126
HIV/AIDS—deaths: fewer than 200 (2009 est.)
country comparison to the world: 108
Drinking water source: *Improved:*
urban: 88% of population
rural: 71% of population
total: 80% of population
Unimproved:
urban: 12% of population
rural: 29% of population
total: 20% of population (2008)
Sanitation facility access: *Improved:*
urban: 51% of population
rural: 39% of population
total: 45% of population
Unimproved:
urban: 49% of population
rural: 61% of population
total: 55% of population (2008)
Nationality: *noun:* Azerbaijani(s)
adjective: Azerbaijani
Ethnic groups: Azeri 90.6%, Dagestani 2.2%, Russian 1.8%, Armenian 1.5%, other 3.9% (1999 census)
note: almost all Armenians live in the separatist Nagorno-Karabakh region
Religions: Muslim 93.4%, Russian Orthodox 2.5%, Armenian Orthodox 2.3%, other 1.8% (1995 est.)
note: religious affiliation is still nominal in Azerbaijan; percentages for actual practicing adherents are much lower
Languages: Azerbaijani (Azeri) (official) 90.3%, Lezgi 2.2%, Russian 1.8%, Armenian 1.5%, other 3.3%, unspecified 1% (1999 census)
Literacy: *definition:* age 15 and over can read and write
total population: 98.8%
male: 99.5%
female: 98.2% (1999 census)
School life expectancy (primary to tertiary education): *total:* 12 years
male: 12 years
female: 12 years (2009)
Education expenditures: 2.8% of GDP (2009)
country comparison to the world: 141

GOVERNMENT

Country name: *conventional long form:* Republic of Azerbaijan
conventional short form: Azerbaijan
local long form: Azarbaycan Respublikasi
local short form: Azarbaycan
former: Azerbaijan Soviet Socialist Republic
Government type: republic
Capital: *name:* Baku (Baki, Baky)
geographic coordinates: 40 23 N, 49 52 E
time difference: UTC+4 (9 hours ahead of Washington, DC during Standard Time)
daylight saving time: +1hr, begins last Sunday in March; ends last Sunday in October
Administrative divisions: 59 rayons (rayonlar; rayon–singular), 11 cities (saharlar; sahar–singular), 1 autonomous republic (muxtar respublika)
rayons: Abseron Rayonu, Agcabadi Rayonu, Agdam Rayonu, Agdas Rayonu, Agstafa Rayonu, Agsu Rayonu, Astara Rayonu, Balakan Rayonu, Barda Rayonu, Beylaqan Rayonu, Bilasuvar Rayonu,

Cabrayil Rayonu, Calilabad Rayonu, Daskasan Rayonu, Davaci Rayonu, Fuzuli Rayonu, Gadabay Rayonu, Goranboy Rayonu, Goycay Rayonu, Haciqabul Rayonu, Imisli Rayonu, Ismayilli Rayonu, Kalbacar Rayonu, Kurdamir Rayonu, Lacin Rayonu, Lankaran Rayonu, Lerik Rayonu, Masalli Rayonu, Neftcala Rayonu, Oguz Rayonu, Qabala Rayonu, Qax Rayonu, Qazax Rayonu, Qobustan Rayonu, Quba Rayonu, Qubadli Rayonu, Qusar Rayonu, Saatli Rayonu, Sabirabad Rayonu, Saki Rayonu, Salyan Rayonu, Samaxi Rayonu, Samkir Rayonu, Samux Rayonu, Siyazan Rayonu, Susa Rayonu, Tartar Rayonu, Tovuz Rayonu, Ucar Rayonu, Xacmaz Rayonu, Xanlar Rayonu, Xizi Rayonu, Xocali Rayonu, Xocavand Rayonu, Yardimli Rayonu, Yevlax Rayonu, Zangilan Rayonu, Zaqatala Rayonu, Zardab Rayonu
cities: Ali Bayramli Sahari, Baki Sahari, Ganca Sahari, Lankaran Sahari, Mingacevir Sahari, Naftalan Sahari, Saki Sahari, Sumqayit Sahari, Susa Sahari, Xankandi Sahari, Yevlax Sahari
autonomous republic: Naxcivan Muxtar Respublikasi (Nakhichevan)
Independence: 30 August 1991 (declared from the Soviet Union); 18 October 1991 (adopted by the Supreme Council of Azerbaijan)
National holiday: Founding of the Democratic Republic of Azerbaijan, 28 May (1918)
Constitution: adopted 12 November 1995; modified by referendum 24 August 2002
Legal system: civil law system
International law organization participation: has not submitted an ICJ jurisdiction declaration; non-party state to the ICCt
Suffrage: 18 years of age; universal
Executive branch: *chief of state:* President Ilham ALIYEV (since 31 October 2003)
head of government: Prime Minister Artur RASIZADE (since 4 November 2003); First Deputy Prime Minister Yaqub EYYUBOV (since June 2006)
cabinet: Council of Ministers appointed by the president and confirmed by the National Assembly
(For more information visit the World Leaders website)
elections: president elected by popular vote for a five-year term (eligible for unlimited terms); election last held on 15 October 2008 (next to be held in October 2013); prime minister and first deputy prime minister appointed by the president and confirmed by the National Assembly
election results: Ilham ALIYEV reelected president; percent of vote–Ilham ALIYEV 89%, Igbal AGHAZADE 2.9%, five other candidates with smaller percentages
note: several political parties boycotted the election due to unfair conditions; OSCE observers concluded that the election did not meet international standards
Legislative branch: unicameral National Assembly or Milli Mejlis (125 seats; members elected by popular vote to serve five-year terms)
elections: last held on 7 November 2010 (next to be held in November 2015)
election results: percent of vote by party–YAP 45.8%, CSP 1.6%, Motherland 1.4%, independents 48.2%, other 3.1%; seats by party–YAP 71, CSP 3, Motherland 2, Democratic Reforms 1, Great Creation 1, Hope Party 1, Social Welfare 1, Civil Unity 1, Whole Azerbaijan Popular Front 1, Justice 1, independents 42
Judicial branch: Constitutional Court the president proposes judges of all the courts to the Parliament which appoints them; Supreme Court; Economic Court
Political parties and leaders: Azerbaijan Democratic Party or ADP [Sardar JALALOGLU]; Civil Solidarity Party or CSP [Sabir RUSTAMKHANLI]; Civil Unity Party [Sabir HACIYEV]; Classic People's Front of Azerbaijan [Mirmahmud MIRALI-OGLU]; Democratic Reform Party [Asim MOLLAZADE]; Great Creation Party [Fazil Gazanfaroglu MUSTAFAYEV]; Hope (Umid) Party [Iqbal AGAZADE]; Justice Party [Ilyas ISMAYILOV]; Liberal Party of Azerbaijan [Lala Shovkat HACIYEVA]; Motherland Party [Fazail AGAMALI]; Musavat (Equality) [Isa GAMBAR, chairman]; Open Society Party [Rasul GULIYEV, in exile in the US]; Social Democratic Party of Azerbaijan or SDP [Araz ALIZADE and Ayaz MUTALIBOV (in exile)]; Social Welfare Party [Hussein KAZIMLI]; United Popular Azerbaijan Front Party or AXCP [Ali KARIMLI]; Whole Azerbaijan Popular Front Party [Gudrat HASANGULIYEV]; Yeni (New) Azerbaijan Party or YAP [President Ilham ALIYEV]
note: opposition parties regularly factionalize and form new parties
Political pressure groups and leaders: Azerbaijan Public Forum [Eldar NAMAZOV]; Karabakh Liberation Organization
International organization participation: ADB, BSEC, CE, CICA, CIS, EAPC, EBRD, ECO, FAO, GCTU, GUAM, IAEA, IBRD, ICAO, ICRM, IDA, IDB, IFAD, IFC, IFRCS, ILO, IMF, IMO, Interpol, IOC, IOM, IPU, ISO, ITSO, ITU, ITUC, MIGA, NAM (observer), OAS (observer), OIC, OPCW, OSCE, PFP, SECI (observer), UN, UNCTAD, UNESCO, UNIDO, UNWTO, UPU, WCO, WFTU, WHO, WIPO, WMO, WTO (observer)
Diplomatic representation in the US: *chief of mission:* Ambassador Yashar ALIYEV
chancery: 2741 34th Street NW, Washington, DC 20008
telephone: [1] (202) 337-3500
FAX: [1] (202) 337-5911
Consulate(s) general: Los Angeles
Diplomatic representation from the US: *chief of mission:* Ambassador Matthew BRYZA
embassy: 83 Azadlig Prospecti, Baku AZ1007
mailing address: American Embassy Baku, US Department of State, 7050 Baku Place, Washington, DC 20521-7050
telephone: [994] (12) 4980-335 through 337
FAX: [994] (12) 4656-671
Flag description: three equal horizontal bands of blue (top), red, and green; a crescent and eight-pointed star in white are centered in the red band; the blue band recalls Azerbaijan's Turkic heritage, red stands for modernization and progress, and green refers to Islam; the crescent moon is an Islamic symbol, while the eight-pointed star represents the eight Turkic peoples of the world
National anthem: *name:* "Azerbaijan Marsi" (March of Azerbaijan)
lyrics/music: Ahmed JAVAD/Uzeyir HAJIBEYOV
note: adopted 1992; although originally written in 1919 during a brief period of independence, "Azerbaijan Marsi" did not become the official anthem until after the dissolution of the Soviet Union

ECONOMY

Economy—overview: Azerbaijan's high economic growth during 2006-08 was attributable to large and growing oil exports, but some non-export sectors also featured double-digit growth, spurred by growth in the construction, banking, and real estate sectors. In 2009, economic growth remained above 9% even as oil prices moderated and growth in the construction sector cooled. In 2010, economic growth slowed to 3.7%, although the impact of the global financial crisis was less severe than in many other countries in the region. The current global economic slowdown presents some challenges for the Azerbaijani economy as oil prices remain below their mid-2008 highs, highlighting Azerbaijan's reliance on energy exports and lackluster attempts to diversify its economy. Azerbaijan's oil production increased dramatically in 1997, when Azerbaijan signed the first production-sharing arrangement (PSA) with the Azerbaijan International Operating Company. Oil exports through the Baku-Tbilisi-Ceyhan Pipeline remain the main economic driver while efforts to boost Azerbaijan's gas production are underway. However, Azerbaijan has made only limited progress on instituting market-based economic reforms. Pervasive public and private sector corruption and structural economic inefficiencies remain a drag on long-term growth, particularly in non-energy sectors. Several other obstacles impede Azerbaijan's economic progress: the need for stepped up foreign investment in the non-energy sector and the continuing conflict with Armenia over the Nagorno-Karabakh region. Trade with Russia and the other former Soviet republics is declining in importance, while trade is building with Turkey and the nations of Europe. Long-term prospects will depend on world oil prices, the location of new oil and gas pipelines in the region, and Azerbaijan's ability to manage its energy wealth to promote sustainable growth in non-energy sectors of the economy and spur employment.
GDP (purchasing power parity): $90.79 billion (2010 est.)
country comparison to the world: 73
$86.47 billion (2009 est.)
$79.11 billion (2008 est.)
note: data are in 2010 US dollars
GDP (official exchange rate): $54.37 billion (2010 est.)
GDP—real growth rate: 5% (2010 est.)
country comparison to the world: 72
9.3% (2009 est.)
10.8% (2008 est.)
GDP—per capita (PPP): $10,900 (2010 est.)
country comparison to the world: 101

$10,500 (2009 est.)
$9,700 (2008 est.)
note: data are in 2010 US dollars
GDP—composition by sector:
agriculture: 5.5%
industry: 61.4%
services: 33.1% (2010 est.)
Labor force: 5.874 million (2010 est.)
country comparison to the world: 65
Labor force—by occupation:
agriculture: 38.3%
industry: 12.1%
services: 49.6% (2008)
Unemployment rate: 0.9% (2010 est.)
country comparison to the world: 3
6% (2009 est.)
Population below poverty line: 11% (2009 est.)
Household income or consumption by percentage share: *lowest 10%:* 6.1%
highest 10%: 17.5% (2005)
Distribution of family income—Gini index: 36.5 (2001)
country comparison to the world: 82
36 (1995)
Investment (gross fixed): 17.3% of GDP (2010 est.)
country comparison to the world: 117
Budget: *revenues:* $14.19 billion
expenditures: $14.64 billion (2010 est.)
Public debt: 4.6% of GDP (2010 est.)
country comparison to the world: 129
6.7% of GDP (2009 est.)
Inflation rate (consumer prices): 5.1% (2010 est.)
country comparison to the world: 143
1.5% (2009 est.)
Central bank discount rate: 2% (31 December 2009)
country comparison to the world: 50
8% (31 December 2008)
note: this is the Refinancing Rate, the key policy rate for the National Bank of Azerbaijan
Commercial bank prime lending rate: 20.03% (31 December 2009 est.)
country comparison to the world: 19
19.76% (31 December 2008 est.)
Stock of narrow money: $7.34 billion (31 December 2010 est.)
country comparison to the world: 77
$6.519 billion (31 December 2009 est.)
Stock of broad money: $11.64 billion (31 December 2010 est.)
country comparison to the world: 96
$10.54 billion (31 December 2009 est.)
Stock of domestic credit: $8.135 billion (31 December 2008 est.)
country comparison to the world: 100
$5.726 billion (31 December 2007 est.)
Market value of publicly traded shares: $NA
Agriculture—products: cotton, grain, rice, grapes, fruit, vegetables, tea, tobacco; cattle, pigs, sheep, goats
Industries: petroleum and natural gas, petroleum products, oilfield equipment; steel, iron ore; cement; chemicals and petrochemicals; textiles
Industrial production growth rate: 3.5% (2010 est.)
country comparison to the world: 95
Electricity—production: 18.6 billion kWh (2007 est.)
country comparison to the world: 71
Electricity—consumption: 18 billion kWh (2007 est.)
country comparison to the world: 72
Electricity—exports: 786 million kWh (2007 est.)
Electricity—imports: 548 million kWh (2007 est.)
Oil—production: 1.011 million bbl/day (2009 est.)
country comparison to the world: 23
Oil—consumption: 136,000 bbl/day (2009 est.)
country comparison to the world: 70
Oil—exports: 528,900 bbl/day (2007 est.)
country comparison to the world: 29
Oil—imports: 2,848 bbl/day (2007 est.)
country comparison to the world: 169
Oil—proved reserves: 7 billion bbl (1 January 2010 est.)
country comparison to the world: 19
Natural gas—production: 23 billion cu m (2009 est.)
country comparison to the world: 30
Natural gas—consumption: 10.12 billion cu m (2008)
country comparison to the world: 48
Natural gas—exports: 5.564 billion cu m (2008 est.)
country comparison to the world: 25
Natural gas—imports: 0 cu m (2008 est.)
country comparison to the world: 143
Natural gas—proved reserves: 849.5 billion cu m (1 January 2010 est.)
country comparison to the world: 28
Current account balance: $15.96 billion (2010 est.)
country comparison to the world: 19
$10.18 billion (2009 est.)
Exports: $28.07 billion (2010 est.)
country comparison to the world: 63
$21.1 billion (2009 est.)
Exports—commodities: oil and gas 90%, machinery, cotton, foodstuffs
Exports—partners: Italy 20.69%, India 10.67%, US 9.24%, France 8.15%, Germany 7.62%, Indonesia 6.63%, Canada 5.13% (2009)
Imports: $7.035 billion (2010 est.)
country comparison to the world: 103
$6.514 billion (2009 est.)
Imports—commodities: machinery and equipment, oil products, foodstuffs, metals, chemicals
Imports—partners: Turkey 18.69%, Russia 16.98%, Germany 7.87%, Ukraine 7.3%, China 6.18%, UK 5.73% (2009)
Reserves of foreign exchange and gold: $6.33 billion (31 December 2010 est.)
country comparison to the world: 66
$5.364 billion (31 December 2009 est.)
Debt—external: $3.221 billion (31 December 2010 est.)
country comparison to the world: 123
$3.44 billion (31 December 2009 est.)
Stock of direct foreign investment—at home: $8.918 billion (31 December 2010 est.)
country comparison to the world: 82
$8.318 billion (31 December 2009 est.)
Stock of direct foreign investment—abroad: $6.058 billion (31 December 2010 est.)
country comparison to the world: 58
$5.558 billion (31 December 2009 est.)
Exchange rates: Azerbaijani manats (AZN) per US dollar—0.8035 (2010), 0.8038 (2009), 0.8219 (2008), 0.8581 (2007), 0.8934 (2006)

COMMUNICATIONS

Telephones—main lines in use: 1.397 million (2009)
country comparison to the world: 69
Telephones—mobile cellular: 7.757 million (2009)
country comparison to the world: 74
Telephone system: *general assessment:* requires considerable expansion and modernization; fixed-line telephony and a broad range of other telecom services are controlled by a state-owned telecommunications monopoly and growth has been stagnant; more competition exists in the mobile-cellular market with four providers in 2009
domestic: teledensity of 17 fixed lines per 100 persons; mobile-cellular teledensity has increased and is rapidly approaching 100 telephones per 100 persons; satellite service connects Baku to a modern switch in its exclave of Nakhchivan
international: country code—994; the Trans-Asia-Europe (TAE) fiber-optic link transits Azerbaijan providing international connectivity to neighboring countries; the old Soviet system of cable and microwave is still serviceable; satellite earth stations—2 (2009)
Broadcast media: 1 state-run and 1 public television channel; 4 domestic commercial TV stations and about 15 regional TV stations; Turkish, Russian, and Iranian TV and radio broadcasts are available, especially in border regions; cable TV services are available in Baku; 1 state-run and 1 public radio network operating; a small number of private commercial radio stations broadcasting; local FM relays of Baku commercial stations are available in many localities; local relays of several international broadcasters had been available until late 2008 when their broadcasts were banned from FM frequencies (2008)
Internet country code: .az
Internet hosts: 22,737 (2010)
country comparison to the world: 105
Internet users: 2.42 million (2009)
country comparison to the world: 70

TRANSPORTATION

Airports: 35 (2010)
country comparison to the world: 109
Airports—with paved runways: *total:* 27
over 3,047 m: 3
2,438 to 3,047 m: 6
1,524 to 2,437 m: 13
914 to 1,523 m: 4
under 914 m: 1 (2010)
Airports—with unpaved runways: *total:* 8
under 914 m: 8 (2010)
Heliports: 1 (2010)
Pipelines: condensate 1 km; gas 3,361 km; oil 1,424 km (2010)
Railways: *total:* 2,918 km
country comparison to the world: 56
broad gauge: 2,918 km 1.520-m gauge (1,278 km electrified) (2009)
Roadways: *total:* 59,141 km
country comparison to the world: 74
paved: 29,210 km
unpaved: 29,931 km (2004)

Merchant marine: *total:* 92
country comparison to the world: 54
by type: cargo 27, passenger 2, passenger/cargo 9, petroleum tanker 48, roll on/roll off 3, specialized tanker 3
foreign-owned: 1 (Turkey 1)
registered in other countries: 2 (Malta 1, Panama 1) (2010)
Ports and terminals: Baku (Baki)

MILITARY

Military branches: Army, Navy, Air and Air Defense Forces (2010)
Military service age and obligation: men between 18 and 35 are liable for military service; 18 years of age for voluntary military service; length of military service is 18 months and 12 months for university graduates (2006)
Manpower available for military service:
males age 16–49: 2,354,249
females age 16–49: 2,334,632 (2010 est.)
Manpower fit for military service: *males age 16–49:* 1,773,993
females age 16–49: 1,964,012 (2010 est.)
Manpower reaching militarily significant age annually: *male:* 76,923
female: 71,024 (2010 est.)
Military expenditures: 2.6% of GDP (2005 est.)
country comparison to the world: 58

TRANSNATIONAL ISSUES

Disputes—international: Azerbaijan, Kazakhstan, and Russia ratified Caspian seabed delimitation treaties based on equidistance, while Iran continues to insist on a one-fifth slice of the lake; the dispute over the break-away Nagorno-Karabakh region and the Armenian military occupation of surrounding lands in Azerbaijan remains the primary focus of regional instability; residents have evacuated the former Soviet-era small ethnic enclaves in Armenia and Azerbaijan; local border forces struggle to control the illegal transit of goods and people across the porous, undemarcated Armenian, Azerbaijani, and Georgian borders; bilateral talks continue with Turkmenistan on dividing the seabed and contested oilfields in the middle of the Caspian
Refugees and internally displaced persons: refugees (country of origin): 2,400 (Russia)
IDPs: 580,000-690,000 (conflict with Armenia over Nagorno-Karabakh) (2007)
Trafficking in persons: current situation: Azerbaijan is primarily a source and transit country for men, women, and children trafficked for the purposes of commercial sexual exploitation and forced labor; women and some children from Azerbaijan are trafficked to Turkey and the UAE for the purpose of sexual exploitation; men and boys are trafficked to Russia for the purpose of forced labor; Azerbaijan serves as a transit country for victims from Uzbekistan, Kyrgyzstan, Kazakhstan, and Moldova trafficked to Turkey and the UAE for sexual exploitation
tier rating: Tier 2 Watch List–Azerbaijan is on the Tier 2 Watch List for its failure to provide evidence of increasing efforts to combat trafficking in persons, particularly efforts to investigate, prosecute, and punish traffickers; to address complicity among law enforcement personnel; and to adequately identify and protect victims in Azerbaijan; the government has yet to develop a much-needed mechanism to identify potential trafficking victims and refer them to safety and care; poor treatment of trafficking victims in courtrooms continues to be a problem (2008)
Illicit drugs: limited illicit cultivation of cannabis and opium poppy, mostly for CIS consumption; small government eradication program; transit point for Southwest Asian opiates bound for Russia and to a lesser extent the rest of Europe

BAHAMAS, THE

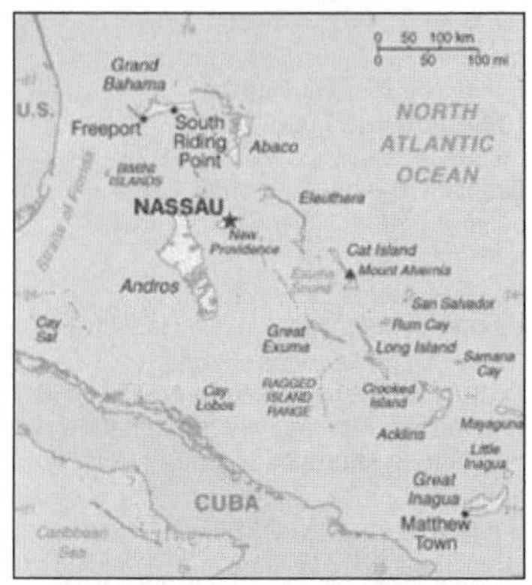

INTRODUCTION

Background: Lucayan Indians inhabited the islands when Christopher COLUMBUS first set foot in the New World on San Salvador in 1492. British settlement of the islands began in 1647; the islands became a colony in 1783. Since attaining independence from the UK in 1973, The Bahamas has prospered through tourism and international banking and investment management. Because of its geography, the country is a major transshipment point for illegal drugs, particularly shipments to the US and Europe, and its territory is used for smuggling illegal migrants into the US.

GEOGRAPHY

Location: Caribbean, chain of islands in the North Atlantic Ocean, southeast of Florida, northeast of Cuba
Geographic coordinates: 24 15 N, 76 00 W
Map references: Central America and the Caribbean
Area: *total:* 13,880 sq km
country comparison to the world: 160
land: 10,010 sq km
water: 3,870 sq km
Area—comparative: slightly smaller than Connecticut
Land boundaries: 0 km
Coastline: 3,542 km
Maritime claims: territorial sea: 12 nm
exclusive economic zone: 200 nm
Climate: tropical marine; moderated by warm waters of Gulf Stream
Terrain: long, flat coral formations with some low rounded hills
Elevation extremes: *lowest point:* Atlantic Ocean 0 m
highest point: Mount Alvernia on Cat Island 63 m
Natural resources: salt, aragonite, timber, arable land
Land use: *arable land:* 0.58%
permanent crops: 0.29%
other: 99.13% (2005)
Irrigated land: 10 sq km (2003)
Total renewable water resources: NA
Natural hazards: hurricanes and other tropical storms cause extensive flood and wind damage
Environment—current issues: coral reef decay; solid waste disposal
Environment—international agreements: *party to:* Biodiversity, Climate Change, Climate Change-Kyoto Protocol, Desertification, Endangered Species, Hazardous Wastes, Law of the Sea, Ozone Layer Protection, Ship Pollution, Wetlands
signed, but not ratified: none of the selected agreements
Geography—note: strategic location adjacent to US and Cuba; extensive island chain of which 30 are inhabited

PEOPLE

Population: 313,312 (July 2011 est.)
country comparison to the world: 177
note: estimates for this country explicitly take into account the effects of excess mortality due to AIDS; this can result in lower life expectancy, higher infant mortality, higher death rates, lower population growth rates, and changes in the distribution of population by age and sex than would otherwise be expected
Age structure: *0–14 years:* 24.4% (male 38,834/female 37,715)
15–64 years: 69.2% (male 106,882/female 110,081)
65 years and over: 6.3% (male 7,578/female 12,222) (2011 est.)
Median age: *total:* 30.2 years
male: 29.1 years
female: 31.3 years (2011 est.)
Population growth rate: 0.922% (2011 est.)
country comparison to the world: 123
Birth rate: 16.1 births/1,000 population (2011 est.)
country comparison to the world: 128
Death rate: 6.88 deaths/1,000 population (July 2011 est.)
country comparison to the world: 141
Net migration rate: 0 migrant(s)/1,000 population (2011 est.)
country comparison to the world: 74
Urbanization: *urban population:* 84% of total population (2010)
rate of urbanization: 1.3% annual rate of change (2010-15 est.)
Major cities—population: NASSAU (capital) 248,000 (2009)
Sex ratio: *at birth:* 1.03 male(s)/female
under 15 years: 1.03 male(s)/female
15–64 years: 0.97 male(s)/female
65 years and over: 0.62 male(s)/female
total population: 0.96 male(s)/female (2011 est.)
Infant mortality rate: *total:* 13.49 deaths/1,000 live births
country comparison to the world: 127
male: 13.29 deaths/1,000 live births
female: 13.69 deaths/1,000 live births (2011 est.)
Life expectancy at birth: *total population:* 71.18 years
country comparison to the world: 140
male: 68.8 years
female: 73.63 years (2011 est.)
Total fertility rate: 1.99 children born/woman (2011 est.)
country comparison to the world: 129
HIV/AIDS—adult prevalence rate: 3.1% (2009 est.)
country comparison to the world: 23
HIV/AIDS—people living with HIV/AIDS: 6,600 (2009 est.)
country comparison to the world: 115
HIV/AIDS—deaths: fewer than 500 (2009 est.)
country comparison to the world: 98
Drinking water source: *Improved:*
urban: 98% of population
rural: 86% of population
total: 96% of population
Unimproved:
urban: 2% of population
rural: 14% of population
total: 4% of population (2000)
Sanitation facility access: *Improved:*
urban: 100% of population
rural: 100% of population
total: 100% of population (2008)
Nationality: *noun:* Bahamian(s)
adjective: Bahamian
Ethnic groups: black 85%, white 12%, Asian and Hispanic 3%
Religions: Baptist 35.4%, Anglican 15.1%, Roman Catholic 13.5%, Pentecostal 8.1%, Church of God 4.8%, Methodist 4.2%, other Christian 15.2%, none or unspecified 2.9%, other 0.8% (2000 census)
Languages: English (official), Creole (among Haitian immigrants)
Literacy: *definition:* age 15 and over can read and write
total population: 95.6%
male: 94.7%
female: 96.5% (2003 est.)
School life expectancy (primary to tertiary education): *total:* 12 years
male: 12 years
female: 12 years (2006)
Education expenditures: NA

GOVERNMENT

Country name: *conventional long form:* Commonwealth of The Bahamas
conventional short form: The Bahamas
Government type: constitutional parliamentary democracy and a Commonwealth realm
Capital: *name:* Nassau
geographic coordinates: 25 05 N, 77 21 W
time difference: UTC-5 (same time as Washington, DC during Standard Time)
daylight saving time: +1hr, begins second Sunday in March; ends first Sunday in November
Administrative divisions: 31 districts; Acklins Islands, Berry Islands, Bimini, Black Point, Cat Island, Central Abaco, Central Andros, Central Eleuthera, City of Freeport, Crooked Island and Long Cay, East Grand Bahama, Exuma, Grand Cay, Harbour Island, Hope Town, Inagua, Long Island, Mangrove Cay, Mayaguana, Moore's Island, North Abaco, North Andros, North Eleuthera, Ragged Island, Rum Cay, San Salvador, South Abaco, South Andros, South Eleuthera, Spanish Wells, West Grand Bahama
Independence: 10 July 1973 (from the UK)
National holiday: Independence Day, 10 July (1973)
Constitution: 10 July 1973

Legal system: common law system based on the English model
International law organization participation: has not submitted an ICJ jurisdiction declaration; non-party state to the ICCt
Suffrage: 18 years of age; universal
Executive branch: *chief of state:* Queen ELIZABETH II (since 6 February 1952); represented by Governor General Sir Arthur A. FOULKES (since 14 April 2010)
head of government: Prime Minister Hubert A. INGRAHAM (since 4 May 2007)
cabinet: Cabinet appointed by the governor general on the prime minister's recommendation
(For more information visit the World Leaders website)
elections: the monarchy is hereditary; governor general appointed by the monarch; following legislative elections, the leader of the majority party or the leader of the majority coalition is usually appointed prime minister by the governor general; the prime minister recommends the deputy prime minister
Legislative branch: bicameral Parliament consists of the Senate (16 seats; members appointed by the governor general upon the advice of the prime minister and the opposition leader to serve five-year terms) and the House of Assembly (41 seats; members elected by direct popular vote to serve five-year terms); the government may dissolve the parliament and call elections at any time
elections: last held on 2 May 2007 (next to be held by May 2012)
election results: percent of vote by party–FNM 49.86%, PLP 47.02%; seats by party–FNM 23, PLP 18
Judicial branch: Privy Council in London; Courts of Appeal; Supreme (lower) Court; Magistrates' Courts
Political parties and leaders: Free National Movement or FNM [Hubert INGRAHAM]; Progressive Liberal Party or PLP [Perry CHRISTIE]
Political pressure groups and leaders: Friends of the Environment
other: trade unions
International organization participation: ACP, AOSIS, C, Caricom, CDB, FAO, G-77, IADB, IBRD, ICAO, ICRM, IDA, IFAD, IFC, IFRCS, ILO, IMF, IMO, IMSO, Interpol, IOC, IOM, ITSO, ITU, LAES, MIGA, NAM, OAS, OPANAL, OPCW, PetroCaribe, UN, UNCTAD, UNESCO, UNIDO, UNWTO, UPU, WCO, WHO, WIPO, WMO, WTO (observer)
Diplomatic representation in the US: *chief of mission:* Ambassador Cornelius A. SMITH
chancery: 2220 Massachusetts Avenue NW, Washington, DC 20008
telephone: [1] (202) 319-2660
FAX: [1] (202) 319-2668
consulate(s) general: Miami, New York
Diplomatic representation from the US: *chief of mission:* Ambassador Nicole A. AVANT
embassy: 42 Queen Street, Nassau, New Providence
mailing address: local or express mail address: P. O. Box N-8197, Nassau; US Department of State, 3370 Nassau Place, Washington, DC 20521-3370
telephone: [1] (242) 322-1181, 328-2206 (after hours)
FAX: [1] (242) 328-2206
Flag description: three equal horizontal bands of aquamarine (top), gold, and aquamarine, with a black equilateral triangle based on the hoist side; the band colors represent the golden beaches of the islands surrounded by the aquamarine sea; black represents the vigor and force of a united people, while the pointing triangle indicates the enterprise and determination of the Bahamian people to develop the rich resources of land and sea
National anthem: *name:* "March On, Bahamaland!"
lyrics/music: Timothy GIBSON
note: adopted 1973; as a Commonwealth country, in addition to the national anthem, "God Save the Queen" serves as the royal anthem (see United Kingdom)

ECONOMY

Economy—overview: The Bahamas is one of the wealthiest Caribbean countries with an economy heavily dependent on tourism and offshore banking. Tourism together with tourism-driven construction and manufacturing accounts for approximately 60% of GDP and directly or indirectly employs half of the archipelago's labor force. Prior to 2006, a steady growth in tourism receipts and a boom in construction of new hotels, resorts, and residences led to solid GDP growth but since then tourism receipts have begun to drop off. The global recession in 2009 took a sizeable toll on the Bahamas, resulting in a contraction in GDP and a widening budget deficit. The decline continued in 2010 as tourism from the US and sector investment lagged. Financial services constitute the second-most important sector of the Bahamian economy and, when combined with business services, account for about 36% of GDP. However, the financial sector currently is smaller than it has been in the past because of the enactment of new and stricter financial regulations in 2000 that caused many international businesses to relocate elsewhere. Manufacturing and agriculture combined contribute approximately a tenth of GDP and show little growth, despite government incentives aimed at those sectors. Overall growth prospects in the short run rest heavily on the fortunes of the tourism sector.
GDP (purchasing power parity): $8.921 billion (2010 est.)
country comparison to the world: 152
$8.877 billion (2009 est.)
$9.275 billion (2008 est.)
note: data are in 2010 US dollars
GDP (official exchange rate): $7.538 billion (2010 est.)
GDP—real growth rate: 0.5% (2010 est.)
country comparison to the world: 181
-4.3% (2009 est.)
-1.7% (2008 est.)
GDP—per capita (PPP): $28,700 (2010 est.)
country comparison to the world: 49
$28,900 (2009 est.)
$30,400 (2008 est.)
note: data are in 2010 US dollars
GDP—composition by sector: *agriculture:* 1.2%
industry: 14.7%
services: 84.1% (2001 est.)
Labor force: 184,000 (2009)
country comparison to the world: 173
Labor force—by occupation: *agriculture:* 5%
industry: 5%
tourism: 50%
other *services:* 40% (2005 est.)
Unemployment rate: 7.6% (2006 est.)
country comparison to the world: 81
Population below poverty line: 9.3% (2004)
Household income or consumption by percentage share: *lowest 10%:* NA%
highest 10%: 27% (2000)
Budget: *revenues:* $1.03 billion
expenditures: $1.03 billion (FY04/05)
Inflation rate (consumer prices): 2.4% (2007 est.)
country comparison to the world: 61
Central bank discount rate: 5.25% (31 December 2009)
country comparison to the world: 78
5.25% (31 December 2008)
Commercial bank prime lending rate: 5.5% (31 December 2009 est.)
country comparison to the world: 139
5.5% (31 December 2008 est.)
Stock of narrow money: $1.284 billion (31 December 2009)
country comparison to the world: 131
$1.275 billion (31 December 2008)
Stock of broad money: $5.991 billion (31 December 2009 est.)
country comparison to the world: 114
$5.893 billion (31 December 2008 est.)
Stock of domestic credit: $7.993 billion (31 December 2009)
country comparison to the world: 101
$7.883 billion (31 December 2008)
Market value of publicly traded shares: $NA
Agriculture—products: citrus, vegetables; poultry
Industries: tourism, banking, cement, oil transshipment, salt, rum, aragonite, pharmaceuticals, spiral-welded steel pipe
Industrial production growth rate: NA%
Electricity—production: 2.045 billion kWh (2007 est.)
country comparison to the world: 133
Electricity—consumption: 1.902 billion kWh (2007 est.)
country comparison to the world: 137
Electricity—exports: 0 kWh (2008 est.)
Electricity—imports: 0 kWh (2008 est.)
Oil—production: 0 bbl/day (2009 est.)
country comparison to the world: 152
Oil—consumption: 36,000 bbl/day (2009 est.)
country comparison to the world: 108
Oil—exports: transshipments of 41,570 bbl/day (2007 est.)
country comparison to the world: 80
Oil—imports: 20,560 bbl/day (2009 est.)
country comparison to the world: 110
Oil—proved reserves: 0 bbl (1 January 2010 est.)
country comparison to the world: 108
Natural gas—production: 0 cu m (2008 est.)
country comparison to the world: 100
Natural gas—consumption: 0 cu m (2008 est.)

country comparison to the world: 148
Natural gas—exports: 0 cu m (2008 est.)
country comparison to the world: 59
Natural gas—imports: 0 cu m (2008 est.)
country comparison to the world: 151
Natural gas—proved reserves: 0 cu m (1 January 2009 est.)
country comparison to the world: 111
Current account balance: $-283.2 million (2009 est.)
country comparison to the world: 96
$-1.442 billion (2007 est.)
Exports: $674 million (2006)
country comparison to the world: 162
Exports—commodities: mineral products and salt, animal products, rum, chemicals, fruit and vegetables
Exports—partners: US 35.99%, Singapore 18.64%, Poland 12.1%, Germany 6.24% (2009)
Imports: $2.401 billion (2006)
country comparison to the world: 148
Imports—commodities: machinery and transport equipment, manufactures, chemicals, mineral fuels; food and live animals
Imports—partners: US 27.23%, South Korea 20.08%, Japan 14.55%, Singapore 5.89%, China 4.75%, Venezuela 4.26%, Italy 4.12% (2009)
Debt—external: $342.6 million (2004 est.)
country comparison to the world: 168
Exchange rates: Bahamian dollars (BSD) per US dollar–1 (2009), 1 (2008), 1 (2007), 1 (2006)

COMMUNICATIONS

Telephones—main lines in use: 129,000 (2009)
country comparison to the world: 141
Telephones—mobile cellular: 358,800 (2009)
country comparison to the world: 166
Telephone system: *general assessment:* modern facilities
domestic: totally automatic system; highly developed; the Bahamas Domestic Submarine Network links 14 of the islands and is designed to satisfy increasing demand for voice and broadband Internet services
international: country code–1-242; landing point for the Americas Region Caribbean Ring System (ARCOS-1) fiber-optic submarine cable that provides links to South and Central America, parts of the Caribbean, and the US; satellite earth stations–2 (2007)
Broadcast media: 2 television stations operated by government-owned, commercially run Broadcasting Corporation of the Bahamas (BCB); multi-channel cable TV subscription service is available; about 15 radio stations operating with BCB operating a multi-channel radio broadcasting network alongside privately-owned radio stations (2007)
Internet country code: .bs
Internet hosts: 21,939 (2010)
country comparison to the world: 107
Internet users: 115,800 (2009)
country comparison to the world: 156

TRANSPORTATION

Airports: 62 (2010)
country comparison to the world: 78
Airports—with paved runways: *total:* 23
over 3,047 m: 2
2,438 to 3,047 m: 3
1,524 to 2,437 m: 13
914 to 1,523 m: 5 (2010)
Airports—with unpaved runways:
total: 39
1,524 to 2,437 m: 5
914 to 1,523 m: 12
under 914 m: 22 (2010)
Heliports: 1 (2010)
Roadways: *total:* 2,717 km
country comparison to the world: 168
paved: 1,560 km
unpaved: 1,157 km (2002)
Merchant marine: *total:* 1,170
country comparison to the world: 10
by type: barge carrier 1, bulk carrier 229, cargo 191, carrier 2, chemical tanker 80, combination ore/oil 8, container 50, liquefied gas 78, passenger 100, passenger/cargo 29, petroleum tanker 222, refrigerated cargo 106, roll on/roll off 12, specialized tanker 2, vehicle carrier 60
foreign-owned: 1,080 (Angola 5, Belgium 9, Bermuda 12, Brazil 1, Canada 102, China 4, Croatia 1, Cyprus 14, Denmark 59, Finland 8, France 19, Germany 39, Greece 209, Guernsey 6, Hong Kong 2, Indonesia 2, Ireland 3, Italy 5, Japan 93, Jordan 2, Kuwait 2, Malaysia 13, Monaco 14, Montenegro 2, Netherlands 22, Nigeria 2, Norway 198, Poland 32, Saudi Arabia 16, Singapore 7, Slovenia 1, Spain 9, Sweden 6, Switzerland 2, Thailand 4, Trinidad and Tobago 1, Turkey 3, UAE 27, UK 24, US 100)
note: this country allows large numbers of ships owned by foreign entities to be registered in its national shipping registry and to fly its flag; these ships operate under the laws of the flag state
registered in other countries: 10 (Bolivia 1, Malta 1, Panama 7, Peru 1) (2010)
Ports and terminals: Freeport, Nassau, South Riding Point

MILITARY

Military branches: Royal Bahamian Defense Force: Land Force, Navy, Air Wing (2011)
Military service age and obligation: 18 years of age; no conscription (2010)
Manpower available for military service:
males age 16–49: 85,568 (2010 est.)
Manpower fit for military service: *males age 16–49:* 63,429
females age 16–49: 64,645 (2010 est.)
Manpower reaching militarily significant age annually: *male:* 2,829
female: 2,750 (2010 est.)
Military expenditures: 0.7% of GDP (2009)
country comparison to the world: 151

TRANSNATIONAL ISSUES

Disputes—international: disagrees with the US on the alignment the northern axis of a potential maritime boundary
Illicit drugs: transshipment point for cocaine and marijuana bound for US and Europe; offshore financial center

BAHRAIN

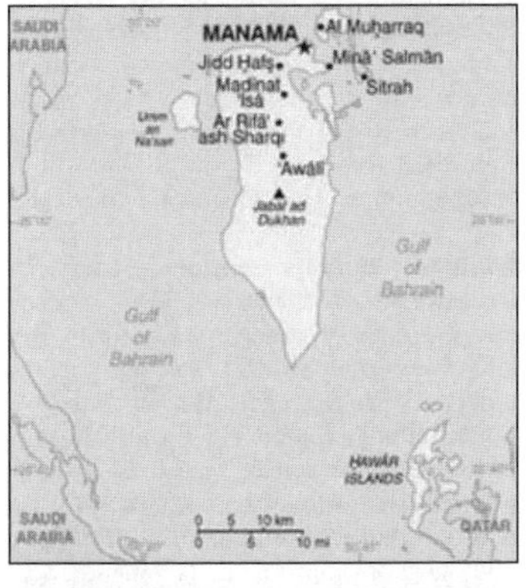

INTRODUCTION

Background: In 1783, the Sunni Al-Khalifa family captured Bahrain from the Persians. In order to secure these holdings, it entered into a series of treaties with the UK during the 19th century that made Bahrain a British protectorate. The archipelago attained its independence in 1971. Facing declining oil reserves, Bahrain has turned to petroleum processing and refining and has transformed itself into an international banking center. Bahrain's small size and central location among Persian Gulf countries require it to play a delicate balancing act in foreign affairs among its larger neighbors. In addition, the Sunni-led government has struggled to manage relations with its approximately 70% Shia-majority population. During the mid-to-late 1990s Shia activists mounted a low-intensity uprising to demand that the Sunni-led government stop systemic economic, social, and political discrimination against Shia Bahrainis. King HAMAD bin Isa Al-Khalifa, after succeeding his late father in 1999, pushed economic and political reforms in part to improve relations with the Shia community. After boycotting the country's first round of democratic elections under the newly-promulgated constitution in 2002, Shia political societies participated in 2006 and 2010 in legislative and municipal elections and Wafiq, the largest Shia political society, won the largest bloc of seats in the elected lower-house of the legislature both times. Nevertheless, Shia discontent has persisted, often manifesting itself in street demonstrations and occasional low-level violence.
In early 2011, Bahrain's fractious opposition sought to ride a rising tide of popular Arab protests to petition for the redress of popular grievances. In mid-February, on the tenth anniversary marking the King's initiation of his democratic reform initia-

tive, a vanguard of hardline activists–who rejected the legitimacy of the Al Khalifa regime and have sometimes instigated low-level violence–organized demonstrations in Shia neighborhoods demanding a new constitution, release of hundreds of Shia prisoners, and an end to discriminations in all sectors of society. Cycles of protestor deaths, funerals, and clashes with security forces ensued, escalating domestic tensions and leading Wifaq legislators to formally resign from the national legislature in protest in late-March 2011. Nearly a month of mostly peaceful opposition demonstrations followed before hardline elements within the Shia opposition undertook new provocative protests. The government's offers of modest political and economic concessions went and the king's "national dialogue" with the opposition–led by his son, the reform-minded Crown Prince–also languished in disagreements over procedure and preconditions. In mid-March 2011, with the backing of Gulf Cooperation Council (GCC) capitals–especially Riyadh and Abu Dhabi–King HAMAD put an end to the mass public gatherings and increasingly disruptive demonstrations by declaring a state of emergency and authorizing the military to take all measures to "protect the safety of the country and its citizens." Manama also welcomed a contingent of mostly Saudi and Emirati forces as part of a GCC deployment intended to help Bahraini security forces maintain order. By mid-April security forces had largely relegated demonstrations to outlying Shia neighborhoods and villages, and negotiations between the government and opposition reached a stalemate. Manama exacted retribution against opposition groups and their supporters through mass firings, arrests, and sectarian incitement.

GEOGRAPHY

Location: Middle East, archipelago in the Persian Gulf, east of Saudi Arabia
Geographic coordinates: 26 00 N, 50 33 E
Map references: Middle East
Area: *total:* 760 sq km
country comparison to the world: 187
land: 760 sq km
water: 0 sq km
Area—comparative: 3.5 times the size of Washington, DC
Land boundaries: 0 km
Coastline: 161 km
Maritime claims: territorial sea: 12 nm
contiguous zone: 24 nm
continental shelf: extending to boundaries to be determined
Climate: arid; mild, pleasant winters; very hot, humid summers
Terrain: mostly low desert plain rising gently to low central escarpment
Elevation extremes: *lowest point:* Persian Gulf 0 m
highest point: Jabal ad Dukhan 122 m
Natural resources: oil, associated and nonassociated natural gas, fish, pearls
Land use: *arable land:* 2.82%
permanent crops: 5.63%
other: 91.55% (2005)
Irrigated land: 40 sq km (2003)
Total renewable water resources: 0.1 cu km (1997)
Freshwater withdrawal (domestic/industrial/agricultural): *total:* 0.3 cu km/yr (40%/3%/57%)
per capita: 411 cu m/yr (2000)
Natural hazards: periodic droughts; dust storms
Environment—current issues: desertification resulting from the degradation of limited arable land, periods of drought, and dust storms; coastal degradation (damage to coastlines, coral reefs, and sea vegetation) resulting from oil spills and other discharges from large tankers, oil refineries, and distribution stations; lack of freshwater resources (groundwater and seawater are the only sources for all water needs)
Environment—international agreements: *party to:* Biodiversity, Climate Change, Climate Change-Kyoto Protocol, Desertification, Hazardous Wastes, Law of the Sea, Ozone Layer Protection, Wetlands
signed, but not ratified: none of the selected agreements
Geography—note: close to primary Middle Eastern petroleum sources; strategic location in Persian Gulf, through which much of the Western world's petroleum must transit to reach open ocean

PEOPLE

Population: 1,214,705
country comparison to the world: 156
note: includes 235,108 non-nationals (July 2011 est.)
Age structure: *0–14 years:* 20.5% (male 126,313/female 122,359)
15–64 years: 77% (male 595,244/female 339,635)
65 years and over: 2.6% (male 14,791/female 16,363) (2011 est.)
Median age: *total:* 30.9 years
male: 32.2 years
female: 28.1 years (2011 est.)
Population growth rate: 2.814% (2011 est.)
country comparison to the world: 15
Birth rate: 14.64 births/1,000 population (2011 est.)
country comparison to the world: 137
Death rate: 2.61 deaths/1,000 population (July 2011 est.)
country comparison to the world: 221
Net migration rate: 16.1 migrant(s)/1,000 population (2011 est.)
country comparison to the world: 4
Urbanization: *urban population:* 89% of total population (2010)
rate of urbanization: 1.8% annual rate of change (2010-15 est.)
Major cities—population: MANAMA (capital) 163,000 (2009)
Sex ratio: *at birth:* 1.028 male(s)/female
under 15 years: 1.02 male(s)/female
15–64 years: 1.33 male(s)/female
65 years and over: 1.13 male(s)/female
total population: 1.24 male(s)/female (2011 est.)
Infant mortality rate: *total:* 10.43 deaths/1,000 live births
country comparison to the world: 146
male: 11.68 deaths/1,000 live births
female: 9.14 deaths/1,000 live births (2011 est.)
Life expectancy at birth: *total population:* 78.15 years
country comparison to the world: 52
male: 76.03 years
female: 80.33 years (2011 est.)
Total fertility rate: 1.88 children born/woman (2011 est.)
country comparison to the world: 143
HIV/AIDS—adult prevalence rate: 0.2% (2001 est.)
country comparison to the world: 93
HIV/AIDS—people living with HIV/AIDS: fewer than 600 (2007 est.)
country comparison to the world: 149
HIV/AIDS—deaths: fewer than 200 (2003 est.)
country comparison to the world: 107
Sanitation facility access: *Improved:*
urban: 100% of population
rural: 100% of population
total: 100% of population (2008)
Nationality: *noun:* Bahraini(s)
adjective: Bahraini
Ethnic groups: Bahraini 62.4%, non-Bahraini 37.6% (2001 census)
Religions: Muslim (Shia and Sunni) 81.2%, Christian 9%, other 9.8% (2001 census)
Languages: Arabic (official), English, Farsi, Urdu
Literacy: *definition:* age 15 and over can read and write
total population: 86.5%
male: 88.6%
female: 83.6% (2001 census)
School life expectancy (primary to tertiary education): *total:* 14 years
male: 13 years
female: 14 years (2006)
Education expenditures: 2.9% of GDP (2008)
country comparison to the world: 136

GOVERNMENT

Country name: *conventional long form:* Kingdom of Bahrain
conventional short form: Bahrain
local long form: Mamlakat al Bahrayn
local short form: Al Bahrayn
former: Dilmun, State of Bahrain
Government type: constitutional monarchy
Capital: *name:* Manama
geographic coordinates: 26 14 N, 50 34 E
time difference: UTC+3 (8 hours ahead of Washington, DC during Standard Time)
Administrative divisions: 5 governorates; Asamah, Janubiyah, Muharraq, Shamaliyah, Wasat
note: each governorate administered by an appointed governor
Independence: 15 August 1971 (from the UK)
National holiday: National Day, 16 December (1971); note–15 August 1971 was the date of independence from the UK, 16 December 1971 was the date of independence from British protection
Constitution: adopted 14 February 2002
Legal system: mixed legal system of Islamic law and English common law
International law organization participation: has not submitted an ICJ jurisdiction declaration; non-party state to the ICCt
Suffrage: 20 years of age; universal; note–Bahraini Cabinet in May 2011 endorsed a draft law lowering eligibility to 18 years

Executive branch: *chief of state:* King HAMAD bin Isa Al-Khalifa (since 6 March 1999); Heir Apparent Crown Prince SALMAN bin Hamad Al-Khalifa (son of the monarch, born 21 October 1969)
head of government: Prime Minister KHALIFA bin Salman Al-Khalifa (since 1971); Deputy Prime Ministers ALI bin Khalifa bin Salman Al-Khalifa, MUHAMMAD bin Mubarak Al-Khalifa, Jawad bin Salim al-ARAIDH
cabinet: Cabinet appointed by the monarch (For more information visit the World Leaders website)
elections: the monarchy is hereditary; prime minister appointed by the monarch
Legislative branch: bicameral legislature consists of the Consultative Council (40 members appointed by the King) and the Council of Representatives or Chamber of Deputies (40 seats; members directly elected to serve four-year terms)
elections: Council of Representatives—last held in two rounds on 23 and 30 October 2010 (next election to be held in 2014)
election results: Council of Representatives—percent of vote by society—NA; seats by society—Wifaq (Shia) 18, Asala (Sunni Salafi) 3, Minbar (Sunni Muslim Brotherhood) 2, independents 17
Judicial branch: High Civil Appeals Court
Political parties and leaders: political parties prohibited but political societies were legalized per a July 2005 law
Political pressure groups and leaders: Shia activists; Sunni Islamist legislators
other: several small leftist and other groups are active
International organization participation: ABEDA, AFESD, AMF, CICA, FAO, G-77, GCC, IAEA, IBRD, ICAO, ICC, ICRM, IDB, IFC, IFRCS, IHO, ILO, IMF, IMO, IMSO, Interpol, IOC, IOM (observer), IPU, ISO, ITSO, ITU, ITUC, LAS, MIGA, NAM, OAPEC, OIC, OPCW, PCA, UN, UNCTAD, UNESCO, UNIDO, UNWTO, UPU, WCO, WFTU, WHO, WIPO, WMO, WTO
Diplomatic representation in the US: *chief of mission:* Ambassador Huda Azra Ibrahim NUNU
chancery: 3502 International Drive NW, Washington, DC 20008
telephone: [1] (202) 342-1111
FAX: [1] (202) 362-2192
consulate(s) general: New York
Diplomatic representation from the US: *chief of mission:* Ambassador (vacant); Charge d'Affaires Stephanie WILLIAMS
embassy: Building #979, Road 3119 (next to Al-Ahli Sports Club), Block 331, Zinj District, Manama
mailing address: PSC 451, Box 660, FPO AE 09834-5100; international mail: American Embassy, Box 26431, Manama
telephone: [973] 1724-2700
FAX: [973] 1727-0547
Flag description: red, the traditional color for flags of Persian Gulf states, with a white serrated band (five white points) on the hoist side; the five points represent the five pillars of Islam
note: until 2002 the flag had eight white points, but this was reduced to five to avoid confusion with the Qatari flag
National anthem: *name:* "Bahrainona" (Our Bahrain)
lyrics/music: unknown
note: adopted 1971; although Mohamed Sudqi AYYASH wrote the original lyrics, they were changed in 2002 following the transformation of Bahrain from an emirate to a kingdom

ECONOMY

Economy—overview: Bahrain is one of the most diversified economies in the Persian Gulf. Highly developed communication and transport facilities make Bahrain home to numerous multinational firms with business in the Gulf. As part of its diversification plans, Bahrain implemented a Free Trade Agreement (FTA) with the US in August 2006, the first FTA between the US and a Gulf state. Bahrain's economy, however, continues to depend heavily on oil. Petroleum production and refining account for more than 60% of Bahrain's export receipts, 70% of government revenues, and 11% of GDP (exclusive of allied industries). Other major economic activities are production of aluminum—Bahrain's second biggest export after oil—finance, and construction. Bahrain competes with Malaysia as a worldwide center for Islamic banking and continues to seek new natural gas supplies as feedstock to support its expanding petrochemical and aluminum industries. Unemployment, especially among the young, is a long-term economic problem Bahrain struggles to address. In 2009, to help lower unemployment among Bahraini nationals, Bahrain reduced sponsorship for expatriate workers, increasing the costs of employing foreign labor. The global financial crisis caused funding for many non-oil projects to dry up and resulted in slower economic growth for Bahrain. Other challenges facing Bahrain include the slow growth of government debt as a result of a large subsidy program, the financing of large government projects, and debt restructuring, such as the bailout of state-owned Gulf Air.
GDP (purchasing power parity): $29.71 billion (2010 est.)
country comparison to the world: 109
$28.55 billion (2009 est.)
$27.69 billion (2008 est.)
note: data are in 2010 US dollars
GDP (official exchange rate): $22.66 billion (2010 est.)
GDP—real growth rate: 4.1% (2010 est.)
country comparison to the world: 95
3.1% (2009 est.)
6.3% (2008 est.)
GDP—per capita (PPP): $40,300 (2010 est.)
country comparison to the world: 21
$39,200 (2009 est.)
$38,500 (2008 est.)
note: data are in 2010 US dollars
GDP—composition by sector:
agriculture: 0.5%
industry: 56.6%
services: 42.9% (2010 est.)
Labor force: 611,000
country comparison to the world: 154
note: 44% of the population in the 15-64 age group is non-national (2010 est.)
Labor force—by occupation:
agriculture: 1%
industry: 79%
services: 20% (1997 est.)
Unemployment rate: 15% (2005 est.)
country comparison to the world: 150
Population below poverty line: NA%
Household income or consumption by percentage share:
lowest 10%: NA%
highest 10%: NA%
Investment (gross fixed): 26.6% of GDP (2010 est.)
country comparison to the world: 33
Budget: *revenues:* $5.933 billion
expenditures: $5.948 billion (2010 est.)
Public debt:
59.2% of GDP (2010 est.)
country comparison to the world: 36
38.5% of GDP (2009 est.)
Inflation rate (consumer prices): 3.3% (2010 est.)
country comparison to the world: 91
2.8% (2009 est.)
Commercial bank prime lending rate: NA% (31 December 2009 est.)
NA% (31 December 2008 est.)
Stock of narrow money: $6.372 billion (31 December 2010 est.)
country comparison to the world: 81
$5.74 billion (31 December 2009 est.)
Stock of broad money: $21.02 billion (31 December 2010 est.)
country comparison to the world: 82
$18.93 billion (31 December 2009 est.)
Stock of domestic credit: $18.46 billion (31 December 2010 est.)
country comparison to the world: 82
$16.34 billion (31 December 2009 est.)
Market value of publicly traded shares: $16.93 billion (31 December 2009)
country comparison to the world: 62
$21.18 billion (31 December 2008)
$28.13 billion (31 December 2007)
Agriculture—products: fruit, vegetables; poultry, dairy products; shrimp, fish
Industries: petroleum processing and refining, aluminum smelting, iron pelletization, fertilizers, Islamic and offshore banking, insurance, ship repairing, tourism
Industrial production growth rate: 1.5% (2010 est.)
country comparison to the world: 138
Electricity—production: 10.25 billion kWh (2007 est.)
country comparison to the world: 89
Electricity—consumption: 10.1 billion kWh (2007 est.)
country comparison to the world: 87
Electricity—exports: 0 kWh (2008 est.)
Electricity—imports: 0 kWh (2008 est.)
Oil—production: 48,560 bbl/day (2009 est.)
country comparison to the world: 63
Oil—consumption: 39,000 bbl/day (2009 est.)
country comparison to the world: 103
Oil—exports: 238,300 bbl/day (2007 est.)
country comparison to the world: 50
Oil—imports: 228,400 bbl/day (2007 est.)
country comparison to the world: 40
Oil—proved reserves: 124.6 million bbl (1 January 2010 est.)
country comparison to the world: 67

Natural gas—production: 12.64 billion cu m (2008 est.)
country comparison to the world: 38
Natural gas—consumption: 12.64 billion cu m (2008 est.)
country comparison to the world: 43
Natural gas—exports: 0 cu m (2008 est.)
country comparison to the world: 54
Natural gas—imports: 0 cu m (2008 est.)
country comparison to the world: 147
Natural gas—proved reserves: 92.03 billion cu m (1 January 2010 est.)
country comparison to the world: 54
Current account balance: $589 million (2010 est.)
country comparison to the world: 52
$560.2 million (2009 est.)
Exports: $15.13 billion (2010 est.)
country comparison to the world: 75
$12.05 billion (2009 est.)
Exports—commodities: petroleum and petroleum products, aluminum, textiles
Exports—partners: India 4.2%, Saudi Arabia 2.78% (2009)
Imports: $12.14 billion (2010 est.)
country comparison to the world: 85
$9.613 billion (2009 est.)
Imports—commodities: crude oil, machinery, chemicals
Imports—partners: Saudi Arabia 22.91%, France 9.76%, US 7.95%, China 6.4%, South Korea 5.26%, Japan 5.19%, Germany 5.01%, UK 4.34% (2009)
Reserves of foreign exchange and gold: $3.766 billion (31 December 2010 est.)
country comparison to the world: 83
$3.54 billion (31 December 2009 est.)
Debt—external: $14.68 billion (31 December 2010 est.)
country comparison to the world: 78
$10.55 billion (31 December 2009 est.)
Stock of direct foreign investment—at home: $15.77 billion (31 December 2010 est.)
country comparison to the world: 74
$15 billion (31 December 2009 est.)
Stock of direct foreign investment—abroad: $8.399 billion (31 December 2010 est.)
country comparison to the world: 51
$7.549 billion (31 December 2009 est.)
Exchange rates: Bahraini dinars (BHD) per US dollar–0.376 (2010), 0.376 (2009), 0.376 (2008), 0.376 (2007), 0.376 (2006)

COMMUNICATIONS

Telephones—main lines in use: 238,400 (2009)
country comparison to the world: 124
Telephones—mobile cellular: 1.578 million (2009)
country comparison to the world: 137
Telephone system: *general assessment:* modern system
domestic: modern fiber-optic integrated services; digital network with rapidly growing use of mobile-cellular telephones
international: country code–973; landing point for the Fiber-Optic Link Around the Globe (FLAG) submarine cable network that provides links to Asia, Middle East, Europe, and US; tropospheric scatter to Qatar and UAE; microwave radio relay to Saudi Arabia; satellite earth station–1 (2007)
Broadcast media: state-run broadcast media; Bahrain Radio and Television Corporation (BRTC) operates 5 terrestrial TV networks; satellite TV systems provide access to international broadcasts; state-run BRTC broadcasts over several radio stations; 1 private FM station directs broadcasts to Indian listeners; radio and TV broadcasts from countries in the region are available (2007)
Internet country code: .bh
Internet hosts: 53,944 (2010)
country comparison to the world: 86
Internet users: 419,500 (2009)
country comparison to the world: 122

TRANSPORTATION

Airports: 4 (2010)
country comparison to the world: 183
Airports—with paved runways: *total:* 4
over 3,047 m: 3
2,438 to 3,047 m: 1 (2010)
Heliports: 1 (2010)
Pipelines: gas 20 km; oil 29 km (2010)
Roadways: *total:* 3,851 km
country comparison to the world: 158
paved: 3,121 km
unpaved: 730 km (2007)
Merchant marine: *total:* 7
country comparison to the world: 125
by type: bulk carrier 2, container 4, petroleum tanker 1
foreign-owned: 5 (Kuwait 5)
registered in other countries: 6 (Honduras 5, Saint Kitts and Nevis 1) (2010)
Ports and terminals: Mina' Salman, Sitrah

MILITARY

Military branches: Bahrain Defense Forces (BDF): Ground Force (includes Air Defense Force), Royal Bahrain Navy (RBN), Royal Bahraini Air Force (RBAF), Bahrain National Guard (BNG) (2011)
Military service age and obligation: 17 years of age for voluntary military service; 15 years of age for NCOs, technicians, and cadets; no conscription (2010)
Manpower available for military service: *males age 16–49:* 508,863
females age 16–49: 290,801 (2010 est.)
Manpower fit for military service: *males age 16–49:* 423,757
females age 16–49: 245,302 (2010 est.)
Manpower reaching militarily significant age annually: *male:* 8,988
female: 8,117 (2010 est.)
Military expenditures: 4.5% of GDP (2006)
country comparison to the world: 21

TRANSNATIONAL ISSUES

Disputes—international: none

BANGLADESH

INTRODUCTION

Background: Europeans began to set up trading posts in the area of Bangladesh in the 16th century; eventually the British came to dominate the region and it became part of British India. In 1947, West Pakistan and East Bengal (both primarily Muslim) separated from India (largely Hindu) and jointly became the new country of Pakistan. East Bengal became East Pakistan in 1955, but the awkward arrangement of a two-part country with its territorial units separated by 1,600 km left the Bengalis marginalized and dissatisfied. East Pakistan seceded from its union with West Pakistan in 1971 and was renamed Bangladesh. A military-backed, emergency caretaker regime suspended parliamentary elections planned for January 2007 in an effort to reform the political system and root out corruption. In contrast to the strikes and violent street rallies that had marked Bangladeshi politics in previous years, the parliamentary elections finally held in late December 2008 were mostly peaceful and Sheikh HASINA Wajed was elected prime minister. About a third of this extremely poor country floods annually during the monsoon rainy season, hampering economic development.

GEOGRAPHY

Location: Southern Asia, bordering the Bay of Bengal, between Burma and India
Geographic coordinates: 24 00 N, 90 00 E
Map references: Asia
Area: *total:* 143,998 sq km
country comparison to the world: 94
land: 130,168 sq km
water: 13,830 sq km
Area—comparative: slightly smaller than Iowa
Land boundaries: *total:* 4,246 km
border countries: Burma 193 km, India 4,053 km
Coastline: 580 km
Maritime claims: territorial sea: 12 nm
contiguous zone: 18 nm
exclusive economic zone: 200 nm
continental shelf: up to the outer limits of the continental margin

Climate: tropical; mild winter (October to March); hot, humid summer (March to June); humid, warm rainy monsoon (June to October)
Terrain: mostly flat alluvial plain; hilly in southeast
Elevation extremes: *lowest point:* Indian Ocean 0 m
highest point: Keokradong 1,230 m
Natural resources: natural gas, arable land, timber, coal
Land use: *arable land:* 55.39%
permanent crops: 3.08%
other: 41.53% (2005)
Irrigated land: 47,250 sq km (2003)
Total renewable water resources: 1,210.6 cu km (1999)
Freshwater withdrawal (domestic/industrial/agricultural): *total:* 79.4 cu km/yr (3%/1%/96%)
per capita: 560 cu m/yr (2000)
Natural hazards: droughts; cyclones; much of the country routinely inundated during the summer monsoon season
Environment—current issues: many people are landless and forced to live on and cultivate flood-prone land; water-borne diseases prevalent in surface water; water pollution, especially of fishing areas, results from the use of commercial pesticides; ground water contaminated by naturally occurring arsenic; intermittent water shortages because of falling water tables in the northern and central parts of the country; soil degradation and erosion; deforestation; severe overpopulation
Environment—international agreements: *party to:* Biodiversity, Climate Change, Climate Change-Kyoto Protocol, Desertification, Endangered Species, Environmental Modification, Hazardous Wastes, Law of the Sea, Ozone Layer Protection, Ship Pollution, Wetlands
signed, but not ratified: none of the selected agreements
Geography—note: most of the country is situated on deltas of large rivers flowing from the Himalayas: the Ganges unites with the Jamuna (main channel of the Brahmaputra) and later joins the Meghna to eventually empty into the Bay of Bengal

PEOPLE

Population: 158,570,535 (July 2011 est.)
country comparison to the world: 7
Age structure: *0–14 years:* 34.3% (male 27,551,594/female 26,776,647)
15–64 years: 61.1% (male 45,956,431/female 50,891,519)
65 years and over: 4.7% (male 3,616,225/female 3,778,119) (2011 est.)
Median age: *total:* 23.3 years
male: 22.7 years
female: 23.7 years (2011 est.)
Population growth rate: 1.566% (2011 est.)
country comparison to the world: 76
Birth rate: 22.98 births/1,000 population (2011 est.)
country comparison to the world: 72
Death rate: 5.75 deaths/1,000 population (July 2011 est.)
country comparison to the world: 172
Net migration rate: -1.57 migrant(s)/1,000 population (2011 est.)
country comparison to the world: 158
Urbanization: *urban population:* 28% of total population (2010)
rate of urbanization: 3.1% annual rate of change (2010-15 est.)
Major cities—population: DHAKA (capital) 14.251 million; Chittagong 4.816 million; Khulna 1.636 million; Rajshahi 853,000 (2009)
Sex ratio: *at birth:* 1.04 male(s)/female
under 15 years: 1.01 male(s)/female
15–64 years: 0.89 male(s)/female
65 years and over: 0.93 male(s)/female
total population: 0.93 male(s)/female (2011 est.)
Infant mortality rate:
total: 50.73 deaths/1,000 live births
country comparison to the world: 47
male: 53.23 deaths/1,000 live births
female: 48.13 deaths/1,000 live births (2011 est.)
Life expectancy at birth: *total population:* 69.75 years
country comparison to the world: 148
male: 67.93 years
female: 71.65 years (2011 est.)
Total fertility rate: 2.6 children born/woman (2011 est.)
country comparison to the world: 81
HIV/AIDS—adult prevalence rate: less than 0.1% (2009 est.)
country comparison to the world: 117
HIV/AIDS—people living with HIV/AIDS: 6,300 (2009 est.)
country comparison to the world: 117
HIV/AIDS—deaths: fewer than 200 (2009 est.)
country comparison to the world: 114
Major infectious diseases: *degree of risk:* high
food or waterborne diseases: bacterial and protozoal diarrhea, hepatitis A and E, and typhoid fever
vectorborne diseases: dengue fever and malaria are high risks in some locations
water contact disease: leptospirosis
animal contact disease: rabies
note: highly pathogenic H5N1 avian influenza has been identified in this country; it poses a negligible risk with extremely rare cases possible among US citizens who have close contact with birds (2009)
Drinking water source: *Improved:*
urban: 85% of population
rural: 78% of population
total: 80% of population
Unimproved: urban: 15% of population
rural: 22% of population
total: 20% of population (2008)
Sanitation facility access: *Improved:*
urban: 56% of population
rural: 52% of population
total: 53% of population
Unimproved:
urban: 44% of population
rural: 48% of population
total: 47% of population (2008)
Nationality: *noun:* Bangladeshi(s)
adjective: Bangladeshi
Ethnic groups: Bengali 98%, other 2% (includes tribal groups, non-Bengali Muslims) (1998)
Religions: Muslim 89.5%, Hindu 9.6%, other 0.9% (2004)
Languages: Bangla (official, also known as Bengali), English
Literacy: *definition:* age 15 and over can read and write
total population: 47.9%
male: 54%
female: 41.4% (2001 Census)
School life expectancy (primary to tertiary education): *total:* 8 years
male: 8 years
female: 8 years (2007)
Education expenditures: 2.4% of GDP (2008)
country comparison to the world: 149

GOVERNMENT

Country name: *conventional long form:* People's Republic of Bangladesh
conventional short form: Bangladesh
local long form: Gana Prajatantri Bangladesh
local short form: former: East Bengal, East Pakistan
Government type: parliamentary democracy
Capital: *name:* Dhaka
geographic coordinates: 23 43 N, 90 24 E
time difference: UTC+6 (11 hours ahead of Washington, DC during Standard Time)
Administrative divisions: 7 divisions; Barisal, Chittagong, Dhaka, Khulna, Rajshahi, Rangpur, Sylhet
Independence: 16 December 1971 (from West Pakistan); note–26 March 1971 is the date of independence from West Pakistan, 16 December 1971 is known as Victory Day and commemorates the official creation of the state of Bangladesh
National holiday: Independence Day, 26 March (1971); note–26 March 1971 is the date of independence from West Pakistan, 16 December 1971 is Victory Day and commemorates the official creation of the state of Bangladesh
Constitution: 4 November 1972; effective 16 December 1972; suspended following coup of 24 March 1982; restored 10 November 1986; amended many times
Legal system: mixed legal system of mostly English common law and Islamic law
International law organization participation: has not submitted an ICJ jurisdiction declaration; accepts ICCt jurisdiction
Suffrage: 18 years of age; universal
Executive branch: *chief of state:* President Zillur RAHMAN (since 12 February 2009)
head of government: Prime Minister Sheikh HASINA Wajed (since 6 January 2009)
cabinet: Cabinet selected by the prime minister and appointed by the president (For more information visit the World Leaders website)
elections: president elected by National Parliament for a five-year term (eligible for a second term); last election held on 11 February 2009 (next to be held in 2014)
election results: Zillur RAHMAN declared president-elect by the Election Commission on 11 February 2009 (sworn in on 12 February); he ran unopposed as president; percent of National Parliament vote–NA
Legislative branch: unicameral National Parliament or Jatiya Sangsad; 300 seats (45 reserved for women) elected by popular vote from single territorial constituencies; members serve five-year terms

elections: last held on 29 December 2008 (next to be held in 2013)
election results: percent of vote by party–AL 49%, BNP 33.2%, JP 7%, JIB 4.6%, other 6.2%; seats by party–AL 230, BNP 30, JP 27, JIB 2, other 11
Judicial branch: Supreme Court (the chief justices and other judges are appointed by the president)
Political parties and leaders: Awami League or AL [Sheikh HASINA]; Communist Party of Bangladesh or CPB [Manjurul A. KHAN]; Bangladesh Nationalist Party or BNP [Khaleda ZIA]; Bikalpa Dhara Bangladesh or BDB [Badrudozza CHOWDHURY]; Islami Oikya Jote or IOJ [multiple leaders]; Jamaat-e-Islami Bangladesh or JIB [Matiur Rahman NIZAMI]; Jatiya Party or JP (Ershad faction) [Hussain Mohammad ERSHAD]; Liberal Democratic Party or LDP [Oli AHMED]
Political pressure groups and leaders: Advocacy to End Gender-based Violence through the MoWCA (Ministry of Women's and Children's Affairs)
other: environmentalists; Islamist groups; religious leaders; teachers; union leaders
International organization participation: ADB, ARF, BIMSTEC, C, CICA (observer), CP, D-8, FAO, G-77, IAEA, IBRD, ICAO, ICC, ICRM, IDA, IDB, IFAD, IFC, IFRCS, IHO, ILO, IMF, IMO, IMSO, Interpol, IOC, IOM, IPU, ISO, ITSO, ITU, ITUC, MIGA, MINURSO, MONUSCO, NAM, OIC, OPCW, SAARC, SACEP, UN, UNAMID, UNCTAD, UNESCO, UNHCR, UNIDO, UNIFIL, UNMIL, UNMIS, UNMIT, UNOCI, UNWTO, UPU, WCO, WFTU, WHO, WIPO, WMO, WTO
Diplomatic representation in the US: *chief of mission:* Ambassador Akramul QADER
chancery: 3510 International Drive NW, Washington, DC 20008
telephone: [1] (202) 244-0183
FAX: [1] (202) 244-7830/2771
consulate(s) general: Los Angeles, New York
Diplomatic representation from the US: *chief of mission:* Ambassador James F. MORIARTY
embassy: Madani Avenue, Baridhara, Dhaka 1212
mailing address: G. P. O. Box 323, Dhaka 1000
telephone: [880] (2) 885-5500
FAX: [880] (2) 882-3744
Flag description: green field with a large red disk shifted slightly to the hoist side of center; the red disk represents the rising sun and the sacrifice to achieve independence; the green field symbolizes the lush vegetation of Bangladesh
National anthem: *name:* "Amar Shonar Bangla" (My Golden Bengal)
lyrics/music: Rabindranath TAGORE
note: adopted 1971; Rabindranath TAGORE, a Nobel laureate, also wrote India's national anthem

ECONOMY

Economy—overview: The economy has grown 5-6% per year since 1996 despite political instability, poor infrastructure, corruption, insufficient power supplies, and slow implementation of economic reforms. Bangladesh remains a poor, overpopulated, and inefficiently-governed nation. Although more than half of GDP is generated through the service sector, 45% of Bangladeshis are employed in the agriculture sector, with rice as the single-most-important product. Bangladesh's growth was resilient during the 2008-09 global financial crisis and recession. Garment exports, totaling $12.3 billion in FY09 and remittances from overseas Bangladeshis, totaling $11 billion in FY10, accounted for almost 25% of GDP.
GDP (purchasing power parity): $258.6 billion (2010 est.)
country comparison to the world: 45
$243.9 billion (2009 est.)
$230.6 billion (2008 est.)
note: data are in 2010 US dollars
GDP (official exchange rate): $104.9 billion (2010 est.)
GDP—real growth rate: 6% (2010 est.)
country comparison to the world: 55
5.8% (2009 est.)
6% (2008 est.)
GDP—per capita (PPP): $1,700 (2010 est.)
country comparison to the world: 196
$1,600 (2009 est.)
$1,500 (2008 est.)
note: data are in 2010 US dollars
GDP—composition by sector: *agriculture:* 18.4%
industry: 28.7%
services: 52.9% (2010 est.)
Labor force: 73.87 million
country comparison to the world: 8
note: extensive export of labor to Saudi Arabia, Kuwait, UAE, Oman, Qatar, and Malaysia; workers' remittances were $10.9 billion in FY09/10 (2010 est.)
Labor force—by occupation:
agriculture: 45%
industry: 30%
services: 25% (2008)
Unemployment rate: 4.8% (2010 est.)
country comparison to the world: 44
5.1% (2009 est.)
note: about 40% of the population is underemployed; many participants in the labor force work only a few hours a week, at low wages
Population below poverty line: 40% (2010 est.)
Household income or consumption by percentage share: *lowest 10%:* NA
highest 10%: 26.6% (2008 est.)
Distribution of family income—Gini index: 33.2 (2005)
country comparison to the world: 95
33.6 (1996)
Investment (gross fixed): 24.4% of GDP (2010 est.)
country comparison to the world: 48
Budget: *revenues:* $11.43 billion
expenditures: $15.9 billion (2010 est.)
Public debt: 39.3% of GDP (2010 est.)
country comparison to the world: 75
39.7% of GDP (2009 est.)
Inflation rate (consumer prices): 8.1% (2010 est.)
country comparison to the world: 185
5.4% (2009 est.)
Central bank discount rate: 5% (31 October 2010)
country comparison to the world: 81
5% (31 December 2008)
Commercial bank prime lending rate: 11.18% (30 September 2010 est.)
country comparison to the world: 46
14.6% (31 December 2009 est.)
Stock of narrow money: $13.98 billion (31 December 2010 est.)
country comparison to the world: 67
$10.92 billion (31 December 2009 est.)
Stock of broad money: $57.21 billion (31 December 2010 est.)
country comparison to the world: 64
$63.03 billion (31 December 2009)
Stock of domestic credit: $62.2 billion (31 December 2010 est.)
country comparison to the world: 60
$53.77 billion (31 December 2009 est.)
Market value of publicly traded shares: $7.068 billion (31 December 2009)
country comparison to the world: 74
$6.671 billion (31 December 2008)
$6.793 billion (31 December 2007)
Agriculture—products: rice, jute, tea, wheat, sugarcane, potatoes, tobacco, pulses, oilseeds, spices, fruit; beef, milk, poultry
Industries: cotton textiles, jute, garments, tea processing, paper newsprint, cement, chemical fertilizer, light engineering, sugar
Industrial production growth rate: 6.4% (2010 est.)
country comparison to the world: 54
Electricity—production: 25.62 billion kWh (2009 est.)
country comparison to the world: 65
Electricity—consumption: 23.94 billion kWh (2009 est.)
country comparison to the world: 65
Electricity—exports: 0 kWh (2008 est.)
Electricity—imports: 0 kWh (2008 est.)
Oil—production: 5,733 bbl/day (2009 est.)
country comparison to the world: 93
Oil—consumption: 82,340 bbl/day (2010)
country comparison to the world: 84
Oil—exports: 2,612 bbl/day (2007 est.)
country comparison to the world: 108
Oil—imports: 77,340 bbl/day (2010 est.)
country comparison to the world: 76
Oil—proved reserves: 28 million bbl (1 January 2010 est.)
country comparison to the world: 83
Natural gas—production: 19.91 billion cu m (2010 est.)
country comparison to the world: 32
Natural gas—consumption: 20.1 billion cu m (2010 est.)
country comparison to the world: 34
Natural gas—exports: 0 cu m (2008 est.)
country comparison to the world: 60
Natural gas—imports: 0 cu m (2008 est.)
country comparison to the world: 152
Natural gas—proved reserves: 195.4 billion cu m (1 January 2010 est.)
country comparison to the world: 46
Current account balance: $3.734 billion (2010 est.)
country comparison to the world: 35
$2.416 billion (2009 est.)
Exports: $16.24 billion (2010 est.)
country comparison to the world: 73
$15.58 billion (2009 est.)

Exports—commodities: garments, frozen fish and seafood, jute and jute goods, leather
Exports—partners: US 22.5%, Germany 14.2%, UK 9.6%, France 7%, Netherlands 6.4% (2009)
Imports: $21.34 billion (2010 est.)
country comparison to the world: 68
$20.3 billion (2009 est.)
Imports—commodities: machinery and equipment, chemicals, iron and steel, textiles, foodstuffs, petroleum products, cement
Imports—partners: China 16.16%, India 12.61%, Singapore 7.55%, Japan 4.63%, Malaysia 4.46% (2009)
Reserves of foreign exchange and gold: $10.79 billion (31 December 2010 est.)
country comparison to the world: 57
$10.34 billion (31 December 2009 est.)
Debt—external: $24.46 billion (31 December 2010 est.)
country comparison to the world: 68
$24.22 billion (31 December 2009 est.)
Stock of direct foreign investment—at home: $6.72 billion (31 December 2010 est.)
country comparison to the world: 85
$5.617 billion (31 December 2009 est.)
Stock of direct foreign investment—abroad: $82 million (31 December 2010 est.)
country comparison to the world: 81
$81 million (31 December 2009 est.)
Exchange rates: taka (BDT) per US dollar–70.59 (2010), 69.04 (2009), 68.554 (2008), 69.893 (2007), 69.031 (2006)

COMMUNICATIONS

Telephones—main lines in use: 1.522 million (2009)
country comparison to the world: 64
Telephones—mobile cellular: 50.4 million (2009)
country comparison to the world: 24
Telephone system: *general assessment:* inadequate for a modern country; introducing digital systems; trunk systems include VHF and UHF microwave radio relay links, and some fiber-optic cable in cities
domestic: fixed-line teledensity remains only about 1 per 100 persons; mobile-cellular telephone subscribership has been increasing rapidly and now exceeds 30 telephones per 100 persons
international: country code–880; landing point for the SEA-ME-WE-4 fiber-optic submarine cable system that provides links to Europe, the Middle East, and Asia; satellite earth stations–6; international radiotelephone communications and landline service to neighboring countries (2009)
Broadcast media: state-owned broadcaster (BTV) operates 1 terrestrial TV station, 3 radio networks, and about 10 local stations; 8 private satellite TV stations and 3 private radio stations also broadcasting; foreign satellite TV stations are gaining audience share in the large cities; several international radio broadcasters are available (2007)
Internet country code: .bd
Internet hosts: 68,224 (2010)
country comparison to the world: 81
Internet users: 617,300 (2009)
country comparison to the world: 112

TRANSPORTATION

Airports: 17 (2010)
country comparison to the world: 141
Airports—with paved runways: *total:* 15
over 3,047 m: 2
2,438 to 3,047 m: 2
1,524 to 2,437 m: 6
914 to 1,523 m: 1
under 914 m: 4 (2010)
Airports—with unpaved runways:
total: 2
1,524 to 2,437 m: 1
under 914 m: 1 (2010)
Pipelines: gas 2,714 km (2010)
Railways: *total:* 2,622 km
country comparison to the world: 63
broad gauge: 946 km 1.676-m gauge
narrow gauge: 1,676 km 1.000-m gauge (2010)
Roadways: *total:* 239,226 km
country comparison to the world: 21
paved: 22,726 km
unpaved: 216,500 km (2003)
Waterways: 8,370 km (includes up to 3,060 km of main cargo routes; the network is reduced to 5,200 km in the dry season) (2007)
country comparison to the world: 17
Merchant marine: *total:* 50
country comparison to the world: 70
by type: bulk carrier 16, cargo 25, container 5, petroleum tanker 4
foreign-owned: 4 (China 1, Singapore 3)
registered in other countries: 9 (Comoros 1, Malta 1, Panama 3, Saint Vincent and the Grenadines 1, Sierra Leone 1, Singapore 2) (2010)
Ports and terminals: Chittagong, Mongla Port
Transportation—note: the International Maritime Bureau reports the territorial waters of Bangladesh remain a high risk for armed robbery against ships; attacks against vessels increased in 2010 for the second consecutive year; 23 commercial vessels were attacked both at anchor and while underway; crews were robbed and stores or cargoes stolen

MILITARY

Military branches: Bangladesh Defense Force: Bangladesh Army (Sena Bahini), Bangladesh Navy (Noh Bahini, BN), Bangladesh Air Force (Biman Bahini, BAF) (2010)
Military service age and obligation: 16 years of age for voluntary enlisted military service (Air Force); 17 years of age (Army and Navy); conscription is by law possible in times of emergency, but has never been implemented (2010)
Manpower available for military service:
males age 16–49: 36,520,491 (2010 est.)
Manpower fit for military service: *males age 16–49:* 30,486,086
females age 16–49: 35,616,093 (2010 est.)
Manpower reaching militarily significant age annually: *male:* 1,606,963
female: 1,689,442 (2010 est.)
Military expenditures: 1.3% of GDP (2009)
country comparison to the world: 113

TRANSNATIONAL ISSUES

Disputes—international: Bangladesh referred its maritime boundary claims with Burma and India to the International Tribunal on the Law of the Sea; discussions with India remain stalled to delimit a small section of river boundary, exchange territory for 51 small Bangladeshi exclaves in India and 111 small Indian exclaves in Bangladesh, allocate divided villages, and stop illegal cross-border trade, migration, violence, and transit of terrorists through the porous border; Bangladesh protests India's fencing and walling-off high-traffic sections of the porous boundary; a joint Bangladesh-India boundary commission agreed to fully demarcate the Bangladesh-India boundary in the Dhubri-Kruigram sector; the Naf river on the border with Burma serves as a smuggling and illegal transit route; Bangladesh struggles to accommodate 29,000 Rohingya, Burmese Muslim minority from Arakan State, living as refugees in Cox's Bazar; Burmese border authorities are constructing a 200 km (124 mi) wire fence designed to deter illegal cross-border transit and tensions from the military build-up along border
Refugees and internally displaced persons: refugees (country of origin): 26,268 (Burma)
IDPs: 65,000 (land conflicts, religious persecution) (2007)
Trafficking in persons: current situation: Bangladesh is a source and transit country for men, women, and children trafficked for the purposes of forced labor and commercial sexual exploitation; a significant share of Bangladesh's trafficking victims are men recruited for work overseas with fraudulent employment offers who are subsequently exploited under conditions of forced labor or debt bondage; children are trafficked within Bangladesh for commercial sexual exploitation, bonded labor, and forced labor; women and children from Bangladesh are also trafficked to India and Pakistan for sexual exploitation
tier rating: Bangladesh is placed on Tier 2 Watch List because it does not fully comply with the minimum standards for the elimination of trafficking; however, it is making significant efforts to do so, including some progress in addressing sex trafficking; the government did not demonstrate sufficient progress in criminally prosecuting and convicting labor trafficking offenders, particularly those responsible for the recruitment of Bangladeshi workers for the purpose of labor trafficking (2009)
Illicit drugs: transit country for illegal drugs produced in neighboring countries

BARBADOS

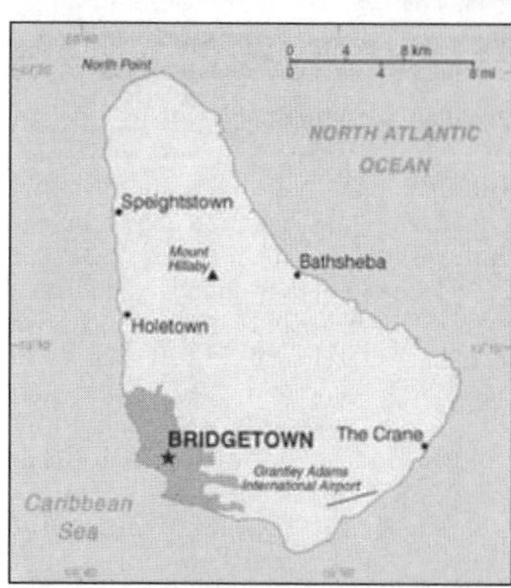

INTRODUCTION

Background: The island was uninhabited when first settled by the British in 1627. Slaves worked the sugar plantations established on the island until 1834 when slavery was abolished. The economy remained heavily dependent on sugar, rum, and molasses production through most of the 20th century. The gradual introduction of social and political reforms in the 1940s and 1950s led to complete independence from the UK in 1966. In the 1990s, tourism and manufacturing surpassed the sugar industry in economic importance.

GEOGRAPHY

Location: Caribbean, island in the North Atlantic Ocean, northeast of Venezuela
Geographic coordinates: 13 10 N, 59 32 W
Map references: Central America and the Caribbean
Area: *total:* 430 sq km
country comparison to the world: 200
land: 430 sq km
water: 0 sq km
Area—comparative: 2.5 times the size of Washington, DC
Land boundaries: 0 km
Coastline: 97 km
Maritime claims: territorial sea: 12 nm
exclusive economic zone: 200 nm
Climate: tropical; rainy season (June to October)
Terrain: relatively flat; rises gently to central highland region
Elevation extremes: *lowest point:* Atlantic Ocean 0 m
highest point: Mount Hillaby 336 m
Natural resources: petroleum, fish, natural gas
Land use: *arable land:* 37.21%
permanent crops: 2.33%
other: 60.46% (2005)
Irrigated land: 50 sq km (2003)
Total renewable water resources: 0.1 cu km (2003)
Freshwater withdrawal (domestic/industrial/agricultural): *total:* 0.09 cu km/yr (33%/44%/22%)
per capita: 333 cu m/yr (2000)
Natural hazards: infrequent hurricanes; periodic landslides
Environment—current issues: pollution of coastal waters from waste disposal by ships; soil erosion; illegal solid waste disposal threatens contamination of aquifers
Environment—international agreements: *party to:* Biodiversity, Climate Change, Climate Change-Kyoto Protocol, Desertification, Endangered Species, Hazardous Wastes, Law of the Sea, Marine Dumping, Ozone Layer Protection, Ship Pollution, Wetlands
signed, but not ratified: none of the selected agreements
Geography—note: easternmost Caribbean island

PEOPLE

Population: 286,705 (July 2011 est.)
country comparison to the world: 180
Age structure: *0–14 years:* 18.9% (male 27,127/female 27,127)
15–64 years: 71.3% (male 100,594/female 103,751)
65 years and over: 9.8% (male 10,982/ female 17,124) (2011 est.)
Median age: *total:* 36.5 years
male: 35.4 years
female: 37.6 years (2011 est.)
Population growth rate: 0.366% (2011 est.)
country comparison to the world: 163
Birth rate: 12.35 births/1,000 population (2011 est.)
country comparison to the world: 158
Death rate: 8.39 deaths/1,000 population (July 2011 est.)
country comparison to the world: 88
Net migration rate: -0.3 migrant(s)/1,000 population (2011 est.)
country comparison to the world: 128
Urbanization: *urban population:* 44% of total population (2010)
rate of urbanization: 1.7% annual rate of change (2010-15 est.)
Major cities—population: BRIDGETOWN (capital) 112,000 (2009)
Sex ratio: *at birth:* 1.013 male(s)/female
under 15 years: 1 male(s)/female
15–64 years: 0.97 male(s)/female
65 years and over: 0.64 male(s)/female
total population: 0.94 male(s)/female (2011 est.)
Infant mortality rate: *total:* 11.86 deaths/1,000 live births
country comparison to the world: 136
male: 13.48 deaths/1,000 live births
female: 10.22 deaths/1,000 live births (2011 est.)
Life expectancy at birth: *total population:* 74.34 years
country comparison to the world: 101
male: 72.07 years
female: 76.64 years (2011 est.)
Total fertility rate: 1.68 children born/ woman (2011 est.)
country comparison to the world: 170
HIV/AIDS—adult prevalence rate: 1.4% (2009 est.)
country comparison to the world: 36
HIV/AIDS—people living with HIV/AIDS: 2,100 (2009 est.)
country comparison to the world: 135
HIV/AIDS—deaths: fewer than 100 (2009 est.)
country comparison to the world: 122
Drinking water source: *Improved:*
urban: 100% of population
rural: 100% of population
total: 100% of population (2008)
Sanitation facility access: *Improved:*
urban: 100% of population
rural: 100% of population
total: 100% of population (2008)
Nationality: *noun:* Barbadian(s) or Bajan (colloquial)
adjective: Barbadian or Bajan (colloquial)
Ethnic groups: black 93%, white 3.2%, mixed 2.6%, East Indian 1%, other 0.2% (2000 census)
Religions: Protestant 63.4% (Anglican 28.3%, Pentecostal 18.7%, Methodist 5.1%, other 11.3%), Roman Catholic 4.2%, other Christian 7%, other 4.8%, none or unspecified 20.6% (2008 est.)
Languages: English
Literacy: *definition:* age 15 and over has ever attended school
total population: 99.7%
male: 99.7%
female: 99.7% (2002 est.)
School life expectancy (primary to tertiary education): *total:* 13 years
male: 13 years
female: 14 years (2001)
Education expenditures: 6.7% of GDP (2008)
country comparison to the world: 19

GOVERNMENT

Country name: *conventional long form:* none
conventional short form: Barbados
Government type: parliamentary democracy and a Commonwealth realm
Capital: *name:* Bridgetown
geographic coordinates: 13 06 N, 59 37 W
time difference: UTC-4 (1 hour ahead of Washington, DC during Standard Time)
Administrative divisions: 11 parishes and 1 city*; Bridgetown*, Christ Church, Saint Andrew, Saint George, Saint James, Saint John, Saint Joseph, Saint Lucy, Saint Michael, Saint Peter, Saint Philip, Saint Thomas
Independence: 30 November 1966 (from the UK)
National holiday: Independence Day, 30 November (1966)
Constitution: 30 November 1966
Legal system: English common law; no judicial review of legislative acts
International law organization participation: accepts compulsory ICJ jurisdiction with reservations; accepts ICCt jurisdiction
Suffrage: 18 years of age; universal
Executive branch: *chief of state:* Queen ELIZABETH II (since 6 February 1952); represented by Governor General Sir Clifford Straughn HUSBANDS (since 1 June 1996)
head of government: Prime Minister Fruendel STUART (since 23 October 2010)
cabinet: Cabinet appointed by the governor general on the advice of the prime minister

(For more information visit the World Leaders website)
elections: the monarchy is hereditary; governor general appointed by the monarch; following legislative elections, the leader of the majority party or the leader of the majority coalition is usually appointed prime minister by the governor general; the prime minister recommends the deputy prime minister
Legislative branch: bicameral Parliament consists of the Senate (21 seats; members appointed by the governor general—12 on the advice of the Prime Minister, 2 on the advice of the opposition leader, and 7 at his discretion) and the House of Assembly (30 seats; members are elected by direct popular vote to serve five-year terms)
elections: House of Assembly–last held on 15 January 2008 (next to be called in 2012)
election results: House of Assembly–percent of vote by party–DLP 52.5%, BLP 47.3%; seats by party–DLP 20, BLP 10
Judicial branch: Supreme Court of Judicature consists of a High Court and a Court of Appeal (judges are appointed by the Service Commissions for the Judicial and Legal Services); Caribbean Court of Justice or CCJ is the highest court of appeal; based in Port of Spain, Trinidad and Tobago
Political parties and leaders: Barbados Labor Party or BLP [Owen ARTHUR]; Democratic Labor Party or DLP [Freundel STUART]; People's Empowerment Party or PEP [David COMISSIONG]
Political pressure groups and leaders: Barbados Secondary Teachers' Union or BSTU [Patrick FROST]; Barbados Union of Teachers or BUT [Herbert GITTENS]; Congress of Trade Unions and Staff Associations of Barbados or CTUSAB, (includes the BWU, NUPW, BUT, and BSTU) [Leroy TROTMAN]; Barbados Workers Union or BWU [Leroy TROTMAN]; Clement Payne Labor Union [David COMISSIONG]; National Union of Public Workers [Joseph GODDARD]
International organization participation: ACP, AOSIS, C, Caricom, CDB, FAO, G-77, IADB, IBRD, ICAO, ICRM, IDA, IFAD, IFC, IFRCS, ILO, IMF, IMO, Interpol, IOC, ISO, ITSO, ITU, ITUC, LAES, MIGA, NAM, OAS, OPANAL, OPCW, UN, UNCTAD, UNESCO, UNIDO, UPU, WCO, WFTU, WHO, WIPO, WMO, WTO
Diplomatic representation in the US: *chief of mission:* Ambassador John BEALE
chancery: 2144 Wyoming Avenue NW, Washington, DC 20008
telephone: [1] (202) 939-9200
FAX: [1] (202) 332-7467
consulate(s) general: Miami, New York
consulate(s): Los Angeles
Diplomatic representation from the US: *chief of mission:* Ambassador (vacant); Charge d' Affaires D. Brent HARDT
embassy: U.S. Embassy, Wildey Business Park, Wildey, St. Michael BB 14006
mailing address: P. O. Box 302, Bridgetown BB 11000; CMR 1014, APO AA 34055
telephone: [1] (246) 227-4399
FAX: [1] (246) 431-0179
Flag description: three equal vertical bands of blue (hoist side), gold, and blue with the head of a black trident centered on the gold band; the band colors represent the blue of the sea and sky and the gold of the beaches; the trident head represents independence and a break with the past (the colonial coat of arms contained a complete trident)
National anthem: *name:* "The National Anthem of Barbados"
lyrics/music: Irving BURGIE/C. Van Roland EDWARDS
note: adopted 1966; the anthem is also known as "In Plenty and In Time of Need"

ECONOMY

Economy—overview: Historically, the Barbadian economy was dependent on sugarcane cultivation and related activities. However, in recent years the economy has diversified into light industry and tourism with about three-quarters of GDP and 80% of exports being attributed to services. Growth has rebounded since 2003, bolstered by increases in construction projects and tourism revenues, reflecting its success in the higher-end segment, but the sector faced declining revenues in 2009 with the global economic downturn. The country enjoys one of the highest per capita incomes in the region. Offshore finance and information services are important foreign exchange earners and thrive from having the same time zone as eastern US financial centers and a relatively highly educated workforce. The government continues its efforts to reduce unemployment, to encourage direct foreign investment, and to privatize remaining state-owned enterprises. The public debt-to-GDP ratio rose to over 100% in 2009, largely because a sharp slowdown in tourism and financial services led to a wide budget deficit.
GDP (purchasing power parity): $6.227 billion (2010 est.)
country comparison to the world: 155
$6.258 billion (2009 est.)
$6.568 billion (2008 est.)
note: data are in 2010 US dollars
GDP (official exchange rate): $3.963 billion (2010 est.)
GDP—real growth rate: -0.5% (2010 est.)
country comparison to the world: 194
-4.7% (2009 est.)
-0.2% (2008 est.)
GDP—per capita (PPP): $21,800 (2010 est.)
country comparison to the world: 60
$22,000 (2009 est.)
$23,200 (2008 est.)
note: data are in 2010 US dollars
GDP—composition by sector:
agriculture: 6%
industry: 16%
services: 78% (2000 est.)
Labor force: 175,000 (2007 est.)
country comparison to the world: 175
Labor force—by occupation:
agriculture: 10%
industry: 15%
services: 75% (1996 est.)
Unemployment rate: 10.7% (2003 est.)
country comparison to the world: 116
Population below poverty line: NA%
Household income or consumption by percentage share: *lowest 10%:* NA%
highest 10%: NA%
Budget: *revenues:* $847 million (including grants)
expenditures: $886 million (2000 est.)
Public debt: NA% of GDP (2009)
Inflation rate (consumer prices): 5.5% (2007 est.)
country comparison to the world: 148
Central bank discount rate: 7% (31 December 2009)
country comparison to the world: 43
10% (31 December 2008)
Commercial bank prime lending rate: 9.25% (31 December 2009 est.)
country comparison to the world: 86
10.03% (31 December 2008 est.)
Stock of narrow money: $1.793 billion (31 December 2009)
country comparison to the world: 123
$1.748 billion (31 December 2008)
Stock of broad money: $4.563 billion (31 December 2009)
country comparison to the world: 124
$4.618 billion (31 December 2008)
Stock of domestic credit: $4.554 billion (31 December 2008 est.)
country comparison to the world: 112
$4.124 billion (31 December 2007 est.)
Market value of publicly traded shares: $NA (31 December 2009)
country comparison to the world: 82
$4.964 billion (31 December 2008)
$5.599 billion (31 December 2007)
Agriculture—products: sugarcane, vegetables, cotton
Industries: tourism, sugar, light manufacturing, component assembly for export
Industrial production growth rate: -3.2% (2000 est.)
country comparison to the world: 162
Electricity—production: 1.003 billion kWh (2007 est.)
country comparison to the world: 146
Electricity—consumption: 939.9 million kWh (2007 est.)
country comparison to the world: 146
Electricity—exports: 0 kWh (2008 est.)
Electricity—imports: 0 kWh (2008 est.)
Oil—production: 765 bbl/day (2009 est.)
country comparison to the world: 107
Oil—consumption: 9,000 bbl/day (2009 est.)
country comparison to the world: 150
Oil—exports: 1,750 bbl/day (2007 est.)
country comparison to the world: 116
Oil—imports: 10,390 bbl/day (2007 est.)
country comparison to the world: 140
Oil—proved reserves: 1.79 million bbl (1 January 2010 est.)
country comparison to the world: 97
Natural gas—production: 29.17 million cu m (2008 est.)
country comparison to the world: 86
Natural gas—consumption: 29.17 million cu m (2008 est.)
country comparison to the world: 110
Natural gas—exports: 0 cu m (2008 est.)
country comparison to the world: 55
Natural gas—imports: 0 cu m (2008 est.)
country comparison to the world: 148
Natural gas—proved reserves: 113.3 million cu m (1 January 2010 est.)
country comparison to the world: 101

Current account balance: $-254 million (2007 est.)
country comparison to the world: 95
Exports: $385 million (2006)
country comparison to the world: 173
Exports—commodities: manufactures, sugar and molasses, rum, other foods and beverages, chemicals, electrical components
Exports—partners: Trinidad and Tobago 17.48%, Jamaica 15.63%, US 8.93%, Saint Lucia 8.13%, UK 5.36%, Saint Vincent and the Grenadines 5.04%, Antigua and Barbuda 4.12% (2009)
Imports: $1.586 billion (2006)
country comparison to the world: 159
Imports—commodities: consumer goods, machinery, foodstuffs, construction materials, chemicals, fuel, electrical components
Imports—partners: Trinidad and Tobago 28.52%, US 27.96%, Colombia 7.13%, China 4.76%, UK 4.39% (2009)
Reserves of foreign exchange and gold: $620 million (2007)
country comparison to the world: 123
Debt—external: $668 million (2003)
country comparison to the world: 156
Exchange rates: Barbadian dollars (BBD) per US dollar–NA (2007), 2 (2006), 2 (2005), 2 (2004), 2 (2003)

COMMUNICATIONS

Telephones—main lines in use: 135,700 (2009)
country comparison to the world: 137
Telephones—mobile cellular: 337,100 (2009)
country comparison to the world: 168
Telephone system: *general assessment:* island-wide automatic telephone system
domestic: fixed-line teledensity of roughly 50 per 100 persons; mobile-cellular telephone density approaching 125 per 100 persons
international: country code–1-246; landing point for the East Caribbean Fiber System (ECFS) submarine cable with links to 13 other islands in the eastern Caribbean extending from the British Virgin Islands to Trinidad; satellite earth stations–1 (Intelsat -Atlantic Ocean); tropospheric scatter to Trinidad and Saint Lucia (2009)
Broadcast media: government-owned Caribbean Broadcasting Corporation (CBC) operates the lone terrestrial television station; CBC also operates a multi-channel cable TV subscription service; roughly a dozen radio stations, consisting of a CBC-operated network alongside privately-owned radio stations, in operation (2007)
Internet country code: .bb
Internet hosts: 1,508 (2010)
country comparison to the world: 159
Internet users: 188,000 (2008)
country comparison to the world: 143

TRANSPORTATION

Airports: 1 (2010)
country comparison to the world: 235
Airports—with paved runways: *total:* 1
over 3,047 m: 1 (2010)
Pipelines: gas 33 km; oil 62 km; refined products 4 km
Roadways: *total:* 1,600 km
country comparison to the world: 176
paved: 1,600 km (2004)
Merchant marine: *total:* 95
country comparison to the world: 52
by type: bulk carrier 19, cargo 55, chemical tanker 9, passenger 1, passenger/cargo 1, petroleum tanker 4, refrigerated cargo 5, roll on/roll off 1
foreign-owned: 89 (Canada 13, Greece 14, Iran 4, Lebanon 2, Norway 41, Sweden 6, Syria 1, Turkey 1, UK 7)
registered in other countries: 1 (unknown 1) (2010)
Ports and terminals: Bridgetown

MILITARY

Military branches: Royal Barbados Defense Force: Troops Command, Barbados Coast Guard (2011)
Military service age and obligation: 18 years of age for voluntary military service (younger volunteers require parental consent); no conscription (2009)
Manpower available for military service:
males age 16–49: 73,820
females age 16–49: 73,835 (2010 est.)
Manpower fit for military service: *males age 16–49:* 58,125
females age 16–49: 58,016 (2010 est.)
Manpower reaching militarily significant age annually: *male:* 1,842
female: 1,849 (2010 est.)
Military expenditures: 0.8% of GDP (2009)
country comparison to the world: 145
Military—note: the Royal Barbados Defense Force includes a land-based Troop Command and a small Coast Guard; the primary role of the land element is to defend the island against external aggression; the Command consists of a single, part-time battalion with a small regular cadre that is deployed throughout the island; it increasingly supports the police in patrolling the coastline to prevent smuggling and other illicit activities (2007)

TRANSNATIONAL ISSUES

Disputes—international: Barbados and Trinidad and Tobago abide by the April 2006 Permanent Court of Arbitration decision delimiting a maritime boundary and limiting catches of flying fish in Trinidad and Tobago's exclusive economic zone; joins other Caribbean states to counter Venezuela's claim that Aves Island sustains human habitation, a criterion under the UN Convention on the Law of the Sea (UNCLOS), which permits Venezuela to extend its EEZ/continental shelf over a large portion of the eastern Caribbean Sea
Illicit drugs: one of many Caribbean transshipment points for narcotics bound for Europe and the US; offshore financial center

BELARUS

INTRODUCTION

Background: After seven decades as a constituent republic of the USSR, Belarus attained its independence in 1991. It has retained closer political and economic ties to Russia than any of the other former Soviet republics. Belarus and Russia signed a treaty on a two-state union on 8 December 1999 envisioning greater political and economic integration. Although Belarus agreed to a framework to carry out the accord, serious implementation has yet to take place. Since his election in July 1994 as the country's first president, Aleksandr LUKASHENKO has steadily consolidated his power through authoritarian means. Government restrictions on freedom of speech and the press, peaceful assembly, and religion remain in place.

GEOGRAPHY

Location: Eastern Europe, east of Poland
Geographic coordinates: 53 00 N, 28 00 E
Map references: Europe
Area: *total:* 207,600 sq km
country comparison to the world: 85
land: 202,900 sq km
water: 4,700 sq km
Area—comparative: slightly smaller than Kansas
Land boundaries: *total:* 3,306 km
border countries: Latvia 171 km, Lithuania 680 km, Poland 605 km, Russia 959 km, Ukraine 891 km
Coastline: 0 km (landlocked)
Maritime claims: none (landlocked)
Climate: cold winters, cool and moist summers; transitional between continental and maritime
Terrain: generally flat and contains much marshland
Elevation extremes: *lowest point:* Nyoman River 90 m
highest point: Dzyarzhynskaya Hara 346 m
Natural resources: timber, peat deposits, small quantities of oil and natural gas, granite, dolomitic limestone, marl, chalk, sand, gravel, clay
Land use:
arable land: 26.77%
permanent crops: 0.6%

other: 72.63% (2005)
Irrigated land: 1,310 sq km (2003)
Total renewable water resources: 58 cu km (1997)
Freshwater withdrawal (domestic/industrial/agricultural): *total*: 2.79 cu km/yr (23%/47%/30%)
per capita: 286 cu m/yr (2000)
Natural hazards: NA
Environment—current issues: soil pollution from pesticide use; southern part of the country contaminated with fallout from 1986 nuclear reactor accident at Chornobyl' in northern Ukraine
Environment—international agreements: *party to*: Air Pollution, Air Pollution-Nitrogen Oxides, Air Pollution-Sulfur 85, Biodiversity, Climate Change, Climate Change-Kyoto Protocol, Desertification, Endangered Species, Environmental Modification, Hazardous Wastes, Law of the Sea, Marine Dumping, Ozone Layer Protection, Ship Pollution, Wetlands
signed, but not ratified: none of the selected agreements
Geography—note: landlocked; glacial scouring accounts for the flatness of Belarusian terrain and for its 11,000 lakes

PEOPLE

Population: 9,577,552 (July 2011 est.)
country comparison to the world: 88
Age structure: *0–14 years*: 14.2% (male 699,048/female 660,130)
15–64 years: 71.7% (male 3,328,548/ female 3,542,359)
65 years and over: 14.1% (male 427,086/ female 920,381) (2011 est.)
Median age: *total*: 39 years
male: 36.1 years
female: 42.1 years (2011 est.)
Population growth rate: -0.363% (2011 est.)
country comparison to the world: 219
Birth rate: 9.76 births/1,000 population (2011 est.)
country comparison to the world: 196
Death rate: 13.77 deaths/1,000 population (July 2011 est.)
country comparison to the world: 17
Net migration rate: 0.38 migrant(s)/1,000 population (2011 est.)
country comparison to the world: 66
Urbanization: *urban population*: 75% of total population (2010)
rate of urbanization: 0.1% annual rate of change (2010-15 est.)
Major cities—population: MINSK (capital) 1.837 million (2009)
Sex ratio: *at birth*: 1.062 male(s)/female
under 15 years: 1.06 male(s)/female
15–64 years: 0.94 male(s)/female
65 years and over: 0.47 male(s)/female
total population: 0.87 male(s)/female (2011 est.)
Infant mortality rate: *total*: 6.25 deaths/1,000 live births
country comparison to the world: 173
male: 7.24 deaths/1,000 live births
female: 5.19 deaths/1,000 live births (2011 est.)
Life expectancy at birth: *total population*: 71.2 years
country comparison to the world: 139
male: 65.57 years
female: 77.18 years (2011 est.)
Total fertility rate: 1.26 children born/ woman (2011 est.)
country comparison to the world: 215
HIV/AIDS—adult prevalence rate: 0.3% (2009 est.)
country comparison to the world: 80
HIV/AIDS—people living with HIV/AIDS: 17,000 (2009 est.)
country comparison to the world: 83
HIV/AIDS—deaths: fewer than 1,000 (2009 est.)
country comparison to the world: 68
Drinking water source: *Improved*:
urban: 100% of population
rural: 100% of population
total: 100% of population (2008)
Sanitation facility access: *Improved*:
urban: 91% of population
rural: 97% of population
total: 93% of population
Unimproved: *urban*: 9% of population
rural: 3% of population
total: 7% of population (2008)
Nationality: *noun*: Belarusian(s)
adjective: Belarusian
Ethnic groups: Belarusian 81.2%, Russian 11.4%, Polish 3.9%, Ukrainian 2.4%, other 1.1% (1999 census)
Religions: Eastern Orthodox 80%, other (including Roman Catholic, Protestant, Jewish, and Muslim) 20% (1997 est.)
Languages: Belarusian (official) 36.7%, Russian (official) 62.8%, other 0.5% (includes small Polish- and Ukrainian-speaking minorities) (1999 census)
Literacy: *definition*: age 15 and over can read and write
total population: 99.6%
male: 99.8%
female: 99.4% (1999 census)
School life expectancy (primary to tertiary education): *total*: 15 years
male: 14 years
female: 15 years (2007)
Education expenditures: 4.5% of GDP (2009)
country comparison to the world: 83

GOVERNMENT

Country name: *conventional long form*: Republic of Belarus
conventional short form: Belarus
local long form: Respublika Byelarus'
local short form: Byelarus'
former: Belorussian (Byelorussian) Soviet Socialist Republic
Government type: republic in name, although in fact a dictatorship
Capital: *name*: Minsk
geographic coordinates: 53 54 N, 27 34 E
time difference: UTC+2 (7 hours ahead of Washington, DC during Standard Time)
daylight saving time: +1hr, begins last Sunday in March and will continue throughout 2011
Administrative divisions: 6 provinces (voblastsi, singular—voblasts') and 1 municipality* (horad); Brest, Homyel' (Gomel), Horad Minsk* (Minsk City), Hrodna (Grodno), Mahilyow (Mogilev), Minsk, Vitsyebsk (Vitebsk)
note: administrative divisions have the same names as their administrative centers; Russian spelling provided for reference when different from Belarusian
Independence: 25 August 1991 (from the Soviet Union)
National holiday: Independence Day, 3 July (1944); note—3 July 1944 was the date Minsk was liberated from German troops, 25 August 1991 was the date of independence from the Soviet Union
Constitution: 15 March 1994; revised by national referendum 24 November 1996 giving the presidency greatly expanded powers; became effective 27 November 1996; revised again 17 October 2004 removing presidential term limits
Legal system: civil law system; note—nearly all major codes (civil, civil procedure, criminal, criminal procedure, family and labor) have been revised and came into force in 1999 or 2000
International law organization participation: has not submitted an ICJ jurisdiction declaration; non-party state to the ICCt
Suffrage: 18 years of age; universal
Executive branch: *chief of state*: President Aleksandr LUKASHENKO (since 20 July 1994)
head of government: Prime Minister Mikhail MYASNIKOVICH (since 28 December 2010); First Deputy Prime Minister Vladimir SEMASHKO (since December 2003)
cabinet: Council of Ministers
(For more information visit the World Leaders website)
elections: president elected by popular vote for a five-year term; first election took place on 23 June and 10 July 1994; according to the 1994 constitution, the next election should have been held in 1999, however, Aleksandr LUKASHENKO extended his term to 2001 via a November 1996 referendum; subsequent election held on 9 September 2001; an October 2004 referendum ended presidential term limits and allowed the president to run in a third (19 March 2006) and fourth election (19 December 2010); prime minister and deputy prime ministers appointed by the president
election results: Aleksandr LUKASHENKO reelected president; percent of vote—Aleksandr LUKASHENKO 79.7%, Andrey SANNIKAU 2.6%, other candidates 17.7%; note—election marred by electoral fraud
Legislative branch: bicameral National Assembly or Natsionalnoye Sobraniye consists of the Council of the Republic or Sovet Respubliki (64 seats; 56 members elected by regional and Minsk city councils and 8 members appointed by the president, to serve four-year terms) and the Chamber of Representatives or Palata Predstaviteley (110 seats; members elected by popular vote to serve four-year terms)
elections: Palata Predstaviteley—last held on 28 September 2008 (next to be held in the spring of 2012); international observers determined that despite minor improvements the election ultimately fell short of democratic standards; pro-LUKASHENKO candidates won every seat
election results: Sovet Respubliki—percent of vote by party—NA; seats by party—NA; Palata Predstaviteley—percent of vote by party—NA; seats by party—KPB 6, AP 1, no affiliation 103
Judicial branch: Supreme Court (judges are appointed by the president); Constitu-

tional Court (half of the judges appointed by the president and half appointed by the Chamber of Representatives)

Political parties and leaders: pro-government parties: Belarusian Agrarian Party or AP [Mikhail SHIMANSKY]; Belarusian Patriotic Movement (Belarusian Patriotic Party) or BPR [Nikolay ULAKHOVICH, chairman]; Communist Party of Belarus or KPB [Tatsyana HOLUBEVA]; Liberal Democratic Party or LDP [Sergey GAYDUKEVICH]; Republican Party of Labor and Justice [Vasiliy ZADNEPRYANYY]

opposition parties: Belarusian Christian Democracy Party [Pavel SEVERINETS] (unregistered); Belarusian Party of Communists or PKB [Sergey KALYAKIN]; Belarusian Party of Labor [Aleksandr BUKHVOSTOV] (unregistered); Belarusian Popular Front or BPF [Aleksey YANUKEVICH]; Belarusian Social-Democratic Hramada [Stanislav SHUSHKEVICH]; Belarusian Social Democratic Party Hramada ("Assembly") or BSDPH [Anatoliy LEVKOVICH]; Belarusian Social Democratic Party People's Assembly ("Narodnaya Hramada") [Nikolay STATKEVICH] (unregistered); Belarusian Women's Party Nadzeya ("Hope") [Yelena YESKOVA, chairperson]; Christian Conservative Party or BPF [Zyanon PAZNIAK]; European Belarus Campaign [Andrey SANNIKOV]; Party of Freedom and Progress [Vladimir NOVOSYAD] (unregistered); "Tell the Truth" Campaign [Vladimir NEKLYAYEV]; United Civic Party or UCP [Anatoliy LEBEDKO]

Political pressure groups and leaders: Assembly of Pro-Democratic NGOs (unregistered) [Sergey MATSKEVICH]; Belarusian Congress of Democratic Trade Unions [Aleksandr YAROSHUK]; Belarusian Association of Journalists [Zhana LITVINA]; Belarusian Helsinki Committee [Aleh HULAK]; Belarusian Independence Bloc (unregistered) and For Freedom movement [Aleksandr MILINKEVICH]; Belarusian Organization of Working Women [Irina ZHIKHAR]; BPF-Youth [Andrus KRECHKA]; Charter 97 (unregistered) [Andrey SANNIKOV]; Perspektiva small business association [Anatol SHUMCHENKO]; Nasha Vyasna (unregistered) ("Our Spring") human rights center; "Tell the Truth" Movement [Vladimir NEKLYAYEV]; Women's Independent Democratic Movement [Ludmila PETINA]; Young Belarus (Malady Belarus) [Zmitser KASPYAROVICH]; Youth Front (Malady Front) [Zmitser DASHKEVICH]

International organization participation: BSEC (observer), CBSS (observer), CEI, CIS, CSTO, EAEC, EAPC, EBRD, FAO, GCTU, IAEA, IBRD, ICAO, ICRM, IDA, IFC, IFRCS, ILO, IMF, IMSO, Interpol, IOC, IOM, IPU, ISO, ITU, ITUC, MIGA, NAM, NSG, OPCW, OSCE, PCA, PFP, SCO (dialogue member), UN, UNCTAD, UNESCO, UNIDO, UNWTO, UPU, WCO, WFTU, WHO, WIPO, WMO, WTO (observer)

Diplomatic representation in the US: *chief of mission:* Ambassador (vacant); Charge d'Affaires Oleg KRAVCHENKO
chancery: 1619 New Hampshire Avenue NW, Washington, DC 20009
telephone: [1] (202) 986-1604
FAX: [1] (202) 986-1805
consulate(s) general: New York

Diplomatic representation from the US: *chief of mission:* Ambassador (vacant); Charge d'Affaires Michael SCANLAN
embassy: 46 Starovilenskaya Street, Minsk 220002
mailing address: PSC 78, Box B Minsk, APO 09723
telephone: [375] (17) 210-12-83, 217-7347 through 7348
FAX: [375] (17) 334-7853

Flag description: red horizontal band (top) and green horizontal band one-half the width of the red band; a white vertical stripe on the hoist side bears Belarusian national ornamentation in red; the red band color recalls past struggles from oppression, the green band represents hope and the many forests of the country

National anthem: *name:* "My, Bielarusy" (We Belarusians)
lyrics/music: Mikhas KLIMKOVICH and Uladzimir KARYZNA/Nester SAKALOUSKI
note: music adopted 1955, lyrics adopted 2002; after the fall of the Soviet Union, Belarus kept the music of its Soviet-era anthem but adopted new lyrics; also known as "Dziarzauny himn Respubliki Bielarus" (State Anthem of the Republic of Belarus)

ECONOMY

Economy—overview: Belarus has seen limited structural reform since 1995, when President LUKASHENKO launched the country on the path of "market socialism." In keeping with this policy, LUKASHENKO reimposed administrative controls over prices and currency exchange rates and expanded the state's right to intervene in the management of private enterprises. Since 2005, the government has re-nationalized a number of private companies. In addition, businesses have been subjected to pressure by central and local governments, including arbitrary changes in regulations, numerous rigorous inspections, retroactive application of new business regulations, and arrests of "disruptive" businessmen and factory owners. Continued state control over economic operations hampers market entry for businesses, both domestic and foreign. Government statistics indicate GDP growth was strong, surpassing 10% in 2008, despite the roadblocks of a tough, centrally directed economy with a high rate of inflation and a low rate of unemployment. However, the global crisis pushed the country into recession in 2009, and GDP grew only 0.2% for the year. Slumping foreign demand hit the industrial sector hard. Minsk has depended on a standby-agreement with the IMF to assist with balance of payments shortfalls. In line with IMF conditions, in 2009, Belarus devalued the ruble more than 40% and tightened some fiscal and monetary policies. On 1 January 2010, Russia, Kazakhstan and Belarus launched a customs union, with unified trade regulations and customs codes still under negotiation. In late January, Russia and Belarus amended their 2007 oil supply agreement. The new terms raised prices for above quota purchases, increasing Belarus' current account deficit. GDP grew 4.8% in 2010, in part, on the strength of renewed export growth. In December 2010, Belarus, Russia and Kazakhstan signed an agreement to form a Common Economic Space and Russia removed all Belarusian oil duties.

GDP (purchasing power parity): $131.2 billion (2010 est.)
country comparison to the world: 61
$121.9 billion (2009 est.)
$121.7 billion (2008 est.)
note: data are in 2010 US dollars

GDP (official exchange rate): $54.71 billion (2010 est.)

GDP—real growth rate: 7.6% (2010 est.)
country comparison to the world: 25
0.2% (2009 est.)
10.2% (2008 est.)

GDP—per capita (PPP): $13,600 (2010 est.)
country comparison to the world: 88
$12,600 (2009 est.)
$12,600 (2008 est.)
note: data are in 2010 US dollars

GDP—composition by sector: *agriculture:* 9%
industry: 42.9%
services: 48.1% (2010 est.)

Labor force: 5 million (2009)
country comparison to the world: 75

Labor force—by occupation: *agriculture:* 14%
industry: 34.7%
services: 51.3% (2003 est.)

Unemployment rate: 1% (2009 est.)
country comparison to the world: 5
1.6% (2005)
note: official registered unemployed; large number of underemployed workers

Population below poverty line: 27.1% (2003 est.)

Household income or consumption by percentage share: *lowest 10%:* 3.6%
highest 10%: 22% (2005)

Distribution of family income—Gini index: 27.9 (2005)
country comparison to the world: 123
21.7 (1998)

Investment (gross fixed): 36% of GDP (2010 est.)
country comparison to the world: 7

Budget: *revenues:* $23.27 billion
expenditures: $24.32 billion (2010 est.)

Inflation rate (consumer prices): 7% (2010 est.)
country comparison to the world: 173
12.9% (2009 est.)

Central bank discount rate: 13.5% (31 December 2009)
country comparison to the world: 31
12% (31 December 2008)

Commercial bank prime lending rate: 11.68% (31 December 2009 est.)
country comparison to the world: 108
8.55% (31 December 2008 est.)

Stock of narrow money: $4.747 billion (31 December 2010 est.)
country comparison to the world: 93
$4.381 billion (31 December 2009 est.)

Stock of broad money: $13.62 billion (31 December 2009)
country comparison to the world: 91

$14.07 billion (31 December 2008)
Stock of domestic credit: $19.99 billion (31 December 2010 est.)
country comparison to the world: 79
$17.15 billion (31 December 2009 est.)
Market value of publicly traded shares: $NA
Agriculture—products: grain, potatoes, vegetables, sugar beets, flax; beef, milk
Industries: metal-cutting machine tools, tractors, trucks, earthmovers, motorcycles, televisions, synthetic fibers, fertilizer, textiles, radios, refrigerators
Industrial production growth rate: 10.5% (2010 est.)
country comparison to the world: 18
Electricity—production: 29.92 billion kWh (2007 est.)
country comparison to the world: 62
Electricity—consumption: 30.54 billion kWh (2007 est.)
country comparison to the world: 59
Electricity—exports: 5.062 billion kWh (2007 est.)
Electricity—imports: 9.406 billion kWh (2007 est.)
Oil—production: 31,400 bbl/day (2009 est.)
country comparison to the world: 70
Oil—consumption: 173,000 bbl/day (2009 est.)
country comparison to the world: 59
Oil—exports: 303,900 bbl/day (2007 est.)
country comparison to the world: 40
Oil—imports: 444,800 bbl/day (2007 est.)
country comparison to the world: 28
Oil—proved reserves: 198 million bbl (1 January 2010 est.)
country comparison to the world: 60
Natural gas—production: 152 million cu m (2008 est.)
country comparison to the world: 76
Natural gas—consumption: 17 billion cu m (2009 est.)
country comparison to the world: 37
Natural gas—exports: 0 cu m (2009)
country comparison to the world: 64
Natural gas—imports: 17.6 billion cu m (2009 est.)
country comparison to the world: 13
Natural gas—proved reserves: 2.832 billion cu m (1 January 2010 est.)
country comparison to the world: 94
Current account balance: $-5.062 billion (2010 est.)
country comparison to the world: 170
$-6.402 billion (2009 est.)
Exports: $24.49 billion (2010 est.)
country comparison to the world: 66
$21.34 billion (2009 est.)
Exports—commodities: machinery and equipment, mineral products, chemicals, metals, textiles, foodstuffs
Exports—partners: Russia 33.6%, Netherlands 13.78%, Ukraine 8.68%, Latvia 6.32%, Poland 4.19%, Germany 4.17% (2009)
Imports: $29.79 billion (2010 est.)
country comparison to the world: 60
$28.31 billion (2009 est.)
Imports—commodities: mineral products, machinery and equipment, chemicals, foodstuffs, metals
Imports—partners: Russia 56.42%, Germany 8.31%, Ukraine 4.79%, China 4.04% (2009)
Reserves of foreign exchange and gold: $5.755 billion (31 December 2010 est.)
country comparison to the world: 67
$4.831 billion (31 December 2009 est.)
Debt—external: $24.8 billion (31 December 2010 est.)
country comparison to the world: 67
$19.74 billion (31 December 2009 est.)
Exchange rates: Belarusian rubles (BYB/BYR) per US dollar–3,019.9 (2010), 2,789.49 (2009), 2,130 (2008), 2,145 (2007), 2,144.6 (2006)

COMMUNICATIONS

Telephones—main lines in use: 3.969 million (2009)
country comparison to the world: 41
Telephones—mobile cellular: 9.686 million (2009)
country comparison to the world: 67
Telephone system: *general assessment:* Belarus lags behind its neighbors in upgrading telecommunications infrastructure; modernization of the network progressing with roughly two-thirds of switching equipment now digital
domestic: state-owned Beltelcom is the sole provider of fixed-line local and long distance service; fixed-line teledensity is improving although rural areas continue to be underserved; multiple GSM mobile-cellular networks are experiencing rapid growth; mobile-cellular teledensity reached 100 telephones per 100 persons in 2009
international: country code–375; Belarus is a member of the Trans-European Line (TEL), Trans-Asia-Europe (TAE) fiber-optic line, and has access to the Trans-Siberia Line (TSL); 3 fiber-optic segments provide connectivity to Latvia, Poland, Russia, and Ukraine; worldwide service is available to Belarus through this infrastructure; additional analog lines to Russia; Intelsat, Eutelsat, and Intersputnik earth stations (2008)
Broadcast media: 4 state-controlled national TV channels; Polish and Russian TV broadcasts are available in some areas; state-run Belarusian Radio operates 3 national networks and an external service; Russian and Polish radio broadcasts are available (2007)
Internet country code: .by
Internet hosts: 147,311 (2010)
country comparison to the world: 71
Internet users: 2.643 million (2009)
country comparison to the world: 69

TRANSPORTATION

Airports: 67 (2010)
country comparison to the world: 74
Airports—with paved runways: *total:* 35
over 3,047 m: 2
2,438 to 3,047 m: 22
1,524 to 2,437 m: 3
914 to 1,523 m: 1
under 914 m: 7 (2010)
Airports—with unpaved runways: *total:* 32
2,438 to 3,047 m: 2
1,524 to 2,437 m: 1
914 to 1,523 m: 2
under 914 m: 27 (2010)
Heliports: 1 (2010)
Pipelines: gas 5,250 km; oil 1,528 km; refined products 1,730 km (2010)
Railways: *total:* 5,537 km
country comparison to the world: 32
broad gauge: 5,512 km 1.520-m gauge (874 km electrified)
standard gauge: 25 km 1.435-m gauge (2010)
Roadways: *total:* 94,797 km
country comparison to the world: 48
paved: 84,028 km
unpaved: 10,769 km (2005)
Waterways: 2,500 km (use limited by its location on the perimeter of the country and by its shallowness) (2003)
country comparison to the world: 35
Ports and terminals: Mazyr

MILITARY

Military branches: Belarus Armed Forces: Land Force, Air and Air Defense Force, Special Operations Force (2011)
Military service age and obligation: 18-27 years of age for compulsory military service; conscript service obligation–12-18 months, depending on academic qualifications (2010)
Manpower available for military service: *males age 16–49:* 2,401,785
females age 16–49: 2,429,653 (2010 est.)
Manpower fit for military service: *males age 16–49:* 1,693,626
females age 16–49: 2,012,401 (2010 est.)
Manpower reaching militarily significant age annually: *male:* 51,855
female: 48,760 (2010 est.)
Military expenditures: 1.4% of GDP (2005 est.)
country comparison to the world: 104

TRANSNATIONAL ISSUES

Disputes—international: boundary demarcated with Latvia and Lithuania; Poland seeks enhanced demarcation and security along this Schengen hard border with financial assistance from the EU
Illicit drugs: limited cultivation of opium poppy and cannabis, mostly for the domestic market; transshipment point for illicit drugs to and via Russia, and to the Baltics and Western Europe; a small and lightly regulated financial center; anti-money-laundering legislation does not meet international standards and was weakened further when know-your-customer requirements were curtailed in 2008; few investigations or prosecutions of money-laundering activities (2008)

BELGIUM

INTRODUCTION

Background: Belgium became independent from the Netherlands in 1830; it was occupied by Germany during World Wars I and II. The country prospered in the past half century as a modern, technologically advanced European state and member of NATO and the EU. Tensions between the Dutch-speaking Flemings of the north and the French-speaking Walloons of the south have led in recent years to constitutional amendments granting these regions formal recognition and autonomy.

GEOGRAPHY

Location: Western Europe, bordering the North Sea, between France and the Netherlands
Geographic coordinates: 50 50 N, 4 00 E
Map references: Europe
Area: *total:* 30,528 sq km
country comparison to the world: 140
land: 30,278 sq km
water: 250 sq km
Area—comparative: about the size of Maryland
Land boundaries: *total:* 1,385 km
border countries: France 620 km, Germany 167 km, Luxembourg 148 km, Netherlands 450 km
Coastline: 66.5 km
Maritime claims: territorial sea: 12 nm
contiguous zone: 24 nm
exclusive economic zone: geographic coordinates define outer limit
continental shelf: median line with neighbors
Climate: temperate; mild winters, cool summers; rainy, humid, cloudy
Terrain: flat coastal plains in northwest, central rolling hills, rugged mountains of Ardennes Forest in southeast
Elevation extremes: *lowest point:* North Sea 0 m
highest point: Botrange 694 m
Natural resources: construction materials, silica sand, carbonates
Land use: *arable land:* 27.42%
permanent crops: 0.69%
other: 71.89%
note: includes Luxembourg (2005)
Irrigated land: 400 sq km (2003)
Total renewable water resources: 20.8 cu km (2005)
Freshwater withdrawal (domestic/industrial/agricultural): *total:* 7.44 cu km/yr (13%/85%/1%)
per capita: 714 cu m/yr (1998)
Natural hazards: flooding is a threat along rivers and in areas of reclaimed coastal land, protected from the sea by concrete dikes
Environment—current issues: the environment is exposed to intense pressures from human activities: urbanization, dense transportation network, industry, extensive animal breeding and crop cultivation; air and water pollution also have repercussions for neighboring countries; uncertainties regarding federal and regional responsibilities (now resolved) had slowed progress in tackling environmental challenges
Environment—international agreements: *party to:* Air Pollution, Air Pollution-Nitrogen Oxides, Air Pollution-Persistent Organic Pollutants, Air Pollution-Sulfur 85, Air Pollution-Sulfur 94, Air Pollution-Volatile Organic Compounds, Antarctic-Environmental Protocol, Antarctic-Marine Living Resources, Antarctic Seals, Antarctic Treaty, Biodiversity, Climate Change, Climate Change-Kyoto Protocol, Desertification, Endangered Species, Environmental Modification, Hazardous Wastes, Law of the Sea, Marine Dumping, Marine Life Conservation, Ozone Layer Protection, Ship Pollution, Tropical Timber 83, Tropical Timber 94, Wetlands, Whaling
signed, but not ratified: none of the selected agreements
Geography—note: crossroads of Western Europe; most West European capitals within 1,000 km of Brussels, the seat of both the European Union and NATO

PEOPLE

Population: 10,431,477 (July 2011 est.)
country comparison to the world: 80
Age structure: *0–14 years:* 15.9% (male 846,706/female 812,486)
15–64 years: 66.1% (male 3,475,404/female 3,416,060)
65 years and over: 18% (male 783,895/female 1,096,926) (2011 est.)
Median age: *total:* 42.3 years
male: 41 years
female: 43.6 years (2011 est.)
Population growth rate: 0.071% (2011 est.)
country comparison to the world: 187
Birth rate: 10.06 births/1,000 population (2011 est.)
country comparison to the world: 191
Death rate: 10.57 deaths/1,000 population (July 2011 est.)
country comparison to the world: 46
Net migration rate: 1.22 migrant(s)/1,000 population (2011 est.)
country comparison to the world: 51
Urbanization: *urban population:* 97% of total population (2010)
rate of urbanization: 0.4% annual rate of change (2010-15 est.)
Major cities—population: BRUSSELS (capital) 1.892 million; Antwerp 961,000 (2009)
Sex ratio:
at birth: 1.045 male(s)/female
under 15 years: 1.04 male(s)/female
15–64 years: 1.02 male(s)/female
65 years and over: 0.71 male(s)/female
total population: 0.96 male(s)/female (2011 est.)
Infant mortality rate:
total: 4.33 deaths/1,000 live births
country comparison to the world: 193
male: 4.86 deaths/1,000 live births
female: 3.78 deaths/1,000 live births (2011 est.)
Life expectancy at birth: *total population:* 79.51 years
country comparison to the world: 37
male: 76.35 years
female: 82.81 years (2011 est.)
Total fertility rate: 1.65 children born/woman (2011 est.)
country comparison to the world: 175
HIV/AIDS—adult prevalence rate: 0.2% (2009 est.)
country comparison to the world: 94
HIV/AIDS—people living with HIV/AIDS: 14,000 (2009 est.)
country comparison to the world: 87
HIV/AIDS—deaths: fewer than 100 (2009 est.)
country comparison to the world: 123
Drinking water source: *Improved:*
urban: 100% of population
rural: 100% of population
total: 100% of population (2008)
Sanitation facility access: *Improved:*
urban: 100% of population
rural: 100% of population
total: 100% of population (2008)
Nationality: *noun:* Belgian(s)
adjective: Belgian
Ethnic groups: Fleming 58%, Walloon 31%, mixed or other 11%
Religions: Roman Catholic 75%, other (includes Protestant) 25%
Languages: Dutch (official) 60%, French (official) 40%, German (official) less than 1%, legally bilingual (Dutch and French)
Literacy: *definition:* age 15 and over can read and write
total population: 99%
male: 99%
female: 99% (2003 est.)
School life expectancy (primary to tertiary education): *total:* 16 years
male: 16 years
female: 16 years (2008)
Education expenditures: 6.01% of GDP (2007)
country comparison to the world: 29

GOVERNMENT

Country name: *conventional long form:* Kingdom of Belgium
conventional short form: Belgium
local long form: Royaume de Belgique/Koninkrijk Belgie
local short form: Belgique/Belgie
Government type: federal parliamentary democracy under a constitutional monarchy

Capital: *name:* Brussels
geographic coordinates: 50 50 N, 4 20 E
time difference: UTC+1 (6 hours ahead of Washington, DC during Standard Time)
daylight saving time: +1hr, begins last Sunday in March; ends last Sunday in October
Administrative divisions: 3 regions (French: regions, singular—region; Dutch: gewesten, singular—gewest); Brussels-Capital Region, also known as Brussels Hoofdstedelijk Gewest (Dutch), Region de Bruxelles-Capitale (French long form), Bruxelles-Capitale (French short form); Flemish Region (Flanders), also known as Vlaams Gewest (Dutch long form), Vlaanderen (Dutch short form), Region Flamande (French long form), Flandre (French short form); Walloon Region (Wallonia), also known as Region Wallone (French long form), Wallonie (French short form), Waals Gewest (Dutch long form), Wallonie (Dutch short form)
note: as a result of the 1993 constitutional revision that furthered devolution into a federal state, there are now three levels of government (federal, regional, and linguistic community) with a complex division of responsibilities
Independence: 4 October 1830 (a provisional government declared independence from the Netherlands); 21 July 1831 (King LEOPOLD I ascended to the throne)
National holiday: 21 July (1831) ascension to the Throne of King LEOPOLD I
Constitution: 7 February 1831; revised 14 July 1993 to create a federal state; amended many times;
Legal system: civil law system based on the French Civil Code; note—Belgian law continues to be modified in conformance with the legislative norms mandated by the European Union; judicial review of legislative acts
International law organization participation: accepts compulsory ICJ jurisdiction with reservations; accepts ICCt jurisdiction
Suffrage: 18 years of age; universal and compulsory
Executive branch: *chief of state:* King ALBERT II (since 9 August 1993); Heir Apparent Prince PHILIPPE, son of the monarch
head of government: Prime Minister Yves LETERME (since 25 November 2009); note—the king accepted the resignation of LETERME on 26 April 2010; LETERME remains as caretaker
cabinet: Council of Ministers are formally appointed by the monarch
(For more information visit the World Leaders website)
elections: the monarchy is hereditary and constitutional; following legislative elections, the leader of the majority party or the leader of the majority coalition usually appointed prime minister by the monarch and then approved by parliament
Legislative branch: bicameral Parliament consists of a Senate or Senaat in Dutch, Senat in French (71 seats; 40 members directly elected by popular vote, 31 indirectly elected; members serve four-year terms) and a Chamber of Deputies or Kamer van Volksvertegenwoordigers in Dutch, Chambre des Representants in French (150 seats; members directly elected by popular vote on the basis of proportional representation to serve four-year terms)
elections: Senate and Chamber of Deputies—last held on 13 June 2010 (next to be held no later than June 2014)
election results: Senate—percent of vote by party—N-VA 19.6%, PS 13.6%, CD&V 10%, sp.a 9.5%, MR 9.3%, Open VLD 8.2%, VB 7.6%, Ecolo 5.5%, CDH 5.1% Groen! 3.9%, other 7.7%; seats by party—N-VA 9, PS 7, CD&V 4, sp.a 4, MR 4, Open VLD 4, VB 3, Ecolo 2, CDH 2, Groen! 1; Chamber of Deputies—percent of vote by party—N-VA 17.4%, PS 13.7%, CD&V 10.9%, MR 9.3%, sp.a 9.2%, Open VLD 8.6%, VB 7.8%, CDH 5.5%, Ecolo 4.8%, Groen! 4.4%, List Dedecker 2.3%, the Popular Party 1.3%, other 4.8%; seats by party—N-VA 27, PS 26, CD&V 17, MR 18, sp.a 13, Open VLD 13, VB 12, CDH 9, Ecolo 8, Groen! 5, List Dedecker 1, the Popular Party 1
note: as a result of the 1993 constitutional revision that furthered devolution into a federal state, there are now three levels of government (federal, regional, and linguistic community) with a complex division of responsibilities; this reality leaves six governments, each with its own legislative assembly
Judicial branch: Constitutional Court (previously Court of Arbitration) (12 judges, 6 Dutch-speaking and 6 French-speaking, appointed by the King); Supreme Court of Justice or Hof van Cassatie (in Dutch) or Cour de Cassation (in French) (judges are appointed for life by the government; candidacies have to be submitted by the High Justice Council)
Political parties and leaders: Flemish parties: Christian Democratic and Flemish or CDV [Wouter BEKE]; Dedecker List or LDD [Lode VEREECK]; Flemish Liberals and Democrats or Open VLD [Alexander DE CROO]; Groen! [Wouter VAN BESIEN] (formerly AGALEV, Flemish Greens); New Flemish Alliance or N-VA [Bart DE WEVER]; Social Progressive Alternative or SP.A [Caroline GENNEZ]; Vlaams Belang (Flemish Interest) or VB [Bruno VALKENIERS]
Francophone parties: Ecolo (Francophone Greens) [Jean-Michel JAVAUX, Sarah TURINE]; Humanist and Democratic Center or CDH [Joelle MILQUET]; Popular Party or PP [Mischael MODRIKAMEN]; Reform Movement or MR [Didier REYNDERS]; Socialist Party or PS [Elio DI RUPO]; other minor parties
Political pressure groups and leaders: Christian, Socialist, and Liberal Trade Unions; Federation of Belgian Industries
other: numerous other associations representing bankers, manufacturers, middle-class artisans, and the legal and medical professions; various organizations represent the cultural interests of Flanders and Wallonia; various peace groups such as Pax Christi and groups representing immigrants
International organization participation: ADB (nonregional members), AfDB (nonregional members), Australia Group, Benelux, BIS, CE, CERN, EAPC, EBRD, EIB, EMU, ESA, EU, FAO, FATF, G-9, G-10, IADB, IAEA, IBRD, ICAO, ICC, ICRM, IDA, IEA, IFAD, IFC, IFRCS, IHO, ILO, IMF, IMO, IMSO, Interpol, IOC, IOM, IPU, ISO, ITSO, ITU, ITUC, MIGA, MONUSCO, NATO, NEA, NSG, OAS (observer), OECD, OIF, OPCW, OSCE, Paris Club, PCA, Schengen Convention, SECI (observer), UN, UNCTAD, UNESCO, UNHCR, UNIDO, UNIFIL, UNMIS, UNRWA, UNTSO, UPU, WCO, WHO, WIPO, WMO, WTO, ZC
Diplomatic representation in the US: *chief of mission:* Ambassador Jan MATTHYSEN
chancery: 3330 Garfield Street NW, Washington, DC 20008
telephone: [1] (202) 333-6900
FAX: [1] (202) 333-3079
consulate(s) general: Atlanta, Los Angeles, New York
Diplomatic representation from the US: *chief of mission:* Ambassador Howard W. GUTMAN
embassy: 27 Boulevard du Regent [Regentlaan], B-1000 Brussels
mailing address: PSC 82, Box 002, APO AE 09710
telephone: [32] (2) 508-2111
FAX: [32] (2) 511-2725
Flag description: three equal vertical bands of black (hoist side), yellow, and red; the vertical design was based on the flag of France; the colors are those of the arms of the duchy of Brabant (yellow lion with red claws and tongue on a black field)
National anthem: *name:* "La Brabanconne" (The Song of Brabant)
lyrics/music: Louis-Alexandre DECHET[French] Victor CEULEMANS [Dutch]/Francois VAN CAMPENHOUT
note: adopted 1830; Louis-Alexandre DECHET was an actor at the theater in which the revolution against the Netherlands began; according to legend, he wrote the lyrics with a group of young people in a Brussels cafe

ECONOMY

Economy—overview: This modern, open, and private-enterprise-based economy has capitalized on its central geographic location, highly developed transport network, and diversified industrial and commercial base. Industry is concentrated mainly in the more heavily-populated region of Flanders in the north. With few natural resources, Belgium imports substantial quantities of raw materials and exports a large volume of manufactures, making its economy vulnerable to volatility in world markets, yet also able to benefit from them. Roughly three-quarters of Belgium's trade is with other EU countries, and Belgium has benefited most from its proximity to Germany. In 2010 Belgian GDP grew by 2.1%, the unemployment rate rose slightly, and the government reduced the budget deficit, which had worsened in 2008 and 2009 because of large-scale bail-outs in the financial sector. Belgium's budget deficit decreased from 6% of GDP to 4.8% in 2010, while public debt was just under 100% of GDP. Belgian banks were severely affected by the international financial crisis with three

major banks receiving capital injections from the government. An ageing population and rising social expenditures are mid- to long-term challenges to public finances.
GDP (purchasing power parity): $394.3 billion (2010 est.)
country comparison to the world: 31
$386.7 billion (2009 est.)
$397.3 billion (2008 est.)
note: data are in 2010 US dollars
GDP (official exchange rate): $465.7 billion (2010 est.)
GDP—real growth rate: 2% (2010 est.)
country comparison to the world: 150
-2.7% (2009 est.)
0.8% (2008 est.)
GDP—per capita (PPP): $37,800 (2010 est.)
country comparison to the world: 26
$37,100 (2009 est.)
$38,200 (2008 est.)
note: data are in 2010 US dollars
GDP—composition by sector: *agriculture:* 0.7%
industry: 22.1%
services: 77.2% (2010 est.)
Labor force: 5.02 million (2010 est.)
country comparison to the world: 74
Labor force—by occupation: *agriculture:* 2%
industry: 25%
services: 73% (2007 est.)
Unemployment rate: 8.5% (2010 est.)
country comparison to the world: 99
7.9% (2009 est.)
Population below poverty line: 15.2% (2007 est.)
Household income or consumption by percentage share: *lowest 10%:* 3.4%
highest 10%: 28.4% (2006)
Distribution of family income—Gini index: 28 (2005)
country comparison to the world: 121
28.7 (1996)
Investment (gross fixed): 20.8% of GDP (2010 est.)
country comparison to the world: 81
Budget: *revenues:* $220.6 billion
expenditures: $242.6 billion (2010 est.)
Public debt: 98.6% of GDP (2010 est.)
country comparison to the world: 10
96.2% of GDP (2009 est.)
Inflation rate (consumer prices): 2.3% (2010 est.)
country comparison to the world: 58
0% (2009 est.)
Central bank discount rate: 1.75% (31 December 2010)
country comparison to the world: 130
1.75% (31 December 2009)
note: this is the European Central Bank's rate on the marginal lending facility, which offers overnight credit to banks in the euro area
Commercial bank prime lending rate: 6.15% (31 December 2009 est.)
country comparison to the world: 126
7.03% (31 December 2008 est.)
Stock of narrow money: $172.9 billion (31 December 2010 est.)
country comparison to the world: 19
$178.7 billion (31 December 2009 est.)
note: see entry for the European Union for money supply in the euro area; the European Central Bank (ECB) controls monetary policy for the 17 members of the Economic and Monetary Union (EMU); individual members of the EMU do not control the quantity of money circulating within their own borders
Stock of broad money: $539.4 billion (31 December 2010 est.)
country comparison to the world: 22
$536.7 billion (31 December 2009 est.)
Stock of domestic credit: $801.1 billion (31 December 2009 est.)
country comparison to the world: 17
$767.1 billion (31 December 2008 est.)
Market value of publicly traded shares: $261.4 billion (31 December 2009)
country comparison to the world: 27
$167.4 billion (31 December 2008)
$386.4 billion (31 December 2007)
Agriculture—products: sugar beets, fresh vegetables, fruits, grain, tobacco; beef, veal, pork, milk
Industries: engineering and metal products, motor vehicle assembly, transportation equipment, scientific instruments, processed food and beverages, chemicals, basic metals, textiles, glass, petroleum
Industrial production growth rate: 4% (2010 est.)
country comparison to the world: 89
Electricity—production: 82.17 billion kWh (2007 est.)
country comparison to the world: 36
Electricity—consumption: 84.88 billion kWh (2007 est.)
country comparison to the world: 34
Electricity—exports: 6.561 billion kWh (2008 est.)
Electricity—imports: 17.16 billion kWh (2008 est.)
Oil—production: 11,220 bbl/day (2009 est.)
country comparison to the world: 82
Oil—consumption: 608,200 bbl/day (2009 est.)
country comparison to the world: 27
Oil—exports: 433,700 bbl/day (2008 est.)
country comparison to the world: 31
Oil—imports: 1.12 million bbl/day (2008 est.)
country comparison to the world: 16
Oil—proved reserves: 0 bbl (1 January 2010 est.)
country comparison to the world: 107
Natural gas—production: 0 cu m (2008 est.)
country comparison to the world: 99
Natural gas—consumption: 16.87 billion cu m (2009 est.)
country comparison to the world: 39
Natural gas—exports: 0 cu m (2008 est.)
country comparison to the world: 58
Natural gas—imports: 16.78 billion cu m (2009 est.)
country comparison to the world: 14
Natural gas—proved reserves: 0 cu m (1 January 2010 est.)
country comparison to the world: 110
Current account balance: $-1.129 billion (2010 est.)
country comparison to the world: 138
$1.251 billion (2009 est.)
Exports: $279.2 billion (2010 est.)
country comparison to the world: 16
$261.1 billion (2009 est.)
Exports—commodities: machinery and equipment, chemicals, finished diamonds, metals and metal products, foodstuffs
Exports—partners: Germany 19.58%, France 17.71%, Netherlands 11.84%, UK 7.21%, US 5.37%, Italy 4.77% (2009)
Imports: $281.7 billion (2010 est.)
country comparison to the world: 17
$261.3 billion (2009 est.)
Imports—commodities: raw materials, machinery and equipment, chemicals, raw diamonds, pharmaceuticals, foodstuffs, transportation equipment, oil products
Imports—partners: Netherlands 17.93%, Germany 17.14%, France 11.69%, Ireland 6.26%, US 5.74%, UK 5.07%, China 4.09% (2009)
Reserves of foreign exchange and gold: $NA (31 December 2010 est.)
$23.98 billion (31 December 2009 est.)
Debt—external: $1.241 trillion (30 June 2010)
country comparison to the world: 12
$1.354 trillion (31 December 2008)
Stock of direct foreign investment—at home: $741.7 billion (31 December 2010 est.)
country comparison to the world: 6
$705.2 billion (31 December 2009 est.)
Stock of direct foreign investment—abroad: $632.8 billion (31 December 2010 est.)
country comparison to the world: 10
$595.8 billion (31 December 2009 est.)
Exchange rates: euros (EUR) per US dollar– 0.755 (2010), 0.72 (2009), 0.6827 (2008), 0.7345 (2007), 0.7964 (2006)

COMMUNICATIONS

Telephones—main lines in use: 4.255 million (2009)
country comparison to the world: 36
Telephones—mobile cellular: 12.419 million (2009)
country comparison to the world: 55
Telephone system: *general assessment:* highly developed, technologically advanced, and completely automated domestic and international telephone and telegraph facilities
domestic: nationwide mobile-cellular telephone system; extensive cable network; limited microwave radio relay network
international: country code–32; landing point for a number of submarine cables that provide links to Europe, the Middle East, and Asia; satellite earth stations–7 (Intelsat–3) (2007)
Broadcast media: a segmented market with the three major communities (Flemish, French, and German-speaking) each having responsibility for their own broadcast media; multiple TV channels exist for each community; additionally, in excess of 90% of households are connected to cable and can access broadcasts of TV stations from neighboring countries; each community has a public radio network co-existing with private broadcasters (2007)
Internet country code: .be
Internet hosts: 4.465 million (2010)
country comparison to the world: 19
Internet users: 8.113 million (2009)
country comparison to the world: 36

TRANSPORTATION

Airports: 43 (2010)
country comparison to the world: 100
Airports—with paved runways: *total:* 27
over 3,047 m: 6
2,438 to 3,047 m: 9
1,524 to 2,437 m: 2
914 to 1,523 m: 1
under 914 m: 9 (2010)

Airports—with unpaved runways: *total:* 16
914 to 1,523 m: 1
under 914 m: 15 (2010)
Heliports: 1 (2010)
Pipelines: gas 2,826 km; oil 154 km; refined products 535 km (2010)
Railways: *total:* 3,233 km
country comparison to the world: 53
standard gauge: 3,233 km 1.435-m gauge (2,950 km electrified) (2010)
Roadways: *total:* 152,256 km
country comparison to the world: 33
paved: 119,079 km (includes 1,763 km of expressways)
unpaved: 33,177 km (2006)
Waterways: 2,043 km (1,528 km in regular commercial use) (2010)
country comparison to the world: 43
Merchant marine: *total:* 81
country comparison to the world: 55
by type: bulk carrier 21, cargo 8, chemical tanker 5, container 4, liquefied gas 23, passenger 2, petroleum tanker 11, roll on/roll off 7
foreign-owned: 13 (Denmark 4, France 5, UK 2, US 2)
registered in other countries: 104 (Bahamas 9, Cambodia 1, Cyprus 2, France 7, Gibraltar 2, Greece 16, Hong Kong 16, Liberia 1, Luxembourg 9, Malta 14, Moldova 2, Mozambique 2, North Korea 1, Panama 2, Portugal 8, Russia 4, Saint Kitts and Nevis 1, Saint Vincent and the Grenadines 6, Vanuatu 1) (2010)
Ports and terminals: cargo ports (tonnage): Antwerp, Gent, Liege, Zeebrugge
container ports (TEUs): Antwerp (8,662,891), Zeebrugge (2,209,715)

MILITARY

Military branches: Belgian Armed Forces: Land Operations Command, Naval Operations Command, Air Operations Commands (2010)
Military service age and obligation: 18 years of age for voluntary military service; conscription suspended (2010)
Manpower available for military service: *males age 16–49:* 2,359,232
females age 16–49: 2,291,689 (2010 est.)
Manpower fit for military service: *males age 16–49:* 1,934,957
females age 16–49: 1,877,268 (2010 est.)
Manpower reaching militarily significant age annually: *male:* 59,665
female: 57,142 (2010 est.)
Military expenditures: 1.3% of GDP (2005 est.)
country comparison to the world: 112

TRANSNATIONAL ISSUES

Disputes—international: none
Illicit drugs: growing producer of synthetic drugs and cannabis; transit point for US-bound ecstasy; source of precursor chemicals for South American cocaine processors; transshipment point for cocaine, heroin, hashish, and marijuana entering Western Europe; despite a strengthening of legislation, the country remains vulnerable to money laundering related to narcotics, automobiles, alcohol, and tobacco; significant domestic consumption of ecstasy

BELIZE

INTRODUCTION

Background: Belize was the site of several Mayan city states until their decline at the end of the first millennium A.D. The British and Spanish disputed the region in the 17th and 18th centuries; it formally became the colony of British Honduras in 1854. Territorial disputes between the UK and Guatemala delayed the independence of Belize until 1981. Guatemala refused to recognize the new nation until 1992 and the two countries are involved in an ongoing border dispute. Guatemala and Belize plan to hold a simultaneous referendum to determine if this dispute will go before the International Court of Justice at The Hague, though they have not yet set a date. Tourism has become the mainstay of the economy. Current concerns include the country's heavy foreign debt burden, high unemployment, growing involvement in the Mexican and South American drug trade, high crime rates, and one of the highest prevalence rates of HIV/AIDS in Central America.

GEOGRAPHY

Location: Central America, bordering the Caribbean Sea, between Guatemala and Mexico
Geographic coordinates: 17 15 N, 88 45 W
Map references: Central America and the Caribbean
Area: *total:* 22,966 sq km
country comparison to the world: 151
land: 22,806 sq km
water: 160 sq km
Area—comparative: slightly smaller than Massachusetts
Land boundaries: *total:* 516 km
border countries: Guatemala 266 km, Mexico 250 km
Coastline: 386 km
Maritime claims: territorial sea: 12 nm in the north, 3 nm in the south; note—from the mouth of the Sarstoon River to Ranguana Cay, Belize's territorial sea is 3 nm; according to Belize's Maritime Areas Act, 1992, the purpose of this limitation is to provide a framework for negotiating a definitive agreement on territorial differences with Guatemala
exclusive economic zone: 200 nm
Climate: tropical; very hot and humid; rainy season (May to November); dry season (February to May)
Terrain: flat, swampy coastal plain; low mountains in south
Elevation extremes: *lowest point:* Caribbean Sea 0 m
highest point: Doyle's Delight 1,160 m
Natural resources: arable land potential, timber, fish, hydropower
Land use: *arable land:* 3.05%
permanent crops: 1.39%
other: 95.56% (2005)
Irrigated land: 30 sq km (2003)
Total renewable water resources: 18.6 cu km (2000)
Freshwater withdrawal (domestic/industrial/agricultural): *total:* 0.15 cu km/yr (7%/73%/20%)
per capita: 556 cu m/yr (2000)
Natural hazards: frequent, devastating hurricanes (June to November) and coastal flooding (especially in south)
Environment—current issues: deforestation; water pollution from sewage, industrial effluents, agricultural runoff; solid and sewage waste disposal
Environment—international agreements: *party to:* Biodiversity, Climate Change, Climate Change-Kyoto Protocol, Desertification, Endangered Species, Hazardous Wastes, Law of the Sea, Ozone Layer Protection, Ship Pollution, Wetlands, Whaling
signed, but not ratified: none of the selected agreements
Geography—note: only country in Central America without a coastline on the North Pacific Ocean

PEOPLE

Population: 321,115 (July 2011 est.)
country comparison to the world: 176
Age structure: *0–14 years:* 36.8% (male 60,327/female 57,933)
15–64 years: 59.6% (male 96,886/female 94,605)
65 years and over: 3.5% (male 5,404/female 5,960) (2011 est.)
Median age: *total:* 21 years
male: 20.8 years
female: 21.2 years (2011 est.)
Population growth rate: 2.056% (2011 est.)
country comparison to the world: 45
Birth rate: 26.43 births/1,000 population (2011 est.)
country comparison to the world: 53

Death rate: 5.87 deaths/1,000 population (July 2011 est.)
country comparison to the world: 170
Net migration rate: 0 migrant(s)/1,000 population (2011 est.)
country comparison to the world: 75
Urbanization:
urban population: 52% of total population (2010)
rate of urbanization: 2.7% annual rate of change (2010-15 est.)
Major cities—population: BELMOPAN (capital) 20,000 (2009)
Sex ratio: *at birth:* 1.05 male(s)/female
under 15 years: 1.04 male(s)/female
15–64 years: 1.02 male(s)/female
65 years and over: 0.91 male(s)/female
total population: 1.03 male(s)/female (2011 est.)
Infant mortality rate: *total:* 21.95 deaths/1,000 live births
country comparison to the world: 91
male: 24.43 deaths/1,000 live births
female: 19.35 deaths/1,000 live births (2011 est.)
Life expectancy at birth: *total population:* 68.23 years
country comparison to the world: 154
male: 66.53 years
female: 70.02 years (2011 est.)
Total fertility rate: 3.21 children born/woman (2011 est.)
country comparison to the world: 52
HIV/AIDS—adult prevalence rate: 2.3% (2009 est.)
country comparison to the world: 27
HIV/AIDS—people living with HIV/AIDS: 4,800 (2009 est.)
country comparison to the world: 120
HIV/AIDS—deaths: fewer than 500 (2009 est.)
country comparison to the world: 96
Major infectious diseases: *degree of risk:* high
food or waterborne diseases: bacterial diarrhea, hepatitis A, and typhoid fever
vectorborne diseases: dengue fever and malaria
water contact disease: leptospirosis (2009)
Drinking water source: *Improved:*
urban: 99% of population
rural: 100% of population
total: 99% of population
Unimproved: urban: 1% of population
rural: 0% of population
total: 1% of population (2008)
Sanitation facility access: *Improved:*
urban: 93% of population
rural: 86% of population
total: 90% of population
Unimproved: urban: 7% of population
rural: 14% of population
total: 10% of population (2008)
Nationality: *noun:* Belizean(s)
adjective: Belizean
Ethnic groups: mestizo 48.7%, Creole 24.9%, Maya 10.6%, Garifuna 6.1%, other 9.7% (2000 census)
Religions: Roman Catholic 49.6%, Protestant 27% (Pentecostal 7.4%, Anglican 5.3%, Seventh-Day Adventist 5.2%, Mennonite 4.1%, Methodist 3.5%, Jehovah's Witnesses 1.5%), other 14%, none 9.4% (2000)
Languages: Spanish 46%, Creole 32.9%, Mayan dialects 8.9%, English 3.9% (official), Garifuna 3.4% (Carib), German 3.3%, other 1.4%, unknown 0.2% (2000 census)
Literacy: *definition:* age 15 and over can read and write
total population: 76.9%
male: 76.7%
female: 77.1% (2000 census)
School life expectancy (primary to tertiary education): *total:* 12 years
male: 12 years
female: 13 years (2009)
Education expenditures: 5.7% of GDP (2008)
country comparison to the world: 36

GOVERNMENT

Country name: *conventional long form:* none
conventional short form: Belize
former: British Honduras
Government type: parliamentary democracy and a Commonwealth realm
Capital: *name:* Belmopan
geographic coordinates: 17 15 N, 88 46 W
time difference: UTC-6 (1 hour behind Washington, DC during Standard Time)
Administrative divisions: 6 districts; Belize, Cayo, Corozal, Orange Walk, Stann Creek, Toledo
Independence: 21 September 1981 (from the UK)
National holiday: Independence Day, 21 September (1981)
Constitution: 21 September 1981
Legal system: English common law
International law organization participation: has not submitted an ICJ jurisdiction declaration; accepts ICCt jurisdiction
Suffrage: 18 years of age; universal
Executive branch: *chief of state:* Queen ELIZABETH II (since 6 February 1952); represented by Governor General Sir Colville YOUNG, Sr. (since 17 November 1993)
head of government: Prime Minister Dean Oliver BARROW (since 8 February 2008); Deputy Prime Minister Gaspar VEGA (since 12 February 2008)
cabinet: Cabinet appointed by the governor general on the advice of the prime minister from the General Assembly
(For more information visit the World Leaders website)
elections: the monarchy is hereditary; governor general appointed by the monarch; following legislative elections, the leader of the majority party or the leader of the majority coalition usually appointed prime minister by the governor general; prime minister recommends the deputy prime minister
Legislative branch: bicameral National Assembly consists of the Senate (12 seats; members appointed by the governor general—6 on the advice of the prime minister, 3 on the advice of the leader of the opposition, and 1 each on the advice of the Belize Council of Churches and Evangelical Association of Churches, the Belize Chamber of Commerce and Industry and the Belize Better Business Bureau, and the National Trade Union Congress and the Civil Society Steering Committee; to serve five-year terms) and the House of Representatives (31 seats; members are elected by direct popular vote to serve five-year terms)
elections: House of Representatives—last held on 6 February 2008 (next to be held in 2013)
election results: percent of vote by party—UDP 56.3%, PUP 40.9%; seats by party—UDP 25, PUP 6
Judicial branch: Supreme Court (the chief justice is appointed by the governor general on the advice of the prime minister); Court of Appeal; Privy Council in the UK; member of the Caribbean Court of Justice (CCJ); Summary Jurisdiction Courts (criminal) and District Courts (civil jurisdiction)
Political parties and leaders: National Alliance for Belizean Rights or NABR; National Reform Party or NRP [Cornelius DUECK]; People's National Party or PNP [Wil MAHEIA]; People's United Party or PUP [John BRICENO]; United Democratic Party or UDP [Dean BARROW]; Vision Inspired by the People or VIP [Paul MORGAN]; We the People Reform Movement or WTP [Hipolito BAUTISTA]
Political pressure groups and leaders: Society for the Promotion of Education and Research or SPEAR [Nicole HAYLOCK]; Association of Concerned Belizeans or ACB [David VASQUEZ]; National Trade Union Congress of Belize or NTUC/B [Rene GOMEZ]
International organization participation: ACP, AOSIS, C, Caricom, CDB, FAO, G-77, IADB, IAEA, IBRD, ICAO, ICRM, IDA, IFAD, IFC, IFRCS, ILO, IMF, IMO, Interpol, IOC, IOM, ITU, ITUC, LAES, MIGA, NAM, OAS, OPANAL, OPCW, PCA, PetroCaribe, RG, SICA, UN, UNCTAD, UNESCO, UNIDO, UPU, WCO, WHO, WIPO, WMO, WTO
Diplomatic representation in the US: *chief of mission:* Ambassador Nestor MENDEZ
chancery: 2535 Massachusetts Avenue NW, Washington, DC 20008
telephone: [1] (202) 332-9636
FAX: [1] (202) 332-6888
consulate(s) general: Los Angeles
Diplomatic representation from the US: *chief of mission:* Ambassador Vinai THUMMALAPALLY
embassy: Floral Park Road, Belmopan City, Cayo District
mailing address: P.O. Box 497, Belmopan City, Cayo District, Belize
telephone: [501] 822-4011
FAX: [501] 822-4012
Flag description: blue with a narrow red stripe along the top and the bottom edges; centered is a large white disk bearing the coat of arms; the coat of arms features a shield flanked by two workers in front of a mahogany tree with the related motto SUB UMBRA FLOREO (I Flourish in the Shade) on a scroll at the bottom, all encircled by a green garland of 50 mahogany leaves; the colors are those of the two main political parties: blue for the PUP and red for the UDP; various elements of the coat of arms—the figures, the tools, the mahogany tree, and the garland of leaves—recall the logging industry that led to British settlement of Belize
note: Belize's flag is the only national flag that depicts human beings; two British

overseas territories, Montserrat and the British Virgin Islands, also depict humans
National anthem: *name:* "Land of the Free"
lyrics/music: Samuel Alfred HAYNES/ Selwyn Walford YOUNG
note: adopted 1981; as a Commonwealth country, in addition to the national anthem, "God Save the Queen" serves as the royal anthem (see United Kingdom)

ECONOMY

Economy—overview: Tourism is the number one foreign exchange earner in this small economy, followed by exports of marine products, citrus, cane sugar, bananas, and garments. The government's expansionary monetary and fiscal policies, initiated in September 1998, led to GDP growth averaging nearly 4% in 1999-2007. Oil discoveries in 2006 bolstered this growth. Exploration efforts have continued and production has increased a small amount. In February 2007, the government restructured nearly all of its public external commercial debt, which helped reduce interest payments and relieved some of the country's liquidity concerns. Growth slipped to 0% in 2009 and 1.5% in 2010 as a result of the global slowdown, natural disasters, and the drop in the price of oil. With weak economic growth and a large public debt burden, fiscal spending is likely to be tight. A key government objective remains the reduction of poverty and inequality with the help of international donors. Although Belize has the second highest per capita income in Central America, the average income figure masks a huge income disparity between rich and poor. The 2010 Poverty Assessment shows that more than 4 out of 10 people live in poverty. The sizable trade deficit and heavy foreign debt burden continue to be major concerns.
GDP (purchasing power parity): $2.651 billion (2010 est.)
country comparison to the world: 181
$2.599 billion (2009 est.)
$2.6 billion (2008 est.)
note: data are in 2010 US dollars
GDP (official exchange rate): $1.396 billion (2010 est.)
GDP—real growth rate: 2% (2010 est.)
country comparison to the world: 149
0% (2009 est.)
3.8% (2008 est.)
GDP—per capita (PPP): $8,400 (2010 est.)
country comparison to the world: 119
$8,400 (2009 est.)
$8,600 (2008 est.)
note: data are in 2010 US dollars
GDP—composition by sector: *agriculture:* 29%
industry: 16.9%
services: 54.1% (2008 est.)
Labor force: 120,500
country comparison to the world: 179
note: shortage of skilled labor and all types of technical personnel (2008 est.)
Labor force—by occupation: *agriculture:* 10.2%
industry: 18.1%
services: 71.7% (2007)
Unemployment rate: 13.1% (2009)
country comparison to the world: 135
8.2% (2008)
Population below poverty line: 43% (2010 est.)
Household income or consumption by percentage share: *lowest 10%:* NA%
highest 10%: NA%
Investment (gross fixed): 26.2% of GDP (2010 est.)
country comparison to the world: 36
Budget: *revenues:* $370.5 million
expenditures: $418 million (2010 est.)
Public debt: 80% of GDP (2010 est.)
country comparison to the world: 17
Inflation rate (consumer prices): 4.1% (2010 est.)
country comparison to the world: 119
-1.1% (2009 est.)
Central bank discount rate: 12% (31 December 2009)
country comparison to the world: 30
12% (31 December 2008)
Commercial bank prime lending rate: 14.08% (31 December 2009 est.)
country comparison to the world: 51
14.14% (31 December 2008 est.)
Stock of narrow money: $389.5 million (31 December 2010 est.)
country comparison to the world: 161
$336.5 million (31 December 2009 est.)
Stock of broad money: $1.351 billion (31 December 2010 est.)
country comparison to the world: 151
$1.084 billion (31 December 2009 est.)
Stock of domestic credit: $1.291 billion (31 December 2010 est.)
country comparison to the world: 141
$1.036 billion (31 December 2009 est.)
Market value of publicly traded shares: $NA
Agriculture—products: bananas, cacao, citrus, sugar; fish, cultured shrimp; lumber
Industries: garment production, food processing, tourism, construction, oil
Industrial production growth rate: 1.4% (2010 est.)
country comparison to the world: 142
Electricity—production: 213.5 million kWh (2007 est.)
country comparison to the world: 176
Electricity—consumption: 198.5 million kWh (2007 est.)
country comparison to the world: 178
Electricity—exports: 0 kWh (2008 est.)
Electricity—imports: 248.4 million kWh (2005)
Oil—production: 3,990 bbl/day (2009 est.)
country comparison to the world: 99
Oil—consumption: 7,000 bbl/day (2009 est.)
country comparison to the world: 157
Oil—exports: 2,260 bbl/day (2007 est.)
country comparison to the world: 110
Oil—imports: 7,204 bbl/day (2007 est.)
country comparison to the world: 147
Oil—proved reserves: 6.7 million bbl (1 January 2010 est.)
country comparison to the world: 94
Natural gas—production: 0 cu m (2008 est.)
country comparison to the world: 101
Natural gas—consumption: 0 cu m (2008 est.)
country comparison to the world: 149
Natural gas—exports: 0 cu m (2008 est.)
country comparison to the world: 61
Natural gas—imports: 0 cu m (2008 est.)
country comparison to the world: 153
Natural gas—proved reserves: 0 cu m (1 January 2010 est.)
country comparison to the world: 112
Current account balance: $-151 million (2010 est.)
country comparison to the world: 85
$-93.3 million (2009 est.)
Exports: $404 million (2010 est.)
country comparison to the world: 172
$381.9 million (2009 est.)
Exports—commodities: sugar, bananas, citrus, clothing, fish products, molasses, wood, crude oil
Exports—partners: US 30.7%, UK 29.77%, Nigeria 4.9%, Cote d'Ivoire 4.45% (2009)
Imports: $740 million (2010 est.)
country comparison to the world: 182
$620.5 million (2009 est.)
Imports—commodities: machinery and transport equipment, manufactured goods; fuels, chemicals, pharmaceuticals; food, beverages, tobacco
Imports—partners: US 33.65%, Mexico 14.17%, Cuba 8.51%, Guatemala 6.75%, Spain 6.07%, China 4.12% (2009)
Reserves of foreign exchange and gold: $219 million (31 December 2010 est.)
country comparison to the world: 132
$213.7 million (31 December 2009 est.)
Debt—external: $1.01 billion (2009 est.)
country comparison to the world: 150
$954.1 million (2008 est.)
Exchange rates: Belizean dollars (BZD) per US dollar–2 (2010), 2 (2009), 2 (2008), 2 (2007), 2 (2006)

COMMUNICATIONS

Telephones—main lines in use: 31,200 (2009)
country comparison to the world: 179
Telephones—mobile cellular: 161,800 (2009)
country comparison to the world: 175
Telephone system: *general assessment:* above-average system; trunk network depends primarily on microwave radio relay
domestic: fixed-line teledensity of 10 per 100 persons; mobile-cellular teledensity roughly 55 per 100 persons
international: country code–501; landing point for the Americas Region Caribbean Ring System (ARCOS-1) fiber-optic telecommunications submarine cable that provides links to South and Central America, parts of the Caribbean, and the US; satellite earth station–8 (Intelsat–2, unknown–6) (2008)
Broadcast media: 8 privately-owned TV stations; multi-channel cable TV provides access to foreign stations; about 25 radio stations broadcasting on roughly 50 different frequencies; state-run radio was privatized in 1998 (2007)
Internet country code: .bz
Internet hosts: 2,880 (2010)
country comparison to the world: 147
Internet users: 36,000 (2009)
country comparison to the world: 178

TRANSPORTATION

Airports: 45 (2010)
country comparison to the world: 96
Airports—with paved runways: *total:* 4
2,438 to 3,047 m: 1

914 to 1,523 m: 1
under 914 m: 2 (2010)
Airports—with unpaved runways: *total:* 41
2,438 to 3,047 m: 1
914 to 1,523 m: 13
under 914 m: 27 (2010)
Roadways: *total:* 3,007 km
country comparison to the world: 165
paved: 575 km
unpaved: 2,432 km (2006)
Waterways: 825 km (navigable only by small craft) (2010)
country comparison to the world: 71
Merchant marine: *total:* 231
country comparison to the world: 33
by type: barge carrier 1, bulk carrier 37, cargo 146, chemical tanker 1, passenger 1, passenger/cargo 1, petroleum tanker 7, refrigerated cargo 27, roll on/roll off 10
foreign-owned: 171 (Chile 1, China 64, Croatia 1, Cyprus 1, Estonia 1, Germany 1, Greece 2, Iceland 2, Italy 3, Japan 1, Latvia 10, Lithuania 2, Netherlands 1, Nigeria 2, Norway 3, Peru 1, Russia 32, Singapore 7, Spain 1, Syria 2, Turkey 18, UAE 5, UK 4, Ukraine 6) (2010)
Ports and terminals: Belize City, Big Creek

MILITARY

Military branches: Belize Defense Force (BDF): Army, BDF Air Wing (includes Special Boat Unit), BDF Volunteer Guard (2011)
Military service age and obligation: 18 years of age for voluntary military service; law allows for conscription only if volunteers are insufficient; conscription has never been implemented; volunteers typically outnumber available positions by 3:1 (2008)
Manpower available for military service: *males age 16–49:* 81,284
females age 16–49: 79,185 (2010 est.)
Manpower fit for military service: *males age 16–49:* 59,431
females age 16–49: 57,221 (2010 est.)
Manpower reaching militarily significant age annually: *male:* 3,723
female: 3,584 (2010 est.)
Military expenditures: 1.4% of GDP (2009)
country comparison to the world: 105

TRANSNATIONAL ISSUES

Disputes—international: Guatemala persists in its territorial claim to half of Belize, but agrees to Line of Adjacency to keep Guatemalan squatters out of Belize's forested interior; Belize and Mexico are working to solve minor border demarcation discrepancies arising from inaccuracies in the 1898 border treaty
Trafficking in persons: current situation: Belize is a source, transit, and destination country for men, women, and children trafficked for the purposes of commercial sexual exploitation and forced labor; the most common form of trafficking in Belize is the internal sex trafficking of minors; some Central American men, women, and children, particularly from Guatemala, Honduras, and El Salvador, migrate voluntarily to Belize in search of work but are subsequently subjected to conditions of forced labor or forced prostitution
tier rating: Belize is placed on Tier 2 Watch List because it does not fully comply with the minimum standards for the elimination of trafficking; however, it is making significant efforts to do so; despite efforts to raise public awareness of human trafficking and provide protection services for trafficking victims, the government did not show evidence of progress in convicting and sentencing trafficking offenders last year (2009)
Illicit drugs: transshipment point for cocaine; small-scale illicit producer of cannabis, primarily for local consumption; offshore sector money-laundering activity related to narcotics trafficking and other crimes (2008)

BENIN

INTRODUCTION

Background: Present day Benin was the site of Dahomey, a prominent West African kingdom that rose in the 15th century. The territory became a French Colony in 1872 and achieved independence on 1 August 1960, as the Republic of Benin. A succession of military governments ended in 1972 with the rise to power of Mathieu KEREKOU and the establishment of a government based on Marxist-Leninist principles. A move to representative government began in 1989. Two years later, free elections ushered in former Prime Minister Nicephore SOGLO as president, marking the first successful transfer of power in Africa from a dictatorship to a democracy. KEREKOU was returned to power by elections held in 1996 and 2001, though some irregularities were alleged. KEREKOU stepped down at the end of his second term in 2006 and was succeeded by Thomas YAYI Boni, a political outsider and independent. YAYI has attempted to stem corruption and has strongly promoted accelerating Benin's economic growth.

GEOGRAPHY

Location: Western Africa, bordering the Bight of Benin, between Nigeria and Togo
Geographic coordinates: 9 30 N, 2 15 E
Map references: Africa
Area: *total:* 112,622 sq km
country comparison to the world: 101
land: 110,622 sq km
water: 2,000 sq km
Area—comparative: slightly smaller than Pennsylvania
Land boundaries: *total:* 1,989 km
border countries: Burkina Faso 306 km, Niger 266 km, Nigeria 773 km, Togo 644 km
Coastline: 121 km
Maritime claims: territorial sea: 200 nm
Climate: tropical; hot, humid in south; semiarid in north
Terrain: mostly flat to undulating plain; some hills and low mountains
Elevation extremes: *lowest point:* Atlantic Ocean 0 m
highest point: Mont Sokbaro 658 m
Natural resources: small offshore oil deposits, limestone, marble, timber
Land use: *arable land:* 23.53%
permanent crops: 2.37%
other: 74.1% (2005)
Irrigated land: 120 sq km (2003)
Total renewable water resources: 25.8 cu km (2001)
Freshwater withdrawal (domestic/industrial/agricultural): *total:* 0.13 cu km/yr (32%/23%/45%)
per capita: 15 cu m/yr (2001)
Natural hazards: hot, dry, dusty harmattan wind may affect north from December to March
Environment—current issues: inadequate supplies of potable water; poaching threatens wildlife populations; deforestation; desertification
Environment—international agreements: *party to:* Biodiversity, Climate Change, Climate Change-Kyoto Protocol, Desertification, Endangered Species, Environmental Modification, Hazardous Wastes, Law of the Sea, Ozone Layer Protection, Ship Pollution, Wetlands, Whaling
signed, but not ratified: none of the selected agreements
Geography—note: sandbanks create difficult access to a coast with no natural harbors, river mouths, or islands

PEOPLE

Population: 9,325,032 (July 2011 est.)
country comparison to the world: 89
note: estimates for this country explicitly take into account the effects of excess mortality due to AIDS; this can result in lower life expectancy, higher infant mortality, higher death rates, lower population growth rates, and changes in the distribution of population by age and sex than would otherwise be expected
Age structure: *0–14 years:* 44.7% (male 2,126,973/female 2,042,340)
15–64 years: 52.6% (male 2,443,370/female 2,461,421)
65 years and over: 2.7% (male 101,640/female 149,288) (2011 est.)
Median age: *total:* 17.4 years
male: 17 years
female: 17.9 years (2011 est.)
Population growth rate: 2.911% (2011 est.)
country comparison to the world: 13
Birth rate: 38.11 births/1,000 population (2011 est.)
country comparison to the world: 15
Death rate: 9 deaths/1,000 population (July 2011 est.)
country comparison to the world: 70
Net migration rate: 0 migrant(s)/1,000 population (2011 est.)
country comparison to the world: 77
Urbanization: *urban population:* 42% of total population (2010)
rate of urbanization: 4% annual rate of change (2010-15 est.)
Major cities—population: COTONOU (seat of government) 815,000; PORTO-NOVO (capital) 276,000 (2009)
Sex ratio: *at birth:* 1.05 male(s)/female
under 15 years: 1.04 male(s)/female
15–64 years: 0.99 male(s)/female
65 years and over: 0.69 male(s)/female
total population: 1 male(s)/female (2011 est.)
Infant mortality rate: *total:* 61.56 deaths/1,000 live births
country comparison to the world: 30
male: 64.89 deaths/1,000 live births
female: 58.07 deaths/1,000 live births (2011 est.)
Life expectancy at birth: *total population:* 59.84 years
country comparison to the world: 187
male: 58.61 years
female: 61.14 years (2011 est.)
Total fertility rate: 5.31 children born/woman (2011 est.)
country comparison to the world: 14
HIV/AIDS—adult prevalence rate: 1.2% (2009 est.)
country comparison to the world: 39
HIV/AIDS—people living with HIV/AIDS: 60,000 (2009 est.)
country comparison to the world: 56
HIV/AIDS—deaths: 2,700 (2009 est.)
country comparison to the world: 48
Major infectious diseases: *degree of risk:* very high
food or waterborne diseases: bacterial and protozoal diarrhea, hepatitis A, and typhoid fever
vectorborne diseases: malaria and yellow fever
respiratory disease: meningococcal meningitis
animal contact disease: rabies (2009)
Drinking water source: *Improved:*
urban: 84% of population
rural: 69% of population
total: 75% of population
Unimproved: urban: 16% of population
rural: 31% of population
total: 25% of population (2008)
Sanitation facility access: *Improved:*
urban: 24% of population
rural: 4% of population
total: 12% of population
Unimproved:
urban: 76% of population
rural: 96% of population
total: 88% of population (2008)
Nationality: *noun:* Beninese (singular and plural)
adjective: Beninese
Ethnic groups: Fon and related 39.2%, Adja and related 15.2%, Yoruba and related 12.3%, Bariba and related 9.2%, Peulh and related 7%, Ottamari and related 6.1%, Yoa-Lokpa and related 4%, Dendi and related 2.5%, other 1.6% (includes Europeans), unspecified 2.9% (2002 census)
Religions: Christian 42.8% (Catholic 27.1%, Celestial 5%, Methodist 3.2%, other Protestant 2.2%, other 5.3%), Muslim 24.4%, Vodoun 17.3%, other 15.5% (2002 census)
Languages: French (official), Fon and Yoruba (most common vernaculars in south), tribal languages (at least six major ones in north)
Literacy: *definition:* age 15 and over can read and write
total population: 34.7%
male: 47.9%
female: 23.3% (2002 census)
School life expectancy (primary to tertiary education): *total:* 9 years
male: 11 years
female: 8 years (2005)
Education expenditures: 3.5% of GDP (2007)
country comparison to the world: 117

GOVERNMENT

Country name: *conventional long form:* Republic of Benin
conventional short form: Benin
local long form: Republique du Benin
local short form: Benin
former: Dahomey
Government type: republic
Capital: *name:* Porto-Novo (official capital)
geographic coordinates: 6 29 N, 2 37 E
time difference: UTC+1 (6 hours ahead of Washington, DC during Standard Time)
note: Cotonou (seat of government)
Administrative divisions: 12 departments; Alibori, Atakora, Atlantique, Borgou, Collines, Kouffo, Donga, Littoral, Mono, Oueme, Plateau, Zou
Independence: 1 August 1960 (from France)
National holiday: National Day, 1 August (1960)
Constitution: adopted by referendum 2 December 1990
Legal system: civil law system modeled largely on the French system and some customary law
International law organization participation: has not submitted an ICJ jurisdiction declaration; accepts ICCt jurisdiction
Suffrage: 18 years of age; universal
Executive branch: *chief of state:* President Thomas YAYI Boni (since 6 April 2006); note–the president is both the chief of state and head of government
head of government: President Thomas YAYI Boni (since 6 April 2006)
cabinet: Council of Ministers appointed by the president
(For more information visit the World Leaders website)
elections: president elected by popular vote for a five-year term (eligible for a second term); last held on 13 March 2011 (next to be held in March 2016)
election results: Thomas YAYI Boni re-elected president; percent of vote–Thomas YAYI Boni 53.1%, Adrien HOUNGBEDJI 35.6%, Abdoulaye Bio TCHANE 6.1%, other 5.2%
Legislative branch: unicameral National Assembly or Assemblee Nationale (83 seats; members are elected by direct popular vote to serve four-year terms)
elections: last held on 30 April 2011 (next to be held in 2015)
election results: percent of vote by party–NA; seats by party–FCBE 41, UN 30, other 12
Judicial branch: Constitutional Court or Cour Constitutionnelle (7 members; 4 appointed by the National Assembly, 3 appointed by the President; appointed for a 5-year term for one term); Supreme Court or Cour Supreme (President of the Supreme Court appointed by the President for a 5-year term); High Court of Justice (composed of members of the Constitutional Court and 6 members appointed by the National Assembly)
Political parties and leaders: African Movement for Democracy and Progress or MADEP [Sefou FAGBOHOUN]; Alliance for Dynamic Democracy or ADD; Alliance of Progress Forces or AFP; Benin Renaissance or RB [Rosine SOGLO]; Democratic Renewal Party or PRD [Adrien HOUNGBEDJI]; Force Cowrie for an Emerging Benin or FCBE; Impulse for Progress and Democracy or IPD [Theophile NATA]; Key Force or FC [Lazare SÈHOUÉTO]; Movement for the People's Alternative or MAP [Olivier CAPO-CHICHI]; Rally for Democracy and Progress or RDP [Dominique HOUNGNINOU]; Social Democrat Party or PSD [Bruno AMOUSSOU]; Union for Democracy and National Solidarity or UDS [Sacca LAFIA]; Union for the Relief or UPR [Issa SALIFOU]
note: approximately 20 additional minor parties
Political pressure groups and leaders: *other:* economic groups; environmentalists; political groups; teachers' unions and other educational groups
International organization participation: ACP, AfDB, AU, ECOWAS, Entente, FAO, FZ, G-77, IAEA, IBRD, ICAO, ICRM, IDA, IDB, IFAD, IFC, IFRCS, ILO, IMF, IMO, Interpol, IOC, IOM, IPU, ISO (correspondent), ITSO, ITU, ITUC, MIGA, MONUSCO, NAM, OAS (observer), OIC, OIF, OPCW, PCA, UN, UNCTAD, UNESCO, UNHCR, UNIDO, UNMIL, UNMIS, UNOCI, UNWTO,

UPU, WAEMU, WCO, WFTU, WHO, WIPO, WMO, WTO
Diplomatic representation in the US: *chief of mission:* Ambassador Cyrille Segbe OGUIN
chancery: 2124 Kalorama Road NW, Washington, DC 20008
telephone: [1] (202) 232-6656
FAX: [1] (202) 265-1996
Diplomatic representation from the US: *chief of mission:* Ambassador James A. KNIGHT
embassy: Rue Caporal Bernard Anani, Cotonou
mailing address: 01 B. P. 2012, Cotonou
telephone: [229] 21-30-06-50
FAX: [229] 21-30-03-84
Flag description: two equal horizontal bands of yellow (top) and red (bottom) with a vertical green band on the hoist side; green symbolizes hope and revival, yellow wealth, and red courage
note: uses the popular Pan-African colors of Ethiopia
National anthem: *name:* "L'Aube Nouvelle" (The Dawn of a New Day)
lyrics/music: Gilbert Jean DAGNON
note: adopted 1960

ECONOMY

Economy—overview: The economy of Benin remains underdeveloped and dependent on subsistence agriculture, cotton production, and regional trade. Growth in real output had averaged about 4% before the global recession, but fell to 2.7% in 2009 and 3% in 2010. Inflation has subsided over the past several years. In order to raise growth, Benin plans to attract more foreign investment, place more emphasis on tourism, facilitate the development of new food processing systems and agricultural products, and encourage new information and communication technology. Specific projects to improve the business climate by reforms to the land tenure system, the commercial justice system, and the financial sector were included in Benin's $307 million Millennium Challenge Account grant signed in February 2006. The 2001 privatization policy continues in telecommunications, water, electricity, and agriculture. As result of these reforms, Benin has become the most competitive country in the West African Economic and Monetary Union, according to the World Economic Forum. The Paris Club and bilateral creditors have eased the external debt situation, with Benin benefiting from a G-8 debt reduction announced in July 2005, while pressing for more rapid structural reforms. An insufficient electrical supply continues to adversely affect Benin's economic growth though the government recently has taken steps to increase domestic power production.
GDP (purchasing power parity): $13.99 billion (2010 est.)
country comparison to the world: 139
$13.66 billion (2009 est.)
$13.3 billion (2008 est.)
note: data are in 2010 US dollars
GDP (official exchange rate): $6.649 billion (2010 est.)
GDP—real growth rate: 2.5% (2010 est.)
country comparison to the world: 138
2.7% (2009 est.)
5% (2008 est.)
GDP—per capita (PPP): $1,500 (2010 est.)
country comparison to the world: 200
$1,600 (2009 est.)
$1,600 (2008 est.)
note: data are in 2010 US dollars
GDP—composition by sector: *agriculture:* 33.2%
industry: 14.5%
services: 52.3% (2007 est.)
Labor force: 3.662 million (2007 est.)
country comparison to the world: 96
Unemployment rate: NA%
Population below poverty line: 37.4% (2007 est.)
Household income or consumption by percentage share: *lowest 10%:* 3.1%
highest 10%: 29% (2003)
Distribution of family income—Gini index: 36.5 (2003)
country comparison to the world: 81
Investment (gross fixed): 18.5% of GDP (2010 est.)
country comparison to the world: 106
Budget: *revenues:* $1.348 billion
expenditures: $1.731 billion (2010 est.)
Inflation rate (consumer prices): 1.6% (2010 est.)
country comparison to the world: 42
2.2% (2009 est.)
Central bank discount rate: 4.25% (31 December 2009)
country comparison to the world: 95
4.75% (31 December 2008)
Commercial bank prime lending rate: NA%
NA%
Stock of narrow money: $1.551 billion (31 December 2010 est.)
country comparison to the world: 125
$1.619 billion (31 December 2009 est.)
Stock of broad money: $2.424 billion (31 December 2010 est.)
country comparison to the world: 137
$2.517 billion (31 December 2009 est.)
Stock of domestic credit: $1.222 billion (31 December 2010 est.)
country comparison to the world: 143
$1.269 billion (31 December 2009 est.)
Market value of publicly traded shares: $NA
Agriculture—products: cotton, corn, cassava (tapioca), yams, beans, palm oil, peanuts, cashews; livestock
Industries: textiles, food processing, construction materials, cement
Industrial production growth rate: 3% (2010 est.)
country comparison to the world: 110
Electricity—production: 124 million kWh (2007 est.)
country comparison to the world: 185
Electricity—consumption: 597 million kWh (2007 est.)
country comparison to the world: 157
Electricity—exports: 0 kWh (2008 est.)
Electricity—imports: 588 million kWh (2007 est.)
Oil—production: 0 bbl/day (2009 est.)
country comparison to the world: 153
Oil—consumption: 23,000 bbl/day (2009 est.)
country comparison to the world: 119
Oil—exports: 8,770 bbl/day (2007 est.)
country comparison to the world: 94
Oil—imports: 28,900 bbl/day (2007 est.)
country comparison to the world: 100
Oil—proved reserves: 8 million bbl (1 January 2010 est.)
country comparison to the world: 93
Natural gas—production: 0 cu m (2008 est.)
country comparison to the world: 103
Natural gas—consumption: 0 cu m (2008 est.)
country comparison to the world: 150
Natural gas—exports: 0 cu m (2008 est.)
country comparison to the world: 63
Natural gas—imports: 0 cu m (2008 est.)
country comparison to the world: 156
Natural gas—proved reserves: 1.133 billion cu m (1 January 2010 est.)
country comparison to the world: 97
Current account balance: $-582 million (2010 est.)
country comparison to the world: 121
$-644 million (2009 est.)
Exports: $1.125 billion (2010 est.)
country comparison to the world: 150
$994 million (2009 est.)
Exports—commodities: cotton, cashews, shea butter, textiles, palm products, seafood
Exports—partners: India 19.72%, China 13.18%, Niger 6.94%, Nigeria 6.56%, Indonesia 5.73%, Togo 5.63%, Namibia 4.17% (2009)
Imports: $1.812 billion (2010 est.)
country comparison to the world: 153
$1.703 billion (2009 est.)
Imports—commodities: foodstuffs, capital goods, petroleum products
Imports—partners: China 35.62%, US 7.51%, France 7.38%, Thailand 6.71%, Malaysia 6.13%, Netherlands 4.83%, Belgium 4.02% (2009)
Reserves of foreign exchange and gold: $1.254 billion (31 December 2010 est.)
country comparison to the world: 113
$1.23 billion (31 December 2009 est.)
Debt—external: $2.894 billion (31 December 2009 est.)
country comparison to the world: 129
$986.2 million (31 December 2008 est.)
Exchange rates: Communaute Financiere Africaine francs (XOF) per US dollar–495.28 (2010), 472.19 (2009), 447.81 (2008), 493.51 (2007), 522.59 (2006)

COMMUNICATIONS

Telephones—main lines in use: 127,100 (2009)
country comparison to the world: 142
Telephones—mobile cellular: 5.033 million (2009)
country comparison to the world: 94
Telephone system: *general assessment:* inadequate system of open-wire, microwave radio relay, and cellular connections; fixed-line network characterized by aging, deteriorating equipment
domestic: fixed-line teledensity only about 2 per 100 persons; spurred by the presence of multiple mobile-cellular providers, cellular telephone subscribership has been increasing rapidly
international: country code—229; landing point for the SAT-3/WASC fiber-optic

submarine cable that provides connectivity to Europe and Asia; long distance fiber-optic links with Togo, Burkina Faso, Niger, and Nigeria; satellite earth stations—7 (Intelsat-Atlantic Ocean) (2008)
Broadcast media: state-run Office de Radiodiffusion et de Television du Benin (ORTB) operates a TV station with multiple channels giving it a wide broadcast reach; several privately-owned TV stations broadcast from Cotonou; satellite TV subscription service is available; state-owned radio, under ORTB control, includes a national station supplemented by a number of regional stations; substantial number of privately-owned radio broadcast stations; transmissions of a few international broadcasters are available on FM in Cotonou (2007)
Internet country code: .bj
Internet hosts: 1,286 (2010)
country comparison to the world: 165
Internet users: 200,100 (2009)
country comparison to the world: 139

TRANSPORTATION

Airports: 5 (2010)
country comparison to the world: 178
Airports—with paved runways: *total:* 1
1,524 to 2,437 m: 1 (2010)
Airports—with unpaved runways: *total:* 4
2,438 to 3,047 m: 1
1,524 to 2,437 m: 1
914 to 1,523 m: 2 (2010)
Railways: *total:* 438 km
country comparison to the world: 113
narrow gauge: 438 km 1.000-m gauge (2010)
Roadways: *total:* 16,000 km
country comparison to the world: 119
paved: 1,400 km
unpaved: 14,600 km (2006)
Waterways: 150 km (seasonal navigation on River Niger along northern border) (2010)
country comparison to the world: 102
Ports and terminals: Cotonou

MILITARY

Military branches: Benin Armed Forces (Forces Armees Beninoises, FAB): Army (l'Arme de Terre), Benin Navy (Forces Navales Beninois, FNB), Benin Air Force (Force Aerienne du Benin, FAB) (2011)
Military service age and obligation: 21 years of age for compulsory and voluntary military service; in practice, volunteers eligible at age 18; both sexes are eligible for military service; conscript tour of duty—18 months (2006)
Manpower available for military service:
males age 16–49: 2,095,373
females age 16–49: 2,038,351 (2010 est.)
Manpower fit for military service: *males age 16–49:* 1,385,065
females age 16–49: 1,400,045 (2010 est.)
Manpower reaching militarily significant age annually: *male:* 108,496
female: 104,526 (2010 est.)
Military expenditures: 1% of GDP (2009)
country comparison to the world: 132

TRANSNATIONAL ISSUES

Disputes—international: talks continue between Benin and Togo on funding the Adjrala hydroelectric dam on the Mona River; Benin retains a border dispute with Burkina Faso around the town of Koualou; location of Benin-Niger-Nigeria tripoint is unresolved
Refugees and internally displaced persons: refugees (country of origin): 9,444 (Togo) (2007)
Illicit drugs: transshipment point used by traffickers for cocaine destined for Western Europe; vulnerable to money laundering due to poorly enforced financial regulations (2008)

BERMUDA

(OVERSEAS TERRITORY OF THE UK)

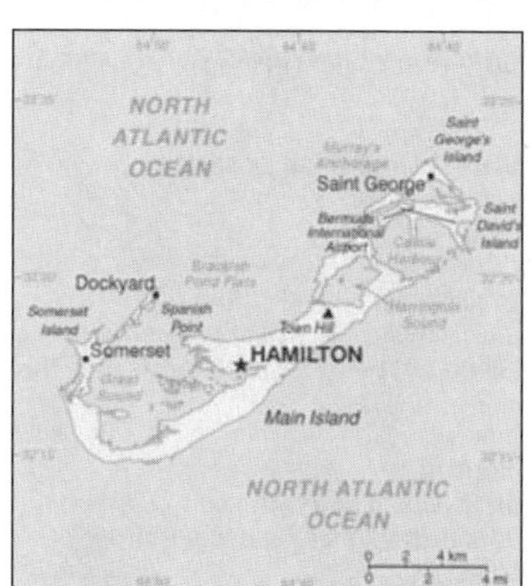

INTRODUCTION

Background: Bermuda was first settled in 1609 by shipwrecked English colonists headed for Virginia. Tourism to the island to escape North American winters first developed in Victorian times. Tourism continues to be important to the island's economy, although international business has overtaken it in recent years. Bermuda has developed into a highly successful offshore financial center. Although a referendum on independence from the UK was soundly defeated in 1995, the present government has reopened debate on the issue.

GEOGRAPHY

Location: North America, group of islands in the North Atlantic Ocean, east of South Carolina (US)
Geographic coordinates: 32 20 N, 64 45 W
Map references: North America
Area: *total:* 54 sq km
country comparison to the world: 230
land: 54 sq km
water: 0 sq km
Area—comparative: about one-third the size of Washington, DC
Land boundaries: 0 km
Coastline: 103 km
Maritime claims: territorial sea: 12 nm
exclusive fishing zone: 200 nm
Climate: subtropical; mild, humid; gales, strong winds common in winter
Terrain: low hills separated by fertile depressions
Elevation extremes: *lowest point:* Atlantic Ocean 0 m
highest point: Town Hill 76 m
Natural resources: limestone, pleasant climate fostering tourism
Land use: *arable land:* 20%
permanent crops: 0%
other: 80% (55% developed, 45% rural/open space) (2005)
Irrigated land: NA
Natural hazards: hurricanes (June to November)
Environment—current issues: sustainable development
Geography—note: consists of about 138 coral islands and islets with ample rainfall, but no rivers or freshwater lakes; some land was leased by the US Government from 1941 to 1995

PEOPLE

Population: 68,679 (July 2011 est.)
country comparison to the world: 200
Age structure: *0–14 years:* 18% (male 6,212/female 6,129)
15–64 years: 67% (male 22,701/female 23,293)
65 years and over: 15.1% (male 4,304/female 6,040) (2011 est.)
Median age: *total:* 42 years
male: 40.4 years
female: 43.5 years (2011 est.)
Population growth rate: 0.594% (2011 est.)
country comparison to the world: 145
Birth rate: 11.42 births/1,000 population (2011 est.)
country comparison to the world: 168
Death rate: 7.57 deaths/1,000 population (July 2011 est.)
country comparison to the world: 115
Net migration rate: 2.1 migrant(s)/1,000 population (2011 est.)
country comparison to the world: 39
Urbanization: *urban population:* 100% of total population (2010)
rate of urbanization: 0.2% annual rate of change (2010-15 est.)
Major cities—population: HAMILTON (capital) 12,000 (2009)
Sex ratio: *at birth:* 1.015 male(s)/female
under 15 years: 1.01 male(s)/female
15–64 years: 0.97 male(s)/female
65 years and over: 0.71 male(s)/female
total population: 0.94 male(s)/female (2011 est.)
Infant mortality rate:
total: 2.47 deaths/1,000 live births
country comparison to the world: 220
male: 2.57 deaths/1,000 live births
female: 2.36 deaths/1,000 live births (2011 est.)

Life expectancy at birth: *total population:* 80.71 years
country comparison to the world: 20
male: 77.49 years
female: 83.99 years (2011 est.)
Total fertility rate: 1.97 children born/ woman (2011 est.)
country comparison to the world: 130
HIV/AIDS—adult prevalence rate: 0.3% (2005)
country comparison to the world: 86
HIV/AIDS—people living with HIV/AIDS: 163 (2005)
country comparison to the world: 161
HIV/AIDS—deaths: 392 (2005)
country comparison to the world: 99
Nationality:
noun: Bermudian(s)
adjective: Bermudian
Ethnic groups: black 54.8%, white 34.1%, mixed 6.4%, other races 4.3%, unspecified 0.4% (2000 census)
Religions: Anglican 23%, Roman Catholic 15%, African Methodist Episcopal 11%, other Protestant 18%, other 12%, unaffiliated 6%, unspecified 1%, none 14% (2000 census)
Languages: English (official), Portuguese
Literacy: *definition:* age 15 and over can read and write
total population: 98%
male: 98%
female: 99% (2005 est.)
School life expectancy (primary to tertiary education): *total:* 12 years
male: 12 years
female: 12 years (2006)
Education expenditures: 2.6% of GDP (2009)
country comparison to the world: 147

GOVERNMENT

Country name: *conventional long form:* none
conventional short form: Bermuda
former: Somers Islands
Dependency status: overseas territory of the UK
Government type: parliamentary; self-governing territory
Capital: *name:* Hamilton
geographic coordinates: 32 17 N, 64 47 W
time difference: UTC-4 (1 hour ahead of Washington, DC during Standard Time)
daylight saving time: +1hr, begins second Sunday in March; ends first Sunday in November
Administrative divisions: 9 parishes and 2 municipalities*; Devonshire, Hamilton, Hamilton*, Paget, Pembroke, Saint George*, Saint George's, Sandys, Smith's, Southampton, Warwick
Independence: none (overseas territory of the UK)
National holiday: Bermuda Day, 24 May
Constitution: 8 June 1968; amended 1989 and 2003
Legal system: English common law
International law organization participation: has not submitted an ICJ jurisdiction declaration; non-party state to the ICCt
Suffrage: 18 years of age; universal
Executive branch: *chief of state:* Queen ELIZABETH II (since 6 February 1952); represented by Governor Sir Richard GOZNEY (since 12 December 2007)
head of government: Premier Paula COX (since 29 October 2010); Deputy Premier Derrick BURGESS
cabinet: Cabinet nominated by the premier, appointed by the governor
(For more information visit the World Leaders website)
elections: the monarchy is hereditary; governor appointed by the monarch; following legislative elections, the leader of the majority party or the leader of the majority coalition usually appointed premier by the governor
Legislative branch: bicameral Parliament consists of the Senate (11 seats; members appointed by the governor, the premier, and the opposition to serve a five-year term) and the House of Assembly (36 seats; members are elected by popular vote to serve up to five-year terms)
elections: last general election held on 18 December 2007 (next to be held not later than 2012)
election results: percent of vote by party—PLP 52.5%, UBP 47.3%; seats by party—PLP 22, UBP 14
Judicial branch: Supreme Court (Chief Justice and other justices appointed by the governor; remain in office until they reach 65 years of age); Court of Appeal (President of the Court of Appeal and other justices appointed by the governor for a specific period laid out in their respective instruments of appointment); Magistrate Courts
Political parties and leaders: Progressive Labor Party or PLP [Ewart BROWN]; United Bermuda Party or UBP [Kim SWAN]
Political pressure groups and leaders: Bermuda Employer's Union [Eddie SAINTS]; Bermuda Industrial Union or BIU [Derrick BURGESS]; Bermuda Public Services Union or BPSU [Ed BALL]; Bermuda Union of Teachers [Michael CHARLES]
International organization participation: Caricom (associate), Interpol (subbureau), IOC, ITUC, UPU, WCO
Diplomatic representation in the US: none (overseas territory of the UK)
Diplomatic representation from the US: *chief of mission:* Consul General Grace W. SHELTON
consulate(s) general: Crown Hill, 16 Middle Road, Devonshire DVO3
mailing address: P. O. Box HM325, Hamilton HMBX; American Consulate General Hamilton, US Department of State, 5300 Hamilton Place, Washington, DC 20520-5300
telephone: [1] (441) 295-1342
FAX: [1] (441) 295-1592, 296-9233
Flag description: red, with the flag of the UK in the upper hoist-side quadrant and the Bermudian coat of arms (a white shield with a red lion standing on a green grassy field holding a scrolled shield showing the sinking of the ship Sea Venture off Bermuda in 1609) centered on the outer half of the flag; it was the shipwreck of the vessel, filled with English colonists originally bound for Virginia, that led to settling of Bermuda
note: the flag is unusual in that it is only British overseas territory that uses a red ensign, all others use blue
National anthem: *name:* "Hail to Bermuda"
lyrics/music: Bette JOHNS
note: serves as a local anthem; as a territory of the United Kingdom, "God Save the Queen" is official (see United Kingdom)

ECONOMY

Economy—overview: Bermuda enjoys the third highest per capita income in the world, more than 50% higher than that of the US; the average cost of a house by the mid-2000s exceeded $1,000,000. Its economy is primarily based on providing financial services for international business and luxury facilities for tourists. A number of reinsurance companies relocated to the island following the 11 September 2001 attacks and again after Hurricane Katrina in August 2005 contributing to the expansion of an already robust international business sector. Bermuda's tourism industry—which derives over 80% of its visitors from the US—continues to struggle but remains the island's number two industry. Most capital equipment and food must be imported. Bermuda's industrial sector is largely focused on construction and agriculture is limited, with only 20% of the land being arable.
GDP (purchasing power parity): $4.5 billion (2004 est.)
country comparison to the world: 166
GDP (official exchange rate): $NA
GDP—real growth rate: 4.6% (2004 est.)
country comparison to the world: 81
GDP—per capita (PPP): $69,900 (2004 est.)
country comparison to the world: 4
GDP—composition by sector:
agriculture: 1%
industry: 10%
services: 89% (2002 est.)
Labor force: 38,360 (2004)
country comparison to the world: 199
Labor force—by occupation: agriculture and fishing: 3%
laborers: 17%
clerical: 19%
professional and technical: 21%
administrative and managerial: 15%
sales: 7%
services: 19% (2004 est.)
Unemployment rate: 2.1% (2004 est.)
country comparison to the world: 15
Population below poverty line: 19% (2000)
Household income or consumption by percentage share: *lowest 10%:* NA%
highest 10%: NA%
Budget: *revenues:* $738 million
expenditures: $665 million (FY04/05)
Inflation rate (consumer prices): 2.8% (November 2005)
country comparison to the world: 75
Market value of publicly traded shares: $1.36 billion (31 December 2009)
country comparison to the world: 99
$1.912 billion (31 December 2008)
$2.731 billion (31 December 2007)
Agriculture—products: bananas, vegetables, citrus, flowers; dairy products, honey
Industries: international business, tourism, light manufacturing

Industrial production growth rate: NA%
Electricity—production: 675.6 million kWh (2007 est.)
country comparison to the world: 152
Electricity—consumption: 628.3 million kWh (2007 est.)
country comparison to the world: 154
Electricity—exports: 0 kWh (2008 est.)
Electricity—imports: 0 kWh (2008 est.)
Oil—production: 0 bbl/day (2009 est.)
country comparison to the world: 151
Oil—consumption: 5,000 bbl/day (2009 est.)
country comparison to the world: 169
Oil—exports: 0 bbl/day (2007 est.)
country comparison to the world: 149
Oil—imports: 4,500 bbl/day (2007 est.)
country comparison to the world: 161
Oil—proved reserves: 0 bbl (1 January 2010 est.)
country comparison to the world: 106
Natural gas—production: 0 cu m (2008 est.)
country comparison to the world: 98
Natural gas—consumption: 0 cu m (2008 est.)
country comparison to the world: 147
Natural gas—exports: 0 cu m (2008 est.)
country comparison to the world: 57
Natural gas—imports: 0 cu m (2008 est.)
country comparison to the world: 150
Natural gas—proved reserves: 0 cu m (1 January 2010 est.)
country comparison to the world: 109
Exports: $763 million (2006)
country comparison to the world: 160
Exports—commodities: reexports of pharmaceuticals
Exports—partners: Spain 16.91%, India 10.15%, Brazil 9.55%, Germany 7.4% (2009)
Imports: $1.162 billion (2006)
country comparison to the world: 169
Imports—commodities: clothing, fuels, machinery and transport equipment, construction materials, chemicals, food and live animals
Imports—partners: US 31.2%, South Korea 26.71%, Brazil 6.77%, Ireland 6.11%, Singapore 5.35% (2009)
Debt—external: $160 million (FY99/00)
country comparison to the world: 177
Stock of direct foreign investment—at home: $NA
Stock of direct foreign investment—abroad: $NA
Exchange rates: Bermudian dollars (BMD) per US dollar–1.000 (fixed rate pegged to the US dollar)

COMMUNICATIONS

Telephones—main lines in use: 57,700 (2009)
country comparison to the world: 158
Telephones—mobile cellular: 85,000 (2009)
country comparison to the world: 187
Telephone system: *general assessment:* a good, fully automatic digital telephone system with fiber-optic trunk lines
domestic: the system has a high fixed-line teledensity coupled with a mobile-cellular teledensity of roughly 125 per 100 persons
international: country code–1-441; landing points for the GlobeNet, Gemini Bermuda, and the Challenger Bermuda-1 (CB-1)submarine cables; satellite earth stations–3 (2009)
Broadcast media: 3 television stations; cable and satellite TV subscription services are available; roughly 10 radio stations operating (2007)
Internet country code: .bm
Internet hosts: 19,855 (2010)
country comparison to the world: 112
Internet users: 54,000 (2009)
country comparison to the world: 172

TRANSPORTATION

Airports: 1 (2010)
country comparison to the world: 212
Airports—with paved runways: *total:* 1
2,438 to 3,047 m: 1 (2010)
Roadways: *total:* 447 km
country comparison to the world: 196
paved: 447 km
note: public roads–225 km; private roads–222 km (2007)
Merchant marine: *total:* 139
country comparison to the world: 43
by type: bulk carrier 22, chemical tanker 3, container 15, liquefied gas 38, passenger 26, passenger/cargo 6, petroleum tanker 20, refrigerated cargo 9
foreign-owned: 114 (China 13, France 1, Germany 15, Greece 2, Hong Kong 5, Ireland 2, Israel 3, Japan 2, Monaco 2, Nigeria 11, Norway 5, Sweden 17, UK 11, US 25)
registered in other countries: 180 (Bahamas 12, Cyprus 1, Greece 3, Hong Kong 12, Isle of Man 7, Liberia 4, Malta 8, Marshall Islands 34, Norway 5, Panama 15, Philippines 43, Saint Vincent and the Grenadines 1, Singapore 21, UK 9, US 5) (2010)
Ports and terminals: Hamilton, Ireland Island, Saint George

MILITARY

Military branches: Bermuda Regiment (2009)
Military service age and obligation: 18-30 years of age for voluntary or compulsory enlistment in the Bermuda Regiment; males must register at age 18; term of service is 38 months (2009)
Manpower available for military service: *males age 16–49:* 15,081 (2010 est.)
Manpower fit for military service: *males age 16–49:* 12,323
females age 16–49: 12,174 (2010 est.)
Manpower reaching militarily significant age annually: *male:* 433
female: 410 (2010 est.)
Military expenditures: 0.11% of GDP (2005 est.)
country comparison to the world: 171
Military—note: defense is the responsibility of the UK

TRANSNATIONAL ISSUES

Disputes—international: none

BHUTAN

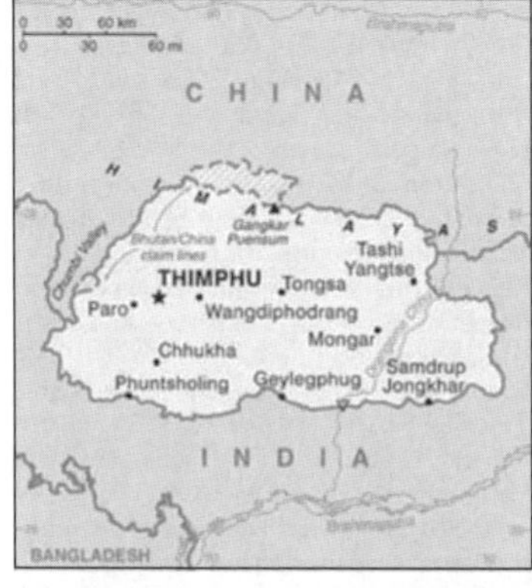

INTRODUCTION

Background: In 1865, Britain and Bhutan signed the Treaty of Sinchulu, under which Bhutan would receive an annual subsidy in exchange for ceding some border land to British India. Under British influence, a monarchy was set up in 1907; three years later, a treaty was signed whereby the British agreed not to interfere in Bhutanese internal affairs and Bhutan allowed Britain to direct its foreign affairs. This role was assumed by independent India after 1947. Two years later, a formal Indo-Bhutanese accord returned the areas of Bhutan annexed by the British, formalized the annual subsidies the country received, and defined India's responsibilities in defense and foreign relations. A refugee issue of over 100,000 Bhutanese in Nepal remains unresolved; 90% of the refugees are housed in seven United Nations Office of the High Commissioner for Refugees (UNHCR) camps. In March 2005, King Jigme Singye WANGCHUCK unveiled the government's draft constitution–which would introduce major democratic reforms–and pledged to hold a national referendum for its approval. In December 2006, the King abdicated the throne to his son, Jigme Khesar Namgyel WANGCHUCK, in order to give him experience as head of state before the democratic transition. In early 2007, India and Bhutan renegotiated their treaty to allow Bhutan greater autonomy in conducting its foreign policy, although Thimphu continues to coordinate policy decisions in this area with New Delhi. In July 2007, seven ministers of Bhutan's 10-member cabinet resigned to join the political process, and the cabinet acted as a caretaker regime until democratic elections for seats to the country's first parliament were completed in March 2008. The king ratified the country's first constitution in July 2008.

GEOGRAPHY

Location: Southern Asia, between China and India
Geographic coordinates: 27 30 N, 90 30 E
Map references: Asia
Area: *total:* 38,394 sq km
country comparison to the world: 136
land: 38,394 sq km
water: 0 sq km
Area—comparative: about one-half the size of Indiana
Land boundaries: *total:* 1,075 km
border countries: China 470 km, India 605 km
Coastline: 0 km (landlocked)
Maritime claims: none (landlocked)
Climate: varies; tropical in southern plains; cool winters and hot summers in central valleys; severe winters and cool summers in Himalayas
Terrain: mostly mountainous with some fertile valleys and savanna
Elevation extremes: *lowest point:* Drangeme Chhu 97 m
highest point: Gangkar Puensum 7,570 m
Natural resources: timber, hydropower, gypsum, calcium carbonate
Land use: *arable land:* 2.3%
permanent crops: 0.43%
other: 97.27% (2005)
Irrigated land: 400 sq km (2003)
Total renewable water resources: 95 cu km (1987)
Freshwater withdrawal (domestic/industrial/agricultural): *total:* 0.43 cu km/yr (5%/1%/94%)
per capita: 199 cu m/yr (2000)
Natural hazards: violent storms from the Himalayas are the source of the country's name, which translates as Land of the Thunder Dragon; frequent landslides during the rainy season
Environment—current issues: soil erosion; limited access to potable water
Environment—international agreements: *party to:* Biodiversity, Climate Change, Climate Change-Kyoto Protocol, Desertification, Endangered Species, Hazardous Wastes, Ozone Layer Protection
signed, but not ratified: Law of the Sea
Geography—note: landlocked; strategic location between China and India; controls several key Himalayan mountain passes

PEOPLE

Population: 708,427 (July 2011 est.)
country comparison to the world: 164
note: the Factbook population estimate is consistent with the first modern census of Bhutan, conducted in 2005; previous Factbook population estimates for this country, which were on the order of three times the total population reported here, were based on Bhutanese government publications that did not include the census
Age structure: *0–14 years:* 28.9% (male 104,622/female 100,383)
15–64 years: 65.3% (male 245,054/female 217,864)
65 years and over: 5.7% (male 21,347/female 19,157) (2011 est.)
Median age:
total: 24.8 years
male: 25.4 years
female: 24.2 years (2011 est.)
Population growth rate: 1.201% (2011 est.)
country comparison to the world: 98
Birth rate: 19.13 births/1,000 population (2011 est.)
country comparison to the world: 103
Death rate: 7.12 deaths/1,000 population (July 2011 est.)
country comparison to the world: 129
Net migration rate: 0 migrant(s)/1,000 population (2011 est.)
country comparison to the world: 78
Urbanization: *urban population:* 35% of total population (2010)
rate of urbanization: 3.7% annual rate of change (2010-15 est.)
Major cities—population: THIMPHU (capital) 89,000 (2009)
Sex ratio: *at birth:* 1.05 male(s)/female
under 15 years: 1.04 male(s)/female
15–64 years: 1.13 male(s)/female
65 years and over: 1.12 male(s)/female
total population: 1.1 male(s)/female (2011 est.)
Infant mortality rate: *total:* 44.48 deaths/1,000 live births
country comparison to the world: 55
male: 45.24 deaths/1,000 live births
female: 43.69 deaths/1,000 live births (2011 est.)
Life expectancy at birth: *total population:* 67.3 years
country comparison to the world: 158
male: 66.46 years
female: 68.19 years (2011 est.)
Total fertility rate: 2.2 children born/woman (2011 est.)
country comparison to the world: 107
HIV/AIDS—adult prevalence rate: 0.2% (2009 est.)
country comparison to the world: 96
HIV/AIDS—people living with HIV/AIDS: fewer than 1,000 (2009 est.)
country comparison to the world: 143
HIV/AIDS—deaths: fewer than 100 (2009 est.)
country comparison to the world: 125
Major infectious diseases: *degree of risk:* intermediate
food or waterborne diseases: bacterial and protozoal diarrhea, hepatitis A, and typhoid fever
vectorborne diseases: dengue fever and malaria
water contact disease: leptospirosis (2009)
Drinking water source: *Improved:*
urban: 99% of population
rural: 88% of population
total: 92% of population
Unimproved: urban: 1% of population
rural: 12% of population
total: 8% of population (2008)
Sanitation facility access: *Improved:*
urban: 87% of population
rural: 54% of population
total: 65% of population
Unimproved:
urban: 13% of population
rural: 46% of population
total: 35% of population (2008)
Nationality: *noun:* Bhutanese (singular and plural)
adjective: Bhutanese
Ethnic groups: Bhote 50%, ethnic Nepalese 35% (includes Lhotsampas—one of several Nepalese ethnic groups), indigenous or migrant tribes 15%
Religions: Lamaistic Buddhist 75%, Indian- and Nepalese-influenced Hinduism 25%
Languages: Sharchhopka 28%, Dzongkha (official) 24%, Lhotshamkha 22%, other 26% (2005 Census)
Literacy: *definition:* age 15 and over can read and write
total population: 47%
male: 60%
female: 34% (2003 est.)
School life expectancy (primary to tertiary education): *total:* 11 years
male: 11 years
female: 11 years (2008)
Education expenditures: 4.8% of GDP (2008)
country comparison to the world: 69

GOVERNMENT

Country name: *conventional long form:* Kingdom of Bhutan
conventional short form: Bhutan
local long form: Druk Gyalkhap
local short form: Druk Yul
Government type: constitutional monarchy
Capital: *name:* Thimphu
geographic coordinates: 27 29 N, 89 36 E
time difference: UTC+6 (11 hours ahead of Washington, DC during Standard Time)
Administrative divisions: 20 districts (dzongkhag, singular and plural); Bumthang, Chhukha, Chirang, Daga, Gasa, Geylegphug, Ha, Lhuntshi, Mongar, Paro, Pemagatsel, Punakha, Samchi, Samdrup Jongkhar, Shemgang, Tashigang, Tashi Yangtse, Thimphu, Tongsa, Wangdi Phodrang
Independence: 1907 (became a unified kingdom under its first hereditary king)
National holiday: National Day (Ugyen WANGCHUCK became first hereditary king), 17 December (1907)
Constitution: ratified 18 July 2008
Legal system: civil law based on Buddhist religious law
International law organization participation: has not submitted an ICJ jurisdiction declaration; non-party state to the ICCt
Suffrage: 18 years of age; universal
Executive branch: *chief of state:* King Jigme Khesar Namgyel WANGCHUCK (since 14 December 2006); note–King Jigme Singye WANGCHUCK abdicated the throne on 14 December 2006 and his son immediately succeeded him; the nearly two-year delay between the former King's abdication and his son's coronation on 6 November 2008 was to ensure an astrologically auspicious coronation date and to give the new king, who had limited experience, deeper administrative expertise under the guidance of his father
head of government: Prime Minister Jigme THINLEY (since 9 April 2008)
cabinet: Council of Ministers (Lhengye Shungtsog) nominated by the monarch, approved by the National Assembly; members serve fixed, five-year terms; note–there is also a Royal Advisory Council (Lodoi Tsokde); members are nominated by the monarch
(For more information visit the World Leaders website)

elections: the monarchy is hereditary, but democratic reforms in July 1998 grant the National Assembly authority to remove the monarch with two-thirds vote; election of a new National Assembly occurred in March 2008; the leader of the majority party nominated as the prime minister

Legislative branch: bicameral Parliament consists of the non-partisan National Council (25 seats; 20 members elected by each of the 20 electoral districts (dzongkhags) for four-year terms and 5 members nominated by the King); and the National Assembly (47 seats; members elected by direct, popular vote for five-year terms)
elections: National Council elections last held on 31 December 2007 and 29 January 2008 (next to be held by December 2012); National Assembly elections last held on 24 March 2008 (next to be held by March 2013)
election results: National Council–NA; National Assembly–percent of vote by party–DPT 67%, PDP 33%; seats by party–DPT 45, PDP 2

Judicial branch: Supreme Court of Appeal (the monarch); High Court (judges appointed by the monarch); note–the draft constitution establishes a Supreme Court that will serve as chief court of appeal

Political parties and leaders: Bhutan Peace and Prosperity Party (Druk Phuensum Tshogpa) or DPT [Jigme THINLEY]; People's Democratic Party or PDP [Tshering TOBGAY]

Political pressure groups and leaders: United Front for Democracy (exiled); Druk National Congress (exiled)
other: Buddhist clergy; ethnic Nepalese organizations leading militant antigovernment campaign; Indian merchant community

International organization participation: ADB, BIMSTEC, CP, FAO, G-77, IBRD, ICAO, IDA, IFAD, IFC, IMF, Interpol, IOC, IOM (observer), ISO (correspondent), ITSO, ITU, NAM, OPCW, SAARC, SACEP, UN, UNCTAD, UNESCO, UNIDO, UNWTO, UPU, WCO, WHO, WIPO, WMO, WTO (observer)

Diplomatic representation in the US: none; note–the Permanent Mission to the UN for Bhutan has consular jurisdiction in the US; the permanent representative to the UN is Daw PENJO; address: 763 First Avenue, New York, NY 10017; telephone [1] (212) 682-2268; FAX [1] (212) 661-0551
consulate(s) general: New York

Diplomatic representation from the US: the US and Bhutan have no formal diplomatic relations, although informal contact is maintained between the Bhutanese and US Embassy in New Delhi (India)

Flag description: divided diagonally from the lower hoist-side corner; the upper triangle is yellow and the lower triangle is orange; centered along the dividing line is a large black and white dragon facing away from the hoist side; the dragon, called the Druk (Thunder Dragon), is the emblem of the nation; its white color stands for purity and the jewels in its claws symbolize wealth; the background colors represent spiritual and secular powers within Bhutan: the orange is associated with Bhuddhism, while the yellow denotes the ruling dynasty

National anthem: *name:* "Druk tsendhen" (The Thunder Dragon Kingdom)
lyrics/music: Gyaldun Dasho Thinley DORJI/Aku TONGMI
note: adopted 1953

ECONOMY

Economy—overview: The economy, one of the world's smallest and least developed, is based on agriculture and forestry, which provide the main livelihood for more than 60% of the population. Agriculture consists largely of subsistence farming and animal husbandry. Rugged mountains dominate the terrain and make the building of roads and other infrastructure difficult and expensive. The economy is closely aligned with India's through strong trade and monetary links and dependence on India's financial assistance. The industrial sector is technologically backward, with most production of the cottage industry type. Most development projects, such as road construction, rely on Indian migrant labor. Model education, social, and environment programs are underway with support from multilateral development organizations. Each economic program takes into account the government's desire to protect the country's environment and cultural traditions. For example, the government, in its cautious expansion of the tourist sector, encourages visits by upscale, environmentally conscientious tourists. Complicated controls and uncertain policies in areas such as industrial licensing, trade, labor, and finance continue to hamper foreign investment. Hydropower exports to India have boosted Bhutan's overall growth. New hydropower projects will be the driving force behind Bhutan's ability to create employment and sustain growth in the coming years.

GDP (purchasing power parity): $3.875 billion (2010 est.)
country comparison to the world: 167
$3.63 billion (2009 est.)
$3.34 billion (2008 est.)
note: data are in 2010 US dollars

GDP (official exchange rate): $1.412 billion (2010 est.)

GDP—real growth rate: 6.7% (2010 est.)
country comparison to the world: 41
8.7% (2009 est.)
3% (2008 est.)

GDP—per capita (PPP): $5,500 (2010 est.)
country comparison to the world: 142
$5,300 (2009 est.)
$4,900 (2008 est.)
note: data are in 2010 US dollars

GDP—composition by sector:
agriculture: 17.6%
industry: 45%
services: 37.4% (2009)

Labor force: 299,900
country comparison to the world: 165
note: major shortage of skilled labor (2008)

Labor force—by occupation: *agriculture:* 43.7%
industry: 39.1%
services: 17.2% (2004 est.)

Unemployment rate: 4% (2009)
country comparison to the world: 36
2.5% (2004)

Population below poverty line: 23.2% (2008)

Household income or consumption by percentage share: *lowest 10%:* 2.3%
highest 10%: 37.6% (2003)

Budget: *revenues:* $302 million
expenditures: $588 million
note: the government of India finances nearly three-fifths of Bhutan's budget expenditures (FY09/10)

Public debt: 57.8% of GDP (2009)
country comparison to the world: 40
81.4% of GDP (2004)

Inflation rate (consumer prices): 4.3% (2008 est.)
country comparison to the world: 123
4.9% (2007 est.)

Central bank discount rate: NA%

Commercial bank prime lending rate: NA%

Stock of narrow money: $335 million (31 December 2008)
country comparison to the world: 164
$381.1 million (31 December 2007)

Stock of broad money: $647.6 million (31 December 2008)

Stock of domestic credit: $169.9 million (31 December 2007)
country comparison to the world: 178

Market value of publicly traded shares: $NA

Agriculture—products: rice, corn, root crops, citrus, foodgrains; dairy products, eggs

Industries: cement, wood products, processed fruits, alcoholic beverages, calcium carbide, tourism

Industrial production growth rate: NA%

Electricity—production: 1.48 billion kWh (2009 est.)
country comparison to the world: 141

Electricity—consumption: 184 million kWh (2009 est.)
country comparison to the world: 179

Electricity—exports: 1.296 billion kWh (2009 est.)

Electricity—imports: 0 kWh (2009 est.)

Oil—production: 0 bbl/day (2009 est.)
country comparison to the world: 155

Oil—consumption: 1,000 bbl/day (2009 est.)
country comparison to the world: 201

Oil—exports: 0 bbl/day (2008 est.)
country comparison to the world: 151

Oil—imports: 1,250 bbl/day (2008 est.)
country comparison to the world: 184

Oil—proved reserves: 0 bbl (1 January 2010 est.)
country comparison to the world: 111

Natural gas—production: 0 cu m (2008 est.)
country comparison to the world: 105

Natural gas—consumption: 0 cu m (2008 est.)
country comparison to the world: 152

Natural gas—exports: 0 cu m (2008 est.)
country comparison to the world: 66

Natural gas—imports: 0 cu m (2008 est.)
country comparison to the world: 158

Natural gas—proved reserves: 0 cu m (1 January 2010 est.)
country comparison to the world: 115

Current account balance: $164 million (2008 est.)
country comparison to the world: 56
$116 million (2007 est.)

Exports: $513 million (2008)
country comparison to the world: 167

$350 million (2006)
Exports—commodities: electricity (to India), ferrosilicon, cement, calcium carbide, copper wire, manganese, vegetable oil
Exports—partners: India 86.3%, Bangladesh 8.1%, Italy 1.5% (2008)
Imports: $533 million (2008)
country comparison to the world: 188
$320 million (2006)
Imports—commodities: fuel and lubricants, passenger cars, machinery and parts, fabrics, rice
Imports—partners: India 63%, Japan 12.3%, China 5.1% (2008)
Debt—external: $836 million (2009)
country comparison to the world: 153
$713.3 million (2006)
Exchange rates: ngultrum (BTN) per US dollar–46.6 (2009), 41.487 (2007), 45.279 (2006), 44.101 (2005), 45.317 (2004)

COMMUNICATIONS

Telephones—main lines in use: 26,300 (2009)
country comparison to the world: 183
Telephones—mobile cellular: 327,100 (2009)
country comparison to the world: 169
Telephone system: *general assessment:* urban towns and district headquarters have telecommunications services
domestic: low teledensity; domestic service is poor especially in rural areas; mobile-cellular service available since 2003
international: country code–975; international telephone and telegraph service via landline and microwave relay through India; satellite earth station–1 Intelsat (2009)
Broadcast media: state-owned TV station established in 1999; cable TV service offers dozens of Indian and other international channels; first radio station, privately launched in 1973, is now state-owned; 1 private radio station began operations in 2006 (2007)
Internet country code: .bt
Internet hosts: 9,147 (2010)
country comparison to the world: 125
Internet users: 50,000 (2009)
country comparison to the world: 173

TRANSPORTATION

Airports: 2 (2010)
country comparison to the world: 196
Airports—with paved runways: *total:* 1
1,524 to 2,437 m: 1 (2010)
Airports—with unpaved runways: *total:* 1
914 to 1,523 m: 1 (2010)
Roadways: *total:* 8,050 km
country comparison to the world: 141
paved: 4,991 km
unpaved: 3,059 km (2003)

MILITARY

Military branches: Royal Bhutan Army (includes Royal Bodyguard and Royal Bhutan Police) (2009)
Military service age and obligation: 18 years of age for voluntary military service; no conscription (2010)
Manpower available for military service: *males age 16–49:* 202,407
females age 16–49: 180,349 (2010 est.)
Manpower fit for military service: *males age 16–49:* 157,664
females age 16–49: 144,861 (2010 est.)
Manpower reaching militarily significant age annually: *male:* 7,363
female: 7,095 (2010 est.)
Military expenditures: 1% of GDP (2005 est.)
country comparison to the world: 131

TRANSNATIONAL ISSUES

Disputes—international: lacking any treaty describing the boundary, Bhutan and China continue negotiations to establish a common boundary alignment to resolve territorial disputes arising from substantial cartographic discrepancies, the largest of which lie in Bhutan's northwest and along the Chumbi salient; Bhutan protests Chinese road construction and other activities on Bhutanese soil; Chinese border soldiers frequently intrude deep into Bhutanese territory; Bhutan cooperates with India to expel Indian Nagaland separatists

BOLIVIA

INTRODUCTION

Background: Bolivia, named after independence fighter Simon BOLIVAR, broke away from Spanish rule in 1825; much of its subsequent history has consisted of a series of nearly 200 coups and countercoups. Democratic civilian rule was established in 1982, but leaders have faced difficult problems of deep-seated poverty, social unrest, and illegal drug production. In December 2005, Bolivians elected Movement Toward Socialism leader Evo MORALES president–by the widest margin of any leader since the restoration of civilian rule in 1982–after he ran on a promise to change the country's traditional political class and empower the nation's poor, indigenous majority. However, since taking office, his controversial strategies have exacerbated racial and economic tensions between the Amerindian populations of the Andean west and the non-indigenous communities of the eastern lowlands. In December 2009, President MORALES easily won reelection, and his party took control of the legislative branch of the government, which will allow him to continue his process of change.

GEOGRAPHY

Location: Central South America, southwest of Brazil
Geographic coordinates: 17 00 S, 65 00 W
Map references: South America
Area: *total:* 1,098,581 sq km
country comparison to the world: 28
land: 1,083,301 sq km
water: 15,280 sq km
Area—comparative: slightly less than three times the size of Montana
Land boundaries: *total:* 6,940 km
border countries: Argentina 832 km, Brazil 3,423 km, Chile 860 km, Paraguay 750 km, Peru 1,075 km
Coastline: 0 km (landlocked)
Maritime claims: none (landlocked)
Climate: varies with altitude; humid and tropical to cold and semiarid
Terrain: rugged Andes Mountains with a highland plateau (Altiplano), hills, lowland plains of the Amazon Basin
Elevation extremes: *lowest point:* Rio Paraguay 90 m
highest point: Nevado Sajama 6,542 m
Natural resources: tin, natural gas, petroleum, zinc, tungsten, antimony, silver, iron, lead, gold, timber, hydropower
Land use: *arable land:* 2.78%
permanent crops: 0.19%
other: 97.03% (2005)
Irrigated land: 1,320 sq km (2003)
Total renewable water resources: 622.5 cu km (2000)
Freshwater withdrawal (domestic/industrial/agricultural): *total:* 1.44 cu km/yr (13%/7%/81%)
per capita: 157 cu m/yr (2000)
Natural hazards: flooding in the northeast (March-April)
volcanism: Bolivia experiences volcanic activity in Andes Mountains on the border with Chile; historically active volcanoes in this region are Irruputuncu (elev. 5,163 m), which last erupted in 1995 and Olca-Paruma
Environment—current issues: the clearing of land for agricultural purposes and the international demand for tropical timber are contributing to deforestation; soil erosion from overgrazing and poor cultivation methods (including slash-and-burn agriculture); desertification; loss of biodiversity; industrial pollution of water supplies used for drinking and irrigation
Environment—international agreements: *party to:* Biodiversity, Climate Change, Climate Change-Kyoto Protocol, Deserti-

fication, Endangered Species, Hazardous Wastes, Law of the Sea, Marine Dumping, Ozone Layer Protection, Ship Pollution, Tropical Timber 83, Tropical Timber 94, Wetlands
signed, but not ratified: Environmental Modification, Marine Life Conservation
Geography—note: landlocked; shares control of Lago Titicaca, world's highest navigable lake (elevation 3,805 m), with Peru

PEOPLE

Population: 10,118,683 (July 2011 est.)
country comparison to the world: 83
Age structure: *0–14 years:* 34.6% (male 1,785,453/female 1,719,173)
15–64 years: 60.7% (male 3,014,419/female 3,129,942)
65 years and over: 4.6% (male 207,792/female 261,904) (2011 est.)
Median age: *total:* 22.5 years
male: 21.8 years
female: 23.2 years (2011 est.)
Population growth rate: 1.694% (2011 est.)
country comparison to the world: 67
Birth rate: 24.71 births/1,000 population (2011 est.)
country comparison to the world: 63
Death rate: 6.85 deaths/1,000 population (July 2011 est.)
country comparison to the world: 143
Net migration rate: -0.92 migrant(s)/1,000 population (2011 est.)
country comparison to the world: 150
Urbanization: *urban population:* 67% of total population (2010)
rate of urbanization: 2.2% annual rate of change (2010-15 est.)
Major cities—population: LA PAZ (capital) 1.642 million; Santa Cruz 1.584 million; Sucre 281,000 (2009)
Sex ratio: *at birth:* 1.05 male(s)/female
under 15 years: 1.04 male(s)/female
15–64 years: 0.96 male(s)/female
65 years and over: 0.79 male(s)/female
total population: 0.98 male(s)/female (2011 est.)
Infant mortality rate: *total:* 42.16 deaths/1,000 live births
country comparison to the world: 60
male: 45.95 deaths/1,000 live births
female: 38.18 deaths/1,000 live births (2011 est.)
Life expectancy at birth: *total population:* 67.57 years
country comparison to the world: 156
male: 64.84 years
female: 70.42 years (2011 est.)
Total fertility rate: 3 children born/woman (2011 est.)
country comparison to the world: 64
HIV/AIDS—adult prevalence rate: 0.2% (2009 est.)
country comparison to the world: 95
HIV/AIDS—people living with HIV/AIDS: 12,000 (2009 est.)
country comparison to the world: 93
HIV/AIDS—deaths: fewer than 1,000 (2009 est.)
country comparison to the world: 67
Major infectious diseases: *degree of risk:* high
food or waterborne diseases: bacterial diarrhea, hepatitis A, and typhoid fever
vectorborne diseases: dengue fever, malaria, and yellow fever
water contact disease: leptospirosis (2009)
Drinking water source: *Improved:*
urban: 96% of population
rural: 67% of population
total: 86% of population
Unimproved: urban: 4% of population
rural: 33% of population
total: 14% of population (2008)
Sanitation facility access: *Improved:*
urban: 34% of population
rural: 9% of population
total: 25% of population
Unimproved: urban: 66% of population
rural: 91% of population
total: 75% of population (2008)
Nationality: *noun:* Bolivian(s)
adjective: Bolivian
Ethnic groups: Quechua 30%, mestizo (mixed white and Amerindian ancestry) 30%, Aymara 25%, white 15%
Religions: Roman Catholic 95%, Protestant (Evangelical Methodist) 5%
Languages: Spanish (official) 60.7%, Quechua (official) 21.2%, Aymara (official) 14.6%, foreign languages 2.4%, other 1.2% (2001 census)
Literacy: *definition:* age 15 and over can read and write
total population: 86.7%
male: 93.1%
female: 80.7% (2001 census)
School life expectancy (primary to tertiary education):
total: 14 years
male: 14 years
female: 14 years (2007)
Education expenditures: 6.3% of GDP (2006)
country comparison to the world: 25

GOVERNMENT

Country name: *conventional long form:* Plurinational State of Bolivia
conventional short form: Bolivia
local long form: Estado Plurinacional de Bolivia
local short form: Bolivia
Government type: republic; note—the new constitution defines Bolivia as a "Social Unitarian State"
Capital: *name:* La Paz (administrative capital)
geographic coordinates: 16 30 S, 68 09 W
time difference: UTC-4 (1 hour ahead of Washington, DC during Standard Time)
note: Sucre (constitutional capital)
Administrative divisions: 9 departments (departamentos, singular—departamento); Beni, Chuquisaca, Cochabamba, La Paz, Oruro, Pando, Potosi, Santa Cruz, Tarija
Independence: 6 August 1825 (from Spain)
National holiday: Independence Day, 6 August (1825)
Constitution: 7 February 2009
Legal system: civil law system with influences from Roman, Spanish, canon (religious), French, and indigenous law
International law organization participation: has not submitted an ICJ jurisdiction declaration; accepts ICCt jurisdiction
Suffrage: 18 years of age, universal and compulsory
Executive branch: *chief of state:* President Juan Evo MORALES Ayma (since 22 January 2006); Vice President Alvaro GARCIA Linera (since 22 January 2006); note—the president is both chief of state and head of government
head of government: President Juan Evo MORALES Ayma (since 22 January 2006); Vice President Alvaro GARCIA Linera (since 22 January 2006)
cabinet: Cabinet appointed by the president (For more information visit the World Leaders website)
elections: president and vice president elected on the same ticket by popular vote for a five-year term and are eligible for re-election; election last held on 6 December 2009 (next to be held in 2014)
election results: Juan Evo MORALES Ayma reelected president; percent of vote—Juan Evo MORALES Ayma 64%; Manfred REYES VILLA 26%; Samuel DORIA MEDINA Arana 6%; Rene JOAQUINO 2%; other 2%
Legislative branch: bicameral Plurinational Legislative Assembly or Asamblea Legislativa Plurinacional consists of Chamber of Senators or Camara de Senadores (36 seats; members are elected by proportional representation from party lists to serve five-year terms) and Chamber of Deputies or Camara de Diputados (130 seats total; 70 uninominal deputies directly elected from a single district, 7 "special" indigenous deputies directly elected from non-contiguous indigenous districts, and 53 plurinominal deputies elected by proportional representation from party lists; all deputies serve five-year terms)
elections: Chamber of Senators and Chamber of Deputies—last held on 6 December 2009 (next to be held in 2014)
election results: Chamber of Senators—percent of vote by party—NA; seats by party—MAS 26, PPB-CN 10; Chamber of Deputies—percent of vote by party—NA; seats by party—MAS 89, PPB-CN 36, UN 3, AS 2
Judicial branch: Supreme Court or Tribunal Supremo de Justicia (judges elected by popular vote from list of candidates pre-selected by Assembly for six-year terms); District Courts (one in each department); Plurinational Constitutional Tribunal (seven primary or titulares and seven alternate or suplente magistrates elected by popular vote from list of candidates pre-selected by Assembly for six-year terms; to rule on constitutional issues (at least two candidates must be indigenous)); Plurinational Electoral Organ (seven members elected by the Assembly and the president; one member must be of indigenous origin to six-year terms); Agro-Environmental Court (judges elected by popular vote from list of candidates pre-selected by Assembly for six-year terms; to run on agro-environmental issues); provincial and local courts (to try minor cases)
Political parties and leaders: Bolivia-National Convergence or PPB-CN [Manfred REYES VILLA]; Fearless Movement or MSM [Juan DE GRANADO Cosio]; Movement Toward Socialism or MAS [Juan Evo MORALES Ayma]; National Unity or UN [Samuel DORIA MEDINA Arana]; People or Gente [Roman LOAYZA]; Social Alliance or AS [Rene JOAQUINO]
Political pressure groups and leaders: Bolivian Workers Central or COR;

Federation of Neighborhood Councils of El Alto or FEJUVE; Landless Movement or MST; National Coordinator for Change or CONALCAM; Sole Confederation of Campesino Workers of Bolivia or CSUTCB
other: Cocalero groups; indigenous organizations (including Confederation of Indigenous Peoples of Eastern Bolivia or CIDOB and National Council of Ayullus and Markas of Quollasuyu or CONAMAQ); labor unions (including the Central Bolivian Workers' Union or COB and Cooperative Miners Federation or FENCOMIN)
International organization participation: CAN, FAO, G-77, IADB, IAEA, IBRD, ICAO, ICC, ICRM, IDA, IFAD, IFC, IFRCS, ILO, IMF, IMO, Interpol, IOC, IOM, IPU, ISO (correspondent), ITSO, ITU, LAES, LAIA, Mercosur (associate), MIGA, MINUSTAH, MONUSCO, NAM, OAS, OPANAL, OPCW, PCA, RG, UN, UNASUR, UNCTAD, UNESCO, UNFICYP, UNIDO, Union Latina, UNMIL, UNMIS, UNOCI, UNWTO, UPU, WCO, WFTU, WHO, WIPO, WMO, WTO
Diplomatic representation in the US: *chief of mission:* Ambassador (vacant); Charge d'Affaires Freddy BERSATTI
chancery: 3014 Massachusetts Avenue NW, Washington, DC 20008
telephone: [1] (202) 483-4410
FAX: [1] (202) 328-3712
consulate(s) general: Los Angeles, Miami, New York, San Francisco
note: as of September 2008, the US has expelled the Bolivian ambassador to the US
Diplomatic representation from the US: *chief of mission:* Ambassador (vacant); Charge d'Affaires John CREAMER
embassy: Avenida Arce 2780, Casilla 425, La Paz
mailing address: P. O. Box 425, La Paz; APO AA 34032
telephone: [591] (2) 216-8000
FAX: [591] (2) 216-8111
note: in September 2008, the Bolivian Government expelled the US Ambassador to Bolivia, and the countries have yet to reinstate ambassadors
Flag description: three equal horizontal bands of red (top), yellow, and green with the coat of arms centered on the yellow band; red stands for bravery and the blood of national heroes, yellow for the nation's mineral resources, and green for the fertility of the land
note: similar to the flag of Ghana, which has a large black five-pointed star centered in the yellow band; in 2009, a presidential decree made it mandatory for a so-called wiphala—a square, multi-colored flag representing the country's indigenous peoples—to be used alongside the traditional flag
National anthem: *name:* "Cancion Patriotica" (Patriotic Song)
lyrics/music: Jose Ignacio de SANJINES/ Leopoldo Benedetto VINCENTI
note: adopted 1852

ECONOMY

Economy—overview: Bolivia is one of the poorest and least developed countries in Latin America. Following a disastrous economic crisis during the early 1980s, reforms spurred private investment, stimulated economic growth, and cut poverty rates in the 1990s. The period 2003-05 was characterized by political instability, racial tensions, and violent protests against plans—subsequently abandoned—to export Bolivia's newly discovered natural gas reserves to large northern hemisphere markets. In 2005, the government passed a controversial hydrocarbons law that imposed significantly higher royalties and required foreign firms then operating under risk-sharing contracts to surrender all production to the state energy company in exchange for a predetermined service fee. After higher prices for mining and hydrocarbons exports produced a fiscal surplus in 2008, the global recession in 2009 slowed growth. Nevertheless, Bolivia recorded the highest growth rate in South America that year. During 2010 an increase in world commodity prices resulted in the biggest trade surplus in history. However, a lack of foreign investment in the key sectors of mining and hydrocarbons and higher food prices pose challenges for the Bolivian economy.
GDP (purchasing power parity): $47.88 billion (2010 est.)
country comparison to the world: 93
$45.96 billion (2009 est.)
$44.47 billion (2008 est.)
note: data are in 2010 US dollars
GDP (official exchange rate): $19.37 billion (2010 est.)
GDP—real growth rate: 4.2% (2010 est.)
country comparison to the world: 90
3.4% (2009 est.)
6.1% (2008 est.)
GDP—per capita (PPP): $4,800 (2010 est.)
country comparison to the world: 152
$4,700 (2009 est.)
$4,600 (2008 est.)
note: data are in 2010 US dollars
GDP—composition by sector:
agriculture: 12%
industry: 38%
services: 50% (2010 est.)
Labor force: 4.186 million (2010 est.)
country comparison to the world: 88
Labor force—by occupation:
agriculture: 40%
industry: 17%
services: 43% (2006 est.)
Unemployment rate: 6.5% (2010 est.)
country comparison to the world: 65
7.7% (2009 est.)
note: data are for urban areas; widespread underemployment
Population below poverty line: 30.3%
note: based on percent of population living on less than the international standard of $2/day (2009 est.)
Household income or consumption by percentage share:
lowest 10%: 0.5%
highest 10%: 44.1% (2005)
Distribution of family income—Gini index: 58.2 (2009)
country comparison to the world: 9
57.9 (1999)
Investment (gross fixed): 20.1% of GDP (2010 est.)
country comparison to the world: 86
Budget: *revenues:* $10.11 billion
expenditures: $10 billion (2010 est.)
Public debt: 39.7% of GDP (2010 est.)
country comparison to the world: 73
42.3% of GDP (2009 est.)
Inflation rate (consumer prices): 7.2% (2010 est.)
country comparison to the world: 178
0.3% (2009 est.)
Central bank discount rate: 3% (31 December 2010)
country comparison to the world: 109
3% (31 December 2009)
Commercial bank prime lending rate: 5.75% (31 December 2010 est.)
country comparison to the world: 130
6.08% (31 December 2009 est.)
Stock of narrow money: $5.367 billion (31 December 2010 est.)
country comparison to the world: 86
$4.47 billion (31 December 2009 est.)
Stock of broad money: $12.16 billion (31 December 2009)
country comparison to the world: 95
$11.04 billion (31 December 2008)
Stock of domestic credit: $7.06 billion (31 December 2010 est.)
country comparison to the world: 106
$5.891 billion (31 December 2009 est.)
Market value of publicly traded shares: $3.915 billion (31 December 2010)
country comparison to the world: 93
$2.792 billion (31 December 2009)
$2.672 billion (31 December 2008)
Agriculture—products: soybeans, coffee, coca, cotton, corn, sugarcane, rice, potatoes; timber
Industries: mining, smelting, petroleum, food and beverages, tobacco, handicrafts, clothing
Industrial production growth rate: 4% (2010 est.)
country comparison to the world: 88
Electricity—production: 6.085 billion kWh (2010 est.)
country comparison to the world: 106
Electricity—consumption: 5.814 billion kWh (2010 est.)
country comparison to the world: 107
Electricity—exports: 0 kWh (2010 est.)
Electricity—imports: 0 kWh (2010 est.)
Oil—production: 43,740 bbl/day (2010 est.)
country comparison to the world: 66
Oil—consumption: 31,070 bbl/day (2010 est.)
country comparison to the world: 111
Oil—exports: 5,621 bbl/day (2010 est.)
country comparison to the world: 102
Oil—imports: 17,330 bbl/day (2010 est.)
country comparison to the world: 116
Oil—proved reserves: 465 million bbl (1 January 2010 est.)
country comparison to the world: 49
Natural gas—production: 14.4 billion cu m (2008 est.)
country comparison to the world: 35
Natural gas—consumption: 2.41 billion cu m (2008 est.)
country comparison to the world: 77
Natural gas—exports: 11.59 billion cu m (2010 est.)
country comparison to the world: 18
Natural gas—imports: 0 cu m (2010 est.)
country comparison to the world: 154

Natural gas—proved reserves: 750.4 billion cu m (1 January 2010 est.)
country comparison to the world: 30
Current account balance: $690.2 million (2010 est.)
country comparison to the world: 47
$757.1 million (2009 est.)
Exports: $6.956 billion (2010 est.)
country comparison to the world: 99
$5.452 billion (2009 est.)
Exports—commodities: natural gas, soybeans and soy products, crude petroleum, zinc ore, tin
Exports—partners: Brazil 41.38%, US 13.87%, Japan 5.62%, Colombia 5.32%, South Korea 4.7%, Peru 4.16% (2009)
Imports: $5.366 billion (2010 est.)
country comparison to the world: 110
$4.466 billion (2009 est.)
Imports—commodities: petroleum products, plastics, paper, aircraft and aircraft parts, prepared foods, automobiles, insecticides, soybeans
Imports—partners: Brazil 27.12%, Argentina 15.69%, US 12.77%, Chile 9.11%, Peru 6.85% (2009)
Reserves of foreign exchange and gold: $9.73 billion (31 December 2010 est.)
country comparison to the world: 59
$8.581 billion (31 December 2009 est.)
Debt—external: $2.864 billion (31 December 2010 est.)
country comparison to the world: 131
$2.594 billion (31 December 2009 est.)
Stock of direct foreign investment—at home: $7.257 billion (31 December 2010)
country comparison to the world: 84
$6.876 billion (31 December 2009)
Stock of direct foreign investment—abroad: $21 million (31 December 2010)
country comparison to the world: 83
$63.8 million (31 December 2008)
Exchange rates: bolivianos (BOB) per US dollar–7.04 (2010), 7.07 (2009), 7.253 (2008), 7.8616 (2007), 8.0159 (2006)

COMMUNICATIONS

Telephones—main lines in use: 810,200 (2009)
country comparison to the world: 88
Telephones—mobile cellular: 7.148 million (2009)
country comparison to the world: 81
Telephone system: *general assessment:* Bolivian National Telecommunications Company (ENTEL) was privatized in 1995 but re-nationalized in 2007; the primary trunk system is being expanded and employs digital microwave radio relay; some areas are served by fiber-optic cable; system operations, reliability, and coverage have steadily improved.
domestic: most telephones are concentrated in La Paz, Santa Cruz, and other capital cities; mobile-cellular telephone use expanding rapidly and, in 2009, teledensity reached about 75 per 100 persons
international: country code–591; satellite earth station–1 Intelsat (Atlantic Ocean) (2009)
Broadcast media: large number of radio and television broadcasting stations with private media outlets dominating; state-owned and private radio and television stations generally operating freely, although both pro-government and anti-government groups have attacked media outlets in response to their reporting (2010)
Internet country code: .bo
Internet hosts: 125,462 (2010)
country comparison to the world: 74
Internet users: 1.103 million (2009)
country comparison to the world: 95

TRANSPORTATION

Airports: 881 (2010)
country comparison to the world: 8
Airports—with paved runways: *total:* 16
over 3,047 m: 3
2,438 to 3,047 m: 4
1,524 to 2,437 m: 4
914 to 1,523 m: 5 (2010)
Airports—with unpaved runways:
total: 865
over 3,047 m: 1
2,438 to 3,047 m: 4
1,524 to 2,437 m: 58
914 to 1,523 m: 187
under 914 m: 615 (2010)
Pipelines: gas 5,330 km; liquid petroleum gas 51 km; oil 2,510 km; refined products 1,627 km (2010)
Railways: *total:* 3,652 km
country comparison to the world: 46
narrow gauge: 3,652 km 1.000-m gauge (2010)
Roadways: *total:* 13,602 km (does not include urban roads)
country comparison to the world: 126
paved: 4,990 km
unpaved: 8,612 km (2004)
Waterways: 10,000 km (commercially navigable almost exclusively in the northern and eastern parts of the country) (2010)
country comparison to the world: 13
Merchant marine: *total:* 22
country comparison to the world: 99
by type: bulk carrier 3, cargo 11, carrier 1, passenger/cargo 1, petroleum tanker 2, roll on/roll off 3, specialized tanker 1
foreign-owned: 7 (Bahamas 1, Ecuador 1, Iran 1, Syria 4) (2010)
Ports and terminals: Puerto Aguirre (inland port on the Paraguay/Parana waterway at the Bolivia/Brazil border); Bolivia has free port privileges in maritime ports in Argentina, Brazil, Chile, and Paraguay

MILITARY

Military branches: Bolivian Armed Forces: Bolivian Army (Ejercito Boliviano, EB), Bolivian Navy (Fuerza Naval Boliviana, FNB; includes marines), Bolivian Air Force (Fuerza Aerea Boliviana, FAB) (2011)
Military service age and obligation: 18-49 years of age for 12-month compulsory male and female military service; when annual number of volunteers falls short of goal, compulsory recruitment is effected, including conscription of boys as young as 14; 15-19 years of age for voluntary premilitary service, provides exemption from further military service (2011)
Manpower available for military service:
males age 16–49: 2,472,490
females age 16–49: 2,535,768 (2010 est.)
Manpower fit for military service: *males age 16–49:* 1,762,260
females age 16–49: 2,013,281 (2010 est.)
Manpower reaching militarily significant age annually: *male:* 108,334
female: 104,945 (2010 est.)
Military expenditures: 1.3% of GDP (2009)
country comparison to the world: 114

TRANSNATIONAL ISSUES

Disputes—international: Chile and Peru rebuff Bolivia's reactivated claim to restore the Atacama corridor, ceded to Chile in 1884, but Chile offers instead unrestricted but not sovereign maritime access through Chile for Bolivian natural gas; contraband smuggling, human trafficking, and illegal narcotic trafficking are problems in the porous areas of the border with Argentina
Illicit drugs: world's third-largest cultivator of coca (after Colombia and Peru) with an estimated 29,500 hectares under cultivation in 2007, increased slightly when compared to 2006; third largest producer of cocaine, estimated at 120 metric tons potential pure cocaine in 2007; transit country for Peruvian and Colombian cocaine destined for Brazil, Argentina, Chile, Paraguay, and Europe; cultivation generally increasing since 2000, despite eradication and alternative crop programs; weak border controls; some money-laundering activity related to narcotics trade; major cocaine consumption (2008)

BOSNIA AND HERZEGOVINA

INTRODUCTION

Background: Bosnia and Herzegovina's declaration of sovereignty in October 1991 was followed by a declaration of independence from the former Yugoslavia on 3 March 1992 after a referendum boycotted by ethnic Serbs. The Bosnian Serbs–supported by neighboring Serbia and Montenegro–responded with armed resistance aimed at partitioning the republic along ethnic lines and joining Serb-held areas to form a "Greater Serbia." In March 1994, Bosniaks and Croats reduced the number of warring factions from three to two by signing an agreement creating a joint Bosniak/Croat Federation of Bosnia and Herzegovina. On 21 November 1995, in Dayton, Ohio, the warring parties initialed a peace agreement that brought to a halt three years of interethnic civil strife (the final agreement was signed in Paris on 14 December 1995). The Dayton Peace Accords retained Bosnia and Herzegovina's international boundaries and created a multi-ethnic and democratic government charged with conducting foreign,

diplomatic, and fiscal policy. Also recognized was a second tier of government composed of two entities roughly equal in size: the Bosniak/Bosnian Croat Federation of Bosnia and Herzegovina and the Bosnian Serb-led Republika Srpska (RS). The Federation and RS governments were charged with overseeing most government functions. The Dayton Accords also established the Office of the High Representative (OHR) to oversee the implementation of the civilian aspects of the agreement. The Peace Implementation Council (PIC) at its conference in Bonn in 1997 also gave the High Representative the authority to impose legislation and remove officials, the so-called "Bonn Powers." In 1995-96, a NATO-led international peacekeeping force (IFOR) of 60,000 troops served in Bosnia to implement and monitor the military aspects of the agreement. IFOR was succeeded by a smaller, NATO-led Stabilization Force (SFOR) whose mission was to deter renewed hostilities. European Union peacekeeping troops (EUFOR) replaced SFOR in December 2004; their mission is to maintain peace and stability throughout the country. EUFOR's mission changed from peacekeeping to civil policing in October 2007, with its presence reduced from nearly 7,000 to less than 2,500 troops. Troop strength at the end of 2010 stood at roughly 1,500. In January 2010, Bosnia and Herzegovina assumed a nonpermanent seat on the UN Security Council for the 2010-11 term.

GEOGRAPHY

Location: Southeastern Europe, bordering the Adriatic Sea and Croatia
Geographic coordinates: 44 00 N, 18 00 E
Map references: Europe
Area: *total:* 51,197 sq km
country comparison to the world: 128
land: 51,187 sq km
water: 10 sq km
Area—comparative: slightly smaller than West Virginia
Land boundaries: *total:* 1,538 km
border countries: Croatia 932 km, Montenegro 249 km, Serbia 357 km
Coastline: 20 km
Maritime claims: no data available
Climate: hot summers and cold winters; areas of high elevation have short, cool summers and long, severe winters; mild, rainy winters along coast
Terrain: mountains and valleys
Elevation extremes: *lowest point:* Adriatic Sea 0 m
highest point: Maglic 2,386 m
Natural resources: coal, iron ore, bauxite, copper, lead, zinc, chromite, cobalt, manganese, nickel, clay, gypsum, salt, sand, timber, hydropower
Land use: *arable land:* 19.61%
permanent crops: 1.89%
other: 78.5% (2005)
Irrigated land: 30 sq km (2003)
Total renewable water resources: 37.5 cu km (2003)
Natural hazards: destructive earthquakes
Environment—current issues: air pollution from metallurgical plants; sites for disposing of urban waste are limited; water shortages and destruction of infrastructure because of the 1992-95 civil strife; deforestation
Environment—international agreements: *party to:* Air Pollution, Biodiversity, Climate Change, Climate Change-Kyoto Protocol, Desertification, Hazardous Wastes, Law of the Sea, Marine Life Conservation, Ozone Layer Protection, Wetlands
signed, but not ratified: none of the selected agreements
Geography—note: within Bosnia and Herzegovina's recognized borders, the country is divided into a joint Bosniak/Croat Federation (about 51% of the territory) and the Bosnian Serb-led Republika Srpska or RS (about 49% of the territory); the region called Herzegovina is contiguous to Croatia and Montenegro, and traditionally has been settled by an ethnic Croat majority in the west and an ethnic Serb majority in the east

PEOPLE

Population: 4,622,163 (July 2011 est.)
country comparison to the world: 120
Age structure: *0–14 years:* 14% (male 333,989/female 313,234)
15–64 years: 71% (male 1,655,669/female 1,625,750)
65 years and over: 15% (male 283,233/female 410,288) (2011 est.)
Median age: *total:* 40.7 years
male: 39.6 years
female: 41.9 years (2011 est.)
Population growth rate: 0.008% (2011 est.)
country comparison to the world: 192
Birth rate: 8.89 births/1,000 population (2011 est.)
country comparison to the world: 212
Death rate: 8.8 deaths/1,000 population (July 2011 est.)
country comparison to the world: 79
Net migration rate: 0 migrant(s)/1,000 population (2011 est.)
country comparison to the world: 76
Urbanization: *urban population:* 49% of total population (2010)
rate of urbanization: 1.1% annual rate of change (2010-15 est.)
Major cities—population: SARAJEVO (capital) 392,000 (2009)
Sex ratio: *at birth:* 1.074 male(s)/female
under 15 years: 1.07 male(s)/female
15–64 years: 1.02 male(s)/female
65 years and over: 0.69 male(s)/female
total population: 0.97 male(s)/female (2011 est.)
Infant mortality rate:
total: 8.67 deaths/1,000 live births
country comparison to the world: 155
male: 9.95 deaths/1,000 live births
female: 7.3 deaths/1,000 live births (2011 est.)
Life expectancy at birth: *total population:* 78.81 years
country comparison to the world: 45
male: 75.25 years
female: 82.63 years (2011 est.)
Total fertility rate: 1.27 children born/woman (2011 est.)
country comparison to the world: 213
HIV/AIDS—adult prevalence rate: less than 0.1% (2007 est.)
country comparison to the world: 118
HIV/AIDS—people living with HIV/AIDS: 900 (2007 est.)
country comparison to the world: 147
HIV/AIDS—deaths: 100 (2001 est.)
country comparison to the world: 124
Drinking water source: *Improved:*
urban: 100% of population
rural: 98% of population
total: 99% of population
Unimproved: urban: 0% of population
rural: 2% of population
total: 1% of population (2008)
Sanitation facility access: *Improved:*
urban: 99% of population
rural: 92% of population
total: 95% of population
Unimproved: urban: 1% of population
rural: 8% of population
total: 5% of population (2008)
Nationality:
noun: Bosnian(s), Herzegovinian(s)
adjective: Bosnian, Herzegovinian
Ethnic groups: Bosniak 48%, Serb 37.1%, Croat 14.3%, other 0.6% (2000)
note: Bosniak has replaced Muslim as an ethnic term in part to avoid confusion with the religious term Muslim–an adherent of Islam
Religions: Muslim 40%, Orthodox 31%, Roman Catholic 15%, other 14%
Languages: Bosnian (official), Croatian (official), Serbian
Literacy: *definition:* age 15 and over can read and write
total population: 96.7%
male: 99%
female: 94.4% (2000 est.)
School life expectancy (primary to tertiary education): *total:* 14 years
male: 13 years
female: 14 years (2009)
Education expenditures: NA

GOVERNMENT

Country name: *conventional long form:* none
conventional short form: Bosnia and Herzegovina
local long form: none
local short form: Bosna i Hercegovina
former: People's Republic of Bosnia and Herzegovina, Socialist Republic of Bosnia and Herzegovina
Government type: emerging federal democratic republic
Capital: *name:* Sarajevo
geographic coordinates: 43 52 N, 18 25 E
time difference: UTC+1 (6 hours ahead of Washington, DC during Standard Time)

daylight saving time: +1hr, begins last Sunday in March; ends last Sunday in October

Administrative divisions: 2 first-order administrative divisions and 1 internationally supervised district*—Brcko district (Brcko Distrikt)*, the Bosniak/Croat Federation of Bosnia and Herzegovina (Federacija Bosna i Hercegovina) and the Bosnian Serb-led Republika Srpska; note—Brcko district is in northeastern Bosnia and is a self-governing administrative unit under the sovereignty of Bosnia and Herzegovina and formally held in condominium between the two entities; the District remains under international supervision

Independence: 1 March 1992 (from Yugoslavia; referendum for independence completed on 1 March 1992; independence declared on 3 March 1992)

National holiday: National Day, 25 November (1943)

Constitution: the Dayton Peace Accords, signed 14 December 1995 in Paris, included a constitution; note—each of the entities also has its own constitution

Legal system: civil law system; Constitutional Court review of legislative acts

International law organization participation: has not submitted an ICJ jurisdiction declaration; accepts ICCt jurisdiction

Suffrage: 18 years of age, 16 if employed; universal

Executive branch: *chief of state:* Chairman of the Presidency Nebojsa RADMANOVIC (chairman of the presidency since 10 November 2010; presidency member since 6 November 2006—Serb); other members of the three-member presidency rotate every eight months: Bakir IZETBEGOVIC (presidency member since 10 November 2010—Bosniak); Zeljko KOMSIC (presidency member since 6 November 2006—Croat)
head of government: Chairman of the Council of Ministers Nikola SPIRIC (since 11 January 2007)
cabinet: Council of Ministers nominated by the council chairman; approved by the state-level House of Representatives
(For more information visit the World Leaders website)
elections: the three members of the presidency (one Bosniak, one Croat, one Serb) elected by popular vote for a four-year term (eligible for a second term, but then ineligible for four years) by constituencies referring to the three ethnic groups; the candidate with the most votes in a constituency is elected; the chairmanship rotates every eight months and resumes where it left off following each general election; election last held on 3 October 2010 (next to be held in October 2014); the chairman of the Council of Ministers appointed by the presidency and confirmed by the state-level House of Representatives
election results: percent of vote—Nebojsa RADMANOVIC with 48.9% of the votes for the Serb seat; Zeljko KOMSIC with 60.6% of the votes for the Croat seat; Bakir IZETBEGOVIC with 34.9% of the votes for the Bosniak seat
note: President of the Federation of Bosnia and Herzegovina: Zivko BUDIMIR (since 17 March 2011); Vice Presidents Spomenka MICIC (since 21 February 2007) and Mirsad KEBO (since 21 February 2007); President of the Republika Srpska: Milorad DODIK (since 15 November 2010)

Legislative branch: bicameral Parliamentary Assembly or Skupstina consists of the House of Peoples or Dom Naroda (15 seats, 5 Bosniak, 5 Croat, 5 Serb; members elected by the Bosniak/Croat Federation's House of Peoples and the Republika Srpska's National Assembly to serve four-year terms); and the state-level House of Representatives or Predstavnicki Dom (42 seats, 28 seats allocated for the Federation of Bosnia and Herzegovina and 14 seats for the Republika Srpska; members elected by popular vote on the basis of proportional representation to serve four-year terms); note—Bosnia's election law specifies four-year terms for the state and first-order administrative division entity legislatures
elections: House of Peoples—last constituted in February 2007 (next to be constituted in 2011); state-level House of Representatives—elections last held on 3 October 2010 (next to be held in October 2014)
election results: House of Peoples—percent of vote by party/coalition—NA; seats by party/coalition—NA; state-level House of Representatives—percent of vote by party/coalition—NA; seats by party/coalition—SDP BiH 8, SNSD 8, SDA 7, SDS 4, SBBBiH 4, HDZ-BiH 3, SBiH 2, HDZ-1990/HSP 2, other 4
note: the Bosniak/Croat Federation has a bicameral legislature that consists of a House of Peoples (58 seats—17 Bosniak, 17 Croat, 17 Serb, 7 other); last constituted February 2007; and a House of Representatives (98 seats; members elected by popular vote to serve four-year terms); elections last held on 3 October 2010 (next to be held in October 2014); percent of vote by party—NA; seats by party/coalition—SDP 28, SDA 23, SBBBiH 13, HDZ-BiH 12, SBiH 9, HDZ-1990/HSP 5, NSRzB 5, other 3; the Republika Srpska has a National Assembly (83 seats; members elected by popular vote to serve four-year terms); elections last held on 3 October 2010 (next to be held in October 2014); percent of vote by party—NA; seats by party/coalition—SNSD 37, SDS 18, PDP 7, DNS 6, SP 4, DP 3, SDP 3, SDA 2, NDS 2 SRS-RS 1; as a result of the 2002 constitutional reform process, a 28-member Republika Srpska Council of Peoples (COP) was established in the Republika Srpska National Assembly including 8 Croats, 8 Bosniaks, 8 Serbs, and 4 members of the smaller communities

Judicial branch: BiH Constitutional Court (consists of nine members: four members are selected by the Bosniak/Croat Federation's House of Representatives, two members by the Republika Srpska's National Assembly, and three non-Bosnian members by the president of the European Court of Human Rights); BiH State Court (consists of 44 national judges and seven international judges and has three divisions—Administrative, Appellate and Criminal—having jurisdiction over cases related to state-level law and cases initiated in the entities that question BiH's sovereignty, political independence, or national security or with economic crimes that have serious repercussions to BiH's economy, beyond that of an entity or Brcko District); a War Crimes Chamber opened in March 2005
note: the entities each have a Supreme Court; each entity also has a number of lower courts; there are 10 cantonal courts in the Federation, plus a number of municipal courts; the Republika Srpska has five district courts and a number of municipal courts

Political parties and leaders: Alliance for a Better Future of BiH or SBB-BiH [Fahrudin RADONCIC]; Alliance of Independent Social Democrats or SNSD [Milorad DODIK]; Bosnian Party or BOSS [Mirnes AJANOVIC]; Bosnian Patriotic Party or BPS [Sefer HALILOVIC]; Civic Democratic Party or GDS [Ibrahim SPAHIC]; Croat Party of Rights or HSP [Zvonko JURISIC]; Croat Peasants' Party-New Croat Initiative or HSS-NHI [Ante COLAK]; Croatian Christian Democratic Union of Bosnia and Herzegovina or HKDU [Ivan MUSA]; Croatian Democratic Union of Bosnia and Herzegovina or HDZ-BiH [Dragan COVIC]; Croatian Democratic Union 1990 or HDZ-1990 [Bozo LJUBIC]; Croatian Peoples Union [Milenko BRKIC]; Democratic National Union or DNZ [Rifat DOLIC]; Democratic Party or DP [Dragan CAVIC]; Democratic Peoples' Alliance or DNS [Marko PAVIC]; Liberal Democratic Party or LDS [Rasim KADIC]; Nasa Stranka or NS [NA; leadership elections late 2010/early 2011]; New Socialist Party or NSP [Zdravko KRSMANOVIC]; Party for Bosnia and Herzegovina or SBiH [Haris SILAJDZIC]; Party of Democratic Action or SDA [Sulejman TIHIC]; Party of Democratic Progress or PDP [Mladen IVANIC]; Peoples' Party of Work for Progress or NSRzB [Mladen IVANKOVIC-LIJANOVIC]; Serb Democratic Party or SDS [Mladen BOSIC]; Serb Radical Party of the Republika Srpska or SRS-RS [Milanko MIHAJLICA]; Serb Radical Party-Dr. Vojislav Seselj or SRS-VS [Mirko BLAGOJEVIC]; Social Democratic Party of BiH or SDP BiH [Zlatko LAGUMDZIJA]; Social Democratic Union or SDU [Nermin PECANAC]; Socialist Party of Republika Srpska or SPRS [Petar DJOKIC]

Political pressure groups and leaders: *other:* war veterans; displaced persons associations; family associations of missing persons; private media

International organization participation: BIS, CE, CEI, EAPC, EBRD, FAO, G-77, IAEA, IBRD, ICAO, ICRM, IDA, IDB, IFAD, IFC, IFRCS, ILO, IMF, IMO, IMSO, Interpol, IOC, IOM, IPU, ISO, ITSO, ITU, ITUC, MIGA, MONUSCO, NAM (observer), OAS (observer), OIC (observer), OPCW, OSCE, PFP, SECI, UN, UN Security Council (temporary), UNCTAD, UNESCO, UNIDO, UNWTO, UPU, WCO, WHO, WIPO, WMO, WTO (observer)

Diplomatic representation in the US: *chief of mission:* Ambassador Mitar KUJUNDZIC

chancery: 2109 E Street NW, Washington, DC 20037
telephone: [1] (202) 337-1500
FAX: [1] (202) 337-1502
consulate(s) general: Chicago, New York
Diplomatic representation from the US: *chief of mission:* Ambassador Patrick S. MOON
embassy: 1 Robert C. Frasure Street, 71000 Sarajevo
mailing address: use embassy street address
telephone: [387] (33) 704-000
FAX: [387] (33) 659-722
branch office(s): Banja Luka, Mostar
Flag description: a wide medium blue vertical band on the fly side with a yellow isosceles triangle abutting the band and the top of the flag; the remainder of the flag is medium blue with seven full five-pointed white stars and two half stars top and bottom along the hypotenuse of the triangle; the triangle approximates the shape of the country and its three points stand for the constituent peoples–Bosniaks, Croats, and Serbs; the stars represent Europe and are meant to be continuous (thus the half stars at top and bottom); the colors (white, blue, and yellow) are often associated with neutrality and peace, and traditionally are linked with Bosnia
National anthem: *name:* "Drzavna himna Bosne i Hercegovine" (The National Anthem of Bosnia and Herzegovina)
lyrics/music: Dusan SESTIC and Benjamin ISOVIC/Dusan SESTIC
note: music adopted 1999; lyrics adopted 2009

ECONOMY

Economy—overview: The interethnic warfare in Bosnia and Herzegovina caused production to plummet by 80% from 1992 to 1995 and unemployment to soar. With an uneasy peace in place, output recovered in 1996-99 at high percentage rates from a low base; but output growth slowed in 2000-02. Part of the lag in output was made up during 2003-08, when GDP growth exceeded 5% per year. However, the country experienced a decline in GDP of more than 3% in 2009 reflecting local effects of the global economic crisis. One of Bosnia's main economic challenges in 2010 has been to reduce spending on public sector wages and social benefits to meet the IMF's criteria for obtaining funding for budget shortfalls. Banking reform accelerated in 2001 as all the Communist-era payments bureaus were shut down; foreign banks, primarily from Austria and Italy, now control most of the banking sector. The konvertibilna marka (convertible mark or BAM)–the national currency introduced in 1998–is pegged to the euro, and confidence in the currency and the banking sector has increased. Bosnia's private sector is growing, but foreign investment has dropped off sharply since 2007. Government spending, at roughly 50% of GDP, remains high because of redundant government offices at the state, entity and municipal level. Privatization of state enterprises has been slow, particularly in the Federation where political division between ethnically-based political parties makes agreement on economic policy more difficult. A sizeable current account deficit and high unemployment rate remain the two most serious macroeconomic problems. Successful implementation of a value-added tax in 2006 provided a predictable source of revenue for the government and helped rein in gray-market activity. National-level statistics have also improved over time but a large share of economic activity remains unofficial and unrecorded. Bosnia and Herzegovina became a full member of the Central European Free Trade Agreement in September 2007. Bosnia and Herzegovina's top economic priorities are: acceleration of integration into the EU; strengthening the fiscal system; public administration reform; World Trade Organization (WTO) membership; and securing economic growth by fostering a dynamic, competitive private sector. The country has received a substantial amount of foreign assistance and will need to demonstrate its ability to implement its economic reform agenda in order to advance its stated goal of EU accession. In 2009, Bosnia and Herzegovina undertook an International Monetary Fund (IMF) standby arrangement, necessitated by sharply increased social spending and a fiscal crisis exacerbated by the global economic downturn. The program aims to reduce recurrent government spending and to strengthen revenue collection.
GDP (purchasing power parity): $30.33 billion (2010 est.)
country comparison to the world: 107
$30.09 billion (2009 est.)
$31.04 billion (2008 est.)
note: data are in 2010 US dollars
GDP (official exchange rate): $16.83 billion (2010 est.)
GDP—real growth rate: 0.8% (2010 est.)
country comparison to the world: 177
-3.1% (2009 est.)
5.7% (2008 est.)
GDP—per capita (PPP): $6,600 (2010 est.)
country comparison to the world: 136
$6,500 (2009 est.)
$6,800 (2008 est.)
note: data are in 2010 US dollars
GDP—composition by sector:
agriculture: 6.5%
industry: 28.4%
services: 65.1% (2010 est.)
Labor force: 2.6 million (2010 est.)
country comparison to the world: 109
Labor force—by occupation: *agriculture:* 20.5%
industry: 32.6%
services: 47% (2008)
Unemployment rate: 43.1% (2010 est.)
country comparison to the world: 188
44.2% (2009 est.)
note: official rate
Population below poverty line: 18.6% (2007 est.)
Household income or consumption by percentage share:
lowest 10%: 2.8%
highest 10%: 27.4% (2004)
Distribution of family income—Gini index: 34.1 (2007)
country comparison to the world: 91
Budget: *revenues:* $7.75 billion
expenditures: $7.82 billion (2010 est.)
Public debt: 39% of GDP (2010 est.)
country comparison to the world: 77
35% of GDP (2009 est.)
Inflation rate (consumer prices): 3.1% (2010 est.)
country comparison to the world: 88
0.8% (2009 est.)
Commercial bank prime lending rate: 7.93% (31 December 2010 est.)
country comparison to the world: 118
7.93% (31 December 2009 est.)
Stock of narrow money: $4.098 billion (31 December 2010 est.)
country comparison to the world: 100
$4.182 billion (31 December 2009 est.)
Stock of broad money: $9.307 billion (31 December 2010 est.)
country comparison to the world: 103
$9.236 billion (31 December 2009 est.)
Stock of domestic credit: $10.09 billion (31 December 2010 est.)
country comparison to the world: 94
$10.01 billion (31 December 2009 est.)
Market value of publicly traded shares: $NA
Agriculture—products: wheat, corn, fruits, vegetables; livestock
Industries: steel, coal, iron ore, lead, zinc, manganese, bauxite, aluminum, vehicle assembly, textiles, tobacco products, wooden furniture, ammunition, domestic appliances, oil refining
Industrial production growth rate: 1.6% (2010 est.)
country comparison to the world: 137
Electricity—production: 14.58 billion kWh (2009 est.)
country comparison to the world: 80
Electricity—consumption: 10.8 billion kWh (2009 est.)
country comparison to the world: 84
Electricity—exports: 3.9 billion kWh (2009 est.)
Electricity—imports: 1.2 billion kWh (2009 est.)
Oil—production: NA bbl/day (2008 est.)
Oil—consumption: NA bbl/day (2009 est.)
Oil—exports: 191.8 bbl/day (2007 est.)
country comparison to the world: 132
Oil—imports: 25,990 bbl/day (2007 est.)
country comparison to the world: 103
Oil—proved reserves: 0 bbl (1 January 2010 est.)
country comparison to the world: 109
Natural gas—production: 0 cu m (2010 est.)
country comparison to the world: 102
Natural gas—consumption: 390 million cu m (2010 est.)
country comparison to the world: 97
Natural gas—exports: 0 cu m (2010 est.)
country comparison to the world: 62
Natural gas—imports: 390 million cu m (2010 est.)
country comparison to the world: 61
Natural gas—proved reserves: 0 cu m (1 January 2010 est.)
country comparison to the world: 113
Current account balance: $-1.175 billion (2010 est.)
country comparison to the world: 140
$-2.667 billion (2009 est.)
Exports: $4.804 billion (2010 est.)
country comparison to the world: 108
$3.929 billion (2009 est.)

Exports—commodities: metals, clothing, wood products
Exports—partners: Croatia 19.07%, Slovenia 18.58%, Italy 16.87%, Germany 13.38%, Austria 10.25% (2009)
Imports: $9.22 billion (2010 est.)
country comparison to the world: 92
$8.773 billion (2009 est.)
Imports—commodities: machinery and equipment, chemicals, fuels, foodstuffs
Imports—partners: Croatia 22.17%, Germany 14.04%, Slovenia 13.45%, Italy 11.89%, Austria 6.61%, Hungary 5.74% (2009)
Reserves of foreign exchange and gold: $4.2 billion (31 December 2010 est.)
country comparison to the world: 75
$3.245 billion (31 December 2009 est.)
Debt—external: $7.996 billion (31 December 2010 est.)
country comparison to the world: 91
$8.048 billion (31 December 2009 est.)
Exchange rates: konvertibilna markas (BAM) per US dollar–1.5088 (2010), 1.4079 (2009), 1.3083 (2008), 1.4419 (2007), 1.5576 (2006)

COMMUNICATIONS

Telephones—main lines in use: 998,600 (2009)
country comparison to the world: 79
Telephones—mobile cellular: 3.257 million (2009)
country comparison to the world: 110
Telephone system: *general assessment:* post-war reconstruction of the telecommunications network, aided by a internationally sponsored program, resulting in sharp increases in the number of fixed telephone lines available
domestic: fixed-line teledensity roughly 22 per 100 persons; mobile-cellular subscribership has been increasing rapidly and, in 2009, reached 70 telephones per 100 persons
international: country code–387; no satellite earth stations (2009)
Broadcast media: 3 public TV broadcasters: Radio and TV of Bosnia and Herzegovina, Federation TV (operating 2 networks), and Republika Srpska Radio-TV; a local commercial network of 5 TV stations; 3 private, near-national TV stations and dozens of small independent TV stations broadcasting; 3 large public radio broadcasters and a large number of private radio stations (2010)
Internet country code: .ba
Internet hosts: 95,234 (2010)
country comparison to the world: 79
Internet users: 1.422 million (2009)
country comparison to the world: 85

TRANSPORTATION

Airports: 25 (2010)
country comparison to the world: 129
Airports—with paved runways: *total:* 7
2,438 to 3,047 m: 4
1,524 to 2,437 m: 1
under 914 m: 2 (2010)
Airports—with unpaved runways:
total: 18
1,524 to 2,437 m: 1
914 to 1,523 m: 6
under 914 m: 11 (2010)
Heliports: 5 (2010)
Pipelines: gas 147 km; oil 9 km
Railways: *total:* 601 km
country comparison to the world: 107
standard gauge: 601 km 1.435-m gauge (392 km electrified) (2009)
Roadways: *total:* 22,926 km
country comparison to the world: 105
paved: 19,426 km (4,652 km of interurban roads)
unpaved: 3,500 km (2010)
Waterways: (Sava River on northern border; open to shipping but use limited) (2009)
Ports and terminals: Bosanska Gradiska, Bosanski Brod, Bosanski Samac, and Brcko (all inland waterway ports on the Sava River), Orasje

MILITARY

Military branches: Armed Forces of Bosnia and Herzegovina (AFBiH): Army of Bosnia and Herzegovina, Air and Air Defense Forces of Bosnia and Herzegovina (Zrakoplovstvo i Protuzracna Obrana, ZPO) (2010)
Military service age and obligation: 18 years of age for voluntary military service; conscription abolished in January 2006; 4-month service obligation; mandatory retirement at age 35 or after 15 years of service (2010)
Manpower available for military service:
males age 16–49: 1,180,829
females age 16–49: 1,143,919 (2010 est.)
Manpower fit for military service: *males age 16–49:* 968,242
females age 16–49: 937,327 (2010 est.)
Manpower reaching militarily significant age annually: *male:* 26,601
female: 24,879 (2010 est.)
Military expenditures: 4.5% of GDP (2005 est.)
country comparison to the world: 19

TRANSNATIONAL ISSUES

Disputes—international: Serbia delimited about half of the boundary with Bosnia and Herzegovina, but sections along the Drina River remain in dispute
Refugees and internally displaced persons: refugees (country of origin): 7,269 (Croatia)
IDPs: 131,600 (Bosnian Croats, Serbs, and Bosniaks displaced in 1992-95 war) (2007)
Illicit drugs: increasingly a transit point for heroin being trafficked to Western Europe; minor transit point for marijuana; remains highly vulnerable to money-laundering activity given a primarily cash-based and unregulated economy, weak law enforcement, and instances of corruption

BOTSWANA

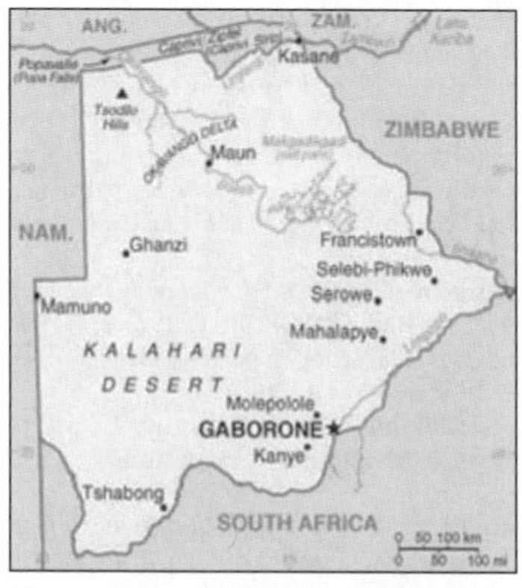

INTRODUCTION

Background: Formerly the British protectorate of Bechuanaland, Botswana adopted its new name upon independence in 1966. Four decades of uninterrupted civilian leadership, progressive social policies, and significant capital investment have created one of the most dynamic economies in Africa. Mineral extraction, principally diamond mining, dominates economic activity, though tourism is a growing sector due to the country's conservation practices and extensive nature preserves. Botswana has one of the world's highest known rates of HIV/AIDS infection, but also one of Africa's most progressive and comprehensive programs for dealing with the disease.

GEOGRAPHY

Location: Southern Africa, north of South Africa
Geographic coordinates: 22 00 S, 24 00 E
Map references: Africa
Area: *total:* 581,730 sq km
country comparison to the world: 47
land: 566,730 sq km
water: 15,000 sq km
Area—comparative: slightly smaller than Texas
Land boundaries: *total:* 4,013 km
border countries: Namibia 1,360 km, South Africa 1,840 km, Zimbabwe 813 km
Coastline: 0 km (landlocked)
Maritime claims: none (landlocked)
Climate: semiarid; warm winters and hot summers
Terrain: predominantly flat to gently rolling tableland; Kalahari Desert in southwest
Elevation extremes: *lowest point:* junction of the Limpopo and Shashe Rivers 513 m
highest point: Tsodilo Hills 1,489 m
Natural resources: diamonds, copper, nickel, salt, soda ash, potash, coal, iron ore, silver
Land use: *arable land:* 0.65%
permanent crops: 0.01%
other: 99.34% (2005)

Irrigated land: 10 sq km (2003)
Total renewable water resources: 14.7 cu km (2001)
Freshwater withdrawal (domestic/industrial/agricultural): *total:* 0.19 cu km/yr (41%/18%/41%)
per capita: 107 cu m/yr (2000)
Natural hazards: periodic droughts; seasonal August winds blow from the west, carrying sand and dust across the country, which can obscure visibility
Environment—current issues: overgrazing; desertification; limited freshwater resources
Environment—international agreements: *party to:* Biodiversity, Climate Change, Climate Change-Kyoto Protocol, Desertification, Endangered Species, Hazardous Wastes, Law of the Sea, Ozone Layer Protection, Wetlands
signed, but not ratified: none of the selected agreements
Geography—note: landlocked; population concentrated in eastern part of the country

PEOPLE

Population: 2,065,398 (July 2011 est.)
country comparison to the world: 144
note: estimates for this country explicitly take into account the effects of excess mortality due to AIDS; this can result in lower life expectancy, higher infant mortality, higher death rates, lower population growth rates, and changes in the distribution of population by age and sex than would otherwise be expected
Age structure: *0–14 years:* 33.9% (male 356,346/female 343,452)
15–64 years: 62.2% (male 649,931/female 634,998)
65 years and over: 3.9% (male 32,542/female 48,129) (2011 est.)
Median age: *total:* 22.3 years
male: 22.2 years
female: 22.4 years (2011 est.)
Population growth rate: 1.656% (2011 est.)
country comparison to the world: 69
Birth rate: 22.31 births/1,000 population (2011 est.)
country comparison to the world: 76
Death rate: 10.57 deaths/1,000 population (July 2011 est.)
country comparison to the world: 47
Net migration rate: 4.82 migrant(s)/1,000 population
country comparison to the world: 19
note: there is an increasing flow of Zimbabweans into South Africa and Botswana in search of better economic opportunities (2011 est.)
Urbanization: *urban population:* 61% of total population (2010)
rate of urbanization: 2.3% annual rate of change (2010-15 est.)
Major cities—population: GABORONE (capital) 196,000 (2009)
Sex ratio: *at birth:* 1.03 male(s)/female
under 15 years: 1.04 male(s)/female
15–64 years: 1.02 male(s)/female
65 years and over: 0.68 male(s)/female
total population: 1.01 male(s)/female (2011 est.)
Infant mortality rate: *total:* 11.14 deaths/1,000 live births
country comparison to the world: 142
male: 11.76 deaths/1,000 live births
female: 10.51 deaths/1,000 live births (2011 est.)
Life expectancy at birth: *total population:* 58.05 years
country comparison to the world: 192
male: 58.78 years
female: 57.3 years (2011 est.)
Total fertility rate: 2.5 children born/woman (2011 est.)
country comparison to the world: 82
HIV/AIDS—adult prevalence rate: 24.8% (2009 est.)
country comparison to the world: 2
HIV/AIDS—people living with HIV/AIDS: 320,000 (2009 est.)
country comparison to the world: 19
HIV/AIDS—deaths: 5,800 (2009 est.)
country comparison to the world: 34
Major infectious diseases:
degree of risk: high
food or waterborne diseases: bacterial diarrhea, hepatitis A, and typhoid fever
vectorborne disease: malaria (2009)
Drinking water source: *Improved:*
urban: 99% of population
rural: 90% of population
total: 95% of population
Unimproved: urban: 1% of population
rural: 10% of population
total: 5% of population (2008)
Sanitation facility access: *Improved:*
urban: 74% of population
rural: 39% of population
total: 60% of population
Unimproved: urban: 26% of population
rural: 61% of population
total: 40% of population (2008)
Nationality: *noun:* Motswana (singular), Batswana (plural)
adjective: Motswana (singular), Batswana (plural)
Ethnic groups: Tswana (or Setswana) 79%, Kalanga 11%, Basarwa 3%, other, including Kgalagadi and white 7%
Religions: Christian 71.6%, Badimo 6%, other 1.4%, unspecified 0.4%, none 20.6% (2001 census)
Languages: Setswana 78.2%, Kalanga 7.9%, Sekgalagadi 2.8%, English (official) 2.1%, other 8.6%, unspecified 0.4% (2001 census)
Literacy: *definition:* age 15 and over can read and write
total population: 81.2%
male: 80.4%
female: 81.8% (2003 est.)
School life expectancy (primary to tertiary education): *total:* 12 years
male: 12 years
female: 12 years (2007)
Education expenditures: 8.9% of GDP (2009)
country comparison to the world: 8

GOVERNMENT

Country name: *conventional long form:* Republic of Botswana
conventional short form: Botswana
local long form: Republic of Botswana
local short form: Botswana
former: Bechuanaland
Government type: parliamentary republic
Capital: *name:* Gaborone
geographic coordinates: 24 45 S, 25 55 E
time difference: UTC+2 (7 hours ahead of Washington, DC during Standard Time)
Administrative divisions: 9 districts and 5 town councils*; Central, Francistown*, Gaborone*, Ghanzi, Jwaneng*, Kgalagadi, Kgatleng, Kweneng, Lobatse*, Northeast, Northwest, Selebi-Pikwe*, Southeast, Southern
Independence: 30 September 1966 (from the UK)
National holiday: Independence Day (Botswana Day), 30 September (1966)
Constitution: March 1965; effective 30 September 1966
Legal system: mixed legal system of civil law influenced by the Roman-Dutch model and also customary and common law
International law organization participation: accepts compulsory ICJ jurisdiction with reservations; accepts ICCt jurisdiction
Suffrage: 18 years of age; universal
Executive branch: *chief of state:* President Seretse Khama Ian KHAMA (since 1 April 2008); Vice President Mompati MERAFHE (since 1 April 2008); note—the president is both the chief of state and head of government
head of government: President Seretse Khama Ian KHAMA (since 1 April 2008); Vice President Mompati MERAFHE (since 1 April 2008)
cabinet: Cabinet appointed by the president
(For more information visit the World Leaders website)
elections: president indirectly elected for a five-year term (eligible for a second term); election last held on 20 October 2009 (next to be held in October 2014); vice president appointed by the president
election results: Seretse Khama Ian KHAMA elected president; percent of National Assembly vote—NA
Legislative branch: bicameral Parliament consists of the House of Chiefs (a largely advisory 15-member body with 8 ex-officio members consisting of the chiefs of the principal tribes, and 7 non-permanent members serving 5-year terms, consisting of 4 elected subchiefs and 3 members selected by the other 12 members) and the National Assembly (63 seats; 57 members directly elected by popular vote, 4 appointed by the majority party, and 2, the President and Attorney General, serve as ex-officio members; members serve five-year terms)
elections: National Assembly elections last held on 16 October 2009 (next to be held in 2014)
election results: percent of vote by party—BDP 53.3%, BNF 21.9%, BCP 19.2%, 2.3%, other 3.3%; seats by party—BDP 45, BNF 6, BCP 4, BAM 1, other 1
Judicial branch: High Court; Court of Appeal; Magistrates' Courts (one in each district)
Political parties and leaders: Botswana Alliance Movement or BAM [Ephraim Lepetu SETSHWAELO]; Botswana Congress Party or BCP [Gilson SALESHANDO]; Botswana Democratic Party or BDP [Daniel KWELAGOBE]; Botswana National Front or BNF [Otswoletse

MOUPO]; Botswana Peoples Party or BPP [Bernard BALIKANI]; MELS Movement of Botswana or MELS [Themba JOINA]; New Democratic Front or NDF [Dick BAYFORD]
note: a number of minor parties joined forces in 1999 to form the BAM but did not capture any parliamentary seats—includes the United Action Party [Ephraim Lepetu SETSHWAELO]; the Independence Freedom Party or IFP [Motsamai MPHO]; the Botswana Progressive Union [D. K. KWELE]
Political pressure groups and leaders: First People of the Kalahari (Bushman organization); Pitso Ya Ba Tswana; Society for the Promotion of Ikalanga Language (Kalanga elites)
other: diamond mining companies
International organization participation: ACP, AfDB, AU, C, FAO, G-77, IAEA, IBRD, ICAO, ICRM, IDA, IFAD, IFC, IFRCS, ILO, IMF, Interpol, IOC, IOM, IPU, ISO, ITSO, ITU, ITUC, MIGA, NAM, OPCW, SACU, SADC, UN, UNCTAD, UNESCO, UNIDO, UNWTO, UPU, WCO, WFTU, WHO, WIPO, WMO, WTO
Diplomatic representation in the US: *chief of mission:* Ambassador Tabelelo Mazile SERETSE
chancery: 1531-1533 New Hampshire Avenue NW, Washington, DC 20036
telephone: [1] (202) 244-4990
FAX: [1] (202) 244-4164
Diplomatic representation from the US: *chief of mission:* Ambassador Stephen J. NOLAN
embassy: Embassy Enclave (off Khama Crescent), Gaborone
mailing address: Embassy Enclave, P. O. Box 90, Gaborone
telephone: [267] 395-3982
FAX: [267] 395-6947
Flag description: light blue with a horizontal white-edged black stripe in the center; the blue symbolizes water in the form of rain, while the black and white bands represent racial harmony
National anthem: *name:* "Fatshe leno la rona" (Our Land)
lyrics/music: Kgalemang Tumedisco MOTSETE
note: adopted 1966

ECONOMY

Economy—overview: Botswana has maintained one of the world's highest economic growth rates since independence in 1966, though growth fell below 5% in 2007-08, and turned sharply negative in 2009, with industry falling nearly 30%. Through fiscal discipline and sound management, Botswana transformed itself from one of the poorest countries in the world to a middle-income country with a per capita GDP of $13,100 in 2010. Two major investment services rank Botswana as the best credit risk in Africa. Diamond mining has fueled much of the expansion and currently accounts for more than one-third of GDP, 70-80% of export earnings, and about half of the government's revenues. Botswana's heavy reliance on a single luxury export was a critical factor in the sharp economic contraction of 2009. Tourism, financial services, subsistence farming, and cattle raising are other key sectors. Although unemployment was 7.5% in 2007 according to official reports, unofficial estimates place it closer to 40%. The prevalence of HIV/AIDS is second highest in the world and threatens Botswana's impressive economic gains. An expected leveling off in diamond mining production within the next two decades overshadows long-term prospects.
GDP (purchasing power parity): $28.49 billion (2010 est.)
country comparison to the world: 110
$26.24 billion (2009 est.)
$27.24 billion (2008 est.)
note: data are in 2010 US dollars
GDP (official exchange rate): $14.03 billion (2010 est.)
GDP—real growth rate:
8.6% (2010 est.)
country comparison to the world: 12
-3.7% (2009 est.)
3.1% (2008 est.)
GDP—per capita (PPP): $14,000 (2010 est.)
country comparison to the world: 84
$13,200 (2009 est.)
$14,000 (2008 est.)
note: data are in 2010 US dollars
GDP—composition by sector: *agriculture:* 2.3%
industry: 45.8%
services: 51.9% (2009 est.)
Labor force: 685,300 formal sector employees (2007)
country comparison to the world: 151
Labor force—by occupation: *agriculture:* NA%
industry: NA%
services: NA%
Unemployment rate: 7.5% (2007 est.)
country comparison to the world: 79
Population below poverty line: 30.3% (2003)
Household income or consumption by percentage share: *lowest 10%:* NA%
highest 10%: NA%
Distribution of family income—Gini index: 63 (1993)
country comparison to the world: 4
Investment (gross fixed): 28.2% of GDP (2010 est.)
country comparison to the world: 23
Budget: *revenues:* $4.165 billion
expenditures: $5.888 billion (2010 est.)
Public debt: 22.6% of GDP (2010 est.)
country comparison to the world: 107
18.6% of GDP (2009 est.)
Inflation rate (consumer prices): 7.1% (2010 est.)
country comparison to the world: 175
8.1% (2009 est.)
Central bank discount rate: 10% (31 December 2009)
country comparison to the world: 18
15% (31 December 2008)
Commercial bank prime lending rate: 13.76% (31 December 2009 est.)
country comparison to the world: 32
16.54% (31 December 2008 est.)
Stock of narrow money: $1.146 billion (31 December 2010 est.)
country comparison to the world: 136
$939.1 million (31 December 2009 est.)
Stock of broad money: $6.679 billion (31 December 2010 est.)
country comparison to the world: 113
$5.357 billion (31 December 2009 est.)
Stock of domestic credit: $1.361 billion (31 December 2008 est.)
country comparison to the world: 139
$2.06 billion (31 December 2007 est.)
Market value of publicly traded shares: $3.991 billion (31 December 2009)
country comparison to the world: 88
$3.556 billion (31 December 2008)
$5.887 billion (31 December 2007)
Agriculture—products: livestock, sorghum, maize, millet, beans, sunflowers, groundnuts
Industries: diamonds, copper, nickel, salt, soda ash, potash, coal, iron ore, silver; livestock processing; textiles
Industrial production growth rate: 6.9% (2010 est.)
country comparison to the world: 48
Electricity—production: 1.052 billion kWh (2007 est.)
country comparison to the world: 144
Electricity—consumption: 2.648 billion kWh (2007 est.)
country comparison to the world: 130
Electricity—exports: 0 kWh (2008 est.)
Electricity—imports: 2.181 billion kWh (2008 est.)
Oil—production: 0 bbl/day (2009 est.)
country comparison to the world: 150
Oil—consumption: 15,000 bbl/day (2009 est.)
country comparison to the world: 136
Oil—exports: 0 bbl/day (2007 est.)
country comparison to the world: 148
Oil—imports: 15,180 bbl/day (2007 est.)
country comparison to the world: 125
Oil—proved reserves: 0 bbl (1 January 2010 est.)
country comparison to the world: 105
Natural gas—production: 0 cu m (2008 est.)
country comparison to the world: 97
Natural gas—consumption: 0 cu m (2008 est.)
country comparison to the world: 146
Natural gas—exports: 0 cu m (2008 est.)
country comparison to the world: 56
Natural gas—imports: 0 cu m (2008 est.)
country comparison to the world: 149
Natural gas—proved reserves: 0 cu m (1 January 2010 est.)
country comparison to the world: 108
Current account balance: $-552 million (2010 est.)
country comparison to the world: 118
$-762 million (2009 est.)
Exports: $4.419 billion (2010 est.)
country comparison to the world: 111
$3.385 billion (2009 est.)
Exports—commodities: diamonds, copper, nickel, soda ash, meat, textiles
Imports: $4.518 billion (2010 est.)
country comparison to the world: 125
$4.243 billion (2009 est.)
Imports—commodities: foodstuffs, machinery, electrical goods, transport equipment, textiles, fuel and petroleum products, wood and paper products, metal and metal products
Reserves of foreign exchange and gold: $7.834 billion (31 December 2010 est.)

country comparison to the world: 62
$8.704 billion (31 December 2009 est.)
Debt—external: $2.222 billion (31 December 2010 est.)
country comparison to the world: 138
$1.681 billion (31 December 2009 est.)
Exchange rates: pulas (BWP) per US dollar–
6.7413 (2010)
7.1551 (2009)
6.7907 (2008)
6.2035 (2007)
5.8447 (2006)

COMMUNICATIONS

Telephones—main lines in use: 144,200 (2009)
country comparison to the world: 134
Telephones—mobile cellular: 1.874 million (2009)
country comparison to the world: 136
Telephone system: *general assessment:* Botswana is participating in regional development efforts; expanding fully digital system with fiber-optic cables linking the major population centers in the east as well as a system of open-wire lines, microwave radio relays links, and radiotelephone communication stations
domestic: fixed-line teledensity has declined in recent years and now stands at roughly 7 telephones per 100 persons; mobile-cellular subscribership is rapidly approaching a teledensity of 100 telephones per 100 persons
international: country code–267; international calls are made via satellite, using international direct dialing; 2 international exchanges; digital microwave radio relay links to Namibia, Zambia, Zimbabwe, and South Africa; satellite earth station–1 Intelsat (Indian Ocean) (2008)
Broadcast media: 2 TV stations–1 state-owned and 1 privately-owned; privately-owned satellite TV subscription service is available; 2 state-owned national radio stations; 3 privately-owned radio stations broadcast locally (2007)
Internet country code: .bw
Internet hosts: 2,739 (2010)
country comparison to the world: 148
Internet users: 120,000 (2009)
country comparison to the world: 154

TRANSPORTATION

Airports: 78 (2010)
country comparison to the world: 71
Airports—with paved runways: *total:* 9
2,438 to 3,047 m: 2
1,524 to 2,437 m: 6
914 to 1,523 m: 1 (2010)
Airports—with unpaved runways: *total:* 69
1,524 to 2,437 m: 4
914 to 1,523 m: 52
under 914 m: 13 (2010)
Railways: *total:* 888 km
country comparison to the world: 93
narrow gauge: 888 km 1.067-m gauge (2010)
Roadways: *total:* 25,798 km
country comparison to the world: 102
paved: 8,410 km
unpaved: 17,388 km (2005)

MILITARY

Military branches: Botswana Defense Force (BDF): Ground Forces Command, Air Arm Command, Logistics Command (2011)
Military service age and obligation: 18 is the apparent age of voluntary military service; official minimum age is unknown (2001)
Manpower available for military service: *males age 16–49:* 557,647
females age 16–49: 531,095 (2010 est.)
Manpower fit for military service: *males age 16–49:* 340,949
females age 16–49: 302,332 (2010 est.)
Manpower reaching militarily significant age annually: *male:* 23,649
female: 23,063 (2010 est.)
Military expenditures: 3.3% of GDP (2006)
country comparison to the world: 38

TRANSNATIONAL ISSUES

Disputes—international: none

BOUVET ISLAND

(TERRITORY OF NORWAY)

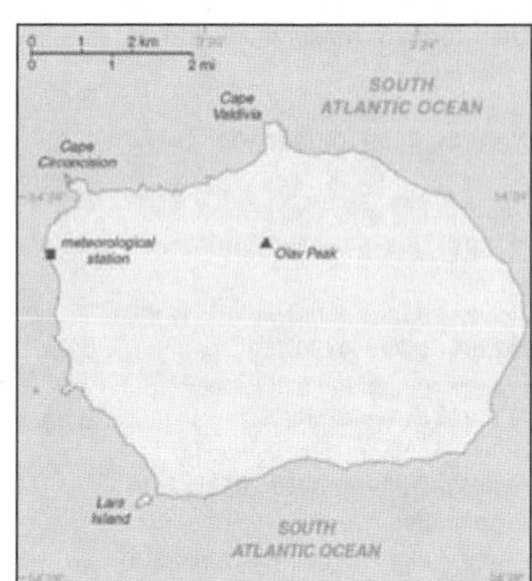

INTRODUCTION

Background: This uninhabited volcanic island is almost entirely covered by glaciers and is difficult to approach. It was discovered in 1739 by a French naval officer after whom the island was named. No claim was made until 1825, when the British flag was raised. In 1928, the UK waived its claim in favor of Norway, which had occupied the island the previous year. In 1971, Norway designated Bouvet Island and the adjacent territorial waters a nature reserve. Since 1977, it has run an automated meteorological station on the island.

GEOGRAPHY

Location: island in the South Atlantic Ocean, southwest of the Cape of Good Hope (South Africa)
Geographic coordinates: 54 26 S, 3 24 E
Map references: Antarctic Region
Area: *total:* 49 sq km
country comparison to the world: 231
land: 49 sq km, *water:* 0 sq km
Area—comparative: about 0.3 times the size of Washington, DC
Land boundaries: 0 km
Coastline: 29.6 km
Maritime claims: territorial sea: 4 nm
Climate: antarctic
Terrain: volcanic; coast is mostly inaccessible
Elevation extremes: *lowest point:* South Atlantic Ocean 0 m
highest point: Olav Peak 935 m
Natural resources: none
Land use: *arable land:* 0%
permanent crops: 0%
other: 100% (93% ice) (2005)
Irrigated land: 0 sq km
Natural hazards: NA
Environment—current issues: NA
Geography—note: covered by glacial ice; declared a nature reserve by Norway

PEOPLE

Population: uninhabited

GOVERNMENT

Country name: *conventional long form:* none
conventional short form: Bouvet Island
Dependency status: territory of Norway; administered by the Polar Department of the Ministry of Justice and Oslo Police
Legal system: the laws of Norway, where applicable, apply
Flag description: the flag of Norway is used

ECONOMY

Economy—overview: no economic activity; declared a nature reserve

COMMUNICATIONS

Internet country code: .bv
Internet hosts: 6 (2010)
country comparison to the world: 225
Communications—note: automatic meteorological station

TRANSPORTATION

Ports and terminals: none; offshore anchorage only

MILITARY

Military—note: defense is the responsibility of Norway

TRANSNATIONAL ISSUES

Disputes—international: none

BRAZIL

INTRODUCTION

Background: Following more than three centuries under Portuguese rule, Brazil gained its independence in 1822, maintaining a monarchical system of government until the abolition of slavery in 1888 and the subsequent proclamation of a republic by the military in 1889. Brazilian coffee exporters politically dominated the country until populist leader Getulio VARGAS rose to power in 1930. By far the largest and most populous country in South America, Brazil underwent more than half a century of populist and military government until 1985, when the military regime peacefully ceded power to civilian rulers. Brazil continues to pursue industrial and agricultural growth and development of its interior. Exploiting vast natural resources and a large labor pool, it is today South America's leading economic power and a regional leader, one of the first in the area to begin an economic recovery. Highly unequal income distribution and crime remain pressing problems. In January 2010, Brazil assumed a nonpermanent seat on the UN Security Council for the 2010-11 term.

GEOGRAPHY

Location: Eastern South America, bordering the Atlantic Ocean
Geographic coordinates: 10 00 S, 55 00 W
Map references: South America
Area: *total:* 8,514,877 sq km
country comparison to the world: 5
land: 8,459,417 sq km
water: 55,460 sq km
note: includes Arquipelago de Fernando de Noronha, Atol das Rocas, Ilha da Trindade, Ilhas Martin Vaz, and Penedos de Sao Pedro e Sao Paulo
Area—comparative: slightly smaller than the US
Land boundaries: *total:* 16,885 km
border countries: Argentina 1,261 km, Bolivia 3,423 km, Colombia 1,644 km, French Guiana 730 km, Guyana 1,606 km, Paraguay 1,365 km, Peru 2,995 km, Suriname 593 km, Uruguay 1,068 km, Venezuela 2,200 km
Coastline: 7,491 km
Maritime claims: territorial sea: 12 nm
contiguous zone: 24 nm
exclusive economic zone: 200 nm
continental shelf: 200 nm or to edge of the continental margin
Climate: mostly tropical, but temperate in south
Terrain: mostly flat to rolling lowlands in north; some plains, hills, mountains, and narrow coastal belt
Elevation extremes: *lowest point:* Atlantic Ocean 0 m
highest point: Pico da Neblina 2,994 m
Natural resources: bauxite, gold, iron ore, manganese, nickel, phosphates, platinum, tin, rare earth elements, uranium, petroleum, hydropower, timber
Land use: *arable land:* 6.93%
permanent crops: 0.89%
other: 92.18% (2005)
Irrigated land: 29,200 sq km (2003)
Total renewable water resources: 8,233 cu km (2000)
Freshwater withdrawal (domestic/industrial/agricultural): *total:* 59.3 cu km/yr (20%/18%/62%)
per capita: 318 cu m/yr (2000)
Natural hazards: recurring droughts in northeast; floods and occasional frost in south
Environment—current issues: deforestation in Amazon Basin destroys the habitat and endangers a multitude of plant and animal species indigenous to the area; there is a lucrative illegal wildlife trade; air and water pollution in Rio de Janeiro, Sao Paulo, and several other large cities; land degradation and water pollution caused by improper mining activities; wetland degradation; severe oil spills
Environment—international agreements: *party to:* Antarctic-Environmental Protocol, Antarctic-Marine Living Resources, Antarctic Seals, Antarctic Treaty, Biodiversity, Climate Change, Climate Change-Kyoto Protocol, Desertification, Endangered Species, Environmental Modification, Hazardous Wastes, Law of the Sea, Marine Dumping, Ozone Layer Protection, Ship Pollution, Tropical Timber 83, Tropical Timber 94, Wetlands, Whaling
signed, but not ratified: none of the selected agreements
Geography—note: largest country in South America; shares common boundaries with every South American country except Chile and Ecuador

PEOPLE

Population: 203,429,773 (July 2011 est.)
country comparison to the world: 5
note: Brazil conducted a census in August 2000, which reported a population of 169,872,855; that figure was about 3.8% lower than projections by the US Census Bureau, and is close to the implied underenumeration of 4.6% for the 1991 census
Age structure: *0–14 years:* 26.2% (male 27,219,651/female 26,180,040)
15–64 years: 67% (male 67,524,642/female 68,809,357)
65 years and over: 6.7% (male 5,796,433/female 7,899,650) (2011 est.)
Median age: *total:* 29.3 years
male: 28.5 years
female: 30.1 years (2011 est.)
Population growth rate: 1.134% (2011 est.)
country comparison to the world: 104
Birth rate: 17.79 births/1,000 population (2011 est.)
country comparison to the world: 109
Death rate: 6.36 deaths/1,000 population (July 2011 est.)
country comparison to the world: 152
Net migration rate: -0.09 migrant(s)/1,000 population (2011 est.)
country comparison to the world: 121
Urbanization: *urban population:* 87% of total population (2010)
rate of urbanization: 1.1% annual rate of change (2010-15 est.)
Major cities—population: Sao Paulo 19.96 million; Rio de Janeiro 11.836 million; Belo Horizonte 5.736 million; Porto Alegre 4.034 million; BRASILIA (capital) 3.789 million (2009)
Sex ratio: *at birth:* 1.05 male(s)/female
under 15 years: 1.04 male(s)/female
15–64 years: 0.98 male(s)/female
65 years and over: 0.73 male(s)/female
total population: 0.98 male(s)/female (2011 est.)
Infant mortality rate:
total: 21.17 deaths/1,000 live births
country comparison to the world: 93
male: 24.63 deaths/1,000 live births
female: 17.53 deaths/1,000 live births (2011 est.)
Life expectancy at birth: *total population:* 72.53 years
country comparison to the world: 124
male: 68.97 years
female: 76.27 years (2011 est.)
Total fertility rate: 2.18 children born/woman (2011 est.)
country comparison to the world: 109
HIV/AIDS—adult prevalence rate: NA
HIV/AIDS—people living with HIV/AIDS: NA
HIV/AIDS—deaths: NA
Drinking water source: *Improved:*
urban: 99% of population
rural: 84% of population
total: 97% of population
Unimproved: urban: 1% of population
rural: 16% of population
total: 3% of population (2008)
Sanitation facility access: *Improved:*
urban: 87% of population
rural: 37% of population
total: 80% of population
Unimproved: urban: 13% of population
rural: 63% of population
total: 20% of population (2008)
Nationality: *noun:* Brazilian(s)
adjective: Brazilian
Ethnic groups: white 53.7%, mulatto (mixed white and black) 38.5%, black 6.2%, other (includes Japanese, Arab, Amerindian) 0.9%, unspecified 0.7% (2000 census)

Religions: Roman Catholic (nominal) 73.6%, Protestant 15.4%, Spiritualist 1.3%, Bantu/voodoo 0.3%, other 1.8%, unspecified 0.2%, none 7.4% (2000 census)
Languages: Portuguese (official and most widely spoken language)
note: less common languages include Spanish (border areas and schools), German, Italian, Japanese, English, and a large number of minor Amerindian languages
Literacy: *definition:* age 15 and over can read and write
total population: 88.6%
male: 88.4%
female: 88.8% (2004 est.)
School life expectancy (primary to tertiary education): *total:* 14 years
male: 14 years
female: 14 years (2008)
Education expenditures: 5.08% of GDP (2007)
country comparison to the world: 55

GOVERNMENT

Country name: *conventional long form:* Federative Republic of Brazil
conventional short form: Brazil
local long form: Republica Federativa do Brasil
local short form: Brasil
Government type: federal republic
Capital: *name:* Brasilia
geographic coordinates: 15 47 S, 47 55 W
time difference: UTC-3 (2 hours ahead of Washington, DC during Standard Time)
daylight saving time: +1hr, begins third Sunday in October; ends last Sunday in February
note: Brazil is divided into three time zones, including one for the Fernando de Noronha Islands
Administrative divisions: 26 states (estados, singular–estado) and 1 federal district* (distrito federal); Acre, Alagoas, Amapa, Amazonas, Bahia, Ceara, Distrito Federal*, Espirito Santo, Goias, Maranhao, Mato Grosso, Mato Grosso do Sul, Minas Gerais, Para, Paraiba, Parana, Pernambuco, Piaui, Rio de Janeiro, Rio Grande do Norte, Rio Grande do Sul, Rondonia, Roraima, Santa Catarina, Sao Paulo, Sergipe, Tocantins
Independence: 7 September 1822 (from Portugal)
National holiday: Independence Day, 7 September (1822)
Constitution: 5 October 1988
Legal system: civil law; note–a new Brazilian civil law code was enacted in 2002 replacing the 1916 code
International law organization participation: has not submitted an ICJ jurisdiction declaration; accepts ICCt jurisdiction
Suffrage: voluntary between 16 and 18 years of age and over 70; compulsory over 18 and under 70 years of age; note–military conscripts do not vote
Executive branch: *chief of state:* President Dilma ROUSSEFF (since 1 January 2011); Vice President Michel TEMER (since 1 January 2011); note–the president is both the chief of state and head of government
head of government: President Dilma ROUSSEFF (since 1 January 2011); Vice President Michel TEMER (since 1 January 2011)
cabinet: Cabinet appointed by the president (For more information visit the World Leaders website)
elections: president and vice president elected on the same ticket by popular vote for a single four-year term; election last held on 3 October 2010 with runoff on 31 October 2010 (next to be held on 5 October 2014 and, if necessary, a runoff election on 2 November 2014)
election results: Dilma ROUSSEFF (PT) elected president in a runoff election; percent of vote–Dilma ROUSSEFF 56.01%, Jose SERRA (PSDB) 43.99%
Legislative branch: bicameral National Congress or Congresso Nacional consists of the Federal Senate or Senado Federal (81 seats; 3 members from each state and federal district elected according to the principle of majority to serve eight-year terms; one-third and two-thirds of members elected every four years, alternately) and the Chamber of Deputies or Camara dos Deputados (513 seats; members are elected by proportional representation to serve four-year terms)
elections: Federal Senate–last held on 3 October 2010 for two-thirds of the Senate (next to be held in October 2014 for one-third of the Senate); Chamber of Deputies–last held on 3 October 2010 (next to be held in October 2014)
election results: Federal Senate–percent of vote by party–NA; seats by party–PMDB 20, PT 13, PSDB 10, DEM (formerly PFL) 7, PTdoB 6, PP 5, PDT 4, PR 4, PSB 4, PPS 1, PRB 1, other 3; Chamber of Deputies–percent of vote by party–NA; seats by party–PT 87, PMDB 80, PSDB 53, DEM (formerly PFL) 43, PP 41, PR 41, PSB 34, PDT 28, PTdoB 21, PSC 17, PCdoB 15, PV 15, PPS 12, other 18
Judicial branch: Supreme Federal Tribunal or STF (11 ministers are appointed for life by the president and confirmed by the Senate); Higher Tribunal of Justice; Regional Federal Tribunals (judges are appointed for life); note–though appointed "for life," judges, like all federal employees, have a mandatory retirement age of 70
Political parties and leaders: Brazilian Democratic Movement Party or PMDB [Federal Deputy Michel TEMER]; Brazilian Labor Party or PTB [Roberto JEFFERSON]; Brazilian Renewal Labor Party or PRTB [Jose Levy FIDELIX da Cruz]; Brazilian Republican Party or PRB [Vitor Paulo Araujo DOS SANTOS]; Brazilian Social Democracy Party or PSDB [Senator Sergio GUERRA]; Brazilian Socialist Party or PSB [Governor Eduardo Henrique Accioly CAMPOS]; Christian Labor Party or PTC [Daniel TOURINHO]; Communist Party of Brazil or PCdoB [Jose Renato RABELO]; Democratic Labor Party or PDT [Carlos Roberto LUPI]; the Democrats or DEM [Federal Deputy Rodrigo MAIA] (formerly Liberal Front Party or PFL); Freedom and Socialism Party or PSOL [Heloisa HELENA]; Green Party or PV [Jose Luiz de Franca PENNA]; Humanist Party of Solidarity or PHS [Paulo Roberto MATOS]; Labor Party of Brazil or PTdoB [Luis Henrique de Oliveira RESENDE]; Liberal Front Party or PFL (now known as the Democrats or DEM); National Mobilization Party or PMN [Oscar Noronha FILHO]; Party of the Republic or PR [Sergio TAMER]; Popular Socialist Party or PPS [Federal Deputy Fernando CORUJA]; Progressive Party or PP [Francisco DORNELLES]; Social Christian Party or PSC [Vitor Jorge Abdala NOSSEIS]; Workers' Party or PT [Jose Eduardo DUTRA]
Political pressure groups and leaders: Landless Workers' Movement or MST
other: labor unions and federations; large farmers' associations; religious groups including evangelical Christian churches and the Catholic Church
International organization participation: AfDB (nonregional member), BIS, CAN (associate), CPLP, FAO, FATF, G-15, G-20, G-24, G-77, IADB, IAEA, IBRD, ICAO, ICC, ICRM, IDA, IFAD, IFC, IFRCS, IHO, ILO, IMF, IMO, IMSO, Interpol, IOC, IOM, IPU, ISO, ITSO, ITU, ITUC, LAES, LAIA, LAS (observer), Mercosur, MIGA, MINURSO, MINUSTAH, NAM (observer), NSG, OAS, OPANAL, OPCW, Paris Club (associate), PCA, RG, SICA (observer), UN, UN Security Council (temporary), UNASUR, UNCTAD, UNESCO, UNFICYP, UNHCR, UNIDO, Union Latina, UNITAR, UNMIL, UNMIS, UNMIT, UNOCI, UNWTO, UPU, WCO, WFTU, WHO, WIPO, WMO, WTO
Diplomatic representation in the US: *chief of mission:* Ambassador Mauro Luiz Iecker VIEIRA
chancery: 3006 Massachusetts Avenue NW, Washington, DC 20008
note: temporary address–1025 Thomas Jefferson St. NW, Suite 300 W, Washington, DC 20007
telephone: [1] (202) 238-2805
FAX: [1] (202) 238-2827
consulate(s) general: Boston, Chicago, Houston, Los Angeles, Miami, New York, San Francisco
Diplomatic representation from the US: *chief of mission:* Ambassador Thomas A. SHANNON
embassy: Avenida das Nacoes, Quadra 801, Lote 3, Distrito Federal Cep 70403-900, Brasilia
mailing address: Unit 7500, DPO, AA 34030
telephone: [55] (61) 3312-7000
FAX: [55] (61) 3225-9136
consulate(s) general: Rio de Janeiro, Sao Paulo
consulate(s): Recife
Flag description: green with a large yellow diamond in the center bearing a blue celestial globe with 27 white five-pointed stars; the globe has a white equatorial band with the motto ORDEM E PROGRESSO (Order and Progress); the current flag was inspired by the banner of the former Empire of Brazil (1822-1889); on the imperial flag, the green represented the House of Braganza of Pedro I, the first Emperor of Brazil, while the yellow stood for the Habsburg Family of his wife; on the modern flag the green represents the forests of the country and the yellow rhombus its mineral wealth; the blue circle and stars, which replaced the coat of arms of the original flag, depict the sky over Rio de Janeiro on the morning of 15 November 1889–the day the Republic of Brazil was declared; the number of stars has changed with the

creation of new states and has risen from an original 21 to the current 27 (one for each state and the Federal District)
National anthem: *name:* "Hino Nacional Brasileiro" (Brazilian National Anthem)
lyrics/music: Joaquim Osorio Duque ESTRADA/Francisco Manoel DA SILVA
note: music adopted 1890, lyrics adopted 1922; the anthem's music, composed in 1822, was used unofficially for many years before it was adopted

ECONOMY

Economy—overview: Characterized by large and well-developed agricultural, mining, manufacturing, and service sectors, Brazil's economy outweighs that of all other South American countries, and Brazil is expanding its presence in world markets. Since 2003, Brazil has steadily improved its macroeconomic stability, building up foreign reserves, and reducing its debt profile by shifting its debt burden toward real denominated and domestically held instruments. In 2008, Brazil became a net external creditor and two ratings agencies awarded investment grade status to its debt. After record growth in 2007 and 2008, the onset of the global financial crisis hit Brazil in September 2008. Brazil experienced two quarters of recession, as global demand for Brazil's commodity-based exports dwindled and external credit dried up. However, Brazil was one of the first emerging markets to begin a recovery. Consumer and investor confidence revived and GDP growth returned to positive in 2010, boosted by an export recovery. Brazil's strong growth and high interest rates make it an attractive destination for foreign investors. Large capital inflows over the past year have contributed to the rapid appreciation of its currency and led the government to raise taxes on some foreign investments. President Dilma ROUSSEFF has pledged to retain the previous administration's commitment to inflation targeting by the Central Bank, a floating exchange rate, and fiscal restraint.
GDP (purchasing power parity): $2.172 trillion (2010 est.)
country comparison to the world: 9
$2.021 trillion (2009 est.)
$2.034 trillion (2008 est.)
note: data are in 2010 US dollars
GDP (official exchange rate): $2.09 trillion (2010 est.)
GDP—real growth rate: 7.5% (2010 est.)
country comparison to the world: 30
-0.6% (2009 est.)
5.2% (2008 est.)
GDP—per capita (PPP): $10,800 (2010 est.)
country comparison to the world: 103
$10,200 (2009 est.)
$10,400 (2008 est.)
note: data are in 2010 US dollars
GDP—composition by sector: *agriculture:* 6.1%
industry: 26.4%
services: 67.5% (2010 est.)
Labor force: 103.6 million (2010 est.)
country comparison to the world: 6
Labor force—by occupation: *agriculture:* 20%
industry: 14%
services: 66% (2003 est.)
Unemployment rate: 7% (2010 est.)
country comparison to the world: 71
8.1% (2009 est.)
Population below poverty line: 26% (2008)
Household income or consumption by percentage share: *lowest 10%:* 1.1%
highest 10%: 43% (2007)
Distribution of family income—Gini index: 56.7 (2005)
country comparison to the world: 10
60.7 (1998)
Investment (gross fixed): 18.5% of GDP (2010 est.)
country comparison to the world: 107
Budget: *revenues:* $464.4 billion
expenditures: $552.6 billion (2010 est.)
Public debt: 60.8% of GDP (2010 est.)
country comparison to the world: 31
59.5% of GDP (2009 est.)
Inflation rate (consumer prices): 4.9% (2010 est.)
country comparison to the world: 138
4.9% (2009 est.)
Central bank discount rate: 15.17% (31 December 2009)
country comparison to the world: 9
20.48% (31 December 2008)
Commercial bank prime lending rate: 44.65% (31 December 2009 est.)
country comparison to the world: 2
47.25% (31 December 2008 est.)
Stock of narrow money: $165.8 billion (31 December 2010 est.)
country comparison to the world: 21
$125.3 billion (31 December 2009 est.)
Stock of broad money: $1.522 trillion (31 December 2009)
country comparison to the world: 11
$972.8 billion (31 December 2008)
Stock of domestic credit: $2.104 trillion (31 December 2010 est.)
country comparison to the world: 11
$1.542 trillion (31 December 2009 est.)
Market value of publicly traded shares: $1.167 trillion (31 December 2009)
country comparison to the world: 16
$589.4 billion (31 December 2008)
$1.37 trillion (31 December 2007)
Agriculture—products: coffee, soybeans, wheat, rice, corn, sugarcane, cocoa, citrus; beef
Industries: textiles, shoes, chemicals, cement, lumber, iron ore, tin, steel, aircraft, motor vehicles and parts, other machinery and equipment
Industrial production growth rate: 11.5% (2010 est.)
country comparison to the world: 15
Electricity—production: 438.8 billion kWh (2007 est.)
country comparison to the world: 10
Electricity—consumption: 404.3 billion kWh (2007 est.)
country comparison to the world: 10
Electricity—exports: 2.034 billion kWh (2007 est.)
Electricity—imports: 42.06 billion kWh; note–supplied by Paraguay (2008 est.)
Oil—production: 2.572 million bbl/day (2009 est.)
country comparison to the world: 9
Oil—consumption: 2.46 million bbl/day (2009 est.)
country comparison to the world: 7
Oil—exports: 570,100 bbl/day (2007 est.)
country comparison to the world: 27
Oil—imports: 632,900 bbl/day (2007 est.)
country comparison to the world: 20
Oil—proved reserves: 13.2 billion bbl (1 January 2010 est.)
country comparison to the world: 17
Natural gas—production: 10.28 billion cu m (2009 est.)
country comparison to the world: 41
Natural gas—consumption: 18.72 billion cu m (2009 est.)
country comparison to the world: 36
Natural gas—exports: NA (2009 est.)
Natural gas—imports: 8.44 billion cu m (2009 est.)
country comparison to the world: 25
Natural gas—proved reserves: 364.2 billion cu m (1 January 2010 est.)
country comparison to the world: 37
Current account balance: $-52.73 billion (2010 est.)
country comparison to the world: 187
$-24.3 billion (2009 est.)
Exports: $199.7 billion (2010 est.)
country comparison to the world: 24
$153 billion (2009 est.)
Exports—commodities: transport equipment, iron ore, soybeans, footwear, coffee, autos
Exports—partners: China 12.49%, US 10.5%, Argentina 8.4%, Netherlands 5.39%, Germany 4.05% (2009)
Imports: $187.7 billion (2010 est.)
country comparison to the world: 22
$127.7 billion (2009 est.)
Imports—commodities: machinery, electrical and transport equipment, chemical products, oil, automotive parts, electronics
Imports—partners: US 16.12%, China 12.61%, Argentina 8.77%, Germany 7.65%, Japan 4.3% (2009)
Reserves of foreign exchange and gold: $290.9 billion (31 December 2010 est.)
country comparison to the world: 6
$238.5 billion (31 December 2009 est.)
Debt—external: $310.8 billion (31 December 2010 est.)
country comparison to the world: 26
$273.7 billion (31 December 2009 est.)
Stock of direct foreign investment—at home: $349.2 billion (31 December 2010 est.)
country comparison to the world: 13
$319.9 billion (31 December 2009 est.)
Stock of direct foreign investment—abroad: $131 billion (31 December 2010 est.)
country comparison to the world: 23
$117.4 billion (31 December 2009 est.)
Exchange rates: reals (BRL) per US dollar–1.77 (2010), 2 (2009), 1.8644 (2008), 1.85 (2007), 2.1761 (2006)

COMMUNICATIONS

Telephones—main lines in use: 41.497 million (2009)
country comparison to the world: 6
Telephones—mobile cellular: 173.959 million (2009)
country comparison to the world: 5
Telephone system: *general assessment:* good working system including an extensive microwave radio relay system and a domestic satellite system with 64 earth stations; mobile-cellular usage has more than tripled in the past 5 years

domestic: fixed-line connections have remained relatively stable in recent years and stand at about 20 per 100 persons; less expensive mobile-cellular technology has been a major driver in expanding telephone service to the lower-income segments of the population with mobile-cellular teledensity approaching 90 per 100 persons in 2009
international: country code–55; landing point for a number of submarine cables, including Americas-1, Americas-2, Atlantis-2, GlobeNet, South Amrica-1, South American Crossing/Latin American Nautilius, and UNISUR that provide direct connectivity to South and Central America, the Caribbean, the US, Africa, and Europe; satellite earth stations–3 Intelsat (Atlantic Ocean), 1 Inmarsat (Atlantic Ocean region east), connected by microwave relay system to Mercosur Brazilsat B3 satellite earth station (2009)
Broadcast media: state-run Radiobras operates a radio and a television network; more than 1,000 radio stations and more than 100 TV channels operating–mostly privately owned; private media ownership highly concentrated (2007)
Internet country code: .br
Internet hosts: 19.316 million (2010)
country comparison to the world: 5
Internet users: 75.982 million (2009)
country comparison to the world: 4

TRANSPORTATION

Airports: 4,072 (2010)
country comparison to the world: 2
Airports—with paved runways: *total:* 726
over 3,047 m: 7
2,438 to 3,047 m: 28
1,524 to 2,437 m: 176
914 to 1,523 m: 460
under 914 m: 55 (2010)
Airports—with unpaved runways: *total:* 3,346
1,524 to 2,437 m: 87
914 to 1,523 m: 1,617
under 914 m: 1,642 (2010)
Heliports: 13 (2010)
Pipelines: condensate/gas 62 km; gas 13,514 km; liquid petroleum gas 352 km; oil 3,729 km; refined products 4,684 km (2010)
Railways: *total:* 28,538 km
country comparison to the world: 10
broad gauge: 5,627 km 1.600-m gauge (467 km electrified)
standard gauge: 194 km 1.440-m gauge
narrow gauge: 22,717 km 1.000-m gauge (2010)
Roadways: *total:* 1,751,868 km
country comparison to the world: 4
paved: 96,353 km
unpaved: 1,655,515 km (2004)
Waterways: 50,000 km (most in areas remote from industry and population) (2010)
country comparison to the world: 3
Merchant marine: *total:* 126
country comparison to the world: 45
by type: bulk carrier 19, cargo 18, chemical tanker 6, container 12, liquefied gas 12, passenger/cargo 10, petroleum tanker 42, roll on/roll off 7
foreign-owned: 26 (Chile 1, Denmark 3, Germany 6, Greece 1, Norway 3, Spain 12)
registered in other countries: 27 (Argentina 1, Bahamas 1, Ghana 1, Liberia 20, Marshall Islands 1, Panama 3) (2010)
Ports and terminals: cargo ports (tonnage): Ilha Grande (Gebig), Paranagua, Rio Grande, Santos, Sao Sebastiao, Tubarao
container ports (TEUs): Santos (2,677,839), Itajai (693,580)
Transportation—note: the International Maritime Bureau reports that the territorial and offshore waters in the Atlantic Ocean remain a significant risk for piracy and armed robbery against ships; 2010 saw an 80% increase in attacks over 2009; numerous commercial vessels were attacked and hijacked both at anchor and while underway; crews were robbed and stores or cargoes stolen

MILITARY

Military branches: Brazilian Army (Exercito Brasileiro, EB), Brazilian Navy (Marinha do Brasil (MB), includes Naval Air and Marine Corps (Corpo de Fuzileiros Navais)), Brazilian Air Force (Forca Aerea Brasileira, FAB) (2011)
Military service age and obligation: 21-45 years of age for compulsory military service; conscript service obligation–9 to 12 months; 17-45 years of age for voluntary service; an increasing percentage of the ranks are "long-service" volunteer professionals; women were allowed to serve in the armed forces beginning in early 1980s when the Brazilian Army became the first army in South America to accept women into career ranks; women serve in Navy and Air Force only in Women's Reserve Corps (2001)
Manpower available for military service: *males age 16–49:* 53,350,703
females age 16–49: 53,433,918 (2010 est.)
Manpower fit for military service: *males age 16–49:* 38,993,989
females age 16–49: 44,841,661 (2010 est.)
Manpower reaching militarily significant age annually: *male:* 1,733,168
female: 1,672,477 (2010 est.)
Military expenditures: 1.7% of GDP (2009)
country comparison to the world: 91

TRANSNATIONAL ISSUES

Disputes—international: uncontested boundary dispute between Brazil and Uruguay over Braziliera/Brasiliera Island in the Quarai/Cuareim River leaves the tripoint with Argentina in question; smuggling of firearms and narcotics continues to be an issue along the Uruguay-Brazil border; Colombian-organized illegal narcotics and paramilitary activities penetrate Brazil's border region with Venezuela
Illicit drugs: second-largest consumer of cocaine in the world; illicit producer of cannabis; trace amounts of coca cultivation in the Amazon region, used for domestic consumption; government has a large-scale eradication program to control cannabis; important transshipment country for Bolivian, Colombian, and Peruvian cocaine headed for Europe; also used by traffickers as a way station for narcotics air transshipments between Peru and Colombia; upsurge in drug-related violence and weapons smuggling; important market for Colombian, Bolivian, and Peruvian cocaine; illicit narcotics proceeds are often laundered through the financial system; significant illicit financial activity in the Tri-Border Area (2008)

BRITISH INDIAN OCEAN TERRITORY

(OVERSEAS TERRITORY OF THE UK)

PEROS BANHOS
SALOMON ISLANDS
INDIAN OCEAN
Nelsons Island
THREE BROTHERS
EAGLE ISLANDS
CHAGOS ARCHIPELAGO
Danger Island
EGMONT ISLANDS
Diego Garcia
0 20 40 km
0 20 40 mi

INTRODUCTION

Background: Formerly administered as part of the British Crown Colony of Mauritius, the British Indian Ocean Territory (BIOT) was established as an overseas territory of the UK in 1965. A number of the islands of the territory were later transferred to the Seychelles when it attained independence in 1976. Subsequently, BIOT has consisted only of the six main island groups comprising the Chagos Archipelago. The largest and most southerly of the islands, Diego Garcia, contains a joint UK-US naval support facility. All of the remaining islands are uninhabited. Between 1967 and 1973, former agricultural workers, earlier residents in the islands, were relocated primarily to Mauritius, but also to the Seychelles. Negotiations between 1971 and 1982 resulted in the establishment of a trust fund by the British Government as compensation for the displaced islanders, known as Chagossians. Beginning in 1998, the islanders pursued a series of lawsuits against the British Government seeking further compensation and the right to return to the territory. In 2006 and 2007, British court rulings invalidated the immigration policies contained in the 2004 BIOT Constitution Order that had excluded the islanders from the archipelago, but upheld the special military status of Diego Garcia. In 2008, the House of Lords, as the final court of appeal in the UK, ruled in favor of the British Government by overturning the lower court rulings and finding no right of return for the Chagossians.

GEOGRAPHY

Location: archipelago in the Indian Ocean, south of India, about halfway between Africa and Indonesia
Geographic coordinates: 6 00 S, 71 30 E; note—Diego Garcia 7 20 S, 72 25 E
Map references: Political Map of the World
Area: *total:* 54,400 sq km
country comparison to the world: 127
land: 60 sq km; Diego Garcia 44 sq km
water: 54,340 sq km
note: includes the entire Chagos Archipelago of 55 islands
Area—comparative: land area is about 0.3 times the size of Washington, DC
Land boundaries: 0 km
Coastline: 698 km
Maritime claims: territorial sea: 3 nm
exclusive fishing zone: 200 nm
Climate: tropical marine; hot, humid, moderated by trade winds
Terrain: flat and low (most areas do not exceed two meters in elevation)
Elevation extremes:
lowest point: Indian Ocean 0 m
highest point: unnamed location on Diego Garcia 15 m
Natural resources: coconuts, fish, sugarcane
Land use: *arable land:* 0%
permanent crops: 0%
other: 100% (2005)
Irrigated land: 0 sq km
Natural hazards: NA
Environment—current issues: NA
Geography—note: archipelago of 55 islands; Diego Garcia, largest and southernmost island, occupies strategic location in central Indian Ocean; island is site of joint US-UK military facility

PEOPLE

Population: no indigenous inhabitants
note: approximately 1,200 former agricultural workers resident in the Chagos Archipelago, often referred to as Chagossians or Ilois, were relocated to Mauritius and the Seychelles in the 1960s and 1970s; in November 2004, approximately 4,000 UK and US military personnel and civilian contractors were living on the island of Diego Garcia
School life expectancy (primary to tertiary education): NA
Education expenditures: NA

GOVERNMENT

Country name: *conventional long form:* British Indian Ocean Territory
conventional short form: none
abbreviation: BIOT
Dependency status: overseas territory of the UK; administered by a commissioner, resident in the Foreign and Commonwealth Office in London
Legal system: the laws of the UK, where applicable, apply
Executive branch:
chief of state: Queen ELIZABETH II (since 6 February 1952)
head of government: Commissioner Colin ROBERTS (since July 2008); Administrator Joanne YEADON (since December 2007); note—both reside in the UK and are represented by the officer commanding British Forces on Diego Garcia
cabinet: NA
(For more information visit the World Leaders website)
elections: none; the monarchy is hereditary; commissioner and administrator appointed by the monarch
Diplomatic representation in the US: none (overseas territory of the UK)
Diplomatic representation from the US: none (overseas territory of the UK)
Flag description: white with six blue wavy horizontal stripes; the flag of the UK is in the upper hoist-side quadrant; the striped section bears a palm tree and yellow crown (the symbols of the territory) centered on the outer half of the flag; the wavy stripes represent the Indian Ocean; although not officially described, the six blue stripes may stand for the six main atolls of the archipelago

ECONOMY

Economy—overview: All economic activity is concentrated on the largest island of Diego Garcia, where a joint UK-US military facility is located. Construction projects and various services needed to support the military installation are performed by military and contract employees from the UK, Mauritius, the Philippines, and the US. There are no industrial or agricultural activities on the islands. The territory earns foreign exchange by selling fishing licenses and postage stamps.
Electricity—production: NA kWh; note—electricity supplied by the US military
Electricity—consumption: NA kWh
Exchange rates: the US dollar is used

COMMUNICATIONS

Telephones—main lines in use: NA
Telephone system: *general assessment:* separate facilities for military and public needs are available
domestic: all commercial telephone services are available, including connection to the Internet
international: country code (Diego Garcia)–246; international telephone service is carried by satellite (2000)
Broadcast media: Armed Forces Radio and Television Service (AFRTS) broadcasts over 3 separate frequencies for US and UK military personnel stationed on the islands (2009)
Internet country code: .io
Internet hosts: 827 (2010)
country comparison to the world: 169

TRANSPORTATION

Airports: 1 (2010)
country comparison to the world: 220
Airports—with paved runways: *total:* 1
over 3,047 m: 1 (2010)
Roadways: note: short section of paved road between port and airfield on Diego Garcia
Ports and terminals: Diego Garcia

MILITARY

Military branches: no regular military forces
Military—note: defense is the responsibility of the UK; the US lease on Diego Garcia expires in 2016

TRANSNATIONAL ISSUES

Disputes—international: Mauritius and Seychelles claim the Chagos Islands; in 2001, the former inhabitants of the archipelago, evicted 1967–1973, were granted UK citizenship and the right of return, followed by Orders in Council in 2004 that banned rehabilitation, a High Court ruling reversing the ban, a Court of Appeal refusal to hear the case, and a Law Lords' decision in 2008 denying the right of return; in addition, the United Kingdom created the world's largest marine protection area around the Chagos islands prohibiting the extraction of any natural resources therein

BRITISH VIRGIN ISLANDS

(OVERSEAS TERRITORY OF THE UK)

INTRODUCTION

Background: First inhabited by Arawak and later by Carib Indians, the Virgin Islands were settled by the Dutch in 1648 and then annexed by the English in 1672. The islands were part of the British colony of the Leeward Islands from 1872-1960; they were granted autonomy in 1967. The economy is closely tied to the larger and more populous US Virgin Islands to the west; the US dollar is the legal currency.

GEOGRAPHY

Location: Caribbean, between the Caribbean Sea and the North Atlantic Ocean, east of Puerto Rico
Geographic coordinates: 18 30 N, 64 30 W
Map references: Central America and the Caribbean
Area: *total:* 151 sq km
country comparison to the world: 218
land: 151 sq km
water: 0 sq km
note: comprised of 16 inhabited and more than 20 uninhabited islands; includes the islands of Tortola, Anegada, Virgin Gorda, Jost van Dyke
Area—comparative: about 0.9 times the size of Washington, DC
Land boundaries: 0 km
Coastline: 80 km
Maritime claims: territorial sea: 3 nm
exclusive fishing zone: 200 nm
Climate: subtropical; humid; temperatures moderated by trade winds

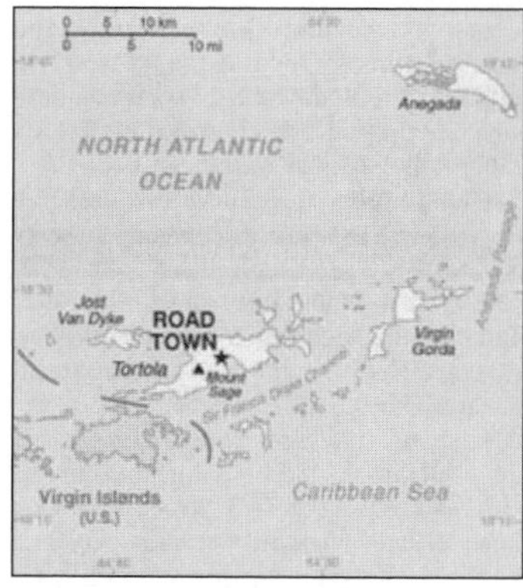

Terrain: coral islands relatively flat; volcanic islands steep, hilly
Elevation extremes:
lowest point: Caribbean Sea 0 m
highest point: Mount Sage 521 m
Natural resources: NEGL
Land use: *arable land:* 20%
permanent crops: 6.67%
other: 73.33% (2005)
Irrigated land: NA
Natural hazards: hurricanes and tropical storms (July to October)
Environment—current issues: limited natural freshwater resources (except for a few seasonal streams and springs on Tortola, most of the islands' water supply comes from wells and rainwater catchments)
Geography—note: strong ties to nearby US Virgin Islands and Puerto Rico

PEOPLE

Population: 25,383 (July 2011 est.)
country comparison to the world: 216
Age structure: *0–14 years:* 19.6% (male 2,526/female 2,457)
15–64 years: 74.1% (male 9,631/female 9,174)
65 years and over: 6.3% (male 827/female 768) (2011 est.)
Median age: *total:* 32.9 years
male: 32.9 years
female: 32.8 years (2011 est.)
Population growth rate: 1.741% (2011 est.)
country comparison to the world: 64
Birth rate: 14.5 births/1,000 population (2011 est.)
country comparison to the world: 142
Death rate: 4.49 deaths/1,000 population (July 2011 est.)
country comparison to the world: 200
Net migration rate: 7.41 migrant(s)/1,000 population (2011 est.)
country comparison to the world: 13
Urbanization: *urban population:* 41% of total population (2010)
rate of urbanization: 1.7% annual rate of change (2010-15 est.)
Major cities—population: ROAD TOWN (capital) 9,000 (2009)
Sex ratio: *at birth:* 1.05 male(s)/female
under 15 years: 1.03 male(s)/female
15–64 years: 1.05 male(s)/female
65 years and over: 1.07 male(s)/female
total population: 1.05 male(s)/female (2011 est.)
Infant mortality rate: *total:* 13.63 deaths/1,000 live births
country comparison to the world: 126
male: 15.47 deaths/1,000 live births
female: 11.71 deaths/1,000 live births (2011 est.)
Life expectancy at birth: *total population:* 77.63 years
country comparison to the world: 59
male: 76.32 years
female: 78.99 years (2011 est.)
Total fertility rate: 1.71 children born/woman (2011 est.)
country comparison to the world: 167
HIV/AIDS—adult prevalence rate: NA
HIV/AIDS—people living with HIV/AIDS: NA
HIV/AIDS—deaths: NA
Drinking water source: *Improved:*
urban: 98% of population
rural: 98% of population
total: 98% of population
Unimproved:
urban: 2% of population
rural: 2% of population
total: 2% of population (2008)
Sanitation facility access: *Improved:*
urban: 100% of population
rural: 100% of population
total: 100% of population (2008)
Nationality: *noun:* British Virgin Islander(s)
adjective: British Virgin Islander
Ethnic groups: black 82%, white 6.8%, other 11.2% (includes Indian and mixed) (2008)
Religions: Protestant 86% (Methodist 33%, Anglican 17%, Church of God 9%, Seventh-Day Adventist 6%, Baptist 4%, Jehovah's Witnesses 2%, other 15%), Roman Catholic 10%, other 2%, none 2% (1991)
Languages: English (official)
Literacy: *definition:* age 15 and over can read and write
total population: 97.8%
male: NA
female: NA (1991 est.)
School life expectancy (primary to tertiary education):
total: 15 years
male: 15 years
female: 16 years (2009)
Education expenditures: 3.2% of GDP (2007)
country comparison to the world: 127

GOVERNMENT

Country name: *conventional long form:* none
conventional short form: British Virgin Islands
abbreviation: BVI
Dependency status: overseas territory of the UK; internal self-governing
Government type: NA
Capital: *name:* Road Town
geographic coordinates: 18 27 N, 64 37 W
time difference: UTC-4 (1 hour ahead of Washington, DC during Standard Time)
Administrative divisions: none (overseas territory of the UK)
Independence: none (overseas territory of the UK)
National holiday: Territory Day, 1 July (1956)
Constitution: 13 June 2007
Legal system: English common law
Suffrage: 18 years of age; universal
Executive branch: *chief of state:* Queen ELIZABETH II (since 6 February 1952); represented by Governor Boyd MCCLEARY (since 20 August 2010)
head of government: Premier Ralph T. O'NEAL (since 23 August 2007)
cabinet: Executive Council appointed by the governor from members of the House of Assembly
(For more information visit the World Leaders website)
elections: the monarchy is hereditary; governor appointed by the monarch; following legislative elections, the leader of the majority party or the leader of the majority coalition usually appointed premier by the governor
Legislative branch: unicameral House of Assembly (13 elected seats and 1 non-voting ex officio member in the attorney general; members are elected by direct popular vote, 1 member from each of nine electoral districts, 4 at-large members; members serve four-year terms)
elections: last held on 20 August 2007 (next to be held in 2011)
election results: percent of vote by party—VIP 45.2%, NDP 39.6%, independent 15.2%; seats by party—VIP 10, NDP 2, independent 1
Judicial branch: Eastern Caribbean Supreme Court, consisting of the High Court of Justice and the Court of Appeal (one judge of the Supreme Court is a resident of the islands and presides over the High Court); Magistrate's Court; Juvenile Court; Court of Summary Jurisdiction
Political parties and leaders: Concerned Citizens Movement or CCM [Ethlyn SMITH]; National Democratic Party or NDP [Orlando SMITH]; United Party or UP [Gregory MADURO]; Virgin Islands Party or VIP [Ralph T. O'NEAL]
Political pressure groups and leaders: The Family Support Network; The Women's Desk
other: environmentalists
International organization participation: Caricom (associate), CDB, Interpol (subbureau), IOC, OECS, UNESCO (associate), UPU
Diplomatic representation in the US: none (overseas territory of the UK)
Diplomatic representation from the US: none (overseas territory of the UK)
Flag description: blue, with the flag of the UK in the upper hoist-side quadrant and the Virgin Islander coat of arms centered in the outer half of the flag; the coat of arms depicts a woman flanked on either side by a vertical column of six oil lamps above a scroll bearing the Latin word VIGILATE (Be Watchful); the islands were named by COLUMBUS in 1493 in honor of Saint Ursula and her 11 virgin followers (some sources say 11,000) who reputedly were martyred by the Huns in the 4th or 5th century; the figure on the banner holding a lamp represents the saint, the other lamps symbolize her followers
National anthem: note: as a territory of the United Kingdom, "God Save the Queen" is official (see United Kingdom)

ECONOMY

Economy—overview: The economy, one of the most stable and prosperous in the Caribbean, is highly dependent on tourism generating an estimated 45% of the national income. More than 934,000 tourists, mainly from the US, visited the islands in 2008. In the mid-1980s, the government began offering offshore registration to companies wishing to incorporate in the islands, and incorporation fees now generate substantial revenues. Roughly 400,000 companies were on the offshore registry by yearend 2000. The adoption of a comprehensive insurance law in late 1994, which provides a blanket of confidentiality with regulated statutory gateways for investigation of criminal offenses, made the British Virgin Islands even more attractive to international business. Livestock raising is the most important agricultural activity; poor soils limit the islands' ability to meet domestic food requirements. Because of traditionally close links with the US Virgin Islands, the British Virgin Islands has used the US dollar as its currency since 1959.
GDP (purchasing power parity): $853.4 million (2004 est.)
country comparison to the world: 204
GDP (official exchange rate): $1.095 billion (2008)
GDP—real growth rate: -0.6% (2008 est.)
country comparison to the world: 195
GDP—per capita (PPP): $38,500 (2004 est.)
country comparison to the world: 24
GDP—composition by sector:
agriculture: 0.9%
industry: 10.7%
services: 88.3% (1996 est.)
Labor force: 12,770 (2004)
country comparison to the world: 213
Labor force—by occupation: *agriculture:* 0.6%
industry: 40%
services: 59.4% (2005)
Unemployment rate: 3.6% (1997)
country comparison to the world: 31
Population below poverty line: NA%
Household income or consumption by percentage share: *lowest 10%:* NA%
highest 10%: NA%
Budget: *revenues:* $204.7 million
expenditures: $180.4 million (2004)
Inflation rate (consumer prices): 7.1% (2008)
country comparison to the world: 176
2% (2005)
Agriculture—products: fruits, vegetables; livestock, poultry; fish
Industries: tourism, light industry, construction, rum, concrete block, offshore financial center
Industrial production growth rate: NA%
Electricity—production: 45 million kWh (2007 est.)
country comparison to the world: 199
Electricity—consumption: 41.85 million kWh (2007 est.)
country comparison to the world: 198
Electricity—exports: 0 kWh (2008 est.)
Electricity—imports: 0 kWh (2008 est.)
Oil—production: 0 bbl/day (2009 est.)
country comparison to the world: 139
Oil—consumption: 1,000 bbl/day (2009 est.)
country comparison to the world: 203
Oil—exports: 0 bbl/day (2007 est.)
country comparison to the world: 139
Oil—imports: 691.4 bbl/day (2007 est.)
country comparison to the world: 193
Oil—proved reserves: 0 bbl (1 January 2010 est.)
country comparison to the world: 201
Natural gas—production: 0 cu m (2008 est.)
country comparison to the world: 198
Natural gas—consumption: 0 cu m (2008 est.)
country comparison to the world: 134
Natural gas—exports: 0 cu m (2008 est.)
country comparison to the world: 197
Natural gas—imports: 0 cu m (2008 est.)
country comparison to the world: 134
Natural gas—proved reserves: 0 cu m (1 January 2010 est.)
country comparison to the world: 200
Current account balance: $134.3 million (1999)
country comparison to the world: 57
Exports: $25.3 million (2002)
country comparison to the world: 204
Exports—commodities: rum, fresh fish, fruits, animals; gravel, sand
Imports: $187 million (2002 est.) (2002 est.)
country comparison to the world: 202
Imports—commodities: building materials, automobiles, foodstuffs, machinery
Debt—external: $36.1 million (1997)
country comparison to the world: 188
Exchange rates: the US dollar is used

COMMUNICATIONS

Telephones—main lines in use: 20,100 (2009)
country comparison to the world: 195
Telephones—mobile cellular: 24,000 (2009)
country comparison to the world: 206
Telephone system: *general assessment:* good overall telephone service
domestic: fixed line connections exceed 80 per 100 persons and mobile cellular subscribership is approaching 100 per 100 persons
international: country code–1-284; connected via submarine cable to Bermuda; the East Caribbean Fiber System (ECFS) submarine cable provides connectivity to 13 other islands in the eastern Caribbean (2009)
Broadcast media: 1 private TV station; multi-channel TV is available from cable and satellite subscription services; about a half dozen private radio stations operating (2007)
Internet country code: .vg
Internet hosts: 497 (2010)
country comparison to the world: 180
Internet users: 4,000 (2002)
country comparison to the world: 206

TRANSPORTATION

Airports: 4 (2010)
country comparison to the world: 188
Airports—with paved runways:
total: 2
914 to 1,523 m: 1
under 914 m: 1 (2010)
Airports—with unpaved runways: *total:* 2
914 to 1,523 m: 2 (2010)
Roadways: *total:* 200 km
country comparison to the world: 207
paved: 200 km (2007)
Merchant marine: registered in other countries: 1 (Panama 1) (2008)
country comparison to the world: 162
Ports and terminals: Road Harbor

MILITARY

Manpower available for military service:
males age 16–49: 7,266 (2010 est.)
Manpower fit for military service: *males age 16–49:* 6,057
females age 16–49: 5,805 (2010 est.)
Manpower reaching militarily significant age annually: *male:* 168
female: 162 (2010 est.)
Military—note: defense is the responsibility of the UK

TRANSNATIONAL ISSUES

Disputes—international: none
Illicit drugs: transshipment point for South American narcotics destined for the US and Europe; large offshore financial center makes it vulnerable to money laundering

BRUNEI

INTRODUCTION

Background: The Sultanate of Brunei's influence peaked between the 15th and 17th centuries when its control extended over coastal areas of northwest Borneo and the southern Philippines. Brunei subsequently entered a period of decline brought on by internal strife over royal succession, colonial expansion of European powers, and piracy. In 1888, Brunei became a British protectorate; independence was achieved in 1984. The same family has ruled Brunei for over six centuries. Brunei benefits from extensive petroleum and natural gas fields, the source of one of the highest per capita GDPs in Asia.

GEOGRAPHY

Location: Southeastern Asia, bordering the South China Sea and Malaysia
Geographic coordinates: 4 30 N, 114 40 E
Map references: Southeast Asia
Area: *total:* 5,765 sq km
country comparison to the world: 172
land: 5,265 sq km

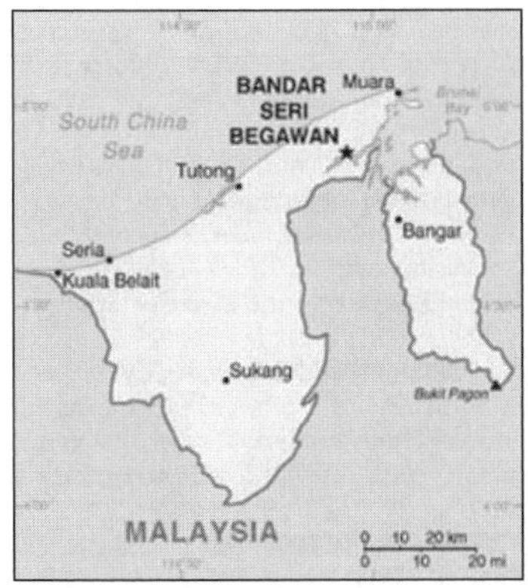

water: 500 sq km
Area—comparative: slightly smaller than Delaware
Land boundaries: *total:* 381 km
border countries: Malaysia 381 km
Coastline: 161 km
Maritime claims: territorial sea: 12 nm
exclusive economic zone: 200 nm or to median line
Climate: tropical; hot, humid, rainy
Terrain: flat coastal plain rises to mountains in east; hilly lowland in west
Elevation extremes: *lowest point:* South China Sea 0 m
highest point: Bukit Pagon 1,850 m
Natural resources: petroleum, natural gas, timber
Land use: *arable land:* 2.08%
permanent crops: 0.87%
other: 97.05% (2005)
Irrigated land: 10 sq km (2003)
Total renewable water resources: 8.5 cu km (1999)
Freshwater withdrawal (domestic/industrial/agricultural): *total:* 0.09
per capita: 243 cu m/yr (1994)
Natural hazards: typhoons, earthquakes, and severe flooding are rare
Environment—current issues: seasonal smoke/haze resulting from forest fires in Indonesia
Environment—international agreements: *party to:* Biodiversity, Climate Change, Desertification, Endangered Species, Hazardous Wastes, Law of the Sea, Ozone Layer Protection, Ship Pollution
signed, but not ratified: none of the selected agreements
Geography—note: close to vital sea lanes through South China Sea linking Indian and Pacific Oceans; two parts physically separated by Malaysia; almost an enclave within Malaysia

PEOPLE

Population: 401,890 (July 2011 est.)
country comparison to the world: 174
Age structure: *0–14 years:* 25.5% (male 52,944/female 49,729)
15–64 years: 70.9% (male 141,121/female 143,977)
65 years and over: 3.5% (male 6,881/female 7,238) (2011 est.)
Median age: *total:* 28.4 years
male: 28.3 years
female: 28.6 years (2011 est.)
Population growth rate: 1.712% (2011 est.)
country comparison to the world: 65
Birth rate: 17.87 births/1,000 population (2011 est.)
country comparison to the world: 107
Death rate: 3.35 deaths/1,000 population (July 2011 est.)
country comparison to the world: 214
Net migration rate: 2.6 migrant(s)/1,000 population (2011 est.)
country comparison to the world: 33
Urbanization: *urban population:* 76% of total population (2010)
rate of urbanization: 2.2% annual rate of change (2010-15 est.)
Major cities—population: BANDAR SERI BEGAWAN (capital) 22,000 (2009)
Sex ratio: *at birth:* 1.047 male(s)/female
under 15 years: 1.06 male(s)/female
15–64 years: 0.99 male(s)/female
65 years and over: 0.94 male(s)/female
total population: 1 male(s)/female (2011 est.)
Infant mortality rate:
total: 11.51 deaths/1,000 live births
country comparison to the world: 140
male: 13.74 deaths/1,000 live births
female: 9.17 deaths/1,000 live births (2011 est.)
Life expectancy at birth: *total population:* 76.17 years
country comparison to the world: 75
male: 73.91 years
female: 78.53 years (2011 est.)
Total fertility rate: 1.86 children born/woman (2011 est.)
country comparison to the world: 149
HIV/AIDS—adult prevalence rate: less than 0.1% (2003 est.)
country comparison to the world: 120
HIV/AIDS—people living with HIV/AIDS: fewer than 200 (2003 est.)
country comparison to the world: 158
HIV/AIDS—deaths: fewer than 200 (2003 est.)
country comparison to the world: 102
Nationality: *noun:* Bruneian(s)
adjective: Bruneian
Ethnicgroups: Malay66.3%,Chinese11.2%, indigenous 3.4%, other 19.1% (2004 est.)
Religions: Muslim (official) 67%, Buddhist 13%, Christian 10%, other (includes indigenous beliefs) 10%
Languages: Malay (official), English, Chinese
Literacy: *definition:* age 15 and over can read and write
total population: 92.7%
male: 95.2%
female: 90.2% (2001 census)
School life expectancy (primary to tertiary education): *total:* 14 years
male: 14 years
female: 14 years (2009)
Education expenditures: NA

GOVERNMENT

Country name: *conventional long form:* Brunei Darussalam
conventional short form: Brunei
local long form: Negara Brunei Darussalam
local short form: Brunei
Government type: constitutional sultanate (locally known as Malay Islamic Monarchy)
Capital: *name:* Bandar Seri Begawan
geographic coordinates: 4 53 N, 114 56 E
time difference: UTC+8 (13 hours ahead of Washington, DC during Standard Time)
Administrative divisions: 4 districts (daerah-daerah, singular—daerah); Belait, Brunei-Muara, Temburong, Tutong
Independence: 1January1984(fromtheUK)
National holiday: National Day, 23 February (1984); note—1 January 1984 was the date of independence from the UK, 23 February 1984 was the date of independence from British protection
Constitution: 29 September 1959 (some provisions suspended under a State of Emergency since December 1962, others since independence on 1 January 1984)
Legal system: mixed legal system based on English common law and Islamic law
International law organization participation: has not submitted an ICJ jurisdiction declaration; non-party state to the ICCt
Suffrage: 18 years of age for village elections; universal
Executive branch: *chief of state:* Sultan and Prime Minister Sir HASSANAL Bolkiah (since 5 October 1967); note—the monarch is both the chief of state and head of government
head of government: Sultan and Prime Minister Sir HASSANAL Bolkiah (since 5 October 1967)
cabinet: Council of Cabinet Ministers appointed and presided over by the monarch; deals with executive matters; note—there is also a Religious Council (members appointed by the monarch) that advises on religious matters, a Privy Council (members appointed by the monarch) that deals with constitutional matters, and the Council of Succession (members appointed by the monarch) that determines the succession to the throne if the need arises (For more information visit the World Leaders website)
elections: none; the monarchy is hereditary
Legislative branch: The Sultan appointed a Legislative Council with 29 members in September 2005; he increased the size of the council to 33 members in June 2011; the council meets annually in March
elections: last held in March 1962 (date of next election NA)
note: The Legislative Council met on 25 September 2004 for first time in 20 years with 21 members appointed by the Sultan; it passed constitutional amendments calling for a 45-seat council with 15 elected members; no timeframe for an election was announced
Judicial branch: Supreme Court—chief justice and judges are sworn in by monarch for three-year terms; Judicial Committee of Privy Council in London is final court of appeal for civil cases; Sharia courts deal with Islamic laws (2006)
Political parties and leaders: National Development Party or NDP [YASSIN Affendi]
note: Brunei National Solidarity Party or PPKB [Abdul LATIF bin Chuchu] and People's Awareness Party or PAKAR [Awang Haji MAIDIN bin Haji Ahmad] were deregistered in 2007; parties are small and have limited activity
Political pressure groups and leaders: NA

International organization participation: ADB, APEC, ARF, ASEAN, C, CP, EAS, G-77, IBRD, ICAO, ICRM, IDB, IFRCS, ILO, IMF, IMO, IMSO, Interpol, IOC, ISO (correspondent), ITSO, ITU, NAM, OIC, OPCW, UN, UNCTAD, UNESCO, UNIFIL, UNWTO, UPU, WCO, WHO, WIPO, WMO, WTO
Diplomatic representation in the US: *chief of mission:* Ambassador Yusoff Abd HAMID
chancery: 3520 International Court NW, Washington, DC 20008
telephone: [1] (202) 237-1838
FAX: [1] (202) 885-0560
Diplomatic representation from the US: *chief of mission:* Ambassador Daniel L. SHIELDS III
embassy: Simpang 336-52-16-9, Jalan Kebangsaan, Bandar Seri Begawan, BC4115
mailing address: Unit 4280, Box 40, FPO AP 96507; P.O. Box 2991, Bandar Seri Begawan BS8675, Negara Brunei Darussalam
telephone: [673] 238-4616
FAX: [673] 238-4604
Flag description: yellow with two diagonal bands of white (top, almost double width) and black starting from the upper hoist side; the national emblem in red is superimposed at the center; yellow is the color of royalty and symbolizes the sultanate; the white and black bands denote Brunei's chief ministers; the emblem includes five main components: a swallow-tailed flag, the royal umbrella representing the monarchy, the wings of four feathers symbolizing justice, tranquility, prosperity, and peace, the two upraised hands signifying the government's pledge to preserve and promote the welfare of the people, and the crescent moon denoting Islam, the state religion; the state motto "Always render service with God's guidance" appears in yellow Arabic script on the crescent; a ribbon below the crescent reads "Brunei, the Abode of Peace"
National anthem: *name:* "Allah Peliharakan Sultan" (God Bless His Majesty)
lyrics/music: Pengiran Haji Mohamed YUSUF bin Abdul Rahim/Awang Haji BESAR bin Sagap
note: adopted 1951

ECONOMY

Economy—overview: Brunei has a small well-to-do economy that encompasses a mixture of foreign and domestic entrepreneurship, government regulation, welfare measures, and village tradition. Crude oil and natural gas production account for just over half of GDP and more than 90% of exports. Per capita GDP is among the highest in Asia, and substantial income from overseas investment supplements income from domestic production. The government provides for all medical services and free education through the university level and subsidizes rice and housing. A new monetary authority was established in January 2011 with responsibilities that include monetary policy, monitoring of financial institutions, and currency trading activities.
GDP (purchasing power parity): $20.38 billion (2010 est.)
country comparison to the world: 123
$19.58 billion (2009 est.)
$19.93 billion (2008 est.)
note: data are in 2010 US dollars
GDP (official exchange rate): $13.02 billion (2010 est.)
GDP—real growth rate: 4.1% (2010 est.)
country comparison to the world: 96
-1.8% (2009 est.)
-1.9% (2008 est.)
GDP—per capita (PPP): $51,600 (2010 est.)
country comparison to the world: 8
$50,400 (2009 est.)
$52,300 (2008 est.)
note: data are in 2010 US dollars
GDP—composition by sector:
agriculture: 0.7%
industry: 74.1%
services: 25.3% (2008 est.)
Labor force: 188,800 (2008)
country comparison to the world: 172
Labor force—by occupation:
agriculture: 4.2%
industry: 62.8%
services: 33% (2008 est.)
Unemployment rate: 3.7% (2008)
country comparison to the world: 32
3.4% (2007)
Population below poverty line: NA%
Household income or consumption by percentage share: *lowest 10%:* NA%
highest 10%: NA%
Budget:
revenues: $6.889 billion
expenditures: $4 billion (2008 est.)
Inflation rate (consumer prices): 2.7% (2008 est.)
country comparison to the world: 73
0.3% (2007 est.)
Commercial bank prime lending rate: 5.5% (31 December 2009 est.)
country comparison to the world: 138
5.5% (31 December 2008 est.)
Stock of narrow money: $3.374 billion (30 March 2009)
country comparison to the world: 107
$3.046 billion (31 December 2008)
Stock of broad money: $8.569 billion (31 December 2009)
country comparison to the world: 106
$7.597 billion (31 December 2008)
Stock of domestic credit: $1.274 billion (31 December 2008 est.)
country comparison to the world: 142
$2.38 billion (31 December 2007 est.)
Market value of publicly traded shares: $NA
Agriculture—products: rice, vegetables, fruits; chickens, water buffalo, cattle, goats, eggs
Industries: petroleum, petroleum refining, liquefied natural gas, construction
Industrial production growth rate: -5.4% (2008 est.)
country comparison to the world: 165
Electricity—production: 3.069 billion kWh (2008)
country comparison to the world: 125
Electricity—consumption: 2.98 billion kWh (2008)
country comparison to the world: 127
Electricity—exports: 0 kWh (2008 est.)
Electricity—imports: 0 kWh (2008 est.)
Oil—production: 146,000 bbl/day (2009 est.)
country comparison to the world: 48
Oil—consumption: 16,000 bbl/day (2009 est.)
country comparison to the world: 131
Oil—exports: 152,900 bbl/day (2007)
country comparison to the world: 59
Oil—imports: 237.6 bbl/day (2007 est.)
country comparison to the world: 201
Oil—proved reserves: 1.1 billion bbl (1 January 2010 est.)
country comparison to the world: 42
Natural gas—production: 13.4 billion cu m (2008 est.)
country comparison to the world: 36
Natural gas—consumption: 4.2 billion cu m (2008 est.)
country comparison to the world: 65
Natural gas—exports: 9.2 billion cu m (2008 est.)
country comparison to the world: 21
Natural gas—imports: 0 cu m (2008 est.)
country comparison to the world: 159
Natural gas—proved reserves: 390.8 billion cu m (1 January 2010 est.)
country comparison to the world: 36
Current account balance: $7.024 billion (2008 est.)
country comparison to the world: 27
$7.101 billion (2007 est.)
Exports: $10.67 billion (2008)
country comparison to the world: 83
$8.25 billion (2007)
Exports—commodities: crude oil, natural gas, garments
Exports—partners: Japan 38.04%, Indonesia 25.95%, South Korea 14.17%, Australia 7.24% (2009)
Imports: $2.61 billion (2008 est.)
country comparison to the world: 145
$2.055 billion (2007 est.)
Imports—commodities: machinery and transport equipment, manufactured goods, food, chemicals
Imports—partners: Singapore 38.4%, Malaysia 18.7%, Japan 7.2%, China 5.42%, Thailand 5.19%, US 4.45%, UK 4.25% (2009)
Debt—external: $0 (2005)
country comparison to the world: 195
Exchange rates: Bruneian dollars (BND) per US dollar–1.36 (2010), 1.45 (2009), 2 (2006), 2 (2005), 2 (2004)

COMMUNICATIONS

Telephones—main lines in use: 80,500 (2009)
country comparison to the world: 150
Telephones—mobile cellular: 425,000 (2009)
country comparison to the world: 163
Telephone system: *general assessment:* service throughout the country is good; international service is good to Southeast Asia, Middle East, Western Europe, and the US
domestic: every service available
international: country code–673; landing point for the SEA-ME-WE-3 optical telecommunications submarine cable that provides links to Asia, the Middle East, and Europe; the Asia-America Gateway submarine cable network, scheduled for completion by late 2008, will provide new links to Asia and the US; satellite earth

stations—2 Intelsat (1 Indian Ocean and 1 Pacific Ocean) (2009)
Broadcast media: state-controlled Radio Television Brunei (RTB) operates 4 channels; 3 Malaysian TV stations are available; foreign TV broadcasts are available via satellite and cable systems; RTB operates 5 radio networks broadcasting on multiple frequencies; British Forces Broadcast Service (BFBS) provides radio broadcasts on 2 FM stations; some radio broadcast stations from Malaysia are available via repeaters (2009)
Internet country code: .bn
Internet hosts: 50,997 (2010)
country comparison to the world: 88
Internet users: 314,900 (2009)
country comparison to the world: 128

TRANSPORTATION

Airports: 2 (2010)
country comparison to the world: 197
Airports—with paved runways: *total:* 2
over 3,047 m: 1
914 to 1,523 m: 1 (2010)
Heliports: 3 (2010)
Pipelines: condensate 33 km; gas 37 km; oil 18 km (2010)
Roadways: *total:* 2,971 km
country comparison to the world: 166
paved: 2,411 km
unpaved: 560 km (2008)
Waterways: 209 km (navigable by craft drawing less than 1.2 m; the Belait, Brunei, and Tutong rivers are major transport links) (2011)
country comparison to the world: 97
Merchant marine: *total:* 9
country comparison to the world: 117
by type: chemical tanker 1, liquefied gas 8 (2010)
Ports and terminals: Lumut, Muara, Seria

MILITARY

Military branches: Royal Brunei Armed Forces: Royal Brunei Land Forces, Royal Brunei Navy, Royal Brunei Air Force (Tentera Udara Diraja Brunei) (2010)
Military service age and obligation: 18 years of age (est.) for voluntary military service; non-Malays are ineligible to serve (2007)
Manpower available for military service:
males age 16–49: 112,688
females age 16–49: 117,536 (2010 est.)
Manpower fit for military service: *males age 16–49:* 95,141
females age 16–49: 99,386 (2010 est.)
Manpower reaching militarily significant age annually: *male:* 3,572
female: 3,465 (2010 est.)
Military expenditures: 4.5% of GDP (2006)
country comparison to the world: 20

TRANSNATIONAL ISSUES

Disputes—international: per Letters of Exchange signed in 2009, Malaysia in 2010 ceded two hydrocarbon concession blocks to Brunei in exchange for Brunei's sultan dropping claims to the Limbang corridor, which divides Brunei; nonetheless, Brunei claims a maritime boundary extending as far as a median with Vietnam, thus asserting an implicit claim to Lousia Reef
Illicit drugs: drug trafficking and illegally importing controlled substances are serious offenses in Brunei and carry a mandatory death penalty

BULGARIA

INTRODUCTION

Background: The Bulgars, a Central Asian Turkic tribe, merged with the local Slavic inhabitants in the late 7th century to form the first Bulgarian state. In succeeding centuries, Bulgaria struggled with the Byzantine Empire to assert its place in the Balkans, but by the end of the 14th century the country was overrun by the Ottoman Turks. Northern Bulgaria attained autonomy in 1878 and all of Bulgaria became independent from the Ottoman Empire in 1908. Having fought on the losing side in both World Wars, Bulgaria fell within the Soviet sphere of influence and became a People's Republic in 1946. Communist domination ended in 1990, when Bulgaria held its first multiparty election since World War II and began the contentious process of moving toward political democracy and a market economy while combating inflation, unemployment, corruption, and crime. The country joined NATO in 2004 and the EU in 2007.

GEOGRAPHY

Location: Southeastern Europe, bordering the Black Sea, between Romania and Turkey
Geographic coordinates: 43 00 N, 25 00 E
Map references: Europe
Area: *total:* 110,879 sq km
country comparison to the world: 104
land: 108,489 sq km
water: 2,390 sq km
Area—comparative: slightly larger than Tennessee
Land boundaries: *total:* 1,808 km
border countries: Greece 494 km, Macedonia 148 km, Romania 608 km, Serbia 318 km, Turkey 240 km
Coastline: 354 km
Maritime claims: territorial sea: 12 nm
contiguous zone: 24 nm
exclusive economic zone: 200 nm
Climate: temperate; cold, damp winters; hot, dry summers
Terrain: mostly mountains with lowlands in north and southeast
Elevation extremes: *lowest point:* Black Sea 0 m
highest point: Musala 2,925 m
Natural resources: bauxite, copper, lead, zinc, coal, timber, arable land
Land use: *arable land:* 29.94%
permanent crops: 1.9%
other: 68.16% (2005)
Irrigated land: 5,880 sq km (2003)
Total renewable water resources: 19.4 cu km (2005)
Freshwater withdrawal (domestic/industrial/agricultural): *total:* 6.92 cu km/yr (3%/78%/19%)
per capita: 895 cu m/yr (2003)
Natural hazards: earthquakes; landslides
Environment—current issues: air pollution from industrial emissions; rivers polluted from raw sewage, heavy metals, detergents; deforestation; forest damage from air pollution and resulting acid rain; soil contamination from heavy metals from metallurgical plants and industrial wastes
Environment—international agreements: *party to:* Air Pollution, Air Pollution-Nitrogen Oxides, Air Pollution-Persistent Organic Pollutants, Air Pollution-Sulfur 85, Air Pollution-Sulfur 94, Air Pollution-Volatile Organic Compounds, Antarctic-Environmental Protocol, Antarctic-Marine Living Resources, Antarctic Treaty, Biodiversity, Climate Change, Climate Change-Kyoto Protocol, Desertification, Endangered Species, Environmental Modification, Hazardous Wastes, Law of the Sea, Marine Dumping, Ozone Layer Protection, Ship Pollution, Wetlands
signed, but not ratified: none of the selected agreements
Geography—note: strategic location near Turkish Straits; controls key land routes from Europe to Middle East and Asia

PEOPLE

Population: 7,093,635 (July 2011 est.)
country comparison to the world: 99
Age structure: *0–14 years:* 13.9% (male 506,403/female 480,935)
15–64 years: 67.9% (male 2,367,680/female 2,446,799)
65 years and over: 18.2% (male 522,343/female 769,475) (2011 est.)
Median age: *total:* 41.9 years
male: 39.6 years
female: 44 years (2011 est.)
Population growth rate: -0.781% (2011 est.)
country comparison to the world: 228
Birth rate: 9.32 births/1,000 population (2011 est.)
country comparison to the world: 204
Death rate: 14.32 deaths/1,000 population (July 2011 est.)

country comparison to the world: 13
Net migration rate: -2.82 migrant(s)/1,000 population (2011 est.)
country comparison to the world: 174
Urbanization: *urban population:* 71% of total population (2010)
rate of urbanization: -0.3% annual rate of change (2010-15 est.)
Major cities—population: SOFIA (capital) 1.192 million (2009)
Sex ratio: *at birth:* 1.06 male(s)/female
under 15 years: 1.05 male(s)/female
15–64 years: 0.97 male(s)/female
65 years and over: 0.68 male(s)/female
total population: 0.92 male(s)/female (2011 est.)
Infant mortality rate: *total:* 16.68 deaths/1,000 live births
country comparison to the world: 106
male: 19.93 deaths/1,000 live births
female: 13.25 deaths/1,000 live births (2011 est.)
Life expectancy at birth: *total population:* 73.59 years
country comparison to the world: 114
male: 69.99 years
female: 77.41 years (2011 est.)
Total fertility rate: 1.42 children born/woman (2011 est.)
country comparison to the world: 196
HIV/AIDS—adult prevalence rate: 0.1% (2009 est.)
country comparison to the world: 119
HIV/AIDS—people living with HIV/AIDS: 3,800 (2009 est.)
country comparison to the world: 123
HIV/AIDS—deaths: fewer than 200 (2009 est.)
country comparison to the world: 103
Drinking water source: *Improved:*
urban: 100% of population
rural: 100% of population
total: 100% of population (2008)
Sanitation facility access: *Improved:*
urban: 100% of population
rural: 100% of population
total: 100% of population (2008)
Nationality: *noun:* Bulgarian(s)
adjective: Bulgarian
Ethnic groups: Bulgarian 83.9%, Turk 9.4%, Roma 4.7%, other 2% (including Macedonian, Armenian, Tatar, Circassian) (2001 census)
Religions: Bulgarian Orthodox 82.6%, Muslim 12.2%, other Christian 1.2%, other 4% (2001 census)
Languages: Bulgarian (official) 84.5%, Turkish 9.6%, Roma 4.1%, other and unspecified 1.8% (2001 census)
Literacy: *definition:* age 15 and over can read and write
total population: 98.2%
male: 98.7%
female: 97.7% (2001 census)
School life expectancy (primary to tertiary education): *total:* 14 years
male: 13 years
female: 14 years (2008)
Education expenditures: 4.1% of GDP (2007)
country comparison to the world: 101

GOVERNMENT

Country name: *conventional long form:* Republic of Bulgaria
conventional short form: Bulgaria
local long form: Republika Balgariya
local short form: Balgariya
Government type: parliamentary democracy
Capital: *name:* Sofia
geographic coordinates: 42 41 N, 23 19 E
time difference: UTC+2 (7 hours ahead of Washington, DC during Standard Time)
daylight saving time: +1hr, begins last Sunday in March; ends last Sunday in October
Administrative divisions: 28 provinces (oblasti, singular—oblast); Blagoevgrad, Burgas, Dobrich, Gabrovo, Khaskovo, Kurdzhali, Kyustendil, Lovech, Montana, Pazardzhik, Pernik, Pleven, Plovdiv, Razgrad, Ruse, Shumen, Silistra, Sliven, Smolyan, Sofiya (Sofia), Sofiya-Grad (Sofia City), Stara Zagora, Turgovishte, Varna, Veliko Turnovo, Vidin, Vratsa, Yambol
Independence: 3 March 1878 (as an autonomous principality within the Ottoman Empire); 22 September 1908 (complete independence from the Ottoman Empire)
National holiday: Liberation Day, 3 March (1878)
Constitution: adopted 12 July 1991
Legal system: civil law
International law organization participation: accepts compulsory ICJ jurisdiction with reservations; accepts ICCt jurisdiction
Suffrage: 18 years of age; universal
Executive branch: *chief of state:* President Georgi PARVANOV (since 22 January 2002); Vice President Angel MARIN (since 22 January 2002)
head of government: Prime Minister Boyko BORISSOV (since 27 July 2009); Deputy Prime Ministers Simeon DJANKOV and Tsvetan TSVETANOV (since 27 July 2009)
cabinet: Council of Ministers nominated by the prime minister and elected by the National Assembly
(For more information visit the World Leaders website)
elections: president and vice president elected on the same ticket by popular vote for a five-year term (eligible for a second term); election last held on 22 and 29 October 2006 (next to be held in 2011); chairman of the Council of Ministers (prime minister) elected by the National Assembly; deputy prime ministers nominated by the prime minister and elected by the National Assembly
election results: Georgi PARVANOV reelected president; percent of vote—Georgi PARVANOV 77.3%, Volen SIDEROV 22.7%; Boyko BORISSOV elected prime minister; result of legislative vote—162 to 77 with 1 abstention
Legislative branch: unicameral National Assembly or Narodno Sabranie (240 seats; members elected by popular vote to serve four-year terms)
elections: last held on 5 July 2009 (next to be held in mid-2013)
election results: percent of vote by party—GERB 39.7%, BSP 17.7%, MRF 14.4%, ATAKA 9.4%, Blue Coalition 6.8%, RZS 4.1%, other 7.9%; seats by party—GERB 117, BSP 40, MRF 37, ATAKA 21, Blue Coalition 15, RZS 8, independents 2
Judicial branch: independent judiciary comprised of judges, prosecutors and investigating magistrates who are appointed, promoted, demoted, and dismissed by a 25-member Supreme Judicial Council (consists of the chairmen of the two Supreme Courts, the Chief Prosecutor, and 22 members, half of whom are elected by the National Assembly and the other half by the bodies of the judiciary for a 5-year term in office); three levels of case review; 182 courts of which two Supreme Courts act as the last instance on civil and criminal cases (the Supreme Court of Cassation) and appeals of government decisions (the Supreme Administrative Court)
Political parties and leaders: Agrarian National Union or ANU [Stefan LICHEV]; ATAKA (Attack party) [Volen SIDEROV]; Blue Coalition [Ivan KOSTOV and Martin DIMITROV] (a coalition of center-right parties dominated by UDF and DSB); Bulgarian New Democracy [Borislav RALCHEV]; Bulgarian Socialist Party or BSP [Sergei STANISHEV]; Citizens for the European Development of Bulgaria or GERB [Boyko BORISSOV]; Coalition for Bulgaria or CfB [Sergei STANISHEV] (coalition of parties dominated by BSP); Democrats for a Strong Bulgaria or DSB [Ivan KOSTOV]; Gergyovden [Petar STOYANOVICH]; Internal Macedonian Revolutionary Organization or IMRO [Krasimir KARAKACHANOV]; Liberal Initiative for Democratic European Development or LIDER [Khristo KOVACHKI]; Movement for Rights and Freedoms or MRF [Ahmed DOGAN]; National Movement for Stability and Progress or NDSV [Hristina HRISTOVA] (formerly National Movement Simeon II or NMS2); New Time [Emil KOSHLUKOV]; Order, Law, Justice or RZS [Yane YANEV]; Union of Democratic Forces or UDF [Martin DIMITROV]; Union of Free Democrats or UFD [Stefan SOFIYANSKI]; United Agrarians [Anastasia MOZER]
Political pressure groups and leaders: Confederation of Independent Trade Unions of Bulgaria or CITUB; Podkrepa Labor Confederation
other: numerous regional, ethnic, and national interest groups with various agendas
International organization participation: Australia Group, BIS, BSEC, CE, CEI, CERN, EAPC, EBRD, EIB, EU, FAO, G-9, IAEA, IBRD, ICAO, ICC, ICRM, IFC, IFRCS, ILO, IMF, IMO, IMSO, Interpol, IOC, IOM, IPU, ISO, ITSO, ITU, ITUC, MIGA, NATO, NSG, OAS (observer), OIF, OPCW, OSCE, PCA, SECI, UN, UNCTAD, UNESCO, UNIDO, UNMIL, UNWTO, UPU, WCO, WFTU, WHO, WIPO, WMO, WTO, ZC
Diplomatic representation in the US: *chief of mission:* Ambassador Elena POPTODOROVA
chancery: 1621 22nd Street NW, Washington, DC 20008
telephone: [1] (202) 387-0174
FAX: [1] (202) 234-7973
consulate(s) general: Chicago, Los Angeles, New York
Diplomatic representation from the US: *chief of mission:* Ambassador James B. WARLICK, Jr

embassy: 16 Kozyak Street, Sofia 1407
mailing address: American Embassy Sofia, US Department of State, 5740 Sofia Place, Washington, DC 20521-5740
telephone: [359] (2) 937-5100
FAX: [359] (2) 937-5320
Flag description: three equal horizontal bands of white (top), green, and red; the pan-Slavic white-blue-red colors were modified by substituting a green band (representing freedom) for the blue
note: the national emblem, formerly on the hoist side of the white stripe, has been removed
National anthem:
name: "Mila Rodino" (Dear Homeland)
lyrics/music: Tsvetan Tsvetkov RADOSLAVOV
note: adopted 1964; the anthem was composed in 1885 by a student en route to fight in the Serbo-Bulgarian War

ECONOMY

Economy—overview: Bulgaria, a former Communist country that entered the EU on 1 January 2007, averaged more than 6% annual growth from 2004 to 2008, driven by significant amounts of foreign direct investment and consumption. Successive governments have demonstrated a commitment to economic reforms and responsible fiscal planning, but the global downturn sharply reduced domestic demand, exports, capital inflows, and industrial production. GDP contracted by approximately 5% in 2009, and stagnated in 2010, despite a significant recovery in exports. The economy is expected to grow modestly in 2011, however. Corruption in the public administration, a weak judiciary, and the presence of organized crime remain significant challenges.
GDP (purchasing power parity): $96.78 billion (2010 est.)
country comparison to the world: 72
$96.63 billion (2009 est.)
$102.2 billion (2008 est.)
note: data are in 2010 US dollars
GDP (official exchange rate): $47.7 billion (2010 est.)
GDP—real growth rate: 0.2% (2010 est.)
country comparison to the world: 185
-5.5% (2009 est.)
6.2% (2008 est.)
GDP—per capita (PPP):
$13,500 (2010 est.)
country comparison to the world: 89
$13,400 (2009 est.)
$14,100 (2008 est.)
note: data are in 2010 US dollars
GDP—composition by sector:
agriculture: 6%
industry: 30.3%
services: 63.7% (2009)
Labor force: 3.4 million (2009 est.)
country comparison to the world: 98
Labor force—by occupation: *agriculture:* 7.1%
industry: 35.2%
services: 57.7% (2009)
Unemployment rate: 9.2% (2010)
country comparison to the world: 102
9.1% (2009 est.)
Population below poverty line: 21.8% (2008)
Household income or consumption by percentage share: *lowest 10%:* 2.9%
highest 10%: 23% (2009)
Distribution of family income—Gini index: 33.5 (2008)
country comparison to the world: 94
26 (2001)
Investment (gross fixed): 22.8% of GDP (2010 est.)
country comparison to the world: 61
Budget: *revenues:* $15.71 billion
expenditures: $17.52 billion (2010 est.)
Public debt: 16.2% of GDP (2010 est.)
country comparison to the world: 117
15.5% of GDP (2009)
Inflation rate (consumer prices): 4.4% (2010)
country comparison to the world: 124
1.6% (2009)
Central bank discount rate: 20% (31 December 2010)
note: Bulgarian National Bank (BNB) has had no independent monetary policy since the introduction of the Currency Board regime in 1997; this is BNB's base interest rate
Commercial bank prime lending rate: 11.34% (31 December 2009 est.)
country comparison to the world: 80
10.86% (31 December 2008 est.)
Stock of narrow money: $12.7 billion (31 December 2010 est.)
country comparison to the world: 68
$13.33 billion (31 December 2009 est.)
Stock of broad money: $35.37 billion (31 December 2010 est.)
country comparison to the world: 73
$35.07 billion (31 December 2009 est.)
Stock of domestic credit: $34.54 billion (31 December 2010 est.)
country comparison to the world: 68
$34.98 billion (31 December 2009)
Market value of publicly traded shares: $6.354 billion (31 December 2010)
country comparison to the world: 72
$7.103 billion (31 December 2009)
$8.858 billion (31 December 2008)
Agriculture—products: vegetables, fruits, tobacco, wine, wheat, barley, sunflowers, sugar beets; livestock
Industries: electricity, gas, water; food, beverages, tobacco; machinery and equipment, base metals, chemical products, coke, refined petroleum, nuclear fuel
Industrial production growth rate: 0.4% (2010 est.)
country comparison to the world: 151
Electricity—production: 4.309 billion kWh (2009)
country comparison to the world: 119
Electricity—consumption: 28.3 billion kWh (2009)
country comparison to the world: 62
Electricity—exports: 7.073 billion kWh (2009)
Electricity—imports: 2.66 billion kWh (2009)
Oil—production: 3,227 bbl/day (2009 est.)
country comparison to the world: 102
Oil—consumption: 125,000 bbl/day (2009 est.)
country comparison to the world: 71
Oil—exports: 76,570 bbl/day (2007 est.)
country comparison to the world: 70
Oil—imports: 189,000 bbl/day (2007 est.)
country comparison to the world: 47
Oil—proved reserves: 15 million bbl (1 January 2010 est.)
country comparison to the world: 86
Natural gas—production: 54 million cu m (2010)
country comparison to the world: 81
Natural gas—consumption: 2.62 billion cu m (2010)
country comparison to the world: 76
Natural gas—exports: 0 cu m (2010)
country comparison to the world: 67
Natural gas—imports: 2.48 billion cu m (2010)
country comparison to the world: 44
Natural gas—proved reserves: 5.663 billion cu m (1 January 2010 est.)
country comparison to the world: 91
Current account balance: $-12.8 million (2010 est.)
country comparison to the world: 64
$-5 billion (2009 est.)
Exports: $19.33 billion (2010 est.)
country comparison to the world: 69
$16.53 billion (2009 est.)
Exports—commodities: clothing, footwear, iron and steel, machinery and equipment, fuels
Exports—partners: Germany 11.21%, Greece 9.43%, Italy 9.24%, Romania 8.52%, Turkey 7.33%, Belgium 5.61%, France 4.44% (2009)
Imports: $22.78 billion (2010 est.)
country comparison to the world: 67
$22.22 billion (2009 est.)
Imports—commodities: machinery and equipment; metals and ores; chemicals and plastics; fuels, minerals, and raw materials
Imports—partners: Russia 13.14%, Germany 12.23%, Italy 7.78%, Greece 6.17%, Romania 5.65%, Turkey 5.48%, Ukraine 4.81%, Austria 4.08% (2009)
Reserves of foreign exchange and gold: $17.27 billion (31 December 2010 est.)
country comparison to the world: 47
$18.53 billion (31 December 2009 est.)
Debt—external: $47.15 billion (30 November 2010 est.)
country comparison to the world: 57
$54.37 billion (31 December 2009)
Stock of direct foreign investment—at home: $51.28 billion (31 December 2010 est.)
country comparison to the world: 54
$49.28 billion (31 December 2009)
Stock of direct foreign investment—abroad: $1.372 billion (31 December 2010 est.)
country comparison to the world: 70
$1.194 billion (31 December 2009)
Exchange rates: leva (BGN) per US dollar–1.5138 (2010), 1.404 (2009), 1.3171 (2008), 1.4366 (2007), 1.5576 (2006)

COMMUNICATIONS

Telephones—main lines in use: 2.164 million (2009)
country comparison to the world: 53
Telephones—mobile cellular: 10.617 million (2009)
country comparison to the world: 62
Telephone system: *general assessment:* inherited an extensive but antiquated telecommunications network from the Soviet era; quality has improved with a

modern digital trunk line now connecting switching centers in most of the regions; remaining areas are connected by digital microwave radio relay
domestic: the Bulgaria Telecommunications Company's fixed-line monopoly terminated in 2005 in an effort to upgrade fixed-line services; mobile-cellular teledensity, fostered by multiple service providers, approached 150 telephones per 100 persons in 2009
international: country code–359; submarine cable provides connectivity to Ukraine and Russia; a combination submarine cable and land fiber-optic system provides connectivity to Italy, Albania, and Macedonia; satellite earth stations–3 (1 Intersputnik in the Atlantic Ocean region, 2 Intelsat in the Atlantic and Indian Ocean regions) (2009)
Broadcast media: 4 national terrestrial television stations with 1 state-owned and 3 privately-owned; a vast array of TV stations are available from cable and satellite TV providers; state-owned national radio broadcasts over 3 networks; large number of private radio stations broadcasting, especially in urban areas (2010)
Internet country code: .bg
Internet hosts: 785,546 (2010)
country comparison to the world: 46
Internet users: 3.395 million (2009)
country comparison to the world: 63

TRANSPORTATION

Airports: 210 (2010)
country comparison to the world: 30
Airports—with paved runways:
total: 130
over 3,047 m: 2
2,438 to 3,047 m: 17
1,524 to 2,437 m: 15
under 914 m: 96 (2010)
Airports—with unpaved runways:
total: 80
1,524 to 2,437 m: 1
914 to 1,523 m: 6
under 914 m: 73 (2010)
Heliports: 3 (2010)
Pipelines: gas 2,844 km; oil 346 km; refined products 156 km (2010)
Railways: *total:* 4,151 km
country comparison to the world: 39
standard gauge: 4,071 km 1.435-m gauge (2,831 km electrified)
narrow gauge: 80 km 0.760-m gauge (2009)
Roadways: *total:* 40,231 km
country comparison to the world: 88
paved: 39,587 km (includes 418 km of expressways)
unpaved: 644 km (2008)
Waterways: 470 km (2009)
country comparison to the world: 84
Merchant marine: *total:* 37
country comparison to the world: 79
by type: bulk carrier 16, cargo 10, chemical tanker 1, liquefied gas 2, passenger/cargo 1, petroleum tanker 2, roll on/roll off 4, specialized tanker 1
foreign-owned: 27 (Germany 25, Russia 2)
registered in other countries: 31 (Comoros 8, Malta 7, Panama 6, Saint Vincent and the Grenadines 10) (2010)
Ports and terminals: Burgas, Varna

MILITARY

Military branches: Bulgarian Armed Forces: Ground Forces, Naval Forces, Bulgarian Air Forces (Bulgarski Voennovazdyshni Sily, BVVS) (2011)
Military service age and obligation: 18-27 years of age for voluntary military service; conscription ended in January 2008; service obligation 6-9 months (2010)
Manpower available for military service:
males age 16–49: 1,637,470
females age 16–49: 1,621,352 (2010 est.)
Manpower fit for military service: *males age 16–49:* 1,320,955
females age 16–49: 1,337,616 (2010 est.)
Manpower reaching militarily significant age annually: *male:* 33,444
female: 32,075 (2010 est.)
Military expenditures: 2.6% of GDP (2005 est.)
country comparison to the world: 57

TRANSNATIONAL ISSUES

Disputes—international: none
Illicit drugs: major European transshipment point for Southwest Asian heroin and, to a lesser degree, South American cocaine for the European market; limited producer of precursor chemicals; vulnerable to money laundering because of corruption, organized crime; some money laundering of drug-related proceeds through financial institutions (2008)

BURKINA FASO

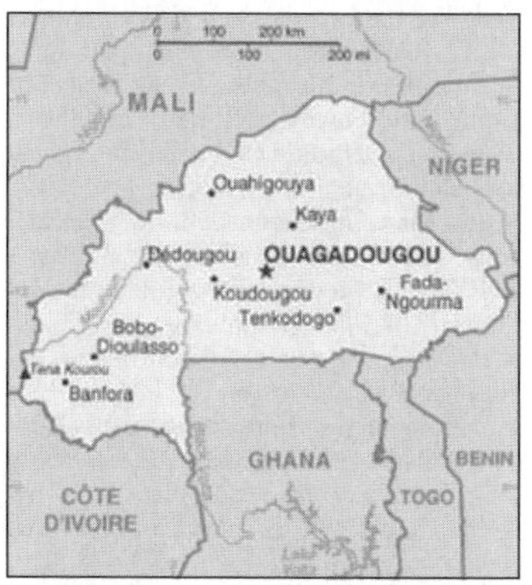

INTRODUCTION

Background: Burkina Faso (formerly Upper Volta) achieved independence from France in 1960. Repeated military coups during the 1970s and 1980s were followed by multiparty elections in the early 1990s. Current President Blaise COMPAORE came to power in a 1987 military coup and has won every election since then. Burkina Faso's high population density and limited natural resources result in poor economic prospects for the majority of its citizens. Recent unrest in Cote d'Ivoire and northern Ghana has hindered the ability of several hundred thousand seasonal Burkinabe farm workers to find employment in neighboring countries.

GEOGRAPHY

Location: Western Africa, north of Ghana
Geographic coordinates: 13 00 N, 2 00 W
Map references: Africa
Area: *total:* 274,200 sq km
country comparison to the world: 74
land: 273,800 sq km
water: 400 sq km
Area—comparative: slightly larger than Colorado
Land boundaries: *total:* 3,193 km
border countries: Benin 306 km, Cote d'Ivoire 584 km, Ghana 549 km, Mali 1,000 km, Niger 628 km, Togo 126 km
Coastline: 0 km (landlocked)
Maritime claims: none (landlocked)
Climate: tropical; warm, dry winters; hot, wet summers
Terrain: mostly flat to dissected, undulating plains; hills in west and southeast
Elevation extremes: *lowest point:* Mouhoun (Black Volta) River 200 m
highest point: Tena Kourou 749 m
Natural resources: manganese, limestone, marble; small deposits of gold, phosphates, pumice, salt
Land use: *arable land:* 17.66%
permanent crops: 0.22%
other: 82.12% (2005)
Irrigated land: 250 sq km (2003)
Total renewable water resources: 17.5 cu km (2001)
Freshwater withdrawal (domestic/industrial/agricultural): *total:* 0.8 cu km/yr (13%/1%/86%)
per capita: 60 cu m/yr (2000)
Natural hazards: recurring droughts
Environment—current issues: recent droughts and desertification severely affecting agricultural activities, population distribution, and the economy; overgrazing; soil degradation; deforestation
Environment—international agreements:
party to: Biodiversity, Climate Change, Climate Change-Kyoto Protocol, Desertification, Endangered Species, Hazardous Wastes, Law of the Sea, Marine Life Conservation, Ozone Layer Protection, Wetlands
signed, but not ratified: none of the selected agreements
Geography—note: landlocked savanna cut by the three principal rivers of the Black, Red, and White Voltas

PEOPLE

Population: 16,751,455 (July 2011 est.)
country comparison to the world: 61
note: estimates for this country explicitly take into account the effects of excess

mortality due to AIDS; this can result in lower life expectancy, higher infant mortality, higher death rates, lower population growth rates, and changes in the distribution of population by age and sex than would otherwise be expected
Age structure: *0–14 years:* 45.8% (male 3,849,350/female 3,828,483)
15–64 years: 51.7% (male 4,320,779/ female 4,334,197)
65 years and over: 2.5% (male 162,157/ female 256,489) (2011 est.)
Median age: *total:* 16.9 years
male: 16.7 years
female: 17.1 years (2011 est.)
Population growth rate: 3.085% (2011 est.)
country comparison to the world: 10
Birth rate: 43.59 births/1,000 population (2011 est.)
country comparison to the world: 5
Death rate: 12.74 deaths/1,000 population (July 2011 est.)
country comparison to the world: 25
Net migration rate: 0 migrant(s)/1,000 population (2011 est.)
country comparison to the world: 110
Urbanization: *urban population:* 26% of total population (2010)
rate of urbanization: 6.2% annual rate of change (2010-15 est.)
Major cities—population: OUAGADOUGOU (capital) 1.777 million (2009)
Sex ratio: *at birth:* 1.03 male(s)/female
under 15 years: 1.01 male(s)/female
15–64 years: 1 male(s)/female
65 years and over: 0.64 male(s)/female
total population: 0.99 male(s)/female (2011 est.)
Infant mortality rate: *total:* 81.4 deaths/1,000 live births
country comparison to the world: 10
male: 88.88 deaths/1,000 live births
female: 73.69 deaths/1,000 live births (2011 est.)
Life expectancy at birth: *total population:* 53.7 years
country comparison to the world: 202
male: 51.75 years
female: 55.71 years (2011 est.)
Total fertility rate: 6.14 children born/ woman (2011 est.)
country comparison to the world: 6
HIV/AIDS—adult prevalence rate: 1.2% (2009 est.)
country comparison to the world: 42
HIV/AIDS—people living with HIV/AIDS: 110,000 (2009 est.)
country comparison to the world: 41
HIV/AIDS—deaths: 7,100 (2009 est.)
country comparison to the world: 29
Major infectious diseases: *degree of risk:* very high
food or waterborne diseases: bacterial and protozoal diarrhea, hepatitis A, and typhoid fever
vectorborne disease: malaria and yellow fever
water contact disease: schistosomiasis
respiratory disease: meningococcal meningitis
animal contact disease: rabies
note: highly pathogenic H5N1 avian influenza has been identified in this country; it poses a negligible risk with extremely rare cases possible among US citizens who have close contact with birds (2009)
Drinking water source: *Improved:*
urban: 95% of population
rural: 72% of population
total: 76% of population
Unimproved: urban: 5% of population
rural: 28% of population
total: 24% of population (2008)
Sanitation facility access: *Improved:*
urban: 33% of population
rural: 6% of population
total: 11% of population
Unimproved: urban: 67% of population
rural: 94% of population
total: 89% of population (2008)
Nationality: *noun:* Burkinabe (singular and plural)
adjective: Burkinabe
Ethnic groups: Mossi over 40%, other approximately 60% (includes Gurunsi, Senufo, Lobi, Bobo, Mande, and Fulani)
Religions: Muslim 50%, indigenous beliefs 40%, Christian (mainly Roman Catholic) 10%
Languages: French (official), native African languages belonging to Sudanic family spoken by 90% of the population
Literacy: *definition:* age 15 and over can read and write
total population: 21.8%
male: 29.4%
female: 15.2% (2003 est.)
School life expectancy (primary to tertiary education): *total:* 6 years
male: 7 years
female: 6 years (2009)
Education expenditures: 4.6% of GDP (2007)
country comparison to the world: 77

GOVERNMENT

Country name: *conventional long form:* none
conventional short form: Burkina Faso
local long form: none
local short form: Burkina Faso
former: Upper Volta, Republic of Upper Volta
Government type: parliamentary republic
Capital: *name:* Ouagadougou
geographic coordinates: 12 22 N, 1 31 W
time difference: UTC 0 (5 hours ahead of Washington, DC during Standard Time)
Administrative divisions: 45 provinces; Bale, Bam, Banwa, Bazega, Bougouriba, Boulgou, Boulkiemde, Comoe, Ganzourgou, Gnagna, Gourma, Houet, Ioba, Kadiogo, Kenedougou, Komondjari, Kompienga, Kossi, Koulpelogo, Kouritenga, Kourweogo, Leraba, Loroum, Mouhoun, Nahouri, Namentenga, Nayala, Noumbiel, Oubritenga, Oudalan, Passore, Poni, Sanguie, Sanmatenga, Seno, Sissili, Soum, Sourou, Tapoa, Tuy, Yagha, Yatenga, Ziro, Zondoma, Zoundweogo
Independence: 5 August 1960 (from France)
National holiday: Republic Day, 11 December (1958); note—commemorates the day that Upper Volta became an autonomous republic in the French Community
Constitution: approved by referendum 2 June 1991; formally adopted 11 June 1991; last amended January 2002
Legal system: civil law based on the French model and customary law
International law organization participation: has not submitted an ICJ jurisdiction declaration; accepts ICCt jurisdiction
Suffrage: 18 years of age; universal
Executive branch: *chief of state:* President BlaiseCOMPAORE(since15October1987)
head of government: Prime Minister Luc-Adolphe TIAO (since 18 April 2011)
cabinet: Council of Ministers appointed by the president on the recommendation of the prime minister
(For more information visit the World Leaders website)
elections: president elected by popular vote for a five-year term (eligible for a second term); election last held on 21 November 2010 (next to be held in 2015); prime minister appointed by the president with the consent of the legislature
election results: Blaise COMPAORE reelected president; percent of popular vote—Blaise COMPAORE 80.2%, Hama Arba DIALLO 8.2%, Benewende Stanislas SANKARA 6.3%, other 5.3%
Legislative branch: unicameral National Assembly or Assemblee Nationale (111 seats; members are elected by popular vote to serve five-year terms)
elections: National Assembly election last held on 6 May 2007 (next to be held in May 2012)
election results: percent of vote by party—NA; seats by party—CDP 73, ADF-RDA 14, UPR 5, UNIR-MS 4, CFD-B 3, UPS 2, PDP-PS 2, RDB 2, PDS 2, PAREN 1, PAI 1, RPC 1, UDPS 1
Judicial branch: Supreme Court of Appeals or Cour de Cassation; Council of State or Conseil d'Etat; Court of Accounts or la Cour des Comptes; Constitutional Council or Conseil Constitutionnel
Political parties and leaders: African Democratic Rally-Alliance for Democracy and Federation or ADF-RDA [Gilbert OUEDRAOGO]; Citizen's Popular Rally or RPC [Antoine QUARE]; Coalition of Democratic Forces of Burkina or CFD-B [Amadou Diemdioda DICKO]; Congress for Democracy and Progress or CDP [Roch Marc-Christian KABORE]; Democratic and Popular Rally or RDP [Nana THIBAUT]; Movement for Tolerance and Progress or MTP [Nayabtigungou Congo KABORE]; Party for African Independence or PAI [Soumane TOURE]; Party for Democracy and Progress-Socialist Party or PDP-PS [Ali LANKOANDE]; Party for Democracy and Socialism or PDS [Felix SOUBEIGA]; Party for National Rebirth or PAREN [Jeanne TRAORE]; Rally for the Development of Burkina or RDB [Antoine KARGOUGOU]; Rally of Ecologists of Burkina Faso or RDEB [Ram OUEDRAGO]; Republican Party for Integration and Solidarity or PARIS; Union for Democracy and Social Progress or UDPS [Fidele HIEN]; Union for Rebirth—Sankarist Movement or UNIR-MS [Benewende STANISLAS]; Union for the Republic or UPR [Toussaint Abel COULIBALY]; Union of Sankarist Parties or UPS [Ernest Nongma OUEDRAOGO]
Political pressure groups and leaders: Burkinabe General Confederation of Labor or CGTB [Tole SAGNON];

Burkinabe Movement for Human Rights or MBDHP [Chrysigone ZOUGMORE]; Group of 14 February [Benewende STANISLAS]; National Confederation of Burkinabe Workers or CNTB [Laurent OUEDRAOGO]; National Organization of Free Unions or ONSL [Paul KABORE]
other: watchdog/political action groups throughout the country in both organizations and communities
International organization participation: ACP, AfDB, AU, ECOWAS, Entente, FAO, FZ, G-77, IAEA, IBRD, ICAO, ICRM, IDA, IDB, IFAD, IFC, IFRCS, ILO, IMF, Interpol, IOC, IOM, IPU, ISO (correspondent), ITSO, ITU, ITUC, MIGA, MONUSCO, NAM, OIC, OIF, OPCW, PCA, UN, UNAMID, UNCTAD, UNESCO, UNIDO, UNITAR, UNMIS, UNWTO, UPU, WADB (regional), WAEMU, WCO, WFTU, WHO, WIPO, WMO, WTO
Diplomatic representation in the US: *chief of mission:* Ambassador Paramanga Ernest YONLI
chancery: 2340 Massachusetts Avenue NW, Washington, DC 20008
telephone: [1] (202) 332-5577
FAX: [1] (202) 667-1882
Diplomatic representation from the US: *chief of mission:* Ambassador Thomas DOUGHERTY
embassy: 602 Avenue Raoul Follereau, Koulouba, Secteur 4
mailing address: 01 B. P. 35, Ouagadougou 01; pouch mail–US Department of State, 2440 Ouagadougou Place, Washington, DC 20521-2440
telephone: [226] 50-30-67-23
FAX: [226] 50-30-38-90
Flag description: two equal horizontal bands of red (top) and green with a yellow five-pointed star in the center; red recalls the country's struggle for independence, green is for hope and abundance, and yellow represents the country's mineral wealth
note: uses the popular Pan-African colors of Ethiopia
National anthem: *name:* "Le Ditanye" (Anthem of Victory)
lyrics/music: Thomas SANKARA
note: adopted 1974; also known as "Une Seule Nuit" (One Single Night), Burkina Faso's anthem was written by the country's president, an avid guitar player

ECONOMY

Economy—overview: Burkina Faso is a poor, landlocked country that relies heavily on cotton and gold exports for revenue. The country has few natural resources and a weak industrial base. About 90% of the population is engaged in subsistence agriculture, which is vulnerable to periodic drought. Cotton is the main cash crop. Since 1998, Burkina Faso has embarked upon a gradual privatization of state-owned enterprises and in 2004 revised its investment code to attract foreign investment. As a result of this new code and other legislation favoring the mining sector, the country has seen an upswing in gold exploration and production. By 2010, gold had become the main source of export revenue.
GDP (purchasing power parity): $19.99 billion (2010 est.)
country comparison to the world: 126
$18.9 billion (2009 est.)
$18.3 billion (2008 est.)
note: data are in 2010 US dollars
GDP (official exchange rate): $8.781 billion (2010 est.)
GDP—real growth rate: 5.8% (2010 est.)
country comparison to the world: 56
3.2% (2009 est.)
5.2% (2008 est.)
GDP—per capita (PPP): $1,200 (2010 est.)
country comparison to the world: 206
$1,200 (2009 est.)
$1,200 (2008 est.)
note: data are in 2010 US dollars
GDP—composition by sector: *agriculture:* 30.1%
industry: 20.7%
services: 49.2% (2009 est.)
Labor force: 6.668 million
country comparison to the world: 64
note: a large part of the male labor force migrates annually to neighboring countries for seasonal employment (2007)
Labor force—by occupation:
agriculture: 90%
industry and *services:* 10% (2000 est.)
Unemployment rate: 77% (2004)
country comparison to the world: 197
Population below poverty line: 46.4% (2004)
Household income or consumption by percentage share: *lowest 10%:* 2.8%
highest 10%: 32.2% (2004)
Distribution of family income—Gini index: 39.5 (2007)
country comparison to the world: 63
48.2 (1994)
Investment (gross fixed): 19.7% of GDP (2010 est.)
country comparison to the world: 93
Budget: *revenues:* $1.87 billion
expenditures: $2.343 billion (2010 est.)
Inflation rate (consumer prices): 1.4% (2010 est.)
country comparison to the world: 34
2.6% (2009 est.)
Central bank discount rate: 4.25% (31 December 2009)
country comparison to the world: 98
4.75% (31 December 2008)
Commercial bank prime lending rate: NA%
Stock of narrow money: $1.416 billion (31 December 2010 est.)
country comparison to the world: 127
$1.303 billion (31 December 2009 est.)
Stock of broad money: $2.406 billion (31 December 2010 est.)
country comparison to the world: 138
$2.22 billion (31 December 2009 est.)
Stock of domestic credit: $1.373 billion (31 December 2010 est.)
country comparison to the world: 138
$1.236 billion (31 December 2009 est.)
Market value of publicly traded shares: $NA
Agriculture—products: cotton, peanuts, shea nuts, sesame, sorghum, millet, corn, rice; livestock
Industries: cotton lint, beverages, agricultural processing, soap, cigarettes, textiles, gold
Industrial production growth rate: 5.5% (2010 est.)
country comparison to the world: 64
Electricity—production: 611.6 million kWh (2007 est.)
country comparison to the world: 155
Electricity—consumption: 568.8 million kWh (2007 est.)
country comparison to the world: 159
Electricity—exports: 0 kWh (2008 est.)
Electricity—imports: 0 kWh (2008 est.)
Oil—production: 0 bbl/day (2009 est.)
country comparison to the world: 137
Oil—consumption: 9,000 bbl/day (2009 est.)
country comparison to the world: 149
Oil—exports: 0 bbl/day (2007 est.)
country comparison to the world: 208
Oil—imports: 8,283 bbl/day (2007 est.)
country comparison to the world: 145
Oil—proved reserves: 0 bbl (1 January 2010 est.)
country comparison to the world: 198
Natural gas—production: 0 cu m (2008 est.)
country comparison to the world: 195
Natural gas—consumption: 0 cu m (2008 est.)
country comparison to the world: 132
Natural gas—exports: 0 cu m (2008 est.)
country comparison to the world: 193
Natural gas—imports: 0 cu m (2008 est.)
country comparison to the world: 131
Natural gas—proved reserves: 0 cu m (1 January 2010 est.)
country comparison to the world: 197
Current account balance: $-486 million (2010 est.)
country comparison to the world: 115
$-330 million (2009 est.)
Exports: $991 million (2010 est.)
country comparison to the world: 151
$772 million (2009 est.)
Exports—commodities: cotton, livestock, gold
Exports—partners: Singapore 16.76%, Belgium 12.78%, China 7.59%, Ghana 6.89%, India 6.36%, Denmark 5.76%, Niger 5.13%, Thailand 4.52% (2009)
Imports: $1.48 billion (2010 est.)
country comparison to the world: 162
$1.186 billion (2009 est.)
Imports—commodities: capital goods, foodstuffs, petroleum
Imports—partners: Cote d'Ivoire 24.31%, France 19.48%, Togo 6.42% (2009)
Reserves of foreign exchange and gold: $1.588 billion (31 December 2010 est.)
country comparison to the world: 110
$1.296 billion (31 December 2009 est.)
Debt—external: $2.002 billion (31 December 2010 est.)
country comparison to the world: 140
$1.784 billion (31 December 2009 est.)
Exchange rates: Communaute Financiere Africaine francs (XOF) per US dollar–495.28 (2010), 472.19 (2009), 447.81 (2008), 493.51 (2007), 522.59 (2006)

COMMUNICATIONS

Telephones—main lines in use: 167,000 (2009)
country comparison to the world: 133
Telephones—mobile cellular: 3.299 million (2009)
country comparison to the world: 109

Telephone system: *general assessment:* system includes microwave radio relay, open-wire, and radiotelephone communication stations; in 2006 the government sold a 51 percent stake in the national telephone company and ultimately plans to retain only a 23 percent stake in the company
domestic: fixed-line connections stand at less than 1 per 100 persons; mobile-cellular usage, fostered by multiple providers, is increasing rapidly from a low base
international: country code–226; satellite earth station–1 Intelsat (Atlantic Ocean) (2009)
Broadcast media: 2 TV stations–1 state-owned and 1 privately-owned; state-owned radio runs a national and regional network; substantial number of privately-owned radio broadcast stations; transmissions of several international broadcasters available in Ouagadougou (2007)
Internet country code: .bf
Internet hosts: 1,877 (2010)
country comparison to the world: 155
Internet users: 178,100 (2009)
country comparison to the world: 144

TRANSPORTATION

Airports: 24 (2010)
country comparison to the world: 132
Airports—with paved runways: *total:* 2
over 3,047 m: 1
2,438 to 3,047 m: 1 (2010)
Airports—with unpaved runways:
total: 22
1,524 to 2,437 m: 4
914 to 1,523 m: 12
under 914 m: 6 (2010)
Railways: *total:* 622 km
country comparison to the world: 106
narrow gauge: 622 km 1.000-m gauge
note: another 660 km of this railway extends into Cote d'Ivoire (2010)
Roadways:
total: 92,495 km
country comparison to the world: 52
paved: 3,857 km
unpaved: 88,638 km (2004)

MILITARY

Military branches: Army, Air Force of Burkina Faso (Force Aerienne de Burkina Faso, FABF), National Gendarmerie (2011)
Military service age and obligation: 18 years of age for voluntary military service; women may serve in supporting roles (2009)
Manpower available for military service: *males age 16–49:* 3,735,735 (2010 est.)
Manpower fit for military service: *males age 16–49:* 2,366,168
females age 16–49: 2,367,673 (2010 est.)
Manpower reaching militarily significant age annually: *male:* 193,905
female: 191,662 (2010 est.)
Military expenditures: 1.2% of GDP (2006)
country comparison to the world: 121

TRANSNATIONAL ISSUES

Disputes—international: adding to illicit cross-border activities, Burkina Faso has issues concerning unresolved boundary alignments with its neighbors; demarcation is currently underway with Mali, the dispute with Niger was referred to the ICJ in 2010, and a dispute over several villages with Benin persists; Benin retains a border dispute with Burkina Faso around the town of Koualou

BURMA

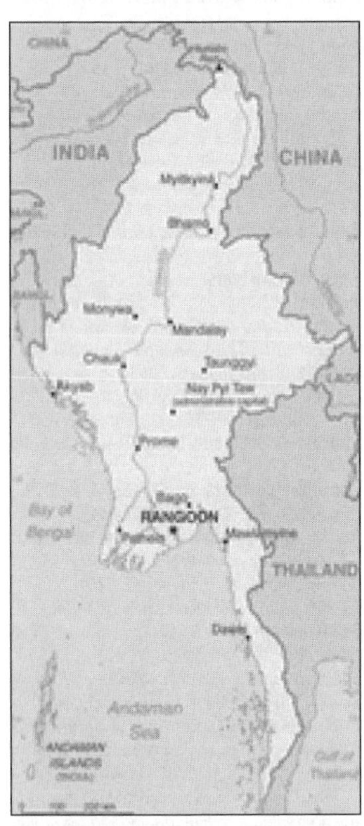

INTRODUCTION

Background: Britain conquered Burma over a period of 62 years (1824-1886) and incorporated it into its Indian Empire. Burma was administered as a province of India until 1937 when it became a separate, self-governing colony; independence from the Commonwealth was attained in 1948. Gen. NE WIN dominated the government from 1962 to 1988, first as military ruler, then as self-appointed president, and later as political kingpin. In September 1988, the military deposed NE WIN and established a new ruling junta. Despite multiparty legislative elections in 1990 that resulted in the main opposition party–the National League for Democracy (NLD)–winning a landslide victory, the junta refused to hand over power. NLD leader and Nobel Peace Prize recipient AUNG SAN SUU KYI, who was under house arrest from 1989 to 1995 and 2000 to 2002, was imprisoned in May 2003 and subsequently transferred to house arrest. She was finally released in November 2010. After the ruling junta in August 2007 unexpectedly increased fuel prices, tens of thousands of Burmese marched in protest, led by prodemocracy activists and Buddhist monks. In late September 2007, the government brutally suppressed the protests, killing at least 13 people and arresting thousands for participating in the demonstrations. Since then, the regime has continued to raid homes and monasteries and arrest persons suspected of participating in the pro-democracy protests. Burma in early May 2008 was struck by Cyclone Nargis, which claimed over 138,000 dead and tens of thousands injured and homeless. Despite this tragedy, the junta proceeded with its May constitutional referendum, the first vote in Burma since 1990. Parliamentary elections held in November 2010, considered flawed by many in the international community, saw the junta's Union Solidarity and Development Party garnering over 75% of the seats. Parliament convened in January 2011 and selected former Prime Minister THEIN SEIN as president. The vast majority of national-level appointees named by THEIN SEIN are former or current military officers.

GEOGRAPHY

Location: Southeastern Asia, bordering the Andaman Sea and the Bay of Bengal, between Bangladesh and Thailand
Geographic coordinates: 22 00 N, 98 00 E
Map references: Southeast Asia
Area:
total: 676,578 sq km
country comparison to the world: 40
land: 653,508 sq km
water: 23,070 sq km
Area—comparative: slightly smaller than Texas
Land boundaries: *total:* 5,876 km
border countries: Bangladesh 193 km, China 2,185 km, India 1,463 km, Laos 235 km, Thailand 1,800 km
Coastline: 1,930 km
Maritime claims: territorial sea: 12 nm
contiguous zone: 24 nm
exclusive economic zone: 200 nm
continental shelf: 200 nm or to the edge of the continental margin
Climate: tropical monsoon; cloudy, rainy, hot, humid summers (southwest monsoon, June to September); less cloudy, scant rainfall, mild temperatures, lower humidity during winter (northeast monsoon, December to April)
Terrain: central lowlands ringed by steep, rugged highlands
Elevation extremes: *lowest point:* Andaman Sea 0 m
highest point: Hkakabo Razi 5,881 m
Natural resources: petroleum, timber, tin, antimony, zinc, copper, tungsten, lead,

coal, marble, limestone, precious stones, natural gas, hydropower
Land use: *arable land:* 14.92%
permanent crops: 1.31%
other: 83.77% (2005)
Irrigated land: 18,700 sq km (2003)
Total renewable water resources: 1,045.6 cu km (1999)
Freshwater withdrawal (domestic/industrial/agricultural): *total:* 33.23 cu km/yr (1%/1%/98%)
per capita: 658 cu m/yr (2000)
Natural hazards: destructive earthquakes and cyclones; flooding and landslides common during rainy season (June to September); periodic droughts
Environment—current issues: deforestation; industrial pollution of air, soil, and water; inadequate sanitation and water treatment contribute to disease
Environment—international agreements: *party to:* Biodiversity, Climate Change, Climate Change-Kyoto Protocol, Desertification, Endangered Species, Law of the Sea, Ozone Layer Protection, Ship Pollution, Tropical Timber 83, Tropical Timber 94
signed, but not ratified: none of the selected agreements
Geography—note: strategic location near major Indian Ocean shipping lanes

PEOPLE

Population: 53,999,804 (July 2011 est.)
country comparison to the world: 24
note: estimates for this country take into account the effects of excess mortality due to AIDS; this can result in lower life expectancy, higher infant mortality, higher death rates, lower population growth rates, and changes in the distribution of population by age and sex than would otherwise be expected
Age structure: *0–14 years:* 27.5% (male 7,560,859/female 7,278,652)
15–64 years: 67.5% (male 18,099,707/female 18,342,696)
65 years and over: 5% (male 1,184,291/female 1,533,599) (2011 est.)
Median age: *total:* 26.9 years
male: 26.3 years
female: 27.5 years (2011 est.)
Population growth rate: 1.084% (2011 est.)
country comparison to the world: 109
Birth rate: 19.31 births/1,000 population (2011 est.)
country comparison to the world: 97
Death rate: 8.16 deaths/1,000 population (July 2011 est.)
country comparison to the world: 97
Net migration rate: -0.31 migrant(s)/1,000 population (2011 est.)
country comparison to the world: 129
Urbanization: *urban population:* 34% of total population (2010)
rate of urbanization: 2.9% annual rate of change (2010-15 est.)
Major cities—population: RANGOON (capital) 4.259 million; Mandalay 1.009 million; Nay Pyi Taw 992,000 (2009)
Sex ratio:
at birth: 1.06 male(s)/female
under 15 years: 1.04 male(s)/female
15–64 years: 0.99 male(s)/female
65 years and over: 0.77 male(s)/female
total population: 0.99 male(s)/female (2011 est.)
Infant mortality rate: *total:* 49.23 deaths/1,000 live births
country comparison to the world: 49
male: 56.16 deaths/1,000 live births
female: 41.88 deaths/1,000 live births (2011 est.)
Life expectancy at birth: *total population:* 64.88 years
country comparison to the world: 168
male: 62.57 years
female: 67.33 years (2011 est.)
Total fertility rate: 2.26 children born/woman (2011 est.)
country comparison to the world: 102
HIV/AIDS—adult prevalence rate: 0.6% (2009 est.)
country comparison to the world: 62
HIV/AIDS—people living with HIV/AIDS: 240,000 (2009 est.)
country comparison to the world: 25
HIV/AIDS—deaths: 18,000 (2009 est.)
country comparison to the world: 17
Major infectious diseases: *degree of risk:* very high
food or waterborne diseases: bacterial and protozoal diarrhea, hepatitis A, and typhoid fever
vectorborne diseases: dengue fever and malaria
water contact disease: leptospirosis
animal contact disease: rabies
note: highly pathogenic H5N1 avian influenza has been identified in this country; it poses a negligible risk with extremely rare cases possible among US citizens who have close contact with birds (2009)
Drinking water source: *Improved:*
urban: 75% of population
rural: 69% of population
total: 71% of population
Unimproved:
urban: 25% of population
rural: 31% of population
total: 29% of population (2008)
Sanitation facility access: *Improved:*
urban: 86% of population
rural: 79% of population
total: 81% of population
Unimproved: urban: 14% of population
rural: 21% of population
total: 19% of population (2008)
Nationality: *noun:* Burmese (singular and plural)
adjective: Burmese
Ethnic groups: Burman 68%, Shan 9%, Karen 7%, Rakhine 4%, Chinese 3%, Indian 2%, Mon 2%, other 5%
Religions: Buddhist 89%, Christian 4% (Baptist 3%, Roman Catholic 1%), Muslim 4%, animist 1%, other 2%
Languages: Burmese (official)
note: minority ethnic groups have their own languages
Literacy: *definition:* age 15 and over can read and write
total population: 89.9%
male: 93.9%
female: 86.4% (2006 est.)
School life expectancy (primary to tertiary education): *total:* 9 years
male: 8 years
female: 8 years (2007)
Education expenditures: NA

GOVERNMENT

Country name: *conventional long form:* Union of Burma
conventional short form: Burma
local long form: Pyidaungzu Myanma Naingngandaw (translated by the US Government as Union of Myanma and by the Burmese as Union of Myanmar)
local short form: Myanma Naingngandaw
former: Socialist Republic of the Union of Burma
note: since 1989 the military authorities in Burma have promoted the name Myanmar as a conventional name for their state; the US Government did not adopt the name, which is a derivative of the Burmese short-form name Myanma Naingngandaw
Government type: military regime (a nominally civilian government has been named, but a formal transfer of power has not yet taken place)
Capital: *name:* Rangoon (Yangon)
geographic coordinates: 16 48 N, 96 09 E
time difference: UTC+6.5 (11.5 hours ahead of Washington, DC during Standard Time)
note: Nay Pyi Taw is the administrative capital
Administrative divisions: 7 divisions (taing-myar, singular–taing) and 7 states* (pyi ne-myar, singular–pyi ne)
divisions: Ayeyarwady, Bago, Magway, Mandalay, Sagaing, Tanintharyi, Yangon
states: Chin, Kachin, Kayah, Kayin, Mon, Rakhine (Arakan), Shan
Independence: 4 January 1948 (from the UK)
National holiday: Independence Day, 4 January (1948); Union Day, 12 February (1947)
Constitution: 3 January 1974; suspended 18 September 1988; a new constitution was to take effect when the bicameral legislature convened 31 January 2011, but no announcement has been made
Legal system: mixed legal system of English common law (as introduced in codifications designed for colonial India) and customary law
International law organization participation: has not submitted an ICJ jurisdiction declaration; non-party state to the ICCt
Suffrage: 18 years of age; universal
Executive branch: *chief of state:* President THEIN SEIN (since 4 February 2011); Vice President SAI MOUK KHAM (since 3 February 2011); Vice President TIN AUNG MYINT OO (since 4 February 2011)
head of government: Prime Minister THEIN SEIN (since 24 October 2007)
cabinet: cabinet is appointed by the president and confirmed by the parliament
(For more information visit the World Leaders website)
elections: THEIN SEIN elected president by the parliament from among three vice presidents; the upper house, the lower house, and military members of the parliament each nominate one vice president (president serves a five-year term)
Legislative branch: bicameral, consists of the House of Nationalities [Amyotha Hluttaw] (224 seats, 168 directly elected and 56 appointed by the military;

members serve five-year terms) and the House of Representatives [Pythu Hluttaw] (440 seats, 330 directly elected and 110 appointed by the military; members serve five-year terms)
elections: last held on 7 November 2010 (next to be held in December 2015)
election results: House of Nationalities–percent of vote by party–USDP 74.8%, others (NUP, SNDP, RNDP, NDF, AMRDP) 25.2%; seats by party–USDP 129, others 39; House of Representatives–percent of vote by party–USDP 79.6%, others (NUP, SNDP, RNDP, NDF, AMRDP) 20.4%; seats by party–USDP 259, others 66
Judicial branch: remnants of the British-era legal system are in place, but there is no guarantee of a fair public trial; the judiciary is not independent of the executive; the 2011 constitution calls for a Supreme Court, a Courts-Martial, and a Constitutional Tribunal of the Union
Political parties and leaders: All Mon Region Democracy Party or AMRDP [NAING NGWE THEIN]; National Democratic Force or NDF [KHIN MAUNG SWE, Dr.THAN NYEIN]; National League for Democracy or NLD [AUNG SHWE, U TIN OO, AUNG SAN SUU KYI]; note–the party was deregisted because it did not register for the 2010 election, but it is still active; National Unity Party or NUP [TUN YE]; Rakhine Nationalities Development Party or RNDP [Dr. AYE MG]; Shan Nationalities Democratic Party [SAI AIKE PAUNG]; Shan Nationalities League for Democracy or SNLD [HKUN HTUN OO]; Union Solidarity and Development Party or USDP [SHWE MANN, HTAY OO]; numerous smaller parties
Political pressure groups and leaders: Thai border: Ethnic Nationalities Council or ENC; Federation of Trade Unions-Burma or FTUB (exile trade union and labor advocates); National Coalition Government of the Union of Burma or NCGUB (self-proclaimed government in exile) ["Prime Minister" Dr. SEIN WIN] consists of individuals, some legitimately elected to the People's Assembly in 1990 (the group fled to a border area and joined insurgents in December 1990 to form a parallel government in exile); National Council-Union of Burma or NCUB (exile coalition of opposition groups)
Inside Burma: Kachin Independence Organization or KIO; Karen National Union or KNU; Karenni National People's Party or KNPP; Union Solidarity and Development Association or USDA (pro-regime, a social and political mass-member organization) [HTAY OO, general secretary] became the Union Solidarity and Development Party in 2010; United Wa State Army or UWSA; 88 Generation Students (pro-democracy movement); several other Shan factions
note: freedom of expression is highly restricted in Burma; political groups, other than parties approved by the government, are limited in number
International organization participation: ADB, ARF, ASEAN, BIMSTEC, CP, EAS, FAO, G-77, IAEA, IBRD, ICAO, ICRM, IDA, IFAD, IFC, IFRCS, IHO, ILO, IMF, IMO, Interpol, IOC, ISO (correspondent), ITU, NAM, OPCW (signatory), SAARC (observer), UN, UNCTAD, UNESCO, UNIDO, UPU, WCO, WHO, WIPO, WMO, WTO
Diplomatic representation in the US: *chief of mission:* Ambassador (vacant); Charge d'Affaires HAN THU; note–Burma does not have an ambassador to the United States
chancery: 2300 S Street NW, Washington, DC 20008
telephone: [1] (202) 332-3344
FAX: [1] (202) 332-4351
consulate(s) general: none; Burma has a Mission to the UN in New York
Diplomatic representation from the US: *chief of mission:* Charge d'Affaires Larry M. DINGER; note–The United States does not have an ambassador to Burma
embassy: 110 University Avenue, Kamayut Township, Rangoon
mailing address: Box B, APO AP 96546
telephone: [95] (1) 536-509, 535-756, 538-038
FAX: [95] (1) 650-306
Flag description: design consists of three equal horizontal stripes of yellow (top), green, and red; centered on the green band is a large white five-pointed star that partially overlaps onto the adjacent colored stripes; the design revives the triband colors used by Burma from 1943-45, during the Japanese occupation
National anthem: *name:* "Kaba Ma Kyei" (Till the End of the World, Myanmar)
lyrics/music: SAYA TIN
note: adopted 1948; Burma is among a handful of non-European nations that have anthems rooted in indigenous traditions; the beginning portion of the anthem is a traditional Burmese anthem before transitioning into a Western-style orchestrated work

ECONOMY

Economy—overview: Burma, a resource-rich country, suffers from pervasive government controls, inefficient economic policies, corruption, and rural poverty. Despite Burma's emergence as a natural gas exporter, socio-economic conditions have deteriorated under the regime's mismanagement, leaving most of the public in poverty, while military leaders and their business cronies exploit the country's ample natural resources. The transfer of state assets, especially real estate, to cronies and military families in 2010 under the guise of a privatization policy further widened the gap between the economic elite and the public. The economy suffers from serious macroeconomic imbalances–including unpredictable inflation, fiscal deficits, multiple official exchange rates that overvalue the Burmese kyat, a distorted interest rate regime, unreliable statistics, and an inability to reconcile national accounts. Burma's poor investment climate hampers the inflow of foreign investment; in recent years, foreign investors have shied away from nearly every sector except for natural gas, power generation, timber, and mining. The exploitation of natural resources does not benefit the population at large. The business climate is widely perceived as opaque, corrupt, and highly inefficient. Over 60% of the FY 2009-10 budget was allocated to state owned enterprises–most operating at a deficit. The most productive sectors will continue to be in extractive industries–especially oil and gas, mining, and timber–with the latter two causing significant environmental degradation. Other areas, such as manufacturing, tourism and services, struggle in the face of inadequate infrastructure, unpredictable trade policies, neglected health and education systems, and endemic corruption. A major banking crisis in 2003 caused 20 private banks to close; private banks still operate under tight restrictions, limiting the private sector's access to credit. The United States, the European Union, and Canada have imposed financial and economic sanctions on Burma. US sanctions, prohibiting most financial transactions with Burmese entities, impose travel bans on senior Burmese military and civilian leaders and others connected to the ruling regime, and ban imports of Burmese products. These sanctions affected the country's fledgling garment industry, isolated the struggling banking sector, and raised the costs of doing business with Burmese companies, particularly firms tied to Burmese regime leaders. The global crisis of 2008-09 caused exports and domestic consumer demand to drop. Remittances from overseas Burmese workers–who had provided significant financial support for their families–slowed or dried up as jobs were lost and migrant workers returned home. Although the Burmese government has good economic relations with its neighbors, significant improvements in economic governance, the business climate, and the political situation are needed to promote serious foreign investment.
GDP (purchasing power parity): $76.47 billion (2010 est.)
country comparison to the world: 80
$72.65 billion (2009 est.)
$69.1 billion (2008 est.)
note: data are in 2010 US dollars
GDP (official exchange rate): $42.95 billion (2010 est.)
GDP—real growth rate: 5.3% (2010 est.)
country comparison to the world: 65
5.1% (2009 est.)
3.6% (2008 est.)
GDP—per capita (PPP): $1,400 (2010 est.)
country comparison to the world: 203
$1,400 (2009 est.)
$1,300 (2008 est.)
note: data are in 2010 US dollars
GDP—composition by sector: *agriculture:* 43.2%
industry: 20%
services: 36.8% (2010 est.)
Labor force: 31.68 million (2010 est.)
country comparison to the world: 18
Labor force—by occupation: *agriculture:* 70%
industry: 7%
services: 23% (2001)
Unemployment rate: 5.7% (2010 est.)
country comparison to the world: 57
4.9% (2009 est.)

Population below poverty line: 32.7% (2007 est.)
Household income or consumption by percentage share: *lowest 10%:* 2.8%
highest 10%: 32.4% (1998)
Investment (gross fixed): 15.1% of GDP (2010 est.)
country comparison to the world: 131
Budget: *revenues:* $1.369 billion
expenditures: $2.951 billion (2010 est.)
Inflation rate (consumer prices): 9.6% (2010 est.)
country comparison to the world: 194
1.5% (2009 est.)
Central bank discount rate: 12% (31 December 2009)
country comparison to the world: 29
12% (31 December 2008)
Commercial bank prime lending rate: 17% (31 December 2009 est.)
country comparison to the world: 30
17% (31 December 2008 est.)
Stock of narrow money: $4.907 billion (31 December 2010 est.)
country comparison to the world: 90
$4.038 billion (31 December 2009 est.)
note: this number reflects the vastly overvalued official exchange rate of 5.38 kyat per dollar in 2007; at the unofficial black market rate of 1,305 kyat per dollar for 2007, the stock of kyats would equal only US$2.465 billion and Burma's velocity of money (the number of times money turns over in the course of a year) would be six, in line with the velocity of money for other countries in the region; in January-February 2011, the unofficial black market rate averaged 890 kyat per dollar.
Stock of broad money: $7.8 billion (31 December 2010 est.)
country comparison to the world: 108
$6.231 billion (31 December 2009 est.)
Stock of domestic credit: $8.552 billion (31 December 2010 est.)
country comparison to the world: 99
$6.858 billion (31 December 2009 est.)
Market value of publicly traded shares: $NA
Agriculture—products: rice, pulses, beans, sesame, groundnuts, sugarcane; hardwood; fish and fish products
Industries: agricultural processing; wood and wood products; copper, tin, tungsten, iron; cement, construction materials; pharmaceuticals; fertilizer; oil and natural gas; garments, jade and gems
Industrial production growth rate: 4.3% (2010 est.)
country comparison to the world: 79
Electricity—production: 6.286 billion kWh (2007 est.)
country comparison to the world: 105
Electricity—consumption: 4.403 billion kWh (2007 est.)
country comparison to the world: 115
Electricity—exports: 0 kWh (2008 est.)
Electricity—imports: 0 kWh (2008 est.)
Oil—production: 18,880 bbl/day (2009 est.)
country comparison to the world: 75
Oil—consumption: 42,000 bbl/day (2009 est.)
country comparison to the world: 100
Oil—exports: 2,200 bbl/day (2007 est.)
country comparison to the world: 111
Oil—imports: 18,250 bbl/day (2007 est.)
country comparison to the world: 113
Oil—proved reserves: 50 million bbl (1 January 2010 est.)
country comparison to the world: 79
Natural gas—production: 12.4 billion cu m (2008 est.)
country comparison to the world: 39
Natural gas—consumption: 3.85 billion cu m (2008 est.)
country comparison to the world: 66
Natural gas—exports: 8.55 billion cu m (2008 est.)
country comparison to the world: 23
Natural gas—imports: 0 cu m (2008 est.)
country comparison to the world: 155
Natural gas—proved reserves: 283.2 billion cu m (1 January 2010 est.)
country comparison to the world: 41
Current account balance: $652 million (2010 est.)
country comparison to the world: 48
$705 million (2009 est.)
Exports: $7.841 billion (2010 est.)
country comparison to the world: 95
$6.862 billion (2009 est.)
note: official export figures are grossly underestimated due to the value of timber, gems, narcotics, rice, and other products smuggled to Thailand, China, and Bangladesh
Exports—commodities: natural gas, wood products, pulses, beans, fish, rice, clothing, jade and gems
Exports—partners: Thailand 46.57%, India 12.99%, China 9.01%, Japan 5.65% (2009)
Imports: $4.532 billion (2010 est.)
country comparison to the world: 124
$4.02 billion (2009 est.)
note: import figures are grossly underestimated due to the value of consumer goods, diesel fuel, and other products smuggled in from Thailand, China, Malaysia, and India
Imports—commodities: fabric, petroleum products, fertilizer, plastics, machinery, transport equipment; cement, construction materials, crude oil; food products, edible oil
Imports—partners: China 33.1%, Thailand 26.28%, Singapore 15.18% (2009)
Reserves of foreign exchange and gold: $3.762 billion (31 December 2010 est.)
country comparison to the world: 84
$3.561 billion (31 December 2009 est.)
Debt—external: $7.145 billion (31 December 2010 est.)
country comparison to the world: 97
$7.079 billion (31 December 2009 est.)
Exchange rates: kyats (MMK) per US dollar–966 (2010), 1,055 (2009), 1,205 (2008), 1,296 (2007), 1,280 (2006)

COMMUNICATIONS

Telephones—main lines in use: 812,000 (2009)
country comparison to the world: 87
Telephones—mobile cellular: 502,000 (2009)
country comparison to the world: 159
Telephone system: *general assessment:* meets minimum requirements for local and intercity service for business and government
domestic: system barely capable of providing basic service; mobile-cellular phone system is grossly underdeveloped
international: country code–95; landing point for the SEA-ME-WE-3 optical telecommunications submarine cable that provides links to Asia, the Middle East, and Europe; satellite earth stations–2, Intelsat (Indian Ocean) and ShinSat (2009)
Broadcast media: government controls all domestic broadcast media; 2 state-controlled television stations with 1 of the stations controlled by the armed forces; a third TV channel, a pay-TV station, is a joint state-private venture; access to satellite TV is limited; 1 state-controlled domestic radio station and 6 FM stations that are joint state-private ventures; transmissions of several international broadcasters are available in parts of Burma; the opposition-backed station Democratic Voice of Burma broadcasts into Burma via shortwave (2009)
Internet country code: .mm
Internet hosts: 172 (2010)
country comparison to the world: 197
Internet users: 110,000 (2009)
country comparison to the world: 158

TRANSPORTATION

Airports: 76 (2010)
country comparison to the world: 72
Airports—with paved runways: *total:* 37
over 3,047 m: 12
2,438 to 3,047 m: 8
1,524 to 2,437 m: 15
914 to 1,523 m: 1
under 914 m: 1 (2010)
Airports—with unpaved runways: *total:* 39
over 3,047 m: 1
1,524 to 2,437 m: 4
914 to 1,523 m: 11
under 914 m: 23 (2010)
Heliports: 6 (2010)
Pipelines: gas 3,046 km; oil 551 km (2010)
Railways: *total:* 5,031 km
country comparison to the world: 36
narrow gauge: 5,031 km 1.000-m gauge (2010)
Roadways: *total:* 27,000 km
country comparison to the world: 100
paved: 3,200 km
unpaved: 23,800 km (2006)
Waterways: 12,800 km (2008)
country comparison to the world: 10
Merchant marine: *total:* 26
country comparison to the world: 91
by type: bulk carrier 1, cargo 19, passenger 2, passenger/cargo 3, specialized tanker 1
foreign-owned: 3 (Cyprus 1, Germany 1, Japan 1)
registered in other countries: 3 (Panama 3) (2010)
Ports and terminals: Moulmein, Rangoon, Sittwe

MILITARY

Military branches: Myanmar Armed Forces (Tatmadaw): Army (Tatmadaw Kyi), Navy (Tatmadaw Yay), Air Force (Tatmadaw Lay) (2011)
Military service age and obligation: 18-35 years of age (men) and 18-27 years of age (women) for compulsory military service; service obligation 2 years; male (ages 18-45) and female (ages 18-35) professionals (including doctors, engineers, mechanics) serve up to 3 years; service terms may be

extended to 5 years in an officially declared emergency; forced conscription of children, although officially prohibited, reportedly continues (2011)

Manpower available for military service: *males age 16–49:* 14,747,845
females age 16–49: 14,710,871 (2010 est.)

Manpower fit for military service: *males age 16–49:* 10,451,515
females age 16–49: 11,181,537 (2010 est.)

Manpower reaching militarily significant age annually: *male:* 522,478
female: 506,388 (2010 est.)

Military expenditures: 2.1% of GDP (2005 est.)
country comparison to the world: 68

TRANSNATIONAL ISSUES

Disputes—international: over half of Burma's population consists of diverse ethnic groups who have substantial numbers of kin in neighboring countries; the Naf river on the border with Bangladesh serves as a smuggling and illegal transit route; Bangladesh struggles to accommodate 29,000 Rohingya, Burmese Muslim minority from Arakan State, living as refugees in Cox's Bazar; Burmese border authorities are constructing a 200 km (124 mi) wire fence designed to deter illegal cross-border transit and tensions from the military build-up along border with Bangladesh in 2010; Bangladesh referred its maritime boundary claims with Burma and India to the International Tribunal on the Law of the Sea; Burmese forces attempting to dig in to the largely autonomous Shan State to rout local militias tied to the drug trade, prompts local residents to periodically flee into neighboring Yunnan Province in China; fencing along the India-Burma international border at Manipur's Moreh town is in progress to check illegal drug trafficking and movement of militants; 140,000 mostly Karen refugees fleeing civil strife, political upheaval and economic stagnation in Burma live in remote camps in Thailand near the border

Refugees and internally displaced persons: *IDPs:* 503,000 (government offensives against ethnic insurgent groups near the eastern borders; most IDPs are ethnic Karen, Karenni, Shan, Tavoyan, and Mon) (2007)

Trafficking in persons: current situation: Burma is a source country for women, children, and men trafficked for the purpose of forced labor and commercial sexual exploitation; Burmese women and children are trafficked to East and Southeast Asia for commercial sexual exploitation, domestic servitude, and forced labor; Burmese children are subjected to conditions of forced labor in Thailand as hawkers and beggars; women are trafficked for commercial sexual exploitation to Malaysia and China; some trafficking victims transit Burma from Bangladesh to Malaysia and from China to Thailand; Burma's internal trafficking remains the most serious concern occurring primarily from villages to urban centers and economic hubs for labor in industrial zones, agricultural estates, and commercial sexual exploitation; the Burmese military continues to engage in the unlawful conscription of child soldiers, and continues to be the main perpetrator of forced labor inside Burma; ethnic insurgent groups also used compulsory labor of adults and unlawful recruitment of children; the regime's widespread use of and lack of accountability in forced labor and recruitment of child soldiers is particularly worrying and represents the top causal factor for Burma's significant trafficking problem
tier rating: Tier 3—serious problems remain in Burma, and in some areas, most notably in the area of forced labor, the Government of Burma is not making significant efforts to comply with the minimum standards for the elimination of trafficking, warranting a ranking of Tier 3; in other areas, particularly with regard to international sex trafficking of women and girls, the Government of Burma is making significant efforts (2010)

Illicit drugs: remains world's second largest producer of illicit opium with an estimated production in 2008 of 340 metric tons, an increase of 26%, and poppy cultivation in 2008 totaled 22,500 hectares, a 4% increase from 2007; production in the United Wa State Army's areas of greatest control remains low; Shan state is the source of 94% of Burma's poppy cultivation; lack of government will to take on major narcotrafficking groups and lack of serious commitment against money laundering continues to hinder the overall antidrug effort; major source of methamphetamine and heroin for regional consumption (2008)

BURUNDI

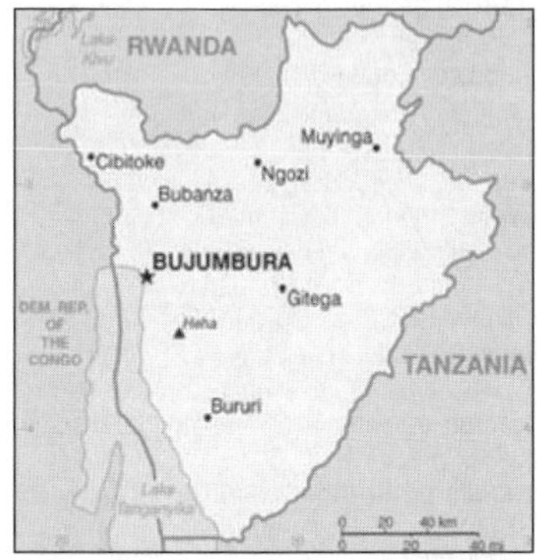

INTRODUCTION

Background: Burundi's first democratically elected president was assassinated in October 1993 after only 100 days in office, triggering widespread ethnic violence between Hutu and Tutsi factions. More than 200,000 Burundians perished during the conflict that spanned almost a dozen years. Hundreds of thousands of Burundians were internally displaced or became refugees in neighboring countries. An internationally brokered power-sharing agreement between the Tutsi-dominated government and the Hutu rebels in 2003 paved the way for a transition process that led to an integrated defense force, established a new constitution in 2005, and elected a majority Hutu government in 2005. The new government, led by President Pierre NKURUNZIZA, signed a South African brokered ceasefire with the country's last rebel group in September of 2006 but still faces many challenges.

GEOGRAPHY

Location: Central Africa, east of Democratic Republic of the Congo

Geographic coordinates: 3 30 S, 30 00 E

Map references: Africa

Area: *total:* 27,830 sq km
country comparison to the world: 146
land: 25,680 sq km
water: 2,150 sq km

Area—comparative: slightly smaller than Maryland

Land boundaries: *total:* 974 km
border countries: Democratic Republic of the Congo 233 km, Rwanda 290 km, Tanzania 451 km

Coastline: 0 km (landlocked)

Maritime claims: none (landlocked)

Climate: equatorial; high plateau with considerable altitude variation (772 m to 2,670 m above sea level); average annual temperature varies with altitude from 23 to 17 degrees centigrade but is generally moderate as the average altitude is about 1,700 m; average annual rainfall is about 150 cm; two wet seasons (February to May and September to November), and two dry seasons (June to August and December to January)

Terrain: hilly and mountainous, dropping to a plateau in east, some plains

Elevation extremes: *lowest point:* Lake Tanganyika 772 m
highest point: Heha 2,670 m

Natural resources: nickel, uranium, rare earth oxides, peat, cobalt, copper, platinum, vanadium, arable land, hydropower, niobium, tantalum, gold, tin, tungsten, kaolin, limestone

Land use: *arable land:* 35.57%
permanent crops: 13.12%
other: 51.31% (2005)

Irrigated land: 210 sq km (2003)

Total renewable water resources: 3.6 cu km (1987)

Freshwater withdrawal (domestic/industrial/agricultural): *total:* 0.29 cu km/yr (17%/6%/77%)
per capita: 38 cu m/yr (2000)

Natural hazards: flooding; landslides; drought

Environment—current issues: soil erosion as a result of overgrazing and the expansion of agriculture into marginal lands; deforestation (little forested land remains because of uncontrolled cutting of trees for fuel); habitat loss threatens wildlife populations
Environment—international agreements: *party to:* Biodiversity, Climate Change, Climate Change-Kyoto Protocol, Desertification, Endangered Species, Hazardous Wastes, Ozone Layer Protection, Wetlands
signed, but not ratified: Law of the Sea
Geography—note: landlocked; straddles crest of the Nile-Congo watershed; the Kagera, which drains into Lake Victoria, is the most remote headstream of the White Nile

PEOPLE

Population: 10,216,190 (July 2011 est.)
country comparison to the world: 81
note: estimates for this country explicitly take into account the effects of excess mortality due to AIDS; this can result in lower life expectancy, higher infant mortality, higher death rates, lower population growth rates, and changes in the distribution of population by age and sex than would otherwise be expected
Age structure: *0–14 years:* 46% (male 2,360,214/female 2,335,541)
15–64 years: 51.6% (male 2,598,011/female 2,669,376)
65 years and over: 2.5% (male 101,207/female 151,841) (2011 est.)
Median age: *total:* 16.9 years
male: 16.6 years
female: 17.2 years (2011 est.)
Population growth rate: 3.462% (2011 est.)
country comparison to the world: 5
Birth rate: 41.01 births/1,000 population (2011 est.)
country comparison to the world: 9
Death rate: 9.61 deaths/1,000 population (July 2011 est.)
country comparison to the world: 60
Net migration rate: 3.22 migrant(s)/1,000 population (2011 est.)
country comparison to the world: 28
Urbanization: *urban population:* 11% of total population (2010)
rate of urbanization: 4.9% annual rate of change (2010-15 est.)
Major cities—population: BUJUMBURA (capital) 455,000 (2009)
Sex ratio: *at birth:* 1.03 male(s)/female
under 15 years: 1.01 male(s)/female
15–64 years: 0.97 male(s)/female
65 years and over: 0.67 male(s)/female
total population: 0.98 male(s)/female (2011 est.)
Infant mortality rate: *total:* 61.82 deaths/1,000 live births
country comparison to the world: 29
male: 66.4 deaths/1,000 live births
female: 57.1 deaths/1,000 live births (2011 est.)
Life expectancy at birth: *total population:* 58.78 years
country comparison to the world: 190
male: 57.09 years
female: 60.52 years (2011 est.)
Total fertility rate: 6.16 children born/woman (2011 est.)
country comparison to the world: 5
HIV/AIDS—adult prevalence rate: 3.3% (2009 est.)
country comparison to the world: 21
HIV/AIDS—people living with HIV/AIDS: 180,000 (2009 est.)
country comparison to the world: 29
HIV/AIDS—deaths: 15,000 (2009 est.)
country comparison to the world: 19
Major infectious diseases: *degree of risk:* very high
food or waterborne diseases: bacterial and protozoal diarrhea, hepatitis A, and typhoid fever
vectorborne disease: malaria
water contact disease: schistosomiasis
animal contact disease: rabies (2009)
Drinking water source: *Improved:*
urban: 83% of population
rural: 71% of population
total: 72% of population
Unimproved: urban: 17% of population
rural: 29% of population
total: 28% of population (2008)
Sanitation facility access: *Improved:*
urban: 49% of population
rural: 46% of population
total: 46% of population
Unimproved: urban: 51% of population
rural: 54% of population
total: 54% of population (2008)
Nationality: *noun:* Burundian(s)
adjective: Burundian
Ethnic groups: Hutu (Bantu) 85%, Tutsi (Hamitic) 14%, Twa (Pygmy) 1%, Europeans 3,000, South Asians 2,000
Religions: Christian 67% (Roman Catholic 62%, Protestant 5%), indigenous beliefs 23%, Muslim 10%
Languages: Kirundi (official), French (official), Swahili (along Lake Tanganyika and in the Bujumbura area)
Literacy: *definition:* age 15 and over can read and write
total population: 59.3%
male: 67.3%
female: 52.2% (2000 est.)
School life expectancy (primary to tertiary education): *total:* 10 years
male: 9 years
female: 7 years (2009)
Education expenditures: 8.3% of GDP (2009)
country comparison to the world: 10

GOVERNMENT

Country name: *conventional long form:* Republic of Burundi
conventional short form: Burundi
local long form: Republique du Burundi/Republika y'u Burundi
local short form: Burundi
former: Urundi
Government type: republic
Capital: *name:* Bujumbura
geographic coordinates: 3 22 S, 29 21 E
time difference: UTC+2 (7 hours ahead of Washington, DC during Standard Time)
Administrative divisions: 17 provinces; Bubanza, Bujumbura Mairie, Bujumbura Rural, Bururi, Cankuzo, Cibitoke, Gitega, Karuzi, Kayanza, Kirundo, Makamba, Muramvya, Muyinga, Mwaro, Ngozi, Rutana, Ruyigi
Independence: 1 July 1962 (from UN trusteeship under Belgian administration)
National holiday: Independence Day, 1 July (1962)
Constitution: ratified by popular referendum 28 February 2005
Legal system: mixed legal system of Belgian civil law and customary law
International law organization participation: has not submitted an ICJ jurisdiction declaration; accepts ICCt jurisdiction
Suffrage: 18 years of age; universal
Executive branch: *chief of state:* President Pierre NKURUNZIZA–Hutu (since 26 August 2005); First Vice President Therence SINUNGURUZA–Tutsi (since 29 August 2010); Second Vice President Gervais RUFYIKIRI–Hutu (since 29 August 2010); note–the president is both the chief of state and head of government
head of government: President Pierre NKURUNZIZA–Hutu (since 26 August 2005); First Vice President Therence SINUNGURUZA–Tutsi (since 29 August 2010); Second Vice President Gervais RUFYIKIRI–Hutu (since 29 August 2010)
cabinet: Council of Ministers appointed by president
(For more information visit the World Leaders website)
elections: the president elected by popular vote for a five-year term (eligible for a second term); elections last held on 28 June 2010 (next to be held in 2015); vice presidents nominated by the president, endorsed by parliament
election results: Pierre NKURUNZIZA elected president by popular vote; Pierre NKURUNZIZA 91.6%, other 8.4%; note–opposition parties withdrew from the election due to alleged government interference in the electoral process
Legislative branch: bicameral Parliament or Parlement, consists of a Senate (54 seats; 34 members elected by indirect vote to serve five-year terms, with remaining seats assigned to ethnic groups and former chiefs of state) and a National Assembly or Assemblee Nationale (minimum 100 seats, 60% Hutu and 40% Tutsi with at least 30% being women; additional seats appointed by a National Independent Electoral Commission to ensure ethnic representation; members are elected by popular vote to serve five-year terms)
elections: last held on 23 July 2010 (next to be held in 2015)
election results: Senate–percent of vote by party–NA%; seats by party–TBD; National Assembly–percent of vote by party–CNDD-FDD 81.2%, UPRONA 11.6%, FRODEBU 5.9%, others 1.3%; seats by party–CNDD-FDD 81, UPRONA 17, FRODEBU 5, other 3
Judicial branch: Supreme Court or Cour Supreme; Constitutional Court; High Court of Justice (composed of the Supreme Court and the Constitutional Court)
Political parties and leaders: governing parties: Burundi Democratic Front or FRODEBU [Leonce NGENDAKUMANA]; National Council for the Defense of Democracy–Front for the Defense of Democracy or CNDD-FDD [Jeremie NGENDAKUMANA]; Unity for National Progress or UPRONA [Bonaventure NIYOYANKANA]

note: a multiparty system was introduced after 1998, included are: National Council for the Defense of Democracy or CNDD [Leonard NYANGOMA]; National Resistance Movement for the Rehabilitation of the Citizen or MRC-Rurenzangemero [Epitace BANYAGANAKANDI]; Party for National Redress or PARENA [Jean-Baptiste BAGAZA]
Political pressure groups and leaders: Forum for the Strengthening of Civil Society or FORSC [Pacifique NININAHAZWE] (civil society umbrella organization); Observatoire de lutte contre la corruption et les malversations economiques or OLUCOME [Gabriel RUFYIRI] (anti-corruption pressure group)
other: Hutu and Tutsi militias (loosely organized)
International organization participation: ACP, AfDB, AU, CEPGL, COMESA, EAC, FAO, G-77, IAEA, IBRD, ICAO, ICRM, IDA, IFAD, IFC, IFRCS, ILO, IMF, Interpol, IOC, IOM, IPU, ISO (subscriber), ITU, ITUC, MIGA, NAM, OIF, OPCW, UN, UNAMID, UNCTAD, UNESCO, UNIDO, UNWTO, UPU, WCO, WHO, WIPO, WMO, WTO
Diplomatic representation in the US: *chief of mission:* Ambassador Angele NIYUHIRE
chancery: Suite 212, 2233 Wisconsin Avenue NW, Washington, DC 20007
telephone: [1] (202) 342-2574
FAX: [1] (202) 342-2578
Diplomatic representation from the US: *chief of mission:* Ambassador Pamela J. H. SLUTZ
embassy: Avenue des Etats-Unis, Bujumbura
mailing address: B. P. 1720, Bujumbura
telephone: [257] 223454
FAX: [257] 222926
Flag description: divided by a white diagonal cross into red panels (top and bottom) and green panels (hoist side and fly side) with a white disk superimposed at the center bearing three red six-pointed stars outlined in green arranged in a triangular design (one star above, two stars below); green symbolizes hope and optimism, white purity and peace, and red the blood shed in the struggle for independence; the three stars in the disk represent the three major ethnic groups: Hutu, Twa, Tutsi, as well as the three elements in the national motto: unity, work, progress
National anthem: *name:* "Burundi Bwacu" (Our Beloved Burundi)
lyrics/music: Jean-Baptiste NTAHOKAJA/ Marc BARENGAYABO
note: adopted 1962

ECONOMY

Economy—overview: Burundi is a landlocked, resource-poor country with an underdeveloped manufacturing sector. The economy is predominantly agricultural which accounts for just over 30% of GDP and employs more than 90% of the population. Burundi's primary exports are coffee and tea, which account for 90% of foreign exchange earnings, though exports are a relatively small share of GDP. Burundi's export earnings–and its ability to pay for imports–rests primarily on weather conditions and international coffee and tea prices. The Tutsi minority, 14% of the population, dominates the coffee trade. An ethnic-based war that lasted for over a decade resulted in more than 200,000 deaths, forced more than 48,000 refugees into Tanzania, and displaced 140,000 others internally. Only one in two children go to school, and approximately one in 15 adults has HIV/AIDS. Food, medicine, and electricity remain in short supply. Less than 2% of the population has electricity in its homes. Burundi's GDP grew around 4% annually in 2006-10. Political stability and the end of the civil war have improved aid flows and economic activity has increased, but underlying weaknesses–a high poverty rate, poor education rates, a weak legal system, a poor transportation network, overburdened utilities, and low administrative capacity–risk undermining planned economic reforms. The purchasing power of most Burundians has decreased as wage increases have not kept up with inflation. Burundi will continue to remain heavily dependent on aid from bilateral and multilateral donors; the delay of funds after a corruption scandal cut off bilateral aid in 2007 reduced government's revenues and its ability to pay salaries. Burundi joined the East African Community, which should boost Burundi's regional trade ties, and received $700 million in debt relief in 2009. Government corruption is also hindering the development of a healthy private sector as companies seek to navigate an environment with ever-changing rules.
GDP (purchasing power parity): $3.397 billion (2010 est.)
country comparison to the world: 172
$3.272 billion (2009 est.)
$3.161 billion (2008 est.)
note: data are in 2010 US dollars
GDP (official exchange rate): $1.489 billion (2010 est.)
GDP—real growth rate: 3.9% (2010 est.)
country comparison to the world: 102
3.5% (2009 est.)
4.5% (2008 est.)
GDP—per capita (PPP): $300 (2010 est.)
country comparison to the world: 228
$300 (2009 est.)
$300 (2008 est.)
note: data are in 2010 US dollars
GDP—composition by sector: *agriculture:* 31.6%
industry: 21.4%
services: 47% (2010 est.)
Labor force: 4.245 million (2007)
country comparison to the world: 87
Labor force—by occupation: *agriculture:* 93.6%
industry: 2.3%
services: 4.1% (2002 est.)
Unemployment rate: NA%
Population below poverty line: 68% (2002 est.)
Household income or consumption by percentage share: *lowest 10%:* 4.1%
highest 10%: 28% (2006)
Distribution of family income—Gini index: 42.4 (1998)
country comparison to the world: 50
Investment (gross fixed): 25.1% of GDP (2010 est.)
country comparison to the world: 42
Budget: *revenues:* $386.3 million
expenditures: $476.2 million (2010 est.)
Inflation rate (consumer prices): 9.8% (2010 est.)
country comparison to the world: 196
28.4% (2009)
Central bank discount rate: 10% (31 December 2009)
country comparison to the world: 41
10.08% (31 December 2008)
Commercial bank prime lending rate: 14.08% (31 December 2009 est.)
country comparison to the world: 33
16.52% (31 December 2008 est.)
Stock of narrow money: $329.3 million (31 December 2010 est.)
country comparison to the world: 166
$293.6 million (31 December 2009 est.)
Stock of broad money: $568.3 million (31 December 2010 est.)
country comparison to the world: 171
$506.7 million (31 December 2009 est.)
Stock of domestic credit: $465.7 million (31 December 2010 est.)
country comparison to the world: 165
$415.2 million (31 December 2009 est.)
Market value of publicly traded shares: $NA
Agriculture—products: coffee, cotton, tea, corn, sorghum, sweet potatoes, bananas, manioc (tapioca); beef, milk, hides
Industries: light consumer goods such as blankets, shoes, soap; assembly of imported components; public works construction; food processing
Industrial production growth rate: 7% (2010 est.)
country comparison to the world: 46
Electricity—production: 92 million kWh (2007 est.)
country comparison to the world: 192
Electricity—consumption: 125.6 million kWh (2007 est.)
country comparison to the world: 185
Electricity—exports: 0 kWh (2008 est.)
Electricity—imports: 40 million kWh; note–supplied by the Democratic Republic of the Congo (2007 est.)
Oil—production: 0 bbl/day (2009 est.)
country comparison to the world: 156
Oil—consumption: 3,000 bbl/day (2009 est.)
country comparison to the world: 177
Oil—exports: 0 bbl/day (2007 est.)
country comparison to the world: 152
Oil—imports: 2,495 bbl/day (2007 est.)
country comparison to the world: 172
Oil—proved reserves: 0 bbl (1 January 2010 est.)
country comparison to the world: 112
Natural gas—production: 0 cu m (2008 est.)
country comparison to the world: 106
Natural gas—consumption: 0 cu m (2008 est.)
country comparison to the world: 153
Natural gas—exports: 0 cu m (2008 est.)
country comparison to the world: 68
Natural gas—imports: 0 cu m (2008 est.)
country comparison to the world: 160
Natural gas—proved reserves: 0 cu m (1 January 2010 est.)
country comparison to the world: 116

Current account balance: $-136 million (2010 est.)
country comparison to the world: 81
$-127 million (2009 est.)
Exports: $71 million (2010 est.)
country comparison to the world: 199
$68 million (2009 est.)
Exports—commodities: coffee, tea, sugar, cotton, hides
Exports—partners: Germany 21.6%, Switzerland 14.86%, Belgium 9.32%, Sweden 8.94%, Pakistan 5.82% (2009)
Imports: $336 million (2010 est.)
country comparison to the world: 192
$275 million (2009 est.)
Imports—commodities: capital goods, petroleum products, foodstuffs
Imports—partners: Saudi Arabia 16.87%, Belgium 11.17%, Uganda 8.62%, Kenya 7.57%, China 5.66%, France 5.35%, Germany 4.46%, India 4.24%, Tanzania 4.21% (2009)
Reserves of foreign exchange and gold: $320 million (31 December 2010 est.)
country comparison to the world: 127
$323 million (31 December 2009 est.)
Debt—external: $1.2 billion (2003)
country comparison to the world: 147
Exchange rates: Burundi francs (BIF) per US dollar–1,250.75 (2010), 1,230.18 (2009), 1,198 (2008), 1,065 (2007), 1,030 (2006)

COMMUNICATIONS

Telephones—main lines in use: 31,500 (2009)
country comparison to the world: 178
Telephones—mobile cellular: 838,400 (2009)
country comparison to the world: 149
Telephone system: *general assessment:* sparse system of open-wire, radiotelephone communications, and low-capacity microwave radio relays
domestic: telephone density one of the lowest in the world; fixed-line connections stand at well less than 1 per 100 persons; mobile-cellular usage is increasing but remains at a meager 10 per 100 persons
international: country code—257; satellite earth station—1 Intelsat (Indian Ocean) (2009)
Broadcast media: state-controlled La Radiodiffusion et Television Nationale de Burundi (RTNB) operates the lone TV broadcast station and the only national radio network; about 10 privately-owned radio broadcast stations; transmissions of several international broadcasters are available in Bujumbura (2007)
Internet country code: .bi
Internet hosts: 201 (2010)
country comparison to the world: 194
Internet users: 157,800 (2009)
country comparison to the world: 147

TRANSPORTATION

Airports: 8 (2010)
country comparison to the world: 161
Airports—with paved runways: *total:* 1
over 3,047 m: 1 (2010)
Airports—with unpaved runways: *total:* 7
914 to 1,523 m: 4
under 914 m: 3 (2010)
Heliports: 1 (2010)
Roadways: *total:* 12,322 km
country comparison to the world: 129
paved: 1,286 km
unpaved: 11,036 km (2004)
Waterways: (mainly on Lake Tanganyika between Bujumbura, Burundi's principal port, and lake ports in Tanzania, Zambia, and the Democratic Republic of Congo) (2010)
Ports and terminals: Bujumbura

MILITARY

Military branches: National Defense Forces (Forces de Defense Nationale, FDN): Army (includes naval detachment, Air Wing, and Coast Guard), National Gendarmerie (2011)
Military service age and obligation: military service is voluntary; the armed forces law of 31 December 2004 did not specify a minimum age for enlistment, but the government had previously said each recruit must have a primary school-leaving certificate; mandatory retirement age 45 (enlisted), 50 (NCOs), and 55 (officers) (2010)
Manpower available for military service: *males age 16–49:* 2,182,327
females age 16–49: 2,202,125 (2010 est.)
Manpower fit for military service: *males age 16–49:* 1,398,769
females age 16–49: 1,481,417 (2010 est.)
Manpower reaching militarily significant age annually: *male:* 117,956
female: 116,956 (2010 est.)
Military expenditures: 5.9% of GDP (2006 est.)
country comparison to the world: 11

TRANSNATIONAL ISSUES

Disputes—international: Burundi and Rwanda dispute two sq km (0.8 sq mi) of Sabanerwa, a farmed area in the Rukurazi Valley where the Akanyaru/Kanyaru River shifted its course southward after heavy rains in 1965; cross-border conflicts among Tutsi, Hutu, other ethnic groups, associated political rebels, armed gangs, and various government forces persist in the Great Lakes region
Refugees and internally displaced persons: refugees (country of origin): 9,849 (Democratic Republic of the Congo)
IDPs: 100,000 (armed conflict between government and rebels; most IDPs in northern and western Burundi) (2007)

CAMBODIA

INTRODUCTION

Background: Most Cambodians consider themselves to be Khmers, descendants of the Angkor Empire that extended over much of Southeast Asia and reached its zenith between the 10th and 13th centuries. Attacks by the Thai and Cham (from present-day Vietnam) weakened the empire, ushering in a long period of decline. The king placed the country under French protection in 1863 and it became part of French Indochina in 1887. Following Japanese occupation in World War II, Cambodia gained full independence from France in 1953. In April 1975, after a five-year struggle, Communist Khmer Rouge forces captured Phnom Penh and evacuated all cities and towns. At least 1.5 million Cambodians died from execution, forced hardships, or starvation during the Khmer Rouge regime under POL POT. A December 1978 Vietnamese invasion drove the Khmer Rouge into the countryside, began a 10-year Vietnamese occupation, and touched off almost 13 years of civil war. The 1991 Paris Peace Accords mandated democratic elections and a ceasefire, which was not fully respected by the Khmer Rouge. UN-sponsored elections in 1993 helped restore some semblance of normalcy under a coalition government. Factional fighting in 1997 ended the first coalition government, but a second round of national elections in 1998 led to the formation of another coalition government and renewed political stability. The remaining elements of the Khmer Rouge surrendered in early 1999. Some of the surviving Khmer Rouge leaders have been tried or are awaiting trial for crimes against humanity by a hybrid UN-Cambodian tribunal supported by international assistance. Elections in July 2003 were relatively peaceful, but it took one year of negotiations between contending political parties before a coalition government was formed. In October 2004, King Norodom SIHANOUK abdicated the throne and his son, Prince Norodom SIHAMONI, was selected to succeed him. Local elections were held in Cambodia in April 2007, with little of the pre-election violence that preceded prior elections. National elections in July 2008 were relatively peaceful.

GEOGRAPHY

Location: Southeastern Asia, bordering the Gulf of Thailand, between Thailand, Vietnam, and Laos

Geographic coordinates: 13 00 N, 105 00 E

Map references: Southeast Asia

Area: *total:* 181,035 sq km
country comparison to the world: 89
land: 176,515 sq km
water: 4,520 sq km

Area—comparative: slightly smaller than Oklahoma

Land boundaries: *total:* 2,572 km
border countries: Laos 541 km, Thailand 803 km, Vietnam 1,228 km

Coastline: 443 km

Maritime claims: territorial sea: 12 nm
contiguous zone: 24 nm
exclusive economic zone: 200 nm
continental shelf: 200 nm

Climate: tropical; rainy, monsoon season (May to November); dry season (December to April); little seasonal temperature variation

Terrain: mostly low, flat plains; mountains in southwest and north

Elevation extremes: *lowest point:* Gulf of Thailand 0 m
highest point: Phnum Aoral 1,810 m

Natural resources: oil and gas, timber, gemstones, iron ore, manganese, phosphates, hydropower potential

Land use: *arable land:* 20.44%
permanent crops: 0.59%
other: 78.97% (2005)

Irrigated land: 2,700 sq km (2003)

Total renewable water resources: 476.1 cu km (1999)

Freshwater withdrawal (domestic/industrial/agricultural): *total:* 4.08 cu km/yr (1%/0%/98%)
per capita: 290 cu m/yr (2000)

Natural hazards: monsoonal rains (June to November); flooding; occasional droughts

Environment—current issues: illegal logging activities throughout the country and strip mining for gems in the western region along the border with Thailand have resulted in habitat loss and declining biodiversity (in particular, destruction of mangrove swamps threatens natural fisheries); soil erosion; in rural areas, most of the population does not have access to potable water; declining fish stocks because of illegal fishing and overfishing

Environment—international agreements: *party to:* Biodiversity, Climate Change, Climate Change-Kyoto Protocol, Desertification, Endangered Species, Hazardous Wastes, Marine Life Conservation, Ozone Layer Protection, Ship Pollution, Tropical Timber 94, Wetlands, Whaling
signed, but not ratified: Law of the Sea

Geography—note: a land of paddies and forests dominated by the Mekong River and Tonle Sap

PEOPLE

Population: 14,701,717 (July 2011 est.)
country comparison to the world: 66
note: estimates for this country take into account the effects of excess mortality due to AIDS; this can result in lower life expectancy, higher infant mortality, higher death rates, lower population growth rates, and changes in the distribution of population by age and sex than would otherwise be expected

Age structure: *0–14 years:* 32.2% (male 2,375,155/female 2,356,305)
15–64 years: 64.1% (male 4,523,030/female 4,893,761)
65 years and over: 3.8% (male 208,473/female 344,993) (2011 est.)

Median age: *total:* 22.9 years
male: 22.2 years
female: 23.7 years (2011 est.)

Population growth rate: 1.698% (2011 est.)
country comparison to the world: 66

Birth rate: 25.4 births/1,000 population (2011 est.)
country comparison to the world: 57

Death rate: 8.07 deaths/1,000 population (July 2011 est.)
country comparison to the world: 102

Net migration rate: -0.34 migrant(s)/1,000 population (2011 est.)
country comparison to the world: 131

Urbanization: *urban population:* 20% of total population (2010)
rate of urbanization: 3.2% annual rate of change (2010-15 est.)

Major cities—population: PHNOM PENH (capital) 1.519 million (2009)

Sex ratio: *at birth:* 1.045 male(s)/female
under 15 years: 1.02 male(s)/female
15–64 years: 0.95 male(s)/female
65 years and over: 0.6 male(s)/female
total population: 0.96 male(s)/female (2011 est.)

Infant mortality rate: *total:* 55.49 deaths/1,000 live births
country comparison to the world: 37
male: 62.54 deaths/1,000 live births
female: 48.13 deaths/1,000 live births (2011 est.)

Life expectancy at birth: *total population:* 62.67 years
country comparison to the world: 178
male: 60.31 years
female: 65.13 years (2011 est.)

Total fertility rate: 2.84 children born/woman (2011 est.)
country comparison to the world: 71

HIV/AIDS—adult prevalence rate: 0.5% (2009 est.)
country comparison to the world: 65

HIV/AIDS—people living with HIV/AIDS: 63,000 (2009 est.)
country comparison to the world: 53

HIV/AIDS—deaths: 3,100 (2009 est.)
country comparison to the world: 45

Major infectious diseases: *degree of risk:* very high
food or waterborne diseases: bacterial and protozoal diarrhea, hepatitis A, and typhoid fever
vectorborne diseases: dengue fever, Japanese encephalitis, and malaria
note: highly pathogenic H5N1 avian influenza has been identified in this country; it poses a negligible risk with extremely rare cases possible among US citizens who have close contact with birds (2009)

Drinking water source: *Improved:*
urban: 81% of population
rural: 56% of population
total: 61% of population
Unimproved:
urban: 19% of population
rural: 44% of population
total: 39% of population (2008)
Sanitation facility access: *Improved:*
urban: 67% of population
rural: 18% of population
total: 29% of population
Unimproved:
urban: 33% of population
rural: 82% of population
total: 71% of population (2008)
Nationality: *noun:* Cambodian(s)
adjective: Cambodian
Ethnic groups: Khmer 90%, Vietnamese 5%, Chinese 1%, other 4%
Religions: Buddhist (official) 96.4%, Muslim 2.1%, other 1.3%, unspecified 0.2% (1998 census)
Languages: Khmer (official) 95%, French, English
Literacy: *definition:* age 15 and over can read and write
total population: 73.6%
male: 84.7%
female: 64.1% (2004 est.)
School life expectancy (primary to tertiary education): *total:* 10 years
male: 10 years
female: 9 years (2007)
Education expenditures: 2.1% of GDP (2009)
country comparison to the world: 154

GOVERNMENT

Country name: *conventional long form:* Kingdom of Cambodia
conventional short form: Cambodia
local long form: Preahreacheanachakr Kampuchea (phonetic pronunciation)
local short form: Kampuchea
former: Khmer Republic, Democratic Kampuchea, People's Republic of Kampuchea, State of Cambodia
Government type: multiparty democracy under a constitutional monarchy
Capital: *name:* Phnom Penh
geographic coordinates: 11 33 N, 104 55 E
time difference: UTC+7 (12 hours ahead of Washington, DC during Standard Time)
Administrative divisions: 23 provinces (khett, singular and plural) and 1 municipality (krong, singular and plural)
provinces: Banteay Mean Choay, Batdambang, Kampong Cham, Kampong Chhnang, Kampong Spoe, Kampong Thum, Kampot, Kandal, Kaoh Kong, Keb, Krachen, Mondol Kiri, Otdar Mean Choay, Pailin, Pouthisat, Preah Seihanu (Sihanoukville), Preah Vihear, Prey Veng, Rotanokiri, Siem Reab, Stoeng Treng, Svay Rieng, Takev
municipalities: Phnum Penh (Phnom Penh)
Independence: 9 November 1953 (from France)
National holiday: Independence Day, 9 November (1953)
Constitution: promulgated 21 September 1993
Legal system: civil law system (influenced by the UN Transitional Authority in Cambodia) customary law, Communist legal theory, and common law
International law organization participation: accepts compulsory ICJ jurisdiction with reservations; accepts ICCt jurisdiction
Suffrage: 18 years of age; universal
Executive branch: *chief of state:* King Norodom SIHAMONI (since 29 October 2004)
head of government: Prime Minister HUN SEN (since 14 January 1985) [co-prime minister from 1993 to 1997]; Permanent Deputy Prime Minister MEN SAM AN (since 25 September 2008); Deputy Prime Ministers SAR KHENG (since 3 February 1992); SOK AN, TEA BANH, HOR NAMHONG, NHEK BUNCHHAY (since 16 July 2004); BIN CHHIN (since 5 September 2007); KEAT CHHON, YIM CHHAI LY (since 24 September 2008); KE KIMYAN (since 12 March 2009)
cabinet: Council of Ministers named by the prime minister and appointed by the monarch
(For more information visit the World Leaders website)
elections: the king chosen by a Royal Throne Council from among all eligible males of royal descent; following legislative elections, a member of the majority party or majority coalition named prime minister by the Chairman of the National Assembly and appointed by the king
Legislative branch: bicameral, consists of the Senate (61 seats; 2 members appointed by the monarch, 2 elected by the National Assembly, and 57 elected by parliamentarians and commune councils; members serve five-year terms) and the National Assembly (123 seats; members elected by popular vote to serve five-year terms)
elections: Senate—last held on 22 January 2006 (next to be held in January 2012); National Assembly—last held on 27 July 2008 (next to be held in July 2013)
election results: Senate—percent of vote by party—CPP 69%, FUNCINPEC 21%, SRP 10%; seats by party—CPP 45, FUNCINPEC 10, SRP 2; National Assembly—percent of vote by party—CPP 58%, SRP 22%, HRP 7%; NRP 6%; FUNCINPEC 5%; others 2%; seats by party—CPP 90, SRP 26, HRP 3, FUNCINPEC 2, NRP 2
Judicial branch: Supreme Council of the Magistracy (provided for in the constitution and formed in December 1997); Supreme Court (and lower courts) exercises judicial authority
Political parties and leaders: Cambodian People's Party or CPP [CHEA SIM]; Human Rights Party or HRP [KHEM SOKHA, also spelled KEM SOKHA]; National United Front for an Independent, Neutral, Peaceful, and Cooperative Cambodia or FUNCINPEC [KEV PUT REAKSMEI]; Nationalist Party or NP [CHHIM SEAK LENG] (formerly the NRP); Sam Rangsi Party or SRP [SAM RANGSI, also spelled SAM RAINSY]
Political pressure groups and leaders: Cambodian Freedom Fighters or CFF; Partnership for Transparency Fund or PTF (anti-corruption organization); Students Movement for Democracy; The Committee for Free and Fair Elections or Comfrel
other: human rights organizations; vendors
International organization participation: ADB, ARF, ASEAN, CICA (observer), EAS, FAO, G-77, IBRD, ICAO, ICRM, IDA, IFAD, IFC, IFRCS, ILO, IMF, IMO, Interpol, IOC, IOM, IPU, ISO (subscriber), ITU, MIGA, NAM, OIF, OPCW, PCA, UN, UNCTAD, UNESCO, UNIDO, UNIFIL, UNMIS, UNWTO, UPU, WCO, WFTU, WHO, WIPO, WMO, WTO
Diplomatic representation in the US: *chief of mission:* Ambassador HENG HEM
chancery: 4530 16th Street NW, Washington, DC 20011
telephone: [1] (202) 726-7742
FAX: [1] (202) 726-8381
Diplomatic representation from the US: *chief of mission:* Ambassador Carol A. RODLEY
embassy: #1, Street 96, Sangkat Wat Phnom, Khan Daun Penh, Phnom Penh
mailing address: Box P, APO AP 96546
telephone: [855] (23) 728-000
FAX: [855] (23) 728-600
Flag description: three horizontal bands of blue (top), red (double width), and blue with a white three-towered temple representing Angkor Wat outlined in black in the center of the red band; red and blue are traditional Cambodian colors
note: only national flag to incorporate an actual building in its design
National anthem: *name:* "Nokoreach" (Royal Kingdom)
lyrics/music: CHUON NAT/F. PERRUCHOT and J. JEKYLL
note: adopted 1941, restored 1993; the anthem, based on a Cambodian folk tune, was restored after the defeat of the Communist regime

ECONOMY

Economy—overview: From 2004 to 2007, the economy grew about 10% per year, driven largely by an expansion in the garment sector, construction, agriculture, and tourism. GDP contracted slightly in 2009 as a result of the global economic slowdown, but climbed more than 4% in 1010, driven by renewed exports. With the January 2005 expiration of a WTO Agreement on Textiles and Clothing, Cambodian textile producers were forced to compete directly with lower-priced countries such as China, India, Vietnam, and Bangladesh. The garment industry currently employs more than 280,000 people—about 5% of the work force—and contributes more than 70% of Cambodia's exports. In 2005, exploitable oil deposits were found beneath Cambodia's territorial waters, representing a new revenue stream for the government if commercial extraction begins. Mining also is attracting significant investor interest, particularly in the northern parts of the country. The government has said opportunities exist for mining bauxite, gold, iron and gems. In 2006, a US-Cambodia bilateral Trade and Investment Framework Agreement (TIFA) was signed, and several rounds of discussions have been held since 2007. Rubber exports increased about 25% in 2009 due to rising global demand. The tourism industry has continued to grow rapidly, with foreign arrivals exceeding 2 million per year in 2007-08; however, economic troubles abroad dampened growth in 2009. The global financial crisis is weakening demand for Cambodian exports, and construction is declining due to a shortage of credit. The long-term development of the economy remains a daunting challenge. The Cambodian government is working with bilateral and multilateral donors, including the World Bank and IMF, to address the country's many pressing needs. The major economic challenge for Cambodia over the next decade will be fashioning an economic environment in which the private sector can create enough jobs to handle Cambodia's demographic imbalance. More than 50% of the population is less than 25 years old. The population lacks education and productive skills, particularly in the poverty-ridden countryside, which suffers from an almost total lack of basic infrastructure.

GDP (purchasing power parity): $30.18 billion (2010 est.)
country comparison to the world: 108
$28.47 billion (2009 est.)
$29.04 billion (2008 est.)
note: data are in 2010 US dollars

GDP (official exchange rate): $11.63 billion (2010 est.)

GDP—real growth rate: 6% (2010 est.)
country comparison to the world: 54
-2% (2009 est.)
6.7% (2008 est.)

GDP—per capita (PPP): $2,100 (2010 est.)
country comparison to the world: 189
$2,000 (2009 est.)
$2,100 (2008 est.)
note: data are in 2010 US dollars

GDP—composition by sector:
agriculture: 33.4%
industry: 21.4%
services: 45.2% (2009 est.)

Labor force: 8.8 million (2010 est.)
country comparison to the world: 53

Labor force—by occupation:
agriculture: 57.6%
industry: 15.9%
services: 26.5% (2009 est.)

Unemployment rate: 3.5% (2007 est.)
country comparison to the world: 29
2.5% (2000 est.)

Population below poverty line: 31% (2007 est.)

Household income or consumption by percentage share:
lowest 10%: 3%
highest 10%: 34.2% (2007)

Distribution of family income—Gini index: 43 (2007 est.)
country comparison to the world: 48
40 (2004 est.)

Investment (gross fixed): 23% of GDP (2010 est.)
country comparison to the world: 58

Budget: *revenues:* $1.413 billion
expenditures: $2.079 billion (2010 est.)

Inflation rate (consumer prices): 4.1% (2010 est.)
country comparison to the world: 117
-0.7% (2009)

Central bank discount rate: NA% (31 December 2008)
country comparison to the world: 79
5.25% (31 December 2007)

Commercial bank prime lending rate: 17% (31 December 2009)
country comparison to the world: 36
16.01% (31 December 2008)

Stock of narrow money: $850.7 million (31 December 2010 est.)
country comparison to the world: 144
$747.2 million (31 December 2009 est.)

Stock of broad money: $4.982 billion (31 December 2010 est.)
country comparison to the world: 119
$3.899 billion (31 December 2009 est.)

Stock of domestic credit: $2.195 billion (31 December 2010 est.)
country comparison to the world: 124
$1.991 billion (31 December 2009 est.)

Market value of publicly traded shares: $NA

Agriculture—products: rice, rubber, corn, vegetables, cashews, tapioca, silk

Industries: tourism, garments, construction, rice milling, fishing, wood and wood products, rubber, cement, gem mining, textiles

Industrial production growth rate: 5.7% (2010 est.)
country comparison to the world: 63

Electricity—production: 1.273 billion kWh (2007 est.)
country comparison to the world: 142

Electricity—consumption: 1.272 billion kWh (2007 est.)
country comparison to the world: 143

Electricity—exports: 0 kWh (2008 est.)

Electricity—imports: 167 million kWh (2007 est.)

Oil—production: 0 bbl/day (2009 est.)
country comparison to the world: 157

Oil—consumption: 4,000 bbl/day (2009 est.)
country comparison to the world: 173

Oil—exports: 0 bbl/day (2007 est.)
country comparison to the world: 153

Oil—imports: 30,970 bbl/day (2007 est.)
country comparison to the world: 97

Oil—proved reserves: 0 bbl (1 January 2010 est.)
country comparison to the world: 113

Natural gas—production: 0 cu m (2008 est.)
country comparison to the world: 107

Natural gas—consumption: 0 cu m (2008 est.)
country comparison to the world: 154

Natural gas—exports: 0 cu m (2008 est.)
country comparison to the world: 69

Natural gas—imports: 0 cu m (2008 est.)
country comparison to the world: 161

Natural gas—proved reserves: 0 cu m (1 January 2010 est.)
country comparison to the world: 117

Current account balance: $-918 million (2010 est.)
country comparison to the world: 130
$-865.7 million (2009 est.)

Exports: $4.687 billion (2010 est.)
country comparison to the world: 109
$4.186 billion (2009 est.)

Exports—commodities: clothing, timber, rubber, rice, fish, tobacco, footwear

Exports—partners: US 45.32%, Singapore 9.46%, Germany 7.52%, UK 7.07%, Canada 6.31%, Vietnam 4.15% (2009)

Imports: $6.005 billion (2010 est.)
country comparison to the world: 106
$5.876 billion (2009 est.)

Imports—commodities: petroleum products, cigarettes, gold, construction materials, machinery, motor vehicles, pharmaceutical products

Imports—partners: Thailand 24.83%, Vietnam 19.73%, China 14.08%, Singapore 11.34%, Hong Kong 7.41%, Taiwan 5.1%, South Korea 4.06% (2009)

Reserves of foreign exchange and gold: $3.84 billion (31 December 2010 est.)
country comparison to the world: 81
$3.289 billion (31 December 2009 est.)

Debt—external: $4.338 billion (31 December 2010 est.)
country comparison to the world: 111
$4.284 billion (31 December 2009 est.)

Exchange rates: riels (KHR) per US dollar–4,145 (2010), 4,139 (2009), 4,070.94 (2008), 4,006 (2007), 4,103 (2006)

COMMUNICATIONS

Telephones—main lines in use: 54,200 (2009)
country comparison to the world: 160

Telephones—mobile cellular: 5.593 million (2009)
country comparison to the world: 91

Telephone system: *general assessment:* adequate fixed-line and/or cellular service in Phnom Penh and other provincial cities; mobile-cellular phone systems are widely used in urban areas to bypass deficiencies in the fixed-line network; mobile-phone coverage is rapidly expanding in rural areas
domestic: fixed-line connections stand at well less than 1 per 100 persons; mobile-cellular usage, aided by increasing competition among service providers, is increasing and stands at 40 per 100 persons
international: country code–855; adequate but expensive landline and cellular service available to all countries from Phnom Penh and major provincial cities; satellite earth station–1 Intersputnik (Indian Ocean region) (2009)

Broadcast media: mixture of state-owned, joint public-private, and privately-owned broadcast media; 9 TV broadcast stations with most operating on multiple channels, including 1 state-operated station broadcasting from multiple locations, 6 stations either jointly operated or privately-owned with some broadcasting from several locations, and 2 TV relay stations–one relaying a French television station and the other relaying a Vietnamese television station; multi-channel cable and satellite systems are available; roughly 50 radio broadcast stations–1 state-owned broadcaster with multiple stations and a large mixture of public and private broadcasters; several international broadcasters are available (2009)

Internet country code: .kh

Internet hosts: 5,452 (2010)
country comparison to the world: 138

Internet users: 78,500 (2009)
country comparison to the world: 166

TRANSPORTATION

Airports: 17 (2010)
country comparison to the world: 140

Airports—with paved runways: *total:* 6
2,438 to 3,047 m: 3
1,524 to 2,437 m: 2
914 to 1,523 m: 1 (2010)

Airports—with unpaved runways: *total:* 11
1,524 to 2,437 m: 1
914 to 1,523 m: 9
under 914 m: 1 (2010)
Heliports: 1 (2010)
Railways: *total:* 690 km
country comparison to the world: 101
narrow gauge: 690 km 1.000-m gauge
note: under restoration (2010)
Roadways: *total:* 38,093 km
country comparison to the world: 90
paved: 2,977 km
unpaved: 35,116 km (2007)
Waterways: 2,400 km (mainly on Mekong River) (2010)
country comparison to the world: 36
Merchant marine: *total:* 620
country comparison to the world: 20
by type: bulk carrier 40, cargo 526, carrier 5, chemical tanker 5, container 5, liquefied gas 1, passenger 1, passenger/cargo 7, petroleum tanker 12, refrigerated cargo 13, roll on/roll off 5
foreign-owned: 426 (Belgium 1, Canada 2, China 203, Cyprus 8, Egypt 12, Estonia 1, French Polynesia 1, Gabon 1, Greece 2, Hong Kong 11, Indonesia 2, Japan 2, Latvia 1, Lebanon 6, Netherlands 1, Romania 1, Russia 60, Singapore 4, South Korea 11, Syria 22, Taiwan 1, Turkey 26, UAE 2, UK 3, Ukraine 37, US 4, Vietnam 1)
note: this country allows large numbers of ships owned by foreign entities to be registered in its national shipping registry and to fly its flag; these ships operate under the laws of the flag state (2010)
Ports and terminals: Phnom Penh, Kampong Saom (Sihanoukville)

MILITARY

Military branches: Royal Cambodian Armed Forces: Royal Cambodian Army, Royal Khmer Navy, Royal Cambodian Air Force (2011)
Military service age and obligation: conscription law of October 2006 requires all males between 18-30 to register for military service; 18-month service obligation (2006)
Manpower available for military service: *males age 16–49:* 3,883,724
females age 16–49: 4,003,585 (2010 est.)
Manpower fit for military service: *males age 16–49:* 2,638,167
females age 16–49: 2,965,328 (2010 est.)
Manpower reaching militarily significant age annually: *male:* 151,143
female: 154,542 (2010 est.)
Military expenditures: 3% of GDP (2005 est.)
country comparison to the world: 46

TRANSNATIONAL ISSUES

Disputes—international: Cambodia is concerned about Laos' extensive upstream dam construction; Cambodia and Thailand dispute sections of boundary; in 2011 Thailand and Cambodia resorted to arms in the dispute over the location of the boundary on the precipice surmounted by Preah Vihear temple ruins, awarded to Cambodia by ICJ decision in 1962 and part of a planned UN World Heritage site; Cambodia accuses Vietnam of a wide variety of illicit cross-border activities; Progress on a joint development area with Vietnam is hampered by an unresolved dispute over sovereignty of offshore islands
Illicit drugs: narcotics-related corruption reportedly involving some in the government, military, and police; limited methamphetamine production; vulnerable to money laundering due to its cash-based economy and porous borders

CAMEROON

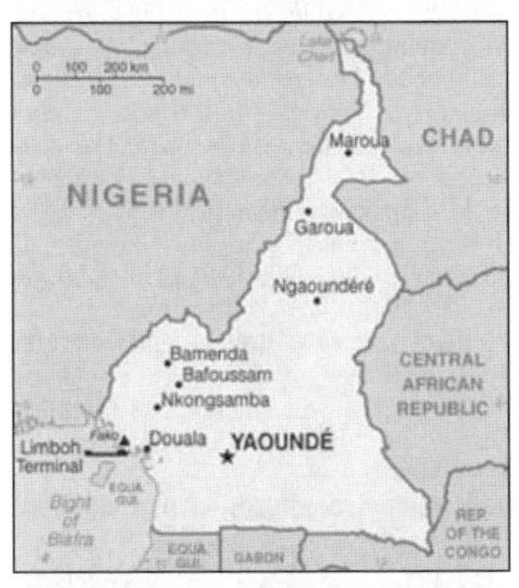

INTRODUCTION

Background: French Cameroon became independent in 1960 as the Republic of Cameroon. The following year the southern portion of neighboring British Cameroon voted to merge with the new country to form the Federal Republic of Cameroon. In 1972, a new constitution replaced the federation with a unitary state, the United Republic of Cameroon. The country has generally enjoyed stability, which has permitted the development of agriculture, roads, and railways, as well as a petroleum industry. Despite slow movement toward democratic reform, political power remains firmly in the hands of President Paul BIYA.

GEOGRAPHY

Location: Western Africa, bordering the Bight of Biafra, between Equatorial Guinea and Nigeria
Geographic coordinates: 6 00 N, 12 00 E
Map references: Africa
Area: *total:* 475,440 sq km
country comparison to the world: 53
land: 472,710 sq km
water: 2,730 sq km
Area—comparative: slightly larger than California
Land boundaries: *total:* 4,591 km
border countries: Central African Republic 797 km, Chad 1,094 km, Republic of the Congo 523 km, Equatorial Guinea 189 km, Gabon 298 km, Nigeria 1,690 km
Coastline: 402 km
Maritime claims:
territorial sea: 12 nm
contiguous zone: 24 nm
Climate: varies with terrain, from tropical along coast to semiarid and hot in north
Terrain: diverse, with coastal plain in southwest, dissected plateau in center, mountains in west, plains in north
Elevation extremes:
lowest point: Atlantic Ocean 0 m
highest point: Fako 4,095 m (on Mt. Cameroon)
Natural resources: petroleum, bauxite, iron ore, timber, hydropower
Land use:
arable land: 12.54%
permanent crops: 2.52%
other: 84.94% (2005)
Irrigated land: 260 sq km (2003)
Total renewable water resources: 285.5 cu km (2003)
Freshwater withdrawal (domestic/industrial/agricultural): *total:* 0.99 cu km/yr (18%/8%/74%)
per capita: 61 cu m/yr (2000)
Natural hazards: volcanic activity with periodic releases of poisonous gases from Lake Nyos and Lake Monoun volcanoes
volcanism: Mt. Cameroon (elev. 4,095 m), which last erupted in 2000, is the most frequently active volcano in West Africa; lakes in Oku volcanic field have released fatal levels of gas on occasion, killing some 1,700 people in 1986
Environment—current issues: waterborne diseases are prevalent; deforestation; overgrazing; desertification; poaching; overfishing
Environment—international agreements: *party to:* Biodiversity, Climate Change, Climate Change-Kyoto Protocol, Desertification, Endangered Species, Hazardous Wastes, Law of the Sea, Ozone Layer Protection, Tropical Timber 83, Tropical Timber 94, Wetlands, Whaling
signed, but not ratified: none of the selected agreements
Geography—note: sometimes referred to as the hinge of Africa; throughout the country there are areas of thermal springs and indications of current or prior volcanic activity; Mount Cameroon, the highest mountain in Sub-Saharan west Africa, is an active volcano

PEOPLE

Population: 19,711,291 (July 2011 est.)
country comparison to the world: 58
note: estimates for this country explicitly take into account the effects of excess mortality due to AIDS; this can result in lower life expectancy, higher infant mortality, higher death rates, lower population growth rates, and changes in the distribution of population by age and sex than would otherwise be expected
Age structure: *0–14 years:* 40.5% (male 4,027,381/female 3,956,219)

15–64 years: 56.2% (male 5,564,570/ female 5,505,857)
65 years and over: 3.3% (male 300,929/ female 356,335) (2011 est.)
Median age: *total:* 19.4 years
male: 19.3 years
female: 19.6 years (2011 est.)
Population growth rate: 2.121% (2011 est.)
country comparison to the world: 41
Birth rate: 33.04 births/1,000 population (2011 est.)
country comparison to the world: 37
Death rate: 11.83 deaths/1,000 population (July 2011 est.)
country comparison to the world: 30
Net migration rate: 0 migrant(s)/1,000 population (2011 est.)
country comparison to the world: 80
Urbanization: *urban population:* 58% of total population (2010)
rate of urbanization: 3.3% annual rate of change (2010-15 est.)
Major cities—population: Douala 2.053 million; YAOUNDE (capital) 1.739 million (2009)
Sex ratio: *at birth:* 1.03 male(s)/female
under 15 years: 1.02 male(s)/female
15–64 years: 1.01 male(s)/female
65 years and over: 0.85 male(s)/female
total population: 1.01 male(s)/female (2011 est.)
Infant mortality rate: *total:* 60.91 deaths/1,000 live births
country comparison to the world: 32
male: 65.48 deaths/1,000 live births
female: 56.2 deaths/1,000 live births (2011 est.)
Life expectancy at birth: *total population:* 54.39 years
country comparison to the world: 201
male: 53.52 years
female: 55.28 years (2011 est.)
Total fertility rate: 4.17 children born/ woman (2011 est.)
country comparison to the world: 39
HIV/AIDS—adult prevalence rate: 5.3% (2009 est.)
country comparison to the world: 13
HIV/AIDS—people living with HIV/AIDS: 610,000 (2009 est.)
country comparison to the world: 15
HIV/AIDS—deaths: 37,000 (2009 est.)
country comparison to the world: 11
Major infectious diseases: *degree of risk:* very high
food or waterborne diseases: bacterial and protozoal diarrhea, hepatitis A and E, and typhoid fever
vectorborne diseases: malaria and yellow fever
water contact disease: schistosomiasis
respiratory disease: meningococcal meningitis
animal contact disease: rabies (2009)
Drinking water source: *Improved:*
urban: 92% of population
rural: 51% of population
total: 74% of population
Unimproved:
urban: 8% of population
rural: 49% of population
total: 26% of population (2008)
Sanitation facility access: *Improved:*
urban: 56% of population
rural: 35% of population
total: 47% of population
Unimproved:
urban: 44% of population
rural: 65% of population
total: 53% of population (2008)
Nationality: *noun:* Cameroonian(s)
adjective: Cameroonian
Ethnic groups: Cameroon Highlanders 31%, Equatorial Bantu 19%, Kirdi 11%, Fulani 10%, Northwestern Bantu 8%, Eastern Nigritic 7%, other African 13%, non-African less than 1%
Religions: indigenous beliefs 40%, Christian 40%, Muslim 20%
Languages: 24 major African language groups, English (official), French (official)
Literacy: *definition:* age 15 and over can read and write
total population: 67.9%
male: 77%
female: 59.8% (2001 est.)
School life expectancy (primary to tertiary education): *total:* 10 years
male: 11 years
female: 9 years (2009)
Education expenditures: 3.7% of GDP (2009)
country comparison to the world: 114

GOVERNMENT

Country name: *conventional long form:* Republic of Cameroon
conventional short form: Cameroon
local long form: Republique du Cameroun/ Republic of Cameroon
local short form: Cameroun/Cameroon
former: French Cameroon, British Cameroon, Federal Republic of Cameroon, United Republic of Cameroon
Government type: republic; multiparty presidential regime
Capital: *name:* Yaounde
geographic coordinates: 3 52 N, 11 31 E
time difference: UTC+1 (6 hours ahead of Washington, DC during Standard Time)
Administrative divisions: 10 regions (regions, singular—region); Adamaoua, Centre, Est, Extreme-Nord, Littoral, Nord, North-West (Nord-Ouest), Ouest, Sud, South-West (Sud-Ouest)
Independence: 1 January 1960 (from French-administered UN trusteeship)
National holiday: Republic Day (National Day), 20 May (1972)
Constitution: approved by referendum 20 May 1972; adopted 2 June 1972; revised January 1996; amended April 2008
Legal system: mixed legal system of English common law, French civil law, and customary law
International law organization participation: accepts compulsory ICJ jurisdiction; non-party state to the ICCt
Suffrage: 20 years of age; universal
Executive branch: *chief of state:* President Paul BIYA (since 6 November 1982)
head of government: Prime Minister Philemon YANG (since 30 June 2009)
cabinet: Cabinet appointed by the president from proposals submitted by the prime minister
(For more information visit the World Leaders website)
elections: president elected by popular vote for a seven-year term (with no term limits per 2008 constitutional amendment); election last held on 11 October 2004 (next to be held by October 2011); prime minister appointed by the president
election results: President Paul BIYA reelected; percent of vote—Paul BIYA 70.9%, John FRU NDI 17.4%, Adamou Ndam NJOYA 4.5%, Garga Haman ADJI 3.7%, other 3.5%
Legislative branch: unicameral National Assembly or Assemblee Nationale (180 seats; members are elected by direct popular vote to serve five-year terms); note—the president can either lengthen or shorten the term of the legislature
elections: last held on 22 July 2007 (next to be held in July 2012)
election results: percent of vote by party—NA; seats by party—CPDM 140, SDF 14, UDC 4, UNDP 4, MP 1, vacant 17
note: the constitution calls for an upper chamber for the legislature, to be called a Senate, but it has yet to be established
Judicial branch: Supreme Court (judges are appointed by the president); High Court of Justice (consists of nine judges and six substitute judges; elected by the National Assembly)
Political parties and leaders: Cameroon People's Democratic Movement or CPDM [Paul BIYA]; Cameroonian Democratic Union or UDC [Adamou Ndam NJOYA]; Movement for the Defense of the Republic or MDR [Dakole DAISSALA]; Movement for the Liberation and Development of Cameroon or MLDC [Marcel YONDO]; National Union for Democracy and Progress or UNDP [Maigari BELLO BOUBA]; Progressive Movement or MP; Social Democratic Front or SDF [John FRU NDI]; Union of Peoples of Cameroon or UPC [Augustin Frederic KODOCK]
Political pressure groups and leaders: Human Rights Defense Group [Albert MUKONG, president]; Southern Cameroon National Council [Ayamba Ette OTUN]
International organization participation: ACP, AfDB, AU, BDEAC, C, CEMAC, FAO, FZ, G-77, IAEA, IBRD, ICAO, ICC, ICRM, IDA, IDB, IFAD, IFC, IFRCS, ILO, IMF, IMO, IMSO, Interpol, IOC, IOM, IPU, ISO, ITSO, ITU, ITUC, MIGA, MONUSCO, NAM, OIC, OIF, OPCW, PCA, UN, UNAMID, UNCTAD, UNESCO, UNIDO, UNWTO, UPU, WCO, WFTU, WHO, WIPO, WMO, WTO
Diplomatic representation in the US: *chief of mission:* Ambassador Joseph FOE-ATANGANA
chancery: 2349 Massachusetts Avenue NW, Washington, DC 20008
telephone: [1] (202) 265-8790
FAX: [1] (202) 387-3826
Diplomatic representation from the US: *chief of mission:* Ambassador Robert P. JACKSON
embassy: Avenue Rosa Parks, Yaounde
mailing address: P. O. Box 817, Yaounde; pouch: American Embassy, US Department of State, Washington, DC 20521-2520
telephone: [237] 2220 15 00; Consular: [237] 2220 16 03

FAX: [237] 2220 16 00 Ext. 4531; Consular *FAX:* [237] 2220 17 52
branch office(s): Douala
Flag description: three equal vertical bands of green (hoist side), red, and yellow, with a yellow five-pointed star centered in the red band; the vertical tricolor recalls the flag of France; red symbolizes unity, yellow the sun, happiness, and the savannahs in the north, and green hope and the forests in the south; the star is referred to as the "star of unity"
note: uses the popular Pan-African colors of Ethiopia
National anthem: *name:* "O Cameroun, Berceau de nos Ancetres" (O Cameroon, Cradle of Our Forefathers)
lyrics/music: Rene Djam AFAME, Samuel Minkio BAMBA, Moise Nyatte NKO'O [French], Benard Nsokika FONLON [English]/Rene Djam AFAME
note: adopted 1957; Cameroon's anthem, also known as "Chant de Ralliement" (The Rallying Song), has been used unofficially since 1948 although officially adopted in 1957; the anthem has French and English versions whose lyrics differ

ECONOMY

Economy—overview: Because of its modest oil resources and favorable agricultural conditions, Cameroon has one of the best-endowed primary commodity economies in sub-Saharan Africa. Still, it faces many of the serious problems confronting other underdeveloped countries, such as stagnant per capita income, a relatively inequitable distribution of income, a top-heavy civil service, endemic corruption, and a generally unfavorable climate for business enterprise. Since 1990, the government has embarked on various IMF and World Bank programs designed to spur business investment, increase efficiency in agriculture, improve trade, and recapitalize the nation's banks. The IMF is pressing for more reforms, including increased budget transparency, privatization, and poverty reduction programs. Weak prices for oil led to the significant slowdown in growth in 2010. The government is under pressure to reduce its budget deficit, which by the government's own forecast will hit 2.8% of GDP, but the presidential election in 2011 may make fiscal austerity difficult.
GDP (purchasing power parity): $44.33 billion (2010 est.)
country comparison to the world: 95
$43.04 billion (2009 est.)
$42.22 billion (2008 est.)
note: data are in 2010 US dollars
GDP (official exchange rate): $22.48 billion (2010 est.)
GDP—real growth rate: 3% (2010 est.)
country comparison to the world: 127
2% (2009 est.)
2.6% (2008 est.)
GDP—per capita (PPP): $2,300 (2010 est.)
country comparison to the world: 184
$2,300 (2009 est.)
$2,300 (2008 est.)
note: data are in 2010 US dollars
GDP—composition by sector:
agriculture: 20%
industry: 30.9%
services: 49.1% (2010 est.)
Labor force: 7.836 million (2010 est.)
country comparison to the world: 59
Labor force—by occupation:
agriculture: 70%
industry: 13%
services: 17% (2001 est.)
Unemployment rate: 30% (2001 est.)
country comparison to the world: 178
Population below poverty line: 48% (2000 est.)
Household income or consumption by percentage share: *lowest 10%:* 2.3%
highest 10%: 35.4% (2001)
Distribution of family income—Gini index: 44.6 (2001)
country comparison to the world: 40
47.7 (1996)
Investment (gross fixed): 21.1% of GDP (2010 est.)
country comparison to the world: 74
Budget: *revenues:* $3.779 billion
expenditures: $4.34 billion (2010 est.)
Public debt: 9.6% of GDP (2010 est.)
country comparison to the world: 122
16.3% of GDP (2009 est.)
Inflation rate (consumer prices): 1.9% (2010 est.)
country comparison to the world: 50
3% (2009 est.)
Central bank discount rate: 4.25% (31 December 2009)
country comparison to the world: 91
4.75% (31 December 2008)
Commercial bank prime lending rate: NA%
Stock of narrow money: $2.888 billion (31 December 2010 est.)
country comparison to the world: 111
$3.074 billion (31 December 2009 est.)
Stock of broad money: $4.831 billion (31 December 2010 est.)
country comparison to the world: 120
$4.921 billion (31 December 2009 est.)
Stock of domestic credit: $848.8 million (31 December 2010 est.)
country comparison to the world: 153
$1.523 billion (31 December 2009 est.)
Market value of publicly traded shares: $NA
Agriculture—products: coffee, cocoa, cotton, rubber, bananas, oilseed, grains, root starches; livestock; timber
Industries: petroleum production and refining, aluminum production, food processing, light consumer goods, textiles, lumber, ship repair
Industrial production growth rate: 4% (2010 est.)
country comparison to the world: 86
Electricity—production: 5.601 billion kWh (2007 est.)
country comparison to the world: 108
Electricity—consumption: 4.801 billion kWh (2007 est.)
country comparison to the world: 112
Electricity—exports: 0 kWh (2008 est.)
Electricity—imports: 0 kWh (2008 est.)
Oil—production: 77,310 bbl/day (2009 est.)
country comparison to the world: 55
Oil—consumption: 26,000 bbl/day (2009 est.)
country comparison to the world: 115
Oil—exports: 107,100 bbl/day (2007 est.)
country comparison to the world: 66
Oil—imports: 45,520 bbl/day (2007 est.)
country comparison to the world: 89
Oil—proved reserves: 200 million bbl (1 January 2010 est.)
country comparison to the world: 58
Natural gas—production: NA cu m (2008 est.)
Natural gas—consumption: NA cu m (2008 est.)
Natural gas—exports: 0 cu m (2008 est.)
country comparison to the world: 76
Natural gas—imports: 0 cu m (2008 est.)
country comparison to the world: 166
Natural gas—proved reserves: 135.1 billion cu m (1 January 2010 est.)
country comparison to the world: 49
Current account balance: $-826 million (2010 est.)
country comparison to the world: 128
$-1.137 billion (2009 est.)
Exports: $4.371 billion (2010 est.)
country comparison to the world: 113
$4.079 billion (2009 est.)
Exports—commodities: crude oil and petroleum products, lumber, cocoa beans, aluminum, coffee, cotton
Exports—partners: Netherlands 13.99%, Spain 12.25%, Italy 11.84%, China 9.14%, US 6.16%, France 5.51%, South Korea 4.66%, Belgium 4.33%, UK 4% (2009)
Imports: $4.869 billion (2010 est.)
country comparison to the world: 119
$4.405 billion (2009 est.)
Imports—commodities: machinery, electrical equipment, transport equipment, fuel, food
Imports—partners: France 21.03%, Nigeria 10.79%, China 10.25%, Belgium 6.62%, US 4.31% (2009)
Reserves of foreign exchange and gold: $4.023 billion (31 December 2010 est.)
country comparison to the world: 79
$3.676 billion (31 December 2009 est.)
Debt—external: $3.344 billion (31 December 2010 est.)
country comparison to the world: 122
$3.231 billion (31 December 2009 est.)
Exchange rates: Cooperation Financiere en Afrique Centrale francs–495.28 (2010), 472.19 (2009), 447.81 (2008), 493.51 (2007), 522.59 (2006)

COMMUNICATIONS

Telephones—main lines in use: 323,800 (2009)
country comparison to the world: 111
Telephones—mobile cellular: 7.397 million (2009)
country comparison to the world: 80
Telephone system: *general assessment:* system includes cable, microwave radio relay, and tropospheric scatter; Camtel, the monopoly provider of fixed-line service, provides connections for only about 1 per 100 persons; equipment is old and outdated, and connections with many parts of the country are unreliable
domestic: mobile-cellular usage, in part a reflection of the poor condition and general inadequacy of the fixed-line network, has increased sharply, reaching a subscribership base of 40 per 100 persons
international: country code–237; landing point for the SAT-3/WASC fiber-optic submarine cable that provides connec-

tivity to Europe and Asia; satellite earth stations–2 Intelsat (Atlantic Ocean) (2009)
Broadcast media: government maintains tight control over broadcast media; state-owned Cameroon Radio Television (CRTV), broadcasting on both a television and radio network, was the only officially recognized and fully licensed broadcaster until August 2007 when the government finally issued licenses to 2 private TV broadcasters and 1 private radio broadcaster; about 70 privately-owned unlicensed radio stations operating but are subject to closure at any time; foreign news services required to partner with state-owned national station (2007)
Internet country code: .cm
Internet hosts: 90 (2010)
country comparison to the world: 205
Internet users: 749,600 (2009)
country comparison to the world: 106

TRANSPORTATION

Airports: 34 (2010)
country comparison to the world: 111
Airports—with paved runways: *total:* 11
over 3,047 m: 2
2,438 to 3,047 m: 5
1,524 to 2,437 m: 3
914 to 1,523 m: 1 (2010)
Airports—with unpaved runways: *total:* 23
1,524 to 2,437 m: 3
914 to 1,523 m: 14
under 914 m: 6 (2010)
Pipelines: oil 886 km (2010)
Railways: *total:* 987 km
country comparison to the world: 88
narrow gauge: 987 km 1.000-m gauge (2010)
Roadways: *total:* 50,000 km
country comparison to the world: 80
paved: 5,000 km
unpaved: 45,000 km (2004)
Waterways: (major rivers in the south, such as the Wouri and the Sanaga, are largely non-navigable; in the north, the Benue, which connects through Nigeria to the Niger River, is navigable in the rainy season only to the port of Garoua) (2010)
Ports and terminals: Douala, Garoua, Limboh Terminal

MILITARY

Military branches: Cameroon Armed Forces (Forces Armees Camerounaises, FAC): Army (L'Armee de Terre), Navy (includes naval infantry), Air Force (Armee de l'Air du Cameroun, AAC), Fire Fighter Corps, Gendarmerie (2011)
Military service age and obligation: 18-23 years of age for male and female voluntary military service; no conscription; high school graduation required; service obligation 4 years; the government periodically calls for volunteers (2010)
Manpower available for military service:
males age 16–49: 4,667,251
females age 16–49: 4,548,909 (2010 est.)
Manpower fit for military service: *males age 16–49:* 2,794,998
females age 16–49: 2,718,110 (2010 est.)
Manpower reaching militarily significant age annually: *male:* 215,248
female: 211,636 (2010 est.)
Military expenditures: 1.3% of GDP (2009)
country comparison to the world: 115

TRANSNATIONAL ISSUES

Disputes—international: Joint Border Commission with Nigeria reviewed 2002 ICJ ruling on the entire boundary and bilaterally resolved differences, including June 2006 Greentree Agreement that immediately ceded sovereignty of the Bakassi Peninsula to Cameroon with a full phase-out of Nigerian control and patriation of residents in 2008; Cameroon and Nigeria agree on maritime delimitation in March 2008; sovereignty dispute between Equatorial Guinea and Cameroon over an island at the mouth of the Ntem River; only Nigeria and Cameroon have heeded the Lake Chad Commission's admonition to ratify the delimitation treaty, which also includes the Chad-Niger and Niger-Nigeria boundaries
Refugees and internally displaced persons: refugees (country of origin): 20,000-30,000 (Chad); 3,000 (Nigeria); 24,000 (Central African Republic) (2007)
Trafficking in persons: current situation: Cameroon is a source, transit, and destination country for women and children trafficked for the purposes of forced labor and commercial sexual exploitation; most victims are children trafficked within country, with girls primarily trafficked for domestic servitude and sexual exploitation; both boys and girls are also trafficked within Cameroon for forced labor in sweatshops, bars, restaurants, and on tea and cocoa plantations; children are trafficked into Cameroon from neighboring states for forced labor in agriculture, fishing, street vending, and spare-parts shops; Cameroon is a transit country for children trafficked between Gabon and Nigeria, and from Nigeria to Saudi Arabia; it is a source country for women transported by sex-trafficking rings to Europe
tier rating: Tier 2 Watch List–Cameroon is on the Tier 2 Watch List for its failure to provide evidence of increasing efforts to combat human trafficking in 2007, particularly in terms of efforts to prosecute and convict trafficking offenders; while Cameroon reported some arrests of traffickers, none of them were prosecuted or punished; the government does not identify trafficking victims among vulnerable populations nor does it monitor the number of victims it intercepts (2008)

CANADA

INTRODUCTION

Background: A land of vast distances and rich natural resources, Canada became a self-governing dominion in 1867 while retaining ties to the British crown. Economically and technologically, the nation has developed in parallel with the US, its neighbor to the south across the World's longest unfortified border. Canada faces the political challenges of meeting public demands for quality improvements in health care, and education, social services, and economic competitiveness, as well as responding to the particular concerns of predominantly francophone Quebec. Canada also aims to develop its diverse energy resources while maintaining its commitment to the environment.

GEOGRAPHY

Location: Northern North America, bordering the North Atlantic Ocean on the east, North Pacific Ocean on the west, and the Arctic Ocean on the north, north of the conterminous US
Geographic coordinates: 60 00 N, 95 00 W
Map references: North America
Area: *total:* 9,984,670 sq km
country comparison to the world: 2
land: 9,093,507 sq km
water: 891,163 sq km
Area—comparative: slightly larger than the US
Land boundaries: *total:* 8,893 km
border countries: US 8,893 km (includes 2,477 km with Alaska)
note: Canada is the World's largest country that borders only one country
Coastline: 202,080 km
Maritime claims: territorial sea: 12 nm

contiguous zone: 24 nm
exclusive economic zone: 200 nm
continental shelf: 200 nm or to the edge of the continental margin
Climate: varies from temperate in south to subarctic and arctic in north
Terrain: mostly plains with mountains in west and lowlands in southeast
Elevation extremes: *lowest point:* Atlantic Ocean 0 m
highest point: Mount Logan 5,959 m
Natural resources: iron ore, nickel, zinc, copper, gold, lead, rare earth elements, molybdenum, potash, diamonds, silver, fish, timber, wildlife, coal, petroleum, natural gas, hydropower
Land use: *arable land:* 4.57%
permanent crops: 0.65%
other: 94.78% (2005)
Irrigated land: 7,850 sq km (2003)
Total renewable water resources: 3,300 cu km (1985)
Freshwater withdrawal (domestic/industrial/agricultural): *total:* 44.72 cu km/yr (20%/69%/12%)
per capita: 1,386 cu m/yr (1996)
Natural hazards: continuous permafrost in north is a serious obstacle to development; cyclonic storms form east of the Rocky Mountains, a result of the mixing of air masses from the Arctic, Pacific, and North American interior, and produce most of the country's rain and snow east of the mountains
volcanism: the vast majority of volcanoes in Western Canada's Coast Mountains remain dormant
Environment—current issues: air pollution and resulting acid rain severely affecting lakes and damaging forests; metal smelting, coal-burning utilities, and vehicle emissions impacting on agricultural and forest productivity; ocean waters becoming contaminated due to agricultural, industrial, mining, and forestry activities
Environment—international agreements: *party to:* Air Pollution, Air Pollution-Nitrogen Oxides, Air Pollution-Persistent Organic Pollutants, Air Pollution-Sulfur 85, Air Pollution-Sulfur 94, Antarctic-Environmental Protocol, Antarctic-Marine Living Resources, Antarctic Seals, Antarctic Treaty, Biodiversity, Climate Change, Climate Change-Kyoto Protocol, Desertification, Endangered Species, Environmental Modification, Hazardous Wastes, Law of the Sea, Marine Dumping, Ozone Layer Protection, Ship Pollution, Tropical Timber 83, Tropical Timber 94, Wetlands
signed, but not ratified: Air Pollution-Volatile Organic Compounds, Marine Life Conservation
Geography—note: second-largest country in world (after Russia); strategic location between Russia and US via north polar route; approximately 90% of the population is concentrated within 160 km of the US border

PEOPLE

Population: 34,030,589 (July 2011 est.)
country comparison to the world: 37
Age structure: *0–14 years:* 15.7% (male 2,736,737/female 2,602,342)
15–64 years: 68.5% (male 11,776,611/female 11,517,972)
65 years and over: 15.9% (male 2,372,356/female 3,024,571) (2011 est.)
Median age: *total:* 41 years
male: 39.8 years
female: 42.1 years (2011 est.)
Population growth rate: 0.794% (2011 est.)
country comparison to the world: 136
Birth rate: 10.28 births/1,000 population (2011 est.)
country comparison to the world: 187
Death rate: 7.98 deaths/1,000 population (July 2011 est.)
country comparison to the world: 105
Net migration rate: 5.65 migrant(s)/1,000 population (2011 est.)
country comparison to the world: 15
Urbanization: *urban population:* 81% of total population (2010)
rate of urbanization: 1.1% annual rate of change (2010-15 est.)
Major cities—population: Toronto 5.377 million; Montreal 3.75 million; Vancouver 2.197 million; OTTAWA (capital) 1.17 million; Calgary 1.16 million (2009)
Sex ratio: *at birth:* 1.056 male(s)/female
under 15 years: 1.05 male(s)/female
15–64 years: 1.02 male(s)/female
65 years and over: 0.78 male(s)/female
total population: 0.98 male(s)/female (2011 est.)
Infant mortality rate: *total:* 4.92 deaths/1,000 live births
country comparison to the world: 183
male: 5.26 deaths/1,000 live births
female: 4.56 deaths/1,000 live births (2011 est.)
Life expectancy at birth: *total population:* 81.38 years
country comparison to the world: 12
male: 78.81 years
female: 84.1 years (2011 est.)
Total fertility rate: 1.58 children born/woman (2011 est.)
country comparison to the world: 178
HIV/AIDS—adult prevalence rate: 0.3% (2009 est.)
country comparison to the world: 91
HIV/AIDS—people living with HIV/AIDS: 68,000 (2009 est.)
country comparison to the world: 50
HIV/AIDS—deaths: fewer than 1,000 (2009 est.)
country comparison to the world: 69
Drinking water source: *Improved:*
urban: 100% of population
rural: 99% of population
total: 100% of population
Unimproved:
urban: 0% of population
rural: 1% of population
total: 0% of population (2008)
Sanitation facility access: *Improved:*
urban: 100% of population
rural: 99% of population
total: 100% of population
Unimproved:
urban: 0% of population
rural: 1% of population
total: 0% of population (2008)
Nationality: *noun:* Canadian(s)
adjective: Canadian
Ethnic groups: British Isles origin 28%, French origin 23%, other European 15%, Amerindian 2%, other, mostly Asian, African, Arab 6%, mixed background 26%
Religions: Roman Catholic 42.6%, Protestant 23.3% (including United Church 9.5%, Anglican 6.8%, Baptist 2.4%, Lutheran 2%), other Christian 4.4%, Muslim 1.9%, other and unspecified 11.8%, none 16% (2001 census)
Languages: English (official) 58.8%, French (official) 21.6%, other 19.6% (2006 Census)
Literacy: *definition:* age 15 and over can read and write
total population: 99%
male: 99%
female: 99% (2003 est.)
School life expectancy (primary to tertiary education): *total:* 17 years
male: 17 years
female: 17 years (2004)
Education expenditures: 4.9% of GDP (2007)
country comparison to the world: 66

GOVERNMENT

Country name: *conventional long form:* none
conventional short form: Canada
Government type: a parliamentary democracy, a federation, and a constitutional monarchy
Capital: *name:* Ottawa
geographic coordinates: 45 25 N, 75 42 W
time difference: UTC-5 (same time as Washington, DC during Standard Time)
daylight saving time: +1hr, begins second Sunday in March; ends first Sunday in November
note: Canada is divided into six time zones
Administrative divisions: 10 provinces and 3 territories*; Alberta, British Columbia, Manitoba, New Brunswick, Newfoundland and Labrador, Northwest Territories*, Nova Scotia, Nunavut*, Ontario, Prince Edward Island, Quebec, Saskatchewan, Yukon Territory*
Independence: 1 July 1867 (union of British North American colonies); 11 December 1931 (recognized by UK per Statute of Westminster)
National holiday: Canada Day, 1 July (1867)
Constitution: made up of unwritten and written acts, customs, judicial decisions, and traditions; the written part of the constitution consists of the Constitution Act of 29 March 1867, which created a federation of four provinces, and the Constitution Act of 17 April 1982, which transferred formal control over the constitution from Britain to Canada, and added a Canadian Charter of Rights and Freedoms as well as procedures for constitutional amendments
Legal system: common law system except in Quebec where civil law based on the French civil code prevails
International law organization participation: accepts compulsory ICJ jurisdiction with reservations; accepts ICCt jurisdiction
Suffrage: 18 years of age; universal
Executive branch: head of state: Queen ELIZABETH II (since 6 February 1952); represented by Governor General David JOHNSTON (since 1 October 2010)

head of government: Prime Minister Stephen Joseph HARPER (since 6 February 2006)
cabinet: Federal Ministry chosen by the prime minister usually from among the members of his own party sitting in Parliament (For more information visit the World Leaders website)
elections: the monarchy is hereditary; governor general appointed by the monarch on the advice of the prime minister for a five-year term; following legislative elections, the leader of the majority party or the leader of the majority coalition in the House of Commons generally designated prime minister by the governor general
Legislative branch: bicameral Parliament or Parlement consists of the Senate or Senat (105 seats; members appointed by the governor general on the advice of the prime minister and serve until 75 years of age) and the House of Commons or Chambre des Communes (308 seats; members elected by direct, popular vote to serve a maximum of four-year terms)
elections: House of Commons–last held on 2 May 2011 (next to be held no later than 19 October 2015)
election results: House of Commons–percent of vote by party–Conservative Party 39.6%, New Democratic Party 30.6%, Liberal Party 18.9%, Bloc Quebecois 6%, Greens 3.9%; seats by party–Conservative Party 166, New Democratic Party 103, Liberal Party 34, Bloc Quebecois 4, Greens 1
Judicial branch: Supreme Court of Canada (judges are appointed by the governor general on the recommendation of the prime minister); Federal Court of Canada; Federal Court of Appeal; Tax Court of Canada; Provincial/Territorial Courts (these are named variously Court of Appeal, Court of Queen's Bench, Superior Court, Supreme Court, and Court of Justice)
Political parties and leaders: Bloc Quebecois [Gilles DUCEPPE]; Conservative Party of Canada [Stephen HARPER]; Green Party [Elizabeth MAY]; Liberal Party [Robert RAE (interim)]; New Democratic Party [Jack LAYTON]
Political pressure groups and leaders: *other:* agricultural sector; automobile industry; business groups; chemical industry; commercial banks; communications sector; energy industry; environmentalists; public administration groups; steel industry; trade unions
International organization participation: ADB (nonregional member), AfDB (nonregional member), APEC, Arctic Council, ARF, ASEAN (dialogue partner), Australia Group, BIS, C, CDB, EAPC, EBRD, FAO, FATF, G-20, G-7, G-8, G-10, IADB, IAEA, IBRD, ICAO, ICC, ICRM, IDA, IEA, IFAD, IFC, IFRCS, IHO, ILO, IMF, IMO, IMSO, Interpol, IOC, IOM, IPU, ISO, ITSO, ITU, ITUC, MIGA, MINUSTAH, MONUSCO, NAFTA, NATO, NEA, NSG, OAS, OECD, OIF, OPCW, OSCE, Paris Club, PCA, PIF (partner), SECI (observer), UN, UNAMID, UNCTAD, UNDOF, UNESCO, UNFICYP, UNHCR, UNMIS, UNRWA, UNTSO, UNWTO, UPU, WCO, WFTU, WHO, WIPO, WMO, WTO, ZC
Diplomatic representation in the US: *chief of mission:* Ambassador Gary DOER
chancery: 501 Pennsylvania Avenue NW, Washington, DC 20001
telephone: [1] (202) 682-1740
FAX: [1] (202) 682-7701
consulate(s) general: Atlanta, Boston, Buffalo, Chicago, Dallas, Denver, Detroit, Los Angeles, Miami, Minneapolis, New York, San Francisco/Silicon Valley,, Seattle
consulate(s): Anchorage, Houston, Philadelphia, Phoenix, Raleigh, San Diego, San Jose (California), Tucson
Diplomatic representation from the US: *chief of mission:* Ambassador David C. JACOBSON
embassy: 490 Sussex Drive, Ottawa, Ontario K1N 1G8
mailing address: P. O. Box 5000, Ogdensburg, NY 13669-0430; P.O. Box 866, Station B, Ottawa, Ontario K1P 5T1
telephone: [1] (613) 688-5335
FAX: [1] (613) 688-3082
consulate(s) general: Calgary, Halifax, Montreal, Quebec, Toronto, Vancouver, Winnipeg
Flag description: two vertical bands of red (hoist and fly side, half width) with white square between them; an 11-pointed red maple leaf is centered in the white square; the maple leaf has long been a Canadian symbol; the official colors of Canada are red and white
National anthem: *name:* "O Canada"
lyrics/music: Adolphe-Basile ROUTHIER [French], Robert Stanley WEIR [English]/ Calixa LAVALLEE
note: adopted 1980; originally written in 1880, "O Canada" served as an unofficial anthem many years before its official adoption; the anthem has French and English versions whose lyrics differ; as a Commonwealth realm, in addition to the national anthem, "God Save the Queen" serves as the royal anthem (see United Kingdom)

ECONOMY

Economy—overview: As an affluent, high-tech industrial society in the trillion-dollar class, Canada resembles the US in its market-oriented economic system, pattern of production, and affluent living standards. Since World War II, the impressive growth of the manufacturing, mining, and service sectors has transformed the nation from a largely rural economy into one primarily industrial and urban. The 1989 US-Canada Free Trade Agreement (FTA) and the 1994 North American Free Trade Agreement (NAFTA) (which includes Mexico) touched off a dramatic increase in trade and economic integration with the US, its principal trading partner. Canada enjoys a substantial trade surplus with the US, which absorbs about three-fourths of Canadian exports each year. Canada is the US's largest foreign supplier of energy, including oil, gas, uranium, and electric power. Given its great natural resources, skilled labor force, and modern capital plant, Canada enjoyed solid economic growth from 1993 through 2007. Buffeted by the global economic crisis, the economy dropped into a sharp recession in the final months of 2008, and Ottawa posted its first fiscal deficit in 2009 after 12 years of surplus. Canada's major banks, however, emerged from the financial crisis of 2008-09 among the strongest in the world, owing to the financial sector's tradition of conservative lending practices and strong capitalization. During 2010, Canada's economy grew only 3%, due to decreased global demand and a highly valued Canadian dollar.
GDP (purchasing power parity): $1.33 trillion (2010 est.)
country comparison to the world: 15
$1.291 trillion (2009 est.)
$1.323 trillion (2008 est.)
note: data are in 2010 US dollars
GDP (official exchange rate): $1.574 trillion (2010 est.)
GDP—real growth rate: 3.1% (2010 est.)
country comparison to the world: 123
-2.5% (2009 est.)
0.5% (2008 est.)
GDP—per capita (PPP): $39,400 (2010 est.)
country comparison to the world: 22
$38,500 (2009 est.)
$39,800 (2008 est.)
note: data are in 2010 US dollars
GDP—composition by sector: *agriculture:* 2%
industry: 20%
services: 78% (2010 est.)
Labor force: 18.59 million (2010 est.)
country comparison to the world: 31
Labor force—by occupation: *agriculture:* 2%
manufacturing: 13%
construction: 6%
services: 76%
other: 3% (2006 est.)
Unemployment rate: 8% (2010 est.)
country comparison to the world: 89
8.3% (2009 est.)
Population below poverty line: 9.4%
note: this figure is the Low Income Cut-Off (LICO), a calculation that results in higher figures than found in many comparable economies; Canada does not have an official poverty line (2008)
Household income or consumption by percentage share: *lowest 10%:* 0.1%
highest 10%: 46% (2005)
Distribution of family income—Gini index: 32.1 (2005)
country comparison to the world: 101
31.5 (1994)
Investment (gross fixed): 22.1% of GDP (2010 est.)
country comparison to the world: 67
Budget: *revenues:* $605.7 billion
expenditures: $677.7 billion (2010 est.)
Public debt: 34% of GDP (2010 est.)
country comparison to the world: 86
29% of GDP (2009 est.)
Inflation rate (consumer prices): 1.6% (2010 est.)
country comparison to the world: 40
0.3% (2009 est.)
Central bank discount rate: 1% (31 December 2010)
country comparison to the world: 139
0.5% (31 December 2009)
Commercial bank prime lending rate: 3% (31 December 2009 est.)
country comparison to the world: 154
2.4% (31 December 2009 est.)
Stock of narrow money: $560.8 billion (31 December 2010 est.)
country comparison to the world: 10

$470.9 billion (31 December 2009 est.)
Stock of broad money: $1.469 trillion (31 December 2010 est.)
country comparison to the world: 12
$1.144 trillion (31 December 2009 est.)
Stock of domestic credit: $2.963 trillion (31 December 2010 est.)
country comparison to the world: 10
$2.606 trillion (31 December 2009 est.)
Market value of publicly traded shares: $1.681 trillion (31 December 2009)
country comparison to the world: 10
$1.002 trillion (31 December 2008)
$2.187 trillion (31 December 2007)
Agriculture—products: wheat, barley, oilseed, tobacco, fruits, vegetables; dairy products; forest products; fish
Industries: transportation equipment, chemicals, processed and unprocessed minerals, food products, wood and paper products, fish products, petroleum and natural gas
Industrial production growth rate: 5.8% (2010 est.)
country comparison to the world: 61
Electricity—production: 620.7 billion kWh (2007 est.)
country comparison to the world: 7
Electricity—consumption: 536.1 billion kWh (2007 est.)
country comparison to the world: 8
Electricity—exports: 55.73 billion kWh (2008 est.)
Electricity—imports: 23.5 billion kWh (2008 est.)
Oil—production: 3.289 million bbl/day (2009 est.)
country comparison to the world: 6
Oil—consumption: 2.151 million bbl/day (2009 est.)
country comparison to the world: 11
Oil—exports: 2.001 million bbl/day (2008 est.)
country comparison to the world: 10
Oil—imports: 1.192 million bbl/day (2008 est.)
country comparison to the world: 15
Oil—proved reserves: 175.2 billion bbl
country comparison to the world: 2
note: includes oil sands (1 January 2010 est.)
Natural gas—production: 161.3 billion cu m (2009 est.)
country comparison to the world: 5
Natural gas—consumption: 94.62 billion cu m (2009 est.)
country comparison to the world: 7
Natural gas—exports: 94.67 billion cu m (2009 est.)
country comparison to the world: 3
Natural gas—imports: 16.59 billion cu m (2009 est.)
country comparison to the world: 16
Natural gas—proved reserves: 1.754 trillion cu m (1 January 2010 est.)
country comparison to the world: 21
Current account balance: $-40.21 billion (2010 est.)
country comparison to the world: 185
$-38.08 billion (2009 est.)
Exports: $406.8 billion (2010 est.)
country comparison to the world: 10
$323.3 billion (2009 est.)
Exports—commodities: motor vehicles and parts, industrial machinery, aircraft, telecommunications equipment; chemicals, plastics, fertilizers; wood pulp, timber, crude petroleum, natural gas, electricity, aluminum
Exports—partners: US 75.02%, UK 3.37%, China 3.09% (2009)
Imports: $406.4 billion (2010 est.)
country comparison to the world: 12
$327.3 billion (2009 est.)
Imports—commodities: machinery and equipment, motor vehicles and parts, crude oil, chemicals, electricity, durable consumer goods
Imports—partners: US 51.1%, China 10.88%, Mexico 4.56% (2009)
Reserves of foreign exchange and gold: $NA (31 December 2010 est.)
$54.36 billion (31 December 2009 est.)
Debt—external: $1.009 trillion (30 June 2010)
country comparison to the world: 15
$781.1 billion (31 December 2008)
Stock of direct foreign investment—at home:
$528.7 billion (31 December 2010 est.)
country comparison to the world: 10
$494.6 billion (31 December 2009 est.)
Stock of direct foreign investment—abroad: $602.5 billion (31 December 2010 est.)
country comparison to the world: 11
$576.2 billion (31 December 2009 est.)
Exchange rates: Canadian dollars (CAD) per US dollar–1.0346 (2010), 1.1431 (2009), 1.0364 (2008), 1.0724 (2007), 1.1334 (2006)

COMMUNICATIONS

Telephones—main lines in use: 18.251 million (2009)
country comparison to the world: 16
Telephones—mobile cellular: 23.081 million (2009)
country comparison to the world: 38
Telephone system: *general assessment:* excellent service provided by modern technology
domestic: domestic satellite system with about 300 earth stations
international: country code–1; submarine cables provide links to the US and Europe; satellite earth stations–7 (5 Intelsat–4 Atlantic Ocean and 1 Pacific Ocean, and 2 Intersputnik–Atlantic Ocean region) (2007)
Broadcast media: 2 public television broadcasting networks each with a large number of network affiliates; several private-commercial networks also with multiple network affiliates; overall, about 150 TV stations; multi-channel satellite and cable systems provide access to a wide range of stations including US stations; mix of public and commercial radio broadcasters with the Canadian Broadcasting Corporation (CBC), the public radio broadcaster, operating 4 radio networks, Radio Canada International, and radio services to indigenous populations in the north; roughly 2,000 licensed radio stations in Canada (2008)
Internet country code: .ca
Internet hosts: 7.77 million (2010)
country comparison to the world: 13
Internet users: 26.96 million (2009)
country comparison to the world: 16

TRANSPORTATION

Airports: 1,404 (2010)
country comparison to the world: 4
Airports—with paved runways: *total:* 514
over 3,047 m: 18
2,438 to 3,047 m: 20
1,524 to 2,437 m: 148
914 to 1,523 m: 249
under 914 m: 79 (2010)
Airports—with unpaved runways: *total:* 890
1,524 to 2,437 m: 73
914 to 1,523 m: 377
under 914 m: 440 (2010)
Heliports: 12 (2010)
Pipelines: gas 835 km; liquid petroleum gas 75,000 km (2010)
Railways: *total:* 46,552 km
country comparison to the world: 5
standard gauge: 46,552 km 1.435-m gauge (2009)
Roadways: *total:* 1,042,300 km
country comparison to the world: 6
paved: 415,600 km (includes 17,000 km of expressways)
unpaved: 626,700 km (2009)
Waterways: 636 km (Saint Lawrence Seaway of 3,769 km, including the Saint Lawrence River of 3,058 km, shared with United States) (2008)
country comparison to the world: 78
Merchant marine: *total:* 184
country comparison to the world: 36
by type: bulk carrier 66, cargo 12, carrier 1, chemical tanker 14, combination ore/oil 1, container 2, passenger 6, passenger/cargo 64, petroleum tanker 12, roll on/roll off 6
foreign-owned: 15 (France 1, Netherlands 1, Norway 4, US 9)
registered in other countries: 223 (Australia 7, Bahamas 102, Barbados 13, Cambodia 2, Cyprus 2, Honduras 1, Hong Kong 70, Liberia 4, Malta 1, Marshall Islands 4, Norway 1, Panama 5, Spain 5, US 1, Vanuatu 5) (2010)
Ports and terminals: Fraser River Port, Halifax, Hamilton, Montreal, Port-Cartier, Quebec City, Saint John (New Brunswick), Sept-Isles, Vancouver

MILITARY

Military branches: Canadian Forces: Land Forces Command (LFC), Maritime Command (MARCOM), Air Command (AIRCOM), Canada Command (homeland security) (2011)
Military service age and obligation: 17 years of age for voluntary male and female military service (with parental consent); 16 years of age for reserve and military college applicants; Canadian citizenship or permanent residence status required; maximum 34 years of age; service obligation 3-9 years (2008)
Manpower available for military service: *males age 16–49:* 8,031,266
females age 16–49: 7,755,550 (2010 est.)
Manpower fit for military service: *males age 16–49:* 6,633,472
females age 16–49: 6,389,669 (2010 est.)
Manpower reaching militarily significant age annually: *male:* 218,069
female: 206,195 (2010 est.)
Military expenditures: 1.1% of GDP (2005 est.)
country comparison to the world: 126

TRANSNATIONAL ISSUES

Disputes—international: managed maritime boundary disputes with the US at Dixon Entrance, Beaufort Sea, Strait of Juan de Fuca, and the Gulf of Maine including the disputed Machias Seal Island and North Rock; Canada and the United States dispute how to divide the Beaufort Sea and the status of the Northwest Passage but continue to work cooperatively to survey the Arctic continental shelf; US works closely with Canada to intensify security measures for monitoring and controlling legal and illegal movement of people, transport, and commodities across the international border; sovereignty dispute with Denmark over Hans Island in the Kennedy Channel between Ellesmere Island and Greenland; commencing the collection of technical evidence for submission to the Commission on the Limits of the Continental Shelf in support of claims for continental shelf beyond 200 nautical miles from its declared baselines in the Arctic, as stipulated in Article 76, paragraph 8, of the United Nations Convention on the Law of the Sea
Illicit drugs: illicit producer of cannabis for the domestic drug market and export to US; use of hydroponics technology permits growers to plant large quantities of high-quality marijuana indoors; increasing ecstasy production, some of which is destined for the US; vulnerable to narcotics money laundering because of its mature financial services sector

CAPE VERDE

INTRODUCTION

Background: The uninhabited islands were discovered and colonized by the Portuguese in the 15th century; Cape Verde subsequently became a trading center for African slaves and later an important coaling and resupply stop for whaling and transatlantic shipping. Following independence in 1975, and a tentative interest in unification with Guinea-Bissau, a one-party system was established and maintained until multi-party elections were held in 1990. Cape Verde continues to exhibit one of Africa's most stable democratic governments. Repeated droughts during the second half of the 20th century caused significant hardship and prompted heavy emigration. As a result, Cape Verde's expatriate population is greater than its domestic one. Most Cape Verdeans have both African and Portuguese antecedents.

GEOGRAPHY

Location: Western Africa, group of islands in the North Atlantic Ocean, west of Senegal
Geographic coordinates: 16 00 N, 24 00 W
Map references: Africa
Area: *total:* 4,033 sq km
country comparison to the world: 175
land: 4,033 sq km
water: 0 sq km
Area—comparative: slightly larger than Rhode Island
Land boundaries: 0 km
Coastline: 965 km
Maritime claims: measured from claimed archipelagic baselines
territorial sea: 12 nm
contiguous zone: 24 nm
exclusive economic zone: 200 nm
Climate: temperate; warm, dry summer; precipitation meager and erratic
Terrain: steep, rugged, rocky, volcanic
Elevation extremes: *lowest point:* Atlantic Ocean 0 m
highest point: Mt. Fogo 2,829 m (a volcano on Fogo Island)
Natural resources: salt, basalt rock, limestone, kaolin, fish, clay, gypsum
Land use: *arable land:* 11.41%
permanent crops: 0.74%
other: 87.85% (2005)
Irrigated land: 30 sq km (2003)
Total renewable water resources: 0.3 cu km (1990)
Freshwater withdrawal (domestic/industrial/agricultural): *total:* 0.02 cu km/yr (7%/2%/91%)
per capita: 39 cu m/yr (2000)
Natural hazards: prolonged droughts; seasonal harmattan wind produces obscuring dust; volcanically and seismically active
volcanism: Fogo (elev. 2,829 m), which last erupted in 1995, is Cape Verde's only active volcano
Environment—current issues: soil erosion; deforestation due to demand for wood used as fuel; water shortages; desertification; environmental damage has threatened several species of birds and reptiles; illegal beach sand extraction; overfishing
Environment—international agreements: *party to:* Biodiversity, Climate Change, Climate Change-Kyoto Protocol, Desertification, Endangered Species, Environmental Modification, Hazardous Wastes, Law of the Sea, Marine Dumping, Ozone Layer Protection, Ship Pollution, Wetlands
signed, but not ratified: none of the selected agreements
Geography—note: strategic location 500 km from west coast of Africa near major north-south sea routes; important communications station; important sea and air refueling site

PEOPLE

Population: 516,100 (July 2011 est.)
country comparison to the world: 169
Age structure: *0–14 years:* 32.6% (male 84,545/female 83,718)
15–64 years: 61.9% (male 154,697/female 164,917)
65 years and over: 5.5% (male 10,648/female 17,575) (2011 est.)
Median age: *total:* 22.7 years
male: 21.9 years
female: 23.5 years (2011 est.)
Population growth rate: 1.446% (2011 est.)
country comparison to the world: 81
Birth rate: 21.47 births/1,000 population (2011 est.)
country comparison to the world: 80
Death rate: 6.34 deaths/1,000 population (July 2011 est.)
country comparison to the world: 155
Net migration rate: -0.66 migrant(s)/1,000 population (2011 est.)
country comparison to the world: 143
Urbanization: *urban population:* 61% of total population (2010)
rate of urbanization: 2.4% annual rate of change (2010-15 est.)
Major cities—population: PRAIA (capital) 125,000 (2009)
Sex ratio: *at birth:* 1.03 male(s)/female
under 15 years: 1.01 male(s)/female
15–64 years: 0.94 male(s)/female
65 years and over: 0.61 male(s)/female
total population: 0.94 male(s)/female (2011 est.)
Infant mortality rate: *total:* 26.94 deaths/1,000 live births
country comparison to the world: 77
male: 30.8 deaths/1,000 live births
female: 22.96 deaths/1,000 live births (2011 est.)
Life expectancy at birth: *total population:* 70.7 years
country comparison to the world: 143
male: 68.51 years
female: 72.96 years (2011 est.)
Total fertility rate: 2.49 children born/woman (2011 est.)
country comparison to the world: 84
HIV/AIDS—adult prevalence rate: 0.04% (2001 est.)
country comparison to the world: 166
HIV/AIDS—people living with HIV/AIDS: 775 (2001)
country comparison to the world: 148
HIV/AIDS—deaths: 225 (as of 2001)
country comparison to the world: 100
Drinking water source: *Improved:*
urban: 85% of population
rural: 82% of population
total: 84% of population
Unimproved:
urban: 15% of population
rural: 17% of population

total: 16% of population (2008)
Sanitation facility access: *Improved:*
urban: 65% of population
rural: 38% of population
total: 54% of population
Unimproved:
urban: 35% of population
rural: 62% of population
total: 46% of population (2008)
Nationality:
noun: Cape Verdean(s)
adjective: Cape Verdean
Ethnic groups: Creole (mulatto) 71%, African 28%, European 1%
Religions: Roman Catholic (infused with indigenous beliefs), Protestant (mostly Church of the Nazarene)
Languages: Portuguese (official), Crioulo (a blend of Portuguese and West African words)
Literacy: *definition:* age 15 and over can read and write
total population: 76.6%
male: 85.8%
female: 69.2% (2003 est.)
School life expectancy (primary to tertiary education): *total:* 12 years
male: 11 years
female: 12 years (2009)
Education expenditures: 5.9% of GDP (2009)
country comparison to the world: 32

GOVERNMENT

Country name: *conventional long form:* Republic of Cape Verde
conventional short form: Cape Verde
local long form: Republica de Cabo Verde
local short form: Cabo Verde
Government type: republic
Capital: *name:* Praia
geographic coordinates: 14 55 N, 23 31 W
time difference: UTC-1 (4 hours ahead of Washington, DC during Standard Time)
Administrative divisions: 17 municipalities (concelhos, singular—concelho); Boa Vista, Brava, Maio, Mosteiros, Paul, Praia, Porto Novo, Ribeira Grande, Sal, Santa Catarina, Santa Cruz, Sao Domingos, Sao Filipe, Sao Miguel, Sao Nicolau, Sao Vicente, Tarrafal
Independence: 5 July 1975 (from Portugal)
National holiday: Independence Day, 5 July (1975)
Constitution: 25 September 1992; a major revision on 23 November 1995 substantially increased the powers of the president; a 1999 revision created the position of national ombudsman (Provedor de Justica)
Legal system: civil law system of Portugal
International law organization participation: has not submitted an ICJ jurisdiction declaration; non-party state to the ICCt
Suffrage: 18 years of age; universal
Executive branch: *chief of state:* President Pedro Verona Rodrigues PIRES (since 22 March 2001)
head of government: Prime Minister Jose Maria Pereira NEVES (since 1 February 2001)
cabinet: Council of Ministers appointed by the president on the recommendation of the prime minister
(For more information visit the World Leaders website)
elections: president elected by popular vote for a five-year term (eligible for a second term); election last held on 12 February 2006 (next to be held on 7 August 2011); prime minister nominated by the National Assembly and appointed by the president
election results: Pedro PIRES reelected president; percent of vote—Pedro PIRES (PAICV) 51.2%, Carlos VIEGA (MPD) 48.8%
Legislative branch: unicameral National Assembly or Assembleia Nacional (72 seats; members elected by popular vote to serve five-year terms)
elections: last held on 6 February 2011 (next to be held by 2016)
election results: percent of vote by party—NA; seats by party—PAICV 38, MPD 32, UCID 2
Judicial branch: Supreme Tribunal of Justice or Supremo Tribunal de Justia; Court of Audit; Military Courts; Fiscal and Customs Courts
Political parties and leaders: African Party for Independence of Cape Verde or PAICV [Jose Maria Pereira NEVES, chairman]; Democratic and Independent Cape Verdean Union or UCID [Antonio MONTEIRO]; Democratic Christian Party or PDC [Manuel RODRIGUES]; Democratic Renovation Party or PRD [Victor FIDALGO]; Movement for Democracy or MPD [Jorge SANTOS]; Party for Democratic Convergence or PCD [Dr. Eurico MONTEIRO]; Party of Work and Solidarity or PTS [Isaias RODRIGUES]; Social Democratic Party or PSD [Joao ALEM]
Political pressure groups and leaders: *other:* environmentalists; political pressure groups
International organization participation: ACP, AfDB, AOSIS, AU, CD, CPLP, ECOWAS, FAO, G-77, IBRD, ICAO, ICRM, IDA, IFAD, IFC, IFRCS, ILO, IMF, IMO, Interpol, IOC, IOM, IPU, ITSO, ITU, ITUC, MIGA, NAM, OIF, OPCW, UN, UNCTAD, UNESCO, UNIDO, Union Latina, UNWTO, UPU, WCO, WHO, WIPO, WMO, WTO
Diplomatic representation in the US: *chief of mission:* Ambassador Fatima Lima VEIGA
chancery: 3415 Massachusetts Avenue NW, Washington, DC 20008
telephone: [1] (202) 965-6820
FAX: [1] (202) 965-1207
consulate(s) general: Boston
Diplomatic representation from the US: *chief of mission:* Ambassador (vacant); Charge d'Affaires Dana BROWN
embassy: Rua Abilio Macedo n6, Praia
mailing address: C. P. 201, Praia
telephone: [238] 2-60-89-00
FAX: [238] 2-61-13-55
Flag description: five unequal horizontal bands; the top-most band of blue—equal to one half the width of the flag—is followed by three bands of white, red, and white, each equal to 1/12 of the width, and a bottom stripe of blue equal to one quarter of the flag width; a circle of 10, yellow, five-pointed stars is centered on the red stripe and positioned 3/8 of the length of the flag from the hoist side; blue stands for the sea and the sky, the circle of stars represents the 10 major islands united into a nation, the stripes symbolize the road to formation of the country through peace (white) and effort (red)
National anthem: *name:* "Cantico da Liberdade" (Song of Freedom)
lyrics/music: Amilcar Spencer LOPES/ Adalberto Higino Tavares SILVA
note: adopted 1996

ECONOMY

Economy—overview: This island economy suffers from a poor natural resource base, including serious water shortages exacerbated by cycles of long-term drought and poor soil for agriculture on several of the islands. The economy is service oriented with commerce, transport, tourism, and public services accounting for about three-fourths of GDP. Although about 40% of the population lives in rural areas, the share of food production in GDP is low. About 82% of food must be imported. The fishing potential, mostly lobster and tuna, is not fully exploited. Cape Verde annually runs a high trade deficit financed by foreign aid and remittances from its large pool of emigrants; remittances supplement GDP by more than 20%. Despite the lack of resources, sound economic management has produced steadily improving incomes. Continued economic reforms are aimed at developing the private sector and attracting foreign investment to diversify the economy. Future prospects depend heavily on the maintenance of aid flows, the encouragement of tourism, remittances, and the momentum of the government's development program. Cape Verde became a member of the WTO in July 2008.
GDP (purchasing power parity): $1.908 billion (2010 est.)
country comparison to the world: 187
$1.81 billion (2009 est.)
$1.747 billion (2008 est.)
note: data are in 2010 US dollars
GDP (official exchange rate): $1.651 billion (2010 est.)
GDP—real growth rate: 5.4% (2010 est.)
country comparison to the world: 63
3.6% (2009 est.)
6.2% (2008 est.)
GDP—per capita (PPP): $3,800 (2010 est.)
country comparison to the world: 159
$3,600 (2009 est.)
$3,500 (2008 est.)
note: data are in 2010 US dollars
GDP—composition by sector: *agriculture:* 9%
industry: 16.2%
services: 74.8% (2010 est.)
Labor force: 196,100 (2007)
country comparison to the world: 170
Unemployment rate: 21% (2000 est.)
country comparison to the world: 170
Population below poverty line: 30% (2000)
Household income or consumption by percentage share: *lowest 10%:* 1.9%
highest 10%: 40.6% (2001)
Investment (gross fixed): 36.9% of GDP (2010 est.)
country comparison to the world: 5
Budget: *revenues:* $520.7 million
expenditures: $680.8 million (2010 est.)
Inflation rate (consumer prices): 2.5% (2010 est.)
country comparison to the world: 64

1% (2009 est.)
Central bank discount rate: 7.5% (31 December 2009)
country comparison to the world: 51
7.5% (31 December 2008)
Commercial bank prime lending rate: 10.98% (31 December 2009 est.)
country comparison to the world: 88
9.99% (31 December 2008 est.)
Stock of narrow money: $585 million (31 December 2010 est.)
country comparison to the world: 157
$628.4 million (31 December 2009 est.)
Stock of broad money: $1.314 billion (31 December 2010 est.)
country comparison to the world: 152
$1.399 billion (31 December 2009 est.)
Stock of domestic credit: $1.179 billion (31 December 2010 est.)
country comparison to the world: 146
$1.256 billion (31 December 2009 est.)
Agriculture—products: bananas, corn, beans, sweet potatoes, sugarcane, coffee, peanuts; fish
Industries: food and beverages, fish processing, shoes and garments, salt mining, ship repair
Industrial production growth rate: 4% (2010 est.)
country comparison to the world: 91
Electricity—production: 250 million kWh (2007 est.)
country comparison to the world: 174
Electricity—consumption: 232.5 million kWh (2007 est.)
country comparison to the world: 173
Electricity—exports: 0 kWh (2008 est.)
Electricity—imports: 0 kWh (2008 est.)
Oil—production: 0 bbl/day (2009 est.)
country comparison to the world: 163
Oil—consumption: 2,000 bbl/day (2009 est.)
country comparison to the world: 186
Oil—exports: 0 bbl/day (2007 est.)
country comparison to the world: 158
Oil—imports: 1,619 bbl/day (2007 est.)
country comparison to the world: 179
Oil—proved reserves: 0 bbl (1 January 2010 est.)
country comparison to the world: 119
Natural gas—production: 0 cu m (2008 est.)
country comparison to the world: 115
Natural gas—consumption: 0 cu m (2008 est.)
country comparison to the world: 162
Natural gas—exports: 0 cu m (2008 est.)
country comparison to the world: 81
Natural gas—imports: 0 cu m (2008 est.)
country comparison to the world: 172
Natural gas—proved reserves: 0 cu m (1 January 2010 est.)
country comparison to the world: 124
Current account balance: $-286 million (2010 est.)
country comparison to the world: 97
$-319 million (2009 est.)
Exports: $114 million (2010 est.)
country comparison to the world: 192
$105 million (2009 est.)
Exports—commodities: fuel, shoes, garments, fish, hides
Exports—partners: Spain 53.98%, Portugal 22.23%, Morocco 7.13% (2009)
Imports: $858 million (2010 est.)
country comparison to the world: 177
$835 million (2009 est.)
Imports—commodities: foodstuffs, industrial products, transport equipment, fuels
Imports—partners: Portugal 44.86%, Netherlands 15.51%, Spain 6.1%, Italy 4.46%, Brazil 4.21% (2009)
Reserves of foreign exchange and gold: $296 million (31 December 2010 est.)
country comparison to the world: 130
$284 million (31 December 2009 est.)
Debt—external: $325 million (2002)
country comparison to the world: 169
Exchange rates: Cape Verdean escudos (CVE) per US dollar–88.58 (2010), 79.38 (2009), 73.84 (2008), 81.235 (2007), 87.946 (2006)

COMMUNICATIONS

Telephones—main lines in use: 72,200 (2009)
country comparison to the world: 154
Telephones—mobile cellular: 392,000 (2009)
country comparison to the world: 165
Telephone system: *general assessment:* effective system, extensive modernization from 1996-2000 following partial privatization in 1995
domestic: major service provider is Cabo Verde Telecom (CVT); fiber-optic ring, completed in 2001, links all islands providing Internet access and ISDN services; cellular service introduced in 1998; broadband services launched in 2004
international: country code–238; landing point for the Atlantis-2 fiber-optic transatlantic telephone cable that provides links to South America, Senegal, and Europe; HF radiotelephone to Senegal and Guinea-Bissau; satellite earth station–1 Intelsat (Atlantic Ocean) (2007)
Broadcast media: state-run TV and radio broadcast network plus a growing number of private broadcasters; Portuguese public TV and radio services for Africa are available; transmissions of a few international broadcasters are obtainable (2007)
Internet country code: .cv
Internet hosts: 26 (2010)
country comparison to the world: 215
Internet users: 150,000 (2009)
country comparison to the world: 148

TRANSPORTATION

Airports: 10 (2010)
country comparison to the world: 157
Airports—with paved runways: *total:* 9
over 3,047 m: 1
1,524 to 2,437 m: 3
914 to 1,523 m: 3
under 914 m: 2 (2010)
Airports—with unpaved runways: *total:* 1
under 914 m: 1 (2010)
Roadways:
total: 1,350 km
country comparison to the world: 178
paved: 932 km
unpaved: 418 km (2000)
Merchant marine: *total:* 13
country comparison to the world: 105
by type: cargo 3, chemical tanker 3, passenger/cargo 7
foreign-owned: 3 (Spain 1, UK 2) (2010)
Ports and terminals: Porto Grande

MILITARY

Military branches: Armed Forces: Army (also called the National Guard), Cape Verde Coast Guard (Guardia Costeira de Cabo Verde, GCCV; includes Air Force (Forca Aerea Caboverdaine), naval infantry) (2011)
Military service age and obligation: 18 years of age (est.) for selective compulsory military service; 14-month conscript service obligation (2006)
Manpower available for military service: *males age 16–49:* 132,087
females age 16–49: 136,956 (2010 est.)
Manpower fit for military service: *males age 16–49:* 106,864
females age 16–49: 117,518 (2010 est.)
Manpower reaching militarily significant age annually: *male:* 6,029
female: 6,026 (2010 est.)
Military expenditures: 0.5% of GDP (2009)
country comparison to the world: 160

TRANSNATIONAL ISSUES

Disputes—international: none
Illicit drugs: used as a transshipment point for Latin American cocaine destined for Western Europe, particularly because of Lusophone links to Brazil, Portugal, and Guinea-Bissau; has taken steps to deter drug money laundering, including a 2002 anti-money laundering reform that criminalizes laundering the proceeds of narcotics trafficking and other crimes and the establishment in 2008 of a Financial Intelligence Unit (2008)

CAYMAN ISLANDS

(OVERSEAS TERRITORY OF THE UK)

INTRODUCTION

Background: The Cayman Islands were colonized from Jamaica by the British during the 18th and 19th centuries and were administered by Jamaica after 1863. In 1959, the islands became a territory within the Federation of the West Indies. When the Federation dissolved in 1962, the Cayman Islands chose to remain a British dependency.

GEOGRAPHY

Location: Caribbean, three-island group (Grand Cayman, Cayman Brac, Little Cayman) in Caribbean Sea, 240 km south of Cuba and 268 km northwest of Jamaica
Geographic coordinates: 19 30 N, 80 30 W
Map references: Central America and the Caribbean
Area: *total:* 264 sq km
country comparison to the world: 209

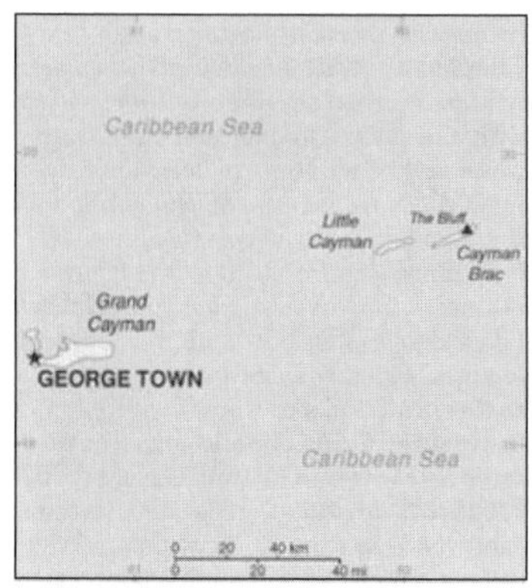

land: 264 sq km
water: 0 sq km
Area—comparative: 1.5 times the size of Washington, DC
Land boundaries: 0 km
Coastline: 160 km
Maritime claims: territorial sea: 12 nm
exclusive fishing zone: 200 nm
Climate: tropical marine; warm, rainy summers (May to October) and cool, relatively dry winters (November to April)
Terrain: low-lying limestone base surrounded by coral reefs
Elevation extremes: *lowest point:* Caribbean Sea 0 m
highest point: The Bluff on Cayman Brac 43 m
Natural resources: fish, climate and beaches that foster tourism
Land use: *arable land:* 3.85%
permanent crops: 0%
other: 96.15% (2005)
Irrigated land: NA
Natural hazards: hurricanes (July to November)
Environment—current issues:
no natural freshwater resources; drinking water supplies must be met by rainwater catchments
Geography—note: important location between Cuba and Central America

PEOPLE

Population: 51,384 (July 2011 est.)
country comparison to the world: 205
note: most of the population lives on Grand Cayman
Age structure: *0–14 years:* 19% (male 4,924/female 4,858)
15–64 years: 71.1% (male 17,766/female 18,743)
65 years and over: 9.9% (male 2,401/female 2,692) (2011 est.)
Median age: *total:* 38.7 years
male: 38.2 years
female: 39.2 years (2011 est.)
Population growth rate: 2.287% (2011 est.)
country comparison to the world: 35
Birth rate: 12.24 births/1,000 population (2011 est.)
country comparison to the world: 163
Death rate: 5.1 deaths/1,000 population (July 2011 est.)
country comparison to the world: 182
Net migration rate: 15.72 migrant(s)/1,000 population
country comparison to the world: 5
note: major destination for Cubans trying to migrate to the US (2011 est.)
Urbanization: *urban population:* 100% of total population (2010)
rate of urbanization: 0.9% annual rate of change (2010-15 est.)
Major cities—population: GEORGE TOWN (capital) 32,000 (2009)
Sex ratio: *at birth:* 1.016 male(s)/female
under 15 years: 1.01 male(s)/female
15–64 years: 0.95 male(s)/female
65 years and over: 0.89 male(s)/female
total population: 0.96 male(s)/female (2011 est.)
Infant mortality rate: *total:* 6.63 deaths/1,000 live births
country comparison to the world: 168
male: 7.6 deaths/1,000 live births
female: 5.65 deaths/1,000 live births (2011 est.)
Life expectancy at birth: *total population:* 80.68 years
country comparison to the world: 21
male: 78.02 years
female: 83.39 years (2011 est.)
Total fertility rate: 1.87 children born/woman (2011 est.)
country comparison to the world: 146
HIV/AIDS—adult prevalence rate: NA
HIV/AIDS—people living with HIV/AIDS: NA
HIV/AIDS—deaths: NA
Drinking water source: *Improved:*
urban: 95% of population
total: 95% of population
Unimproved:
urban: 5% of population
total: 5% of population (2008)
Sanitation facility access: *Improved:*
urban: 96% of population
total: 96% of population
Unimproved:
urban: 4% of population
total: 4% of population (2008)
Nationality: *noun:* Caymanian(s)
adjective: Caymanian
Ethnic groups: mixed 40%, white 20%, black 20%, expatriates of various ethnic groups 20%
Religions: Church of God 25.5%, Roman Catholic 12.6%, Presbyterian/United Church 9.2%, Seventh Day Adventist 8.4%, Baptist 8.3%, Pentecostal 6.7%, Anglican 3.9%, other religions 4%, non-denominational 5.7%, other 6.5%, none 6.1%, unspecified 3.2% (2007)
Languages: English (official) 95%, Spanish 3.2%, other 1.8% (1999 census)
Literacy: *definition:* age 15 and over has ever attended school
total population: 98%
male: 98%
female: 98% (1970 est.)
School life expectancy (primary to tertiary education): *total:* 12 years
male: 11 years
female: 12 years (2008)
Education expenditures: 2.6% of GDP (2006)
country comparison to the world: 146

GOVERNMENT

Country name: *conventional long form:* none
conventional short form: Cayman Islands
Dependency status: overseas territory of the UK
Government type: parliamentary democracy
Capital: *name:* George Town (on Grand Cayman)
geographic coordinates: 19 18 N, 81 23 W
time difference: UTC-5 (same time as Washington, DC during Standard Time)
Administrative divisions: 8 districts; Creek, Eastern, Midland, South Town, Spot Bay, Stake Bay, West End, Western
Independence: none (overseas territory of the UK)
National holiday: Constitution Day, first Monday in July
Constitution: The Cayman Islands Constitution Order 2009, 6 November 2009
Legal system: English common law and local statutes
Suffrage: 18 years of age; universal
Executive branch: *chief of state:* Queen ELIZABETH II (since 6 February 1952); represented by Governor Duncan TAYLOR (since 15 January 2010)
head of government: Premier McKeeva BUSH (since 6 November 2009)
cabinet: The Cabinet (six members are appointed by the governor on the advice of the premier, selected from among the elected members of the Legislative Assembly)
(For more information visit the World Leaders website)
elections: the monarchy is hereditary; the governor appointed by the monarch; following legislative elections, the leader of the majority party or coalition appointed by the governor as premier
Legislative branch: unicameral Legislative Assembly (20 seats; 18 members elected by popular vote and 2 ex officio members from The Cabinet; to serve four-year terms)
elections: last held on 20 May 2009 (next to be held not later than May 2013)
election results: percent of vote by party–NA; seats by party–UDP 9, PPM 5, independent 1
Judicial branch: Grand Court; Cayman Islands Court of Appeal; Summary Court
Political parties and leaders: People's Progressive Movement or PPM [Kurt TIBBETTS]; United Democratic Party or UDP [McKeeva BUSH]
Political pressure groups and leaders: National Trust
other: environmentalists
International organization participation: Caricom (associate), CDB, Interpol (subbureau), IOC, UNESCO (associate), UPU
Diplomatic representation in the US: none (overseas territory of the UK)
Diplomatic representation from the US: none (overseas territory of the UK); consular services provided through the US Embassy in Jamaica
Flag description: a blue field, with the flag of the UK in the upper hoist-side quadrant and the Caymanian coat of arms centered on the outer half of the flag; the coat of arms includes a crest with a pineapple, representing the connection with Jamaica, and a turtle, representing Cayman's seafaring tradition, above a shield bearing a golden lion, symbolizing Great Britain, below which are three green stars (representing the three islands) surmounting white and blue wavy lines representing the sea and a scroll at the bottom bearing the motto HE HATH FOUNDED IT UPON THE SEAS

National anthem: *name:* "Beloved Isle Cayman"
lyrics/music: Leila E. ROSS
note: adopted 1993; served as an unofficial anthem since 1930; as a territory of the United Kingdom, in addition to the local anthem, "God Save the Queen" is official (see United Kingdom)

ECONOMY

Economy—overview: With no direct taxation, the islands are a thriving offshore financial center. More than 93,000 companies were registered in the Cayman Islands as of 2008, including almost 300 banks, 800 insurers, and 10,000 mutual funds. A stock exchange was opened in 1997. Tourism is also a mainstay, accounting for about 70% of GDP and 75% of foreign currency earnings. The tourist industry is aimed at the luxury market and caters mainly to visitors from North America. Total tourist arrivals exceeded 1.9 million in 2008, with about half from the US. About 90% of the islands' food and consumer goods must be imported. The Caymanians enjoy a standard of living roughly equal to that of Switzerland.
GDP (purchasing power parity): $2.25 billion (2008 est.)
country comparison to the world: 183
$2.23 billion (2003 est.)
GDP (official exchange rate): $2.25 billion (2008 est.)
GDP—real growth rate: 1.1% (2008 est.)
country comparison to the world: 169
0.9% (2004 est.)
GDP—per capita (PPP): $43,800 (2004 est.)
country comparison to the world: 15
GDP—composition by sector:
agriculture: 1.4%
industry: 3.2%
services: 95.4% (1994 est.)
Labor force: 39,000
country comparison to the world: 197
note: nearly 55% are non-nationals (2007)
Labor force—by occupation:
agriculture: 1.9%
industry: 19.1%
services: 79% (2008 est.)
Unemployment rate: 4% (2008)
country comparison to the world: 37
4.4% (2004)
Population below poverty line: NA%
Household income or consumption by percentage share:
lowest 10%: NA%
highest 10%: NA%
Budget: *revenues:* $423.8 million
expenditures: $392.6 million (2004)
Inflation rate (consumer prices): 4.1% (2008)
country comparison to the world: 118
4.4% (2004)
Stock of narrow money: $334.3 million (31 December 2008)
country comparison to the world: 165
Stock of broad money: $5.564 billion (31 December 2008 est.)
country comparison to the world: 117
Market value of publicly traded shares: $NA (31 December 2008)
country comparison to the world: 114
$183.5 million (31 December 2007)
$188.4 million (31 December 2006)
Agriculture—products: vegetables, fruit; livestock; turtle farming
Industries: tourism, banking, insurance and finance, construction, construction materials, furniture
Industrial production growth rate: NA%
Electricity—production: 546 million kWh (2007 est.)
country comparison to the world: 156
Electricity—consumption: 507.8 million kWh (2007 est.)
country comparison to the world: 162
Electricity—exports: 0 kWh (2008 est.)
Electricity—imports: 0 kWh (2008 est.)
Oil—production: 0 bbl/day (2009 est.)
country comparison to the world: 159
Oil—consumption: 3,000 bbl/day (2009 est.)
country comparison to the world: 178
Oil—exports: 0 bbl/day (2007 est.)
country comparison to the world: 154
Oil—imports: 3,294 bbl/day (2007 est.)
country comparison to the world: 168
Oil—proved reserves: 0 bbl (1 January 2010 est.)
country comparison to the world: 115
Natural gas—production: 0 cu m (2008 est.)
country comparison to the world: 111
Natural gas—consumption: 0 cu m (2008 est.)
country comparison to the world: 158
Natural gas—exports: 0 cu m (2008 est.)
country comparison to the world: 75
Natural gas—imports: 0 cu m (2008 est.)
country comparison to the world: 165
Natural gas—proved reserves: 0 cu m (1 January 2010 est.)
country comparison to the world: 120
Exports: $13.8 million (2008)
country comparison to the world: 211
$2.52 million (2004)
Exports—commodities: turtle products, manufactured consumer goods
Imports: $876.5 million (2008)
country comparison to the world: 175
$866.9 million (2004)
Imports—commodities: foodstuffs, manufactured goods, fuels
Debt—external: $70 million (1996)
country comparison to the world: 184
Stock of direct foreign investment—at home: $NA
Stock of direct foreign investment—abroad: $NA
Exchange rates: Caymanian dollars (KYD) per US dollar–NA (2007), 0.8496 (2006)

COMMUNICATIONS

Telephones—main lines in use: 38,000 (2009)
country comparison to the world: 171
Telephones—mobile cellular: 33,800 (2004)
country comparison to the world: 200
Telephone system: *general assessment:* reasonably good overall telephone system with a high fixed-line teledensity
domestic: liberalization of telecom market in 2003; introduction of competition in the mobile-cellular market in 2004
international: country code–1-345; landing points for the MAYA-1, Eastern Caribbean Fiber System (ECFS), and the Cayman-Jamaica Fiber System submarine cables that provide links to the US and parts of Central and South America; satellite earth station–1 Intelsat (Atlantic Ocean) (2007)
Broadcast media: 4 television stations; cable and satellite subscription services offer a variety of international programming; government-owned Radio Cayman operates 2 networks broadcasting on 5 stations; 10 privately-owned radio stations operate alongside Radio Cayman (2007)
Internet country code: .ky
Internet hosts: 21,910 (2010)
country comparison to the world: 108
Internet users: 23,000 (2008)
country comparison to the world: 189

TRANSPORTATION

Airports: 3 (2010)
country comparison to the world: 193
Airports—with paved runways: *total:* 2
1,524 to 2,437 m: 2 (2010)
Airports—with unpaved runways: *total:* 1
914 to 1,523 m: 1 (2010)
Roadways:
total: 785 km
country comparison to the world: 186
paved: 785 km (2007)
Merchant marine: *total:* 113
country comparison to the world: 46
by type: bulk carrier 20, cargo 3, chemical tanker 56, liquefied gas 1, petroleum tanker 8, refrigerated cargo 10, vehicle carrier 15
foreign-owned: 99 (Germany 6, Greece 11, Italy 6, Japan 19, Switzerland 1, UK 2, US 54) (2010)
Ports and terminals: Cayman Brac, George Town

MILITARY

Military branches: no regular military forces; Royal Cayman Islands Police Force (2010)
Manpower available for military service: *males age 16–49:* 12,238 (2010 est.)
Manpower fit for military service: *males age 16–49:* 9,981
females age 16–49: 10,417 (2010 est.)
Manpower reaching militarily significant age annually: *male:* 333
female: 342 (2010 est.)
Military—note: defense is the responsibility of the UK

TRANSNATIONAL ISSUES

Disputes—international: none
Illicit drugs: major offshore financial center; vulnerable to drug transshipment to the US and Europe (2008)

CENTRAL AFRICAN REPUBLIC

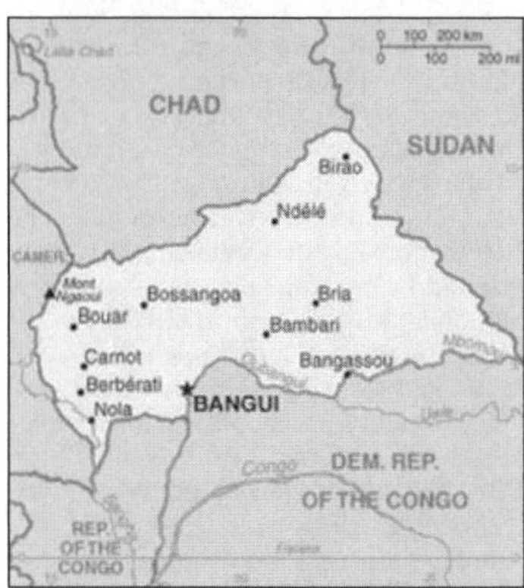

INTRODUCTION

Background: The former French colony of Ubangi-Shari became the Central African Republic upon independence in 1960. After three tumultuous decades of misrule–mostly by military governments–civilian rule was established in 1993 and lasted for one decade. President Ange-Felix PATASSE's civilian government was plagued by unrest, and in March 2003 he was deposed in a military coup led by General Francois BOZIZE, who established a transitional government. Though the government has the tacit support of civil society groups and the main parties, a wide field of candidates contested the municipal, legislative, and presidential elections held in March and May of 2005 in which General BOZIZE was affirmed as president. The government still does not fully control the countryside, where pockets of lawlessness persist. Unrest in the neighboring nations of Chad, Sudan, and the DRC continues to affect stability in the Central African Republic as well.

GEOGRAPHY

Location: Central Africa, north of Democratic Republic of the Congo
Geographic coordinates: 7 00 N, 21 00 E
Map references: Africa
Area: *total:* 622,984 sq km
country comparison to the world: 44
land: 622,984 sq km
water: 0 sq km
Area—comparative: slightly smaller than Texas
Land boundaries: *total:* 5,203 km
border countries: Cameroon 797 km, Chad 1,197 km, Democratic Republic of the Congo 1,577 km, Republic of the Congo 467 km, Sudan 1,165 km
Coastline: 0 km (landlocked)
Maritime claims: none (landlocked)
Climate: tropical; hot, dry winters; mild to hot, wet summers
Terrain: vast, flat to rolling, monotonous plateau; scattered hills in northeast and southwest
Elevation extremes: *lowest point:* Oubangui River 335 m
highest point: Mont Ngaoui 1,420 m
Natural resources: diamonds, uranium, timber, gold, oil, hydropower
Land use: *arable land:* 3.1%
permanent crops: 0.15%
other: 96.75% (2005)
Irrigated land: 20 sq km (2003)
Total renewable water resources: 144.4 cu km (2003)
Freshwater withdrawal (domestic/industrial/agricultural): *total:* 0.03 cu km/yr (80%/16%/4%)
per capita: 7 cu m/yr (2000)
Natural hazards: hot, dry, dusty harmattan winds affect northern areas; floods are common
Environment—current issues: tap water is not potable; poaching has diminished the country's reputation as one of the last great wildlife refuges; desertification; deforestation
Environment—international agreements: *party to:* Biodiversity, Climate Change, Climate Change-Kyoto Protocol, Desertification, Endangered Species, Hazardous Wastes, Ozone Layer Protection, Tropical Timber 94, Wetlands
signed, but not ratified: Law of the Sea
Geography—note: landlocked; almost the precise center of Africa

PEOPLE

Population: 4,950,027 (July 2011 est.)
country comparison to the world: 116
note: estimates for this country explicitly take into account the effects of excess mortality due to AIDS; this can result in lower life expectancy, higher infant mortality, higher death rates, lower population growth rates, and changes in the distribution of population by age and sex than would otherwise be expected
Age structure: *0–14 years:* 41% (male 1,021,144/female 1,007,819)
15–64 years: 55.3% (male 1,353,600/female 1,382,291)
65 years and over: 3.7% (male 73,977/female 111,196) (2011 est.)
Median age: *total:* 19.2 years
male: 18.8 years
female: 19.6 years (2011 est.)
Population growth rate: 2.146% (2011 est.)
country comparison to the world: 40
Birth rate: 36.46 births/1,000 population (2011 est.)
country comparison to the world: 24
Death rate: 15.01 deaths/1,000 population (July 2011 est.)
country comparison to the world: 10
Net migration rate: 0 migrant(s)/1,000 population (2011 est.)
country comparison to the world: 82
Urbanization: *urban population:* 39% of total population (2010)
rate of urbanization: 2.5% annual rate of change (2010-15 est.)
Major cities—population: BANGUI (capital) 702,000 (2009)
Sex ratio: *at birth:* 1.03 male(s)/female
under 15 years: 1.01 male(s)/female
15–64 years: 0.98 male(s)/female
65 years and over: 0.67 male(s)/female
total population: 0.98 male(s)/female (2011 est.)
Infant mortality rate: *total:* 99.38 deaths/1,000 live births
country comparison to the world: 6
male: 107.34 deaths/1,000 live births
female: 91.17 deaths/1,000 live births (2011 est.)
Life expectancy at birth: *total population:* 50.07 years
country comparison to the world: 214
male: 48.84 years
female: 51.35 years (2011 est.)
Total fertility rate: 4.63 children born/woman (2011 est.)
country comparison to the world: 31
HIV/AIDS—adult prevalence rate: 4.7% (2009 est.)
country comparison to the world: 16
HIV/AIDS—people living with HIV/AIDS: 130,000 (2009 est.)
country comparison to the world: 37
HIV/AIDS—deaths: 11,000 (2009 est.)
country comparison to the world: 26
Major infectious diseases: *degree of risk:* very high
food or waterborne diseases: bacterial and protozoal diarrhea, hepatitis A, and typhoid fever
vectorborne disease: malaria
respiratory disease: meningococcal meningitis
water contact disease: schistosomiasis
animal contact disease: rabies (2009)
Drinking water source: *Improved:*
urban: 92% of population
rural: 51% of population
total: 67% of population
Unimproved:
urban: 8% of population
rural: 49% of population
total: 33% of population (2008)
Sanitation facility access: *Improved:*
urban: 43% of population
rural: 28% of population
total: 34% of population
Unimproved:
urban: 57% of population
rural: 72% of population
total: 66% of population (2008)
Nationality: *noun:* Central African(s)
adjective: Central African
Ethnic groups: Baya 33%, Banda 27%, Mandjia 13%, Sara 10%, Mboum 7%, M'Baka 4%, Yakoma 4%, other 2%
Religions: indigenous beliefs 35%, Protestant 25%, Roman Catholic 25%, Muslim 15%
note: animistic beliefs and practices strongly influence the Christian majority
Languages: French (official), Sangho (lingua franca and national language), tribal languages
Literacy: *definition:* age 15 and over can read and write
total population: 48.6%
male: 64.8%
female: 33.5% (2000 est.)
School life expectancy (primary to tertiary education): *total:* 7 years
male: 8 years
female: 5 years (2009)
Education expenditures: 1.3% of GDP (2009)

country comparison to the world: 161

GOVERNMENT

Country name: *conventional long form:* Central African Republic
conventional short form: none
local long form: Republique Centrafricaine
local short form: none
former: Ubangi-Shari, Central African Empire
abbreviation: CAR
Government type: republic
Capital: *name:* Bangui
geographic coordinates: 4 22 N, 18 35 E
time difference: UTC+1 (6 hours ahead of Washington, DC during Standard Time)
Administrative divisions: 14 prefectures (prefectures, singular—prefecture), 2 economic prefectures* (prefectures economiques, singular—prefecture economique), and 1 commune**; Bamingui-Bangoran, Bangui**, Basse-Kotto, Haute-Kotto, Haut-Mbomou, Kemo, Lobaye, Mambere-Kadei, Mbomou, Nana-Grebizi*, Nana-Mambere, Ombella-Mpoko, Ouaka, Ouham, Ouham-Pende, Sangha-Mbaere*, Vakaga
Independence: 13 August 1960 (from France)
National holiday: Republic Day, 1 December (1958)
Constitution: ratified by popular referendum 5 December 2004; effective 27 December 2004
Legal system: civil law system based on the French model
International law organization participation: has not submitted an ICJ jurisdiction declaration; accepts ICCt jurisdiction
Suffrage: 18 years of age; universal
Executive branch: *chief of state:* President Francois BOZIZE (since 15 March 2003 coup)
head of government: Prime Minister Faustin-Archange TOUADERA (since 22 January 2008)
cabinet: Council of Ministers
(For more information visit the World Leaders website)
elections: president elected for a five-year term (eligible for a second term); elections last held on 23 January 2011 (next to be held in 2016); prime minister appointed by the president
election results: Francois BOZIZE elected to a second term as president; percent of vote—Francois BOZIZE (KNK) 64.4%, Ange-Felix PATASSE 21.4%, Martin ZIGUELE (MLPC) 6.8%, Emile Gros Raymond NAKOMBO (RDC) 4.6%, Jean-Jacques DEMAFOUTH (NAP) 2.8%
Legislative branch: unicameral National Assembly or Assemblee Nationale (105 seats; members are elected by popular vote to serve five-year terms)
elections: last held on 23 January 2011 and 27 March 2011 (next to be held in 2016)
election results: percent of vote by party—NA; seats by party—KNK 62, independents 26, MLPC 2, other 15
Judicial branch: Supreme Court or Cour Supreme; Constitutional Court (three judges appointed by the president, three by the president of the National Assembly, and three by fellow judges); Court of Appeal; Criminal Courts; Inferior Courts
Political parties and leaders: Alliance for Democracy and Progress or ADP [Jacques MBOLIEDAS]; Central African Democratic Rally or RDC [Andre KOLINGBA]; Civic Forum or FC [Gen. Timothee MALENDOMA]; Democratic Forum for Modernity or FODEM [Charles MASSI]; Liberal Democratic Party or PLD [Nestor KOMBO-NAGUEMON]; Londo Association or LONDO; Movement for Democracy and Development or MDD [David DACKO]; Movement for the Liberation of the Central African People or MLPC [Ange-Felix PATASSE] (the party of deposed president); National Convergence or KNK; National Unity Party or PUN [Jean-Paul NGOUPANDE]; New Alliance for Progress or NAP [Jean-Jacques DEMAFOUTH]; Patriotic Front for Progress or FPP [Abel GOUMBA]; People's Union for the Republic or UPR [Pierre Sammy MAKFOY]; Social Democratic Party or PSD [Enoch LAKOUE]
Political pressure groups and leaders: Monam (combating gender-base violence)
International organization participation: ACP, AfDB, AU, BDEAC, CEMAC, FAO, FZ, G-77, IAEA, IBRD, ICAO, ICRM, IDA, IFAD, IFC, IFRCS, ILO, IMF, Interpol, IOC, IOM, ISO (subscriber), ITSO, ITU, ITUC, MIGA, NAM, OIC, OIF, OPCW, UN, UNCTAD, UNESCO, UNIDO, UNWTO, UPU, WCO, WHO, WIPO, WMO, WTO
Diplomatic representation in the US: *chief of mission:* Ambassador Stanislas MOUSSA-KEMBE
chancery: 1618 22nd Street NW, Washington, DC 20008
telephone: [1] (202) 483-7800
FAX: [1] (202) 332-9893
Diplomatic representation from the US: *chief of mission:* Ambassador Frederick B. COOK
embassy: Avenue David Dacko, Bangui
mailing address: B. P. 924, Bangui
telephone: [236] 61 02 00
FAX: [236] 61 44 94
note: the embassy is currently operating with a minimal staff
Flag description: four equal horizontal bands of blue (top), white, green, and yellow with a vertical red band in center; a yellow five-pointed star to the hoist side of the blue band; banner combines the Pan-African and French flag colors; red symbolizes the blood spilled in the struggle for independence, blue represents the sky and freedom, white peace and dignity, green hope and faith, and yellow tolerance; the star represents aspiration towards a vibrant future
National anthem: *name:* "Le Renaissance" (The Renaissance)
lyrics/music: Barthelemy BOGANDA/Herbert PEPPER
note: adopted 1960; Barthelemy BOGANDA, who wrote the anthem's lyrics, was the first prime minister of the autonomous French territory

ECONOMY

Economy—overview: Subsistence agriculture, together with forestry, remains the backbone of the economy of the Central African Republic (CAR), with about 60% of the population living in outlying areas. The agricultural sector generates more than half of GDP. Timber has accounted for about 16% of export earnings and the diamond industry, for 40%. Important constraints to economic development include the CAR's landlocked position, a poor transportation system, a largely unskilled work force, and a legacy of misdirected macroeconomic policies. Factional fighting between the government and its opponents remains a drag on economic revitalization. Distribution of income is extraordinarily unequal. Grants from France and the international community can only partially meet humanitarian needs.
GDP (purchasing power parity): $3.446 billion (2010 est.)
country comparison to the world: 171
$3.337 billion (2009 est.)
$3.281 billion (2008 est.)
note: data are in 2010 US dollars
GDP (official exchange rate): $2.018 billion (2010 est.)
GDP—real growth rate: 3.3% (2010 est.)
country comparison to the world: 115
1.7% (2009 est.)
2% (2008 est.)
GDP—per capita (PPP): $700 (2010 est.)
country comparison to the world: 221
$700 (2009 est.)
$700 (2008 est.)
note: data are in 2010 US dollars
GDP—composition by sector:
agriculture: 55%
industry: 20%
services: 25% (2001 est.)
Labor force: 1.926 million (2007)
country comparison to the world: 122
Unemployment rate: 8% (2001 est.)
country comparison to the world: 87
note: 23% unemployment for Bangui
Population below poverty line: NA%
Household income or consumption by percentage share:
lowest 10%: 2.1%
highest 10%: 33% (2003)
Distribution of family income—Gini index: 61.3 (1993)
country comparison to the world: 6
Budget: *revenues:* $334 million
expenditures: $362 million (2009 est.)
Inflation rate (consumer prices): 0.9% (2007 est.)
country comparison to the world: 17
Central bank discount rate: 4.25% (31 December 2009)
country comparison to the world: 90
4.75% (31 December 2008)
Commercial bank prime lending rate: NA%
Stock of narrow money: $288.8 million (31 December 2009)
country comparison to the world: 167
$241.3 million (31 December 2008)
Stock of broad money: $343.4 million (31 December 2009 est.)
country comparison to the world: 178
$292.9 million (31 December 2008 est.)
Stock of domestic credit: $357.6 million (31 December 2009)
country comparison to the world: 168
$339.1 million (31 December 2008)
Market value of publicly traded shares: $NA

Agriculture—products: timber, cotton, coffee, tobacco, manioc (tapioca), yams, millet, corn, bananas; timber
Industries: gold and diamond mining, logging, brewing, textiles, footwear, assembly of bicycles and motorcycles
Industrial production growth rate: 3% (2002)
country comparison to the world: 112
Electricity—production: 115 million kWh (2007 est.)
country comparison to the world: 187
Electricity—consumption: 107 million kWh (2007 est.)
country comparison to the world: 188
Electricity—exports: 0 kWh (2008 est.)
Electricity—imports: 0 kWh (2008 est.)
Oil—production: 0 bbl/day (2009 est.)
country comparison to the world: 162
Oil—consumption: 2,000 bbl/day (2009 est.)
country comparison to the world: 181
Oil—exports: 0 bbl/day (2007 est.)
country comparison to the world: 156
Oil—imports: 2,203 bbl/day (2007 est.)
country comparison to the world: 174
Oil—proved reserves: 0 bbl (1 January 2010 est.)
country comparison to the world: 118
Natural gas—production: 0 cu m (2008 est.)
country comparison to the world: 114
Natural gas—consumption: 0 cu m (2008 est.)
country comparison to the world: 161
Natural gas—exports: 0 cu m (2008 est.)
country comparison to the world: 79
Natural gas—imports: 0 cu m (2008 est.)
country comparison to the world: 170
Natural gas—proved reserves: 0 cu m (1 January 2010 est.)
country comparison to the world: 123
Current account balance: $-77 million (2007 est.)
country comparison to the world: 75
Exports: $146.7 million (2007 est.)
country comparison to the world: 186
Exports—commodities: diamonds, timber, cotton, coffee, tobacco
Exports—partners: Belgium 32.57%, China 10.49%, Indonesia 10.36%, Morocco 10.24%, Democratic Republic of the Congo 6.87%, France 5.79% (2009)
Imports: $237.3 million (2007 est.)
country comparison to the world: 198
Imports—commodities: food, textiles, petroleum products, machinery, electrical equipment, motor vehicles, chemicals, pharmaceuticals
Imports—partners: South Korea 19.29%, France 11.95%, US 7.78%, Cameroon 7.39%, Netherlands 6.77% (2009)
Debt—external: $1.153 billion (2007 est.)
country comparison to the world: 148
Exchange rates: Cooperation Financiere en Afrique Centrale francs (XAF) per US dollar–495.28 (2010), 472.19 (2009), 481.8 (2007), 522.59 (2006)

COMMUNICATIONS

Telephones—main lines in use: 12,000 (2009)
country comparison to the world: 200
Telephones—mobile cellular: 168,000 (2009)
country comparison to the world: 174
Telephone system: *general assessment:* network consists principally of microwave radio relay and low-capacity, low-powered radiotelephone communication
domestic: limited telephone service with less than 1 fixed-line connection per 100 persons; spurred by the presence of multiple mobile-cellular service providers, cellular usage is increasing from a low base; most fixed-line and mobile-cellular telephone services are concentrated in Bangui
international: country code–236; satellite earth station–1 Intelsat (Atlantic Ocean) (2008)
Broadcast media: government-owned network, Radiodiffusion Television Centrafricaine, provides domestic TV broadcasting; licenses for 2 private TV stations are pending; state-owned radio network is supplemented by a small number of privately-owned broadcast stations as well as a few community radio stations; transmissions of at least 2 international broadcasters are available (2007)
Internet country code: .cf
Internet hosts: 20 (2010)
country comparison to the world: 217
Internet users: 22,600 (2009)
country comparison to the world: 191

TRANSPORTATION

Airports: 37 (2010)
country comparison to the world: 106
Airports—with paved runways: *total:* 2
2,438 to 3,047 m: 1
1,524 to 2,437 m: 1 (2010)
Airports—with unpaved runways: *total:* 35
2,438 to 3,047 m: 1
1,524 to 2,437 m: 12
914 to 1,523 m: 16
under 914 m: 6 (2010)
Roadways: *total:* 24,307 km (2000)
country comparison to the world: 104
Waterways: 2,800 km (the primary navigable river is the Ubangi, which joins the River Congo; it was the traditional route for the export of products because it connected with the Congo-Ocean railway at Brazzaville; because of the warfare on both sides of the River Congo from 1997, however, routes through Cameroon became preferred by importers and exporters) (2010)
country comparison to the world: 34
Ports and terminals: Bangui, Nola, Salo, Nzinga

MILITARY

Military branches: Central African Armed Forces (Forces Armees Centrafricaines, FACA): Ground Forces (includes Military Air Service), General Directorate of Gendarmerie Inspection (DGIG), National Police (2011)
Military service age and obligation: 18 years of age for selective military service; 2-year conscript service obligation (2010)
Manpower available for military service: *males age 16–49:* 1,149,856
females age 16–49: 1,145,897 (2010 est.)
Manpower fit for military service: *males age 16–49:* 655,875
females age 16–49: 661,308 (2010 est.)
Manpower reaching militarily significant age annually: *male:* 54,843
female: 53,999 (2010 est.)
Military expenditures: 0.9% of GDP (2009)
country comparison to the world: 143

TRANSNATIONAL ISSUES

Disputes—international: periodic skirmishes over water and grazing rights among related pastoral populations along the border with southern Sudan persist
Refugees and internally displaced persons: refugees (country of origin): 7,900 (Sudan); 3,700 (Democratic Republic of the Congo); note–UNHCR resumed repatriation of Southern Sudanese refugees in 2006
IDPs: 197,000 (ongoing unrest following coup in 2003) (2007)
Trafficking in persons: current situation: Central African Republic is a source, transit, and destination country for men, women, and children trafficked for the purposes of forced labor and sexual exploitation; the majority of victims are children trafficked within the country for sexual exploitation, domestic servitude, street vending, and forced agricultural, mine, market and restaurant labor; to a lesser extent, children are trafficked from the Central African Republic to Cameroon, Nigeria, and the Democratic Republic of Congo; rebels conscript children into armed forces within the country
tier rating: Tier 2 Watch List–Central African Republic is on the Tier 2 Watch List for the third consecutive year for its failure to show evidence of increasing efforts to combat trafficking in 2007; efforts to address trafficking through vigorous law enforcement measures and victim protection efforts were minimal, though awareness about trafficking appeared to be increasing in the country; the government does not actively investigate cases, work to identify trafficking victims among vulnerable populations, or rescue and provide care to victims; the government has not taken measures to reduce demand for commercial sex acts (2008)

CHAD

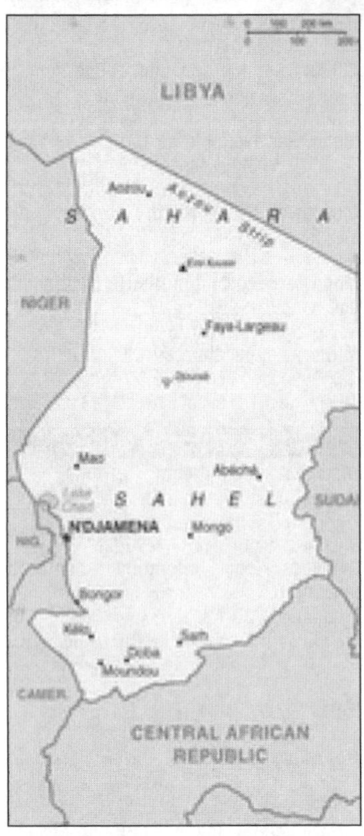

INTRODUCTION

Background: Chad, part of France's African holdings until 1960, endured three decades of civil warfare as well as invasions by Libya before a semblance of peace was finally restored in 1990. The government eventually drafted a democratic constitution and held flawed presidential elections in 1996 and 2001. In 1998, a rebellion broke out in northern Chad, which has sporadically flared up despite several peace agreements between the government and the rebels. In 2005, new rebel groups emerged in western Sudan and made probing attacks into eastern Chad despite signing peace agreements in December 2006 and October 2007. Power remains in the hands of an ethnic minority. In June 2005, President Idriss DEBY held a referendum successfully removing constitutional term limits and won another controversial election in 2006. Sporadic rebel campaigns continued throughout 2006 and 2007. The capital experienced a significant rebel threat in early 2008.

GEOGRAPHY

Location: Central Africa, south of Libya
Geographic coordinates: 15 00 N, 19 00 E
Map references: Africa
Area: *total:* 1.284 million sq km
country comparison to the world: 21
land: 1,259,200 sq km
water: 24,800 sq km
Area—comparative: slightly more than three times the size of California
Land boundaries: *total:* 5,968 km
border countries: Cameroon 1,094 km, Central African Republic 1,197 km, Libya 1,055 km, Niger 1,175 km, Nigeria 87 km, Sudan 1,360 km
Coastline: 0 km (landlocked)
Maritime claims: none (landlocked)
Climate: tropical in south, desert in north
Terrain: broad, arid plains in center, desert in north, mountains in northwest, lowlands in south
Elevation extremes: *lowest point:* Djourab 160 m
highest point: Emi Koussi 3,415 m
Natural resources: petroleum, uranium, natron, kaolin, fish (Lake Chad), gold, limestone, sand and gravel, salt
Land use: *arable land:* 2.8%
permanent crops: 0.02%
other: 97.18% (2005)
Irrigated land: 300 sq km (2003)
Total renewable water resources: 43 cu km (1987)
Freshwater withdrawal (domestic/industrial/agricultural): *total:* 0.23 cu km/yr (17%/0%/83%)
per capita: 24 cu m/yr (2000)
Natural hazards: hot, dry, dusty harmattan winds occur in north; periodic droughts; locust plagues
Environment—current issues: inadequate supplies of potable water; improper waste disposal in rural areas contributes to soil and water pollution; desertification
Environment—international agreements: *party to:* Biodiversity, Climate Change, Desertification, Endangered Species, Hazardous Wastes, Ozone Layer Protection, Wetlands
signed, but not ratified: Law of the Sea, Marine Dumping

Geography—note: landlocked; Lake Chad is the most significant water body in the Sahel

PEOPLE

Population: 10,758,945 (July 2011 est.)
country comparison to the world: 77
Age structure: *0–14 years:* 46% (male 2,510,656/female 2,441,780)
15–64 years: 51% (male 2,531,896/female 2,960,406)
65 years and over: 2.9% (male 131,805/female 182,402) (2011 est.)
Median age:
total: 16.8 years
male: 15.6 years
female: 17.9 years (2011 est.)
Population growth rate: 2.009% (2011 est.)
country comparison to the world: 49
Birth rate: 39.4 births/1,000 population (2011 est.)
country comparison to the world: 13
Death rate: 15.47 deaths/1,000 population (July 2011 est.)
country comparison to the world: 7
Net migration rate: -3.84 migrant(s)/1,000 population (2011 est.)
country comparison to the world: 188
Urbanization: *urban population:* 28% of total population (2010)
rate of urbanization: 4.6% annual rate of change (2010-15 est.)
Major cities—population: N'DJAMENA (capital) 808,000 (2009)
Sex ratio: *at birth:* 1.04 male(s)/female
under 15 years: 1.03 male(s)/female
15–64 years: 0.85 male(s)/female
65 years and over: 0.73 male(s)/female
total population: 0.92 male(s)/female (2011 est.)
Infant mortality rate: *total:* 95.31 deaths/1,000 live births
country comparison to the world: 8
male: 101.18 deaths/1,000 live births
female: 89.22 deaths/1,000 live births (2011 est.)
Life expectancy at birth: *total population:* 48.33 years
country comparison to the world: 219
male: 47.28 years
female: 49.43 years (2011 est.)
Total fertility rate: 5.05 children born/woman (2011 est.)
country comparison to the world: 19
HIV/AIDS—adult prevalence rate: 3.4% (2009 est.)
country comparison to the world: 19
HIV/AIDS—people living with HIV/AIDS: 210,000 (2009 est.)
country comparison to the world: 27
HIV/AIDS—deaths: 11,000 (2009 est.)
country comparison to the world: 24
Major infectious diseases: *degree of risk:* very high
food or waterborne diseases: bacterial and protozoal diarrhea, hepatitis A, and typhoid fever
vectorborne disease: malaria
water contact disease: schistosomiasis
respiratory disease: meningococcal meningitis
animal contact disease: rabies (2009)
Drinking water source: *Improved:*
urban: 67% of population
rural: 44% of population
total: 50% of population
Unimproved:
urban: 33% of population
rural: 56% of population
total: 50% of population (2008)
Sanitation facility access: *Improved:*
urban: 23% of population
rural: 4% of population
total: 9% of population
Unimproved:
urban: 77% of population
rural: 96% of population
total: 91% of population (2008)
Nationality:
noun: Chadian(s)
adjective: Chadian
Ethnic groups: Sara 27.7%, Arab 12.3%, Mayo-Kebbi 11.5%, Kanem-Bornou 9%, Ouaddai 8.7%, Hadjarai 6.7%, Tandjile 6.5%, Gorane 6.3%, Fitri-Batha 4.7%, other 6.4%, unknown 0.3% (1993 census)
Religions: Muslim 53.1%, Catholic 20.1%, Protestant 14.2%, animist 7.3%, other 0.5%, unknown 1.7%, atheist 3.1% (1993 census)

Languages: French (official), Arabic (official), Sara (in south), more than 120 different languages and dialects
Literacy: *definition:* age 15 and over can read and write French or Arabic
total population: 25.7%
male: 40.8%
female: 12.8% (2000 est.)
School life expectancy (primary to tertiary education): *total:* 7 years
male: 9 years
female: 5 years (2009)
Education expenditures: 3.2% of GDP (2009)
country comparison to the world: 125

GOVERNMENT

Country name: *conventional long form:* Republic of Chad
conventional short form: Chad
local long form: Republique du Tchad/ Jumhuriyat Tshad
local short form: Tchad/Tshad
Government type: republic
Capital: *name:* N'Djamena
geographic coordinates: 12 06 N, 15 02 E
time difference: UTC+1 (6 hours ahead of Washington, DC during Standard Time)
Administrative divisions: 22 regions (regions, singular–region); Barh el Gazel, Batha, Borkou, Chari-Baguirmi, Ennedi, Guera, Hadjer-Lamis, Kanem, Lac, Logone Occidental, Logone Oriental, Mandoul, Mayo-Kebbi Est, Mayo-Kebbi Ouest, Moyen-Chari, Ouaddai, Salamat, Sila, Tandjile, Tibesti, Ville de N'Djamena, Wadi Fira
Independence: 11 August 1960 (from France)
National holiday: Independence Day, 11 August (1960)
Constitution: passed by referendum 31 March 1996; a June 2005 referendum removed constitutional term limits
Legal system: mixed legal system of civil and customary law
International law organization participation: has not submitted an ICJ jurisdiction declaration; accepts ICCt jurisdiction
Suffrage: 18 years of age; universal
Executive branch: *chief of state:* President Lt. Gen. Idriss DEBY Itno (since 4 December 1990)
head of government: Prime Minister Emmanuel NADINGAR (since 5 March 2010)
cabinet: Council of State; members are appointed by the president on the recommendation of the prime minister
(For more information visit the World Leaders website)
elections: president elected by popular vote for a five-year term; if no candidate receives at least 50% of the total vote, the two candidates receiving the most votes must stand for a second round of voting; last election held on 25 April 2011 (next to be held by 2016); prime minister appointed by the president
election results: Lt. Gen. Idriss DEBY Itno reelected president; percent of vote–Lt. Gen. Idriss DEBY 83.6%, Albert Pahimi PADACKE 8.6%, Nadji Madou 7.8%
Legislative branch: unicameral National Assembly (188 seats; members elected by popular vote to serve four-year terms)
elections: National Assembly–last held on 13 February 2011 (next to be held by 2015); note–legislative elections, originally scheduled for 2006, were first delayed by National Assembly action and subsequently by an accord, signed in August 2007, between government and opposition parties
election results: percent of vote by party–NA; seats by party–ART 133, UNDR 11, others 44
Judicial branch: Supreme Court; Constitutional Council; High Court of Justice; Court of Appeal; Criminal Courts; Magistrate Courts
Political parties and leaders: Alliance for the Renaissance of Chad or ART, an alliance among the ruling MPS, RDP, and Viva-RNDP; Federation Action for the Republic or FAR [Ngarledjy YORONGAR]; National Rally for Development and Progress or Viva-RNDP [Delwa Kassire KOUMAKOYE]; National Union for Democracy and Renewal or UNDR [Saleh KEBZABO]; Party for Liberty and Development or PLD [Ibni Oumar Mahamat SALEH]; Patriotic Salvation Movement or MPS [Mahamat Saleh AHMAT, chairman]; Rally for Democracy and Progress or RDP [Lol Mahamat CHOUA]; Union for Renewal and Democracy or URD [Gen. Wadal Abdelkader KAMOUGUE]
Political pressure groups and leaders: rebel groups
International organization participation: ACP, AfDB, AU, BDEAC, CEMAC, FAO, FZ, G-77, IAEA, IBRD, ICAO, ICRM, IDA, IDB, IFAD, IFC, IFRCS, ILO, IMF, Interpol, IOC, ITSO, ITU, ITUC, MIGA, NAM, OIC, OIF, OPCW, UN, UNCTAD, UNESCO, UNIDO, UNOCI, UNWTO, UPU, WCO, WHO, WIPO, WMO, WTO
Diplomatic representation in the US: *chief of mission:* Ambassador Mahamoud Adam BECHIR
chancery: 2401 Massachusetts Avenue NW, Washington, DC 20008
telephone: [1] (202) 462-4009
FAX: [1] (202) 265-1937
Diplomatic representation from the US: *chief of mission:* Ambassador Louis NIGRO
embassy: Avenue Felix Eboue, N'Djamena
mailing address: B. P. 413, N'Djamena
telephone: [235] 251-62-11, 251-70-09, 251-77-59
FAX: [235] 251-56-54
Flag description: three equal vertical bands of blue (hoist side), yellow, and red; the flag combines the blue and red French (former colonial) colors with the red and yellow of the Pan-African colors; blue symbolizes the sky, hope, and the south of the country, which is relatively well-watered; yellow represents the sun, as well as the desert in the north of the country; red stands for progress, unity, and sacrifice
note: similar to the flag of Romania; also similar to the flags of Andorra and Moldova, both of which have a national coat of arms centered in the yellow band; design was based on the flag of France
National anthem: *name:* "La Tchadienne" (The Chadian)
lyrics/music: Louis GIDROL and his students/Paul VILLARD
note: adopted 1960

ECONOMY

Economy—overview: Chad's primarily agricultural economy will continue to be boosted by major foreign direct investment projects in the oil sector that began in 2000. At least 80% of Chad's population relies on subsistence farming and livestock raising for its livelihood. Chad's economy has long been handicapped by its landlocked position, high energy costs, and a history of instability. Chad relies on foreign assistance and foreign capital for most public and private sector investment projects. A consortium led by two US companies has been investing $3.7 billion to develop oil reserves–estimated at 1 billion barrels–in southern Chad. Chinese companies are also expanding exploration efforts and are currently building a 300-km pipeline and the country's first refinery. The nation's total oil reserves are estimated at 1.5 billion barrels. Oil production came on stream in late 2003. Chad began to export oil in 2004. Cotton, cattle, and gum arabic provide the bulk of Chad's non-oil export earnings.
GDP (purchasing power parity): $17.36 billion (2010 est.)
country comparison to the world: 131
$16.52 billion (2009 est.)
$16.48 billion (2008 est.)
note: data are in 2010 US dollars
GDP (official exchange rate): $7.848 billion (2010 est.)
GDP—real growth rate: 5.1% (2010 est.)
country comparison to the world: 69
0.3% (2009 est.)
-0.4% (2008 est.)
GDP—per capita (PPP): $1,600 (2010 est.)
country comparison to the world: 198
$1,600 (2009 est.)
$1,600 (2008 est.)
note: data are in 2010 US dollars
GDP—composition by sector: *agriculture:* 50.5%
industry: 7%
services: 42.5% (2010 est.)
Labor force: 4.293 million (2007)
country comparison to the world: 85
Labor force—by occupation: *agriculture:* 80% (subsistence farming, herding, and fishing)
industry and *services:* 20% (2006 est.)
Unemployment rate: NA%
Population below poverty line: 80% (2001 est.)
Household income or consumption by percentage share: *lowest 10%:* 2.6%
highest 10%: 30.8% (2003)
Investment (gross fixed): 14.8% of GDP (2010 est.)
country comparison to the world: 133
Budget: *revenues:* $1.972 billion
expenditures: $2.859 billion (2010 est.)
Inflation rate (consumer prices): 4% (2010 est.)
country comparison to the world: 113
10% (2009 est.)
Central bank discount rate: 4.25% (31 December 2009)
country comparison to the world: 94
4.75% (31 December 2008)

Commercial bank prime lending rate: NA%
Stock of narrow money: $920.9 million (31 December 2010 est.)
country comparison to the world: 141
$937.8 million (31 December 2009 est.)
Stock of broad money: $1.257 billion (31 December 2010 est.)
country comparison to the world: 155
$1.008 billion (31 December 2009 est.)
Stock of domestic credit: $943.8 million (31 December 2010 est.)
country comparison to the world: 152
$566.9 million (31 December 2009 est.)
Market value of publicly traded shares: $NA
Agriculture—products: cotton, sorghum, millet, peanuts, rice, potatoes, manioc (tapioca); cattle, sheep, goats, camels
Industries: oil, cotton textiles, meatpacking, brewing, natron (sodium carbonate), soap, cigarettes, construction materials
Industrial production growth rate: 3% (2010 est.)
country comparison to the world: 113
Electricity—production: 100 million kWh (2007 est.)
country comparison to the world: 191
Electricity—consumption: 93 million kWh (2007 est.)
country comparison to the world: 191
Electricity—exports: 0 kWh (2008 est.)
Electricity—imports: 0 kWh (2008 est.)
Oil—production: 115,000 bbl/day (2009 est.)
country comparison to the world: 51
Oil—consumption: 1,000 bbl/day (2009 est.)
country comparison to the world: 200
Oil—exports: 157,900 bbl/day (2007 est.)
country comparison to the world: 55
Oil—imports: 1,571 bbl/day (2007 est.)
country comparison to the world: 180
Oil—proved reserves: 1.5 billion bbl (1 January 2010 est.)
country comparison to the world: 39
Natural gas—production: 0 cu m (2008 est.)
country comparison to the world: 108
Natural gas—consumption: 0 cu m (2008 est.)
country comparison to the world: 155
Natural gas—exports: 0 cu m (2008 est.)
country comparison to the world: 70
Natural gas—imports: 0 cu m (2008 est.)
country comparison to the world: 162
Natural gas—proved reserves: 0 cu m (1 January 2010 est.)
country comparison to the world: 118
Current account balance: $-2.6 billion (2010 est.)
country comparison to the world: 164
$-2.305 billion (2009 est.)
Exports: $3.036 billion (2010 est.)
country comparison to the world: 122
$2.709 billion (2009 est.)
Exports—commodities: oil, cattle, cotton, gum arabic
Exports—partners: US 90.06%, France 4.81%, China 1.6% (2009)
Imports: $2.631 billion (2010 est.)
country comparison to the world: 144
$2.539 billion (2009 est.)
Imports—commodities: machinery and transportation equipment, industrial goods, foodstuffs, textiles
Imports—partners: France 17.74%, Cameroon 12.7%, China 11.23%, US 7.59%, Italy 6.54%, Ukraine 5.33%, Netherlands 4.37% (2009)
Reserves of foreign exchange and gold: $868 million (31 December 2010 est.)
country comparison to the world: 118
$685 million (31 December 2009 est.)
Debt—external: $NA (31 December 2010 est.)
$1.749 billion (31 December 2008 est.)
Stock of direct foreign investment—at home: $NA (31 December 2010)
$4.5 billion (2006 est.)
Stock of direct foreign investment—abroad: $NA
Exchange rates: Cooperation Financiere en Afrique Centrale francs (XAF) per US dollar–495.28 (2010), 472.19 (2009), 447.81 (2008), 480.1 (2007), 522.59 (2006)

COMMUNICATIONS

Telephones—main lines in use: 13,000 (2009)
country comparison to the world: 199
Telephones—mobile cellular: 2.686 million (2009)
country comparison to the world: 119
Telephone system: *general assessment:* inadequate system of radiotelephone communication stations with high costs and low telephone density
domestic: fixed-line connections for only about 1 per 1000 persons coupled with mobile-cellular subscribership base of only about 25 per 100 persons
international: country code–235; satellite earth station–1 Intelsat (Atlantic Ocean) (2009)
Broadcast media: 1 state-owned TV broadcast station; state-owned radio network, Radiodiffusion Nationale Tchadienne (RNT), operates national and regional stations; about 10 private radio stations; some stations rebroadcast programs from international broadcasters (2007)
Internet country code: .td
Internet hosts: 5 (2010)
country comparison to the world: 226
Internet users: 168,100 (2009)
country comparison to the world: 145

TRANSPORTATION

Airports: 56 (2010)
country comparison to the world: 83
Airports—with paved runways: *total:* 8
over 3,047 m: 2
2,438 to 3,047 m: 3
1,524 to 2,437 m: 2
under 914 m: 1 (2010)
Airports—with unpaved runways: *total:* 48
2,438 to 3,047 m: 2
1,524 to 2,437 m: 15
914 to 1,523 m: 21
under 914 m: 10 (2010)
Pipelines: oil 265 km (2010)
Roadways: *total:* 33,400 km
country comparison to the world: 95
paved: 267 km
unpaved: 33,133 km (2002)
Waterways: (Chari and Legone rivers are navigable only in wet season) (2010)

MILITARY

Military branches: Armed Forces: Chadian National Army (Armee Nationale du Tchad, ANT), Chadian Air Force (Force Aerienne Tchadienne, FAT), Gendarmerie (2008)
Military service age and obligation: 20 years of age for conscripts, with 3-year service obligation; 18 years of age for volunteers; no minimum age restriction for volunteers with consent from a parent or guardian; women are subject to 1 year of compulsory military or civic service at age of 21 (2004)
Manpower available for military service:
males age 16–49: 2,090,244
females age 16–49: 2,441,321 (2010 est.)
Manpower fit for military service: *males age 16–49:* 1,183,242
females age 16–49: 1,395,811 (2010 est.)
Manpower reaching militarily significant age annually:
male: 128,723
female: 128,244 (2010 est.)
Military expenditures: 1.7% of GDP (2009)
country comparison to the world: 90

TRANSNATIONAL ISSUES

Disputes—international: since 2003, Janjawid armed militia and the Sudanese military have driven hundreds of thousands of Darfur residents into Chad; Chad remains an important mediator in the Sudanese civil conflict, reducing tensions with Sudan arising from cross-border banditry; Chadian Aozou rebels reside in southern Libya; only Nigeria and Cameroon have heeded the Lake Chad Commission's admonition to ratify the delimitation treaty, which also includes the Chad-Niger and Niger-Nigeria boundaries
Refugees and internally displaced persons: refugees (country of origin): 234,000 (Sudan); 54,200 (Central African Republic)
IDPs: 178,918 (2007)
Trafficking in persons: current situation: Chad is a source, transit, and destination country for children trafficked for the purposes of forced labor and commercial sexual exploitation; the majority of children are trafficked within Chad for involuntary domestic servitude, forced cattle herding, forced begging, forced labor in petty commerce or the fishing industry, or for commercial sexual exploitation; to a lesser extent, Chadian children are also trafficked to Cameroon, the Central African Republic, and Nigeria for cattle herding; children may also be trafficked from Cameroon and the Central African Republic to Chad's oil producing regions for sexual exploitation
tier rating: Tier 2 Watch List–the Government of Chad does not fully comply with the minimum standards for the elimination of trafficking and is not making any significant efforts to do so; although facing resource constraints, the government has the capacity to conduct basic anti-trafficking law enforcement efforts, yet did not do so during the last year; it showed no results in enforcing government policy prohibiting the recruitment of child soldiers; Chad has not ratified the 2000 UN TIP Protocol (2009)

CHILE

INTRODUCTION

Background: Prior to the arrival of the Spanish in the 16th century, the Inca ruled northern Chile while the indigenous Mapuche inhabited central and southern Chile. Although Chile declared its independence in 1810, decisive victory over the Spanish was not achieved until 1818. In the War of the Pacific (1879-83), Chile defeated Peru and Bolivia and won its present northern regions. It was not until the 1880s that the Mapuche Indians were completely subjugated. After a series of elected governments, the three-year-old Marxist government of Salvador ALLENDE was overthrown in 1973 by a military coup led by Augusto PINOCHET, who ruled until a freely elected president was installed in 1990. Sound economic policies, maintained consistently since the 1980s, have contributed to steady growth, reduced poverty rates by over half, and have helped secure the country's commitment to democratic and representative government. Chile has increasingly assumed regional and international leadership roles befitting its status as a stable, democratic nation.

GEOGRAPHY

Location: Southern South America, bordering the South Pacific Ocean, between Argentina and Peru
Geographic coordinates: 30 00 S, 71 00 W
Map references: South America
Area: *total:* 756,102 sq km
country comparison to the world: 38
land: 743,812 sq km
water: 12,290 sq km
note: includes Easter Island (Isla de Pascua) and Isla Sala y Gomez
Area—comparative: slightly smaller than twice the size of Montana
Land boundaries: *total:* 6,339 km
border countries: Argentina 5,308 km, Bolivia 860 km, Peru 171 km
Coastline: 6,435 km
Maritime claims: territorial sea: 12 nm
contiguous zone: 24 nm
exclusive economic zone: 200 nm
continental shelf: 200/350 nm
Climate: temperate; desert in north; Mediterranean in central region; cool and damp in south
Terrain: low coastal mountains; fertile central valley; rugged Andes in east
Elevation extremes: *lowest point:* Pacific Ocean 0 m
highest point: Nevado Ojos del Salado 6,880 m
Natural resources: copper, timber, iron ore, nitrates, precious metals, molybdenum, hydropower
Land use: *arable land:* 2.62%
permanent crops: 0.43%
other: 96.95% (2005)
Irrigated land: 19,000 sq km (2003)
Total renewable water resources: 922 cu km (2000)
Freshwater withdrawal (domestic/industrial/agricultural): *total:* 12.55 cu km/yr (11%/25%/64%)
per capita: 770 cu m/yr (2000)
Natural hazards: severe earthquakes; active volcanism; tsunamis
volcanism: Chile experiences significant volcanic activity due to the more than three-dozen active volcanoes situated within the Andes Mountains; Lascar (elev. 5,592 m), which last erupted in 2007, is the most active volcano in the northern Chilean Andes; Llaima (elev. 3,125 m) in central Chile, which last erupted in 2009, is another of the country's most active; Chaiten's 2008 eruption forced major evacuations; other notable historically active volcanoes include Cerro Hudson, Copahue, Guallatiri, Llullaillaco, Nevados de Chillan, Puyehue, San Pedro, and Villarrica
Environment—current issues: widespread deforestation and mining threaten natural resources; air pollution from industrial and vehicle emissions; water pollution from raw sewage
Environment—international agreements: *party to:* Antarctic-Environmental Protocol, Antarctic-Marine Living Resources, Antarctic Seals, Antarctic Treaty, Biodiversity, Climate Change, Climate Change-Kyoto Protocol, Desertification, Endangered Species, Environmental Modification, Hazardous Wastes, Law of the Sea, Marine Dumping, Ozone Layer Protection, Ship Pollution, Wetlands, Whaling
signed, but not ratified: none of the selected agreements
Geography—note: strategic location relative to sea lanes between Atlantic and Pacific Oceans (Strait of Magellan, Beagle Channel, Drake Passage); Atacama Desert is one of world's driest regions; the crater lake of Ojos del Salado is the worlds highest lake (at 6,390m)

PEOPLE

Population: 16,888,760 (July 2011 est.)
country comparison to the world: 59
Age structure: *0–14 years:* 22.3% (male 1,928,210/female 1,840,839)
15–64 years: 68.1% (male 5,751,091/female 5,744,014)
65 years and over: 9.6% (male 680,450/female 944,156) (2011 est.)
Median age: *total:* 32.1 years
male: 31.1 years
female: 33.1 years (2011 est.)
Population growth rate: 0.836% (2011 est.)
country comparison to the world: 131
Birth rate: 14.33 births/1,000 population (2011 est.)
country comparison to the world: 144
Death rate: 5.97 deaths/1,000 population (July 2011 est.)
country comparison to the world: 163
Net migration rate: 0 migrant(s)/1,000 population (2011 est.)
country comparison to the world: 79
Urbanization: *urban population:* 89% of total population (2010)
rate of urbanization: 1.1% annual rate of change (2010-15 est.)
Major cities—population: SANTIAGO (capital) 5.883 million; Valparaiso 865,000 (2009)
Sex ratio: *at birth:* 1.05 male(s)/female
under 15 years: 1.05 male(s)/female
15–64 years: 1 male(s)/female
65 years and over: 0.72 male(s)/female
total population: 0.98 male(s)/female (2011 est.)
Infant mortality rate: *total:* 7.34 deaths/1,000 live births
country comparison to the world: 163
male: 8.1 deaths/1,000 live births
female: 6.55 deaths/1,000 live births (2011 est.)
Life expectancy at birth: *total population:* 77.7 years
country comparison to the world: 56
male: 74.44 years
female: 81.13 years (2011 est.)
Total fertility rate: 1.88 children born/woman (2011 est.)
country comparison to the world: 144
HIV/AIDS—adult prevalence rate: 0.4% (2009 est.)
country comparison to the world: 73
HIV/AIDS—people living with HIV/AIDS: 40,000 (2009 est.)
country comparison to the world: 62
HIV/AIDS—deaths: NA
Drinking water source: *Improved:*
urban: 99% of population
rural: 75% of population
total: 96% of population
Unimproved:
urban: 1% of population
rural: 25% of population
total: 4% of population (2008)
Sanitation facility access: *Improved:*

urban: 98% of population
rural: 83% of population
total: 96% of population
Unimproved:
urban: 2% of population
rural: 17% of population
total: 4% of population (2008)
Nationality: *noun:* Chilean(s)
adjective: Chilean
Ethnic groups: white and white-Amerindian 95.4%, Mapuche 4%, other indigenous groups 0.6% (2002 census)
Religions: Roman Catholic 70%, Evangelical 15.1%, Jehovah's Witness 1.1%, other Christian 1%, other 4.6%, none 8.3% (2002 census)
Languages: Spanish (official), Mapudungun, German, English
Literacy: *definition:* age 15 and over can read and write
total population: 95.7%
male: 95.8%
female: 95.6% (2002 census)
School life expectancy (primary to tertiary education): *total:* 15 years
male: 15 years
female: 15 years (2008)
Education expenditures: 4% of GDP (2008)
country comparison to the world: 104

GOVERNMENT

Country name: *conventional long form:* Republic of Chile
conventional short form: Chile
local long form: Republica de Chile
local short form: Chile
Government type: republic
Capital: *name:* Santiago
geographic coordinates: 33 27 S, 70 40 W
time difference: UTC-4 (1 hour ahead of Washington, DC during Standard Time)
daylight saving time: +1hr, begins third Sunday in August; ends second Sunday in May; note—the end of DST was delayed until 8 May 2011 due to the ongoing energy crisis
note: Valparaiso is the seat of the national legislature
Administrative divisions: 15 regions (regiones, singular—region); Aisen del General Carlos Ibanez del Campo, Antofagasta, Araucania, Arica y Parinacota, Atacama, Biobio, Coquimbo, Libertador General Bernardo O'Higgins, Los Lagos, Los Rios, Magallanes y de la Antartica Chilena, Maule, Region Metropolitana (Santiago), Tarapaca, Valparaiso
note: the US does not recognize claims to Antarctica
Independence: 18 September 1810 (from Spain)
National holiday: Independence Day, 18 September (1810)
Constitution: 11 September 1980, effective 11 March 1981; amended several times
Legal system: civil law system influenced by several West European civil legal systems; judicial review of legislative acts in the Supreme Court
International law organization participation: has not submitted an ICJ jurisdiction declaration; accepts ICCt jurisdiction
Suffrage: 18 years of age; universal and compulsory
Executive branch: *chief of state:* President Sebastian PINERA Echenique (since 11 March 2010); note—the president is both the chief of state and head of government
head of government: President Sebastian PINERA Echenique (since 11 March 2010)
cabinet: Cabinet appointed by the president
(For more information visit the World Leaders website)
elections: president elected by popular vote for a single four-year term; election last held on 13 December 2009 with runoff election held on 17 January 2010 (next to be held in December 2013)
election results: Sebastian PINERA Echenique elected president; percent of vote—Sebastian PINERA Echenique 51.6%; Eduardo FREI 48.4%
Legislative branch: bicameral National Congress or Congreso Nacional consists of the Senate or Senado (38 seats; members elected by popular vote to serve eight-year terms; one-half elected every four years) and the Chamber of Deputies or Camara de Diputados (120 seats; members are elected by popular vote to serve four-year terms)
elections: Senate—last held on 13 December 2009 (next to be held in December 2013); Chamber of Deputies—last held on 13 December 2009 (next to be held in December 2013)
election results: Senate—percent of vote by party—NA; seats by party—CPD 9 (PDC 4, PPD 3, PS 2), APC 9 (RN 6, UDI 3); Chamber of Deputies—percent of vote by party—NA; seats by party—APC 58 (UDI 37, RN 18, other 3), CPD 57 (PDC 19, PPD 18, PS 11, PRSD 5, PC 3, other 1), PRI 3, independent 2
Judicial branch: Supreme Court or Corte Suprema (judges are appointed by the president and ratified by the Senate from lists of candidates provided by the court itself; the president of the Supreme Court is elected every three years by the 20-member court); Constitutional Tribunal (eight-members—two each from the Senate, Chamber of Deputies, Supreme Court, and National Security Council—review the constitutionality of laws approved by Congress)
Political parties and leaders: Broad Social Movement or MAS; Clean Chile Vote Happy or CLVF (including Broad Social Movement, Country Force, and Regionalist Party of Independents or PRI); Coalition for Change or CC (formerly known as the Alliance for Chile (Alianza) or APC) (including National Renewal or RN [Carlos LARRAIN Pena], Independent Democratic Union or UDI [Juan Antonio COLOMA Correa], and Chile First [Vlado MIROSEVIC]); Coalition of Parties for Democracy (Concertacion) or CPD (including Christian Democratic Party or PDC [Ignacio WALKER], Party for Democracy or PPD [Carolina TOHA Morales], Radical Social Democratic Party or PRSD [Jose Antonio GOMEZ Urrutia], and Socialist Party or PS [Osvaldo ANDRADE]); Partido Ecologista del Sur; Together We Can Do More (including Communist Party or PC [Guillermo TEILLIER del Valle], and Humanist Party or PH [Danilo MONTEVERDE])
Political pressure groups and leaders: Roman Catholic Church, particularly conservative groups such as Opus Dei; United Labor Central or CUT includes trade unionists from the country's five largest labor confederations
other: revitalized university student federations at all major universities
International organization participation: APEC, BIS, CAN (associate), CD, FAO, G-15, G-77, IADB, IAEA, IBRD, ICAO, ICC, ICRM, IDA, IFAD, IFC, IFRCS, IHO, ILO, IMF, IMO, IMSO, Interpol, IOC, IOM, IPU, ISO, ITSO, ITU, ITUC, LAES, LAIA, Mercosur (associate), MIGA, MINUSTAH, NAM, OAS, OECD, OPANAL, OPCW, PCA, RG, SICA (observer), UN, UNASUR, UNCTAD, UNESCO, UNFICYP, UNHCR, UNIDO, Union Latina, UNMOGIP, UNTSO, UNWTO, UPU, WCO, WFTU, WHO, WIPO, WMO, WTO
Diplomatic representation in the US: *chief of mission:* Ambassador Arturo FERMANDOIS Vohringer
chancery: 1732 Massachusetts Avenue NW, Washington, DC 20036
telephone: [1] (202) 785-1746
FAX: [1] (202) 887-5579
consulate(s) general: Chicago, Houston, Los Angeles, Miami, New York, Philadelphia, San Francisco, San Juan (Puerto Rico)
Diplomatic representation from the US: *chief of mission:* Ambassador Alejandro D. WOLFF
embassy: Avenida Andres Bello 2800, Las Condes, Santiago
mailing address: APO AA 34033
telephone: [56] (2) 330-3000
FAX: [56] (2) 330-3710, 330-3160
Flag description: two equal horizontal bands of white (top) and red; a blue square the same height as the white band at the hoist-side end of the white band; the square bears a white five-pointed star in the center representing a guide to progress and honor; blue symbolizes the sky, white is for the snow-covered Andes, and red represents the blood spilled to achieve independence
note: design was influenced by the US flag
National anthem: *name:* "Himno Nacional de Chile" (National Anthem of Chile)
lyrics/music: Eusebio LILLO Robles and Bernardo DE VERA y Pintado/Ramon CARNICER y Battle
note: music adopted 1828, original lyrics adopted 1818, adapted lyrics adopted 1847; under Augusto PINOCHET"s military rule, a verse glorifying the army was added; however, as a protest, some citizens refused to sing this verse; it was removed when democracy was restored in 1990

ECONOMY

Economy—overview: Chile has a market-oriented economy characterized by a high level of foreign trade and a reputation for strong financial institutions and sound policy that have given it the strongest sovereign bond rating in South America. Exports account for more than one-fourth of GDP, with commodities making up some three-quarters of total exports. Copper alone provides one-third of government revenue. During the early 1990s, Chile's reputation as a role model for economic reform was strengthened when the democratic government of Patricio AYLWIN–which took over from the military in 1990–deepened the economic reform initiated by the military government. Since 1999, growth has averaged 4% per year. Chile deepened its longstanding commitment to trade liberalization with the signing of a free trade agreement with the US, which took effect on 1 January 2004. Chile claims to have more bilateral or regional trade agreements than any other country. It has 57 such agreements (not all of them full free trade agreements), including with the European Union, Mercosur, China, India, South Korea, and Mexico. Over the past seven years, foreign direct investment inflows have quadrupled to some $15 billion in 2010, but FDI had dropped to about $7 billion in 2009 in the face of diminished investment throughout the world. The Chilean government conducts a rule-based countercyclical fiscal policy, accumulating surpluses in sovereign wealth funds during periods of high copper prices and economic growth, and allowing deficit spending only during periods of low copper prices and growth. As of September 2008, those sovereign wealth funds–kept mostly outside the country and separate from Central Bank reserves–amounted to more than $20 billion. Chile used $4 billion from this fund to finance a fiscal stimulus package to fend off recession. In December 2009, the OECD invited Chile to become a full member, after a two year period of compliance with organization mandates, and in May 2010 Chile signed the OECD Convention, becoming the first South American country to join the OECD. The economy started to show signs of a rebound in the fourth quarter of 2009, and GDP grew more than 5% in 2010. Chile achieved this growth despite the magnitude 8.8 earthquake that struck Chile in February 2010, which was one of the top ten strongest earthquakes on record. The earthquake and subsequent tsunamis it generated caused considerable damage near the epicenter, located about 70 miles from Concepcion–and about 200 miles southwest of Santiago. The Chilean Ministry of Finance estimates the total immediate losses were close to 17% of GDP.

GDP (purchasing power parity): $257.9 billion (2010 est.)
country comparison to the world: 46
$245 billion (2009 est.)
$249.2 billion (2008 est.)
note: data are in 2010 US dollars

GDP (official exchange rate): $203.3 billion (2010 est.)

GDP—real growth rate: 5.3% (2010 est.)
country comparison to the world: 64
-1.7% (2009 est.)
3.7% (2008 est.)

GDP—per capita (PPP): $15,400 (2010 est.)
country comparison to the world: 72
$14,800 (2009 est.)
$15,100 (2008 est.)
note: data are in 2010 US dollars

GDP—composition by sector:
agriculture: 5.6%
industry: 40.5%
services: 53.9% (2009 est.)

Labor force: 7.58 million (2010 est.)
country comparison to the world: 61

Labor force—by occupation:
agriculture: 13.2%
industry: 23%
services: 63.9% (2005)

Unemployment rate: 8.7% (2010 est.)
country comparison to the world: 100
9.6% (2009 est.)

Population below poverty line: 11.5% (2009)

Household income or consumption by percentage share: *lowest 10%:* 1.6%
highest 10%: 41.7% (2006)

Distribution of family income—Gini index: 52.4 (2009)
country comparison to the world: 16
57.1 (2000)

Investment (gross fixed): 23.5% of GDP (2010 est.)
country comparison to the world: 53

Budget: *revenues:* $40.97 billion
expenditures: $45.07 billion (2010 est.)

Public debt: 6.2% of GDP (2010 est.)
country comparison to the world: 127
6.1% of GDP (2009 est.)

Inflation rate (consumer prices): 1.7% (2010 est.)
country comparison to the world: 46
1.5% (2009 est.)

Central bank discount rate: 0.5% (31 December 2009)
country comparison to the world: 49
8.25% (31 December 2008)

Commercial bank prime lending rate: 7.25% (31 December 2009 est.)
country comparison to the world: 57
13.26% (31 December 2008 est.)

Stock of narrow money: $29.81 billion (31 December 2010 est.)
country comparison to the world: 58
$23.68 billion (31 December 2009 est.)

Stock of broad money: $160.3 billion (31 December 2009)
country comparison to the world: 45
$127.5 billion (31 December 2008)

Stock of domestic credit: $153.6 billion (31 December 2010 est.)
country comparison to the world: 41
$133.7 billion (31 December 2009 est.)

Market value of publicly traded shares: $209.5 billion (31 December 2009)
country comparison to the world: 31
$132.4 billion (31 December 2008)
$212.9 billion (31 December 2007)

Agriculture—products: grapes, apples, pears, onions, wheat, corn, oats, peaches, garlic, asparagus, beans; beef, poultry, wool; fish; timber

Industries: copper, lithium, other minerals, foodstuffs, fish processing, iron and steel, wood and wood products, transport equipment, cement, textiles

Industrial production growth rate: 3.2% (2010 est.)
country comparison to the world: 101

Electricity—production: 60.6 billion kWh (2007 est.)
country comparison to the world: 43

Electricity—consumption: 57.29 billion kWh (2007 est.)
country comparison to the world: 42

Electricity—exports: 0 kWh (2008 est.)

Electricity—imports: 1.628 billion kWh (2007 est.)

Oil—production: 10,850 bbl/day (2009 est.)
country comparison to the world: 85

Oil—consumption: 277,000 bbl/day (2009 est.)
country comparison to the world: 47

Oil—exports: 49,250 bbl/day (2007 est.)
country comparison to the world: 78

Oil—imports: 311,200 bbl/day (2007 est.)
country comparison to the world: 36

Oil—proved reserves: 150 million bbl (1 January 2010 est.)
country comparison to the world: 65

Natural gas—production: 1.65 billion cu m (2008 est.)
country comparison to the world: 59

Natural gas—consumption: 2.34 billion cu m (2008 est.)
country comparison to the world: 79

Natural gas—exports: 0 cu m (2008 est.)
country comparison to the world: 74

Natural gas—imports: 690 million cu m (2008 est.)
country comparison to the world: 58

Natural gas—proved reserves: 97.97 billion cu m (1 January 2010 est.)
country comparison to the world: 53

Current account balance: $1.033 billion (2010 est.)
country comparison to the world: 46
$4.217 billion (2009 est.)

Exports: $64.28 billion (2010 est.)
country comparison to the world: 45
$53.74 billion (2009 est.)

Exports—commodities: copper, fruit, fish products, paper and pulp, chemicals, wine

Exports—partners: China 16.46%, US 11.31%, Japan 9.06%, South Korea 6.49%, Brazil 4.64%, Mexico 4.09% (2009)

Imports: $54.23 billion (2010 est.)
country comparison to the world: 47
$39.75 billion (2009 est.)

Imports—commodities: petroleum and petroleum products, chemicals, electrical and telecommunications equipment, industrial machinery, vehicles, natural gas

Imports—partners: US 21.77%, China 12.76%, Argentina 9.55%, Brazil 6.46%, South Korea 5.35% (2009)

Reserves of foreign exchange and gold: $26.08 billion (31 December 2010 est.)
country comparison to the world: 39
$25.29 billion (31 December 2009 est.)

Debt—external: $84.51 billion (31 December 2010 est.)
country comparison to the world: 41
$72.76 billion (31 December 2009 est.)

Stock of direct foreign investment—at home: $136.3 billion (31 December 2010 est.)
country comparison to the world: 26
$121.6 billion (31 December 2009 est.)

Stock of direct foreign investment—abroad: $51.15 billion (31 December 2010 est.)
country comparison to the world: 33
$41.2 billion (31 December 2009 est.)

Exchange rates: Chilean pesos (CLP) per US dollar–525.34 (2010), 560.86 (2009), 509.02 (2008), 526.25 (2007), 530.29 (2006)

COMMUNICATIONS

Telephones—main lines in use: 3.575 million (2009)
country comparison to the world: 44
Telephones—mobile cellular: 16.45 million (2009)
country comparison to the world: 44
Telephone system: *general assessment:* privatization begun in 1988; most advanced telecommunications infrastructure in South America; modern system based on extensive microwave radio relay facilities; domestic satellite system with 3 earth stations
domestic: number of fixed-line connections have stagnated in recent years as mobile-cellular usage continues to increase, reaching a level of 100 telephones per 100 persons
international: country code–56; landing points for the Pan American, South America-1, and South American Crossing/Latin America Nautilius submarine cables providing links to the US and to Central and South America; satellite earth stations–2 Intelsat (Atlantic Ocean) (2009)
Broadcast media: national and local terrestrial television channels, coupled with extensive cable TV networks; the state-owned Television Nacional de Chile (TVN) network is self-financed through commercial advertising revenues and is not under direct government control; large number of privately-owned TV stations; about 250 radio stations (2007)
Internet country code: .cl
Internet hosts: 1.056 million (2010)
country comparison to the world: 43
Internet users: 7.009 million (2009)
country comparison to the world: 39

TRANSPORTATION

Airports: 366 (2010)
country comparison to the world: 22
Airports—with paved runways: *total:* 84
over 3,047 m: 5
2,438 to 3,047 m: 8
1,524 to 2,437 m: 23
914 to 1,523 m: 24
under 914 m: 24 (2010)
Airports—with unpaved runways:
total: 282
2,438 to 3,047 m: 3
1,524 to 2,437 m: 12
914 to 1,523 m: 50
under 914 m: 217 (2010)
Pipelines: gas 3,064 km; liquid petroleum gas 517 km; oil 895 km; refined products 768 km (2010)
Railways: *total:* 7,082 km
country comparison to the world: 28
broad gauge: 3,435 km 1.676-m gauge (850 km electrified)
narrow gauge: 3,647 km 1.000-m gauge (2010)
Roadways: *total:* 80,505 km
country comparison to the world: 59
paved: 16,745 km (includes 2,414 km of expressways)
unpaved: 63,760 km (2004)
Merchant marine: *total:* 48
country comparison to the world: 71
by type: bulk carrier 11, cargo 10, chemical tanker 8, container 1, liquefied gas 2, passenger 4, passenger/cargo 3, petroleum tanker 8, roll on/roll off 1
foreign-owned: 1 (Norway 1)
registered in other countries: 48 (Argentina 6, Belize 1, Brazil 1, Cyprus 1, Isle of Man 8, Liberia 7, Panama 17, Singapore 7) (2010)
Ports and terminals: Coronel, Huasco, Lirquen, Puerto Ventanas, San Antonio, San Vicente, Valparaiso

MILITARY

Military branches: Army of the Nation, Chilean Navy (Armada de Chile, includes Naval Aviation, Marine Corps, and Maritime Territory and Merchant Marine Directorate (Directemar)), Chilean Air Force (Fuerza Aerea de Chile, FACh), Carabineros Corps (Cuerpo de Carabineros) (2011)
Military service age and obligation: 18-45 years of age for voluntary male and female military service, although the right to compulsory recruitment is retained; service obligation–12 months for Army, 22 months for Navy and Air Force (2008)
Manpower available for military service:
males age 16–49: 4,324,732
females age 16–49: 4,251,954 (2010 est.)
Manpower fit for military service: *males age 16–49:* 3,621,475
females age 16–49: 3,561,099 (2010 est.)
Manpower reaching militarily significant age annually:
male: 141,500
female: 135,709 (2010 est.)
Military expenditures: 2.7% of GDP (2006)
country comparison to the world: 53

TRANSNATIONAL ISSUES

Disputes—international: Chile and Peru rebuff Bolivia's reactivated claim to restore the Atacama corridor, ceded to Chile in 1884, but Chile has offered instead unrestricted but not sovereign maritime access through Chile to Bolivian natural gas; Chile rejects Peru's unilateral legislation to change its latitudinal maritime boundary with Chile to an equidistance line with a southwestern axis favoring Peru, in October 2007, Peru took its maritime complaint with Chile to the ICJ; territorial claim in Antarctica (Chilean Antarctic Territory) partially overlaps Argentine and British claims; the joint boundary commission, established by Chile and Argentina in 2001, has yet to map and demarcate the delimited boundary in the inhospitable Andean Southern Ice Field (Campo de Hielo Sur)
Illicit drugs: transshipment country for cocaine destined for Europe and the region; some money laundering activity, especially through the Iquique Free Trade Zone; imported precursors passed on to Bolivia; domestic cocaine consumption is rising, making Chile a significant consumer of cocaine (2008)

CHINA

(ALSO SEE SEPARATE HONG KONG, MACAU, AND TAIWAN ENTRIES)

INTRODUCTION

Background: For centuries China stood as a leading civilization, outpacing the rest of the world in the arts and sciences, but in the 19th and early 20th centuries, the country was beset by civil unrest, major famines, military defeats, and foreign occupation. After World War II, the Communists under MAO Zedong established an autocratic socialist system that, while ensuring China's sovereignty, imposed strict controls over everyday life and cost the lives of tens of millions of people. After 1978, MAO's successor DENG Xiaoping and other leaders focused on market-oriented economic development and by 2000 output had quadrupled. For much of the population, living standards have improved dramatically and the room for personal choice has expanded, yet political controls remain tight. China since the early 1990s has increased its global outreach and participation in international organizations.

GEOGRAPHY

Location: Eastern Asia, bordering the East China Sea, Korea Bay, Yellow Sea, and South China Sea, between North Korea and Vietnam
Geographic coordinates: 35 00 N, 105 00 E
Map references: Asia
Area: *total:* 9,596,961 sq km
country comparison to the world: 4
land: 9,569,901 sq km
water: 27,060 sq km
Area—comparative: slightly smaller than the US
Land boundaries: *total:* 22,117 km
border countries: Afghanistan 76 km, Bhutan 470 km, Burma 2,185 km, India 3,380 km, Kazakhstan 1,533 km, North Korea 1,416 km, Kyrgyzstan 858 km, Laos 423 km, Mongolia 4,677 km, Nepal 1,236 km, Pakistan 523 km, Russia (northeast) 3,605 km, Russia (northwest) 40 km, Tajikistan 414 km, Vietnam 1,281 km
regional borders: Hong Kong 30 km, Macau 0.34 km
Coastline: 14,500 km
Maritime claims: territorial sea: 12 nm
contiguous zone: 24 nm
exclusive economic zone: 200 nm
continental shelf: 200 nm or to the edge of the continental margin
Climate: extremely diverse; tropical in south to subarctic in north
Terrain: mostly mountains, high plateaus, deserts in west; plains, deltas, and hills in east
Elevation extremes: *lowest point:* Turpan Pendi -154 m
highest point: Mount Everest 8,850 m
Natural resources: coal, iron ore, petroleum, natural gas, mercury, tin, tungsten, antimony, manganese, molybdenum, vanadium, magnetite, aluminum, lead, zinc, rare earth elements, uranium, hydropower potential (world's largest)
Land use: *arable land:* 14.86%
permanent crops: 1.27%
other: 83.87% (2005)
Irrigated land: 545,960 sq km (2003)
Total renewable water resources: 2,829.6 cu km (1999)
Freshwater withdrawal (domestic/industrial/agricultural): *total:* 549.76 cu km/yr (7%/26%/68%)
per capita: 415 cu m/yr (2000)
Natural hazards: frequent typhoons (about five per year along southern and eastern coasts); damaging floods; tsunamis; earthquakes; droughts; land subsidence
volcanism: China contains some historically active volcanoes including Changbaishan (also known as Baitoushan, Baegdu, or P'aektu-san), Hainan Dao, and Kunlun although most have been relatively inactive in recent centuries
Environment—current issues: air pollution (greenhouse gases, sulfur dioxide particulates) from reliance on coal produces acid rain; water shortages, particularly in the north; water pollution from untreated wastes; deforestation; estimated loss of one-fifth of agricultural land since 1949 to soil erosion and economic development; desertification; trade in endangered species
Environment—international agreements: *party to:* Antarctic-Environmental Protocol, Antarctic Treaty, Biodiversity, Climate Change, Climate Change-Kyoto Protocol, Desertification, Endangered Species, Environmental Modification, Hazardous Wastes, Law of the Sea, Marine Dumping, Ozone Layer Protection, Ship Pollution, Tropical Timber 83, Tropical Timber 94, Wetlands, Whaling
signed, but not ratified: none of the selected agreements
Geography—note: world's fourth largest country (after Russia, Canada, and US); Mount Everest on the border with Nepal is the world's tallest peak

PEOPLE

Population: 1,336,718,015 (July 2011 est.)
country comparison to the world: 1
Age structure: *0–14 years:* 17.6% (male 126,634,384/female 108,463,142)
15–64 years: 73.6% (male 505,326,577/female 477,953,883)
65 years and over: 8.9% (male 56,823,028/female 61,517,001) (2011 est.)
Median age:
total: 35.5 years
male: 34.9 years
female: 36.2 years (2011 est.)
Population growth rate: 0.493% (2011 est.)
country comparison to the world: 152
Birth rate: 12.29 births/1,000 population (2011 est.)
country comparison to the world: 160
Death rate: 7.03 deaths/1,000 population (July 2011 est.)
country comparison to the world: 132
Net migration rate: -0.33 migrant(s)/1,000 population (2011 est.)
country comparison to the world: 130
Urbanization: *urban population:* 47% of total population (2010)
rate of urbanization: 2.3% annual rate of change (2010-15 est.)
Major cities—population: Shanghai 16.575 million; BEIJING (capital) 12.214 million; Chongqing 9.401 million; Shenzhen 9.005 million; Guangzhou 8.884 million (2009)
Sex ratio: *at birth:* 1.133 male(s)/female
under 15 years: 1.17 male(s)/female
15–64 years: 1.06 male(s)/female
65 years and over: 0.93 male(s)/female
total population: 1.06 male(s)/female (2011 est.)
Infant mortality rate: *total:* 16.06 deaths/1,000 live births
country comparison to the world: 112
male: 15.61 deaths/1,000 live births
female: 16.57 deaths/1,000 live births (2011 est.)
Life expectancy at birth: *total population:* 74.68 years
country comparison to the world: 95
male: 72.68 years
female: 76.94 years (2011 est.)
Total fertility rate: 1.54 children born/woman (2011 est.)
country comparison to the world: 182
HIV/AIDS—adult prevalence rate: 0.1% (2009 est.)
country comparison to the world: 122
HIV/AIDS—people living with HIV/AIDS: 740,000 (2009 est.)
country comparison to the world: 14
HIV/AIDS—deaths: 26,000 (2009 est.)
country comparison to the world: 14
Major infectious diseases: *degree of risk:* intermediate
food or waterborne diseases: bacterial diarrhea, hepatitis A, and typhoid fever
vectorborne diseases: Japanese encephalitis and dengue fever
soil contact disease: hantaviral hemorrhagic fever with renal syndrome (HFRS)
animal contact disease: rabies
note: highly pathogenic H5N1 avian influenza has been identified in this country; it poses a negligible risk with extremely rare cases possible among US citizens who have close contact with birds (2009)
Drinking water source: *Improved:*
urban: 98% of population
rural: 82% of population
total: 89% of population
Unimproved:
urban: 2% of population
rural: 18% of population
total: 11% of population (2008)
Sanitation facility access: *Improved:*
urban: 58% of population
rural: 52% of population
total: 55% of population
Unimproved:
urban: 42% of population
rural: 48% of population
total: 45% of population (2008)
Nationality: *noun:* Chinese (singular and plural)
adjective: Chinese
Ethnic groups: Han Chinese 91.5%, Zhuang, Manchu, Hui, Miao, Uighur, Tujia, Yi, Mongol, Tibetan, Buyi, Dong, Yao, Korean, and other nationalities 8.5% (2000 census)
Religions: Daoist (Taoist), Buddhist, Christian 3%-4%, Muslim 1%-2%
note: officially atheist (2002 est.)
Languages: Standard Chinese or Mandarin (Putonghua, based on the Beijing dialect), Yue (Cantonese), Wu (Shanghainese), Minbei (Fuzhou), Minnan (Hokkien-Taiwanese), Xiang, Gan, Hakka dialects, minority languages (see Ethnic groups entry)
note: Mongolian is official in Nei Mongol, Uighur is official in Xinjiang Uygur, and Tibetan is official in Xizang (Tibet)
Literacy: *definition:* age 15 and over can read and write
total population: 91.6%
male: 95.7%
female: 87.6% (2007)
School life expectancy (primary to tertiary education):
total: 12 years
male: 11 years
female: 12 years (2009)
Education expenditures: NA

GOVERNMENT

Country name: *conventional long form:* People's Republic of China
conventional short form: China
local long form: Zhonghua Renmin Gongheguo
local short form: Zhongguo

abbreviation: PRC
Government type: Communist state
Capital: *name:* Beijing
geographic coordinates: 39 55 N, 116 23 E
time difference: UTC+8 (13 hours ahead of Washington, DC during Standard Time)
note: despite its size, all of China falls within one time zone; many people in Xinjiang Province observe an unofficial "Xinjiang timezone" of UTC+6, two hours behind Beijing
Administrative divisions: 23 provinces (sheng, singular and plural), 5 autonomous regions (zizhiqu, singular and plural), and 4 municipalities (shi, singular and plural)
provinces: Anhui, Fujian, Gansu, Guangdong, Guizhou, Hainan, Hebei, Heilongjiang, Henan, Hubei, Hunan, Jiangsu, Jiangxi, Jilin, Liaoning, Qinghai, Shaanxi, Shandong, Shanxi, Sichuan, Yunnan, Zhejiang; (see note on Taiwan)
autonomous regions: Guangxi, Nei Mongol, Ningxia, Xinjiang Uygur, Xizang (Tibet)
municipalities: Beijing, Chongqing, Shanghai, Tianjin
note: China considers Taiwan its 23rd province; see separate entries for the special administrative regions of Hong Kong and Macau
Independence: 1 October 1949 (People's Republic of China established); notable earlier dates: 221 BC (unification under the Qin Dynasty); 1 January 1912 (Qing Dynasty replaced by the Republic of China)
National holiday: Anniversary of the founding of the People's Republic of China, 1 October (1949)
Constitution: most recent promulgation 4 December 1982; amended several times
Legal system: civil law influenced by Soviet and continental European civil law systems; legislature retains power to interpret statutes; constitution ambiguous on judicial review of legislation
International law organization participation: has not submitted an ICJ jurisdiction declaration; non-party state to the ICCt
Suffrage: 18 years of age; universal
Executive branch: *chief of state:* President HU Jintao (since 15 March 2003); Vice President XI Jinping (since 15 March 2008)
head of government: Premier WEN Jiabao (since 16 March 2003); Executive Vice Premier LI Keqiang (17 March 2008), Vice Premier HUI Liangyu (since 17 March 2003), Vice Premier ZHANG Dejiang (since 17 March 2008), and Vice Premier WANG Qishan (since 17 March 2008)
cabinet: State Council appointed by National People's Congress
(For more information visit the World Leaders website)
elections: president and vice president elected by National People's Congress for a five-year term (eligible for a second term); elections last held on 15-17 March 2008 (next to be held in mid-March 2013); premier nominated by president, confirmed by National People's Congress
election results: HU Jintao elected president by National People's Congress with a total of 2,963 votes; XI Jinping elected vice president with a total of 2,919 votes
Legislative branch: unicameral National People's Congress or Quanguo Renmin Daibiao Dahui (2,987 seats; members elected by municipal, regional, and provincial people's congresses, and People's Liberation Army to serve five-year terms)
elections: last held in December 2007-February 2008 (date of next election to be held in late 2012 to early 2013)
election results: percent of vote–NA; seats–2,987
note: only members of the CCP, its eight allied parties, and sympathetic independent candidates are elected
Judicial branch: Supreme People's Court (judges appointed by the National People's Congress); Local People's Courts (comprise higher, intermediate, and basic courts); Special People's Courts (primarily military, maritime, railway transportation, and forestry courts)
Political parties and leaders: Chinese Communist Party or CCP [HU Jintao]; eight registered small parties controlled by CCP
Political pressure groups and leaders: no substantial political opposition groups exist
International organization participation: ADB, AfDB (nonregional member), APEC, ARF, ASEAN (dialogue partner), BIS, CDB, CICA, EAS, FAO, FATF, G-20, G-24 (observer), G-77, IADB, IAEA, IBRD, ICAO, ICC, ICRM, IDA, IFAD, IFC, IFRCS, IHO, ILO, IMF, IMO, IMSO, Interpol, IOC, IOM (observer), IPU, ISO, ITSO, ITU, LAIA (observer), MIGA, MINURSO, MONUSCO, NAM (observer), NSG, OAS (observer), OPCW, PCA, PIF (partner), SAARC (observer), SCO, SICA (observer), UN, UN Security Council, UNAMID, UNCTAD, UNESCO, UNHCR, UNIDO, UNIFIL, UNITAR, UNMIL, UNMIS, UNMIT, UNOCI, UNTSO, UNWTO, UPU, WCO, WHO, WIPO, WMO, WTO, ZC
Diplomatic representation in the US: *chief of mission:* Ambassador ZHANG Yesui
chancery: 3505 International Place NW, Washington, DC 20008
telephone: [1] (202) 495-2266
FAX: [1] (202) 495-2190
consulate(s) general: Chicago, Houston, Los Angeles, New York, San Francisco
Diplomatic representation from the US: *chief of mission:* Ambassador (vacant); Charge d'Affaires Robert S. WONG
embassy: 55 An Jia Lou Lu, 100600 Beijing
mailing address: PSC 461, Box 50, FPO AP 96521-0002
telephone: [86] (10) 8531-3000
FAX: [86] (10) 8531-3300
consulate(s) general: Chengdu, Guangzhou, Shanghai, Shenyang, Wuhan
Flag description: red with a large yellow five-pointed star and four smaller yellow five-pointed stars (arranged in a vertical arc toward the middle of the flag) in the upper hoist-side corner; the color red represents revolution, while the stars symbolize the four social classes–the working class, the peasantry, the urban petty bourgeoisie, and the national bourgeoisie (capitalists)–united under the Communist Party of China
National anthem: *name:* "Yiyongjun Jinxingqu" (The March of the Volunteers)
lyrics/music: TIAN Han/NIE Er
note: adopted 1949; the anthem, though banned during the Cultural Revolution, is more commonly known as "Zhongguo Guoge" (Chinese National Song); it was originally the theme song to the 1935 Chinese movie, "Sons and Daughters in a Time of Storm"

ECONOMY

Economy—overview: Since the late 1970s China has moved from a closed, centrally planned system to a more market-oriented one that plays a major global role–in 2010 China became the world's largest exporter. Reforms began with the phasing out of collectivized agriculture, and expanded to include the gradual liberalization of prices, fiscal decentralization, increased autonomy for state enterprises, creation of a diversified banking system, development of stock markets, rapid growth of the private sector, and opening to foreign trade and investment. China has implemented reforms in a gradualist fashion. In recent years, China has renewed its support for state-owned enterprises in sectors it considers important to "economic security," explicitly looking to foster globally competitive national champions. After keeping its currency tightly linked to the US dollar for years, in July 2005 China revalued its currency by 2.1% against the US dollar and moved to an exchange rate system that references a basket of currencies. From mid 2005 to late 2008 cumulative appreciation of the renminbi against the US dollar was more than 20%, but the exchange rate remained virtually pegged to the dollar from the onset of the global financial crisis until June 2010, when Beijing allowed resumption of a gradual appreciation. The restructuring of the economy and resulting efficiency gains have contributed to a more than tenfold increase in GDP since 1978. Measured on a purchasing power parity (PPP) basis that adjusts for price differences, China in 2010 stood as the second-largest economy in the world after the US, having surpassed Japan in 2001. The dollar values of China's agricultural and industrial output each exceed those of the US; China is second to the US in the value of services it produces. Still, per capita income is below the world average. The Chinese government faces numerous economic challenges, including: (a) reducing its high domestic savings rate and correspondingly low domestic demand; (b) sustaining adequate job growth for tens of millions of migrants and new entrants to the work force; (c) reducing corruption and other economic crimes; and (d) containing environmental damage and social strife related to the economy's rapid transformation. Economic development has progressed further in coastal provinces than in the interior, and approximately 200 million rural laborers and their dependents have relocated to urban areas to find work. One consequence of the "one child" policy is that China is now one of the most rapidly aging countries in the world. Deterioration in the environment–notably air pollution, soil erosion, and the steady fall of the water table, especially in the north–is another long-term problem. China continues to lose arable land because of erosion and economic development. The Chinese government is seeking to add energy production capacity from sources other than coal and oil, focusing on nuclear and alternative energy development. In 2009, the global economic downturn reduced foreign demand for Chinese exports for the first time in many years, but China rebounded quickly, outperforming all other major economies in 2010 with GDP growth around 10%. The economy appears set to remain on a strong growth trajectory in 2011, lending credibility to the stimulus policies the regime rolled out during the global financial crisis. The government vows, in the 12th Five-Year Plan adopted in March 2011, to continue reforming the economy and emphasizes the need to increase domestic consumption in order to make the economy less dependent on exports for GDP growth in the future. However, China likely will make only marginal progress toward these rebalancing goals in 2011. Two economic problems China currently faces are inflation–which, late in 2010, surpassed the government's target of 3%–and local government debt, which swelled as a result of stimulus policies, and is largely off-the-books and potentially low-quality.

GDP (purchasing power parity): $10.09 trillion (2010 est.)
country comparison to the world: 3
$9.144 trillion (2009 est.)
$8.374 trillion (2008 est.)
note: data are in 2010 US dollars

GDP (official exchange rate): $5.878 trillion
note: because China's exchange rate is determine by fiat, rather than by market forces, the official exchange rate measure of GDP is not an accurate measure of China's output; GDP at the official exchange rate substantially understates the actual level of China's output vis-a-vis the rest of the world; in China's situation, GDP at purchasing power parity provides the best measure for comparing output across countries (2010 est.)

GDP—real growth rate: 10.3% (2010 est.)
country comparison to the world: 6
9.2% (2009 est.)
9.6% (2008 est.)

GDP—per capita (PPP): $7,600 (2010 est.)
country comparison to the world: 126
$6,900 (2009 est.)
$6,400 (2008 est.)
note: data are in 2010 US dollars

GDP—composition by sector:
agriculture: 9.6%
industry: 46.8%
services: 43.6% (2010 est.)

Labor force: 780 million (2010 est.)
country comparison to the world: 1

Labor force—by occupation:
agriculture: 38.1%
industry: 27.8%
services: 34.1% (2008 est.)

Unemployment rate: 4.3% (September 2009 est.)
country comparison to the world: 41
4.2% (December 2008 est.)
note: official data for urban areas only; including migrants may boost total unemployment to 9%; substantial unemployment and underemployment in rural areas

Population below poverty line: 2.8%
note: 21.5 million rural population live below the official "absolute poverty" line (approximately $90 per year); an additional 35.5 million rural population live above that level but below the official "low income" line (approximately $125 per year) (2007)

Household income or consumption by percentage share: *lowest 10%:* 3.5%
highest 10%: 15%
note: data are for urban households only (2008)

Distribution of family income—Gini index: 41.5 (2007)
country comparison to the world: 52
40 (2001)

Investment (gross fixed): 47.8% of GDP (2010 est.)
country comparison to the world: 1

Budget: *revenues:* $1.149 trillion
expenditures: $1.27 trillion (2010 est.)

Public debt: 17.5% of GDP (2010 est.)
country comparison to the world: 113
16.9% of GDP (2009 est.)

Inflation rate (consumer prices): 5% (2010 est.)
country comparison to the world: 140
-0.7% (2009 est.)

Central bank discount rate: 2.79% (31 December 2009)
country comparison to the world: 110
2.79% (31 December 2008)

Commercial bank prime lending rate: 5.81% (31 December 2010 est.)
country comparison to the world: 142
5.31% (31 December 2009 est.)

Stock of narrow money: $3.838 trillion (31 December 2010 est.)
country comparison to the world: 4
$3.242 trillion (31 December 2009 est.)

Stock of broad money: $10.08 trillion (31 December 2010 est.)
country comparison to the world: 5
$8.933 trillion (31 December 2009 est.)

Stock of domestic credit: $8.156 trillion (31 December 2010 est.)
country comparison to the world: 4
$7.24 trillion (31 December 2009 est.)

Market value of publicly traded shares: $5.008 trillion (31 December 2009 est.)
country comparison to the world: 4
$2.794 trillion (31 December 2008)
$6.226 trillion (31 December 2007 est.)

Agriculture—products: world leader in gross value of agricultural output; rice, wheat, potatoes, corn, peanuts, tea, millet, barley, apples, cotton, oilseed; pork; fish

Industries: world leader in gross value of industrial output; mining and ore processing, iron, steel, aluminum, and other metals, coal; machine building; armaments; textiles and apparel; petroleum; cement; chemicals; fertilizers; consumer products, including footwear, toys, and electronics; food processing; transportation equipment, including automobiles, rail cars and locomotives, ships, and aircraft; telecommunications equipment, commercial space launch vehicles, satellites

Industrial production growth rate: 11% (2010 est.)
country comparison to the world: 17

Electricity—production: 3.451 trillion kWh (2008 est.)
country comparison to the world: 2

Electricity—consumption: 3.438 trillion kWh (2008 est.)
country comparison to the world: 2

Electricity—exports: 16.64 billion kWh (2008)

Electricity—imports: 3.842 billion kWh (2008)

Oil—production: 3.991 million bbl/day (2009 est.)
country comparison to the world: 5

Oil—consumption: 8.2 million bbl/day (2009 est.)
country comparison to the world: 3

Oil—exports: 388,000 bbl/day (2008 est.)
country comparison to the world: 33

Oil—imports: 4.393 million bbl/day (2008)
country comparison to the world: 4

Oil—proved reserves: 20.35 billion bbl (1 January 2010 est.)
country comparison to the world: 13

Natural gas—production: 82.94 billion cu m (2009)
country comparison to the world: 9

Natural gas—consumption: 87.08 billion cu m (2009)
country comparison to the world: 9

Natural gas—exports: 3.32 billion cu m (2009)
country comparison to the world: 32

Natural gas—imports: 7.462 billion cu m (2009)
country comparison to the world: 27

Natural gas—proved reserves: 3.03 trillion cu m (1 January 2010 est.)
country comparison to the world: 13

Current account balance: $272.5 billion (2010 est.)
country comparison to the world: 1
$297.1 billion (2009 est.)

Exports: $1.506 trillion (2010 est.)
country comparison to the world: 2
$1.204 trillion (2009 est.)

Exports—commodities: electrical and other machinery, including data processing equipment, apparel, textiles, iron and steel, optical and medical equipment

Exports—partners: US 20.03%, Hong Kong 12.03%, Japan 8.32%, South Korea 4.55%, Germany 4.27% (2009)

Imports: $1.307 trillion (2010 est.)
country comparison to the world: 3
$954.3 billion (2009 est.)

Imports—commodities: electrical and other machinery, oil and mineral fuels, optical and medical equipment, metal ores, plastics, organic chemicals

Imports—partners: Japan 12.27%, Hong Kong 10.06%, South Korea 9.04%, US 7.66%, Taiwan 6.84%, Germany 5.54% (2009)

Reserves of foreign exchange and gold: $2.622 trillion (31 December 2010 est.)
country comparison to the world: 1
$2.426 trillion (31 December 2009 est.)

Debt—external: $406.6 billion (31 December 2010 est.)
country comparison to the world: 23
$349.3 billion (31 December 2009 est.)

Stock of direct foreign investment—at home: $574.3 billion (31 December 2010 est.)
country comparison to the world: 9
$473.1 billion (31 December 2009 est.)
Stock of direct foreign investment—abroad: $278.9 billion (31 December 2010 est.)
country comparison to the world: 15
$229.6 billion (31 December 2009 est.)
Exchange rates: Renminbi yuan (RMB) per US dollar–6.7852 (2010), 6.8314 (2009), 6.9385 (2008), 7.61 (2007), 7.97 (2006)

COMMUNICATIONS

Telephones—main lines in use: 313.68 million (2009)
country comparison to the world: 1
Telephones—mobile cellular: 747 million (2009)
country comparison to the world: 1
Telephone system: *general assessment:* domestic and international services are increasingly available for private use; unevenly distributed domestic system serves principal cities, industrial centers, and many towns; China continues to develop its telecommunications infrastructure, and is partnering with foreign providers to expand its global reach; China in the summer of 2008 began a major restructuring of its telecommunications industry, resulting in the consolidation of its six telecom service operators to three, China Telecom, China Mobile and China Unicom, each providing both fixed-line and mobile services
domestic: interprovincial fiber-optic trunk lines and cellular telephone systems have been installed; mobile-cellular subscribership is increasing rapidly; the number of Internet users exceeded 250 million by summer 2008; a domestic satellite system with 55 earth stations is in place
international: country code–86; a number of submarine cables provide connectivity to Asia, the Middle East, Europe, and the US; satellite earth stations–7 (5 Intelsat–4 Pacific Ocean and 1 Indian Ocean; 1 Intersputnik–Indian Ocean region; and 1 Inmarsat–Pacific and Indian Ocean regions) (2008)
Broadcast media: all broadcast media are owned by, or affiliated with, the Communist Party of China or a government agency; no privately-owned television or radio stations with state-run Chinese Central TV, provincial, and municipal stations offering more than 2,000 channels; the Central Propaganda Department lists subjects that are off limits to domestic broadcast media with the government maintaining authority to approve all programming; foreign-made TV programs must be approved prior to broadcast (2008)
Internet country code: .cn
Internet hosts: 15.251 million (2010)
country comparison to the world: 6
Internet users: 389 million (2009)
country comparison to the world: 1

TRANSPORTATION

Airports: 502 (2010)
country comparison to the world: 15
Airports—with paved runways: *total:* 442
over 3,047 m: 63
2,438 to 3,047 m: 137
1,524 to 2,437 m: 132
914 to 1,523 m: 27
under 914 m: 83 (2010)
Airports—with unpaved runways: *total:* 60
over 3,047 m: 4
2,438 to 3,047 m: 7
1,524 to 2,437 m: 9
914 to 1,523 m: 13
under 914 m: 27 (2010)
Heliports: 48 (2010)
Pipelines: gas 38,566 km; oil 23,470 km; refined products 13,706 km (2010)
Railways: *total:* 86,000 km
country comparison to the world: 3
standard gauge: 86,000 km 1.435-m gauge (36,000 km electrified) (2009)
Roadways: *total:* 3,860,800 km
country comparison to the world: 2
paved: 3,056,300 km (includes 65,000 km of expressways)
unpaved: 804,500 km (2009)
Waterways: 110,000 km (navigable waterways) (2010)
country comparison to the world: 1
Merchant marine: *total:* 2,010
country comparison to the world: 3
by type: barge carrier 6, bulk carrier 571, cargo 639, carrier 5, chemical tanker 98, container 204, liquefied gas 55, passenger 9, passenger/cargo 83, petroleum tanker 271, refrigerated cargo 35, roll on/roll off 9, specialized tanker 1, vehicle carrier 24
foreign-owned: 18 (Germany 1, Hong Kong 15, Japan 2)
registered in other countries: 1,623 (Bahamas 4, Bangladesh 1, Belize 64, Bermuda 13, Cambodia 203, Comoros 1, Cyprus 6, France 5, Georgia 11, Germany 2, Honduras 2, Hong Kong 432, India 1, Indonesia 1, Kiribati 28, Liberia 10, Malta 11, Marshall Islands 16, North Korea 1, Norway 25, Panama 574, Philippines 4, Saint Kitts and Nevis 1, Saint Vincent and the Grenadines 82, Sierra Leone 12, Singapore 26, South Korea 9, Thailand 1, Togo 2, Tuvalu 9, UK 7, unknown 59) (2010)
Ports and terminals: Dalian, Guangzhou, Ningbo, Qingdao, Qinhuangdao, Shanghai, Shenzhen, Tianjin

MILITARY

Military branches: People's Liberation Army (PLA): Ground Forces, Navy (includes marines and naval aviation), Air Force (Zhongguo Renmin Jiefangjun Kongjun, PLAAF; includes Airborne Forces), and Second Artillery Corps (strategic missile force); People's Armed Police (PAP); PLA Reserve Force (2010)
Military service age and obligation: 18-22 years of age for selective compulsory military service, with 24-month service obligation; no minimum age for voluntary service (all officers are volunteers); 18-19 years of age for women high school graduates who meet requirements for specific military jobs; in 2010, a decision was made to allow women in combat roles (2010)
Manpower available for military service: *males age 16–49:* 385,821,101
females age 16–49: 363,789,674 (2010 est.)
Manpower fit for military service: *males age 16–49:* 318,265,016
females age 16–49: 300,323,611 (2010 est.)
Manpower reaching militarily significant age annually: *male:* 10,406,544
female: 9,131,990 (2010 est.)
Military expenditures: 4.3% of GDP (2006)
country comparison to the world: 23

TRANSNATIONAL ISSUES

Disputes—international: continuing talks and confidence-building measures work toward reducing tensions over Kashmir that nonetheless remains militarized with portions under the de facto administration of China (Aksai Chin), India (Jammu and Kashmir), and Pakistan (Azad Kashmir and Northern Areas); India does not recognize Pakistan's ceding historic Kashmir lands to China in 1964; China and India continue their security and foreign policy dialogue started in 2005 related to the dispute over most of their rugged, militarized boundary, regional nuclear proliferation, and other matters; China claims most of India's Arunachal Pradesh to the base of the Himalayas; lacking any treaty describing the boundary, Bhutan and China continue negotiations to establish a common boundary alignment to resolve territorial disputes arising from substantial cartographic discrepancies, the largest of which lie in Bhutan's northwest and along the Chumbi salient; Bhutan protests Chinese road construction and other activities on Bhutanese soil; Chinese border soldiers frequently intrude deep into Bhutanese territory; Burmese forces attempting to dig in to the largely autonomous Shan State to rout local militias tied to the drug trade, prompts local residents to periodically flee into neighboring Yunnan Province in China; Chinese maps show an international boundary symbol off the coasts of the littoral states of the South China Seas, where China has interrupted Vietnamese hydrocarbon exploration; China asserts sovereignty over Scarborough Reef along with the Philippines and Taiwan, and over the Spratly Islands together with Malaysia, the Philippines, Taiwan, Vietnam, and Brunei; the 2002 "Declaration on the Conduct of Parties in the South China Sea" eased tensions in the Spratly's but is not the legally binding "code of conduct" sought by some parties; Vietnam and China continue to expand construction of facilities in the Spratly's and in March 2005, the national oil companies of China, the Philippines, and Vietnam signed a joint accord on marine seismic activities in the Spratly Islands; China occupies some of the Paracel Islands also claimed by Vietnam and Taiwan; China and Taiwan continue to reject both Japan's claims to the uninhabited islands of Senkaku-shoto (Diaoyu Tai) and Japan's unilaterally declared equidistance line in the East China Sea, the site of intensive hydrocarbon exploration and exploitation; certain islands in the Yalu and Tumen rivers are in dispute with North Korea; North Korea and China seek to stem illegal migration to China by North Koreans, fleeing privations and oppression, by building a fence along portions of the border and imprisoning North

Koreans deported by China; China and Russia have demarcated the once disputed islands at the Amur and Ussuri confluence and in the Argun River in accordance with their 2004 Agreement; China and Tajikistan have begun demarcating the revised boundary agreed to in the delimitation of 2002; the decade-long demarcation of the China-Vietnam land boundary was completed in 2009; citing environmental, cultural, and social concerns, China has reconsidered construction of 13 dams on the Salween River, but energy-starved Burma, with backing from Thailand, remains intent on building five hydro-electric dams downstream despite regional and international protests; Chinese and Hong Kong authorities met in March 2008 to resolve ownership and use of lands recovered in Shenzhen River channelization, including 96-hectare Lok Ma Chau Loop; Hong Kong developing plans to reduce 2,000 out of 2,800 hectares of its restricted Closed Area by 2010
Refugees and internally displaced persons: refugees (country of origin): 300,897 (Vietnam); estimated 30,000-50,000 (North Korea)
IDPs: 90,000 (2007)
Trafficking in persons: current situation: China is a source, transit, and destination country for men, women, and children trafficked for the purposes of sexual exploitation and forced labor; the majority of trafficking in China occurs within the country's borders, but there is also considerable international trafficking of Chinese citizens to Africa, Asia, Europe, Latin America, the Middle East, and North America; Chinese women are lured abroad through false promises of legitimate employment, only to be forced into commercial sexual exploitation, largely in Taiwan, Thailand, Malaysia, and Japan; women and children are trafficked to China from Mongolia, Burma, North Korea, Russia, and Vietnam for forced labor, marriage, and prostitution; some North Korean women and children seeking to leave their country voluntarily cross the border into China and are then sold into prostitution, marriage, or forced labor
tier rating: Tier 2 Watch List–China is on the Tier 2 Watch List for the fourth consecutive year for its failure to provide evidence of increasing efforts to combat human trafficking, particularly in terms of punishment of trafficking crimes and the protection of Chinese and foreign victims of trafficking; victims are sometimes punished for unlawful acts that were committed as a direct result of their being trafficked, such as violations of prostitution or immigration/emigration controls; the Chinese Government continued to treat North Korean victims of trafficking solely as economic migrants, routinely deporting them back to horrendous conditions in North Korea; additional challenges facing the Chinese Government include the enormous size of its trafficking problem and the significant level of corruption and complicity in trafficking by some local government officials (2008)
Illicit drugs: major transshipment point for heroin produced in the Golden Triangle region of Southeast Asia; growing domestic consumption of synthetic drugs, and heroin from Southeast and Southwest Asia; source country for methamphetamine and heroin chemical precursors, despite new regulations on its large chemical industry (2008)

CHRISTMAS ISLAND

(TERRITORY OF AUSTRALIA)

INTRODUCTION

Background: Named in 1643 for the day of its discovery, the island was annexed and settlement began by the UK in 1888. Phosphate mining began in the 1890s. The UK transferred sovereignty to Australia in 1958. Almost two-thirds of the island has been declared a national park.

GEOGRAPHY

Location: Southeastern Asia, island in the Indian Ocean, south of Indonesia
Geographic coordinates: 10 30 S, 105 40 E
Map references: Oceania
Area: *total:* 135 sq km
country comparison to the world: 220
land: 135 sq km
water: 0 sq km
Area—comparative: about three-quarters the size of Washington, DC
Land boundaries: 0 km
Coastline: 138.9 km
Maritime claims: territorial sea: 12 nm
contiguous zone: 12 nm
exclusive fishing zone: 200 nm
Climate: tropical with a wet season (December to April) and dry season; heat and humidity moderated by trade winds
Terrain: steep cliffs along coast rise abruptly to central plateau
Elevation extremes: *lowest point:* Indian Ocean 0 m
highest point: Murray Hill 361 m
Natural resources: phosphate, beaches
Land use: *arable land:* 0%
permanent crops: 0%
other: 100% (mainly tropical rainforest; 63% of the island is a national park) (2005)
Irrigated land: NA
Natural hazards: the narrow fringing reef surrounding the island can be a maritime hazard
Environment—current issues: loss of rainforest; impact of phosphate mining
Geography—note: located along major sea lanes of Indian Ocean

PEOPLE

Population: 1,402 (July 2010 est.)
country comparison to the world: 232
Age structure: *0–14 years:* NA
15–64 years: NA
65 years and over: NA
Population growth rate: 0% (2010 est.)
country comparison to the world: 197
Birth rate: NA
Death rate: NA
Net migration rate: NA
Sex ratio: NA (2009 est.)
Infant mortality rate: *total:* NA
male: NA
female: NA
Life expectancy at birth: *total population:* NA
male: NA
female: NA
Total fertility rate: NA
HIV/AIDS—adult prevalence rate: NA
HIV/AIDS—people living with HIV/AIDS: NA
HIV/AIDS—deaths: NA
Nationality: *noun:* Christmas Islander(s)
adjective: Christmas Island
Ethnic groups: Chinese 70%, European 20%, Malay 10%
note: no indigenous population (2001)
Religions: Buddhist 36%, Muslim 25%, Christian 18%, other 21% (1997)
Languages: English (official), Chinese, Malay
Literacy: NA
School life expectancy (primary to tertiary education): NA
Education expenditures: NA

GOVERNMENT

Country name: *conventional long form:* Territory of Christmas Island
conventional short form: Christmas Island
Dependency status: non-self governing territory of Australia; administered from Canberra by the Department of Regional Australia, Regional Development and Local Government
Government type: NA
Capital: *name:* The Settlement
geographic coordinates: 10 25 S, 105 43 E
time difference: UTC+7 (12 hours ahead of Washington, DC during Standard Time)
Administrative divisions: none (territory of Australia)
Independence: none (territory of Australia)
National holiday: Australia Day, 26 January (1788)

Constitution: Christmas Island Act of 1958-59 (1 October 1958) as amended by the Territories Law Reform Act of 1992

Legal system: legal system is under the authority of the governor general of Australia and Australian law

Suffrage: 18 years of age

Executive branch: *chief of state:* Queen ELIZABETH II (since 6 February 1952) represented by the Australian governor general
head of government: Administrator Brian LACY (since 5 October 2009)
elections: the monarchy is hereditary; governor general appointed by the monarch on the recommendation of the Australian prime minister; administrator appointed by the governor general of Australia and represents the monarch and Australia

Legislative branch: unicameral Christmas Island Shire Council (9 seats; members elected by popular vote to serve four-year terms)
elections: held every two years with half the members standing for election; last held on 17 October 2009 (next to be held in 2011)
election results: percent of vote–NA; seats–independents 9

Judicial branch: Supreme Court; District Court; Magistrate's Court

Political parties and leaders: none

Political pressure groups and leaders: none

International organization participation: none

Diplomatic representation in the US: none (territory of Australia)

Diplomatic representation from the US: none (territory of Australia)

Flag description: territorial flag; divided diagonally from upper hoist to lower fly; the upper triangle is green with a yellow image of the Golden Bosun Bird superimposed, the lower triangle is blue with the Southern Cross constellation, representing Australia, superimposed; a centered yellow disk displays a green map of the island
note: the flag of Australia is used for official purposes

National anthem: note: as a territory of Australia, "Advance Australia Fair" remains official as the national anthem, while "God Save the Queen" serves as the royal anthem (see Australia)

ECONOMY

Economy—overview: Phosphate mining had been the only significant economic activity, but in December 1987 the Australian government closed the mine. In 1991, the mine was reopened. With the support of the government, a $34 million casino opened in 1993, but closed in 1998.

GDP (purchasing power parity): $NA

Labor force: NA

Budget: *revenues:* $NA
expenditures: $NA

Agriculture—products: NA

Industries: tourism, phosphate extraction (near depletion)

Exports: $NA

Exports—commodities: phosphate

Imports: $NA

Imports—commodities: consumer goods

Exchange rates: Australian dollars (AUD) per US dollar–1.0902 (2010), 1.2822 (2009), 1.2059 (2008), 1.2137 (2007), 1.3285 (2006)

COMMUNICATIONS

Telephones—main lines in use: NA

Telephone system: *general assessment:* service provided by the Australian network
domestic: GSM mobile-cellular telephone service replaced older analog system in February 2005
international: country code–61-8; satellite earth station–1 (Intelsat provides telephone and telex service) (2005)

Broadcast media: 1 community radio station; broadcasts of several Australian radio and television stations are received via satellite (2009)

Internet country code: .cx

Internet hosts: 2,542 (2010)
country comparison to the world: 149

Internet users: 464 (2001)
country comparison to the world: 216

TRANSPORTATION

Airports: 1 (2010)
country comparison to the world: 223

Airports—with paved runways: *total:* 1
1,524 to 2,437 m: 1 (2010)

Railways: *total:* 18 km
country comparison to the world: 133

standard gauge: 18 km 1.435-m (not in operation) (2010)

Roadways: *total:* 140 km
country comparison to the world: 210
paved: 30 km
unpaved: 110 km (2007)

Ports and terminals: Flying Fish Cove

MILITARY

Military—note: defense is the responsibility of Australia

TRANSNATIONAL ISSUES

Disputes—international: none

CLIPPERTON ISLAND

(POSSESSION OF FRANCE)

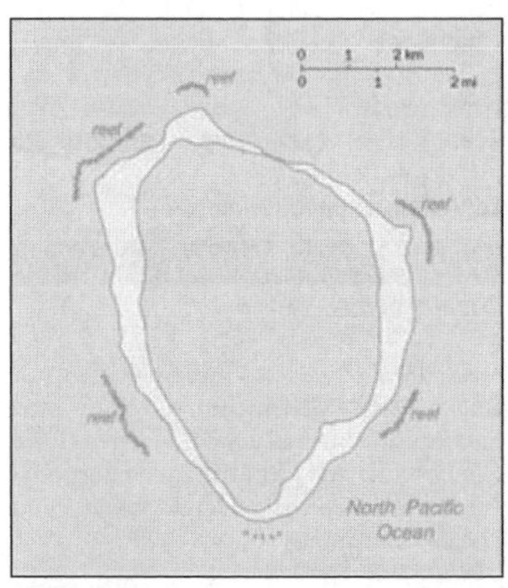

INTRODUCTION

Background: This isolated island was named for John CLIPPERTON, a pirate who made it his hideout early in the 18th century. Annexed by France in 1855, it was seized by Mexico in 1897. Arbitration eventually awarded the island to France, which took possession in 1935.

GEOGRAPHY

Location: Middle America, atoll in the North Pacific Ocean, 1,120 km southwest of Mexico

Geographic coordinates: 10 17 N, 109 13 W

Map references: Political Map of the World

Area:
total: 6 sq km
country comparison to the world: 243
land: 6 sq km
water: 0 sq km

Area—comparative: about 12 times the size of The Mall in Washington, DC

Land boundaries: 0 km

Coastline: 11.1 km

Maritime claims: territorial sea: 12 nm

exclusive economic zone: 200 nm

Climate: tropical; humid, average temperature 20-32 degrees C, wet season (May to October)

Terrain: coral atoll

Elevation extremes: *lowest point:* Pacific Ocean 0 m
highest point: Rocher Clipperton 29 m

Natural resources: fish

Land use: *arable land:* 0%
permanent crops: 0%
other: 100% (all coral) (2005)

Irrigated land: 0 sq km

Natural hazards: NA

Environment—current issues: NA

Geography—note: reef 12 km in circumference

PEOPLE

Population: uninhabited

GOVERNMENT

Country name: *conventional long form:* none
conventional short form: Clipperton Island
local long form: none
local short form: Ile Clipperton
former: sometimes called Ile de la Passion

Dependency status: possession of France; administered directly by the Minister of Overseas France

Legal system: the laws of France, where applicable, apply

Flag description: the flag of France is used

ECONOMY

Economy—overview: Although 115 species of fish have been identified in the territorial waters of Clipperton Island, the only economic activity is tuna fishing.

TRANSPORTATION

Ports and terminals: none; offshore anchorage only

MILITARY

Military—note: defense is the responsibility of France

TRANSNATIONAL ISSUES

Disputes—international: none

COCOS (KEELING) ISLANDS

(TERRITORY OF AUSTRALIA)

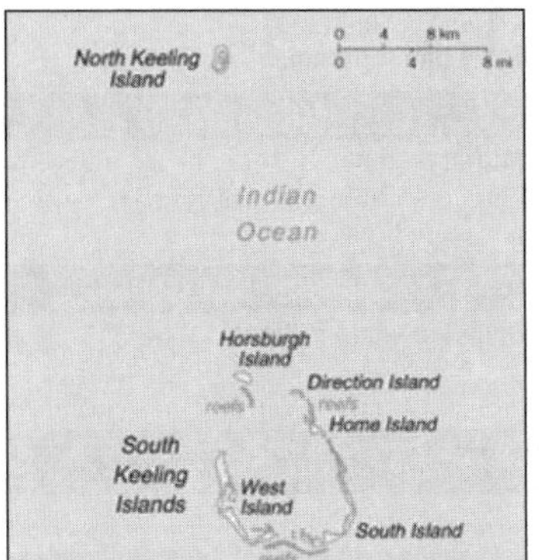

INTRODUCTION

Background: There are 27 coral islands in the group. Captain William KEELING discovered the islands in 1609, but they remained uninhabited until the 19th century. From the 1820s to 1978, members of the CLUNIE-ROSS family controlled the islands and the copra produced from local coconuts. Annexed by the UK in 1857, the Cocos Islands were transferred to the Australian Government in 1955. The population on the two inhabited islands generally is split between the ethnic Europeans on West Island and the ethnic Malays on Home Island.

GEOGRAPHY

Location: Southeastern Asia, group of islands in the Indian Ocean, southwest of Indonesia, about halfway between Australia and Sri Lanka
Geographic coordinates: 12 30 S, 96 50 E
Map references: Oceania
Area: *total:* 14 sq km
country comparison to the world: 239
land: 14 sq km
water: 0 sq km
note: includes the two main islands of West Island and Home Island
Area—comparative: about 24 times the size of The Mall in Washington, DC
Land boundaries: 0 km
Coastline: 26 km
Maritime claims: territorial sea: 12 nm
exclusive fishing zone: 200 nm
Climate: tropical with high humidity, moderated by the southeast trade winds for about nine months of the year
Terrain: flat, low-lying coral atolls
Elevation extremes: *lowest point:* Indian Ocean 0 m
highest point: unnamed location 5 m
Natural resources: fish
Land use: *arable land:* 0%
permanent crops: 0%
other: 100% (2005)
Irrigated land: NA
Natural hazards: cyclone season is October to April
Environment—current issues: freshwater resources are limited to rainwater accumulations in natural underground reservoirs
Geography—note: islands are thickly covered with coconut palms and other vegetation; site of a World War I naval battle in November 1914 between the Australian light cruiser HMAS Sydney and the German raider SMS Emden; after being heavily damaged in the engagement, the Emden was beached by her captain on North Keeling Island

PEOPLE

Population: 596 (July 2010 est.)
country comparison to the world: 236
Age structure: *0–14 years:* NA
15–64 years: NA
65 years and over: NA
Population growth rate: 0% (2010 est.)
country comparison to the world: 195
Birth rate: NA
Death rate: NA
Net migration rate: NA
Infant mortality rate: *total:* NA
male: NA
female: NA
Life expectancy at birth: *total population:* NA
male: NA
female: NA
Total fertility rate: NA
HIV/AIDS—adult prevalence rate: NA
HIV/AIDS—people living with HIV/AIDS: NA
HIV/AIDS—deaths: NA
Nationality: *noun:* Cocos Islander(s)
adjective: Cocos Islander
Ethnic groups: Europeans, Cocos Malays
Religions: Sunni Muslim 80%, other 20% (2002 est.)
Languages: Malay (Cocos dialect), English
Literacy: NA
School life expectancy (primary to tertiary education): NA
Education expenditures: NA

GOVERNMENT

Country name: *conventional long form:* Territory of Cocos (Keeling) Islands
conventional short form: Cocos (Keeling) Islands
Dependency status: non-self governing territory of Australia; administered from Canberra by the Department of Regional Australia, Regional Development and Local Government
Government type: NA
Capital: *name:* West Island
geographic coordinates: 12 10 S, 96 50 E
time difference: UTC+6.5 (11.5 hours ahead of Washington, DC during Standard Time)
Administrative divisions: none (territory of Australia)
Independence: none (territory of Australia)
National holiday: Australia Day, 26 January (1788)
Constitution: Cocos (Keeling) Islands Act of 1955 (23 November 1955) as amended by the Territories Law Reform Act of 1992
Legal system: common law based on the Australian model
Suffrage: 18 years of age
Executive branch: *chief of state:* Queen ELIZABETH II (since 6 February 1952); represented by the Australian governor general
head of government: Administrator (nonresident) Brian LACY (since 5 October 2009)
cabinet: NA
(For more information visit the World Leaders website)
elections: the monarchy is hereditary; governor general appointed by the monarch on the recommendation of the Australian prime minister; administrator appointed by the governor general of Australia and represents the monarch and Australia
Legislative branch: unicameral Cocos (Keeling) Islands Shire Council (7 seats)
elections: held every two years with half the members standing for election; last held in October 2009 (next to be held in Ocotber 2011)
Judicial branch: Supreme Court; Magistrate's Court
Political parties and leaders: none
Political pressure groups and leaders: The Cocos Islands Youth Support Centre
International organization participation: none
Diplomatic representation in the US: none (territory of Australia)
Diplomatic representation from the US: none (territory of Australia)
Flag description: the flag of Australia is used
National anthem: note: as a territory of Australia, "Advance Australia Fair" remains official as the national anthem, while "God Save the Queen" serves as the royal anthem (see Australia)

ECONOMY

Economy—overview: Coconuts, grown throughout the islands, are the sole cash crop. Small local gardens and fishing contribute to the food supply, but additional food and most other necessities must be imported from Australia. There is a small tourist industry.
GDP (purchasing power parity): $NA
GDP—real growth rate: 1% (2003)
country comparison to the world: 174
Labor force: NA
Labor force—by occupation: note: the Cocos Islands Cooperative Society Ltd.

employs construction workers, stevedores, and lighterage workers; tourism employs others
Unemployment rate: 60% (2000 est.)
country comparison to the world: 195
Budget: *revenues:* $NA
expenditures: $NA
Agriculture—products: vegetables, bananas, pawpaws, coconuts
Industries: copra products and tourism
Exports: $NA
Exports—commodities: copra
Imports: $NA
Imports—commodities: foodstuffs
Exchange rates: Australian dollars (AUD) per US dollar–1.0902 (2010), 1.2822 (2009), 1.2059 (2008), 1.2137 (2007), 1.3285 (2006)

COMMUNICATIONS

Telephones—main lines in use: 287 (1992)
country comparison to the world: 228
Telephone system: *general assessment:* connected within Australia's telecommunication system; a local mobile-cellular network is in operation
domestic: NA
international: country code–61; telephone, telex, and facsimile communications with Australia and elsewhere via satellite; satellite earth station–1 (Intelsat) (2001)
Broadcast media: 1 local radio station staffed by community volunteers; broadcasts of several Australian radio and TV stations are received via satellite (2009)
Internet country code: .cc
Internet hosts: 35,312 (2010)
country comparison to the world: 96

TRANSPORTATION

Airports: 1 (2010)
country comparison to the world: 214
Airports—with paved runways:
total: 1
1,524 to 2,437 m: 1 (2010)
Roadways: *total:* 22 km
country comparison to the world: 220
paved: 10 km
unpaved: 12 km (2007)
Ports and terminals: Port Refuge

MILITARY

Military—note: defense is the responsibility of Australia; the territory has a five-person police force

TRANSNATIONAL ISSUES

Disputes—international: none

COLOMBIA

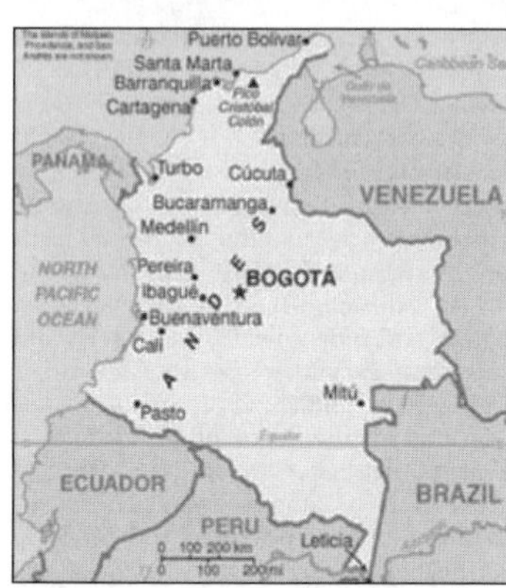

INTRODUCTION

Background: Colombia was one of the three countries that emerged from the collapse of Gran Colombia in 1830 (the others are Ecuador and Venezuela). A four-decade long conflict between government forces and anti-government insurgent groups, principally the Revolutionary Armed Forces of Colombia (FARC) heavily funded by the drug trade, escalated during the 1990s. The insurgents lack the military or popular support necessary to overthrow the government, and violence has been decreasing since about 2002. However, insurgents continue attacks against civilians and large areas of the countryside are under guerrilla influence or are contested by security forces. More than 31,000 former paramilitaries had demobilized by the end of 2006 and the United Self Defense Forces of Colombia (AUC) as a formal organization had ceased to function. In the wake of the paramilitary demobilization, emerging criminal groups arose, whose members include some former paramilitaries. The Colombian Government has stepped up efforts to reassert government control throughout the country, and now has a presence in every one of its administrative departments. However, neighboring countries worry about the violence spilling over their borders. In January 2011, Colombia assumed a nonpermanent seat on the UN Security Council for the 2011-12 term.

GEOGRAPHY

Location: Northern South America, bordering the Caribbean Sea, between Panama and Venezuela, and bordering the North Pacific Ocean, between Ecuador and Panama
Geographic coordinates: 4 00 N, 72 00 W
Map references: South America
Area: *total:* 1,138,910 sq km
country comparison to the world: 26
land: 1,038,700 sq km
water: 100,210 sq km
note: includes Isla de Malpelo, Roncador Cay, and Serrana Bank
Area—comparative: slightly less than twice the size of Texas
Land boundaries: *total:* 6,309 km
border countries: Brazil 1,644 km, Ecuador 590 km, Panama 225 km, Peru 1,800 km, Venezuela 2,050 km
Coastline: 3,208 km (Caribbean Sea 1,760 km, North Pacific Ocean 1,448 km)
Maritime claims: territorial sea: 12 nm
exclusive economic zone: 200 nm
continental shelf: 200 m depth or to the depth of exploitation
Climate: tropical along coast and eastern plains; cooler in highlands
Terrain: flat coastal lowlands, central highlands, high Andes Mountains, eastern lowland plains
Elevation extremes: *lowest point:* Pacific Ocean 0 m
highest point: Pico Cristobal Colon 5,775 m
note: nearby Pico Simon Bolivar also has the same elevation
Natural resources: petroleum, natural gas, coal, iron ore, nickel, gold, copper, emeralds, hydropower
Land use: *arable land:* 2.01%
permanent crops: 1.37%
other: 96.62% (2005)
Irrigated land: 9,000 sq km (2003)
Total renewable water resources: 2,132 cu km (2000)
Freshwater withdrawal (domestic/industrial/agricultural): *total:* 10.71 cu km/yr (50%/4%/46%)
per capita: 235 cu m/yr (2000)
Natural hazards: highlands subject to volcanic eruptions; occasional earthquakes; periodic droughts
volcanism: Galeras (elev. 4,276 m) is one of Colombia's most active volcanoes, having erupted in 2009 and 2010 causing major evacuations; it has been deemed a "Decade Volcano" by the International Association of Volcanology and Chemistry of the Earth's Interior, worthy of study due to its explosive history and close proximity to human populations; Nevado del Ruiz (elev. 5,321 m), 129 km (80 mi) west of Bogota, erupted in 1985 producing lahars that killed 23,000 people; the volcano last erupted in 1991; additionally, after 500 years of dormancy, Nevado del Huila reawakened in 2007 and has experienced frequent eruptions since then; other historically active volcanoes include Cumbal, Dona Juana, Nevado del Tolima, and Purace
Environment—current issues: deforestation; soil and water quality damage from overuse of pesticides; air pollution, especially in Bogota, from vehicle emissions
Environment—international agreements: *party to:* Antarctic Treaty, Biodiversity, Climate Change, Climate Change-Kyoto Protocol, Desertification, Endangered Species, Hazardous Wastes, Marine Life Conservation, Ozone Layer Protection, Ship Pollution, Tropical Timber 83, Tropical Timber 94, Wetlands
signed, but not ratified: Law of the Sea
Geography—note: only South American country with coastlines on both the North Pacific Ocean and Caribbean Sea

PEOPLE

Population: 44,725,543 (July 2011 est.)
country comparison to the world: 30
Age structure: *0–14 years:* 26.7% (male 6,109,495/female 5,834,273)
15–64 years: 67.2% (male 14,826,008/female 15,208,799)

65 years and over: 6.1% (male 1,159,691/ female 1,587,277) (2011 est.)
Median age: *total:* 28 years
male: 27 years
female: 28.9 years (2011 est.)
Population growth rate: 1.156% (2011 est.)
country comparison to the world: 100
Birth rate: 17.49 births/1,000 population (2011 est.)
country comparison to the world: 113
Death rate: 5.26 deaths/1,000 population (July 2011 est.)
country comparison to the world: 180
Net migration rate: -0.67 migrant(s)/1,000 population (2011 est.)
country comparison to the world: 144
Urbanization: *urban population:* 75% of total population (2010)
rate of urbanization: 1.7% annual rate of change (2010-15 est.)
Major cities—population: BOGOTA (capital) 8.262 million; Medellin 3.497 million; Cali 2.352 million; Barranquilla 1.836 million; Bucaramanga 1.065 million (2009)
Sex ratio: *at birth:* 1.06 male(s)/female
under 15 years: 1.05 male(s)/female
15–64 years: 0.97 male(s)/female
65 years and over: 0.74 male(s)/female
total population: 0.98 male(s)/female (2011 est.)
Infant mortality rate: *total:* 16.39 deaths/1,000 live births
country comparison to the world: 109
male: 19.92 deaths/1,000 live births
female: 12.65 deaths/1,000 live births (2011 est.)
Life expectancy at birth: *total population:* 74.55 years
country comparison to the world: 97
male: 71.27 years
female: 78.03 years (2011 est.)
Total fertility rate: 2.15 children born/ woman (2011 est.)
country comparison to the world: 112
HIV/AIDS—adult prevalence rate: 0.5% (2009 est.)
country comparison to the world: 69
HIV/AIDS—people living with HIV/AIDS: 160,000 (2009 est.)
country comparison to the world: 33
HIV/AIDS—deaths: 14,000 (2009 est.)
country comparison to the world: 20
Major infectious diseases: *degree of risk:* high
food or waterborne diseases: bacterial diarrhea
vectorborne diseases: dengue fever, malaria, and yellow fever
water contact disease: leptospirosis (2009)
Drinking water source: *Improved:*
urban: 99% of population
rural: 73% of population
total: 92% of population
Unimproved:
urban: 1% of population
rural: 27% of population
total: 8% of population (2008)
Sanitation facility access: *Improved:*
urban: 81% of population
rural: 55% of population
total: 74% of population
Unimproved:
urban: 19% of population
rural: 45% of population
total: 26% of population (2008)
Nationality: *noun:* Colombian(s)
adjective: Colombian
Ethnic groups: mestizo 58%, white 20%, mulatto 14%, black 4%, mixed black-Amerindian 3%, Amerindian 1%
Religions: Roman Catholic 90%, other 10%
Languages: Spanish (official)
Literacy: *definition:* age 15 and over can read and write
total population: 90.4%
male: 90.1%
female: 90.7% (2005 census)
School life expectancy (primary to tertiary education): *total:* 14 years
male: 13 years
female: 14 years (2009)
Education expenditures: 4.8% of GDP (2009)
country comparison to the world: 68

GOVERNMENT

Country name: *conventional long form:* Republic of Colombia
conventional short form: Colombia
local long form: Republica de Colombia
local short form: Colombia
Government type: republic; executive branch dominates government structure
Capital: *name:* Bogota
geographic coordinates: 4 36 N, 74 05 W
time difference: UTC-5 (same time as Washington, DC during Standard Time)
Administrative divisions: 32 departments (departamentos, singular—departamento) and 1 capital district* (distrito capital); Amazonas, Antioquia, Arauca, Atlantico, Bogota*, Bolivar, Boyaca, Caldas, Caqueta, Casanare, Cauca, Cesar, Choco, Cordoba, Cundinamarca, Guainia, Guaviare, Huila, La Guajira, Magdalena, Meta, Narino, Norte de Santander, Putumayo, Quindio, Risaralda, San Andres y Providencia, Santander, Sucre, Tolima, Valle del Cauca, Vaupes, Vichada
Independence: 20 July 1810 (from Spain)
National holiday: Independence Day, 20 July (1810)
Constitution: 5 July 1991; amended many times
Legal system: civil law system influenced by the Spanish and French civil codes
International law organization participation: has not submitted an ICJ jurisdiction declaration; accepts ICCt jurisdiction
Suffrage: 18 years of age; universal
Executive branch: *chief of state:* President Juan Manuel SANTOS Calderon (since 7 August 2010); Vice President Angelino GARZON (since 7 August 2010); note—the president is both the chief of state and head of government
head of government: President Juan Manuel SANTOS Calderon (since 7 August 2010); Vice President Angelino GARZON (since 7 August 2010)
cabinet: Cabinet appointed by the president
(For more information visit the World Leaders website)
elections: president and vice president elected by popular vote for a four-year term (eligible for a second term); election last held on 30 May 2010 with a runoff election 20 June 2010 (next to be held in May 2014)
election results: Juan Manuel SANTOS Calderon elected president in runoff election; percent of vote—Juan Manuel SANTOS Calderon 69.06%, Antanas MOCKUS 27.52%
Legislative branch: bicameral Congress or Congreso consists of the Senate or Senado (102 seats; members elected by popular vote to serve four-year terms) and the Chamber of Representatives or Camara de Representantes (166 seats; members elected by popular vote to serve four-year terms)
elections: Senate—last held on 14 March 2010 (next to be held in March 2014); Chamber of Representatives—last held on 14 March 2010 (next to be held in March 2014)
election results: Senate—percent of vote by party—NA; seats by party—U Party 28, PC 22, PL 16, PIN 9, CR 8, PDA 8, Green Party 5, other parties 5; Chamber of Representatives—percent of vote by party—NA; seats by party—U Party 47, PC 37, PL 36, CR 16, PIN 12, PDA 4, Green Party 3, other parties 10; note—as of 1 January 2011, the Senate currently has 101 seats after one seat became vacant due to a PL senator losing their seat for illegal collusion with the FARC; the Chamber of Representatives also has one seat vacant after only 165 of the 166 candidates were credentialed
Judicial branch: four roughly coequal, supreme judicial organs; Supreme Court of Justice or Corte Suprema de Justicia (highest court of criminal law; judges are selected by their peers from the nominees of the Superior Judicial Council for eight-year terms); Council of State (highest court of administrative law; judges are selected from the nominees of the Superior Judicial Council for eight-year terms); Constitutional Court (guards integrity and supremacy of the constitution; rules on constitutionality of laws, amendments to the constitution, and international treaties); Superior Judicial Council (administers and disciplines the civilian judiciary; resolves jurisdictional conflicts arising between other courts; members are elected by three sister courts and Congress for eight-year terms)
Political parties and leaders: Alternative Democratic Pole or PDA [Clara LOPEZ]; Conservative Party or PC [Fernando ARAUJO]; Green Party [Luis GARZON]; Liberal Party or PL [Rafael PARDO]; National Integration Party or PIN [Angel ALIRIO Moreno]; Radical Change or CR [German VARGAS Lleras]; Social National Unity Party or U Party [Juan Francisco LOZANO Ramirez]
note: Colombia has seven major political parties, and numerous smaller movements
Political pressure groups and leaders: Central Union of Workers or CUT; Colombian Confederation of Workers or CTC; General Confederation of Workers or CGT; National Liberation Army or ELN; Revolutionary Armed Forces of Colombia or FARC
note: FARC and ELN are the two largest insurgent groups active in Colombia

International organization participation: BCIE, CAN, Caricom (observer), CDB, FAO, G-3, G-24, G-77, IADB, IAEA, IBRD, ICAO, ICC, ICRM, IDA, IFAD, IFC, IFRCS, IHO, ILO, IMF, IMO, IMSO, Interpol, IOC, IOM, IPU, ISO, ITSO, ITU, ITUC, LAES, LAIA, Mercosur (associate), MIGA, NAM, OAS, OPANAL, OPCW, PCA, RG, UN, UN Security Council (temporary), UNASUR, UNCTAD, UNESCO, UNHCR, UNIDO, Union Latina, UNWTO, UPU, WCO, WFTU, WHO, WIPO, WMO, WTO

Diplomatic representation in the US:
chief of mission: Ambassador Gabriel SILVA Lujan
chancery: 2118 Leroy Place NW, Washington, DC 20008
telephone: [1] (202) 387-8338
FAX: [1] (202) 232-8643
consulate(s) general: Atlanta, Boston, Chicago, Houston, Los Angeles, Miami, New York, San Francisco, San Juan (Puerto Rico), Washington, DC

Diplomatic representation from the US: *chief of mission:* Ambassador Michael MCKINLEY
embassy: Calle 24 Bis No. 48-50, Bogota, D.C.
mailing address: Carrera 45 No. 24B-27, Bogota, D.C.
telephone: [57] (1) 315-0811
FAX: [57] (1) 315-2197

Flag description: three horizontal bands of yellow (top, double-width), blue, and red; the flag retains the three main colors of the banner of Gran Columbia, the short-lived South American republic that broke up in 1830; various interpretations of the colors exist and include: yellow for the gold in Colombia's land, blue for the seas on its shores, and red for the blood spilled in attaining freedom; alternatively, the colors have been described as representing more elemental concepts such as sovereignty and justice (yellow), loyalty and vigilance (blue), and valor and generosity (red); or simply the principles of liberty, equality, and fraternity
note: similar to the flag of Ecuador, which is longer and bears the Ecuadorian coat of arms superimposed in the center

National anthem: *name:* "Himno Nacional de la Republica de Colombia" (National Anthem of the Republic of Colombia)
lyrics/music: Rafael NUNEZ/Oreste SINDICI
note: adopted 1920; the anthem was created from an inspirational poem written by President Rafael NUNEZ

ECONOMY

Economy—overview: The SANTOS administration has highlighted five "locomotives" to stimulate economic growth: extractive industries; agriculture; infrastructure; housing; and innovation. Colombia is third largest exporter of oil to the United States. President SANTOS, inaugurated in August 2010, introduced unprecedented legislation to better distribute extractive industry royalties and compensate Colombians who lost their land due to decades of violence. He also seeks to build on improvements in domestic security and on President URIBE's promarket economic policies. Foreign direct investment reached a record $10 billion in 2008, but dropped to $7.2 billion in 2009, before beginning to recover in 2010, notably in the oil sector. Pro-business reforms in the oil and gas sectors and export-led growth, fueled mainly by the Andean Trade Promotion and Drug Eradication Act, have enhanced Colombia's investment climate. Inequality, underemployment, and narcotrafficking remain significant challenges, and Colombia's infrastructure requires major improvements to sustain economic expansion. Because of the global financial crisis and weakening demand for Colombia's exports, Colombia's economy grew only 2.7% in 2008, and 0.8% in 2009 but rebounded to around 4.4% in 2010. In late 2010, Colombia experienced its most severe flooding in decades, with damages estimated to exceed $6 billion. The government has encouraged exporters to diversify their customer base beyond the United States and Venezuela, traditionally Colombia's largest trading partners; the SANTOS administration continues to pursue free trade agreements with Asian and South American partners and a trade accord with Canada is expected to go into effect in 2011, while a negotiated trade agreement with the EU has yet to be approved by the EU parliament. Improved relations with Venezuela have eased worries about restrictions on bilateral trade, but the business sector remains concerned about the pending US Congressional approval of the US-Colombia Trade Promotion Agreement.

GDP (purchasing power parity): $435.4 billion (2010 est.)
country comparison to the world: 29
$417.4 billion (2009 est.)
$411.4 billion (2008 est.)
note: data are in 2010 US dollars

GDP (official exchange rate): $285.5 billion (2010 est.)

GDP—real growth rate: 4.3% (2010 est.)
country comparison to the world: 88
1.5% (2009 est.)
3.5% (2008 est.)

GDP—per capita (PPP): $9,800 (2010 est.)
country comparison to the world: 111
$9,600 (2009 est.)
$9,500 (2008 est.)
note: data are in 2010 US dollars

GDP—composition by sector:
agriculture: 9.3%
industry: 38%
services: 52.7% (2010 est.)

Labor force: 21.27 million (2010 est.)
country comparison to the world: 30

Labor force—by occupation:
agriculture: 18%
industry: 13%
services: 68% (2010 est.)

Unemployment rate: 11.8% (2010 est.)
country comparison to the world: 125
12% (2009 est.)

Population below poverty line: 45.5% (2009)

Household income or consumption by percentage share:
lowest 10%: 0.8%
highest 10%: 45% (2008)

Distribution of family income—Gini index: 58.5 (2009)
country comparison to the world: 8
53.8 (1996)

Investment (gross fixed): 22.8% of GDP (2010 est.)
country comparison to the world: 62

Budget: *revenues:* $74.2 billion
expenditures: $83.9 billion (2011 est.)

Public debt: 44.8% of GDP (2010 est.)
country comparison to the world: 60
45.3% of GDP (2009 est.)

Inflation rate (consumer prices): 3.1% (2010 est.)
country comparison to the world: 87
4% (2009 est.)

Central bank discount rate: 3% (31 October 2010)
country comparison to the world: 74
5.5% (31 December 2009)

Commercial bank prime lending rate: 12.98% (31 December 2009 est.)
country comparison to the world: 28
17.18% (31 December 2008 est.)

Stock of narrow money: $31.83 billion (31 December 2010 est.)
country comparison to the world: 54
$24.41 billion (31 December 2009 est.)

Stock of broad money: $104.9 billion (31 December 2010 est.)
country comparison to the world: 52
$82.39 billion (31 December 2009 est.)

Stock of domestic credit: $123 billion (31 December 2010 est.)
country comparison to the world: 45
$96.66 billion (31 December 2009 est.)

Market value of publicly traded shares: $217.3 billion (31 December 2010)
country comparison to the world: 30
$133.3 billion (31 December 2009)
$87.03 billion (31 December 2008)

Agriculture—products: coffee, cut flowers, bananas, rice, tobacco, corn, sugarcane, cocoa beans, oilseed, vegetables; forest products; shrimp

Industries: textiles, food processing, oil, clothing and footwear, beverages, chemicals, cement; gold, coal, emeralds

Industrial production growth rate: 5.5% (2010 est.)
country comparison to the world: 65

Electricity—production: 50.58 billion kWh (2007)
country comparison to the world: 48

Electricity—consumption: 38.59 billion kWh (2007)
country comparison to the world: 54

Electricity—exports: 876.7 million kWh (2007)

Electricity—imports: 39.4 million kWh (2007)

Oil—production: 785,000 bbl/day (2010 est.)
country comparison to the world: 27

Oil—consumption: 288,000 bbl/day (2009 est.)
country comparison to the world: 43

Oil—exports: 294,000 bbl/day (2008 est.)
country comparison to the world: 43

Oil—imports: 16,540 bbl/day (2007 est.)
country comparison to the world: 121

Oil—proved reserves: 1.9 billion bbl (1 January 2010 est.)
country comparison to the world: 36

Natural gas—production: 9 billion cu m (2008 est.)

country comparison to the world: 43
Natural gas—consumption: 8.1 billion cu m (2008 est.)
country comparison to the world: 53
Natural gas—exports: 900 million cu m (2008 est.)
country comparison to the world: 37
Natural gas—imports: 0 cu m (2008 est.)
country comparison to the world: 168
Natural gas—proved reserves: 112 billion cu m (1 January 2010 est.)
country comparison to the world: 51
Current account balance: $-5.946 billion (2010 est.)
country comparison to the world: 171
$-4.991 billion (2009 est.)
Exports: $40.24 billion (2010 est.)
country comparison to the world: 59
$32.08 billion (2009)
Exports—commodities: petroleum, coffee, coal, nickel, emeralds, apparel, bananas, cut flowers
Exports—partners: US 42%, EU 12.6%, China 5.2%, Ecuador 4.5% (2010 est.)
Imports: $36.26 billion (2010 est.)
country comparison to the world: 53
$32.49 billion (2009)
Imports—commodities: industrial equipment, transportation equipment, consumer goods, chemicals, paper products, fuels, electricity
Imports—partners: US 25.5%, China 13.4%, Mexico 9.4%, Brazil 5.9%, Germany 4.1% (2010 est.)
Reserves of foreign exchange and gold: $28.5 billion (31 December 2010 est.)
country comparison to the world: 38
$25.35 billion (31 December 2009 est.)
Debt—external: $57.74 billion (31 December 2010 est.)
country comparison to the world: 51
$52.9 billion (31 December 2009 est.)
Stock of direct foreign investment—at home: $84.62 billion (31 December 2010 est.)
country comparison to the world: 38
$75.22 billion (31 December 2009 est.)
Stock of direct foreign investment—abroad: $19.2 billion (31 December 2010 est.)
country comparison to the world: 43
$16.2 billion (31 December 2009 est.)
Exchange rates: Colombian pesos (COP) per US dollar–1,869.9 (2010), 2,157.6 (2009), 2,243.6 (2008), 2,013.8 (2007), 2,358.6 (2006)

COMMUNICATIONS

Telephones—main lines in use: 7.5 million (2009)
country comparison to the world: 25
Telephones—mobile cellular: 42.16 million (2009)
country comparison to the world: 29
Telephone system: *general assessment:* modern system in many respects with a nationwide microwave radio relay system, a domestic satellite system with 41 earth stations, and a fiber-optic network linking 50 cities; telecommunications sector liberalized during the 1990s; multiple providers of both fixed-line and mobile-cellular services
domestic: fixed-line connections stand at about 15 per 100 persons; mobile cellular telephone subscribership is about 90 per 100 persons; competition among cellular service providers is resulting in falling local and international calling rates and contributing to the steep decline in the market share of fixed line services
international: country code–57; landing points for the ARCOS, Colombia-Florida Subsea Fiber (CFX-1), Maya-1, Pan American, and the South America-1 submarine cables providing links to the US, parts of the Caribbean, and Central and South America; satellite earth stations–10 (6 Intelsat, 1 Inmarsat, 3 fully digitalized international switching centers) (2009)
Broadcast media: combination of state-owned and privately-owned broadcast media provide service; more than 500 radio stations and large number of national, regional, and local TV stations (2007)
Internet country code: .co
Internet hosts: 2.527 million (2010)
country comparison to the world: 32
Internet users: 22.538 million (2009)
country comparison to the world: 18

TRANSPORTATION

Airports: 990 (2010)
country comparison to the world: 7
Airports—with paved runways: *total:* 116
over 3,047 m: 2
2,438 to 3,047 m: 8
1,524 to 2,437 m: 41
914 to 1,523 m: 50
under 914 m: 15 (2010)
Airports—with unpaved runways: *total:* 874
over 3,047 m: 1
1,524 to 2,437 m: 35
914 to 1,523 m: 228
under 914 m: 610 (2010)
Heliports: 2 (2010)
Pipelines: gas 4,801 km; oil 6,334 km; refined products 3,309 km (2010)
Railways: *total:* 874 km
country comparison to the world: 95
standard gauge: 150 km 1.435-m gauge
narrow gauge: 498 km 0.950-m gauge; 226 km 0.914-m gauge (2010)
Roadways: *total:* 141,374 km (2010)
country comparison to the world: 34
Waterways: 18,000 km (2010)
country comparison to the world: 6
Merchant marine: *total:* 13
country comparison to the world: 104
by type: cargo 11, petroleum tanker 1, specialized tanker 1
registered in other countries: 3 (Antigua and Barbuda 1, Panama 2) (2010)
Ports and terminals: Barranquilla, Buenaventura, Cartagena, Puerto Bolivar, Santa Marta, Turbo

MILITARY

Military branches: National Army (Ejercito Nacional), Republic of Colombia Navy (Armada Republica de Colombia, includes Naval Aviation, Naval Infantry (Infanteria de Marina, IM), and Coast Guard), Colombian Air Force (Fuerza Aerea de Colombia, FAC) (2011)
Military service age and obligation: 18-24 years of age for compulsory and voluntary military service; service obligation–18 months (2004)
Manpower available for military service: *males age 16–49:* 11,692,647
females age 16–49: 11,727,625 (2010 est.)
Manpower fit for military service: *males age 16–49:* 9,150,400
females age 16–49: 9,861,760 (2010 est.)
Manpower reaching militarily significant age annually: *male:* 430,634
female: 413,974 (2010 est.)
Military expenditures: 3.4% of GDP (2005 est.)
country comparison to the world: 34

TRANSNATIONAL ISSUES

Disputes—international: in December 2007, ICJ allocates San Andres, Providencia, and Santa Catalina islands to Colombia under 1928 Treaty but does not rule on 82 degrees W meridian as maritime boundary with Nicaragua; managed dispute with Venezuela over maritime boundary and Venezuelan-administered Los Monjes Islands near the Gulf of Venezuela; Colombian-organized illegal narcotics, guerrilla, and paramilitary activities penetrate all neighboring borders and have caused Colombian citizens to flee mostly into neighboring countries; Colombia, Honduras, Nicaragua, Jamaica, and the US assert various claims to Bajo Nuevo and Serranilla Bank
Refugees and internally displaced persons: *IDPs:* 1.8-3.5 million (conflict between government and illegal armed groups and drug traffickers) (2007)
Illicit drugs: illicit producer of coca, opium poppy, and cannabis; world's leading coca cultivator with 167,000 hectares in coca cultivation in 2007, a 6% increase over 2006, producing a potential of 535 mt of pure cocaine; the world's largest producer of coca derivatives; supplies cocaine to nearly all of the US market and the great majority of other international drug markets; in 2005, aerial eradication dispensed herbicide to treat over 130,000 hectares but aggressive replanting on the part of coca growers means Colombia remains a key producer; a significant portion of narcotics proceeds are either laundered or invested in Colombia through the black market peso exchange; important supplier of heroin to the US market; opium poppy cultivation is estimated to have fallen 25% between 2006 and 2007; most Colombian heroin is destined for the US market (2008)

COMOROS

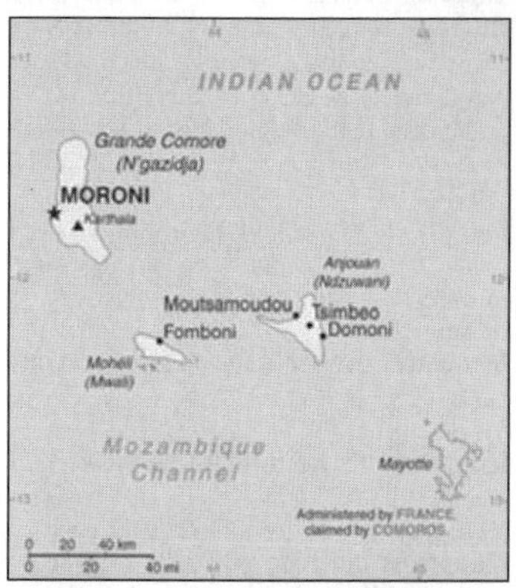

INTRODUCTION

Background: Comoros has endured more than 20 coups or attempted coups since gaining independence from France in 1975. In 1997, the islands of Anjouan and Moheli declared independence from Comoros. In 1999, military chief Col. AZALI seized power in a bloodless coup, and helped negotiate the 2000 Fomboni Accords power-sharing agreement in which the federal presidency rotates among the three islands, and each island maintains its own local government. AZALI won the 2002 presidential election, and each island in the archipelago elected its own president. AZALI stepped down in 2006 and President SAMBI was elected to office. In 2007, Mohamed BACAR effected Anjouan's de-facto secession from the Union, refusing to step down in favor of fresh Anjouanais elections when Comoros' other islands held legitimate elections in July. The African Union (AU) initially attempted to resolve the political crisis by applying sanctions and a naval blockade on Anjouan, but in March 2008, AU and Comoran soldiers seized the island. The move was generally welcomed by the island's inhabitants.

GEOGRAPHY

Location: Southern Africa, group of islands at the northern mouth of the Mozambique Channel, about two-thirds of the way between northern Madagascar and northern Mozambique
Geographic coordinates: 12 10 S, 44 15 E
Map references: Africa
Area: *total:* 2,235 sq km
country comparison to the world: 179
land: 2,235 sq km
water: 0 sq km
Area—comparative: slightly more than 12 times the size of Washington, DC
Land boundaries: 0 km
Coastline: 340 km
Maritime claims: territorial sea: 12 nm
exclusive economic zone: 200 nm
Climate: tropical marine; rainy season (November to May)
Terrain: volcanic islands, interiors vary from steep mountains to low hills
Elevation extremes: *lowest point:* Indian Ocean 0 m
highest point: Karthala 2,360 m
Natural resources: NEGL
Land use: *arable land:* 35.87%
permanent crops: 23.32%
other: 40.81% (2005)
Irrigated land: NA
Total renewable water resources: 1.2 cu km (2003)
Freshwater withdrawal (domestic/industrial/agricultural): *total:* 0.01 cu km/yr (48%/5%/47%)
per capita: 13 cu m/yr (1999)
Natural hazards: cyclones possible during rainy season (December to April); volcanic activity on Grand Comore
volcanism: Karthala (elev. 2,361 m) on Grand Comore Island last erupted in 2007; a 2005 eruption forced thousands of people to be evacuated and produced a large ash cloud
Environment—current issues: soil degradation and erosion results from crop cultivation on slopes without proper terracing; deforestation
Environment—international agreements: *party to:* Biodiversity, Climate Change, Climate Change-Kyoto Protocol, Desertification, Endangered Species, Hazardous Wastes, Law of the Sea, Ozone Layer Protection, Ship Pollution, Wetlands
signed, but not ratified: none of the selected agreements
Geography—note: important location at northern end of Mozambique Channel

PEOPLE

Population: 794,683 (July 2011 est.)
country comparison to the world: 161
Age structure: *0–14 years:* 41.6% (male 166,141/female 164,788)
15–64 years: 55.3% (male 217,046/female 222,093)
65 years and over: 3.1% (male 11,053/female 13,562) (2011 est.)
Median age: *total:* 19 years
male: 18.7 years
female: 19.3 years (2011 est.)
Population growth rate: 2.696% (2011 est.)
country comparison to the world: 19
Birth rate: 34.19 births/1,000 population (2011 est.)
country comparison to the world: 33
Death rate: 7.23 deaths/1,000 population (July 2011 est.)
country comparison to the world: 124
Net migration rate: 0 migrant(s)/1,000 population (2011 est.)
country comparison to the world: 81
Urbanization: *urban population:* 28% of total population (2010)
rate of urbanization: 2.8% annual rate of change (2010-15 est.)
Major cities—population: MORONI (capital) 49,000 (2009)
Sex ratio: *at birth:* 1.03 male(s)/female
under 15 years: 1.01 male(s)/female
15–64 years: 0.98 male(s)/female
65 years and over: 0.83 male(s)/female
total population: 0.98 male(s)/female (2011 est.)
Infant mortality rate: *total:* 62.63 deaths/1,000 live births
country comparison to the world: 27
male: 70.3 deaths/1,000 live births
female: 54.73 deaths/1,000 live births (2011 est.)
Life expectancy at birth: *total population:* 64.2 years
country comparison to the world: 172
male: 61.76 years
female: 66.72 years (2011 est.)
Total fertility rate: 4.72 children born/woman (2011 est.)
country comparison to the world: 28
HIV/AIDS—adult prevalence rate: 0.1% (2009 est.)
country comparison to the world: 123
HIV/AIDS—people living with HIV/AIDS: fewer than 500 (2009 est.)
country comparison to the world: 153
HIV/AIDS—deaths: fewer than 100 (2009 est.)
country comparison to the world: 126
Drinking water source: *Improved:*
urban: 91% of population
rural: 97% of population
total: 95% of population
Unimproved:
urban: 9% of population
rural: 3% of population
total: 5% of population (2008)
Sanitation facility access: *Improved:*
urban: 50% of population
rural: 30% of population
total: 36% of population
Unimproved:
urban: 50% of population
rural: 70% of population
total: 64% of population (2008)
Nationality: *noun:* Comoran(s)
adjective: Comoran
Ethnic groups: Antalote, Cafre, Makoa, Oimatsaha, Sakalava
Religions: Sunni Muslim 98%, Roman Catholic 2%
Languages: Arabic (official), French (official), Shikomoro (a blend of Swahili and Arabic)
Literacy: *definition:* age 15 and over can read and write
total population: 56.5%
male: 63.6%
female: 49.3% (2003 est.)
School life expectancy (primary to tertiary education): *total:* 11 years
male: 12 years
female: 10 years (2005)
Education expenditures: 7.6% of GDP (2008)
country comparison to the world: 13

GOVERNMENT

Country name: *conventional long form:* Union of the Comoros
conventional short form: Comoros
local long form: Udzima wa Komori (Comorian); Union des Comores (French); Jumhuriyat al Qamar al Muttahidah (Arabic)
local short form: Komori (Comorian); Comores (French); Juzur al Qamar (Arabic)
Government type: republic
Capital: *name:* Moroni
geographic coordinates: 11 42 S, 43 14 E

time difference: UTC+3 (8 hours ahead of Washington, DC during Standard Time)

Administrative divisions: 3 islands and 4 municipalities*; Grande Comore (N'gazidja), Anjouan (Ndzuwani), Domoni*, Fomboni*, Moheli (Mwali), Moroni*, Moutsamoudou*

Independence: 6 July 1975 (from France)

National holiday: Independence Day, 6 July (1975)

Constitution: 23 December 2001

Legal system: mixed legal system of Islamic religious law, the French civil code of 1975, and customary law

International law organization participation: has not submitted an ICJ jurisdiction declaration; accepts ICCt jurisdiction

Suffrage: 18 years of age; universal

Executive branch: *chief of state:* President Ikililou DHOININE (since 26 May 2011)
head of government: President Ikililou DHOININE (since 26 May 2011)
cabinet: Council of Ministers appointed by the president
(For more information visit the World Leaders website)
elections: as defined by the 2001 constitution, the presidency rotates every four years among the elected presidents from the three main islands in the Union; election last held on 7 November and 26 December 2010 (next to be held in 2014)
election results: Ikililou DHOININE elected president; percent of vote–Ikililou DHOININE 61.1%, Mohamed Said FAZUL 32.7%, Abdou DJABIR 6.2%

Legislative branch: unicameral Assembly of the Union (33 seats; 15 deputies are selected by the individual islands' local assemblies and 18 by universal suffrage to serve for five years);
elections: last held on 6 and 20 December 2009 (next to be held in 2014)
election results: percent of vote by party–NA; seats by party–pro-union coalition 19, autonomous coalition 4, independents 1; note–9 additional seats are filled by deputies from local island assemblies

Judicial branch: Supreme Court or Cour Supremes (two members appointed by the president, two members elected by the Federal Assembly, one elected by the Council of each island, and others are former presidents of the republic)

Political parties and leaders: Camp of the Autonomous Islands or CdIA (a coalition of parties organized by the islands' presidents in opposition to the Union President); Convention for the Renewal of the Comoros or CRC [AZALI Assowmani]; Front National pour la Justice or FNJ [Ahmed RACHID] (Islamic party in opposition); Mouvement pour la Democratie et le Progress or MDP-NGDC [Abbas DJOUSSOUF]; Parti Comorien pour la Democratie et le Progress or PCDP [Ali MROUDJAE]; Rassemblement National pour le Development or RND [Omar TAMOU, Abdoulhamid AFFRAITANE]

Political pressure groups and leaders: *other:* environmentalists

International organization participation: ACP, AfDB, AMF, AOSIS, AU, COMESA, FAO, FZ, G-77, IBRD, ICAO, ICRM, IDA, IDB, IFAD, IFC, IFRCS, ILO, IMF, IMO, IMSO, InOC, Interpol, IOC, IPU, ITSO, ITU, ITUC, LAS, NAM, OIC, OIF, OPCW, UN, UNCTAD, UNESCO, UNIDO, UPU, WCO, WHO, WIPO, WMO, WTO (observer)

Diplomatic representation in the US: *chief of mission:* Representative to the UN and Ambassador to the US Mohamed TOIHIRI
chancery: Mission to the US, 866 United Nations Plaza, Suite 418, New York, NY 10017
telephone: [1] (212) 750-1637

Diplomatic representation from the US: the US does not have an embassy in Comoros; the ambassador to Madagascar is accredited to Comoros

Flag description: four equal horizontal bands of yellow (top), white, red, and blue, with a green isosceles triangle based on the hoist; centered within the triangle is a white crescent with the convex side facing the hoist and four white, five-pointed stars placed vertically in a line between the points of the crescent; the horizontal bands and the four stars represent the four main islands of the archipelago–Mwali, N'gazidja, Nzwani, and Mahore (Mayotte–territorial collectivity of France, but claimed by Comoros)
note: the crescent, stars, and color green are traditional symbols of Islam

National anthem: *name:* "Udzima wa ya Masiwa" (The Union of the Great Islands)
lyrics/music: Said Hachim SIDI ABDEREMANE/Said Hachim SIDI ABDEREMANE and Kamildine ABDALLAH
note: adopted 1978

ECONOMY

Economy—overview: One of the world's poorest countries, Comoros is made up of three islands that have inadequate transportation links, a young and rapidly increasing population, and few natural resources. The low educational level of the labor force contributes to a subsistence level of economic activity, high unemployment, and a heavy dependence on foreign grants and technical assistance. Agriculture, including fishing, hunting, and forestry, contributes 40% to GDP, employs 80% of the labor force, and provides most of the exports. Export income is heavily reliant on the three main crops of vanilla, cloves, and ylang-ylang and Comoros' export earnings are easily disrupted by disasters such as fires. The country is not self-sufficient in food production; rice, the main staple, accounts for the bulk of imports. The government–which is hampered by internal political disputes–lacks a comprehensive strategy to attract foreign investment and is struggling to upgrade education and technical training, privatize commercial and industrial enterprises, improve health services, diversify exports, promote tourism, and reduce the high population growth rate. Political problems have inhibited growth, which has averaged only about 1% in 2006-09. Remittances from 150,000 Comorans abroad help supplement GDP. In September 2009 the IMF approved Comoros for a three-year $21 million loan. The IMF gave generally positive reports of the country's program performance as of October 2010. The African Development Bank approved a $34.6 million debt-relief package loan for Comoros in September 2010, and Comoros will attempt to qualify for debt relief in 2012 under the IMF and World Bank's Heavily Indebted Poor Countries (HIPC) initiative.

GDP (purchasing power parity): $800 million (2010 est.)
country comparison to the world: 205
$783.4 million (2009 est.)
$769.2 million (2008 est.)
note: data are in 2010 US dollars

GDP (official exchange rate): $534 million (2010 est.)

GDP—real growth rate: 2.1% (2010 est.)
country comparison to the world: 144
1.8% (2009 est.)
1% (2008 est.)

GDP—per capita (PPP): $1,000 (2010 est.)
country comparison to the world: 212
$1,000 (2009 est.)
$1,100 (2008 est.)
note: data are in 2010 US dollars

GDP—composition by sector:
agriculture: 40%
industry: 4%
services: 56% (2001 est.)

Labor force: 268,500 (2007 est.)
country comparison to the world: 166

Labor force—by occupation:
agriculture: 80%
industry and *services:* 20% (1996 est.)

Unemployment rate: 20% (1996 est.)
country comparison to the world: 165

Population below poverty line: 60% (2002 est.)

Household income or consumption by percentage share: *lowest 10%:* 0.9%
highest 10%: 55.2% (2004)

Budget: *revenues:* $27.6 million
expenditures: $NA (2001 est.)

Inflation rate (consumer prices): 3% (2007 est.)
country comparison to the world: 83

Central bank discount rate: 2.21% (31 December 2009)
country comparison to the world: 76
5.36% (31 December 2008)

Commercial bank prime lending rate: 10.5% (31 December 2009 est.)
country comparison to the world: 81
10.5% (31 December 2008 est.)

Stock of narrow money: $104.7 million (31 December 2009)
country comparison to the world: 179
$98.36 million (31 December 2008)

Stock of broad money: $168.6 million (31 December 2009)
country comparison to the world: 183
$143.7 million (31 December 2008)

Stock of domestic credit: $79.75 million (31 December 2008 est.)
country comparison to the world: 181
$60.57 million (31 December 2007 est.)

Agriculture—products: vanilla, cloves, ylang-ylang (perfume essence), copra, coconuts, bananas, cassava (tapioca)

Industries: fishing, tourism, perfume distillation

Industrial production growth rate: NA%

Electricity—production: 22 million kWh (2007 est.)
country comparison to the world: 205

Electricity—consumption: 20.46 million kWh (2007 est.)
country comparison to the world: 204
Electricity—exports: 0 kWh (2008 est.)
Electricity—imports: 0 kWh (2008 est.)
Oil—production: 0 bbl/day (2009 est.)
country comparison to the world: 160
Oil—consumption: 1,000 bbl/day (2009 est.)
country comparison to the world: 199
Oil—exports: 0 bbl/day (2007 est.)
country comparison to the world: 155
Oil—imports: 766.2 bbl/day (2007 est.)
country comparison to the world: 191
Oil—proved reserves: 0 bbl (1 January 2010 est.)
country comparison to the world: 116
Natural gas—production: 0 cu m (2008 est.)
country comparison to the world: 112
Natural gas—consumption: 0 cu m (2008 est.)
country comparison to the world: 159
Natural gas—exports: 0 cu m (2008 est.)
country comparison to the world: 77
Natural gas—imports: 0 cu m (2008 est.)
country comparison to the world: 167
Natural gas—proved reserves: 0 cu m (1 January 2010 est.)
country comparison to the world: 121
Current account balance: $8 million (2007 est.)
country comparison to the world: 61
Exports: $32 million (2006)
country comparison to the world: 203
Exports—commodities: vanilla, ylang-ylang (perfume essence), cloves, copra
Exports—partners: Turkey 25.2%, France 20.44%, Singapore 17.44%, Algeria 8.02%, Italy 6.09%, Saudi Arabia 5% (2009)
Imports: $143 million (2006)
country comparison to the world: 206
Imports—commodities: rice and other foodstuffs, consumer goods, petroleum products, cement, transport equipment
Imports—partners: France 15.5%, China 14.66%, India 10.55%, UAE 7.88%, Pakistan 5.69%, Kenya 4.51% (2009)
Debt—external: $232 million (2000 est.)
country comparison to the world: 173
Exchange rates: Comoran francs (KMF) per US dollar–361.4 (2007), 391.8 (2006), 395.6 (2005), 396.21 (2004), 435.9 (2003)

COMMUNICATIONS

Telephones—main lines in use: 25,400 (2009)
country comparison to the world: 184
Telephones—mobile cellular: 100,000 (2009)
country comparison to the world: 185
Telephone system: *general assessment:* sparse system of microwave radio relay and HF radiotelephone communication stations
domestic: fixed-line connections only about 3 per 100 persons; mobile cellular usage about 15 per 100 persons
international: country code–269; HF radiotelephone communications to Madagascar and Reunion
Broadcast media: national state-owned TV station and a TV station run by Anjouan regional government; national state-owned radio; regional governments on the islands of Grande Comore and Anjouan each operate a radio station; a few independent and small community radio stations operate on the islands of Grande Comore and Moheli, and these two islands have access to Mayotte Radio and French TV (2007)
Internet country code: .km
Internet hosts: 14 (2010)
country comparison to the world: 219
Internet users: 24,300 (2009)
country comparison to the world: 186

TRANSPORTATION

Airports: 4 (2010)
country comparison to the world: 184
Airports—with paved runways: *total:* 4
2,438 to 3,047 m: 1
914 to 1,523 m: 3 (2010)
Roadways: *total:* 880 km
country comparison to the world: 184
paved: 673 km
unpaved: 207 km (2002)
Merchant marine: *total:* 177
country comparison to the world: 37
by type: bulk carrier 19, cargo 102, carrier 5, chemical tanker 6, container 2, passenger 3, passenger/cargo 1, petroleum tanker 15, refrigerated cargo 12, roll on/roll off 12
foreign-owned: 98 (Bangladesh 1, Bulgaria 8, China 1, Cyprus 2, Greece 3, Kenya 1, Kuwait 1, Latvia 1, Lebanon 3, Lithuania 3, Monaco 1, Nigeria 1, Norway 2, Pakistan 3, Russia 21, Syria 6, Turkey 16, UAE 11, UK 1, Ukraine 10, US 2) (2010)
Ports and terminals: Mayotte, Mutsamudu

MILITARY

Military branches: Army of National Development (AND): Comoran Security Force, Comoran Coast Guard, Comoran Federal Police (2011)
Military service age and obligation: 18 years of age for 2-year voluntary military service; no conscription; women first inducted into the Army in 2004 (2011)
Manpower available for military service: *males age 16–49:* 184,236
females age 16–49: 183,363 (2010 est.)
Manpower fit for military service: *males age 16–49:* 134,562
females age 16–49: 145,797 (2010 est.)
Manpower reaching militarily significant age annually: *male:* 8,831
female: 8,809 (2010 est.)
Military expenditures: 2.8% of GDP (2006)
country comparison to the world: 50

TRANSNATIONAL ISSUES

Disputes—international: claims French-administered Mayotte and challenges France's and Madagascar's claims to Banc du Geyser, a drying reef in the Mozambique Channel; in May 2008, African Union forces are called in to assist the Comoros military recapture Anjouan Island from rebels who seized it in 2001

CONGO, DEMOCRATIC REPUBLIC OF THE

INTRODUCTION

Background: Established as a Belgian colony in 1908, the then-Republic of the Congo gained its independence in 1960, but its early years were marred by political and social instability. Col. Joseph MOBUTU seized power and declared himself president in a November 1965 coup. He subsequently changed his name–to MOBUTU Sese Seko–as well as that of the country–to Zaire. MOBUTU retained his position for 32 years through several sham elections, as well as through brutal force. Ethnic strife and civil war, touched off by a massive inflow of refugees in 1994 from fighting in Rwanda and Burundi, led in May 1997 to the toppling of the MOBUTU regime by a rebellion backed by Rwanda and Uganda and fronted by Laurent KABILA. He renamed the country the Democratic Republic of the Congo (DRC), but in August 1998 his regime was itself challenged by a second insurrection again backed by Rwanda and Uganda. Troops from Angola, Chad, Namibia, Sudan, and Zimbabwe intervened to support KABILA's regime. A cease-fire was signed in July 1999 by the DRC, Congolese armed rebel groups, Angola, Namibia, Rwanda, Uganda, and Zimbabwe but sporadic fighting continued. Laurent KABILA was assassinated in January 2001 and his son, Joseph KABILA, was named head of state. In October 2002, the new president was successful in negotiating the withdrawal of Rwandan forces occupying eastern Congo; two months later, the Pretoria Accord was signed by all remaining warring parties to end the fighting and establish a government of national unity. A transitional government was set up in July 2003. Joseph KABILA as president and four vice presidents represented the former government, former rebel groups, the political opposition, and civil society. The transitional government held a successful constitutional referendum in December 2005 and elections for the presidency, National Assembly, and provincial legislatures in 2006. The National Assembly was installed in September 2006 and KABILA was inaugurated president in December 2006. Provincial assemblies were constituted in early 2007, and elected governors and national senators in January 2007. The next national elections are scheduled for November 2011.

GEOGRAPHY

Location: Central Africa, northeast of Angola
Geographic coordinates: 0 00 N, 25 00 E
Map references: Africa
Area: *total:* 2,344,858 sq km
country comparison to the world: 12
land: 2,267,048 sq km
water: 77,810 sq km
Area—comparative: slightly less than one-fourth the size of the US
Land boundaries: *total:* 10,730 km
border countries: Angola 2,511 km (of which 225 km is the boundary of Angola's discontiguous Cabinda Province), Burundi 233 km, Central African Republic 1,577 km, Republic of the Congo 2,410 km, Rwanda 217 km, Sudan 628 km, Tanzania 459 km, Uganda 765 km, Zambia 1,930 km
Coastline: 37 km
Maritime claims: territorial sea: 12 nm
exclusive economic zone: boundaries with neighbors
Climate: tropical; hot and humid in equatorial river basin; cooler and drier in southern highlands; cooler and wetter in eastern highlands; north of Equator–wet season (April to October), dry season (December to February); south of Equator–wet season (November to March), dry season (April to October)
Terrain: vast central basin is a low-lying plateau; mountains in east
Elevation extremes: *lowest point:* Atlantic Ocean 0 m
highest point: Pic Marguerite on Mont Ngaliema (Mount Stanley) 5,110 m
Natural resources: cobalt, copper, niobium, tantalum, petroleum, industrial and gem diamonds, gold, silver, zinc, manganese, tin, uranium, coal, hydropower, timber
Land use:
arable land: 2.86%
permanent crops: 0.47%
other: 96.67% (2005)
Irrigated land: 110 sq km (2003)
Total renewable water resources: 1,283 cu km (2001)
Freshwater withdrawal (domestic/industrial/agricultural): *total:* 0.36 cu km/yr (53%/17%/31%)
per capita: 6 cu m/yr (2000)
Natural hazards: periodic droughts in south; Congo River floods (seasonal); in the east, in the Great Rift Valley, there are active volcanoes
volcanism: Nyiragongo (elev. 3,470 m), which erupted in 2002 and is experiencing ongoing activity, poses a major threat to the city of Goma, home to a quarter of a million people; the volcano produces unusually fast-moving lava, known to travel up to 100 km/hr; Nyiragongo has been deemed a "Decade Volcano" by the International Association of Volcanology and Chemistry of the Earth's Interior, worthy of study due to its explosive history and close proximity to human populations; its neighbor, Nyamuragira, which erupted in 2010, is Africa's most active volcano; Visoke is the only other historically active volcano
Environment—current issues: poaching threatens wildlife populations; water pollution; deforestation; refugees responsible for significant deforestation, soil erosion, and wildlife poaching; mining of minerals (coltan–a mineral used in creating capacitors, diamonds, and gold) causing environmental damage
Environment—international agreements: *party to:* Biodiversity, Climate Change, Climate Change-Kyoto Protocol, Desertification, Endangered Species, Hazardous Wastes, Law of the Sea, Marine Dumping, Ozone Layer Protection, Tropical Timber 83, Tropical Timber 94, Wetlands
signed, but not ratified: Environmental Modification
Geography—note: straddles equator; has narrow strip of land that controls the lower Congo River and is only outlet to South Atlantic Ocean; dense tropical rain forest in central river basin and eastern highlands

PEOPLE

Population: 71,712,867 (July 2011 est.)
country comparison to the world: 19
note: estimates for this country explicitly take into account the effects of excess mortality due to AIDS; this can result in lower life expectancy, higher infant mortality, higher death rates, lower population growth rates, and changes in the distribution of population by age and sex than would otherwise be expected
Age structure: *0–14 years:* 44.4% (male 16,031,347/female 15,811,818)
15–64 years: 53% (male 18,919,942/female 19,116,204)
65 years and over: 2.6% (male 767,119/female 1,066,437) (2011 est.)
Median age:
total: 17.4 years
male: 17.2 years
female: 17.6 years (2011 est.)
Population growth rate: 2.614% (2011 est.)
country comparison to the world: 24
Birth rate: 37.74 births/1,000 population (2011 est.)
country comparison to the world: 18
Death rate: 11.06 deaths/1,000 population (July 2011 est.)
country comparison to the world: 36
Net migration rate: -0.54 migrant(s)/1,000 population (2011 est.)
country comparison to the world: 140
Urbanization: *urban population:* 35% of total population (2010)
rate of urbanization: 4.5% annual rate of change (2010-15 est.)
Major cities—population: KINSHASA (capital) 8.401 million; Lubumbashi 1.543 million; Mbuji-Mayi 1.488 million; Kananga 878,000; Kisangani 812,000 (2009)
Sex ratio: *at birth:* 1.03 male(s)/female
under 15 years: 1.01 male(s)/female
15–64 years: 0.99 male(s)/female
65 years and over: 0.69 male(s)/female
total population: 0.99 male(s)/female (2011 est.)
Infant mortality rate: *total:* 78.43 deaths/1,000 live births
country comparison to the world: 13
male: 82.2 deaths/1,000 live births
female: 74.55 deaths/1,000 live births (2011 est.)
Life expectancy at birth: *total population:* 55.33 years
country comparison to the world: 199
male: 53.9 years
female: 56.8 years (2011 est.)
Total fertility rate: 5.24 children born/woman (2011 est.)
country comparison to the world: 15
HIV/AIDS—adult prevalence rate: NA
HIV/AIDS—people living with HIV/AIDS: NA
HIV/AIDS—deaths: NA
Major infectious diseases: *degree of risk:* very high
food or waterborne diseases: bacterial and protozoal diarrhea, hepatitis A, and typhoid fever
vectorborne diseases: malaria, plague, and African trypanosomiasis (sleeping sickness)
water contact disease: schistosomiasis
animal contact disease: rabies (2009)
Drinking water source: *Improved:*
urban: 80% of population
rural: 28% of population
total: 46% of population
Unimproved:
urban: 20% of population
rural: 72% of population
total: 54% of population (2008)
Sanitation facility access: *Improved:*
urban: 23% of population
rural: 23% of population
total: 23% of population
Unimproved:
urban: 67% of population
rural: 67% of population
total: 67% of population (2008)
Nationality: *noun:* Congolese (singular and plural)
adjective: Congolese or Congo
Ethnic groups: over 200 African ethnic groups of which the majority are Bantu; the four largest tribes–Mongo, Luba, Kongo (all Bantu), and the Mangbetu-Azande (Hamitic) make up about 45% of the population
Religions: Roman Catholic 50%, Protestant 20%, Kimbanguist 10%, Muslim 10%, other (includes syncretic sects and indigenous beliefs) 10%
Languages: French (official), Lingala (a lingua franca trade language), Kingwana (a dialect of Kiswahili or Swahili), Kikongo, Tshiluba
Literacy: *definition:* age 15 and over can read and write French, Lingala, Kingwana, or Tshiluba
total population: 67.2%
male: 80.9%
female: 54.1% (2001 est.)
School life expectancy (primary to tertiary education): *total:* 8 years
male: 9 years
female: 7 years (2009)
Education expenditures: NA

GOVERNMENT

Country name: *conventional long form:* Democratic Republic of the Congo

conventional short form: DRC
local long form: Republique Democratique du Congo
local short form: RDC
former: Congo Free State, Belgian Congo, Congo/Leopoldville, Congo/Kinshasa, Zaire
abbreviation: DRC
Government type: republic
Capital: *name:* Kinshasa
geographic coordinates: 4 19 S, 15 18 E
time difference: UTC+1 (6 hours ahead of Washington, DC during Standard Time)
Administrative divisions: 10 provinces (provinces, singular–province) and 1 city* (ville); Bandundu, Bas-Congo, Equateur, Kasai-Occidental, Kasai-Oriental, Katanga, Kinshasa*, Maniema, Nord-Kivu, Orientale, Sud-Kivu
note: according to the Constitution adopted in December 2005, the current administrative divisions were to be subdivided into 26 new provinces by 2009 but this has yet to be implemented
Independence: 30 June 1960 (from Belgium)
National holiday: Independence Day, 30 June (1960)
Constitution: 18 February 2006
Legal system: civil legal system based on Belgian version of French civil law
International law organization participation: accepts compulsory ICJ jurisdiction with reservations; accepts ICCt jurisdiction
Suffrage: 18 years of age; universal and compulsory
Executive branch: *chief of state:* President Joseph KABILA (since 17 January 2001)
head of government: Prime Minister Adolphe MUZITO (since 10 October 2008)
cabinet: Ministers of State appointed by the president
(For more information visit the World Leaders website)
elections: under the new constitution the president elected by popular vote for a five-year term (eligible for a second term); elections last held on 30 July 2006 and on 29 October 2006 (next to be held on 27 November 2011); prime minister appointed by the president
election results: Joseph KABILA elected president; percent of vote (second round)–Joseph KABILA 58%, Jean-Pierre BEMBA Gombo 42%
note: Joseph KABILA succeeded his father, Laurent Desire KABILA, following the latter's assassination in January 2001; negotiations with rebel leaders led to the establishment of a transitional government in July 2003 with free elections held on 30 July 2006 and a run-off on 29 October 2006 confirming Joseph KABILA as president
Legislative branch: bicameral legislature consists of a Senate (108 seats; members elected by provincial assemblies to serve five-year terms) and a National Assembly (500 seats; 61 members elected by majority vote in single-member constituencies, 439 members elected by open list proportional-representation in multi-member constituencies to serve five-year terms)
elections: Senate–last held on 19 January 2007 (next to be held on 13 June 2012); National Assembly–last held on 30 July 2006 (next to be held on 27 November 2011)
election results: Senate–percent of vote by party–NA; seats by party–PPRD 22, MLC 14, FR 7, RCD 7, PDC 6, CDC 3, MSR 3, PALU 2, independents 26, others 18 (political parties that won a single seat); National Assembly–percent of vote by party–NA; seats by party–PPRD 111, MLC 64, PALU 34, MSR 27, FR 26, RCD 15, independents 63, others 160 (includes 63 political parties that won 10 or fewer seats)
Judicial branch: Constitutional Court; Appeals Court or Cour de Cassation; Council of State; High Military Court; plus civil and military courts and tribunals
Political parties and leaders: Christian Democrat Party or PDC [Jose ENDUNDO]; Congolese Rally for Democracy or RCD [Azarias RUBERWA]; Convention of Christian Democrats or CDC; Forces of Renewal or FR [Mbusa NYAMWISI]; Movement for the Liberation of the Congo or MLC [Jean-Pierre BEMBA]; People's Party for Reconstruction and Democracy or PPRD [Joseph KABILA]; Social Movement for Renewal or MSR [Pierre LUMBI]; Unified Lumumbist Party or PALU [Antoine GIZENGA]; Union for the Congolese Nation or UNC [Vital KAMERHE]; Union for Democracy and Social Progress or UDPS [Etienne TSHISEKEDI]; Union of Mobutuist Democrats or UDEMO [MOBUTU Nzanga]
Political pressure groups and leaders: MONUSCO–UN peacekeeping force; FARDC (Forces Armées de la République Démocratique du Congo)–Army of the Democratic Republic of the Congo which commits atrocities on citizens; FDLR (Forces Democratiques de Liberation du Rwanda)–Rwandan militia group made up of some of the perpetrators of Rwanda's Genocide in 1994; CNDP (National Congress for the Defense of the People)–mainly Congolese Tutsis who want refugees returned and more representation in government
International organization participation: ACP, AfDB, AU, CEPGL, COMESA, FAO, G-24, G-77, IAEA, IBRD, ICAO, ICRM, IDA, IFAD, IFC, IFRCS, IHO, ILO, IMF, IMO, Interpol, IOC, IOM, IPU, ISO, ITSO, ITU, ITUC, MIGA, NAM, OIF, OPCW, PCA, SADC, UN, UNCTAD, UNESCO, UNHCR, UNIDO, UNWTO, UPU, WCO, WFTU, WHO, WIPO, WMO, WTO
Diplomatic representation in the US: *chief of mission:* Ambassador Faida MITIFU
chancery: Suite 601, 1726 M Street, NW, Washington, DC, 20036
telephone: [1] (202) 234-7690 through 7691
FAX: [1] (202) 234-2609
Diplomatic representation from the US: *chief of mission:* Ambassador James F. ENTWISTLE
embassy: 310 Avenue des Aviateurs, Kinshasa
mailing address: Unit 31550, APO AE 09828
telephone: [243] (81) 225-5872
FAX: [243] (81) 301-0561
Flag description: sky blue field divided diagonally from the lower hoist corner to upper fly corner by a red stripe bordered by two narrow yellow stripes; a yellow, five-pointed star appears in the upper hoist corner; blue represents peace and hope, red the blood of the country's martyrs, and yellow the country's wealth and prosperity; the star symbolizes unity and the brilliant future for the country
National anthem: *name:* "Debout Congolaise" (Arise Congolese)
lyrics/music: Joseph LUTUMBA/Simon-Pierre BOKA di Mpasi Londi
note: adopted 1960; the anthem was replaced during the period in which the country was known as Zaire, but was readopted in 1997

ECONOMY

Economy—overview: The economy of the Democratic Republic of the Congo–a nation endowed with vast potential wealth–is slowly recovering from decades of decline. Systemic corruption since independence in 1960 and conflict that began in May 1997 has dramatically reduced national output and government revenue, increased external debt, and resulted in the deaths of more than 5 million people from violence, famine, and disease. Foreign businesses curtailed operations due to uncertainty about the outcome of the conflict, lack of infrastructure, and the difficult operating environment. Conditions began to improve in late 2002 with the withdrawal of a large portion of the invading foreign troops. The transitional government reopened relations with international financial institutions and international donors, and President KABILA began implementing reforms. Progress has been slow and the International Monetary Fund curtailed their program for the DRC at the end of March 2006 because of fiscal overruns. Much economic activity still occurs in the informal sector, and is not reflected in GDP data. Renewed activity in the mining sector, the source of most export income, boosted Kinshasa's fiscal position and GDP growth from 2006-2008, however, the government's review of mining contracts that began in 2006, combined with a fall in world market prices for the DRC's key mineral exports temporarily weakened output in 2009, leading to a balance of payments crisis. The recovery in mineral prices beginning in mid 2009 boosted mineral exports, and emergency funds from the IMF boosted foreign reserves. An uncertain legal framework, corruption, and a lack of transparency in government policy are long-term problems for the mining sector and for the economy as a whole. The global recession cut economic growth in 2009 to less than half its 2008 level, but growth returned to 6% in 2010. The DRC signed a Poverty Reduction and Growth Facility with the IMF in 2009 and received $12 billion in multilateral and bilateral debt relief in 2010.
GDP (purchasing power parity): $23.12 billion (2010 est.)
country comparison to the world: 119
$21.56 billion (2009 est.)
$20.96 billion (2008 est.)
note: data are in 2010 US dollars

GDP (official exchange rate): $13.13 billion (2010 est.)
GDP—real growth rate: 7.2% (2010 est.)
country comparison to the world: 32
2.8% (2009 est.)
6.2% (2008 est.)
GDP—per capita (PPP): $300 (2010 est.)
country comparison to the world: 227
$300 (2009 est.)
$300 (2008 est.)
note: data are in 2010 US dollars
GDP—composition by sector:
agriculture: 37.4%
industry: 26%
services: 36.6% (2008 est.)
Labor force: 23.53 million (2007 est.)
country comparison to the world: 26
Labor force—by occupation:
agriculture: NA%
industry: NA%
services: NA%
Unemployment rate: NA%
Population below poverty line: 71% (2006 est.)
Household income or consumption by percentage share: *lowest 10%:* 2.3%
highest 10%: 34.7% (2006)
Budget: *revenues:* $700 million
expenditures: $2 billion (2006 est.)
Inflation rate (consumer prices): 26.2% (2010 est.)
country comparison to the world: 222
53.4% (2009 est.)
Central bank discount rate: 70% (31 December 2009)
country comparison to the world: 2
40% (31 December 2008)
Commercial bank prime lending rate: 65.42% (31 December 2009 est.)
country comparison to the world: 4
43.15% (31 December 2008 est.)
Stock of narrow money: $613.9 million (31 December 2008)
country comparison to the world: 153
$597 million (31 December 2007)
Stock of broad money: $1.562 billion (31 December 2008 est.)
country comparison to the world: 148
$1.275 billion (31 December 2007 est.)
Stock of domestic credit: $NA (31 December 2010)
$928.5 million (31 December 2008)
Market value of publicly traded shares: $NA
Agriculture—products: coffee, sugar, palm oil, rubber, tea, cotton, cocoa, quinine, cassava (tapioca), manioc, bananas, plantains, peanuts, root crops, corn, fruits; wood products
Industries: mining (diamonds, gold, copper, cobalt, coltan, zinc, tin, diamonds), mineral processing, consumer products (including textiles, plastics, footwear, cigarettes, metal products, processed foods and beverages), timber, cement, commercial ship repair
Industrial production growth rate: NA%
Electricity—production: 8.217 billion kWh (2007 est.)
country comparison to the world: 97
Electricity—consumption: 5.997 billion kWh (2007 est.)
country comparison to the world: 106
Electricity—exports: 1.916 billion kWh (2007 est.)
Electricity—imports: 6 million kWh (2007 est.)
Oil—production: 16,360 bbl/day (2009 est.)
country comparison to the world: 78
Oil—consumption: 10,000 bbl/day (2009 est.)
country comparison to the world: 146
Oil—exports: 20,090 bbl/day (2007 est.)
country comparison to the world: 88
Oil—imports: 11,350 bbl/day (2007 est.)
country comparison to the world: 138
Oil—proved reserves: 180 million bbl (1 January 2010 est.)
country comparison to the world: 61
Natural gas—production: 0 cu m (2008 est.)
country comparison to the world: 110
Natural gas—consumption: 0 cu m (2008 est.)
country comparison to the world: 157
Natural gas—exports: 0 cu m (2008 est.)
country comparison to the world: 73
Natural gas—imports: 0 cu m (2008 est.)
country comparison to the world: 164
Natural gas—proved reserves: 991.1 million cu m (1 January 2010 est.)
country comparison to the world: 98
Current account balance: $-1.47 billion (2010 est.)
country comparison to the world: 148
$-402 million (2007 est.)
Exports: $3.8 billion (2009 est.)
country comparison to the world: 117
$6.6 billion (2008 est.)
Exports—commodities: diamonds, gold, copper, cobalt, wood products, crude oil, coffee
Exports—partners: China 46.75%, US 15.35%, Belgium 10.68%, Zambia 5.78%, Finland 4.38% (2009)
Imports: $5.3 billion (2009 est.)
country comparison to the world: 112
$6.7 billion (2008 est.)
Imports—commodities: foodstuffs, mining and other machinery, transport equipment, fuels
Imports—partners: South Africa 18.22%, Belgium 10.2%, China 8.34%, Zambia 7.77%, France 7.28%, Zimbabwe 6.52%, Kenya 5.48%, Netherlands 4.13%, Italy 3.96% (2009)
Reserves of foreign exchange and gold: $1.01 billion (March 2010 est.)
country comparison to the world: 116
$1 billion (December 2009 est.)
Debt—external: $13.5 billion (2009 est.)
country comparison to the world: 80
$12.7 billion (2008 est.)
Exchange rates: Congolese francs (CDF) per US dollar–495.28 (2010), 472.19 (2009), 559 (2008), 516 (2007), 464.69 (2006)

COMMUNICATIONS

Telephones—main lines in use: 40,000 (2009)
country comparison to the world: 168
Telephones—mobile cellular: 10.163 million (2009)
country comparison to the world: 63
Telephone system: *general assessment:* barely adequate wire and microwave radio relay service in and between urban areas; domestic satellite system with 14 earth stations; inadequate fixed line infrastructure
domestic: state-owned operator providing less than 1 fixed-line connection per 1000 persons; given the backdrop of a wholly inadequate fixed-line infrastructure, the use of mobile-cellular services has surged and subscribership in 2009 exceeded 10 million–roughly 15 per 100 persons
international: country code–243; satellite earth station–1 Intelsat (Atlantic Ocean) (2009)
Broadcast media: state-owned TV broadcast station with near national coverage; more than a dozen privately-owned TV stations with 2 having near national coverage; 2 state-owned radio stations are supplemented by more than 100 private radio stations; transmissions of at least 2 international broadcasters are available (2007)
Internet country code: .cd
Internet hosts: 3,006 (2010)
country comparison to the world: 146
Internet users: 290,000 (2008)
country comparison to the world: 131

TRANSPORTATION

Airports: 198 (2010)
country comparison to the world: 31
Airports—with paved runways: *total:* 26
over 3,047 m: 4
2,438 to 3,047 m: 2
1,524 to 2,437 m: 17
914 to 1,523 m: 2
under 914 m: 1 (2010)
Airports—with unpaved runways: *total:* 172
1,524 to 2,437 m: 20
914 to 1,523 m: 91
under 914 m: 61 (2010)
Pipelines: gas 37 km; oil 39 km; refined products 756 km (2010)
Railways: *total:* 4,007 km
country comparison to the world: 43
narrow gauge: 3,882 km 1.067-m gauge (858 km electrified); 125 km 1.000-m gauge (2010)
Roadways: *total:* 153,497 km
country comparison to the world: 32
paved: 2,794 km
unpaved: 150,703 km (2004)
Waterways: 15,000 km (including the Congo, its tributaries, and unconnected lakes) (2009)
country comparison to the world: 8
Merchant marine: *total:* 1
country comparison to the world: 163
by type: petroleum tanker 1
foreign-owned: 1 (Republic of the Congo 1) (2010)
Ports and terminals: Banana, Boma, Bukavu, Bumba, Goma, Kalemie, Kindu, Kinshasa, Kisangani, Matadi, Mbandaka

MILITARY

Military branches: Armed Forces of the Democratic Republic of the Congo (Forces d'Armees de la Republique Democratique du Congo, FARDC): Army, National Navy (La Marine Nationale), Congolese Air Force (Force Aerienne Congolaise, FAC) (2011)
Military service age and obligation: 18-45 years of age for voluntary military service (2009)
Manpower available for military service: *males age 16–49:* 15,980,106 (2010 est.)

Manpower fit for military service: *males age 16–49:* 10,168,258
females age 16–49: 10,331,693 (2010 est.)
Manpower reaching militarily significant age annually:
male: 877,684
female: 871,880 (2010 est.)
Military expenditures: 2.5% of GDP (2006)
country comparison to the world: 61

TRANSNATIONAL ISSUES

Disputes—international: heads of the Great Lakes states and UN pledged in 2004 to abate tribal, rebel, and militia fighting in the region, including northeast Congo, where the UN Organization Mission in the Democratic Republic of the Congo (MONUC), organized in 1999, maintains over 16,500 uniformed peacekeepers; members of Uganda's Lords Resistance Army forces continue to seek refuge in Congo's Garamba National Park as peace talks with the Uganda government evolve; the location of the boundary in the broad Congo River with the Republic of the Congo is indefinite except in the Pool Malebo/Stanley Pool area; Uganda and DRC dispute Rukwanzi island in Lake Albert and other areas on the Semliki River with hydrocarbon potential; boundary commission continues discussions over Congolese-administered triangle of land on the right bank of the Lunkinda river claimed by Zambia near the DRC village of Pweto; DRC accuses Angola of shifting monuments
Refugees and internally displaced persons: refugees (country of origin): 132,295 (Angola); 37,313 (Rwanda); 17,777 (Burundi); 13,904 (Uganda); 6,181 (Sudan); 5,243 (Republic of Congo)
IDPs: 1.4 million (fighting between government forces and rebels since mid-1990s; most IDPs are in eastern provinces) (2007)
Trafficking in persons: current situation: Democratic Republic of the Congo is a source and destination country for men, women, and children subjected to trafficking for the purposes of forced labor and forced prostitution; the majority of this trafficking is internal, and much of it is perpetrated by armed groups and government forces outside government control within the country's unstable eastern provinces
tier rating: Tier 3—Government of the Democratic Republic of the Congo does not fully comply with the minimum standards for the elimination of trafficking and is not making significant efforts to do so; the government did not show evidence of progress in prosecuting and punishing labor or sex trafficking offenders, including members of its own armed forces; providing protective services for the vast majority of trafficking victims; or raising public awareness of human trafficking; in addition, the government's anti-trafficking law enforcement efforts decreased during the reporting period (2010)
Illicit drugs: one of Africa's biggest producers of cannabis, but mostly for domestic consumption; traffickers exploit lax shipping controls to transit pseudoephedrine through the capital; while rampant corruption and inadequate supervision leaves the banking system vulnerable to money laundering, the lack of a well-developed financial system limits the country's utility as a money-laundering center (2008)

COOK ISLANDS

(SELF-GOVERNING IN FREE ASSOCIATION WITH NEW ZEALAND)

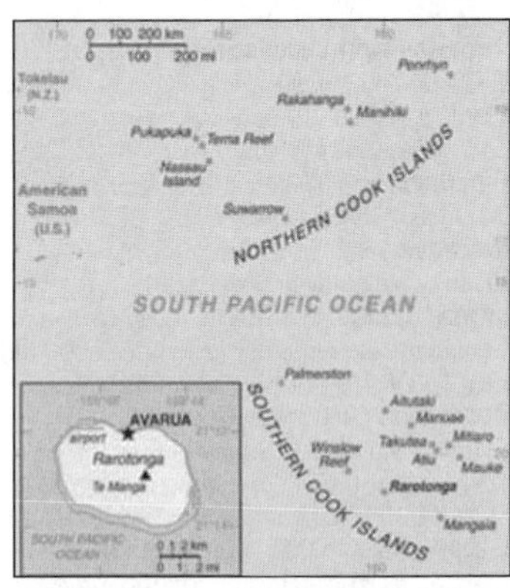

INTRODUCTION

Background: Named after Captain COOK, who sighted them in 1770, the islands became a British protectorate in 1888. By 1900, administrative control was transferred to New Zealand; in 1965, residents chose self-government in free association with New Zealand. The emigration of skilled workers to New Zealand and government deficits are continuing problems.

GEOGRAPHY

Location: Oceania, group of islands in the South Pacific Ocean, about half way between Hawaii and New Zealand
Geographic coordinates: 21 14 S, 159 46 W
Map references: Oceania
Area: *total:* 236 sq km
country comparison to the world: 213
land: 236 sq km
water: 0 sq km
Area—comparative: 1.3 times the size of Washington, DC
Land boundaries: 0 km
Coastline: 120 km
Maritime claims: territorial sea: 12 nm
exclusive economic zone: 200 nm
continental shelf: 200 nm or to the edge of the continental margin
Climate: tropical oceanic; moderated by trade winds; a dry season from April to November and a more humid season from December to March
Terrain: low coral atolls in north; volcanic, hilly islands in south
Elevation extremes: *lowest point:* Pacific Ocean 0 m
highest point: Te Manga 652 m
Natural resources: NEGL
Land use: *arable land:* 16.67%
permanent crops: 8.33%
other: 75% (2005)
Irrigated land: NA
Natural hazards: typhoons (November to March)
Environment—current issues: NA
Environment—international agreements: *party to:* Biodiversity, Climate Change, Climate Change-Kyoto Protocol, Desertification, Hazardous Wastes, Law of the Sea, Ozone Layer Protection
Geography—note: the northern Cook Islands are seven low-lying, sparsely populated, coral atolls; the southern Cook Islands, where most of the population lives, consist of eight elevated, fertile, volcanic isles, including the largest, Rarotonga, at 67 sq km

PEOPLE

Population: 11,124 (July 2011 est.)
country comparison to the world: 222
Age structure: *0–14 years:* 25.1% (male 1,479/female 1,308)
15–64 years: 65% (male 3,737/female 3,499)
65 years and over: 9.9% (male 538/female 563) (2011 est.)
Median age: *total:* 32 years
male: 31.3 years
female: 32.7 years (2011 est.)
Population growth rate: -3.2% (2011 est.)
country comparison to the world: 230
Birth rate: 15.37 births/1,000 population (2011 est.)
country comparison to the world: 133
Death rate: 7.37 deaths/1,000 population NA (July 2011 est.)
country comparison to the world: 122
Urbanization: *urban population:* 75% of total population (2008)
rate of urbanization: 1.4% annual rate of change (2010-15 est.)
Sex ratio: *at birth:* 1.048 male(s)/female
under 15 years: 1.13 male(s)/female
15–64 years: 1.07 male(s)/female
65 years and over: 0.96 male(s)/female
total population: 1.07 male(s)/female (2011 est.)
Infant mortality rate: *total:* 15.81 deaths/1,000 live births
country comparison to the world: 114
male: 19.24 deaths/1,000 live births
female: 12.21 deaths/1,000 live births (2011 est.)
Life expectancy at birth: *total population:* 74.7 years
country comparison to the world: 93
male: 71.91 years
female: 77.62 years (2011 est.)
Total fertility rate: 2.39 children born/woman (2011 est.)
country comparison to the world: 95
HIV/AIDS—adult prevalence rate: NA
HIV/AIDS—people living with HIV/AIDS: NA

HIV/AIDS—deaths: NA
Drinking water source: *Improved:*
urban: 99% of population
rural: 87% of population
total: 95% of population
Unimproved:
urban: 1% of population
rural: 13% of population
total: 5% of population (2000)
Sanitation facility access: *Improved:*
urban: 100% of population
rural: 100% of population
total: 100% of population (2008)
Nationality: *noun:* Cook Islander(s)
adjective: Cook Islander
Ethnic groups: Cook Island Maori (Polynesian) 87.7%, part Cook Island Maori 5.8%, other 6.5% (2001 census)
Religions: Cook Islands Christian Church 55.9%, Roman Catholic 16.8%, Seventh-Day Adventists 7.9%, Church of Latter Day Saints 3.8%, other Protestant 5.8%, other 4.2%, unspecified 2.6%, none 3% (2001 census)
Languages: English (official), Maori
Literacy: *definition:* age 15 and over can read and write
total population: 95%
male: NA
female: NA
School life expectancy (primary to tertiary education): *total:* 12 years
male: 12 years
female: 13 years (2010)
Education expenditures: NA
People—note: 2001 census counted a resident population of 15,017

GOVERNMENT

Country name: *conventional long form:* none
conventional short form: Cook Islands
former: Harvey Islands
Dependency status: self-governing in free association with New Zealand; Cook Islands is fully responsible for internal affairs; New Zealand retains responsibility for external affairs and defense in consultation with the Cook Islands
Government type: self-governing parliamentary democracy
Capital: *name:* Avarua
geographic coordinates: 21 12 S, 159 46 W
time difference: UTC-10 (5 hours behind Washington, DC during Standard Time)
Administrative divisions: none
Independence: none (became self-governing in free association with New Zealand on 4 August 1965 and has the right at any time to move to full independence by unilateral action)
National holiday: Constitution Day, first Monday in August (1965)
Constitution: 4 August 1965
Legal system: common law similar to New Zealand common law
International law organization participation: has not submitted an ICJ jurisdiction declaration (New Zealand normally retains responsibility for external affairs); accepts ICCt jurisdiction
Suffrage: 18 years of age; universal
Executive branch: *chief of state:* Queen ELIZABETH II (since 6 February 1952) represented by Sir Frederick GOODWIN (since 9 February 2001); New Zealand High Commissioner Linda TE PUNI (since 3 June 2010)
head of government: Prime Minister Henry PUNA (since 30 November 2010)
cabinet: Cabinet chosen by the prime minister; collectively responsible to Parliament
(For more information visit the World Leaders website)
elections: the monarchy is hereditary; the UK representative appointed by the monarch; the New Zealand high commissioner appointed by the New Zealand Government; following legislative elections, the leader of the majority party or the leader of the majority coalition usually becomes prime minister
Legislative branch: bicameral Parliament consists of a House of Ariki, or upper house, made up of traditional leaders and a Legislative Assembly, or lower house, (24 seats; members elected by popular vote to serve four-year terms)
note: the House of Ariki advises on traditional matters and maintains considerable influence but has no legislative powers
elections: last held on 17 November 2010 (next to be held by 2014)
election results: percent of vote by party—NA; seats by party—CIP 16, Demo 8
Judicial branch: High Court
Political parties and leaders: Cook Islands Party or CIP [Henry PUNA]; Democratic Party or Demo [Dr. Terepai MAOATE]
Political pressure groups and leaders: Reform Conference (lobby for political system changes)
other: various groups lobbying for political change
International organization participation: ACP, ADB, AOSIS, FAO, ICAO, ICRM, IFAD, IFRCS, IMO, IMSO, IOC, ITUC, OPCW, PIF, Sparteca, SPC, UNESCO, UPU, WHO, WMO
Diplomatic representation in the US: none (self-governing in free association with New Zealand)
Diplomatic representation from the US: none (self-governing in free association with New Zealand)
Flag description: blue, with the flag of the UK in the upper hoist-side quadrant and a large circle of 15 white five-pointed stars (one for every island) centered in the outer half of the flag
National anthem: *name:* "Te Atua Mou E" (To God Almighty)
lyrics/music: Tepaeru Te RITO/Thomas DAVIS
note: adopted 1982; as prime minister, Sir Thomas DAVIS composed the anthem; his wife, a tribal chief, wrote the lyrics

ECONOMY

Economy—overview: Like many other South Pacific island nations, the Cook Islands' economic development is hindered by the isolation of the country from foreign markets, the limited size of domestic markets, lack of natural resources, periodic devastation from natural disasters, and inadequate infrastructure. Agriculture, employing more than one-quarter of the working population, provides the economic base with major exports made up of copra and citrus fruit. Black pearls are the Cook Islands' leading export. Manufacturing activities are limited to fruit processing, clothing, and handicrafts. Trade deficits are offset by remittances from emigrants and by foreign aid overwhelmingly from New Zealand. In the 1980s and 1990s, the country lived beyond its means, maintaining a bloated public service and accumulating a large foreign debt. Subsequent reforms, including the sale of state assets, the strengthening of economic management, the encouragement of tourism, and a debt restructuring agreement, have rekindled investment and growth.
GDP (purchasing power parity): $183.2 million (2005 est.)
country comparison to the world: 215
GDP (official exchange rate): $183.2 million (2005 est.)
GDP—real growth rate: 0.1% (2005 est.)
country comparison to the world: 188
GDP—per capita (PPP): $9,100 (2005 est.)
country comparison to the world: 116
GDP—composition by sector:
agriculture: 15.1%
industry: 9.6%
services: 75.3% (2004)
Labor force: 6,820 (2001)
country comparison to the world: 217
Labor force—by occupation:
agriculture: 29%
industry: 15%
services: 56% (1995)
Unemployment rate: 13.1% (2005)
country comparison to the world: 136
Population below poverty line: NA%
Household income or consumption by percentage share:
lowest 10%: NA%
highest 10%: NA%
Budget: *revenues:* $70.95 million
expenditures: $69.05 million (FY05/06)
Inflation rate (consumer prices): 2.1% (2005 est.)
country comparison to the world: 52
Agriculture—products: copra, citrus, pineapples, tomatoes, beans, pawpaws, bananas, yams, taro, coffee; pigs, poultry
Industries: fruit processing, tourism, fishing, clothing, handicrafts
Industrial production growth rate: 1% (2002)
country comparison to the world: 143
Electricity—production: 31 million kWh (2007 est.)
country comparison to the world: 202
Electricity—consumption: 28.83 million kWh (2007 est.)
country comparison to the world: 201
Electricity—exports: 0 kWh (2008 est.)
Electricity—imports: 0 kWh (2008 est.)
Oil—production: 0 bbl/day (2009 est.)
country comparison to the world: 164
Oil—consumption: 1,000 bbl/day (2009 est.)
country comparison to the world: 198
Oil—exports: 0 bbl/day (2007 est.)
country comparison to the world: 159
Oil—imports: 495 bbl/day (2007 est.)
country comparison to the world: 198
Oil—proved reserves: 0 bbl (1 January 2010 est.)

country comparison to the world: 120
Natural gas—production: 0 cu m (2008 est.)
country comparison to the world: 116
Natural gas—consumption: 0 cu m (2008 est.)
country comparison to the world: 163
Natural gas—exports: 0 cu m (2008 est.)
country comparison to the world: 82
Natural gas—imports: 0 cu m (2008 est.)
country comparison to the world: 173
Natural gas—proved reserves: 0 cu m (1 January 2010 est.)
country comparison to the world: 125
Current account balance: $26.67 million (2005)
country comparison to the world: 58
Exports: $5.222 million (2005)
country comparison to the world: 216
Exports—commodities: copra, papayas, fresh and canned citrus fruit, coffee; fish; pearls and pearl shells; clothing
Imports: $81.04 million (2005)
country comparison to the world: 212
Imports—commodities: foodstuffs, textiles, fuels, timber, capital goods
Debt—external: $141 million (1996 est.)
country comparison to the world: 178
Exchange rates: NZ dollars (NZD) per US dollar–1.3874 (2010), 1.6002 (2009), 1.3811 (2007), 1.5408 (2006)

COMMUNICATIONS

Telephones—main lines in use: 6,900 (2009)
country comparison to the world: 208
Telephones—mobile cellular: 7,000 (2009)
country comparison to the world: 210
Telephone system: *general assessment:* Telecom Cook Islands offers international direct dialing, Internet, email, fax, and Telex
domestic: individual islands are connected by a combination of satellite earth stations, microwave systems, and VHF and HF radiotelephone; within the islands, service is provided by small exchanges connected to subscribers by open-wire, cable, and fiber-optic cable
international: country code–682; satellite earth station–1 Intelsat (Pacific Ocean)
Broadcast media: 1 privately-owned TV station broadcasts from Rarotonga providing a mix of local news and overseas-sourced programs; a satellite program package is available; 6 radio stations broadcast with 1 reportedly reaching all of the islands (2009)
Internet country code: .ck
Internet hosts: 2,521 (2010)
country comparison to the world: 150
Internet users: 6,000 (2009)
country comparison to the world: 204

TRANSPORTATION

Airports: 10 (2010)
country comparison to the world: 156
Airports—with paved runways: *total:* 1
1,524 to 2,437 m: 1 (2010)
Airports—with unpaved runways: *total:* 9
1,524 to 2,437 m: 2
914 to 1,523 m: 5
under 914 m: 2 (2010)
Roadways: *total:* 320 km
country comparison to the world: 201
paved: 33 km
unpaved: 287 km (2003)
Merchant marine:
total: 34
country comparison to the world: 82
by type: bulk carrier 1, cargo 27, passenger 1, refrigerated cargo 5
foreign-owned: 23 (Egypt 1, Germany 1, Latvia 1, Lithuania 2, former Netherlands Antilles 1, Norway 6, NZ 1, Russia 1, Sweden 3, Turkey 4, UK 2) (2010)
Ports and terminals: Avatiu

MILITARY

Military branches: no regular military forces; National Police Department (2009)
Manpower fit for military service: *males age 16–49:* 2,198
females age 16–49: 2,156 (2010 est.)
Manpower reaching militarily significant age annually: *male:* 127
female: 107 (2010 est.)
Military—note: defense is the responsibility of New Zealand in consultation with the Cook Islands and at its request

TRANSNATIONAL ISSUES

Disputes—international: none

CORAL SEA ISLANDS

(TERRITORY OF AUSTRALIA)

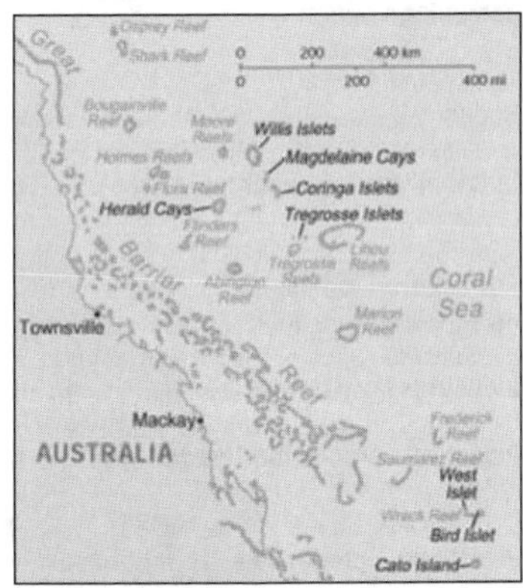

INTRODUCTION

Background: Scattered over more than three-quarters of a million square kilometers of ocean, the Coral Sea Islands were declared a territory of Australia in 1969. They are uninhabited except for a small meteorological staff on the Willis Islets. Automated weather stations, beacons, and a lighthouse occupy many other islands and reefs.

GEOGRAPHY

Location: Oceania, islands in the Coral Sea, northeast of Australia
Geographic coordinates: 18 00 S, 152 00 E
Map references: Oceania
Area: *total:* less than 3 sq km
country comparison to the world: 247
land: less than 3 sq km
water: 0 sq km
note: includes numerous small islands and reefs scattered over a sea area of about 780,000 sq km with the Willis Islets the most important
Area—comparative: NA
Land boundaries: 0 km
Coastline: 3,095 km
Maritime claims: territorial sea: 3 nm
exclusive fishing zone: 200 nm
Climate: tropical
Terrain: sand and coral reefs and islands (or cays)
Elevation extremes: *lowest point:* Pacific Ocean 0 m
highest point: unnamed location on Cato Island 6 m
Natural resources: NEGL
Land use:
arable land: 0%
permanent crops: 0%
other: 100% (mostly grass or scrub cover) (2005)
Irrigated land: 0 sq km
Natural hazards: occasional tropical cyclones
Environment—current issues: no permanent freshwater resources
Geography—note: important nesting area for birds and turtles

PEOPLE

Population: no indigenous inhabitants
note: there is a staff of three to four at the meteorological station on Willis Island (July 2007 est.)

GOVERNMENT

Country name: *conventional long form:* Coral Sea Islands Territory
conventional short form: Coral Sea Islands
Dependency status: territory of Australia; administered from Canberra by the Department of Regional Australia, Regional Development and Local Government
Legal system: the legal system of Australia, where applicable, applies
Diplomatic representation in the US: none (territory of Australia)
Diplomatic representation from the US: none (territory of Australia)
Flag description: the flag of Australia is used

ECONOMY

Economy—overview: no economic activity

COMMUNICATIONS

Communications—note: there are automatic weather stations on many of the isles and reefs relaying data to the mainland

TRANSPORTATION

Ports and terminals: none; offshore anchorage only

MILITARY

Military—note: defense is the responsibility of Australia

TRANSNATIONAL ISSUES

Disputes—international: none

COSTA RICA

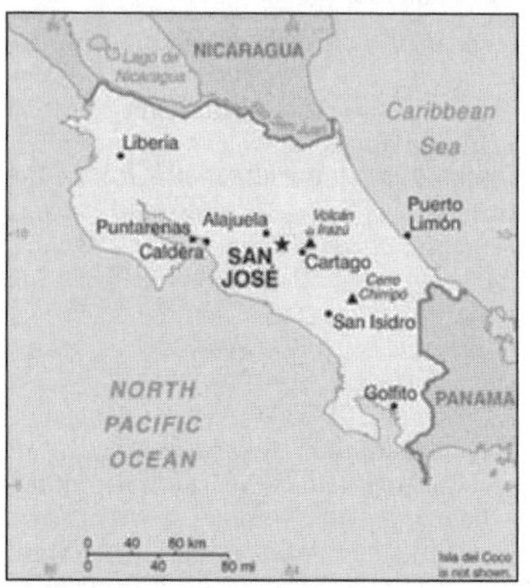

INTRODUCTION

Background: Although explored by the Spanish early in the 16th century, initial attempts at colonizing Costa Rica proved unsuccessful due to a combination of factors, including: disease from mosquito-infested swamps, brutal heat, resistance by natives, and pirate raids. It was not until 1563 that a permanent settlement of Cartago was established in the cooler, fertile central highlands. The area remained a colony for some two and a half centuries. In 1821, Costa Rica became one of several Central American provinces that jointly declared their independence from Spain. Two years later it joined the United Provinces of Central America, but this federation disintegrated in 1838, at which time Costa Rica proclaimed its sovereignty and independence. Since the late 19th century, only two brief periods of violence have marred the country's democratic development. In 1949, Costa Rica dissolved its armed forces. Although it still maintains a large agricultural sector, Costa Rica has expanded its economy to include strong technology and tourism industries. The standard of living is relatively high. Land ownership is widespread.

GEOGRAPHY

Location: Central America, bordering both the Caribbean Sea and the North Pacific Ocean, between Nicaragua and Panama
Geographic coordinates: 10 00 N, 84 00 W
Map references: Central America and the Caribbean
Area: *total:* 51,100 sq km
country comparison to the world: 129
land: 51,060 sq km
water: 40 sq km
note: includes Isla del Coco
Area—comparative: slightly smaller than West Virginia
Land boundaries: *total:* 639 km
border countries: Nicaragua 309 km, Panama 330 km
Coastline: 1,290 km
Maritime claims: territorial sea: 12 nm
exclusive economic zone: 200 nm
continental shelf: 200 nm
Climate: tropical and subtropical; dry season (December to April); rainy season (May to November); cooler in highlands
Terrain: coastal plains separated by rugged mountains including over 100 volcanic cones, of which several are major volcanoes
Elevation extremes: *lowest point:* Pacific Ocean 0 m
highest point: Cerro Chirripo 3,810 m
Natural resources: hydropower
Land use: *arable land:* 4.4%
permanent crops: 5.87%
other: 89.73% (2005)
Irrigated land: 1,080 sq km (2003)
Total renewable water resources: 112.4 cu km (2000)
Freshwater withdrawal (domestic/industrial/agricultural): *total:* 2.68 cu km/yr (29%/17%/53%)
per capita: 619 cu m/yr (2000)
Natural hazards: occasional earthquakes, hurricanes along Atlantic coast; frequent flooding of lowlands at onset of rainy season and landslides; active volcanoes
volcanism: Arenal (elev. 1,670 m), which erupted in 2010, is the most active volcano in Costa Rica; a 1968 eruption destroyed the town of Tabacon; Irazu (elev. 3,432 m), situated just east of San Jose, has the potential to spew ash over the capital city as it did between 1963 and 1965; other historically active volcanoes include Miravalles, Poas, Rincon de la Vieja, and Turrialba
Environment—current issues: deforestation and land use change, largely a result of the clearing of land for cattle ranching and agriculture; soil erosion; coastal marine pollution; fisheries protection; solid waste management; air pollution
Environment—international agreements: *party to:* Biodiversity, Climate Change, Climate Change-Kyoto Protocol, Desertification, Endangered Species, Environmental Modification, Hazardous Wastes, Law of the Sea, Marine Dumping, Ozone Layer Protection, Wetlands, Whaling
signed, but not ratified: Marine Life Conservation
Geography—note: four volcanoes, two of them active, rise near the capital of San Jose in the center of the country; one of the volcanoes, Irazu, erupted destructively in 1963-65

PEOPLE

Population: 4,576,562 (July 2011 est.)
country comparison to the world: 122
Age structure: *0–14 years:* 24.6% (male 574,876/female 549,664)
15–64 years: 69.1% (male 1,588,940/female 1,571,573)
65 years and over: 6.4% (male 135,017/female 156,492) (2011 est.)
Median age: *total:* 28.8 years
male: 28.4 years
female: 29.2 years (2011 est.)
Population growth rate: 1.308% (2011 est.)
country comparison to the world: 90
Birth rate: 16.54 births/1,000 population (2011 est.)
country comparison to the world: 123
Death rate: 4.33 deaths/1,000 population (July 2011 est.)
country comparison to the world: 205
Net migration rate: 0.87 migrant(s)/1,000 population (2011 est.)
country comparison to the world: 55
Urbanization: *urban population:* 64% of total population (2010)
rate of urbanization: 2.1% annual rate of change (2010-15 est.)
Major cities—population: SAN JOSE (capital) 1.416 million (2009)
Sex ratio: *at birth:* 1.05 male(s)/female
under 15 years: 1.05 male(s)/female
15–64 years: 1.01 male(s)/female
65 years and over: 0.86 male(s)/female
total population: 1.01 male(s)/female (2011 est.)
Infant mortality rate: *total:* 9.45 deaths/1,000 live births
country comparison to the world: 153
male: 10.3 deaths/1,000 live births
female: 8.56 deaths/1,000 live births (2011 est.)
Life expectancy at birth: *total population:* 77.72 years
country comparison to the world: 55
male: 75.1 years
female: 80.46 years (2011 est.)
Total fertility rate: 1.93 children born/woman (2011 est.)
country comparison to the world: 135
HIV/AIDS—adult prevalence rate: 0.3% (2009 est.)
country comparison to the world: 84
HIV/AIDS—people living with HIV/AIDS: 9,800 (2009 est.)
country comparison to the world: 99
HIV/AIDS—deaths: fewer than 500 (2009 est.)
country comparison to the world: 95
Major infectious diseases: *degree of risk:* intermediate
food or waterborne diseases: bacterial diarrhea
vectorborne diseases: dengue fever (2009)
Drinking water source: *Improved:*
urban: 100% of population
rural: 91% of population
total: 97% of population
Unimproved:
urban: 0% of population
rural: 9% of population
total: 3% of population (2008)
Sanitation facility access: *Improved:*
urban: 95% of population
rural: 96% of population
total: 95% of population
Unimproved:

urban: 5% of population
rural: 4% of population
total: 5% of population (2008)
Nationality: *noun:* Costa Rican(s)
adjective: Costa Rican
Ethnic groups: white (including mestizo) 94%, black 3%, Amerindian 1%, Chinese 1%, other 1%
Religions: Roman Catholic 76.3%, Evangelical 13.7%, Jehovah's Witnesses 1.3%, other Protestant 0.7%, other 4.8%, none 3.2%
Languages: Spanish (official), English
Literacy: *definition:* age 15 and over can read and write
total population: 94.9%
male: 94.7%
female: 95.1% (2000 census)
School life expectancy (primary to tertiary education):
total: 12 years
male: 12 years
female: 12 years (2005)
Education expenditures: 6.3% of GDP (2009)
country comparison to the world: 24

GOVERNMENT

Country name: *conventional long form:* Republic of Costa Rica
conventional short form: Costa Rica
local long form: Republica de Costa Rica
local short form: Costa Rica
Government type: democratic republic
Capital: *name:* San Jose
geographic coordinates: 9 56 N, 84 05 W
time difference: UTC-6 (1 hour behind Washington, DC during Standard Time)
Administrative divisions: 7 provinces (provincias, singular—provincia); Alajuela, Cartago, Guanacaste, Heredia, Limon, Puntarenas, San Jose
Independence: 15 September 1821 (from Spain)
National holiday: Independence Day, 15 September (1821)
Constitution: 7 November 1949
Legal system: civil law system based on Spanish civil code; judicial review of legislative acts in the Supreme Court
International law organization participation: accepts compulsory ICJ jurisdiction; accepts ICCt jurisdiction
Suffrage: 18 years of age; universal and compulsory
Executive branch: *chief of state:* President Laura CHINCHILLA Miranda (since 8 May 2010); First Vice President Alfio PIVA Mesen (since 8 May 2010); Second Vice President Luis LIBERMAN Ginsburg (since 8 May 2010); note—the president is both the chief of state and head of government
head of government: President Laura CHINCHILLA Miranda (since 8 May 2010); First Vice President Alfio PIVA Mesen (since 8 May 2010); Second Vice President Luis LIBERMAN Ginsburg (since 8 May 2010)
cabinet: Cabinet selected by the president
(For more information visit the World Leaders website)
elections: president and vice presidents elected on the same ticket by popular vote for a single four-year term; election last held on 7 February 2010 (next to be held in February 2014)
election results: Laura CHINCHILLA Miranda elected president; percent of vote—Laura CHINCHILLA Miranda (PLN) 46.7%; Otton SOLIS (PAC) 25.1%, Otto GUEVARA Guth (ML) 20.8%
Legislative branch: unicameral Legislative Assembly or Asamblea Legislativa (57 seats; members elected by direct, popular vote to serve four-year terms)
elections: last held on 7 February 2010 (next to be held in February 2014)
election results: percent of vote by party—NA; seats by party—PLN 23, PAC 10, ML 9, PUSC 6, PASE 4, other 5
Judicial branch: Supreme Court or Corte Suprema (22 justices are elected for renewable eight-year terms by the Legislative Assembly)
Political parties and leaders: Accessibility Without Exclusion or PASE [Oscar Andres LOPEZ Arias]; Citizen Action Party or PAC [Alberto CANAS Escalante]; Costa Rican Renovation Party or PRC [Gerardo Justo OROZCO Alvarez]; Democratic Force Party or PFD [Marco GONZALEZ Nunez]; Frente Amplio [Jose MERINO del Rio]; Homeland First or PP (Patria Primero) [Juan Jose VARGAS Fallas]; Libertarian Movement Party or PML [Otto GUEVARA Guth]; National Democratic Alliance or ADN [Jose Miguel VILLALOBOS Umana]; National Integration Party or PIN [Walter MUNOZ Cespedes]; National Liberation Party or PLN [Francisco Antonio PACHECO Fernandez]; National Rescue Party or PRN [Fabio Enrique DELGADO Hernandez]; National Union Party or PUN [Arturo ACOSTA Mora]; Patriotic Alliance [Mariano FIGUERES Olsen]; Patriotic Union or UP [Jose Miguel CORRALES Bolanos]; Popular Vanguard [Trino BARRANTES Araya]; Social Christian Unity Party or PUSC [Luis FISHMAN Zonzinski]; Union for Change Party or UPC [Antonio ALVAREZ Desanti]
Political pressure groups and leaders: Authentic Confederation of Democratic Workers or CATD (Communist Party affiliate); Chamber of Coffee Growers; Confederated Union of Workers or CUT (Communist Party affiliate); Costa Rican Confederation of Democratic Workers or CCTD (Liberation Party affiliate); Costa Rican Exporter's Chamber or CADEXCO; Costa Rican Solidarity Movement; Costa Rican Union of Private Sector Enterprises or UCCAEP [Rafael CARRILLO]; Federation of Public Service Workers or FTSP; National Association for Economic Development or ANFE; National Association of Educators or ANDE; National Association of Public and Private Employees or ANEP [Albino VARGAS]; Rerum Novarum or CTRN (PLN affiliate) [Gilbert BROWN]
International organization participation: BCIE, CACM, FAO, G-77, IADB, IAEA, IBRD, ICAO, ICC, ICRM, IDA, IFAD, IFC, IFRCS, ILO, IMF, IMO, IMSO, Interpol, IOC, IOM, IPU, ISO, ITSO, ITU, ITUC, LAES, LAIA (observer), MIGA, NAM (observer), OAS, OPANAL, OPCW, PCA, RG, SICA, UN, UNCTAD, UNESCO, UNHCR, UNIDO, Union Latina, UNWTO, UPU, WCO, WFTU, WHO, WIPO, WMO, WTO
Diplomatic representation in the US: *chief of mission:* Ambassador Muni FIGUERES Boggs
chancery: 2114 S Street NW, Washington, DC 20008
telephone: [1] (202) 234-2945 or 2946
FAX: [1] (202) 265-4795
consulate(s) general: Atlanta, Chicago, Houston, Los Angeles, Miami, New Orleans, New York, San Francisco, San Juan (Puerto Rico)
Diplomatic representation from the US: *chief of mission:* Ambassador Anne Slaughter ANDREW
embassy: Calle 120 Avenida O, Pavas, San Jose
mailing address: APO AA 34020
telephone: [506] 2519-2000
FAX: [506] 2519-2305
Flag description: five horizontal bands of blue (top), white, red (double width), white, and blue, with the coat of arms in a white elliptical disk toward the hoist side of the red band; Costa Rica retained the earlier blue-white-blue flag of Central America until 1848 when, in response to revolutionary activity in Europe, it was decided to incorporate the French colors into the national flag and a central red stripe was added; today the blue color is said to stand for the sky, opportunity, and perseverance, white denotes peace, happiness, and wisdom, while red represents the blood shed for freedom, as well as the generosity and vibrancy of the people
note: somewhat resembles the flag of North Korea; similar to the flag of Thailand but with the blue and red colors reversed
National anthem: *name:* "Himno Nacional de Costa Rica" (National Anthem of Costa Rica)
lyrics/music: Jose Maria ZELEDON Brenes/ Manuel Maria GUTIERREZ
note: adopted 1949; the anthem's music was originally written for an 1853 welcome ceremony for diplomatic missions from the United States and United Kingdom; the lyrics were added in 1903

ECONOMY

Economy—overview: Prior to the global economic crisis, Costa Rica enjoyed stable economic growth. The economy contracted 0.7% in 2009, but resumed growth at more than 3% in 2010. While the traditional agricultural exports of bananas, coffee, sugar, and beef are still the backbone of commodity export trade, a variety of industrial and specialized agricultural products have broadened export trade in recent years. High value added goods and services, including microchips, have further bolstered exports. Tourism continues to bring in foreign exchange, as Costa Rica's impressive biodiversity makes it a key destination for ecotourism. Foreign investors remain attracted by the country's political stability and relatively high education levels, as well as the fiscal incentives offered in the free-trade zones; and Costa Rica has attracted one of the highest levels of foreign direct investment per capita in Latin America. However, many business impediments, such as high levels of bureaucracy, difficulty of enforcing contracts, and weak investor protection, remain. Poverty has remained around 15-20% for nearly 20 years, and the strong social safety net that had been put into place by the government has eroded due to increased financial constraints on government expenditures. Unlike the rest of Central America, Costa Rica is not highly dependent on remittances as they only represent about 2% of GDP. Immigration from Nicaragua has increasingly become a concern for the government. The estimated 300,000-500,000 Nicaraguans in Costa Rica legally and illegally are an important source of–mostly unskilled–labor, but also place heavy demands on the social welfare system. The US-Central American-Dominican Republic Free Trade Agreement (CAFTA-DR) entered into force on 1 January 2009, after significant delays within the Costa Rican legislature. CAFTA-DR will likely lead to increased foreign direct investment in key sectors of the economy, including the insurance and telecommunications sectors recently opened to private investors. President CHINCHILLA is likely to push for fiscal reform in the coming year, seeking to boost revenue, possibly through revised tax legislation, to fund an increase in security services and education.

GDP (purchasing power parity): $51.17 billion (2010 est.)
country comparison to the world: 91
$49.12 billion (2009 est.)
$49.76 billion (2008 est.)
note: data are in 2010 US dollars

GDP (official exchange rate): $35.78 billion (2010 est.)

GDP—real growth rate: 4.2% (2010 est.)
country comparison to the world: 92
-1.3% (2009 est.)
2.7% (2008 est.)

GDP—per capita (PPP): $11,300 (2010 est.)
country comparison to the world: 98
$11,000 (2009 est.)
$11,300 (2008 est.)
note: data are in 2010 US dollars

GDP—composition by sector:
agriculture: 6.3%
industry: 22.9%
services: 70.8% (2010 est.)

Labor force: 2.17 million
country comparison to the world: 117
note: this official estimate excludes Nicaraguans living in Costa Rica (2010 est.)

Labor force—by occupation:
agriculture: 14%
industry: 22%
services: 64% (2006 est.)

Unemployment rate: 7.3% (2010 est.)
country comparison to the world: 77
7.8% (2009 est.)

Population below poverty line:
16% (2006 est.)

Household income or consumption by percentage share: *lowest 10%:* 1.5%
highest 10%: 35.5% (2005)

Distribution of family income—Gini index: 48 (2008)
country comparison to the world: 28
45.9 (1997)

Investment (gross fixed): 20.8% of GDP (2010 est.)
country comparison to the world: 82

Budget: *revenues:* $5.085 billion
expenditures: $6.921 billion (2010 est.)

Public debt: 42.4% of GDP (2010 est.)
country comparison to the world: 62
42% of GDP (2009 est.)

Inflation rate (consumer prices): 5.8% (2010 est.)
country comparison to the world: 154
7.8% (2009 est.)

Central bank discount rate: 23% (31 December 2009)
country comparison to the world: 5
25% (31 December 2008)

Commercial bank prime lending rate: 17.4% (31 December 2009 est.)
country comparison to the world: 37
15.83% (31 December 2008 est.)

Stock of narrow money: $4.504 billion (31 December 2010 est.)
country comparison to the world: 95
$3.992 billion (31 December 2009 est.)

Stock of broad money: $16.81 billion (31 December 2009)
country comparison to the world: 87
$15.84 billion (31 December 2008)

Stock of domestic credit: $15.82 billion (31 December 2010 est.)
country comparison to the world: 87
$14.74 billion (31 December 2009 est.)

Market value of publicly traded shares: $1.452 billion (31 December 2009)
country comparison to the world: 100
$1.887 billion (31 December 2008)
$2.035 billion (31 December 2007)

Agriculture—products: bananas, pineapples, coffee, melons, ornamental plants, sugar, corn, rice, beans, potatoes; beef, poultry, dairy; timber

Industries: microprocessors, food processing, medical equipment, textiles and clothing, construction materials, fertilizer, plastic products

Industrial production growth rate: 3% (2010 est.)
country comparison to the world: 109

Electricity—production: 9.29 billion kWh (2008 est.)
country comparison to the world: 93

Electricity—consumption: 8.25 billion kWh (2008 est.)
country comparison to the world: 94

Electricity—exports: 77.16 million kWh (2008 est.)

Electricity—imports: 203.2 million kWh (2007 est.)

Oil—production: 0 bbl/day (2008 est.)
country comparison to the world: 161

Oil—consumption: 44,000 bbl/day (2009 est.)
country comparison to the world: 98

Oil—exports: 2,117 bbl/day (2007 est.)
country comparison to the world: 112

Oil—imports: 46,260 bbl/day (2009 est.)
country comparison to the world: 87

Oil—proved reserves: 0 bbl (1 January 2010 est.)
country comparison to the world: 117

Natural gas—production: 0 cu m (2008 est.)
country comparison to the world: 113

Natural gas—consumption: 0 cu m (2008 est.)
country comparison to the world: 160

Natural gas—exports: 0 cu m (2008 est.)
country comparison to the world: 78

Natural gas—imports: 0 cu m (2008 est.)
country comparison to the world: 169

Natural gas—proved reserves: 0 cu m (1 January 2010 est.)
country comparison to the world: 122

Current account balance: $-1.349 billion (2010 est.)
country comparison to the world: 143
$-537 million (2009 est.)

Exports: $10.01 billion (2010 est.)
country comparison to the world: 86
$8.847 billion (2009 est.)

Exports—commodities: bananas, pineapples, coffee, melons, ornamental plants, sugar; beef; seafood; electronic components, medical equipment

Exports—partners: US 32.61%, Netherlands 12.82%, China 11.81%, Mexico 4.2% (2009)

Imports: $13.32 billion (2010 est.)
country comparison to the world: 81
$10.87 billion (2009 est.)

Imports—commodities: raw materials, consumer goods, capital equipment, petroleum, construction materials

Imports—partners: US 44.72%, Mexico 7.65%, Venezuela 5.56%, China 5.15%, Japan 4.36% (2009)

Reserves of foreign exchange and gold: $4.584 billion (31 December 2010 est.)
country comparison to the world: 73
$4.066 billion (31 December 2009 est.)

Debt—external: $8.55 billion (31 December 2010 est.)
country comparison to the world: 90
$7.972 billion (31 December 2009 est.)

Stock of direct foreign investment—at home: $13.92 billion (31 December 2010 est.)
country comparison to the world: 77
$12.17 billion (31 December 2009 est.)

Stock of direct foreign investment—abroad: $547 million (31 December 2010 est.)
country comparison to the world: 75
$539 million (31 December 2009 est.)

Exchange rates: Costa Rican colones (CRC) per US dollar–513 (2010), 573.29 (2009), 530.41 (2008), 519.53 (2007), 511.3 (2006)

COMMUNICATIONS

Telephones—main lines in use: 1.493 million (2009)
country comparison to the world: 65

Telephones—mobile cellular: 1.95 million (2009)
country comparison to the world: 134

Telephone system: *general assessment:* good domestic telephone service in terms of breadth of coverage; under the terms of CAFTA-DR, the state-run telecommunications monopoly scheduled to be opened to competition from domestic and international firms, has been delayed by the nation's telecommunications regulator.
domestic: point-to-point and point-to-multipoint microwave, fiber-optic, and coaxial cable link rural areas; Internet service is available
international: country code–506; landing points for the Americas Region Caribbean Ring System (ARCOS-1), MAYA-1, and the Pan American Crossing submarine cables that provide links to South and Central America, parts of the Caribbean, and the

US; connected to Central American Microwave System; satellite earth stations—2 Intelsat (Atlantic Ocean) (2009)
Broadcast media: multiple privately-owned television stations and 1 publicly-owned television station; cable network services are widely available; more than 100 privately-owned radio stations and a public radio network (2007)
Internet country code: .cr
Internet hosts: 34,024 (2010)
country comparison to the world: 97
Internet users: 1.485 million (2009)
country comparison to the world: 82

TRANSPORTATION

Airports: 151 (2010)
country comparison to the world: 36
Airports—with paved runways: *total:* 39
2,438 to 3,047 m: 2
1,524 to 2,437 m: 2
914 to 1,523 m: 23
under 914 m: 12 (2010)
Airports—with unpaved runways:
total: 112
914 to 1,523 m: 18
under 914 m: 94 (2010)
Pipelines: refined products 662 km (2010)
Railways: *total:* 278 km
country comparison to the world: 122
narrow gauge: 278 km 1.067-m gauge
note: none of the railway network is in use (2010)
Roadways: *total:* 35,330 km
country comparison to the world: 94
paved: 8,621 km
unpaved: 26,709 km (2004)
Waterways: 730 km (seasonally navigable by small craft) (2010)
country comparison to the world: 75
Merchant marine: *total:* 1
country comparison to the world: 151
by type: passenger/cargo 1 (2010)
Ports and terminals: Caldera, Puerto Limon

MILITARY

Military branches: no regular military forces; Ministry of Public Security, Government, and Police (2011)
Manpower available for military service:
males age 16–49: 1,255,798
females age 16–49: 1,230,202 (2010 est.)
Manpower fit for military service: *males age 16–49:* 1,058,419
females age 16–49: 1,037,053 (2010 est.)
Manpower reaching militarily significant age annually: *male:* 42,201
female: 40,444 (2010 est.)
Military expenditures: 0.6% of GDP (2009)
country comparison to the world: 155

TRANSNATIONAL ISSUES

Disputes—international: the ICJ has given Costa Rica until January 2008 to reply and Nicaragua until July 2008 to rejoin before rendering its decision on the navigation, security, and commercial rights of Costa Rican vessels on the Rio San Juan over which Nicaragua retains sovereignty
Refugees and internally displaced persons: refugees (country of origin): 9,699-11,500 (Colombia) (2007)
Illicit drugs: transshipment country for cocaine and heroin from South America; illicit production of cannabis in remote areas; domestic cocaine consumption, particularly crack cocaine, is rising; significant consumption of amphetamines; seizures of smuggled cash in Costa Rica and at the main border crossing to enter Costa Rica from Nicaragua have risen in recent years (2008)

COTE D'IVOIRE

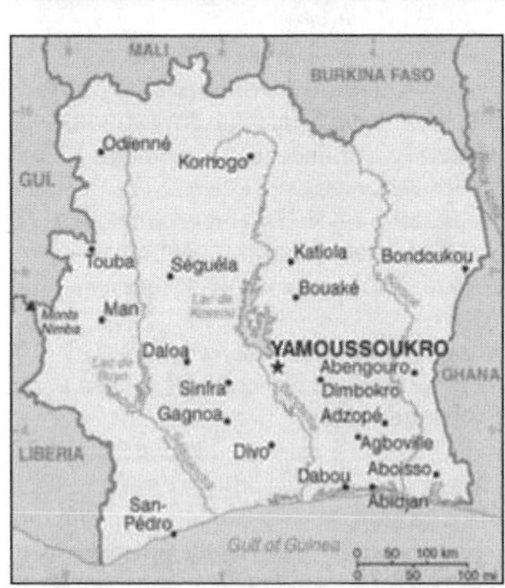

INTRODUCTION

Background: Close ties to France since independence in 1960, the development of cocoa production for export, and foreign investment made Cote d'Ivoire one of the most prosperous of the West African states, but did not protect it from political turmoil. In December 1999, a military coup—the first ever in Cote d'Ivoire's history—overthrew the government. Junta leader Robert GUEI blatantly rigged elections held in late 2000 and declared himself the winner. Popular protest forced him to step aside and brought Laurent GBAGBO into power. Ivorian dissidents and disaffected members of the military launched a failed coup attempt in September 2002. Rebel forces claimed the northern half of the country, and in January 2003 were granted ministerial positions in a unity government under the auspices of the Linas-Marcoussis Peace Accord. President GBAGBO and rebel forces resumed implementation of the peace accord in December 2003 after a three-month stalemate, but issues that sparked the civil war, such as land reform and grounds for citizenship, remained unresolved. In March 2007 President GBAGBO and former New Force rebel leader Guillaume SORO signed the Ouagadougou Political Agreement. As a result of the agreement, SORO joined GBAGBO's government as Prime Minister and the two agreed to reunite the country by dismantling the zone of confidence separating North from South, integrate rebel forces into the national armed forces, and hold elections. Disarmament, demobilization, and reintegration of rebel forces have been problematic as rebels seek to enter the armed forces. Citizen identification and voter registration pose election difficulties, and balloting planned for November 2009 was postponed to 2010. On 28 November 2010, Alassane Dramane OUATTARA won the presidential election, defeating then President Laurent GBAGBO. GBAGBO refused to hand over power, resulting in a 6-month stand-off. In April 2011, after widespread fighting, GBAGBO was formally forced from office by OUATTARA supporters with the support of UN and French forces. Several thousand UN troops and several hundred French remain in Cote d'Ivoire to support the transition process.

GEOGRAPHY

Location: Western Africa, bordering the North Atlantic Ocean, between Ghana and Liberia
Geographic coordinates: 8 00 N, 5 00 W
Map references: Africa
Area: *total:* 322,463 sq km
country comparison to the world: 68
land: 318,003 sq km
water: 4,460 sq km
Area—comparative: slightly larger than New Mexico
Land boundaries: *total:* 3,110 km
border countries: Burkina Faso 584 km, Ghana 668 km, Guinea 610 km, Liberia 716 km, Mali 532 km
Coastline: 515 km
Maritime claims: territorial sea: 12 nm
exclusive economic zone: 200 nm
continental shelf: 200 nm
Climate: tropical along coast, semiarid in far north; three seasons—warm and dry (November to March), hot and dry (March to May), hot and wet (June to October)
Terrain: mostly flat to undulating plains; mountains in northwest
Elevation extremes: *lowest point:* Gulf of Guinea 0 m
highest point: Monts Nimba 1,752 m
Natural resources: petroleum, natural gas, diamonds, manganese, iron ore, cobalt, bauxite, copper, gold, nickel, tantalum, silica sand, clay, cocoa beans, coffee, palm oil, hydropower
Land use: *arable land:* 10.23%
permanent crops: 11.16%
other: 78.61% (2005)
Irrigated land: 730 sq km (2003)
Total renewable water resources: 81 cu km (2001)
Freshwater withdrawal (domestic/industrial/agricultural): *total:* 0.93 cu km/yr (24%/12%/65%)
per capita: 51 cu m/yr (2000)
Natural hazards: coast has heavy surf and no natural harbors; during the rainy season torrential flooding is possible
Environment—current issues: deforestation (most of the country's forests—once the

largest in West Africa—have been heavily logged); water pollution from sewage and industrial and agricultural effluents
Environment—international agreements: *party to:* Biodiversity, Climate Change, Climate Change-Kyoto Protocol, Desertification, Endangered Species, Hazardous Wastes, Law of the Sea, Marine Dumping, Ozone Layer Protection, Ship Pollution, Tropical Timber 83, Tropical Timber 94, Wetlands, Whaling
signed, but not ratified: none of the selected agreements
Geography—note: most of the inhabitants live along the sandy coastal region; apart from the capital area, the forested interior is sparsely populated

PEOPLE

Population: 21,504,162 (July 2011 est.)
country comparison to the world: 56
note: estimates for this country explicitly take into account the effects of excess mortality due to AIDS; this can result in lower life expectancy, higher infant mortality, higher death rates, lower population growth rates, and changes in the distribution of population by age and sex than would otherwise be expected
Age structure: *0–14 years:* 39.8% (male 4,312,133/female 4,240,500)
15–64 years: 57.2% (male 6,262,802/ female 6,039,458)
65 years and over: 3% (male 320,396/female 328,873) (2011 est.)
Median age: *total:* 19.6 years
male: 19.7 years
female: 19.5 years (2011 est.)
Population growth rate: 2.078% (2011 est.)
country comparison to the world: 43
Birth rate: 30.95 births/1,000 population (2011 est.)
country comparison to the world: 42
Death rate: 10.16 deaths/1,000 population (July 2011 est.)
country comparison to the world: 53
Net migration rate: 0 migrant(s)/1,000 population NA (2011 est.)
country comparison to the world: 88
Urbanization: *urban population:* 51% of total population (2010)
rate of urbanization: 3.7% annual rate of change (2010-15 est.)
Major cities—population: ABIDJAN (seat of government) 4.009 million; YAMOUSSOUKRO (capital) 808,000 (2009)
Sex ratio: *at birth:* 1.03 male(s)/female
under 15 years: 1.02 male(s)/female
15–64 years: 1.04 male(s)/female
65 years and over: 0.99 male(s)/female
total population: 1.03 male(s)/female (2011 est.)
Infant mortality rate: *total:* 64.78 deaths/1,000 live births
country comparison to the world: 23
male: 71.54 deaths/1,000 live births
female: 57.83 deaths/1,000 live births (2011 est.)
Life expectancy at birth: *total population:* 56.78 years
country comparison to the world: 195
male: 55.79 years
female: 57.81 years (2011 est.)
Total fertility rate: 3.92 children born/ woman (2011 est.)
country comparison to the world: 41
HIV/AIDS—adult prevalence rate: 3.4% (2009 est.)
country comparison to the world: 20
HIV/AIDS—people living with HIV/AIDS: 450,000 (2009 est.)
country comparison to the world: 17
HIV/AIDS—deaths: 36,000 (2009 est.)
country comparison to the world: 12
Major infectious diseases: *degree of risk:* very high
food or waterborne diseases: bacterial diarrhea, hepatitis A, and typhoid fever
vectorborne diseases: malaria and yellow fever
water contact: schistosomiasis
animal contact disease: rabies
note: highly pathogenic H5N1 avian influenza has been identified in this country; it poses a negligible risk with extremely rare cases possible among US citizens who have close contact with birds (2009)
Drinking water source: *Improved:*
urban: 93% of population
rural: 68% of population
total: 80% of population
Unimproved:
urban: 7% of population
rural: 32% of population
total: 20% of population (2008)
Sanitation facility access: *Improved:*
urban: 36% of population
rural: 11% of population
total: 23% of population
Unimproved:
urban: 64% of population
rural: 89% of population
total: 77% of population (2008)
Nationality:
noun: Ivoirian(s)
adjective: Ivoirian
Ethnic groups: Akan 42.1%, Voltaiques or Gur 17.6%, Northern Mandes 16.5%, Krous 11%, Southern Mandes 10%, other 2.8% (includes 130,000 Lebanese and 14,000 French) (1998)
Religions: Muslim 38.6%, Christian 32.8%, indigenous 11.9%, none 16.7% (2008 est.)
note: the majority of foreigners (migratory workers) are Muslim (70%) and Christian (20%)
Languages: French (official), 60 native dialects of which Dioula is the most widely spoken
Literacy: *definition:* age 15 and over can read and write
total population: 48.7%
male: 60.8%
female: 38.6% (2000 est.)
School life expectancy (primary to tertiary education): *total:* 6 years
male: 8 years
female: 5 years (2000)
Education expenditures: 4.6% of GDP (2008)
country comparison to the world: 76

GOVERNMENT

Country name: *conventional long form:* Republic of Cote d'Ivoire
conventional short form: Cote d'Ivoire
local long form: Republique de Cote d'Ivoire
local short form: Cote d'Ivoire
note: pronounced coat-div-whar
former: Ivory Coast
Government type: republic; multiparty presidential regime established 1960
note: the government is currently disputed as of 31 January 2011, with both candidates in the runoff claiming victory
Capital: *name:* Yamoussoukro
geographic coordinates: 6 49 N, 5 17 W
time difference: UTC 0 (5 hours ahead of Washington, DC during Standard Time)
note: although Yamoussoukro has been the official capital since 1983, Abidjan remains the commercial and administrative center; the US, like other countries, maintains its Embassy in Abidjan
Administrative divisions: 19 regions; Agneby, Bafing, Bas-Sassandra, Denguele, Dix-Huit Montagnes, Fromager, Haut-Sassandra, Lacs, Lagunes, Marahoue, Moyen-Cavally, Moyen-Comoe, N'zi-Comoe, Savanes, Sud-Bandama, Sud-Comoe, Vallee du Bandama, Worodougou, Zanzan
Independence: 7 August 1960 (from France)
National holiday: Independence Day, 7 August (1960)
Constitution: approved by referendum 23 July 2000
Legal system: civil law system based on the French civil code; judicial review in the Constitutional Chamber of the Supreme Court
International law organization participation: accepts compulsory ICJ jurisdiction with reservations; accepts ICCt jurisdiction under Article 12(3)of the Rome Statute
Suffrage: 18 years of age; universal
Executive branch: *chief of state:* President Alassane OUATTARA (since 4 December 2010);
head of government: Prime Minister Guillaume SORO (since 4 April 2007);
cabinet: Council of Ministers appointed by the president
(For more information visit the World Leaders website)
elections: president elected by popular vote for a five-year term (no term limits); election last held on 31 October and 28 November 2010 (next to be held in 2015); prime minister appointed by the president
election results: Alassane OUATTARA elected president; percent of vote—Alassane OUATTARA 54.1%, Laurent GBAGBO 45.9%; note—President OUATTARA was declared winner by the election commission and took the oath of office on 4 December, Prime Minister SORO resigned from the incumbent administration and was subsequently appointed to the same position by OUATTARA; former president GBAGBO refused to cede resulting in a 6-month stand-off, he was finally forced to stand down in April 2011
Legislative branch: unicameral National Assembly or Assemblee Nationale (225 seats; members elected in single- and multi-district elections by direct popular vote to serve five-year terms)
elections: elections last held on 10 December 2000 with by-elections on 14 January 2001 (elections originally scheduled for 2005 have been repeatedly postponed by the government)

election results: percent of vote by party–NA; seats by party–FPI 96, PDCI-RDA 94, RDR 5, PIT 4, other 2, independents 22, vacant 2
note: a Senate was scheduled to be created in October 2006 elections that never took place
Judicial branch: Supreme Court or Cour Supreme consists of four chambers: Judicial Chamber for criminal cases, Audit Chamber for financial cases, Constitutional Chamber for judicial review cases, and Administrative Chamber for civil cases; there is no legal limit to the number of members
Political parties and leaders: Citizen's Democratic Union or UDCY [Theodore MEL EG]; Democratic Party of Cote d'Ivoire or PDCI [Henri Konan BEDIE]; Ivorian Popular Front or FPI [Pascale Affi N'GUESSAN]; Ivorian Worker's Party or PIT [Francis WODIE]; Opposition Movement of the Future or MFA [Innocent Augustin ANAKY]; Rally of the Republicans or RDR [Alassane OUATTARA]; Union for Democracy and Peace in Cote d'Ivoire or UDPCI [Toikeuse MABRI]; over 144 smaller registered parties
Political pressure groups and leaders: Federation of University and High School Students of Cote d'Ivoire or FESCI [Serges KOFFI]; Rally of Houphouetists for Democracy and Peace or RHDP [Alphonse DJEDJE MADY]; Young Patriots [Charles BLE GOUDE]
International organization participation: ACP, AfDB, AU, ECOWAS (suspended), Entente, FAO, FZ, G-24, G-77, IAEA, IBRD, ICAO, ICRM, IDA, IDB, IFAD, IFC, IFRCS, ILO, IMF, IMO, Interpol, IOC, IOM, IPU, ISO, ITSO, ITU, ITUC, MIGA, NAM, OIC, OIF, OPCW, UN, UNCTAD, UNESCO, UNHCR, UNIDO, Union Latina, UNWTO, UPU, WADB (regional), WAEMU, WCO, WFTU, WHO, WIPO, WMO, WTO
Diplomatic representation in the US:
chief of mission: Ambassador Daouda DIABATE
chancery: 2424 Massachusetts Avenue NW, Washington, DC 20008
telephone: [1] (202) 797-0300
FAX: [1] (202) 244-3088
Diplomatic representation from the US:
chief of mission: Ambassador Wanda L. NESBITT
embassy: Cocody Riviera Golf 01, Abidjan
mailing address: B. P. 1712, Abidjan 01
telephone: [225] 22 49 40 00
FAX: [225] 22 49 43 32
Flag description: three equal vertical bands of orange (hoist side), white, and green; orange symbolizes the land (savannah) of the north and fertility, white stands for peace and unity, green represents the forests of the south and the hope for a bright future
note: similar to the flag of Ireland, which is longer and has the colors reversed–green (hoist side), white, and orange; also similar to the flag of Italy, which is green (hoist side), white, and red; design was based on the flag of France
National anthem: *name:* "L'Abidjanaise" (Song of Abidjan)
lyrics/music: Mathieu EKRA, Joachim BONY, and Pierre Marie COTY/Pierre Marie COTY and Pierre Michel PANGO
note: adopted 1960; although the nation's capital city moved from Abidjan to Yamoussoukro in 1983, the anthem still owes its name to the former capital

ECONOMY

Economy—overview: Cote d'Ivoire is heavily dependent on agriculture and related activities, which engage roughly 68% of the population. Cote d'Ivoire is the world's largest producer and exporter of cocoa beans and a significant producer and exporter of coffee and palm oil. Consequently, the economy is highly sensitive to fluctuations in international prices for these products, and, to a lesser extent, in climatic conditions. Cocoa, oil, and coffee are the country's top export revenue earners, but the country is also producing gold. Since the end of the civil war in 2003, political turmoil has continued to damage the economy, resulting in the loss of foreign investment and slow economic growth. GDP grew by more than 2% in 2008 and around 4% per year in 2009-10. Per capita income has declined by 15% since 1999, but registered a slight improvement in 2009-10. Power cuts caused by a turbine failure in early 2010 slowed economic activity. Cote d'Ivoire in 2010 signed agreements to restructure its Paris Club bilateral, other bilateral, and London Club debt. Cote d'Ivoire's long term challenges include political instability and degrading infrastructure.
GDP (purchasing power parity): $37.02 billion (2010 est.)
country comparison to the world: 100
$36.09 billion (2009 est.)
$34.79 billion (2008 est.)
note: data are in 2010 US dollars
GDP (official exchange rate): $22.82 billion (2010 est.)
GDP—real growth rate: 2.6% (2010 est.)
country comparison to the world: 136
3.8% (2009 est.)
2.3% (2008 est.)
GDP—per capita (PPP): $1,800 (2010 est.)
country comparison to the world: 195
$1,800 (2009 est.)
$1,700 (2008 est.)
note: data are in 2010 US dollars
GDP—composition by sector:
agriculture: 28.2%
industry: 21.3%
services: 50.6% (2010 est.)
Labor force: 7.617 million (2010 est.)
country comparison to the world: 60
Labor force—by occupation:
agriculture: 68%
industry and *services:* NA (2007 est.)
Unemployment rate: NA%
note: unemployment may have climbed to 40-50% as a result of the civil war
Population below poverty line: 42% (2006 est.)
Household income or consumption by percentage share: *lowest 10%:* 2%
highest 10%: 34% (2002)
Distribution of family income—Gini index: 44.6 (2002)
country comparison to the world: 41
36.7 (1995)
Investment (gross fixed): 9.7% of GDP (2010 est.)
country comparison to the world: 150
Budget: *revenues:* $4.755 billion
expenditures: $5.158 billion (2010 est.)
Public debt: 63.3% of GDP (2010 est.)
country comparison to the world: 28
66.5% of GDP (2009 est.)
Inflation rate (consumer prices): 1.4% (2010 est.)
country comparison to the world: 30
0.9% (2009 est.)
Central bank discount rate: 4.25% (31 December 2009)
country comparison to the world: 97
4.75% (31 December 2008)
Commercial bank prime lending rate: NA%
Stock of narrow money: $5.094 billion (31 December 2010 est.)
country comparison to the world: 88
$4.959 billion (31 December 2009 est.)
Stock of broad money: $7.653 billion (31 December 2010 est.)
country comparison to the world: 109
$7.437 billion (31 December 2009 est.)
Stock of domestic credit: $5.448 billion (31 December 2010 est.)
country comparison to the world: 109
$5.308 billion (31 December 2009 est.)
Market value of publicly traded shares: $6.141 billion (31 December 2009)
country comparison to the world: 73
$7.071 billion (31 December 2008)
$8.353 billion (31 December 2007)
Agriculture—products: coffee, cocoa beans, bananas, palm kernels, corn, rice, manioc (tapioca), sweet potatoes, sugar, cotton, rubber; timber
Industries: foodstuffs, beverages; wood products, oil refining, truck and bus assembly, textiles, fertilizer, building materials, electricity, ship construction and repair
Industrial production growth rate: 4.5% (2010 est.)
country comparison to the world: 77
Electricity—production: 5.275 billion kWh (2007 est.)
country comparison to the world: 111
Electricity—consumption: 3.231 billion kWh (2007 est.)
country comparison to the world: 122
Electricity—exports: 772 million kWh (2007 est.)
Electricity—imports: 0 kWh (2008 est.)
Oil—production: 58,950 bbl/day (2009 est.)
country comparison to the world: 60
Oil—consumption: 24,000 bbl/day (2009 est.)
country comparison to the world: 116
Oil—exports: 115,700 bbl/day (2007 est.)
country comparison to the world: 64
Oil—imports: 80,960 bbl/day (2007 est.)
country comparison to the world: 70
Oil—proved reserves: 250 million bbl (1 January 2010 est.)
country comparison to the world: 57
Natural gas—production: 1.3 billion cu m (2008 est.)
country comparison to the world: 61
Natural gas—consumption: 1.3 billion cu m (2008 est.)

country comparison to the world: 84
Natural gas—exports: 0 cu m (2008 est.)
country comparison to the world: 115
Natural gas—imports: 0 cu m (2008 est.)
country comparison to the world: 201
Natural gas—proved reserves: 28.32 billion cu m (1 January 2010 est.)
country comparison to the world: 71
Current account balance: $534 million (2010 est.)
country comparison to the world: 53
$1.67 billion (2009 est.)
Exports: $10.25 billion (2010 est.)
country comparison to the world: 84
$10.5 billion (2009 est.)
Exports—commodities: cocoa, coffee, timber, petroleum, cotton, bananas, pineapples, palm oil, fish
Exports—partners: Netherlands 13.92%, France 10.75%, US 7.79%, Germany 7.2%, Nigeria 6.99%, Ghana 5.56% (2009)
Imports: $7.015 billion (2010 est.)
country comparison to the world: 104
$6.318 billion (2009 est.)
Imports—commodities: fuel, capital equipment, foodstuffs
Imports—partners: Nigeria 20.75%, France 14.19%, China 7.18%, Thailand 5.09% (2009)
Reserves of foreign exchange and gold: $3.985 billion (31 December 2010 est.)
country comparison to the world: 80
$3.267 billion (31 December 2009 est.)
Debt—external: $11.6 billion (31 December 2010 est.)
country comparison to the world: 86
$11.34 billion (31 December 2009 est.)
Stock of direct foreign investment—at home: $NA
Stock of direct foreign investment—abroad: $NA
Exchange rates: Communaute Financiere Africaine francs (XOF) per US dollar– 495.28 (2010), 472.19 (2009), 447.81 (2008), 481.83 (2007), 522.89 (2006)

COMMUNICATIONS

Telephones—main lines in use: 282,100 (2009)
country comparison to the world: 118
Telephones—mobile cellular: 13.346 million (2009)
country comparison to the world: 53
Telephone system: *general assessment:* well developed by African standards; telecommunications sector privatized in late 1990s and operational fixed-lines have increased since that time with two fixed-line providers operating over open-wire lines, microwave radio relay, and fiber-optics; 90% digitalized
domestic: with multiple mobile-cellular service providers competing in the market, usage has increased sharply to roughly 65 per 100 persons
international: country code–225; landing point for the SAT-3/WASC fiber-optic submarine cable that provides connectivity to Europe and Asia; satellite earth stations–2 Intelsat (1 Atlantic Ocean and 1 Indian Ocean) (2009)
Broadcast media: state-owned television operates 2 stations; no private terrestrial TV stations, but satellite TV subscription service is available; state-owned radio operates 2 stations; some private radio stations; transmissions of several international broadcasters are available (2007)
Internet country code: .ci
Internet hosts: 9,865 (2010)
country comparison to the world: 121
Internet users: 967,300 (2009)
country comparison to the world: 103

TRANSPORTATION

Airports: 27 (2010)
country comparison to the world: 121
Airports—with paved runways: *total:* 7
over 3,047 m: 1
2,438 to 3,047 m: 2
1,524 to 2,437 m: 4 (2010)
Airports—with unpaved runways: *total:* 20
1,524 to 2,437 m: 6
914 to 1,523 m: 11
under 914 m: 3 (2010)
Pipelines: condensate 86 km; gas 180 km; oil 92 km (2010)
Railways: *total:* 660 km
country comparison to the world: 104
narrow gauge: 660 km 1.000-m gauge
note: an additional 622 km of this railroad extends into Burkina Faso (2010)
Roadways: *total:* 80,000 km
country comparison to the world: 60
paved: 6,500 km
unpaved: 73,500 km
note: includes intercity and urban roads; another 20,000 km of dirt roads are in poor condition and 150,000 km of dirt roads are impassable (2006)
Waterways: 980 km (navigable rivers, canals, and numerous coastal lagoons) (2009)
country comparison to the world: 67
Ports and terminals: Abidjan, Espoir, San-Pedro

MILITARY

Military branches: Republican Forces of Cote d'Ivoire (Force Republiques de Cote d'Ivoire, FRCI): Army, Navy, Cote d'Ivoire Air Force (Force Aerienne de la Cote d'Ivoire)
note: FRCI is the former Armed Forces of the New Forces (FAFN) (2011)
Military service age and obligation: 18-25 years of age for compulsory and voluntary male and female military service; voluntary recruitment of former rebels into the new national army is restricted to ages 22-29 (2011)
Manpower available for military service: *males age 16–49:* 5,247,522
females age 16–49: 5,047,901 (2010 est.)
Manpower fit for military service: *males age 16–49:* 3,360,087
females age 16–49: 3,196,033 (2010 est.)
Manpower reaching militarily significant age annually: *male:* 247,011
female: 242,958 (2010 est.)
Military expenditures: 1.5% of GDP (2009)
country comparison to the world: 97

TRANSNATIONAL ISSUES

Disputes—international: despite the presence of over 9,000 UN forces (UNOCI) in Cote d'Ivoire since 2004, ethnic conflict still leaves displaced hundreds of thousands of Ivoirians in and out of the country as well as driven out migrants from neighboring states who worked in Ivorian cocoa plantations; the March 2007 peace deal between Ivorian rebels and the government brought significant numbers of rebels out of hiding in neighboring states
Refugees and internally displaced persons: refugees (country of origin): 25,615 (Liberia)
IDPs: 709,000 (2002 coup; most IDPs are in western regions) (2007)
Trafficking in persons: Cote d'Ivoire is a source, transit, and destination country for women and children trafficked for forced labor and commercial sexual exploitation; trafficking within the country is more prevalent than international trafficking and the majority of victims are children; women and girls are trafficked from northern areas to southern cities for domestic servitude, restaurant labor, and sexual exploitation; boys are trafficked internally for agricultural and service labor and transnationally for forced labor in agriculture, mining, construction, and in the fishing industry; women and girls are trafficked to and from other West and Central African countries for domestic servitude and forced street vending
tier rating: Tier 2 Watch List–Cote d'Ivoire is on the Tier 2 Watch List for its failure to provide evidence of increasing efforts to eliminate trafficking in 2007, particularly with regard to its law enforcement efforts and protection of sex trafficking victims; in addition, Ivoirian law does not prohibit all forms of trafficking, and Cote d'Ivoire has not ratified the 2000 UN TIP Protocol (2008)
Illicit drugs: illicit producer of cannabis, mostly for local consumption; utility as a narcotic transshipment point to Europe reduced by ongoing political instability; while rampant corruption and inadequate supervision leave the banking system vulnerable to money laundering, the lack of a developed financial system limits the country's utility as a major money-laundering center (2008)

CROATIA

INTRODUCTION

Background: The lands that today comprise Croatia were part of the Austro-Hungarian Empire until the close of World War I. In 1918, the Croats, Serbs, and Slovenes formed a kingdom known after 1929 as Yugoslavia. Following World War II, Yugoslavia became a federal independent Communist state under the strong hand of Marshal TITO. Although Croatia declared its independence from Yugoslavia in 1991, it took four years of sporadic, but often bitter, fighting before occupying Serb armies were mostly cleared from Croatian lands. Under UN supervision, the last Serb-held enclave in eastern Slavonia was returned to Croatia in 1998. In April 2009, Croatia joined NATO; it is a candidate for eventual EU accession.

GEOGRAPHY

Location: Southeastern Europe, bordering the Adriatic Sea, between Bosnia and Herzegovina and Slovenia
Geographic coordinates: 45 10 N, 15 30 E
Map references: Europe
Area: *total:* 56,594 sq km
country comparison to the world: 126
land: 55,974 sq km
water: 620 sq km
Area—comparative: slightly smaller than West Virginia
Land boundaries:
total: 1,982 km
border countries: Bosnia and Herzegovina 932 km, Hungary 329 km, Serbia 241 km, Montenegro 25 km, Slovenia 455 km
Coastline: 5,835 km (mainland 1,777 km, islands 4,058 km)
Maritime claims: territorial sea: 12 nm
continental shelf: 200 m depth or to the depth of exploitation
Climate: Mediterranean and continental; continental climate predominant with hot summers and cold winters; mild winters, dry summers along coast
Terrain: geographically diverse; flat plains along Hungarian border, low mountains and highlands near Adriatic coastline and islands
Elevation extremes: *lowest point:* Adriatic Sea 0 m
highest point: Dinara 1,831 m
Natural resources: oil, some coal, bauxite, low-grade iron ore, calcium, gypsum, natural asphalt, silica, mica, clays, salt, hydropower
Land use: *arable land:* 25.82%
permanent crops: 2.19%
other: 71.99% (2005)
Irrigated land: 110 sq km (2003)
Total renewable water resources: 105.5 cu km (1998)
Natural hazards: destructive earthquakes
Environment—current issues: air pollution (from metallurgical plants) and resulting acid rain is damaging the forests; coastal pollution from industrial and domestic waste; landmine removal and reconstruction of infrastructure consequent to 1992-95 civil strife
Environment—international agreements: *party to:* Air Pollution, Air Pollution-Nitrogen Oxides, Air Pollution-Persistent Organic Pollutants, Air Pollution-Sulfur 94, Air Pollution-Volatile Organic Compounds, Biodiversity, Climate Change, Climate Change-Kyoto Protocol, Desertification, Endangered Species, Hazardous Wastes, Law of the Sea, Marine Dumping, Ozone Layer Protection, Ship Pollution, Wetlands, Whaling
signed, but not ratified: none of the selected agreements
Geography—note: controls most land routes from Western Europe to Aegean Sea and Turkish Straits; most Adriatic Sea islands lie off the coast of Croatia–some 1,200 islands, islets, ridges, and rocks

PEOPLE

Population: 4,483,804 (July 2011 est.)
country comparison to the world: 123
Age structure: *0–14 years:* 15.1% (male 346,553/female 328,677)
15–64 years: 68.1% (male 1,516,884/female 1,536,065)
65 years and over: 16.9% (male 296,268/female 459,357) (2011 est.)
Median age: *total:* 41.4 years
male: 39.5 years
female: 43.3 years (2011 est.)
Population growth rate: -0.076% (2011 est.)
country comparison to the world: 203
Birth rate: 9.6 births/1,000 population (2011 est.)
country comparison to the world: 201
Death rate: 11.91 deaths/1,000 population (July 2011 est.)
country comparison to the world: 29
Net migration rate: 1.55 migrant(s)/1,000 population (2011 est.)
country comparison to the world: 45
Urbanization: *urban population:* 58% of total population (2010)
rate of urbanization: 0.4% annual rate of change (2010-15 est.)
Major cities—population: ZAGREB (capital) 685,000 (2009)
Sex ratio: *at birth:* 1.055 male(s)/female
under 15 years: 1.06 male(s)/female
15–64 years: 0.99 male(s)/female
65 years and over: 0.64 male(s)/female
total population: 0.93 male(s)/female (2011 est.)
Infant mortality rate: *total:* 6.16 deaths/1,000 live births
country comparison to the world: 174
male: 6.24 deaths/1,000 live births
female: 6.08 deaths/1,000 live births (2011 est.)
Life expectancy at birth: *total population:* 75.79 years
country comparison to the world: 80
male: 72.17 years
female: 79.6 years (2011 est.)
Total fertility rate: 1.43 children born/woman (2011 est.)
country comparison to the world: 195
HIV/AIDS—adult prevalence rate: less than 0.1% (2009 est.)
country comparison to the world: 134
HIV/AIDS—people living with HIV/AIDS: fewer than 1,000 (2009 est.)
country comparison to the world: 145
HIV/AIDS—deaths: fewer than 100 (2009 est.)
country comparison to the world: 134
Major infectious diseases: *degree of risk:* intermediate
food or waterborne diseases: bacterial diarrhea
vectorborne diseases: tickborne encephalitis
note: highly pathogenic H5N1 avian influenza has been identified in this country; it poses a negligible risk with extremely rare cases possible among US citizens who have close contact with birds (2009)
Drinking water source: *Improved:*
urban: 100% of population
rural: 97% of population
total: 99% of population
Unimproved:
urban: 0% of population
rural: 3% of population
total: 1% of population (2008)
Sanitation facility access: *Improved:*
urban: 99% of population
rural: 98% of population
total: 99% of population
Unimproved:
urban: 1% of population
rural: 2% of population
total: 1% of population (2008)
Nationality: *noun:* Croat(s), Croatian(s)
adjective: Croatian
Ethnic groups: Croat 89.6%, Serb 4.5%, other 5.9% (including Bosniak, Hungarian, Slovene, Czech, and Roma) (2001 census)
Religions: Roman Catholic 87.8%, Orthodox 4.4%, other Christian 0.4%, Muslim 1.3%, other and unspecified 0.9%, none 5.2% (2001 census)
Languages: Croatian (official) 96.1%, Serbian 1%, other and undesignated (including Italian, Hungarian, Czech, Slovak, and German) 2.9% (2001 census)
Literacy: *definition:* age 15 and over can read and write
total population: 98.1%
male: 99.3%
female: 97.1% (2001 census)
School life expectancy (primary to tertiary education): *total:* 14 years
male: 13 years
female: 14 years (2008)

Education expenditures: 4.6% of GDP (2009)
country comparison to the world: 75

GOVERNMENT

Country name: *conventional long form:* Republic of Croatia
conventional short form: Croatia
local long form: Republika Hrvatska
local short form: Hrvatska
former: People's Republic of Croatia, Socialist Republic of Croatia
Government type: presidential/parliamentary democracy
Capital: *name:* Zagreb
geographic coordinates: 45 48 N, 16 00 E
time difference: UTC+1 (6 hours ahead of Washington, DC during Standard Time)
daylight saving time: +1hr, begins last Sunday in March; ends last Sunday in October
Administrative divisions: 20 counties (zupanije, zupanija—singular) and 1 city* (grad—singular); Bjelovarsko-Bilogorska, Brodsko-Posavska, Dubrovacko-Neretvanska (Dubrovnik-Neretva), Istarska (Istria), Karlovacka, Koprivnicko-Krizevacka, Krapinsko-Zagorska, Licko-Senjska (Lika-Senj), Medimurska, Osjecko-Baranjska, Pozesko-Slavonska (Pozega-Slavonia), Primorsko-Goranska, Sibensko-Kninska, Sisacko-Moslavacka, Splitsko-Dalmatinska (Split-Dalmatia), Varazdinska, Viroviticko-Podravska, Vukovarsko-Srijemska, Zadarska, Zagreb*, Zagrebacka
Independence: 25 June 1991 (from Yugoslavia)
National holiday: Independence Day, 8 October (1991); note—25 June 1991 was the day the Croatian parliament voted for independence; following a three-month moratorium to allow the European Community to solve the Yugoslav crisis peacefully, Parliament adopted a decision on 8 October 1991 to sever constitutional relations with Yugoslavia
Constitution: adopted 22 December 1990; revised 2000, 2001
Legal system: civil law system based on Yugoslav civil codes; note—Croatian legislation is changing the former Yugoslav legal model
International law organization participation: has not submitted an ICJ jurisdiction declaration; accepts ICCt jurisdiction
Suffrage: 18 years of age, 16 if employed; universal
Executive branch: *chief of state:* President Ivo JOSIPOVIC (since 18 February 2010)
head of government: Prime Minister Jadranka KOSOR (since 6 July 2009); Deputy Prime Ministers Bozidar PANKRETIC (since 6 July 2009), Darko MILINOVIC (since 13 November 2009), Domagoj Ivan MILOSEVIC (since 29 December 2010), Petar COBANKOVIC (since 29 December 2010), Slobodan UZELAC (since 12 January 2008), Gordan JANDROKOVIC (since 29 December 2010)
cabinet: Council of Ministers named by the prime minister and approved by the parliamentary assembly
(For more information visit the World Leaders website)
elections: president elected by popular vote for a five-year term (eligible for a second term); election last held on 10 January 2010 (next to be held in December 2015); the leader of the majority party or the leader of the majority coalition usually appointed prime minister by the president and then approved by the assembly
election results: Ivo JOSIPOVIC elected president; percent of vote in the second round—Ivo JOSIPOVIC 60%, Milan BANDIC 40%
Legislative branch: unicameral Assembly or Sabor (153 seats; members elected from party lists by popular vote to serve four-year terms)
elections: last held on 25 November 2007 (next to be held by November 2011)
election results: percent of vote by party—NA; number of seats by party—HDZ 66, SDP 57, HNS 6, HSS 6, HDSSB 3, IDS 3, SDSS 3, other 9
Judicial branch: Supreme Court; Constitutional Court; judges for both courts are appointed for eight-year terms by the Judicial Council of the Republic, which is elected by the Assembly
Political parties and leaders: Croatian Democratic Congress of Slavonia and Baranja or HDSSB [Vladimir SISLJAGIC]; Croatian Democratic Union or HDZ [Jadranka KOSOR]; Croatian Party of the Right or HSP [Anto DJAPIC]; Croatian Peasant Party or HSS [Josip FRISCIC]; Croatian Pensioner Party or HSU [Silvano HRELJA]; Croatian People's Party or HNS [Radimir CACIC]; Croatian Social Liberal Party or HSLS [Darinko KOSOR]; Independent Democratic Serb Party or SDSS [Vojislav STANIMIROVIC]; Istrian Democratic Assembly or IDS [Ivan JAKOVCIC]; Social Democratic Party of Croatia or SDP [Zoran MILANOVIC]
Political pressure groups and leaders: *other:* human rights groups
International organization participation: Australia Group, BIS, BSEC (observer), CE, CEI, EAPC, EBRD, EU (candidate country), FAO, G-11, IADB, IAEA, IBRD, ICAO, ICC, ICRM, IDA, IFAD, IFC, IFRCS, IHO, ILO, IMF, IMO, IMSO, Interpol, IOC, IOM, IPU, ISO, ITSO, ITU, ITUC, MIGA, MINURSO, MINUSTAH, NAM (observer), NATO, NSG, OAS (observer), OIF (observer), OPCW, OSCE, PCA, SECI, UN, UNCTAD, UNDOF, UNESCO, UNFICYP, UNIDO, UNIFIL, UNMIL, UNMIS, UNMOGIP, UNWTO, UPU, WCO, WHO, WIPO, WMO, WTO, ZC
Diplomatic representation in the US: *chief of mission:* Ambassador Kolinda GRABAR-KITAROVIC
chancery: 2343 Massachusetts Avenue NW, Washington, DC 20008
telephone: [1] (202) 588-5899
FAX: [1] (202) 588-8936
consulate(s) general: Chicago, Los Angeles, New York
Diplomatic representation from the US: *chief of mission:* Ambassador James B. FOLEY
embassy: 2 Thomas Jefferson Street, 10010 Zagreb
mailing address: use street address
telephone: [385] (1) 661-2200
FAX: [385] (1) 661-2373
Flag description: three equal horizontal bands of red (top), white, and blue—the Pan-Slav colors—superimposed by the Croatian coat of arms; the coat of arms consists of one main shield (a checkerboard of 13 red and 12 silver (white) fields) surmounted by five smaller shields that form a crown over the main shield; the five small shields represent five historic regions, they are (from left to right): Croatia, Dubrovnik, Dalmatia, Istria, and Slavonia
note: the Pan-Slav colors were inspired by the 19th-century flag of Russia
National anthem: *name:* "Lijepa nasa domovino" (Our Beautiful Homeland)
lyrics/music: Antun MIHANOVIC/Josip RUNJANIN
note: adopted 1972; "Lijepa nasa domovino," whose lyrics were written in 1835, served as an unofficial anthem beginning in 1891

ECONOMY

Economy—overview: Once one of the wealthiest of the Yugoslav republics, Croatia's economy suffered badly during the 1991-95 war as output collapsed and the country missed the early waves of investment in Central and Eastern Europe that followed the fall of the Berlin Wall. Between 2000 and 2007, however, Croatia's economic fortunes began to improve slowly, with moderate but steady GDP growth between 4% and 6% led by a rebound in tourism and credit-driven consumer spending. Inflation over the same period has remained tame and the currency, the kuna, stable. Nevertheless, difficult problems still remain, including a stubbornly high unemployment rate, a growing trade deficit and uneven regional development. The state retains a large role in the economy, as privatization efforts often meet stiff public and political resistance. While macroeconomic stabilization has largely been achieved, structural reforms lag because of deep resistance on the part of the public and lack of strong support from politicians. The EU accession process should accelerate fiscal and structural reform. While long term growth prospects for the economy remain strong, Croatia will face significant pressure as a result of the global financial crisis. Croatia's high foreign debt, anemic export sector, strained state budget, and over-reliance on tourism revenue will result in higher risk to economic stability over the medium term.
GDP (purchasing power parity): $78.09 billion (2010 est.)
country comparison to the world: 79
$79.18 billion (2009 est.)
$84.06 billion (2008 est.)
note: data are in 2010 US dollars
GDP (official exchange rate): $60.59 billion (2010 est.)
GDP—real growth rate: -1.4% (2010 est.)
country comparison to the world: 203
-5.8% (2009 est.)
2.4% (2008 est.)
GDP—per capita (PPP): $17,400 (2010 est.)
country comparison to the world: 67

$17,600 (2009 est.)
$18,700 (2008 est.)
note: data are in 2010 US dollars
GDP—composition by sector:
agriculture: 6.8%
industry: 27.2%
services: 66% (2010 est.)
Labor force: 1.762 million (2010 est.)
country comparison to the world: 124
Labor force—by occupation:
agriculture: 5%
industry: 31.3%
services: 63.6% (2008)
Unemployment rate: 17.6% (2010 est.)
country comparison to the world: 159
16.1% (2009 est.)
Population below poverty line: 17% (2008)
Household income or consumption by percentage share: *lowest 10%:* 3.6%
highest 10%: 23.1% (2005 est.)
Distribution of family income—Gini index: 29 (2008)
country comparison to the world: 117
29 (1998)
Investment (gross fixed): 22.4% of GDP (2010 est.)
country comparison to the world: 65
Budget: *revenues:* $22 billion
expenditures: $24.29 billion (2010 est.)
Public debt: 55% of GDP (2010 est.)
country comparison to the world: 47
46.4% of GDP (2009 est.)
Inflation rate (consumer prices): 1.3% (2010 est.)
country comparison to the world: 29
2.4% (2009 est.)
Central bank discount rate: 9% (31 December 2009)
country comparison to the world: 48
9% (31 December 2008)
Commercial bank prime lending rate: 11.55% (31 December 2009 est.)
country comparison to the world: 85
10.07% (31 December 2008 est.)
Stock of narrow money: $8.72 billion (31 December 2010 est.)
country comparison to the world: 75
$8.964 billion (31 December 2009 est.)
Stock of broad money: $40.8 billion (31 December 2010 est.)
country comparison to the world: 69
$42.59 billion (31 December 2009 est.)
Stock of domestic credit: $48.62 billion (31 December 2010 est.)
country comparison to the world: 64
$48.6 billion (31 December 2009 est.)
Market value of publicly traded shares: $25.64 billion (31 December 2009)
country comparison to the world: 56
$26.79 billion (31 December 2008)
$65.98 billion (31 December 2007)
Agriculture—products: wheat, corn, sugar beets, sunflower seed, barley, alfalfa, clover, olives, citrus, grapes, soybeans, potatoes; livestock, dairy products
Industries: chemicals and plastics, machine tools, fabricated metal, electronics, pig iron and rolled steel products, aluminum, paper, wood products, construction materials, textiles, shipbuilding, petroleum and petroleum refining, food and beverages, tourism
Industrial production growth rate: -0.9% (2010 est.)
country comparison to the world: 156
Electricity—production: 11.49 billion kWh (2008 est.)
country comparison to the world: 86
Electricity—consumption: 18 billion kWh (2008 est.)
country comparison to the world: 71
Electricity—exports: 5.668 billion kWh (2008 est.)
Electricity—imports: 12.24 billion kWh (2008 est.)
Oil—production: 23,960 bbl/day (2009 est.)
country comparison to the world: 72
Oil—consumption: 106,000 bbl/day (2009 est.)
country comparison to the world: 75
Oil—exports: 43,750 bbl/day (2007 est.)
country comparison to the world: 79
Oil—imports: 122,100 bbl/day (2007 est.)
country comparison to the world: 58
Oil—proved reserves: 73.35 million bbl (1 January 2010 est.)
country comparison to the world: 77
Natural gas—production: 2.847 billion cu m (2009 est.)
country comparison to the world: 56
Natural gas—consumption: 3.205 billion cu m (2009 est.)
country comparison to the world: 71
Natural gas—exports: 695.5 million cu m (2009 est.)
country comparison to the world: 40
Natural gas—imports: 1.22 billion cu m (2009 est.)
country comparison to the world: 53
Natural gas—proved reserves: 30.58 billion cu m (1 January 2010 est.)
country comparison to the world: 69
Current account balance: $-2.312 billion (2010 est.)
country comparison to the world: 158
$-3.247 billion (2009 est.)
Exports: $11.51 billion (2010 est.)
country comparison to the world: 81
$10.72 billion (2009 est.)
Exports—commodities: transport equipment, machinery, textiles, chemicals, foodstuffs, fuels
Exports—partners: Italy 19.1%, Bosnia and Herzegovina 12.98%, Germany 11.06%, Slovenia 7.47%, Austria 5.44%, Serbia 5.41% (2009)
Imports: $20.93 billion (2010 est.)
country comparison to the world: 69
$21 billion (2009 est.)
Imports—commodities: machinery, transport and electrical equipment; chemicals, fuels and lubricants; foodstuffs
Imports—partners: Italy 15.46%, Germany 13.57%, Russia 9.29%, China 6.83%, Slovenia 5.75%, Austria 5.04% (2009)
Reserves of foreign exchange and gold: $13.79 billion (31 December 2010 est.)
country comparison to the world: 52
$14.89 billion (31 December 2009 est.)
Debt—external: $59.7 billion (31 December 2010 est.)
country comparison to the world: 49
$62.41 billion (31 December 2009 est.)
Stock of direct foreign investment—at home: $34.63 billion (31 December 2010 est.)
country comparison to the world: 59
$32.13 billion (31 December 2009 est.)
Stock of direct foreign investment—abroad: $6.334 billion (31 December 2010 est.)
country comparison to the world: 56
$5.934 billion (31 December 2009 est.)
Exchange rates: kuna (HRK) per US dollar–5.6356 (2010), 5.2692 (2009), 4.98 (2008), 5.3735 (2007), 5.8625 (2006)

COMMUNICATIONS

Telephones—main lines in use: 1.859 million (2009)
country comparison to the world: 60
Telephones—mobile cellular: 6.035 million (2009)
country comparison to the world: 84
Telephone system: *general assessment:* the telecommunications network has improved steadily since the mid-1990s; local lines are digital
domestic: fixed-line teledensity holding steady at about 40 per 100 persons; mobile-cellular telephone subscriptions exceed the population
international: country code–385; digital international service is provided through the main switch in Zagreb; Croatia participates in the Trans-Asia-Europe (TEL) fiber-optic project, which consists of 2 fiber-optic trunk connections with Slovenia and a fiber-optic trunk line from Rijeka to Split and Dubrovnik; the ADRIA-1 submarine cable provides connectivity to Albania and Greece (2009)
Broadcast media: the national state-owned public broadcaster, Croatian Radiotelevision (HRT), operates 2 terrestrial TV networks, a satellite channel that rebroadcasts programs for Croatians living abroad, and 6 regional TV centers; 2 private broadcasters operate national terrestrial networks; about 15 privately-owned regional TV stations; multi-channel cable and satellite TV subscription services are available; state-owned public broadcaster operates 3 national radio networks and a number of regional radio stations; 2 privately-owned national radio networks and a large number of regional, county, city, and community radio stations (2007)
Internet country code: .hr
Internet hosts: 1.287 million (2010)
country comparison to the world: 38
Internet users: 2.234 million (2009)
country comparison to the world: 73

TRANSPORTATION

Airports: 69 (2010)
country comparison to the world: 73
Airports—with paved runways: *total:* 23
over 3,047 m: 2
2,438 to 3,047 m: 6
1,524 to 2,437 m: 3
914 to 1,523 m: 3
under 914 m: 9 (2010)
Airports—with unpaved runways: *total:* 46
1,524 to 2,437 m: 1
914 to 1,523 m: 7
under 914 m: 38 (2010)
Heliports: 1 (2010)
Pipelines: gas 1,686 km; oil 532 km (2010)
Railways: *total:* 2,722 km
country comparison to the world: 60
standard gauge: 2,722 km 1.435-m gauge (985 km electrified) (2009)
Roadways: *total:* 29,343 km (includes 1,047 km of expressways) (2009)
country comparison to the world: 98

Waterways: 785 km (2009)
country comparison to the world: 74
Merchant marine: *total:* 75
country comparison to the world: 56
by type: bulk carrier 24, cargo 7, chemical tanker 6, passenger/cargo 27, petroleum tanker 10, refrigerated cargo 1
foreign-owned: 2 (Norway 2)
registered in other countries: 33 (Bahamas 1, Belize 1, Liberia 2, Malta 7, Marshall Islands 12, Panama 2, Saint Vincent and the Grenadines 8) (2010)
Ports and terminals: Omisalj, Ploce, Rijeka, Sibernik, Split, Vukovar (on Danube River)

MILITARY

Military branches: Armed Forces of the Republic of Croatia (Oruzane Snage Republike Hrvatske, OSRH), consists of five major commands directly subordinate to a General Staff: Ground Forces (Hrvatska Kopnena Vojska, HKoV), Naval Forces (Hrvatska Ratna Mornarica, HRM; includes coast guard), Air Force and Air Defense Command, Joint Education and Training Command, Logistics Command; Military Police Force supports each of the three Croatian military forces (2010)
Military service age and obligation: 18-27 years of age for voluntary military service; 16 years of age with parental consent; 6-month service obligation; conscription abolished 1 January 2008 (2010)
Manpower available for military service: *males age 16–49:* 1,016,234
females age 16–49: 1,017,355 (2010 est.)
Manpower fit for military service: *males age 16–49:* 770,710
females age 16–49: 839,732 (2010 est.)
Manpower reaching militarily significant age annually: *male:* 28,334
female: 27,015 (2010 est.)
Military expenditures: 2.39% of GDP (2005 est.)
country comparison to the world: 64

TRANSNATIONAL ISSUES

Disputes—international: dispute remains with Bosnia and Herzegovina over several small sections of the boundary related to maritime access that hinders ratification of the 1999 border agreement; the Croatia-Slovenia land and maritime boundary agreement, which would have ceded most of Pirin Bay and maritime access to Slovenia and several villages to Croatia, remains unratified and in dispute; Slovenia also protests Croatia's 2003 claim to an exclusive economic zone in the Adriatic; as a European Union peripheral state, Slovenia imposed a hard border Schengen regime with non-member Croatia in December 2007
Refugees and internally displaced persons: *IDPs:* 2,900-7,000 (Croats and Serbs displaced in 1992-95 war) (2007)
Illicit drugs: transit point along the Balkan route for Southwest Asian heroin to Western Europe; has been used as a transit point for maritime shipments of South American cocaine bound for Western Europe (2008)

CUBA

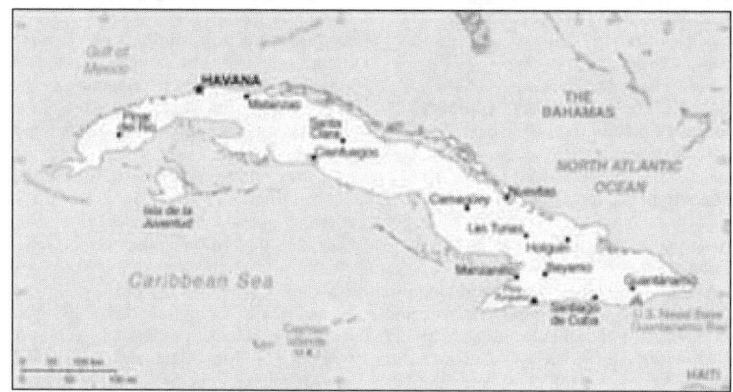

INTRODUCTION

Background: The native Amerindian population of Cuba began to decline after the European discovery of the island by Christopher COLUMBUS in 1492 and following its development as a Spanish colony during the next several centuries. Large numbers of African slaves were imported to work the coffee and sugar plantations, and Havana became the launching point for the annual treasure fleets bound for Spain from Mexico and Peru. Spanish rule eventually provoked an independence movement and occasional rebellions that were harshly suppressed. US intervention during the Spanish-American War in 1898 assisted the Cubans in overthrowing Spanish rule. The Treaty of Paris established Cuban independence from the US in 1902 after which the island experienced a string of governments mostly dominated by the military and corrupt politicians. Fidel CASTRO led a rebel army to victory in 1959; his iron rule held the subsequent regime together for nearly five decades. He stepped down as president in February 2008 in favor of his younger brother Raul CASTRO. Cuba's Communist revolution, with Soviet support, was exported throughout Latin America and Africa during the 1960s, 1970s, and 1980s. The country faced a severe economic downturn in 1990 following the withdrawal of former Soviet subsidies worth $4 billion to $6 billion annually. Cuba at times portrays the US embargo, in place since 1961, as the source if its difficulties. Illicit migration to the US–using homemade rafts, alien smugglers, air flights, or via the southwest border–is a continuing problem. The US Coast Guard intercepted 982 individuals attempting to cross the Straits of Florida in fiscal year 2009.

GEOGRAPHY

Location: Caribbean, island between the Caribbean Sea and the North Atlantic Ocean, 150 km south of Key West, Florida
Geographic coordinates: 21 30 N, 80 00 W
Map references: Central America and the Caribbean
Area: *total:* 110,860 sq km
country comparison to the world: 105
land: 109,820 sq km
water: 1,040 sq km
Area—comparative: slightly smaller than Pennsylvania
Land boundaries: *total:* 29 km
border countries: US Naval Base at Guantanamo Bay 29 km
note: Guantanamo Naval Base is leased by the US and remains part of Cuba
Coastline: 3,735 km
Maritime claims: territorial sea: 12 nm
contiguous zone: 24 nm
exclusive economic zone: 200 nm
Climate: tropical; moderated by trade winds; dry season (November to April); rainy season (May to October)
Terrain: mostly flat to rolling plains, with rugged hills and mountains in the southeast
Elevation extremes: *lowest point:* Caribbean Sea 0 m
highest point: Pico Turquino 2,005 m
Natural resources: cobalt, nickel, iron ore, chromium, copper, salt, timber, silica, petroleum, arable land
Land use:
arable land: 27.63%
permanent crops: 6.54%
other: 65.83% (2005)
Irrigated land: 8,700 sq km (2003)
Total renewable water resources: 38.1 cu km (2000)
Freshwater withdrawal (domestic/industrial/agricultural): *total:* 8.2 cu km/yr (19%/12%/69%)
per capita: 728 cu m/yr (2000)
Natural hazards: the east coast is subject to hurricanes from August to November (in general, the country averages about one hurricane every other year); droughts are common
Environment—current issues: air and water pollution; biodiversity loss; deforestation

Environment—international agreements: *party to:* Antarctic Treaty, Biodiversity, Climate Change, Climate Change-Kyoto Protocol, Desertification, Endangered Species, Environmental Modification, Hazardous Wastes, Law of the Sea, Marine Dumping, Ozone Layer Protection, Ship Pollution, Wetlands
signed, but not ratified: Marine Life Conservation
Geography—note: largest country in Caribbean and westernmost island of the Greater Antilles

PEOPLE

Population: 11,087,330 (July 2011 est.)
country comparison to the world: 74
Age structure: *0–14 years:* 17.3% (male 984,607/female 931,167)
15–64 years: 71.1% (male 3,947,047/female 3,932,128)
65 years and over: 11.7% (male 583,757/female 708,624) (2011 est.)
Median age: *total:* 38.4 years
male: 37.6 years
female: 39.2 years (2011 est.)
Population growth rate: -0.104% (2011 est.)
country comparison to the world: 206
Birth rate: 9.99 births/1,000 population (2011 est.)
country comparison to the world: 193
Death rate: 7.47 deaths/1,000 population (July 2011 est.)
country comparison to the world: 118
Net migration rate: -3.56 migrant(s)/1,000 population (2011 est.)
country comparison to the world: 183
Urbanization: *urban population:* 75% of total population (2010)
rate of urbanization: 0% annual rate of change (2010-15 est.)
Major cities—population: HAVANA (capital) 2.14 million (2009)
Sex ratio: *at birth:* 1.06 male(s)/female
under 15 years: 1.06 male(s)/female
15–64 years: 1 male(s)/female
65 years and over: 0.83 male(s)/female
total population: 0.99 male(s)/female (2011 est.)
Infant mortality rate: *total:* 4.9 deaths/1,000 live births
country comparison to the world: 184
male: 5.27 deaths/1,000 live births
female: 4.52 deaths/1,000 live births (2011 est.)
Life expectancy at birth: *total population:* 77.7 years
country comparison to the world: 57
male: 75.46 years
female: 80.08 years (2011 est.)
Total fertility rate: 1.44 children born/woman (2011 est.)
country comparison to the world: 193
HIV/AIDS—adult prevalence rate: 0.1% (2009 est.)
country comparison to the world: 124
HIV/AIDS—people living with HIV/AIDS: 7,100 (2009 est.)
country comparison to the world: 112
HIV/AIDS—deaths: fewer than 100 (2009 est.)
country comparison to the world: 127
Major infectious diseases: *degree of risk:* intermediate
food or waterborne diseases: bacterial diarrhea and hepatitis A
vectorborne diseases: dengue fever (2009)
Drinking water source: *Improved:*
urban: 96% of population
rural: 89% of population
total: 94% of population
Unimproved:
urban: 4% of population
rural: 11% of population
total: 6% of population (2008)
Sanitation facility access: *Improved:*
urban: 94% of population
rural: 81% of population
total: 91% of population
Unimproved:
urban: 6% of population
rural: 19% of population
total: 9% of population (2008)
Nationality: *noun:* Cuban(s)
adjective: Cuban
Ethnic groups: white 65.1%, mulatto and mestizo 24.8%, black 10.1% (2002 census)
Religions: nominally 85% Roman Catholic prior to CASTRO assuming power; Protestants, Jehovah's Witnesses, Jews, and Santeria are also represented
Languages: Spanish (official)
Literacy: *definition:* age 15 and over can read and write
total population: 99.8%
male: 99.8%
female: 99.8% (2002 census)
School life expectancy (primary to tertiary education): *total:* 18 years
male: 16 years
female: 19 years (2009)
Education expenditures: 13.6% of GDP (2008)
country comparison to the world: 2
People—note: illicit emigration is a continuing problem; Cubans attempt to depart the island and enter the US using homemade rafts, alien smugglers, direct flights, or falsified visas; Cubans also use non-maritime routes to enter the US including direct flights to Miami and overland via the southwest border

GOVERNMENT

Country name: *conventional long form:* Republic of Cuba
conventional short form: Cuba
local long form: Republica de Cuba
local short form: Cuba
Government type: Communist state
Capital: *name:* Havana
geographic coordinates: 23 07 N, 82 21 W
time difference: UTC-5 (same time as Washington, DC during Standard Time)
daylight saving time: +1hr, begins third Sunday in March; ends last Sunday in October
Administrative divisions: 15 provinces (provincias, singular–provincia) and 1 special municipality* (municipio especial); Artemisa, Camaguey, Ciego de Avila, Cienfuegos, Granma, Guantanamo, Holguin, Isla de la Juventud*, La Habana, Las Tunas, Matanzas, Mayabeque, Pinar del Rio, Sancti Spiritus, Santiago de Cuba, Villa Clara
Independence: 20 May 1902 (from Spain 10 December 1898; administered by the US from 1898 to 1902); not acknowledged by the Cuban Government as a day of independence
National holiday: Triumph of the Revolution, 1 January (1959)
Constitution: 24 February 1976; amended July 1992 and June 2002
Legal system: civil law system based on Spanish civil code
International law organization participation: has not submitted an ICJ jurisdiction declaration; non-party state to the ICCt
Suffrage: 16 years of age; universal
Executive branch: *chief of state:* President of the Council of State and President of the Council of Ministers Gen. Raul CASTRO Ruz (president since 24 February 2008); First Vice President of the Council of State and First Vice President of the Council of Ministers Jose Ramon MACHADO Ventura (since 24 February 2008); note–the president is both the chief of state and head of government
head of government: President of the Council of State and President of the Council of Ministers Gen. Raul CASTRO Ruz (president since 24 February 2008); First Vice President of the Council of State and First Vice President of the Council of Ministers Jose Ramon MACHADO Ventura (since 24 February 2008)
cabinet: Council of Ministers proposed by the president of the Council of State and appointed by the National Assembly or the 31-member Council of State, elected by the assembly to act on its behalf when it is not in session
(For more information visit the World Leaders website)
elections: president and vice presidents elected by the National Assembly for a five-year term; election last held on 24 February 2008 (next to be held in 2013)
election results: Gen. Raul CASTRO Ruz elected president; percent of legislative vote–100%; Jose Ramon MACHADO Ventura elected vice president; percent of legislative vote–100%
Legislative branch: unicameral National Assembly of People's Power or Asemblea Nacional del Poder Popular (number of seats in the National Assembly is based on population; 614 seats; members elected directly from slates approved by special candidacy commissions to serve five-year terms)
elections: last held on 20 January 2008 (next to be held in January 2013)
election results: Cuba's Communist Party is the only legal party, and officially sanctioned candidates run unopposed
Judicial branch: People's Supreme Court or Tribunal Supremo Popular (president, vice presidents, and other judges are elected by the National Assembly)
Political parties and leaders: Cuban Communist Party or PCC [Fidel CASTRO Ruz, first secretary]
Political pressure groups and leaders: Human Rights Watch; National Association of Small Farmers
International organization participation: ACP, AOSIS, FAO, G-77, IAEA, ICAO, ICC, ICRM, IFAD, IFRCS, IHO, ILO, IMO, IMSO, Interpol, IOC, IOM

(observer), IPU, ISO, ITSO, ITU, LAES, LAIA, NAM, OAS (excluded from formal participation since 1962), OPANAL, OPCW, PCA, PetroCaribe, RG, UN, UNCTAD, UNESCO, UNIDO, Union Latina, UNWTO, UPU, WCO, WFTU, WHO, WIPO, WMO, WTO

Diplomatic representation in the US: none; note–Cuba has an Interests Section in the Swiss Embassy, headed by Principal Officer Jorge BOLANOS Suarez; address: Cuban Interests Section, Swiss Embassy, 2630 16th Street NW, Washington, DC 20009; *telephone:* [1] (202) 797-8518; *FAX:* [1] (202) 797-8521

Diplomatic representation from the US: none; note–the US has an Interests Section in the Swiss Embassy, headed by Chief of Mission Jonathan D. FARRAR; address: USINT, Swiss Embassy, Calzada between L and M Streets, Vedado, Havana; *telephone:* [53] (7) 833-3551 through 3559 (operator assistance required); *FAX:* [53] (7) 833-1653; protecting power in Cuba is Switzerland

Flag description: five equal horizontal bands of blue (top, center, and bottom) alternating with white; a red equilateral triangle based on the hoist side bears a white, five-pointed star in the center; the blue bands refer to the three old divisions of the island: central, occidental, and oriental; the white bands describe the purity of the independence ideal; the triangle symbolizes liberty, equality, and fraternity, while the red color stands for the blood shed in the independence struggle; the white star, called La Estrella Solitaria (the Lone Star) lights the way to freedom and was taken from the flag of Texas

note: design similar to the Puerto Rican flag, with the colors of the bands and triangle reversed

National anthem: *name:* "La Bayamesa" (The Bayamo Song)

lyrics/music: Pedro FIGUEREDO

note: adopted 1940; Pedro FIGUEREDO first performed "La Bayamesa" in 1868 during the Ten Years War against the Spanish; a leading figure in the uprising, FIGUEREDO was captured in 1870 and executed in front of a firing squad; just prior to the fusillade he is reputed to have shouted, "Morir por la Patria es vivir" (To die for the country is to live), a line from the anthem

ECONOMY

Economy—overview: The government continues to balance the need for economic loosening against a desire for firm political control. The government announced it would eliminate 500,000 state jobs by March 2011 and has expanded opportunities for self-employment. President Raul CASTRO said such changes were needed to update the economic model to ensure the survival of socialism. The government has introduced limited reforms, some initially implemented in the 1990s, to increase enterprise efficiency and alleviate serious shortages of food, consumer goods, and services. The average Cuban's standard of living remains at a lower level than before the downturn of the 1990s, which was caused by the loss of Soviet aid and domestic inefficiencies. Since late 2000, Venezuela has been providing oil on preferential terms, and it currently supplies about 100,000 barrels per day of petroleum products. Cuba has been paying for the oil, in part, with the services of Cuban personnel in Venezuela including some 30,000 medical professionals.

GDP (purchasing power parity): $114.1 billion (2010 est.)
country comparison to the world: 65
$112.4 billion (2009 est.)
$110.8 billion (2008 est.)
note: data are in 2010 US dollars

GDP (official exchange rate): $57.49 billion (2010 est.)

GDP—real growth rate: 1.5% (2010 est.)
country comparison to the world: 160
1.4% (2009 est.)
4.1% (2008 est.)

GDP—per capita (PPP): $9,900 (2010 est.)
country comparison to the world: 110
$9,800 (2009 est.)
$9,700 (2008 est.)
note: data are in 2010 US dollars

GDP—composition by sector:
agriculture: 4.2%
industry: 22.7%
services: 72.9% (2010 est.)

Labor force: 5.164 million
country comparison to the world: 72
note: state sector 78%, non-state sector 22% (2010 est.)

Labor force—by occupation:
agriculture: 20%
industry: 19.4%
services: 60.6% (2005)

Unemployment rate: 2% (2010 est.)
country comparison to the world: 12
1.7% (2009 est.)

Population below poverty line: NA%

Household income or consumption by percentage share:
lowest 10%: NA%
highest 10%: NA%

Investment (gross fixed): 10.5% of GDP (2010 est.)
country comparison to the world: 148

Budget: *revenues:* $46.51 billion
expenditures: $48.89 billion (2010 est.)

Public debt: 34.4% of GDP (2010 est.)
country comparison to the world: 84
34.7% of GDP (2009 est.)

Inflation rate (consumer prices): 0.7% (2010 est.)
country comparison to the world: 14
-0.5% (2009 est.)

Central bank discount rate: NA%

Commercial bank prime lending rate: NA%

Stock of narrow money: $11.57 billion (31 December 2010 est.)
country comparison to the world: 69
$11.74 billion (31 December 2009 est.)

Stock of broad money: $35.92 billion (31 December 2010 est.)
country comparison to the world: 71
$35.61 billion (31 December 2009 est.)

Stock of domestic credit: $NA

Agriculture—products: sugar, tobacco, citrus, coffee, rice, potatoes, beans; livestock

Industries: sugar, petroleum, tobacco, construction, nickel, steel, cement, agricultural machinery, pharmaceuticals

Industrial production growth rate: 0.8% (2010 est.)
country comparison to the world: 149

Electricity—production: 16.89 billion kWh (2007 est.)
country comparison to the world: 72

Electricity—consumption: 13.93 billion kWh (2007 est.)
country comparison to the world: 77

Electricity—exports: 0 kWh (2008 est.)

Electricity—imports: 0 kWh (2008 est.)

Oil—production: 48,340 bbl/day (2009 est.)
country comparison to the world: 64

Oil—consumption: 169,000 bbl/day (2009 est.)
country comparison to the world: 60

Oil—exports: 0 bbl/day (2007 est.)
country comparison to the world: 157

Oil—imports: 104,800 bbl/day (2007 est.)
country comparison to the world: 63

Oil—proved reserves: 178.9 million bbl (1 January 2010 est.)
country comparison to the world: 62

Natural gas—production: 400 million cu m (2008 est.)
country comparison to the world: 68

Natural gas—consumption: 400 million cu m (2008 est.)
country comparison to the world: 96

Natural gas—exports: 0 cu m (2008 est.)
country comparison to the world: 80

Natural gas—imports: 0 cu m (2008 est.)
country comparison to the world: 171

Natural gas—proved reserves: 70.79 billion cu m (1 January 2010 est.)
country comparison to the world: 57

Current account balance: $-87 million (2010 est.)
country comparison to the world: 76
$539 million (2009 est.)

Exports: $3.311 billion (2010 est.)
country comparison to the world: 118
$2.879 billion (2009 est.)

Exports—commodities: sugar, nickel, tobacco, fish, medical products, citrus, coffee

Exports—partners: China 25.68%, Canada 20.31%, Spain 6.79%, Netherlands 4.53% (2009)

Imports: $10.25 billion (2010 est.)
country comparison to the world: 88
$8.91 billion (2009 est.)

Imports—commodities: petroleum, food, machinery and equipment, chemicals

Imports—partners: Venezuela 30.51%, China 15.48%, Spain 8.3%, US 6.87% (2009)

Reserves of foreign exchange and gold: $4.847 billion (31 December 2010 est.)
country comparison to the world: 71
$4.647 billion (31 December 2009 est.)

Debt—external: $19.75 billion (31 December 2010 est.)
country comparison to the world: 71
$19.42 billion (31 December 2009 est.)

Stock of direct foreign investment—at home: $NA (31 December 2009 est.)

Stock of direct foreign investment—abroad: $4.138 billion (2006 est.)
country comparison to the world: 61

Exchange rates: Cuban pesos (CUP) per US dollar–0.9259 (2010), 0.9259 (2009),

0.9259 (2008), 0.9259 (2007), 0.9231 (2006)

COMMUNICATIONS

Telephones—main lines in use: 1.168 million (2009)
country comparison to the world: 72
Telephones—mobile cellular: 443,000 (2009)
country comparison to the world: 162
Telephone system: *general assessment:* greater investment beginning in 1994 and the establishment of a new Ministry of Information Technology and Communications in 2000 has resulted in improvements in the system; national fiber-optic system under development; 95% of switches digitized by end of 2006; mobile-cellular telephone service is expensive and must be paid in convertible pesos, which effectively limits subscribership
domestic: fixed-line density remains low at less than 10 per 100 inhabitants; mobile-cellular service expanding but remains less than 5 per 100 persons
international: country code–53; fiber-optic cable laid to but not linked to US network; satellite earth station–1 Intersputnik (Atlantic Ocean region) (2009)
Broadcast media: government owns and controls all broadcast media with private ownership of electronic media prohibited; government operates 4 national TV networks and many local TV stations; government operates 6 national radio networks, an international station, and many local radio stations; Radio-TV Marti is beamed from the US (2007)
Internet country code: .cu
Internet hosts: 3,025 (2010)
country comparison to the world: 145
Internet users: 1.606 million
country comparison to the world: 79
note: private citizens are prohibited from buying computers or accessing the Internet without special authorization; foreigners may access the Internet in large hotels but are subject to firewalls; some Cubans buy illegal passwords on the black market or take advantage of public outlets to access limited email and the government-controlled "intranet" (2009)

TRANSPORTATION

Airports: 136 (2010)
country comparison to the world: 43
Airports—with paved runways: *total:* 65
over 3,047 m: 7
2,438 to 3,047 m: 9
1,524 to 2,437 m: 17
914 to 1,523 m: 5
under 914 m: 27 (2010)
Airports—with unpaved runways: *total:* 71
914 to 1,523 m: 13
under 914 m: 58 (2010)
Pipelines: gas 41 km; oil 230 km (2010)
Railways: *total:* 8,598 km
country comparison to the world: 25
standard gauge: 8,322 km 1.435-m gauge (124 km electrified)
narrow gauge: 276 km 1.000-m gauge
note: 4,533 km of the track is used by sugar plantations; 4,257 km is standard gauge; 276 km is narrow gauge (2009)
Roadways: *total:* 60,858 km
country comparison to the world: 73
paved: 29,820 km (includes 638 km of expressway)
unpaved: 31,038 km (2000)
Waterways: 240 km (almost all navigable inland waterways are near the mouths of rivers) (2010)
country comparison to the world: 95
Merchant marine: *total:* 5
country comparison to the world: 130
by type: cargo 2, passenger 1, refrigerated cargo 2
registered in other countries: 6 (Cyprus 1, former Netherlands Antilles 1, Panama 4) (2010)
Ports and terminals: Antilla, Cienfuegos, Guantanamo, Havana, Matanzas, Mariel, Nuevitas Bay, Santiago de Cuba, Tanamo

MILITARY

Military branches: Revolutionary Armed Forces (Fuerzas Armadas Revolucionarias, FAR): Revolutionary Army (Ejercito Revolucionario, ER, includes Territorial Militia Troops (Milicia de Tropas de Territoriales, MTT)); Revolutionary Navy (Marina de Guerra Revolucionaria, MGR, includes Marine Corps); Revolutionary Air and Air Defense Forces (DAAFAR), Youth Labor Army (Ejercito Juvenil del Trabajo, EJT) (2011)
Military service age and obligation: 17-28 years of age for compulsory military service; 2-year service obligation; both sexes subject to military service (2006)
Manpower available for military service: *males age 16–49:* 2,998,201
females age 16–49: 2,919,107 (2010 est.)
Manpower fit for military service: *males age 16–49:* 2,446,131
females age 16–49: 2,375,590 (2010 est.)
Manpower reaching militarily significant age annually:
male: 72,823
female: 69,108 (2010 est.)
Military expenditures: 3.8% of GDP (2006 est.)
country comparison to the world: 30
Military—note: the collapse of the Soviet Union deprived the Cuban military of its major economic and logistic support and had a significant impact on the state of Cuban equipment; the army remains well trained and professional in nature; while the lack of replacement parts for its existing equipment has increasingly affected operational capabilities, Cuba remains able to offer considerable resistance to any regional power (2010)

TRANSNATIONAL ISSUES

Disputes—international: US Naval Base at Guantanamo Bay is leased to US and only mutual agreement or US abandonment of the facility can terminate the lease
Trafficking in persons: current situation: Cuba is principally a source country for children subjected to trafficking in persons, specifically commercial sexual exploitation within the country; the scope of trafficking within Cuba is difficult to gauge due to the closed nature of the government and sparse non-governmental or independent reporting
tier rating: Tier 3–Cuba does not fully comply with the minimum standards for the elimination of trafficking and is not making significant efforts to do so; in a positive step, the Government of Cuba shared information about human trafficking and its efforts to address the issue; the government did not prohibit all forms of trafficking during the reporting period, nor did it provide specific evidence that it prosecuted and punished trafficking offenders, protected victims of all forms of trafficking, or implemented victim protection policies or programs to prevent human trafficking (2010)
Illicit drugs: territorial waters and air space serve as transshipment zone for US- and European-bound drugs; established the death penalty for certain drug-related crimes in 1999 (2008)

CURACAO

(PART OF THE KINGDOM OF THE NETHERLANDS)

INTRODUCTION

Background: Originally settled by Arawak Indians, Curacao was seized by the Dutch in 1634 along with the neighboring island of Bonaire. Once the center of the Caribbean slave trade, Curacao was hard hit by the abolition of slavery in 1863. Its prosperity (and that of neighboring Aruba) was restored in the early 20th century with the construction of the Isla Refineria to service the newly discovered Venezuelan oil fields. In 1954, Curacao and several other Dutch Caribbean possessions were reorganized as the Netherlands Antilles, part of the Kingdom of the Netherlands. In referenda in 2005 and 2009, the citizens of Curacao voted to become a self-governing country within the Kingdom of the Netherlands. The change in status became effective in October of 2010 with the dissolution of the Netherlands Antilles.

GEOGRAPHY

Location: Caribbean, an island in the Caribbean Sea–55 km off the coast of Venezuela
Geographic coordinates: 12 10 N, 69 00 W
Map references: Central America and the Caribbean
Area:
total: 444 sq km
country comparison to the world: 198
land: 444 sq km
water: 0 sq km
Area—comparative: more than two times the size of Washington, DC
Land boundaries: 0 km
Coastline: 364 km
Maritime claims: territorial sea: 12 nm

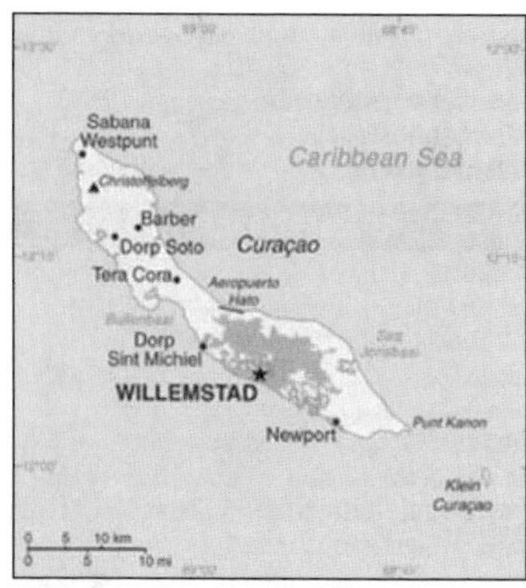

exclusive fishing zone: 12 nm
Climate: tropical marine climate, ameliorated by northeast trade winds, results in mild temperatures; semi-arid with average rainfall of 600 mm/year
Terrain: generally low, hilly terrain
Elevation extremes: *lowest point:* Caribbean Sea 0 m
highest point: Mt. Christoffel, 372m
Natural resources: calcium phosphates, aloes, sorghum, peanuts, vegetables, tropical fruit
Land use: *arable land:* 10%
permanent crops: 0%
other: 90%
Irrigated land: NA
Total renewable water resources: NA
Freshwater withdrawal (domestic/industrial/agricultural): NA
Natural hazards: Curacao is south of the Caribbean hurricane belt and is rarely threatened
Environment—current issues: NA
Geography—note: Curacao is a part of the Windward Islands (southern) group

PEOPLE

Population: 142,180 (est. January 2010)
country comparison to the world: 186
Age structure: *0–14 years:* 21.1% (males 15,337/females 14,589)
15–64 years: 66.7% (males 42,896/females 51,998)
65 years and over: 12.2% (males 6,972/females 10,388) (2010)
Population growth rate: NA
Birth rate: NA
Death rate: 8 deaths/1,000 population (2009)
country comparison to the world: 104
Net migration rate: 1.27 migrant(s)/1,000 population (2008)
country comparison to the world: 49
Sex ratio: *at birth:* 1.15 male(s)/female
under 15 years: 1.05 male(s)/female
15–64 years: 0.82 male(s)/female
65 years and over: 0.67 male(s)/female
total population: 0.85 male(s)/female (2010)
Life expectancy at birth: *total:* NA
males: 72.4 years
females: 80.1 years (2009)
Total fertility rate: 2.1 children born/woman (2009)
country comparison to the world: 118
HIV/AIDS—adult prevalence rate: NA
HIV/AIDS—people living with HIV/AIDS: NA
HIV/AIDS—deaths: NA
Religions: Roman Catholic 80.1%, Protestant 5.5%, none 4.6%, Pentecostal 3.5%, Seventh Day Adventist 2.2%, Jehovah's Witnesses 1.7%, Jewish 0.8%, other 1.3%, not reported 0.3% (2001 census)
Languages: Papiamentu (a Spanish-Portuguese-Dutch-English dialect) 81.2%, Dutch (official) 8%, Spanish 4%, English 2.9%, other 3.9% (2001 census)
School life expectancy (primary to tertiary education): NA
Education expenditures: NA

GOVERNMENT

Country name: Dutch long form: Land Curacao
Dutch short form: Curacao
Papiamentu long form: Pais Korsou
Papiamentu short form: Korsou
former: Netherlands Antilles; Curacao and Dependencies
Dependency status: constituent country within the Kingdom of the Netherlands; full autonomy in internal affairs granted in 2010; Dutch Government responsible for defense and foreign affairs
Government type: parliamentary
Capital: *name:* Willemstad
geographic coordinates: 12 06 N, 68 55 W
time difference: UTC-4 (1 hour ahead of Washington, DC during Standard Time)
Administrative divisions: none (part of the Kingdom of the Netherlands)
Independence: none (part of the Kingdom of the Netherlands)
National holiday: Queen's Day (Birthday of Queen-Mother JULIANA and accession to the throne of her oldest daughter BEATRIX), 30 April (1909 and 1980)
Constitution: Staatsregeling, 10 October 2010; revised Kingdom Charter pending
Legal system: based on Dutch civil law system with some English common law influence
Suffrage: 18 years of age; universal
Executive branch: *chief of state:* Queen BEATRIX of the Netherlands (since 30 April 1980); represented by Governor General Frits GOEDGEDRAG (since 10 October 2010)
head of government: Prime Minister Gerrit SCHOTTE (since 10 October 2010)
cabinet: Executive Council
(For more information visit the World Leaders website)
elections: the monarch is hereditary; governor general appointed by the monarch; following legislative elections, the leader of the majority party is usually elected prime minister by the parliament
Legislative branch: unicameral parliament or Staten (21 seats; members elected by popular vote for four year terms)
elections: last held 27 August 2010 (next to be held in 2014)
election results: percent of vote by party—PAR 30%, MFK 21%, PS 19%, MAN 9%, FOL 7%, PNP 6%; seats by party—PAR 8, MFK 5, PS 4, MAN 2, FOL 1, PNP 1
Judicial branch: Common Court of Justice, Joint High Court of Justice (judges appointed by the monarch)
Political parties and leaders: Frente Obrero Liberashon (Workers' Liberation Front) or FOL [Anthony GODETT]; Movimentu Antiyas Nobo (New Antilles Movement) or MAN [Charles COOPER]; Movementu Futuro Korsou or MFK [Gerrit SCHOTTE]; Partido Antia Restruktura or PAR [Emily DE JONGH-ELHAGE]; People's National Party or PNP [Ersilia DE LANNOOY]; Pueblo Soberano or PS [Herman WIELS]
Diplomatic representation in the US: none (represented by the Kingdom of the Netherlands)
Diplomatic representation from the US: *chief of mission:* Consul General Valerie BELON
consulate(s) general: J. B. Gorsiraweg #1, Willemstad, Curacao
mailing address: P. O. Box 158, Willemstad, Curacao
telephone: [599] (9) 4613066
FAX: [599] (9) 4616489
Flag description: on a blue field a horizontal yellow band somewhat below the center divides the flag into proportions of 5:1:2; two five-pointed white stars—the smaller above and to the left of the larger—appear in the canton; the blue of the upper and lower sections symbolizes the sky and sea respectively; yellow represents the sun; the stars symbolize Curacao and its uninhabited smaller sister island of Klein Curacao; the five star points signify the five continents from which Curacao's people derive
National anthem: *name:* Himmo di Korsou (Anthem of Curacao)
lyrics/music: Guillermo ROSARIO, Mae HENRIQUEZ, Enrique MULLER, Betty DORAN/Frater Candidus NOWENS, Errol "El Toro" COLINA
note: adapted 1978; the lyrics, originally written in 1899, were rewritten in 1978 to make them less colonial in nature

ECONOMY

Economy—overview: Tourism, petroleum refining, and offshore finance are the mainstays of this small economy, which is closely tied to the outside world. Although GDP grew slightly during the past decade, the island enjoys a high per capita income and a well-developed infrastructure compared with other countries in the region. Curacao has an excellent natural harbor that can accommodate large oil tankers. The Venezuelan state oil company leases the single refinery on the island from the government; most of the oil for the refinery is imported from Venezuela; most of the refined products are exported to the US. Almost all consumer and capital goods are imported, with the US, Brazil, Italy, and Mexico being the major suppliers. The government is attempting to diversify its industry and trade and has signed an Association Agreement with the EU to expand business there. Poor soils and inadequate water supplies hamper the development of agriculture. Budgetary problems complicate reform of the health and pension systems for an aging population.
GDP (purchasing power parity): $2.838 billion (2008 est.)
country comparison to the world: 177
$2.606 billion (2007 est.)
$2.452 billion (2006 est.)

note: data are in 2008 US dollars
GDP (official exchange rate): $5.08 billion (2008 est.)
GDP—real growth rate: 3.5% (2008)
country comparison to the world: 109
2.2% (2007)
GDP—percapita(PPP):$15,000(2004est.)
country comparison to the world: 75
GDP—composition by sector:
agriculture: 1%
industry: 15%
services: 84% (2000 est.)
Labor force: 63,000 (2008 est.)
country comparison to the world: 186
Labor force—by occupation:
agriculture: 1.2%
industry: 16.9%
services: 81.8% (2008 est.)
Unemployment rate: 10.3% (2008 est.)
country comparison to the world: 112
Inflation rate (consumer prices): 1.7% (2009 est.)
country comparison to the world: 44
6.8% (2008 est.)
Agriculture—products: aloe, sorghum, peanuts, vegetables, tropical fruit
Industries: tourism, petroleum refining, petroleum transshipment facilities, light manufacturing
Industrial production growth rate: NA%
Exports: $876 million (2008 est.)
country comparison to the world: 153
note: excludes oil
Exports—commodities: petroleum products
Exports—partners: US 13.1%, Guatemala 10.8%, Singapore 10.7%, Dominican Republic 9.6%, Haiti 7.6%, The Bahamas 6.1%, Honduras 4.5%, Mexico 4.2% (2009 est.)
Imports: $1.34 billion (2008 est.)
country comparison to the world: 166
Imports—commodities: crude petroleum, food, manufactures
Imports—partners: Venezuela 57.3%, US 19.2%, Brazil 8.1% (2009 est.)
Exchange rates: Netherlands Antillean guilders (ANG) per US dollar–1.79 (2010),
1.79 (2009), 1.79 (2008), 1.79 (2007), 1.79 (2006)

COMMUNICATIONS

Telephones—main lines in use: NA
Telephones—mobile cellular: NA
Telephone system: *general assessment:* NA
domestic: NA
international: country code–599
Broadcast media: government-run Telecuracao operates a TV station and a radio station; several privately-owned radio stations
Internet country code: .cw
Internet hosts: NA
Internet users: NA

TRANSPORTATION

Airports: 1
country comparison to the world: 213
Airports—with paved runways:
total: 1
over 3,047 m: 1 (2010)
Roadways: *total:* 550 km
country comparison to the world: 191
Ports and terminals: Bullen Baai, Fuik Bay, Willemstad

MILITARY

Military branches: the Royal Netherlands Navy maintains a permanent and active presence in the region from its main operating base on Curacao; other local security forces include a coast guard, paramilitary National Guard (Vrijwilligers Korps Curacao), and Police Force (2010)
Military service age and obligation: no conscription (2010)
Military—note: defense is the responsibility of the Kingdom of the Netherlands

CYPRUS

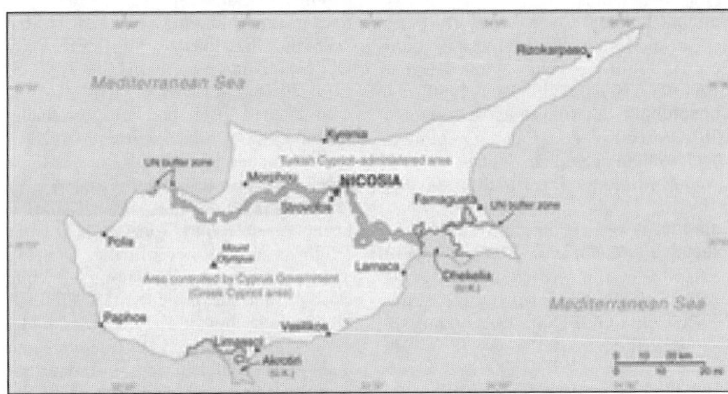

INTRODUCTION

Background: A former British colony, Cyprus became independent in 1960 following years of resistance to British rule. Tensions between the Greek Cypriot majority and Turkish Cypriot minority came to a head in December 1963, when violence broke out in the capital of Nicosia. Despite the deployment of UN peacekeepers in 1964, sporadic intercommunal violence continued forcing most Turkish Cypriots into enclaves throughout the island. In 1974, a Greek Government-sponsored attempt to seize control of Cyprus was met by military intervention from Turkey, which soon controlled more than a third of the island. In 1983, the Turkish Cypriot-occupied area declared itself the "Turkish Republic of Northern Cyprus" ("TRNC"), but it is recognized only by Turkey. The election of a new Cypriot president in 2008 served as the impetus for the UN to encourage both the Greek Cypriot and Turkish Cypriot communities to reopen unification negotiations. In September 2008, the leaders of the two communities began negotiations under UN auspices aimed at reuniting the divided island; the talks remain ongoing. The entire island entered the EU on 1 May 2004, although the EU acquis–the body of common rights and obligations–applies only to the areas under the internationally recognized government, and is suspended in the areas administered by Turkish Cypriots. However, individual Turkish Cypriots able to document their eligibility for Republic of Cyprus citizenship legally enjoy the same rights accorded to other citizens of European Union states.

GEOGRAPHY

Location: Middle East, island in the Mediterranean Sea, south of Turkey
Geographic coordinates:
35 00 N, 33 00 E
Map references: Europe
Area: *total:* 9,251 sq km (of which 3,355 sq km are in north Cyprus)
country comparison to the world: 170
land: 9,241 sq km
water: 10 sq km
Area—comparative: about 0.6 times the size of Connecticut
Land boundaries: *total:* 150.4 km (approximately)
border sovereign base areas: Akrotiri 47.4 km, Dhekelia 103 km (approximately)
Coastline: 648 km
Maritime claims: territorial sea: 12 nm
contiguous zone: 24 nm
continental shelf: 200 m depth or to the depth of exploitation
Climate: temperate; Mediterranean with hot, dry summers and cool winters
Terrain: central plain with mountains to north and south; scattered but significant plains along southern coast
Elevation extremes: *lowest point:* Mediterranean Sea 0 m
highest point: Mount Olympus 1,951 m
Natural resources: copper, pyrites, asbestos, gypsum, timber, salt, marble, clay earth pigment
Land use: *arable land:* 10.81%
permanent crops: 4.32%
other: 84.87% (2005)
Irrigated land: 400 sq km (2003)
Total renewable water resources: 0.4 cu km (2005)
Freshwater withdrawal (domestic/industrial/agricultural): *total:* 0.21 cu km/yr (27%/1%/71%)
per capita: 250 cu m/yr (2000)

Natural hazards: moderate earthquake activity; droughts
Environment—current issues: water resource problems (no natural reservoir catchments, seasonal disparity in rainfall, sea water intrusion to island's largest aquifer, increased salination in the north); water pollution from sewage and industrial wastes; coastal degradation; loss of wildlife habitats from urbanization
Environment—international agreements: *party to:* Air Pollution, Air Pollution-Nitrogen Oxides, Air Pollution-Persistent Organic Pollutants, Air Pollution-Sulfur 94, Biodiversity, Climate Change, Climate Change-Kyoto Protocol, Desertification, Endangered Species, Environmental Modification, Hazardous Wastes, Law of the Sea, Marine Dumping, Ozone Layer Protection, Ship Pollution, Wetlands
signed, but not ratified: none of the selected agreements
Geography—note: the third largest island in the Mediterranean Sea (after Sicily and Sardinia)

PEOPLE

Population: 1,120,489 (July 2011 est.)
country comparison to the world: 158
Age structure: *0–14 years:* 16.2% (male 93,280/female 88,022)
15–64 years: 73.4% (male 427,752/female 394,578)
65 years and over: 10.4% (male 50,761/female 66,096) (2011 est.)
Median age:
total: 34.8 years
male: 33.5 years
female: 36.6 years (2011 est.)
Population growth rate: 1.617% (2011 est.)
country comparison to the world: 70
Birth rate: 11.41 births/1,000 population (2011 est.)
country comparison to the world: 170
Death rate: 6.45 deaths/1,000 population (July 2011 est.)
country comparison to the world: 151
Net migration rate: 11.21 migrant(s)/1,000 population (2011 est.)
country comparison to the world: 8
Urbanization: *urban population:* 70% of total population (2010)
rate of urbanization: 1.3% annual rate of change (2010-15 est.)
Major cities—population: NICOSIA (capital) 240,000 (2009)
Sex ratio: *at birth:* 1.05 male(s)/female
under 15 years: 1.06 male(s)/female
15–64 years: 1.08 male(s)/female
65 years and over: 0.77 male(s)/female
total population: 1.04 male(s)/female (2011 est.)
Infant mortality rate: *total:* 9.38 deaths/1,000 live births
country comparison to the world: 154
male: 11.23 deaths/1,000 live births
female: 7.43 deaths/1,000 live births (2011 est.)
Life expectancy at birth: *total population:* 77.82 years
country comparison to the world: 53
male: 75.04 years
female: 80.74 years (2011 est.)
Total fertility rate: 1.45 children born/woman (2011 est.)
country comparison to the world: 192
HIV/AIDS—adult prevalence rate: 0.1% (2003 est.)
country comparison to the world: 125
HIV/AIDS—people living with HIV/AIDS: fewer than 1,000 (2007 est.)
country comparison to the world: 140
HIV/AIDS—deaths: NA
Drinking water source: *Improved:*
urban: 100% of population
rural: 100% of population
total: 100% of population (2008)
Sanitation facility access: *Improved:*
urban: 100% of population
rural: 100% of population
total: 100% of population (2008)
Nationality: *noun:* Cypriot(s)
adjective: Cypriot
Ethnic groups: Greek 77%, Turkish 18%, other 5% (2001)
Religions: Greek Orthodox 78%, Muslim 18%, other (includes Maronite and Armenian Apostolic) 4%
Languages: Greek (official), Turkish (official), English
Literacy: *definition:* age 15 and over can read and write
total population: 97.6%
male: 98.9%
female: 96.3% (2001 census)
School life expectancy (primary to tertiary education): *total:* 14 years
male: 14 years
female: 14 years (2008)
Education expenditures: 4.1% of GDP (2007)
country comparison to the world: 100

GOVERNMENT

Country name: *conventional long form:* Republic of Cyprus
conventional short form: Cyprus
local long form: Kypriaki Dimokratia/Kibris Cumhuriyeti
local short form: Kypros/Kibris
note: the Turkish Cypriot community, which administers the northern part of the island, refers to itself as the "Turkish Republic of Northern Cyprus" ("TRNC")
Government type: republic
note: a separation of the two ethnic communities inhabiting the island began following the outbreak of communal strife in 1963; this separation was further solidified after the Turkish intervention in July 1974, following a Greek military-junta-supported coup attempt that gave the Turkish Cypriots de facto control in the north; Greek Cypriots control the only internationally recognized government; on 15 November 1983 Turkish Cypriot "President" Rauf DENKTASH declared independence and the formation of a "Turkish Republic of Northern Cyprus" ("TRNC"), which is recognized only by Turkey
Capital: *name:* Nicosia (Lefkosia)
geographic coordinates: 35 10 N, 33 22 E
time difference: UTC+2 (7 hours ahead of Washington, DC during Standard Time)
daylight saving time: +1hr, begins last Sunday in March; ends last Sunday in October
Administrative divisions: 6 districts; Famagusta, Kyrenia, Larnaca, Limassol, Nicosia, Paphos; note—Turkish Cypriot area's administrative divisions include Kyrenia, all but a small part of Famagusta, and small parts of Nicosia (Lefkosia) and Larnaca
Independence: 16 August 1960 (from the UK); note—Turkish Cypriots proclaimed self-rule on 13 February 1975 and independence in 1983, but these proclamations are only recognized by Turkey
National holiday: Independence Day, 1 October (1960); note—Turkish Cypriots celebrate 15 November (1983) as "Independence Day"
Constitution: 16 August 1960
note: from December 1963, the Turkish Cypriots no longer participated in the government; negotiations to create the basis for a new or revised constitution to govern the island and for better relations between Greek and Turkish Cypriots have been held intermittently since the mid-1960s; in 1975, following the 1974 Turkish intervention, Turkish Cypriots created their own constitution and governing bodies within the "Turkish Federated State of Cyprus," which they then called the "Turkish Republic of Northern Cyprus (TRNC)" when the Turkish Cypriots declared independence in 1983; a new constitution for the "TRNC" passed by referendum on 5 May 1985, although the "TRNC" remains unrecognized by any country other than Turkey
Legal system: mixed legal system of English common law and civil law with Greek Orthodox religious law influence
International law organization participation: accepts compulsory ICJ jurisdiction with reservations; accepts ICCt jurisdiction
Suffrage: 18 years of age; universal
Executive branch: *chief of state:* President Demetris CHRISTOFIAS (since 28 February 2008); note—the president is both the chief of state and head of government; post of vice president is currently vacant; under the 1960 constitution, the post is reserved for a Turkish Cypriot
head of government: President Demetris CHRISTOFIAS (since 28 February 2008)
cabinet: Council of Ministers appointed jointly by the president and vice president (For more information visit the World Leaders website)
elections: president elected by popular vote for a five-year term; election last held on 17 and 24 February 2008 (next to be held in February 2013)
election results: Demetris CHRISTOFIAS elected president; percent of vote (first round)—Ioannis KASOULIDES 33.5%, Demetris CHRISTOFIAS 33.3%, Tassos PAPADOPOULOS 31.8%, other 1.4%; (second round) Demetris CHRISTOFIAS 53.4%, Ioannis KASOULIDES 46.6%
note: Dervis EROGLU became "president" of the "TRNC" on 23 April 2010 after "presidential" elections on 18 April 2010; results—Dervis EROGLU 50.4%, Mehmet Ali TALAT 42.9%; Irsen KUCUK is "TRNC acting prime minister"
Legislative branch: unicameral—area under government control: House of Representatives or Vouli Antiprosopon (80 seats; 56 assigned to the Greek Cypriots, 24 to Turkish Cypriots; note—

only those assigned to Greek Cypriots are filled; members are elected by popular vote to serve five-year terms); area administered by Turkish Cypriots: Assembly of the Republic or Cumhuriyet Meclisi (50 seats; members elected by popular vote to serve five-year terms)
elections: area under government control: last held on 22 May 2011 (next to be held in May 2016); area administered by Turkish Cypriots: last held on 19 April 2009 (next to be held in 2014)
election results: area under government control: House of Representatives–percent of vote by party–DISY 34.3%, AKEL 32.7%, DIKO 15.8%, EDEK 8.9%, EURO.KO 3.9%, other 4.5%; seats by party–DISY 20, AKEL 19, DIKO 9, EDEK 5, EURO.KO 2, other 1; area administered by Turkish Cypriots: Assembly of the Republic–percent of vote by party–UBP 44.1%, CTP 29.3%, DP 10.6%, other 16%; seats by party–UBP 26, CTP 15, DP 5, other 4
Judicial branch: Supreme Court (judges are appointed jointly by the president and vice president); subordinate courts
note: there is also a "Supreme Court" in the area administered by Turkish Cypriots
Political parties and leaders: area under government control: Democratic Party or DIKO [Marios KAROYIAN]; Democratic Rally or DISY [Nikos ANASTASIADES]; European Party or EURO.KO [Demetris SYLLOURIS]; Fighting Democratic Movement or ADIK [Dinos MIKHAILIDES]; Green Party of Cyprus [George PERDIKIS]; Movement for Social Democrats or EDEK [Yiannakis OMIROU]; Progressive Party of the Working People or AKEL (Communist Party) [Andros KYPRIANOU]; United Democrats or EDI [Praxoula ANTONIADOU]
area administered by Turkish Cypriots: Communal Democracy Party or TDP [Mehmet CAKICI]; Cyprus Socialist Party or KSP [Yusuf ALKIM]; Democratic Party or DP [Serdaer DENKTASH]; Freedom and Reform Party or ORP [Turgay AVCI]; National Unity Party or UBP [Irsen KUCUK]; Nationalist Justice Party or MAP [Ata TEPE]; New Cyprus Party or YKP [Murat KANATLI]; Politics for the People Party or HIS [Ahmet YONLUER]; Republican Turkish Party or CTP [Ferdi Sabit SOYER]; United Cyprus Party or BKP [Izzet IZCAN]
Political pressure groups and leaders: Confederation of Cypriot Workers or SEK (pro-West); Confederation of Revolutionary Labor Unions or Dev-Is; Federation of Turkish Cypriot Labor Unions or Turk-Sen; Pan-Cyprian Labor Federation or PEO (Communist controlled)
International organization participation: Australia Group, C, CE, EBRD, EIB, EMU, EU, FAO, IAEA, IBRD, ICAO, ICC, IDA, IFAD, IFC, IHO, ILO, IMF, IMO, IMSO, Interpol, IOC, IOM, IPU, ISO, ITSO, ITU, ITUC, MIGA, NSG, OAS (observer), OIF (associate member), OPCW, OSCE, PCA, UN, UNCTAD, UNESCO, UNHCR, UNIDO, UNIFIL, UNWTO, UPU, WCO, WFTU, WHO, WIPO, WMO, WTO
Diplomatic representation in the US: *chief of mission:* Ambassador Pavlos ANASTASIADES
chancery: 2211 R Street NW, Washington, DC 20008
telephone: [1] (202) 462-5772, 462-0873
FAX: [1] (202) 483-6710
consulate(s) general: New York
note: representative of the Turkish Cypriot community in the US is Dilek Yavuz YANIK; office at 1667 K Street NW, Washington, DC; telephone [1] (202) 887-6198
Diplomatic representation from the US: *chief of mission:* Ambassador Frank C. URBANCIC, Jr.
embassy: corner of Metochiou and Ploutarchou Streets, 2407 Engomi, Nicosia
mailing address: P. O. Box 24536, 1385 Nicosia
telephone: [357] (22) 393939
FAX: [357] (22) 780944
Flag description: white with a copper-colored silhouette of the island (the name Cyprus is derived from the Greek word for copper) above two green crossed olive branches in the center of the flag; the branches symbolize the hope for peace and reconciliation between the Greek and Turkish communities
note: the "Turkish Republic of Northern Cyprus" flag retains the white field of the Cyprus national flag but displays narrow horizontal red stripes positioned a small distance from the top and bottom edges between which are centered a red crescent and a red five-pointed star; the banner is modeled after the Turkish national flag but with the colors reversed
National anthem: *name:* "Ymnos eis tin Eleftherian" (Hymn to Liberty)
lyrics/music: Dionysios SOLOMOS/ Nikolaos MANTZAROS
note: adopted 1960; Cyprus adopted the Greek national anthem as its own; the Turkish community in Cyprus uses the anthem of Turkey

ECONOMY

Economy—overview: The area of the Republic of Cyprus under government control has a market economy dominated by the service sector, which accounts for nearly four-fifths of GDP. Tourism, financial services, and real estate are the most important sectors. Erratic growth rates over the past decade reflect the economy's reliance on tourism, the profitability of which often fluctuates with political instability in the region and economic conditions in Western Europe. Nevertheless, the economy in the area under government control has grown at a rate well above the EU average since 2000. Cyprus joined the European Exchange Rate Mechanism (ERM2) in May 2005 and adopted the euro as its national currency on 1 January 2008. An aggressive austerity program in the preceding years, aimed at paving the way for the euro, helped turn a soaring fiscal deficit (6.3% in 2003) into a surplus of 1.2% in 2008, and reduced inflation to 4.7%. This prosperity came under pressure in 2009, as construction and tourism slowed in the face of reduced foreign demand triggered by the ongoing global financial crisis. Although Cyprus lagged behind its EU peers in showing signs of stress from the global crisis, the economy tipped into recession in 2009, contracting by 1.8%, and has been slow to bounce back since, posting an anemic growth rate of 0.6% in 2010. In addition, the budget deficit is on the rise and reached 5.7% of GDP in 2010, a violation of the EU's budget deficit criteria of no more than 3% of GDP. In response to the country's deteriorating finances, Nicosia is promising to implement measures to cut the cost of the state payroll, curb tax evasion, and revamp social benefits. However, it has been slow to act, lacking a consensus in parliament and among the social partners for its proposed measures.
GDP (purchasing power parity): $23.19 billion (2010 est.)
country comparison to the world: 118
$22.95 billion (2009 est.)
$23.34 billion (2008 est.)
note: data are in 2010 US dollars
GDP (official exchange rate): $23.17 billion (2010 est.)
GDP—real growth rate: 1% (2010 est.)
country comparison to the world: 171
-1.7% (2009 est.)
3.6% (2008 est.)
GDP—per capita (PPP): $21,000 (2010 est.)
country comparison to the world: 62
$21,200 (2009 est.)
$21,900 (2008 est.)
note: data are in 2010 US dollars
GDP—composition by sector:
agriculture: 2.1%
industry: 18.6%
services: 79.3% (2010 est.)
Labor force: 400,000 (2010 est.)
country comparison to the world: 158
Labor force—by occupation:
agriculture: 8.5%
industry: 20.5%
services: 71% (2006 est.)
Unemployment rate: 5.6% (2010 est.)
country comparison to the world: 56
4.3% (2009 est.)
Population below poverty line: NA%
Household income or consumption by percentage share:
lowest 10%: NA%
highest 10%: NA%
Distribution of family income—Gini index: 29 (2005)
country comparison to the world: 119
Investment (gross fixed): 19.7% of GDP (2010 est.)
country comparison to the world: 92
Budget: *revenues:*: $9.308 billion
expenditures:: $10.61 billion (2010 est.)
Public debt: 61.1% of GDP (2010 est.)
country comparison to the world: 30
58% of GDP (2009)
Inflation rate (consumer prices): 2.4% (2010 est.)
country comparison to the world: 59
0.3% (2009)
Central bank discount rate: 1.75% (31 December 2010)
country comparison to the world: 117
1.75% (31 December 2009)
note: this is the European Central Bank's rate on the marginal lending facility,

which offers overnight credit to banks in the euro area
Commercial bank prime lending rate: 7.49% (31 December 2009 est.)
country comparison to the world: 122
7.19% (31 December 2008 est.)
Stock of narrow money: $4.341 billion (31 December 2010 est.)
country comparison to the world: 98
$4.602 billion (31 December 2009 est.)
note: see entry for the European Union for money supply in the euro area; the European Central Bank (ECB) controls monetary policy for the 17 members of the EMU; individual members of the EMU do not control the quantity of money circulating within their own borders
Stock of broad money: $50.5 billion (31 December 2010 est.)
country comparison to the world: 67
$53.46 billion (31 December 2009 est.)
Stock of domestic credit: $101.2 billion (31 December 2009 est.)
country comparison to the world: 50
$80.68 billion (31 December 2008 est.)
Market value of publicly traded shares: $4.993 billion (31 December 2009)
country comparison to the world: 70
$7.955 billion (31 December 2008)
$29.48 billion (31 December 2007)
Agriculture—products: citrus, vegetables, barley, grapes, olives, vegetables; poultry, pork, lamb; dairy, cheese
Industries: tourism, food and beverage processing, cement and gypsum production, ship repair and refurbishment, textiles, light chemicals, metal products, wood, paper, stone, and clay products
Industrial production growth rate: 0.1% (2010 est.)
country comparison to the world: 153
Electricity—production: 4.502 billion kWh (2007 est.)
country comparison to the world: 116
Electricity—consumption: 4.277 billion kWh (2007 est.)
country comparison to the world: 117
Electricity—exports: 0 kWh (2010 est.)
Electricity—imports: 0 kWh (2010 est.)
Oil—production: 0 bbl/day (2010 est.)
country comparison to the world: 165
Oil—consumption: 59,000 bbl/day (2009 est.)
country comparison to the world: 92
Oil—exports: 0 bbl/day (2010 est.)
country comparison to the world: 160
Oil—imports: 58,930 bbl/day (2007 est.)
country comparison to the world: 80
Oil—proved reserves: 0 bbl (1 January 2010 est.)
country comparison to the world: 121
Natural gas—production: 0 cu m (2010 est.)
country comparison to the world: 117
Natural gas—consumption: 0 cu m (2008 est.)
country comparison to the world: 164
Natural gas—exports: 0 cu m (2010 est.)
country comparison to the world: 83
Natural gas—imports: 0 cu m (2010 est.)
country comparison to the world: 174
Natural gas—proved reserves: 0 cu m (1 January 2010 est.)
country comparison to the world: 126
Current account balance: $-2.5 billion (2010 est.)
country comparison to the world: 161
$-1.915 billion (2009 est.)
Exports: $2.232 billion (2010 est.)
country comparison to the world: 130
$2.065 billion (2009 est.)
Exports—commodities: citrus, potatoes, pharmaceuticals, cement, clothing
Exports—partners: Greece 23.83%, Germany 9.2%, UK 8.78% (2009)
Imports: $7.962 billion (2010 est.)
country comparison to the world: 100
$7.973 billion (2009 est.)
Imports—commodities: consumer goods, petroleum and lubricants, machinery, transport equipment
Imports—partners: Greece 20.18%, Italy 10.67%, UK 8.95%, Germany 8.79%, Israel 6.99%, China 5.52%, Netherlands 4.85%, France 4.01% (2009)
Reserves of foreign exchange and gold: $NA (31 December 2010 est.)
$1.289 billion (31 December 2009 est.)
Debt—external: $NA (31 December 2010 est.)
$32.61 billion (31 December 2008 est.)
Stock of direct foreign investment—at home: $29.36 billion (31 December 2010 est.)
country comparison to the world: 62
$26.61 billion (31 December 2009 est.)
Stock of direct foreign investment—abroad: $16.57 billion (31 December 2010 est.)
country comparison to the world: 46
$15.79 billion (31 December 2009 est.)
Exchange rates: euros (EUR) per US dollar–0.755 (2010), 0.7198 (2009), 0.6827 (2008), 0.4286 (2007), 0.4586 (2006)
Economy of the area administered by Turkish Cypriots: Economy–overview: The Turkish Cypriot economy has roughly half the per capita GDP of the south, and economic growth tends to be volatile, given the north's relative isolation, bloated public sector, reliance on the Turkish lira, and small market size. The Turkish Cypriots are heavily dependent on transfers from the Turkish Government. Ankara directly finances about one-third of the Turkish Cypriot "administration's" budget. Aid from Turkey has exceeded $400 million annually in recent years. The Turkish Cypriot economy experienced a sharp slowdown in 2008-09 due to the global financial crisis and to its reliance on British and Turkish tourism, both of which declined due to the recession. The Turkish Cypriot budget deficit also deteriorated in 2009 due to decreased state revenues and increased government expenditures on public sector salaries and social services. The Turkish Cypriot economy declined about 0.6% in 2010.
GDP (purchasing power parity): $1.829 billion (2007 est.)
GDP—real growth rate: -0.6% (2010 est.)
GDP–*per capita:* $11,700 (2007 est.)
GDP–composition by sector: *agriculture:* 8.6%, *industry:* 22.5%, *services:* 69.1% (2006 est.)
Labor force: 95,030 (2007 est.)
Labor force–by occupation: *agriculture:* 14.5%, *industry:* 29%, *services:* 56.5% (2004)
Unemployment rate: 9.4% (2005 est.)
Population below poverty line: %NA
Inflation rate: 11.4% (2006)
Budget: *revenues:* $2.5 billion, expenditures: $2.5 billion (2006)
Agriculture—products: citrus fruit, dairy, potatoes, grapes, olives, poultry, lamb
Industries: foodstuffs, textiles, clothing, ship repair, clay, gypsum, copper, furniture
Industrial production growth rate: -0.3% (2007 est.)
Electricity production: 998.9 million kWh (2005)
Electricity consumption: 797.9 million kWh (2005)
Exports: $68.1 million, f.o.b. (2007 est.)
Export—commodities: citrus, dairy, potatoes, textiles
Export—partners: Turkey 40%; direct trade between the area administered by Turkish Cypriots and the area under government control remains limited
Imports: $1.2 billion, f.o.b. (2007 est.)
Import—commodities: vehicles, fuel, cigarettes, food, minerals, chemicals, machinery
Import—partners: Turkey 60%; direct trade between the area administered by Turkish Cypriots and the area under government control remains limited
Reserves of foreign exchange and gold: $NA
Debt—external: $NA
Currency (code): Turkish new lira (YTL)
Exchange rates: Turkish new lira per US dollar: 1.5181 (2010) 1.319 (2007) 1.4286 (2006) 1.3436 (2005)

COMMUNICATIONS

Telephones—main lines in use: area under government control: 414,500 (2009); area administered by Turkish Cypriots: 86,228 (2002)
country comparison to the world: 102
Telephones—mobile cellular: area under government control: 977,500 (2009); area administered by Turkish Cypriots: 147,522 (2002)
country comparison to the world: 146
Telephone system: *general assessment:* excellent in both area under government control and area administered by Turkish Cypriots
domestic: open-wire, fiber-optic cable, and microwave radio relay
international: country code–357 (area administered by Turkish Cypriots uses the country code of Turkey–90); a number of submarine cables, including the SEA-ME-WE-3, combine to provide connectivity to Western Europe, the Middle East, and Asia; tropospheric scatter; satellite earth stations–8 (3 Intelsat–1 Atlantic Ocean and 2 Indian Ocean, 2 Eutelsat, 2 Intersputnik, and 1 Arabsat)
Broadcast media: mixture of state and privately-run television and radio services; the public broadcaster operates 2 TV channels and 4 radio stations; 6 private TV broadcasters, satellite and cable TV services including telecasts from Greece and Turkey, and a number of private radio stations are available; in areas administered by Turkish Cypriots, there are 2 public TV stations, 4 public radio stations, and

privately-owned TV and radio broadcast stations (2007)
Internet country code: .cy
Internet hosts: 187,881 (2010)
country comparison to the world: 66
Internet users: 433,900 (2009)
country comparison to the world: 120

TRANSPORTATION

Airports: 15 (2010)
country comparison to the world: 144
Airports—with paved runways: *total:* 13
2,438 to 3,047 m: 6
1,524 to 2,437 m: 3
914 to 1,523 m: 3
under 914 m: 1 (2010)
Airports—with unpaved runways: *total:* 2
under 914 m: 2 (2010)
Heliports: 9 (2010)
Pipelines: oil 0 km
Roadways: *total:* 14,671 km
country comparison to the world: 121
12,321 km under government control (includes 257 km of expressways),
2,350 km administered by Turkish Cypriots (2008)
Merchant marine: *total:* 839
country comparison to the world: 13
by type: bulk carrier 267, cargo 173, chemical tanker 77, container 193, liquefied gas 10, passenger 3, passenger/cargo 24, petroleum tanker 69, refrigerated cargo 6, roll on/roll off 13, vehicle carrier 4
foreign-owned: 637 (Austria 1, Belgium 2, Bermuda 1, Canada 2, Chile 1, China 6, Cuba 1, Denmark 6, Estonia 7, France 16, Germany 189, Greece 216, Hong Kong 2, India 2, Iran 10, Ireland 3, Israel 1, Italy 6, Japan 19, Monaco 1, Netherlands 24, Norway 12, Philippines 1, Poland 20, Portugal 2, Russia 47, Singapore 1, Slovenia 4, Spain 7, Sweden 5, Syria 1, UAE 5, UK 7, Ukraine 2, US 7)
note: this country allows large numbers of ships owned by foreign entities to be registered in its national shipping registry and to fly its flag; these ships operate under the laws of the flag state
registered in other countries: 138 (Bahamas 14, Belize 1, Burma 1, Cambodia 8, Comoros 2, Finland 1, Gibraltar 1, Greece 4, Hong Kong 3, Liberia 7, Malta 29, Marshall Islands 38, Norway 1, Panama 8, Russia 11, Saint Vincent and the Grenadines 2, Sierra Leone 1, Singapore 3, unknown 3) (2010)
Ports and terminals: area under government control: Larnaca, Limassol, Vasilikos; area administered by Turkish Cypriots: Famagusta, Kyrenia

MILITARY

Military branches: Republic of Cyprus: Greek Cypriot National Guard (Ethniki Forea, EF; includes naval and air elements); northern Cyprus: Turkish Cypriot Security Force (GKK) (2009)
Military service age and obligation: Greek Cypriot National Guard (GCNG): 18-50 years of age for compulsory military service for all Greek Cypriot males; 17 years of age for voluntary service; women may volunteer for a 3-year term; length of service is 25 months (2009)
Manpower available for military service: Greek Cypriot National Guard (GCNG): *males age 16–49:* 327,875
females age 16–49: 287,891 (2010 est.)
Manpower fit for military service: Greek Cypriot National Guard (GCNG): *males age 16–49:* 275,842
females age 16–49: 239,862 (2010 est.)
Manpower reaching militarily significant age annually:
male: 8,167
female: 7,398 (2010 est.)
Military expenditures: 3.8% of GDP (2005 est.) (U)
country comparison to the world: 29

TRANSNATIONAL ISSUES

Disputes—international: hostilities in 1974 divided the island into two de facto autonomous entities, the internationally recognized Cypriot Government and a Turkish-Cypriot community (north Cyprus); the 1,000-strong UN Peacekeeping Force in Cyprus (UNFICYP) has served in Cyprus since 1964 and maintains the buffer zone between north and south; on 1 May 2004, Cyprus entered the European Union still divided, with the EU's body of legislation and standards (acquis communitaire) suspended in the north; Turkey protests Cypriot Government creating hydrocarbon blocks and maritime boundary with Lebanon in March 2007
Refugees and internally displaced persons: *IDPs:* 210,000 (both Turkish and Greek Cypriots; many displaced for over 30 years) (2007)
Illicit drugs: minor transit point for heroin and hashish via air routes and container traffic to Europe, especially from Lebanon and Turkey; some cocaine transits as well; despite a strengthening of anti-money-laundering legislation, remains vulnerable to money laundering; reporting of suspicious transactions in offshore sector remains weak (2008)

CZECH REPUBLIC

INTRODUCTION

Background: At the close of World War I, the Czechs and Slovaks of the former Austro-Hungarian Empire merged to form Czechoslovakia. During the interwar years, having rejected a federal system, the new country's leaders were frequently preoccupied with meeting the demands of other ethnic minorities within the republic, most notably the Sudeten Germans and the Ruthenians (Ukrainians). On the eve of World War II, the Czech part of the country was forcibly annexed to the Third Reich, and the Slovaks declared independence as a fascist ally of Nazi Germany. After the war, a reunited but truncated Czechoslovakia (less Ruthenia) fell within the Soviet sphere of influence. In 1968, an invasion by Warsaw Pact troops ended the efforts of the country's leaders to liberalize Communist Party rule and create "socialism with a human face." Anti-Soviet demonstrations the following year ushered in a period of harsh repression known as "normalization." With the collapse of Soviet-backed authority in 1989, Czechoslovakia regained its freedom through a peaceful "Velvet Revolution." On 1 January 1993, the country underwent a "velvet divorce" into its two national components, the Czech Republic and Slovakia. The Czech Republic joined NATO in 1999 and the European Union in 2004.

GEOGRAPHY

Location: Central Europe, between Germany, Poland, Slovakia, and Austria
Geographic coordinates: 49 45 N, 15 30 E
Map references: Europe
Area: *total:* 78,867 sq km
country comparison to the world: 115
land: 77,247 sq km
water: 1,620 sq km
Area—comparative: slightly smaller than South Carolina
Land boundaries: *total:* 1,989 km
border countries: Austria 362 km, Germany 815 km, Poland 615 km, Slovakia 197 km
Coastline: 0 km (landlocked)
Maritime claims: none (landlocked)
Climate: temperate; cool summers; cold, cloudy, humid winters
Terrain: Bohemia in the west consists of rolling plains, hills, and plateaus surrounded by low mountains; Moravia in the east consists of very hilly country
Elevation extremes: *lowest point:* Elbe River 115 m
highest point: Snezka 1,602 m
Natural resources: hard coal, soft coal, kaolin, clay, graphite, timber
Land use: *arable land:* 38.82%
permanent crops: 3%
other: 58.18% (2005)
Irrigated land: 240 sq km (2003)
Total renewable water resources: 16 cu km (2005)

Freshwater withdrawal (domestic/industrial/agricultural):
total: 1.91 cu km/yr (41%/57%/2%)
per capita: 187 cu m/yr (2002)
Natural hazards: flooding
Environment—current issues: air and water pollution in areas of northwest Bohemia and in northern Moravia around Ostrava present health risks; acid rain damaging forests; efforts to bring industry up to EU code should improve domestic pollution
Environment—international agreements:
party to: Air Pollution, Air Pollution-Nitrogen Oxides, Air Pollution-Persistent Organic Pollutants, Air Pollution-Sulfur 85, Air Pollution-Sulfur 94, Air Pollution-Volatile Organic Compounds, Antarctic-Environmental Protocol, Antarctic Treaty, Biodiversity, Climate Change, Climate Change-Kyoto Protocol, Desertification, Endangered Species, Environmental Modification, Hazardous Wastes, Law of the Sea, Ozone Layer Protection, Ship Pollution, Wetlands, Whaling
signed, but not ratified: none of the selected agreements
Geography—note: landlocked; strategically located astride some of oldest and most significant land routes in Europe; Moravian Gate is a traditional military corridor between the North European Plain and the Danube in central Europe

PEOPLE

Population: 10,190,213 (July 2011 est.)
country comparison to the world: 82
Age structure: *0–14 years:* 13.5% (male 704,495/female 666,191)
15–64 years: 70.2% (male 3,599,774/female 3,554,158)
65 years and over: 16.3% (male 663,982/female 1,001,613) (2011 est.)
Median age: *total:* 40.8 years
male: 39.2 years
female: 42.5 years (2011 est.)
Population growth rate: -0.12% (2011 est.)
country comparison to the world: 207
Birth rate: 8.7 births/1,000 population (2011 est.)
country comparison to the world: 214
Death rate: 10.86 deaths/1,000 population (July 2011 est.)
country comparison to the world: 41
Net migration rate: 0.97 migrant(s)/1,000 population (2011 est.)
country comparison to the world: 54
Urbanization: *urban population:* 74% of total population (2010)
rate of urbanization: 0.3% annual rate of change (2010-15 est.)
Major cities—population: PRAGUE (capital) 1.162 million (2009)
Sex ratio: *at birth:* 1.059 male(s)/female
under 15 years: 1.06 male(s)/female
15–64 years: 1.01 male(s)/female
65 years and over: 0.66 male(s)/female
total population: 0.95 male(s)/female (2011 est.)
Infant mortality rate: *total:* 3.73 deaths/1,000 live births
country comparison to the world: 205
male: 4.06 deaths/1,000 live births
female: 3.38 deaths/1,000 live births (2011 est.)
Life expectancy at birth: *total population:* 77.19 years
country comparison to the world: 63
male: 73.93 years
female: 80.66 years (2011 est.)
Total fertility rate: 1.26 children born/woman (2011 est.)
country comparison to the world: 214
HIV/AIDS—adult prevalence rate: less than 0.1% (2009 est.)
country comparison to the world: 127
HIV/AIDS—people living with HIV/AIDS: 2,000 (2009 est.)
country comparison to the world: 136
HIV/AIDS—deaths: fewer than 100 (2009 est.)
country comparison to the world: 130
Drinking water source: *Improved:*
urban: 100% of population
rural: 100% of population
total: 100% of population (2008)
Sanitation facility access: *Improved:*
urban: 99% of population
rural: 97% of population
total: 98% of population
Unimproved:
urban: 1% of population
rural: 3% of population
total: 2% of population (2008)
Nationality: *noun:* Czech(s)
adjective: Czech
Ethnic groups: Czech 90.4%, Moravian 3.7%, Slovak 1.9%, other 4% (2001 census)
Religions: Roman Catholic 26.8%, Protestant 2.1%, other 3.3%, unspecified 8.8%, unaffiliated 59% (2001 census)
Languages: Czech 94.9%, Slovak 2%, other 2.3%, unidentified 0.8% (2001 census)
Literacy: *definition:* NA
total population: 99%
male: 99%
female: 99% (2003 est.)
School life expectancy (primary to tertiary education): *total:* 15 years
male: 15 years
female: 16 years (2008)
Education expenditures: 4.2% of GDP (2007)
country comparison to the world: 97

GOVERNMENT

Country name: *conventional long form:* Czech Republic
conventional short form: Czech Republic
local long form: Ceska Republika
local short form: Cesko
Government type: parliamentary democracy
Capital: *name:* Prague
geographic coordinates: 50 05 N, 14 28 E
time difference: UTC+1 (6 hours ahead of Washington, DC during Standard Time)
daylight saving time: +1hr, begins last Sunday in March; ends last Sunday in October
Administrative divisions: 13 regions (kraje, singular—kraj) and 1 capital city* (hlavni mesto); Jihocesky (South Bohemia), Jihomoravsky (South Moravia), Karlovarsky, Kralovehradecky, Liberecky, Moravskoslezsky (Moravia-Silesia), Olomoucky, Pardubicky, Plzensky (Pilsen), Praha (Prague)*, Stredocesky (Central Bohemia), Ustecky, Vysocina, Zlinsky
Independence: 1 January 1993 (Czechoslovakia split into the Czech Republic and Slovakia); note—although 1 January is the day the Czech Republic came into being, the Czechs generally consider 28 October 1918, the day the former Czechoslovakia declared its independence from the Austro-Hungarian Empire, as their independence day
National holiday: Czechoslovak Founding Day, 28 October (1918)
Constitution: ratified 16 December 1992, effective 1 January 1993; amended several times
Legal system: civil law system based on former Austro-Hungarian civil codes and socialist theory; note—legislation is actively modernizing the legal system
International law organization participation: has not submitted an ICJ jurisdiction declaration; accepts ICCt jurisdiction
Suffrage: 18 years of age; universal
Executive branch: *chief of state:* President Vaclav KLAUS (since 7 March 2003)
head of government: Prime Minister Petr NECAS (since 28 June 2010); First Deputy Prime Minister Karel SCHWARZENBERG (since 13 July 2010), Deputy Prime Minister Radek JOHN (since 13 July 2010)
cabinet: Cabinet appointed by the president on the recommendation of the prime minister
(For more information visit the World Leaders website)
elections: president elected by Parliament for a five-year term (eligible for a second term); last successful election held on 15 February 2008 (after earlier elections held 8 and 9 February 2008 were inconclusive; next election to be held in 2013); prime minister appointed by the president
election results: Vaclav KLAUS reelected president on 15 February 2008; Vaclav KLAUS 141 votes, Jan SVEJNAR 111 votes (third round; combined votes of both chambers of parliament)
Legislative branch: bicameral Parliament or Parlament consists of the Senate or Senat (81 seats; members elected by popular vote to serve six-year terms; one-third elected every two years) and the Chamber of Deputies or Poslanecka Snemovna (200 seats; members are elected by popular vote to serve four-year terms)
elections: Senate—last held in two rounds on 15-16 and 22-23 October 2010 (next to be held by October 2012); Chamber of Deputies—last held on 28-29 May 2010 (next to be held by 2014)
election results: Senate—percent of vote by party—NA; seats by party—CSSD 41, ODS 25, KDU-CSL 6, TOP 09 5, others 4; Chamber of Deputies—percent of vote by party—CSSD 22.1%, ODS 20.2%, TOP 09 16.7%, KSCM 11.3%, VV 10.9%; seats by party—CSSD 56, ODS 53, TOP 09 41, KSCM 26, VV 24
Judicial branch: Supreme Court; judges are appointed by the president for an unlimited term; Constitutional Court; 15 judges are appointed by the president and confirmed by the Senate for a ten-year term; Supreme Administrative Court; chairman and deputy chairmen are appointed by the

president for a 10-year term; judges are appointed by the president for an unlimited term
Political parties and leaders: Association of Independent Candidates-European Democrats or SNK-ED [Zdenka MARKOVA]; Christian Democratic Union-Czechoslovak People's Party or KDU-CSL [Pavel BELOBRADEK]; Civic Democratic Party or ODS [Petr NECAS]; Communist Party of Bohemia and Moravia or KSCM [Vojtech FILIP]; Czech Social Democratic Party or CSSD [Bohuslav SOBOTKA (acting)]; Green Party [Ondrej LISKA]; Public Affairs or VV [Radek JOHN]; Tradice Odpovednost Prosperita 09 or TOP 09 [Karel SCHWARZENBERG]; Union of Freedom-Democratic Union or US-DEU [Jan CERNY]
Political pressure groups and leaders: Czech-Moravian Confederation of Trade Unions or CMKOS [Jaroslav ZAVADIL]
International organization participation: Australia Group, BIS, BSEC (observer), CD, CE, CEI, CERN, EAPC, EBRD, EIB, ESA, EU, FAO, IAEA, IBRD, ICAO, ICC, ICRM, IDA, IEA, IFC, IFRCS, ILO, IMF, IMO, IMSO, Interpol, IOC, IOM, IPU, ISO, ITSO, ITU, ITUC, MIGA, MONUSCO, NATO, NEA, NSG, OAS (observer), OECD, OIF (observer), OPCW, OSCE, PCA, Schengen Convention, SECI (observer), UN, UNCTAD, UNESCO, UNIDO, UNWTO, UPU, WCO, WFTU, WHO, WIPO, WMO, WTO, ZC
Diplomatic representation in the US: *chief of mission:* Ambassador Norman EISEN
chancery: 3900 Spring of Freedom Street NW, Washington, DC 20008
telephone: [1] (202) 274-9100
FAX: [1] (202) 966-8540
consulate(s) general: Chicago, Los Angeles, New York
Diplomatic representation from the US: *chief of mission:* Ambassador Norman EISEN
embassy: Trziste 15, 118 01 Prague 1
mailing address: use embassy street address
telephone: [420] 257 022 000
FAX: [420] 257 022 809
Flag description: two equal horizontal bands of white (top) and red with a blue isosceles triangle based on the hoist side
note: is identical to the flag of the former Czechoslovakia
National anthem: *name:* "Kde domov muj?" (Where is My Home?)
lyrics/music: Josef Kajetan TYL/Frantisek Jan SKROUP
note: adopted 1993; the anthem is a verse from the former Czechoslovak anthem originally written as part of the opera "Fidlovacka"

ECONOMY

Economy—overview: The Czech Republic is a stable and prosperous market economy, which harmonized its laws and regulations with those of the EU prior to its EU accession in 2004. While the conservative, inward looking Czech financial system has remained relative healthy, the small, open, export-driven Czech economy remains very sensitive to changes in the economic performance of its main export markets, especially Germany. When Western Europe and Germany fell into recession in late 2008, demand for Czech goods plunged, leading to double digit drops in industrial production and exports. As a result, real GDP fell 4.1% in 2009, with most of the decline occurring during the first quarter. Real GDP, however, has slowly recovered with positive quarter-on-quarter growth starting in the second half of 2009 and continuing throughout 2010. The auto industry remains the largest single industry and, together with its suppliers, accounts for as much as 20% of Czech manufacturing. The Czech Republic produced more than a million cars for the first time in 2010, over 80% of which were exported. Foreign and domestic businesses alike voice concerns about corruption, especially in public procurement. Other long term challenges include dealing with a rapidly aging population, funding an unsustainable pension and health care system, and diversifying away from manufacturing and toward a more high-tech, services-based, knowledge economy.
GDP (purchasing power parity): $261.3 billion (2010 est.)
country comparison to the world: 44
$255.4 billion (2009 est.)
$266.4 billion (2008 est.)
note: data are in 2010 US dollars
GDP (official exchange rate): $192.2 billion (2010 est.)
GDP—real growth rate: 2.3% (2010 est.)
country comparison to the world: 141
-4.1% (2009 est.)
2.5% (2008 est.)
GDP—per capita (PPP): $25,600 (2010 est.)
country comparison to the world: 54
$25,000 (2009 est.)
$26,100 (2008 est.)
note: data are in 2010 US dollars
GDP—composition by sector:
agriculture: 2.2%
industry: 38.3%
services: 59.5% (2010 est.)
Labor force: 5.37 million (2010 est.)
country comparison to the world: 71
Labor force—by occupation:
agriculture: 3.1%
industry: 38.6%
services: 58.3% (2009)
Unemployment rate: 7.1% (November 2010 est.)
country comparison to the world: 73
8% (2009 est.)
Population below poverty line: NA%
Household income or consumption by percentage share: *lowest 10%:* 1.5%
highest 10%: NA (2009)
Distribution of family income—Gini index: 26 (2005)
country comparison to the world: 130
25.4 (1996)
Investment (gross fixed): 22.5% of GDP (2010 est.)
country comparison to the world: 64
Budget: *revenues:* $77.9 billion
expenditures: $87.87 billion (2010 est.)
Public debt: 40% of GDP (2010 est.)
country comparison to the world: 72
32.5% of GDP (2009)
Inflation rate (consumer prices): 1.5% (2010 est.)
country comparison to the world: 38
1% (2009 est.)
Central bank discount rate: 0.25% (31 December 2010)
country comparison to the world: 133
1% (31 December 2009)
note: the two-week repo rate was 0.75% on 31 December 2010; this is the main rate CNB uses
Commercial bank prime lending rate: 3.9% (31 December 2010 est.)
country comparison to the world: 149
4.5% (31 December 2009 est.)
Stock of narrow money: $96.82 billion (31 December 2010 est.)
country comparison to the world: 33
$92.95 billion (31 December 2009 est.)
Stock of broad money: $138.6 billion (31 December 2010 est.)
country comparison to the world: 46
$139 billion (31 December 2009 est.)
Stock of domestic credit: $119.5 billion (31 December 2010 est.)
country comparison to the world: 46
$118.8 billion (31 December 2009 est.)
Market value of publicly traded shares: $73.1 billion (31 December 2010)
country comparison to the world: 45
$70.26 billion (31 December 2009)
$57.8 billion (31 December 2008)
Agriculture—products: wheat, potatoes, sugar beets, hops, fruit; pigs, poultry
Industries: motor vehicles, metallurgy, machinery and equipment, glass, armaments
Industrial production growth rate: 15.9% (2010 est.)
country comparison to the world: 7
Electricity—production: 82.25 billion kWh (2009 est.)
country comparison to the world: 35
Electricity—consumption: 53.42 billion kWh (2009 est.)
country comparison to the world: 44
Electricity—exports: 22.23 billion kWh (2009 est.)
Electricity—imports: 8.58 billion kWh (2009 est.)
Oil—production: 10,970 bbl/day (2009 est.)
country comparison to the world: 83
Oil—consumption: 207,600 bbl/day (2009 est.)
country comparison to the world: 54
Oil—exports: 29,670 bbl/day (2008 est.)
country comparison to the world: 85
Oil—imports: 219,900 bbl/day (2008 est.)
country comparison to the world: 42
Oil—proved reserves: 15 million bbl (1 January 2010 est.)
country comparison to the world: 85
Natural gas—production: 178 million cu m (2009 est.)
country comparison to the world: 75
Natural gas—consumption: 8.164 billion cu m (2009 est.)
country comparison to the world: 51
Natural gas—exports: 1.111 billion cu m (2009 est.)
country comparison to the world: 35
Natural gas—imports: 9.683 billion cu m (2009 est.)
country comparison to the world: 21

Natural gas—proved reserves: 3.072 billion cu m (1 January 2010 est.)
country comparison to the world: 92
Current account balance: $-5.956 billion (2010 est.)
country comparison to the world: 172
$-2.146 billion (2009 est.)
Exports: $116.5 billion (2010 est.)
country comparison to the world: 33
$112.6 billion (2009 est.)
Exports—commodities: machinery and transport equipment, raw materials and fuel, chemicals
Exports—partners: Germany 31.7%, Slovakia 8.7%, Poland 6.2%, France 5.5%, UK 4.9%, Austria 4.7%, Italy 4.5% (2010 est.)
Imports: $109.2 billion (2010 est.)
country comparison to the world: 31
$103.1 billion (2009 est.)
Imports—commodities: machinery and transport equipment, raw materials and fuels, chemicals
Imports—partners: Germany 25.6%, China 11.9%, Poland 6.5%, Russia 5.4%, Slovakia 5.2% (2010 est.)
Reserves of foreign exchange and gold: $42.34 billion (31 December 2010 est.)
country comparison to the world: 30
$41.61 billion (31 December 2009 est.)
Debt—external: $86.79 billion (31 December 2010 est.)
country comparison to the world: 40
$86.55 billion (31 December 2009)
Stock of direct foreign investment—at home: $130.4 billion (30 September 2010 est.)
country comparison to the world: 28
$121.9 billion (31 December 2009 est.)
Stock of direct foreign investment—abroad: $14.67 billion (30 September 2010 est.)
country comparison to the world: 49
$14.35 billion (31 December 2009 est.)
Exchange rates: koruny (CZK) per US dollar–19.111 (2010), 19.063 (2009), 17.064 (2008), 20.53 (2007), 22.596 (2006)

COMMUNICATIONS

Telephones—main lines in use: 2.092 million (2009)
country comparison to the world: 54
Telephones—mobile cellular: 14.258 million (2009)
country comparison to the world: 51
Telephone system: *general assessment:* privatization and modernization of the Czech telecommunication system got a late start but is advancing steadily; virtually all exchanges now digital; existing copper subscriber systems enhanced with Asymmetric Digital Subscriber Line (ADSL) equipment to accommodate Internet and other digital signals; trunk systems include fiber-optic cable and microwave radio relay
domestic: access to the fixed-line telephone network expanded throughout the 1990s but the number of fixed line connections has been dropping since then; mobile telephone usage increased sharply beginning in the mid-1990s and the number of cellular telephone subscriptions now greatly exceeds the population
international: country code–420; satellite earth stations–6 (2 Intersputnik–Atlantic and Indian Ocean regions, 1 Intelsat, 1 Eutelsat, 1 Inmarsat, 1 Globalstar) (2009)
Broadcast media: roughly 130 television broadcasters operating some 350 television channels with 4 publicly operated and the remainder in private hands; 13 television stations have national coverage with 4 being publicly operated; cable and satellite TV subscription services are available; about 70 radio broadcasters are registered operating roughly 85 radio stations with 15 stations publicly operated; 16 radio stations provide national coverage with the remainder local or regional (2008)
Internet country code: .cz
Internet hosts: 3.494 million (2010)
country comparison to the world: 25
Internet users: 6.681 million (2009)
country comparison to the world: 40

TRANSPORTATION

Airports: 122 (2010)
country comparison to the world: 50
Airports—with paved runways: *total:* 44
over 3,047 m: 2
2,438 to 3,047 m: 9
1,524 to 2,437 m: 12
914 to 1,523 m: 3
under 914 m: 18 (2010)
Airports—with unpaved runways: *total:* 78
1,524 to 2,437 m: 1
914 to 1,523 m: 27
under 914 m: 50 (2010)
Heliports: 1 (2010)
Pipelines: gas 7,010 km; oil 547 km; refined products 94 km (2010)
Railways: *total:* 9,632 km
country comparison to the world: 22
standard gauge: 9,530 km 1.435-m gauge (3,165 km electrified)
narrow gauge: 102 km 0.750-m gauge (2010)
Roadways: *total:* 127,719 km
country comparison to the world: 36
paved: 127,719 km (includes 729 km of expressways) (2009)
Waterways: 664 km (principally on Elbe, Vltava, Oder, and other navigable rivers, lakes, and canals) (2010)
country comparison to the world: 77
Merchant marine: registered in other countries: 1 (Saint Vincent and the Grenadines 1) (2010)
country comparison to the world: 153
Ports and terminals: Decin, Prague, Usti nad Labem

MILITARY

Military branches: Army of the Czech Republic (ACR): Joint Forces Command (includes Land Forces and Air Forces), Support and Training Forces Command (2011)
Military service age and obligation: 18-28 years of age for male and female voluntary military service; no conscription (2010)
Manpower available for military service: *males age 16–49:* 2,506,826
females age 16–49: 2,407,634 (2010 est.)
Manpower fit for military service: *males age 16–49:* 2,072,267
females age 16–49: 1,988,839 (2010 est.)
Manpower reaching militarily significant age annually: *male:* 49,999
female: 47,501 (2010 est.)
Military expenditures: 1.46% of GDP (2007 est.)
country comparison to the world: 103

TRANSNATIONAL ISSUES

Disputes—international: while threats of international legal action never materialized in 2007, 915,220 Austrians, with the support of the popular Freedom Party, signed a petition in January 2008, demanding that Austria block the Czech Republic's accession to the EU unless Prague closes its controversial Soviet-style nuclear plant in Temelin, bordering Austria
Illicit drugs: transshipment point for Southwest Asian heroin and minor transit point for Latin American cocaine to Western Europe; producer of synthetic drugs for local and regional markets; susceptible to money laundering related to drug trafficking, organized crime; significant consumer of ecstasy (2008)

DENMARK

INTRODUCTION

Background: Once the seat of Viking raiders and later a major north European power, Denmark has evolved into a modern, prosperous nation that is participating in the general political and economic integration of Europe. It joined NATO in 1949 and the EEC (now the EU) in 1973. However, the country has opted out of certain elements of the European Union's Maastricht Treaty, including the European Economic and Monetary Union (EMU), European defense cooperation, and issues concerning certain justice and home affairs.

GEOGRAPHY

Location: Northern Europe, bordering the Baltic Sea and the North Sea, on a peninsula north of Germany (Jutland); also includes several major islands (Sjaelland, Fyn, and Bornholm)
Geographic coordinates: 56 00 N, 10 00 E
Map references: Europe
Area:
total: 43,094 sq km
country comparison to the world: 133
land: 42,434 sq km
water: 660 sq km
note: includes the island of Bornholm in the Baltic Sea and the rest of metropolitan Denmark (the Jutland Peninsula, and the major islands of Sjaelland and Fyn), but excludes the Faroe Islands and Greenland
Area—comparative: slightly less than twice the size of Massachusetts
Land boundaries: *total:* 68 km
border countries: Germany 68 km
Coastline: 7,314 km
Maritime claims: territorial sea: 12 nm
contiguous zone: 24 nm
exclusive economic zone: 200 nm
continental shelf: 200 m depth or to the depth of exploitation
Climate: temperate; humid and overcast; mild, windy winters and cool summers
Terrain: low and flat to gently rolling plains
Elevation extremes: *lowest point:* Lammefjord -7 m
highest point: Mollehoj/Ejer Bavnehoj 171 m
Natural resources: petroleum, natural gas, fish, salt, limestone, chalk, stone, gravel and sand
Land use: *arable land:* 52.59%
permanent crops: 0.19%
other: 47.22% (2005)
Irrigated land: 4,490 sq km (2003)
Total renewable water resources: 6.1 cu km (2003)
Freshwater withdrawal (domestic/industrial/agricultural): *total:* 0.67 cu km/yr (32%/26%/42%)
per capita: 123 cu m/yr (2002)
Natural hazards: flooding is a threat in some areas of the country (e.g., parts of Jutland, along the southern coast of the island of Lolland) that are protected from the sea by a system of dikes
Environment—current issues: air pollution, principally from vehicle and power plant emissions; nitrogen and phosphorus pollution of the North Sea; drinking and surface water becoming polluted from animal wastes and pesticides
Environment—international agreements: *party to:* Air Pollution, Air Pollution-Nitrogen Oxides, Air Pollution-Persistent Organic Pollutants, Air Pollution-Sulfur 85, Air Pollution-Sulfur 94, Air Pollution-Volatile Organic Compounds, Antarctic Treaty, Biodiversity, Climate Change, Climate Change-Kyoto Protocol, Desertification, Endangered Species, Environmental Modification, Hazardous Wastes, Law of the Sea, Marine Dumping, Marine Life Conservation, Ozone Layer Protection, Ship Pollution, Tropical Timber 83, Tropical Timber 94, Wetlands, Whaling
signed, but not ratified: none of the selected agreements
Geography—note: controls Danish Straits (Skagerrak and Kattegat) linking Baltic and North Seas; about one-quarter of the population lives in greater Copenhagen

PEOPLE

Population: 5,529,888 (July 2011 est.)
country comparison to the world: 110
Age structure: *0–14 years:* 17.6% (male 500,265/female 474,829)
15–64 years: 65.3% (male 1,811,198/female 1,798,507)
65 years and over: 17.1% (male 417,957/female 527,132) (2011 est.)
Median age: *total:* 40.9 years
male: 40 years
female: 41.8 years (2011 est.)
Population growth rate: 0.251% (2011 est.)
country comparison to the world: 173
Birth rate: 10.29 births/1,000 population (2011 est.)
country comparison to the world: 186
Death rate: 10.19 deaths/1,000 population (July 2011 est.)
country comparison to the world: 51
Net migration rate: 2.41 migrant(s)/1,000 population (2011 est.)
country comparison to the world: 34
Urbanization: *urban population:* 87% of total population (2010)
rate of urbanization: 0.4% annual rate of change (2010-15 est.)
Major cities—population: COPENHAGEN (capital) 1.174 million (2009)
Sex ratio: *at birth:* 1.055 male(s)/female
under 15 years: 1.05 male(s)/female
15–64 years: 1.01 male(s)/female
65 years and over: 0.78 male(s)/female
total population: 0.98 male(s)/female (2011 est.)
Infant mortality rate: *total:* 4.24 deaths/1,000 live births
country comparison to the world: 196
male: 4.3 deaths/1,000 live births
female: 4.18 deaths/1,000 live births (2011 est.)
Life expectancy at birth: *total population:* 78.63 years
country comparison to the world: 48
male: 76.25 years
female: 81.14 years (2011 est.)
Total fertility rate: 1.74 children born/woman (2011 est.)
country comparison to the world: 163
HIV/AIDS—adult prevalence rate: 0.2% (2009 est.)
country comparison to the world: 97
HIV/AIDS—people living with HIV/AIDS: 5,300 (2009 est.)
country comparison to the world: 119
HIV/AIDS—deaths: fewer than 100 (2009 est.)
country comparison to the world: 128
Drinking water source: *Improved:*
urban: 100% of population
rural: 100% of population
total: 100% of population (2008)
Sanitation facility access: *Improved:*
urban: 100% of population
rural: 100% of population
total: 100% of population (2008)
Nationality: *noun:* Dane(s)
adjective: Danish
Ethnic groups: Scandinavian, Inuit, Faroese, German, Turkish, Iranian, Somali
Religions: Evangelical Lutheran (official) 95%, other Christian (includes Protestant and Roman Catholic) 3%, Muslim 2%
Languages: Danish, Faroese, Greenlandic (an Inuit dialect), German (small minority)
note: English is the predominant second language
Literacy: *definition:* age 15 and over can read and write
total population: 99%
male: 99%
female: 99% (2003 est.)
School life expectancy (primary to tertiary education): *total:* 17 years
male: 16 years
female: 18 years (2008)
Education expenditures: 7.8% of GDP (2007)
country comparison to the world: 11

GOVERNMENT

Country name: *conventional long form:* Kingdom of Denmark
conventional short form: Denmark
local long form: Kongeriget Danmark
local short form: Danmark
Government type: constitutional monarchy
Capital: *name:* Copenhagen
geographic coordinates: 55 40 N, 12 35 E

time difference: UTC+1 (6 hours ahead of Washington, DC during Standard Time)
daylight saving time: +1hr, begins last Sunday in March; ends last Sunday in October
note: applies to continental Denmark only, not to its North Atlantic components
Administrative divisions: metropolitan Denmark—5 regions (regioner, singular—region); Hovedstaden, Midtjylland, Nordjylland, Sjaelland, Syddanmark
note: an extensive local government reform merged 271 municipalities into 98 and 13 counties into five regions, effective 1 January 2007
Independence: ca. 965 (unified and Christianized under HARALD I Gormson); 5 June 1849 (became a constitutional monarchy)
National holiday: none designated; Constitution Day, 5 June (1849) is generally viewed as the National Day
Constitution: 5 June 1953; note—constitution allowed for a unicameral legislature and a female chief of state
Legal system: civil law; judicial review of legislative acts
International law organization participation: accepts compulsory ICJ jurisdiction with reservations; accepts ICCt jurisdiction
Suffrage: 18 years of age; universal
Executive branch: *chief of state:* Queen MARGRETHE II (since 14 January 1972); Heir Apparent Crown Prince FREDERIK, elder son of the monarch (born on 26 May 1968)
head of government: Prime Minister Lars Loekke RASMUSSEN (since 5 April 2009)
cabinet: Council of State appointed by the monarch
(For more information visit the World Leaders website)
elections: the monarchy is hereditary; following legislative elections, the leader of the majority party or the leader of the majority coalition usually appointed prime minister by the monarch
Legislative branch: unicameral People's Assembly or Folketing (179 seats, including 2 from Greenland and 2 from the Faroe Islands; members elected by popular vote on the basis of proportional representation to serve four-year terms unless the Folketing is dissolved earlier)
elections: last held on 13 November 2007 (next to be held by November 2011)
election results: percent of vote by party—Liberal Party 26.2%, Social Democrats 25.5%, Danish People's Party 13.9%, Socialist People's Party 13.0%, Conservative People's Party 10.4%, Social Liberal Party 5.1%, New Alliance 2.8%, Red-Green Unity List 2.2%, other 0.9%; seats by party—Liberal Party 46, Social Democrats 45, Danish People's Party 25, Socialist People's Party 23, Conservative People's Party 18, Social Liberal Party 9, New Alliance 5, Red-Green Alliance 4; note—does not include the two seats from Greenland and the two seats from the Faroe Islands
Judicial branch: Supreme Court (judges are appointed for life by the monarch)
Political parties and leaders: Christian Democrats [Bjarne Hartung KIRKEGAARD] (was Christian People's Party); Conservative People's Party [Lars BARFOED]; Danish People's Party [Pia KJAERSGAARD]; Liberal Alliance [Anders SAMUELSEN]; Liberal Party [Lars Loekke RASMUSSEN]; Red-Green Unity List (Alliance) [collective leadership] (bloc includes Left Socialist Party, Communist Party of Denmark, Socialist Workers' Party); Social Democratic Party [Helle THORNING-SCHMIDT]; Social Liberal Party [Margrethe VESTAGER]; Socialist People's Party [Villy SOEVNDAL]
Political pressure groups and leaders: Confederation of Danish Employers or DA [President Jorn Neergaard LARSEN]; Principal DA member organizations: Confederation of Danish Industries [CEO Karsten DYBVAD]; Confederation of Danish Labor Unions [President Harald BORSTING]; Danish Bankers Association [CEO Joergen HORWITZ]; DaneAge Association [President Bjarne HASTRUP]; Danish Society for Nature Conservation [President Ella Maria BISSCHOP-LARSEN]
other: humanitarian relief; development assistance; human rights NGOs
International organization participation: ADB (nonregional member), AfDB (nonregional member), Arctic Council, Australia Group, BIS, CBSS, CE, CERN, EAPC, EBRD, EIB, ESA, EU, FAO, FATF, G-9, IADB, IAEA, IBRD, ICAO, ICC, ICRM, IDA, IEA, IFAD, IFC, IFRCS, IHO, ILO, IMF, IMO, IMSO, Interpol, IOC, IOM, IPU, ISO, ITSO, ITU, ITUC, MIGA, MONUSCO, NATO, NC, NEA, NIB, NSG, OAS (observer), OECD, OPCW, OSCE, Paris Club, PCA, Schengen Convention, UN, UNCTAD, UNESCO, UNHCR, UNIDO, UNIFIL, UNMIL, UNMIS, UNRWA, UNTSO, UPU, WCO, WHO, WIPO, WMO, WTO, ZC
Diplomatic representation in the US: *chief of mission:* Ambassador Peter TAKSOE-JENSEN
chancery: 3200 Whitehaven Street NW, Washington, DC 20008
telephone: [1] (202) 234-4300
FAX: [1] (202) 328-1470
consulate(s) general: Chicago, New York
Diplomatic representation from the US: *chief of mission:* Ambassador Laurie S. FULTON
embassy: Dag Hammarskjolds Alle 24, 2100 Copenhagen
mailing address: PSC 73, APO AE 09716
telephone: [45] 33 41 71 00
FAX: [45] 35 43 02 23
Flag description: red with a white cross that extends to the edges of the flag; the vertical part of the cross is shifted to the hoist side; the banner is referred to as the Dannebrog (Danish flag) and is one of the oldest national flags in the world; traditions as to the origin of the flag design vary, but the best known is a legend that the banner fell from the sky during an early-13th century battle; caught up by the Danish king before it ever touched the earth, this heavenly talisman inspired the royal army to victory; in actuality, the flag may derive from a crusade banner or ensign
note: the shifted design element was subsequently adopted by the other Nordic countries of Finland, Iceland, Norway, and Sweden
National anthem: *name:* "Der er et yndigt land" (There is a Lovely Land); "Kong Christian" (King Christian)
lyrics/music: Adam Gottlob OEHLENSCHLAGER/Hans Ernst KROYER; Johannes EWALD/unknown
note: Denmark has two national anthems with equal status; "Der er et yndigt land," adopted 1844, is a national anthem, while "Kong Christian," adopted 1780, serves as both a national and royal anthem; "Kong Christian" is also known as "Kong Christian stod ved hojen mast" (King Christian Stood by the Lofty Mast) and "Kongesangen" (The King's Anthem); within Denmark, the royal anthem is played only when royalty is present and is usually followed by the national anthem; when royalty is not present, only the national anthem is performed; outside Denmark, the royal anthem is played, unless the national anthem is requested

ECONOMY

Economy—overview: This thoroughly modern market economy features a high-tech agricultural sector, state-of-the-art industry with world-leading firms in pharmaceuticals, maritime shipping and renewable energy, and a high dependence on foreign trade. Denmark is a member of the European Union (EU); Danish legislation and regulations conform to EU standards on almost all issues. Danes enjoy among the highest standards of living in the world and the Danish economy is characterized by extensive government welfare measures and an equitable distribution of income. Denmark is a net exporter of food and energy and enjoys a comfortable balance of payments surplus, but depends on imports of raw materials for the manufacturing sector. Within the EU, Denmark is among the strongest supporters of trade liberalization. After a long consumption-driven upswing, Denmark's economy began slowing in 2007 with the end of a housing boom. Housing prices dropped markedly in 2008-09. The global financial crisis has exacerbated this cyclical slowdown through increased borrowing costs and lower export demand, consumer confidence, and investment. The global financial crises cut Danish GDP by 0.9% in 2008 and 5.2% in 2009. Historically low levels of unemployment rose sharply with the recession but remain below 5%, based on the national measure, about half the level of the EU; harmonized to OECD standards the unemployment rate was about 8% at the end of 2010. Denmark made a modest recovery in 2010 in part because of increased government spending. An impending decline in the ratio of workers to retirees will be a major long-term issue. Denmark maintained a healthy budget surplus for many years up to 2008, but the budget balance

swung into deficit during 2009-10. Nonetheless, Denmark's fiscal position remains among the strongest in the EU. Despite previously meeting the criteria to join the European Economic and Monetary Union (EMU), so far Denmark has decided not to join, although the Danish krone remains pegged to the euro.
GDP (purchasing power parity): $201.7 billion (2010 est.)
country comparison to the world: 53
$197.7 billion (2009 est.)
$208.5 billion (2008 est.)
note: data are in 2010 US dollars
GDP (official exchange rate): $310.8 billion (2010 est.)
GDP—real growth rate: 2.1% (2010 est.)
country comparison to the world: 145
-5.2% (2009 est.)
-1.1% (2008 est.)
GDP—per capita (PPP): $36,600 (2010 est.)
country comparison to the world: 29
$35,900 (2009 est.)
$38,000 (2008 est.)
note: data are in 2010 US dollars
GDP—composition by sector:
agriculture: 1.1%
industry: 22.8%
services: 76.1% (2010 est.)
Labor force: 2.82 million (2010 est.)
country comparison to the world: 106
Labor force—by occupation:
agriculture: 2.5%
industry: 20.2%
services: 77.3% (2005 est.)
Unemployment rate: 4.2% (2010 est.)
country comparison to the world: 39
4.3% (2009 est.)
Population below poverty line: 12.1% (2007)
Household income or consumption by percentage share: *lowest 10%:* 1.9%
highest 10%: 28.7% (2007)
Distribution of family income—Gini index: 29 (2007)
country comparison to the world: 118
24.7 (1992)
Investment (gross fixed): 17.5% of GDP (2010 est.)
country comparison to the world: 115
Budget: *revenues:* $160.3 billion
expenditures: $175.9 billion (2010 est.)
Public debt: 46.6% of GDP (2010 est.)
country comparison to the world: 56
41.5% of GDP (2009 est.)
Inflation rate (consumer prices): 2.6% (2010 est.)
country comparison to the world: 68
1.3% (2009 est.)
Central bank discount rate: 1.75% (31 December 2010)
country comparison to the world: 116
1.75% (31 December 2009)
Commercial bank prime lending rate: NA%
Stock of narrow money: $148.1 billion (31 December 2010 est.)
country comparison to the world: 23
$153.1 billion (31 December 2009 est.)
Stock of broad money: $209 billion (31 December 2010 est.)
country comparison to the world: 39
$226.8 billion (31 December 2009 est.)
Stock of domestic credit: $636.5 billion (31 December 2010 est.)
country comparison to the world: 22
$671.7 billion (31 December 2009 est.)
Market value of publicly traded shares: $186.9 billion (31 December 2009)
country comparison to the world: 32
$131.5 billion (31 December 2008)
$277.7 billion (31 December 2007)
Agriculture—products: barley, wheat, potatoes, sugar beets; pork, dairy products; fish
Industries: iron, steel, nonferrous metals, chemicals, food processing, machinery and transportation equipment, textiles and clothing, electronics, construction, furniture and other wood products, shipbuilding and refurbishment, windmills, pharmaceuticals, medical equipment
Industrial production growth rate: 4% (2010 est.)
country comparison to the world: 82
Electricity—production: 36.4 billion kWh (2008 est.)
country comparison to the world: 58
Electricity—consumption: 34.3 billion kWh (2008 est.)
country comparison to the world: 56
Electricity—exports: 11.36 billion kWh (2008)
Electricity—imports: 12.82 billion kWh (2008)
Oil—production: 262,100 bbl/day (2009 est.)
country comparison to the world: 40
Oil—consumption: 166,500 bbl/day (2009 est.)
country comparison to the world: 61
Oil—exports: 268,500 bbl/day (2008 est.)
country comparison to the world: 46
Oil—imports: 173,100 bbl/day (2008 est.)
country comparison to the world: 50
Oil—proved reserves: 1.06 billion bbl (1 January 2010 est.)
country comparison to the world: 43
Natural gas—production: 8.398 billion cu m (2009)
country comparison to the world: 44
Natural gas—consumption: 4.41 billion cu m (2009)
country comparison to the world: 61
Natural gas—exports: 3.98 billion cu m (2009)
country comparison to the world: 29
Natural gas—imports: 0 cu m (2008)
country comparison to the world: 175
Natural gas—proved reserves: 61.3 billion cu m (1 January 2010 est.)
country comparison to the world: 62
Current account balance: $14.35 billion (2010 est.)
country comparison to the world: 21
$12.43 billion (2009 est.)
Exports: $99.37 billion (2010 est.)
country comparison to the world: 35
$91.51 billion (2009 est.)
Exports—commodities: machinery and instruments, meat and meat products, dairy products, fish, pharmaceuticals, furniture, windmills
Exports—partners: Germany 17.53%, Sweden 12.68%, UK 8.49%, US 6.05%, Norway 6.01%, Netherlands 4.84%, France 4.57% (2009)
Imports: $90.83 billion (2010 est.)
country comparison to the world: 33
$84.46 billion (2009 est.)
Imports—commodities: machinery and equipment, raw materials and semimanufactures for industry, chemicals, grain and foodstuffs, consumer goods
Imports—partners: Germany 21.07%, Sweden 13.18%, Norway 7%, Netherlands 6.97%, China 6.22%, UK 5.53% (2009)
Reserves of foreign exchange and gold: $NA (31 December 2010 est.)
$76.65 billion (31 December 2009 est.)
Debt—external: $559.5 billion (30 June 2010)
country comparison to the world: 19
$588.8 billion (31 December 2008)
Stock of direct foreign investment—at home: $149.6 billion (31 December 2010 est.)
country comparison to the world: 25
$144.6 billion (31 December 2009 est.)
Stock of direct foreign investment—abroad: $199.8 billion (31 December 2010 est.)
country comparison to the world: 19
$186.6 billion (31 December 2009 est.)
Exchange rates: Danish kroner (DKK) per US dollar–5.624 (2010), 5.361 (2009), 5.0236 (2008), 5.4797 (2007), 5.9468 (2006)

COMMUNICATIONS

Telephones—main lines in use: 2.062 million (2009)
country comparison to the world: 56
Telephones—mobile cellular: 7.406 million (2009)
country comparison to the world: 79
Telephone system: *general assessment:* excellent telephone and telegraph services
domestic: buried and submarine cables and microwave radio relay form trunk network, multiple cellular mobile communications systems
international: country code–45; a series of fiber-optic submarine cables link Denmark with Canada, Faroe Islands, Germany, Iceland, Netherlands, Norway, Poland, Russia, Sweden, and UK; satellite earth stations–18 (6 Intelsat, 10 Eutelsat, 1 Orion, 1 Inmarsat (Blaavand-Atlantic-East)); note–the Nordic countries (Denmark, Finland, Iceland, Norway, and Sweden) share the Danish earth station and the Eik, Norway, station for worldwide Inmarsat access (2008)
Broadcast media: strong public-sector television presence with state-owned Danmarks Radio (DR) operating 4 channels and publicly-owned TV2 operating roughly a half dozen channels; broadcasts of privately-owned stations are available via satellite and cable feed; DR operates 4 nationwide FM radio stations, 15 digital audio broadcasting stations, and about 15 web-based radio stations; approximately 250 commercial and community radio stations are operational (2007)
Internet country code: .dk
Internet hosts: 4.145 million (2010)
country comparison to the world: 22
Internet users: 4.75 million (2009)
country comparison to the world: 48

TRANSPORTATION

Airports: 92 (2010)
country comparison to the world: 65

Airports—with paved runways: *total:* 28
over 3,047 m: 2
2,438 to 3,047 m: 7
1,524 to 2,437 m: 4
914 to 1,523 m: 12
under 914 m: 3 (2010)
Airports—with unpaved runways: *total:* 64
914 to 1,523 m: 3
under 914 m: 61 (2010)
Pipelines: gas 2,858 km; oil 107 km (2010)
Railways: *total:* 2,667 km
country comparison to the world: 61
standard gauge: 2,667 km 1.435-m gauge (640 km electrified) (2008)
Roadways: *total:* 73,197 km
country comparison to the world: 64
paved: 73,197 km (includes 1,111 km of expressways) (2008)
Waterways: 400 km (2010)
country comparison to the world: 88
Merchant marine: *total:* 347
country comparison to the world: 28
by type: bulk carrier 4, cargo 56, carrier 1, chemical tanker 104, container 87, liquefied gas 4, passenger/cargo 40, petroleum tanker 38, refrigerated cargo 4, roll on/roll off 6, specialized tanker 3
foreign-owned: 32 (Germany 10, Greece 1, Iceland 3, Norway 2, Sweden 16)
registered in other countries: 592 (Antigua and Barbuda 20, Bahamas 59, Belgium 4, Brazil 3, Cyprus 6, Egypt 1, France 12, Georgia 1, Gibraltar 6, Hong Kong 41, Isle of Man 26, Italy 4, Jamaica 1, Liberia 4, Lithuania 8, Malaysia 1, Malta 41, Marshall Islands 7, Mexico 2, Netherlands 36, former Netherlands Antilles 1, Norway 11, Panama 46, Portugal 4, Saint Vincent and the Grenadines 19, Singapore 125, South Africa 1, Spain 2, Sweden 15, UK 46, Uruguay 1, US 34, Venezuela 1, unknown 3) (2010)
Ports and terminals: Aalborg, Aarhus, Copenhagen, Ensted, Esbjerg, Fredericia, Kalundborg

MILITARY

Military branches: Defense Command: Army Operational Command, Admiral Danish Fleet, Arctic Command, Tactical Air Command, Home Guard (2010)
Military service age and obligation: 18 years of age for compulsory and voluntary military service; conscripts serve an initial training period that varies from 4 to 12 months according to specialization; reservists are assigned to mobilization units following completion of their conscript service; women eligible to volunteer for military service (2004)
Manpower available for military service: *males age 16–49:* 1,236,337
females age 16–49: 1,224,182 (2010 est.)
Manpower fit for military service: *males age 16–49:* 1,014,560
females age 16–49: 1,003,921 (2010 est.)
Manpower reaching militarily significant age annually:
male: 37,913
female: 35,865 (2010 est.)
Military expenditures: 1.3% of GDP (2007 est.)
country comparison to the world: 116

TRANSNATIONAL ISSUES

Disputes—international: Iceland, the UK, and Ireland dispute Denmark's claim that the Faroe Islands' continental shelf extends beyond 200 nm; Faroese continue to study proposals for full independence; sovereignty dispute with Canada over Hans Island in the Kennedy Channel between Ellesmere Island and Greenland; Denmark (Greenland) and Norway have made submissions to the Commission on the Limits of the Continental shelf (CLCS) and Russia is collecting additional data to augment its 2001 CLCS submission

DHEKELIA

(UK SOVEREIGN BASE AREA)

INTRODUCTION

Background: By terms of the 1960 Treaty of Establishment that created the independent Republic of Cyprus, the UK retained full sovereignty and jurisdiction over two areas of almost 254 square kilometers–Akrotiri and Dhekelia. The larger of these is the Dhekelia Sovereign Base Area, which is also referred to as the Eastern Sovereign Base Area.

GEOGRAPHY

Location: Eastern Mediterranean, on the southeast coast of Cyprus near Famagusta
Geographic coordinates: 34 59 N, 33 45 E
Map references: Europe
Area: *total:* 130.8 sq km
country comparison to the world: 221
note: area surrounds three Cypriot enclaves
Area—comparative: about three-quarters the size of Washington, DC
Land boundaries: *total:* 103 km (approximately)
border countries: Cyprus 103 km (approximately)
Coastline: 27.5 km
Climate: temperate; Mediterranean with hot, dry summers and cool winters
Environment—current issues: netting and trapping of small migrant songbirds in the spring and autumn
Geography—note: British extraterritorial rights also extended to several small off-post sites scattered across Cyprus; of the Sovereign Base Area land 60% is privately owned and farmed, 20% is owned by the Ministry of Defense, and 20% is SBA Crown land

PEOPLE

Population: approximately 15,700 live on the Sovereign Base Areas of Akrotiri and Dhekelia including 7,700 Cypriots, 3,600 service and UK based contract personnel, and 4,400 dependents
country comparison to the world: 219
Languages: English, Greek
School life expectancy (primary to tertiary education): NA
Education expenditures: NA

GOVERNMENT

Country name: *conventional long form:* none
conventional short form: Dhekelia
Dependency status: a special form of UK overseas territory; administered by an administrator who is also the Commander, British Forces Cyprus
Capital: *name:* Episkopi Cantonment (base administrative center for Akrotiri and Dhekelia); located in Akrotiri
geographic coordinates: 34 40 N, 32 51 E
time difference: UTC+2 (7 hours ahead of Washington, DC during Standard Time)
daylight saving time: +1hr, begins last Sunday in March; ends last Sunday in October
Constitution: Sovereign Base Areas of Akrotiri and Dhekelia Order in Council 1960, effective 16 August 1960, functions as a basic legal document
Legal system: the Sovereign Base Area Administration has its own court system to deal with civil and criminal matters; laws applicable to the Cypriot population are, as far as possible, the same as the laws of the Republic of Cyprus
Executive branch: *chief of state:* Queen ELIZABETH II (since 6 February 1952)
head of government: Administrator Air Vice Marshall Graham STACEY (since 4 November 2010); note–reports to the British Ministry of Defense
elections: none; the monarchy is hereditary; the administrator appointed by the monarch
Judicial branch: see Akrotiri
Diplomatic representation in the US: none (overseas territory of the UK)
Diplomatic representation from the US: none (overseas territory of the UK)
Flag description: the flag of the UK is used
National anthem: note: as a United Kingdom area of special sovereignty, "God Save the Queen" is official (see United Kingdom)

ECONOMY

Economy—overview: Economic activity is limited to providing services to the military and their families located in Dhekelia. All food and manufactured goods must be imported.
Industries: none

Exchange rates: note: uses the euro

COMMUNICATIONS

Broadcast media: British Forces Broadcast Service (BFBS) provides multi-channel satellite TV service as well as BFBS radio broadcasts to the Dhekelia Sovereign Base (2009)

MILITARY

Military—note: includes Dhekelia Garrison and Ayios Nikolaos Station connected by a roadway

DJIBOUTI

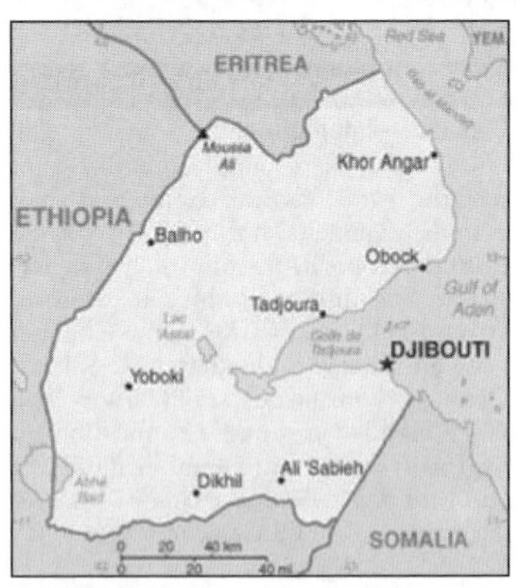

INTRODUCTION

Background: The French Territory of the Afars and the Issas became Djibouti in 1977. Hassan Gouled APTIDON installed an authoritarian one-party state and proceeded to serve as president until 1999. Unrest among the Afars minority during the 1990s led to a civil war that ended in 2001 following the conclusion of a peace accord between Afar rebels and the Issa-dominated government. In 1999, Djibouti's first multi-party presidential elections resulted in the election of Ismail Omar GUELLEH; he was re-elected to a second term in 2005. Djibouti occupies a strategic geographic location at the mouth of the Red Sea and serves as an important transshipment location for goods entering and leaving the east African highlands. The present leadership favors close ties to France, which maintains a significant military presence in the country, but also has strong ties with the US. Djibouti hosts the only US military base in sub-Saharan Africa.

GEOGRAPHY

Location: Eastern Africa, bordering the Gulf of Aden and the Red Sea, between Eritrea and Somalia
Geographic coordinates: 11 30 N, 43 00 E
Map references: Africa
Area: *total:* 23,200 sq km
country comparison to the world: 150
land: 23,180 sq km
water: 20 sq km
Area—comparative: slightly smaller than Massachusetts
Land boundaries: *total:* 516 km
border countries: Eritrea 109 km, Ethiopia 349 km, Somalia 58 km
Coastline: 314 km
Maritime claims: territorial sea: 12 nm
contiguous zone: 24 nm
exclusive economic zone: 200 nm
Climate: desert; torrid, dry
Terrain: coastal plain and plateau separated by central mountains
Elevation extremes: *lowest point:* Lac Assal -155 m
highest point: Moussa Ali 2,028 m
Natural resources: potential geothermal power, gold, clay, granite, limestone, marble, salt, diatomite, gypsum, pumice, petroleum
Land use: *arable land:* 0.04%
permanent crops: 0%
other: 99.96% (2005)
Irrigated land: 10 sq km (2003)
Total renewable water resources: 0.3 cu km (1997)
Freshwater withdrawal (domestic/industrial/agricultural): *total:* 0.02 cu km/yr (84%/0%/16%)
per capita: 25 cu m/yr (2000)
Natural hazards: earthquakes; droughts; occasional cyclonic disturbances from the Indian Ocean bring heavy rains and flash floods
volcanism: Djibouti experiences limited volcanic activity; Ardoukoba (elev. 298 m) last erupted in 1978; Manda-Inakir, located along the Ethiopian border, is also historically active
Environment—current issues: inadequate supplies of potable water; limited arable land; desertification; endangered species
Environment—international agreements: *party to:* Biodiversity, Climate Change, Climate Change-Kyoto Protocol, Desertification, Endangered Species, Hazardous Wastes, Law of the Sea, Ozone Layer Protection, Ship Pollution, Wetlands
signed, but not ratified: none of the selected agreements
Geography—note: strategic location near world's busiest shipping lanes and close to Arabian oilfields; terminus of rail traffic into Ethiopia; mostly wasteland; Lac Assal (Lake Assal) is the lowest point in Africa

PEOPLE

Population: 757,074 (July 2011 est.)
country comparison to the world: 162
Age structure: *0–14 years:* 35% (male 132,592/female 132,114)
15–64 years: 61.7% (male 206,323/female 260,772)
65 years and over: 3.3% (male 11,349/female 13,924) (2011 est.)
Median age: *total:* 21.8 years
male: 20.2 years
female: 23.1 years (2011 est.)
Population growth rate: 2.237% (2011 est.)
country comparison to the world: 37
Birth rate: 25.27 births/1,000 population (2011 est.)
country comparison to the world: 59
Death rate: 8.23 deaths/1,000 population (July 2011 est.)
country comparison to the world: 93
Net migration rate: 5.33 migrant(s)/1,000 population (2011 est.)
country comparison to the world: 16
Urbanization: *urban population:* 76% of total population (2010)
rate of urbanization: 1.8% annual rate of change (2010-15 est.)
Major cities—population: DJIBOUTI (capital) 567,000 (2009)
Sex ratio: *at birth:* 1.03 male(s)/female
under 15 years: 1 male(s)/female
15–64 years: 0.8 male(s)/female
65 years and over: 0.81 male(s)/female
total population: 0.86 male(s)/female (2011 est.)
Infant mortality rate: *total:* 54.94 deaths/1,000 live births
country comparison to the world: 40
male: 62.63 deaths/1,000 live births
female: 47.02 deaths/1,000 live births (2011 est.)
Life expectancy at birth: *total population:* 61.14 years
country comparison to the world: 183
male: 58.69 years
female: 63.66 years (2011 est.)
Total fertility rate: 2.71 children born/woman (2011 est.)
country comparison to the world: 74
HIV/AIDS—adult prevalence rate: 2.5% (2009 est.)
country comparison to the world: 25
HIV/AIDS—people living with HIV/AIDS: 14,000 (2009 est.)
country comparison to the world: 88
HIV/AIDS—deaths: 1,000 (2009 est.)
country comparison to the world: 70
Major infectious diseases: *degree of risk:* high
food or waterborne diseases: bacterial and protozoal diarrhea, hepatitis A and E, and typhoid fever
vectorborne disease: malaria
note: highly pathogenic H5N1 avian influenza has been identified in this country; it poses a negligible risk with extremely rare cases possible among US citizens who have close contact with birds (2009)
Drinking water source: *Improved:*
urban: 98% of population
rural: 52% of population
total: 92% of population
Unimproved:
urban: 2% of population
rural: 48% of population
total: 8% of population (2008)
Sanitation facility access: *Improved:*
urban: 63% of population
rural: 10% of population
total: 56% of population

Unimproved:
urban: 37% of population
rural: 90% of population
total: 44% of population (2008)
Nationality:
noun: Djiboutian(s)
adjective: Djiboutian
Ethnic groups: Somali 60%, Afar 35%, other 5% (includes French, Arab, Ethiopian, and Italian)
Religions: Muslim 94%, Christian 6%
Languages: French (official), Arabic (official), Somali, Afar
Literacy: *definition:* age 15 and over can read and write
total population: 67.9%
male: 78%
female: 58.4% (2003 est.)
School life expectancy (primary to tertiary education): *total:* 5 years
male: 6 years
female: 5 years (2009)
Education expenditures: 8.4% of GDP (2007)
country comparison to the world: 9

GOVERNMENT

Country name: *conventional long form:* Republic of Djibouti
conventional short form: Djibouti
local long form: Republique de Djibouti/Jumhuriyat Jibuti
local short form: Djibouti/Jibuti
former: French Territory of the Afars and Issas, French Somaliland
Government type: republic
Capital: *name:* Djibouti
geographic coordinates: 11 35 N, 43 09 E
time difference: UTC+3 (8 hours ahead of Washington, DC during Standard Time)
Administrative divisions: 6 districts (cercles, singular—cercle); Ali Sabieh, Arta, Dikhil, Djibouti, Obock, Tadjourah
Independence: 27 June 1977 (from France)
National holiday: Independence Day, 27 June (1977)
Constitution: approved by referendum 4 September 1992; note—constitution allows for multiparties
Legal system: mixed legal system based primarily on the French civil code (as it existed in 1997) and Islamic religious law (in matters of family law and successions), and customary law
International law organization participation: accepts compulsory ICJ jurisdiction with reservations; accepts ICCt jurisdiction
Suffrage: 18 years of age; universal
Executive branch: *chief of state:* President Ismail Omar GUELLEH (since 8 May 1999)
head of government: Prime Minister Mohamed Dileita DILEITA (since 4 March 2001)
cabinet: Council of Ministers responsible to the president
(For more information visit the World Leaders website)
elections: president elected by popular vote for a five-year term; president is eligible to hold office until age 75; election last held on 8 April 2011 (next to be held by 2016); prime minister appointed by the president
election results: Ismail Omar GUELLEH reelected president; percent of vote—Ismail Omar GUELLEH 80.6%, Mohamed Warsama RAGUEH 19.4%
Legislative branch: unicameral Chamber of Deputies or Chambre des Deputes (65 seats; members elected by popular vote to serve five-year terms); note—constitutional amendments in 2010 provided for the establishment of a senate
elections: last held on 8 February 2008 (next to be held in 2013)
election results: percent of vote by party—NA; seats—UMP (coalition of parties associated with President Ismail Omar GUELLAH) 65
Judicial branch: Supreme Court or Cour Supreme; Constitutional Court
Political parties and leaders: Democratic National Party or PND [ADEN Robleh Awaleh]; Democratic Renewal Party or PRD [Abdillahi HAMARITEH]; Djibouti Development Party or PDD [Mohamed Daoud CHEHEM]; Front pour la Restauration de l'Unite Democratique or FRUD [Ali Mohamed DAOUD]; People's Progress Assembly or RPP [Ismail Omar GUELLEH] (governing party); Peoples Social Democratic Party or PPSD [Moumin Bahdon FARAH]; Republican Alliance for Democracy or ARD [Ahmed YOUSSOUF]; Union for a Presidential Majority or UMP [Mohamed Dileita DILEITA] (a coalition of parties including RPP, FRUD, PND, and PPSD); Union for Democracy and Justice or UDJ
Political pressure groups and leaders: NA
International organization participation: ACP, AfDB, AFESD, AMF, AU, COMESA, FAO, G-77, IBRD, ICAO, ICRM, IDA, IDB, IFAD, IFC, IFRCS, IGAD, ILO, IMF, IMO, Interpol, IOC, IPU, ITU, ITUC, LAS, MIGA, MINURSO, NAM, OIC, OIF, OPCW, UN, UNCTAD, UNESCO, UNHCR, UNIDO, UNWTO, UPU, WCO, WFTU, WHO, WIPO, WMO, WTO
Diplomatic representation in the US: *chief of mission:* Ambassador Roble OLHAYE Oudine
chancery: Suite 515, 1156 15th Street NW, Washington, DC 20005
telephone: [1] (202) 331-0270
FAX: [1] (202) 331-0302
Diplomatic representation from the US: *chief of mission:* Ambassador James C. SWAN
embassy: Plateau du Serpent, Boulevard Marechal Joffre, Djibouti
mailing address: B. P. 185, Djibouti
telephone: [253] 35 39 95
FAX: [253] 35 39 40
Flag description: two equal horizontal bands of light blue (top) and light green with a white isosceles triangle based on the hoist side bearing a red five-pointed star in the center; blue stands for sea and sky and the Issa Somali people; green symbolizes earth and the Afar people; white represents peace; the red star recalls the struggle for independence and stands for unity
National anthem: *name:* "Jabuuti" (Djibouti)
lyrics/music: Aden ELMI/Abdi ROBLEH
note: adopted 1977

ECONOMY

Economy—overview: The economy is based on service activities connected with the country's strategic location and status as a free trade zone in the Horn of Africa. Two-thirds of Djibouti's inhabitants live in the capital city; the remainder are mostly nomadic herders. Scanty rainfall limits crop production to fruits and vegetables, and most food must be imported. Djibouti provides services as both a transit port for the region and an international transshipment and refueling center. Imports and exports from landlocked neighbor Ethiopia represent 70% of port activity at Djibouti's container terminal. Djibouti has few natural resources and little industry. The nation is, therefore, heavily dependent on foreign assistance to help support its balance of payments and to finance development projects. An unemployment rate of nearly 60% in urban areas continues to be a major problem. While inflation is not a concern, due to the fixed tie of the Djiboutian franc to the US dollar, the artificially high value of the Djiboutian franc adversely affects Djibouti's balance of payments. Per capita consumption dropped an estimated 35% between 1999 and 2006 because of recession, civil war, and a high population growth rate (including immigrants and refugees). Djibouti has experienced relatively minimal impact from the global economic downturn, but its reliance on diesel-generated electricity and imported food leave average consumers vulnerable to global price shocks.
GDP (purchasing power parity): $2.105 billion (2010 est.)
country comparison to the world: 184
$2.014 billion (2009 est.)
$1.918 billion (2008 est.)
note: data are in 2010 US dollars
GDP (official exchange rate): $1.14 billion (2010 est.)
GDP—real growth rate: 4.5% (2010 est.)
country comparison to the world: 82
5% (2009 est.)
5.8% (2008 est.)
GDP—per capita (PPP): $2,800 (2010 est.)
country comparison to the world: 172
$2,800 (2009 est.)
$2,700 (2008 est.)
note: data are in 2010 US dollars
GDP—composition by sector:
agriculture: 3.2%
industry: 14.9%
services: 81.9% (2006 est.)
Labor force: 351,700 (2007)
country comparison to the world: 159
Labor force—by occupation:
agriculture: NA%
industry: NA%
services: NA%
Unemployment rate: 59% (2007 est.)
country comparison to the world: 194
note: data are for urban areas, 83% in rural areas
Population below poverty line: 42% (2007 est.)
Household income or consumption by percentage share:
lowest 10%: 2.4%
highest 10%: 30.9% (2002)

Budget: *revenues:* $135 million
expenditures: $182 million (1999 est.)
Inflation rate (consumer prices): 6% (2009 est.)
country comparison to the world: 157
5% (2007 est.)
Commercial bank prime lending rate: NA% (31 December 2009 est.)
country comparison to the world: 74
11.56% (31 December 2008 est.)
Stock of narrow money: $577.8 million (31 December 2009)
country comparison to the world: 158
$462.7 million (31 December 2008)
Stock of broad money: $940.8 million (31 December 2009 est.)
country comparison to the world: 166
$800.8 million (31 December 2008 est.)
Stock of domestic credit: $339 million (31 December 2009)
country comparison to the world: 169
$269.9 million (31 December 2008)
Agriculture—products: fruits, vegetables; goats, sheep, camels, animal hides
Industries: construction, agricultural processing
Electricity—production: 280 million kWh (2007 est.)
country comparison to the world: 170
Electricity—consumption: 260.4 million kWh (2007 est.)
country comparison to the world: 171
Electricity—exports: 0 kWh (2008 est.)
Electricity—imports: 0 kWh (2008 est.)
Oil—production: 0 bbl/day (2009 est.)
country comparison to the world: 166
Oil—consumption: 12,000 bbl/day (2009 est.)
country comparison to the world: 143
Oil—exports: 19.18 bbl/day (2007 est.)
country comparison to the world: 137
Oil—imports: 8,476 bbl/day (2007 est.)
country comparison to the world: 143
Oil—proved reserves: 0 bbl (1 January 2010 est.)
country comparison to the world: 122
Natural gas—production: 0 cu m (2008 est.)
country comparison to the world: 118
Natural gas—consumption: 0 cu m (2008 est.)
country comparison to the world: 165
Natural gas—exports: 0 cu m (2008 est.)
country comparison to the world: 84
Natural gas—imports: 0 cu m (2008 est.)
country comparison to the world: 176
Natural gas—proved reserves: 0 cu m (1 January 2010 est.)
country comparison to the world: 127
Current account balance: $-352 million (2009 est.)
country comparison to the world: 105
$-212 million (2007 est.)
Exports: $100 million (2009 est.)
country comparison to the world: 194
$340 million (2006 est.)
Exports—commodities: reexports, hides and skins, coffee (in transit)
Exports—partners: Somalia 76.68%, France 4.89%, UAE 4.22% (2009)
Imports: $644 million (2009 est.)
country comparison to the world: 184
$1.555 billion (2006)
Imports—commodities: foods, beverages, transport equipment, chemicals, petroleum products
Imports—partners: Saudi Arabia 16.26%, India 16.03%, China 14.26%, US 9.57%, Malaysia 6.63%, Japan 4.74% (2009)
Debt—external: $428 million (2006)
country comparison to the world: 164
Exchange rates: Djiboutian francs (DJF) per US dollar–177.71 (2007), 174.75 (2006), 177.72 (2005), 177.72 (2004), 177.72 (2003)

COMMUNICATIONS

Telephones—main lines in use: 16,800 (2009)
country comparison to the world: 198
Telephones—mobile cellular: 128,800 (2009)
country comparison to the world: 179
Telephone system: *general assessment:* telephone facilities in the city of Djibouti are adequate, as are the microwave radio relay connections to outlying areas of the country
domestic: Djibouti Telecom is the sole provider of telecommunications services and utilizes mostly a microwave radio relay network; fiber-optic cable is installed in the capital; rural areas connected via wireless local loop radio systems; mobile cellular coverage is primarily limited to the area in and around Djibouti city
international: country code–253; landing point for the SEA-ME-WE-3 optical telecommunications submarine cable with links to Asia, the Middle East, and Europe; satellite earth stations–2 (1 Intelsat–Indian Ocean and 1 Arabsat); Medarabtel regional microwave radio relay telephone network (2009)
Broadcast media: maintains restrictions on the licensing and operation of broadcast media; state-owned Radiodiffusion-Television de Djibouti (RTD) operates the sole terrestrial TV station as well as the only 2 domestic radio networks; no private TV or radio stations; transmissions of several international broadcasters are available (2007)
Internet country code: .dj
Internet hosts: 195 (2010)
country comparison to the world: 196
Internet users: 25,900 (2009)
country comparison to the world: 184

TRANSPORTATION

Airports: 13 (2010)
country comparison to the world: 152
Airports—with paved runways: *total:* 3
over 3,047 m: 1
2,438 to 3,047 m: 1
1,524 to 2,437 m: 1 (2010)
Airports—with unpaved runways: *total:* 10
1,524 to 2,437 m: 1
914 to 1,523 m: 7
under 914 m: 2 (2010)
Railways: *total:* 100 km (Djibouti segment of the 781 km Addis Ababa-Djibouti railway)
country comparison to the world: 126
narrow gauge: 100 km 1.000-m gauge
note: railway is under joint control of Djibouti and Ethiopia but is largely inoperable (2010)
Roadways: *total:* 3,065 km
country comparison to the world: 164
paved: 1,226 km
unpaved: 1,839 km (2000)
Ports and terminals: Djibouti
Transportation—note: the International Maritime Bureau reports offshore waters in the Gulf of Aden are high risk for piracy; numerous vessels, including commercial shipping and pleasure craft, have been attacked and hijacked both at anchor and while underway; crew, passengers, and cargo are held for ransom; the presence of several naval task forces in the Gulf of Aden and additional anti-piracy measures on the part of ship operators reduced the incidence of piracy in that body of water by more than half in 2010

MILITARY

Military branches: Djibouti Armed Forces (Forces Armees Djiboutiennes, FAD): Djibouti National Army (includes Coastal Navy, Djiboutian Air Force (Force Aerienne Djiboutienne, FAD), National Gendarmerie (GN)) (2011)
Military service age and obligation: 18 years of age for voluntary military service; 16-25 years of age for voluntary military training; no conscription (2008)
Manpower available for military service: *males age 16–49:* 170,386
females age 16–49: 221,411 (2010 est.)
Manpower fit for military service: *males age 16–49:* 114,557
females age 16–49: 154,173 (2010 est.)
Manpower reaching militarily significant age annually:
male: 8,360
female: 8,602 (2010 est.)
Military expenditures: 3.8% of GDP (2006)
country comparison to the world: 28

TRANSNATIONAL ISSUES

Disputes—international: Djibouti maintains economic ties and border accords with "Somaliland" leadership while maintaining some political ties to various factions in Somalia; Kuwait is chief investor in the 2008 restoration and upgrade of the Ethiopian-Djibouti rail link; in 2008, Eritrean troops move across the border on Ras Doumera peninsula and occupy Doumera Island with undefined sovereignty in the Red Sea
Refugees and internally displaced persons: refugees (country of origin): 8,642 (Somalia) (2007)

DOMINICA

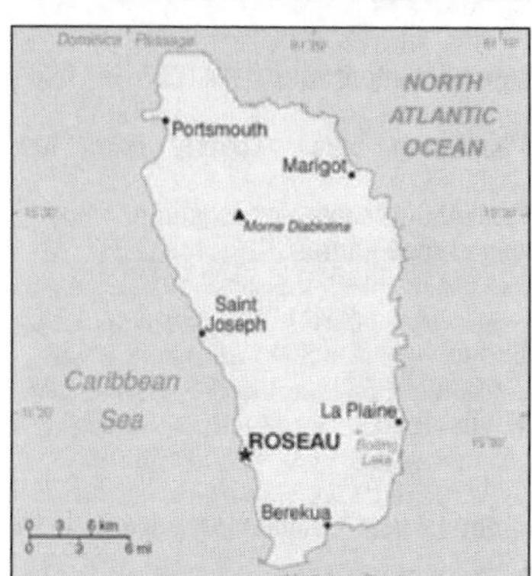

INTRODUCTION

Background: Dominica was the last of the Caribbean islands to be colonized by Europeans due chiefly to the fierce resistance of the native Caribs. France ceded possession to Great Britain in 1763, which made the island a colony in 1805. In 1980, two years after independence, Dominica's fortunes improved when a corrupt and tyrannical administration was replaced by that of Mary Eugenia CHARLES, the first female prime minister in the Caribbean, who remained in office for 15 years. Some 3,000 Carib Indians still living on Dominica are the only pre-Columbian population remaining in the eastern Caribbean.

GEOGRAPHY

Location: Caribbean, island between the Caribbean Sea and the North Atlantic Ocean, about half way between Puerto Rico and Trinidad and Tobago
Geographic coordinates: 15 25 N, 61 20 W
Map references: Central America and the Caribbean
Area: *total:* 751 sq km
country comparison to the world: 188
land: 751 sq km
water: 0 sq km
Area—comparative: slightly more than four times the size of Washington, DC
Land boundaries: 0 km
Coastline: 148 km
Maritime claims: territorial sea: 12 nm
contiguous zone: 24 nm
exclusive economic zone: 200 nm
Climate: tropical; moderated by northeast trade winds; heavy rainfall
Terrain: rugged mountains of volcanic origin
Elevation extremes: *lowest point:* Caribbean Sea 0 m
highest point: Morne Diablotins 1,447 m
Natural resources: timber, hydropower, arable land
Land use: *arable land:* 6.67%
permanent crops: 21.33%
other: 72% (2005)
Irrigated land: NA
Total renewable water resources: NA
Freshwater withdrawal (domestic/industrial/agricultural): *total:* 0.02 cu km/yr
per capita: 213 cu m/yr (1996)
Natural hazards: flash floods are a constant threat; destructive hurricanes can be expected during the late summer months
Environment—current issues: NA
Environment—international agreements: *party to:* Biodiversity, Climate Change, Climate Change-Kyoto Protocol, Desertification, Endangered Species, Environmental Modification, Hazardous Wastes, Law of the Sea, Ozone Layer Protection, Ship Pollution, Whaling
signed, but not ratified: none of the selected agreements
Geography—note: known as "The Nature Island of the Caribbean" due to its spectacular, lush, and varied flora and fauna, which are protected by an extensive natural park system; the most mountainous of the Lesser Antilles, its volcanic peaks are cones of lava craters and include Boiling Lake, the second-largest, thermally active lake in the world

PEOPLE

Population: 72,969 (July 2011 est.)
country comparison to the world: 199
Age structure: *0–14 years:* 22.9% (male 8,551/female 8,188)
15–64 years: 66.8% (male 25,007/female 23,730)
65 years and over: 10.3% (male 3,246/female 4,247) (2011 est.)
Median age: *total:* 30.8 years
male: 30.4 years
female: 31.3 years (2011 est.)
Population growth rate: 0.214% (2011 est.)
country comparison to the world: 179
Birth rate: 15.62 births/1,000 population (2011 est.)
country comparison to the world: 130
Death rate: 8.06 deaths/1,000 population (July 2011 est.)
country comparison to the world: 103
Net migration rate: -5.43 migrant(s)/1,000 population (2011 est.)
country comparison to the world: 196
Urbanization: *urban population:* 67% of total population (2010)
rate of urbanization: 0.3% annual rate of change (2010-15 est.)
Major cities—population: ROSEAU (capital) 14,000 (2009)
Sex ratio: *at birth:* 1.05 male(s)/female
under 15 years: 1.04 male(s)/female
15–64 years: 1.05 male(s)/female
65 years and over: 0.76 male(s)/female
total population: 1.02 male(s)/female (2011 est.)
Infant mortality rate: *total:* 12.78 deaths/1,000 live births
country comparison to the world: 129
male: 17.11 deaths/1,000 live births
female: 8.23 deaths/1,000 live births (2011 est.)
Life expectancy at birth: *total population:* 75.98 years
country comparison to the world: 77
male: 73.03 years
female: 79.08 years (2011 est.)
Total fertility rate: 2.07 children born/woman (2011 est.)
country comparison to the world: 122
HIV/AIDS—adult prevalence rate: NA
HIV/AIDS—people living with HIV/AIDS: NA
HIV/AIDS—deaths: NA
Drinking water source: *Improved:*
urban: 96% of population
rural: 92% of population
total: 95% of population
Unimproved:
urban: 4% of population
rural: 8% of population
total: 5% of population (2000)
Sanitation facility access: *Improved:*
urban: 80% of population
rural: 84% of population
total: 81% of population
Unimproved:
urban: 20% of population
rural: 16% of population
total: 19% of population (2000)
Nationality: *noun:* Dominican(s)
adjective: Dominican
Ethnic groups: black 86.8%, mixed 8.9%, Carib Amerindian 2.9%, white 0.8%, other 0.7% (2001 census)
Religions: Roman Catholic 61.4%, Seventh Day Adventist 6%, Pentecostal 5.6%, Baptist 4.1%, Methodist 3.7%, Church of God 1.2%, Jehovah's Witnesses 1.2%, other Christian 7.7%, Rastafarian 1.3%, other or unspecified 1.6%, none 6.1% (2001 census)
Languages: English (official), French patois
Literacy: *definition:* age 15 and over has ever attended school
total population: 94%
male: 94%
female: 94% (2003 est.)
School life expectancy (primary to tertiary education): *total:* 13 years
male: 13 years
female: 13 years (2008)
Education expenditures: 4.7% of GDP (2008)
country comparison to the world: 71

GOVERNMENT

Country name: *conventional long form:* Commonwealth of Dominica
conventional short form: Dominica
Government type: parliamentary democracy
Capital: *name:* Roseau
geographic coordinates: 15 18 N, 61 24 W
time difference: UTC-4 (1 hour ahead of Washington, DC during Standard Time)
Administrative divisions: 10 parishes; Saint Andrew, Saint David, Saint George, Saint John, Saint Joseph, Saint Luke, Saint Mark, Saint Patrick, Saint Paul, Saint Peter
Independence: 3 November 1978 (from the UK)
National holiday: Independence Day, 3 November (1978)
Constitution: 3 November 1978
Legal system: common law based on the English model

International law organization participation: accepts compulsory ICJ jurisdiction; accepts ICCt jurisdiction
Suffrage: 18 years of age; universal
Executive branch: *chief of state:* President Nicholas J. O. LIVERPOOL (since October 2003)
head of government: Prime Minister Roosevelt SKERRIT (since 8 January 2004)
cabinet: Cabinet appointed by the president on the advice of the prime minister
(For more information visit the World Leaders website)
elections: president elected by the House of Assembly for a five-year term; election last held on 1 October 2003 (next to be held in 2013); prime minister appointed by the president
election results: in the absence of an opposition candidate, Nicholas LIVERPOOL consented to a second term in 2008 at the request of the prime minister and leader of the opposition and no formal election was held in 2008
Legislative branch: unicameral House of Assembly (30 seats; 9 members appointed, 21 elected by popular vote to serve five-year terms)
elections: last held on 18 December 2009 (next to be held in 2015); note—tradition dictates that the election will be held within five years of the last election, but technically it is five years from the first seating of parliament (12 May 2005) plus a 90-day grace period
election results: percent of vote by party—DLP 61.2%, UWP 34.9%; seats by party—DLP 18, UWP 3
Judicial branch: Eastern Caribbean Supreme Court, consisting of the Court of Appeal and the High Court (located in Saint Lucia; one of the six judges must reside in Dominica and preside over the Court of Summary Jurisdiction)
Political parties and leaders: Dominica Freedom Party or DFP [Charles SAVARIN]; Dominica Labor Party or DLP [Roosevelt SKERRIT]; Dominica United Workers Party or UWP [Earl WILLIAMS]
Political pressure groups and leaders: Dominica Liberation Movement or DLM (a small leftist party)
International organization participation: ACP, AOSIS, C, Caricom, CDB, FAO, G-77, IBRD, ICRM, IDA, IFAD, IFC, IFRCS, ILO, IMF, IMO, Interpol, IOC, ISO (subscriber), ITU, ITUC, MIGA, NAM, OAS, OECS, OIF, OPANAL, OPCW, PetroCaribe, UN, UNCTAD, UNESCO, UNIDO, UPU, WHO, WIPO, WMO, WTO
Diplomatic representation in the US: *chief of mission:* Ambassador Hubert J. CHARLES
chancery: 3216 New Mexico Avenue NW, Washington, DC 20016
telephone: [1] (202) 364-6781
FAX: [1] (202) 364-6791
consulate(s) general: New York
Diplomatic representation from the US: the US does not have an embassy in Dominica; the US Ambassador to Barbados is accredited to Dominica
Flag description: green, with a centered cross of three equal bands—the vertical part is yellow (hoist side), black, and white and the horizontal part is yellow (top), black, and white; superimposed in the center of the cross is a red disk bearing a Sisserou Parrot, unique to Dominica, encircled by 10 green, five-pointed stars edged in yellow; the 10 stars represent the 10 administrative divisions (parishes); green symbolizes the island's lush vegetation; the triple-colored cross represents the Christian Trinity; the yellow color denotes sunshine, the main agricultural products (citrus and bananas), and the native Carib Indians; black is for the rich soil and the African heritage of most citizens; white signifies rivers, waterfalls, and the purity of aspirations; the red disc stands for social justice
National anthem: *name:* "Isle of Beauty, Isle of Splendor"
lyrics/music: Wilfred Oscar Morgan POND/Lemuel McPherson CHRISTIAN
note: adopted 1967

ECONOMY

Economy—overview: The Dominican economy has been dependent on agriculture—primarily bananas—in years past, but increasingly has been driven by tourism as the government seeks to promote Dominica as an "ecotourism" destination. In order to diversify the island's production base, the government also is attempting to develop an offshore financial sector and has signed an agreement with the EU to develop geothermal energy resources. In 2003, the government began a comprehensive restructuring of the economy—including elimination of price controls, privatization of the state banana company, and tax increases—to address an economic and financial crisis and to meet IMF requirements. This restructuring paved the way for an economic recovery—real growth for 2006 reached a two-decade high—and helped to reduce the debt burden, which remains at about 85% of GDP. Hurricane Dean struck the island in August 2007 causing damages equivalent to 20% of GDP. In 2009, growth slowed as a result of the global recession; it picked up only slightly in 2010.
GDP (purchasing power parity): $758 million (2010 est.)
country comparison to the world: 207
$751.1 million (2009 est.)
$752.9 million (2008 est.)
note: data are in 2010 US dollars
GDP (official exchange rate): $376 million (2010 est.)
GDP—real growth rate: 1% (2010 est.)
country comparison to the world: 172
-0.3% (2009 est.)
3.2% (2008 est.)
GDP—per capita (PPP): $10,400 (2010 est.)
country comparison to the world: 106
$10,300 (2009 est.)
$10,400 (2008 est.)
note: data are in 2010 US dollars
GDP—composition by sector:
agriculture: 17.7%
industry: 32.8%
services: 49.5% (2004 est.)
Labor force: 25,000 (2000 est.)
country comparison to the world: 206
Labor force—by occupation:
agriculture: 40%
industry: 32%
services: 28% (2000 est.)
Unemployment rate: 23% (2000 est.)
country comparison to the world: 173
Population below poverty line: 30% (2002 est.)
Household income or consumption by percentage share:
lowest 10%: NA%
highest 10%: NA%
Budget: *revenues:* $343 million
expenditures: $277 million (2009)
Public debt: 78% of GDP (2009 est.)
country comparison to the world: 21
85% of GDP (2006 est.)
Inflation rate (consumer prices): 0.1% (2009 est.)
country comparison to the world: 9
2.7% (2007 est.)
Central bank discount rate: 6.5% (31 December 2009)
country comparison to the world: 58
6.5% (31 December 2008)
Commercial bank prime lending rate: 10.02% (31 December 2009 est.)
country comparison to the world: 99
9.06% (31 December 2008 est.)
Stock of narrow money: $74.84 million (31 December 2009)
country comparison to the world: 183
$67.94 million (31 December 2008)
Stock of broad money: $398.5 million (31 December 2009)
country comparison to the world: 177
$362 million (31 December 2008)
Stock of domestic credit: $213.6 million (31 December 2008 est.)
country comparison to the world: 175
$193.1 million (31 December 2007 est.)
Agriculture—products: bananas, citrus, mangos, root crops, coconuts, cocoa
note: forest and fishery potential not exploited
Industries: soap, coconut oil, tourism, copra, furniture, cement blocks, shoes
Industrial production growth rate: NA%
Electricity—production: 85 million kWh (2007 est.)
country comparison to the world: 194
Electricity—consumption: 79.05 million kWh (2007 est.)
country comparison to the world: 193
Electricity—exports: 0 kWh (2008 est.)
Electricity—imports: 0 kWh (2008 est.)
Oil—production: 0 bbl/day (2009 est.)
country comparison to the world: 167
Oil—consumption: 1,000 bbl/day (2009 est.)
country comparison to the world: 197
Oil—exports: 0 bbl/day (2007 est.)
country comparison to the world: 161
Oil—imports: 838.2 bbl/day (2007 est.)
country comparison to the world: 190
Oil—proved reserves: 0 bbl (1 January 2010 est.)
country comparison to the world: 123
Natural gas—production: 0 cu m (2008 est.)
country comparison to the world: 119
Natural gas—consumption: 0 cu m (2008 est.)
country comparison to the world: 166
Natural gas—exports: 0 cu m (2008 est.)
country comparison to the world: 85

Natural gas—imports: 0 cu m (2008 est.)
country comparison to the world: 177
Natural gas—proved reserves: 0 cu m (1 January 2010 est.)
country comparison to the world: 128
Current account balance: $-72 million (2007 est.)
country comparison to the world: 73
Exports: $94 million (2006)
country comparison to the world: 196
Exports—commodities: bananas, soap, bay oil, vegetables, grapefruit, oranges
Exports—partners: Japan 28.62%, UK 19.81%, Antigua and Barbuda 7.7%, Guyana 6.52%, Jamaica 5.4%, Trinidad and Tobago 4.2% (2009)
Imports: $296 million (2006)
country comparison to the world: 196
Imports—commodities: manufactured goods, machinery and equipment, food, chemicals
Imports—partners: Japan 31.29%, US 19.73%, Trinidad and Tobago 11.8%, China 11.58% (2009)
Debt—external: $213 million (2004)
country comparison to the world: 174
Exchange rates: East Caribbean dollars (XCD) per US dollar–2.7 (2010), 2.7 (2009), 2.7 (2005), 2.7 (2004), 2.7 (2003)

COMMUNICATIONS

Telephones—main lines in use: 17,500 (2009)
country comparison to the world: 197
Telephones—mobile cellular: 106,000 (2009)
country comparison to the world: 184
Telephone system: *general assessment:* fully automatic network
domestic: Fixed-line connections continued to decline slowly with the two active operators providing about 25 fixed-line connections per 100 persons; subscribership among the three mobile-cellular providers continued to increase with teledensity approaching 150 per 100 persons in 2009
international: country code–1-767; landing points for the East Caribbean Fiber Optic System (ECFS) and the Global Caribbean Network (GCN) submarine cables providing connectivity to other islands in the eastern Caribbean extending from the British Virgin Islands to Trinidad; microwave radio relay and SHF radiotelephone links to Martinique and Guadeloupe; VHF and UHF radiotelephone links to Saint Lucia
Broadcast media: no terrestrial television service available; subscription cable TV provider offers some locally produced programming plus channels from the US, Latin America, and the Caribbean; state-operated radio broadcasts on 6 stations; privately-owned radio broadcasts on about 15 stations (2007)
Internet country code: .dm
Internet hosts: 718 (2010)
country comparison to the world: 174
Internet users: 28,000 (2009)
country comparison to the world: 182

TRANSPORTATION

Airports: 2 (2010)
country comparison to the world: 198
Airports—with paved runways: *total:* 2
1,524 to 2,437 m: 1
914 to 1,523 m: 1 (2010)
Roadways: *total:* 780 km
country comparison to the world: 187
paved: 393 km
unpaved: 387 km (2000)
Merchant marine: *total:* 40
country comparison to the world: 78
by type: bulk carrier 11, cargo 20, chemical tanker 2, petroleum tanker 4, refrigerated cargo 2, roll on/roll off 1
foreign-owned: 37 (Australia 1, Estonia 6, Germany 2, Greece 9, India 2, Latvia 1, Norway 1, Russia 6, Saudi Arabia 3, Singapore 1, Syria 2, Turkey 1, Ukraine 2)
registered in other countries: 1 (Saint Vincent and the Grenadines 1) (2010)
Ports and terminals: Portsmouth, Roseau

MILITARY

Military branches: no regular military forces; Commonwealth of Dominica Police Force (includes Coast Guard) (2011)
Manpower available for military service: *males age 16–49:* 19,075 (2010 est.)
Manpower fit for military service: *males age 16–49:* 16,035
females age 16–49: 15,499 (2010 est.)
Manpower reaching militarily significant age annually:
male: 675
female: 636 (2010 est.)
Military expenditures: NA

TRANSNATIONAL ISSUES

Disputes—international: Dominica is the only Caribbean state to challenge Venezuela's sovereignty claim over Aves Island and joins the other island nations in challenging whether the feature sustains human habitation, a criterion under the UN Convention on the Law of the Sea (UNCLOS), which permits Venezuela to extend its Exclusive Economic Zone (EEZ) and continental shelf claims over a large portion of the eastern Caribbean Sea
Illicit drugs: transshipment point for narcotics bound for the US and Europe; minor cannabis producer (2008)

DOMINICAN REPUBLIC

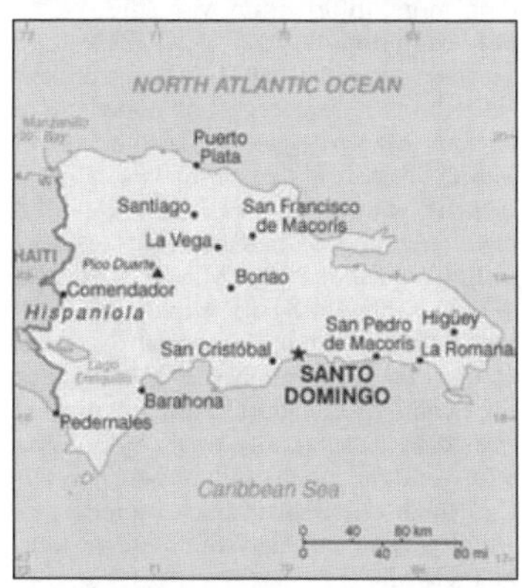

INTRODUCTION

Background: The Taino–indigenous inhabitants of Hispaniola prior to the arrival of the Europeans–divided the island into five chiefdoms and territories. Christopher COLUMBUS explored and claimed the island on his first voyage in 1492; it became a springboard for Spanish conquest of the Caribbean and the American mainland. In 1697, Spain recognized French dominion over the western third of the island, which in 1804 became Haiti. The remainder of the island, by then known as Santo Domingo, sought to gain its own independence in 1821 but was conquered and ruled by the Haitians for 22 years; it finally attained independence as the Dominican Republic in 1844. In 1861, the Dominicans voluntarily returned to the Spanish Empire, but two years later they launched a war that restored independence in 1865. A legacy of unsettled, mostly non-representative rule followed, capped by the dictatorship of Rafael Leonidas TRUJILLO from 1930-61. Juan BOSCH was elected president in 1962 but was deposed in a military coup in 1963. In 1965, the United States led an intervention in the midst of a civil war sparked by an uprising to restore BOSCH. In 1966, Joaquin BALAGUER defeated BOSCH in an election to become president. BALAGUER maintained a tight grip on power for most of the next 30 years when international reaction to flawed elections forced him to curtail his term in 1996. Since then, regular competitive elections have been held in which opposition candidates have won the presidency. Former President (1996-2000) Leonel FERNANDEZ Reyna won election to a new term in 2004 following a constitutional amendment allowing presidents to serve more than one term, and was since reelected to a second consecutive term.

GEOGRAPHY

Location: Caribbean, eastern two-thirds of the island of Hispaniola, between the Caribbean Sea and the North Atlantic Ocean, east of Haiti
Geographic coordinates: 19 00 N, 70 40 W
Map references: Central America and the Caribbean
Area: *total:* 48,670 sq km
country comparison to the world: 131
land: 48,320 sq km
water: 350 sq km
Area—comparative: slightly more than twice the size of New Hampshire
Land boundaries: *total:* 360 km
border countries: Haiti 360 km
Coastline: 1,288 km
Maritime claims: measured from claimed archipelagic straight baselines

territorial sea: 6 nm
contiguous zone: 24 nm
exclusive economic zone: 200 nm
continental shelf: 200 nm or to the edge of the continental margin
Climate: tropical maritime; little seasonal temperature variation; seasonal variation in rainfall
Terrain: rugged highlands and mountains with fertile valleys interspersed
Elevation extremes: *lowest point:* Lago Enriquillo -46 m
highest point: Pico Duarte 3,175 m
Natural resources: nickel, bauxite, gold, silver
Land use: *arable land:* 22.49%
permanent crops: 10.26%
other: 67.25% (2005)
Irrigated land: 2,750 sq km (2003)
Total renewable water resources: 21 cu km (2000)
Freshwater withdrawal (domestic/industrial/agricultural): *total:* 3.39 cu km/yr (32%/2%/66%)
per capita: 381 cu m/yr (2000)
Natural hazards: lies in the middle of the hurricane belt and subject to severe storms from June to October; occasional flooding; periodic droughts
Environment—current issues: water shortages; soil eroding into the sea damages coral reefs; deforestation
Environment—international agreements: *party to:* Biodiversity, Climate Change, Climate Change-Kyoto Protocol, Desertification, Endangered Species, Hazardous Wastes, Marine Dumping, Marine Life Conservation, Ozone Layer Protection, Ship Pollution, Wetlands
signed, but not ratified: Law of the Sea
Geography—note: shares island of Hispaniola with Haiti

PEOPLE

Population: 9,956,648 (July 2011 est.)
country comparison to the world: 85
Age structure: *0–14 years:* 29.5% (male 1,493,251/female 1,441,735)
15–64 years: 64% (male 3,251,419/female 3,120,540)
65 years and over: 6.5% (male 300,245/female 349,458) (2011 est.)
Median age: *total:* 26.1 years
male: 26 years
female: 26.3 years (2011 est.)
Population growth rate: 1.331% (2011 est.)
country comparison to the world: 88
Birth rate: 19.67 births/1,000 population (2011 est.)
country comparison to the world: 90
Death rate: 4.35 deaths/1,000 population (July 2011 est.)
country comparison to the world: 203
Net migration rate: -2.01 migrant(s)/1,000 population (2011 est.)
country comparison to the world: 164
Urbanization: *urban population:* 69% of total population (2010)
rate of urbanization: 2.1% annual rate of change (2010-15 est.)
Major cities—population: SANTO DOMINGO (capital) 2.138 million (2009)
Sex ratio: *at birth:* 1.04 male(s)/female
under 15 years: 1.04 male(s)/female
15–64 years: 1.04 male(s)/female
65 years and over: 0.86 male(s)/female
total population: 1.03 male(s)/female (2011 est.)
Infant mortality rate: *total:* 22.22 deaths/1,000 live births
country comparison to the world: 89
male: 24.21 deaths/1,000 live births
female: 20.14 deaths/1,000 live births (2011 est.)
Life expectancy at birth: *total population:* 77.31 years
country comparison to the world: 61
male: 75.16 years
female: 79.55 years (2011 est.)
Total fertility rate: 2.44 children born/woman (2011 est.)
country comparison to the world: 88
HIV/AIDS—adult prevalence rate: 0.9% (2009 est.)
country comparison to the world: 49
HIV/AIDS—people living with HIV/AIDS: 57,000 (2009 est.)
country comparison to the world: 58
HIV/AIDS—deaths: 2,300 (2009 est.)
country comparison to the world: 53
Major infectious diseases: *degree of risk:* high
food or waterborne diseases: bacterial diarrhea, hepatitis A, and typhoid fever
vectorborne diseases: dengue fever and malaria
water contact disease: leptospirosis (2009)
Drinking water source: *Improved:*
urban: 87% of population
rural: 84% of population
total: 86% of population
Unimproved:
urban: 13% of population
rural: 16% of population
total: 14% of population (2008)
Sanitation facility access: *Improved:*
urban: 87% of population
rural: 74% of population
total: 83% of population
Unimproved:
urban: 13% of population
rural: 26% of population
total: 17% of population (2008)
Nationality: *noun:* Dominican(s)
adjective: Dominican
Ethnic groups: mixed 73%, white 16%, black 11%
Religions: Roman Catholic 95%, other 5%
Languages: Spanish (official)
Literacy: *definition:* age 15 and over can read and write
total population: 87%
male: 86.8%
female: 87.2% (2002 census)
School life expectancy (primary to tertiary education): *total:* 12 years
male: 11 years
female: 13 years (2004)
Education expenditures: 2.3% of GDP (2009)
country comparison to the world: 152

GOVERNMENT

Country name: *conventional long form:* Dominican Republic
conventional short form: The Dominican
local long form: Republica Dominicana
local short form: La Dominicana
Government type: democratic republic
Capital: *name:* Santo Domingo
geographic coordinates: 18 28 N, 69 54 W
time difference: UTC-4 (1 hour ahead of Washington, DC during Standard Time)
Administrative divisions: 31 provinces (provincias, singular—provincia) and 1 district* (distrito); Azua, Bahoruco, Barahona, Dajabon, Distrito Nacional*, Duarte, El Seibo, Elias Pina, Espaillat, Hato Mayor, Independencia, La Altagracia, La Romana, La Vega, Maria Trinidad Sanchez, Monsenor Nouel, Monte Cristi, Monte Plata, Pedernales, Peravia, Puerto Plata, Salcedo, Samana, San Cristobal, San Jose de Ocoa, San Juan, San Pedro de Macoris, Sanchez Ramirez, Santiago, Santiago Rodriguez, Santo Domingo, Valverde
Independence: 27 February 1844 (from Haiti)
National holiday: Independence Day, 27 February (1844)
Constitution: 28 November 1966; amended 25 July 2002 and January 2010
Legal system: civil law system based on the French civil code; Criminal Procedures Code modified in 2004 to include important elements of an accusatory system
International law organization participation: accepts compulsory ICJ jurisdiction; accepts ICCt jurisdiction
Suffrage: 18 years of age, universal and compulsory; married persons regardless of age; note—members of the armed forces and national police cannot vote
Executive branch: *chief of state:* President Leonel FERNANDEZ Reyna (since 16 August 2004); Vice President Rafael ALBURQUERQUE de Castro (since 16 August 2004); note—the president is both the chief of state and head of government
head of government: President Leonel FERNANDEZ Reyna (since 16 August 2004); Vice President Rafael ALBURQUERQUE de Castro (since 16 August 2004)
cabinet: Cabinet nominated by the president
(For more information visit the World Leaders website)
elections: president and vice president elected on the same ticket by popular vote for four-year terms; election last held on 16 May 2008 (next to be held in May 2012)
election results: Leonel FERNANDEZ reelected president; percent of vote—Leonel FERNANDEZ 53.6%, Miguel VARGAS 41%, Amable ARISTY less than 5%
Legislative branch: bicameral National Congress or Congreso Nacional consists of the Senate or Senado (32 seats; members elected by popular vote to serve four-year terms) and the House of Representatives or Camara de Diputados (183 seats; members are elected by popular vote to serve four-year terms)
elections: Senate—last held on 16 May 2010 (next to be held in May 2016); House of Representatives—last held on 16 May 2010 (next to be held in May 2016); in order to synchronize presidential, legislative, and local elections for 2016, those members elected in 2010 will actually serve terms of six years
election results: Senate—percent of vote by party—NA; seats by party—PLD 31, PRD 1; House of Representatives—percent of vote

by party–NA; seats by party–PLD 105, PRD 75, PRSC 3
Judicial branch: Supreme Court or Corte Suprema (judges are appointed by the National Judicial Council comprised of the president, the leaders of both chambers of congress, the president of the Supreme Court, and an additional non-governing party congressional representative)
Political parties and leaders: Dominican Liberation Party or PLD [Leonel FERNANDEZ Reyna]; Dominican Revolutionary Party or PRD [Ramon ALBURQUERQUE]; National Progressive Front [Vincent CASTILLO, Pelegrin CASTILLO]; Social Christian Reformist Party or PRSC [Enrique ANTUN]
Political pressure groups and leaders: Citizen Participation Group (Participacion Ciudadania); Collective of Popular Organizations or COP; Foundation for Institution-Building and Justice (FINJUS)
International organization participation: ACP, AOSIS, BCIE, Caricom (observer), FAO, G-77, IADB, IAEA, IBRD, ICAO, ICC, ICRM, IDA, IFAD, IFC, IFRCS, IHO, ILO, IMF, IMO, Interpol, IOC, IOM, IPU, ISO (correspondent), ITSO, ITU, ITUC, LAES, LAIA (observer), MIGA, NAM, OAS, OPANAL, OPCW, PCA, PetroCaribe, RG, SICA (associated member), UN, UNCTAD, UNESCO, UNIDO, Union Latina, UNWTO, UPU, WCO, WFTU, WHO, WIPO, WMO, WTO
Diplomatic representation in the US:
chief of mission: Ambassador Roberto B. SALADIN Selin
chancery: 1715 22nd Street NW, Washington, DC 20008
telephone: [1] (202) 332-6280
FAX: [1] (202) 265-8057
consulate(s) general: Anchorage, Boston, Chicago, Mayaguez (Puerto Rico), Miami, New Orleans, New York, San Juan (Puerto Rico), Sun Valley (California)
Diplomatic representation from the US: *chief of mission:* Ambassador Raul H. YZAGUIRRE
embassy: corner of Calle Cesar Nicolas Penson and Calle Leopoldo Navarro, Santo Domingo
mailing address: Unit 5500, APO AA 34041-5500
telephone: [1] (809) 221-2171
FAX: [1] (809) 686-7437
Flag description: a centered white cross that extends to the edges divides the flag into four rectangles–the top ones are blue (hoist side) and red, and the bottom ones are red (hoist side) and blue; a small coat of arms featuring a shield supported by a laurel branch (left) and a palm branch (right) is at the center of the cross; above the shield a blue ribbon displays the motto, DIOS, PATRIA, LIBERTAD (God, Fatherland, Liberty), and below the shield, REPUBLICA DOMINICANA appears on a red ribbon; in the shield a bible is opened to a verse that reads "Y la verdad nos hara libre" (And the truth shall set you free); blue stands for liberty, white for salvation, and red for the blood of heroes
National anthem: *name:* "Himno Nacional" (National Anthem)
lyrics/music: Emilio PRUD"HOMME/Jose REYES
note: adopted 1934; also known as "Quisqueyanos valientes" (Valient Sons of Quisqueye); the anthem never refers to the people as Dominican but rather calls them "Quisqueyanos," a reference to the indigenous name of the island

ECONOMY

Economy—overview: The Dominican Republic has long been viewed primarily as an exporter of sugar, coffee, and tobacco, but in recent years the service sector has overtaken agriculture as the economy's largest employer, due to growth in telecommunications, tourism, and free trade zones. The economy is highly dependent upon the US, the destination for nearly 60% of exports. Remittances from the US amount to about a tenth of GDP, equivalent to almost half of exports and three-quarters of tourism receipts. The country suffers from marked income inequality; the poorest half of the population receives less than one-fifth of GDP, while the richest 10% enjoys nearly 40% of GDP. High unemployment and underemployment remains an important long-term challenge. The Central America-Dominican Republic Free Trade Agreement (CAFTA-DR) came into force in March 2007, boosting investment and exports and reducing losses to the Asian garment industry. The growth of the Dominican Republic's economy rebounded in 2010 from the global recession, and remains one of the fastest growing in the region.
GDP (purchasing power parity): $87.25 billion (2010 est.)
country comparison to the world: 75
$80.97 billion (2009 est.)
$78.27 billion (2008 est.)
note: data are in 2010 US dollars
GDP (official exchange rate): $51.63 billion (2010 est.)
GDP—real growth rate: 7.8% (2010 est.)
country comparison to the world: 22
3.5% (2009 est.)
5.3% (2008 est.)
GDP—per capita (PPP): $8,900 (2010 est.)
country comparison to the world: 117
$8,400 (2009 est.)
$8,200 (2008 est.)
note: data are in 2010 US dollars
GDP—composition by sector:
agriculture: 11.5%
industry: 21%
services: 67.5% (2010 est.)
Labor force: 4.498 million (2010 est.)
country comparison to the world: 81
Labor force—by occupation:
agriculture: 14.6%
industry: 22.3%
services: 63.1% (2005)
Unemployment rate: 14.2% (2010 est.)
country comparison to the world: 144
14.9% (2009 est.)
Population below poverty line: 42.2% (2004)
Household income or consumption by percentage share: *lowest 10%:* 1.5%
highest 10%: 38.7% (2005)
Distribution of family income—Gini index: 49.9 (2005)
country comparison to the world: 24
47.4 (1998)
Investment (gross fixed): 15.4% of GDP (2010 est.)
country comparison to the world: 129
Budget: *revenues:* $7.11 billion
expenditures: $8.634 billion (2010 est.)
Public debt: 41.7% of GDP (2010 est.)
country comparison to the world: 64
40.8% of GDP (2009 est.)
Inflation rate (consumer prices): 6.3% (2010 est.)
country comparison to the world: 164
1.4% (2009 est.)
Commercial bank prime lending rate: 18.14% (31 December 2009 est.)
country comparison to the world: 17
19.95% (31 December 2008 est.)
Stock of narrow money: $4.734 billion (31 December 2010 est.)
country comparison to the world: 94
$4.079 billion (31 December 2009 est.)
Stock of broad money: $15.71 billion (31 December 2010 est.)
country comparison to the world: 88
$14 billion (31 December 2009 est.)
Stock of domestic credit: $21.63 billion (31 December 2010 est.)
country comparison to the world: 77
$18.91 billion (31 December 2009 est.)
Market value of publicly traded shares: $NA
Agriculture—products: sugarcane, coffee, cotton, cocoa, tobacco, rice, beans, potatoes, corn, bananas; cattle, pigs, dairy products, beef, eggs
Industries: tourism, sugar processing, ferronickel and gold mining, textiles, cement, tobacco
Industrial production growth rate: 1.5% (2010 est.)
country comparison to the world: 139
Electricity—production: 14.02 billion kWh (2007 est.)
country comparison to the world: 81
Electricity—consumption: 12.7 billion kWh (2007 est.)
country comparison to the world: 80
Electricity—exports: 0 kWh (2008 est.)
Electricity—imports: 0 kWh (2008 est.)
Oil—production: 0 bbl/day (2008 est.)
country comparison to the world: 168
Oil—consumption: 118,000 bbl/day (2009 est.)
country comparison to the world: 73
Oil—exports: 0 bbl/day (2007 est.)
country comparison to the world: 162
Oil—imports: 116,200 bbl/day (2007 est.)
country comparison to the world: 60
Oil—proved reserves: 0 bbl (1 January 2010 est.)
country comparison to the world: 124
Natural gas—production: 0 cu m (2008 est.)
country comparison to the world: 120
Natural gas—consumption: 470 million cu m (2008 est.)
country comparison to the world: 95
Natural gas—exports: 0 cu m (2008 est.)
country comparison to the world: 86
Natural gas—imports: 470 million cu m (2008 est.)
country comparison to the world: 60

Natural gas—proved reserves: 0 cu m (1 January 2010 est.)
country comparison to the world: 129
Current account balance: $-3.862 billion (2010 est.)
country comparison to the world: 168
$-2.328 billion (2009 est.)
Exports: $6.161 billion (2010 est.)
country comparison to the world: 103
$5.462 billion (2009 est.)
Exports—commodities: ferronickel, sugar, gold, silver, coffee, cocoa, tobacco, meats, consumer goods
Exports—partners: US 54.08%, Haiti 9.78% (2009)
Imports: $14.53 billion (2010 est.)
country comparison to the world: 79
$12.28 billion (2009 est.)
Imports—commodities: foodstuffs, petroleum, cotton and fabrics, chemicals and pharmaceuticals
Imports—partners: US 42.79%, Venezuela 7.04%, Mexico 6.17%, Colombia 5.59% (2009)
Reserves of foreign exchange and gold: $2.705 billion (31 December 2010 est.)
country comparison to the world: 92
$2.905 billion (31 December 2009 est.)
Debt—external: $13.09 billion (31 December 2010 est.)
country comparison to the world: 82
$11.04 billion (31 December 2009 est.)
Stock of direct foreign investment—at home: $19.45 billion (31 December 2010 est.)
country comparison to the world: 68
$17.95 billion (31 December 2009 est.)
Stock of direct foreign investment—abroad: $NA (31 December 2010 est.)
$59 million (31 December 2009 est.)
Exchange rates: Dominican pesos (DOP) per US dollar–36.92 (2010), 36.03 (2009), 34.775 (2008), 33.113 (2007), 33.406 (2006)

COMMUNICATIONS

Telephones—main lines in use: 965,400 (2009)
country comparison to the world: 81
Telephones—mobile cellular: 8.63 million (2009)
country comparison to the world: 71
Telephone system: *general assessment:* relatively efficient system based on island-wide microwave radio relay network
domestic: fixed-line teledensity is about 10 per 100 persons; multiple providers of mobile-cellular service with a subscribership of roughly 75 per 100 persons
international: country code–1-809; landing point for the Americas Region Caribbean Ring System (ARCOS-1), Antillas 1, and the Fibralink submarine cables that provide links to South and Central America, parts of the Caribbean, and US; satellite earth station–1 Intelsat (Atlantic Ocean) (2009)
Broadcast media: combination of state-owned and privately-owned broadcast media; 1 state-owned television network and a number of private TV networks; networks operate repeaters to extend signals throughout country; combination of state-owned and privately-owned radio stations; more than 300 radio stations operating (2007)
Internet country code: .do
Internet hosts: 283,298 (2010)
country comparison to the world: 60
Internet users: 2.701 million (2009)
country comparison to the world: 68

TRANSPORTATION

Airports: 35 (2010)
country comparison to the world: 110
Airports—with paved runways: *total:* 16
over 3,047 m: 3
2,438 to 3,047 m: 4
1,524 to 2,437 m: 4
914 to 1,523 m: 4
under 914 m: 1 (2010)
Airports—with unpaved runways: *total:* 19
1,524 to 2,437 m: 1
914 to 1,523 m: 1
under 914 m: 17 (2010)
Pipelines: oil 99 km
Railways:
total: 142 km
country comparison to the world: 125
standard gauge: 142 km 1.435-m gauge (2010)
Roadways: *total:* 19,705 km
country comparison to the world: 109
paved: 9,872 km
unpaved: 9,833 km (2002)
Merchant marine: *total:* 1
country comparison to the world: 152
by type: cargo 1
registered in other countries: 1 (Panama 1) (2008)
Ports and terminals: Andres (Boca Chica), Puerto Haina, Puerto Plata, Santo Domingo

MILITARY

Military branches: Army, Navy (Marina de Guerra), Air Force (Fuerza Aerea Dominicana, FAD) (2011)
Military service age and obligation: 16-21 years of age for compulsory military service; recruits must be Dominican Republic citizens; women may volunteer (2010)
Manpower available for military service:
males age 16–49: 2,580,083
females age 16–49: 2,464,698 (2010 est.)
Manpower fit for military service: *males age 16–49:* 2,188,358
females age 16–49: 2,090,180 (2010 est.)
Manpower reaching militarily significant age annually:
male: 100,047
female: 96,302 (2010 est.)
Military expenditures: 0.7% of GDP (2009)
country comparison to the world: 152

TRANSNATIONAL ISSUES

Disputes—international: Haitian migrants cross the porous border into the Dominican Republic to find work; illegal migrants from the Dominican Republic cross the Mona Passage each year to Puerto Rico to find better work
Trafficking in persons: current situation: the Dominican Republic is a source, transit, and destination country for men, women, and children trafficked for the purposes of commercial sexual exploitation and forced labor; a large number of Dominican women are trafficked into prostitution and sexual exploitation in Western Europe, Australia, Central and South America, and Caribbean destinations; a significant number of women, boys, and girls are trafficked within the country for sexual exploitation and domestic servitude
tier rating: Tier 3–for its failure to show evidence of increasing efforts to combat human trafficking, particularly in terms of not adequately investigating and prosecuting public officials who may be complicit with trafficking activity, and inadequate government efforts to protect trafficking victims; the government has taken measures to reduce demand for commercial sex acts with children through criminal prosecutions (2008)
Illicit drugs: transshipment point for South American drugs destined for the US and Europe; has become a transshipment point for ecstasy from the Netherlands and Belgium destined for US and Canada; substantial money laundering activity in particular by Colombian narcotics traffickers; significant amphetamine consumption (2008)

ECUADOR

INTRODUCTION

Background: What is now Ecuador formed part of the northern Inca Empire until the Spanish conquest in 1533. Quito became a seat of Spanish colonial government in 1563 and part of the Viceroyalty of New Granada in 1717. The territories of the Viceroyalty–New Granada (Colombia), Venezuela, and Quito–gained their independence between 1819 and 1822 and formed a federation known as Gran Colombia. When Quito withdrew in 1830, the traditional name was changed in favor of the "Republic of the Equator." Between 1904 and 1942, Ecuador lost territories in a series of conflicts with its neighbors. A border war with Peru that flared in 1995 was resolved in 1999. Although Ecuador marked 30 years of civilian governance in 2004, the period was marred by political instability. Protests in Quito contributed to the mid-term ouster of three of Ecuador's last four democratically elected Presidents. In September 2008, voters approved a new constitution, Ecuador's 20th since gaining independence. General elections, under the new constitutional framework, were held in April 2009, and voters re-elected President Rafael CORREA.

GEOGRAPHY

Location: Western South America, bordering the Pacific Ocean at the Equator, between Colombia and Peru
Geographic coordinates: 2 00 S, 77 30 W
Map references: South America
Area: *total:* 283,561 sq km
country comparison to the world: 73
land: 276,841 sq km
water: 6,720 sq km
note: includes Galapagos Islands
Area—comparative: slightly smaller than Nevada
Land boundaries: *total:* 2,010 km
border countries: Colombia 590 km, Peru 1,420 km
Coastline: 2,237 km
Maritime claims: *territorial sea:* 200 nm
continental shelf: 100 nm from 2,500-m isobath
Climate: tropical along coast, becoming cooler inland at higher elevations; tropical in Amazonian jungle lowlands
Terrain: coastal plain (costa), inter-Andean central highlands (sierra), and flat to rolling eastern jungle (oriente)
Elevation extremes: *lowest point:* Pacific Ocean 0 m
highest point: Chimborazo 6,267 m
note: due to the fact that the earth is not a perfect sphere and has an equatorial bulge, the highest point on the planet furthest from its center is Mount Chimborazo not Mount Everest, which is merely the highest peak above sea-level
Natural resources: petroleum, fish, timber, hydropower
Land use: *arable land:* 5.71%
permanent crops: 4.81%
other: 89.48% (2005)
Irrigated land: 8,650 sq km (2003)
Total renewable water resources: 432 cu km (2000)
Freshwater withdrawal (domestic/industrial/agricultural): *total:* 16.98 cu km/yr (12%/5%/82%)
per capita: 1,283 cu m/yr (2000)
Natural hazards: frequent earthquakes; landslides; volcanic activity; floods; periodic droughts
volcanism: Ecuador experiences volcanic activity in the Andes Mountains; Sangay (elev. 5,230 m), which erupted in 2010, is mainland Ecuador's most active volcano; other historically active volcanoes in the Andes include Antisana, Cayambe, Chacana, Cotopaxi, Guagua Pichincha, Reventador, Sumaco, and Tungurahua; Fernandina (elev. 1,476 m), a shield volcano that last erupted in 2009, is the most active of the many Galapagos volcanoes; other historically active Galapagos volcanoes include Wolf, Sierra Negra, Cerro Azul, Pinta, Marchena, and Santiago
Environment—current issues: deforestation; soil erosion; desertification; water pollution; pollution from oil production wastes in ecologically sensitive areas of the Amazon Basin and Galapagos Islands
Environment—international agreements: *party to:* Antarctic-Environmental Protocol, Antarctic Treaty, Biodiversity, Climate Change, Climate Change-Kyoto Protocol, Desertification, Endangered Species, Hazardous Wastes, Ozone Layer Protection, Ship Pollution, Tropical Timber 83, Tropical Timber 94, Wetlands
signed, but not ratified: none of the selected agreements
Geography—note: Cotopaxi in Andes is highest active volcano in world

PEOPLE

Population: 15,007,343 (July 2011 est.)
country comparison to the world: 65
Age structure: *0–14 years:* 30.1% (male 2,301,840/female 2,209,971)
15–64 years: 63.5% (male 4,699,548/female 4,831,521)
65 years and over: 6.4% (male 463,481/female 500,982) (2011 est.)
Median age: *total:* 25.7 years
male: 25 years
female: 26.3 years (2011 est.)
Population growth rate: 1.443% (2011 est.)
country comparison to the world: 82
Birth rate: 19.96 births/1,000 population (2011 est.)
country comparison to the world: 89
Death rate: 5 deaths/1,000 population (July 2011 est.)
country comparison to the world: 186
Net migration rate: -0.52 migrant(s)/1,000 population (2011 est.)
country comparison to the world: 138
Urbanization: *urban population:* 67% of total population (2010)
rate of urbanization: 2% annual rate of change (2010-15 est.)
Major cities—population: Guayaquil 2.634 million; QUITO (capital) 1.801 million (2009)
Sex ratio: *at birth:* 1.05 male(s)/female
under 15 years: 1.04 male(s)/female
15–64 years: 0.97 male(s)/female
65 years and over: 0.93 male(s)/female
total population: 0.99 male(s)/female (2011 est.)
Infant mortality rate: *total:* 19.65 deaths/1,000 live births
country comparison to the world: 99
male: 23.02 deaths/1,000 live births
female: 16.11 deaths/1,000 live births (2011 est.)
Life expectancy at birth: *total population:* 75.73 years
country comparison to the world: 82
male: 72.79 years
female: 78.82 years (2011 est.)
Total fertility rate: 2.42 children born/woman (2011 est.)
country comparison to the world: 91
HIV/AIDS—adult prevalence rate: 0.4% (2009 est.)
country comparison to the world: 72
HIV/AIDS—people living with HIV/AIDS: 37,000 (2009 est.)
country comparison to the world: 64
HIV/AIDS—deaths: 2,200 (2009 est.)
country comparison to the world: 54
Major infectious diseases: *degree of risk:* high
food or waterborne diseases: bacterial diarrhea, hepatitis A, and typhoid fever
vectorborne diseases: dengue fever and malaria
water contact disease: leptospirosis (2009)
Drinking water source: *Improved:*
urban: 97% of population
rural: 88% of population
total: 94% of population
Unimproved: urban: 3% of population
rural: 12% of population
total: 6% of population (2008)
Sanitation facility access: *Improved:*
urban: 96% of population
rural: 84% of population
total: 92% of population
Unimproved: urban: 4% of population
rural: 16% of population
total: 8% of population (2008)
Nationality: *noun:* Ecuadorian(s)
adjective: Ecuadorian
Ethnic groups: mestizo (mixed Amerindian and white) 65%, Amerindian 25%, Spanish and others 7%, black 3%
Religions: Roman Catholic 95%, other 5%

Languages: Spanish (official), Amerindian languages (especially Quechua)
Literacy: *definition:* age 15 and over can read and write
total population: 91%
male: 92.3%
female: 89.7% (2001 census)
School life expectancy (primary to tertiary education): *total:* 14 years
male: 13 years
female: 14 years (2008)
Education expenditures: NA

GOVERNMENT

Country name: *conventional long form:* Republic of Ecuador
conventional short form: Ecuador
local long form: Republica del Ecuador
local short form: Ecuador
Government type: republic
Capital: *name:* Quito
geographic coordinates: 0 13 S, 78 30 W
time difference: UTC-5 (same time as Washington, DC during Standard Time)
Administrative divisions: 24 provinces (provincias, singular—provincia); Azuay, Bolivar, Canar, Carchi, Chimborazo, Cotopaxi, El Oro, Esmeraldas, Galapagos, Guayas, Imbabura, Loja, Los Rios, Manabi, Morona-Santiago, Napo, Orellana, Pastaza, Pichincha, Santa Elena, Santo Domingo de los Tsachilas, Sucumbios, Tungurahua, Zamora-Chinchipe
Independence: 24 May 1822 (from Spain)
National holiday: Independence Day (independence of Quito), 10 August (1809)
Constitution: 20 October 2008
Legal system: civil law based on the Chilean civil code with modifications
International law organization participation: has not submitted an ICJ jurisdiction declaration; accepts ICCt jurisdiction
Suffrage: 16 years of age; universal, compulsory for persons ages 18-65, optional for other eligible voters
Executive branch: *chief of state:* President Rafael CORREA Delgado (since 15 January 2007); Vice President Lenin MORENO Garces (since 15 January 2007); note—the president is both the chief of state and head of government
head of government: President Rafael CORREA Delgado (since 15 January 2007); Vice President Lenin MORENO Garces (since 15 January 2007)
cabinet: Cabinet appointed by the president (For more information visit the World Leaders website)
elections: the president and vice president elected on the same ticket by popular vote for a four-year term and can be re-elected for another consecutive term; election last held on 26 April 2009 (next to be held in 2013)
election results: President Rafael CORREA Delgado reelected president; percent of vote—Rafael CORREA Delgado 52%; Lucio GUTIERREZ 28.2%; Alvaro NOBOA 11.4%; other 8.4%
Legislative branch: unicameral National Assembly or Asamblea Nacional (124 seats; members are elected through a party-list proportional representation system to serve four-year terms)
elections: last held on 26 April 2009 (next to be held in 2013)
election results: percent of vote by party—NA; seats by party—PAIS 59, PSP 19, PSC 11, PRIAN 7, MPD 5, PRE 3, other 20; note—defections by members of National Assembly are commonplace, resulting in frequent changes in the numbers of seats held by the various parties
Judicial branch: National Court of Justice or Corte Nacional de Justicia (according to the Constitution, justices are elected through a procedure overseen by the Judiciary Council); Constitutional Court or Corte Constitucional (Constitutional Court justices are appointed by a commission composed of two delegates each from the Executive, Legislative, and Transparency branches of government)
Political parties and leaders: Alianza PAIS movement [Rafael Vicente CORREA Delgado]; Democratic Left or ID [Dalton BACIGALUPO]; Ethical and Democratic Network or RED [Martha ROLDOS]; Institutional Renewal and National Action Party or PRIAN [Vicente TAIANO]; Pachakutik Plurinational Unity Movement—New Country or MUPP-NP [Rafael ANTUNI]; Patriotic Society Party or PSP [Lucio GUTIERREZ Borbua]; Popular Democratic Movement or MPD [Luis VILLACIS]; Roldosist Party or PRE [Abdala BUCARAM Pulley, director]; Social Christian Party or PSC [Pascual DEL CIOPPO]; Socialist Party—Broad Front or PS-FA [Rafael QUINTERO]; Warrior's Spirit Movement [Jaime NEBOT]
Political pressure groups and leaders: Confederation of Indigenous Nationalities of Ecuador or CONAIE [Marlon SANTI, president]; Federation of Indigenous Evangelists of Ecuador or FEINE [Manuel CHUGCHILAN, president]; National Federation of Indigenous Afro-Ecuatorianos and Peasants or FENOCIN [Luis Alberto ANDRANGO Cadena, president]; National Teacher's Union or UNE [Mariana PALLASCO]
International organization participation: CAN, FAO, G-11, G-77, IADB, IAEA, IBRD, ICAO, ICC, ICRM, IDA, IFAD, IFC, IFRCS, IHO, ILO, IMF, IMO, Interpol, IOC, IOM, IPU, ISO, ITSO, ITU, ITUC, LAES, LAIA, Mercosur (associate), MIGA, MINUSTAH, NAM, OAS, OPANAL, OPCW, OPEC, PCA, RG, UN, UNASUR, UNCTAD, UNESCO, UNHCR, UNIDO, Union Latina, UNMIL, UNMIS, UNOCI, UNWTO, UPU, WCO, WFTU, WHO, WIPO, WMO, WTO
Diplomatic representation in the US: *chief of mission:* Ambassador Luis Benigno GALLEGOS Chiriboga
chancery: 2535 15th Street NW, Washington, DC 20009
telephone: [1] (202) 234-7200
FAX: [1] (202) 667-3482
consulate(s) general: Atlanta, Boston, Chicago, Dallas, Houston, Las Vegas, Los Angeles, Miami, Minneapolis, New Haven, New Orleans, New York, Newark (New Jersey), Phoenix, San Francisco, San Juan (Puerto Rico)
Diplomatic representation from the US: *chief of mission:* Ambassador Heather HODGES
embassy: Avenida Avigiras E12-170 y Avenida Eloy Alfaro, Quito
mailing address: Avenida Guayacanes N52-205 y Avenida Avigiras
telephone: [593] (2) 398-5000
FAX: [593] (2) 398-5100
consulate(s) general: Guayaquil
Flag description: three horizontal bands of yellow (top, double width), blue, and red with the coat of arms superimposed at the center of the flag; the flag retains the three main colors of the banner of Gran Columbia, the South American republic that broke up in 1830; the yellow color represents sunshine, grain, and mineral wealth, blue the sky, sea, and rivers, and red the blood of patriots spilled in the struggle for freedom and justice
note: similar to the flag of Colombia, which is shorter and does not bear a coat of arms
National anthem: *name:* "Salve, Oh Patria!" (We Salute You Our Homeland)
lyrics/music: Juan Leon MERA/Antonio NEUMANE
note: adopted 1948; Juan Leon MERA wrote the lyrics in 1865; only the chorus and second verse are sung

ECONOMY

Economy—overview: Ecuador is substantially dependent on its petroleum resources, which have accounted for more than half of the country's export earnings and approximately one-third of public sector revenues in recent years. In 1999/2000, Ecuador suffered a severe economic crisis, with GDP contracting by 5.3%. Poverty increased significantly, the banking system collapsed, and Ecuador defaulted on its external debt. In March 2000, the Congress approved a series of structural reforms that also provided for the adoption of the US dollar as legal tender. Dollarization stabilized the economy, and positive growth returned in the years that followed, helped by high oil prices, remittances, and increased non-traditional exports. From 2002-06 the economy grew an average of 5.2% per year, the highest five-year average in 25 years. After moderate growth in 2007, the economy reached a growth rate of 7.2% in 2008, in large part due to high global petroleum prices and increased public sector investment. President Rafael CORREA, who took office in January 2007, defaulted in December 2008 on Ecuador's sovereign debt, which, with a total face value of approximately US$3.2 billion, represented about 80% of Ecuador's private external debt. In May 2009, Ecuador bought back 91% of its "defaulted" bonds via an international auction. Economic policies under the CORREA administration—including an announcement in late 2009 of its intention to terminate 13 bilateral investment treaties, including one with the United States—have generated economic uncertainty and discouraged private investment. The Ecuadorian economy contracted 0.4% in 2009 due to the global financial crisis and to the sharp decline in world oil prices and remittance flows. Growth picked up to a 3.7% rate in 2010, according to Ecuadorian government estimates.

GDP (purchasing power parity): $115 billion (2010 est.)
country comparison to the world: 64
$111.4 billion (2009 est.)
$111 billion (2008 est.)
note: data are in 2010 US dollars

GDP (official exchange rate): $58.91 billion (2010 est.)

GDP—real growth rate: 3.2% (2010 est.)
country comparison to the world: 120
0.4% (2009 est.)
7.2% (2008 est.)

GDP—per capita (PPP): $7,800 (2010 est.)
country comparison to the world: 125
$7,600 (2009 est.)
$7,700 (2008 est.)
note: data are in 2010 US dollars

GDP—composition by sector:
agriculture: 6.4%
industry: 35.9%
services: 57.7% (2010 est.)

Labor force: 4.59 million (urban) (2010 est.)
country comparison to the world: 80

Labor force—by occupation: *agriculture:* 8.3%
industry: 21.2%
services: 70.4% (2005)

Unemployment rate: 5% (2010 est.)
country comparison to the world: 46
6.5% (2009 est.)

Population below poverty line: 33.1% (June 2010)

Household income or consumption by percentage share: *lowest 10%:* 1%
highest 10%: 35.3%
note: data for urban households only (June 2010)

Distribution of family income—Gini index: 46.9 (June 2010)
country comparison to the world: 34
50.5 (2006)
note: data are for urban households

Investment (gross fixed): 23.7% of GDP (2010 est.)
country comparison to the world: 51

Budget: *revenues:* $14.48 billion
expenditures: $18.21 billion (2011 est.)

Public debt: 23.2% of GDP (2010 est.)
country comparison to the world: 105
19.9% of GDP (2009 est.)

Inflation rate (consumer prices): 3.3% (2010 est.)
country comparison to the world: 90
4.3% (2009 est.)

Central bank discount rate: 8.68% (31 December 2010)
country comparison to the world: 47
9.19% (31 December 2009)

Commercial bank prime lending rate: 19% (31 December 2009)
country comparison to the world: 98
9.14% (31 December 2008)

Stock of narrow money: $6.198 billion (31 December 2010 est.)
country comparison to the world: 84
$5.201 billion (31 December 2009 est.)

Stock of broad money: $21.22 billion (31 December 2010 est.)
country comparison to the world: 81
$18.83 billion (31 December 2009 est.)

Stock of domestic credit: $17.47 billion (31 December 2010 est.)
country comparison to the world: 85
$14.56 billion (31 December 2009 est.)

Market value of publicly traded shares: $4.248 billion (31 December 2009)
country comparison to the world: 84
$4.562 billion (31 December 2008)
$4.266 billion (31 December 2007)

Agriculture—products: bananas, coffee, cocoa, rice, potatoes, manioc (tapioca), plantains, sugarcane; cattle, sheep, pigs, beef, pork, dairy products; balsa wood; fish, shrimp

Industries: petroleum, food processing, textiles, wood products, chemicals

Industrial production growth rate: 3.6% (2010 est.)
country comparison to the world: 93

Electricity—production: 16.42 billion kWh (2007 est.)
country comparison to the world: 74

Electricity—consumption: 15.81 billion kWh (2007 est.)
country comparison to the world: 75

Electricity—exports: 20.68 million kWh (2007 est.)

Electricity—imports: 1.12 billion kWh (2007 est.)

Oil—production: 485,700 bbl/day (2009 est.)
country comparison to the world: 32

Oil—consumption: 181,000 bbl/day (2009 est.)
country comparison to the world: 58

Oil—exports: 338,000 bbl/day (2010 est.)
country comparison to the world: 36

Oil—imports: 80,500 bbl/day (2007 est.)
country comparison to the world: 72

Oil—proved reserves: 6.542 billion bbl (1 January 2010 est.)
country comparison to the world: 22

Natural gas—production: 283.2 million cu m (2009 est.)
country comparison to the world: 72

Natural gas—consumption: 283.2 million cu m (2009 est.)
country comparison to the world: 98

Natural gas—exports: 0 cu m (2009 est.)
country comparison to the world: 87

Natural gas—imports: 0 cu m (2009 est.)
country comparison to the world: 178

Natural gas—proved reserves: 7.985 billion cu m (1 January 2010 est.)
country comparison to the world: 82

Current account balance: $-692 million (2010 est.)
country comparison to the world: 124
$-337.4 million (2009 est.)

Exports: $17.37 billion (2010 est.)
country comparison to the world: 72
$14.35 billion (2009 est.)

Exports—commodities: petroleum, bananas, cut flowers, shrimp, cacao, coffee, wood, fish

Exports—partners: US 33.5%, Peru 6.8%, Chile 6.5%, Columbia 4.9% (2009 est.)

Imports: $17.65 billion (2010 est.)
country comparison to the world: 76
$14.27 billion (2009 est.)

Imports—commodities: industrial materials, fuels and lubricants, nondurable consumer goods

Imports—partners: US 25.4%, Columbia 10.6%, Venezuela 6.5%, Brazil 4.5% (2009 est.)

Reserves of foreign exchange and gold: $3.59 billion (31 December 2010 est.)
country comparison to the world: 88
$3.792 billion (31 December 2009 est.)

Debt—external: $14.71 billion (31 December 2010 est.)
country comparison to the world: 77
$13.48 billion (31 December 2009 est.)

Stock of direct foreign investment—at home: $12.3 billion (31 December 2010 est.)
country comparison to the world: 78
$11.95 billion (31 December 2009 est.)

Stock of direct foreign investment—abroad: $NA (31 December 2010 est.)
$8.019 billion (31 December 2009 est.)

Exchange rates: the US dollar became El Salvador's currency in 2001

COMMUNICATIONS

Telephones—main lines in use: 2.004 million (2009)
country comparison to the world: 57

Telephones—mobile cellular: 13.635 million (2009)
country comparison to the world: 52

Telephone system: *general assessment:* generally elementary but being expanded
domestic: fixed-line services provided by multiple telecommunications operators; fixed-line teledensity stands at about 14 per 100 persons; mobile-cellular use has surged and subscribership reached about 95 per 100 persons in 2009
international: country code–593; landing points for the PAN-AM and South America-1 submarine cables that provide links to the west coast of South America, Panama, Colombia, Venezuela, and extending onward to Aruba and the US Virgin Islands in the Caribbean; satellite earth station–1 Intelsat (Atlantic Ocean) (2009)

Broadcast media: many TV and radio stations are privately-owned; the government owns and runs one national television station and controls two others, as well as multiple radio stations; Ecuador has multiple television networks and TV channels, and a large number of local channels; more than 400 radio stations; broadcast media required by law to give the government free air time to broadcast programs produced by the state (2007)

Internet country code: .ec

Internet hosts: 67,975 (2010)
country comparison to the world: 82

Internet users: 3.352 million (2009)
country comparison to the world: 64

TRANSPORTATION

Airports: 428 (2010)
country comparison to the world: 18

Airports—with paved runways:
total: 105
over 3,047 m: 3
2,438 to 3,047 m: 5
1,524 to 2,437 m: 17
914 to 1,523 m: 25
under 914 m: 55 (2010)

Airports—with unpaved runways:
total: 323

914 to 1,523 m: 39
under 914 m: 284 (2010)
Heliports: 2 (2010)
Pipelines: extra heavy crude 434 km; gas 5 km; oil 1,378 km; refined products 1,262 km (2010)
Railways:
total: 965 km
country comparison to the world: 90
narrow gauge: 965 km 1.067-m gauge (2010)
Roadways:
total: 43,670 km
country comparison to the world: 85
paved: 6,472 km
unpaved: 37,198 km (2006)
Waterways: 1,500 km (most inaccessible) (2010)
country comparison to the world: 53
Merchant marine: *total:* 41
country comparison to the world: 77
by type: cargo 1, chemical tanker 3, liquefied gas 1, passenger 9, petroleum tanker 26, refrigerated cargo 1
registered in other countries: 7 (Bolivia 1, Panama 6) (2010)
Ports and terminals: Esmeraldas, Guayaquil, Manta, Puerto Bolivar

MILITARY

Military branches: Ecuadorian Armed Forces: Ecuadorian Land Force (Fuerza Terrestre Ecuatoriana, FTE), Ecuadorian Navy (Fuerza Naval del Ecuador (FNE), includes Naval Infantry, Naval Aviation, Coast Guard), Ecuadorian Air Force (Fuerza Aerea Ecuatoriana, FAE) (2011)
Military service age and obligation: 20 years of age for selective conscript military service; 12-month service obligation (2008)
Manpower available for military service: *males age 16–49:* 3,728,906
females age 16–49: 3,844,918 (2010 est.)
Manpower fit for military service: *males age 16–49:* 2,834,213
females age 16–49: 3,269,535 (2010 est.)
Manpower reaching militarily significant age annually: *male:* 152,593
female: 147,143 (2010 est.)
Military expenditures: 0.9% of GDP (2009)
country comparison to the world: 135

TRANSNATIONAL ISSUES

Disputes—international: organized illegal narcotics operations in Colombia penetrate across Ecuador's shared border, which thousands of Colombians also cross to escape the violence in their home country
Refugees and internally displaced persons:
refugees (country of origin): 11,526 (Colombia); note–UNHCR estimates as many as 250,000 Columbians are seeking asylum in Ecuador, many of whom do not register as refugees for fear of deportation (2007)
Illicit drugs: significant transit country for cocaine originating in Colombia and Peru, with much of the US-bound cocaine passing through Ecuadorian Pacific waters; importer of precursor chemicals used in production of illicit narcotics; attractive location for cash-placement by drug traffickers laundering money because of dollarization and weak anti-money-laundering regime; increased activity on the northern frontier by trafficking groups and Colombian insurgents (2008)

EGYPT

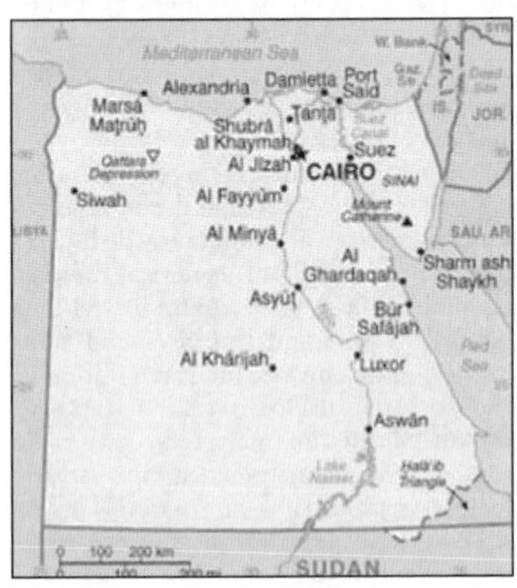

INTRODUCTION

Background: The regularity and richness of the annual Nile River flood, coupled with semi-isolation provided by deserts to the east and west, allowed for the development of one of the world's great civilizations. A unified kingdom arose circa 3200 B.C., and a series of dynasties ruled in Egypt for the next three millennia. The last native dynasty fell to the Persians in 341 B.C., who in turn were replaced by the Greeks, Romans, and Byzantines. It was the Arabs who introduced Islam and the Arabic language in the 7th century and who ruled for the next six centuries. A local military caste, the Mamluks took control about 1250 and continued to govern after the conquest of Egypt by the Ottoman Turks in 1517. Following the completion of the Suez Canal in 1869, Egypt became an important world transportation hub, but also fell heavily into debt. Ostensibly to protect its investments, Britain seized control of Egypt's government in 1882, but nominal allegiance to the Ottoman Empire continued until 1914. Partially independent from the UK in 1922, Egypt acquired full sovereignty with the overthrow of the British-backed monarchy in 1952. The completion of the Aswan High Dam in 1971 and the resultant Lake Nasser have altered the time-honored place of the Nile River in the agriculture and ecology of Egypt. A rapidly growing population (the largest in the Arab world), limited arable land, and dependence on the Nile all continue to overtax resources and stress society. The government has struggled to meet the demands of Egypt's growing population through economic reform and massive investment in communications and physical infrastructure.

GEOGRAPHY

Location: Northern Africa, bordering the Mediterranean Sea, between Libya and the Gaza Strip, and the Red Sea north of Sudan, and includes the Asian Sinai Peninsula
Geographic coordinates: 27 00 N, 30 00 E
Map references: Africa
Area: *total:* 1,001,450 sq km
country comparison to the world: 30
land: 995,450 sq km
water: 6,000 sq km
Area—comparative: slightly more than three times the size of New Mexico
Land boundaries: *total:* 2,665 km
border countries: Gaza Strip 11 km, Israel 266 km, Libya 1,115 km, Sudan 1,273 km
Coastline: 2,450 km
Maritime claims: *territorial sea:* 12 nm
contiguous zone: 24 nm
exclusive economic zone: 200 nm
continental shelf: 200 m depth or to the depth of exploitation
Climate: desert; hot, dry summers with moderate winters
Terrain: vast desert plateau interrupted by Nile valley and delta
Elevation extremes: *lowest point:* Qattara Depression -133 m
highest point: Mount Catherine 2,629 m
Natural resources: petroleum, natural gas, iron ore, phosphates, manganese, limestone, gypsum, talc, asbestos, lead, rare earth elements, zinc
Land use: *arable land:* 2.92%
permanent crops: 0.5%
other: 96.58% (2005)
Irrigated land: 34,220 sq km (2003)
Total renewable water resources: 86.8 cu km (1997)
Freshwater withdrawal (domestic/industrial/agricultural): *total:* 68.3 cu km/yr (8%/6%/86%)
per capita: 923 cu m/yr (2000)
Natural hazards: periodic droughts; frequent earthquakes; flash floods; landslides; hot, driving windstorm called khamsin occurs in spring; dust storms; sandstorms
Environment—current issues: agricultural land being lost to urbanization and windblown sands; increasing soil salination below Aswan High Dam; desertification; oil pollution threatening coral reefs, beaches, and marine habitats; other water pollution from agricultural pesticides, raw sewage, and industrial effluents; limited natural freshwater resources away from the Nile, which is the only perennial water source; rapid growth in population overstraining the Nile and natural resources

Environment—international agreements: *party to:* Biodiversity, Climate Change, Climate Change-Kyoto Protocol, Desertification, Endangered Species, Environmental Modification, Hazardous Wastes, Law of the Sea, Marine Dumping, Ozone Layer Protection, Ship Pollution, Tropical Timber 83, Tropical Timber 94, Wetlands
signed, but not ratified: none of the selected agreements
Geography—note: controls Sinai Peninsula, only land bridge between Africa and remainder of Eastern Hemisphere; controls Suez Canal, a sea link between Indian Ocean and Mediterranean Sea; size, and juxtaposition to Israel, establish its major role in Middle Eastern geopolitics; dependence on upstream neighbors; dominance of Nile basin issues; prone to influxes of refugees from Sudan and the Palestinian territories

PEOPLE

Population: 82,079,636 (July 2011 est.)
country comparison to the world: 15
Age structure: *0–14 years:* 32.7% (male 13,725,282/female 13,112,157)
15–64 years: 62.8% (male 26,187,921/female 25,353,947)
65 years and over: 4.5% (male 1,669,313/female 2,031,016) (2011 est.)
Median age:
total: 24.3 years
male: 24 years
female: 24.6 years (2011 est.)
Population growth rate: 1.96% (2011 est.)
country comparison to the world: 57
Birth rate: 24.63 births/1,000 population (2011 est.)
country comparison to the world: 64
Death rate: 4.82 deaths/1,000 population (July 2011 est.)
country comparison to the world: 193
Net migration rate: -0.21 migrant(s)/1,000 population (2011 est.)
country comparison to the world: 124
Urbanization: *urban population:* 43.4% of total population (2010)
rate of urbanization: 2.1% annual rate of change (2010-15 est.)
Major cities—population: CAIRO (capital) 10.902 million; Alexandria 4.387 million (2009)
Sex ratio: *at birth:* 1.05 male(s)/female
under 15 years: 1.05 male(s)/female
15–64 years: 1.03 male(s)/female
65 years and over: 0.83 male(s)/female
total population: 1.03 male(s)/female (2011 est.)
Infant mortality rate:
total: 25.2 deaths/1,000 live births
country comparison to the world: 81
male: 26.8 deaths/1,000 live births
female: 23.52 deaths/1,000 live births (2011 est.)
Life expectancy at birth: *total population:* 72.66 years
country comparison to the world: 123
male: 70.07 years
female: 75.38 years (2011 est.)
Total fertility rate: 2.97 children born/woman (2011 est.)
country comparison to the world: 65
HIV/AIDS—adult prevalence rate: less than 0.1% (2009 est.)
country comparison to the world: 126
HIV/AIDS—people living with HIV/AIDS: 11,000 (2009 est.)
country comparison to the world: 95
HIV/AIDS—deaths: fewer than 500 (2009 est.)
country comparison to the world: 94
Major infectious diseases: *degree of risk:* intermediate
food or waterborne diseases: bacterial diarrhea, hepatitis A, and typhoid fever
vectorborne disease: Rift Valley fever
water contact disease: schistosomiasis
note: highly pathogenic H5N1 avian influenza has been identified in this country; it poses a negligible risk with extremely rare cases possible among US citizens who have close contact with birds (2009)
Drinking water source: *Improved:*
urban: 100% of population
rural: 98% of population
total: 99% of population
Unimproved: urban: 0% of population
rural: 4% of population
total: 1% of population (2008)
Sanitation facility access: *Improved:*
urban: 97% of population
rural: 92% of population
total: 94% of population
Unimproved: urban: 3% of population
rural: 8% of population
total: 6% of population (2008)
Nationality: *noun:* Egyptian(s)
adjective: Egyptian
Ethnic groups: Egyptian 99.6%, other 0.4% (2006 census)
Religions: Muslim (mostly Sunni) 90%, Coptic 9%, other Christian 1%
Languages: Arabic (official), English and French widely understood by educated classes
Literacy: *definition:* age 15 and over can read and write
total population: 71.4%
male: 83%
female: 59.4% (2005 est.)
School life expectancy (primary to tertiary education): *total:* 11 years
male: 11 years
female: 11 years (2004)
Education expenditures: 3.8% of GDP (2008)
country comparison to the world: 110

GOVERNMENT

Country name: *conventional long form:* Arab Republic of Egypt
conventional short form: Egypt
local long form: Jumhuriyat Misr al-Arabiyah
local short form: Misr
former: United Arab Republic (with Syria)
Government type: republic
Capital: *name:* Cairo
geographic coordinates: 30 03 N, 31 15 E
time difference: UTC+2 (7 hours ahead of Washington, DC during Standard Time)
Administrative divisions: 29 governorates (muhafazat, singular—muhafazat); Ad Daqahliyah, Al Bahr al Ahmar (Red Sea), Al Buhayrah (El Beheira), Al Fayyum (El Faiyum), Al Gharbiyah, Al Iskandariyah (Alexandria), Al Isma'iliyah (Ismailia), Al Jizah (Giza), Al Minufiyah (El Monofia), Al Minya, Al Qahirah (Cairo), Al Qalyubiyah, Al Uqsur (Luxor), Al Wadi al Jadid (New Valley), As Suways (Suez), Ash Sharqiyah, Aswan, Asyut, Bani Suwayf (Beni Suef), Bur Sa'id (Port Said), Dumyat (Damietta), Helwan, Janub Sina' (South Sinai), Kafr ash Shaykh, Matruh (Western Desert), Qina (Qena), Shamal Sina' (North Sinai), Sittah Uktubar, Suhaj (Sohag)
Independence: 28 February 1922 (from UK protectorate status; the revolution that began on 23 July 1952 led to a republic being declared on 18 June 1953 and all British troops withdrawn on 18 June 1956); note—it was ca. 3200 B.C. that the Two Lands of Upper (southern) and Lower (northern) Egypt were first united politically
National holiday: Revolution Day, 23 July (1952)
Constitution: 11 September 1971; amended 22 May 1980, 25 May 2005, and 26 March 2007; note—constitution dissolved by the military caretaker government 13 February 2011
Legal system: mixed legal system based on Napoleonic civil law and Islamic religious law; judicial review by Supreme Court and Council of State (oversees validity of administrative decisions)
International law organization participation: accepts compulsory ICJ jurisdiction with reservations; non-party state to the ICCt
Suffrage: 18 years of age; universal and compulsory
Executive branch: *chief of state:* President (vacant); Vice President (vacant); note—following the resignation of President Mohamed Hosni MUBARAK in February 2011, the Supreme Council of the Armed Forces, headed by Defense Minister Muhammad Hussein TANTAWI, assumed control of the government
head of government: Prime Minister Essam Abdel Aziz SHARAF (since 4 March 2011); Deputy Prime Minister Yehia El-GAMAL (since 24 February 2011)
cabinet: a new cabinet was sworn in on 7 March 2011
(For more information visit the World Leaders website)
elections: president elected by popular vote for a six-year term (no term limits)
election results: Hosni MUBARAK reelected president; percent of vote—Hosni MUBARAK 88.6%, Ayman NOUR 7.6%, Noman GOMAA 2.9%
Legislative branch: bicameral system consists of the Advisory Council or Majlis al-Shura (Shura Council) that traditionally functions mostly in a consultative role (264 seats; 176 members elected by popular vote, 88 appointed by the president; members serve six-year terms; mid-term elections for half of the elected members) and the People's Assembly or Majlis al-Sha'b (518 seats; 508 members elected by popular vote, 64 seats reserved for women, 10 appointed by the president; members serve five-year terms)
elections: Advisory Council—last held in June 2010 (next to be held in 2013); People's Assembly—last held in November-December 2010 in one round of voting and one run-off election (next to be held in 2015); note—on 13 February 2011 the ruling military council dissolved the parliament

election results: Advisory Council—percent of vote by party—NA; seats by party—NDP 80, Al-Geel 1, Nasserist 1, NWP 1, Tagammu 1, Tomorrow Party 1, independents 3; People's Assembly—percent of vote by party—NA; seats by party—NDP 419, NWP 6, Tagammu 5, Democratic Peace Party 1, Social Justice Party 1, Tomorrow Party 1, independents 71, seats undecided 4, seats appointed by president 10

Judicial branch: Supreme Constitutional Court

Political parties and leaders: Al-Geel; Democratic Peace Party; Nasserist Party [Ahmed HASSAN]; National Democratic Party or NDP (governing party) [Mohamed Hosni MUBARAK]; National Progressive Unionist Grouping or Tagammu [Rifaat EL-SAID]; New Wafd Party or NWP [Sayed EL-BEDAWY]; Social Justice Party [Mohamed Abdel Al HASAN]; Tomorrow Party [Ayman NOUR]
note: formation of political parties must be approved by the government; only parties with representation in elected bodies are listed

Political pressure groups and leaders: Muslim Brotherhood (technically illegal)
note: despite a constitutional ban against religious-based parties and political activity, the technically illegal Muslim Brotherhood constitutes Egypt's most potentially significant political opposition; President MUBARAK has alternated between tolerating limited political activity by the Brotherhood and blocking its influence (its members compete as independents in elections but do not currently hold any seats in the legislature); civic society groups are sanctioned, but constrained in practical terms; only trade unions and professional associations affiliated with the government are officially sanctioned; Internet social networking groups and bloggers

International organization participation: ABEDA, AfDB, AFESD, AMF, AU, BSEC (observer), CAEU, CICA, COMESA, D-8, EBRD, FAO, G-15, G-24, G-77, IAEA, IBRD, ICAO, ICC, ICRM, IDA, IDB, IFAD, IFC, IFRCS, IHO, ILO, IMF, IMO, IMSO, Interpol, IOC, IOM, IPU, ISO, ITSO, ITU, LAS, MIGA, MINURSO, MONUSCO, NAM, OAPEC, OAS (observer), OIC, OIF, OSCE (partner), PCA, UN, UNAMID, UNCTAD, UNESCO, UNHCR, UNIDO, UNMIL, UNMIS, UNOCI, UNRWA, UNWTO, UPU, WCO, WFTU, WHO, WIPO, WMO, WTO

Diplomatic representation in the US:
chief of mission: Ambassador Sameh Hassan SHOUKRY
chancery: 3521 International Court NW, Washington, DC 20008
telephone: [1] (202) 895-5400
FAX: [1] (202) 244-4319
consulate(s) general: Chicago, Houston, New York, San Francisco

Diplomatic representation from the US:
chief of mission: Ambassador Margaret SCOBEY
embassy: 8 Kamal El Din Salah St., Garden City, Cairo
mailing address: Unit 64900, Box 15, APO AE 09839-4900; 5 Tawfik Diab Street, Garden City, Cairo
telephone: [20] (2) 2797-3300
FAX: [20] (2) 2797-3200

Flag description: three equal horizontal bands of red (top), white, and black; the national emblem (a gold Eagle of Saladin facing the hoist side with a shield superimposed on its chest above a scroll bearing the name of the country in Arabic) centered in the white band; the band colors derive from the Arab Liberation flag and represent oppression (black), overcome through bloody struggle (red), to be replaced by a bright future (white)
note: similar to the flag of Syria, which has two green stars in the white band, Iraq, which has an Arabic inscription centered in the white band, and Yemen, which has a plain white band

National anthem: *name:* "Bilady, Bilady, Bilady" (My Homeland, My Homeland, My Homeland)
lyrics/music: Younis-al QADI/Sayed DARWISH
note: adopted 1979; after the signing of the 1979 peace with Israel, Egypt sought to create an anthem less militaristic than its previous one; Sayed DARWISH, commonly considered the father of modern Egyptian music, composed the anthem

ECONOMY

Economy—overview: Occupying the northeast corner of the African continent, Egypt is bisected by the highly fertile Nile valley, where most economic activity takes place. Egypt's economy was highly centralized during the rule of former President Gamal Abdel NASSER but opened up considerably under former Presidents Anwar EL-SADAT and Mohamed Hosni MUBARAK. Cairo from 2004 to 2008 aggressively pursued economic reforms to attract foreign investment and facilitate GDP growth. The global financial crisis slowed the reform efforts. The budget deficit climbed to over 8% of GDP and Egypt's GDP growth slowed to 4.6% in 2009, predominately due to reduced growth in export-oriented sectors, including manufacturing and tourism, and Suez Canal revenues. In 2010, the government spent more on infrastructure and public projects, and exports drove GDP growth to more than 5%, but GDP growth in 2011 is unlikely to bounce back to pre-global financial recession levels, when it stood at 7%. Despite the relatively high levels of economic growth over the past few years, living conditions for the average Egyptian remain poor.

GDP (purchasing power parity): $497.8 billion (2010 est.)
country comparison to the world: 27
$473.4 billion (2009 est.)
$452.3 billion (2008 est.)
note: data are in 2010 US dollars

GDP (official exchange rate): $218.5 billion (2010 est.)

GDP—real growth rate: 5.1% (2010 est.)
country comparison to the world: 70
4.7% (2009 est.)
7.2% (2008 est.)

GDP—per capita (PPP): $6,200 (2010 est.)
country comparison to the world: 138
$6,000 (2009 est.)
$5,900 (2008 est.)
note: data are in 2010 US dollars

GDP—composition by sector: *agriculture:* 13.5%
industry: 37.9%
services: 48.6% (2010 est.)

Labor force: 26.1 million (2010 est.)
country comparison to the world: 21

Labor force—by occupation:
agriculture: 32%
industry: 17%
services: 51% (2001 est.)

Unemployment rate: 9.7% (2010 est.)
country comparison to the world: 107
9.4% (2009 est.)

Population below poverty line: 20% (2005 est.)

Household income or consumption by percentage share: *lowest 10%:* 3.9%
highest 10%: 27.6% (2005)

Distribution of family income—Gini index: 34.4 (2001)
country comparison to the world: 90

Investment (gross fixed): 18.4% of GDP (2010 est.)
country comparison to the world: 108

Budget: *revenues:* $46.82 billion
expenditures: $64.19 billion (2010 est.)

Public debt: 80.5% of GDP (2010 est.)
country comparison to the world: 16
80.9% of GDP (2009 est.)

Inflation rate (consumer prices): 12.8% (2010 est.)
country comparison to the world: 208
11.9% (2009 est.)

Central bank discount rate: 8.5% (31 December 2009)
country comparison to the world: 34
11.5% (31 December 2008)

Commercial bank prime lending rate: 11.98% (31 December 2009 est.)
country comparison to the world: 65
12.33% (31 December 2008 est.)

Stock of narrow money: $37.8 billion (31 December 2010 est.)
country comparison to the world: 49
$33.42 billion (31 December 2009 est.)

Stock of broad money: $166.2 billion (31 December 2010 est.)
country comparison to the world: 42
$146.7 billion (31 December 2009 est.)

Stock of domestic credit: $145.6 billion (31 December 2010 est.)
country comparison to the world: 42
$131.5 billion (31 December 2009 est.)

Market value of publicly traded shares: $89.95 billion (31 December 2009)
country comparison to the world: 43
$85.89 billion (31 December 2008)
$139.3 billion (31 December 2007)

Agriculture—products: cotton, rice, corn, wheat, beans, fruits, vegetables; cattle, water buffalo, sheep, goats

Industries: textiles, food processing, tourism, chemicals, pharmaceuticals, hydrocarbons, construction, cement, metals, light manufactures

Industrial production growth rate: 5.5% (2010 est.)
country comparison to the world: 66

Electricity—production: 118.4 billion kWh (2007 est.)

country comparison to the world: 28
Electricity—consumption: 104.1 billion kWh (2007 est.)
country comparison to the world: 29
Electricity—exports: 814 million kWh (2007 est.)
Electricity—imports: 251 million kWh (2007 est.)
Oil—production: 680,500 bbl/day (2009 est.)
country comparison to the world: 29
Oil—consumption: 683,000 bbl/day (2009 est.)
country comparison to the world: 25
Oil—exports: 89,300 bbl/day (2009 est.)
country comparison to the world: 68
Oil—imports: 48,450 bbl/day (2009 est.)
country comparison to the world: 85
Oil—proved reserves: 4.3 billion bbl (1 January 2010 est.)
country comparison to the world: 27
Natural gas—production: 62.7 billion cu m (2009 est.)
country comparison to the world: 14
Natural gas—consumption: 42.5 billion cu m (2009 est.)
country comparison to the world: 21
Natural gas—exports: 8.55 billion cu m (2009 est.)
country comparison to the world: 22
Natural gas—imports: 0 cu m (2009 est.)
country comparison to the world: 179
Natural gas—proved reserves: 1.656 trillion cu m (1 January 2010 est.)
country comparison to the world: 22
Current account balance: $270 million (2010 est.)
country comparison to the world: 54
$-3.195 billion (2009 est.)
Exports:
$25.34 billion (2010 est.)
country comparison to the world: 64
$23.09 billion (2009 est.)
Exports—commodities: crude oil and petroleum products, cotton, textiles, metal products, chemicals, processed food
Exports—partners: US 7.95%, Italy 7.26%, Spain 6.78%, India 6.69%, Saudi Arabia 5.53%, Syria 5.3%, France 4.39%, South Korea 4.27% (2009)
Imports: $46.52 billion (2010 est.)
country comparison to the world: 49
$45.56 billion (2009 est.)
Imports—commodities: machinery and equipment, foodstuffs, chemicals, wood products, fuels
Imports—partners: US 9.92%, China 9.63%, Germany 6.98%, Italy 6.88%, Turkey 4.94% (2009)
Reserves of foreign exchange and gold: $35.72 billion (31 December 2010 est.)
country comparison to the world: 34
$33.93 billion (31 December 2009 est.)
Debt—external: $30.61 billion (31 December 2010 est.)
country comparison to the world: 64
$29.66 billion (31 December 2009 est.)
Stock of direct foreign investment—at home: $72.41 billion (31 December 2010 est.)
country comparison to the world: 48
$66.71 billion (31 December 2009 est.)
Stock of direct foreign investment—abroad: $4.9 billion (31 December 2010 est.)
country comparison to the world: 59
$4.272 billion (31 December 2009 est.)
Exchange rates: Egyptian pounds (EGP) per US dollar–5.6124 (2010), 5.545 (2009), 5.4 (2008), 5.67 (2007), 5.725 (2006)

COMMUNICATIONS

Telephones—main lines in use: 10.313 million (2009)
country comparison to the world: 21
Telephones—mobile cellular: 55.352 million (2009)
country comparison to the world: 19
Telephone system: *general assessment:* underwent extensive upgrading during 1990s; principal centers at Alexandria, Cairo, Al Mansurah, Ismailia, Suez, and Tanta are connected by coaxial cable and microwave radio relay
domestic: largest fixed-line system in the region; as of 2010 there were three mobile-cellular networks with a total of more than 55 million subscribers
international: country code–20; landing point for Aletar, the SEA-ME-WE-3 and SEA-ME-WE-4 submarine cable networks, Link Around the Globe (FLAG) Falcon and FLAG FEA; satellite earth stations–4 (2 Intelsat–Atlantic Ocean and Indian Ocean, 1 Arabsat, and 1 Inmarsat); tropospheric scatter to Sudan; microwave radio relay to Israel; a participant in Medarabtel (2009)
Broadcast media: mix of state-run and private broadcast media; state-run TV operates 2 national and 6 regional terrestrial networks as well as a few satellite channels; about 20 private satellite channels and a large number of Arabic satellite channels are available via subscription; state-run radio operates about 70 stations belonging to 8 networks; 2 privately-owned radio stations operational (2008)
Internet country code: .eg
Internet hosts: 187,197 (2010)
country comparison to the world: 67
Internet users: 20.136 million (2009)
country comparison to the world: 21

TRANSPORTATION

Airports: 86 (2010)
country comparison to the world: 66
Airports—with paved runways: *total:* 73
over 3,047 m: 15
2,438 to 3,047 m: 36
1,524 to 2,437 m: 15
914 to 1,523 m: 2
under 914 m: 5 (2010)
Airports—with unpaved runways: *total:* 13
2,438 to 3,047 m: 1
1,524 to 2,437 m: 3
914 to 1,523 m: 5
under 914 m: 4 (2010)
Heliports: 6 (2010)
Pipelines: condensate 320 km; condensate/gas 13 km; gas 6,628 km; liquid petroleum gas 956 km; oil 4,332 km; oil/gas/water 3 km; refined products 895 km; water 13 km (2010)
Railways: *total:* 5,083 km
country comparison to the world: 34
standard gauge: 5,083 km 1.435-m gauge (62 km electrified) (2010)
Roadways: *total:* 65,050 km
country comparison to the world: 70
paved: 47,500 km
unpaved: 17,550 km (2009)
Waterways: 3,500 km (includes the Nile River, Lake Nasser, Alexandria-Cairo Waterway, and numerous smaller canals in Nile Delta; the Suez Canal (193.5 km including approaches) is navigable by oceangoing vessels drawing up to 17.68 m) (2010)
country comparison to the world: 29
Merchant marine: *total:* 66
country comparison to the world: 63
by type: bulk carrier 11, cargo 24, container 3, passenger/cargo 7, petroleum tanker 12, roll on/roll off 9
foreign-owned: 13 (Denmark 1, France 1, Greece 8, Jordan 2, Lebanon 1)
registered in other countries: 52 (Cambodia 12, Cook Islands 1, Georgia 11, Honduras 2, Malta 1, Marshall Islands 1, Moldova 5, Panama 11, Saint Vincent and the Grenadines 4, Saudi Arabia 1, Sierra Leone 2, unknown 1) (2010)
Ports and terminals: Ayn Sukhnah, Alexandria, Damietta, El Dekheila, Port Said, Sidi Kurayr, Suez

MILITARY

Military branches: Army, Navy, Air Force, Air Defense Command
Military service age and obligation: 18-30 years of age for male conscript military service; service obligation 12-36 months, followed by a 9-year reserve obligation (2008)
Manpower available for military service: *males age 16–49:* 21,012,199
females age 16–49: 20,145,021 (2010 est.)
Manpower fit for military service: *males age 16–49:* 18,060,543
females age 16–49: 17,244,838 (2010 est.)
Manpower reaching militarily significant age annually: *male:* 783,405
female: 748,647 (2010 est.)
Military expenditures: 3.4% of GDP (2005 est.)
country comparison to the world: 35

TRANSNATIONAL ISSUES

Disputes—international: Sudan claims but Egypt de facto administers security and economic development of Halaib region north of the 22nd parallel boundary; Egypt no longer shows its administration of the Bir Tawil trapezoid in Sudan on its maps; Gazan breaches in the security wall with Egypt in January 2008 highlight difficulties in monitoring the Sinai border; Saudi Arabia claims Egyptian-administered islands of Tiran and Sanafir
Refugees and internally displaced persons: refugees (country of origin): 60,000–80,000 (Iraq); 70,198 (Palestinian Territories); 12,157 (Sudan) (2007)
Illicit drugs: transit point for cannabis, heroin, and opium moving to Europe, Israel, and North Africa; transit stop for Nigerian drug couriers; concern as money laundering site due to lax enforcement of financial regulations

EL SALVADOR

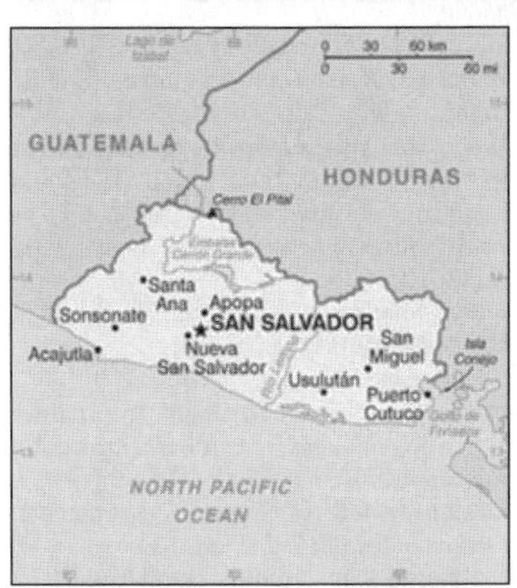

INTRODUCTION

Background: El Salvador achieved independence from Spain in 1821 and from the Central American Federation in 1839. A 12-year civil war, which cost about 75,000 lives, was brought to a close in 1992 when the government and leftist rebels signed a treaty that provided for military and political reforms.

GEOGRAPHY

Location: Central America, bordering the North Pacific Ocean, between Guatemala and Honduras
Geographic coordinates: 13 50 N, 88 55 W
Map references: Central America and the Caribbean
Area: *total:* 21,041 sq km
country comparison to the world: 152
land: 20,721 sq km
water: 320 sq km
Area—comparative: slightly smaller than Massachusetts
Land boundaries: *total:* 545 km
border countries: Guatemala 203 km, Honduras 342 km
Coastline: 307 km
Maritime claims: *territorial sea:* 12 nm
contiguous zone: 24 nm
exclusive economic zone: 200 nm
Climate: tropical; rainy season (May to October); dry season (November to April); tropical on coast; temperate in uplands
Terrain: mostly mountains with narrow coastal belt and central plateau
Elevation extremes: *lowest point:* Pacific Ocean 0 m
highest point: Cerro El Pital 2,730 m
Natural resources: hydropower, geothermal power, petroleum, arable land
Land use: *arable land:* 31.37%
permanent crops: 11.88%
other: 56.75% (2005)
Irrigated land: 450 sq km (2003)
Total renewable water resources: 25.2 cu km (2001)
Freshwater withdrawal (domestic/industrial/agricultural): *total:* 1.28 cu km/yr (25%/16%/59%)
per capita: 186 cu m/yr (2000)
Natural hazards: known as the Land of Volcanoes; frequent and sometimes destructive earthquakes and volcanic activity; extremely susceptible to hurricanes
volcanism: El Salvador experiences significant volcanic activity; San Salvador (elev. 1,893 m), which last erupted in 1917, has the potential to cause major harm to the country's capital, which lies just below the volcano's slopes; San Miguel (elev. 2,130 m), which last erupted in 2002, is one of the most active volcanoes in the country; other historically active volcanoes include Conchaguita, Ilopango, Izalco, and Santa Ana
Environment—current issues: deforestation; soil erosion; water pollution; contamination of soils from disposal of toxic wastes
Environment—international agreements: *party to:* Biodiversity, Climate Change, Climate Change-Kyoto Protocol, Desertification, Endangered Species, Hazardous Wastes, Ozone Layer Protection, Wetlands
signed, but not ratified: Law of the Sea
Geography—note: smallest Central American country and only one without a coastline on Caribbean Sea

PEOPLE

Population: 6,071,774 (July 2011 est.)
country comparison to the world: 106
Age structure: *0–14 years:* 30.6% (male 953,842/female 905,688)
15–64 years: 63% (male 1,802,113/female 2,021,191)
65 years and over: 6.4% (male 173,363/female 215,577) (2011 est.)
Median age: *total:* 24.3 years
male: 22.9 years
female: 25.7 years (2011 est.)
Population growth rate: 0.318% (2011 est.)
country comparison to the world: 169
Birth rate: 17.75 births/1,000 population (2011 est.)
country comparison to the world: 111
Death rate: 5.62 deaths/1,000 population (July 2011 est.)
country comparison to the world: 174
Net migration rate: -8.95 migrant(s)/1,000 population (2011 est.)
country comparison to the world: 207
Urbanization: *urban population:* 64% of total population (2010)
rate of urbanization: 1.4% annual rate of change (2010-15 est.)
Major cities—population: SAN SALVADOR (capital) 1.534 million (2009)
Sex ratio: *at birth:* 1.05 male(s)/female
under 15 years: 1.05 male(s)/female
15–64 years: 0.89 male(s)/female
65 years and over: 0.81 male(s)/female
total population: 0.93 male(s)/female (2011 est.)
Infant mortality rate:
total: 20.3 deaths/1,000 live births
country comparison to the world: 97
male: 22.36 deaths/1,000 live births
female: 18.15 deaths/1,000 live births (2011 est.)
Life expectancy at birth: *total population:* 73.44 years
country comparison to the world: 117
male: 70.16 years
female: 76.87 years (2011 est.)
Total fertility rate: 2.08 children born/woman (2011 est.)
country comparison to the world: 120
HIV/AIDS—adult prevalence rate: 0.8% (2009 est.)
country comparison to the world: 55
HIV/AIDS—people living with HIV/AIDS: 34,000 (2009 est.)
country comparison to the world: 68
HIV/AIDS—deaths: 1,400 (2009 est.)
country comparison to the world: 61
Major infectious diseases: *degree of risk:* high
food or waterborne diseases: bacterial diarrhea, hepatitis A, and typhoid fever
vectorborne diseases: dengue fever
water contact disease: leptospirosis (2009)
Drinking water source: *Improved:*
urban: 94% of population
rural: 76% of population
total: 87% of population
Unimproved: urban: 6% of population
rural: 24% of population
total: 13% of population (2008)
Sanitation facility access: *Improved:*
urban: 89% of population
rural: 83% of population
total: 87% of population
Unimproved: urban: 89% of population
rural: 83% of population
total: 87% of population (2008)
Nationality: *noun:* Salvadoran(s)
adjective: Salvadoran
Ethnic groups: mestizo 90%, white 9%, Amerindian 1%
Religions: Roman Catholic 57.1%, Protestant 21.2%, Jehovah's Witnesses 1.9%, Mormon 0.7%, other religions 2.3%, none 16.8% (2003 est.)
Languages: Spanish (official), Nahua (among some Amerindians)
Literacy: *definition:* age 15 and over can read and write
total population: 81.1%
male: 82.8%
female: 79.6% (2007 census)
School life expectancy (primary to tertiary education): *total:* 12 years
male: 12 years
female: 12 years (2008)
Education expenditures: 3.6% of GDP (2008)
country comparison to the world: 116

GOVERNMENT

Country name: *conventional long form:* Republic of El Salvador
conventional short form: El Salvador
local long form: Republica de El Salvador
local short form: El Salvador
Government type: republic
Capital: *name:* San Salvador
geographic coordinates: 13 42 N, 89 12 W
time difference: UTC-6 (1 hour behind Washington, DC during Standard Time)
daylight saving time: none scheduled for 2011
Administrative divisions: 14 departments (departamentos, singular—departamento); Ahuachapan, Cabanas, Chalatenango,

Cuscatlan, La Libertad, La Paz, La Union, Morazan, San Miguel, San Salvador, San Vicente, Santa Ana, Sonsonate, Usulutan
Independence: 15 September 1821 (from Spain)
National holiday: Independence Day, 15 September (1821)
Constitution: 20 December 1983
Legal system: civil law system with minor common law influence; judicial review of legislative acts in the Supreme Court
International law organization participation: has not submitted an ICJ jurisdiction declaration; non-party state to the ICCt
Suffrage: 18 years of age; universal
Executive branch: *chief of state:* President Mauricio FUNES Cartagena (since 1 June 2009); Vice President Salvador SANCHEZ CEREN (since 1 June 2009); note—the president is both the chief of state and head of government
head of government: President Mauricio FUNES Cartagena (since 1 June 2009); Vice President Salvador SANCHEZ CEREN (since 1 June 2009)
cabinet: Council of Ministers selected by the president
(For more information visit the World Leaders website)
elections: president and vice president elected on the same ticket by popular vote for a single five-year term; election last held on 15 March 2009 (next to be held in March 2014)
election results: Mauricio FUNES Cartagena elected president; percent of vote—Mauricio FUNES Cartagena 51.3%, Rodrigo AVILA 48.7%
Legislative branch: unicameral Legislative Assembly or Asamblea Legislativa (84 seats; members elected by direct, popular vote to serve three-year terms)
elections: last held on 18 January 2009 (next to be held in March 2012)
election results: percent of vote by party—NA; seats by party—FMLN 35, ARENA 32, PCN 11, PDC 5, CD 1; note—as of 1 January 2011, the current composition of the legislature by seats is as follows: FMLN 35, ARENA 19, GANA 16, PCN 10, PDC 2, CD 1, independent 1
Judicial branch: Supreme Court or Corte Suprema (15 judges are selected by the Legislative Assembly; the 15 judges are assigned to four Supreme Court chambers—constitutional, civil, penal, and administrative conflict)
Political parties and leaders: Christian Democratic Party or PDC [Rodolfo PARKER]; Democratic Convergence or CD [Oscar KATTAN] (formerly United Democratic Center or CDU); Farabundo Marti National Liberation Front or FMLN [Medardo GONZALEZ]; National Conciliation Party or PCN [Ciro CRUZ ZEPEDA]; Nationalist Republican Alliance or ARENA [Alfredo CRISTIANI]; Great Alliance for National Unity or GANA [Andres ROVIRA]
Political pressure groups and leaders: labor organizations—Electrical Industry Union of El Salvador or SIES; Federation of the Construction Industry, Similar Transport and other activities, or FESINCONTRANS; National Confederation of Salvadoran Workers or CNTS; National Union of Salvadoran Workers or UNTS; Port Industry Union of El Salvador or SIPES; Salvadoran Union of Ex-Petrolleros and Peasant Workers or USEPOC; Salvadoran Workers Central or CTS; Workers Union of Electrical Corporation or STCEL; business organizations—National Association of Small Enterprise or ANEP; Salvadoran Assembly Industry Association or ASIC; Salvadoran Industrial Association or ASI
International organization participation: BCIE, CACM, CD, FAO, G-11, G-77, IADB, IAEA, IBRD, ICAO, ICC, ICRM, IDA, IFAD, IFC, IFRCS, ILO, IMF, IMO, Interpol, IOC, IOM, IPU, ISO (correspondent), ITSO, ITU, ITUC, LAES, LAIA (observer), MIGA, MINURSO, NAM (observer), OAS, OPANAL, OPCW, PCA, RG, SICA, UN, UNCTAD, UNESCO, UNIDO, UNIFIL, Union Latina, UNMIL, UNMIS, UNOCI, UNWTO, UPU, WCO, WFTU, WHO, WIPO, WMO, WTO
Diplomatic representation in the US: *chief of mission:* Ambassador Francisco Robert ALTSCHUL Fuentes
chancery: Suite 100, 1400 16th Street, Washington, DC 20036
telephone: [1] (202) 265-9671
FAX: [1] (202) 234-3763
consulate(s) general: Chicago, Dallas, Duluth (Georgia), Houston, Las Vegas, Los Angeles, Miami, New York (2), Nogales (Arizona), Santa Ana (California), San Francisco, Washington (DC), Woodbridge (Virginia)
consulate(s): Boston, Elizabeth (New Jersey)
Diplomatic representation from the US: *chief of mission:* Ambassador (vacant); Charge d'Affaires Robert BLAU
embassy: Final Boulevard Santa Elena Sur, Antiguo Cuscatlan, La Libertad, San Salvador
mailing address: Unit 3450, APO AA 34023; 3450 San Salvador Place, Washington, DC 20521-3450
telephone: [503] 2501-2999
FAX: [503] 2501-2150
Flag description: three equal horizontal bands of blue (top), white, and blue with the national coat of arms centered in the white band; the coat of arms features a round emblem encircled by the words REPUBLICA DE EL SALVADOR EN LA AMERICA CENTRAL; the banner is based on the former blue-white-blue flag of the Federal Republic of Central America; the blue bands symbolize the Pacific Ocean and the Caribbean Sea, while the white band represents the land between the two bodies of water, as well as peace and prosperity
note: similar to the flag of Nicaragua, which has a different coat of arms centered in the white band—it features a triangle encircled by the words REPUBLICA DE NICARAGUA on top and AMERICA CENTRAL on the bottom; also similar to the flag of Honduras, which has five blue stars arranged in an X pattern centered in the white band
National anthem: *name:* "Himno Nacional de El Salvador" (National Anthem of El Salvador)
lyrics/music: Juan Jose CANAS/Juan ABERLE
note: officially adopted 1953, in use since 1879; the anthem of El Salvador is one of the world's longest

ECONOMY

Economy—overview: Despite being the smallest country geographically in Central America, El Salvador has the third largest economy in the region. The economy took a hit from the global recession and real GDP contracted by 3.5% in 2009. The economy began a slow recovery in 2010 on the back of improved export and remittances figures. Remittances accounted for 16% of GDP in 2009, and about a third of all households receive these transfers. In 2006 El Salvador was the first country to ratify the Dominican Republic-Central American Free Trade Agreement (CAFTA-DR), which has bolstered the export of processed foods, sugar, and ethanol, and supported investment in the apparel sector amid increased Asian competition and the expiration of the Multi-Fiber Agreement in 2005. El Salvador has promoted an open trade and investment environment, and has embarked on a wave of privatizations extending to telecom, electricity distribution, banking, and pension funds. In late 2006, the government and the Millennium Challenge Corporation signed a five-year, $461 million compact to stimulate economic growth and reduce poverty in the country's northern region, the primary conflict zone during the civil war, through investments in education, public services, enterprise development, and transportation infrastructure. With the adoption of the US dollar as its currency in 2001, El Salvador lost control over monetary policy. Any counter-cyclical policy response to the downturn must be through fiscal policy, which is constrained by legislative requirements for a two-thirds majority to approve any international financing, and by already high levels of debt.
GDP (purchasing power parity): $43.57 billion (2010 est.)
country comparison to the world: 96
$43.24 billion (2009 est.)
$44.83 billion (2008 est.)
note: data are in 2010 US dollars
GDP (official exchange rate): $21.7 billion (2010 est.)
GDP—real growth rate: 0.7% (2010 est.)
country comparison to the world: 180
-3.5% (2009 est.)
2.4% (2008 est.)
GDP—per capita (PPP): $7,200 (2010 est.)
country comparison to the world: 129
$7,200 (2009 est.)
$7,500 (2008 est.)
note: data are in 2010 US dollars
GDP—composition by sector:
agriculture: 11%
industry: 29.1%
services: 59.9% (2010 est.)
Labor force: 2.94 million (2010 est.)
country comparison to the world: 105

Labor force—by occupation: *agriculture:* 19%
industry: 23%
services: 58% (2006 est.)
Unemployment rate: 7% (2010 est.)
country comparison to the world: 72
7.2% (2009 est.)
note: data are official rates; but the economy has much underemployment
Population below poverty line: 37.8% (2009 est.)
Household income or consumption by percentage share: *lowest 10%:* 1%
highest 10%: 37% (2005)
Distribution of family income—Gini index: 52.4 (2002)
country comparison to the world: 15
52.5 (2001)
Investment (gross fixed): 13.7% of GDP (2010 est.)
country comparison to the world: 140
Budget: *revenues:* $3.894 billion
expenditures: $4.915 billion (2010 est.)
Public debt: 55% of GDP (2010 est.)
country comparison to the world: 46
52.3% of GDP (2009 est.)
Inflation rate (consumer prices): 0.8% (2010 est.)
country comparison to the world: 15
-0.2% (2009)
Commercial bank prime lending rate: 6.59% (31 December 2010)
country comparison to the world: 109
8.42% (31 December 2009)
Stock of narrow money: $2.534 billion (31 December 2010 est.)
country comparison to the world: 117
$2.153 billion (31 December 2009 est.)
Stock of broad money: $9.666 billion (31 December 2010 est.)
country comparison to the world: 100
$9.011 billion (31 December 2009 est.)
Stock of domestic credit: $10.01 billion (31 December 2010 est.)
country comparison to the world: 95
$9.867 billion (31 December 2009 est.)
Market value of publicly traded shares: $4.432 billion (31 December 2009)
country comparison to the world: 83
$4.656 billion (31 December 2008)
$6.743 billion (31 December 2007)
Agriculture—products: coffee, sugar, corn, rice, beans, oilseed, cotton, sorghum; beef, dairy products
Industries: food processing, beverages, petroleum, chemicals, fertilizer, textiles, furniture, light metals
Industrial production growth rate: 0.9% (2010 est.)
country comparison to the world: 147
Electricity—production: 5.445 billion kWh (2009 est.)
country comparison to the world: 110
Electricity—consumption: 4.524 billion kWh (2009 est.)
country comparison to the world: 114
Electricity—exports: 78.7 million kWh (2009 est.)
Electricity—imports: 208.4 million kWh (2009 est.)
Oil—production: 0 bbl/day (2008 est.)
country comparison to the world: 171
Oil—consumption: 46,000 bbl/day (2009 est.)
country comparison to the world: 97
Oil—exports: 1,927 bbl/day (2007 est.)
country comparison to the world: 113
Oil—imports: 46,310 bbl/day (2007 est.)
country comparison to the world: 86
Oil—proved reserves: 0 bbl (1 January 2010 est.)
country comparison to the world: 128
Natural gas—production: 0 cu m (2008 est.)
country comparison to the world: 123
Natural gas—consumption: 0 cu m (2008 est.)
country comparison to the world: 168
Natural gas—exports: 0 cu m (2008 est.)
country comparison to the world: 91
Natural gas—imports: 0 cu m (2008 est.)
country comparison to the world: 182
Natural gas—proved reserves: 0 cu m (1 January 2010 est.)
country comparison to the world: 132
Current account balance: $-907 million (2010 est.)
country comparison to the world: 129
$-374 million (2009)
Exports: $4.377 billion (2010 est.)
country comparison to the world: 112
$3.797 billion (2009)
Exports—commodities: offshore assembly exports, coffee, sugar, textiles and apparel, gold, ethanol, chemicals, electricity, iron and steel manufactures
Exports—partners: US 43.86%, Guatemala 13.92%, Honduras 13.22%, Nicaragua 5.65% (2009)
Imports: $7.98 billion (2010 est.)
country comparison to the world: 99
$7.255 billion (2009)
Imports—commodities: raw materials, consumer goods, capital goods, fuels, foodstuffs, petroleum, electricity
Imports—partners: US 29.79%, Mexico 10.26%, Guatemala 9.7%, China 4.5%, Honduras 4.4% (2009)
Reserves of foreign exchange and gold: $2.882 billion (31 December 2010 est.)
country comparison to the world: 91
$2.985 billion (31 December 2009)
Debt—external: $11.45 billion (31 December 2010 est.)
country comparison to the world: 87
$10.83 billion (31 December 2009 est.)
Stock of direct foreign investment—at home: $7.522 billion (31 December 2010 est.)
country comparison to the world: 83
$7.132 billion (31 December 2009 est.)
Stock of direct foreign investment—abroad: $273 million (31 December 2010 est.)
country comparison to the world: 78
$333 million (31 December 2009 est.)
Exchange rates: Communications
Telephones—main lines in use: 1.099 million (2009)
country comparison to the world: 76
Telephones—mobile cellular: 7.566 million (2009)
country comparison to the world: 78
Telephone system: *general assessment:* multiple mobile-cellular providers are expanding services rapidly and in 2009 teledensity exceeded 100 per 100 persons; growth in fixed-line services has slowed in the face of mobile-cellular competition
domestic: nationwide microwave radio relay system
international: country code–503; satellite earth station–1 Intelsat (Atlantic Ocean); connected to Central American Microwave System (2009)
Broadcast media: multiple privately-owned national terrestrial television networks, supplemented by cable TV networks that carry international channels; hundreds of commercial radio broadcast stations and 1 government-owned radio broadcast station (2007)
Internet country code: .sv
Internet hosts: 13,849 (2010)
country comparison to the world: 119
Internet users: 746,000 (2009)
country comparison to the world: 107

TRANSPORTATION

Airports: 65 (2010)
country comparison to the world: 77
Airports—with paved runways: *total:* 4
over 3,047 m: 1
1,524 to 2,437 m: 1
914 to 1,523 m: 2 (2010)
Airports—with unpaved runways:
total: 61
1,524 to 2,437 m: 1
914 to 1,523 m: 13
under 914 m: 47 (2010)
Heliports: 1 (2010)
Railways: *total:* 283 km
country comparison to the world: 121
narrow gauge: 283 km 0.600-m gauge
note: railways have been inoperable since 2005 because of disuse and high costs that led to a lack of maintenance (2010)
Roadways: *total:* 10,886 km
country comparison to the world: 134
paved: 2,827 km (includes 327 km of expressways)
unpaved: 8,059 km (2000)
Waterways: (Rio Lempa is partially navigable for small craft) (2010)
Ports and terminals: Acajutla, Puerto Cutuco

MILITARY

Military branches: Salvadoran Army (ES), Salvadoran Navy (FNES), Salvadoran Air Force (Fuerza Aerea Salvadorena, FAS) (2011)
Military service age and obligation: 18 years of age for selective compulsory military service; 16-22 years of age for voluntary male or female service; service obligation–12 months, with 11 months for officers and NCOs (2009)
Manpower available for military service: *males age 16–49:* 1,449,214
females age 16–49: 1,611,248 (2010 est.)
Manpower fit for military service: *males age 16–49:* 1,079,038
females age 16–49: 1,373,368 (2010 est.)
Manpower reaching militarily significant age annually: *male:* 71,530
female: 68,971 (2010 est.)
Military expenditures: 0.6% of GDP (2009)
country comparison to the world: 158

TRANSNATIONAL ISSUES

Disputes—international: International Court of Justice (ICJ) ruled on the delimitation of "bolsones" (disputed areas) along the El Salvador-Honduras boundary, in 1992, with final agreement by the parties in 2006 after an Organization of American States (OAS) survey and a further ICJ ruling in 2003; the 1992 ICJ ruling advised a tripartite resolution to a maritime boundary in the Gulf of Fonseca advocating Honduran access to the Pacific; El Salvador continues to claim tiny Conejo Island, not identified in the ICJ decision, off Honduras in the Gulf of Fonseca

Illicit drugs: transshipment point for cocaine; small amounts of marijuana produced for local consumption; significant use of cocaine

EQUATORIAL GUINEA

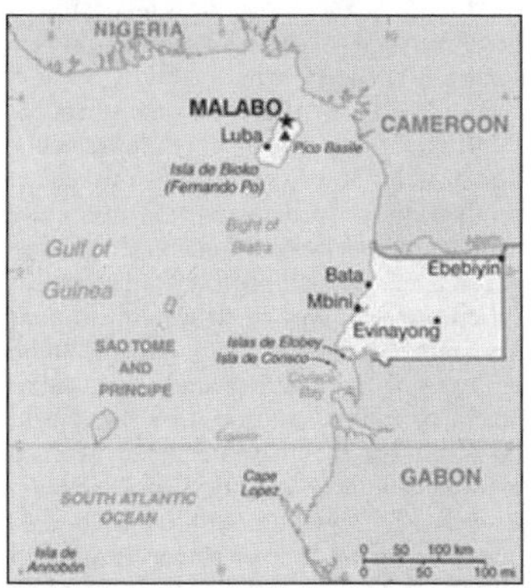

INTRODUCTION

Background: Equatorial Guinea gained independence in 1968 after 190 years of Spanish rule. This tiny country, composed of a mainland portion plus five inhabited islands, is one of the smallest on the African continent. President Teodoro OBIANG NGUEMA MBASOGO has ruled the country since 1979 when he seized power in a coup. Although nominally a constitutional democracy since 1991, the 1996, 2002, and 2009 presidential elections–as well as the 1999, 2004, and 2008 legislative elections–were widely seen as flawed. The president exerts almost total control over the political system and has discouraged political opposition. Equatorial Guinea has experienced rapid economic growth due to the discovery of large offshore oil reserves, and in the last decade has become Sub-Saharan Africa's third largest oil exporter. Despite the country's economic windfall from oil production resulting in a massive increase in government revenue in recent years, improvements in the population's living standards have been slow to develop.

GEOGRAPHY

Location: Western Africa, bordering the Bight of Biafra, between Cameroon and Gabon
Geographic coordinates: 2 00 N, 10 00 E
Map references: Africa
Area:
total: 28,051 sq km
country comparison to the world: 145
land: 28,051 sq km
water: 0 sq km
Area—comparative: slightly smaller than Maryland
Land boundaries: *total:* 539 km
border countries: Cameroon 189 km, Gabon 350 km
Coastline: 296 km
Maritime claims: *territorial sea:* 12 nm
exclusive economic zone: 200 nm
Climate: tropical; always hot, humid
Terrain: coastal plains rise to interior hills; islands are volcanic
Elevation extremes: *lowest point:* Atlantic Ocean 0 m
highest point: Pico Basile 3,008 m
Natural resources: petroleum, natural gas, timber, gold, bauxite, diamonds, tantalum, sand and gravel, clay
Land use: *arable land:* 4.63%
permanent crops: 3.57%
other: 91.8% (2005)
Irrigated land: NA
Total renewable water resources: 26 cu km (2001)
Freshwater withdrawal (domestic/industrial/agricultural): *total:* 0.11 cu km/yr (83%/16%/1%)
per capita: 220 cu m/yr (2000)
Natural hazards: violent windstorms; flash floods
volcanism: Santa Isabel (elev. 3,007 m), which last erupted in 1923, is the country's only historically active volcano; Santa Isabel, along with two dormant volcanoes, forms Bioko Island in the Gulf of Guinea
Environment—current issues: tap water is not potable; deforestation
Environment—international agreements: *party to:* Biodiversity, Climate Change, Climate Change-Kyoto Protocol, Desertification, Endangered Species, Hazardous Wastes, Law of the Sea, Marine Dumping, Ozone Layer Protection, Ship Pollution, Wetlands
signed, but not ratified: none of the selected agreements
Geography—note: insular and continental regions widely separated

PEOPLE

Population: 668,225 (July 2011 est.)
country comparison to the world: 165
Age structure: *0–14 years:* 41.5% (male 140,946/female 136,294)
15–64 years: 54.4% (male 179,141/female 184,358)
65 years and over: 4.1% (male 11,880/female 15,606) (2011 est.)
Median age:
total: 19.1 years
male: 18.5 years
female: 19.7 years (2011 est.)
Population growth rate: 2.641% (2011 est.)
country comparison to the world: 23
Birth rate: 35.43 births/1,000 population (2011 est.)
country comparison to the world: 28
Death rate: 9.03 deaths/1,000 population (July 2011 est.)
country comparison to the world: 69
Net migration rate: 0 migrant(s)/1,000 population (2011 est.)
country comparison to the world: 83
Urbanization: *urban population:* 40% of total population (2010)
rate of urbanization: 3.1% annual rate of change (2010-15 est.)
Major cities—population: MALABO (capital) 128,000 (2009)
Sex ratio: *at birth:* 1.03 male(s)/female
under 15 years: 1.03 male(s)/female
15–64 years: 0.97 male(s)/female
65 years and over: 0.78 male(s)/female
total population: 0.99 male(s)/female (2011 est.)
Infant mortality rate: *total:* 77.3 deaths/1,000 live births
country comparison to the world: 15
male: 78.37 deaths/1,000 live births
female: 76.19 deaths/1,000 live births (2011 est.)
Life expectancy at birth:
total population: 62.37 years
country comparison to the world: 181
male: 61.4 years
female: 63.36 years (2011 est.)
Total fertility rate: 4.91 children born/woman (2011 est.)
country comparison to the world: 22
HIV/AIDS—adult prevalence rate: 5% (2009 est.)
country comparison to the world: 15
HIV/AIDS—people living with HIV/AIDS: 20,000 (2009 est.)
country comparison to the world: 77
HIV/AIDS—deaths: fewer than 1,000 (2009 est.)
country comparison to the world: 71
Major infectious diseases: *degree of risk:* very high
food or waterborne diseases: bacterial and protozoal diarrhea, hepatitis A, and typhoid fever
vectorborne disease: malaria and yellow fever
animal contact disease: rabies (2009)
Drinking water source: *Improved:* *urban:* 45% of population
rural: 42% of population
total: 43% of population
Unimproved: urban: 55% of population
rural: 58% of population
total: 57% of population (2000)
Nationality: *noun:* Equatorial Guinean(s) or Equatoguinean(s)
adjective: Equatorial Guinean or Equatoguinean
Ethnic groups: Fang 85.7%, Bubi 6.5%, Mdowe 3.6%, Annobon 1.6%, Bujeba 1.1%, other 1.4% (1994 census)
Religions: nominally Christian and predominantly Roman Catholic, pagan practices

Languages: Spanish (official) 67.6%, other (includes French (official), Fang, Bubi) 32.4% (1994 census)
Literacy: *definition:* age 15 and over can read and write
total population: 87%
male: 93.4%
female: 80.5% (2000 est.)
School life expectancy (primary to tertiary education): *total:* 8 years
male: 9 years
female: 7 years (2002)
Education expenditures: 0.6% of GDP (2003)
country comparison to the world: 164

GOVERNMENT

Country name: *conventional long form:* Republic of Equatorial Guinea
conventional short form: Equatorial Guinea
local long form: Republica de Guinea Ecuatorial/Republique de Guinee equatoriale
local short form: Guinea Ecuatorial/Guinee equatoriale
former: Spanish Guinea
Government type: republic
Capital: *name:* Malabo
geographic coordinates: 3 45 N, 8 47 E
time difference: UTC+1 (6 hours ahead of Washington, DC during Standard Time)
Administrative divisions: 7 provinces (provincias, singular–provincia); Annobon, Bioko Norte, Bioko Sur, Centro Sur, Kie-Ntem, Litoral, Wele-Nzas
Independence: 12 October 1968 (from Spain)
National holiday: Independence Day, 12 October (1968)
Constitution: approved by national referendum 17 November 1991; amended January 1995
Legal system: mixed system of civil and customary law
International law organization participation: has not submitted an ICJ jurisdiction declaration; non-party state to the ICCt
Suffrage: 18 years of age; universal
Executive branch: *chief of state:* President Brig. Gen. (Ret.) Teodoro OBIANG NGUEMA MBASOGO (since 3 August 1979 when he seized power in a military coup)
head of government: Prime Minister Ignacio MILAM Tang (since 8 July 2008)
cabinet: Council of Ministers appointed by the president
(For more information visit the World Leaders website)
elections: president elected by popular vote for a seven-year term (no term limits); election last held on 29 November 2009 (next to be held in 2016); prime minister and deputy prime ministers appointed by the president
election results: Teodoro OBIANG NGUEMA MBASOGO reelected president; percent of vote–Teodoro OBIANG NGUEMA MBASOGO 95.8%, Placido Mico ABOGO 3.6%
Legislative branch: unicameral House of People's Representatives or Camara de Representantes del Pueblo (100 seats; members directly elected by popular vote to serve five-year terms)
elections: last held on 4 May 2008 (next to be held in 2012)
election results: percent of vote by party–NA; seats by party–PDGE 89, EC 10, CPDS 1
note: Parliament has little power since the constitution vests all executive authority in the president
Judicial branch:
Supreme Tribunal
Political parties and leaders: Convergence Party for Social Democracy or CPDS [Placido MICO Abogo]; Democratic Party for Equatorial Guinea or PDGE [Teodoro OBIANG NGUEMA MBASOGO] (ruling party); Electoral Coalition or EC; Party for Progress of Equatorial Guinea or PPGE [Severo MOTO]; Popular Action of Equatorial Guinea or APGE [Avelino MOCACHE]; Popular Union or UP [Daniel MARTINEZ Ayecaba]
Political pressure groups and leaders: ASODEGUE (Madrid-based pressure group for democratic reform); EG Justice (US-based anti-corruption group)
International organization participation: ACP, AfDB, AU, BDEAC, CEMAC, CPLP (associate), FAO, FZ, G-77, IBRD, ICAO, ICRM, IDA, IFAD, IFC, IFRCS, ILO, IMF, IMO, Interpol, IOC, ITSO, ITU, MIGA, NAM, OAS (observer), OIF, OPCW, UN, UNCTAD, UNESCO, UNIDO, UNWTO, UPU, WHO, WIPO, WTO (observer)
Diplomatic representation in the US: *chief of mission:* Ambassador Purificacion ANGUE ONDO
chancery: 2020 16th Street NW, Washington, DC 20009
telephone: [1] (202) 518-5700
FAX: [1] (202) 518-5252
Diplomatic representation from the US: *chief of mission:* Ambassador Alberto M. FERNANDEZ
embassy: KM-3, Carreterade de Aeropuerto (El Paraiso), Apartado 95, Malabo note–relocated embassy is opened for limited functions; inquiries should continue to be directed to the US Embassy in Yaounde, Cameroon
mailing address: B.P. 817, Yaounde, Cameroon; US Embassy Yaounde, US Department of State, Washington, DC 20521-2520
telephone: [237] 2220-1500
FAX: [237] 2220-1572
Flag description: three equal horizontal bands of green (top), white, and red, with a blue isosceles triangle based on the hoist side and the coat of arms centered in the white band; the coat of arms has six yellow six-pointed stars (representing the mainland and five offshore islands) above a gray shield bearing a silk-cotton tree and below which is a scroll with the motto UNIDAD, PAZ, JUSTICIA (Unity, Peace, Justice); green symbolizes the jungle and natural resources, blue represents the sea that connects the mainland to the islands, white stands for peace, and red recalls the fight for independence
National anthem: *name:* "Caminemos pisando la senda" (Let Us Tread the Path)
lyrics/music: Atanasio Ndongo MIYONO/Atanasio Ndongo MIYONO or Ramiro Sanchez LOPEZ (disputed)
note: adopted 1968

ECONOMY

Economy—overview: The discovery and exploitation of large oil and gas reserves have contributed to dramatic economic growth but fluctuating oil prices have produced huge swings in GDP growth in recent years. Forestry and farming are also minor components of GDP. Subsistence farming is the dominate form of livelihood. Although pre-independence Equatorial Guinea counted on cocoa production for hard currency earnings, the neglect of the rural economy under successive regimes has diminished potential for agriculture-led growth (the government has stated its intention to reinvest some oil revenue into agriculture). A number of aid programs sponsored by the World Bank and the IMF have been cut off since 1993 because of corruption and mismanagement. The government has been widely criticized for its lack of transparency and misuse of oil revenues; however, in 2010, under Equatorial Guinea's candidacy in the Extractive Industries Transparency Initiative, the government published oil revenue figures for the first time. Undeveloped natural resources include gold, zinc, diamonds, columbite-tantalite, and other base metals. Growth remained strong in 2008, when oil production peaked, but slowed in 2009-10, as the price of oil and the production level fell.
GDP (purchasing power parity): $23.82 billion (2010 est.)
country comparison to the world: 116
$24.02 billion (2009 est.)
$22.71 billion (2008 est.)
note: data are in 2010 US dollars
GDP (official exchange rate): $14.49 billion (2010 est.)
GDP—real growth rate: -0.8% (2010 est.)
country comparison to the world: 196
5.7% (2009 est.)
10.7% (2008 est.)
GDP—per capita (PPP): $36,600 (2010 est.)
country comparison to the world: 28
$37,900 (2009 est.)
$36,800 (2008 est.)
note: data are in 2010 US dollars; population figures are uncertain for Equatorial Guinea; these per capita income figures are based on a estimated population of less than 700,000; some estimates put the figure as high as 1.2 million people; if true, the per capita GDP figures would be significantly lower
GDP—composition by sector:
agriculture: 2.2%
industry: 93.9%
services: 3.8% (2010 est.)
Labor force: 195,200 (2007)
country comparison to the world: 171
Unemployment rate: 22.3% (2009 est.)
country comparison to the world: 172
Population below poverty line: NA%
Household income or consumption by percentage share: *lowest 10%:* NA%
highest 10%: NA%

Investment (gross fixed): 29.1% of GDP (2010 est.)
country comparison to the world: 20
Budget: *revenues:* $6.739 billion
expenditures: $6.984 billion (2010 est.)
Public debt: 4.1% of GDP (2010 est.)
country comparison to the world: 131
5.4% of GDP (2009 est.)
Inflation rate (consumer prices): 8.2% (2010 est.)
country comparison to the world: 188
7.1% (2009 est.)
Central bank discount rate: 4.25% (31 December 2009)
country comparison to the world: 86
4.75% (31 December 2008)
Commercial bank prime lending rate: NA%
NA%
Stock of narrow money: $1.86 billion (31 December 2010 est.)
country comparison to the world: 121
$1.295 billion (31 December 2009 est.)
Stock of broad money: $2.207 billion (31 December 2010 est.)
country comparison to the world: 139
$1.473 billion (31 December 2009 est.)
Stock of domestic credit: $1.534 billion (31 December 2009)
country comparison to the world: 135
$3.579 billion (31 December 2008)
Agriculture—products: coffee, cocoa, rice, yams, cassava (tapioca), bananas, palm oil nuts; livestock; timber
Industries: petroleum, natural gas, sawmilling
Industrial production growth rate: 1.8% (2010 est.)
country comparison to the world: 134
Electricity—production: 28 million kWh (2007 est.)
country comparison to the world: 204
Electricity—consumption: 26.04 million kWh (2007 est.)
country comparison to the world: 203
Electricity—exports: 0 kWh (2008 est.)
Electricity—imports: 0 kWh (2008 est.)
Oil—production: 346,000 bbl/day (2009 est.)
country comparison to the world: 35
Oil—consumption: 1,000 bbl/day (2009 est.)
country comparison to the world: 196
Oil—exports: 362,900 bbl/day (2007 est.)
country comparison to the world: 35
Oil—imports: 1,114 bbl/day (2007 est.)
country comparison to the world: 187
Oil—proved reserves: 1.1 billion bbl (1 January 2010 est.)
country comparison to the world: 41
Natural gas—production: 6.67 billion cu m (2008 est.)
country comparison to the world: 47
Natural gas—consumption: 1.5 billion cu m (2008 est.)
country comparison to the world: 83
Natural gas—exports: 5.17 billion cu m (2008 est.)
country comparison to the world: 27
Natural gas—imports: 0 cu m (2008 est.)
country comparison to the world: 180
Natural gas—proved reserves: 36.81 billion cu m (1 January 2010 est.)
country comparison to the world: 67
Current account balance: $-1.477 billion (2010 est.)
country comparison to the world: 149
$-1.883 billion (2009 est.)
Exports: $10.24 billion (2010 est.)
country comparison to the world: 85
$8.495 billion (2009 est.)
Exports—commodities: petroleum products, timber
Exports—partners: US 30.31%, China 12.54%, Japan 9.21%, Spain 7.5%, South Korea 7.01%, Taiwan 5.63%, Italy 5.38%, Netherlands 4.09% (2009)
Imports: $5.743 billion (2010 est.)
country comparison to the world: 107
$5.258 billion (2009 est.)
Imports—commodities: petroleum sector equipment, other equipment, construction materials, vehicles
Imports—partners: China 19.97%, US 17.28%, Spain 14.94%, France 9.49%, Cote d'Ivoire 6.34%, Italy 5.02% (2009)
Reserves of foreign exchange and gold: $4.086 billion (31 December 2010 est.)
country comparison to the world: 78
$3.252 billion (31 December 2009 est.)
Debt—external: $832 million (31 December 2010 est.)
country comparison to the world: 154
$766 million (31 December 2009 est.)
Exchange rates: Cooperation Financiere en Afrique Centrale francs per US dollar–495.28 (2010), 472.19 (2009), 447.81 (2008), 481.83 (2007), 522.4 (2006)

COMMUNICATIONS

Telephones—main lines in use: 10,000 (2009)
country comparison to the world: 202
Telephones—mobile cellular: 445,000 (2009)
country comparison to the world: 161
Telephone system: *general assessment:* digital fixed-line network in most major urban areas and good mobile coverage
domestic: fixed-line density is about 2 per 100 persons; mobile-cellular subscribership has been increasing and in 2009 stood at about 70 percent of the population
international: country code–240; international communications from Bata and Malabo to African and European countries; satellite earth station–1 Intelsat (Indian Ocean) (2009)
Broadcast media: state maintains control of broadcast media with domestic broadcast media limited to 1 state-owned TV station, 1 state-owned radio station, and 1 private radio station owned by the president's eldest son; satellite TV service is available; transmissions of multiple international broadcasters are accessible (2007)
Internet country code: .gq
Internet hosts: 9 (2010)
country comparison to the world: 222
Internet users: 14,400 (2009)
country comparison to the world: 199

TRANSPORTATION

Airports: 7 (2010)
country comparison to the world: 166
Airports—with paved runways: *total:* 6
2,438 to 3,047 m: 2
1,524 to 2,437 m: 1
914 to 1,523 m: 1
under 914 m: 2 (2010)
Airports—with unpaved runways: *total:* 1
2,438 to 3,047 m: 1 (2010)
Pipelines: gas 37 km (2010)
Roadways: *total:* 2,880 km (2000)
country comparison to the world: 167
Merchant marine: *total:* 4
country comparison to the world: 131
by type: cargo 1, chemical tanker 1, petroleum tanker 2
foreign-owned: 1 (Norway 1) (2010)
Ports and terminals: Bata, Luba, Malabo (2010)

MILITARY

Military branches: National Guard (Guardia Nacional de Guinea Ecuatoria, GNGE (Army), with Coast Guard (Navy) and Air Wing) (2010)
Military service age and obligation: 18 years of age for selective compulsory military service; service obligation 2 years; women hold only administrative positions in the Coast Guard (2011)
Manpower available for military service: *males age 16–49:* 151,147
females age 16–49: 150,345 (2010 est.)
Manpower fit for military service: *males age 16–49:* 113,277
females age 16–49: 115,320 (2010 est.)
Manpower reaching militarily significant age annually: *male:* 7,398
female: 7,126 (2010 est.)
Military expenditures: 0.1% of GDP (2009)
country comparison to the world: 172

TRANSNATIONAL ISSUES

Disputes—international: in 2002, ICJ ruled on an equidistance settlement of Cameroon-Equatorial Guinea-Nigeria maritime boundary in the Gulf of Guinea, but a dispute between Equatorial Guinea and Cameroon over an island at the mouth of the Ntem River and imprecisely defined maritime coordinates in the ICJ decision delay final delimitation; UN urges Equatorial Guinea and Gabon to resolve the sovereignty dispute over Gabon-occupied Mbane and lesser islands and to create a maritime boundary in the hydrocarbon-rich Corisco Bay
Trafficking in persons: current situation: Equatorial Guinea is primarily a destination country for children trafficked for the purpose of forced labor and possibly for the purpose of sexual exploitation; children have been trafficked from nearby countries for domestic servitude, market labor, ambulant vending, and possibly sexual exploitation; women may also be trafficked to Equatorial Guinea from Cameroon, Benin, other neighboring countries, and China for sexual exploitation
tier rating: Tier 2 Watch List–Equatorial Guinea is on the Tier 2 Watch List for its failure to provide evidence of increasing efforts to eliminate trafficking, particularly in the areas of prosecuting and convicting trafficking offenders and failing to formalize mechanisms to provide assistance to victims; although the government made some effort to enforce laws against child labor exploitation, it failed to report any trafficking prosecutions or convictions in 2007; the government continued to lack shelters or formal procedures for providing care to victims (2008)

ERITREA

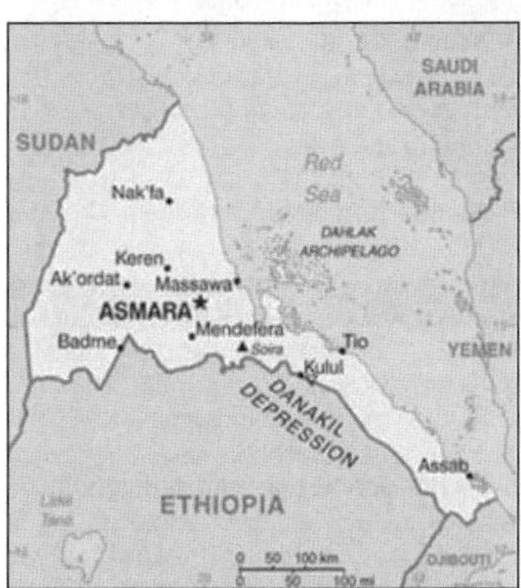

INTRODUCTION

Background: The UN awarded Eritrea to Ethiopia in 1952 as part of a federation. Ethiopia's annexation of Eritrea as a province 10 years later sparked a 30-year struggle for independence that ended in 1991 with Eritrean rebels defeating governmental forces; independence was overwhelmingly approved in a 1993 referendum. A two-and-a-half-year border war with Ethiopia that erupted in 1998 ended under UN auspices in December 2000. Eritrea hosted a UN peacekeeping operation that monitored a 25 km-wide Temporary Security Zone (TSZ) on the border with Ethiopia. Eritrea's denial of fuel to the mission caused the UN to withdraw the mission and terminate its mandate 31 July 2008. An international commission, organized to resolve the border dispute, posted its findings in 2002. However, both parties have been unable to reach agreement on implementing the decision. On 30 November 2007, the Eritrea-Ethiopia Boundary Commission remotely demarcated the border by coordinates and dissolved itself, leaving Ethiopia still occupying several tracts of disputed territory, including the town of Badme. Eritrea accepted the EEBC's "virtual demarcation" decision and called on Ethiopia to remove its troops from the TSZ that it states is Eritrean territory. Ethiopia has not accepted the virtual demarcation decision. In 2009 the UN imposed sanctions on Eritrea after accusing it of backing anti-Ethiopian Islamist insurgents in Somalia.

GEOGRAPHY

Location: Eastern Africa, bordering the Red Sea, between Djibouti and Sudan
Geographic coordinates: 15 00 N, 39 00 E
Map references: Africa
Area: *total:* 117,600 sq km
country comparison to the world: 100
land: 101,000 sq km
water: 16,600 sq km
Area—comparative: slightly larger than Pennsylvania
Land boundaries: *total:* 1,626 km
border countries: Djibouti 109 km, Ethiopia 912 km, Sudan 605 km
Coastline: 2,234 km (mainland on Red Sea 1,151 km, islands in Red Sea 1,083 km)
Maritime claims: *territorial sea:* 12 nm
Climate: hot, dry desert strip along Red Sea coast; cooler and wetter in the central highlands (up to 61 cm of rainfall annually, heaviest June to September); semiarid in western hills and lowlands
Terrain: dominated by extension of Ethiopian north-south trending highlands, descending on the east to a coastal desert plain, on the northwest to hilly terrain and on the southwest to flat-to-rolling plains
Elevation extremes: *lowest point:* near Kulul within the Danakil Depression -75 m
highest point: Soira 3,018 m
Natural resources: gold, potash, zinc, copper, salt, possibly oil and natural gas, fish
Land use: *arable land:* 4.78%
permanent crops: 0.03%
other: 95.19% (2005)
Irrigated land: 210 sq km (2003)
Total renewable water resources: 6.3 cu km (2001)
Freshwater withdrawal (domestic/industrial/agricultural): *total:* 0.3 cu km/yr (3%/0%/97%)
per capita: 68 cu m/yr (2000)
Natural hazards: frequent droughts; locust swarms
volcanism: Dubbi (elev. 1,625 m), which last erupted in 1861, is the country's only historically active volcano
Environment—current issues: deforestation; desertification; soil erosion; overgrazing; loss of infrastructure from civil warfare
Environment—international agreements: *party to:* Biodiversity, Climate Change, Climate Change-Kyoto Protocol, Desertification, Endangered Species, Hazardous Wastes, Ozone Layer Protection
signed, but not ratified: none of the selected agreements
Geography—note: strategic geopolitical position along world's busiest shipping lanes; Eritrea retained the entire coastline of Ethiopia along the Red Sea upon de jure independence from Ethiopia on 24 May 1993

PEOPLE

Population: 5,939,484 (July 2011 est.)
country comparison to the world: 107
Age structure: *0–14 years:* 42.1% (male 1,256,384/female 1,244,569)
15–64 years: 54.3% (male 1,580,535/female 1,641,911)
65 years and over: 3.6% (male 96,627/female 119,458) (2011 est.)
Median age: *total:* 18.7 years
male: 18.3 years
female: 19.1 years (2011 est.)
Population growth rate: 2.472% (2011 est.)
country comparison to the world: 28
Birth rate: 32.8 births/1,000 population (2011 est.)
country comparison to the world: 38
Death rate: 8.08 deaths/1,000 population (July 2011 est.)
country comparison to the world: 101
Net migration rate: 0 migrant(s)/1,000 population (2011 est.)
country comparison to the world: 84
Urbanization: *urban population:* 22% of total population (2010)
rate of urbanization: 5.2% annual rate of change (2010-15 est.)
Major cities—population: ASMARA (capital) 649,000 (2009)
Sex ratio: *at birth:* 1.03 male(s)/female
under 15 years: 1.01 male(s)/female
15–64 years: 0.96 male(s)/female
65 years and over: 0.82 male(s)/female
total population: 0.98 male(s)/female (2011 est.)
Infant mortality rate: *total:* 41.33 deaths/1,000 live births
country comparison to the world: 63
male: 46.77 deaths/1,000 live births
female: 35.72 deaths/1,000 live births (2011 est.)
Life expectancy at birth: *total population:* 62.52 years
country comparison to the world: 179
male: 60.4 years
female: 64.69 years (2011 est.)
Total fertility rate: 4.48 children born/woman (2011 est.)
country comparison to the world: 34
HIV/AIDS—adult prevalence rate: 0.8% (2009 est.)
country comparison to the world: 56
HIV/AIDS—people living with HIV/AIDS: 25,000 (2009 est.)
country comparison to the world: 73
HIV/AIDS—deaths: 1,700 (2009 est.)
country comparison to the world: 57
Major infectious diseases: *degree of risk:* high
food or waterborne diseases: bacterial diarrhea, hepatitis A, and typhoid fever
vectorborne disease: malaria (2009)
Drinking water source: *Improved:*
urban: 74% of population
rural: 57% of population
total: 61% of population
Unimproved: urban: 26% of population
rural: 43% of population
total: 39% of population (2008)
Sanitation facility access: *Improved:*
urban: 52% of population
rural: 4% of population
total: 14% of population
Unimproved: urban: 48% of population
rural: 96% of population
total: 86% of population (2008)
Nationality: *noun:* Eritrean(s)
adjective: Eritrean
Ethnic groups: nine recognized ethnic groups: Tigrinya 55%, Tigre 30%, Saho 4%, Kunama 2%, Rashaida 2%, Bilen 2%, other (Afar, Beni Amir, Nera) 5% (2010 est.)
Religions: Muslim, Coptic Christian, Roman Catholic, Protestant
Languages: Tigrinya (official), Arabic (official), English (official), Tigre, Kunama, Afar, other Cushitic languages
Literacy: *definition:* age 15 and over can read and write
total population: 58.6%
male: 69.9%
female: 47.6% (2003 est.)
School life expectancy (primary to tertiary education): *total:* 5 years
male: 6 years

female: 4 years (2009)
Education expenditures: 2% of GDP (2006)
country comparison to the world: 156

GOVERNMENT

Country name: *conventional long form:* State of Eritrea
conventional short form: Eritrea
local long form: Hagere Ertra
local short form: Ertra
former: Eritrea Autonomous Region in Ethiopia
Government type: transitional government
note: following a successful referendum on independence for the Autonomous Region of Eritrea on 23-25 April 1993, a National Assembly, composed entirely of the People's Front for Democracy and Justice or PFDJ, was established as a transitional legislature; a Constitutional Commission was also established to draft a constitution; ISAIAS Afworki was elected president by the transitional legislature; the constitution, ratified in May 1997, did not enter into effect, pending parliamentary and presidential elections; parliamentary elections were scheduled in December 2001 but were postponed indefinitely; currently the sole legal party is the People's Front for Democracy and Justice (PFDJ)
Capital: *name:* Asmara (Asmera)
geographic coordinates: 15 20 N, 38 56 E
time difference: UTC+3 (8 hours ahead of Washington, DC during Standard Time)
Administrative divisions: 6 regions (zobatat, singular–zoba); Anseba, Debub (South), Debubawi K'eyih Bahri (Southern Red Sea), Gash Barka, Ma'akel (Central), Semenawi Keyih Bahri (Northern Red Sea)
Independence: 24 May 1993 (from Ethiopia)
National holiday: Independence Day, 24 May (1993)
Constitution: adopted 23 May 1997, but has not yet been fully implemented
Legal system: mixed legal system of civil, customary, and Islamic religious law
International law organization participation: has not submitted an ICJ jurisdiction declaration; non-party state to the ICCt
Suffrage: 18 years of age; universal
Executive branch: *chief of state:* President ISAIAS Afworki (since 8 June 1993); note–the president is both the chief of state and head of government and is head of the State Council and National Assembly
head of government: President ISAIAS Afworki (since 8 June 1993)
cabinet: State Council the collective is executive authority; members appointed by the president
(For more information visit the World Leaders website)
elections: president elected by the National Assembly for a five-year term (eligible for a second term); the most recent and only election was held on 8 June 1993 (next election date uncertain as the National Assembly did not hold a presidential election in December 2001 as anticipated)
election results: ISAIAS Afworki elected president; percent of National Assembly vote–ISAIAS Afworki 95%, other 5%
Legislative branch: unicameral National Assembly (150 seats; members elected by direct popular vote to serve five-year terms)
elections: in May 1997, following the adoption of the new constitution, 75 members of the PFDJ Central Committee (the old Central Committee of the EPLF), 60 members of the 527-member Constituent Assembly, which had been established in 1997 to discuss and ratify the new constitution, and 15 representatives of Eritreans living abroad were formed into a Transitional National Assembly to serve as the country's legislative body until country-wide elections to a National Assembly were held; although only 75 of 150 members of the Transitional National Assembly were elected, the constitution stipulates that once past the transition stage, all members of the National Assembly will be elected by secret ballot of all eligible voters; National Assembly elections scheduled for December 2001 were postponed indefinitely
Judicial branch: Supreme Court; Regional, subregional, and village courts
Political parties and leaders: People's Front for Democracy and Justice or PFDJ [ISAIAS Afworki] (the only party recognized by the government); note–a National Assembly committee drafted a law on political parties in January 2001, but the full National Assembly has yet to debate or vote on it
Political pressure groups and leaders: Eritrean Democratic Party (EDP) [HAGOS, Mesfin]; Eritrean Islamic Jihad or EIJ (includes Eritrean Islamic Jihad Movement or EIJM also known as the Abu Sihel Movement); Eritrean Islamic Salvation or EIS (also known as the Arafa Movement); Eritrean Liberation Front or ELF [ABDULLAH Muhammed]; Eritrean National Alliance or ENA (a coalition including EIJ, EIS, ELF, and a number of ELF factions) [HERUY Tedla Biru]; Eritrean Public Forum or EPF [ARADOM Iyob]
International organization participation: ACP, AfDB, AU, COMESA, FAO, G-77, IAEA, IBRD, ICAO, IDA, IFAD, IFC, IFRCS (observer), ILO, IMF, IMO, Interpol, IOC, ISO (subscriber), ITU, ITUC, LAS (observer), MIGA, NAM, OPCW, PCA, UN, UNCTAD, UNESCO, UNIDO, UNWTO, UPU, WCO, WFTU, WHO, WIPO, WMO
Diplomatic representation in the US: *chief of mission:* Ambassador (vacant); Charge d'Affaires Berhane Gebrehiwet SOLOMON
chancery: 1708 New Hampshire Avenue NW, Washington, DC 20009
telephone: [1] (202) 319-1991
FAX: [1] (202) 319-1304
consulate(s) general: Oakland (California)
Diplomatic representation from the US: *chief of mission:* Ambassador (vacant); Charge d'Affaires Joel REIFMAN
embassy: 179 Ala Street, Asmara
mailing address: P. O. Box 211, Asmara
telephone: [291] (1) 120004
FAX: [291] (1) 127584
Flag description: red isosceles triangle (based on the hoist side) dividing the flag into two right triangles; the upper triangle is green, the lower one is blue; a gold wreath encircling a gold olive branch is centered on the hoist side of the red triangle; green stands for the country's agriculture economy, red signifies the blood shed in the fight for freedom, and blue symbolizes the bounty of the sea; the wreath-olive branch symbol is similar to that on the first flag of Eritrea from 1952; the shape of the red triangle broadly mimics the shape of the country
National anthem: *name:* "Ertra, Ertra, Ertra" (Eritrea, Eritrea, Eritrea)
lyrics/music: SOLOMON Tsehaye Beraki/ Isaac Abraham MEHAREZGI and ARON Tekle Tesfatsion
note: adopted 1993; upon independence from Ethiopia in 1993, Eritrea adopted its own national anthem

ECONOMY

Economy—overview: Since independence from Ethiopia in 1993, Eritrea has faced the economic problems of a small, desperately poor country, accentuated by the recent implementation of restrictive economic policies. Eritrea has a command economy under the control of the sole political party, the People's Front for Democracy and Justice (PFDJ). Like the economies of many African nations, a large share of the population–nearly 80%–is engaged in subsistence agriculture, but they produce only a small share of total output. Since the conclusion of the Ethiopian-Eritrea war in 2000, the government has maintained a firm grip on the economy, expanding the use of the military and party-owned businesses to complete Eritrea's development agenda. The government strictly controls the use of foreign currency by limiting access and availability. Few private enterprises remain in Eritrea. Eritrea's economy depends heavily on taxes paid by members of the diaspora. Erratic rainfall and the delayed demobilization of agriculturalists from the military continue to interfere with agricultural production, and Eritrea's recent harvests have been unable to meet the food needs of the country. The Government continues to place its hope for additional revenue on the development of several international mining projects. Despite difficulties for international companies in working with the Eritrean Government, a Canadian mining company signed a contract with the government in 2007 and began mineral extraction in 2010. Eritrea's economic future depends upon its ability to master social problems such as illiteracy, unemployment, and low skills, and more importantly, on the government's willingness to support a true market economy.
GDP (purchasing power parity): $3.625 billion (2010 est.)
country comparison to the world: 169
$3.548 billion (2009 est.)
$3.415 billion (2008 est.)
note: data are in 2010 US dollars
GDP (official exchange rate): $2.117 billion (2010 est.)
GDP—real growth rate: 2.2% (2010 est.)
country comparison to the world: 143

3.9% (2009 est.)
-9.8% (2008 est.)
GDP—per capita (PPP): $600 (2010 est.)
country comparison to the world: 223
$600 (2009 est.)
$600 (2008 est.)
note: data are in 2010 US dollars
GDP—composition by sector: *agriculture:* 11.8%
industry: 20.4%
services: 67.7% (2010 est.)
Labor force: 1.935 million (2007)
country comparison to the world: 121
Labor force—by occupation: *agriculture:* 80%
industry and *services:* 20% (2004 est.)
Unemployment rate: NA%
Population below poverty line: 50% (2004 est.)
Household income or consumption by percentage share: *lowest 10%:* NA%
highest 10%: NA%
Investment (gross fixed): 10.3% of GDP (2010 est.)
country comparison to the world: 149
Budget: *revenues:* $463.4 million
expenditures: $920.1 million (2010 est.)
Inflation rate (consumer prices): 20% (2010 est.)
country comparison to the world: 220
20% (2009 est.)
Commercial bank prime lending rate: NA%
Stock of narrow money: $1.382 billion (31 December 2010 est.)
country comparison to the world: 128
$1.007 billion (31 December 2009 est.)
Stock of broad money: $2.872 billion (31 December 2010 est.)
country comparison to the world: 133
$2.171 billion (31 December 2009 est.)
Stock of domestic credit: $2.919 billion (31 December 2010 est.)
country comparison to the world: 121
$2.206 billion (31 December 2009 est.)
Agriculture—products: sorghum, lentils, vegetables, corn, cotton, tobacco, sisal; livestock, goats; fish
Industries: food processing, beverages, clothing and textiles, light manufacturing, salt, cement
Industrial production growth rate: 8% (2010 est.)
country comparison to the world: 34
Electricity—production: 271 million kWh (2007 est.)
country comparison to the world: 172
Electricity—consumption: 228 million kWh (2007 est.)
country comparison to the world: 177
Electricity—exports: 0 kWh (2008 est.)
Electricity—imports: 0 kWh (2008 est.)
Oil—production: 0 bbl/day (2009 est.)
country comparison to the world: 170
Oil—consumption: 5,000 bbl/day (2009 est.)
country comparison to the world: 168
Oil—exports: 0 bbl/day (2007 est.)
country comparison to the world: 163
Oil—imports: 4,790 bbl/day (2007 est.)
country comparison to the world: 159
Oil—proved reserves: 0 bbl (1 January 2010 est.)
country comparison to the world: 127
Natural gas—production: 0 cu m (2008 est.)
country comparison to the world: 122
Natural gas—consumption: 0 cu m (2008 est.)
country comparison to the world: 167
Natural gas—exports: 0 cu m (2008 est.)
country comparison to the world: 90
Natural gas—imports: 0 cu m (2008 est.)
country comparison to the world: 181
Natural gas—proved reserves: 0 cu m (1 January 2010 est.)
country comparison to the world: 131
Current account balance: $-212 million (2010 est.)
country comparison to the world: 93
$-188 million (2009 est.)
Exports: $25 million (2010 est.)
country comparison to the world: 205
$20 million (2009 est.)
Exports—commodities: livestock, sorghum, textiles, food, small manufactures
Exports—partners: India 25.3%, Italy 20.7%, Sudan 14.1%, China 12.9%, France 5.5%, Saudi Arabia 5.4% (2008)
Imports: $738 million (2010 est.)
country comparison to the world: 183
$682 million (2009 est.)
Imports—commodities: machinery, petroleum products, food, manufactured goods
Imports—partners: Saudi Arabia 20.7%, India 13.6%, Italy 12.6%, China 9.9%, US 5.1%, Germany 4.6% (2008)
Reserves of foreign exchange and gold: $104 million (31 December 2010 est.)
country comparison to the world: 135
$88 million (31 December 2009 est.)
Debt—external: $NA (31 December 2010 est.)
$961.9 million (31 December 2008 est.)
Exchange rates: nakfa (ERN) per US dollar–15.375 (2010), 15.375 (2009), 15.38 (2008), 15.5 (2007), 15.4 (2006)

COMMUNICATIONS

Telephones—main lines in use: 48,500 (2009)
country comparison to the world: 164
Telephones—mobile cellular: 141,100 (2009)
country comparison to the world: 177
Telephone system: *general assessment:* inadequate; most telephones are in Asmara; government is seeking international tenders to improve the system (2002)
domestic: combined fixed-line and mobile-cellular subscribership is only about 3 per 100 persons (2009)
international: country code–291; note–international connections exist
Broadcast media: government controls broadcast media with private ownership prohibited; 1 state-owned TV station; state-owned radio operates 2 networks; purchases of satellite dishes and subscriptions to international broadcast media are permitted (2007)
Internet country code: .er
Internet hosts: 1,241 (2010)
country comparison to the world: 166
Internet users: 200,000 (2008)
country comparison to the world: 140

TRANSPORTATION

Airports: 13 (2010)
country comparison to the world: 151
Airports—with paved runways: *total:* 4
over 3,047 m: 2
2,438 to 3,047 m: 2 (2010)
Airports—with unpaved runways:
total: 9
over 3,047 m: 1
2,438 to 3,047 m: 1
1,524 to 2,437 m: 5
914 to 1,523 m: 2 (2010)
Heliports: 1 (2010)
Railways: *total:* 306 km
country comparison to the world: 119
narrow gauge: 306 km 0.950-m gauge (2010)
Roadways: *total:* 4,010 km
country comparison to the world: 157
paved: 874 km
unpaved: 3,136 km (2000)
Merchant marine: *total:* 4
country comparison to the world: 132
by type: cargo 2, petroleum tanker 1, roll on/roll off 1 (2010)
Ports and terminals: Assab, Massawa

MILITARY

Military branches: Eritrean Armed Forces: Ground Forces, Navy, Air Force (2011)
Military service age and obligation: 18-40 years of age for male and female voluntary and compulsory military service; 16-month conscript service obligation (2006)
Manpower available for military service: *males age 16–49:* 1,350,446
females age 16–49: 1,362,575 (2010 est.)
Manpower fit for military service: *males age 16–49:* 896,096
females age 16–49: 953,757 (2010 est.)
Manpower reaching militarily significant age annually: *male:* 66,829
female: 66,731 (2010 est.)
Military expenditures: 6.3% of GDP (2006 est.)
country comparison to the world: 8

TRANSNATIONAL ISSUES

Disputes—international: Eritrea and Ethiopia agreed to abide by 2002 Ethiopia-Eritrea Boundary Commission's (EEBC) delimitation decision but, neither party responded to the revised line detailed in the November 2006 EEBC Demarcation Statement; Sudan accuses Eritrea of supporting eastern Sudanese rebel groups; in 2008 Eritrean troops move across the border on Ras Doumera peninsula and occupy Doumera Island with undefined sovereignty in the Red Sea
Refugees and internally displaced persons: *IDPs:* 32,000 (border war with Ethiopia from 1998-2000; most IDPs are near the central border region) (2007)
Trafficking in persons: current situation: Eritrea is a source country for men, women, and children trafficked for the purposes of forced labor and commercial sexual exploitation; each year, large numbers of migrant workers depart Eritrea in search of work, particularly in the Gulf States, where some likely become victims of forced labor, including in domestic servitude, or commercial sexual exploitation; thousands of Eritreans flee the country illegally, mostly to Sudan, Ethiopia, and Kenya where their illegal status makes them vulnerable to situations of human trafficking; the govern-

ment remains complicit in conscripting children into military service
tier rating: Tier 3–the Government of Eritrea does not fully comply with the minimum standards for the elimination of trafficking and is not making significant efforts to do so; the Eritrean government does not operate with transparency and published neither data nor statistics regarding its efforts to combat human trafficking; it did not respond to requests to provide information for this report; the government made no known progress in prosecuting and punishing trafficking crimes over the reporting period and did not appear to provide any significant assistance to victims of trafficking during the reporting period (2009)

ESTONIA

INTRODUCTION

Background: After centuries of Danish, Swedish, German, and Russian rule, Estonia attained independence in 1918. Forcibly incorporated into the USSR in 1940–an action never recognized by the US–it regained its freedom in 1991 with the collapse of the Soviet Union. Since the last Russian troops left in 1994, Estonia has been free to promote economic and political ties with the West. It joined both NATO and the EU in the spring of 2004.

GEOGRAPHY

Location: Eastern Europe, bordering the Baltic Sea and Gulf of Finland, between Latvia and Russia
Geographic coordinates: 59 00 N, 26 00 E
Map references: Europe
Area: *total:* 45,228 sq km
country comparison to the world: 132
land: 42,388 sq km
water: 2,840 sq km
note: includes 1,520 islands in the Baltic Sea
Area—comparative: slightly smaller than New Hampshire and Vermont combined
Land boundaries: *total:* 633 km
border countries: Latvia 343 km, Russia 290 km
Coastline: 3,794 km
Maritime claims: *territorial sea:* 12 nm
exclusive economic zone: limits fixed in coordination with neighboring states
Climate: maritime; wet, moderate winters, cool summers
Terrain: marshy, lowlands; flat in the north, hilly in the south
Elevation extremes:
lowest point: Baltic Sea 0 m
highest point: Suur Munamagi 318 m
Natural resources: oil shale, peat, rare earth elements, phosphorite, clay, limestone, sand, dolomite, arable land, sea mud
Land use: *arable land:* 12.05%
permanent crops: 0.35%
other: 87.6% (2005)
Irrigated land: 40 sq km (2003)
Total renewable water resources: 21.1 cu km (2005)
Freshwater withdrawal (domestic/industrial/agricultural): *total:* 1.41 cu km/yr (56%/39%/5%)
per capita: 1,060 cu m/yr (2002)
Natural hazards: sometimes flooding occurs in the spring
Environment—current issues: air polluted with sulfur dioxide from oil-shale burning power plants in northeast; however, the amount of pollutants emitted to the air have fallen steadily, the emissions of 2000 were 80% less than in 1980; the amount of unpurified wastewater discharged to water bodies in 2000 was 1/20 the level of 1980; in connection with the start-up of new water purification plants, the pollution load of wastewater decreased; Estonia has more than 1,400 natural and manmade lakes, the smaller of which in agricultural areas need to be monitored; coastal seawater is polluted in certain locations
Environment—international agreements: *party to:* Air Pollution, Air Pollution-Nitrogen Oxides, Air Pollution-Persistent Organic Pollutants, Air Pollution-Sulfur 85, Air Pollution-Volatile Organic Compounds, Antarctic Treaty, Biodiversity, Climate Change, Climate Change-Kyoto Protocol, Endangered Species, Hazardous Wastes, Law of the Sea, Ozone Layer Protection, Ship Pollution, Wetlands
signed, but not ratified: none of the selected agreements
Geography—note: the mainland terrain is flat, boggy, and partly wooded; offshore lie more than 1,500 islands

PEOPLE

Population: 1,282,963 (July 2011 est.)
country comparison to the world: 154
Age structure: *0–14 years:* 15.1% (male 99,919/female 94,066)
15–64 years: 67.2% (male 410,132/female 451,736)
65 years and over: 17.7% (male 74,803/female 152,307) (2011 est.)
Median age: *total:* 40.5 years
male: 37 years
female: 43.9 years (2011 est.)
Population growth rate: -0.641% (2011 est.)
country comparison to the world: 226
Birth rate: 10.45 births/1,000 population (2011 est.)
country comparison to the world: 183
Death rate: 13.55 deaths/1,000 population (July 2011 est.)
country comparison to the world: 20
Net migration rate: -3.31 migrant(s)/1,000 population (2011 est.)
country comparison to the world: 179
Urbanization: *urban population:* 69% of total population (2010)
rate of urbanization: 0.1% annual rate of change (2010-15 est.)
Major cities—population: TALLINN (capital) 399,000 (2009)
Sex ratio: *at birth:* 1.063 male(s)/female
under 15 years: 1.06 male(s)/female
15–64 years: 0.91 male(s)/female
65 years and over: 0.49 male(s)/female
total population: 0.84 male(s)/female (2011 est.)
Infant mortality rate:
total: 7.06 deaths/1,000 live births
country comparison to the world: 166
male: 8.21 deaths/1,000 live births
female: 5.85 deaths/1,000 live births (2011 est.)
Life expectancy at birth: *total population:* 73.33 years
country comparison to the world: 118
male: 68.02 years
female: 78.97 years (2011 est.)
Total fertility rate: 1.44 children born/woman (2011 est.)
country comparison to the world: 194
HIV/AIDS—adult prevalence rate: 1.2% (2009 est.)
country comparison to the world: 40
HIV/AIDS—people living with HIV/AIDS: 9,900 (2009 est.)
country comparison to the world: 98
HIV/AIDS—deaths: fewer than 500 (2009 est.)
country comparison to the world: 93
Major infectious diseases:
degree of risk: intermediate
food or waterborne diseases: bacterial diarrhea
vectorborne disease: tickborne encephalitis (2009)
Drinking water source: *Improved:*
urban: 99% of population
rural: 97% of population
total: 98% of population
Unimproved: urban: 1% of population
rural: 3% of population
total: 2% of population (2008)
Sanitation facility access: *Improved:*
urban: 96% of population
rural: 94% of population
total: 95% of population
Unimproved: urban: 4% of population
rural: 6% of population
total: 5% of population (2008)
Nationality: *noun:* Estonian(s)
adjective: Estonian
Ethnic groups: Estonian 68.7%, Russian 25.6%, Ukrainian 2.1%, Belarusian 1.2%, Finn 0.8%, other 1.6% (2008 census)
Religions: Evangelical Lutheran 13.6%, Orthodox 12.8%, other Christian (including Methodist, Seventh-Day

Adventist, Roman Catholic, Pentecostal) 1.4%, unaffiliated 34.1%, other and unspecified 32%, none 6.1% (2000 census)
Languages: Estonian (official) 67.3%, Russian 29.7%, other 2.3%, unknown 0.7% (2000 census)
Literacy: *definition:* age 15 and over can read and write
total population: 99.8%
male: 99.8%
female: 99.8% (2000 census)
School life expectancy (primary to tertiary education): *total:* 16 years
male: 15 years
female: 17 years (2008)
Education expenditures: 4.9% of GDP (2007)
country comparison to the world: 64

GOVERNMENT

Country name: *conventional long form:* Republic of Estonia
conventional short form: Estonia
local long form: Eesti Vabariik
local short form: Eesti
former: Estonian Soviet Socialist Republic
Government type: parliamentary republic
Capital: *name:* Tallinn
geographic coordinates: 59 26 N, 24 43 E
time difference: UTC+2 (7 hours ahead of Washington, DC during Standard Time)
daylight saving time: +1hr, begins last Sunday in March; ends last Sunday in October
Administrative divisions: 15 counties (maakonnad, singular—maakond); Harjumaa (Tallinn), Hiiumaa (Kardla), Ida-Virumaa (Johvi), Jarvamaa (Paide), Jogevamaa (Jogeva), Laanemaa (Haapsalu), Laane-Virumaa (Rakvere), Parnumaa (Parnu), Polvamaa (Polva), Raplamaa (Rapla), Saaremaa (Kuressaare), Tartumaa (Tartu), Valgamaa (Valga), Viljandimaa (Viljandi), Vorumaa (Voru)
note: counties have the administrative center name following in parentheses
Independence: 20 August 1991 (declared); 6 September 1991 (recognized by the Soviet Union)
National holiday: Independence Day, 24 February (1918); note—24 February 1918 was the date Estonia declared its independence from Soviet Russia and established its statehood; 20 August 1991 was the date it declared its independence from the Soviet Union
Constitution: adopted 28 June 1992
Legal system: civil law system
International law organization participation: accepts compulsory ICJ jurisdiction with reservations; accepts ICCt jurisdiction
Suffrage: 18 years of age; universal for all Estonian citizens
Executive branch: *chief of state:* President Toomas Hendrik ILVES (since 9 October 2006)
head of government: Prime Minister Andrus ANSIP (since 12 April 2005)
cabinet: Ministers appointed by the prime minister, approved by Parliament
(For more information visit the World Leaders website)
elections: president elected by Parliament for a five-year term (eligible for a second term); if a candidate does not secure two-thirds of the votes after three rounds of balloting in the Parliament, then an electoral assembly (made up of Parliament plus members of local councils) elects the president, choosing between the two candidates with the largest number of votes; election last held on 23 September 2006 (next to be held in the fall of 2011); prime minister nominated by the president and approved by Parliament
election results: Toomas Hendrik ILVES elected president on 23 September 2006 by a 345-member electoral assembly; ILVES received 174 votes to incumbent Arnold RUUTEL's 162; remaining 9 ballots left blank or invalid
Legislative branch: unicameral Parliament or Riigikogu (101 seats; members elected by popular vote to serve four-year terms)
elections: last held on 6 March 2011 (next to be held in March 2015)
election results: percent of vote by party—Estonian Reform Party 28.6%, Center Party of Estonia 23.3%, Union of Pro Patria and Res Publica 20.5%, Social Democratic Party 17.1%, Estonian Greens 3.8%, Estonian People's Union 2.1%, other 4.6%; seats by party—Estonian Reform Party 33, Center Party 26, Union of Pro Patria and Res Publica 23, Social Democratic Party 19
Judicial branch: Supreme Court (chairman appointed for life by Parliament)
Political parties and leaders: Center Party of Estonia (Keskerakond) [Edgar SAVISAAR]; Estonian Greens (Rohelised) [Marek STRANDBERG]; Estonian People's Union (Rahvaliit) [Andrus BLOK]; Estonian Reform Party (Reformierakond) [Andrus ANSIP]; Social Democratic Party [Sven MIKSER]; Union of Pro Patria and Res Publica (Isamaa je Res Publica Liit) [Mart LAAR]
International organization participation: Australia Group, BA, BIS, CBSS, CE, EAPC, EBRD, EIB, EMU, ESA (cooperating state), EU, FAO, IAEA, IBRD, ICAO, ICRM, IDA, IFC, IFRCS, IHO, ILO, IMF, IMO, Interpol, IOC, IOM, IPU, ISO, ITSO, ITU, ITUC, MIGA, NATO, NIB, NSG, OAS (observer), OECD, OPCW, OSCE, PCA, Schengen Convention, UN, UNCTAD, UNESCO, UNHCR, UNTSO, UPU, WCO, WHO, WIPO, WMO, WTO
Diplomatic representation in the US: *chief of mission:* Ambassador Vaino REINART
chancery: 2131 Massachusetts Avenue NW, Washington, DC 20008
telephone: [1] (202) 588-0101
FAX: [1] (202) 588-0108
consulate(s) general: New York
Diplomatic representation from the US: *chief of mission:* Ambassador Michael C. POLT
embassy: Kentmanni 20, 15099 Tallinn
mailing address: use embassy street address
telephone: [372] 668-8100
FAX: [372] 668-8265
Flag description: three equal horizontal bands of blue (top), black, and white; various interpretations are linked to the flag colors; blue represents faith, loyalty, and devotion, while also reminiscent of the sky, sea, and lakes of the country; black symbolizes the soil of the country and the dark past and suffering endured by the Estonian people; white refers to the striving towards enlightenment and virtue, and is the color of birch bark and snow, as well as summer nights illuminated by the midnight sun
National anthem: *name:* "Mu isamaa, mu onn ja room" (My Native Land, My Pride and Joy)
lyrics/music: Johann Voldemar JANNSEN/Fredrik PACIUS
note: adopted 1920, though banned between 1940 and 1990 under Soviet occupation; the anthem, used in Estonia since 1869, shares the same melody with that of Finland but has different lyrics

ECONOMY

Economy—overview: Estonia, a 2004 European Union entrant, has a modern market-based economy and one of the higher per capita income levels in Central Europe and the Baltic region. Estonia's successive governments have pursued a free market, pro-business economic agenda and have wavered little in their commitment to pro-market reforms. The current government has followed relatively sound fiscal policies that have resulted in balanced budgets and very low public debt. The economy benefits from strong electronics and telecommunications sectors and strong trade ties with Finland, Sweden, and Germany. Tallinn's priority has been to sustain high growth rates—on average 8% per year from 2003 to 2007. Estonia's economy slowed down markedly and fell sharply into recession in mid-2008, primarily as a result of an investment and consumption slump following the bursting of the real estate market bubble. GDP dropped nearly 14% in 2009, among the world's highest rates of contraction. Rising exports to Sweden and Finland lead an economic recovery in 2010, but unemployment stands above 17%. Estonia joined the OECD in December 2010 and adopted the euro in January 2011.
GDP (purchasing power parity): $24.69 billion (2010 est.)
country comparison to the world: 113
$23.95 billion (2009 est.)
$27.81 billion (2008 est.)
note: data are in 2010 US dollars
GDP (official exchange rate): $19.78 billion (2010 est.)
GDP—real growth rate: 3.1% (2010 est.)
country comparison to the world: 121
-13.9% (2009 est.)
-5.1% (2008 est.)
GDP—per capita (PPP): $19,100 (2010 est.)
country comparison to the world: 63
$18,400 (2009 est.)
$21,300 (2008 est.)
note: data are in 2010 US dollars
GDP—composition by sector: *agriculture:* 2.5%
industry: 28.7%
services: 68.8% (2010 est.)
Labor force: 688,000 (2010 est.)
country comparison to the world: 150

Labor force—by occupation:
agriculture: 2.8%
industry: 22.7%
services: 74.5% (2008)
Unemployment rate: 17.5% (2010 est.)
country comparison to the world: 158
13.8% (2009 est.)
Population below poverty line: 19.7% (2008)
Household income or consumption by percentage share: *lowest 10%:* 2.7%
highest 10%: 27.7% (2004)
Distribution of family income—Gini index: 31.4 (2009)
country comparison to the world: 105
37 (1999)
Investment (gross fixed): 22.5% of GDP (2010 est.)
country comparison to the world: 63
Budget: *revenues:* $7.851 billion
expenditures: $8.21 billion (2010 est.)
Public debt: 7.7% of GDP (2010 est.)
country comparison to the world: 125
7.1% of GDP (2009 est.)
Inflation rate (consumer prices): 2.4% (2010 est.)
country comparison to the world: 62
0.2% (2009)
Commercial bank prime lending rate: 9.39% (31 December 2009 est.)
country comparison to the world: 107
8.55% (31 December 2008 est.)
Stock of narrow money: $5.345 billion (31 December 2010 est.)
country comparison to the world: 87
$5.822 billion (31 December 2009 est.)
note: this figure represents the US dollar value of Estonian kroon in circulation prior to Estonia's joining the Economic and Monetary Union (EMU); see entry for the European Union for money supply in the euro area; the European Central Bank (ECB) controls monetary policy for the 17 members of the EMU; individual members of the EMU do not control the quantity of money circulating within their own borders
Stock of broad money: $10.7 billion (31 December 2010 est.)
country comparison to the world: 98
$11.37 billion (31 December 2009 est.)
Stock of domestic credit: $18.94 billion (31 December 2010 est.)
country comparison to the world: 80
$20.32 billion (31 December 2009 est.)
Market value of publicly traded shares: $2.654 billion (31 December 2009)
country comparison to the world: 98
$1.95 billion (31 December 2008)
$6.037 billion (31 December 2007)
Agriculture—products: grain, potatoes, vegetables; livestock and dairy products; fish
Industries: engineering, electronics, wood and wood products, textiles; information technology, telecommunications
Industrial production growth rate: 10% (2010 est.)
country comparison to the world: 20
Electricity—production: 8.779 billion kWh (2009 est.)
country comparison to the world: 95
Electricity—consumption: 7.08 billion kWh (2009 est.)
country comparison to the world: 98
Electricity—exports: 2.943 billion kWh (2009 est.)
Electricity—imports: 3.025 billion kWh (2009 est.)
Oil—production: 7,600 bbl/day (2009 est.)
country comparison to the world: 88
Oil—consumption: 30,000 bbl/day (2009 est.)
country comparison to the world: 112
Oil—exports: 7,280 bbl/day (2007 est.)
country comparison to the world: 96
Oil—imports: 30,590 bbl/day (2007 est.)
country comparison to the world: 98
Oil—proved reserves: 0 bbl (1 January 2010 est.)
country comparison to the world: 126
Natural gas—production: 0 cu m (2009 est.)
country comparison to the world: 121
Natural gas—consumption: 1.02 billion cu m (2009 est.)
country comparison to the world: 89
Natural gas—exports: 0 cu m (2009 est.)
country comparison to the world: 89
Natural gas—imports: 1.02 billion cu m (2009 est.)
country comparison to the world: 55
Natural gas—proved reserves: 0 cu m (1 January 2010 est.)
country comparison to the world: 130
Current account balance: $265 million (2010 est.)
country comparison to the world: 55
$898.7 million (2009 est.)
Exports: $11.5 billion (2010 est.)
country comparison to the world: 82
$9.05 billion (2009)
Exports—commodities: machinery and electrical equipment 21%, wood and wood products 9%, metals 9%, furniture 7%, vehicles and parts 5%, food products and beverages 4%, textiles 4%, plastics 3%
Exports—partners: Finland 18.57%, Sweden 12.52%, Latvia 9.51%, Russia 9.33%, Germany 6.09%, Lithuania 4.76%, US 4.26% (2009)
Imports: $12.17 billion (2010 est.)
country comparison to the world: 84
$10.16 billion (2009)
Imports—commodities: machinery and electrical equipment 22%, mineral fuels 18%, chemical products 3%, foodstuffs 6%, plastics 6%, textiles 5%
Imports—partners: Finland 14.52%, Lithuania 10.84%, Latvia 10.47%, Germany 10.33%, Russia 8.59%, Sweden 8.34%, Poland 5.63% (2009)
Reserves of foreign exchange and gold: $3.641 billion (31 December 2010 est.)
country comparison to the world: 87
$3.981 billion (31 December 2009 est.)
Debt—external: $25.13 billion (31 December 2010 est.)
country comparison to the world: 66
$25.56 billion (31 December 2009 est.)
Stock of direct foreign investment—at home: $17.53 billion (31 December 2010 est.)
country comparison to the world: 71
$16.23 billion (31 December 2009 est.)
Stock of direct foreign investment—abroad: $7.134 billion (31 December 2010 est.)
country comparison to the world: 54
$6.534 billion (31 December 2009 est.)
Exchange rates: kroon (EEK) per US dollar–11.8 (2010), 11.23 (2009), 10.7 (2008), 11.535 (2007), 12.473 (2006)

COMMUNICATIONS

Telephones—main lines in use: 492,800 (2009)
country comparison to the world: 99
Telephones—mobile cellular: 2.72 million (2009)
country comparison to the world: 117
Telephone system: *general assessment:* foreign investment in the form of joint business ventures greatly improved telephone service with a wide range of high quality voice, data, and Internet services available
domestic: substantial fiber-optic cable systems carry telephone, TV, and radio traffic in the digital mode; Internet services are widely available; schools and libraries are connected to the Internet, a large percentage of the population files income-tax returns online, and online voting was used for the first time in the 2005 local elections
international: country code–372; fiber-optic cables to Finland, Sweden, Latvia, and Russia provide worldwide packet-switched service; 2 international switches are located in Tallinn (2008)
Broadcast media: the publicly-owned broadcaster, Eesti Rahvusringhaaling (ERR), operates 2 television channels; national private TV channels expanding service; a range of channels are aimed at Russian-speaking viewers; high penetration rate for cable TV services with more than half of Estonian households connected; publicly-owned broadcaster, ERR, operates 4 radio networks and there are a growing number of private commercial radio stations broadcasting nationally, regionally, and locally (2008)
Internet country code: .ee
Internet hosts: 729,534 (2010)
country comparison to the world: 48
Internet users: 971,700 (2009)
country comparison to the world: 102

TRANSPORTATION

Airports: 19 (2010)
country comparison to the world: 137
Airports—with paved runways: *total:* 13
over 3,047 m: 2
2,438 to 3,047 m: 7
1,524 to 2,437 m: 2
914 to 1,523 m: 2 (2010)
Airports—with unpaved runways: *total:* 6
1,524 to 2,437 m: 2
914 to 1,523 m: 1
under 914 m: 3 (2010)
Heliports: 1 (2010)
Pipelines: gas 859 km (2010)
Railways: *total:* 1,196 km
country comparison to the world: 84
broad gauge: 1,196 km 1.520-m and 1.524-m gauge (131 km electrified) (2010)
Roadways: *total:* 58,034 km
country comparison to the world: 76
paved: 34,936 km (includes 104 km of expressways)
unpaved: 23,098 km (2009)
Waterways: 335 km (320 km are navigable year round) (2010)

country comparison to the world: 91
Merchant marine: *total:* 24
country comparison to the world: 95
by type: cargo 4, chemical tanker 1, passenger/cargo 17, petroleum tanker 2
foreign-owned: 3 (Germany 1, Norway 2)
registered in other countries: 77 (Antigua and Barbuda 20, Belize 1, Cambodia 1, Cyprus 7, Dominica 6, Finland 2, Latvia 4, Malta 16, former Netherlands Antilles 1, Norway 1, Saint Kitts and Nevis 3, Saint Vincent and the Grenadines 10, Sierra Leone 1, Sweden 3, Venezuela 1) (2010)
Ports and terminals: Kuivastu, Kunda, Muuga, Parnu Reid, Sillamae, Tallinn

MILITARY

Military branches: Estonian Defense Forces: Land Force, Navy, Air Force (Eesti Ohuvagi), Defense League (Kaitseliit, KL) (2011)
Military service age and obligation: obligation for compulsory service ages 16-60, with conscription "likely" ages 18-27; service requirement 8-11 months (2009)
Manpower available for military service: *males age 16–49:* 291,801
females age 16–49: 302,696 (2010 est.)
Manpower fit for military service: *males age 16–49:* 210,854
females age 16–49: 251,185 (2010 est.)
Manpower reaching militarily significant age annually: *male:* 6,668
female: 6,309 (2010 est.)
Military expenditures: 2% of GDP (2005 est.)
country comparison to the world: 70

TRANSNATIONAL ISSUES

Disputes—international: Russia recalled its signature to the 1996 technical border agreement with Estonia in 2005, rather than concede to Estonia's appending a prepared unilateral declaration referencing Soviet occupation and territorial losses; Russia demands better accommodation of Russian-speaking population in Estonia; Estonian citizen groups continue to press for realignment of the boundary based on the 1920 Tartu Peace Treaty that would bring the now divided ethnic Setu people and parts of the Narva region within Estonia; as a member state that forms part of the EU's external border, Estonia must implement the strict Schengen border rules with Russia
Illicit drugs: growing producer of synthetic drugs; increasingly important transshipment zone for cannabis, cocaine, opiates, and synthetic drugs since joining the European Union and the Schengen Accord; potential money laundering related to organized crime and drug trafficking is a concern, as is possible use of the gambling sector to launder funds; major use of opiates and ecstasy

ETHIOPIA

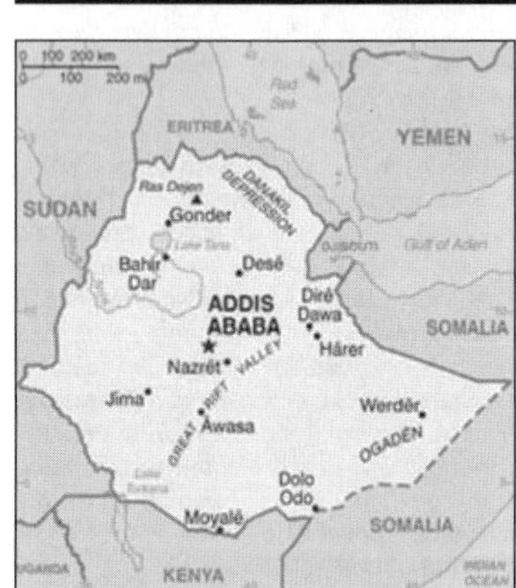

INTRODUCTION

Background: Unique among African countries, the ancient Ethiopian monarchy maintained its freedom from colonial rule with the exception of a short-lived Italian occupation from 1936-41. In 1974, a military junta, the Derg, deposed Emperor Haile SELASSIE (who had ruled since 1930) and established a socialist state. Torn by bloody coups, uprisings, wide-scale drought, and massive refugee problems, the regime was finally toppled in 1991 by a coalition of rebel forces, the Ethiopian People's Revolutionary Democratic Front (EPRDF). A constitution was adopted in 1994, and Ethiopia's first multiparty elections were held in 1995. A border war with Eritrea late in the 1990s ended with a peace treaty in December 2000. In November 2007, the Eritrea-Ethiopia Border Commission remotely demarcated the border by geographical coordinates, but final demarcation of the boundary on the ground is currently on hold because of Ethiopian objections to an international commission's finding requiring it to surrender territory considered sensitive to Ethiopia.

GEOGRAPHY

Location: Eastern Africa, west of Somalia
Geographic coordinates: 8 00 N, 38 00 E
Map references: Africa
Area: *total:* 1,104,300 sq km
country comparison to the world: 27
land: 1 million sq km
water: 104,300 sq km
Area—comparative: slightly less than twice the size of Texas
Land boundaries: *total:* 5,328 km
border countries: Djibouti 349 km, Eritrea 912 km, Kenya 861 km, Somalia 1,600 km, Sudan 1,606 km
Coastline: 0 km (landlocked)
Maritime claims: none (landlocked)
Climate: tropical monsoon with wide topographic-induced variation
Terrain: high plateau with central mountain range divided by Great Rift Valley
Elevation extremes: *lowest point:* Danakil Depression -125 m
highest point: Ras Dejen 4,533 m
Natural resources: small reserves of gold, platinum, copper, potash, natural gas, hydropower
Land use: *arable land:* 10.01%
permanent crops: 0.65%
other: 89.34% (2005)
Irrigated land: 2,900 sq km (2003)
Total renewable water resources: 110 cu km (1987)
Freshwater withdrawal (domestic/industrial/agricultural): *total:* 5.56 cu km/yr (6%/0%/94%)
per capita: 72 cu m/yr (2002)
Natural hazards: geologically active Great Rift Valley susceptible to earthquakes, volcanic eruptions; frequent droughts
volcanism: Ethiopia experiences volcanic activity in the Great Rift Valley; Erta Ale (elev. 613 m), which has caused frequent lava flows in recent years, is the country's most active volcano; Dabbahu became active in 2005, causing evacuations; other historically active volcanoes include Alayta, Dalaffilla, Dallol, Dama Ali, Fentale, Kone, Manda Hararo, and Manda-Inakir
Environment—current issues: deforestation; overgrazing; soil erosion; desertification; water shortages in some areas from water-intensive farming and poor management
Environment—international agreements: *party to:* Biodiversity, Climate Change, Climate Change-Kyoto Protocol, Desertification, Endangered Species, Hazardous Wastes, Ozone Layer Protection
signed, but not ratified: Environmental Modification, Law of the Sea
Geography—note: landlocked—entire coastline along the Red Sea was lost with the de jure independence of Eritrea on 24 May 1993; the Blue Nile, the chief headstream of the Nile by water volume, rises in T'ana Hayk (Lake Tana) in northwest Ethiopia; three major crops are believed to have originated in Ethiopia: coffee, grain sorghum, and castor bean

PEOPLE

Population: 90,873,739 (July 2011 est.)
country comparison to the world: 13
note: estimates for this country explicitly take into account the effects of excess mortality due to AIDS; this can result in lower life expectancy, higher infant mortality, higher death rates, lower population growth rates, and changes in the distribution of population by age and sex than would otherwise be expected
Age structure: *0–14 years:* 46.3% (male 20,990,369/female 21,067,961)
15–64 years: 51% (male 22,707,235/female 23,682,385)
65 years and over: 2.7% (male 1,037,488/female 1,388,301) (2011 est.)
Median age: *total:* 16.8 years
male: 16.5 years
female: 17.1 years (2011 est.)
Population growth rate: 3.194% (2011 est.)
country comparison to the world: 8

Birth rate: 42.99 births/1,000 population (2011 est.)
country comparison to the world: 6
Death rate: 11.04 deaths/1,000 population (July 2011 est.)
country comparison to the world: 37
Net migration rate: -0.01 migrant(s)/1,000 population
country comparison to the world: 115
note: repatriation of Ethiopian refugees residing in Sudan is expected to continue for several years; some Sudanese, Somali, and Eritrean refugees, who fled to Ethiopia from the fighting or famine in their own countries, continue to return to their homes (2011 est.)
Urbanization: *urban population:* 17% of total population (2010)
rate of urbanization: 3.8% annual rate of change (2010-15 est.)
Major cities—population: ADDIS ABABA (capital) 2.863 million (2009)
Sex ratio: *at birth:* 1.03 male(s)/female
under 15 years: 1 male(s)/female
15–64 years: 0.96 male(s)/female
65 years and over: 0.75 male(s)/female
total population: 0.97 male(s)/female (2011 est.)
Infant mortality rate:
total: 77.12 deaths/1,000 live births
country comparison to the world: 16
male: 88.03 deaths/1,000 live births
female: 65.88 deaths/1,000 live births (2011 est.)
Life expectancy at birth: *total population:* 56.19 years
country comparison to the world: 196
male: 53.64 years
female: 58.81 years (2011 est.)
Total fertility rate: 6.02 children born/woman (2011 est.)
country comparison to the world: 7
HIV/AIDS—adult prevalence rate: NA
HIV/AIDS—people living with HIV/AIDS: NA
HIV/AIDS—deaths: NA
Major infectious diseases: *degree of risk:* high
food or waterborne diseases: bacterial and protozoal diarrhea, hepatitis A and E, and typhoid fever
vectorborne diseases: malaria
respiratory disease: meningococcal meningitis
animal contact disease: rabies
water contact disease: schistosomiasis (2009)
Drinking water source: *Improved:*
urban: 98% of population
rural: 26% of population
total: 38% of population
Unimproved:
urban: 2% of population
rural: 74% of population
total: 62% of population (2008)
Sanitation facility access: *Improved:*
urban: 29% of population
rural: 8% of population
total: 12% of population
Unimproved: urban: 71% of population
rural: 92% of population
total: 88% of population (2008)
Nationality: *noun:* Ethiopian(s)
adjective: Ethiopian
Ethnic groups: Oromo 34.5%, Amara 26.9%, Somalie 6.2%, Tigraway 6.1%, Sidama 4%, Guragie 2.5%, Welaita 2.3%, Hadiya 1.7%, Affar 1.7%, Gamo 1.5%, Gedeo 1.3%, other 11.3% (2007 Census)
Religions: Orthodox 43.5%, Muslim 33.9%, Protestant 18.6%, traditional 2.6%, Catholic 0.7%, other 0.7% (2007 Census)
Languages: Amarigna (Amharic) (official) 32.7%, Oromigna (official regional) 31.6%, Tigrigna (official regional) 6.1%, Somaligna 6%, Guaragigna 3.5%, Sidamigna 3.5%, Hadiyigna 1.7%, other 14.8%, English (official) (major foreign language taught in schools), Arabic (official) (1994 census)
Literacy: *definition:* age 15 and over can read and write
total population: 42.7%
male: 50.3%
female: 35.1% (2003 est.)
School life expectancy (primary to tertiary education): *total:* 8 years
male: 9 years
female: 8 years (2008)
Education expenditures: 5.5% of GDP (2007)
country comparison to the world: 42

GOVERNMENT

Country name: *conventional long form:* Federal Democratic Republic of Ethiopia
conventional short form: Ethiopia
local long form: Ityop'iya Federalawi Demokrasiyawi Ripeblik
local short form: Ityop'iya
former: Abyssinia, Italian East Africa
abbreviation: FDRE
Government type: federal republic
Capital: *name:* Addis Ababa
geographic coordinates: 9 02 N, 38 42 E
time difference: UTC+3 (8 hours ahead of Washington, DC during Standard Time)
Administrative divisions: 9 ethnically based states (kililoch, singular—kilil) and 2 self-governing administrations* (astedaderoch, singular—astedader); Adis Abeba* (Addis Ababa), Afar, Amara (Amhara), Binshangul Gumuz, Dire Dawa*, Gambela Hizboch (Gambela Peoples), Hareri Hizb (Harari People), Oromiya (Oromia), Sumale (Somali), Tigray, Ye Debub Biheroch Bihereseboch na Hizboch (Southern Nations, Nationalities and Peoples)
Independence: oldest independent country in Africa and one of the oldest in the world—at least 2,000 years (may be traced to the Aksumite Kingdom, which coalesced in the first century B.C.)
National holiday: National Day (defeat of MENGISTU regime), 28 May (1991)
Constitution: ratified 8 December 1994, effective 22 August 1995
Legal system: civil law system
International law organization participation: has not submitted an ICJ jurisdiction declaration; non-party state to the ICCt
Suffrage: 18 years of age; universal
Executive branch: *chief of state:* President GIRMA Woldegiorgis (since 8 October 2001)
head of government: Prime Minister MELES Zenawi (since August 1995)
cabinet: Council of Ministers ministers selected by the prime minister and approved by the House of People's Representatives
(For more information visit the World Leaders website)
elections: president elected by both chambers of Parliament for a six-year term (eligible for a second term); election last held on 9 October 2007 (next to be held in October 2013); prime minister designated by the party in power following legislative elections
election results: GIRMA Woldegiorgis elected president; percent of vote by the House of People's Representatives—79%
Legislative branch: bicameral Parliament consists of the House of Federation (or upper chamber responsible for interpreting the constitution and federal-regional issues) (108 seats; members chosen by state assemblies to serve five-year terms) and the House of People's Representatives (or lower chamber responsible for passing legislation) (547 seats; members directly elected by popular vote from single-member districts to serve five-year terms)
elections: last held on 23 May 2010 (next to be held in 2015)
election results: percent of vote—NA; seats by party—EPRDF 499, SPDP 24, BGPDP 9, ANDP 8, GPUDM 3, HNL 1, FORUM 1, APDO 1, independent 1
Judicial branch: Federal Supreme Court (the president and vice president of the Federal Supreme Court are recommended by the prime minister and appointed by the House of People's Representatives; for other federal judges, the prime minister submits to the House of People's Representatives for appointment candidates selected by the Federal Judicial Administrative Council)
Political parties and leaders: Afar National Democratic Party or ANDP [Mohammed KEDIR]; Arena Tigray; Argoba People's Democratic Organization or APDO; Benishangul Gumuz People's Democratic Party or BGPDP [Mulualem BESSE]; Coalition for Unity and Democratic Party or CUDP; Ethiopian Federal Democratic Forum or FORUM (a UDJ-led 8-party alliance established for the 2010 parliamentary elections); Ethiopian People's Revolutionary Democratic Front or EPRDF; Gambella Peoples Unity Democratic Movement or GPUDM [Umod UBONG]; Gurage Nationalities' Democratic Movement or GNDM; Harari National League or HNL [Murad ABDULHADI]; Oromo Federalist Democratic Movement or OFDM [BULCHA Demeksa]; Oromo People's Congress or OPC [IMERERA Gudina]; Somali Democratic Alliance Forces or SODAF; Somali People's Democratic Party or SPDP; United Ethiopian Democratic Forces or UEDF [BEYENE Petros]; Unity for Democracy and Justice or UDJ [Birtukan MEDEKSA, currently imprisoned]
Political pressure groups and leaders: Ethiopian People's Patriotic Front or EPPF; Ogaden National Liberation Front

or ONLF; Oromo Liberation Front or OLF [DAOUD Ibsa]
International organization participation: ACP, AfDB, AU, COMESA, FAO, G-24, G-77, IAEA, IBRD, ICAO, ICRM, IDA, IFAD, IFC, IFRCS, IGAD, ILO, IMF, IMO, Interpol, IOC, IOM (observer), IPU, ISO, ITSO, ITU, ITUC, MIGA, NAM, OPCW, PCA, UN, UNAMID, UNCTAD, UNESCO, UNHCR, UNIDO, UNMIL, UNOCI, UNWTO, UPU, WCO, WFTU, WHO, WIPO, WMO, WTO (observer)
Diplomatic representation in the US: *chief of mission:* Ambassador GIRMA Birru
chancery: 3506 International Drive NW, Washington, DC 20008
telephone: [1] (202) 364-1200
FAX: [1] (202) 587-0195
consulate(s) general: Los Angeles
consulate(s): New York
Diplomatic representation from the US: *chief of mission:* Ambassador Donald E. BOOTH
embassy: Entoto Street, Addis Ababa
mailing address: P. O. Box 1014, Addis Ababa
telephone: [251] 11-517-40-00
FAX: [251] 11-517-40-01
Flag description: three equal horizontal bands of green (top), yellow, and red, with a yellow pentagram and single yellow rays emanating from the angles between the points on a light blue disk centered on the three bands; green represents hope and the fertility of the land, yellow symbolizes justice and harmony, while red stands for sacrifice and heroism in the defense of the land; the blue of the disk symbolizes peace and the pentagram represents the unity and equality of the nationalities and peoples of Ethiopia
note: Ethiopia is the oldest independent country in Africa, and the three main colors of her flag (adopted ca. 1895) were so often adopted by other African countries upon independence that they became known as the Pan-African colors; the emblem in the center of the current flag was added in 1996
National anthem: *name:* "Whedefit Gesgeshi Woude Henate Ethiopia" (March Forward, Dear Mother Ethiopia)
lyrics/music: DEREJE Melaku Mengesha/ SOLOMON Lulu
note: adopted 1992

ECONOMY

Economy—overview: Ethiopia's poverty-stricken economy is based on agriculture, accounting for almost 45% of GDP, and 85% of total employment. The agricultural sector suffers from frequent drought and poor cultivation practices. Coffee is critical to the Ethiopian economy with exports of some $350 million in 2006, but historically low prices have seen many farmers switching to qat to supplement income. Under Ethiopia's constitution, the state owns all land and provides long-term leases to the tenants; the system continues to hamper growth in the industrial sector as entrepreneurs are unable to use land as collateral for loans. In November 2001, Ethiopia qualified for debt relief from the Highly Indebted Poor Countries (HIPC) initiative, and in December 2005 the IMF forgave Ethiopia's debt. The global economic downturn led to balance of payments pressures, partially alleviated by recent emergency funding from the IMF. While GDP growth has remained high, per capita income is among the lowest in the world.
GDP (purchasing power parity): $86.12 billion (2010 est.)
country comparison to the world: 76
$79.74 billion (2009 est.)
$72.48 billion (2008 est.)
note: data are in 2010 US dollars
GDP (official exchange rate): $29.72 billion (2010 est.)
GDP—real growth rate: 8% (2010 est.)
country comparison to the world: 19
10% (2009 est.)
11.2% (2008 est.)
GDP—per capita (PPP): $1,000 (2010 est.)
country comparison to the world: 213
$900 (2009 est.)
$900 (2008 est.)
note: data are in 2010 US dollars
GDP—composition by sector:
agriculture: 42.9%
industry: 13.7%
services: 43.4% (2010 est.)
Labor force: 37.9 million (2007)
country comparison to the world: 17
Labor force—by occupation:
agriculture: 85%
industry: 5%
services: 10% (2009 est.)
Unemployment rate: NA%
Population below poverty line: 38.7% (FY05/06 est.)
Household income or consumption by percentage share: *lowest 10%:* 4.1%
highest 10%: 25.6% (2005)
Distribution of family income—Gini index: 30 (2000)
country comparison to the world: 113
40 (1995)
Investment (gross fixed): 25.2% of GDP (2010 est.)
country comparison to the world: 40
Budget: *revenues:* $4.36 billion
expenditures: $5.098 billion (2010 est.)
Public debt: 39.3% of GDP (2010 est.)
country comparison to the world: 74
35.4% of GDP (2009 est.)
Inflation rate (consumer prices): 7% (2010 est.)
country comparison to the world: 174
8.5% (2009 est.)
Central bank discount rate: NA%
Commercial bank prime lending rate: 8% (31 December 2008)
country comparison to the world: 127
7% (31 December 2006)
Stock of narrow money: $4.764 billion (31 December 2010 est.)
country comparison to the world: 92
$4.972 billion (31 December 2009 est.)
Stock of broad money: $8.248 billion (31 December 2010 est.)
country comparison to the world: 107
$8.641 billion (31 December 2009 est.)
Stock of domestic credit: $8.661 billion (31 December 2010 est.)
country comparison to the world: 98
$9.292 billion (31 December 2009 est.)
Market value of publicly traded shares: $NA
Agriculture—products: cereals, pulses, coffee, oilseed, cotton, sugarcane, potatoes, qat, cut flowers; hides, cattle, sheep, goats; fish
Industries: food processing, beverages, textiles, leather, chemicals, metals processing, cement
Industrial production growth rate: 9.5% (2010 est.)
country comparison to the world: 22
Electricity—production: 3.46 billion kWh (2007 est.)
country comparison to the world: 123
Electricity—consumption: 3.13 billion kWh (2007 est.)
country comparison to the world: 125
Electricity—exports: 0 kWh (2008 est.)
Electricity—imports: 0 kWh (2008 est.)
Oil—production: 0 bbl/day (2009 est.)
country comparison to the world: 172
Oil—consumption: 38,000 bbl/day (2009 est.)
country comparison to the world: 105
Oil—exports: 0 bbl/day (2007 est.)
country comparison to the world: 164
Oil—imports: 33,590 bbl/day (2007 est.)
country comparison to the world: 96
Oil—proved reserves: 430,000 bbl (1 January 2010 est.)
country comparison to the world: 99
Natural gas—production: 0 cu m (2008 est.)
country comparison to the world: 124
Natural gas—consumption: 0 cu m (2008 est.)
country comparison to the world: 169
Natural gas—exports: 0 cu m (2008 est.)
country comparison to the world: 92
Natural gas—imports: 0 cu m (2008 est.)
country comparison to the world: 183
Natural gas—proved reserves: 24.92 billion cu m (1 January 2010 est.)
country comparison to the world: 74
Current account balance: $-2.232 billion (2010 est.)
country comparison to the world: 157
$-1.996 billion (2009 est.)
Exports: $1.729 billion (2010 est.)
country comparison to the world: 135
$1.636 billion (2009 est.)
Exports—commodities: coffee, qat, gold, leather products, live animals, oilseeds
Exports—partners: China 10.87%, Germany 9.75%, Saudi Arabia 7.39%, US 7.21%, Netherlands 6.38%, Switzerland 5.33%, Sudan 4.35%, Belgium 4% (2009)
Imports: $7.517 billion (2010 est.)
country comparison to the world: 101
$6.946 billion (2009 est.)
Imports—commodities: food and live animals, petroleum and petroleum products, chemicals, machinery, motor vehicles, cereals, textiles
Imports—partners: China 14.73%, Saudi Arabia 8.41%, India 7.65%, US 4.3% (2009)
Reserves of foreign exchange and gold: $1.88 billion (31 December 2010 est.)
country comparison to the world: 106
$1.781 billion (31 December 2009 est.)
Debt—external: $4.289 billion (31 December 2010 est.)
country comparison to the world: 113

$3.621 billion (31 December 2009 est.)
Exchange rates: birr (ETB) per US dollar–14.4 (2010), 11.78 (2009), 9.57 (2008), 8.96 (2007), 8.69 (2006)

COMMUNICATIONS

Telephones—main lines in use: 915,100 (2009)
country comparison to the world: 83
Telephones—mobile cellular: 4.052 million (2009)
country comparison to the world: 104
Telephone system: *general assessment:* inadequate telephone system with the Ethiopian Telecommunications Corporation (ETC) maintaining a monopoly over telecommunication services; open-wire, microwave radio relay; radio communication in the HF, VHF, and UHF frequencies; 2 domestic satellites provide the national trunk service
domestic: the number of fixed lines and mobile telephones is increasing from a small base; combined fixed and mobile-cellular teledensity is only about 5 per 100 persons
international: country code–251; open-wire to Sudan and Djibouti; microwave radio relay to Kenya and Djibouti; satellite earth stations–3 Intelsat (1 Atlantic Ocean and 2 Pacific Ocean) (2009)
Broadcast media: 1 public TV broadcast station broadcasting nationally and 1 public radio broadcaster with stations in each of the 13 administrative districts; a few commercial radio stations and roughly a dozen community radio stations (2009)
Internet country code: .et
Internet hosts: 151 (2010)
country comparison to the world: 200
Internet users: 447,300 (2009)
country comparison to the world: 119

TRANSPORTATION

Airports: 61 (2010)
country comparison to the world: 79
Airports—with paved runways: *total:* 17
over 3,047 m: 3
2,438 to 3,047 m: 8
1,524 to 2,437 m: 4
914 to 1,523 m: 1
under 914 m: 1 (2010)
Airports—with unpaved runways: *total:* 44
2,438 to 3,047 m: 3
1,524 to 2,437 m: 12
914 to 1,523 m: 22
under 914 m: 7 (2010)
Railways: *total:* 681 km (Ethiopian segment of the 781 km Addis Ababa-Djibouti railroad)
country comparison to the world: 102
narrow gauge: 681 km 1.000-m gauge
note: railway is under joint control of Djibouti and Ethiopia but is largely inoperable (2008)
Roadways: *total:* 36,469 km
country comparison to the world: 93
paved: 6,980 km
unpaved: 29,489 km (2004)
Merchant marine: *total:* 9
country comparison to the world: 120
by type: cargo 8, roll on/roll off 1 (2010)
Ports and terminals: Ethiopia is landlocked and uses ports of Djibouti in Djibouti and Berbera in Somalia

MILITARY

Military branches: Ethiopian National Defense Force (ENDF): Ground Forces, Ethiopian Air Force (ETAF) (2011)
note: Ethiopia is landlocked and has no navy; following the secession of Eritrea (1993), Ethiopian naval facilities remained in Eritrean possession
Military service age and obligation: 18 years of age for voluntary military service; no compulsory military service, but the military can conduct callups when necessary and compliance is compulsory (2011)
Manpower available for military service: *males age 16–49:* 19,067,499
females age 16–49: 19,726,816 (2010 est.)
Manpower fit for military service: *males age 16–49:* 11,868,084
females age 16–49: 12,889,260 (2010 est.)
Manpower reaching militarily significant age annually: *male:* 967,411
female: 981,714 (2010 est.)
Military expenditures: 1.2% of GDP (2009)
country comparison to the world: 123

TRANSNATIONAL ISSUES

Disputes—international: Eritrea and Ethiopia agreed to abide by the 2002 Eritrea-Ethiopia Boundary Commission's (EEBC) delimitation decision, but neither party responded to the revised line detailed in the November 2006 EEBC Demarcation Statement; the undemarcated former British administrative line has little meaning as a political separation to rival clans within Ethiopia's Ogaden and southern Somalia's Oromo region; Ethiopian forces invaded southern Somalia and routed Islamist Courts from Mogadishu in January 2007; "Somaliland" secessionists provide port facilities in Berbera and trade ties to landlocked Ethiopia; civil unrest in eastern Sudan has hampered efforts to demarcate the porous boundary with Ethiopia
Refugees and internally displaced persons: refugees (country of origin): 66,980 (Sudan); 16,576 (Somalia); 13,078 (Eritrea)
IDPs: 200,000 (border war with Eritrea from 1998-2000, ethnic clashes in Gambela, and ongoing Ethiopian military counterinsurgency in Somali region; most IDPs are in Tigray and Gambela Provinces) (2007)
Illicit drugs: transit hub for heroin originating in Southwest and Southeast Asia and destined for Europe, as well as cocaine destined for markets in southern Africa; cultivates qat (khat) for local use and regional export, principally to Djibouti and Somalia (legal in all three countries); the lack of a well-developed financial system limits the country's utility as a money laundering center

EUROPEAN UNION

INTRODUCTION

Preliminary statement: The evolution of what is today the European Union (EU) from a regional economic agreement among six neighboring states in 1951 to today's hybrid intergovernmental and supranational organization of 27 countries across the European continent stands as an unprecedented phenomenon in the annals of history. Dynastic unions for territorial consolidation were long the norm in Europe; on a few occasions even country-level unions were arranged–the Polish-Lithuanian Commonwealth and the Austro-Hungarian Empire were examples. But for such a large number of nation-states to cede some of their sovereignty to an overarching entity is unique.
Although the EU is not a federation in the strict sense, it is far more than a free-trade association such as ASEAN, NAFTA, or Mercosur, and it has certain attributes associated with independent nations: its own flag, currency (for some members), and law-making abilities, as well as diplomatic representation and a common foreign and security policy in its dealings with external partners.
Thus, inclusion of basic intelligence on the EU has been deemed appropriate as a new, separate entity in The World Factbook. However, because of the EU's special status, this description is placed after the regular country entries.
Background: Following the two devastating World Wars in the first half of the 20th century, a number of European leaders in the late 1940s became convinced that the only way to establish a lasting peace was to unite the two chief belligerent nations–France and Germany–both economically and politically. In 1950, the French Foreign Minister Robert SCHUMAN proposed an eventual union of all Europe, the first step of which would be the integration of the coal and steel industries of Western Europe. The following year the European Coal and Steel Community (ECSC) was set up when six members, Belgium, France, West Germany, Italy, Luxembourg, and the Netherlands, signed the Treaty of Paris.

The ECSC was so successful that within a few years the decision was made to integrate other elements of the countries' economies. In 1957, envisioning an "ever closer union," the Treaties of Rome created the European Economic Community (EEC) and the European Atomic Energy Community (Euratom), and the six member states undertook to eliminate trade barriers among themselves by forming a common market. In 1967, the institutions of all three communities were formally merged into the European Community (EC), creating a single Commission, a single Council of Ministers, and the body known today as the European Parliament. Members of the European Parliament were initially selected by national parliaments, but in 1979 the first direct elections were undertaken and they have been held every five years since.
In 1973, the first enlargement of the EC took place with the addition of Denmark, Ireland, and the United Kingdom. The 1980s saw further membership expansion with Greece joining in 1981 and Spain and Portugal in 1986. The 1992 Treaty of Maastricht laid the basis for further forms of cooperation in foreign and defense policy, in judicial and internal affairs, and in the creation of an economic and monetary union–including a common currency. This further integration created the European Union (EU), at the time standing alongside the European Community. In 1995, Austria, Finland, and Sweden joined the EU/EC, raising the membership total to 15.
A new currency, the euro, was launched in world money markets on 1 January 1999; it became the unit of exchange for all EU member states except the United Kingdom, Sweden, and Denmark. In 2002, citizens of those 12 countries began using euro banknotes and coins. Ten new countries joined the EU in 2004–Cyprus, the Czech Republic, Estonia, Hungary, Latvia, Lithuania, Malta, Poland, Slovakia, and Slovenia–and in 2007 Bulgaria and Romania joined, bringing the membership to 27, where it stands today.
In an effort to ensure that the EU could function efficiently with an expanded membership, the Treaty of Nice (signed in 2000) set forth rules aimed at streamlining the size and procedures of EU institutions. An effort to establish a "Constitution for Europe," growing out of a Convention held in 2002-2003, foundered when it was rejected in referenda in France and the Netherlands in 2005. A subsequent effort in 2007 incorporated many of the features of the rejected Constitution while also making a number of substantive and symbolic changes. The new treaty, initially known as the Reform Treaty but subsequently referred to as the Treaty of Lisbon, sought to amend existing treaties rather than replace them. The treaty was approved at the EU intergovernmental conference of the 27 member states held in Lisbon in December 2007, after which the process of national ratifications began. In October 2009, an Irish referendum approved the Lisbon Treaty (overturning a previous rejection) and cleared the way for an ultimate unanimous endorsement. Poland and the Czech Republic signed on soon after. The Lisbon Treaty, again invoking the idea of an "ever closer union," came into force on 1 December 2009 and the European Union officially replaced and succeeded the European Community.

GEOGRAPHY

Location: Europe between the North Atlantic Ocean in the west and Russia, Belarus, and Ukraine to the east
Map references: Europe
Area: *total:* 4,324,782 sq km
Area—comparative: less than one-half the size of the US
Land boundaries: *total:* 12,440.8 km
border countries: Albania 282 km, Andorra 120.3 km, Belarus 1,050 km, Croatia 999 km, Holy See 3.2 km, Liechtenstein 34.9 km, Macedonia 394 km, Moldova 450 km, Monaco 4.4 km, Norway 2,348 km, Russia 2,257 km, San Marino 39 km, Serbia 945 km, Switzerland 1,811 km, Turkey 446 km, Ukraine 1,257 km
note: data for European Continent only
Coastline: 65,992.9 km
Maritime claims: NA
Climate: cold temperate; potentially subarctic in the north to temperate; mild wet winters; hot dry summers in the south
Terrain: fairly flat along the Baltic and Atlantic coast; mountainous in the central and southern areas
Elevation extremes: *lowest point:* Lammefjord, Denmark -7 m; Zuidplaspolder, Netherlands -7 m
highest point: Mont Blanc 4,807 m; note–situated on the border between France and Italy
Natural resources: iron ore, natural gas, petroleum, coal, copper, lead, zinc, bauxite, uranium, potash, salt, hydropower, arable land, timber, fish
Land use: *arable land:* NA
permanent crops: NA
other: NA
Irrigated land: 168,050 sq km (2003 est.)
Natural hazards: flooding along coasts; avalanches in mountainous area; earthquakes in the south; volcanic eruptions in Italy; periodic droughts in Spain; ice floes in the Baltic
Environment—current issues: NA
Environment—international agreements:
party to: Air Pollution, Air Pollution-Nitrogen Oxides, Air Pollution-Persistent Organic Pollutants, Air Pollution-Sulphur 94, Antarctic-Marine Living Resources, Biodiversity, Climate Change, Climate Change-Kyoto Protocol, Desertification, Hazardous Wastes, Law of the Sea, Ozone Layer Protection, Tropical Timber 83, Tropical Timber 94
signed but not ratified: Air Pollution-Volatile Organic Compounds

PEOPLE

Population: 492,387,344 (July 2010 est.)
Age structure: *0–14 years:* 15.44% (male 38,992,677/female 36,940,450)
15–64 years: 67.23% (male 166,412,403/female 164,295,636)
65 years and over: 17.33% (male 35,376,333/female 49,853,361) (2009 est.)
Median age: note–see individual country entries of member states (2009 est.)
Population growth rate: 0.098 % (2010 est.)
Birth rate: 9.83 births/1,000 population (2010 est.)
Death rate: 10.33 deaths/1,000 population (July 2010 est.)
Net migration rate: 1.48 migrant(s)/1,000 population (2010 est.)
Sex ratio:
at birth: 1.06 male(s)/female
under 15 years: 1.05 male(s)/female
15–64 years: 1 male(s)/female
65 years and over: 0.73 male(s)/female
total population: 0.95 male(s)/female (2011 est.)
Infant mortality rate:
total: 5.61 deaths/1,000 live births
country comparison to the world: 179
male: 6.26 deaths/1,000 live births
female: 4.93 deaths/1,000 live births (2010 est.)
Life expectancy at birth: *total population:* 78.82 years
country comparison to the world: 44
male: 75.7 years
female: 82.13 years (2010 est.)
Total fertility rate: 1.51 children born/woman (2010 est.)
HIV/AIDS—adult prevalence rate: note–see individual country entries of member states
HIV/AIDS—people living with HIV/AIDS: note–see individual country entries of member states
HIV/AIDS—deaths: note–see individual country entries of member states
Religions: Roman Catholic, Protestant, Orthodox, Muslim, Jewish
Languages: Bulgarian, Czech, Danish, Dutch, English, Estonian, Finnish, French, Gaelic, German, Greek, Hungarian, Italian, Latvian, Lithuanian, Maltese, Polish, Portuguese, Romanian, Slovak, Slovene, Spanish, Swedish
note: only official languages are listed; German, the major language of Germany, Austria, and Switzerland, is the most widely spoken mother tongue–over 19% of the EU population; English is the most widely spoken language–about 49% of the EU population is conversant with it (2007)
School life expectancy (primary to tertiary education): NA

GOVERNMENT

Union name:
conventional long form: European Union
abbreviation: EU
Political structure: a hybrid intergovernmental and supranational organization
Capital: *name:* Brussels (Belgium), Strasbourg (France), Luxembourg
geographic coordinates: (Brussels) 50 50 N, 4 20 E
time difference: UTC+1 (6 hours ahead of Washington, DC during Standard Time)
daylight saving time: +1hr, begins last Sunday in March; ends last Sunday in October

note: the Council of the European Union meets in Brussels, Belgium; the European Parliament meets in Brussels and Strasbourg, France; the Court of Justice of the European Union meets in Luxembourg

Member states: 27 countries: Austria, Belgium, Bulgaria, Cyprus, Czech Republic, Denmark, Estonia, Finland, France, Germany, Greece, Hungary, Ireland, Italy, Latvia, Lithuania, Luxembourg, Malta, Netherlands, Poland, Portugal, Romania, Slovakia, Slovenia, Spain, Sweden, UK; note—candidate countries: Croatia, Iceland, Macedonia, Montenegro, Turkey

Independence: 7 February 1992 (Maastricht Treaty signed establishing the EU); 1 November 1993 (Maastricht Treaty entered into force)

note: Treaties of Rome, which were signed on 25 March 1957 and entered into force on 1 January 1958, created the European Economic Community and the European Atomic Energy Community; the Treaty of Lisbon, which was signed on 13 December 2007 and entered into force on 1 December 2009, replaced and succeeded the European Community with the European Union

National holiday: Europe Day 9 May (1950); note—the day in 1950 that Robert SCHUMAN proposed the creation of what became the European Coal and Steel Community, the progenitor of today's European Union, with the aim of achieving a united Europe

Constitution: none

note: the EU legal order, although based on a series of treaties, has often been described as "constitutional" in nature; the Treaty on European Union (TEU), as modified by the Lisbon Treaty, states in Article 1 that "the HIGH CONTRACTING PARTIES establish among themselves a EUROPEAN UNION ... on which the Member States confer competences to attain objectives they have in common"; Article 1 of the TEU states further that the EU is "founded on the present Treaty and on the Treaty on the Functioning of the European Union (hereinafter referred to as 'the Treaties')," both possessing the same legal value; Article 6 of the TEU provides that a separately adopted Charter of Fundamental Rights of the European Union "shall have the same legal value as the Treaties"

Legal system: unique supranational law system in which, according to an interpretive declaration of member-state governments appended to the Treaty of Lisbon, "the Treaties and the law adopted by the Union on the basis of the Treaties have primacy over the law of Member States" under conditions laid down in the case law of the Court of Justice; key principles of EU law include fundamental rights as guaranteed by the European Convention for the Protection of Human Rights and Fundamental Freedoms and as resulting from constitutional traditions common to the EU's states

Suffrage: voting for the European Parliament is permitted in each member state at 18 years of age; universal

Executive branch: under the EU treaties there are three distinct institutions each of which conducts functions that may be regarded as executive in nature: the European Council: brings together heads of state and government, along with the president of the European Commission, and meets at least four times a year; its aim is to provide the impetus for the development of the Union and to issue general policy guidelines; leaders of the EU member states appointed former Belgian Prime Minister Herman VAN ROMPUY to be the first full-time president of the European Council in November 2009; he took office on 1 December 2009 and will serve a two-and-one-half-year term, renewable once; his core responsibilities include chairing the summits each year and providing policy and organizational continuity

the Council: consists of ministers of each EU member state and meets regularly in different configurations depending on the subject matter; it carries out policy-making and coordinating functions (also legislative functions); although the name is similar, the "Council" is an institution distinct from the head of state-level "European Council"; ministers of EU member states chair meetings of the Council based on a six-month rotating presidency

the European Commission: is comprised of 27 members, one from each member country; each commissioner is responsible for one or more policy areas; its responsibilities include promoting the general interest of the EU, acting as "guardian of the Treaties," executing the budget and managing programs, ensuring the Union's external representation, and additional duties; its president is Jose Manuel BARROSO (since 2004); the president of the European Commission is designated by member state governments and confirmed by the European Parliament; working from member state recommendations, the Commission president then assembles a "college" of Commission members; the European Parliament confirms the entire Commission for a five-year term; the next confirmation process will likely be held in January 2015

note: for external representation and foreign policy making, leaders of the EU member states appointed UK Baroness Catherine ASHTON to be the first High Representative of the Union for Foreign Affairs and Security Policy; ASHTON took office on 1 December 2009; her concurrent appointment as Vice President of the European Commission endows her position with the policymaking influence of the Council of the EU and the budgetary influence of the European Commission; the High Representative helps develop and implement the EU's common foreign and security policy, represents and acts for the Union in many international contexts, and oversees the new diplomatic corps of the EU, formally established on 1 December 2010, the European External Action Service

Legislative branch: two legislative bodies consisting of the Council of the European Union (27 member-state ministers having 345 votes; the number of votes is roughly proportional to member-states' population) and the European Parliament (736 seats; seats allocated among member states in proportion to population; members elected by direct universal suffrage for a five-year term); note—the Council is the main decision-making body of the EU, although the Commission proposes most EU legislative acts

elections: last held on 4-7 June 2009 (next to be held in June 2014)

election results: percent of vote—EPP 36%, S&D 25%, ALDE 11.4%, Greens/EFA 7.5%, ECR 7.3%, GUE/NGL 4.8%, EFD 4.3%, independents 3.7%; seats by party—EPP 265, S&D 184, ALDE 84, Greens/EFA 55, ECR 54, GUE/NGL 35, EFD 32, independents 27

Judicial branch: Court of Justice of the European Union (ensures that the treaties are interpreted and applied uniformly throughout the EU, resolves disputed issues among the EU institutions, issues opinions on questions of EU law referred by member state courts)—27 judges (one from each member state) appointed for a six-year term; note—the court can sit in chambers, in a "Grand Chamber" of 13 judges, or as the full court; General Court (a court below the Court of Justice)—27 judges appointed for a six-year term; Civil Service Tribunal—7 judges appointed for a three-year term

Political parties and leaders: Confederal Group of the European United Left-Nordic Green Left or GUE/NGL [Lothar BISKY]; Europe of Freedom and Democracy Group or EFD [Nigel FARAGE and Francesco SPERONI]; European Conservatives and Reformists Group or ECR [Michael KAMINSKI]; Group of Greens/European Free Alliance or Greens/EFA [Rebecca HARMS and Daniel COHN-BENDIT]; Group of the Alliance of Liberals and Democrats for Europe or ALDE [Guy VERHOFSTADT]; Group of the European People's Party or EPP [Joseph DAUL]; Group of the Progressive Alliance of Socialists and Democrats in the European Parliament or S&D [Martin SCHULZ]

International organization participation: ARF (dialogue member), ASEAN (dialogue member), Australian Group, BIS, CBSS, CERN, EBRD, FAO, FATF, G-8, G-10, G-20, IDA, IEA, LAIA (observer), NSG (observer), OAS (observer), OECD, PIF (partner), SAARC (observer), UN (observer), UNRWA (observer), WCO, WTO, ZC (observer)

Diplomatic representation in the US: *chief of mission:* Ambassador Joao P. Castanheira do VALE DE ALMEIDA

chancery: 2175 K Street, NW, Washington, DC 20037

telephone: [1] (202) 862-9500

FAX: [1] (202) 429-1766

Diplomatic representation from the US: *chief of mission:* Ambassador William E. KENNARD

embassy: 13 Zinnerstraat/Rue Zinner, B-1000 Brussels

mailing address: same as above
telephone: [32] (2) 508-2111
FAX: [32] (2) 508-2063
Flag description: a blue field with 12 five-pointed gold stars arranged in a circle in the center; blue represents the sky of the Western world, the stars are the peoples of Europe in a circle, a symbol of unity; the number of stars is fixed
National anthem: *name:* "Ode to Joy""
lyrics/music: none/Ludwig VON BEETHOVEN, arranged by Herbert VON KARAJAN
note: adopted 1972, not in use until 1986; according to the European Union, the song is meant to represent all of Europe rather than just the organization; the song also serves as the anthem for the Council of Europe

ECONOMY

Economy—overview: Internally, the EU has abolished trade barriers, adopted a common currency, and is striving toward convergence of living standards. Internationally, the EU aims to bolster Europe's trade position and its political and economic power. Because of the great differences in per capita income among member states (from $7,000 to $78,000) and in national attitudes toward issues like inflation, debt, and foreign trade, the EU faces difficulties in devising and enforcing common policies. Eleven established EU member states, under the auspices of the European Economic and Monetary Union (EMU), introduced the euro as their common currency on 1 January 1999 (Greece did so two years later), but the UK and Denmark have 'opt-outs' that allow them to keep their national currencies, and Sweden has not taken the steps needed to participate. Between 2004 and 2007, the EU admitted 12 countries that are, in general, less advanced economically than the other 15. Of the 12 most recent member states, only Slovenia (1 January 2007), Cyprus and Malta (1 January 2008), Slovakia (1 January 2009), and Estonia (1 January 2011) have adopted the euro; the remaining states other than the UK and Denmark are legally required to adopt the currency upon meeting EU's fiscal and monetary convergence criteria. The EU has recovered from the global financial crisis faster than expected, with business investment growing by an estimated 2% in 2010, but with public investment and housing development lagging. Strong corporate profits should enable this recovery to continue in 2011. Nevertheless, significant risks to growth remain, including, high official debts and deficits, aging populations, over-regulation of non-financial businesses, and doubts about the sustainability of the EMU. In June 2010, prompted by the Greek financial crisis, the EU and the IMF set up a $1 trillion bailout fund to rescue any EMU member in danger of default, but it has not calmed market jitters that have diminished the value of the euro. Discussions are currently under way to create a permanent European Stabilization Mechanism (ESM) in 2013, when the existing European Financial Stability Facility expires.
GDP (purchasing power parity): $14.82 trillion (2010 est.)
country comparison to the world: 1
$14.56 trillion (2009 est.)
$15.18 trillion (2008 est.)
note: data are in 2010 US dollars
GDP (official exchange rate): $16.07 trillion (2010 est.)
GDP—real growth rate: 1.8% (2010 est.)
country comparison to the world: 154
-4.1% (2009 est.)
0.6% (2008 est.)
GDP—per capita (PPP): $32,700 (2010 est.)
country comparison to the world: 42
$32,200 (2009 est.)
$33,700 (2008 est.)
note: data are in 2010 US dollars
GDP—composition by sector:
agriculture: 1.8%
industry: 25%
services: 73.2% (2010 est.)
Labor force: 225.3 million (2010 est.)
country comparison to the world: 3
Labor force—by occupation: *agriculture:* 5.6%
industry: 27.7%
services: 66.7% (2007 est.)
Unemployment rate: 9.5% (2010 est.)
country comparison to the world: 105
9% (2009 est.)
Population below poverty line: note—see individual country entries of member states
Household income or consumption by percentage share: *lowest 10%:* 2.8%
highest 10%: 25.2% (1915 est.)
Distribution of family income—Gini index: 30.4 (2009 est.)
country comparison to the world: 111
31.2 (1996 est.)
Investment (gross fixed): 18.6% of GDP (2010 est.)
country comparison to the world: 105
Inflation rate (consumer prices): 1.8% (2010 est.)
country comparison to the world: 47
1.8% (2009 est.)
Central bank discount rate: 1.75% (31 December 2010)
country comparison to the world: 113
1.75% (31 December 2009)
note: this is the European Central Bank's rate on the marginal lending facility, which offers overnight credit to banks in the euro area
Commercial bank prime lending rate: 7.52% (31 December 2009 est.)
country comparison to the world: 106
8.58% (31 December 2008 est.)
Stock of narrow money: $5.542 trillion (31 December 2008)
country comparison to the world: 3
$5.649 trillion (31 December 2007)
note: this is the quantity of money, M1, for the euro area, converted into US dollars at the exchange rate for the date indicated; it excludes the stock of money carried by non-euro-area members of the European Union
Stock of broad money: $11.17 trillion (31 December 2008 est.)
country comparison to the world: 4
$10.83 trillion (31 December 2007 est.)
note: this is the quantity of broad money for the euro area, converted into US dollars at the exchange rate for the date indicated; it excludes the stock of broad money carried by non-euro-area members of the European Union
Stock of domestic credit: $22.65 trillion (31 December 2009 est.)
country comparison to the world: 2
$21.24 trillion (31 December 2008 est.)
note: this figure refers to the euro area only; it excludes credit data for non-euro-area members of the EU
Market value of publicly traded shares: $9.903 trillion (31 December 2009 est.)
country comparison to the world: 2
$7.644 trillion (31 December 2008)
$15.27 trillion (31 December 2007 est.)
Agriculture—products: wheat, barley, oilseeds, sugar beets, wine, grapes; dairy products, cattle, sheep, pigs, poultry; fish
Industries: among the world's largest and most technologically advanced, the EU industrial base includes: ferrous and non-ferrous metal production and processing, metal products, petroleum, coal, cement, chemicals, pharmaceuticals, aerospace, rail transportation equipment, passenger and commercial vehicles, construction equipment, industrial equipment, shipbuilding, electrical power equipment, machine tools and automated manufacturing systems, electronics and telecommunications equipment, fishing, food and beverage processing, furniture, paper, textiles, tourism
Industrial production growth rate: 4.1% (2010 est.)
country comparison to the world: 81
Electricity—production: 3.078 trillion kWh (2007 est.)
country comparison to the world: 3
Electricity—consumption: 2.901 trillion kWh (2007 est.)
country comparison to the world: 3
Electricity—exports: NA kWh
Electricity—imports: NA kWh
Oil—production: 2.365 million bbl/day (2009 est.)
country comparison to the world: 13
Oil—consumption: 13.63 million bbl/day (2009 est.)
country comparison to the world: 2
Oil—exports: 2.196 million bbl/day (2008 est.)
country comparison to the world: 7
Oil—imports: 8.613 million bbl/day (2008 est.)
country comparison to the world: 2
Oil—proved reserves: 5.453 billion bbl (1 January 2010 est.)
country comparison to the world: 25
Natural gas—production: 181.6 billion cu m (2009 est.)
country comparison to the world: 4
Natural gas—consumption: 487.9 billion cu m (2009 est.)
country comparison to the world: 2
Natural gas—exports: NA cu m
Natural gas—imports: NA cu m
Natural gas—proved reserves: 2.25 trillion cu m (1 January 2010 est.)
country comparison to the world: 18
Current account balance: $NA (2010)

$51.4 billion (2009 est.)
Exports: $1.952 trillion (2007)
country comparison to the world: 1
$1.33 trillion (2005)
note: external exports, excluding intra-EU trade
Exports—commodities: machinery, motor vehicles, aircraft, plastics, pharmaceuticals and other chemicals, fuels, iron and steel, nonferrous metals, wood pulp and paper products, textiles, meat, dairy products, fish, alcoholic beverages
Imports:
$1.69 trillion (2007)
country comparison to the world: 2
$1.466 trillion (2005)
note: external imports, excluding intra-EU trade
Imports—commodities: machinery, vehicles, aircraft, plastics, crude oil, chemicals, textiles, metals, foodstuffs, clothing
Reserves of foreign exchange and gold: $NA
Debt—external: $13.72 trillion (30 June 2010)
country comparison to the world: 2
note: this is the external debt for the euro area only; it excludes the external debt of the non-euro-area members of the EU
Stock of direct foreign investment—at home: $NA
Exchange rates: euros per US dollar–0.755 (2010), 0.7198 (2009), 0.6827 (2008), 0.7345 (2007), 0.7964 (2006)

COMMUNICATIONS

Telephones—main lines in use: 238 million (2005)
Telephones—mobile cellular: 466 million (2005)
Telephone system: note–see individual country entries of member states
Internet country code: .eu; note–see country entries of member states for individual country codes
Internet hosts: 140,277; note–this sum reflects the number of Internet hosts assigned the .eu Internet country code (2010)
Internet users: 247 million (2006)

TRANSPORTATION

Airports: 3,383 (2010)
Airports—with paved runways:
total: 1,992
over 3,047 m: 116
2,438 to 3,047 m: 340
1,524 to 2,437 m: 546
914 to 1,523 m: 422
under 914 m: 568 (2010)
Airports—with unpaved runways:
total: 1,391
over 3,047 m: 2
2,438 to 3,047 m: 1
1,524 to 2,437 m: 22
914 to 1,523 m: 254
under 914 m: 1,112 (2010)
Heliports: 99 (2010)
Railways: *total:* 230,237 km (2010)
Roadways: *total:* 5,814,080 km (2010)
Waterways: 44,103 km (2010)
Ports and terminals: Antwerp (Belgium), Barcelona (Spain), Braila (Romania), Bremen (Germany), Burgas (Bulgaria), Constanta (Romania), Copenhagen (Denmark), Galati (Romania), Gdansk (Poland), Hamburg (Germany), Helsinki (Finland), Las Palmas (Canary Islands, Spain), Le Havre (France), Lisbon (Portugal), London (UK), Marseille (France), Naples (Italy), Peiraiefs or Piraeus (Greece), Riga (Latvia), Rotterdam (Netherlands), Stockholm (Sweden), Talinn (Estonia), Tulcea (Romania), Varna (Bulgaria)

MILITARY

Military—note: the five-nation Eurocorps–created in 1992 by France, Germany, Belgium, Spain, and Luxembourg–has deployed troops and police on peacekeeping missions to Bosnia-Herzegovina, Macedonia, and the Democratic Republic of the Congo and assumed command of the ISAF in Afghanistan in August 2004; Eurocorps directly commands the 5,000-man Franco-German Brigade, the Multinational Command Support Brigade, and EUFOR in Bosnia and Herzegovina; in November 2004, the EU Council of Ministers formally committed to creating 13 1,500-man battle groups by the end of 2007, to respond to international crises on a rotating basis; 22 of the EU's 27 nations have agreed to supply troops; France, Italy, and the UK formed the first of three battle groups in 2005; Norway, Sweden, Estonia, and Finland established the Nordic Battle Group effective 1 January 2008; nine other groups are to be formed; a rapid-reaction naval EU Maritime Task Group was stood up in March 2007 (2007)

TRANSNATIONAL ISSUES

Disputes—international: as a political union, the EU has no border disputes with neighboring countries, but Estonia has no land boundary agreements with Russia, Slovenia disputes its land and maritime boundaries with Croatia, and Spain has territorial and maritime disputes with Morocco and with the UK over Gibraltar; the EU has set up a Schengen area–consisting of 22 EU member states that have signed the convention implementing the Schengen agreements or "acquis" (1985 and 1990) on the free movement of persons and the harmonization of border controls in Europe; these agreements became incorporated into EU law with the implementation of the 1997 Treaty of Amsterdam on 1 May 1999; in addition, non-EU states Iceland and Norway (as part of the Nordic Union) have been included in the Schengen area since 1996 (full members in 2001), and Switzerland since 2008 bringing the total current membership to 25; the UK (since 2000) and Ireland (since 2002) take part in only some aspects of the Schengen area, especially with respect to police and criminal matters; nine of the 12 new member states that joined the EU since 2004 joined Schengen on 21 December 2007; of the three remaining EU states, Romania and Bulgaria may join by late 2011, while Cyprus' entry is held up by the ongoing Cyprus dispute

FALKLAND ISLANDS (ISLAS MALVINAS)

(OVERSEAS TERRITORY OF THE UK; ALSO CLAIMED BY ARGENTINA)

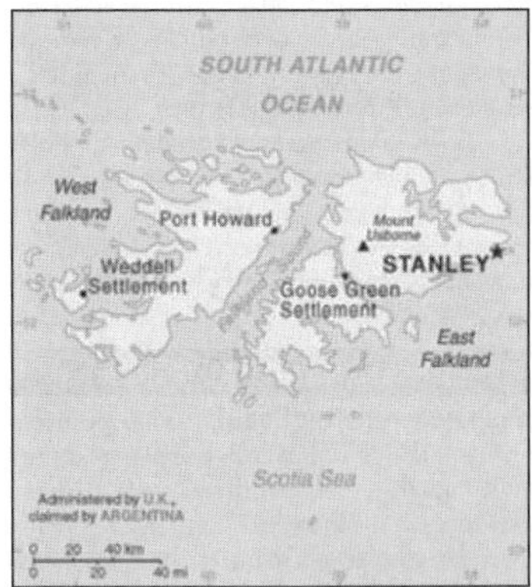

INTRODUCTION

Background: Although first sighted by an English navigator in 1592, the first landing (English) did not occur until almost a century later in 1690, and the first settlement (French) was not established until 1764. The colony was turned over to Spain two years later and the islands have since been the subject of a territorial dispute, first between Britain and Spain, then between Britain and Argentina. The UK asserted its claim to the islands by establishing a naval garrison there in 1833. Argentina invaded the islands on 2 April 1982. The British responded with an expeditionary force that landed seven weeks later and after fierce fighting forced an Argentine surrender on 14 June 1982.

GEOGRAPHY

Location: Southern South America, islands in the South Atlantic Ocean, east of southern Argentina
Geographic coordinates: 51 45 S, 59 00 W
Map references: South America
Area: *total:* 12,173 sq km
country comparison to the world: 164
land: 12,173 sq km
water: 0 sq km
note: includes the two main islands of East and West Falkland and about 200 small islands
Area—comparative: slightly smaller than Connecticut
Land boundaries: 0 km
Coastline: 1,288 km
Maritime claims: *territorial sea:* 12 nm
continental shelf: 200 nm
exclusive fishing zone: 200 nm
Climate: cold marine; strong westerly winds, cloudy, humid; rain occurs on more than half of days in year; average annual rainfall is 24 inches in Stanley; occasional snow all year, except in January and February, but typically does not accumulate
Terrain: rocky, hilly, mountainous with some boggy, undulating plains
Elevation extremes: *lowest point:* Atlantic Ocean 0 m
highest point: Mount Usborne 705 m
Natural resources: fish, squid, wildlife, calcified seaweed, sphagnum moss
Land use: *arable land:* 0%
permanent crops: 0%
other: 100% (99% permanent pastures, 1% other) (2005)
Irrigated land: NA
Natural hazards: strong winds persist throughout the year
Environment—current issues: overfishing by unlicensed vessels is a problem; reindeer were introduced to the islands in 2001 for commercial reasons; this is the only commercial reindeer herd in the world unaffected by the 1986 Chornobyl disaster
Geography—note: deeply indented coast provides good natural harbors; short growing season

PEOPLE

Population: 3,140 (July 2008 est.)
country comparison to the world: 229
Age structure: *0–14 years:* NA
15–64 years: NA
65 years and over: NA
Population growth rate: 0.011% (2009 est.)
country comparison to the world: 191
Birth rate: NA
Death rate: NA
Net migration rate: NA
Urbanization: *urban population:* 74% of total population (2010)
rate of urbanization: 0.9% annual rate of change (2010-15 est.)
Major cities—population: STANLEY (capital) 2,000 (2009)
Infant mortality rate: *total:* NA
male: NA
female: NA
Life expectancy at birth: *total population:* NA
male: NA
female: NA
Total fertility rate: NA
HIV/AIDS—adult prevalence rate: NA
HIV/AIDS—people living with HIV/AIDS: NA
HIV/AIDS—deaths: NA
Nationality: *noun:* Falkland Islander(s)
adjective: Falkland Island
Ethnic groups: British
Religions: Christian 67.2%, none 31.5%, other 1.3% (2006 census)
Languages: English
Literacy: NA
School life expectancy (primary to tertiary education): NA
Education expenditures: NA

GOVERNMENT

Country name: *conventional long form:* none
conventional short form: Falkland Islands (Islas Malvinas)
Dependency status: overseas territory of the UK; also claimed by Argentina
Government type: NA
Capital: *name:* Stanley
geographic coordinates: 51 42 S, 57 51 W
time difference: UTC-4 (1 hour ahead of Washington, DC during Standard Time)
daylight saving time: +1hr, begins first Sunday in September; ends third Sunday in April; note–DST will be observed throughout 2011 on a trial basis
Administrative divisions: none (overseas territory of the UK; also claimed by Argentina)
Independence: none (overseas territory of the UK; also claimed by Argentina)
National holiday: Liberation Day, 14 June (1982)
Constitution: 1 January 2009
Legal system: English common law and local statutes
Suffrage: 18 years of age; universal
Executive branch: *chief of state:* Queen ELIZABETH II (since 6 February 1952)
head of government: Governor Nigel HAYWOOD (since 16 October 2010) is the Queen's representative; Chief Executive Dr. Tim THOROGOOD (since 3 January 2008)
cabinet: Executive Council; three members elected by the Legislative Council, two ex officio members (chief executive and the financial secretary), and the governor; the governor must obey the rulings of the Executive Council on domestic affairs
(For more information visit the World Leaders website)
elections: the monarchy is hereditary; governor appointed by the monarch; chief executive appointed by the governor
Legislative branch: unicameral Legislative Assembly (10 seats; 2 members are ex officio and 8 are elected by popular vote; members to serve four-year terms); presided over by the governor
elections: last held on 5 November 2009 (next to be held in November 2013)
election results: percent of vote–NA; seats–independents 8
Judicial branch: Supreme Court (chief justice is a nonresident); Magistrates Court (senior magistrate presides over civil and criminal divisions); Court of Summary Jurisdiction
Political parties and leaders: none; all independents
Political pressure groups and leaders: Falkland Islands Association (supports freedom of the people from external causes)
International organization participation: UPU
Diplomatic representation in the US: none (overseas territory of the UK; also claimed by Argentina)
Diplomatic representation from the US: none (overseas territory of the UK; also claimed by Argentina)
Flag description: blue with the flag of the UK in the upper hoist-side quadrant and the Falkland Island coat of arms centered on the outer half of the flag; the coat of arms contains a white ram (sheep raising was once the major economic activity) above the sailing ship Desire (whose crew discovered the islands) with a scroll at the bottom bearing the motto DESIRE THE RIGHT
National anthem: *name:* "Song of the Falklands""
lyrics/music: Christopher LANHAM
note: adopted 1930s; the song is the local unofficial anthem; as a territory of the

United Kingdom, "God Save the Queen" is official (see United Kingdom)

ECONOMY

Economy—overview: The economy was formerly based on agriculture, mainly sheep farming, but today fishing contributes the bulk of economic activity. In 1987, the government began selling fishing licenses to foreign trawlers operating within the Falkland Islands' exclusive fishing zone. These license fees total more than $40 million per year, which help support the island's health, education, and welfare system. Squid accounts for 75% of the fish taken. Dairy farming supports domestic consumption; crops furnish winter fodder. Foreign exchange earnings come from shipments of high-grade wool to the UK and the sale of postage stamps and coins. The islands are now self-financing except for defense. The British Geological Survey announced a 200-mile oil exploration zone around the islands in 1993, and early seismic surveys suggest substantial reserves capable of producing 500,000 barrels per day; to date, no exploitable site has been identified. An agreement between Argentina and the UK in 1995 seeks to defuse licensing and sovereignty conflicts that would dampen foreign interest in exploiting potential oil reserves. Political tensions between the UK and Argentina rose in early 2010 after a UK company began oil drilling activities in the waters around the Falkland Islands but abated somewhat when the drilling operation failed to discover commercially exploitable oil reserves. Tourism, especially eco-tourism, is increasing rapidly, with about 30,000 visitors in 2001. Another large source of income is interest paid on money the government has in the bank. The British military presence also provides a sizeable economic boost.
GDP (purchasing power parity): $105.1 million (2002 est.)
country comparison to the world: 219
GDP (official exchange rate): $105.1 million (2002 est.)
GDP—per capita (PPP): $35,400 (2002 est.)
country comparison to the world: 35
GDP—composition by sector: *agriculture:* 95%
industry: NA%
services: NA% (1996)
Labor force: 1,724 (1996)
country comparison to the world: 224
Labor force—by occupation: *agriculture:* 95% (mostly sheepherding and fishing)
industry and *services:* 5% (1996)
Unemployment rate: NA%
Population below poverty line: NA%
Household income or consumption by percentage share: *lowest 10%:* NA%
highest 10%: NA%
Budget: *revenues:* $66.2 million
expenditures: $67.9 million (FY98/99 est.)
Inflation rate (consumer prices): 3.6% (1998)
country comparison to the world: 103
Agriculture—products: fodder and vegetable crops; sheep, dairy products; fish, squid
Industries: fish and wool processing; tourism
Industrial production growth rate: NA%
Electricity—production: 16 million kWh (2007 est.)
country comparison to the world: 208
Electricity—consumption: 14.88 million kWh (2007 est.)
country comparison to the world: 208
Electricity—exports: 0 kWh (2008 est.)
Electricity—imports: 0 kWh (2008 est.)
Oil—production: 0 bbl/day (2009 est.)
country comparison to the world: 174
Oil—consumption: 0 bbl/day (2009 est.)
country comparison to the world: 204
Oil—exports: 0 bbl/day (2007 est.)
country comparison to the world: 165
Oil—imports: 270.9 bbl/day (2007 est.)
country comparison to the world: 199
Oil—proved reserves: 0 bbl (1 January 2010 est.)
country comparison to the world: 131
Natural gas—production: 0 cu m (2008 est.)
country comparison to the world: 126
Natural gas—consumption: 0 cu m (2008 est.)
country comparison to the world: 171
Natural gas—exports: 0 cu m (2008 est.)
country comparison to the world: 95
Natural gas—imports: 0 cu m (2008 est.)
country comparison to the world: 185
Natural gas—proved reserves: 0 cu m (1 January 2010 est.)
country comparison to the world: 135
Exports: $125 million (2004 est.)
country comparison to the world: 189
Exports—commodities: wool, hides, meat, fish, squid
Imports: $90 million (2004 est.)
country comparison to the world: 211
Imports—commodities: fuel, food and drink, building materials, clothing
Debt—external: $NA
Exchange rates: Falkland pounds (FKP) per US dollar–0.6388 (2010), 0.6175 (2009), 0.5418 (2006), 0.5493 (2005), 0.5462 (2004)

COMMUNICATIONS

Telephones—main lines in use: 2,000 (2009)
country comparison to the world: 223
Telephones—mobile cellular: 3,300 (2009)
country comparison to the world: 211
Telephone system: *general assessment:* NA
domestic: government-operated radiotelephone and private VHF/CB radiotelephone networks provide effective service to almost all points on both islands
international: country code–500; satellite earth station–1 Intelsat (Atlantic Ocean) with links through London to other countries
Broadcast media: television service provided by a multi-channel service provider; radio services provided by the public broadcaster Falkland Islands Radio Service (FIRS), broadcasting on both AM and FM frequencies, and by the British Forces Broadcasting Service (BFBS) (2007)
Internet country code: .fk
Internet hosts: 91 (2010)
country comparison to the world: 203
Internet users: 2,900 (2009)
country comparison to the world: 208

TRANSPORTATION

Airports: 7 (2010)
country comparison to the world: 169
Airports—with paved runways: *total:* 2
2,438 to 3,047 m: 1
914 to 1,523 m: 1 (2010)
Airports—with unpaved runways: *total:* 5
under 914 m: 5 (2010)
Roadways: *total:* 440 km
country comparison to the world: 197
paved: 50 km
unpaved: 390 km (2008)
Ports and terminals: Stanley

MILITARY

Military branches: no regular military forces
Military expenditures: NA
Military—note: defense is the responsibility of the UK

TRANSNATIONAL ISSUES

Disputes—international: Argentina, which claims the islands in its constitution and briefly occupied them by force in 1982, agreed in 1995 to no longer seek settlement by force; UK continues to reject Argentine requests for sovereignty talks

FAROE ISLANDS

(PART OF THE KINGDOM OF DENMARK)

INTRODUCTION

Background: The population of the Faroe Islands is largely descended from Viking settlers who arrived in the 9th century. The islands have been connected politically to Denmark since the 14th century. A high degree of self government was granted the Faroese in 1948, who have autonomy over most internal affairs while Denmark is responsible for justice, defense, and foreign affairs. The Faroe Islands are not part of the European Union.

GEOGRAPHY

Location: Northern Europe, island group between the Norwegian Sea and the North Atlantic Ocean, about half way between Iceland and Norway
Geographic coordinates: 62 00 N, 7 00 W
Map references: Europe
Area: *total:* 1,393 sq km
country comparison to the world: 182
land: 1,393 sq km
water: 0 sq km (some lakes and streams)

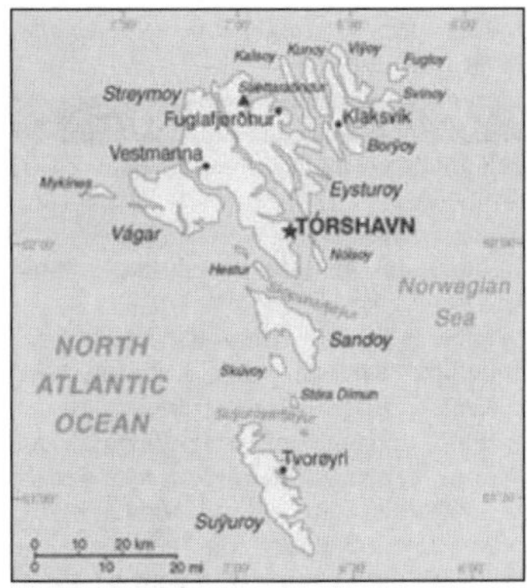

Area—comparative: eight times the size of Washington, DC
Land boundaries: 0 km
Coastline: 1,117 km
Maritime claims: *territorial sea:* 3 nm
continental shelf: 200 nm or agreed boundaries or median line
exclusive fishing zone: 200 nm or agreed boundaries or median line
Climate: mild winters, cool summers; usually overcast; foggy, windy
Terrain: rugged, rocky, some low peaks; cliffs along most of coast
Elevation extremes: *lowest point:* Atlantic Ocean 0 m
highest point: Slaettaratindur 882 m
Natural resources: fish, whales, hydropower, possible oil and gas
Land use: *arable land:* 2.14%
permanent crops: 0%
other: 97.86% (2005)
Irrigated land: 0 sq km
Natural hazards: NA
Environment—current issues: NA
Environment—international agreements: *party to:* Marine Dumping–associate member to the London Convention and Ship Pollution
Geography—note: archipelago of 17 inhabited islands and one uninhabited island, and a few uninhabited islets; strategically located along important sea lanes in northeastern Atlantic; precipitous terrain limits habitation to small coastal lowlands

PEOPLE

Population: 49,267 (July 2011 est.)
country comparison to the world: 207
Age structure: *0–14 years:* 21% (male 5,362/female 4,975)
15–64 years: 64.2% (male 16,837/female 14,788)
65 years and over: 14.8% (male 3,487/female 3,818) (2011 est.)
Median age: *total:* 37.3 years
male: 36.6 years
female: 38.1 years (2011 est.)
Population growth rate: 0.428% (2011 est.)
country comparison to the world: 156
Birth rate: 12.95 births/1,000 population (2011 est.)
country comparison to the world: 154
Death rate: 8.67 deaths/1,000 population (July 2011 est.)
country comparison to the world: 83
Net migration rate: 0 migrant(s)/1,000 population (2011 est.)
country comparison to the world: 85
Urbanization: *urban population:* 40% of total population (2010)
rate of urbanization: 0.9% annual rate of change (2010-15 est.)
Sex ratio: *at birth:* 1.071 male(s)/female
under 15 years: 1.07 male(s)/female
15–64 years: 1.15 male(s)/female
65 years and over: 0.9 male(s)/female
total population: 1.09 male(s)/female (2011 est.)
Infant mortality rate: *total:* 6.06 deaths/1,000 live births
country comparison to the world: 175
male: 6.31 deaths/1,000 live births
female: 5.79 deaths/1,000 live births (2011 est.)
Life expectancy at birth: *total population:* 79.72 years
country comparison to the world: 33
male: 77.25 years
female: 82.35 years (2011 est.)
Total fertility rate: 2.42 children born/woman (2011 est.)
country comparison to the world: 90
HIV/AIDS—adult prevalence rate: NA
HIV/AIDS—people living with HIV/AIDS: NA
HIV/AIDS—deaths: NA
Nationality: *noun:* Faroese (singular and plural)
adjective: Faroese
Ethnic groups: Scandinavian
Religions: Evangelical Lutheran 83.8%, other and unspecified 16.2% (2006 census)
Languages: Faroese (derived from Old Norse), Danish
Literacy: NA; note–probably 99%, the same as Denmark proper
School life expectancy (primary to tertiary education): NA
Education expenditures: NA

GOVERNMENT

Country name: *conventional long form:* none
conventional short form: Faroe Islands
local long form: none
local short form: Foroyar
Dependency status: part of the Kingdom of Denmark; self-governing overseas administrative division of Denmark since 1948
Government type: NA
Capital: *name:* Torshavn
geographic coordinates: 62 01 N, 6 46 W
time difference: UTC 0 (5 hours ahead of Washington, DC during Standard Time)
daylight saving time: +1hr, begins last Sunday in March; ends last Sunday in October
Administrative divisions: none (part of the Kingdom of Denmark; self-governing overseas administrative division of Denmark); there are no first-order administrative divisions as defined by the US Government, but there are 34 municipalities
Independence: none (part of the Kingdom of Denmark; self-governing overseas administrative division of Denmark)
National holiday: Olaifest (Olavsoka), 29 July
Constitution: 5 June 1953 (Danish Constitution)
Legal system: the laws of Denmark, where applicable, apply
Suffrage: 18 years of age; universal
Executive branch: *chief of state:* Queen MARGRETHE II of Denmark (since 14 January 1972), represented by High Commissioner Dan Michael KNUDSEN, chief administrative officer (since 2008)
head of government: Prime Minister Kaj Leo JOHANNESSEN (since 26 September 2008)
cabinet: Landsstyri appointed by the prime minister
(For more information visit the World Leaders website)
elections: the monarchy is hereditary; high commissioner appointed by the monarch; following legislative elections, the leader of the majority party or the leader of the majority coalition is usually elected prime minister by the Faroese Parliament; election last held on 19 January 2008 (next to be held no later than January 2012)
election results: Joannes EIDESGAARD elected prime minister in 2008; governing coalition collapses in September 2008, Kaj Leo JOHANNESSEN becomes prime minister
Legislative branch: unicameral Faroese Parliament or Logting (33 seats; members elected by popular vote on a proportional basis from the seven constituencies to serve four-year terms)
elections: last held on 19 January 2008 (next to be held no later than January 2012)
election results: percent of vote by party–Republican Party 23.3%, Union Party 21%, People's Party 20.1%, Social Democratic Party 19.3%, Center Party 8.4%, Independence Party 7.2%, other 0.7%; seats by party–Republican Party 8, Union Party 7, People's Party 7, Social Democratic Party 6, Center Party 3, Independence Party 2
note: election of two seats to the Danish Parliament was last held on 13 November 2007 (next to be held no later than November 2011); results–percent of vote by party–NA; seats by party–Republican Party 1, Union Party 1
Judicial branch: none
Political parties and leaders: Center Party [Jenis av RANA]; Independence Party [Kari P. HOJGAARD]; People's Party [Jorgen NICLASEN]; Republican Party [Hogni HOYDAL]; Social Democratic Party [Joannes EIDESGAARD]; Union Party [Kaj Leo JOHANNESEN]
Political pressure groups and leaders: conservationists
International organization participation: Arctic Council, IMO (associate), NC, NIB, UNESCO (associate), UPU
Diplomatic representation in the US: none (self-governing overseas administrative division of Denmark)
Diplomatic representation from the US: none (self-governing overseas administrative division of Denmark)
Flag description: white with a red cross outlined in blue extending to the edges of the flag; the vertical part of the cross is shifted toward the hoist side in the style of the Dannebrog (Danish flag); referred to as Merkid, meaning "the banner" or "the mark," the flag resembles those of neighboring Iceland and Norway, and uses the same three colors–but in a different

sequence; white represents the clear Faroese sky as well as the foam of the waves; red and blue are traditional Faroese colors

National anthem: *name:* "Mitt alfagra land" (My Fairest Land)
lyrics/music: Simun av SKAROI/Peter ALBERG
note: adopted 1948; the anthem is also known as "Tu alfagra land mitt" (Thou Fairest Land of Mine); as an autonomous overseas division of Denmark, the Faroe Islands are permitted their own national anthem

ECONOMY

Economy—overview: The Faroese economy is dependent on fishing, which makes the economy vulnerable to price swings. The sector accounts for about 95% of exports and nearly half of GDP. In early 2008 the Faroese economy began to slow as a result of smaller catches and historically high oil prices that continue to trouble the economy. Reduced catches, especially of cod and haddock, have continued to strain the Faroese economy. GDP grew 0.5% in 2008-09. The slowdown in the Faroese economy followed a strong performance since the mid-1990s with annual growth rates averaging close to 6%, mostly a result of increased fish landings and salmon farming, and high export prices. Unemployment reached its lowest level in the first half of 2008, but increased to 3.9% in 2009 and is rising. The Faroese Home Rule Government produced increasing budget surpluses that helped to reduce the large public debt, most of it to Denmark. However, total dependence on fishing and salmon farming make the Faroese economy very vulnerable to fluctuations in world demand. In addition, budget surpluses turned to deficits in 2008-09, and the economy at both the country and local level is running large deficits. Initial discoveries of oil in the Faroese area give hope for eventual oil production, which may provide a foundation for a more diversified economy and less dependence on Danish economic assistance. Aided by an annual subsidy from Denmark amounting to about 6% of Faroese GDP, the Faroese have a standard of living almost equal to that of Denmark and Greenland.

GDP (purchasing power parity): $1.59 billion (2008 est.)
country comparison to the world: 192

GDP (official exchange rate): $2.45 billion (2008 est.)

GDP—real growth rate: 0.5% (2008 est.)
country comparison to the world: 182

GDP—per capita (PPP): $32,900 (2008 est.)
country comparison to the world: 41

GDP—composition by sector:
agriculture: 16%
industry: 29%
services: 55% (2007 est.)

Labor force: 34,680 (November 2008)
country comparison to the world: 201

Labor force—by occupation:
agriculture: 10.2%
industry: 20.5%
services: 69.2% (2008)

Unemployment rate: 3.9% (2009)
country comparison to the world: 34
1.2% (2008)

Population below poverty line: NA%

Household income or consumption by percentage share: *lowest 10%:* NA%
highest 10%: NA%

Budget: *revenues:* $1.163 billion
expenditures: $1.139 billion
note: Denmark supplies the Faroe Islands with almost one-third of their public funds (2006)

Inflation rate (consumer prices): -1.1% (2009)
country comparison to the world: 5
6.4% (2008)

Agriculture—products: milk, potatoes, vegetables; sheep; salmon, other fish

Industries: fishing, fish processing, small ship repair and refurbishment, handicrafts

Industrial production growth rate: 8% (2007 est.)
country comparison to the world: 33

Electricity—production: 275.8 million kWh (2008 est.)
country comparison to the world: 171

Electricity—consumption: 264.4 million kWh (2008 est.)
country comparison to the world: 170

Electricity—exports: 0 kWh (2008)

Electricity—imports: 0 kWh (2008)

Oil—production: 0 bbl/day (2009 est.)
country comparison to the world: 175

Oil—consumption: 5,000 bbl/day (2009 est.)
country comparison to the world: 167

Oil—exports: 0 bbl/day (2008)
country comparison to the world: 166

Oil—imports: 4,922 bbl/day (2008)
country comparison to the world: 158

Oil—proved reserves: 0 bbl (1 January 2010 est.)
country comparison to the world: 132

Natural gas—production: 0 cu m (2008)
country comparison to the world: 127

Natural gas—consumption: 0 cu m (2008)
country comparison to the world: 172

Natural gas—exports: 0 cu m (2008)
country comparison to the world: 96

Natural gas—imports: 0 cu m (2008)
country comparison to the world: 186

Natural gas—proved reserves: 0 cu m (1 January 2010 est.)
country comparison to the world: 136

Exports: $848 million (2008)
country comparison to the world: 157
$634 million (2006)

Exports—commodities: fish and fish products 94%, stamps, ships

Exports—partners: Hungary 36.26%, Denmark 21.36%, UK 12.21%, Nigeria 7.72%, US 6.49%, Norway 5.46% (2009)

Imports: $983 million (2008)
country comparison to the world: 172
$751 million (2006)

Imports—commodities: consumer goods 36%, raw materials and semi-manufactures 32%, machinery and transport equipment 29%, fuels, fish, salt

Imports—partners: Denmark 54.42%, Norway 20.76%, Sweden 4.79% (2009)

Debt—external: $68.1 million (2006)
country comparison to the world: 185

Exchange rates: Danish kroner (DKK) per US dollar–5.624 (2009), 5.0236 (2008), 5.4797 (2007), 5.9468 (2006)

COMMUNICATIONS

Telephones—main lines in use: 20,900 (2009)
country comparison to the world: 193

Telephones—mobile cellular: 57,000 (2009)
country comparison to the world: 194

Telephone system: *general assessment:* good international communications; good domestic facilities
domestic: conversion to digital system completed in 1998; both NMT (analog) and GSM (digital) mobile telephone systems are installed
international: country code–298; satellite earth stations–1 Orion; 1 fiber-optic submarine cable to the Shetland Islands, linking the Faroe Islands with Denmark and Iceland; fiber-optic submarine cable connection to Canada-Europe cable

Broadcast media: 1 publicly-owned TV station; the Faroese telecommunications company distributes local and international channels through its digital terrestrial network; publicly-owned radio station supplemented by 2 privately-owned stations broadcasting over multiple frequencies (2008)

Internet country code: .fo

Internet hosts: 8,936 (2010)
country comparison to the world: 128

Internet users: 37,500 (2009)
country comparison to the world: 175

TRANSPORTATION

Airports: 1 (2010)
country comparison to the world: 215

Airports—with paved runways: *total:* 1
914 to 1,523 m: 1 (2010)

Roadways: *total:* 463 km (2006)
country comparison to the world: 194

Merchant marine: *total:* 26
country comparison to the world: 90

by type: cargo 11, chemical tanker 6, container 2, passenger 1, passenger/cargo 3, refrigerated cargo 2, roll on/roll off 1

foreign-owned: 11 (Norway 6, Sweden 5) (2010)

Ports and terminals: Fuglafjordur, Torshavn, Vagur

MILITARY

Military branches: no regular military forces

Manpower available for military service: *males age 16–49:* 11,831 (2010 est.)

Manpower fit for military service: *males age 16–49:* 9,827
females age 16–49: 8,418 (2010 est.)

Manpower reaching militarily significant age annually: *male:* 372
female: 373 (2010 est.)

Military expenditures: NA

Military—note: defense is the responsibility of Denmark

TRANSNATIONAL ISSUES

Disputes—international: because anticipated offshore hydrocarbon resources have not been realized, earlier Faroese proposals for full independence have been deferred; Iceland, the UK, and Ireland dispute Denmark's claim that the Faroe Islands' continental shelf extends beyond 200 nm

FIJI

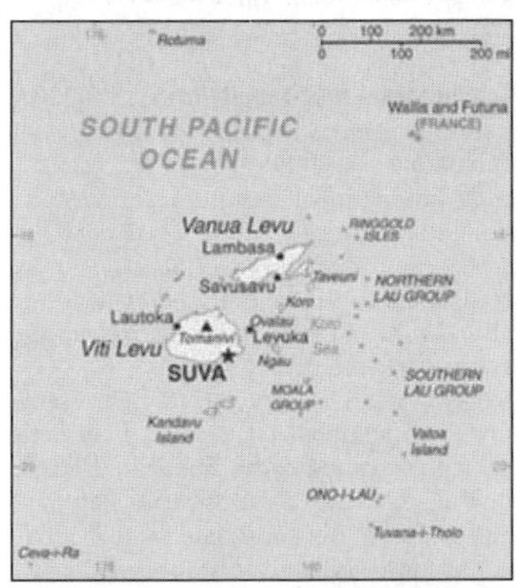

INTRODUCTION

Background: Fiji became independent in 1970 after nearly a century as a British colony. Democratic rule was interrupted by two military coups in 1987 caused by concern over a government perceived as dominated by the Indian community (descendants of contract laborers brought to the islands by the British in the 19th century). The coups and a 1990 constitution that cemented native Melanesian control of Fiji led to heavy Indian emigration; the population loss resulted in economic difficulties, but ensured that Melanesians became the majority. A new constitution enacted in 1997 was more equitable. Free and peaceful elections in 1999 resulted in a government led by an Indo-Fijian, but a civilian-led coup in May 2000 ushered in a prolonged period of political turmoil. Parliamentary elections held in August 2001 provided Fiji with a democratically elected government led by Prime Minister Laisenia QARASE. Re-elected in May 2006, QARASE was ousted in a December 2006 military coup led by Commodore Voreqe BAINIMARAMA, who initially appointed himself acting president but in January 2007 became interim prime minister. Since taking power BAINIMARAMA has neutralized his opponents, crippled Fiji's democratic institutions, and refused to hold elections.

GEOGRAPHY

Location: Oceania, island group in the South Pacific Ocean, about two-thirds of the way from Hawaii to New Zealand
Geographic coordinates: 18 00 S, 175 00 E
Map references: Oceania
Area: *total:* 18,274 sq km
country comparison to the world: 156
land: 18,274 sq km
water: 0 sq km
Area—comparative: slightly smaller than New Jersey
Land boundaries: 0 km
Coastline: 1,129 km
Maritime claims: measured from claimed archipelagic straight baselines
territorial sea: 12 nm
exclusive economic zone: 200 nm
continental shelf: 200 m depth or to the depth of exploitation; rectilinear shelf claim added
Climate: tropical marine; only slight seasonal temperature variation
Terrain: mostly mountains of volcanic origin
Elevation extremes: *lowest point:* Pacific Ocean 0 m
highest point: Tomanivi 1,324 m
Natural resources: timber, fish, gold, copper, offshore oil potential, hydropower
Land use: *arable land:* 10.95%
permanent crops: 4.65%
other: 84.4% (2005)
Irrigated land: 30 sq km (2003)
Total renewable water resources: 28.6 cu km (1987)
Freshwater withdrawal (domestic/industrial/agricultural): *total:* 0.07 cu km/yr (14%/14%/71%)
per capita: 82 cu m/yr (2000)
Natural hazards: cyclonic storms can occur from November to January
Environment—current issues: deforestation; soil erosion
Environment—international agreements: *party to:* Biodiversity, Climate Change, Climate Change-Kyoto Protocol, Desertification, Endangered Species, Law of the Sea, Marine Life Conservation, Ozone Layer Protection, Tropical Timber 83, Tropical Timber 94, Wetlands
signed, but not ratified: none of the selected agreements
Geography—note: includes 332 islands; approximately 110 are inhabited

PEOPLE

Population: 883,125 (July 2011 est.)
country comparison to the world: 159
Age structure: *0–14 years:* 28.9% (male 130,409/female 124,870)
15–64 years: 65.9% (male 297,071/female 284,643)
65 years and over: 5.2% (male 21,187/female 24,945) (2011 est.)
Median age: *total:* 26.9 years
male: 26.7 years
female: 27.1 years (2011 est.)
Population growth rate: 0.798% (2011 est.)
country comparison to the world: 135
Birth rate: 21.11 births/1,000 population (2011 est.)
country comparison to the world: 82
Death rate: 5.9 deaths/1,000 population (July 2011 est.)
country comparison to the world: 168
Net migration rate: -7.24 migrant(s)/1,000 population (2011 est.)
country comparison to the world: 204
Urbanization: *urban population:* 52% of total population (2010)
rate of urbanization: 1.3% annual rate of change (2010-15 est.)
Major cities—population: SUVA (capital) 174,000 (2009)
Sex ratio: *at birth:* 1.05 male(s)/female
under 15 years: 1.04 male(s)/female
15–64 years: 1 male(s)/female
65 years and over: 0.81 male(s)/female
total population: 1 male(s)/female (2011 est.)
Infant mortality rate: *total:* 11 deaths/1,000 live births
country comparison to the world: 144
male: 12.12 deaths/1,000 live births
female: 9.82 deaths/1,000 live births (2011 est.)
Life expectancy at birth: *total population:* 71.31 years
country comparison to the world: 138
male: 68.73 years
female: 74.03 years (2011 est.)
Total fertility rate: 2.61 children born/woman (2011 est.)
country comparison to the world: 80
HIV/AIDS—adult prevalence rate: 0.1% (2009 est.)
country comparison to the world: 129
HIV/AIDS—people living with HIV/AIDS: fewer than 1,000 (2009 est.)
country comparison to the world: 142
HIV/AIDS—deaths: fewer than 100 (2009 est.)
country comparison to the world: 132
Nationality: *noun:* Fijian(s)
adjective: Fijian
Ethnic groups: Fijian 57.3% (predominantly Melanesian with a Polynesian admixture), Indian 37.6%, Rotuman 1.2%, other 3.9% (European, other Pacific Islanders, Chinese) (2007 census)
Religions: Christian 64.5% (Methodist 34.6%, Roman Catholic 9.1%, Assembly of God 5.7%, Seventh Day Adventist 3.9%, Anglican 0.8%, other 10.4%), Hindu 27.9%, Muslim 6.3%, Sikh 0.3%, other or unspecified 0.3%, none 0.7% (2007 census)
Languages: English (official), Fijian (official), Hindustani
Literacy: *definition:* age 15 and over can read and write
total population: 93.7%
male: 95.5%
female: 91.9% (2003 est.)
School life expectancy (primary to tertiary education): *total:* 13 years
male: 13 years
female: 13 years (2005)
Education expenditures: 6.2% of GDP (2004)
country comparison to the world: 26

GOVERNMENT

Country name: *conventional long form:* Republic of Fiji
conventional short form: Fiji
local long form: Republic of Fiji/Matanitu ko Viti
local short form: Fiji/Viti
Government type: republic
Capital: *name:* Suva (on Viti Levu)
geographic coordinates: 18 08 S, 178 25 E
time difference: UTC+12 (17 hours ahead of Washington, DC during Standard Time)
daylight saving time: +1hr, begins fourth Sunday in October; ends first Sunday in March
Administrative divisions: 4 divisions and 1 dependency*; Central, Eastern, Northern, Rotuma*, Western
Independence: 10 October 1970 (from the UK)
National holiday: Independence Day, second Monday of October (1970)
Constitution: enacted 25 July 1997; effective 28 July 1998; note—constitution

encourages multiculturalism and makes multiparty government mandatory
Legal system: common law system based on the English model
International law organization participation: has not submitted an ICJ jurisdiction declaration; accepts ICCt jurisdiction
Suffrage: 21 years of age; universal
Executive branch: *chief of state:* President Ratu Epeli NAILATIKAU (since 30 July 2009)
head of government: Prime Minister Laisenia QARASE (since 10 September 2000); note—although QARASE is still the legal prime minister, he has been confined to his home island; former President ILOILOVATU appointed Commodore Voreqe BAINIMARAMA interim prime minister under the military regime
cabinet: Cabinet appointed by the prime minister from among the members of Parliament and responsible to Parliament; note—coup leader Commodore Voreqe BAINIMARAMA has appointed an interim cabinet
(For more information visit the World Leaders website)
elections: under the constitution, president elected by the Great Council of Chiefs for a five-year term (eligible for a second term); in 2007 the Great Council of Chiefs was suspended from its role in electing the president; prime minister appointed by the president
election results: Ratu Epeli NAILATIKAU was appointed by Chief Justice Anthony GATES
Legislative branch: bicameral Parliament consists of the Senate (32 seats; 14 members appointed by the president on the advice of the Great Council of Chiefs, 9 appointed by the president on the advice of the Prime Minister, 8 on the advice of the opposition leader, and 1 appointed on the advice of the council of Rotuma) and the House of Representatives (71 seats; 23 members reserved for ethnic Fijians, 19 reserved for ethnic Indians, 3 reserved for other ethnic groups, 1 reserved for the council of Rotuma constituency encompassing the whole of Fiji, and 25 open seats; members serve five-year terms)
elections: House of Representatives—last held on 6-13 May 2006 (next to be held in 2011)
election results: House of Representatives—percent of vote by party—SDL 44.6%, FLP 39.2%, UPP 0.8%, independents 4.9%, other 10.5%; seats by party—SDL 36, FLP 31, UPP 2, independents 2
Judicial branch: Supreme Court (judges are appointed by the president); Court of Appeal; High Court; Magistrates' Courts
Political parties and leaders: Dodonu Ni Taukei Party or DNT [Fereti S. DEWA]; Fiji Democratic Party or FDP [Filipe BOLE] (a merger of the Christian Democrat Alliance or VLV [Poesci Waqalevu BUNE], Fijian Association Party or FAP, Fijian Political Party or SVT [Sitiveni RABUKA] (primarily Fijian), and New Labor Unity Party or NLUP [Ofa SWANN]); Fiji Labor Party or FLP [Mahendra CHAUDHRY]; General Voters Party or GVP (became part of United General Party); Girmit Heritage Party or GHP; Justice and Freedom Party or AIM; Lio 'On Famor Rotuma Party or LFR; National Federation Party or NFP [Pramond RAE] (primarily Indian); Nationalist Vanua Takolavo Party or NVTLP [Saula TELAWA]; Party of National Unity or PANU [Ponipate LESAVUA]; Party of the Truth or POTT; United Fiji Party/Sogosogo Duavata ni Lewenivanua or SDL [Laisenia QARASE]; United Peoples Party or UPP [Millis Mick BEDDOES]
Political pressure groups and leaders: Group Against Racial Discrimination or GARD [Dr. Anirudk SINGH] (for restoration of a democratic government); Viti Landowners Association
International organization participation: ACP, ADB, AOSIS, C (suspended), CP, FAO, G-77, IBRD, ICAO, ICRM, IDA, IFAD, IFC, IFRCS, IHO, ILO, IMF, IMO, Interpol, IOC, ISO, ITSO, ITU, ITUC, MIGA, OPCW, PCA, PIF, Sparteca, SPC, UN, UNCTAD, UNESCO, UNIDO, UNMIS, UNMIT, UNWTO, UPU, WCO, WFTU, WHO, WIPO, WMO, WTO
Diplomatic representation in the US: *chief of mission:* Ambassador Winston THOMPSON
chancery: 2000 M Street, NW, Suite 710, Washington, DC 20036
telephone: [1] (202) 466-8320
FAX: [1] (202) 466-8325
Diplomatic representation from the US: *chief of mission:* Ambassador C. Steven MCGANN
embassy: 31 Loftus Street, Suva
mailing address: P. O. Box 218, Suva
telephone: [679] 331-4466
FAX: [679] 330-0081
Flag description: light blue with the flag of the UK in the upper hoist-side quadrant and the Fijian shield centered on the outer half of the flag; the blue symbolizes the Pacific ocean and the Union Jack reflects the links with Great Britain; the shield—taken from Fiji's coat of arms—depicts a yellow lion above a white field quartered by the cross of Saint George; the four quarters depict stalks of sugarcane, a palm tree, bananas, and a white dove
National anthem: *name:* "God Bless Fiji"
lyrics/music: Michael Francis Alexander PRESCOTT/C. Austin MILES (adapted by Michael Francis Alexander PRESCOTT)
note: adopted 1970; the anthem is known in Fijian as "Meda Dau Doka" (Let Us Show Pride); adapted from the hymn, "Dwelling in Beulah Land," the anthem's English lyrics are generally sung, although they differ in meaning from the official Fijian lyrics

ECONOMY

Economy—overview: Fiji, endowed with forest, mineral, and fish resources, is one of the most developed of the Pacific island economies though still with a large subsistence sector. Sugar exports, remittances from Fijians working abroad, and a growing tourist industry—with 400,000 to 500,000 tourists annually—are the major sources of foreign exchange. Fiji's sugar has special access to European Union markets but will be harmed by the EU's decision to cut sugar subsidies. Sugar processing makes up one-third of industrial activity but is not efficient. Fiji's tourism industry was damaged by the December 2006 coup and is facing an uncertain recovery time. In 2007 tourist arrivals were down almost 6%, with substantial job losses in the service sector, and GDP dipped. The coup has created a difficult business climate. The EU has suspended all aid until the interim government takes steps toward new elections. Long-term problems include low investment, uncertain land ownership rights, and the government's inability to manage its budget. Overseas remittances from Fijians working in Kuwait and Iraq have decreased significantly. Fiji's current account deficit reached 23% of GDP in 2006.
GDP (purchasing power parity): $3.869 billion (2010 est.)
country comparison to the world: 168
$3.864 billion (2009 est.)
$3.983 billion (2008 est.)
note: data are in 2010 US dollars
GDP (official exchange rate): $3.131 billion (2010 est.)
GDP—real growth rate: 0.1% (2010 est.)
country comparison to the world: 187
-3% (2009 est.)
-0.1% (2008 est.)
GDP—per capita (PPP): $4,400 (2010 est.)
country comparison to the world: 155
$4,400 (2009 est.)
$4,600 (2008 est.)
note: data are in 2010 US dollars
GDP—composition by sector:
agriculture: 8.9%
industry: 13.5%
services: 77.6% (2004 est.)
Labor force: 335,000 (2007 est.)
country comparison to the world: 161
Labor force—by occupation: *agriculture:* 70%
industry and *services:* 30% (2001 est.)
Unemployment rate: 7.6% (1999)
country comparison to the world: 83
Population below poverty line: 25.5% (FY90/91)
Household income or consumption by percentage share: *lowest 10%:* NA%
highest 10%: NA%
Budget: *revenues:* $1.363 billion
expenditures: $1.376 billion (2006)
Inflation rate (consumer prices): 4.8% (2007)
country comparison to the world: 135
Central bank discount rate: 3% (31 December 2009)
country comparison to the world: 66
6.32% (31 December 2008)
Commercial bank prime lending rate: 7.85% (31 December 2009 est.)
country comparison to the world: 117
7.97% (31 December 2008 est.)
Stock of narrow money: $748 million (31 December 2008)
country comparison to the world: 148
$1.042 billion (31 December 2007)
Stock of broad money: $1.76 billion (31 December 2008)
country comparison to the world: 147
Stock of domestic credit: $1.799 billion
country comparison to the world: 130
Market value of publicly traded shares: $NA (31 December 2009)

country comparison to the world: 107
$568.2 million (31 December 2008)
$522.2 million (31 December 2007)
Agriculture—products: sugarcane, coconuts, cassava (tapioca), rice, sweet potatoes, bananas; cattle, pigs, horses, goats; fish
Industries: tourism, sugar, clothing, copra, gold, silver, lumber, small cottage industries
Industrial production growth rate: NA%
Electricity—production: 928 million kWh (2007 est.)
country comparison to the world: 147
Electricity—consumption: 863 million kWh (2007 est.)
country comparison to the world: 147
Electricity—exports: 0 kWh (2008 est.)
Electricity—imports: 0 kWh (2008 est.)
Oil—production: 0 bbl/day (2009 est.)
country comparison to the world: 173
Oil—consumption: 11,000 bbl/day (2009 est.)
country comparison to the world: 144
Oil—exports: 2,455 bbl/day (2007 est.)
country comparison to the world: 109
Oil—imports: 20,340 bbl/day (2007 est.)
country comparison to the world: 111
Oil—proved reserves: 0 bbl (1 January 2010 est.)
country comparison to the world: 130
Natural gas—production: 0 cu m (2008 est.)
country comparison to the world: 125
Natural gas—consumption: 0 cu m (2008 est.)
country comparison to the world: 170
Natural gas—exports: 0 cu m (2008 est.)
country comparison to the world: 94
Natural gas—imports: 0 cu m (2008 est.)
country comparison to the world: 184
Natural gas—proved reserves: 0 cu m (1 January 2010 est.)
country comparison to the world: 134
Current account balance: $-507 million (2007 est.)
country comparison to the world: 117
Exports: $1.202 billion (2006)
country comparison to the world: 147
Exports—commodities: sugar, garments, gold, timber, fish, molasses, coconut oil
Exports—partners: US 15.21%, Australia 12.11%, UK 11.23%, Samoa 5.39%, Tonga 4.74%, Japan 4.44% (2009)
Imports: $3.12 billion (2006)
country comparison to the world: 138
Imports—commodities: manufactured goods, machinery and transport equipment, petroleum products, food, chemicals
Imports—partners: Singapore 27.27%, Australia 19.36%, NZ 15.15%, China 6.92%, India 5.23%, Thailand 4.25% (2009)
Debt—external: $127 million (2004 est.)
country comparison to the world: 179
Stock of direct foreign investment—at home: $NA
Stock of direct foreign investment—abroad: $NA
Exchange rates: Fijian dollars (FJD) per US dollar–NA (2007), 1.7313 (2006), 1.691 (2005), 1.7331 (2004), 1.8958 (2003)

COMMUNICATIONS

Telephones—main lines in use: 136,800 (2009)
country comparison to the world: 136
Telephones—mobile cellular: 640,000 (2009)
country comparison to the world: 156
Telephone system: *general assessment:* modern local, interisland, and international (wire/radio integrated) public and special-purpose telephone, telegraph, and teleprinter facilities; regional radio communications center
domestic: telephone or radio telephone links to almost all inhabited islands; most towns and large villages have automatic telephone exchanges and direct dialing; combined fixed and mobile-cellular teledensity is about 80 per 100 persons
international: country code–679; access to important cable links between US and Canada as well as between NZ and Australia; satellite earth stations–2 Inmarsat (Pacific Ocean) (2009)
Broadcast media: Fiji TV, a publicly-traded company, operates a free-to-air channel as well as the Sky Fiji and Sky Pacific multi-channel pay-TV services; state-owned commercial company, Fiji Broadcasting Corporation, Ltd, operates 6 radio stations–2 public broadcasters and 4 commercial broadcasters with multiple repeaters; 5 radio stations with repeaters operated by Communications Fiji, Ltd; transmissions of multiple international broadcasters are available (2009)
Internet country code: .fj
Internet hosts: 17,088 (2010)
country comparison to the world: 113
Internet users: 114,200 (2009)
country comparison to the world: 157

TRANSPORTATION

Airports: 28 (2010)
country comparison to the world: 118
Airports—with paved runways: *total:* 4
over 3,047 m: 1
1,524 to 2,437 m: 1
914 to 1,523 m: 2 (2010)
Airports—with unpaved runways: *total:* 24
914 to 1,523 m: 5
under 914 m: 19 (2010)
Railways: *total:* 597 km
country comparison to the world: 108
narrow gauge: 597 km 0.600-m gauge
note: belongs to the government-owned Fiji Sugar Corporation; used to haul sugarcane during the harvest season, which runs from May to December (2008)
Roadways: *total:* 3,440 km
country comparison to the world: 162
paved: 1,692 km
unpaved: 1,748 km (2000)
Waterways: 203 km (122 km are navigable by motorized craft and 200-metric-ton barges) (2010)
country comparison to the world: 98
Merchant marine: *total:* 10
country comparison to the world: 112
by type: passenger 4, passenger/cargo 4, roll on/roll off 2
foreign-owned: 2 (Australia 2) (2010)
Ports and terminals: Lautoka, Levuka, Suva

MILITARY

Military branches: Republic of Fiji Military Forces (RFMF): Land Forces, Naval Forces (2011)
Military service age and obligation: 18 years of age for voluntary military service (2010)
Manpower available for military service: *males age 16–49:* 233,240
females age 16–49: 222,587 (2010 est.)
Manpower fit for military service: *males age 16–49:* 183,730
females age 16–49: 188,325 (2010 est.)
Manpower reaching militarily significant age annually: *male:* 8,403
female: 8,039 (2010 est.)
Military expenditures: 1.9% of GDP (2009)
country comparison to the world: 75

TRANSNATIONAL ISSUES

Disputes—international: none
Trafficking in persons: current situation: Fiji is a source country for children trafficked for the purpose of commercial sexual exploitation and a destination country for a small number of women from China and India trafficked for the purposes of forced labor and commercial sexual exploitation
tier rating: Tier 2 Watch List–Fiji does not fully comply with the minimum standards for the elimination of trafficking and is not making significant efforts to do so; the government has demonstrated no action to investigate or prosecute traffickers, assist victims, take steps to reduce the demand for commercial sex acts, or support any anti-trafficking information or education campaigns; Fiji has not ratified the 2000 UN TIP Protocol (2009)

FINLAND

INTRODUCTION

Background: Finland was a province and then a grand duchy under Sweden from the 12th to the 19th centuries, and an autonomous grand duchy of Russia after 1809. It won its complete independence in 1917. During World War II, it was able to successfully defend its freedom and resist invasions by the Soviet Union–albeit with some loss of territory. In the subsequent half century, the Finns made a remarkable transformation from a farm/forest economy to a diversified modern industrial economy; per capita income is now among the highest in Western Europe. A member of the European Union since 1995, Finland was the only Nordic state to join the euro system at its initiation in January 1999. In the 21st century, the key features of Finland's modern welfare state are a high standard of education, equality promotion, and national social security system–currently challenged by an aging population and the fluctuations of an export-driven economy.

GEOGRAPHY

Location: Northern Europe, bordering the Baltic Sea, Gulf of Bothnia, and Gulf of Finland, between Sweden and Russia
Geographic coordinates: 64 00 N, 26 00 E
Map references: Europe
Area: *total:* 338,145 sq km
country comparison to the world: 64
land: 303,815 sq km
water: 34,330 sq km
Area—comparative: slightly smaller than Montana
Land boundaries: *total:* 2,654 km
border countries: Norway 727 km, Sweden 614 km, Russia 1,313 km
Coastline: 1,250 km
Maritime claims: *territorial sea:* 12 nm (in the Gulf of Finland–3 nm)
contiguous zone: 24 nm
exclusive fishing zone: 12 nm; extends to continental shelf boundary with Sweden
continental shelf: 200 m depth or to the depth of exploitation
Climate: cold temperate; potentially subarctic but comparatively mild because of moderating influence of the North Atlantic Current, Baltic Sea, and more than 60,000 lakes
Terrain: mostly low, flat to rolling plains interspersed with lakes and low hills
Elevation extremes:
lowest point: Baltic Sea 0 m
highest point: Haltiatunturi 1,328 m
Natural resources: timber, iron ore, copper, lead, zinc, chromite, nickel, gold, silver, limestone
Land use: *arable land:* 6.54%
permanent crops: 0.02%
other: 93.44% (2005)
Irrigated land: 640 sq km (2003)
Total renewable water resources: 110 cu km (2005)
Freshwater withdrawal (domestic/industrial/agricultural): *total:* 2.33 cu km/yr (14%/84%/3%)
per capita: 444 cu m/yr (1999)
Natural hazards: NA
Environment—current issues: air pollution from manufacturing and power plants contributing to acid rain; water pollution from industrial wastes, agricultural chemicals; habitat loss threatens wildlife populations
Environment—international agreements: *party to:* Air Pollution, Air Pollution-Nitrogen Oxides, Air Pollution-Persistent Organic Pollutants, Air Pollution-Sulfur 85, Air Pollution-Sulfur 94, Air Pollution-Volatile Organic Compounds, Antarctic-Environmental Protocol, Antarctic-Marine Living Resources, Antarctic Treaty, Biodiversity, Climate Change, Climate Change-Kyoto Protocol, Desertification, Endangered Species, Environmental Modification, Hazardous Wastes, Law of the Sea, Marine Dumping, Marine Life Conservation, Ozone Layer Protection, Ship Pollution, Tropical Timber 83, Tropical Timber 94, Wetlands, Whaling
signed, but not ratified: none of the selected agreements
Geography—note: long boundary with Russia; Helsinki is northernmost national capital on European continent; population concentrated on small southwestern coastal plain

PEOPLE

Population: 5,259,250 (July 2011 est.)
country comparison to the world: 113
Age structure: *0–14 years:* 16% (male 429,450/female 414,570)
15–64 years: 66.1% (male 1,759,059/female 1,719,173)
65 years and over: 17.8% (male 385,671/female 551,327) (2011 est.)
Median age: *total:* 42.5 years
male: 40.8 years
female: 44.3 years (2011 est.)
Population growth rate: 0.075% (2011 est.)
country comparison to the world: 186
Birth rate: 10.37 births/1,000 population (2011 est.)
country comparison to the world: 184
Death rate: 10.24 deaths/1,000 population (July 2011 est.)
country comparison to the world: 49
Net migration rate: 0.62 migrant(s)/1,000 population (2011 est.)
country comparison to the world: 60
Urbanization: *urban population:* 85% of total population (2010)
rate of urbanization: 0.6% annual rate of change (2010-15 est.)
Major cities—population: HELSINKI (capital) 1.107 million (2009)
Sex ratio: *at birth:* 1.04 male(s)/female
under 15 years: 1.04 male(s)/female
15–64 years: 1.02 male(s)/female
65 years and over: 0.69 male(s)/female
total population: 0.96 male(s)/female (2011 est.)
Infant mortality rate:
total: 3.43 deaths/1,000 live births
country comparison to the world: 211
male: 3.73 deaths/1,000 live births
female: 3.11 deaths/1,000 live births (2011 est.)
Life expectancy at birth: *total population:* 79.27 years
country comparison to the world: 39
male: 75.79 years
female: 82.89 years (2011 est.)
Total fertility rate: 1.73 children born/woman (2011 est.)
country comparison to the world: 165
HIV/AIDS—adult prevalence rate: 0.1% (2009 est.)
country comparison to the world: 128
HIV/AIDS—people living with HIV/AIDS: 2,600 (2009 est.)
country comparison to the world: 132
HIV/AIDS—deaths: fewer than 100 (2009 est.)
country comparison to the world: 131
Drinking water source: *Improved:*
urban: 100% of population
rural: 100% of population
total: 100% of population (2008)
Sanitation facility access:
Improved:
urban: 100% of population
rural: 100% of population
total: 100% of population (2008)
Nationality: *noun:* Finn(s)
adjective: Finnish
Ethnic groups: Finn 93.4%, Swede 5.6%, Russian 0.5%, Estonian 0.3%, Roma (Gypsy) 0.1%, Sami 0.1% (2006)
Religions: Lutheran Church of Finland 82.5%, Orthodox Church 1.1%, other Christian 1.1%, other 0.1%, none 15.1% (2006)
Languages: Finnish (official) 91.2%, Swedish (official) 5.5%, other (small Sami- and Russian-speaking minorities) 3.3% (2007)
Literacy: *definition:* age 15 and over can read and write
total population: 100%
male: 100%
female: 100% (2000 est.)
School life expectancy (primary to tertiary education): *total:* 17 years
male: 16 years
female: 18 years (2008)
Education expenditures: 5.9% of GDP (2007)
country comparison to the world: 33

GOVERNMENT

Country name: *conventional long form:* Republic of Finland
conventional short form: Finland
local long form: Suomen tasavalta/Republiken Finland
local short form: Suomi/Finland
Government type: republic
Capital: *name:* Helsinki
geographic coordinates: 60 10 N, 24 56 E
time difference: UTC+2 (7 hours ahead of Washington, DC during Standard Time)
daylight saving time: +1hr, begins last Sunday in March; ends last Sunday in October
Administrative divisions: 19 regions (maakunnat, singular–maakunta (Finnish); landskapen, singular–landskapet (Swedish)); Aland (Swedish), Ahvenanmaa (Finnish); Etela-Karjala (Finnish), Sodra Karelen (Swedish) [South Karelia]; Etela-Pohjanmaa (Finnish), Sodra Osterbotten (Swedish) [South Ostrobothnia]; Etela-Savo (Finnish), Sodra Savolax (Swedish) [South Savo]; Kanta-Hame (Finnish), Egentliga Tavastland (Swedish); Kainuu (Finnish), Kajanaland (Swedish); Keski-Pohjanmaa (Finnish), Mellersta Osterbotten (Swedish) [Central Ostrobothnia]; Keski-Suomi (Finnish), Mellersta Finland (Swedish) [Central Finland]; Kymenlaakso (Finnish), Kymmenedalen (Swedish); Lappi (Finnish), Lappland (Swedish); Paijat-Hame (Finnish), Paijanne-Tavastland (Swedish); Pirkanmaa (Finnish), Birkaland (Swedish) [Tampere]; Osterbotten (Swedish), Pohjanmaa (Finnish) [Ostrobothnia]; Pohjois-Karjala (Finnish), Norra Karelen (Swedish) [North Karelia]; Pohjois-Pohjanmaa (Finnish), Norra Osterbotten (Swedish) [North Ostrobothnia]; Pohjois-Savo (Finnish), Norra Savolax (Swedish) [North Savo]; Satakunta (Finnish and Swedish); Uusimaa (Finnish), Nyland (Swedish) [Newland]; Varsinais-Suomi (Finnish), Egentliga Finland (Swedish) [Southwest Finland]
Independence: 6 December 1917 (from Russia)
National holiday: Independence Day, 6 December (1917)
Constitution: 1 March 2000
Legal system: civil law system based on the Swedish model; note–the president

may request the Supreme Court to review laws

International law organization participation: accepts compulsory ICJ jurisdiction with reservations; accepts ICCt jurisdiction

Suffrage: 18 years of age; universal

Executive branch: *chief of state:* President Tarja HALONEN (since 1 March 2000)
head of government: Prime Minister Mari KIVINIEMI (since 22 June 2010); Deputy Prime Minister Jyrki KATAINEN (since 19 April 2007)
cabinet: Council of State or Valtioneuvosto appointed by the president, responsible to parliament
(For more information visit the World Leaders website)
elections: president elected by popular vote for a six-year term (eligible for a second term); election last held on 15 January 2006 (next to be held in January 2012); the parliament elects a prime minister who is then appointed to office by the president; Prime Minister KIVINIEMI elected on 22 June 2010
election results: percent of vote–Tarja HALONEN (SDP) 46.3%, Sauli NIINISTO (Kok) 24.1%, Matti VANHANEN (Kesk) 18.6%, Heidi HAUTALA (VIHR) 3.5% other 10.5%; a runoff election between HALONEN and NIINISTO was held 29 January 2006–HALONEN 51.8%, NIINISTO 48.2%; Mari KIVINIEMI elected prime minister; election results 115-56
note: government coalition–Kesk, KOK, VIHR, and SFP

Legislative branch: unicameral Parliament or Eduskunta (200 seats; members elected by popular vote on a proportional basis to serve four-year terms)
elections: last held on 17 April 2011 (next to be held in April 2015)
election results: percent of vote by party–Kok 20.4%, SDP 19.1%, True Finns 19%, Kesk 15.8%, VAS 8.1%, VIHR 7.2%, SFP 4.3%, KD 4%, other 2%; seats by party–Kok 44, SDP 42, True Finns 39, Kesk 35, VAS 14, VIHR 10, SFP 9, KD 6, other 1 (the constituency of Aland)

Judicial branch: general courts–deal with criminal and civil cases (include district courts, Courts of Appeal, and the Supreme Court or Korkein Oikeus, whose judges are appointed by the president); administrative courts

Political parties and leaders: Center Party or Kesk [Mari KIVINIEMI]; Christian Democrats or KD [Paivi RASANEN]; Green Party or VIHR [Anni SINNEMAKI]; Left Alliance or VAS [Paavo ARHINMAKI]; National Coalition Party or Kok [Jyrki KATAINEN]; Social Democratic Party or SDP [Jutta URPILAINEN]; Swedish People's Party or SFP [Stefan WALLIN]; True Finns or PS [Timo SOINI]

International organization participation: ADB (nonregional member), AfDB (nonregional member), Arctic Council, Australia Group, BIS, CBSS, CE, CERN, EAPC, EBRD, EIB, EMU, ESA, EU, FAO, FATF, G-9, IADB, IAEA, IBRD, ICAO, ICC, ICRM, IDA, IEA, IFAD, IFC, IFRCS, IHO, ILO, IMF, IMO, IMSO, Interpol, IOC, IOM, IPU, ISO, ITSO, ITU, ITUC, MIGA, NC, NEA, NIB, NSG, OAS (observer), OECD, OPCW, OSCE, Paris Club, PCA, PFP, Schengen Convention, UN, UNCTAD, UNESCO, UNHCR, UNIDO, UNMIL, UNMIS, UNMOGIP, UNRWA, UNTSO, UPU, WCO, WFTU, WHO, WIPO, WMO, WTO, ZC

Diplomatic representation in the US: *chief of mission:* Ambassador Pekka LINTU
chancery: 3301 Massachusetts Avenue NW, Washington, DC 20008
telephone: [1] (202) 298-5800
FAX: [1] (202) 298-6030
consulate(s) general: Los Angeles, New York

Diplomatic representation from the US: *chief of mission:* Ambassador Bruce J. ORECK
embassy: Itainen Puistotie 14B, 00140 Helsinki
mailing address: APO AE 09723
telephone: [358] (9) 616250
FAX: [358] (9) 6162 5800

Flag description: white with a blue cross extending to the edges of the flag; the vertical part of the cross is shifted to the hoist side in the style of the Dannebrog (Danish flag); the blue represents the thousands of lakes scattered across the country, while the white is for the snow that covers the land in winter

National anthem: *name:* "Maamme" (Our Land)
lyrics/music: Johan Ludvig RUNEBERG/ Fredrik PACIUS
note: in use since 1848; although never officially adopted by law, the anthem has been popular since it was first sung by a student group in 1848; Estonia's anthem uses the same melody as that of Finland

ECONOMY

Economy—overview: Finland has a highly industrialized, largely free-market economy with per capita output roughly that of Austria, Belgium, the Netherlands, and Sweden. Trade is important with exports accounting for over one third of GDP in recent years. Finland is strongly competitive in manufacturing–principally the wood, metals, engineering, telecommunications, and electronics industries. Finland excels in high-tech exports such as mobile phones. Except for timber and several minerals, Finland depends on imports of raw materials, energy, and some components for manufactured goods. Because of the climate, agricultural development is limited to maintaining self-sufficiency in basic products. Forestry, an important export earner, provides a secondary occupation for the rural population. Finland had been one of the best performing economies within the EU in recent years and its banks and financial markets avoided the worst of global financial crisis. However, the world slowdown hit exports and domestic demand hard in 2009, with Finland experiencing one of the deepest contractions in the euro zone. A recovery of exports, domestic trade, and household consumption stimulated economic growth in 2010. The recession left a deep mark on general government finances and the debt ratio, turning previously strong budget surpluses into deficits. Despite good growth prospects, general government finances will remain in deficit during the next few years. The great challenge of economic policy will be to implement a post-recession exit strategy in which measures supporting growth will be combined with general government adjustment measures. Longer-term, Finland must address a rapidly aging population and decreasing productivity that threaten competitiveness, fiscal sustainability, and economic growth.

GDP (purchasing power parity): $186 billion (2010 est.)
country comparison to the world: 56
$180.3 billion (2009 est.)
$196.5 billion (2008 est.)
note: data are in 2010 US dollars

GDP (official exchange rate): $239.2 billion (2010 est.)

GDP—real growth rate: 3.1% (2010 est.)
country comparison to the world: 122
-8.2% (2009 est.)
0.9% (2008 est.)

GDP—per capita (PPP): $35,400 (2010 est.)
country comparison to the world: 34
$34,400 (2009 est.)
$37,500 (2008 est.)
note: data are in 2010 US dollars

GDP—composition by sector:
agriculture: 2.6%
industry: 29.1%
services: 68.2% (2010 est.)

Labor force: 2.68 million (2010 est.)
country comparison to the world: 107

Labor force—by occupation: agriculture and forestry: 4.9%
industry: 16.7%
construction: 7.1%
commerce: 19.4%
finance, insurance, and business *services:* 12.8%
transport and communications: 6.3%
public services: 32.8% (2009)

Unemployment rate: 8.4% (2010 est.)
country comparison to the world: 97
8.2% (2009 est.)

Population below poverty line: NA%

Household income or consumption by percentage share: *lowest 10%:* 3.6%
highest 10%: 24.7% (2007)

Distribution of family income—Gini index: 26.8 (2008)
country comparison to the world: 125
25.6 (1991)

Investment (gross fixed): 18.7% of GDP (2010 est.)
country comparison to the world: 102

Budget: *revenues:* $66.58 billion
expenditures: $65.33 billion
note: Central Government Budget (2010 est.)

Public debt: 45.4% of GDP (2010 est.)
country comparison to the world: 59
44% of GDP (2009 est.)

Inflation rate (consumer prices): 1.2% (2010 est.)
country comparison to the world: 26
0% (2009 est.)

Central bank discount rate: 1.75% (31 December 2010)
country comparison to the world: 123
1.75% (31 December 2009)
note: this is the European Central Bank's rate on the marginal lending facility, which

offers overnight credit to banks in the euro area
Commercial bank prime lending rate: 3.51% (31 December 2009 est.)
country comparison to the world: 135
5.79% (31 December 2008 est.)
Stock of narrow money: $108 billion (31 December 2010 est.)
country comparison to the world: 30
$110.4 billion (31 December 2009 est.)
note: see entry for the European Union for money supply in the euro area; the European Central Bank (ECB) controls monetary policy for the 17 members of the Economic and Monetary Union (EMU); individual members of the EMU do not control the quantity of money circulating within their own borders
Stock of broad money: $160.4 billion (31 December 2010 est.)
country comparison to the world: 44
$168.5 billion (31 December 2009 est.)
Stock of domestic credit: $259.2 billion (31 December 2009 est.)
country comparison to the world: 36
$241.6 billion (31 December 2008 est.)
Market value of publicly traded shares: $91.02 billion (31 December 2009)
country comparison to the world: 28
$154.4 billion (31 December 2008)
$369.2 billion (31 December 2007)
Agriculture—products: barley, wheat, sugar beets, potatoes; dairy cattle; fish
Industries: metals and metal products, electronics, machinery and scientific instruments, shipbuilding, pulp and paper, foodstuffs, chemicals, textiles, clothing
Industrial production growth rate: 6% (2010 est.)
country comparison to the world: 60
Electricity—production: 77.44 billion kWh (2008 est.)
country comparison to the world: 38
Electricity—consumption: 87.25 billion kWh (2008)
country comparison to the world: 32
Electricity—exports: 3.335 billion kWh (2008)
Electricity—imports: 16.11 billion kWh (2008)
Oil—production: 8,718 bbl/day (2009 est.)
country comparison to the world: 87
Oil—consumption: 206,200 bbl/day (2009 est.)
country comparison to the world: 55
Oil—exports: 130,500 bbl/day (2009 est.)
country comparison to the world: 62
Oil—imports: 337,900 bbl/day (2009 est.)
country comparison to the world: 31
Oil—proved reserves: 0 bbl (1 January 2010 est.)
country comparison to the world: 129
Natural gas—production: NA
Natural gas—consumption: 4.289 billion cu m (2009)
country comparison to the world: 63
Natural gas—exports: 0 cu m (2008 est.)
country comparison to the world: 93
Natural gas—imports: 4.289 billion cu m (2009)
country comparison to the world: 34
Natural gas—proved reserves: 0 cu m (1 January 2010 est.)
country comparison to the world: 133
Current account balance: $4.696 billion (2010 est.)
country comparison to the world: 33
$3.343 billion (2009)
Exports: $73.53 billion (2010 est.)
country comparison to the world: 40
$62.54 billion (2009)
Exports—commodities: electrical and optical equipment, machinery, transport equipment, paper and pulp, chemicals, basic metals; timber
Exports—partners: Germany 10.32%, Sweden 9.79%, Russia 9%, US 7.85%, Netherlands 5.9%, UK 5.24%, China 4.1% (2009)
Imports: $69.11 billion (2010 est.)
country comparison to the world: 39
$57.68 billion (2009)
Imports—commodities: foodstuffs, petroleum and petroleum products, chemicals, transport equipment, iron and steel, machinery, textile yarn and fabrics, grains
Imports—partners: Russia 16.28%, Germany 15.76%, Sweden 14.65%, Netherlands 6.99%, China 5.29%, France 4.22% (2009)
Reserves of foreign exchange and gold: $9.128 billion (31 December 2010 est.)
country comparison to the world: 61
$11.45 billion (31 December 2009 est.)
Debt—external: $370.8 billion (30 June 2010)
country comparison to the world: 24
$339.5 billion (31 December 2008)
Stock of direct foreign investment—at home: $87.99 billion (31 December 2010 est.)
country comparison to the world: 36
$81.77 billion (31 December 2009)
Stock of direct foreign investment—abroad: $122.2 billion (31 December 2010 est.)
country comparison to the world: 24
$122.6 billion (31 December 2009)
Exchange rates: euros (EUR) per US dollar–0.755 (2010), 0.7198 (2009), 0.6827 (2008), 0.7345 (2007), 0.7964 (2006)

COMMUNICATIONS

Telephones—main lines in use: 1.43 million (2009)
country comparison to the world: 66
Telephones—mobile cellular: 7.7 million (2009)
country comparison to the world: 76
Telephone system: *general assessment:* modern system with excellent service
domestic: digital fiber-optic fixed-line network and an extensive mobile-cellular network provide domestic needs
international: country code–358; submarine cables provide links to Estonia and Sweden; satellite earth stations–access to Intelsat transmission service via a Swedish satellite earth station, 1 Inmarsat (Atlantic and Indian Ocean regions); note–Finland shares the Inmarsat earth station with the other Nordic countries (Denmark, Iceland, Norway, and Sweden)
Broadcast media: a mix of publicly-operated TV stations and privately-owned TV stations; the 2 publicly-owned TV stations recently expanded services and the largest private TV station has introduced several special-interest pay-TV channels; cable and satellite multi-channel subscription services are available; all TV signals have been broadcast digitally since September 2007; analog broadcasts via cable networks were terminated in February 2008; public broadcasting maintains a network of 13 national and 25 regional radio stations; a large number of private radio broadcasters (2008)
Internet country code: .fi; note–Aland Islands assigned .ax
Internet hosts: 4.394 million (2010)
country comparison to the world: 21
Internet users: 4.393 million (2009)
country comparison to the world: 55

TRANSPORTATION

Airports: 148 (2010)
country comparison to the world: 38
Airports—with paved runways: *total:* 75
over 3,047 m: 3
2,438 to 3,047 m: 26
1,524 to 2,437 m: 10
914 to 1,523 m: 22
under 914 m: 14 (2010)
Airports—with unpaved runways: *total:* 73
914 to 1,523 m: 3
under 914 m: 70 (2010)
Pipelines: gas 694 km (2010)
Railways: *total:* 5,919 km
country comparison to the world: 31
broad gauge: 5,919 km 1.524-m gauge (3,067 km electrified) (2009)
Roadways:
total: 78,141 km
country comparison to the world: 62
paved: 50,914 km (includes 739 km of expressways)
unpaved: 27,227 km (2009)
Waterways: 7,842 km (includes Saimaa Canal system of 3,577 km; southern part leased from Russia; water transport is used frequently in the summer and is widely replaced with sledges on the ice in winter; there are 187,888 lakes in Finland that cover 31,500 km) (2010)
country comparison to the world: 18
Merchant marine: *total:* 93
country comparison to the world: 53
by type: bulk carrier 1, cargo 26, carrier 1, chemical tanker 6, container 3, passenger 4, passenger/cargo 16, petroleum tanker 5, roll on/roll off 28, vehicle carrier 3
foreign-owned: 6 (Cyprus 1, Estonia 2, Iceland 1, Norway 2)
registered in other countries: 52 (Bahamas 8, Germany 5, Gibraltar 2, Liberia 2, Malta 2, Netherlands 14, Norway 1, Panama 2, Sweden 16) (2010)
Ports and terminals: Helsinki, Kotka, Naantali, Porvoo, Raahe, Rauma

MILITARY

Military branches: Finnish Defense Forces (FDF): Army, Navy (includes Coastal Defense Forces), Air Force (Suomen Ilmavoimat) (2007)
Military service age and obligation: 18 years of age for male voluntary and compulsory–and female voluntary–national military and nonmilitary service; service obligation 6-12 months; mandatory retirement at age 60 (2010)
Manpower available for military service:
males age 16–49: 1,155,368
females age 16–49: 1,106,193 (2010 est.)

Manpower fit for military service: *males age 16–49:* 955,151
females age 16–49: 912,983 (2010 est.)
Manpower reaching militarily significant age annually: *male:* 32,599
female: 31,416 (2010 est.)
Military expenditures: 2% of GDP (2005 est.)
country comparison to the world: 73

TRANSNATIONAL ISSUES

Disputes—international: various groups in Finland advocate restoration of Karelia and other areas ceded to the Soviet Union, but the Finnish Government asserts no territorial demands

FRANCE

INTRODUCTION

Background: Although ultimately a victor in World Wars I and II, France suffered extensive losses in its empire, wealth, manpower, and rank as a dominant nation-state. Nevertheless, France today is one of the most modern countries in the world and is a leader among European nations. Since 1958, it has constructed a hybrid presidential-parliamentary governing system resistant to the instabilities experienced in earlier more purely parliamentary administrations. In recent decades, its reconciliation and cooperation with Germany have proved central to the economic integration of Europe, including the introduction of a common exchange currency, the euro, in January 1999. In the early 21st century, five French overseas entities–French Guiana, Guadeloupe, Martinique, Mayotte, and Reunion–became French regions and were made part of France proper.

GEOGRAPHY

Location: metropolitan France: Western Europe, bordering the Bay of Biscay and English Channel, between Belgium and Spain, southeast of the UK; bordering the Mediterranean Sea, between Italy and Spain
French Guiana: Northern South America, bordering the North Atlantic Ocean, between Brazil and Suriname
Guadeloupe: Caribbean, islands between the Caribbean Sea and the North Atlantic Ocean, southeast of Puerto Rico
Martinique: Caribbean, island between the Caribbean Sea and North Atlantic Ocean, north of Trinidad and Tobago
Mayotte: Southern Indian Ocean, island in the Mozambique Channel, about half way between northern Madagascar and northern Mozambique
Reunion: Southern Africa, island in the Indian Ocean, east of Madagascar
Geographic coordinates:
metropolitan France: 46 00 N, 2 00 E
French Guiana: 4 00 N, 53 00 W
Guadeloupe: 16 15 N, 61 35 W
Martinique: 14 40 N, 61 00 W
Mayotte: 12 50 S, 45 10 E
Reunion: 21 06 S, 55 36 E
Map references: metropolitan France: Europe
French Guiana:
South America
Guadeloupe: Central America and the Caribbean
Martinique: Central America and the Caribbean
Mayotte: Africa
Reunion: World
Area: *total:* 643,801 sq km; 551,500 sq km (metropolitan France)
country comparison to the world: 42
land: 640,427 sq km; 549,970 sq km (metropolitan France)
water: 3,374 sq km; 1,530 sq km (metropolitan France)
note: the first numbers include the overseas regions of French Guiana, Guadeloupe, Martinique, Mayotte, and Reunion
Area—comparative: slightly less than the size of Texas
Land boundaries: metropolitan France–*total:* 2,889 km
border countries: Andorra 56.6 km, Belgium 620 km, Germany 451 km, Italy 488 km, Luxembourg 73 km, Monaco 4.4 km, Spain 623 km, Switzerland 573 km
French Guiana—total: 1,183 km
border countries: Brazil 673 km, Suriname 510 km
Coastline: *total:* 4,853 km
metropolitan France: 3,427 km
Maritime claims: *territorial sea:* 12 nm
contiguous zone: 24 nm
exclusive economic zone: 200 nm (does not apply to the Mediterranean)
continental shelf: 200 m depth or to the depth of exploitation
Climate: metropolitan France: generally cool winters and mild summers, but mild winters and hot summers along the Mediterranean; occasional strong, cold, dry, north-to-northwesterly wind known as mistral
French Guiana: tropical; hot, humid; little seasonal temperature variation
Guadeloupe and Martinique: subtropical tempered by trade winds; moderately high humidity; rainy season (June to October); vulnerable to devastating cyclones (hurricanes) every eight years on average
Mayotte: tropical; marine; hot, humid, rainy season during northeastern monsoon (November to May); dry season is cooler (May to November)
Reunion: tropical, but temperature moderates with elevation; cool and dry (May to November), hot and rainy (November to April)
Terrain: metropolitan France: mostly flat plains or gently rolling hills in north and west; remainder is mountainous, especially Pyrenees in south, Alps in east
French Guiana: low-lying coastal plains rising to hills and small mountains
Guadeloupe: Basse-Terre is volcanic in origin with interior mountains; Grande-Terre is low limestone formation; most of the seven other islands are volcanic in origin
Martinique: mountainous with indented coastline; dormant volcano
Mayotte: generally undulating, with deep ravines and ancient volcanic peaks
Reunion: mostly rugged and mountainous; fertile lowlands along coast
Elevation extremes: *lowest point:* Rhone River delta -2 m
highest point: Mont Blanc 4,807 m
note: in order to assess the possible effects of climate change on the ice and snow cap of Mont Blanc, its surface and peak have been extensively measured in recent years; these new peak measurements have exceeded the traditional height of 4,807 m and have varied between 4,808 m and 4,811 m; the actual rock summit is 4,792 m and is 40 m away from the ice-covered summit
Natural resources: metropolitan France: coal, iron ore, bauxite, zinc, uranium, antimony, arsenic, potash, feldspar, fluorspar, gypsum, timber, fish
French Guiana: gold deposits, petroleum, kaolin, niobium, tantalum, clay
Land use: *arable land:* 33.46%
permanent crops: 2.03%
other: 64.51%
note: French Guiana–arable land 0.13%, permanent crops 0.04%, other 99.83% (90% forest, 10% other); Guadeloupe–arable land 11.70%, permanent crops 2.92%, other 85.38%; Martinique–arable land 9.09%, permanent crops 10.0%, other 80.91%; Reunion–arable land 13.94%, permanent crops 1.59%, other 84.47% (2005)
Irrigated land: *total:* 26,190 sq km;
metropolitan France: 26,000 sq km (2003)
Total renewable water resources: 189 cu km (2005)
Freshwater withdrawal (domestic/industrial/agricultural): *total:* 33.16 cu km/yr (16%/74%/10%)
per capita: 548 cu m/yr (2000)
Natural hazards: metropolitan France: flooding; avalanches; midwinter windstorms; drought; forest fires in south near the Mediterranean

overseas departments: hurricanes (cyclones); flooding; volcanic activity (Guadeloupe, Martinique, Reunion)
Environment—current issues: some forest damage from acid rain; air pollution from industrial and vehicle emissions; water pollution from urban wastes, agricultural runoff
Environment—international agreements: *party to:* Air Pollution, Air Pollution-Nitrogen Oxides, Air Pollution-Persistent Organic Pollutants, Air Pollution-Sulfur 85, Air Pollution-Sulfur 94, Air Pollution-Volatile Organic Compounds, Antarctic-Environmental Protocol, Antarctic-Marine Living Resources, Antarctic Seals, Antarctic Treaty, Biodiversity, Climate Change, Climate Change-Kyoto Protocol, Desertification, Endangered Species, Hazardous Wastes, Law of the Sea, Marine Dumping, Marine Life Conservation, Ozone Layer Protection, Ship Pollution, Tropical Timber 83, Tropical Timber 94, Wetlands, Whaling
signed, but not ratified: none of the selected agreements
Geography—note: largest West European nation

PEOPLE

Population: 65,312,249 (July 2011 est.)
country comparison to the world: 21
note: the above figure is for metropolitan France and five overseas regions; the metropolitan France population is 62,814,233
Age structure: *0–14 years:* 18.5% (male 6,180,905/female 5,886,849)
15–64 years: 64.7% (male 21,082,175/female 21,045,867)
65 years and over: 16.8% (male 4,578,089/female 6,328,834) (2011 est.)
Median age: *total:* 39.9 years
male: 38.4 years
female: 41.5 years (2011 est.)
Population growth rate: 0.5% (2011 est.)
country comparison to the world: 151
Birth rate: 12.29 births/1,000 population (2011 est.)
country comparison to the world: 161
Death rate: 8.76 deaths/1,000 population (July 2011 est.)
country comparison to the world: 80
Net migration rate: 1.46 migrant(s)/1,000 population (2011 est.)
country comparison to the world: 46
Urbanization: *urban population:* 85% of total population (2010)
rate of urbanization: 1% annual rate of change (2010-15 est.)
Major cities—population: PARIS (capital) 10.41 million; Marseille-Aix-en-Provence 1.457 million; Lyon 1.456 million; Lille 1.028 million; Nice-Cannes 977,000 (2009)
Sex ratio: *at birth:* 1.051 male(s)/female
under 15 years: 1.05 male(s)/female
15–64 years: 1 male(s)/female
65 years and over: 0.72 male(s)/female
total population: 0.96 male(s)/female (2011 est.)
Infant mortality rate:
total: 3.29 deaths/1,000 live births
country comparison to the world: 214
male: 3.61 deaths/1,000 live births
female: 2.96 deaths/1,000 live births (2011 est.)
Life expectancy at birth: *total population:* 81.19 years
country comparison to the world: 13
male: 78.02 years
female: 84.54 years (2011 est.)
Total fertility rate: 1.96 children born/woman (2011 est.)
country comparison to the world: 133
HIV/AIDS—adult prevalence rate: 0.4% (2009 est.)
country comparison to the world: 71
HIV/AIDS—people living with HIV/AIDS: 150,000 (2009 est.)
country comparison to the world: 34
HIV/AIDS—deaths: 1,700 (2009 est.)
country comparison to the world: 56
Drinking water source: *Improved:*
urban: 100% of population
rural: 100% of population
total: 100% of population (2008)
Sanitation facility access: *Improved:*
urban: 100% of population
rural: 100% of population
total: 100% of population (2008)
Nationality: *noun:* Frenchman(men), Frenchwoman(women)
adjective: French
Ethnic groups: Celtic and Latin with Teutonic, Slavic, North African, Indochinese, Basque minorities
overseas departments: black, white, mulatto, East Indian, Chinese, Amerindian
Religions: Roman Catholic 83%-88%, Protestant 2%, Jewish 1%, Muslim 5%-10%, unaffiliated 4%
overseas departments: Roman Catholic, Protestant, Hindu, Muslim, Buddhist, pagan
Languages: French (official) 100%, rapidly declining regional dialects and languages (Provencal, Breton, Alsatian, Corsican, Catalan, Basque, Flemish)
overseas departments: French, Creole patois, Mahorian (a Swahili dialect)
Literacy: *definition:* age 15 and over can read and write
total population: 99%
male: 99%
female: 99% (2003 est.)
School life expectancy (primary to tertiary education): *total:* 16 years
male: 16 years
female: 16 years (2008)
Education expenditures: 5.6% of GDP (2007)
country comparison to the world: 38

GOVERNMENT

Country name: *conventional long form:* French Republic
conventional short form: France
local long form: Republique francaise
local short form: France
Government type: republic
Capital: *name:* Paris
geographic coordinates: 48 52 N, 2 20 E
time difference: UTC+1 (6 hours ahead of Washington, DC during Standard Time)
daylight saving time: +1hr, begins last Sunday in March; ends last Sunday in October
note: applies to metropolitan France only, not to its overseas departments, collectivities, or territories
Administrative divisions: 27 regions (regions, singular–region); Alsace, Aquitaine, Auvergne, Basse-Normandie (Lower Normandy), Bourgogne (Burgundy), Bretagne (Brittany), Centre, Champagne-Ardenne, Corse (Corsica), Franche-Comte, Guadeloupe, Guyane (French Guiana), Haute-Normandie (Upper Normandy), Ile-de-France, Languedoc-Roussillon, Limousin, Lorraine, Martinique, Mayotte, Midi-Pyrenees, Nord-Pas-de-Calais, Pays de la Loire, Picardie, Poitou-Charentes, Provence-Alpes-Cote d'Azur, Reunion, Rhone-Alpes
note: France is divided into 22 metropolitan regions (including the "territorial collectivity" of Corse or Corsica) and 5 overseas regions (French Guiana, Guadeloupe, Martinique, Mayotte, and Reunion) and is subdivided into 96 metropolitan departments and 5 overseas departments (which are the same as the overseas regions)
Dependent areas: Clipperton Island, French Polynesia, French Southern and Antarctic Lands, New Caledonia, Saint Barthelemy, Saint Martin, Saint Pierre and Miquelon, Wallis and Futuna
note: the US does not recognize claims to Antarctica; New Caledonia has been considered a "sui generis" collectivity of France since 1998, a unique status falling between that of an independent country and a French overseas department
Independence: no official date of independence: 486 (Frankish tribes unified under Merovingian kingship); 10 August 843 (Western Francia established from the division of the Carolingian Empire); 14 July 1789 (French monarchy overthrown); 22 September 1792 (First French Republic founded); 4 October 1958 (Fifth French Republic established)
National holiday: Fete de la Federation, 14 July (1790); note–although often incorrectly referred to as Bastille Day, the celebration actually commemorates the holiday held on the first anniversary of the storming of the Bastille (on 14 July 1789) and the establishment of a constitutional monarchy; other names for the holiday are Fete Nationale (National Holiday) and quatorze juillet (14th of July)
Constitution: adopted by referendum 28 September 1958; effective 4 October 1958; amended many times
note: amended in 1962 concerning election of president; amended to comply with provisions of 1992 EC Maastricht Treaty, 1997 Amsterdam Treaty, 2003 Treaty of Nice; amended in 1993 to tighten immigration laws; amended in 2000 to change the seven-year presidential term to a five-year term; amended in 2005 to make the EU constitutional treaty compatible with the Constitution of France and to ensure that the decision to ratify EU accession treaties would be made by referendum
Legal system: civil law; review of administrative but not legislative acts
International law organization participation: has not submitted an ICJ jurisdiction declaration; accepts ICCt jurisdiction
Suffrage: 18 years of age; universal

Executive branch: *chief of state:* President Nicolas SARKOZY (since 16 May 2007)
head of government: Prime Minister Francois FILLON (since 17 May 2007)
cabinet: Council of Ministers appointed by the president at the suggestion of the prime minister
(For more information visit the World Leaders website)
elections: president elected by popular vote for a five-year term (eligible for a second term); election last held on 22 April and 6 May 2007 (next to be held in the spring of 2012); prime minister appointed by the president
election results: Nicolas SARKOZY elected; first round: percent of vote—Nicolas SARKOZY 31.2%, Segolene ROYAL 25.9%, Francois BAYROU 18.6%, Jean-Marie LE PEN 10.4%, others 13.9%; second round: SARKOZY 53.1%, ROYAL 46.9%

Legislative branch: bicameral Parliament or Parlement consists of the Senate or Senat (343 seats; 323 for metropolitan France and overseas departments, 2 for New Caledonia, 2 for French Polynesia, 1 for Saint-Pierre and Miquelon, 1 for Saint-Barthelemy, 1 for Saint-Martin, 1 for Wallis and Futuna, and 12 for French nationals abroad; members indirectly elected by an electoral college to serve six-year terms; one third elected every three years); note—between 2006 and 2011, 15 new seats will be added to the Senate for a total of 348 seats—328 for metropolitan France and overseas departments, 2 for New Caledonia, 2 for French Polynesia, 1 for Saint-Pierre and Miquelon, 1 for Saint-Barthelemy, 1 for Saint-Martin, 1 for Wallis and Futuna, and 12 for French nationals abroad; Mayotte's previously held 2 seats as an overseas collectivity are now included in the total as an overseas department; starting in 2008, members will be indirectly elected by an electoral college to serve six-year terms with one-half elected every three years; and the National Assembly or Assemblee Nationale (577 seats; 555 for metropolitan France, 15 for overseas departments, 7 for overseas dependencies; members elected by popular vote under a single-member majority system to serve five-year terms)
elections: Senate—last held on 21 September 2008 (next to be held in September 2011); National Assembly—last held on 10 and 17 June 2007 (next to be held in June 2012)
election results: Senate—percent of vote by party—NA; seats by party—UMP 151, PS 102, PCF 22, MoDem 11, NC 11, Greens 5, PG 2, other 39; National Assembly—percent of vote by party—UMP 46.4%, PS 42.2%, miscellaneous left wing parties 2.5%, PCF 2.3%, NC 2.1%, PRG 1.6%, miscellaneous right wing parties 1.2%, the Greens 0.4%, other 1.2%; seats by party—UMP 313, PS 186, NC 22, miscellaneous left wing parties 15, PCF 16, miscellaneous right wing parties 9, PRG 7, the Greens 3, other 6

Judicial branch: Supreme Court of Appeals or Cour de Cassation (judges are appointed by the president from nominations of the High Council of the Judiciary); Constitutional Council or Conseil Constitutionnel (three members appointed by the president, three appointed by the president of the National Assembly, and three appointed by the president of the Senate); Council of State or Conseil d'Etat

Political parties and leaders: Democratic Movement or MoDem [Francois BAYROU] (previously Union for French Democracy or UDF); French Communist Party or PCF [Pierre LAURENT]; Greens [Cecile DUFLOT]; Left Party or PG [Jean-Luc MELENCHON]; Left Radical Party or PRG [Jean-Michel BAYLET] (previously Radical Socialist Party or PRS and the Left Radical Movement or MRG); Movement for France or MPF [Philippe DE VILLIERS]; National Front or FN [Jean-Marie LE PEN]; New Anticapitalist Party or NPA [Olivier BESANCENOT]; New Center or NC [Herve MORIN]; Radical Party [Jean-Louis BORLOO]; Rally for France or RPF [Charles PASQUA]; Republican and Citizen Movement or MRC [Jean Pierre CHEVENEMENT]; Socialist Party or PS [Martine AUBRY]; Union for a Popular Movement or UMP [Jean-Francois COPE]; Worker's Struggle or LO [Nathalie ARTHAUD]

Political pressure groups and leaders: Confederation francaise democratique du travail or CFDT, left-leaning labor union with approximately 803,000 members; Confederation francaise de l'encadrement—Confederation generale des cadres or CFE-CGC, independent white-collar union with 196,000 members; Confederation francaise des travailleurs chretiens of CFTC, independent labor union founded by Catholic workers that claims 132,000 members; Confederation generale du travail or CGT, historically communist labor union with approximately 700,000 members; Confederation generale du travail—Force ouvriere or FO, independent labor union with an estimated 300,000 members; Mouvement des entreprises de France or MEDEF, employers' union with 750,000 companies as members (claimed)
French Guiana: conservationists; gold mining pressure groups; hunting pressure groups
Guadeloupe: Christian Movement for the Liberation of Guadeloupe or KLPG; General Federation of Guadeloupe Workers or CGT-G; General Union of Guadeloupe Workers or UGTG; Movement for an Independent Guadeloupe or MPGI; The Socialist Renewal Movement
Martinique: Caribbean Revolutionary Alliance or ARC; Central Union for Martinique Workers or CSTM; Frantz Fanon Circle; League of Workers and Peasants; Proletarian Action Group or GAP
Reunion: NA

International organization participation: ADB (nonregional member), AfDB (nonregional member), Arctic Council (observer), Australia Group, BDEAC, BIS, BSEC (observer), CBSS (observer), CE, CERN, EAPC, EBRD, EIB, EMU, ESA, EU, FAO, FATF, FZ, G-20, G-5, G-7, G-8, G-10, IADB, IAEA, IBRD, ICAO, ICC, ICRM, IDA, IEA, IFAD, IFC, IFRCS, IHO, ILO, IMF, IMO, IMSO, InOC, Interpol, IOC, IOM, IPU, ISO, ITSO, ITU, ITUC, MIGA, MINURSO, MINUSTAH, MONUSCO, NATO, NEA, NSG, OAS (observer), OECD, OIF, OPCW, OSCE, Paris Club, PCA, PIF (partner), Schengen Convention, SECI (observer), SPC, UN, UN Security Council, UNCTAD, UNESCO, UNHCR, UNIDO, UNIFIL, Union Latina, UNITAR, UNMIL, UNOCI, UNRWA, UNTSO, UNWTO, UPU, WCO, WFTU, WHO, WIPO, WMO, WTO, ZC

Diplomatic representation in the US: *chief of mission:* Ambassador Francois M. DELATTRE
chancery: 4101 Reservoir Road NW, Washington, DC 20007
telephone: [1] (202) 944-6000
FAX: [1] (202) 944-6166
consulate(s) general: Atlanta, Boston, Chicago, Houston, Los Angeles, Miami, New Orleans, New York, San Francisco

Diplomatic representation from the US: *chief of mission:* Ambassador Charles H. RIVKIN
embassy: 2 Avenue Gabriel, 75382 Paris Cedex 08
mailing address: PSC 116, APO AE 09777
telephone: [33] (1) 43-12-22-22
FAX: [33] (1) 42 66 97 83
consulate(s) general: Marseille, Strasbourg

Flag description: three equal vertical bands of blue (hoist side), white, and red; known as the "Le drapeau tricolore" (French Tricolor), the origin of the flag dates to 1790 and the French Revolution when the "ancient French color" of white was combined with the blue and red colors of the Parisian militia; the official flag for all French dependent areas
note: the design and/or colors are similar to a number of other flags, including those of Belgium, Chad, Cote d'Ivoire, Ireland, Italy, Luxembourg, and Netherlands

National anthem: *name:* "La Marseillaise" (The Song of Marseille)
lyrics/music: Claude-Joseph ROUGET de Lisle
note: adopted 1795, restored 1870; originally known as "Chant de Guerre pour l'Armee du Rhin" (War Song for the Army of the Rhine), the National Guard of Marseille made the song famous by singing it while marching into Paris in 1792 during the French Revolutionary Wars

ECONOMY

Economy—overview: France is in the midst of transition from a well-to-do modern economy that has featured extensive government ownership and intervention to one that relies more on market mechanisms. The government has partially or fully privatized many large companies, banks, and insurers, and has ceded stakes in such leading firms as Air France, France Telecom, Renault, and Thales. It maintains a strong presence in some sectors, particularly power, public transport, and defense industries. With at least 75 million foreign tourists per year, France is the most visited country in the world and maintains the third largest income in the world from tourism. France's leaders remain committed to a capitalism in which they maintain social equity by means of laws, tax policies, and social spending that reduce income disparity and the impact of free markets on public health and welfare. France has weathered the global economic crisis better than most other big EU economies because of the relative resilience of domestic consumer spending, a large public sector, and less exposure to the downturn in global demand than in some other countries. Nonetheless, France's real GDP contracted 2.5% in 2009, but recovered somewhat in 2010, while the unemployment rate increased from 7.4% in 2008 to 9.5% in 2010. The government pursuit of aggressive stimulus and investment measures in response to the economic crisis, however, are contributing to a deterioration of France's public finances. The government budget deficit rose sharply from 3.4% of GDP in 2008 to 7.8% of GDP in 2010, while France's public debt rose from 68% of GDP to 84% over the same period. Paris is terminating stimulus measures, eliminating tax credits, and freezing most government spending to bring the budget deficit under the 3% euro-zone ceiling by 2013, and to highlight France's commitment to fiscal discipline at a time of intense financial market scrutiny of euro zone debt levels. President SARKOZY–who secured passage of pension reform in 2010–is expected to seek passage of some tax reforms in 2011, but he may delay additional, more costly, reforms until after the 2012 election.

GDP (purchasing power parity): $2.145 trillion (2010 est.)
country comparison to the world: 10
$2.114 trillion (2009 est.)
$2.169 trillion (2008 est.)
note: data are in 2010 US dollars

GDP (official exchange rate): $2.583 trillion (2010 est.)

GDP—real growth rate: 1.5% (2010 est.)
country comparison to the world: 161
-2.5% (2009 est.)
0.1% (2008 est.)

GDP—per capita (PPP): $33,100 (2010 est.)
country comparison to the world: 39
$32,800 (2009 est.)
$33,900 (2008 est.)
note: data are in 2010 US dollars

GDP—composition by sector:
agriculture: 1.8%
industry: 19.2%
services: 79% (2010 est.)

Labor force: 28.21 million (2010 est.)
country comparison to the world: 20

Labor force—by occupation: *agriculture:* 3.8%
industry: 24.3%
services: 71.8% (2005)

Unemployment rate: 9.5% (2010 est.)
country comparison to the world: 104
9.1% (2009 est.)

Population below poverty line: 6.2% (2004)

Household income or consumption by percentage share:
lowest 10%: 3%
highest 10%: 24.8% (2004)

Distribution of family income—Gini index: 32.7 (2008)
country comparison to the world: 98
32.7 (1995)

Investment (gross fixed): 19.9% of GDP (2010 est.)
country comparison to the world: 89

Budget: *revenues:* $1.241 trillion
expenditures: $1.441 trillion (2010 est.)

Public debt: 83.5% of GDP (2010 est.)
country comparison to the world: 14
77.6% of GDP (2009 est.)

Inflation rate (consumer prices): 1.5% (2010 est.)
country comparison to the world: 37
0.1% (2009 est.)

Central bank discount rate: 1.75% (31 December 2010)
country comparison to the world: 125
1.75% (31 December 2009)
note: this is the European Central Bank's rate on the marginal lending facility, which offers overnight credit to banks in the euro area

Commercial bank prime lending rate: 7.46% (31 December 2009 est.)
country comparison to the world: 113
8.13% (31 December 2008 est.)

Stock of narrow money: $858.6 billion (31 December 2010 est.)
country comparison to the world: 8
$862.3 billion (31 December 2009 est.)
note: see entry for the European Union for money supply in the euro area; the European Central Bank (ECB) controls monetary policy for the 17 members of the Economic and Monetary Union (EMU); individual members of the EMU do not control the quantity of money circulating within their own borders

Stock of broad money: $2.292 trillion (31 December 2010 est.)
country comparison to the world: 8
$2.306 trillion (31 December 2009 est.)

Stock of domestic credit: $4.319 trillion (31 December 2009 est.)
country comparison to the world: 7
$4.121 trillion (31 December 2008 est.)

Market value of publicly traded shares: $1.972 trillion (31 December 2009)
country comparison to the world: 7
$1.492 trillion (31 December 2008)
$2.771 trillion (31 December 2007)

Agriculture—products: wheat, cereals, sugar beets, potatoes, wine grapes; beef, dairy products; fish

Industries: machinery, chemicals, automobiles, metallurgy, aircraft, electronics; textiles, food processing; tourism

Industrial production growth rate: 3.5% (2010 est.)
country comparison to the world: 97

Electricity—production: 535.7 billion kWh (2007 est.)
country comparison to the world: 9

Electricity—consumption: 447.2 billion kWh (2007 est.)
country comparison to the world: 9

Electricity—exports: 58.69 billion kWh (2008 est.)

Electricity—imports: 10.68 billion kWh (2008 est.)

Oil—production: 70,820 bbl/day (2009 est.)
country comparison to the world: 57

Oil—consumption: 1.875 million bbl/day (2009 est.)
country comparison to the world: 13

Oil—exports: 597,800 bbl/day (2008 est.)
country comparison to the world: 24

Oil—imports: 2.386 million bbl/day (2008 est.)
country comparison to the world: 9

Oil—proved reserves: 101.2 million bbl (1 January 2010 est.)
country comparison to the world: 68

Natural gas—production: 877 million cu m (2009 est.)
country comparison to the world: 64

Natural gas—consumption: 44.84 billion cu m (2009 est.)
country comparison to the world: 19

Natural gas—exports: 1.931 billion cu m (2009 est.)
country comparison to the world: 34

Natural gas—imports: 45.85 billion cu m (2009 est.)
country comparison to the world: 5

Natural gas—proved reserves: 7.079 billion cu m (1 January 2010 est.)
country comparison to the world: 83

Current account balance: $-53.29 billion (2010 est.)
country comparison to the world: 188
$-51.86 billion (2009 est.)

Exports: $508.7 billion (2010 est.)
country comparison to the world: 6
$473.9 billion (2009 est.)

Exports—commodities: machinery and transportation equipment, aircraft, plastics, chemicals, pharmaceutical products, iron and steel, beverages

Exports—partners: Germany 15.88%, Italy 8.16%, Spain 7.8%, Belgium 7.44%, UK 7.04%, US 5.65%, Netherlands 3.99% (2009)

Imports: $577.7 billion (2010 est.)
country comparison to the world: 6
$535.8 billion (2009 est.)

Imports—commodities: machinery and equipment, vehicles, crude oil, aircraft, plastics, chemicals

Imports—partners: Germany 19.41%, Belgium 11.61%, Italy 7.97%, Netherlands 7.15%, Spain 6.68%, UK 4.9%, US 4.72%, China 4.44% (2009)

Reserves of foreign exchange and gold: $NA (31 December 2010 est.)
$133.1 billion (31 December 2009 est.)

Debt—external: $4.698 trillion (30 June 2010)
country comparison to the world: 5
$4.935 trillion (31 December 2008)

Stock of direct foreign investment—at home: $1.207 trillion (31 December 2010 est.)
country comparison to the world: 2
$1.151 trillion (31 December 2009 est.)

Stock of direct foreign investment—abroad: $1.837 trillion (31 December 2010 est.)
country comparison to the world: 2
$1.711 trillion (31 December 2009 est.)

Exchange rates: euros (EUR) per US dollar–0.755 (2010), 0.7198 (2009), 0.6827 (2008), 0.7345 (2007), 0.7964 (2006)

COMMUNICATIONS

Telephones—main lines in use: 36.441 million; 35.5 million (metropolitan France) (2009)
country comparison to the world: 7

Telephones—mobile cellular: 60.95 million; 59.543 million (metropolitan France) (2009)
country comparison to the world: 18
Telephone system: *general assessment:* highly developed
domestic: extensive cable and microwave radio relay; extensive use of fiber-optic cable; domestic satellite system
international: country code–33; numerous submarine cables provide links throughout Europe, Asia, Australia, the Middle East, and US; satellite earth stations–more than 3 (2 Intelsat (with total of 5 antennas–2 for Indian Ocean and 3 for Atlantic Ocean), NA Eutelsat, 1 Inmarsat–Atlantic Ocean region); HF radiotelephone communications with more than 20 countries
overseas departments: country codes: French Guiana–594; Guadeloupe–590; Martinique–596; Mayotte–262; Reunion–262
Broadcast media: a mix of both publicly-operated and privately-owned TV stations; state-owned France Televisions operates 4 networks, one of which is a network of regional stations, and has part-interest in several thematic cable/satellite channels and international channels; a large number of privately-owned regional and local TV stations; multi-channel satellite and cable services provide a large number of channels; public broadcaster Radio France operates 7 national networks, a series of regional networks, and operates services for overseas territories and foreign audiences; Radio France Internationale (RFI), under the Ministry of Foreign Affairs, is a leading international broadcaster; a large number of commercial FM stations, with many of them consolidating into commercial networks (2008)
Internet country code: metropolitan France–.fr; French Guiana–.gf; Guadeloupe–.gp; Martinique–.mq; Mayotte–.yt; Reunion–.re
Internet hosts: 15,182,001; 15.161 million (metropolitan France) (2010)
country comparison to the world: 7
Internet users: 45.262 million; 44.625 million (metropolitan France) (2009)
country comparison to the world: 8

TRANSPORTATION

Airports: 475 (2010)
country comparison to the world: 16
Airports—with paved runways: *total:* 297
over 3,047 m: 14
2,438 to 3,047 m: 27
1,524 to 2,437 m: 98
914 to 1,523 m: 83
under 914 m: 76 (2010)
Airports—with unpaved runways:
total: 177
914 to 1,523 m: 69
under 914 m: 108 (2010)
Heliports: 1 (2010)
Pipelines: gas 15,276 km; oil 2,939 km; refined products 5,084 km (2010)
Railways: *total:* 29,640 km
country comparison to the world: 9
standard gauge: 29,473 km 1.435-m gauge (15,361 km electrified)
narrow gauge: 167 km 1.000-m gauge (63 km electrified) (2009)
Roadways: *total:* 1,027,183 km (metropolitan France; includes 10,958 km of expressways)
country comparison to the world: 7
note: there are another 5,100 km of roadways in overseas departments (2007)
Waterways: metropolitan France: 8,501 km (1,621 km accessible to craft of 3,000 metric tons) (2010)
country comparison to the world: 16
Merchant marine: *total:* 167
country comparison to the world: 38
by type: bulk carrier 2, cargo 8, chemical tanker 36, container 25, liquefied gas 12, passenger 11, passenger/cargo 44, petroleum tanker 17, refrigerated cargo 1, roll on/roll off 11
foreign-owned: 57 (Belgium 7, China 5, Denmark 12, French Polynesia 12, Germany 1, New Caledonia 3, Norway 1, NZ 1, Singapore 3, Spain 1, Sweden 6, Switzerland 5)
registered in other countries: 146 (Bahamas 19, Belgium 5, Bermuda 1, Canada 1, Cyprus 16, Egypt 1, Hong Kong 3, Indonesia 1, Italy 2, Luxembourg 16, Malta 13, Morocco 4, Netherlands 2, Norway 4, Panama 13, Saint Vincent and the Grenadines 2, Singapore 3, South Korea 1, Taiwan 1, UK 33, US 4, unknown 1) (2010)
Ports and terminals: Calais, Dunkerque, Le Havre, Marseille, Nantes, Paris, Rouen

MILITARY

Military branches: Army (Armee de Terre; includes Marines, Foreign Legion, Army Light Aviation), Navy (Marine Nationale, includes Naval Air, Maritime Gendarmerie (Coast Guard)), Air Force (Armee de l'Air (AdlA), includes Air Defense), National Gendarmerie (2011)
Military service age and obligation: 17-40 years of age for male and female voluntary military service (with parental consent); no conscription; 12-month service obligation; women serve in noncombat posts (2010)
Manpower available for military service:
males age 16–49: 14,563,662
females age 16–49: 14,238,434 (2010 est.)
Manpower fit for military service: *males age 16–49:* 12,025,341
females age 16–49: 11,721,827 (2010 est.)
Manpower reaching militarily significant age annually:
male: 396,050
female: 377,839 (2010 est.)
Military expenditures: 2.6% of GDP (2005 est.)
country comparison to the world: 56

TRANSNATIONAL ISSUES

Disputes—international: Madagascar claims the French territories of Bassas da India, Europa Island, Glorioso Islands, and Juan de Nova Island; Comoros claims Mayotte; Mauritius claims Tromelin Island; territorial dispute between Suriname and the French overseas department of French Guiana; France asserts a territorial claim in Antarctica (Adelie Land); France and Vanuatu claim Matthew and Hunter Islands, east of New Caledonia
Illicit drugs: metropolitan France: transshipment point for South American cocaine, Southwest Asian heroin, and European synthetics
French Guiana: small amount of marijuana grown for local consumption; minor transshipment point to Europe
Martinique: transshipment point for cocaine and marijuana bound for the US and Europe

FRENCH POLYNESIA

(OVERSEAS LANDS OF FRANCE)

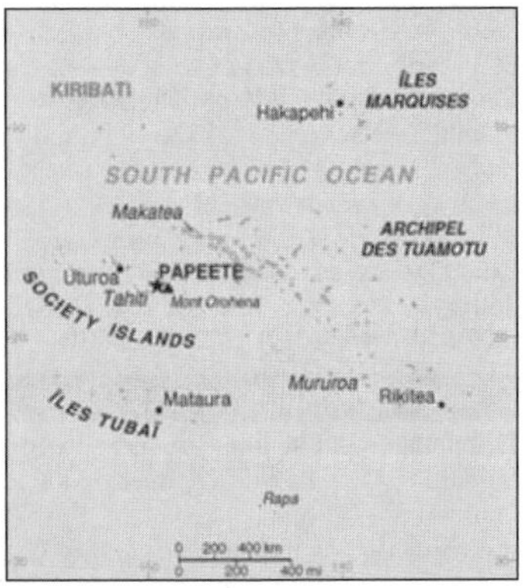

INTRODUCTION

Background: The French annexed various Polynesian island groups during the 19th century. In September 1995, France stirred up widespread protests by resuming nuclear testing on the Mururoa atoll after a three-year moratorium. The tests were suspended in January 1996. In recent years, French Polynesia's autonomy has been considerably expanded.

GEOGRAPHY

Location: Oceania, archipelagoes in the South Pacific Ocean about half way between South America and Australia
Geographic coordinates: 15 00 S, 140 00 W
Map references: Oceania
Area: *total:* 4,167 sq km (118 islands and atolls)
country comparison to the world: 174
land: 3,827 sq km
water: 340 sq km
Area—comparative: slightly less than one-third the size of Connecticut
Land boundaries: 0 km
Coastline: 2,525 km
Maritime claims: *territorial sea:* 12 nm
exclusive economic zone: 200 nm
Climate: tropical, but moderate
Terrain: mixture of rugged high islands and low islands with reefs
Elevation extremes: *lowest point:* Pacific Ocean 0 m

highest point: Mont Orohena 2,241 m
Natural resources: timber, fish, cobalt, hydropower
Land use: *arable land:* 0.75%
permanent crops: 5.5%
other: 93.75% (2005)
Irrigated land: 10 sq km (2003)
Natural hazards: occasional cyclonic storms in January
Environment—current issues: NA
Geography—note: includes five archipelagoes (four volcanic, one coral); Makatea in French Polynesia is one of the three great phosphate rock islands in the Pacific Ocean–the others are Banaba (Ocean Island) in Kiribati and Nauru

PEOPLE

Population: 294,935 (July 2011 est.)
country comparison to the world: 179
Age structure: *0–14 years:* 23.5% (male 35,376/female 33,840)
15–64 years: 69.3% (male 105,823/female 98,597)
65 years and over: 7.2% (male 10,742/female 10,557) (2011 est.)
Median age: *total:* 29.9 years
male: 30.2 years
female: 29.6 years (2011 est.)
Population growth rate: 1.331% (2011 est.)
country comparison to the world: 89
Birth rate: 15.53 births/1,000 population (2011 est.)
country comparison to the world: 131
Death rate: 4.87 deaths/1,000 population (July 2011 est.)
country comparison to the world: 191
Net migration rate: 2.65 migrant(s)/1,000 population (2011 est.)
country comparison to the world: 31
Urbanization: *urban population:* 51% of total population (2010)
rate of urbanization: 1.3% annual rate of change (2010-15 est.)
Major cities—population: PAPEETE (capital) 133,000 (2009)
Sex ratio: *at birth:* 1.05 male(s)/female
under 15 years: 1.04 male(s)/female
15–64 years: 1.07 male(s)/female
65 years and over: 1.02 male(s)/female
total population: 1.06 male(s)/female (2011 est.)
Infant mortality rate:
total: 7.27 deaths/1,000 live births
country comparison to the world: 164
male: 8.35 deaths/1,000 live births
female: 6.15 deaths/1,000 live births (2011 est.)
Life expectancy at birth:
total population: 77.1 years
country comparison to the world: 65
male: 74.62 years
female: 79.7 years (2011 est.)
Total fertility rate: 1.87 children born/woman (2011 est.)
country comparison to the world: 148
HIV/AIDS—adult prevalence rate: NA
HIV/AIDS—people living with HIV/AIDS: NA
HIV/AIDS—deaths: NA
Drinking water source: *Improved:*
urban: 100% of population
rural: 100% of population
total: 100% of population (2008)
Sanitation facility access: *Improved:*
urban: 99% of population
rural: 97% of population
total: 98% of population
Unimproved: urban: 1% of population
rural: 3% of population
total: 2% of population (2008)
Nationality: *noun:* French Polynesian(s)
adjective: French Polynesian
Ethnic groups: Polynesian 78%, Chinese 12%, local French 6%, metropolitan French 4%
Religions: Protestant 54%, Roman Catholic 30%, other 10%, no religion 6%
Languages: French (official) 61.1%, Polynesian (official) 31.4%, Asian languages 1.2%, other 0.3%, unspecified 6% (2002 census)
Literacy: *definition:* age 14 and over can read and write
total population: 98%
male: 98%
female: 98% (1977 est.)
School life expectancy (primary to tertiary education): NA
Education expenditures: NA

GOVERNMENT

Country name: *conventional long form:* Overseas Lands of French Polynesia
conventional short form: French Polynesia
local long form: Pays d'outre-mer de la Polynesie Francaise
local short form: Polynesie Francaise
former: French Colony of Oceania
Dependency status: overseas lands of France; overseas territory of France from 1946-2003; overseas collectivity of France since 2003, though it is often referred to as an overseas country due to its degree of autonomy
Government type: NA
Capital: *name:* Papeete
geographic coordinates: 17 32 S, 149 34 W
time difference: UTC-10 (5 hours behind Washington, DC during Standard Time)
Administrative divisions: none (overseas lands of France); there are no first-order administrative divisions as defined by the US Government, but there are five archipelagic divisions named Archipel des Marquises, Archipel des Tuamotu, Archipel des Tubuai, Iles du Vent, Iles Sous-le-Vent
Independence: none (overseas lands of France)
National holiday: Bastille Day, 14 July (1789); note–the local holiday is Internal Autonomy Day, 29 June (1880)
Constitution: 4 October 1958 (French Constitution)
Legal system: the laws of France, where applicable, apply
Suffrage: 18 years of age; universal
Executive branch: *chief of state:* President Nicolas SARKOZY (since 16 May 2007), represented by High Commissioner of the Republic Richard DIDIER (since 24 January 2011)
head of government: President of French Polynesia Oscar TEMARU (since 1 April 2011); President of the Assembly of French Polynesia Jacqui DROLLET (since 14 April 2011)
cabinet: Council of Ministers; president submits a list of members of the Assembly for approval by them to serve as ministers (For more information visit the World Leaders website)
elections: French president elected by popular vote for a five-year term; high commissioner appointed by the French president on the advice of the French Ministry of Interior; president of the French Polynesia government and the president of the Assembly of French Polynesia elected by the members of the assembly for five-year terms (no term limits)
Legislative branch: unicameral Assembly of French Polynesia or Assemblee de la Polynesia francaise (57 seats; members elected by popular vote to serve five-year terms)
elections: last held on 27 January 2008 (first round) and 10 February 2008 (second round) (next to be held in 2013)
election results: percent of vote by party–Our Home alliance 45.2%, Union for Democracy alliance 37.2%, Popular Rally (Tahoeraa Huiraatira) 17.2% other 0.5%; seats by party–Our Home alliance 27, Union for Democracy alliance 20, Popular Rally 10
note: two seats were elected to the French Senate on 21 September 2008 (next to be held in September 2014); results–percent of vote by party–NA; seats by party–UMP 1, independent 1; two seats were elected to the French National Assembly on 10-17 June 2007 (next to be held in 2012); results–percent of vote by party–NA; seats by party–UMP 2
Judicial branch: Court of Appeal or Cour d'Appel; Court of the First Instance or Tribunal de Premiere Instance; Court of Administrative Law or Tribunal Administratif
Political parties and leaders: Alliance for a New Democracy or ADN (includes the parties The New Star and This Country is Yours); New Fatherland Party (Ai'a Api); Our Home alliance; People's Servant Party (Tavini Huiraatira); Popular Rally (Tahoeraa Huiraatira); Union for Democracy alliance or UPD
Political pressure groups and leaders: NA
International organization participation: ITUC, PIF (associate member), SPC, UPU
Diplomatic representation in the US: none (overseas lands of France)
Diplomatic representation from the US: none (overseas lands of France)
Flag description: two red horizontal bands encase a wide white band in a 1:2:1 ratio; centered on the white band is a disk with a blue and white wave pattern depicting the sea on the lower half and a gold and white ray pattern depicting the sun on the upper half; a Polynesian canoe rides on the wave pattern; the canoe has a crew of five represented by five stars that symbolize the five island groups; red and white are traditional Polynesian colors
note: similar to the red-white-red flag of Tahiti, the largest of the islands in French Polynesia, which has no emblem in the white band; the flag of France is used for official occasions
National anthem: *name:* "Ia Ora 'O Tahiti Nui" (Long Live Tahiti Nui)
lyrics/music: Maeva BOUGES, Irmine TEHEI, Angele TEROROTUA, Johanna

NOUVEAU, Patrick AMARU, Louis MAMATUI and Jean-Pierre CELESTIN
note: adopted 1993; serves as a local anthem; as a territory of France, "La Marseillaise" is official (see France)
Government—note: under certain acts of France, French Polynesia has acquired autonomy in all areas except those relating to police and justice, monetary policy, tertiary education, immigration, and defense and foreign affairs; the duties of its president are fashioned after those of the French prime minister

ECONOMY

Economy—overview: Since 1962, when France stationed military personnel in the region, French Polynesia has changed from a subsistence agricultural economy to one in which a high proportion of the work force is either employed by the military or supports the tourist industry. With the halt of French nuclear testing in 1996, the military contribution to the economy fell sharply. Tourism accounts for about one-fourth of GDP and is a primary source of hard currency earnings. Other sources of income are pearl farming and deep-sea commercial fishing. The small manufacturing sector primarily processes agricultural products. The territory benefits substantially from development agreements with France aimed principally at creating new businesses and strengthening social services.
GDP (purchasing power parity): $4.718 billion (2004 est.)
country comparison to the world: 164
$4.58 billion (2003 est.)
GDP (official exchange rate): $6.1 billion (2004)
GDP—real growth rate: 2.7% (2005)
country comparison to the world: 132
5.1% (2002)
GDP—per capita (PPP): $18,000 (2004 est.)
country comparison to the world: 66
$17,500 (2003 est.)
GDP—composition by sector:
agriculture: 3.5%
industry: 20.4%
services: 76.1% (2005)
Labor force: 116,000 (2007)
country comparison to the world: 180
Labor force—by occupation:
agriculture: 13%
industry: 19%
services: 68% (2002)
Unemployment rate: 11.7% (2005)
country comparison to the world: 124
Population below poverty line: NA%
Household income or consumption by percentage share: *lowest 10%:* NA%
highest 10%: NA%
Budget: *revenues:* $865 million
expenditures: $644.1 million (1999)
Inflation rate (consumer prices): 1.1% (2007)
country comparison to the world: 25
1.1% (2006 est.)
Market value of publicly traded shares: $NA
Agriculture—products: fish; coconuts, vanilla, vegetables, fruits, coffee; poultry, beef, dairy products
Industries: tourism, pearls, agricultural processing, handicrafts, phosphates
Industrial production growth rate: NA%
Electricity—production: 650 million kWh (2007 est.)
country comparison to the world: 153
Electricity—consumption: 604.5 million kWh (2007 est.)
country comparison to the world: 155
Electricity—exports: 0 kWh (2008 est.)
Electricity—imports: 0 kWh (2008 est.)
Oil—production: 0 bbl/day (2009 est.)
country comparison to the world: 176
Oil—consumption: 7,000 bbl/day (2009 est.)
country comparison to the world: 156
Oil—exports: 0 bbl/day (2007 est.)
country comparison to the world: 167
Oil—imports: 6,701 bbl/day (2007 est.)
country comparison to the world: 149
Oil—proved reserves: 0 bbl (1 January 2010 est.)
country comparison to the world: 133
Natural gas—production: 0 cu m (2008 est.)
country comparison to the world: 128
Natural gas—consumption: 0 cu m (2008 est.)
country comparison to the world: 173
Natural gas—exports: 0 cu m (2008 est.)
country comparison to the world: 97
Natural gas—imports: 0 cu m (2008 est.)
country comparison to the world: 187
Natural gas—proved reserves: 0 cu m (1 January 2010 est.)
country comparison to the world: 137
Exports: $211 million (2005 est.)
country comparison to the world: 181
Exports—commodities: cultured pearls, coconut products, mother-of-pearl, vanilla, shark meat
Imports: $1.706 billion (2005 est.)
country comparison to the world: 156
Imports—commodities: fuels, foodstuffs, machinery and equipment
Debt—external: $NA
Exchange rates: Comptoirs Francais du Pacifique francs (XPF) per US dollar—87.59 (2007), 94.97 (2006), 95.89 (2005), 96.04 (2004), 105.66 (2003)

COMMUNICATIONS

Telephones—main lines in use: 54,300 (2009)
country comparison to the world: 159
Telephones—mobile cellular: 208,300 (2009)
country comparison to the world: 171
Telephone system: *general assessment:* NA
domestic: combined fixed and mobile-cellular density is roughly 90 per 100 persons
international: country code—689; satellite earth station—1 Intelsat (Pacific Ocean) (2009)
Broadcast media: the publicly-owned French Overseas Network (RFO), which operates in France's overseas departments and territories, broadcasts on 2 television channels and 1 radio station; a government-owned TV station is operating; a small number of privately-owned radio stations also broadcast (2008)
Internet country code: .pf
Internet hosts: 36,056 (2010)
country comparison to the world: 95
Internet users: 120,000 (2009)
country comparison to the world: 153

TRANSPORTATION

Airports: 53 (2010)
country comparison to the world: 89
Airports—with paved runways:
total: 46
over 3,047 m: 2
1,524 to 2,437 m: 4
914 to 1,523 m: 33
under 914 m: 7 (2010)
Airports—with unpaved runways:
total: 7
914 to 1,523 m: 3
under 914 m: 4 (2010)
Heliports: 1 (2010)
Roadways: *total:* 2,590 km
country comparison to the world: 169
paved: 1,735 km
unpaved: 855 km (1999)
Merchant marine: registered in other countries: 13 (Cambodia 1, France 12) (2010)
country comparison to the world: 109
Ports and terminals: Papeete

MILITARY

Military branches: no regular military forces (2011)
Manpower available for military service: *males age 16–49:* 82,722 (2010 est.)
Manpower fit for military service: *males age 16–49:* 67,363
females age 16–49: 66,053 (2010 est.)
Manpower reaching militarily significant age annually: *male:* 2,498
female: 2,390 (2010 est.)
Military—note: defense is the responsibility of France

TRANSNATIONAL ISSUES

Disputes—international: none

FRENCH SOUTHERN AND ANTARCTIC LANDS

(OVERSEAS TERRITORY OF FRANCE)

INTRODUCTION

Background: In February 2007, the Iles Eparses became an integral part of the French Southern and Antarctic Lands (TAAF). The Southern Lands are now divided into five administrative districts, two of which are archipelagos, Iles Crozet and Iles Kerguelen; the third is a district composed of two volcanic islands, Ile Saint-Paul and Ile Amsterdam; the fourth, Iles Eparses, consists of five scattered tropical islands around Madagascar. They contain no permanent inhabitants and are visited only by researchers studying the native fauna, scientists at the various scientific stations, fishermen, and military personnel. The fifth district

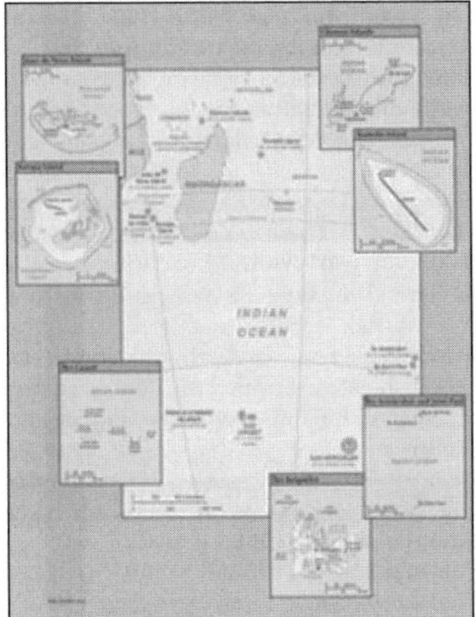

is the Antarctic portion, which consists of "Adelie Land," a thin slice of the Antarctic continent discovered and claimed by the French in 1840.

Ile Amsterdam: Discovered but not named in 1522 by the Spanish, the island subsequently received the appellation of Nieuw Amsterdam from a Dutchman; it was claimed by France in 1843. A short-lived attempt at cattle farming began in 1871. A French meteorological station established on the island in 1949 is still in use.

Ile Saint Paul: Claimed by France since 1893, the island was a fishing industry center from 1843 to 1914. In 1928, a spiny-lobster cannery was established, but when the company went bankrupt in 1931, seven workers were abandoned. Only two survived until 1934 when rescue finally arrived.

Iles Crozet: A large archipelago formed from the Crozet Plateau, Iles Crozet is divided into two main groups: L'Occidental (the West), which includes Ile aux Cochons, Ilots des Apotres, Ile des Pingouins, and the reefs Brisants de l'Heroine; and L'Oriental (the east), which includes Ile d'Est and Ile de la Possession (the largest island of the Crozets). Discovered and claimed by France in 1772, the islands were used for seal hunting and as a base for whaling. Originally administered as a dependency of Madagascar, they became part of the TAAF in 1955.

Iles Kerguelen: This island group, discovered in 1772, is made up of one large island (Ile Kerguelen) and about 300 smaller islands. A permanent group of 50 to 100 scientists resides at the main base at Port-aux-Francais.

Adelie Land: The only non-insular district of the TAAF is the Antarctic claim known as "Adelie Land." The US Government does not recognize it as a French dependency.

Bassas da India: A French possession since 1897, this atoll is a volcanic rock surrounded by reefs and is awash at high tide.

Europa Island: This heavily wooded island has been a French possession since 1897; it is the site of a small military garrison that staffs a weather station.

Glorioso Islands: A French possession since 1892, the Glorioso Islands are composed of two lushly vegetated coral islands (Ile Glorieuse and Ile du Lys) and three rock islets. A military garrison operates a weather and radio station on Ile Glorieuse.

Juan de Nova Island: Named after a famous 15th century Spanish navigator and explorer, the island has been a French possession since 1897. It has been exploited for its guano and phosphate. Presently a small military garrison oversees a meteorological station.

Tromelin Island: First explored by the French in 1776, the island came under the jurisdiction of Reunion in 1814. At present, it serves as a sea turtle sanctuary and is the site of an important meteorological station.

GEOGRAPHY

Location: southeast and east of Africa, islands in the southern Indian Ocean, some near Madagascar and others about equidistant between Africa, Antarctica, and Australia; note—French Southern and Antarctic Lands include Ile Amsterdam, Ile Saint-Paul, Iles Crozet, Iles Kerguelen, Bassas da India, Europa Island, Glorioso Islands, Juan de Nova Island, and Tromelin Island in the southern Indian Ocean, along with the French-claimed sector of Antarctica, "Adelie Land"; the US does not recognize the French claim to "Adelie Land"

Geographic coordinates: Ile Amsterdam (Ile Amsterdam et Ile Saint-Paul): 37 50 S, 77 32 E
Ile Saint-Paul (Ile Amsterdam et Ile Saint-Paul): 38 72 S, 77 53 E
Iles Crozet: 46 25 S, 51 00 E
Iles Kerguelen: 49 15 S, 69 35 E
Bassas da India (Iles Eparses): 21 30 S, 39 50 E
Europa Island (Iles Eparses): 22 20 S, 40 22 E
Glorioso Islands (Iles Eparses): 11 30 S, 47 20 E
Juan de Nova Island (Iles Eparses): 17 03 S, 42 45 E
Tromelin Island (Iles Eparses): 15 52 S, 54 25 E

Map references: Antarctic Region

Area: Ile Amsterdam (Ile Amsterdam et Ile Saint-Paul): total—55 sq km; land—55 sq km; water—0 sq km
country comparison to the world: 228
Ile Saint-Paul (Ile Amsterdam et Ile Saint-Paul): total—7 sq km; land—7 sq km; water—0 sq km
Iles Crozet: total—352 sq km; land—352 sq km; water—0 sq km
Iles Kerguelen: total—7,215 sq km; land—7,215 sq km; water—0 sq km
Bassas da India (Iles Eparses): total—80 sq km; land—0.2 sq km; water—79.8 sq km (lagoon)
Europa Island (Iles Eparses): total—28 sq km; land—28 sq km; water—0 sq km
Glorioso Islands (Iles Eparses): total—5 sq km; land—5 sq km; water—0 sq km
Juan de Nova Island (Iles Eparses): total—4.4 sq km; land—4.4 sq km; water—0 sq km
Tromelin Island (Iles Eparses): total—1 sq km; land—1 sq km; water—0 sq km
note: excludes "Adelie Land" claim of about 500,000 sq km in Antarctica that is not recognized by the US

Area—comparative: Ile Amsterdam (Ile Amsterdam et Ile Saint-Paul): less than one-half the size of Washington, DC
Ile Saint-Paul (Ile Amsterdam et Ile Saint-Paul): more than 10 times the size of The Mall in Washington, DC
Iles Crozet: about twice the size of Washington, DC
Iles Kerguelen: slightly larger than Delaware
Bassas da India (Iles Eparses): land area about one-third the size of The Mall in Washington, DC
Europa Island (Iles Eparses): about one-sixth the size of Washington, DC
Glorioso Islands (Iles Eparses): about eight times the size of The Mall in Washington, DC
Juan de Nova Island (Iles Eparses): about seven times the size of The Mall in Washington, DC
Tromelin Island (Iles Eparses): about 1.7 times the size of The Mall in Washington, DC

Land boundaries: 0 km

Coastline: Ile Amsterdam (Ile Amsterdam et Ile Saint-Paul): 28 km
Ile Saint-Paul (Ile Amsterdam et Ile Saint-Paul): Iles Kerguelen: 2,800 km
Bassas da India (Iles Eparses): 35.2 km
Europa Island (Iles Eparses): 22.2 km
Glorioso Islands (Iles Eparses): 35.2 km
Juan de Nova Island (Iles Eparses): 24.1 km
Tromelin Island (Iles Eparses): 3.7 km

Maritime claims: *territorial sea:* 12 nm
exclusive economic zone: 200 nm from Iles Kerguelen and Iles Eparses (does not include the rest of French Southern and Antarctic Lands); Juan de Nova Island and Tromelin Island claim a continental shelf of 200-m depth or to the depth of exploitation

Climate: Ile Amsterdam et Ile Saint-Paul: oceanic with persistent westerly winds and high humidity

Iles Crozet: windy, cold, wet, and cloudy

Iles Kerguelen: oceanic, cold, overcast, windy

Iles Eparses: tropical

Terrain: Ile Amsterdam (Ile Amsterdam et Ile Saint-Paul): a volcanic island with steep coastal cliffs; the center floor of the volcano is a large plateau
Ile Saint-Paul (Ile Amsterdam et Ile Saint-Paul): triangular in shape, the island is the top of a volcano, rocky with steep cliffs on the eastern side; has active thermal springs
Iles Crozet: a large archipelago formed from the Crozet Plateau is divided into two groups of islands
Iles Kerguelen: the interior of the large island of Ile Kerguelen is composed of rugged terrain of high mountains, hills, valleys, and plains with a number of peninsulas stretching off its coasts
Bassas da India (Iles Eparses): atoll, awash at high tide; shallow (15 m) lagoon
Europa Island, Glorioso Islands, Juan de Nova Island: low, flat, and sandy
Tromelin Island (Iles Eparses): low, flat, sandy; likely volcanic seamount

Elevation extremes: *lowest point:* Indian Ocean 0 m
highest point: Mont de la Dives on Ile Amsterdam (Ile Amsterdam et Ile Saint-Paul) 867 m; unnamed location on Ile Saint-Paul (Ile Amsterdam et Ile Saint-Paul) 272 m; Pic Marion-Dufresne in Iles Crozet 1,090 m; Mont Ross in Iles Kerguelen 1,850 m; unnamed location on Bassas de India (Iles Eparses) 2.4 m; unnamed location on Europa Island (Iles Eparses) 24 m; unnamed location on Glorioso Islands (Iles Eparses) 12 m; unnamed location on Juan de Nova Island (Iles Eparses) 10 m; unnamed location on Tromelin Island (Iles Eparses) 7 m

Natural resources: fish, crayfish
note: Glorioso Islands and Tromelin Island (Iles Eparses) have guano, phosphates, and coconuts
Land use: Ile Amsterdam (Ile Amsterdam et Ile Saint-Paul)—100% trees, grasses, ferns, and moss; Ile Saint-Paul (Ile Amsterdam et Ile Saint-Paul)—100% grass, ferns, and moss; Iles Crozet—100% tossock grass, heath, and fern; Iles Kerguelen—100% tossock grass and Kerguelen cabbage; Bassas da India (Iles Eparses)—100% rock, coral reef, and sand; Europa Island (Iles Eparses)—100% mangrove swamp and dry woodlands; Glorioso Islands (Iles Eparses)—100% lush vegetation and coconut palms; Juan de Nova Island (Iles Eparses)—90% forest, 10% other; Tromelin Island (Iles Eparses)—100% grasses and scattered brush (2005)
Irrigated land: 0 sq km
Natural hazards: Ile Amsterdam and Ile Saint-Paul are inactive volcanoes; Iles Eparses subject to periodic cyclones; Bassas da India is a maritime hazard since it is under water for a period of three hours prior to and following the high tide and surrounded by reefs
volcanism: Reunion Island—Piton de la Fournaise (elev. 2,632 m,), which has erupted many times in recent years, including 2010, is one of the world's most active volcanoes; although rare, eruptions outside the volcano's caldera could threaten nearby cities
Environment—current issues: introduction of foreign species on Iles Crozet has caused severe damage to the original ecosystem; overfishing of Patagonian toothfish around Iles Crozet and Iles Kerguelen
Geography—note: islands component is widely scattered across remote locations in the southern Indian Ocean
Bassas da India (Iles Eparses): the atoll is a circular reef that sits atop a long-extinct, submerged volcano
Europa Island and Juan de Nova Island (Iles Eparses): wildlife sanctuary for seabirds and sea turtles
Glorioso Island (Iles Eparses): the islands and rocks are surrounded by an extensive reef system
Tromelin Island (Iles Eparses): climatologically important location for forecasting cyclones in the western Indian Ocean; wildlife sanctuary (seabirds, tortoises)

PEOPLE

Population: no indigenous inhabitants
Ile Amsterdam (Ile Amsterdam et Ile Saint-Paul): has no permanent residents but has a meteorological station
Ile Saint-Paul (Ile Amsterdam et Ile Saint-Paul): is uninhabited but is frequently visited by fishermen and has a scientific research cabin for short stays
Iles Crozet: are uninhabited except for 18 to 30 people staffing the Alfred Faure research station on Ile del la Possession
Iles Kerguelen: 50 to 100 scientists are located at the main base at Port-aux-Francais on Ile Kerguelen
Bassas da India (Iles Eparses): uninhabitable
Europa Island, Glorioso Islands, Juan de Nova Island (Iles Eparses): a small French military garrison and a few meteorologists on each possession; visited by scientists
Tromelin Island (Iles Eparses): uninhabited, except for visits by scientists

GOVERNMENT

Country name: *conventional long form:* Territory of the French Southern and Antarctic Lands
conventional short form: French Southern and Antarctic Lands
local long form: Territoire des Terres Australes et Antarctiques Francaises
local short form: Terres Australes et Antarctiques Francaises
abbreviation: TAAF
Dependency status: overseas territory of France since 1955
Administrative divisions: none (overseas territory of France); there are no first-order administrative divisions as defined by the US Government, but there are five administrative districts named Iles Crozet, Iles Eparses, Iles Kerguelen, Ile Saint-Paul et Ile Amsterdam; the fifth district is the "Adelie Land" claim in Antarctica that is not recognized by the US
Legal system: the laws of France, where applicable, apply
Executive branch: *chief of state:* President Nicolas SARKOZY (since 16 May 2007), represented by Senior Administrator Christian GAUDIN (since 4 November 2010)
International organization participation: UPU
Diplomatic representation in the US: none (overseas territory of France)
Diplomatic representation from the US: none (overseas territory of France)
Flag description: the flag of France is used
National anthem: note: as a territory of France, "La Marseillaise" is official (see France)

ECONOMY

Economy—overview: Economic activity is limited to servicing meteorological and geophysical research stations, military bases, and French and other fishing fleets. The fish catches landed on Iles Kerguelen by foreign ships are exported to France and Reunion.

COMMUNICATIONS

Internet country code: .tf
Internet hosts: 44 (2010)
country comparison to the world: 212
Communications—note: one or more meteorological stations on each possession

TRANSPORTATION

Airports: 4; note—one each on Europa Island, Glorioso Islands, Juan de Nova Island, and Tromelin Island in the Iles Eparses district (2010)
country comparison to the world: 185
Ports and terminals: none; offshore anchorage only
Transportation—note: aids to navigation—lighthouses: Europa Island 18 m; Juan de Nova Island (W side) 37 m; Tromelin Island (NW point) 11m (all in the Iles Eparses district)

MILITARY

Military—note: defense is the responsibility of France

TRANSNATIONAL ISSUES

Disputes—international: French claim to "Adelie Land" in Antarctica is not recognized by the US
Bassas da India, Europa Island, Glorioso Islands, Juan de Nova Island (Iles Eparses): claimed by Madagascar; the vegetated drying cays of Banc du Geyser, which were claimed by Madagascar in 1976, also fall within the EEZ claims of the Comoros and France (Glorioso Islands)
Tromelin Island (Iles Eparses): claimed by Mauritius

GABON

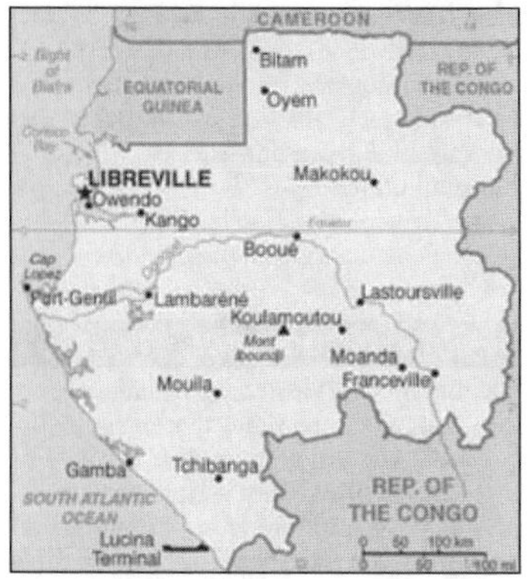

INTRODUCTION

Background: Until recently, only two autocratic presidents had ruled Gabon since its independence from France in 1960. The recent president of Gabon, El Hadj Omar BONGO Ondimba–one of the longest-serving heads of state in the world–had dominated the country's political scene for four decades. President BONGO introduced a nominal multiparty system and a new constitution in the early 1990s. However, allegations of electoral fraud during local elections in 2002-03 and the presidential elections in 2005 exposed the weaknesses of formal political structures in Gabon. President BONGO died in June 2009. New elections in August 2009 brought Ali Ben BONGO, son of the former president, to power. Despite political conditions, a small population, abundant natural resources, and considerable foreign support have helped make Gabon one of the more prosperous and stable African countries. In January 2010, Gabon assumed a nonpermanent seat on the UN Security Council for the 2010-11 term.

GEOGRAPHY

Location: Western Africa, bordering the Atlantic Ocean at the Equator, between Republic of the Congo and Equatorial Guinea
Geographic coordinates: 1 00 S, 11 45 E
Map references: Africa
Area: *total:* 267,667 sq km
country comparison to the world: 76
land: 257,667 sq km
water: 10,000 sq km
Area—comparative: slightly smaller than Colorado
Land boundaries: *total:* 2,551 km
border countries: Cameroon 298 km, Republic of the Congo 1,903 km, Equatorial Guinea 350 km
Coastline: 885 km
Maritime claims:
territorial sea: 12 nm
contiguous zone: 24 nm
exclusive economic zone: 200 nm
Climate: tropical; always hot, humid
Terrain: narrow coastal plain; hilly interior; savanna in east and south
Elevation extremes: *lowest point:* Atlantic Ocean 0 m
highest point: Mont Iboundji 1,575 m
Natural resources: petroleum, natural gas, diamond, niobium, manganese, uranium, gold, timber, iron ore, hydropower
Land use: *arable land:* 1.21%
permanent crops: 0.64%
other: 98.15% (2005)
Irrigated land: 70 sq km (2003)
Total renewable water resources: 164 cu km (1987)
Freshwater withdrawal (domestic/industrial/agricultural): *total:* 0.12 cu km/yr (50%/8%/42%)
per capita: 87 cu m/yr (2000)
Natural hazards: NA
Environment—current issues: deforestation; poaching
Environment—international agreements: *party to:* Biodiversity, Climate Change, Climate Change-Kyoto Protocol, Desertification, Endangered Species, Hazardous Wastes, Law of the Sea, Marine Dumping, Ozone Layer Protection, Ship Pollution, Tropical Timber 83, Tropical Timber 94, Wetlands, Whaling
signed, but not ratified: none of the selected agreements
Geography—note: a small population and oil and mineral reserves have helped Gabon become one of Africa's wealthier countries; in general, these circumstances have allowed the country to maintain and conserve its pristine rain forest and rich biodiversity

PEOPLE

Population: 1,576,665 (July 2011 est.)
country comparison to the world: 151
note: estimates for this country explicitly take into account the effects of excess mortality due to AIDS; this can result in lower life expectancy, higher infant mortality, higher death rates, lower population growth rates, and changes in the distribution of population by age and sex than would otherwise be expected
Age structure: *0–14 years:* 42.2% (male 333,746/female 330,959)
15–64 years: 54% (male 424,392/female 426,478)
65 years and over: 3.9% (male 25,687/female 35,403) (2011 est.)
Median age: *total:* 18.6 years
male: 18.4 years
female: 18.9 years (2011 est.)
Population growth rate: 1.999% (2011 est.)
country comparison to the world: 51
Birth rate: 35.19 births/1,000 population (2011 est.)
country comparison to the world: 30
Death rate: 13 deaths/1,000 population (July 2011 est.)
country comparison to the world: 23
Net migration rate: -2.2 migrant(s)/1,000 population (2011 est.)
country comparison to the world: 169
Urbanization: *urban population:* 86% of total population (2010)
rate of urbanization: 2.1% annual rate of change (2010-15 est.)
Major cities—population: LIBREVILLE (capital) 619,000 (2009)
Sex ratio: *at birth:* 1.03 male(s)/female
under 15 years: 1.01 male(s)/female
15–64 years: 1 male(s)/female
65 years and over: 0.72 male(s)/female
total population: 0.99 male(s)/female (2011 est.)
Infant mortality rate:
total: 49.95 deaths/1,000 live births
country comparison to the world: 48
male: 57.87 deaths/1,000 live births
female: 41.8 deaths/1,000 live births (2011 est.)
Life expectancy at birth: *total population:* 52.49 years
country comparison to the world: 207
male: 51.78 years
female: 53.22 years (2011 est.)
Total fertility rate: 4.59 children born/woman (2011 est.)
country comparison to the world: 32
HIV/AIDS—adult prevalence rate: 5.2% (2009 est.)
country comparison to the world: 14
HIV/AIDS—people living with HIV/AIDS: 46,000 (2009 est.)
country comparison to the world: 60
HIV/AIDS—deaths: 2,400 (2009 est.)
country comparison to the world: 52
Major infectious diseases: *degree of risk:* very high
food or waterborne diseases: bacterial diarrhea, hepatitis A, and typhoid fever
vectorborne disease: malaria and chikungunya
water contact disease: schistosomiasis
animal contact disease: rabies (2009)
Drinking water source: *Improved:*
urban: 95% of population
rural: 41% of population
total: 87% of population
Unimproved: urban: 5% of population
rural: 59% of population
total: 13% of population (2008)
Sanitation facility access: *Improved:*
urban: 33% of population
rural: 30% of population
total: 33% of population
Unimproved:
urban: 27% of population
rural: 30% of population
total: 27% of population (2008)
Nationality: *noun:* Gabonese (singular and plural)
adjective: Gabonese
Ethnic groups: Bantu tribes, including four major tribal groupings (Fang, Bapounou, Nzebi, Obamba); other Africans and Europeans, 154,000, including 10,700 French and 11,000 persons of dual nationality
Religions: Christian 55%-75%, animist, Muslim less than 1%
Languages: French (official), Fang, Myene, Nzebi, Bapounou/Eschira, Bandjabi
Literacy: *definition:* age 15 and over can read and write
total population: 63.2%

male: 73.7%
female: 53.3% (1995 est.)
School life expectancy (primary to tertiary education): *total:* 13 years
male: 12 years
female: 12 years (2002)
Education expenditures: NA

GOVERNMENT

Country name: *conventional long form:* Gabonese Republic
conventional short form: Gabon
local long form: Republique Gabonaise
local short form: Gabon
Government type: republic; multiparty presidential regime
Capital: *name:* Libreville
geographic coordinates: 0 23 N, 9 27 E
time difference: UTC+1 (6 hours ahead of Washington, DC during Standard Time)
Administrative divisions: 9 provinces; Estuaire, Haut-Ogooue, Moyen-Ogooue, Ngounie, Nyanga, Ogooue-Ivindo, Ogooue-Lolo, Ogooue-Maritime, Woleu-Ntem
Independence: 17 August 1960 (from France)
National holiday: Independence Day, 17 August (1960)
Constitution: adopted 14 March 1991
Legal system: mixed legal system of French civil law and customary law
International law organization participation: has not submitted an ICJ jurisdiction declaration; accepts ICCt jurisdiction
Suffrage: 18 years of age; universal
Executive branch: *chief of state:* President Ali Ben BONGO Ondimba (since 16 October 2009)
head of government: Prime Minister Paul BIYOGHE MBA (since 15 July 2009)
cabinet: Council of Ministers appointed by the prime minister in consultation with the president
(For more information visit the World Leaders website)
elections: president elected by popular vote for a seven-year term (no term limits); election last held on 30 August 2009 (next to be held in 2016); prime minister appointed by the president
election results: President Ali Ben BONGO Ondimba elected; percent of vote–Ali Ben BONGO Ondimba 41.7%, Andre MBA OBAME 25.9%, Pierre MAMBOUNDOU 25.2%, Zacharie MYBOTO 3.9%, other 3.3%
note: President BONGO died on 8 June 2009 after serving as president for 32 years; in accordance with the constitution he was replaced on an interim basis by the president of the Senate, Rose Francine ROGOMBE on 10 June 2009; new elections were held on 30 August 2009 and the son of the former president, Ali Ben BONGO Ondimba, was elected president
Legislative branch: bicameral legislature consists of the Senate (102 seats; members elected by members of municipal councils and departmental assemblies to serve six-year terms) and the National Assembly or Assemblee Nationale (120 seats; members are elected by direct, popular vote to serve five-year terms)
elections: Senate–last held on 18 January 2009 (next to be held in January 2015); National Assembly–last held on 17 and 24 December 2006 (next to be held in December 2011)
election results: Senate–percent of vote by party–NA; seats by party–PDG 75, RPG 6, UGDD 3, CLR 2, PGCI 2, PSD 2, UPG 2, ADERE 1, independents 9; National Assembly–percent of vote by party–NA; seats by party–PDG 82, RPG 8, UPG 8, UGDD 4, ADERE 3, CLR 2, PGP-Ndaot 2, PSD 2, independents 4, others 5
Judicial branch: Supreme Court or Cour Supreme consisting of three chambers–Judicial, Administrative, and Accounts; Constitutional Court; Courts of Appeal; Court of State Security; county courts
Political parties and leaders: Circle of Liberal Reformers or CLR [General Jean Boniface ASSELE]; Congress for Democracy and Justice or CDJ [Jules Aristide Bourdes OGOULIGUENDE]; Democratic and Republican Alliance or ADERE [Divungui-di-Ndinge DIDJOB]; Gabonese Democratic Party or PDG [Simplice Nguedet MANZELA] (former sole party); Gabonese Party for Progress or PGP [Benoit Mouity NZAMBA]; Gabonese Union for Democracy and Development or UGDD [Zacherie MYBOTO]; National Rally of Woodcutters or RNB; National Rally of Woodcutters-Rally for Gabon or RNB-RPG (Bucherons) [Fr. Paul M'BA-ABESSOLE]; Party of Development and Social Solidarity or PDS [Seraphin Ndoat REMBOGO]; People's Unity Party or PUP [Louis Gaston MAYILA]; Social Democratic Party or PSD [Pierre Claver MAGANGA-MOUSSAVOU]; Union for Democracy and Social Integration or UDIS; Union of Gabonese Patriots or UPG [Pierre MAMBOUNDOU]
Political pressure groups and leaders: NA
International organization participation: ACP, AfDB, AU, BDEAC, CEMAC, FAO, FZ, G-24, G-77, IAEA, IBRD, ICAO, ICRM, IDA, IDB, IFAD, IFC, IFRCS, ILO, IMF, IMO, IMSO, Interpol, IOC, IOM, IPU, ISO (correspondent), ITSO, ITU, ITUC, MIGA, NAM, OIC, OIF, OPCW, UN, UN Security Council (temporary), UNCTAD, UNESCO, UNIDO, UNWTO, UPU, WCO, WHO, WIPO, WMO, WTO
Diplomatic representation in the US: *chief of mission:* Ambassador Carlos Victor BOUNGOU
chancery: Suite 200, 2034 20th Street NW, Washington, DC 20009
telephone: [1] (202) 797-1000
FAX: [1] (202) 332-0668
consulate(s): New York
Diplomatic representation from the US: *chief of mission:* Ambassador Erik D. BENJAMINSON
embassy: Boulevard du Bord de Mer, Libreville
mailing address: Centre Ville, B. P. 4000, Libreville; pouch:2270 Libreville Place, Washington, DC 20521-2270
telephone: [241] 76 20 03 through 76 20 04, after hours–07380171
FAX: [241] 74 55 07
Flag description: three equal horizontal bands of green (top), yellow, and blue; green represents the country's forests and natural resources, gold represents the equator (which transects Gabon) as well as the sun, blue represents the sea
National anthem:
name: "La Concorde" (The Concorde)
lyrics/music: Georges Aleka DAMAS
note: adopted 1960

ECONOMY

Economy—overview: Gabon enjoys a per capita income four times that of most sub-Saharan African nations, but because of high income inequality, a large proportion of the population remains poor. Gabon depended on timber and manganese until oil was discovered offshore in the early 1970s. The oil sector now accounts for more than 50% of GDP although the industry is in decline as fields pass their peak production. Gabon continues to face fluctuating prices for its oil, timber, and manganese exports and the global recession led to a GDP contraction of 1.4% in 2009. Despite the abundance of natural wealth, poor fiscal management hobbles the economy. In 1997, an IMF mission to Gabon criticized the government for overspending on off-budget items, overborrowing from the central bank, and slipping on its schedule for privatization and administrative reform. The rebound of oil prices from 1999 to 2008 helped growth, but drops in production have hampered Gabon from fully realizing potential gains. Gabon signed a 14-month Stand-By Arrangement with the IMF in May 2007, and later that year issued a $1 billion sovereign bond to buy back a sizable portion of its Paris Club debt.
GDP (purchasing power parity): $22.48 billion (2010 est.)
country comparison to the world: 120
$21.27 billion (2009 est.)
$21.57 billion (2008 est.)
note: data are in 2010 US dollars
GDP (official exchange rate): $13.06 billion (2010 est.)
GDP—real growth rate: 5.7% (2010 est.)
country comparison to the world: 57
-1.4% (2009 est.)
2.3% (2008 est.)
GDP—per capita (PPP): $14,500 (2010 est.)
country comparison to the world: 80
$14,000 (2009 est.)
$14,500 (2008 est.)
note: data are in 2010 US dollars
GDP—composition by sector:
agriculture: 4.5%
industry: 62.7%
services: 32.8% (2010 est.)
Labor force: 712,000 (2010 est.)
country comparison to the world: 148
Labor force—by occupation:
agriculture: 60%
industry: 15%
services: 25% (2000 est.)
Unemployment rate: 21% (2006 est.)
country comparison to the world: 169
Population below poverty line: NA%

Household income or consumption by percentage share: *lowest 10%:* 2.5%
highest 10%: 32.7% (2005)
Investment (gross fixed): 28.8% of GDP (2010 est.)
country comparison to the world: 21
Budget: *revenues:* $3.557 billion
expenditures: $2.945 billion (2010 est.)
Public debt: 25.8% of GDP (2010 est.)
country comparison to the world: 96
27.6% of GDP (2009 est.)
Inflation rate (consumer prices): -1.3% (2010 est.)
country comparison to the world: 3
1.9% (2009 est.)
Central bank discount rate: 4.25% (31 December 2009)
country comparison to the world: 99
4.75% (31 December 2008)
Commercial bank prime lending rate: NA%
Stock of narrow money: $1.835 billion (31 December 2010 est.)
country comparison to the world: 122
$1.623 billion (31 December 2009 est.)
Stock of broad money: $2.764 billion (31 December 2010 est.)
country comparison to the world: 134
$2.468 billion (31 December 2009 est.)
Stock of domestic credit: $1.074 billion (31 December 2010 est.)
country comparison to the world: 148
$826.8 million (31 December 2009 est.)
Market value of publicly traded shares: $NA
Agriculture—products: cocoa, coffee, sugar, palm oil, rubber; cattle; okoume (a tropical softwood); fish
Industries: petroleum extraction and refining; manganese, gold; chemicals, ship repair, food and beverages, textiles, lumbering and plywood, cement
Industrial production growth rate: 4.8% (2010 est.)
country comparison to the world: 72
Electricity—production: 1.774 billion kWh (2007 est.)
country comparison to the world: 136
Electricity—consumption: 1.446 billion kWh (2007 est.)
country comparison to the world: 141
Electricity—exports: 0 kWh (2008 est.)
Electricity—imports: 0 kWh (2008 est.)
Oil—production: 241,700 bbl/day (2009 est.)
country comparison to the world: 41
Oil—consumption: 14,000 bbl/day (2009 est.)
country comparison to the world: 138
Oil—exports: 227,300 bbl/day (2007 est.)
country comparison to the world: 52
Oil—imports: 4,185 bbl/day (2007 est.)
country comparison to the world: 165
Oil—proved reserves: 2 billion bbl (1 January 2010 est.)
country comparison to the world: 35
Natural gas—production: 90 million cu m (2008 est.)
country comparison to the world: 79
Natural gas—consumption: 90 million cu m (2008 est.)
country comparison to the world: 105
Natural gas—exports: 0 cu m (2008 est.)
country comparison to the world: 99
Natural gas—imports: 0 cu m (2008 est.)
country comparison to the world: 189
Natural gas—proved reserves: 28.32 billion cu m (1 January 2010 est.)
country comparison to the world: 72
Current account balance: $591 million (2010 est.)
country comparison to the world: 51
$887 million (2009 est.)
Exports: $6.803 billion (2010 est.)
country comparison to the world: 100
$6.04 billion (2009 est.)
Exports—commodities: crude oil 70%, timber, manganese, uranium
Exports—partners: Russia 30.62%, US 16.56%, China 15.87%, France 4.28% (2009)
Imports: $2.433 billion (2010 est.)
country comparison to the world: 147
$2.298 billion (2009 est.)
Imports—commodities: machinery and equipment, foodstuffs, chemicals, construction materials
Imports—partners: France 32.21%, US 7.92%, China 7.02%, Belgium 4.99%, Italy 4.81%, Cameroon 4.56%, Netherlands 4.35% (2009)
Reserves of foreign exchange and gold: $2.602 billion (31 December 2010 est.)
country comparison to the world: 93
$1.993 billion (31 December 2009 est.)
Debt—external: $2.374 billion (31 December 2010 est.)
country comparison to the world: 136
$2.352 billion (31 December 2009 est.)
Exchange rates: Cooperation Financiere en Afrique Centrale francs per US dollar— 495.28 (2010), 472.19 (2009), 447.81 (2008), 481.83 (2007), 522.89 (2006)

COMMUNICATIONS

Telephones—main lines in use: 26,500 (2009)
country comparison to the world: 182
Telephones—mobile cellular: 1.373 million (2009)
country comparison to the world: 141
Telephone system: *general assessment:* adequate system of cable, microwave radio relay, tropospheric scatter, radiotelephone communication stations, and a domestic satellite system with 12 earth stations
domestic: a growing mobile-cellular network with multiple providers is making telephone service more widely available; subscribership reached 90 per 100 persons in 2009
international: country code—241; landing point for the SAT-3/WASC fiber-optic submarine cable that provides connectivity to Europe and Asia; satellite earth stations—3 Intelsat (Atlantic Ocean) (2009)
Broadcast media: state owns and operates 2 TV stations and 2 radio broadcast stations; a few private radio and TV stations are operational; transmissions of at least 2 international broadcasters are accessible; satellite service subscriptions are available (2007)
Internet country code: .ga
Internet hosts: 90 (2010)
country comparison to the world: 204
Internet users: 98,800 (2009)
country comparison to the world: 160

TRANSPORTATION

Airports: 44 (2010)
country comparison to the world: 97
Airports—with paved runways: *total:* 13
over 3,047 m: 1
2,438 to 3,047 m: 1
1,524 to 2,437 m: 9
914 to 1,523 m: 1
under 914 m: 1 (2010)
Airports—with unpaved runways: *total:* 31
1,524 to 2,437 m: 6
914 to 1,523 m: 11
under 914 m: 14 (2010)
Pipelines: gas 294 km; oil 893 km (2010)
Railways: *total:* 649 km
country comparison to the world: 105
standard gauge: 649 km 1.435-m gauge (2009)
Roadways: *total:* 9,170 km
country comparison to the world: 138
paved: 937 km
unpaved: 8,233 km (2004)
Waterways: 1,600 km (310 km on Ogooue River) (2010)
country comparison to the world: 51
Merchant marine: registered in other countries: 2 (Cambodia 1, Panama 1) (2010)
country comparison to the world: 142
Ports and terminals: Gamba, Libreville, Lucinda, Owendo, Port-Gentil

MILITARY

Military branches: Army, Navy, Air Force, National Gendarmerie, National Police
Military service age and obligation: 20 years of age for voluntary military service; no conscription (2009)
Manpower available for military service: *males age 16–49:* 350,640
females age 16–49: 351,718 (2010 est.)
Manpower fit for military service: *males age 16–49:* 202,404
females age 16–49: 195,389 (2010 est.)
Manpower reaching militarily significant age annually: *male:* 17,638
female: 17,614 (2010 est.)
Military expenditures: 0.9% of GDP (2009)
country comparison to the world: 142

TRANSNATIONAL ISSUES

Disputes—international: UN urges Equatorial Guinea and Gabon to resolve the sovereignty dispute over Gabon-occupied Mbane Island and lesser islands and to establish a maritime boundary in hydrocarbon-rich Corisco Bay
Refugees and internally displaced persons: refugees (country of origin): 7,178 (Republic of Congo) (2007)
Trafficking in persons: current situation: Gabon is predominantly a destination country for children trafficked from other African countries for the purpose of forced labor; girls are primarily trafficked for domestic servitude, forced market vending, forced restaurant labor, and sexual exploitation, while boys are trafficked for forced street hawking and forced labor in small workshops

tier rating: Tier 2 Watch List–Gabon is on the Tier 2 Watch List for its failure to provide evidence of increasing efforts to combat human trafficking in 2007, particularly in terms of efforts to convict and punish trafficking offenders; the government has not reported the convictions or sentences of any trafficking offenders; the government did not take steps to reduce demand for commercial sex acts (2008)

GAMBIA, THE

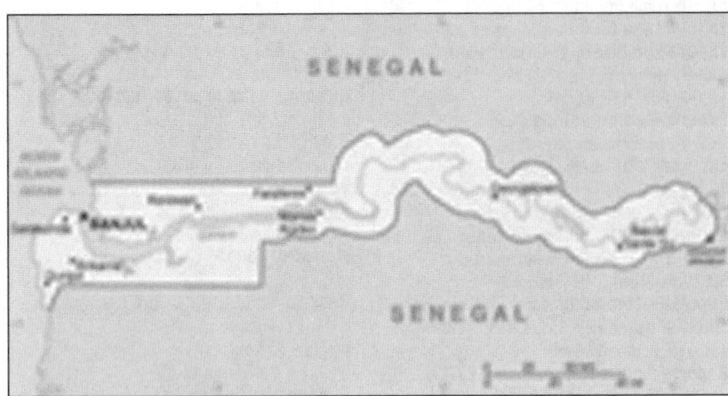

INTRODUCTION

Background: The Gambia gained its independence from the UK in 1965. Geographically surrounded by Senegal, it formed a short-lived federation of Senegambia between 1982 and 1989. In 1991 the two nations signed a friendship and cooperation treaty, but tensions have flared up intermittently since then. Yahya JAMMEH led a military coup in 1994 that overthrew the president and banned political activity. A new constitution and presidential elections in 1996, followed by parliamentary balloting in 1997, completed a nominal return to civilian rule. JAMMEH has been elected president in all subsequent elections including most recently in late 2006.

GEOGRAPHY

Location: Western Africa, bordering the North Atlantic Ocean and Senegal
Geographic coordinates: 13 28 N, 16 34 W
Map references: Africa
Area:
total: 11,295 sq km
country comparison to the world: 166
land: 10,000 sq km
water: 1,295 sq km
Area—comparative: slightly less than twice the size of Delaware
Land boundaries: *total:* 740 km
border countries: Senegal 740 km
Coastline: 80 km
Maritime claims: *territorial sea:* 12 nm
contiguous zone: 18 nm
exclusive fishing zone: 200 nm
continental shelf: extent not specified
Climate: tropical; hot, rainy season (June to November); cooler, dry season (November to May)
Terrain: flood plain of the Gambia River flanked by some low hills
Elevation extremes: *lowest point:* Atlantic Ocean 0 m
highest point: unnamed elevation 53 m
Natural resources: fish, clay, silica sand, titanium (rutile and ilmenite), tin, zircon
Land use: *arable land:* 27.88%
permanent crops: 0.44%
other: 71.68% (2005)
Irrigated land: 20 sq km (2003)
Total renewable water resources: 8 cu km (1982)
Freshwater withdrawal (domestic/industrial/agricultural): *total:* 0.03 cu km/yr (23%/12%/65%)
per capita: 20 cu m/yr (2000)
Natural hazards: drought (rainfall has dropped by 30% in the last 30 years)
Environment—current issues: deforestation; desertification; water-borne diseases prevalent
Environment—international agreements: *party to:* Biodiversity, Climate Change, Climate Change-Kyoto Protocol, Desertification, Endangered Species, Hazardous Wastes, Law of the Sea, Ozone Layer Protection, Ship Pollution, Wetlands, Whaling *signed, but not ratified:* none of the selected agreements
Geography—note: almost an enclave of Senegal; smallest country on the continent of Africa

PEOPLE

Population: 1,797,860 (July 2011 est.)
country comparison to the world: 148
Age structure: *0–14 years:* 40% (male 360,732/female 358,440)
15–64 years: 56.9% (male 501,946/female 520,826)
65 years and over: 3.1% (male 26,645/female 29,271) (2011 est.)
Median age: *total:* 19.4 years
male: 19.2 years
female: 19.7 years (2011 est.)
Population growth rate: 2.396% (2011 est.)
country comparison to the world: 32
Birth rate: 34.19 births/1,000 population (2011 est.)
country comparison to the world: 32
Death rate: 7.65 deaths/1,000 population (July 2011 est.)
country comparison to the world: 113
Net migration rate: -2.58 migrant(s)/1,000 population (2011 est.)
country comparison to the world: 171
Urbanization: *urban population:* 58% of total population (2010)
rate of urbanization: 3.7% annual rate of change (2010-15 est.)
Major cities—population: BANJUL (capital) 436,000 (2009)
Sex ratio: *at birth:* 1.03 male(s)/female
under 15 years: 1.01 male(s)/female
15–64 years: 0.98 male(s)/female
65 years and over: 0.98 male(s)/female
total population: 1 male(s)/female (2011 est.)
Infant mortality rate:
total: 71.67 deaths/1,000 live births
country comparison to the world: 19
male: 77.3 deaths/1,000 live births
female: 65.87 deaths/1,000 live births (2011 est.)
Life expectancy at birth: *total population:* 63.51 years
country comparison to the world: 175
male: 61.23 years
female: 65.86 years (2011 est.)
Total fertility rate: 4.23 children born/woman (2011 est.)
country comparison to the world: 37
HIV/AIDS—adult prevalence rate: 2% (2009 est.)
country comparison to the world: 28
HIV/AIDS—people living with HIV/AIDS: 18,000 (2009 est.)
country comparison to the world: 81
HIV/AIDS—deaths: fewer than 1,000 (2009 est.)
country comparison to the world: 72
Major infectious diseases: *degree of risk:* very high
food or waterborne diseases: bacterial and protozoal diarrhea, hepatitis A, and typhoid fever
vectorborne diseases: malaria
water contact disease: schistosomiasis
respiratory disease: meningococcal meningitis
animal contact disease: rabies (2009)
Drinking water source: *Improved:*
urban: 96% of population
rural: 86% of population
total: 92% of population
Unimproved: urban: 4% of population
rural: 14% of population
total: 8% of population (2008)
Sanitation facility access: *Improved:*
urban: 68% of population
rural: 65% of population
total: 67% of population
Unimproved: urban: 32% of population
rural: 35% of population

total: 33% of population (2008)
Nationality: *noun:* Gambian(s)
adjective: Gambian
Ethnic groups: African 99% (Mandinka 42%, Fula 18%, Wolof 16%, Jola 10%, Serahuli 9%, other 4%), non-African 1% (2003 census)
Religions: Muslim 90%, Christian 8%, indigenous beliefs 2%
Languages: English (official), Mandinka, Wolof, Fula, other indigenous vernaculars
Literacy: *definition:* age 15 and over can read and write
total population: 40.1%
male: 47.8%
female: 32.8% (2003 est.)
School life expectancy (primary to tertiary education):
total: 9 years
male: 9 years
female: 9 years (2008)
Education expenditures: 2% of GDP (2004)
country comparison to the world: 155

GOVERNMENT

Country name: *conventional long form:* Republic of The Gambia
conventional short form: The Gambia
Government type: republic
Capital: *name:* Banjul
geographic coordinates: 13 27 N, 16 34 W
time difference: UTC 0 (5 hours ahead of Washington, DC during Standard Time)
Administrative divisions: 5 divisions and 1 city*; Banjul*, Central River, Lower River, North Bank, Upper River, Western
Independence: 18 February 1965 (from the UK)
National holiday: Independence Day, 18 February (1965)
Constitution: approved by national referendum 8 August 1996; effective 16 January 1997
Legal system: mixed legal system of English common law, Islamic law, and customary law
International law organization participation: accepts compulsory ICJ jurisdiction with reservations; accepts ICCt jurisdiction
Suffrage: 18 years of age; universal
Executive branch: *chief of state:* President Yahya JAMMEH (since 18 October 1996); note—from 1994 to 1996 he was chairman of the junta; Vice President Isatou NJIE-SAIDY (since 20 March 1997); note—the president is both the chief of state and head of government
head of government: President Yahya JAMMEH (since 18 October 1996); Vice President Isatou NJIE-SAIDY (since 20 March 1997)
cabinet: Cabinet appointed by the president (For more information visit the World Leaders website)
elections: president elected by popular vote for a five-year term (no term limits); election last held on 22 September 2006 (next to be held in 2011)
election results: Yahya JAMMEH reelected president; percent of vote—Yahya JAMMEH 67.3%, Ousainou DARBOE 26.6%, Halifa SALLAH 6%
Legislative branch: unicameral National Assembly (53 seats; 48 members elected by popular vote, 5 appointed by the president; members to serve five-year terms)
elections: last held on 25 January 2007 (next to be held in 2012)
election results: percent of vote by party—NA; seats by party—APRC 47, UDP 4, NADD 1, independent 1
Judicial branch: Supreme Court
Political parties and leaders: Alliance for Patriotic Reorientation and Construction or APRC [Yahya A. J. J. JAMMEH] (the ruling party); Gambia People's Democratic Party or GPDP [Henry GOMEZ]; National Alliance for Democracy and Development or NADD [Halifa SALLAH]; National Convention Party or NCP [Sheriff DIBBA]; National Reconciliation Party or NRP [Hamat N. K. BAH]; People's Democratic Organization for Independence and Socialism or PDOIS [Halifa SALLAH]; United Democratic Party or UDP [Ousainou DARBOE]
Political pressure groups and leaders: National Environment Agency or NEA; West African Peace Building Network-Gambian Chapter or WANEB-GAMBIA; Youth Employment Network Gambia or YENGambia
other: special needs group advocates; teachers and principals
International organization participation: ACP, AfDB, AU, C, ECOWAS, FAO, G-77, IBRD, ICAO, ICRM, IDA, IDB, IFAD, IFC, IFRCS, ILO, IMF, IMO, Interpol, IOC, IOM, IPU, ISO (correspondent), ITSO, ITU, ITUC, MIGA, NAM, OIC, OPCW, UN, UNAMID, UNCTAD, UNESCO, UNIDO, UNMIL, UNOCI, UNWTO, UPU, WCO, WFTU, WHO, WIPO, WMO, WTO
Diplomatic representation in the US: *chief of mission:* Ambassador Alieu Momodou NGUM
chancery: Suite 240, Georgetown Plaza, 2233 Wisconsin Avenue NW, Washington, DC 20007
telephone: [1] (202) 785-1379, 1399, 1425
FAX: [1] (202) 785-1430
Diplomatic representation from the US: *chief of mission:* Ambassador (vacant); Charge d'Affaires Cindy GREGG
embassy: Kairaba Avenue, Fajara, Banjul
mailing address: P. M. B. No. 19, Banjul
telephone: [220] 439-2856, 437-6169, 437-6170
FAX: [220] 439-2475
Flag description: three equal horizontal bands of red (top), blue with white edges, and green; red stands for the sun and the savannah, blue represents the Gambia River, and green symbolizes forests and agriculture; the white stripes denote unity and peace
National anthem: *name:* "For The Gambia, Our Homeland"
lyrics/music: Virginia Julie HOWE/adapted by Jeremy Frederick HOWE
note: adopted 1965; the music is an adaptation of the traditional Mandinka song "Foday Kaba Dumbuya"

ECONOMY

Economy—overview: The Gambia has sparse natural resource deposits and a limited agricultural base, and relies in part on remittances from workers overseas and tourist receipts. About three-quarters of the population depends on the agricultural sector for its livelihood. Small-scale manufacturing activity features the processing of peanuts, fish, and hides. The Gambia's natural beauty and proximity to Europe has made it one of the larger markets for tourism in West Africa, boosted by government and private sector investments in eco-tourism and upscale facilities. In the past few years, The Gambia's re-export trade—traditionally a major segment of economic activity—has declined, but its banking sector has grown rapidly. Unemployment and underemployment rates remain high; economic progress depends on sustained bilateral and multilateral aid, on responsible government economic management, and on continued technical assistance from multilateral and bilateral donors. The quality of fiscal management, however, is weak. The government has promised to raise civil service wages over the next two years and the deficit is projected to worsen.
GDP (purchasing power parity): $3.494 billion (2010 est.)
country comparison to the world: 170
$3.304 billion (2009 est.)
$3.098 billion (2008 est.)
note: data are in 2010 US dollars
GDP (official exchange rate): $1.067 billion (2010 est.)
GDP—real growth rate: 5.7% (2010 est.)
country comparison to the world: 59
6.7% (2009 est.)
6.3% (2008 est.)
GDP—per capita (PPP): $1,900 (2010 est.)
country comparison to the world: 191
$1,900 (2009 est.)
$1,800 (2008 est.)
note: data are in 2010 US dollars
GDP—composition by sector:
agriculture: 30.1%
industry: 16.3%
services: 53.6% (2010 est.)
Labor force: 777,100 (2007)
country comparison to the world: 146
Labor force—by occupation:
agriculture: 75%
industry: 19%
services: 6% (1996)
Unemployment rate: NA%
Population below poverty line: NA%
Household income or consumption by percentage share: *lowest 10%:* 2%
highest 10%: 36.9% (2003)
Distribution of family income—Gini index: 50.2 (1998)
country comparison to the world: 22
Investment (gross fixed): 28% of GDP (2010 est.)
country comparison to the world: 24
Budget: *revenues:* $183.9 million
expenditures: $202.5 million (2010 est.)
Inflation rate (consumer prices): 5.5% (2010 est.)
country comparison to the world: 149
4.6% (2009 est.)
Central bank discount rate: 9% (31 December 2009)
country comparison to the world: 37

11% (31 December 2008)
Commercial bank prime lending rate: 27% (31 December 2009 est.)
country comparison to the world: 6
27.92% (31 December 2007)
Stock of narrow money: $222.9 million (31 December 2010 est.)
country comparison to the world: 172
$210.2 million (31 December 2009 est.)
Stock of broad money: $453.9 million (31 December 2010 est.)
country comparison to the world: 173
$438.9 million (31 December 2009 est.)
Stock of domestic credit: $293.5 million (31 December 2010 est.)
country comparison to the world: 170
$283.7 million (31 December 2009 est.)
Market value of publicly traded shares: $NA
Agriculture—products: rice, millet, sorghum, peanuts, corn, sesame, cassava (tapioca), palm kernels; cattle, sheep, goats
Industries: processing peanuts, fish, and hides; tourism, beverages, agricultural machinery assembly, woodworking, metalworking, clothing
Industrial production growth rate: 8.9%
country comparison to the world: 26
note: although The Gambia had the highest industrial growth rate in the world in 2009, this growth is from a tiny industrial base (2010 est.)
Electricity—production: 160 million kWh (2007 est.)
country comparison to the world: 180
Electricity—consumption: 148.8 million kWh (2007 est.)
country comparison to the world: 183
Electricity—exports: 0 kWh (2008 est.)
Electricity—imports: 0 kWh (2008 est.)
Oil—production: 0 bbl/day (2009 est.)
country comparison to the world: 177
Oil—consumption: 2,000 bbl/day (2009 est.)
country comparison to the world: 187
Oil—exports: 41.62 bbl/day (2007 est.)
country comparison to the world: 134
Oil—imports: 2,266 bbl/day (2007 est.)
country comparison to the world: 173
Oil—proved reserves: 0 bbl (1 January 2010 est.)
country comparison to the world: 134
Natural gas—production: 0 cu m (2008 est.)
country comparison to the world: 129
Natural gas—consumption: 0 cu m (2008 est.)
country comparison to the world: 174
Natural gas—exports: 0 cu m (2008 est.)
country comparison to the world: 98
Natural gas—imports: 0 cu m (2008 est.)
country comparison to the world: 188
Natural gas—proved reserves: 0 cu m (1 January 2010 est.)
country comparison to the world: 138
Current account balance: $-90 million (2010 est.)
country comparison to the world: 77
$-81 million (2009 est.)
Exports: $107 million (2010 est.)
country comparison to the world: 193
$95 million (2009 est.)
Exports—commodities: peanut products, fish, cotton lint, palm kernels
Exports—partners: India 42.06%, France 15.34%, UK 9.03%, China 7.38%, Hong Kong 4.55%, Belgium 3.97% (2009)
Imports: $306 million (2010 est.)
country comparison to the world: 195
$280 million (2009 est.)
Imports—commodities: foodstuffs, manufactures, fuel, machinery and transport equipment
Imports—partners: China 20.45%, Senegal 11.97%, Brazil 8.48%, Cote d'Ivoire 4.71%, Netherlands 4.68%, US 4.49% (2009)
Reserves of foreign exchange and gold: $203 million (31 December 2010 est.)
country comparison to the world: 133
$224 million (31 December 2009 est.)
Debt—external: $530 million (31 December 2010 est.)
country comparison to the world: 159
$489 million (31 December 2009 est.)
Exchange rates: dalasis (GMD) per US dollar–28.5193 (2010), 26.6444 (2009), 22.75 (2008), 27.79 (2007), 28.066 (2006)

COMMUNICATIONS

Telephones—main lines in use: 49,000 (2009)
country comparison to the world: 163
Telephones—mobile cellular: 1.433 million (2009)
country comparison to the world: 140
Telephone system: *general assessment:* adequate microwave radio relay and open-wire network; state-owned Gambia Telecommunications partially privatized in 2007
domestic: combined fixed-line and mobile-cellular teledensity, aided by multiple mobile-cellular providers, approached 85 per 100 persons in 2009
international: country code–220; microwave radio relay links to Senegal and Guinea-Bissau; a landing station for the Africa Coast to Europe (ACE) undersea fiber-optic cable is scheduled for completion in 2011; satellite earth station–1 Intelsat (Atlantic Ocean) (2009)
Broadcast media: state-owned, single-channel TV service; state-owned radio station and 4 privately-owned radio stations; transmissions of multiple international broadcasters are available, some via shortwave radio; foreign cable and satellite TV subscription services are obtainable in some parts of the country (2007)
Internet country code: .gm
Internet hosts: 1,453 (2010)
country comparison to the world: 162
Internet users: 130,100 (2009)
country comparison to the world: 150

TRANSPORTATION

Airports: 1 (2010)
country comparison to the world: 216
Airports—with paved runways: *total:* 1
over 3,047 m: 1 (2010)
Roadways: *total:* 3,742 km
country comparison to the world: 159
paved: 723 km
unpaved: 3,019 km (2004)
Waterways: 390 km (on River Gambia; small ocean-going vessels can reach 190 km) (2010)
country comparison to the world: 89
Merchant marine: *total:* 5
country comparison to the world: 129
by type: passenger/cargo 4, petroleum tanker 1 (2010)
Ports and terminals: Banjul

MILITARY

Military branches: Office of the Chief of Defense Staff: Gambian National Army (GNA), Gambian Navy (GN) (2010)
Military service age and obligation: 18 years of age for male and female voluntary military service; no conscription (2010)
Manpower available for military service: *males age 16–49:* 423,306
females age 16–49: 438,641 (2010 est.)
Manpower fit for military service: *males age 16–49:* 315,176
females age 16–49: 347,017 (2010 est.)
Manpower reaching militarily significant age annually: *male:* 20,508
female: 20,853 (2010 est.)
Military expenditures: 0.9% of GDP (2009)
country comparison to the world: 139

TRANSNATIONAL ISSUES

Disputes—international: attempts to stem refugees, cross-border raids, arms smuggling, and other illegal activities by separatists from southern Senegal's Casamance region, as well as from conflicts in other west African states
Refugees and internally displaced persons: refugees (country of origin): 5,955 (Sierra Leone) (2007)

GAZA STRIP

INTRODUCTION

Background: The September 1993 Israel-PLO Declaration of Principles on Interim Self-Government Arrangements provided for a transitional period of Palestinian self-rule in the West Bank and Gaza Strip. Under a series of agreements signed between May 1994 and September 1999, Israel transferred to the Palestinian Authority (PA) security and civilian responsibility for many Palestinian-populated areas of the West Bank and Gaza Strip. Negotiations to determine the permanent status of the West Bank and Gaza Strip stalled following the outbreak of an intifada in September 2000. In April 2003, the Quartet (US, EU, UN, and Russia) presented a roadmap to a final settlement of the conflict by 2005 based on reciprocal steps by the two parties leading to

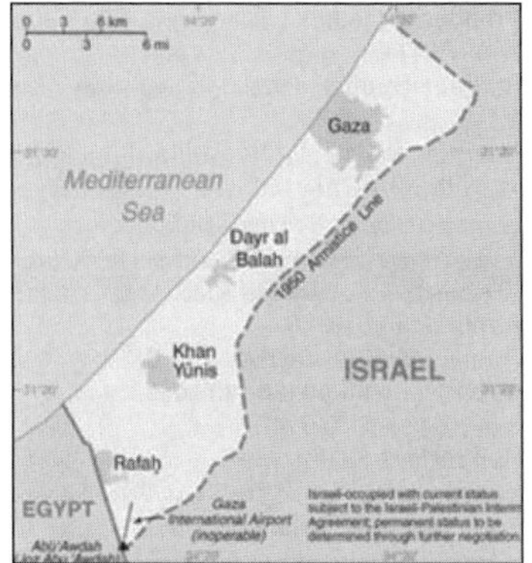

two states, Israel and a democratic Palestine. Following Palestinian leader Yasir ARAFAT's death in late 2004, Mahmud ABBAS was elected PA president in January 2005. A month later, Israel and the PA agreed to the Sharm el-Sheikh Commitments in an effort to move the peace process forward. In September 2005, Israel unilaterally withdrew all its settlers and soldiers and dismantled its military facilities in the Gaza Strip and withdrew settlers and redeployed soldiers from four small northern West Bank settlements. Nonetheless, Israel still controls maritime, airspace, and most access to the Gaza Strip; and it enforces a restricted zone along the border inside Gaza. In January 2006, the Islamic Resistance Movement, HAMAS, won control of the Palestinian Legislative Council (PLC). HAMAS took control of the PA government in March 2006, but President ABBAS had little success negotiating with HAMAS to present a political platform acceptable to the international community so as to lift economic sanctions on Palestinians. Violent clashes between Fatah and HAMAS supporters in the Gaza Strip in 2006 and early 2007 resulted in numerous Palestinian deaths and injuries. In February 2007, ABBAS and HAMAS Political Bureau Chief MISHAL signed the Mecca Agreement in Saudi Arabia that resulted in the formation of a Palestinian National Unity Government (NUG) headed by HAMAS member Ismail HANIYA. However, fighting continued in the Gaza Strip, and in June 2007, HAMAS militants succeeded in a violent takeover of all military and governmental institutions in the Gaza Strip. ABBAS dismissed the NUG and through a series of presidential decrees formed a PA government in the West Bank led by independent Salam FAYYAD. HAMAS rejected the NUG's dismissal, and despite multiple rounds of Egyptian-brokered reconciliation negotiations, the two groups have failed to bridge their differences. Late November 2007 through June 2008 witnessed a substantial increase in Israeli-Palestinian violence. An Egyptian-brokered truce in June 2008 between Israel and HAMAS brought about a five-month pause in hostilities, but spiraling end-of-year violence resulted in the deaths of an estimated 1,100 to 1,400 Palestinians and left tens of thousands of people homeless. International donors pledged $4.5 billion in aid to rebuild the Gaza Strip, but by the end of 2010 large-scale reconstruction had not begun.

GEOGRAPHY

Location: Middle East, bordering the Mediterranean Sea, between Egypt and Israel
Geographic coordinates: 31 25 N, 34 20 E
Map references: Middle East
Area:
total: 360 sq km
country comparison to the world: 204
land: 360 sq km
water: 0 sq km
Area—comparative: slightly more than twice the size of Washington, DC
Land boundaries: *total:* 62 km
border countries: Egypt 11 km, Israel 51 km
Coastline: 40 km
Maritime claims: Israeli-occupied with current status subject to the Israeli-Palestinian Interim Agreement; permanent status to be determined through further negotiation
Climate: temperate, mild winters, dry and warm to hot summers
Terrain: flat to rolling, sand- and dune-covered coastal plain
Elevation extremes: *lowest point:* Mediterranean Sea 0 m
highest point: Abu 'Awdah (Joz Abu 'Awdah) 105 m
Natural resources: arable land, natural gas
Land use: *arable land:* 29%
permanent crops: 21%
other: 50% (2002)
Irrigated land: 155 sq km; (note—includes West Bank) (2003)
Natural hazards: droughts
Environment—current issues: desertification; salination of fresh water; sewage treatment; water-borne disease; soil degradation; depletion and contamination of underground water resources
Geography—note: strategic strip of land along Mideast-North African trade routes has experienced an incredibly turbulent history; the town of Gaza itself has been besieged countless times in its history

PEOPLE

Population: 1,657,155 (July 2011 est.)
country comparison to the world: 149
Age structure: *0–14 years:* 43.9% (male 374,110/female 354,088)
15–64 years: 53.5% (male 453,253/female 432,855)
65 years and over: 2.6% (male 17,326/female 25,523) (2011 est.)
Median age: *total:* 17.7 years
male: 17.5 years
female: 17.9 years (2011 est.)
Population growth rate: 3.201% (2011 est.)
country comparison to the world: 7
Birth rate: 35.3 births/1,000 population (2011 est.)
country comparison to the world: 29
Death rate: 3.29 deaths/1,000 population (July 2011 est.)
country comparison to the world: 216
Net migration rate: 0 migrant(s)/1,000 population (2011 est.)
country comparison to the world: 87
Urbanization: *urban population:* 72% of total population (2008)
rate of urbanization: 3.3% annual rate of change (2005-10 est.)
Sex ratio: *at birth:* 1.06 male(s)/female
under 15 years: 1.06 male(s)/female
15–64 years: 1.05 male(s)/female
65 years and over: 0.68 male(s)/female
total population: 1.04 male(s)/female (2011 est.)
Infant mortality rate: *total:* 17.12 deaths/1,000 live births
country comparison to the world: 105
male: 18.25 deaths/1,000 live births
female: 15.92 deaths/1,000 live births (2011 est.)
Life expectancy at birth: *total population:* 73.92 years
country comparison to the world: 111
male: 72.27 years
female: 75.68 years (2011 est.)
Total fertility rate: 4.74 children born/woman (2011 est.)
country comparison to the world: 26
HIV/AIDS—adult prevalence rate: NA
HIV/AIDS—people living with HIV/AIDS: NA
HIV/AIDS—deaths: NA
Nationality: *noun:* NA
adjective: NA
Ethnic groups: Palestinian Arab
Religions: Muslim (predominantly Sunni) 99.3%, Christian 0.7%
Languages: Arabic, Hebrew (spoken by many Palestinians), English (widely understood)
Literacy: *definition:* age 15 and over can read and write
total population: 92.4%
male: 96.7%
female: 88% (2004 est.)
School life expectancy (primary to tertiary education): *total:* 14 years
male: 13 years
female: 14 years (2006)
Education expenditures: NA

GOVERNMENT

Country name: *conventional long form:* none
conventional short form: Gaza Strip
local long form: none
local short form: Qita' Ghazzah

ECONOMY

Economy—overview: High population density, limited land and sea access, continuing isolation, and strict internal and external security controls have degraded economic conditions in the Gaza Strip—the smaller of the two areas in the Palestinian Territories. Israeli-imposed crossings closures, which became more restrictive after HAMAS violently took over the territory in June 2007, and fighting between HAMAS and Israel during December 2008-January 2009, resulted in the near collapse of most of the private sector, extremely high unemployment, and high poverty rates. Shortages of goods are met through large-scale humanitarian assistance—led by UNRWA—and the HAMAS-regulated black market tunnel trade that flourishes under the Gaza Strip's border with Egypt.

However, changes to the blockade in 2010 included moving from a white list–in which only approved items were allowed into Gaza through the crossings–to a black list, where all but non-approved items were allowed into Gaza through the crossings. Israeli authorities have recently signaled that exports from the territory might be possible in the future, but currently regular exports from Gaza are not permitted.
GDP (purchasing power parity): see entry for West Bank
GDP—real growth rate: see entry for West Bank
GDP—per capita (PPP): see entry for West Bank
GDP—composition by sector: see entry for West Bank
Labor force: 339,000 (2009)
country comparison to the world: 160
Labor force—by occupation: *agriculture:* 12%
industry: 5%
services: 83% (June 2008)
Unemployment rate: 40% (2010 est.)
country comparison to the world: 184
40% (2009 est.)
Population below poverty line: 70% (2009 est.)
Budget: see entry for West Bank
Inflation rate (consumer prices): 3.5% (2010 est.)
country comparison to the world: 101
9.9% (2009 est.)
note: includes West Bank
Commercial bank prime lending rate: see entry for West Bank
Stock of domestic credit: note: see entry for West Bank
Agriculture—products: olives, fruit, vegetables, flowers; beef, dairy products
Industries: textiles, food processing
Industrial production growth rate: see entry for West Bank
Electricity—production: 65,000 kWh (2009)
country comparison to the world: 213
Electricity—consumption: 202,000 kWh (2009)
country comparison to the world: 213
Electricity—exports: 0 kWh (2008 est.)
Electricity—imports: 120,000 kWh; note–from Israeli Electric Company (2009)
Oil—production: see entry for West Bank
Oil—consumption: see entry for West Bank
Oil—exports: see entry for West Bank
Oil—imports: see entry for West Bank
Oil—proved reserves: 0 bbl
country comparison to the world: 140
Exports—commodities: strawberries, carnations
Imports: see entry for West Bank
Imports—commodities: food, consumer goods
note: Israel permits limited imports through crossings with Gaza, but many "dual use" goods, such as construction materials, are smuggled through tunnels beneath Gaza's border with Egypt
Debt—external: see entry for West Bank
Exchange rates: new Israeli shekels (ILS) per US dollar–3.739 (2010), 3.9323 (2009), 3.56 (2008), 4.14 (2007), 4.4565 (2006)

COMMUNICATIONS

Telephones—main lines in use: 360,400 (includes West Bank) (2010)
country comparison to the world: 109
Telephones—mobile cellular: 2.405 million (includes West Bank) (2010)
country comparison to the world: 124
Telephone system: *general assessment:* Gaza continues to repair the damage to its telecommunications infrastructure caused by fighting in 2009
domestic: Israeli company BEZEK and the Palestinian company PALTEL are responsible for fixed line services; the Palestinian JAWWAL company provides cellular services
international: country code–970 (2009)
Broadcast media: 1 television station and about 10 radio stations (2008)
Internet country code: .ps; note–same as West Bank
Internet users: 1.379 million (includes West Bank) (2009)
country comparison to the world: 87

TRANSPORTATION

Airports: 1 (2010)
country comparison to the world: 218
Airports—with paved runways: *total:* 1
over 3,047 m: 1 (2010)
Heliports: 1 (2010)
Roadways: note: see entry for West Bank
Ports and terminals: Gaza

MILITARY

Military branches: Palestinian Authority security forces have operated only in the West Bank, not in the Gaza Strip, since HAMAS seized power in June 2007; law and order and other security functions are performed by HAMAS security organizations (2008)
Manpower available for military service: *males age 16–49:* 385,961 (2010 est.)
Manpower fit for military service: *males age 16–49:* 335,820
females age 16–49: 319,847 (2010 est.)
Manpower reaching militarily significant age annually: *male:* 18,805
female: 17,903 (2010 est.)
Military expenditures: NA

TRANSNATIONAL ISSUES

Disputes—international: West Bank and Gaza Strip are Israeli-occupied with current status subject to the Israeli-Palestinian Interim Agreement–permanent status to be determined through further negotiation; Israel removed settlers and military personnel from the Gaza Strip in August 2005
Refugees and internally displaced persons: refugees (country of origin): 1.017 million (Palestinian Refugees (UNRWA)) (2007)

GEORGIA

INTRODUCTION

Background: The region of present day Georgia contained the ancient kingdoms of Colchis and Kartli-Iberia. The area came under Roman influence in the first centuries A.D. and Christianity became the state religion in the 330s. Domination by Persians, Arabs, and Turks was followed by a Georgian golden age (11th-13th centuries) that was cut short by the Mongol invasion of 1236. Subsequently, the Ottoman and Persian empires competed for influence in the region. Georgia was absorbed into the Russian Empire in the 19th century. Independent for three years (1918-1921) following the Russian revolution, it was forcibly incorporated into the USSR until the Soviet Union dissolved in 1991. An attempt by the incumbent Georgian government to manipulate national legislative elections in November 2003 touched off widespread protests that led to the resignation of Eduard SHEVARDNADZE, president since 1995. New elections in early 2004 swept Mikheil SAAKASHVILI into power along with his United National Movement party. Progress on market reforms and democratization has been made in the years since independence, but this progress has been complicated by Russian assistance and support to the breakaway regions of Abkhazia and South Ossetia. After a series of Russian and separatist provocations in summer 2008, Georgian military action in South Ossetia in early August led to a Russian military response that not only occupied the breakaway areas, but large portions of Georgia proper as well. Russian troops pulled back from most occupied Georgian territory, but in late August 2008 Russia unilaterally recognized the independence of Abkhazia and South Ossetia. This action was strongly condemned by most of the world's nations and international organizations.

GEOGRAPHY

Location: Southwestern Asia, bordering the Black Sea, between Turkey and Russia, with a sliver of land north of the Caucasus extending into Europe
Geographic coordinates: 42 00 N, 43 30 E
Map references: Middle East
Area: *total:* 69,700 sq km
country comparison to the world: 120
land: 69,700 sq km
water: 0 sq km
Area—comparative: slightly smaller than South Carolina
Land boundaries: *total:* 1,461 km
border countries: Armenia 164 km, Azerbaijan 322 km, Russia 723 km, Turkey 252 km
Coastline: 310 km
Maritime claims: *territorial sea:* 12 nm
exclusive economic zone: 200 nm
Climate: warm and pleasant; Mediterranean-like on Black Sea coast
Terrain: largely mountainous with Great Caucasus Mountains in the north and Lesser Caucasus Mountains in the south; Kolkhet'is Dablobi (Kolkhida Lowland) opens to the Black Sea in the west; Mtkvari River Basin in the east; good soils in river valley flood plains, foothills of Kolkhida Lowland
Elevation extremes:
lowest point: Black Sea 0 m
highest point: Mt'a Shkhara 5,201 m
Natural resources: timber, hydropower, manganese deposits, iron ore, copper, minor coal and oil deposits; coastal climate and soils allow for important tea and citrus growth
Land use: *arable land:* 11.51%
permanent crops: 3.79%
other: 84.7% (2005)
Irrigated land: 4,690 sq km (2003)
Total renewable water resources: 63.3 cu km (1997)
Freshwater withdrawal (domestic/industrial/agricultural): *total:* 3.61 cu km/yr (20%/21%/59%)
per capita: 808 cu m/yr (2000)
Natural hazards: earthquakes
Environment—current issues: air pollution, particularly in Rust'avi; heavy pollution of Mtkvari River and the Black Sea; inadequate supplies of potable water; soil pollution from toxic chemicals
Environment—international agreements: *party to:* Air Pollution, Biodiversity, Climate Change, Climate Change-Kyoto Protocol, Desertification, Endangered Species, Hazardous Wastes, Law of the Sea, Ozone Layer Protection, Ship Pollution, Wetlands
signed, but not ratified: none of the selected agreements
Geography—note: strategically located east of the Black Sea; Georgia controls much of the Caucasus Mountains and the routes through them

PEOPLE

Population: 4,585,874 (July 2011 est.)
country comparison to the world: 121
Age structure: *0–14 years:* 15.6% (male 383,856/female 333,617)
15–64 years: 68.3% (male 1,511,844/female 1,620,727)
65 years and over: 16% (male 293,143/female 442,687) (2011 est.)
Median age:
total: 39.1 years
male: 36.6 years
female: 41.6 years (2011 est.)
Population growth rate: -0.326% (2011 est.)
country comparison to the world: 217
Birth rate: 10.73 births/1,000 population (2011 est.)
country comparison to the world: 179
Death rate: 9.92 deaths/1,000 population (July 2011 est.)
country comparison to the world: 56
Net migration rate: -4.06 migrant(s)/1,000 population (2011 est.)
country comparison to the world: 189
Urbanization: *urban population:* 53% of total population (2010)
rate of urbanization: -0.4% annual rate of change (2010-15 est.)
Major cities—population: TBILISI (capital) 1.115 million (2009)
Sex ratio: *at birth:* 1.113 male(s)/female
under 15 years: 1.15 male(s)/female
15–64 years: 0.93 male(s)/female
65 years and over: 0.66 male(s)/female
total population: 0.91 male(s)/female (2011 est.)
Infant mortality rate:
total: 15.17 deaths/1,000 live births
country comparison to the world: 118
male: 17.1 deaths/1,000 live births
female: 13.02 deaths/1,000 live births (2011 est.)
Life expectancy at birth:
total population: 77.12 years
country comparison to the world: 64
male: 73.8 years
female: 80.82 years (2011 est.)
Total fertility rate: 1.45 children born/woman (2011 est.)
country comparison to the world: 191
HIV/AIDS—adult prevalence rate: 0.1% (2009 est.)
country comparison to the world: 130
HIV/AIDS—people living with HIV/AIDS: 3,500 (2009 est.)
country comparison to the world: 127
HIV/AIDS—deaths: fewer than 100 (2009 est.)
country comparison to the world: 133
Drinking water source: *Improved:*
urban: 100% of population
rural: 96% of population
total: 98% of population
Unimproved: urban: 0% of population
rural: 4% of population
total: 2% of population (2008)
Sanitation facility access: *Improved:*
urban: 96% of population
rural: 93% of population
total: 95% of population
Unimproved: urban: 4% of population
rural: 7% of population
total: 5% of population (2008)
Nationality: *noun:* Georgian(s)
adjective: Georgian
Ethnic groups: Georgian 83.8%, Azeri 6.5%, Armenian 5.7%, Russian 1.5%, other 2.5% (2002 census)
Religions: Orthodox Christian (official) 83.9%, Muslim 9.9%, Armenian-Gregorian 3.9%, Catholic 0.8%, other 0.8%, none 0.7% (2002 census)
Languages: Georgian (official) 71%, Russian 9%, Armenian 7%, Azeri 6%, other 7%
note: Abkhaz is the official language in Abkhazia
Literacy: *definition:* age 15 and over can read and write
total population: 100%
male: 100%
female: 100% (2004 est.)
School life expectancy (primary to tertiary education): *total:* 13 years
male: 13 years
female: 13 years (2009)
Education expenditures: 3.2% of GDP (2009)
country comparison to the world: 129

GOVERNMENT

Country name: *conventional long form:* none
conventional short form: Georgia
local long form: none
local short form: Sak'art'velo
former: Georgian Soviet Socialist Republic
Government type: republic
Capital: *name:* T'bilisi
geographic coordinates: 41 43 N, 44 47 E
time difference: UTC+4 (9 hours ahead of Washington, DC during Standard Time)
Administrative divisions: 9 regions (mkharebi, singular—mkhare), 1 city (k'alak'i), and 2 autonomous republics (avtomnoy respubliki, singular—avtom respublika)
regions: Guria, Imereti, Kakheti, Kvemo Kartli, Mtskheta-Mtianeti, Racha-Lechkhumi and Kvemo Svaneti, Samegrelo and Zemo Svaneti, Samtskhe-Javakheti, Shida Kartli
city: Tbilisi
autonomous republics: Abkhazia or Ap'khazet'is Avtonomiuri Respublika (Sokhumi), Ajaria or Acharis Avtonomiuri Respublika (Bat'umi)
note: the administrative centers of the two autonomous republics are shown in parentheses
Independence: 9 April 1991 (from the Soviet Union); notable earlier date: A.D. 1008 (Georgia unified under BAGRAT III)
National holiday: Independence Day, 26 May (1918); note—26 May 1918 was the date of independence from Soviet Russia, 9 April 1991 was the date of independence from the Soviet Union
Constitution: adopted 24 August 1995

Legal system: civil law system
International law organization participation: accepts compulsory ICJ jurisdiction; accepts ICCt jurisdiction
Suffrage: 18 years of age; universal
Executive branch: *chief of state:* President Mikheil SAAKASHVILI (since 25 January 2004); the president is the chief of state and serves as head of government for the power ministries of internal affairs and defense
head of government: Prime Minister Nikoloz GILAURI (since 6 February 2009); the prime minister is head of government for all the ministries of government except the power ministries of internal affairs and defense
cabinet: Cabinet of Ministers
(For more information visit the World Leaders website)
elections: president elected by popular vote for a five-year term (eligible for a second term); election last held on 5 January 2008 (next to be held in January 2013)
election results: Mikheil SAAKASHVILI reelected president; percent of vote—Mikheil SAAKASHVILI 53.5%, Levan GACHECHILADZE 25.7%, Badri PATARKATSISHVILI 7.1%, other 13.7%
Legislative branch: unicameral Parliament or Parlamenti (also known as Supreme Council or Umaghlesi Sabcho) (150 seats; 75 members elected by proportional representation, 75 from single-seat constituencies; members to serve four-year terms)
elections: last held on 21 May 2008 (next to be held in the spring of 2012)
election results: percent of vote by party—United National Movement 59.2%, National Council-New Rights (a Joint Opposition, nine-party bloc) 17.7%, Christian Democratic Movement 8.8%, Labor Party 7.4%, Republican Party 3.8%; seats by party—United National Movement 120, National Council-New Rights 16, Christian Democratic Movement 6, Labor Party 6, Republican Party 2
Judicial branch: Supreme Court (judges elected by the Supreme Council on the president's or chairman of the Supreme Court's recommendation); Constitutional Court; first and second instance courts
Political parties and leaders: Christian Democratic Movement [Giorgi TARGAMADZE]; Conservative Party [Kakha KUKAVA]; Democratic Movement United Georgia [Nino BURJANADZE]; For Fair Georgia [Zurab NOGAIDELI]; Georgian Party [Sozar SUBARI]; Georgian People's Front [Nodar NATADZE]; Greens [Giorgi GACHECHILADZE]; Industry Will Save Georgia (Industrialists) or IWSG [Georgi TOPADZE]; Labor Party [Shalva NATELASHVILI]; National Democratic Party or NDP [Bachuki KARDAVA]; National Forum [Kakhaber SHARTAVA]; New Rights [David GAMKRELIDZE]; Our Georgia-Free Democrats (OGFD) [Irakli ALASANIA]; People's Party [Koba DAVITASHVILI; Republican Party [David USUPASHVILI]; Socialist Party or SPG [Irakli MINDELI]; Traditionalists [Akaki ASATIANI]; United National Movement or UNM [Mikheil SAAKASHVILI]
Political pressure groups and leaders: separatists in the breakaway regions of Abkhazia and South Ossetia
International organization participation: ADB, BSEC, CE, EAPC, EBRD, FAO, G-11, GCTU, GUAM, IAEA, IBRD, ICAO, ICC, ICRM, IDA, IFAD, IFC, IFRCS, ILO, IMF, IMO, Interpol, IOC, IOM, IPU, ISO (correspondent), ITSO, ITU, ITUC, MIGA, OAS (observer), OIF (observer), OPCW, OSCE, PFP, SECI (observer), UN, UNCTAD, UNESCO, UNIDO, UNWTO, UPU, WCO, WHO, WIPO, WMO, WTO
Diplomatic representation in the US: *chief of mission:* Ambassador Temuri YAKOBASHVILI
chancery: 2209 Massachusetts Avenue NW, Washington, DC 20008
telephone: [1] (202) 387-2390
FAX: [1] (202) 393-4537
consulate(s) general: New York
Diplomatic representation from the US: *chief of mission:* Ambassador John BASS
embassy: 11 George Balanchine Street, T'bilisi 0131
mailing address: 7060 T'bilisi Place, Washington, DC 20521-7060
telephone: [995] (32) 27-70-00
FAX: [995] (32) 53-23-10
Flag description: white rectangle with a central red cross extending to all four sides of the flag; each of the four quadrants displays a small red bolnur-katskhuri cross; although adopted as the official Georgian flag in 2004, the five-cross flag design appears to date back to the 14th century
National anthem: *name:* "Tavisupleba" (Liberty)
lyrics/music: Dawit MAGRADSE/ Zakaria PALIASHVILI (adapted by Joseb KETSCHAKMADSE)
note: adopted 2004; after the Rose Revolution, a new anthem with music based on the operas "Abesalom da Eteri" and "Daisi" was adopted

ECONOMY

Economy—overview: Georgia's economy sustained GDP growth of more than 10% in 2006-07, based on strong inflows of foreign investment and robust government spending. However, GDP growth slowed in 2008 following the August 2008 conflict with Russia, and turned negative in 2009 as foreign direct investment and workers' remittances declined in the wake of the global financial crisis, but rebounded in 2010. Georgia's main economic activities include the cultivation of agricultural products such as grapes, citrus fruits, and hazelnuts; mining of manganese and copper; and output of a small industrial sector producing alcoholic and nonalcoholic beverages, metals, machinery, aircraft and chemicals. Areas of recent improvement include growth in the construction, banking services, and mining sectors, but reduced availability of external investment and the slowing regional economy are emerging risks. The country imports nearly all its needed supplies of natural gas and oil products. It has sizeable hydropower capacity, a growing component of its energy supplies. Georgia has overcome the chronic energy shortages and gas supply interruptions of the past by renovating hydropower plants and by increasingly relying on natural gas imports from Azerbaijan instead of from Russia. The construction on the Baku-T'bilisi-Ceyhan oil pipeline, the Baku-T'bilisi-Erzerum gas pipeline, and the Kars-Akhalkalaki Railroad are part of a strategy to capitalize on Georgia's strategic location between Europe and Asia and develop its role as a transit point for gas, oil and other goods. Georgia has historically suffered from a chronic failure to collect tax revenues; however, the government, since coming to power in 2004, has simplified the tax code, improved tax administration, increased tax enforcement, and cracked down on petty corruption. However, the economic downturn of 2008-09 eroded the tax base and led to a decline in the budget surplus and an increase in public borrowing needs. The country is pinning its hopes for renewed growth on a determined effort to continue to liberalize the economy by reducing regulation, taxes, and corruption in order to attract foreign investment, but the economy faces a more difficult investment climate both domestically and internationally.
GDP (purchasing power parity): $22.44 billion (2010 est.)
country comparison to the world: 121
$21.1 billion (2009 est.)
$21.93 billion (2008 est.)
note: data are in 2010 US dollars
GDP (official exchange rate): $11.67 billion (2010 est.)
GDP—real growth rate: 6.4% (2010 est.)
country comparison to the world: 46
-3.8% (2009 est.)
2.4% (2008 est.)
GDP—per capita (PPP): $4,900 (2010 est.)
country comparison to the world: 150
$4,600 (2009 est.)
$4,700 (2008 est.)
note: data are in 2010 US dollars
GDP—composition by sector: *agriculture:* 11%
industry: 27.1%
services: 62% (2010 est.)
Labor force: 1.918 million (2007 est.)
country comparison to the world: 123
Labor force—by occupation: *agriculture:* 55.6%
industry: 8.9%
services: 35.5% (2006 est.)
Unemployment rate: 16.4% (2009 est.)
country comparison to the world: 155
13.6% (2006 est.)
Population below poverty line: 31% (2006)
Household income or consumption by percentage share: *lowest 10%:* 1.9%
highest 10%: 30.6% (2008)
Distribution of family income—Gini index: 40.8 (2009)
country comparison to the world: 59
37.1 (1996)

Investment (gross fixed): 14.5% of GDP (2010 est.)
country comparison to the world: 136
Budget: *revenues:* $3.172 billion
expenditures: $3.915 billion (2010 est.)
Inflation rate (consumer prices): 5.7% (2010 est.)
country comparison to the world: 151
1.7% (2009 est.)
Central bank discount rate: 8% (25 December 2008)
NA% (31 December 2007)
note: this is the Refinancing Rate, the key monetary policy rate of the Georgian National Bank
Commercial bank prime lending rate: 25.52% (31 December 2009 est.)
country comparison to the world: 14
21.24% (31 December 2008 est.)
Stock of narrow money: $1.175 billion (31 December 2010 est.)
country comparison to the world: 135
$1.122 billion (31 December 2009 est.)
Stock of broad money: $2.146 billion (31 December 2010 est.)
country comparison to the world: 140
$1.28 billion (31 December 2009 est.)
Stock of domestic credit: $3.243 billion (31 December 2010 est.)
country comparison to the world: 118
$3.569 billion (31 December 2009 est.)
Market value of publicly traded shares: $733.3 million (31 December 2009)
country comparison to the world: 110
$327.3 million (31 December 2008)
$1.389 billion (31 December 2007)
Agriculture—products: citrus, grapes, tea, hazelnuts, vegetables; livestock
Industries: steel, aircraft, machine tools, electrical appliances, mining (manganese and copper), chemicals, wood products, wine
Industrial production growth rate: 4% (2010 est.)
country comparison to the world: 85
Electricity—production: 7.97 billion kWh (2008 est.)
country comparison to the world: 98
Electricity—consumption: 6.902 billion kWh (2008 est.)
country comparison to the world: 100
Electricity—exports: 628 million kWh (2007 est.)
Electricity—imports: 430 million kWh (2007 est.)
Oil—production: 994.9 bbl/day (2009 est.)
country comparison to the world: 105
Oil—consumption: 13,000 bbl/day (2009 est.)
country comparison to the world: 139
Oil—exports: 1,486 bbl/day (2008 est.)
country comparison to the world: 118
Oil—imports: 16,590 bbl/day (2008 est.)
country comparison to the world: 120
Oil—proved reserves: 35 million bbl (1 January 2010 est.)
country comparison to the world: 82
Natural gas—production: 8 million cu m (2008 est.)
country comparison to the world: 90
Natural gas—consumption: 1.73 billion cu m (2008 est.)
country comparison to the world: 81
Natural gas—exports: 0 cu m (2008 est.)
country comparison to the world: 100
Natural gas—imports: 1.72 billion cu m (2008 est.)
country comparison to the world: 49
Natural gas—proved reserves: 8.495 billion cu m (1 January 2010 est.)
country comparison to the world: 80
Current account balance: $-1.404 billion (2010 est.)
country comparison to the world: 146
$-1.259 billion (2009 est.)
Exports: $2.29 billion (2010 est.)
country comparison to the world: 129
$1.893 billion (2009 est.)
Exports—commodities: scrap metal, wine, mineral water, ores, vehicles, fruits and nuts
Exports—partners: Turkey 17.87%, Azerbaijan 12.3%, Bulgaria 9.6%, Canada 8.78%, UK 7.49%, Ukraine 6.82%, Spain 5.27%, US 4.99% (2009)
Imports:
$4.828 billion (2010 est.)
country comparison to the world: 120
$4.293 billion (2009 est.)
Imports—commodities: fuels, vehicles, machinery and parts, grain and other foods, pharmaceuticals
Imports—partners: Turkey 16.81%, Azerbaijan 9.72%, Ukraine 9.17%, Russia 7.39%, US 6.63%, Germany 6.22% (2009)
Reserves of foreign exchange and gold: $2.35 billion (31 December 2010 est.)
country comparison to the world: 95
$2.11 billion (31 December 2009 est.)
Debt—external: $3.381 billion (31 December 2009)
country comparison to the world: 121
$7.711 billion (31 December 2008)
Exchange rates: laris (GEL) per US dollar–1.8009 (2010), 1.6705 (2009), 1.47 (2008), 1.7 (2007), 1.78 (2006)

COMMUNICATIONS

Telephones—main lines in use: 620,000 (2009)
country comparison to the world: 94
Telephones—mobile cellular: 2.837 million (2009)
country comparison to the world: 115
Telephone system: *general assessment:* fixed-line telecommunications network has only limited coverage outside Tbilisi; long list of people waiting for fixed line connections; multiple mobile-cellular providers provide services to an increasing subscribership throughout the country
domestic: cellular telephone networks cover the entire country; mobile-cellular teledensity roughly 60 per 100 people; urban fixed-line telephone density is about 20 per 100 people; rural telephone density is about 4 per 100 people; intercity facilities include a fiber-optic line between T'bilisi and K'ut'aisi; nationwide pager service is available
international: country code—995; the Georgia-Russia fiber optic submarine cable provides connectivity to Russia; international service is available by microwave, landline, and satellite through the Moscow switch; international electronic mail and telex service are available
Broadcast media: 1 state-owned public television station in Tbilisi and 8 privately-owned TV stations; state-run public broadcaster operates 2 networks; dozens of cable TV operators and several major commercial TV stations are operating; state-owned public radio broadcaster operates 2 networks; several dozen private stations broadcast (2008)
Internet country code: .ge
Internet hosts: 110,680 (2010)
country comparison to the world: 76
Internet users: 1.3 million (2009)
country comparison to the world: 90

TRANSPORTATION

Airports: 22 (2010)
country comparison to the world: 133
Airports—with paved runways: *total:* 18
over 3,047 m: 1
2,438 to 3,047 m: 7
1,524 to 2,437 m: 4
914 to 1,523 m: 4
under 914 m: 2 (2010)
Airports—with unpaved runways: *total:* 4
914 to 1,523 m: 2
under 914 m: 2 (2010)
Heliports: 3 (2010)
Pipelines: gas 1,596 km; oil 1,258 km (2010)
Railways: *total:* 1,612 km
country comparison to the world: 78
broad gauge: 1,575 km 1.520-m gauge (1,575 electrified)
narrow gauge: 37 km 0.912-m gauge (37 electrified) (2009)
Roadways: *total:* 20,329 km
country comparison to the world: 108
paved: 7,854 km (includes 13 km of expressways)
unpaved: 12,475 km (2006)
Merchant marine: *total:* 193
country comparison to the world: 34
by type: bulk carrier 18, cargo 151, carrier 1, chemical tanker 3, container 2, liquefied gas 1, passenger/cargo 3, petroleum tanker 3, refrigerated cargo 2, roll on/roll off 7, vehicle carrier 2
foreign-owned: 132 (China 11, Denmark 1, Egypt 11, Germany 4, Greece 3, Hong Kong 4, Israel 1, Italy 2, Latvia 1, Lebanon 1, Pakistan 1, Romania 7, Russia 7, Syria 35, Turkey 22, UAE 1, UK 4, Ukraine 15, US 1)
registered in other countries: 1 (unknown 1) (2010)
Ports and terminals: Bat'umi, P'ot'i
Transportation—note: large parts of transportation network are in poor condition because of lack of maintenance and repair

MILITARY

Military branches: Georgian Armed Forces: Land Forces (include Air and Air Defense Forces); separatist Abkhazia Armed Forces: Ground Forces, Air Forces; separatist South Ossetia Armed Forces (2011)
Military service age and obligation: 18 to 34 years of age for compulsory and volun-

tary active duty military service; conscript service obligation—18 months (2005)
Manpower available for military service: *males age 16–49:* 1,080,840
females age 16–49: 1,122,031 (2010 est.)
Manpower fit for military service: *males age 16–49:* 893,003
females age 16–49: 931,683 (2010 est.)
Manpower reaching militarily significant age annually: *male:* 29,723
female: 27,242 (2010 est.)
Military expenditures: 1.9% of GDP (2010 est.)
country comparison to the world: 76

TRANSNATIONAL ISSUES

Disputes—international: Russia's military support and subsequent recognition of Abkhazia and South Ossetia independence in 2008 continue to sour relations with Georgia; Georgia continues to restrain the return of Meshkhetian Turks dispersed by Stalin; ethnic Armenian groups in Javakheti region of Georgia seek greater autonomy from the Georgian government; local border forces struggle to control the illegal transit of goods and people across the porous, undemarcated Armenian, Azerbaijani, and Georgian borders
Refugees and internally displaced persons: refugees (country of origin): 1,100 (Russia)
IDPs: 220,000-240,000 (displaced from Abkhazia and South Ossetia) (2007)
Illicit drugs: limited cultivation of cannabis and opium poppy, mostly for domestic consumption; used as transshipment point for opiates via Central Asia to Western Europe and Russia

GERMANY

INTRODUCTION

Background: As Europe's largest economy and second most populous nation (after Russia), Germany is a key member of the continent's economic, political, and defense organizations. European power struggles immersed Germany in two devastating World Wars in the first half of the 20th century and left the country occupied by the victorious Allied powers of the US, UK, France, and the Soviet Union in 1945. With the advent of the Cold War, two German states were formed in 1949: the western Federal Republic of Germany (FRG) and the eastern German Democratic Republic (GDR). The democratic FRG embedded itself in key Western economic and security organizations, the EC, which became the EU, and NATO, while the Communist GDR was on the front line of the Soviet-led Warsaw Pact. The decline of the USSR and the end of the Cold War allowed for German unification in 1990. Since then, Germany has expended considerable funds to bring Eastern productivity and wages up to Western standards. In January 1999, Germany and 10 other EU countries introduced a common European exchange currency, the euro. In January 2011, Germany assumed a nonpermanent seat on the UN Security Council for the 2011-12 term.

GEOGRAPHY

Location: Central Europe, bordering the Baltic Sea and the North Sea, between the Netherlands and Poland, south of Denmark
Geographic coordinates: 51 00 N, 9 00 E
Map references: Europe
Area: *total:* 357,022 sq km
country comparison to the world: 62
land: 348,672 sq km
water: 8,350 sq km
Area—comparative: slightly smaller than Montana
Land boundaries: *total:* 3,621 km
border countries: Austria 784 km, Belgium 167 km, Czech Republic 646 km, Denmark 68 km, France 451 km, Luxembourg 138 km, Netherlands 577 km, Poland 456 km, Switzerland 334 km
Coastline: 2,389 km
Maritime claims: *territorial sea:* 12 nm
exclusive economic zone: 200 nm
continental shelf: 200 m depth or to the depth of exploitation
Climate: temperate and marine; cool, cloudy, wet winters and summers; occasional warm mountain (foehn) wind
Terrain: lowlands in north, uplands in center, Bavarian Alps in south
Elevation extremes: *lowest point:* Neuendorf bei Wilster -3.54 m
highest point: Zugspitze 2,963 m
Natural resources: coal, lignite, natural gas, iron ore, copper, nickel, uranium, potash, salt, construction materials, timber, arable land
Land use: *arable land:* 33.13%
permanent crops: 0.6%
other: 66.27% (2005)
Irrigated land: 4,850 sq km (2003)
Total renewable water resources: 188 cu km (2005)
Freshwater withdrawal (domestic/industrial/agricultural): *total:* 38.01 cu km/yr (12%/68%/20%)
per capita: 460 cu m/yr (2001)
Natural hazards: flooding
Environment—current issues: emissions from coal-burning utilities and industries contribute to air pollution; acid rain, resulting from sulfur dioxide emissions, is damaging forests; pollution in the Baltic Sea from raw sewage and industrial effluents from rivers in eastern Germany; hazardous waste disposal; government established a mechanism for ending the use of nuclear power over the next 15 years; government working to meet EU commitment to identify nature preservation areas in line with the EU's Flora, Fauna, and Habitat directive
Environment—international agreements: *party to:* Air Pollution, Air Pollution-Nitrogen Oxides, Air Pollution-Persistent Organic Pollutants, Air Pollution-Sulfur 85, Air Pollution-Sulfur 94, Air Pollution-Volatile Organic Compounds, Antarctic-Environmental Protocol, Antarctic-Marine Living Resources, Antarctic Seals, Antarctic Treaty, Biodiversity, Climate Change, Climate Change-Kyoto Protocol, Desertification, Endangered Species, Environmental Modification, Hazardous Wastes, Law of the Sea, Marine Dumping, Ozone Layer Protection, Ship Pollution, Tropical Timber 83, Tropical Timber 94, Wetlands, Whaling
signed, but not ratified: none of the selected agreements
Geography—note: strategic location on North European Plain and along the entrance to the Baltic Sea

PEOPLE

Population: 81,471,834 (July 2011 est.)
country comparison to the world: 16
Age structure: *0–14 years:* 13.3% (male 5,569,390/female 5,282,245)
15–64 years: 66.1% (male 27,227,487/female 26,617,915)
65 years and over: 20.6% (male 7,217,163/female 9,557,634) (2011 est.)
Median age: *total:* 44.9 years
male: 43.7 years
female: 46 years (2011 est.)
Population growth rate: -0.208% (2011 est.)
country comparison to the world: 212
Birth rate: 8.3 births/1,000 population (2011 est.)
country comparison to the world: 219
Death rate: 10.92 deaths/1,000 population (July 2011 est.)
country comparison to the world: 39
Net migration rate: 0.54 migrant(s)/1,000 population (2011 est.)
country comparison to the world: 62
Urbanization: *urban population:* 74% of total population (2010)
rate of urbanization: 0% annual rate of change (2010-15 est.)
Major cities—population: BERLIN (capital) 3.438 million; Hamburg 1.786 million; Munich 1.349 million; Cologne 1.001 million (2009)

Sex ratio: *at birth:* 1.055 male(s)/female
under 15 years: 1.05 male(s)/female
15–64 years: 1.04 male(s)/female
65 years and over: 0.72 male(s)/female
total population: 0.97 male(s)/female (2011 est.)
Infant mortality rate: *total:* 3.54 deaths/1,000 live births
country comparison to the world: 208
male: 3.84 deaths/1,000 live births
female: 3.21 deaths/1,000 live births (2011 est.)
Life expectancy at birth: *total population:* 80.07 years
country comparison to the world: 27
male: 77.82 years
female: 82.44 years (2011 est.)
Total fertility rate: 1.41 children born/woman (2011 est.)
country comparison to the world: 198
HIV/AIDS—adult prevalence rate: 0.1% (2009 est.)
country comparison to the world: 131
HIV/AIDS—people living with HIV/AIDS: 67,000 (2009 est.)
country comparison to the world: 51
HIV/AIDS—deaths: fewer than 1,000 (2009 est.)
country comparison to the world: 73
Drinking water source: *Improved:*
urban: 100% of population
rural: 100% of population
total: 100% of population (2008)
Sanitation facility access: *Improved:*
urban: 100% of population
rural: 100% of population
total: 100% of population (2008)
Nationality: *noun:* German(s)
adjective: German
Ethnic groups: German 91.5%, Turkish 2.4%, other 6.1% (made up largely of Greek, Italian, Polish, Russian, Serbo-Croatian, Spanish)
Religions: Protestant 34%, Roman Catholic 34%, Muslim 3.7%, unaffiliated or other 28.3%
Languages: German
Literacy: *definition:* age 15 and over can read and write
total population: 99%
male: 99%
female: 99% (2003 est.)
School life expectancy (primary to tertiary education): *total:* 16 years
male: 16 years
female: 16 years (2006)
Education expenditures: 4.5% of GDP (2007)
country comparison to the world: 82
People—note: second most populous country in Europe after Russia

GOVERNMENT

Country name: *conventional long form:* Federal Republic of Germany
conventional short form: Germany
local long form: Bundesrepublik Deutschland
local short form: Deutschland
former: German Empire, German Republic, German Reich
Government type: federal republic
Capital: *name:* Berlin
geographic coordinates: 52 31 N, 13 24 E
time difference: UTC+1 (6 hours ahead of Washington, DC during Standard Time)
daylight saving time: +1hr, begins last Sunday in March; ends last Sunday in October
Administrative divisions: 16 states (Laender, singular—Land); Baden-Wurttemberg, Bayern (Bavaria), Berlin, Brandenburg, Bremen, Hamburg, Hessen (Hesse), Mecklenburg-Vorpommern (Mecklenburg-Western Pomerania), Niedersachsen (Lower Saxony), Nordrhein-Westfalen (North Rhine-Westphalia), Rheinland-Pfalz (Rhineland-Palatinate), Saarland, Sachsen (Saxony), Sachsen-Anhalt (Saxony-Anhalt), Schleswig-Holstein, Thueringen (Thuringia); note—Bayern, Sachsen, and Thueringen refer to themselves as free states (Freistaaten, singular—Freistaat)
Independence: 18 January 1871 (German Empire unification); divided into four zones of occupation (UK, US, USSR, and France) in 1945 following World War II; Federal Republic of Germany (FRG or West Germany) proclaimed on 23 May 1949 and included the former UK, US, and French zones; German Democratic Republic (GDR or East Germany) proclaimed on 7 October 1949 and included the former USSR zone; West Germany and East Germany unified on 3 October 1990; all four powers formally relinquished rights on 15 March 1991; notable earlier dates: 10 August 843 (Eastern Francia established from the division of the Carolingian Empire); 2 February 962 (crowning of OTTO I, recognized as the first Holy Roman Emperor)
National holiday: Unity Day, 3 October (1990)
Constitution: 23 May 1949, known as Basic Law; became constitution of the united Germany 3 October 1990
Legal system: civil law system
International law organization participation: accepts compulsory ICJ jurisdiction with reservations; accepts ICCt jurisdiction
Suffrage: 18 years of age; universal
Executive branch: *chief of state:* President Christian WULFF (since 30 June 2010)
head of government: Chancellor Angela MERKEL (since 22 November 2005)
cabinet: Cabinet or Bundesminister (Federal Ministers) appointed by the president on the recommendation of the chancellor
(For more information visit the World Leaders website)
elections: president elected for a five-year term (eligible for a second term) by a Federal Assembly, including all members of the Federal Diet and an equal number of delegates elected by the state parliaments; election last held on 30 June 2010 (next to be held by June 2015); chancellor elected by an absolute majority of the Federal Diet for a four-year term; Bundestag vote for Chancellor last held after 27 September 2009 (next to follow the legislative election to be held no later than 2013)
election results: Christian WULFF elected president; received 625 votes of the Federal Assembly against 494 for GAUCK and 121 abstentions; Angela MERKEL reelected chancellor; vote by Federal Diet 323 to 285 with four abstentions
Legislative branch: bicameral legislature consists of the Federal Council or Bundesrat (69 votes; state governments sit in the Council; each has three to six votes in proportion to population and is required to vote as a block) and the Federal Diet or Bundestag (622 seats; members elected by popular vote for a four-year term under a system of personalized proportional representation; a party must win 5% of the national vote or three direct mandates to gain proportional representation and caucus recognition)
elections: Bundestag—last held on 27 September 2009 (next to be held no later than autumn 2013); note—there are no elections for the Bundesrat; composition is determined by the composition of the state-level governments; the composition of the Bundesrat has the potential to change any time one of the 16 states holds an election
election results: Bundestag—percent of vote by party—CDU/CSU 33.8%, SPD 23%, FDP 14.6%, Left 11.9%, Greens 10.7%, other 6%; seats by party—CDU/CSU 239, SPD 146, FDP 93, Left 76, Greens 68
Judicial branch: Federal Constitutional Court or Bundesverfassungsgericht (half the judges are elected by the Bundestag and half by the Bundesrat); Federal Court of Justice; Federal Administrative Court
Political parties and leaders: Alliance '90/Greens [Claudia ROTH and Cem OZDEMIR]; Christian Democratic Union or CDU [Angela MERKEL]; Christian Social Union or CSU [Horst SEEHOFER]; Free Democratic Party or FDP [Guido WESTERWELLE]; Left Party or Die Linke [Klaus ERNST and Gesine LOETZSCH]; Social Democratic Party or SPD [Sigmar GABRIEL]
Political pressure groups and leaders: business associations and employers' organizations; trade unions; religious, immigrant, expellee, and veterans groups
International organization participation: ADB (nonregional member), AfDB (nonregional member), Arctic Council (observer), Australia Group, BIS, BSEC (observer), CBSS, CDB, CE, CERN, EAPC, EBRD, EIB, EMU, ESA, EU, FAO, FATF, G-20, G-5, G-7, G-8, G-10, IADB, IAEA, IBRD, ICAO, ICC, ICRM, IDA, IEA, IFAD, IFC, IFRCS, IHO, ILO, IMF, IMO, IMSO, Interpol, IOC, IOM, IPU, ISO, ITSO, ITU, ITUC, MIGA, NATO, NEA, NSG, OAS (observer), OECD, OPCW, OSCE, Paris Club, PCA, Schengen Convention, SECI (observer), SICA (observer), UN, UN Security Council (temporary), UNAMID, UNCTAD, UNESCO, UNHCR, UNIDO, UNIFIL, UNMIL, UNMIS, UNRWA, UNWTO, UPU, WCO, WHO, WIPO, WMO, WTO, ZC
Diplomatic representation in the US: *chief of mission:* Ambassador Klaus SCHARIOTH
chancery: 4645 Reservoir Road NW, Washington, DC 20007

telephone: [1] (202) 298-4000
FAX: [1] (202) 298-4249
consulate(s) general: Atlanta, Boston, Chicago, Houston, Los Angeles, Miami, New York, San Francisco

Diplomatic representation from the US:
chief of mission: Ambassador Philip D. MURPHY
embassy: Pariser Platz 2, 14191 Berlin; note—new embassy opened 4 July 2008
mailing address: PSC 120, Box 1000, APO AE 09265, Clayallee 170, 14195 Berlin
telephone: [49] (030) 2385174
FAX: [49] (030) 8305-1215
consulate(s) general: Duesseldorf, Frankfurt am Main, Hamburg, Leipzig, Munich

Flag description: three equal horizontal bands of black (top), red, and gold; these colors have played an important role in German history and can be traced back to the medieval banner of the Holy Roman Emperor—a black eagle with red claws and beak on a gold field

National anthem: *name:* "Lied der Deutschen" (Song of the Germans)
lyrics/music: August Heinrich HOFFMANN VON FALLERSLEBE/Franz Joseph HAYDN
note: adopted 1922, restored 1990; the anthem, also known as "Deutschlandlied" (Song of Germany), was abolished in 1945 because of the Nazi's use of the first verse, specifically the phrase, "Deutschland, Deutschland uber alles" (Germany, Germany above all) to promote nationalism; since restoration in 1990, only the third verse is sung

ECONOMY

Economy—overview: The German economy—the fifth largest economy in the world in PPP terms and Europe's largest—is a leading exporter of machinery, vehicles, chemicals, and household equipment and benefits from a highly skilled labor force. Like its western European neighbors, Germany faces significant demographic challenges to sustained long-term growth. Low fertility rates and declining net immigration are increasing pressure on the country's social welfare system and necessitate structural reforms. The modernization and integration of the eastern German economy—where unemployment can exceed 20% in some municipalities—continues to be a costly long-term process, with annual transfers from west to east amounting in 2008 alone to roughly $12 billion. Reforms launched by the government of Chancellor Gerhard SCHROEDER (1998-2005), deemed necessary to address chronically high unemployment and low average growth, contributed to strong growth in 2006 and 2007 and falling unemployment. These advances, as well as a government subsidized, reduced working hour scheme, help explain the relatively modest increase in unemployment during the 2008-09 recession—the deepest since World War II—and its decrease to 7.4%in 2010. GDP contracted 4.7% in 2009 but grew by 3.6% in 2010. In its annual projection for 2011, the Federal Government expects the upswing to continue, with GDP forecast to grow this year at a real rate of 2.3%. The recovery was attributable primarily to rebounding manufacturing orders and exports—increasingly outside the Euro Zone. Domestic demand, however, is becoming more significant driver of Germany's economic expansion. Stimulus and stabilization efforts initiated in 2008 and 2009 and tax cuts introduced in Chancellor Angela MERKEL's second term increased Germany's budget deficit to 3.5% in 2010. The Bundesbank expects the deficit to drop to about 2.5% in 2011, below the EU's 3% limit. A constitutional amendment approved in 2009 likewise limits the federal government to structural deficits of no more than 0.35% of GDP per annum as of 2016.

GDP (purchasing power parity): $2.94 trillion (2010 est.)
country comparison to the world: 6
$2.841 trillion (2009 est.)
$2.98 trillion (2008 est.)
note: data are in 2010 US dollars

GDP (official exchange rate): $3.316 trillion (2010 est.)

GDP—real growth rate: 3.5% (2010 est.)
country comparison to the world: 111
-4.7% (2009 est.)
0.7% (2008 est.)

GDP—per capita (PPP): $35,700 (2010 est.)
country comparison to the world: 33
$34,500 (2009 est.)
$36,200 (2008 est.)
note: data are in 2010 US dollars

GDP—composition by sector:
agriculture: 0.8%
industry: 27.9%
services: 71.3% (2010 est.)

Labor force: 43.35 million (2010 est.)
country comparison to the world: 14

Labor force—by occupation:
agriculture: 2.4%
industry: 29.7%
services: 67.8% (2005)

Unemployment rate: 7.4% (2010 est.)
country comparison to the world: 78
7.5% (2009 est.)
note: this is the International Labor Organization's estimated rate for international comparisons; Germany's Federal Employment Agency estimated a seasonally adjusted rate of 10.8%

Population below poverty line: 15.5% (2010 est.)

Household income or consumption by percentage share: *lowest 10%:* 3.6%
highest 10%: 24% (2000)

Distribution of family income—Gini index: 27 (2006)
country comparison to the world: 124
30 (1994)

Investment (gross fixed): 18% of GDP (2010 est.)
country comparison to the world: 112

Budget: *revenues:* $1.396 trillion
expenditures: $1.516 trillion (2010 est.)

Public debt: 78.8% of GDP (2010 est.)
country comparison to the world: 19
72.5% of GDP (2009 est.)

Inflation rate (consumer prices): 1.1% (2010 est.)
country comparison to the world: 24
0.4% (2009 est.)

Central bank discount rate: 1.75% (31 December 2010)
country comparison to the world: 129
1.75% (31 December 2009)
note: this is the European Central Bank's rate on the marginal lending facility, which offers overnight credit to banks in the euro area

Commercial bank prime lending rate: 4.96% (31 December 2009 est.)
country comparison to the world: 133
5.97% (31 December 2008 est.)

Stock of narrow money: $1.627 trillion (31 December 2010 est.)
country comparison to the world: 6
$1.681 trillion (31 December 2009 est.)
note: see entry for the European Union for money supply in the euro area; the European Central Bank (ECB) controls monetary policy for the 17 members of the Economic and Monetary Union (EMU); individual members of the EMU do not control the quantity of money circulating within their own borders

Stock of broad money: $4.288 trillion (31 December 2010 est.)
country comparison to the world: 6
$4.202 trillion (31 December 2009 est.)

Stock of domestic credit: $5.2 trillion (31 December 2009 est.)
country comparison to the world: 5
$5.019 trillion (31 December 2008 est.)

Market value of publicly traded shares: $1.298 trillion (31 December 2009)
country comparison to the world: 9
$1.108 trillion (31 December 2008)
$2.106 trillion (31 December 2007)

Agriculture—products: potatoes, wheat, barley, sugar beets, fruit, cabbages; cattle, pigs, poultry

Industries: among the world's largest and most technologically advanced producers of iron, steel, coal, cement, chemicals, machinery, vehicles, machine tools, electronics, food and beverages, shipbuilding, textiles

Industrial production growth rate: 9% (2010 est.)
country comparison to the world: 24

Electricity—production: 593.4 billion kWh (2007 est.)
country comparison to the world: 8

Electricity—consumption: 547.3 billion kWh (2007 est.)
country comparison to the world: 7

Electricity—exports: 61.7 billion kWh (2008 est.)

Electricity—imports: 41.67 billion kWh (2008 est.)

Oil—production: 156,800 bbl/day (2009 est.)
country comparison to the world: 45

Oil—consumption: 2.437 million bbl/day (2009 est.)
country comparison to the world: 8

Oil—exports: 536,600 bbl/day (2008 est.)
country comparison to the world: 28

Oil—imports: 2.862 million bbl/day (2008 est.)
country comparison to the world: 7

Oil—proved reserves: 276 million bbl (1 January 2010 est.)

country comparison to the world: 55
Natural gas—production: 15.29 billion cu m (2009 est.)
country comparison to the world: 34
Natural gas—consumption: 96.26 billion cu m (2009 est.)
country comparison to the world: 5
Natural gas—exports: 12.64 billion cu m (2009 est.)
country comparison to the world: 16
Natural gas—imports: 94.57 billion cu m (2009 est.)
country comparison to the world: 2
Natural gas—proved reserves: 175.6 billion cu m (1 January 2010 est.)
country comparison to the world: 47
Current account balance: $162.3 billion (2010 est.)
country comparison to the world: 3
$168.1 billion (2009 est.)
Exports: $1.337 trillion (2010 est.)
country comparison to the world: 3
$1.145 trillion (2009 est.)
Exports—commodities: machinery, vehicles, chemicals, metals and manufactures, foodstuffs, textiles
Exports—partners: France 10.2%, US 6.7%, Netherlands 6.7%, UK 6.6%, Italy 6.3%, Austria 6%, China 4.5%, Switzerland 4.4% (2009 est.)
Imports: $1.12 trillion (2010 est.)
country comparison to the world: 4
$956.7 billion (2009 est.)
Imports—commodities: machinery, vehicles, chemicals, foodstuffs, textiles, metals
Imports—partners: Netherlands 8.5%, China 8.2%, France 8.2%, US 5.9%, Italy 5.9%, UK 4.9%, Belgium 4.3%, Austria 4.3%, Switzerland 4.2% (2009 est.)
Reserves of foreign exchange and gold: $NA (31 December 2010 est.)
$180.8 billion (31 December 2009 est.)
Debt—external: $4.713 trillion (30 June 2010)
country comparison to the world: 4
$5.158 trillion (31 December 2008)
Stock of direct foreign investment—at home: $1.057 trillion (31 December 2010 est.)
country comparison to the world: 4
$1.054 trillion (31 December 2009 est.)
Stock of direct foreign investment—abroad: $1.484 trillion (31 December 2010 est.)
country comparison to the world: 4
$1.46 trillion (31 December 2009 est.)
Exchange rates: euros (EUR) per US dollar–0.755 (2010), 0.7198 (2009), 0.6827 (2008), 0.7345 (2007), 0.7964 (2006)

COMMUNICATIONS

Telephones—main lines in use: 48.7 million (2009)
country comparison to the world: 3
Telephones—mobile cellular: 105 million (2009)
country comparison to the world: 8
Telephone system: *general assessment:* Germany has one of the world's most technologically advanced telecommunications systems; as a result of intensive capital expenditures since reunification, the formerly backward system of the eastern part of the country, dating back to World War II, has been modernized and integrated with that of the western part
domestic: Germany is served by an extensive system of automatic telephone exchanges connected by modern networks of fiber-optic cable, coaxial cable, microwave radio relay, and a domestic satellite system; cellular telephone service is widely available, expanding rapidly, and includes roaming service to many foreign countries
international: country code–49; Germany's international service is excellent worldwide, consisting of extensive land and undersea cable facilities as well as earth stations in the Inmarsat, Intelsat, Eutelsat, and Intersputnik satellite systems (2001)
Broadcast media: a mixture of publicly-operated and privately-owned TV and radio stations; national and regional public broadcasters compete with nearly 400 privately-owned national and regional TV stations; more than 90% of households have cable or satellite TV; hundreds of radio stations broadcasting including multiple national radio networks, regional radio networks, and a large number of local radio stations (2008)
Internet country code: .de
Internet hosts: 21.729 million (2010)
country comparison to the world: 4
Internet users: 65.125 million (2009)
country comparison to the world: 5

TRANSPORTATION

Airports: 549 (2010)
country comparison to the world: 13
Airports—with paved runways: *total:* 330
over 3,047 m: 13
2,438 to 3,047 m: 53
1,524 to 2,437 m: 59
914 to 1,523 m: 70
under 914 m: 135 (2010)
Airports—with unpaved runways: *total:* 219
1,524 to 2,437 m: 2
914 to 1,523 m: 33
under 914 m: 184 (2010)
Heliports: 25 (2010)
Pipelines: gas 24,688 km; oil 3,687 km; refined products 4,875 km (2010)
Railways: *total:* 41,981 km
country comparison to the world: 6
standard gauge: 41,722 km 1.435-m gauge (20,053 km electrified)
narrow gauge: 220 km 1.000-m gauge (75 km electrified); 39 km 0.750-m gauge (24 km electrified) (2009)
Roadways: *total:* 644,480 km
country comparison to the world: 11
paved: 644,480 km (includes 12,800 km of expressways)
note: includes local roads (2010)
Waterways: 7,467 km (Rhine River carries most goods; Main-Danube Canal links North Sea and Black Sea) (2010)
country comparison to the world: 19
Merchant marine: *total:* 421
country comparison to the world: 25
by type: barge carrier 2, bulk carrier 7, cargo 44, carrier 1, chemical tanker 15, container 293, liquefied gas 7, passenger 4, passenger/cargo 27, petroleum tanker 10, refrigerated cargo 1, roll on/roll off 9, vehicle carrier 1
foreign-owned: 10 (China 2, Finland 5, Greece 1, Sweden 1, Switzerland 1)
registered in other countries: 3,287 (Antigua and Barbuda 1050, Australia 2, Bahamas 39, Belize 1, Bermuda 15, Brazil 6, Bulgaria 25, Burma 1, Cayman Islands 6, China 1, Cook Islands 1, Cyprus 189, Denmark 10, Dominica 2, Estonia 1, France 1, Georgia 4, Gibraltar 125, Hong Kong 10, Isle of Man 56, Italy 1, Jamaica 10, Liberia 1049, Luxembourg 9, Malta 127, Marshall Islands 247, Morocco 2, Netherlands 92, former Netherlands Antilles 32, NZ 2, Panama 27, Portugal 13, Saint Vincent and the Grenadines 2, Singapore 30, Slovakia 4, Spain 5, Sri Lanka 5, Sweden 3, Turkey 1, UK 77, US 3, Venezuela 1) (2010)
Ports and terminals: Bremen, Bremerhaven, Duisburg, Hamburg, Karlsruhe, Lubeck, Neuss-Dusseldorf, Rostock, Wilhemshaven

MILITARY

Military branches: Federal Armed Forces (Bundeswehr): Army (Heer), Navy (Deutsche Marine, includes naval air arm), Air Force (Luftwaffe), Joint Support Services (Streitkraeftbasis), Central Medical Service (Zentraler Sanitaetsdienst) (2010)
Military service age and obligation: 18 years of age (conscripts serve a 9-month tour of compulsory military service) (2004)
Manpower available for military service:
males age 16–49: 18,529,299
females age 16–49: 17,888,543 (2010 est.)
Manpower fit for military service: *males age 16–49:* 15,027,886
females age 16–49: 14,510,527 (2010 est.)
Manpower reaching militarily significant age annually: *male:* 405,438
female: 384,930 (2010 est.)
Military expenditures: 1.5% of GDP (2005 est.)
country comparison to the world: 99

TRANSNATIONAL ISSUES

Disputes—international: none
Illicit drugs: source of precursor chemicals for South American cocaine processors; transshipment point for and consumer of Southwest Asian heroin, Latin American cocaine, and European-produced synthetic drugs; major financial center

GHANA

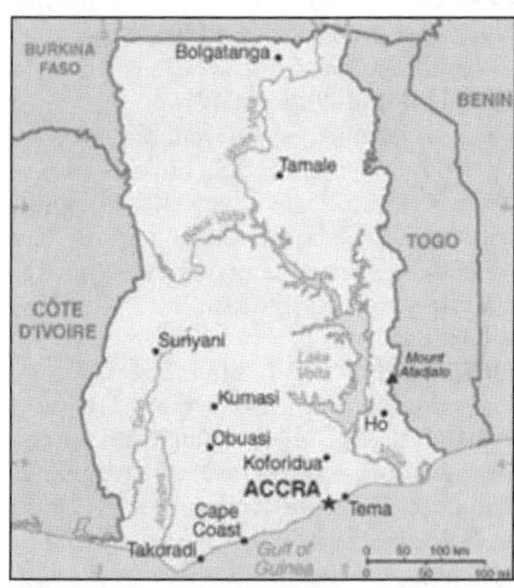

INTRODUCTION

Background: Formed from the merger of the British colony of the Gold Coast and the Togoland trust territory, Ghana in 1957 became the first sub-Saharan country in colonial Africa to gain its independence. Ghana endured a long series of coups before Lt. Jerry RAWLINGS took power in 1981 and banned political parties. After approving a new constitution and restoring multiparty politics in 1992, RAWLINGS won presidential elections in 1992 and 1996, but was constitutionally prevented from running for a third term in 2000. John KUFUOR succeeded him and was reelected in 2004. John Atta MILLS took over as head of state in early 2009.

GEOGRAPHY

Location: Western Africa, bordering the Gulf of Guinea, between Cote d'Ivoire and Togo
Geographic coordinates: 8 00 N, 2 00 W
Map references: Africa
Area: *total:* 238,533 sq km
country comparison to the world: 81
land: 227,533 sq km
water: 11,000 sq km
Area—comparative: slightly smaller than Oregon
Land boundaries: *total:* 2,094 km
border countries: Burkina Faso 549 km, Cote d'Ivoire 668 km, Togo 877 km
Coastline: 539 km
Maritime claims: *territorial sea:* 12 nm
contiguous zone: 24 nm
exclusive economic zone: 200 nm
continental shelf: 200 nm
Climate: tropical; warm and comparatively dry along southeast coast; hot and humid in southwest; hot and dry in north
Terrain: mostly low plains with dissected plateau in south-central area
Elevation extremes: *lowest point:* Atlantic Ocean 0 m
highest point: Mount Afadjato 885 m
Natural resources: gold, timber, industrial diamonds, bauxite, manganese, fish, rubber, hydropower, petroleum, silver, salt, limestone
Land use: *arable land:* 17.54%
permanent crops: 9.22%
other: 73.24% (2005)
Irrigated land: 310 sq km (2003)
Total renewable water resources: 53.2 cu km (2001)
Freshwater withdrawal (domestic/industrial/agricultural): *total:* 0.98 cu km/yr (24%/10%/66%)
per capita: 44 cu m/yr (2000)
Natural hazards: dry, dusty, northeastern harmattan winds occur from January to March; droughts
Environment—current issues: recurrent drought in north severely affects agricultural activities; deforestation; overgrazing; soil erosion; poaching and habitat destruction threatens wildlife populations; water pollution; inadequate supplies of potable water
Environment—international agreements: *party to:* Biodiversity, Climate Change, Climate Change-Kyoto Protocol, Desertification, Endangered Species, Environmental Modification, Hazardous Wastes, Law of the Sea, Ozone Layer Protection, Ship Pollution, Tropical Timber 83, Tropical Timber 94, Wetlands
signed, but not ratified: Marine Life Conservation
Geography—note: Lake Volta is the world's largest artificial lake

PEOPLE

Population: 24,791,073 (July 2011 est.)
country comparison to the world: 47
note: estimates for this country explicitly take into account the effects of excess mortality due to AIDS; this can result in lower life expectancy, higher infant mortality, higher death rates, lower population growth rates, and changes in the distribution of population by age and sex than would otherwise be expected
Age structure: *0–14 years:* 36.5% (male 4,568,273/female 4,468,939)
15–64 years: 60% (male 7,435,449/female 7,436,204)
65 years and over: 3.6% (male 399,737/female 482,471) (2011 est.)
Median age: *total:* 21.4 years
male: 21.1 years
female: 21.6 years (2011 est.)
Population growth rate: 1.822% (2011 est.)
country comparison to the world: 63
Birth rate: 27.55 births/1,000 population (2011 est.)
country comparison to the world: 47
Death rate: 8.75 deaths/1,000 population (July 2011 est.)
country comparison to the world: 81
Net migration rate: -0.58 migrant(s)/1,000 population (2011 est.)
country comparison to the world: 141
Urbanization: *urban population:* 51% of total population (2010)
rate of urbanization: 3.4% annual rate of change (2010-15 est.)
Major cities—population: ACCRA (capital) 2.269 million; Kumasi 1.773 million (2009)
Sex ratio: *at birth:* 1.03 male(s)/female
under 15 years: 1.02 male(s)/female
15–64 years: 1 male(s)/female
65 years and over: 0.84 male(s)/female
total population: 1 male(s)/female (2011 est.)
Infant mortality rate: *total:* 48.55 deaths/1,000 live births
country comparison to the world: 50
male: 51.99 deaths/1,000 live births
female: 45.01 deaths/1,000 live births (2011 est.)
Life expectancy at birth: *total population:* 61 years
country comparison to the world: 186
male: 59.78 years
female: 62.25 years (2011 est.)
Total fertility rate: 3.48 children born/woman (2011 est.)
country comparison to the world: 46
HIV/AIDS—adult prevalence rate: 1.8% (2009 est.)
country comparison to the world: 31
HIV/AIDS—people living with HIV/AIDS: 260,000 (2009 est.)
country comparison to the world: 24
HIV/AIDS—deaths: 18,000 (2009 est.)
country comparison to the world: 16
Major infectious diseases: *degree of risk:* very high
food or waterborne diseases: bacterial and protozoal diarrhea, hepatitis A, and typhoid fever
vectorborne diseases: malaria
water contact disease: schistosomiasis
respiratory disease: meningococcal meningitis
animal contact disease: rabies
note: highly pathogenic H5N1 avian influenza has been identified in this country; it poses a negligible risk with extremely rare cases possible among US citizens who have close contact with birds (2009)
Drinking water source: *Improved:*
urban: 90% of population
rural: 74% of population
total: 82% of population
Unimproved: urban: 10% of population
rural: 26% of population
total: 18% of population (2008)
Sanitation facility access: *Improved:*
urban: 18% of population
rural: 7% of population
total: 13% of population
Unimproved: urban: 82% of population
rural: 93% of population
total: 87% of population (2008)
Nationality: *noun:* Ghanaian(s)
adjective: Ghanaian
Ethnic groups: Akan 45.3%, Mole-Dagbon 15.2%, Ewe 11.7%, Ga-Dangme 7.3%, Guan 4%, Gurma 3.6%, Grusi 2.6%, Mande-Busanga 1%, other tribes 1.4%, other 7.8% (2000 census)
Religions: Christian 68.8% (Pentecostal/Charismatic 24.1%, Protestant 18.6%, Catholic 15.1%, other 11%), Muslim 15.9%, traditional 8.5%, other 0.7%, none 6.1% (2000 census)
Languages: Asante 14.8%, Ewe 12.7%, Fante 9.9%, Boron (Brong) 4.6%, Dagomba 4.3%, Dangme 4.3%, Dagarte (Dagaba) 3.7%, Akyem 3.4%, Ga 3.4%,

Akuapem 2.9%, other (includes English (official)) 36.1% (2000 census)

Literacy:
definition: age 15 and over can read and write
total population: 57.9%
male: 66.4%
female: 49.8% (2000 census)

School life expectancy (primary to tertiary education):
total: 10 years
male: 11 years
female: 10 years (2009)

Education expenditures: 5.4% of GDP (2005)
country comparison to the world: 46

GOVERNMENT

Country name: *conventional long form:* Republic of Ghana
conventional short form: Ghana
former: Gold Coast

Government type:
constitutional democracy

Capital: *name:* Accra
geographic coordinates: 5 33 N, 0 13 W
time difference: UTC 0 (5 hours ahead of Washington, DC during Standard Time)

Administrative divisions: 10 regions; Ashanti, Brong-Ahafo, Central, Eastern, Greater Accra, Northern, Upper East, Upper West, Volta, Western

Independence: 6 March 1957 (from the UK)

National holiday: Independence Day, 6 March (1957)

Constitution: approved 28 April 1992

Legal system: mixed system of English common law and customary law

International law organization participation: has not submitted an ICJ jurisdiction declaration; accepts ICCt jurisdiction

Suffrage: 18 years of age; universal

Executive branch: *chief of state:* President John Evans Atta MILLS (since 7 January 2009); Vice President John Dramani MAHAMA (since 7 January 2009); note–the president is both the chief of state and head of government
head of government: President John Evans Atta MILLS (since 7 January 2009); Vice President John Dramani MAHAMA (since 7 January 2009)
cabinet: Council of Ministers; president nominates members subject to approval by Parliament
(For more information visit the World Leaders website)
elections: president and vice president elected on the same ticket by popular vote for four-year terms (eligible for a second term); election last held on 7 and 28 December 2008 (next to be held on 7 December 2012)
election results: John Evans Atta MILLS elected president in run-off election; percent of vote–John Evans Atta MILLS 50.23%, Nana Addo Dankwa AKUFO-ADDO 49.77%

Legislative branch: unicameral Parliament (230 seats; members elected by direct, popular vote in single-seat constituencies to serve four-year terms)
elections: last held on 7 December 2008 (next to be held on 7 December 2012)
election results: percent of vote by party–NA; seats by party–NDC 114, NPP 107, PNC 2, CPP 1, independent 4, other 2

Judicial branch: Supreme Court; High Court; Court of Appeal; regional tribunals

Political parties and leaders: Convention People's Party or CPP [Ladi NYLANDER]; Democratic Freedom Party or DFP [Alhaji Abudu Rahman ISSAKAH]; Every Ghanaian Living Everywhere or EGLE; Great Consolidated Popular Party or GCPP [Dan LARTEY]; National Democratic Congress or NDC [Dr. Kwabena ADJEI]; New Patriotic Party or NPP [Jake OBETSEBI-LAMPEY]; People's National Convention or PNC [Alhaji Amed RAMADAN]; Reform Party [Kyeretwie OPUKU]; United Renaissance Party or URP [Charles WAYO]

Political pressure groups and leaders: Christian Aid (water rights); Committee for Joint Action or CJA (education reform); National Coalition Against the Privatization of Water or CAP (water rights); Oxfam (water rights); Public Citizen (water rights); Students Coalition Against EPA [Kwabena Ososukene OKAI] (education reform); Third World Network (education reform)

International organization participation: ACP, AfDB, AU, C, ECOWAS, FAO, G-24, G-77, IAEA, IBRD, ICAO, ICC, ICRM, IDA, IFAD, IFC, IFRCS, ILO, IMF, IMO, IMSO, Interpol, IOC, IOM, IPU, ISO, ITSO, ITU, ITUC, MIGA, MINURSO, MONUSCO, NAM, OAS (observer), OIF (associate member), OPCW, UN, UNAMID, UNCTAD, UNESCO, UNHCR, UNIDO, UNIFIL, UNMIL, UNOCI, UNWTO, UPU, WCO, WFTU, WHO, WIPO, WMO, WTO

Diplomatic representation in the US:
chief of mission: Ambassador Daniel Ohene AGYEKUM
chancery: 3512 International Drive NW, Washington, DC 20008
telephone: [1] (202) 686-4520
FAX: [1] (202) 686-4527
consulate(s) general: New York

Diplomatic representation from the US:
chief of mission: Ambassador Donald G. TEITELBAUM
embassy: 24 4th Circular Rd. Cantonments, Accra
mailing address: P. O. Box 194, Accra
telephone: [233] (21) 741-000
FAX: [233] (21) 741-389

Flag description: three equal horizontal bands of red (top), yellow, and green, with a large black five-pointed star centered in the yellow band; red symbolizes the blood shed for independence, yellow represents the country's mineral wealth, while green stands for its forests and natural wealth; the black star is said to be the lodestar of African freedom
note: uses the popular Pan-African colors of Ethiopia; similar to the flag of Bolivia, which has a coat of arms centered in the yellow band

National anthem: *name:* "God Bless Our Homeland Ghana"
lyrics/music: unknown/Philip GBEHO
note: music adopted 1957, lyrics adopted 1966; the lyrics were changed twice, once when a republic was declared in 1960 and again after a 1966 coup

ECONOMY

Economy—overview: Ghana is well endowed with natural resources and agriculture accounts for roughly one-third of GDP and employs more than half of the workforce, mainly small landholders. The services sector accounts for 40% of GDP. Gold and cocoa production and individual remittances are major sources of foreign exchange. Oil production at Ghana's offshore Jubilee field began in mid-December and is expected to boost economic growth. Ghana signed a Millennium Challenge Corporation (MCC) Compact in 2006, which aims to assist in transforming Ghana's agricultural sector. Ghana opted for debt relief under the Heavily Indebted Poor Country (HIPC) program in 2002, and is also benefiting from the Multilateral Debt Relief Initiative that took effect in 2006. In 2009 Ghana signed a three-year Poverty Reduction and Growth Facility with the IMF to improve macroeconomic stability, private sector competitiveness, human resource development, and good governance and civic responsibility. Sound macroeconomic management along with high prices for gold and cocoa helped sustain GDP growth in 2008-10. In early 2010 President John Atta MILLS targeted recovery from high inflation and current account and budget deficits as his priorities.

GDP (purchasing power parity): $61.97 billion (2010 est.)
country comparison to the world: 86
$58.61 billion (2009 est.)
$56 billion (2008 est.)
note: data are in 2010 US dollars

GDP (official exchange rate): $31.08 billion (2010 est.)

GDP—real growth rate: 5.7% (2010 est.)
country comparison to the world: 58
4.7% (2009 est.)
8.4% (2008 est.)

GDP—per capita (PPP): $2,500 (2010 est.)
country comparison to the world: 175
$2,500 (2009 est.)
$2,400 (2008 est.)
note: data are in 2010 US dollars

GDP—composition by sector: *agriculture:* 33.7%
industry: 24.7%
services: 41.6% (2010 est.)

Labor force: 10.56 million (2010 est.)
country comparison to the world: 48

Labor force—by occupation: *agriculture:* 56%
industry: 15%
services: 29% (2005 est.)

Unemployment rate: 11% (2000 est.)
country comparison to the world: 122

Population below poverty line: 28.5% (2007 est.)

Household income or consumption by percentage share: *lowest 10%:* 2%
highest 10%: 32.8% (2006)
Distribution of family income—Gini index: 39.4 (2005-06)
country comparison to the world: 64
40.7 (1999)
Investment (gross fixed): 39.8% of GDP (2010 est.)
country comparison to the world: 4
Budget: *revenues:* $5.518 billion
expenditures: $7.025 billion (2010 est.)
Public debt: 59.9% of GDP (2010 est.)
country comparison to the world: 33
55.2% of GDP (2009 est.)
Inflation rate (consumer prices): 10.9% (2010 est.)
country comparison to the world: 198
19.3% (2009 est.)
Central bank discount rate: 18% (31 December 2009)
country comparison to the world: 14
17% (31 December 2008)
Commercial bank prime lending rate: NA%
Stock of narrow money: $6.26 billion (31 December 2010 est.)
country comparison to the world: 83
$5.203 billion (31 December 2009 est.)
Stock of broad money: $9.583 billion (31 December 2010 est.)
country comparison to the world: 102
$7.823 billion (31 December 2009 est.)
Stock of domestic credit: $7.155 billion (31 December 2010 est.)
country comparison to the world: 105
$6.987 billion (31 December 2009 est.)
Market value of publicly traded shares: $2.508 billion (31 December 2009)
country comparison to the world: 90
$3.394 billion (31 December 2008)
$2.38 billion (31 December 2007)
Agriculture—products: cocoa, rice, cassava (tapioca), peanuts, corn, shea nuts, bananas; timber
Industries: mining, lumbering, light manufacturing, aluminum smelting, food processing, cement, small commercial ship building
Industrial production growth rate: 5% (2010 est.)
country comparison to the world: 68
Electricity—production: 6.746 billion kWh (2007 est.)
country comparison to the world: 102
Electricity—consumption: 5.702 billion kWh (2007 est.)
country comparison to the world: 109
Electricity—exports: 249 million kWh (2007 est.)
Electricity—imports: 435 million kWh (2007 est.)
Oil—production: 7,081 bbl/day (2009 est.)
country comparison to the world: 89
Oil—consumption: 57,000 bbl/day (2009 est.)
country comparison to the world: 93
Oil—exports: 4,843 bbl/day (2007 est.)
country comparison to the world: 104
Oil—imports: 45,380 bbl/day (2007 est.)
country comparison to the world: 90
Oil—proved reserves: 15 million bbl (1 January 2010 est.)
country comparison to the world: 88
Natural gas—production: 0 cu m (2008 est.)
country comparison to the world: 130
Natural gas—consumption: 0 cu m (2008 est.)
country comparison to the world: 175
Natural gas—exports: 0 cu m (2008 est.)
country comparison to the world: 101
Natural gas—imports: 0 cu m (2008 est.)
country comparison to the world: 190
Natural gas—proved reserves: 22.65 billion cu m (1 January 2010 est.)
country comparison to the world: 76
Current account balance: $-1.871 billion (2010 est.)
country comparison to the world: 153
$-1.199 billion (2009 est.)
Exports: $7.326 billion (2010 est.)
country comparison to the world: 98
$5.84 billion (2009 est.)
Exports—commodities: gold, cocoa, timber, tuna, bauxite, aluminum, manganese ore, diamonds, horticulture
Exports—partners: Netherlands 13.45%, UK 7.87%, France 5.85%, Ukraine 5.84%, Malaysia 3.97% (2009)
Imports: $10.18 billion (2010 est.)
country comparison to the world: 89
$8.046 billion (2009 est.)
Imports—commodities: capital equipment, petroleum, foodstuffs
Imports—partners: China 16.8%, Nigeria 11.88%, US 6.63%, Cote d'Ivoire 5.99%, India 5.57%, France 5.09%, UK 4.23% (2009)
Reserves of foreign exchange and gold: $3.8 billion (31 December 2010 est.)
country comparison to the world: 82
$3.165 billion (31 December 2009 est.)
Debt—external: $6.483 billion (31 December 2010 est.)
country comparison to the world: 98
$5.427 billion (31 December 2009 est.)
Stock of direct foreign investment—at home: $NA
Stock of direct foreign investment—abroad: $NA
Exchange rates: cedis (GHC) per US dollar–1.429 (2010), 1.409 (2009), 1.1 (2008), 0.95 (2007), 9,174.8 (2006)

COMMUNICATIONS

Telephones—main lines in use: 267,400 (2009)
country comparison to the world: 121
Telephones—mobile cellular: 15.109 million (2009)
country comparison to the world: 49
Telephone system: *general assessment:* primarily microwave radio relay; wireless local loop has been installed; outdated and unreliable fixed-line infrastructure heavily concentrated in Accra
domestic: competition among multiple mobile-cellular providers has spurred growth with a subscribership of more than 60 per 100 persons and rising
international: country code–233; landing point for the SAT-3/WASC, Main One, and GLO-1 fiber-optic submarine cables that provide connectivity to South Africa, Europe, and Asia; satellite earth stations–4 Intelsat (Atlantic Ocean); microwave radio relay link to Panaftel system connects Ghana to its neighbors (2009)
Broadcast media: state-owned TV station, 2 state-owned radio networks; several privately-owned TV stations and a large number of privately-owned radio stations; transmissions of multiple international broadcasters are accessible; several cable and satellite TV subscription services are obtainable (2007)
Internet country code: .gh
Internet hosts: 41,082 (2010)
country comparison to the world: 93
Internet users: 1.297 million (2009)
country comparison to the world: 93

TRANSPORTATION

Airports: 11 (2010)
country comparison to the world: 155
Airports—with paved runways: *total:* 7
over 3,047 m: 1
2,438 to 3,047 m: 1
1,524 to 2,437 m: 3
914 to 1,523 m: 2 (2010)
Airports—with unpaved runways: *total:* 4
914 to 1,523 m: 3
under 914 m: 1 (2010)
Pipelines: gas 1 km; oil 5 km; refined products 312 km (2010)
Railways: *total:* 947 km
country comparison to the world: 91
narrow gauge: 947 km 1.067-m gauge (2009)
Roadways: *total:* 62,221 km
country comparison to the world: 72
paved: 9,955 km
unpaved: 52,266 km (2006)
Waterways: 1,293 km (168 km for launches and lighters on Volta, Ankobra, and Tano rivers; 1,125 km of arterial and feeder waterways on Lake Volta) (2011)
country comparison to the world: 58
Merchant marine: *total:* 4
country comparison to the world: 135
by type: petroleum tanker 1, refrigerated cargo 3
foreign-owned: 2 (Brazil 1, South Korea 1) (2010)
Ports and terminals: Takoradi, Tema

MILITARY

Military branches: Ghana Army, Ghana Navy, Ghana Air Force (2011)
Military service age and obligation: 18 years of age for voluntary military service, with basic education certificate; no conscription (2010)
Manpower available for military service: *males age 16–49:* 6,268,191
females age 16–49: 6,194,339 (2010 est.)
Manpower fit for military service: *males age 16–49:* 4,136,406
females age 16–49: 4,220,761 (2010 est.)
Manpower reaching militarily significant age annually: *male:* 267,896
female: 260,992 (2010 est.)
Military expenditures: 1.7% of GDP (2009)
country comparison to the world: 89

TRANSNATIONAL ISSUES

Disputes—international: Ghana struggles to accommodate returning nationals who worked in the cocoa plantations and escaped fighting in Cote d'Ivoire

Refugees and internally displaced persons: refugees (country of origin): 35,653 (Liberia); 8,517 (Togo) (2007)
Illicit drugs: illicit producer of cannabis for the international drug trade; major transit hub for Southwest and Southeast Asian heroin and, to a lesser extent, South American cocaine destined for Europe and the US; widespread crime and money laundering problem, but the lack of a well developed financial infrastructure limits the country's utility as a money laundering center; significant domestic cocaine and cannabis use

GIBRALTAR

(OVERSEAS TERRITORY OF THE UK)

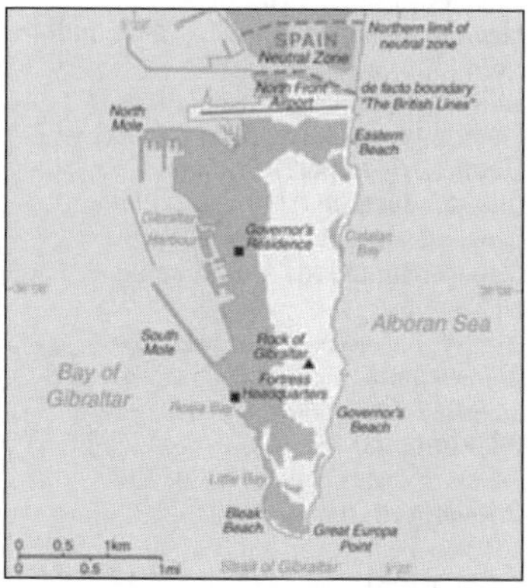

INTRODUCTION

Background: Strategically important, Gibraltar was reluctantly ceded to Great Britain by Spain in the 1713 Treaty of Utrecht; the British garrison was formally declared a colony in 1830. In a referendum held in 1967, Gibraltarians voted overwhelmingly to remain a British dependency. The subsequent granting of autonomy in 1969 by the UK led to Spain closing the border and severing all communication links. A series of talks were held by the UK and Spain between 1997 and 2002 on establishing temporary joint sovereignty over Gibraltar. In response to these talks, the Gibraltar Government called a referendum in late 2002 in which the majority of citizens voted overwhelmingly against any sharing of sovereignty with Spain. Since late 2004, tripartite talks among Spain, the UK, and Gibraltar have been held with the aim of cooperatively resolving problems that affect the local population, and work continues on cooperation agreements in areas such as taxation and financial services; communications and maritime security; policy, legal and customs services; environmental protection; and education and visa services. Throughout 2009, a dispute over Gibraltar's claim to territorial waters extending out three miles gave rise to periodic non-violent maritime confrontations between Spanish and UK naval patrols. A new noncolonial constitution came into effect in 2007, and the European Court of First Instance recognized Gibraltar's right to regulate its own tax regime in December 2008, but the UK retains responsibility for defense, foreign relations, internal security, and financial stability.

GEOGRAPHY

Location: Southwestern Europe, bordering the Strait of Gibraltar, which links the Mediterranean Sea and the North Atlantic Ocean, on the southern coast of Spain
Geographic coordinates: 36 08 N, 5 21 W
Map references: Europe
Area: *total:* 6.5 sq km
country comparison to the world: 242
land: 6.5 sq km
water: 0 sq km
Area—comparative: more than 10 times the size of The National Mall in Washington, D.C.
Land boundaries: *total:* 1.2 km
border countries: Spain 1.2 km
Coastline: 12 km
Maritime claims: *territorial sea:* 3 nm
Climate: Mediterranean with mild winters and warm summers
Terrain: a narrow coastal lowland borders the Rock of Gibraltar
Elevation extremes: *lowest point:* Mediterranean Sea 0 m
highest point: Rock of Gibraltar 426 m
Natural resources: none
Land use: *arable land:* 0%
permanent crops: 0%
other: 100% (2005)
Irrigated land: NA
Natural hazards: NA
Environment—current issues: limited natural freshwater resources: large concrete or natural rock water catchments collect rainwater (no longer used for drinking water) and adequate desalination plant
Geography—note: strategic location on Strait of Gibraltar that links the North Atlantic Ocean and Mediterranean Sea

PEOPLE

Population: 28,956 (July 2011 est.)
country comparison to the world: 215
Age structure: *0–14 years:* 20.4% (male 3,040/female 2,862)
15–64 years: 65.8% (male 9,607/female 9,451)
65 years and over: 13.8% (male 1,934/female 2,062) (2011 est.)
Median age: *total:* 33.3 years
male: 32.4 years
female: 34.3 years (2011 est.)
Population growth rate: 0.273% (2011 est.)
country comparison to the world: 170
Birth rate: 14.23 births/1,000 population (2011 est.)
country comparison to the world: 145
Death rate: 8.18 deaths/1,000 population (July 2011 est.)
country comparison to the world: 96
Net migration rate: -3.32 migrant(s)/1,000 population (2011 est.)
country comparison to the world: 180
Urbanization: *urban population:* 100% of total population (2010)
rate of urbanization: 0.2% annual rate of change (2010-15 est.)
Sex ratio: *at birth:* 1.07 male(s)/female
under 15 years: 1.06 male(s)/female
15–64 years: 1.02 male(s)/female
65 years and over: 0.93 male(s)/female
total population: 1.01 male(s)/female (2011 est.)
Infant mortality rate: *total:* 6.69 deaths/1,000 live births
country comparison to the world: 167
male: 7.44 deaths/1,000 live births
female: 5.87 deaths/1,000 live births (2011 est.)
Life expectancy at birth: *total population:* 78.68 years
country comparison to the world: 47
male: 75.84 years
female: 81.72 years (2011 est.)
Total fertility rate: 1.96 children born/woman (2011 est.)
country comparison to the world: 132
HIV/AIDS—adult prevalence rate: NA
HIV/AIDS—people living with HIV/AIDS: NA
HIV/AIDS—deaths: NA
Nationality: *noun:* Gibraltarian(s)
adjective: Gibraltar
Ethnic groups: Spanish, Italian, English, Maltese, Portuguese, German, North Africans
Religions: Roman Catholic 78.1%, Church of England 7%, other Christian 3.2%, Muslim 4%, Jewish 2.1%, Hindu 1.8%, other or unspecified 0.9%, none 2.9% (2001 census)
Languages: English (used in schools and for official purposes), Spanish, Italian, Portuguese
Literacy: *definition:* NA
total population: above 80%
male: NA
female: NA
School life expectancy (primary to tertiary education): NA
Education expenditures: NA

GOVERNMENT

Country name: *conventional long form:* none
conventional short form: Gibraltar
Dependency status: overseas territory of the UK
Government type: NA
Capital: *name:* Gibraltar

geographic coordinates: 36 08 N, 5 21 W
time difference: UTC+1 (6 hours ahead of Washington, DC during Standard Time)
daylight saving time: +1hr, begins last Sunday in March; ends last Sunday in October
Administrative divisions: none (overseas territory of the UK)
Independence: none (overseas territory of the UK)
National holiday: National Day, 10 September (1967); note–day of the national referendum to decide whether to remain with the UK or join Spain
Constitution: 5 June 2006; came into force 2 January 2007
Legal system: the laws of the UK, where applicable, apply
Suffrage: 18 years of age; universal; and British citizens who have been residents six months or more
Executive branch: *chief of state:* Queen ELIZABETH II (since 6 February 1952); represented by Governor Vice Admiral Sir Adrian JOHNS (since 26 October 2009)
head of government: Chief Minister Peter CARUANA (since 17 May 1996)
cabinet: Council of Ministers appointed from among the 17 elected members of the Parliament by the governor in consultation with the chief minister
(For more information visit the World Leaders website)
elections: the monarchy is hereditary; governor appointed by the monarch; following legislative elections, the leader of the majority party or the leader of the majority coalition is usually appointed chief minister by the governor
Legislative branch: unicameral Parliament (18 seats: 17 members elected by popular vote, 1 for the speaker appointed by Parliament; members serve four-year terms)
elections: last held on 11 October 2007 (next to be held not later than October 2011)
election results: percent of vote by party–GSD 49.3%, GSLP 31.8%, Gibraltar Liberal Party 13.6%; seats by party–GSD 10, GSLP 4, Gibraltar Liberal Party 3
Judicial branch: Supreme Court; Court of Appeal
Political parties and leaders: Gibraltar Liberal Party [Joseph GARCIA]; Gibraltar Social Democrats or GSD [Peter CARUANA]; Gibraltar Socialist Labor Party or GSLP [Joseph John BOSSANO]
Political pressure groups and leaders: Chamber of Commerce; Gibraltar Representatives Organization; Women's Association
International organization participation: Interpol (subbureau), UPU
Diplomatic representation in the US: none (overseas territory of the UK)
Diplomatic representation from the US: none (overseas territory of the UK)
Flag description: two horizontal bands of white (top, double width) and red with a three-towered red castle in the center of the white band; hanging from the castle gate is a gold key centered in the red band; the design is that of Gibraltar's coat of arms granted on 10 July 1502 by King Ferdinand and Queen Isabella of Spain; the castle symbolizes Gibraltar as a fortress, while the key represents Gibraltar's strategic importance–the key to the Mediterranean
National anthem: *name:* "Gibraltar Anthem"
lyrics/music: Peter EMBERLEY
note: adopted 1994; serves as a local anthem; as a territory of the United Kingdom, "God Save the Queen" remains official (see United Kingdom)

ECONOMY

Economy—overview: Self-sufficient Gibraltar benefits from an extensive shipping trade, offshore banking, and its position as an international conference center. Tax rates are low to attract foreign investment. The British military presence has been sharply reduced and now contributes about 7% to the local economy, compared with 60% in 1984. The financial sector, tourism (almost 5 million visitors in 1998), gaming revenues, shipping services fees, and duties on consumer goods also generate revenue. The financial sector, tourism, and the shipping sector contribute 30%, 30%, and 25%, respectively, of GDP. Telecommunications, e-commerce, and e-gaming account for the remaining 15%. In recent years, Gibraltar has seen major structural change from a public to a private sector economy, but changes in government spending still have a major impact on the level of employment.
GDP (purchasing power parity): $1.275 billion (2008)
country comparison to the world: 195
$1.203 billion (2007 est.)
$1.106 billion (2006 est.)
GDP (official exchange rate): $1.106 billion (2006 est.)
GDP—real growth rate: 6% (2008)
country comparison to the world: 53
8.8% (2007)
0% (2006 est.)
GDP—per capita (PPP): $43,000 (2006 est.)
country comparison to the world: 16
$41,200 (2007 est.)
$38,400 (2006 est.)
GDP—composition by sector:
agriculture: 0%
industry: 0%
services: 100% (2008 est.)
Labor force: 12,690 (including non-Gibraltar laborers) (2001)
country comparison to the world: 214
Labor force—by occupation: *agriculture:* negligible
industry: 40%
services: 60% (2001)
Unemployment rate: 3% (2005 est.)
country comparison to the world: 25
Population below poverty line: NA%
Household income or consumption by percentage share: *lowest 10%:* NA%
highest 10%: NA%
Budget: *revenues:* $475.8 million
expenditures: $452.3 million (2008 est.)
Public debt: 7.5% of GDP (2008 est.)
country comparison to the world: 126
13.5% of GDP (2006 est.)
Inflation rate (consumer prices): 2.8% (2008)
country comparison to the world: 79
2.6% (2006)
Agriculture—products: none
Industries: tourism, banking and finance, ship repairing, tobacco
Industrial production growth rate: NA%
Electricity—production: 146 million kWh (2007 est.)
country comparison to the world: 182
Electricity—consumption: 146 million kWh (2007 est.)
country comparison to the world: 184
Electricity—exports: 0 kWh (2008 est.)
Electricity—imports: 0 kWh (2008 est.)
Oil—production: 0 bbl/day (2009 est.)
country comparison to the world: 178
Oil—consumption: 21,000 bbl/day (2009 est.)
country comparison to the world: 121
Oil—exports: 0 bbl/day (2007 est.)
country comparison to the world: 168
Oil—imports: 25,610 bbl/day (2007 est.)
country comparison to the world: 104
Oil—proved reserves: 0 bbl (1 January 2010 est.)
country comparison to the world: 135
Natural gas—production: 0 cu m (2008 est.)
country comparison to the world: 131
Natural gas—consumption: 0 cu m (2008 est.)
country comparison to the world: 176
Natural gas—exports: 0 cu m (2008 est.)
country comparison to the world: 102
Natural gas—imports: 0 cu m (2008 est.)
country comparison to the world: 191
Natural gas—proved reserves: 0 cu m (1 January 2010 est.)
country comparison to the world: 139
Exports: $271 million (2004 est.)
country comparison to the world: 177
Exports—commodities: (principally reexports) petroleum 51%, manufactured goods
Imports: $2.967 billion (2004 est.)
country comparison to the world: 142
Imports—commodities: fuels, manufactured goods, and foodstuffs
Debt—external: $NA
Exchange rates: Gibraltar pounds (GIP) per US dollar–0.6388 (2010), 0.6175 (2009), 0.4993 (2007), 0.5418 (2006)

COMMUNICATIONS

Telephones—main lines in use: 24,000 (2009)
country comparison to the world: 187
Telephones—mobile cellular: 28,600 (2009)
country comparison to the world: 202
Telephone system: *general assessment:* adequate, automatic domestic system and adequate international facilities
domestic: automatic exchange facilities
international: country code–350; radiotelephone; microwave radio relay; satellite earth station–1 Intelsat (Atlantic Ocean)
Broadcast media: Gibraltar Broadcasting Corporation (GBC) provides television and radio broadcasting services via 1 television station and 4 radio stations; British Forces Broadcasting Service (BFBS)

operates 1 radio station; broadcasts from Spanish radio and TV stations are accessible (2008)
Internet country code: .gi
Internet hosts: 2,053 (2010)
country comparison to the world: 154
Internet users: 20,200 (2009)
country comparison to the world: 192

TRANSPORTATION

Airports: 1 (2010)
country comparison to the world: 217
Airports—with paved runways: *total:* 1
1,524 to 2,437 m: 1 (2010)
Roadways: *total:* 29 km
country comparison to the world: 218
paved: 29 km (2007)
Merchant marine: *total:* 265
country comparison to the world: 32
by type: bulk carrier 1, cargo 139, chemical tanker 65, container 35, liquefied gas 2, petroleum tanker 11, roll on/roll off 4, vehicle carrier 8
foreign-owned: 250 (Belgium 2, Cyprus 1, Denmark 6, Finland 2, Germany 125, Greece 7, Iceland 1, Italy 4, Jersey 1, Morocco 4, Netherlands 33, Norway 42, Singapore 1, Sweden 12, UAE 5, UK 4)
note: this country allows large numbers of ships owned by foreign entities to be registered in its national shipping registry and to fly its flag; these ships operate under the laws of the flag state
registered in other countries: 6 (Liberia 5, Panama 1) (2010)
Ports and terminals: Gibraltar

MILITARY

Military branches: Royal Gibraltar Regiment (2009)
Manpower available for military service: *males age 16–49:* 7,037 (2010 est.)
Manpower fit for military service: *males age 16–49:* 6,017
females age 16–49: 5,706 (2010 est.)
Manpower reaching militarily significant age annually: *male:* 228
female: 220 (2010 est.)
Military—note: defense is the responsibility of the UK; the Royal Gibraltar Regiment replaced the last British regular infantry forces in 1992

TRANSNATIONAL ISSUES

Disputes—international: in 2002, Gibraltar residents voted overwhelmingly by referendum to reject any "shared sovereignty" arrangement; the government of Gibraltar insists on equal participation in talks between the UK and Spain; Spain disapproves of UK plans to grant Gibraltar even greater autonomy

GREECE

INTRODUCTION

Background: Greece achieved independence from the Ottoman Empire in 1829. During the second half of the 19th century and the first half of the 20th century, it gradually added neighboring islands and territories, most with Greek-speaking populations. In World War II, Greece was first invaded by Italy (1940) and subsequently occupied by Germany (1941-44); fighting endured in a protracted civil war between supporters of the king and other anti-Communists and Communist rebels. Following the latter's defeat in 1949, Greece joined NATO in 1952. In 1967, a group of military officers seized power, establishing a military dictatorship that suspended many political liberties and forced the king to flee the country. In 1974, democratic elections and a referendum created a parliamentary republic and abolished the monarchy. In 1981, Greece joined the EC (now the EU); it became the 12th member of the European Economic and Monetary Union in 2001. In 2010, the prospect of a Greek default on its euro-denominated debt created severe strains within the EMU and raised the question of whether a member country might voluntarily leave the common currency or be removed.

GEOGRAPHY

Location: Southern Europe, bordering the Aegean Sea, Ionian Sea, and the Mediterranean Sea, between Albania and Turkey
Geographic coordinates: 39 00 N, 22 00 E
Map references: Europe
Area: *total:* 131,957 sq km
country comparison to the world: 96
land: 130,647 sq km
water: 1,310 sq km
Area—comparative: slightly smaller than Alabama
Land boundaries: *total:* 1,228 km
border countries: Albania 282 km, Bulgaria 494 km, Turkey 206 km, Macedonia 246 km
Coastline: 13,676 km
Maritime claims: *territorial sea:* 12 nm
continental shelf: 200 m depth or to the depth of exploitation
Climate: temperate; mild, wet winters; hot, dry summers
Terrain: mostly mountains with ranges extending into the sea as peninsulas or chains of islands
Elevation extremes: *lowest point:* Mediterranean Sea 0 m
highest point: Mount Olympus 2,917 m
Natural resources: lignite, petroleum, iron ore, bauxite, lead, zinc, nickel, magnesite, marble, salt, hydropower potential
Land use: *arable land:* 20.45%
permanent crops: 8.59%
other: 70.96% (2005)
Irrigated land: 14,530 sq km (2003)
Total renewable water resources: 72 cu km (2005)
Freshwater withdrawal (domestic/industrial/agricultural): *total:* 8.7 cu km/yr (16%/3%/81%)
per capita: 782 cu m/yr (1997)
Natural hazards: severe earthquakes
volcanism: Santorini (elev. 367 m) has been deemed a "Decade Volcano" by the International Association of Volcanology and Chemistry of the Earth's Interior, worthy of study due to its explosive history and close proximity to human populations; although there have been very few eruptions in recent centuries, Methana and Nisyros in the Aegean are classified as historically active
Environment—current issues: air pollution; water pollution
Environment—international agreements: *party to:* Air Pollution, Air Pollution-Nitrogen Oxides, Air Pollution-Sulfur 94, Antarctic-Environmental Protocol, Antarctic-Marine Living Resources, Antarctic Treaty, Biodiversity, Climate Change, Climate Change-Kyoto Protocol, Desertification, Endangered Species, Environmental Modification, Hazardous Wastes, Law of the Sea, Marine Dumping, Ozone Layer Protection, Ship Pollution, Tropical Timber 83, Tropical Timber 94, Wetlands
signed, but not ratified: Air Pollution-Persistent Organic Pollutants, Air Pollution-Volatile Organic Compounds
Geography—note: strategic location dominating the Aegean Sea and southern approach to Turkish Straits; a peninsular country, possessing an archipelago of about 2,000 islands

PEOPLE

Population: 10,760,136 (July 2011 est.)
country comparison to the world: 76
Age structure: *0–14 years:* 14.2% (male 787,143/female 741,356)
15–64 years: 66.2% (male 3,555,447/female 3,567,383)
65 years and over: 19.6% (male 923,177/female 1,185,630) (2011 est.)
Median age: *total:* 42.5 years
male: 41.4 years
female: 43.6 years (2011 est.)
Population growth rate: 0.083% (2011 est.)
country comparison to the world: 185
Birth rate: 9.21 births/1,000 population (2011 est.)
country comparison to the world: 206
Death rate: 10.7 deaths/1,000 population (July 2011 est.)
country comparison to the world: 44

Net migration rate: 2.32 migrant(s)/1,000 population (2011 est.)
country comparison to the world: 36
Urbanization: *urban population:* 61% of total population (2010)
rate of urbanization: 0.6% annual rate of change (2010-15 est.)
Major cities—population: ATHENS (capital) 3.252 million; Thessaloniki 834,000 (2009)
Sex ratio: *at birth:* 1.064 male(s)/female
under 15 years: 1.06 male(s)/female
15–64 years: 1 male(s)/female
65 years and over: 0.78 male(s)/female
total population: 0.96 male(s)/female (2011 est.)
Infant mortality rate: *total:* 5 deaths/1,000 live births
country comparison to the world: 182
male: 5.49 deaths/1,000 live births
female: 4.48 deaths/1,000 live births (2011 est.)
Life expectancy at birth: *total population:* 79.92 years
country comparison to the world: 30
male: 77.36 years
female: 82.65 years (2011 est.)
Total fertility rate: 1.38 children born/woman (2011 est.)
country comparison to the world: 203
HIV/AIDS—adult prevalence rate: 0.1% (2009 est.)
country comparison to the world: 132
HIV/AIDS—people living with HIV/AIDS: 8,800 (2009 est.)
country comparison to the world: 104
HIV/AIDS—deaths: fewer than 500 (2009 est.)
country comparison to the world: 92
Drinking water source: *Improved:*
urban: 100% of population
rural: 99% of population
total: 100% of population
Unimproved: urban: 0% of population
rural: 1% of population
total: 0% of population (2008)
Sanitation facility access: *Improved:*
urban: 99% of population
rural: 97% of population
total: 98% of population
Unimproved: urban: 1% of population
rural: 3% of population
total: 2% of population (2008)
Nationality: *noun:* Greek(s)
adjective: Greek
Ethnic groups: population: Greek 93%, other (foreign citizens) 7% (2001 census)
note: percents represent citizenship, since Greece does not collect data on ethnicity
Religions: Greek Orthodox (official) 98%, Muslim 1.3%, other 0.7%
Languages: Greek (official) 99%, other (includes English and French) 1%
Literacy: *definition:* age 15 and over can read and write
total population: 96%
male: 97.8%
female: 94.2% (2001 census)
School life expectancy (primary to tertiary education): *total:* 17 years
male: 16 years
female: 17 years (2007)
Education expenditures: 4% of GDP (2005)
country comparison to the world: 105

GOVERNMENT

Country name: *conventional long form:* Hellenic Republic
conventional short form: Greece
local long form: Elliniki Dhimokratia
local short form: Ellas or Ellada
former: Kingdom of Greece
Government type: parliamentary republic
Capital: *name:* Athens
geographic coordinates: 37 59 N, 23 44 E
time difference: UTC+2 (7 hours ahead of Washington, DC during Standard Time)
daylight saving time: +1hr, begins last Sunday in March; ends last Sunday in October
Administrative divisions: 51 prefectures (nomoi, singular—nomos) and 1 autonomous region*; Achaia, Agion Oros* (Mount Athos), Aitolia kai Akarnania, Argolis, Arkadia, Arta, Attiki, Chalkidiki, Chania, Chios, Dodekanisos, Drama, Evros, Evrytania, Evvoia, Florina, Fokidos, Fthiotis, Grevena, Ileia, Imathia, Ioannina, Irakleion, Karditsa, Kastoria, Kavala, Kefallinia, Kerkyra, Kilkis, Korinthia, Kozani, Kyklades, Lakonia, Larisa, Lasithi, Lefkada, Lesvos, Magnisia, Messinia, Pella, Pieria, Preveza, Rethymnis, Rodopi, Samos, Serres, Thesprotia, Thessaloniki, Trikala, Voiotia, Xanthi, Zakynthos
Independence: 1829 (from the Ottoman Empire)
National holiday: Independence Day, 25 March (1821)
Constitution: 11 June 1975; amended March 1986 and April 2001
Legal system: civil legal system based on Roman law
International law organization participation: accepts compulsory ICJ jurisdiction with reservations; accepts ICCt jurisdiction
Suffrage: 18 years of age; universal and compulsory
Executive branch: *chief of state:* President Karolos PAPOULIAS (since 12 March 2005)
head of government: Prime Minister Georgios Andreas PAPANDREOU (since 6 October 2009)
cabinet: Cabinet appointed by the president on the recommendation of the prime minister
(For more information visit the World Leaders website)
elections: president elected by parliament for a five-year term (eligible for a second term); election last held on 3 February 2010 (next to be held by February 2015); president appoints leader of the party securing plurality of vote in election to become prime minister and form a government
election results: Karolos PAPOULIAS reelected president; number of parliamentary votes, 266 out of 300
Legislative branch: unicameral Parliament or Vouli ton Ellinon (300 seats; members elected by direct popular vote to serve four-year terms)
elections: last held on 4 October 2009 (next to be held by 2013)
election results: percent of vote by party—PASOK 43.9%, ND 33.5%, KKE 7.5%, LAOS 5.6%, SYRIZA 4.6%, other 4.9%; seats by party—PASOK 160, ND 91, KKE 21, LAOS 15, SYRIZA 13; note—seats by party as of 15 December 2010—PASOK 156, ND 86, KKE 21 LAOS 15, SYRIZA 9, DISY 5, Democratic Left 4, independents 4 (DISY and Democratic Left entered parliament as members of ND and SYRIZA, respectively, and the independents entered parliament as members of PASOK); only parties surpassing a 3% threshold are entitled to parliamentary seats; parties need 10 seats to become formal parliamentary groups, but can retain that status if the party participated in the last election and received the minimum 3% threshold
Judicial branch: Supreme Civil and Criminal Court; all judges are appointed for life by the president after consultation with a judicial council; Supreme Administrative Court and Court of Auditors; Courts of Appeal; Courts of First Instance
Political parties and leaders: Anticapitalist Left Cooperation for the Overthrow or ANTARSYA [Petros KONSTANTINOU]; Coalition of the Radical Left or SYRIZA [Alexis TSIPRAS]; Communist Party of Greece or KKE [Aleka PAPARIGA]; Democratic Left [Fotis KOUVELIS]; Democratic Alliance or DISY [Theodora BAKOGIANNI]; Ecologist Greens [Nikos CHRYSOGELOS]; Golden Dawn [Nikolaos MICHALOLIAKOS]; New Democracy or ND [Antonis SAMARAS]; Panhellenic Socialist Movement or PASOK [Georgios PAPANDREOU]; Popular Orthodox Rally or LAOS [Georgios KARATZAFERIS]
Political pressure groups and leaders: Civil Servants Confederation or ADEDY [Spyros PAPASPYROS]; Federation of Greek Industries or SEV [Dimitris DASKALOPOULOS]; General Confederation of Greek Workers or GSEE [Ioannis PANAGOPOULOS]
International organization participation: Australia Group, BIS, BSEC, CE, CERN, EAPC, EBRD, EIB, EMU, ESA, EU, FAO, FATF, IAEA, IBRD, ICAO, ICC, ICRM, IDA, IEA, IFAD, IFC, IFRCS, IHO, ILO, IMF, IMO, IMSO, Interpol, IOC, IOM, IPU, ISO, ITSO, ITU, ITUC, MIGA, MINURSO, NATO, NEA, NSG, OAS (observer), OECD, OIF, OPCW, OSCE, PCA, Schengen Convention, SECI, UN, UNCTAD, UNESCO, UNHCR, UNIDO, UNIFIL, UNMIS, UNWTO, UPU, WCO, WFTU, WHO, WIPO, WMO, WTO, ZC
Diplomatic representation in the US: *chief of mission:* Ambassador Vassilis KASKARELIS
chancery: 2217 Massachusetts Avenue NW, Washington, DC 20008
telephone: [1] (202) 939-1300
FAX: [1] (202) 939-1324
consulate(s) general: Boston, Chicago, Los Angeles, New York, San Francisco, Tampa
consulate(s): Atlanta, Houston, New Orleans
Diplomatic representation from the US: *chief of mission:* Ambassador Daniel Bennett SMITH

embassy: 91 Vasilisis Sophias Avenue, 10160 Athens
mailing address: PSC 108, APO AE 09842-0108
telephone: [30] (210) 721-2951
FAX: [30] (210) 645-6282
consulate(s) general: Thessaloniki

Flag description: nine equal horizontal stripes of blue alternating with white; a blue square bearing a white cross appears in the upper hoist-side corner; the cross symbolizes Greek Orthodoxy, the established religion of the country; there is no agreed upon meaning for the nine stripes or for the colors; the exact shade of blue has never been set by law and has varied from a light to a dark blue over time

National anthem: *name:* "Ymnos eis tin Eleftherian" (Hymn to Liberty)
lyrics/music: Dionysios SOLOMOS/ Nikolaos MANTZAROS
note: adopted 1864; the anthem is based on a 158 verse poem by the same name, which was inspired by the Greek Revolution of 1821 against the Ottomans; Cyprus also uses "Hymn to Liberty" as its anthem

ECONOMY

Economy—overview: Greece has a capitalist economy with the public sector accounting for about 40% of GDP and with per capita GDP about two-thirds that of the leading euro-zone economies. Tourism provides 15% of GDP. Immigrants make up nearly one-fifth of the work force, mainly in agricultural and unskilled jobs. Greece is a major beneficiary of EU aid, equal to about 3.3% of annual GDP. The Greek economy grew by nearly 4.0% per year between 2003 and 2007, due partly to infrastructural spending related to the 2004 Athens Olympic Games, and in part to an increased availability of credit, which has sustained record levels of consumer spending. But the economy went into recession in 2009 as a result of the world financial crisis, tightening credit conditions, and Athens' failure to address a growing budget deficit, which was triggered by falling state revenues, and increased government expenditures. The economy contracted by 2% in 2009, and 4.8% in 2010. Greece violated the EU's Growth and Stability Pact budget deficit criterion of no more than 3% of GDP from 2001 to 2006, but finally met that criterion in 2007-08, before exceeding it again in 2009, with the deficit reaching 15.4% of GDP. Austerity measures reduced the deficit to 9.4% of GDP in 2010. Public debt, inflation, and unemployment are above the euro-zone average while per capita income is below; unemployment rose to 12% in 2010. Eroding public finances, a credibility gap stemming from inaccurate and misreported statistics, and consistent underperformance on following through with reforms prompted major credit rating agencies in late 2009 to downgrade Greece's international debt rating, and has led the country into a financial crisis. Under intense pressure by the EU and international market participants, the government has adopted a medium-term austerity program that includes cutting government spending, reducing the size of the public sector, decreasing tax evasion, reforming the health care and pension systems, and improving competitiveness through structural reforms to the labor and product markets. Athens, however, faces long-term challenges to push through unpopular reforms in the face of often vocal opposition from the country's powerful labor unions and the general public. Greek labor unions are striking over new austerity measures, but the strikes so far have had a limited impact on the government's will to adopt reforms. An uptick in widespread unrest, however, could challenge the government's ability to implement reforms and meet budget targets, and could also lead to rioting or violence. In April 2010 a leading credit agency assigned Greek debt its lowest possible credit rating; in May, the International Monetary Fund and Eurozone governments provided Greece emergency short- and medium-term loans worth $147 billion so that the country could make debt repayments to creditors. In exchange for the largest bailout ever assembled, the government announced combined spending cuts and tax increases totaling $40 billion over three years, on top of the tough austerity measures already taken. Greece, however, struggled to boost revenues and cut spending to meet 2010 targets set by the EU and the IMF, especially after Eurostat–the EU's statistical office–revised upward Greece's deficit and debt numbers for 2009 and 2010. Greece's lenders are calling on Athens to step up efforts in 2011 to increase tax collection, shore up public enterprises, and rein in health spending, and are planning to give Greece more time to repay its EU-IMF loan. Greece responded by introducing major structural reforms, but investors still question whether Greece can sustain fiscal efforts in the face of a bleak economic outlook and public discontent.

GDP (purchasing power parity): $318.1 billion (2010 est.)
country comparison to the world: 39
$333.2 billion (2009 est.)
$340.1 billion (2008 est.)
note: data are in 2010 US dollars

GDP (official exchange rate): $305.4 billion (2010 est.)

GDP—real growth rate: -4.5% (2010 est.)
country comparison to the world: 212
-2% (2009 est.)
1% (2008 est.)

GDP—per capita (PPP): $29,600 (2010 est.)
country comparison to the world: 47
$31,000 (2009 est.)
$31,700 (2008 est.)
note: data are in 2010 US dollars

GDP—composition by sector: *agriculture:* 4%
industry: 17.6%
services: 78.5% (2010 est.)

Labor force: 5.05 million (2010 est.)
country comparison to the world: 73

Labor force—by occupation: *agriculture:* 12.4%
industry: 22.4%
services: 65.1% (2005 est.)

Unemployment rate: 12% (2010 est.)
country comparison to the world: 128
9.4% (2009 est.)

Population below poverty line: 20% (2009 est.)

Household income or consumption by percentage share: *lowest 10%:* 2.5%
highest 10%: 26% (2000 est.)

Distribution of family income—Gini index: 33 (2005)
country comparison to the world: 96
35.4 (1998)

Investment (gross fixed): 14.8% of GDP (2010 est.)
country comparison to the world: 134

Budget: *revenues:* $114.5 billion
expenditures: $142.9 billion (2010 est.)

Public debt: 144% of GDP (2010 est.)
country comparison to the world: 5
126.8% of GDP (2009 est.)

Inflation rate (consumer prices): 4.5% (2010 est.)
country comparison to the world: 128
1.2% (2009 est.)

Central bank discount rate: 1.75% (31 December 2010)
country comparison to the world: 127
1.75% (31 December 2009)
note: this is the European Central Bank's rate on the marginal lending facility, which offers overnight credit to banks in the euro area

Commercial bank prime lending rate: 8.59% (31 December 2009 est.)
country comparison to the world: 105
8.65% (31 December 2008 est.)

Stock of narrow money: $152.8 billion (31 December 2010 est.)
country comparison to the world: 22
$172.8 billion (31 December 2009 est.)
note: see entry for the European Union for money supply in the euro area; the European Central Bank (ECB) controls monetary policy for the 17 members of the Economic and Monetary Union (EMU); individual members of the EMU do not control the quantity of money circulating within their own borders

Stock of broad money: $335.9 billion (31 December 2010 est.)
country comparison to the world: 26
$368.4 billion (31 December 2009 est.)

Stock of domestic credit: $419.9 billion (31 December 2009 est.)
country comparison to the world: 25
$394.6 billion (31 December 2008 est.)

Market value of publicly traded shares: $54.72 billion (31 December 2009)
country comparison to the world: 41
$90.4 billion (31 December 2008)
$264.9 billion (31 December 2007)

Agriculture—products: wheat, corn, barley, sugar beets, olives, tomatoes, wine, tobacco, potatoes; beef, dairy products

Industries: tourism, food and tobacco processing, textiles, chemicals, metal products; mining, petroleum

Industrial production growth rate: 3.2% (2010 est.)
country comparison to the world: 102

Electricity—production: 58.79 billion kWh (2007 est.)

country comparison to the world: 44
Electricity—consumption: 58.28 billion kWh (2007 est.)
country comparison to the world: 40
Electricity—exports: 1.962 billion kWh (2008 est.)
Electricity—imports: 7.575 billion kWh (2008 est.)
Oil—production: 6,779 bbl/day (2009 est.)
country comparison to the world: 90
Oil—consumption: 414,400 bbl/day (2009 est.)
country comparison to the world: 34
Oil—exports: 153,000 bbl/day (2008 est.)
country comparison to the world: 58
Oil—imports: 520,900 bbl/day (2008 est.)
country comparison to the world: 24
Oil—proved reserves: 10 million bbl (1 January 2010 est.)
country comparison to the world: 91
Natural gas—production: 9 million cu m (2009 est.)
country comparison to the world: 89
Natural gas—consumption: 3.528 billion cu m (2009 est.)
country comparison to the world: 70
Natural gas—exports: 0 cu m (2008 est.)
country comparison to the world: 105
Natural gas—imports: 3.556 billion cu m (2009 est.)
country comparison to the world: 37
Natural gas—proved reserves: 991.1 million cu m (1 January 2010 est.)
country comparison to the world: 99
Current account balance: $-17.1 billion (2010 est.)
country comparison to the world: 180
$-34.43 billion (2009 est.)
Exports: $21.14 billion (2010 est.)
country comparison to the world: 67
$21.34 billion (2009 est.)
Exports—commodities: food and beverages, manufactured goods, petroleum products, chemicals, textiles
Exports—partners: Germany 11.11%, Italy 11.05%, Cyprus 7.28%, Bulgaria 6.74%, US 4.95%, UK 4.4%, Turkey 4.23% (2009)
Imports: $44.9 billion (2010 est.)
country comparison to the world: 50
$64.2 billion (2009 est.)
Imports—commodities: machinery, transport equipment, fuels, chemicals
Imports—partners: Germany 13.73%, Italy 12.71%, China 7.08%, France 6.1%, Netherlands 6.02%, South Korea 5.68%, Belgium 4.34%, Spain 4.08% (2009)
Reserves of foreign exchange and gold: $NA (31 December 2010 est.)
$5.546 billion (31 December 2009 est.)
Debt—external: $532.9 billion (30 June 2010)
country comparison to the world: 20
$504.6 billion (31 December 2008)
Stock of direct foreign investment—at home: $48.1 billion (31 December 2010 est.)
country comparison to the world: 55
$44.93 billion (31 December 2009 est.)
Stock of direct foreign investment—abroad: $38.66 billion (31 December 2010 est.)
country comparison to the world: 35
$40.45 billion (31 December 2009 est.)
Exchange rates: euros (EUR) per US dollar–, 0.7715 (2010), 0.7179 (2009), 0.6827 (2008), 0.7345 (2007), 0.7964 (2006)

COMMUNICATIONS

Telephones—main lines in use: 5.93 million (2009)
country comparison to the world: 30
Telephones—mobile cellular: 13.295 million (2009)
country comparison to the world: 54
Telephone system: *general assessment:* adequate, modern networks reach all areas; good mobile telephone and international service
domestic: microwave radio relay trunk system; extensive open-wire connections; submarine cable to offshore islands
international: country code–30; landing point for the SEA-ME-WE-3 optical telecommunications submarine cable that provides links to Europe, Middle East, and Asia; a number of smaller submarine cables provide connectivity to various parts of Europe, the Middle East, and Cyprus; tropospheric scatter; satellite earth stations–4 (2 Intelsat–1 Atlantic Ocean and 1 Indian Ocean, 1 Eutelsat, and 1 Inmarsat–Indian Ocean region)
Broadcast media: broadcast media dominated by the private sector; roughly 150 private TV channels, about a dozen of the private channels broadcast at the national or regional level; 3 publicly-owned terrestrial TV channels with national coverage, 1 publicly-owned satellite channel, and 3 stations designed for digital terrestrial transmissions; multi-channel satellite and cable TV services obtainable; upwards of 1,500 radio stations broadcasting, nearly all of them privately-owned; state-run broadcaster has 7 national stations, 2 international stations, and 19 regional stations (2007)
Internet country code: .gr
Internet hosts: 2.574 million (2010)
country comparison to the world: 31
Internet users: 4.971 million (2009)
country comparison to the world: 46

TRANSPORTATION

Airports: 81 (2010)
country comparison to the world: 68
Airports—with paved runways:
total: 67
over 3,047 m: 6
2,438 to 3,047 m: 14
1,524 to 2,437 m: 20
914 to 1,523 m: 18
under 914 m: 9 (2010)
Airports—with unpaved runways:
total: 14
914 to 1,523 m: 2
under 914 m: 12 (2010)
Heliports: 9 (2010)
Pipelines: gas 1,240 km; oil 75 km (2010)
Railways: *total:* 2,548 km
country comparison to the world: 64
standard gauge: 1,565 km 1.435-m gauge (764 km electrified)
narrow gauge: 961 km 1.000-m gauge; 22 km 0.750-m gauge (2009)
Roadways:
total: 117,533 km
country comparison to the world: 37
paved: 107,895 km (includes 880 km of expressways)
unpaved: 9,638 km (2005)
Waterways: 6 km (the 6 km long Corinth Canal crosses the Isthmus of Corinth; it shortens a sea voyage by 325 km) (2010)
country comparison to the world: 107
Merchant marine: *total:* 886
country comparison to the world: 12
by type: bulk carrier 263, cargo 53, carrier 1, chemical tanker 72, container 34, liquefied gas 13, passenger 8, passenger/cargo 116, petroleum tanker 312, roll on/roll off 13, specialized tanker 1
foreign-owned: 62 (Belgium 16, Bermuda 3, Cyprus 4, Italy 5, UK 27, US 7)
registered in other countries: 2,391 (Antigua and Barbuda 5, Bahamas 209, Barbados 14, Belize 2, Bermuda 2, Brazil 1, Cambodia 2, Cayman Islands 11, Comoros 3, Cyprus 216, Denmark 1, Dominica 9, Egypt 8, Georgia 3, Germany 1, Gibraltar 7, Honduras 4, Hong Kong 22, Indonesia 1, Isle of Man 57, Italy 8, Jamaica 8, Liberia 454, Malta 458, Marshall Islands 358, Mexico 1, Moldova 4, Panama 402, Philippines 4, Portugal 5, Saint Vincent and the Grenadines 63, Sao Tome and Principe 1, Saudi Arabia 4, Singapore 19, Slovakia 1, Togo 1, UAE 3, UK 1, Uruguay 1, Vanuatu 4, Venezuela 4, unknown 8) (2010)
Ports and terminals: Agioi Theodoroi, Aspropyrgos, Pachi, Piraeus, Thessaloniki

MILITARY

Military branches: Hellenic Army (Ellinikos Stratos, ES), Hellenic Navy (Ellinikos Polemiko Navtiko, EPN), Hellenic Air Force (Elliniki Polemiki Aeroporia, EPA) (2011)
Military service age and obligation: 19-45 years of age for compulsory military service; during wartime the law allows for recruitment beginning January of the year of inductee's 18th birthday, thus including 17 year olds; 17 years of age for volunteers; conscript service obligation–1 year for all services; women are eligible for voluntary military service (2008)
Manpower available for military service:
males age 16–49: 2,485,389
females age 16–49: 2,469,854 (2010 est.)
Manpower fit for military service: *males age 16–49:* 2,032,378
females age 16–49: 2,016,552 (2010 est.)
Manpower reaching militarily significant age annually: *male:* 52,754
female: 49,485 (2010 est.)
Military expenditures: 4.3% of GDP (2005 est.)
country comparison to the world: 22

TRANSNATIONAL ISSUES

Disputes—international: Greece and Turkey continue discussions to resolve their complex maritime, air, territorial, and boundary disputes in the Aegean Sea; Cyprus question with Turkey; Greece rejects the use of the name Macedonia or Republic of Macedonia; the mass migration of unemployed Albanians still remains a problem for developed countries, chiefly Greece and Italy

Illicit drugs: a gateway to Europe for traffickers smuggling cannabis and heroin from the Middle East and Southwest Asia to the West and precursor chemicals to the East; some South American cocaine transits or is consumed in Greece; money laundering related to drug trafficking and organized crime

GREENLAND

(PART OF THE KINGDOM OF DENMARK)

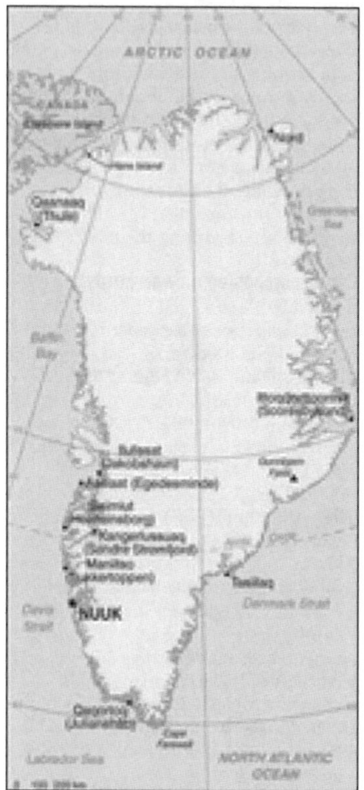

INTRODUCTION

Background: Greenland, the world's largest island, is about 81% ice capped. Vikings reached the island in the 10th century from Iceland; Danish colonization began in the 18th century, and Greenland was made an integral part of Denmark in 1953. It joined the European Community (now the EU) with Denmark in 1973 but withdrew in 1985 over a dispute centered on stringent fishing quotas. Greenland was granted self-government in 1979 by the Danish parliament; the law went into effect the following year. Greenland voted in favor of increased self-rule in November 2008 and acquired greater responsibility for internal affairs in June 2009. Denmark, however, continues to exercise control of Greenland's foreign affairs, security, and financial policy in consultation with Greenland's Home Rule Government.

GEOGRAPHY

Location: Northern North America, island between the Arctic Ocean and the North Atlantic Ocean, northeast of Canada
Geographic coordinates: 72 00 N, 40 00 W
Map references: North America
Area: *total:* 2,166,086 sq km
country comparison to the world: 13
land: 2,166,086 sq km (410,449 sq km ice-free, 1,755,637 sq km ice-covered)
Area—comparative: slightly more than three times the size of Texas
Land boundaries: 0 km
Coastline: 44,087 km
Maritime claims: *territorial sea:* 3 nm
exclusive fishing zone: 200 nm or agreed boundaries or median line
continental shelf: 200 nm or agreed boundaries or median line
Climate: arctic to subarctic; cool summers, cold winters
Terrain: flat to gradually sloping icecap covers all but a narrow, mountainous, barren, rocky coast
Elevation extremes: *lowest point:* Atlantic Ocean 0 m
highest point: Gunnbjorn Fjeld 3,700 m
Natural resources: coal, iron ore, lead, zinc, molybdenum, diamonds, gold, platinum, niobium, tantalite, uranium, fish, seals, whales, hydropower, possible oil and gas
Land use: *arable land:* 0%
permanent crops: 0%
other: 100% (2005)
Irrigated land: NA
Natural hazards: continuous permafrost over northern two-thirds of the island
Environment—current issues: protection of the arctic environment; preservation of the Inuit traditional way of life, including whaling and seal hunting
Geography—note: dominates North Atlantic Ocean between North America and Europe; sparse population confined to small settlements along coast; close to one-quarter of the population lives in the capital, Nuuk; world's second largest ice cap

PEOPLE

Population: 57,670 (July 2011 est.)
country comparison to the world: 204
Age structure: *0–14 years:* 22.3% (male 6,514/female 6,330)
15–64 years: 70.2% (male 21,599/female 18,861)
65 years and over: 7.6% (male 2,269/female 2,097) (2011 est.)
Median age: *total:* 33.6 years
male: 35 years
female: 32.1 years (2011 est.)
Population growth rate: 0.05% (2011 est.)
country comparison to the world: 189
Birth rate: 14.6 births/1,000 population (2011 est.)
country comparison to the world: 140
Death rate: 8.12 deaths/1,000 population (July 2011 est.)
country comparison to the world: 99
Net migration rate: -5.98 migrant(s)/1,000 population (2011 est.)
country comparison to the world: 198
Urbanization: *urban population:* 84% of total population (2010)
rate of urbanization: 0.3% annual rate of change (2010-15 est.)
Major cities—population: NUUK (capital) 15,000 (2009)
Sex ratio: *at birth:* 1.051 male(s)/female
under 15 years: 1.03 male(s)/female
15–64 years: 1.15 male(s)/female
65 years and over: 1.05 male(s)/female
total population: 1.12 male(s)/female (2011 est.)
Infant mortality rate:
total: 10.05 deaths/1,000 live births
country comparison to the world: 148
male: 11.47 deaths/1,000 live births
female: 8.55 deaths/1,000 live births (2011 est.)
Life expectancy at birth: *total population:* 70.96 years
country comparison to the world: 141
male: 68.33 years
female: 73.74 years (2011 est.)
Total fertility rate: 2.13 children born/woman (2011 est.)
country comparison to the world: 115
HIV/AIDS—adult prevalence rate: NA
HIV/AIDS—people living with HIV/AIDS: 100 (1999)
country comparison to the world: 163
HIV/AIDS—deaths: NA
Nationality:
noun: Greenlander(s)
adjective: Greenlandic
Ethnic groups: Inuit 89%, Danish and other 11% (2009)
Religions: Evangelical Lutheran, traditional Inuit spiritual beliefs
Languages: Greenlandic (East Inuit) (official), Danish (official), English
Literacy: *definition:* age 15 and over can read and write
total population: 100%
male: 100%
female: 100% (2001 est.)
School life expectancy (primary to tertiary education): NA
Education expenditures: NA

GOVERNMENT

Country name: *conventional long form:* none
conventional short form: Greenland
local long form: none
local short form: Kalaallit Nunaat
Dependency status: part of the Kingdom of Denmark; self-governing overseas administrative division of Denmark since 1979

Government type: parliamentary democracy within a constitutional monarchy
Capital: *name:* Nuuk (Godthab)
geographic coordinates: 64 11 N, 51 45 W
time difference: UTC-3 (2 hours ahead of Washington, DC during Standard Time)
daylight saving time: +1hr, begins last Sunday in March; ends last Sunday in October
note: Greenland is divided into four time zones
Administrative divisions: 4 municipalities (kommuner, singular kommune); Kujalleq, Qaasuitsup, Qeqqata, Sermersooq
note: the North and East Greenland National Park (Avannaarsuani Tunumilu Nuna Allanngutsaaliugaq) and the Thule Air Base in Pituffik (in northwest Greenland) are two unincorporated areas; the national park's 972,000 sq km—about 46% of the island—make it the largest national park in the world and also the most northerly
Independence: none (extensive self-rule as part of the Kingdom of Denmark; foreign affairs is the responsibility of Denmark, but Greenland actively participates in international agreements relating to Greenland)
National holiday: June 21 (longest day)
Constitution: (November 2008) Act on Greenland Self Government
Legal system: the laws of Denmark, where applicable, apply
Suffrage: 18 years of age; universal
Executive branch: *chief of state:* Queen MARGRETHE II of Denmark (since 14 January 1972), represented by High Commissioner Soeren Hald MOELLER (since April 2005)
head of government: Prime Minister Kuupik KLEIST (since 12 June 2009)
cabinet: Home Rule Government elected by the Parliament (Landsting) on the basis of the strength of parties
(For more information visit the World Leaders website)
elections: the monarchy is hereditary; high commissioner appointed by the monarch; prime minister elected by parliament (usually the leader of the majority party)
election results: Kuupik KLEIST elected prime minister
Legislative branch: unicameral Parliament or Inatsisartut (Landsting) (31 seats; members elected by popular vote on the basis of proportional representation to serve four-year terms)
elections: last held on 2 June 2009 (next to be held by 2014)
election results: percent of vote by party—IA 43.7%, Siumut 26.5%, Demokratiit 12.7%, Atassut 10.9%; Kattusseqatigiit 3.8%, other 2.4%; seats by party—IA 14, Siumut 9, Demokraatiit 4, Atassut 3, Kattusseqatigiit 1
note: two representatives were elected to the Danish Parliament or Folketing on 13 November 2007 (next to be held by November 2011); percent of vote by party—NA; seats by party—Siumut 1, Inuit Ataqatigiit 1
Judicial branch: High Court or Landsret (appeals can be made to the Ostre Landsret or Eastern Division of the High Court or Supreme Court in Copenhagen)
Political parties and leaders: Atassut Party (Solidarity) [Gerhardt PETERSEN] (a conservative party favoring continuing close relations with Denmark); Demokratiit [Jens B. FREDERIKSEN]; Inuit Ataqatigiit or IA (Inuit Community) [Kuupik KLEIST] (a leftist party favoring complete independence from Denmark rather than home rule); Kattusseqatigiit (Candidate List) [Anthon FREDERIKSEN] (an independent right-of-center party with no official platform); Siumut (Forward Party) [Alega HAMMOND] (a social democratic party advocating more distinct Greenlandic identity and greater autonomy from Denmark)
Political pressure groups and leaders: *other:* conservationists; environmentalists
International organization participation: Arctic Council, NC, NIB, UPU
Diplomatic representation in the US: none (self-governing overseas administrative division of Denmark)
Diplomatic representation from the US: none (self-governing overseas administrative division of Denmark)
Flag description: two equal horizontal bands of white (top) and red with a large disk slightly to the hoist side of center—the top half of the disk is red, the bottom half is white; the design represents the sun reflecting off a field of ice; the colors are the same as those of the Danish flag and symbolize Greenland's links to the Kingdom of Denmark
National anthem: *name:* "Nunarput utoqqarsuanngoravit" ("Our Country, Who's Become So Old" also translated as "You Our Ancient Land")
lyrics/music: Henrik LUND/Jonathan PETERSEN
note: adopted 1916; the government also recognizes "Nuna asiilasooq" as a secondary anthem

ECONOMY

Economy—overview: The economy remains critically dependent on exports of shrimp and fish and on a substantial subsidy—about $650 million in 2009—from the Danish Government, which supplies nearly 60% of government revenues. The public sector, including publicly owned enterprises and the municipalities, plays the dominant role in Greenland's economy. Greenland's GDP contracted about 2% in 2009 as a result of the global economic slowdown. Budget surpluses turned to deficits beginning in 2007 and unemployment has risen. During the last decade the Greenland Home Rule Government (GHRG) pursued conservative fiscal and monetary policies, but public pressure has increased for better schools, health care and retirement systems. The Greenlandic economy has benefited from increasing catches and exports of shrimp, Greenland halibut and, more recently, crabs. Due to Greenland's continued dependence on exports of fish—which account for 82% of exports—the economy remains very sensitive to foreign developments. International consortia are increasingly active in exploring for hydrocarbon resources off Greenland's western coast, and international studies indicate the potential for oil and gas fields in northern and northeastern Greenland. In May 2007 a US aluminum producer concluded a memorandum of understanding with the Greenland Home Rule Government to build an aluminum smelter and a power generation facility, which takes advantage of Greenland's abundant hydropower potential. Within the area of mining, olivine sand continues to be produced and gold production has resumed in south Greenland. Tourism also offers another avenue of economic growth for Greenland, with increasing numbers of cruise lines now operating in Greenland's western and southern waters during the peak summer tourism season.
GDP (purchasing power parity): $1.989 billion (2009 est.)
country comparison to the world: 186
$2.03 billion (2008 est.)
GDP (official exchange rate): $2.03 billion (2008)
GDP—real growth rate: -2% (2009 est.)
country comparison to the world: 207
1.5% (2008 est.)
4% (2007 est.)
GDP—per capita (PPP): $36,500 (2008 est.)
country comparison to the world: 30
$35,900 (2007 est.)
GDP—composition by sector: *agriculture:* 4.9%
industry: 31.9%
services: 63.2% (2007 est.)
Labor force: 28,240 (January 2009)
country comparison to the world: 205
Labor force—by occupation: *agriculture:* 4.9%
industry: 31.9%
services: 63.2% (2007 est.)
Unemployment rate: 6.8% (2007 est.)
country comparison to the world: 68
7.3% (2006 est.)
Population below poverty line: 9.2% (2007 est.)
Household income or consumption by percentage share: *lowest 10%:* NA%
highest 10%: NA%
Budget: *revenues:* $1.47 billion
expenditures: $1.51 billion (2007)
Inflation rate (consumer prices): 9.4% (2008 est.)
country comparison to the world: 192
1% (2005 est.)
Agriculture—products: forage crops, garden and greenhouse vegetables; sheep, reindeer; fish
Industries: fish processing (mainly shrimp and Greenland halibut); gold, niobium, tantalite, uranium, iron and diamond mining; handicrafts, hides and skins, small shipyards
Industrial production growth rate: NA%
Electricity—production: 310.3 million kWh (2008 est.)
country comparison to the world: 166
Electricity—consumption: 285.6 million kWh (2008 est.)
country comparison to the world: 169

Electricity—exports: 0 kWh (2008)
Electricity—imports: 0 kWh (2008)
Oil—production: 0 bbl/day
country comparison to the world: 180
Oil—consumption: 4,000 bbl/day (2009 est.)
country comparison to the world: 170
Oil—exports: 1,183 bbl/day (2008)
country comparison to the world: 120
Oil—imports: 5,172 bbl/day (2008 est.)
country comparison to the world: 157
Oil—proved reserves: 0 bbl (1 January 2010 est.)
country comparison to the world: 137
Natural gas—production: 0 cu m (2008)
country comparison to the world: 133
Natural gas—consumption: 0 cu m (2008)
country comparison to the world: 178
Natural gas—exports: 0 cu m (2008)
country comparison to the world: 104
Natural gas—imports: 0 cu m (2008)
country comparison to the world: 193
Natural gas—proved reserves: 0 cu m (1 January 2010 est.)
country comparison to the world: 141
Exports: $485 million (2008)
country comparison to the world: 168
$428 million (2007)
Exports—commodities: fish and fish products 72%, metals 10% (2008)
Exports—partners: Denmark 61.13%, Japan 13.69%, China 6.15%, Sweden 5.21% (2009)
Imports: $867 million (2008)
country comparison to the world: 176
$669 million (2007)
Imports—commodities: machinery and transport equipment, manufactured goods, food, petroleum products
Imports—partners: Denmark 74.93%, Sweden 11.73%, Norway 2.29% (2009)
Debt—external: $58 million (2009)
country comparison to the world: 187
$25 million (1999)
Exchange rates: Danish kroner (DKK) per US dollar–5.624 (2010), 5.361 (2009), 5.4797 (2007), 5.9468 (2006)

COMMUNICATIONS

Telephones—main lines in use: 22,000 (2009)
country comparison to the world: 190
Telephones—mobile cellular: 53,500 (2009)
country comparison to the world: 195
Telephone system: *general assessment:* adequate domestic and international service provided by satellite, cables and microwave radio relay; totally digital since 1995
domestic: microwave radio relay and satellite
international: country code–299; satellite earth stations–15 (12 Intelsat, 1 Eutelsat, 2 Americom GE-2 (all Atlantic Ocean)) (2000)
Broadcast media: the Greenland Broadcasting Company provides public radio and television services throughout the island with a broadcast station and a series of repeaters; a few private local television and radio stations broadcast; Danish public radio rebroadcasts are available (2007)
Internet country code: .gl
Internet hosts: 15,668 (2010)
country comparison to the world: 116
Internet users: 36,000 (2009)
country comparison to the world: 177

TRANSPORTATION

Airports: 15 (2010)
country comparison to the world: 147
Airports—with paved runways: *total:* 10
2,438 to 3,047 m: 2
1,524 to 2,437 m: 1
914 to 1,523 m: 1
under 914 m: 6 (2010)
Airports—with unpaved runways: *total:* 5
1,524 to 2,437 m: 1
914 to 1,523 m: 2
under 914 m: 2 (2010)
Roadways: note: although there are short roads in towns, there are no roads between towns; inter-urban transport takes place either by sea or air (2005)
Merchant marine: *total:* 1
country comparison to the world: 154
by type: passenger 1 (2010)
Ports and terminals: Sisimiut

MILITARY

Military branches: no regular military forces
Manpower available for military service: *males age 16–49:* 15,280 (2010 est.)
Manpower fit for military service: *males age 16–49:* 10,765
females age 16–49: 11,399 (2010 est.)
Manpower reaching militarily significant age annually: *male:* 488
female: 478 (2010 est.)
Military—note: defense is the responsibility of Denmark

TRANSNATIONAL ISSUES

Disputes—international: managed dispute between Canada and Denmark over Hans Island in the Kennedy Channel between Canada's Ellesmere Island and Greenland; Denmark (Greenland) and Norway have made submissions to the Commission on the Limits of the Continental shelf (CLCS) and Russia is collecting additional data to augment its 2001 CLCS submission

GRENADA

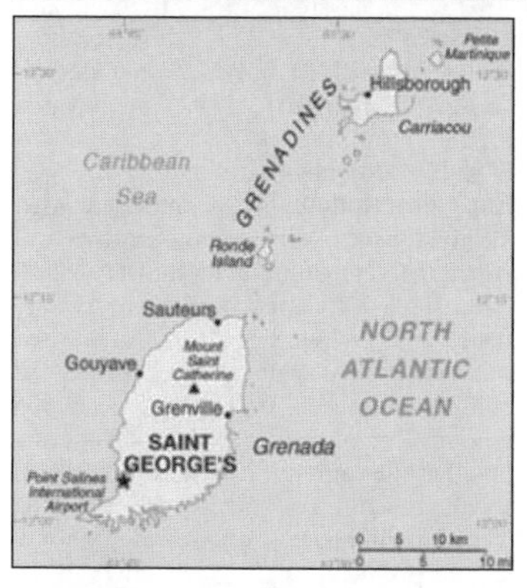

INTRODUCTION

Background: Carib Indians inhabited Grenada when COLUMBUS discovered the island in 1498, but it remained uncolonized for more than a century. The French settled Grenada in the 17th century, established sugar estates, and imported large numbers of African slaves. Britain took the island in 1762 and vigorously expanded sugar production. In the 19th century, cacao eventually surpassed sugar as the main export crop; in the 20th century, nutmeg became the leading export. In 1967, Britain gave Grenada autonomy over its internal affairs. Full independence was attained in 1974 making Grenada one of the smallest independent countries in the Western Hemisphere. Grenada was seized by a Marxist military council on 19 October 1983. Six days later the island was invaded by US forces and those of six other Caribbean nations, which quickly captured the ringleaders and their hundreds of Cuban advisers. Free elections were reinstituted the following year and have continued since that time. Hurricane Ivan struck Grenada in September of 2004 causing severe damage.

GEOGRAPHY

Location: Caribbean, island between the Caribbean Sea and Atlantic Ocean, north of Trinidad and Tobago
Geographic coordinates: 12 07 N, 61 40 W
Map references: Central America and the Caribbean
Area: *total:* 344 sq km
country comparison to the world: 205
land: 344 sq km
water: 0 sq km
Area—comparative: twice the size of Washington, DC
Land boundaries: 0 km
Coastline: 121 km
Maritime claims: *territorial sea:* 12 nm
exclusive economic zone: 200 nm
Climate: tropical; tempered by northeast trade winds
Terrain: volcanic in origin with central mountains
Elevation extremes: *lowest point:* Caribbean Sea 0 m
highest point: Mount Saint Catherine 840 m
Natural resources: timber, tropical fruit, deepwater harbors
Land use: *arable land:* 5.88%
permanent crops: 29.41%
other: 64.71% (2005)
Irrigated land: NA
Total renewable water resources: NA
Natural hazards: lies on edge of hurricane belt; hurricane season lasts from June to November
Environment—current issues: NA
Environment—international agreements: *party to:* Biodiversity, Climate Change, Climate Change-Kyoto Protocol, Deserti-

fication, Endangered Species, Law of the Sea, Ozone Layer Protection, Whaling
signed, but not ratified: none of the selected agreements
Geography—note: the administration of the islands of the Grenadines group is divided between Saint Vincent and the Grenadines and Grenada

PEOPLE

Population: 108,419 (July 2011 est.)
country comparison to the world: 188
Age structure: *0–14 years:* 25.4% (male 14,152/female 13,390)
15–64 years: 65.7% (male 36,245/female 34,960)
65 years and over: 8.9% (male 4,372/female 5,300) (2011 est.)
Median age: *total:* 28.6 years
male: 28.6 years
female: 28.6 years (2011 est.)
Population growth rate: 0.551% (2011 est.)
country comparison to the world: 149
Birth rate: 17.01 births/1,000 population (2011 est.)
country comparison to the world: 120
Death rate: 7.94 deaths/1,000 population (July 2011 est.)
country comparison to the world: 108
Net migration rate: -3.56 migrant(s)/1,000 population (2011 est.)
country comparison to the world: 184
Urbanization: *urban population:* 39% of total population (2010)
rate of urbanization: 1.6% annual rate of change (2010-15 est.)
Major cities—population: SAINT GEORGE'S (capital) 40,000 (2009)
Sex ratio: *at birth:* 1.098 male(s)/female
under 15 years: 1.05 male(s)/female
15–64 years: 1.04 male(s)/female
65 years and over: 0.82 male(s)/female
total population: 1.02 male(s)/female (2011 est.)
Infant mortality rate: *total:* 11.43 deaths/1,000 live births
country comparison to the world: 141
male: 10.54 deaths/1,000 live births
female: 12.41 deaths/1,000 live births (2011 est.)
Life expectancy at birth: *total population:* 73.04 years
country comparison to the world: 121
male: 70.51 years
female: 75.82 years (2011 est.)
Total fertility rate: 2.18 children born/woman (2011 est.)
country comparison to the world: 108
HIV/AIDS—adult prevalence rate: NA
HIV/AIDS—people living with HIV/AIDS: NA
HIV/AIDS—deaths: NA
Drinking water source: *Improved:*
urban: 97% of population
rural: 93% of population
total: 94% of population
Unimproved: urban: 3% of population
rural: 7% of population
total: 6% of population (2000)
Sanitation facility access: *Improved:*
urban: 96% of population
rural: 97% of population
total: 97% of population
Unimproved: urban: 4% of population
rural: 3% of population
total: 3% of population (2008)
Nationality: *noun:* Grenadian(s)
adjective: Grenadian
Ethnic groups: black 82%, mixed black and European 13%, European and East Indian 5%, and trace of Arawak/Carib Amerindian
Religions: Roman Catholic 53%, Anglican 13.8%, other Protestant 33.2%
Languages: English (official), French patois
Literacy: *definition:* age 15 and over can read and write
total population: 96%
male: NA
female: NA (2003 est.)
School life expectancy (primary to tertiary education): *total:* 16 years
male: 15 years
female: 16 years (2009)
Education expenditures: 4.9% of GDP (2003)
country comparison to the world: 63

GOVERNMENT

Country name: *conventional long form:* none
conventional short form: Grenada
Government type: parliamentary democracy and a Commonwealth realm
Capital: *name:* Saint George's
geographic coordinates: 12 03 N, 61 45 W
time difference: UTC-4 (1 hour ahead of Washington, DC during Standard Time)
Administrative divisions: 6 parishes and 1 dependency*; Carriacou and Petite Martinique*, Saint Andrew, Saint David, Saint George, Saint John, Saint Mark, Saint Patrick
Independence: 7 February 1974 (from the UK)
National holiday: Independence Day, 7 February (1974)
Constitution: 19 December 1973
Legal system: common law based on English model
International law organization participation: has not submitted an ICJ jurisdiction declaration; non-party state to the ICCt
Suffrage: 18 years of age; universal
Executive branch: *chief of state:* Queen ELIZABETH II (since 6 February 1952); represented by Governor General Carlyle Arnold GLEAN (since 27 November 2008)
head of government: Prime Minister Tillman THOMAS (since 9 July 2008)
cabinet: Cabinet appointed by the governor general on the advice of the prime minister
(For more information visit the World Leaders website)
elections: the monarchy is hereditary; governor general appointed by the monarch; following legislative elections, the leader of the majority party or the leader of the majority coalition is usually appointed prime minister by the governor general
Legislative branch: bicameral Parliament consists of the Senate (13 seats, 10 members appointed by the government and 3 by the leader of the opposition) and the House of Representatives (15 seats; members elected by popular vote to serve five-year terms)
elections: last held on 8 July 2008 (next to be held in 2013)
election results: House of Representatives—percent of vote by party—NA; seats by party—NDC 11, NNP 4
Judicial branch: Eastern Caribbean Supreme Court, consisting of a court of Appeal and a High Court of Justice (two High Court judges are assigned to and reside in Grenada); Itinerant Court of Appeal three judges; member of the Caribbean Court of Justice (CCJ)
Political parties and leaders: Grenada United Labor Party or GULP [Gloria Payne BANFIELD]; National Democratic Congress or NDC [Tillman THOMAS]; New National Party or NNP [Keith MITCHELL]
Political pressure groups and leaders: Committee for Human Rights in Grenada or CHRG; New Jewel Movement Support Group; The British Grenada Friendship Society; The New Jewel 19 Committee
International organization participation: ACP, AOSIS, C, Caricom, CDB, FAO, G-77, IBRD, ICAO, ICRM, IDA, IFAD, IFC, IFRCS, ILO, IMF, IMO, Interpol, IOC, ITU, ITUC, LAES, MIGA, NAM, OAS, OECS, OPANAL, OPCW, PetroCaribe, UN, UNCTAD, UNESCO, UNIDO, UPU, WHO, WIPO, WTO
Diplomatic representation in the US: *chief of mission:* Ambassador Gillian M.S. BRISTOL
chancery: 1701 New Hampshire Avenue NW, Washington, DC 20009
telephone: [1] (202) 265-2561
FAX: [1] (202) 265-2468
consulate(s) general: New York
Diplomatic representation from the US: *chief of mission:* the US Ambassador to Barbados is accredited to Grenada
embassy: Lance-aux-Epines Stretch, Saint George's
mailing address: P. O. Box 54, Saint George's
telephone: [1] (473) 444-1173 through 1177
FAX: [1] (473) 444-4820
Flag description: a rectangle divided diagonally into yellow triangles (top and bottom) and green triangles (hoist side and outer side), with a red border around the flag; there are seven yellow, five-pointed stars with three centered in the top red border, three centered in the bottom red border, and one on a red disk superimposed at the center of the flag; there is also a symbolic nutmeg pod on the hoist-side triangle (Grenada is the world's second-largest producer of nutmeg, after Indonesia); the seven stars stand for the seven administrative divisions, with the central star denoting the capital, St. George; yellow represents the sun and the warmth of the people, green stands for vegetation and agriculture, and red symbolizes harmony, unity, and courage
National anthem: *name:* "Hail Grenada"
lyrics/music: Irva Merle BAPTISTE/Louis Arnold MASANTO
note: adopted 1974

ECONOMY

Economy—overview: Grenada relies on tourism as its main source of foreign exchange especially since the construction of an international airport in 1985. Hurricanes Ivan (2004) and Emily (2005) severely damaged the agricultural sector—particularly nutmeg and cocoa cultivation—which had been a key driver of economic growth. Grenada has rebounded from the devastating effects of the hurricanes but is now saddled with the debt burden from the rebuilding process. Public debt-to-GDP is nearly 110%, leaving the THOMAS administration limited room to engage in public investments and social spending. Strong performances in construction and manufacturing, together with the development of tourism and an offshore financial industry, have also contributed to growth in national output; however, economic growth was stagnant in 2010 after a sizeable contraction in 2009, because of the global economic slowdown's effects on tourism and remittances.
GDP (purchasing power parity): $1.098 billion (2010 est.)
country comparison to the world: 198
$1.114 billion (2009 est.)
$1.204 billion (2008 est.)
note: data are in 2010 US dollars
GDP (official exchange rate): $674 million (2010 est.)
GDP—real growth rate: -1.4% (2010 est.)
country comparison to the world: 204
-7.6% (2009 est.)
2.2% (2008 est.)
GDP—per capita (PPP): $10,200 (2010 est.)
country comparison to the world: 108
$10,400 (2009 est.)
$11,300 (2008 est.)
note: data are in 2010 US dollars
GDP—composition by sector: *agriculture:* 5.4%
industry: 18%
services: 76.6% (2003)
Labor force: 42,300 (1996)
country comparison to the world: 192
Labor force—by occupation: *agriculture:* 24%
industry: 14%
services: 62% (1999 est.)
Unemployment rate: 12.5% (2000)
country comparison to the world: 132
Population below poverty line: 32% (2000)
Household income or consumption by percentage share: *lowest 10%:* NA%
highest 10%: NA%
Budget: *revenues:* $175.3 million
expenditures: $215.9 million (2009 est.)
Inflation rate (consumer prices): 3.7% (2007 est.)
country comparison to the world: 104
Central bank discount rate: 6.5% (31 December 2009)
country comparison to the world: 57
6.5% (31 December 2008)
Commercial bank prime lending rate: 11.06% (31 December 2009 est.)
country comparison to the world: 92
9.53% (31 December 2008 est.)
Stock of narrow money: $123.1 million (31 December 2009)
country comparison to the world: 178
$131.7 million (31 December 2008)
Stock of broad money: $743.5 million (31 December 2009)
country comparison to the world: 169
$719.5 million (31 December 2008)
Stock of domestic credit: $658 million (31 December 2008 est.)
country comparison to the world: 161
$575.8 million (31 December 2007 est.)
Market value of publicly traded shares: $NA
Agriculture—products: bananas, cocoa, nutmeg, mace, citrus, avocados, root crops, sugarcane, corn, vegetables
Industries: food and beverages, textiles, light assembly operations, tourism, construction
Electricity—production: 178.7 million kWh (2007 est.)
country comparison to the world: 179
Electricity—consumption: 155.7 million kWh (2007 est.)
country comparison to the world: 182
Electricity—exports: 0 kWh (2008 est.)
Electricity—imports: 0 kWh (2008 est.)
Oil—production: 0 bbl/day (2009 est.)
country comparison to the world: 179
Oil—consumption: 3,000 bbl/day (2009 est.)
country comparison to the world: 176
Oil—exports: 0 bbl/day (2007 est.)
country comparison to the world: 169
Oil—imports: 1,923 bbl/day (2007 est.)
country comparison to the world: 176
Oil—proved reserves: 0 bbl (1 January 2010 est.)
country comparison to the world: 136
Natural gas—production: 0 cu m (2008 est.)
country comparison to the world: 132
Natural gas—consumption: 0 cu m (2008 est.)
country comparison to the world: 177
Natural gas—exports: 0 cu m (2008 est.)
country comparison to the world: 103
Natural gas—imports: 0 cu m (2008 est.)
country comparison to the world: 192
Natural gas—proved reserves: 0 cu m (1 January 2010 est.)
country comparison to the world: 140
Current account balance: $-138 million (2007 est.)
country comparison to the world: 82
Exports: $38 million (2006)
country comparison to the world: 202
Exports—commodities: bananas, cocoa, nutmeg, fruit and vegetables, clothing, mace
Exports—partners: Saint Lucia 19.73%, Antigua and Barbuda 13.41%, US 12.21%, Saint Kitts and Nevis 12.03%, Dominica 12% (2009)
Imports: $343 million (2006)
country comparison to the world: 191
Imports—commodities: food, manufactured goods, machinery, chemicals, fuel
Imports—partners: Trinidad and Tobago 39.76%, US 18.11% (2009)
Debt—external: $347 million (2004)
country comparison to the world: 167
Exchange rates: East Caribbean dollars (XCD) per US dollar–2.7 (2010), 2.7 (2009), 2.7 (2005), 2.7 (2004), 2.7 (2003)

COMMUNICATIONS

Telephones—main lines in use: 28,600 (2009)
country comparison to the world: 181
Telephones—mobile cellular: 64,000 (2009)
country comparison to the world: 192
Telephone system: *general assessment:* automatic, island-wide telephone system
domestic: interisland VHF and UHF radiotelephone links
international: country code–1-473; landing point for the East Caribbean Fiber Optic System (ECFS) submarine cable with links to 13 other islands in the eastern Caribbean extending from the British Virgin Islands to Trinidad; SHF radiotelephone links to Trinidad and Tobago and Saint Vincent; VHF and UHF radio links to Trinidad
Broadcast media: the Grenada Broadcasting Network, jointly owned by the government and the Caribbean Communications Network of Trinidad and Tobago, operates a television station and 2 radio stations; multi-channel cable TV subscription service is available; a dozen private radio stations also broadcast (2007)
Internet country code: .gd
Internet hosts: 52 (2010)
country comparison to the world: 209
Internet users: 25,000 (2009)
country comparison to the world: 185

TRANSPORTATION

Airports: 3 (2010)
country comparison to the world: 194
Airports—with paved runways: *total:* 3
2,438 to 3,047 m: 1
1,524 to 2,437 m: 1
under 914 m: 1 (2010)
Roadways: *total:* 1,127 km
country comparison to the world: 182
paved: 687 km
unpaved: 440 km (2000)
Ports and terminals: Saint George's

MILITARY

Military branches: no regular military forces; Royal Grenada Police Force (includes Coast Guard) (2010)
Manpower available for military service: *males age 16–49:* 27,468 (2010 est.)
Manpower fit for military service: *males age 16–49:* 22,596
females age 16–49: 22,588 (2010 est.)
Manpower reaching militarily significant age annually: *male:* 995
female: 1,002 (2010 est.)
Military expenditures: NA

TRANSNATIONAL ISSUES

Disputes—international: none
Illicit drugs: small-scale cannabis cultivation; lesser transshipment point for marijuana and cocaine to US

GUAM

(TERRITORY OF THE US)

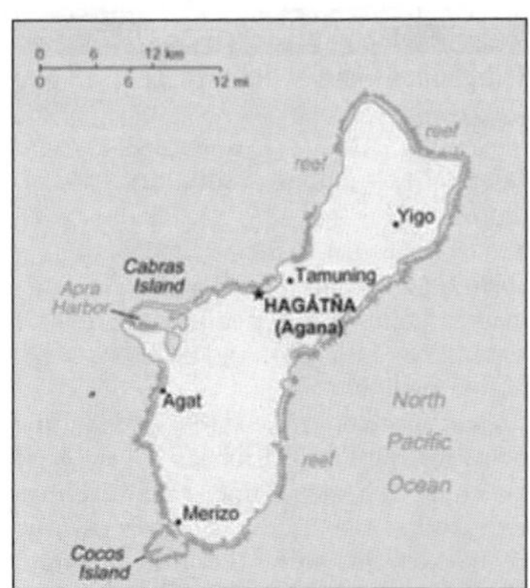

INTRODUCTION

Background: Guam was ceded to the US by Spain in 1898. Captured by the Japanese in 1941, it was retaken by the US three years later. The military installation on the island is one of the most strategically important US bases in the Pacific.

GEOGRAPHY

Location: Oceania, island in the North Pacific Ocean, about three-quarters of the way from Hawaii to the Philippines
Geographic coordinates: 13 28 N, 144 47 E
Map references: Oceania
Area: *total:* 544 sq km
land: 544 sq km
water: 0 sq km
Area—comparative: three times the size of Washington, DC
Land boundaries: 0 km
Coastline: 125.5 km
Maritime claims: *territorial sea:* 12 nm
exclusive economic zone: 200 nm
Climate: tropical marine; generally warm and humid, moderated by northeast trade winds; dry season (January to June), rainy season (July to December); little seasonal temperature variation
Terrain: volcanic origin, surrounded by coral reefs; relatively flat coralline limestone plateau (source of most fresh water), with steep coastal cliffs and narrow coastal plains in north, low hills in center, mountains in south
Elevation extremes:
lowest point: Pacific Ocean 0 m
highest point: Mount Lamlam 406 m
Natural resources: aquatic wildlife (supporting tourism), fishing (largely undeveloped)
Land use: *arable land:* 3.64%
permanent crops: 18.18%
other: 78.18% (2005)
Irrigated land: NA
Natural hazards: frequent squalls during rainy season; relatively rare but potentially destructive typhoons (June–December)
Environment—current issues: extirpation of native bird population by the rapid proliferation of the brown tree snake, an exotic, invasive species
Geography—note: largest and southernmost island in the Mariana Islands archipelago; strategic location in western North Pacific Ocean

PEOPLE

Population: 183,286 (July 2011 est.)
Age structure: *0–14 years:* 27% (male 25,577/female 23,836)
15–64 years: 65.5% (male 61,237/female 58,891)
65 years and over: 7.5% (male 6,287/female 7,458) (2011 est.)
Median age:
total: 29.4 years
male: 29 years
female: 29.8 years (2011 est.)
Population growth rate: 1.314% (2011 est.)
Birth rate: 17.85 births/1,000 population (2011 est.)
Death rate: 4.71 deaths/1,000 population (July 2011 est.)
Net migration rate: 0 migrant(s)/1,000 population (2011 est.)
Urbanization: *urban population:* 93% of total population (2010)
rate of urbanization: 1.2% annual rate of change (2010-15 est.)
Major cities—population: HAGATNA (capital) 153,000 (2009)
Sex ratio: *at birth:* 1.06 male(s)/female
under 15 years: 1.07 male(s)/female
15–64 years: 1.04 male(s)/female
65 years and over: 0.85 male(s)/female
total population: 1.03 male(s)/female (2011 est.)
Infant mortality rate:
total: 5.83 deaths/1,000 live births
male: 6.24 deaths/1,000 live births
female: 5.38 deaths/1,000 live births (2011 est.)
Life expectancy at birth: *total population:* 78.34 years
male: 75.3 years
female: 81.57 years (2011 est.)
Total fertility rate: 2.48 children born/woman (2011 est.)
HIV/AIDS—adult prevalence rate: NA
HIV/AIDS—people living with HIV/AIDS: NA
HIV/AIDS—deaths: NA
Drinking water source: *Improved:*
urban: 100% of population
rural: 100% of population
total: 100% of population (2008)
Sanitation facility access: *Improved:*
urban: 99% of population
rural: 98% of population
total: 99% of population
Unimproved: urban: 1% of population
rural: 2% of population
total: 1% of population (2008)
Nationality:
noun: Guamanian(s) (US citizens)
adjective: Guamanian
Ethnic groups: Chamorro 37.1%, Filipino 26.3%, other Pacific islander 11.3%, white 6.9%, other Asian 6.3%, other ethnic origin or race 2.3%, mixed 9.8% (2000 census)
Religions: Roman Catholic 85%, other 15% (1999 est.)
Languages: English 38.3%, Chamorro 22.2%, Philippine languages 22.2%, other Pacific island languages 6.8%, Asian languages 7%, other languages 3.5% (2000 census)
Literacy: *definition:* age 15 and over can read and write
total population: 99%
male: 99%
female: 99% (1990 est.)
School life expectancy (primary to tertiary education): NA
Education expenditures: NA

GOVERNMENT

Country name: *conventional long form:* Territory of Guam
conventional short form: Guam
local long form: Guahan
local short form: Guahan
Dependency status: organized, unincorporated territory of the US with policy relations between Guam and the US under the jurisdiction of the Office of Insular Affairs, US Department of the Interior
Government type: NA
Capital: *name:* Hagatna (Agana)
geographic coordinates: 13 28 N, 144 44 E
time difference: UTC+10 (15 hours ahead of Washington, DC during Standard Time)
Administrative divisions: none (territory of the US)
Independence: none (territory of the US)
National holiday: Discovery Day, first Monday in March (1521)
Constitution: Organic Act of Guam, 1 August 1950
Legal system: common law modeled on US system; US federal laws apply
Suffrage: 18 years of age; universal; US citizens but do not vote in US presidential elections
Executive branch: *chief of state:* President Barack H. OBAMA (since 20 January 2009); Vice President Joseph R. BIDEN (since 20 January 2009)
head of government: Governor Eddie CALVO (since 3 January 2011); Lieutenant Governor Ray TENORIO (since 3 January 2011)
cabinet: heads of executive departments; appointed by the governor with the consent of the Guam legislature
(For more information visit the World Leaders website)
elections: under the US Constitution, residents of unincorporated territories, such as Guam, do not vote in elections for US president and vice president; however, they may vote in Democratic and Republican presidential primary elections; governor and lieutenant governor elected on the same ticket by popular vote for a four-year term (can serve two consecutive terms, then must wait a full term before running again); election last held on 2 November 2010 (next to be held in November 2014)

election results: Eddie CALVO elected governor with 50.6% percent of vote against 49.4% for Carl GUTIERREZ; Ray TENORIO elected lieutenant governor
Legislative branch: unicameral Legislature (15 seats; members elected by popular vote to serve two-year terms)
elections: last held on 2 November 2010 (next to be held in November 2012)
election results: percent of vote by party–NA; seats by party–Democratic Party 9, Republican Party 6
note: Guam elects one nonvoting delegate to the US House of Representatives; election last held on 2 November 2010 (next to be held in November 2012); results–percent of vote by party–NA; seats by party–Democratic Party 1
Judicial branch: Federal District Court (judge is appointed by the president); Supreme Court of Guam (hears appeals from Superior Court–judges appointed by governor); Territorial Superior Court (judges appointed for eight-year terms by the governor)
Political parties and leaders: Democratic Party [Michael PHILLIPS]; Republican Party [Philip J. FLORES] (controls the legislature)
Political pressure groups and leaders: Guam Federation of Teachers' Union; Guam Waterworks Authority Workers
other: activists; indigenous groups
International organization participation: IOC, SPC, UPU
Diplomatic representation in the US: none (territory of the US)
Diplomatic representation from the US: none (territory of the US)
Flag description: territorial flag is dark blue with a narrow red border on all four sides; centered is a red-bordered, pointed, vertical ellipse containing a beach scene, a proa or outrigger canoe with sail, and a palm tree with the word GUAM superimposed in bold red letters; the proa is sailing in Agana Bay with the promontory of Punta Dos Amantes, near the capital, in the background; blue represents the sea and red the blood shed in the struggle against oppression
note: the US flag is the national flag
National anthem: *name:* "Fanohge Chamoru" (Stand Ye Guamanians)
lyrics/music: Ramon Manalisay SABLAN [English], Lagrimas UNTALAN [Chamoru]/Ramon Manalisay SABLAN
note: adopted 1919; the local anthem is also known as "Guam Hymn"; as a territory of the United States, "The Star-Spangled Banner," which generally follows the playing of "Stand Ye Guamanians," is official (see United States)

ECONOMY

Economy—overview: The economy depends largely on US military spending and tourism. Total US grants, wage payments, and procurement outlays amounted to $1.3 billion in 2004. Over the past 30 years, the tourist industry has grown to become the largest income source following national defense. The Guam economy continues to experience expansion in both its tourism and military sectors.
GDP (purchasing power parity): $2.5 billion (2005 est.)
GDP (official exchange rate): $2.773 billion (2001)
GDP—real growth rate: NA%
GDP—per capita (PPP): $15,000 (2005 est.)
GDP—composition by sector:
agriculture: NA%
industry: NA%
services: NA%
Labor force: 82,950 (2007 est.)
Labor force—by occupation: *agriculture:* 26%
industry: 10%
services: 64% (2004 est.)
Unemployment rate: 11.4% (2002 est.)
Population below poverty line: 23% (2001 est.)
Household income or consumption by percentage share: *lowest 10%:* NA%
highest 10%: NA%
Budget: *revenues:* $319.6 million
expenditures: $427.8 million (2002 est.)
Inflation rate (consumer prices): 2.5% (2005 est.)
Agriculture—products: fruits, copra, vegetables; eggs, pork, poultry, beef
Industries: US military, tourism, construction, transshipment services, concrete products, printing and publishing, food processing, textiles
Industrial production growth rate: NA%
Electricity—production: 1.767 billion kWh (2007 est.)
Electricity—consumption: 1.644 billion kWh (2007 est.)
Electricity—exports: 0 kWh (2008 est.)
Electricity—imports: 0 kWh (2008 est.)
Oil—production: 0 bbl/day (2009 est.)
Oil—consumption: 10,620 bbl/day (2009 est.)
Oil—exports: 0 bbl/day (2007 est.)
Oil—imports: 14,230 bbl/day (2007 est.)
Oil—proved reserves: 0 bbl (1 January 2010 est.)
Natural gas—production: 0 cu m (2008 est.)
Natural gas—consumption: 0 cu m (2008 est.)
Natural gas—exports: 0 cu m (2008 est.)
Natural gas—imports: 0 cu m (2008 est.)
Natural gas—proved reserves: 0 cu m (1 January 2010 est.)
Exports: $45 million (2004 est.)
Exports—commodities: transshipments of refined petroleum products, construction materials, fish, food and beverage products
Imports: $701 million (2004 est.)
Imports—commodities: petroleum and petroleum products, food, manufactured goods
Debt—external: $NA
Exchange rates: the US dollar is used

COMMUNICATIONS

Telephones—main lines in use: 65,500 (2009)
Telephones—mobile cellular: 98,000 (2004)
Telephone system: *general assessment:* modern system, integrated with US facilities for direct dialing, including free use of 800 numbers
domestic: digital system, including mobile-cellular service and local access to the Internet
international: country code–1-671; major landing point for submarine cables between Asia and the US (Guam is a trans-Pacific communications hub for major carriers linking the US and Asia); satellite earth stations–2 Intelsat (Pacific Ocean)
Broadcast media: about a dozen TV broadcast channels, including digital channels; multi-channel cable TV services are available; roughly 20 radio stations broadcasting (2009)
Internet country code: .gu
Internet hosts: 24 (2010)
Internet users: 90,000 (2009)

TRANSPORTATION

Airports: 5; note–2 serviceable (2010)
Airports—with paved runways: *total:* 4
over 3,047 m: 2
2,438 to 3,047 m: 1
914 to 1,523 m: 1 (2010)
Airports—with unpaved runways: *total:* 1
under 914 m: 1 (2010)
Roadways: *total:* 1,045 km (2008)
Ports and terminals: Apra Harbor

MILITARY

Manpower fit for military service: *males age 16–49:* 38,358
females age 16–49: 36,869 (2010 est.)
Manpower reaching militarily significant age annually: *male:* 1,701
female: 1,608 (2010 est.)
Military—note: defense is the responsibility of the US

TRANSNATIONAL ISSUES

Disputes—international: none

GUATEMALA

INTRODUCTION

Background: The Mayan civilization flourished in Guatemala and surrounding regions during the first millennium A.D. After almost three centuries as a Spanish colony, Guatemala won its independence in 1821. During the second half of the 20th century, it experienced a variety of military and civilian governments, as well as a 36-year guerrilla war. In 1996, the government signed a peace agreement formally ending the conflict, which had left more than 100,000 people dead and had created, by some estimates, some 1 million refugees.

GEOGRAPHY

Location: Central America, bordering the North Pacific Ocean, between El Salvador and Mexico, and bordering the Gulf of Honduras (Caribbean Sea) between Honduras and Belize
Geographic coordinates: 15 30 N, 90 15 W
Map references: Central America and the Caribbean
Area: *total:* 108,889 sq km
country comparison to the world: 106
land: 107,159 sq km
water: 1,730 sq km
Area—comparative: slightly smaller than Tennessee
Land boundaries: *total:* 1,687 km
border countries: Belize 266 km, El Salvador 203 km, Honduras 256 km, Mexico 962 km
Coastline: 400 km
Maritime claims: *territorial sea:* 12 nm
exclusive economic zone: 200 nm
continental shelf: 200 m depth or to the depth of exploitation
Climate: tropical; hot, humid in lowlands; cooler in highlands
Terrain: mostly mountains with narrow coastal plains and rolling limestone plateau
Elevation extremes: *lowest point:* Pacific Ocean 0 m
highest point: Volcan Tajumulco 4,211 m
note: highest point in Central America
Natural resources: petroleum, nickel, rare woods, fish, chicle, hydropower
Land use: *arable land:* 13.22%
permanent crops: 5.6%
other: 81.18% (2005)
Irrigated land: 1,300 sq km (2003)
Total renewable water resources: 111.3 cu km (2000)
Freshwater withdrawal (domestic/industrial/agricultural): *total:* 2.01 cu km/yr (6%/13%/80%)
per capita: 160 cu m/yr (2000)
Natural hazards: numerous volcanoes in mountains, with occasional violent earthquakes; Caribbean coast extremely susceptible to hurricanes and other tropical storms
volcanism: Guatemala experiences significant volcanic activity in the Sierra Madre range; Santa Maria (elev. 3,772 m) has been deemed a "Decade Volcano" by the International Association of Volcanology and Chemistry of the Earth's Interior, worthy of study due to its explosive history and close proximity to human populations; Pacaya (elev. 2,552 m), which erupted in May 2010 causing an ashfall on Guatemala City and prompting evacuations, is one of the country's most active volcanoes; the volcano has frequently been in eruption since 1965; other historically active volcanoes include Acatenango, Almolonga, Atitlan, Fuego, and Tacana
Environment—current issues: deforestation in the Peten rainforest; soil erosion; water pollution
Environment—international agreements: *party to:* Antarctic Treaty, Biodiversity, Climate Change, Climate Change-Kyoto Protocol, Desertification, Endangered Species, Environmental Modification, Hazardous Wastes, Law of the Sea, Marine Dumping, Ozone Layer Protection, Ship Pollution, Wetlands, Whaling
signed, but not ratified: none of the selected agreements
Geography—note: no natural harbors on west coast

PEOPLE

Population: 13,824,463 (July 2011 est.)
country comparison to the world: 69
Age structure: *0–14 years:* 38.1% (male 2,678,340/female 2,582,472)
15–64 years: 58% (male 3,889,573/female 4,130,698)
65 years and over: 3.9% (male 252,108/female 291,272) (2011 est.)
Median age: *total:* 20 years
male: 19.4 years
female: 20.7 years (2011 est.)
Population growth rate: 1.986% (2011 est.)
country comparison to the world: 54
Birth rate: 26.96 births/1,000 population (2011 est.)
country comparison to the world: 48
Death rate: 4.98 deaths/1,000 population (July 2011 est.)
country comparison to the world: 187
Net migration rate: -2.12 migrant(s)/1,000 population (2011 est.)
country comparison to the world: 165
Urbanization: *urban population:* 49% of total population (2010)
rate of urbanization: 3.4% annual rate of change (2010-15 est.)
Major cities—population: GUATEMALA CITY (capital) 1.075 million (2009)
Sex ratio: *at birth:* 1.05 male(s)/female
under 15 years: 1.04 male(s)/female
15–64 years: 0.94 male(s)/female
65 years and over: 0.86 male(s)/female
total population: 0.97 male(s)/female (2011 est.)
Infant mortality rate: *total:* 26.02 deaths/1,000 live births
country comparison to the world: 78
male: 28.26 deaths/1,000 live births
female: 23.67 deaths/1,000 live births (2011 est.)
Life expectancy at birth: *total population:* 70.88 years
country comparison to the world: 142
male: 69.03 years
female: 72.83 years (2011 est.)
Total fertility rate: 3.27 children born/woman (2011 est.)
country comparison to the world: 50
HIV/AIDS—adult prevalence rate: 0.8% (2009 est.)
country comparison to the world: 54
HIV/AIDS—people living with HIV/AIDS: 62,000 (2009 est.)
country comparison to the world: 54
HIV/AIDS—deaths: 2,600 (2009 est.)
country comparison to the world: 49
Major infectious diseases: *degree of risk:* high
food or waterborne diseases: bacterial diarrhea, hepatitis A, and typhoid fever
vectorborne disease: dengue fever and malaria
water contact disease: leptospirosis (2009)
Drinking water source: *Improved:*
urban: 98% of population
rural: 90% of population
total: 94% of population
Unimproved: urban: 2% of population
rural: 10% of population
total: 6% of population (2008)
Sanitation facility access: *Improved:*
urban: 89% of population
rural: 73% of population
total: 81% of population
Unimproved: urban: 11% of population
rural: 27% of population
total: 19% of population (2008)
Nationality: *noun:* Guatemalan(s)
adjective: Guatemalan
Ethnic groups: Mestizo (mixed Amerindian-Spanish–in local Spanish called Ladino) and European 59.4%, K'iche 9.1%, Kaqchikel 8.4%, Mam 7.9%, Q'eqchi 6.3%, other Mayan 8.6%, indigenous non-Mayan 0.2%, other 0.1% (2001 census)
Religions: Roman Catholic, Protestant, indigenous Mayan beliefs
Languages: Spanish (official) 60%, Amerindian languages 40%
note: there are 23 officially recognized Amerindian languages, including Quiche, Cakchiquel, Kekchi, Mam, Garifuna, and Xinca
Literacy: *definition:* age 15 and over can read and write
total population: 69.1%
male: 75.4%
female: 63.3% (2002 census)
School life expectancy (primary to tertiary education): *total:* 11 years
male: 11 years
female: 10 years (2007)
Education expenditures: 3.2% of GDP (2008)
country comparison to the world: 126

GOVERNMENT

Country name: *conventional long form:* Republic of Guatemala
conventional short form: Guatemala
local long form: Republica de Guatemala
local short form: Guatemala
Government type: constitutional democratic republic
Capital: *name:* Guatemala City
geographic coordinates: 14 37 N, 90 31 W
time difference: UTC-6 (1 hour behind Washington, DC during Standard Time)

Administrative divisions: 22 departments (departamentos, singular–departamento); Alta Verapaz, Baja Verapaz, Chimaltenango, Chiquimula, El Progreso, Escuintla, Guatemala, Huehuetenango, Izabal, Jalapa, Jutiapa, Peten, Quetzaltenango, Quiche, Retalhuleu, Sacatepequez, San Marcos, Santa Rosa, Solola, Suchitepequez, Totonicapan, Zacapa

Independence: 15 September 1821 (from Spain)

National holiday: Independence Day, 15 September (1821)

Constitution: 31 May 1985, effective 14 January 1986; suspended 25 May 1993; reinstated 5 June 1993; amended November 1993

Legal system: civil law system; judicial review of legislative acts

International law organization participation: has not submitted an ICJ jurisdiction declaration; non-party state to the ICCt

Suffrage: 18 years of age; universal; note–active duty members of the armed forces may not vote and are restricted to their barracks on election day

Executive branch: *chief of state:* President Alvaro COLOM Caballeros (since 14 January 2008); Vice President Jose Rafael ESPADA (since 14 January 2008); note–the president is both the chief of state and head of government
head of government: President Alvaro COLOM Caballeros (since 14 January 2008); Vice President Jose Rafael ESPADA (since 14 January 2008)
cabinet: Council of Ministers appointed by the president
(For more information visit the World Leaders website)
elections: president and vice president elected on the same ticket by popular vote for a four-year term (may not serve consecutive terms); election last held on 9 September 2007; runoff held on 4 November 2007 (next to be held in September 2011)
election results: Alvaro COLOM Caballeros elected president; percent of vote–Alvaro COLOM Caballeros 52.8%, Otto PEREZ Molina 47.2%

Legislative branch: unicameral Congress of the Republic or Congreso de la Republica (158 seats; members elected through a party list proportional representation system)
elections: last held on 9 September 2007 (next to be held in September 2011)
election results: percent of vote by party–UNE 30.4%, GANA 23.4%, PP 18.9%, FRG 9.5%, PU 5.1%, other 12.7%; seats by party–UNE 48, GANA 37, PP 30, FRG 15, PU 8, CASA 5, EG 4, PAN 4, UCN 4, URNG 2, UD 1

Judicial branch: Constitutional Court or Corte de Constitucionalidad is Guatemala's highest court (five judges are elected by Congress for concurrent five-year terms); Supreme Court of Justice or Corte Suprema de Justicia (13 members are elected by Congress to serve concurrent five-year terms and elect a president of the Court each year from among their number; the president of the Supreme Court of Justice also supervises trial judges around the country, who are named to five-year terms)

Political parties and leaders: Center of Social Action or CASA [Feliz Adolfo RUANO de Leon]; Democracy Front or FRENTE [Alfonso CABRERA]; Democratic Union or UD [Edwin Armando MARTINEZ Herrera]; Encounter for Guatemala or EG [Nineth MONTENGRO]; Grand National Alliance or GANA [Jaime Antonio MARTINEZ Lohayza]; Guatemalan National Revolutionary Unity or URNG [Hector Alfredo NUILA Ericastilla]; Guatemalan Republican Front or FRG [Luis Fernando PEREZ]; Independent Bloc Guatemala or BG [Macario Efrain OLIVA Muralles]; Independent Democratic Freedom Renewed or LIDER [Manuel BALDIZON]; National Advancement Party or PAN [Juan GUTIERREZ]; National Unity for Hope or UNE [Roberto KESTLER Velasquez]; Nationalist Change Union or UCN [Mario ESTRADA]; Patriot Party or PP [Ingrid Roxana BALDETTI Elias]; Unionista Party or PU [Alvaro ARZU Irigoyen]

Political pressure groups and leaders: Agrarian Owners Group or UNAGRO; Alliance Against Impunity or AAI; Committee for Campesino Unity or CUC; Coordinating Committee of Agricultural, Commercial, Industrial, and Financial Associations or CACIF; International Commission Against Impunity in Guatemala or CICIG; Mutual Support Group or GAM

International organization participation: BCIE, CACM, FAO, G-24, G-77, IADB, IAEA, IBRD, ICAO, ICC, ICRM, IDA, IFAD, IFC, IFRCS, IHO, ILO, IMF, IMO, Interpol, IOC, IOM, IPU, ISO (correspondent), ITSO, ITU, ITUC, LAES, LAIA (observer), MIGA, MINUSTAH, MONUSCO, NAM, OAS, OPANAL, OPCW, PCA, PetroCaribe, RG, SICA, UN, UNCTAD, UNESCO, UNIDO, UNIFIL, Union Latina, UNITAR, UNMIS, UNOCI, UNWTO, UPU, WCO, WFTU, WHO, WIPO, WMO, WTO

Diplomatic representation in the US: *chief of mission:* Ambassador Francisco VILLAGRAN de Leon
chancery: 2220 R Street NW, Washington, DC 20008
telephone: [1] (202) 745-4952
FAX: [1] (202) 745-1908
consulate(s) general: Atlanta, Chicago, Denver, Houston, Los Angeles, Miami, New York, Phoenix, Providence, San Francisco

Diplomatic representation from the US: *chief of mission:* Ambassador Stephen G. MCFARLAND
embassy: 7-01 Avenida Reforma, Zone 10, Guatemala City
mailing address: APO AA 34024
telephone: [502] 2326-4000
FAX: [502] 2326-4654

Flag description: three equal vertical bands of light blue (hoist side), white, and light blue, with the coat of arms centered in the white band; the coat of arms includes a green and red quetzal (the national bird) representing liberty and a scroll bearing the inscription LIBERTAD 15 DE SEPTIEMBRE DE 1821 (the original date of independence from Spain) all superimposed on a pair of crossed rifles signifying Guatemala's willingness to defend itself and a pair of crossed swords representing honor and framed by a laurel wreath symbolizing victory; the blue bands stand for the Pacific Ocean and the Caribbean Sea and the sea and sky; the white band denotes peace and purity

National anthem: *name:* "Himno Nacional de Guatemala" (National Anthem of Guatemala)
lyrics/music: Jose Joaquin PALMA/Rafael Alvarez OVALLE
note: adopted 1897, modified lyrics adopted 1934; Cuban poet Jose Joaquin PALMA anonymously submitted lyrics to a public contest calling for a national anthem; his authorship was not discovered until 1911

ECONOMY

Economy—overview: Guatemala is the most populous country in Central America with a GDP per capita roughly one-half that of the average for Latin America and the Caribbean. The agricultural sector accounts for nearly 15% of GDP and half of the labor force; key agricultural exports include coffee, sugar, and bananas. The 1996 peace accords, which ended 36 years of civil war, removed a major obstacle to foreign investment, and since then Guatemala has pursued important reforms and macroeconomic stabilization. The Dominican Republic-Central American Free Trade Agreement (CAFTA-DR) entered into force in July 2006 spurring increased investment and diversification of exports, with the largest increases in ethanol and non-traditional agricultural exports. While CAFTA-DR has helped improve the investment climate, concerns over security, the lack of skilled workers and poor infrastructure continue to hamper foreign direct investment. The distribution of income remains highly unequal with the richest 10% of the population accounting for more than 40% of Guatemala's overall consumption. More than half of the population is below the national poverty line and 15% lives in extreme poverty. Poverty among indigenous groups, which make up 38% of the population, averages 76% and extreme poverty rises to 28%. 43% of children under five are chronically malnourished, one of the highest malnutrition rates in the world. President COLOM entered into office with the promise to increase education, healthcare, and rural development, and in April 2008 he inaugurated a conditional cash transfer program, modeled after programs in Brazil and Mexico, that provide financial incentives for poor families to keep their children in school and get regular health check-ups. Given Guatemala's large expatriate community in the United States, it is the top remittance recipient in Central

America, with inflows serving as a primary source of foreign income equivalent to nearly two-thirds of exports or one-tenth of GDP. Economic growth fell in 2009 as export demand from US and other Central American markets fell and foreign investment slowed amid the global recession, but the economy recovered gradually in 2010 and will likely return to more normal growth rates by 2012. President COLOM, in his last year in office, will likely face opposition to economic reform, particularly over a long-delayed tax reform and an IMF-recommended reform to strengthen the banking sector.

GDP (purchasing power parity): $70.15 billion (2010 est.)
country comparison to the world: 82
$68.36 billion (2009 est.)
$68 billion (2008 est.)
note: data are in 2010 US dollars
GDP (official exchange rate): $41.47 billion (2010 est.)
GDP—real growth rate: 2.6% (2010 est.)
country comparison to the world: 137
0.5% (2009 est.)
3.3% (2008 est.)
GDP—per capita (PPP): $5,200 (2010 est.)
country comparison to the world: 146
$5,100 (2009 est.)
$5,200 (2008 est.)
note: data are in 2010 US dollars
GDP—composition by sector: *agriculture:* 13.3%
industry: 24.4%
services: 62.3% (2010 est.)
Labor force: 4.26 million (2010 est.)
country comparison to the world: 86
Labor force—by occupation: *agriculture:* 50%
industry: 15%
services: 35% (1999 est.)
Unemployment rate: 3.2% (2005 est.)
country comparison to the world: 26
Population below poverty line: 56.2% (2004 est.)
Household income or consumption by percentage share: *lowest 10%:* 1.3%
highest 10%: 42.4% (2006)
Distribution of family income—Gini index: 55.1 (2007)
country comparison to the world: 11
55.8 (1998)
Investment (gross fixed): 13.9% of GDP (2010 est.)
country comparison to the world: 138
Budget: *revenues:* $4.897 billion
expenditures: $6.124 billion (2010 est.)
Public debt: 29.6% of GDP (2010 est.)
country comparison to the world: 91
27.9% of GDP (2009 est.)
Inflation rate (consumer prices): 3.9% (2010 est.)
country comparison to the world: 112
1.9% (2009 est.)
Central bank discount rate: NA%
Commercial bank prime lending rate: 13.85% (31 December 2009 est.)
country comparison to the world: 55
13.39% (31 December 2008 est.)
Stock of narrow money: $6.6 billion (31 December 2010 est.)
country comparison to the world: 80
$6.13 billion (31 December 2009 est.)
Stock of broad money: $25.4 billion (31 December 2010 est.)
country comparison to the world: 76
$22.9 billion (31 December 2009 est.)
Stock of domestic credit: $15.58 billion (31 December 2010 est.)
country comparison to the world: 88
$14.8 billion (31 December 2009 est.)
Market value of publicly traded shares: $NA
Agriculture—products: sugarcane, corn, bananas, coffee, beans, cardamom; cattle, sheep, pigs, chickens
Industries: sugar, textiles and clothing, furniture, chemicals, petroleum, metals, rubber, tourism
Industrial production growth rate: 2.6% (2010 est.)
country comparison to the world: 119
Electricity—production: 8.425 billion kWh (2007 est.)
country comparison to the world: 96
Electricity—consumption: 7.115 billion kWh (2007 est.)
country comparison to the world: 97
Electricity—exports: 131.9 million kWh (2007 est.)
Electricity—imports: 8.11 million kWh (2007 est.)
Oil—production: 13,530 bbl/day (2009 est.)
country comparison to the world: 80
Oil—consumption: 79,000 bbl/day (2009 est.)
country comparison to the world: 86
Oil—exports: 21,850 bbl/day (2007 est.)
country comparison to the world: 87
Oil—imports: 72,440 bbl/day (2007 est.)
country comparison to the world: 77
Oil—proved reserves: 83.07 million bbl (1 January 2010 est.)
country comparison to the world: 74
Natural gas—production: 0 cu m (2008 est.)
country comparison to the world: 134
Natural gas—consumption: 0 cu m (2008 est.)
country comparison to the world: 179
Natural gas—exports: 0 cu m (2008 est.)
country comparison to the world: 106
Natural gas—imports: 0 cu m (2008 est.)
country comparison to the world: 194
Natural gas—proved reserves: 2.96 billion cu m (1 January 2006 est.)
country comparison to the world: 93
Current account balance: $-1.345 billion (2010 est.)
country comparison to the world: 142
$-267.4 million (2009 est.)
Exports: $8.47 billion (2010 est.)
country comparison to the world: 91
$7.214 billion (2009)
Exports—commodities: coffee, sugar, petroleum, apparel, bananas, fruits and vegetables, cardamom
Exports—partners: US 40.41%, El Salvador 11.2%, Honduras 8.48%, Mexico 5.86% (2009)
Imports: $12.65 billion (2010 est.)
country comparison to the world: 83
$11.52 billion (2009)
Imports—commodities: fuels, machinery and transport equipment, construction materials, grain, fertilizers, electricity
Imports—partners: US 36.46%, Mexico 10.49%, China 5.88%, El Salvador 5.14% (2009)
Reserves of foreign exchange and gold: $5.709 billion (31 December 2010 est.)
country comparison to the world: 69
$4.973 billion (31 December 2009 est.)
Debt—external: $17.47 billion (31 December 2010 est.)
country comparison to the world: 76
$16.04 billion (31 December 2009 est.)
Exchange rates: quetzales (GTQ) per US dollar–8.0798 (2010), 8.1616 (2009), 7.5895 (2008), 7.6833 (2007), 7.6026 (2006)

COMMUNICATIONS

Telephones—main lines in use: 1.413 million (2009)
country comparison to the world: 68
Telephones—mobile cellular: 17.308 million (2009)
country comparison to the world: 43
Telephone system: *general assessment:* fairly modern network centered in the city of Guatemala
domestic: state-owned telecommunications company privatized in the late 1990s opening the way for competition; fixed-line teledensity roughly 10 per 100 persons; fixed-line investments are being concentrated on improving rural connectivity; mobile-cellular teledensity exceeds 100 per 100 persons
international: country code–502; landing point for both the Americas Region Caribbean Ring System (ARCOS-1) and the SAM-1 fiber optic submarine cable system that together provide connectivity to South and Central America, parts of the Caribbean, and the US; connected to Central American Microwave System; satellite earth station–1 Intelsat (Atlantic Ocean) (2008)
Broadcast media: 4 privately-owned national terrestrial TV channels dominate TV broadcasting; multi-channel satellite and cable services are available; 1 government-owned radio station and hundreds of privately-owned radio stations (2007)
Internet country code: .gt
Internet hosts: 196,870 (2010)
country comparison to the world: 65
Internet users: 2.279 million (2009)
country comparison to the world: 72

TRANSPORTATION

Airports: 372 (2010)
country comparison to the world: 21
Airports—with paved runways: *total:* 13
2,438 to 3,047 m: 3
1,524 to 2,437 m: 3
914 to 1,523 m: 4
under 914 m: 3 (2010)
Airports—with unpaved runways: *total:* 359
2,438 to 3,047 m: 1
1,524 to 2,437 m: 3
914 to 1,523 m: 84
under 914 m: 271 (2010)
Pipelines: oil 480 km (2010)
Railways: *total:* 332 km
country comparison to the world: 118
narrow gauge: 332 km 0.914-m gauge (2009)

Roadways:
total: 14,095 km
country comparison to the world: 123
paved: 4,863 km (includes 75 km of expressways)
unpaved: 9,232 km (2000)
Waterways: 990 km (260 km navigable year round; additional 730 km navigable during high-water season) (2010)
country comparison to the world: 66
Ports and terminals: Puerto Quetzal, Santo Tomas de Castilla

MILITARY

Military branches: National Army of Guatemala (Ejercito Nacional de Guatemala, ENG), Guatemalan Navy (Marina Nacional, includes Marines), Guatemalan Air Force (Fuerza Aerea Guatemalteca, FAG) (2009)
Military service age and obligation: all male citizens between the ages of 18 and 50 are liable for military service; conscript service obligation varies from 12 to 24 months; women can serve as officers (2009)
Manpower available for military service:
males age 16–49: 3,165,870
females age 16–49: 3,371,217 (2010 est.)
Manpower fit for military service: *males age 16–49:* 2,590,843
females age 16–49: 2,926,544 (2010 est.)
Manpower reaching militarily significant age annually: *male:* 171,092
female: 168,151 (2010 est.)
Military expenditures: 0.4% of GDP (2009)
country comparison to the world: 166

TRANSNATIONAL ISSUES

Disputes—international: annual ministerial meetings under the OAS-initiated Agreement on the Framework for Negotiations and Confidence Building Measures continue to address Guatemalan land and maritime claims in Belize and the Caribbean Sea; Guatemala persists in its territorial claim to half of Belize, but agrees to Line of Adjacency to keep Guatemalan squatters out of Belize's forested interior; Mexico must deal with thousands of impoverished Guatemalans and other Central Americans who cross the porous border looking for work in Mexico and the United States
Refugees and internally displaced persons: *IDPs:* undetermined (the UN does not estimate there are any IDPs, although some NGOs estimate over 200,000 IDPs as a result of over three decades of internal conflict that ended in 1996) (2007)
Trafficking in persons: current situation: Guatemala is a source, transit, and destination country for Guatemalans and Central Americans trafficked for the purposes of commercial sexual exploitation and forced labor; human trafficking is a significant and growing problem in the country; Guatemalan women and children are trafficked within the country for commercial sexual exploitation, primarily to Mexico and the United States; Guatemalan men, women, and children are also trafficked within the country, and to Mexico and the United States, for forced labor
tier rating: Tier 2 Watch List—for a second consecutive year, Guatemala is on the Tier 2 Watch List for its failure to provide evidence of increasing efforts to combat trafficking in persons, particularly with respect to ensuring that trafficking offenders are appropriately prosecuted for their crimes; while prosecutors initiated trafficking prosecutions, they continued to face problems in court with application of Guatemala's comprehensive anti-trafficking law; the government made modest improvements to its protection efforts, but assistance remained inadequate overall in 2007 (2008)
Illicit drugs: major transit country for cocaine and heroin; in 2005, cultivated 100 hectares of opium poppy after reemerging as a potential source of opium in 2004; potential production of less than 1 metric ton of pure heroin; marijuana cultivation for mostly domestic consumption; proximity to Mexico makes Guatemala a major staging area for drugs (particularly for cocaine); money laundering is a serious problem; corruption is a major problem

GUERNSEY

(BRITISH CROWN DEPENDENCY)

INTRODUCTION

Background: Guernsey and the other Channel Islands represent the last remnants of the medieval Dukedom of Normandy, which held sway in both France and England. The islands were the only British soil occupied by German troops in World War II. Guernsey is a British crown dependency but is not part of the UK or of the European Union. However, the UK Government is constitutionally responsible for its defense and international representation.

GEOGRAPHY

Location: Western Europe, islands in the English Channel, northwest of France
Geographic coordinates: 49 28 N, 2 35 W
Map references: Europe
Area: *total:* 78 sq km
country comparison to the world: 226
land: 78 sq km
water: 0 sq km
note: includes Alderney, Guernsey, Herm, Sark, and some other smaller islands
Area—comparative: about one-half the size of Washington, DC
Land boundaries: 0 km
Coastline: 50 km
Maritime claims: *territorial sea:* 3 nm
exclusive fishing zone: 12 nm
Climate: temperate with mild winters and cool summers; about 50% of days are overcast
Terrain: mostly level with low hills in southwest
Elevation extremes: *lowest point:* Atlantic Ocean 0 m
highest point: unnamed elevation on Sark 114 m
Natural resources: cropland
Land use: *arable land:* NA
permanent crops: NA
other: NA
Irrigated land: NA
Natural hazards: NA
Environment—current issues: NA
Geography—note: large, deepwater harbor at Saint Peter Port

PEOPLE

Population: 65,068 (July 2011 est.)
country comparison to the world: 203
Age structure: *0–14 years:* 14.9% (male 5,036/female 4,670)
15–64 years: 68% (male 22,195/female 22,049)
65 years and over: 17.1% (male 4,952/female 6,166) (2011 est.)
Median age: *total:* 42.2 years
male: 41.1 years
female: 43.2 years (2011 est.)
Population growth rate: 0.438% (2011 est.)
country comparison to the world: 155
Birth rate: 10.13 births/1,000 population (2011 est.)
country comparison to the world: 190
Death rate: 8.44 deaths/1,000 population (July 2011 est.)
country comparison to the world: 87
Net migration rate: 2.69 migrant(s)/1,000 population (2011 est.)
country comparison to the world: 30
Urbanization: *urban population:* 31% of total population (2010)
rate of urbanization: 0.8% annual rate of change (2010-15 est.)
Sex ratio: *at birth:* 1.05 male(s)/female
under 15 years: 1.03 male(s)/female
15–64 years: 0.98 male(s)/female
65 years and over: 0.77 male(s)/female
total population: 0.95 male(s)/female (2011 est.)

Infant mortality rate:
total: 3.55 deaths/1,000 live births
country comparison to the world: 207
male: 3.87 deaths/1,000 live births
female: 3.21 deaths/1,000 live births (2011 est.)
Life expectancy at birth: *total population:* 82.16 years
country comparison to the world: 6
male: 79.5 years
female: 84.95 years (2011 est.)
Total fertility rate: 1.54 children born/woman (2011 est.)
country comparison to the world: 181
HIV/AIDS—adult prevalence rate: NA
HIV/AIDS—people living with HIV/AIDS: NA
HIV/AIDS—deaths: NA
Nationality: *noun:* Channel Islander(s)
adjective: Channel Islander
Ethnic groups: British and Norman-French descent with small percentages from other European countries
Religions: Anglican, Roman Catholic, Presbyterian, Baptist, Congregational, Methodist
Languages: English, French, Norman-French dialect spoken in country districts
Literacy: NA
School life expectancy (primary to tertiary education): NA
Education expenditures: NA

GOVERNMENT

Country name: *conventional long form:* Bailiwick of Guernsey
conventional short form: Guernsey
Dependency status: British crown dependency
Government type: parliamentary democracy
Capital: *name:* Saint Peter Port
geographic coordinates: 49 27 N, 2 32 W
time difference: UTC 0 (5 hours ahead of Washington, DC during Standard Time)
daylight saving time: +1hr, begins last Sunday in March; ends last Sunday in October
Administrative divisions: none (British crown dependency); there are no first-order administrative divisions as defined by the US Government, but there are 10 parishes: Castel, Forest, Saint Andrew, Saint Martin, Saint Peter Port, Saint Pierre du Bois, Saint Sampson, Saint Saviour, Torteval, Vale
Independence: none (British crown dependency)
National holiday: Liberation Day, 9 May (1945)
Constitution: unwritten; partly statutes, partly common law and practice
Legal system: customary legal system based on Norman customary law, and includes elements of the French Civil Code and English common law
Suffrage: 16 years of age; universal
Executive branch: *chief of state:* Queen ELIZABETH II (since 6 February 1952), represented by Lieutenant Governor Air Marshall Peter WALKER (since 15 April 2011)
head of government: Chief Minister Lyndon TROTT (since 1 May 2008); Bailiff Sir Geoffrey ROWLAND (since June 2005)
cabinet: Policy Council elected by the States of Deliberation
(For more information visit the World Leaders website)
elections: the monarchy is hereditary; lieutenant governor and bailiff appointed by the monarch; chief minister elected by States of Deliberation
election results: Lyndon TROTT elected chief minister, percent of vote of the States of Deliberation NA
Legislative branch: unicameral States of Deliberation (45 seats; members elected by popular vote to serve four-year terms; note—there are also 10 Douzaine representatives—one from each parish, 2 representatives from Alderney and the appointed attorney general and soliciter general); note—Alderney and Sark have parliaments
elections: last held on 23 April 2008 (next to be held in 2012)
election results: percent of vote—NA; seats—all independents
Judicial branch: Royal Court (judges elected by an electoral college and the bailiff)
Political parties and leaders: none; all independents
Political pressure groups and leaders: Stop Traffic Endangering Pedestrian Safety or STEPS; No More Masts [Colin FALLAIZE]
International organization participation: UPU
Diplomatic representation in the US: none (British crown dependency)
Diplomatic representation from the US: none (British crown dependency)
Flag description: white with the red cross of Saint George (patron saint of England) extending to the edges of the flag and a yellow equal-armed cross of William the Conqueror superimposed on the Saint George cross; the red cross represents the old ties with England and the fact that Guernsey is a British Crown dependency; the gold cross is a replica of the one used by Duke William of Normandy at the Battle of Hastings
National anthem: *name:* "Sarnia Cherie" (Guernsey Dear)
lyrics/music: George DEIGHTON/Domencio SANTANGELO
note: adopted 1911; serves as a local anthem; as a British crown dependency, "God Save the Queen" remains official (see United Kingdom)

ECONOMY

Economy—overview: Financial services—banking, fund management, insurance—account for about 23% of employment and about 55% of total income in this tiny, prosperous Channel Island economy. Tourism, manufacturing, and horticulture, mainly tomatoes and cut flowers, have been declining. Financial services, construction, retail, and the public sector have been growing. Light tax and death duties make Guernsey a popular tax haven. The evolving economic integration of the EU nations is changing the environment under which Guernsey operates.
GDP (purchasing power parity): $2.742 billion (2005)
country comparison to the world: 178
GDP (official exchange rate): $2.742 billion (2005)
GDP—real growth rate: 3% (2005 est.)
country comparison to the world: 126
GDP—per capita (PPP): $44,600 (2005)
country comparison to the world: 14
GDP—composition by sector:
agriculture: 3%
industry: 10%
services: 87% (2000)
Labor force: 31,470 (March 2006)
country comparison to the world: 203
Unemployment rate: 0.9% (March 2006 est.)
country comparison to the world: 4
Population below poverty line: NA%
Household income or consumption by percentage share: *lowest 10%:* NA%
highest 10%: NA%
Budget: *revenues:* $563.6 million
expenditures: $530.9 million (2005)
Inflation rate (consumer prices): 3.4% (June 2006)
country comparison to the world: 97
Agriculture—products: tomatoes, greenhouse flowers, sweet peppers, eggplant, fruit; Guernsey cattle
Industries: tourism, banking
Industrial production growth rate: NA%
Electricity—production: NA kWh
Electricity—consumption: NA kWh
Electricity—exports: 0 kWh (2002)
Electricity—imports: 0 kWh (2002)
Exports: $NA
Exports—commodities: tomatoes, flowers and ferns, sweet peppers, eggplant, other vegetables
Imports: $NA
Imports—commodities: coal, gasoline, oil, machinery and equipment
Debt—external: $NA
Exchange rates: Guernsey pound
0.6388 (2010), 0.6175 (2009), 0.4993 (2007), 0.5418 (2006)

COMMUNICATIONS

Telephones—main lines in use: 45,100 (2009)
country comparison to the world: 165
Telephones—mobile cellular: 43,800 (2004)
country comparison to the world: 197
Telephone system: *general assessment:* NA
domestic: fixed-line and mobile-cellular services widely available; combined fixed and mobile-cellular teledensity exceeds 100 per 100 persons
international: country code—44; 1 submarine cable
Broadcast media: multiple UK terrestrial television broadcasts—received via a transmitter in Jersey with relays in Jersey, Guernsey, and Alderney—will begin switching from analog to digital broadcasts in November 2010; satellite packages are available; BBC Radio Guernsey and 1 other radio station operating (2009)
Internet country code: .gg
Internet hosts: 197 (2010)
country comparison to the world: 195
Internet users: 48,300 (2009)
country comparison to the world: 174

TRANSPORTATION

Airports: 2 (2010)
country comparison to the world: 199
Airports—with paved runways: *total:* 2
914 to 1,523 m: 1
under 914 m: 1 (2010)
Ports and terminals: Braye Bay, Saint Peter Port

MILITARY

Manpower fit for military service: *males age 16–49:* 12,493
females age 16–49: 12,272 (2010 est.)
Manpower reaching militarily significant age annually: *male:* 354
female: 342 (2010 est.)
Military—note: defense is the responsibility of the UK

TRANSNATIONAL ISSUES

Disputes—international: none

GUINEA

INTRODUCTION

Background: Guinea has had a history of authoritarian rule since gaining its independence from France in 1958. Lansana CONTE came to power in 1984 when the military seized the government after the death of the first president, Sekou TOURE. Guinea did not hold democratic elections until 1993 when Gen. CONTE (head of the military government) was elected president of the civilian government. He was reelected in 1998 and again in 2003, though all the polls were marred by irregularities. History repeated itself in December 2008 when following President CONTE's death, Capt. Moussa Dadis CAMARA led a military coup, seizing power and suspending the constitution. His unwillingness to yield to domestic and international pressure to step down led to heightened political tensions that culminated in September 2009 when presidential guards opened fire on an opposition rally killing more than 150 people, and in early December 2009 when CAMARA was wounded in an assassination attempt and evacuated to Morocco and subsequently to Burkina Faso. A transitional government led by General Sekouba KONATE held democratic elections in 2010 and Alpha CONDE was elected president in the country's first free and fair elections since independence.

GEOGRAPHY

Location: Western Africa, bordering the North Atlantic Ocean, between Guinea-Bissau and Sierra Leone
Geographic coordinates: 11 00 N, 10 00 W
Map references: Africa
Area: *total:* 245,857 sq km
country comparison to the world: 78
land: 245,717 sq km
water: 140 sq km
Area—comparative: slightly smaller than Oregon
Land boundaries: *total:* 3,399 km
border countries: Cote d'Ivoire 610 km, Guinea-Bissau 386 km, Liberia 563 km, Mali 858 km, Senegal 330 km, Sierra Leone 652 km
Coastline: 320 km
Maritime claims: *territorial sea:* 12 nm
exclusive economic zone: 200 nm
Climate: generally hot and humid; monsoonal-type rainy season (June to November) with southwesterly winds; dry season (December to May) with northeasterly harmattan winds
Terrain: generally flat coastal plain, hilly to mountainous interior
Elevation extremes: *lowest point:* Atlantic Ocean 0 m
highest point: Mont Nimba 1,752 m
Natural resources: bauxite, iron ore, diamonds, gold, uranium, hydropower, fish, salt
Land use: *arable land:* 4.47%
permanent crops: 2.64%
other: 92.89% (2005)
Irrigated land: 950 sq km (2003)
Total renewable water resources: 226 cu km (1987)
Freshwater withdrawal (domestic/industrial/agricultural): *total:* 1.51 cu km/yr (8%/2%/90%)
per capita: 161 cu m/yr (2000)
Natural hazards: hot, dry, dusty harmattan haze may reduce visibility during dry season
Environment—current issues: deforestation; inadequate supplies of potable water; desertification; soil contamination and erosion; overfishing, overpopulation in forest region; poor mining practices have led to environmental damage
Environment—international agreements: *party to:* Biodiversity, Climate Change, Climate Change-Kyoto Protocol, Desertification, Endangered Species, Hazardous Wastes, Law of the Sea, Ozone Layer Protection, Ship Pollution, Wetlands, Whaling
signed, but not ratified: none of the selected agreements
Geography—note: the Niger and its important tributary the Milo have their sources in the Guinean highlands

PEOPLE

Population: 10,601,009 (July 2011 est.)
country comparison to the world: 79
Age structure: *0–14 years:* 42.5% (male 2,278,048/female 2,229,602)
15–64 years: 54% (male 2,860,845/female 2,860,004)
65 years and over: 3.5% (male 164,051/female 208,459) (2011 est.)
Median age: *total:* 18.6 years
male: 18.3 years
female: 18.8 years (2011 est.)
Population growth rate: 2.645% (2011 est.)
country comparison to the world: 22
Birth rate: 36.9 births/1,000 population (2011 est.)
country comparison to the world: 21
Death rate: 10.45 deaths/1,000 population (July 2011 est.)
country comparison to the world: 48
Net migration rate: 0 migrant(s)/1,000 population (2011 est.)
country comparison to the world: 86
Urbanization: *urban population:* 35% of total population (2010)
rate of urbanization: 4.3% annual rate of change (2010-15 est.)
Major cities—population: CONAKRY (capital) 1.597 million (2009)
Sex ratio: *at birth:* 1.03 male(s)/female
under 15 years: 1.02 male(s)/female
15–64 years: 1 male(s)/female
65 years and over: 0.78 male(s)/female
total population: 1 male(s)/female (2011 est.)
Infant mortality rate: *total:* 61.03 deaths/1,000 live births
country comparison to the world: 31
male: 64.29 deaths/1,000 live births
female: 57.68 deaths/1,000 live births (2011 est.)
Life expectancy at birth: *total population:* 58.11 years
country comparison to the world: 191
male: 56.63 years
female: 59.64 years (2011 est.)
Total fertility rate: 5.1 children born/woman (2011 est.)
country comparison to the world: 17
HIV/AIDS—adult prevalence rate: 1.3% (2009 est.)
country comparison to the world: 37
HIV/AIDS—people living with HIV/AIDS: 79,000 (2009 est.)
country comparison to the world: 46
HIV/AIDS—deaths: 4,700 (2009 est.)
country comparison to the world: 39
Major infectious diseases: *degree of risk:* very high
food or waterborne diseases: bacterial and protozoal diarrhea, hepatitis A, and typhoid fever
vectorborne diseases: malaria and yellow fever
water contact disease: schistosomiasis

aerosolized dust or soil contact disease: Lassa fever
animal contact disease: rabies (2009)
Drinking water source: *Improved:*
urban: 89% of population
rural: 61% of population
total: 71% of population
Unimproved: urban: 11% of population
rural: 39% of population
total: 29% of population (2008)
Sanitation facility access: *Improved:*
urban: 34% of population
rural: 11% of population
total: 19% of population
Unimproved: urban: 66% of population
rural: 89% of population
total: 81% of population (2008)
Nationality: *noun:* Guinean(s)
adjective: Guinean
Ethnic groups: Peuhl 40%, Malinke 30%, Soussou 20%, smaller ethnic groups 10%
Religions: Muslim 85%, Christian 8%, indigenous beliefs 7%
Languages: French (official)
note: each ethnic group has its own language
Literacy: *definition:* age 15 and over can read and write
total population: 29.5%
male: 42.6%
female: 18.1% (2003 est.)
School life expectancy (primary to tertiary education): *total:* 9 years
male: 10 years
female: 7 years (2009)
Education expenditures: 2.4% of GDP (2008)
country comparison to the world: 150

GOVERNMENT

Country name: *conventional long form:* Republic of Guinea
conventional short form: Guinea
local long form: Republique de Guinee
local short form: Guinee
former: French Guinea
Government type: republic
Capital: *name:* Conakry
geographic coordinates: 9 33 N, 13 42 W
time difference: UTC 0 (5 hours ahead of Washington, DC during Standard Time)
Administrative divisions: 33 prefectures and 1 special zone (zone special)*; Beyla, Boffa, Boke, Conakry*, Coyah, Dabola, Dalaba, Dinguiraye, Dubreka, Faranah, Forecariah, Fria, Gaoual, Gueckedou, Kankan, Kerouane, Kindia, Kissidougou, Koubia, Koundara, Kouroussa, Labe, Lelouma, Lola, Macenta, Mali, Mamou, Mandiana, Nzerekore, Pita, Siguiri, Telimele, Tougue, Yomou
Independence: 2 October 1958 (from France)
National holiday: Independence Day, 2 October (1958)
Constitution: 7 May 2010 (Loi Fundamentale)
Legal system: civil law system based on the French model
International law organization participation: accepts compulsory ICJ jurisdiction with reservations; accepts ICCt jurisdiction
Suffrage: 18 years of age; universal
Executive branch: *chief of state:* President Alpha CONDE (since 21 December 2010)
head of government: Prime Minister Mohamed Said FOFANA (since 24 December 2010)
cabinet: Council of Ministers appointed by the president
(For more information visit the World Leaders website)
elections: president elected by popular vote for a five-year term (eligible for a second term); candidate must receive a majority of the votes cast to be elected president; election last held on 27 June 2010 with a runoff election held on 7 November 2010
election results: Alpha CONDE elected president in a runoff election; percent of vote Alpha CONDE 52.5%, Cellou Dalein DIALLO 47.5%
Legislative branch: the legislature was dissolved by junta leader Moussa Dadis CAMARA in December 2008 and in February 2010, the Transition Government appointed a 155 member National Transition Council (CNT) that has since acted in the legislature's place
elections: last held on 30 June 2002
Judicial branch: Constitutional Court; Court of First Instance or Tribunal de Premiere Instance; Court of Appeal or Cour d'Appel; Supreme Court or Cour Supreme
Political parties and leaders: Rally for the Guinean People or RPG [Alpha CONDE]; Union of Democratic Forces of Guinea or UFDG [Cellou Dalein DIALLO]; Union of Republican Forces or UFR [Sidya TOURE]
note: Listed are the three most popular parties in first round voting for president in 2010; overall, there are more than 130 registered parties
Political pressure groups and leaders: National Confederation of Guinean Workers-Labor Union of Guinean Workers or CNTG-USTG Alliance (includes National Confederation of Guinean Workers or CNTG and Labor Union of Guinean Workers or USTG); Syndicate of Guinean Teachers and Researchers or SLECG
International organization participation: ACP, AfDB, AU (suspended), ECOWAS (suspended), FAO, G-77, IBRD, ICAO, ICRM, IDA, IDB, IFAD, IFC, IFRCS, ILO, IMF, IMO, Interpol, IOC, IOM, ISO (correspondent), ITSO, ITU, ITUC, MIGA, MINURSO, NAM, OIC, OIF, OPCW, UN, UNCTAD, UNESCO, UNHCR, UNIDO, UNMIS, UNOCI, UNWTO, UPU, WCO, WFTU, WHO, WIPO, WMO, WTO
Diplomatic representation in the US: *chief of mission:* Ambassador Mory Karamoko KABA
chancery: 2112 Leroy Place NW, Washington, DC 20008
telephone: [1] (202) 986-4300
FAX: [1] (202) 483-8688
Diplomatic representation from the US: *chief of mission:* Ambassador Patricia Newton MOLLER
embassy: Koloma, Conakry, east of Hamdallaye Circle
mailing address: B. P. 603, Transversale No. 2, Centre Administratif de Koloma, Commune de Ratoma, Conakry
telephone: [224] 65-10-40-00
FAX: [224] 65-10-42-97
Flag description: three equal vertical bands of red (hoist side), yellow, and green; red represents the people's sacrifice for liberation and work; yellow stands for the sun, for the riches of the earth, and for justice; green symbolizes the country's vegetation and unity
note: uses the popular Pan-African colors of Ethiopia; the colors from left to right are the reverse of those on the flags of neighboring Mali and Senegal
National anthem: *name:* "Liberte" (Liberty)
lyrics/music: unknown/Fodeba KEITA
note: adopted 1958

ECONOMY

Economy—overview: Guinea is a poor country that possesses major mineral, hydropower, and agricultural resources. The country has almost half of the world's bauxite reserves and significant iron ore, gold, and diamond reserves. However, Guinea has been unable to profit from this potential, as rampant corruption, dilapidated electricity and other degraded infrastructure, and political uncertainty have drained investor confidence. In the time since a 2008 coup following the death of long-term President Lansana CONTE, international donors, including the G-8, the IMF, and the World Bank, have significantly curtailed their development programs. Throughout 2009, policies of the ruling military junta severely weakened the economy. The junta leaders spent and printed money at an accelerated rate, driving inflation and debt to perilously high levels. In early 2010, the junta collapsed and was replaced by a Transition Government, which ceded power in December 2010 to the country's first-ever democratically elected president, Alpha CONDE. International assistance and investment are expected to return to Guinea, but the levels will depend upon the ability of the new government to combat corruption and reform its banking system. IMF and World Bank programs will be especially critical as Guinea attempts to gain debt relief. Since the 2009 global economic downturn, the price and value of bauxite and alumina exports has steadily risen. Export levels will likely continue to grow as investor confidence returns. International investors have expressed keen interest in Guinea's vast iron ore reserves, which could further propel the country's growth.
GDP (purchasing power parity): $10.81 billion (2010 est.)
country comparison to the world: 150
$10.6 billion (2009 est.)
$10.63 billion (2008 est.)
note: data are in 2010 US dollars
GDP (official exchange rate): $4.633 billion (2010 est.)

GDP—real growth rate: 1.9% (2010 est.)
country comparison to the world: 152
-0.3% (2009 est.)
4.9% (2008 est.)
GDP—per capita (PPP): $1,000 (2010 est.)
country comparison to the world: 214
$1,100 (2009 est.)
$1,100 (2008 est.)
note: data are in 2010 US dollars
GDP—composition by sector:
agriculture: 25.8%
industry: 45.7%
services: 28.5% (2010 est.)
Labor force: 4.392 million (2007 est.)
country comparison to the world: 83
Labor force—by occupation:
agriculture: 76%
industry and *services:* 24% (2006 est.)
Unemployment rate: NA%
Population below poverty line: 47% (2006 est.)
Household income or consumption by percentage share: *lowest 10%:* 1.9%
highest 10%: 41% (2006)
Distribution of family income—Gini index: 38.1 (2006)
country comparison to the world: 71
40.3 (1994)
Investment (gross fixed): 14.6% of GDP (2010 est.)
country comparison to the world: 135
Budget: *revenues:* $574.1 million
expenditures: $875.4 million (2010 est.)
Inflation rate (consumer prices): 15% (2010 est.)
country comparison to the world: 218
9% (2009 est.)
Central bank discount rate: NA%
country comparison to the world: 7
22.25% (31 December 2005)
Commercial bank prime lending rate: NA%
Stock of narrow money: $496.2 million (31 December 2010 est.)
country comparison to the world: 160
$459.7 million (31 December 2009 est.)
Stock of broad money: $830 million (31 December 2010 est.)
country comparison to the world: 167
$761.9 million (31 December 2009 est.)
Stock of domestic credit: $734.4 million (31 December 2010 est.)
country comparison to the world: 158
$674.2 million (31 December 2009 est.)
Market value of publicly traded shares: $NA
Agriculture—products: rice, coffee, pineapples, palm kernels, cassava (tapioca), bananas, sweet potatoes; cattle, sheep, goats; timber
Industries: bauxite, gold, diamonds, iron; alumina refining; light manufacturing, and agricultural processing
Industrial production growth rate: 3% (2010 est.)
country comparison to the world: 111
Electricity—production: 850 million kWh
country comparison to the world: 148
note: excludes electricity generated at interior mining sites (2007 est.)
Electricity—consumption: 790.5 million kWh (2007 est.)
country comparison to the world: 149
Electricity—exports: 0 kWh (2008 est.)
Electricity—imports: 0 kWh (2008 est.)
Oil—production: 0 bbl/day (2009 est.)
country comparison to the world: 181
Oil—consumption: 9,000 bbl/day (2009 est.)
country comparison to the world: 151
Oil—exports: 0 bbl/day (2007 est.)
country comparison to the world: 170
Oil—imports: 8,674 bbl/day (2007 est.)
country comparison to the world: 142
Oil—proved reserves: 0 bbl (1 January 2010 est.)
country comparison to the world: 138
Natural gas—production: 0 cu m (2008 est.)
country comparison to the world: 135
Natural gas—consumption: 0 cu m (2008 est.)
country comparison to the world: 180
Natural gas—exports: 0 cu m (2008 est.)
country comparison to the world: 107
Natural gas—imports: 0 cu m (2008 est.)
country comparison to the world: 195
Natural gas—proved reserves: 0 cu m (1 January 2010 est.)
country comparison to the world: 142
Current account balance: $-434 million (2010 est.)
country comparison to the world: 111
$-538 million (2009 est.)
Exports: $1.468 billion (2010 est.)
country comparison to the world: 139
$1.18 billion (2009 est.)
Exports—commodities: bauxite, alumina, gold, diamonds, coffee, fish, agricultural products
Exports—partners: India 19.68%, Spain 13.18%, Russia 7.24%, Germany 6.86%, Ireland 5.87%, US 5.71%, Ukraine 5.6% (2009)
Imports: $1.551 billion (2010 est.)
country comparison to the world: 160
$1.236 billion (2009 est.)
Imports—commodities: petroleum products, metals, machinery, transport equipment, textiles, grain and other foodstuffs
Imports—partners: China 8.67%, Netherlands 6.67%, France 4.33%, UK 4.22% (2009)
Reserves of foreign exchange and gold: $NA (31 December 2010 est.)
$51 million (31 December 2009 est.)
Debt—external: $3.072 billion (31 December 2009 est.)
country comparison to the world: 126
$3.222 billion (31 December 2008 est.)
Exchange rates: Guinean francs (GNF) per US dollar–6,100 (2010), 5,500 (2009), 5,500 (2008), 4,122.8 (2007), 5,350 (2006)

COMMUNICATIONS

Telephones—main lines in use: 22,000 (2009)
country comparison to the world: 191
Telephones—mobile cellular: 5.607 million (2009)
country comparison to the world: 90
Telephone system: *general assessment:* inadequate system of open-wire lines, small radiotelephone communication stations, and new microwave radio relay system
domestic: Conakry reasonably well served; coverage elsewhere remains inadequate and large companies tend to rely on their own systems for nationwide links; fixed-line teledensity less than 1 per 100 persons; mobile-cellular subscribership is expanding and exceeded 50 per 100 persons in 2009
international: country code–224; satellite earth station–1 Intelsat (Atlantic Ocean)
Broadcast media: government maintains marginal control over broadcast media; single state-run TV station; state-run radio broadcast station also operates several stations in rural areas; a steadily increasing number of privately-owned radio stations, nearly all in Conakry, and about a dozen community radio stations; foreign television programming available via satellite and cable subscription services (2011)
Internet country code: .gn
Internet hosts: 14 (2010)
country comparison to the world: 220
Internet users: 95,000 (2009)
country comparison to the world: 161

TRANSPORTATION

Airports: 16 (2010)
country comparison to the world: 143
Airports—with paved runways: *total:* 4
over 3,047 m: 1
1,524 to 2,437 m: 3 (2010)
Airports—with unpaved runways: *total:* 12
1,524 to 2,437 m: 7
914 to 1,523 m: 3
under 914 m: 2 (2010)
Railways: *total:* 1,185 km
country comparison to the world: 86
standard gauge: 238 km 1.435-m gauge
narrow gauge: 947 km 1.000-m gauge (2009)
Roadways: *total:* 44,348 km
country comparison to the world: 83
paved: 4,342 km
unpaved: 40,006 km (2003)
Waterways: 1,300 km (navigable by shallow-draft native craft in the northern part of the Niger system) (2009)
country comparison to the world: 55
Ports and terminals: Conakry, Kamsar

MILITARY

Military branches: National Armed Forces: Army, Navy (Armee de Mer or Marine Guineenne, includes Marines), Guinean Air Force (Force Aerienne de Guinee) (2009)
Military service age and obligation: 18-25 years of age for compulsory and voluntary military service; 18-month conscript service obligation (2009)
Manpower available for military service:
males age 16–49: 2,359,203
females age 16–49: 2,329,784 (2010 est.)
Manpower fit for military service: *males age 16–49:* 1,493,991
females age 16–49: 1,535,418 (2010 est.)
Manpower reaching militarily significant age annually: *male:* 118,443
female: 115,901 (2010 est.)
Military expenditures: 1.1% of GDP (2009)
country comparison to the world: 125

TRANSNATIONAL ISSUES

Disputes—international: conflicts among rebel groups, warlords, and youth gangs in neighboring states have spilled over into Guinea resulting in domestic instability; Sierra Leone considers Guinea's definition of the flood plain limits to define the left bank boundary of the Makona and Moa rivers excessive and protests Guinea's continued occupation of these lands, including the hamlet of Yenga, occupied since 1998

Refugees and internally displaced persons: refugees (country of origin): 21,856 (Liberia); 5,259 (Sierra Leone); 3,900 (Cote d'Ivoire)
IDPs: 19,000 (cross-border incursions from Cote d'Ivoire, Liberia, Sierra Leone) (2007)

Trafficking in persons: current situation: Guinea is a source, transit, and destination country for men, women, and children trafficked for the purposes of forced labor and sexual exploitation; the majority of victims are children, and internal trafficking is more prevalent than transnational trafficking; within the country, girls are trafficked primarily for domestic servitude and sexual exploitation, while boys are trafficked for forced agricultural labor, and as forced beggars, street vendors, shoe shiners, and laborers in gold and diamond mines; some Guinean men are also trafficked for agricultural labor within Guinea; transnationally, girls are trafficked into Guinea for domestic servitude and likely also for sexual exploitation
tier rating: Tier 2 Watch List–Guinea is on the Tier 2 Watch List for its failure to provide evidence of increasing efforts to eliminate trafficking over 2006; Guinea demonstrated minimal law enforcement efforts for a second year in a row, while protection efforts diminished over efforts in 2006; the government did not report any trafficking convictions in 2007; due to a lack of resources, the government does not provide shelter services for trafficking victims; the government took no measures to reduce the demand for commercial sexual exploitation (2008)

GUINEA-BISSAU

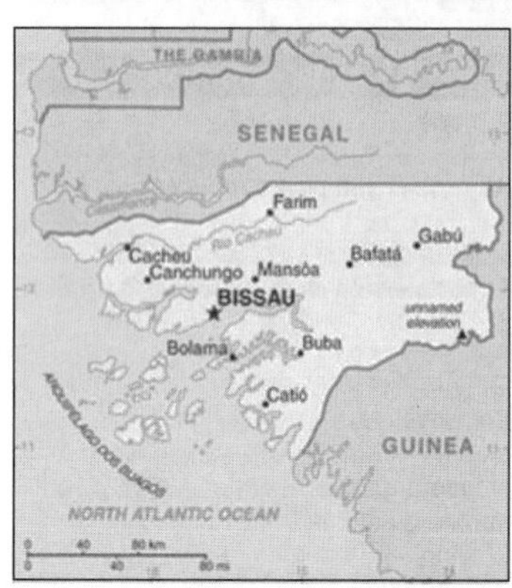

INTRODUCTION

Background: Since independence from Portugal in 1974, Guinea-Bissau has experienced considerable political and military upheaval. In 1980, a military coup established authoritarian dictator Joao Bernardo 'Nino' VIEIRA as president. Despite setting a path to a market economy and multiparty system, VIEIRA's regime was characterized by the suppression of political opposition and the purging of political rivals. Several coup attempts through the 1980s and early 1990s failed to unseat him. In 1994 VIEIRA was elected president in the country's first free elections. A military mutiny and resulting civil war in 1998 eventually led to VIEIRA's ouster in May 1999. In February 2000, a transitional government turned over power to opposition leader Kumba YALA after he was elected president in transparent polling. In September 2003, after only three years in office, YALA was ousted by the military in a bloodless coup, and businessman Henrique ROSA was sworn in as interim president. In 2005, former President VIEIRA was re-elected president pledging to pursue economic development and national reconciliation; he was assassinated in March 2009. Malam Bacai SANHA was elected in an emergency election held in June 2009.

GEOGRAPHY

Location: Western Africa, bordering the North Atlantic Ocean, between Guinea and Senegal
Geographic coordinates: 12 00 N, 15 00 W
Map references: Africa
Area: *total:* 36,125 sq km
country comparison to the world: 137
land: 28,120 sq km
water: 8,005 sq km
Area—comparative: slightly less than three times the size of Connecticut
Land boundaries: *total:* 724 km
border countries: Guinea 386 km, Senegal 338 km
Coastline: 350 km
Maritime claims: *territorial sea:* 12 nm
exclusive economic zone: 200 nm
Climate: tropical; generally hot and humid; monsoonal-type rainy season (June to November) with southwesterly winds; dry season (December to May) with northeasterly harmattan winds
Terrain: mostly low coastal plain rising to savanna in east
Elevation extremes: *lowest point:* Atlantic Ocean 0 m
highest point: unnamed elevation in the eastern part of the country 300 m
Natural resources: fish, timber, phosphates, bauxite, clay, granite, limestone, unexploited deposits of petroleum
Land use: *arable land:* 8.31%
permanent crops: 6.92%
other: 84.77% (2005)
Irrigated land: 250 sq km (2003)
Total renewable water resources: 31 cu km (2003)
Freshwater withdrawal (domestic/industrial/agricultural): *total:* 0.18 cu km/yr (13%/5%/82%)
per capita: 113 cu m/yr (2000)
Natural hazards: hot, dry, dusty harmattan haze may reduce visibility during dry season; brush fires
Environment—current issues: deforestation; soil erosion; overgrazing; overfishing
Environment—international agreements: *party to:* Biodiversity, Climate Change, Climate Change-Kyoto Protocol, Desertification, Endangered Species, Hazardous Wastes, Law of the Sea, Ozone Layer Protection, Wetlands
signed, but not ratified: none of the selected agreements
Geography—note: this small country is swampy along its western coast and low-lying inland

PEOPLE

Population: 1,596,677 (July 2011 est.)
country comparison to the world: 150
Age structure: *0–14 years:* 40.4% (male 321,889/female 323,202)
15–64 years: 56.4% (male 435,986/female 465,117)
65 years and over: 3.2% (male 19,975/female 30,508) (2011 est.)
Median age: *total:* 19.5 years
male: 18.9 years
female: 20 years (2011 est.)
Population growth rate: 1.988% (2011 est.)
country comparison to the world: 52
Birth rate: 35.15 births/1,000 population (2011 est.)
country comparison to the world: 31
Death rate: 15.27 deaths/1,000 population (July 2011 est.)
country comparison to the world: 8
Net migration rate: 0 migrant(s)/1,000 population (2011 est.)
country comparison to the world: 104
Urbanization: *urban population:* 30% of total population (2010)
rate of urbanization: 3% annual rate of change (2010-15 est.)
Major cities—population: BISSAU (capital) 302,000 (2009)
Sex ratio: *at birth:* 1.03 male(s)/female
under 15 years: 1 male(s)/female
15–64 years: 0.93 male(s)/female
65 years and over: 0.66 male(s)/female
total population: 0.95 male(s)/female (2011 est.)
Infant mortality rate: *total:* 96.23 deaths/1,000 live births
country comparison to the world: 7
male: 106.11 deaths/1,000 live births
female: 86.06 deaths/1,000 live births (2011 est.)
Life expectancy at birth: *total population:* 48.7 years

country comparison to the world: 217
male: 46.8 years
female: 50.67 years (2011 est.)
Total fertility rate: 4.51 children born/woman (2011 est.)
country comparison to the world: 33
HIV/AIDS—adult prevalence rate: 2.5% (2009 est.)
country comparison to the world: 26
HIV/AIDS—people living with HIV/AIDS: 22,000 (2009 est.)
country comparison to the world: 76
HIV/AIDS—deaths: 1,200 (2009 est.)
country comparison to the world: 65
Major infectious diseases: *degree of risk:* very high
food or waterborne diseases: bacterial and protozoal diarrhea, hepatitis A, and typhoid fever
vectorborne diseases: malaria and yellow fever
water contact disease: schistosomiasis
animal contact disease: rabies (2009)
Drinking water source: *Improved:*
urban: 83% of population
rural: 51% of population
total: 61% of population
Unimproved: urban: 17% of population
rural: 49% of population
total: 39% of population (2008)
Sanitation facility access: *Improved:*
urban: 49% of population
rural: 9% of population
total: 21% of population
Unimproved: urban: 51% of population
rural: 91% of population
total: 79% of population (2008)
Nationality: *noun:* Guinean(s)
adjective: Guinean
Ethnic groups: African 99% (includes Balanta 30%, Fula 20%, Manjaca 14%, Mandinga 13%, Papel 7%), European and mulatto less than 1%
Religions: Muslim 50%, indigenous beliefs 40%, Christian 10%
Languages: Portuguese (official), Crioulo, African languages
Literacy: *definition:* age 15 and over can read and write
total population: 42.4%
male: 58.1%
female: 27.4% (2003 est.)
School life expectancy (primary to tertiary education): *total:* 9 years (2006)
Education expenditures: NA

GOVERNMENT

Country name: *conventional long form:* Republic of Guinea-Bissau
conventional short form: Guinea-Bissau
local long form: Republica da Guine-Bissau
local short form: Guine-Bissau
former: Portuguese Guinea
Government type: republic
Capital: *name:* Bissau
geographic coordinates: 11 51 N, 15 35 W
time difference: UTC 0 (5 hours ahead of Washington, DC during Standard Time)
Administrative divisions: 9 regions (regioes, singular–regiao); Bafata, Biombo, Bissau, Bolama, Cacheu, Gabu, Oio, Quinara, Tombali; note–Bolama may have been renamed Bolama/Bijagos
Independence: 24 September 1973 (declared); 10 September 1974 (from Portugal)
National holiday: Independence Day, 24 September (1973)
Constitution: 16 May 1984; amended several times
Legal system: mixed legal system of civil law (influenced by the early French Civil Code) and customary law
International law organization participation: accepts compulsory ICJ jurisdiction; non-party state to the ICCt
Suffrage: 18 years of age; universal
Executive branch: *chief of state:* President Malam Bacai SANHA (since 8 September 2009)
head of government: Prime Minister Carlos GOMES Junior (since 25 December 2008)
cabinet: NA
(For more information visit the World Leaders website)
elections: president elected by popular vote for a five-year term (no term limits); election last held on 28 June 2009 with a runoff between the two leading candidates held on 26 July 2009 (next to be held by 2014); prime minister appointed by the president after consultation with party leaders in the legislature
election results: Malam Bacai SANHA elected president; percent of vote, second ballot–Malam Bacai SANHA 63.5%, Kumba YALA 36.5%
Legislative branch: unicameral National People's Assembly or Assembleia Nacional Popular (100 seats; members elected by popular vote to serve four-year terms)
elections: last held on 16 November 2008 (next to be held in 2012)
election results: percent of vote by party–PAIGC 49.8%, PRS 25.3%, PRID 7.5%, PND 2.4%, AD 1.4%, other parties 13.6%; seats by party–PAIGC 67, PRS 28, PRID 3, PND 1, AD 1
Judicial branch: Supreme Court or Supremo Tribunal da Justica (consists of nine justices appointed by the president and serve at his pleasure; final court of appeals in criminal and civil cases); Regional Courts (one in each of nine regions; first court of appeals for Sectoral Court decisions; hear all felony cases and civil cases valued at more than $1,000); 24 Sectoral Courts (judges are not necessarily trained lawyers; they hear civil cases valued at less than $1,000 and misdemeanor criminal cases)
Political parties and leaders: African Party for the Independence of Guinea-Bissau and Cape Verde or PAIGC [Carlos GOMES Junior]; Democratic Alliance or AD [Victor MANDINGA]; Democratic Social Front or FDS [Rafael BARBOSA]; Electoral Union or UE [Joaquim BALDE]; Guinea-Bissau Civic Forum/Social Democracy or FCGSD [Antonieta Rosa GOMES]; Guinea-Bissau Democratic Party or PDG; Guinea-Bissau Socialist Democratic Party or PDSG [Serifo BALDE]; Labor and Solidarity Party or PST [Lancuba INDJAI]; New Democracy Party or PND; Party for Democratic Convergence or PCD [Victor MANDINGA]; Party for Renewal and Progress or PRP; Party for Social Renewal or PRS [Kumba YALA]; Progress Party or PP; Republican Party for Independence and Development or PRID [Aristides GOMES]; Union of Guinean Patriots or UPG [Francisca VAZ]; Union for Change or UM [Amine SAAD]; United Platform or UP (coalition formed by PCD, FDS, FLING, and RGB-MB); United Popular Alliance or APU; United Social Democratic Party or PUSD [Francisco FADUL]
Political pressure groups and leaders: NA
International organization participation: ACP, AfDB, AOSIS, AU, CPLP, ECOWAS, FAO, FZ, G-77, IBRD, ICAO, ICRM, IDA, IDB, IFAD, IFC, IFRCS, ILO, IMF, IMO, Interpol, IOC, IOM, IPU, ITSO, ITU, ITUC, MIGA, NAM, OIC, OIF, OPCW, UN, UNCTAD, UNESCO, UNIDO, Union Latina, UNWTO, UPU, WADB (regional), WAEMU, WCO, WFTU, WHO, WIPO, WMO, WTO
Diplomatic representation in the US: *chief of mission:* none; note–Guinea-Bissau does not have official representation in Washington, DC
Diplomatic representation from the US: the US Embassy suspended operations on 14 June 1998 in the midst of violent conflict between forces loyal to then President VIEIRA and military-led junta; the US Ambassador to Senegal is accredited to Guinea-Bissau
Flag description: two equal horizontal bands of yellow (top) and green with a vertical red band on the hoist side; there is a black five-pointed star centered in the red band; yellow symbolizes the sun; green denotes hope; red represents blood shed during the struggle for independence; the black star stands for African unity
note: uses the popular Pan-African colors of Ethiopia; the flag design was heavily influenced by the Ghanaian flag
National anthem: *name:* "Esta e a Nossa Patria Bem Amada" (This Is Our Beloved Country)
lyrics/music: Amilcar Lopes CABRAL/XIAO He
note: adopted 1974; a delegation from Portuguese Guinea visited China in 1963 and heard music by XIAO He; Amilcar Lopes CABRA, the leader of Guinea-Bissau's independence movement, asked the composer to create a piece that would inspire his people to struggle for independence
Economy-Bissau
Economy—overview: One of the poorest countries in the world, Guinea-Bissau's legal economy depends mainly on farming and fishing, but trafficking narcotics is probably the most lucrative trade. Cashew crops have increased remarkably in recent years. Guinea-Bissau exports fish and seafood along with small amounts of peanuts, palm kernels, and timber. Rice is the major crop and staple food. However, intermittent fighting between Senegalese-backed government troops and a military junta destroyed much of the country's infrastructure and caused widespread damage to the economy

in 1998; the civil war led to a 28% drop in GDP that year, with partial recovery in 1999-2002. In December 2003, the World Bank, IMF, and UNDP were forced to step in to provide emergency budgetary support in the amount of $107 million for 2004, representing over 80% of the total national budget. The combination of limited economic prospects, a weak and faction-ridden government, and favorable geography have made this West African country a way station for drugs bound for Europe.
GDP (purchasing power parity): $1.784 billion (2010 est.)
country comparison to the world: 189
$1.724 billion (2009 est.)
$1.674 billion (2008 est.)
note: data are in 2010 US dollars
GDP (official exchange rate): $837 million (2010 est.)
GDP—real growth rate: 3.5% (2010 est.)
country comparison to the world: 112
3% (2009 est.)
3.2% (2008 est.)
GDP—per capita (PPP): $1,100 (2010 est.)
country comparison to the world: 210
$1,100 (2009 est.)
$1,100 (2008 est.)
note: data are in 2010 US dollars
GDP—composition by sector:
agriculture: 62%
industry: 12%
services: 26% (1999 est.)
Labor force: 632,700 (2007)
country comparison to the world: 152
Labor force—by occupation: *agriculture:* 82%
industry and *services:* 18% (2000 est.)
Unemployment rate: NA%
Population below poverty line: NA%
Household income or consumption by percentage share: *lowest 10%:* 2.9%
highest 10%: 28% (2002)
Budget: *revenues:* $NA
expenditures: $NA
Inflation rate (consumer prices): 3.8% (2007 est.)
country comparison to the world: 108
Central bank discount rate: 4.25% (31 December 2009)
country comparison to the world: 89
4.75% (31 December 2008)
Commercial bank prime lending rate: NA%
Stock of narrow money: $192.1 million (31 December 2009)
country comparison to the world: 175
$171.2 million (31 December 2008)
Stock of broad money: $209.3 million (31 December 2009 est.)
country comparison to the world: 182
$189.2 million (31 December 2008 est.)
Stock of domestic credit: $42.56 million (31 December 2009)
country comparison to the world: 183
$58.87 million (31 December 2008)
Market value of publicly traded shares: $NA
Agriculture—products: rice, corn, beans, cassava (tapioca), cashew nuts, peanuts, palm kernels, cotton; timber; fish
Industries: agricultural products processing, beer, soft drinks
Industrial production growth rate: 4.7% (2003 est.)
country comparison to the world: 75
Electricity—production: 65 million kWh (2007 est.)
country comparison to the world: 197
Electricity—consumption: 60.45 million kWh (2007 est.)
country comparison to the world: 196
Electricity—exports: 0 kWh (2008 est.)
Electricity—imports: 0 kWh (2008 est.)
Oil—production: 0 bbl/day (2009 est.)
country comparison to the world: 123
Oil—consumption: 3,000 bbl/day (2009 est.)
country comparison to the world: 179
Oil—exports: 0 bbl/day (2007 est.)
country comparison to the world: 196
Oil—imports: 2,545 bbl/day (2007 est.)
country comparison to the world: 171
Oil—proved reserves: 0 bbl (1 January 2010 est.)
country comparison to the world: 178
Natural gas—production: 0 cu m (2008 est.)
country comparison to the world: 174
Natural gas—consumption: 0 cu m (2008 est.)
country comparison to the world: 116
Natural gas—exports: 0 cu m (2008 est.)
country comparison to the world: 160
Natural gas—imports: 0 cu m (2008 est.)
country comparison to the world: 108
Natural gas—proved reserves: 0 cu m (1 January 2010 est.)
country comparison to the world: 180
Current account balance: $-6 million (2007 est.)
country comparison to the world: 62
Exports: $133 million (2006)
country comparison to the world: 187
Exports—commodities: fish, shrimp; cashew nuts, peanuts, palm kernels, sawn lumber
Exports—partners: India 62.21%, Nigeria 31.28%, Portugal 1.48% (2009)
Imports: $200 million (2006)
country comparison to the world: 201
Imports—commodities: foodstuffs, machinery and transport equipment, petroleum products
Imports—partners: Portugal 17.33%, Senegal 13.66%, Netherlands 9.27%, India 9.11%, Thailand 5.2%, Brazil 4.49% (2009)
Debt—external: $941.5 million (2000 est.)
country comparison to the world: 152
Exchange rates: Communaute Financiere Africaine francs (XOF) per US dollar–495.28 (2010), 472.19 (2009), 493.51 (2007), 522.59 (2006)

COMMUNICATIONS

Telephones—main lines in use: 4,800 (2009)
country comparison to the world: 213
Telephones—mobile cellular: 560,300 (2009)
country comparison to the world: 158
domestic: fixed-line teledensity less than 1 per 100 persons; mobile-cellular teledensity reached 35 per 100 in 2009
international: country code–245 (2008)
Broadcast media: 1 state-owned TV station and a second station, RTP Africa, is operated by Portuguese public broadcaster Radio e Televisao de Portugal (RTP); 1 state-owned radio station, several private radio stations, and some community radio stations; multiple international broadcasters are available (2007)
Internet country code: .gw
Internet hosts: 82 (2010)
country comparison to the world: 206
Internet users: 37,100 (2009)
country comparison to the world: 176

TRANSPORTATION

Airports: 9 (2010)
country comparison to the world: 158
Airports—with paved runways:
total: 2
over 3,047 m: 1
1,524 to 2,437 m: 1 (2010)
Airports—with unpaved runways:
total: 7
1,524 to 2,437 m: 1
914 to 1,523 m: 3
under 914 m: 3 (2010)
Roadways:
total: 3,455 km
country comparison to the world: 161
paved: 965 km
unpaved: 2,490 km (2002)
Waterways: (rivers are navigable for some distance; many inlets and creeks give shallow-water access to much of interior) (2009)
Ports and terminals: Bissau, Buba, Cacheu, Farim

MILITARY

Military branches: People's Revolutionary Armed Force (FARP): Army, Navy, National Air Force (Forca Aerea Nacional); Presidential Guard (2011)
Military service age and obligation: 18-25 years of age for selective compulsory military service (Air Force service is voluntary); 16 years of age or younger with parental consent, for voluntary service (2010)
Manpower available for military service:
males age 16–49: 370,790
females age 16–49: 372,171 (2010 est.)
Manpower fit for military service:
males age 16–49: 205,460
females age 16–49: 212,277 (2010 est.)
Manpower reaching militarily significant age annually: *male:* 17,639
female: 17,865 (2010 est.)
Military expenditures: 3.1% of GDP (2005 est.)
country comparison to the world: 41

TRANSNATIONAL ISSUES

Disputes—international: in 2006, political instability within Senegal's Casamance region resulted in thousands of Senegalese refugees, cross-border raids, and arms smuggling into Guinea-Bissau
Refugees and internally displaced persons: refugees (country of origin): 7,454 (Senegal) (2007)
Trafficking in persons: current situation: Guinea-Bissau is a source country for children trafficked primarily for forced begging

and forced agricultural labor to other West African countries
tier rating: Tier 2 Watch List—for the second year in a row, Guinea-Bissau is on the Tier 2 Watch List for its failure to combat severe forms of trafficking in persons, as evidenced by the continued failure to pass an anti-trafficking law and inadequate efforts to investigate or prosecute trafficking crimes or convict and punish trafficking offenders (2008)
Illicit drugs: increasingly important transit country for South American cocaine enroute to Europe; enabling environment for trafficker operations thanks to pervasive corruption; archipelago-like geography around the capital facilitates drug smuggling

GUYANA

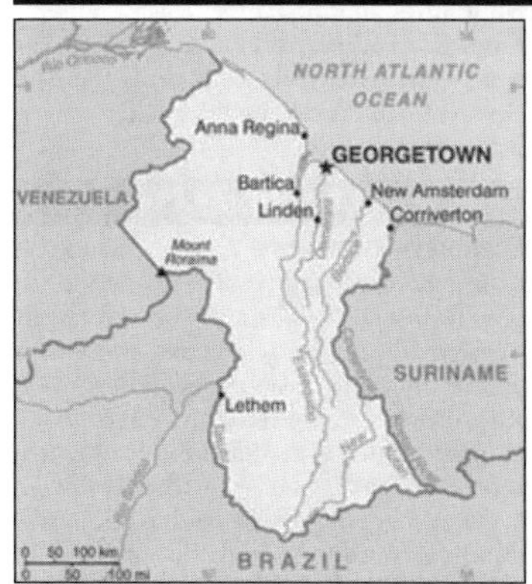

INTRODUCTION

Background: Originally a Dutch colony in the 17th century, by 1815 Guyana had become a British possession. The abolition of slavery led to black settlement of urban areas and the importation of indentured servants from India to work the sugar plantations. This ethnocultural divide has persisted and has led to turbulent politics. Guyana achieved independence from the UK in 1966, and since then it has been ruled mostly by socialist-oriented governments. In 1992, Cheddi JAGAN was elected president in what is considered the country's first free and fair election since independence. After his death five years later, his wife, Janet JAGAN, became president but resigned in 1999 due to poor health. Her successor, Bharrat JAGDEO, was reelected in 2001 and again in 2006.

GEOGRAPHY

Location: Northern South America, bordering the North Atlantic Ocean, between Suriname and Venezuela
Geographic coordinates: 5 00 N, 59 00 W
Map references: South America
Area: *total:* 214,969 sq km
country comparison to the world: 84
land: 196,849 sq km
water: 18,120 sq km
Area—comparative: slightly smaller than Idaho
Land boundaries: *total:* 2,949 km
border countries: Brazil 1,606 km, Suriname 600 km, Venezuela 743 km
Coastline: 459 km
Maritime claims:
territorial sea: 12 nm
exclusive economic zone: 200 nm
continental shelf: 200 nm or to the outer edge of the continental margin
Climate: tropical; hot, humid, moderated by northeast trade winds; two rainy seasons (May to August, November to January)
Terrain: mostly rolling highlands; low coastal plain; savanna in south
Elevation extremes: *lowest point:* Atlantic Ocean 0 m
highest point: Mount Roraima 2,835 m
Natural resources: bauxite, gold, diamonds, hardwood timber, shrimp, fish
Land use: *arable land:* 2.23%
permanent crops: 0.14%
other: 97.63% (2005)
Irrigated land: 1,500 sq km (2003)
Total renewable water resources: 241 cu km (2000)
Freshwater withdrawal (domestic/industrial/agricultural): *total:* 1.64 cu km/yr (2%/1%/98%)
per capita: 2,187 cu m/yr (2000)
Natural hazards: flash floods are a constant threat during rainy seasons
Environment—current issues: water pollution from sewage and agricultural and industrial chemicals; deforestation
Environment—international agreements: *party to:* Biodiversity, Climate Change, Climate Change-Kyoto Protocol, Desertification, Endangered Species, Hazardous Wastes, Law of the Sea, Ozone Layer Protection, Ship Pollution, Tropical Timber 83, Tropical Timber 94
signed, but not ratified: none of the selected agreements
Geography—note: the third-smallest country in South America after Suriname and Uruguay; substantial portions of its western and eastern territories are claimed by Venezuela and Suriname respectively

PEOPLE

Population: 744,768 (July 2011 est.)
country comparison to the world: 163
note: estimates for this country explicitly take into account the effects of excess mortality due to AIDS; this can result in lower life expectancy, higher infant mortality, higher death rates, lower population growth rates, and changes in the distribution of population by age and sex than would otherwise be expected
Age structure: *0–14 years:* 31.9% (male 120,981/female 116,654)
15–64 years: 63.3% (male 235,566/female 235,717)
65 years and over: 4.8% (male 14,801/female 21,049) (2011 est.)
Median age: *total:* 23.9 years
male: 23.2 years
female: 24.6 years (2011 est.)
Population growth rate: -0.44% (2011 est.)
country comparison to the world: 221
Birth rate: 17.12 births/1,000 population (2011 est.)
country comparison to the world: 118
Death rate: 7.2 deaths/1,000 population (July 2011 est.)
country comparison to the world: 125
Net migration rate: -14.32 migrant(s)/1,000 population (2011 est.)
country comparison to the world: 216
Urbanization: *urban population:* 29% of total population (2010)
rate of urbanization: 0.5% annual rate of change (2010-15 est.)
Major cities—population: GEORGETOWN (capital) 132,000 (2009)
Sex ratio: *at birth:* 1.05 male(s)/female
under 15 years: 1.04 male(s)/female
15–64 years: 1 male(s)/female
65 years and over: 0.71 male(s)/female
total population: 1 male(s)/female (2011 est.)
Infant mortality rate:
total: 36.76 deaths/1,000 live births
country comparison to the world: 68
male: 41.09 deaths/1,000 live births
female: 32.2 deaths/1,000 live births (2011 est.)
Life expectancy at birth: *total population:* 67.08 years
country comparison to the world: 159
male: 63.27 years
female: 71.07 years (2011 est.)
Total fertility rate: 2.34 children born/woman (2011 est.)
country comparison to the world: 96
HIV/AIDS—adult prevalence rate: 1.2% (2009 est.)
country comparison to the world: 41
HIV/AIDS—people living with HIV/AIDS: 5,900 (2009 est.)
country comparison to the world: 118
HIV/AIDS—deaths: fewer than 500 (2009 est.)
country comparison to the world: 91
Major infectious diseases: *degree of risk:* high
food or waterborne diseases: bacterial and protozoal diarrhea, hepatitis A, and typhoid fever
vectorborne diseases: dengue fever and malaria
water contact disease: leptospirosis (2009)
Drinking water source: *Improved:*
urban: 98% of population
rural: 93% of population
total: 94% of population
Unimproved: urban: 2% of population
rural: 7% of population
total: 6% of population (2008)
Sanitation facility access: *Improved:*
urban: 85% of population
rural: 80% of population
total: 81% of population
Unimproved: urban: 15% of population
rural: 20% of population
total: 19% of population (2008)
Nationality: *noun:* Guyanese (singular and plural)

adjective: Guyanese
Ethnic groups: East Indian 43.5%, black (African) 30.2%, mixed 16.7%, Amerindian 9.1%, other 0.5% (2002 census)
Religions: Hindu 28.4%, Pentecostal 16.9%, Roman Catholic 8.1%, Anglican 6.9%, Seventh Day Adventist 5%, Methodist 1.7%, Jehovah Witness 1.1%, other Christian 17.7%, Muslim 7.2%, other 4.3%, none 4.3% (2002 census)
Languages: English, Amerindian dialects, Creole, Caribbean Hindustani (a dialect of Hindi), Urdu
Literacy: *definition:* age 15 and over has ever attended school
total population: 91.8%
male: 92%
female: 91.6% (2002 Census)
School life expectancy (primary to tertiary education): *total:* 12 years
male: 12 years
female: 12 years (2009)
Education expenditures: 6.1% of GDP (2007)
country comparison to the world: 28

GOVERNMENT

Country name: *conventional long form:* Cooperative Republic of Guyana
conventional short form: Guyana
former: British Guiana
Government type: republic
Capital: *name:* Georgetown
geographic coordinates: 6 48 N, 58 10 W
time difference: UTC-4 (1 hour ahead of Washington, DC during Standard Time)
Administrative divisions: 10 regions; Barima-Waini, Cuyuni-Mazaruni, Demerara-Mahaica, East Berbice-Corentyne, Essequibo Islands-West Demerara, Mahaica-Berbice, Pomeroon-Supenaam, Potaro-Siparuni, Upper Demerara-Berbice, Upper Takutu-Upper Essequibo
Independence: 26 May 1966 (from the UK)
National holiday: Republic Day, 23 February (1970)
Constitution: 6 October 1980
Legal system: common law system, based on the English model, with some Roman-Dutch civil law influence
International law organization participation: has not submitted an ICJ jurisdiction declaration; accepts ICCt jurisdiction
Suffrage: 18 years of age; universal
Executive branch: *chief of state:* President Bharrat JAGDEO (since 11 August 1999); note—assumed presidency after resignation of President Janet JAGAN and was reelected in 2001, and again in 2006
head of government: Prime Minister Samuel HINDS (since October 1992, except for a period as chief of state after the death of President Cheddi JAGAN on 6 March 1997)
cabinet: Cabinet of Ministers appointed by the president, responsible to the legislature
(For more information visit the World Leaders website)
elections: president elected by popular vote as leader of a party list in parliamentary elections, which must be held at least every five years (no term limits); elections last held on 28 August 2006 (next to be held by August 2011); prime minister appointed by the president
election results: President Bharrat JAGDEO reelected; percent of vote 54.6%
Legislative branch: unicameral National Assembly (65 seats; members elected by popular vote, also not more than 4 non-elected non-voting ministers and 2 non-elected non-voting parliamentary secretaries appointed by the president; members to serve five-year terms)
elections: last held on 28 August 2006 (next to be held by August 2011)
election results: percent of vote by party—PPP/C 54.6%, PNC/R 34%, AFC 8.1%, other 3.3%; seats by party—PPP/C 36, PNC/R 22, AFC 5, other 2
Judicial branch: Supreme Court of Judicature, consisting of the High Court and the Court of Appeal, with right of final appeal to the Caribbean Court of Justice (CCJ)
Political parties and leaders: Alliance for Change or AFC [Raphael TROTMAN and Khemraj RAMJATTAN]; Guyana Action Party or GAP [Paul HARDY]; Justice for All Party [C.N. SHARMA]; People's National Congress/Reform or PNC/R [Robert Herman Orlando CORBIN]; People's Progressive Party/Civic or PPP/C [Bharrat JAGDEO]; Rise, Organize, and Rebuild or ROAR [Ravi DEV]; The United Force or TUF [Manzoor NADIR]; The Unity Party [Joey JAGAN]; Vision Guyana [Peter RAMSAROOP]; Working People's Alliance or WPA [Rupert ROOPNARAINE]
Political pressure groups and leaders: Amerindian People's Association; Guyana Bar Association; Guyana Citizens Initiative; Guyana Human Rights Association; Guyana Public Service Union or GPSU; Private Sector Commission; Trades Union Congress
International organization participation: ACP, AOSIS, C, Caricom, CDB, FAO, G-77, IADB, IBRD, ICAO, ICRM, IDA, IFAD, IFC, IFRCS, ILO, IMF, IMO, Interpol, IOC, IOM (observer), ISO (subscriber), ITU, ITUC, LAES, MIGA, NAM, OAS, OIC, OPANAL, OPCW, PCA, PetroCaribe, RG, UN, UNASUR, UNCTAD, UNESCO, UNIDO, UPU, WCO, WFTU, WHO, WIPO, WMO, WTO
Diplomatic representation in the US: *chief of mission:* Ambassador Bayney KARRAN
chancery: 2490 Tracy Place NW, Washington, DC 20008
telephone: [1] (202) 265-6900
FAX: [1] (202) 232-1297
consulate(s) general: New York
Diplomatic representation from the US: *chief of mission:* Ambassador (vacant); Charge d'Affaires Karen L. WILLIAMS
embassy: US Embassy, 100 Young and Duke Streets, Kingston, Georgetown
mailing address: P. O. Box 10507, Georgetown; US Embassy, 3170 Georgetown Place, Washington DC 20521-3170
telephone: [592] 225-4900 through 4909
FAX: [592] 225-8497
Flag description: green, with a red isosceles triangle (based on the hoist side) superimposed on a long, yellow arrowhead; there is a narrow, black border between the red and yellow, and a narrow, white border between the yellow and the green; green represents forest and foliage; yellow stands for mineral resources and a bright future; white symbolizes Guyana's rivers; red signifies zeal and the sacrifice of the people; black indicates perseverance
National anthem: *name:* "Dear Land of Guyana, of Rivers and Plains"
lyrics/music: Archibald Leonard LUKERL/ Robert Cyril Gladstone POTTER
note: adopted 1966

ECONOMY

Economy—overview: The Guyanese economy exhibited moderate economic growth in recent years and is based largely on agriculture and extractive industries. The economy is heavily dependent upon the export of six commodities—sugar, gold, bauxite, shrimp, timber, and rice—which represent nearly 60% of the country's GDP and are highly susceptible to adverse weather conditions and fluctuations in commodity prices. Guyana's entrance into the Caricom Single Market and Economy (CSME) in January 2006 has broadened the country's export market, primarily in the raw materials sector. Economic recovery since a 2005 flood-related contraction was buoyed by increases in remittances and foreign direct investment in the sugar and rice industries as well as the mining sector. Chronic problems include a shortage of skilled labor and a deficient infrastructure. The government is juggling a sizable external debt against the urgent need for expanded public investment. In March 2007, the Inter-American Development Bank, Guyana's principal donor, canceled Guyana's nearly $470 million debt, equivalent to nearly 48% of GDP, which along with other Highly Indebted Poor Country (HIPC) debt forgiveness brought the debt-to-GDP ratio down from 183% in 2006 to 120% in 2007. Guyana became heavily indebted as a result of the inward-looking, state-led development model pursued in the 1970s and 1980s. Growth slowed in 2009-10 as a result of the world recession. The slowdown in the domestic economy and lower import costs helped to narrow the country's current account deficit, despite generally lower earnings from exports.
GDP (purchasing power parity): $5.379 billion (2010 est.)
country comparison to the world: 160
$5.19 billion (2009 est.)
$5.024 billion (2008 est.)
note: data are in 2010 US dollars
GDP (official exchange rate): $2.215 billion (2010 est.)
GDP—real growth rate: 3.6% (2010 est.)
country comparison to the world: 108
3.3% (2009 est.)
2% (2008 est.)
GDP—per capita (PPP): $7,200 (2010 est.)
country comparison to the world: 130
$6,900 (2009 est.)
$6,600 (2008 est.)

note: data are in 2010 US dollars
GDP—composition by sector: *agriculture:* 24.3%
industry: 24.7%
services: 51% (2010 est.)
Labor force: 333,900 (2007 est.)
country comparison to the world: 162
Labor force—by occupation: *agriculture:* NA%
industry: NA%
services: NA%
Unemployment rate: 11% (2007)
country comparison to the world: 120
Population below poverty line: NA%
Household income or consumption by percentage share: *lowest 10%:* 1.3%
highest 10%: 33.8% (1999)
Distribution of family income—Gini index: 43.2 (1999)
country comparison to the world: 45
Investment (gross fixed): 34.1% of GDP (2010 est.)
country comparison to the world: 11
Budget: *revenues:* $619.5 million
expenditures: $655.7 million (2010 est.)
Public debt: 57% of GDP (2010 est.)
country comparison to the world: 41
Inflation rate (consumer prices): 6.8% (2010 est.)
country comparison to the world: 170
2.9% (2009 est.)
Central bank discount rate: 6.75% (31 December 2009)
country comparison to the world: 54
6.75% (31 December 2008)
Commercial bank prime lending rate: 14.54% (31 December 2009 est.)
country comparison to the world: 47
14.58% (31 December 2008 est.)
Stock of narrow money: $386.9 million (31 December 2010 est.)
country comparison to the world: 162
$252.9 million (31 December 2009 est.)
Stock of broad money: $1.303 billion (31 December 2010 est.)
country comparison to the world: 153
$905.6 million (31 December 2009 est.)
Stock of domestic credit: $754 million (31 December 2010 est.)
country comparison to the world: 157
$524 million (31 December 2009 est.)
Market value of publicly traded shares: $NA (31 December 2009)
country comparison to the world: 111
$289.9 million (31 December 2008)
$262.4 million (31 December 2007)
Agriculture—products: sugarcane, rice, edible oils; shrimp, fish, beef, pork, poultry
Industries: bauxite, sugar, rice milling, timber, textiles, gold mining
Industrial production growth rate: 2.5% (2010 est.)
country comparison to the world: 123
Electricity—production: 821 million kWh (2007 est.)
country comparison to the world: 150
Electricity—consumption: 667 million kWh (2007 est.)
country comparison to the world: 151
Electricity—exports: 0 kWh (2008 est.)
Electricity—imports: 0 kWh (2008 est.)
Oil—production: 0 bbl/day (2009 est.)
country comparison to the world: 182
Oil—consumption: 10,000 bbl/day (2009 est.)
country comparison to the world: 148
Oil—exports: 0 bbl/day (2007 est.)
country comparison to the world: 171
Oil—imports: 10,550 bbl/day (2007 est.)
country comparison to the world: 139
Oil—proved reserves: 0 bbl (1 January 2010 est.)
country comparison to the world: 139
Natural gas—production: 0 cu m (2008 est.)
country comparison to the world: 136
Natural gas—consumption: 0 cu m (2008 est.)
country comparison to the world: 181
Natural gas—exports: 0 cu m (2008 est.)
country comparison to the world: 108
Natural gas—imports: 0 cu m (2008 est.)
country comparison to the world: 196
Natural gas—proved reserves: 0 cu m (1 January 2010 est.)
country comparison to the world: 143
Current account balance: $-311 million (2010 est.)
country comparison to the world: 98
$-265 million (2009 est.)
Exports: $814 million (2010 est.)
country comparison to the world: 159
$763 million (2009 est.)
Exports—commodities: sugar, gold, bauxite, alumina, rice, shrimp, molasses, rum, timber
Exports—partners: Canada 27.52%, US 16.93%, UK 10.84%, Ukraine 5.54%, Netherlands 5%, Trinidad and Tobago 4.33%, Jamaica 4.12% (2009)
Imports:
$1.366 billion (2010 est.)
country comparison to the world: 165
$1.161 billion (2009 est.)
Imports—commodities: manufactures, machinery, petroleum, food
Imports—partners: US 25.23%, Trinidad and Tobago 23.23%, Cuba 6.41%, China 6.05% (2009)
Reserves of foreign exchange and gold: $506 million (31 December 2010 est.)
country comparison to the world: 125
$631.4 million (31 December 2009 est.)
Debt—external: $804.3 million (30 September 2008)
country comparison to the world: 155
$1.2 billion (2002)
Exchange rates: Guyanese dollars (GYD) per US dollar–204.07 (2010), 203.95 (2009), 203.86 (2008), 201.89 (2007), 200.28 (2006)

COMMUNICATIONS

Telephones—main lines in use: 130,000 (2009)
country comparison to the world: 139
Telephones—mobile cellular: 281,400 (2005)
country comparison to the world: 170
Telephone system: *general assessment:* fair system for long-distance service; microwave radio relay network for trunk lines; many areas still lack fixed-line telephone services
domestic: fixed-line teledensity is about 15 per 100 persons; mobile-cellular teledensity about 35 per 100 persons in 2005
international: country code–592; tropospheric scatter to Trinidad; satellite earth station–1 Intelsat (Atlantic Ocean)
Broadcast media: government-dominated broadcast media; the National Communications Network (NCN) TV is state-owned; a few private TV stations relay satellite services; the state owns and operates 2 radio stations broadcasting on multiple frequencies capable of reaching the entire country; government limits on licensing of new private radio stations continue to constrain competition in broadcast media (2007)
Internet country code: .gy
Internet hosts: 8,840 (2010)
country comparison to the world: 132
Internet users: 189,600 (2009)
country comparison to the world: 142

TRANSPORTATION

Airports: 96 (2010)
country comparison to the world: 63
Airports—with paved runways: *total:* 10
1,524 to 2,437 m: 2
914 to 1,523 m: 1
under 914 m: 7 (2010)
Airports—with unpaved runways: *total:* 86
914 to 1,523 m: 13
under 914 m: 73 (2010)
Roadways: *total:* 7,970 km
country comparison to the world: 142
paved: 590 km
unpaved: 7,380 km (2000)
Waterways: 330 km (the Berbice, Demerara, and Essequibo rivers are navigable by oceangoing vessels for 150 km, 100 km, and 80 km respectively) (2010)
country comparison to the world: 92
Merchant marine: *total:* 8
country comparison to the world: 121
by type: cargo 6, petroleum tanker 1, refrigerated cargo 1
registered in other countries: 3 (Saint Vincent and the Grenadines 2, unknown 1) (2010)
Ports and terminals: Georgetown

MILITARY

Military branches: Guyana Defense Force: Army (includes Coast Guard, Air Corps) (2009)
Military service age and obligation: 18-25 years of age for voluntary military service; no conscription (2008)
Manpower available for military service: *males age 16–49:* 189,840 (2010 est.)
Manpower fit for military service: *males age 16–49:* 133,239
females age 16–49: 147,719 (2010 est.)
Manpower reaching militarily significant age annually: *male:* 8,849
female: 8,460 (2010 est.)
Military expenditures: 1.8% of GDP (2006)
country comparison to the world: 83

TRANSNATIONAL ISSUES

Disputes—international: all of the area west of the Essequibo River is claimed by Venezuela preventing any discussion of a maritime boundary; Guyana has expressed its intention to join Barbados in asserting claims before UNCLOS that Trinidad and Tobago's maritime boundary with Venezuela extends into their waters; Suriname claims a triangle of land between the New and Kutari/Koetari rivers in a historic dispute over the headwaters of the Courantyne; Guyana seeks arbitration under provisions of the UN Convention on the Law of the Sea (UNCLOS) to resolve the long-standing dispute with Suriname over the axis of the territorial sea boundary in potentially oil-rich waters

Trafficking in persons: current situation: Guyana is a source, transit, and destination country for men, women, and children trafficked for the purposes of commercial sexual exploitation and forced labor; most trafficking appears to take place in remote mining camps in the country's interior; some women and girls are trafficked from northern Brazil; reporting from other nations suggests Guyanese women and girls are trafficked for sexual exploitation to neighboring countries and Guyanese men and boys are subject to labor exploitation in construction and agriculture; trafficking victims from Suriname, Brazil, and Venezuela transit Guyana en route to Caribbean destinations

tier rating: Tier 2 Watch List—for a second consecutive year, Guyana is on the Tier 2 Watch List for failing to provide evidence of increasing efforts to combat trafficking, particularly in the area of law enforcement actions against trafficking offenders; the government has yet to produce an anti-trafficking conviction under the comprehensive Combating of Trafficking in Persons Act, which became law in 2005; the government operates no shelters for trafficking victims, but did include limited funding for anti-trafficking NGOs in its 2008 budget; the government did not make any effort to reduce demand for commercial sex acts during 2007 (2008)

Illicit drugs: transshipment point for narcotics from South America—primarily Venezuela—to Europe and the US; producer of cannabis; rising money laundering related to drug trafficking and human smuggling

HAITI

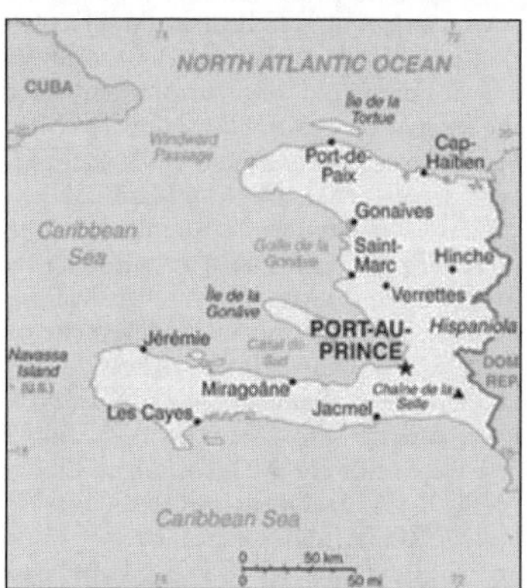

INTRODUCTION

Background: The native Taino Amerindians–who inhabited the island of Hispaniola when it was discovered by COLUMBUS in 1492–were virtually annihilated by Spanish settlers within 25 years. In the early 17th century, the French established a presence on Hispaniola. In 1697, Spain ceded to the French the western third of the island, which later became Haiti. The French colony, based on forestry and sugar-related industries, became one of the wealthiest in the Caribbean but only through the heavy importation of African slaves and considerable environmental degradation. In the late 18th century, Haiti's nearly half million slaves revolted under Toussaint L'OUVERTURE. After a prolonged struggle, Haiti became the first black republic to declare independence in 1804. The poorest country in the Western Hemisphere, Haiti has been plagued by political violence for most of its history. After an armed rebellion led to the forced resignation and exile of President Jean-Bertrand ARISTIDE in February 2004, an interim government took office to organize new elections under the auspices of the United Nations Stabilization Mission in Haiti (MINUSTAH). Continued violence and technical delays prompted repeated postponements, but Haiti finally did inaugurate a democratically elected president and parliament in May of 2006. A massive magnitude 7.0 earthquake struck Haiti in January 2010 with an epicenter about 15 km southwest of the capital, Port-au-Prince. An estimated 2 million people lived within the zone of heavy to moderate structural damage. The earthquake was assessed as the worst in this region over the last 200 years and massive international assistance will be required to help the country recover.

GEOGRAPHY

Location: Caribbean, western one-third of the island of Hispaniola, between the Caribbean Sea and the North Atlantic Ocean, west of the Dominican Republic

Geographic coordinates: 19 00 N, 72 25 W

Map references: Central America and the Caribbean

Area: *total:* 27,750 sq km
country comparison to the world: 147
land: 27,560 sq km
water: 190 sq km

Area—comparative: slightly smaller than Maryland

Land boundaries: *total:* 360 km
border countries: Dominican Republic 360 km

Coastline: 1,771 km

Maritime claims: *territorial sea:* 12 nm
contiguous zone: 24 nm
exclusive economic zone: 200 nm
continental shelf: to depth of exploitation

Climate: tropical; semiarid where mountains in east cut off trade winds

Terrain: mostly rough and mountainous

Elevation extremes: *lowest point:* Caribbean Sea 0 m
highest point: Chaine de la Selle 2,680 m

Natural resources: bauxite, copper, calcium carbonate, gold, marble, hydropower

Land use: *arable land:* 28.11%
permanent crops: 11.53%
other: 60.36% (2005)

Irrigated land: 920 sq km (2003)

Total renewable water resources: 14 cu km (2000)

Freshwater withdrawal (domestic/industrial/agricultural): *total:* 0.99 cu km/yr (5%/1%/94%)
per capita: 116 cu m/yr (2000)

Natural hazards: lies in the middle of the hurricane belt and subject to severe storms from June to October; occasional flooding and earthquakes; periodic droughts

Environment—current issues: extensive deforestation (much of the remaining forested land is being cleared for agriculture and used as fuel); soil erosion; inadequate supplies of potable water

Environment—international agreements: *party to:* Biodiversity, Climate Change, Climate Change-Kyoto Protocol, Desertification, Law of the Sea, Marine Dumping, Marine Life Conservation, Ozone Layer Protection
signed, but not ratified: Hazardous Wastes

Geography—note: shares island of Hispaniola with Dominican Republic (western one-third is Haiti, eastern two-thirds is the Dominican Republic)

PEOPLE

Population: 9,719,932 (July 2011 est.)
country comparison to the world: 87
note: estimates for this country explicitly take into account the effects of excess mortality due to AIDS; this can result in lower life expectancy, higher infant mortality, higher death rates, lower population growth rates, and changes in the distribution of population by age and sex than would otherwise be expected

Age structure: *0–14 years:* 35.9% (male 1,748,677/female 1,742,199)
15–64 years: 60.1% (male 2,898,251/female 2,947,272)
65 years and over: 3.9% (male 170,584/female 212,949) (2011 est.)

Median age: *total:* 21.4 years
male: 21.1 years
female: 21.6 years (2011 est.)

Population growth rate: 0.787%
country comparison to the world: 137
note: the preliminary 2011 numbers differ significantly from those of 2010, which were strongly influenced by the demographic effect of the January 2010 earthquake; the latest figures more closely correspond to those of 2009 (2011 est.)

Birth rate: 24.4 births/1,000 population (2011 est.)
country comparison to the world: 66

Death rate: 8.21 deaths/1,000 population
country comparison to the world: 94
note: the preliminary 2011 numbers differ significantly from those of 2010, which were strongly influenced by the demographic effect of the January 2010 earthquake; the latest figures more closely correspond to those of 2009 (July 2011 est.)

Net migration rate: -8.32 migrant(s)/1,000 population (2011 est.)
country comparison to the world: 205

Urbanization: *urban population:* 52% of total population (2010)
rate of urbanization: 3.9% annual rate of change (2010-15 est.)

Major cities—population: PORT-AU-PRINCE (capital) 2.143 million (2010)

Sex ratio: *at birth:* 1.011 male(s)/female
under 15 years: 1.02 male(s)/female
15–64 years: 0.99 male(s)/female
65 years and over: 0.62 male(s)/female
total population: 0.98 male(s)/female (2011 est.)

Infant mortality rate:
total: 54.02 deaths/1,000 live births
country comparison to the world: 41
male: 58.16 deaths/1,000 live births
female: 49.83 deaths/1,000 live births
note: the preliminary 2011 numbers differ significantly from those of 2010, which were strongly influenced by the demographic effect of the January 2010 earthquake; the latest figures more closely correspond to those of 2009 (2011 est.)

Life expectancy at birth: *total population:* 62.17 years
country comparison to the world: 182
male: 60.84 years
female: 63.53 years
note: the preliminary 2011 numbers differ significantly from those of 2010, which were strongly influenced by the demographic effect of the January 2010 earthquake; the latest figures more closely correspond to those of 2009 (2011 est.)

Total fertility rate: 3.07 children born/woman (2011 est.)
country comparison to the world: 62

HIV/AIDS—adult prevalence rate: 1.9% (2009 est.)
country comparison to the world: 30

HIV/AIDS—people living with HIV/AIDS: 120,000 (2009 est.)
country comparison to the world: 38

HIV/AIDS—deaths: 7,100 (2009 est.)
country comparison to the world: 30

Major infectious diseases: *degree of risk:* high

food or waterborne diseases: bacterial and protozoal diarrhea, hepatitis A and E, and typhoid fever
vectorborne diseases: dengue fever and malaria
water contact disease: leptospirosis (2009)
Drinking water source: *Improved:*
urban: 71% of population
rural: 55% of population
total: 63% of population
Unimproved: urban: 29% of population
rural: 45% of population
total: 37% of population (2008)
Sanitation facility access: *Improved:*
urban: 24% of population
rural: 10% of population
total: 17% of population
Unimproved: urban: 76% of population
rural: 90% of population
total: 83% of population (2008)
Nationality: *noun:* Haitian(s)
adjective: Haitian
Ethnic groups: black 95%, mulatto and white 5%
Religions: Roman Catholic 80%, Protestant 16% (Baptist 10%, Pentecostal 4%, Adventist 1%, other 1%), none 1%, other 3%
note: roughly half of the population practices voodoo
Languages: French (official), Creole (official)
Literacy: *definition:* age 15 and over can read and write
total population: 52.9%
male: 54.8%
female: 51.2% (2003 est.)
School life expectancy (primary to tertiary education): NA
Education expenditures: NA

GOVERNMENT

Country name: *conventional long form:* Republic of Haiti
conventional short form: Haiti
local long form: Republique d'Haiti/ Repiblik d' Ayiti
local short form: Haiti/Ayiti
Government type: republic
Capital: *name:* Port-au-Prince
geographic coordinates: 18 32 N, 72 20 W
time difference: UTC-5 (same time as Washington, DC during Standard Time)
daylight saving time: no DST planned for 2011
Administrative divisions: 10 departments (departements, singular–departement); Artibonite, Centre, Grand'Anse, Nippes, Nord, Nord-Est, Nord-Ouest, Ouest, Sud, Sud-Est
Independence: 1 January 1804 (from France)
National holiday: Independence Day, 1 January (1804)
Constitution: approved March 1987
note: suspended June 1988 with most articles reinstated March 1989; constitutional government ousted in a military coup in September 1991, although in October 1991 military government claimed to be observing the constitution; returned to constitutional rule in October 1994; constitution, while technically in force between 2004-2006, was not enforced; returned to constitutional rule in May 2006
Legal system: civil law system strongly influenced by Napoleonic Code
International law organization participation: accepts compulsory ICJ jurisdiction; non-party state to the ICCt
Suffrage: 18 years of age; universal
Executive branch: *chief of state:* President Michel MARTELLY (since 14 May 2011)
head of government: Prime Minister Jean-Max BELLERIVE (since 7 November 2009); note–submitted his resignation on 13 May 2011
cabinet: Cabinet chosen by the prime minister in consultation with the president
(For more information visit the World Leaders website)
elections: president elected by popular vote for a five-year term (may not serve consecutive terms); election last held on 28 November 2010; runoff scheduled for 16 January 2011 (next to be held in 2015); prime minister appointed by the president, ratified by the National Assembly
election results: Michel MARTELLY wins the runoff election held on 20 March 2011 with 67.6% of the vote against 31.7% for Mirlande MANIGAT
Legislative branch: bicameral National Assembly or Assemblee Nationale consists of the Senate (30 seats; members elected by popular vote to serve six-year terms; one-third elected every two years) and the Chamber of Deputies (99 seats; members elected by popular vote to serve four-year terms); note–in reestablishing the Senate in 2006, the candidate in each department receiving the most votes in the last election serves six years, the candidate with the second most votes serves four years, and the candidate with the third most votes serves two years
elections: Senate–last held on 28 November 2010 with run-off elections scheduled for 16 January 2011 (next regular election, for one third of seats, to be held in 2012); Chamber of Deputies–last held on 28 November 2010 with run-off elections schedule for 16 January 2011 (next regular election to be held in 2014)
election results: 2010 election results are not final; 2006 Senate–percent of vote by party–NA; seats by party–L'ESPWA 11, FUSION 5, OPL 4, FL 3, LAAA 2, UNCRH 2, PONT 2, ALYANS 1; 2006 Chamber of Deputies–percent of vote by party–NA; seats by party–L'ESPWA 23, FUSION 17, FRN 12, OPL 10, ALYANS 10, LAAA 5, MPH 3, MOCHRENA 3, other 10; results for six other seats contested on 3 December 2006 remain unknown
Judicial branch: Supreme Court or Cour de Cassation
Political parties and leaders: Assembly of Progressive National Democrats or RDNP [Mirlande MANIGAT]; Christian and Citizen For Haiti's Reconstruction or ACCRHA [Chavannes JEUNE]; Convention for Democratic Unity or KID [Evans PAUL]; Cooperative Action to Rebuild Haiti or KONBA [Jean William JEANTY]; December 16 Platform or Platfom 16 Desanm [Dr. Gerard BLOT]; Democratic Alliance or ALYANS [Evans PAUL] (coalition composed of KID and PPRH); Effort and Solidarity to Create an Alternative for the People or ESKAMP [Joseph JASME]; Fanmi Lavalas or FL [Maryse NARCISSE]; For Us All or PONT [Jean-Marie CHERESTAL]; Grouping of Citizens for Hope or RESPE [Charles-Henri BAKER]; Haiti in Action or AAA [Youri LATORTUE]; Haitian Youth Democratic Movement or MODEJHA [Jean Hector ANACACIS]; Haitians for Haiti [Yvon NEPTUNE]; Independent Movement for National Reconstruction or MIRN [Luc FLEURINORD]; Lavni Organization or LAVNI [Yves CRISTALIN]; Liberal Party of Haiti or PLH [Jean Andre VICTOR]; Love Haiti or Renmen Ayiti [Jean-Henry CEANT and Camille LEBLANC]; Merging of Haitian Social Democratics or FUSION [Victor BENOIT] (coalition of Ayiti Capable, Haitian National Revolutionary Party, and National Congress of Democratic Movements); Mobilization for National Development or MDN [Hubert de RONCERAY]; Mobilization for Progress in Haiti or MPH [Samir MOURRA]; National Coalition of Nonaligned Political Parties or CONACED [Osner FEVRY]; National Front for the Reconstruction of Haiti or FRN [Guy PHILIPPE]; New Christian Movement for a New Haiti or MOCHRENA [Luc MESADIEU]; Open the Gate Party or PLB [Anes LUBIN]; Peasant's Response or Repons Peyizan [Michel MARTELLY]; Platform Alternative for Progress and Democracy or ALTENATIV [Victor BENOIT and Evans PAUL]; Platform of Haitian Patriots or PLAPH [Dejean BELISAIRE and Himler REBU]; Popular Party for the Renewal of Haiti or PPRH [Claude ROMAIN]; Strength in Unity or Ansanm Nou Fo [Leslie VOLTAIRE]; Struggling People's Organization or OPL [Harry MARSAN]; Union [Chavannes JEUNE]; Union of Haitian Citizens for Democracy, Development, and Education or UCADDE [Jeantel JOSEPH]; Union of Nationalist and Progressive Haitians or UNPH [Edouard FRANCISQUE]; Unity or Inite [Rene PREVAL] (coalition that includes Front for Hope or L'ESPWA); Vigilance or Veye Yo [Lavarice GAUDIN]; Youth for People's Power or JPP [Rene CIVIL]
Political pressure groups and leaders: Autonomous Organizations of Haitian Workers or CATH [Fignole ST-CYR]; Confederation of Haitian Workers or CTH; Economic Forum of the Private Sector or EF [Reginald BOULOS]; Federation of Workers Trade Unions or FOS; General Organization of Independent Haitian Workers [Patrick NUMAS]; Grand-Anse Resistance Committee, or KOREGA; The Haitian Association of Industries or ADIH [Georges SASSINE]; National Popular Assembly or APN; Papaye Peasants Movement or MPP [Chavannes JEAN-BAPTISTE]; Popular Organizations Gathering Power or PROP; Protestant Federation of Haiti; Roman Catholic Church

International organization participation: ACP, AOSIS, Caricom, CDB, FAO, G-77, IADB, IAEA, IBRD, ICAO, ICRM, IDA, IFAD, IFC, IFRCS, ILO, IMF, IMO, Interpol, IOC, IOM, ITSO, ITU, ITUC, LAES, MIGA, NAM, OAS, OIF, OPANAL, OPCW, PCA, PetroCaribe, RG, UN, UNCTAD, UNESCO, UNIDO, Union Latina, UNWTO, UPU, WCO, WFTU, WHO, WIPO, WMO, WTO

Diplomatic representation in the US:
chief of mission: Ambassador Louis Harold JOSEPH
chancery: 2311 Massachusetts Avenue NW, Washington, DC 20008
telephone: [1] (202) 332-4090
FAX: [1] (202) 745-7215
consulate(s) general: Boston, Chicago, Miami, New York, San Juan (Puerto Rico)
consulate(s): Orlando (Florida)

Diplomatic representation from the US:
chief of mission: Ambassador Kenneth H. MERTEN
embassy: Tabarre 41, Route de Tabarre, Port-au-Prince
mailing address: use mailing address
telephone: [509] 229-8000
FAX: [509] 229-8028

Flag description: two equal horizontal bands of blue (top) and red with a centered white rectangle bearing the coat of arms, which contains a palm tree flanked by flags and two cannons above a scroll bearing the motto L'UNION FAIT LA FORCE (Union Makes Strength); the colors are taken from the French Tricolor and represent the union of blacks and mulattoes

National anthem: *name:* "La Dessalinienne" (The Dessalines Song)
lyrics/music: Justin LHERISSON/Nicolas GEFFRARD
note: adopted 1904; the anthem is named for Jean-Jacques DESSALINES, a leader in the Haitian Revolution and first ruler of an independent Haiti

ECONOMY

Economy—overview: Haiti is a free market economy that enjoys the advantages of low labor costs and tariff-free access to the US for many of its exports. Poverty, corruption, and poor access to education for much of the population are among Haiti's most serious disadvantages. Over the longer term, Haiti needs to create jobs for its young workforce and to build institutional capacity. Haiti's economy suffered a severe setback when a 7.0 magnitude earthquake destroyed much of its capital city, Port-au-Prince, and neighboring areas in January 2010. Already the poorest country in the Western Hemisphere with 80% of the population living under the poverty line and 54% in abject poverty, the damage to Port-au-Prince caused the country's GDP to contract an estimated 5.1% in 2010. Two-thirds of all Haitians depend on the agricultural sector, mainly small-scale subsistence farming, and remain vulnerable to damage from frequent natural disasters, exacerbated by the country's widespread deforestation. US economic engagement under the Haitian Hemispheric Opportunity through Partnership Encouragement (HOPE) Act, passed in December 2006, has boosted apparel exports and investment by providing duty-free access to the US. Congress voted in 2010 to extend the legislation until 2020 under the Haitian Economic Lift Act (HELP); the apparel sector accounts for three-quarters of Haitian exports and nearly one-tenth of GDP. Remittances are the primary source of foreign exchange, equaling nearly 20% of GDP and more than twice the earnings from exports. Haiti suffers from a lack of investment, partly because of limited infrastructure and a lack of security. In 2005, Haiti paid its arrears to the World Bank, paving the way for reengagement with the Bank. Haiti received debt forgiveness for over $1 billion through the Highly-Indebted Poor Country (HIPC) initiative in mis-2009. The remainder of its outstanding external debt was cancelled by donor countries in early 2010 but has since risen to about $400 million. The government relies on formal international economic assistance for fiscal sustainability, with over half of its annual budget coming from outside sources.

GDP (purchasing power parity): $11.48 billion (2010 est.)
country comparison to the world: 146
$12.09 billion (2009 est.)
$11.75 billion (2008 est.)
note: data are in 2010 US dollars

GDP (official exchange rate): $6.632 billion (2010 est.)

GDP—real growth rate: -5.1% (2010 est.)
country comparison to the world: 213
2.9% (2009 est.)
0.8% (2008 est.)

GDP—per capita (PPP): $1,200 (2010 est.)
country comparison to the world: 207
$1,200 (2009 est.)
$1,200 (2008 est.)
note: data are in 2010 US dollars

GDP—composition by sector:
agriculture: 25%
industry: 16%
services: 59% (2010 est.)

Labor force: 4.81 million
country comparison to the world: 77
note: shortage of skilled labor, unskilled labor abundant (2010 est.)

Labor force—by occupation: *agriculture:* 38.1%
industry: 11.5%
services: 50.4% (2010)

Unemployment rate: 40.6% (2010 est.)
country comparison to the world: 187
note: widespread unemployment and underemployment; more than two-thirds of the labor force do not have formal jobs

Population below poverty line: 80% (2003 est.)

Household income or consumption by percentage share: *lowest 10%:* 0.7%
highest 10%: 47.7% (2001)

Distribution of family income—Gini index: 59.2 (2001)
country comparison to the world: 7

Investment (gross fixed): 25% of GDP (2010 est.)
country comparison to the world: 44

Budget: *revenues:* $900 million
expenditures: $2.6 billion (2010 est.)

Inflation rate (consumer prices): 4.7% (2010 est.)
country comparison to the world: 134
-4.7% (2009 est.)

Commercial bank prime lending rate: 8% (31 December 2010 est.)
country comparison to the world: 116
8% (31 December 2009 est.)

Stock of narrow money: $970 million (31 December 2010 est.)
country comparison to the world: 139
$709 million (31 December 2009 est.)

Stock of broad money: $3.249 billion (31 December 2009)
country comparison to the world: 131
$2.49 billion (31 December 2008)

Stock of domestic credit: $960.1 million (31 December 2010 est.)
country comparison to the world: 151
$1.273 billion (31 December 2009 est.)

Market value of publicly traded shares: $NA

Agriculture—products: coffee, mangoes, sugarcane, rice, corn, sorghum; wood

Industries: textiles, sugar refining, flour milling, cement, light assembly based on imported parts

Industrial production growth rate: -4.8% (2010 est.)
country comparison to the world: 163

Electricity—production: 650 million kWh (2010 est.)
country comparison to the world: 154

Electricity—consumption: 309 million kWh (2010 est.)
country comparison to the world: 167

Electricity—exports: NA kWh (2010 est.)

Electricity—imports: 0 kWh (2010 est.)

Oil—production: 0 bbl/day (2009 est.)
country comparison to the world: 183

Oil—consumption: 12,000 bbl/day (2009 est.)
country comparison to the world: 142

Oil—exports: 0 bbl/day (2007 est.)
country comparison to the world: 172

Oil—imports: 12,280 bbl/day (2007 est.)
country comparison to the world: 136

Oil—proved reserves: 0 bbl (1 January 2010 est.)
country comparison to the world: 141

Natural gas—production: 0 cu m (2008 est.)
country comparison to the world: 137

Natural gas—consumption: 0 cu m (2008 est.)
country comparison to the world: 182

Natural gas—exports: 0 cu m (2008 est.)
country comparison to the world: 109

Natural gas—imports: 0 cu m (2008 est.)
country comparison to the world: 197

Natural gas—proved reserves: 0 cu m (1 January 2010 est.)
country comparison to the world: 144

Current account balance: $-781 million (2010 est.)
country comparison to the world: 125
$-627 million (2009 est.)

Exports: $530.2 million (2010 est.)
country comparison to the world: 164
$550 million (2009 est.)

Exports—commodities: apparel, manufactures, oils, cocoa, mangoes, coffee

Exports—partners: US 79.76%, Dominican Republic 7.24%, Canada 2.96% (2009)

Imports: $2.727 billion (2010 est.)

country comparison to the world: 143
$1.996 billion (2009 est.)
Imports—commodities: food, manufactured goods, machinery and transport equipment, fuels, raw materials
Imports—partners: US 33.11%, Dominican Republic 23.53%, Netherlands Antilles 10.75%, China 5.36% (2009)
Reserves of foreign exchange and gold: $1.587 billion (31 December 2010 est.)
country comparison to the world: 111
$615 million (31 December 2009 est.)
Debt—external: $350 million (31 December 2010 est.)
country comparison to the world: 166
$1.362 billion (31 December 2009 est.)
Exchange rates: gourdes (HTG) per US dollar–40.15 (2010), 42.02 (2009), 39.216 (2008), 37.138 (2007), 40.232 (2006)

COMMUNICATIONS

Telephones—main lines in use: 108,300 (2009)
country comparison to the world: 143
Telephones—mobile cellular: 3.648 million (2009)
country comparison to the world: 108
Telephone system: *general assessment:* telecommunications infrastructure is among the least developed in Latin America and the Caribbean; domestic facilities barely adequate; international facilities slightly better
domestic: mobile-cellular telephone services are expanding rapidly due, in part, to the introduction of low-cost GSM phones; mobile-cellular teledensity reached 40 per 100 persons in 2009
international: country code–509; satellite earth station–1 Intelsat (Atlantic Ocean)
Broadcast media: several television stations, including 1 government-owned; cable TV subscription service is available; government-owned radio network; more than 250 private and community radio stations operating with about 50 FM stations in Port-au-Prince alone (2007)
Internet country code: .ht
Internet hosts: 273 (2010)
country comparison to the world: 185
Internet users: 1 million (2009)
country comparison to the world: 99

TRANSPORTATION

Airports: 14 (2010)
country comparison to the world: 148
Airports—with paved runways: *total:* 4
2,438 to 3,047 m: 1
914 to 1,523 m: 3 (2010)
Airports—with unpaved runways: *total:* 10
914 to 1,523 m: 2
under 914 m: 8 (2010)
Roadways: *total:* 4,160 km
country comparison to the world: 155
paved: 1,011 km
unpaved: 3,149 km (2000)
Ports and terminals: Cap-Haitien, Gonaives, Jacmel, Port-au-Prince

MILITARY

Military branches: no regular military forces–small Coast Guard; the regular Haitian Armed Forces (FAdH)–Army, Navy, and Air Force–have been demobilized but still exist on paper until or unless they are constitutionally abolished (2009)
Manpower available for military service: *males age 16–49:* 2,398,804
females age 16–49: 2,415,039 (2010 est.)
Manpower fit for military service: *males age 16–49:* 1,666,324
females age 16–49: 1,704,364 (2010 est.)
Manpower reaching militarily significant age annually: *male:* 115,246
female: 115,282 (2010 est.)
Military expenditures: 0.4% of GDP (2006)
country comparison to the world: 167

TRANSNATIONAL ISSUES

Disputes—international: since 2004, about 8,000 peacekeepers from the UN Stabilization Mission in Haiti (MINUSTAH) maintain civil order in Haiti; despite efforts to control illegal migration, Haitians cross into the Dominican Republic and sail to neighboring countries; Haiti claims US-administered Navassa Island
Illicit drugs: Caribbean transshipment point for cocaine en route to the US and Europe; substantial bulk cash smuggling activity; Colombian narcotics traffickers favor Haiti for illicit financial transactions; pervasive corruption; significant consumer of cannabis

HEARD ISLAND AND MCDONALD ISLANDS

(TERRITORY OF AUSTRALIA)

INTRODUCTION

Background: These uninhabited, barren, sub-Antarctic islands were transferred from the UK to Australia in 1947. Populated by large numbers of seal and bird species, the islands have been designated a nature preserve.

GEOGRAPHY

Location: islands in the Indian Ocean, about two-thirds of the way from Madagascar to Antarctica
Geographic coordinates: 53 06 S, 72 31 E
Map references: Antarctic Region
Area: *total:* 412 sq km
country comparison to the world: 201
land: 412 sq km
water: 0 sq km
Area—comparative: slightly more than two times the size of Washington, DC
Land boundaries: 0 km
Coastline: 101.9 km
Maritime claims: *territorial sea:* 12 nm
exclusive fishing zone: 200 nm
Climate: antarctic
Terrain: Heard Island–80% ice-covered, bleak and mountainous, dominated by a large massif (Big Ben) and an active volcano (Mawson Peak); McDonald Islands–small and rocky
Elevation extremes: *lowest point:* Indian Ocean 0 m
highest point: Mawson Peak on Big Ben volcano 2,745 m
Natural resources: fish
Land use: *arable land:* 0%
permanent crops: 0%
other: 100% (2005)
Irrigated land: 0 sq km
Natural hazards: Mawson Peak, an active volcano, is on Heard Island
Environment—current issues: NA
Geography—note: Mawson Peak on Heard Island is the highest Australian mountain (at 2,745 meters, it is taller than Mt. Kosciuszko in Australia proper), and one of only two active volcanoes located in Australian territory, the other being McDonald Island; in 1992, McDonald Island broke its dormancy and began erupting; it has erupted several times since, the most recent being in 2005

PEOPLE

Population: uninhabited

GOVERNMENT

Country name: *conventional long form:* Territory of Heard Island and McDonald Islands
conventional short form: Heard Island and McDonald Islands
abbreviation: HIMI
Dependency status: territory of Australia; administered from Canberra by the Department of Sustainability, Environment, Water, Population and Communities
Legal system: the laws of Australia, where applicable, apply
Diplomatic representation in the US: none (territory of Australia)
Diplomatic representation from the US: none (territory of Australia)
Flag description: the flag of Australia is used

ECONOMY

Economy—overview: The islands have no indigenous economic activity, but the Australian Government allows limited fishing in the surrounding waters.

COMMUNICATIONS

Internet country code: .hm

TRANSPORTATION

Ports and terminals: none; offshore anchorage only

MILITARY

Military—note: defense is the responsibility of Australia; Australia conducts fisheries patrols

TRANSNATIONAL ISSUES

Disputes—international: none

HOLY SEE (VATICAN CITY)

INTRODUCTION

Background: Popes in their secular role ruled portions of the Italian peninsula for more than a thousand years until the mid 19th century, when many of the Papal States were seized by the newly united Kingdom of Italy. In 1870, the pope's holdings were further circumscribed when Rome itself was annexed. Disputes between a series of "prisoner" popes and Italy were resolved in 1929 by three Lateran Treaties, which established the independent state of Vatican City and granted Roman Catholicism special status in Italy. In 1984, a concordat between the Holy See and Italy modified certain of the earlier treaty provisions, including the primacy of Roman Catholicism as the Italian state religion. Present concerns of the Holy See include religious freedom, international development, the environment, the Middle East, China, the decline of religion in Europe, terrorism, interreligious dialogue and reconciliation, and the application of church doctrine in an era of rapid change and globalization. About 1 billion people worldwide profess the Catholic faith.

GEOGRAPHY

Location: Southern Europe, an enclave of Rome (Italy)
Geographic coordinates: 41 54 N, 12 27 E
Map references: Europe
Area: *total:* 0.44 sq km
country comparison to the world: 249
land: 0.44 sq km
water: 0 sq km
Area—comparative: about 0.7 times the size of The National Mall in Washington, DC
Land boundaries: *total:* 3.2 km
border countries: Italy 3.2 km
Coastline: 0 km (landlocked)
Maritime claims: none (landlocked)
Climate: temperate; mild, rainy winters (September to May) with hot, dry summers (May to September)
Terrain: urban; low hill
Elevation extremes: *lowest point:* unnamed location 19 m
highest point: unnamed elevation 75 m
Natural resources: none
Land use: *arable land:* 0%
permanent crops: 0%
other: 100% (urban area) (2005)
Irrigated land: 0 sq km
Natural hazards: NA
Environment—current issues: NA
Environment—international agreements: *party to:* Ozone Layer Protection
signed, but not ratified: Air Pollution, Environmental Modification
Geography—note: landlocked; enclave in Rome, Italy; world's smallest state; beyond the territorial boundary of Vatican City, the Lateran Treaty of 1929 grants the Holy See extraterritorial authority over 23 sites in Rome and five outside of Rome, including the Pontifical Palace at Castel Gandolfo (the Pope's summer residence)

PEOPLE

Population: 832 (July 2011 est.)
country comparison to the world: 235
Population growth rate: 0.004% (2011 est.)
country comparison to the world: 194
Urbanization: *urban population:* 100% of total population (2010)
rate of urbanization: 0.1% annual rate of change (2010-15 est.)
HIV/AIDS—adult prevalence rate: NA
HIV/AIDS—people living with HIV/AIDS: NA
HIV/AIDS—deaths: NA
Nationality: *noun:* none
adjective: none
Ethnic groups: Italians, Swiss, other
Religions: Roman Catholic
Languages: Italian, Latin, French, various other languages
Literacy: [*definition:* age 15 and over can read and write
total population: 100%
male: 100%
female: 100%

GOVERNMENT

Country name: *conventional long form:* The Holy See (State of the Vatican City)
conventional short form: Holy See (Vatican City)
local long form: Santa Sede (Stato della Citta del Vaticano)
local short form: Santa Sede (Citta del Vaticano)
Government type: ecclesiastical
Capital: *name:* Vatican City
geographic coordinates: 41 54 N, 12 27 E
time difference: UTC+1 (6 hours ahead of Washington, DC during Standard Time)
daylight saving time: +1hr, begins last Sunday in March; ends last Sunday in October
Administrative divisions: none
Independence: 11 February 1929 (from Italy); note–the three treaties signed with Italy on 11 February 1929 acknowledged, among other things, the full sovereignty of the Vatican and established its territorial extent; however, the origin of the Papal States, which over centuries varied considerably in extent, may be traced back to 754
National holiday: Election Day of Pope BENEDICT XVI, 19 April (2005)
Constitution: Fundamental Law promulgated by Pope JOHN PAUL II 26 November 2000, effective 22 February 2001 (replaced the first Fundamental Law of 1929)
Legal system: religious legal system based on canon (religious) law
International law organization participation: has not submitted an ICJ jurisdiction declaration; non-party state to the ICCt
Suffrage: limited to cardinals less than 80 years old
Executive branch: *chief of state:* Pope BENEDICT XVI (since 19 April 2005)
head of government: Secretary of State Cardinal Tarcisio BERTONE (since 15 September 2006)
cabinet: Pontifical Commission for the State of Vatican City appointed by the pope
(For more information visit the World Leaders website)
elections: pope elected for life by the College of Cardinals; election last held on 19 April 2005 (next to be held after the death of the current pope); secretary of state appointed by the pope
election results: Joseph RATZINGER elected Pope BENEDICT XVI
Legislative branch: unicameral Pontifical Commission for Vatican City State
Judicial branch: there are three tribunals responsible for civil and criminal matters within Vatican City; three other tribunals rule on issues pertaining to the Holy See
note: judicial duties were established by the Motu Proprio, papal directive, of Pope PIUS XII on 1 May 1946
Political parties and leaders: none
Political pressure groups and leaders: none (exclusive of influence exercised by church officers)

International organization participation: IAEA, Interpol, IOM (observer), ITSO, ITU, ITUC, OAS (observer), OPCW, OSCE, Schengen Convention (de facto member), UN (observer), UNCTAD, UNHCR, Union Latina (observer), UNWTO (observer), UPU, WIPO, WTO (observer)

Diplomatic representation in the US: *chief of mission:* Apostolic Nuncio Archbishop Pietro SAMBI
chancery: 3339 Massachusetts Avenue NW, Washington, DC 20008
telephone: [1] (202) 333-7121
FAX: [1] (202) 337-4036

Diplomatic representation from the US: *chief of mission:* Ambassador Miguel Humberto DIAZ
embassy: Villa Domiziana, Via delle Terme Deciane 26, 00153 Rome
mailing address: PSC 833, Box 66, APO AE 09624
telephone: [39] (06) 4674-3428
FAX: [39] (06) 575-3411

Flag description: two vertical bands of yellow (hoist side) and white with the arms of the Holy See, consisting of the crossed keys of Saint Peter surmounted by the three-tiered papal tiara, centered in the white band; the yellow color represents the pope's spiritual power, the white his worldly power

National anthem: *name:* "Inno e Marcia Pontificale" (Hymn and Pontifical March)
lyrics/music: Raffaello LAVAGNA/Charles-Francois GOUNOD
note: adopted 1950; although used as such, "Inno e Marcia Pontificale" is not officially a national anthem but rather a hymn meant to appeal to Roman Catholics throughout the world

ECONOMY

Economy—overview: The Holy See is supported financially by a variety of sources, including investments, real estate income, and donations from Catholic individuals, dioceses, and institutions; these help fund the Roman Curia (Vatican bureaucracy), diplomatic missions, and media outlets. The separate Vatican City State budget includes the Vatican museums and post office and is supported financially by the sale of stamps, coins, medals, and tourist mementos; by fees for admission to museums; and by publications sales. Moreover, an annual collection taken up in dioceses and direct donations go to a non-budgetary fund known as Peter's Pence, which is used directly by the Pope for charity, disaster relief, and aid to churches in developing nations. The incomes and living standards of lay workers are comparable to those of counterparts who work in the city of Rome.

GDP (purchasing power parity): $NA

Labor force: NA

Labor force—by occupation: note: essentially services with a small amount of industry; nearly all dignitaries, priests, nuns, guards, and the approximately 3,000 lay workers live outside the Vatican

Population below poverty line: NA%

Budget: *revenues:* $355.5 million
expenditures: $356.8 million (2008)

Industries: printing; production of coins, medals, postage stamps; mosaics and staff uniforms; worldwide banking and financial activities

Electricity—production: NA kWh

Electricity—consumption: NA kWh

Electricity—imports: NA kWh; note—electricity supplied by Italy; a small portion of electricity is self-produced from solar panels

Exchange rates: euros (EUR) per US dollar—0.755 (2010), 0.7198 (2009), 0.6827 (2008), 0.7345 (2007), 0.7964 (2006)

COMMUNICATIONS

Telephones—main lines in use: 5,120 (2005)
country comparison to the world: 211

Telephone system: *general assessment:* automatic digital exchange
domestic: connected via fiber optic cable to Telecom Italia network
international: country code—39; uses Italian system

Broadcast media: the Vatican Television Center (CTV) transmits live broadcasts of the Pope's Sunday and Wednesday audiences, as well as the Pope's public celebrations; CTV also produces documentaries; Vatican Radio is the Holy See's official broadcasting service broadcasting via shortwave, AM and FM frequencies, and via satellite and Internet connections (2008)

Internet country code: .va

Internet hosts: 68 (2010)
country comparison to the world: 208

MILITARY

Military branches: Pontifical Swiss Guard Corps (Corpo della Guardia Svizzera Pontificia) (2010)

Military—note: defense is the responsibility of Italy; ceremonial and limited security duties performed by Pontifical Swiss Guard

TRANSNATIONAL ISSUES

Disputes—international: none

HONDURAS

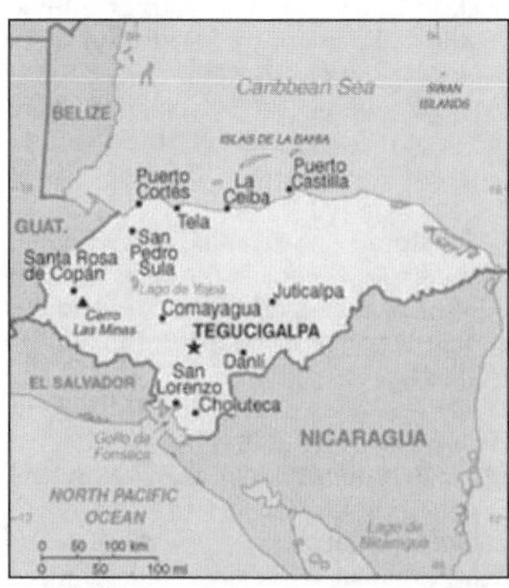

INTRODUCTION

Background: Once part of Spain's vast empire in the New World, Honduras became an independent nation in 1821. After two and a half decades of mostly military rule, a freely elected civilian government came to power in 1982. During the 1980s, Honduras proved a haven for anti-Sandinista contras fighting the Marxist Nicaraguan Government and an ally to Salvadoran Government forces fighting leftist guerrillas. The country was devastated by Hurricane Mitch in 1998, which killed about 5,600 people and caused approximately $2 billion in damage. Since then, the economy has slowly rebounded.

GEOGRAPHY

Location: Central America, bordering the Caribbean Sea, between Guatemala and Nicaragua and bordering the Gulf of Fonseca (North Pacific Ocean), between El Salvador and Nicaragua

Geographic coordinates: 15 00 N, 86 30 W

Map references: Central America and the Caribbean

Area: *total:* 112,090 sq km
country comparison to the world: 102
land: 111,890 sq km
water: 200 sq km

Area—comparative: slightly larger than Tennessee

Land boundaries: *total:* 1,520 km
border countries: Guatemala 256 km, El Salvador 342 km, Nicaragua 922 km

Coastline: Caribbean Sea 669 km; Gulf of Fonseca 163 km

Maritime claims: *territorial sea:* 12 nm
contiguous zone: 24 nm
exclusive economic zone: 200 nm
continental shelf: natural extension of territory or to 200 nm

Climate: subtropical in lowlands, temperate in mountains

Terrain: mostly mountains in interior, narrow coastal plains

Elevation extremes: *lowest point:* Caribbean Sea 0 m
highest point: Cerro Las Minas 2,870 m

Natural resources: timber, gold, silver, copper, lead, zinc, iron ore, antimony, coal, fish, hydropower

Land use: *arable land:* 9.53%
permanent crops: 3.21%
other: 87.26% (2005)

Irrigated land: 800 sq km (2003)

Total renewable water resources: 95.9 cu km (2000)

Freshwater withdrawal (domestic/industrial/agricultural): *total:* 0.86 cu km/yr (8%/12%/80%)
per capita: 119 cu m/yr (2000)

Natural hazards: frequent, but generally mild, earthquakes; extremely susceptible to damaging hurricanes and floods along the Caribbean coast
Environment—current issues: urban population expanding; deforestation results from logging and the clearing of land for agricultural purposes; further land degradation and soil erosion hastened by uncontrolled development and improper land use practices such as farming of marginal lands; mining activities polluting Lago de Yojoa (the country's largest source of fresh water), as well as several rivers and streams, with heavy metals
Environment—international agreements: *party to:* Biodiversity, Climate Change, Climate Change-Kyoto Protocol, Desertification, Endangered Species, Hazardous Wastes, Law of the Sea, Marine Dumping, Ozone Layer Protection, Ship Pollution, Tropical Timber 83, Tropical Timber 94, Wetlands
signed, but not ratified: none of the selected agreements
Geography—note: has only a short Pacific coast but a long Caribbean shoreline, including the virtually uninhabited eastern Mosquito Coast

PEOPLE

Population: 8,143,564 (July 2011 est.)
country comparison to the world: 93
note: estimates for this country explicitly take into account the effects of excess mortality due to AIDS; this can result in lower life expectancy, higher infant mortality, higher death rates, lower population growth rates, and changes in the distribution of population by age and sex than would otherwise be expected
Age structure: *0–14 years:* 36.7% (male 1,528,271/female 1,464,428)
15–64 years: 59.5% (male 2,431,607/female 2,412,951)
65 years and over: 3.8% (male 136,035/female 170,272) (2011 est.)
Median age:
total: 21 years
male: 20.6 years
female: 21.4 years (2011 est.)
Population growth rate: 1.888% (2011 est.)
country comparison to the world: 61
Birth rate: 25.14 births/1,000 population (2011 est.)
country comparison to the world: 61
Death rate: 5.02 deaths/1,000 population (July 2011 est.)
country comparison to the world: 184
Net migration rate: -1.25 migrant(s)/1,000 population (2011 est.)
country comparison to the world: 156
Urbanization: *urban population:* 52% of total population (2010)
rate of urbanization: 3.1% annual rate of change (2010-15 est.)
Major cities—population: TEGUCIGALPA (capital) 1 million (2009)
Sex ratio: *at birth:* 1.05 male(s)/female
under 15 years: 1.04 male(s)/female
15–64 years: 1.01 male(s)/female
65 years and over: 0.81 male(s)/female
total population: 1.01 male(s)/female (2011 est.)
Infant mortality rate: *total:* 20.44 deaths/1,000 live births
country comparison to the world: 96
male: 23.14 deaths/1,000 live births
female: 17.61 deaths/1,000 live births (2011 est.)
Life expectancy at birth: *total population:* 70.61 years
country comparison to the world: 144
male: 68.93 years
female: 72.37 years (2011 est.)
Total fertility rate: 3.09 children born/woman (2011 est.)
country comparison to the world: 60
HIV/AIDS—adult prevalence rate: 0.8% (2009 est.)
country comparison to the world: 53
HIV/AIDS—people living with HIV/AIDS: 39,000 (2009 est.)
country comparison to the world: 63
HIV/AIDS—deaths: 2,500 (2009 est.)
country comparison to the world: 51
Major infectious diseases: *degree of risk:* high
food or waterborne diseases: bacterial diarrhea, hepatitis A, and typhoid fever
vectorborne diseases: dengue fever and malaria
water contact disease: leptospirosis (2009)
Drinking water source: *Improved:*
urban: 95% of population
rural: 77% of population
total: 86% of population
Unimproved: urban: 5% of population
rural: 23% of population
total: 14% of population (2008)
Sanitation facility access: *Improved:*
urban: 80% of population
rural: 62% of population
total: 71% of population
Unimproved: urban: 20% of population
rural: 38% of population
total: 29% of population (2008)
Nationality: *noun:* Honduran(s)
adjective: Honduran
Ethnic groups: mestizo (mixed Amerindian and European) 90%, Amerindian 7%, black 2%, white 1%
Religions: Roman Catholic 97%, Protestant 3%
Languages: Spanish (official), Amerindian dialects
Literacy: *definition:* age 15 and over can read and write
total population: 80%
male: 79.8%
female: 80.2% (2001 census)
School life expectancy (primary to tertiary education): *total:* 11 years
male: 11 years
female: 12 years (2008)
Education expenditures: NA

GOVERNMENT

Country name: *conventional long form:* Republic of Honduras
conventional short form: Honduras
local long form: Republica de Honduras
local short form: Honduras
Government type: democratic constitutional republic
Capital: *name:* Tegucigalpa
geographic coordinates: 14 06 N, 87 13 W
time difference: UTC-6 (1 hour behind Washington, DC during Standard Time)
daylight saving time: none scheduled for 2011
Administrative divisions: 18 departments (departamentos, singular—departamento); Atlantida, Choluteca, Colon, Comayagua, Copan, Cortes, El Paraiso, Francisco Morazan, Gracias a Dios, Intibuca, Islas de la Bahia, La Paz, Lempira, Ocotepeque, Olancho, Santa Barbara, Valle, Yoro
Independence: 15 September 1821 (from Spain)
National holiday: Independence Day, 15 September (1821)
Constitution: 11 January 1982, effective 20 January 1982; amended many times
Legal system: civil law system
International law organization participation: accepts compulsory ICJ jurisdiction with reservations; accepts ICCt jurisdiction
Suffrage: 18 years of age; universal and compulsory
Executive branch: *chief of state:* President Porfirio LOBO Sosa (since 27 January 2010); Vice President Maria Antonieta Guillen de BOGRAN (since 27 January 2010); note—the president is both the chief of state and head of government
head of government: President Porfirio LOBO Sosa (since 27 January 2010); Vice President Maria Antonieta Guillen de BOGRAN (since 27 January 2010)
cabinet: Cabinet appointed by president
(For more information visit the World Leaders website)
elections: president elected by popular vote for a four-year term; election last held on 29 November 2009 (next to be held in November 2013)
election results: Porfirio "Pepe" LOBO Sosa elected president; percent of vote—Porfirio "Pepe" LOBO Sosa 56.3%, Elvin SANTOS Lozano 38.1%, other 5.6%
Legislative branch: unicameral National Congress or Congreso Nacional (128 seats; members elected proportionally by department to serve four-year terms)
elections: last held on 29 November 2009 (next to be held in November 2013)
election results: percent of vote by party—NA; seats by party—PNH 71, PL 45, PDC 5, PUD 4, PINU 3
Judicial branch: Supreme Court of Justice or Corte Suprema de Justicia (15 judges are elected for seven-year terms by the National Congress)
Political parties and leaders: Christian Democratic Party or PDC [Felicito AVILA Ordonez]; Democratic Unification Party or PUD [Cesar HAM]; Liberal Party or PL [Roberto MICHELETTI Bain]; National Party or PN [Antonio ALVAREZ Arias]; Social Democratic Innovation and Unity Party or PINU [Jorge Rafael AGUILAR Paredes]
Political pressure groups and leaders: Beverage and Related Industries Syndicate or STIBYS; Committee for the Defense of Human Rights in Honduras or CODEH; Confederation of Honduran Workers or CTH; Coordinating Committee of Popular Organizations or CCOP;

General Workers Confederation or CGT; Honduran Council of Private Enterprise or COHEP; National Association of Honduran Campesinos or ANACH; National Union of Campesinos or UNC; Popular Bloc or BP; United Confederation of Honduran Workers or CUTH; United Farm Workers' Movement of the Aguan (MUCA)

International organization participation: BCIE, CACM, FAO, G-11, G-77, IADB, IAEA, IBRD, ICAO, ICRM, IDA, IFAD, IFC, IFRCS, ILO, IMF, IMO, Interpol, IOC (suspended), IOM, ISO (subscriber), ITSO, ITU, ITUC, LAES, LAIA (observer), MIGA, MINURSO, NAM, OAS (suspended), OPANAL, OPCW, PCA, PetroCaribe, RG (suspended), SICA, UN, UNCTAD, UNESCO, UNIDO, Union Latina, UNWTO, UPU, WCO (suspended), WFTU, WHO, WIPO, WMO, WTO

Diplomatic representation in the US: *chief of mission:* Ambassador Jorge Ramon HERNANDEZ Alcerro
chancery: Suite 4-M, 3007 Tilden Street NW, Washington, DC 20008
telephone: [1] (202) 966-2604
FAX: [1] (202) 966-9751
consulate(s) general: Atlanta, Chicago, Houston, Los Angeles, Miami, New Orleans, New York, Phoenix, San Francisco
honorary *consulate(s):* Jacksonville

Diplomatic representation from the US: *chief of mission:* Ambassador Hugo LLORENS
embassy: Avenida La Paz, Apartado Postal No. 3453, Tegucigalpa
mailing address: American Embassy, APO AA 34022, Tegucigalpa
telephone: [504] 236-9320, 238-5114
FAX: [504] 238-4357

Flag description: three equal horizontal bands of blue (top), white, and blue, with five blue, five-pointed stars arranged in an X pattern centered in the white band; the stars represent the members of the former Federal Republic of Central America–Costa Rica, El Salvador, Guatemala, Honduras, and Nicaragua; the blue bands symbolize the Pacific Ocean and the Caribbean Sea; the white band represents the land between the two bodies of water and the peace and prosperity of its people
note: similar to the flag of El Salvador, which features a round emblem encircled by the words REPUBLICA DE EL SALVADOR EN LA AMERICA CENTRAL centered in the white band; also similar to the flag of Nicaragua, which features a triangle encircled by the words REPUBLICA DE NICARAGUA on top and AMERICA CENTRAL on the bottom, centered in the white band

National anthem: *name:* "Himno Nacional de Honduras" (National Anthem of Honduras)
lyrics/music: Augusto Constancio COELLO/Carlos HARTLING
note: adopted 1915; the anthem's seven verses chronicle Honduran history; on official occasions, only the chorus and last verse are sung

ECONOMY

Economy—overview: Honduras, the second poorest country in Central America, suffers from extraordinarily unequal distribution of income, as well as high underemployment. While historically dependent on the export of bananas and coffee, Honduras has diversified its export base to include apparel and automobile wire harnessing. Nearly half of Honduras's economic activity is directly tied to the US, with exports to the US accounting for 30% of GDP and remittances for another 20%. The US-Central America Free Trade Agreement (CAFTA) came into force in 2006 and has helped foster foreign direct investment, but physical and political insecurity, as well as crime and perceptions of corruption, may deter potential investors; about 70% of FDI is from US firms. The economy registered sluggish economic growth in 2010, insufficient to improve living standards for the nearly 60% of the population in poverty. The LOBO administration inherited a difficult fiscal position with off-budget debts accrued in previous administrations and government salaries nearly equivalent to tax collections. His government has displayed a commitment to improving tax collection and cutting expenditures, and attracting foreign investment. This enabled Tegucigalpa to secure an IMF Precautionary Stand-By agreement in October 2010. The IMF agreement has helped renew multilateral and bilateral donor confidence in Honduras following the ZELAYA administration's economic mismanagement and the 2009 coup.

GDP (purchasing power parity): $33.63 billion (2010 est.)
country comparison to the world: 104
$32.72 billion (2009 est.)
$33.44 billion (2008 est.)
note: data are in 2010 US dollars

GDP (official exchange rate): $15.35 billion (2010 est.)

GDP—real growth rate: 2.8% (2010 est.)
country comparison to the world: 130
-2.1% (2009 est.)
4.1% (2008 est.)

GDP—per capita (PPP): $4,200 (2010 est.)
country comparison to the world: 156
$4,200 (2009 est.)
$4,400 (2008 est.)
note: data are in 2010 US dollars

GDP—composition by sector: *agriculture:* 12.4%
industry: 26.9%
services: 60.8% (2010 est.)

Labor force: 3.394 million (2010 est.)
country comparison to the world: 99

Labor force—by occupation: *agriculture:* 39.2%
industry: 20.9%
services: 39.8% (2005 est.)

Unemployment rate: 5.1% (2010 est.)
country comparison to the world: 49
3.2% (2009 est.)
note: about one-third of the people are underemployed

Population below poverty line: 65% (2010)

Household income or consumption by percentage share: *lowest 10%:* 0.7%
highest 10%: 42.2% (2006)

Distribution of family income—Gini index: 53.8 (2003)
country comparison to the world: 12
56.3 (1998)

Investment (gross fixed): 23.3% of GDP (2010 est.)
country comparison to the world: 55

Budget: *revenues:* $2.923 billion
expenditures: $3.651 billion (2010 est.)

Public debt: 26.1% of GDP (2010 est.)
country comparison to the world: 95
23.7% of GDP (2009 est.)

Inflation rate (consumer prices): 4.6% (2010 est.)
country comparison to the world: 132
5.5% (2009 est.)

Central bank discount rate: NA%

Commercial bank prime lending rate: 19.16% (31 December 2009 est.)
country comparison to the world: 26
17.94% (31 December 2008 est.)

Stock of narrow money: $1.296 billion (31 December 2010 est.)
country comparison to the world: 129
$1.564 billion (31 December 2009 est.)

Stock of broad money: $7.618 billion (31 December 2010 est.)
country comparison to the world: 110
$7.064 billion (31 December 2009 est.)

Stock of domestic credit: $7.581 billion (31 December 2010 est.)
country comparison to the world: 104
$7.029 billion (31 December 2009 est.)

Market value of publicly traded shares: $NA

Agriculture—products: bananas, coffee, citrus, corn, African palm; beef; timber; shrimp, tilapia, lobster

Industries: sugar, coffee, woven and knit apparel, wood products, cigars

Industrial production growth rate: 2.4% (2010 est.)
country comparison to the world: 125

Electricity—production: 6.58 billion kWh (2009 est.)
country comparison to the world: 103

Electricity—consumption: 6.54 billion kWh
country comparison to the world: 103
note: approximately 1.5 billion kWh in transmission and distribution losses (2009 est.)

Electricity—exports: 0 kWh (2008 est.)

Electricity—imports: 11.8 million kWh (2007 est.)

Oil—production: 0 bbl/day (2009 est.)
country comparison to the world: 185

Oil—consumption: 56,000 bbl/day (2009 est.)
country comparison to the world: 94

Oil—exports: 0 bbl/day (2007 est.)
country comparison to the world: 173

Oil—imports: 46,130 bbl/day (2007 est.)
country comparison to the world: 88

Oil—proved reserves: 0 bbl (1 January 2010 est.)
country comparison to the world: 143

Natural gas—production: 0 cu m (2008 est.)
country comparison to the world: 139

Natural gas—consumption: 0 cu m (2008 est.)
country comparison to the world: 183

Natural gas—exports: 0 cu m (2008 est.)

country comparison to the world: 111
Natural gas—imports: 0 cu m (2008 est.)
country comparison to the world: 198
Natural gas—proved reserves: 0 cu m (1 January 2010 est.)
country comparison to the world: 146
Current account balance: $-1.048 billion (2010 est.)
country comparison to the world: 136
$-1.327 billion (2009 est.)
Exports: $5.879 billion (2010 est.)
country comparison to the world: 105
$5.09 billion (2009 est.)
Exports—commodities: apparel, coffee, shrimp, wire harnesses, cigars, bananas, gold, palm oil, fruit, lobster, lumber
Exports—partners: US 59.6%, El Salvador 5.61%, Guatemala 5.28%, Mexico 4.19%, Germany 4.04% (2009)
Imports:
$8.878 billion (2010 est.)
country comparison to the world: 94
$5.924 billion (2009 est.)
Imports—commodities: machinery and transport equipment, industrial raw materials, chemical products, fuels, foodstuffs
Imports—partners: US 46.81%, Guatemala 8.92%, El Salvador 7.13%, Mexico 5.54%, Costa Rica 4.91% (2009)
Reserves of foreign exchange and gold: $2.302 billion (31 December 2010 est.)
country comparison to the world: 96
$2.127 billion (31 December 2009 est.)
Debt—external: $3.54 billion (31 December 2010 est.)
country comparison to the world: 119
$3.311 billion (31 December 2009 est.)
Exchange rates: lempiras (HNL) per US dollar–18.9 (2010), 18.9 (2009), 18.983 (2008), 18.9 (2007), 18.895 (2006)

COMMUNICATIONS

Telephones—main lines in use: 830,000 (2009)
country comparison to the world: 85
Telephones—mobile cellular: 7.714 million (2009)
country comparison to the world: 75
Telephone system: *general assessment:* the number of fixed-line connections are increasing but still limited; competition among multiple providers of mobile-cellular services is contributing to a sharp increase in the number of subscribers
domestic: beginning in 2003, private sub-operators allowed to provide fixed-lines in order to expand telephone coverage contributing to an increase in fixed-line teledensity to roughly 10 per 100 persons; mobile-cellular subscribership reached 100 per 100 persons in 2009
international: country code–504; landing point for both the Americas Region Caribbean Ring System (ARCOS-1) and the MAYA-1 fiber optic submarine cable system that together provide connectivity to South and Central America, parts of the Caribbean, and the US; satellite earth stations–2 Intelsat (Atlantic Ocean); connected to Central American Microwave System
Broadcast media: multiple privately-owned terrestrial television networks, supplemented by multiple cable TV networks; Radio Honduras is the lone government-owned radio network; roughly 300 privately-owned radio stations (2007)
Internet country code: .hn
Internet hosts: 16,075 (2010)
country comparison to the world: 115
Internet users: 731,700 (2009)
country comparison to the world: 108

TRANSPORTATION

Airports: 104 (2010)
country comparison to the world: 58
Airports—with paved runways: *total:* 12
2,438 to 3,047 m: 3
1,524 to 2,437 m: 2
914 to 1,523 m: 4
under 914 m: 3 (2010)
Airports—with unpaved runways: *total:* 92
1,524 to 2,437 m: 2
914 to 1,523 m: 16
under 914 m: 74 (2010)
Railways:
total: 75 km
country comparison to the world: 128
narrow gauge: 75 km 1.067-m gauge (2009)
Roadways: *total:* 14,239 km
country comparison to the world: 122
paved: 3,159 km
unpaved: 11,080 km (1,420 km summer only) (2009)
Waterways: 465 km (most navigable only by small craft) (2010)
country comparison to the world: 85
Merchant marine: *total:* 104
country comparison to the world: 49
by type: bulk carrier 8, cargo 50, carrier 2, chemical tanker 7, container 1, passenger 3, passenger/cargo 2, petroleum tanker 22, refrigerated cargo 6, roll on/roll off 3
foreign-owned: 49 (Bahrain 5, Canada 1, China 2, Egypt 2, Greece 4, Hong Kong 1, Israel 1, Japan 4, Lebanon 2, Mexico 1, Montenegro 2, Panama 1, Singapore 12, South Korea 6, Taiwan 2, Tanzania 1, UK 1, Vietnam 1) (2010)
Ports and terminals: La Ceiba, Puerto Cortes, San Lorenzo, Tela

MILITARY

Military branches: Army, Navy (includes Naval Infantry), Honduran Air Force (Fuerza Aerea Hondurena, FAH) (2008)
Military service age and obligation: 18 years of age for voluntary 2 to 3 year military service (2004)
Manpower available for military service:
males age 16–49: 2,045,914
females age 16–49: 1,991,418 (2010 est.)
Manpower fit for military service: *males age 16–49:* 1,525,578
females age 16–49: 1,539,688 (2010 est.)
Manpower reaching militarily significant age annually: *male:* 95,895
female: 92,087 (2010 est.)
Military expenditures: 0.6% of GDP (2006 est.)
country comparison to the world: 156

TRANSNATIONAL ISSUES

Disputes—international: International Court of Justice (ICJ) ruled on the delimitation of "bolsones" (disputed areas) along the El Salvador-Honduras border in 1992 with final settlement by the parties in 2006 after an Organization of American States (OAS) survey and a further ICJ ruling in 2003; the 1992 ICJ ruling advised a tripartite resolution to a maritime boundary in the Gulf of Fonseca with consideration of Honduran access to the Pacific; El Salvador continues to claim tiny Conejo Island, not mentioned in the ICJ ruling, off Honduras in the Gulf of Fonseca; Honduras claims the Belizean-administered Sapodilla Cays off the coast of Belize in its constitution, but agreed to a joint ecological park around the cays should Guatemala consent to a maritime corridor in the Caribbean under the OAS-sponsored 2002 Belize-Guatemala Differendum; memorials and countermemorials were filed by the parties in Nicaragua's 1999 and 2001 proceedings against Honduras and Colombia at the ICJ over the maritime boundary and territorial claims in the western Caribbean Sea–final public hearings are scheduled for 2007
Illicit drugs: transshipment point for drugs and narcotics; illicit producer of cannabis, cultivated on small plots and used principally for local consumption; corruption is a major problem; some money-laundering activity

HONG KONG

(SPECIAL ADMINISTRATIVE REGION OF CHINA)

INTRODUCTION

Background: Occupied by the UK in 1841, Hong Kong was formally ceded by China the following year; various adjacent lands were added later in the 19th century. Pursuant to an agreement signed by China and the UK on 19 December 1984, Hong Kong became the Hong Kong Special Administrative Region (SAR) of the People's Republic of China on 1 July 1997. In this agreement, China promised that, under its "one country, two systems" formula, China's socialist economic system would not be imposed on Hong Kong and that Hong Kong would enjoy a high degree of autonomy in all matters except foreign and defense affairs for the next 50 years.

GEOGRAPHY

Location: Eastern Asia, bordering the South China Sea and China
Geographic coordinates: 22 15 N, 114 10 E
Map references: Southeast Asia

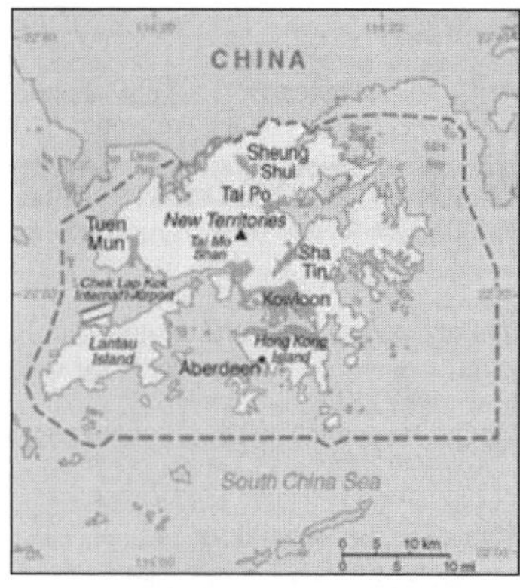

Area: *total:* 1,104 sq km
country comparison to the world: 183
land: 1,054 sq km
water: 50 sq km
Area—comparative: six times the size of Washington, DC
Land boundaries: *total:* 30 km
regional border: China 30 km
Coastline: 733 km
Maritime claims: *territorial sea:* 3 nm
Climate: subtropical monsoon; cool and humid in winter, hot and rainy from spring through summer, warm and sunny in fall
Terrain: hilly to mountainous with steep slopes; lowlands in north
Elevation extremes: *lowest point:* South China Sea 0 m
highest point: Tai Mo Shan 958 m
Natural resources: outstanding deepwater harbor, feldspar
Land use: *arable land:* 5.05%
permanent crops: 1.01%
other: 93.94% (2001)
Irrigated land: 20 sq km (1998 est.)
Natural hazards: occasional typhoons
Environment—current issues: air and water pollution from rapid urbanization
Environment—international agreements: *party to:* Marine Dumping (associate member), Ship Pollution (associate member)
Geography—note: composed of more than 200 islands

PEOPLE

Population: 7,122,508 (July 2011 est.)
country comparison to the world: 98
Age structure: *0–14 years:* 11.6% (male 431,728/female 394,898)
15–64 years: 74.8% (male 2,573,929/female 2,757,095)
65 years and over: 13.5% (male 452,278/female 512,580) (2011 est.)
Median age: *total:* 43.4 years
male: 42.8 years
female: 43.8 years (2011 est.)
Population growth rate: 0.448% (2011 est.)
country comparison to the world: 153
Birth rate: 7.49 births/1,000 population (2011 est.)
country comparison to the world: 220
Death rate: 7.07 deaths/1,000 population (July 2011 est.)
country comparison to the world: 131
Net migration rate: 4.06 migrant(s)/1,000 population (2011 est.)
country comparison to the world: 24
Urbanization: *urban population:* 100% of total population (2010)
rate of urbanization: 0.9% annual rate of change (2010-15 est.)
Sex ratio: *at birth:* 1.075 male(s)/female
under 15 years: 1.09 male(s)/female
15–64 years: 0.94 male(s)/female
65 years and over: 0.88 male(s)/female
total population: 0.95 male(s)/female (2011 est.)
Infant mortality rate:
total: 2.9 deaths/1,000 live births
country comparison to the world: 217
male: 3.08 deaths/1,000 live births
female: 2.71 deaths/1,000 live births (2011 est.)
Life expectancy at birth: *total population:* 82.04 years
country comparison to the world: 8
male: 79.32 years
female: 84.97 years (2011 est.)
Total fertility rate: 1.07 children born/woman (2011 est.)
country comparison to the world: 222
HIV/AIDS—adult prevalence rate: 0.1% (2003 est.)
country comparison to the world: 133
HIV/AIDS—people living with HIV/AIDS: 2,600 (2003 est.)
country comparison to the world: 131
HIV/AIDS—deaths: fewer than 200 (2003 est.)
country comparison to the world: 112
Nationality: *noun:* Chinese/Hong Konger
adjective: Chinese/Hong Kong
Ethnic groups: Chinese 95%, Filipino 1.6%, Indonesian 1.3%, other 2.1% (2006 census)
Religions: eclectic mixture of local religions 90%, Christian 10%
Languages: Cantonese (official) 90.8%, English (official) 2.8%, Putonghua (Mandarin) 0.9%, other Chinese dialects 4.4%, other 1.1% (2006 census)
Literacy: *definition:* age 15 and over has ever attended school
total population: 93.5%
male: 96.9%
female: 89.6% (2002)
School life expectancy (primary to tertiary education): *total:* 16 years
male: 15 years
female: 16 years (2009)
Education expenditures: 4.5% of GDP (2009)
country comparison to the world: 80

GOVERNMENT

Country name: *conventional long form:* Hong Kong Special Administrative Region
conventional short form: Hong Kong
official long form: Xianggang Tebie Xingzhengqu
official short form: Xianggang
abbreviation: HK
Dependency status: special administrative region of China
Government type: limited democracy
Administrative divisions: none (special administrative region of China)
Independence: none (special administrative region of China)
National holiday: National Day (Anniversary of the Founding of the People's Republic of China), 1 October (1949); note—1 July 1997 is celebrated as Hong Kong Special Administrative Region Establishment Day
Constitution: Basic Law, approved March 1990 by China's National People's Congress, is Hong Kong's charter
Legal system: mixed legal system of common law based on the English model and Chinese customary law (in matters of family and land tenure)
Suffrage: direct election—18 years of age for half the legislature and a majority of seats in 18 district councils; universal for permanent residents living in the territory of Hong Kong for the past seven years
indirect election—limited to about 220,000 members of functional constituencies for the other half of the legislature and an 800-member election committee for the chief executive drawn from broad sectoral groupings, central government bodies, and municipal organizations
Executive branch: *chief of state:* President of China HU Jintao (since 15 March 2003)
head of government: Chief Executive Donald TSANG Yam-kuen (since 24 June 2005)
cabinet: Executive Council or ExCo consists of 15 official members and 14 non-official members
(For more information visit the World Leaders website)
elections: chief executive elected for five-year term by 800-member electoral committee; election last held on 25 March 2007 (next to be held in 2012)
note: the LegCo voted in June 2010 to expand the electoral committee to 1,200 seats for the next election
election results: Donald TSANG elected chief executive receiving 84.1% of the vote of the election committee; Alan LEONG Kah-kit received 15.9%
Legislative branch: unicameral Legislative Council or LegCo (60 seats; 30 members indirectly elected by functional constituencies, 30 elected by popular vote; members serve four-year terms)
note: the LegCo voted in June 2010 to expand to 70 seats for the next election; the measure was approved by the National People's Congress Standing Committee in August 2010; the 10 new seats will be elected by popular vote
elections: last held on 7 September 2008 (next to be held in September 2012)
election results: percent of vote by party—pro-democracy 57%; pro-Beijing 40%, independent 3%; seats by parties—(pro-Beijing 35) DAB 13, Liberal Party 7, FTU 1, others 14; (pro-democracy 23) Democratic Party 8, Civic Party 5, CTU 3, League of Social Democrats 3, ADPL 2, The Frontier 1, NWSC 1; others 11; independents and non-voting LegCo president 2
Judicial branch: Court of Final Appeal in the Hong Kong Special Administrative Region
Political parties and leaders: parties: Association for Democracy and People's Livelihood or ADPL [LIU Sung Lee]; Civic Party [Audrey EU Yuet-mee]; Democratic Alliance for the Betterment and Progress

of Hong Kong or DAB [TAM Yiu Cheng]; Democratic Party [Albert HO Chun-yan]; League of Social Democrats [Raymond WONG Yuk-man]; Liberal Party [Miriam LAU Kin-yee]; The Frontier (disbanded)
others: Confederation of Trade Unions or CTU; Federation of Trade Unions or FTU; Neighborhood and Workers Service Center or NWSC
note: political blocs include: pro-democracy–ADPL, Civic Party, Democratic Party, League of Social Democrats; pro-Beijing–DAB, Liberal Party, The Professional Forum (an informal group of three generally pro-government and pro-business LegCo members from functional constituencies and one independent elected from a geographic constituency), and Economic Synergy; there is no political party ordinance, so there are no registered political parties; politically active groups register as societies or companies

Political pressure groups and leaders: Chinese General Chamber of Commerce (pro-China); Chinese Manufacturers' Association of Hong Kong; Confederation of Trade Unions or CTU (pro-democracy) [LEE Cheuk-yan, general secretary]; Federation of Hong Kong Industries; Federation of Trade Unions or FTU (pro-China) [CHENG Yiu-tong, executive councilor]; Hong Kong Alliance in Support of the Patriotic Democratic Movement in China [LEE Cheuk-yan, chairman]; Hong Kong and Kowloon Trade Union Council (pro-Taiwan); Hong Kong General Chamber of Commerce; Hong Kong Professional Teachers' Union [FUNG Wai-wah, president]; Neighborhood and Workers' Service Center or NWSC [LEUNG Yiu-chung, LegCo member] (pro-democracy); Civic Act-up [Cyd HO Sau-lan, LegCo member] (pro-democracy)

International organization participation: ADB, APEC, BIS, FATF, ICC, IHO, IMF, IMO (associate), Interpol (subbureau), IOC, ISO (correspondent), ITUC, UNWTO (associate), UPU, WCO, WTO

Diplomatic representation in the US: none (special administrative region of China); Hong Kong Economic and Trade Office carries out normal liaison and communication with the US Government and other US entities
representative: Donald TONG
office: 1520 18th Street NW, Washington, DC 20036
telephone: [1] 202 331-8947
FAX: [1] 202 331-0318
NKETO offices: New York, San Francisco

Diplomatic representation from the US:
chief of mission: Consul General Stephen M. YOUNG
consulate(s) general: 26 Garden Road, Hong Kong
mailing address: PSC 461, Box 1, FPO AP 96521-0006
telephone: [852] 2523-9011
FAX: [852] 2845-1598

Flag description: red with a stylized, white, five-petal Bauhinia flower in the center; each petal contains a small, red, five-pointed star in its middle; the red color is the same as that on the Chinese flag and represents the motherland; the fragrant Bauhinia–developed in Hong Kong the late 19th century–has come to symbolize the region; the five stars echo those on the flag of China

National anthem: note: as a Special Administrative Region of China, "Yiyonggjun Jinxingqu" is official (see China)

ECONOMY

Economy—overview: Hong Kong has a free market economy highly dependent on international trade and finance–the value of goods and services trade, including the sizable share of re-exports, is about four times GDP. Hong Kong's open economy left it exposed to the global economic slowdown, but its increasing integration with China, through trade, tourism, and financial links, helped it recover more quickly than many observers anticipated. The Hong Kong government is promoting the Special Administrative Region (SAR) as the site for Chinese renminbi (RMB) internationalization. Hong Kong residents are allowed to establish RMB-denominated savings accounts; RMB-denominated corporate and Chinese government bonds have been issued in Hong Kong; and RMB trade settlement is allowed. The territory far exceeded the RMB conversion quota set by Beijing for trade settlements in 2010 due to the growth of earnings from exports to the mainland. RMB deposits grew to roughly 4.6% of total system deposits in Hong Kong by the end of 2010, an increase of over 392% since the beginning of the year. The government is pursuing efforts to introduce additional use of RMB in Hong Kong financial markets and is seeking to expand the RMB quota for 2011. The mainland has long been Hong Kong's largest trading partner, accounting for about half of Hong Kong's exports by value. Hong Kong's natural resources are limited, and food and raw materials must be imported. As a result of China's easing of travel restrictions, the number of mainland tourists to the territory has surged from 4.5 million in 2001 to 22.5 million in 2010, outnumbering visitors from all other countries combined. Hong Kong has also established itself as the premier stock market for Chinese firms seeking to list abroad. In 2010 mainland Chinese companies constituted about 19% of the firms listed on the Hong Kong Stock Exchange and accounted for 62% of the Exchange's market capitalization. During the past decade, as Hong Kong's manufacturing industry moved to the mainland, its service industry has grown rapidly and in 2009 accounted for more than 90% of the territory's GDP. GDP growth averaged a strong 3.8% from 1989 to 2010. Hong Kong's GDP fell in 2009 as a result of the global financial crisis, but a recovery began in third quarter 2009, and the economy grew nearly 6.8% in 2010. The Hong Kong government adopted several temporary fiscal policy support measures in response to the crisis that it may discontinue if strong growth is sustained. Credit expansion and tight housing supply conditions caused Hong Kong property prices to rise rapidly in 2010, and some lower income segments of the population are increasingly unable to afford adequate housing. Hong Kong continues to link its currency closely to the US dollar, maintaining an arrangement established in 1983.

GDP (purchasing power parity): $325.8 billion (2010 est.)
country comparison to the world: 37
$305 billion (2009 est.)
$313.3 billion (2008 est.)
note: data are in 2010 US dollars

GDP (official exchange rate): $225 billion (2010 est.)

GDP—real growth rate: 6.8% (2010 est.)
country comparison to the world: 40
-2.7% (2009 est.)
2.3% (2008 est.)

GDP—per capita (PPP): $45,900 (2010 est.)
country comparison to the world: 13
$43,200 (2009 est.)
$44,600 (2008 est.)
note: data are in 2010 US dollars

GDP—composition by sector:
agriculture: 0.1%
industry: 7.6%
services: 92.3% (2010 est.)

Labor force: 3.7 million (2010 est.)
country comparison to the world: 93

Labor force—by occupation: manufacturing: 4.7%
construction: 2.2%
wholesale and retail trade, restaurants, and hotels: 41.7%
financing, insurance, and real estate: 12%
transport and communications: 6.3%
community and social *services:* 17%
note: above data exclude public sector (2010 est.)

Unemployment rate: 4.3% (2010 est.)
country comparison to the world: 40
5.4% (2009 est.)

Population below poverty line: NA%

Household income or consumption by percentage share:
lowest 10%: NA%
highest 10%: NA%

Distribution of family income—Gini index: 53.3 (2007)
country comparison to the world: 13

Investment (gross fixed): 21.4% of GDP (2010 est.)
country comparison to the world: 72

Budget: *revenues:* $48.1 billion
expenditures: $38.9 billion (2010 est.)

Public debt: 18.2% of GDP (2010 est.)
country comparison to the world: 112
37.4% of GDP (2009 est.)

Inflation rate (consumer prices): 4.5% (2010 est.)
country comparison to the world: 126
2.4% (2009 est.)

Central bank discount rate: 0.5% (31 December 2010)
country comparison to the world: 138
0.5% (31 December 2009)

Commercial bank prime lending rate: 5% (31 December 2010 est.)
country comparison to the world: 147
5% (31 December 2009 est.)

Stock of narrow money: $130.4 billion (31 December 2010 est.)

country comparison to the world: 26
$115.6 billion (31 December 2009 est.)
Stock of broad money: $914.9 billion (31 December 2009)
country comparison to the world: 18
$846.5 billion (31 December 2008)
Stock of domestic credit: $390.1 billion (31 December 2010 est.)
country comparison to the world: 29
$317.4 billion (31 December 2009 est.)
Market value of publicly traded shares: $2.702 trillion (31 December 2010)
country comparison to the world: 5
$2.292 trillion (31 December 2009)
$1.32 trillion (31 December 2008 est.)
Agriculture—products: fresh vegetables; poultry, pork; fish
Industries: textiles, clothing, tourism, banking, shipping, electronics, plastics, toys, watches, clocks
Industrial production growth rate: -0.3% (2010 est.)
country comparison to the world: 154
Electricity—production: 38.23 billion kWh (2010 est.)
country comparison to the world: 55
Electricity—consumption: 42.64 billion kWh (2010 est.)
country comparison to the world: 50
Electricity—exports: 2.23 billion kWh (2010 est.)
Electricity—imports: 12.26 billion kWh (2010 est.)
Oil—production: 0 bbl/day (2010 est.)
country comparison to the world: 184
Oil—consumption: 418,200 bbl/day (2010 est.)
country comparison to the world: 33
Oil—exports: 10,020 bbl/day (2010)
country comparison to the world: 93
Oil—imports: 428,200 bbl/day (2010)
country comparison to the world: 29
Oil—proved reserves: 0 bbl (1 January 2011 est.)
country comparison to the world: 142
Natural gas—production: 0 cu m (2010 est.)
country comparison to the world: 138
Natural gas—consumption: 3.7 billion cu m (2010 est.)
country comparison to the world: 67
Natural gas—exports: 0 cu m (2010 est.)
country comparison to the world: 110
Natural gas—imports: 3.7 billion cu m (2010 est.)
country comparison to the world: 36
Natural gas—proved reserves: 0 cu m (1 January 2011 est.)
country comparison to the world: 145
Current account balance: $18.07 billion (2010 est.)
country comparison to the world: 18
$17.3 billion (2009 est.)
Exports: $388.6 billion (2010 est.)
country comparison to the world: 12
$316.6 billion (2009)
Exports—commodities: electrical machinery and appliances, textiles, apparel, footwear, watches and clocks, toys, plastics, precious stones, printed material
Exports—partners: China 51.2%, US 11.6%, Japan 4.4% (2009 est.)
Imports: $431.4 billion (2010 est.)
country comparison to the world: 9
$345.2 billion (2009 est.)
Imports—commodities: raw materials and semi-manufactures, consumer goods, capital goods, foodstuffs, fuel (most is reexported)
Imports—partners: China 46.4%, Japan 8.8%, Taiwan 6.5%, Singapore 6.5%, US 5.3% (2009 est.)
Reserves of foreign exchange and gold: $268.9 billion (31 December 2010 est.)
country comparison to the world: 9
$255.8 billion (31 December 2009 est.)
Debt—external: $750.8 billion (31 December 2010 est.)
country comparison to the world: 18
$664.7 billion (31 December 2009)
Stock of direct foreign investment—at home: $962.2 billion (31 December 2010 est.)
country comparison to the world: 5
$931 billion (31 December 2009 est.)
Stock of direct foreign investment—abroad: $873.1 billion (31 December 2010 est.)
country comparison to the world: 6
$827.4 billion (31 December 2009 est.)
Exchange rates: Hong Kong dollars (HKD) per US dollar–7.78 (2010), 7.75 (2009), 7.751 (2008), 7.802 (2007), 7.7678 (2006)

COMMUNICATIONS

Telephones—main lines in use: 4.188 million (2009)
country comparison to the world: 37
Telephones—mobile cellular: 12.207 million (2009)
country comparison to the world: 56
Telephone system: *general assessment:* modern facilities provide excellent domestic and international services
domestic: microwave radio relay links and extensive fiber-optic network
international: country code–852; multiple international submarine cables provide connections to Asia, US, Australia, the Middle East, and Western Europe; satellite earth stations–3 Intelsat (1 Pacific Ocean and 2 Indian Ocean); coaxial cable to Guangzhou, China
Broadcast media: 2 commercial terrestrial television networks each with multiple stations; multi-channel satellite and cable TV systems are available; 3 radio networks, one of which is government-funded, operate about 15 radio stations (2010)
Internet country code: .hk
Internet hosts: 817,701 (2010)
country comparison to the world: 45
Internet users: 4.873 million (2009)
country comparison to the world: 47

TRANSPORTATION

Airports: 2 (2010)
country comparison to the world: 200
Airports—with paved runways: *total:* 2
over 3,047 m: 1
1,524 to 2,437 m: 1 (2010)
Heliports: 9 (2010)
Roadways: *total:* 2,067 km
country comparison to the world: 172
paved: 2,067 km (2010)
Merchant marine: *total:* 1,429
country comparison to the world: 5
by type: barge carrier 2, bulk carrier 629, cargo 177, carrier 11, chemical tanker 134, container 274, liquefied gas 37, passenger 4, passenger/cargo 9, petroleum tanker 139, roll on/roll off 5, vehicle carrier 8
foreign-owned: 855 (Belgium 16, Bermuda 12, Canada 70, China 432, Cyprus 3, Denmark 41, France 3, Germany 10, Greece 22, Indonesia 8, Iran 1, Japan 84, Libya 1, Norway 49, Russia 1, Singapore 13, South Korea 3, Taiwan 26, UAE 2, UK 27, US 31)
note: this country allows large numbers of ships owned by foreign entities to be registered in its national shipping registry and to fly its flag; these ships operate under the laws of the flag state
registered in other countries: 297 (Bahamas 2, Bermuda 5, Cambodia 11, China 15, Cyprus 2, Georgia 4, Honduras 1, India 1, Kiribati 1, Liberia 47, Malaysia 8, Malta 2, Marshall Islands 3, former Netherlands Antilles 1, NZ 1, Panama 125, Saint Vincent and the Grenadines 4, Seychelles 1, Sierra Leone 4, Singapore 38, Thailand 1, Tuvalu 1, UK 8, unknown 11) (2010)
Ports and terminals: Hong Kong

MILITARY

Military branches: no regular indigenous military forces; Hong Kong garrison of China's People's Liberation Army (PLA) includes elements of the PLA Ground Forces, PLA Navy, and PLA Air Force; these forces are under the direct leadership of the Central Military Commission in Beijing and under administrative control of the adjacent Guangzhou Military Region (2009)
Manpower available for military service: *males age 16–49:* 1,704,090
females age 16–49: 1,873,175 (2010 est.)
Manpower fit for military service: *males age 16–49:* 1,387,213
females age 16–49: 1,505,875 (2010 est.)
Manpower reaching militarily significant age annually:
male: 39,579
female: 36,554 (2010 est.)
Military expenditures: NA
Military—note: defense is the responsibility of China

TRANSNATIONAL ISSUES

Disputes—international: none
Illicit drugs: despite strenuous law enforcement efforts, faces difficult challenges in controlling transit of heroin and methamphetamine to regional and world markets; modern banking system provides conduit for money laundering; rising indigenous use of synthetic drugs, especially among young people

HUNGARY

INTRODUCTION

Background: Hungary became a Christian kingdom in A.D. 1000 and for many centuries served as a bulwark against Ottoman Turkish expansion in Europe. The kingdom eventually became part of the polyglot Austro-Hungarian Empire, which collapsed during World War I. The country fell under Communist rule following World War II. In 1956, a revolt and an announced withdrawal from the Warsaw Pact were met with a massive military intervention by Moscow. Under the leadership of Janos KADAR in 1968, Hungary began liberalizing its economy, introducing so-called "Goulash Communism." Hungary held its first multiparty elections in 1990 and initiated a free market economy. It joined NATO in 1999 and the EU five years later. In 2011, Hungary assumed the six-month rotating presidency of the EU for the first time.

GEOGRAPHY

Location: Central Europe, northwest of Romania
Geographic coordinates: 47 00 N, 20 00 E
Map references: Europe
Area: *total:* 93,028 sq km
country comparison to the world: 109
land: 89,608 sq km
water: 3,420 sq km
Area—comparative: slightly smaller than Indiana
Land boundaries:
total: 2,185 km
border countries: Austria 366 km, Croatia 329 km, Romania 443 km, Serbia 166 km, Slovakia 676 km, Slovenia 102 km, Ukraine 103 km
Coastline: 0 km (landlocked)
Maritime claims: none (landlocked)
Climate: temperate; cold, cloudy, humid winters; warm summers
Terrain: mostly flat to rolling plains; hills and low mountains on the Slovakian border
Elevation extremes: *lowest point:* Tisza River 78 m
highest point: Kekes 1,014 m
Natural resources: bauxite, coal, natural gas, fertile soils, arable land
Land use: *arable land:* 49.58%
permanent crops: 2.06%
other: 48.36% (2005)
Irrigated land: 2,300 sq km (2003)
Total renewable water resources: 120 cu km (2005)
Freshwater withdrawal (domestic/industrial/agricultural): *total:* 21.03 cu km/yr (9%/59%/32%)
per capita: 2,082 cu m/yr (2001)
Environment—current issues: the upgrading of Hungary's standards in waste management, energy efficiency, and air, soil, and water pollution to meet EU requirements will require large investments
Environment—international agreements: *party to:* Air Pollution, Air Pollution-Nitrogen Oxides, Air Pollution-Persistent Organic Pollutants, Air Pollution-Sulfur 85, Air Pollution-Sulfur 94, Air Pollution-Volatile Organic Compounds, Antarctic Treaty, Biodiversity, Climate Change, Climate Change-Kyoto Protocol, Desertification, Endangered Species, Environmental Modification, Hazardous Wastes, Law of the Sea, Marine Dumping, Ozone Layer Protection, Ship Pollution, Wetlands, Whaling
signed, but not ratified: none of the selected agreements
Geography—note: landlocked; strategic location astride main land routes between Western Europe and Balkan Peninsula as well as between Ukraine and Mediterranean basin; the north-south flowing Duna (Danube) and Tisza Rivers divide the country into three large regions

PEOPLE

Population: 9,976,062 (July 2011 est.)
country comparison to the world: 84
Age structure: *0–14 years:* 14.9% (male 767,824/female 721,242)
15–64 years: 68.2% (male 3,361,538/female 3,444,450)
65 years and over: 16.9% (male 622,426/female 1,058,582) (2011 est.)
Median age: *total:* 40.2 years
male: 38.1 years
female: 42.8 years (2011 est.)
Population growth rate: -0.17% (2011 est.)
country comparison to the world: 211
Birth rate: 9.6 births/1,000 population (2011 est.)
country comparison to the world: 200
Death rate: 12.68 deaths/1,000 population (July 2011 est.)
country comparison to the world: 26
Net migration rate: 1.39 migrant(s)/1,000 population (2011 est.)
country comparison to the world: 47
Urbanization: *urban population:* 68% of total population (2010)
rate of urbanization: 0.3% annual rate of change (2010-15 est.)
Major cities—population: BUDAPEST (capital) 1.705 million (2009)
Sex ratio: *at birth:* 1.057 male(s)/female
under 15 years: 1.06 male(s)/female
15–64 years: 0.98 male(s)/female
65 years and over: 0.57 male(s)/female
total population: 0.91 male(s)/female (2011 est.)
Infant mortality rate:
total: 5.31 deaths/1,000 live births
country comparison to the world: 180
male: 5.57 deaths/1,000 live births
female: 5.04 deaths/1,000 live births (2011 est.)
Life expectancy at birth: *total population:* 74.79 years
country comparison to the world: 92
male: 71.04 years
female: 78.76 years (2011 est.)
Total fertility rate: 1.4 children born/woman (2011 est.)
country comparison to the world: 199
HIV/AIDS—adult prevalence rate: less than 0.1% (2009 est.)
country comparison to the world: 135
HIV/AIDS—people living with HIV/AIDS: 3,000 (2009 est.)
country comparison to the world: 129
HIV/AIDS—deaths: fewer than 200 (2009 est.)
country comparison to the world: 113
Major infectious diseases: *degree of risk:* intermediate
food or waterborne diseases: bacterial diarrhea and hepatitis A
vectorborne diseases: tickborne encephalitis (2009)
Drinking water source:
Improved:
urban: 100% of population
rural: 100% of population
total: 100% of population (2008)
Sanitation facility access: *Improved:*
urban: 100% of population
rural: 100% of population
total: 100% of population (2008)
Nationality: *noun:* Hungarian(s)
adjective: Hungarian
Ethnic groups: Hungarian 92.3%, Roma 1.9%, other or unknown 5.8% (2001 census)
Religions: Roman Catholic 51.9%, Calvinist 15.9%, Lutheran 3%, Greek Catholic 2.6%, other Christian 1%, other or unspecified 11.1%, unaffiliated 14.5% (2001 census)
Languages: Hungarian 93.6%, other or unspecified 6.4% (2001 census)
Literacy: *definition:* age 15 and over can read and write

total population: 99.4%
male: 99.5%
female: 99.3% (2003 est.)
School life expectancy (primary to tertiary education): *total:* 15 years
male: 15 years
female: 16 years (2008)
Education expenditures: 5.2% of GDP (2007)
country comparison to the world: 52

GOVERNMENT

Country name: *conventional long form:* Republic of Hungary
conventional short form: Hungary
local long form: Magyar Koztarsasag
local short form: Magyarorszag
Government type: parliamentary democracy
Capital: *name:* Budapest
geographic coordinates: 47 30 N, 19 05 E
time difference: UTC+1 (6 hours ahead of Washington, DC during Standard Time)
daylight saving time: +1hr, begins last Sunday in March; ends last Sunday in October
Administrative divisions: 19 counties (megyek, singular—megye), 23 urban counties (singular—megyei varos), and 1 capital city (fovaros)
counties: Bacs-Kiskun, Baranya, Bekes, Borsod-Abauj-Zemplen, Csongrad, Fejer, Gyor-Moson-Sopron, Hajdu-Bihar, Heves, Jasz-Nagykun-Szolnok, Komarom-Esztergom, Nograd, Pest, Somogy, Szabolcs-Szatmar-Bereg, Tolna, Vas, Veszprem, Zala
urban counties: Bekescsaba, Debrecen, Dunaujvaros, Eger, Erd, Gyor, Hodmezovasarhely, Kaposvar, Kecskemet, Miskolc, Nagykanizsa, Nyiregyhaza, Pecs, Salgotarjan, Sopron, Szeged, Szekesfehervar, Szekszard, Szolnok, Szombathely, Tatabanya, Veszprem, Zalaegerszeg
capital city: Budapest
Independence: 16 November 1918 (republic proclaimed); notable earlier dates: 25 December 1000 (crowning of King STEPHEN I, traditional founding date); 30 March 1867 (Austro-Hungarian dual monarchy established)
National holiday: Saint Stephen's Day, 20 August; note—commemorates the date when his remains were transferred to Buda (now Budapest)
Constitution: 18 August 1949, effective 20 August 1949; revised 19 April 1972; 18 October 1989; and 1997
note: 18 October 1989 revision ensured legal rights for individuals and constitutional checks on the authority of the prime minister and also established the principle of parliamentary oversight; 1997 amendment streamlined the judicial system
Legal system: civil legal system influenced by the German model
International law organization participation: accepts compulsory ICJ jurisdiction with reservations; accepts ICCt jurisdiction
Suffrage: 18 years of age; universal
Executive branch: *chief of state:* President Pal SCHMITT (since 6 August 2010)
head of government: Prime Minister Viktor ORBAN (since 29 May 2010)
cabinet: Council of Ministers prime minister elected by the National Assembly on the recommendation of the president; other ministers proposed by the prime minister and appointed and relieved of their duties by the president
(For more information visit the World Leaders website)
elections: president elected by the National Assembly for a five-year term (eligible for a second term); election last held on 29 June 2010 (next to be held by June 2015); prime minister elected by the National Assembly on the recommendation of the president; election last held 29 May 2010
election results: Pal SCHMITT elected president; National Assembly vote—Pal SCHMITT 263, Andras BALOGH 58; Viktor ORBAN was elected prime minister; National Assembly vote—261 to 107
note: to be elected, the president must win two-thirds of legislative vote in the first two rounds or a simple majority in the third round
Legislative branch: unicameral National Assembly or Orszaggyules (386 seats; members elected by popular vote under a system of proportional and direct representation to serve four-year terms)
elections: last held on 11 and 25 April 2010 (next to be held in April 2014)
election results: percent of vote by party (5% or more of the vote required for parliamentary representation in the first round)—Fidesz 52.7%, MSzP 19.3%, Jobbik 16.7%, LMP 7.5%; seats by party—Fidesz 263, MSzP 59, Jobbik 47, LMP 16, independent 1
Judicial branch: Constitutional Court (judges are elected by the National Assembly for nine-year terms)
Political parties and leaders: Christian Democratic People's Party or KDNP [Zsolt SEMJEN]; Hungarian Civic Alliance or Fidesz [Viktor ORBAN, chairman]; Hungarian Socialist Party or MSzP [Attila MESTERHAZY]; Movement for a Better Hungary or Jobbik [Gabor VONA]; Politics Can Be Different or LMP [13-member leadership]
Political pressure groups and leaders: Air Work Group (works to reduce air pollution in towns and cities); Danube Circle (protests the building of the Gabchikovo-Nagymaros dam); Green Future (protests the impact of lead contamination of local factory on health of the people); Hungarian Civil Liberties Union (Tarsasag a Szabadsagjogokert) or TASZ (freedom of expression, information privacy); Hungarian Helsinki Committee (asylum seekers' rights, human rights in law enforcement and the judicial system); environmentalists: Hungarian Ornithological and Nature Conservation Society (Magyar Madartani Egyesulet)or MME; Green Alternative (Zold Alternativa)
International organization participation: Australia Group, BIS, CE, CEI, CERN, EAPC, EBRD, EIB, ESA (cooperating state), EU, FAO, G-9, IAEA, IBRD, ICAO, ICC, ICRM, IDA, IEA, IFC, IFRCS, ILO, IMF, IMO, IMSO, Interpol, IOC, IOM, IPU, ISO, ITSO, ITU, ITUC, MIGA, MINURSO, NATO, NEA, NSG, OAS (observer), OECD, OIF (observer), OPCW, OSCE, PCA, Schengen Convention, SECI, UN, UNCTAD, UNESCO, UNFICYP, UNHCR, UNIDO, UNIFIL, UNWTO, UPU, WCO, WFTU, WHO, WIPO, WMO, WTO, ZC
Diplomatic representation in the US: *chief of mission:* Ambassador Gyorgy SZAPARY
chancery: 3910 Shoemaker Street NW, Washington, DC 20008
telephone: [1] (202) 362-6730
FAX: [1] (202) 966-8135
consulate(s) general: Chicago, Los Angeles, New York
Diplomatic representation from the US: *chief of mission:* Ambassador Eleni Tsakopoulos KOUNALAKIS
embassy: Szabadsag ter 12, H-1054 Budapest
mailing address: pouch: American Embassy Budapest, 5270 Budapest Place, US Department of State, Washington, DC 20521-5270
telephone: [36] (1) 475-4400
FAX: [36] (1) 475-4764
Flag description: three equal horizontal bands of red (top), white, and green; the flag dates to the national movement of the 18th and 19th centuries, and fuses the medieval colors of the Hungarian coat of arms with the revolutionary tricolor form of the French flag; folklore attributes virtues to the colors: red for strength, white for faithfulness, and green for hope; alternatively, the red is seen as being for the blood spilled in defense of the land, white for freedom, and green for the pasturelands that make up so much of the country
National anthem:
name: "Himnusz" (Hymn)
lyrics/music: Ferenc KOLCSEY/Ferenc ERKEL
note: adopted 1844; the anthem is also known as "Isten, aldd meg a magyart" (God, Bless the Hungarians)

ECONOMY

Economy—overview: Hungary has made the transition from a centrally planned to a market economy, with a per capita income nearly two-thirds that of the EU-25 average. The private sector accounts for more than 80% of GDP. Foreign ownership of and investment in Hungarian firms are widespread, with cumulative foreign direct investment worth more than $70 billion. The government's austerity measures, imposed since late 2006, have reduced the budget deficit from over 9% of GDP in 2006 to 3.2% in 2010, with a target of less than 3% in 2011. Hungary's impending inability to service its short-term debt—brought on by the global financial crisis in late 2008—led Budapest to obtain an IMF/EU/World Bank-arranged financial assistance package worth over $25 billion. The global economic downturn, declining exports, and low domestic consumption and fixed asset accumulation, dampened by government austerity measures, resulted in an economic contraction of 6.3% in 2009. In 2010 the new government implemented a number of changes including cutting business and personal income taxes, but imposed "crisis taxes" on financial institutions, energy and telecom companies, and retailers. The economy rebounded in 2010 with a big boost from exports, especially to Germany, and growth of more than 2.5% is expected in 2011. Unemployment remained high, at more than 10% in 2010.

GDP (purchasing power parity): $187.6 billion (2010 est.)
country comparison to the world: 55
$185.4 billion (2009 est.)
$198.7 billion (2008 est.)
note: data are in 2010 US dollars

GDP (official exchange rate): $129 billion (2010 est.)

GDP—real growth rate: 1.2% (2010 est.)
country comparison to the world: 168
-6.7% (2009 est.)
0.8% (2008 est.)

GDP—per capita (PPP): $18,800 (2010 est.)
country comparison to the world: 64
$18,500 (2009 est.)
$19,800 (2008 est.)
note: data are in 2010 US dollars

GDP—composition by sector:
agriculture: 3.3%
industry: 30.8%
services: 65.9% (2010 est.)

Labor force: 4.3 million (2010 est.)
country comparison to the world: 84

Labor force—by occupation:
agriculture: 4.7%
industry: 30.9%
services: 64.4% (2010)

Unemployment rate: 10.7% (2010 est.)
country comparison to the world: 115
11.4% (2009 est.)

Population below poverty line: 13.9% (2010)

Household income or consumption by percentage share: *lowest 10%:* 3.1%
highest 10%: 22.6% (2009)

Distribution of family income—Gini index: 24.7 (2009)
country comparison to the world: 135
24.4 (1998)

Investment (gross fixed): 19.4% of GDP (2010 est.)
country comparison to the world: 97

Budget: *revenues:* $63.1 billion
expenditures: $67.31 billion (2010 est.)

Public debt: 79.6% of GDP (2010 est.)
country comparison to the world: 18
78.8% of GDP (2009 est.)

Inflation rate (consumer prices): 4.9% (2010 est.)
country comparison to the world: 137
4.2% (2009)

Central bank discount rate: 5.75% (31 December 2010)
country comparison to the world: 68
6.25% (31 December 2009)

Commercial bank prime lending rate: 7.1% (30 November 2010 est.)
country comparison to the world: 104
8.68% (31 December 2009 est.)

Stock of narrow money: $31.81 billion (31 December 2010 est.)
country comparison to the world: 55
$32.5 billion (31 December 2009 est.)

Stock of broad money: $68.87 billion (31 December 2010 est.)
country comparison to the world: 60
$76.38 billion (31 December 2009 est.)

Stock of domestic credit: $108.1 billion (31 December 2010 est.)
country comparison to the world: 49
$116.7 billion (31 December 2009 est.)

Market value of publicly traded shares: $27.88 billion (31 December 2010)
country comparison to the world: 55
$30.2 billion (31 December 2009)
$18.58 billion (31 December 2008)

Agriculture—products: wheat, corn, sunflower seed, potatoes, sugar beets; pigs, cattle, poultry, dairy products

Industries: mining, metallurgy, construction materials, processed foods, textiles, chemicals (especially pharmaceuticals), motor vehicles

Industrial production growth rate: 11% (2010 est.)
country comparison to the world: 16

Electricity—production: 37.55 billion kWh (2010 est.)
country comparison to the world: 56

Electricity—consumption: 42.7 billion kWh (2010 est.)
country comparison to the world: 49

Electricity—exports: 4.703 billion kWh (2010 est.)

Electricity—imports: 9.879 billion kWh (2010 est.)

Oil—production: 21,430 bbl/day (2010 est.)
country comparison to the world: 74

Oil—consumption: 137,300 bbl/day (2010 est.)
country comparison to the world: 69

Oil—exports: 0 bbl/day (2010 est.)
country comparison to the world: 174

Oil—imports: 171,600 bbl/day (2010 est.)
country comparison to the world: 51

Oil—proved reserves: 26.57 million bbl (1 January 2010 est.)
country comparison to the world: 84

Natural gas—production: 2.609 billion cu m (2009)
country comparison to the world: 57

Natural gas—consumption: 11.05 billion cu m (2009)
country comparison to the world: 46

Natural gas—exports: 227 million cu m (2009)
country comparison to the world: 42

Natural gas—imports: 9.63 billion cu m (2009)
country comparison to the world: 22

Natural gas—proved reserves: 8.098 billion cu m (1 January 2010 est.)
country comparison to the world: 81

Current account balance: $-2.128 billion (2010 est.)
country comparison to the world: 155
$592.3 million (2009 est.)

Exports: $93.74 billion (2010 est.)
country comparison to the world: 36
$82.1 billion (2009 est.)

Exports—commodities: machinery and equipment 61.1%, other manufactures 28.7%, food products 6.5%, raw materials 2%, fuels and electricity 1.6% (2009 est.)

Exports—partners: Germany 25.5%, Italy 5.5%, UK 5.4%, Romania 5.3%, Slovakia 5.1%, France 4.9%, Austria 4.7% (2010 est.)

Imports: $87.44 billion (2010 est.)
country comparison to the world: 34
$76.45 billion (2009 est.)

Imports—commodities: machinery and equipment 50%, fuels and electricity 11%, food products, raw materials

Imports—partners: Germany 26.1%, Russia 7.7%, China 6.8%, Austria 5.9%, Netherlands 4.4%, Poland 4.3%, Italy 4.2% (2010 est.)

Reserves of foreign exchange and gold: $44.99 billion (31 December 2010 est.)
country comparison to the world: 27
$44.18 billion (31 December 2009 est.)

Debt—external: $148.4 billion (31 December 2010 est.)
country comparison to the world: 33
$149.8 billion (31 December 2009 est.)

Stock of direct foreign investment—at home: $82.07 billion (31 December 2010 est.)
country comparison to the world: 42
$70.41 billion (31 December 2009 est.)

Stock of direct foreign investment—abroad: $19.8 billion (31 December 2010 est.)
country comparison to the world: 41
$19.41 billion (31 December 2009 est.)

Exchange rates: forints (HUF) per US dollar–206.15 (2010), 202.34 (2009), 171.8 (2008), 183.83 (2007), 210.39 (2006)

COMMUNICATIONS

Telephones—main lines in use: 3.069 million (2009)
country comparison to the world: 50

Telephones—mobile cellular: 11.793 million (2009)
country comparison to the world: 58

Telephone system: *general assessment:* the telephone system has been modernized; the system is digital and highly automated; trunk services are carried by fiber-optic cable and digital microwave radio relay; a program for fiber-optic subscriber connections was initiated in 1996
domestic: competition among mobile-cellular service providers has led to a sharp increase in the use of mobile-cellular phones since 2000 and a decrease in the number of fixed-line connections
international: country code–36; Hungary has fiber-optic cable connections with all neighboring countries; the international switch is in Budapest; satellite earth stations–2 Intelsat (Atlantic Ocean and Indian Ocean regions), 1 Inmarsat, 1 very small aperture terminal (VSAT) system of ground terminals

Broadcast media: mixed system of state-supported public service broadcast media and private broadcasters; the 3 publicly-owned TV channels and the 2 main privately-owned TV stations are the major national broadcasters; a large number of special interest channels have emerged; highly developed market for satellite and cable TV services with about two-thirds of viewers utilizing multi-channel services; 3 state-supported public-service radio networks and 2 major national commercial stations; a large number of local stations including commercial, public service, nonprofit, and community radio stations; digital transition postponed to the end of 2012 (2007)

Internet country code: .hu

Internet hosts: 2.655 million (2010)

country comparison to the world: 30
Internet users: 6.176 million (2009)
country comparison to the world: 41

TRANSPORTATION

Airports: 43 (2010)
country comparison to the world: 99
Airports—with paved runways:
total: 22
over 3,047 m: 2
2,438 to 3,047 m: 7
1,524 to 2,437 m: 5
914 to 1,523 m: 6
under 914 m: 2 (2010)
Airports—with unpaved runways:
total: 21
1,524 to 2,437 m: 2
914 to 1,523 m: 8
under 914 m: 11 (2010)
Heliports: 5 (2010)
Pipelines: gas 4,716 km; oil 984 km; refined products 361 km (2010)
Railways: *total:* 9,208 km
country comparison to the world: 23
broad gauge: 36 km 1.524-m gauge
standard gauge: 7,802 km 1.435-m gauge (2,911 km electrified)
narrow gauge: 219 km 0.760-m gauge (2009)
Roadways: *total:* 197,519 km
country comparison to the world: 25
paved: 74,993 km (43,898 km of interurban roads including 911 km of expressways)
unpaved: 112,526 km (2010)
Waterways: 1,622 km (most on Danube River) (2010)
country comparison to the world: 48
Ports and terminals: Budapest, Dunaujvaros, Gyor-Gonyu, Csepel, Baja, Mohacs

MILITARY

Military branches: Land Forces, Hungarian Air Force (Magyar Legiero, ML) (2011)
Military service age and obligation: 18-25 years of age for voluntary military service; no conscription; 6-month service obligation (2010)
Manpower available for military service:
males age 16–49: 2,349,948
females age 16–49: 2,290,568 (2010 est.)
Manpower fit for military service: *males age 16–49:* 1,902,639
females age 16–49: 1,897,378 (2010 est.)
Manpower reaching militarily significant age annually: *male:* 59,237
female: 55,533 (2010 est.)
Military expenditures: 1.75% of GDP (2005 est.)
country comparison to the world: 85

TRANSNATIONAL ISSUES

Disputes—international: bilateral government, legal, technical and economic working group negotiations continue in 2006 with Slovakia over Hungary's failure to complete its portion of the Gabcikovo-Nagymaros hydroelectric dam project along the Danube; as a member state that forms part of the EU's external border, Hungary has implemented the strict Schengen border rules
Illicit drugs: transshipment point for Southwest Asian heroin and cannabis and for South American cocaine destined for Western Europe; limited producer of precursor chemicals, particularly for amphetamine and methamphetamine; efforts to counter money laundering, related to organized crime and drug trafficking are improving but remain vulnerable; significant consumer of ecstasy

ICELAND

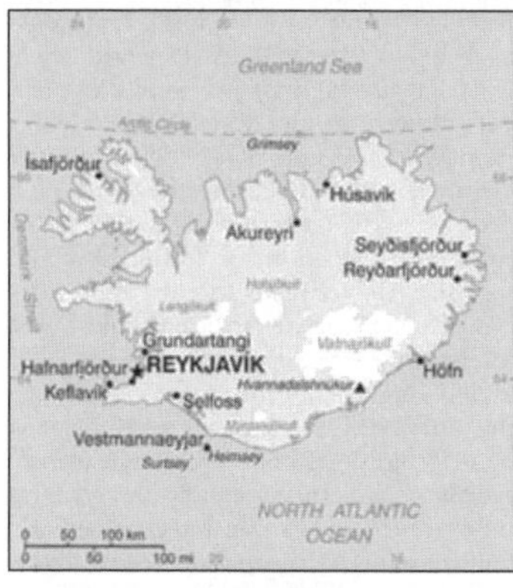

INTRODUCTION

Background: Settled by Norwegian and Celtic (Scottish and Irish) immigrants during the late 9th and 10th centuries A.D., Iceland boasts the world's oldest functioning legislative assembly, the Althing, established in 930. Independent for over 300 years, Iceland was subsequently ruled by Norway and Denmark. Fallout from the Askja volcano of 1875 devastated the Icelandic economy and caused widespread famine. Over the next quarter century, 20% of the island's population emigrated, mostly to Canada and the US. Limited home rule from Denmark was granted in 1874 and complete independence attained in 1944. The second half of the 20th century saw substantial economic growth driven primarily by the fishing industry. The economy diversified greatly after the country joined the European Economic Area in 1994, but Iceland was especially hard hit by the global financial crisis in the years following 2008. Literacy, longevity, and social cohesion are first rate by world standards.

GEOGRAPHY

Location: Northern Europe, island between the Greenland Sea and the North Atlantic Ocean, northwest of the United Kingdom
Geographic coordinates: 65 00 N, 18 00 W
Map references: Europe
Area: *total:* 103,000 sq km
country comparison to the world: 107
land: 100,250 sq km
water: 2,750 sq km
Area—comparative: slightly smaller than Kentucky
Land boundaries: 0 km
Coastline: 4,970 km
Maritime claims: *territorial sea:* 12 nm
exclusive economic zone: 200 nm
continental shelf: 200 nm or to the edge of the continental margin
Climate: temperate; moderated by North Atlantic Current; mild, windy winters; damp, cool summers
Terrain: mostly plateau interspersed with mountain peaks, icefields; coast deeply indented by bays and fiords
Elevation extremes: *lowest point:* Atlantic Ocean 0 m
highest point: Hvannadalshnukur 2,110 m (at Vatnajokull glacier)
Natural resources: fish, hydropower, geothermal power, diatomite
Land use: *arable land:* 0.07%
permanent crops: 0%
other: 99.93% (2005)
Irrigated land: NA
Total renewable water resources: 170 cu km (2005)
Freshwater withdrawal (domestic/industrial/agricultural): *total:* 0.17 cu km/yr (34%/66%/0%)
per capita: 567 cu m/yr (2003)
Natural hazards: earthquakes and volcanic activity
volcanism: Iceland, situated on top of a hotspot, experiences severe volcanic activity; Eyjafjallajokull (elev. 1,666 m) erupted in 2010, sending ash high into the atmosphere and seriously disrupting European air traffic; scientists continue to monitor nearby Katla (elev. 1,512 m), which has a high probability of eruption in the very near future, potentially disrupting air traffic; Grimsvotn and Hekla are Iceland's most frequently active volcanoes; other historically active volcanoes include Askja, Bardarbunga, Brennisteinsfjoll, Esjufjoll, Hengill, Krafla, Krisuvik, Kverkfjoll, Oraefajokull, Reykjanes, Torfajokull, and Vestmannaeyjar
Environment—current issues: water pollution from fertilizer runoff; inadequate wastewater treatment
Environment—international agreements: *party to:* Air Pollution, Air Pollution-Persistent Organic Pollutants, Biodiversity, Climate Change, Climate Change-Kyoto Protocol, Desertification, Endangered Species, Hazardous Wastes, Kyoto Protocol, Law of the Sea, Marine Dumping, Ozone Layer Protection, Ship Pollution, Transboundary Air Pollution, Wetlands, Whaling
signed, but not ratified: Environmental Modification, Marine Life Conservation
Geography—note: strategic location between Greenland and Europe; westernmost European country; Reykjavik is the northernmost national capital in the world; more land covered by glaciers than in all of continental Europe

PEOPLE

Population: 311,058 (July 2011 est.)
country comparison to the world: 178
Age structure: *0–14 years:* 20.2% (male 31,929/female 31,034)
15–64 years: 67.1% (male 105,541/female 103,202)
65 years and over: 12.7% (male 17,974/female 21,378) (2011 est.)
Median age: *total:* 35.6 years
male: 35.2 years
female: 36.1 years (2011 est.)
Population growth rate: 0.687% (2011 est.)
country comparison to the world: 141
Birth rate: 13.29 births/1,000 population (2011 est.)
country comparison to the world: 152
Death rate: 6.96 deaths/1,000 population (July 2011 est.)
country comparison to the world: 137
Net migration rate: 0.53 migrant(s)/1,000 population (2011 est.)
country comparison to the world: 63
Urbanization: *urban population:* 93% of total population (2010)
rate of urbanization: 1.5% annual rate of change (2010-15 est.)
Major cities—population: REYKJAVIK (capital) 198,000 (2009)
Sex ratio: *at birth:* 1.04 male(s)/female
under 15 years: 1.03 male(s)/female
15–64 years: 1.02 male(s)/female
65 years and over: 0.83 male(s)/female
total population: 1 male(s)/female (2011 est.)
Infant mortality rate: *total:* 3.2 deaths/1,000 live births
country comparison to the world: 215
male: 3.34 deaths/1,000 live births
female: 3.05 deaths/1,000 live births (2011 est.)
Life expectancy at birth: *total population:* 80.9 years
country comparison to the world: 18
male: 78.72 years
female: 83.17 years (2011 est.)
Total fertility rate: 1.89 children born/woman (2011 est.)
country comparison to the world: 142
HIV/AIDS—adult prevalence rate: 0.3% (2009 est.)
country comparison to the world: 83
HIV/AIDS—people living with HIV/AIDS: fewer than 1,000 (2009 est.)
country comparison to the world: 144
HIV/AIDS—deaths: fewer than 100 (2009 est.)
country comparison to the world: 135
Drinking water source: *Improved:*
urban: 100% of population
rural: 100% of population
total: 100% of population (2008)
Sanitation facility access: *Improved:*
urban: 100% of population
rural: 100% of population
total: 100% of population (2008)
Nationality: *noun:* Icelander(s)
adjective: Icelandic
Ethnic groups: homogeneous mixture of descendants of Norse and Celts 94%, population of foreign origin 6%
Religions: Lutheran Church of Iceland 80.7%, Roman Catholic Church 2.5%, Reykjavik Free Church 2.4%, Hafnarfjorour Free Church 1.6%, other religions 3.6%, unaffiliated 3%, other or unspecified 6.2% (2006 est.)
Languages: Icelandic, English, Nordic languages, German widely spoken
Literacy: *definition:* age 15 and over can read and write
total population: 99%
male: 99%
female: 99% (2003 est.)
School life expectancy (primary to tertiary education): *total:* 18 years
male: 17 years
female: 20 years (2008)
Education expenditures: 7.4% of GDP (2007)

country comparison to the world: 14

GOVERNMENT

Country name:
conventional long form: Republic of Iceland
conventional short form: Iceland
local long form: Lydveldid Island
local short form: Island
Government type: constitutional republic
Capital: *name:* Reykjavik
geographic coordinates: 64 09 N, 21 57 W
time difference: UTC 0 (5 hours ahead of Washington, DC during Standard Time)
Administrative divisions: 8 regions; Austurland, Hofudhborgarsvaedhi, Nordhurland Eystra, Nordhurland Vestra, Sudhurland, Sudhurnes, Vestfirdhir, Vesturland
Independence: 1 December 1918 (became a sovereign state under the Danish Crown); 17 June 1944 (from Denmark)
National holiday: Independence Day, 17 June (1944)
Constitution: 16 June 1944, effective 17 June 1944; amended many times
Legal system: civil law system influenced by the Danish model
International law organization participation: has not submitted an ICJ jurisdiction declaration; accepts ICCt jurisdiction
Suffrage: 18 years of age; universal
Executive branch:
chief of state: President Olafur Ragnar GRIMSSON (since 1 August 1996)
head of government: Prime Minister Johanna SIGURDARDOTTIR (since 1 February 2009);
cabinet: Cabinet appointed by the prime minister
(For more information visit the World Leaders website)
elections: president, a largely ceremonial post, elected by popular vote for a four-year term (no term limits); election last held on 28 June 2004 (next to be held in June 2012); note—the presidential election of 28 June 2008 was not held because Olafur Ragnar GRIMSSON had no challengers; he was sworn in on 1 August 2008; following legislative elections, the leader of the majority party or the leader of the majority coalition usually the prime minister
election results: Olafur Ragnar GRIMSSON elected president; percent of vote—Olafur Ragnar GRIMSSON 85.6%, Baldur AGUSTSSON 12.5%, Astthor MAGNUSSON 1.9%;
Legislative branch: unicameral Althingi (parliament) (63 seats; members elected by popular vote to serve four-year terms)
elections: last held on 25 April 2009 (next to be held in 2013)
election results: percent of vote by party—Social Democratic Alliance 29.8%, Independence Party 23.7%, Left-Green Movement 21.7%, Progressive Party 14.8%, Citizens' Movement 7.2%, other 2.8%; seats by party—Social Democratic Alliance 20, Independence Party 16, Left-Green Alliance 14, Progressive Party 9, Citizens' Movement 4
note: the Citizens' Movement disintegrated in September 2009; three of its former MPs are now represented under the banner of The Movement and the fourth former MP is an independent
Judicial branch: Supreme Court or Haestirettur (justices are appointed for life by the Minister of Justice); eight district courts (justices are appointed for life by the Minister of Justice)
Political parties and leaders: Independence Party or IP [Bjarni BENEDIKTSSON]; Left-Green Movement or LGM [Steingrimur SIGFUSSON]; Progressive Party or PP [Sigmundur David GUNNLAUGSSON]; Social Democratic Alliance or SDA [Johanna SIGURDARDOTTIR]; The Movement [Birgitta JONSDOTTIR]
International organization participation: Arctic Council, Australia Group, BIS, CBSS, CE, EAPC, EBRD, EFTA, EU (candidate country), FAO, FATF, IAEA, IBRD, ICAO, ICC, ICRM, IDA, IFAD, IFC, IFRCS, IHO, ILO, IMF, IMO, IMSO, Interpol, IOC, IPU, ISO, ITSO, ITU, ITUC, MIGA, NATO, NC, NEA, NIB, NSG, OAS (observer), OECD, OPCW, OSCE, PCA, Schengen Convention, UN, UNCTAD, UNESCO, UPU, WCO, WHO, WIPO, WMO, WTO
Diplomatic representation in the US:
chief of mission: Ambassador Hjalmar W. HANNESSON
chancery: House of Sweden, 2900 K Street NW #509, Washington, DC 20007
telephone: [1] (202) 265-6653
FAX: [1] (202) 265-6656
consulate(s) general: New York
Diplomatic representation from the US: *chief of mission:* Ambassador Luis E. ARREAGA
embassy: Laufasvegur 21, 101 Reykjavik
mailing address: US Department of State, 5640 Reykjavik Place, Washington, D.C. 20521-5640
telephone: [354] 562-9100
FAX: [354] 562-9118
Flag description: blue with a red cross outlined in white extending to the edges of the flag; the vertical part of the cross is shifted to the hoist side in the style of the Dannebrog (Danish flag); the colors represent three of the elements that make up the island: red is for the island's volcanic fires, white recalls the snow and ice fields of the island, and blue is for the surrounding ocean
National anthem:
name: "Lofsongur" (Song of Praise)
lyrics/music: Matthias JOCHUMSSON/Sveinbjorn SVEINBJORNSSON
note: adopted 1944; the anthem, also known as "O, Guo vors Lands" (O, God of Our Land), was originally written and performed in 1874

ECONOMY

Economy—overview: Iceland's Scandinavian-type social-market economy combines a capitalist structure and free-market principles with an extensive welfare system. Prior to the 2008 crisis, Iceland had achieved high growth, low unemployment, and a remarkably even distribution of income. The economy depends heavily on the fishing industry, which provides 40% of export earnings, more than 12% of GDP, and employs 7% of the work force. It remains sensitive to declining fish stocks as well as to fluctuations in world prices for its main exports: fish and fish products, aluminum, and ferrosilicon. Iceland's economy has been diversifying into manufacturing and service industries in the last decade, particularly within the fields of software production, biotechnology, and tourism. Abundant geothermal and hydropower sources have attracted substantial foreign investment in the aluminum sector and boosted economic growth, although the financial crisis has put several investment projects on hold. Much of Iceland's economic growth in recent years came as the result of a boom in domestic demand following the rapid expansion of the country's financial sector. Domestic banks expanded aggressively in foreign markets, and consumers and businesses borrowed heavily in foreign currencies, following the privatization of the banking sector in the early 2000s. Worsening global financial conditions throughout 2008 resulted in a sharp depreciation of the krona vis-a-vis other major currencies. The foreign exposure of Icelandic banks, whose loans and other assets totaled more than 10 times the country's GDP, became unsustainable. Iceland's three largest banks collapsed in late 2008. The country secured over $10 billion in loans from the IMF and other countries to stabilize its currency and financial sector, and to back government guarantees for foreign deposits in Icelandic banks. GDP fell 6.8% in 2009, and unemployment peaked at 9.4% in February 2009. GDP fell 3.4% in 2010. Since the collapse of Iceland's financial sector, government economic priorities have included: stabilizing the krona, reducing Iceland's high budget deficit, containing inflation, restructuring the financial sector, and diversifying the economy. Three new banks were established to take over the domestic assets of the collapsed banks. Two of them have foreign majority ownership, while the State holds a majority of the shares of the third. British and Dutch authorities have pressed claims totaling over $5 billion against Iceland to compensate their citizens for losses suffered on deposits held in the failed Icelandic bank, Landsbanki Islands. Iceland agreed to new terms with the UK and the Netherlands to compensate British and Dutch depositors, but the agreement must first be approved by the Icelandic President. Iceland began accession negotiations with the EU in July 2010; however, public support has dropped substantially because of concern about losing control over fishing resources and in reaction to measures taken by Brussels during the ongoing Eurozone crisis.
GDP (purchasing power parity): $11.82 billion (2010 est.)
country comparison to the world: 145
$12.24 billion (2009 est.)
$13.15 billion (2008 est.)
note: data are in 2010 US dollars
GDP (official exchange rate): $12.59 billion (2010 est.)

GDP—real growth rate: -3.5% (2010 est.)
country comparison to the world: 210
-6.9% (2009 est.)
1.4% (2008 est.)
GDP—per capita (PPP): $38,300 (2010 est.)
country comparison to the world: 25
$39,900 (2009 est.)
$43,200 (2008 est.)
note: data are in 2010 US dollars
GDP—composition by sector:
agriculture: 5.5%
industry: 24.7%
services: 69.9% (2010 est.)
Labor force: 178,800 (2010)
country comparison to the world: 174
Labor force—by occupation:
agriculture: 4.8%
industry: 22.2%
services: 73% (2008)
Unemployment rate:
8.3% (2010 est.)
country comparison to the world: 93
8.6% (2009 est.)
Population below poverty line: NA%
Household income or consumption by percentage share: *lowest 10%:* NA%
highest 10%: NA%
Distribution of family income—Gini index: 28 (2006)
country comparison to the world: 122
25 (2005)
Investment (gross fixed): 12.4% of GDP (2010 est.)
country comparison to the world: 144
Budget: *revenues:* $4.81 billion
expenditures: $5.673 billion (2010 est.)
Public debt: 123.8% of GDP (2010 est.)
country comparison to the world: 6
113.9% of GDP (2009 est.)
Inflation rate (consumer prices): 5.5% (2010 est.)
country comparison to the world: 147
12% (2009 est.)
Central bank discount rate: 14.55% (31 December 2009)
country comparison to the world: 8
22% (31 December 2008)
Commercial bank prime lending rate: 18.99% (31 December 2009 est.)
country comparison to the world: 21
19.29% (31 December 2007)
Stock of narrow money: $4.413 billion (31 December 2010 est.)
country comparison to the world: 96
$4.438 billion (31 December 2009 est.)
Stock of broad money: $19.97 billion (31 December 2010 est.)
country comparison to the world: 83
$24.28 billion (31 December 2009 est.)
Stock of domestic credit: $46.03 billion (31 December 2010 est.)
country comparison to the world: 65
$54.65 billion (31 December 2009 est.)
Market value of publicly traded shares: $1.128 billion (31 December 2009)
country comparison to the world: 78
$5.557 billion (31 December 2008)
$40.56 billion (31 December 2007)
Agriculture—products: potatoes, green vegetables; mutton, chicken, pork, beef, dairy products; fish
Industries: fish processing; aluminum smelting, ferrosilicon production; geothermal power, hydropower, tourism
Industrial production growth rate: -1% (2010 est.)
country comparison to the world: 157
Electricity—production: 16.84 billion kWh (2009 est.)
country comparison to the world: 73
Electricity—consumption: 16.48 billion kWh (2009 est.)
country comparison to the world: 74
Electricity—exports: 0 kWh (2008 est.)
Electricity—imports:
0 kWh (2008 est.)
Oil—production: 0 bbl/day (2009 est.)
country comparison to the world: 186
Oil—consumption: 18,900 bbl/day (2009 est.)
country comparison to the world: 127
Oil—exports: 1,915 bbl/day (2008 est.)
country comparison to the world: 114
Oil—imports: 16,390 bbl/day (2008 est.)
country comparison to the world: 122
Oil—proved reserves: 0 bbl
(1 January 2010 est.)
country comparison to the world: 144
Natural gas—production: 0 cu m (2008 est.)
country comparison to the world: 140
Natural gas—consumption: 0 cu m (2008 est.)
country comparison to the world: 184
Natural gas—exports: 0 cu m (2008 est.)
country comparison to the world: 112
Natural gas—imports: 0 cu m (2008 est.)
country comparison to the world: 199
Natural gas—proved reserves: 0 cu m (1 January 2010 est.)
country comparison to the world: 147
Current account balance: $-42 million (2010 est.)
country comparison to the world: 69
$-440 million (2009 est.)
Exports:
$4.619 billion (2010 est.)
country comparison to the world: 110
$4.05 billion (2009 est.)
Exports—commodities: fish and fish products 40%, aluminum, animal products, ferrosilicon, diatomite
Exports—partners: Netherlands 30.71%, UK 12.73%, Germany 11.21%, Norway 5.75%, Spain 4.82% (2009)
Imports: $3.677 billion (2010 est.)
country comparison to the world: 131
$3.318 billion (2009 est.)
Imports—commodities: machinery and equipment, petroleum products, foodstuffs, textiles
Imports—partners: Norway 12.97%, Netherlands 8.62%, Germany 8.3%, Sweden 8.03%, Denmark 7.27%, US 6.94%, China 4.98%, UK 4.55%, Brazil 4.09% (2009)
Reserves of foreign exchange and gold: $4.206 billion (31 December 2010 est.)
country comparison to the world: 74
$3.883 billion (31 December 2009 est.)
Debt—external: $3.073 billion (2002)
country comparison to the world: 125
Stock of direct foreign investment—at home: $NA (31 December 2010)
$9.2 billion (31 December 2008)
Stock of direct foreign investment—abroad: $NA
$8.8 billion (31 December 2008)
Exchange rates: Icelandic kronur (ISK) per US dollar–139.32 (2010), 123.64 (2009), 85.619 (2008), 63.391 (2007), 70.195 (2006)

COMMUNICATIONS

Telephones—main lines in use: 185,200 (2009)
country comparison to the world: 127
Telephones—mobile cellular: 349,000 (2009)
country comparison to the world: 167
Telephone system: *general assessment:* telecommunications infrastructure is modern and fully digitized, with satellite-earth stations, fiber-optic cables, and an extensive broadband network
domestic: liberalization of the telecommunications sector beginning in the late 1990s has led to increased competition especially in the mobile services segment of the market
international: country code–354; the CANTAT-3 and FARICE-1 submarine cable systems provide connectivity to Canada, the Faroe Islands, UK, Denmark, and Germany; a planned new section of the Hibernia-Atlantic submarine cable will provide additional connectivity to Canada, US, and Ireland; satellite earth stations–2 Intelsat (Atlantic Ocean), 1 Inmarsat (Atlantic and Indian Ocean regions); note–Iceland shares the Inmarsat earth station with the other Nordic countries (Denmark, Finland, Norway, and Sweden)
Broadcast media: state-owned public television broadcaster operates 1 TV channel nationally; several privately-owned TV stations broadcast nationally and roughly another half-dozen operate locally; about half the households utilize multi-channel cable or satellite TV services; state-owned public radio broadcaster operates 2 national networks and 4 regional stations; 2 privately-owned radio stations operate nationally and another 15 provide more limited coverage (2007)
Internet country code: .is
Internet hosts: 344,748 (2010)
country comparison to the world: 55
Internet users: 301,600 (2009)
country comparison to the world: 129

TRANSPORTATION

Airports: 99 (2010)
country comparison to the world: 59
Airports—with paved runways: *total:* 6
over 3,047 m: 1
1,524 to 2,437 m: 3
914 to 1,523 m: 2 (2010)
Airports—with unpaved runways: *total:* 93
1,524 to 2,437 m: 3
914 to 1,523 m: 27
under 914 m: 63 (2010)
Roadways: *total:* 12,869 km
country comparison to the world: 128
paved/oiled gravel: 4,438 km (does not include urban roads)
unpaved: 8,431 km (2009)
Merchant marine: *total:* 2
country comparison to the world: 143
by type: passenger/cargo 2
registered in other countries: 19 (Antigua and Barbuda 9, Belize 2, Denmark 3, Finland 1, Gibraltar 1, Norway 3) (2010)

Ports and terminals: Grundartangi, Hafnarfjordur, Reykjavik

MILITARY

Military branches: no regular military forces; Icelandic National Police (2008)
Manpower available for military service: *males age 16–49:* 75,337 (2010 est.)
Manpower fit for military service: *males age 16–49:* 62,781
females age 16–49: 61,511 (2010 est.)
Manpower reaching militarily significant age annually: *male:* 2,277
female: 2,200 (2010 est.)
Military expenditures: 0% of GDP (2005 est.)
country comparison to the world: 173
Military—note: Iceland has no standing military force; under a 1951 bilateral agreement–still valid–its defense was provided by the US-manned Icelandic Defense Force (IDF) headquartered at Keflavik; however, all US military forces in Iceland were withdrawn as of October 2006; although wartime defense of Iceland remains a NATO commitment, in April 2007, Iceland and Norway signed a bilateral agreement providing for Norwegian aerial surveillance and defense of Icelandic airspace (2008)

TRANSNATIONAL ISSUES

Disputes—international: Iceland, the UK, and Ireland dispute Denmark's claim that the Faroe Islands' continental shelf extends beyond 200 nm

INDIA

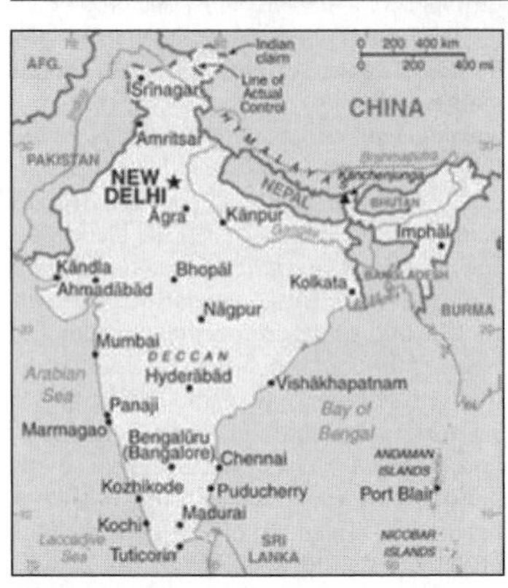

INTRODUCTION

Background: The Indus Valley civilization, one of the world's oldest, flourished during the 3rd and 2nd millennia B.C. and extended into northwestern India. Aryan tribes from the northwest infiltrated onto the Indian subcontinent about 1500 B.C.; their merger with the earlier Dravidian inhabitants created the classical Indian culture. The Maurya Empire of the 4th and 3rd centuries B.C.–which reached its zenith under ASHOKA–united much of South Asia. The Golden Age ushered in by the Gupta dynasty (4th to 6th centuries A.D.) saw a flowering of Indian science, art, and culture. Islam spread across the subcontinent over a period of 700 years. In the 10th and 11th centuries, Turks and Afghans invaded India and established the Delhi Sultanate. In the early 16th century, the Emperor BABUR established the Mughal Dynasty which ruled India for more than three centuries. European explorers began establishing footholds in India during the 16th century. By the 19th century, Great Britain had become the dominant political power on the subcontinent. The British Indian Army played a vital role in both World Wars. Nonviolent resistance to British rule, led by Mohandas GANDHI and Jawaharlal NEHRU, eventually brought about independence in 1947. Communal violence led to the subcontinent's bloody partition, which resulted in the creation of two separate states, India and Pakistan. The two countries have fought three wars since independence, the last of which in 1971 resulted in East Pakistan becoming the separate nation of Bangladesh. India's nuclear weapons tests in 1998 caused Pakistan to conduct its own tests that same year. In November 2008, terrorists allegedly originating from Pakistan conducted a series of coordinated attacks in Mumbai, India's financial capital. Despite pressing problems such as significant overpopulation, environmental degradation, extensive poverty, and widespread corruption, rapid economic development is fueling India's rise on the world stage. In January 2011, India assumed a nonpermanent seat in the UN Security Council for the 2011-12 term.

GEOGRAPHY

Location: Southern Asia, bordering the Arabian Sea and the Bay of Bengal, between Burma and Pakistan
Geographic coordinates: 20 00 N, 77 00 E
Map references: Asia
Area: *total:* 3,287,263 sq km
country comparison to the world: 7
land: 2,973,193 sq km
water: 314,070 sq km
Area—comparative: slightly more than one-third the size of the US
Land boundaries: *total:* 14,103 km
border countries: Bangladesh 4,053 km, Bhutan 605 km, Burma 1,463 km, China 3,380 km, Nepal 1,690 km, Pakistan 2,912 km
Coastline: 7,000 km
Maritime claims: *territorial sea:* 12 nm
contiguous zone: 24 nm
exclusive economic zone: 200 nm
continental shelf: 200 nm or to the edge of the continental margin
Climate: varies from tropical monsoon in south to temperate in north
Terrain: upland plain (Deccan Plateau) in south, flat to rolling plain along the Ganges, deserts in west, Himalayas in north
Elevation extremes: *lowest point:* Indian Ocean 0 m
highest point: Kanchenjunga 8,598 m
Natural resources: coal (fourth-largest reserves in the world), iron ore, manganese, mica, bauxite, rare earth elements, titanium ore, chromite, natural gas, diamonds, petroleum, limestone, arable land
Land use: *arable land:* 48.83%
permanent crops: 2.8%
other: 48.37% (2005)
Irrigated land: 558,080 sq km (2003)
Total renewable water resources: 1,907.8 cu km (1999)
Freshwater withdrawal (domestic/industrial/agricultural): *total:* 645.84 cu km/yr (8%/5%/86%)
per capita: 585 cu m/yr (2000)
Natural hazards: droughts; flash floods, as well as widespread and destructive flooding from monsoonal rains; severe thunderstorms; earthquakes
volcanism: Barren Island (elev. 354 m) in the Andaman Sea has been active in recent years
Environment—current issues: deforestation; soil erosion; overgrazing; desertification; air pollution from industrial effluents and vehicle emissions; water pollution from raw sewage and runoff of agricultural pesticides; tap water is not potable throughout the country; huge and growing population is overstraining natural resources
Environment—international agreements: *party to:* Antarctic-Environmental Protocol, Antarctic-Marine Living Resources, Antarctic Treaty, Biodiversity, Climate Change, Climate Change-Kyoto Protocol, Desertification, Endangered Species, Environmental Modification, Hazardous Wastes, Law of the Sea, Ozone Layer Protection, Ship Pollution, Tropical Timber 83, Tropical Timber 94, Wetlands, Whaling
signed, but not ratified: none of the selected agreements
Geography—note: dominates South Asian subcontinent; near important Indian Ocean trade routes; Kanchenjunga, third tallest mountain in the world, lies on the border with Nepal

PEOPLE

Population: 1,189,172,906 (July 2011 est.)
country comparison to the world: 2
Age structure: *0–14 years:* 29.7% (male 187,450,635/female 165,415,758)
15–64 years: 64.9% (male 398,757,331/female 372,719,379)
65 years and over: 5.5% (male 30,831,190/female 33,998,613) (2011 est.)
Median age: *total:* 26.2 years
male: 25.6 years
female: 26.9 years (2011 est.)
Population growth rate: 1.344% (2011 est.)

country comparison to the world: 86
Birth rate: 20.97 births/1,000 population (2011 est.)
country comparison to the world: 84
Death rate: 7.48 deaths/1,000 population (July 2011 est.)
country comparison to the world: 117
Net migration rate: -0.05 migrant(s)/1,000 population (2011 est.)
country comparison to the world: 118
Urbanization: *urban population:* 30% of total population (2010)
rate of urbanization: 2.4% annual rate of change (2010-15 est.)
Major cities—population: NEW DELHI (capital) 21.72 million; Mumbai 19.695 million; Kolkata 15.294 million; Chennai 7.416 million; Bangalore 7.079 million (2009)
Sex ratio: *at birth:* 1.12 male(s)/female
under 15 years: 1.13 male(s)/female
15–64 years: 1.07 male(s)/female
65 years and over: 0.91 male(s)/female
total population: 1.08 male(s)/female (2011 est.)
Infant mortality rate: *total:* 47.57 deaths/1,000 live births
country comparison to the world: 51
male: 46.18 deaths/1,000 live births
female: 49.14 deaths/1,000 live births (2011 est.)
Life expectancy at birth: *total population:* 66.8 years
country comparison to the world: 161
male: 65.77 years
female: 67.95 years (2011 est.)
Total fertility rate: 2.62 children born/woman (2011 est.)
country comparison to the world: 79
HIV/AIDS—adult prevalence rate: 0.3% (2009 est.)
country comparison to the world: 82
HIV/AIDS—people living with HIV/AIDS: 2.4 million (2009 est.)
country comparison to the world: 4
HIV/AIDS—deaths: 170,000 (2009 est.)
country comparison to the world: 3
Major infectious diseases: *degree of risk:* high
food or waterborne diseases: bacterial diarrhea, hepatitis A and E, and typhoid fever
vectorborne diseases: chikungunya, dengue fever, Japanese encephalitis, and malaria
animal contact disease: rabies
water contact disease: leptospirosis
note: highly pathogenic H5N1 avian influenza has been identified in this country; it poses a negligible risk with extremely rare cases possible among US citizens who have close contact with birds (2009)
Drinking water source: *Improved:*
urban: 96% of population
rural: 84% of population
total: 88% of population
Unimproved: urban: 4% of population
rural: 16% of population
total: 12% of population (2008)
Sanitation facility access: *Improved:*
urban: 54% of population
rural: 21% of population
total: 31% of population
Unimproved: urban: 46% of population
rural: 79% of population
total: 69% of population (2008)

Nationality: *noun:* Indian(s)
adjective: Indian
Ethnic groups: Indo-Aryan 72%, Dravidian 25%, Mongoloid and other 3% (2000)
Religions: Hindu 80.5%, Muslim 13.4%, Christian 2.3%, Sikh 1.9%, other 1.8%, unspecified 0.1% (2001 census)
Languages: Hindi 41%, Bengali 8.1%, Telugu 7.2%, Marathi 7%, Tamil 5.9%, Urdu 5%, Gujarati 4.5%, Kannada 3.7%, Malayalam 3.2%, Oriya 3.2%, Punjabi 2.8%, Assamese 1.3%, Maithili 1.2%, other 5.9%
note: English enjoys the status of subsidiary official language but is the most important language for national, political, and commercial communication; Hindi is the most widely spoken language and primary tongue of 41% of the people; there are 14 other official languages: Bengali, Telugu, Marathi, Tamil, Urdu, Gujarati, Malayalam, Kannada, Oriya, Punjabi, Assamese, Kashmiri, Sindhi, and Sanskrit; Hindustani is a popular variant of Hindi/Urdu spoken widely throughout northern India but is not an official language (2001 census)
Literacy: *definition:* age 15 and over can read and write
total population: 61%
male: 73.4%
female: 47.8% (2001 census)
School life expectancy (primary to tertiary education): *total:* 10 years
male: 11 years
female: 10 years (2007)
Education expenditures: 3.1% of GDP (2006)
country comparison to the world: 130

GOVERNMENT

Country name: *conventional long form:* Republic of India
conventional short form: India
local long form: Republic of India/Bharatiya Ganarajya
local short form: India/Bharat
Government type: federal republic
Capital: *name:* New Delhi
geographic coordinates: 28 36 N, 77 12 E
time difference: UTC+5.5 (10.5 hours ahead of Washington, DC during Standard Time)
Administrative divisions: 28 states and 7 union territories*; Andaman and Nicobar Islands*, Andhra Pradesh, Arunachal Pradesh, Assam, Bihar, Chandigarh*, Chhattisgarh, Dadra and Nagar Haveli*, Daman and Diu*, Delhi*, Goa, Gujarat, Haryana, Himachal Pradesh, Jammu and Kashmir, Jharkhand, Karnataka, Kerala, Lakshadweep*, Madhya Pradesh, Maharashtra, Manipur, Meghalaya, Mizoram, Nagaland, Orissa, Puducherry*, Punjab, Rajasthan, Sikkim, Tamil Nadu, Tripura, Uttar Pradesh, Uttarakhand, West Bengal
Independence: 15 August 1947 (from the UK)
National holiday: Republic Day, 26 January (1950)
Constitution: 26 January 1950; amended many times
Legal system: common law system based on the English model; separate personal law codes apply to Muslims, Christians, and Hindus; judicial review of legislative acts
International law organization participation: accepts compulsory ICJ jurisdiction with reservations; non-party state to the ICCt
Suffrage: 18 years of age; universal
Executive branch: *chief of state:* President Pratibha Devisingh PATIL (since 25 July 2007); Vice President Mohammad Hamid ANSARI (since 11 August 2007)
head of government: Prime Minister Manmohan SINGH (since 22 May 2004)
cabinet: Cabinet appointed by the president on the recommendation of the prime minister
(For more information visit the World Leaders website)
elections: president elected by an electoral college consisting of elected members of both houses of Parliament and the legislatures of the states for a five-year term (no term limits); election last held in July 2007 (next to be held in July 2012); vice president elected by both houses of Parliament for a five-year term; election last held in August 2007 (next to be held August 2012); prime minister chosen by parliamentary members of the majority party following legislative elections; election last held April–May 2009 (next to be held no later than May 2014)
election results: Pratibha PATIL elected president; percent of vote–Pratibha PATIL 65.8%, Bhairon Singh SHEKHAWAT–34.2%
Legislative branch: bicameral Parliament or Sansad consists of the Council of States or Rajya Sabha (a body consisting of not more than 250 members up to 12 of whom are appointed by the president, the remainder are chosen by the elected members of the state and territorial assemblies; members serve six-year terms) and the People's Assembly or Lok Sabha (545 seats; 543 members elected by popular vote, 2 appointed by the president; members serve five-year terms)
elections: People's Assembly–last held in five phases on 16, 22-23, 30 April and 7, 13 May 2009 (next must be held by May 2014)
election results: People's Assembly–percent of vote by party–NA; seats by party–INC 206, BJP 116, SP 23, BSP 21, JD (U) 20, AITC 19, DMK 18, CPI-M 16, BJD 14, SS 11, AIADMK 9, NCP 9, other 61, vacant 2; note–seats by party as November 2009–INC 207, BJP 116, SP 22, BSP 21, JD (U) 20, AITC 19, DMK 18, CPI-M 16, BJD 14, SS 11, AIADMK 9, NCP 9, other 61, vacant 2
Judicial branch: Supreme Court (one chief justice and 25 associate justices are appointed by the president and remain in office until they reach the age of 65 or are removed for "proved misbehavior")
Political parties and leaders: All India Anna Dravida Munnetra Kazhagam or AIADMK [J. JAYALALITHAA]; All India Trinamool Congress or AITC [Mamata BANERJEE]; Bahujan Samaj Party or BSP [MAYAWATI]; Bharatiya Janata Party or BJP [Nitin GADKARI]; Biju Janata Dal or BJD [Naveen PATNAIK]; Communist Party of

India or CPI [B. BARDHAN]; Communist Party of India-Marxist or CPI-M [Prakash KARAT]; Dravida Munnetra Kazhagam or DMK [Kalaignar M.KARUNANIDHI]; Indian National Congress or INC [Sonia GANDHI]; Janata Dal (United) or JD(U) [Sharad YADAV]; Left Front (an alliance of Indian leftist parties); Nationalist Congress Party or NCP [Sharad PAWAR]; Rashtriya Lok Dal or RLD [Ajit SINGH]; Samajwadi Party or SP [Mulayam Singh YADAV]; Shiromani Akali Dal or SAD [Parkash Singh BADAL]; Shiv Sena or SS [Bal THACKERAY]; Telugu Desam Party or TDP [Chandrababu NAIDU]; note–India has dozens of national and regional political parties; only parties or coalitions with four or more seats in the People's Assembly are listed

Political pressure groups and leaders: All Parties Hurriyat Conference in the Kashmir Valley (separatist group); Bajrang Dal (religious organization); National Socialist Council of Nagaland in the northeast (separatist group); Rashtriya Swayamsevak Sangh [Mohan BHAGWAT] (religious organization); Vishwa Hindu Parishad [Ashok SINGHAL] (religious organization)
other: numerous religious or militant/chauvinistic organizations; various separatist groups seeking greater communal and/or regional autonomy

International organization participation: ADB, AfDB (nonregional member), ARF, ASEAN (dialogue partner), BIMSTEC, BIS, C, CD, CERN (observer), CICA, CP, EAS, FAO, FATF, G-15, G-20, G-24, G-77, IAEA, IBRD, ICAO, ICC, ICRM, IDA, IFAD, IFC, IFRCS, IHO, ILO, IMF, IMO, IMSO, Interpol, IOC, IOM, IPU, ISO, ITSO, ITU, ITUC, LAS (observer), MIGA, MONUSCO, NAM, OAS (observer), OPCW, PCA, PIF (partner), SAARC, SACEP, SCO (observer), UN, UN Security Council (temporary), UNCTAD, UNDOF, UNESCO, UNHCR, UNIDO, UNIFIL, UNITAR, UNMIS, UNMIT, UNOCI, UNWTO, UPU, WCO, WFTU, WHO, WIPO, WMO, WTO

Diplomatic representation in the US: *chief of mission:* Ambassador Meera SHANKAR
chancery: 2107 Massachusetts Avenue NW, Washington, DC 20008; note–Consular Wing located at 2536 Massachusetts Avenue NW, Washington, DC 20008
telephone: [1] (202) 939-7000
FAX: [1] (202) 265-4351
consulate(s) general: Chicago, Houston, New York, San Francisco

Diplomatic representation from the US: *chief of mission:* Ambassador Timothy J. ROEMER
embassy: Shantipath, Chanakyapuri, New Delhi 110021
mailing address: use embassy street address
telephone: [91] (011) 2419-8000
FAX: [91] (11) 2419-0017
consulate(s) general: Chennai (Madras), Hyderabad; Kolkata (Calcutta), Mumbai (Bombay)

Flag description: three equal horizontal bands of saffron (subdued orange) (top), white, and green, with a blue chakra (24-spoked wheel) centered in the white band; saffron represents courage, sacrifice, and the spirit of renunciation; white signifies purity and truth; green stands for faith and fertility; the blue chakra symbolizes the wheel of life in movement and death in stagnation
note: similar to the flag of Niger, which has a small orange disk centered in the white band

National anthem: *name:* "Jana-Gana-Mana" (Thou Art the Ruler of the Minds of All People)
lyrics/music: Rabindranath TAGORE
note: adopted 1950; Rabindranath TAGORE, a Nobel laureate, also wrote Bangladesh's national anthem

ECONOMY

Economy—overview: India is developing into an open-market economy, yet traces of its past autarkic policies remain. Economic liberalization, including industrial deregulation, privatization of state-owned enterprises, and reduced controls on foreign trade and investment, began in the early 1990s and has served to accelerate the country's growth, which has averaged more than 7% per year since 1997. India's diverse economy encompasses traditional village farming, modern agriculture, handicrafts, a wide range of modern industries, and a multitude of services. Slightly more than half of the work force is in agriculture, but services are the major source of economic growth, accounting for more than half of India's output, with only one-third of its labor force. India has capitalized on its large educated English-speaking population to become a major exporter of information technology services and software workers. In 2010, the Indian economy rebounded robustly from the global financial crisis–in large part because of strong domestic demand–and growth exceeded 8% year-on-year in real terms. Merchandise exports, which account for about 15% of GDP, returned to pre-financial crisis levels. An industrial expansion and high food prices, resulting from the combined effects of the weak 2009 monsoon and inefficiencies in the government's food distribution system, fueled inflation which peaked at about 11% in the first half of 2010, but has gradually decreased to single digits following a series of central bank interest rate hikes. In 2010 New Delhi reduced subsidies for fuel and fertilizers, sold a small percentage of its shares in some state-owned enterprises and auctioned off rights to radio bandwidth for 3G telecommunications in part to lower the government's deficit. The Indian Government seeks to reduce its budget deficit to 5.5% of GDP in FY 2010-11, down from 6.8% in the previous fiscal year. India's long term challenges include widespread poverty, inadequate physical and social infrastructure, limited non-agricultural employment opportunities, insufficient access to quality basic and higher education, and accommodating rural-to-urban migration.

GDP (purchasing power parity): $4.06 trillion (2010 est.)
country comparison to the world: 5
$3.679 trillion (2009 est.)
$3.447 trillion (2008 est.)
note: data are in 2010 US dollars

GDP (official exchange rate): $1.538 trillion (2010 est.)

GDP—real growth rate: 10.4% (2010 est.)
country comparison to the world: 5
6.8% (2009 est.)
6.2% (2008 est.)

GDP—per capita (PPP): $3,500 (2010 est.)
country comparison to the world: 163
$3,200 (2009 est.)
$3,000 (2008 est.)
note: data are in 2010 US dollars

GDP—composition by sector: *agriculture:* 16.1%
industry: 28.6%
services: 55.3% (2010 est.)

Labor force: 478.3 million (2010 est.)
country comparison to the world: 2

Labor force—by occupation:
agriculture: 52%
industry: 14%
services: 34% (2009 est.)

Unemployment rate: 10.8% (2010 est.)
country comparison to the world: 118
10.7% (2009 est.)

Population below poverty line: 25% (2007 est.)

Household income or consumption by percentage share: *lowest 10%:* 3.6%
highest 10%: 31.1% (2005)

Distribution of family income—Gini index: 36.8 (2004)
country comparison to the world: 79
37.8 (1997)

Investment (gross fixed): 32% of GDP (2010 est.)
country comparison to the world: 15

Budget: *revenues:* $170.7 billion
expenditures: $268 billion (2010 est.)

Public debt: 55.9% of GDP (2010 est.)
country comparison to the world: 45
57.3% of GDP (2009 est.)

Inflation rate (consumer prices): 11.7% (2010 est.)
country comparison to the world: 201
10.9% (2009 est.)

Central bank discount rate: 6% (31 December 2009)
country comparison to the world: 73
6% (31 December 2008)

Commercial bank prime lending rate: 12.19% (31 December 2009 est.)
country comparison to the world: 56
13.31% (31 December 2008 est.)

Stock of narrow money: $328.4 billion (31 December 2010 est.)
country comparison to the world: 15
$268.4 billion (31 December 2009 est.)

Stock of broad money: $1.29 trillion (31 December 2010 est.)
country comparison to the world: 14
$1.04 trillion (31 December 2009 est.)

Stock of domestic credit: $1.164 trillion (31 December 2010 est.)
country comparison to the world: 14
$938.8 billion (31 December 2009 est.)

Market value of publicly traded shares: $1.179 trillion (31 December 2009)
country comparison to the world: 15
$645.5 billion (31 December 2008)
$1.819 trillion (31 December 2007)

Agriculture—products: rice, wheat, oilseed, cotton, jute, tea, sugarcane, lentils, onions, potatoes; dairy products, sheep, goats, poultry; fish
Industries: textiles, chemicals, food processing, steel, transportation equipment, cement, mining, petroleum, machinery, software, pharmaceuticals
Industrial production growth rate: 9.7% (2010 est.)
country comparison to the world: 21
Electricity—production: 723.8 billion kWh (2009 est.)
country comparison to the world: 6
Electricity—consumption: 568 billion kWh (2007 est.)
country comparison to the world: 6
Electricity—exports: 810 million kWh (2009 est.)
Electricity—imports: 5.27 billion kWh (2009 est.)
Oil—production: 878,700 bbl/day (2009 est.)
country comparison to the world: 24
Oil—consumption: 2.98 million bbl/day (2009 est.)
country comparison to the world: 5
Oil—exports: 738,600 bbl/day (2007 est.)
country comparison to the world: 23
Oil—imports: 2.9 million bbl/day (2007 est.)
country comparison to the world: 6
Oil—proved reserves: 5.8 billion bbl (1 January 2010 est.)
country comparison to the world: 23
Natural gas—production: 38.65 billion cu m (2009 est.)
country comparison to the world: 22
Natural gas—consumption: 51.27 billion cu m (2009 est.)
country comparison to the world: 16
Natural gas—exports: 0 cu m (2008 est.)
country comparison to the world: 113
Natural gas—imports: 12.62 billion cu m (2009 est.)
country comparison to the world: 17
Natural gas—proved reserves: 1.075 trillion cu m (1 January 2010 est.)
country comparison to the world: 26
Current account balance: $-26.91 billion (2010 est.)
country comparison to the world: 182
$-26.63 billion (2009 est.)
Exports: $201 billion (2010 est.)
country comparison to the world: 23
$168.2 billion (2009 est.)
Exports—commodities: petroleum products, precious stones, machinery, iron and steel, chemicals, vehicles, apparel
Exports—partners: UAE 12.87%, US 12.59%, China 5.59% (2009)
Imports: $327 billion (2010 est.)
country comparison to the world: 13
$274.3 billion (2009 est.)
Imports—commodities: crude oil, precious stones, machinery, fertilizer, iron and steel, chemicals
Imports—partners: China 10.94%, US 7.16%, Saudi Arabia 5.36%, UAE 5.18%, Australia 5.02%, Germany 4.86%, Singapore 4.02% (2009)
Reserves of foreign exchange and gold: $284.1 billion (31 December 2010 est.)
country comparison to the world: 7
$274.7 billion (31 December 2009 est.)
Debt—external: $237.1 billion (31 December 2010 est.)
country comparison to the world: 29
$221.3 billion (31 December 2009 est.)
Stock of direct foreign investment—at home: $191.1 billion (31 December 2010 est.)
country comparison to the world: 23
$157.9 billion (31 December 2009 est.)
Stock of direct foreign investment—abroad: $89.04 billion (31 December 2010 est.)
country comparison to the world: 26
$76.62 billion (31 December 2009 est.)
Exchange rates: Indian rupees (INR) per US dollar–46.163 (2010), 48.405 (2009), 43.319 (2008), 41.487 (2007), 45.3 (2006)

COMMUNICATIONS

Telephones—main lines in use: 35.77 million (2010)
country comparison to the world: 8
Telephones—mobile cellular: 670 million (2010)
country comparison to the world: 2
Telephone system: *general assessment:* supported by recent deregulation and liberalization of telecommunications laws and policies, India has emerged as one of the fastest growing telecom markets in the world; total telephone subscribership base reached 700 million, an overall teledensity of 60%, and subscribership is currently growing more than 15 million per month; urban teledensity has reached 100% and rural teledensity is about 20% and steadily growing
domestic: mobile cellular service introduced in 1994 and organized nationwide into four metropolitan areas and 19 telecom circles each with multiple private service providers and one or more state-owned service providers; in recent years significant trunk capacity added in the form of fiber-optic cable and one of the world's largest domestic satellite systems, the Indian National Satellite system (INSAT), with 6 satellites supporting 33,000 very small aperture terminals (VSAT)
international: country code–91; a number of major international submarine cable systems, including Sea-Me-We-3 with landing sites at Cochin and Mumbai (Bombay), Sea-Me-We-4 with a landing site at Chennai, Fiber-Optic Link Around the Globe (FLAG) with a landing site at Mumbai (Bombay), South Africa–Far East (SAFE) with a landing site at Cochin, the i2i cable network linking to Singapore with landing sites at Mumbai (Bombay) and Chennai (Madras), and Tata Indicom linking Singapore and Chennai (Madras), provide a significant increase in the bandwidth available for both voice and data traffic; satellite earth stations–8 Intelsat (Indian Ocean) and 1 Inmarsat (Indian Ocean region); 9 gateway exchanges operating from Mumbai (Bombay), New Delhi, Kolkata (Calcutta), Chennai (Madras), Jalandhar, Kanpur, Gandhinagar, Hyderabad, and Ernakulam (2010)
Broadcast media: Doordarshan, India's public TV network, operates about 20 national, regional, and local services; large number of privately-owned TV stations are distributed by cable and satellite service providers; government controls AM radio with All India Radio operating domestic and external networks; news broadcasts via radio are limited to the All India Radio Network; since 2000, privately-owned FM stations are permitted but limited to broadcasting entertainment and educational content (2007)
Internet country code: .in
Internet hosts: 4.536 million (2010)
country comparison to the world: 18
Internet users: 61.338 million (2009)
country comparison to the world: 6

TRANSPORTATION

Airports: 352 (2010)
country comparison to the world: 23
Airports—with paved runways: *total:* 249
over 3,047 m: 21
2,438 to 3,047 m: 57
1,524 to 2,437 m: 75
914 to 1,523 m: 81
under 914 m: 15 (2010)
Airports—with unpaved runways: *total:* 103
over 3,047 m: 1
2,438 to 3,047 m: 3
1,524 to 2,437 m: 8
914 to 1,523 m: 43
under 914 m: 48 (2010)
Heliports: 40 (2010)
Pipelines: condensate/gas 2 km; gas 9,596 km; liquid petroleum gas 2,152 km; oil 7,448 km; refined products 10,486 km (2010)
Railways: *total:* 63,974 km
country comparison to the world: 4
broad gauge: 54,257 km 1.676-m gauge (18,927 km electrified)
narrow gauge: 7,180 km 1.000-m gauge; 2,537 km 0.762-m gauge and 0.610-m gauge (2010)
Roadways: *total:* 3,320,410 km (includes 200 km of expressways) (2009)
country comparison to the world: 3
Waterways: 14,500 km (5,200 km on major rivers and 485 km on canals suitable for mechanized vessels) (2008)
country comparison to the world: 9
Merchant marine: *total:* 324
country comparison to the world: 29
by type: bulk carrier 94, cargo 78, chemical tanker 23, container 15, liquefied gas 11, passenger 4, passenger/cargo 12, petroleum tanker 87
foreign-owned: 8 (China 1, Hong Kong 1, Jersey 1, Malaysia 1, UAE 4)
registered in other countries: 56 (Cyprus 2, Dominica 2, Liberia 1, Malta 4, Marshall Islands 8, Nigeria 1, Panama 17, Singapore 19, unknown 2) (2010)
Ports and terminals: Chennai, Jawaharal Nehru, Kandla, Kolkata (Calcutta), Mumbai (Bombay), Sikka, Vishakhapatnam

MILITARY

Military branches: Army, Navy (includes naval air arm), Air Force, Coast Guard (2011)
Military service age and obligation: 17 years 6 months of age for voluntary

military service; no conscription; women may join as officers, but for noncombat roles only (2010)
Manpower available for military service: *males age 16–49:* 319,129,420
females age 16–49: 296,071,637 (2010 est.)
Manpower fit for military service: *males age 16–49:* 249,531,562
females age 16–49: 240,039,958 (2010 est.)
Manpower reaching militarily significant age annually: *male:* 12,151,065
female: 10,745,891 (2010 est.)
Military expenditures: 2.5% of GDP (2006)
country comparison to the world: 62

TRANSNATIONAL ISSUES

Disputes—international: since China and India launched a security and foreign policy dialogue in 2005, consolidated discussions related to the dispute over most of their rugged, militarized boundary, regional nuclear proliferation, Indian claims that China transferred missiles to Pakistan, and other matters continue; various talks and confidence-building measures have cautiously begun to defuse tensions over Kashmir, particularly since the October 2005 earthquake in the region; Kashmir nevertheless remains the site of the world's largest and most militarized territorial dispute with portions under the de facto administration of China (Aksai Chin), India (Jammu and Kashmir), and Pakistan (Azad Kashmir and Northern Areas); India and Pakistan have maintained the 2004 cease fire in Kashmir and initiated discussions on defusing the armed stand-off in the Siachen glacier region; Pakistan protests India's fencing the highly militarized Line of Control and construction of the Baglihar Dam on the Chenab River in Jammu and Kashmir, which is part of the larger dispute on water sharing of the Indus River and its tributaries; UN Military Observer Group in India and Pakistan (UNMOGIP) has maintained a small group of peacekeepers since 1949; India does not recognize Pakistan's ceding historic Kashmir lands to China in 1964; to defuse tensions and prepare for discussions on a maritime boundary, India and Pakistan seek technical resolution of the disputed boundary in Sir Creek estuary at the mouth of the Rann of Kutch in the Arabian Sea; Pakistani maps continue to show its Junagadh claim in Indian Gujarat State; discussions with Bangladesh remain stalled to delimit a small section of river boundary, to exchange territory for 51 Bangladeshi exclaves in India and 111 Indian exclaves in Bangladesh, to allocate divided villages, and to stop illegal cross-border trade, migration, violence, and transit of terrorists through the porous border; Bangladesh protests India's fencing and walling-off high-traffic sections of the porous boundary; a joint Bangladesh-India boundary commission agreed to fully demarcate the Bangladesh-India boundary in the Dhubri-Kruigram sector; Bangladesh referred its maritime boundary claims with Burma and India to the International Tribunal on the Law of the Sea; fencing along the India-Burma international border at Manipur's Moreh town is in progress to check illegal drug trafficking and movement of militants; Bhutan cooperates with India to expel Indian Nagaland separatists; Joint Border Committee with Nepal continues to examine contested boundary sections, including the 400 square kilometer dispute over the source of the Kalapani River; India maintains a strict border regime to keep out Maoist insurgents and control illegal cross-border activities from Nepal
Refugees and internally displaced persons: refugees (country of origin): 77,200 (Tibet/China); 69,609 (Sri Lanka); 9,472 (Afghanistan)
IDPs: at least 600,000 (about half are Kashmiri Pandits from Jammu and Kashmir) (2007)
Trafficking in persons: current situation: India is a source, destination, and transit country for men, women, and children trafficked for the purposes of forced labor and commercial sexual exploitation; internal forced labor may constitute India's largest trafficking problem; men, women, and children are held in debt bondage and face forced labor working in brick kilns, rice mills, agriculture, and embroidery factories; women and girls are trafficked within the country for the purposes of commercial sexual exploitation and forced marriage; children are subjected to forced labor as factory workers, domestic servants, beggars, and agriculture workers, and have been used as armed combatants by some terrorist and insurgent groups; India is also a destination for women and girls from Nepal and Bangladesh trafficked for the purpose of commercial sexual exploitation; Indian women are trafficked to the Middle East for commercial sexual exploitation; men and women from Bangladesh and Nepal are trafficked through India for forced labor and commercial sexual exploitation in the Middle East
tier rating: Tier 2 Watch List–India is on the Tier 2 Watch List for a fifth consecutive year for its failure to provide evidence of increasing efforts to combat human trafficking in 2007; despite the reported extent of the trafficking crisis in India, government authorities made uneven efforts to prosecute traffickers and protect trafficking victims; government authorities continued to rescue victims of commercial sexual exploitation and forced child labor and child armed combatants, and began to show progress in law enforcement against these forms of trafficking; a critical challenge overall is the lack of punishment for traffickers, effectively resulting in impunity for acts of human trafficking; India has not ratified the 2000 UN TIP Protocol (2008)
Illicit drugs: world's largest producer of licit opium for the pharmaceutical trade, but an undetermined quantity of opium is diverted to illicit international drug markets; transit point for illicit narcotics produced in neighboring countries and throughout Southwest Asia; illicit producer of methaqualone; vulnerable to narcotics money laundering through the hawala system; licit ketamine and precursor production

INDIAN OCEAN

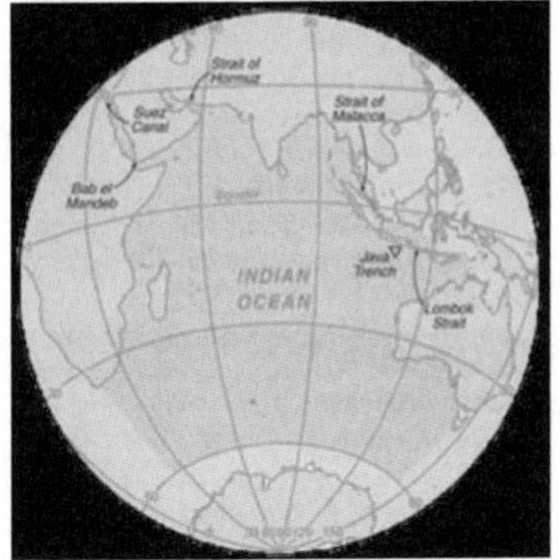

INTRODUCTION

Background: The Indian Ocean is the third largest of the world's five oceans (after the Pacific Ocean and Atlantic Ocean, but larger than the Southern Ocean and Arctic Ocean). Four critically important access waterways are the Suez Canal (Egypt), Bab el Mandeb (Djibouti-Yemen), Strait of Hormuz (Iran-Oman), and Strait of Malacca (Indonesia-Malaysia). The decision by the International Hydrographic Organization in the spring of 2000 to delimit a fifth ocean, the Southern Ocean, removed the portion of the Indian Ocean south of 60 degrees south latitude.

GEOGRAPHY

Location: body of water between Africa, the Southern Ocean, Asia, and Australia
Geographic coordinates: 20 00 S, 80 00 E
Map references: Political Map of the World
Area: *total:* 68.556 million sq km
note: includes Andaman Sea, Arabian Sea, Bay of Bengal, Flores Sea, Great Australian Bight, Gulf of Aden, Gulf of Oman, Java Sea, Mozambique Channel, Persian Gulf, Red Sea, Savu Sea, Strait of Malacca, Timor Sea, and other tributary water bodies
Area—comparative: about 5.5 times the size of the US
Coastline: 66,526 km

Climate: northeast monsoon (December to April), southwest monsoon (June to October); tropical cyclones occur during May/June and October/November in the northern Indian Ocean and January/February in the southern Indian Ocean
Terrain: surface dominated by counterclockwise gyre (broad, circular system of currents) in the southern Indian Ocean; unique reversal of surface currents in the northern Indian Ocean; low atmospheric pressure over southwest Asia from hot, rising, summer air results in the southwest monsoon and southwest-to-northeast winds and currents, while high pressure over northern Asia from cold, falling, winter air results in the northeast monsoon and northeast-to-southwest winds and currents; ocean floor is dominated by the Mid-Indian Ocean Ridge and subdivided by the Southeast Indian Ocean Ridge, Southwest Indian Ocean Ridge, and Ninetyeast Ridge
Elevation extremes: *lowest point:* Java Trench -7,258 m
highest point: sea level 0 m
Natural resources: oil and gas fields, fish, shrimp, sand and gravel aggregates, placer deposits, polymetallic nodules
Natural hazards: occasional icebergs pose navigational hazard in southern reaches
Environment—current issues: endangered marine species include the dugong, seals, turtles, and whales; oil pollution in the Arabian Sea, Persian Gulf, and Red Sea
Geography—note: major chokepoints include Bab el Mandeb, Strait of Hormuz, Strait of Malacca, southern access to the Suez Canal, and the Lombok Strait

ECONOMY

Economy—overview: The Indian Ocean provides major sea routes connecting the Middle East, Africa, and East Asia with Europe and the Americas. It carries a particularly heavy traffic of petroleum and petroleum products from the oilfields of the Persian Gulf and Indonesia. Its fish are of great and growing importance to the bordering countries for domestic consumption and export. Fishing fleets from Russia, Japan, South Korea, and Taiwan also exploit the Indian Ocean, mainly for shrimp and tuna. Large reserves of hydrocarbons are being tapped in the offshore areas of Saudi Arabia, Iran, India, and western Australia. An estimated 40% of the world's offshore oil production comes from the Indian Ocean. Beach sands rich in heavy minerals and offshore placer deposits are actively exploited by bordering countries, particularly India, South Africa, Indonesia, Sri Lanka, and Thailand.

TRANSPORTATION

Ports and terminals: Chennai (Madras, India); Colombo (Sri Lanka); Durban (South Africa); Jakarta (Indonesia); Kolkata (Calcutta, India); Melbourne (Australia); Mumbai (Bombay, India); Richards Bay (South Africa)
Transportation—note: the International Maritime Bureau reports the territorial waters of littoral states and offshore waters as high risk for piracy and armed robbery against ships, particularly in the Gulf of Aden, along the east coast of Africa, the Bay of Bengal, and the Strait of Malacca; numerous vessels, including commercial shipping and pleasure craft, have been attacked and hijacked both at anchor and while underway; hijacked vessels are often disguised and cargoes stolen; crew and passengers are often held for ransom, murdered, or cast adrift; the presence of several naval task forces in the Gulf of Aden and additional anti-piracy measures on the part of ship operators have reduced the piracy incidents in that body of water by more than half in 2010; in response, Somali-based pirates, using hijacked fishing trawlers as "mother ships" to extend their range, shifted operations as far south as the Mozambique Channel and eastward to the vicinity of the Maldives

TRANSNATIONAL ISSUES

Disputes—international: some maritime disputes (see littoral states)

INDONESIA

INTRODUCTION

Background: The Dutch began to colonize Indonesia in the early 17th century; Japan occupied the islands from 1942 to 1945. Indonesia declared its independence after Japan's surrender, but it required four years of intermittent negotiations, recurring hostilities, and UN mediation before the Netherlands agreed to transfer sovereignty in 1949. Free and fair legislative elections took place in 1999 after decades of repressive rule. Indonesia is now the world's third most populous democracy, the world's largest archipelagic state, and home to the world's largest Muslim population. Current issues include: alleviating poverty, improving education, preventing terrorism, consolidating democracy after four decades of authoritarianism, implementing economic and financial reforms, stemming corruption, holding the military and police accountable for human rights violations, addressing climate change, and controlling infectious diseases, particularly those of global and regional importance. In 2005, Indonesia reached a historic peace agreement with armed separatists in Aceh, which led to democratic elections in Aceh in December 2006. Indonesia continues to face low intensity armed resistance by the separatist Free Papua Movement.

GEOGRAPHY

Location: Southeastern Asia, archipelago between the Indian Ocean and the Pacific Ocean
Geographic coordinates: 5 00 S, 120 00 E
Map references: Southeast Asia
Area:
total: 1,904,569 sq km
country comparison to the world: 16
land: 1,811,569 sq km
water: 93,000 sq km
Area—comparative: slightly less than three times the size of Texas
Land boundaries: *total:* 2,830 km
border countries: Timor-Leste 228 km, Malaysia 1,782 km, Papua New Guinea 820 km
Coastline: 54,716 km
Maritime claims: measured from claimed archipelagic straight baselines
territorial sea: 12 nm
exclusive economic zone: 200 nm
Climate: tropical; hot, humid; more moderate in highlands
Terrain: mostly coastal lowlands; larger islands have interior mountains
Elevation extremes: *lowest point:* Indian Ocean 0 m
highest point: Puncak Jaya 5,030 m
Natural resources: petroleum, tin, natural gas, nickel, timber, bauxite, copper, fertile soils, coal, gold, silver
Land use: *arable land:* 11.03%
permanent crops: 7.04%
other: 81.93% (2005)
Irrigated land: 45,000 sq km (2003)
Total renewable water resources: 2,838 cu km (1999)
Freshwater withdrawal (domestic/industrial/agricultural): *total:* 82.78 cu km/yr (8%/1%/91%)

per capita: 372 cu m/yr (2000)
Natural hazards: occasional floods; severe droughts; tsunamis; earthquakes; volcanoes; forest fires
volcanism: Indonesia contains the most volcanoes of any country in the world–some 76 are historically active; significant volcanic activity occurs on Java, western Sumatra, the Sunda Islands, Halmahera Island, Sulawesi Island, Sangihe Island, and in the Banda Sea; Merapi (elev. 2,968 m), Indonesia's most active volcano and in eruption since 2010, has been deemed a "Decade Volcano" by the International Association of Volcanology and Chemistry of the Earth's Interior, worthy of study due to its explosive history and close proximity to human populations; other notable historically active volcanoes include Agung, Awu, Karangetang, Krakatau (Krakatoa), Makian, Raung, and Tambora
Environment—current issues: deforestation; water pollution from industrial wastes, sewage; air pollution in urban areas; smoke and haze from forest fires
Environment—international agreements: *party to:* Biodiversity, Climate Change, Climate Change-Kyoto Protocol, Desertification, Endangered Species, Hazardous Wastes, Law of the Sea, Ozone Layer Protection, Ship Pollution, Tropical Timber 83, Tropical Timber 94, Wetlands
signed, but not ratified: Marine Life Conservation
Geography—note: archipelago of 17,508 islands (6,000 inhabited); straddles equator; strategic location astride or along major sea lanes from Indian Ocean to Pacific Ocean

PEOPLE

Population: 245,613,043 (July 2011 est.)
country comparison to the world: 4
Age structure: *0–14 years:* 27.3% (male 34,165,213/female 32,978,841)
15–64 years: 66.5% (male 82,104,636/female 81,263,055)
65 years and over: 6.1% (male 6,654,695/female 8,446,603) (2011 est.)
Median age: *total:* 28.2 years
male: 27.7 years
female: 28.7 years (2011 est.)
Population growth rate: 1.069% (2011 est.)
country comparison to the world: 111
Birth rate: 18.1 births/1,000 population (2011 est.)
country comparison to the world: 105
Death rate: 6.26 deaths/1,000 population (July 2011 est.)
country comparison to the world: 156
Net migration rate: -1.15 migrant(s)/1,000 population (2011 est.)
country comparison to the world: 153
Urbanization: *urban population:* 44% of total population (2010)
rate of urbanization: 1.7% annual rate of change (2010-15 est.)
Major cities—population: JAKARTA (capital) 9.121 million; Surabaya 2.509 million; Bandung 2.412 million; Medan 2.131 million; Semarang 1.296 million (2009)
Sex ratio: *at birth:* 1.05 male(s)/female
under 15 years: 1.04 male(s)/female
15–64 years: 1.01 male(s)/female
65 years and over: 0.79 male(s)/female
total population: 1 male(s)/female (2011 est.)
Infant mortality rate:
total: 27.95 deaths/1,000 live births
country comparison to the world: 72
male: 32.63 deaths/1,000 live births
female: 23.03 deaths/1,000 live births (2011 est.)
Life expectancy at birth: *total population:* 71.33 years
country comparison to the world: 137
male: 68.8 years
female: 73.99 years (2011 est.)
Total fertility rate: 2.25 children born/woman (2011 est.)
country comparison to the world: 103
HIV/AIDS—adult prevalence rate: 0.2% (2009 est.)
country comparison to the world: 99
HIV/AIDS—people living with HIV/AIDS: 310,000 (2009 est.)
country comparison to the world: 20
HIV/AIDS—deaths: 8,300 (2009 est.)
country comparison to the world: 27
Major infectious diseases: *degree of risk:* high
food or waterborne diseases: bacterial diarrhea, hepatitis A and E, and typhoid fever
vectorborne diseases: chikungunya, dengue fever, and malaria
note: highly pathogenic H5N1 avian influenza has been identified in this country; it poses a negligible risk with extremely rare cases possible among US citizens who have close contact with birds (2009)
Drinking water source: *Improved:*
urban: 89% of population
rural: 71% of population
total: 80% of population
Unimproved: urban: 11% of population
rural: 29% of population
total: 20% of population (2008)
Sanitation facility access: *Improved:*
urban: 67% of population
rural: 36% of population
total: 52% of population
Unimproved: urban: 33% of population
rural: 64% of population
total: 48% of population (2008)
Nationality: *noun:* Indonesian(s)
adjective: Indonesian
Ethnic groups: Javanese 40.6%, Sundanese 15%, Madurese 3.3%, Minangkabau 2.7%, Betawi 2.4%, Bugis 2.4%, Banten 2%, Banjar 1.7%, other or unspecified 29.9% (2000 census)
Religions: Muslim 86.1%, Protestant 5.7%, Roman Catholic 3%, Hindu 1.8%, other or unspecified 3.4% (2000 census)
Languages: Bahasa Indonesia (official, modified form of Malay), English, Dutch, local dialects (of which the most widely spoken is Javanese)
Literacy: *definition:* age 15 and over can read and write
total population: 90.4%
male: 94%
female: 86.8% (2004 est.)
School life expectancy (primary to tertiary education): *total:* 13 years
male: 13 years
female: 13 years (2009)
Education expenditures: 2.8% of GDP (2008)
country comparison to the world: 140

GOVERNMENT

Country name: *conventional long form:* Republic of Indonesia
conventional short form: Indonesia
local long form: Republik Indonesia
local short form: Indonesia
former: Netherlands East Indies, Dutch East Indies
Government type: republic
Capital: *name:* Jakarta
geographic coordinates: 6 10 S, 106 49 E
time difference: UTC+7 (12 hours ahead of Washington, DC during Standard Time)
note: Indonesia is divided into three time zones
Administrative divisions: 30 provinces (provinsi-provinsi, singular–provinsi), 2 special regions* (daerah-daerah istimewa, singular–daerah istimewa), and 1 special capital city district** (daerah khusus ibukota); Aceh*, Bali, Banten, Bengkulu, Gorontalo, Jakarta Raya**, Jambi, Jawa Barat (West Java), Jawa Tengah (Central Java), Jawa Timur (East Java), Kalimantan Barat (West Kalimantan), Kalimantan Selatan (South Kalimantan), Kalimantan Tengah (Central Kalimantan), Kalimantan Timur (East Kalimantan), Kepulauan Bangka Belitung (Bangka Belitung Islands), Kepulauan Riau (Riau Islands), Lampung, Maluku, Maluku Utara (North Maluku), Nusa Tenggara Barat (West Nusa Tenggara), Nusa Tenggara Timur (East Nusa Tenggara), Papua, Papua Barat (West Papua), Riau, Sulawesi Barat (West Sulawesi), Sulawesi Selatan (South Sulawesi), Sulawesi Tengah (Central Sulawesi), Sulawesi Tenggara (Southeast Sulawesi), Sulawesi Utara (North Sulawesi), Sumatera Barat (West Sumatra), Sumatera Selatan (South Sumatra), Sumatera Utara (North Sumatra), Yogyakarta*
note: following the implementation of decentralization beginning on 1 January 2001, regencies and municipalities have become the key administrative units responsible for providing most government services
Independence: 17 August 1945 (declared); 27 December 1949 (recognized by the Netherlands); note–in August 2005 the Netherlands announced that it had recognized de facto Indonesian independence on 17 August 1945
National holiday: Independence Day, 17 August (1945)
Constitution: August 1945; abrogated by Federal Constitution of 1949 and Provisional Constitution of 1950, restored 5 July 1959; series of amendments concluded in 2002
Legal system: civil law system based on the Roman-Dutch model and influenced by customary law
International law organization participation: has not submitted an ICJ jurisdiction declaration; non-party state to the ICCt
Suffrage: 17 years of age; universal and married persons regardless of age
Executive branch: *chief of state:* President Susilo Bambang YUDHOYONO (since 20 October 2004); Vice President BOEDIONO (since 20 October 2009); note–the

president is both the chief of state and head of government
head of government: President Susilo Bambang YUDHOYONO (since 20 October 2004); Vice President BOEDIONO (since 20 October 2009)
cabinet: Cabinet appointed by the president (For more information visit the World Leaders website)
elections: president and vice president elected for five-year terms (eligible for a second term) by direct vote of the citizenry; election last held on 8 July 2009 (next to be held in 2014)
election results: Susilo Bambang YUDHOYONO elected president; percent of vote–Susilo Bambang YUDHOYONO 60.8%, MEGAWATI Sukarnoputri 26.8%, Jusuf KALLA 12.4%

Legislative branch: People's Consultative Assembly (Majelis Permusyawaratan Rakyat or MPR) is the upper house; it consists of members of the DPR and DPD and has role in inaugurating and impeaching the president and in amending the constitution but does not formulate national policy; House of Representatives or Dewan Perwakilan Rakyat (DPR) (560 seats, members elected to serve five-year terms), formulates and passes legislation at the national level; House of Regional Representatives (Dewan Perwakilan Daerah or DPD), constitutionally mandated role includes providing legislative input to DPR on issues affecting regions (132 members, four from each of Indonesia's 30 provinces, two special regions, and one special capital city district)
elections: last held on 9 April 2009 (next to be held in 2014)
election results: percent of vote by party–PD 20.9%, GOLKAR 14.5%, PDI-P 14.0%, PKS 7.9%, PAN 6.0%, PPP 5.3%, PKB 4.9%, GERINDRA 4.5%, HANURA 3.8%, others 18.2%; seats by party–PD 148, GOLKAR 107, PDI-P 94, PKS 57, PAN 46, PPP 37, PKB 28, GERINDRA 26, HANURA 17
note: 29 other parties received less than 2.5% of the vote so did not obtain any seats; because of election rules, the number of seats won does not always follow the percentage of votes received by parties

Judicial branch: Supreme Court or Mahkamah Agung is the final court of appeal but does not have the power of judicial review (justices are appointed by the president from a list of candidates selected by the legislature); in March 2004 the Supreme Court assumed administrative and financial responsibility for the lower court system from the Ministry of Justice and Human Rights; Constitutional Court or Mahkamah Konstitusi (invested by the president on 16 August 2003) has the power of judicial review, jurisdiction over the results of a general election, and reviews actions to dismiss a president from office; Labor Court under supervision of Supreme Court began functioning in January 2006; the Anti-Corruption Court has jurisdiction over corruption cases brought by the independent Corruption Eradication Commission

Political parties and leaders: Democrat Party or PD [Anas URANINGRUM]; Functional Groups Party or GOLKAR [Aburizal BAKRIE]; Great Indonesia Movement Party or GERINDRA [SUHARDI]; Indonesia Democratic Party-Struggle or PDI-P [MEGAWATI Sukarnoputri]; National Awakening Party or PKB [Muhaiman ISKANDAR]; National Mandate Party or PAN [Hatta RAJASA]; People's Conscience Party or HANURA [WIRANTO]; Prosperous Justice Party or PKS [Luthfi Hasan ISHAQ]; United Development Party or PPP [Suryadharma ALI]

Political pressure groups and leaders: Commission for the "Disappeared" and Victims of Violence or KontraS; Indonesia Corruption Watch or ICW; Indonesian Forum for the Environment or WALHI; Islamic Defenders Front or FPI; People's Democracy Fortress or Bendera

International organization participation: ADB, APEC, ARF, ASEAN, BIS, CICA (observer), CP, D-8, EAS, FAO, G-11, G-15, G-20, G-77, IAEA, IBRD, ICAO, ICC, ICRM, IDA, IDB, IFAD, IFC, IFRCS, IHO, ILO, IMF, IMO, IMSO, Interpol, IOC, IOM (observer), IPU, ISO, ITSO, ITU, ITUC, MIGA, MONUSCO, NAM, OIC, OPCW, PIF (partner), UN, UNAMID, UNCTAD, UNESCO, UNIDO, UNIFIL, UNMIL, UNMIS, UNWTO, UPU, WCO, WFTU, WHO, WIPO, WMO, WTO

Diplomatic representation in the US:
chief of mission: Ambassador Dino Patti DJALAL
chancery: 2020 Massachusetts Avenue NW, Washington, DC 20036
telephone: [1] (202) 775-5200
FAX: [1] (202) 775-5365
consulate(s) general: Chicago, Houston, Los Angeles, New York, San Francisco

Diplomatic representation from the US: *chief of mission:* Ambassador Scot A. MARCIEL
embassy: Jalan 1 Medan Merdeka Selatan 4-5, Jakarta 10110
mailing address: Unit 8129, Box 1, FPO AP 96520
telephone: [62] (21) 3435-9000
FAX: [62] (21) 3435-9922
consulate(s) general: Surabaya

Flag description: two equal horizontal bands of red (top) and white; the colors derive from the banner of the Majapahit Empire of the 13th-15th centuries; red symbolizes courage, white represents purity
note: similar to the flag of Monaco, which is shorter; also similar to the flag of Poland, which is white (top) and red

National anthem: *name:* "Indonesia Raya" (Great Indonesia)
lyrics/music: Wage Rudolf SOEPRATMAN
note: adopted 1945

ECONOMY

Economy—overview: Indonesia, a vast polyglot nation, has weathered the global financial crisis relatively smoothly because of its heavy reliance on domestic consumption as the driver of economic growth. Increasing investment by both local and foreign investors is also supporting solid growth. Although the economy slowed to 4.5% growth in 2009 from the 6%-plus growth rate recorded in 2007 and 2008, by 2010 growth returned to a 6% rate. During the recession, Indonesia outperformed most of its regional neighbors. The government made economic advances under the first administration of President YUDHOYONO, introducing significant reforms in the financial sector, including tax and customs reforms, the use of Treasury bills, and capital market development and supervision. Indonesia's debt-to-GDP ratio in recent years has declined steadily because of increasingly robust GDP growth and sound fiscal stewardship, leading two of the three leading credit agencies to upgrade credit ratings for Indonesia's sovereign debt to one notch below investment grade. Indonesia still struggles with poverty and unemployment, inadequate infrastructure, corruption, a complex regulatory environment, and unequal resource distribution among regions. YUDHOYONO and his vice president, respected economist BOEDIONO, have maintained broad continuity of economic policy, although the economic reform agenda has been slowed during the first year of their term by corruption scandals and the departure of an internationally respected finance minister. In late 2010, increasing inflation, driven by higher and volatile food prices, posed an increasing challenge to economic policymakers and threatened to push millions of the near-poor below the poverty line. The government in 2011 faces the ongoing challenge of improving Indonesia's infrastructure to remove impediments to growth, while addressing climate change concerns, particularly with regard to conserving Indonesia's forests and peatlands, the focus of a potentially trailblazing $1 billion REDD+ pilot project.

GDP (purchasing power parity): $1.03 trillion (2010 est.)
country comparison to the world: 16
$970.6 billion (2009 est.)
$928.2 billion (2008 est.)
note: data are in 2010 US dollars

GDP (official exchange rate): $706.7 billion (2010 est.)

GDP—real growth rate: 6.1% (2010 est.)
country comparison to the world: 50
4.6% (2009 est.)
6% (2008 est.)

GDP—per capita (PPP): $4,200 (2010 est.)
country comparison to the world: 157
$4,000 (2009 est.)
$3,900 (2008 est.)
note: data are in 2010 US dollars

GDP—composition by sector:
agriculture: 16.5%
industry: 46.4%
services: 37.1% (3rd quarter, 2010 est.)

Labor force: 116.5 million (2010 est.)
country comparison to the world: 5

Labor force—by occupation: *agriculture:* 38.3%
industry: 12.8%
services: 48.9% (2010 est.)

Unemployment rate: 7.1% (2010 est.)
country comparison to the world: 75

7.9% (2009 est.)
Population below poverty line: 13.33% (2010)
Household income or consumption by percentage share: *lowest 10%:* 3%
highest 10%: 32.3% (2006)
Distribution of family income—Gini index: 37 (2009)
country comparison to the world: 78
39.4 (2005)
Investment (gross fixed): 32.5% of GDP (2010 est.)
country comparison to the world: 14
Budget: *revenues:* $119.5 billion
expenditures: $132.9 billion (2011 est.)
Public debt: 26.4% of GDP (2010 est.)
country comparison to the world: 93
27.4% of GDP (2009 est.)
Inflation rate (consumer prices): 5.1% (2010 est.)
country comparison to the world: 144
4.8% (2009 est.)
Central bank discount rate: 6.37% (31 December 2010)
country comparison to the world: 65
6.46% (31 December 2009)
note: this figure represents the 3-month SBI rate; BI has not employed the one-month SBI since September 2010
Commercial bank prime lending rate: 13.29% (30 November 2010 est.)
country comparison to the world: 48
14.5% (31 December 2009 est.)
note: these figures represent the average annualized rate on working capital loans
Stock of narrow money: $62.27 billion (31 December 2010 est.)
country comparison to the world: 42
$49.63 billion (31 December 2009 est.)
Stock of broad money: $243.1 billion (31 December 2010 est.)
country comparison to the world: 37
$205.8 billion (31 December 2009 est.)
Stock of domestic credit: $253.1 billion (31 December 2010 est.)
country comparison to the world: 37
$192.3 billion (31 December 2009 est.)
Market value of publicly traded shares: $361.2 billion (31 December 2010)
country comparison to the world: 26
$178.2 billion (31 December 2009)
$98.76 billion (31 December 2008)
Agriculture—products: rice, cassava (tapioca), peanuts, rubber, cocoa, coffee, palm oil, copra; poultry, beef, pork, eggs
Industries: petroleum and natural gas, textiles, apparel, footwear, mining, cement, chemical fertilizers, plywood, rubber, food, tourism
Industrial production growth rate: 3.6% (2010 est.)
country comparison to the world: 94
Electricity—production: 129 billion kWh (2008 est.)
country comparison to the world: 27
Electricity—consumption: 119.3 billion kWh (2007 est.)
country comparison to the world: 28
Electricity—exports: 0 kWh (2009 est.)
Electricity—imports: 0 kWh (2009 est.)
Oil—production: 1.023 million bbl/day (2009 est.)
country comparison to the world: 22
Oil—consumption: 1.115 million bbl/day (2009 est.)
country comparison to the world: 18
Oil—exports: 322,000 bbl/day (2009 est.)
country comparison to the world: 37
Oil—imports: 456,700 bbl/day (2009 est.)
country comparison to the world: 27
Oil—proved reserves: 4.05 billion bbl (1 January 2010 est.)
country comparison to the world: 28
Natural gas—production: 85.7 billion cu m (2009 est.)
country comparison to the world: 8
Natural gas—consumption: 45.2 billion cu m (2008)
country comparison to the world: 18
Natural gas—exports: 33.5 billion cu m (2008 est.)
country comparison to the world: 7
Natural gas—imports: 0 cu m (2008 est.)
country comparison to the world: 200
Natural gas—proved reserves: 3.001 trillion cu m (1 January 2010 est.)
country comparison to the world: 14
Current account balance: $8.532 billion (2010 est.)
country comparison to the world: 26
$10.75 billion (2009 est.)
Exports: $146.3 billion (2010 est.)
country comparison to the world: 30
$119.5 billion (2009 est.)
Exports—commodities: oil and gas, electrical appliances, plywood, textiles, rubber
Exports—partners: Japan 17.28%, Singapore 11.29%, US 10.81%, China 7.62%, South Korea 5.53%, India 4.35%, Taiwan 4.11%, Malaysia 4.07% (2009)
Imports:
$111.1 billion (2010 est.)
country comparison to the world: 30
$84.35 billion (2009 est.)
Imports—commodities: machinery and equipment, chemicals, fuels, foodstuffs
Imports—partners: Singapore 24.96%, China 12.52%, Japan 8.92%, Malaysia 5.88%, South Korea 5.64%, US 4.88%, Thailand 4.45% (2009)
Reserves of foreign exchange and gold: $96.21 billion (31 December 2010 est.)
country comparison to the world: 18
$66.12 billion (31 December 2009 est.)
Debt—external: $196.1 billion (31 December 2010 est.)
country comparison to the world: 31
$172.9 billion (31 December 2009 est.)
Stock of direct foreign investment—at home: $81.21 billion (31 December 2010 est.)
country comparison to the world: 43
$72.84 billion (31 December 2009 est.)
Stock of direct foreign investment—abroad: $33.71 billion (31 December 2010 est.)
country comparison to the world: 36
$30.18 billion (31 December 2009 est.)
Exchange rates: Indonesian rupiah (IDR) per US dollar–9,169.5 (2010), 10,389.9 (2009), 9,698.9 (2008), 9,143 (2007), 9,159.3 (2006)

COMMUNICATIONS

Telephones—main lines in use: 33.958 million (2009)
country comparison to the world: 9
Telephones—mobile cellular: 159.248 million (2009)
country comparison to the world: 6
Telephone system: *general assessment:* domestic service includes an interisland microwave system, an HF radio police net, and a domestic satellite communications system; international service good
domestic: coverage provided by existing network has been expanded by use of over 200,000 telephone kiosks many located in remote areas; mobile-cellular subscribership growing rapidly
international: country code–62; landing point for both the SEA-ME-WE-3 and SEA-ME-WE-4 submarine cable networks that provide links throughout Asia, the Middle East, and Europe; satellite earth stations–2 Intelsat (1 Indian Ocean and 1 Pacific Ocean)
Broadcast media: mixture of about a dozen national television networks–2 public broadcasters, the remainder private broadcasters–each with multiple transmitters; more than 100 local TV stations operating; widespread use of satellite and cable TV systems; public radio broadcaster operates 6 national networks as well as regional and local stations; overall, more than 700 radio stations operating with more than 650 privately-operated (2008)
Internet country code: .id
Internet hosts: 1.269 million (2010)
country comparison to the world: 39
Internet users: 20 million (2009)
country comparison to the world: 22

TRANSPORTATION

Airports: 684 (2010)
country comparison to the world: 10
Airports—with paved runways: *total:* 171
over 3,047 m: 4
2,438 to 3,047 m: 19
1,524 to 2,437 m: 50
914 to 1,523 m: 64
under 914 m: 34 (2010)
Airports—with unpaved runways: *total:* 513
1,524 to 2,437 m: 4
914 to 1,523 m: 25
under 914 m: 484 (2010)
Heliports: 64 (2010)
Pipelines: condensate 812 km; condensate/gas 73 km; gas 7,165 km; oil 5,984 km; oil/gas/water 12 km; refined products 617 km; water 44 km (2010)
Railways: *total:* 5,042 km
country comparison to the world: 35
narrow gauge: 5,042 km 1.067-m gauge (565 km electrified) (2009)
Roadways: *total:* 437,759 km
country comparison to the world: 14
paved: 258,744 km
unpaved: 179,015 km (2008)
Waterways: 21,579 km (2011)
country comparison to the world: 5
Merchant marine: *total:* 1,244
country comparison to the world: 8
by type: bulk carrier 95, cargo 601, chemical tanker 57, container 112, liquefied gas 17, passenger 47, passenger/cargo 76, petroleum tanker 214, refrigerated cargo 4, roll on/roll off 12, specialized tanker 1, vehicle carrier 8
foreign-owned: 61 (China 1, France 1, Greece 1, Japan 7, Malaysia 1, Norway 4, Singapore 42, South Korea 1, Taiwan 1, US 2)
registered in other countries: 87 (Bahamas 2, Cambodia 2, Hong Kong 8, Liberia 4,

Mongolia 2, Panama 14, Singapore 53, unknown 2) (2010)

Ports and terminals: Banjarmasin, Belawan, Kotabaru, Krueg Geukueh, Palembang, Panjang, Sungai Pakning, Tanjung Perak, Tanjung Priok

Transportation—note: the International Maritime Bureau reports the territorial and offshore waters in the Strait of Malacca and South China Sea as high risk for piracy and armed robbery against ships; 2010 saw the highest levels of armed robbery against ships since 2007; 40 commercial vessels were attacked, boarded, or hijacked both at anchor or while underway; hijacked vessels are often disguised and cargo diverted to ports in East Asia; crews have been murdered or cast adrift

MILITARY

Military branches: Indonesian Armed Forces (Tentara Nasional Indonesia, TNI): Army (TNI-Angkatan Darat (TNI-AD)), Navy (TNI-Angkatan Laut (TNI-AL); includes marines (Korps Marinir, KorMar), naval air arm), Air Force (TNI-Angkatan Udara (TNI-AU)), National Air Defense Command (Kommando Pertahanan Udara Nasional (Kohanudnas)) (2011)

Military service age and obligation: 18 years of age for selective compulsory and voluntary military service; 2-year conscript service obligation, with reserve obligation to age 45 (officers); Indonesian citizens only (2008)

Manpower available for military service: *males age 16–49:* 65,847,171
females age 16–49: 63,228,017 (2010 est.)

Manpower fit for military service: *males age 16–49:* 54,264,299
females age 16–49: 53,274,361 (2010 est.)

Manpower reaching militarily significant age annually: *male:* 2,263,892
female: 2,191,267 (2010 est.)

Military expenditures: 3% of GDP (2005 est.)
country comparison to the world: 47

TRANSNATIONAL ISSUES

Disputes—international: Indonesia has a stated foreign policy objective of establishing stable fixed land and maritime boundaries with all of its neighbors; some sections of border along Timor-Leste's Oecussi exclave and maritime boundaries with Timor-Leste remain unresolved; many refugees from Timor-Leste who left in 2003 still reside in Indonesia and refuse repatriation; a 1997 treaty between Indonesia and Australia settled some parts of their maritime boundary but outstanding issues remain; ICJ's award of Sipadan and Ligitan islands to Malaysia in 2002 left the sovereignty of Unarang rock and the maritime boundary in the Ambalat oil block in the Celebes Sea in dispute; the ICJ decision has prompted Indonesia to assert claims to and to establish a presence on its smaller outer islands; Indonesia and Singapore continue to work on finalization of their 1973 maritime boundary agreement by defining unresolved areas north of Indonesia's Batam Island; Indonesian secessionists, squatters, and illegal migrants create repatriation problems for Papua New Guinea; maritime delimitation talks continue with Palau; Indonesian groups challenge Australia's claim to Ashmore Reef; Australia has closed parts of the Ashmore and Cartier Reserve to Indonesian traditional fishing and placed restrictions on certain catches

Refugees and internally displaced persons: *IDPs:* 200,000-350,000 (government offensives against rebels in Aceh; most IDPs in Aceh, Central Kalimantan, Central Sulawesi Provinces, and Maluku) (2007)

Illicit drugs: illicit producer of cannabis largely for domestic use; producer of methamphetamine and ecstasy

IRAN

INTRODUCTION

Background: Known as Persia until 1935, Iran became an Islamic republic in 1979 after the ruling monarchy was overthrown and Shah Mohammad Reza PAHLAVI was forced into exile. Conservative clerical forces established a theocratic system of government with ultimate political authority vested in a learned religious scholar referred to commonly as the Supreme Leader who, according to the constitution, is accountable only to the Assembly of Experts–a popularly elected 86-member body of clerics. US-Iranian relations have been strained since a group of Iranian students seized the US Embassy in Tehran on 4 November 1979 and held it until 20 January 1981. During 1980-88, Iran fought a bloody, indecisive war with Iraq that eventually expanded into the Persian Gulf and led to clashes between US Navy and Iranian military forces between 1987 and 1988. Iran has been designated a state sponsor of terrorism for its activities in Lebanon and elsewhere in the world and remains subject to US, UN, and EU economic sanctions and export controls because of its continued involvement in terrorism and its nuclear weapons ambitions. Following the election of reformer Hojjat ol-Eslam Mohammad KHATAMI as president in 1997 and a reformist Majles (legislature) in 2000, a campaign to foster political reform in response to popular dissatisfaction was initiated. The movement floundered as conservative politicians, through the control of unelected institutions, prevented reform measures from being enacted and increased repressive measures. Starting with nationwide municipal elections in 2003 and continuing through Majles elections in 2004, conservatives reestablished control over Iran's elected government institutions, which culminated with the August 2005 inauguration of hardliner Mahmud AHMADI-NEJAD as president. His controversial reelection in June 2009 sparked nationwide protests over allegations of electoral fraud. The UN Security Council has passed a number of resolutions (1696 in July 2006, 1737 in December 2006, 1747 in March 2007, 1803 in March 2008, and 1835 in September 2008 and 1929 in June 2010) calling for Iran to suspend its uranium enrichment and reprocessing activities and comply with its IAEA obligations and responsibilities. Resolutions 1737, 1477, 1803 and 1929 subject a number of Iranian individuals and entities involved in Iran's nuclear and ballistic missile programs to sanctions. Additionally, several Iranian entities are subject to US sanctions under Executive Order 13382 designations for proliferation activities and EO 13224 designations for support of terrorism. In mid-February 2011, opposition activists conducted the largest antiregime rallies since December 2009, spurred by the success of uprisings in Tunisia and Egypt. Protester turnout probably was at most tens of thousands and security forces were deployed to disperse protesters. Additional protests in March 2011 failed to elicit significant participation largely because of the robust security response, although discontent still smolders.

GEOGRAPHY

Location: Middle East, bordering the Gulf of Oman, the Persian Gulf, and the Caspian Sea, between Iraq and Pakistan

Geographic coordinates: 32 00 N, 53 00 E

Map references: Middle East

Area: *total:* 1,648,195 sq km
country comparison to the world: 18
land: 1,531,595 sq km
water: 116,600 sq km

Area—comparative: slightly smaller than Alaska

Land boundaries: *total:* 5,440 km
border countries: Afghanistan 936 km, Armenia 35 km, Azerbaijan-proper 432 km, Azerbaijan-Naxcivan exclave 179 km,

Iraq 1,458 km, Pakistan 909 km, Turkey 499 km, Turkmenistan 992 km
Coastline: 2,440 km; note–Iran also borders the Caspian Sea (740 km)
Maritime claims: *territorial sea:* 12 nm
contiguous zone: 24 nm
exclusive economic zone: bilateral agreements or median lines in the Persian Gulf
continental shelf: natural prolongation
Climate: mostly arid or semiarid, subtropical along Caspian coast
Terrain: rugged, mountainous rim; high, central basin with deserts, mountains; small, discontinuous plains along both coasts
Elevation extremes:
lowest point: Caspian Sea -28 m
highest point: Kuh-e Damavand 5,671 m
Natural resources: petroleum, natural gas, coal, chromium, copper, iron ore, lead, manganese, zinc, sulfur
Land use: *arable land:* 9.78%
permanent crops: 1.29%
other: 88.93% (2005)
Irrigated land: 76,500 sq km (2003)
Total renewable water resources: 137.5 cu km (1997)
Freshwater withdrawal (domestic/industrial/agricultural): *total:* 72.88 cu km/yr (7%/2%/91%)
per capita: 1,048 cu m/yr (2000)
Natural hazards: periodic droughts, floods; dust storms, sandstorms; earthquakes
Environment—current issues: air pollution, especially in urban areas, from vehicle emissions, refinery operations, and industrial effluents; deforestation; overgrazing; desertification; oil pollution in the Persian Gulf; wetland losses from drought; soil degradation (salination); inadequate supplies of potable water; water pollution from raw sewage and industrial waste; urbanization
Environment—international agreements: *party to:* Biodiversity, Climate Change, Climate Change-Kyoto Protocol, Desertification, Endangered Species, Hazardous Wastes, Marine Dumping, Ozone Layer Protection, Ship Pollution, Wetlands
signed, but not ratified: Environmental Modification, Law of the Sea, Marine Life Conservation
Geography—note: strategic location on the Persian Gulf and Strait of Hormuz, which are vital maritime pathways for crude oil transport

PEOPLE

Population: 77,891,220 (July 2011 est.)
country comparison to the world: 18
Age structure: *0–14 years:* 24.1% (male 9,608,342/female 9,128,427)
15–64 years: 70.9% (male 28,083,193/female 27,170,445)
65 years and over: 5% (male 1,844,967/female 2,055,846) (2011 est.)
Median age: *total:* 26.8 years
male: 26.6 years
female: 27.1 years (2011 est.)
Population growth rate: 1.248% (2011 est.)
country comparison to the world: 94
Birth rate: 18.55 births/1,000 population (2011 est.)
country comparison to the world: 104
Death rate: 5.94 deaths/1,000 population (July 2011 est.)
country comparison to the world: 165
Net migration rate: -0.13 migrant(s)/1,000 population (2011 est.)
country comparison to the world: 123
Urbanization: *urban population:* 71% of total population (2010)
rate of urbanization: 1.9% annual rate of change (2010-15 est.)
Major cities—population: TEHRAN (capital) 7.19 million; Mashhad 2.592 million; Esfahan 1.704 million; Karaj 1.531 million; Tabriz 1.459 million (2009)
Sex ratio: *at birth:* 1.05 male(s)/female
under 15 years: 1.05 male(s)/female
15–64 years: 1.02 male(s)/female
65 years and over: 0.91 male(s)/female
total population: 1.02 male(s)/female (2011 est.)
Infant mortality rate: *total:* 42.26 deaths/1,000 live births
country comparison to the world: 59
male: 42.75 deaths/1,000 live births
female: 41.75 deaths/1,000 live births (2011 est.)
Life expectancy at birth: *total population:* 70.06 years
country comparison to the world: 146
male: 68.58 years
female: 71.61 years (2011 est.)
Total fertility rate: 1.88 children born/woman (2011 est.)
country comparison to the world: 145
HIV/AIDS—adult prevalence rate: 0.2% (2009 est.)
country comparison to the world: 100
HIV/AIDS—people living with HIV/AIDS: 92,000 (2009 est.)
country comparison to the world: 44
HIV/AIDS—deaths: 6,400 (2009 est.)
country comparison to the world: 33
Major infectious diseases: *degree of risk:* intermediate
food or waterborne diseases: bacterial diarrhea
vectorborne diseases: Crimean Congo hemorrhagic fever and malaria
note: highly pathogenic H5N1 avian influenza has been identified in this country; it poses a negligible risk with extremely rare cases possible among US citizens who have close contact with birds (2009)
Drinking water source: *Improved:*
urban: 98% of population
rural: 83% of population
total: 93% of population
Unimproved: urban: 2% of population
rural: 17% of population
total: 7% of population (2000)
Sanitation facility access: *Improved:*
urban: 86% of population
rural: 78% of population
total: 83% of population
Unimproved: urban: 14% of population
rural: 22% of population
total: 17% of population (2000)
Nationality: *noun:* Iranian(s)
adjective: Iranian
Ethnic groups: Persian 51%, Azeri 24%, Gilaki and Mazandarani 8%, Kurd 7%, Arab 3%, Lur 2%, Baloch 2%, Turkmen 2%, other 1%
Religions: Muslim 98% (Shia 89%, Sunni 9%), other (includes Zoroastrian, Jewish, Christian, and Baha'i) 2%
Languages: Persian and Persian dialects 58%, Turkic and Turkic dialects 26%, Kurdish 9%, Luri 2%, Balochi 1%, Arabic 1%, Turkish 1%, other 2%
Literacy: *definition:* age 15 and over can read and write
total population: 77%
male: 83.5%
female: 70.4% (2002 est.)
School life expectancy (primary to tertiary education): *total:* 13 years
male: 13 years
female: 13 years (2009)
Education expenditures: 4.7% of GDP (2009)
country comparison to the world: 72

GOVERNMENT

Country name: *conventional long form:* Islamic Republic of Iran
conventional short form: Iran
local long form: Jomhuri-ye Eslami-ye Iran
local short form: Iran
former: Persia
Government type: theocratic republic
Capital: *name:* Tehran
geographic coordinates: 35 40 N, 51 25 E
time difference: UTC+3.5 (8.5 hours ahead of Washington, DC during Standard Time)
daylight saving time: +1hr, begins fourth Tuesday in March; ends fourth Thursday in September
Administrative divisions: 31 provinces (ostanha, singular–ostan); Alborz, Ardabil, Azarbayjan-e Gharbi (West Azerbaijan), Azarbayjan-e Sharqi (East Azerbaijan), Bushehr, Chahar Mahal va Bakhtiari, Esfahan, Fars, Gilan, Golestan, Hamadan, Hormozgan, Ilam, Kerman, Kermanshah, Khorasan-e Jonubi (South Khorasan), Khorasan-e Razavi (Razavi Khorasan), Khorasan-e Shomali (North Khorasan), Khuzestan, Kohgiluyeh va Bowyer Ahmad, Kordestan, Lorestan, Markazi, Mazandaran, Qazvin, Qom, Semnan, Sistan va Baluchestan, Tehran, Yazd, Zanjan
Independence: 1 April 1979 (Islamic Republic of Iran proclaimed); notable earlier dates: ca. 625 B.C. (unification of Iran under the Medes); ca. A.D. 1501 (Iran reunified under the Safavids); 12 December 1925 (modern Iran established under the Pahlavis)
National holiday: Republic Day, 1 April (1979)
Constitution: 2-3 December 1979; revised 1989
note: the revision in 1989 expanded powers of the presidency and eliminated the prime ministership
Legal system: religious legal system based on sharia law
International law organization participation: has not submitted an ICJ jurisdiction declaration; non-party state to the ICCt
Suffrage: 18 years of age; universal
Executive branch: *chief of state:* Supreme Leader Ali Hoseini-KHAMENEI (since 4 June 1989)
head of government: President Mahmud AHMADI-NEJAD (since 3 August 2005);

First Vice President Mohammad Reza RAHIMI (since 13 September 2009)
cabinet: Council of Ministers selected by the president with legislative approval; the Supreme Leader has some control over appointments to the more sensitive ministries
(For more information visit the World Leaders website)
note: also considered part of the Executive branch of government are three oversight bodies: 1) Assembly of Experts (Majles-Khebregan), a popularly elected body charged with determining the succession of the Supreme Leader, reviewing his performance, and deposing him if deemed necessary; 2) Expediency Council or the Council for the Discernment of Expediency (Majma-e-Tashkhis-e-Maslahat-e-Nezam) exerts supervisory authority over the executive, judicial, and legislative branches and resolves legislative issues on which the Majles and the Council of Guardians disagree and since 1989 has been used to advise national religious leaders on matters of national policy; in 2005 the Council's powers were expanded to act as a supervisory body for the government; 3) Council of Guardians of the Constitution or Council of Guardians or Guardians Council (Shora-ye Negban-e Qanon-e Asassi) determines whether proposed legislation is both constitutional and faithful to Islamic law, vets candidates in popular elections for suitability, and supervises national elections
elections: Supreme Leader appointed for life by the Assembly of Experts; president elected by popular vote for a four-year term (eligible for a second term and third nonconsecutive term); election last held on 12 June 2009 (next presidential election slated for June 2013)
election results: Mahmud AHMADI-NEJAD reelected president; percent of vote—Mahmud AHMADI-NEJAD 62.6%, Mir-Hosein MUSAVI-Khamenei 33.8%, other 3.6%; voter turnout 85% (according to official figures published by the government)

Legislative branch: unicameral Islamic Consultative Assembly or Majles-e-Shura-ye-Eslami or Majles (290 seats; members elected by popular vote to serve four-year terms)
elections: last held on 14 March 2008 with a runoff held on 25 April 2008 (next to be held in 2012)
election results: percent of vote—NA; seats by party—conservatives/Islamists 167, reformers 39, independents 74, religious minorities 5, other 5

Judicial branch: The Supreme Court (Qeveh Qazaieh) and the four-member High Council of the Judiciary have a single head and overlapping responsibilities; together they supervise the enforcement of all laws and establish judicial and legal policies; lower courts include a special clerical court, a revolutionary court, and a special administrative court

Political parties and leaders: formal political parties are a relatively new phenomenon in Iran and most conservatives still prefer to work through political pressure groups rather than parties; often political parties or coalitions are formed prior to elections and disbanded soon thereafter; a loose pro-reform coalition called the 2nd Khordad Front, which includes political parties as well as less formal groups and organizations, achieved considerable success in elections for the sixth Majles in early 2000; groups in the coalition included the Islamic Iran Participation Front (IIPF), Executives of Construction Party (Kargozaran), Solidarity Party, Islamic Labor Party, Mardom Salari, Mojahedin of the Islamic Revolution Organization (MIRO), and Militant Clerics Society (Ruhaniyun); the coalition participated in the seventh Majles elections in early 2004; following his defeat in the 2005 presidential elections, former MCS Secretary General and sixth Majles Speaker Mehdi KARUBI formed the National Trust Party; a new conservative group, Islamic Iran Developers Coalition (Abadgaran), took a leading position in the new Majles after winning a majority of the seats in February 2004; following the 2004 Majles elections, traditional and hardline conservatives have attempted to close ranks under the United Front of Principalists and the Broad Popular Coalition of Principalists; several reformist groups, such as the Mujahadin of the Islamic Revolution, came together as a reformist coalition in advance of the 2008 Majles elections; the IIPF has repeatedly complained that the overwhelming majority of its candidates have been unfairly disqualified from the 2008 elections

Political pressure groups and leaders: groups that generally support the Islamic Republic: Ansar-e Hizballah-Islamic Coalition Party (Motalefeh); Followers of the Line of the Imam and the Leader; Islamic Engineers Society; Tehran Militant Clergy Association (Ruhaniyat); active pro-reform student group: Office of Strengthening Unity (OSU); opposition groups: Baluchistan People's Party (BPP); Freedom Movement of Iran; Green Path movement [Mehdi KARUBI, Mir-Hosein MUSAVI]; Marz-e Por Gohar; National Front; and various ethnic and Monarchist organizations; armed political groups that have been repressed by the government: Democratic Party of Iranian Kurdistan (KDPI); Jundallah; Komala; Mujahedin-e Khalq Organization (MEK or MKO); People's Fedayeen; People's Free Life Party of Kurdistan (PJAK)

International organization participation: CICA, CP, D-8, ECO, FAO, G-15, G-24, G-77, IAEA, IBRD, ICAO, ICC, ICRM, IDA, IDB, IFAD, IFC, IFRCS, IHO, ILO, IMF, IMO, IMSO, Interpol, IOC, IOM, IPU, ISO, ITSO, ITU, MIGA, NAM, OIC, OPCW, OPEC, PCA, SAARC (observer), SCO (observer), UN, UNCTAD, UNESCO, UNHCR, UNIDO, UNITAR, UNMIS, UNWTO, UPU, WCO, WFTU, WHO, WIPO, WMO, WTO (observer)
Diplomatic representation in the US: none; note—Iran has an Interests Section in the Pakistani Embassy; address: Iranian Interests Section, Pakistani Embassy, 2209 Wisconsin Avenue NW, Washington, DC 20007; *telephone:* [1] (202) 965-4990; FAX [1] (202) 965-1073

Diplomatic representation from the US: none; note—the US Interests Section is located in the Embassy of Switzerland No. 39 Shahid Mousavi (Golestan 5th), Pasdaran Ave., Tehran, Iran; telephone [98] 21 2254 2178/2256 5273; FAX [98] 21 2258 0432

Flag description: three equal horizontal bands of green (top), white, and red; the national emblem (a stylized representation of the word Allah in the shape of a tulip, a symbol of martyrdom) in red is centered in the white band; ALLAH AKBAR (God is Great) in white Arabic script is repeated 11 times along the bottom edge of the green band and 11 times along the top edge of the red band; green is the color of Islam and also represents growth, white symbolizes honesty and peace, red stands for bravery and martyrdom

National anthem: *name:* "Soroud-e Melli-e Jomhouri-e Eslami-e Iran" (National Anthem of the Islamic Republic of Iran)
lyrics/music: multiple authors/Hassan RIAHI
note: adopted 1990

ECONOMY

Economy—overview: Iran's economy is marked by an inefficient state sector, reliance on the oil sector, which provides the majority of government revenues, and statist policies, which create major distortions throughout the system. Private sector activity is typically limited to small-scale workshops, farming, and services. Price controls, subsidies, and other rigidities weigh down the economy, undermining the potential for private-sector-led growth. Significant informal market activity flourishes. The legislature in late 2009 passed President Mahmud AHMADI-NEJAD's bill to reduce subsidies, particularly on food and energy. The bill would phase out subsidies—which benefit Iran's upper and middle classes the most—over three to five years and replace them with cash payments to Iran's lower classes. However, the start of the program was delayed repeatedly throughout 2010 over fears of public reaction to higher prices. This is the most extensive economic reform since the government implemented gasoline rationing in 2007. The recovery of world oil prices in the last year increased Iran's oil export revenue by at least $10 billion over 2009, easing some of the financial impact of the newest round of international sanctions. Although inflation has fallen substantially since the mid-2000s, Iran continues to suffer from double-digit unemployment and underemployment. Underemployment among Iran's educated youth has convinced many to seek jobs overseas, resulting in a significant "brain drain."

GDP (purchasing power parity): $818.7 billion (2010 est.)
country comparison to the world: 20
$810.3 billion (2009 est.)
$809.8 billion (2008 est.)
note: data are in 2010 US dollars

GDP (official exchange rate): $357.2 billion (2010 est.)
GDP—real growth rate: 1% (2010 est.)
country comparison to the world: 175
0.1% (2009 est.)
1% (2008 est.)
GDP—per capita (PPP): $10,600 (2010 est.)
country comparison to the world: 105
$10,700 (2009 est.)
$10,800 (2008 est.)
note: data are in 2010 US dollars
GDP—composition by sector:
agriculture: 11%
industry: 45.9%
services: 43.1% (2010 est.)
Labor force: 25.7 million
country comparison to the world: 22
note: shortage of skilled labor (2010 est.)
Labor force—by occupation:
agriculture: 25%
industry: 31%
services: 45% (June 2007)
Unemployment rate: 14.6% (2010 est.)
country comparison to the world: 147
10.3% (2008 est.)
note: data are according to the Iranian Government
Population below poverty line: 18% (2007 est.)
Household income or consumption by percentage share: *lowest 10%:* 2.6%
highest 10%: 29.6% (2005)
Distribution of family income—Gini index: 44.5 (2006)
country comparison to the world: 42
Investment (gross fixed): 27.6% of GDP (2010 est.)
country comparison to the world: 27
Budget: *revenues:* $105.7 billion
expenditures: $98.83 billion (2010 est.)
Public debt: 16.2% of GDP (2010 est.)
country comparison to the world: 115
16.8% of GDP (2009 est.)
Inflation rate (consumer prices): 11.8% (2010 est.)
country comparison to the world: 204
13.5% (2009 est.)
note: official Iranian estimate
Central bank discount rate: NA%
Commercial bank prime lending rate: 12% (31 December 2009 est.)
country comparison to the world: 69
12% (31 December 2008 est.)
Stock of narrow money: $50.37 billion (31 December 2010 est.)
country comparison to the world: 45
$48.74 billion (31 December 2009 est.)
Stock of broad money: $167.4 billion (31 December 2010 est.)
country comparison to the world: 41
$147.2 billion (31 December 2009 est.)
Stock of domestic credit: $132.2 billion (31 December 2010 est.)
country comparison to the world: 43
$120.2 billion (31 December 2009 est.)
Market value of publicly traded shares: $63.3 billion (31 December 2009)
country comparison to the world: 52
$49.04 billion (31 December 2008)
$45.57 billion (31 December 2007)
Agriculture—products: wheat, rice, other grains, sugar beets, sugar cane, fruits, nuts, cotton; dairy products, wool; caviar
Industries: petroleum, petrochemicals, fertilizers, caustic soda, textiles, cement and other construction materials, food processing (particularly sugar refining and vegetable oil production), ferrous and non-ferrous metal fabrication, armaments
Industrial production growth rate: 4.3% excluding oil (2010 est.)
country comparison to the world: 80
Electricity—production: 212.8 billion kWh (2009 est.)
country comparison to the world: 19
Electricity—consumption: 206.7 billion kWh (2009 est.)
country comparison to the world: 18
Electricity—exports: 6.15 billion kWh (2009 est.)
Electricity—imports: 2.06 billion kWh (2009 est.)
Oil—production: 4.172 million bbl/day (2009 est.)
country comparison to the world: 4
Oil—consumption: 1.809 million bbl/day (2009 est.)
country comparison to the world: 14
Oil—exports: 2.4 million bbl/day (2010 est.)
country comparison to the world: 4
Oil—imports: 168,000 bbl/day (2009 est.)
country comparison to the world: 53
Oil—proved reserves: 137.6 billion bbl based on Iranian claims
country comparison to the world: 3
note: Iran has about 10% of world reserves (1 January 2010 est.)
Natural gas—production: 200 billion cu m (2009 est.)
country comparison to the world: 3
Natural gas—consumption: 140 billion cu m
country comparison to the world: 4
note: excludes injection and flaring (2009 est.)
Natural gas—exports: 5.4 billion cu m (2009 est.)
country comparison to the world: 26
Natural gas—imports: 5.2 billion cu m (2009 est.)
country comparison to the world: 31
Natural gas—proved reserves: 29.61 trillion cu m (1 January 2010 est.)
country comparison to the world: 2
Current account balance: $9.76 billion (2010 est.)
country comparison to the world: 24
$1.913 billion (2009 est.)
Exports: $78.69 billion (2010 est.)
country comparison to the world: 37
$69.04 billion (2009 est.)
Exports—commodities: petroleum 80%, chemical and petrochemical products, fruits and nuts, carpets
Exports—partners: China 16.58%, Japan 11.9%, India 10.54%, South Korea 7.54%, Turkey 4.36% (2009)
Imports: $58.97 billion (2010 est.)
country comparison to the world: 44
$58.97 billion (2009 est.)
Imports—commodities: industrial supplies, capital goods, foodstuffs and other consumer goods, technical services
Imports—partners: UAE 15.14%, China 13.48%, Germany 9.66%, South Korea 7.16%, Italy 5.27%, Russia 4.81%, India 4.12% (2009)
Reserves of foreign exchange and gold: $75.06 billion (31 December 2010 est.)
country comparison to the world: 20
$81.31 billion (31 December 2009 est.)
Debt—external: $12.84 billion (31 December 2010 est.)
country comparison to the world: 83
$12.63 billion (31 December 2009 est.)
Stock of direct foreign investment—at home: $16.82 billion (31 December 2010 est.)
country comparison to the world: 72
$15.13 billion (31 December 2009 est.)
Stock of direct foreign investment—abroad: $2.075 billion (31 December 2010 est.)
country comparison to the world: 67
$1.825 billion (31 December 2009 est.)
Exchange rates: Iranian rials (IRR) per US dollar–10,308.2 (2010), 9,864.3 (2009), 9,142.8 (2008), 9,407.5 (2007), 9,227.1 (2006)

COMMUNICATIONS

Telephones—main lines in use: 25.804 million (2009)
country comparison to the world: 11
Telephones—mobile cellular: 52.555 million (2009)
country comparison to the world: 21
Telephone system: *general assessment:* currently being modernized and expanded with the goal of not only improving the efficiency and increasing the volume of the urban service but also bringing telephone service to several thousand villages, not presently connected
domestic: the addition of new fiber cables and modern switching and exchange systems installed by Iran's state-owned telecom company have improved and expanded the fixed-line network greatly; fixed-line availability has more than doubled to nearly 26 million lines since 2000; additionally, mobile-cellular service has increased dramatically serving more than 50 million subscribers in 2009; combined fixed and mobile-cellular subscribership now exceeds 100 per 100 persons
international: country code–98; submarine fiber-optic cable to UAE with access to Fiber-Optic Link Around the Globe (FLAG); Trans-Asia-Europe (TAE) fiber-optic line runs from Azerbaijan through the northern portion of Iran to Turkmenistan with expansion to Georgia and Azerbaijan; HF radio and microwave radio relay to Turkey, Azerbaijan, Pakistan, Afghanistan, Turkmenistan, Syria, Kuwait, Tajikistan, and Uzbekistan; satellite earth stations–13 (9 Intelsat and 4 Inmarsat) (2009)
Broadcast media: state-run broadcast media with no private, independent broadcasters; Islamic Republic of Iran Broadcasting (IRIB), the state-run TV broadcaster, operates 5 nationwide channels, a news channel, about 30 provincial channels, and several international channels; about 20 foreign Persian-language TV stations broadcasting on satellite TV are capable of being seen in Iran; satellite dishes are illegal and, while their use had been tolerated, authorities began confiscating satellite dishes following the unrest stemming from the 2009 presidential election; IRIB operates 8 nationwide radio networks, a number

of provincial stations, and an external service; most major international broadcasters transmit to Iran (2009)
Internet country code: .ir
Internet hosts: 119,947 (2010)
country comparison to the world: 75
Internet users: 8.214 million (2009)
country comparison to the world: 35

TRANSPORTATION

Airports: 319 (2010)
country comparison to the world: 24
Airports—with paved runways: *total:* 133
over 3,047 m: 42
2,438 to 3,047 m: 27
1,524 to 2,437 m: 24
914 to 1,523 m: 34
under 914 m: 6 (2010)
Airports—with unpaved runways:
total: 186
over 3,047 m: 1
2,438 to 3,047 m: 1
1,524 to 2,437 m: 9
914 to 1,523 m: 142
under 914 m: 33 (2010)
Heliports: 19 (2010)
Pipelines: condensate 7 km; condensate/gas 12 km; gas 20,155 km; liquid petroleum gas 570 km; oil 7,123 km; refined products 7,937 km (2010)
Railways: *total:* 8,442 km
country comparison to the world: 26
broad gauge: 94 km 1.676-m gauge
standard gauge: 8,348 km 1.435-m gauge (148 km electrified) (2008)
Roadways: *total:* 172,927 km
country comparison to the world: 28
paved: 125,908 km (includes 1,429 km of expressways)
unpaved: 47,019 km (2006)
Waterways: 850 km (on Karun River; some navigation on Lake Urmia) (2009)
country comparison to the world: 70
Merchant marine: *total:* 74
country comparison to the world: 57
by type: bulk carrier 11, cargo 40, chemical tanker 5, container 9, liquefied gas 1, passenger/cargo 3, petroleum tanker 1, refrigerated cargo 2, roll on/roll off 2
foreign-owned: 1 (UAE 1)
registered in other countries: 78 (Barbados 4, Bolivia 1, Cyprus 10, Hong Kong 1, Malta 56, Panama 5, Ukraine 1) (2010)
Ports and terminals: Assaluyeh, Bandar Abbas, Bandar-e-Eman Khomeyni

MILITARY

Military branches: Islamic Republic of Iran Regular Forces (Artesh): Ground Forces, Navy, Air Force (IRIAF), Khatemolanbia Air Defense Headquarters; Islamic Revolutionary Guard Corps (Sepah-e Pasdaran-e Enqelab-e Eslami, IRGC): Ground Resistance Forces, Navy, Aerospace Force, Qods Force (special operations); Law Enforcement Forces (2011)
Military service age and obligation: 19 years of age for compulsory military service; 16 years of age for volunteers; 17 years of age for Law Enforcement Forces; 15 years of age for Basij Forces (Popular Mobilization Army); conscript military service obligation—18 months; women exempt from military service (2008)
Manpower available for military service:
males age 16–49: 23,619,215
females age 16–49: 22,628,341 (2010 est.)
Manpower fit for military service: *males age 16–49:* 20,149,222
females age 16–49: 19,417,275 (2010 est.)
Manpower reaching militarily significant age annually: *male:* 715,111
female: 677,372 (2010 est.)
Military expenditures: 2.5% of GDP (2006)
country comparison to the world: 59

TRANSNATIONAL ISSUES

Disputes—international: Iran protests Afghanistan's limiting flow of dammed Helmand River tributaries during drought; Iraq's lack of a maritime boundary with Iran prompts jurisdiction disputes beyond the mouth of the Shatt al Arab in the Persian Gulf; Iran and UAE dispute Tunb Islands and Abu Musa Island, which are occupied by Iran; Azerbaijan, Kazakhstan, and Russia ratified Caspian seabed delimitation treaties based on equidistance, while Iran continues to insist on a one-fifth slice of the lake; Afghan and Iranian commissioners have discussed boundary monument densification and resurvey
Refugees and internally displaced persons: refugees (country of origin): 914,268 (Afghanistan); 54,024 (Iraq) (2007)
Trafficking in persons: current situation: Iran is a source, transit, and destination country for men, women, and children trafficked for the purposes of sexual exploitation and involuntary servitude; Iranian women are trafficked internally for the purpose of forced prostitution and for forced marriages to settle debts; Iranian and Afghan children living in Iran are trafficked internally for the purpose of forced marriages, commercial sexual exploitation and involuntary servitude as beggars or laborers to pay debts, provide income or support drug addiction of their families; press reports indicate that criminal organizations play a significant role in human trafficking to and from Iran, in connection with smuggling of migrants, drugs, and arms
tier rating: Tier 3–Iran did not provide evidence of law enforcement activities against trafficking, and credible reports indicate that Iranian authorities' response is not sufficient to penalize offenders, protect victims, and eliminate trafficking; some aspects of Iranian law and policy hinder efforts to combat trafficking including punishment of victims and legal obstacles to punishing offenders; Iran has not ratified the 2000 UN TIP Protocol (2009)
Illicit drugs: despite substantial interdiction efforts and considerable control measures along the border with Afghanistan, Iran remains one of the primary transshipment routes for Southwest Asian heroin to Europe; suffers one of the highest opiate addiction rates in the world, and has an increasing problem with synthetic drugs; lacks anti-money laundering laws; has reached out to neighboring countries to share counter-drug intelligence

IRAQ

INTRODUCTION

Background: Formerly part of the Ottoman Empire, Iraq was occupied by Britain during the course of World War I; in 1920, it was declared a League of Nations mandate under UK administration. In stages over the next dozen years, Iraq attained its independence as a kingdom in 1932. A "republic" was proclaimed in 1958, but in actuality a series of strongmen ruled the country until 2003. The last was SADDAM Husayn. Territorial disputes with Iran led to an inconclusive and costly eight-year war (1980-88). In August 1990, Iraq seized Kuwait but was expelled by US-led, UN coalition forces during the Gulf War of January-February 1991. Following Kuwait's liberation, the UN Security Council (UNSC) required Iraq to scrap all weapons of mass destruction and long-range missiles and to allow UN verification inspections. Continued Iraqi noncompliance with UNSC resolutions over a period of 12 years led to the US-led invasion of Iraq in March 2003 and the ouster of the SADDAM Husayn regime. US forces remained in Iraq under a UNSC mandate through 2009 and under a bilateral security agreement thereafter, helping to provide security and to train and mentor Iraqi security forces. In October 2005, Iraqis approved a constitution in a national referendum and, pursuant to this document, elected a 275-member Council of Representatives (CoR) in December 2005. The CoR approved most cabinet ministers in May 2006, marking the transition to Iraq's first constitutional government in nearly a half century. In January 2009, Iraq held elections for provincial councils in all provinces except for the three provinces comprising the Kurdistan Regional Government and Kirkuk province. Iraq held a national legislative election in March 2010, and after nine months of deadlock the CoR approved the new government in December 2010.

GEOGRAPHY

Location: Middle East, bordering the Persian Gulf, between Iran and Kuwait
Geographic coordinates: 33 00 N, 44 00 E
Map references: Middle East
Area: *total:* 438,317 sq km
country comparison to the world: 58
land: 437,367 sq km
water: 950 sq km
Area—comparative: slightly more than twice the size of Idaho
Land boundaries: *total:* 3,650 km
border countries: Iran 1,458 km, Jordan 181 km, Kuwait 240 km, Saudi Arabia 814 km, Syria 605 km, Turkey 352 km
Coastline: 58 km
Maritime claims: *territorial sea:* 12 nm
continental shelf: not specified
Climate: mostly desert; mild to cool winters with dry, hot, cloudless summers; northern mountainous regions along Iranian and Turkish borders experience cold winters with occasionally heavy snows that melt in early spring, sometimes causing extensive flooding in central and southern Iraq
Terrain: mostly broad plains; reedy marshes along Iranian border in south with large flooded areas; mountains along borders with Iran and Turkey
Elevation extremes: *lowest point:* Persian Gulf 0 m
highest point: unnamed peak; 3,611 m; note–this peak is neither Gundah Zhur 3,607 m nor Kuh-e Hajji-Ebrahim 3,595 m
Natural resources: petroleum, natural gas, phosphates, sulfur
Land use: *arable land:* 13.12%
permanent crops: 0.61%
other: 86.27% (2005)
Irrigated land: 35,250 sq km (2003)
Total renewable water resources: 96.4 cu km (1997)
Freshwater withdrawal (domestic/industrial/agricultural): *total:* 42.7 cu km/yr (3%/5%/92%)
per capita: 1,482 cu m/yr (2000)
Natural hazards: dust storms; sandstorms; floods
Environment—current issues: government water control projects have drained most of the inhabited marsh areas east of An Nasiriyah by drying up or diverting the feeder streams and rivers; a once sizable population of Marsh Arabs, who inhabited these areas for thousands of years, has been displaced; furthermore, the destruction of the natural habitat poses serious threats to the area's wildlife populations; inadequate supplies of potable water; development of the Tigris and Euphrates rivers system contingent upon agreements with upstream riparian Turkey; air and water pollution; soil degradation (salination) and erosion; desertification
Environment—international agreements: *party to:* Biodiversity, Law of the Sea, Ozone Layer Protection
signed, but not ratified: Environmental Modification
Geography—note: strategic location on Shatt al Arab waterway and at the head of the Persian Gulf

PEOPLE

Population: 30,399,572 (July 2011 est.)
country comparison to the world: 39
Age structure: *0–14 years:* 38% (male 5,882,682/female 5,678,741)
15–64 years: 58.9% (male 9,076,558/female 8,826,545)
65 years and over: 3.1% (male 435,908/female 499,138) (2011 est.)
Median age: *total:* 20.9 years
male: 20.8 years
female: 21 years (2011 est.)
Population growth rate: 2.399% (2011 est.)
country comparison to the world: 31
Birth rate: 28.81 births/1,000 population (2011 est.)
country comparison to the world: 44
Death rate: 4.82 deaths/1,000 population (July 2011 est.)
country comparison to the world: 194
Net migration rate: 0 migrant(s)/1,000 population (2011 est.)
country comparison to the world: 89
Urbanization: *urban population:* 66% of total population (2010)
rate of urbanization: 2.6% annual rate of change (2010-15 est.)
Major cities—population: BAGHDAD (capital) 5.751 million; Mosul 1.447 million; Erbil 1.009 million; Basra 923,000; As Sulaymaniyah 836,000 (2009)
Sex ratio: *at birth:* 1.05 male(s)/female
under 15 years: 1.04 male(s)/female
15–64 years: 1.03 male(s)/female
65 years and over: 0.88 male(s)/female
total population: 1.03 male(s)/female (2011 est.)
Infant mortality rate: *total:* 41.68 deaths/1,000 live births
country comparison to the world: 61
male: 45.93 deaths/1,000 live births
female: 37.21 deaths/1,000 live births (2011 est.)
Life expectancy at birth: *total population:* 70.55 years
country comparison to the world: 145
male: 69.15 years
female: 72.02 years (2011 est.)
Total fertility rate: 3.67 children born/woman (2011 est.)
country comparison to the world: 42
HIV/AIDS—adult prevalence rate: less than 0.1% (2001 est.)
country comparison to the world: 136
HIV/AIDS—people living with HIV/AIDS: fewer than 500 (2003 est.)
country comparison to the world: 152
HIV/AIDS—deaths: NA
Major infectious diseases: *degree of risk:* intermediate
food or waterborne diseases: bacterial diarrhea, hepatitis A, and typhoid fever
note: highly pathogenic H5N1 avian influenza has been identified in this country; it poses a negligible risk with extremely rare cases possible among US citizens who have close contact with birds (2009)
Drinking water source: *Improved:*
urban: 91% of population
rural: 55% of population
total: 79% of population
Unimproved: urban: 9% of population
rural: 45% of population
total: 21% of population (2008)
Sanitation facility access: *Improved:*
urban: 76% of population
rural: 66% of population
total: 73% of population
Unimproved:
urban: 24% of population
rural: 34% of population
total: 27% of population (2008)
Nationality: *noun:* Iraqi(s)
adjective: Iraqi
Ethnic groups: Arab 75%-80%, Kurdish 15%-20%, Turkoman, Assyrian, or other 5%
Religions: Muslim 97% (Shia 60%-65%, Sunni 32%-37%), Christian or other 3%
note: while there has been voluntary relocation of many Christian families to northern Iraq, recent reporting indicates that the overall Christian population may have dropped by as much as 50 percent since the fall of the Saddam HUSSEIN regime in 2003, with many fleeing to Syria, Jordan, and Lebanon
Languages: Arabic (official), Kurdish (official in Kurdish regions), Turkoman (a Turkish dialect), Assyrian (Neo-Aramaic), Armenian
Literacy: *definition:* age 15 and over can read and write
total population: 74.1%
male: 84.1%
female: 64.2% (2000 est.)
School life expectancy (primary to tertiary education): *total:* 10 years
male: 11 years
female: 8 years (2005)
Education expenditures: NA

GOVERNMENT

Country name: *conventional long form:* Republic of Iraq
conventional short form: Iraq
local long form: Jumhuriyat al-Iraq
local short form: Al Iraq
Government type: parliamentary democracy
Capital: *name:* Baghdad
geographic coordinates: 33 20 N, 44 23 E
time difference: UTC+3 (8 hours ahead of Washington, DC during Standard Time)
Administrative divisions: 18 governorates (muhafazat, singular–muhafazah) and 1 region*; Al Anbar, Al Basrah, Al Muthanna, Al Qadisiyah (Ad Diwaniyah), An Najaf, Arbil (Erbil), As Sulaymaniyah, Babil, Baghdad, Dahuk, Dhi Qar, Diyala, Karbala', Kirkuk, Kurdistan Regional Government*, Maysan, Ninawa, Salah ad Din, Wasit
Independence: 3 October 1932 (from League of Nations mandate under British administration); note–on 28 June 2004 the Coalition Provisional Authority transferred sovereignty to the Iraqi Interim Government

National holiday: Republic Day, July 14 (1958); note–the Government of Iraq has yet to declare an official national holiday but still observes Republic Day
Constitution: ratified 15 October 2005 (subject to review by the Constitutional Review Committee and a possible public referendum)
Legal system: mixed legal system of civil and Islamic law
International law organization participation: has not submitted an ICJ jurisdiction declaration; non-party state to the ICCt
Suffrage: 18 years of age; universal
Executive branch: *chief of state:* President Jalal TALABANI (since 6 April 2005)
head of government: Prime Minister Nuri al-MALIKI (since 20 May 2006)
cabinet: Council of Ministers consists of ministers appointed by the Presidency Council plus the Prime Minister and Deputy Prime Ministers
(For more information visit the World Leaders website)
elections: president elected by Council of Representatives (parliament) to serve a four-year term (eligible for a second term);election last held on 11 November 2010 (next to be held in 2014)
election results: President Jalal TALABANI reelected on 11 November 2010; parliamentary vote count on second ballot–195 votes; Nuri al-MALIKI reselected prime minister
Legislative branch: unicameral Council of Representatives (325 seats consisting of 317 members elected by an optional open-list, proportional representation system and 8 seats reserved for minorities; members serve four-year terms); note–Iraq's Constitution calls for the establishment of an upper house, the Federation Council
elections: last held on 7 March 2010 for an enlarged 325-seat parliament (next to be held in 2014)
election results: Council of Representatives–percent of vote by coalition–Iraqi National Movement 25.9%, State of Law coalition 25.8%, Iraqi National Alliance 19.4%, Kurdistan Alliance 15.3%, Goran (Change) List 4.4%, Tawafuq Front 2.7%, Iraqi Unity Alliance 2.9%, Kurdistan Islamic Union 2.3%, Kurdistan Islamic Group 1.4%; seats by coalition–Iraqi National Movement 91, State of Law Coalition 89, Iraqi National Alliance 70, Kurdistan Alliance 43, Goran (Change) List 8, Tawafuq Front 6, Iraqi Unity Alliance 4, Kurdistan Islamic Union 4, Kurdistan Islamic Group 2, seats reserved for minorities 8
Judicial branch: the Iraq Constitution calls for the federal judicial power to be comprised of the Higher Judicial Council, Federal Supreme Court, Federal Court of Cassation, Public Prosecution Department, Judiciary Oversight Commission and other federal courts that are regulated in accordance with the law
Political parties and leaders: Badr Organization [Hadi al-AMIRI]; Da'wa al-Islamiya Party [Prime Minister Nuri al-MALIKI]; Da'wa Tanzim [Hashim al-MUSAWI branch]; Da-wa Tanzim [Abd al-Karim al-ANZI branch]; Fadilah Party [Hashim al-HASHIMI]; Hadba Gathering [Athil al-NUJAYFI]; Iraqi Charter Assembly [Ahmad Abd al-Ghafur al-SAMARRAI]; Iraqi Constitutional Party [Jawad al-BULANI]; Iraqi Front for National Dialogue [Salih al-MUTLAQ]; Iraqi Islamic Party or IIP [Usama al-TIKRITI]; Iraqi Justice and Reform Movement [Shaykh Abdallah al-YAWR]; Iraqi National Congress or INC [Ahmad CHALABI]; Iraqi National Accord or INA [former Prime Minister Ayad ALLAWI]; Islamic Supreme Council of Iraq or ISCI [Ammar al-HAKIM]; Kurdistan Democratic Party or KDP [Kurdistan Regional Government President Masud BARZANI]; National Gathering [Deputy Prime Minister Rafi al-ISSAWI]; National Movement for Reform and Development [Jamal al-KARBULI]; National Reform Trend [former Prime Minister Ibrahim al-JAFARI]; Patriotic Union of Kurdistan or PUK [Jalal TALABANI]; Renewal List [Vice President Tariq al-HASHIMI]; Sadrist Trend [Muqtada al-SADR]; Sahawa al-Iraq [Ahmad al-RISHAWI]; Tawafuq Front
note: numerous smaller local, tribal, and minority parties
Political pressure groups and leaders: Sunni militias; Shia militias, some associated with political parties
International organization participation: ABEDA, AFESD, AMF, CAEU, CICA, FAO, G-77, IAEA, IBRD, ICAO, ICRM, IDA, IDB, IFAD, IFC, IFRCS, ILO, IMF, IMO, IMSO, Interpol, IOC, IPU, ISO, ITSO, ITU, LAS, MIGA, NAM, OAPEC, OIC, OPCW, OPEC, PCA, UN, UNCTAD, UNESCO, UNIDO, UNWTO, UPU, WCO, WFTU, WHO, WIPO, WMO, WTO (observer)
Diplomatic representation in the US: *chief of mission:* Ambassador Samir Shakir al-SUMAYDI
chancery: 3421 Massachusetts Ave, NW, Washington, DC 20007
telephone: [1] (202) 742-1600
FAX: [1] (202) 333-1129
Diplomatic representation from the US: *chief of mission:* Ambassador James F. JEFFREY
embassy: Baghdad
mailing address: APO AE 09316
telephone: 1-240-553-0589 ext. 5340 or 5635; note–Consular Section
FAX: NA
Flag description: three equal horizontal bands of red (top), white, and black; the Takbir (Arabic expression meaning "God is great") in green Arabic script is centered in the white band; the band colors derive from the Arab Liberation flag and represent oppression (black), overcome through bloody struggle (red), to be replaced by a bright future (white); the Council of Representatives approved this flag in 2008 as a compromise temporary replacement for the Ba'athist Saddam-era flag
note: similar to the flag of Syria, which has two stars but no script, Yemen, which has a plain white band, and that of Egypt, which has a gold Eagle of Saladin centered in the white band
National anthem: *name:* "Mawtini" (My Homeland)
lyrics/music: Ibrahim TOUQAN/ Mohammad FLAYFEL
note: adopted 2004; following the ousting of Saddam HUSSEIN, Iraq adopted "Mawtini," a popular folk song throughout the Arab world, which also serves as an unofficial anthem of the Palestinian people

ECONOMY

Economy—overview: An improved security environment and an initial wave of foreign investment are helping to spur economic activity, particularly in the energy, construction, and retail sectors. Broader economic improvement, long-term fiscal health, and sustained increases in the standard of living still depend on the government passing major policy reforms and on continued development of Iraq's massive oil reserves. Although foreign investors viewed Iraq with increasing interest in 2010, most are still hampered by difficulties in acquiring land for projects and by other regulatory impediments. Iraq's economy is dominated by the oil sector, which provides over 90% of government revenue and 80% of foreign exchange earnings. Since mid-2009, oil export earnings have returned to levels seen before Operation Iraqi Freedom and government revenues have rebounded, along with global oil prices. In 2011 Baghdad probably will increase oil exports above the current level of 1.9 million barrels per day (bbl/day) as a result of new contracts with international oil companies, but is likely to fall short of the 2.4 million bbl/day it is forecasting in its budget. Iraq is making modest progress in building the institutions needed to implement economic policy. In 2010, Bagdad signed a new agreement with both the IMF and World Bank for conditional aid programs that will help strengthen Iraq's economic institutions. Some reform-minded leaders within the Iraqi government are seeking to pass laws to strengthen the economy. This legislation includes a package of laws to establish a modern legal framework for the oil sector and a mechanism to equitably divide oil revenues within the nation, although these and other important reforms are still under contentious and sporadic negotiation. Iraq's recent contracts with major oil companies have the potential to greatly expand oil revenues, but Iraq will need to upgrade its oil processing, pipeline, and export infrastructure to enable these deals to reach their potential. The Government of Iraq is pursuing a strategy to gain additional foreign investment in Iraq's economy. This includes an amendment to the National Investment Law, multiple international trade and investment events, as well as potential participation in joint ventures with state-owned enterprises. Provincial Councils also are using their own budgets to promote and facilitate investment at the local level. However, widespread corruption, inadequate infrastructure, insufficient essential services, and antiquated commercial laws and regulations stifle investment and continue to constrain the growth of private, non-energy sectors. The Central Bank has successfully held the exchange rate at approximately

1,170 Iraqi dinar/US dollar since January 2009. Inflation has decreased consistently since 2006 as the security situation has improved. However, Iraqi leaders remain hard pressed to translate macroeconomic gains into improved lives for ordinary Iraqis. Unemployment remains a problem throughout the country. Reducing corruption and implementing reforms–such as bank restructuring and developing the private sector–would be important steps in this direction.
GDP (purchasing power parity): $113.4 billion (2010 est.)
country comparison to the world: 66
$112.4 billion (2009 est.)
$107.9 billion (2008 est.)
note: data are in 2010 US dollars
GDP (official exchange rate): $82.15 billion (2010 est.)
GDP—real growth rate: 0.8% (2010 est.)
country comparison to the world: 176
4.2% (2009 est.)
9.5% (2008 est.)
GDP—per capita (PPP): $3,800 (2010 est.)
country comparison to the world: 161
$3,900 (2009 est.)
$3,800 (2008 est.)
note: data are in 2010 US dollars
GDP—composition by sector: *agriculture:* 9.7%
industry: 63%
services: 27.3% (2010 est.)
Labor force: 8.5 million (2009 est.)
country comparison to the world: 55
Labor force—by occupation: *agriculture:* 21.6%
industry: 18.7%
services: 59.8% (2008 est.)
Unemployment rate: 15.3% (2009 est.)
country comparison to the world: 154
15.2% (2008 est.)
Population below poverty line: 25% (2008 est.)
Household income or consumption by percentage share: *lowest 10%:* NA%
highest 10%: NA%
Budget: *revenues:* $52.8 billion
expenditures: $72.4 billion (2010 est.)
Inflation rate (consumer prices): 4.2% (2010 est.)
country comparison to the world: 122
6.8% (2009 est.)
Central bank discount rate: 8.83% (31 December 2009)
country comparison to the world: 15
16.75% (31 December 2008)
Commercial bank prime lending rate: 15.64% (31 December 2009 est.)
country comparison to the world: 20
19.5% (31 December 2008 est.)
Stock of narrow money: $35.69 billion (31 December 2010 est.)
country comparison to the world: 50
$30.02 billion (31 December 2009 est.)
Stock of broad money: $46.01 billion (31 December 2010 est.)
country comparison to the world: 68
$37.9 billion (31 December 2009 est.)
Stock of domestic credit: $21.94 billion (31 December 2008 est.)
country comparison to the world: 76
$10.16 billion (31 December 2007 est.)
Market value of publicly traded shares: $2.6 billion (31 July 2010)
country comparison to the world: 97
$2 billion (31 July 2009)
$1.878 billion (31 March 2008)
Agriculture—products: wheat, barley, rice, vegetables, dates, cotton; cattle, sheep, poultry
Industries: petroleum, chemicals, textiles, leather, construction materials, food processing, fertilizer, metal fabrication/processing
Industrial production growth rate: 4.8% (2010 est.)
country comparison to the world: 74
Electricity—production: 46.39 billion kWh (2009 est.)
country comparison to the world: 49
Electricity—consumption: 52 billion kWh (2009 est.)
country comparison to the world: 45
Electricity—exports: 0 kWh (2009 est.)
Electricity—imports: 5.6 billion kWh (2009 est.)
Oil—production: 2.399 million bbl/day (2009 est.)
country comparison to the world: 12
Oil—consumption: 687,000 bbl/day (2009 est.)
country comparison to the world: 24
Oil—exports: 1.91 million bbl/day (2009 est.)
country comparison to the world: 11
Oil—imports: 116,900 bbl/day (2009 est.)
country comparison to the world: 59
Oil—proved reserves: 115 billion bbl (1 January 2010 est.)
country comparison to the world: 4
Natural gas—production: 1.88 billion cu m (2008 est.)
country comparison to the world: 58
Natural gas—consumption: 9.454 billion cu m
country comparison to the world: 49
note: 1.48 billion cu m were flared (2008 est.)
Natural gas—exports: 0 cu m (2008 est.)
country comparison to the world: 116
Natural gas—imports: 0 cu m (2008 est.)
country comparison to the world: 202
Natural gas—proved reserves: 3.17 trillion cu m (1 January 2010 est.)
country comparison to the world: 11
Current account balance: $2.715 billion (2010 est.)
country comparison to the world: 41
$-19.9 billion (2009 est.)
Exports: $49.1 billion (2010 est.)
country comparison to the world: 55
$40.86 billion (2009 est.)
Exports—commodities: crude oil 84%, crude materials excluding fuels, food and live animals
Exports—partners: US 27.62%, India 14.45%, Italy 10.14%, South Korea 8.62%, Taiwan 5.61%, China 4.23%, Netherlands 4.13%, Japan 3.99% (2009)
Imports:
$42.56 billion (2010 est.)
country comparison to the world: 51
$50 billion (2008 est.)
Imports—commodities: food, medicine, manufactures
Imports—partners: Turkey 24.99%, Syria 17.36%, US 8.66%, China 6.79%, Jordan 4.17%, Italy 3.98%, Germany 3.97% (2009)
Reserves of foreign exchange and gold: $45.68 billion (31 December 2010 est.)
country comparison to the world: 25
$44.38 billion (31 December 2009 est.)
Debt—external: $52.58 billion (31 December 2010 est.)
country comparison to the world: 55
$73 billion (31 December 2009 est.)
Exchange rates: Iraqi dinars (IQD) per US dollar–1,170 (2010), 1,170 (2009), 1,176 (2008), 1,255 (2007), 1,466 (2006)

COMMUNICATIONS

Telephones—main lines in use: 1.108 million (2009)
country comparison to the world: 74
Telephones—mobile cellular: 19.722 million (2009)
country comparison to the world: 40
Telephone system: *general assessment:* the 2003 liberation of Iraq severely disrupted telecommunications throughout Iraq including international connections; widespread government efforts to rebuild domestic and international communications through fiber optic links are in progress; the mobile cellular market has expanded rapidly and its subscribership base is expected to continue increasing rapidly
domestic: repairs to switches and lines destroyed during 2003 continue; additional switching capacity is improving access; mobile-cellular service is available and centered on 3 GSM networks which are being expanded beyond their regional roots, improving country-wide connectivity; wireless local loop is available in some metropolitan areas and additional licenses have been issued with the hope of overcoming the lack of fixed-line infrastructure
international: country code–964; satellite earth stations–4 (2 Intelsat–1 Atlantic Ocean and 1 Indian Ocean, 1 Intersputnik–Atlantic Ocean region, and 1 Arabsat (inoperative)); local microwave radio relay connects border regions to Jordan, Kuwait, Syria, and Turkey; international terrestrial fiber-optic connections have been established with Saudi Arabia, Turkey, and Kuwait with planned connections to Iran and Jordan; a link to the Fiber-Optic Link Around the Globe (FLAG) submarine fiber-optic cable is planned (2009)
Broadcast media: the number of private radio and television stations has increased rapidly since 2003; government-owned TV and radio stations are operated by the publicly-funded Iraqi Public Broadcasting Service; private broadcast media are mostly linked to political, ethnic, or religious groups; satellite TV is available to an estimated 70% of viewers and many of the broadcasters are based abroad; transmissions of multiple international radio broadcasters are accessible (2007)
Internet country code: .iq
Internet hosts: 9 (2010)
country comparison to the world: 223
Internet users: 325,900 (2009)
country comparison to the world: 126

TRANSPORTATION

Airports: 104 (2010)

country comparison to the world: 57
Airports—with paved runways: *total:* 75
over 3,047 m: 20
2,438 to 3,047 m: 36
1,524 to 2,437 m: 5
914 to 1,523 m: 6
under 914 m: 8 (2010)
Airports—with unpaved runways:
total: 29
over 3,047 m: 3
2,438 to 3,047 m: 4
1,524 to 2,437 m: 3
914 to 1,523 m: 13
under 914 m: 6 (2010)
Heliports: 21 (2010)
Pipelines: gas 2,447 km; liquid petroleum gas 918 km; oil 5,104 km; refined products 1,637 km (2010)
Railways: *total:* 2,272 km
country comparison to the world: 66
standard gauge: 2,272 km 1.435-m gauge (2008)
Roadways: *total:* 44,900 km
country comparison to the world: 82
paved: 37,851 km
unpaved: 7,049 km (2002)
Waterways: 5,279 km (the Euphrates River (2,815 km), Tigris River (1,899 km), and Third River (565 km) are the principal waterways) (2010)
country comparison to the world: 23
Merchant marine: *total:* 2
country comparison to the world: 148
by type: petroleum tanker 2
registered in other countries: 2 (Marshall Islands 2) (2010)
Ports and terminals: Al Basrah, Khawr az Zubayr, Umm Qasr

MILITARY

Military branches: Counterterrorism Service Forces: Counterterrorism Command; Iraqi Special Operations Forces (ISOF); Ministry of Defense Forces: Iraqi Army (includes Army Aviation Directorate, former National Guard Iraqi Intervention Forces, and Strategic Infrastructure Battalions), Iraqi Navy (former Iraqi Coastal Defense Force, includes Iraq Marine Force), Iraqi Air Force (Al-Quwwat al-Jawwiya al-Iraqiya) (2011)
Military service age and obligation: 18-40 years of age for voluntary military service (2010)
Manpower available for military service: *males age 16–49:* 7,767,329
females age 16–49: 7,461,766 (2010 est.)
Manpower fit for military service: *males age 16–49:* 6,591,185
females age 16–49: 6,421,717 (2010 est.)
Manpower reaching militarily significant age annually: *male:* 332,194
female: 322,010 (2010 est.)
Military expenditures: 8.6% of GDP (2006)
country comparison to the world: 5

TRANSNATIONAL ISSUES

Disputes—international: coalition forces assist Iraqis in monitoring internal and cross-border security; approximately two million Iraqis have fled the conflict in Iraq, with the majority taking refuge in Syria and Jordan, and lesser numbers to Egypt, Lebanon, Iran, and Turkey; Iraq's lack of a maritime boundary with Iran prompts jurisdiction disputes beyond the mouth of the Shatt al Arab in the Persian Gulf; Turkey has expressed concern over the autonomous status of Kurds in Iraq
Refugees and internally displaced persons: refugees (country of origin): 10,000-15,000 (Palestinian Territories); 11,773 (Iran); 16,832 (Turkey)
IDPs: 2.4 million (ongoing US-led war and ethno-sectarian violence) (2007)

IRELAND

INTRODUCTION

Background: Celtic tribes arrived on the island between 600-150 B.C. Invasions by Norsemen that began in the late 8th century were finally ended when King Brian BORU defeated the Danes in 1014. English invasions began in the 12th century and set off more than seven centuries of Anglo-Irish struggle marked by fierce rebellions and harsh repressions. A failed 1916 Easter Monday Rebellion touched off several years of guerrilla warfare that in 1921 resulted in independence from the UK for 26 southern counties; six northern (Ulster) counties remained part of the UK. In 1949, Ireland withdrew from the British Commonwealth; it joined the European Community in 1973. Irish governments have sought the peaceful unification of Ireland and have cooperated with Britain against terrorist groups. A peace settlement for Northern Ireland is gradually being implemented despite some difficulties. In 2006, the Irish and British governments developed and began to implement the St. Andrews Agreement, building on the Good Friday Agreement approved in 1998.

GEOGRAPHY

Location: Western Europe, occupying five-sixths of the island of Ireland in the North Atlantic Ocean, west of Great Britain
Geographic coordinates: 53 00 N, 8 00 W
Map references: Europe
Area:
total: 70,273 sq km
country comparison to the world: 119
land: 68,883 sq km
water: 1,390 sq km
Area—comparative: slightly larger than West Virginia
Land boundaries: *total:* 360 km
border countries: UK 360 km
Coastline: 1,448 km
Maritime claims:
territorial sea: 12 nm
exclusive fishing zone: 200 nm
Climate: temperate maritime; modified by North Atlantic Current; mild winters, cool summers; consistently humid; overcast about half the time
Terrain: mostly level to rolling interior plain surrounded by rugged hills and low mountains; sea cliffs on west coast
Elevation extremes:
lowest point: Atlantic Ocean 0 m
highest point: Carrauntoohil 1,041 m
Natural resources: natural gas, peat, copper, lead, zinc, silver, barite, gypsum, limestone, dolomite
Land use: *arable land:* 16.82%
permanent crops: 0.03%
other: 83.15% (2005)
Irrigated land: NA
Total renewable water resources: 46.8 cu km (2003)
Freshwater withdrawal (domestic/industrial/agricultural): *total:* 1.18 cu km/yr (23%/77%/0%)
per capita: 284 cu m/yr (1994)
Natural hazards: NA
Environment—current issues: water pollution, especially of lakes, from agricultural runoff
Environment—international agreements: *party to:* Air Pollution, Air Pollution-Nitrogen Oxides, Air Pollution-Sulfur 94, Biodiversity, Climate Change, Climate Change-Kyoto Protocol, Desertification, Endangered Species, Environmental Modification, Hazardous Wastes, Law of the Sea, Marine Dumping, Ozone Layer Protection, Ship Pollution, Tropical Timber 83, Tropical Timber 94, Wetlands, Whaling
signed, but not ratified: Air Pollution-Persistent Organic Pollutants, Marine Life Conservation
Geography—note: strategic location on major air and sea routes between North America and northern Europe; over 40% of the population resides within 100 km of Dublin

PEOPLE

Population: 4,670,976 (July 2011 est.)
country comparison to the world: 119
Age structure: *0–14 years:* 21.1% (male 503,921/female 483,454)
15–64 years: 67.3% (male 1,581,959/female 1,560,238)
65 years and over: 11.6% (male 246,212/female 295,192) (2011 est.)

Median age: *total:* 34.8 years
male: 34.5 years
female: 35.1 years (2011 est.)
Population growth rate: 1.061% (2011 est.)
country comparison to the world: 113
Birth rate: 16.1 births/1,000 population (2011 est.)
country comparison to the world: 127
Death rate: 6.34 deaths/1,000 population (July 2011 est.)
country comparison to the world: 154
Net migration rate: 0.86 migrant(s)/1,000 population (2011 est.)
country comparison to the world: 56
Urbanization: *urban population:* 62% of total population (2010)
rate of urbanization: 1.8% annual rate of change (2010-15 est.)
Major cities—population: DUBLIN (capital) 1.084 million (2009)
Sex ratio: *at birth:* 1.057 male(s)/female
under 15 years: 1.07 male(s)/female
15–64 years: 1 male(s)/female
65 years and over: 0.81 male(s)/female
total population: 0.99 male(s)/female (2011 est.)
Infant mortality rate:
total: 3.85 deaths/1,000 live births
country comparison to the world: 203
male: 4.24 deaths/1,000 live births
female: 3.44 deaths/1,000 live births (2011 est.)
Life expectancy at birth: *total population:* 80.19 years
country comparison to the world: 26
male: 77.96 years
female: 82.55 years (2011 est.)
Total fertility rate: 2.02 children born/woman (2011 est.)
country comparison to the world: 127
HIV/AIDS—adult prevalence rate: 0.2% (2009 est.)
country comparison to the world: 98
HIV/AIDS—people living with HIV/AIDS: 6,900 (2009 est.)
country comparison to the world: 113
HIV/AIDS—deaths: fewer than 100 (2009 est.)
country comparison to the world: 129
Drinking water source: *Improved:*
urban: 100% of population
rural: 100% of population
total: 100% of population (2008)
Sanitation facility access: *Improved:*
urban: 100% of population
rural: 98% of population
total: 99% of population
Unimproved: urban: 0% of population
rural: 2% of population
total: 1% of population (2008)
Nationality: *noun:* Irishman(men), Irishwoman(women), Irish (collective plural)
adjective: Irish
Ethnic groups: Irish 87.4%, other white 7.5%, Asian 1.3%, black 1.1%, mixed 1.1%, unspecified 1.6% (2006 census)
Religions: Roman Catholic 87.4%, Church of Ireland 2.9%, other Christian 1.9%, other 2.1%, unspecified 1.5%, none 4.2% (2006 census)
Languages: English (official,the language generally used), Irish (Gaelic or Gaeilge) (official, spoken mainly in areas along the western coast
Literacy: *definition:* age 15 and over can read and write
total population: 99%
male: 99%
female: 99% (2003 est.)
School life expectancy (primary to tertiary education): *total:* 18 years
male: 18 years
female: 18 years (2008)
Education expenditures: 4.9% of GDP (2007)
country comparison to the world: 65

GOVERNMENT

Country name: *conventional long form:* none
conventional short form: Ireland
local long form: none
local short form: Eire
Government type: republic, parliamentary democracy
Capital:
name: Dublin
geographic coordinates: 53 19 N, 6 14 W
time difference: UTC 0 (5 hours ahead of Washington, DC during Standard Time)
daylight saving time: +1hr, begins last Sunday in March; ends last Sunday in October
Administrative divisions: 29 counties and 5 cities*; Carlow, Cavan, Clare, Cork, Cork*, Donegal, Dublin*, Dun Laoghaire-Rathdown, Fingal, Galway, Galway*, Kerry, Kildare, Kilkenny, Laois, Leitrim, Limerick, Limerick*, Longford, Louth, Mayo, Meath, Monaghan, North Tipperary, Offaly, Roscommon, Sligo, South Dublin, South Tipperary, Waterford, Waterford*, Westmeath, Wexford, Wicklow
Independence: 6 December 1921 (from the UK by treaty)
National holiday: Saint Patrick's Day, 17 March
Constitution: adopted 1 July 1937 by plebiscite; effective 29 December 1937
Legal system: common law system based on the English model but substantially modified by customary law; judicial review of legislative acts in Supreme Court
International law organization participation: has not submitted an ICJ jurisdiction declaration; accepts ICCt jurisdiction
Suffrage: 18 years of age; universal
Executive branch: *chief of state:* President Mary MCALEESE (since 11 November 1997)
head of government: Taoiseach (Prime Minister) Enda KENNY (since 9 March 2011)
cabinet: Cabinet appointed by the president with previous nomination by the prime minister and approval of the House of Representatives
(For more information visit the World Leaders website)
elections: president elected by popular vote for a seven-year term (eligible for a second term); election last held on 31 October 1997 (next scheduled for March 2011); note—Mary MCALEESE was appointed to a second term when no other candidate qualified for the 2004 presidential election; prime minister (taoiseach) nominated by the House of Representatives (Dail Eireann) and appointed by the president
election results: Mary MCALEESE elected president; percent of vote—Mary MCALEESE 44.8%, Mary BANOTTI 29.6%
Legislative branch: bicameral Parliament or Oireachtas consists of the Senate or Seanad Eireann (60 seats; 49 members elected by the universities and from candidates put forward by five vocational panels, 11 are nominated by the prime minister; members serve five-year terms) and the lower house of Parliament or Dail Eireann (166 seats; members elected by popular vote on the basis of proportional representation to serve five-year terms)
elections: Senate—last held in July 2007 (next to be held 27 April 2011); House of Representatives—last held on 25 February 2011 (next to be held probably in 2016)
election results: Senate—percent of vote by party—NA; seats by party—Fianna Fail 28, Fine Gael 14, Labor Party 6, Progressive Democrats 2, Green Party 2, Sinn Fein 1, independents 7; House of Representatives—percent of vote by party—Fine Gael 45.8%, Labor Party 22.3%, Fianna Fail 12.0%, Sinn Fein 8.4%, United Left Alliance 3.0%, New Vision 0.6%, independents 7.8%; seats by party—Fine Gael 76, Labor Party 37, Fianna Fail 20, Sinn Fein 14, United Left Alliance 5, New Vision 1, independents 13; note—after November 2009 disbandment of the Progressive Democrats, the two members of the Senate continued as independent DPs
note: on 8 November 2008, delegates voted to disband the Progressive Democrats, and in November 2009 it officially stopped operating as a political party
Judicial branch: Supreme Court (judges appointed by the president on the advice of the prime minister and cabinet)
Political parties and leaders: Fianna Fail [Michael MARTIN]; Fine Gael [Enda KENNY]; Green Party [John GORMLEY]; Labor Party [Eamon GILMORE]; New Vision; Progressive Democrats or PD [Noel GREALISH] (formerly dissolved on 20 November 2009); Sinn Fein [Gerry ADAMS]; Socialist Party [Joe HIGGINS]; The Workers' Party [Michael FINNEGAN]; United Left Alliance
Political pressure groups and leaders: Families Acting for Innocent Relatives or FAIR [Brian MCCONNELL] (seek compensation for victims of violence); Families Against Intimidation and Terror or FAIT (oppose terrorism); Gaeltacht Civil Rights Campaign (Coiste Cearta Sibhialta na Gaeilge) or CCSG (encourages the use of the Irish language and campaigns for greater civil rights in Irish speaking areas); Iona Institute [David QUINN] (a conservative Catholic think tank); Irish Anti-War Movement [Richard Boyd BARRETT] (campaigns against wars around the world); Irish Republican Army or IRA (terrorist group); Keep Ireland Open (environmental group); Midland Railway Action Group or MRAG [Willie ALLEN] (transportation promoters); Peace and Neutrality Alliance [Roger COLE] (campaigns to protect Irish neutrality); Rail Users Ireland (formerly the Platform 11—transportation promoters); 32 Country Sovereignty Move-

ment or 32CSM (supports a fully sovereign Ireland); Ulster Defence Association or UDA (terrorist group)

International organization participation: ADB (nonregional member), Australia Group, BIS, CE, EAPC, EBRD, EIB, EMU, ESA, EU, FAO, FATF, IAEA, IBRD, ICAO, ICC, ICRM, IDA, IEA, IFAD, IFC, IFRCS, IHO, ILO, IMF, IMO, Interpol, IOC, IOM, IPU, ISO, ITSO, ITU, ITUC, MIGA, MINURSO, MONUSCO, NEA, NSG, OAS (observer), OECD, OPCW, OSCE, Paris Club, PCA, PFP, UN, UNCTAD, UNESCO, UNHCR, UNIDO, UNIFIL, UNOCI, UNRWA, UNTSO, UPU, WCO, WHO, WIPO, WMO, WTO, ZC

Diplomatic representation in the US: *chief of mission:* Ambassador Michael COLLINS
chancery: 2234 Massachusetts Avenue NW, Washington, DC 20008
telephone: [1] (202) 462-3939
FAX: [1] (202) 232-5993
consulate(s) general: Atlanta, Boston, Chicago, New York, San Francisco

Diplomatic representation from the US: *chief of mission:* Ambassador Daniel ROONEY
embassy: 42 Elgin Road, Ballsbridge, Dublin 4
mailing address: use embassy street address
telephone: [353] (1) 668-8777
FAX: [353] (1) 668-9946

Flag description: three equal vertical bands of green (hoist side), white, and orange; officially the flag colors have no meaning, but a common interpretation is that the green represents the Irish nationalist (Gaelic) tradition of Ireland; orange represents the Orange tradition (minority supporters of William of Orange); white symbolizes peace (or a lasting truce) between the green and the orange
note: similar to the flag of Cote d'Ivoire, which is shorter and has the colors reversed–orange (hoist side), white, and green; also similar to the flag of Italy, which is shorter and has colors of green (hoist side), white, and red

National anthem: *name:* "Amhran na bhFiann" (The Soldier's Song)
lyrics/music: Peadar KEARNEY [English], Liam O RINN [Irish]/Patrick HEENEY and Peadar KEARNEY
note: adopted 1926; instead of "Amhran na bhFiann," the song "Ireland's Call" is often used in athletic events where citizens of the Republic of Ireland and Northern Ireland compete as a unified team

ECONOMY

Economy—overview: Ireland is a small, modern, trade-dependent economy. Ireland was among the initial group of 12 EU nations that began circulating the euro on 1 January 2002. GDP growth averaged 6% in 1995-2007, but economic activity has dropped sharply since the onset of the world financial crisis, with GDP falling by over 3% in 2008, nearly 8% in 2009, and 1% in 2010. Ireland entered into a recession in 2008 for the first time in more than a decade, with the subsequent collapse of its domestic property and construction markets. Property prices rose more rapidly in Ireland in the decade up to 2007 than in any other developed economy. Since their 2007 peak, average house prices have fallen 50%. In the wake of the collapse of the construction sector and the downturn in consumer spending and business investment, the export sector, dominated by foreign multinationals, has become a key component of Ireland's economy. Agriculture, once the most important sector, is now dwarfed by industry and services. In 2008 the COWEN government moved to guarantee all bank deposits, recapitalize the banking system, and establish partly-public venture capital funds in response to the country's economic downturn. In 2009, in continued efforts to stabilize the banking sector, the Irish Government established the National Asset Management Agency (NAMA) to acquire problem commercial property and development loans from Irish banks. Faced with sharply reduced revenues and a burgeoning budget deficit, the Irish Government introduced the first in a series of draconian budgets in 2009. In addition to across-the-board cuts in spending, the 2009 budget included wage reductions for all public servants. These measures were not sufficient. The budget deficit reached nearly 32% of GDP in 2010 because of additional government support for the banking sector. In late 2010, the COWEN Government agreed to a $112 billion loan package from the EU and IMF to help Dublin further increase the capitalization of its banking sector and avoid defaulting on its sovereign debt. The government also initiated a four-year austerity plan to cut an additional $20 billion from its budget. A return to modest growth is expected in 2011.

GDP (purchasing power parity): $172.3 billion (2010 est.)
country comparison to the world: 57
$174.2 billion (2009 est.)
$188.4 billion (2008 est.)
note: data are in 2010 US dollars

GDP (official exchange rate): $204.3 billion (2010 est.)

GDP—real growth rate: -1% (2010 est.)
country comparison to the world: 199
-7.6% (2009 est.)
-3.5% (2008 est.)

GDP—per capita (PPP): $37,300 (2010 est.)
country comparison to the world: 27
$38,000 (2009 est.)
$41,700 (2008 est.)
note: data are in 2010 US dollars

GDP—composition by sector: *agriculture:* 2%
industry: 29%
services: 70% (2009 est.)

Labor force: 2.15 million (2010 est.)
country comparison to the world: 119

Labor force—by occupation:
agriculture: 5%
industry: 20%
services: 76% (2010 est.)

Unemployment rate: 13.7% (2010 est.)
country comparison to the world: 140
12.4% (2009)

Population below poverty line: 5.5% (2009 est.)

Household income or consumption by percentage share: *lowest 10%:* 2.9%
highest 10%: 27.2% (2000)

Distribution of family income—Gini index: 29.3 (2009)
country comparison to the world: 116
35.9 (1987)

Investment (gross fixed): 16.5% of GDP (2010 est.)
country comparison to the world: 123

Budget: *revenues:* $68.7 billion
expenditures: $135.1 billion (2010 est.)

Public debt: 94.2% of GDP (2010 est.)
country comparison to the world: 11
65.5% of GDP (2009 est.)

Inflation rate (consumer prices): -1.6% (2010 est.)
country comparison to the world: 2
-1.7% (2009 est.)

Central bank discount rate: 1.75% (31 December 2010)
country comparison to the world: 121
1.75% (31 December 2009)
note: this is the European Central Bank's rate on the marginal lending facility, which offers overnight credit to banks in the euro area

Commercial bank prime lending rate: 2.1% (31 December 2010 est.)
country comparison to the world: 153
2.5% (31 December 2009 est.)

Stock of narrow money: $127.7 billion (31 December 2010 est.)
country comparison to the world: 27
$141 billion (31 December 2009 est.)
note: see entry for the European Union for money supply in the euro area; the European Central Bank (ECB) controls monetary policy for the 17 members of the Economic and Monetary Union (EMU); individual members of the EMU do not control the quantity of money circulating within their own borders

Stock of broad money: $257.1 billion (31 December 2010 est.)
country comparison to the world: 31
$275.9 billion (31 December 2009 est.)

Stock of domestic credit: $745.7 billion (31 December 2009 est.)
country comparison to the world: 19
$738.5 billion (31 December 2008 est.)

Market value of publicly traded shares: $63.1 billion (31 December 2010)
country comparison to the world: 49
$61.7 billion (31 December 2009)
$49.4 billion (31 December 2008)

Agriculture—products: beef, dairy products, barley, potatoes, wheat

Industries: pharmaceuticals, chemicals, computer hardware and software, food products, beverages and brewing; medical devices

Industrial production growth rate: 5% (2010 est.)
country comparison to the world: 70

Electricity—production: 27.28 billion kWh (2010 est.)
country comparison to the world: 63

Electricity—consumption: 26.99 billion kWh (2010 est.)
country comparison to the world: 64

Electricity—exports: 290 million kWh (2010 est.)

Electricity—imports: 756 million kWh (2010 est.)

Oil—production: 0 bbl/day (2009 est.)
country comparison to the world: 169

Oil—consumption: 160,900 bbl/day (2009 est.)
country comparison to the world: 63
Oil—exports: 19,270 bbl/day (2009 est.)
country comparison to the world: 89
Oil—imports: 181,600 bbl/day (2009 est.)
country comparison to the world: 49
Oil—proved reserves: 0 bbl (1 January 2010)
country comparison to the world: 125
Natural gas—production: 392 million cu m (2009)
country comparison to the world: 69
Natural gas—consumption: 4.999 billion cu m (2009)
country comparison to the world: 58
Natural gas—exports: 0 cu m (2009)
country comparison to the world: 88
Natural gas—imports: 4.628 billion cu m (2009)
country comparison to the world: 33
Natural gas—proved reserves: 9.911 billion cu m (1 January 2010 est.)
country comparison to the world: 79
Current account balance: $-3.191 billion (2010 est.)
country comparison to the world: 167
$-6.769 billion (2009)
Exports: $115.7 billion (2010 est.)
country comparison to the world: 34
$117.5 billion (2009)
Exports—commodities: machinery and equipment, computers, chemicals, pharmaceuticals; live animals, animal products
Exports—partners: US 20.52%, Belgium 17.78%, UK 16.31%, Germany 5.66%, France 5.56%, Spain 4.19% (2009)
Imports: $70.36 billion (2010 est.)
country comparison to the world: 38
$62.85 billion (2009)
Imports—commodities: data processing equipment, other machinery and equipment, chemicals, petroleum and petroleum products, textiles, clothing
Imports—partners: UK 35.28%, US 16.87%, Germany 6.76%, Netherlands 5.86%, France 4.76% (2009)
Reserves of foreign exchange and gold: $2.104 billion (31 December 2010)
country comparison to the world: 101
$2.087 billion (31 December 2009)
Debt—external: $2.253 trillion (30 September 2010)
country comparison to the world: 7
$2.087 trillion (31 December 2009)
Stock of direct foreign investment—at home: $228 billion (31 September 2010 est.)
country comparison to the world: 19
$236.2 billion (31 December 2009)
Stock of direct foreign investment—abroad: $286.2 billion (31 September 2010)
country comparison to the world: 14
$264.6 billion (31 December 2009)
Exchange rates: euros (EUR) per US dollar–0.755 (2010), 0.7198 (2009), 0.6827 (2008), 0.7345 (2007), 0.7964 (2006)

COMMUNICATIONS

Telephones—main lines in use: 2.08 million (2009)
country comparison to the world: 55
Telephones—mobile cellular: 4.871 million (2009)
country comparison to the world: 98
Telephone system: *general assessment:* modern digital system using cable and microwave radio relay
domestic: system privatized but dominated by former state monopoly operator; increasing levels of broadband access particularly in urban areas
international: country code–353; landing point for the Hibernia-Atlantic submarine cable with links to the US, Canada, and UK; satellite earth station–1 Intelsat (Atlantic Ocean)
Broadcast media: publicly-owned broadcaster Radio Telefis Eireann (RTE) operates 2 TV stations; commercial television stations are available; about 75% of households utilize multi-channel satellite and TV services that provide access to a wide range of stations; RTE operates 4 national radio stations and has launched digital audio broadcasts on several stations; a number of commercial broadcast stations operate at the national, regional, and local levels (2007)
Internet country code: .ie
Internet hosts: 1.339 million (2010)
country comparison to the world: 36
Internet users: 3.042 million (2009)
country comparison to the world: 67

TRANSPORTATION

Airports: 39 (2010)
country comparison to the world: 105
Airports—with paved runways: *total:* 16
over 3,047 m: 1
2,438 to 3,047 m: 1
1,524 to 2,437 m: 4
914 to 1,523 m: 5
under 914 m: 5 (2010)
Airports—with unpaved runways: *total:* 23
914 to 1,523 m: 2
under 914 m: 21 (2010)
Pipelines: gas 1,888 km (2010)
Railways: *total:* 3,237 km
country comparison to the world: 52
broad gauge: 1,872 km 1.600-m gauge (37 km electrified)
narrow gauge: 1,365 km 0.914-m gauge (operated by the Irish Peat Board to transport peat to power stations and briquetting plants) (2008)
Roadways: *total:* 96,036 km
country comparison to the world: 47
paved: 96,036 km (includes 896 km of expressways) (2010)
Waterways: 956 km (pleasure craft only) (2010)
country comparison to the world: 68
Merchant marine: *total:* 28
country comparison to the world: 88
by type: cargo 25, chemical tanker 2, container 1
foreign-owned: 5 (Norway 3, US 2)
registered in other countries: 21 (Bahamas 3, Bermuda 2, Cyprus 3, Isle of Man 1, Kazakhstan 1, Malta 1, Netherlands 7, Panama 1, Slovakia 1, Sweden 1) (2010)
Ports and terminals: Cork, Dublin, Shannon Foynes, Waterford

MILITARY

Military branches: Irish Defense Forces (IDF; Oglaigh na h-Eireann): Army, Naval Service, Air Corps (2011)
Military service age and obligation: 17-25 years of age for male and female voluntary military service (17-27 years of age for the Naval Service); enlistees 16 years of age can be recruited for apprentice specialist positions; 17-35 years of age for the Reserve Defense Forces (RDF); maximum obligation 12 years (5 years IDF, 7 years RDF); EU citizenship or 5-year residence in Ireland required (2010)
Manpower available for military service: *males age 16–49:* 1,179,125
females age 16–49: 1,163,728 (2010 est.)
Manpower fit for military service: *males age 16–49:* 977,631
females age 16–49: 965,900 (2010 est.)
Manpower reaching militarily significant age annually: *male:* 28,564
female: 27,197 (2010 est.)
Military expenditures: 0.9% of GDP (2005 est.)
country comparison to the world: 136

TRANSNATIONAL ISSUES

Disputes—international: Ireland, Iceland, and the UK dispute Denmark's claim that the Faroe Islands' continental shelf extends beyond 200 nm
Illicit drugs: transshipment point for and consumer of hashish from North Africa to the UK and Netherlands and of European-produced synthetic drugs; increasing consumption of South American cocaine; minor transshipment point for heroin and cocaine destined for Western Europe; despite recent legislation, narcotics-related money laundering–using bureaux de change, trusts, and shell companies involving the offshore financial community–remains a concern

ISLE OF MAN

(BRITISH CROWN DEPENDENCY)

INTRODUCTION

Background: Part of the Norwegian Kingdom of the Hebrides until the 13th century when it was ceded to Scotland, the isle came under the British crown in 1765. Current concerns include reviving the almost extinct Manx Gaelic language. Isle of Man is a British crown dependency but is not part of the UK or of the European Union. However, the UK Government remains constitutionally responsible for its defense and international representation.

GEOGRAPHY

Location: Western Europe, island in the Irish Sea, between Great Britain and Ireland
Geographic coordinates: 54 15 N, 4 30 W
Map references: Europe
Area: *total:* 572 sq km

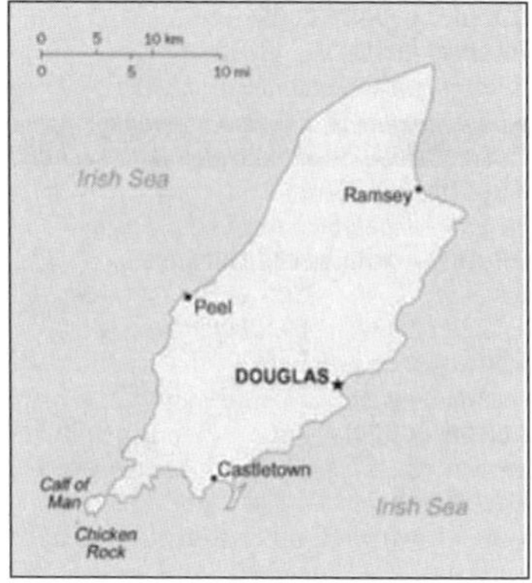

country comparison to the world: 193
land: 572 sq km
water: 0 sq km
Area—comparative: slightly more than three times the size of Washington, DC
Land boundaries: 0 km
Coastline: 160 km
Maritime claims: *territorial sea:* 12 nm
exclusive fishing zone: 12 nm
Climate: temperate; cool summers and mild winters; overcast about a third of the time
Terrain: hills in north and south bisected by central valley
Elevation extremes: *lowest point:* Irish Sea 0 m
highest point: Snaefell 621 m
Natural resources: none
Land use: *arable land:* 9%
permanent crops: 0%
other: 91% (permanent pastures, forests, mountain, and heathland) (2002)
Irrigated land: 0 sq km
Natural hazards: NA
Environment—current issues: waste disposal (both household and industrial); transboundary air pollution
Geography—note: one small islet, the Calf of Man, lies to the southwest and is a bird sanctuary

PEOPLE

Population: 84,655 (July 2011 est.)
country comparison to the world: 198
Age structure: *0–14 years:* 16.6% (male 7,362/female 6,664)
15–64 years: 65.2% (male 27,752/female 27,466)
65 years and over: 18.2% (male 7,003/female 8,408) (2011 est.)
Median age: *total:* 42.6 years
male: 41.8 years
female: 43.3 years (2011 est.)
Population growth rate: 0.921% (2011 est.)
country comparison to the world: 124
Birth rate: 11.42 births/1,000 population (2011 est.)
country comparison to the world: 169
Death rate: 9.92 deaths/1,000 population (July 2011 est.)
country comparison to the world: 57
Net migration rate: 7.71 migrant(s)/1,000 population (2011 est.)
country comparison to the world: 12
Urbanization: *urban population:* 51% of total population (2010)
rate of urbanization: 0% annual rate of change (2010-15 est.)
Major cities—population: DOUGLAS (capital) 26,000 (2009)
Sex ratio: *at birth:* 1.08 male(s)/female
under 15 years: 1.05 male(s)/female
15–64 years: 1.01 male(s)/female
65 years and over: 0.71 male(s)/female
total population: 0.96 male(s)/female (2011 est.)
Infant mortality rate:
total: 4.32 deaths/1,000 live births
country comparison to the world: 194
male: 4.25 deaths/1,000 live births
female: 4.4 deaths/1,000 live births (2011 est.)
Life expectancy at birth: *total population:* 80.64 years
country comparison to the world: 22
male: 79.09 years
female: 82.32 years (2011 est.)
Total fertility rate: 1.96 children born/woman (2011 est.)
country comparison to the world: 131
HIV/AIDS—adult prevalence rate: NA
HIV/AIDS—people living with HIV/AIDS: NA
HIV/AIDS—deaths: NA
Nationality: *noun:* Manxman (men), Manxwoman (women)
adjective: Manx
Ethnic groups: Manx (Norse-Celtic descent), Britons
Religions: Anglican, Roman Catholic, Methodist, Baptist, Presbyterian, Society of Friends
Languages: English, Manx Gaelic (about 2% of the population has some knowledge)
Literacy: NA
School life expectancy (primary to tertiary education): NA
Education expenditures: NA

GOVERNMENT

Country name: *conventional long form:* none
conventional short form: Isle of Man
abbreviation: I.O.M.
Dependency status: British crown dependency
Government type: parliamentary democracy
Capital: *name:* Douglas
geographic coordinates: 54 09 N, 4 29 W
time difference: UTC 0 (5 hours ahead of Washington, DC during Standard Time)
daylight saving time: +1hr, begins last Sunday in March; ends last Sunday in October
Administrative divisions: none; there are no first-order administrative divisions as defined by the US Government, but there are 24 local authorities each with its own elections
Independence: none (British crown dependency)
National holiday: Tynwald Day, 5 July
Constitution: unwritten; note—The Isle of Man Constitution Act of 1961 does not embody the unwritten Manx Constitution
Legal system: the laws of the UK where applicable apply and include Manx statutes
Suffrage: 16 years of age; universal
Executive branch: *chief of state:* Lord of Mann Queen ELIZABETH II (since 6 February 1952); represented by Lieutenant Governor Adam WOOD (since 7 April 2011)
head of government: Chief Minister Tony BROWN (since 14 December 2006)
cabinet: Council of Ministers
(For more information visit the World Leaders website)
elections: the monarchy is hereditary; lieutenant governor appointed by the monarch; the chief minister elected by the Tynwald for a five-year term; election last held on 14 December 2006 (next to be held in December 2011)
election results: House of Keys speaker Tony BROWN elected chief minister by the Tynwald
Legislative branch: bicameral Tynwald consists of the Legislative Council (11 seats; members composed of the President of Tynwald, the Lord Bishop of Sodor and Man, a nonvoting attorney general, and 8 others named by the House of Keys) and the House of Keys (24 seats; members elected by popular vote to serve five-year terms)
elections: House of Keys—last held on 23 November 2006 (next to be held in November 2011)
election results: House of Keys—percent of vote by party—NA; seats by party—Liberal Vannin Party 2, Manx Labor Party 1, independents 21
Judicial branch: High Court of Justice (justices are appointed by the Lord Chancellor of England on the nomination of the lieutenant governor)
Political parties and leaders: Alliance for Progressive Government; Liberal Vannin Party [Peter KARRAN]; Manx Labor Party; Manx Nationalist Party (Mec Vannin) [Bernard MOFFATT]
note: most members sit as independents
Political pressure groups and leaders: Alliance for Progressive Government or APG (a government watchdog); Mec Vannin (political party advocating a sovereign state and environment policies); note—has only had one member elected to the Tynwald
International organization participation: UPU
Diplomatic representation in the US: none (British crown dependency)
Diplomatic representation from the US: none (British crown dependency)
Flag description: red with the Three Legs of Man emblem (triskelion), in the center; the three legs are joined at the thigh and bent at the knee; in order to have the toes pointing clockwise on both sides of the flag, a two-sided emblem is used; the flag is based on the coat-of-arms of the last recognized Norse King of Mann, Magnus III (r. 1252-1265); the triskelion has its roots in an early Celtic sun symbol
National anthem: *name:* "Arrane Ashoonagh dy Vannin" (O Land of Our Birth)
lyrics/music: William Henry GILL [English], John J. KNEEN [Manx]/traditional
note: adopted 2003, in use since 1907; serves as a local anthem; as a British crown dependency, "God Save the Queen" is official (see United Kingdom) and is played when the sovereign, members of the royal family, or the lieutenant governor are present

ECONOMY

Economy—overview: Offshore banking, manufacturing, and tourism are key sectors of the economy. The government offers low taxes and other incentives to high-technology companies and financial institutions to locate on the island; this has paid off in expanding employment opportunities in high-income industries. As a result, agriculture and fishing, once the mainstays of the economy, have declined in their contributions to GDP. The Isle of Man also attracts online gambling sites and the film industry. Trade is mostly with the UK. The Isle of Man enjoys free access to EU markets.
GDP (purchasing power parity): $2.719 billion (2005 est.)
country comparison to the world: 180
GDP (official exchange rate): $2.719 billion (2005 est.)
GDP—real growth rate: 5.2% (2005)
country comparison to the world: 67
GDP—per capita (PPP): $35,000 (2005 est.)
country comparison to the world: 36
GDP—composition by sector:
agriculture: 1%
industry: 13%
services: 86% (2000 est.)
Labor force: 39,690 (2001)
country comparison to the world: 195
Labor force—by occupation: agriculture, forestry, and fishing: 3%
manufacturing: 11%
construction: 10%
transport and communication: 8%
wholesale and retail distribution: 11%
professional and scientific *services:* 18%
public administration: 6%
banking and finance: 18%
tourism: 2%
entertainment and catering: 3%
miscellaneous *services:* 10% (2001)
Unemployment rate: 1.8% (October 2010 est.)
country comparison to the world: 11
1.5% (December 2006 est.)
Population below poverty line: NA%
Household income or consumption by percentage share: *lowest 10%:* NA%
highest 10%: NA%
Budget: *revenues:* $965 million
expenditures: $943 million (FY05/06 est.)
Inflation rate (consumer prices): 3.1% (2006 est.)
country comparison to the world: 86
Market value of publicly traded shares: $NA
Agriculture—products: cereals, vegetables; cattle, sheep, pigs, poultry
Industries: financial services, light manufacturing, tourism
Exports: $NA
Exports—commodities: tweeds, herring, processed shellfish, beef, lamb
Imports: $NA
Imports—commodities: timber, fertilizers, fish
Debt—external: $NA
Exchange rates: Manx pounds (IMP) per US dollar–0.6388 (2010), 0.6175 (2009), 0.5302 (2008), 0.4993 (2007), 0.5418 (2006)

COMMUNICATIONS

Telephones—main lines in use: 51,000 (1999)
country comparison to the world: 161
Telephone system: *general assessment:* NA
domestic: landline, telefax, mobile cellular telephone system
international: country code–44; fiber-optic cable, microwave radio relay, satellite earth station, submarine cable
Broadcast media: national public radio broadcasts over 3 FM stations and 1 AM station; 2 commercial broadcasters operating with 1 having multiple FM stations; receives radio and TV services via relays from British TV and radio broadcasters (2008)
Internet country code: .im
Internet hosts: 765 (2010)
country comparison to the world: 173

TRANSPORTATION

Airports: 1 (2010)
country comparison to the world: 219
Airports—with paved runways:
total: 1
1,524 to 2,437 m: 1 (2010)
Railways: *total:* 63 km
country comparison to the world: 129
narrow gauge: 6 km 1.076-m gauge (6 km electrified); 57 km 0.914-m gauge (29 km electrified)
note: primarily summer tourist attractions (2008)
Roadways: *total:* 500 km (2008)
country comparison to the world: 193
Merchant marine: *total:* 292
country comparison to the world: 31
by type: bulk carrier 45, cargo 49, chemical tanker 48, container 6, liquefied gas 41, passenger/cargo 2, petroleum tanker 91, roll on/roll off 5, vehicle carrier 5
foreign-owned: 200 (Bermuda 7, Chile 8, Denmark 26, Germany 56, Greece 57, Ireland 1, Japan 15, Norway 26, Singapore 1, Sweden 1, US 2) (2010)
Ports and terminals: Douglas, Ramsey

MILITARY

Manpower fit for military service: *males age 16–49:* 15,206
females age 16–49: 15,127 (2010 est.)
Manpower reaching militarily significant age annually:
male: 507
female: 494 (2010 est.)
Military—note: defense is the responsibility of the UK

TRANSNATIONAL ISSUES

Disputes—international: none

ISRAEL

(ALSO SEE SEPARATE GAZA STRIP AND WEST BANK ENTRIES)

INTRODUCTION

Background: Following World War II, the British withdrew from their mandate of Palestine, and the UN partitioned the area into Arab and Jewish states, an arrangement rejected by the Arabs. Subsequently, the Israelis defeated the Arabs in a series of wars without ending the deep tensions between the two sides. The territories Israel occupied since the 1967 war are not included in the Israel country profile, unless otherwise noted. On 25 April 1982, Israel withdrew from the Sinai pursuant to the 1979 Israel-Egypt Peace Treaty. In keeping with the framework established at the Madrid Conference in October 1991, bilateral negotiations were conducted between Israel and Palestinian representatives and Syria to achieve a permanent settlement. Israel and Palestinian officials signed on 13 September 1993 a Declaration of Principles (also known as the "Oslo Accords") guiding an interim period of Palestinian self-rule. Outstanding territorial and other disputes with Jordan were resolved in the 26 October 1994 Israel-Jordan Treaty of Peace. In addition, on 25 May 2000, Israel withdrew unilaterally from southern Lebanon, which it had occupied since 1982. In April 2003, US President BUSH, working in conjunction with the EU, UN, and Russia–the "Quartet"–took the lead in laying out a roadmap to a final settlement of the conflict by 2005, based on reciprocal steps by the two parties leading to two states, Israel and a democratic Palestine. However, progress toward a permanent status agreement was undermined by Israeli-Palestinian violence between September 2003 and February 2005. In the summer of 2005, Israel unilaterally disengaged from the Gaza Strip, evacuating settlers and its military while retaining control over most points of entry into the Gaza Strip. The election of HAMAS to head the Palestinian Legislative Council froze relations between Israel and the Palestinian Authority (PA). Ehud OLMERT became prime minister in March 2006 and presided over a 34-day conflict with Hizballah in Lebanon in June-August 2006 and a 23-day conflict with HAMAS in the Gaza Strip during December 2008 and January 2009. OLMERT, who in June 2007 resumed talks with PA President Mahmoud ABBAS, resigned in September 2008. Prime Minister Binyamin NETANYAHU formed a coalition in March 2009 following a February 2009 general election. Direct talks launched in September 2010 collapsed following the expiration of Israel's 10-month partial settlement construction moratorium in the West Bank. Diplomatic

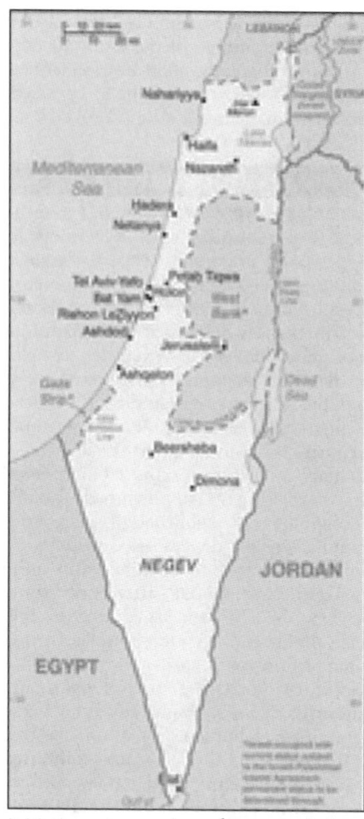

initiatives to revive the negotiations through proximity talks began at the end of 2010.

GEOGRAPHY

Location: Middle East, bordering the Mediterranean Sea, between Egypt and Lebanon
Geographic coordinates: 31 30 N, 34 45 E
Map references: Middle East
Area: *total:* 20,770 sq km
country comparison to the world: 153
land: 20,330 sq km
water: 440 sq km
Area—comparative: slightly larger than New Jersey
Land boundaries: *total:* 1,017 km
border countries: Egypt 266 km, Gaza Strip 51 km, Jordan 238 km, Lebanon 79 km, Syria 76 km, West Bank 307 km
Coastline: 273 km
Maritime claims: *territorial sea:* 12 nm
continental shelf: to depth of exploitation
Climate: temperate; hot and dry in southern and eastern desert areas
Terrain: Negev desert in the south; low coastal plain; central mountains; Jordan Rift Valley
Elevation extremes: *lowest point:* Dead Sea -408 m
highest point: Har Meron 1,208 m
Natural resources: timber, potash, copper ore, natural gas, phosphate rock, magnesium bromide, clays, sand
Land use: *arable land:* 15.45%
permanent crops: 3.88%
other: 80.67% (2005)
Irrigated land: 1,940 sq km (2003)
Total renewable water resources: 1.7 cu km (2001)
Freshwater withdrawal (domestic/industrial/agricultural): *total:* 2.05 cu km/yr (31%/7%/62%)
per capita: 305 cu m/yr (2000)
Natural hazards: sandstorms may occur during spring and summer; droughts; periodic earthquakes
Environment—current issues: limited arable land and natural freshwater resources pose serious constraints; desertification; air pollution from industrial and vehicle emissions; groundwater pollution from industrial and domestic waste, chemical fertilizers, and pesticides
Environment—international agreements: *party to:* Biodiversity, Climate Change, Climate Change-Kyoto Protocol, Desertification, Endangered Species, Hazardous Wastes, Ozone Layer Protection, Ship Pollution, Wetlands, Whaling
signed, but not ratified: Marine Life Conservation
Geography—note: Lake Tiberias (Sea of Galilee) is an important freshwater source; there are about 355 Israeli civilian sites including about 145 small outpost communities in the West Bank, 41 sites in the Golan Heights, and 32 in East Jerusalem (2010 est.)

PEOPLE

Population: 7,473,052 (July 2010 est.)
country comparison to the world: 96
note: approximately 296,700 Israeli settlers live in the West Bank (2009 est.); approximately 19,100 Israeli settlers live in the Golan Heights (2008 est.); approximately 192,800 Israeli settlers live in East Jerusalem (2008 est.)
Age structure: *0–14 years:* 27.6% (male 1,057,113/female 1,008,978)
15–64 years: 62.2% (male 2,358,858/female 2,292,281)
65 years and over: 10.1% (male 331,034/female 424,788) (2011 est.)
Median age: *total:* 29.4 years
male: 28.7 years
female: 30.1 years (2011 est.)
Population growth rate: 1.584% (2011 est.)
country comparison to the world: 73
Birth rate: 19.24 births/1,000 population (2011 est.)
country comparison to the world: 98
Death rate: 5.47 deaths/1,000 population (July 2011 est.)
country comparison to the world: 176
Net migration rate: 2.08 migrant(s)/1,000 population (2011 est.)
country comparison to the world: 40
Urbanization: *urban population:* 92% of total population (2010)
rate of urbanization: 1.5% annual rate of change (2010-15 est.)
Major cities—population: Tel Aviv-Yafo 3.219 million; Haifa 1.027 million; JERUSALEM (capital) 768,000 (2009)
Sex ratio: *at birth:* 1.05 male(s)/female
under 15 years: 1.05 male(s)/female
15–64 years: 1.03 male(s)/female
65 years and over: 0.78 male(s)/female
total population: 1 male(s)/female (2011 est.)
Infant mortality rate: *total:* 4.12 deaths/1,000 live births
country comparison to the world: 200
male: 4.3 deaths/1,000 live births
female: 3.94 deaths/1,000 live births (2011 est.)
Life expectancy at birth: *total population:* 80.96 years
country comparison to the world: 17
male: 78.79 years
female: 83.24 years (2011 est.)
Total fertility rate: 2.7 children born/woman (2011 est.)
country comparison to the world: 75
HIV/AIDS—adult prevalence rate: 0.2% (2009 est.)
country comparison to the world: 101
HIV/AIDS—people living with HIV/AIDS: 7,500 (2009 est.)
country comparison to the world: 110
HIV/AIDS—deaths: fewer than 100 (2009 est.)
country comparison to the world: 136
Drinking water source: *Improved:*
urban: 100% of population
rural: 100% of population
total: 100% of population (2008)
Sanitation facility access: *Improved:*
urban: 100% of population
rural: 100% of population
total: 100% of population (2008)
Nationality: *noun:* Israeli(s)
adjective: Israeli
Ethnic groups: Jewish 76.4% (of which Israel-born 67.1%, Europe/America-born 22.6%, Africa-born 5.9%, Asia-born 4.2%), non-Jewish 23.6% (mostly Arab) (2004)
Religions: Jewish 75.6%, Muslim 16.9%, Christian 2%, Druze 1.7%, other 3.8% (2008 census)
Languages: Hebrew (official), Arabic (used officially for Arab minority), English (most commonly used foreign language)
Literacy: *definition:* age 15 and over can read and write
total population: 97.1%
male: 98.5%
female: 95.9% (2004 est.)
School life expectancy (primary to tertiary education): *total:* 15 years
male: 15 years
female: 16 years (2008)
Education expenditures: 5.9% of GDP (2007)
country comparison to the world: 30

GOVERNMENT

Country name: *conventional long form:* State of Israel
conventional short form: Israel
local long form: Medinat Yisra'el
local short form: Yisra'el
Government type: parliamentary democracy
Capital: *name:* Jerusalem
geographic coordinates: 31 46 N, 35 14 E
time difference: UTC+2 (7 hours ahead of Washington, DC during Standard Time)
daylight saving time: +1hr, begins first Friday in April; ends the Sunday between the holidays of Rosh Hashana and Yom Kippur
note: Israel proclaimed Jerusalem as its capital in 1950, but the US, like all other countries, maintains its Embassy in Tel Aviv
Administrative divisions: 6 districts (mehozot, singular—mehoz); Central,

Haifa, Jerusalem, Northern, Southern, Tel Aviv

Independence: 14 May 1948 (from League of Nations mandate under British administration)

National holiday: Independence Day, 14 May (1948); note–Israel declared independence on 14 May 1948, but the Jewish calendar is lunar and the holiday may occur in April or May

Constitution: no formal constitution; some of the functions of a constitution are filled by the Declaration of Establishment (1948), the Basic Laws of the parliament (Knesset), and the Israeli citizenship law; note–since May 2003 the Constitution, Law, and Justice Committee of the Knesset has been working on a draft constitution

Legal system: mixed legal system of English common law, British Mandate regulations, and Jewish, Christian, and Muslim religious laws

International law organization participation: has not submitted an ICJ jurisdiction declaration; withdrew acceptance of ICCt jurisdiction in 2002

Suffrage: 18 years of age; universal

Executive branch: *chief of state:* President Shimon PERES (since 15 July 2007)
head of government: Prime Minister Binyamin NETANYAHU (since 31 March 2009)
cabinet: Cabinet selected by prime minister and approved by the Knesset
(For more information visit the World Leaders website)
elections: president largely a ceremonial role and is elected by the Knesset for a seven-year term (one-term limit); election last held 13 June 2007 (next to be held in 2014 but can be called earlier); following legislative elections, the president, in consultation with party leaders, assigns the task of forming a governing coalition to a Knesset member who he or she determines is most likely to accomplish that task
election results: Shimon PERES elected president; number of votes in first round–Shimon PERES 58, Reuven RIVLIN 37, Colette AVITAL 21; PERES elected president in second round with 86 votes (unopposed)

Legislative branch: unicameral Knesset (120 seats; political parties are elected by popular vote and assigned seats for members on a proportional basis; members serve four-year terms)
elections: last held on 10 February 2009 (next scheduled election to be held in 2013)
election results: percent of vote by party–Kadima 23.2%, Likud-Ahi 22.3%, YB 12.1%, Labor 10.2%, SHAS 8.8%, United Torah Judaism 4.5%, United Arab List 3.5%, National Union 3.4%, Hadash 3.4%, The Jewish Home 3%, The New Movement-Meretz 3%, Balad 2.6%; seats by party–Kadima 28, Likud-Ahi 27, YB 15, Labor 13, SHAS 11, United Torah Judaism 5, United Arab List 4, National Union 4, HADASH 4, The Jewish Home 3, The New Movement-Meretz 3, Balad 3

Judicial branch: Supreme Court (justices appointed by Judicial Selection Committee–made up of all three branches of the government; mandatory retirement age is 70)

Political parties and leaders: Balad [Jamal ZAHALKA]; Democratic Front for Peace and Equality (HADASH) [Muhammad BARAKEH]; Independence [Ehud BARAK]; Kadima [Tzipora "Tzipi" LIVNI]; Labor Party [Eitan CABEL]; Likud [Binyamin NETANYAHU]; National Union [Yaakov KATZ]; SHAS [Eliyahu YISHAI]; The Jewish Home (HaBayit HaYehudi) [Daniel HERSCHKOWITZ]; The New Movement-Meretz [Haim ORON]; United Arab List-Ta'al [Ibrahim SARSUR]; United Torah Judaism or UTJ [Yaakov LITZMAN]; Yisrael Beiteinu or YB [Avigdor LIEBERMAN]

Political pressure groups and leaders: B'Tselem [Jessica MONTELL, Executive Director] monitors human rights abuses; Peace Now [Yariv OPPENHEIMER, Secretary General] supports territorial concessions in the West Bank and Gaza Strip; YESHA Council of Settlements [Danny DAYAN, Chairman] promotes settler interests and opposes territorial compromise; Breaking the Silence [Yehuda SHAUL, Executive Director] collects testimonies from soldiers who served in the West Bank and Gaza Strip

International organization participation: BIS, BSEC (observer), CERN (observer), CICA, EBRD, FAO, IADB, IAEA, IBRD, ICAO, ICC, ICRM, IDA, IFAD, IFC, IFRCS, ILO, IMF, IMO, IMSO, Interpol, IOC, IOM, IPU, ISO, ITSO, ITU, ITUC, MIGA, OAS (observer), OECD, OPCW (signatory), OSCE (partner), Paris Club (associate), PCA, SECI (observer), UN, UNCTAD, UNESCO, UNHCR, UNIDO, UNWTO, UPU, WCO, WHO, WIPO, WMO, WTO

Diplomatic representation in the US: *chief of mission:* Ambassador Michael OREN
chancery: 3514 International Drive NW, Washington, DC 20008
telephone: [1] (202) 364-5500
FAX: [1] (202) 364-5607
consulate(s) general: Atlanta, Boston, Chicago, Houston, Los Angeles, Miami, New York, Philadelphia, San Francisco

Diplomatic representation from the US: *chief of mission:* Ambassador James B. CUNNINGHAM
embassy: 71 Hayarkon Street, Tel Aviv 63903
telephone: [972] (3) 519-7575
FAX: [972] (3) 516-4390
consulate(s) general: Jerusalem; note–an independent US mission, established in 1928, whose members are not accredited to a foreign government

Flag description: white with a blue hexagram (six-pointed linear star) known as the Magen David (Shield of David) centered between two equal horizontal blue bands near the top and bottom edges of the flag; the basic design resembles a Jewish prayer shawl (tallit), which is white with blue stripes; the hexagram as a Jewish symbol dates back to medieval times

National anthem: *name:* "Hatikvah" (The Hope)
lyrics/music: Naftali Herz IMBER/traditional, arranged by Samuel COHEN
note: adopted 2004, unofficial since 1948; used as the anthem of the Zionist movement since 1897; the 1888 arrangement by Shmuel COHEN is thought to be based on the Romanian folk song "Carul cu boi" (The Ox Driven Cart)

ECONOMY

Economy—overview: Israel has a technologically advanced market economy. It depends on imports of crude oil, grains, raw materials, and military equipment. Despite limited natural resources, Israel has intensively developed its agricultural and industrial sectors over the past 20 years. Cut diamonds, high-technology equipment, and agricultural products (fruits and vegetables) are the leading exports. Israel usually posts sizable trade deficits, which are covered by large transfer payments from abroad and by foreign loans. Roughly half of the government's external debt is owed to the US, its major source of economic and military aid. Israel's GDP, after contracting slightly in 2001 and 2002 due to the Palestinian conflict and troubles in the high-technology sector, grew about 5% per year from 2004-07. The global financial crisis of 2008-09 spurred a brief recession in Israel, but the country entered the crisis with solid fundamentals–following years of prudent fiscal policy and a series of liberalizing reforms–and a resilient banking sector, and the economy has shown signs of an early recovery. Following GDP growth of 4% in 2008, Israel's GDP slipped to 0.2% in 2009, but reached 3.4% in 2010, as exports rebounded. The global economic downturn affected Israel's economy primarily through reduced demand for Israel's exports in the United States and EU, Israel's top trading partners. Exports of goods and services account for about 40% of the country's GDP. The Israeli Government responded to the recession by implementing a modest fiscal stimulus package and an aggressive expansionary monetary policy–including cutting interest rates to record lows, purchasing government bonds, and intervening in the foreign currency market. The Bank of Israel began raising interest rates in the summer of 2009 when inflation rose above the upper end of the Bank's target and the economy began to show signs of recovery.

GDP (purchasing power parity): $219.4 billion (2010 est.)
country comparison to the world: 52
$209.8 billion (2009 est.)
$208.1 billion (2008 est.)
note: data are in 2010 US dollars

GDP (official exchange rate): $213.1 billion (2010 est.)

GDP—real growth rate: 4.6% (2010 est.)
country comparison to the world: 80
0.8% (2009 est.)
4.2% (2008 est.)

GDP—per capita (PPP): $29,800 (2010 est.)
country comparison to the world: 46
$29,000 (2009 est.)

$29,300 (2008 est.)
note: data are in 2010 US dollars
GDP—composition by sector: *agriculture:* 2.4%
industry: 32.6%
services: 65% (2010 est.)
Labor force: 3.08 million (2010 est.)
country comparison to the world: 101
Labor force—by occupation: *agriculture:* 2%
industry: 16%
services: 82% (September 2008)
Unemployment rate: 6.4% (2010 est.)
country comparison to the world: 61
7.6% (2009 est.)
Population below poverty line: 23.6%
note: Israel's poverty line is $7.30 per person per day (2007)
Household income or consumption by percentage share: *lowest 10%:* 2.5%
highest 10%: 24.3% (2008)
Distribution of family income—Gini index: 39.2 (2008)
country comparison to the world: 65
35.5 (2001)
Investment (gross fixed): 16.7% of GDP (2010 est.)
country comparison to the world: 121
Budget: *revenues:* $60.59 billion
expenditures: $68.68 billion (2010 est.)
Public debt: 77.3% of GDP (2010 est.)
country comparison to the world: 22
77.7% of GDP (2009 est.)
Inflation rate (consumer prices): 2.6% (2010 est.)
country comparison to the world: 67
3.3% (2009 est.)
Central bank discount rate: 1% (31 December 2009)
country comparison to the world: 111
2.5% (31 December 2008)
Commercial bank prime lending rate: 3.73% (31 December 2009 est.)
country comparison to the world: 131
6.06% (31 December 2008 est.)
Stock of narrow money: $27.58 billion (31 December 2010 est.)
country comparison to the world: 59
$25.16 billion (31 December 2009 est.)
Stock of broad money: $208.8 billion (31 December 2009 est.)
country comparison to the world: 40
$195.7 billion (31 December 2008 est.)
Stock of domestic credit: $169.9 billion (31 December 2010 est.)
country comparison to the world: 40
$148.5 billion (31 December 2009 est.)
Market value of publicly traded shares: $182.1 billion (31 December 2009)
country comparison to the world: 29
$134.5 billion (31 December 2008)
$236.4 billion (31 December 2007)
Agriculture—products: citrus, vegetables, cotton; beef, poultry, dairy products
Industries: high-technology products (including aviation, communications, computer-aided design and manufactures, medical electronics, fiber optics), wood and paper products, potash and phosphates, food, beverages, and tobacco, caustic soda, cement, construction, metals products, chemical products, plastics, diamond cutting, textiles, footwear
Industrial production growth rate: 5.7% (2010 est.)
country comparison to the world: 62
Electricity—production: 54.5 billion kWh (2008 est.)
country comparison to the world: 46
Electricity—consumption: 46.38 billion kWh (2007 est.)
country comparison to the world: 48
Electricity—exports: 2.081 billion kWh (2007)
Electricity—imports: 0 kWh (2008)
Oil—production: 3,806 bbl/day (2009 est.)
country comparison to the world: 100
Oil—consumption: 231,000 bbl/day (2009 est.)
country comparison to the world: 52
Oil—exports: 69,580 bbl/day (2007 est.)
country comparison to the world: 72
Oil—imports: 318,900 bbl/day (2007 est.)
country comparison to the world: 34
Oil—proved reserves: 1.94 million bbl (1 January 2010 est.)
country comparison to the world: 96
Natural gas—production: 1.19 billion cu m (2008 est.)
country comparison to the world: 62
Natural gas—consumption: 1.19 billion cu m (2008 est.)
country comparison to the world: 87
Natural gas—exports: 0 cu m (2008 est.)
country comparison to the world: 114
Natural gas—imports: NA
Natural gas—proved reserves: 30.44 billion cu m (1 January 2010 est.)
country comparison to the world: 70
Current account balance: $6.269 billion (2010 est.)
country comparison to the world: 30
$7.637 billion (2009 est.)
Exports: $54.31 billion (2010 est.)
country comparison to the world: 49
$45.9 billion (est.)
Exports—commodities: machinery and equipment, software, cut diamonds, agricultural products, chemicals, textiles and apparel
Exports—partners: US 35.05%, Hong Kong 6.02%, Belgium 4.95% (2009)
Imports:
$55.6 billion (2010 est.)
country comparison to the world: 46
$45.99 billion (2009 est.)
Imports—commodities: raw materials, military equipment, investment goods, rough diamonds, fuels, grain, consumer goods
Imports—partners: US 12.35%, China 7.43%, Germany 7.1%, Switzerland 6.94%, Belgium 5.42%, Italy 4.49%, UK 4.03%, Netherlands 3.98% (2009)
Reserves of foreign exchange and gold: $66.98 billion (31 December 2010 est.)
country comparison to the world: 21
$60.61 billion (31 December 2009 est.)
Debt—external: $89.68 billion (31 December 2010 est.)
country comparison to the world: 39
$86.78 billion (31 December 2009 est.)
Stock of direct foreign investment—at home: $64.82 billion (31 December 2010 est.)
country comparison to the world: 51
$58.82 billion (31 December 2009 est.)
Stock of direct foreign investment—abroad: $58.42 billion (31 December 2010 est.)
country comparison to the world: 30
$55.02 billion (31 December 2009 est.)
Exchange rates: new Israeli shekels (ILS) per US dollar—3.739 (2010), 3.93 (2009) 3.588 (2008), 4.14 (2007), 4.4565 (2006)

COMMUNICATIONS

Telephones—main lines in use: 3.25 million (2009)
country comparison to the world: 48
Telephones—mobile cellular: 9.022 million (2009)
country comparison to the world: 70
Telephone system: *general assessment:* most highly developed system in the Middle East although not the largest
domestic: good system of coaxial cable and microwave radio relay; all systems are digital; four privately-owned mobile-cellular service providers with countrywide coverage
international: country code—972; submarine cables provide links to Europe, Cyprus, and parts of the Middle East; satellite earth stations—3 Intelsat (2 Atlantic Ocean and 1 Indian Ocean) (2008)
Broadcast media: state broadcasting network, operated by the Israel Broadcasting Authority (IBA), broadcasts on 2 channels, one in Hebrew and the other in Arabic; 5 commercial channels including a channel broadcasting in Russian, a channel broadcasting Knesset proceedings, and a music channel supervised by a public body; multi-channel satellite and cable TV packages provide access to foreign channels; IBA broadcasts on 8 radio networks with multiple repeaters and Israel Defense Forces Radio broadcasts over multiple stations; about 15 privately-owned radio stations; overall more than 100 stations and repeater stations operating (2008)
Internet country code: .il
Internet hosts: 1.689 million (2010)
country comparison to the world: 35
Internet users: 4.525 million (2009)
country comparison to the world: 51

TRANSPORTATION

Airports: 48 (2010)
country comparison to the world: 92
Airports—with paved runways: *total:* 30
over 3,047 m: 2
2,438 to 3,047 m: 5
1,524 to 2,437 m: 6
914 to 1,523 m: 11
under 914 m: 6 (2010)
Airports—with unpaved runways: *total:* 18
1,524 to 2,437 m: 1
914 to 1,523 m: 3
under 914 m: 14 (2010)
Heliports: 3 (2010)
Pipelines: gas 211 km; oil 442 km; refined products 261 km (2010)
Railways: *total:* 975 km
country comparison to the world: 89
standard gauge: 975 km 1.435-m gauge (2010)
Roadways: *total:* 18,290 km
country comparison to the world: 115
paved: 18,290 km (includes 146 km of expressways) (2009)
Merchant marine: *total:* 10
country comparison to the world: 113
by type: cargo 2, container 8

registered in other countries: 51 (Bermuda 3, Cyprus 1, Georgia 1, Honduras 1, Liberia 31, Malta 5, Marshall Islands 1, Moldova 4, Panama 1, Saint Vincent and the Grenadines 3) (2010)
Ports and terminals: Ashdod, Elat (Eilat), Hadera, Haifa

MILITARY

Military branches: Israel Defense Forces (IDF), Israel Naval Forces (IN), Israel Air Force (IAF) (2010)
Military service age and obligation: 18 years of age for compulsory (Jews, Druzes) and voluntary (Christians, Muslims, Circassians) military service; both sexes are obligated to military service; conscript service obligation–36 months for enlisted men, 21 months for enlisted women, 48 months for officers; pilots commit to 9 years service; reserve obligation to age 41-51 (men), 24 (women) (2010)
Manpower available for military service: *males age 16–49:* 1,797,960
females age 16–49: 1,713,230 (2010 est.)
Manpower fit for military service: *males age 16–49:* 1,517,510
females age 16–49: 1,446,132 (2010 est.)
Manpower reaching militarily significant age annually: *male:* 62,304
female: 59,418 (2010 est.)
Military expenditures: 7.3% of GDP (2006)
country comparison to the world: 6

TRANSNATIONAL ISSUES

Disputes—international: West Bank and Gaza Strip are Israeli-occupied with current status subject to the Israeli-Palestinian Interim Agreement–permanent status to be determined through further negotiation; Israel continues construction of a "seam line" separation barrier along parts of the Green Line and within the West Bank; Israel withdrew its settlers and military from the Gaza Strip and from four settlements in the West Bank in August 2005; Golan Heights is Israeli-occupied (Lebanon claims the Shab'a Farms area of Golan Heights); since 1948, about 350 peacekeepers from the UN Truce Supervision Organization (UNTSO) headquartered in Jerusalem monitor ceasefires, supervise armistice agreements, prevent isolated incidents from escalating, and assist other UN personnel in the region
Refugees and internally displaced persons: *IDPs:* 150,000-420,000 (Arab villagers displaced from homes in northern Israel) (2007)
Illicit drugs: increasingly concerned about ecstasy, cocaine, and heroin abuse; drugs arrive in country from Lebanon and, increasingly, from Jordan; money-laundering center

ITALY

INTRODUCTION

Background: Italy became a nation-state in 1861 when the regional states of the peninsula, along with Sardinia and Sicily, were united under King Victor EMMANUEL II. An era of parliamentary government came to a close in the early 1920s when Benito MUSSOLINI established a Fascist dictatorship. His alliance with Nazi Germany led to Italy's defeat in World War II. A democratic republic replaced the monarchy in 1946 and economic revival followed. Italy was a charter member of NATO and the European Economic Community (EEC). It has been at the forefront of European economic and political unification, joining the Economic and Monetary Union in 1999. Persistent problems include illegal immigration, organized crime, corruption, high unemployment, sluggish economic growth, and the low incomes and technical standards of southern Italy compared with the prosperous north.

GEOGRAPHY

Location: Southern Europe, a peninsula extending into the central Mediterranean Sea, northeast of Tunisia
Geographic coordinates: 42 50 N, 12 50 E
Map references: Europe
Area: *total:* 301,340 sq km
country comparison to the world: 71
land: 294,140 sq km
water: 7,200 sq km
note: includes Sardinia and Sicily
Area—comparative: slightly larger than Arizona
Land boundaries: *total:* 1,899.2 km
border countries: Austria 430 km, France 488 km, Holy See (Vatican City) 3.2 km, San Marino 39 km, Slovenia 199 km, Switzerland 740 km
Coastline: 7,600 km
Maritime claims: *territorial sea:* 12 nm
continental shelf: 200 m depth or to the depth of exploitation
Climate: predominantly Mediterranean; Alpine in far north; hot, dry in south
Terrain: mostly rugged and mountainous; some plains, coastal lowlands
Elevation extremes: *lowest point:* Mediterranean Sea 0 m
highest point: Mont Blanc (Monte Bianco) de Courmayeur 4,748 m (a secondary peak of Mont Blanc)
Natural resources: coal, mercury, zinc, potash, marble, barite, asbestos, pumice, fluorspar, feldspar, pyrite (sulfur), natural gas and crude oil reserves, fish, arable land
Land use: *arable land:* 26.41%
permanent crops: 9.09%
other: 64.5% (2005)
Irrigated land: 27,500 sq km (2003)
Total renewable water resources: 175 cu km (2005)
Freshwater withdrawal (domestic/industrial/agricultural): *total:* 41.98 cu km/yr (18%/37%/45%)
per capita: 723 cu m/yr (1998)
Natural hazards: regional risks include landslides, mudflows, avalanches, earthquakes, volcanic eruptions, flooding; land subsidence in Venice
volcanism: Italy experiences significant volcanic activity; Etna (elev. 3,330 m), which is in eruption as of 2010, is Europe's most active volcano; flank eruptions pose a threat to nearby Sicilian villages; Etna, along with the famous Vesuvius, which remains a threat to the millions of nearby residents in the Bay of Naples area, have both been deemed "Decade Volcanoes" by the International Association of Volcanology and Chemistry of the Earth's Interior, worthy of study due to their explosive history and close proximity to human populations; Stromboli, on its namesake island, has also been continuously active with moderate volcanic activity; other historically active volcanoes include Campi Flegrei, Ischia, Larderello, Pantelleria, Vulcano, and Vulsini
Environment—current issues: air pollution from industrial emissions such as sulfur dioxide; coastal and inland rivers polluted from industrial and agricultural effluents; acid rain damaging lakes; inadequate industrial waste treatment and disposal facilities
Environment—international agreements: *party to:* Air Pollution, Air Pollution-Nitrogen Oxides, Air Pollution-Persistent Organic Pollutants, Air Pollution-Sulfur 85, Air Pollution-Sulfur 94, Air Pollution-Volatile Organic Compounds, Antarctic-Environmental Protocol, Antarctic-Marine Living Resources, Antarctic Seals, Antarctic Treaty, Biodiversity, Climate Change, Climate Change-Kyoto Protocol, Desertification, Endangered Species, Environmental Modification, Hazardous Wastes, Law of the Sea, Marine Dumping, Ozone Layer Protection, Ship Pollution, Tropical Timber 83, Tropical Timber 94, Wetlands, Whaling
signed, but not ratified: none of the selected agreements
Geography—note: strategic location dominating central Mediterranean as well as southern sea and air approaches to Western Europe

PEOPLE

Population: 61,016,804 (July 2011 est.)
country comparison to the world: 23
Age structure: *0–14 years:* 13.8% (male 4,315,292/female 4,124,624)
15–64 years: 65.9% (male 19,888,901/ female 20,330,495)
65 years and over: 20.3% (male 5,248,418/ female 7,109,074) (2011 est.)
Median age: *total:* 43.5 years
male: 42.4 years
female: 44.7 years (2011 est.)
Population growth rate: 0.42% (2011 est.)
country comparison to the world: 157
Birth rate: 9.18 births/1,000 population (2011 est.)
country comparison to the world: 208
Death rate: 9.84 deaths/1,000 population (July 2011 est.)
country comparison to the world: 59
Net migration rate: 4.86 migrant(s)/1,000 population (2011 est.)
country comparison to the world: 18
Urbanization: *urban population:* 68% of total population (2010)
rate of urbanization: 0.5% annual rate of change (2010-15 est.)
Major cities—population: ROME (capital) 3.357 million; Milan 2.962 million; Naples 2.27 million; Turin 1.662 million; Palermo 872,000 (2009)
Sex ratio: *at birth:* 1.059 male(s)/female
under 15 years: 1.06 male(s)/female
15–64 years: 1.03 male(s)/female
65 years and over: 0.72 male(s)/female
total population: 0.96 male(s)/female (2011 est.)
Infant mortality rate: *total:* 3.38 deaths/1,000 live births
country comparison to the world: 213
male: 3.59 deaths/1,000 live births
female: 3.16 deaths/1,000 live births (2011 est.)
Life expectancy at birth: *total population:* 81.77 years
country comparison to the world: 10
male: 79.16 years
female: 84.53 years (2011 est.)
Total fertility rate: 1.39 children born/ woman (2011 est.)
country comparison to the world: 202
HIV/AIDS—adult prevalence rate: 0.3% (2009 est.)
country comparison to the world: 81
HIV/AIDS—people living with HIV/AIDS: 140,000 (2009 est.)
country comparison to the world: 35
HIV/AIDS—deaths: fewer than 1,000 (2009 est.)
country comparison to the world: 74
Drinking water source: *Improved:*
urban: 100% of population
rural: 100% of population
total: 100% of population (2008)
Nationality: *noun:* Italian(s)
adjective: Italian
Ethnic groups: Italian (includes small clusters of German-, French-, and Slovene-Italians in the north and Albanian-Italians and Greek-Italians in the south)
Religions: Roman Catholic 90% (approximately; about one-third practicing), other 10% (includes mature Protestant and Jewish communities and a growing Muslim immigrant community)
Languages: Italian (official), German (parts of Trentino-Alto Adige region are predominantly German speaking), French (small French-speaking minority in Valle d'Aosta region), Slovene (Slovene-speaking minority in the Trieste-Gorizia area)
Literacy: *definition:* age 15 and over can read and write
total population: 98.4%
male: 98.8%
female: 98% (2001 census)
School life expectancy (primary to tertiary education): *total:* 16 years
male: 16 years
female: 17 years (2008)
Education expenditures: 4.3% of GDP (2007)
country comparison to the world: 91

GOVERNMENT

Country name: *conventional long form:* Italian Republic
conventional short form: Italy
local long form: Repubblica Italiana
local short form: Italia
former: Kingdom of Italy
Government type: republic
Capital: *name:* Rome
geographic coordinates: 41 54 N, 12 29 E
time difference: UTC+1 (6 hours ahead of Washington, DC during Standard Time)
daylight saving time: +1hr, begins last Sunday in March; ends last Sunday in October
Administrative divisions: 15 regions (regioni, singular—regione) and 5 autonomous regions (regioni autonome, singular—regione autonoma)
regions: Abruzzo, Basilicata, Calabria, Campania, Emilia-Romagna, Lazio (Latium), Liguria, Lombardia, Marche, Molise, Piemonte (Piedmont), Puglia (Apulia), Toscana (Tuscany), Umbria, Veneto (Venetia)
autonomous regions: Friuli-Venezia Giulia; Sardegna (Sardinia); Sicilia (Sicily); Trentino-Alto Adige (Trentino-South Tyrol) or Trentino-Suedtirol (German); Valle d'Aosta (Aosta Valley) or Vallee d'Aoste (French)
Independence: 17 March 1861 (Kingdom of Italy proclaimed; Italy was not finally unified until 1870)
National holiday: Republic Day, 2 June (1946)
Constitution: passed 11 December 1947, effective 1 January 1948; amended many times
Legal system: civil law system; judicial review under certain conditions in Constitutional Court
International law organization participation: has not submitted an ICJ jurisdiction declaration; accepts ICCt jurisdiction
Suffrage: 18 years of age; universal (except in senatorial elections, where minimum age is 25)
Executive branch: *chief of state:* President Giorgio NAPOLITANO (since 15 May 2006)
head of government: Prime Minister Silvio BERLUSCONI (since 8 May 2008); note—in Italy the prime minister is referred to as the president of the Council of Ministers
cabinet: Council of Ministers proposed by the prime minister and nominated by the president
(For more information visit the World Leaders website)
elections: president elected by an electoral college consisting of both houses of parliament and 58 regional representatives for a seven-year term (no term limits); election last held on 10 May 2006 (next to be held in May 2013); prime minister appointed by the president and confirmed by parliament
election results: Giorgio NAPOLITANO elected president on the fourth round of voting; electoral college vote—543
Legislative branch: bicameral Parliament or Parlamento consists of the Senate or Senato della Repubblica (315 seats; members elected by proportional vote with the winning coalition in each region receiving 55% of seats from that region; members to serve five-year terms; and up to 5 senators for life appointed by the president of the Republic) and the Chamber of Deputies or Camera dei Deputati (630 seats; members elected by popular vote with the winning national coalition receiving 54% of chamber seats; members to serve five-year terms); note—it has not been clarified if each president has the power to designate up to five senators or if five is the number of senators for life who might sit in the Senate
elections: Senate—last held on 13-14 April 2008 (next to be held in April 2013); Chamber of Deputies—last held on 13-14 April 2008 (next to be held in April 2013)
election results: Senate—percent of vote by party—NA; seats by party—S. BERLUSCONI coalition 174 (PdL 147, LN 25, MpA 2), W. VELTRONI coalition 132 (PD 118, IdV 3), UdC 3, other 6; Chamber of Deputies—percent of vote by party—NA; seats by party—S. BERLUSCONI coalition 344 (PdL 276, LN 60, MpA 8), W. VELTRONI coalition 246 (PD 217, IdV 29), UdC 36, other 4
Judicial branch: Constitutional Court or Corte Costituzionale (composed of 15 judges: one-third appointed by the president, one-third elected by parliament, one-third elected by the ordinary and administrative Supreme Courts)
Political parties and leaders: Center-Right coalition: Lega Nord or LN [Umberto BOSSI]; Movement for Autonomy or MpA [Raffaele LOMBARDO]; People of Freedom or PdL [Silvio BERLUSCONI]
Center-Left coalition: Democratic Party or PD [Pier Luigi BERSANI]; Italy of Values or IdV [Antonio DI PIETRO]
other non-allied parties: Future and Liberty Party or FLI [Gianfranco FINI]; Union of the Center or UdC [Pier Ferdinando CASINI]
Political pressure groups and leaders: manufacturers and merchants associations—Confcommercio; Confindustria; organized farm groups—Confcoltivatori; Confagricoltura; Roman Catholic Church; three major trade union confederations—Confederazione Generale Italiana del

Lavoro or CGIL [Guglielmo EPIFANI] which is left wing; Confederazione Italiana dei Sindacati Lavoratori or CISL [Raffaele BONANNO], which is Roman Catholic centrist; Unione Italiana del Lavoro or UIL [Luigi ANGELETTI] which is lay centrist)

International organization participation: ADB (nonregional member), AfDB (nonregional member), Australia Group, BIS, BSEC (observer), CBSS (observer), CD, CDB, CE, CEI, CERN, EAPC, EBRD, EIB, EMU, ESA, EU, FAO, FATF, G-20, G-7, G-8, G-10, IADB, IAEA, IBRD, ICAO, ICC, ICRM, IDA, IEA, IFAD, IFC, IFRCS, IHO, ILO, IMF, IMO, IMSO, Interpol, IOC, IOM, IPU, ISO, ITSO, ITU, ITUC, LAIA (observer), MIGA, MINURSO, NATO, NEA, NSG, OAS (observer), OECD, OPCW, OSCE, Paris Club, PCA, PIF (partner), Schengen Convention, SECI (observer), SICA (observer), UN, UNAMID, UNCTAD, UNESCO, UNHCR, UNIDO, UNIFIL, Union Latina, UNMOGIP, UNRWA, UNTSO, UNWTO, UPU, WCO, WHO, WIPO, WMO, WTO, ZC

Diplomatic representation in the US: *chief of mission:* Ambassador Giulio TERZI di Sant' Agata
chancery: 3000 Whitehaven Street NW, Washington, DC 20008
telephone: [1] (202) 612-4400
FAX: [1] (202) 518-2151
consulate(s) general: Boston, Chicago, Houston, Miami, New York, Los Angeles, Philadelphia, San Francisco
consulate(s): Detroit

Diplomatic representation from the US: *chief of mission:* Ambassador David THORNE
embassy: Via Vittorio Veneto 121, 00187-Rome
mailing address: PSC 59, Box 100, APO AE 09624
telephone: [39] (06) 46741
FAX: [39] (06) 488-2672, 4674-2356
consulate(s) general: Florence, Milan, Naples

Flag description: three equal vertical bands of green (hoist side), white, and red; design inspired by the French flag brought to Italy by Napoleon in 1797; colors are those of Milan (red and white) combined with the green uniform color of the Milanese civic guard
note: similar to the flag of Mexico, which is longer, uses darker shades of red and green, and has its coat of arms centered on the white band; Ireland, which is longer and is green (hoist side), white, and orange; also similar to the flag of the Cote d'Ivoire, which has the colors reversed–orange (hoist side), white, and green

National anthem: *name:* "Il Canto degli Italiani" (The Song of the Italians)
lyrics/music: Goffredo MAMELI/Michele NOVARO
note: adopted 1946; the anthem, originally written in 1847, is also known as "L'Inno di Mameli" (Mameli's Hymn), and "Fratelli D'Italia" (Brothers of Italy)

ECONOMY

Economy—overview: Italy has a diversified industrial economy, which is divided into a developed industrial north, dominated by private companies, and a less-developed, welfare-dependent, agricultural south, with high unemployment. The Italian economy is driven in large part by the manufacture of high-quality consumer goods produced by small and medium-sized enterprises, many of them family owned. Italy also has a sizable underground economy, which by some estimates accounts for as much as 15% of GDP. These activities are most common within the agriculture, construction, and service sectors. Italy has moved slowly on implementing needed structural reforms, such as reducing graft, overhauling costly entitlement programs, and increasing employment opportunities for young workers, particularly women. The international financial crisis worsened conditions in Italy's labor market, with unemployment rising from 6.2% in 2007 to 8.4% in 2010, but in the longer-term Italy's low fertility rate and quota-driven immigration policies will increasingly strain its economy. A rise in exports and investment driven by the global economic recovery nevertheless helped the economy grow by about 1% in 2010 following a 5% contraction in 2009. The Italian government has struggled to limit government spending, but Italy's exceedingly high public debt remains above 115% of GDP, and its fiscal deficit–just 1.5% of GDP in 2007–exceeded 5% in 2009 and 2010, as the costs of servicing the country's debt rose.

GDP (purchasing power parity): $1.774 trillion (2010 est.)
country comparison to the world: 11
$1.751 trillion (2009 est.)
$1.847 trillion (2008 est.)
note: data are in 2010 US dollars

GDP (official exchange rate): $2.055 trillion (2010 est.)

GDP—real growth rate: 1.3% (2010 est.)
country comparison to the world: 165
-5.2% (2009 est.)
-1.3% (2008 est.)

GDP—per capita (PPP): $30,500 (2010 est.)
country comparison to the world: 43
$30,100 (2009 est.)
$31,800 (2008 est.)
note: data are in 2010 US dollars

GDP—composition by sector: *agriculture:* 1.8%
industry: 24.9%
services: 73.3% (2010 est.)

Labor force: 25.05 million (2010 est.)
country comparison to the world: 23

Labor force—by occupation:
agriculture: 4.2%
industry: 30.7%
services: 65.1% (2005)

Unemployment rate: 8.4% (2010 est.)
country comparison to the world: 96
7.8% (2009 est.)

Population below poverty line: NA%

Household income or consumption by percentage share:
lowest 10%: 2.3%
highest 10%: 26.8% (2000)

Distribution of family income—Gini index: 32 (2006)
country comparison to the world: 102
27.3 (1995)

Investment (gross fixed): 19.1% of GDP (2010 est.)
country comparison to the world: 99

Budget: *revenues:* $940.3 billion
expenditures: $1.042 trillion (2010 est.)

Public debt: 118.1% of GDP (2010 est.)
country comparison to the world: 8
115.8% of GDP (2009 est.)

Inflation rate (consumer prices): 1.4% (2010 est.)
country comparison to the world: 33
0.8% (2009 est.)

Central bank discount rate: 1.75% (31 December 2010)
country comparison to the world: 128
1.75% (31 December 2009)
note: this is the European Central Bank's rate on the marginal lending facility, which offers overnight credit to banks in the euro area

Commercial bank prime lending rate: 10.26% (31 December 2009 est.)
country comparison to the world: 75
11.31% (31 December 2008 est.)

Stock of narrow money: $1.234 trillion (31 December 2010 est.)
country comparison to the world: 7
$1.267 trillion (31 December 2009 est.)
note: see entry for the European Union for money supply in the euro area; the European Central Bank (ECB) controls monetary policy for the 17 members of the Economic and Monetary Union (EMU); individual members of the EMU do not control the quantity of money circulating within their own borders

Stock of broad money: $1.884 trillion (31 December 2010 est.)
country comparison to the world: 10
$1.846 trillion (31 December 2009 est.)

Stock of domestic credit: $3.274 trillion (31 December 2009 est.)
country comparison to the world: 9
$3.047 trillion (31 December 2008 est.)

Market value of publicly traded shares: $317.3 billion (31 December 2009)
country comparison to the world: 17
$520.9 billion (31 December 2008)
$1.073 trillion (31 December 2007)

Agriculture—products: fruits, vegetables, grapes, potatoes, sugar beets, soybeans, grain, olives; beef, dairy products; fish

Industries: tourism, machinery, iron and steel, chemicals, food processing, textiles, motor vehicles, clothing, footwear, ceramics

Industrial production growth rate: 0.5% (2010 est.)
country comparison to the world: 150

Electricity—production: 289.7 billion kWh (2007 est.)
country comparison to the world: 14

Electricity—consumption: 315 billion kWh (2007 est.)
country comparison to the world: 13

Electricity—exports: 3.431 billion kWh (2008 est.)

Electricity—imports: 43 billion kWh (2008 est.)

Oil—production: 146,500 bbl/day (2009 est.)
country comparison to the world: 47

Oil—consumption: 1.537 million bbl/day (2009 est.)
country comparison to the world: 16

Oil—exports: 586,900 bbl/day (2008 est.)
country comparison to the world: 26
Oil—imports: 1.911 million bbl/day (2008 est.)
country comparison to the world: 10
Oil—proved reserves: 423.7 million bbl (1 January 2010 est.)
country comparison to the world: 53
Natural gas—production: 8.119 billion cu m (2009 est.)
country comparison to the world: 45
Natural gas—consumption: 78.12 billion cu m (2009 est.)
country comparison to the world: 10
Natural gas—exports: 124 million cu m (2009 est.)
country comparison to the world: 43
Natural gas—imports: 69.24 billion cu m (2009 est.)
country comparison to the world: 4
Natural gas—proved reserves: 69.83 billion cu m (1 January 2010 est.)
country comparison to the world: 58
Current account balance: $-61.98 billion (2010 est.)
country comparison to the world: 189
$-66.2 billion (2009 est.)
Exports: $458.4 billion (2010 est.)
country comparison to the world: 8
$407.2 billion (2009 est.)
Exports—commodities: engineering products, textiles and clothing, production machinery, motor vehicles, transport equipment, chemicals; food, beverages and tobacco; minerals, and nonferrous metals
Exports—partners: Germany 12.6%, France 11.57%, US 5.92%, Spain 5.69%, UK 5.13%, Switzerland 4.69% (2009)
Imports: $459.7 billion (2010 est.)
country comparison to the world: 8
$403.9 billion (2009 est.)
Imports—commodities: engineering products, chemicals, transport equipment, energy products, minerals and nonferrous metals, textiles and clothing; food, beverages, and tobacco
Imports—partners: Germany 16.68%, France 8.82%, China 6.53%, Netherlands 5.63%, Spain 4.3%, Russia 4.12%, Belgium 4.08% (2009)
Reserves of foreign exchange and gold: $NA (31 December 2010 est.)
$132.8 billion (31 December 2009 est.)
Debt—external: $2.223 trillion (30 June 2010 est.)
country comparison to the world: 9
$2.328 trillion (31 December 2008)
Stock of direct foreign investment—at home: $405.1 billion (31 December 2010 est.)
country comparison to the world: 12
$368.9 billion (31 December 2009 est.)
Stock of direct foreign investment—abroad: $601.1 billion (31 December 2010 est.)
country comparison to the world: 12
$555.2 billion (31 December 2009 est.)
Exchange rates: euros (EUR) per US dollar–0.7715 (2010), 0.7179 (2009), 0.6827 (2008), 0.7345 (2007), 0.7964 (2006)

COMMUNICATIONS

Telephones—main lines in use: 21.3 million (2009)
country comparison to the world: 12
Telephones—mobile cellular: 90.613 million (2009)
country comparison to the world: 12
Telephone system: *general assessment:* modern, well developed, fast; fully automated telephone, telex, and data services
domestic: high-capacity cable and microwave radio relay trunks
international: country code–39; a series of submarine cables provide links to Asia, Middle East, Europe, North Africa, and US; satellite earth stations–3 Intelsat (with a total of 5 antennas–3 for Atlantic Ocean and 2 for Indian Ocean), 1 Inmarsat (Atlantic Ocean region), and NA Eutelsat
Broadcast media: two Italian media giants–the publicly-owned Radiotelevisione Italiana (RAI) with 3 national terrestrial stations and privately-owned Mediaset with 3 national terrestrial stations–dominate; additional broadcasts by a large number of private stations and Sky Italia–a satellite TV network; RAI operates 3 AM/FM nationwide radio stations; some 1,300 commercial radio stations (2007)
Internet country code: .it
Internet hosts: 23.16 million (2010)
country comparison to the world: 3
Internet users: 29.235 million (2009)
country comparison to the world: 13

TRANSPORTATION

Airports: 132 (2010)
country comparison to the world: 44
Airports—with paved runways: *total:* 101
over 3,047 m: 9
2,438 to 3,047 m: 30
1,524 to 2,437 m: 18
914 to 1,523 m: 31
under 914 m: 13 (2010)
Airports—with unpaved runways: *total:* 31
1,524 to 2,437 m: 1
914 to 1,523 m: 11
under 914 m: 19 (2010)
Heliports: 6 (2010)
Pipelines: gas 18,348 km; oil 1,241 km (2010)
Railways: *total:* 20,254 km
country comparison to the world: 15
standard gauge: 18,611 km 1.435-m gauge (12,662 km electrified)
narrow gauge: 123 km 1.000-m gauge (123 km electrified); 1,290 km 0.950-m gauge (151 km electrified); 231 km 0.850-m gauge (2010)
Roadways: *total:* 487,700 km
country comparison to the world: 13
paved: 487,700 km (includes 6,700 km of expressways) (2007)
Waterways: 2,400 km (used for commercial traffic; of limited overall value compared to road and rail) (2009)
country comparison to the world: 37
Merchant marine: *total:* 667
country comparison to the world: 17
by type: bulk carrier 81, cargo 47, carrier 1, chemical tanker 169, container 22, liquefied gas 25, passenger 23, passenger/cargo 160, petroleum tanker 56, refrigerated cargo 4, roll on/roll off 34, specialized tanker 11, vehicle carrier 34
foreign-owned: 78 (Denmark 4, France 2, Germany 1, Greece 8, Luxembourg 12, Nigeria 1, Norway 6, Sweden 1, Switzerland 6, Taiwan 11, Turkey 3, UK 2, US 21)
registered in other countries: 213 (Bahamas 5, Belize 3, Cayman Islands 6, Cyprus 6, Georgia 2, Gibraltar 4, Greece 5, Kiribati 1, Liberia 48, Malta 52, Marshall Islands 1, Netherlands 9, Norway 3, Panama 23, Portugal 10, Russia 9, Saint Kitts and Nevis 1, Saint Vincent and the Grenadines 5, Singapore 3, Slovakia 2, Spain 1, Sweden 5, Turkey 2, UK 4, unknown 3) (2010)
Ports and terminals: Augusta, Cagliari, Genoa, Livorno, Santa Panagia (Melilli), Taranto, Trieste, Venice

MILITARY

Military branches: Italian Armed Forces: Army (Esercito Italiano, EI), Navy (Marina Militare Italiana, MMI), Italian Air Force (Aeronautica Militare Italiana, AMI), Carabinieri Corps (Arma dei Carabinieri, CC) (2011)
Military service age and obligation: 18-27 year of age for voluntary military service; conscription abolished January 2005; women may serve in any military branch; 10-month service obligation, with a reserve obligation to age 45 (Army and Air Force) or 39 (Navy) (2006)
Manpower available for military service: *males age 16–49:* 13,865,688
females age 16–49: 14,003,755 (2010 est.)
Manpower fit for military service: *males age 16–49:* 11,247,446
females age 16–49: 11,348,695 (2010 est.)
Manpower reaching militarily significant age annually: *male:* 288,188
female: 281,671 (2010 est.)
Military expenditures: 1.8% of GDP (2005 est.)
country comparison to the world: 81

TRANSNATIONAL ISSUES

Disputes—international: Italy's long coastline and developed economy entices tens of thousands of illegal immigrants from southeastern Europe and northern Africa
Illicit drugs: important gateway for and consumer of Latin American cocaine and Southwest Asian heroin entering the European market; money laundering by organized crime and from smuggling

J

JAMAICA

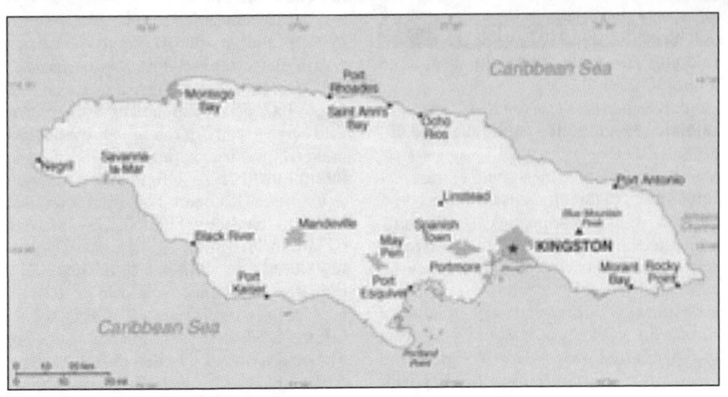

INTRODUCTION

Background: The island–discovered by Christopher COLUMBUS in 1494–was settled by the Spanish early in the 16th century. The native Taino Indians, who had inhabited Jamaica for centuries, were gradually exterminated and replaced by African slaves. England seized the island in 1655 and established a plantation economy based on sugar, cocoa, and coffee. The abolition of slavery in 1834 freed a quarter million slaves, many of whom became small farmers. Jamaica gradually obtained increasing independence from Britain. In 1958 it joined other British Caribbean colonies in forming the Federation of the West Indies. Jamaica gained full independence when it withdrew from the Federation in 1962. Deteriorating economic conditions during the 1970s led to recurrent violence as rival gangs affiliated with the major political parties evolved into powerful organized crime networks involved in international drug smuggling and money laundering. Violent crime, drug trafficking, and poverty pose significant challenges to the government today. Nonetheless, many rural and resort areas remain relatively safe and contribute substantially to the economy.

GEOGRAPHY

Location: Caribbean, island in the Caribbean Sea, south of Cuba
Geographic coordinates: 18 15 N, 77 30 W
Map references: Central America and the Caribbean
Area: *total:* 10,991 sq km
country comparison to the world: 167
land: 10,831 sq km
water: 160 sq km
Area—comparative: slightly smaller than Connecticut
Land boundaries: 0 km
Coastline: 1,022 km
Maritime claims: measured from claimed archipelagic straight baselines
territorial sea: 12 nm
contiguous zone: 24 nm
exclusive economic zone: 200 nm
continental shelf: 200 nm or to edge of the continental margin
Climate: tropical; hot, humid; temperate interior
Terrain: mostly mountains, with narrow, discontinuous coastal plain
Elevation extremes: *lowest point:* Caribbean Sea 0 m
highest point: Blue Mountain Peak 2,256 m
Natural resources: bauxite, gypsum, limestone
Land use: *arable land:* 15.83%
permanent crops: 10.01%
other: 74.16% (2005)
Irrigated land: 250 sq km (2002)
Total renewable water resources: 9.4 cu km (2000)
Freshwater withdrawal (domestic/industrial/agricultural): *total:* 0.41 cu km/yr (34%/17%/49%)
per capita: 155 cu m/yr (2000)
Natural hazards: hurricanes (especially July to November)
Environment—current issues: heavy rates of deforestation; coastal waters polluted by industrial waste, sewage, and oil spills; damage to coral reefs; air pollution in Kingston results from vehicle emissions
Environment—international agreements: *party to:* Biodiversity, Climate Change, Climate Change-Kyoto Protocol, Desertification, Endangered Species, Hazardous Wastes, Law of the Sea, Marine Dumping, Marine Life Conservation, Ozone Layer Protection, Ship Pollution, Wetlands
signed, but not ratified: none of the selected agreements
Geography—note: strategic location between Cayman Trench and Jamaica Channel, the main sea lanes for the Panama Canal

PEOPLE

Population: 2,868,380 (July 2011 est.)
country comparison to the world: 138
Age structure: *0–14 years:* 30.1% (male 438,888/female 424,383)
15–64 years: 62.3% (male 882,548/female 904,242)
65 years and over: 7.6% (male 97,717/female 120,602) (2011 est.)
Median age: *total:* 24.2 years
male: 23.7 years
female: 24.7 years (2011 est.)
Population growth rate: 0.733% (2011 est.)
country comparison to the world: 138
Birth rate: 19.2 births/1,000 population (2011 est.)
country comparison to the world: 99
Death rate: 6.54 deaths/1,000 population (July 2011 est.)
country comparison to the world: 149
Net migration rate: -5.34 migrant(s)/1,000 population (2011 est.)
country comparison to the world: 195
Urbanization: *urban population:* 52% of total population (2010)
rate of urbanization: 0.6% annual rate of change (2010-15 est.)
Major cities—population: KINGSTON (capital) 580,000 (2009)
Sex ratio: *at birth:* 1.05 male(s)/female
under 15 years: 1.03 male(s)/female
15–64 years: 0.97 male(s)/female
65 years and over: 0.81 male(s)/female
total population: 0.98 male(s)/female (2011 est.)
Infant mortality rate: *total:* 14.6 deaths/1,000 live births
country comparison to the world: 123
male: 15.18 deaths/1,000 live births
female: 14 deaths/1,000 live births (2011 est.)
Life expectancy at birth: *total population:* 73.45 years
country comparison to the world: 116
male: 71.79 years
female: 75.19 years (2011 est.)
Total fertility rate: 2.17 children born/woman (2011 est.)
country comparison to the world: 110
HIV/AIDS—adult prevalence rate: 1.7% (2009 est.)
country comparison to the world: 32
HIV/AIDS—people living with HIV/AIDS: 32,000 (2009 est.)
country comparison to the world: 69
HIV/AIDS—deaths: 1,200 (2009 est.)
country comparison to the world: 64
Drinking water source: *Improved:*
urban: 98% of population
rural: 89% of population
total: 94% of population
Unimproved: urban: 2% of population
rural: 11% of population
total: 6% of population (2008)
Sanitation facility access: *Improved:*
urban: 82% of population
rural: 84% of population
total: 83% of population
Unimproved: urban: 18% of population
rural: 16% of population
total: 17% of population (2008)
Nationality: *noun:* Jamaican(s)
adjective: Jamaican
Ethnic groups: black 91.2%, mixed 6.2%, other or unknown 2.6% (2001 census)
Religions: Protestant 62.5% (Seventh-Day Adventist 10.8%, Pentecostal 9.5%, Other Church of God 8.3%, Baptist 7.2%, New Testament Church of God 6.3%, Church of God in Jamaica 4.8%, Church of God of Prophecy 4.3%, Anglican 3.6%, other Christian 7.7%), Roman Catholic 2.6%, other or unspecified 14.2%, none 20.9%, (2001 census)

Languages: English, English patois
Literacy: *definition:* age 15 and over has ever attended school
total population: 87.9%
male: 84.1%
female: 91.6% (2003 est.)
School life expectancy (primary to tertiary education): *total:* 14 years
male: 13 years
female: 15 years (2008)
Education expenditures: 5.8% of GDP (2009)
country comparison to the world: 35

GOVERNMENT

Country name: *conventional long form:* none
conventional short form: Jamaica
Government type: constitutional parliamentary democracy and a Commonwealth realm
Capital: *name:* Kingston
geographic coordinates: 18 00 N, 76 48 W
time difference: UTC-5 (same time as Washington, DC during Standard Time)
Administrative divisions: 14 parishes; Clarendon, Hanover, Kingston, Manchester, Portland, Saint Andrew, Saint Ann, Saint Catherine, Saint Elizabeth, Saint James, Saint Mary, Saint Thomas, Trelawny, Westmoreland
note: for local government purposes, Kingston and Saint Andrew were amalgamated in 1923 into the present single corporate body known as the Kingston and Saint Andrew Corporation
Independence: 6 August 1962 (from the UK)
National holiday: Independence Day, 6 August (1962)
Constitution: 6 August 1962
Legal system: common law system based on the English model
International law organization participation: has not submitted an ICJ jurisdiction declaration; non-party state to the ICCt
Suffrage: 18 years of age; universal
Executive branch: *chief of state:* Queen ELIZABETH II (since 6 February 1952); represented by Governor General Dr. Patrick L. ALLEN (since 26 February 2009)
head of government: Prime Minister Bruce GOLDING (since 11 September 2007)
cabinet: Cabinet is appointed by the governor general on the advice of the prime minister
(For more information visit the World Leaders website)
elections: the monarchy is hereditary; governor general appointed by the monarch on the recommendation of the prime minister; following legislative elections, the leader of the majority party or the leader of the majority coalition in the House of Representatives is appointed prime minister by the governor general
Legislative branch: bicameral Parliament consists of the Senate (a 21-member body appointed by the governor general on the recommendations of the prime minister and the leader of the opposition; ruling party is allocated 13 seats, and the opposition is allocated 8 seats) and the House of Representatives (60 seats; members elected by popular vote to serve five-year terms)
elections: last held on 3 September 2007 (next to be held no later than October 2012)
election results: percent of vote by party—JLP 50.1%, PNP 49.8%; seats by party—JLP 33, PNP 27
Judicial branch: Supreme Court (judges appointed by the governor general on the advice of the prime minister); Court of Appeal; Privy Council in UK; member of the Caribbean Court of Justice (CCJ)
Political parties and leaders: Jamaica Labor Party or JLP [Bruce GOLDING]; People's National Party or PNP [Portia SIMPSON-MILLER]; National Democratic Movement or NDM [Michael WILLIAMS]
Political pressure groups and leaders: New Beginnings Movement or NBM; Rastafarians (black religious/racial cultists, pan-Africanists)
International organization participation: ACP, AOSIS, C, Caricom, CDB, FAO, G-15, G-77, IADB, IAEA, IBRD, ICAO, ICRM, IFAD, IFC, IFRCS, IHO, ILO, IMF, IMO, Interpol, IOC, IOM, ISO, ITSO, ITU, LAES, MIGA, NAM, OAS, OPANAL, OPCW, PetroCaribe, RG, UN, UNCTAD, UNESCO, UNIDO, UNITAR, UNWTO, UPU, WCO, WFTU, WHO, WIPO, WMO, WTO
Diplomatic representation in the US: *chief of mission:* Ambassador Audrey P. MARKS
chancery: 1520 New Hampshire Avenue NW, Washington, DC 20036
telephone: [1] (202) 452-0660
FAX: [1] (202) 452-0081
consulate(s) general: Miami, New York
Diplomatic representation from the US: *chief of mission:* Ambassador Pamela BRIDGEWATER
embassy: 142 Old Hope Road, Kingston 6
mailing address: P.O. Box 541, Kingston 5
telephone: [1] (876) 702-6000
FAX: [1] (876) 702-6001
Flag description: diagonal yellow cross divides the flag into four triangles—green (top and bottom) and black (hoist side and outer side); green represents hope, vegetation, and agriculture, black reflects hardships overcome and to be faced, and yellow recalls golden sunshine and the island's natural resources
National anthem: *name:* "Jamaica, Land We Love"
lyrics/music: Hugh Braham SHERLOCK/ Robert Charles LIGHTBOURNE
note: adopted 1962

ECONOMY

Economy—overview: The Jamaican economy is heavily dependent on services, which now account for more than 60% of GDP. The country continues to derive most of its foreign exchange from tourism, remittances, and bauxite/alumina. Remittances account for nearly 15% of GDP and exports of bauxite and alumina make up about 10%. The bauxite/alumina sector was most affected by the global downturn while the tourism industry was resilient, experiencing an increase of 4% in tourist arrivals. Tourism revenues account for roughly 10% of GDP, and both arrivals and revenues grew in 2010, up 4% and 6% respectively. The Economic growth faces many challenges: high crime and corruption, large-scale unemployment and underemployment, and a debt-to-GDP ratio of more than 120%. Jamaica's onerous public debt burden—the fourth highest in the world on a per capita basis—is the result of government bailouts to ailing sectors of the economy, most notably to the financial sector in the mid-to-late 1990s. In early 2010, the Jamaican government created the Jamaica Debt Exchange (JDX) in order to retire high-priced domestic bonds and significantly reduce annual debt servicing. The Government of Jamaica signed a $1.27 billion, 27-month Standby Agreement with the International Monetary Fund for balance of payment support in February 2010. Other multilaterals have also provided millions of dollars in loans and grants. Despite the improvement, debt servicing costs still hinder the government's ability to spend on infrastructure and social programs, particularly as job losses rise in a shrinking economy. The GOLDING administration faces the difficult prospect of having to achieve fiscal discipline in order to maintain debt payments, while simultaneously attacking a serious crime problem that is hampering economic growth. High unemployment exacerbates the crime problem, including gang violence that is fueled by the drug trade.
GDP (purchasing power parity): $23.72 billion (2010 est.)
country comparison to the world: 117
$23.99 billion (2009 est.)
$24.74 billion (2008 est.)
note: data are in 2010 US dollars
GDP (official exchange rate): $13.69 billion (2010 est.)
GDP—real growth rate: -1.1% (2010 est.)
country comparison to the world: 200
-3% (2009 est.)
-0.9% (2008 est.)
GDP—per capita (PPP): $8,300 (2010 est.)
country comparison to the world: 120
$8,500 (2009 est.)
$8,800 (2008 est.)
note: data are in 2010 US dollars
GDP—composition by sector: *agriculture:* 5.7%
industry: 29.7%
services: 64.6% (2010 est.)
Labor force: 1.317 million (2010 est.)
country comparison to the world: 136
Labor force—by occupation: *agriculture:* 17%
industry: 19%
services: 64% (2006)
Unemployment rate: 12.9% (2010 est.)
country comparison to the world: 133
11.4% (2009 est.)
Population below poverty line: 16.5% (2009 est.)
Household income or consumption by percentage share: *lowest 10%:* 2.1%
highest 10%: 35.8% (2004)
Distribution of family income—Gini index: 45.5 (2004)
country comparison to the world: 38
37.9 (2000)
Investment (gross fixed): 25.1% of GDP (2010 est.)
country comparison to the world: 43
Budget: *revenues:* $3.611 billion
expenditures: $4.644 billion (2010 est.)

Public debt: 123.2% of GDP (2010 est.)
country comparison to the world: 7
124.4% of GDP (2009 est.)
Inflation rate (consumer prices): 13% (2010 est.)
country comparison to the world: 210
9.6% (2009 est.)
Central bank discount rate: NA%
Commercial bank prime lending rate: 16.43% (31 December 2009 est.)
country comparison to the world: 31
16.83% (31 December 2008 est.)
Stock of narrow money: $1.432 billion (31 December 2010 est.)
country comparison to the world: 126
$1.371 billion (31 December 2009 est.)
Stock of broad money: $5.782 billion (31 December 2010 est.)
country comparison to the world: 115
$5.472 billion (31 December 2009 est.)
Stock of domestic credit: $7.922 billion (31 December 2010 est.)
country comparison to the world: 102
$7.282 billion (31 December 2009 est.)
Market value of publicly traded shares: $6.201 billion (31 December 2009)
country comparison to the world: 71
$7.513 billion (31 December 2008)
$12.33 billion (31 December 2007)
Agriculture—products: sugarcane, bananas, coffee, citrus, yams, ackees, vegetables; poultry, goats, milk; crustaceans, mollusks
Industries: tourism, bauxite/alumina, agro processing, light manufactures, rum, cement, metal, paper, chemical products, telecommunications
Industrial production growth rate: -2% (2010 est.)
country comparison to the world: 160
Electricity—production: 7.324 billion kWh (2007 est.)
country comparison to the world: 99
Electricity—consumption: 6.345 billion kWh (2007 est.)
country comparison to the world: 105
Electricity—exports: 0 kWh (2008 est.)
Electricity—imports: 0 kWh (2008 est.)
Oil—production: 0 bbl/day (2008 est.)
country comparison to the world: 187
Oil—consumption: 77,000 bbl/day (2009 est.)
country comparison to the world: 87
Oil—exports: 0 bbl/day (2007 est.)
country comparison to the world: 175
Oil—imports: 77,720 bbl/day (2007 est.)
country comparison to the world: 75
Oil—proved reserves: 0 bbl (1 January 2010 est.)
country comparison to the world: 145
Natural gas—production: 0 cu m (2008 est.)
country comparison to the world: 141
Natural gas—consumption: 0 cu m (2008 est.)
country comparison to the world: 185
Natural gas—exports: 0 cu m (2008 est.)
country comparison to the world: 118
Natural gas—imports: 0 cu m (2008 est.)
country comparison to the world: 203
Natural gas—proved reserves: 0 cu m (1 January 2010 est.)
country comparison to the world: 148
Current account balance: $-1.382 billion (2010 est.)
country comparison to the world: 144
$-876 million (2009 est.)
Exports:
$1.487 billion (2010 est.)
country comparison to the world: 138
$1.263 billion (2009 est.)
Exports—commodities: alumina, bauxite, sugar, rum, coffee, yams, beverages, chemicals, wearing apparel, mineral fuels
Exports—partners: US 38.19%, Canada 12.2%, UK 10.79%, Norway 4.89%, Netherlands 4.69% (2009)
Imports: $5.378 billion (2010 est.)
country comparison to the world: 109
$4.581 billion (2009 est.)
Imports—commodities: food and other consumer goods, industrial supplies, fuel, parts and accessories of capital goods, machinery and transport equipment, construction materials
Imports—partners: US 28.32%, Trinidad and Tobago 22.98%, Venezuela 12.14%, China 4.61%, Brazil 4.18% (2009)
Reserves of foreign exchange and gold: $1.85 billion (31 December 2010 est.)
country comparison to the world: 107
$2.081 billion (31 December 2009 est.)
Debt—external: $12.66 billion (31 December 2010 est.)
country comparison to the world: 84
$10.56 billion (31 December 2009 est.)
Exchange rates: Jamaican dollars (JMD) per US dollar–87.41 (2010), 87.89 (2009), 72.236 (2008), 69.034 (2007), 65.768 (2006)

COMMUNICATIONS

Telephones—main lines in use: 302,300 (2009)
country comparison to the world: 114
Telephones—mobile cellular: 2.971 million (2009)
country comparison to the world: 114
Telephone system: *general assessment:* fully automatic domestic telephone network
domestic: the 1999 agreement to open the market for telecommunications services resulted in rapid growth in mobile-cellular telephone usage while the number of fixed-lines in use has declined; combined mobile-cellular teledensity exceeded 110 per 100 persons in 2009
international: country code–1-876; the Fibralink submarine cable network provides enhanced delivery of business and broadband traffic and is linked to the Americas Region Caribbean Ring System (ARCOS-1) submarine cable in the Dominican Republic; the link to ARCOS-1 provides seamless connectivity to US, parts of the Caribbean, Central America, and South America; satellite earth stations–2 Intelsat (Atlantic Ocean) (2008)
Broadcast media: privately-owned Radio Jamaica Limited and its subsidiaries operate multiple television stations, subscription cable services, and radio stations; 2 other privately-owned television stations broadcast; roughly 70 radio stations (2007)
Internet country code: .jm
Internet hosts: 3,099 (2010)
country comparison to the world: 143
Internet users: 1.581 million (2009)
country comparison to the world: 80

TRANSPORTATION

Airports: 27 (2010)
country comparison to the world: 122
Airports—with paved runways: *total:* 12
2,438 to 3,047 m: 2
914 to 1,523 m: 3
under 914 m: 7 (2010)
Airports—with unpaved runways: *total:* 15
under 914 m: 15 (2010)
Roadways: *total:* 21,552 km
country comparison to the world: 107
paved: 15,937 km (includes 33 km of expressways)
unpaved: 5,615 km (2005)
Merchant marine: *total:* 19
country comparison to the world: 100
by type: bulk carrier 9, cargo 5, container 4, roll on/roll off 1
foreign-owned: 19 (Denmark 1, Germany 10, Greece 8) (2010)
Ports and terminals: Discovery Bay (Port Rhoades), Kingston, Montego Bay, Port Antonio, Port Esquivel, Port Kaiser, Rocky Point

MILITARY

Military branches: Jamaica Defense Force: Ground Forces, Coast Guard, Air Wing (2010)
Military service age and obligation: 18 years of age for voluntary military service; younger recruits may be conscripted with parental consent (2001)
Manpower available for military service: *males age 16–49:* 726,263
females age 16–49: 742,958 (2010 est.)
Manpower fit for military service: *males age 16–49:* 590,673
females age 16–49: 596,414 (2010 est.)
Manpower reaching militarily significant age annually: *male:* 33,369
female: 32,702 (2010 est.)
Military expenditures: 0.6% of GDP (2006 est.)
country comparison to the world: 154

TRANSNATIONAL ISSUES

Disputes—international: none
Illicit drugs: transshipment point for cocaine from South America to North America and Europe; illicit cultivation and consumption of cannabis; government has an active manual cannabis eradication program; corruption is a major concern; substantial money-laundering activity; Colombian narcotics traffickers favor Jamaica for illicit financial transactions

JAN MAYEN

(TERRITORY OF NORWAY)

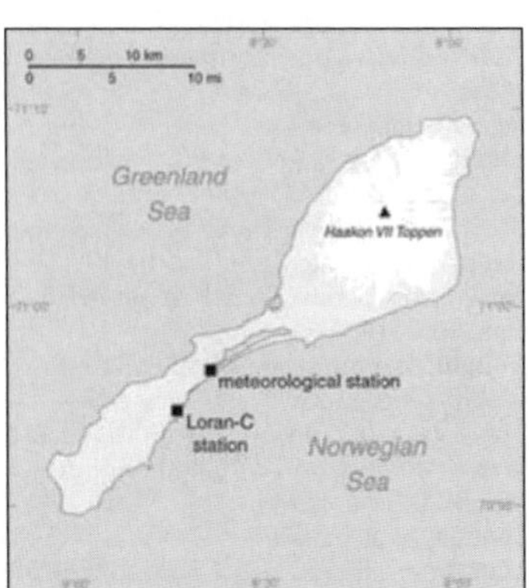

INTRODUCTION

Background: This desolate, arctic, mountainous island was named after a Dutch whaling captain who indisputably discovered it in 1614 (earlier claims are inconclusive). Visited only occasionally by seal hunters and trappers over the following centuries, the island came under Norwegian sovereignty in 1929. The long dormant Beerenberg volcano resumed activity in 1970; the most recent eruption occurred in 1985. It is the northernmost active volcano on earth.

GEOGRAPHY

Location: Northern Europe, island between the Greenland Sea and the Norwegian Sea, northeast of Iceland
Geographic coordinates: 71 00 N, 8 00 W
Map references: Europe
Area: *total:* 377 sq km
country comparison to the world: 203
land: 377 sq km
water: 0 sq km
Area—comparative: slightly more than twice the size of Washington, DC
Land boundaries: 0 km
Coastline: 124.1 km
Maritime claims: *territorial sea:* 4 nm
contiguous zone: 10 nm
exclusive economic zone: 200 nm
continental shelf: 200 m depth or to the depth of exploitation
Climate: arctic maritime with frequent storms and persistent fog
Terrain: volcanic island, partly covered by glaciers
Elevation extremes: *lowest point:* Norwegian Sea 0 m
highest point: Haakon VII Toppen on Beerenberg 2,277 m
note: Beerenberg volcano has numerous peaks; the highest point on the volcano rim is named Haakon VII Toppen, after Norway's first king following the reestablishment of Norwegian independence in 1905
Natural resources: none
Land use: *arable land:* 0%
permanent crops: 0%
other: 100% (2005)
Irrigated land: 0 sq km
Natural hazards: dominated by the volcano Beerenberg
volcanism: Beerenberg (elev. 2,227 m) is Norway's only active volcano; volcanic activity resumed in 1970; the most recent eruption occurred in 1985
Environment—current issues: NA
Geography—note: barren volcanic island with some moss and grass

PEOPLE

Population: no indigenous inhabitants
note: personnel operate the Long Range Navigation (Loran-C) base and the weather and coastal services radio station

GOVERNMENT

Country name: *conventional long form:* none
conventional short form: Jan Mayen
Dependency status: territory of Norway; since August 1994, administered from Oslo through the county governor (fylkesmann) of Nordland; however, authority has been delegated to a station commander of the Norwegian Defense Communication Service
Legal system: the laws of Norway, where applicable, apply
Flag description: the flag of Norway is used

ECONOMY

Economy—overview: Jan Mayen is a volcanic island with no exploitable natural resources, although surrounding waters contain substantial fish stocks and potential untapped petroleum resources. Economic activity is limited to providing services for employees of Norway's radio and meteorological stations on the island.

COMMUNICATIONS

Broadcast media: a coastal radio station has been remotely operated since 1994 (2008)

TRANSPORTATION

Airports: 1 (2010)
country comparison to the world: 222
Airports—with unpaved runways: *total:* 1
1,524 to 2,437 m: 1 (2010)
Ports and terminals: none; offshore anchorage only

MILITARY

Military—note: defense is the responsibility of Norway

TRANSNATIONAL ISSUES

Disputes—international: none

JAPAN

INTRODUCTION

Background: In 1603, after decades of civil warfare, the Tokugawa shogunate (a military-led, dynastic government) ushered in a long period of relative political stability and isolation from foreign influence. For more than two centuries this policy enabled Japan to enjoy a flowering of its indigenous culture. Japan opened its ports after signing the Treaty of Kanagawa with the US in 1854 and began to intensively modernize and industrialize. During the late 19th and early 20th centuries, Japan became a regional power that was able to defeat the forces of both China and Russia. It occupied Korea, Formosa (Taiwan), and southern Sakhalin Island. In 1931-32 Japan occupied Manchuria, and in 1937 it launched a full-scale invasion of China. Japan attacked US forces in 1941–triggering America's entry into World War II–and soon occupied much of East and Southeast Asia. After its defeat in World War II, Japan recovered to become an economic power and an ally of the US. While the emperor retains his throne as a symbol of national unity, elected politicians hold actual decision-making power. Following three decades of unprecedented growth, Japan's economy experienced a major slowdown starting in the 1990s, but the country remains a major economic power. In March 2011, Japan's strongest-ever earthquake, and an accompanying tsunami, devastated the northeast part of Honshu island, killing thousands and damaging several nuclear power plants. The catastrophe hobbled the country's economy and its energy infrastructure, and severely strained its capacity to deal with the humanitarian disaster.

GEOGRAPHY

Location: Eastern Asia, island chain between the North Pacific Ocean and the Sea of Japan, east of the Korean Peninsula
Geographic coordinates: 36 00 N, 138 00 E

Map references: Asia
Area: *total:* 377,915 sq km
country comparison to the world: 61
land: 364,485 sq km
water: 13,430 sq km
note: includes Bonin Islands (Ogasawara-gunto), Daito-shoto, Minami-jima, Okino-tori-shima, Ryukyu Islands (Nansei-shoto), and Volcano Islands (Kazan-retto)
Area—comparative: slightly smaller than California
Land boundaries: 0 km
Coastline: 29,751 km
Maritime claims: *territorial sea:* 12 nm; between 3 nm and 12 nm in the international straits—La Perouse or Soya, Tsugaru, Osumi, and Eastern and Western Channels of the Korea or Tsushima Strait
contiguous zone: 24 nm
exclusive economic zone: 200 nm
Climate: varies from tropical in south to cool temperate in north
Terrain: mostly rugged and mountainous
Elevation extremes: *lowest point:* Hachiro-gata -4 m
highest point: Fujiyama 3,776 m
Natural resources: negligible mineral resources, fish
note: with virtually no energy natural resources, Japan is the world's largest importer of coal and liquefied natural gas, as well as the second largest importer of oil
Land use: *arable land:* 11.64%
permanent crops: 0.9%
other: 87.46% (2005)
Irrigated land: 25,920 sq km (2003)
Total renewable water resources: 430 cu km (1999)
Freshwater withdrawal (domestic/industrial/agricultural): *total:* 88.43 cu km/yr (20%/18%/62%)
per capita: 690 cu m/yr (2000)
Natural hazards: many dormant and some active volcanoes; about 1,500 seismic occurrences (mostly tremors but occasional severe earthquakes) every year; tsunamis; typhoons
volcanism: both Unzen (elev. 1,500 m) and Sakura-jima (elev. 1,117 m), which lies near the densely populated city of Kagoshima, have been deemed "Decade Volcanoes" by the International Association of Volcanology and Chemistry of the Earth's Interior, worthy of study due to their explosive history and close proximity to human populations; other notable historically active volcanoes include Asama, Honshu Island's most active volcano, Aso, Bandai, Fuji, Iwo-Jima, Kikai, Kirishima, Komaga-take, Oshima, Suwanosejima, Tokachi, Yake-dake, and Usu
Environment—current issues: air pollution from power plant emissions results in acid rain; acidification of lakes and reservoirs degrading water quality and threatening aquatic life; Japan is one of the largest consumers of fish and tropical timber, contributing to the depletion of these resources in Asia and elsewhere
Environment—international agreements: *party to:* Antarctic-Environmental Protocol, Antarctic-Marine Living Resources, Antarctic Seals, Antarctic Treaty, Biodiversity, Climate Change, Climate Change-Kyoto Protocol, Desertification, Endangered Species, Environmental Modification, Hazardous Wastes, Law of the Sea, Marine Dumping, Ozone Layer Protection, Ship Pollution, Tropical Timber 83, Tropical Timber 94, Wetlands, Whaling
signed, but not ratified: none of the selected agreements
Geography—note: strategic location in northeast Asia

PEOPLE

Population: 126,475,664 (July 2011 est.)
country comparison to the world: 10
Age structure: *0–14 years:* 13.1% (male 8,521,571/female 8,076,173)
15–64 years: 64% (male 40,815,840/female 40,128,235)
65 years and over: 22.9% (male 12,275,829/female 16,658,016) (2011 est.)
Median age: *total:* 44.8 years
male: 43.2 years
female: 46.7 years (2011 est.)
Population growth rate: -0.278% (2011 est.)
country comparison to the world: 215
Birth rate: 7.31 births/1,000 population (2011 est.)
country comparison to the world: 221
Death rate: 10.09 deaths/1,000 population (July 2011 est.)
country comparison to the world: 55
Net migration rate: 0 migrant(s)/1,000 population (2011 est.)
country comparison to the world: 90
Urbanization: *urban population:* 67% of total population (2010)
rate of urbanization: 0.2% annual rate of change (2010-15 est.)
Major cities—population: TOKYO (capital) 36.507 million; Osaka-Kobe 11.325 million; Nagoya 3.257 million; Fukuoka-Kitakyushu 2.809 million; Sapporo 2.673 million (2009)
Sex ratio: *at birth:* 1.056 male(s)/female
under 15 years: 1.06 male(s)/female
15–64 years: 1.02 male(s)/female
65 years and over: 0.74 male(s)/female
total population: 0.95 male(s)/female (2011 est.)
Infant mortality rate: *total:* 2.78 deaths/1,000 live births
country comparison to the world: 218
male: 2.98 deaths/1,000 live births
female: 2.58 deaths/1,000 live births (2011 est.)
Life expectancy at birth: *total population:* 82.25 years
country comparison to the world: 5
male: 78.96 years
female: 85.72 years (2011 est.)
Total fertility rate: 1.21 children born/woman (2011 est.)
country comparison to the world: 219
HIV/AIDS—adult prevalence rate: less than 0.1% (2009 est.)
country comparison to the world: 137
HIV/AIDS—people living with HIV/AIDS: 8,100 (2009 est.)
country comparison to the world: 109
HIV/AIDS—deaths: fewer than 100 (2009 est.)
country comparison to the world: 137
Drinking water source: *Improved:*
urban: 100% of population
rural: 100% of population
total: 100% of population (2008)
Sanitation facility access: *Improved:*
urban: 100% of population
rural: 100% of population
total: 100% of population (2008)
Nationality: *noun:* Japanese (singular and plural)
adjective: Japanese
Ethnic groups: Japanese 98.5%, Koreans 0.5%, Chinese 0.4%, other 0.6%
note: up to 230,000 Brazilians of Japanese origin migrated to Japan in the 1990s to work in industries; some have returned to Brazil (2004)
Religions: Shintoism 83.9%, Buddhism 71.4%, Christianity 2%, other 7.8%
note: total adherents exceeds 100% because many people belong to both Shintoism and Buddhism (2005)
Languages: Japanese
Literacy: *definition:* age 15 and over can read and write
total population: 99%
male: 99%
female: 99% (2002)
School life expectancy (primary to tertiary education): *total:* 15 years
male: 15 years
female: 15 years (2008)
Education expenditures: 3.5% of GDP (2007)
country comparison to the world: 118

GOVERNMENT

Country name: *conventional long form:* none
conventional short form: Japan
local long form: Nihon-koku/Nippon-koku
local short form: Nihon/Nippon
Government type: a parliamentary government with a constitutional monarchy
Capital: *name:* Tokyo
geographic coordinates: 35 41 N, 139 45 E
time difference: UTC+9 (14 hours ahead of Washington, DC during Standard Time)
Administrative divisions: 47 prefectures; Aichi, Akita, Aomori, Chiba, Ehime, Fukui, Fukuoka, Fukushima, Gifu, Gunma, Hiroshima, Hokkaido, Hyogo, Ibaraki, Ishikawa, Iwate, Kagawa, Kagoshima, Kanagawa, Kochi, Kumamoto, Kyoto, Mie, Miyagi, Miyazaki, Nagano, Nagasaki, Nara, Niigata, Oita, Okayama, Okinawa, Osaka, Saga, Saitama, Shiga, Shimane, Shizuoka, Tochigi, Tokushima, Tokyo, Tottori, Toyama, Wakayama, Yamagata, Yamaguchi, Yamanashi
Independence: 3 May 1947 (current constitution adopted as amendment to Meiji Constitution); notable earlier dates: 660 B.C. (traditional date of the founding of the nation by Emperor JIMMU); 29 November 1890 (Meiji Constitution provides for constitutional monarchy)
National holiday: Birthday of Emperor AKIHITO, 23 December (1933)
Constitution: 3 May 1947
Legal system: civil law system based on German model; system also reflects Anglo-American influence and Japanese traditions; judicial review of legislative acts in the Supreme Court
International law organization participation: accepts compulsory ICJ jurisdiction with reservations; accepts ICCt jurisdiction

Suffrage: 20 years of age; universal

Executive branch: *chief of state:* Emperor AKIHITO (since 7 January 1989)

head of government: Prime Minister Naoto KAN (since 8 June 2010)

cabinet: Cabinet is appointed by the prime minister

(For more information visit the World Leaders website)

elections: Diet designates the prime minister; constitution requires that the prime minister commands parliamentary majority; following legislative elections, the leader of majority party or leader of majority coalition in House of Representatives usually becomes prime minister; the monarchy is hereditary

Legislative branch: bicameral Diet or Kokkai consists of the House of Councillors or Sangi-in (242 seats—members elected for fixed six-year terms; half reelected every three years; 146 members in multi-seat constituencies and 96 by proportional representation) and the House of Representatives or Shugi-in (480 seats—members elected for maximum four-year terms; 300 in single-seat constituencies; 180 members by proportional representation in 11 regional blocs); the prime minister has the right to dissolve the House of Representatives at any time with the concurrence of the cabinet

elections: House of Councillors—last held on 11 July 2010 (next to be held in July 2013); House of Representatives—last held on 30 August 2009 (next to be held by August 2013)

election results: House of Councillors—percent of vote by party—DPJ 31.6%, LDP 24.1%, YP 13.6%, NK 13.1%, JCP 6.1%, SDP 3.8%, others 7.7%; seats by party—DPJ 106, LDP 84, NK 19, YP 11, JCP 6, SDP 4, others 12

House of Representatives—percent of vote by party (by proportional representation)—DPJ 42.4%, LDP 26.7%, NK 11.5%, JCP 7.0%, SDP 4.3%, others 8.1%; seats by party—DPJ 308, LDP 119, NK 21, JCP 9, SDP 7, others 16 (2009)

Judicial branch: Supreme Court (chief justice is appointed by the monarch after designation by the cabinet; all other justices are appointed by the cabinet)

Political parties and leaders: Democratic Party of Japan or DPJ [Naoto KAN]; Japan Communist Party or JCP [Kazuo SHII]; Liberal Democratic Party or LDP [Sadakazu TANIGAKI]; New Komeito or NK [Natsuo YAMAGUCHI]; People's New Party or PNP [Shizuka KAMEI]; Social Democratic Party or SDP [Mizuho FUKUSHIMA]; Your Party or YP [Yoshimi WATANABE]

Political pressure groups and leaders: *other:* business groups; trade unions

International organization participation: ADB, AfDB (nonregional member), APEC, ARF, ASEAN (dialogue partner), Australia Group, BIS, CERN (observer), CICA (observer), CP, EAS, EBRD, FAO, FATF, G-20, G-5, G-7, G-8, G-10, IADB, IAEA, IBRD, ICAO, ICC, ICRM, IDA, IEA, IFAD, IFC, IFRCS, IHO, ILO, IMF, IMO, IMSO, Interpol, IOC, IOM, IPU, ISO, ITSO, ITU, ITUC, LAIA (observer), MIGA, NEA, NSG, OAS (observer), OECD, OPCW, OSCE (partner), Paris Club, PCA, PIF (partner), SAARC (observer), SECI (observer), UN, UNCTAD, UNDOF, UNESCO, UNHCR, UNIDO, UNMIS, UNRWA, UNWTO, UPU, WCO, WFTU, WHO, WIPO, WMO, WTO, ZC

Diplomatic representation in the US: *chief of mission:* Ambassador Ichiro FUJISAKI

chancery: 2520 Massachusetts Avenue NW, Washington, DC 20008

telephone: [1] (202) 238-6700

FAX: [1] (202) 328-2187

consulate(s) general: Atlanta, Boston, Chicago, Denver, Detroit, Agana (Guam), Honolulu, Houston, Los Angeles, Miami, New York, Portland (Oregon), San Francisco, Seattle

consulate(s): Anchorage, Nashville

Diplomatic representation from the US: *chief of mission:* Ambassador John V. ROOS

embassy: 1-10-5 Akasaka, Minato-ku, Tokyo 107-8420

mailing address: Unit 9800, Box 300, APO AP 96303-0300

telephone: [81] (03) 3224-5000

FAX: [81] (03) 3505-1862

consulate(s) general: Naha (Okinawa), Osaka-Kobe, Sapporo

consulate(s): Fukuoka, Nagoya

Flag description: white with a large red disk (representing the sun without rays) in the center

National anthem: *name:* "Kimigayo" (The Emperor"s Reign)

lyrics/music: unknown/Hiromori HAYASHI

note: adopted 1999; in use as unofficial national anthem since 1883; oldest anthem lyrics in the world, dating to the 10th century or earlier; there is some opposition to the anthem because of its association with militarism and worship of the emperor

ECONOMY

Economy—overview: In the years following World War II, government-industry cooperation, a strong work ethic, mastery of high technology, and a comparatively small defense allocation (1% of GDP) helped Japan develop a technologically advanced economy. Two notable characteristics of the post-war economy were the close interlocking structures of manufacturers, suppliers, and distributors, known as keiretsu, and the guarantee of lifetime employment for a substantial portion of the urban labor force. Both features are now eroding under the dual pressures of global competition and domestic demographic change. Japan's industrial sector is heavily dependent on imported raw materials and fuels. A tiny agricultural sector is highly subsidized and protected, with crop yields among the highest in the world. Usually self sufficient in rice, Japan imports about 60% of its food on a caloric basis. Japan maintains one of the world's largest fishing fleets and accounts for nearly 15% of the global catch. For three decades, overall real economic growth had been spectacular—a 10% average in the 1960s, a 5% average in the 1970s, and a 4% average in the 1980s. Growth slowed markedly in the 1990s, averaging just 1.7%, largely because of the after effects of inefficient investment and an asset price bubble in the late 1980s that required a protracted period of time for firms to reduce excess debt, capital, and labor. Measured on a purchasing power parity (PPP) basis that adjusts for price differences, Japan in 2010 stood as the third-largest economy in the world after China, which surpassed Japan in 2001. The Japanese financial sector was not heavily exposed to sub-prime mortgages or their derivative instruments and weathered the initial effect of the recent global credit crunch, but a sharp downturn in business investment and global demand for Japan's exports in late 2008 pushed Japan further into recession. Government stimulus spending helped the economy recover in late 2009 and 2010. Prime Minister KAN's government has proposed opening the agricultural and services sectors to greater foreign competition and boosting exports through free-trade agreements, but debate continues on restructuring the economy and funding new stimulus programs in the face of a tight fiscal situation. Japan's huge government debt, which exceeds 200% of GDP, persistent deflation, reliance on exports to drive growth, and an aging and shrinking population are major long-term challenges for the economy. A 9.0-magnitude earthquake and an ensuing tsunami devastated the northeast coast of Honshu Island on 11 March 2011, washing away buildings and infrastructure as much as 6 miles inland, killing thousands, severely damaging several nuclear power plants, displacing and leaving homeless more than 320,000 people, and leaving a million households without running water. Radiation leaks at the Fukushima Daiichai nuclear power plant prompted mass evacuations and the declaration of a no-fly zone—initially for people and planes within 12.5 miles of the plant but later expanded to 19 miles. Radioactive iodine-131 has been found as far as 100 miles from the plant in samples of water, milk, fish, beef, and certain vegetables, at levels that make these foods unfit for consumption and create uncertainty regarding possible long-term contamination of the area. Energy-cutting efforts by electric companies and train lines slowed the pace of business throughout Honshu Island, and the stock market gyrated, dropping as much as 10% in a single day. In order to stabilize financial markets and retard appreciation of the yen, the Bank of Japan injected more than $325 billion in yen into the economy. Estimates of the direct costs of the damage—rebuilding homes and factories—range from $235 billion to $310 billion. Some economic forecasters, who previously had anticipated slower growth for Japan in 2011, now believe GDP may decline as much as 1% for the year.

GDP (purchasing power parity): $4.31 trillion (2010 est.)

country comparison to the world: 4

$4.146 trillion (2009 est.)

$4.424 trillion (2008 est.)

note: data are in 2010 US dollars
GDP (official exchange rate): $5.459 trillion (2010 est.)
GDP—real growth rate: 3.9% (2010 est.)
country comparison to the world: 101
-6.3% (2009 est.)
-1.2% (2008 est.)
GDP—per capita (PPP): $34,000 (2010 est.)
country comparison to the world: 38
$32,600 (2009 est.)
$34,800 (2008 est.)
note: data are in 2010 US dollars
GDP—composition by sector:
agriculture: 1.1%
industry: 23%
services: 75.9% (2010 est.)
Labor force: 65.7 million (2010 est.)
country comparison to the world: 9
Labor force—by occupation:
agriculture: 3.9%
industry: 26.2%
services: 69.8% (2010 est.)
Unemployment rate: 5.1% (2010 est.)
country comparison to the world: 48
5.1% (2009 est.)
Population below poverty line: 15.7% (2007)
note: Ministry of Health, Labor and Welfare (MHLW) press release, 20 October 2009
Household income or consumption by percentage share:
lowest 10%: 1.9%
highest 10%: 27.5% (2008)
Distribution of family income—Gini index: 37.6 (2008)
country comparison to the world: 74
24.9 (1993)
Investment (gross fixed): 20.3% of GDP (2010 est.)
country comparison to the world: 84
Budget:
revenues: $1.638 trillion
expenditures: $2.16 trillion (2010 est.)
Public debt: 225.8% of GDP (2010 est.)
country comparison to the world: 1
217.6% of GDP (2009 est.)
Inflation rate (consumer prices): -0.7% (2010 est.)
country comparison to the world: 7
-1.4% (2009 est.)
Central bank discount rate: 0.3% (31 December 2009)
country comparison to the world: 140
0.3% (31 December 2008)
Commercial bank prime lending rate: 1.6% (31 December 2010 est.)
country comparison to the world: 156
1.65% (31 December 2009 est.)
Stock of narrow money: $5.712 trillion (31 December 2010 est.)
country comparison to the world: 2
$5.206 trillion (31 December 2009 est.)
Stock of broad money: $16.46 trillion (31 December 2009)
country comparison to the world: 2
$15.43 trillion (31 December 2008)
Stock of domestic credit: $16.39 trillion (31 December 2008 est.)
country comparison to the world: 3
$13.32 trillion (31 December 2007 est.)
Market value of publicly traded shares: $3.537 trillion (31 December 2010)
country comparison to the world: 3
$3.29 trillion (31 December 2009)
$3.22 trillion (31 December 2008)
Agriculture—products: rice, sugar beets, vegetables, fruit; pork, poultry, dairy products, eggs; fish
Industries: among world's largest and technologically advanced producers of motor vehicles, electronic equipment, machine tools, steel and nonferrous metals, ships, chemicals, textiles, processed foods
Industrial production growth rate: 15.5% (2010 est.)
country comparison to the world: 8
Electricity—production: 956.5 billion kWh (2009 est.)
country comparison to the world: 4
Electricity—consumption: 858.5 billion kWh (2009 est.)
country comparison to the world: 4
Electricity—exports: 0 kWh (2009 est.)
Electricity—imports: 0 kWh (2009 est.)
Oil—production: 132,700 bbl/day (2009 est.)
country comparison to the world: 49
Oil—consumption: 4.363 million bbl/day (2009 est.)
country comparison to the world: 4
Oil—exports: 380,900 bbl/day (2008 est.)
country comparison to the world: 34
Oil—imports: 5.033 million bbl/day (2008 est.)
country comparison to the world: 3
Oil—proved reserves: 44.12 million bbl (1 January 2010 est.)
country comparison to the world: 80
Natural gas—production: 3.539 billion cu m (2009 est.)
country comparison to the world: 51
Natural gas—consumption: 94.67 billion cu m (2009 est.)
country comparison to the world: 6
Natural gas—exports: 0 cu m (2008 est.)
country comparison to the world: 117
Natural gas—imports: 90.29 billion cu m (2009 est.)
country comparison to the world: 3
Natural gas—proved reserves: 20.9 billion cu m (1 January 2010 est.)
country comparison to the world: 77
Current account balance: $166.5 billion (2010 est.)
country comparison to the world: 2
$141.8 billion (2009 est.)
Exports: $765.2 billion (2010 est.)
country comparison to the world: 5
$580.8 billion (2009 est.)
Exports—commodities: transport equipment, motor vehicles, semiconductors, electrical machinery, chemicals
Exports—partners: China 18.88%, US 16.42%, South Korea 8.13%, Taiwan 6.27%, Hong Kong 5.49% (2009)
Imports:
$636.8 billion (2010 est.)
country comparison to the world: 5
$501.6 billion (2009 est.)
Imports—commodities: machinery and equipment, fuels, foodstuffs, chemicals, textiles, raw materials
Imports—partners: China 22.2%, US 10.96%, Australia 6.29%, Saudi Arabia 5.29%, UAE 4.12%, South Korea 3.98%, Indonesia 3.95% (2009)
Reserves of foreign exchange and gold: $1.096 trillion (31 December 2010 est.)
country comparison to the world: 2
$1.049 trillion (31 December 2009 est.)
Debt—external: $2.441 trillion (30 September 2010)
country comparison to the world: 6
$2.053 trillion (31 December 2009)
Stock of direct foreign investment—at home: $199.4 billion (31 December 2010 est.)
country comparison to the world: 21
$199.9 billion (31 December 2009 est.)
Stock of direct foreign investment—abroad: $719.9 billion (31 December 2010 est.)
country comparison to the world: 8
$740.4 billion (31 December 2009 est.)
Exchange rates: yen (JPY) per US dollar–87.78 (2010), 93.57 (2009), 103.58 (2008), 117.99 (2007), 116.18 (2006)

COMMUNICATIONS

Telephones—main lines in use: 44.364 million (2009)
country comparison to the world: 5
Telephones—mobile cellular: 114.917 million (2009)
country comparison to the world: 7
Telephone system: *general assessment:* excellent domestic and international service
domestic: high level of modern technology and excellent service of every kind
international: country code–81; numerous submarine cables provide links throughout Asia, Australia, the Middle East, Europe, and US; satellite earth stations–7 Intelsat (Pacific and Indian Oceans), 1 Intersputnik (Indian Ocean region), 3 Inmarsat (Pacific and Indian Ocean regions), and 8 SkyPerfect JSAT (2008)
Broadcast media: a mixture of public and commercial broadcast TV and radio stations; 5 national terrestrial television networks including 1 public broadcaster; the large number of radio and TV stations available provide a wide range of choices; satellite and cable services provide access to international channels (2008)
Internet country code: .jp
Internet hosts: 54.846 million (2010)
country comparison to the world: 2
Internet users: 99.182 million (2009)
country comparison to the world: 3

TRANSPORTATION

Airports: 176 (2010)
country comparison to the world: 34
Airports—with paved runways:
total: 144
over 3,047 m: 7
2,438 to 3,047 m: 44
1,524 to 2,437 m: 38
914 to 1,523 m: 28
under 914 m: 27 (2010)
Airports—with unpaved runways: *total:* 32
914 to 1,523 m: 4
under 914 m: 28 (2010)
Heliports: 15 (2010)
Pipelines: gas 4,135 km; oil 171 km; oil/gas/water 53 km (2010)
Railways: *total:* 26,435 km
country comparison to the world: 11
standard gauge: 3,978 km 1.435-m gauge (3,978 km electrified)
narrow gauge: 96 km 1.372-m gauge (96 km electrified); 22,313 km 1.067-m gauge (15,235 km electrified); 48 km 0.762-m gauge (48 km electrified) (2009)

Roadways: *total:* 1,203,777 km
country comparison to the world: 5
paved: 961,366 km (includes 7,560 km of expressways)
unpaved: 242,411 km (2008)
Waterways: 1,770 km (seagoing vessels use inland seas) (2010)
country comparison to the world: 46
Merchant marine: *total:* 673
country comparison to the world: 16
by type: bulk carrier 152, cargo 31, carrier 3, chemical tanker 28, container 2, liquefied gas 63, passenger 12, passenger/cargo 120, petroleum tanker 152, refrigerated cargo 4, roll on/roll off 52, vehicle carrier 54
foreign-owned: 1 (Norway 1)
registered in other countries: 3,064 (Bahamas 93, Belize 1, Bermuda 2, Burma 1, Cambodia 2, Cayman Islands 19, China 2, Cyprus 19, Honduras 4, Hong Kong 84, Indonesia 7, Isle of Man 15, Liberia 102, Malaysia 4, Malta 5, Marshall Islands 41, Netherlands 1, Panama 2347, Philippines 82, Portugal 9, Saint Kitts and Nevis 3, Saint Vincent and the Grenadines 3, Sierra Leone 3, Singapore 146, South Korea 15, Thailand 2, UK 4, Vanuatu 44, unknown 4) (2010)
Ports and terminals: Chiba, Kawasaki, Kobe, Mizushima, Moji, Nagoya, Osaka, Tokyo, Tomakomai, Yokohama

MILITARY

Military branches: Japanese Ministry of Defense (MOD): Ground Self-Defense Force (Rikujou Jietai, GSDF), Maritime Self-Defense Force (Kaijou Jietai, MSDF), Air Self-Defense Force (Koukuu Jieitai, ASDF) (2011)
Military service age and obligation: 18 years of age for voluntary military service; Maritime Self-Defense Force mandatory retirement at age 54 (2011)
Manpower available for military service:
males age 16–49: 27,301,443
females age 16–49: 26,307,003 (2010 est.)
Manpower fit for military service: *males age 16–49:* 22,390,431
females age 16–49: 21,540,322 (2010 est.)
Manpower reaching militarily significant age annually: *male:* 623,365
female: 591,253 (2010 est.)
Military expenditures: 0.8% of GDP (2006)
country comparison to the world: 150

TRANSNATIONAL ISSUES

Disputes—international: the sovereignty dispute over the islands of Etorofu, Kunashiri, and Shikotan, and the Habomai group, known in Japan as the "Northern Territories" and in Russia as the "Southern Kuril Islands," occupied by the Soviet Union in 1945, now administered by Russia and claimed by Japan, remains the primary sticking point to signing a peace treaty formally ending World War II hostilities; Japan and South Korea claim Liancourt Rocks (Take-shima/Tok-do) occupied by South Korea since 1954; China and Taiwan dispute both Japan's claims to the uninhabited islands of the Senkaku-shoto (Diaoyu Tai) and Japan's unilaterally declared exclusive economic zone in the East China Sea, the site of intensive hydrocarbon prospecting

JERSEY

(BRITISH CROWN DEPENDENCY)

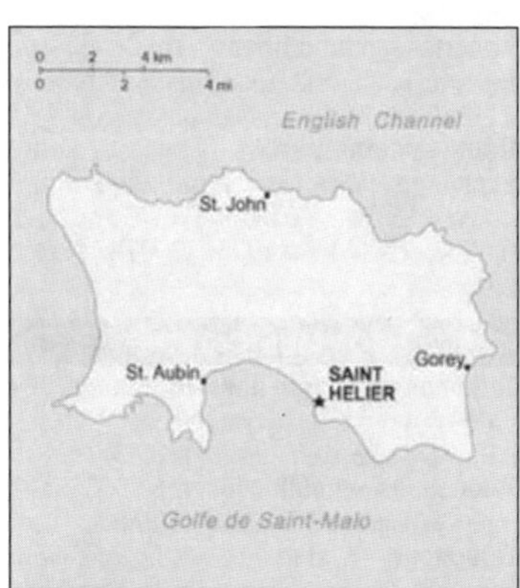

INTRODUCTION

Background: Jersey and the other Channel Islands represent the last remnants of the medieval Dukedom of Normandy that held sway in both France and England. These islands were the only British soil occupied by German troops in World War II. Jersey is a British crown dependency but is not part of the UK or of the European Union. However, the UK Government is constitutionally responsible for its defense and international representation.

GEOGRAPHY

Location: Western Europe, island in the English Channel, northwest of France
Geographic coordinates: 49 15 N, 2 10 W
Map references: Europe
Area: *total:* 116 sq km
country comparison to the world: 223
land: 116 sq km
water: 0 sq km
Area—comparative: about two-thirds the size of Washington, DC
Land boundaries: 0 km
Coastline: 70 km
Maritime claims: *territorial sea:* 3 nm
exclusive fishing zone: 12 nm
Climate: temperate; mild winters and cool summers
Terrain: gently rolling plain with low, rugged hills along north coast
Elevation extremes: *lowest point:* Atlantic Ocean 0 m
highest point: unnamed elevation 143 m
Natural resources: arable land
Land use:
arable land: 0%
permanent crops: 0%
other: 100% (2005)
Irrigated land: NA
Natural hazards: NA
Environment—current issues: NA
Geography—note: largest and southernmost of Channel Islands; about 30% of population concentrated in Saint Helier

PEOPLE

Population: 94,161 (July 2011 est.)
country comparison to the world: 194
Age structure: *0–14 years:* 16.5% (male 8,067/female 7,507)
15–64 years: 68.7% (male 32,187/female 32,485)
65 years and over: 14.8% (male 5,953/female 7,962) (2011 est.)
Median age: *total:* 40.1 years
male: 38.3 years
female: 41.3 years (2011 est.)
Population growth rate: 0.841% (2011 est.)
country comparison to the world: 130
Birth rate: 10.9 births/1,000 population (2011 est.)
country comparison to the world: 176
Death rate: 7.52 deaths/1,000 population (July 2011 est.)
country comparison to the world: 116
Net migration rate: 5.03 migrant(s)/1,000 population (2011 est.)
country comparison to the world: 17
Urbanization: *urban population:* 31% of total population (2010)
rate of urbanization: 0.8% annual rate of change (2010-15 est.)
Sex ratio: *at birth:* 1.06 male(s)/female
under 15 years: 1.08 male(s)/female
15–64 years: 1 male(s)/female
65 years and over: 0.8 male(s)/female
total population: 0.97 male(s)/female (2011 est.)
Infant mortality rate: *total:* 3.98 deaths/1,000 live births
country comparison to the world: 202
male: 4.2 deaths/1,000 live births
female: 3.75 deaths/1,000 live births (2011 est.)
Life expectancy at birth: *total population:* 81.38 years
country comparison to the world: 11
male: 78.96 years
female: 83.94 years (2011 est.)
Total fertility rate: 1.66 children born/woman (2011 est.)
country comparison to the world: 173
HIV/AIDS—adult prevalence rate: NA
HIV/AIDS—people living with HIV/AIDS: NA
HIV/AIDS—deaths: NA
Nationality: *noun:* Channel Islander(s)
adjective: Channel Islander
Ethnic groups: Jersey 51.1%, Britons 34.8%, Irish, French, and other white 6.6%, Portuguese/Madeiran 6.4%, other 1.1% (2001 census)

Religions: Anglican, Roman Catholic, Baptist, Congregational New Church, Methodist, Presbyterian
Languages: English 94.5% (official), Portuguese 4.6%, other 0.9% (2001 census)
Literacy: NA
School life expectancy (primary to tertiary education): NA
Education expenditures: NA

GOVERNMENT

Country name: *conventional long form:* Bailiwick of Jersey
conventional short form: Jersey
Dependency status: British crown dependency
Government type: parliamentary democracy
Capital: *name:* Saint Helier
geographic coordinates: 49 11 N, 2 06 W
time difference: UTC 0 (5 hours ahead of Washington, DC during Standard Time)
daylight saving time: +1hr, begins last Sunday in March; ends last Sunday in October
Administrative divisions: none (British crown dependency); there are no first-order administrative divisions as defined by the US Government, but there are 12 parishes: Grouville, Saint Brelade, Saint Clement, Saint Helier, Saint John, Saint Lawrence, Saint Martin, Saint Mary, Saint Ouen, Saint Peter, Saint Saviour, and Trinity
Independence: none (British crown dependency)
National holiday: Liberation Day, 9 May (1945)
Constitution: unwritten; partly statutes, partly common law and practice
Legal system: the laws of the UK, where applicable, apply; local statutes
Suffrage: 16 years of age; universal
Executive branch: *chief of state:* Queen ELIZABETH II (since 6 February 1952); represented by Lieutenant Governor Andrew RIDGEWAY (since 14 June 2006)
head of government: Chief Minister Terry LE SUEUR (12 December 2008); Bailiff Michael BIRT (since 9 July 2009)
cabinet: Cabinet (since December 2005)
(For more information visit the World Leaders website)
elections: ministers of the Cabinet including the chief minister are elected by the Assembly of States; the monarchy is hereditary; lieutenant governor and bailiff appointed by the monarch
Legislative branch: unicameral Assembly of the States of Jersey (58 seats; 55 are voting members, of which 12 are senators elected for six-year terms, 12 are constables or heads of parishes elected for three-year terms, 29 are deputies elected for three-year terms, the bailiff and the deputy bailiff, and 3 non-voting members include the Dean of Jersey, the Attorney General, and the Solicitor General appointed by the monarch)
elections: last held on 15 October 2008 for senators and 26 November 2008 for deputies (next to be held in 2011)
election results: percent of vote—NA; seats—independents 55
Judicial branch: Royal Court (judges elected by an electoral college and the bailiff)
Political parties and leaders: two declared parties: Centre Party; Jersey Democratic Alliance
note: all senators and deputies elected in 2008 were independents
Political pressure groups and leaders: Institute of Directors, Jersey branch (provides business support); Jersey Hospitality Association [Robert JONES] (trade association); Jersey Rights Association [David ROTHERHAM] (human rights); La Societe Jersiaise (education and conservation group); Progress Jersey [Daren O'TOOLE, Gino RISOLI] (human rights); Royal Jersey Agriculture and Horticultural Society or RJA&HS (development and management of the Jersey breed of cattle); Save Jersey's Heritage (protects heritage through building preservation)
Diplomatic representation in the US: none (British crown dependency)
Diplomatic representation from the US: none (British crown dependency)
Flag description: white with a diagonal red cross extending to the corners of the flag; in the upper quadrant, surmounted by a yellow crown, a red shield with three lions in yellow; according to tradition, the ships of Jersey—in an attempt to differentiate themselves from English ships flying the horizontal cross of St. George—rotated the cross to the "X" (saltire) configuration; because this arrangement still resembled the Irish cross of St. Patrick, the yellow Plantagenet crown and Jersey coat of arms were added
National anthem: *name:* "Isle de Siez Nous" (Island Home)
lyrics/music: Gerard LE FEUVRE
note: adopted 2008; serves as a local anthem; as a British crown dependency, "God Save the Queen" is official (see United Kingdom)

ECONOMY

Economy—overview: Jersey's economy is based on international financial services, agriculture, and tourism. In 2005 the finance sector accounted for about 50% of the island's output. Potatoes, cauliflower, tomatoes, and especially flowers are important export crops, shipped mostly to the UK. The Jersey breed of dairy cattle is known worldwide and represents an important export income earner. Milk products go to the UK and other EU countries. Tourism accounts for one-quarter of GDP. In recent years, the government has encouraged light industry to locate in Jersey with the result that an electronics industry has developed, displacing more traditional industries. All raw material and energy requirements are imported as well as a large share of Jersey's food needs. Light taxes and death duties make the island a popular tax haven. Living standards come close to those of the UK.
GDP (purchasing power parity): $5.1 billion (2005 est.)
country comparison to the world: 161
GDP (official exchange rate): $5.1 billion (2005 est.)
GDP—real growth rate: NA%
GDP—per capita (PPP): $57,000 (2005 est.)
country comparison to the world: 6
GDP—composition by sector: *agriculture:* 1%
industry: 2%
services: 97% (2005)
Labor force: 53,560 (June 2006)
country comparison to the world: 188
Unemployment rate: 2.2% (2006 est.)
country comparison to the world: 17
Population below poverty line: NA%
Household income or consumption by percentage share: *lowest 10%:* NA%
highest 10%: NA%
Budget: *revenues:* $829 million
expenditures: $851 million (2005)
Inflation rate (consumer prices): 3.7% (2006)
country comparison to the world: 105
Market value of publicly traded shares: $NA
Agriculture—products: potatoes, cauliflower, tomatoes; beef, dairy products
Industries: tourism, banking and finance, dairy, electronics
Industrial production growth rate: NA%
Electricity—consumption: 630.1 million kWh (2004 est.)
country comparison to the world: 153
Electricity—imports: NA kWh; note—electricity supplied by France
Exports: $NA
Exports—commodities: light industrial and electrical goods, dairy cattle, foodstuffs, textiles, flowers
Imports: $NA
Imports—commodities: machinery and transport equipment, manufactured goods, foodstuffs, mineral fuels, chemicals
Debt—external: $NA
Exchange rates: Jersey pounds per US dollar—0.6388 (2010), 0.6175 (2009), 0.5302 (2008), 0.4993 (2007), 0.5418 (2006)

COMMUNICATIONS

Telephones—main lines in use: 73,900 (2009)
country comparison to the world: 153
Telephones—mobile cellular: 83,900 (2004)
country comparison to the world: 188
Telephone system: *general assessment:* state-owned, partially-competitive market; increasingly modern, with some broadband access
domestic: digital telephone system launch announced in 2006 and currently being implemented; fixed-line and mobile-cellular services widely available; combined fixed and mobile-cellular density exceeds 100 per 100 persons
international: country code—44; submarine cable connectivity to Guernsey, the UK, and France (2008)
Broadcast media: multiple UK terrestrial television broadcasts—received via a transmitter in Jersey with relays in Jersey, Guernsey, and Alderney—will begin switching from analog to digital broadcasts in November 2010; satellite packages available; BBC Radio Jersey and 1 other radio station operating (2009)
Internet country code: .je
Internet hosts: 237 (2010)
country comparison to the world: 191
Internet users: 29,500 (2009)
country comparison to the world: 181

TRANSPORTATION

Airports: 1 (2010)
country comparison to the world: 221
Airports—with paved runways:
total: 1
1,524 to 2,437 m: 1 (2010)
Roadways:
total: 576 km (2010)
country comparison to the world: 190
Merchant marine: registered in other countries: 11 (Gibraltar 1, India 1, Marshall Islands 9) (2010)
country comparison to the world: 111
Ports and terminals: Gorey, Saint Aubin, Saint Helier

MILITARY

Manpower fit for military service: *males age 16–49:* 18,688
females age 16–49: 18,615 (2010 est.)
Manpower reaching militarily significant age annually: *male:* 664
female: 590 (2010 est.)
Military—note: defense is the responsibility of the UK

TRANSNATIONAL ISSUES

Disputes—international: none

JORDAN

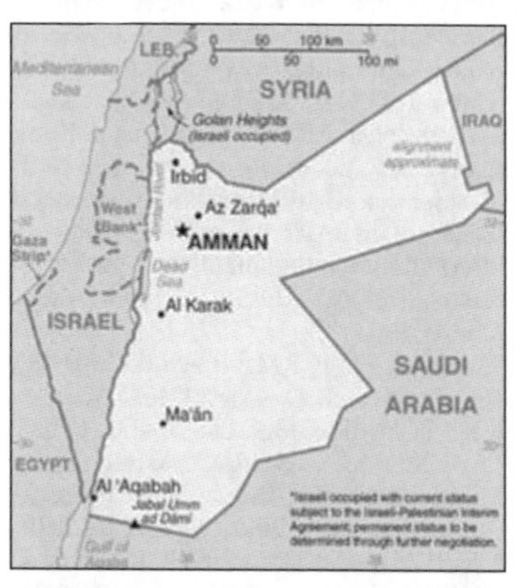

INTRODUCTION

Background: Following World War I and the dissolution of the Ottoman Empire, the UK received a mandate to govern much of the Middle East. Britain separated out a semi-autonomous region of Transjordan from Palestine in the early 1920s, and the area gained its independence in 1946; it adopted the name of Jordan in 1950. The country's long-time ruler was King HUSSEIN (1953-99). A pragmatic leader, he successfully navigated competing pressures from the major powers (US, USSR, and UK), various Arab states, Israel, and a large internal Palestinian population. Jordan lost the West Bank to Israel in the 1967 war and barely managed to defeat Palestinian rebels who attempted to overthrow the monarchy in 1970. King HUSSEIN in 1988 permanently relinquished Jordanian claims to the West Bank. In 1989, he reinstituted parliamentary elections and initiated a gradual political liberalization; political parties were legalized in 1992. In 1994, he signed a peace treaty with Israel. King ABDALLAH II, the son of King HUSSEIN, assumed the throne following his father's death in February 1999. Since then, he has consolidated his power and undertaken an aggressive economic reform program. Jordan acceded to the World Trade Organization in 2000, and began to participate in the European Free Trade Association in 2001. In 2003, Jordan staunchly supported the Coalition ouster of SADDAM in Iraq and, following the outbreak of insurgent violence in Iraq, absorbed thousands of displaced Iraqis. Municipal elections were held in July 2007 under a system in which 20% of seats in all municipal councils were reserved by quota for women. Parliamentary elections were held in November 2010 and saw independent pro-government candidates win the vast majority of seats. Beginning in January 2011 in the wake of unrest in Tunisia and Egypt, several thousand Jordanians staged weekly demonstrations and marches in Amman and other cities throughout Jordan to protest government corruption, rising prices, rampant poverty, and high unemployment. In response, King Abdallah replaced his prime minister and formed a National Dialogue Commission with a reform mandate. Some opposition groups also called for sweeping political and constitutional reforms, particularly on a controversial election law.

GEOGRAPHY

Location: Middle East, northwest of Saudi Arabia, between Israel (to the west) and Iraq
Geographic coordinates: 31 00 N, 36 00 E
Map references: Middle East
Area: *total:* 89,342 sq km
country comparison to the world: 111
land: 88,802 sq km
water: 540 sq km
Area—comparative: slightly smaller than Indiana
Land boundaries: *total:* 1,635 km
border countries: Iraq 181 km, Israel 238 km, Saudi Arabia 744 km, Syria 375 km, West Bank 97 km
Coastline: 26 km
Maritime claims: *territorial sea:* 3 nm
Climate: mostly arid desert; rainy season in west (November to April)
Terrain: mostly desert plateau in east, highland area in west; Great Rift Valley separates East and West Banks of the Jordan River
Elevation extremes: *lowest point:* Dead Sea -408 m
highest point: Jabal Umm ad Dami 1,854 m
Natural resources: phosphates, potash, shale oil
Land use:
arable land: 3.32%
permanent crops: 1.18%
other: 95.5% (2005)
Irrigated land: 750 sq km (2003)
Total renewable water resources: 0.9 cu km (1997)
Freshwater withdrawal (domestic/industrial/agricultural): *total:* 1.01 cu km/yr (21%/4%/75%)
per capita: 177 cu m/yr (2000)
Natural hazards: droughts; periodic earthquakes
Environment—current issues: limited natural freshwater resources; deforestation; overgrazing; soil erosion; desertification
Environment—international agreements: *party to:* Biodiversity, Climate Change, Climate Change-Kyoto Protocol, Desertification, Endangered Species, Hazardous Wastes, Law of the Sea, Marine Dumping, Ozone Layer Protection, Wetlands
signed, but not ratified: none of the selected agreements
Geography—note: strategic location at the head of the Gulf of Aqaba and as the Arab country that shares the longest border with Israel and the occupied West Bank

PEOPLE

Population: 6,508,271 (July 2011 est.)
country comparison to the world: 102
Age structure: *0–14 years:* 35.3% (male 1,180,595/female 1,114,533)
15–64 years: 59.9% (male 1,977,075/female 1,921,504)
65 years and over: 4.8% (male 153,918/female 160,646) (2011 est.)
Median age: *total:* 22.1 years
male: 21.8 years
female: 22.4 years (2011 est.)
Population growth rate: 0.984% (2011 est.)
country comparison to the world: 117
Birth rate: 26.79 births/1,000 population (2011 est.)
country comparison to the world: 50
Death rate: 2.69 deaths/1,000 population (July 2011 est.)
country comparison to the world: 220
Net migration rate: -14.26 migrant(s)/1,000 population (2011 est.)
country comparison to the world: 215
Urbanization: *urban population:* 79% of total population (2010)
rate of urbanization: 1.6% annual rate of change (2010-15 est.)
Major cities—population: AMMAN (capital) 1.088 million (2009)
Sex ratio: *at birth:* 1.06 male(s)/female
under 15 years: 1.06 male(s)/female
15–64 years: 1.03 male(s)/female
65 years and over: 0.97 male(s)/female
total population: 1.04 male(s)/female (2011 est.)
Infant mortality rate: *total:* 16.42 deaths/1,000 live births
country comparison to the world: 107
male: 16.98 deaths/1,000 live births
female: 15.83 deaths/1,000 live births (2011 est.)

Life expectancy at birth: *total population:* 80.05 years
country comparison to the world: 29
male: 78.73 years
female: 81.45 years (2011 est.)
Total fertility rate: 3.39 children born/woman (2011 est.)
country comparison to the world: 49
HIV/AIDS—adult prevalence rate: less than 0.1% (2001 est.)
country comparison to the world: 138
HIV/AIDS—people living with HIV/AIDS: 600 (2007 est.)
country comparison to the world: 150
HIV/AIDS—deaths: fewer than 500 (2003 est.)
country comparison to the world: 90
Drinking water source: *Improved:*
urban: 98% of population
rural: 91% of population
total: 96% of population
Unimproved: urban: 2% of population
rural: 9% of population
total: 4% of population (2008)
Sanitation facility access:
Improved:
urban: 98% of population
rural: 97% of population
total: 98% of population
Unimproved:
urban: 2% of population
rural: 3% of population
total: 2% of population (2008)
Nationality: *noun:* Jordanian(s)
adjective: Jordanian
Ethnic groups: Arab 98%, Circassian 1%, Armenian 1%
Religions: Sunni Muslim 92%, Christian 6% (majority Greek Orthodox, but some Greek and Roman Catholics, Syrian Orthodox, Coptic Orthodox, Armenian Orthodox, and Protestant denominations), other 2% (several small Shia Muslim and Druze populations) (2001 est.)
Languages: Arabic (official), English (widely understood among upper and middle classes)
Literacy: *definition:* age 15 and over can read and write
total population: 89.9%
male: 95.1%
female: 84.7% (2003 est.)
School life expectancy (primary to tertiary education): *total:* 13 years
male: 13 years
female: 13 years (2008)
Education expenditures: NA

GOVERNMENT

Country name: *conventional long form:* Hashemite Kingdom of Jordan
conventional short form: Jordan
local long form: Al Mamlakah al Urduniyah al Hashimiyah
local short form: Al Urdun
former: Transjordan
Government type: constitutional monarchy
Capital: *name:* Amman
geographic coordinates: 31 57 N, 35 56 E
time difference: UTC+2 (7 hours ahead of Washington, DC during Standard Time)
daylight saving time: +1hr, begins first Friday in April; ends last Friday in October
Administrative divisions: 12 governorates (muhafazat, singular—muhafazah); Ajlun, Al 'Aqabah, Al Balqa', Al Karak, Al Mafraq, 'Amman, At Tafilah, Az Zarqa', Irbid, Jarash, Ma'an, Madaba
Independence: 25 May 1946 (from League of Nations mandate under British administration)
National holiday: Independence Day, 25 May (1946)
Constitution: 1 January 1952; amended many times
Legal system: mixed legal system of civil law and Islamic religious law; judicial review of legislative acts in a specially provided High Tribunal
International law organization participation: has not submitted an ICJ jurisdiction declaration; accepts ICCt jurisdiction
Suffrage: 18 years of age; universal
Executive branch: *chief of state:* King ABDALLAH II (since 7 February 1999); Crown Prince HUSSEIN (born 28 June 1994), eldest son of King ABDALLAH II
head of government: Prime Minister Marouf al-BAKHIT (since 1 February 2011); Deputy Prime Minister Saad Hayel SROUR
cabinet: Cabinet appointed by the prime minister in consultation with the monarch; note—on 1 February 2011 the King dismissed the cabinet and designated Marouf al-BAKHIT the new prime minister
(For more information visit the World Leaders website)
elections: the monarchy is hereditary; prime minister appointed by the monarch
Legislative branch: bicameral National Assembly or Majlis al-'Umma consists of the Senate, also called the House of Notables or Majlis al-Ayan (60 seats; members appointed by the monarch to serve four-year terms) and the Chamber of Deputies, also called the House of Representatives or Majlis al-Nuwaab (120 seats; members elected using a single, non-transferable vote system in multi-member districts to serve four-year terms); note—the new electoral law enacted in May 2010 allocated an additional 10 seats (6 seats added to the number reserved for women, bringing the total to 12; 2 additional seats for Amman; and 1 seat each for the cities of Zarqa and Irbid; unchanged are 9 seats reserved for Christian candidates, 9 for Bedouin candidates, and 3 for Jordanians of Chechen or Circassian descent
elections: Chamber of Deputies—last held on 9 November 2010 (next scheduled in 2014); note—the King dissolved the previous Chamber of Deputies in November 2009, midway through the parliamentary term
election results: Chamber of Deputies—percent of vote by party—NA; seats by party—independents and other 120 (includes 12 seats filled by women's quota and 1 woman was directly elected); note—the IAF boycotted the election
Judicial branch: Court of Cassation (Supreme Court)
Political parties and leaders: Arab Ba'ath Socialist Party [Fuad DABBOUR]; Ba'ath Arab Progressive Party [Tayseer al-HAMSI]; Call Party [Mohammed Abu BAKR]; Democratic People's Party [Ablah al-ULBAH]; Democratic Popular Unity Party [Sa'ed DIAB]; Islamic Action Front or IAF [Hamzeh MANSOUR]; Islamic Center Party [Marwan al-FA'OURI; Jordanian Communist Party [Munir HAMARNEH]; Jordanian National Party [Mona Abu BAKR]; Jordanian United Front [Amjad al-MAJALI]; Life Party [Thaher 'AMROU]; Message Party [Hazem QASHOU]; National Constitution Party [Ahmed al-SHUNAQ]; National Current Party [Abd al-Hadi al-MAJALI]; National Movement for Direct Democracy [Mohammed al-QAQ]
Political pressure groups and leaders: Anti-Normalization Committee [Hamzeh MANSOUR, chairman]; Higher Coordination Committee of Opposition Parties [Hamzeh MANZOUR]; Jordan Bar Association [Saleh al-ARMUTI, chairman]; Jordanian Press Association [Sayf al-SHARIF, president]; Jordanian Muslim Brotherhood [Dr. Hamam SAID, controller general]
International organization participation: ABEDA, AFESD, AMF, CAEU, CICA, FAO, G-11, G-77, IAEA, IBRD, ICAO, ICC, ICRM, IDA, IDB, IFAD, IFC, IFRCS, ILO, IMF, IMO, Interpol, IOC, IOM, IPU, ISO, ITSO, ITU, ITUC, LAS, MIGA, MINURSO, MINUSTAH, MONUSCO, NAM, OIC, OPCW, OSCE (partner), PCA, UN, UNAMID, UNCTAD, UNESCO, UNHCR, UNIDO, UNMIL, UNMIS, UNOCI, UNRWA, UNWTO, UPU, WCO, WFTU, WHO, WIPO, WMO, WTO
Diplomatic representation in the US: *chief of mission:* Ambassador Alia Hatough BOURAN
chancery: 3504 International Drive NW, Washington, DC 20008
telephone: [1] (202) 966-2664
FAX: [1] (202) 966-3110
Diplomatic representation from the US: *chief of mission:* Ambassador Robert S. BEECROFT
embassy: Abdoun, Amman
mailing address: P. O. Box 354, Amman 11118 Jordan; Unit 70200, Box 5, DPO AE 09892-0200
telephone: [962] (6) 590-6000
FAX: [962] (6) 592-0121
Flag description: three equal horizontal bands of black (top), representing the Abbassid Caliphate, white, representing the Ummayyad Caliphate, and green, representing the Fatimid Caliphate; a red isosceles triangle on the hoist side, representing the Great Arab Revolt of 1916, and bearing a small white seven-pointed star symbolizing the seven verses of the opening Sura (Al-Fatiha) of the Holy Koran; the seven points on the star represent faith in One God, humanity, national spirit, humility, social justice, virtue, and aspirations; design is based on the Arab Revolt flag of World War I
National anthem: *name:* "As-salam al-malaki al-urdoni" (Long Live the King of Jordan)
lyrics/music: Abdul-Mone'm al-RIFAI'/Abdul-Qader al-TANEER
note: adopted 1946; the shortened version of the anthem is used most commonly,

while the full version is reserved for special occasions

ECONOMY

Economy—overview: Jordan's economy is among the smallest in the Middle East, with insufficient supplies of water, oil, and other natural resources, underlying the government's heavy reliance on foreign assistance. Other economic challenges for the government include chronic high rates of poverty, unemployment, inflation, and a large budget deficit. Since assuming the throne in 1999, King ABDALLAH has implemented significant economic reforms, such as opening the trade regime, privatizing state-owned companies, and eliminating most fuel subsidies, which in the past few years have spurred economic growth by attracting foreign investment and creating some jobs. The global economic slowdown, however, has depressed Jordan's GDP growth. Export-oriented sectors such as manufacturing, mining, and the transport of re-exports have been hit the hardest. The Government approved two supplementary budgets in 2010, but sweeping tax cuts planned for 2010 did not materialize because of Amman's need for additional revenue to cover excess spending. The budget deficit is likely to remain high, at 5-6% of GDP, and Amman likely will continue to depend heavily on foreign assistance to finance the deficit in 2011. Jordan's financial sector has been relatively isolated from the international financial crisis because of its limited exposure to overseas capital markets. Jordan is currently exploring nuclear power generation to forestall energy shortfalls.
GDP (purchasing power parity): $34.53 billion (2010 est.)
country comparison to the world: 103
$33.49 billion (2009 est.)
$32.73 billion (2008 est.)
note: data are in 2010 US dollars
GDP (official exchange rate): $27.53 billion (2010 est.)
GDP—real growth rate: 3.1% (2010 est.)
country comparison to the world: 124
2.3% (2009 est.)
7.6% (2008 est.)
GDP—per capita (PPP): $5,400 (2010 est.)
country comparison to the world: 144
$5,300 (2009 est.)
$5,300 (2008 est.)
note: data are in 2010 US dollars
GDP—composition by sector: *agriculture:* 3.4%
industry: 30.3%
services: 66.2% (2010 est.)
Labor force: 1.719 million (2010 est.)
country comparison to the world: 126
Labor force—by occupation:
agriculture: 2.7%
industry: 20%
services: 77.4% (2007 est.)
Unemployment rate: 13.4% (2010 est.)
country comparison to the world: 137
12.9% (2009 est.)
note: official rate; unofficial rate is approximately 30%
Population below poverty line: 14.2% (2002)
Household income or consumption by percentage share: *lowest 10%:* 3%
highest 10%: 30.7% (2006)
Distribution of family income—Gini index: 39.7 (2007)
country comparison to the world: 62
36.4 (1997)
Investment (gross fixed): 30.1% of GDP (2010 est.)
country comparison to the world: 19
Budget: *revenues:* $6.269 billion
expenditures: $8.701 billion (2010 est.)
Public debt: 61.4% of GDP (2010 est.)
country comparison to the world: 29
64.7% of GDP (2009 est.)
Inflation rate (consumer prices): 4.4% (2010 est.)
country comparison to the world: 125
-0.7% (2009 est.)
Central bank discount rate: 4.75% (31 December 2009)
country comparison to the world: 67
6.25% (31 December 2008)
Commercial bank prime lending rate: 9.25% (31 December 2009 est.)
country comparison to the world: 100
9.03% (31 December 2008 est.)
Stock of narrow money: $9.386 billion (31 December 2010 est.)
country comparison to the world: 72
$8.437 billion (31 December 2009 est.)
Stock of broad money: $35.53 billion (31 December 2010 est.)
country comparison to the world: 72
$33.38 billion (31 December 2009 est.)
Stock of domestic credit: $26.85 billion (31 December 2010 est.)
country comparison to the world: 72
$25.14 billion (31 December 2009 est.)
Market value of publicly traded shares: $31.86 billion (31 December 2009)
country comparison to the world: 53
$35.85 billion (31 December 2008)
$41.22 billion (31 December 2007)
Agriculture—products: citrus, tomatoes, cucumbers, olives, strawberries, stone fruits; sheep, poultry, dairy
Industries: clothing, fertilizers, potash, phosphate mining, pharmaceuticals, petroleum refining, cement, inorganic chemicals, light manufacturing, tourism
Industrial production growth rate: 2.7% (2010 est.)
country comparison to the world: 118
Electricity—production: 12.21 billion kWh (2007 est.)
country comparison to the world: 84
Electricity—consumption: 10.4 billion kWh (2007 est.)
country comparison to the world: 85
Electricity—exports: 176 million kWh (2007 est.)
Electricity—imports: 200 million kWh (2007 est.)
Oil—production: 0 bbl/day (2008 est.)
country comparison to the world: 188
Oil—consumption: 108,000 bbl/day (2009 est.)
country comparison to the world: 74
Oil—exports: 0 bbl/day (2007 est.)
country comparison to the world: 176
Oil—imports: 108,200 bbl/day (2007 est.)
country comparison to the world: 61
Oil—proved reserves: 1 million bbl (1 January 2010 est.)
country comparison to the world: 98
Natural gas—production: 250 million cu m (2008 est.)
country comparison to the world: 73
Natural gas—consumption: 2.97 billion cu m (2008 est.)
country comparison to the world: 74
Natural gas—exports: 0 cu m (2008 est.)
country comparison to the world: 119
Natural gas—imports: 2.72 billion cu m (2008 est.)
country comparison to the world: 42
Natural gas—proved reserves: 6.031 billion cu m (1 January 2010 est.)
country comparison to the world: 87
Current account balance: $-975 million (2010 est.)
country comparison to the world: 132
$-1.27 billion (2009 est.)
Exports: $7.333 billion (2010 est.)
country comparison to the world: 97
$6.366 billion (2009 est.)
Exports—commodities: clothing, fertilizers, potash, phosphates, vegetables, pharmaceuticals
Exports—partners: US 17.13%, Iraq 17%, India 13.59%, Saudi Arabia 10.56%, Syria 4.18%, UAE 4.09% (2009)
Imports: $12.97 billion (2010 est.)
country comparison to the world: 82
$12.5 billion (2009 est.)
Imports—commodities: crude oil, machinery, transport equipment, iron, cereals
Imports—partners: Saudi Arabia 17.3%, China 10.95%, US 6.94%, Germany 6.29%, Egypt 6.1% (2009)
Reserves of foreign exchange and gold: $12.64 billion (31 December 2010 est.)
country comparison to the world: 54
$12.14 billion (31 December 2009 est.)
Debt—external: $5.522 billion (31 December 2010 est.)
country comparison to the world: 102
$6.766 billion (31 December 2009 est.)
Stock of direct foreign investment—at home: $22.19 billion (31 December 2010 est.)
country comparison to the world: 66
$19.76 billion (31 December 2009 est.)
Stock of direct foreign investment—abroad: $NA
Exchange rates: Jordanian dinars (JOD) per US dollar–0.709 (2010), 0.709 (2009), 0.709 (2008), 0.709 (2007), 0.709 (2006)

COMMUNICATIONS

Telephones—main lines in use: 501,200 (2009)
country comparison to the world: 97
Telephones—mobile cellular: 6.014 million (2009)
country comparison to the world: 85
Telephone system: *general assessment:* service has improved recently with increased use of digital switching equipment; microwave radio relay transmission and coaxial and fiber-optic cable are employed on trunk lines; growing mobile-cellular usage in both urban and rural areas is reducing use of fixed-line services; Internet penetration remains modest and slow-growing
domestic: 1995 telecommunications law opened all non-fixed-line services to

private competition; in 2005, monopoly over fixed-line services terminated and the entire telecommunications sector was opened to competition; currently multiple mobile-cellular providers with subscribership rapidly approaching 100 per 100 persons
international: country code–962; landing point for the Fiber-Optic Link Around the Globe (FLAG) FEA and FLAG Falcon submarine cable networks; satellite earth stations–33 (3 Intelsat, 1 Arabsat, and 29 land and maritime Inmarsat terminals); fiber-optic cable to Saudi Arabia and microwave radio relay link with Egypt and Syria; participant in Medarabtel (2010)
Broadcast media: radio and TV dominated by the government-owned Jordan Radio and Television Corporation (JRTV) that operates a main network, a sports network, a film network, and a satellite channel; first independent TV broadcaster aired in 2007; international satellite TV and Israeli and Syrian TV broadcasts are available; roughly 30 radio stations operational with JRTV operating the main government-owned station; transmissions of multiple international radio broadcasters are available (2007)
Internet country code: .jo
Internet hosts: 42,412 (2010)
country comparison to the world: 92
Internet users: 1.642 million (2009)
country comparison to the world: 78

TRANSPORTATION

Airports: 18 (2010)
country comparison to the world: 138
Airports—with paved runways: *total:* 16
over 3,047 m: 8
2,438 to 3,047 m: 5
1,524 to 2,437 m: 1
914 to 1,523 m: 1
under 914 m: 1 (2010)
Airports—with unpaved runways: *total:* 2
under 914 m: 2 (2010)
Heliports: 1 (2010)
Pipelines: gas 439 km; oil 49 km (2010)
Railways: *total:* 507 km
country comparison to the world: 111
narrow gauge: 507 km 1.050-m gauge (2010)
Roadways: *total:* 7,891 km
country comparison to the world: 143
paved: 7,891 km (2009)
Merchant marine: *total:* 13
country comparison to the world: 107
by type: cargo 5, passenger/cargo 6, petroleum tanker 1, roll on/roll off 1
foreign-owned: 7 (UAE 7)
registered in other countries: 20 (Bahamas 2, Egypt 2, Panama 13, Syria 2, unknown 1) (2010)
Ports and terminals: Al 'Aqabah

MILITARY

Military branches: Jordanian Armed Forces (JAF): Royal Jordanian Land Force (RJLF), Royal Jordanian Navy, Royal Jordanian Air Force (Al-Quwwat al-Jawwiya al-Malakiya al-Urduniya, RJAF), Special Operations Command (Socom); Public Security Directorate (normally falls under Ministry of Interior, but comes under JAF in wartime or crisis) (2011)
Military service age and obligation: 17 years of age for voluntary military service; conscription for males at age 18 was suspended in 1999, but reinstated in July 2007 in order to provide youth training necessary for job market needs; all males under age 37 are required to register; women not subject to conscription, but can volunteer to serve in non-combat military positions in the Royal Jordanian Arab Army Women's Corps (2010)
Manpower available for military service: *males age 16–49:* 1,674,260
females age 16–49: 1,611,315 (2010 est.)
Manpower fit for military service: *males age 16–49:* 1,439,192
females age 16–49: 1,384,500 (2010 est.)
Manpower reaching militarily significant age annually: *male:* 73,574
female: 69,420 (2010 est.)
Military expenditures: 8.6% of GDP (2006)
country comparison to the world: 4

TRANSNATIONAL ISSUES

Disputes—international: approximately two million Iraqis have fled the conflict in Iraq, with the majority taking refuge in Syria and Jordan; 2004 Agreement settles border dispute with Syria pending demarcation
Refugees and internally displaced persons: refugees (country of origin): 1,835,704 (Palestinian Refugees (UNRWA)); 500,000 (Iraq)
IDPs: 160,000 (1967 Arab-Israeli War) (2007)

KAZAKHSTAN

INTRODUCTION

Background: Ethnic Kazakhs, a mix of Turkic and Mongol nomadic tribes who migrated into the region in the 13th century, were rarely united as a single nation. The area was conquered by Russia in the 18th century, and Kazakhstan became a Soviet Republic in 1936. During the 1950s and 1960s agricultural "Virgin Lands" program, Soviet citizens were encouraged to help cultivate Kazakhstan's northern pastures. This influx of immigrants (mostly Russians, but also some other deported nationalities) skewed the ethnic mixture and enabled non-ethnic Kazakhs to outnumber natives. Independence in 1991 drove many of these newcomers to emigrate. Kazakhstan's economy is larger than those of all the other Central Asian states largely due to the country's vast natural resources. Current issues include: developing a cohesive national identity; expanding the development of the country's vast energy resources and exporting them to world markets; diversifying the economy outside the oil, gas, and mining sectors; enhancing Kazakhstan's economic competitiveness; developing a multiparty parliament and advancing political and social reform; and strengthening relations with neighboring states and other foreign powers.

GEOGRAPHY

Location: Central Asia, northwest of China; a small portion west of the Ural (Zhayyq) River in eastern-most Europe
Geographic coordinates: 48 00 N, 68 00 E
Map references: Asia
Area: *total:* 2,724,900 sq km
country comparison to the world: 9
land: 2,699,700 sq km
water: 25,200 sq km
Area—comparative: slightly less than four times the size of Texas
Land boundaries: *total:* 12,185 km
border countries: China 1,533 km, Kyrgyzstan 1,224 km, Russia 6,846 km, Turkmenistan 379 km, Uzbekistan 2,203 km
Coastline: 0 km (landlocked); note–Kazakhstan borders the Aral Sea, now split into two bodies of water (1,070 km), and the Caspian Sea (1,894 km)
Maritime claims: none (landlocked)
Climate: continental, cold winters and hot summers, arid and semiarid
Terrain: vast flat steppe extending from the Volga in the west to the Altai Mountains in the east and from the plains of western Siberia in the north to oases and deserts of Central Asia in the south
Elevation extremes: *lowest point:* Vpadina Kaundy -132 m
highest point: Khan Tangiri Shyngy (Pik Khan-Tengri) 6,995 m
Natural resources: major deposits of petroleum, natural gas, coal, iron ore, manganese, chrome ore, nickel, cobalt, copper, molybdenum, lead, zinc, bauxite, gold, uranium
Land use: *arable land:* 8.28%
permanent crops: 0.05%
other: 91.67% (2005)
Irrigated land: 35,560 sq km (2003)
Total renewable water resources: 109.6 cu km (1997)
Freshwater withdrawal (domestic/industrial/agricultural): *total:* 35 cu km/yr (2%/17%/82%)
per capita: 2,360 cu m/yr (2000)
Natural hazards: earthquakes in the south; mudslides around Almaty
Environment—current issues: radioactive or toxic chemical sites associated with former defense industries and test ranges scattered throughout the country pose health risks for humans and animals; industrial pollution is severe in some cities; because the two main rivers that flowed into the Aral Sea have been diverted for irrigation, it is drying up and leaving behind a harmful layer of chemical pesticides and natural salts; these substances are then picked up by the wind and blown into noxious dust storms; pollution in the Caspian Sea; soil pollution from overuse of agricultural chemicals and salination from poor infrastructure and wasteful irrigation practices
Environment—international agreements: *party to:* Air Pollution, Biodiversity, Climate Change, Desertification, Endangered Species, Environmental Modification, Hazardous Wastes, Ozone Layer Protection, Ship Pollution, Wetlands
signed, but not ratified: Climate Change-Kyoto Protocol
Geography—note: landlocked; Russia leases approximately 6,000 sq km of territory enclosing the Baykonur Cosmodrome; in January 2004, Kazakhstan and Russia extended the lease to 2050

PEOPLE

Population: 15,522,373 (July 2011 est.)
country comparison to the world: 64
Age structure: *0–14 years:* 21.6% (male 1,709,929/female 1,637,132)
15–64 years: 71% (male 5,373,755/female 5,654,461)
65 years and over: 7.4% (male 392,689/female 754,407) (2011 est.)
Median age:
total: 30.2 years
male: 28.7 years
female: 31.9 years (2011 est.)
Population growth rate: 0.4% (2011 est.)
country comparison to the world: 158
Birth rate: 16.65 births/1,000 population (2011 est.)
country comparison to the world: 122
Death rate: 9.38 deaths/1,000 population (July 2011 est.)
country comparison to the world: 63
Net migration rate: -3.27 migrant(s)/1,000 population (2011 est.)
country comparison to the world: 178
Urbanization: *urban population:* 59% of total population (2010)
rate of urbanization: 1.3% annual rate of change (2010-15 est.)
Major cities—population: Almaty 1.383 million; ASTANA (capital) 650,000 (2009)
Sex ratio: *at birth:* 1.058 male(s)/female
under 15 years: 1.04 male(s)/female
15–64 years: 0.95 male(s)/female
65 years and over: 0.53 male(s)/female
total population: 0.93 male(s)/female (2011 est.)
Infant mortality rate: *total:* 24.15 deaths/1,000 live births
country comparison to the world: 83
male: 28.44 deaths/1,000 live births
female: 19.62 deaths/1,000 live births (2011 est.)
Life expectancy at birth: *total population:* 68.51 years
country comparison to the world: 152
male: 63.24 years
female: 74.08 years (2011 est.)
Total fertility rate: 1.87 children born/woman (2011 est.)
country comparison to the world: 147
HIV/AIDS–adult prevalence rate: 0.1% (2009 est.)
country comparison to the world: 141
HIV/AIDS–people living with HIV/AIDS: 13,000 (2009 est.)
country comparison to the world: 90
HIV/AIDS–deaths: fewer than 500 (2009 est.)
country comparison to the world: 87
Drinking water source: *Improved:*
urban: 99% of population
rural: 90% of population
total: 95% of population
Unimproved: urban: 1% of population
rural: 10% of population

total: 5% of population (2008)
Sanitation facility access: *Improved: urban:* 97% of population
rural: 98% of population
total: 97% of population
Unimproved: urban: 3% of population
rural: 2% of population
total: 3% of population (2008)
Nationality: *noun:* Kazakhstani(s)
adjective: Kazakhstani
Ethnic groups: Kazakh (Qazaq) 63.1%, Russian 23.7%, Uzbek 2.8%, Ukrainian 2.1%, Uighur 1.4%, Tatar 1.3%, German 1.1%, other 4.5% (2009 census)
Religions: Muslim 47%, Russian Orthodox 44%, Protestant 2%, other 7%
Languages: Kazakh (Qazaq, state language) 64.4%, Russian (official, used in everyday business, designated the "language of interethnic communication") 95% (2001 est.)
Literacy: *definition:* age 15 and over can read and write
total population: 99.5%
male: 99.8%
female: 99.3% (1999 est.)
School life expectancy (primary to tertiary education): *total:* 15 years
male: 15 years
female: 16 years (2010)
Education expenditures: 2.8% of GDP (2007)
country comparison to the world: 138

GOVERNMENT

Country name: *conventional long form:* Republic of Kazakhstan
conventional short form: Kazakhstan
local long form: Qazaqstan Respublikasy
local short form: Qazaqstan
former: Kazakh Soviet Socialist Republic
Government type: republic; authoritarian presidential rule, with little power outside the executive branch
Capital:
name: Astana
geographic coordinates: 51 10 N, 71 25 E
time difference: UTC+6 (11 hours ahead of Washington, DC during Standard Time)
note: Kazakhstan is divided into two time zones
Administrative divisions: 14 provinces (oblystar, singular–oblys) and 3 cities* (qalalar, singular–qala); Almaty Oblysy, Almaty Qalasy*, Aqmola Oblysy (Astana), Aqtobe Oblysy, Astana Qalasy*, Atyrau Oblysy, Batys Qazaqstan Oblysy [West Kazakhstan] (Oral), Bayqongyr Qalasy [Baykonur]*, Mangghystau Oblysy (Aqtau), Ongtustik Qazaqstan Oblysy [South Kazakhstan] (Shymkent), Pavlodar Oblysy, Qaraghandy Oblysy, Qostanay Oblysy, Qyzylorda Oblysy, Shyghys Qazaqstan Oblysy [East Kazakhstan] (Oskemen), Soltustik Qazaqstan Oblysy (Petropavlovsk), Zhambyl Oblysy (Taraz)
note: administrative divisions have the same names as their administrative centers (exceptions have the administrative center name following in parentheses); in 1995, the Governments of Kazakhstan and Russia entered into an agreement whereby Russia would lease for a period of 20 years an area of 6,000 sq km enclosing the Baykonur space launch facilities and the city of Bayqongyr (Baykonur, formerly Leninsk); in 2004, a new agreement extended the lease to 2050
Independence: 16 December 1991 (from the Soviet Union)
National holiday: Independence Day, 16 December (1991)
Constitution: first post-independence constitution adopted 28 January 1993; new constitution adopted by national referendum 30 August 1995
Legal system: civil law system influenced by Roman-Germanic law and by the theory and practice of the Russian Federation
International law organization participation: has not submitted an ICJ jurisdiction declaration; non-party state to the ICCt
Suffrage: 18 years of age; universal
Executive branch: *chief of state:* President Nursultan A. NAZARBAYEV (chairman of the Supreme Soviet from 22 February 1990, elected president 1 December 1991)
head of government: Prime Minister Karim MASIMOV (since 10 January 2007); First Deputy Prime Minister Umirzak SHUKEYEV (since 3 March 2009), Deputy Prime Ministers Yerbol ORYNBAYEV (since 29 October 2007), Aset ISEKESHEV (since 12 March 2010)
cabinet: Council of Ministers appointed by the president
(For more information visit the World Leaders website)
elections: president elected by popular vote for a five-year term; election last held on 3 April 2011 (next to be held December 2016); prime minister and deputy prime ministers appointed by the president, with Mazhilis approval; note–constitutional amendments of May 2007 shortened the presidential term from seven years to five years and established a two-consecutive-term limit; changes will take effect after NAZARBAYEV's term ends; he, and only he, is allowed to run for president indefinitely
note: constitutional amendments of January 2011 moved election date from 2012 to April 2011 but kept five-year term; subsequent election to take place December 2016
election results: Nursultan A. NAZARBAYEV reelected president; percent of vote–Nursultan A. NAZARBAYEV 95.5%, other 4.5%
Legislative branch: bicameral Parliament consists of the Senate (47 seats; 15 members are appointed by the president; 32 members elected by local assemblies; members serve six-year terms, but elections are staggered with half of the members up for re-election every three years) and the Mazhilis (107 seats; 9 out of the 107 Mazhilis members elected by the Assembly of the People of Kazakhstan, a presidentially appointed advisory body designed to represent the country's ethnic minorities; non-appointed members are popularly elected to serve five-year terms)
elections: Senate–(indirect) last held in October 2008 (next to be held in 2011); Mazhilis–last held on 18 August 2007 (next to be held in 2012)
election results: Senate–percent of vote by party–NA; seats by party–Nur Otan 16; Mazhilis–percent of vote by party–Nur-Otan 88.1%, NSDP 4.6%, Ak Zhol 3.3%, Auyl 1.6%, Communist People's Party 1.3%, Patriots Party 0.8% Ruhaniyat 0.4%; seats by party–Nur-Otan 98; note–parties had to achieve a threshold of 7% of the electorate to qualify for seats in the Mazhilis; changes to electoral legislation enacted since the 2007 election now ensure that the second-placed party will enter the Majilis at the next parliamentary election, even if it does not clear the 7% threshold
Judicial branch: Supreme Court (44 members); Constitutional Council (seven members)
Political parties and leaders: Adilet (Justice) [Maksut NARIKBAYEV, Zeynulla ALSHIMBAYEV, Serik ABDRAHMANOV, Bakhytbek AKHMETZHAN, Yerkin ONGARBAYEV, Tolegan SYDYKOV] (formerly Democratic Party of Kazakhstan); Agrarian and Industrial Union of Workers Block or AIST (Agrarian Party and Civic Party); Ak Zhol Party (Bright Path) [Alikhan BAYMENOV]; Alga [Vladimir KOZLOV] (unregistered); Auyl (Village) [Gani KALIYEV]; Azat (Freedom) Party [Bolat ABILOV] (formerly True Ak Zhol Party); Azat NSDP [co-chaired by Bolat ABILOV and Zharmakhan TUYAKBAY]; Azat and NSDP united in 2009, but the authorities have refused to register Azat NSDP as a single party; Communist Party of Kazakhstan or KPK [Serikbolsyn ABDILDIN]; Communist People's Party of Kazakhstan [Vladislav KOSAREV]; National Social Democratic Party or NSDP [Zharmakhan TUYAKBAY]; Nur-Otan [Bakhytzhan ZHUMAGULOV] (the Agrarian, Asar, and Civic parties merged with Otan); Patriots' Party [Gani KASYMOV]; Rukhaniyat (Spirituality) [Serikzhan MAMBETALIN]
Political pressure groups and leaders: Adil-Soz [Tamara KALEYEVA]; Almaty Helsinki Committee [Ninel FOKINA]; Confederation of Free Trade Unions [Sergei BELKIN]; For Fair Elections [Yevgeniy ZHOVTIS (jailed), Sabit ZHUSUPOV, Sergey DUVANOV, Ibrash NUSUPBAYEV]; Kazakhstan International Bureau on Human Rights [Yevgeniy ZHOVTIS, executive director]; Pan-National Social Democratic Party of Kazakhstan [Zharmakhan TUYAKBAY]; Pensioners Movement or Pokoleniye [Irina SAVOSTINA, chairwoman]; Republican Network of International Monitors [Dos KUSHIM]; Transparency International [Sergey ZLOTNIKOV]
International organization participation: ADB, CICA, CIS, CSTO, EAEC, EAPC, EBRD, ECO, FAO, GCTU, IAEA, IBRD, ICAO, ICRM, IDA, IDB, IFAD, IFC, IFRCS, ILO, IMF, IMO, Interpol, IOC, IOM, IPU, ISO, ITSO, ITU, MIGA, NAM (observer), NSG, OAS (observer), OIC, OPCW, OSCE, PFP, SCO, UN, UNCTAD, UNESCO, UNIDO, UNWTO, UPU, WCO, WFTU, WHO, WIPO, WMO, WTO (observer), ZC

Diplomatic representation in the US: *chief of mission:* Ambassador Yerlan IDRISSOV
chancery: 1401 16th Street NW, Washington, DC 20036
telephone: [1] (202) 232-5488
FAX: [1] (202) 232-5845
consulate(s) general: Los Angeles
consulate(s): New York

Diplomatic representation from the US: *chief of mission:* Ambassador Richard E. HOAGLAND
embassy: Ak Bulak 4, Str. 23-22, Building #3, Astana 010010
mailing address: use embassy street address
telephone: [7] (7172) 70-21-00
FAX: [7] (7172) 34-08-90

Flag description: a gold sun with 32 rays above a soaring golden steppe eagle, both centered on a sky blue background; the hoist side displays a national ornamental pattern "koshkar-muiz" (the horns of the ram) in gold; the blue color is of religious significance to the Turkic peoples of the country, and so symbolizes cultural and ethnic unity; it also represents the endless sky as well as water; the sun, a source of life and energy, exemplifies wealth and plenitude; the sun's rays are shaped like grain, which is the basis of abundance and prosperity; the eagle has appeared on the flags of Kazakh tribes for centuries and represents freedom, power, and the flight to the future

National anthem: *name:* "Menin Qazaqstanim" (My Kazakhstan)
lyrics/music: Zhumeken NAZHIMEDENOV and Nursultan NAZARBAYEV/Shamshi KALDAYAKOV
note: adopted 2006; President Nursultan NAZARBAYEV played a role in revising the lyrics

ECONOMY

Economy—overview: Kazakhstan, geographically the largest of the former Soviet republics, excluding Russia, possesses enormous fossil fuel reserves and plentiful supplies of other minerals and metals, such as uranium, copper, and zinc. It also has a large agricultural sector featuring livestock and grain. In 2002 Kazakhstan became the first country in the former Soviet Union to receive an investment-grade credit rating, and from 2000 through 2007, Kazakhstan's economy grew more than 9% per year. Extractive industries, particularly hydrocarbons and mining, have been the engines of this growth. However, geographic limitations and decaying infrastructure present serious obstacles. Landlocked, with restricted access to the high seas, Kazakhstan relies on its neighbors to export its products, especially oil and gas. Although its Caspian Sea ports and rail lines carrying oil have been upgraded, civil aviation has been neglected. Telecoms are improving, but require considerable investment, as does the information technology base. Supply and distribution of electricity can be erratic. At the end of 2007, global financial markets froze up and the loss of capital inflows to Kazakhstani banks caused a credit crunch. The subsequent and sharp fall of oil and commodity prices in 2008 aggravated the economic situation, and Kazakhstan plunged into recession. While the global financial crisis took a significant toll on Kazakhstan's economy, it has rebounded well. In response to the crisis, Kazakhstan's government devalued the tenge (Kazakhstan's currency) to stabilize market pressures and injected $19 billion in economic stimulus. Rising commodity prices have helped revive Kazakhstan's economy, which registered 7% growth in 2010. Barring a dramatic decline in oil prices, strong growth is expected to continue in 2011. Despite solid macroeconomic indicators, the government realizes that its economy suffers from an overreliance on oil and extractive industries, the so-called "Dutch disease." In response, Kazakhstan has embarked on an ambitious diversification program, aimed at developing targeted sectors like transport, pharmaceuticals, telecommunications, petrochemicals and food processing.

GDP (purchasing power parity): $196.4 billion (2010 est.)
country comparison to the world: 54
$183.6 billion (2009 est.)
$181.4 billion (2008 est.)
note: data are in 2010 US dollars

GDP (official exchange rate): $138.4 billion (2010 est.)

GDP—real growth rate: 7% (2010 est.)
country comparison to the world: 35
1.2% (2009 est.)
3.2% (2008 est.)

GDP—per capita (PPP): $12,700 (2010 est.)
country comparison to the world: 92
$11,900 (2009 est.)
$11,800 (2008 est.)
note: data are in 2010 US dollars

GDP—composition by sector: *agriculture:* 5.4%
industry: 42.8%
services: 51.8% (2010 est.)

Labor force: 8.718 million (2010 est.)
country comparison to the world: 54

Labor force—by occupation: *agriculture:* 28.2%
industry: 18.2%
services: 53.6% (2010)

Unemployment rate: 5.5% (2010 est.)
country comparison to the world: 53
6.3% (2009 est.)

Population below poverty line: 8.2% (2009)

Household income or consumption by percentage share: *lowest 10%:* 3.3%
highest 10%: 26.5% (2004 est.)

Distribution of family income—Gini index: 26.7 (2009)
country comparison to the world: 127
31.5 (2003)

Investment (gross fixed): 24.7% of GDP (2010 est.)
country comparison to the world: 46

Budget: *revenues:* $27.5 billion
expenditures: $31.6 billion (2010 est.)

Public debt: 16.2% of GDP (2010 est.)
country comparison to the world: 118
14.2% of GDP (2009 est.)

Inflation rate (consumer prices): 7.8% (2010 est.)
country comparison to the world: 182
7.3% (2009 est.)

Central bank discount rate: 7% (31 December 2010)
country comparison to the world: 40
10.5% (31 December 2008)

Commercial bank prime lending rate: NA%

Stock of narrow money: $15 billion (31 December 2010 est.)
country comparison to the world: 66
$16.66 billion (31 December 2009 est.)

Stock of broad money: $58 billion (31 December 2010 est.)
country comparison to the world: 63
$52.83 billion (31 December 2009 est.)

Stock of domestic credit: $44.53 billion (31 December 2010 est.)
country comparison to the world: 66
$39.72 billion (31 December 2009 est.)

Market value of publicly traded shares: $57.66 billion (31 December 2009)
country comparison to the world: 54
$31.08 billion (31 December 2008)
$41.38 billion (31 December 2007)

Agriculture—products: grain (mostly spring wheat), cotton; livestock

Industries: oil, coal, iron ore, manganese, chromite, lead, zinc, copper, titanium, bauxite, gold, silver, phosphates, sulfur, uranium, iron and steel; tractors and other agricultural machinery, electric motors, construction materials

Industrial production growth rate: 7.3% (2010 est.)
country comparison to the world: 42
Electricity–production: 78.4 billion kWh (2009 est.)
country comparison to the world: 37

Electricity—consumption: 77.9 billion kWh (2009 est.)
country comparison to the world: 36
Electricity–exports: 3.617 billion kWh (2007 est.)

Electricity—imports: 1.94 billion kWh (2009 est.)

Oil—production: 1.54 million bbl/day (2009 est.)
country comparison to the world: 19

Oil—consumption: 241,000 bbl/day (2009 est.)
country comparison to the world: 51

Oil—exports: 1.345 million bbl/day (2009 est.)
country comparison to the world: 19

Oil—imports: 164,000 bbl/day (2007 est.)
country comparison to the world: 54

Oil—proved reserves: 30 billion bbl (1 January 2010 est.)
country comparison to the world: 11

Natural gas—production: 35.61 billion cu m (2009 est.)
country comparison to the world: 24

Natural gas—consumption: 33.68 billion cu m (2008 est.)
country comparison to the world: 27

Natural gas—exports: 17.66 billion cu m (2008 est.)
country comparison to the world: 12

Natural gas—imports: 3.72 billion cu m (2009 est.)
country comparison to the world: 35

Natural gas—proved reserves: 2.407 trillion cu m (1 January 2010 est.)
country comparison to the world: 15

Current account balance: $6.993 billion (2010 est.)
country comparison to the world: 28
$-3.405 billion (2009 est.)
Exports: $59.23 billion (2010 est.)
country comparison to the world: 47
$43.84 billion (2009 est.)
Exports—commodities: oil and oil products 59%, ferrous metals 19%, chemicals 5%, machinery 3%, grain, wool, meat, coal
Exports—partners: China 16.34%, France 9.23%, Germany 8.32%, Russia 6.9%, Ukraine 5.52%, Romania 5.25%, Italy 5.12%, US 4.34% (2009)
Imports: $30.11 billion (2010 est.)
country comparison to the world: 59
$28.77 billion (2009 est.)
Imports—commodities: machinery and equipment, metal products, foodstuffs
Imports—partners: Russia 28.5%, China 26.72%, Germany 6.59%, Italy 5.58%, Ukraine 4.8% (2009)
Reserves of foreign exchange and gold: $32.44 billion (31 December 2010 est.)
country comparison to the world: 36
$23.22 billion (31 December 2009 est.)
Debt—external: $94.44 billion (31 December 2010 est.)
country comparison to the world: 37
$106.3 billion (31 December 2009 est.)
Stock of direct foreign investment—at home: $83.3 billion (31 December 2010 est.)
country comparison to the world: 40
$69.46 billion (31 December 2009 est.)
Stock of direct foreign investment—abroad: $7.208 billion (31 December 2010 est.)
country comparison to the world: 53
$5.708 billion (31 December 2009 est.)
Exchange rates: tenge (KZT) per US dollar–147.28 (2010), 147.5 (2009), 120.25 (2008), 122.55 (2007), 126.09 (2006)

COMMUNICATIONS

Telephones—main lines in use: 3.763 million (2009)
country comparison to the world: 43
Telephones—mobile cellular: 14.995 million (2009)
country comparison to the world: 50
Telephone system: *general assessment:* inherited an outdated telecommunications network from the Soviet era requiring modernization
domestic: intercity by landline and microwave radio relay; number of fixed-line connections is gradually increasing and fixed-line teledensity now roughly 25 per 100 persons; mobile-cellular usage is increasing and the subscriber base now is roughly 100 per 100 persons
international: country code–7; international traffic with other former Soviet republics and China carried by landline and microwave radio relay and with other countries by satellite and by the Trans-Asia-Europe (TAE) fiber-optic cable; satellite earth stations–2 Intelsat (2008)
Broadcast media: state owns nearly all radio and TV transmission facilities and operates national TV and radio networks; nearly all nationwide TV networks are wholly or partly owned by the government; some former state-owned media outlets have been privatized and are controlled by the president's daughter, who heads the Khabar Agency that runs multiple TV and radio stations; a number of privately-owned TV stations; households with satellite dishes have access to foreign media; a small number of commercial radio stations operating along with state-run radio stations (2008)
Internet country code: .kz
Internet hosts: 53,984 (2010)
country comparison to the world: 85
Internet users: 5.299 million (2009)
country comparison to the world: 44

TRANSPORTATION

Airports: 97 (2010)
country comparison to the world: 62
Airports—with paved runways: *total:* 65
over 3,047 m: 10
2,438 to 3,047 m: 26
1,524 to 2,437 m: 16
914 to 1,523 m: 5
under 914 m: 8 (2010)
Airports—with unpaved runways: *total:* 32
over 3,047 m: 5
2,438 to 3,047 m: 6
1,524 to 2,437 m: 3
914 to 1,523 m: 5
under 914 m: 13 (2010)
Heliports: 3 (2010)
Pipelines: condensate 658 km; gas 12,317 km; oil 11,201 km; refined products 1,095 km; water 1,465 km (2010)
Railways: *total:* 15,079 km
country comparison to the world: 19
broad gauge: 15,079 km 1.520-m gauge (4,000 km electrified) (2010)
Roadways: *total:* 93,612 km
country comparison to the world: 50
paved: 84,100 km
unpaved: 9,512 km (2008)
Waterways: 4,000 km (on the Ertis (Irtysh) River (80%) and Syr Darya (Syrdariya) River) (2010)
country comparison to the world: 27
Merchant marine: *total:* 8
country comparison to the world: 124
by type: petroleum tanker 6, refrigerated cargo 1, specialized tanker 1
foreign-owned: 1 (Ireland 1) (2010)
Ports and terminals: Aqtau (Shevchenko), Atyrau (Gur'yev), Oskemen (Ust-Kamenogorsk), Pavlodar, Semey (Semipalatinsk)

MILITARY

Military branches: Kazakhstan Armed Forces: Ground Forces, Navy, Air Mobile Forces, Air Defense Forces (2010)
Military service age and obligation: 18 years of age for compulsory military service; conscript service obligation–2 years; minimum age for volunteers NA (2004)
Manpower available for military service: *males age 16–49:* 4,163,629
females age 16–49: 4,179,051 (2010 est.)
Manpower fit for military service: *males age 16–49:* 2,909,999
females age 16–49: 3,528,169 (2010 est.)
Manpower reaching militarily significant age annually: *male:* 125,322
female: 119,541 (2010 est.)
Military expenditures: 1.1% of GDP (2010)
country comparison to the world: 127
Transnational Issues
Disputes—international: Kyrgyzstan has yet to ratify the 2001 boundary delimitation with Kazakhstan; field demarcation of the boundaries with Turkmenistan commenced in 2005, and with Uzbekistan in 2004; demarcation is scheduled to get underway with Russia in 2007; demarcation with China was completed in 2002; creation of a seabed boundary with Turkmenistan in the Caspian Sea remains under discussion; Azerbaijan, Kazakhstan, and Russia ratified Caspian seabed delimitation treaties based on equidistance, while Iran continues to insist on a one-fifth slice of the lake
Refugees and internally displaced persons: refugees (country of origin): 3,700 (Russia); 508 (Afghanistan) (2007)
Illicit drugs: significant illicit cultivation of cannabis for CIS markets, as well as limited cultivation of opium poppy and ephedra (for the drug ephedrine); limited government eradication of illicit crops; transit point for Southwest Asian narcotics bound for Russia and the rest of Europe; significant consumer of opiates

KENYA

INTRODUCTION

Background: Founding president and liberation struggle icon Jomo KENYATTA led Kenya from independence in 1963 until his death in 1978, when President Daniel Toroitich arap MOI took power in a constitutional succession. The country was a de facto one-party state from 1969 until 1982 when the ruling Kenya African National Union (KANU) made itself the sole legal party in Kenya. MOI acceded to internal and external pressure for political liberalization in late 1991. The ethnically fractured opposition failed to dislodge KANU from power in elections in 1992 and 1997, which were marred by violence and fraud, but were viewed as having generally reflected the will of the Kenyan people. President MOI stepped down in December 2002 following fair and peaceful elections. Mwai KIBAKI, running as the candidate of the multiethnic, united opposition group, the National Rainbow Coalition (NARC), defeated KANU candidate Uhuru KENYATTA and assumed the presidency following a campaign centered on an anticorruption platform. KIBAKI's NARC coalition splintered in 2005 over a constitutional review process. Government defectors joined with KANU to form a new opposition coalition, the Orange Democratic Movement, which defeated the government's draft constitution in a popular referendum in November 2005. KIBAKI's reelection in December 2007 brought charges of vote rigging from ODM candidate Raila ODINGA and unleashed two months of violence in which as many as 1,500 people died. UN-sponsored talks in late February produced a powersharing accord bringing ODINGA into the government in the restored position of prime minister. Kenya in August 2010 adopted a new constitution that eliminates the role of prime minister after the next presidential election.

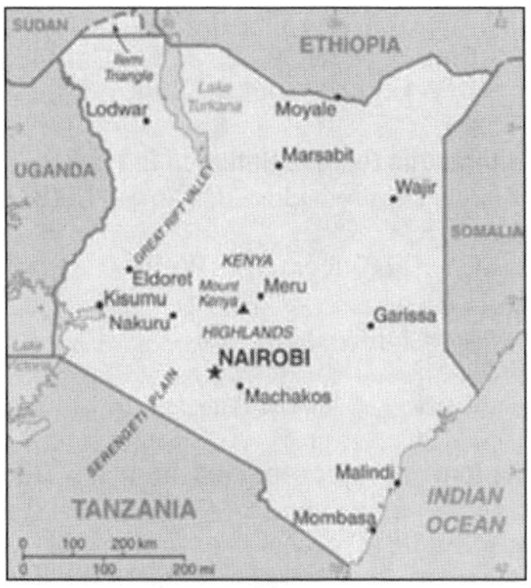

GEOGRAPHY

Location: Eastern Africa, bordering the Indian Ocean, between Somalia and Tanzania
Geographic coordinates: 1 00 N, 38 00 E
Map references: Africa
Area: *total:* 580,367 sq km
country comparison to the world: 48
land: 569,140 sq km
water: 11,227 sq km
Area—comparative: slightly more than twice the size of Nevada
Land boundaries: *total:* 3,477 km
border countries: Ethiopia 861 km, Somalia 682 km, Sudan 232 km, Tanzania 769 km, Uganda 933 km
Coastline: 536 km
Maritime claims: *territorial sea:* 12 nm
exclusive economic zone: 200 nm
continental shelf: 200 m depth or to the depth of exploitation
Climate: varies from tropical along coast to arid in interior
Terrain: low plains rise to central highlands bisected by Great Rift Valley; fertile plateau in west
Elevation extremes: *lowest point:* Indian Ocean 0 m
highest point: Mount Kenya 5,199 m
Natural resources: limestone, soda ash, salt, gemstones, fluorspar, zinc, diatomite, gypsum, wildlife, hydropower
Land use: *arable land:* 8.01%
permanent crops: 0.97%
other: 91.02% (2005)
Irrigated land: 1,030 sq km (2003)
Total renewable water resources: 30.2 cu km (1990)
Freshwater withdrawal (domestic/industrial/agricultural): *total:* 1.58 cu km/yr (30%/6%/64%)
per capita: 46 cu m/yr (2000)
Natural hazards: recurring drought; flooding during rainy seasons
volcanism: Kenya experiences limited volcanic activity; the Barrier (elev. 1,032 m) last erupted in 1921; South Island is the only other historically active volcano
Environment—current issues: water pollution from urban and industrial wastes; degradation of water quality from increased use of pesticides and fertilizers; water hyacinth infestation in Lake Victoria; deforestation; soil erosion; desertification; poaching
Environment—international agreements: *party to:* Biodiversity, Climate Change, Climate Change-Kyoto Protocol, Desertification, Endangered Species, Hazardous Wastes, Law of the Sea, Marine Dumping, Marine Life Conservation, Ozone Layer Protection, Ship Pollution, Wetlands, Whaling
signed, but not ratified: none of the selected agreements
Geography—note: the Kenyan Highlands comprise one of the most successful agricultural production regions in Africa; glaciers are found on Mount Kenya, Africa's second highest peak; unique physiography supports abundant and varied wildlife of scientific and economic value

PEOPLE

Population: 41,070,934 (July 2011 est.)
country comparison to the world: 33
note: estimates for this country explicitly take into account the effects of excess mortality due to AIDS; this can result in lower life expectancy, higher infant mortality, higher death rates, lower population growth rates, and changes in the distribution of population by age and sex than would otherwise be expected
Age structure: *0–14 years:* 42.2% (male 8,730,845/female 8,603,270)
15–64 years: 55.1% (male 11,373,997/female 11,260,402)
65 years and over: 2.7% (male 497,389/female 605,031) (2011 est.)
Median age:
total: 18.9 years
male: 18.8 years
female: 19 years (2011 est.)
Population growth rate: 2.462% (2011 est.)
country comparison to the world: 29
Birth rate: 33.54 births/1,000 population (2011 est.)
country comparison to the world: 34
Death rate: 8.93 deaths/1,000 population (July 2011 est.)
country comparison to the world: 72
Net migration rate: 0 migrant(s)/1,000 population (2011 est.)
country comparison to the world: 91
Urbanization: *urban population:* 22% of total population (2010)
rate of urbanization: 4.2% annual rate of change (2010-15 est.)
Major cities—population: NAIROBI (capital) 3.375 million; Mombassa 966,000 (2009)
Sex ratio: *at birth:* 1.02 male(s)/female
under 15 years: 1.01 male(s)/female
15–64 years: 1.01 male(s)/female
65 years and over: 0.83 male(s)/female
total population: 1.01 male(s)/female (2011 est.)
Infant mortality rate: *total:* 52.29 deaths/1,000 live births
country comparison to the world: 43
male: 55.03 deaths/1,000 live births
female: 49.49 deaths/1,000 live births (2011 est.)
Life expectancy at birth: *total population:* 59.48 years
country comparison to the world: 189
male: 58.91 years
female: 60.07 years (2011 est.)
Total fertility rate: 4.19 children born/woman (2011 est.)
country comparison to the world: 38
HIV/AIDS–adult prevalence rate: 6.3% (2009 est.)
country comparison to the world: 11
HIV/AIDS–people living with HIV/AIDS: 1.5 million (2009 est.)
country comparison to the world: 5
HIV/AIDS–deaths: 80,000 (2009 est.)
country comparison to the world: 6
Major infectious diseases: *degree of risk:* high
food or waterborne diseases: bacterial and protozoal diarrhea, hepatitis A, and typhoid fever
vectorborne disease: malaria and Rift Valley fever
water contact disease: schistosomiasis
animal contact disease: rabies (2009)
Drinking water source: *Improved:*
urban: 83% of population
rural: 52% of population
total: 59% of population
Unimproved: urban: 17% of population
rural: 48% of population
total: 41% of population (2008)
Sanitation facility access: *Improved:*
urban: 27% of population
rural: 32% of population
total: 31% of population
Unimproved: urban: 73% of population
rural: 68% of population
total: 69% of population (2008)
Nationality: *noun:* Kenyan(s)
adjective: Kenyan
Ethnic groups: Kikuyu 22%, Luhya 14%, Luo 13%, Kalenjin 12%, Kamba 11%, Kisii 6%, Meru 6%, other African 15%, non-African (Asian, European, and Arab) 1%
Religions: Protestant 45%, Roman Catholic 33%, Muslim 10%, indigenous beliefs 10%, other 2%
note: a large majority of Kenyans are Christian, but estimates for the percentage of the population that adheres to Islam or indigenous beliefs vary widely
Languages: English (official), Kiswahili (official), numerous indigenous languages
Literacy: *definition:* age 15 and over can read and write
total population: 85.1%
male: 90.6%
female: 79.7% (2003 est.)
School life expectancy (primary to tertiary education): *total:* 11 years
male: 11 years
female: 11 years (2009)
Education expenditures: 7% of GDP (2006)
country comparison to the world: 16

GOVERNMENT

Country name: *conventional long form:* Republic of Kenya
conventional short form: Kenya
local long form: Republic of Kenya/Jamhuri ya Kenya
local short form: Kenya
former: British East Africa
Government type: republic
Capital: *name:* Nairobi
geographic coordinates: 1 17 S, 36 49 E
time difference: UTC+3 (8 hours ahead of Washington, DC during Standard Time)
Administrative divisions: 7 provinces and 1 area*; Central, Coast, Eastern, Nairobi Area*, North Eastern, Nyanza, Rift Valley, Western; note—the constitution promulgated in August 2010 designates 47 yet-to-be-defined counties as first-order administrative units
Independence: 12 December 1963 (from the UK)
National holiday: Independence Day, 12 December (1963); Madaraka Day, 1 June; Mashujaa Day, 20 October
Constitution: 27 August 2010; the new constitution abolishes the position of prime minister and establishes a bicameral legislature; many details have yet to be finalized and will require significant legislative action
Legal system: mixed legal system of English common law, Islamic law, and customary law; judicial review in High Court
International law organization participation: accepts compulsory ICJ jurisdiction with reservations; accepts ICCt jurisdiction
Suffrage: 18 years of age; universal
Executive branch: *chief of state:* President Mwai KIBAKI (since 30 December 2002); Vice President Stephen Kalonzo MUSYOKA (since 10 January 2008);
head of government: President Mwai KIBAKI (since 30 December 2002); Vice President Stephen Kalonzo MUSYOKA (since 10 January 2008); Prime Minister Raila Amolo ODINGA (since 17 April 2008); note—according to the 2008 powersharing agreement the role of the prime minister was not well defined; constitutionally, the president remains chief of state and head of government, but the prime minister is charged with coordinating government business
cabinet: Cabinet appointed by the president and chaired by the prime minister, who is the leader of the largest party in parliament
(For more information visit the World Leaders website)
elections: president elected by popular vote for a five-year term (eligible for a second term); in addition to receiving the largest number of votes in absolute terms, the presidential candidate must also win 25% or more of the vote in at least five of Kenya's seven provinces and one area to avoid a runoff; election last held on 27 December 2007 (next to be held in December 2012); vice president appointed by the president; note—the new constitution sets elections for August 2011 but this date is expected to slip
election results: President Mwai KIBAKI reelected; percent of vote—Mwai KIBAKI 46%, Raila ODINGA 44%, Kalonzo MUSYOKA 9%, other 3.4%
Legislative branch: unicameral National Assembly or Bunge usually referred to as Parliament (224 seats; 210 members elected by popular vote to serve five-year terms, 12 nominated members appointed by the president but selected by the parties in proportion to their parliamentary vote totals, 2 ex-officio members); note—the constitution promulgated in August 2010 changes the legislature to a bicameral parliament consisting of a 290 member National Assembly and a 94 member Senate; parliament members will serve five year terms
elections: last held on 27 December 2007 (next to be held in December 2012)
election results: percent of vote by party—NA; seats by party—ODM 99, PNU 46, ODM-K 16, KANU 14 other 35; ex-officio 2; seats appointed by the president—ODM 6, PNU 3, ODM-K 2, KANU 1
Judicial branch: Court of Appeal (chief justice is appointed by the president); High Court; note—the constitution promulgated in August 2010 specifies three superior courts consisting of a Supreme Court, Court of Appeals, and High Court, and three subordinate courts consisting of Magistrate courts, Kadhis courts (sentences according to Muslim law), and Courts Martial
Political parties and leaders: Forum for the Restoration of Democracy-Kenya or FORD-Kenya [Musikari KOMBO]; Forum for the Restoration of Democracy-People or FORD-People [Reuben OYONDI]; Kenya African National Union or KANU [Uhuru KENYATTA]; National Rainbow Coalition-Kenya or NARC-Kenya [Martha KARUA]; Orange Democratic Movement or ODM [Raila ODINGA]; Orange Democratic Movement-Kenya or ODM-K [Kalonzo MUSYOKA]; Party of National Unity or PNU [Mwai KIBAKI]; Shirikisho Party of Kenya or SPK [Chirau Ali MWAKWERE]
Political pressure groups and leaders: Council of Islamic Preachers of Kenya or CIPK [Sheikh Idris MOHAMMED]; Kenya Human Rights Commission [L. Muthoni WANYEKI]; Muslim Human Rights Forum [Ali-Amin KIMATHI]; National Muslim Leaders Forum or NAMLEF [Abdullahi ABDI]; Protestant National Council of Churches of Kenya or NCCK [Canon Peter Karanja MWANGI]; Roman Catholic and other Christian churches; Supreme Council of Kenya Muslims or SUPKEM [Shaykh Abdul Gafur al-BUSAIDY]
other: labor unions
International organization participation: ACP, AfDB, AU, C, COMESA, EAC, EADB, FAO, G-15, G-77, IAEA, IBRD, ICAO, ICRM, IDA, IFAD, IFC, IFRCS, IGAD, ILO, IMF, IMO, IMSO, Interpol, IOC, IOM, IPU, ISO, ITSO, ITU, ITUC, MIGA, MONUSCO, NAM, OPCW, PCA, UN, UNAMID, UNCTAD, UNESCO, UNHCR, UNIDO, UNMIS, UNWTO, UPU, WCO, WHO, WIPO, WMO, WTO
Diplomatic representation in the US: *chief of mission:* Ambassador Elkanah ODEMBO Absalom
chancery: 2249 R Street NW, Washington, DC 20008
telephone: [1] (202) 387-6101
FAX: [1] (202) 462-3829
consulate(s) general: Los Angeles
consulate(s): New York
Diplomatic representation from the US: *chief of mission:* Ambassador Michael E. RANNEBERGER
embassy: US Embassy, United Nations Avenue, Nairobi; P. O. Box 606 Village Market, Nairobi 00621
mailing address: Box 21A, Unit 64100, APO AE 09831
telephone: [254] (20) 363-6000
FAX: [254] (20) 363-410
Flag description: three equal horizontal bands of black (top), red, and green; the red band is edged in white; a large Maasai warrior's shield covering crossed spears is superimposed at the center; black symbolizes the majority population, red the blood shed in the struggle for freedom, green stands for natural wealth, and white for peace; the shield and crossed spears symbolize the defense of freedom
National anthem: *name:* "Ee Mungu Nguvu Yetu" (Oh God of All Creation)
lyrics/music: Graham HYSLOP, Thomas KALUME, Peter KIBUKOSYA, Washington OMONDI, and George W. SENOGA-ZAKE/traditional, adapted by Graham HYSLOP, Thomas KALUME, Peter KIBUKOSYA, Washington OMONDI, and George W. SENOGA-ZAKE
note: adopted 1963; the anthem is based on a traditional Kenyan folk song

ECONOMY

Economy—overview: Although the regional hub for trade and finance in East Africa, Kenya has been hampered by corruption and by reliance upon several primary goods whose prices have remained low. In 1997, the IMF suspended Kenya's Enhanced Structural Adjustment Program due to the government's failure to maintain reforms and curb corruption. The IMF, which had resumed loans in 2000 to help Kenya through a drought, again halted lending in 2001 when the government failed to institute several anticorruption measures. In the key December 2002 elections, Daniel Arap MOI's 24-year-old reign ended, and a new opposition government took on the formidable economic problems facing the nation. After some early progress in rooting out corruption and encouraging donor support, the KIBAKI government was rocked by high-level graft scandals in 2005 and 2006. In 2006, the World Bank and IMF delayed loans pending action by the government on corruption. The international financial institutions and donors have since resumed lending, despite little action on the government's part to deal

with corruption. Post-election violence in early 2008, coupled with the effects of the global financial crisis on remittance and exports, reduced GDP growth to 1.7 in 2008, but the economy rebounded in 2009-10.
GDP (purchasing power parity): $66.03 billion (2010 est.)
country comparison to the world: 83
$62.9 billion (2009 est.)
$61.31 billion (2008 est.)
note: data are in 2010 US dollars
GDP (official exchange rate): $32.16 billion (2010 est.)
GDP—real growth rate: 5% (2010 est.)
country comparison to the world: 73
2.6% (2009 est.)
1.6% (2008 est.)
GDP—per capita (PPP): $1,600 (2010 est.)
country comparison to the world: 199
$1,600 (2009 est.)
$1,600 (2008 est.)
note: data are in 2010 US dollars
GDP—composition by sector:
agriculture: 22%
industry: 16%
services: 62% (2010 est.)
Labor force:
17.94 million (2010 est.)
country comparison to the world: 33
Labor force—by occupation:
agriculture: 75%
industry and *services:* 25% (2007 est.)
Unemployment rate: 40% (2008 est.)
country comparison to the world: 185
40% (2001 est.)
Population below poverty line: 50% (2000 est.)
Household income or consumption by percentage share: *lowest 10%:* 1.8%
highest 10%: 37.8% (2005)
Distribution of family income—Gini index: 42.5 (2008 est.)
country comparison to the world: 49
44.9 (1997)
Investment (gross fixed): 21.3% of GDP (2010 est.)
country comparison to the world: 73
Budget: *revenues:* $7.017 billion
expenditures: $9.045 billion (2010 est.)
Public debt: 50.9% of GDP (2010 est.)
country comparison to the world: 50
46.3% of GDP (2009 est.)
Inflation rate (consumer prices): 4.2% (2010 est.)
country comparison to the world: 121
9.3% (2009 est.)
Central bank discount rate: NA%
Commercial bank prime lending rate: 14.8% (31 December 2009 est.)
country comparison to the world: 53
14.02% (31 December 2008 est.)
Stock of narrow money: $6.333 billion (31 December 2010 est.)
country comparison to the world: 82
$5.717 billion (31 December 2009 est.)
Stock of broad money: $15.38 billion (31 December 2010 est.)
country comparison to the world: 89
$13.5 billion (31 December 2009 est.)
Stock of domestic credit: $14.11 billion (31 December 2010 est.)
country comparison to the world: 89
$13.17 billion (31 December 2009 est.)
Market value of publicly traded shares: $10.76 billion (31 December 2009)
country comparison to the world: 67
$10.92 billion (31 December 2008)
$13.39 billion (31 December 2007)
Agriculture—products: tea, coffee, corn, wheat, sugarcane, fruit, vegetables; dairy products, beef, pork, poultry, eggs
Industries: small-scale consumer goods (plastic, furniture, batteries, textiles, clothing, soap, cigarettes, flour), agricultural products, horticulture, oil refining; aluminum, steel, lead; cement, commercial ship repair, tourism
Industrial production growth rate: 4% (2010 est.)
country comparison to the world: 83
Electricity—production: 5.223 billion kWh (2008 est.)
country comparison to the world: 112
Electricity—consumption: 4.863 billion kWh (2008 est.)
country comparison to the world: 111
Electricity—exports: 58.3 million kWh (2007 est.)
Electricity—imports: 22.5 million kWh (2007 est.)
Oil—production: 0 bbl/day (2008 est.)
country comparison to the world: 189
Oil—consumption: 76,000 bbl/day (2009 est.)
country comparison to the world: 88
Oil—exports: 7,270 bbl/day (2007 est.)
country comparison to the world: 97
Oil—imports: 80,530 bbl/day (2007 est.)
country comparison to the world: 71
Oil—proved reserves: 0 bbl
(1 January 2010 est.)
country comparison to the world: 146
Natural gas—production: 0 cu m (2008 est.)
country comparison to the world: 142
Natural gas—consumption: 0 cu m (2008 est.)
country comparison to the world: 186
Natural gas—exports: 0 cu m (2008 est.)
country comparison to the world: 120
Natural gas—imports: 0 cu m (2008 est.)
country comparison to the world: 76
Natural gas—proved reserves: 0 cu m (1 January 2010 est.)
country comparison to the world: 149
Current account balance: $-1.414 billion (2010 est.)
country comparison to the world: 147
$-1.611 billion (2009 est.)
Exports: $5.141 billion (2010 est.)
country comparison to the world: 107
$4.459 billion (2009 est.)
Exports—commodities: tea, horticultural products, coffee, petroleum products, fish, cement
Exports—partners: UK 11.31%, Netherlands 9.81%, Uganda 9.07%, Tanzania 8.83%, US 5.93%, Pakistan 5.63% (2009)
Imports: $10.4 billion (2010 est.)
country comparison to the world: 87
$9.715 billion (2009 est.)
Imports—commodities: machinery and transportation equipment, petroleum products, motor vehicles, iron and steel, resins and plastics
Imports—partners: India 11.67%, China 10.58%, UAE 9.32%, South Africa 8.36%, Saudi Arabia 6.53%, US 6.25%, Japan 5.1% (2009)
Reserves of foreign exchange and gold: $4.585 billion (31 December 2010 est.)
country comparison to the world: 72
$3.85 billion (31 December 2009 est.)
Debt—external: $7.935 billion (31 December 2010 est.)
country comparison to the world: 92
$7.795 billion (31 December 2009 est.)
Stock of direct foreign investment—at home: $2.337 billion (31 December 2010 est.)
country comparison to the world: 88
$2.129 billion (31 December 2009 est.)
Stock of direct foreign investment—abroad: $338 million (31 December 2010 est.)
country comparison to the world: 76
$288 million (31 December 2009 est.)
Exchange rates: Kenyan shillings (KES) per US dollar–79.217 (2010), 77.352 (2009), 68.358 (2008), 68.309 (2007), 72.101 (2006)

COMMUNICATIONS

Telephones—main lines in use: 664,100 (2009)
country comparison to the world: 91
Telephones—mobile cellular: 19.365 million (2009)
country comparison to the world: 41
Telephone system: *general assessment:* inadequate; fixed-line telephone system is small and inefficient; trunks are primarily microwave radio relay; business data commonly transferred by a very small aperture terminal (VSAT) system
domestic: sole fixed-line provider, Telkom Kenya, is slated for privatization; multiple providers in the mobile-cellular segment of the market fostering a boom in mobile-cellular telephone usage with teledensity reaching 50 per 100 persons in 2009
international: country code–254; The East Africa Marine System (TEAMS) and the SEACOM undersea fiber-optic cable systems; satellite earth stations–4 Intelsat
Broadcast media: about a half-dozen privately-owned TV stations and a state-owned television broadcaster that operates 2 channels; satellite and cable TV subscription services are available; state-owned radio broadcaster operates 2 national radio channels and provides regional and local radio services in multiple languages; a large number of private radio broadcasters, including provincial stations broadcasting in local languages; transmissions of several international broadcasters are available (2007)
Internet country code: .ke
Internet hosts: 47,676 (2010)
country comparison to the world: 90
Internet users: 3.996 million (2009)
country comparison to the world: 59

TRANSPORTATION

Airports: 191 (2010)
country comparison to the world: 33
Airports—with paved runways: *total:* 17
over 3,047 m: 4

2,438 to 3,047 m: 2
1,524 to 2,437 m: 4
914 to 1,523 m: 6
under 914 m: 1 (2010)
Airports—with unpaved runways: *total:* 174
1,524 to 2,437 m: 12
914 to 1,523 m: 107
under 914 m: 55 (2010)
Pipelines: oil 4 km; refined products 928 km (2010)
Railways: *total:* 2,066 km
country comparison to the world: 71
narrow gauge: 2,066 km 1.000-m gauge (2010)
Roadways: *total:* 160,886 km
country comparison to the world: 31
paved: 11,197 km
unpaved: 149,689 km (2008)
Waterways: (the only significant inland waterway in the country is the part of Lake Victoria within the boundaries of Kenya; Kisumu is the main port and has ferry connections to Uganda and Tanzania) (2010)
Merchant marine: *total:* 1
country comparison to the world: 155
by type: petroleum tanker 1
registered in other countries: 5 (Comoros 1, Saint Vincent and the Grenadines 2, Tuvalu 1, unknown 1) (2010)
Ports and terminals: Kisumu, Mombasa

MILITARY

Military branches: Kenya Armed Forces: Kenya Army, Kenya Navy, Kenya Air Force (2010)
Military service age and obligation: 18-26 years of age for voluntary service (under 18 with parental consent), with a 9-year obligation (7 years for Kenyan Navy); applicants must be Kenyan citizens and provide a national identity card (obtained at age 18) and a school-leaving certificate (2010)
Manpower available for military service: *males age 16–49:* 9,768,140
females age 16–49: 9,466,257 (2010 est.)
Manpower fit for military service: *males age 16–49:* 6,361,268
females age 16–49: 6,106,870 (2010 est.)
Manpower reaching militarily significant age annually: *male:* 422,104
female: 416,927 (2010 est.)
Military expenditures: 2.8% of GDP (2006)
country comparison to the world: 49
Transnational Issues
Disputes—international: Kenya served as an important mediator in brokering Sudan's north-south separation in February 2005; Kenya provides shelter to almost a quarter of a million refugees, including Ugandans who flee across the border periodically to seek protection from Lord's Resistance Army (LRA) rebels; Kenya works hard to prevent the clan and militia fighting in Somalia from spreading across the border, which has long been open to nomadic pastoralists; the boundary that separates Kenya's and Sudan's sovereignty is unclear in the "Ilemi Triangle," which Kenya has administered since colonial times
Refugees and internally displaced persons: refugees (country of origin): 173,702 (Somalia); 73,004 (Sudan); 16,428 (Ethiopia)
IDPs: 250,000-400,000 (2007 post-election violence; KANU attacks on opposition tribal groups in 1990s) (2007)
Illicit drugs: widespread harvesting of small plots of marijuana; transit country for South Asian heroin destined for Europe and North America; Indian methaqualone also transits on way to South Africa; significant potential for money-laundering activity given the country's status as a regional financial center; massive corruption, and relatively high levels of narcotics-associated activities

KIRIBATI

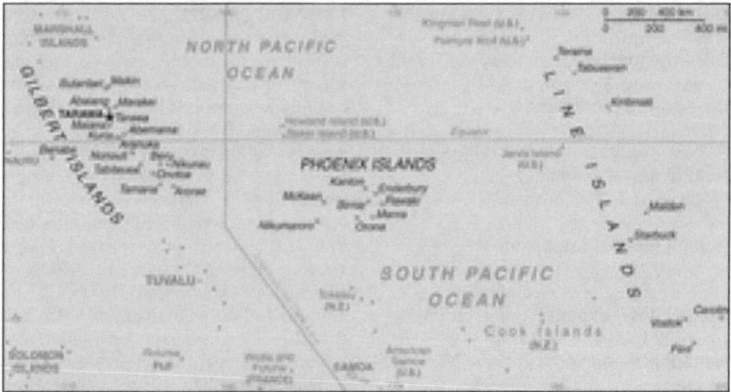

INTRODUCTION

Background: The Gilbert Islands became a British protectorate in 1892 and a colony in 1915; they were captured by the Japanese in the Pacific War in 1941. The islands of Makin and Tarawa were the sites of major US amphibious victories over entrenched Japanese garrisons in 1943. The Gilbert Islands were granted self-rule by the UK in 1971 and complete independence in 1979 under the new name of Kiribati. The US relinquished all claims to the sparsely inhabited Phoenix and Line Island groups in a 1979 treaty of friendship with Kiribati.

GEOGRAPHY

Location: Oceania, group of 33 coral atolls in the Pacific Ocean, straddling the Equator, as well as the International Date Line; the capital Tarawa is about half way between Hawaii and Australia
Geographic coordinates: 1 25 N, 173 00 E
Map references: Oceania
Area: *total:* 811 sq km
country comparison to the world: 186
land: 811 sq km
water: 0 sq km
note: includes three island groups–Gilbert Islands, Line Islands, Phoenix Islands
Area—comparative: four times the size of Washington, DC
Land boundaries: 0 km
Coastline: 1,143 km
Maritime claims: *territorial sea:* 12 nm
exclusive economic zone: 200 nm
Climate: tropical; marine, hot and humid, moderated by trade winds
Terrain: mostly low-lying coral atolls surrounded by extensive reefs
Elevation extremes: *lowest point:* Pacific Ocean 0 m
highest point: unnamed elevation on Banaba 81 m
Natural resources: phosphate (production discontinued in 1979)
Land use: *arable land:* 2.74%
permanent crops: 47.95%
other: 49.31% (2005)
Irrigated land: NA
Natural hazards: typhoons can occur any time, but usually November to March; occasional tornadoes; low level of some of the islands make them sensitive to changes in sea level
Environment—current issues: heavy pollution in lagoon of south Tarawa atoll due to heavy migration mixed with traditional practices such as lagoon latrines and open-pit dumping; ground water at risk
Environment—international agreements: *party to:* Biodiversity, Climate Change, Climate Change-Kyoto Protocol, Desertification, Hazardous Wastes, Law of the Sea, Marine Dumping, Ozone Layer Protection, Whaling
signed, but not ratified: none of the selected agreements
Geography—note: 21 of the 33 islands are inhabited; Banaba (Ocean Island) in Kiribati is one of the three great phosphate rock islands in the Pacific Ocean–the others are Makatea in French Polynesia, and Nauru; Kiribati is the only country in the world to fall into all four hemispheres (northern, southern, eastern, and western)

PEOPLE

Population: 100,743 (July 2011 est.)
country comparison to the world: 193
Age structure: *0–14 years:* 33.9% (male 17,385/female 16,750)
15–64 years: 62.4% (male 30,598/female 32,259)

65 years and over: 3.7% (male 1,461/female 2,290) (2011 est.)
Median age: *total:* 22.5 years
male: 21.7 years
female: 23.4 years (2011 est.)
Population growth rate: 1.249% (2011 est.)
country comparison to the world: 93
Birth rate: 22.73 births/1,000 population (2011 est.)
country comparison to the world: 74
Death rate: 7.4 deaths/1,000 population (July 2011 est.)
country comparison to the world: 120
Net migration rate: -2.85 migrant(s)/1,000 population (2011 est.)
country comparison to the world: 175
Urbanization: *urban population:* 44% of total population (2010)
rate of urbanization: 1.9% annual rate of change (2010-15 est.)
Major cities—population: TARAWA (capital) 43,000 (2009)
Sex ratio: *at birth:* 1.05 male(s)/female
under 15 years: 1.04 male(s)/female
15–64 years: 0.95 male(s)/female
65 years and over: 0.64 male(s)/female
total population: 0.97 male(s)/female (2011 est.)
Infant mortality rate:
total: 38.89 deaths/1,000 live births
country comparison to the world: 64
male: 40.13 deaths/1,000 live births
female: 37.58 deaths/1,000 live births (2011 est.)
Life expectancy at birth: *total population:* 64.39 years
country comparison to the world: 171
male: 62.03 years
female: 66.88 years (2011 est.)
Total fertility rate: 2.78 children born/woman (2011 est.)
country comparison to the world: 72
HIV/AIDS–adult prevalence rate: NA
HIV/AIDS—people living with HIV/AIDS: NA
HIV/AIDS—deaths: NA
Drinking water source: *Improved:*
urban: 77% of population
rural: 50% of population
total: 62% of population
Unimproved: urban: 23% of population
rural: 50% of population
total: 38% of population (2000)
Sanitation facility access: *Improved:*
urban: 47% of population
rural: 22% of population
total: 33% of population
Unimproved: urban: 53% of population
rural: 78% of population
total: 67% of population (2000)
Nationality: *noun:* I-Kiribati (singular and plural)
adjective: I-Kiribati
Ethnic groups: Micronesian 98.8%, other 1.2% (2000 census)
Religions: Roman Catholic 55%, Protestant 36%, Mormon 3.1%, Bahai 2.2%, Seventh-Day Adventist 1.9%, other 1.8% (2005 census)
Languages: I-Kiribati, English (official)
Literacy: NA
School life expectancy (primary to tertiary education): *total:* 12 years
male: 12 years
female: 13 years (2008)
Education expenditures: NA

GOVERNMENT

Country name: *conventional long form:* Republic of Kiribati
conventional short form: Kiribati
local long form: Republic of Kiribati
local short form: Kiribati
note: pronounced keer-ree-bahss
former: Gilbert Islands
Government type: republic
Capital: *name:* Tarawa
geographic coordinates: 1 19 N, 172 58 E
time difference: UTC+12 (17 hours ahead of Washington, DC during Standard Time)
note: on 1 January 1995, Kiribati proclaimed that all of its territory was in the same time zone as its Gilbert Islands group (UTC +12) even though the Phoenix Islands and the Line Islands under its jurisdiction were on the other side of the International Date Line
Administrative divisions: 3 units; Gilbert Islands, Line Islands, Phoenix Islands; note–in addition, there are 6 districts (Banaba, Central Gilberts, Line Islands, Northern Gilberts, Southern Gilberts, Tarawa) and 21 island councils–one for each of the inhabited islands (Abaiang, Abemama, Aranuka, Arorae, Banaba, Beru, Butaritari, Kanton, Kiritimati, Kuria, Maiana, Makin, Marakei, Nikunau, Nonouti, Onotoa, Tabiteuea, Tabuaeran, Tamana, Tarawa, Teraina)
Independence: 12 July 1979 (from the UK)
National holiday: Independence Day, 12 July (1979)
Constitution: 12 July 1979
Legal system: English common law supplemented by customary law
International law organization participation: has not submitted an ICJ jurisdiction declaration; non-party state to the ICCt
Suffrage: 18 years of age; universal
Executive branch: *chief of state:* President Anote TONG (since 10 July 2003); Vice President Teima ONORIO; note–the president is both the chief of state and head of government
head of government: President Anote TONG (since 10 July 2003); Vice President Teima ONORIO
cabinet: 12-member cabinet appointed by the president from among the members of the House of Parliament
(For more information visit the World Leaders website)
elections: the House of Parliament chooses the presidential candidates from among its members and then those candidates compete in a general election; president elected by popular vote for a four-year term (eligible for two more terms); election last held on 17 October 2007 (next to be held in 2011); vice president appointed by the president
election results: Anote TONG 63.7%, Nabuti MWEMWENIKARAWA 32.9%
Legislative branch: unicameral House of Parliament or Maneaba Ni Maungatabu (46 seats; 44 members elected by popular vote, 1 ex officio member–the attorney general, 1 nominated by the Rabi Council of Leaders (representing Banaba Island); members serve four-year terms)
elections: legislative elections were held in two rounds–the first round on 22 August 2007 and the second round on 30 August 2007 (next to be held in 2011)
election results: percent of vote by party–NA; seats by party–NA, other 2 (includes attorney general)
Judicial branch: Court of Appeal; High Court; 26 Magistrates' courts; judges at all levels are appointed by the president
Political parties and leaders: Boutokaan Te Koaua Party or BTK [Taberannang TIMEON]; Maneaban Te Mauri Party or MTM [Teburoro TITO]; Maurin Kiribati Pati or MKP; National Progressive Party or NPP [Dr. Harry TONG]
note: there is no tradition of formally organized political parties in Kiribati; they more closely resemble factions or interest groups because they have no party headquarters, formal platforms, or party structures
Political pressure groups and leaders: NA
International organization participation: ACP, ADB, AOSIS, C, FAO, IBRD, ICAO, ICRM, IDA, IFAD, IFC, IFRCS, ILO, IMF, IMO, IOC, ITU, ITUC, OPCW, PIF, Sparteca, SPC, UN, UNCTAD, UNESCO, UPU, WHO, WMO
Diplomatic representation in the US: Kiribati does not have an embassy in the US; there is an honorary consulate in Honolulu
Diplomatic representation from the US: the US does not have an embassy in Kiribati; the US ambassador to Fiji is accredited to Kiribati
Flag description: the upper half is red with a yellow frigate bird flying over a yellow rising sun, and the lower half is blue with three horizontal wavy white stripes to represent the Pacific ocean; the white stripes represent the three island groups–the Gilbert, Line, and Phoenix Islands; the 17 rays of the sun represent the 16 Gilbert Islands and Banaba (formerly Ocean Island); the frigate bird symbolizes authority and freedom
National anthem: *name:* "Teirake kaini Kiribati" (Stand Up, Kiribati)
lyrics/music: Urium Tamuera IOTEBA
note: adopted 1979

ECONOMY

Economy—overview: A remote country of 33 scattered coral atolls, Kiribati has few natural resources and is one of the least developed Pacific Islands. Commercially viable phosphate deposits were exhausted at the time of independence from the UK in 1979. Copra and fish now represent the bulk of production and exports. The economy has fluctuated widely in recent years. Economic development is constrained by a shortage of skilled workers, weak infrastructure, and remoteness from international markets. Tourism provides more than one-fifth of GDP. Private sector initiatives and a financial sector are in the early stages of development. Foreign financial aid from the EU, UK, US, Japan, Australia, New Zealand, Canada, UN agencies, and Taiwan

accounts for 20-25% of GDP. Remittances from seamen on merchant ships abroad account for more than $5 million each year. Kiribati receives around $15 million annually for the government budget from an Australian trust fund.
GDP (purchasing power parity): $618 million (2010 est.)
country comparison to the world: 210
$606.7 million (2009 est.)
$611.3 million (2008 est.)
note: data are in 2010 US dollars
GDP (official exchange rate): $147 million (2010 est.)
GDP—real growth rate: 1.8% (2010 est.)
country comparison to the world: 155
-0.7% (2009 est.)
-1.1% (2008 est.)
GDP—per capita (PPP): $6,200 (2010 est.)
country comparison to the world: 137
$6,200 (2009 est.)
$6,300 (2008 est.)
note: data are in 2010 US dollars
GDP—composition by sector: *agriculture:* 8.9%
industry: 24.2%
services: 66.8% (2004)
Labor force: 7,870 economically active, not including subsistence farmers (2001 est.)
country comparison to the world: 216
Labor force—by occupation:
agriculture: 2.7%
industry: 32%
services: 65.3% (2000)
Unemployment rate: 2% (1992 est.)
country comparison to the world: 13
Population below poverty line: NA%
Household income or consumption by percentage share: *lowest 10%:* NA%
highest 10%: NA%
Budget: *revenues:* $55.52 million
expenditures: $59.71 million (FY05)
Inflation rate (consumer prices): 0.2% (2007 est.)
country comparison to the world: 10
Market value of publicly traded shares: $NA
Agriculture—products: copra, taro, breadfruit, sweet potatoes, vegetables; fish
Industries: fishing, handicrafts
Industrial production growth rate: NA%
Electricity—production: 14 million kWh (2007 est.)
country comparison to the world: 209
Electricity—consumption: 13.02 million kWh (2007 est.)
country comparison to the world: 209
Electricity—exports: 0 kWh (2008 est.)
Electricity—imports: 0 kWh (2008 est.)
Oil—production: 0 bbl/day (2009 est.)
country comparison to the world: 190
Oil—consumption: 0 bbl/day
country comparison to the world: 205
Oil—exports: 0 bbl/day (2007 est.)
country comparison to the world: 178
Oil—imports: 260.8 bbl/day (2007 est.)
country comparison to the world: 200
Oil—proved reserves: 0 bbl (1 January 2010 est.)
country comparison to the world: 148
Natural gas—production: 0 cu m (2008 est.)
country comparison to the world: 144
Natural gas—consumption: 0 cu m (2008 est.)
country comparison to the world: 188
Natural gas—exports: 0 cu m (2008 est.)
country comparison to the world: 123
Natural gas—imports: 0 cu m (2008 est.)
country comparison to the world: 78
Natural gas—proved reserves: 0 cu m (1 January 2010 est.)
country comparison to the world: 151
Current account balance: $-21 million (2007 est.)
country comparison to the world: 65
Exports: $17 million (2004 est.)
country comparison to the world: 209
Exports—commodities: copra 62%, coconuts, seaweed, fish
Imports: $62 million (2004 est.)
country comparison to the world: 215
Imports—commodities: foodstuffs, machinery and equipment, miscellaneous manufactured goods, fuel
Debt—external: $10 million (1999 est.)
country comparison to the world: 190
Exchange rates: Australian dollars (AUD) per US dollar–1.0902 (2010), 1.2822 (2009), 1.2059 (2008), 1.2137 (2007), 1.3285 (2006)

COMMUNICATIONS

Telephones—main lines in use: 4,000 (2009)
country comparison to the world: 215
Telephones—mobilecellular: 1,000(2009)
country comparison to the world: 216
Telephone system: *general assessment:* generally good quality national and international service
domestic: wire line service available on Tarawa and Kiritimati (Christmas Island); connections to outer islands by HF/VHF radiotelephone; wireless service available in Tarawa since 1999
international: country code–686; Kiribati is being linked to the Pacific Ocean Cooperative Telecommunications Network, which should improve telephone service; satellite earth station–1 Intelsat (Pacific Ocean)
Broadcast media: 1 television broadcast station that provides about 1 hour of local programming Monday-Friday; multichannel TV packages provide access to Australian and US stations; 1 government-operated radio station broadcasting on AM, FM, and shortwave (2009)
Internet country code: .ki
Internet hosts: 31 (2010)
country comparison to the world: 214
Internet users: 7,800 (2009)
country comparison to the world: 203

TRANSPORTATION

Airports: 19 (2010)
country comparison to the world: 136
Airports—with paved runways: *total:* 4
1,524 to 2,437 m: 4 (2010)
Airports—with unpaved runways: *total:* 15
914 to 1,523 m: 11
under 914 m: 4 (2010)
Roadways: *total:* 670 km (2000)
country comparison to the world: 189
Waterways: 5 km (small network of canals in Line Islands) (2007)
country comparison to the world: 108
Merchant marine: *total:* 71
country comparison to the world: 61
by type: bulk carrier 6, cargo 32, chemical tanker 6, passenger/cargo 1, petroleum tanker 11, refrigerated cargo 15
foreign-owned: 51 (China 28, Hong Kong 1, Italy 1, Singapore 11, South Korea 2, Taiwan 5, Turkey 3) (2010)
Ports and terminals: Betio (Tarawa Atoll), Canton Island, English Harbor

MILITARY

Military branches: no regular military forces (establishment prevented by the constitution); Police Force (2011)
Manpower available for military service: *males age 16–49:* 25,190 (2010 est.)
Manpower fit for military service: *males age 16–49:* 18,364
females age 16–49: 20,302 (2010 est.)
Manpower reaching militarily significant age annually: *male:* 1,132
female: 1,120 (2010 est.)
Military expenditures: NA
Military—note: Kiribati does not have military forces; defense assistance is provided by Australia and NZ

TRANSNATIONAL ISSUES

Disputes—international: none

KOREA, NORTH

INTRODUCTION

Background: An independent kingdom for much of its long history, Korea was occupied by Japan beginning in 1905 following the Russo-Japanese War. Five years later, Japan formally annexed the entire peninsula. Following World War II, Korea was split with the northern half coming under Soviet-sponsored Communist control. After failing in the Korean War (1950-53) to conquer the US-backed Republic of Korea (ROK) in the southern portion by force, North Korea (DPRK), under its founder President KIM Il Sung, adopted a policy of ostensible diplomatic and economic "self-reliance" as a check against outside influence. The DPRK demonized the US as the ultimate threat to its social system through state-funded propaganda, and molded political, economic, and military policies around the core ideological objective of eventual unification of Korea under Pyongyang's control. KIM Il Sung's son, the current ruler KIM Jong Il, was officially designated as his father's successor in 1980, assuming a growing political and managerial role until the elder KIM's death in 1994. In 2010, KIM Jong Il began the process of preparing the way for his youngest son, KIM Jong Un, to succeed

him in power. After decades of economic mismanagement and resource misallocation, the DPRK since the mid-1990s has relied heavily on international aid to feed its population. North Korea's history of regional military provocations, proliferation of military-related items, long-range missile development, WMD programs including tests of nuclear devices in 2006 and 2009, and massive conventional armed forces are of major concern to the international community. The regime has marked 2012, the centenary of KIM Il Sung's birth, a banner year; to that end, the country has been focused on development of the economy.

GEOGRAPHY

Location: Eastern Asia, northern half of the Korean Peninsula bordering the Korea Bay and the Sea of Japan, between China and South Korea
Geographic coordinates: 40 00 N, 127 00 E
Map references: Asia
Area: *total:* 120,538 sq km
country comparison to the world: 98
land: 120,408 sq km
water: 130 sq km
Area—comparative: slightly smaller than Mississippi
Land boundaries: *total:* 1,673 km
border countries: China 1,416 km, South Korea 238 km, Russia 19 km
Coastline: 2,495 km
Maritime claims: *territorial sea:* 12 nm
exclusive economic zone: 200 nm
note: military boundary line 50 nm in the Sea of Japan and the exclusive economic zone limit in the Yellow Sea where all foreign vessels and aircraft without permission are banned
Climate: temperate with rainfall concentrated in summer
Terrain: mostly hills and mountains separated by deep, narrow valleys; coastal plains wide in west, discontinuous in east
Elevation extremes: *lowest point:* Sea of Japan 0 m
highest point: Paektu-san 2,744 m
Natural resources: coal, lead, tungsten, zinc, graphite, magnesite, iron ore, copper, gold, pyrites, salt, fluorspar, hydropower
Land use: *arable land:* 22.4%
permanent crops: 1.66%
other: 75.94% (2005)
Irrigated land: 14,600 sq km (2003)
Total renewable water resources: 77.1 cu km (1999)
Freshwater withdrawal (domestic/industrial/agricultural): *total:* 9.02 cu km/yr (20%/25%/55%)
per capita: 401 cu m/yr (2000)
Natural hazards: late spring droughts often followed by severe flooding; occasional typhoons during the early fall
volcanism: Changbaishan (elev. 2,744 m) (also known as Baitoushan, Baegdu or P'aektu-san), on the Chinese border, is considered historically active
Environment—current issues: water pollution; inadequate supplies of potable water; waterborne disease; deforestation; soil erosion and degradation
Environment—international agreements: *party to:* Antarctic Treaty, Biodiversity, Climate Change, Climate Change-Kyoto Protocol, Desertification, Environmental Modification, Hazardous Wastes, Ozone Layer Protection, Ship Pollution
signed, but not ratified: Law of the Sea
Geography—note: strategic location bordering China, South Korea, and Russia; mountainous interior is isolated and sparsely populated

PEOPLE

Population: 24,457,492 (July 2011 est.)
country comparison to the world: 48
Age structure: *0–14 years:* 22.4% (male 2,766,006/female 2,700,378)
15–64 years: 68.6% (male 8,345,737/ female 8,423,482)
65 years and over: 9.1% (male 738,693/ female 1,483,196) (2011 est.)
Median age:
total: 32.9 years
male: 31.2 years
female: 34.6 years (2011 est.)
Population growth rate: 0.538% (2011 est.)
country comparison to the world: 150
Birth rate: 14.51 births/1,000 population (2011 est.)
country comparison to the world: 141
Death rate: 9.08 deaths/1,000 population (July 2011 est.)
country comparison to the world: 68
Net migration rate: -0.04 migrant(s)/1,000 population (2011 est.)
country comparison to the world: 117
Urbanization: *urban population:* 60% of total population (2010)
rate of urbanization: 0.6% annual rate of change (2010-15 est.)
Sex ratio: *at birth:* 1.047 male(s)/female
under 15 years: 1.03 male(s)/female
15–64 years: 0.98 male(s)/female
65 years and over: 0.64 male(s)/female
total population: 0.95 male(s)/female (2011 est.)
Infant mortality rate: *total:* 27.11 deaths/1,000 live births
country comparison to the world: 76
male: 30.04 deaths/1,000 live births
female: 24.05 deaths/1,000 live births (2011 est.)
Life expectancy at birth: *total population:* 68.89 years
country comparison to the world: 149
male: 65.03 years
female: 72.93 years (2011 est.)
Total fertility rate: 2.02 children born/ woman (2011 est.)
country comparison to the world: 126
HIV/AIDS–adult prevalence rate: NA
HIV/AIDS—people living with HIV/AIDS: NA
HIV/AIDS—deaths: NA
Drinking water source: *Improved:*
urban: 100% of population
rural: 100% of population
total: 100% of population (2008)
Sanitation facility access: *Improved:*
urban: 58% of population
rural: 60% of population
total: 59% of population
Unimproved: urban: 42% of population
rural: 40% of population
total: 41% of population (2000)
Nationality: *noun:* Korean(s)
adjective: Korean
Ethnic groups: racially homogeneous; there is a small Chinese community and a few ethnic Japanese
Religions: traditionally Buddhist and Confucianist, some Christian and syncretic Chondogyo (Religion of the Heavenly Way)
note: autonomous religious activities now almost nonexistent; government-sponsored religious groups exist to provide illusion of religious freedom
Languages: Korean
Literacy: *definition:* age 15 and over can read and write
total population: 99%
male: 99%
female: 99% (1991 est.)
School life expectancy (primary to tertiary education): NA
Education expenditures: NA

GOVERNMENT

Country name: *conventional long form:* Democratic People's Republic of Korea
conventional short form: North Korea
local long form: Choson-minjujuui-inmin-konghwaguk
local short form: Choson
abbreviation: DPRK
Government type: Communist state one-man dictatorship
Capital: *name:* Pyongyang
geographic coordinates: 39 01 N, 125 45 E
time difference: UTC+9 (14 hours ahead of Washington, DC during Standard Time)
Administrative divisions: 9 provinces (do, singular and plural) and 2 municipalities (si, singular and plural)
provinces: Chagang-do (Chagang), Hamgyong-bukto (North Hamgyong), Hamgyong-namdo (South Hamgyong), Hwanghae-bukto (North Hwanghae), Hwanghae-namdo (South Hwanghae), Kangwon-do (Kangwon), P'yongan-bukto (North P'yongan), P'yongan-namdo (South P'yongan), Yanggang-do (Yanggang)
municipalities: Nason-si, P'yongyang-si (Pyongyang)
Independence: 15 August 1945 (from Japan)
National holiday: Founding of the Democratic People's Republic of Korea (DPRK), 9 September (1948)
Constitution: adopted 1948; revised several times most recently in 2009

Legal system: civil law system based on the Prussian model; system influenced by Japanese traditions and Communist legal theory

International law organization participation: has not submitted an ICJ jurisdiction declaration; non-party state to the ICCt

Suffrage: 17 years of age; universal

Executive branch: *chief of state:* KIM Jong Il (since July 1994); note—on 9 April 2009, rubberstamp Supreme People's Assembly (SPA) reelected KIM Jong Il chairman of the National Defense Commission, a position accorded nation's "highest administrative authority"; SPA reelected KIM Yong Nam in 2009 president of its Presidium also with responsibility of representing state and receiving diplomatic credentials

head of government: Premier CHOE Yong Rim (since 7 June 2010); Vice Premier HAN Kwang Bok (since 7 June 2010), Vice Premier JO Pyong Ju (since 7 June 2010), Vice Premier JON Ha Chol (since 7 June 2010), Vice Premier KANG Nung Su (since 7 June 2010), Vice Premier KIM Rak Hui (since 7 June 2010), Vice Premier PAK Su Gil (since 18 September 2009), Vice Premier RI Mu Yong (since 31 May 2011); Vice Premier RO Tu Chol (since 3 September 2003)

cabinet: Naegak (cabinet) members, except for Minister of People's Armed Forces, are appointed by SPA

(For more information visit the World Leaders website)

elections: last election held in September 2003; date of next election NA

election results: KIM Jong Il and KIM Yong Nam were only nominees for positions and ran unopposed

Legislative branch: unicameral Supreme People's Assembly or Ch'oego Inmin Hoeui (687 seats; members elected by popular vote to serve five-year terms)

elections: last held on 8 March 2009 (next to be held in March 2014)

election results: percent of vote by party—NA; seats by party—NA; ruling party approves a list of candidates who are elected without opposition; a token number of seats are reserved for minor parties

Judicial branch: Central Court (judges are elected by the Supreme People's Assembly)

Political parties and leaders: major party—Korean Workers' Party or KWP [KIM Jong Il]; minor parties—Chondoist Chongu Party [RYU Mi Yong] (under KWP control), Social Democratic Party [KIM Yong Dae] (under KWP control)

Political pressure groups and leaders: none

International organization participation: ARF, FAO, G-77, ICAO, ICRM, IFAD, IFRCS, IHO, IMO, IOC, IPU, ISO, ITSO, ITU, NAM, UN, UNCTAD, UNESCO, UNIDO, UNWTO, UPU, WFTU, WHO, WIPO, WMO

Diplomatic representation in the US: none; North Korea has a Permanent Mission to the UN in New York

Diplomatic representation from the US: none; note—Swedish Embassy in Pyongyang represents the US as consular protecting power

Flag description: three horizontal bands of blue (top), red (triple width), and blue; the red band is edged in white; on the hoist side of the red band is a white disk with a red five-pointed star; the broad red band symbolizes revolutionary traditions; the narrow white bands stands for purity, strength, and dignity; the blue bands signify sovereignty, peace, and friendship; the red star represents socialism

National anthem: *name:* "Aegukka" (Patriotic Song)

lyrics/music: PAK Se Yong/KIM Won Gyun

note: adopted 1947; both North Korea and South Korea's anthems share the same name and have a vaguely similar melody but have different lyrics; the North Korean anthem is also known as "Ach'imun pinnara" (Let Morning Shine)

ECONOMY

Economy—overview: North Korea, one of the world's most centrally directed and least open economies, faces chronic economic problems. Industrial capital stock is nearly beyond repair as a result of years of underinvestment, shortages of spare parts, and poor maintenance. Large-scale military spending draws off resources needed for investment and civilian consumption. Industrial and power output have stagnated for years at a fraction of pre-1990 levels. Frequent weather-related crop failures aggravated chronic food shortages caused by on-going systemic problems, including a lack of arable land, collective farming practices, poor soil quality, insufficient fertilization, and persistent shortages of tractors and fuel. Large-scale international food aid deliveries have allowed the people of North Korea to escape widespread starvation since famine threatened in 1995, but the population continues to suffer from prolonged malnutrition and poor living conditions. Since 2002, the government has allowed private "farmers' markets" to begin selling a wider range of goods. It also permitted some private farming—on an experimental basis—in an effort to boost agricultural output. In October 2005, the government tried to reverse some of these policies by forbidding private sales of grains and reinstituting a centralized food rationing system. By December 2005, the government terminated most international humanitarian assistance operations in North Korea (calling instead for developmental assistance only) and restricted the activities of remaining international and non-governmental aid organizations. In mid-2008, North Korea began receiving food aid under a US program to deliver 500,000 metric tons of food via the World Food Program and US nongovernmental organizations; but Pyongyang stopped accepting the aid in March 2009. In December 2009, North Korea carried out a redenomination of its currency, capping the amount of North Korean won that could be exchanged for the new notes, and limiting the exchange to a one-week window. A concurrent crackdown on markets and foreign currency use yielded severe shortages and inflation, forcing Pyongyang to ease the restrictions by February 2010. In response to the sinking of the South Korean destroyer Cheonan and the shelling of Yeonpyong Island, South Korea's government cut off most aid, trade, and bilateral cooperation activities, with the exception of operations at the Kaesong Industrial Complex. The year 2012 will be the 100th anniversary of Kim Il-sung's birthday. The North Korean government often highlights its 2012 goal of becoming a "strong and prosperous" nation. Attracting foreign investment, especially from neighboring China, will be a key factor for improving the overall standard of living. Nevertheless, firm political control remains the government's overriding concern, which likely will inhibit changes to North Korea's current economic system.

GDP (purchasing power parity): $40 billion (2009 est.)

country comparison to the world: 99

$40 billion (2009 est.)

$40 billion (2008 est.)

note: data are in 2010 US dollars;

North Korea does not publish reliable National Income Accounts data; the data shown here are derived from purchasing power parity (PPP) GDP estimates for North Korea that were made by Angus MADDISON in a study conducted for the OECD; his figure for 1999 was extrapolated to 2009 using estimated real growth rates for North Korea's GDP and an inflation factor based on the US GDP deflator; the results were rounded to the nearest $10 billion.

GDP (official exchange rate): $28 billion (2009 est.)

GDP—real growth rate: -0.9% (2009 est.)

country comparison to the world: 197

3.7% (2008 est.)

GDP—per capita (PPP): $1,800 (2009 est.)

country comparison to the world: 193

$1,800 (2009 est.)

$1,900 (2008 est.)

note: data are in 2010 US dollars

GDP—composition by sector: *agriculture:* 20.9%

industry: 46.9%

services: 32.1% (2002 est.)

Labor force: 12.2 million

country comparison to the world: 43

note: estimates vary widely (2009 est.)

Labor force—by occupation:

agriculture: 35%

industry and *services:* 65% (2008 est.)

Unemployment rate: NA%

Population below poverty line: NA%

Household income or consumption by percentage share: *lowest 10%:* NA%

highest 10%: NA%

Budget: *revenues:* $3.2 billion

expenditures: $3.3 billion (2007 est.)

Inflation rate (consumer prices): NA%

Agriculture—products: rice, corn, potatoes, soybeans, pulses; cattle, pigs, pork, eggs

Industries: military products; machine building, electric power, chemicals; mining (coal, iron ore, limestone, magnesite,

graphite, copper, zinc, lead, and precious metals), metallurgy; textiles, food processing; tourism
Industrial production growth rate: NA%
Electricity—production: 22.5 billion kWh (2008 est.)
country comparison to the world: 68
Electricity—consumption: 18.8 billion kWh (2008 est.)
country comparison to the world: 70
Electricity—exports: 0 kWh (2008 est.)
Electricity—imports: 0 kWh (2008 est.)
Oil—production: 118 bbl/day (2009 est.)
country comparison to the world: 110
Oil—consumption: 16,000 bbl/day (2009 est.)
country comparison to the world: 133
Oil—exports: 0 bbl/day (2007 est.)
country comparison to the world: 177
Oil—imports: 13,890 bbl/day (2007 est.)
country comparison to the world: 131
Oil—proved reserves: 0 bbl (1 January 2010 est.)
country comparison to the world: 147
Natural gas—production: 0 cu m (2008 est.)
country comparison to the world: 143
Natural gas—consumption: 0 cu m (2008 est.)
country comparison to the world: 187
Natural gas—exports: 0 cu m (2008 est.)
country comparison to the world: 122
Natural gas—imports: 0 cu m (2008 est.)
country comparison to the world: 77
Natural gas—proved reserves: 0 cu m (1 January 2010 est.)
country comparison to the world: 150
Exports: $1.997 billion (2009)
country comparison to the world: 133
$2.062 billion (2008)
Exports—commodities: minerals, metallurgical products, manufactures (including armaments), textiles, agricultural and fishery products
Exports—partners: South Korea 47%, China 40%, Hong Kong 2% (2009 est.)
Imports: $3.096 billion (2009)
country comparison to the world: 139
$3.574 billion (2008)
Imports—commodities: petroleum, coking coal, machinery and equipment, textiles, grain
Imports—partners: China 61%, South Korea 24%, Singapore 2%, India 2% (2009 est.)
Debt—external: $12.5 billion (2001 est.)
country comparison to the world: 85
Exchange rates: North Korean won (KPW) per US dollar (market rate), 1,800 (December 2010), 3,630 (December 2008) 140 (2007), 141 (2006)

COMMUNICATIONS

Telephones—main lines in use: 1.18 million (2008)
country comparison to the world: 71
Telephone system: *general assessment:* adequate system; nationwide fiber-optic network; mobile-cellular service expanding beyond Pyongyang
domestic: fiber-optic links installed down to the county level; telephone directories unavailable; GSM mobile-cellular service initiated in 2002 but suspended in 2004; Orascom Telecom Holding, an Egyptian company, launched W-CDMA mobile service on December 15, 2008 for the Pyongyang area and has expanded service to several large cities
international: country code—850; satellite earth stations—2 (1 Intelsat—Indian Ocean, 1 Russian—Indian Ocean region); other international connections through Moscow and Beijing (2009)
Broadcast media: no independent media; radios and televisions are pre-tuned to government stations; 4 government-owned television stations; the Korean Workers' Party owns and operates the Korean Central Broadcasting Station, and the state-run Voice of Korea operates an external broadcast service; the government prohibits listening to and jams foreign broadcasts (2008)
Internet country code: .kp
Internet hosts: 3 (2010)
country comparison to the world: 230

TRANSPORTATION

Airports: 79 (2010)
country comparison to the world: 70
Airports—with paved runways: *total:* 37
over 3,047 m: 2
2,438 to 3,047 m: 23
1,524 to 2,437 m: 7
914 to 1,523 m: 1
under 914 m: 4 (2010)
Airports—with unpaved runways: *total:* 42
2,438 to 3,047 m: 2
1,524 to 2,437 m: 18
914 to 1,523 m: 14
under 914 m: 8 (2010)
Heliports: 22 (2010)
Pipelines: oil 154 km (2010)
Railways: *total:* 5,242 km
country comparison to the world: 33
standard gauge: 5,242 km 1.435-m gauge (3,500 km electrified) (2009)
Roadways: *total:* 25,554 km
country comparison to the world: 103
paved: 724 km
unpaved: 24,830 km (2006)
Waterways: 2,250 km (most navigable only by small craft) (2010)
country comparison to the world: 38
Merchant marine: *total:* 158
country comparison to the world: 42
by type: bulk carrier 8, cargo 129, carrier 1, container 3, passenger/cargo 1, petroleum tanker 11, refrigerated cargo 3, roll on/roll off 2
foreign-owned: 19 (Belgium 1, China 1, Nigeria 1, Romania 1, Singapore 2, South Korea 1, Syria 6, UAE 6)
registered in other countries: 5 (Mongolia 1, Sierra Leone 1, unknown 3) (2010)
Ports and terminals: Ch'ongjin, Haeju, Hungnam (Hamhung), Namp'o, Senbong, Songnim, Sonbong (formerly Unggi), Wonsan

MILITARY

Military branches: North Korean People's Army: Ground Forces, Navy, Air Force; civil security forces (2005)
Military service age and obligation: 17 years of age (2004)
Manpower available for military service: *males age 16–49:* 6,515,279
females age 16–49: 6,418,693 (2010 est.)
Manpower fit for military service: *males age 16–49:* 4,836,567
females age 16–49: 5,230,137 (2010 est.)
Manpower reaching militarily significant age annually: *male:* 207,737
female: 204,553 (2010 est.)
Military expenditures: NA

TRANSNATIONAL ISSUES

Disputes—international: risking arrest, imprisonment, and deportation, tens of thousands of North Koreans cross into China to escape famine, economic privation, and political oppression; North Korea and China dispute the sovereignty of certain islands in Yalu and Tumen rivers; Military Demarcation Line within the 4-km wide Demilitarized Zone has separated North from South Korea since 1953; periodic incidents in the Yellow Sea with South Korea which claims the Northern Limiting Line as a maritime boundary; North Korea supports South Korea in rejecting Japan's claim to Liancourt Rocks (Tok-do/Take-shima)
Refugees and internally displaced persons: *IDPs:* undetermined (flooding in mid-2007 and famine during mid-1990s) (2007)
Trafficking in persons: current situation: North Korea is a source country for men, women, and children trafficked for the purposes of forced labor and commercial sexual exploitation; the most common form of trafficking involves North Korean women and girls who cross the border into China voluntarily; additionally, North Korean women and girls are lured out of North Korea to escape poor social and economic conditions by the promise of food, jobs, and freedom, only to be forced into prostitution, marriage, or exploitative labor arrangements once in China
tier rating: Tier 3—North Korea does not fully comply with minimum standards for the elimination of trafficking and is not making significant efforts to do so; the government does not acknowledge the existence of human rights abuses in the country or recognize trafficking, either within the country or transnationally; North Korea has not ratified the 2000 UN TIP Protocol (2008)
Illicit drugs: for years, from the 1970s into the 2000s, citizens of the Democratic People's Republic of (North) Korea (DPRK), many of them diplomatic employees of the government, were apprehended abroad while trafficking in narcotics, including two in Turkey in December 2004; police investigations in Taiwan and Japan in recent years have linked North Korea to large illicit shipments of heroin and methamphetamine, including an attempt by the North Korean merchant ship Pong Su to deliver 150 kg of heroin to Australia in April 2003

KOREA, SOUTH

INTRODUCTION

Background: An independent Korean state or collection of states has existed almost continuously for several millennia. Between its initial unification in the 7th century—from three predecessor Korean states—until the 20th century, Korea existed as a single independent country. In 1905, following the Russo-Japanese War, Korea became a protectorate of imperial Japan, and in 1910 it was annexed as a colony. Korea regained its independence following Japan's surrender to the United States in 1945. After World War II, a Republic of Korea (ROK) was set up in the southern half of the Korean Peninsula while a Communist-style government was installed in the north (the DPRK). During the Korean War (1950-53), US troops and UN forces fought alongside soldiers from the ROK to defend South Korea from DPRK attacks supported by China and the Soviet Union. An armistice was signed in 1953, splitting the peninsula along a demilitarized zone at about the 38th parallel. Thereafter, South Korea achieved rapid economic growth with per capita income rising to roughly 17 times the level of North Korea. In 1993, KIM Young-sam became South Korea's first civilian president following 32 years of military rule. South Korea today is a fully functioning modern democracy. President LEE Myung-bak has pursued a policy of global engagement since taking office in February 2008, highlighted by Seoul's hosting of the G-20 summit in November 2010. Serious tensions with North Korea have punctuated inter-Korean relations in recent years, including the North's sinking of the South Korean warship Cheonan in March 2010 and its artillery attack on South Korean soldiers and citizens in November 2010.

GEOGRAPHY

Location: Eastern Asia, southern half of the Korean Peninsula bordering the Sea of Japan and the Yellow Sea
Geographic coordinates: 37 00 N, 127 30 E
Map references: Asia
Area: *total:* 99,720 sq km
country comparison to the world: 108
land: 96,920 sq km
water: 2,800 sq km
Area—comparative: slightly larger than Indiana
Land boundaries: *total:* 238 km
border countries: North Korea 238 km
Coastline: 2,413 km
Maritime claims: *territorial sea:* 12 nm; between 3 nm and 12 nm in the Korea Strait
contiguous zone: 24 nm
exclusive economic zone: 200 nm
continental shelf: not specified
Climate: temperate, with rainfall heavier in summer than winter
Terrain: mostly hills and mountains; wide coastal plains in west and south
Elevation extremes: *lowest point:* Sea of Japan 0 m
highest point: Halla-san 1,950 m
Natural resources: coal, tungsten, graphite, molybdenum, lead, hydropower potential
Land use: *arable land:* 16.58%
permanent crops: 2.01%
other: 81.41% (2005)
Irrigated land: 8,780 sq km (2003)
Total renewable water resources: 69.7 cu km (1999)
Freshwater withdrawal (domestic/industrial/agricultural): *total:* 18.59 cu km/yr (36%/16%/48%)
per capita: 389 cu m/yr (2000)
Natural hazards: occasional typhoons bring high winds and floods; low-level seismic activity common in southwest
volcanism: Halla (elev. 1,950 m) is considered historically active although it has not erupted in many centuries
Environment—current issues: air pollution in large cities; acid rain; water pollution from the discharge of sewage and industrial effluents; drift net fishing
Environment—international agreements: *party to:* Antarctic-Environmental Protocol, Antarctic-Marine Living Resources, Antarctic Treaty, Biodiversity, Climate Change, Climate Change-Kyoto Protocol, Desertification, Endangered Species, Environmental Modification, Hazardous Wastes, Law of the Sea, Marine Dumping, Ozone Layer Protection, Ship Pollution, Tropical Timber 83, Tropical Timber 94, Wetlands, Whaling
signed, but not ratified: none of the selected agreements
Geography—note: strategic location on Korea Strait

PEOPLE

Population: 48,754,657 (July 2011 est.)
country comparison to the world: 26
Age structure: *0–14 years:* 15.7% (male 3,980,541/female 3,650,631)
15–64 years: 72.9% (male 18,151,023/female 17,400,809)
65 years and over: 11.4% (male 2,259,621/female 3,312,032) (2011 est.)
Median age: *total:* 38.4 years
male: 37 years
female: 39.8 years (2011 est.)
Population growth rate: 0.23% (2011 est.)
country comparison to the world: 178
Birth rate: 8.55 births/1,000 population (2011 est.)
country comparison to the world: 216
Death rate: 6.26 deaths/1,000 population (July 2011 est.)
country comparison to the world: 157
Net migration rate: 0 migrant(s)/1,000 population (2011 est.)
country comparison to the world: 92
Urbanization: *urban population:* 83% of total population (2010)
rate of urbanization: 0.6% annual rate of change (2010-15 est.)
Major cities—population: SEOUL (capital) 9.778 million; Busan (Pusan) 3.439 million; Incheon (Inch'on) 2.572 million; Daegu (Taegu) 2.458 million; Daejon (Taejon) 1.497 million (2009)
Sex ratio: *at birth:* 1.069 male(s)/female
under 15 years: 1.1 male(s)/female
15–64 years: 1.04 male(s)/female
65 years and over: 0.67 male(s)/female
total population: 1 male(s)/female (2011 est.)
Infant mortality rate: *total:* 4.16 deaths/1,000 live births
country comparison to the world: 198
male: 4.37 deaths/1,000 live births
female: 3.93 deaths/1,000 live births (2011 est.)
Life expectancy at birth: *total population:* 79.05 years
country comparison to the world: 41
male: 75.84 years
female: 82.49 years (2011 est.)
Total fertility rate: 1.23 children born/woman (2011 est.)
country comparison to the world: 218
HIV/AIDS—adult prevalence rate: less than 0.1% (2009 est.)
country comparison to the world: 139
HIV/AIDS—people living with HIV/AIDS: 9,500 (2009 est.)
country comparison to the world: 101
HIV/AIDS—deaths: fewer than 500 (2009 est.)
country comparison to the world: 88
Drinking water source: *Improved:*
urban: 100% of population
rural: 88% of population
total: 98% of population
Unimproved:
urban: 0% of population
rural: 12% of population
total: 2% of population (2008)
Sanitation facility access: *Improved:*
urban: 100% of population
rural: 100% of population
total: 100% of population (2008)
Nationality: *noun:* Korean(s)
adjective: Korean
Ethnic groups: homogeneous (except for about 20,000 Chinese)
Religions: Christian 26.3% (Protestant 19.7%, Roman Catholic 6.6%), Buddhist 23.2%, other or unknown 1.3%, none 49.3% (1995 census)
Languages: Korean, English (widely taught in junior high and high school)
Literacy: *definition:* age 15 and over can read and write

total population: 97.9%
male: 99.2%
female: 96.6% (2002)
School life expectancy (primary to tertiary education): *total:* 17 years
male: 18 years
female: 16 years (2008)
Education expenditures: 4.2% of GDP (2007)
country comparison to the world: 96

GOVERNMENT

Country name: *conventional long form:* Republic of Korea
conventional short form: South Korea
local long form: Taehan-min'guk
local short form: Han'guk
abbreviation: ROK
Government type: republic
Capital: *name:* Seoul
geographic coordinates: 37 33 N, 126 59 E
time difference: UTC+9 (14 hours ahead of Washington, DC during Standard Time)
Administrative divisions: 9 provinces (do, singular and plural) and 7 metropolitan cities (gwangyoksi, singular and plural)
provinces: Chungcheong-bukto (North Chungcheong), Chungcheong-namdo (South Chungcheong), Gangwon, Gyeonggi, Gyeongsang-bukto (North Gyeongsang), Gyeongsang-namdo (South Gyeongsang), Jeju, Jeolla-bukto (North Jeolla), Jeolla-namdo (South Jeolla)
metropolitan cities: Busan (Pusan), Daegu (Taegu), Daejon (Taejon), Gwangju (Kwangju), Incheon (Inch'on), Seoul, Ulsan
Independence: 15 August 1945 (from Japan)
National holiday: Liberation Day, 15 August (1945)
Constitution: 17 July 1948; note–amended or rewritten many times; current constitution approved 29 October 1987
Legal system: mixed legal system combining European civil law, Anglo-American law, and Chinese classical thought
International law organization participation: has not submitted an ICJ jurisdiction declaration; accepts ICCt jurisdiction
Suffrage: 19 years of age; universal
Executive branch: *chief of state:* President LEE Myung-bak (since 25 February 2008)
head of government: Prime Minister KIM Hwang-sik (since 1 October 2010)
cabinet: State Council appointed by the president on the prime minister's recommendation
(For more information visit the World Leaders website)
elections: president elected by popular vote for a single five-year term; election last held on 19 December 2007 (next to be held in December 2012); prime minister appointed by president with consent of National Assembly
election results: LEE Myung-bak elected president on 19 December 2007; percent of vote–LEE Myung-bak (GNP) 48.7%; CHUNG Dong-young (UNDP) 26.1%; LEE Hoi-chang (independent) 15.1; others 10.1%
Legislative branch: unicameral National Assembly or Kukhoe (299 seats; 245 members elected in single-seat constituencies, 54 elected by proportional representation; members serve four-year terms)
elections: last held on 9 April 2008 (next to be held in April 2012)
election results: percent of vote by party–NA; seats by party–GNP 172, UDP 83, LFP 20, PPA 8, DLP 5, RKP 1, independents 9
Judicial branch: Supreme Court (justices appointed by the president with consent of National Assembly); Constitutional Court (justices appointed by the president based partly on nominations by National Assembly and Chief Justice of the court)
Political parties and leaders: Democratic Party or DP [CHUNG Sye-kyun] (formerly the United Democratic Party or UDP); Democratic Labor Party or DLP [KANG Ki-kap]; Grand National Party or GNP [AHN Sang-soo]; Liberty Forward Party or LFP [LEE Hoi-chang]; New Progressive Party or NPP [ROH Hoe-chan]; Pro-Park Alliance or PPA [SUH Choung-won]; Renewal Korea Party or RKP [SONG Yong-o]
Political pressure groups and leaders: Federation of Korean Industries; Federation of Korean Trade Unions; Korean Confederation of Trade Unions; Korean National Council of Churches; Korean Traders Association; Korean Veterans' Association; National Council of Labor Unions; National Democratic Alliance of Korea; National Federation of Farmers' Associations; National Federation of Student Associations
International organization participation: ADB, AfDB (nonregional member), APEC, ARF, ASEAN (dialogue partner), Australia Group, BIS, CD, CICA, CP, EAS, EBRD, FAO, FATF, G-20, IADB, IAEA, IBRD, ICAO, ICC, ICRM, IDA, IEA, IFAD, IFC, IFRCS, IHO, ILO, IMF, IMO, IMSO, Interpol, IOC, IOM, IPU, ISO, ITSO, ITU, ITUC, LAIA (observer), MIGA, MINURSO, NEA, NSG, OAS (observer), OECD, OPCW, OSCE (partner), Paris Club (associate), PCA, PIF (partner), SAARC (observer), UN, UNAMID, UNCTAD, UNESCO, UNHCR, UNIDO, UNIFIL, UNMIL, UNMOGIP, UNOCI, UNWTO, UPU, WCO, WHO, WIPO, WMO, WTO, ZC
Diplomatic representation in the US: *chief of mission:* Ambassador HAN Duck-soo
chancery: 2450 Massachusetts Avenue NW, Washington, DC 20008
telephone: [1] (202) 939-5600
FAX: [1] (202) 387-0205
consulate(s) general: Agana (Guam), Atlanta, Boston, Chicago, Honolulu, Houston, Los Angeles, New York, San Francisco, Seattle
Diplomatic representation from the US: *chief of mission:* Ambassador Kathleen STEPHENS
embassy: 32 Sejongno, Jongno-gu, Seoul 110-710
mailing address: US Embassy Seoul, APO AP 96205-5550
telephone: [82] (2) 397-4114
FAX: [82] (2) 738-8845
Flag description: white with a red (top) and blue yin-yang symbol in the center; there is a different black trigram from the ancient I Ching (Book of Changes) in each corner of the white field; the Korean national flag is called Taegukki; white is a traditional Korean color and represents peace and purity; the blue section represents the negative cosmic forces of the yin, while the red symbolizes the opposite positive forces of the yang; each trigram (kwae) denotes one of the four universal elements, which together express the principle of movement and harmony
National anthem: *name:* "Aegukga" (Patriotic Song)
lyrics/music: YUN Ch'i-Ho or AN Ch'ang-Ho/AHN Eaktay
note: adopted 1948, well known by 1910; both North Korea and South Korea's anthems share the same name and have a vaguely similar melody but have different lyrics

ECONOMY

Economy—overview: Since the 1960s, South Korea has achieved an incredible record of growth and global integration to become a high-tech industrialized economy. Four decades ago, GDP per capita was comparable with levels in the poorer countries of Africa and Asia. In 2004, South Korea joined the trillion dollar club of world economies, and currently is among the world's 20 largest economies. Initially, a system of close government and business ties, including directed credit and import restrictions, made this success possible. The government promoted the import of raw materials and technology at the expense of consumer goods, and encouraged savings and investment over consumption. The Asian financial crisis of 1997-98 exposed longstanding weaknesses in South Korea's development model including high debt/equity ratios and massive short-term foreign borrowing. GDP plunged by 6.9% in 1998, and then recovered by 9% in 1999-2000. Korea adopted numerous economic reforms following the crisis, including greater openness to foreign investment and imports. Growth moderated to about 4-5% annually between 2003 and 2007. With the global economic downturn in late 2008, South Korean GDP growth slowed to 0.2% in 2009. In the third quarter of 2009, the economy began to recover, in large part due to export growth, low interest rates, and an expansionary fiscal policy, and growth exceeded 6% in 2010. The South Korean economy's long term challenges include a rapidly aging population, inflexible labor market, and overdependence on manufacturing exports to drive economic growth.
GDP (purchasing power parity): $1.459 trillion (2010 est.)
country comparison to the world: 13
$1.375 trillion (2009 est.)
$1.373 trillion (2008 est.)
note: data are in 2010 US dollars
GDP (official exchange rate): $1.007 trillion (2010 est.)
GDP—real growth rate: 6.1% (2010 est.)
country comparison to the world: 49
0.2% (2009 est.)
2.3% (2008 est.)
GDP—per capita (PPP): $30,000 (2010 est.)
country comparison to the world: 45
$28,300 (2009 est.)
$28,400 (2008 est.)
note: data are in 2010 US dollars

GDP—composition by sector:
agriculture: 3%
industry: 39.4%
services: 57.6% (2008 est.)
Labor force: 24.62 million (2010 est.)
country comparison to the world: 25
Labor force—by occupation:
agriculture: 7.3%
industry: 24.3%
services: 68.4% (2010 est.)
Unemployment rate: 3.3% (2010 est.)
country comparison to the world: 27
3.7% (2009 est.)
Population below poverty line: 15% (2006 est.)
Household income or consumption by percentage share: *lowest 10%:* 2.7%
highest 10%: 24.2% (2007)
Distribution of family income—Gini index: 31.4 (2009)
country comparison to the world: 106
35.8 (2000)
Investment (gross fixed): 28.7% of GDP (2010 est.)
country comparison to the world: 22
Budget: *revenues:* $248.3 billion
expenditures: $267.3 billion (2010 est.)
Public debt: 23.7% of GDP (2010 est.)
country comparison to the world: 103
23.5% of GDP (2009 est.)
Inflation rate (consumer prices): 3% (2010 est.)
country comparison to the world: 85
2.8% (2009 est.)
Central bank discount rate: 1.25% (31 December 2009)
country comparison to the world: 126
1.75% (31 December 2008)
Commercial bank prime lending rate: 5.65% (31 December 2009 est.)
country comparison to the world: 124
7.17% (31 December 2008 est.)
Stock of narrow money: $101.9 billion (31 December 2010 est.)
country comparison to the world: 31
$82.54 billion (31 December 2009 est.)
Stock of broad money: $1.346 trillion (31 December 2009)
country comparison to the world: 13
$1.132 trillion (31 December 2008)
Stock of domestic credit: $1.057 trillion (31 December 2010 est.)
country comparison to the world: 15
$935.4 billion (31 December 2009 est.)
Market value of publicly traded shares: $836.5 billion (31 December 2009)
country comparison to the world: 18
$494.6 billion (31 December 2008)
$1.124 trillion (31 December 2007)
Agriculture—products: rice, root crops, barley, vegetables, fruit; cattle, pigs, chickens, milk, eggs; fish
Industries: electronics, telecommunications, automobile production, chemicals, shipbuilding, steel
Industrial production growth rate: 12.1% (2010 est.)
country comparison to the world: 11
Electricity–production: 417 billion kWh (2009 est.)
country comparison to the world: 11
Electricity—consumption: 402 billion kWh (2009 est.)
country comparison to the world: 11
Electricity–exports: 0 kWh (2010)
Electricity—imports: 0 kWh (2010)
Oil—production: 48,180 bbl/day (2010 est.)
country comparison to the world: 65
Oil—consumption: 2.185 million bbl/day (2010 est.)
country comparison to the world: 10
Oil—exports: 907,100 bbl/day
country comparison to the world: 21
note: exports consist of oil derivatives (gasoline, light oil, and diesel), not crude oil (2009)
Oil—imports: 3.074 million bbl/day (2009)
country comparison to the world: 5
Oil—proved reserves: 0 bbl (1 January 2010 est.)
country comparison to the world: 149
Natural gas—production: 651 million cu m (2009 est.)
country comparison to the world: 66
Natural gas—consumption: 34.09 billion cu m (2009 est.)
country comparison to the world: 25
Natural gas—exports: 0 cu m (2010 est.)
country comparison to the world: 124
Natural gas—imports: 32.69 billion cu m (2009 est.)
country comparison to the world: 10
Natural gas—proved reserves: 50 billion cu m (1 January 2008 est.)
country comparison to the world: 64
Current account balance: $36.35 billion (2010 est.)
country comparison to the world: 12
$42.67 billion (2009 est.)
Exports: $466.3 billion (2010 est.)
country comparison to the world: 7
$373.6 billion (2009 est.)
Exports—commodities: semiconductors, wireless telecommunications equipment, motor vehicles, computers, steel, ships, petrochemicals
Exports—partners: China 23.2%, US 10.1%, Japan 5.8%, Hong Kong 5.3% (2009 est.)
Imports: $417.9 billion (2010 est.)
country comparison to the world: 10
$317.5 billion (2009 est.)
Imports—commodities: machinery, electronics and electronic equipment, oil, steel, transport equipment, organic chemicals, plastics
Imports—partners: China 16.8%, Japan 15.3%, US 9%, Saudi Arabia 6.1%, Australia 4.6% (2009 est.)
Reserves of foreign exchange and gold: $274.6 billion (31 December 2010 est.)
country comparison to the world: 8
$270 billion (31 December 2009 est.)
Debt—external: $370.1 billion (31 December 2010 est.)
country comparison to the world: 25
$370.8 billion (31 December 2009 est.)
Stock of direct foreign investment—at home: $112.1 billion (31 December 2010 est.)
country comparison to the world: 30
$110.8 billion (31 December 2009 est.)
Stock of direct foreign investment—abroad: $115.6 billion (31 December 2009)
country comparison to the world: 25
$74.6 billion (30 June 2008)
Exchange rates: South Korean won (KRW) per US dollar–1,153.77 (2010), 1,276.93 (2009), 1,101.7 (2008), 929.2 (2007), 954.8 (2006)

COMMUNICATIONS

Telephones—main lines in use: 19.289 million (2009)
country comparison to the world: 15
Telephones—mobile cellular: 47.944 million (2009)
country comparison to the world: 25
Telephone system: *general assessment:* excellent domestic and international services featuring rapid incorporation of new technologies
domestic: fixed-line and mobile-cellular services widely available with a combined telephone subscribership of roughly 140 per 100 persons; rapid assimilation of a full range of telecommunications technologies leading to a boom in e-commerce
international: country code–82; numerous submarine cables provide links throughout Asia, Australia, the Middle East, Europe, and US; satellite earth stations–66
Broadcast media: multiple national television networks with 2 of the 3 largest networks publicly operated; the largest privately-owned network, Seoul Broadcasting Service (SBS), has ties with other commercial TV networks; cable and satellite TV subscription services are available; publicly-operated radio broadcast networks and a large number of privately-owned radio broadcasting networks, each with multiple affiliates, and independent local stations (2010)
Internet country code: .kr
Internet hosts: 291,329 (2010)
country comparison to the world: 58
Internet users: 39.4 million (2009)
country comparison to the world: 11

TRANSPORTATION

Airports: 116 (2010)
country comparison to the world: 53
Airports—with paved runways: *total:* 72
over 3,047 m: 4
2,438 to 3,047 m: 21
1,524 to 2,437 m: 13
914 to 1,523 m: 12
under 914 m: 22 (2010)
Airports—with unpaved runways: *total:* 44
914 to 1,523 m: 2
under 914 m: 42 (2010)
Heliports: 510 (2010)
Pipelines: gas 2,139 km; refined products 864 km (2010)
Railways: *total:* 3,381 km
country comparison to the world: 50
standard gauge: 3,381 km 1.435-m gauge (1,843 km electrified) (2008)
Roadways: *total:* 103,029 km
country comparison to the world: 40
paved: 80,642 km (includes 3,367 km of expressways)
unpaved: 22,387 km (2008)
Waterways: 1,608 km (most navigable only by small craft) (2010)
country comparison to the world: 49
Merchant marine: *total:* 819
country comparison to the world: 14
by type: bulk carrier 201, cargo 246, carrier 5, chemical tanker 132, container 69, liquefied gas 40, passenger 5, passenger/cargo 21, petroleum tanker 67, refrigerated cargo 15, roll on/roll off 9, vehicle carrier 9
foreign-owned: 33 (China 9, France 1, Japan 15, US 8)

registered in other countries: 438 (Cambodia 11, Ghana 1, Honduras 6, Hong Kong 3, Indonesia 1, Kiribati 2, Liberia 1, Malta 3, Marshall Islands 25, North Korea 1, Panama 366, Philippines 1, Russia 1, Singapore 9, Tuvalu 1, unknown 6) (2010)
Ports and terminals: Incheon (Inch'on), Pohang (P'ohang), Busan (Pusan), Ulsan, Yeosu (Yosu)

MILITARY

Military branches: Republic of Korea Army, Navy (includes Marine Corps), Air Force (2011)
Military service age and obligation: 20-30 years of age for compulsory military service, with middle school education required; conscript service obligation—21 months (Army, Marines), 23 months (Navy), 24 months (Air Force); 18-26 years of age for voluntary military service; women, in service since 1950, admitted to 7 service branches, including infantry, but excluded from artillery, armor, anti-air, and chaplaincy corps; some 4,000 women serve as commissioned and noncommissioned officers, approx. 2.3% of all officers; HIV-positive individuals are exempt from military service (2010)
Manpower available for military service: *males age 16–49:* 13,185,794
females age 16–49: 12,423,496 (2010 est.)
Manpower fit for military service: *males age 16–49:* 10,864,566
females age 16–49: 10,168,709 (2010 est.)
Manpower reaching militarily significant age annually: *male:* 365,760
female: 321,225 (2010 est.)
Military expenditures: 2.7% of GDP (2006)
country comparison to the world: 52
Transnational Issues
Disputes—international: Military Demarcation Line within the 4-km wide Demilitarized Zone has separated North from South Korea since 1953; periodic incidents with North Korea in the Yellow Sea over the Northern Limit Line, which South Korea claims as a maritime boundary; South Korea and Japan claim Liancourt Rocks (Tok-do/Take-shima), occupied by South Korea since 1954

KOSOVO

INTRODUCTION

Background: The central Balkans were part of the Roman and Byzantine Empires before ethnic Serbs migrated to the territories of modern Kosovo in the 7th century. During the medieval period, Kosovo became the center of a Serbian Empire and saw the construction of many important Serb religious sites, including many architecturally significant Serbian Orthodox monasteries. The defeat of Serbian forces at the Battle of Kosovo in 1389 led to five centuries of Ottoman rule during which large numbers of Turks and Albanians moved to Kosovo. By the end of the 19th century, Albanians replaced the Serbs as the dominant ethnic group in Kosovo. Serbia reacquired control over Kosovo from the Ottoman Empire during the First Balkan War of 1912. After World War II, Kosovo became an autonomous province of Serbia in the Socialist Federal Republic of Yugoslavia (S.F.R.Y.) with status almost equivalent to that of a republic under the 1974 S.F.R.Y. constitution. Despite legislative concessions, Albanian nationalism increased in the 1980s, which led to riots and calls for Kosovo's independence. At the same time, Serb nationalist leaders, such as Slobodan MILOSEVIC, exploited Kosovo Serb claims of maltreatment to secure votes from supporters, many of whom viewed Kosovo as their cultural heartland. Under MILOSEVIC's leadership, Serbia instituted a new constitution in 1989 that revoked Kosovo's status as an autonomous province of Serbia. Kosovo Albanian leaders responded in 1991 by organizing a referendum that declared Kosovo independent. Under MILOSEVIC, Serbia carried out repressive measures against the Albanians in the early 1990s as the unofficial Kosovo government, led by Ibrahim RUGOVA, used passive resistance in an attempt to try to gain international assistance and recognition of an independent Kosovo. Albanians dissatisfied with RUGOVA's passive strategy in the 1990s created the Kosovo Liberation Army and launched an insurgency. Starting in 1998, Serbian military, police, and paramilitary forces under MILOSEVIC conducted a brutal counterinsurgency campaign that resulted in massacres and massive expulsions of ethnic Albanians. Approximately 800,000 Albanians were forced from their homes in Kosovo during this time. International attempts to mediate the conflict failed, and MILOSEVIC's rejection of a proposed settlement led to a three-month NATO military operation against Serbia beginning in March 1999 that forced Serbia to agree to withdraw its military and police forces from Kosovo. UN Security Council Resolution 1244 (1999) placed Kosovo under a transitional administration, the UN Interim Administration Mission in Kosovo (UNMIK), pending a determination of Kosovo's future status. A UN-led process began in late 2005 to determine Kosovo's final status. The negotiations ran in stages between 2006 and 2007, but ended without agreement between Belgrade and Pristina. On 17 February 2008, the Kosovo Assembly declared Kosovo independent. Since then, over 70 countries have recognized Kosovo, and it has joined the International Monetary Fund and World Bank. Serbia continues to reject Kosovo's independence and in October 2008, it sought an advisory opinion from the International Court of Justice (ICJ) on the legality under international law of Kosovo's declaration of independence. The ICJ released the advisory opinion in July 2010 affirming that Kosovo's declaration of independence did not violate general principles of international law, UN Security Council Resolution 1244, or the Constitutive Framework. The opinion was closely tailored to Kosovo's unique history and circumstances.

GEOGRAPHY

Location: Southeast Europe, between Serbia and Macedonia
Geographic coordinates: 42 35 N, 21 00 E
Map references: Europe
Area: *total:* 10,887 sq km
country comparison to the world: 168
land: 10,887 sq km
water: 0 sq km
Area—comparative: slightly larger than Delaware
Land boundaries: *total:* 702 km
border countries: Albania 112 km, Macedonia 159 km, Montenegro 79 km, Serbia 352 km
Coastline: 0 km (landlocked)
Maritime claims: none (landlocked)
Climate: influenced by continental air masses resulting in relatively cold winters with heavy snowfall and hot, dry summers and autumns; Mediterranean and alpine influences create regional variation; maximum rainfall between October and December
Terrain: flat fluvial basin with an elevation of 400-700 m above sea level surrounded by several high mountain ranges with elevations of 2,000 to 2,500 m
Elevation extremes: *lowest point:* Drini i Bardhe/Beli Drim 297 m (located on the border with Albania)
highest point: Gjeravica/Deravica 2,656 m
Natural resources: nickel, lead, zinc, magnesium, lignite, kaolin, chrome, bauxite

PEOPLE

Population: 1,825,632 (July 2011 est.)
country comparison to the world: 147
Age structure: *0–14 years:* 27.2% (male 258,078/female 237,987)
15–64 years: 66.1% (male 630,350/female 576,946)
65 years and over: 6.7% (male 51,668/female 70,603) (2011 est.)
Median age: *total:* 26.7 years
male: 26.3 years
female: 27.2 years (2011 est.)
Sex ratio: *at birth:* 1.085 male(s)/female

under 15 years: 1.09 male(s)/female
15–64 years: 1.09 male(s)/female
65 years and over: 0.74 male(s)/female
total population: 1.06 male(s)/female (2011 est.)

Nationality: *noun:* Kosovar (Albanian), Kosovac (Serbian)
adjective: Kosovar (Albanian), Kosovski (Serbian)
note: Kosovan, a neutral term, is sometimes also used as a noun or adjective

Ethnic groups: Albanians 92%, other (Serb, Bosniak, Gorani, Roma, Turk, Ashkali, Egyptian) 8% (2008)

Religions: Muslim, Serbian Orthodox, Roman Catholic

Languages: Albanian (official), Serbian (official), Bosnian, Turkish, Roma

Literacy: *definition:* age 15 and over can read and write
total population: 91.9%
male: 96.6%
female: 87.5% (2007 Census)

School life expectancy (primary to tertiary education): NA

Education expenditures: 4.3% of GDP (2008)
country comparison to the world: 92

GOVERNMENT

Country name: *conventional long form:* Republic of Kosovo
conventional short form: Kosovo
local long form: Republika e Kosoves (Republika Kosovo)
local short form: Kosova (Kosovo)

Government type: republic

Capital: *name:* Pristina (Prishtine, Prishtina)
geographic coordinates: 42 40 N, 21 10 E
time difference: UTC+1 (6 hours ahead of Washington, DC during Standard Time)

daylight saving time: +1hr, begins last Sunday in March; ends last Sunday in October

Administrative divisions: 30 municipalities (komunat, singular–komuna in Albanian; opstine, singular–opstina in Serbian); Decan (Decani), Dragash (Dragas), Ferizaj (Urosevac), Fushe Kosove (Kosovo Polje), Gjakove (Dakovica), Gjilan (Gnjilane), Gllogovc/Drenas (Glogovac), Istog (Istok), Kacanik, Kamenice/Dardana (Kamenica), Kline (Klina), Leposaviq (Leposavic), Lipjan (Lipljan), Malisheve (Malisevo), Mitrovice (Mitrovica), Novoberde (Novo Brdo), Obiliq (Obilic), Peje (Pec), Podujeve (Podujevo), Prishtine (Pristina), Prizren, Rahovec (Orahovac), Shterpce (Strpce), Shtime (Stimlje), Skenderaj (Srbica), Suhareke (Suva Reka), Viti (Vitina), Vushtrri (Vucitrn), Zubin Potok, Zvecan
note–the Government of Kosovo has announced the establishment of 8 additional municipalities in accordance with UN Special Envoy AHTISAARI's mandated decentralization process; the boundaries of several municipalities are pending final approval; the municipalities are: Gracanice (Gracanica), Hani i Elezit (Dzeneral Jankovic), Junik, Kllokot-Verboc (Klokot-Vrbovac), Mamushe (Mamusa), Partes, Ranillug (Ranilug); in addition, the current Mitrovice (Mitrovica) municipality is to be split into Mitrovice (Mitrovica) North and Mitrovice (Mitrovica) South

Independence: 17 February 2008 (from Serbia)

National holiday: Independence Day, 17 February (2008)

Constitution: adopted by the Kosovo Assembly 9 April 2008; effective 15 June 2008

Legal system: evolving legal system; mixture of applicable Kosovo law, UNMIK laws and regulations, and applicable laws of the Former Socialist Republic of Yugoslavia that were in effect in Kosovo as of 22 March 1989

International law organization participation: has not submitted an ICJ jurisdiction declaration; non-party state to the ICCt

Suffrage: 18 years of age; universal

Executive branch: *chief of state:* President Atifete JAHJAGA (since 7 April 2011);
head of government: Prime Minister Hashim THACI (since 9 January 2008)
cabinet: ministers; elected by the Kosovo Assembly
(For more information visit the World Leaders website)
elections: the president is elected for a five-year term by the Kosovo Assembly; election last held on 7 April 2011; note–the prime minister elected by the Kosovo Assembly
election results: Atifete JAHJAGA elected in one round (JAHJAGA 80, Suzana NOVOBERDALIU 10); Hashim THACI elected prime minister by the Assembly

Legislative branch: unicameral national Assembly (120 seats; 100 seats directly elected, 10 seats guaranteed for ethnic Serbs, 10 seats guaranteed for other ethnic minorities; members to serve four-year terms)
elections: last held on 12 December 2010 with runoff elections in a few municipalities in January 2011 (next expected to be held in 2015)
election results: percent of vote by party–NA; seats by party–NA; note–2010 extraordinary assembly election results were announced by the Central Elections Commission 30 January 2011; certification of the results was still pending as of 31 January

Judicial branch: Supreme Court; Appellate Court; basic courts
note: the Law on Courts, which went into effect on 1 January 2011, provided for a reorganization of the court system; the Kosovo Constitution dictates that the Supreme Court of Kosovo is the highest judicial authority, and provides for a Kosovo Judicial Council (KJC) that proposes to the president candidates for appointment or reappointment as judges and prosecutors; the KJC is also responsible for decisions on the promotion and transfer of judges and disciplinary proceedings against judges; at least 15 percent of Supreme Court and district court judges shall be from non-majority communities

Political parties and leaders: Albanian Christian Democratic Party of Kosovo or PShDK [Marjan DEMAJ]; Alliance for a New Kosovo or AKR [Behgjet PACOLLI]; Alliance for the Future of Kosovo or AAK [Ramush HARADINAJ]; Alliance of Independent Social Democrats of Kosovo and Metohija or SDSKIM [Ljubisa ZIVIC]; Bosniak Vakat Coalition or DSV [Sadik IDRIZI]; Citizens' Initiative of Gora or GIG [Murselj HALJILJI]; Democratic Action Party or SDA [Numan BALIC]; Democratic League of Dardania or LDD [Nexhat DACI]; Democratic League of Kosovo or LDK [Isa MUSTAFA]; Democratic Party of Ashkali of Kosovo or PDAK [Berat QERIMI]; Democratic Party of Bosniaks [Dzezair MURATI]; Democratic Party of Kosovo or PDK [Hashim THACI]; Independent Liberal Party or SLS [Slobadan PETROVIC]; Kosovo Democratic Turkish Party of KDTP [Mahir YAGCILAR]; Movement for Self-Determination (Vetevendosje) [Albin KKURTI]; New Democratic Initiative of Kosovo or IRDK [Xhevdet NEZIRAJ]; New Democratic Party or ND [Predrag JOVIC]; New Spirit or FER [Shpend AHMETI]; Serb National Party or SNS [Mihailo SCEPANOVIC]; Serbian Democratic Party of Kosovo and Metohija or SDS KiM [Slavisa PETKOVIC]; Serbian Kosovo and Metohija Party or SKMS [Dragisa MIRIC]; Serbian National Council of Northern Kosovo and Metohija or SNV [Milan IVANOVIC]; Social Democratic Party of Kosovo or PSDK [Agim CEKU]; Socialist Party of Kosovo or PSK [Emrush XHEMAJLI]; United Roma Party of Kosovo or PREBK [Ilaz KADOLLI]

Political pressure groups and leaders: Council for the Defense of Human Rights and Freedom (human rights); Organization for Democracy, Anti-Corruption and Dignity Rise! [Avni ZOGIANI]; Serb National Council (SNV); The Speak Up Movement [Ramadan ILAZI]

International organization participation: IBRD, IDA, IFC, IMF, ITUC, MIGA

Diplomatic representation in the US: *chief of mission:* Ambassador Avni SPAHIU
chancery: 1101 30th Street NW, Suites 330/340, Washington, DC 20007
telephone: 202-380-3581
FAX: 202-380-3628
consulate(s) general: New York

Diplomatic representation from the US: *chief of mission:* Ambassador Christopher William DELL
embassy: Arberia/Dragodan, Nazim Hikmet 30, Pristina, Kosovo
mailing address: use embassy street address
telephone: [381] 38 59 59 3000
FAX: [381] 38 549 890

Flag description: centered on a dark blue field is the geographical shape of Kosovo in a gold color surmounted by six white, five-pointed stars arrayed in a slight arc; each star represents one of the major ethnic groups of Kosovo: Albanians, Serbs, Turks, Gorani, Roma, and Bosniaks

National anthem: *name:* "Europe"
lyrics/music: none/Mendi MENGJIQI
note: adopted 2008; Kosovo chose to not include lyrics in its anthem so as not to offend minority ethnic groups in the country

ECONOMY

Economy—overview: Over the past few years Kosovo's economy has shown significant progress in transitioning to a market-based system and maintaining macroeconomic stability, but it is still highly dependent on the international community and the diaspora for financial and technical assistance. Remittances from the diaspora–located mainly in Germany, Switzerland, and the Nordic countries–are estimated to account for about 13-15% of GDP, and donor-financed activities and aid for another 7.5%. Kosovo's citizens are the poorest in Europe with an average annual per capita income of only $2,800. Unemployment, around 40% of the population, is a significant problem that encourages outward migration and black market activity. Most of Kosovo's population lives in rural towns outside of the capital, Pristina. Inefficient, near-subsistence farming is common–the result of small plots, limited mechanization, and lack of technical expertise. With international assistance, Kosovo has been able to privatize 50% of its state-owned enterprises (SOEs) by number, and over 90% of SOEs by value. Minerals and metals–including lignite, lead, zinc, nickel, chrome, aluminum, magnesium, and a wide variety of construction materials–once formed the backbone of industry, but output has declined because of ageing equipment and insufficient investment. A limited and unreliable electricity supply due to technical and financial problems is a major impediment to economic development, but Kosovo has received technical assistance to help improve accounting and controls. The US Government is cooperating with the Ministry for Energy and Mines and the World Bank to prepare a commercial tender for a project to include construction of a new power plant and the development of a coal mine to supply the new power plant as well as two existing plants. Privatization of the distribution and supply divisions of Kosovo Energy Corporation is also planned. The official currency of Kosovo is the euro, but the Serbian dinar is also used in Serb enclaves. Kosovo's tie to the euro has helped keep core inflation low. Kosovo has one of the most open economies in the region, and continues to work with the international community on measures to improve the business environment and attract foreign investment. Kosovo has maintained a budget surplus as a result of efficient value added tax (VAT) collection at the borders and inefficient budget execution. In order to help integrate Kosovo into regional economic structures, UNMIK signed (on behalf of Kosovo) its accession to the Central Europe Free Trade Area (CEFTA) in 2006. However, Serbia and Bosnia have refused to recognize Kosovo's customs stamp or extend reduced tariff privileges for Kosovo products under CEFTA. In July 2008, Kosovo received pledges of $1.9 billion from 37 countries in support of its reform priorities. In June 2009, Kosovo joined the World Bank and International Monetary Fund, and Kosovo began servicing its share of the former Yugoslavia's debt.
GDP (purchasing power parity): $11.97 billion (2010 est.)
country comparison to the world: 144
$11.51 billion (2009 est.)
$11.19 billion (2008 est.)
GDP (official exchange rate): $5.601 billion (2010 est.)
GDP—real growth rate: 4% (2010 est.)
country comparison to the world: 98
2.9% (2009 est.)
6.9% (2008 est.)
GDP—per capita (PPP): $6,600 (2010 est.)
country comparison to the world: 135
$6,400 (2009 est.)
$5,300 (2008 est.)
GDP—composition by sector: *agriculture:* 12.9%
industry: 22.6%
services: 64.5% (2009 est.)
Labor force: 310,000 (2009 est.)
country comparison to the world: 164
Labor force—by occupation:
agriculture: 23.6%
industry: NA
services: NA (2010)
Unemployment rate: 45% (2009 est.)
country comparison to the world: 189
Population below poverty line: 30% (2010 est.)
Distribution of family income—Gini index: 30 (FY05/06)
country comparison to the world: 115
Investment (gross fixed): 35% of GDP (2010 est.)
country comparison to the world: 9
Budget: *revenues:* $1.458 billion
expenditures: $1.581 billion (2010 est.)
Public debt: NA% of GDP (2010)
7% of GDP (2009 est.)
Inflation rate (consumer prices): 3.5% (2010 est.)
country comparison to the world: 100
Commercial bank prime lending rate: 14.4% (30 June 2010 est.)
country comparison to the world: 52
14.09% (31 December 2009 est.)
Agriculture—products: wheat, corn, berries, potatoes, peppers
Industries: mineral mining, construction materials, base metals, leather, machinery, appliances
Electricity—production: 4.777 billion kWh (2009)
country comparison to the world: 114
Electricity—consumption: 5.388 billion kWh (2009)
country comparison to the world: 110
Oil–production: 0 bbl/day (2007)
country comparison to the world: 191
Oil—consumption: NA bbl/day
Oil—proved reserves: NA bbl
Natural gas—production: 0 cu m (2007)
country comparison to the world: 145
Natural gas—consumption: 0 cu m (2007)
country comparison to the world: 189
Natural gas—proved reserves: NA cu m
Current account balance: $-2.716 billion (2010 est.)
country comparison to the world: 166
$-2.77 billion (2009 est.)
Exports: $527 million (2007 est.)
country comparison to the world: 166
Exports–commodities: mining and processed metal products, scrap metals, leather products, machinery, appliances
Imports: $2.6 billion (2007 est.) (2007 est.)
country comparison to the world: 146
Imports–commodities: foodstuffs, wood, petroleum, chemicals, machinery and electrical equipment
Reserves of foreign exchange and gold: $NA
Debt—external: $NA
Stock of direct foreign investment—at home: $21.2 billion (31 December 2010 est.)
country comparison to the world: 67
$21.32 billion (31 December 2009 est.)
Exchange rates: euros (EUR) per US dollar–0.755 (2010), 0.7198 (2009), 0.6827 (2008), 0.7345 (2007)

COMMUNICATIONS

Telephones—main lines in use: 106,300 (2006)
country comparison to the world: 144
Telephones—mobile cellular: 562,000 (2007)
country comparison to the world: 157

TRANSPORTATION

Airports: 8 (2010)
country comparison to the world: 162
Airports—with paved runways: *total:* 4
2,438 to 3,047 m: 1
1,524 to 2,437 m: 1
under 914 m: 2 (2010)
Airports—with unpaved runways:
total: 4
under 914 m: 4 (2010)
Heliports: 2 (2010)
Railways: *total:* 430 km
country comparison to the world: 114
standard gauge: 430 km 1.435-m gauge (2007)
Roadways: *total:* 1,926 km
country comparison to the world: 175
paved: 1,668 km
unpaved: 258 km (2009)

MILITARY

Military branches: Kosovo Security Force (2010)
Manpower fit for military service: *males age 16–49:* 430,926
females age 16–49: 389,614 (2010 est.)

TRANSNATIONAL ISSUES

Disputes—international: Serbia with several other states protest the US and other states' recognition of Kosovo's declaring itself as a sovereign and independent state in February 2008; ethnic Serbian municipalities along Kosovo's northern border challenge final status of Kosovo-Serbia boundary; several thousand NATO-led KFOR peacekeepers under UNMIK authority continue to keep the peace within Kosovo between the ethnic Albanian majority and the Serb minority in Kosovo; Kosovo and Macedonia completed demarcation of their boundary in September 2008
Refugees and internally displaced persons: IDP's: 21,000 (2007)

KUWAIT

INTRODUCTION

Background: Britain oversaw foreign relations and defense for the ruling Kuwaiti AL-SABAH dynasty from 1899 until independence in 1961. Kuwait was attacked and overrun by Iraq on 2 August 1990. Following several weeks of aerial bombardment, a US-led, UN coalition began a ground assault on 23 February 1991 that liberated Kuwait in four days. Kuwait spent more than $5 billion to repair oil infrastructure damaged during 1990-91. The AL-SABAH family has ruled since returning to power in 1991 and reestablished an elected legislature that in recent years has become increasingly assertive. The country witnessed the historic election in May 2009 of four women to its National Assembly. Amid the 2010-11 uprisings and protests across the Arab world, stateless Arabs, known as bidoon, staged small protests in February and March 2011 demanding citizenship, jobs, and other benefits available to Kuwaiti nationals. Youth activist groups–supported by opposition legislators and the prime minister's rivals within the ruling family–in March of 2011 rallied for an end to corruption and the prime minister's ouster.

GEOGRAPHY

Location: Middle East, bordering the Persian Gulf, between Iraq and Saudi Arabia
Geographic coordinates: 29 30 N, 45 45 E
Map references: Middle East
Area: *total:* 17,818 sq km
country comparison to the world: 157
land: 17,818 sq km
water: 0 sq km
Area—comparative: slightly smaller than New Jersey
Land boundaries: *total:* 462 km
border countries: Iraq 240 km, Saudi Arabia 222 km
Coastline: 499 km
Maritime claims: *territorial sea:* 12 nm
Climate: dry desert; intensely hot summers; short, cool winters
Terrain: flat to slightly undulating desert plain
Elevation extremes: *lowest point:* Persian Gulf 0 m
highest point: unnamed elevation 306 m
Natural resources: petroleum, fish, shrimp, natural gas
Land use: *arable land:* 0.84%
permanent crops: 0.17%
other: 98.99% (2005)
Irrigated land: 130 sq km (2003)
Total renewable water resources: 0.02 cu km (1997)
Freshwater withdrawal (domestic/industrial/agricultural): *total:* 0.44 cu km/yr (45%/2%/52%)
per capita: 164 cu m/yr (2000)
Natural hazards: sudden cloudbursts are common from October to April and bring heavy rain, which can damage roads and houses; sandstorms and dust storms occur throughout the year but are most common between March and August
Environment—current issues: limited natural freshwater resources; some of world's largest and most sophisticated desalination facilities provide much of the water; air and water pollution; desertification
Environment—international agreements: *party to:* Biodiversity, Climate Change, Climate Change-Kyoto Protocol, Desertification, Endangered Species, Environmental Modification, Hazardous Wastes, Law of the Sea, Ozone Layer Protection
signed, but not ratified: Marine Dumping
Geography—note: strategic location at head of Persian Gulf

PEOPLE

Population: 2,595,628 (July 2011 est.)
country comparison to the world: 139
note: includes 1,291,354 non-nationals
Age structure: *0–14 years:* 25.8% (male 348,816/female 321,565)
15–64 years: 72.2% (male 1,153,433/female 720,392)
65 years and over: 2% (male 25,443/female 25,979) (2011 est.)
Median age: *total:* 28.5 years
male: 29.8 years
female: 26.3 years (2011 est.)
Population growth rate: 1.986%
country comparison to the world: 53
note: this rate reflects a return to pre-Gulf crisis immigration of expatriates (2011 est.)
Birth rate: 21.32 births/1,000 population (2011 est.)
country comparison to the world: 81
Death rate: 2.11 deaths/1,000 population (July 2011 est.)
country comparison to the world: 223
Net migration rate: 0.65 migrant(s)/1,000 population (2011 est.)
country comparison to the world: 59
Urbanization: *urban population:* 98% of total population (2010)
rate of urbanization: 2.1% annual rate of change (2010-15 est.)
Major cities—population: KUWAIT (capital) 2.23 million (2009)
Sex ratio: *at birth:* 1.047 male(s)/female
under 15 years: 1.04 male(s)/female
15–64 years: 1.79 male(s)/female
65 years and over: 1.65 male(s)/female
total population: 1.54 male(s)/female (2011 est.)
Infant mortality rate:
total: 8.07 deaths/1,000 live births
country comparison to the world: 160
male: 7.76 deaths/1,000 live births
female: 8.39 deaths/1,000 live births (2011 est.)
Life expectancy at birth: *total population:* 77.09 years
country comparison to the world: 66
male: 75.95 years
female: 78.3 years (2011 est.)
Total fertility rate: 2.64 children born/woman (2011 est.)
country comparison to the world: 77
HIV/AIDS–adult prevalence rate: 0.1% (2001 est.)
country comparison to the world: 140
HIV/AIDS–people living with HIV/AIDS: NA (2007 est.)
HIV/AIDS—deaths: NA
Drinking water source: *Improved:*
urban: 99% of population
rural: 99% of population
total: 99% of population
Unimproved: urban: 1% of population
rural: 1% of population
total: 1% of population (2008)
Sanitation facility access: *Improved:*
urban: 100% of population
rural: 100% of population
total: 100% of population (2008)
Nationality: *noun:* Kuwaiti(s)
adjective: Kuwaiti
Ethnic groups: Kuwaiti 45%, other Arab 35%, South Asian 9%, Iranian 4%, other 7%
Religions: Muslim 85% (Sunni 70%, Shia 30%), other (includes Christian, Hindu, Parsi) 15%
Languages: Arabic (official), English widely spoken
Literacy: *definition:* age 15 and over can read and write
total population: 93.3%
male: 94.4%
female: 91% (2005 census)
School life expectancy (primary to tertiary education): *total:* 12 years
male: 12 years
female: 13 years (2006)
Education expenditures: 3.8% of GDP (2006)
country comparison to the world: 112

GOVERNMENT

Country name: *conventional long form:* State of Kuwait
conventional short form: Kuwait
local long form: Dawlat al Kuwayt
local short form: Al Kuwayt
Government type: constitutional emirate
Capital: *name:* Kuwait City
geographic coordinates: 29 22 N, 47 58 E
time difference: UTC+3 (8 hours ahead of Washington, DC during Standard Time)
Administrative divisions: 6 governorates (muhafazat, singular–muhafazah); Al

Ahmadi, Al 'Asimah, Al Farwaniyah, Al Jahra', Hawalli, Mubarak al Kabir
Independence: 19 June 1961 (from the UK)
National holiday: National Day, 25 February (1950)
Constitution: approved and promulgated 11 November 1962
Legal system: mixed legal system consisting of English common law, French civil law, and Islamic religious law
International law organization participation: has not submitted an ICJ jurisdiction declaration; non-party state to the ICCt
Suffrage: 21 years of age; universal; note—males in the military or police are not allowed to vote; adult females were allowed to vote as of 16 May 2005; all voters must have been citizens for 20 years
Executive branch: *chief of state:* Amir SABAH al-Ahmad al-Jabir al-Sabah (since 29 January 2006); Crown Prince NAWAF al-Ahmad al-Jabir al-Sabah (since 7 February 2006)
head of government: Prime Minister NASIR AL-MUHAMMAD al-Ahmad al-Sabah (since 3 April 2007); First Deputy Prime Minister JABIR AL_MUBARAK al-Hamad al-Sabah; Deputy Prime Ministers MUHAMMAD AL_SABAH al-Salim al-Sabah,Muhammad Muhsin al-AFASI First Deputy Prime Minister JABIR AL-MUBAREK al-Hamad al-Sabah (since 9 February 2006); Deputy Prime Minister MUHAMMAD AL-SABAH al-Salim al-Sabah (since 9 February 2006)
cabinet: Council of Ministers appointed by the prime minister and approved by the amir; the cabinet of Prime Minister NASIR AL-MUHAMMAD al-Ahmad al-Sabah resigned on 31 March 2011
(For more information visit the World Leaders website)
elections: none; the amir is hereditary; the amir appoints the prime minister and deputy prime ministers
Legislative branch: unicameral National Assembly or Majlis al-Umma (50 seats; members elected by popular vote to serve four-year terms; all cabinet ministers are also ex officio voting members of the National Assembly)
elections: last held on 16 May 2009 (next election to be held in 2013)
election results: percent of vote by bloc–NA; seats by bloc–tribal MPs 25 (all Sunni Muslims, and represented primarily by the Al-Mutairi, Al-Azmi, Al-Ajmi, and Al-Rasheedi tribes), Shia Muslims 9, liberals 7, independents 6, Salafi (Sunni) Islamists 3
Judicial branch: High Court of Appeal
Political parties and leaders: none; formation of political parties is in practice illegal but is not forbidden by law
Political pressure groups and leaders: *other:* Islamists; merchants; political groups; secular liberals and pro-governmental deputies; Shia activists; tribal groups
International organization participation: ABEDA, AfDB (nonregional member), AFESD, AMF, BDEAC, CAEU, FAO, G-77, GCC, IAEA, IBRD, ICAO, ICC, ICRM, IDA, IDB, IFAD, IFC, IFRCS, IHO, ILO, IMF, IMO, IMSO, Interpol, IOC, IPU, ISO, ITSO, ITU, ITUC, LAS, MIGA, NAM, OAPEC, OIC, OPCW, OPEC, Paris Club (associate), PCA, UN, UNCTAD, UNESCO, UNIDO, UNWTO, UPU, WCO, WFTU, WHO, WIPO, WMO, WTO
Diplomatic representation in the US: *chief of mission:* Ambassador SALIM al-Abdallah al-Jabir al-Sabah
chancery: 2940 Tilden Street NW, Washington, DC 20008
telephone: [1] (202) 966-0702
FAX: [1] (202) 364-2868
consulate(s) general: Los Angeles
Diplomatic representation from the US: *chief of mission:* Ambassador Deborah K. JONES
embassy: Bayan 36302, Block 13, Al-Masjed Al-Aqsa Street (near the Bayan palace), Kuwait City
mailing address: P. O. Box 77 Safat 13001 Kuwait; or PSC 1280 APO AE 09880-9000
telephone: [965] 2259-1001
FAX: [965] 2538-0282
Flag description: three equal horizontal bands of green (top), white, and red with a black trapezoid based on the hoist side; colors and design are based on the Arab Revolt flag of World War I; green represents fertile fields, white stands for purity, red denotes blood on Kuwaiti swords, black signifies the defeat of the enemy
National anthem: *name:* "Al-Nasheed Al-Watani" (National Anthem)
lyrics/music: Ahmad MUSHARI al-Adwani/Ibrahim Nasir al-SOULA
note: adopted 1978; the anthem is only used on formal occasions

ECONOMY

Economy—overview: Kuwait has a geographically small, but wealthy, relatively open economy with self-reported crude oil reserves of about 102 billion barrels–about 9% of world reserves. Petroleum accounts for nearly half of GDP, 95% of export revenues, and 95% of government income. Kuwaiti officials have committed to increasing oil production to 4 million barrels per day by 2020. The rise in global oil prices throughout 2010 is reviving government consumption and economic growth as Kuwait experiences a 20% increase in government budget revenue. Kuwait has done little to diversify its economy, in part, because of this positive fiscal situation, and, in part, due to the poor business climate and the acrimonious relationship between the National Assembly and the executive branch, which has stymied most movement on economic reforms. Nonetheless, the government in May 2010 passed a privatization bill that allows the government to sell assets to private investors, and in January passed an economic development plan that pledges to spend up to $130 billion in five years to diversify the economy away from oil, attract more investment, and boost private sector participation in the economy. Increasing government expenditures by so large an amount during the planned time frame may be difficult to accomplish.
GDP (purchasing power parity): $136.5 billion (2010 est.)
country comparison to the world: 60
$133.9 billion (2009 est.)
$141.2 billion (2008 est.)
note: data are in 2010 US dollars
GDP (official exchange rate): $131.3 billion (2010 est.)
GDP—real growth rate: 2% (2010 est.)
country comparison to the world: 148
-5.2% (2009 est.)
5% (2008 est.)
GDP—per capita (PPP): $48,900 (2010 est.)
country comparison to the world: 10
$49,700 (2009 est.)
$54,300 (2008 est.)
note: data are in 2010 US dollars
GDP—composition by sector:
agriculture: 0.3%
industry: 48.1%
services: 51.6% (2010 est.)
Labor force: 2.154 million
country comparison to the world: 118
note: non-Kuwaitis represent about 60% of the labor force (2010 est.)
Labor force—by occupation:
agriculture: NA%
industry: NA%
services: NA%
Unemployment rate:
2.2% (2004 est.)
country comparison to the world: 19
Population below poverty line: NA%
Household income or consumption by percentage share: *lowest 10%:* NA%
highest 10%: NA%
Investment (gross fixed): 13.8% of GDP (2010 est.)
country comparison to the world: 139
Budget: *revenues:* $64.81 billion
expenditures: $38.12 billion (2010 est.)
Public debt: 12.6% of GDP (2010 est.)
country comparison to the world: 120
13.1% of GDP (2009 est.)
Inflation rate (consumer prices):
3.8% (2010 est.)
country comparison to the world: 107
4% (2009 est.)
Central bank discount rate: 3% (31 December 2009)
country comparison to the world: 104
3.75% (31 December 2008)
Commercial bank prime lending rate: 5.9% (31 December 2009)
country comparison to the world: 119
7.61% (31 December 2008 est.)
Stock of narrow money: $18.12 billion (31 December 2010 est.)
country comparison to the world: 63
$16.38 billion (31 December 2009 est.)
Stock of broad money: $88.71 billion (31 December 2010 est.)
country comparison to the world: 56
$86.53 billion (31 December 2009 est.)
Stock of domestic credit: $96.71 billion (31 December 2010 est.)
country comparison to the world: 52
$90.71 billion (31 December 2009 est.)
Market value of publicly traded shares: $95.94 billion (31 December 2009)
country comparison to the world: 38
$107.2 billion (31 December 2008)
$188 billion (31 December 2007)
Agriculture—products: fish

Industries: petroleum, petrochemicals, cement, shipbuilding and repair, water desalination, food processing, construction materials
Industrial production growth rate: 2.1% (2010 est.)
country comparison to the world: 126
Electricity–production: 45.83 billion kWh (2007 est.)
country comparison to the world: 50
Electricity—consumption: 40.21 billion kWh (2007 est.)
country comparison to the world: 51
Electricity–exports: 0 kWh (2008 est.)
Electricity—imports: 0 kWh (2008 est.)
Oil—production: 2.494 million bbl/day (2009 est.)
country comparison to the world: 10
Oil—consumption: 320,000 bbl/day (2009 est.)
country comparison to the world: 40
Oil—exports: 2.349 million bbl/day (2007 est.)
country comparison to the world: 5
Oil—imports: 0 bbl/day (2007 est.)
country comparison to the world: 204
Oil—proved reserves: 104 billion bbl (1 January 2010 est.)
country comparison to the world: 5
Natural gas—production: 12.7 billion cu m (2008 est.)
country comparison to the world: 37
Natural gas—consumption: 12.7 billion cu m (2008 est.)
country comparison to the world: 42
Natural gas—exports: 0 cu m (2008 est.)
country comparison to the world: 125
Natural gas—imports: 300 million cu m (2009 est.) (2009 est.)
country comparison to the world: 63
Natural gas—proved reserves: 1.798 trillion cu m (1 January 2010 est.)
country comparison to the world: 20
Current account balance: $38.2 billion (2010 est.)
country comparison to the world: 11
$28.61 billion (2009 est.)
Exports:
$65.03 billion (2010 est.)
country comparison to the world: 43
$50.34 billion (2009 est.)
Exports—commodities: oil and refined products, fertilizers
Exports—partners: Japan 17.9%, South Korea 17.31%, India 12.43%, Taiwan 9.07%, US 7.9%, China 7.55%, Singapore 5.48% (2009)
Imports: $20.36 billion (2010 est.)
country comparison to the world: 70
$17.08 billion (2009 est.)
Imports—commodities: food, construction materials, vehicles and parts, clothing
Imports—partners: US 11.18%, China 9.07%, Germany 7.63%, Japan 7.14%, Saudi Arabia 6.24%, Italy 5%, France 4.77%, India 4.09%, UK 4.02% (2009)
Reserves of foreign exchange and gold: $22.42 billion (31 December 2010 est.)
country comparison to the world: 42
$20.38 billion (31 December 2009 est.)
Debt—external: $56.81 billion (31 December 2010 est.)
country comparison to the world: 53
$55.23 billion (31 December 2009 est.)
Stock of direct foreign investment—at home:$1.281billion(31December2010est.)
country comparison to the world: 89
$1.081 billion (31 December 2009 est.)
Stock of direct foreign investment—abroad: $44.31 billion (31 December 2010 est.)
country comparison to the world: 34
$34.73 billion (31 December 2009 est.)
Exchange rates: Kuwaiti dinars (KD) per US dollar–0.2888 (2010), 0.2877 (2009), 0.2679 (2008), 0.2844 (2007), 0.29 (2006)

COMMUNICATIONS

Telephones—main lines in use: 553,500 (2009)
country comparison to the world: 95
Telephones—mobile cellular: 3.876 million (2009)
country comparison to the world: 106
Telephone system: *general assessment:* the quality of service is excellent
domestic: new telephone exchanges provide a large capacity for new subscribers; trunk traffic is carried by microwave radio relay, coaxial cable, and open-wire and fiber-optic cable; a mobile-cellular telephone system operates throughout Kuwait, and the country is well supplied with pay telephones
international: country code–965; linked to international submarine cable Fiber-Optic Link Around the Globe (FLAG); linked to Bahrain, Qatar, UAE via the Fiber-Optic Gulf (FOG) cable; coaxial cable and microwave radio relay to Saudi Arabia; satellite earth stations–6 (3 Intelsat–1 Atlantic Ocean and 2 Indian Ocean, 1 Inmarsat–Atlantic Ocean, and 2 Arabsat)
Broadcast media: state-owned TV broadcaster operates 4 networks and a satellite channel; several private TV broadcasters have emerged since 2003; satellite TV is available with pan-Arab TV stations especially popular; state-owned Radio Kuwait broadcasts on a number of channels in Arabic and English; first private radio station emerged in 2005; transmissions of at least 2 international radio broadcasters are available (2007)
Internet country code: .kw
Internet hosts: 2,485 (2010)
country comparison to the world: 151
Internet users: 1.1 million (2009)
country comparison to the world: 96

TRANSPORTATION

Airports: 7 (2010)
country comparison to the world: 168
Airports—with paved runways: *total:* 4
over 3,047 m: 1
2,438 to 3,047 m: 2
1,524 to 2,437 m: 1 (2010)
Airports—with unpaved runways: *total:* 3
1,524 to 2,437 m: 1
under 914 m: 2 (2010)
Heliports: 4 (2010)
Pipelines: gas 269 km; oil 540 km; refined products 57 km (2010)
Roadways: *total:* 5,749 km
country comparison to the world: 150
paved: 4,887 km
unpaved: 862 km (2004)
Merchant marine: *total:* 30
country comparison to the world: 84
by type: bulk carrier 1, carrier 3, container 6, liquefied gas 4, petroleum tanker 16
registered in other countries: 47 (Bahamas 2, Bahrain 5, Comoros 1, Libya 1, Malta 2, Panama 12, Qatar 7, Saint Kitts and Nevis 3, Saudi Arabia 4, UAE 10) (2010)
Ports and terminals: Ash Shu'aybah, Ash Shuwaykh, Az Zawr (Mina' Sa'ud), Mina' 'Abd Allah, Mina' al Ahmadi

MILITARY

Military branches: Kuwaiti Land Forces (KLF), Kuwaiti Navy, Kuwaiti Air Force (Al-Quwwat al-Jawwiya al-Kuwaitiya), Kuwaiti National Guard (KNG) (2009)
Military service age and obligation: 18-30 years of age for compulsory and 18-25 years of age for voluntary military service; women age 18-30 may be subject to compulsory military service; conscription suspended in 2001 (2009)
Manpower available for military service: *males age 16–49:* 1,002,480
females age 16–49: 616,958 (2010 est.)
Manpower fit for military service: *males age 16–49:* 840,912
females age 16–49: 523,206 (2010 est.)
Manpower reaching militarily significant age annually: *male:* 17,653
female: 16,232 (2010 est.)
Military expenditures: 5.3% of GDP (2006)
country comparison to the world: 14
Transnational Issues
Disputes—international: Kuwait and Saudi Arabia continue negotiating a joint maritime boundary with Iran; no maritime boundary exists with Iraq in the Persian Gulf
Trafficking in persons: current situation: Kuwait is a destination country for men and women who migrate legally from South and Southeast Asia for domestic or low-skilled labor, but are subjected to conditions of involuntary servitude by employers in Kuwait including conditions of physical and sexual abuse, non-payment of wages, confinement to the home, and withholding of passports to restrict their freedom of movement; Kuwait is reportedly a transit point for South and East Asian workers recruited for low-skilled work in Iraq; some of these workers are deceived as to the true location and nature of this work, and others are subjected to conditions of involuntary servitude in Iraq
tier rating: Tier 3–Kuwaiti government has shown an inability to define trafficking and has demonstrated insufficient political will to address human trafficking adequately; much of the human trafficking found in Kuwait involves domestic workers in private residences and the government is reluctant to prosecute Kuwaiti citizens; the government has not enacted legislation targeting human trafficking nor established a permanent shelter for victims of trafficking (2009)

KYRGYZSTAN

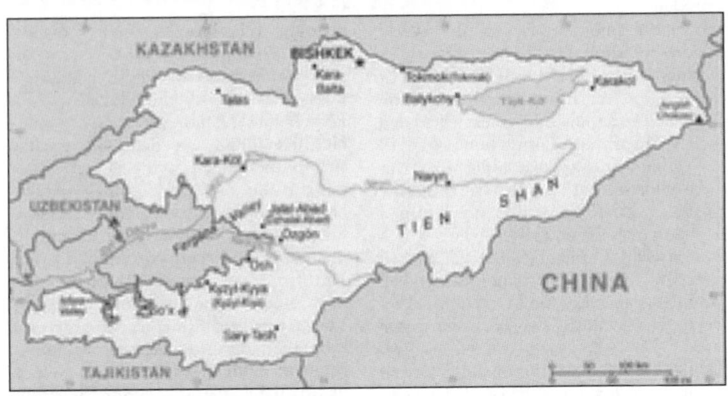

INTRODUCTION

Background: A Central Asian country of incredible natural beauty and proud nomadic traditions, most of Kyrgyzstan was formally annexed to Russia in 1876. The Kyrgyz staged a major revolt against the Tsarist Empire in 1916 in which almost one-sixth of the Kyrgyz population was killed. Kyrgyzstan became a Soviet republic in 1936 and achieved independence in 1991 when the USSR dissolved. Nationwide demonstrations in the spring of 2005 resulted in the ouster of President Askar AKAEV, who had run the country since 1990. Subsequent presidential elections in July 2005 were won overwhelmingly by former prime minister Kurmanbek BAKIEV. Over the next few years, the new president manipulated the parliament to accrue new powers for himself. In July 2009, after months of harassment against his opponents and media critics, BAKIEV won re-election in a presidential campaign that the international community deemed flawed. In April 2010, nationwide protests led to the resignation and expulsion of BAKIEV. He was replaced by President Roza OTUNBAEVA who will serve as president until 31 December 2011 according to a presidential decree issued 19 May 2010. Presidential elections are scheduled to be held in October 2011. Continuing concerns include: endemic corruption, poor interethnic relations, and terrorism.

GEOGRAPHY

Location: Central Asia, west of China, south of Kazakhstan
Geographic coordinates: 41 00 N, 75 00 E
Map references: Asia
Area: *total:* 199,951 sq km
country comparison to the world: 86
land: 191,801 sq km
water: 8,150 sq km
Area—comparative: slightly smaller than South Dakota
Land boundaries: *total:* 3,051 km
border countries: China 858 km, Kazakhstan 1,224 km, Tajikistan 870 km, Uzbekistan 1,099 km
Coastline: 0 km (landlocked)
Maritime claims: none (landlocked)
Climate: dry continental to polar in high Tien Shan Mountains; subtropical in southwest (Fergana Valley); temperate in northern foothill zone
Terrain: peaks of Tien Shan and associated valleys and basins encompass entire nation
Elevation extremes: *lowest point:* Kara-Daryya (Karadar'ya) 132 m
highest point: Jengish Chokusu (Pik Pobedy) 7,439 m
Natural resources: abundant hydropower; significant deposits of gold and rare earth metals; locally exploitable coal, oil, and natural gas; other deposits of nepheline, mercury, bismuth, lead, and zinc
Land use: *arable land:* 6.55%
permanent crops: 0.28%
other: 93.17%
note: Kyrgyzstan has the world's largest natural-growth walnut forest (2005)
Irrigated land: 10,720 sq km (2003)
Total renewable water resources: 46.5 cu km (1997)
Freshwater withdrawal (domestic/industrial/agricultural): *total:* 10.08 cu km/yr (3%/3%/94%)
per capita: 1,916 cu m/yr (2000)
Natural hazards: NA
Environment—current issues: water pollution; many people get their water directly from contaminated streams and wells; as a result, water-borne diseases are prevalent; increasing soil salinity from faulty irrigation practices
Environment—international agreements: *party to:* Air Pollution, Biodiversity, Climate Change, Climate Change-Kyoto Protocol, Desertification, Hazardous Wastes, Ozone Layer Protection, Wetlands
signed, but not ratified: none of the selected agreements
Geography—note: landlocked; entirely mountainous, dominated by the Tien Shan range; 94% of the country is 1,000 m above sea level with an average elevation of 2,750 m; many tall peaks, glaciers, and high-altitude lakes

PEOPLE

Population: 5,587,443 (July 2011 est.)
country comparison to the world: 109
Age structure: *0–14 years:* 29.3% (male 834,024/female 801,750)
15–64 years: 65.4% (male 1,790,534/female 1,865,521)
65 years and over: 5.3% (male 114,200/female 181,414) (2011 est.)
Median age: *total:* 25 years
male: 24.1 years
female: 26 years (2011 est.)
Population growth rate: 1.427% (2011 est.)
country comparison to the world: 85
Birth rate: 23.66 births/1,000 population (2011 est.)
country comparison to the world: 70
Death rate: 6.79 deaths/1,000 population (July 2011 est.)
country comparison to the world: 145
Net migration rate: -2.6 migrant(s)/1,000 population (2011 est.)
country comparison to the world: 172
Urbanization: *urban population:* 35% of total population (2010)
rate of urbanization: 1.3% annual rate of change (2010-15 est.)
Major cities—population: BISHKEK (capital) 854,000 (2009)
Sex ratio: *at birth:* 1.053 male(s)/female
under 15 years: 1.04 male(s)/female
15–64 years: 0.96 male(s)/female
65 years and over: 0.64 male(s)/female
total population: 0.96 male(s)/female (2011 est.)
Infant mortality rate: *total:* 29.27 deaths/1,000 live births
country comparison to the world: 71
male: 34.01 deaths/1,000 live births
female: 24.28 deaths/1,000 live births (2011 est.)
Life expectancy at birth: *total population:* 70.04 years
country comparison to the world: 147
male: 66.04 years
female: 74.24 years (2011 est.)
Total fertility rate: 2.63 children born/woman (2011 est.)
country comparison to the world: 78
HIV/AIDS–adult prevalence rate: 0.3% (2009 est.)
country comparison to the world: 88
HIV/AIDS–people living with HIV/AIDS: 9,800 (2009 est.)
country comparison to the world: 100
HIV/AIDS–deaths: fewer than 500 (2009 est.)
country comparison to the world: 89
Drinking water source: *Improved:*
urban: 99% of population
rural: 85% of population
total: 90% of population
Unimproved: urban: 1% of population
rural: 15% of population
total: 10% of population (2008)
Sanitation facility access: *Improved:*
urban: 94% of population
rural: 93% of population
total: 93% of population
Unimproved: urban: 6% of population
rural: 7% of population
total: 7% of population (2008)
Nationality: *noun:* Kyrgyzstani(s)
adjective: Kyrgyzstani
Ethnic groups: Kyrgyz 64.9%, Uzbek 13.8%, Russian 12.5%, Dungan 1.1%,

Ukrainian 1%, Uighur 1%, other 5.7% (1999 census)
Religions: Muslim 75%, Russian Orthodox 20%, other 5%
Languages: Kyrgyz (official) 64.7%, Uzbek 13.6%, Russian (official) 12.5%, Dungun 1%, other 8.2% (1999 census)
Literacy: *definition:* age 15 and over can read and write
total population: 98.7%
male: 99.3%
female: 98.1% (1999 census)
School life expectancy (primary to tertiary education): *total:* 12 years
male: 12 years
female: 13 years (2009)
Education expenditures: 5.9% of GDP (2008)
country comparison to the world: 31

GOVERNMENT

Country name: *conventional long form:* Kyrgyz Republic
conventional short form: Kyrgyzstan
local long form: Kyrgyz Respublikasy
local short form: Kyrgyzstan
former: Kirghiz Soviet Socialist Republic
Government type: republic
Capital: *name:* Bishkek
geographic coordinates: 42 52 N, 74 36 E
time difference: UTC+6 (11 hours ahead of Washington, DC during Standard Time)
Administrative divisions: 7 provinces (oblastlar, singular–oblasty) and 1 city* (shaar); Batken Oblasty, Bishkek Shaary*, Chuy Oblasty (Bishkek), Jalal-Abad Oblasty, Naryn Oblasty, Osh Oblasty, Talas Oblasty, Ysyk-Kol Oblasty (Karakol)
note: administrative divisions have the same names as their administrative centers (exceptions have the administrative center name following in parentheses)
Independence: 31 August 1991 (from the Soviet Union)
National holiday: Independence Day, 31 August (1991)
Constitution: 27 June 2010
Legal system: civil law system which includes features of French civil law and Russian Federation laws
International law organization participation: has not submitted an ICJ jurisdiction declaration; non-party state to the ICCt
Suffrage: 18 years of age; universal
Executive branch: *chief of state:* President Roza OTUNBAEVA (since 19 May 2010); note–OTUNBAEVA became acting president on 7 April 2010 following the early April 2010 riots that overthrew President Kurmanbek BAKIEV; she was appointed president through 31 December 2011 by a 19 May 2010 decree of the provisional government, which also prohibited her from running in the next presidential election; she was officially sworn in on 3 July 2010
head of government: Prime Minister Almazbek ATAMBAEV (since 17 December 2010); First Deputy Prime Minister (vacant); Deputy Prime Ministers–Shamil ATAKHANOV, Ibragim JUNUSOV (since 17 December 2010)
cabinet: Cabinet of Ministers proposed by the prime minister, appointed by the president; ministers in charge of defense and security, are appointed solely by the president
(For more information visit the World Leaders website)
elections: Kurmanbek BAKIEV reelected by popular vote for a five-year term; election last held on 23 July 2009 (next scheduled for 2011); prime minister nominated by the parliamentary party holding more than 50% of the seats; if no such party exists, the president selects the party that will form a coalition majority and government
election results: Kurmanbek BAKIEV elected president; percent of vote–Kurmanbek BAKIEV 76.1%, Almaz ATAMBAEV 8.4%, Temir SARIEV 6.7%, other candidates 8.8%
Legislative branch: unicameral Supreme Council or Jogorku Kengesh (120 seats; members elected by popular vote to serve five-year terms)
elections: last held on 10 October 2010 (next to be held in 2015)
election results: Supreme Council–percent of vote by party–NA; seats by party–Ata-Jurt 28, SDPK 26, Ar-Namys 25, Respublika 23, Ata-Meken 18
Judicial branch: Supreme Court; Constitutional Court (judges of both the Supreme and Constitutional Courts are appointed for 10-year terms by the Jogorku Kengesh on the recommendation of the president; their mandatory retirement age is 70 years); Higher Court of Arbitration; Local Courts (judges appointed by the president on the recommendation of the National Council on Legal Affairs for a probationary period of five years, then 10 years)
Political parties and leaders: Ar-Namys (Dignity) Party [Feliks KULOV]; Ata-Jurt (Homeland) [Kamchybek TASHIEV, Akhmat KELDIBEKOV]; Ata-Meken (Fatherland) [Omurbek TEKEBAEV]; Butun Kyrgyzstan (All Kyrgyzstan) [Adakhan MADUMAROV, Miroslav NIYAZOV]; Respublika [Omurbek BABANOV]; Social-Democratic Party of Kyrgyzstan (SDPK) [Almazbek ATAMBAEV]
Political pressure groups and leaders: Adilet Legal Clinic [Cholpon JAKUPOVA]; Coalition for Democracy and Civil Society [Dinara OSHURAKHUNOVA]; Inter-bilim [Asiya SASYKBAEVA]
International organization participation: ADB, CICA, CIS, CSTO, EAEC, EAPC, EBRD, ECO, FAO, GCTU, IAEA, IBRD, ICAO, ICRM, IDA, IDB, IFAD, IFC, IFRCS, ILO, IMF, Interpol, IOC, IOM, IPU, ISO (correspondent), ITSO, ITU, MIGA, NAM (observer), OIC, OPCW, OSCE, PCA, PFP, SCO, UN, UNCTAD, UNESCO, UNIDO, UNMIL, UNMIS, UNWTO, UPU, WCO, WFTU, WHO, WIPO, WMO, WTO
Diplomatic representation in the US: *chief of mission:* Ambassador Mukhtar JUMALIEV
chancery: 2360 Massachusetts Ave. NW, Washington, DC 20008
telephone: [1] (202) 449-9822
FAX: [1] (202) 386-7550
consulate(s): New York
Diplomatic representation from the US: *chief of mission:* Ambassador Tatiana GFOELLER
embassy: 171 Prospect Mira, Bishkek 720016
mailing address: use embassy street address
telephone: [996] (312) 551-241, (517) 777-217
FAX: [996] (312) 551-264
Flag description: red field with a yellow sun in the center having 40 rays representing the 40 Kyrgyz tribes; on the obverse side the rays run counterclockwise, on the reverse, clockwise; in the center of the sun is a red ring crossed by two sets of three lines, a stylized representation of a "tunduk"–the crown of a traditional Kyrgyz yurt; red symbolizes bravery and valor, the sun evinces peace and wealth
National anthem: *name:* "Kyrgyz Respublikasynyn Mamlekettik Gimni" (National Anthem of the Kyrgyz Republic)
lyrics/music: Djamil SADYKOV and Eshmambet KULUEV/Nasyr DAVLESOV and Kalyi MOLDOBASANOV
note: adopted 1992

ECONOMY

Economy—overview: Kyrgyzstan is a poor, mountainous country with a dominant agricultural sector. Cotton, tobacco, wool, and meat are the main agricultural products, although only tobacco and cotton are exported in any quantity. Industrial exports include gold, mercury, uranium, natural gas, and electricity. The economy depends heavily on gold exports–mainly from output at the Kumtor gold mine. Following independence, Kyrgyzstan was progressive in carrying out market reforms, such as an improved regulatory system and land reform. Kyrgyzstan was the first Commonwealth of Independent States (CIS) country to be accepted into the World Trade Organization. Much of the government's stock in enterprises has been sold. Drops in production had been severe after the breakup of the Soviet Union in December 1991, but by mid-1995, production began to recover and exports began to increase. In 2005, the BAKIEV government and international financial institutions initiated a comprehensive medium-term poverty reduction and economic growth strategy. Bishkek agreed to pursue much needed tax reform and, in 2006, became eligible for the heavily indebted poor countries (HIPC) initiative. The government made steady strides in controlling its substantial fiscal deficit, nearly closing the gap between revenues and expenditures in 2006, before boosting expenditures more than 20% in 2007-08. GDP grew about 8% annually in 2007-08, partly due to higher gold prices internationally, but slowed to 2.3% in 2009. The overthrow of President BAKIEV in April, 2010 and subsequent ethnic clashes left hundreds dead and damaged infrastructure. Shrinking trade and agricultural production, as well as political instability, caused GDP to contract about 3.5% in 2010. The fiscal deficit widened to 11% of GDP, reflecting significant increases in crisis-related spending, including both rehabilitation of damaged infrastructure and bank

recapitalization. Progress in reconstruction, fighting corruption, restructuring domestic industry, and attracting foreign aid and investment are key to future growth.
GDP (purchasing power parity): $12.02 billion (2010 est.)
country comparison to the world: 143
$12.18 billion (2009 est.)
$11.84 billion (2008 est.)
note: data are in 2010 US dollars
GDP (official exchange rate): $4.615 billion (2010 est.)
GDP—real growth rate: -1.4% (2010 est.)
country comparison to the world: 202
2.9% (2009 est.)
7.6% (2008 est.)
GDP—per capita (PPP): $2,200 (2010 est.)
country comparison to the world: 186
$2,200 (2009 est.)
$2,200 (2008 est.)
note: data are in 2010 US dollars
GDP—composition by sector:
agriculture: 24.6%
industry: 25%
services: 50.4% (2010 est.)
Labor force: 2.344 million (2007)
country comparison to the world: 112
Labor force—by occupation: *agriculture:* 48%
industry: 12.5%
services: 39.5% (2005 est.)
Unemployment rate: 18% (2004 est.)
country comparison to the world: 161
Population below poverty line: 40% (2004 est.)
Household income or consumption by percentage share: *lowest 10%:* 3.6%
highest 10%: 25.9% (2004)
Distribution of family income—Gini index: 30.3 (2003)
country comparison to the world: 112
29 (2001)
Investment (gross fixed): 26.4% of GDP (2010 est.)
country comparison to the world: 34
Budget: *revenues:* $980 million
expenditures: $1.46 billion (2010 est.)
Inflation rate (consumer prices): 4.8% (2010 est.)
country comparison to the world: 136
6.9% (2009 est.)
Central bank discount rate: 9.07% (31 December 2009)
country comparison to the world: 17
15.11% (31 December 2008)
Commercial bank prime lending rate: 23.03% (31 December 2009 est.)
country comparison to the world: 18
19.86% (31 December 2008 est.)
Stock of narrow money: $714.9 million (31 December 2010 est.)
country comparison to the world: 149
$826.4 million (31 December 2009 est.)
Stock of broad money: $1.1 billion (31 December 2010 est.)
country comparison to the world: 160
$1.247 billion (31 December 2009 est.)
Stock of domestic credit: $505.4 million (31 December 2010 est.)
country comparison to the world: 164
$572.9 million (31 December 2009 est.)
Market value of publicly traded shares: $71.84 million (31 December 2009)
country comparison to the world: 117
$93.79 million (31 December 2008)
$121 million (31 December 2007)
Agriculture—products: tobacco, cotton, potatoes, vegetables, grapes, fruits and berries; sheep, goats, cattle, wool
Industries: small machinery, textiles, food processing, cement, shoes, sawn logs, refrigerators, furniture, electric motors, gold, rare earth metals
Industrial production growth rate: 6% (2010 est.)
country comparison to the world: 59
Electricity–production: 15.96 billion kWh (2007 est.)
country comparison to the world: 76
Electricity—consumption: 9 billion kWh (2007 est.)
country comparison to the world: 90
Electricity–exports: 2.379 billion kWh (2007 est.)
Electricity—imports: 0 kWh (2008 est.)
Oil—production: 979.1 bbl/day (2009 est.)
country comparison to the world: 106
Oil—consumption: 15,000 bbl/day (2009 est.)
country comparison to the world: 135
Oil—exports: 1,890 bbl/day (2007 est.)
country comparison to the world: 115
Oil—imports: 12,850 bbl/day (2007 est.)
country comparison to the world: 135
Oil—proved reserves: 40 million bbl (1 January 2010 est.)
country comparison to the world: 81
Natural gas—production: 30 million cu m (2008 est.)
country comparison to the world: 85
Natural gas—consumption: 750 million cu m (2008 est.)
country comparison to the world: 91
Natural gas—exports: 0 cu m (2008 est.)
country comparison to the world: 121
Natural gas—imports: 720 million cu m (2008 est.)
country comparison to the world: 57
Natural gas—proved reserves: 5.663 billion cu m (1 January 2010 est.)
country comparison to the world: 90
Current account balance: $-210 million (2010 est.)
country comparison to the world: 91
$184 million (2009 est.)
Exports: $1.682 billion (2010 est.)
country comparison to the world: 136
$1.726 billion (2009 est.)
Exports—commodities: cotton, wool, meat, tobacco; gold, mercury, uranium, natural gas, hydropower; machinery; shoes
Exports—partners: Switzerland 25.96%, Russia 25.88%, Uzbekistan 15.72%, Kazakhstan 12.47% (2009)
Imports: $3.075 billion (2010 est.)
country comparison to the world: 140
$2.987 billion (2009 est.)
Imports—commodities: oil and gas, machinery and equipment, chemicals, foodstuffs
Imports—partners: China 57.03%, Russia 19.34%, Kazakhstan 5.9% (2009)
Reserves of foreign exchange and gold: $1.615 billion (31 December 2010 est.)
country comparison to the world: 109
$1.585 billion (31 December 2009 est.)
Debt—external: $3.738 billion (30 June 2010)
country comparison to the world: 118
$3.467 billion (31 December 2008)
Stock of direct foreign investment—at home: $NA (31 December 2009 est.)
Stock of direct foreign investment—abroad: $NA
Exchange rates: soms (KGS) per US dollar–46.337 (2010), 42.905 (2009), 36.108 (2008), 37.746 (2007), 40.149 (2006)

COMMUNICATIONS

Telephones—main lines in use: 498,300 (2009)
country comparison to the world: 98
Telephones—mobile cellular: 4.487 million (2009)
country comparison to the world: 100
Telephone system: *general assessment:* telecommunications infrastructure is being upgraded; loans from the European Bank for Reconstruction and Development (EBRD) are being used to install a digital network, digital radio-relay stations, and fiber-optic links
domestic: fixed-line penetration remains low and concentrated in urban areas; multiple mobile-cellular service providers with growing coverage; mobile-cellular subscribership exceeded 80 per 100 persons in 2009
international: country code–996; connections with other CIS countries by landline or microwave radio relay and with other countries by leased connections with Moscow international gateway switch and by satellite; satellite earth stations–2 (1 Intersputnik, 1 Intelsat); connected internationally by the Trans-Asia-Europe (TAE) fiber-optic line
Broadcast media: state-run television broadcaster operates 2 nationwide networks and 6 regional stations; roughly 20 private TV stations operating with most rebroadcasting other channels; state-run radio broadcaster operates 2 networks; about 20 private radio stations operating (2007)
Internet country code: .kg
Internet hosts: 97,976 (2010)
country comparison to the world: 78
Internet users: 2.195 million (2009)
country comparison to the world: 74

TRANSPORTATION

Airports: 28 (2010)
country comparison to the world: 120
Airports—with paved runways: *total:* 18
over 3,047 m: 1
2,438 to 3,047 m: 3
1,524 to 2,437 m: 11
under 914 m: 3 (2010)
Airports—with unpaved runways: *total:* 10
1,524 to 2,437 m: 1
914 to 1,523 m: 1
under 914 m: 8 (2010)
Pipelines: gas 480 km; oil 16 km (2010)
Railways: *total:* 470 km
country comparison to the world: 112
broad gauge: 470 km 1.520-m gauge (2010)
Roadways: *total:* 18,500 km
country comparison to the world: 114
paved: 16,909 km (includes 140 km of expressways)
unpaved: 1,591 km (2003)
Waterways: 600 km (2010)
country comparison to the world: 79
Ports and terminals: Balykchy (Ysyk-Kol or Rybach'ye)

MILITARY

Military branches: Ground Forces, Air Force (includes Air Defense Forces), National Guard (2010)
Military service age and obligation: 18-27 years of age for compulsory male military service in the armed forces or Interior Ministry; service obligation—1 year, with optional fee-based 3-year service in the callup mobilization reserve; women may volunteer at age 19; 16-17 years of age for military cadets, who cannot take part in military operations (2011)
Manpower available for military service: *males age 16–49:* 1,456,881
females age 16–49: 1,470,317 (2010 est.)
Manpower fit for military service: *males age 16–49:* 1,119,224
females age 16–49: 1,257,263 (2010 est.)
Manpower reaching militarily significant age annually: *male:* 56,606
female: 54,056 (2010 est.)
Military expenditures: 0.5% of GDP (2009)
country comparison to the world: 164
Transnational Issues
Disputes—international: Kyrgyzstan has yet to ratify the 2001 boundary delimitation with Kazakhstan; disputes in Isfara Valley delay completion of delimitation with Tajikistan; delimitation of 130 km of border with Uzbekistan is hampered by serious disputes around enclaves and other areas
Illicit drugs: limited illicit cultivation of cannabis and opium poppy for CIS markets; limited government eradication of illicit crops; transit point for Southwest Asian narcotics bound for Russia and the rest of Europe; major consumer of opiates

LAOS

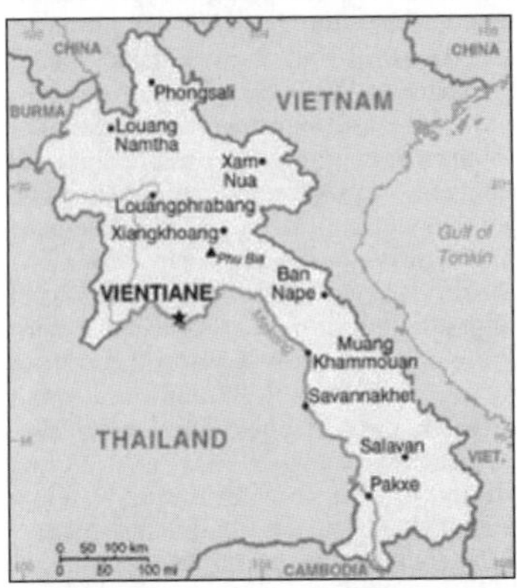

INTRODUCTION

Background: Modern-day Laos has its roots in the ancient Lao kingdom of Lan Xang, established in the 14th Century under King FA NGUM. For 300 years Lan Xang had influence reaching into present-day Cambodia and Thailand, as well as over all of what is now Laos. After centuries of gradual decline, Laos came under the domination of Siam (Thailand) from the late 18th century until the late 19th century when it became part of French Indochina. The Franco-Siamese Treaty of 1907 defined the current Lao border with Thailand. In 1975, the Communist Pathet Lao took control of the government ending a six-century-old monarchy and instituting a strict socialist regime closely aligned to Vietnam. A gradual, limited return to private enterprise and the liberalization of foreign investment laws began in 1988. Laos became a member of ASEAN in 1997.

GEOGRAPHY

Location: Southeastern Asia, northeast of Thailand, west of Vietnam
Geographic coordinates: 18 00 N, 105 00 E
Map references: Southeast Asia
Area: *total:* 236,800 sq km
country comparison to the world: 83
land: 230,800 sq km
water: 6,000 sq km
Area—comparative: slightly larger than Utah
Land boundaries: total: 5,083 km
border countries: Burma 235 km, Cambodia 541 km, China 423 km, Thailand 1,754 km, Vietnam 2,130 km
Coastline: 0 km (landlocked)
Maritime claims: none (landlocked)
Climate: tropical monsoon; rainy season (May to November); dry season (December to April)
Terrain: mostly rugged mountains; some plains and plateaus
Elevation extremes: *lowest point:* Mekong River 70 m
highest point: Phu Bia 2,817 m
Natural resources: timber, hydropower, gypsum, tin, gold, gemstones
Land use: arable land: 4.01%
permanent crops: 0.34%
other: 95.65% (2005)
Irrigated land: 1,750 sq km (2003)
Total renewable water resources: 333.6 cu km (2003)
Freshwater withdrawal (domestic/industrial/agricultural): *total:* 3 cu km/yr (4%/6%/90%)
per capita: 507 cu m/yr (2000)
Natural hazards: floods, droughts
Environment—current issues: unexploded ordnance; deforestation; soil erosion; most of the population does not have access to potable water
Environment—international agreements: *party to:* Biodiversity, Climate Change, Climate Change-Kyoto Protocol, Desertification, Endangered Species, Environmental Modification, Law of the Sea, Ozone Layer Protection
signed, but not ratified: none of the selected agreements
Geography—note: landlocked; most of the country is mountainous and thickly forested; the Mekong River forms a large part of the western boundary with Thailand

PEOPLE

Population: 6,477,211 (July 2011 est.)
country comparison to the world: 103
Age structure: *0–14 years:* 36.7% (male 1,197,579/female 1,181,523)
15–64 years: 59.6% (male 1,908,176/female 1,950,544)
65 years and over: 3.7% (male 107,876/female 131,513) (2011 est.)
Median age: *total:* 21 years
male: 20.7 years
female: 21.3 years (2011 est.)
Population growth rate: 1.684% (2011 est.)
country comparison to the world: 68
Birth rate: 26.13 births/1,000 population (2011 est.)
country comparison to the world: 55
Death rate: 8.13 deaths/1,000 population (July 2011 est.)
country comparison to the world: 98
Net migration rate: -1.16 migrant(s)/1,000 population (2011 est.)
country comparison to the world: 154
Urbanization: *urban population:* 33% of total population (2010)
rate of urbanization: 4.9% annual rate of change (2010-15 est.)
Major cities—population: VIENTIANE (capital) 799,000 (2009)
Sex ratio: *at birth:* 1.04 male(s)/female
under 15 years: 1.01 male(s)/female
15–64 years: 0.98 male(s)/female
65 years and over: 0.75 male(s)/female
total population: 0.98 male(s)/female (2011 est.)
Infant mortality rate:
total: 59.46 deaths/1,000 live births
country comparison to the world: 35
male: 65.49 deaths/1,000 live births
female: 53.18 deaths/1,000 live births (2011 est.)
Life expectancy at birth:
total population: 62.39 years
country comparison to the world: 180
male: 60.5 years
female: 64.36 years (2011 est.)
Total fertility rate: 3.14 children born/woman (2011 est.)
country comparison to the world: 56
HIV/AIDS—adult prevalence rate: 0.2% (2009 est.)
country comparison to the world: 102
HIV/AIDS—people living with HIV/AIDS: 8,500 (2009 est.)
country comparison to the world: 107
HIV/AIDS—deaths: fewer than 200 (2009 est.)
country comparison to the world: 106
Major infectious diseases: degree of risk: very high
food or waterborne diseases: bacterial and protozoal diarrhea, hepatitis A, and typhoid fever
vectorborne diseases: dengue fever and malaria
note: highly pathogenic H5N1 avian influenza has been identified in this country; it poses a negligible risk with extremely rare cases possible among US citizens who have close contact with birds (2008) (2009)
Drinking water source: *Improved:*
urban: 72% of population
rural: 51% of population
total: 57% of population
Unimproved: urban: 28% of population
rural: 49% of population
total: 43% of population (2008)
Sanitation facility access: *Improved:*
urban: 86% of population
rural: 38% of population
total: 53% of population
Unimproved:
urban: 14% of population
rural: 62% of population
total: 47% of population (2008)
Nationality: noun: Lao(s) or Laotian(s)
adjective: Lao or Laotian
Ethnic groups: Lao 55%, Khmou 11%, Hmong 8%, other (over 100 minor ethnic groups) 26% (2005 census)
Religions: Buddhist 67%, Christian 1.5%, other and unspecified 31.5% (2005 census)
Languages: Lao (official), French, English, various ethnic languages
Literacy: *definition:* age 15 and over can read and write
total population: 73%
male: 83%
female: 63% (2005 Census)
School life expectancy (primary to tertiary education): *total:* 9 years
male: 10 years
female: 9 years (2008)
Education expenditures: 2.3% of GDP (2008)
country comparison to the world: 151

GOVERNMENT

Country name: *conventional long form:* Lao People's Democratic Republic
conventional short form: Laos
local long form: Sathalanalat Paxathipatai Paxaxon Lao
local short form: Pathet Lao (unofficial)
Government type: Communist state
Capital: *name:* Vientiane (Viangchan)
geographic coordinates: 17 58 N, 102 36 E
time difference: UTC+7 (12 hours ahead of Washington, DC during Standard Time)

Administrative divisions: 16 provinces (khoueng, singular and plural) and 1 capital city* (nakhon luang, singular and plural); Attapu, Bokeo, Bolikhamxai, Champasak, Houaphan, Khammouan, Louangnamtha, Louangphrabang, Oudomxai, Phongsali, Salavan, Savannakhet, Viangchan (Vientiane)*, Viangchan, Xaignabouli, Xekong, Xiangkhoang
Independence: 19 July 1949 (from France)
National holiday: Republic Day, 2 December (1975)
Constitution: promulgated 14 August 1991; amended in 2003
Legal system: civil law system similar in form to the French system
International law organization participation: has not submitted an ICJ jurisdiction declaration; non-party state to the ICCt
Suffrage: 18 years of age; universal
Executive branch: *chief of state:* President Lt. Gen. CHOUMMALI Saignason (since 8 June 2006); Vice President BOUN-GNANG Volachit (since 8 June 2006)
head of government: Prime Minister THONGSING Thammavong (since 24 December 2010); Deputy Prime Ministers Maj. Gen. ASANG Laoli (since May 2002), Lt. Gen. DOUANGCHAI Phichit (since 8 June 2006), SOMSAVAT Lengsavat (since 26 February 1998), and THONGLOUN Sisoulit (since 27 March 2001)
cabinet: Ministers appointed by president, approved by National Assembly
(For more information visit the World Leaders website)
elections: president and vice president elected by National Assembly for five-year terms; election last held on 8 June 2006 (next to be held in 2011); prime minister nominated by the president and elected by the National Assembly for five-year term
election results: CHOUMMALI Saignason elected president; BOUN-GNANG Volachit elected vice president; percent of National Assembly vote–100%; BOUASONE Bouphavanh elected prime minister; percent of National Assembly vote–97%
Legislative branch: unicameral National Assembly (132 seats; members elected by popular vote from a list of candidates selected by the Lao People's Revolutionary Party to serve five-year terms)
elections: last held on 30 April 2011 (next to be held in 2016)
election results: percent of vote by party–NA; seats by party–LPRP 128, independents 4
Judicial branch: People's Supreme Court (the president of the People's Supreme Court is elected by the National Assembly on the recommendation of the National Assembly Standing Committee; the vice president of the People's Supreme Court and the judges are appointed by the National Assembly Standing Committee)
Political parties and leaders: Lao People's Revolutionary Party or LPRP [CHOUMMALI Saignason]; other parties proscribed
Political pressure groups and leaders: NA
International organization participation: ADB, ARF, ASEAN, CP, EAS, FAO, G-77, IBRD, ICAO, ICRM, IDA, IFAD, IFC, IFRCS, ILO, IMF, Interpol, IOC, IPU, ISO (subscriber), ITU, MIGA, NAM, OIF, OPCW, PCA, UN, UNCTAD, UNESCO, UNIDO, UNWTO, UPU, WCO, WFTU, WHO, WIPO, WMO, WTO (observer)
Diplomatic representation in the US: *chief of mission:* Ambassador SENG Soukhathivong
chancery: 2222 S Street NW, Washington, DC 20008
telephone: [1] (202) 332-6416
FAX: [1] (202) 332-4923
Diplomatic representation from the US: *chief of mission:* Ambassador Karen B. STEWART
embassy: 19 Rue Bartholonie, That Dam, Vientiane
mailing address: American Embassy Vientiane, APO AP 96546
telephone: [856] 21-26-7000
FAX: [856] 21-26-7190
Flag description: three horizontal bands of red (top), blue (double width), and red with a large white disk centered in the blue band; the red bands recall the blood shed for liberation; the blue band represents the Mekong River and prosperity; the white disk symbolizes the full moon against the Mekong River, but also signifies the unity of the people under the Pathet Lao, as well as the country's bright future
National anthem: *name:* "Pheng Xat Lao" (Hymn of the Lao People)
lyrics/music: SISANA Sisane/THONGDY Sounthonevichit
note: music adopted 1945, lyrics adopted 1975; the anthem's lyrics were changed following the 1975 Communist revolution that overthrew the monarchy

ECONOMY

Economy—overview: The government of Laos, one of the few remaining one-party Communist states, began decentralizing control and encouraging private enterprise in 1986. The results, starting from an extremely low base, were striking–growth averaged 6% per year from 1988-2008 except during the short-lived drop caused by the Asian financial crisis that began in 1997. Despite this high growth rate, Laos remains a country with an underdeveloped infrastructure, particularly in rural areas. It has a rudimentary, but improving, road system, and limited external and internal telecommunications. China has signed a deal with the Lao to build a high speed rail system in the country. Construction on the $7 billion project is slated to begin in April 2011 and will take five years. Electricity is available in urban areas and in many rural districts. Subsistence agriculture, dominated by rice cultivation in lowland areas, accounts for about 30% of GDP and 75% of total employment. The government in FY09/10 received $586 million from international donors. Economic growth has reduced official poverty rates from 46% in 1992 to 26% in 2010. The economy has benefited from high foreign investment in hydropower, mining, and construction. Laos gained Normal Trade Relations status with the US in 2004, and is taking steps required to join the World Trade Organization, such as reforming import licensing. Related trade policy reforms will improve the business environment. On the fiscal side, Laos initiated a VAT tax system in 2010. Simplified investment procedures and expanded bank credits for small farmers and small entrepreneurs will improve Lao's economic prospects. The government appears committed to raising the country's profile among investors. The World Bank has declared that Laos's goal of graduating from the UN Development Program's list of least-developed countries by 2020 is achievable. According Laotian officials, the 7th Socio-Economic Development Plan for 2011-15 will outline efforts to achieve Millennium Development Goals.
GDP (purchasing power parity): $15.69 billion (2010 est.)
country comparison to the world: 135
$14.56 billion (2009 est.)
$13.54 billion (2008 est.)
note: data are in 2010 US dollars
GDP (official exchange rate): $6.341 billion (2010 est.)
GDP—real growth rate: 7.7% (2010 est.)
country comparison to the world: 23
7.6% (2009 est.)
7.8% (2008 est.)
GDP—per capita (PPP): $2,500 (2010 est.)
country comparison to the world: 176
$2,300 (2009 est.)
$2,200 (2008 est.)
note: data are in 2010 US dollars
GDP—composition by sector: *agriculture:* 29.8%
industry: 31.7%
services: 38.5% (2010 est.)
Labor force: 3.69 million (2010 est.)
country comparison to the world: 95
Labor force—by occupation: *agriculture:* 75.1%
industry and services: NA (2010 est.)
Unemployment rate: 2.5% (2009 est.)
country comparison to the world: 21
2.4% (2005 est.)
Population below poverty line: 26% (2010 est.)
Household income or consumption by percentage share: *lowest 10%:* 3.4%
highest 10%: 28.5% (2002)
Distribution of family income—Gini index: 34.6 (2002)
country comparison to the world: 89
37 (1997)
Budget: *revenues:* $1.137 billion
expenditures: $1.328 billion (2010 est.)
Inflation rate (consumer prices): 6% (2010 est.)
country comparison to the world: 158
0% (2009 est.)
Central bank discount rate: 4.3% (31 December 2010)
country comparison to the world: 103
4% (31 December 2009)
Commercial bank prime lending rate: 26% (31 December 2010)
country comparison to the world: 79
11% (30 November 2009)
Stock of narrow money: $630 million (31 December 2010 est.)
country comparison to the world: 151
$691.1 million (31 December 2009)

Stock of broad money: $1.818 billion (31 December 2010 est.)
country comparison to the world: 144
$1.549 billion (31 December 2009 est.)
Stock of domestic credit: $1.562 billion (31 December 2010 est.)
country comparison to the world: 133
$1.095 billion (31 December 2009 est.)
Agriculture—products: sweet potatoes, vegetables, corn, coffee, sugarcane, tobacco, cotton, tea, peanuts, rice; water buffalo, pigs, cattle, poultry
Industries: copper, tin, gold, and gypsum mining; timber, electric power, agricultural processing, construction, garments, cement, tourism
Industrial production growth rate: 17.7% (2010 est.)
country comparison to the world: 4
Electricity—production: 1.553 billion kWh (2010 est.)
country comparison to the world: 139
Electricity—consumption: 2.23 billion kWh (2010 est.)
country comparison to the world: 133
Electricity—exports: 341 million kWh (2010 est.)
Electricity—imports: 999 million kWh (2010 est.)
Oil—production: 0 bbl/day (2009 est.)
country comparison to the world: 192
Oil—consumption: 1,918 bbl/day (2010 est.)
country comparison to the world: 188
Oil—exports: 0 bbl/day (2007 est.)
country comparison to the world: 179
Oil—imports: 1,918 bbl/day (2010 est.)
country comparison to the world: 177
Oil—proved reserves: 0 bbl
country comparison to the world: 150
Natural gas—production: 0 cu m (2008 est.)
country comparison to the world: 146
Natural gas—consumption: NA cu m (2008 est.)
Natural gas—exports: 0 cu m (2008 est.)
country comparison to the world: 126
Natural gas—imports: NA cu m (2008 est.)
Natural gas—proved reserves: 0 cu m (1 January 2010 est.)
country comparison to the world: 152
Current account balance: $-195 million (2010 est.)
country comparison to the world: 89
$-395 million (2009 est.)
Exports: $1.95 billion (2010 est.)
country comparison to the world: 134
$1.147 billion (2009 est.)
Exports—commodities: wood products, coffee, electricity, tin, copper, gold
Exports—partners: Thailand 29.18%, China 15.04%, Vietnam 14.96%, UK 4.29% (2009)
Imports: $1.504 billion (2010 est.)
country comparison to the world: 161
$1.308 billion (2009 est.)
Imports—commodities: machinery and equipment, vehicles, fuel, consumer goods
Imports—partners: Thailand 66.2%, China 11.45%, Vietnam 5.3% (2009)
Reserves of foreign exchange and gold: $756 million (31 December 2010 est.)
country comparison to the world: 120
$712.4 million (31 December 2009 est.)
Debt—external: $5.797 billion (2010 est.)
country comparison to the world: 101
$5.548 billion (2009 est.)
Exchange rates: kips (LAK) per US dollar– 8,320.27 (2010), 8,516.04 (2009), 8,760.69 (2008), 9,658 (2007), 10,235 (2006)

COMMUNICATIONS

Telephones—main lines in use: 132,200 (2009)
country comparison to the world: 138
Telephones—mobile cellular: 3.235 million (2009)
country comparison to the world: 111
Telephone system: general assessment: service to general public is poor but improving; the government relies on a radiotelephone network to communicate with remote areas
domestic: multiple service providers; mobile cellular usage growing very rapidly
international: country code–856; satellite earth station–1 Intersputnik (Indian Ocean region) and a second to be developed by China (2008)
Broadcast media: 2 television stations operating out of Vientiane–1 government-operated and the other jointly-owned by the government and a Thai company; roughly 15 provincial stations operating with nearly all programming relayed via satellite from the government-operated station in Vientiane; relays from Hanoi provide access to a Vietnamese television station; broadcasts available from stations in Thailand and Vietnam in border areas; multi-channel satellite and cable TV systems provide access to a wide range of foreign stations; state-controlled radio with state-operated Lao National Radio (LNR) broadcasting on 5 frequencies–1 AM, 2 SW, and 2 FM; LNR's AM and FM programs are relayed via satellite constituting a large part of the programming schedules of the provincial radio stations; Thai radio broadcasts available in border areas and transmissions of multiple international broadcasters are also accessible (2008)
Internet country code: .la
Internet hosts: 1,468 (2010)
country comparison to the world: 161
Internet users: 300,000 (2009)
country comparison to the world: 130

TRANSPORTATION

Airports: 41 (2010)
country comparison to the world: 103
Airports—with paved runways: *total:* 9
2,438 to 3,047 m: 2
1,524 to 2,437 m: 4
914 to 1,523 m: 3 (2010)
Airports—with unpaved runways: *total:* 32
1,524 to 2,437 m: 2
914 to 1,523 m: 9
under 914 m: 21 (2010)
Pipelines: refined products 540 km (2010)
Roadways: *total:* 36,831 km
country comparison to the world: 92
paved: 4,811 km
unpaved: 32,020 km (2007)
Waterways: 4,600 km (primarily on the Mekong River and its tributaries; 2,900 additional km are intermittently navigable by craft drawing less than 0.5 m) (2010)
country comparison to the world: 24
Merchant marine: *total:* 1
country comparison to the world: 156
by type: cargo 1 (2008)

MILITARY

Military branches: Lao People's Armed Forces (LPAF): Lao People's Army (LPA; includes Riverine Force), Air Force (2011)
Military service age and obligation: 18 years of age for compulsory military service; minimum 18-month service obligation (2010)
Manpower available for military service: *males age 16–49:* 1,574,362
females age 16–49: 1,607,856 (2010 est.)
Manpower fit for military service: *males age 16–49:* 1,111,629
females age 16–49: 1,190,035 (2010 est.)
Manpower reaching militarily significant age annually: *male:* 71,400
female: 73,038 (2010 est.)
Military expenditures: 0.5% of GDP (2006)
country comparison to the world: 163
Military—note: serving one of the world's least developed countries, the Lao People's Armed Forces (LPAF) is small, poorly funded, and ineffectively resourced; its mission focus is border and internal security, primarily in countering ethnic Hmong insurgent groups; together with the Lao People's Revolutionary Party and the government, the Lao People's Army (LPA) is the third pillar of state machinery, and as such is expected to suppress political and civil unrest and similar national emergencies, but the LPA also has upgraded skills to respond to avian influenza outbreaks; there is no perceived external threat to the state and the LPA maintains strong ties with the neighboring Vietnamese military (2008)

TRANSNATIONAL ISSUES

Disputes—international: Southeast Asian states have enhanced border surveillance to check the spread of avian flu; talks continue on completion of demarcation with Thailand but disputes remain over islands in the Mekong River; concern among Mekong Commission members that China's construction of dams on the Mekong River will affect water levels; Cambodia is concerned about Laos' extensive upstream dam construction
Illicit drugs: estimated opium poppy cultivation in 2008 was 1,900 hectares, about a 73% increase from 2007; estimated potential opium production in 2008 more than tripled to 17 metric tons; unsubstantiated reports of domestic methamphetamine production; growing domestic methamphetamine problem (2007)

LATVIA

INTRODUCTION

Background: The name "Latvia" originates from the ancient Latgalians, one of four eastern Baltic tribes that formed the ethnic core of the Latvian people (ca. 8th-12th centuries A.D.). The region subsequently came under the control of Germans, Poles, Swedes, and finally, Russians. A Latvian republic emerged following World War I, but it was annexed by the USSR in 1940–an action never recognized by the US and many other countries. Latvia reestablished its independence in 1991 following the breakup of the Soviet Union. Although the last Russian troops left in 1994, the status of the Russian minority (some 30% of the population) remains of concern to Moscow. Latvia joined both NATO and the EU in the spring of 2004.

GEOGRAPHY

Location: Eastern Europe, bordering the Baltic Sea, between Estonia and Lithuania
Geographic coordinates: 57 00 N, 25 00 E
Map references: Europe
Area: *total:* 64,589 sq km
country comparison to the world: 123
land: 62,249 sq km
water: 2,340 sq km
Area—comparative: slightly larger than West Virginia
Land boundaries: total: 1,382 km
border countries: Belarus 171 km, Estonia 343 km, Lithuania 576 km, Russia 292 km
Coastline: 498 km
Maritime claims: *territorial sea:* 12 nm
exclusive economic zone: 200 nm
continental shelf: 200 m depth or to the depth of exploitation
Climate: maritime; wet, moderate winters
Terrain: low plain
Elevation extremes:
lowest point: Baltic Sea 0 m
highest point: Gaizina Kalns 312 m
Natural resources: peat, limestone, dolomite, amber, hydropower, timber, arable land
Land use: arable land: 28.19%
permanent crops: 0.45%
other: 71.36% (2005)
Irrigated land: 200 sq km
note: land in Latvia is often too wet and in need of drainage not irrigation; approximately 16,000 sq km or 85% of agricultural land has been improved by drainage (2003)
Total renewable water resources: 49.9 cu km (2005)
Freshwater withdrawal (domestic/industrial/agricultural): *total:* 0.25 cu km/yr (55%/33%/12%)
per capita: 108 cu m/yr (2003)
Natural hazards: NA
Environment—current issues: Latvia's environment has benefited from a shift to service industries after the country regained independence; the main environmental priorities are improvement of drinking water quality and sewage system, household, and hazardous waste management, as well as reduction of air pollution; in 2001, Latvia closed the EU accession negotiation chapter on environment committing to full enforcement of EU environmental directives by 2010
Environment—international agreements: *party to:* Air Pollution, Air Pollution-Persistent Organic Pollutants, Biodiversity, Climate Change, Climate Change-Kyoto Protocol, Desertification, Endangered Species, Hazardous Wastes, Law of the Sea, Ozone Layer Protection, Ship Pollution, Wetlands
signed, but not ratified: none of the selected agreements
Geography—note: most of the country is composed of fertile low-lying plains with some hills in the east

PEOPLE

Population: 2,204,708 (July 2011 est.)
country comparison to the world: 141
Age structure: *0–14 years:* 13.5% (male 152,706/female 145,756)
15–64 years: 69.5% (male 747,044/female 785,521)
65 years and over: 16.9% (male 121,570/female 252,111) (2011 est.)
Median age: *total:* 40.6 years
male: 37.6 years
female: 43.7 years (2011 est.)
Population growth rate: -0.597% (2011 est.)
country comparison to the world: 224
Birth rate: 9.96 births/1,000 population (2011 est.)
country comparison to the world: 194
Death rate: 13.6 deaths/1,000 population (July 2011 est.)
country comparison to the world: 18
Net migration rate: -2.33 migrant(s)/1,000 population (2011 est.)
country comparison to the world: 170
Urbanization: *urban population:* 68% of total population (2010)
rate of urbanization: -0.4% annual rate of change (2010-15 est.)
Major cities—population: RIGA (capital) 711,000 (2009)
Sex ratio: *at birth:* 1.054 male(s)/female
under 15 years: 1.05 male(s)/female
15–64 years: 0.95 male(s)/female
65 years and over: 0.48 male(s)/female
total population: 0.86 male(s)/female (2011 est.)
Infant mortality rate:
total: 8.42 deaths/1,000 live births
country comparison to the world: 159
male: 10.2 deaths/1,000 live births
female: 6.53 deaths/1,000 live births (2011 est.)
Life expectancy at birth: *total population:* 72.68 years
country comparison to the world: 122
male: 67.56 years
female: 78.07 years (2011 est.)
Total fertility rate: 1.32 children born/woman (2011 est.)
country comparison to the world: 207
HIV/AIDS—adult prevalence rate: 0.7% (2009 est.)
country comparison to the world: 61
HIV/AIDS—people living with HIV/AIDS: 8,600 (2009 est.)
country comparison to the world: 106
HIV/AIDS—deaths: fewer than 1,000 (2009 est.)
country comparison to the world: 75
Major infectious diseases: degree of risk: intermediate
food or waterborne diseases: bacterial diarrhea
vectorborne diseases: tickborne encephalitis (2009)
Drinking water source: *Improved:*
urban: 100% of population
rural: 96% of population
total: 99% of population
Unimproved: urban: 0% of population
rural: 4% of population
total: 1% of population (2008)
Sanitation facility access: *Improved:*
urban: 82% of population
rural: 71% of population
total: 78% of population
Unimproved: urban: 18% of population
rural: 29% of population
total: 22% of population (2008)
Nationality: noun: Latvian(s)
adjective: Latvian
Ethnic groups: Latvian 59.3%, Russian 27.8%, Belarusian 3.6%, Ukrainian 2.5%, Polish 2.4%, Lithuanian 1.3%, other 3.1% (2009)
Religions: Lutheran 19.6%, Orthodox 15.3%, other Christian 1%, other 0.4%, unspecified 63.7% (2006)
Languages: Latvian (official) 58.2%, Russian 37.5%, Lithuanian and other 4.3% (2000 census)
Literacy: *definition:* age 15 and over can read and write
total population: 99.7%
male: 99.8%
female: 99.7% (2000 census)
School life expectancy (primary to tertiary education): *total:* 15 years
male: 14 years
female: 17 years (2008)
Education expenditures: 5% of GDP (2007)
country comparison to the world: 57

GOVERNMENT

Country name: *conventional long form:* Republic of Latvia
conventional short form: Latvia
local long form: Latvijas Republika
local short form: Latvija
former: Latvian Soviet Socialist Republic
Government type: parliamentary democracy

Capital: *name:* Riga
geographic coordinates: 56 57 N, 24 06 E
time difference: UTC+2 (7 hours ahead of Washington, DC during Standard Time)
daylight saving time: +1hr, begins last Sunday in March; ends last Sunday in October
Administrative divisions: 109 municipalities (novadi, singular-novads) and 9 cities
municipalities: Adazu Novads, Aglonas Novads, Aizkraukles Novads, Aizputes Novads, Aknistes Novads, Alojas Novads, Alsungas Novads, Aluksnes Novads, Amatas Novads, Apes Novads, Auces Novads, Babites Novads, Baldones Novads, Baltinavas Novads, Balvu Novads, Bauskas Novads, Beverinas Novads, Brocenu Novads, Burtnieku Novads, Carnikavas Novads, Cesu Novads, Cesvaines Novads, Ciblas Novads, Dagdas Novads, Daugavpils Novads, Dobeles Novads, Dundagas Novads, Durbes Novads, Engures Novads, Erglu Novads, Garkalnes Novads, Grobinas Novads, Gulbenes Novads, Iecavas Novads, Ikskiles Novads, Ilukstes Novads, Incukalna Novads, Jaunjelgavas Novads, Juanpiebalgas Novads, Jaunpils Novads, Jekabpils Novads, Jelgavas Novads, Kandavas Novads, Karsavas Novads, Keguma Novads, Kekavas Novads, Kocenu Novads, Kokneses Novads, Kraslavas Novads, Krimuldas Novads, Krustpils Novads, Kuldigas Novads, Lielvardes Novads, Ligatnes Novads, Limbazu Novads, Livanu Novads, Lubanas Novads, Ludzas Novads, Madonas Novads, Malpils Novads, Marupes Novads, Mazsalacas Novads, Nauksenu Novads, Neretas Novads, Nicas Novads, Ogres Novads, Olaines Novads, Ozolnieku Novads, Pargaujas Novads, Pavilostas Novads, Plavinu Novads, Preilu Novads, Priekules Novads, Priekulu Novads, Raunas Novads, Rezeknes Novads, Riebinu Novads, Rojas Novads, Ropazu Novads, Rucavas Novads, Rugaju Novads, Rujienas Novads, Rundales Novads, Salacgrivas Novads, Salas Novads, Salaspils Novads, Saldus Novads, Saulkrastu Novads, Sejas Novads, Siguldas Novads, Skriveru Novads, Skrundas Novads, Smiltenes Novads, Stopinu Novads, Strencu Novads, Talsu Novads, Tervetes Novads, Tukuma Novads, Vainodes Novads, Valkas Novads, Varaklanu Novads, Varkavas Novads, Vecpiebalgas Novads, Vecumnieku Novads, Ventspils Novads, Viesites Novads, Vilakas Novads, Vilanu Novads, Zilupes Novads
cities: Daugavpils, Jekabpils, Jelgava, Jurmala, Liepaja, Rezekne, Riga, Valmiera, Ventspils
Independence: 4 May 1990 (declared); 6 September 1991 (recognized by the Soviet Union)
National holiday: Independence Day, 18 November (1918); note—18 November 1918 was the date Latvia declared independence from Soviet Russia and established its statehood; 4 May 1990 was the date it declared its independence from the Soviet Union
Constitution: 15 February 1922; restored to force by the Constitutional Law of the Republic of Latvia adopted by the Supreme Council 21 August 1991; multiple amendments since
Legal system: civil law system with traces of socialist legal traditions and practices
International law organization participation: has not submitted an ICJ jurisdiction declaration; accepts ICCt jurisdiction
Suffrage: 18 years of age; universal for Latvian citizens
Executive branch: *chief of state:* President Valdis ZATLERS (since 8 July 2007)
head of government: Prime Minister Valdis DOMBROVSKIS (since 12 March 2009)
cabinet: Cabinet of Ministers nominated by the prime minister and appointed by Parliament
(For more information visit the World Leaders website)
elections: president elected by Parliament for a four-year term (eligible for a second term); election last held on 31 May 2007 (next to be held in 2011); prime minister appointed by the president, confirmed by Parliament
election results: Valdis ZATLERS elected president; parliamentary vote—Valdis ZATLERS 58, Aivars ENDZINS 39
Legislative branch: unicameral Parliament or Saeima (100 seats; members elected by proportional representation from party lists by popular vote to serve four-year terms)
elections: last held on 2 October 2010 (next to be held in October 2014)
election results: percent of vote by party—Unity bloc 31.2%, SC 26%, ZZS 19.7%, National Alliance 7.7%, For a Good Latvia bloc 7.7%; seats by party—Unity bloc 33, SC 29, ZZS 22, National Alliance 8, For a Good Latvia 8
Judicial branch: Supreme Court (judges' appointments are confirmed by parliament); Constitutional Court (judges' appointments are confirmed by parliament)
Political parties and leaders: All For Latvia! [Irnants PARADNIEKS, Raivis DZINTARS]; Civic Union [Sandra KALNIETE, Girts Valdis KRISTOVSKIS]; First Party of Latvia/Latvia's Way or LPP/LC [Ainars SLESERS]; For a Good Latvia (alliance of TP, LPP/LC); For Human Rights in a United Latvia or PCTVL [Jakovs PLINERS, Tatjana ZDANOKA]; For the Fatherland and Freedom/Latvian National Independence Movement or TB/LNNK [Roberts ZILE, Maris GRINBLATS]; Harmony Center or SC [Nils USAKOVS, Janis URBANOVICS]; National Alliance (alliance of TB/LNNK, All For Latvia!); New Era Party or JL [Solvita ABOLTINA, Dzintars ZAKIS]; People's Party or TP [Andris SKELE]; Society for Different Politics or SCP [Aigars STOKENBERGS; Artis PABRIKS]; The Union of Latvian Greens and Farmers Party or ZZS [Augusts BRIGMANIS]; Unity bloc (alliance of Civic Union, New Era, SCP)
Political pressure groups and leaders: Free Trade Union Confederation of Latvia [Peteris KRIGERS], Employers' Confederation of Latvia [Elina EGLE], Farmers' Parliament [Juris LAZDINS]
International organization participation: Australia Group, BA, BIS, CBSS, CE, EAPC, EBRD, EIB, EU, FAO, IAEA, IBRD, ICAO, ICRM, IDA, IFC, IFRCS, IHO, ILO, IMF, IMO, IMSO, Interpol, IOC, IOM, IPU, ISO (correspondent), ITU, ITUC, MIGA, NATO, NIB, NSG, OAS (observer), OIF (observer), OPCW, OSCE, PCA, Schengen Convention, UN, UNCTAD, UNESCO, UNWTO, UPU, WCO, WHO, WIPO, WMO, WTO
Diplomatic representation in the US: *chief of mission:* Ambassador Andrejs PILDEGOVICS
chancery: 2306 Massachusetts Ave. NW, Washington, DC 20008
telephone: [1] (202) 328-2840
FAX: [1] (202) 328-2860
Diplomatic representation from the US: *chief of mission:* Ambassador Judith G. GARBER
embassy: 7 Raina Boulevard, Riga LV-1510
mailing address: American Embassy Riga, US Department of State, 4520 Riga Place, Washington, DC 20520-4520
telephone: [371] 670-36200
FAX: [371] 678-20047
Flag description: three horizontal bands of maroon (top), white (half-width), and maroon; the flag is one of the older banners in the world; a medieval chronicle mentions a red standard with a white stripe being used by Latvian tribes in about 1280
National anthem: *name:* "Dievs, sveti Latviju!" (God Bless Latvia)
lyrics/music: Karlis BAUMANIS
note: adopted 1920, restored 1990; the song was first performed in 1873 while Latvia was a part of Russia; the anthem was banned during the Soviet occupation from 1940 to 1990

ECONOMY

Economy—overview: Latvia is a small, open economy with exports contributing significantly to its GDP. Due to its geographical location, transit services are highly-developed, along with timber and wood-processing, agriculture and food products, and manufacturing of machinery and electronic devices. The bulk of the country's economic activity, however, is in the services sector. Corruption continues to be an impediment to attracting FDI flows and Latvia's low birth rate and decreasing population are major challenges to its long-term economic vitality. Latvia's economy experienced GDP growth of more than 10% per year during 2006-07, but entered a severe recession in 2008 as a result of an unsustainable current account deficit and large debt exposure amid the softening world economy. GDP plunged 18% in 2009—the three Baltic states had the world's worst declines that year. Thanks to strong export growth in 2009 and 2010, the economy experienced its first real quarterly GDP growth in over two years (2.9%) in the third quarter of 2010. The IMF, EU, and other international donors provided substantial financial assistance to Latvia as part of an agreement to defend the currency's peg to the euro. This agreement calls for reduction of Latvia's fiscal deficit to below 3% of GDP by 2012, in order to meet the Maastricht Treaty criteria for euro adoption. DOMBROVSKIS' government enacted major spending cuts to reduce the fiscal deficit to a maximum of 8.5% of GDP in 2010, and Latvia has approved a 2011 budget

with a projected deficit of 5.4% of GDP. The majority of companies, banks, and real estate have been privatized, although the state still holds sizable stakes in a few large enterprises. Latvia officially joined the World Trade Organization in February, 1999. EU membership, a top foreign policy goal, came in May 2004. Latvia's current major financial policy goal, entrance into the euro zone, is targeted for 2014.
GDP (purchasing power parity): $32.51 billion (2010 est.)
country comparison to the world: 106
$32.62 billion (2009 est.)
$39.76 billion (2008 est.)
note: data are in 2010 US dollars
GDP (official exchange rate): $24.05 billion (2010 est.)
GDP—real growth rate: -0.3% (2010 est.)
country comparison to the world: 192
-18% (2009 est.)
-4.2% (2008 est.)
GDP—per capita (PPP): $14,700 (2010 est.)
country comparison to the world: 77
$14,600 (2009 est.)
$17,700 (2008 est.)
note: data are in 2010 US dollars
GDP—composition by sector:
agriculture: 4.2%
industry: 20.6%
services: 75.2% (2010 est.)
Labor force: 1.178 million (2010 est.)
country comparison to the world: 139
Labor force—by occupation:
agriculture: 12.1%
industry: 25.8%
services: 61.8% (2005 est.)
Unemployment rate: 14.3% (2010 est.)
country comparison to the world: 145
16% (2009 est.)
Population below poverty line: NA%
Household income or consumption by percentage share: *lowest 10%:* 2.7%
highest 10%: 27.4% (2004)
Distribution of family income—Gini index: 36 (2005)
country comparison to the world: 84
32 (1999)
Investment (gross fixed): 15.7% of GDP (2010 est.)
country comparison to the world: 128
Budget: *revenues:* $8.028 billion
expenditures: $9.863 billion (2010 est.)
Public debt: 46.2% of GDP (2010 est.)
country comparison to the world: 58
36.6% of GDP (2009 est.)
Inflation rate (consumer prices): -1.2% (2010 est.)
country comparison to the world: 4
3.5% (2009 est.)
Central bank discount rate: 4% (31 December 2009)
country comparison to the world: 70
6% (31 December 2008)
Commercial bank prime lending rate: 16.23% (31 December 2009 est.)
country comparison to the world: 70
11.85% (31 December 2008 est.)
Stock of narrow money: $5.769 billion (31 December 2010 est.)
country comparison to the world: 85
$5.893 billion (31 December 2009 est.)
Stock of broad money: $11.17 billion (31 December 2010 est.)
country comparison to the world: 97
$11.46 billion (31 December 2009 est.)
Stock of domestic credit: $27.59 billion (31 December 2010 est.)
country comparison to the world: 70
$27.76 billion (31 December 2009 est.)
Market value of publicly traded shares: $1.824 billion (31 December 2009)
country comparison to the world: 101
$1.609 billion (31 December 2008)
$3.111 billion (31 December 2007)
Agriculture—products: grain, rapeseed, potatoes, vegetables; pork, poultry, milk, eggs; fish
Industries: processed foods, processed wood products, textiles, processed metals, pharmaceuticals, railroad cars, synthetic fibers, electronics
Industrial production growth rate: -1.8% (2010 est.)
country comparison to the world: 159
Electricity—production: 4.62 billion kWh (2007 est.)
country comparison to the world: 115
Electricity—consumption: 6.822 billion kWh (2007 est.)
country comparison to the world: 101
Electricity—exports: 2.123 billion kWh (2008 est.)
Electricity—imports: 4.643 billion kWh (2008 est.)
Oil—production: 0 bbl/day (2009 est.)
country comparison to the world: 194
Oil—consumption: 40,000 bbl/day (2009 est.)
country comparison to the world: 101
Oil—exports: 5,873 bbl/day (2007 est.)
country comparison to the world: 99
Oil—imports: 43,400 bbl/day (2007 est.)
country comparison to the world: 92
Oil—proved reserves: 0 bbl (1 January 2010 est.)
country comparison to the world: 152
Natural gas—production: 0 cu m (2008 est.)
country comparison to the world: 148
Natural gas—consumption: 1.527 billion cu m (2009 est.)
country comparison to the world: 82
Natural gas—exports: 0 cu m (2008 est.)
country comparison to the world: 128
Natural gas—imports: 1.743 billion cu m (2009 est.)
country comparison to the world: 48
Current account balance: $1.62 billion (2010 est.)
country comparison to the world: 43
$2.53 billion (2009 est.)
Exports:
$7.894 billion (2010 est.)
country comparison to the world: 94
$7.223 billion (2009 est.)
Exports—commodities: food products, wood and wood products, metals, machinery and equipment, textiles
Exports—partners: Lithuania 15.19%, Estonia 13.57%, Russia 13.17%, Germany 8.13%, Sweden 5.7% (2009)
Imports: $9.153 billion (2010 est.)
country comparison to the world: 93
$8.906 billion (2009 est.)
Imports—commodities: machinery and equipment, consumer goods, chemicals, fuels, vehicles
Imports—partners: Lithuania 16.36%, Germany 11.34%, Russia 10.68%, Poland 8.11%, Estonia 7.69% (2009)
Reserves of foreign exchange and gold: $7.17 billion (31 December 2010 est.)
country comparison to the world: 64
$6.907 billion (31 December 2009 est.)
Debt—external: $37.28 billion (31 December 2010 est.)
country comparison to the world: 59
$41.58 billion (31 December 2009 est.)
Stock of direct foreign investment—at home: $11.71 billion (31 December 2010 est.)
country comparison to the world: 80
$11.61 billion (31 December 2009 est.)
Stock of direct foreign investment—abroad: $1.097 billion (31 December 2010 est.)
country comparison to the world: 72
$1.037 billion (31 December 2009 est.)
Exchange rates: lati (LVL) per US dollar–0.5422 (2010), 0.5056 (2009), 0.4701 (2008), 0.5162 (2007), 0.5597 (2006)

COMMUNICATIONS

Telephones—main lines in use: 644,000 (2009)
country comparison to the world: 92
Telephones—mobile cellular: 2.243 million (2009)
country comparison to the world: 128
Telephone system: general assessment: recent efforts focused on bringing competition to the telecommunications sector; the number of fixed lines is decreasing as mobile-cellular telephone service expands
domestic: number of telecommunications operators has grown rapidly since the fixed-line market opened to competition in 2003; combined fixed-line and mobile-cellular subscribership exceeds 125 per 100 persons
international: country code–371; the Latvian network is now connected via fiber optic cable to Estonia, Finland, and Sweden (2008)
Broadcast media: several national and regional commercial TV stations are foreign-owned, 2 national TV stations are publicly-owned; system supplemented by privately-owned regional and local TV stations; cable and satellite multi-channel TV services with domestic and foreign broadcasts are available; publicly-owned broadcaster operates 4 radio networks with dozens of stations throughout the country; dozens of private broadcasters also operate radio stations (2007)
Internet country code: .lv
Internet hosts: 289,478 (2010)
country comparison to the world: 59
Internet users: 1.504 million (2009)
country comparison to the world: 81

TRANSPORTATION

Airports: 42 (2010)
country comparison to the world: 101
Airports—with paved runways: *total:* 19
over 3,047 m: 1
2,438 to 3,047 m: 3
1,524 to 2,437 m: 5
914 to 1,523 m: 3
under 914 m: 7 (2010)
Airports—with unpaved runways: *total:* 23
under 914 m: 23 (2010)
Pipelines: gas 948 km; refined products 415 km (2010)

Railways: *total:* 2,239 km
country comparison to the world: 67
broad gauge: 2,206 km 1.520-m gauge
narrow gauge: 33 km 0.750-m gauge (2010)
Roadways: *total:* 73,074 km
country comparison to the world: 65
paved: 14,459 km
unpaved: 58,615 km (2010)
Waterways: 300 km (navigable year round) (2010)
country comparison to the world: 94
Merchant marine: *total:* 13
country comparison to the world: 108
by type: cargo 3, chemical tanker 1, passenger/cargo 5, petroleum tanker 3, roll on/roll off 1
foreign-owned: 4 (Estonia 4)
registered in other countries: 90 (Antigua and Barbuda 16, Belize 10, Cambodia 1, Comoros 1, Cook Islands 1, Dominica 1, Georgia 1, Liberia 9, Malta 11, Marshall Islands 18, Panama 4, Saint Kitts and Nevis 2, Saint Vincent and the Grenadines 15) (2010)
Ports and terminals: Riga, Ventspils

MILITARY

Military branches: National Armed Forces (Nacionalo Brunoto Speku): Land Forces, Navy (Latvijas Juras Speki; includes Coast Guard (Latvijas Kara Flotes)), Latvian Air Force (Latvijas Gaisa Speki), Latvian Home Guard (Latvijas Zemessardze) (2011)
Military service age and obligation: 18 years of age for voluntary male and female military service; conscription abolished January 2007; under current law, every citizen is entitled to serve in the armed forces for life (2009)
Manpower available for military service: *males age 16–49:* 546,090
females age 16–49: 540,810 (2010 est.)
Manpower fit for military service: *males age 16–49:* 401,691
females age 16–49: 447,638 (2010 est.)
Manpower reaching militarily significant age annually: *male:* 10,482
female: 9,858 (2010 est.)
Military expenditures: 1.2% of GDP (2005 est.)
country comparison to the world: 120

TRANSNATIONAL ISSUES

Disputes—international: Russia demands better Latvian treatment of ethnic Russians in Latvia; boundary demarcated with Latvia and Lithuania; the Latvian parliament has not ratified its 1998 maritime boundary treaty with Lithuania, primarily due to concerns over oil exploration rights; as a member state that forms part of the EU's external border, Latvia has implemented the strict Schengen border rules with Russia
Illicit drugs: transshipment and destination point for cocaine, synthetic drugs, opiates, and cannabis from Southwest Asia, Western Europe, Latin America, and neighboring Balkan countries; despite improved legislation, vulnerable to money laundering due to nascent enforcement capabilities and comparatively weak regulation of offshore companies and the gaming industry; CIS organized crime (including counterfeiting, corruption, extortion, stolen cars, and prostitution) accounts for most laundered proceeds

LEBANON

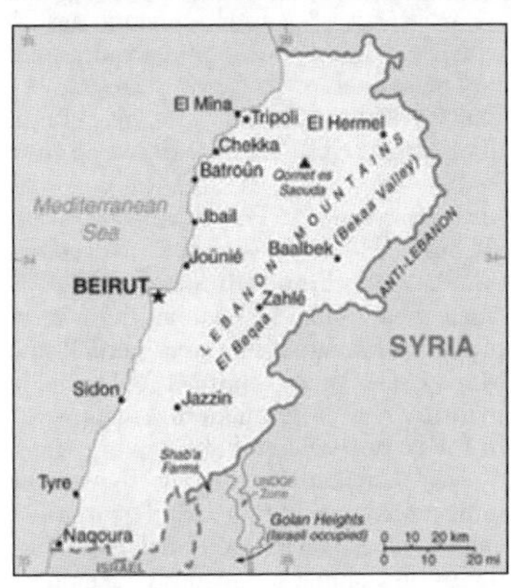

INTRODUCTION

Background: Following World War I, France acquired a mandate over the northern portion of the former Ottoman Empire province of Syria. The French separated out the region of Lebanon in 1920, and granted this area independence in 1943. A lengthy civil war (1975-90) devastated the country, but Lebanon has since made progress toward rebuilding its political institutions. Under the Ta'if Accord–the blueprint for national reconciliation–the Lebanese established a more equitable political system, particularly by giving Muslims a greater voice in the political process while institutionalizing sectarian divisions in the government. Since the end of the war, Lebanon has conducted several successful elections. Most militias have been reduced or disbanded, with the exception of Hizballah, designated by the US State Department as a Foreign Terrorist Organization, and Palestinian militant groups. During Lebanon's civil war, the Arab League legitimized in the Ta'if Accord Syria's troop deployment, numbering about 16,000 based mainly east of Beirut and in the Bekaa Valley. Israel's withdrawal from southern Lebanon in May 2000 and the passage in September 2004 of UNSCR 1559–a resolution calling for Syria to withdraw from Lebanon and end its interference in Lebanese affairs–encouraged some Lebanese groups to demand that Syria withdraw its forces as well. The assassination of former Prime Minister Rafiq HARIRI and 22 others in February 2005 led to massive demonstrations in Beirut against the Syrian presence ("the Cedar Revolution"), and Syria withdrew the remainder of its military forces in April 2005. In May-June 2005, Lebanon held its first legislative elections since the end of the civil war free of foreign interference, handing a majority to the bloc led by Sa'ad HARIRI, the slain prime minister's son. In July 2006, Hizballah kidnapped two Israeli soldiers leading to a 34-day conflict with Israel in which approximately 1,200 Lebanese civilians were killed. UNSCR 1701 ended the war in August 2006, and Lebanese Armed Forces (LAF) deployed throughout the country for the first time in decades, charged with securing Lebanon's borders against weapons smuggling and maintaining a weapons-free zone in south Lebanon with the help of the UN Interim Force in Lebanon (UNIFIL). The LAF in May-September 2007 battled Sunni extremist group Fatah al-Islam in the Nahr al-Barid Palestinian refugee camp, winning a decisive victory, but destroying the camp and displacing 30,000 Palestinian residents. Lebanese politicians in November 2007 were unable to agree on a successor to Emile LAHUD when he stepped down as president, creating a political vacuum until the election of LAF Commander Gen. Michel SULAYMAN in May 2008 and the formation of a new unity government in July 2008. Legislative elections in June 2009 again produced victory for the bloc led by Sa'ad HARIRI, but a period of prolonged negotiation over the composition of the cabinet ensued. A national unity government was finally formed in November 2009 and approved by the National Assembly the following month. In January 2010, Lebanon assumed a nonpermanent seat on the UN Security Council for the 2010-11 term. Inspired by the popular revolts that began in late 2010 against dictatorships across the Middle East and North Africa, marches and demonstrations in Lebanon were directed instead against sectarian politics. Protesters saw the country's religious sectarian politics as the primary cause of Lebanon's anemic government. The first protests in late February 2011, although limited in size, gained some traction.

GEOGRAPHY

Location: Middle East, bordering the Mediterranean Sea, between Israel and Syria
Geographic coordinates: 33 50 N, 35 50 E
Map references: Middle East
Area: *total:* 10,400 sq km
country comparison to the world: 169
land: 10,230 sq km
water: 170 sq km
Area—comparative: about 0.7 times the size of Connecticut
Land boundaries: total: 454 km
border countries: Israel 79 km, Syria 375 km
Coastline: 225 km
Maritime claims: *territorial sea:* 12 nm
Climate: Mediterranean; mild to cool, wet winters with hot, dry summers; Lebanon mountains experience heavy winter snows
Terrain: narrow coastal plain; El Beqaa (Bekaa Valley) separates Lebanon and Anti-Lebanon Mountains
Elevation extremes: *lowest point:* Mediterranean Sea 0 m

highest point: Qornet es Saouda 3,088 m
Natural resources: limestone, iron ore, salt, water-surplus state in a water-deficit region, arable land
Land use: arable land: 16.35%
permanent crops: 13.75%
other: 69.9% (2005)
Irrigated land: 1,040 sq km (2003)
Total renewable water resources: 4.8 cu km (1997)
Freshwater withdrawal (domestic/industrial/agricultural): *total:* 1.38 cu km/yr (33%/1%/67%)
per capita: 385 cu m/yr (2000)
Natural hazards: dust storms, sandstorms
Environment—current issues: deforestation; soil erosion; desertification; air pollution in Beirut from vehicular traffic and the burning of industrial wastes; pollution of coastal waters from raw sewage and oil spills
Environment—international agreements: *party to:* Biodiversity, Climate Change, Climate Change-Kyoto Protocol, Desertification, Hazardous Wastes, Law of the Sea, Ozone Layer Protection, Ship Pollution, Wetlands
signed, but not ratified: Environmental Modification, Marine Life Conservation
Geography—note: Nahr el Litani is the only major river in Near East not crossing an international boundary; rugged terrain historically helped isolate, protect, and develop numerous factional groups based on religion, clan, and ethnicity

PEOPLE

Population: 4,143,101 (July 2011 est.)
country comparison to the world: 127
Age structure: *0–14 years:* 23% (male 487,930/female 464,678)
15–64 years: 68% (male 1,370,628/female 1,446,173)
65 years and over: 9% (male 173,073/female 200,619) (2011 est.)
Median age: *total:* 29.8 years
male: 28.7 years
female: 31 years (2011 est.)
Population growth rate: 0.244% (2011 est.)
country comparison to the world: 175
Birth rate: 15.02 births/1,000 population (2011 est.)
country comparison to the world: 135
Death rate: 6.54 deaths/1,000 population (July 2011 est.)
country comparison to the world: 150
Net migration rate: -6.04 migrant(s)/1,000 population (2011 est.)
country comparison to the world: 199
Urbanization: *urban population:* 87% of total population (2010)
rate of urbanization: 0.9% annual rate of change (2010-15 est.)
Major cities—population: BEIRUT (capital) 1.909 million (2009)
Sex ratio: *at birth:* 1.05 male(s)/female
under 15 years: 1.05 male(s)/female
15–64 years: 0.95 male(s)/female
65 years and over: 0.87 male(s)/female
total population: 0.96 male(s)/female (2011 est.)
Infant mortality rate: *total:* 15.85 deaths/1,000 live births
country comparison to the world: 113
male: 15.99 deaths/1,000 live births
female: 15.71 deaths/1,000 live births (2011 est.)
Life expectancy at birth:
total population: 75.01 years
country comparison to the world: 90
male: 73.48 years
female: 76.62 years (2011 est.)
Total fertility rate: 1.77 children born/woman (2011 est.)
country comparison to the world: 158
HIV/AIDS—adult prevalence rate: 0.1% (2009 est.)
country comparison to the world: 142
HIV/AIDS—people living with HIV/AIDS: 3,600 (2009 est.)
country comparison to the world: 125
HIV/AIDS—deaths: fewer than 500 (2009 est.)
country comparison to the world: 86
Drinking water source: *Improved:*
urban: 100% of population
rural: 100% of population
total: 100% of population (2008)
Sanitation facility access: *Improved:*
urban: 100% of population
rural: 87% of population
total: 98% of population
Unimproved: urban: 0% of population
rural: 13% of population
total: 2% of population (2000)
Nationality: noun: Lebanese (singular and plural)
adjective: Lebanese
Ethnic groups: Arab 95%, Armenian 4%, other 1%
note: many Christian Lebanese do not identify themselves as Arab but rather as descendents of the ancient Canaanites and prefer to be called Phoenicians
Religions: Muslim 59.7% (Shia, Sunni, Druze, Isma'ilite, Alawite or Nusayri), Christian 39% (Maronite Catholic, Greek Orthodox, Melkite Catholic, Armenian Orthodox, Syrian Catholic, Armenian Catholic, Syrian Orthodox, Roman Catholic, Chaldean, Assyrian, Copt, Protestant), other 1.3%
note: 17 religious sects recognized
Languages: Arabic (official), French, English, Armenian
Literacy: *definition:* age 15 and over can read and write
total population: 87.4%
male: 93.1%
female: 82.2% (2003 est.)
School life expectancy (primary to tertiary education): *total:* 14 years
male: 13 years
female: 14 years (2009)
Education expenditures: 1.8% of GDP (2009)
country comparison to the world: 159

GOVERNMENT

Country name: *conventional long form:* Lebanese Republic
conventional short form: Lebanon
local long form: Al Jumhuriyah al Lubnaniyah
local short form: Lubnan
former: Greater Lebanon
Government type: republic
Capital: *name:* Beirut
geographic coordinates: 33 52 N, 35 30 E
time difference: UTC+2 (7 hours ahead of Washington, DC during Standard Time)
daylight saving time: +1hr, begins last Sunday in March; ends last Sunday in October
Administrative divisions: 6 governorates (mohafazat, singular–mohafazah); Beqaa, Beyrouth (Beirut), Liban-Nord, Liban-Sud, Mont-Liban, Nabatiye
note: two new governorates–Aakar and Baalbek-Hermel–have been legislated but not yet implemented
Independence: 22 November 1943 (from League of Nations mandate under French administration)
National holiday: Independence Day, 22 November (1943)
Constitution: 23 May 1926; amended a number of times, most recently in 1990 to include changes necessitated by the Charter of Lebanese National Reconciliation (Ta'if Accord) of October 1989
Legal system: mixed legal system of civil law based on the French civil code and religious laws covering personal status, marriage, divorce, and other family relations of the Jewish, Islamic, and Christian communities
International law organization participation: has not submitted an ICJ jurisdiction declaration; non-party state to the ICCt
Suffrage: 21 years of age; compulsory for all males; authorized for women at age 21 with elementary education; excludes military personnel
Executive branch: *chief of state:* President Michel SULAYMAN (since 25 May 2008)
head of government: Prime Minister Sa'ad al-Din al-HARIRI (since 9 November 2009), Deputy Prime Minister Elias MURR (since 9 November 2009; note–the government is in a caretaker status until Prime Minister-Designate Najib MIQATI is able to form a new government that is approved by the National Assembly
cabinet: Cabinet chosen by the prime minister in consultation with the president and members of the National Assembly; note–the Cabinet resigned on 12 January 2010 following the resignation of over a third of the ministers
(For more information visit the World Leaders website)
elections: president elected by the National Assembly for a six-year term (may not serve consecutive terms); election last held on 25 May 2008 (next to be held in 2014); the prime minister and deputy prime minister appointed by the president in consultation with the National Assembly
election results: Michel SULAYMAN elected president; National Assembly vote–118 for, 6 abstentions, 3 invalidated; 1 seat unfilled due to death of incumbent
Legislative branch: unicameral National Assembly or Majlis al-Nuwab (Arabic) or Assemblee Nationale (French) (128 seats; members elected by popular vote on the basis of sectarian proportional representation to serve four-year terms)
elections: last held on 7 June 2009 (next to be held in 2013)
election results: percent of vote by group–March 8 Coalition 54.7%, March 14 Coalition 45.3%; seats by group–March 14 Coalition 71; March 8 Coalition 57

Judicial branch: four Courts of Cassation (three courts for civil and commercial cases and one court for criminal cases); Constitutional Council (called for in Ta'if Accord–rules on constitutionality of laws); Supreme Council (hears charges against the president and prime minister as needed)

Political parties and leaders: 14 March Coalition: Democratic Left [Ilyas ATALLAH]; Democratic Renewal Movement [Nassib LAHUD]; Future Movement Bloc [Sa'ad al-HARIRI]; Kataeb Party [Amine GEMAYEL]; Lebanese Forces [Samir JA'JA]; Tripoli Independent Bloc
8 March Coalition: Development and Resistance Bloc [Nabih BERRI, leader of Amal Movement]; Free Patriotic Movement [Michel AWN]; Loyalty to the Resistance Bloc [Mohammad RA'AD] (includes Hizballah [Hassan NASRALLAH]); Nasserite Popular Movement [Usama SAAD]; Popular Bloc [Elias SKAFF]; Syrian Ba'th Party [Sayez SHUKR]; Syrian Social Nationalist Party [Ali QANSO]; Tashnaq [Hovig MEKHITIRIAN]

Independent: Democratic Gathering Bloc [Walid JUNBLATT, leader of Progressive Socialist Party]; Metn Bloc [Michel MURR]

Political pressure groups and leaders: Maronite Church [Patriarch Nasrallah SFAYR]
other: note–most sects retain militias and a number of militant groups operate in Palestinian refugee camps

International organization participation: ABEDA, AFESD, AMF, FAO, G-24, G-77, IAEA, IBRD, ICAO, ICC, ICRM, IDA, IDB, IFAD, IFC, IFRCS, ILO, IMF, IMO, IMSO, Interpol, IOC, IPU, ISO, ITSO, ITU, LAS, MIGA, NAM, OAS (observer), OIC, OIF, OPCW, PCA, UN, UN Security Council (temporary), UNCTAD, UNESCO, UNHCR, UNIDO, UNRWA, UNWTO, UPU, WCO, WFTU, WHO, WIPO, WMO, WTO (observer)

Diplomatic representation in the US: *chief of mission:* Ambassador Antoine CHEDID
chancery: 2560 28th Street NW, Washington, DC 20008
telephone: [1] (202) 939-6300
FAX: [1] (202) 939-6324
consulate(s) general: Detroit, New York, Los Angeles

Diplomatic representation from the US: *chief of mission:* Ambassador Maura CONNELLY
embassy: Awkar, Lebanon (Awkar facing the Municipality)
mailing address: P. O. Box 70-840, Antelias, Lebanon; from US: US Embassy Beirut, 6070 Beirut Place, Washington, DC 20521-6070
telephone: [961] (4) 542600, 543600
FAX: [961] (4) 544136

Flag description: three horizontal bands consisting of red (top), white (middle, double width), and red (bottom) with a green cedar tree centered in the white band; the red bands symbolize blood shed for liberation, the white band denotes peace, the snow of the mountains, and purity; the green cedar tree is the symbol of Lebanon and represents eternity, steadiness, happiness, and prosperity

National anthem: *name:* "Kulluna lil-watan" (All Of Us, For Our Country!)
lyrics/music: Rachid NAKHLE/Wadih SABRA
note: adopted 1927; the anthem was chosen following a nationwide competition

ECONOMY

Economy—overview: Lebanon has a free-market economy and a strong laissez-faire commercial tradition. The government does not restrict foreign investment; however, the investment climate suffers from red tape, corruption, arbitrary licensing decisions, high taxes, tariffs, and fees, archaic legislation, and weak intellectual property rights. The Lebanese economy is service-oriented; main growth sectors include banking and tourism. The 1975-90 civil war seriously damaged Lebanon's economic infrastructure, cut national output by half, and all but ended Lebanon's position as a Middle Eastern entrepot and banking hub. In the years since, Lebanon has rebuilt much of its war-torn physical and financial infrastructure by borrowing heavily–mostly from domestic banks. In an attempt to reduce the ballooning national debt, the Rafiq HARIRI government in 2000 began an austerity program, reining in government expenditures, increasing revenue collection, and passing legislation to privatize state enterprises, but economic and financial reform initiatives stalled and public debt continued to grow despite receipt of more than $2 billion in bilateral assistance at the 2002 Paris II Donors Conference. The Israeli-Hizballah conflict in July-August 2006 caused an estimated $3.6 billion in infrastructure damage, and prompted international donors to pledge nearly $1 billion in recovery and reconstruction assistance. Donors met again in January 2007 at the Paris III Donor Conference and pledged more than $7.5 billion to Lebanon for development projects and budget support, conditioned on progress on Beirut's fiscal reform and privatization program. An 18-month political stalemate and sporadic sectarian and political violence hampered economic activity, particularly tourism, retail sales, and investment, until the new government was formed in July 2008. Political stability following the Doha Accord of May 2008 helped boost tourism and, together with a strong banking sector, enabled real GDP growth of 7% per year in 2009-10 despite a slowdown in the region.

GDP (purchasing power parity): $59.37 billion (2010 est.)
country comparison to the world: 87
$55.23 billion (2009 est.)
$50.9 billion (2008 est.)
note: data are in 2010 US dollars

GDP (official exchange rate): $39.25 billion (2010 est.)

GDP—real growth rate: 7.5% (2010 est.)
country comparison to the world: 29
8.5% (2009 est.)
9.3% (2008 est.)

GDP—per capita (PPP): $14,400 (2010 est.)
country comparison to the world: 81
$13,500 (2009 est.)
$12,600 (2008 est.)
note: data are in 2010 US dollars

GDP—composition by sector: *agriculture:* 5.1%
industry: 15.9%
services: 79% (2010 est.)

Labor force: 1.481 million
country comparison to the world: 131
note: in addition, there are as many as 1 million foreign workers (2007 est.)

Labor force—by occupation: *agriculture:* NA%
industry: NA%
services: NA%

Unemployment rate: NA%

Population below poverty line: 28% (1999 est.)

Household income or consumption by percentage share: *lowest 10%:* NA%
highest 10%: NA%

Investment (gross fixed): 30.8% of GDP (2010 est.)
country comparison to the world: 16

Budget: *revenues:* $9.001 billion
expenditures: $10.95 billion (2010 est.)

Public debt: 150.7% of GDP (2010 est.)
country comparison to the world: 3
154.8% of GDP (2009 est.)

Inflation rate (consumer prices): 3.7% (2010 est.)
country comparison to the world: 106
1.2% (2009 est.)

Central bank discount rate: 10% (31 December 2009)
country comparison to the world: 33
12% (31 December 2008)

Commercial bank prime lending rate: 9.57% (31 December 2009 est.)
country comparison to the world: 89
9.96% (31 December 2008 est.)

Stock of narrow money: $3.692 billion (31 December 2010 est.)
country comparison to the world: 104
$3.21 billion (31 December 2009 est.)

Stock of broad money: $92.01 billion (31 December 2010 est.)
country comparison to the world: 54
$82.07 billion (31 December 2009 est.)

Stock of domestic credit: $62.68 billion (31 December 2010 est.)
country comparison to the world: 59
$56.98 billion (31 December 2009 est.)

Market value of publicly traded shares: $12.89 billion (31 December 2009)
country comparison to the world: 68
$9.641 billion (31 December 2008)
$10.86 billion (31 December 2007)

Agriculture—products: citrus, grapes, tomatoes, apples, vegetables, potatoes, olives, tobacco; sheep, goats

Industries: banking, tourism, food processing, wine, jewelry, cement, textiles, mineral and chemical products, wood and furniture products, oil refining, metal fabricating

Industrial production growth rate: 2.1% (2010 est.)
country comparison to the world: 127

Electricity—production: 10.41 billion kWh (2009)
country comparison to the world: 88

Electricity—consumption: 9.793 billion kWh (2009)
country comparison to the world: 88

Electricity—exports: 0 kWh (2009 est.)

Electricity—imports: 1.114 billion kWh (2009 est.)

Oil—production: 0 bbl/day (2009 est.)
country comparison to the world: 193
Oil—consumption: 90,000 bbl/day (2009 est.)
country comparison to the world: 78
Oil—exports: 0 bbl/day (2009)
country comparison to the world: 180
Oil—imports: 86,750 bbl/day (2007 est.)
country comparison to the world: 69
Oil—proved reserves: 0 bbl (1 January 2010 est.)
country comparison to the world: 151
Natural gas—production: 0 cu m (2009 est.)
country comparison to the world: 147
Natural gas—consumption: 0 cu m (2008 est.)
country comparison to the world: 190
Natural gas—exports: 0 cu m (2009 est.)
country comparison to the world: 127
Natural gas—imports: 0 cu m (2008 est.)
country comparison to the world: 79
Natural gas—proved reserves: 0 cu m (1 January 2010 est.)
country comparison to the world: 153
Current account balance: $-6.972 billion (2010 est.)
country comparison to the world: 173
$-7.555 billion (2009 est.)
Exports: $5.187 billion (2010 est.)
country comparison to the world: 106
$4.716 billion (2009 est.)
Exports—commodities: jewelry, base metals, chemicals, miscellaneous consumer goods, fruit and vegetables, tobacco, construction minerals, electric power machinery and switchgear, textile fibers, paper
Exports—partners: Switzerland 22%, UAE 10%, Iraq 8%, Saudi Arabia 7%, Syria 6% (2009 est.)
Imports: $17.97 billion (2010 est.)
country comparison to the world: 75
$15.9 billion (2009 est.)
Imports—commodities: petroleum products, cars, medicinal products, clothing, meat and live animals, consumer goods, paper, textile fabrics, tobacco, electrical machinery and equipment, chemicals
Imports—partners: US 11%, France 10%, China 9%, Italy 8%, Germany 8%, Turkey 4% (2009 est.)
Reserves of foreign exchange and gold: $41.57 billion (31 December 2010 est.)
country comparison to the world: 31
$39.16 billion (31 December 2009 est.)
Debt—external: $34.45 billion (31 December 2010 est.)
country comparison to the world: 60
$31.89 billion (31 December 2009 est.)
Stock of direct foreign investment—at home: $NA
Stock of direct foreign investment—abroad: $NA
Exchange rates: Lebanese pounds (LBP) per US dollar–1,507.5 (2010), 1,507.5 (2009), 1,507.5 (2008), 1,507.5 (2007), 1,507.5 (2006)

COMMUNICATIONS

Telephones—main lines in use: 750,000 (2009)
country comparison to the world: 89
Telephones—mobile cellular: 1.526 million (2009)
country comparison to the world: 138
Telephone system: general assessment: repair of the telecommunications system, severely damaged during the civil war, now complete
domestic: two mobile-cellular networks provide good service; combined fixed-line and mobile-cellular subscribership exceeds 55 per 100 persons
international: country code–961; submarine cable links to Cyprus, Egypt, and Syria; satellite earth stations–2 Intelsat (1 Indian Ocean and 1 Atlantic Ocean); coaxial cable to Syria (2009)
Broadcast media: 7 TV stations in operation, 1 of which is state-owned; more than 30 radio stations, 1 of which is state-owned; satellite and cable TV services are available; transmissions of at least 2 international broadcasters are accessible through partner stations (2007)
Internet country code: .lb
Internet hosts: 51,451 (2010)
country comparison to the world: 87
Internet users: 1 million (2009)
country comparison to the world: 101

TRANSPORTATION

Airports: 7 (2010)
country comparison to the world: 167
Airports—with paved runways: *total:* 5
over 3,047 m: 1
2,438 to 3,047 m: 2
914 to 1,523 m: 1
under 914 m: 1 (2010)
Airports—with unpaved runways:
total: 2
914 to 1,523 m: 2 (2010)
Pipelines: gas 102 km (2010)
Railways: *total:* 401 km
country comparison to the world: 116
standard gauge: 319 km 1.435-m gauge
narrow gauge: 82 km 1.050-m gauge
note: rail system unusable because of the damage done during fighting in the 1980s and in 2006 (2008)
Roadways: *total:* 6,970 km (includes 170 km of expressways) (2005)
country comparison to the world: 148
Merchant marine: *total:* 29
country comparison to the world: 86
by type: bulk carrier 3, cargo 12, carrier 11, refrigerated cargo 1, vehicle carrier 2
foreign-owned: 3 (Syria 3)
registered in other countries: 40 (Barbados 2, Cambodia 6, Comoros 3, Egypt 1, Georgia 1, Honduras 2, Liberia 1, Malta 7, Moldova 1, Panama 2, Saint Vincent and the Grenadines 4, Syria 2, Togo 6, unknown 2) (2010)
Ports and terminals: Beirut, Tripoli

MILITARY

Military branches: Lebanese Armed Forces (LAF): Army ((Al Jaysh al Lubnaniya) includes Navy (Al Quwwat al Bahiriyya al Lubnaniya), Air Force (Al Quwwat al Jawwiya al Lubnaniya)) (2010)
Military service age and obligation: 18-30 years of age for voluntary military service; no conscription (2007)
Manpower available for military service:
males age 16–49: 1,081,016
females age 16–49: 1,115,349 (2010 est.)
Manpower fit for military service: *males age 16–49:* 920,825
females age 16–49: 941,806 (2010 est.)
Manpower reaching militarily significant age annually:
male: 36,856
female: 35,121 (2010 est.)
Military expenditures: 3.1% of GDP (2005 est.)
country comparison to the world: 40

TRANSNATIONAL ISSUES

Disputes—international: lacking a treaty or other documentation describing the boundary, portions of the Lebanon-Syria boundary are unclear with several sections in dispute; since 2000, Lebanon has claimed Shab'a Farms area in the Israeli-occupied Golan Heights; the roughly 2,000-strong UN Interim Force in Lebanon (UNIFIL) has been in place since 1978
Refugees and internally displaced persons: refugees (country of origin): 405,425 (Palestinian refugees (UNRWA)); 50,000-60,000 (Iraq)
IDPs: 17,000 (1975-90 civil war, Israeli invasions); 200,000 (July-August 2006 war) (2007)
Illicit drugs: cannabis cultivation dramatically reduced to 2,500 hectares in 2002 despite continued significant cannabis consumption; opium poppy cultivation minimal; small amounts of Latin American cocaine and Southwest Asian heroin transit country on way to European markets and for Middle Eastern consumption; money laundering of drug proceeds fuels concern that extremists are benefiting from drug trafficking

LESOTHO

INTRODUCTION

Background: Basutoland was renamed the Kingdom of Lesotho upon independence from the UK in 1966. The Basuto National Party ruled for the first two decades. King MOSHOESHOE was exiled in 1990, but returned to Lesotho in 1992 and was reinstated in 1995 and subsequently succeeded by his son, King LETSIE III, in 1996. Constitutional government was restored in 1993 after seven years of military rule. In 1998, violent protests and a military mutiny following a contentious election prompted a brief but bloody intervention by South African and Botswana military forces under the aegis of the Southern African Development Community. Subsequent constitutional reforms restored relative political stability. Peaceful parliamentary elections were held in 2002, but the National Assembly elections of February 2007 were hotly contested and aggrieved parties continue to dispute how the

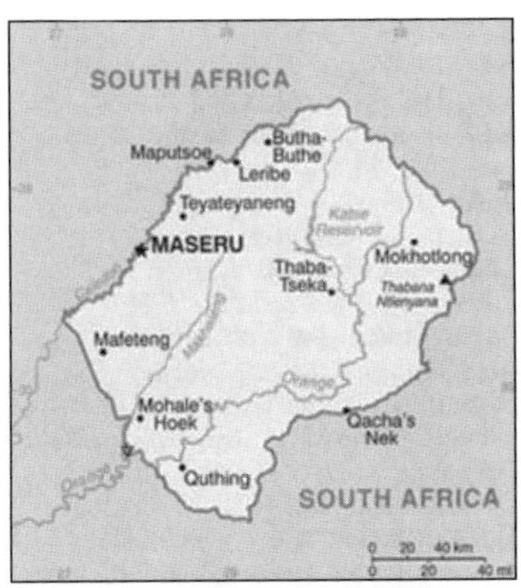

electoral law was applied to award proportional seats in the Assembly.

GEOGRAPHY

Location: Southern Africa, an enclave of South Africa
Geographic coordinates: 29 30 S, 28 30 E
Map references: Africa
Area: *total:* 30,355 sq km
country comparison to the world: 141
land: 30,355 sq km
water: 0 sq km
Area—comparative: slightly smaller than Maryland
Land boundaries: total: 909 km
border countries: South Africa 909 km
Coastline: 0 km (landlocked)
Maritime claims: none (landlocked)
Climate: temperate; cool to cold, dry winters; hot, wet summers
Terrain: mostly highland with plateaus, hills, and mountains
Elevation extremes: *lowest point:* junction of the Orange and Makhaleng Rivers 1,400 m
highest point: Thabana Ntlenyana 3,482 m
Natural resources: water, agricultural and grazing land, diamonds, sand, clay, building stone
Land use: arable land: 10.87%
permanent crops: 0.13%
other: 89% (2005)
Irrigated land: 30 sq km (2003)
Total renewable water resources: 5.2 cu km (1987)
Freshwater withdrawal (domestic/industrial/agricultural): *total:* 0.05 cu km/yr (40%/40%/20%)
per capita: 28 cu m/yr (2000)
Natural hazards: periodic droughts
Environment—current issues: population pressure forcing settlement in marginal areas results in overgrazing, severe soil erosion, and soil exhaustion; desertification; Highlands Water Project controls, stores, and redirects water to South Africa
Environment—international agreements: *party to:* Biodiversity, Climate Change, Climate Change-Kyoto Protocol, Desertification, Endangered Species, Hazardous Wastes, Law of the Sea, Marine Life Conservation, Ozone Layer Protection, Wetlands
signed, but not ratified: none of the selected agreements
Geography—note: landlocked, completely surrounded by South Africa; mountainous, more than 80% of the country is 1,800 m above sea level

PEOPLE

Population: 1,924,886 (July 2011 est.)
country comparison to the world: 146
note: estimates for this country explicitly take into account the effects of excess mortality due to AIDS; this can result in lower life expectancy, higher infant mortality, higher death rates, lower population growth rates, and changes in the distribution of population by age and sex than would otherwise be expected
Age structure: *0–14 years:* 33.5% (male 323,934/female 321,727)
15–64 years: 61.1% (male 573,773/female 602,443)
65 years and over: 5.4% (male 50,956/female 52,053) (2011 est.)
Median age: *total:* 22.9 years
male: 22.8 years
female: 22.9 years (2011 est.)
Population growth rate: 0.332% (2011 est.)
country comparison to the world: 166
Birth rate: 26.93 births/1,000 population (2011 est.)
country comparison to the world: 49
Death rate: 15.19 deaths/1,000 population (July 2011 est.)
country comparison to the world: 9
Net migration rate: -8.42 migrant(s)/1,000 population (2011 est.)
country comparison to the world: 206
Urbanization: *urban population:* 27% of total population (2010)
rate of urbanization: 3.4% annual rate of change (2010-15 est.)
Major cities—population: MASERU (capital) 220,000 (2009)
Sex ratio: *at birth:* 1.03 male(s)/female
under 15 years: 1.01 male(s)/female
15–64 years: 0.95 male(s)/female
65 years and over: 0.96 male(s)/female
total population: 0.97 male(s)/female (2011 est.)
Infant mortality rate: *total:* 55.04 deaths/1,000 live births
country comparison to the world: 39
male: 59.28 deaths/1,000 live births
female: 50.67 deaths/1,000 live births (2011 est.)
Life expectancy at birth: *total population:* 51.63 years
country comparison to the world: 212
male: 51.51 years
female: 51.76 years (2011 est.)
Total fertility rate: 2.94 children born/woman (2011 est.)
country comparison to the world: 68
HIV/AIDS—adult prevalence rate: 23.6% (2009 est.)
country comparison to the world: 3
HIV/AIDS—people living with HIV/AIDS: 290,000 (2009 est.)
country comparison to the world: 21
HIV/AIDS—deaths: 14,000 (2009 est.)
country comparison to the world: 21
Drinking water source: *Improved:*
urban: 97% of population
rural: 81% of population
total: 85% of population
Unimproved: urban: 3% of population
rural: 19% of population
total: 15% of population (2008)
Sanitation facility access: *Improved:*
urban: 40% of population
rural: 25% of population
total: 29% of population
Unimproved: urban: 60% of population
rural: 75% of population
total: 71% of population (2008)
Nationality: noun: Mosotho (singular), Basotho (plural)
adjective: Basotho
Ethnic groups: Sotho 99.7%, Europeans, Asians, and other 0.3%,
Religions: Christian 80%, indigenous beliefs 20%
Languages: Sesotho (southern Sotho), English (official), Zulu, Xhosa
Literacy: *definition:* age 15 and over can read and write
total population: 84.8%
male: 74.5%
female: 94.5% (2003 est.)
School life expectancy (primary to tertiary education): *total:* 10 years
male: 10 years
female: 10 years (2008)
Education expenditures: 12.4% of GDP (2008)
country comparison to the world: 3

GOVERNMENT

Country name: *conventional long form:* Kingdom of Lesotho
conventional short form: Lesotho
local long form: Kingdom of Lesotho
local short form: Lesotho
former: Basutoland
Government type: parliamentary constitutional monarchy
Capital: *name:* Maseru
geographic coordinates: 29 19 S, 27 29 E
time difference: UTC+2 (7 hours ahead of Washington, DC during Standard Time)
Administrative divisions: 10 districts; Berea, Butha-Buthe, Leribe, Mafeteng, Maseru, Mohale's Hoek, Mokhotlong, Qacha's Nek, Quthing, Thaba-Tseka
Independence: 4 October 1966 (from the UK)
National holiday: Independence Day, 4 October (1966)
Constitution: 2 April 1993
Legal system: mixed legal system of English common law and Roman-Dutch law; judicial review of legislative acts in High Court and Court of Appeal
International law organization participation: accepts compulsory ICJ jurisdiction with reservations; accepts ICCt jurisdiction
Suffrage: 18 years of age; universal
Executive branch: *chief of state:* King LETSIE III (since 7 February 1996); note—King LETSIE III formerly occupied the throne from November 1990 to February 1995 while his father was in exile
head of government: Prime Minister Pakalitha MOSISILI (since 23 May 1998)
cabinet: Cabinet
(For more information visit the World Leaders website)
elections: according to the constitution, the leader of the majority party in the Assembly automatically becomes prime minister; the monarchy is hereditary, but, under the terms of the constitution that came into effect after the March 1993 election, the monarch is a "living symbol of national unity" with no executive or legis-

lative powers; under traditional law the college of chiefs has the power to depose the monarch, determine who is next in the line of succession, or who shall serve as regent in the event that the successor is not of mature age
Legislative branch: bicameral Parliament consists of the Senate (33 members–22 principal chiefs and 11 other members appointed by the ruling party) and the Assembly (120 seats, 80 by popular vote and 40 by proportional vote; members elected by popular vote to serve five-year terms)
elections: last held on 17 February 2007 (next to be held in 2012)
election results: percent of vote by party–NA; seats by party–LCD 61, NIP 21, ABC 17, LWP 10, ACP 4, BNP 3, other 4
Judicial branch: High Court (chief justice appointed by the monarch acting on the advice of the prime minister); Court of Appeal; Magistrate Courts; customary or traditional court
Political parties and leaders: Alliance of Congress Parties or ACP (including the Lesotho People's Congress or LCP [Kelebone MAOPE], the Basotholand African Congress or BAC [Khauhelo RALITAPOLE], and a faction of the Basotho Congress Party or BCP [Ntsukunyane MPHANYA]); All Basotho Convention or ABC [Thomas THABANE]; Basotho Batho Democratic Party or BBDP; Basotho Congress Party or BCP; Basotho Democratic National Party or BDNP [Thabang NYEOE]; Basotho National Party or BNP [vacant]; Basotholand African National Congress or BANC; Christian Democratic Party or CDP [Enerst RAMOKOENA]; Lesotho Congress for Democracy or LCD [Pakalitha MOSISILI] (the governing party); Lesotho Workers Party or LWP [Macaefa BILLY]; National Independent Party or NIP [Anthony MANYELI]
Political pressure groups and leaders: Media Institute of Southern Africa, Lesotho chapter [Thabang MATJAMA] (pushes for media freedom)
International organization participation: ACP, AfDB, AU, C, FAO, G-77, IAEA, IBRD, ICAO, ICRM, IDA, IFAD, IFC, IFRCS, ILO, IMF, Interpol, IOC, IOM, IPU, ISO (correspondent), ITU, MIGA, NAM, OPCW, SACU, SADC, UN, UNCTAD, UNESCO, UNHCR, UNIDO, UNWTO, UPU, WCO, WFTU, WHO, WIPO, WMO, WTO
Diplomatic representation in the US: *chief of mission:* Ambassador (vacant); Charge d'Affaires Molefi Christopher NYAKA
chancery: 2511 Massachusetts Avenue NW, Washington, DC 20008
telephone: [1] (202) 797-5533
FAX: [1] (202) 234-6815
Diplomatic representation from the US: *chief of mission:* Ambassador Michele T. BOND
embassy: 254 Kingsway, Maseru West (Consular Section)
mailing address: P. O. Box 333, Maseru 100, Lesotho
telephone: [266] 22 312666
FAX: [266] 22 310116
Flag description: three horizontal stripes of blue (top), white, and green in the proportions of 3:4:3; the colors represent rain, peace, and prosperity respectively; centered in the white stripe is a black Basotho hat representing the indigenous people; the flag was unfurled in October 2006 to celebrate 40 years of independence
National anthem: *name:* "Lesotho fatse la bo ntat'a rona" (Lesotho, Land of Our Fathers)
lyrics/music: Francois COILLARD/Ferdinand-Samuel LAUR
note: adopted 1967; the anthem's music derives from an 1823 Swiss songbook

ECONOMY

Economy—overview: Small, landlocked, and mountainous, Lesotho relies on remittances from Basotho employed in South Africa, customs duties from the Southern Africa Customs Union (SACU), and export revenue for the majority of government revenue. However, the government has recently strengthened its tax system to reduce dependency on customs duties. Completion of a major hydropower facility in January 1998 permitted the sale of water to South Africa and generated royalties for Lesotho. Lesotho produces about 90% of its own electrical power needs. As the number of mineworkers has declined steadily over the past several years, a small manufacturing base has developed based on farm products that support the milling, canning, leather, and jute industries, as well as an apparel-assembly sector. Despite Lesotho's market-based economy being heavily tied to its neighbor South Africa, the US is an important trade partner because of the export sector's heavy dependence on apparel exports. Exports have grown significantly because of the trade benefits contained in the Africa Growth and Opportunity Act. Most of the labor force is engaged in subsistence agriculture, especially livestock herding, although drought has decreased agricultural activity. The extreme inequality in the distribution of income remains a major drawback. Lesotho has signed an Interim Poverty Reduction and Growth Facility with the IMF. In July 2007, Lesotho signed a Millennium Challenge Account Compact with the US worth $362.5 million. Economic growth dropped in 2009, due mainly to the effects of the global economic crisis as demand for the country's exports declined and SACU revenue fell precipitously when South Africa–the primary contributor to the SACU revenue pool–went into recession, but growth returned to 3.5% in 2010.
GDP (purchasing power parity): $3.303 billion (2010 est.)
country comparison to the world: 173
$3.223 billion (2009 est.)
$3.129 billion (2008 est.)
note: data are in 2010 US dollars
GDP (official exchange rate): $2.127 billion (2010 est.)
GDP—real growth rate: 2.4% (2010 est.)
country comparison to the world: 139
3% (2009 est.)
4.7% (2008 est.)
GDP—per capita (PPP): $1,700 (2010 est.)
country comparison to the world: 197
$1,700 (2009 est.)
$1,600 (2008 est.)
note: data are in 2010 US dollars
GDP—composition by sector:
agriculture: 7.1%
industry: 34.6%
services: 58.2% (2010 est.)
Labor force: 854,600 (2007 est.)
country comparison to the world: 145
Labor force—by occupation: *agriculture:* 86% of resident population engaged in subsistence agriculture; roughly 35% of the active male wage earners work in South Africa
industry and services: 14% (2002 est.)
Unemployment rate: 45% (2002)
country comparison to the world: 190
Population below poverty line: 49% (1999)
Household income or consumption by percentage share: *lowest 10%:* 1%
highest 10%: 39.4% (2003)
Distribution of family income—Gini index: 63.2 (1995)
country comparison to the world: 3
56 (1986-87)
Investment (gross fixed): 21.9% of GDP (2010 est.)
country comparison to the world: 69
Budget: *revenues:* $968.4 million
expenditures: $1.193 billion (2010 est.)
Inflation rate (consumer prices): 6.1% (2010 est.)
country comparison to the world: 162
7.2% (2009 est.)
Central bank discount rate: 10.66% (31 December 2009)
country comparison to the world: 23
14.05% (31 December 2008)
Commercial bank prime lending rate: 13% (31 December 2009 est.)
country comparison to the world: 35
16.19% (31 December 2008 est.)
Stock of narrow money: $653.3 million (31 December 2010 est.)
country comparison to the world: 150
$509.5 million (31 December 2009 est.)
Stock of broad money: $1.057 billion (31 December 2010 est.)
country comparison to the world: 163
$876 million (31 December 2009 est.)
Stock of domestic credit: $177.7 million (31 December 2010 est.)
country comparison to the world: 177
$147.3 million (31 December 2009 est.)
Agriculture—products: corn, wheat, pulses, sorghum, barley; livestock
Industries: food, beverages, textiles, apparel assembly, handicrafts, construction, tourism
Industrial production growth rate: 3% (2010 est.)
country comparison to the world: 115
Electricity—production: 502 million kWh
country comparison to the world: 159
note: electricity supplied by South Africa (2007 est.)
Electricity—consumption: 516.9 million kWh (2007 est.)
country comparison to the world: 161
Electricity—exports: 0 kWh (2008 est.)
Electricity—imports: 50 million kWh; note–electricity supplied by South Africa (2008 est.)
Oil—production: 0 bbl/day (2009 est.)
country comparison to the world: 196

Oil—consumption: 2,000 bbl/day (2009 est.)
country comparison to the world: 184
Oil—exports: 0 bbl/day (2007 est.)
country comparison to the world: 181
Oil—imports: 1,553 bbl/day (2007 est.)
country comparison to the world: 181
Oil—proved reserves: 0 bbl (1 January 2010 est.)
country comparison to the world: 154
Natural gas—production: 0 cu m (2008 est.)
country comparison to the world: 151
Natural gas—consumption: 0 cu m (2008 est.)
country comparison to the world: 192
Natural gas—exports: 0 cu m (2008 est.)
country comparison to the world: 131
Natural gas—imports: 0 cu m (2008 est.)
country comparison to the world: 81
Natural gas—proved reserves: 0 cu m (1 January 2010 est.)
country comparison to the world: 156
Current account balance: $-125 million (2010 est.)
country comparison to the world: 80
$194 million (2009 est.)
Exports: $985 million (2010 est.)
country comparison to the world: 152
$821 million (2009 est.)
Exports—commodities: manufactures 75% (clothing, footwear, road vehicles), wool and mohair, food and live animals
Exports—partners: US 58.9%, Belgium 37%, Madagascar 1.2% (2008)
Imports: $1.766 billion (2010 est.)
country comparison to the world: 155
$1.572 billion (2009 est.)
Imports—commodities: food; building materials, vehicles, machinery, medicines, petroleum products
Imports—partners: China 26.3%, Taiwan 20.1%, Hong Kong 16.4%, South Korea 14.1%, India 9.2% (2008)
Reserves of foreign exchange and gold: $893 million (31 December 2010 est.)
country comparison to the world: 117
$988 million (31 December 2009 est.)
Debt—external: $647 million (31 December 2010 est.)
country comparison to the world: 158
$671 million (31 December 2009 est.)
Exchange rates: maloti (LSL) per US dollar–7.9 (2010), 8.47 (2009), 7.75 (2008) 7.25 (2007), 6.85 (2006)

COMMUNICATIONS

Telephones—main lines in use: 40,000 (2009)
country comparison to the world: 169
Telephones—mobile cellular: 661,000 (2009)
country comparison to the world: 153
Telephone system: general assessment: rudimentary system consisting of a modest number of landlines, a small microwave radio relay system, and a small radiotelephone communication system; mobile-cellular telephone system is expanding
domestic: privatized in 2001, Telecom Lesotho was tasked with providing an additional 50,000 fixed-line connections within five years, a target not met; mobile-cellular service dominates the market and is expanding with a subscribership exceeding 30 per 100 persons in 2009; rural services are scant
international: country code–266; satellite earth station–1 Intelsat (Atlantic Ocean) (2009)
Broadcast media: 1 state-owned TV station and 2 state-owned radio stations; government controls most private broadcast media; satellite TV subscription service is available; transmissions of multiple international broadcasters are obtainable (2008)
Internet country code: .ls
Internet hosts: 632 (2010)
country comparison to the world: 175
Internet users: 76,800 (2009)
country comparison to the world: 167

TRANSPORTATION

Airports: 26 (2010)
country comparison to the world: 128
Airports—with paved runways: *total:* 3
over 3,047 m: 1
914 to 1,523 m: 1
under 914 m: 1 (2010)
Airports—with unpaved runways:
total: 23
914 to 1,523 m: 5
under 914 m: 18 (2010)
Roadways:
total: 7,091 km
country comparison to the world: 147
paved: 1,404 km
unpaved: 5,687 km (2003)

MILITARY

Military branches: Lesotho Defense Force (LDF): Army (includes Air Wing) (2010)
Military service age and obligation: 18-24 years of age for voluntary military service; no conscription; women serve as commissioned officers (2009)
Manpower available for military service:
males age 16–49: 472,456
females age 16–49: 508,953 (2010 est.)
Manpower fit for military service: *males age 16–49:* 270,184
females age 16–49: 275,734 (2010 est.)
Manpower reaching militarily significant age annually:
male: 19,110
female: 20,037 (2010 est.)
Military expenditures:
2.6% of GDP (2006)
country comparison to the world: 55
Military—note: Lesotho's declared policy is maintenance of its independent sovereignty and preservation of internal security; in practice, external security is guaranteed by South Africa; restructuring of the Lesotho Defense Force (LDF) and Ministry of Defense and Public Service over the past five years has focused on subordinating the defense apparatus to civilian control and restoring the LDF's cohesion; the restructuring has considerably improved capabilities and professionalism, but the LDF is disproportionately large for a small, poor country; the government has outlined a reduction to a planned 1,500-man strength, but these plans have met with vociferous resistance from the political opposition and from inside the LDF (2008)

TRANSNATIONAL ISSUES

Disputes—international: none

LIBERIA

INTRODUCTION

Background: Settlement of freed slaves from the US in what is today Liberia began in 1822; by 1847, the Americo-Liberians were able to establish a republic. William TUBMAN, president from 1944-71, did much to promote foreign investment and to bridge the economic, social, and political gaps between the descendents of the original settlers and the inhabitants of the interior. In 1980, a military coup led by Samuel DOE ushered in a decade of authoritarian rule. In December 1989, Charles TAYLOR launched a rebellion against DOE's regime that led to a prolonged civil war in which DOE himself was killed. A period of relative peace in 1997 allowed for elections that brought TAYLOR to power, but major fighting resumed in 2000. An August 2003 peace agreement ended the war and prompted the resignation of former president Charles TAYLOR, who faces war crimes charges in The Hague related to his involvement in Sierra Leone's civil war. After two years of rule by a transitional government, democratic elections in late 2005 brought President Ellen JOHNSON SIRLEAF to power. The UN Mission in Liberia (UNMIL) maintains a strong presence throughout the country, but the security situation is still fragile and the process of rebuilding the social and economic structure of this war-torn country continues.

GEOGRAPHY

Location: Western Africa, bordering the North Atlantic Ocean, between Cote d'Ivoire and Sierra Leone

Geographic coordinates: 6 30 N, 9 30 W
Map references: Africa
Area: *total:* 111,369 sq km
country comparison to the world: 103
land: 96,320 sq km
water: 15,049 sq km
Area—comparative: slightly larger than Tennessee
Land boundaries: total: 1,585 km
border countries: Guinea 563 km, Cote d'Ivoire 716 km, Sierra Leone 306 km
Coastline: 579 km
Maritime claims: *territorial sea:* 200 nm
Climate: tropical; hot, humid; dry winters with hot days and cool to cold nights; wet, cloudy summers with frequent heavy showers
Terrain: mostly flat to rolling coastal plains rising to rolling plateau and low mountains in northeast
Elevation extremes: *lowest point:* Atlantic Ocean 0 m
highest point: Mount Wuteve 1,380 m
Natural resources: iron ore, timber, diamonds, gold, hydropower
Land use: arable land: 3.43%
permanent crops: 1.98%
other: 94.59% (2005)
Irrigated land: 30 sq km (2003)
Total renewable water resources: 232 cu km (1987)
Freshwater withdrawal (domestic/industrial/agricultural): *total:* 0.11 cu km/yr (27%/18%/55%)
per capita: 34 cu m/yr (2000)
Natural hazards: dust-laden harmattan winds blow from the Sahara (December to March)
Environment—current issues: tropical rain forest deforestation; soil erosion; loss of biodiversity; pollution of coastal waters from oil residue and raw sewage
Environment—international agreements: *party to:* Biodiversity, Climate Change, Climate Change-Kyoto Protocol, Desertification, Endangered Species, Hazardous Wastes, Law of the Sea, Ozone Layer Protection, Ship Pollution, Tropical Timber 83, Tropical Timber 94, Wetlands
signed, but not ratified: Environmental Modification, Marine Life Conservation
Geography—note: facing the Atlantic Ocean, the coastline is characterized by lagoons, mangrove swamps, and river-deposited sandbars; the inland grassy plateau supports limited agriculture

PEOPLE

Population: 3,786,764 (July 2011 est.)
country comparison to the world: 129
Age structure: *0–14 years:* 44.3% (male 843,182/female 834,922)
15–64 years: 52.7% (male 989,623/female 1,007,577)
65 years and over: 2.9% (male 56,189/female 55,271) (2011 est.)
Median age: *total:* 18.3 years
male: 18.2 years
female: 18.3 years (2011 est.)
Population growth rate: 2.663% (2011 est.)
country comparison to the world: 20
Birth rate: 37.25 births/1,000 population (2011 est.)
country comparison to the world: 20
Death rate: 10.62 deaths/1,000 population (July 2011 est.)
country comparison to the world: 45
Net migration rate: 0 migrant(s)/1,000 population (2011 est.)
country comparison to the world: 93
Urbanization: *urban population:* 48% of total population (2010)
rate of urbanization: 3.4% annual rate of change (2010-15 est.)
Major cities—population: MONROVIA (capital) 882,000 (2009)
Sex ratio: *at birth:* 1.03 male(s)/female
under 15 years: 1.01 male(s)/female
15–64 years: 0.98 male(s)/female
65 years and over: 1.03 male(s)/female
total population: 1 male(s)/female (2011 est.)
Infant mortality rate:
total: 74.52 deaths/1,000 live births
country comparison to the world: 18
male: 78.96 deaths/1,000 live births
female: 69.95 deaths/1,000 live births (2011 est.)
Life expectancy at birth:
total population: 57 years
country comparison to the world: 194
male: 55.44 years
female: 58.6 years (2011 est.)
Total fertility rate: 5.13 children born/woman (2011 est.)
country comparison to the world: 16
HIV/AIDS—adult prevalence rate: 1.5% (2009 est.)
country comparison to the world: 34
HIV/AIDS—people living with HIV/AIDS: 37,000 (2009 est.)
country comparison to the world: 65
HIV/AIDS—deaths: 3,600 (2009 est.)
country comparison to the world: 44
Major infectious diseases: degree of risk: very high
food or waterborne diseases: bacterial and protozoal diarrhea, hepatitis A, and typhoid fever
vectorborne diseases: malaria and yellow fever
water contact disease: schistosomiasis
aerosolized dust or soil contact disease: Lassa fever
animal contact disease: rabies (2009)
Drinking water source: *Improved:*
urban: 79% of population
rural: 51% of population
total: 68% of population
Unimproved: urban: 21% of population
rural: 49% of population
total: 32% of population (2008)
Sanitation facility access: *Improved:*
urban: 25% of population
rural: 4% of population
total: 17% of population
Unimproved: urban: 75% of population
rural: 96% of population
total: 83% of population (2008)
Nationality: noun: Liberian(s)
adjective: Liberian
Ethnic groups: Kpelle 20.3%, Bassa 13.4%, Grebo 10%, Gio 8%, Mano 7.9%, Kru 6%, Lorma 5.1%, Kissi 4.8%, Gola 4.4%, other 20.1% (2008 census)
Religions: Christian 85.6%, Muslim 12.2%, Traditional 0.6%, other 0.2%, none 1.4% (2008 census)
Languages: English 20% (official), some 20 ethnic group languages few of which can be written or used in correspondence
Literacy: *definition:* age 15 and over can read and write
total population: 57.5%
male: 73.3%
female: 41.6% (2003 est.)
School life expectancy (primary to tertiary education): *total:* 11 years
male: 13 years
female: 9 years (2000)
Education expenditures: 2.7% of GDP (2008)
country comparison to the world: 145

GOVERNMENT

Country name: *conventional long form:* Republic of Liberia
conventional short form: Liberia
Government type: republic
Capital: *name:* Monrovia
geographic coordinates: 6 18 N, 10 48 W
time difference: UTC 0 (5 hours ahead of Washington, DC during Standard Time)
Administrative divisions: 15 counties; Bomi, Bong, Gbarpolu, Grand Bassa, Grand Cape Mount, Grand Gedeh, Grand Kru, Lofa, Margibi, Maryland, Montserrado, Nimba, River Cess, River Gee, Sinoe
Independence: 26 July 1847
National holiday: Independence Day, 26 July (1847)
Constitution: 6 January 1986
Legal system: mixed legal system of common law (based on Anglo-American law) and customary law
International law organization participation: accepts compulsory ICJ jurisdiction with reservations; accepts ICCt jurisdiction
Suffrage: 18 years of age; universal
Executive branch: *chief of state:* President Ellen JOHNSON SIRLEAF (since 16 January 2006); note–the President is both the chief of state and head of government
head of government: President Ellen JOHNSON SIRLEAF (since 16 January 2006)
cabinet: Cabinet appointed by the president and confirmed by the Senate
(For more information visit the World Leaders website)
elections: president elected by popular vote for a six-year term (eligible for a second term); election last held on 8 November 2005 (next to be held on 11 October 2011)
election results: Ellen JOHNSON SIRLEAF elected president; percent of vote, second round–Ellen JOHNSON SIRLEAF 59.6%, George WEAH 40.4%
Legislative branch: bicameral National Assembly consists of the Senate (30 seats; note–number of seats changed in 11 October 2005 elections; members elected by popular vote to serve nine-year terms) and the House of Representatives (64 seats; members elected by popular vote to serve six-year terms)
elections: Senate–last held on 11 October 2005 (next to be held in October 2011); House of Representatives–last held on 11 October 2005 (next to be held on 11 October 2011)
election results: Senate–percent of vote by party–NA; seats by party–COTOL 7, NPP 4, CDC 3, LP 3, UP 3, APD 3, other 7; House of Representatives–percent of vote by party–NA; seats by party–CDC 15, LP

9, COTOL 8, UP 8, APD 5, NPP 4, other 15; note—the UP now holds 13 out of 30 senate seats and 16 out of 64 house seats following a merger with several smaller parties in 2009
note: junior senators—those who received the second most votes in each county in the 11 October 2005 election—will only serve a six-year first term because the Liberian constitution mandates staggered Senate elections to ensure continuity of government; all senators will be eligible for nine-year terms thereafter
Judicial branch: Supreme Court
Political parties and leaders: Alliance for Peace and Democracy or APD [Togba-na TIPOTEH]; Congress for Democratic Change or CDC [George WEAH]; Liberty Party or LP [Charles BRUMSKINE]; National Patriotic Party or NPP [Roland MASSAQUOI]; Unity Party or UP [Varney SHERMAN]
Political pressure groups and leaders: *other:* demobilized former military officers
International organization participation: ACP, AfDB, AU, ECOWAS, FAO, G-77, IAEA, IBRD, ICAO, ICRM, IDA, IFAD, IFC, IFRCS, ILO, IMF, IMO, IMSO, Interpol, IOC, IOM, IPU, ISO (correspondent), ITU, ITUC, MIGA, NAM, OPCW, UN, UNCTAD, UNESCO, UNIDO, UPU, WCO, WFTU, WHO, WIPO, WMO, WTO (observer)
Diplomatic representation in the US: *chief of mission:* Ambassador (vacant); Charge d'Affaires William V. BULL
chancery: 5201 16th Street NW, Washington, DC 20011
telephone: [1] (202) 723-0437
FAX: [1] (202) 723-0436
consulate(s) general: New York
Diplomatic representation from the US: *chief of mission:* Ambassador Linda THOMAS-GREENFIELD
embassy: 111 United Nations Drive, P. O. Box 98, Mamba Point, 1000 Monrovia, 10
mailing address: use embassy street address
telephone: [231] 7-705-4826
FAX: [231] 7-701-0370
Flag description: 11 equal horizontal stripes of red (top and bottom) alternating with white; a white five-pointed star appears on a blue square in the upper hoist-side corner; the stripes symbolize the signatories of the Liberian Declaration of Independence; the blue square represents the African mainland, and the star represents the freedom granted to the ex-slaves; according to the constitution, the blue color signifies liberty, justice, and fidelity, the white color purity, cleanliness, and guilelessness, and the red color steadfastness, valor, and fervor
note: the design is based on the US flag
National anthem: *name:* "All Hail, Liberia Hail!"
lyrics/music: Daniel Bashiel WARNER/ Olmstead LUCA
note: lyrics adopted 1847, music adopted 1860; the anthem's author would become the third president of Liberia

ECONOMY

Economy—overview: Liberia is a low income country heavily reliant on foreign assistance for revenue. Civil war and government mismanagement destroyed much of Liberia's economy, especially the infrastructure in and around the capital, Monrovia. Many businesses fled the country, taking capital and expertise with them, but with the conclusion of fighting and the installation of a democratically-elected government in 2006, several have returned. Liberia has the distinction of having the highest ratio of direct foreign investment to GDP in the world. Richly endowed with water, mineral resources, forests, and a climate favorable to agriculture, Liberia had been a producer and exporter of basic products, primarily raw timber and rubber and is reviving those sectors. Local manufacturing, mainly foreign owned, had been small in scope. President JOHNSON SIRLEAF, a Harvard-trained banker and administrator, has taken steps to reduce corruption, build support from international donors, and encourage private investment. Embargos on timber and diamond exports have been lifted, opening new sources of revenue for the government and Liberia shipped its first major timber exports to Europe in 2010. The country reached its Heavily Indebted Poor Countries initiative completion point in 2010 and nearly $5 billion of international debt was permanently eliminated. This new status will enable Liberia to establish a sovereign credit rating and issue bonds. Liberia's Paris Club creditors agreed to cancel Liberia's debt as well. Rebuilding infrastructure and raising incomes will depend on generous financial and technical assistance from donor countries and foreign investment in key sectors, such as infrastructure and power generation.
GDP (purchasing power parity): $1.691 billion (2010 est.)
country comparison to the world: 190
$1.608 billion (2009 est.)
$1.537 billion (2008 est.)
note: data are in 2010 US dollars
GDP (official exchange rate): $974 million (2010 est.)
GDP—real growth rate: 5.1% (2010 est.)
country comparison to the world: 68
4.6% (2009 est.)
7.1% (2008 est.)
GDP—per capita (PPP): $500 (2010 est.)
country comparison to the world: 226
$400 (2009 est.)
$400 (2008 est.)
note: data are in 2010 US dollars
GDP—composition by sector:
agriculture: 76.9%
industry: 5.4%
services: 17.7% (2002 est.)
Labor force: 1.372 million (2007)
country comparison to the world: 134
Labor force—by occupation:
agriculture: 70%
industry: 8%
services: 22% (2000 est.)
Unemployment rate: 85% (2003 est.)
country comparison to the world: 198
Population below poverty line: 80% (2000 est.)
Household income or consumption by percentage share: *lowest 10%:* 2.4%
highest 10%: 30.1% (2007)
Budget: *revenues:* $NA
expenditures: $NA
Inflation rate (consumer prices): 11.2% (2007 est.)
country comparison to the world: 199
Commercial bank prime lending rate: 14.4% (31 December 2008)
country comparison to the world: 41
15.05% (31 December 2007)
Stock of narrow money: $206.9 million (31 December 2008)
country comparison to the world: 174
$145.6 million (31 December 2007)
Stock of broad money: $NA
Stock of domestic credit: $1.202 billion (31 December 2008)
country comparison to the world: 145
$1.157 billion (31 December 2007)
Market value of publicly traded shares: $NA
Agriculture—products: rubber, coffee, cocoa, rice, cassava (tapioca), palm oil, sugarcane, bananas; sheep, goats; timber
Industries: rubber processing, palm oil processing, timber, diamonds
Industrial production growth rate: NA%
Electricity—production: 350 million kWh (2007 est.)
country comparison to the world: 164
Electricity—consumption: 325.5 million kWh (2007 est.)
country comparison to the world: 166
Electricity—exports: 0 kWh (2008 est.)
Electricity—imports: 0 kWh (2008 est.)
Oil—production: 0 bbl/day (2009 est.)
country comparison to the world: 195
Oil—consumption: 4,000 bbl/day (2009 est.)
country comparison to the world: 171
Oil—exports: 23.37 bbl/day (2007 est.)
country comparison to the world: 136
Oil—imports: 4,263 bbl/day (2007 est.)
country comparison to the world: 164
Oil—proved reserves: 0 bbl (1 January 2010 est.)
country comparison to the world: 153
Natural gas—production: 0 cu m (2008 est.)
country comparison to the world: 150
Natural gas—consumption: 0 cu m (2008 est.)
country comparison to the world: 191
Natural gas—exports: 0 cu m (2008 est.)
country comparison to the world: 130
Natural gas—imports: 0 cu m (2008 est.)
country comparison to the world: 80
Natural gas—proved reserves: 0 cu m (1 January 2010 est.)
country comparison to the world: 155
Current account balance: $-224 million (2007)
country comparison to the world: 94
Exports: $1.197 billion (2006)
country comparison to the world: 148
Exports—commodities: rubber, timber, iron, diamonds, cocoa, coffee
Exports—partners: Germany 27.92%, Poland 17.12%, South Africa 15.83%, India 10.48%, Greece 7.09%, US 6.23%, Norway 5.24% (2009)
Imports: $7.143 billion (2006)
country comparison to the world: 102
Imports—commodities: fuels, chemicals, machinery, transportation equipment, manufactured goods; foodstuffs
Imports—partners: South Korea 28.29%, Singapore 19.06%, Japan 17.06%, China 14.58%, Taiwan 4.02% (2009)

Debt—external: $3.2 billion (2005 est.)
country comparison to the world: 124
Stock of direct foreign investment—at home: $NA
Stock of direct foreign investment—abroad: $NA
Exchange rates: Liberian dollars (LRD) per US dollar–NA (2007), 59.43 (2006), 53.098(2005),54.906(2004),59.379(2003)

COMMUNICATIONS

Telephones—main lines in use: 2,000 (2009)
country comparison to the world: 222
Telephones—mobile cellular: 842,000 (2009)
country comparison to the world: 148
Telephone system: general assessment: the limited services available are found almost exclusively in the capital Monrovia; fixed-line service stagnant and extremely limited; telephone coverage extended to a number of other towns and rural areas by four mobile-cellular network operators
domestic: mobile-cellular subscription base growing and teledensity reached 25 per 100 persons in 2009
international: country code–231; satellite earth station–1 Intelsat (Atlantic Ocean) (2009)
Broadcast media: 3 private TV stations; satellite TV service is available; 1 state-owned radio station; about 15 independent radio stations broadcasting in Monrovia, with another 25 local stations operating in other areas; transmissions of 2 international broadcasters are available (2007)
Internet country code: .lr
Internet hosts: 8 (2010)
country comparison to the world: 224
Internet users: 20,000 (2009)
country comparison to the world: 193

TRANSPORTATION

Airports: 29 (2010)
country comparison to the world: 115
Airports—with paved runways: *total:* 2
over 3,047 m: 1
1,524 to 2,437 m: 1 (2010)
Airports—with unpaved runways: *total:* 27
1,524 to 2,437 m: 5
914 to 1,523 m: 8
under 914 m: 14 (2010)
Pipelines: oil 4 km
Railways: *total:* 429 km
country comparison to the world: 115
standard gauge: 345 km 1.435-m gauge
narrow gauge: 84 km 1.067-m gauge
note: most sections of the railways were inoperable because of damage suffered during the civil wars from 1980 to 2003, but many are being rebuilt (2010)
Roadways: *total:* 10,600 km
country comparison to the world: 135
paved: 657 km
unpaved: 9,943 km (2000)
Merchant marine: *total:* 2,512
country comparison to the world: 2
by type: barge carrier 3, bulk carrier 507, cargo 136, carrier 1, chemical tanker 232, combination ore/oil 6, container 875, liquefied gas 93, passenger 2, passenger/cargo 2, petroleum tanker 509, refrigerated cargo 109, roll on/roll off 2, specialized tanker 10, vehicle carrier 25
foreign-owned: 2,356 (Angola 1, Argentina 3, Australia 2, Belgium 1, Bermuda 4, Brazil 20, Canada 4, Chile 7, China 10, Croatia 2, Cyprus 7, Denmark 4, Finland 2, Germany 1049, Gibraltar 5, Greece 454, Hong Kong 47, India 1, Indonesia 4, Isle of Man 19, Israel 31, Italy 48, Japan 102, Latvia 9, Lebanon 1, Monaco 10, Netherlands 35, Nigeria 4, Norway 42, Poland 13, Qatar 5, Romania 3, Russia 108, Saudi Arabia 24, Singapore 27, Slovenia 5, South Korea 1, Sweden 10, Switzerland 17, Syria 1, Taiwan 88, Turkey 15, UAE 27, UK 25, Ukraine 16, Uruguay 1, US 39, Vietnam 3)
note: this country allows large numbers of ships owned by foreign entities to be registered in its national shipping registry and to fly its flag; these ships operate under the laws of the flag state (2010)
Ports and terminals: Buchanan, Monrovia

MILITARY

Military branches: Armed Forces of Liberia (AFL): Army, Navy, Air Force
Military service age and obligation: 18 years of age for voluntary military service; no conscription (2010)
Manpower available for military service:
males age 16–49: 815,826
females age 16–49: 828,484 (2010 est.)
Manpower fit for military service: *males age 16–49:* 524,243
females age 16–49: 544,349 (2010 est.)
Manpower reaching militarily significant age annually: *male:* 36,585
female: 38,516 (2010 est.)
Military expenditures: 1.3% of GDP (2006 est.)
country comparison to the world: 117

TRANSNATIONAL ISSUES

Disputes—international: although civil unrest continues to abate with the assistance of 18,000 UN Mission in Liberia (UNMIL) peacekeepers, as of January 2007, Liberian refugees still remain in Guinea, Cote d'Ivoire, Sierra Leone, and Ghana; Liberia, in turn, shelters refugees fleeing turmoil in Cote d'Ivoire; despite the presence of over 9,000 UN forces (UNOCI) in Cote d'Ivoire since 2004, ethnic conflict continues to spread into neighboring states who can no longer send their migrant workers to Ivorian cocoa plantations; UN sanctions ban Liberia from exporting diamonds and timber
Refugees and internally displaced persons: refugees (country of origin): 12,600 (Cote d'Ivoire)
IDPs: 13,000 (civil war from 1990-2004; IDP resettlement began in November 2004) (2007)
Illicit drugs: transshipment point for Southeast and Southwest Asian heroin and South American cocaine for the European and US markets; corruption, criminal activity, arms-dealing, and diamond trade provide significant potential for money laundering, but the lack of well-developed financial system limits the country's utility as a major money-laundering center.

LIBYA

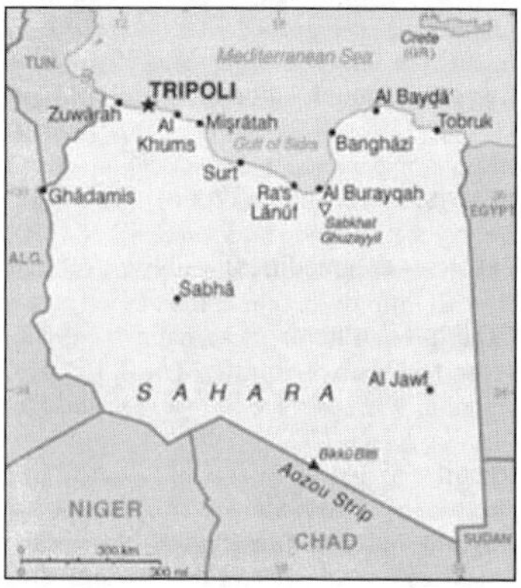

INTRODUCTION

Background: The Italians supplanted the Ottoman Turks in the area around Tripoli in 1911 and did not relinquish their hold until 1943 when defeated in World War II. Libya then passed to UN administration and achieved independence in 1951. Following a 1969 military coup, Col. Muammar Abu Minyar al-QADHAFI began to espouse his own political system, the Third Universal Theory. The system is a combination of socialism and Islam derived in part from tribal practices and is supposed to be implemented by the Libyan people themselves in a unique form of "direct democracy." QADHAFI has always seen himself as a revolutionary and visionary leader. He used oil funds during the 1970s and 1980s to promote his ideology outside Libya, supporting subversives and terrorists abroad to hasten the end of Marxism and capitalism. In addition, beginning in 1973, he engaged in military operations in northern Chad's Aozou Strip–to gain access to minerals and to use as a base of influence in Chadian politics–but was forced to retreat in 1987. UN sanctions in 1992 isolated QADHAFI politically following the downing of Pan AM Flight 103 over Lockerbie, Scotland. During the 1990s, QADHAFI began to rebuild his relationships with Europe. UN sanctions were suspended in April 1999 and finally lifted in September 2003 after Libya accepted responsibility for the Lockerbie bombing. In December 2003, Libya announced that it had agreed to reveal and end its programs to develop weapons of mass destruction and to renounce terrorism. QADHAFI subsequently made significant strides in normalizing relations with Western nations. He hosted various Western European leaders as well as many working-level and commercial delegations, and made his first trip to Western Europe in 15 years when he traveled to Brussels in April 2004. The US rescinded Libya's designation as a state sponsor of terrorism in June 2006. In August 2008, the US and Libya signed a bilateral comprehensive claims settlement agreement to compensate claimants in both countries who allege injury or death at the hands of the other country, including the Lockerbie bombing, the LaBelle disco bombing, and the UTA 772 bombing. In October 2008, the US Government received $1.5 billion pursuant to the agreement to distribute to US national claimants, and as a result effectively normalized its bilateral relationship with Libya. The two countries then exchanged ambassadors for the first time since 1973 in January 2009. Libya in May 2010 was elected to its first three-year seat on the UN Human Rights Council, prompting protests from international non-governmental organizations and human rights campaigners. Unrest that began in several Near Eastern and North African countries in late December 2010 spread to several Libyan cities in February 2011. In response to QADHAFI's harsh military crackdown on protesters, the UN Security Council adopted Resolution 1973, which demanded an immediate ceasefire and authorized the international community to establish a no-fly zone over Libya.

GEOGRAPHY

Location: Northern Africa, bordering the Mediterranean Sea, between Egypt and Tunisia
Geographic coordinates: 25 00 N, 17 00 E
Map references: Africa
Area: *total:* 1,759,540 sq km
country comparison to the world: 17
land: 1,759,540 sq km
water: 0 sq km
Area—comparative: slightly larger than Alaska
Land boundaries: total: 4,348 km
border countries: Algeria 982 km, Chad 1,055 km, Egypt 1,115 km, Niger 354 km, Sudan 383 km, Tunisia 459 km
Coastline: 1,770 km
Maritime claims:
territorial sea: 12 nm
note: Gulf of Sidra closing line—32 degrees, 30 minutes north
exclusive fishing zone: 62 nm
Climate: Mediterranean along coast; dry, extreme desert interior
Terrain: mostly barren, flat to undulating plains, plateaus, depressions
Elevation extremes: *lowest point:* Sabkhat Ghuzayyil -47 m
highest point: Bikku Bitti 2,267 m
Natural resources: petroleum, natural gas, gypsum
Land use: arable land: 1.03%
permanent crops: 0.19%
other: 98.78% (2005)
Irrigated land: 4,700 sq km (2003)
Total renewable water resources: 0.6 cu km (1997)
Freshwater withdrawal (domestic/industrial/agricultural): *total:* 4.27 cu km/yr (14%/3%/83%)
per capita: 730 cu m/yr (2000)
Natural hazards: hot, dry, dust-laden ghibli is a southern wind lasting one to four days in spring and fall; dust storms, sandstorms
Environment—current issues: desertification; limited natural freshwater resources; the Great Manmade River Project, the largest water development scheme in the world, brings water from large aquifers under the Sahara to coastal cities
Environment—international agreements: *party to:* Biodiversity, Climate Change, Climate Change-Kyoto Protocol, Desertification, Endangered Species, Hazardous Wastes, Marine Dumping, Ozone Layer Protection, Ship Pollution, Wetlands
signed, but not ratified: Law of the Sea
Geography—note: more than 90% of the country is desert or semidesert

PEOPLE

Population: 6,597,960 (July 2011 est.)
country comparison to the world: 101
note: includes 166,510 non-nationals
Age structure: *0–14 years:* 32.8% (male 1,104,590/female 1,057,359)
15–64 years: 62.7% (male 2,124,053/ female 2,011,226)
65 years and over: 4.6% (male 146,956/ female 153,776) (2011 est.)
Median age: *total:* 24.5 years
male: 24.5 years
female: 24.4 years (2011 est.)
Population growth rate: 2.064% (2011 est.)
country comparison to the world: 44
Birth rate: 24.04 births/1,000 population (2011 est.)
country comparison to the world: 68
Death rate: 3.4 deaths/1,000 population (July 2011 est.)
country comparison to the world: 213
Net migration rate: 0 migrant(s)/1,000 population (2011 est.)
country comparison to the world: 94
Urbanization: *urban population:* 78% of total population (2010)
rate of urbanization: 2.1% annual rate of change (2010-15 est.)
Major cities—population: TRIPOLI (capital) 1.095 million (2009)
Sex ratio: *at birth:* 1.05 male(s)/female
under 15 years: 1.04 male(s)/female
15–64 years: 1.06 male(s)/female
65 years and over: 0.96 male(s)/female
total population: 1.05 male(s)/female (2011 est.)
Infant mortality rate:
total: 20.09 deaths/1,000 live births
country comparison to the world: 98
male: 22.06 deaths/1,000 live births
female: 18.02 deaths/1,000 live births (2011 est.)
Life expectancy at birth: *total population:* 77.65 years
country comparison to the world: 58
male: 75.34 years
female: 80.08 years (2011 est.)
Total fertility rate: 2.96 children born/ woman (2011 est.)
country comparison to the world: 66
HIV/AIDS—adult prevalence rate: 0.3% (2001 est.)
country comparison to the world: 85
HIV/AIDS—people living with HIV/AIDS: 10,000 (2001 est.)
country comparison to the world: 96
HIV/AIDS—deaths: NA
Drinking water source: *Improved:*
urban: 54% of population
rural: 55% of population
total: 54% of population
Unimproved: urban: 46% of population
rural: 45% of population
total: 46% of population (2000)
Sanitation facility access: *Improved:*
urban: 97% of population
rural: 96% of population
total: 97% of population
Unimproved: urban: 3% of population
rural: 4% of population
total: 3% of population (2008)
Nationality: noun: Libyan(s)
adjective: Libyan
Ethnic groups: Berber and Arab 97%, other 3% (includes Greeks, Maltese, Italians, Egyptians, Pakistanis, Turks, Indians, and Tunisians)
Religions: Sunni Muslim 97%, other 3%
Languages: Arabic, Italian, English,
note: all are widely understood in the major cities
Literacy: *definition:* age 15 and over can read and write
total population: 82.6%
male: 92.4%
female: 72% (2003 est.)
School life expectancy (primary to tertiary education): *total:* 17 years
male: 16 years
female: 17 years (2003)
Education expenditures: NA

GOVERNMENT

Country name: *conventional long form:* Great Socialist People's Libyan Arab Jamahiriya
conventional short form: Libya
local long form: Al Jamahiriyah al Arabiyah al Libiyah ash Shabiyah al Ishtirakiyah al Uthma
local short form: none
Government type: Jamahiriya (a state of the masses) in theory, governed by the populace through local councils; in practice, an authoritarian state
Capital: *name:* Tripoli (Tarabulus)
geographic coordinates: 32 53 N, 13 10 E
time difference: UTC+2 (7 hours ahead of Washington, DC during Standard Time)
Administrative divisions: 22 districts (shabiyat, singular–shabiyat); Al Butnan, Al Jabal al Akhdar, Al Jabal al Gharbi, Al Jafarah, Al Jufrah, Al Kufrah, Al Marj, Al Marqab, Al Wahat, An Nuqat al Khams, Az Zawiyah, Banghazi, Darnah, Ghat, Misratah, Murzuq, Nalut, Sabha, Surt, Tarabulus, Wadi al Hayat, Wadi ash Shati
Independence: 24 December 1951 (from UN trusteeship)
National holiday: Revolution Day, 1 September (1969)
Constitution: none; note–following the September 1969 military overthrow of the Libyan government, the Revolutionary Command Council replaced the existing constitution with the Constitutional Proclamation in December 1969; in March 1977, Libya adopted the Declaration of the Establishment of the People's Authority
Legal system: mixed system of civil and Islamic law

International law organization participation: has not submitted an ICJ jurisdiction declaration; non-party state to the ICCt
Suffrage: 18 years of age; universal and technically compulsory
Executive branch: *chief of state:* Revolutionary Leader Col. Muammar Abu Minyar al-QADHAFI (since 1 September 1969); note—holds no official title, but is de facto chief of state
head of government: Secretary of the General People's Committee (Prime Minister) al-Baghdadi Ali al-MAHMUDI (since 5 March 2006)
cabinet: General People's Committee established by the General People's Congress
(For more information visit the World Leaders website)
elections: national elections are indirect through a hierarchy of people's committees; head of government elected by the General People's Congress; election last held in March 2010 (next elections expected in early 2011)
election results: NA
Legislative branch: unicameral General People's Congress (760 seats; members elected indirectly through a hierarchy of people's committees)
Judicial branch: Supreme Court
Political parties and leaders: none
Political pressure groups and leaders: *other:* anti-QADHAFI Libyan exile movement; Islamic elements
International organization participation: ABEDA, AfDB, AFESD, AMF, AMU, AU, CAEU, COMESA, FAO, G-77, IAEA, IBRD, ICAO, ICRM, IDA, IDB, IFAD, IFC, IFRCS, ILO, IMF, IMO, IMSO, Interpol, IOC, IOM, IPU, ISO, ITSO, ITU, LAS, MIGA, NAM, OAPEC, OIC, OPCW, OPEC, PCA, UN, UNCTAD, UNESCO, UNIDO, UNWTO, UPU, WCO, WFTU, WHO, WIPO, WMO, WTO (observer)
Diplomatic representation in the US: *chief of mission:* Ambassador (vacant)
chancery: 2600 Virginia Avenue NW, Suite 705, Washington, DC 20037
telephone: [1] (202) 944-9601
FAX: [1] (202) 944-9060
Diplomatic representation from the US: *chief of mission:* Ambassador Gene A. CRETZ
embassy: off Jaraba Street, behind the Libyan-Swiss clinic, Ben Ashour
mailing address: US Embassy, 8850 Tripoli Place, Washington, DC 20521-8850
telephone: [218] 91-220-3239
Flag description: plain green; green is the traditional color of Islam (the state religion)
National anthem: *name:* "Allahu Akbar" (God Is Greatest)
lyrics/music: Mahmoud el-SHERIF/Abdalla Shams el-DIN
note: adopted 1969; the anthem was originally a battle song for the Egyptian Army in the 1956 Suez War

ECONOMY

Economy—overview: The Libyan economy depends primarily upon revenues from the oil sector, which contribute about 95% of export earnings, 25% of GDP, and 80% of government revenue. The weakness in world hydrocarbon prices in 2009 reduced Libyan government tax income and constrained economic growth. Substantial revenues from the energy sector coupled with a small population give Libya one of the highest per capita GDPs in Africa, but little of this income flows down to the lower orders of society. Libyan officials in the past five years have made progress on economic reforms as part of a broader campaign to reintegrate the country into the international fold. This effort picked up steam after UN sanctions were lifted in September 2003 and as Libya announced in December 2003 that it would abandon programs to build weapons of mass destruction. The process of lifting US unilateral sanctions began in the spring of 2004; all sanctions were removed by June 2006, helping Libya attract greater foreign direct investment, especially in the energy sector. Libyan oil and gas licensing rounds continue to draw high international interest; the National Oil Corporation (NOC) set a goal of nearly doubling oil production to 3 million bbl/day by 2012. In November 2009, the NOC announced that that target may slip to as late as 2017. Libya faces a long road ahead in liberalizing the socialist-oriented economy, but initial steps—including applying for WTO membership, reducing some subsidies, and announcing plans for privatization—are laying the groundwork for a transition to a more market-based economy. The non-oil manufacturing and construction sectors, which account for more than 20% of GDP, have expanded from processing mostly agricultural products to include the production of petrochemicals, iron, steel, and aluminum. Climatic conditions and poor soils severely limit agricultural output, and Libya imports about 75% of its food. Libya's primary agricultural water source remains the Great Manmade River Project, but significant resources are being invested in desalinization research to meet growing water demands.
GDP (purchasing power parity): $90.57 billion (2010 est.)
country comparison to the world: 74
$86.95 billion (2009 est.)
$89.01 billion (2008 est.)
note: data are in 2010 US dollars
GDP (official exchange rate): $74.23 billion (2010 est.)
GDP—real growth rate: 4.2% (2010 est.)
country comparison to the world: 94
-2.3% (2009 est.)
2.3% (2008 est.)
GDP—per capita (PPP): $14,000 (2010 est.)
country comparison to the world: 83
$13,700 (2009 est.)
$14,400 (2008 est.)
note: data are in 2010 US dollars
GDP—composition by sector: *agriculture:* 2.6%
industry: 63.8%
services: 33.6% (2010 est.)
Labor force: 1.729 million (2010 est.)
country comparison to the world: 125
Labor force—by occupation:
agriculture: 17%
industry: 23%
services: 59% (2004 est.)
Unemployment rate: 30% (2004 est.)
country comparison to the world: 179
Population below poverty line: NA
note: About one-third of Libyans live at or below the national poverty line
Household income or consumption by percentage share: *lowest 10%:* NA%
highest 10%: NA%
Investment (gross fixed): 13.2% of GDP (2010 est.)
country comparison to the world: 141
Budget: *revenues:* $42.31 billion
expenditures: $38.92 billion (2010 est.)
Public debt: 3.3% of GDP (2010 est.)
country comparison to the world: 132
3.9% of GDP (2009 est.)
Inflation rate (consumer prices): 3% (2010 est.)
country comparison to the world: 84
2.4% (2009 est.)
Central bank discount rate: 4% (31 December 2009)
country comparison to the world: 85
5% (31 December 2008)
Commercial bank prime lending rate: 8.41% (31 December 2008)
country comparison to the world: 132
6% (31 December 2007)
Stock of narrow money: $29.85 billion (31 December 2010 est.)
country comparison to the world: 57
$29.82 billion (31 December 2009 est.)
Stock of broad money: $35.98 billion (31 December 2010 est.)
country comparison to the world: 70
$36.2 billion (31 December 2009 est.)
Stock of domestic credit: $55.03 billion (31 December 2010 est.)
country comparison to the world: 61
$41.13 billion (31 December 2009 est.)
Market value of publicly traded shares: $NA
Agriculture—products: wheat, barley, olives, dates, citrus, vegetables, peanuts, soybeans; cattle
Industries: petroleum, petrochemicals, aluminum, iron and steel, food processing, textiles, handicrafts, cement
Industrial production growth rate: 2.7% (2010 est.)
country comparison to the world: 117
Electricity—production: 23.98 billion kWh (2007 est.)
country comparison to the world: 66
Electricity—consumption: 22.17 billion kWh (2007 est.)
country comparison to the world: 66
Electricity—exports: 104 million kWh (2007 est.)
Electricity—imports: 77 million kWh (2007 est.)
Oil—production: 1.79 million bbl/day (2009 est.)
country comparison to the world: 18
Oil—consumption: 280,000 bbl/day (2009 est.)
country comparison to the world: 45
Oil—exports: 1.542 million bbl/day (2007 est.)
country comparison to the world: 15
Oil—imports: 575.3 bbl/day (2007 est.)
country comparison to the world: 195
Oil—proved reserves: 47 billion bbl (1 January 2010 est.)

country comparison to the world: 9
Natural gas—production: 15.9 billion cu m (2008 est.)
country comparison to the world: 33
Natural gas—consumption: 5.5 billion cu m (2008 est.)
country comparison to the world: 57
Natural gas—exports: 10.4 billion cu m (2008 est.)
country comparison to the world: 20
Natural gas—imports: 0 cu m (2008 est.)
country comparison to the world: 82
Natural gas—proved reserves: 1.539 trillion cu m (1 January 2010 est.)
country comparison to the world: 23
Current account balance: $15.53 billion (2010 est.)
country comparison to the world: 20
$10.06 billion (2009 est.)
Exports: $44.89 billion (2010 est.)
country comparison to the world: 58
$37.16 billion (2009 est.)
Exports—commodities: crude oil, refined petroleum products, natural gas, chemicals
Exports—partners: Italy 37.65%, Germany 10.11%, France 8.44%, Spain 7.94%, Switzerland 5.93%, US 5.27% (2009)
Imports: $24.47 billion (2010 est.)
country comparison to the world: 64
$22.01 billion (2009 est.)
Imports—commodities: machinery, semi-finished goods, food, transport equipment, consumer products
Imports—partners: Italy 18.9%, China 10.54%, Turkey 9.92%, Germany 9.78%, France 5.63%, Tunisia 5.25%, South Korea 4.02% (2009)
Reserves of foreign exchange and gold: $107.3 billion (31 December 2010 est.)
country comparison to the world: 15
$104.2 billion (31 December 2009 est.)
Debt—external: $6.378 billion (31 December 2010 est.)
country comparison to the world: 99
$5.891 billion (31 December 2009 est.)
Stock of direct foreign investment—at home: $18.64 billion (31 December 2010 est.)
country comparison to the world: 70
$15.56 billion (31 December 2009 est.)
Stock of direct foreign investment—abroad: $15.32 billion (31 December 2010 est.)
country comparison to the world: 48
$13.92 billion (31 December 2009 est.)
Exchange rates: Libyan dinars (LYD) per US dollar–1.2648 (2010), 1.2535 (2009), 1.2112 (2008), 1.2604 (2007), 1.3108 (2006)

COMMUNICATIONS

Telephones—main lines in use: 1.101 million (2009)
country comparison to the world: 75
Telephones—mobile cellular: 5.004 million (2009)
country comparison to the world: 95
Telephone system: general assessment: telecommunications system is state-owned and service is poor, but investment is being made to upgrade; state retains monopoly in fixed-line services; mobile-cellular telephone system became operational in 1996
domestic: multiple providers for a mobile telephone system that is growing rapidly; combined fixed-line and mobile-cellular teledensity is approaching 100 telephones per 100 persons
international: country code–218; satellite earth stations–4 Intelsat, NA Arabsat, and NA Intersputnik; submarine cable to France and Italy; microwave radio relay to Tunisia and Egypt; tropospheric scatter to Greece; participant in Medarabtel (2009)
Broadcast media: state controls broadcast media; state-owned terrestrial TV station and about a half-dozen state-owned satellite stations broadcast; some provinces operate local TV stations; a single, non-state-owned TV station launched in 2007; pan-Arab satellite TV stations are available; state-owned radio broadcasts on a number of frequencies, some of which carry regional programming; Voice of Africa, Libya's external radio service, can also be heard; a single, non-state-owned radio station broadcasting (2007)
Internet country code: .ly
Internet hosts: 12,432 (2010)
country comparison to the world: 120
Internet users: 353,900 (2009)
country comparison to the world: 124

TRANSPORTATION

Airports: 137 (2010)
country comparison to the world: 42
Airports—with paved runways: *total:* 59
over 3,047 m: 24
2,438 to 3,047 m: 5
1,524 to 2,437 m: 23
914 to 1,523 m: 6
under 914 m: 1 (2010)
Airports—with unpaved runways: *total:* 78
over 3,047 m: 3
2,438 to 3,047 m: 2
1,524 to 2,437 m: 14
914 to 1,523 m: 42
under 914 m: 17 (2010)
Heliports: 2 (2010)
Pipelines: condensate 776 km; gas 3,216 km; oil 6,960 km (2010)
Roadways: *total:* 100,024 km
country comparison to the world: 42
paved: 57,214 km
unpaved: 42,810 km (2003)
Merchant marine: *total:* 27
country comparison to the world: 89
by type: cargo 5, chemical tanker 4, liquefied gas 3, petroleum tanker 13, roll on/roll off 2
foreign-owned: 5 (Kuwait 1, Norway 1, Syria 2, UK 1)
registered in other countries: 5 (Hong Kong 1, Malta 4) (2010)
Ports and terminals: Az Zawiyah, Marsa al Burayqah (Marsa el Brega), Ra's Lanuf, Tripoli

MILITARY

Military branches: Armed Peoples on Duty (APOD, Army), Libyan Arab Navy, Libyan Arab Air Force (Al-Quwwat al-Jawwiya al-Jamahiriya al-Arabia al-Libyya, LAAF), Libyan Coast Guard (2008)
Military service age and obligation: 17 years of age (2004)
Manpower available for military service: *males age 16–49:* 1,775,078
females age 16–49: 1,714,194 (2010 est.)
Manpower fit for military service: *males age 16–49:* 1,511,144
females age 16–49: 1,458,934 (2010 est.)
Manpower reaching militarily significant age annually: *male:* 59,547
female: 57,070 (2010 est.)
Military expenditures: 3.9% of GDP (2005 est.)
country comparison to the world: 25

TRANSNATIONAL ISSUES

Disputes—international: Liechtenstein dormant disputes include Libyan claims of about 32,000 sq km still reflected on its maps of southeastern Algeria and the FLN's assertions of a claim to Chirac Pastures in southeastern Morocco; various Chadian rebels from the Aozou region reside in southern Libya
Refugees and internally displaced persons: refugees (country of origin): 8,000 (Palestinian Territories) (2007)
Trafficking in persons: current situation: Libya is a transit and destination country for men and women from sub-Saharan Africa and Asia trafficked for the purposes of forced labor and commercial sexual exploitation
tier rating: Tier 2 Watch List–Libya is on the Tier 2 Watch List for its failure to provide evidence of increasing efforts to address trafficking in persons in 2007 when compared to 2006, particularly in the area of investigating and prosecuting trafficking offenses; Libya did not publicly release any data on investigations or punishment of any trafficking offenses (2008)

LIECHTENSTEIN

INTRODUCTION

Background: The Principality of Liechtenstein was established within the Holy Roman Empire in 1719. Occupied by both French and Russian troops during the Napoleonic wars, it became a sovereign state in 1806 and joined the Germanic Confederation in 1815. Liechtenstein became fully independent in 1866 when the Confederation dissolved. Until the end of World War I, it was closely tied to Austria, but the economic devastation caused by that conflict forced Liechtenstein to enter into a customs and monetary union with Switzerland. Since World War II (in which Liechtenstein remained neutral), the country's low taxes have spurred outstanding economic growth. In 2000, shortcomings in banking regulatory oversight resulted in concerns about the use of financial institutions for money

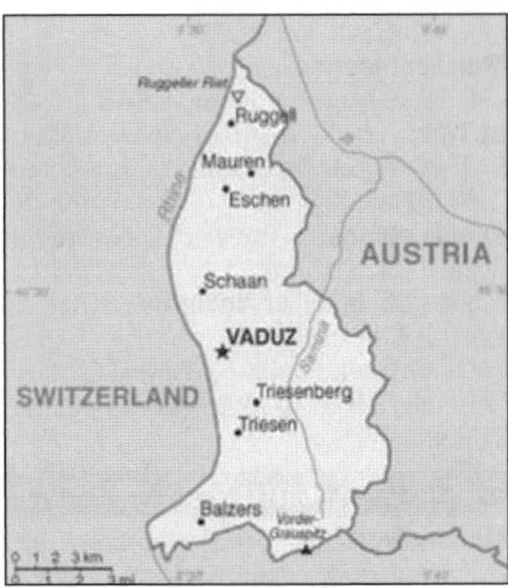

laundering. However, Liechtenstein implemented anti-money-laundering legislation and a Mutual Legal Assistance Treaty with the US that went into effect in 2003.

GEOGRAPHY

Location: Central Europe, between Austria and Switzerland
Geographic coordinates: 47 16 N, 9 32 E
Map references: Europe
Area: *total:* 160 sq km
country comparison to the world: 217
land: 160 sq km
water: 0 sq km
Area—comparative: about 0.9 times the size of Washington, DC
Land boundaries: total: 76 km
border countries: Austria 34.9 km, Switzerland 41.1 km
Coastline: 0 km (doubly landlocked)
Maritime claims: none (landlocked)
Climate: continental; cold, cloudy winters with frequent snow or rain; cool to moderately warm, cloudy, humid summers
Terrain: mostly mountainous (Alps) with Rhine Valley in western third
Elevation extremes: *lowest point:* Ruggeller Riet 430 m
highest point: Vorder-Grauspitz 2,599 m
Natural resources: hydroelectric potential, arable land
Land use: arable land: 25%
permanent crops: 0%
other: 75% (2005)
Irrigated land: NA
Natural hazards: NA
Environment—current issues: NA
Environment—international agreements: *party to:* Air Pollution, Air Pollution-Nitrogen Oxides, Air Pollution-Persistent Organic Pollutants, Air Pollution-Sulfur 85, Air Pollution-Sulfur 94, Air Pollution-Volatile Organic Compounds, Biodiversity, Climate Change, Climate Change-Kyoto Protocol, Desertification, Endangered Species, Hazardous Wastes, Ozone Layer Protection, Wetlands
signed, but not ratified: Law of the Sea
Geography—note: along with Uzbekistan, one of only two doubly landlocked countries in the world; variety of microclimatic variations based on elevation

PEOPLE

Population: 35,236 (July 2011 est.)
country comparison to the world: 211
Age structure: *0–14 years:* 16.1% (male 2,809/female 2,856)
15–64 years: 69% (male 11,970/female 12,326)
65 years and over: 15% (male 2,304/female 2,971) (2011 est.)
Median age: *total:* 41.8 years
male: 41.2 years
female: 42.3 years (2011 est.)
Population growth rate: 0.653% (2011 est.)
country comparison to the world: 142
Birth rate: 9.65 births/1,000 population (2011 est.)
country comparison to the world: 198
Death rate: 7.61 deaths/1,000 population (July 2011 est.)
country comparison to the world: 114
Net migration rate: 4.48 migrant(s)/1,000 population (2011 est.)
country comparison to the world: 21
Urbanization: *urban population:* 14% of total population (2010)
rate of urbanization: 0.9% annual rate of change (2010-15 est.)
Major cities—population: VADUZ (capital) 5,000 (2009)
Sex ratio: *at birth:* 1 male(s)/female
under 15 years: 0.98 male(s)/female
15–64 years: 0.97 male(s)/female
65 years and over: 0.77 male(s)/female
total population: 0.94 male(s)/female (2011 est.)
Infant mortality rate:
total: 4.15 deaths/1,000 live births
country comparison to the world: 199
male: 5.37 deaths/1,000 live births
female: 2.92 deaths/1,000 live births (2011 est.)
Life expectancy at birth:
total population: 80.31 years
country comparison to the world: 24
male: 76.86 years
female: 83.77 years (2011 est.)
Total fertility rate: 1.53 children born/woman (2011 est.)
country comparison to the world: 183
HIV/AIDS—adult prevalence rate: NA
HIV/AIDS—people living with HIV/AIDS: NA
HIV/AIDS—deaths: NA
Nationality: noun: Liechtensteiner(s)
adjective: Liechtenstein
Ethnic groups: Liechtensteiner 65.6%, other 34.4% (2000 census)
Religions: Roman Catholic 76.2%, Protestant 7%, unknown 10.6%, other 6.2% (June 2002)
Languages: German (official), Alemannic dialect
Literacy: *definition:* age 10 and over can read and write
total population: 100%
male: 100%
female: 100%
School life expectancy (primary to tertiary education): *total:* 12 years
male: 13 years
female: 12 years (2008)
Education expenditures: 2% of GDP (2007)
country comparison to the world: 157

GOVERNMENT

Country name: *conventional long form:* Principality of Liechtenstein
conventional short form: Liechtenstein
local long form: Fuerstentum Liechtenstein
local short form: Liechtenstein
Government type: constitutional monarchy
Capital: *name:* Vaduz
geographic coordinates: 47 08 N, 9 31 E
time difference: UTC+1 (6 hours ahead of Washington, DC during Standard Time)
daylight saving time: +1hr, begins last Sunday in March; ends last Sunday in October
Administrative divisions: 11 communes (Gemeinden, singular–Gemeinde); Balzers, Eschen, Gamprin, Mauren, Planken, Ruggell, Schaan, Schellenberg, Triesen, Triesenberg, Vaduz
Independence: 23 January 1719 (Principality of Liechtenstein established); 12 July 1806 (independence from the Holy Roman Empire)
National holiday: Assumption Day, 15 August
Constitution: 5 October 1921; amended 15 September 2003
Legal system: civil law system influenced by Swiss, Austrian, and German law
International law organization participation: accepts compulsory ICJ jurisdiction with reservations; accepts ICCt jurisdiction
Suffrage: 18 years of age; universal
Executive branch: *chief of state:* Prince HANS ADAM II (since 13 November 1989, assumed executive powers on 26 August 1984); Heir Apparent Prince ALOIS, son of the monarch (born 11 June 1968); note–on 15 August 2004, HANS ADAM transferred the official duties of the ruling prince to ALOIS, but HANS ADAM retains status of chief of state
head of government: Head of Government Klaus TSCHUETSCHER (since 25 March 2009)
cabinet: Cabinet elected by the Parliament, confirmed by the monarch
(For more information visit the World Leaders website)
elections: the monarchy is hereditary; following legislative elections, the leader of the majority party in the Landtag usually appointed the head of government by the monarch and the leader of the largest minority party in the Landtag usually appointed the deputy head of government by the monarch if there is a coalition government
Legislative branch: unicameral Parliament or Landtag (25 seats; members elected by popular vote under proportional representation to serve four-year terms)
elections: last held on 8 February 2009 (next to be held in February 2013)
election results: percent of vote by party–VU 47.6%, FBP 43.5%, FL 8.9%; seats by party–VU 13, FBP 11, FL 1
Judicial branch: Supreme Court or Oberster Gerichtshof; Court of Appeal or Obergericht
Political parties and leaders: Die Freie Liste (The Free List) or FL [Wolfgang MARXER]; Fortschrittliche Buergerpartei (Progressive Citizens' Party) or FBP [Alexander BATLINER]; Vaterlaendische Union (Fatherland Union) or VU [Adolf HEEB]
Political pressure groups and leaders: NA
International organization participation: CE, EBRD, EFTA, IAEA, ICRM, IFRCS, Interpol, IOC, IPU, ITSO, ITU, ITUC,

OPCW, OSCE, PCA, Schengen Convention (de facto member), UN, UNCTAD, UPU, WIPO, WTO

Diplomatic representation in the US: *chief of mission:* Ambassador Claudia FRITSCHE
chancery: 2900 K Street, NW, Suite 602B, Washington, DC 20007
telephone: [1] (202) 331-0590
FAX: [1] (202) 331-3221

Diplomatic representation from the US: the US does not have an embassy in Liechtenstein; the US Ambassador to Switzerland is accredited to Liechtenstein

Flag description: two equal horizontal bands of blue (top) and red with a gold crown on the hoist side of the blue band; the colors may derive from the blue and red livery design used in the principality's household in the 18th century; the prince's crown was introduced in 1937 to distinguish the flag from that of Haiti

National anthem: *name:* "Oben am jungen Rhein" (High Above the Young Rhine)
lyrics/music: Jakob Joseph JAUCH/ unknown
note: adopted 1850, revised 1963; the anthem uses the tune of "God Save the Queen"

ECONOMY

Economy—overview: Despite its small size and limited natural resources, Liechtenstein has developed into a prosperous, highly industrialized, free-enterprise economy with a vital financial service sector and likely the second highest per capita income in the world. The Liechtenstein economy is widely diversified with a large number of small businesses. Low business taxes—the maximum tax rate is 20%—and easy incorporation rules have induced many holding companies to establish nominal offices in Liechtenstein providing 30% of state revenues. The country participates in a customs union with Switzerland and uses the Swiss franc as its national currency. It imports more than 90% of its energy requirements. Liechtenstein has been a member of the European Economic Area (an organization serving as a bridge between the European Free Trade Association (EFTA) and the EU) since May 1995. The government is working to harmonize its economic policies with those of an integrated Europe. In 2008, Liechtenstein came under renewed international pressure—particularly from Germany—to improve transparency in its banking and tax systems. In December 2008, Liechtenstein signed a Tax Information Exchange Agreement with the US. Upon Liechtenstein's conclusion of 12 bilateral information-sharing agreements, the OECD in October 2009 removed the principality from its "grey list" of countries that had yet to implement the organization's Model Tax Convention. By the end of 2010, Liechtenstein had signed 25 Tax Information Exchange Agreements or Double Tax Agreements.

GDP (purchasing power parity): $5.028 billion (2008)
country comparison to the world: 162
$4.16 billion (2007)
$4.035 billion (2006 est.)

GDP (official exchange rate): $5.08 billion (2008)

GDP—real growth rate: 1.8% (2008 est.)
country comparison to the world: 153
3.1% (2007 est.)

GDP—per capita (PPP): $141,100 (2008 est.)
country comparison to the world: 2
$122,100 (2007 est.)

GDP—composition by sector:
agriculture: 6%
industry: 36%
services: 58% (2008)

Labor force: 32,880 (2009)
country comparison to the world: 202
note: 51% of people at work in Liechtenstein commute daily from Austria, Switzerland, and Germany (2009)

Labor force—by occupation:
agriculture: 1.5%
industry: 43.5%
services: 55% (December 2006)

Unemployment rate: 1.5% (2007)
country comparison to the world: 8
1.3% (September 2002)

Population below poverty line: NA%

Household income or consumption by percentage share: *lowest 10%:* NA%
highest 10%: NA%

Budget: *revenues:* $943 million
expenditures: $820 million (2008 est.)

Inflation rate (consumer prices): 0.7% (2010)
country comparison to the world: 12
0.3% (2009)

Market value of publicly traded shares: $NA

Agriculture—products: wheat, barley, corn, potatoes; livestock, dairy products

Industries: electronics, metal manufacturing, dental products, ceramics, pharmaceuticals, food products, precision instruments, tourism, optical instruments

Industrial production growth rate: NA%

Exports: $2.83 billion (2009)
country comparison to the world: 125
$3.92 billion (2008)

Exports—commodities: small specialty machinery, connectors for audio and video, parts for motor vehicles, dental products, hardware, prepared foodstuffs, electronic equipment, optical products

Imports: $1.77 billion (2009)
country comparison to the world: 154
$2.59 billion (2008)

Imports—commodities: agricultural products, raw materials, energy products, machinery, metal goods, textiles, foodstuffs, motor vehicles

Debt—external: $0 (2001)
country comparison to the world: 198

Exchange rates: Swiss francs (CHF) per US dollar—1.0429 (2010), 1.0881 (2009), 1.0774 (2008), 1.1973 (2007), 1.2539 (2006)

COMMUNICATIONS

Telephones—main lines in use: 19,600 (2009)
country comparison to the world: 196

Telephones—mobile cellular: 35,000 (2009)
country comparison to the world: 199

Telephone system: general assessment: automatic telephone system
domestic: fixed-line and mobile-cellular services widely available; combined telephone service subscribership exceeds 150 per 100 persons
international: country code—423; linked to Swiss networks by cable and microwave radio relay (2008)

Broadcast media: relies on foreign terrestrial and satellite broadcasters for most broadcast media services; first Liechtenstein-based television station established August 2008; Radio Liechtenstein operates multiple radio stations; a Swiss-based broadcaster operates several radio stations in Liechtenstein (2008)

Internet country code: .li

Internet hosts: 9,418 (2010)
country comparison to the world: 124

Internet users: 23,000 (2009)
country comparison to the world: 190

TRANSPORTATION

Pipelines: gas 20 km (2010)

Railways: *total:* 9 km
country comparison to the world: 134
standard gauge: 9 km 1.435-m gauge (electrified)
note: belongs to the Austrian Railway System connecting Austria and Switzerland (2011)

Roadways: *total:* 380 km
country comparison to the world: 200
paved: 380 km (2010)

Waterways: 28 km (2010)
country comparison to the world: 106

MILITARY

Military branches: no regular military forces (constitutionally prohibited); Principality of Liechtenstein National Police (Landespolizei, LP) (2010)

Manpower available for military service: *males age 16–49:* 8,009 (2010 est.)

Manpower fit for military service: *males age 16–49:* 6,538
females age 16–49: 6,746 (2010 est.)

Manpower reaching militarily significant age annually: *male:* 219
female: 211 (2010 est.)

Military—note: Liechtenstein has no military forces but is interested in European security policy and is an active member of the Organization for Security and Cooperation in Europe (OSCE)

TRANSNATIONAL ISSUES

Disputes—international: none

Illicit drugs: has strengthened money laundering controls, but money laundering remains a concern due to Liechtenstein's sophisticated offshore financial services sector

LITHUANIA

INTRODUCTION

Background: Lithuanian lands were united under MINDAUGAS in 1236; over the next century, through alliances and conquest, Lithuania extended its territory to include most of present-day Belarus and Ukraine. By the end of the 14th century Lithuania was the largest state in Europe. An alliance with Poland in 1386 led the two countries into a union through the person of a common ruler. In 1569, Lithuania and Poland formally united into a single dual state, the Polish-Lithuanian Commonwealth. This entity survived until 1795 when its remnants were partitioned by surrounding countries. Lithuania regained its independence following World War I but was annexed by the USSR in 1940–an action never recognized by the US and many other countries. On 11 March 1990, Lithuania became the first of the Soviet republics to declare its independence, but Moscow did not recognize this proclamation until September of 1991 (following the abortive coup in Moscow). The last Russian troops withdrew in 1993. Lithuania subsequently restructured its economy for integration into Western European institutions; it joined both NATO and the EU in the spring of 2004.

GEOGRAPHY

Location: Eastern Europe, bordering the Baltic Sea, between Latvia and Russia
Geographic coordinates: 56 00 N, 24 00 E
Map references: Europe
Area: *total:* 65,300 sq km
country comparison to the world: 122
land: 62,680 sq km
water: 2,620 sq km
Area—comparative: slightly larger than West Virginia
Land boundaries: total: 1,574 km
border countries: Belarus 680 km, Latvia 576 km, Poland 91 km, Russia (Kaliningrad) 227 km
Coastline: 90 km
Maritime claims: *territorial sea:* 12 nm
Climate: transitional, between maritime and continental; wet, moderate winters and summers
Terrain: lowland, many scattered small lakes, fertile soil
Elevation extremes:
lowest point: Baltic Sea 0 m
highest point: Aukstojas 294 m
Natural resources: peat, arable land, amber
Land use: arable land: 44.81%
permanent crops: 0.9%
other: 54.29% (2005)
Irrigated land: 70 sq km (2003)
Total renewable water resources: 24.5 cu km (2005)
Freshwater withdrawal (domestic/industrial/agricultural): *total:* 3.33 cu km/yr (78%/15%/7%)
per capita: 971 cu m/yr (2003)
Natural hazards: NA
Environment—current issues: contamination of soil and groundwater with petroleum products and chemicals at military bases
Environment—international agreements: *party to:* Air Pollution, Air Pollution-Nitrogen Oxides, Air Pollution-Persistent Organic Pollutants, Air Pollution-Sulphur 85, Air Pollution-Sulphur 94, Air Pollution-Volatile Organic Compounds, Biodiversity, Climate Change, Climate Change-Kyoto Protocol, Desertification, Endangered Species, Environmental Modification, Hazardous Wastes, Law of the Sea, Ozone Layer Protection, Ship Pollution, Wetlands
signed, but not ratified: none of the selected agreements
Geography—note: fertile central plains are separated by hilly uplands that are ancient glacial deposits

PEOPLE

Population: 3,535,547 (July 2011 est.)
country comparison to the world: 130
Age structure: *0–14 years:* 13.8% (male 250,146/female 236,984)
15–64 years: 69.7% (male 1,211,707/female 1,254,195)
65 years and over: 16.5% (male 201,358/female 381,157) (2011 est.)
Median age: *total:* 40.1 years
male: 37.5 years
female: 42.7 years (2011 est.)
Population growth rate: -0.276% (2011 est.)
country comparison to the world: 214
Birth rate: 9.29 births/1,000 population (2011 est.)
country comparison to the world: 205
Death rate: 11.33 deaths/1,000 population (July 2011 est.)
country comparison to the world: 35
Net migration rate: -0.72 migrant(s)/1,000 population (2011 est.)
country comparison to the world: 146
Urbanization: *urban population:* 67% of total population (2010)
rate of urbanization: -0.5% annual rate of change (2010-15 est.)
Major cities—population: VILNIUS (capital) 546,000 (2009)
Sex ratio: *at birth:* 1.057 male(s)/female
under 15 years: 1.06 male(s)/female
15–64 years: 0.96 male(s)/female
65 years and over: 0.53 male(s)/female
total population: 0.89 male(s)/female (2011 est.)
Infant mortality rate:
total: 6.27 deaths/1,000 live births
country comparison to the world: 172
male: 7.49 deaths/1,000 live births
female: 4.99 deaths/1,000 live births (2011 est.)
Life expectancy at birth:
total population: 75.34 years
country comparison to the world: 86
male: 70.48 years
female: 80.48 years (2011 est.)
Total fertility rate: 1.25 children born/woman (2011 est.)
country comparison to the world: 217
HIV/AIDS—adult prevalence rate: 0.1% (2009 est.)
country comparison to the world: 143
HIV/AIDS—people living with HIV/AIDS: 1,200 (2009 est.)
country comparison to the world: 138
HIV/AIDS—deaths: fewer than 100 (2009 est.)
country comparison to the world: 138
Major infectious diseases:
degree of risk: intermediate
food or waterborne diseases: bacterial diarrhea
vectorborne diseases: tickborne encephalitis (2009)
Nationality:
noun: Lithuanian(s)
adjective: Lithuanian
Ethnic groups: Lithuanian 84%, Polish 6.1%, Russian 4.9%, Belarusian 1.1%, other or unspecified 3.9% (2009)
Religions: Roman Catholic 79%, Russian Orthodox 4.1%, Protestant (including Lutheran and Evangelical Christian Baptist) 1.9%, other or unspecified 5.5%, none 9.5% (2001 census)
Languages: Lithuanian (official) 82%, Russian 8%, Polish 5.6%, other and unspecified 4.4% (2001 census)
Literacy: *definition:* age 15 and over can read and write
total population: 99.6%
male: 99.6%
female: 99.6% (2001 census)
School life expectancy (primary to tertiary education): *total:* 16 years
male: 15 years
female: 17 years (2008)
Education expenditures: 4.7% of GDP (2007)
country comparison to the world: 74

GOVERNMENT

Country name: *conventional long form:* Republic of Lithuania
conventional short form: Lithuania
local long form: Lietuvos Respublika
local short form: Lietuva
former: Lithuanian Soviet Socialist Republic
Government type: parliamentary democracy
Capital: *name:* Vilnius
geographic coordinates: 54 41 N, 25 19 E
time difference: UTC+2 (7 hours ahead of Washington, DC during Standard Time)
daylight saving time: +1hr, begins last Sunday in March; ends last Sunday in October

Administrative divisions: 10 counties (apskritys, singular–apskritis); Alytaus, Kauno, Klaipedos, Marijampoles, Panevezio, Siauliu, Taurages, Telsiu, Utenos, Vilniaus
Independence: 11 March 1990 (declared); 6 September 1991 (recognized by the Soviet Union); notable earlier dates: 6 July 1253 (coronation of Mindaugas, traditional founding date), 1 July 1569 (Polish-Lithuanian Commonwealth created)
National holiday: Independence Day, 16 February (1918); note–16 February 1918 was the date Lithuania declared its independence from Soviet Russia and established its statehood; 11 March 1990 was the date it declared its independence from the Soviet Union
Constitution: adopted 25 October 1992; last amended 13 July 2004
Legal system: civil law system; legislative acts can be appealed to the constitutional court
International law organization participation: has not submitted an ICJ jurisdiction declaration; accepts ICCt jurisdiction
Suffrage: 18 years of age; universal
Executive branch: *chief of state:* President Dalia GRYBAUSKAITE (since 12 July 2009)
head of government: Prime Minister Andrius KUBILIUS (since 27 November 2008)
cabinet: Council of Ministers appointed by the president on the nomination of the prime minister
(For more information visit the World Leaders website)
elections: president elected by popular vote for a five-year term (eligible for a second term); election last held on 17 May 2009 (next to be held in May 2014); prime minister appointed by the president on the approval of the Parliament
election results: Dalia GRYBAUSKAITE elected president; percent of vote–Dalia GRYBAUSKAITE 69.1%, Algirdas BUTKEVICIUS 11.8%, Valentinas MAZURONIS 6.2%, others 12.9%; Andrius KUBILIUS' government approved by Parliament 83-40 with 5 abstentions
Legislative branch: unicameral Parliament or Seimas (141 seats; 71 members elected by popular vote, 70 elected by proportional representation; members to serve four-year terms)
elections: last held on 12 and 26 October 2008 (next to be held in October 2012)
election results: percent of vote by party–TS-LKD 19.7%, TPP 15.1%, TT 12.7%, LSDP 11.7%, DP+J 9%, LRLS 5.7%, LCS 5.3%, LLRA 4.8%, LVLS 3.7%, NS 3.6%, other 8.7%; seats by faction–TS-LKD 44, LSDP 26, TPP 16, TT 15, LRLS 11, DP+J 10, LCS 8, LLRA 3, LVLS 3, NS 1, independent 4; note–seats by faction as of 25 January 2011–TS-LKD 45, LSDP 24, TT 18, LCS and TPP 13, LRLS 13, Christian Party 10, DP 10, unaffiliated 7, vacant 1; note–TS-LKD, LRLS, LCS and TPP form the ruling coalition
Judicial branch: Constitutional Court; Supreme Court; Court of Appeal; judges for all courts appointed by the president
Political parties and leaders: Christian party [Gediminas VAGNORIUS]; Civil Democracy Party or PDP [Algimantas MATULEVICIUS]; Electoral Action of Lithuanian Poles or LLRA [Valdemar TOMASZEVSKI]; Homeland Union–Lithuanian Christian Democrats or TS-LKD [Andrius KUBILIUS]; Labor Party or DP [Viktor USPASKICH]; Liberal and Center Union or LCS [Gintautas BABRAVICIUS]; Liberal Movement or LS or LRLS [Eligijus MASIULIS]; Lithuanian Farmers' Union or LVLS or VLS [Ramunas KARBAUSKIS]; Lithuanian People's Party (not yet officially established) [Kazimiera PRUNSKIENE]; National Revival or TPP [Arunas VALINSKAS]; New Union (Social Liberal) or NS [Arturas PAULAUSKAS]; Order and Justice Party or TT [Rolandas PAKSAS]; Social Democratic Party or LSDP [Algirdas BUTKEVICIUS]
International organization participation: Australia Group, BA, BIS, CBSS, CD, CE, EAPC, EBRD, EIB, EU, FAO, IAEA, IBRD, ICAO, ICC, ICRM, IFC, IFRCS, ILO, IMF, IMO, Interpol, IOC, IOM, IPU, ISO, ITU, ITUC, MIGA, NATO, NIB, NSG, OAS (observer), OIF (observer), OPCW, OSCE, PCA, Schengen Convention, UN, UNCTAD, UNESCO, UNIDO, UNWTO, UPU, WCO, WHO, WIPO, WMO, WTO
Diplomatic representation in the US: *chief of mission:* Ambassador Zygimantas PAVILIONIS
chancery: 2622 16th Street NW, Washington, DC 20009
telephone: [1] (202) 234-5860
FAX: [1] (202) 328-0466
consulate(s) general: Chicago, New York
Diplomatic representation from the US: *chief of mission:* Ambassador Anne E. DERSE
embassy: Akmenu gatve 6, Vilnius, LT-03106
mailing address: American Embassy, Akmenu Gatve 6, Vilnius LT-03106
telephone: [370] (5) 266 5500
FAX: [370] (5) 266 5510
Flag description: three equal horizontal bands of yellow (top), green, and red; yellow symbolizes golden fields, as well as the sun, light, and goodness; green represents the forests of the countryside, in addition to nature, freedom, and hope; red stands for courage and the blood spilled in defense of the homeland
National anthem: *name:* "Tautiska giesme" (The National Song)
lyrics/music: Vincas KUDIRKA
note: adopted 1918, restored 1990; the anthem was written in 1898 while Lithuania was a part of Russia; it was banned during the Soviet occupation from 1940 to 1990

ECONOMY

Economy—overview: Lithuania gained membership in the World Trade Organization and joined the EU in May 2004. Despite Lithuania's EU accession, Lithuania's trade with its Central and Eastern European neighbors, and Russia in particular, accounts for a growing percentage of total trade. Privatization of the large, state-owned utilities is nearly complete. Foreign government and business support have helped in the transition from the old command economy to a market economy. Lithuania's economy grew on average 8% per year for the four years prior to 2008 driven by exports and domestic demand. However, GDP plunged nearly 15% in 2009–during the 2008-09 crisis the three former Soviet Baltic republics had the world's worst economic declines. In 2009, the government launched a high-profile campaign, led by Prime Minister KUBILIUS, to attract foreign investment and to develop export markets. The current account deficit, which had risen to roughly 15% of GDP in 2007-08, recovered to a surplus of 4% 2009 and 3.4% in 2010 in the wake of a cutback in imports to almost half the 2008 level. Nevertheless, economic growth was flat and unemployment continued upward to 17.9% in 2010.
GDP (purchasing power parity): $56.59 billion (2010 est.)
country comparison to the world: 89
$55.84 billion (2009 est.)
$65.5 billion (2008 est.)
note: data are in 2010 US dollars
GDP (official exchange rate): $36.36 billion (2010 est.)
GDP—real growth rate: 1.3% (2010 est.)
country comparison to the world: 166
-14.7% (2009 est.)
2.9% (2008 est.)
GDP—per capita (PPP): $16,000 (2010 est.)
country comparison to the world: 70
$15,700 (2009 est.)
$18,400 (2008 est.)
note: data are in 2010 US dollars
GDP—composition by sector:
agriculture: 4.3%
industry: 27.6%
services: 68.2% (2010 est.)
Labor force: 1.633 million (2010 est.)
country comparison to the world: 128
Labor force—by occupation:
agriculture: 14%
industry: 29.1%
services: 56.9% (2005)
Unemployment rate: 17.9% (2010 est.)
country comparison to the world: 160
13.7% (2009 est.)
Population below poverty line: 4% (2003)
Household income or consumption by percentage share: *lowest 10%:* 2.7%
highest 10%: 27.4% (2004)
Distribution of family income—Gini index: 36 (2005)
country comparison to the world: 85
34 (1999)
Investment (gross fixed): 15.2% of GDP (2010 est.)
country comparison to the world: 130
Budget: *revenues:* $11.26 billion
expenditures: $13.48 billion (2010 est.)
Public debt: 36.7% of GDP (2010 est.)
country comparison to the world: 81
29.5% of GDP (2009 est.)
Inflation rate (consumer prices): 0.9% (2010 est.)
country comparison to the world: 18
4.5% (2009 est.)
Central bank discount rate: 1.75% (February 2010)
country comparison to the world: 100
4.73% (31 December 2008)

Commercial bank prime lending rate: 8.39% (31 December 2009 est.)
country comparison to the world: 110
8.41% (31 December 2008 est.)
Stock of narrow money: $8.917 billion (31 December 2010 est.)
country comparison to the world: 73
$8.896 billion (31 December 2009 est.)
Stock of broad money: $17.26 billion (31 December 2010 est.)
country comparison to the world: 86
$17.6 billion (31 December 2009 est.)
Stock of domestic credit: $25.35 billion (31 December 2010 est.)
country comparison to the world: 73
$25.85 billion (31 December 2009 est.)
Market value of publicly traded shares: $4.477 billion (31 December 2009)
country comparison to the world: 86
$3.625 billion (31 December 2008)
$10.13 billion (31 December 2007)
Agriculture—products: grain, potatoes, sugar beets, flax, vegetables; beef, milk, eggs; fish
Industries: metal-cutting machine tools, electric motors, television sets, refrigerators and freezers, petroleum refining, shipbuilding (small ships), furniture making, textiles, food processing, fertilizers, agricultural machinery, optical equipment, electronic components, computers, amber jewelry
Industrial production growth rate: 2.5% (2010 est.)
country comparison to the world: 122
Electricity—production: 12.09 billion kWh (2007 est.)
country comparison to the world: 85
Electricity—consumption: 9.612 billion kWh (2007 est.)
country comparison to the world: 89
Electricity—exports: 6.606 billion kWh (2008 est.)
Electricity—imports: 5.649 billion kWh (2008 est.)
Oil—production: 6,333 bbl/day (2009 est.)
country comparison to the world: 91
Oil—consumption: 74,000 bbl/day (2009 est.)
country comparison to the world: 89
Oil—exports: 137,200 bbl/day (2007 est.)
country comparison to the world: 60
Oil—imports: 204,000 bbl/day (2007 est.)
country comparison to the world: 44
Oil—proved reserves: 12 million bbl (1 January 2010 est.)
country comparison to the world: 90
Natural gas—production: 0 cu m (2008 est.)
country comparison to the world: 149
Natural gas—consumption: 3.53 billion cu m (2008 est.)
country comparison to the world: 69
Natural gas—exports: 0 cu m (2008 est.)
country comparison to the world: 129
Natural gas—imports: 3.53 billion cu m (2008 est.)
country comparison to the world: 38
Natural gas—proved reserves: 0 cu m (1 January 2010 est.)
country comparison to the world: 154
Current account balance: $1.231 billion (2010 est.)
country comparison to the world: 44
$1.492 billion (2009 est.)
Exports: $19.29 billion (2010 est.)
country comparison to the world: 70
$16.48 billion (2009 est.)
Exports—commodities: mineral products 22%, machinery and equipment 10%, chemicals 9%, textiles 7%, foodstuffs 7%, plastics 7%
Exports—partners: Russia 13.2%, Latvia 10%, Germany 9.6%, Poland 7.1%, Estonia 7.1%, Belarus 4.7%, UK 4.3% (2009 est.)
Imports: $20.34 billion (2010 est.)
country comparison to the world: 71
$17.56 billion (2009 est.)
Imports—commodities: mineral products, machinery and equipment, transport equipment, chemicals, textiles and clothing, metals
Imports—partners: Russia 30.1%, Germany 11.1%, Poland 9.9%, Latvia 6.3% (2009 est.)
Reserves of foreign exchange and gold: $6.418 billion (31 December 2010 est.)
country comparison to the world: 65
$6.66 billion (31 December 2009 est.)
Debt—external: $27.6 billion (31 December 2010 est.)
country comparison to the world: 65
$28.69 billion (31 December 2009 est.)
Stock of direct foreign investment—at home: $14.11 billion (31 December 2010 est.)
country comparison to the world: 76
$13.81 billion (31 December 2009 est.)
Stock of direct foreign investment—abroad: $2.507 billion (31 December 2010 est.)
country comparison to the world: 64
$2.307 billion (31 December 2009 est.)
Exchange rates: litai (LTL) per US dollar—2.6637 (2010), 2.4787 (2009), 2.3251 (2008), 2.5362 (2007), 2.7498 (2006)

COMMUNICATIONS

Telephones—main lines in use: 747,400 (2009)
country comparison to the world: 90
Telephones—mobile cellular: 4.962 million (2009)
country comparison to the world: 96
Telephone system: general assessment: adequate; being modernized to provide improved international capability and better residential access
domestic: rapid expansion of mobile-cellular services has resulted in a steady decline in the number of fixed-line connections; mobile-cellular teledensity stands at about 140 per 100 persons
international: country code—370; major international connections to Denmark, Sweden, and Norway by submarine cable for further transmission by satellite; landline connections to Latvia and Poland (2008)
Broadcast media: public broadcaster operates 3 channels with the third channel—a satellite channel—introduced in 2007; various privately-owned commercial TV broadcasters operate national and multiple regional channels; large number of privately-owned local TV stations; multi-channel cable and satellite TV services are available; publicly-owned broadcaster operates 3 radio networks; large number of privately-owned commercial broadcasters, many with repeater stations in various regions throughout the country (2007)
Internet country code: .lt
Internet hosts: 1.17 million (2010)
country comparison to the world: 40
Internet users: 1.964 million (2009)
country comparison to the world: 75

TRANSPORTATION

Airports: 81 (2010)
country comparison to the world: 69
Airports—with paved runways: *total:* 26
over 3,047 m: 3
2,438 to 3,047 m: 1
1,524 to 2,437 m: 7
914 to 1,523 m: 2
under 914 m: 13 (2010)
Airports—with unpaved runways: *total:* 55
over 3,047 m: 1
914 to 1,523 m: 3
under 914 m: 51 (2010)
Pipelines: gas 1,695 km; refined products 114 km (2010)
Railways: *total:* 1,767 km
country comparison to the world: 76
broad gauge: 1,745 km 1.524-m gauge (122 km electrified)
standard gauge: 22 km 1.435-m gauge (2010)
Roadways: *total:* 81,030 km
country comparison to the world: 58
paved: 71,563 km (includes 309 km of expressways)
unpaved: 9,467 km (2008)
Waterways: 441 km (navigable year round) (2007)
country comparison to the world: 87
Merchant marine: *total:* 42
country comparison to the world: 75
by type: cargo 22, container 1, passenger/cargo 6, refrigerated cargo 11, roll on/roll off 2
foreign-owned: 8 (Denmark 8)
registered in other countries: 29 (Antigua and Barbuda 4, Belize 2, Comoros 3, Cook Islands 2, Norway 1, Panama 4, Saint Vincent and the Grenadines 10, unknown 3) (2010)
Ports and terminals: Butinge, Klaipeda

MILITARY

Military branches: Ground Forces, Naval Forces, Air Forces (Karines Oro Pajegos, KOP), National Defense Volunteer Forces (2010)
Military service age and obligation: 19-26 years of age for compulsory military service; 18 years of age for volunteers; 12-month conscript service obligation; male registration required at age 16 (2009)
Manpower available for military service: *males age 16–49:* 890,074
females age 16–49: 875,780 (2010 est.)
Manpower fit for military service: *males age 16–49:* 669,111
females age 16–49: 724,803 (2010 est.)
Manpower reaching militarily significant age annually: *male:* 20,425
female: 19,527 (2010 est.)
Military expenditures: 1.2% of GDP (2007 est.)
country comparison to the world: 119

TRANSNATIONAL ISSUES

Disputes—international: Lithuania and Russia committed to demarcating their boundary in 2006 in accordance with the land and maritime treaty ratified by Russia in May 2003 and by Lithuania in 1999; Lithuania operates a simplified transit regime for Russian nationals traveling from the Kaliningrad coastal exclave into Russia, while still conforming,

as a EU member state having an external border with a non-EU member, to strict Schengen border rules; boundary demarcated with Latvia and Lithuania; as of January 2007, ground demarcation of the boundary with Belarus was complete and mapped with final ratification documents in preparation
Illicit drugs: transshipment and destination point for cannabis, cocaine, ecstasy, and opiates from Southwest Asia, Latin America, Western Europe, and neighboring Baltic countries; growing production of high-quality amphetamines, but limited production of cannabis, methamphetamines; susceptible to money laundering despite changes to banking legislation

LUXEMBOURG

INTRODUCTION

Background: Founded in 963, Luxembourg became a grand duchy in 1815 and an independent state under the Netherlands. It lost more than half of its territory to Belgium in 1839 but gained a larger measure of autonomy. Full independence was attained in 1867. Overrun by Germany in both world wars, it ended its neutrality in 1948 when it entered into the Benelux Customs Union and when it joined NATO the following year. In 1957, Luxembourg became one of the six founding countries of the European Economic Community (later the European Union), and in 1999 it joined the euro currency area.

GEOGRAPHY

Location: Western Europe, between France and Germany
Geographic coordinates: 49 45 N, 6 10 E
Map references: Europe
Area: *total:* 2,586 sq km
country comparison to the world: 178
land: 2,586 sq km
water: 0 sq km
Area—comparative: slightly smaller than Rhode Island
Land boundaries: total: 359 km
border countries: Belgium 148 km, France 73 km, Germany 138 km
Coastline: 0 km (landlocked)
Maritime claims: none (landlocked)
Climate: modified continental with mild winters, cool summers
Terrain: mostly gently rolling uplands with broad, shallow valleys; uplands to slightly mountainous in the north; steep slope down to Moselle flood plain in the southeast
Elevation extremes: *lowest point:* Moselle River 133 m
highest point: Buurgplaatz 559 m
Natural resources: iron ore (no longer exploited), arable land
Land use: arable land: 27.42%
permanent crops: 0.69%
other: 71.89% (includes Belgium) (2005)
Irrigated land: NA
Total renewable water resources: 1.6 cu km (2005)
Freshwater withdrawal (domestic/industrial/agricultural): *total:* 0.06 cu km/yr (42%/45%/13%)
per capita: 121 cu m/yr (1999)
Natural hazards: NA
Environment—current issues: air and water pollution in urban areas, soil pollution of farmland
Environment—international agreements: *party to:* Air Pollution, Air Pollution-Nitrogen Oxides, Air Pollution-Persistent Organic Pollutants, Air Pollution-Sulfur 85, Air Pollution-Sulfur 94, Air Pollution-Volatile Organic Compounds, Biodiversity, Climate Change, Climate Change-Kyoto Protocol, Desertification, Endangered Species, Hazardous Wastes, Law of the Sea, Marine Dumping, Ozone Layer Protection, Ship Pollution, Tropical Timber 83, Tropical Timber 94, Wetlands
signed, but not ratified: Environmental Modification
Geography—note: landlocked; the only Grand Duchy in the world

PEOPLE

Population: 503,302 (July 2011 est.)
country comparison to the world: 171
Age structure: *0–14 years:* 18.2% (male 47,274/female 44,366)
15–64 years: 66.9% (male 169,343/female 167,211)
65 years and over: 14.9% (male 31,086/female 44,022) (2011 est.)
Median age: *total:* 39.4 years
male: 38.4 years
female: 40.4 years (2011 est.)
Population growth rate: 1.145% (2011 est.)
country comparison to the world: 102
Birth rate: 11.69 births/1,000 population (2011 est.)
country comparison to the world: 166
Death rate: 8.48 deaths/1,000 population (July 2011 est.)
country comparison to the world: 85
Net migration rate: 8.24 migrant(s)/1,000 population (2011 est.)
country comparison to the world: 11
Urbanization: *urban population:* 85% of total population (2010)
rate of urbanization: 1.4% annual rate of change (2010-15 est.)
Major cities—population: LUXEMBOURG (capital) 90,000 (2009)
Sex ratio: *at birth:* 1.066 male(s)/female
under 15 years: 1.07 male(s)/female
15–64 years: 1.01 male(s)/female
65 years and over: 0.7 male(s)/female
total population: 0.97 male(s)/female (2011 est.)
Infant mortality rate:
total: 4.44 deaths/1,000 live births
country comparison to the world: 192
male: 4.46 deaths/1,000 live births
female: 4.42 deaths/1,000 live births (2011 est.)
Life expectancy at birth: *total population:* 79.61 years
country comparison to the world: 36
male: 76.36 years
female: 83.08 years (2011 est.)
Total fertility rate: 1.77 children born/woman (2011 est.)
country comparison to the world: 159
HIV/AIDS—adult prevalence rate: 0.3% (2009 est.)
country comparison to the world: 89
HIV/AIDS—people living with HIV/AIDS: fewer than 1,000 (2009 est.)
country comparison to the world: 146
HIV/AIDS—deaths: fewer than 100 (2009 est.)
country comparison to the world: 140
Drinking water source: *Improved:*
urban: 100% of population
rural: 100% of population
total: 100% of population (2008)
Sanitation facility access: *Improved:*
urban: 100% of population
rural: 100% of population
total: 100% of population (2008)
Nationality: noun: Luxembourger(s)
adjective: Luxembourg
Ethnic groups: Luxembourger 63.1%, Portuguese 13.3%, French 4.5%, Italian 4.3%, German 2.3%, other EU 7.3%, other 5.2% (2000 census)
Religions: Roman Catholic 87%, other (includes Protestant, Jewish, and Muslim) 13% (2000)
Languages: Luxembourgish (national language), German (administrative language), French (administrative language)
Literacy: *definition:* age 15 and over can read and write
total population: 100%
male: 100%
female: 100% (2000 est.)
School life expectancy (primary to tertiary education): *total:* 13 years
male: 13 years
female: 13 years (2006)
Education expenditures: NA

GOVERNMENT

Country name: *conventional long form:* Grand Duchy of Luxembourg
conventional short form: Luxembourg
local long form: Grand Duche de Luxembourg
local short form: Luxembourg
Government type: constitutional monarchy

Capital: *name:* Luxembourg
geographic coordinates: 49 36 N, 6 07 E
time difference: UTC+1 (6 hours ahead of Washington, DC during Standard Time)
daylight saving time: +1hr, begins last Sunday in March; ends last Sunday in October
Administrative divisions: 3 districts; Diekirch, Grevenmacher, Luxembourg
Independence: 1839 (from the Netherlands)
National holiday: National Day (Birthday of Grand Duchess Charlotte) 23 June; note—the actual date of birth was 23 January 1896, but the festivities were shifted by five months to allow observance during a more favorable time of year
Constitution: 17 October 1868; occasional revisions
Legal system: civil law system
International law organization participation: accepts compulsory ICJ jurisdiction; accepts ICCt jurisdiction
Suffrage: 18 years of age; universal and compulsory
Executive branch: *chief of state:* Grand Duke HENRI (since 7 October 2000); Heir Apparent Prince GUILLAUME (son of the monarch, born 11 November 1981)
head of government: Prime Minister Jean-Claude JUNCKER (since 20 January 1995); Deputy Prime Minister Jean ASSELBORN (since 31 July 2004)
cabinet: Council of Ministers recommended by the prime minister and appointed by the monarch
(For more information visit the World Leaders website)
elections: the monarchy is hereditary; following popular elections to the Chamber of Deputies, the leader of the majority party or the leader of the majority coalition usually appointed prime minister by the monarch; the deputy prime minister appointed by the monarch; they are responsible to the Chamber of Deputies
note: government coalition—CSV and LSAP
Legislative branch: unicameral Chamber of Deputies or Chambre des Deputes (60 seats; members elected by popular vote to serve five-year terms)
elections: last held on 7 June 2009 (next to be held by June 2014)
election results: percent of vote by party—CSV 38%, LSAP 21.6%, DP 15%, Green Party 11.7%, ADR 8.1%, The Left 3.3%, other 2.3%; seats by party—CSV 26, LSAP 13, DP 9, Green Party 7, ADR 4, The Left 1
note: there is also a Council of State that serves as an advisory body to the Chamber of Deputies; the Council of State has 21 members appointed by the Grand Duke on the advice of the prime minister
Judicial branch: judicial courts and tribunals (three Justices of the Peace, two district courts, and one Supreme Court of Appeals); administrative courts and tribunals (State Prosecutor's Office, administrative courts and tribunals, and the Constitutional Court); judges for all courts are appointed for life by the monarch
Political parties and leaders: Alternative Democratic Reform Party or ADR [Robert MEHLEN]; Christian Social People's Party or CSV [Michel WOLTER]; dei Lenk/la Gauche (the Left); Democratic Party or DP [Claude MEISCH]; Green Party [Francois BAUSCH]; Luxembourg Socialist Workers' Party or LSAP [Alex BODRY]; other minor parties
Political pressure groups and leaders: ABBL (bankers' association); ALEBA (financial sector trade union); Centrale Paysanne (federation of agricultural producers); CEP (professional sector chamber); CGFP (trade union representing civil service); Chambre de Commerce (Chamber of Commerce); Chambre des Metiers (Chamber of Artisans); FEDIL (federation of industrialists); Greenpeace (environment protection); LCGP (center-right trade union); Mouvement Ecologique (protection of ecology); OGBL (center-left trade union)
International organization participation: ADB (nonregional member), Australia Group, Benelux, CE, EAPC, EBRD, EIB, EMU, ESA, EU, FAO, FATF, IAEA, IBRD, ICAO, ICC, ICRM, IDA, IEA, IFAD, IFC, IFRCS, ILO, IMF, IMO, Interpol, IOC, IOM, IPU, ISO, ITSO, ITU, ITUC, MIGA, NATO, NEA, NSG, OAS (observer), OECD, OIF, OPCW, OSCE, PCA, Schengen Convention, UN, UNCTAD, UNESCO, UNHCR, UNIDO, UPU, WCO, WHO, WIPO, WMO, WTO, ZC
Diplomatic representation in the US: *chief of mission:* Ambassador Jean-Paul SENNINGER
chancery: 2200 Massachusetts Avenue NW, Washington, DC 20008
telephone: [1] (202) 265-4171 through 72
FAX: [1] (202) 328-8270
consulate(s) general: New York, San Francisco
Diplomatic representation from the US: *chief of mission:* Ambassador Cynthia STROUM
embassy: 22 Boulevard Emmanuel Servais, L-2535 Luxembourg City
mailing address: American Embassy Luxembourg, Unit 1410, APO AE 09126-1410 (official mail); American Embassy Luxembourg, PSC 9, Box 9500, APO AE 09123 (personal mail)
telephone: [352] 46 01 23
FAX: [352] 46 14 01
Flag description: three equal horizontal bands of red (top), white, and light blue; similar to the flag of the Netherlands, which uses a darker blue and is shorter; the coloring is derived from the Grand Duke's coat of arms (a red lion on a white and blue striped field)
National anthem: *name:* "Ons Heemecht" (Our Motherland); "De Wilhelmus" (The William)
lyrics/music: Michel LENTZ/Jean-Antoine ZINNEN; Nikolaus WELTER/unknown
note: "Ons Heemecht," adopted 1864, is the national anthem, while "De Wilhelmus," adopted 1919, serves as a royal anthem for use when members of the grand ducal family enter or exit a ceremony in Luxembourg

ECONOMY

Economy—overview: This small, stable, high-income economy—benefiting from its proximity to France, Belgium, and Germany—has historically featured solid growth, low inflation, and low unemployment. The industrial sector, initially dominated by steel, has become increasingly diversified to include chemicals, rubber, and other products. Growth in the financial sector, which now accounts for about 28% of GDP, has more than compensated for the decline in steel. Most banks are foreign owned and have extensive foreign dealings, but Luxembourg has lost some of its advantages as a tax haven because of OECD and EU pressure. The economy depends on foreign and cross-border workers for about 60% of its labor force. Luxembourg, like all EU members, suffered from the global economic crisis that began in late 2008, but unemployment has trended below the EU average. Following strong expansion from 2004 to 2007, Luxembourg's economy contracted and 3.7% in 2009, but rebounded 3.2% in 2010. The country continues to enjoy an extraordinarily high standard of living—GDP per capita ranks third in the world, after Liechtenstein and Qatar, and is the highest in the EU. Turmoil in the world financial markets and lower global demand during 2008-09 prompted the government to inject capital into the banking sector and implement stimulus measures to boost the economy. Government stimulus measures and support for the banking sector, however, led to a 5% government budget deficit in 2009. Nevertheless, the deficit was cut below 3% in 2010.
GDP (purchasing power parity): $41.09 billion (2010 est.)
country comparison to the world: 98
$39.74 billion (2009 est.)
$41.25 billion (2008 est.)
note: data are in 2010 US dollars
GDP (official exchange rate): $54.95 billion (2010 est.)
GDP—real growth rate: 3.4% (2010 est.)
country comparison to the world: 114
-3.7% (2009 est.)
1.4% (2008 est.)
GDP—per capita (PPP): $82,600 (2010 est.)
country comparison to the world: 3
$80,800 (2009 est.)
$84,900 (2008 est.)
note: data are in 2010 US dollars
GDP—composition by sector: *agriculture:* 0.4%
industry: 13.6%
services: 86% (2007 est.)
Labor force: 206,000
country comparison to the world: 168
note: 125,400 workers commute daily from France, Belgium, and Germany (2010 est.)
Labor force—by occupation: *agriculture:* 2.2%
industry: 17.2%
services: 80.6% (2007 est.)
Unemployment rate: 5.5% (2010 est.)
country comparison to the world: 52
5.7% (2009 est.)
Population below poverty line: NA%
Household income or consumption by percentage share: *lowest 10%:* 3.5%
highest 10%: 23.8% (2000)
Distribution of family income—Gini index: 26 (2005)

country comparison to the world: 132
Investment (gross fixed): 16.8% of GDP (2010 est.)
country comparison to the world: 120
Budget: *revenues:* $20.88 billion
expenditures: $22.11 billion (2010 est.)
Public debt: 16.2% of GDP (2010 est.)
country comparison to the world: 116
14.6% of GDP (2009 est.)
Inflation rate (consumer prices): 2.1% (2010 est.)
country comparison to the world: 54
0.4% (2009 est.)
Central bank discount rate: 1.75% (31 December 2010)
country comparison to the world: 124
1.75% (31 December 2009)
note: this is the European Central Bank's rate on the marginal lending facility, which offers overnight credit to banks in the euro area
Stock of narrow money: $120.8 billion (31 December 2010 est.)
country comparison to the world: 29
$121 billion (31 December 2009 est.)
note: see entry for the European Union for money supply in the euro area; the European Central Bank (ECB) controls monetary policy for the 17 members of the Economic and Monetary Union (EMU); individual members of the EMU do not control the quantity of money circulating within their own borders
Stock of broad money: $255.5 billion (31 December 2010 est.)
country comparison to the world: 35
$231.7 billion (31 December 2009 est.)
Stock of domestic credit: $395.1 billion (31 December 2009 est.)
country comparison to the world: 28
$369.6 billion (31 December 2008 est.)
Market value of publicly traded shares: $105.6 billion (31 December 2009)
country comparison to the world: 47
$66.46 billion (31 December 2008)
$166.1 billion (31 December 2007)
Agriculture—products: grapes, barley, oats, potatoes, wheat, fruits; dairy and livestock products
Industries: banking and financial services, iron and steel, information technology, telecommunications, cargo transportation, food processing, chemicals, metal products, engineering, tires, glass, aluminum, tourism
Industrial production growth rate: 1.7% (2009 est.)
country comparison to the world: 135
Electricity—production: 2.696 billion kWh (2007 est.)
country comparison to the world: 128
Electricity—consumption: 6.525 billion kWh (2007 est.)
country comparison to the world: 104
Electricity—exports: 2.483 billion kWh (2008 est.)
Electricity—imports: 6.83 billion kWh (2008 est.)
Oil—production: 0 bbl/day (2009 est.)
country comparison to the world: 197
Oil—consumption: 50,720 bbl/day (2009 est.)
country comparison to the world: 95
Oil—exports: 63 bbl/day (2008 est.)
country comparison to the world: 133
Oil—imports: 59,210 bbl/day (2008 est.)
country comparison to the world: 79
Oil—proved reserves: 0 bbl (1 January 2010 est.)
country comparison to the world: 155
Natural gas—production: 0 cu m (2008 est.)
country comparison to the world: 152
Natural gas—consumption: 1.268 billion cu m (2009 est.)
country comparison to the world: 85
Natural gas—exports: 0 cu m (2008 est.)
country comparison to the world: 132
Natural gas—imports: 1.263 billion cu m (2009 est.)
country comparison to the world: 50
Natural gas—proved reserves: 0 cu m (1 January 2010 est.)
country comparison to the world: 157
Current account balance: $3.396 billion (2010 est.)
country comparison to the world: 37
$2.985 billion (2009 est.)
Exports: $17.82 billion (2010 est.)
country comparison to the world: 71
$15.5 billion (2009 est.)
Exports—commodities: machinery and equipment, steel products, chemicals, rubber products, glass
Exports—partners: Germany 19.78%, France 15.87%, Belgium 11.07%, UK 7.96%, Italy 7.49%, Netherlands 4.31% (2009)
Imports: $23.67 billion (2010 est.)
country comparison to the world: 65
$19.76 billion (2009 est.)
Imports—commodities: minerals, metals, foodstuffs, quality consumer goods
Imports—partners: Belgium 27.22%, Germany 23.14%, China 18.62%, France 8.85%, Netherlands 5.06% (2009)
Reserves of foreign exchange and gold: $NA (31 December 2010 est.)
$810 million (31 December 2009 est.)
Debt—external: $1.892 trillion (30 June 2010)
country comparison to the world: 11
$2.02 trillion (31 December 2008)
Stock of direct foreign investment—at home: $NA (31 December 2009 est.)
$11.21 billion (31 December 2008 est.)
Stock of direct foreign investment—abroad: $NA
Exchange rates: euros (EUR) per US dollar–0.755 (2010), 0.7198 (2009), 0.6827 (2008), 0.7345 (2007), 0.7964 (2006)

COMMUNICATIONS

Telephones—main lines in use: 273,600 (2009)
country comparison to the world: 120
Telephones—mobile cellular: 719,000 (2009)
country comparison to the world: 152
Telephone system: general assessment: highly developed, completely automated and efficient system, mainly buried cables
domestic: fixed line teledensity over 50 per 100 persons; nationwide mobile-cellular telephone system with market for mobile-cellular phones virtually saturated
international: country code–352 (2008)
Broadcast media: Luxembourg has a long tradition of operating radio and TV services to pan-European audiences and is home to Europe's largest privately-owned broadcast media group, the RTL group, which operates 45 television stations and 31 radio stations in Europe; also home to Europe's largest satellite operator, Societe Europeenne des Satellites (SES); domestically, the RTL group operates TV and radio networks; other domestic private radio and TV operators and French and German stations are available; satellite and cable TV services are accessible (2008)
Internet country code: .lu
Internet hosts: 244,225 (2010)
country comparison to the world: 63
Internet users: 424,500 (2009)
country comparison to the world: 121

TRANSPORTATION

Airports: 2 (2010)
country comparison to the world: 201
Airports—with paved runways: *total:* 1
over 3,047 m: 1 (2010)
Airports—with unpaved runways: *total:* 1
under 914 m: 1 (2010)
Heliports: 1 (2010)
Pipelines: gas 142 km; refined products 27 km (2010)
Railways: *total:* 275 km
country comparison to the world: 123
standard gauge: 275 km 1.435-m gauge (243 km electrified) (2010)
Roadways: *total:* 5,227 km
country comparison to the world: 152
paved: 5,227 km (includes 147 km of expressways) (2008)
Waterways: 37 km (on Moselle River) (2010)
country comparison to the world: 105
Merchant marine: *total:* 47
country comparison to the world: 72
by type: bulk carrier 3, cargo 3, chemical tanker 16, container 10, passenger 3, petroleum tanker 2, roll on/roll off 10
foreign-owned: 45 (Belgium 9, France 16, Germany 9, Netherlands 2, Switzerland 1, UK 5, US 3)
registered in other countries: 16 (Italy 12, Malta 3, Panama 1) (2010)
Ports and terminals: Mertert

MILITARY

Military branches: Army (2010)
Military service age and obligation: 17-25 years of age for male and female voluntary military service; soldiers under 18 are not deployed into combat or with peacekeeping missions; no conscription; Luxembourg citizen or EU citizen with 3-year residence in Luxembourg (2010)
Manpower available for military service:
males age 16–49: 118,665
females age 16–49: 117,456 (2010 est.)
Manpower fit for military service: *males age 16–49:* 97,290
females age 16–49: 96,361 (2010 est.)
Manpower reaching militarily significant age annually: *male:* 3,263
female: 3,084 (2010 est.)
Military expenditures: 0.9% of GDP (2005 est.)
country comparison to the world: 140

TRANSNATIONAL ISSUES

Disputes—international: none

MACAU

INTRODUCTION

Background: Colonized by the Portuguese in the 16th century, Macau was the first European settlement in the Far East. Pursuant to an agreement signed by China and Portugal on 13 April 1987, Macau became the Macau Special Administrative Region (SAR) of the People's Republic of China on 20 December 1999. In this agreement, China promised that, under its "one country, two systems" formula, China's socialist economic system would not be practiced in Macau, and that Macau would enjoy a high degree of autonomy in all matters except foreign and defense affairs for the next 50 years.

GEOGRAPHY

Location: Eastern Asia, bordering the South China Sea and China
Geographic coordinates: 22 10 N, 113 33 E
Map references: Southeast Asia
Area: *total:* 28.2 sq km
country comparison to the world: 235
land: 28.2 sq km
water: 0 sq km
Area—comparative: less than one-sixth the size of Washington, DC
Land boundaries: *total:* 0.34 km
regional border: China 0.34 km
Coastline: 41 km
Maritime claims: not specified
Climate: subtropical; marine with cool winters, warm summers
Terrain: generally flat
Elevation extremes: *lowest point:* South China Sea 0 m
highest point: Coloane Alto 172 m
Natural resources: NEGL
Land use: *arable land:* 0%
permanent crops: 0%
other: 100% (2005)
Irrigated land: NA
Natural hazards: typhoons
Environment—current issues: NA
Environment—international agreements: *party to:* Marine Dumping (associate member), Ship Pollution (associate member)
Geography—note: essentially urban; an area of land reclaimed from the sea measuring 5.2 sq km and known as Cotai now connects the islands of Coloane and Taipa; the island area is connected to the mainland peninsula by three bridges

PEOPLE

Population: 573,003 (July 2011 est.)
country comparison to the world: 167
Age structure: *0–14 years:* 15% (male 45,635/female 40,523)
15–64 years: 76.8% (male 205,998/female 233,820)
65 years and over: 8.2% (male 22,043/female 24,984) (2011 est.)
Median age: *total:* 36.2 years
male: 36.8 years
female: 35.6 years (2011 est.)
Population growth rate: 0.879% (2011 est.)
country comparison to the world: 127
Birth rate: 9.03 births/1,000 population (2011 est.)
country comparison to the world: 209
Death rate: 3.72 deaths/1,000 population (July 2011 est.)
country comparison to the world: 208
Net migration rate: 3.49 migrant(s)/1,000 population (2011 est.)
country comparison to the world: 26
Urbanization: *urban population:* 100% of total population (2010)
rate of urbanization: 0.7% annual rate of change (2010-15 est.)
Sex ratio: *at birth:* 1.05 male(s)/female
under 15 years: 1.14 male(s)/female
15–64 years: 0.88 male(s)/female
65 years and over: 0.88 male(s)/female
total population: 0.92 male(s)/female (2011 est.)
Infant mortality rate: *total:* 3.18 deaths/1,000 live births
country comparison to the world: 216
male: 3.34 deaths/1,000 live births
female: 3.02 deaths/1,000 live births (2011 est.)
Life expectancy at birth: *total population:* 84.41 years
country comparison to the world: 2
male: 81.45 years
female: 87.52 years (2011 est.)
Total fertility rate: 0.92 children born/woman (2011 est.)
country comparison to the world: 223
HIV/AIDS—adult prevalence rate: NA
HIV/AIDS—people living with HIV/AIDS: NA
HIV/AIDS—deaths: NA
Nationality: *noun:* Chinese
adjective: Chinese
Ethnic groups: Chinese 94.3%, other 5.7% (includes Macanese–mixed Portuguese and Asian ancestry) (2006 census)
Religions: Buddhist 50%, Roman Catholic 15%, none or other 35% (1997 est.)
Languages: Cantonese 85.7%, Hokkien 4%, Mandarin 3.2%, other Chinese dialects 2.7%, English 1.5%, Tagalog 1.3%, other 1.6% (2001 census)
Literacy: *definition:* age 15 and over can read and write
total population: 91.3%
male: 95.3%
female: 87.8% (2001 census)
School life expectancy (primary to tertiary education): *total:* 14 years
male: 15 years
female: 14 years (2009)
Education expenditures: 2.2% of GDP (2008)
country comparison to the world: 153

GOVERNMENT

Country name: *conventional long form:* Macau Special Administrative Region
conventional short form: Macau
official long form: Aomen Tebie Xingzhengqu (Chinese); Regiao Administrativa Especial de Macau (Portuguese)
official short form: Aomen (Chinese); Macau (Portuguese)
Dependency status: special administrative region of China
Government type: limited democracy
Administrative divisions: none (special administrative region of the People's Republic of China)
Independence: none (special administrative region of China)
National holiday: National Day (Anniversary of the Founding of the People's Republic of China), 1 October (1949); note—20 December 1999 is celebrated as Macau Special Administrative Region Establishment Day
Constitution: Basic Law, approved 31 March 1993 by China's National People's Congress, is Macau's charter
Legal system: civil law system based on the Portuguese model
Suffrage: direct election 18 years of age for some non-executive positions, universal for permanent residents living in Macau for the past seven years; indirect election limited to organizations registered as "corporate voters" (257 are currently registered) and a 300-member Election Committee drawn from broad regional groupings, municipal organizations, and central government bodies
Executive branch: *chief of state:* President of China HU Jintao (since 15 March 2003)
head of government: Chief Executive Fernando CHUI Sai-on (since 20 December 2009)
cabinet: Executive Council consists of 1 government secretary, 3 legislators, 4 businessmen, 1 pro-Beijing unionist, and 1 pro-Beijing educator
(For more information visit the World Leaders website)
elections: chief executive chosen by a 300-member Election Committee for a five-year term (eligible for a second term); election last held on 26 July 2009 (next to be held in July 2014)
election results: Fernando CHUI Sai-on elected in 2009 with 282 votes, took office on 20 December 2009
Legislative branch: unicameral Legislative Assembly (29 seats; 12 members elected by popular vote, 10 by indirect vote, and 7 appointed by the chief executive; members to serve four-year terms)
elections: last held on 20 September 2009 (next to be held in September 2013)
election results: percent of vote—UPD 14.9%, ACUM 12%, APMD 11.6%, NUDM 9.9%, UPP 9.9%, ANMD 7.8%, UMG 7.3%, MUDAR 5.5%, others 21.1%; seats by

political group—UPD 2, ACUM 2, APMD 2, NUMD 1, UPP 1, ANMD 1, UMG 1, MUDAR 1; 10 seats filled by professional and business groups; 7 members appointed by the chief executive
Judicial branch: Court of Final Appeal in Macau Special Administrative Region
Political parties and leaders: Alliance for Change or MUDAR; Macau Development Alliance or NUDM [Angela LEONG On-kei]; Macau-Guangdong Union or UNG; Macau United Citizens' Association or ACUM [CHAN Meng-kam]; New Democratic Macau Association or APMD [Antonio NG Kuok-cheong]; New Hope or NE [Jose Maria Pereira COUTINHO]; Union for Promoting Progress or UPP [LEONG Heng-teng]
note: there is no political party ordinance, so there are no registered political parties; politically active groups register as societies or companies
Political pressure groups and leaders: Civic Power [Agnes LAM Iok-fong]; Macau New Chinese Youth Association [LEONG Sin-man]; Macau Society of Tourism and Entertainment or STDM [Stanley HO]; Macau Worker's Union [HO Heng-kuok]; Union for Democracy Development [Antonio NG Kuok-cheong]
International organization participation: IHO, IMF, IMO (associate), Interpol (subbureau), ISO (correspondent), UNESCO (associate), UNWTO (associate), UPU, WCO, WTO
Diplomatic representation in the US: none (special administrative region of China)
Diplomatic representation from the US: the US has no offices in Macau; US Consulate General in Hong Kong is accredited to Macau
Flag description: green with a lotus flower above a stylized bridge and water in white, beneath an arc of five gold, five-pointed stars: one large in the center of the arc and two smaller on either side; the lotus is the floral emblem of Macau, the three petals represent the peninsula and two islands that make up Macau; the five stars echo those on the flag of China
National anthem: *note:* as a Special Administrative Region of China, "Yiyonggjun Jinxingqu" is official (see China)

ECONOMY

Economy—overview: Macau's economy slowed dramatically in 2009 as a result of the global economic slowdown, but strong growth resumed in 2010, largely on the back of strong tourism and gaming sectors. After opening up its locally-controlled casino industry to foreign competition in 2001, the territory attracted tens of billions of dollars in foreign investment, transforming Macau into one of the world's largest gaming center. Macau's gaming and tourism businesses were fueled by China's decision to relax travel restrictions on Chinese citizens wishing to visit Macau. By 2006, Macau's gaming revenue surpassed that of the Las Vegas strip, and gaming-related taxes accounted for more than 70% of total government revenue. In 2008, Macau introduced measures to cool the rapidly developing sector. This city of nearly 552,300 hosted nearly 25 million visitors in 2010. Almost 53% came from mainland China. Macau's traditional manufacturing industry has virtually disappeared since the termination of the Multi-Fiber Agreement in 2005. In 2010, total exports were less than US$900 million, while gaming receipts were almost US$24 billion, a 58% increase over 2009. The Macau government plans to tighten control over the opening of new casinos and strengthen supervision of local casino operations in 2011 and has introduced measures to diversify the economy. The Closer Economic Partnership Agreement (CEPA) between Macau and mainland China that came into effect on 1 January 2004 offers Macau-made products tariff-free access to the mainland; nevertheless, China is Macau's second largest goods export market, behind Hong Kong, and followed by the United States. Macau's currency, the pataca, is closely tied to the Hong Kong dollar, which is also freely accepted in the territory.
GDP (purchasing power parity): $18.47 billion (2009 est.)
country comparison to the world: 128
$18.14 billion (2008 est.)
$14.4 billion (2006)
note: data are in 2010 US dollars
GDP (official exchange rate): $22.1 billion (2009 est.)
GDP—real growth rate: 1% (2009 est.)
country comparison to the world: 173
12.9% (2008)
26% (2007)
GDP—per capita (PPP): $33,000 (2009)
country comparison to the world: 40
$31,800 (2008)
$28,400 (2006)
GDP—composition by sector:
agriculture: 0.1%
industry: 2.8%
services: 97.1% (2009 est.)
Labor force: 330,900 (2010 est.)
country comparison to the world: 163
Labor force—by occupation:
manufacturing: 4.3%
construction: 8.7%
transport and communications: 5.5%
wholesale and retail trade: 13.3%
restaurants and hotels: 12.7%
gambling: 13.3%
public sector: 6.7%
financial services: 2.4%
other services and agriculture: 33.2% (2010)
Unemployment rate: 2.9% (2010)
country comparison to the world: 22
3.6% (2009)
Population below poverty line: NA%
Household income or consumption by percentage share:
lowest 10%: NA%
highest 10%: NA%
Budget: *revenues:* $9.95 billion
expenditures: $7.18 billion (2010)
Inflation rate (consumer prices): 2.8% (2010)
country comparison to the world: 77
1.2% (2009)
Commercial bank prime lending rate: 5.25% (31 December 2010 est.)
country comparison to the world: 144
5.25% (31 December 2009 est.)
Stock of narrow money: $4.34 billion (31 December 2009)
country comparison to the world: 99
$3.831 billion (31 December 2008)
Stock of broad money: $30.41 billion (31 December 2009 est.)
country comparison to the world: 74
$26.56 billion (31 December 2008)
Stock of domestic credit: $16.33 billion (31 December 2010 est.)
country comparison to the world: 86
$12.64 billion (31 December 2009 est.)
Market value of publicly traded shares: $46.1 billion (31 February 2011 est.)
country comparison to the world: 95
$2.3 billion (31 December 2008)
$413.1 million (2004 est.)
Agriculture—products: only 2% of land area is cultivated, mainly by vegetable growers; fishing, mostly for crustaceans, is important; some of the catch is exported to Hong Kong
Industries: tourism, gambling, clothing, textiles, electronics, footwear, toys
Industrial production growth rate: -23.7%
country comparison to the world: 167
Electricity—production: 1.1 billion kWh (2010 est.)
country comparison to the world: 143
Electricity—consumption: 3.66 billion kWh (2010 est.)
country comparison to the world: 119
Electricity—exports: 0 kWh (2010 est.)
Electricity—imports: 2.79 billion kWh (2010 est.)
Oil—production: 0 bbl/day (2010 est.)
country comparison to the world: 199
Oil—consumption: 6,490 bbl/day (2010 est.)
country comparison to the world: 158
Oil—exports: 0 bbl/day (2010 est.)
country comparison to the world: 182
Oil—imports: 6,284 bbl/day (2010 est.)
country comparison to the world: 152
Oil—proved reserves: 0 bbl (1 January 2011 est.)
country comparison to the world: 157
Natural gas—production: 0 cu m (2010 est.)
country comparison to the world: 154
Natural gas—consumption: 154.7 million cu m (2010)
country comparison to the world: 101
Natural gas—exports: 0 cu m (2010 est.)
country comparison to the world: 134
Natural gas—imports: 154.5 million cu m (2010 est.)
country comparison to the world: 65
Natural gas—proved reserves: 174,000 cu m (2010 est.)
country comparison to the world: 103
Current account balance: $6.23 billion (2009)
country comparison to the world: 31
Exports: $870 million (2010 est.)
country comparison to the world: 154
$950 million (2009 est.)
note: includes reexports
Exports—commodities: clothing, textiles, footwear, toys, electronics, machinery and parts

Exports—partners: Hong Kong 38.7%, US 17.9%, China 14.4%, Germany 4% (2009 est.)
Imports: $5.5 billion (2010 est.)
country comparison to the world: 108
$4.5 billion (2009 est.)
Imports—commodities: raw materials and semi-manufactured goods, consumer goods (foodstuffs, beverages, tobacco), capital goods, mineral fuels and oils
Imports—partners: China 31.1%, Hong Kong 10.8%, Japan 8.1%, France 8%, US 6.2% (2009 est.)
Reserves of foreign exchange and gold: $23.73 billion (2010)
country comparison to the world: 41
Debt—external: $0 (2010)
country comparison to the world: 196
Stock of direct foreign investment—at home: $12.1 billion (2008 est.)
country comparison to the world: 79
$10.5 billion (#REF! est.)
Stock of direct foreign investment—abroad: $240 million (2009 est.)
country comparison to the world: 80
$964 million (2008)
Exchange rates: patacas (MOP) per US dollar–8.002 (2010) 7.983 (2008) 8.011 (2007) 8.0015 (2006)

COMMUNICATIONS

Telephones—main lines in use: 168,903 (2010)
country comparison to the world: 132
Telephones—mobile cellular: 1.109 million (2010)
country comparison to the world: 144
Telephone system: *general assessment:* fairly modern communication facilities maintainedfordomesticandinternationalservices
domestic: termination of monopoly over mobile-cellular telephone services in 2001 spurred sharp increase in subscriptions with mobile-cellular teledensity approaching 200 per 100 persons in 2010; fixed-line subscribership appears to have peaked and is now in decline
international: country code–853; landing point for the SEA-ME-WE-3 submarine cable network that provides links to Asia, the Middle East, and Europe; HF radiotelephone communication facility; satellite earth station–1 Intelsat (Indian Ocean) (2010)
Broadcast media: local government dominates broadcast media; 2 television stations operated by the government with one broadcasting in Portuguese and the other in Cantonese and Mandarin; cable and satellite TV services are available; 3 radio stations broadcasting, of which 2 are government-operated (2010)
Internet country code: .mo
Internet hosts: 252 (2010)
country comparison to the world: 189
Internet users: 270,200 (2009)
country comparison to the world: 134

TRANSPORTATION

Airports: 1 (2010)
country comparison to the world: 224
Airports—with paved runways: *total:* 1
over 3,047 m: 1 (2010)
Heliports: 2 (2010)
Roadways: *total:* 413 km
country comparison to the world: 198
paved: 413 km (2009)
Ports and terminals: Macau

MILITARY

Military branches: no regular military forces
Manpower available for military service: *males age 16–49:* 150,780 (2010 est.)
Manpower fit for military service: *males age 16–49:* 124,189
females age 16–49: 149,514 (2010 est.)
Manpower reaching militarily significant age annually: *male:* 4,274
female: 3,674 (2010 est.)
Military—note: defense is the responsibility of China

TRANSNATIONAL ISSUES

Disputes—international: none
Illicit drugs: transshipment point for drugs going into mainland China; consumer of opiates and amphetamines

MACEDONIA

INTRODUCTION

Background: Macedonia gained its independence peacefully from Yugoslavia in 1991. Greece's objection to the new state's use of what it considered a Hellenic name and symbols delayed international recognition, which occurred under the provisional designation of "the Former Yugoslav Republic of Macedonia." In 1995, Greece lifted a 20-month trade embargo and the two countries agreed to normalize relations. The United States began referring to Macedonia by its constitutional name, Republic of Macedonia, in 2004 and negotiations continue between Greece and Macedonia to resolve the name issue. Some ethnic Albanians, angered by perceived political and economic inequities, launched an insurgency in 2001 that eventually won the support of the majority of Macedonia's Albanian population and led to the internationally-brokered Ohrid Framework Agreement, which ended the fighting by establishing a set of new laws enhancing the rights of minorities. Fully implementing the Framework Agreement and stimulating economic growth and development continue to be challenges for Macedonia, although progress has been made on both fronts over the past several years.

GEOGRAPHY

Location: Southeastern Europe, north of Greece
Geographic coordinates: 41 50 N, 22 00 E
Map references: Europe
Area: *total:* 25,713 sq km
country comparison to the world: 149
land: 25,433 sq km
water: 280 sq km
Area—comparative: slightly larger than Vermont
Land boundaries: *total:* 766 km
border countries: Albania 151 km, Bulgaria 148 km, Greece 246 km, Kosovo 159 km, Serbia 62 km
Coastline: 0 km (landlocked)
Maritime claims: none (landlocked)
Climate: warm, dry summers and autumns; relatively cold winters with heavy snowfall
Terrain: mountainous territory covered with deep basins and valleys; three large lakes, each divided by a frontier line; country bisected by the Vardar River
Elevation extremes: *lowest point:* Vardar River 50 m
highest point: Golem Korab (Maja e Korabit) 2,764 m
Natural resources: low-grade iron ore, copper, lead, zinc, chromite, manganese, nickel, tungsten, gold, silver, asbestos, gypsum, timber, arable land
Land use: *arable land:* 22.01%
permanent crops: 1.79%
other: 76.2% (2005)
Irrigated land:
550 sq km (2003)
Total renewable water resources: 6.4 cu km (2001)
Freshwater withdrawal (domestic/industrial/agricultural): *total:* 2.27
per capita: 1,118 cu m/yr (2000)
Natural hazards: high seismic risks
Environment—current issues: air pollution from metallurgical plants
Environment—international agreements: *party to:* Air Pollution, Biodiversity, Climate Change, Climate Change-Kyoto Protocol, Desertification, Endangered Species, Hazardous Wastes, Law of the Sea, Ozone Layer Protection, Wetlands
signed, but not ratified: none of the selected agreements
Geography—note: landlocked; major transportation corridor from Western and Central Europe to Aegean Sea and Southern Europe to Western Europe

PEOPLE

Population: 2,077,328 (July 2011 est.)
country comparison to the world: 143
Age structure: *0–14 years:* 18.5% (male 198,643/female 184,775)

15–64 years: 70% (male 733,601/female 720,103)
65 years and over: 11.6% (male 103,620/female 136,586) (2011 est.)
Median age: *total:* 35.8 years
male: 34.8 years
female: 36.9 years (2011 est.)
Population growth rate: 0.248% (2011 est.)
country comparison to the world: 174
Birth rate: 11.87 births/1,000 population (2011 est.)
country comparison to the world: 165
Death rate: 8.91 deaths/1,000 population (July 2011 est.)
country comparison to the world: 73
Net migration rate: -0.48 migrant(s)/1,000 population (2011 est.)
country comparison to the world: 136
Urbanization: *urban population:* 59% of total population (2010)
rate of urbanization: 0.3% annual rate of change (2010-15 est.)
Major cities—population: SKOPJE (capital) 480,000 (2009)
Sex ratio: *at birth:* 1.077 male(s)/female
under 15 years: 1.08 male(s)/female
15–64 years: 1.02 male(s)/female
65 years and over: 0.76 male(s)/female
total population: 1 male(s)/female (2011 est.)
Infant mortality rate: *total:* 8.54 deaths/1,000 live births
country comparison to the world: 157
male: 8.76 deaths/1,000 live births
female: 8.3 deaths/1,000 live births (2011 est.)
Life expectancy at birth: *total population:* 75.14 years
country comparison to the world: 88
male: 72.61 years
female: 77.87 years (2011 est.)
Total fertility rate: 1.58 children born/woman (2011 est.)
country comparison to the world: 177
HIV/AIDS—adult prevalence rate: less than 0.1% (2007 est.)
country comparison to the world: 146
HIV/AIDS—people living with HIV/AIDS: fewer than 200 (2007 est.)
country comparison to the world: 159
HIV/AIDS—deaths: fewer than 100 (2003 est.)
country comparison to the world: 142
Drinking water source: *Improved:*
urban: 100% of population
rural: 99% of population
total: 100% of population
Unimproved:
urban: 0% of population
rural: 1% of population
total: 0% of population (2008)
Sanitation facility access: *Improved:*
urban: 92% of population
rural: 82% of population
total: 89% of population
Unimproved:
urban: 8% of population
rural: 18% of population
total: 11% of population (2008)
Nationality: *noun:* Macedonian(s)
adjective: Macedonian
Ethnic groups: Macedonian 64.2%, Albanian 25.2%, Turkish 3.9%, Roma (Gypsy) 2.7%, Serb 1.8%, other 2.2% (2002 census)
Religions: Macedonian Orthodox 64.7%, Muslim 33.3%, other Christian 0.37%, other and unspecified 1.63% (2002 census)
Languages: Macedonian (official) 66.5%, Albanian (official) 25.1%, Turkish 3.5%, Roma 1.9%, Serbian 1.2%, other 1.8% (2002 census)
Literacy: *definition:* age 15 and over can read and write
total population: 96.1%
male: 98.2%
female: 94.1% (2002 census)
School life expectancy (primary to tertiary education): *total:* 13 years
male: 13 years
female: 13 years (2008)
Education expenditures: NA

GOVERNMENT

Country name: *conventional long form:* Republic of Macedonia
conventional short form: Macedonia
local long form: Republika Makedonija
local short form: Makedonija
note: the provisional designation used by the UN, EU, and NATO is the "former Yugoslav Republic of Macedonia" (FYROM)
former: People's Republic of Macedonia, Socialist Republic of Macedonia
Government type: parliamentary democracy
Capital: *name:* Skopje
geographic coordinates: 42 00 N, 21 26 E
time difference: UTC+1 (6 hours ahead of Washington, DC during Standard Time)
daylight saving time: +1hr, begins last Sunday in March; ends last Sunday in October
Administrative divisions: 84 municipalities (opstini, singular–opstina); Aerodrom (Skopje), Aracinovo, Berovo, Bitola, Bogdanci, Bogovinje, Bosilovo, Brvenica, Butel (Skopje), Cair (Skopje), Caska, Centar (Skopje), Centar Zupa, Cesinovo, Cucer Sandevo, Debar, Debarca, Delcevo, Demir Hisar, Demir Kapija, Dojran, Dolneni, Dorce Petrov (Gjorce Petrov) (Skopje), Drugovo, Gazi Baba (Skopje), Gevgelija, Gostivar, Gradsko, Ilinden, Jegunovce, Karbinci, Karpos (Skopje), Kavadarci, Kicevo, Kisela Voda (Skopje), Kocani, Konce, Kratovo, Kriva Palanka, Krivogastani, Krusevo, Kumanovo, Lipkovo, Lozovo, Makedonska Kamenica, Makedonski Brod, Mavrovo i Rostusa, Mogila, Negotino, Novaci, Novo Selo, Ohrid, Oslomej, Pehcevo, Petrovec, Plasnica, Prilep, Probistip, Radovis, Rankovce, Resen, Rosoman, Saraj (Skopje), Sopiste, Staro Nagoricane, Stip, Struga, Strumica, Studenicani, Suto Orizari (Skopje), Sveti Nikole, Tearce, Tetovo, Valandovo, Vasilevo, Veles, Vevcani, Vinica, Vranestica, Vrapciste, Zajas, Zelenikovo, Zelino, Zrnovci
note: the 10 municipalities followed by Skopje in parentheses collectively constitute the larger Skopje Municipality
Independence: 8 September 1991 (referendum by registered voters endorsed independence from Yugoslavia)
National holiday: Independence Day, 8 September (1991); also known as National Day
Constitution: adopted 17 November 1991, effective 20 November 1991; amended November 2001, 2005 and in 2009
note: amended November 2001 by a series of new constitutional amendments strengthening minority rights, in 2005 with amendments related to the judiciary, and in 2009 with amendments related to the threshold required to elect the president
Legal system: civil law system; judicial review of legislative acts
International law organization participation: has not submitted an ICJ jurisdiction declaration; accepts ICCt jurisdiction
Suffrage: 18 years of age; universal
Executive branch: *chief of state:* President Gjorge IVANOV (since 12 May 2009)
head of government: Prime Minister Nikola GRUEVSKI (since 26 August 2006)
cabinet: Council of Ministers elected by the majority vote of all the deputies in the Assembly; note–current cabinet formed by the government coalition parties VMRO-DPMNE, BDI/DUI, and several small parties
(For more information visit the World Leaders website)
elections: president elected by popular vote for a five-year term (eligible for a second term); two-round election: first round held on 22 March 2009, second round held on 5 April 2009 (next to be held in March 2014); prime minister elected by the Assembly following legislative elections
election results: Gjorge IVANOV elected president on second-round ballot; percent of vote–Gjorge IVANOV 63.1%, Ljubomir FRCKOSKI 36.9%
Legislative branch: unicameral Assembly or Sobranie (120 seats; members elected by popular vote from party lists based on the percentage of the overall vote the parties gain in each of six electoral districts; members serve four-year terms)
elections: last held on 1 June and 15 June 2008 (next to be held by July 2012)
election results: percent of vote by party–VMRO-DPMNE-led block 49%, SDSM-led block 24%, BDI/DUI 13%, PDSh/DPA 8%, other 6%; seats by party–VMRO-DPMNE-led block 63, SDSM-led block 27, BDI/DUI 18, PDSh/DPA 11, PEI 1
Judicial branch: Supreme Court; Constitutional Court; Judicial Council
note: the Judicial Council appoints the judges
Political parties and leaders: Democratic Alliance or DS [Pavle TRAJANOV]; Democratic Party of Serbs in Macedonia [Ivan STOILJKOVIC]; Democratic Party of the Albanians or PDSh/DPA [Menduh THACI]; Democratic Party of Turks in Macedonia [Kenan HASIPI]; Democratic Union for Integration or BDI/DUI [Ali AHMETI]; Internal Macedonian Revolutionary Organization–Democratic Party for Macedonian National Unity (VMRO-DPMNE) [Nikola GRUEVSKI]; Liberal Democratic Party or LDP; Liberal Party [Borce STOJANOVSKI]; Movement for

Reconstruction of Macedonia or DOM [Liljana POPOVSKA]; New Alternative [Gjorgji OROVCANEC]; New Democracy or DR [Imer SELMANI]; New Social-Democratic Party or NSDP [Tito PETKOVSKI]; Party for Democratic Action in Macedonia or SDAM [Avdija PEPIC]; Party for European Future or PEI [Fijat CANOSKI]; Social-Democratic Union of Macedonia or SDSM [Branko CRVENKOVSKI]; Socialist Party or SP [Ljubisav IVANOV-DZINGO]; Union of Roma of Macedonia [Amdi BAJRAM]; United for Macedonia or OM [Ljube BOSKOVSKI]; VMRO-Macedonian [Borislav STOJMENOV]
Political pressure groups and leaders: Federation of Free Trade Unions [Rasko MISHKOSKI]; Federation of Trade Unions [Zivko MITREVSKI]; Trade Union of Education, Science and Culture [Yakim NEDELKOV]
International organization participation: BIS, CE, CEI, EAPC, EBRD, EU (candidate country), FAO, IAEA, IBRD, ICAO, ICRM, IDA, IFAD, IFC, IFRCS, ILO, IMF, IMO, Interpol, IOC, IOM (observer), IPU, ISO, ITU, ITUC, MIGA, OIF, OPCW, OSCE, PCA, PFP, SECI, UN, UNCTAD, UNESCO, UNHCR, UNIDO, UNIFIL, UNWTO, UPU, WCO, WHO, WIPO, WMO, WTO
Diplomatic representation in the US: *chief of mission:* Ambassador Zoran JOLEVSKI
chancery: 2129 Wyoming Avenue NW, Washington, DC 20008
telephone: [1] (202) 667-0501
FAX: [1] (202) 667-2131
consulate(s) general: Southfield (Michigan), Chicago
Diplomatic representation from the US: *chief of mission:* Ambassador Philip T. REEKER
embassy: Str. Samolilova, Nr. 21, 1000 Skopje
mailing address: American Embassy Skopje, US Department of State, 7120 Skopje Place, Washington, DC 20521-7120 (pouch)
telephone: [389] 2 310-2000
FAX: [389] 2 310-2499
Flag description: a yellow sun (the Sun of Liberty) with eight broadening rays extending to the edges of the red field; the red and yellow colors have long been associated with Macedonia
National anthem: *name:* "Denes Nad Makedonija" (Today Over Macedonia)
lyrics/music: Vlado MALESKI/Todor SKALOVSKI
note: adopted 1991; the song, written in 1943, previously served as the anthem of the Socialist Republic of Macedonia while part of Yugoslavia

ECONOMY

Economy—overview: Having a small, open economy makes Macedonia vulnerable to economic developments in Europe and dependent on regional integration and progress toward EU membership for continued economic growth. At independence in September 1991, Macedonia was the least developed of the Yugoslav republics, producing a mere 5% of the total federal output of goods and services. The collapse of Yugoslavia ended transfer payments from the central government and eliminated advantages from inclusion in a de facto free trade area. An absence of infrastructure, UN sanctions on the downsized Yugoslavia, and a Greek economic embargo over a dispute about the country's constitutional name and flag hindered economic growth until 1996. Since then, Macedonia has maintained macroeconomic stability with low inflation, but it has so far lagged the region in attracting foreign investment and creating jobs, despite making extensive fiscal and business sector reforms. Official unemployment remains high at 31.7%, but may be overstated based on the existence of an extensive gray market, estimated to be more than 20% of GDP, that is not captured by official statistics. In the wake of the global economic downturn, Macedonia has experienced decreased foreign direct investment, lowered credit, and a large trade deficit. However, as a result of conservative fiscal policies and a sound financial system, in 2010 the country received slightly improved credit ratings. Macroeconomic stability also was maintained by a prudent monetary policy, which kept the domestic currency at the pegged level against the euro, while interest rates were falling. As a result, GDP growth was modest, but positive, in 2010.
GDP (purchasing power parity): $20 billion (2010 est.)
country comparison to the world: 125
$19.86 billion (2009 est.)
$20.04 billion (2008 est.)
note: data are in 2010 US dollars; Macedonia has a large informal sector
GDP (official exchange rate): $9.108 billion (2010 est.)
GDP—real growth rate: 0.7% (2010 est.)
country comparison to the world: 179
-0.9% (2009 est.)
5% (2008 est.)
GDP—per capita (PPP): $9,700 (2010 est.)
country comparison to the world: 112
$9,600 (2009 est.)
$9,700 (2008 est.)
note: data are in 2010 US dollars
GDP—composition by sector: *agriculture:* 8.7%
industry: 22.1%
services: 69.2% (2010 est.)
Labor force: 949,300 (2010 est.)
country comparison to the world: 143
Labor force—by occupation:
agriculture: 19.9%
industry: 22.1%
services: 58% (September 2010)
Unemployment rate: 31.7% (3rd quarter, 2010 est.)
country comparison to the world: 180
32.2% (2009)
Population below poverty line: 28.7% (2008)
Household income or consumption by percentage share: *lowest 10%:* 2.4%
highest 10%: 29.6% (2003)
Distribution of family income—Gini index: 39 (2003)
country comparison to the world: 68
Investment (gross fixed): 23.2% of GDP (2010 est.)
country comparison to the world: 56
Budget: *revenues:* $2.772 billion
expenditures: $3.011 billion (2010 est.)
Public debt: 34.2% of GDP (November 2010 est.)
country comparison to the world: 85
31.6% of GDP (2009 est.)
Inflation rate (consumer prices): 1.6% (2010 est.)
country comparison to the world: 43
-0.8% (2009 est.)
Central bank discount rate: 6.5% (31 December 2009)
country comparison to the world: 55
6.5% (31 December 2008)
note: series discontinued in January 2010. Discount rate was replaced by a referent rate for calculating the penalty rate: 4.0% (31 December 2010)
Commercial bank prime lending rate: 8.2% (30 November 2010 est.)
country comparison to the world: 95
9.3% (31 December 2009 est.)
Stock of narrow money: $1.146 billion (31 December 2010 est.)
country comparison to the world: 137
$1.224 billion (31 December 2009 est.)
Stock of broad money: $4.134 billion (31 December 2010 est.)
country comparison to the world: 126
$4.858 billion (31 December 2009 est.)
Stock of domestic credit: $4.001 billion (31 December 2010 est.)
country comparison to the world: 115
$4.055 billion (31 December 2009 est.)
Market value of publicly traded shares: $2.647 billion (31 December 2010)
country comparison to the world: 92
$2.859 billion (31 December 2009)
$823.5 million (31 December 2008)
Agriculture—products: grapes, tobacco, vegetables, fruits; milk, eggs
Industries: food processing, beverages, textiles, chemicals, iron, steel, cement, energy, pharmaceuticals
Industrial production growth rate: 4% (November 2010 est.)
country comparison to the world: 84
Electricity—production: 6.819 billion kWh (2010 est.)
country comparison to the world: 101
Electricity—consumption: 8.189 billion kWh (2010 est.)
country comparison to the world: 95
Electricity—exports: 0 kWh (2010 est.)
Electricity—imports: 1.37 billion kWh (2010 est.)
Oil—production: 0 bbl/day (2010)
country comparison to the world: 204
Oil—consumption: 18,200 bbl/day (2010)
country comparison to the world: 128
Oil—exports: 8,594 bbl/day (2010)
country comparison to the world: 95
Oil—imports: 18,200 bbl/day (2009)
country comparison to the world: 114
Oil—proved reserves: 0 bbl (1 January 2011 est.)
country comparison to the world: 162
Natural gas—production: 0 cu m (2010 est.)
country comparison to the world: 158

Natural gas—consumption: 117.4 million cu m (2010)
country comparison to the world: 102
Natural gas—exports: 0 cu m (2010 est.)
country comparison to the world: 139
Natural gas—imports: 117.4 million cu m (2010)
country comparison to the world: 67
Natural gas—proved reserves: 0 cu m (1 January 2011 est.)
country comparison to the world: 164
Current account balance: $-328 million (2010 est.)
country comparison to the world: 101
$-645.6 million (2009 est.)
Exports: $3.171 billion (2010 est.)
country comparison to the world: 120
$2.686 billion (2009 est.)
Exports—commodities: food, beverages, tobacco; textiles, miscellaneous manufactures, iron and steel
Exports—partners: Germany 20.31%, Greece 13.09%, Italy 11.08%, Bulgaria 10.61%, Croatia 7.74% (2009)
Imports: $5.113 billion (2010 est.)
country comparison to the world: 116
$4.842 billion (2009 est.)
Imports—commodities: machinery and equipment, automobiles, chemicals, fuels, food products
Imports—partners: Germany 15.11%, Greece 14.88%, Bulgaria 9.08%, Italy 7.68%, Turkey 7.59%, Slovenia 6.26%, Hungary 4.31% (2009)
Reserves of foreign exchange and gold: $2.217 billion (30 November 2010 est.)
country comparison to the world: 99
$2.292 billion (31 December 2009 est.)
Debt—external: $5.485 billion (30 September 2010 est.)
country comparison to the world: 103
$5.5 billion (31 December 2009 est.)
Stock of direct foreign investment—at home: $3.739 billion (31 October 2010 est.)
country comparison to the world: 86
$3.554 billion (31 December 2009 est.)
Stock of direct foreign investment—abroad: $NA (31 December 2010)
$564 million (31 December 2009 est.)
Exchange rates: Macedonian denars (MKD) per US dollar–46.434 (2010) 44.1 (2009) 41.414 (2008) 44.732 (2007) 48.978 (2006)

COMMUNICATIONS

Telephones—main lines in use: 442,200 (2009)
country comparison to the world: 101
Telephones—mobile cellular: 1.943 million (2009)
country comparison to the world: 135
Telephone system: general assessment: competition from the mobile-cellular segment of the telecommunications market has led to a drop in fixed-line telephone subscriptions
domestic: combined fixed-line and mobile-cellular telephone subscribership about 115 per 100 persons
international: country code–389 (2009)
Broadcast media: public television broadcaster operates 3 national channels and a satellite network; 5 privately-owned TV channels broadcast nationally using terrestrial transmitters and about 15 broadcast on national level via satellite; roughly 75 local commercial TV stations broadcasting; large number of cable operators offering domestic and international programming; public radio broadcaster operates over multiple stations; 3 privately-owned radio stations broadcast nationally; about 70 local commercial radio stations functioning (2010)
Internet country code: .mk
Internet hosts: 60,533 (2010)
country comparison to the world: 84
Internet users: 1.057 million (2009)
country comparison to the world: 97

TRANSPORTATION

Airports: 14 (2010)
country comparison to the world: 150
Airports—with paved runways: *total:* 10
2,438 to 3,047 m: 2
under 914 m: 8 (2010)
Airports—with unpaved runways: *total:* 4
914 to 1,523 m: 1
under 914 m: 3 (2010)
Pipelines: gas 268 km; oil 120 km (2010)
Railways: *total:* 699 km
country comparison to the world: 100
standard gauge: 699 km 1.435-m gauge (234 km electrified) (2010)
Roadways: *total:* 13,736 km (includes 216 km of expressways) (2010)
country comparison to the world: 125

MILITARY

Military branches: Army of the Republic of Macedonia (ARM): Joint Operational Command, with subordinate Air Wing (Makedonsko Voeno Vozduhoplovstvo, MVV); Special Operations Regiment; Logistic Support Command; Training Command (2010)
Military service age and obligation: 18 years of age for voluntary military service; no conscription (2010)
Manpower available for military service: *males age 16–49:* 532,196
females age 16–49: 511,964 (2010 est.)
Manpower fit for military service: *males age 16–49:* 443,843
females age 16–49: 426,251 (2010 est.)
Manpower reaching militarily significant age annually: *male:* 16,144
female: 14,920 (2010 est.)
Military expenditures: 6% of GDP (2005 est.)
country comparison to the world: 9

TRANSNATIONAL ISSUES

Disputes—international: Kosovo and Macedonia completed demarcation of their boundary in September 2008; Greece continues to reject the use of the name Macedonia or Republic of Macedonia
Refugees and internally displaced persons: *IDPs:* fewer than 1,000 (ethnic conflict in 2001) (2007)
Illicit drugs: major transshipment point for Southwest Asian heroin and hashish; minor transit point for South American cocaine destined for Europe; although not a financial center and most criminal activity is thought to be domestic, money laundering is a problem due to a mostly cash-based economy and weak enforcement

MADAGASCAR

INTRODUCTION

Background: Formerly an independent kingdom, Madagascar became a French colony in 1896 but regained independence in 1960. During 1992-93, free presidential and National Assembly elections were held ending 17 years of single-party rule. In 1997, in the second presidential race, Didier RATSIRAKA, the leader during the 1970s and 1980s, was returned to the presidency. The 2001 presiden tial election was contested between the followers of Didier RATSIRAKA and Marc RAVALOMANANA, nearly causing secession of half of the country. In April 2002, the High Constitutional Court announced RAVALOMANANA the winner. RAVALOMANANA achieved a second term following a landslide victory in the generally free and fair presidential elections of 2006. In early 2009, protests over increasing restrictions on opposition press and activities resulted in RAVALOMANANA stepping down and the presidency was conferred to the mayor of Antananarivo, Andry RAJOELINA. Following negotiations in July and August of 2009, a power-sharing agreement with a 15-month transitional period was established, but has not yet been implemented.

GEOGRAPHY

Location: Southern Africa, island in the Indian Ocean, east of Mozambique
Geographic coordinates: 20 00 S, 47 00 E
Map references: Africa
Area: *total:* 587,041 sq km
country comparison to the world: 46
land: 581,540 sq km
water: 5,501 sq km
Area—comparative: slightly less than twice the size of Arizona
Land boundaries: 0 km
Coastline: 4,828 km
Maritime claims: *territorial sea:* 12 nm
contiguous zone: 24 nm
exclusive economic zone: 200 nm
continental shelf: 200 nm or 100 nm from the 2,500-m isobath
Climate: tropical along coast, temperate inland, arid in south
Terrain: narrow coastal plain, high plateau and mountains in center
Elevation extremes: *lowest point:* Indian Ocean 0 m

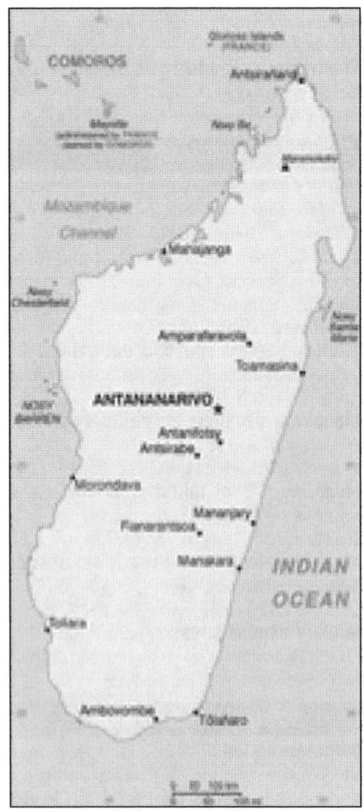

highest point: Maromokotro 2,876 m
Natural resources: graphite, chromite, coal, bauxite, rare earth elements, salt, quartz, tar sands, semiprecious stones, mica, fish, hydropower
Land use: *arable land:* 5.03%
permanent crops: 1.02%
other: 93.95% (2005)
Irrigated land: 10,860 sq km (2003)
Total renewable water resources: 337 cu km (1984)
Freshwater withdrawal (domestic/industrial/agricultural): *total:* 14.96 cu km/yr (3%/2%/96%)
per capita: 804 cu m/yr (2000)
Natural hazards: periodic cyclones; drought; and locust infestation
volcanism: Madagascar's volcanoes have not erupted in historical times
Environment—current issues: soil erosion results from deforestation and overgrazing; desertification; surface water contaminated with raw sewage and other organic wastes; several endangered species of flora and fauna unique to the island
Environment—international agreements: *party to:* Biodiversity, Climate Change, Climate Change-Kyoto Protocol, Desertification, Endangered Species, Hazardous Wastes, Law of the Sea, Marine Life Conservation, Ozone Layer Protection, Ship Pollution, Wetlands
signed, but not ratified: none of the selected agreements
Geography—note: world's fourth-largest island; strategic location along Mozambique Channel

PEOPLE

Population: 21,926,221 (July 2011 est.)
country comparison to the world: 53
Age structure: *0–14 years:* 43.1% (male 4,762,589/female 4,693,259)
15–64 years: 53.8% (male 5,864,520/female 5,938,029)
65 years and over: 3% (male 295,409/female 372,415) (2011 est.)
Median age: *total:* 18.2 years
male: 17.9 years
female: 18.4 years (2011 est.)
Population growth rate: 2.973% (2011 est.)
country comparison to the world: 12
Birth rate: 37.51 births/1,000 population (2011 est.)
country comparison to the world: 19
Death rate: 7.79 deaths/1,000 population (July 2011 est.)
country comparison to the world: 112
Net migration rate: 0 migrant(s)/1,000 population (2011 est.)
country comparison to the world: 95
Urbanization: *urban population:* 30% of total population (2010)
rate of urbanization: 3.9% annual rate of change (2010-15 est.)
Major cities—population: ANTANANARIVO (capital) 1.816 million (2009)
Sex ratio: *at birth:* 1.03 male(s)/female
under 15 years: 1.01 male(s)/female
15–64 years: 0.99 male(s)/female
65 years and over: 0.8 male(s)/female
total population: 0.99 male(s)/female (2011 est.)
Infant mortality rate: *total:* 51.45 deaths/1,000 live births
country comparison to the world: 45
male: 56.23 deaths/1,000 live births
female: 46.52 deaths/1,000 live births (2011 est.)
Life expectancy at birth: *total population:* 63.63 years
country comparison to the world: 174
male: 61.62 years
female: 65.7 years (2011 est.)
Total fertility rate: 5.02 children born/woman (2011 est.)
country comparison to the world: 20
HIV/AIDS—adult prevalence rate: 0.2% (2009 est.)
country comparison to the world: 103
HIV/AIDS—people living with HIV/AIDS: 24,000 (2009 est.)
country comparison to the world: 74
HIV/AIDS—deaths: 1,700 (2009 est.)
country comparison to the world: 55
Major infectious diseases: degree of risk: very high
food or waterborne diseases: bacterial and protozoal diarrhea, hepatitis A, and typhoid fever
vectorborne diseases: chikungunya, malaria, and plague
water contact disease: schistosomiasis (2009)
Drinking water source: *Improved:*
urban: 71% of population
rural: 29% of population
total: 41% of population
Unimproved:
urban: 29% of population
rural: 71% of population
total: 59% of population (2008)
Sanitation facility access: *Improved:*
urban: 15% of population
rural: 10% of population
total: 11% of population
Unimproved:
urban: 85% of population
rural: 90% of population
total: 89% of population (2008)
Nationality: *noun:* Malagasy (singular and plural)
adjective: Malagasy
Ethnic groups: Malayo-Indonesian (Merina and related Betsileo), Cotiers (mixed African, Malayo-Indonesian, and Arab ancestry—Betsimisaraka, Tsimihety, Antaisaka, Sakalava), French, Indian, Creole, Comoran
Religions: indigenous beliefs 52%, Christian 41%, Muslim 7%
Languages: French (official), Malagasy (official), English
Literacy: *definition:* age 15 and over can read and write
total population: 68.9%
male: 75.5%
female: 62.5% (2003 est.)
School life expectancy (primary to tertiary education): *total:* 11 years
male: 11 years
female: 11 years (2009)
Education expenditures: 3% of GDP (2009)
country comparison to the world: 133

GOVERNMENT

Country name: *conventional long form:* Republic of Madagascar
conventional short form: Madagascar
local long form: Republique de Madagascar/Repoblikan'i Madagasikara
local short form: Madagascar/Madagasikara
former: Malagasy Republic
Government type: republic
Capital: *name:* Antananarivo
geographic coordinates: 18 55 S, 47 31 E
time difference: UTC+3 (8 hours ahead of Washington, DC during Standard Time)
Administrative divisions: 6 provinces (faritany); Antananarivo, Antsiranana, Fianarantsoa, Mahajanga, Toamasina, Toliara
Independence: 26 June 1960 (from France)
National holiday: Independence Day, 26 June (1960)
Constitution: passed by referendum 17 November 2010; promulgated 11 December 2010 (2010)
Legal system: civil law system based on the old French civil code and customary law in matters of marriage, family, and obligation
International law organization participation: accepts compulsory ICJ jurisdiction with reservations; accepts ICCt jurisdiction
Suffrage: 18 years of age; universal
Executive branch: *chief of state:* President Andry RAJOELINA (since 18 March 2009)
head of government: Prime Minister Albert Camille VITAL (since 18 December 2009)
cabinet: Council of Ministers appointed by the prime minister
(For more information visit the World Leaders website)
elections: president elected by popular vote for a five-year term (eligible for a second term); election last held on 3 December

2006 (next to be held in September 2011); prime minister appointed by the president; note—a power-sharing agreement in the summer of 2009 established a 15-month transition, concluding in general elections now scheduled for September 2011
election results: percent of vote—Marc RAVALOMANANA 54.8%, Jean LAHINIRIKO 11.7%, Roland RATSIRAKA 10.1%, Herizo RAZAFIMAHALEO 9.1%, Norbert RATSIRAHONANA 4.2%, Ny Hasina ANDRIAMANJATO 4.2%, Elia RAVELOMANANTSOA 2.6%, Pety RAKOTONIAINA 1.7%, other 1.6%; note—RAVALOMANANA stepped down on 17 March 2009
note: on 17 March 2009, democratically elected President Marc RAVALOMANANA stepped down handing the government over to the military, which in turn conferred the presidency on opposition leader and Antananarivo mayor Andry RAJOELINA, who will head the High Transition Authority; a power-sharing agreement reached in August 2009 established a 15-month transition period, concluding in general elections in 2010; as of December 2010 the agreement had not been fully implemented
Legislative branch: bicameral legislature consists of a Senate or Senat (100 seats; two-thirds of the members appointed by regional assemblies; the remaining one-third appointed by the president; members to serve four-year terms) and a National Assembly or Assemblee Nationale (127 seats—reduced from 160 seats by an April 2007 national referendum; members elected by popular vote to serve four-year terms)
elections: National Assembly—last held on 23 September 2007 (next to be held in September 2011); note—a power-sharing agreement in the summer of 2009 established a 15-month transition, concluding in general elections now scheduled for September 2011
election results: National Assembly—percent of vote by party—NA; seats by party—TIM 106, LEADER/Fanilo 1, independents 20
Judicial branch: Supreme Court or Cour Supreme; High Constitutional Court or Haute Cour Constitutionnelle
Political parties and leaders: Association for the Rebirth of Madagascar or AREMA [Pierrot RAJAONARIVELO]; Democratic Party for Union in Madagascar or PSDUM [Jean LAHINIRIKO]; Economic Liberalism and Democratic Action for National Recovery or LEADER/Fanilo [Herizo RAZAFIMAHALEO]; Fihaonana Party or FP [Guy-Willy RAZANAMASY]; I Love Madagascar or TIM [Marc RAVALOMANANA]; Renewal of the Social Democratic Party or RPSD [Evariste MARSON]
Political pressure groups and leaders: Committee for the Defense of Truth and Justice or KMMR; Committee for National Reconciliation or CRN [Albert Zafy]; National Council of Christian Churches or FFKM
International organization participation: ACP, AfDB, AU, COMESA, FAO, G-77, IAEA, IBRD, ICAO, ICC, ICRM, IDA, IFAD, IFC, IFRCS, ILO, IMF, IMO, InOC, Interpol, IOC, IOM, ISO (correspondent), ITSO, ITU, ITUC, MIGA, NAM, OIF, OPCW, PCA, SADC, UN, UNCTAD, UNESCO, UNHCR, UNIDO, UNWTO, UPU, WCO, WFTU, WHO, WIPO, WMO, WTO
Diplomatic representation in the US: *chief of mission:* Ambassador (vacant); Charge d'Affaires Eulalie N. RAVELOSOA
chancery: 2374 Massachusetts Avenue NW, Washington, DC 20008
telephone: [1] (202) 265-5525 through 5526
FAX: [1] (202) 265-3034
consulate(s) general: Los Angeles, New York
Diplomatic representation from the US: *chief of mission:* Ambassador R. Niels MARQUARDT
embassy: 14-16 Rue Rainitovo, Antsahavola, Antananarivo 101
mailing address: B. P. 620, Antsahavola, Antananarivo
telephone: [261] (20) 22-212-57, 22-212-73, 22-209-56
FAX: [261] (20) 22-345-39
Flag description: two equal horizontal bands of red (top) and green with a vertical white band of the same width on hoist side; by tradition, red stands for sovereignty, green for hope, white for purity
National anthem: *name:* "Ry Tanindraza nay malala o" (Oh, Our Beloved Fatherland)
lyrics/music: Pasteur RAHAJASON/ Norbert RAHARISOA
note: adopted 1959

ECONOMY

Economy—overview: After discarding socialist economic policies in the mid-1990s, Madagascar followed a World Bank- and IMF-led policy of privatization and liberalization that has been undermined since the start of the political crisis. This strategy placed the country on a slow and steady growth path from an extremely low level. Agriculture, including fishing and forestry, is a mainstay of the economy, accounting for more than one-fourth of GDP and employing 80% of the population. Exports of apparel have boomed in recent years primarily due to duty-free access to the US. However, Madagascar's failure to comply with the requirements of the African Growth and Opportunity Act (AGOA) led to the termination of the country's duty-free access in January 2010. Deforestation and erosion, aggravated by the use of firewood as the primary source of fuel, are serious concerns. Former President RAVALOMANANA worked aggressively to revive the economy following the 2002 political crisis, which triggered a 12% drop in GDP that year. The current political crisis which began in early 2009 has dealt additional blows to the economy. Tourism dropped more than 50% in 2009, compared with the previous year, and many investors are wary of entering the uncertain investment environment.
GDP (purchasing power parity): $19.41 billion (2010 est.)
country comparison to the world: 127
$19.8 billion (2009 est.)
$20.55 billion (2008 est.)
note: data are in 2010 US dollars
GDP (official exchange rate): $8.345 billion (2010 est.)
GDP—real growth rate: -2% (2010 est.)
country comparison to the world: 208
-3.7% (2009 est.)
7.1% (2008 est.)
GDP—per capita (PPP): $900 (2010 est.)
country comparison to the world: 218
$1,000 (2009 est.)
$1,000 (2008 est.)
note: data are in 2010 US dollars
GDP—composition by sector: *agriculture:* 26.5%
industry: 16.7%
services: 56.8% (2010 est.)
Labor force: 9.504 million (2007)
country comparison to the world: 51
Population below poverty line: 50% (2004 est.)
Household income or consumption by percentage share: *lowest 10%:* 2.6%
highest 10%: 41.5% (2005)
Distribution of family income—Gini index: 47.5 (2001)
country comparison to the world: 30
38.1 (1999)
Investment (gross fixed): 34.6% of GDP (2010 est.)
country comparison to the world: 10
Budget: *revenues:* $896.9 million
expenditures: $1.547 billion (2010 est.)
Inflation rate (consumer prices): 8.1% (2010 est.)
country comparison to the world: 186
9% (2009 est.)
Central bank discount rate: NA%
Commercial bank prime lending rate: 45% (31 December 2009 est.)
country comparison to the world: 3
45% (31 December 2008 est.)
Stock of narrow money: $1.233 billion (31 December 2010 est.)
country comparison to the world: 134
$1.228 billion (31 December 2009 est.)
Stock of broad money: $2.012 billion (31 December 2010 est.)
country comparison to the world: 142
$1.994 billion (31 December 2009 est.)
Stock of domestic credit: $1.02 billion (31 December 2010 est.)
country comparison to the world: 149
$997.6 million (31 December 2009 est.)
Market value of publicly traded shares: $NA
Agriculture—products: coffee, vanilla, sugarcane, cloves, cocoa, rice, cassava (tapioca), beans, bananas, peanuts; livestock products
Industries: meat processing, seafood, soap, breweries, tanneries, sugar, textiles, glassware, cement, automobile assembly plant, paper, petroleum, tourism
Industrial production growth rate: 2% (2010 est.)
country comparison to the world: 129
Electricity—production: 1.045 billion kWh (2007 est.)
country comparison to the world: 145
Electricity—consumption: 971.4 million kWh (2007 est.)
country comparison to the world: 145
Electricity—exports: 0 kWh (2008 est.)
Electricity—imports: 0 kWh (2008 est.)
Oil—production: 0 bbl/day (2009 est.)
country comparison to the world: 198

Oil—consumption: 21,000 bbl/day (2009 est.)
country comparison to the world: 122
Oil—exports: 364.9 bbl/day (2007 est.)
country comparison to the world: 126
Oil—imports: 16,940 bbl/day (2007 est.)
country comparison to the world: 118
Oil—proved reserves: 0 bbl (1 January 2010 est.)
country comparison to the world: 156
Natural gas—production: 0 cu m (2008 est.)
country comparison to the world: 153
Natural gas—consumption: 0 cu m (2008 est.)
country comparison to the world: 193
Natural gas—exports: 0 cu m (2008 est.)
country comparison to the world: 133
Natural gas—imports: 0 cu m (2008 est.)
country comparison to the world: 83
Natural gas—proved reserves: 0 cu m (1 January 2006 est.)
country comparison to the world: 158
Current account balance: $-600 million (2010 est.)
country comparison to the world: 123
$-561 million (2009 est.)
Exports: $1.412 billion (2010 est.)
country comparison to the world: 142
$1.309 billion (2009 est.)
Exports—commodities: coffee, vanilla, shellfish, sugar, cotton cloth, chromite, petroleum products
Exports—partners: France 28.9%, US 20.49%, Germany 5.89%, China 4.36% (2009)
Imports: $1.958 billion (2010 est.)
country comparison to the world: 152
$1.893 billion (2009 est.)
Imports—commodities: capital goods, petroleum, consumer goods, food
Imports—partners: China 12.99%, Thailand 11.93%, Bahrain 7.1%, France 6.89%, US 4.13% (2009)
Reserves of foreign exchange and gold: $1.038 billion (31 December 2010 est.)
country comparison to the world: 115
$1.136 billion (31 December 2009 est.)
Debt—external: $2.973 billion (31 December 2010 est.)
country comparison to the world: 128
$2.261 billion (31 December 2009 est.)
Stock of direct foreign investment—at home: $NA
Stock of direct foreign investment—abroad: $NA
Exchange rates: Malagasy ariary (MGA) per US dollar–2,062.5 (2010) 1,956.2 (2009) 1,654.78 (2008) 1,880 (2007) 2,161.4 (2006)

COMMUNICATIONS

Telephones—main lines in use: 181,200 (2009)
country comparison to the world: 128
Telephones—mobile cellular: 5.997 million (2009)
country comparison to the world: 86
Telephone system: general assessment: system is above average for the region; Antananarivo's main telephone exchange modernized in the late 1990s, but the rest of the analogue-based telephone system is poorly developed; have been adding fixed line connections since 2005
domestic: combined fixed-line and mobile-cellular teledensity about 30 per 100 persons
international: country code–261; SEACOM undersea fiber-optic cable and the Lion undersea cable connecting to Reunion and Mauritius; satellite earth stations–2 (1 Intelsat–Indian Ocean, 1 Intersputnik–Atlantic Ocean region) (2009)
Broadcast media: state-owned Radio Nationale Malagasy (RNM) and Television Malagasy (TVM) have an extensive national network reach; privately-owned radio and TV broadcasters in cities and major towns; state-run radio predominates in rural areas; relays of 2 international broadcasters are available in Antananarivo (2007)
Internet country code: .mg
Internet hosts: 27,606 (2010)
country comparison to the world: 99
Internet users: 319,900 (2009)
country comparison to the world: 127

TRANSPORTATION

Airports: 84 (2010)
country comparison to the world: 67
Airports—with paved runways:
total: 27
over 3,047 m: 1
2,438 to 3,047 m: 2
1,524 to 2,437 m: 6
914 to 1,523 m: 17
under 914 m: 1 (2010)
Airports—with unpaved runways: *total:* 57
1,524 to 2,437 m: 2
914 to 1,523 m: 35
under 914 m: 20 (2010)
Railways: *total:* 854 km
country comparison to the world: 97
narrow gauge: 854 km 1.000-m gauge (2010)
Roadways: *total:* 65,663 km
country comparison to the world: 69
paved: 7,617 km
unpaved: 58,046 km (2003)
Waterways: 600 km (432 km navigable) (2010)
country comparison to the world: 80
Merchant marine: *total:* 8
country comparison to the world: 123
by type: cargo 4, passenger/cargo 2, petroleum tanker 2 (2010)
Ports and terminals: Antsiranana (Diego Suarez), Mahajanga, Toamasina, Toliara (Tulear)

MILITARY

Military branches: People's Armed Forces: Intervention Force, Development Force, and Aeronaval Force (navy and air); National Gendarmerie
Military service age and obligation: 18-25 years of age for male-only voluntary military service; no conscription; service obligation–18 months (either military or equivalent civil service); 20-30 years of age for National Gendarmerie recruits (35 years of age for those with military experience) (2010)
Manpower available for military service:
males age 16–49: 4,900,729
females age 16–49: 4,909,061 (2010 est.)
Manpower fit for military service: *males age 16–49:* 3,390,071
females age 16–49: 3,682,180 (2010 est.)
Manpower reaching militarily significant age annually: *male:* 248,184
female: 246,769 (2010 est.)
Military expenditures:
1% of GDP (2006)
country comparison to the world: 130

TRANSNATIONAL ISSUES

Disputes—international: claims Bassas da India, Europa Island, Glorioso Islands, and Juan de Nova Island (all administered by France); the vegetated drying cays of Banc du Geyser, which were claimed by Madagascar in 1976, also fall within the EEZ claims of the Comoros and France (Glorioso Islands, part of the French Southern and Antarctic Lands)
Illicit drugs: illicit producer of cannabis (cultivated and wild varieties) used mostly for domestic consumption; transshipment point for heroin

MALAWI

INTROD\UCTION

Background: Established in 1891, the British protectorate of Nyasaland became the independent nation of Malawi in 1964. After three decades of one-party rule under President Hastings Kamuzu BANDA the country held multiparty elections in 1994, under a provisional constitution that came into full effect the following year. Current President Bingu wa MUTHARIKA, elected in May 2004 after a failed attempt by the previous president to amend the constitution to permit another term, struggled to assert his authority against his predecessor and subsequently started his own party, the Democratic Progressive Party (DPP) in 2005. As president, MUTHARIKA has overseen some economic improvement. Population growth, increasing pressure on agricultural lands, corruption, and the spread of HIV/AIDS pose major problems for Malawi. MUTHARIKA was reelected to a second term in May 2009.

GEOGRAPHY

Location: Southern Africa, east of Zambia, west and north of Mozambique
Geographic coordinates: 13 30 S, 34 00 E
Map references: Africa
Area:
total: 118,484 sq km
country comparison to the world: 99
land: 94,080 sq km
water: 24,404 sq km
Area—comparative: slightly smaller than Pennsylvania

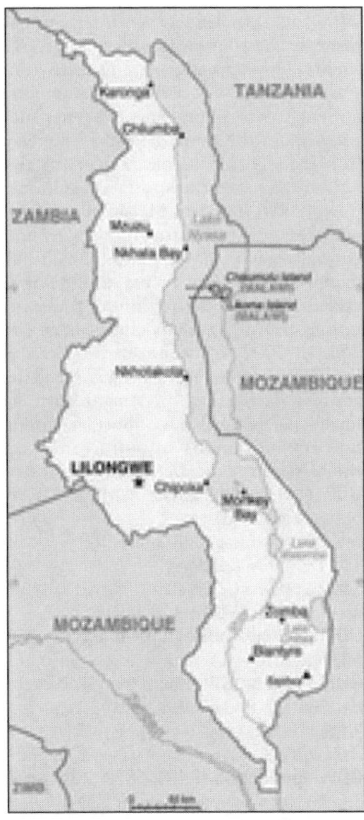

Land boundaries: *total:* 2,881 km
border countries: Mozambique 1,569 km, Tanzania 475 km, Zambia 837 km
Coastline: 0 km (landlocked)
Maritime claims: none (landlocked)
Climate: sub-tropical; rainy season (November to May); dry season (May to November)
Terrain: narrow elongated plateau with rolling plains, rounded hills, some mountains
Elevation extremes: *lowest point:* junction of the Shire River and international boundary with Mozambique 37 m
highest point: Sapitwa (Mount Mlanje) 3,002 m
Natural resources: limestone, arable land, hydropower, unexploited deposits of uranium, coal, and bauxite
Land use: *arable land:* 20.68%
permanent crops: 1.18%
other: 78.14% (2005)
Irrigated land: 560 sq km (2003)
Total renewable water resources: 17.3 cu km (2001)
Freshwater withdrawal (domestic/industrial/agricultural): *total:* 1.01 cu km/yr (15%/5%/80%)
per capita: 78 cu m/yr (2000)
Natural hazards: NA
Environment—current issues: deforestation; land degradation; water pollution from agricultural runoff, sewage, industrial wastes; siltation of spawning grounds endangers fish populations
Environment—international agreements: *party to:* Biodiversity, Climate Change, Climate Change-Kyoto Protocol, Desertification, Endangered Species, Environmental Modification, Hazardous Wastes, Marine Life Conservation, Ozone Layer Protection, Ship Pollution, Wetlands
signed, but not ratified: Law of the Sea
Geography—note: landlocked; Lake Nyasa, some 580 km long, is the country's most prominent physical feature

PEOPLE

Population: 15,879,252 (July 2011 est.)
country comparison to the world: 63
note: estimates for this country explicitly take into account the effects of excess mortality due to AIDS; this can result in lower life expectancy, higher infant mortality, higher death rates, lower population growth rates, and changes in the distribution of population by age and sex than would otherwise be expected
Age structure: *0–14 years:* 45.1% (male 3,586,696/female 3,571,298)
15–64 years: 52.2% (male 4,140,874/female 4,155,015)
65 years and over: 2.7% (male 182,304/female 243,065) (2011 est.)
Median age: *total:* 17.2 years
male: 17.1 years
female: 17.4 years (2011 est.)
Population growth rate: 2.763% (2011 est.)
country comparison to the world: 17
Birth rate: 40.85 births/1,000 population (2011 est.)
country comparison to the world: 10
Death rate: 13.22 deaths/1,000 population (July 2011 est.)
country comparison to the world: 21
Net migration rate: 0 migrant(s)/1,000 population (2011 est.)
country comparison to the world: 98
Urbanization: *urban population:* 20% of total population (2010)
rate of urbanization: 5.3% annual rate of change (2010-15 est.)
Major cities—population: Blantyre 856,000; LILONGWE (capital) 821,000 (2009)
Sex ratio:
at birth: 1.015 male(s)/female
under 15 years: 1 male(s)/female
15–64 years: 1 male(s)/female
65 years and over: 0.76 male(s)/female
total population: 0.99 male(s)/female (2011 est.)
Infant mortality rate: *total:* 81.04 deaths/1,000 live births
country comparison to the world: 11
male: 85.11 deaths/1,000 live births
female: 76.9 deaths/1,000 live births (2011 est.)
Life expectancy at birth: *total population:* 51.7 years
country comparison to the world: 211
male: 50.93 years
female: 52.48 years (2011 est.)
Total fertility rate: 5.43 children born/woman (2011 est.)
country comparison to the world: 12
HIV/AIDS—adult prevalence rate: 11% (2009 est.)
country comparison to the world: 9
HIV/AIDS—people living with HIV/AIDS: 920,000 (2009 est.)
country comparison to the world: 13
HIV/AIDS—deaths: 51,000 (2009 est.)
country comparison to the world: 9
Major infectious diseases: degree of risk: very high
food or waterborne diseases: bacterial and protozoal diarrhea, hepatitis A, and typhoid fever
vectorborne diseases: malaria and plague
water contact disease: schistosomiasis
animal contact disease: rabies (2009)
Drinking water source: *Improved:*
urban: 95% of population
rural: 77% of population
total: 80% of population
Unimproved:
urban: 5% of population
rural: 23% of population
total: 20% of population (2008)
Sanitation facility access: *Improved:*
urban: 51% of population
rural: 57% of population
total: 56% of population
Unimproved:
urban: 49% of population
rural: 43% of population
total: 44% of population (2008)
Nationality: *noun:* Malawian(s)
adjective: Malawian
Ethnic groups: Chewa, Nyanja, Tumbuka, Yao, Lomwe, Sena, Tonga, Ngoni, Ngonde, Asian, European
Religions: Christian 79.9%, Muslim 12.8%, other 3%, none 4.3% (1998 census)
Languages: Chichewa (official) 57.2%, Chinyanja 12.8%, Chiyao 10.1%, Chitumbuka 9.5%, Chisena 2.7%, Chilomwe 2.4%, Chitonga 1.7%, other 3.6% (1998 census)
Literacy: *definition:* age 15 and over can read and write
total population: 62.7%
male: 76.1%
female: 49.8% (2003 est.)
School life expectancy (primary to tertiary education): *total:* 9 years
male: 9 years
female: 9 years (2007)
Education expenditures: 4.2% of GDP (2003)
country comparison to the world: 98

GOVERNMENT

Country name: *conventional long form:* Republic of Malawi
conventional short form: Malawi
local long form: Dziko la Malawi
local short form: Malawi
former: British Central African Protectorate, Nyasaland Protectorate, Nyasaland
Government type: multiparty democracy
Capital: *name:* Lilongwe
geographic coordinates: 13 59 S, 33 47 E
time difference: UTC+2 (7 hours ahead of Washington, DC during Standard Time)
Administrative divisions: 28 districts; Balaka, Blantyre, Chikwawa, Chiradzulu, Chitipa, Dedza, Dowa, Karonga, Kasungu, Likoma, Lilongwe, Machinga (Kasupe), Mangochi, Mchinji, Mulanje, Mwanza, Mzimba, Neno, Ntcheu, Nkhata Bay, Nkhotakota, Nsanje, Ntchisi, Phalombe, Rumphi, Salima, Thyolo, Zomba
Independence: 6 July 1964 (from the UK)
National holiday: Independence Day (Republic Day), 6 July (1964)
Constitution: 18 May 1994
Legal system: mixed legal system of English common law and customary law;

judicial review of legislative acts in the Supreme Court of Appeal
International law organization participation: accepts compulsory ICJ jurisdiction with reservations; accepts ICCt jurisdiction
Suffrage: 18 years of age; universal
Executive branch: *chief of state:* President Bingu wa MUTHARIKA (since 24 May 2004); note–the president is both the chief of state and head of government
head of government: President Bingu wa MUTHARIKA (since 24 May 2004)
cabinet: 46-member Cabinet named by the president
(For more information visit the World Leaders website)
elections: president elected by popular vote for a five-year term (eligible for a second term); election last held on 19 May 2009 (next to be held in May 2014)
election results: Bingu wa MUTHARIKA elected president; percent of vote–Bingu wa MUTHARIKA 66%, John TEMBO 30.7%, other 3.3%
Legislative branch: unicameral National Assembly (193 seats; members elected by popular vote to serve five-year terms)
elections: last held on 19 May 2009 (next to be held in May 2014)
election results: percent of vote by party–NA; seats by party–DPP 114, MCP 26, UDF 17, independents 32, other 4
Judicial branch: Supreme Court of Appeal; High Court (chief justice appointed by the president, puisne judges appointed on the advice of the Judicial Service Commission); magistrate's courts
Political parties and leaders: Alliance for Democracy or AFORD [Dindi NYASULU]; Congress of Democrats or CODE [Ralph KASAMBARA]; Democratic Progressive Party or DPP [Bingu wa MUTHARIKA]; Malawi Congress Party or MCP [John TEMBO]; Malawi Democratic Party or MDP [Kampelo KALUA]; Malawi Forum for Unity and Development or MAFUNDE [George MNESA]; Maravi People's Party [Uladi MUSSA]; National Unity Party or NUP [Harry CHIUME]; New Rainbow Coalition Party [Beatrice MWALE]; New Republican Party [Gwanda CHAKUWAMBA]; People's Progressive Movement or PPM [Aleke BANDA]; People's Transformation Movement or PETRA [Kamuzu CHIBAMBO]; Republican Party or RP [Stanley MASAULI]; United Democratic Front or UDF [Bakili MULUZI]; United Democratic Party [Kenedy KALAMBO]
Political pressure groups and leaders: Agri-Ecology Media (agriculture and environmental group); Council for NGOs in Malawi or CONGOMA (human rights, democracy, and development); Human Rights Consultative Committee or HRCC (human rights); Malawi Law Society (human rights and law reform); Malawi Movement for the Restoration of Democracy or MMRD (acts to restore and maintain democracy); Public Affairs Committee or PAC (promotes democracy, development, peace and unity)
International organization participation: ACP, AfDB, AU, C, COMESA, FAO, G-77, IAEA, IBRD, ICAO, ICRM, IDA, IFAD, IFC, IFRCS, ILO, IMF, IMO, Interpol, IOC, IPU, ISO (correspondent), ITSO, ITU, ITUC, MIGA, MONUSCO, NAM, OPCW, SADC, UN, UNAMID, UNCTAD, UNESCO, UNIDO, UNWTO, UPU, WCO, WFTU, WHO, WIPO, WMO, WTO
Diplomatic representation in the US: *chief of mission:* Ambassador Stephen D. Tennyson MATENJE
chancery: 2408 Massachusetts Avenue NW, Washington, DC 20008
telephone: [1] (202) 721-0270
FAX: [1] (202) 721-0288
Diplomatic representation from the US: *chief of mission:* Ambassador [vacant]; Charge d'Affaires Lisa VICKERS
embassy: 16 Jomo Kenyatta Road, Lilongwe 3
mailing address: P. O. Box 30016, Lilongwe 3, Malawi
telephone: [265] (1) 773 166
FAX: [265] (1) 770 471
Flag description: three equal horizontal bands of red (top), black, and green; a white sun disc is centered on the black band, its surrounding 45 white rays extend partially into the red and green bands; black represents the native peoples, red the blood shed in their struggle for freedom, and green the color of nature; the sun represents Malawi's economic progress since attaining independence
National anthem: *name:* "Mulungu dalitsa Malawi" (Oh God Bless Our Land of Malawi)
lyrics/music: Michael-Fredrick Paul SAUKA
note: adopted 1964

ECONOMY

Economy—overview: Landlocked Malawi ranks among the world's most densely populated and least developed countries. The economy is predominately agricultural with about 80% of the population living in rural areas. Agriculture, which has benefited from fertilizer subsidies since 2006, accounts for more than one-third of GDP and 90% of export revenues. The performance of the tobacco sector is key to short-term growth as tobacco accounts for more than half of exports. The economy depends on substantial inflows of economic assistance from the IMF, the World Bank, and individual donor nations. In 2006, Malawi was approved for relief under the Heavily Indebted Poor Countries (HIPC) program. In December 2007, the US granted Malawi eligibility status to receive financial support within the Millennium Challenge Corporation (MCC) initiative. The government faces many challenges including developing a market economy, improving educational facilities, facing up to environmental problems, dealing with the rapidly growing problem of HIV/AIDS, and satisfying foreign donors that fiscal discipline is being tightened. Since 2005 President MUTHARIKA'S government has exhibited improved financial discipline under the guidance of Finance Minister Goodall GONDWE and signed a three year Poverty Reduction and Growth Facility worth $56 million with the IMF. Improved relations with the IMF lead other international donors to resume aid as well. The government has announced infrastructure projects that could yield improvements, such as a new oil pipeline, for better fuel access, and the potential for a waterway link through Mozambican rivers to the ocean, for better transportation options. Since 2009, however, Malawi has experienced some setbacks, including a general shortage of foreign exchange, which has damaged its ability to pay for imports, and fuel shortages that hinder transportation and productivity. Investment fell 23% in 2009, and continued to decline in 2010. The government has failed to address barriers to investment such as unreliable power, water shortages, poor telecommunications infrastructure, and the high costs of services.
GDP (purchasing power parity): $12.98 billion (2010 est.)
country comparison to the world: 140
$12.18 billion (2009 est.)
$11.32 billion (2008 est.)
note: data are in 2010 US dollars
GDP (official exchange rate): $5.053 billion (2010 est.)
GDP—real growth rate: 6.6% (2010 est.)
country comparison to the world: 42
7.6% (2009 est.)
8.6% (2008 est.)
GDP—per capita (PPP): $800 (2010 est.)
country comparison to the world: 220
$800 (2009 est.)
$800 (2008 est.)
note: data are in 2010 US dollars
GDP—composition by sector:
agriculture: 33.4%
industry: 21.7%
services: 44.9% (2010 est.)
Labor force: 5.747 million (2007 est.)
country comparison to the world: 66
Labor force—by occupation:
agriculture: 90%
industry and *services:* 10% (2003 est.)
Unemployment rate: NA%
Population below poverty line: 53% (2004)
Household income or consumption by percentage share:
lowest 10%: 3%
highest 10%: 31.9% (2004)
Distribution of family income—Gini index: 39 (2004)
country comparison to the world: 69
Investment (gross fixed): 27.7% of GDP (2010 est.)
country comparison to the world: 26
Budget: *revenues:* $1.735 billion
expenditures: $1.769 billion (2010 est.)
Public debt: 40.4% of GDP (2010 est.)
country comparison to the world: 70
44.6% of GDP (2009 est.)
Inflation rate (consumer prices): 8% (2010 est.)
country comparison to the world: 184
8.4% (2009 est.)
Central bank discount rate: 15% (31 December 2009)
country comparison to the world: 19
15% (31 December 2008)
Commercial bank prime lending rate: 25.25% (31 December 2009 est.)
country comparison to the world: 8

25.28% (31 December 2008 est.)
Stock of narrow money: $626.5 million (31 December 2010 est.)
country comparison to the world: 152
$580.3 million (31 December 2009 est.)
Stock of broad money: $1.434 billion (31 December 2010 est.)
country comparison to the world: 149
$1.233 billion (31 December 2009 est.)
Stock of domestic credit: $1.72 billion (31 December 2010 est.)
country comparison to the world: 131
$1.515 billion (31 December 2009 est.)
Market value of publicly traded shares: $1.771 billion (31 December 2008)
country comparison to the world: 106
$587.2 million (31 December 2006)
Agriculture—products: tobacco, sugarcane, cotton, tea, corn, potatoes, cassava (tapioca), sorghum, pulses, groundnuts, Macadamia nuts; cattle, goats
Industries: tobacco, tea, sugar, sawmill products, cement, consumer goods
Industrial production growth rate: 17.3% (2010 est.)
country comparison to the world: 5
Electricity—production: 1.69 billion kWh (2007 est.)
country comparison to the world: 137
Electricity—consumption: 1.572 billion kWh (2007 est.)
country comparison to the world: 139
Electricity—exports: 0 kWh (2008 est.)
Electricity—imports: 0 kWh (2008 est.)
Oil—production: 0 bbl/day (2009 est.)
country comparison to the world: 202
Oil—consumption: 8,000 bbl/day (2009 est.)
country comparison to the world: 153
Oil—exports: 0 bbl/day (2007 est.)
country comparison to the world: 184
Oil—imports: 6,960 bbl/day (2007 est.)
country comparison to the world: 148
Oil—proved reserves: 0 bbl (1 January 2010 est.)
country comparison to the world: 160
Natural gas—production: 0 cu m (2008 est.)
country comparison to the world: 157
Natural gas—consumption: 0 cu m (2008 est.)
country comparison to the world: 196
Natural gas—exports: 0 cu m (2008 est.)
country comparison to the world: 138
Natural gas—imports: 0 cu m (2008 est.)
country comparison to the world: 85
Natural gas—proved reserves: 0 cu m (1 January 2010 est.)
country comparison to the world: 162
Current account balance: $-315 million (2010 est.)
country comparison to the world: 99
$-332 million (2009 est.)
Exports: $1.189 billion (2010 est.)
country comparison to the world: 149
$912 million (2009 est.)
Exports—commodities: tobacco 53%, tea, sugar, cotton, coffee, peanuts, wood products, apparel
Exports—partners: Germany 12.37%, Egypt 8.52%, South Africa 7.67%, Zimbabwe 7.55%, US 7.4%, Russia 6.79%, Netherlands 6.64%, Japan 4.1% (2009)
Imports: $1.675 billion (2010 est.)
country comparison to the world: 157
$1.502 billion (2009 est.)
Imports—commodities: food, petroleum products, semi-manufactures, consumer goods, transportation equipment
Imports—partners: South Africa 40.15%, China 6.79%, India 6.73%, France 5.03%, Tanzania 4.81%, Mozambique 4.03% (2009)
Reserves of foreign exchange and gold: $301 million (31 December 2010 est.)
country comparison to the world: 129
$163.4 million (31 December 2009 est.)
Debt—external: $1.213 billion (31 December 2010 est.)
country comparison to the world: 146
$1.166 billion (31 December 2009 est.)
Stock of direct foreign investment—at home: $NA
Stock of direct foreign investment—abroad: $NA
Exchange rates: Malawian kwachas (MWK) per US dollar–151.65 (2010) 141.14 (2009) 142.41 (2008) 141.12 (2007) 135.96 (2006)

COMMUNICATIONS

Telephones—main lines in use: 175,000 (2009)
country comparison to the world: 130
Telephones—mobile cellular: 2.4 million (2009)
country comparison to the world: 126
Telephone system: *general assessment:* rudimentary; privatization of Malawi Telecommunications (MTL), a necessary step in bringing improvement to telecommunications services, completed in 2006
domestic: limited fixed-line subscribership of about 1 per 100 persons; mobile-cellular services are expanding but network coverage is limited and is based around the main urban areas; mobile-cellular subscribership about 15 per 100 persons
international: country code–265; satellite earth stations–2 Intelsat (1 Indian Ocean, 1 Atlantic Ocean) (2009)
Broadcast media: radio is the main broadcast medium; state-run radio has the widest geographic broadcasting reach, but about a dozen privately-owned radio stations broadcast in major urban areas; the single television network is government-owned; relays of multiple international broadcasters are available (2007)
Internet country code: .mw
Internet hosts: 870 (2010)
country comparison to the world: 167
Internet users: 716,400 (2009)
country comparison to the world: 109

TRANSPORTATION

Airports: 32 (2010)
country comparison to the world: 113
Airports—with paved runways: *total:* 6
over 3,047 m: 1
1,524 to 2,437 m: 1
914 to 1,523 m: 4 (2010)
Airports—with unpaved runways:
total: 26
1,524 to 2,437 m: 1
914 to 1,523 m: 13
under 914 m: 12 (2010)
Railways: *total:* 797 km
country comparison to the world: 99
narrow gauge: 797 km 1.067-m gauge (2010)
Roadways: *total:* 15,451 km
country comparison to the world: 120
paved: 6,956 km
unpaved: 8,495 km (2003)
Waterways: 700 km (on Lake Nyasa [Lake Malawi] and Shire River) (2010)
country comparison to the world: 76
Ports and terminals: Chipoka, Monkey Bay, Nkhata Bay, Nkhotakota, Chilumba

MILITARY

Military branches: Malawi Defense Forces (MDF): Army (includes Air Wing, Naval Detachment) (2011)
Military service age and obligation: 18 years of age for voluntary military service; standard obligation is 2 years of active duty and 5 years reserve service (2007)
Manpower available for military service: *males age 16–49:* 3,514,809 (2010 est.)
Manpower fit for military service: *males age 16–49:* 2,132,909
females age 16–49: 2,043,925 (2010 est.)
Manpower reaching militarily significant age annually:
male: 183,683
female: 183,028 (2010 est.)
Military expenditures: 1.3% of GDP (2006)
country comparison to the world: 111

TRANSNATIONAL ISSUES

Disputes—international: disputes with Tanzania over the boundary in Lake Nyasa (Lake Malawi) and the meandering Songwe River remain dormant

MALAYSIA

INTRODUCTION

Background: During the late 18th and 19th centuries, Great Britain established colonies and protectorates in the area of current Malaysia; these were occupied by Japan from 1942 to 1945. In 1948, the British-ruled territories on the Malay Peninsula formed the Federation of Malaya, which became independent in 1957. Malaysia was formed in 1963 when the former British colonies of Singapore and the East Malaysian states of Sabah and Sarawak on the northern coast of Borneo joined the Federation. The first several years of the country's history were marred by a Communist insurgency, Indonesian confrontation with Malaysia, Philippine claims to Sabah, and Singapore's secession from the Federation in 1965. During the 22-year term of Prime Minister MAHATHIR bin Mohamad (1981-2003),

Malaysia was successful in diversifying its economy from dependence on exports of raw materials to expansion in manufacturing, services, and tourism. Current Prime Minister Mohamed NAJIB bin Abdul Razak (in office since April 2009) has continued these pro-business policies.

GEOGRAPHY

Location: Southeastern Asia, peninsula bordering Thailand and northern one-third of the island of Borneo, bordering Indonesia, Brunei, and the South China Sea, south of Vietnam
Geographic coordinates: 2 30 N, 112 30 E
Map references: Southeast Asia
Area: *total:* 329,847 sq km
country comparison to the world: 66
land: 328,657 sq km
water: 1,190 sq km
Area—comparative: slightly larger than New Mexico
Land boundaries: *total:* 2,669 km
border countries: Brunei 381 km, Indonesia 1,782 km, Thailand 506 km
Coastline: 4,675 km (Peninsular Malaysia 2,068 km, East Malaysia 2,607 km)
Maritime claims: *territorial sea:* 12 nm
exclusive economic zone: 200 nm
continental shelf: 200 m depth or to the depth of exploitation; specified boundary in the South China Sea
Climate: tropical; annual southwest (April to October) and northeast (October to February) monsoons
Terrain: coastal plains rising to hills and mountains
Elevation extremes: *lowest point:* Indian Ocean 0 m
highest point: Gunung Kinabalu 4,100 m
Natural resources: tin, petroleum, timber, copper, iron ore, natural gas, bauxite
Land use: *arable land:* 5.46%
permanent crops: 17.54%
other: 77% (2005)
Irrigated land: 3,650 sq km (2003)
Total renewable water resources: 580 cu km (1999)
Freshwater withdrawal (domestic/industrial/agricultural): *total:* 9.02 cu km/yr (17%/21%/62%)
per capita: 356 cu m/yr (2000)
Natural hazards: flooding; landslides; forest fires
Environment—current issues: air pollution from industrial and vehicular emissions; water pollution from raw sewage; deforestation; smoke/haze from Indonesian forest fires
Environment—international agreements: *party to:* Biodiversity, Climate Change, Climate Change-Kyoto Protocol, Desertification, Endangered Species, Hazardous Wastes, Law of the Sea, Marine Life Conservation, Ozone Layer Protection, Ship Pollution, Tropical Timber 83, Tropical Timber 94, Wetlands
signed, but not ratified: none of the selected agreements
Geography—note: strategic location along Strait of Malacca and southern South China Sea

PEOPLE

Population: 28,728,607 (July 2011 est.)
country comparison to the world: 43
Age structure: *0–14 years:* 29.6% (male 4,374,495/female 4,132,009)
15–64 years: 65.4% (male 9,539,972/female 9,253,574)
65 years and over: 5% (male 672,581/female 755,976) (2011 est.)
Median age: *total:* 26.8 years
male: 26.7 years
female: 27 years (2011 est.)
Population growth rate: 1.576% (2011 est.)
country comparison to the world: 74
Birth rate: 21.08 births/1,000 population (2011 est.)
country comparison to the world: 83
Death rate: 4.93 deaths/1,000 population (July 2011 est.)
country comparison to the world: 189
Net migration rate: -0.39 migrant(s)/1,000 population
country comparison to the world: 133
note: does not reflect net flow of an unknown number of illegal immigrants from other countries in the region (2011 est.)
Urbanization: *urban population:* 72% of total population (2010)
rate of urbanization: 2.4% annual rate of change (2010-15 est.)
Major cities—population: KUALA LUMPUR (capital) 1.493 million; Klang 1.071 million; Johor Bahru 958,000 (2009)
Sex ratio:
at birth: 1.069 male(s)/female
under 15 years: 1.06 male(s)/female
15–64 years: 1.01 male(s)/female
65 years and over: 0.79 male(s)/female
total population: 1.01 male(s)/female (2011 est.)
Infant mortality rate: *total:* 15.02 deaths/1,000 live births
country comparison to the world: 119
male: 17.37 deaths/1,000 live births
female: 12.52 deaths/1,000 live births (2011 est.)
Life expectancy at birth: *total population:* 73.79 years
country comparison to the world: 112
male: 71.05 years
female: 76.73 years (2011 est.)
Total fertility rate: 2.67 children born/woman (2011 est.)
country comparison to the world: 76
HIV/AIDS—adult prevalence rate: 0.5% (2009 est.)
country comparison to the world: 67
HIV/AIDS—people living with HIV/AIDS: 100,000 (2009 est.)
country comparison to the world: 42
HIV/AIDS—deaths: 5,800 (2009 est.)
country comparison to the world: 36
Major infectious diseases: degree of risk: high
food or waterborne diseases: bacterial diarrhea
vectorborne diseases: dengue fever and malaria
note: highly pathogenic H5N1 avian influenza has been identified in this country; it poses a negligible risk with extremely rare cases possible among US citizens who have close contact with birds (2009)
Drinking water source: *Improved:*
urban: 100% of population
rural: 99% of population
total: 100% of population
Unimproved:
urban: 0% of population
rural: 1% of population
total: 0% of population (2008)
Sanitation facility access: *Improved:*
urban: 96% of population
rural: 95% of population
total: 96% of population
Unimproved:
urban: 4% of population
rural: 5% of population
total: 4% of population (2008)
Nationality: *noun:* Malaysian(s)
adjective: Malaysian
Ethnic groups: Malay 50.4%, Chinese 23.7%, indigenous 11%, Indian 7.1%, others 7.8% (2004 est.)
Religions: Muslim 60.4%, Buddhist 19.2%, Christian 9.1%, Hindu 6.3%, Confucianism, Taoism, other traditional Chinese religions 2.6%, other or unknown 1.5%, none 0.8% (2000 census)
Languages: Bahasa Malaysia (official), English, Chinese (Cantonese, Mandarin, Hokkien, Hakka, Hainan, Foochow), Tamil, Telugu, Malayalam, Panjabi, Thai
note: in East Malaysia there are several indigenous languages; most widely spoken are Iban and Kadazan
Literacy: *definition:* age 15 and over can read and write
total population: 88.7%
male: 92%
female: 85.4% (2000 census)

School life expectancy (primary to tertiary education):
total: 13 years
male: 12 years
female: 13 years (2008)
Education expenditures: 4.1% of GDP (2008)
country comparison to the world: 99

GOVERNMENT

Country name: *conventional long form:* none
conventional short form: Malaysia
local long form: none
local short form: Malaysia
former: Federation of Malaya
Government type: constitutional monarchy
note: nominally headed by paramount ruler (commonly referred to as the King) and a bicameral Parliament consisting of a nonelected upper house and an elected lower house; all Peninsular Malaysian states have hereditary rulers (commonly referred to as sultans) except Melaka and Pulau Pinang (Penang); those two states along with Sabah and Sarawak in East Malaysia have governors appointed by government; powers of state governments are limited by federal constitution; under terms of federation, Sabah and Sarawak retain certain constitutional prerogatives (e.g., right to maintain their own immigration controls)
Capital: *name:* Kuala Lumpur
geographic coordinates: 3 10 N, 101 42 E
time difference: UTC+8 (13 hours ahead of Washington, DC during Standard Time)
note: Putrajaya is referred to as administrative center not capital; Parliament meets in Kuala Lumpur
Administrative divisions: 13 states (negeri-negeri, singular—negeri) Johor, Kedah, Kelantan, Melaka, Negeri Sembilan, Pahang, Perak, Perlis, Pulau Pinang, Sabah, Sarawak, Selangor, Terengganu; and 1 federal territory (Wilayah Persekutuan) with three components, city of Kuala Lumpur, Labuan, and Putrajaya
Independence: 31 August 1957 (from the UK)
National holiday: Independence Day 31 August (1957) (independence of Malaya); Malaysia Day 16 September (1963) (formation of Malaysia)
Constitution: 31 August 1957; amended many times the latest in 2007
Legal system: mixed legal system of English common law, Islamic law, and customary law; judicial review of legislative acts in the Supreme Court at request of supreme head of the federation
International law organization participation: has not submitted an ICJ jurisdiction declaration; non-party state to the ICCt
Suffrage: 21 years of age; universal
Executive branch: *chief of state:* King—Sultan MIZAN Zainal Abidin (since 13 December 2006); the position of the king is primarily ceremonial
head of government: Prime Minister Mohamed NAJIB bin Abdul Razak (since 3 April 2009); Deputy Prime Minister MUHYIDDIN bin Mohamed Yassin (since 9 April 2009)
cabinet: Cabinet appointed by the prime minister from among the members of Parliament with consent of the king
(For more information visit the World Leaders website)
elections: kings elected by and from the hereditary rulers of nine of the states for five-year terms; selection based on principle of rotation among rulers of states; election last held on 3 November 2006 (next to be held in 2011); prime minister designated from among the members of the House of Representatives; following legislative elections, the leader who commands the support of the majority of members in the House becomes prime minister (since independence this has been the leader of the UMNO party)
election results: Sultan MIZAN Zainal Abidin elected king
Legislative branch: bicameral Parliament or Parlimen consists of Senate or Dewan Negara (70 seats; 44 members appointed by the king, 26 elected by 13 state legislatures to serve three-year terms with a two term limit) and House of Representatives or Dewan Rakyat (222 seats; members elected by popular vote to serve up to five-year terms)
elections: House of Representatives—last held on 8 March 2008 (next to be held by June 2013)
election results: House of Representatives—percent of vote—BN coalition 50.3%, opposition parties 46.8%, others 2.9%; seats—BN coalition 140, opposition parties 82; (seats by party as of March 2011—BN coalition 137, opposition parties 76, independents 9)
Judicial branch: civil courts include Federal Court, Court of Appeal, High Court of Malaya on peninsula Malaysia, and High Court of Sabah and Sarawak in states of Borneo (judges are appointed by the king on the advice of the prime minister); sharia courts include Sharia Appeal Court, Sharia High Court, and Sharia Subordinate Courts at state-level and deal with religious and family matters such as custody, divorce, and inheritance only for Muslims; decisions of sharia courts cannot be appealed to civil courts
Political parties and leaders: *National Front (Barisan Nasional) or BN (ruling coalition) consists of the following parties:* Gerakan Rakyat Malaysia Party or PGRM [KOH Tsu Koon]; Liberal Democratic Party (Parti Liberal Demokratik—Sabah) or LDP [LIEW Vui Keong]; Malaysian Chinese Association (Persatuan China Malaysia) or MCA [CHUA Soi Lek]; Malaysian Indian Congress (Kongres India Malaysia) or MIC [Govindasamy PALANIVEL]; Parti Bersatu Rakyat Sabah or PBRS [Joseph KURUP]; Parti Bersatu Sabah or PBS [Joseph PAIRIN Kitingan]; Parti Pesaka Bumiputera Bersatu or PBB [Abdul TAIB Mahmud]; Parti Rakyat Sarawak or PRS [James MASING]; Sarawak United People's Party (Parti Bersatu Rakyat Sarawak) or SUPP [George CHAN Hong Nam]; United Malays National Organization or UMNO [NAJIB bin Abdul Razak]; United Pasokmomogun Kadazandusun Murut Organization (Pertubuhan Pasko Momogun Kadazan Dusun Bersatu) or UPKO [Bernard DOMPOK]; People's Progressive Party (Parti Progresif Penduduk Malaysia) or PPP [M.Kayveas]; Sarawak Progressive Democratic Party or SPDP [William MAWAN])
People's Alliance (Pakatan Rakyat) or PR (opposition coalition) consists of the following parties: Democratic Action Party (Parti Tindakan Demokratik) or DAP [KARPAL Singh]; Islamic Party of Malaysia (Parti Islam se Malaysia) or PAS [Abdul HADI Awang]; People's Justice Party (Parti Keadilan Rakyat) or PKR [WAN AZIZAH Wan Ismail]; Sarawak National Party or SNAP [Edwin DUNDANG]
independent party: Sabah Progressive Party (Parti Progresif Saban) or SAPP [YONG Teck Lee]
Political pressure groups and leaders: Bar Council; BERSIH (electoral reform coalition); PEMBELA (Muslim NGO coalition); PERKASA (defense of Malay rights)
other: religious groups; women's groups; youth groups
International organization participation: ADB, APEC, ARF, ASEAN, BIS, C, CICA (observer), CP, D-8, EAS, FAO, G-15, G-77, IAEA, IBRD, ICAO, ICC, ICRM, IDA, IDB, IFAD, IFC, IFRCS, IHO, ILO, IMF, IMO, IMSO, Interpol, IOC, IPU, ISO, ITSO, ITU, ITUC, MIGA, MINURSO, MONUSCO, NAM, OIC, OPCW, PCA, PIF (partner), UN, UNAMID, UNCTAD, UNESCO, UNIDO, UNIFIL, UNMIL, UNMIS, UNMIT, UNWTO, UPU, WCO, WFTU, WHO, WIPO, WMO, WTO
Diplomatic representation in the US: *chief of mission:* Ambassador JAMALUDDIN Jarjis
chancery: 3516 International Court NW, Washington, DC 20008
telephone: [1] (202) 572-9700
FAX: [1] (202) 572-9882
consulate(s) general: Los Angeles, New York
Diplomatic representation from the US: *chief of mission:* Ambassador Paul W. JONES
embassy: 376 Jalan Tun Razak, 50400 Kuala Lumpur
mailing address: US Embassy Kuala Lumpur, APO AP 96535-8152
telephone: [60] (3) 2168-5000
FAX: [60] (3) 2142-2207
Flag description: 14 equal horizontal stripes of red (top) alternating with white (bottom); there is a blue rectangle in the upper hoist-side corner bearing a yellow crescent and a yellow 14-pointed star; the flag is often referred to as Jalur Gemilang (Stripes of Glory); the 14 stripes stand for the equal status in the federation of the 13 member states and the federal government; the 14 points on the star represent the unity between these entities; the crescent is a traditional symbol of Islam; blue symbolizes the unity of the Malay people and yellow is the royal color of Malay rulers
note: the design is based on the flag of the US
National anthem: *name:* "Negaraku" (My Country)
lyrics/music: collective, led by Tunku ABDUL RAHMAN/Pierre Jean DE BERANGER

note: adopted 1957; the full version is only performed in the presence of the king; the tune, which was adopted from a popular French melody titled "La Rosalie," was originally the anthem of the state of Perak

ECONOMY

Economy—overview: Malaysia, a middle-income country, has transformed itself since the 1970s from a producer of raw materials into an emerging multi-sector economy. Under current Prime Minister NAJIB, Malaysia is attempting to achieve high-income status by 2020 and to move farther up the value-added production chain by attracting investments in Islamic finance, high technology industries, biotechnology, and services. The NAJIB administration also is continuing efforts to boost domestic demand and reduce the economy's dependence on exports. Nevertheless, exports–particularly of electronics, oil and gas, palm oil and rubber–remain a significant driver of the economy. As an oil and gas exporter, Malaysia has profited from higher world energy prices, although the rising cost of domestic gasoline and diesel fuel, combined with strained government finances, has forced Kuala Lumpur begin to reduce government subsidies. The government is also trying to lessen its dependence on state oil producer Petronas, which supplies more than 40% of government revenue. The central bank maintains healthy foreign exchange reserves and its well-developed regulatory regime has limited Malaysia's exposure to riskier financial instruments and the global financial crisis. Nevertheless, decreasing worldwide demand for consumer goods hurt Malaysia's exports and economic growth in 2009, although both showed signs of recovery in 2010. In order to attract increased investment, NAJIB has raised possible revisions to the special economic and social preferences accorded to ethnic Malays under the New Economic Policy of 1970, but he has encountered significant opposition, especially from Malay nationalists and other vested interests.

GDP (purchasing power parity): $414.4 billion (2010 est.)
country comparison to the world: 30
$386.8 billion (2009 est.)
$393.5 billion (2008 est.)
note: data are in 2010 US dollars

GDP (official exchange rate): $238 billion (2010 est.)

GDP—real growth rate: 7.2% (2010 est.)
country comparison to the world: 33
-1.7% (2009 est.)
4.7% (2008 est.)

GDP—per capita (PPP): $14,700 (2010 est.)
country comparison to the world: 76
$13,900 (2009 est.)
$14,400 (2008 est.)
note: data are in 2010 US dollars

GDP—composition by sector:
agriculture: 9.1%
industry: 41.6%
services: 49.3% (2010 est.)

Labor force: 12.2 million (2010 est.)
country comparison to the world: 42

Labor force—by occupation:
agriculture: 13%
industry: 36%
services: 51% (2005 est.)

Unemployment rate: 3.5% (2010 est.)
country comparison to the world: 28
3.7% (2009 est.)

Population below poverty line: 3.6% (2007 est.)

Household income or consumption by percentage share: *lowest 10%*: 2.6%
highest 10%: 28.5% (2005 est.)

Distribution of family income—Gini index: 44.1 (2009)
country comparison to the world: 43
49.2 (1997)

Investment (gross fixed): 20.1% of GDP (2010 est.)
country comparison to the world: 87

Budget: *revenues*: $46.78 billion
expenditures: $46.34 billion (2010 est.)

Public debt: 53.1% of GDP (2010 est.)
country comparison to the world: 49
53.3% of GDP (2009 est.)

Inflation rate (consumer prices): 1.7% (2010 est.)
country comparison to the world: 45
0.6% (2009 est.)
note: approximately 30% of goods are price-controlled

Central bank discount rate: 2.83% (31 December 2010)
country comparison to the world: 134
1% (31 December 2009)

Commercial bank prime lending rate: 6.27% (31 December 2010 est.)
country comparison to the world: 137
5.51% (31 December 2009 est.)

Stock of narrow money: $73.8 billion (31 December 2010 est.)
country comparison to the world: 37
$64.8 billion (31 December 2009 est.)

Stock of broad money: $358.1 billion (31 December 2010 est.)
country comparison to the world: 24
$319.2 billion (31 December 2009 est.)

Stock of domestic credit: $314.7 billion (31 December 2010 est.)
country comparison to the world: 33
$265.2 billion (31 December 2009 est.)

Market value of publicly traded shares: $256 billion (31 December 2009)
country comparison to the world: 25
$187.1 billion (31 December 2008)
$325.7 billion (31 December 2007)

Agriculture—products: Peninsular Malaysia–rubber, palm oil, cocoa, rice; Sabah–subsistence crops, coconuts, rice; rubber, timber; Sarawak–rubber, timber; pepper

Industries: Peninsular Malaysia–rubber and oil palm processing and manufacturing, light manufacturing, pharmaceuticals, medical technology, electronics, tin mining and smelting, logging, timber processing; Sabah–logging, petroleum production; Sarawak–agriculture processing, petroleum production and refining, logging

Industrial production growth rate: 7.5% (2010 est.)
country comparison to the world: 37

Electricity—production: 107.4 billion kWh (2009 est.)
country comparison to the world: 32

Electricity—consumption: 93.8 billion kWh (2009 est.)
country comparison to the world: 31

Electricity—exports: 91.7 million kWh (2009 est.)

Electricity—imports: 0 kWh (2009 est.)

Oil—production: 693,700 bbl/day (2009 est.)
country comparison to the world: 28

Oil—consumption: 536,000 bbl/day (2009 est.)
country comparison to the world: 31

Oil—exports: 511,900 bbl/day (2007 est.)
country comparison to the world: 30

Oil—imports: 314,600 bbl/day (2007 est.)
country comparison to the world: 35

Oil—proved reserves: 2.9 billion bbl (1 January 2010 est.)
country comparison to the world: 32

Natural gas—production: 57.3 billion cu m (2008 est.)
country comparison to the world: 17

Natural gas—consumption: 26.27 billion cu m (2008 est.)
country comparison to the world: 30

Natural gas—exports: 31.03 billion cu m (2008 est.)
country comparison to the world: 8

Natural gas—imports: 0 cu m (2008 est.)
country comparison to the world: 91

Natural gas—proved reserves: 2.35 trillion cu m (1 January 2010 est.)
country comparison to the world: 16

Current account balance: $34.14 billion (2010 est.)
country comparison to the world: 13
$33.24 billion (2009 est.)

Exports: $210.3 billion (2010 est.)
country comparison to the world: 22
$163.2 billion (2009 est.)

Exports—commodities: electronic equipment, petroleum and liquefied natural gas, wood and wood products, palm oil, rubber, textiles, chemicals

Exports—partners: Singapore 13.4%, China 12.6%, Japan 10.4%, US 9.5%, Thailand 5.3%, Hong Kong 5.1% (2010 est.)

Imports: $174.3 billion (2010 est.)
country comparison to the world: 23
$128.3 billion (2009 est.)

Imports—commodities: electronics, machinery, petroleum products, plastics, vehicles, iron and steel products, chemicals

Imports—partners: China 12.6%, Japan 12.6%, Singapore 11.4%, US 10.7%, Thailand 6.2%, Indonesia 5.6% (2010 est.)

Reserves of foreign exchange and gold: $106.5 billion (31 December 2010 est.)
country comparison to the world: 16
$96.71 billion (31 December 2009 est.)

Debt—external: $72.6 billion (31 December 2010 est.)
country comparison to the world: 45
$67.4 billion (31 December 2009 est.)

Stock of direct foreign investment—at home: $77.44 billion (31 December 2010 est.)
country comparison to the world: 46
$74.64 billion (31 December 2009 est.)

Stock of direct foreign investment—abroad: $82.65 billion (31 December 2010 est.)

country comparison to the world: 27
$75.62 billion (31 December 2009 est.)
Exchange rates: ringgits (MYR) per US dollar–3.04 (2010) 3.52 (2009) 3.33 (2008) 3.46 (2007) 3.6683 (2006)

COMMUNICATIONS

Telephones—main lines in use: 4.312 million (2009)
country comparison to the world: 35
Telephones—mobile cellular: 30.379 million (2009)
country comparison to the world: 31
Telephone system: general assessment: modern system featuring good intercity service on Peninsular Malaysia provided mainly by microwave radio relay and an adequate intercity microwave radio relay network between Sabah and Sarawak via Brunei; international service excellent
domestic: domestic satellite system with 2 earth stations; combined fixed-line and mobile-cellular teledensity 135 per 100 persons
international: country code–60; landing point for several major international submarine cable networks that provide connectivity to Asia, Middle East, and Europe; satellite earth stations–2 Intelsat (1 Indian Ocean, 1 Pacific Ocean) (2008)
Broadcast media: state-owned television broadcaster operates 2 TV networks with relays throughout the country, and the leading private commercial media group operates 4 TV stations with numerous relays throughout the country; satellite TV subscription service is available; state-owned radio broadcaster operates multiple national networks as well as regional and local stations; large number of private commercial radio broadcasters and some subscription satellite radio services are available; about 400 radio stations overall (2008)
Internet country code: .my
Internet hosts: 344,452 (2010)
country comparison to the world: 56
Internet users: 15.355 million (2009)
country comparison to the world: 26

TRANSPORTATION

Airports: 118 (2010)
country comparison to the world: 51
Airports—with paved runways: *total:* 38
over 3,047 m: 7
2,438 to 3,047 m: 10
1,524 to 2,437 m: 6
914 to 1,523 m: 8
under 914 m: 7 (2010)
Airports—with unpaved runways: *total:* 80
914 to 1,523 m: 7
under 914 m: 73 (2010)
Heliports: 3 (2010)
Pipelines: condensate 3 km; gas 1,757 km; liquid petroleum gas 155 km; oil 30 km; refined products 114 km (2010)
Railways: *total:* 1,849 km
country comparison to the world: 75
standard gauge: 57 km 1.435-m gauge (57 km electrified)
narrow gauge: 1,792 km 1.000-m gauge (150 km electrified) (2010)
Roadways: *total:* 98,721 km
country comparison to the world: 43
paved: 80,280 km (includes 1,821 km of expressways)
unpaved: 18,441 km (2004)
Waterways: 7,200 km (Peninsular Malaysia 3,200 km; Sabah 1,500 km; Sarawak 2,500 km) (2011)
country comparison to the world: 20
Merchant marine: *total:* 321
country comparison to the world: 30
by type: bulk carrier 9, cargo 97, carrier 2, chemical tanker 45, container 44, liquefied gas 35, passenger/cargo 4, petroleum tanker 79, roll on/roll off 2, vehicle carrier 4
foreign-owned: 35 (Denmark 1, Hong Kong 8, Japan 4, Nigeria 1, Russia 2, Singapore 19)
registered in other countries: 79 (Bahamas 13, India 1, Indonesia 1, Malta 1, Marshall Islands 11, Panama 12, Papua New Guinea 1, Philippines 1, Saint Kitts and Nevis 1, Sierra Leone 1, Singapore 27, Thailand 3, Tuvalu 1, US 2, unknown 3) (2010)
Ports and terminals: Bintulu, Johor Bahru, George Town (Penang), Port Kelang (Port Klang), Tanjung Pelepas
Transportation—note: the International Maritime Bureau reports that the territorial and offshore waters in the Strait of Malacca and South China Sea remain high risk for piracy and armed robbery against ships; in the past, commercial vessels have been attacked and hijacked both at anchor and while underway; hijacked vessels are often disguised and cargo diverted to ports in East Asia; crews have been murdered or cast adrift; increased naval patrols since 2005 in the Strait of Malacca resulted in no reported incidents in 2010

MILITARY

Military branches: Malaysian Armed Forces (Angkatan Tentera Malaysia, ATM): Malaysian Army (Tentera Darat Malaysia), Royal Malaysian Navy (Tentera Laut Diraja Malaysia, TLDM), Royal Malaysian Air Force (Tentera Udara Diraja Malaysia, TUDM) (2010)
Military service age and obligation: 18 years of age for voluntary military service (2005)
Manpower available for military service: *males age 16–49:* 7,501,518
females age 16–49: 7,315,999 (2010 est.)
Manpower fit for military service: *males age 16–49:* 6,247,306
females age 16–49: 6,175,274 (2010 est.)
Manpower reaching militarily significant age annually: *male:* 265,008
female: 254,812 (2010 est.)
Military expenditures: 2.03% of GDP (2005 est.)
country comparison to the world: 69

TRANSNATIONAL ISSUES

Disputes—international: while the 2002 "Declaration on the Conduct of Parties in the South China Sea" has eased tensions over the Spratly Islands, it is not the legally binding "code of conduct" sought by some parties; Malaysia was not party to the March 2005 joint accord among the national oil companies of China, the Philippines, and Vietnam on conducting marine seismic activities in the Spratly Islands; disputes continue over deliveries of fresh water to Singapore, Singapore's land reclamation, bridge construction, and maritime boundaries in the Johor and Singapore Straits; in 2008, ICJ awards sovereignty of Pedra Branca (Pulau Batu Puteh/Horsburgh Island) to Singapore, and Middle Rocks to Malaysia, but does not rule on maritime regimes, boundaries, or disposition of South Ledge; ICJ awarded Ligitan and Sipadan islands, also claimed by Indonesia and Philippines, to Malaysia but left maritime boundary and sovereignty of Unarang rock in the hydrocarbon-rich Celebes Sea in dispute; separatist violence in Thailand's predominantly Muslim southern provinces prompts measures to close and monitor border with Malaysia to stem terrorist activities; Philippines retains a dormant claim to Malaysia's Sabah State in northern Borneo; Per Letters of Exchange signed in 2009, Malaysia in 2010 ceded two hydrocarbon concession blocks to Brunei in exchange for Brunei's sultan dropping claims to the Limbang corridor, which divides Brunei; piracy remains a problem in the Malacca Strait
Refugees and internally displaced persons: refugees (country of origin): 15,174 (Indonesia); 21,544 (Burma) (2007)
Trafficking in persons: current situation: Malaysia is a destination and, to a lesser extent, a source and transit country for women and children trafficked for the purpose of commercial sexual exploitation, and men, women, and children for forced labor; Malaysia is mainly a destination country for men, women, and children who migrate willingly from South and Southeast Asia to work, some of whom are subjected to conditions of involuntary servitude by Malaysian employers in the domestic, agricultural, construction, plantation, and industrial sectors; to a lesser extent, some Malaysian women, primarily of Chinese ethnicity, are trafficked abroad for commercial sexual exploitation
tier rating: Tier 2 Watch List–the Government of Malaysia does not fully comply with the minimum standards for the elimination of trafficking and is not making significant efforts to do so, despite some progress in enforcing the 2007 comprehensive anti-trafficking law; it has yet to fully address labor trafficking in Malaysia; there are credible allegations of involvement of Malaysian immigration officials in trafficking and extorting Burmese refugees; the government did not develop mechanisms to effectively screen victims of trafficking in vulnerable groups and condones the confiscation of passports of migrant workers by employers (2009)
Illicit drugs: drug trafficking prosecuted vigorously and carries severe penalties; heroin still primary drug of abuse, but synthetic drug demand remains strong; continued ecstasy and methamphetamine producer for domestic users and, to a lesser extent, the regional drug market

MALDIVES

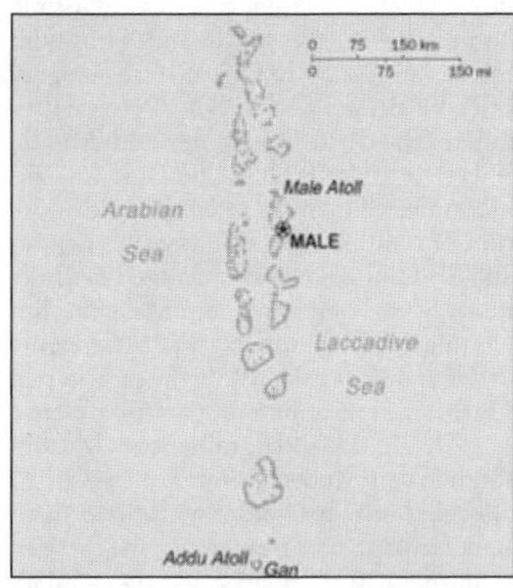

INTRODUCTION

Background: Maldives was long a sultanate, first under Dutch and then under British protection. It became a republic in 1968, three years after independence. President Maumoon Abdul GAYOOM dominated the islands' political scene for 30 years, elected to six successive terms by single-party referendums. Following riots in the capital Male in August 2004, the president and his government pledged to embark upon democratic reforms including a more representative political system and expanded political freedoms. Progress was sluggish, however, and many promised reforms were slow to be realized. Nonetheless, political parties were legalized in 2005. In June 2008, a constituent assembly–termed the "Special Majlis"–finalized a new constitution, which was ratified by the president in August. The first-ever presidential elections under a multi-candidate, multi-party system were held in October 2008. GAYOOM was defeated in a runoff poll by Mohamed NASHEED, a political activist who had been jailed several years earlier by the former regime. Challenges facing President NASHEED include strengthening democracy and combating poverty and drug abuse. Maldives officials have played a prominent role in the international climate change discussion (due to the islands' low elevation and the threat from sea-level rise) and on the United Nations Human Rights Council.

GEOGRAPHY

Location: Southern Asia, group of atolls in the Indian Ocean, south-southwest of India
Geographic coordinates: 3 15 N, 73 00 E
Map references: Asia
Area: *total:* 298 sq km
country comparison to the world: 208
land: 298 sq km
water: 0 sq km
Area—comparative: about 1.7 times the size of Washington, DC
Land boundaries: 0 km
Coastline: 644 km
Maritime claims: measured from claimed archipelagic straight baselines
territorial sea: 12 nm
contiguous zone: 24 nm
exclusive economic zone: 200 nm
Climate: tropical; hot, humid; dry, northeast monsoon (November to March); rainy, southwest monsoon (June to August)
Terrain: flat, with white sandy beaches
Elevation extremes: *lowest point:* Indian Ocean 0 m
highest point: unnamed location on Viligili in the Addu Atholhu 2.4 m
Natural resources: fish
Land use: *arable land:* 13.33%
permanent crops: 30%
other: 56.67% (2005)
Irrigated land: NA
Total renewable water resources: 0.03 cu km (1999)
Freshwater withdrawal (domestic/industrial/agricultural): *total:* 0.003 cu km/yr (98%/2%/0%)
per capita: 9 cu m/yr (1987)
Natural hazards: tsunamis; low elevation of islands makes them sensitive to sea level rise
Environment—current issues: depletion of freshwater aquifers threatens water supplies; global warming and sea level rise; coral reef bleaching
Environment—international agreements: *party to:* Biodiversity, Climate Change, Climate Change-Kyoto Protocol, Desertification, Hazardous Wastes, Law of the Sea, Ozone Layer Protection, Ship Pollution
signed, but not ratified: none of the selected agreements
Geography—note: 1,190 coral islands grouped into 26 atolls (200 inhabited islands, plus 80 islands with tourist resorts); archipelago with strategic location astride and along major sea lanes in Indian Ocean

PEOPLE

Population: 394,999 (July 2011 est.)
country comparison to the world: 175
Age structure: *0–14 years:* 21.5% (male 43,332/female 41,642)
15–64 years: 74.4% (male 177,365/female 116,552)
65 years and over: 4.1% (male 7,888/female 8,220) (2011 est.)
Median age: *total:* 26.2 years
male: 26.8 years
female: 25.2 years (2011 est.)
Population growth rate: -0.151% (2011 est.)
country comparison to the world: 209
Birth rate: 14.83 births/1,000 population (2011 est.)
country comparison to the world: 136
Death rate: 3.71 deaths/1,000 population (July 2011 est.)
country comparison to the world: 209
Net migration rate: -12.62 migrant(s)/1,000 population (2011 est.)
country comparison to the world: 214
Urbanization: *urban population:* 40% of total population (2010)
rate of urbanization: 4.2% annual rate of change (2010-15 est.)
Major cities—population: MALE (capital) 120,000 (2009)
Sex ratio: *at birth:* 1.05 male(s)/female
under 15 years: 1.04 male(s)/female
15–64 years: 1.57 male(s)/female
65 years and over: 0.98 male(s)/female
total population: 1.4 male(s)/female (2011 est.)
Infant mortality rate: *total:* 27.45 deaths/1,000 live births
country comparison to the world: 75
male: 29.93 deaths/1,000 live births
female: 24.84 deaths/1,000 live births (2011 est.)
Life expectancy at birth: *total population:* 74.45 years
country comparison to the world: 100
male: 72.22 years
female: 76.8 years (2011 est.)
Total fertility rate: 1.81 children born/woman (2011 est.)
country comparison to the world: 152
HIV/AIDS—adult prevalence rate: less than 0.1% (2009 est.)
country comparison to the world: 150
HIV/AIDS—people living with HIV/AIDS: fewer than 100 (2009 est.)
country comparison to the world: 162
HIV/AIDS—deaths: fewer than 100 (2009 est.)
country comparison to the world: 145
Drinking water source: *Improved:*
urban: 99% of population
rural: 86% of population
total: 91% of population
Unimproved:
urban: 1% of population
rural: 14% of population
total: 9% of population (2008)
Sanitation facility access: *Improved:*
urban: 100% of population
rural: 96% of population
total: 98% of population
Unimproved:
urban: 0% of population
rural: 4% of population
total: 2% of population (2008)
Nationality: *noun:* Maldivian(s)
adjective: Maldivian
Ethnic groups: South Indians, Sinhalese, Arabs
Religions: Sunni Muslim
Languages: Dhivehi (official, dialect of Sinhala, script derived from Arabic), English (spoken by most government officials)
Literacy: *definition:* age 15 and over can read and write
total population: 93.8%
male: 93%
female: 94.7% (2006 Census)
School life expectancy (primary to tertiary education): *total:* 12 years
male: 13 years
female: 12 years (2006)
Education expenditures: 11.2% of GDP (2009)
country comparison to the world: 5

GOVERNMENT

Country name: *conventional long form:* Republic of Maldives
conventional short form: Maldives
local long form: Dhivehi Raajjeyge Jumhooriyyaa

local short form: Dhivehi Raajje
Government type: republic
Capital: *name:* Male
geographic coordinates: 4 10 N, 73 30 E
time difference: UTC+5 (10 hours ahead of Washington, DC during Standard Time)
Administrative divisions: 19 atolls (atholhu, singular and plural) and the capital city*; Alifu, Baa, Dhaalu, Faafu, Gaafu Alifu, Gaafu Dhaalu, Gnaviyani, Haa Alifu, Haa Dhaalu, Kaafu, Laamu, Lhaviyani, Maale (Male)*, Meemu, Noonu, Raa, Seenu, Shaviyani, Thaa, Vaavu
Independence: 26 July 1965 (from the UK)
National holiday: Independence Day, 26 July (1965)
Constitution: new constitution ratified 7 August 2008
Legal system: Islamic religious legal system with English common law influences, primarily in commercial matters
International law organization participation: has not submitted an ICJ jurisdiction declaration; non-party state to the ICCt
Suffrage: 18 years of age; universal
Executive branch: *chief of state:* President Mohamed "Anni" NASHEED (since 11 November 2008); Vice President Mohamed WAHEED Hassan Maniku (since 11 November 2008); note—the president is both the chief of state and head of government
head of government: President Mohamed "Anni" NASHEED (since 11 November 2008); Vice President Mohamed WAHEED Hassan Maniku (since 11 November 2008)
cabinet: Cabinet of Ministers is appointed by the president
(For more information visit the World Leaders website)
elections: under the new constitution, the president elected by direct vote; president elected for a five-year term (eligible for a second term); election last held on 8 and 28 October 2008 (next to be held in 2013)
election results: Mohamed NASHEED elected president; percent of vote—Mohamed NASHEED 54.3%, Maumoon Abdul GAYOOM 45.7%
Legislative branch: unicameral People's Council or People's Majlis (77 seats; members elected by direct vote to serve five-year terms); note—the Majlis in February 2009 passed legislation that increased the number of seats to 77 from 50
elections: last held on 9 May 2009 (next to be held in 2014)
election results: percent of vote—DRP 36.4%, MDP 33.8 %, PA 9.1%, DQP 2.6% Republican Party 1.2%, independents 16.9%; seats by party as of 23 February 2011—DRP 27, MDP 33, PA 7, DQP 1, Republican Party 2, independents 7
Judicial branch: Supreme Court; Supreme Court judges are appointed by the president with approval of voting members of the People's Council; High Court; Trial Courts; all lower court judges are appointed by the Judicial Service Commission
Political parties and leaders: Adhaalath (Justice) Party or AP [Shaykh Hussein RASHEED Ahmed]; Dhivehi Quamee Party or DQP [Hassan SAEED]; Dhivehi Rayyithunge Party (Maldivian People's Party) or DRP [THASMEEN Ali]; Gaumii Ithihaad (National Alliance) or GI [Mohamed WAHEED]; Islamic Democratic Party or IDP; Maldivian Democratic Party or MDP [Mariya Ahmed DIDI]; Maldives National Congress or MNC; Maldives Social Democratic Party or MSDP; People's Alliance or PA [Abdullah YAMEEN]; People's Party or PP; Poverty Alleviation Party or PAP; Republican (Jumhooree) Party or JP [Gasim IBRAHIM]; Social Liberal Party or SLP [Ibrahim ISMAIL]
Political pressure groups and leaders: *other:* various unregistered political parties
International organization participation: ADB, AOSIS, C, CP, FAO, G-77, IBRD, ICAO, IDA, IDB, IFAD, IFC, IFRCS (observer), ILO, IMF, IMO, Interpol, IOC, IPU, ITU, MIGA, NAM, OIC, OPCW, SAARC, SACEP, UN, UNCTAD, UNESCO, UNIDO, UNWTO, UPU, WCO, WHO, WIPO, WMO, WTO
Diplomatic representation in the US: *chief of mission:* Ambassador Abdul GHAFOOR Mohamed
chancery: 800 2nd Avenue, Suite 400E, New York, NY 10017
telephone: [1] (212) 599-6195
FAX: [1] (212) 661-6405
Diplomatic representation from the US: the US does not have an embassy in Maldives; the US Ambassador to Sri Lanka, Ambassador Patricia A. BUTENIS, is accredited to Maldives and makes periodic visits
Flag description: red with a large green rectangle in the center bearing a vertical white crescent moon; the closed side of the crescent is on the hoist side of the flag; red recalls those who have sacrificed their lives in defense of their country, the green rectangle represents peace and prosperity, and the white crescent signifies Islam
National anthem: *name:* "Gaumee Salaam" (National Salute)
lyrics/music: Mohamed Jameel DIDI/Wannakuwattawaduge DON AMARADEVA
note: lyrics adopted 1948, music adopted 1972; between 1948 and 1972, the lyrics were sung to the tune of "Auld Lang Syne"

ECONOMY

Economy—overview: Tourism, Maldives' largest economic activity, accounts for 28% of GDP and more than 60% of foreign exchange receipts. Over 90% of government tax revenue comes from import duties and tourism-related taxes. Fishing is the second leading sector, but the fish catch has dropped sharply in recent years. Agriculture and manufacturing continue to play a lesser role in the economy, constrained by the limited availability of cultivable land and the shortage of domestic labor. Most staple foods must be imported. In the last decade, real GDP growth averaged around 6% per year except for 2005, when GDP declined following the Indian Ocean tsunami, and in 2009, when GDP shrank by 3% as tourist arrivals declined and capital flows plunged in the wake of the global financial crisis. Falling tourist arrivals and fish exports, combined with high government spending on social needs, subsidies, and civil servant salaries contributed to a balance of payments crisis, which was eased with a December 2009, $79.3 million dollar IMF standby agreement. However, after the first two disbursements, the IMF withheld subsequent disbursements due to concerns over Maldives' growing budget deficit. Maldives has had chronic budget deficits in recent years and the government's plans to cut expenditures have not progressed well. A new Goods and Services Tax on Tourism (GST) was introduced in January 2011 and a new Business Profit Tax is to be introduced during the year. These taxes are expected to increase government revenue by about 25%. The government has privatized the main airport and is partially privatizing the energy sector. Tourism will remain the engine of the economy. The Government of the Maldives has aggressively promoted building new island resorts. Due to increasing tourist arrivals, the government expects GDP growth around 4.0% in 2011. Diversifying the economy beyond tourism and fishing, reforming public finance, and increasing employment opportunities are major challenges facing the government. Over the longer term Maldivian authorities worry about the impact of erosion and possible global warming on their low-lying country; 80% of the area is 1 meter or less above sea level.
GDP (purchasing power parity): $2.734 billion (2010 est.)
country comparison to the world: 179
$2.532 billion (2009 est.)
$2.658 billion (2008 est.)
note: data are in 2010 US dollars
GDP (official exchange rate): $1.87 billion (2010 est.)
GDP—real growth rate: 8% (2010 est.)
country comparison to the world: 18
-4.8% (2009 est.)
12.8% (2008 est.)
GDP—per capita (PPP): $6,900 (2010 est.)
country comparison to the world: 133
$6,400 (2009 est.)
$6,900 (2008 est.)
note: data are in 2010 US dollars
GDP—composition by sector:
agriculture: 5.6%
industry: 16.9%
services: 77.5% (2009 est.)
Labor force: 110,000 (2010)
country comparison to the world: 182
Labor force—by occupation:
agriculture: 11%
industry: 23%
services: 65% (2006 est.)
Unemployment rate: 14.5% (2010 est.)
country comparison to the world: 146
14.4% (2006 est.)
Population below poverty line: 16% (2008)
Household income or consumption by percentage share: *lowest 10%:* NA%
highest 10%: NA%
Budget: *revenues:* $476 million

expenditures: $758 million (2010 est.)
Inflation rate (consumer prices): 6% (2010 est.)
country comparison to the world: 159
7.3% (2009 est.)
Central bank discount rate: 13% (31 December 2009)
country comparison to the world: 26
13% (31 December 2008)
Commercial bank prime lending rate: 13% (31 December 2009 est.)
country comparison to the world: 60
13% (31 December 2008 est.)
Stock of narrow money: $588 million (31 October 2009)
country comparison to the world: 156
$581 million (31 December 2008)
Stock of broad money: $1.239 billion (31 December 2009)
country comparison to the world: 157
$1.064 billion (31 December 2008)
Stock of domestic credit: $1.548 billion (31 December 2008 est.)
country comparison to the world: 134
$1.08 billion (31 December 2007 est.)
Market value of publicly traded shares: $NA
Agriculture—products: coconuts, corn, sweet potatoes; fish
Industries: tourism, fish processing, shipping, boat building, coconut processing, garments, woven mats, rope, handicrafts, coral and sand mining
Industrial production growth rate: -0.9% (2004 est.)
country comparison to the world: 155
Electricity—production: 542 million kWh (2009 est.)
country comparison to the world: 157
Electricity—consumption: 542 million kWh (2009 est.)
country comparison to the world: 160
Electricity—exports: 0 kWh (2009 est.)
Electricity—imports: 0 kWh (2009 est.)
Oil—production: 0 bbl/day (2010 est.)
country comparison to the world: 208
Oil—consumption: 6,000 bbl/day (2009 est.)
country comparison to the world: 161
Oil—exports: 0 bbl/day (2010 est.)
country comparison to the world: 188
Oil—imports: 5,490 bbl/day (2008 est.)
country comparison to the world: 155
Oil—proved reserves: 0 bbl (1 January 2010 est.)
country comparison to the world: 166
Natural gas—production: 0 cu m (2010 est.)
country comparison to the world: 163
Natural gas—consumption: 0 cu m (2010 est.)
country comparison to the world: 201
Natural gas—exports: 0 cu m (2010 est.)
country comparison to the world: 145
Natural gas—imports: 0 cu m (2010 est.)
country comparison to the world: 90
Natural gas—proved reserves: 0 cu m (1 January 2010 est.)
country comparison to the world: 168
Current account balance: $-463 million (2010 est.)
country comparison to the world: 114
$-419 million (2009 est.)
Exports: $163 million (2009 est.)
country comparison to the world: 185
$331 million (2008 est.)
Exports—commodities: fish
Exports—partners: France 17.01%, Thailand 15.16%, Italy 13.49%, UK 13.13%, Sri Lanka 12.38% (2009)
Imports: $967 million (2009 est.)
country comparison to the world: 173
$1.388 billion (2008 est.)
Imports—commodities: petroleum products, ships, foodstuffs, clothing, intermediate and capital goods
Imports—partners: Singapore 24.62%, UAE 15.7%, India 11.02%, Malaysia 8.98%, Sri Lanka 5.4%, Thailand 5.36% (2009)
Debt—external: $943 million (2010 est.)
country comparison to the world: 151
$933 million (2009 est.)
Exchange rates: rufiyaa (MVR) per US dollar–12.8 (2010) 12.8 (2008) 12.8 (2007) 12.8 (2006)

COMMUNICATIONS

Telephones—main lines in use: 49,913 (2009)
country comparison to the world: 162
Telephones—mobile cellular: 461,149 (2009)
country comparison to the world: 160
Telephone system: *general assessment*: telephone services have improved; interatoll communication through microwave links; all inhabited islands and resorts are connected with telephone and fax service
domestic: each island now has at least 1 public telephone, and there are mobile-cellular networks with a rapidly expanding subscribership that exceeds 100 per 100 persons
international: country code–960; linked to international submarine cable Fiber-Optic LinkAroundtheGlobe(FLAG);satelliteearth station–3 Intelsat (Indian Ocean) (2009)
Broadcast media: state-owned radio and television monopoly until recently; state-owned TV operates 2 channels; 2 privately-owned TV stations; state owns Voice of Maldives and operates both an entertainment and a music-based station; there are 5 privately-owned radio broadcast stations operating (2009)
Internet country code: .mv
Internet hosts: 2,164 (2010)
country comparison to the world: 153
Internet users: 86,400 (2009)
country comparison to the world: 163

TRANSPORTATION

Airports: 5 (2010)
country comparison to the world: 179
Airports—with paved runways: *total*: 3
over 3,047 m: 1
2,438 to 3,047 m: 1
914 to 1,523 m: 1 (2010)
Airports—with unpaved runways: *total*: 2
914 to 1,523 m: 2 (2010)
Roadways: *total*: 88 km
country comparison to the world: 214
paved roads: 88 km–60 km in Male; 14 km on Addu Atolis; 14 km on Laamu
note: village roads are mainly compacted coral (2006)
Merchant marine: *total*: 24
country comparison to the world: 94
by type: bulk carrier 1, cargo 20, petroleum tanker 1, refrigerated cargo 2
registered in other countries: 4 (Panama 3, Tuvalu 1) (2010)
Ports and terminals: Male

MILITARY

Military branches: Maldives National Defense Force (MNDF): Marine Corps, Security Protection Group, Coast Guard (2010)
Military service age and obligation: 18-28 years of age for voluntary military service; no conscription (2010)
Manpower available for military service: *males age 16–49*: 156,319
females age 16–49: 98,815 (2010 est.)
Manpower fit for military service: *males age 16–49*: 135,374
females age 16–49: 85,181 (2010 est.)
Manpower reaching militarily significant age annually: *male*: 4,167
female: 3,595 (2010 est.)
Military expenditures: 5.5% of GDP (2005 est.)
country comparison to the world: 12
Military—note: the Maldives National Defense Force (MNDF), with its small size and with little serviceable equipment, is inadequate to prevent external aggression and is primarily tasked to reinforce the Maldives Police Service (MPS) and ensure security in the exclusive economic zone (2008)

TRANSNATIONAL ISSUES

Disputes—international: none
Refugees and internally displaced persons: *IDPs*: 1,000-10,000 (December 2004 tsunami victims) (2007)

MALI

INTRODUCTION

Background: The Sudanese Republic and Senegal became independent of France in 1960 as the Mali Federation. When Senegal withdrew after only a few months, what formerly made up the Sudanese Republic was renamed Mali. Rule by dictatorship was brought to a close in 1991 by a military coup–led by the current president Amadou TOURE–enabling Mali's emergence as one of the strongest democracies on the continent. President Alpha KONARE won Mali's first democratic presidential election in 1992 and was reelected in 1997. In keeping with Mali's two-term constitutional limit, KONARE stepped down in 2002 and was succeeded by Amadou TOURE, who was subsequently elected to a second term in 2007. The elections were widely judged to be free and fair.

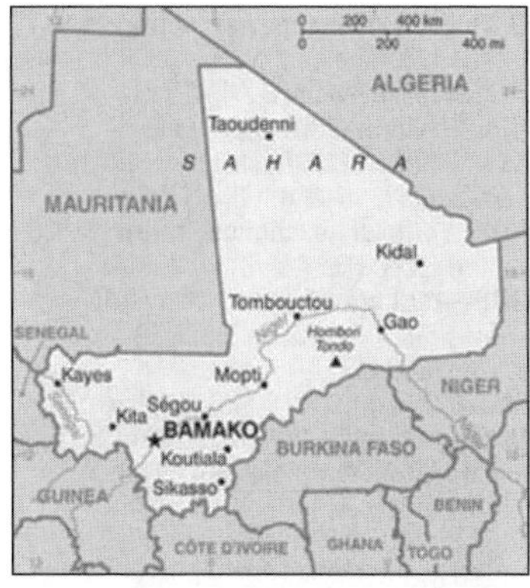

GEOGRAPHY

Location: interior Western Africa, southwest of Algeria, north of Guinea, Cote d'Ivoire, and Burkina Faso, west of Niger
Geographic coordinates: 17 00 N, 4 00 W
Map references: Africa
Area: *total:* 1,240,192 sq km
country comparison to the world: 24
land: 1,220,190 sq km
water: 20,002 sq km
Area—comparative: slightly less than twice the size of Texas
Land boundaries: *total:* 7,243 km
border countries: Algeria 1,376 km, Burkina Faso 1,000 km, Guinea 858 km, Cote d'Ivoire 532 km, Mauritania 2,237 km, Niger 821 km, Senegal 419 km
Coastline: 0 km (landlocked)
Maritime claims: none (landlocked)
Climate: subtropical to arid; hot and dry (February to June); rainy, humid, and mild (June to November); cool and dry (November to February)
Terrain: mostly flat to rolling northern plains covered by sand; savanna in south, rugged hills in northeast
Elevation extremes: *lowest point:* Senegal River 23 m
highest point: Hombori Tondo 1,155 m
Natural resources: gold, phosphates, kaolin, salt, limestone, uranium, gypsum, granite, hydropower
note: bauxite, iron ore, manganese, tin, and copper deposits are known but not exploited
Land use: *arable land:* 3.76%
permanent crops: 0.03%
other: 96.21% (2005)
Irrigated land: 2,360 sq km (2003)
Total renewable water resources: 100 cu km (2001)
Freshwater withdrawal (domestic/industrial/agricultural): *total:* 6.55 cu km/yr (9%/1%/90%)
per capita: 484 cu m/yr (2000)
Natural hazards: hot, dust-laden harmattan haze common during dry seasons; recurring droughts; occasional Niger River flooding
Environment—current issues: deforestation; soil erosion; desertification; inadequate supplies of potable water; poaching
Environment—international agreements: *party to:* Biodiversity, Climate Change, Climate Change-Kyoto Protocol, Desertification, Endangered Species, Hazardous Wastes, Law of the Sea, Ozone Layer Protection, Wetlands, Whaling
signed, but not ratified: none of the selected agreements
Geography—note: landlocked; divided into three natural zones: the southern, cultivated Sudanese; the central, semiarid Sahelian; and the northern, arid Saharan

PEOPLE

Population: 14,159,904 (July 2011 est.)
country comparison to the world: 67
Age structure: *0–14 years:* 47.3% (male 3,372,717/female 3,325,188)
15–64 years: 49.7% (male 3,438,687/female 3,605,143)
65 years and over: 3% (male 199,862/female 218,307) (2011 est.)
Median age: *total:* 16.3 years
male: 15.9 years
female: 16.7 years (2011 est.)
Population growth rate: 2.61% (2011 est.)
country comparison to the world: 25
Birth rate: 45.62 births/1,000 population (2011 est.)
country comparison to the world: 3
Death rate: 14.29 deaths/1,000 population (July 2011 est.)
country comparison to the world: 14
Net migration rate: -5.23 migrant(s)/1,000 population (2011 est.)
country comparison to the world: 194
Urbanization: *urban population:* 36% of total population (2010)
rate of urbanization: 4.4% annual rate of change (2010-15 est.)
Major cities—population: BAMAKO (capital) 1.628 million (2009)
Sex ratio: *at birth:* 1.03 male(s)/female
under 15 years: 1.01 male(s)/female
15–64 years: 0.95 male(s)/female
65 years and over: 0.92 male(s)/female
total population: 0.98 male(s)/female (2011 est.)
Infant mortality rate: *total:* 111.35 deaths/1,000 live births
country comparison to the world: 4
male: 118.15 deaths/1,000 live births
female: 104.34 deaths/1,000 live births (2011 est.)
Life expectancy at birth: *total population:* 52.61 years
country comparison to the world: 206
male: 51.01 years
female: 54.26 years (2011 est.)
Total fertility rate: 6.44 children born/woman (2011 est.)
country comparison to the world: 3
HIV/AIDS—adult prevalence rate: 1% (2009 est.)
country comparison to the world: 45
HIV/AIDS—people living with HIV/AIDS: 76,000 (2009 est.)
country comparison to the world: 48
HIV/AIDS—deaths: 4,400 (2009 est.)
country comparison to the world: 41
Major infectious diseases: *degree of risk:* very high
food or waterborne diseases: bacterial and protozoal diarrhea, hepatitis A, and typhoid fever
vectorborne disease: malaria
water contact disease: schistosomiasis
respiratory disease: meningococcal meningitis (2009)
Drinking water source: *Improved:*
urban: 81% of population
rural: 44% of population
total: 56% of population
Unimproved:
urban: 19% of population
rural: 56% of population
total: 44% of population (2008)
Sanitation facility access: *Improved:*
urban: 54% of population
rural: 32% of population
total: 36% of population
Unimproved:
urban: 46% of population
rural: 68% of population
total: 54% of population (2008)
Nationality: *noun:* Malian(s)
adjective: Malian
Ethnic groups: Mande 50% (Bambara, Malinke, Soninke), Peul 17%, Voltaic 12%, Songhai 6%, Tuareg and Moor 10%, other 5%
Religions: Muslim 90%, Christian 1%, indigenous beliefs 9%
Languages: French (official), Bambara 80%, numerous African languages
Literacy: *definition:* age 15 and over can read and write
total population: 46.4%
male: 53.5%
female: 39.6% (2003 est.)
School life expectancy (primary to tertiary education): *total:* 8 years
male: 9 years
female: 7 years (2009)
Education expenditures: 4.4% of GDP (2009)
country comparison to the world: 88

GOVERNMENT

Country name: *conventional long form:* Republic of Mali
conventional short form: Mali
local long form: Republique de Mali
local short form: Mali
former: French Sudan and Sudanese Republic
Government type: republic
Capital: *name:* Bamako
geographic coordinates: 12 39 N, 8 00 W
time difference: UTC 0 (5 hours ahead of Washington, DC during Standard Time)
Administrative divisions: 8 regions (regions, singular–region); Gao, Kayes, Kidal, Koulikoro, Mopti, Segou, Sikasso, Tombouctou
Independence: 22 September 1960 (from France)
National holiday: Independence Day, 22 September (1960)
Constitution: adopted 12 January 1992
Legal system: civil law system based on the French civil law model and influenced by customary law; judicial review of legislative acts in Constitutional Court
International law organization participation: has not submitted an ICJ jurisdiction declaration; accepts ICCt jurisdiction
Suffrage: 18 years of age; universal
Executive branch: *chief of state:* President Amadou Toumani TOURE (since 8 June 2002)

head of government: Prime Minister CISSE Mariam Kaidama Sidibe (since 3 April 2011)
cabinet: Council of Ministers appointed by the prime minister
(For more information visit the World Leaders website)
elections: president elected by popular vote for a five-year term (eligible for a second term); election last held on 29 April 2007 (next to be held in April 2012); prime minister appointed by the president
election results: Amadou Toumani TOURE reelected president; percent of vote—Amadou Toumani TOURE 71.2%, Ibrahim Boubacar KEITA 19.2%, other 9.6%
Legislative branch: unicameral National Assembly or Assemblee Nationale (147 seats; members elected by popular vote to serve five-year terms)
elections: last held on 1 and 22 July 2007 (next to be held in July 2012)
election results: percent of vote by party—NA; seats by party—ADP coalition 113 (ADEMA 51, URD 34, MPR 8, CNID 7, UDD 3, and other 10), FDR coalition 15 (RPM 11, PARENA 4), SADI 4, independent 15
Judicial branch: Supreme Court or Cour Supreme
Political parties and leaders: African Solidarity for Democracy and Independence or SADI [Oumar MARIKO, secretary general]; Alliance for Democracy or ADEMA [Diounconda TRAORE]; Alliance for Democracy and Progress or ADP (a coalition of political parties including ADEMA and URD formed in December 2006 to support the presidential candidacy of Amadou TOURE); Alliance for Democratic Change (political group comprised mainly of Tuareg from Mali's northern region); Convergence 2007 [Soumeylou Boubeye MAIGA]; Front for Democracy and the Republic or FDR (a coalition of political parties including RPM and PARENA formed to oppose the presidential candidacy of Amadou TOURE); National Congress for Democratic Initiative or CNID [Mountaga TALL]; Party for Democracy and Progress or PDP [Mady KONATE]; Party for National Renewal or PARENA [Tiebile DRAME]; Patriotic Movement for Renewal or MPR [Choguel MAIGA]; Rally for Democracy and Labor or RDT [Amadou Ali NIANGADOU]; Rally for Mali or RPM [Ibrahim Boubacar KEITA]; Sudanese Union/African Democratic Rally or US/RDA [Mamadou Basir GOLOGO]; Union for Democracy and Development or UDD [Moussa Balla COULIBALY]; Union for Republic and Democracy or URD [Soumaila CISSE]
Political pressure groups and leaders: *other:* the army; Islamic authorities; rebels in the northern region; state-run cotton company CMDT; tuaregs
International organization participation: ACP, AfDB, AU, CD, ECOWAS, FAO, FZ, G-77, IAEA, IBRD, ICAO, ICRM, IDA, IDB, IFAD, IFC, IFRCS, ILO, IMF, Interpol, IOC, IOM, IPU, ISO, ITSO, ITU, ITUC, MIGA, MONUSCO, NAM, OIC, OIF, OPCW, UN, UNAMID, UNCTAD, UNESCO, UNIDO, UNMIL, UNWTO, UPU, WADB (regional), WAEMU, WCO, WFTU, WHO, WIPO, WMO, WTO
Diplomatic representation in the US: *chief of mission:* Ambassador Mamadou TRAORE
chancery: 2130 R Street NW, Washington, DC 20008
telephone: [1] (202) 332-2249, 939-8950
FAX: [1] (202) 332-6603
Diplomatic representation from the US: *chief of mission:* Ambassador Gillian A. MILOVANOVIC
embassy: located just off the Roi Bin Fahad Aziz Bridge just west of the Bamako central district
mailing address: ACI 2000, Rue 243, Porte 297, Bamako
telephone: [223] 270-2300
FAX: [223] 270-2479
Flag description: three equal vertical bands of green (hoist side), yellow, and red
note: uses the popular Pan-African colors of Ethiopia; the colors from left to right are the same as those of neighboring Senegal (which has an additional green central star) and the reverse of those on the flag of neighboring Guinea
National anthem: *name:* "Le Mali" (Mali)
lyrics/music: Seydou Badian KOUYATE/ Banzoumana SISSOKO
note: adopted 1962; the anthem is also known as "Pour L'Afrique et pour toi, Mali" (For Africa and for You, Mali) and "A ton appel Mali" (At Your Call, Mali)

ECONOMY

Economy—overview: Among the 25 poorest countries in the world, Mali is a landlocked country highly dependent on gold mining and agricultural exports for revenue. The country's fiscal status fluctuates with gold and agricultural commodity prices and the harvest. Mali remains dependent on foreign aid. Economic activity is largely confined to the riverine area irrigated by the Niger River and about 65% of its land area is desert or semidesert. About 10% of the population is nomadic and some 80% of the labor force is engaged in farming and fishing. Industrial activity is concentrated on processing farm commodities. The government has continued an IMF-recommended structural adjustment program that has helped the economy grow, diversify, and attract foreign investment. Mali is developing its cotton and iron ore extraction industries to diversify its revenue sources because gold production has started to fall. Mali has invested in tourism but security issues are hurting the industry. Mali's adherence to economic reform and the 50% devaluation of the CFA franc in January 1994 have pushed up economic growth to a 5% average in 1996-2010. Worker remittances and external trade routes for the landlocked country have been jeopardized by continued unrest in neighboring Cote d'Ivoire. However, Mali is building a road network that will connect it to all adjacent countries and it has a railway line to Senegal. In 2010, Mali experienced a regional drought that hurt livestock and livelihoods.
GDP (purchasing power parity): $16.77 billion (2010 est.)
country comparison to the world: 134
$16.06 billion (2009 est.)
$15.37 billion (2008 est.)
note: data are in 2010 US dollars
GDP (official exchange rate): $9.268 billion (2010 est.)
GDP—real growth rate: 4.5% (2010 est.)
country comparison to the world: 83
4.5% (2009 est.)
5% (2008 est.)
GDP—per capita (PPP): $1,200 (2010 est.)
country comparison to the world: 208
$1,200 (2009 est.)
$1,200 (2008 est.)
note: data are in 2010 US dollars
GDP—composition by sector: *agriculture:* 45%
industry: 17%
services: 38% (2001 est.)
Labor force: 3.241 million (2007 est.)
country comparison to the world: 100
Labor force—by occupation: *agriculture:* 80%
industry and *services:* 20% (2005 est.)
Unemployment rate: 30% (2004 est.)
country comparison to the world: 177
Population below poverty line: 36.1% (2005 est.)
Household income or consumption by percentage share: *lowest 10%:* 2.7%
highest 10%: 30.5% (2006)
Distribution of family income—Gini index: 40.1 (2001)
country comparison to the world: 60
50.5 (1994)
Budget: *revenues:* $1.5 billion
expenditures: $1.8 billion (2006 est.)
Inflation rate (consumer prices): 2.5% (2007 est.)
country comparison to the world: 65
Central bank discount rate: 4.25% (31 December 2009)
country comparison to the world: 87
4.75% (31 December 2008)
Commercial bank prime lending rate: NA%
Stock of narrow money: $1.758 billion (31 December 2009)
country comparison to the world: 124
$1.559 billion (31 December 2008)
Stock of broad money: $2.514 billion (31 December 2009 est.)
country comparison to the world: 136
$2.12 billion (31 December 2008 est.)
Stock of domestic credit: $994.9 million (31 December 2009)
country comparison to the world: 150
$1.095 billion (31 December 2008)
Market value of publicly traded shares: $NA
Agriculture—products: cotton, millet, rice, corn, vegetables, peanuts; cattle, sheep, goats
Industries: food processing; construction; phosphate and gold mining
Industrial production growth rate: NA%
Electricity—production: 515 million kWh (2007 est.)
country comparison to the world: 158
Electricity—consumption: 479 million kWh (2007 est.)
country comparison to the world: 163

Electricity—exports: NA kWh
note: Mali may be providing electricity to Senegal and Mauritania (2008 est.)
Electricity—imports: 0 kWh (2008 est.)
Oil—production: 0 bbl/day (2009 est.)
country comparison to the world: 205
Oil—consumption: 6,000 bbl/day (2009 est.)
country comparison to the world: 162
Oil—exports: 0 bbl/day (2007 est.)
country comparison to the world: 185
Oil—imports: 4,402 bbl/day (2007 est.)
country comparison to the world: 163
Oil—proved reserves: 0 bbl (1 January 2010 est.)
country comparison to the world: 163
Natural gas—production: 0 cu m (2008 est.)
country comparison to the world: 159
Natural gas—consumption: 0 cu m (2008 est.)
country comparison to the world: 197
Natural gas—exports: 0 cu m (2008 est.)
country comparison to the world: 140
Natural gas—imports: 0 cu m (2008 est.)
country comparison to the world: 86
Natural gas—proved reserves: 0 cu m (1 January 2010 est.)
country comparison to the world: 165
Current account balance: $-446 million (2007 est.)
country comparison to the world: 112
Exports: $294 million (2006)
country comparison to the world: 175
Exports—commodities: cotton, gold, livestock
Exports—partners: China 14.61%, Thailand 8.28%, Pakistan 6.74%, Morocco 6.48%, Burkina Faso 4.67%, France 4.6%, India 4.45% (2009)
Imports: $2.358 billion (2006)
country comparison to the world: 149
Imports—commodities: petroleum, machinery and equipment, construction materials, foodstuffs, textiles
Imports—partners: Senegal 12.21%, France 11.57%, Cote d'Ivoire 10.05%, China 5.89% (2009)
Debt—external: $2.8 billion (2002)
country comparison to the world: 133
Exchange rates: Communaute Financiere Africaine francs (XOF) per US dollar–495.28 (2010) 472.19 (2009) 493.51 (2007) 522.59 (2006)

COMMUNICATIONS

Telephones—main lines in use: 81,000 (2009)
country comparison to the world: 149
Telephones—mobile cellular: 3.742 million (2009)
country comparison to the world: 107
Telephone system: general assessment: domestic system unreliable but improving; increasing use of local radio loops to extend network coverage to remote areas
domestic: fixed-line subscribership remains less than 1 per 100 persons; mobile-cellular subscribership has increased sharply to nearly 30 per 100 persons
international: country code–223; satellite communications center and fiber-optic links to neighboring countries; satellite earth stations–2 Intelsat (1 Atlantic Ocean, 1 Indian Ocean) (2008)
Broadcast media: national public TV broadcaster; 2 privately-owned companies provide subscription services to foreign multi-channel TV packages; national public radio broadcaster supplemented by a large number of privately-owned and community broadcast stations; transmissions of multiple international broadcasters are available (2007)
Internet country code: .ml
Internet hosts: 524 (2010)
country comparison to the world: 179
Internet users: 249,800 (2009)
country comparison to the world: 135

TRANSPORTATION

Airports: 20 (2010)
country comparison to the world: 135
Airports—with paved runways: *total:* 8
2,438 to 3,047 m: 4
1,524 to 2,437 m: 3
914 to 1,523 m: 1 (2010)
Airports—with unpaved runways:
total: 12
1,524 to 2,437 m: 4
914 to 1,523 m: 5
under 914 m: 3 (2010)
Railways: *total:* 593 km
country comparison to the world: 109
narrow gauge: 593 km 1.000-m gauge (2010)
Roadways: *total:* 18,709 km
country comparison to the world: 113
paved: 3,368 km
unpaved: 15,341 km (2004)
Waterways: 1,800 km (downstream of Koulikoro; low water levels on the River Niger cause problems in dry years; in the months before the rainy season the river is not navigable by commercial vessels) (2010)
country comparison to the world: 45
Ports and terminals: Koulikoro

MILITARY

Military branches: *Malian Armed Forces:* Army, Republic of Mali Air Force (Force Aerienne de la Republique du Mali, FARM), National Guard (2008)
Military service age and obligation: 18 years of age for compulsory and voluntary military service; conscript service obligation–2 years (2010)
Manpower available for military service:
males age 16–49: 2,848,412
females age 16–49: 2,981,106 (2010 est.)
Manpower fit for military service: *males age 16–49:* 1,825,779
females age 16–49: 1,968,563 (2010 est.)
Manpower reaching militarily significant age annually: *male:* 158,031
female: 159,733 (2010 est.)
Military expenditures: 1.9% of GDP (2006)
country comparison to the world: 74

TRANSNATIONAL ISSUES

Disputes—international: demarcation is currently underway with Burkina Faso
Refugees and internally displaced persons: refugees (country of origin): 6,300 (Mauritania) (2007)

MALTA

INTRODUCTION

Background: Great Britain formally acquired possession of Malta in 1814. The island staunchly supported the UK through both world wars and remained in the Commonwealth when it became independent in 1964. A decade later Malta became a republic. Since about the mid-1980s, the island has transformed itself into a freight transshipment point, a financial center, and a tourist destination. Malta became an EU member in May 2004 and began using the euro as currency in 2008.

GEOGRAPHY

Location: Southern Europe, islands in the Mediterranean Sea, south of Sicily (Italy)
Geographic coordinates: 35 50 N, 14 35 E
Map references: Europe
Area: *total:* 316 sq km
country comparison to the world: 206
land: 316 sq km
water: 0 sq km
Area—comparative: slightly less than twice the size of Washington, DC
Land boundaries: 0 km
Coastline: 196.8 km (excludes 56.01 km for the island of Gozo)
Maritime claims: *territorial sea:* 12 nm
contiguous zone: 24 nm
continental shelf: 200 m depth or to the depth of exploitation
exclusive fishing zone: 25 nm
Climate: Mediterranean; mild, rainy winters; hot, dry summers
Terrain: mostly low, rocky, flat to dissected plains; many coastal cliffs
Elevation extremes: *lowest point:* Mediterranean Sea 0 m
highest point: Ta'Dmejrek 253 m (near Dingli)
Natural resources: limestone, salt, arable land

Land use: *arable land:* 31.25%
permanent crops: 3.13%
other: 65.62% (2005)
Irrigated land: 20 sq km (2003)
Total renewable water resources: 0.07 cu km (2005)
Freshwater withdrawal (domestic/industrial/agricultural): *total:* 0.02 cu km/yr (74%/1%/25%)
per capita: 50 cu m/yr (2000)
Natural hazards: NA
Environment—current issues: limited natural freshwater resources; increasing reliance on desalination
Environment—international agreements: *party to:* Air Pollution, Biodiversity, Climate Change, Climate Change-Kyoto Protocol, Desertification, Endangered Species, Hazardous Wastes, Law of the Sea, Marine Dumping, Ozone Layer Protection, Ship Pollution, Wetlands
signed, but not ratified: none of the selected agreements
Geography—note: the country comprises an archipelago, with only the three largest islands (Malta, Ghawdex or Gozo, and Kemmuna or Comino) being inhabited; numerous bays provide good harbors; Malta and Tunisia are discussing the commercial exploitation of the continental shelf between their countries, particularly for oil exploration

PEOPLE

Population: 408,333 (July 2011 est.)
country comparison to the world: 173
Age structure: *0–14 years:* 15.7% (male 32,829/female 31,198)
15–64 years: 68.5% (male 142,006/female 137,803)
65 years and over: 15.8% (male 28,305/female 36,192) (2011 est.)
Median age: *total:* 40 years
male: 38.8 years
female: 41.4 years (2011 est.)
Population growth rate: 0.375% (2011 est.)
country comparison to the world: 161
Birth rate: 10.35 births/1,000 population (2011 est.)
country comparison to the world: 185
Death rate: 8.6 deaths/1,000 population (July 2011 est.)
country comparison to the world: 84
Net migration rate: 2.01 migrant(s)/1,000 population (2011 est.)
country comparison to the world: 41
Urbanization: *urban population:* 95% of total population (2010)
rate of urbanization: 0.5% annual rate of change (2010-15 est.)
Major cities—population: VALLETTA (capital) 199,000 (2009)
Sex ratio: *at birth:* 1.058 male(s)/female
under 15 years: 1.05 male(s)/female
15–64 years: 1.03 male(s)/female
65 years and over: 0.77 male(s)/female
total population: 0.99 male(s)/female (2011 est.)
Infant mortality rate: *total:* 3.69 deaths/1,000 live births
country comparison to the world: 206
male: 4.12 deaths/1,000 live births
female: 3.22 deaths/1,000 live births (2011 est.)
Life expectancy at birth: *total population:* 79.72 years
country comparison to the world: 34
male: 77.45 years
female: 82.12 years (2011 est.)
Total fertility rate: 1.52 children born/woman (2011 est.)
country comparison to the world: 184
HIV/AIDS—adult prevalence rate: 0.1% (2009 est.)
country comparison to the world: 148
HIV/AIDS—people living with HIV/AIDS: fewer than 500 (2009 est.)
country comparison to the world: 155
HIV/AIDS—deaths: fewer than 100 (2009 est.)
country comparison to the world: 143
Drinking water source: *Improved:*
urban: 100% of population
rural: 100% of population
total: 100% of population (2008)
Sanitation facility access: *Improved:*
urban: 100% of population
rural: 100% of population
total: 100% of population (2008)
Nationality: *noun:* Maltese (singular and plural)
adjective: Maltese
Ethnic groups: Maltese (descendants of ancient Carthaginians and Phoenicians with strong elements of Italian and other Mediterranean stock)
Religions: Roman Catholic 98%
Languages: Maltese (official) 90.2%, English (official) 6%, multilingual 3%, other 0.8% (2005 census)
Literacy: *definition:* age 10 and over can read and write
total population: 92.8%
male: 91.7%
female: 93.9% (2005 Census)
School life expectancy (primary to tertiary education): *total:* 14 years
male: 14 years
female: 15 years (2008)
Education expenditures: 6.4% of GDP (2007)
country comparison to the world: 22

GOVERNMENT

Country name: *conventional long form:* Republic of Malta
conventional short form: Malta
local long form: Repubblika ta' Malta
local short form: Malta
Government type: republic
Capital: *name:* Valletta
geographic coordinates: 35 53 N, 14 30 E
time difference: UTC+1 (6 hours ahead of Washington, DC during Standard Time)
daylight saving time: +1hr, begins last Sunday in March; ends last Sunday in October
Administrative divisions: none (administered directly from Valletta); note–local councils carry out administrative orders and have some responsibility for local road and other public maintenance
Independence: 21 September 1964 (from the UK)
National holiday: Independence Day, 21 September (1964); Republic Day, 13 December (1974)
Constitution: 1964; amended many times
Legal system: mixed legal system of English common law and civil law (based on the Roman and Napoleonic civil codes)
International law organization participation: accepts compulsory ICJ jurisdiction with reservations; accepts ICCt jurisdiction
Suffrage: 18 years of age; universal
Executive branch: *chief of state:* President George ABELA (since 4 April 2009)
head of government: Prime Minister Lawrence GONZI (since 23 March 2004)
cabinet: Cabinet appointed by the president on the advice of the prime minister
(For more information visit the World Leaders website)
elections: president elected by a resolution of the House of Representatives for a five-year term; election last held on 12 January 2009 (next to be held by April 2014); following legislative elections, the leader of the majority party or leader of a majority coalition usually appointed prime minister by the president for a five-year term; the deputy prime minister appointed by the president on the advice of the prime minister
election results: George ABELA elected president by the House of Representatives
Legislative branch: unicameral House of Representatives (normally 65 seats; members are elected by popular vote on the basis of proportional representation by the Single Transferrable Vote (STV) to serve five-year terms; note–the parliament elected in 2008 is composed of 69 seats; when the political party winning an absolute majority of first-count votes (or a plurality of first-count votes in an election where only two parties are represented in parliament) does not win an absolute majority of seats, the constitution provides for the winning party to be awarded additional number of seats in parliament to guarantee it an absolute majority; in the event that more than two parties are repressented in parliament, with none acquiring the absolute majority of votes, the party winning the majority of seats prevails
elections: last held on 8 March 2008 (next to be held by March 2013)
election results: percent of vote by party–PN 49.3%, PL 48.8%, other 1.9%; seats by party–PN 35, PL 34
Judicial branch: Constitutional Court; Court of First Instance; Court of Appeal
note: magistrates and judges for the courts are appointed by the president on the advice of the prime minister
Political parties and leaders: Alternativa Demokratika/Alliance for Social Justice or AD [Michael BRIGUGLIO]; Labor Party or PL [Joseph MUSCAT]; Nationalist Party or PN [Lawrence GONZI]; The Malta Communist Party [Victor DEGIOVANNI]
Political pressure groups and leaders: Alleanza Liberal-Demokratika Maltra or ALDM (for divorce, abortion, gay marriage, the rights existent in other EU member states); Alleanza Nazzionali Repubblikana or ANR (for traditional values, anti-immigration); Alternattiva Demokratika (pro-environment); Flimkien Ghal-Ambjent Ahjar (pro-environment); Ghazda tal-Konsumaturi (consumer rights)

other: environmentalists
International organization participation: Australia Group, C, CE, EAPC, EBRD, EIB, EMU, EU, FAO, IAEA, IBRD, ICAO, ICRM, IFAD, IFC, IFRCS, ILO, IMF, IMO, IMSO, Interpol, IOC, IOM, IPU, ISO, ITSO, ITU, ITUC, MIGA, NSG, OPCW, OSCE, PCA, PFP, Schengen Convention, UN, UNCTAD, UNESCO, UNIDO, Union Latina (observer), UNWTO, UPU, WCO, WHO, WIPO, WMO, WTO
Diplomatic representation in the US: *chief of mission:* Ambassador Mark MICELI-FARRUGIA
chancery: 2017 Connecticut Avenue NW, Washington, DC 20008
telephone: [1] (202) 462-3611 through 3612
FAX: [1] (202) 387-5470
consulate(s): New York
Diplomatic representation from the US: *chief of mission:* Ambassador Douglas W. KMIEC
embassy: 3rd Floor, Development House, Saint Anne Street, Floriana, FRN 9010
mailing address: P. O. Box 535, Valletta, VLT1000
telephone: [356] 2561 4000
FAX: [356] 2124 3229
Flag description: two equal vertical bands of white (hoist side) and red; in the upper hoist-side corner is a representation of the George Cross, edged in red; according to legend, the colors are taken from the red and white checkered banner of Count Roger of Sicily who removed a bi-colored corner and granted it to Malta in 1091; an uncontested explanation is that the colors are those of the Knights of Saint John who ruled Malta from 1530 to 1798; in 1942, King George VI of the United Kingdom awarded the George Cross to the islanders for their exceptional bravery and gallantry in World War II; since independence in 1964, the George Cross bordered in red has appeared directly on the white field
National anthem: *name:* "L-Innu Malti" (The Maltese Anthem)
lyrics/music: Dun Karm PSAILA/Robert SAMMUT
note: adopted 1945; the anthem is written in the form of a prayer

ECONOMY

Economy—overview: Malta produces only about 20% of its food needs, has limited fresh water supplies, and has few domestic energy sources. Malta's geographic position between the EU and Africa makes it a target for illegal immigration, which has strained Malta's political and economic resources. Malta adopted the euro on 1 January 2008. Malta's financial services industry has grown in recent years and in 2008-09 it escaped significant damage from the international financial crisis, largely because the sector is centered on the indigenous real estate market and is not highly leveraged. Locally, the restricted damage from the financial crisis has been attributed to the stability of the Maltese banking system and to its prudent risk-management practices. The global economic downturn and high electricity and water prices hurt Malta's real economy, which is dependent on foreign trade, manufacturing–especially electronics and pharmaceuticals–and tourism, but growth bounced back as the global economy recovered in 2010. Following a 1.2% contraction in 2009, GDP grew 2% in 2010. In early 2011, the EU ended excessive deficit procedures against Malta, after Malta had taken measures to correct an excessive deficit in 2010 and appeared likely to reach its deficit target of 2.8% of GDP in 2011.
GDP (purchasing power parity): $10.41 billion (2010 est.)
country comparison to the world: 151
$10.04 billion (2009 est.)
$10.39 billion (2008 est.)
note: data are in 2010 US dollars
GDP (official exchange rate): $8.288 billion (2010 est.)
GDP—real growth rate: 3.7% (2010 est.)
country comparison to the world: 105
-3.4% (2009 est.)
5.3% (2008 est.)
GDP—per capita (PPP): $25,600 (2010 est.)
country comparison to the world: 52
$24,800 (2009 est.)
$25,800 (2008 est.)
note: data are in 2010 US dollars
GDP—composition by sector: *agriculture:* 1.9%
industry: 17.2%
services: 80.9% (2010 est.)
Labor force: 163,100 (2010)
country comparison to the world: 177
Labor force—by occupation:
agriculture: 1.3%
industry: 24.8%
services: 73.9% (2010)
Unemployment rate: 6.9% (2009)
country comparison to the world: 70
6.1% (2008)
Population below poverty line: NA%
Household income or consumption by percentage share: *lowest 10%:* NA%
highest 10%: NA%
Distribution of family income—Gini index: 26 (2007)
country comparison to the world: 131
Investment (gross fixed): 8.8% of GDP (2010 est.)
country comparison to the world: 151
Budget: *revenues:* $4.455 billion
expenditures: $3.322 billion (2010 est.)
Public debt: 69.1% of GDP (2010 est.)
country comparison to the world: 25
68.8% of GDP (2009 est.)
Inflation rate (consumer prices): 3.3% (2010 est.)
country comparison to the world: 94
-0.7% (2009 est.)
Central bank discount rate: 1.75% (31 December 2010)
country comparison to the world: 122
1.75% (31 December 2009)
note: this is the European Central Bank's rate on the marginal lending facility, which offers overnight credit to banks in the euro area
Commercial bank prime lending rate: 4.47% (31 December 2010)
country comparison to the world: 150
4.47% (31 December 2009)
Stock of narrow money: $7.046 billion (31 December 2010 est.)
country comparison to the world: 79
$6.147 billion (31 December 2009 est.)
note: see entry for the European Union for money supply in the euro area; the European Central Bank (ECB) controls monetary policy for the 17 members of the EMU; individual members of the EMU do not control the quantity of money circulating within their own borders
Stock of broad money: $13.21 billion (31 December 2010 est.)
country comparison to the world: 93
$12.53 billion (31 December 2009 est.)
Stock of domestic credit: $18.66 billion (31 December 2010)
country comparison to the world: 81
$20.4 billion (31 December 2009)
Market value of publicly traded shares: $1.982 billion (31 December 2009)
country comparison to the world: 87
$3.572 billion (31 December 2008)
$5.633 billion (31 December 2007)
Agriculture—products: potatoes, cauliflower, grapes, wheat, barley, tomatoes, citrus, cut flowers, green peppers; pork, milk, poultry, eggs
Industries: tourism, electronics, ship building and repair, construction, food and beverages, pharmaceuticals, footwear, clothing, tobacco, aviation services, financial services, information technology services
Industrial production growth rate: NA%
Electricity—production: 2.113 billion kWh (2010)
country comparison to the world: 132
Electricity—consumption: 1.991 billion kWh (2010)
country comparison to the world: 136
Electricity—exports: 0 kWh (2009)
Electricity—imports: 0 kWh (2009)
Oil—production: 0 bbl/day (2009 est.)
country comparison to the world: 207
Oil—consumption: 19,000 bbl/day (2009 est.)
country comparison to the world: 126
Oil—exports: 0 bbl/day (2009 est.)
country comparison to the world: 187
Oil—imports: 17,910 bbl/day (2007 est.)
country comparison to the world: 115
Oil—proved reserves: 0 bbl (1 January 2010 est.)
country comparison to the world: 165
Natural gas—production: 0 cu m (2009 est.)
country comparison to the world: 162
Natural gas—consumption: 0 cu m (2009 est.)
country comparison to the world: 200
Natural gas—exports: 0 cu m (2009 est.)
country comparison to the world: 144
Natural gas—imports: 0 cu m (2009 est.)
country comparison to the world: 89
Natural gas—proved reserves: 0 cu m (1 January 2010 est.)
country comparison to the world: 167
Current account balance: $-362.8 million (2010)
country comparison to the world: 106
$-574.6 million (2009)
Exports: $3.124 billion (2010)
country comparison to the world: 121
$2.347 billion (2009)
Exports—commodities:
electrical machinery, mechanical appliances, fish and crustaceans, pharmaceutical products, printed material

Exports—partners: Germany 13.3%, Singapore 12.5%, France 11.4%, US 9.4%, Hong Kong 6.5%, UK 5.9%, Italy 4.8% (2009 est.)
Imports: $5.159 billion (2010)
country comparison to the world: 114
$4.334 billion (2009)
Imports—commodities: mineral fuels and oils, electrical machinery, non-electrical machinery, aircraft and other transport equipment, plastic and other semi-manufactured goods; food, drink, tobacco
Imports—partners: Italy 24.4%, UK 11.7%, Germany 9.3%, France 7.6%, China 4.2% (2009 est.)
Reserves of foreign exchange and gold: $522 million (31 December 2010 est.)
country comparison to the world: 124
$871 million (31 December 2009)
Debt—external: $5.978 billion (31 December 2010)
country comparison to the world: 100
$5.578 billion (2009)
Stock of direct foreign investment—at home: $16.63 billion (31 December 2010)
country comparison to the world: 73
Stock of direct foreign investment—abroad: $1.213 billion
country comparison to the world: 71
Exchange rates: euros (EUR) per US dollar–0.755 (2010) 0.72 (2009) 0.6827 (2008) 0.3106 (2007) 0.37 (2006)

COMMUNICATIONS

Telephones—main lines in use: 252,700 (2009)
country comparison to the world: 123
Telephones—mobile cellular: 422,100 (2009)
country comparison to the world: 164
Telephone system: *general assessment:* automatic system featuring submarine cable and microwave radio relay between islands
domestic: combined fixed-line and mobile-cellular subscribership exceeds 165 per 100 persons
international: country code–356; submarine cable connects to Italy; satellite earth station–1 Intelsat (Atlantic Ocean) (2008)
Broadcast media: 2 publicly-owned television stations, Television Malta (TVM) broadcasting nationally and an educational channel; several privately-owned national television stations, two of which are owned by political parties; Italian and British broadcast programs are available; multi-channel cable and satellite TV services are obtainable; publicly-owned radio broadcaster operates 1 station; roughly 20 commercial radio stations operating (2011)
Internet country code: .mt
Internet hosts: 24,941 (2010)
country comparison to the world: 102
Internet users: 240,600 (2009)
country comparison to the world: 137

TRANSPORTATION

Airports: 1 (2010)
country comparison to the world: 225
Airports—with paved runways: *total:* 1
over 3,047 m: 1 (2010)
Roadways: *total:* 2,227 km
country comparison to the world: 171
paved: 2,014 km
unpaved: 213 km (2005)
Merchant marine: *total:* 1,571
country comparison to the world: 4
by type: bulk carrier 522, cargo 377, carrier 1, chemical tanker 280, container 91, liquefied gas 31, passenger 45, passenger/cargo 22, petroleum tanker 141, refrigerated cargo 14, roll on/roll off 30, specialized tanker 2, vehicle carrier 15
foreign-owned: 1,401 (Angola 7, Austria 1, Azerbaijan 1, Bahamas 1, Bangladesh 1, Belgium 14, Bermuda 8, Bulgaria 7, Canada 1, China 11, Croatia 7, Cyprus 29, Denmark 41, Egypt 1, Estonia 16, Finland 2, France 13, Germany 127, Greece 458, Hong Kong 2, India 4, Iran 56, Ireland 1, Israel 5, Italy 52, Japan 5, Kuwait 2, Latvia 11, Lebanon 7, Libya 4, Luxembourg 3, Malaysia 1, Netherlands 2, Nigeria 1, Norway 84, Poland 22, Portugal 3, Romania 8, Russia 47, Singapore 3, Slovenia 4, South Korea 3, Spain 10, Sweden 3, Switzerland 14, Syria 5, Turkey 211, UAE 1, UK 16, Ukraine 30, US 35)
note: this country allows large numbers of ships owned by foreign entities to be registered in its national shipping registry and to fly its flag; these ships operate under the laws of the flag state
registered in other countries: 2 (Panama 2) (2010)
Ports and terminals: Marsaxlokk (Malta Freeport), Valletta

MILITARY

Military branches: Armed Forces of Malta (AFM; includes air and maritime elements) (2010)
Military service age and obligation: 17 years 6 months of age for voluntary military service; no conscription (2010)
Manpower available for military service: *males age 16–49:* 95,499
females age 16–49: 90,919 (2010 est.)
Manpower fit for military service: *males age 16–49:* 79,645
females age 16–49: 75,684 (2010 est.)
Manpower reaching militarily significant age annually: *male:* 2,554
female: 2,385 (2010 est.)
Military expenditures: 0.7% of GDP (2006 est.)
country comparison to the world: 153

TRANSNATIONAL ISSUES

Disputes—international: none
Illicit drugs: minor transshipment point for hashish from North Africa to Western Europe

MARSHALL ISLANDS

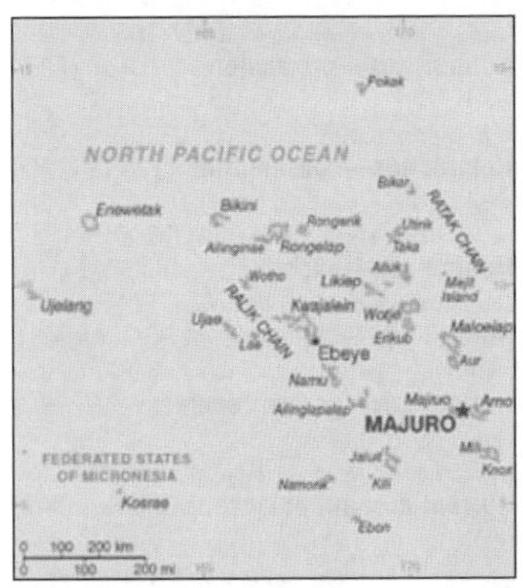

INTRODUCTION

Background: After almost four decades under US administration as the easternmost part of the UN Trust Territory of the Pacific Islands, the Marshall Islands attained independence in 1986 under a Compact of Free Association. Compensation claims continue as a result of US nuclear testing on some of the atolls between 1947 and 1962. The Marshall Islands hosts the US Army Kwajalein Atoll (USAKA) Reagan Missile Test Site, a key installation in the US missile defense network.

GEOGRAPHY

Location: Oceania, two archipelagic island chains of 29 atolls, each made up of many small islets, and five single islands in the North Pacific Ocean, about half way between Hawaii and Australia
Geographic coordinates: 9 00 N, 168 00 E
Map references: Oceania
Area: *total:* 181 sq km
country comparison to the world: 215
land: 181 sq km
water: 0 sq km
note: the archipelago includes 11,673 sq km of lagoon waters and includes the atolls of Bikini, Enewetak, Kwajalein, Majuro, Rongelap, and Utirik
Area—comparative: about the size of Washington, DC
Land boundaries: 0 km
Coastline: 370.4 km
Maritime claims: *territorial sea:* 12 nm
contiguous zone: 24 nm
exclusive economic zone: 200 nm
Climate: tropical; hot and humid; wet season May to November; islands border typhoon belt
Terrain: low coral limestone and sand islands
Elevation extremes: *lowest point:* Pacific Ocean 0 m
highest point: unnamed location on Likiep 10 m
Natural resources: coconut products, marine products, deep seabed minerals
Land use: *arable land:* 11.11%
permanent crops: 44.44%
other: 44.45% (2005)
Irrigated land: 0 sq km
Natural hazards: infrequent typhoons
Environment—current issues: inadequate supplies of potable water; pollution of

Majuro lagoon from household waste and discharges from fishing vessels
Environment—international agreements: *party to:* Biodiversity, Climate Change, Climate Change-Kyoto Protocol, Desertification, Hazardous Wastes, Law of the Sea, Ozone Layer Protection, Ship Pollution, Wetlands, Whaling
signed, but not ratified: none of the selected agreements
Geography—note: the islands of Bikini and Enewetak are former US nuclear test sites; Kwajalein atoll, famous as a World War II battleground, surrounds the world's largest lagoon and is used as a US missile test range; the island city of Ebeye is the second largest settlement in the Marshall Islands, after the capital of Majuro, and one of the most densely populated locations in the Pacific

PEOPLE

Population: 67,182 (July 2011 est.)
country comparison to the world: 202
Age structure: *0–14 years:* 38.2% (male 13,062/female 12,576)
15–64 years: 58.8% (male 20,171/female 19,340)
65 years and over: 3% (male 988/female 1,045) (2011 est.)
Median age: *total:* 21.8 years
male: 21.8 years
female: 21.8 years (2011 est.)
Population growth rate: 1.954% (2011 est.)
country comparison to the world: 58
Birth rate: 29.11 births/1,000 population (2011 est.)
country comparison to the world: 43
Death rate: 4.38 deaths/1,000 population (July 2011 est.)
country comparison to the world: 202
Net migration rate: -5.19 migrant(s)/1,000 population (2011 est.)
country comparison to the world: 193
Urbanization: *urban population:* 72% of total population (2010)
rate of urbanization: 2.3% annual rate of change (2010-15 est.)
Major cities—population: MAJURO (capital) 30,000 (2009)
Sex ratio: *at birth:* 1.05 male(s)/female
under 15 years: 1.04 male(s)/female
15–64 years: 1.04 male(s)/female
65 years and over: 0.94 male(s)/female
total population: 1.04 male(s)/female (2011 est.)
Infant mortality rate: *total:* 23.74 deaths/1,000 live births
country comparison to the world: 85
male: 26.69 deaths/1,000 live births
female: 20.64 deaths/1,000 live births (2011 est.)
Life expectancy at birth: *total population:* 71.76 years
country comparison to the world: 132
male: 69.67 years
female: 73.95 years (2011 est.)
Total fertility rate: 3.44 children born/woman (2011 est.)
country comparison to the world: 48
HIV/AIDS—adult prevalence rate: NA
HIV/AIDS—people living with HIV/AIDS: NA
HIV/AIDS—deaths: NA
Drinking water source: *Improved:*
urban: 92% of population
rural: 99% of population
total: 94% of population
Unimproved:
urban: 8% of population
rural: 1% of population
total: 6% of population (2008)
Sanitation facility access: *Improved:*
urban: 83% of population
rural: 53% of population
total: 73% of population
Unimproved:
urban: 17% of population
rural: 47% of population
total: 27% of population (2008)
Nationality: *noun:* Marshallese (singular and plural)
adjective: Marshallese
Ethnic groups: Marshallese 92.1%, mixed Marshallese 5.9%, other 2% (2006)
Religions: Protestant 54.8%, Assembly of God 25.8%, Roman Catholic 8.4%, Bukot nan Jesus 2.8%, Mormon 2.1%, other Christian 3.6%, other 1%, none 1.5% (1999 census)
Languages: Marshallese (official) 98.2%, other languages 1.8% (1999 census)
note: English (official), widely spoken as a second language
Literacy: *definition:* age 15 and over can read and write
total population: 93.7%
male: 93.6%
female: 93.7% (1999)
School life expectancy (primary to tertiary education): *total:* 13 years
male: 11 years
female: 11 years (2003)
Education expenditures: 12% of GDP (2004)
country comparison to the world: 4

GOVERNMENT

Country name: *conventional long form:* Republic of the Marshall Islands
conventional short form: Marshall Islands
local long form: Republic of the Marshall Islands
local short form: Marshall Islands
abbreviation: RMI
former: Trust Territory of the Pacific Islands, Marshall Islands District
Government type: constitutional government in free association with the US; the Compact of Free Association entered into force on 21 October 1986 and the Amended Compact entered into force in May 2004
Capital: *name:* Majuro
geographic coordinates: 7 06 N, 171 23 E
time difference: UTC+12 (17 hours ahead of Washington, DC during Standard Time)
Administrative divisions: 33 municipalities; Ailinginae, Ailinglaplap, Ailuk, Arno, Aur, Bikar, Bikini, Bokak, Ebon, Enewetak, Erikub, Jabat, Jaluit, Jemo, Kili, Kwajalein, Lae, Lib, Likiep, Majuro, Maloelap, Mejit, Mili, Namorik, Namu, Rongelap, Rongrik, Toke, Ujae, Ujelang, Utirik, Wotho, Wotje
Independence: 21 October 1986 (from the US-administered UN trusteeship)
National holiday: Constitution Day, 1 May (1979)
Constitution: 1 May 1979
Legal system: mixed legal system of US and English common law, customary law, and local statutes
International law organization participation: has not submitted an ICJ jurisdiction declaration; accepts ICCt jurisdiction
Suffrage: 18 years of age; universal
Executive branch: *chief of state:* President Jurelang ZEDKAIA (since 2 November 2009); note—the president is both the chief of state and head of government
head of government: President Jurelang ZEDKAIA (since 2 November 2009)
cabinet: Cabinet selected by the president from among the members of the legislature (For more information visit the World Leaders website)
elections: president elected by Nitijela (legislature) from among its members for a four-year term; election last held on 7 January 2008 (next to be held in 2012)
election results: Litokwa TOMEING removed as president by no confidence vote on 21 October 2009; legislature elects ZEDKAIA president on 26 October 2009
Legislative branch: unicameral legislature or Nitijela (33 seats; members elected by popular vote to serve four-year terms)
elections: last held on 19 November 2007 (next to be held by November 2011)
election results: percent of vote by party—NA; seats by party—independents 33
note: the Council of Chiefs or Ironij is a 12-member body comprised of tribal chiefs that advises on matters affecting customary law and practice
Judicial branch: Supreme Court; High Court; Traditional Rights Court
Political parties and leaders: traditionally there have been no formally organized political parties; what has existed more closely resembles factions or interest groups because they do not have party headquarters, formal platforms, or party structures; the following two "groupings" have competed in legislative balloting in recent years—Aelon Kein Ad Party [Michael KABUA] and United Democratic Party or UDP [Litokwa TOMEING]
Political pressure groups and leaders: NA
International organization participation: ACP, ADB, AOSIS, FAO, G-77, IAEA, IBRD, ICAO, IDA, IFAD, IFC, ILO, IMF, IMO, IMSO, Interpol, IOC, ITU, OPCW, PIF, Sparteca, SPC, UN, UNCTAD, UNESCO, WHO
Diplomatic representation in the US: *chief of mission:* Ambassador Banny DEBRUM
chancery: 2433 Massachusetts Avenue NW, Washington, DC 20008
telephone: [1] (202) 234-5414
FAX: [1] (202) 232-3236
consulate(s) general: Honolulu
Diplomatic representation from the US: *chief of mission:* Ambassador Martha L. CAMPBELL
embassy: Oceanside, Mejen Weto, Long Island, Majuro
mailing address: P. O. Box 1379, Majuro, Republic of the Marshall Islands 96960-1379
telephone: [692] 247-4011
FAX: [692] 247-4012

Flag description: blue with two stripes radiating from the lower hoist-side corner–orange (top) and white; a white star with four large rays and 20 small rays appears on the hoist side above the two stripes; blue represents the Pacific Ocean, the orange stripe signifies the Ralik Chain or sunset and courage, while the white stripe signifies the Ratak Chain or sunrise and peace; the star symbolizes the cross of Christianity, each of the 24 rays designates one of the electoral districts in the country and the four larger rays highlight the principal cultural centers of Majuro, Jaluit, Wotje, and Ebeye; the rising diagonal band can also be interpreted as representing the equator, with the star showing the archipelago's position just to the north
National anthem: *name:* "Forever Marshall Islands"
lyrics/music: Amata KABUA
note: adopted 1981

ECONOMY

Economy—overview: US Government assistance is the mainstay of this tiny island economy. The Marshall Islands received more than $1 billion in aid from the US from 1986-2002. Agricultural production, primarily subsistence, is concentrated on small farms; the most important commercial crops are coconuts and breadfruit. Small-scale industry is limited to handicrafts, tuna processing, and copra. The tourist industry, now a small source of foreign exchange employing less than 10% of the labor force, remains the best hope for future added income. The islands have few natural resources, and imports far exceed exports. Under the terms of the Amended Compact of Free Association, the US will provide millions of dollars per year to the Marshall Islands (RMI) through 2023, at which time a Trust Fund made up of US and RMI contributions will begin perpetual annual payouts. Government downsizing, drought, a drop in construction, the decline in tourism, and less income from the renewal of fishing vessel licenses have held GDP growth to an average of 1% over the past decade.
GDP (purchasing power parity): $133.5 million (2008 est.)
country comparison to the world: 218
$115 million (2001 est.)
note: data are in 2010 US dollars
GDP (official exchange rate): $161.7 million (2008 est.)
GDP—real growth rate: -0.3% (2008 est.)
country comparison to the world: 193
3.5% (2005 est.)
GDP—per capita (PPP): $2,500 (2008 est.)
country comparison to the world: 181
$2,900 (2005 est.)
GDP—composition by sector: *agriculture:* 31.7%
industry: 14.9%
services: 53.4% (2004 est.)
Labor force: 14,680 (2000)
country comparison to the world: 212
Labor force—by occupation: *agriculture:* 21.4%
industry: 20.9%
services: 57.7% (2000)
Unemployment rate: 36% (2006 est.)
country comparison to the world: 183
30.9% (2000 est.)
Population below poverty line: NA%
Household income or consumption by percentage share: *lowest 10%:* NA%
highest 10%: NA%
Budget: *revenues:* $123.3 million
expenditures: $1.213 billion (2008)
Inflation rate (consumer prices): 12.9% (2008 est.)
country comparison to the world: 209
3% (2005 est.)
Agriculture—products: coconuts, tomatoes, melons, taro, breadfruit, fruits; pigs, chickens
Industries: copra, tuna processing, tourism, craft items (from seashells, wood, and pearls)
Industrial production growth rate: NA%
Exports: $19.4 million (2008 est.)
country comparison to the world: 207
$9.1 million (2000 est.)
Exports—commodities: copra cake, coconut oil, handicrafts, fish
Imports: $79.4 million (2008 est.)
country comparison to the world: 213
$54.7 million (2000 est.)
Imports—commodities: foodstuffs, machinery and equipment, fuels, beverages and tobacco
Debt—external: $87 million (2008 est.)
country comparison to the world: 180
$86.5 million (FY99/00 est.)
Exchange rates: the US dollar is used

COMMUNICATIONS

Telephones—main lines in use: 4,400 (2009)
country comparison to the world: 214
Telephones—mobile cellular: 1,000 (2009)
country comparison to the world: 217
Telephone system: general assessment: digital switching equipment; modern services include telex, cellular, Internet, international calling, caller ID, and leased data circuits
domestic: Majuro Atoll and Ebeye and Kwajalein islands have regular, seven-digit, direct-dial telephones; other islands interconnected by high frequency radiotelephone (used mostly for government purposes) and mini-satellite telephones
international: country code–692; satellite earth stations–2 Intelsat (Pacific Ocean); US Government satellite communications system on Kwajalein (2005)
Broadcast media: no television broadcast station; a cable network is available on Majuro with programming via videotape replay and satellite relays; 4 radio broadcast stations; American Armed Forces Radio and Television Service (AFRTS) provides satellite radio and television service to Kwajalein Atoll (2009)
Internet country code: .mh
Internet hosts: 3 (2010)
country comparison to the world: 228
Internet users: 2,200 (2009)
country comparison to the world: 209

TRANSPORTATION

Airports: 15 (2010)
country comparison to the world: 146
Airports—with paved runways: *total:* 4
1,524 to 2,437 m: 3
914 to 1,523 m: 1 (2010)
Airports—with unpaved runways: *total:* 11
914 to 1,523 m: 10
under 914 m: 1 (2010)
Roadways: *total:* 2,028 km (includes 75 km of expressways) (2007)
country comparison to the world: 174
Merchant marine: *total:* 1,381
country comparison to the world: 7
by type: barge carrier 1, bulk carrier 415, cargo 63, chemical tanker 314, combination ore/oil 2, container 206, liquefied gas 83, passenger 7, passenger/cargo 1, petroleum tanker 259, refrigerated cargo 14, roll on/roll off 9, vehicle carrier 7
foreign-owned: 1,284 (Australia 1, Bermuda 34, Brazil 1, Canada 4, China 16, Croatia 12, Cyprus 38, Denmark 7, Egypt 1, Germany 247, Greece 358, Hong Kong 3, India 8, Iraq 2, Isle of Man 2, Israel 1, Italy 1, Japan 41, Jersey 9, Latvia 18, Malaysia 11, Mexico 4, Monaco 21, Netherlands 16, Norway 57, Pakistan 1, Qatar 24, Romania 2, Russia 6, Singapore 28, Slovenia 6, South Korea 25, Switzerland 12, Taiwan 2, Turkey 72, UAE 17, UK 7, Ukraine 1, US 168)
note: this country allows large numbers of ships owned by foreign entities to be registered in its national shipping registry and to fly its flag; these ships operate under the laws of the flag state (2010)
Ports and terminals: Enitwetak Island, Kwajalein, Majuro

MILITARY

Military branches: no regular military forces; under the 1983 Compact of Free Association, the US has full authority and responsibility for security and defense of the Marshall Islands; Marshall Islands Police (2009)
Manpower available for military service: *males age 16–49:* 16,446 (2010 est.)
Manpower fit for military service: *males age 16–49:* 13,568
females age 16–49: 13,606 (2010 est.)
Manpower reaching militarily significant age annually: *male:* 653
female: 631 (2010 est.)
Military expenditures: NA
Military—note: defense is the responsibility of the US

TRANSNATIONAL ISSUES

Disputes—international: claims US territory of Wake Island

MAURITANIA

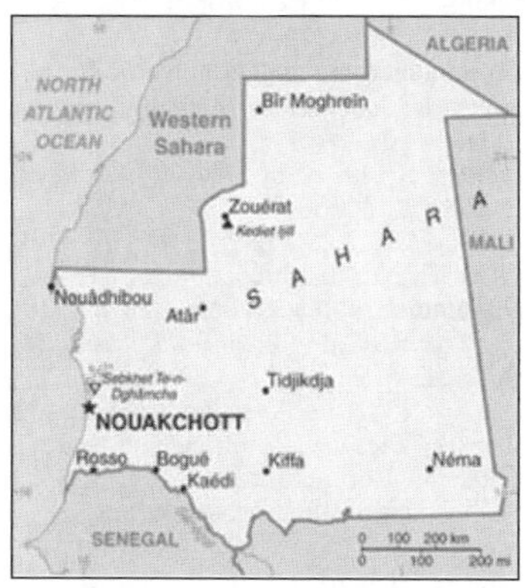

INTRODUCTION

Background: Independent from France in 1960, Mauritania annexed the southern third of the former Spanish Sahara (now Western Sahara) in 1976 but relinquished it after three years of raids by the Polisario guerrilla front seeking independence for the territory. Maaouya Ould Sid Ahmed TAYA seized power in a coup in 1984 and ruled Mauritania with a heavy hand for more than two decades. A series of presidential elections that he held were widely seen as flawed. A bloodless coup in August 2005 deposed President TAYA and ushered in a military council that oversaw a transition to democratic rule. Independent candidate Sidi Ould Cheikh ABDALLAHI was inaugurated in April 2007 as Mauritania's first freely and fairly elected president. His term ended prematurely in August 2008 when a military junta led by General Mohamed Ould Abdel AZIZ deposed him and ushered in a military council government. AZIZ was subsequently elected president in July 2009 and sworn in the following month. The country continues to experience ethnic tensions among its black population (Afro-Mauritanians) and white and black Moor (Arab-Berber) communities, and is having to confront a growing terrorism threat by al-Qa'ida in the Islamic Maghreb (AQIM).

GEOGRAPHY

Location: Northern Africa, bordering the North Atlantic Ocean, between Senegal and Western Sahara
Geographic coordinates: 20 00 N, 12 00 W
Map references: Africa
Area: *total:* 1,030,700 sq km
country comparison to the world: 29
land: 1,030,700 sq km
water: 0 sq km
Area—comparative: slightly larger than three times the size of New Mexico
Land boundaries: *total:* 5,074 km
border countries: Algeria 463 km, Mali 2,237 km, Senegal 813 km, Western Sahara 1,561 km
Coastline: 754 km
Maritime claims: *territorial sea:* 12 nm
contiguous zone: 24 nm
exclusive economic zone: 200 nm
continental shelf: 200 nm or to the edge of the continental margin
Climate: desert; constantly hot, dry, dusty
Terrain: mostly barren, flat plains of the Sahara; some central hills
Elevation extremes: *lowest point:* Sebkhet Te-n-Dghamcha -5 m
highest point: Kediet Ijill 915 m
Natural resources: iron ore, gypsum, copper, phosphate, diamonds, gold, oil, fish
Land use: *arable land:* 0.2%
permanent crops: 0.01%
other: 99.79% (2005)
Irrigated land: 490 sq km (2002)
Total renewable water resources: 11.4 cu km (1997)
Freshwater withdrawal (domestic/industrial/agricultural): *total:* 1.7 cu km/yr (9%/3%/88%)
per capita: 554 cu m/yr (2000)
Natural hazards: hot, dry, dust/sand-laden sirocco wind blows primarily in March and April; periodic droughts
Environment—current issues: overgrazing, deforestation, and soil erosion aggravated by drought are contributing to desertification; limited natural freshwater resources away from the Senegal, which is the only perennial river; locust infestation
Environment—international agreements: *party to:* Biodiversity, Climate Change, Climate Change-Kyoto Protocol, Desertification, Endangered Species, Hazardous Wastes, Law of the Sea, Ozone Layer Protection, Ship Pollution, Wetlands, Whaling
signed, but not ratified: none of the selected agreements
Geography—note: most of the population is concentrated in the cities of Nouakchott and Nouadhibou and along the Senegal River in the southern part of the country

PEOPLE

Population: 3,281,634 (July 2011 est.)
country comparison to the world: 133
Age structure: *0–14 years:* 40.4% (male 665,314/female 660,352)
15–64 years: 56.2% (male 866,859/female 975,821)
65 years and over: 3.5% (male 48,075/female 65,213) (2011 est.)
Median age: *total:* 19.5 years
male: 18.6 years
female: 20.4 years (2011 est.)
Population growth rate: 2.349% (2011 est.)
country comparison to the world: 34
Birth rate: 33.23 births/1,000 population (2011 est.)
country comparison to the world: 36
Death rate: 8.83 deaths/1,000 population (July 2011 est.)
country comparison to the world: 76
Net migration rate: -0.91 migrant(s)/1,000 population (2011 est.)
country comparison to the world: 149
Urbanization: *urban population:* 41% of total population (2010)
rate of urbanization: 2.9% annual rate of change (2010-15 est.)
Major cities—population: NOUAKCHOTT (capital) 709,000 (2009)
Sex ratio: *at birth:* 1.03 male(s)/female
under 15 years: 1.01 male(s)/female
15–64 years: 0.89 male(s)/female
65 years and over: 0.74 male(s)/female
total population: 0.93 male(s)/female (2011 est.)
Infant mortality rate: *total:* 60.42 deaths/1,000 live births
country comparison to the world: 34
male: 65.55 deaths/1,000 live births
female: 55.13 deaths/1,000 live births (2011 est.)
Life expectancy at birth: *total population:* 61.14 years
country comparison to the world: 184
male: 58.94 years
female: 63.41 years (2011 est.)
Total fertility rate: 4.3 children born/woman (2011 est.)
country comparison to the world: 35
HIV/AIDS—adult prevalence rate: 0.7% (2009 est.)
country comparison to the world: 60
HIV/AIDS—people living with HIV/AIDS: 14,000 (2009 est.)
country comparison to the world: 89
HIV/AIDS—deaths: fewer than 1,000 (2009 est.)
country comparison to the world: 77
Major infectious diseases: degree of risk: high
food or waterborne diseases: bacterial and protozoal diarrhea, hepatitis A, and typhoid fever
vectorborne diseases: malaria and Rift Valley fever
respiratory disease: meningococcal meningitis
animal contact disease: rabies (2009)
Drinking water source: *Improved:*
urban: 52% of population
rural: 47% of population
total: 49% of population
Unimproved:
urban: 48% of population
rural: 53% of population
total: 51% of population (2008)
Sanitation facility access: *Improved:*
urban: 50% of population
rural: 9% of population
total: 26% of population
Unimproved:
urban: 50% of population
rural: 91% of population
total: 74% of population (2008)
Nationality: *noun:* Mauritanian(s)
adjective: Mauritanian
Ethnic groups: mixed Moor/black 40%, Moor 30%, black 30%
Religions: Muslim 100%
Languages: Arabic (official and national), Pulaar, Soninke, Wolof (all national languages), French, Hassaniya
Literacy: *definition:* age 15 and over can read and write
total population: 51.2%
male: 59.5%

female: 43.4% (2000 census)
School life expectancy (primary to tertiary education): *total:* 8 years
male: 8 years
female: 8 years (2007)
Education expenditures: 4.4% of GDP (2008)
country comparison to the world: 86

GOVERNMENT

Country name: *conventional long form:* Islamic Republic of Mauritania
conventional short form: Mauritania
local long form: Al Jumhuriyah al Islamiyah al Muritaniyah
local short form: Muritaniyah
Government type: military junta
Capital: *name:* Nouakchott
geographic coordinates: 18 07 N, 16 02 W
time difference: UTC 0 (5 hours ahead of Washington, DC during Standard Time)
Administrative divisions: 13 regions (wilayas, singular–wilaya); Adrar, Assaba, Brakna, Dakhlet Nouadhibou, Gorgol, Guidimaka, Hodh ech Chargui, Hodh El Gharbi, Inchiri, Nouakchott, Tagant, Tiris Zemmour, Trarza
Independence: 28 November 1960 (from France)
National holiday: Independence Day, 28 November (1960)
Constitution: 12 July 1991
Legal system: mixed legal system of Islamic and French civil law
International law organization participation: has not submitted an ICJ jurisdiction declaration; non-party state to the ICCt
Suffrage: 18 years of age; universal
Executive branch: *chief of state:* President Mohamed Ould Abdel AZIZ (since 5 August 2009); note–AZIZ, who deposed democratically elected President Sidi Ould Cheikh ABDELLAHI in a coup and installed himself as President of the High State Council on 6 August 2008, retired from the military and stepped down from the Presidency in April 2009 to run for president; he was elected president in an election held on 18 July 2009
head of government: Prime Minister Moulaye Ould Mohamed LAGHDAF (since 14 August 2008)
cabinet: Council of Ministers
(For more information visit the World Leaders website)
elections: following the August 2008 coup, the High State Council planned to hold a new presidential election in June 2009; the election was subsequently rescheduled to 18 July 2009 following the Dakar Accords, which brought Mauritania back to constitutional rule; under Mauritania's constitution, the president elected by popular vote for a five-year term; election last held on 18 July 2009 (next to be held by 2014)
election results: percent of vote–Mohamed Ould Abdel AZIZ 52.6%, Messaoud Ould BOULKHEIR 16.3%, Ahmed Ould DADDAH 13.7%, other 17.4%
Legislative branch: bicameral legislature consists of the Senate or Majlis al-Shuyukh (56 seats; 53 members elected by municipal leaders and 3 members elected for Mauritanians abroad to serve six-year terms; a portion of seats up for election every two years) and the National Assembly or Al Jamiya Al Wataniya (95 seats; members elected by popular vote to serve five-year terms)
elections: Senate–last held in November 2009; National Assembly–last held on 19 November and 3 December 2006 (next to be held in 2011)
election results: Senate–percent of vote by party–NA; seats by party–CPM (Coalition of Majority Parties) 45, COD 7, RNRD-TAWASSOUL 4; National Assembly–percent of vote by party–NA; seats by party–CPM 63 (UPR 50, PRDR 7, UDP 3, HATEM-PMUC 2, RD 1), COD 27 (RFD 9, UFP 6, APP 6, PNDD-ADIL 6), RNRD-TAWASSOUL 4, FP 1
Judicial branch: Supreme Court or Cour Supreme; Court of Appeals; lower courts
Political parties and leaders: Alternative or El-Badil [Mohamed Yahdhi Ould MOCTAR HACEN]; Coalition of Majority Parties or CPM (parties supporting the regime including PRDR, UPR, RD, HATEM-PMUC, UCD); Coordination of Democratic Opposition or COD (coalition of opposition political parties opposed to the government including APP, RFD, UFP, PNDD-ADIL, Alternative or El-Badil); Democratic Renewal or RD [Moustapha Ould ABDEIDARRAHMANE]; Mauritanian Party for Unity and Change or HATEM-PMUC [Saleh Ould HANENA]; National Pact for Democracy and Development or PNDD-ADIL [Yahya Ould Ahmed Ould WAGHEF] (independents formerly supporting President Abdellahi); National Rally for Freedom, Democracy and Equality or RNDLE; National Rally for Reform and Development or RNRD-TAWASSOUL [Mohamed Jamil MANSOUR] (moderate Islamists); Popular Front or FP [Ch'bih Ould CHEIKH MALAININE]; Popular Progressive Alliance or APP [Messoud Ould BOULKHEIR]; Rally of Democratic Forces or RFD [Ahmed Ould DADDAH]; Republican Party for Democracy and Renewal or PRDR [Mintata Mint HDEID]; Socialist and Democratic Unity Party or PUDS; Union for Democracy and Progress or UDP [Naha Mint MOUKNASS]; Union for the Republic or UPR; Union of Democratic Center or UCD [Cheikh Sid'Ahmed Ould BABA]; Union of the Forces for Progress or UFP [Mohamed Ould MAOULOUD];
Political pressure groups and leaders: General Confederation of Mauritanian Workers or CGTM [Abdallahi Ould MOHAMED, secretary general]; Independent Confederation of Mauritanian Workers or CLTM [Samory Ould BEYE]; Mauritanian Workers Union or UTM [Mohamed Ely Ould BRAHIM, secretary general]
other: Arab nationalists; Ba'thists; Islamists
International organization participation: ABEDA, ACP, AfDB, AFESD, AMF, AMU, AU, CAEU, FAO, G-77, IAEA, IBRD, ICAO, ICRM, IDA, IDB, IFAD, IFC, IFRCS, ILO, IMF, IMO, Interpol, IOC, IOM, IPU, ISO (correspondent), ITSO, ITU, ITUC, LAS, MIGA, NAM, OIC, OIF, OPCW, UN, UNCTAD, UNESCO, UNIDO, UNWTO, UPU, WCO, WHO, WIPO, WMO, WTO
Diplomatic representation in the US: *chief of mission:* Ambassador Mohamed Lemine El HAYCEN
chancery: 2129 Leroy Place NW, Washington, DC 20008
telephone: [1] (202) 232-5700 through 5701
FAX: [1] (202) 319-2623
Diplomatic representation from the US: *chief of mission:* Ambassador Mark M. BOULWARE
embassy: 288 Rue Abdallaye, Rue 42-100 (between Presidency building and Spanish Embassy), Nouakchott
mailing address: BP 222, Nouakchott
telephone: [222] 525-2660 through 2663
FAX: [222] 525-1592
Flag description: green with a yellow five-pointed star above a yellow, horizontal crescent; the closed side of the crescent is down; the crescent, star, and color green are traditional symbols of Islam; the gold color stands for the sands of the Sahara
National anthem: *name:* "Hymne National de la Republique Islamique de Mauritanie" (National Anthem of the Islamic Republic of Mauritania)
lyrics/music: Baba Ould CHEIKH/ traditional, arranged by Tolia NIKIPROWETZKY
note: adopted 1960; the unique rhythm of the Mauritanian anthem makes it particularly challenging to sing

ECONOMY

Economy—overview: Half the population still depends on agriculture and livestock for a livelihood, even though many of the nomads and subsistence farmers were forced into the cities by recurrent droughts in the 1970s and 1980s. Mauritania has extensive deposits of iron ore, which account for nearly 40% of total exports. The nation's coastal waters are among the richest fishing areas in the world but overexploitation by foreigners threatens this key source of revenue. The country's first deepwater port opened near Nouakchott in 1986. Before 2000, drought and economic mismanagement resulted in a buildup of foreign debt. In February 2000, Mauritania qualified for debt relief under the Heavily Indebted Poor Countries (HIPC) initiative and nearly all of its foreign debt has since been forgiven. A new investment code approved in December 2001 improved the opportunities for direct foreign investment. Mauritania and the IMF agreed to a three-year Poverty Reduction and Growth Facility (PRGF) arrangement in 2006. Mauritania made satisfactory progress, but the IMF, World Bank, and other international actors suspended assistance and investment in Mauritania after the August 2008 coup. Since the presidential election in July 2009, donors have resumed assistance. Oil prospects, while initially promising, have largely failed to materialize, and the government has placed a priority on attracting private investment to spur economic growth. The

Government also emphasizes reduction of poverty, improvement of health and education, and privatization of the economy.
GDP (purchasing power parity): $6.655 billion (2010 est.)
country comparison to the world: 154
$6.358 billion (2009 est.)
$6.437 billion (2008 est.)
note: data are in 2010 US dollars
GDP (official exchange rate): $3.799 billion (2010 est.)
GDP—real growth rate: 4.7% (2010 est.)
country comparison to the world: 78
-1.2% (2009 est.)
3.5% (2008 est.)
GDP—per capita (PPP): $2,100 (2010 est.)
country comparison to the world: 188
$2,000 (2009 est.)
$2,100 (2008 est.)
note: data are in 2010 US dollars
GDP—composition by sector: *agriculture:* 12.5%
industry: 46.7%
services: 40.7% (2008 est.)
Labor force: 1.318 million (2007)
country comparison to the world: 135
Labor force—by occupation: *agriculture:* 50%
industry: 10%
services: 40% (2001 est.)
Unemployment rate: 30% (2008 est.)
country comparison to the world: 176
20% (2004 est.)
Population below poverty line: 40% (2004 est.)
Household income or consumption by percentage share: *lowest 10%:* 2.5%
highest 10%: 29.5% (2000)
Distribution of family income—Gini index: 39 (2000)
country comparison to the world: 66
37.3 (1995)
Budget: *revenues:* $770 million
expenditures: $770 million (2007 est.)
Inflation rate (consumer prices): 7.3% (2007 est.)
country comparison to the world: 181
Central bank discount rate: NA% (31 December 2009)
country comparison to the world: 28
12% (31 December 2007)
Commercial bank prime lending rate: NA%
Stock of domestic credit: $NA
Market value of publicly traded shares: $NA
Agriculture—products: dates, millet, sorghum, rice, corn; cattle, sheep
Industries: fish processing, oil production, mining of iron ore, gold, and copper
note: gypsum deposits have never been exploited
Industrial production growth rate: 2% (2000 est.)
country comparison to the world: 130
Electricity—production: 415.3 million kWh (2007 est.)
country comparison to the world: 162
Electricity—consumption: 386.2 million kWh (2007 est.)
country comparison to the world: 165
Electricity—exports: 0 kWh (2008 est.)
Electricity—imports: 0 kWh (2008 est.)
Oil—production: 16,510 bbl/day (2009 est.)
country comparison to the world: 77
Oil—consumption: 20,000 bbl/day (2009 est.)
country comparison to the world: 124
Oil—exports: 30,620 bbl/day (2007 est.)
country comparison to the world: 83
Oil—imports: 20,610 bbl/day (2007 est.)
country comparison to the world: 109
Oil—proved reserves: 100 million bbl (1 January 2010 est.)
country comparison to the world: 70
Natural gas—production: 0 cu m (2008 est.)
country comparison to the world: 161
Natural gas—consumption: 0 cu m (2008 est.)
country comparison to the world: 199
Natural gas—exports: 0 cu m (2008 est.)
country comparison to the world: 143
Natural gas—imports: 0 cu m (2008 est.)
country comparison to the world: 88
Natural gas—proved reserves: 28.32 billion cu m (1 January 2010 est.)
country comparison to the world: 73
Current account balance: $-184 million (2007 est.)
country comparison to the world: 87
Exports: $1.395 billion (2006)
country comparison to the world: 143
Exports—commodities: iron ore, fish and fish products, gold, copper, petroleum
Exports—partners: China 42.06%, Italy 9.71%, Japan 7.57%, Cote d'Ivoire 6.16%, Spain 5.63%, Netherlands 4.32% (2009)
Imports: $1.475 billion (2006)
country comparison to the world: 163
Imports—commodities: machinery and equipment, petroleum products, capital goods, foodstuffs, consumer goods
Imports—partners: France 14.3%, Netherlands 10.33%, China 9.94%, Brazil 5.58%, Belgium 4.87%, Germany 4.04%, Spain 4.02% (2009)
Debt—external: $NA
Exchange rates: ouguiyas (MRO) per US dollar–261.5 (2010 est.) 262.4 (2009) 238.2 (2008) 258.6 (2007) 271.3 (2006)

COMMUNICATIONS

Telephones—main lines in use: 74,500 (2009)
country comparison to the world: 152
Telephones—mobile cellular: 2.182 million (2009)
country comparison to the world: 130
Telephone system: *general assessment:* limited system of cable and open-wire lines, minor microwave radio relay links, and radiotelephone communications stations; mobile-cellular services expanding rapidly
domestic: Mauritel, the national telecommunications company, was privatized in 2001 but remains the monopoly provider of fixed-line services; fixed-line teledensity 2 per 100 persons; mobile-cellular network coverage extends mainly to urban areas with a teledensity of 70 per 100 persons; mostly cable and open-wire lines; a domestic satellite telecommunications system links Nouakchott with regional capitals
international: country code–222; satellite earth stations–3 (1 Intelsat–Atlantic Ocean, 2 Arabsat); fiber-optic and Asymmetric Digital Subscriber Line (ADSL) cables for Internet access (2008)
Broadcast media: broadcast media state-owned; 1 state-run TV and 1 state-run radio network; Television de Mauritanie, the state-run TV station, has an additional 6 regional TV stations that provide local programming (2008)
Internet country code: .mr
Internet hosts: 23 (2010)
country comparison to the world: 216
Internet users: 75,000 (2009)
country comparison to the world: 169

TRANSPORTATION

Airports: 28 (2010)
country comparison to the world: 119
Airports—with paved runways: *total:* 9
2,438 to 3,047 m: 5
1,524 to 2,437 m: 4 (2010)
Airports—with unpaved runways:
total: 19
1,524 to 2,437 m: 9
914 to 1,523 m: 8
under 914 m: 2 (2010)
Railways: 728 km
standard gauge: 728 km 1.435-m gauge (2010)
Roadways: *total:* 11,066 km
country comparison to the world: 133
paved: 2,966 km
unpaved: 8,100 km (2006)
Waterways: (some is navigation possible on the Senegal River) (2010)
Ports and terminals: Nouadhibou, Nouakchott

MILITARY

Military branches: Mauritanian Armed Forces: Army, Mauritanian Navy (Marine Mauritanienne; includes naval infantry), Islamic Air Force of Mauritania (Force Aerienne Islamique de Mauritanie, FAIM) (2010)
Military service age and obligation: 18 years of age (est.); conscript service obligation–2 years; majority of servicemen believed to be volunteers; service in Air Force and Navy is voluntary (2006)
Manpower available for military service: *males age 16–49:* 718,713
females age 16–49: 804,622 (2010 est.)
Manpower fit for military service: *males age 16–49:* 480,042
females age 16–49: 581,473 (2010 est.)
Manpower reaching militarily significant age annually: *male:* 36,116
female: 36,826 (2010 est.)
Military expenditures: 5.5% of GDP (2006)
country comparison to the world: 13

TRANSNATIONAL ISSUES

Disputes—international: Mauritanian claims to Western Sahara remain dormant
Trafficking in persons: current situation: Mauritania is a source and destination country for children trafficked for forced labor and sexual exploitation; slavery-related practices, rooted in ancestral master-slave relationships, continue to exist in isolated parts of the country;

Mauritanian boys called talibe are trafficked within the country by religious teachers for forced begging; children are also trafficked by street gangs within the country that force them to steal, beg, and sell drugs; girls are trafficked internally for domestic servitude and sexual exploitation; women and children from neighboring states are trafficked into Mauritania for purposes of forced begging, domestic servitude, and sexual exploitation
tier rating: Tier 3—the Government of Mauritania does not fully comply with the minimum standards for the elimination of trafficking and is not making significant efforts to do so; the government did not show evidence of overall progress in prosecuting and punishing trafficking offenders, protecting trafficking victims, or preventing new incidents of trafficking during the past year; progress that the previous government demonstrated in 2007 through enactment of strengthened anti-slavery legislation and deepened political will to eliminate slavery and trafficking has stalled; law enforcement efforts to address human trafficking including traditional slavery practices decreased (2009)

MAURITIUS

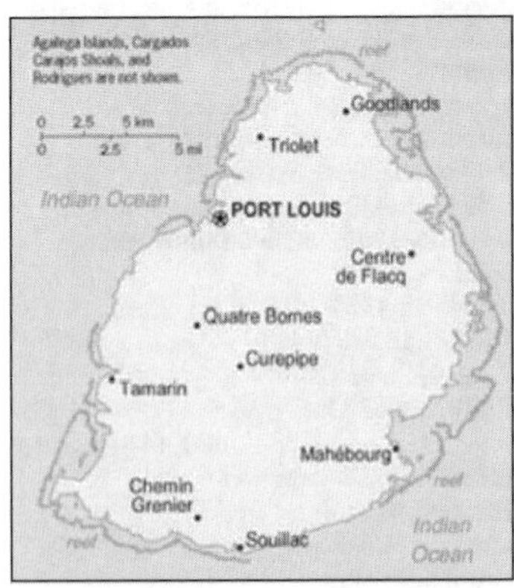

INTRODUCTION

Background: Although known to Arab and Malay sailors as early as the 10th century, Mauritius was first explored by the Portuguese in the 16th century and subsequently settled by the Dutch—who named it in honor of Prince Maurits van NASSAU—in the 17th century. The French assumed control in 1715, developing the island into an important naval base overseeing Indian Ocean trade, and establishing a plantation economy of sugar cane. The British captured the island in 1810, during the Napoleonic Wars. Mauritius remained a strategically important British naval base, and later an air station, playing an important role during World War II for anti-submarine and convoy operations, as well as the collection of signals intelligence. Independence from the UK was attained in 1968. A stable democracy with regular free elections and a positive human rights record, the country has attracted considerable foreign investment and has earned one of Africa's highest per capita incomes. Recent poor weather, declining sugar prices, and declining textile and apparel production, have slowed economic growth, leading to some protests over standards of living in the Creole community.

GEOGRAPHY

Location: Southern Africa, island in the Indian Ocean, east of Madagascar
Geographic coordinates: 20 17 S, 57 33 E
Map references: Africa
Area: *total:* 2,040 sq km
country comparison to the world: 180
land: 2,030 sq km
water: 10 sq km
note: includes Agalega Islands, Cargados Carajos Shoals (Saint Brandon), and Rodrigues
Area—comparative: almost 11 times the size of Washington, DC
Land boundaries: 0 km
Coastline: 177 km
Maritime claims: measured from claimed archipelagic straight baselines
territorial sea: 12 nm
exclusive economic zone: 200 nm
continental shelf: 200 nm or to the edge of the continental margin
Climate: tropical, modified by southeast trade winds; warm, dry winter (May to November); hot, wet, humid summer (November to May)
Terrain: small coastal plain rising to discontinuous mountains encircling central plateau
Elevation extremes: *lowest point:* Indian Ocean 0 m
highest point: Mont Piton 828 m
Natural resources: arable land, fish
Land use: *arable land:* 49.02%
permanent crops: 2.94%
other: 48.04% (2005)
Irrigated land: 220 sq km (2003)
Total renewable water resources: 2.2 cu km (2001)
Freshwater withdrawal (domestic/industrial/agricultural): *total:* 0.61 cu km/yr (25%/14%/60%)
per capita: 488 cu m/yr (2000)
Natural hazards: cyclones (November to April); almost completely surrounded by reefs that may pose maritime hazards
Environment—current issues: water pollution, degradation of coral reefs
Environment—international agreements: *party to:* Antarctic-Marine Living Resources, Biodiversity, Climate Change, Climate Change-Kyoto Protocol, Desertification, Endangered Species, Environmental Modification, Hazardous Wastes, Law of the Sea, Marine Life Conservation, Ozone Layer Protection, Ship Pollution, Wetlands
signed, but not ratified: none of the selected agreements
Geography—note: the main island, from which the country derives its name, is of volcanic origin and is almost entirely surrounded by coral reefs; home of the dodo, a large flightless bird related to pigeons, driven to extinction by the end of the 17th century through a combination of hunting and the introduction of predatory species

PEOPLE

Population: 1,303,717 (July 2011 est.)
country comparison to the world: 153
Age structure: *0–14 years:* 21.8% (male 145,185/female 139,579)
15–64 years: 70.7% (male 457,743/female 463,875)
65 years and over: 7.5% (male 38,944/female 58,391) (2011 est.)
Median age: *total:* 32.7 years
male: 31.9 years
female: 33.6 years (2011 est.)
Population growth rate: 0.729% (2011 est.)
country comparison to the world: 139
Birth rate: 13.97 births/1,000 population (2011 est.)
country comparison to the world: 147
Death rate: 6.68 deaths/1,000 population (July 2011 est.)
country comparison to the world: 146
Net migration rate: 0 migrant(s)/1,000 population (2011 est.)
country comparison to the world: 99
Urbanization: *urban population:* 42% of total population (2010)
rate of urbanization: 0.8% annual rate of change (2010-15 est.)
Major cities—population: PORT LOUIS (capital) 149,000 (2009)
Sex ratio: *at birth:* 1.05 male(s)/female
under 15 years: 1.04 male(s)/female
15–64 years: 0.99 male(s)/female
65 years and over: 0.67 male(s)/female
total population: 0.97 male(s)/female (2011 est.)
Infant mortality rate: *total:* 11.52 deaths/1,000 live births
country comparison to the world: 139
male: 13.7 deaths/1,000 live births
female: 9.23 deaths/1,000 live births (2011 est.)
Life expectancy at birth: *total population:* 74.48 years
country comparison to the world: 99
male: 71.01 years
female: 78.12 years (2011 est.)
Total fertility rate: 1.79 children born/woman (2011 est.)
country comparison to the world: 155
HIV/AIDS—adult prevalence rate: 1% (2009 est.)
country comparison to the world: 47
HIV/AIDS—people living with HIV/AIDS: 8,800 (2009 est.)
country comparison to the world: 103
HIV/AIDS—deaths: fewer than 500 (2009 est.)

country comparison to the world: 85
Drinking water source: *Improved:*
urban: 100% of population
rural: 99% of population
total: 99% of population
Unimproved:
urban: 0% of population
rural: 1% of population
total: 1% of population (2008)
Sanitation facility access: *Improved:*
urban: 93% of population
rural: 90% of population
total: 91% of population
Unimproved:
urban: 7% of population
rural: 10% of population
total: 9% of population (2008)
Nationality: *noun:* Mauritian(s)
adjective: Mauritian
Ethnic groups: Indo-Mauritian 68%, Creole 27%, Sino-Mauritian 3%, Franco-Mauritian 2%
Religions: Hindu 48%, Roman Catholic 23.6%, Muslim 16.6%, other Christian 8.6%, other 2.5%, unspecified 0.3%, none 0.4% (2000 census)
Languages: Creole 80.5%, Bhojpuri 12.1%, French 3.4%, English (official; spoken by less than 1% of the population), other 3.7%, unspecified 0.3% (2000 census)
Literacy: *definition:* age 15 and over can read and write
total population: 84.4%
male: 88.4%
female: 80.5% (2000 census)
School life expectancy (primary to tertiary education):
total: 14 years
male: 13 years
female: 14 years (2008)
Education expenditures: 3.2% of GDP (2009)
country comparison to the world: 128

GOVERNMENT

Country name: *conventional long form:* Republic of Mauritius
conventional short form: Mauritius
local long form: Republic of Mauritius
local short form: Mauritius
Government type: parliamentary democracy
Capital: *name:* Port Louis
geographic coordinates: 20 09 S, 57 29 E
time difference: UTC+4 (9 hours ahead of Washington, DC during Standard Time)
Administrative divisions: 9 districts and 3 dependencies*; Agalega Islands*, Black River, Cargados Carajos Shoals*, Flacq, Grand Port, Moka, Pamplemousses, Plaines Wilhems, Port Louis, Riviere du Rempart, Rodrigues*, Savanne
Independence: 12 March 1968 (from the UK)
National holiday: Independence Day, 12 March (1968)
Constitution: 12 March 1968; amended 12 March 1992
Legal system: civil legal system based on French civil law with some elements of English common law
International law organization participation: accepts compulsory ICJ jurisdiction with reservations; accepts ICCt jurisdiction
Suffrage: 18 years of age; universal
Executive branch: *chief of state:* President Sir Anerood JUGNAUTH (since 7 October 2003); Vice President Monique OHSAN-BELLEPEAU (since 13 November 2010)
head of government: Prime Minister Navinchandra RAMGOOLAM (since 5 July 2005)
cabinet: Council of Ministers appointed by the president on the recommendation of the prime minister
(For more information visit the World Leaders website)
elections: president and vice president elected by the National Assembly for five-year terms (eligible for a second term); elections last held on 19 September 2008 (next to be held in 2013); prime minister and deputy prime minister appointed by the president, responsible to the National Assembly
election results: Sir Anerood JUGNAUTH reelected president by unanimous vote; percent of vote by the National Assembly—NA
Legislative branch: unicameral National Assembly (70 seats; 62 members elected by popular vote, 8 appointed by the election commission to give representation to various ethnic minorities; members to serve five-year terms)
elections: last held on 5 May 2010 (next to be held in 2015)
election results: percent of vote by party—NA; seats by party—AF 41, MMM 18, MR 2, MSF 1; appointed seats—to be assigned 8
Judicial branch: Supreme Court
Political parties and leaders: Alliance of the Future or AF [Navinchandra RAMGOOLAM] (governing coalition—includes MLD, MMSM, MR, MSD, PMXD); Mauritian Labor Party or MLP [Navinchandra RAMGOOLAM]; Mauritian Militant Movement or MMM [Paul BERENGER]; Mauritian Militant Socialist Movement or MMSM [Pravind JUGNAUTH]; Mauritian Socialist Militant Movement or MSMM [Madan DULLOO]; Mauritian Solidarity Front of FSM [Cehl FAKEERMEEAH]; Mouvement Republicain or MR [Jayarama VALAYDEN]; Parti Mauricien Xavier Duval or PMXD [Xavier Luc DUVAL]; Rodrigues Movement or MR [Joseph (Nicholas) Von MALLY]; Rodrigues Peoples Organization or OPR [Serge CLAIR]
Political pressure groups and leaders: *other:* various labor unions
International organization participation: ACP, AfDB, AOSIS, AU, C, COMESA, CPLP (associate), FAO, G-77, IAEA, IBRD, ICAO, ICRM, IDA, IFAD, IFC, IFRCS, IHO, ILO, IMF, IMO, IMSO, InOC, Interpol, IOC, IOM, IPU, ISO, ITSO, ITU, ITUC, MIGA, NAM, OIF, OPCW, PCA, SAARC (observer), SADC, UN, UNCTAD, UNESCO, UNIDO, UNWTO, UPU, WCO, WFTU, WHO, WIPO, WMO, WTO
Diplomatic representation in the US: *chief of mission:* Ambassador Somduth SOBORUN
chancery: 1709 N Street NW, Washington, DC 20036
telephone: [1] (202) 244-1491 through 1492
FAX: [1] (202) 966-0983
Diplomatic representation from the US: *chief of mission:* Ambassador Mary Jo WILLS
embassy: 4th Floor, Rogers House, John Kennedy Street, Port Louis
mailing address: international mail: P. O. Box 544, Port Louis; US mail: American Embassy, Port Louis, US Department of State, Washington, DC 20521-2450
telephone: [230] 202-4400
FAX: [230] 208-9534
Flag description: four equal horizontal bands of red (top), blue, yellow, and green; red represents the blood shed for independence, blue the Indian Ocean surrounding the island, yellow has been interpreted as the new light of independence, golden sunshine, or the bright future, and green can symbolize either agriculture or the lush vegetation of the island
National anthem: *name:* "Motherland"
lyrics/music: Jean Georges PROSPER/ Philippe GENTIL
note: adopted 1968

ECONOMY

Economy—overview: Since independence in 1968, Mauritius has developed from a low-income, agriculturally based economy to a middle-income diversified economy with growing industrial, financial, and tourist sectors. For most of the period, annual growth has been in the order of 5% to 6%. This remarkable achievement has been reflected in more equitable income distribution, increased life expectancy, lowered infant mortality, and a much-improved infrastructure. The economy rests on sugar, tourism, textiles and apparel, and financial services, and is expanding into fish processing, information and communications technology, and hospitality and property development. Sugarcane is grown on about 90% of the cultivated land area and accounts for 15% of export earnings. The government's development strategy centers on creating vertical and horizontal clusters of development in these sectors. Mauritius has attracted more than 32,000 offshore entities, many aimed at commerce in India, South Africa, and China. Investment in the banking sector alone has reached over $1 billion. Mauritius, with its strong textile sector, has been well poised to take advantage of the Africa Growth and Opportunity Act (AGOA). Mauritius' sound economic policies and prudent banking practices helped to mitigate negative effects from the global financial crisis in 2008-09. GDP grew 3.6% in 2010 and the country continues to expand its trade and investment outreach around the globe.
GDP (purchasing power parity): $18.06 billion (2010 est.)
country comparison to the world: 129
$17.36 billion (2009 est.)
$16.85 billion (2008 est.)

note: data are in 2010 US dollars
GDP (official exchange rate): $9.729 billion (2010 est.)
GDP—real growth rate: 4% (2010 est.)
country comparison to the world: 97
3% (2009 est.)
5.5% (2008 est.)
GDP—per capita (PPP): $14,000 (2010 est.)
country comparison to the world: 82
$13,500 (2009 est.)
$13,200 (2008 est.)
note: data are in 2010 US dollars
GDP—composition by sector: *agriculture:* 4.8%
industry: 24.6%
services: 70.5% (2010 est.)
Labor force: 597,000 (2010 est.)
country comparison to the world: 155
Labor force—by occupation: *agriculture and fishing:* 9%
construction and industry: 30%
transportation and communication: 7%
trade, restaurants, hotels: 22%
finance: 6%
other *services:* 25% (2007)
Unemployment rate: 7.5% (2010 est.)
country comparison to the world: 80
7.3% (2009 est.)
Population below poverty line: 8% (2006 est.)
Household income or consumption by percentage share: *lowest 10%:* NA%
highest 10%: NA%
Distribution of family income—Gini index: 39 (2006 est.)
country comparison to the world: 67
37 (1987 est.)
Investment (gross fixed): 23.8% of GDP (2010 est.)
country comparison to the world: 50
Budget: *evenues:* $2.114 billion
expenditures: $2.583 billion (2010 est.)
Public debt: 60.5% of GDP (2010 est.)
country comparison to the world: 32
62.4% of GDP (2009 est.)
Inflation rate (consumer prices): 2.9% (2010 est.)
country comparison to the world: 81
2.5% (2009 est.)
Central bank discount rate: NA%
Commercial bank prime lending rate: 19.25% (31 December 2009 est.)
country comparison to the world: 13
21.54% (31 December 2008 est.)
Stock of narrow money: $1.889 billion (31 December 2010 est.)
country comparison to the world: 120
$1.906 billion (31 December 2009 est.)
Stock of broad money: $9.605 billion (31 December 2010 est.)
country comparison to the world: 101
$9.277 billion (31 December 2009 est.)
Stock of domestic credit: $10.23 billion (31 December 2010 est.)
country comparison to the world: 92
$9.423 billion (31 December 2009 est.)
Market value of publicly traded shares: $4.74 billion (31 December 2009)
country comparison to the world: 89
$3.443 billion (31 December 2008)
$5.666 billion (31 December 2007)
Agriculture—products: sugarcane, tea, corn, potatoes, bananas, pulses; cattle, goats; fish
Industries: food processing (largely sugar milling), textiles, clothing, mining, chemicals, metal products, transport equipment, nonelectrical machinery, tourism
Industrial production growth rate: 3.3% (2010 est.)
country comparison to the world: 99
Electricity—production: 2.321 billion kWh (2007 est.)
country comparison to the world: 130
Electricity—consumption: 2.158 billion kWh (2007 est.)
country comparison to the world: 134
Electricity—exports: 0 kWh (2008 est.)
Electricity—imports: 0 kWh (2008 est.)
Oil—production: 0 bbl/day (2009 est.)
country comparison to the world: 206
Oil—consumption: 23,000 bbl/day (2009 est.)
country comparison to the world: 118
Oil—exports: 0 bbl/day (2007 est.)
country comparison to the world: 186
Oil—imports: 22,200 bbl/day (2007 est.)
country comparison to the world: 107
Oil—proved reserves: 0 bbl (1 January 2010 est.)
country comparison to the world: 164
Natural gas—production: 0 cu m (2008 est.)
country comparison to the world: 160
Natural gas—consumption: 0 cu m (2008 est.)
country comparison to the world: 198
Natural gas—exports: 0 cu m (2008 est.)
country comparison to the world: 142
Natural gas—imports: 0 cu m (2008 est.)
country comparison to the world: 87
Natural gas—proved reserves: 0 cu m (1 January 2010 est.)
country comparison to the world: 166
Current account balance: $-949 million (2010 est.)
country comparison to the world: 131
$-674.6 million (2009 est.)
Exports: $2.041 billion (2010 est.)
country comparison to the world: 132
$1.942 billion (2009 est.)
Exports—commodities: clothing and textiles, sugar, cut flowers, molasses, fish
Exports—partners: UK 25.55%, France 16.89%, US 9.51%, Italy 5.68%, UAE 5.47%, Belgium 4.93%, Madagascar 4.11% (2009)
Imports: $3.935 billion (2010 est.)
country comparison to the world: 129
$3.499 billion (2009 est.)
Imports—commodities: manufactured goods, capital equipment, foodstuffs, petroleum products, chemicals
Imports—partners: India 24.5%, France 14.02%, South Africa 8.55%, China 8.17% (2009)
Reserves of foreign exchange and gold: $2.36 billion (31 December 2010 est.)
country comparison to the world: 94
$2.304 billion (31 December 2009 est.)
Debt—external: $5.043 billion (31 December 2010 est.)
country comparison to the world: 105
$4.474 billion (31 December 2009 est.)
Stock of direct foreign investment—at home: $NA
Stock of direct foreign investment—abroad: $NA
Exchange rates: Mauritian rupees (MUR) per US dollar—30.991 (2010) 31.96 (2009) 27.973 (2008) 31.798 (2007) 31.656 (2006)

COMMUNICATIONS

Telephones—main lines in use: 379,100 (2009)
country comparison to the world: 105
Telephones—mobile cellular: 1.087 million (2009)
country comparison to the world: 145
Telephone system: *general assessment:* small system with good service
domestic: monopoly over fixed-line services terminated in 2005; fixed-line teledensity roughly 30 per 100 persons; mobile-cellular services launched in 1989 with teledensity in 2009 reaching 85 per 100 persons
international: country code—230; landing point for the SAFE submarine cable that provides links to Asia and South Africa where it connects to the SAT-3/WASC submarine cable that provides further links to parts of East Africa, and Europe; satellite earth station—1 Intelsat (Indian Ocean); new microwave link to Reunion; HF radiotelephone links to several countries (2009)
Broadcast media: the government maintains control over TV broadcasting through the Mauritius Broadcasting Corporation (MBC), which operates 3 analog and 10 digital TV stations; MBC is a shareholder in a local company that operates 2 pay TV stations; the state retains the largest radio broadcast network with multiple stations; several private radio broadcasters have entered the market since 2001; transmissions of at least 2 international broadcasters are available (2007)
Internet country code: .mu
Internet hosts: 36,653 (2010)
country comparison to the world: 94
Internet users: 290,000 (2009)
country comparison to the world: 132

TRANSPORTATION

Airports: 5 (2010)
country comparison to the world: 181
Airports—with paved runways: *total:* 2
over 3,047 m: 1
914 to 1,523 m: 1 (2010)
Airports—with unpaved runways: *total:* 3
914 to 1,523 m: 2
under 914 m: 1 (2010)
Roadways: *total:* 2,066 km
country comparison to the world: 173
paved: 2,066 km (includes 75 km of expressways) (2009)
Merchant marine: *total:* 3
country comparison to the world: 137
by type: passenger/cargo 2, refrigerated cargo 1 (2010)
Ports and terminals: Port Louis

MILITARY

Military branches: no regular military forces; Mauritius Police Force, Special Mobile Force, National Coast Guard (2009)
Manpower available for military service: *males age 16–49:* 343,628 (2010 est.)
Manpower fit for military service: *males age 16–49:* 280,596

females age 16–49: 283,317 (2010 est.)
Manpower reaching militarily significant age annually: *male:* 10,193
female: 10,104 (2010 est.)
Military expenditures: 0.3% of GDP (2006 est.)
country comparison to the world: 169

TRANSNATIONAL ISSUES

Disputes—international: Mauritius and Seychelles claim the Chagos Islands; claims French-administered Tromelin Island
Illicit drugs: consumer and transshipment point for heroin from South Asia; small amounts of cannabis produced and consumed locally; significant offshore financial industry creates potential for money laundering, but corruption levels are relatively low and the government appears generally to be committed to regulating its banking industry

MEXICO

INTRODUCTION

Background: The site of advanced Amerindian civilizations, Mexico came under Spanish rule for three centuries before achieving independence early in the 19th century. A devaluation of the peso in late 1994 threw Mexico into economic turmoil, triggering the worst recession in over half a century. The global financial crisis beginning in late 2008 caused another massive economic downturn the following year. As the economy recovers, ongoing economic and social concerns include low real wages, underemployment for a large segment of the population, inequitable income distribution, and few advancement opportunities for the largely Amerindian population in the impoverished southern states. The elections held in 2000 marked the first time since the 1910 Mexican Revolution that an opposition candidate–Vicente FOX of the National Action Party (PAN)–defeated the party in government, the Institutional Revolutionary Party (PRI). He was succeeded in 2006 by another PAN candidate Felipe CALDERON. National elections, including the presidential election, are scheduled for July 2012. Since 2007, Mexico's powerful drug-trafficking organizations have engaged in bloody fueding, resulting in tens of thousands of drug-related homicides.

GEOGRAPHY

Location: Middle America, bordering the Caribbean Sea and the Gulf of Mexico, between Belize and the United States and bordering the North Pacific Ocean, between Guatemala and the United States
Geographic coordinates: 23 00 N, 102 00 W
Map references: North America
Area: *total:* 1,964,375 sq km
country comparison to the world: 15
land: 1,943,945 sq km
water: 20,430 sq km
Area—comparative: slightly less than three times the size of Texas
Land boundaries: *total:* 4,353 km
border countries: Belize 250 km, Guatemala 962 km, US 3,141 km
Coastline: 9,330 km
Maritime claims: *territorial sea:* 12 nm
contiguous zone: 24 nm
exclusive economic zone: 200 nm
continental shelf: 200 nm or to the edge of the continental margin
Climate: varies from tropical to desert
Terrain: high, rugged mountains; low coastal plains; high plateaus; desert
Elevation extremes: *lowest point:* Laguna Salada -10 m
highest point: Volcan Pico de Orizaba 5,700 m
Natural resources: petroleum, silver, copper, gold, lead, zinc, natural gas, timber
Land use: *arable land:* 12.66%
permanent crops: 1.28%
other: 86.06% (2005)
Irrigated land: 63,200 sq km (2003)
Total renewable water resources: 457.2 cu km (2000)
Freshwater withdrawal (domestic/industrial/agricultural): *total:* 78.22 cu km/yr (17%/5%/77%)
per capita: 731 cu m/yr (2000)
Natural hazards: tsunamis along the Pacific coast, volcanoes and destructive earthquakes in the center and south, and hurricanes on the Pacific, Gulf of Mexico, and Caribbean coasts
volcanism: Mexico experiences volcanic activity in the central-southern part of the country; the volcanoes in Baja California are mostly dormant; Colima (elev. 3,850 m), which erupted in 2010, is Mexico's most active volcano and is responsible for causing periodic evacuations of nearby villagers; it has been deemed a "Decade Volcano" by the International Association of Volcanology and Chemistry of the Earth's Interior, worthy of study due to its explosive history and close proximity to human populations; Popocatepetl (elev. 5,426 m) poses a threat to Mexico City; other historically active volcanoes include Barcena, Ceboruco, El Chichon, Michoacan-Guanajuato, Pico de Orizaba, San Martin, Socorro, and Tacana
Environment—current issues: scarcity of hazardous waste disposal facilities; rural to urban migration; natural freshwater resources scarce and polluted in north, inaccessible and poor quality in center and extreme southeast; raw sewage and industrial effluents polluting rivers in urban areas; deforestation; widespread erosion; desertification; deteriorating agricultural lands; serious air and water pollution in the national capital and urban centers along US-Mexico border; land subsidence in Valley of Mexico caused by groundwater depletion
note: the government considers the lack of clean water and deforestation national security issues
Environment—international agreements: *party to:* Biodiversity, Climate Change, Climate Change-Kyoto Protocol, Desertification, Endangered Species, Hazardous Wastes, Law of the Sea, Marine Dumping, Marine Life Conservation, Ozone Layer Protection, Ship Pollution, Wetlands, Whaling
signed, but not ratified: none of the selected agreements
Geography—note: strategic location on southern border of US; corn (maize), one of the world's major grain crops, is thought to have originated in Mexico

PEOPLE

Population: 113,724,226 (July 2011 est.)
country comparison to the world: 11
Age structure: *0–14 years:* 28.2% (male 16,395,974/female 15,714,182)
15–64 years: 65.2% (male 35,842,495/female 38,309,528)
65 years and over: 6.6% (male 3,348,495/female 4,113,552) (2011 est.)
Median age: *total:* 27.1 years
male: 26 years
female: 28.1 years (2011 est.)
Population growth rate: 1.102% (2011 est.)
country comparison to the world: 105
Birth rate: 19.13 births/1,000 population (2011 est.)
country comparison to the world: 102

Death rate: 4.86 deaths/1,000 population (July 2011 est.)
country comparison to the world: 192
Net migration rate: -3.24 migrant(s)/1,000 population (2011 est.)
country comparison to the world: 177
Urbanization: *urban population:* 78% of total population (2010)
rate of urbanization: 1.2% annual rate of change (2010-15 est.)
note: Mexico City is the second-largest urban agglomeration in the Western Hemisphere, after Sao Paulo (Brazil), but before New York-Newark (US)
Major cities—population: MEXICO CITY (capital) 19.319 million; Guadalajara 4.338 million; Monterrey 3.838 million; Puebla 2.278 million; Tijuana 1.629 million (2009)
Sex ratio: *at birth:* 1.05 male(s)/female
under 15 years: 1.04 male(s)/female
15–64 years: 0.94 male(s)/female
65 years and over: 0.82 male(s)/female
total population: 0.96 male(s)/female (2011 est.)
Infant mortality rate: *total:* 17.29 deaths/1,000 live births
country comparison to the world: 104
male: 19.14 deaths/1,000 live births
female: 15.36 deaths/1,000 live births (2011 est.)
Life expectancy at birth: *total population:* 76.47 years
country comparison to the world: 72
male: 73.65 years
female: 79.43 years (2011 est.)
Total fertility rate: 2.29 children born/woman (2011 est.)
country comparison to the world: 101
HIV/AIDS—adult prevalence rate: 0.3% (2009 est.)
country comparison to the world: 90
HIV/AIDS—people living with HIV/AIDS: 220,000 (2009 est.)
country comparison to the world: 26
HIV/AIDS—deaths: NA
Major infectious diseases: degree of risk: intermediate
food or waterborne diseases: bacterial diarrhea, hepatitis A, and typhoid fever
vectorborne disease: dengue fever
water contact disease: leptospirosis (2009)
Drinking water source: *Improved:*
urban: 96% of population
rural: 87% of population
total: 94% of population
Unimproved:
urban: 4% of population
rural: 13% of population
total: 6% of population (2008)
Sanitation facility access: *Improved:*
urban: 90% of population
rural: 68% of population
total: 85% of population
Unimproved:
urban: 10% of population
rural: 32% of population
total: 15% of population (2008)
Nationality: *noun:* Mexican(s)
adjective: Mexican
Ethnic groups: mestizo (Amerindian-Spanish) 60%, Amerindian or predominantly Amerindian 30%, white 9%, other 1%
Religions: Roman Catholic 76.5%, Protestant 6.3% (Pentecostal 1.4%, Jehovah's Witnesses 1.1%, other 3.8%), other 0.3%, unspecified 13.8%, none 3.1% (2000 census)
Languages: Spanish only 92.7%, Spanish and indigenous languages 5.7%, indigenous only 0.8%, unspecified 0.8%
note: indigenous languages include various Mayan, Nahuatl, and other regional languages (2005)
Literacy: *definition:* age 15 and over can read and write
total population: 86.1%
male: 86.9%
female: 85.3% (2005 Census)
School life expectancy (primary to tertiary education): *total:* 14 years
male: 14 years
female: 14 years (2008)
Education expenditures: 4.8% of GDP (2007)
country comparison to the world: 70

GOVERNMENT

Country name: *conventional long form:* United Mexican States
conventional short form: Mexico
local long form: Estados Unidos Mexicanos
local short form: Mexico
Government type: federal republic
Capital: *name:* Mexico City (Distrito Federal)
geographic coordinates: 19 26 N, 99 08 W
time difference: UTC-6 (1 hour behind Washington, DC during Standard Time)
daylight saving time: +1hr, begins first Sunday in April; ends last Sunday in October
note: Mexico is divided into three time zones
Administrative divisions: 31 states (estados, singular–estado) and 1 federal district* (distrito federal); Aguascalientes, Baja California, Baja California Sur, Campeche, Chiapas, Chihuahua, Coahuila de Zaragoza, Colima, Distrito Federal*, Durango, Guanajuato, Guerrero, Hidalgo, Jalisco, Mexico, Michoacan de Ocampo, Morelos, Nayarit, Nuevo Leon, Oaxaca, Puebla, Queretaro de Arteaga, Quintana Roo, San Luis Potosi, Sinaloa, Sonora, Tabasco, Tamaulipas, Tlaxcala, Veracruz-Llave, Yucatan, Zacatecas
Independence: 16 September 1810 (declared); 27 September 1821 (recognized by Spain)
National holiday: Independence Day, 16 September (1810)
Constitution: 5 February 1917
Legal system: civil law system with US constitutional law theory influence; judicial review of legislative acts
International law organization participation: accepts compulsory ICJ jurisdiction with reservations; accepts ICCt jurisdiction
Suffrage: 18 years of age; universal and compulsory (but not enforced)
Executive branch: *chief of state:* President Felipe de Jesus CALDERON Hinojosa (since 1 December 2006); note–the president is both the chief of state and head of government
head of government: President Felipe de Jesus CALDERON Hinojosa (since 1 December 2006)
cabinet: Cabinet appointed by the president; note–appointment of attorney general, the head of the Bank of Mexico, and senior treasury officials require consent of the Senate
(For more information visit the World Leaders website)
elections: president elected by popular vote for a single six-year term; election last held on 2 July 2006 (next to be held 1 July 2012)
election results: Felipe CALDERON elected president; percent of vote–Felipe CALDERON 35.9%, Andres Manuel LOPEZ OBRADOR 35.3%, Roberto MADRAZO 22.3%, other 6.5%
Legislative branch: bicameral National Congress or Congreso de la Union consists of the Senate or Camara de Senadores (128 seats; 96 members elected by popular vote to serve six-year terms, and 32 seats allocated on the basis of each party's popular vote) and the Chamber of Deputies or Camara de Diputados (500 seats; 300 members are elected by popular vote; remaining 200 members are allocated on the basis of each party's popular vote; members to serve three-year terms)
elections: Senate–last held on 2 July 2006 for all of the seats (next to be held on 1 July 2012); Chamber of Deputies–last held on 5 July 2009 (next to be held on 1 July 2012)
election results: Senate–percent of vote by party–NA; seats by party–PAN 52, PRI 33, PRD 26, PVEM 6, CD 5, PT 5, independent 1; Chamber of Deputies–percent of vote by party–NA; seats by party–PRI 237, PAN 143, PRD 72, PVEM 21, PT 13, CD 6, other 8; note–as of 1 January 2011, the current composition of the Senate is: PAN 50, PRI 33, PRD 25, PVEM 6, CD 6, PT 5, independent 3; the current composition of the Chamber of Deputies is: PRI 237, PAN 142, PRD 69, PVEM 21, PT 13, CD 8, other 10
Judicial branch: Supreme Court of Justice or Suprema Corte de Justicia de la Nacion (justices or ministros are appointed by the president with consent of the Senate)
Political parties and leaders: Convergence for Democracy or CD [Luis WALTON Aburto]; Institutional Revolutionary Party or PRI [Humberto MOREIRA Valdes]; Labor Party or PT [Alberto ANAYA Gutierrez]; Mexican Green Ecological Party or PVEM [Jorge Emilio GONZALEZ Martinez]; National Action Party (Partido Accion Nacional) or PAN [Gustavo MADERO Munoz]; New Alliance Party (Partido Nueva Alianza) or PNA/PANAL [Jorge Antonio KAHWAGI Macari]; Party of the Democratic Revolution (Partido de la Revolucion Democratica) or PRD [Jesus ORTEGA Martinez]
Political pressure groups and leaders: Businessmen's Coordinating Council or CCE; Confederation of Employers of the Mexican Republic or COPARMEX; Confederation of Industrial Chambers

or CONCAMIN; Confederation of Mexican Workers or CTM; Confederation of National Chambers of Commerce or CONCANACO; Coordinator for Foreign Trade Business Organizations or COECE; Federation of Unions Providing Goods and Services or FESEBES; National Chamber of Transformation Industries or CANACINTRA; National Peasant Confederation or CNC; National Small Business Chamber or CANACOPE; National Syndicate of Education Workers or SNTE; National Union of Workers or UNT; Popular Assembly of the People of Oaxaca or APPO; Roman Catholic Church

International organization participation: APEC, BCIE, BIS, CAN (observer), Caricom (observer), CD, CDB, CSN (observer), EBRD, FAO, FATF, G-20, G-3, G-15, G-24, IADB, IAEA, IBRD, ICAO, ICC, ICRM, IDA, IFAD, IFC, IFRCS, IHO, ILO, IMF, IMO, IMSO, Interpol, IOC, IOM, IPU, ISO, ITSO, ITU, ITUC, LAES, LAIA, MIGA, NAFTA, NAM (observer), NEA, OAS, OECD, OPANAL, OPCW, Paris Club (associate), PCA, RG, SICA (observer), UN, UNASUR (observer), UNCTAD, UNESCO, UNHCR, UNIDO, UNWTO, UPU, WCO, WFTU, WHO, WIPO, WMO, WTO

Diplomatic representation in the US: *chief of mission:* Ambassador Arturo SARUKHAN Casamitjana
chancery: 1911 Pennsylvania Avenue NW, Washington, DC 20006
telephone: [1] (202) 728-1600
FAX: [1] (202) 728-1698
consulate(s) general: Atlanta, Austin, Boston, Chicago, Dallas, Denver, El Paso, Houston, Laredo (Texas), Los Angeles, Miami, New York, Nogales (Arizona), Phoenix, Sacramento, San Antonio, San Diego, San Francisco, San Jose, San Juan (Puerto Rico)
consulate(s): Albuquerque, Anchorage (Alaska), Boise (Idaho), Brownsville (Texas), Calexico (California), Del Rio (Texas), Detroit, Douglas (Arizona), Eagle Pass (Texas), Fresno (California), Indianapolis (Indiana), Kansas City (Missouri), Las Vegas, Little Rock (Arkansas), McAllen (Texas), Midland (Texas), New Orleans, Omaha, Orlando, Oxnard (California), Philadelphia, Portland (Oregon), Presidio (Texas), Raleigh (North Carolina), Saint Paul, Salt Lake City, San Bernardino, Santa Ana (California), Seattle, Tucson, Washington DC, Yuma (Arizona); note–Washington DC Consular Section located in a separate building from the Mexican Embassy and has jurisdiction over DC, parts of Virginia, Maryland, and West Virginia

Diplomatic representation from the US: *chief of mission:* Ambassador Carlos PASCUAL
embassy: Paseo de la Reforma 305, Colonia Cuauhtemoc, 06500 Mexico, Distrito Federal
mailing address: P. O. Box 9000, Brownsville, TX 78520-9000
telephone: [52] (55) 5080-2000
FAX: [52] (55) 5511-9980
consulate(s) general: Ciudad Juarez, Guadalajara, Hermosillo, Matamoros, Monterrey, Nuevo Laredo, Tijuana
consulate(s): Merida, Nogales

Flag description: three equal vertical bands of green (hoist side), white, and red; Mexico's coat of arms (an eagle with a snake in its beak perched on a cactus) is centered in the white band; green signifies hope, joy, and love; white represents peace and honesty; red stands for hardiness, bravery, strength, and valor; the coat of arms is derived from a legend that the wandering Aztec people were to settle at a location where they would see an eagle on a cactus eating a snake; the city they founded, Tenochtitlan, is now Mexico City
note: similar to the flag of Italy, which is shorter, uses lighter shades of red and green, and does not have anything in its white band

National anthem: *name:* "Himno Nacional Mexicano" (National Anthem of Mexico)
lyrics/music: Francisco Gonzalez BOCANEGRA/Jaime Nuno ROCA
note: adopted 1943, in use since 1854; the anthem is also known as "Mexicanos, al grito de Guerra" (Mexicans, to the War Cry); according to tradition, Francisco Gonzalez BOCANEGRA, an accomplished poet, was uninterested in submitting lyrics to a national anthem contest; his fiancee locked him in a room and refused to release him until the lyrics were completed

ECONOMY

Economy—overview: Mexico has a free market economy in the trillion dollar class. It contains a mixture of modern and outmoded industry and agriculture, increasingly dominated by the private sector. Recent administrations have expanded competition in seaports, railroads, telecommunications, electricity generation, natural gas distribution, and airports. Per capita income is roughly one-third that of the US; income distribution remains highly unequal. Since the implementation of the North American Free Trade Agreement (NAFTA) in 1994, Mexico's share of US imports has increased from 7% to 12%, and its share of Canadian imports has doubled to 5%. Mexico has free trade agreements with over 50 countries including, Guatemala, Honduras, El Salvador, the European Free Trade Area, and Japan, putting more than 90% of trade under free trade agreements. In 2007, during its first year in office, the Felipe CALDERON administration was able to garner support from the opposition to successfully pass pension and fiscal reforms. The administration passed an energy reform measure in 2008 and another fiscal reform in 2009. Mexico's GDP plunged 6.5% in 2009 as world demand for exports dropped, asset prices tumbled, and remittances and investment declined. GDP posted positive growth of 5% in 2010, with exports–particularly to the United States–leading the way, while domestic consumption and investment lagged. The administration continues to face many economic challenges, including improving the public education system, upgrading infrastructure, modernizing labor laws, and fostering private investment in the energy sector. CALDERON has stated that his top economic priorities remain reducing poverty and creating jobs.

GDP (purchasing power parity): $1.567 trillion (2010 est.)
country comparison to the world: 12
$1.486 trillion (2009 est.)
$1.582 trillion (2008 est.)
note: data are in 2010 US dollars

GDP (official exchange rate): $1.039 trillion (2010 est.)

GDP—real growth rate: 5.5% (2010 est.)
country comparison to the world: 61
-6.1% (2009 est.)
1.5% (2008 est.)

GDP—per capita (PPP): $13,900 (2010 est.)
country comparison to the world: 85
$13,400 (2009 est.)
$14,400 (2008 est.)
note: data are in 2010 US dollars

GDP—composition by sector: *agriculture:* 4.2%
industry: 33.3%
services: 62.5% (2010 est.)

Labor force: 46.99 million (2010 est.)
country comparison to the world: 12

Labor force—by occupation: *agriculture:* 13.7%
industry: 23.4%
services: 62.9% (2005)

Unemployment rate: 5.6% (2010 est.)
country comparison to the world: 55
5.5% (2009 est.)
note: underemployment may be as high as 25%

Population below poverty line: 18.2%
note: based on food-based definition of poverty; asset based poverty amounted to more than 47% (2008)

Household income or consumption by percentage share: *lowest 10%:* 1.7%
highest 10%: 36.3% (2008)

Distribution of family income—Gini index: 48.2 (2008)
country comparison to the world: 27
53.1 (1998)

Investment (gross fixed): 21.1% of GDP (2010 est.)
country comparison to the world: 75

Budget: *revenues:* $237 billion
expenditures: $267 billion (2010 est.)

Public debt: 41.5% of GDP (2010 est.)
country comparison to the world: 66
39.1% of GDP (2009 est.)

Inflation rate (consumer prices): 4.1% (2010 est.)
country comparison to the world: 120
3.6% (2009)

Central bank discount rate: NA%

Commercial bank prime lending rate: 7.07% (31 December 2009 est.)
country comparison to the world: 103
8.71% (31 December 2008 est.)

Stock of narrow money: $135.7 billion (31 December 2010 est.)
country comparison to the world: 25
$119.5 billion (31 December 2009 est.)

Stock of broad money: $583.8 billion (31 December 2010 est.)
country comparison to the world: 21
$493 billion (31 December 2009 est.)

Stock of domestic credit: $342.4 billion (31 December 2010 est.)
country comparison to the world: 30

$288.5 billion (31 December 2009 est.)
Market value of publicly traded shares: $340.6 billion (31 December 2009)
country comparison to the world: 24
$232.6 billion (31 December 2008)
$397.7 billion (31 December 2007)
Agriculture—products: corn, wheat, soybeans, rice, beans, cotton, coffee, fruit, tomatoes; beef, poultry, dairy products; wood products
Industries: food and beverages, tobacco, chemicals, iron and steel, petroleum, mining, textiles, clothing, motor vehicles, consumer durables, tourism
Industrial production growth rate: 6% (2010 est.)
country comparison to the world: 58
Electricity—production: 245 billion kWh (2008 est.)
country comparison to the world: 15
Electricity—consumption: 181.5 billion kWh (2009 est.)
country comparison to the world: 20
Electricity—exports: 1.288 billion kWh (2008 est.)
Electricity—imports: 584 million kWh (2008 est.)
Oil—production: 3.001 million bbl/day (2009 est.)
country comparison to the world: 7
Oil—consumption: 2.078 million bbl/day (2009 est.)
country comparison to the world: 12
Oil—exports: 1.225 million bbl/day (2009 est.)
country comparison to the world: 20
Oil—imports: 521,100 bbl/day (2008 est.)
country comparison to the world: 23
Oil—proved reserves: 12.42 billion bbl (1 January 2010 est.)
country comparison to the world: 18
Natural gas—production: 60.35 billion cu m (2009 est.)
country comparison to the world: 15
Natural gas—consumption: 59.8 billion cu m (2009 est.)
country comparison to the world: 12
Natural gas—exports: 688 million cu m (2009 est.)
country comparison to the world: 41
Natural gas—imports: 11.84 billion cu m (2009 est.)
country comparison to the world: 18
Natural gas—proved reserves: 359.7 billion cu m (1 January 2010 est.)
country comparison to the world: 38
Current account balance: $-7 billion (2010 est.)
country comparison to the world: 174
$-6.23 billion (2009 est.)
Exports: $303 billion (2010 est.)
country comparison to the world: 15
$229.8 billion (2009 est.)
Exports—commodities: manufactured goods, oil and oil products, silver, fruits, vegetables, coffee, cotton
Exports—partners: US 80.5%, Canada 3.6%, Germany 1.4% (2009 est.)
Imports:
$306 billion (2010 est.)
country comparison to the world: 16
$234.4 billion (2009 est.)
Imports—commodities: metalworking machines, steel mill products, agricultural machinery, electrical equipment, car parts for assembly, repair parts for motor vehicles, aircraft, and aircraft parts
Imports—partners: US 48%, China 13.5%, Japan 4.8%, South Korea 4.6%, Germany 4.1% (2009 est.)
Reserves of foreign exchange and gold: $116.4 billion (31 December 2010 est.)
country comparison to the world: 14
$99.86 billion (31 December 2009 est.)
Debt—external: $212.5 billion (31 December 2010 est.)
country comparison to the world: 30
$204.5 billion (31 December 2009 est.)
Stock of direct foreign investment—at home: $328.4 billion (31 December 2010 est.)
country comparison to the world: 15
$308.4 billion (31 December 2009 est.)
Stock of direct foreign investment—abroad: $62.93 billion (31 December 2010 est.)
country comparison to the world: 29
$53.46 billion (31 December 2009 est.)
Exchange rates: Mexican pesos (MXN) per US dollar—12.687 (2010) 13.514 (2009) 11.016 (2008) 10.8 (2007) 10.899 (2006)

COMMUNICATIONS

Telephones—main lines in use: 19.425 million (2009)
country comparison to the world: 14
Telephones—mobile cellular: 83.528 million (2009)
country comparison to the world: 13
Telephone system: *general assessment:* adequate telephone service for business and government; improving quality and increasing mobile cellular availability, with mobile subscribers far outnumbering fixed-line subscribers; domestic satellite system with 120 earth stations; extensive microwave radio relay network; considerable use of fiber-optic cable and coaxial cable
domestic: despite the opening to competition in January 1997, Telmex remains dominant; Fixed-line teledensity is less than 20 per 100 persons; mobile-cellular teledensity reached 75 per 100 persons in 2009
international: country code—52; Columbus-2 fiber-optic submarine cable with access to the US, Virgin Islands, Canary Islands, Spain, and Italy; the Americas Region Caribbean Ring System (ARCOS-1) and the MAYA-1 submarine cable system together provide access to Central America, parts of South America and the Caribbean, and the US; satellite earth stations—120 (32 Intelsat, 2 Solidaridad (giving Mexico improved access to South America, Central America, and much of the US as well as enhancing domestic communications), 1 Panamsat, numerous Inmarsat mobile earth stations); linked to Central American Microwave System of trunk connections (2009)
Broadcast media: large number of television stations and more than 1,400 radio stations, most are privately owned; the Televisa group once had a virtual monopoly in TV broadcasting, but new broadcasting groups and foreign satellite and cable operators are now available (2007)
Internet country code: .mx
Internet hosts: 12.854 million (2010)
country comparison to the world: 9
Internet users: 31.02 million (2009)
country comparison to the world: 12

TRANSPORTATION

Airports: 1,819 (2010)
country comparison to the world: 3
Airports—with paved runways: *total:* 250
over 3,047 m: 12
2,438 to 3,047 m: 30
1,524 to 2,437 m: 85
914 to 1,523 m: 83
under 914 m: 40 (2010)
Airports—with unpaved runways:
total: 1,569
over 3,047 m: 1
2,438 to 3,047 m: 1
1,524 to 2,437 m: 66
914 to 1,523 m: 438
under 914 m: 1,063 (2010)
Heliports: 1 (2010)
Pipelines: gas 16,594 km; liquid petroleum gas 2,152 km; oil 7,499 km; oil/gas/water 4 km; refined products 7,264 km; water 33 km (2010)
Railways: *total:* 17,166 km
country comparison to the world: 16
standard gauge: 17,166 km 1.435-m gauge (22 km electrified) (2010)
Roadways: *total:* 366,095 km
country comparison to the world: 17
paved: 132,289 km (includes 6,279 km of expressways)
unpaved: 233,806 km (2008)
Waterways: 2,900 km (navigable rivers and coastal canals mostly connected with ports on the country's east coast) (2010)
country comparison to the world: 33
Merchant marine: *total:* 60
country comparison to the world: 65
by type: bulk carrier 4, cargo 3, chemical tanker 12, liquefied gas 4, passenger/cargo 11, petroleum tanker 22, roll on/roll off 4
foreign-owned: 5 (Denmark 2, Greece 1, South Africa 1, UAE 1)
registered in other countries: 18 (Antigua and Barbuda 2, Honduras 1, Marshall Islands 4, Panama 6, Portugal 1, Spain 2, Venezuela 1, unknown 1) (2010)
Ports and terminals: Altamira, Coatzacoalcos, Lazaro Cardenas, Manzanillo, Salina Cruz, Veracruz

MILITARY

Military branches: Secretariat of National Defense (Secretaria de Defensa Nacional, Sedena): Army (Ejercito), Mexican Air Force (Fuerza Aerea Mexicana, FAM); Secretariat of the Navy (Secretaria de Marina, Semar): Mexican Navy (Armada de Mexico (ARM), includes Naval Air Force (FAN), Mexican Marine Corps (Cuerpo de Infanteria de Marina, Mexmar or CIM)) (2011)
Military service age and obligation: 18 years of age for compulsory military service, conscript service obligation—12 months; 16 years of age with consent for voluntary enlistment; conscripts serve only in the Army; Navy and Air Force service is all voluntary; women are eligible for voluntary military service (2007)

Manpower available for military service: *males age 16–49:* 28,815,506
females age 16–49: 30,363,558 (2010 est.)
Manpower fit for military service: *males age 16–49:* 23,239,866
females age 16–49: 25,642,549 (2010 est.)
Manpower reaching militarily significant age annually: *male:* 1,105,371
female: 1,067,007 (2010 est.)
Military expenditures: 0.5% of GDP (2006 est.)
country comparison to the world: 161

TRANSNATIONAL ISSUES

Disputes—international: abundant rainfall in recent years along much of the Mexico-US border region has ameliorated periodically strained water-sharing arrangements; the US has intensified security measures to monitor and control legal and illegal personnel, transport, and commodities across its border with Mexico; Mexico must deal with thousands of impoverished Guatemalans and other Central Americans who cross the porous border looking for work in Mexico and the United States; Belize and Mexico are working to solve minor border demarcation discrepancies arising from inaccuracies in the 1898 border treaty
Refugees and internally displaced persons: *IDPs:* 5,500-10,000 (government's quashing of Zapatista uprising in 1994 in eastern Chiapas Region) (2007)
Illicit drugs: major drug-producing nation; cultivation of opium poppy in 2007 rose to 6,900 hectares yielding a potential production of 18 metric tons of pure heroin, or 50 metric tons of "black tar" heroin, the dominant form of Mexican heroin in the western United States; marijuana cultivation increased to 8,900 hectares in 2007 and yielded a potential production of 15,800 metric tons; government conducts the largest independent illicit-crop eradication program in the world; continues as the primary transshipment country for US-bound cocaine from South America, with an estimated 90% of annual cocaine movements toward the US stopping in Mexico; major drug syndicates control the majority of drug trafficking throughout the country; producer and distributor of ecstasy; significant money-laundering center; major supplier of heroin and largest foreign supplier of marijuana and methamphetamine to the US market (2007)

MICRONESIA, FEDERATED STATES OF

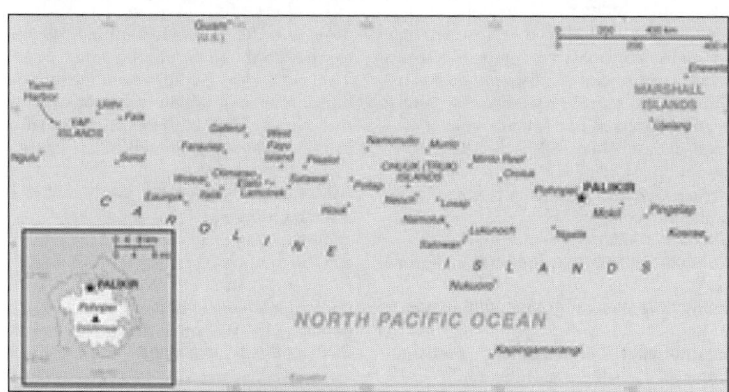

INTRODUCTION

Background: In 1979 the Federated States of Micronesia, a UN Trust Territory under US administration, adopted a constitution. In 1986 independence was attained under a Compact of Free Association with the US, which was amended and renewed in 2004. Present concerns include large-scale unemployment, overfishing, and overdependence on US aid.

GEOGRAPHY

Location: Oceania, island group in the North Pacific Ocean, about three-quarters of the way from Hawaii to Indonesia
Geographic coordinates: 6 55 N, 158 15 E
Map references: Oceania
Area: *total:* 702 sq km
country comparison to the world: 190
land: 702 sq km
water: 0 sq km (fresh water only)
note: includes Pohnpei (Ponape), Chuuk (Truk) Islands, Yap Islands, and Kosrae (Kosaie)
Area—comparative: four times the size of Washington, DC (land area only)
Land boundaries: 0 km
Coastline: 6,112 km
Maritime claims: *territorial sea:* 12 nm
exclusive economic zone: 200 nm
Climate: tropical; heavy year-round rainfall, especially in the eastern islands; located on southern edge of the typhoon belt with occasionally severe damage
Terrain: islands vary geologically from high mountainous islands to low, coral atolls; volcanic outcroppings on Pohnpei, Kosrae, and Chuuk
Elevation extremes: *lowest point:* Pacific Ocean 0 m
highest point: Dolohmwar (Totolom) 791 m
Natural resources: timber, marine products, deep-seabed minerals, phosphate
Land use: *arable land:* 5.71%
permanent crops: 45.71%
other: 48.58% (2005)
Irrigated land: NA
Natural hazards: typhoons (June to December)
Environment—current issues: overfishing, climate change, pollution
Environment—international agreements: *party to:* Biodiversity, Climate Change, Climate Change-Kyoto Protocol, Desertification, Hazardous Wastes, Law of the Sea, Ozone Layer Protection
signed, but not ratified: none of the selected agreements
Geography—note: four major island groups totaling 607 islands

PEOPLE

Population: 106,836 (July 2011 est.)
country comparison to the world: 189
Age structure: *0–14 years:* 33.6% (male 18,223/female 17,639)
15–64 years: 63.4% (male 33,566/female 34,215)
65 years and over: 3% (male 1,375/female 1,818) (2011 est.)
Median age: *total:* 22.7 years
male: 22.2 years
female: 23.3 years (2011 est.)
Population growth rate: -0.313% (2011 est.)
country comparison to the world: 216
Birth rate: 22.22 births/1,000 population (2011 est.)
country comparison to the world: 77
Death rate: 4.35 deaths/1,000 population (July 2011 est.)
country comparison to the world: 204
Net migration rate: -20.99 migrant(s)/1,000 population (2011 est.)
country comparison to the world: 219
Urbanization: *urban population:* 23% of total population (2010)
rate of urbanization: 1.3% annual rate of change (2010-15 est.)
Major cities—population: PALIKIR (capital) 7,000 (2009)
Sex ratio: *at birth:* 1.05 male(s)/female
under 15 years: 1.03 male(s)/female
15–64 years: 0.99 male(s)/female
65 years and over: 0.74 male(s)/female
total population: 0.99 male(s)/female (2011 est.)
Infant mortality rate: *total:* 24.34 deaths/1,000 live births
country comparison to the world: 82
male: 26.89 deaths/1,000 live births
female: 21.66 deaths/1,000 live births (2011 est.)
Life expectancy at birth: *total population:* 71.52 years
country comparison to the world: 134
male: 69.58 years
female: 73.55 years (2011 est.)
Total fertility rate: 2.74 children born/woman (2011 est.)
country comparison to the world: 73
HIV/AIDS—adult prevalence rate: NA
HIV/AIDS—people living with HIV/AIDS: NA
HIV/AIDS—deaths: NA

Drinking water source: *Improved:*
urban: 94% of population
rural: 92% of population
total: 92% of population
Unimproved:
urban: 6% of population
rural: 8% of population
total: 8% of population (2000)
Sanitation facility access:
Improved:
urban: 59% of population
rural: 16% of population
total: 26% of population
Unimproved:
urban: 41% of population
rural: 84% of population
total: 74% of population (2000)
Nationality: *noun:* Micronesian(s)
adjective: Micronesian; Chuukese, Kosraen(s), Pohnpeian(s), Yapese
Ethnic groups: Chuukese 48.8%, Pohnpeian 24.2%, Kosraean 6.2%, Yapese 5.2%, Yap outer islands 4.5%, Asian 1.8%, Polynesian 1.5%, other 6.4%, unknown 1.4% (2000 census)
Religions: Roman Catholic 52.7%, Congregational 40.1%, Baptist 0.9%, Seventh-Day Adventist 0.7%, other 3.8%, none or unspecified 0.8% (2000 Census)
Languages: English (official and common language), Chuukese, Kosrean, Pohnpeian, Yapese, Ulithian, Woleaian, Nukuoro, Kapingamarangi
Literacy: *definition:* age 15 and over can read and write
total population: 89%
male: 91%
female: 88% (1980 est.)
School life expectancy (primary to tertiary education): NA
Education expenditures: NA

GOVERNMENT

Country name: *conventional long form:* Federated States of Micronesia
conventional short form: none
local long form: Federated States of Micronesia
local short form: none
former: Trust Territory of the Pacific Islands, Ponape, Truk, and Yap Districts
abbreviation: FSM
Government type: constitutional government in free association with the US; the Compact of Free Association entered into force on 3 November 1986 and the Amended Compact entered into force in May 2004
Capital: *name:* Palikir
geographic coordinates: 6 55 N, 158 09 E
time difference: UTC+11 (16 hours ahead of Washington, DC during Standard Time)
Administrative divisions: 4 states; Chuuk (Truk), Kosrae (Kosaie), Pohnpei (Ponape), Yap
Independence: 3 November 1986 (from the US-administered UN trusteeship)
National holiday: Constitution Day, 10 May (1979)
Constitution: 10 May 1979
Legal system: mixed legal system of common and customary law
International law organization participation: has not submitted an ICJ jurisdiction declaration; non-party state to the ICCt
Suffrage: 18 years of age; universal
Executive branch: *chief of state:* President Emanuel MORI (since 11 May 2007); Vice President Alik L. ALIK (since 11 May 2007); note—the president is both the chief of state and head of government
head of government: President Emanuel MORI (since 11 May 2007); Vice President Alik L. ALIK (since 11 May 2007)
cabinet: Cabinet includes the vice president and the heads of the 8 executive departments (For more information visit the World Leaders website)
elections: president and vice president elected by Congress from among the four senators at large for a four-year term (eligible for a second term); election last held on 11 May 2011 (next to be held in May 2015); note—a proposed constitutional amendment to establish popular elections for president and vice president failed
election results: Emanuel MORI reelected president by Congress unopposed; Alik L. ALIK reelected vice president
Legislative branch: unicameral Congress (14 seats; 4—one elected from each state to serve four-year terms and 10—elected from single-member districts delineated by population to serve two-year terms; members elected by popular vote)
elections: last held on 8 March 2011 (next to be held in March 2013)
election results: percent of vote—NA%; seats—independents 14
Judicial branch: Supreme Court
Political parties and leaders: no formal parties
Political pressure groups and leaders: NA
International organization participation: ACP, ADB, AOSIS, FAO, G-77, IBRD, ICAO, ICRM, IDA, IFC, IFRCS, IMF, IOC, ITSO, ITU, MIGA, OPCW, PIF, Sparteca, SPC, UN, UNCTAD, UNESCO, WHO, WMO
Diplomatic representation in the US: *chief of mission:* Ambassador Yosiwo GEORGE
chancery: 1725 N Street NW, Washington, DC 20036
telephone: [1] (202) 223-4383
FAX: [1] (202) 223-4391
consulate(s) general: Honolulu, Tamuning (Guam)
Diplomatic representation from the US: *chief of mission:* Ambassador Peter A. PRAHAR
embassy: 101 Upper Pics Road, Kolonia
mailing address: P. O. Box 1286, Kolonia, Pohnpei, 96941
telephone: [691] 320-2187
FAX: [691] 320-2186
Flag description: light blue with four white five-pointed stars centered; the stars are arranged in a diamond pattern; blue symbolizes the Pacific Ocean, the stars represent the four island groups of Chuuk, Kosrae, Pohnpei, and Yap
National anthem: *name:* "Patriots of Micronesia"
lyrics/music: unknown
note: adopted 1991; the anthem is also known as "Across All Micronesia;" the music is based on the 1820 German patriotic song "Ich hab mich ergeben," which was the West German national anthem from 1949-1950; variants of this tune are used in Johannes Brahms' "Festival Overture" and Gustav Mahler's "Third Symphony"

ECONOMY

Economy—overview: Economic activity consists primarily of subsistence farming and fishing. The islands have few mineral deposits worth exploiting, except for high-grade phosphate. The potential for a tourist industry exists, but the remote location, a lack of adequate facilities, and limited air connections hinder development. Under the original terms of the Compact of Free Association, the US provided $1.3 billion in grant aid during the period 1986-2001; the level of aid has been subsequently reduced. The Amended Compact of Free Association with the US guarantees the Federated States of Micronesia (FSM) millions of dollars in annual aid through 2023, and establishes a Trust Fund into which the US and the FSM make annual contributions in order to provide annual payouts to the FSM in perpetuity after 2023. The country's medium-term economic outlook appears fragile due not only to the reduction in US assistance but also to the current slow growth of the private sector.
GDP (purchasing power parity): $238.1 million (2008 est.)
country comparison to the world: 213
$277 million (2002 est.)
note: data are in 2008 US dollars
GDP supplemented by grant aid, averaging perhaps $100 million annually
GDP (official exchange rate): $238.1 million (2008)
GDP—real growth rate: NA%
0.3% (2005 est.)
GDP—per capita (PPP): $2,200 (2008 est.)
country comparison to the world: 187
$2,300 (2005 est.)
note: data are in 2008 US dollars
GDP—composition by sector: *agriculture:* 28.9%
industry: 15.2%
services: 55.9% (2004 est.)
Labor force: 16,360 (2008)
country comparison to the world: 211
Labor force—by occupation:
agriculture: 0.9%
industry: 34.4%
services: 64.7%
note: two-thirds of the labor force are government employees (FY05 est.)
Unemployment rate: 22% (2000 est.)
country comparison to the world: 171
Population below poverty line: 26.7% (2000)
Household income or consumption by percentage share: *lowest 10%:* NA%
highest 10%: NA%
BudLget:
revenues: $166 million (FY07 est.)
expenditures: $152.7 million (FY07 est.)
Inflation rate (consumer prices): 2.2% (2005)
country comparison to the world: 57
Commercial bank prime lending rate: 15.38% (31 December 2009 est.)

country comparison to the world: 50
14.38% (31 December 2008 est.)
Stock of narrow money: $29.02 million (31 December 2009)
country comparison to the world: 185
$21.21 million (31 December 2008)
Stock of broad money: $114 million (31 December 2009 est.)
country comparison to the world: 185
$98 million (31 December 2008 est.)
Stock of domestic credit: $65.68 million (31 December 2009)
country comparison to the world: 182
$43.75 million (31 December 2008)
Agriculture—products: black pepper, tropical fruits and vegetables, coconuts, bananas, cassava (tapioca), sakau (kava), Kosraen citrus, betel nuts, sweet potatoes; pigs, chickens; fish
Industries: tourism, construction; fish processing, specialized aquaculture; craft items (from shell, wood, and pearls)
Industrial production growth rate: NA%
Electricity—production: 192 million kWh (2002)
country comparison to the world: 177
Electricity—consumption: 178.6 million kWh (2002)
country comparison to the world: 180
Electricity—exports: 0 kWh (2002)
Electricity—imports: 0 kWh (2002)
Current account balance: $-34.3 million (FY05 est.)
country comparison to the world: 68
Exports: $14 million (2004 est.)
country comparison to the world: 210
Exports—commodities: fish, garments, bananas, black pepper, sakau (kava), betel nut
Imports: $132.7 million (2004)
country comparison to the world: 208
Imports—commodities: food, manufactured goods, machinery and equipment, beverages
Debt—external: $60.8 million (FY05 est.)
country comparison to the world: 186
Exchange rates: the US dollar is used

COMMUNICATIONS

Telephones—main lines in use: 8,700 (2009)
country comparison to the world: 203
Telephones—mobile cellular: 38,000 (2009)
country comparison to the world: 198
Telephone system: *general assessment:* adequate system
domestic: islands interconnected by shortwave radiotelephone (used mostly for government purposes), satellite (Intelsat) ground stations, and some coaxial and fiber-optic cable; mobile-cellular service available on Kosrae, Pohnpei, and Yap
international: country code–691; satellite earth stations–5 Intelsat (Pacific Ocean) (2002)
Broadcast media: no television broadcast stations; each state has a multi-channel cable service with television transmissions carrying roughly 95% imported programming and 5% local programming; about a half dozen radio stations in operation (2009)
Internet country code: .fm
Internet hosts: 3,097 (2010)
country comparison to the world: 144
Internet users: 17,000 (2009)
country comparison to the world: 197

TRANSPORTATION

Airports: 6 (2010)
country comparison to the world: 170
Airports—with paved runways: *total:* 6
1,524 to 2,437 m: 4
914 to 1,523 m: 2 (2010)
Roadways: *total:* 240 km
country comparison to the world: 206
paved: 42 km
unpaved: 198 km (2000)
Merchant marine: *total:* 3
country comparison to the world: 136
by type: cargo 1, passenger/cargo 2 (2010)
Ports and terminals: Colonia (Tomil Harbor), Lele Harbor, Pohnepi Harbor

MILITARY

Military branches: no regular military forces
Manpower available for military service: *males age 16–49:* 26,712 (2010 est.)
Manpower fit for military service: *males age 16–49:* 22,008
females age 16–49: 23,501 (2010 est.)
Manpower reaching militarily significant age annually: *male:* 1,276
female: 1,253 (2010 est.)
Military—note: defense is the responsibility of the US

TRANSNATIONAL ISSUES

Disputes—international: none
Illicit drugs: major consumer of cannabis

MOLDOVA

INTRODUCTION

Background: Part of Romania during the interwar period, Moldova was incorporated into the Soviet Union at the close of World War II. Although the country has been independent from the USSR since 1991, Russian forces have remained on Moldovan territory east of the Dniester River supporting the Slavic majority population, mostly Ukrainians and Russians, who have proclaimed a "Transnistria" republic. One of the poorest nations in Europe, Moldova became the first former Soviet state to elect a Communist, Vladimir VORONIN, as its president in 2001. VORONIN served as Moldova's president until he resigned in September 2009, following the opposition's gain of a narrow majority in July parliamentary elections and the Communist Party's (PCRM) subsequent inability to attract the three-fifths of parliamentary votes required to elect a president. Moldova's four opposition parties formed a new coalition, the Alliance for European Integration (AIE), which acted as Moldova's governing coalition until December 2010. Moldova experienced significant political uncertainty in 2009 and 2010, holding three general elections (in April 2009, July 2009, and November 2010) and four presidential ballots in parliament, all of which failed to secure a president. Following the November 2010 parliamentary elections, a reconstituted AIE-coalition of three parties formed a government, but remains two votes short of the three-fifths majority required to elect a president.

GEOGRAPHY

Location: Eastern Europe, northeast of Romania
Geographic coordinates: 47 00 N, 29 00 E
Map references: Europe
Area: *total:* 33,851 sq km
country comparison to the world: 139
land: 32,891 sq km
water: 960 sq km
Area—comparative: slightly larger than Maryland
Land boundaries: *total:* 1,390 km
border countries: Romania 450 km, Ukraine 940 km
Coastline: 0 km (landlocked)
Maritime claims: none (landlocked)
Climate: moderate winters, warm summers
Terrain: rolling steppe, gradual slope south to Black Sea
Elevation extremes: *lowest point:* Dniester (Nistru) 2 m

highest point: Dealul Balanesti 430 m
Natural resources: lignite, phosphorites, gypsum, arable land, limestone
Land use: *arable land:* 54.52%
permanent crops: 8.81%
other: 36.67% (2005)
Irrigated land: 3,000 sq km (2003)
Total renewable water resources: 11.7 cu km (1997)
Freshwater withdrawal (domestic/industrial/agricultural): *total:* 2.31 cu km/yr (10%/58%/33%)
per capita: 549 cu m/yr (2000)
Natural hazards: landslides
Environment—current issues: heavy use of agricultural chemicals, including banned pesticides such as DDT, has contaminated soil and groundwater; extensive soil erosion from poor farming methods
Environment—international agreements: *party to:* Air Pollution, Air Pollution-Persistent Organic Pollutants, Biodiversity, Climate Change, Climate Change-Kyoto Protocol, Desertification, Endangered Species, Hazardous Wastes, Ozone Layer Protection, Ship Pollution, Wetlands
signed, but not ratified: none of the selected agreements
Geography—note: landlocked; well endowed with various sedimentary rocks and minerals including sand, gravel, gypsum, and limestone

PEOPLE

Population: 4,314,377 (July 2011 est.)
country comparison to the world: 124
Age structure: *0–14 years:* 15.5% (male 344,101/female 325,995)
15–64 years: 74% (male 1,550,386/female 1,643,108)
65 years and over: 10.4% (male 164,512/female 286,275) (2011 est.)
Median age: *total:* 35.4 years
male: 33.5 years
female: 37.4 years (2011 est.)
Population growth rate: -0.072% (2011 est.)
country comparison to the world: 202
Birth rate: 11.16 births/1,000 population (2011 est.)
country comparison to the world: 173
Death rate: 10.74 deaths/1,000 population (July 2011 est.)
country comparison to the world: 43
Net migration rate: -1.13 migrant(s)/1,000 population (2011 est.)
country comparison to the world: 151
Urbanization: *urban population:* 47% of total population (2010)
rate of urbanization: 0.9% annual rate of change (2010-15 est.)
Major cities—population: CHISINAU (capital) 650,000 (2009)
Sex ratio: *at birth:* 1.059 male(s)/female
under 15 years: 1.06 male(s)/female
15–64 years: 0.94 male(s)/female
65 years and over: 0.58 male(s)/female
total population: 0.91 male(s)/female (2011 est.)
Infant mortality rate: *total:* 12.43 deaths/1,000 live births
country comparison to the world: 131
male: 13.85 deaths/1,000 live births
female: 10.92 deaths/1,000 live births (2011 est.)
Life expectancy at birth: *total population:* 71.37 years
country comparison to the world: 136
male: 67.68 years
female: 75.28 years (2011 est.)
Total fertility rate: 1.29 children born/woman (2011 est.)
country comparison to the world: 210
HIV/AIDS—adult prevalence rate: 0.4% (2009 est.)
country comparison to the world: 70
HIV/AIDS—people living with HIV/AIDS: 12,000 (2009 est.)
country comparison to the world: 94
HIV/AIDS—deaths: fewer than 1,000 (2009 est.)
country comparison to the world: 76
Drinking water source: *Improved:*
urban: 96% of population
rural: 85% of population
total: 90% of population
Unimproved:
urban: 4% of population
rural: 15% of population
total: 10% of population (2008)
Sanitation facility access: *Improved:*
urban: 85% of population
rural: 74% of population
total: 79% of population
Unimproved:
urban: 15% of population
rural: 26% of population
total: 21% of population (2008)
Nationality: *noun:* Moldovan(s)
adjective: Moldovan
Ethnic groups: Moldovan/Romanian 78.2%, Ukrainian 8.4%, Russian 5.8%, Gagauz 4.4%, Bulgarian 1.9%, other 1.3% (2004 census)
note: internal disputes with ethnic Slavs in the Transnistrian region
Religions: Eastern Orthodox 98%, Jewish 1.5%, Baptist and other 0.5% (2000)
Languages: Moldovan (official, virtually the same as the Romanian language), Russian, Gagauz (a Turkish dialect)
Literacy: *definition:* age 15 and over can read and write
total population: 99.1%
male: 99.7%
female: 98.6% (2005 est.)
School life expectancy (primary to tertiary education): *total:* 12 years
male: 12 years
female: 12 years (2009)
Education expenditures: 9.6% of GDP (2009)
country comparison to the world: 7

GOVERNMENT

Country name: *conventional long form:* Republic of Moldova
conventional short form: Moldova
local long form: Republica Moldova
local short form: Moldova
former: Moldavian Soviet Socialist Republic, Moldovan Soviet Socialist Republic
Government type: republic
Capital: *name:* Chisinau in Romanian (Kishinev in Russian)
note: pronounced KEE-shee-now (KIH-shi-nyev)
geographic coordinates: 47 00 N, 28 51 E
time difference: UTC+2 (7 hours ahead of Washington, DC during Standard Time)
daylight saving time: +1hr, begins last Sunday in March; ends last Sunday in October
Administrative divisions: 32 raions (raioane, singular–raion), 3 municipalities (municipii, singular–municipiul), 1 autonomous territorial unit (unitatea teritoriala autonoma), and 1 territorial unit (unitatea teritoriala)
raions: Anenii Noi, Basarabeasca, Briceni, Cahul, Cantemir, Calarasi, Causeni, Cimislia, Criuleni, Donduseni, Drochia, Dubasari, Edinet, Falesti, Floresti, Glodeni, Hincesti, Ialoveni, Leova, Nisporeni, Ocnita, Orhei, Rezina, Riscani, Singerei, Soldanesti, Soroca, Stefan-Voda, Straseni, Taraclia, Telenesti, Ungheni
municipalities: Balti, Bender, Chisinau
autonomous territorial unit: Gagauzia
territorial unit: Stinga Nistrului (Transnistria)
Independence: 27 August 1991 (from the Soviet Union)
National holiday: Independence Day, 27 August (1991)
Constitution: adopted 29 July 1994; effective 27 August 1994; note–replaced 1979 Soviet Constitution
Legal system: civil law system with Germanic law influences; Constitutional Court review of legislative acts
International law organization participation: has not submitted an ICJ jurisdiction declaration; accepts ICCt jurisdiction
Suffrage: 18 years of age; universal
Executive branch: *chief of state:* Acting President Marian LUPU (since 30 December 2010)
note: Vladimir VORONIN, president since 4 April 2001, resigned on 11 September 2009; during the first AEI government, Speaker of Parliament Mihai GHIMPU served as acting president; Marian LUPU, the Speaker of Parliament, is serving as acting president until new elections can be held
head of government: Prime Minister Vladimir FILAT (since 25 September 2009); reelected/confirmed on 14 January 2011
note: Vladimir Filat resigned on 27 December 2010, but was reappointed on 31 December 2010
cabinet: Cabinet selected by president, subject to approval of Parliament
(For more information visit the World Leaders website)
elections: president elected by Parliament for a four-year term (eligible for a second term); last successful election held on 4 April 2005, most recent (failed) election held on 10 December 2009); note–prime minister designated by the president upon consultation with Parliament; within 15 days from designation, the prime minister-designate must request a vote of confidence from the Parliament regarding his/her work program and entire cabinet; prime minister (re)designated on 31 December 2010; the prime minister and cabinet

received a vote of confidence 14 January 2011
election results: Vladimir VORONIN reelected president (2005); parliamentary votes–Vladimir VORONIN 75, Gheorghe DUCA 1; Vladimir FILAT (re)designated prime minister; parliamentary votes of confidence–59 of 101
Legislative branch: unicameral Parliament or Parlamentul (101 seats; members elected on an at-large basis by popular vote to serve four-year terms)
elections: last held on 28 November 2010 (next to be held in 2014, unless Parliament fails to elect a president); note–this was the third parliamentary election in less than two years; the earlier parliaments (elected 5 April 2009 and 29 July 2009) were dissolved after they could not agree on a presidential candidate
election results: percent of vote by party–PCRM 39.3%, PLDM 29.4%, PD 12.7%, PL 10%, other 8.6%; seats by party–PCRM 42, PLDM 32, PD 15, PL 12; note–the PLDM, PD, and PL governing coalition, termed the Alliance for European Integration, has 59 seats; it remains 2 votes short of the 61 needed to elect a new president
Judicial branch: Supreme Court; Constitutional Court (the sole authority for constitutional judicature)
Political parties and leaders: represented in Parliament: Communist Party of the Republic of Moldova or PCRM [Vladimir VORONIN]; Democratic Party or PD [Marian LUPU]; Liberal Democratic Party or PLDM [Vladimir FILAT]; Liberal Party or PL [Mihai GHIMPU]; Alliance for European Integration or AIE (coalition of the PD, PLDM, and PL)
not represented in Parliament: Christian Democratic People's Party or PPCD [Iurie ROSCA]; Conservative Party or PC [Natalia NIRCA]; Ecological Party of Moldova "Green Alliance" or PEMAVE [Vladimir BRAGA]; European Action Movement or MAE [Veaceslav UNTILA]; For Nation and Country Party or PpNT [Sergiu MOCANU]; Humanist Party of Moldova or PUM [Valeriu PASAT]; Labor Party or PM [Gheorghe SIMA]; National Liberal Party or PNL [Vitalia PAVLICENKO]; Our Moldova Alliance or AMN [Serafim URECHEAN]; Patriots of Moldova Party or PPM [Mihail GARBUZ]; Popular Republican Party or PPR [Nicolae ANDRONIC]; Republican Party of Moldova or PRM [Andrei STRATAN]; Roma Social Political Movement of the Republic of Moldova or MRRM [Ion BUCUR]; Social Democratic Party or PSD [Victor SELIN]; Social Political Movement "Equality" or MR [Valeriy KLIMENCO]; United Moldova Party or PMUEM [Vladimir TURCAN]
Political pressure groups and leaders: NA
International organization participation: BSEC, CE, CEI, CIS, EAEC (observer), EAPC, EBRD, FAO, GCTU, GUAM, IAEA, IBRD, ICAO, ICRM, IDA, IFAD, IFC, IFRCS, ILO, IMF, IMO, Interpol, IOC, IOM, IPU, ISO (correspondent), ITU, MIGA, OIF, OPCW, OSCE, PFP, SECI, UN, UNCTAD, UNESCO, UNHCR, UNIDO, Union Latina, UNMIL, UNMIS, UNOCI, UNWTO, UPU, WCO, WHO, WIPO, WMO, WTO
Diplomatic representation in the US: *chief of mission:* Ambassador Igor MUNTEANU
chancery: 2101 S Street NW, Washington, DC 20008
telephone: [1] (202) 667-1130
FAX: [1] (202) 667-1204
Diplomatic representation from the US: *chief of mission:* Ambassador Asif J. CHAUDHRY
embassy: 103 Mateevici Street, Chisinau MD-2009
mailing address: use embassy street address
telephone: [373] (22) 40-8300
FAX: [373] (22) 23-3044
Flag description: three equal vertical bands of blue (hoist side), yellow, and red; emblem in center of flag is of a Roman eagle of gold outlined in black with a red beak and talons carrying a yellow cross in its beak and a green olive branch in its right talons and a yellow scepter in its left talons; on its breast is a shield divided horizontally red over blue with a stylized aurochs head, star, rose, and crescent all in black-outlined yellow; based on the color scheme of the flag of Romania–with which Moldova shares a history and culture–but Moldova's blue band is lighter; the reverse of the flag does not display any coat of arms
note: one of only three national flags that differ on their obverse and reverse sides–the others are Paraguay and Saudi Arabia
National anthem: *name:* "Limba noastra" (Our Language)
lyrics/music: Alexei MATEEVICI/Alexandru CRISTEA
note: adopted 1994

ECONOMY

Economy—overview: Moldova remains one of the poorest countries in Europe despite recent progress from its small economic base. It enjoys a favorable climate and good farmland but has no major mineral deposits. As a result, the economy depends heavily on agriculture, featuring fruits, vegetables, wine, and tobacco. Moldova must import almost all of its energy supplies. Moldova's dependence on Russian energy was underscored at the end of 2005, when a Russian-owned electrical station in Moldova's separatist Transnistria region cut off power to Moldova and Russia's Gazprom cut off natural gas in disputes over pricing. In January 2009, gas supplies were cut during a dispute between Russia and Ukraine. Russia's decision to ban Moldovan wine and agricultural products, coupled with its decision to double the price Moldova paid for Russian natural gas, have hurt growth. The onset of the global financial crisis and poor economic conditions in Moldova's main foreign markets caused GDP to fall 6% in 2009. Unemployment almost doubled and inflation disappeared–at -0.1%, a record low. Moldova's IMF agreement expired in May 2009. In fall 2009, the IMF allocated $186 million to Moldova to cover its immediate budgetary needs, and the government signed a new agreement with the IMF in January 2010 for a program worth $574 million. In 2010, an upturn in the world economy boosted GDP growth to 6.5% and inflation to 7.3%. Economic reforms have been slow because of corruption and strong political forces backing government controls. Nevertheless, the government's primary goal of EU integration has resulted in some market-oriented progress. The granting of EU trade preferences and increased exports to Russia will encourage higher growth rates, but the agreements are unlikely to serve as a panacea, given the extent to which export success depends on higher quality standards and other factors. The economy has made a modest recovery, but remains vulnerable to political uncertainty, weak administrative capacity, vested bureaucratic interests, higher fuel prices, poor agricultural weather, and the skepticism of foreign investors as well as the presence of an illegal separatist regime in Moldova's Transnistria region.
GDP (purchasing power parity): $10.99 billion (2010 est.)
country comparison to the world: 149
$10.28 billion (2009 est.)
$10.93 billion (2008 est.)
note: data are in 2010 US dollars
GDP (official exchange rate): $5.81 billion (2010 est.)
GDP—real growth rate: 6.9% (2010 est.)
country comparison to the world: 38
-6% (2009 est.)
7.8% (2008 est.)
GDP—per capita (PPP): $2,500 (2010 est.)
country comparison to the world: 177
$2,400 (2009 est.)
$2,500 (2008 est.)
note: data are in 2010 US dollars
GDP—composition by sector: *agriculture:* 16.3%
industry: 20.1%
services: 63.6% (2010 est.)
Labor force: 1.203 million (2010 est.)
country comparison to the world: 138
Labor force—by occupation: *agriculture:* 40.6%
industry: 16%
services: 43.3% (2005)
Unemployment rate: 6.5% (3rd quarter, 2010 est.)
country comparison to the world: 64
3.1% (2009 est.)
Population below poverty line: 26.3% (2009)
Household income or consumption by percentage share: *lowest 10%:* 3%
highest 10%: 28.2% (2004)
Distribution of family income—Gini index: 37.4 (2007)
country comparison to the world: 75
33.2 (2003)
Investment (gross fixed): 21.7% of GDP (2010 est.)
country comparison to the world: 71
Budget: *revenues:* $2.164 billion
expenditures: $2.462 billion (2010 est.)
Public debt: 25% of GDP (2010 est.)
country comparison to the world: 100
28.9% of GDP (2009)
Inflation rate (consumer prices): 7.3% (2010 est.)

country comparison to the world: 180
-0.1% (2009 est.)
Commercial bank prime lending rate: 20.54% (31 December 2009 est.)
country comparison to the world: 15
21.06% (31 December 2008 est.)
Stock of narrow money: $1.293 billion (31 December 2010 est.)
country comparison to the world: 130
$1.073 billion (31 December 2009 est.)
Stock of broad money: $2.038 billion (31 December 2010 est.)
country comparison to the world: 141
$1.702 billion (31 December 2009 est.)
Stock of domestic credit: $2.097 billion (31 December 2010)
country comparison to the world: 125
$1.823 billion (31 December 2009)
Market value of publicly traded shares: $573.9 million (2004)
Agriculture—products: vegetables, fruits, grapes, grain, sugar beets, sunflower seed, tobacco; beef, milk; wine
Industries: sugar, vegetable oil, food processing, agricultural machinery; foundry equipment, refrigerators and freezers, washing machines; hosiery, shoes, textiles
Industrial production growth rate: 7% (2010)
country comparison to the world: 43
Electricity—production: 3.617 billion kWh (2007 est.)
country comparison to the world: 122
Electricity—consumption: 4.37 billion kWh (2007 est.)
country comparison to the world: 116
Electricity—exports: 240 million kWh (2007 est.)
Electricity—imports: 2.931 billion kWh (2007 est.)
Oil—production: 0 bbl/day (2009 est.)
country comparison to the world: 200
Oil—consumption: 19,000 bbl/day (2009 est.)
country comparison to the world: 125
Oil—exports: 36.49 bbl/day (2007 est.)
country comparison to the world: 135
Oil—imports: 14,230 bbl/day (2007 est.)
country comparison to the world: 130
Oil—proved reserves: 0 bbl (1 January 2010 est.)
country comparison to the world: 158
Natural gas—production: 60,000 cu m (2010 est.)
country comparison to the world: 92
Natural gas—consumption: 3.176 billion cu m (2010 est.)
country comparison to the world: 72
Natural gas—exports: 0 cu m (2010 est.)
country comparison to the world: 135
Natural gas—imports: 3.176 billion cu m (2008 est.)
country comparison to the world: 40
Natural gas—proved reserves: 0 cu m (1 January 2010 est.)
country comparison to the world: 159
Current account balance: $-565 million (2010 est.)
country comparison to the world: 119
$-439.3 million (2009)
Exports: $1.45 billion (2010 est.)
country comparison to the world: 140
$1.297 billion (2009)
Exports—commodities: foodstuffs, textiles, machinery
Exports—partners: Russia 23.77%, Italy 14.11%, Romania 12.74%, Germany 6.92%, Turkey 6.08%, Belarus 5.38% (2009)
Imports: $3.66 billion (2010 est.)
country comparison to the world: 132
$3.278 billion (2009 est.)
Imports—commodities: mineral products and fuel, machinery and equipment, chemicals, textiles
Imports—partners: Ukraine 19.9%, Romania 15.1%, Russia 14.52%, Germany 8.69%, Italy 5.7%, Belarus 4.38% (2009)
Reserves of foreign exchange and gold: $1.71 billion (31 December 2010 est.)
country comparison to the world: 108
$1.48 billion (31 December 2009 est.)
Debt—external: $4.618 billion (30 September 2010 est.)
country comparison to the world: 109
$4.364 billion (31 December 2009 est.)
Stock of direct foreign investment—at home: $2.649 billion (1 January 2010 est.)
country comparison to the world: 87
$1.813 billion (2008)
Stock of direct foreign investment—abroad: $62.44 million (1 January 2010)
country comparison to the world: 82
Exchange rates: Moldovan lei (MDL) per US dollar–12.37 (2010) 11.11 (2009) 10.326 (2008) 12.177 (2007) 13.131 (2006)

COMMUNICATIONS

Telephones—main lines in use: 1.139 million (2009)
country comparison to the world: 73
Telephones—mobile cellular: 2.785 million (2009)
country comparison to the world: 116
Telephone system: *general assessment:* poor service outside Chisinau; some modernization is under way
domestic: depending on location, new subscribers may face long wait for service; multiple private operators of GSM mobile-cellular telephone service are operating; GPRS system is being introduced; a CDMA mobile telephone network began operations in 2007; combined fixed-line and mobile-cellular teledensity 90 per 100 persons
international: country code–373; service through Romania and Russia via landline; satellite earth stations–at least 3 (Intelsat, Eutelsat, and Intersputnik) (2009)
Broadcast media: state-owned national radio-TV broadcaster operates 2 television and 2 radio stations; a total of nearly 40 terrestrial TV channels and some 50 radio stations are in operation; Russian and Romanian channels also are available (2007)
Internet country code: .md
Internet hosts: 492,181 (2010)
country comparison to the world: 50
Internet users: 1.333 million (2009)
country comparison to the world: 89

TRANSPORTATION

Airports: 11 (2010)
country comparison to the world: 154
Airports—with paved runways: *total:* 5
over 3,047 m: 1
2,438 to 3,047 m: 2
1,524 to 2,437 m: 2 (2010)
Airports—with unpaved runways: *total:* 6
1,524 to 2,437 m: 3
under 914 m: 3 (2010)
Pipelines: gas 1,906 km (2010)
Railways: *total:* 1,190 km
country comparison to the world: 85
broad gauge: 1,176 km 1.520-m gauge
standard gauge: 14 km 1.435-m gauge (2010)
Roadways: *total:* 9,343 km
country comparison to the world: 137
paved: 8,810 km
unpaved: 533 km (2008)
Waterways: 558 km (in public use on Danube, Dniester and Prut rivers) (2008)
country comparison to the world: 83
Merchant marine: *total:* 107
country comparison to the world: 48
by type: bulk carrier 7, cargo 89, chemical tanker 2, passenger/cargo 1, petroleum tanker 1, refrigerated cargo 1, roll on/roll off 6
foreign-owned: 63 (Belgium 2, Egypt 5, Greece 4, Israel 4, Lebanon 1, Romania 2, Russia 5, Syria 3, Turkey 18, UK 6, Ukraine 12, Yemen 1) (2010)

MILITARY

Military branches: National Army: Land Forces Command (includes special forces), Air Forces Command (includes air defense unit), Logistics Command (2010)
Military service age and obligation: 18 years of age for compulsory military service; 17 years of age for voluntary service; male registration required at age 16; 12-month service obligation (2009)
Manpower available for military service: *males age 16–49:* 1,143,440
females age 16–49: 1,156,958 (2010 est.)
Manpower fit for military service: *males age 16–49:* 875,224
females age 16–49: 969,903 (2010 est.)
Manpower reaching militarily significant age annually: *male:* 28,213
female: 26,614 (2010 est.)
Military expenditures: 0.4% of GDP (2005 est.)
country comparison to the world: 165

TRANSNATIONAL ISSUES

Disputes—international: Moldova and Ukraine operate joint customs posts to monitor the transit of people and commodities through Moldova's break-away Transnistria region, which remains under OSCE supervision
Trafficking in persons: current situation: Moldova is a major source and, to a lesser extent, a transit country for women and girls trafficked for the purpose of commercial sexual exploitation; Moldovan women are trafficked to the Middle East, Eastern Europe, and Western Europe; girls and young women are trafficked within the country from rural areas to Chisinau; children are also trafficked to neighboring countries for forced labor and begging; labor trafficking of men to work in the construction, agriculture, and service sectors of Russia is increasingly a problem; according to an ILO report, Moldova's national Bureau of Statistics

estimated that there were likely over 25,000 Moldovan victims of trafficking for forced labor in 2008
tier rating: Tier 2 Watch List–The Government of Moldova does not fully comply with the minimum standards for the elimination of trafficking; however, it is making significant efforts to do so; despite initial efforts to combat trafficking-related complicity since the government's reassessment on the Tier 2 Watch List in September 2008, and increased victim assistance, the government did not demonstrate sufficiently meaningful efforts to curb trafficking-related corruption, which is a government-acknowledged problem in Moldova; the government improved victim protection efforts, deployed more law-enforcement officers in the effort and contributed direct financial assistance toward victim protection and assistance for the first time (2010)
Illicit drugs: limited cultivation of opium poppy and cannabis, mostly for CIS consumption; transshipment point for illicit drugs from Southwest Asia via Central Asia to Russia, Western Europe, and possibly the US; widespread crime and underground economic activity

MONACO

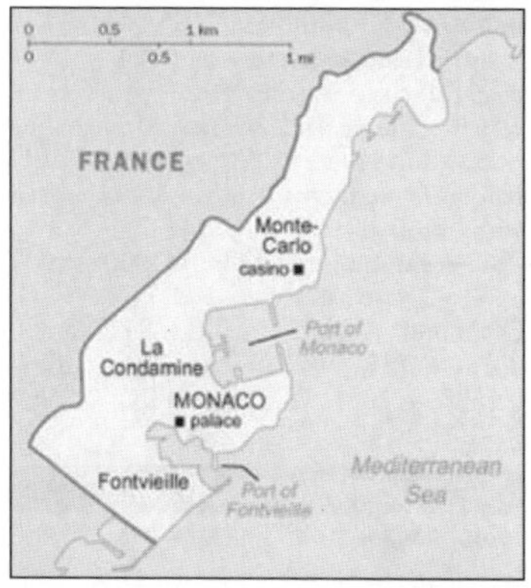

INTRODUCTION

Background: The Genoese built a fortress on the site of present day Monaco in 1215. The current ruling Grimaldi family first seized temporary control in 1297, and again in 1331, but were not able to permanently secure their holding until 1419. Economic development was spurred in the late 19th century with a railroad linkup to France and the opening of a casino. Since then, the principality's mild climate, splendid scenery, and gambling facilities have made Monaco world famous as a tourist and recreation center.

GEOGRAPHY

Location: Western Europe, bordering the Mediterranean Sea on the southern coast of France, near the border with Italy
Geographic coordinates: 43 44 N, 7 24 E
Map references: Europe
Area: *total:* 2 sq km
country comparison to the world: 248
land: 2 sq km
water: 0 sq km
Area—comparative: about three times the size of The Mall in Washington, DC
Land boundaries: *total:* 4.4 km
border countries: France 4.4 km
Coastline: 4.1 km
Maritime claims: *territorial sea:* 12 nm
exclusive economic zone: 12 nm
Climate: Mediterranean with mild, wet winters and hot, dry summers
Terrain: hilly, rugged, rocky
Elevation extremes: *lowest point:* Mediterranean Sea 0 m
highest point: Mont Agel 140 m
Natural resources: none
Land use: *arable land:* 0%
permanent crops: 0%
other: 100% (urban area) (2005)
Irrigated land: NA
Natural hazards: NA
Environment—current issues: NA
Environment—international agreements: *party to:* Air Pollution, Air Pollution-Sulfur 94, Air Pollution-Volatile Organic Compounds, Biodiversity, Climate Change, Climate Change-Kyoto Protocol, Desertification, Endangered Species, Hazardous Wastes, Law of the Sea, Marine Dumping, Ozone Layer Protection, Ship Pollution, Wetlands, Whaling
signed, but not ratified: none of the selected agreements
Geography—note: second-smallest independent state in the world (after Holy See); almost entirely urban

PEOPLE

Population: 30,539 (July 2011 est.)
country comparison to the world: 214
Age structure: *0–14 years:* 12.3% (male 1,930/female 1,841)
15–64 years: 60.8% (male 9,317/female 9,249)
65 years and over: 26.9% (male 3,640/female 4,562) (2011 est.)
Median age: *total:* 49.4 years
male: 48.4 years
female: 50.5 years (2011 est.)
Population growth rate: -0.124% (2011 est.)
country comparison to the world: 208
Birth rate: 6.94 births/1,000 population (2011 est.)
country comparison to the world: 222
Death rate: 8.28 deaths/1,000 population (July 2011 est.)
country comparison to the world: 91
Net migration rate: 0.1 migrant(s)/1,000 population (2011 est.)
country comparison to the world: 70
Urbanization: *urban population:* 100% of total population (2010)
rate of urbanization: 0.3% annual rate of change (2010-15 est.)
Sex ratio: *at birth:* 1.039 male(s)/female
under 15 years: 1.05 male(s)/female
15–64 years: 1 male(s)/female
65 years and over: 0.8 male(s)/female
total population: 0.95 male(s)/female (2011 est.)
Infant mortality rate: *total:* 1.79 deaths/1,000 live births
country comparison to the world: 222
male: 2.04 deaths/1,000 live births
female: 1.54 deaths/1,000 live births (2011 est.)
Life expectancy at birth: *total population:* 89.73 years
country comparison to the world: 1
male: 85.77 years
female: 93.84 years (2011 est.)
Total fertility rate: 1.5 children born/woman (2011 est.)
country comparison to the world: 186
HIV/AIDS—adult prevalence rate: NA
HIV/AIDS—people living with HIV/AIDS: NA
HIV/AIDS—deaths: NA
Drinking water source: *Improved:*
urban: 100% of population
total: 100% of population (2008)
Sanitation facility access: *Improved:*
urban: 100% of population
total: 100% of population (2008)
Nationality: *noun:* Monegasque(s) or Monacan(s)
adjective: Monegasque or Monacan
Ethnic groups: French 47%, Monegasque 16%, Italian 16%, other 21%
Religions: Roman Catholic 90%, other 10%
Languages: French (official), English, Italian, Monegasque
Literacy: *definition:* age 15 and over can read and write
total population: 99%
male: 99%
female: 99% (2003 est.)
School life expectancy (primary to tertiary education): *total:* 18 years
male: 18 years
female: 17 years (2009)
Education expenditures: 1.2% of GDP (2004)
country comparison to the world: 163

GOVERNMENT

Country name: *conventional long form:* Principality of Monaco
conventional short form: Monaco
local long form: Principaute de Monaco
local short form: Monaco
Government type: constitutional monarchy
Capital: *name:* Monaco
geographic coordinates: 43 44 N, 7 25 E
time difference: UTC+1 (6 hours ahead of Washington, DC during Standard Time)
daylight saving time: +1hr, begins last Sunday in March; ends last Sunday in October
Administrative divisions: none; there are no first-order administrative divisions as defined by the US Government, but there are four quarters (quartiers, singular—quartier); Fontvieille, La Condamine, Monaco-Ville, Monte-Carlo

Independence: 1419 (beginning of permanent rule by the House of Grimaldi)
National holiday: National Day (Saint Rainier's Day), 19 November (1857)
Constitution: 17 December 1962; modified 2 April 2002
Legal system: civil law system influenced by French legal tradition
International law organization participation: has not submitted an ICJ jurisdiction declaration; non-party state to the ICCt
Suffrage: 18 years of age; universal
Executive branch: *chief of state:* Prince ALBERT II (since 6 April 2005)
head of government: Minister of State Michel ROGER (since 29 March 2010)
cabinet: Council of Government under the authority of the monarch
(For more information visit the World Leaders website)
elections: the monarchy is hereditary; minister of state appointed by the monarch from a list of three French national candidates presented by the French Government
Legislative branch: unicameral National Council or Conseil National (24 seats; 16 members elected by list majority system, 8 by proportional representation to serve five-year terms)
elections: last held on 3 February 2008 (next to be held in February 2013)
election results: percent of vote by party—UPM 52.2%, REM 40.5%, Monaco Together 7.3%; seats by party—UPM 21, REM 3
Judicial branch: Supreme Court or Tribunal Supreme (judges appointed by the monarch on the basis of nominations by the National Council)
Political parties and leaders: Monaco Together; Rally and Issues for Monaco or REM; Union for Monaco or UPM (including National Union for the Future of Monaco or UNAM)
Political pressure groups and leaders: NA
International organization participation: CE, FAO, IAEA, ICAO, ICC, ICRM, IFRCS, IHO, IMO, IMSO, Interpol, IOC, IPU, ITSO, ITU, OAS (observer), OIF, OPCW, OSCE, Schengen Convention (de facto member), UN, UNCTAD, UNESCO, UNIDO, Union Latina, UNWTO, UPU, WHO, WIPO, WMO
Diplomatic representation in the US: *chief of mission:* Ambassador Gilles NOGHES
chancery: 3400 International Drive NW, Suite 2K-100, Washington, DC 20008
telephone: (202) 234-1530
FAX: (202) 244-7656
consulate(s) general: New York
Diplomatic representation from the US: the US does not have an embassy in Monaco; the US Ambassador to France is accredited to Monaco; the US Consul General in Marseille (France), under the authority of the US ambassador to France, handles routine diplomatic and consular matters concerning Monaco
Flag description: two equal horizontal bands of red (top) and white; the colors are those of the ruling House of Grimaldi and have been in use since 1339, making the flag one of the world's oldest national banners
note: similar to the flag of Indonesia which is longer and the flag of Poland which is white (top) and red
National anthem: *name:* "A Marcia de Muneghu" (The March of Monaco)
lyrics/music: Louis NOTARI/Charles ALBRECHT
note: music adopted 1867, lyrics adopted 1931; although French is much more commonly spoken, only the Monegasque lyrics are official; the French version is known as "Hymne Monegasque" (Monegasque Anthem); the words are generally only sung on official occasions

ECONOMY

Economy—overview: Monaco, bordering France on the Mediterranean coast, is a popular resort, attracting tourists to its casino and pleasant climate. The principality also is a major banking center and has successfully sought to diversify into services and small, high-value-added, nonpolluting industries. The state has no income tax and low business taxes and thrives as a tax haven both for individuals who have established residence and for foreign companies that have set up businesses and offices. Monaco, however, is not a tax-free shelter; it charges nearly 20% value-added tax, collects stamp duties, and companies face a 33% tax on profits unless they can show that three-quarters of profits are generated within the principality. Monaco was formally removed from the OECD's "grey list" of uncooperative tax jurisdictions in late 2009, but continues to face international pressure to abandon its banking secrecy laws and help combat tax evasion. The state retains monopolies in a number of sectors, including tobacco, the telephone network, and the postal service. Living standards are high, roughly comparable to those in prosperous French metropolitan areas.
GDP (purchasing power parity): $976.3 million (2006 est.)
country comparison to the world: 201
note: data are in 2010 US dollars
GDP (official exchange rate): $NA
GDP—real growth rate: NA%
GDP—per capita (PPP): $30,000 (2006 est.)
country comparison to the world: 44
GDP—composition by sector: *agriculture:* 0%
industry: 4.9%
services: 95.1% (2005)
Labor force: 44,000
country comparison to the world: 191
note: includes workers from all foreign countries (2005 est.)
Unemployment rate: 0% (2005)
country comparison to the world: 1
Population below poverty line: NA%
Household income or consumption by percentage share: *lowest 10%:* NA%
highest 10%: NA%
Budget: *revenues:* $863 million
expenditures: $920.6 million (2005 est.)
Inflation rate (consumer prices): 1.9% (2000)
country comparison to the world: 51
Market value of publicly traded shares: $NA
Agriculture—products: none
Industries: tourism, construction, small-scale industrial and consumer products
Industrial production growth rate: NA%
Electricity—consumption: NA kWh
Electricity—imports: NA kWh; note—electricity supplied by France
Exports: $716.3 million (2005)
country comparison to the world: 161
note: full customs integration with France, which collects and rebates Monegasque trade duties; also participates in EU market system through customs union with France
Imports: $916.1 million (2005)
country comparison to the world: 174
note: full customs integration with France, which collects and rebates Monegasque trade duties; also participates in EU market system through customs union with France
Debt—external: $18 billion (2000 est.)
country comparison to the world: 73
Exchange rates: euros (EUR) per US dollar—0.755 (2010) 0.7198 (2009) 0.6827 (2008) 0.7345 (2007) 0.7964 (2006)

COMMUNICATIONS

Telephones—main lines in use: 35,400 (2009)
country comparison to the world: 174
Telephones—mobile cellular: 23,000 (2009)
country comparison to the world: 207
Telephone system: *general assessment:* modern automatic telephone system; the country's sole fixed line operator offers a full range of services to residential and business customers
domestic: combined fixed-line and mobile-cellular teledensity exceeds 100%
international: country code—377; no satellite earth stations; connected by cable into the French communications system
Broadcast media: TV Monte-Carlo (TMC) operates a TV network; Radio Monte-Carlo has both an Italian-language and a French-language network; a few private radio stations operating (2008)
Internet country code: .mc
Internet hosts: 23,621 (2010)
country comparison to the world: 104
Internet users: 23,000 (2009)
country comparison to the world: 188

TRANSPORTATION

Heliports: 1 (2010)
Roadways: *total:* 50 km
country comparison to the world: 217
paved: 50 km (2007)
Merchant marine: *registered in other countries:* 68 (Bahamas 14, Bermuda 2, Comoros 1, Cyprus 1, Liberia 10, Marshall Islands 21, Norway 1, Panama 14, Saint Vincent and the Grenadines 3, Vanuatu 1) (2010)
country comparison to the world: 62
Ports and terminals: Monaco

MILITARY

Military branches: no regular military forces; the Palace Guard performs ceremonial duties

Manpower available for military service: *males age 16–49:* 5,749 (2010 est.)
Manpower fit for military service: *males age 16–49:* 4,629
females age 16–49: 4,597 (2010 est.)
Manpower reaching militarily significant age annually: *male:* 153
female: 141 (2010 est.)
Military—note: defense is the responsibility of France

TRANSNATIONAL ISSUES

Disputes—international: none

MONGOLIA

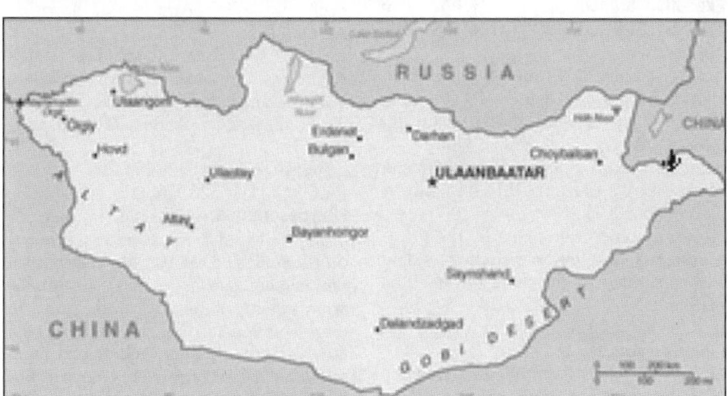

INTRODUCTION

Background: The Mongols gained fame in the 13th century when under Chinggis KHAAN they established a huge Eurasian empire through conquest. After his death the empire was divided into several powerful Mongol states, but these broke apart in the 14th century. The Mongols eventually retired to their original steppe homelands and in the late 17th century came under Chinese rule. Mongolia won its independence in 1921 with Soviet backing and a Communist regime was installed in 1924. The modern country of Mongolia, however, represents only part of the Mongols' historical homeland; more ethnic Mongolians live in the Inner Mongolia Autonomous Region in the People's Republic of China than in Mongolia. Following a peaceful democratic revolution, the ex-Communist Mongolian People's Revolutionary Party (MPRP) won elections in 1990 and 1992, but was defeated by the Democratic Union Coalition (DUC) in the 1996 parliamentary election. The MPRP won an overwhelming majority in the 2000 parliamentary election, but the party lost seats in the 2004 election and shared power with democratic coalition parties from 2004-08. The MPRP regained a solid majority in the 2008 parliamentary elections but nevertheless formed a coalition government with the Democratic Party. In 2010 the MPRP voted to retake the name of the Mongolian People's Party (MPP), a name it used in the early 1920s. The prime minister and most cabinet members are MPP members.

GEOGRAPHY

Location: Northern Asia, between China and Russia
Geographic coordinates: 46 00 N, 105 00 E
Map references: Asia
Area: *total:* 1,564,116 sq km
country comparison to the world: 19
land: 1,553,556 sq km
water: 10,560 sq km
Area—comparative: slightly smaller than Alaska
Land boundaries: *total:* 8,220 km
border countries: China 4,677 km, Russia 3,543 km
Coastline: 0 km (landlocked)
Maritime claims: none (landlocked)
Climate: desert; continental (large daily and seasonal temperature ranges)
Terrain: vast semidesert and desert plains, grassy steppe, mountains in west and southwest; Gobi Desert in south-central
Elevation extremes: *lowest point:* Hoh Nuur 560 m
highest point: Nayramadlin Orgil (Huyten Orgil) 4,374 m
Natural resources: oil, coal, copper, molybdenum, tungsten, phosphates, tin, nickel, zinc, fluorspar, gold, silver, iron
Land use: *arable land:* 0.76%
permanent crops: 0%
other: 99.24% (2005)
Irrigated land: 840 sq km (2003)
Total renewable water resources: 34.8 cu km (1999)
Freshwater withdrawal (domestic/industrial/agricultural): *total:* 0.44 cu km/yr (20%/27%/52%)
per capita: 166 cu m/yr (2000)
Natural hazards: dust storms; grassland and forest fires; drought; "zud," which is harsh winter conditions
Environment—current issues: limited natural freshwater resources in some areas; the policies of former Communist regimes promoted rapid urbanization and industrial growth that had negative effects on the environment; the burning of soft coal in power plants and the lack of enforcement of environmental laws severely polluted the air in Ulaanbaatar; deforestation, overgrazing, and the converting of virgin land to agricultural production increased soil erosion from wind and rain; desertification and mining activities had a deleterious effect on the environment
Environment—international agreements: *party to:* Biodiversity, Climate Change, Climate Change-Kyoto Protocol, Desertification, Endangered Species, Environmental Modification, Hazardous Wastes, Law of the Sea, Ozone Layer Protection, Ship Pollution, Wetlands, Whaling
signed, but not ratified: none of the selected agreements
Geography—note: landlocked; strategic location between China and Russia

PEOPLE

Population: 3,133,318 (July 2011 est.)
country comparison to the world: 134
Age structure: *0–14 years:* 27.3% (male 437,241/female 419,693)
15–64 years: 68.7% (male 1,074,949/female 1,076,455)
65 years and over: 4% (male 54,415/female 70,565) (2011 est.)
Median age: *total:* 26.2 years
male: 25.8 years
female: 26.6 years (2011 est.)
Population growth rate: 1.489% (2011 est.)
country comparison to the world: 80
Birth rate: 20.93 births/1,000 population (2011 est.)
country comparison to the world: 85
Death rate: 6.04 deaths/1,000 population (July 2011 est.)
country comparison to the world: 162
Net migration rate: 0 migrant(s)/1,000 population (2011 est.)
country comparison to the world: 96
Urbanization: *urban population:* 62% of total population (2010)
rate of urbanization: 1.9% annual rate of change (2010-15 est.)
Major cities—population: ULAANBAATAR (capital) 949,000 (2009)
Sex ratio: *at birth:* 1.05 male(s)/female
under 15 years: 1.04 male(s)/female
15–64 years: 1 male(s)/female
65 years and over: 0.77 male(s)/female
total population: 1 male(s)/female (2011 est.)
Infant mortality rate: *total:* 37.26 deaths/1,000 live births
country comparison to the world: 67
male: 40.26 deaths/1,000 live births
female: 34.11 deaths/1,000 live births (2011 est.)
Life expectancy at birth: *total population:* 68.31 years
country comparison to the world: 153
male: 65.85 years
female: 70.89 years (2011 est.)
Total fertility rate: 2.21 children born/woman (2011 est.)

country comparison to the world: 104
HIV/AIDS—adult prevalence rate: less than 0.1% (2009 est.)
country comparison to the world: 145
HIV/AIDS—people living with HIV/AIDS: fewer than 500 (2009 est.)
country comparison to the world: 156
HIV/AIDS—deaths: fewer than 100 (2009 est.)
country comparison to the world: 141
Drinking water source: *Improved:*
urban: 97% of population
rural: 49% of population
total: 76% of population
Unimproved:
urban: 3% of population
rural: 51% of population
total: 24% of population (2008)
Sanitation facility access: *Improved:*
urban: 64% of population
rural: 32% of population
total: 50% of population
Unimproved:
urban: 46% of population
rural: 68% of population
total: 50% of population (2008)
Nationality: *noun:* Mongolian(s)
adjective: Mongolian
Ethnic groups: Mongol (mostly Khalkha) 94.9%, Turkic (mostly Kazakh) 5%, other (including Chinese and Russian) 0.1% (2000)
Religions: Buddhist Lamaist 50%, Shamanist and Christian 6%, Muslim 4%, none 40% (2004)
Languages: Khalkha Mongol 90%, Turkic, Russian (1999)
Literacy: *definition:* age 15 and over can read and write
total population: 97.8%
male: 98%
female: 97.5% (2000 census)
School life expectancy (primary to tertiary education): *total:* 14 years
male: 13 years
female: 15 years (2009)
Education expenditures: 5.6% of GDP (2009)
country comparison to the world: 39

GOVERNMENT

Country name: *conventional long form:* none
conventional short form: Mongolia
local long form: none
local short form: Mongol Uls
former: Outer Mongolia
Government type: parliamentary
Capital: *name:* Ulaanbaatar
geographic coordinates: 47 55 N, 106 55 E
time difference: UTC+8 (13 hours ahead of Washington, DC during Standard Time)
Administrative divisions: 21 provinces (aymguud, singular—aymag) and 1 municipality* (singular—hot); Arhangay, Bayanhongor, Bayan-Olgiy, Bulgan, Darhan-Uul, Dornod, Dornogovi, Dundgovi, Dzavhan (Zavkhan), Govi-Altay, Govisumber, Hentiy, Hovd, Hovsgol, Omnogovi, Orhon, Ovorhangay, Selenge, Suhbaatar, Tov, Ulaanbaatar*, Uvs
Independence: 11 July 1921 (from China)
National holiday: Independence Day/Revolution Day, 11 July (1921)
Constitution: 13 January 1992
Legal system: civil law system influenced by Soviet and Romano-Germanic legal systems; constitution ambiguous on judicial review of legislative acts
International law organization participation: has not submitted an ICJ jurisdiction declaration; accepts ICCt jurisdiction
Suffrage: 18 years of age; universal
Executive branch: *chief of state:* President Tsakhia ELBEGDORJ (since 18 June 2009)
head of government: Prime Minister Sukhbaatar BATBOLD (since 29 October 2009); First Deputy Prime Minister (Norov ALTANKHUYAG (since 20 September 2008); Deputy Prime Minister Miegombyn ENKHBOLD (since 6 December 2007)
cabinet: Cabinet nominated by the prime minister in consultation with the president and confirmed by the State Great Hural (parliament)
(For more information visit the World Leaders website)
elections: presidential candidates nominated by political parties represented in State Great Hural and elected by popular vote for a four-year term (eligible for a second term); election last held on 24 May 2009 (next to be held in May 2013); following legislative elections, leader of majority party or majority coalition usually elected prime minister by State Great Hural
election results: in elections in May 2009, Tsakhia ELBEGDORJ elected president; percent of vote—Tsakhia ELBEGDORJ 51.2%, Nambar ENKHBAYAR 47.4%, others 1.3%
Legislative branch: unicameral State Great Hural 76 seats; members elected by popular vote to serve four-year terms
elections: last held on 29 June 2008 (next to be held in June 2012)
election results: percent of vote by party—NA; seats by party—MPP 46, DP 27, others 3
Judicial branch: Supreme Court (serves as appeals court for people's and provincial courts but rarely overturns verdicts of lower courts; judges are nominated by the General Council of Courts and approved by the president)
Political parties and leaders: Civil Will-Green Party or CWGP [Dangaasuren EHKHBAT]; Democratic Party or DP [Norov ALTANHUYAG]; Mongolian People's Party or MPP [Sukhbaatar BATBOLD]; Mongolian People's Revolutionary Party or MPRP [Nambaryn ENKHBAYAR]
Political pressure groups and leaders: *other:* human rights groups; women's groups
International organization participation: ADB, ARF, CD, CICA, CP, EBRD, FAO, G-77, IAEA, IBRD, ICAO, ICRM, IDA, IFAD, IFC, IFRCS, ILO, IMF, IMO, Interpol, IOC, IOM, IPU, ISO, ITSO, ITU, ITUC, MIGA, MINURSO, MONUSCO, NAM, OPCW, OSCE (partner), SCO (observer), UN, UNCTAD, UNESCO, UNIDO, UNMIL, UNMIS, UNWTO, UPU, WCO, WHO, WIPO, WMO, WTO
Diplomatic representation in the US: *chief of mission:* Ambassador Khasbazar BEKHBAT
chancery: 2833 M Street NW, Washington, DC 20007
telephone: [1] (202) 333-7117
FAX: [1] (202) 298-9227
consulate(s) general: New York
Diplomatic representation from the US: *chief of mission:* Ambassador Jonathan ADDLETON
embassy: Big Ring Road, 11th Micro Region, Ulaanbaatar, 14171 Mongolia
mailing address: PSC 461, Box 300, FPO AP 96521-0002; P.O. Box 1021, Ulaanbaatar-13
telephone: [976] (11) 329-095
FAX: [976] (11) 320-776
Flag description: three equal, vertical bands of red (hoist side), blue, and red; centered on the hoist-side red band in yellow is the national emblem ("soyombo"—a columnar arrangement of abstract and geometric representation for fire, sun, moon, earth, water, and the yin-yang symbol); blue represents the sky, red symbolizes progress and prosperity
National anthem: *name:* "Mongol ulsyn toriin duulal" (National Anthem of Mongolia)
lyrics/music: Tsendiin DAMDINSUREN/Bilegiin DAMDINSUREN and Luvsanjamts MURJORJ
note: music adopted 1950, lyrics adopted 2006; the anthem's lyrics have been altered on numerous occasions

ECONOMY

Economy—overview: Economic activity in Mongolia has traditionally been based on herding and agriculture—Mongolia's extensive mineral deposits, however, have attracted foreign investors. The country holds copper, gold, coal, molybdenum, fluorspar, uranium, tin, and tungsten deposits, which account for a large part of foreign direct investment and government revenues. Soviet assistance, at its height one-third of GDP, disappeared almost overnight in 1990 and 1991 at the time of the dismantlement of the USSR. The following decade saw Mongolia endure both deep recession, because of political inaction and natural disasters, as well as economic growth, because of reform-embracing, free-market economics and extensive privatization of the formerly state-run economy. Severe winters and summer droughts in 2000-02 resulted in massive livestock die-off and zero or negative GDP growth. This was compounded by falling prices for Mongolia's primary sector exports and widespread opposition to privatization. Growth averaged nearly 9% per year in 2004-08 largely because of high copper prices and new gold production. In 2008 Mongolia experienced a soaring inflation rate with year-to-year inflation reaching nearly 30%—the highest inflation rate in over a decade. By late 2008, as the country began to feel the effects of the global financial crisis, falling commodity prices helped lower inflation, but also reduced government revenues and forced cuts in spending. In early 2009, the International Monetary Fund reached a $236 million Stand-by Arrangement with Mongolia and the country has started to move out of the crisis. Although the banking sector remains unstable, the government is now enforcing stricter supervision regulations. In October 2009, the government passed long-awaited legislation on an investment agreement to develop Mongolia's Oyu Tolgoi mine, considered to be one of the world's largest untapped copper deposits. The economy grew 6.1% in 2010, largely on the strength of exports to nearby countries, and international reserves reached $1.6 billion in September, an all time high for Mongolia. Mongolia's economy continues to be heavily influenced by its neighbors. Mongolia purchases 95% of its petroleum products and a substantial amount of electric power from Russia, leaving it vulnerable to price increases. Trade with China represents more than half of Mongolia's total external trade–China receives more than three-fourths of Mongolia's exports. Remittances from Mongolians working abroad are sizable, but have fallen due to the economic crisis; money laundering is a growing concern. Mongolia joined the World Trade Organization in 1997 and seeks to expand its participation in regional economic and trade regimes.

GDP (purchasing power parity): $11.02 billion (2010 est.)
country comparison to the world: 148
$10.38 billion (2009 est.)
$10.51 billion (2008 est.)
note: data are in 2010 US dollars
GDP (official exchange rate): $6.125 billion (2010 est.)
GDP—real growth rate: 6.1% (2010 est.)
country comparison to the world: 52
-1.3% (2009 est.)
8.9% (2008 est.)
GDP—per capita (PPP): $3,600 (2010 est.)
country comparison to the world: 162
$3,400 (2009 est.)
$3,500 (2008 est.)
note: data are in 2010 US dollars
GDP—composition by sector: *agriculture:* 15%
industry: 31%
services: 54% (2010 est.)
Labor force: 1.068 million (2008)
country comparison to the world: 140
Labor force—by occupation: *agriculture:* 34%
industry: 5%
services: 61% (2008)
Unemployment rate: 11.5% (2009)
country comparison to the world: 123
2.8% (2008)
Population below poverty line: 36.1% (2004)
Household income or consumption by percentage share: *lowest 10%:* 2.9%
highest 10%: 24.9% (2005)
Distribution of family income—Gini index: 32.8 (2002)
country comparison to the world: 97
44 (1998)
Budget: *revenues:* $2.26 billion
expenditures: $2.26 billion (2010)
Inflation rate (consumer prices): 13% (2010 est.)
country comparison to the world: 211
4.2% (2009 est.)
Central bank discount rate: 10.99% (31 December 2010)
country comparison to the world: 38
10.82% (31 December 2009)
Commercial bank prime lending rate: 19.97% (31 December 2010 est.)
country comparison to the world: 12
21.67% (31 December 2009 est.)
Stock of narrow money: $921.4 million (31 December 2009)
country comparison to the world: 140
$451.4 million (31 December 2008)
Stock of broad money: $1.996 billion (31 December 2009)
country comparison to the world: 143
$1.791 billion (31 December 2008)
Stock of domestic credit: $1.466 billion (31 December 2010 est.)
country comparison to the world: 136
$1.011 billion (31 December 2009 est.)
Market value of publicly traded shares: $1.093 billion (31 December 2010)
country comparison to the world: 108
$430.2 million (31 December 2009)
$407 million (31 December 2008)
Agriculture—products: wheat, barley, vegetables, forage crops; sheep, goats, cattle, camels, horses
Industries: construction and construction materials; mining (coal, copper, molybdenum, fluorspar, tin, tungsten, and gold); oil; food and beverages; processing of animal products, cashmere and natural fiber manufacturing
Industrial production growth rate: 3% (2006 est.)
country comparison to the world: 108
Electricity—production: 4.5 billion kWh (2010)
country comparison to the world: 117
Electricity—consumption: 3.023 billion kWh (2010)
country comparison to the world: 126
Electricity—exports: 20.7 million kWh (2010)
Electricity—imports: 214.1 million kWh (2010)
Oil—production: 5,975 bbl/day (2010)
country comparison to the world: 92
Oil—consumption: 16,000 bbl/day (2009 est.)
country comparison to the world: 134
Oil—exports: 5,834 bbl/day (2010 est.)
country comparison to the world: 100
Oil—imports: 15,730 bbl/day (2010)
country comparison to the world: 123
Oil—proved reserves: NA bbl (1 January 2010 est.)
Natural gas—production: 0 cu m (2010 est.)
country comparison to the world: 155
Natural gas—consumption: 0 cu m (2010 est.)
country comparison to the world: 194
Natural gas—exports: 0 cu m (2010 est.)
country comparison to the world: 136
Natural gas—imports: 11,790 cu m (2010 est.)
country comparison to the world: 70
Natural gas—proved reserves: 0 cu m (1 January 2010 est.)
country comparison to the world: 160
Current account balance: $-378.8 million (2010 est.)
country comparison to the world: 109
$-228.7 million (2009 est.)
Exports: $2.899 billion (2010)
country comparison to the world: 124
$1.902 billion (2009)
Exports—commodities: copper, apparel, livestock, animal products, cashmere, wool, hides, fluorspar, other nonferrous metals, coal
Exports—partners: China 78.52%, Canada 9.46%, Russia 3.02% (2009)
Imports: $3.3 billion (2010)
country comparison to the world: 137
$2.131 billion (2009)
Imports—commodities: machinery and equipment, fuel, cars, food products, industrial consumer goods, chemicals, building materials, sugar, tea
Imports—partners: China 35.99%, Russia 31.56%, South Korea 7.08%, Japan 4.8% (2009)
Debt—external: $1.86 billion (2009)
country comparison to the world: 142
$1.6 billion (2008)
Stock of direct foreign investment—at home: $NA
Stock of direct foreign investment—abroad: $NA
Exchange rates: togrog/tugriks (MNT) per US dollar–1,357.5 (2010) 1,442.8 (2009) 1,170 (2007) 1,165 (2006)

COMMUNICATIONS

Telephones—main lines in use: 188,900 (2009)
country comparison to the world: 126
Telephones—mobile cellular: 2.249 million (2009)
country comparison to the world: 127
Telephone system: *general assessment:* network is improving with international direct dialing available in many areas; a fiber-optic network has been installed that is improving broadband and communication services between major urban centers with multiple companies providing inter-city fiber-optic cable services
domestic: very low fixed-line teledensity; there are multiple mobile-cellular providers and subscribership is increasing rapidly;
international: country code–976; satellite earth stations–7
Broadcast media: following a law passed in 2005, Mongolia's state-run radio and TV provider converted to a public service provider; also available are private radio and TV broadcasters, as well as multichannel satellite and cable TV providers; more than 100 radio stations, including some 20 via repeaters for the public broadcaster; transmissions of multiple international broadcasters are available (2008)
Internet country code: .mn
Internet hosts: 7,942 (2010)
country comparison to the world: 134
Internet users: 330,000 (2008)
country comparison to the world: 125

TRANSPORTATION

Airports: 46 (2010)
country comparison to the world: 95

Airports—with paved runways: *total:* 14
over 3,047 m: 1
2,438 to 3,047 m: 10
1,524 to 2,437 m: 3 (2010)
Airports—with unpaved runways: *total:* 32
over 3,047 m: 1
2,438 to 3,047 m: 4
1,524 to 2,437 m: 25
914 to 1,523 m: 1
under 914 m: 1 (2010)
Heliports: 1 (2010)
Railways: *total:* 1,908 km
country comparison to the world: 74
broad gauge: 1,908 km 1.520-m gauge
note: the railway is 50 percent owned by the Russian State Railway (2010)
Roadways: *total:* 49,249 km
country comparison to the world: 81
paved: 3,015 km
unpaved: 46,234 km (2010)
Waterways: 580 km (the only waterway in operation is Lake Hovsgol (135 km); Selenge River (270 km) and Orhon River (175 km) are navigable but carry little traffic; lakes and rivers freeze in winter, they are open from May to September) (2010)
country comparison to the world: 82
Merchant marine: *total:* 58
country comparison to the world: 67
by type: bulk carrier 20, cargo 29, chemical tanker 2, liquefied gas 2, passenger/cargo 1, roll on/roll off 3, vehicle carrier 1
foreign-owned: 44 (Indonesia 2, North Korea 1, Russia 4, Singapore 1, Turkey 1, Ukraine 1, Vietnam 34) (2010)

MILITARY

Military branches: Mongolian Armed Forces: Mongolian Army, Mongolian Air Force; there is no navy (2010)
Military service age and obligation: 18-25 years of age for compulsory military service; conscript service obligation–12 months in land or air defense forces or police; a small portion of Mongolian land forces (2.5 percent) is comprised of contract soldiers; women cannot be deployed overseas for military operations (2006)
Manpower available for military service: *males age 16–49:* 898,546
females age 16–49: 891,192 (2010 est.)
Manpower fit for military service: *males age 16–49:* 726,199
females age 16–49: 756,628 (2010 est.)
Manpower reaching militarily significant age annually: *male:* 30,829
female: 29,648 (2010 est.)
Military expenditures: 1.4% of GDP (2006)
country comparison to the world: 110

TRANSNATIONAL ISSUES

Disputes—international: none

MONTENEGRO

INTRODUCTION

Background: The use of the name Montenegro began in the 15th century when the Crnojevic dynasty began to rule the Serbian principality of Zeta; over subsequent centuries Montenegro was able to maintain its independence from the Ottoman Empire. From the 16th to 19th centuries, Montenegro became a theocracy ruled by a series of bishop princes; in 1852, it was transformed into a secular principality. After World War I, Montenegro was absorbed by the Kingdom of Serbs, Croats, and Slovenes, which became the Kingdom of Yugoslavia in 1929; at the conclusion of World War II, it became a constituent republic of the Socialist Federal Republic of Yugoslavia. When the latter dissolved in 1992, Montenegro federated with Serbia, first as the Federal Republic of Yugoslavia and, after 2003, in a looser union of Serbia and Montenegro. In May 2006, Montenegro invoked its right under the Constitutional Charter of Serbia and Montenegro to hold a referendum on independence from the state union. The vote for severing ties with Serbia exceeded 55%–the threshold set by the EU–allowing Montenegro to formally declare its independence on 3 June 2006.

GEOGRAPHY

Location: Southeastern Europe, between the Adriatic Sea and Serbia
Geographic coordinates: 42 30 N, 19 18 E
Map references: Europe
Area: *total:* 13,812 sq km
country comparison to the world: 161
land: 13,452 sq km
water: 360 sq km
Area—comparative: slightly smaller than Connecticut
Land boundaries: *total:* 625 km
border countries: Albania 172 km, Bosnia and Herzegovina 225 km, Croatia 25 km, Kosovo 79 km, Serbia 124 km
Coastline: 293.5 km
Maritime claims: *territorial sea:* 12 nm
continental shelf: defined by treaty
Climate: Mediterranean climate, hot dry summers and autumns and relatively cold winters with heavy snowfalls inland
Terrain: highly indented coastline with narrow coastal plain backed by rugged high limestone mountains and plateaus
Elevation extremes: *lowest point:* Adriatic Sea 0 m
highest point: Bobotov Kuk 2,522 m
Natural resources: bauxite, hydroelectricity
Land use: *arable land:* 13.7%
permanent crops: 1%
other: 85.3%
Irrigated land: NA
Natural hazards: destructive earthquakes
Environment—current issues: pollution of coastal waters from sewage outlets, especially in tourist-related areas such as Kotor
Environment—international agreements: *party to:* Air Pollution, Biodiversity, Climate Change, Climate Change-Kyoto Protocol, Desertification, Hazardous Wastes, Law of the Sea, Marine Dumping, Marine Life Conservation, Ozone Layer Protection, Ship Pollution
signed, but not ratified: none of the selected agreements
Geography—note: strategic location along the Adriatic coast

PEOPLE

Population: 661,807 (July 2011 est.)
country comparison to the world: 166
Age structure: *0–14 years:* 15.5% (male 50,060/female 52,823)
15–64 years: 71% (male 244,057/female 225,620)
65 years and over: 13.5% (male 35,551/female 53,696) (2011 est.)
Median age: *total:* 37.8 years
male: 36.5 years
female: 39.2 years (2011 est.)
Population growth rate: -0.705% (2011 est.)
country comparison to the world: 227
Birth rate: 11 births/1,000 population (2011 est.)
country comparison to the world: 175
Death rate: 8.89 deaths/1,000 population (July 2011 est.)
country comparison to the world: 74
Urbanization: *urban population:* 61% of total population (2010)
rate of urbanization: 0.1% annual rate of change (2010-15 est.)
Major cities—population: PODGORICA (capital) 144,000 (2009)
Sex ratio: *at birth:* 1.072 male(s)/female
under 15 years: 0.95 male(s)/female
15–64 years: 1.08 male(s)/female
65 years and over: 0.67 male(s)/female
total population: 0.99 male(s)/female (2011 est.)
Major infectious diseases: degree of risk: intermediate
food or waterborne diseases: bacterial diarrhea
vectorborne disease: Crimean Congo hemorrhagic fever (2009)
Drinking water source: *Improved:*
urban: 100% of population
rural: 96% of population
total: 98% of population

Unimproved:
urban: 0% of population
rural: 4% of population
total: 2% of population (2008)

Sanitation facility access: *Improved:*
urban: 96% of population
rural: 86% of population
total: 92% of population
Unimproved:
urban: 4% of population
rural: 14% of population
total: 8% of population (2008)

Nationality: *noun:* Montenegrin(s)
adjective: Montenegrin

Ethnic groups: Montenegrin 43%, Serbian 32%, Bosniak 8%, Albanian 5%, other (Muslims, Croats, Roma (Gypsy)) 12% (2003 census)

Religions: Orthodox 74.2%, Muslim 17.7%, Catholic 3.5%, other 0.6%, unspecified 3%, atheist 1% (2003 census)

Languages: Serbian 63.6%, Montenegrin (official) 22%, Bosnian 5.5%, Albanian 5.3%, unspecified 3.7% (2003 census)

School life expectancy (primary to tertiary education): NA

Education expenditures: NA

GOVERNMENT

Country name: *conventional long form:* none
conventional short form: Montenegro
local long form: none
local short form: Crna Gora
former: People's Republic of Montenegro, Socialist Republic of Montenegro, Republic of Montenegro

Government type: republic

Capital: *name:* Podgorica
geographic coordinates: 42 26 N, 19 16 E
time difference: UTC+1 (6 hours ahead of Washington, DC during Standard Time)
daylight saving time: +1 hr, begins last Sunday in March; ends last Sunday in October

Administrative divisions: 21 municipalities (opstine, singular–opstina); Andrijevica, Bar, Berane, Bijelo Polje, Budva, Cetinje, Danilovgrad, Herceg Novi, Kolasin, Kotor, Mojkovac, Niksic, Plav, Pljevlja, Pluzine, Podgorica, Rozaje, Savnik, Tivat, Ulcinj, Zabljak

Independence: 3 June 2006 (from Serbia and Montenegro)

National holiday: National Day, 13 July (1878)

Constitution: approved 19 October 2007 (by the Assembly)

Legal system: civil law

International law organization participation: has not submitted an ICJ jurisdiction declaration; accepts ICCt jurisdiction

Suffrage: 18 years of age; universal

Executive branch: *chief of state:* President Filip VUJANOVIC (since 6 April 2008)
head of government: Prime Minister Igor LUKSIC (since 29 December 2010)
cabinet: Ministries act as cabinet
(For more information visit the World Leaders website)
elections: president elected by direct vote for five-year term (eligible for a second term); election last held on 6 April 2008 (next to be held in 2013); prime minister proposed by president, accepted by Assembly
election results: Filip VUJANOVIC reelected president; Filip VUJANOVIC 51.9%, Andrija MANDIC 19.6%, Nebojsa MEDOJEVIC 16.6%, Srdan MILIC 11.9%

Legislative branch: unicameral Assembly (81 seats; members elected by direct vote to serve four-year terms; note—seats increased from 74 seats in 2006)
elections: last held on 29 March 2009 (next to be held in 2013)
election results: percent of vote by party—Coalition for European Montenegro 51.94%, SNP 16.83%, NOVA 9.22%, PZP 6.03%, other (including Albanian minority parties) 15.98%; seats by party—Coalition for European Montenegro 48, SNP 16, NOVA 8, PZP 5, Albanian minority parties 4

Judicial branch: Constitutional Court (five judges serve nine-year terms); Supreme Court (judges have life tenure)

Political parties and leaders: Albanian Alternative or AA [Vesel SINISHTAJ]; Coalition for European Montenegro (bloc) [Milo DJUKANOVIC] (includes Democratic Party of Socialists or DPS [Milo DJUKANOVIC], Social Democratic Party or SDP [Ranko KRIVOKAPIC], Bosniak Party of BS [Rafet HUSOVIC], and Croatian Civic Initiative or HGI [Marija VUCINOVIC); Coalition SNP-NS-DSS (bloc) (includes Socialist People's Party or SNP [Srdjan MILIC], People's Party of Montenegro or NS [Predrag POPOVIC], and Democratic Serbian Party of Montenegro or DSS [Ranko KADIC]); Democratic League-Party of Democratic Prosperity or SPP [Mehmet BARDHIJ]; Democratic Union of Albanians or DUA [Ferhat DINOSHA]; For a Different Montenegro (bloc) [Goran BATRICEVIC] (includes Democratic Center or DC [Goran BATRICEVIC] and Liberal Party of Montenegro or LP [Miodrag ZIVKOVIC]); FORCA [Nazif CUNGU]; Movement for Changes or PZP [Nebojsa MEDOJEVIC]; National Coalition (includes People's Party of Montenegro or NS [Predrag POPOVIC] and Democratic Serbian Party of Montenegro or DSS [Ranko KADIC]); New Serb Democracy or NOVA [Andrija MANDIC]; Socialist People's Party of Montenegro or SNP [Srdjan MILIC]

International organization participation: CE, CEI, EAPC, EBRD, FAO, IAEA, IBRD, ICAO, ICRM, IDA, IFC, IFRCS, ILO, IMF, IMO, IMSO, Interpol, IOC, IOM, IPU, ISO (correspondent), ITSO, ITU, ITUC, MIGA, NAM (observer), OPCW, OSCE, PCA, PFP, SECI, UN, UNCTAD, UNESCO, UNHCR, UNIDO, UNMIL, UNWTO, UPU, WCO, WHO, WIPO, WMO, WTO (observer)

Diplomatic representation in the US: *chief of mission:* Ambassador Srdjan DARMANOVIC
chancery: 1610 New Hampshire Avenue NW, Washington, DC, 20009
telephone: [1] (202) 234-6108
FAX: [1] (202) 234-6109
consulate(s) general: New York

Diplomatic representation from the US: *chief of mission:* Ambassador (vacant); Charge d'Affaires Benjamin LOWENTHAL
embassy: Ljubljanska bb, 81000 Podgorica, Montenegro
mailing address: use embassy street address
telephone: [382] 81 225 417
FAX: [382] 81 241 358

Flag description: a red field bordered by a narrow golden-yellow stripe with the Montenegrin coat of arms centered; the arms consist of a double-headed golden eagle—symbolizing the unity of church and state—surmounted by a crown; the eagle holds a golden scepter in its right claw and a blue orb in its left; the breast shield over the eagle shows a golden lion passant on a green field in front of a blue sky; the lion is symbol of episcopal authority and harks back to the three and a half centuries that Montenegro was ruled as a theocracy

National anthem: *name:* "Oj, svijetla majska zoro" (Oh, Bright Dawn of May)
lyrics/music: Sekula DRLJEVIC/unknown, arranged by Zarko MIKOVIC
note: adopted 2004; the anthem's music is based on a Montenegrin folk song

ECONOMY

Economy—overview: Montenegro severed its economy from federal control and from Serbia during the MILOSEVIC era and maintained its own central bank, adopted the Deutchmark, then the euro—rather than the Yugoslav dinar—as official currency, collected customs tariffs, and managed its own budget. The dissolution of the loose political union between Serbia and Montenegro in 2006 led to separate membership in several international financial institutions, such as the European Bank for Reconstruction and Development. In January 2007, Montenegro joined the World Bank and IMF. Montenegro is pursuing its own membership in the World Trade Organization and signed a Stabilization and Association agreement with the European Union in October 2007. The European Council granted candidate country status to Montenegro at the December 2010 session. Unemployment and regional disparities in development are key political and economic problems. Montenegro has privatized its large aluminum complex—the dominant industry—as well as most of its financial sector, and has begun to attract foreign direct investment in the tourism sector. The global financial crisis has had a significant negative impact on the economy, due to the ongoing credit crunch, a decline in the real estate sector, and a fall in aluminum exports.

GDP (purchasing power parity): $6.724 billion (2010 est.)
country comparison to the world: 153
$6.653 billion (2009 est.)
$7.055 billion (2008 est.)
note: data are in 2010 US dollars

GDP (official exchange rate): $4.017 billion (2010 est.)

GDP—real growth rate: 1.1% (2010 est.)
country comparison to the world: 170
-5.7% (2009 est.)
6.9% (2008 est.)

GDP—per capita (PPP): $10,100 (2010 est.)

country comparison to the world: 109
$9,900 (2009 est.)
$10,400 (2008 est.)
note: data are in 2010 US dollars
GDP—composition by sector: *agriculture:* NA%
industry: NA%
services: NA%
Labor force: 259,100 (2004)
country comparison to the world: 167
Labor force—by occupation: *agriculture:* 2%
industry: 30%
services: 68% (2004 est.)
Unemployment rate: 14.7% (2007 est.)
country comparison to the world: 148
Population below poverty line: 7% (2007 est.)
Distribution of family income—Gini index: 30 (2003)
country comparison to the world: 114
Investment (gross fixed): 30.5% of GDP (2006 est.)
country comparison to the world: 17
Budget: *revenues:* $NA
expenditures: $NA
Public debt: 38% of GDP (2006)
country comparison to the world: 80
Inflation rate (consumer prices): 3.4% (2007)
country comparison to the world: 96
Commercial bank prime lending rate: 9.36% (31 December 2009 est.)
country comparison to the world: 96
9.24% (31 December 2008 est.)
Stock of narrow money: $816.8 million (31 December 2008)
country comparison to the world: 145
$1.172 billion (31 December 2007)
Stock of broad money: $1.406 billion (31 December 2008)
country comparison to the world: 150
$1.446 billion (31 December 2007)
Stock of domestic credit: $3.29 billion (31 December 2009)
country comparison to the world: 117
$3.771 billion (31 December 2008)
Market value of publicly traded shares: $4.289 billion (31 December 2009)
country comparison to the world: 91
$2.863 billion (31 December 2008)
$3.699 billion (31 December 2007)
Agriculture—products: tobacco, potatoes, citrus fruits, olives, grapes; sheep
Industries: steelmaking, aluminum, agricultural processing, consumer goods, tourism
Electricity—production: 2.864 billion kWh (2005 est.)
country comparison to the world: 127
Electricity—consumption: 18.6 million kWh (2005)
country comparison to the world: 206
Electricity—exports: 0 kWh (2005)
Electricity—imports: NA kWh (2005)
Oil—production: 0 bbl/day (2009 est.)
country comparison to the world: 203
Oil—consumption: 5,000 bbl/day (2009 est.)
country comparison to the world: 165
Oil—exports: 313.6 bbl/day (2005)
country comparison to the world: 128
Oil—imports: 6,093 bbl/day (2005)
country comparison to the world: 153
Oil—proved reserves: 0 bbl (1 January 2010 est.)
country comparison to the world: 161
Natural gas—consumption: NA cu m
Natural gas—proved reserves: 0 cu m (1 January 2010 est.)
country comparison to the world: 163
Current account balance: $-1.102 billion (2007 est.)
country comparison to the world: 137
Exports: $171.3 million (2003)
country comparison to the world: 183
Exports—partners: Italy 29.52%, Greece 22.65%, Slovenia 11.83%, Hungary 8.96%, US 7.93% (2009)
Imports: $601.7 million (2003)
country comparison to the world: 185
Imports—partners: Italy 17.54%, Slovenia 14.62%, Germany 10.5%, Austria 7.82%, China 7.82%, Russia 4.4%, Hungary 4.11%, Greece 4.11%, Netherlands 4% (2009)
Reserves of foreign exchange and gold: $NA
Debt—external: $650 million (2006)
country comparison to the world: 157
Exchange rates: euros (EUR) per US dollar–0.755 (2010) 0.72 (2009) 0.6827 (2008) 0.7345 (2007) 0.7964 (2006)

COMMUNICATIONS

Telephones—main lines in use: 366,600 (2009)
country comparison to the world: 107
Telephones—mobile cellular: 752,000 (2009)
country comparison to the world: 151
Telephone system: *general assessment:* modern telecommunications system with access to European satellites
domestic: GSM mobile-cellular service, available through multiple providers with national coverage, is growing
international: country code–382; 2 international switches connect the national system
Broadcast media: state-owned national radio-TV broadcaster operates 2 terrestrial television networks, 1 satellite TV channel, and 2 radio networks; roughly a dozen privately-owned TV broadcasters operate networks nationally, regionally, and locally; in addition to the 2 state-owned national radio networks, roughly 50 privately-owned radio stations and networks broadcast (2007)
Internet country code: .me
Internet hosts: 6,247 (2010)
country comparison to the world: 137
Internet users: 280,000 (2009)
country comparison to the world: 133

TRANSPORTATION

Airports: 5 (2010)
country comparison to the world: 180
Airports—with paved runways: *total:* 4
2,438 to 3,047 m: 2
1,524 to 2,437 m: 1
under 914 m: 1 (2010)
Airports—with unpaved runways: *total:* 1
914 to 1,523 m: 1 (2010)
Heliports: 1 (2010)
Railways: *total:* 250 km
country comparison to the world: 124
standard gauge: 250 km 1.435-m gauge (169 km electrified) (2010)
Roadways: *total:* 7,624 km
country comparison to the world: 145
paved: 5,097 km
unpaved: 2,527 km (2010)
Merchant marine: *total:* 2
country comparison to the world: 145
by type: cargo 1, passenger/cargo 1
registered in other countries: 5 (Bahamas 2, Honduras 2, Slovakia 1) (2010)
Ports and terminals: Bar

MILITARY

Military branches: Armed Forces of the Republic of Montenegro: Army of Montenegro (includes Montenegrin Navy (Mornarica Crne Gore, MCG)), Air Force (2011)
Military service age and obligation: compulsory national military service abolished August 2006
Manpower fit for military service: *males age 16–49:* 149,159
females age 16–49: 131,823 (2010 est.)
Manpower reaching militarily significant age annually: *male:* 3,120
female: 3,677 (2010 est.)

TRANSNATIONAL ISSUES

Disputes—international:

MONTSERRAT

OVERSEAS TERRITORY OF THE UK)

INTRODUCTION

Background: English and Irish colonists from St. Kitts first settled on Montserrat in 1632; the first African slaves arrived three decades later. The British and French fought for possession of the island for most of the 18th century, but it finally was confirmed as a British possession in 1783. The island's sugar plantation economy was converted to small farm landholdings in the mid 19th century. Much of this island was devastated and two-thirds of the population fled abroad because of the eruption of the Soufriere Hills Volcano that began on 18 July 1995. Montserrat has endured volcanic activity since, with the last eruption occurring in July 2003.

GEOGRAPHY

Location: Caribbean, island in the Caribbean Sea, southeast of Puerto Rico
Geographic coordinates: 16 45 N, 62 12 W

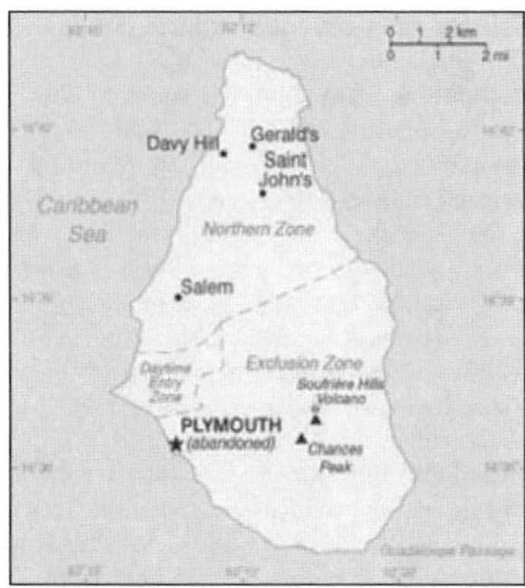

Map references: Central America and the Caribbean
Area: *total:* 102 sq km
country comparison to the world: 224
land: 102 sq km
water: 0 sq km
Area—comparative: about 0.6 times the size of Washington, DC
Land boundaries: 0 km
Coastline: 40 km
Maritime claims: *territorial sea:* 3 nm
exclusive fishing zone: 200 nm
Climate: tropical; little daily or seasonal temperature variation
Terrain: volcanic island, mostly mountainous, with small coastal lowland
Elevation extremes: *lowest point:* Caribbean Sea 0 m
highest point: lava dome in English's Crater (in the Soufriere Hills volcanic complex) estimated at over 930 m (2006)
Natural resources: NEGL
Land use: *arable land:* 20%
permanent crops: 0%
other: 80% (2005)
Irrigated land: NA
Natural hazards: severe hurricanes (June to November); volcanic eruptions
volcanism: Soufriere Hills volcano, at a height of 915 m (3,010 ft), has erupted continuously since 1995; a massive eruption in 1997 destroyed most of the capital, Plymouth, and resulted in approximately half of the island becoming uninhabitable
Environment—current issues: land erosion occurs on slopes that have been cleared for cultivation
Geography—note: the island is entirely volcanic in origin and comprised of three major volcanic centers of differing ages

PEOPLE

Population: 5,140 (July 2011 est.)
country comparison to the world: 228
note: an estimated 8,000 refugees left the island following the resumption of volcanic activity in July 1995; some have returned
Age structure: *0–14 years:* 26.9% (male 717/female 665)
15–64 years: 66.6% (male 1,648/female 1,777)
65 years and over: 6.5% (male 208/female 125) (2011 est.)
Median age: *total:* 29.6 years
male: 29.2 years
female: 30.1 years (2011 est.)
Population growth rate: 0.447% (2011 est.)
country comparison to the world: 154
Birth rate: 11.67 births/1,000 population (2011 est.)
country comparison to the world: 167
Death rate: 7.2 deaths/1,000 population (July 2011 est.)
country comparison to the world: 126
Net migration rate: 0 migrant(s)/1,000 population (2011 est.)
country comparison to the world: 97
Urbanization: *urban population:* 14% of total population (2010)
rate of urbanization: 2.4% annual rate of change (2010-15 est.)
Sex ratio: *at birth:* 1.033 male(s)/female
under 15 years: 1.09 male(s)/female
15–64 years: 0.91 male(s)/female
65 years and over: 2.03 male(s)/female
total population: 1.01 male(s)/female (2011 est.)
Infant mortality rate: *total:* 15.23 deaths/1,000 live births
country comparison to the world: 117
male: 11.54 deaths/1,000 live births
female: 19.1 deaths/1,000 live births (2011 est.)
Life expectancy at birth: *total population:* 73.16 years
country comparison to the world: 120
male: 74.99 years
female: 71.24 years (2011 est.)
Total fertility rate: 1.26 children born/woman (2011 est.)
country comparison to the world: 216
HIV/AIDS—adult prevalence rate: NA
HIV/AIDS—people living with HIV/AIDS: NA
HIV/AIDS—deaths: NA
Drinking water source: *Improved:*
urban: 100% of population
rural: 100% of population
total: 100% of population (2008)
Sanitation facility access: *Improved:*
urban: 96% of population
rural: 96% of population
total: 96% of population
Unimproved: urban: 4% of population
rural: 4% of population
total: 4% of population (2008)
Nationality: *noun:* Montserratian(s)
adjective: Montserratian
Ethnic groups: black, white
Religions: Anglican, Methodist, Roman Catholic, Pentecostal, Seventh-Day Adventist, other Christian denominations
Languages: English
Literacy: *definition:* age 15 and over has ever attended school
total population: 97%
male: 97%
female: 97% (1970 est.)
School life expectancy (primary to tertiary education): *total:* 15 years
male: 14 years
female: 17 years (2007)
Education expenditures: 3.3% of GDP (2004)
country comparison to the world: 122

GOVERNMENT

Country name: *conventional long form:* none
conventional short form: Montserrat
Dependency status: overseas territory of the UK
Government type: NA
Capital: *name:* Plymouth
geographic coordinates: 16 42 N, 62 13 W
time difference: UTC-4 (1 hour ahead of Washington, DC during Standard Time)
note: Plymouth was abandoned in 1997 because of volcanic activity; interim government buildings have been built at Brades Estate in the Carr's Bay/Little Bay vicinity at the northwest end of Montserrat
Administrative divisions: 3 parishes; Saint Anthony, Saint Georges, Saint Peter
Independence: none (overseas territory of the UK)
National holiday: Birthday of Queen ELIZABETH II, second Saturday in June (1926)
Constitution: effective 19 December 1989
Legal system: English common law
Suffrage: 18 years of age; universal
Executive branch: *chief of state:* Queen ELIZABETH II (since 6 February 1952); represented by Governor Adrian DAVIS (since 8 April 2011)
head of government: Chief Minister Rueben MEADE (since 10 September 2009)
cabinet: Executive Council consists of the governor, the chief minister, 3 other ministers, the attorney general, and the finance secretary
(For more information visit the World Leaders website)
elections: the monarchy is hereditary; governor appointed by the monarch; following legislative elections, the leader of the majority party usually becomes chief minister
Legislative branch: unicameral Legislative Council (11 seats; 9 members popularly elected to serve five-year terms; the attorney general and financial secretary sit as ex-officio members)
elections: last held on 8 September 2009 (next to be held by 2014)
election results: percent of vote by party—NA; seats by party—MCAP 6, independents 3
Judicial branch: Eastern Caribbean Supreme Court (based in Saint Lucia, one judge of the Supreme Court is a resident of the islands and presides over the High Court)
Political parties and leaders: Montserrat Democratic Party or MDP [Lowell LEWIS]; Movement for Change and Prosperity or MCAP [Roselyn CASSELL-SEALY]; New People's Liberation Movement or NPLM [John A. OSBORNE]
Political pressure groups and leaders: NA
International organization participation: Caricom, CDB, Interpol (subbureau), OECS, UPU
Diplomatic representation in the US: none (overseas territory of the UK)
Diplomatic representation from the US: none (overseas territory of the UK)
Flag description: blue, with the flag of the UK in the upper hoist-side quadrant and the Montserratian coat of arms centered in the outer half of the flag; the arms feature a woman in green dress, Erin, the female personification of Ireland, standing beside a yellow harp and embracing a large dark cross with her right arm; Erin and the harp are symbols of Ireland reflecting the territory's Irish ancestry; blue represents

awareness, trustworthiness, determination, and righteousness
National anthem: note: as a territory of the United Kingdom, "God Save the Queen" is official (see United Kingdom)

ECONOMY

Economy—overview: Severe volcanic activity, which began in July 1995, has put a damper on this small, open economy. A catastrophic eruption in June 1997 closed the airports and seaports, causing further economic and social dislocation. Two-thirds of the 12,000 inhabitants fled the island. Some began to return in 1998 but lack of housing limited the number. The agriculture sector continued to be affected by the lack of suitable land for farming and the destruction of crops. Prospects for the economy depend largely on developments in relation to the volcanic activity and on public sector construction activity. The UK has launched a three-year $122.8 million aid program to help reconstruct the economy. Half of the island is expected to remain uninhabitable for another decade.
GDP (purchasing power parity): $29 million (2002 est.)
country comparison to the world: 224
GDP (official exchange rate): $NA
GDP—real growth rate: -1% (2002 est.)
country comparison to the world: 198
GDP—per capita (PPP): $3,400 (2002 est.)
country comparison to the world: 165
GDP—composition by sector:
agriculture: 1.2%
industry: 23.1%
services: 75.7% (1999 est.)
Labor force: NA
Unemployment rate: 6% (1998 est.)
country comparison to the world: 59
Population below poverty line: NA%
Household income or consumption by percentage share: *lowest 10%:* NA%
highest 10%: NA%
Budget: *revenues:* $31.4 million
expenditures: $31.6 million (1997 est.)
Inflation rate (consumer prices): 2.6% (2002 est.)
country comparison to the world: 69
Central bank discount rate: 6.5% (31 December 2009)
country comparison to the world: 56
6.5% (31 December 2008)
Commercial bank prime lending rate: 9.04% (31 December 2009 est.)
country comparison to the world: 90
9.89% (31 December 2008 est.)
Stock of narrow money: $14.13 million (31 December 2009)
country comparison to the world: 188
$14.51 million (31 December 2008)
Stock of broad money: $69.63 million (31 December 2009)
country comparison to the world: 187
$62.13 million (31 December 2008)
Stock of domestic credit: $9.93 million (31 December 2008 est.)
country comparison to the world: 185
$5.537 million (31 December 2007 est.)
Agriculture—products: cabbages, carrots, cucumbers, tomatoes, onions, peppers; livestock products
Industries: tourism, rum, textiles, electronic appliances
Industrial production growth rate: NA%
Electricity—production: 22 million kWh (2007 est.)
country comparison to the world: 206
Electricity—consumption: 20.46 million kWh (2007 est.)
country comparison to the world: 205
Electricity—exports: 0 kWh (2008 est.)
Electricity—imports: 0 kWh (2008 est.)
Oil—production: 0 bbl/day (2009 est.)
country comparison to the world: 201
Oil—consumption: 1,000 bbl/day (2009 est.)
country comparison to the world: 195
Oil—exports: 0 bbl/day (2007 est.)
country comparison to the world: 183
Oil—imports: 520.6 bbl/day (2007 est.)
country comparison to the world: 197
Oil—proved reserves: 0 bbl (1 January 2010 est.)
country comparison to the world: 159
Natural gas—production: 0 cu m (2008 est.)
country comparison to the world: 156
Natural gas—consumption: 0 cu m (2008 est.)
country comparison to the world: 195
Natural gas—exports: 0 cu m (2008 est.)
country comparison to the world: 137
Natural gas—imports: 0 cu m (2008 est.)
country comparison to the world: 84
Natural gas—proved reserves: 0 cu m (1 January 2010 est.)
country comparison to the world: 161
Exports: $700,000 (2001 est.)
country comparison to the world: 218
Exports—commodities: electronic components, plastic bags, apparel; hot peppers, limes, live plants; cattle
Imports: $17 million (2001 est.)
country comparison to the world: 219
Imports—commodities: machinery and transportation equipment, foodstuffs, manufactured goods, fuels, lubricants, and related materials
Debt—external: $8.9 million (1997)
country comparison to the world: 191
Exchange rates: East Caribbean dollars (XCD) per US dollar–2.7 (2010) 2.7 (2009) 2.7 (2005) 2.7 (2004) 2.7 (2003)

COMMUNICATIONS

Telephones—main lines in use: 2,700 (2009)
country comparison to the world: 219
Telephones—mobile cellular: 3,000 (2008)
country comparison to the world: 212
Telephone system: *general assessment:* modern and fully digitalized
domestic: combined fixed-line and mobile-cellular teledensity exceeds 100 per 100 persons
international: country code–1-664; landing point for the East Caribbean Fiber System (ECFS) optic submarine cable with links to 13 other islands in the eastern Caribbean extending from the British Virgin Islands to Trinidad
Broadcast media: Radio Montserrat, a public radio broadcaster, transmits on 1 station and has a repeater transmission to a second station; repeater transmissions from the GEM Radio Network of Trinidad and Tobago provide another 2 radio stations; cable and satellite TV are obtainable (2007)
Internet country code: .ms
Internet hosts: 552 (2010)
country comparison to the world: 177
Internet users: 1,200 (2009)
country comparison to the world: 212

TRANSPORTATION

Airports: 2 (2010)
country comparison to the world: 209
Airports—with paved runways: *total:* 2
under 914 m: 2 (2010)
Roadways: note: volcanic eruptions that began in 1995 destroyed most of the 227 km road system; a new road infrastructure has been built on the north end of the island (2008)
Ports and terminals: Little Bay, Plymouth

MILITARY

Military branches: no regular military forces; Royal Montserrat Police Force (2010)
Manpower available for military service: *males age 16–49:* 1,353 (2010 est.)
Manpower fit for military service:
males age 16–49: 1,135
females age 16–49: 1,223 (2010 est.)
Manpower reaching militarily significant age annually: *male:* 35
female: 34 (2010 est.)
Military—note: defense is the responsibility of the UK

TRANSNATIONAL ISSUES

Disputes—international: none
Illicit drugs: transshipment point for South American narcotics destined for the US and Europe
none
Refugees and internally displaced persons: refugees (country of origin): 7,000 (Kosovo); note–mostly ethnic Serbs and Roma who fled Kosovo in 1999
IDPs: 16,192 (ethnic conflict in 1999 and riots in 2004) (2007)

MOROCCO

INTRODUCTION

Background: In 788, about a century after the Arab conquest of North Africa, successive Moorish dynasties began to rule in Morocco. In the 16th century, the Sa'adi monarchy, particularly under Ahmad AL-MANSUR (1578-1603), repelled foreign invaders and inaugurated a golden age. The Alaouite dynasty, to which the current Moroccan royal family belongs, established a sultanate in Morocco beginning in the 17th

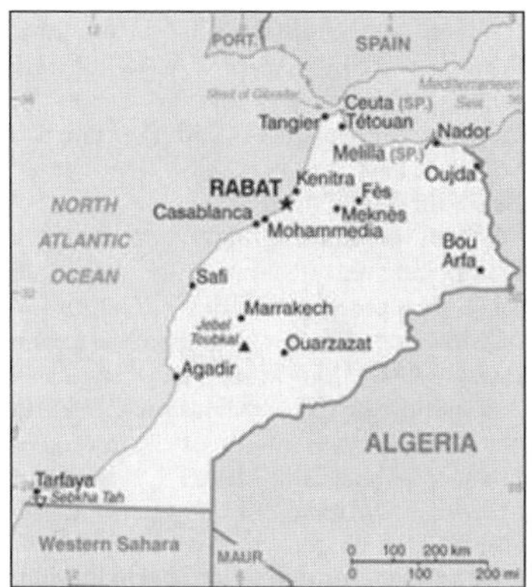

century. In 1860, Spain occupied northern Morocco and ushered in a half century of trade rivalry among European powers that saw Morocco's sovereignty steadily erode; in 1912, the French imposed a protectorate over the country. A protracted independence struggle with France ended successfully in 1956. The internationalized city of Tangier and most Spanish possessions were turned over to the new country that same year. Sultan MOHAMMED V, the current monarch's grandfather, organized the new state as a constitutional monarchy and in 1957 assumed the title of king. Morocco annexed Western Sahara during the late 1970s, but final resolution on the status of the territory remains unresolved. Gradual political reforms in the 1990s resulted in the establishment of a bicameral legislature, which first met in 1997. Under King MOHAMMED VI–who in 1999 succeeded his father to the throne–human rights have improved. Morocco enjoys a moderately free press, but the government occasionally takes action against journalists who report on three broad subjects considered to be taboo: the monarchy, Islam, and the status of Western Sahara. Despite the continuing reforms, ultimate authority remains in the hands of the monarch. Influenced by protests elsewhere in the Middle East and North Africa, thousands of Moroccans in February 2011 rallied in Rabat and several other major cities to demand constitutional reform and more democracy and to protest government corruption and high food prices. A number of similar demonstrations and marches continued through March. Police response to most of the protests was subdued compared to the violence noted in protests elsewhere in the region. In early March, in response to demonstrator demands, King MOHAMMED VI announced the formation of a commission to reform the country's constitution. A draft constitution is to be presented by June 2011, followed by a popular referendum.

GEOGRAPHY

Location: Northern Africa, bordering the North Atlantic Ocean and the Mediterranean Sea, between Algeria and Western Sahara
Geographic coordinates: 32 00 N, 5 00 W
Map references: Africa
Area: *total:* 446,550 sq km
country comparison to the world: 57
land: 446,300 sq km
water: 250 sq km
Area—comparative: slightly larger than California
Land boundaries: *total:* 2,017.9 km
border countries: Algeria 1,559 km, Western Sahara 443 km, Spain (Ceuta) 6.3 km, Spain (Melilla) 9.6 km
Coastline: 1,835 km
Maritime claims: *territorial sea:* 12 nm
contiguous zone: 24 nm
exclusive economic zone: 200 nm
continental shelf: 200 m depth or to the depth of exploitation
Climate: Mediterranean, becoming more extreme in the interior
Terrain: northern coast and interior are mountainous with large areas of bordering plateaus, intermontane valleys, and rich coastal plains
Elevation extremes: *lowest point:* Sebkha Tah -55 m
highest point: Jebel Toubkal 4,165 m
Natural resources: phosphates, iron ore, manganese, lead, zinc, fish, salt
Land use: *arable land:* 19%
permanent crops: 2%
other: 79% (2005)
Irrigated land: 14,450 sq km (2003)
Total renewable water resources: 29 cu km (2003)
Freshwater withdrawal (domestic/industrial/agricultural): *total:* 12.6 cu km/yr (10%/3%/87%)
per capita: 400 cu m/yr (2000)
Natural hazards: northern mountains geologically unstable and subject to earthquakes; periodic droughts
Environment—current issues: land degradation/desertification (soil erosion resulting from farming of marginal areas, overgrazing, destruction of vegetation); water supplies contaminated by raw sewage; siltation of reservoirs; oil pollution of coastal waters
Environment—international agreements: *party to:* Biodiversity, Climate Change, Climate Change-Kyoto Protocol, Desertification, Endangered Species, Hazardous Wastes, Law of the Sea, Marine Dumping, Ozone Layer Protection, Ship Pollution, Wetlands, Whaling
signed, but not ratified: Environmental Modification
Geography—note: strategic location along Strait of Gibraltar

PEOPLE

Population: 31,968,361 (July 2011 est.)
country comparison to the world: 38
Age structure: *0–14 years:* 27.8% (male 4,514,623/female 4,382,487)
15–64 years: 66.1% (male 10,335,931/ female 10,785,380)
65 years and over: 6.1% (male 881,622/ female 1,068,318) (2011 est.)
Median age:
total: 26.9 years
male: 26.3 years
female: 27.4 years (2011 est.)
Population growth rate: 1.067% (2011 est.)
country comparison to the world: 112
Birth rate: 19.19 births/1,000 population (2011 est.)
country comparison to the world: 100
Death rate: 4.75 deaths/1,000 population (July 2011 est.)
country comparison to the world: 195
Net migration rate: -3.77 migrant(s)/1,000 population (2011 est.)
country comparison to the world: 187
Urbanization: *urban population:* 58% of total population (2010)
rate of urbanization: 2.1% annual rate of change (2010-15 est.)
Major cities—population: Casablanca 3.245 million; RABAT (capital) 1.77 million; Fes 1.044 million; Marrakech 909,000; Tangier 768,000 (2009)
Sex ratio: *at birth:* 1.05 male(s)/female
under 15 years: 1.03 male(s)/female
15–64 years: 0.96 male(s)/female
65 years and over: 0.83 male(s)/female
total population: 0.97 male(s)/female (2011 est.)
Infant mortality rate: *total:* 27.53 deaths/1,000 live births
country comparison to the world: 74
male: 32.32 deaths/1,000 live births
female: 22.51 deaths/1,000 live births (2011 est.)
Life expectancy at birth: *total population:* 75.9 years
country comparison to the world: 78
male: 72.84 years
female: 79.11 years (2011 est.)
Total fertility rate: 2.21 children born/ woman (2011 est.)
country comparison to the world: 105
HIV/AIDS—adult prevalence rate: 0.1% (2009 est.)
country comparison to the world: 147
HIV/AIDS—people living with HIV/AIDS: 26,000 (2009 est.)
country comparison to the world: 72
HIV/AIDS—deaths: 1,200 (2009 est.)
country comparison to the world: 63
Drinking water source: *Improved:*
urban: 98% of population
rural: 60% of population
total: 81% of population
Unimproved:
urban: 2% of population
rural: 40% of population
total: 19% of population (2008)
Sanitation facility access: *Improved:*
urban: 83% of population
rural: 52% of population
total: 69% of population
Unimproved:
urban: 17% of population
rural: 48% of population
total: 31% of population (2008)
Nationality: *noun:* Moroccan(s)
adjective: Moroccan
Ethnic groups: Arab-Berber 99.1%, other 0.7%, Jewish 0.2%
Religions: Muslim 98.7%, Christian 1.1%, Jewish 0.2%
Languages: Arabic (official), Berber dialects, French (often the language of business, government, and diplomacy)
Literacy: *definition:* age 15 and over can read and write
total population: 52.3%
male: 65.7%

female: 39.6% (2004 census)
School life expectancy (primary to tertiary education): *total:* 10 years
male: 11 years
female: 10 years (2007)
Education expenditures: 5.6% of GDP (2008)
country comparison to the world: 41

GOVERNMENT

Country name: *conventional long form:* Kingdom of Morocco
conventional short form: Morocco
local long form: Al Mamlakah al Maghribiyah
local short form: Al Maghrib
Government type: constitutional monarchy
Capital: *name:* Rabat
geographic coordinates: 34 01 N, 6 49 W
time difference: UTC 0 (5 hours ahead of Washington, DC during Standard Time)
daylight saving time: +1 hr, begins first Sunday in April; ends last Sunday in July
Administrative divisions: 15 regions; Grand Casablanca, Chaouia-Ouardigha, Doukkala-Abda, Fes-Boulemane, Gharb-Chrarda-Beni Hssen, Guelmim-Es Smara, Laayoune-Boujdour-Sakia El Hamra, Marrakech-Tensift-Al Haouz, Meknes-Tafilalet, Oriental, Rabat-Sale-Zemmour-Zaer, Souss-Massa-Draa, Tadla-Azilal, Tanger-Tetouan, Taza-Al Hoceima-Taounate
note: Morocco claims the territory of Western Sahara, the political status of which is considered undetermined by the US Government; portions of the regions Guelmim-Es Smara and Laayoune-Boujdour-Sakia El Hamra as claimed by Morocco lie within Western Sahara; Morocco also claims Oued Eddahab-Lagouira, another region that falls entirely within Western Sahara
Independence: 2 March 1956 (from France)
National holiday: Throne Day (accession of King MOHAMMED VI to the throne), 30 July (1999)
Constitution: 10 March 1972; revised 4 September 1992, amended September 1996
note: the amendment of September 1996 created a bicameral legislature
Legal system: mixed legal system of civil law based on French law and Islamic law; judicial review of legislative acts by Supreme Court
International law organization participation: has not submitted an ICJ jurisdiction declaration; non-party state to the ICCt
Suffrage: 18 years of age; universal
Executive branch: *chief of state:* King MOHAMMED VI (since 30 July 1999)
head of government: Prime Minister Abbas EL FASSI (since 19 September 2007)
cabinet: Council of Ministers appointed by the monarch
(For more information visit the World Leaders website)
elections: the monarchy is hereditary; prime minister appointed by the monarch following legislative elections
Legislative branch: bicameral Parliament consists of the Chamber of Counselors (or upper house) (270 seats; members elected indirectly by local councils, professional organizations, and labor syndicates to serve nine-year terms; one-third of the members are elected every three years) and Chamber of Representatives (or lower house) (325 seats; 295 members elected by multi-seat constituencies and 30 from national lists of women; members elected by popular vote to serve five-year terms)
elections: Chamber of Counselors—last held on 3 October 2009 (next to be held in 2012); Chamber of Representatives—last held on 7 September 2007 (next to be held in 2012)
election results: Chamber of Counselors—percent of vote by party—NA; seats by party—NA; Chamber of Representatives—percent of vote by party—NA; seats by party—PI 52, PJD 46, MP 41, RNI 39, USFP 38, UC 27, PPS 17, FFD 9, MDS 9, Al Ahd 8, other 39
Judicial branch: Supreme Court (judges are appointed on the recommendation of the Supreme Council of the Judiciary, presided over by the monarch)
Political parties and leaders: Action Party or PA [Mohammed EL IDRISSI]; Al Ahd (The Covenant) Party [Najib EL OUAZZANI]; Alliance des Libert'es (Alliance of Liberty) or ADL [Ali BELHAJ]; An-Nahj Ad-Dimocrati or An-Nahj [Abdellah EL HARIF]; Authenticity and Modernity Party or PAM [Mohamed Cheikh BIADILLAH, secretary general]; Choura et Istiqlal (Consultation and Independence) Party or PCI [Abdelwahed MAACH]; Citizens' Forces or FC [Abderrahman LAHJOUJI]; Citizenship and Development Initiative or ICD [Mohamed BENHAMOU]; Constitutional Union Party or UC [Mohammed ABIED]; Democratic and Social Movement or MDS [Mahmoud ARCHANE]; Democratic Forces Front or FFD [Touhami EL KHIARI]; Democratic Socialist Vanguard Party or PADS [Ahmed BENJELLOUN]; Democratic Society Party or PSD [Zhor CHEKKAFI]; Democratic Union or UD [Bouazza IKKEN]; Environment and Development Party or PED [Ahmed EL ALAMI]; Istiqlal (Independence) Party or PI [Abbas EL FASSI]; Justice and Development Party or PJD [Abdelilah BENKIRANE]; Labor Party or PT [Abdelkrim BENATIK]; Moroccan Liberal Party or PML [Mohamed ZIANE]; National Democratic Party or PND [Abdallah KADIRI]; National Ittihadi Congress Party or CNI [Abdelmajid BOUZOUBAA]; National Popular Movement or MNP [Mahjoubi AHERDANE]; National Rally of Independents or RNI [Mustapha EL MANSOURI]; National Union of Popular Forces or UNFP [Abdellah IBRAHIM]; Popular Movement or MP [Mohamed LAENSER]; Progress and Socialism Party or PPS [Ismail ALAOUI]; Reform and Development Party or PRD [Abderrahmane EL KOUHEN]; Renaissance and Virtue Party or PRV [Mohamed KHALIDI]; Renewal and Equity Party or PRE [Chakir ACHABAR]; Social Center Party or PSC [Lahcen MADIH]; Socialist Democratic Party or PSD [Aissa OUARDIGHI]; Socialist Union of Popular Forces or USFP [Abdelwahed RADI]; Unified Socialist Left Party or PGSU [Mohamed Ben Said AIT IDDER]
Political pressure groups and leaders: Democratic Confederation of Labor or CDT [Noubir AMAOUI]; General Union of Moroccan Workers or UGTM [Abderrazzak AFILAL]; Moroccan Employers Association or CGEM [Hassan CHAMI]; National Labor Union of Morocco or UNMT [Abdelslam MAATI]; Union of Moroccan Workers or UMT [Mahjoub BENSEDDIK]
International organization participation: ABEDA, AfDB, AFESD, AMF, AMU, CD, EBRD, FAO, G-11, G-77, IAEA, IBRD, ICAO, ICC, ICRM, IDA, IDB, IFAD, IFC, IFRCS, IHO, ILO, IMF, IMO, IMSO, Interpol, IOC, IOM, IPU, ISO, ITSO, ITU, ITUC, LAS, MIGA, MONUSCO, NAM, OAS (observer), OIC, OIF, OPCW, OSCE (partner), Paris Club (associate), PCA, UN, UNCTAD, UNESCO, UNHCR, UNIDO, UNMIS, UNOCI, UNWTO, UPU, WCO, WHO, WIPO, WMO, WTO
Diplomatic representation in the US: *chief of mission:* Ambassador Aziz MEKOUAR
chancery: 1601 21st Street NW, Washington, DC 20009
telephone: [1] (202) 462-7979
FAX: [1] (202) 265-0161
consulate(s) general: New York
Diplomatic representation from the US: *chief of mission:* Ambassador Samuel L. KAPLAN
embassy: 2 Avenue de Mohamed El Fassi, Rabat
mailing address: PSC 74, Box 021, APO AE 09718
telephone: [212] (37) 76 22 65
FAX: [212] (37) 76 56 61
consulate(s) general: Casablanca
Flag description: red with a green pentacle (five-pointed, linear star) known as Sulayman's (Solomon's) seal in the center of the flag; red and green are traditional colors in Arab flags, although the use of red is more commonly associated with the Arab states of the Persian gulf; the pentacle represents the five pillars of Islam and signifies the association between God and the nation; design dates to 1912
National anthem: *name:* "Hymne Cherifien" (Hymn of the Sharif)
lyrics/music: Ali Squalli HOUSSAINI/Leo MORGAN
note: music adopted 1956, lyrics adopted 1970

ECONOMY

Economy—overview: Morocco's market economy benefits from the country's relatively low labor costs and proximity to Europe, which aid key areas of the economy such as agriculture, light manufacturing, tourism, and remittances. Morocco is also the world's largest exporter of phosphate, which has long provided a source of export earnings and economic stability. Economic policies pursued since 2003 by King MOHAMMED VI have brought macroeconomic stability to the country with generally low inflation, improved financial performance, and steady progress in developing the service and industrial sectors. In 2006, Morocco entered a Free Trade Agreement (FTA) with the US, and in 2008 entered into an advanced status in its 2000 Association Agreement with the EU. However, poverty, illiteracy, and unemployment rates remain high. In response to these challenges, King MOHAMMED in 2005 launched a National Initiative for Human Development, a $2 billion program aimed at alleviating poverty and underdevelopment by expanding electricity to rural areas and replacing urban slums with public and subsidized housing, among other policies. Morocco's trade and budget deficits widened in 2010, and reducing government spending and adapting to sluggish economic growth in Europe will be challenges in 2011. Morocco's long-term challenges include improving education and job prospects for young Moroccans, closing the disparity in wealth between the rich and the poor, confronting corruption, and expanding and diversifying exports beyond phosphates and low-value-added products.

GDP (purchasing power parity): $151.4 billion (2010 est.)
country comparison to the world: 58
$146.8 billion (2009 est.)
$139.9 billion (2008 est.)
note: data are in 2010 US dollars

GDP (official exchange rate): $103.5 billion (2010 est.)

GDP—real growth rate: 3.2% (2010 est.)
country comparison to the world: 119
4.9% (2009 est.)
5.6% (2008 est.)

GDP—per capita (PPP): $4,800 (2010 est.)
country comparison to the world: 151
$4,700 (2009 est.)
$4,500 (2008 est.)
note: data are in 2010 US dollars

GDP—composition by sector:
agriculture: 17.1%
industry: 31.6%
services: 51.4% (2010 est.)

Labor force: 11.63 million (2010 est.)
country comparison to the world: 45

Labor force—by occupation:
agriculture: 44.6%
industry: 19.8%
services: 35.5% (2006 est.)

Unemployment rate: 9.8% (2010 est.)
country comparison to the world: 108
9.1% (2009 est.)

Population below poverty line: 15% (2007 est.)

Household income or consumption by percentage share: *lowest 10%:* 2.7%
highest 10%: 33.2% (2007)

Distribution of family income—Gini index: 40.9 (2005 est.)
country comparison to the world: 57
39.5 (1999 est.)

Investment (gross fixed): 30.2% of GDP (2010 est.)
country comparison to the world: 18

Budget: *revenues:* $23.42 billion
expenditures: $27.08 billion (2010 est.)

Public debt: 58.2% of GDP (2010 est.)
country comparison to the world: 39
56.9% of GDP (2009 est.)

Inflation rate (consumer prices): 2.5% (2010 est.)
country comparison to the world: 66
1% (2009 est.)

Central bank discount rate: 3.31% (31 December 2009)
country comparison to the world: 107
3.32% (31 December 2008)

Commercial bank prime lending rate: 6.5% (31 December 2008)

Stock of narrow money: $67.33 billion (31 December 2010 est.)
country comparison to the world: 40
$64.58 billion (31 December 2009 est.)

Stock of broad money: $108.7 billion (31 December 2009)
country comparison to the world: 51
$99.5 billion (31 December 2008)

Stock of domestic credit: $93.21 billion (31 December 2010 est.)
country comparison to the world: 53
$91.83 billion (31 December 2009 est.)

Market value of publicly traded shares: $62.91 billion (31 December 2009)
country comparison to the world: 48
$65.75 billion (31 December 2008)
$75.49 billion (31 December 2007)

Agriculture—products: barley, wheat, citrus fruits, grapes, vegetables, olives; livestock; wine

Industries: phosphate rock mining and processing, food processing, leather goods, textiles, construction, energy, tourism

Industrial production growth rate: 4.4% (2010 est.)
country comparison to the world: 78

Electricity—production: 19.78 billion kWh (2008 est.)
country comparison to the world: 70

Electricity—consumption: 20.78 billion kWh (2007 est.)
country comparison to the world: 68

Electricity—exports: 0 kWh (2008 est.)

Electricity—imports: 3.429 billion kWh (2009 est.)

Oil—production: 4,053 bbl/day (2009 est.)
country comparison to the world: 98

Oil—consumption: 187,000 bbl/day (2009 est.)
country comparison to the world: 57

Oil—exports: 17,420 bbl/day (2007 est.)
country comparison to the world: 90

Oil—imports: 195,800 bbl/day (2007 est.)
country comparison to the world: 45

Oil—proved reserves: 100 million bbl (1 January 2010 est.)
country comparison to the world: 69

Natural gas—production: 60 million cu m (2008 est.)
country comparison to the world: 80

Natural gas—consumption: 560 million cu m (2008 est.)
country comparison to the world: 94

Natural gas—exports: 0 cu m (2008 est.)
country comparison to the world: 141

Natural gas—imports: 500 million cu m (2008 est.)
country comparison to the world: 59

Natural gas—proved reserves: 1.501 billion cu m (1 January 2010 est.)
country comparison to the world: 96

Current account balance: $-7.922 billion (2010 est.)
country comparison to the world: 175
$-4.958 billion (2009 est.)

Exports: $14.49 billion (2010 est.)
country comparison to the world: 76
$13.92 billion (2009 est.)

Exports—commodities: clothing and textiles, electric components, inorganic chemicals, transistors, crude minerals, fertilizers (including phosphates), petroleum products, citrus fruits, vegetables, fish

Exports—partners: Spain 22.02%, France 20.22%, India 4.91%, Italy 4% (2009)

Imports: $34.19 billion (2010 est.)
country comparison to the world: 54
$30.55 billion (2009 est.)

Imports—commodities: crude petroleum, textile fabric, telecommunications equipment, wheat, gas and electricity, transistors, plastics

Imports—partners: France 16.95%, Spain 14.72%, China 7.1%, Italy 6.76%, Germany 6.28%, US 5.66%, Saudi Arabia 5.11% (2009)

Reserves of foreign exchange and gold: $24.57 billion (31 December 2010 est.)
country comparison to the world: 40
$23.58 billion (31 December 2009 est.)

Debt—external: $22.69 billion (31 December 2010 est.)
country comparison to the world: 69
$21.12 billion (31 December 2009 est.)

Stock of direct foreign investment—at home: $42.19 billion (31 December 2010 est.)
country comparison to the world: 57
$40.72 billion (31 December 2009 est.)

Stock of direct foreign investment—abroad: $1.047 billion (31 December 2010 est.)
country comparison to the world: 74
$1.333 billion (31 December 2009 est.)

Exchange rates: Moroccan dirhams (MAD) per US dollar–8.3619 (2010) 8.0571 (2009) 7.526 (2008) 8.3563 (2007) 8.7722 (2006)

COMMUNICATIONS

Telephones—main lines in use: 3.516 million (2009)
country comparison to the world: 46

Telephones—mobile cellular: 25.311 million (2009)
country comparison to the world: 35

Telephone system: *general assessment:* good system composed of open-wire lines, cables, and microwave radio relay links; principal switching centers are Casablanca and Rabat; national network nearly 100% digital using fiber-optic links; improved rural service employs microwave radio relay; Internet available but expensive
domestic: fixed-line teledensity is roughly 10 per 100 persons; mobile-cellular subscribership approached 75 per 100 persons in 2009
international: country code–212; landing point for the Atlas Offshore, Estepona-Tetouan, Euroafrica, Spain-Morocco, and SEA-ME-WE-3 fiber-optic telecommunications undersea cables that provide connectivity to Asia, the Middle East, and Europe; satellite earth stations–2 Intelsat (Atlantic

Ocean) and 1 Arabsat; microwave radio relay to Gibraltar, Spain, and Western Sahara; coaxial cable and microwave radio relay to Algeria; participant in Medarabtel; fiber-optic cable link from Agadir to Algeria and Tunisia (2009)
Broadcast media: 2 television broadcast networks with state-run Radio-Television Marocaine (RTM) operating one network and the state partially owning the other; foreign TV broadcasts are available via satellite dish; 3 radio broadcast networks with RTM operating one; the government-owned network includes 10 regional radio channels in addition to its national service (2007)
Internet country code: .ma
Internet hosts: 277,793 (2010)
country comparison to the world: 61
Internet users: 13.213 million (2009)
country comparison to the world: 29

TRANSPORTATION

Airports: 58 (2010)
country comparison to the world: 82
Airports—with paved runways: *total:* 32
over 3,047 m: 11
2,438 to 3,047 m: 7
1,524 to 2,437 m: 10
914 to 1,523 m: 4 (2010)
Airports—with unpaved runways: *total:* 26
2,438 to 3,047 m: 1
1,524 to 2,437 m: 7
914 to 1,523 m: 10
under 914 m: 8 (2010)
Heliports: 1 (2010)
Pipelines: gas 830 km; oil 439 km (2010)
Railways: *total:* 2,067 km
country comparison to the world: 70
standard gauge: 2,067 km 1.435-m gauge (1,022 km electrified) (2010)
Roadways: *total:* 57,625 km
country comparison to the world: 77
paved: 35,664 km (includes 639 km of expressways)
unpaved: 21,961 km (2006)
Merchant marine: *total:* 30
country comparison to the world: 83
by type: cargo 2, chemical tanker 2, container 7, passenger/cargo 15, petroleum tanker 1, roll on/roll off 3
foreign-owned: 6 (France 4, Germany 2)
registered in other countries: 5 (Gibraltar 4, Panama 1) (2010)
Ports and terminals: Casablanca, Jorf Lasfar, Mohammedia, Safi, Tangier

MILITARY

Military branches: Royal Armed Forces (Forces Armees Royales, FAR): Royal Moroccan Army (includes Air Defense), Royal Moroccan Navy (includes Coast Guard, Marines), Royal Moroccan Air Force (Al Quwwat al Jawyiya al Malakiya Marakishiya; Force Aerienne Royale Marocaine) (2010)
Military service age and obligation: 18 years of age for voluntary military service; service obligation–18 months (2010)
Manpower available for military service: *males age 16–49:* 8,252,682
females age 16–49: 8,691,419 (2010 est.)
Manpower fit for military service: *males age 16–49:* 7,026,016
females age 16–49: 7,377,045 (2010 est.)
Manpower reaching militarily significant age annually: *male:* 300,327
female: 298,366 (2010 est.)
Military expenditures: 5% of GDP (2003 est.)
country comparison to the world: 16

TRANSNATIONAL ISSUES

Disputes—international: claims and administers Western Sahara whose sovereignty remains unresolved; Morocco protests Spain's control over the coastal enclaves of Ceuta, Melilla, and Penon de Velez de la Gomera, the islands of Penon de Alhucemas and Islas Chafarinas, and surrounding waters; both countries claim Isla Perejil (Leila Island); discussions have not progressed on a comprehensive maritime delimitation, setting limits on resource exploration and refugee interdiction, since Morocco's 2002 rejection of Spain's unilateral designation of a median line from the Canary Islands; Morocco serves as one of the primary launching areas of illegal migration into Spain from North Africa; Algeria's border with Morocco remains an irritant to bilateral relations, each nation accusing the other of harboring militants and arms smuggling; the FLN's assertions of a claim to Chirac Pastures in southeastern Morocco is a dormant dispute
Illicit drugs: one of the world's largest producers of illicit hashish; shipments of hashish mostly directed to Western Europe; transit point for cocaine from South America destined for Western Europe; significant consumer of cannabis

MOZAMBIQUE

INTRODUCTION

Background: Almost five centuries as a Portuguese colony came to a close with independence in 1975. Large-scale emigration, economic dependence on South Africa, a severe drought, and a prolonged civil war hindered the country's development until the mid 1990s. The ruling Front for the Liberation of Mozambique (Frelimo) party formally abandoned Marxism in 1989, and a new constitution the following year provided for multiparty elections and a free market economy. A UN-negotiated peace agreement between Frelimo and rebel Mozambique National Resistance (Renamo) forces ended the fighting in 1992. In December 2004, Mozambique underwent a delicate transition as Joaquim CHISSANO stepped down after 18 years in office. His elected successor, Armando Emilio GUEBUZA, promised to continue the sound economic policies that have encouraged foreign investment. President GUEBUZA was reelected to a second term in October 2009. However, the elections were flawed by voter fraud, questionable disqualification of candidates, and Frelimo use of government resources during the campaign. As a result, Freedom House removed Mozambique from its list of electoral democracies.

GEOGRAPHY

Location: Southeastern Africa, bordering the Mozambique Channel, between South Africa and Tanzania
Geographic coordinates: 18 15 S, 35 00 E
Map references: Africa
Area: *total:* 799,380 sq km
country comparison to the world: 35
land: 786,380 sq km
water: 13,000 sq km
Area—comparative: slightly less than twice the size of California
Land boundaries: *total:* 4,571 km
border countries: Malawi 1,569 km, South Africa 491 km, Swaziland 105 km, Tanzania 756 km, Zambia 419 km, Zimbabwe 1,231 km
Coastline: 2,470 km
Maritime claims: *territorial sea:* 12 nm
exclusive economic zone: 200 nm
Climate: tropical to subtropical
Terrain: mostly coastal lowlands, uplands in center, high plateaus in northwest, mountains in west
Elevation extremes: *lowest point:* Indian Ocean 0 m
highest point: Monte Binga 2,436 m
Natural resources: coal, titanium, natural gas, hydropower, tantalum, graphite
Land use: *arable land:* 5.43%
permanent crops: 0.29%
other: 94.28% (2005)
Irrigated land: 1,180 sq km (2003)
Total renewable water resources: 216 cu km (1992)
Freshwater withdrawal (domestic/industrial/agricultural): *total:* 0.63 cu km/yr (11%/2%/87%)
per capita: 32 cu m/yr (2000)
Natural hazards: severe droughts; devastating cyclones and floods in central and southern provinces
Environment—current issues: a long civil war and recurrent drought in the hinterlands have resulted in increased migration of the population to urban and coastal areas with adverse environmental consequences; desertification; pollution of surface and coastal waters; elephant poaching for ivory is a problem
Environment—international agreements: *party to:* Biodiversity, Climate Change, Climate Change-Kyoto Protocol, Desertification, Endangered Species, Hazardous Wastes, Law of the Sea, Ozone Layer Protection, Ship Pollution, Wetlands
signed, but not ratified: none of the selected agreements
Geography—note: the Zambezi flows through the north-central and most fertile part of the country

PEOPLE

Population: 22,948,858 (July 2011 est.)
country comparison to the world: 51
note: estimates for this country explicitly take into account the effects of excess mortality due to AIDS; this can result in lower life expectancy, higher infant mortality, higher death rates, lower population growth rates, and changes in the distribution of population by age and sex than would otherwise be expected; the 1997 Mozambican census reported a population of 16,099,246
Age structure: *0–14 years:* 45.9% (male 5,295,776/female 5,245,485)
15–64 years: 51.1% (male 5,550,501/female 6,174,668)
65 years and over: 3% (male 313,892/female 368,536) (2011 est.)
Median age: *total:* 16.8 years
male: 16.1 years
female: 17.4 years (2011 est.)
Population growth rate: 2.444% (2011 est.)
country comparison to the world: 30
Birth rate: 39.62 births/1,000 population (2011 est.)
country comparison to the world: 12
Death rate: 13 deaths/1,000 population (July 2011 est.)
country comparison to the world: 22
Net migration rate: -2.18 migrant(s)/1,000 population (2011 est.)
country comparison to the world: 168
Urbanization: *urban population:* 38% of total population (2010)
rate of urbanization: 4% annual rate of change (2010-15 est.)
Major cities—population: MAPUTO (capital) 1.589 million; Matola 761,000 (2009)
Sex ratio: *at birth:* 1.017 male(s)/female
under 15 years: 1.01 male(s)/female
15–64 years: 0.96 male(s)/female
65 years and over: 0.71 male(s)/female
total population: 0.98 male(s)/female (2011 est.)
Infant mortality rate: *total:* 78.95 deaths/1,000 live births
country comparison to the world: 12
male: 81.18 deaths/1,000 live births
female: 76.68 deaths/1,000 live births (2011 est.)
Life expectancy at birth: *total population:* 51.78 years
country comparison to the world: 210
male: 51.01 years
female: 52.57 years (2011 est.)
Total fertility rate: 5.46 children born/woman (2011 est.)
country comparison to the world: 11
HIV/AIDS—adult prevalence rate: 11.5% (2009 est.)
country comparison to the world: 8
HIV/AIDS—people living with HIV/AIDS: 1.4 million (2009 est.)
country comparison to the world: 6
HIV/AIDS—deaths: 74,000 (2009 est.)
country comparison to the world: 7
Major infectious diseases: *degree of risk:* very high
food or waterborne diseases: bacterial and protozoal diarrhea, hepatitis A, and typhoid fever
vectorborne diseases: malaria and plague
water contact disease: schistosomiasis
animal contact disease: rabies (2009)
Drinking water source: *Improved:*
urban: 77% of population
rural: 29% of population
total: 47% of population
Unimproved:
urban: 23% of population
rural: 71% of population
total: 53% of population (2008)
Sanitation facility access:
Improved:
urban: 38% of population
rural: 4% of population
total: 17% of population
Unimproved:
urban: 62% of population
rural: 96% of population
total: 83% of population (2008)
Nationality: *noun:* Mozambican(s)
adjective: Mozambican
Ethnic groups: African 99.66% (Makhuwa, Tsonga, Lomwe, Sena, and others), Europeans 0.06%, Euro-Africans 0.2%, Indians 0.08%
Religions: Catholic 23.8%, Muslim 17.8%, Zionist Christian 17.5%, other 17.8%, none 23.1% (1997 census)
Languages: Emakhuwa 26.1%, Xichangana 11.3%, Portuguese 8.8% (official, spoken by 27% of population as a second language), Elomwe 7.6%, Cisena 6.8%, Echuwabo 5.8%, other Mozambican languages 32%, other foreign languages 0.3%, unspecified 1.3% (1997 census)
Literacy: *definition:* age 15 and over can read and write
total population: 47.8%
male: 63.5%
female: 32.7% (2003 est.)
School life expectancy (primary to tertiary education): *total:* 9 years
male: 10 years
female: 8 years (2007)
Education expenditures: 5% of GDP (2006)
country comparison to the world: 59

GOVERNMENT

Country name: *conventional long form:* Republic of Mozambique
conventional short form: Mozambique
local long form: Republica de Mocambique
local short form: Mocambique
former: Portuguese East Africa
Government type: republic
Capital: *name:* Maputo
geographic coordinates: 25 57 S, 32 35 E
time difference: UTC+2 (7 hours ahead of Washington, DC during Standard Time)
Administrative divisions: 10 provinces (provincias, singular–provincia), 1 city (cidade)*; Cabo Delgado, Gaza, Inhambane, Manica, Maputo, Cidade de Maputo*, Nampula, Niassa, Sofala, Tete, Zambezia
Independence: 25 June 1975 (from Portugal)
National holiday: Independence Day, 25 June (1975)
Constitution: 30 November 1990
Legal system: mixed legal system of Portuguese civil law, Islamic law, and customary law
International law organization participation: has not submitted an ICJ jurisdiction declaration; non-party state to the ICCt
Suffrage: 18 years of age; universal
Executive branch: *chief of state:* President Armando GUEBUZA (since 2 February 2005)
head of government: Prime Minister Aires Bonifacio ALI (since 16 January 2010)
cabinet: Cabinet
(For more information visit the World Leaders website)
elections: president elected by popular vote for a five-year term (eligible for a second term); election last held on 28 October 2009 (next to be held in 2014); prime minister appointed by the president
election results: Armando GUEBUZA reelected president; percent of vote—Armando GUEBUZA 76.3%, Afonso DHLAKAMA 14.9%, Daviz SIMANGO 8.8%
Legislative branch: unicameral Assembly of the Republic or Assembleia da Republica (250 seats; members directly elected by popular vote to serve five-year terms)
elections: last held on 28 October 2009 (next to be held in 2014)
election results: percent of vote by party—FRELIMO 74.7%, RENAMO 17.7%, MDM 3.9%, other 3.7%; seats by party—FRELIMO 191, RENAMO 51, MDM 8
Judicial branch: Supreme Court (the court of final appeal; some of its professional judges are appointed by the president, and some are elected by the Assembly); other courts include an Administrative Court, Constitutional Court, customs courts,

maritime courts, courts marshal, labor courts

Political parties and leaders: Democratic Movement of Mozambique (Movimento Democratico de Mocambique) or MDM [Daviz SIMANGO]; Front for the Liberation of Mozambique (Frente de Liberatacao de Mocambique) or FRELIMO [Armando Emilio GUEBUZA]; Mozambique National Resistance (Resistencia Nacional Mocambicana) or RENAMO [Afonso DHLAKAMA]

Political pressure groups and leaders: Mozambican League of Human Rights (Liga Mocambicana dos Direitos Humanos) or LDH [Alice MABOTE, president]

International organization participation: ACP, AfDB, AU, C, CPLP, FAO, G-77, IAEA, IBRD, ICAO, ICRM, IDA, IDB, IFAD, IFC, IFRCS, IHO, ILO, IMF, IMO, IMSO, Interpol, IOC, IOM (observer), IPU, ISO (correspondent), ITSO, ITU, ITUC, MIGA, MONUSCO, NAM, OIC, OIF (observer), OPCW, SADC, UN, UNCTAD, UNESCO, UNHCR, UNIDO, Union Latina, UNWTO, UPU, WCO, WFTU, WHO, WIPO, WMO, WTO

Diplomatic representation in the US: *chief of mission:* Ambassador Amelia Matos SUMBANA
chancery: 1525 New Hampshire Avenue NW, Washington, DC 20036
telephone: [1] (202) 293-7146
FAX: [1] (202) 835-0245

Diplomatic representation from the US: *chief of mission:* Ambassador Leslie V. ROWE
embassy: Avenida Kenneth Kuanda 193, Maputo
mailing address: P. O. Box 783, Maputo
telephone: [258] (21) 492797
FAX: [258] (21) 490114

Flag description: three equal horizontal bands of green (top), black, and yellow with a red isosceles triangle based on the hoist side; the black band is edged in white; centered in the triangle is a yellow five-pointed star bearing a crossed rifle and hoe in black superimposed on an open white book; green represents the riches of the land, white peace, black the African continent, yellow the country's minerals, and red the struggle for independence; the rifle symbolizes defense and vigilance, the hoe refers to the country's agriculture, the open book stresses the importance of education, and the star represents Marxism and internationalism

National anthem: *name:* "Patria Amada" (Lovely Fatherland)
lyrics/music: Salomao J. MANHICA/ unknown
note: adopted 2002

ECONOMY

Economy—overview: At independence in 1975, Mozambique was one of the world's poorest countries. Socialist mismanagement and a brutal civil war from 1977-92 exacerbated the situation. In 1987, the government embarked on a series of macroeconomic reforms designed to stabilize the economy. These steps, combined with donor assistance and with political stability since the multi-party elections in 1994, have led to dramatic improvements in the country's growth rate. Fiscal reforms, including the introduction of a value-added tax and reform of the customs service, have improved the government's revenue collection abilities. In spite of these gains, Mozambique remains dependent upon foreign assistance for more than half of its annual budget, and the majority of the population remains below the poverty line. Subsistence agriculture continues to employ the vast majority of the country's work force and smallholder agricultural productivity and productivity growth is weak. A substantial trade imbalance persists although the opening of the Mozal aluminum smelter, the country's largest foreign investment project to date, has increased export earnings. At the end of 2007, and after years of negotiations, the government took over Portugal's majority share of the Cahora Bassa Hydroelectricity (HCB) company, a dam that was not transferred to Mozambique at independence because of the ensuing civil war and unpaid debts. More electrical power capacity is needed for additional investment projects in titanium extraction and processing and garment manufacturing that could further close the import/export gap. Mozambique's once substantial foreign debt has been reduced through forgiveness and rescheduling under the IMF's Heavily Indebted Poor Countries (HIPC) and Enhanced HIPC initiatives, and is now at a manageable level. In July 2007 the Millennium Challenge Corporation (MCC) signed a compact with Mozambique; the compact entered into force in September 2008 and will continue for five years. Compact projects will focus on improving sanitation, roads, agriculture, and the business regulation environment in an effort to spur economic growth in the four northern provinces of the country. Mozambique grew at an average annual rate of 9% in the decade up to 2007, one of Africa's strongest performances. However, heavy reliance on aluminum, which accounts for about one-third of exports, subjects the economy to volatile international prices. The sharp decline in aluminum prices during the global economic crisis lowered GDP growth by several percentage points. Despite 8.3% GDP growth in 2010, the increasing cost of living prompted citizens to riot in September 2010, after fuel, water, electricity, and bread price increases were announced. In an attempt to contain the cost of living, the government implemented subsidies, decreased taxes and tariffs, and instituted other fiscal measures.

GDP (purchasing power parity): $21.81 billion (2010 est.)
country comparison to the world: 122
$20.38 billion (2009 est.)
$19.17 billion (2008 est.)
note: data are in 2010 US dollars

GDP (official exchange rate): $9.893 billion (2010 est.)

GDP—real growth rate: 7% (2010 est.)
country comparison to the world: 34
6.3% (2009 est.)
6.8% (2008 est.)

GDP—per capita (PPP): $1,000 (2010 est.)
country comparison to the world: 211
$900 (2009 est.)
$900 (2008 est.)
note: data are in 2010 US dollars

GDP—composition by sector: *agriculture:* 28.8%
industry: 26%
services: 45.2% (2010 est.)

Labor force: 9.87 million (2010 est.)
country comparison to the world: 50

Labor force—by occupation: *agriculture:* 81%
industry: 6%
services: 13% (1997 est.)

Unemployment rate: 21% (1997 est.)
country comparison to the world: 168

Population below poverty line: 70% (2001 est.)

Household income or consumption by percentage share: *lowest 10%:* 2.1%
highest 10%: 39.2% (2003)

Distribution of family income—Gini index: 47.3 (2002)
country comparison to the world: 31
39.6 (1997)

Investment (gross fixed): 17.2% of GDP (2010 est.)
country comparison to the world: 119

Budget: *revenues:* $2.346 billion
expenditures: $2.898 billion (2010 est.)

Public debt: 40.8% of GDP (2010 est.)
country comparison to the world: 68
33.4% of GDP (2009 est.)

Inflation rate (consumer prices): 13.5% (2010 est.)
country comparison to the world: 216
3.3% (2009 est.)

Central bank discount rate: 9.95% (31 December 2009)
country comparison to the world: 44
9.95% (31 December 2008)

Commercial bank prime lending rate: 15.68% (31 December 2009 est.)
country comparison to the world: 24
18.31% (31 December 2008 est.)

Stock of narrow money: $2.657 billion (31 December 2010 est.)
country comparison to the world: 113
$2.812 billion (31 December 2009 est.)

Stock of broad money: $3.803 billion (31 December 2010 est.)
country comparison to the world: 128
$4.074 billion (31 December 2009 est.)

Stock of domestic credit: $2.74 billion (31 December 2010 est.)
country comparison to the world: 123
$2.311 billion (31 December 2009 est.)

Market value of publicly traded shares: $NA

Agriculture—products: cotton, cashew nuts, sugarcane, tea, cassava (tapioca), corn, coconuts, sisal, citrus and tropical fruits, potatoes, sunflowers; beef, poultry

Industries: food, beverages, chemicals (fertilizer, soap, paints), aluminum, petroleum products, textiles, cement, glass, asbestos, tobacco

Industrial production growth rate: 8% (2010 est.)
country comparison to the world: 36

Electricity—production: 15.91 billion kWh (2007 est.)
country comparison to the world: 77

Electricity—consumption: 10.16 billion kWh (2007 est.)
country comparison to the world: 86
Electricity—exports: 11.82 billion kWh (2007 est.)
Electricity—imports: 8.278 billion kWh (2007 est.)
Oil—production: 0 bbl/day (2009 est.)
country comparison to the world: 209
Oil—consumption: 18,000 bbl/day (2009 est.)
country comparison to the world: 129
Oil—exports: 0 bbl/day (2007 est.)
country comparison to the world: 189
Oil—imports: 13,760 bbl/day (2007 est.)
country comparison to the world: 133
Oil—proved reserves: 0 bbl (1 January 2010 est.)
country comparison to the world: 167
Natural gas—production: 3.3 billion cu m (2008 est.)
country comparison to the world: 52
Natural gas—consumption: 100 million cu m (2008 est.)
country comparison to the world: 103
Natural gas—exports: 3.2 billion cu m (2008 est.)
country comparison to the world: 33
Natural gas—imports: 0 cu m (2008 est.)
country comparison to the world: 92
Natural gas—proved reserves: 127.4 billion cu m (1 January 2010 est.)
country comparison to the world: 50
Current account balance: $-1.028 billion (2010 est.)
country comparison to the world: 133
$-866 million (2009 est.)
Exports:
$2.517 billion (2010 est.)
country comparison to the world: 127
$1.947 billion (2009 est.)
Exports—commodities: aluminum, prawns, cashews, cotton, sugar, citrus, timber; bulk electricity
Exports—partners: Netherlands 47.62%, South Africa 11.6% (2009)
Imports: $3.527 billion (2010 est.)
country comparison to the world: 135
$3.059 billion (2009 est.)
Imports—commodities: machinery and equipment, vehicles, fuel, chemicals, metal products, foodstuffs, textiles
Imports—partners: South Africa 33.54%, Netherlands 8.42%, India 5.93%, China 4.24% (2009)
Reserves of foreign exchange and gold: $1.982 billion (31 December 2010 est.)
country comparison to the world: 104
$1.829 billion (31 December 2009 est.)
Debt—external: $4.99 billion (31 December 2010 est.)
country comparison to the world: 108
$4.246 billion (31 December 2009 est.)
Exchange rates: meticais (MZM) per US dollar–35 (2010) 26.28 (2009) 24.125 (2008) 26.264 (2007) 25.4 (2006)

COMMUNICATIONS

Telephones—main lines in use: 82,400 (2009)
country comparison to the world: 148
Telephones—mobile cellular: 5.971 million (2009)
country comparison to the world: 87
Telephone system: *general assessment:* a fair telecommunications system that is shackled with a heavy state presence, lack of competition, and high operating costs and charges
domestic: stagnation in the fixed-line network contrasts with rapid growth in the mobile-cellular network; mobile-cellular coverage now includes all the main cities and key roads, including those from Maputo to the South African and Swaziland borders, the national highway through Gaza and Inhambane provinces, the Beira corridor, and from Nampula to Nacala; extremely low fixed-line teledensity; despite significant growth in mobile-cellular services, teledensity remains low at about 25 per 100 persons
international: country code–258; satellite earth stations–5 Intelsat (2 Atlantic Ocean and 3 Indian Ocean); landing point for the SEACOM fiber-optic cable
Broadcast media: 1 state-run TV station supplemented by private TV station; Portuguese state TV's African service, RTP Africa, and Brazilian-owned TV Miramar are available; state-run radio provides nearly 100% territorial coverage and broadcasts in multiple languages; a number of privately-owned and community-operated stations also broadcast; transmissions of multiple international broadcasters are available (2007)
Internet country code: .mz
Internet hosts: 21,172 (2010)
country comparison to the world: 109
Internet users: 613,600 (2009)
country comparison to the world: 113

TRANSPORTATION

Airports: 106 (2010)
country comparison to the world: 54
Airports—with paved runways: *total:* 23
over 3,047 m: 1
2,438 to 3,047 m: 3
1,524 to 2,437 m: 10
914 to 1,523 m: 4
under 914 m: 5 (2010)
Airports—with unpaved runways: *total:* 83
2,438 to 3,047 m: 1
1,524 to 2,437 m: 9
914 to 1,523 m: 34
under 914 m: 39 (2010)
Pipelines: gas 918 km; refined products 278 km (2010)
Railways: *total:* 4,787 km
country comparison to the world: 38
narrow gauge: 4,787 km 1.067-m gauge (2010)
Roadways: *total:* 30,400 km
country comparison to the world: 96
paved: 5,685 km
unpaved: 24,715 km (2000)
Waterways: 460 km (Zambezi River navigable to Tete and along Cahora Bassa Lake) (2010)
country comparison to the world: 86
Merchant marine: *total:* 2
country comparison to the world: 146
by type: cargo 2
foreign-owned: 2 (Belgium 2) (2010)
Ports and terminals: Beira, Maputo, Nacala

MILITARY

Military branches: Mozambique Armed Defense Forces (Forcas Armadas de Defesa de Mocambique, FADM): Mozambique Army, Mozambique Navy (Marinha de Guerra de Mocambique, MGM), Mozambique Air Force (Forca Aerea de Mocambique, FAM) (2011)
Military service age and obligation: registration for military service is mandatory for all males and females at 18 years of age; 18-35 years of age for selective compulsory military service; 18 years of age for voluntary service; 2-year service obligation; women may serve as officers or enlisted (2010)
Manpower available for military service: *males age 16–49:* 4,613,367 (2010 est.)
Manpower fit for military service: *males age 16–49:* 2,677,473
females age 16–49: 2,941,073 (2010 est.)
Manpower reaching militarily significant age annually: *male:* 274,602
female: 280,008 (2010 est.)
Military expenditures: 0.8% of GDP (2006)
country comparison to the world: 149

TRANSNATIONAL ISSUES

Disputes—international: none
Illicit drugs: southern African transit point for South Asian hashish and heroin, and South American cocaine probably destined for the European and South African markets; producer of cannabis (for local consumption) and methaqualone (for export to South Africa); corruption and poor regulatory capability make the banking system vulnerable to money laundering, but the lack of a well-developed financial infrastructure limits the country's utility as a money-laundering center

NAMIBIA

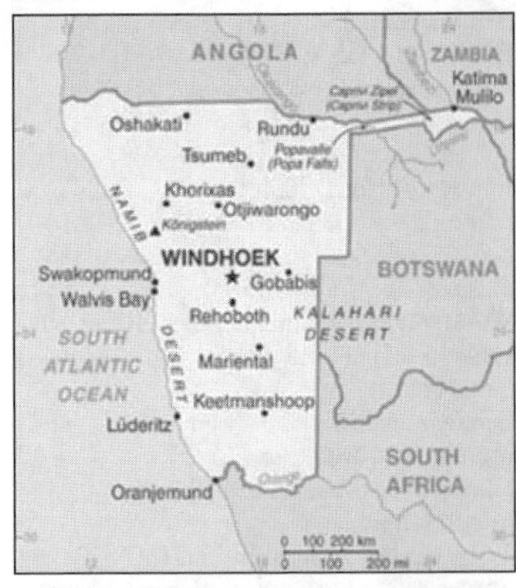

INTRODUCTION

Background: South Africa occupied the German colony of South-West Africa during World War I and administered it as a mandate until after World War II, when it annexed the territory. In 1966 the Marxist South-West Africa People's Organization (SWAPO) guerrilla group launched a war of independence for the area that became Namibia, but it was not until 1988 that South Africa agreed to end its administration in accordance with a UN peace plan for the entire region. Namibia has been governed by SWAPO since the country won independence in 1990. Hifikepunye POHAMBA was elected president in November 2004 in a landslide victory replacing Sam NUJOMA who led the country during its first 14 years of self rule. POHAMBA was reelected in November 2009.

GEOGRAPHY

Location: Southern Africa, bordering the South Atlantic Ocean, between Angola and South Africa
Geographic coordinates: 22 00 S, 17 00 E
Map references: Africa
Area: *total:* 824,292 sq km
country comparison to the world: 34
land: 823,290 sq km
water: 1,002 sq km
Area—comparative: slightly more than half the size of Alaska
Land boundaries: *total:* 3,936 km
border countries: Angola 1,376 km, Botswana 1,360 km, South Africa 967 km, Zambia 233 km
Coastline: 1,572 km
Maritime claims: *territorial sea:* 12 nm
contiguous zone: 24 nm
exclusive economic zone: 200 nm
Climate: desert; hot, dry; rainfall sparse and erratic
Terrain: mostly high plateau; Namib Desert along coast; Kalahari Desert in east
Elevation extremes: *lowest point:* Atlantic Ocean 0 m
highest point: Konigstein 2,606 m
Natural resources: diamonds, copper, uranium, gold, silver, lead, tin, lithium, cadmium, tungsten, zinc, salt, hydropower, fish
note: suspected deposits of oil, coal, and iron ore
Land use: *arable land:* 0.99%
permanent crops: 0.01%
other: 99% (2005)
Irrigated land: 80 sq km (2003)
Total renewable water resources: 45.5 cu km (1991)
Freshwater withdrawal (domestic/industrial/agricultural): *total:* 0.3 cu km/yr (24%/5%/71%)
per capita: 148 cu m/yr (2000)
Natural hazards: prolonged periods of drought
Environment—current issues: limited natural freshwater resources; desertification; wildlife poaching; land degradation has led to few conservation areas
Environment—international agreements: *party to:* Antarctic-Marine Living Resources, Biodiversity, Climate Change, Climate Change-Kyoto Protocol, Desertification, Endangered Species, Hazardous Wastes, Law of the Sea, Ozone Layer Protection, Wetlands
signed, but not ratified: none of the selected agreements
Geography—note: first country in the world to incorporate the protection of the environment into its constitution; some 14% of the land is protected, including virtually the entire Namib Desert coastal strip

PEOPLE

Population: 2,147,585 (July 2011 est.)
country comparison to the world: 142
note: estimates for this country explicitly take into account the effects of excess mortality due to AIDS; this can result in lower life expectancy, higher infant mortality, higher death rates, lower population growth rates, and changes in the distribution of population by age and sex than would otherwise be expected
Age structure: *0–14 years:* 34.2% (male 371,078/female 364,232)
15–64 years: 61.7% (male 671,853/female 652,414)
65 years and over: 4.1% (male 38,851/female 49,157) (2011 est.)
Median age: *total:* 21.7 years
male: 21.7 years
female: 21.8 years (2011 est.)
Population growth rate: 0.873% (2011 est.)
country comparison to the world: 128
Birth rate: 21.48 births/1,000 population (2011 est.)
country comparison to the world: 79
Death rate: 12.95 deaths/1,000 population (July 2011 est.)
country comparison to the world: 24
Net migration rate: 0.2 migrant(s)/1,000 population (2011 est.)
country comparison to the world: 69
Urbanization: *urban population:* 38% of total population (2010)
rate of urbanization: 3.3% annual rate of change (2010-15 est.)
Major cities—population: WINDHOEK (capital) 342,000 (2009)
Sex ratio: *at birth:* 1.03 male(s)/female
under 15 years: 1.02 male(s)/female
15–64 years: 1.03 male(s)/female
65 years and over: 0.8 male(s)/female
total population: 1.01 male(s)/female (2011 est.)
Infant mortality rate: *total:* 45.59 deaths/1,000 live births
country comparison to the world: 53
male: 48.86 deaths/1,000 live births
female: 42.21 deaths/1,000 live births (2011 est.)
Life expectancy at birth: *total population:* 52.19 years
country comparison to the world: 209
male: 52.48 years
female: 51.89 years (2011 est.)
Total fertility rate: 2.49 children born/woman (2011 est.)
country comparison to the world: 83
HIV/AIDS—adult prevalence rate: 13.1% (2009 est.)
country comparison to the world: 7
HIV/AIDS—people living with HIV/AIDS: 180,000 (2009 est.)
country comparison to the world: 31
HIV/AIDS—deaths: 6,700 (2009 est.)
country comparison to the world: 32
Major infectious diseases: *degree of risk:* high
food or waterborne diseases: bacterial diarrhea, hepatitis A, and typhoid fever
vectorborne disease: malaria
water contact disease: schistosomiasis (2009)
Drinking water source: *Improved:*
urban: 99% of population
rural: 88% of population
total: 92% of population
Unimproved:
urban: 1% of population
rural: 12% of population
total: 8% of population (2008)
Sanitation facility access: *Improved:*
urban: 60% of population
rural: 17% of population
total: 33% of population
Unimproved:
urban: 40% of population
rural: 83% of population
total: 67% of population (2008)
Nationality: *noun:* Namibian(s)
adjective: Namibian
Ethnic groups: black 87.5%, white 6%, mixed 6.5%
note: about 50% of the population belong to the Ovambo tribe and 9% to the Kavangos tribe; other ethnic groups include Herero 7%, Damara 7%, Nama 5%, Caprivian 4%, Bushmen 3%, Baster 2%, Tswana 0.5%
Religions: Christian 80% to 90% (Lutheran 50% at least), indigenous beliefs 10% to 20%
Languages: English (official) 7%, Afrikaans (common language of most of the population and about 60% of the white population), German 32%, indigenous languages (includes Oshivambo, Herero, Nama) 1%
Literacy: *definition:* age 15 and over can read and write
total population: 85%
male: 86.8%
female: 83.5% (2001 census)
School life expectancy (primary to tertiary education): *total:* 12 years

male: 12 years
female: 12 years (2008)
Education expenditures: 6.4% of GDP (2008)
country comparison to the world: 23

GOVERNMENT

Country name: *conventional long form:* Republic of Namibia
conventional short form: Namibia
local long form: Republic of Namibia
local short form: Namibia
former: German South-West Africa (Sued-West Afrika), South-West Africa
Government type: republic
Capital: *name:* Windhoek
geographic coordinates: 22 34 S, 17 05 E
time difference: UTC+1 (6 hours ahead of Washington, DC during Standard Time)
daylight saving time: +1hr, begins first Sunday in September; ends first Sunday in April
Administrative divisions: 13 regions; Caprivi, Erongo, Hardap, Karas, Khomas, Kunene, Ohangwena, Okavango, Omaheke, Omusati, Oshana, Oshikoto, Otjozondjupa
Independence: 21 March 1990 (from South African mandate)
National holiday: Independence Day, 21 March (1990)
Constitution: ratified 9 February 1990, effective 12 March 1990
Legal system: mixed legal system of uncodified civil law based on Roman-Dutch law and customary law
International law organization participation: has not submitted an ICJ jurisdiction declaration; accepts ICCt jurisdiction
Suffrage: 18 years of age; universal
Executive branch: *chief of state:* President Hifikepunye POHAMBA (since 21 March 2005)
head of government: Prime Minister Nahas ANGULA (since 21 March 2005)
cabinet: Cabinet appointed by the president from among the members of the National Assembly
(For more information visit the World Leaders website)
elections: president elected by popular vote for a five-year term (eligible for a second term); election last held on 27-28 November 2009 (next to be held in 2014)
election results: Hifikepunye POHAMBA reelected president; percent of vote—Hifikepunye POHAMBA 76.4%, Hidipo HAMUTENYA 11.0%, Katuutire KAURA 3.0%, Kuaima RIRUAKO 2.9%, Justus GAROEB 2.4%, Ignatius SHIXWAMENI 1.3%, Hendrick MUDGE 1.2%, other 1.8%
Legislative branch: bicameral legislature consists of the National Council, primarily an advisory body (26 seats; two members chosen from each regional council to serve six-year terms), and the National Assembly (72 seats; members elected by popular vote to serve five-year terms)
elections: National Council—elections for regional councils to determine members of the National Council held on 26-27 November 2010 (next to be held in 2016); National Assembly—last held on 26-27 November 2009 (next to be held in November 2014)
election results: National Council—percent of vote by party—NA; seats by party—SWAPO 24, UDF 1, DTA 1; National Assembly—percent of vote by party—SWAPO 75.3%, RDP 11.3%, DTA 3.1%, NUDO 3.0%, UDF 2.4%, APP 1.4%, RP 0.8%, COD 0.7%, SWANU 0.6%, other 1.3%; seats by party—SWAPO 54, RDP 8, DTA 2, NUDO 2, UDF 2, APP 1, COD 1, RP 1, SWANU 1
Judicial branch: Supreme Court (judges appointed by the president on the recommendation of the Judicial Service Commission)
Political parties and leaders: All People's Party or APP [Ignatius SHIXWAMENI]; Congress of Democrats or COD [Benjamin ULENGA]; Democratic Turnhalle Alliance of Namibia or DTA [Katuutire KAURA]; Monitor Action Group or MAG [Jurie VILJOEN]; National Democratic Movement for Change or NamDMC; National Unity Democratic Organization or NUDO [Kuaima RIRUAKO]; Rally for Democracy and Progress or RDP [Hidipo HAMUTENYA]; Republican Party or RP [Hendrick MUDGE]; South West Africa National Union or SWANU [Usutuaije MAAMBERUA]; South West Africa People's Organization or SWAPO [Hifikepunye POHAMBA]; United Democratic Front or UDF [Justus GAROEB]
Political pressure groups and leaders: National Society for Human Rights or NSHR (NAMRIGHTS as of 2010); various labor unions
International organization participation: ACP, AfDB, AU, C, FAO, G-77, IAEA, IBRD, ICAO, ICRM, IFAD, IFC, IFRCS, ILO, IMF, IMO, Interpol, IOC, IOM, IPU, ISO, ITSO, ITU, ITUC, MIGA, NAM, OPCW, SACU, SADC, UN, UNAMID, UNCTAD, UNESCO, UNHCR, UNIDO, UNMIL, UNMIS, UNOCI, UNWTO, UPU, WCO, WHO, WIPO, WMO, WTO
Diplomatic representation in the US: *chief of mission:* Ambassador Martin ANDJABA
chancery: 1605 New Hampshire Avenue NW, Washington, DC 20009
telephone: [1] (202) 986-0540
FAX: [1] (202) 986-0443
Diplomatic representation from the US: *chief of mission:* Ambassador Wanda L. NESBITT
embassy: 14 Lossen Street, Windhoek
mailing address: Private Bag 12029 Ausspannplatz, Windhoek
telephone: [264] (61) 295-8500
FAX: [264] (61) 295-8603
Flag description: a wide red stripe edged by narrow white stripes divides the flag diagonally from lower hoist corner to upper fly corner; the upper hoist-side triangle is blue and charged with a yellow, 12-rayed sunburst; the lower fly-side triangle is green; red signifies the heroism of the people and their determination to build a future of equal opportunity for all; white stands for peace, unity, tranquility, and harmony; blue represents the Namibian sky and the Atlantic Ocean, the country's precious water resources and rain; the yellow sun denotes power and existence; green symbolizes vegetation and agricultural resources
National anthem: *name:* "Namibia, Land of the Brave"
lyrics/music: Axali DOESEB
note: adopted 1991

ECONOMY

Economy—overview: The economy is heavily dependent on the extraction and processing of minerals for export. Mining accounts for 8% of GDP, but provides more than 50% of foreign exchange earnings. Rich alluvial diamond deposits make Namibia a primary source for gem-quality diamonds. Namibia is the world's fourth-largest producer of uranium. It also produces large quantities of zinc and is a small producer of gold and other minerals. The mining sector employs only about 3% of the population while about 35-40% of the population depends on subsistence agriculture for its livelihood. Namibia normally imports about 50% of its cereal requirements; in drought years food shortages are a major problem in rural areas. A high per capita GDP, relative to the region, hides one of the world's most unequal income distributions, as shown by Namibia's 70.7 GINI coefficient. The Namibian economy is closely linked to South Africa with the Namibian dollar pegged one-to-one to the South African rand. Until 2010, Namibia drew 40% of its budget revenues from the Southern African Customs Union (SACU). Increased payments from SACU put Namibia's budget into surplus in 2007 for the first time since independence. SACU allotments to Namibia increased in 2009, but will drop for 2010 and 2011 because South Africa went into recession during the global economic crisis, reducing overall SACU income. Increased fish production and mining of zinc, copper, and uranium spurred growth in 2003-08, but growth in recent years was undercut by poor fish catches, a dramatic decline in demand for diamonds, higher costs of producing metals, and the global recession. A rebound in diamond and uranium prices in 2010 provided a significant boost to Namibia's mining sector. Copper mines, which closed in 2008, are slated to reopen in 2011.
GDP (purchasing power parity): $14.6 billion (2010 est.)
country comparison to the world: 138
$13.98 billion (2009 est.)
$14.1 billion (2008 est.)
note: data are in 2010 US dollars
GDP (official exchange rate): $11.87 billion (2010 est.)
GDP—real growth rate: 4.4% (2010 est.)
country comparison to the world: 86
-0.8% (2009 est.)
4.3% (2008 est.)
GDP—per capita (PPP): $6,900 (2010 est.)
country comparison to the world: 132
$6,600 (2009 est.)
$6,700 (2008 est.)
note: data are in 2010 US dollars
GDP—composition by sector: *agriculture:* 9%
industry: 32.7%

services: 58.2% (2010 est.)
Labor force: 729,000 (2010 est.)
country comparison to the world: 147
Labor force—by occupation: *agriculture:* 16.3%
industry: 22.4%
services: 61.3%
note: statistics are for the formal sector only; about half of Namibia's people are unemployed while about two-thirds live in rural areas; roughly two-thirds of rural dwellers rely on subsistence agriculture (2008 est.)
Unemployment rate: 51.2% (2008 est.)
country comparison to the world: 193
36.7% (2004 est.)
Population below poverty line: 55.8%
note: the UNDP's 2005 Human Development Report indicated that 34.9% of the population live on $1 per day and 55.8% live on $2 per day (2005 est.)
Household income or consumption by percentage share: *lowest 10%:* 1.1%
highest 10%: 53% (2008)
Distribution of family income—Gini index: 70.7 (2003)
country comparison to the world: 1
Investment (gross fixed): 24% of GDP (2010 est.)
country comparison to the world: 49
Budget: *revenues:* $2.977 billion
expenditures: $3.817 billion (2010 est.)
Public debt: 20% of GDP (2010 est.)
country comparison to the world: 111
15.1% of GDP (2009 est.)
Inflation rate (consumer prices): 4.6% (2010 est.)
country comparison to the world: 131
8.8% (2009 est.)
Central bank discount rate: 7% (31 December 2009)
country comparison to the world: 42
10% (31 December 2008)
Commercial bank prime lending rate: 11.12% (31 December 2009 est.)
country comparison to the world: 54
13.74% (31 December 2008 est.)
Stock of narrow money: $3.049 billion (31 December 2010 est.)
country comparison to the world: 109
$2.495 billion (31 December 2009 est.)
Stock of broad money: $4.756 billion (31 December 2010 est.)
country comparison to the world: 121
$3.691 billion (31 December 2009 est.)
Stock of domestic credit: $5.122 billion (31 December 2010 est.)
country comparison to the world: 111
$4.041 billion (31 December 2009 est.)
Market value of publicly traded shares: $846.3 million (31 December 2009)
country comparison to the world: 104
$618.7 million (31 December 2008)
$702 million (31 December 2007)
Agriculture—products: millet, sorghum, peanuts, grapes; livestock; fish
Industries: meatpacking, fish processing, dairy products; mining (diamonds, lead, zinc, tin, silver, tungsten, uranium, copper)
Industrial production growth rate: 6.5% (2010 est.)
country comparison to the world: 53
Electricity—production: 1.491 billion kWh (2009 est.)
country comparison to the world: 140
Electricity—consumption: 2.845 billion kWh (2009 est.)
country comparison to the world: 128
Electricity—exports: 40 million kWh (2007 est.)
Electricity—imports: 2.045 billion kWh (2007 est.)
Oil—production: 0 bbl/day (2009 est.)
country comparison to the world: 140
Oil—consumption: 22,000 bbl/day (2009 est.)
country comparison to the world: 120
Oil—exports: 0 bbl/day (2007 est.)
country comparison to the world: 140
Oil—imports: 19,120 bbl/day (2007 est.)
country comparison to the world: 112
Oil—proved reserves: 0 bbl (1 January 2010 est.)
country comparison to the world: 203
Natural gas—production: 0 cu m (2008 est.)
country comparison to the world: 200
Natural gas—consumption: 0 cu m (2008 est.)
country comparison to the world: 136
Natural gas—exports: 0 cu m (2008 est.)
country comparison to the world: 200
Natural gas—imports: 0 cu m (2008 est.)
country comparison to the world: 136
Natural gas—proved reserves: 62.29 billion cu m (1 January 2010 est.)
country comparison to the world: 61
Current account balance: $-187 million (2010 est.)
country comparison to the world: 88
$-160.9 million (2009 est.)
Exports: $4.277 billion (2010 est.)
country comparison to the world: 114
$3.535 billion (2009 est.)
Exports—commodities: diamonds, copper, gold, zinc, lead, uranium; cattle, processed fish, karakul skins
Imports: $5.152 billion (2010 est.)
country comparison to the world: 115
$4.519 billion (2009 est.)
Imports—commodities: foodstuffs; petroleum products and fuel, machinery and equipment, chemicals
Reserves of foreign exchange and gold: $1.961 billion (31 December 2010 est.)
country comparison to the world: 105
$2.051 billion (31 December 2009 est.)
Debt—external: $2.373 billion (31 December 2010 est.)
country comparison to the world: 137
$2.175 billion (31 December 2009 est.)
Stock of direct foreign investment—at home: $NA
Stock of direct foreign investment—abroad: $NA
Exchange rates: Namibian dollars (NAD) per US dollar–7.57 (2010) 8.42 (2009) 7.75 (2008) 7.18 (2007) 6.7649 (2006)

COMMUNICATIONS

Telephones—main lines in use: 142,100 (2009)
country comparison to the world: 135
Telephones—mobile cellular: 1.217 million (2009)
country comparison to the world: 142
Telephone system: *general assessment:* good system; core fiber-optic network links most centers and connections are now digital
domestic: multiple mobile-cellular providers with a combined subscribership of nearly 60 telephones per 100 persons; combined fixed-line and mobile-cellular teledensity about 65 per 100 persons
international: country code–264; fiber-optic cable to South Africa, microwave radio relay link to Botswana, direct links to other neighboring countries; connected to the South African Far East (SAFE) submarine cable through South Africa; satellite earth stations–4 Intelsat (2008)
Broadcast media: 1 private and 1 state-run television station; satellite and cable TV service is available; state-run radio service broadcasts in multiple languages; about a dozen private radio stations operating; transmissions of multiple international broadcasters are available (2007)
Internet country code: .na
Internet hosts: 76,020 (2010)
country comparison to the world: 80
Internet users: 127,500 (2009)
country comparison to the world: 151

TRANSPORTATION

Airports: 129 (2010)
country comparison to the world: 47
Airports—with paved runways: *total:* 21
over 3,047 m: 3
2,438 to 3,047 m: 2
1,524 to 2,437 m: 13
914 to 1,523 m: 3 (2010)
Airports—with unpaved runways: *total:* 108
2,438 to 3,047 m: 1
1,524 to 2,437 m: 25
914 to 1,523 m: 71
under 914 m: 11 (2010)
Railways: *total:* 2,626 km
country comparison to the world: 62
narrow gauge: 2,626 km 1.067-m gauge (2010)
Roadways: *total:* 64,189 km
country comparison to the world: 71
paved: 5,477 km
unpaved: 58,712 km (2010)
Merchant marine: *total:* 1
country comparison to the world: 150
by type: cargo 1 (2010)
Ports and terminals: Luderitz, Walvis Bay

MILITARY

Military branches: Namibian Defense Force (NDF): Army, Navy, Air Force (2010)
Military service age and obligation: 18 years of age for voluntary military service; no conscription (2010)
Manpower available for military service: *males age 16–49:* 568,231 (2010 est.)
Manpower fit for military service: *males age 16–49:* 351,431
females age 16–49: 311,513 (2010 est.)
Manpower reaching militarily significant age annually: *male:* 26,413
female: 26,038 (2010 est.)
Military expenditures: 3.7% of GDP (2006)
country comparison to the world: 31

TRANSNATIONAL ISSUES

Disputes—international: concerns from international experts and local populations over the Okavango Delta ecology in Botswana and human displacement scuttled Namibian plans to construct a hydroelectric dam on Popa Falls along the Angola-Namibia border; managed dispute with South Africa over the location of the boundary in the Orange River; Namibia has supported, and in 2004 Zimbabwe dropped objections to, plans between Botswana and Zambia to build a bridge over the Zambezi River, thereby de facto recognizing a short, but not clearly delimited, Botswana-Zambia boundary in the river

Refugees and internally displaced persons: refugees (country of origin): 4,700 (Angola) (2007)

NAURU

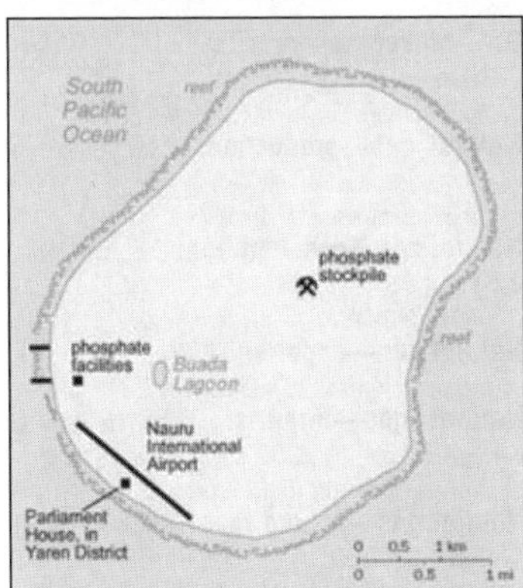

INTRODUCTION

Background: The exact origins of the Nauruans are unclear since their language does not resemble any other in the Pacific. Germany annexed the island in 1888. A German-British consortium began mining the island's phosphate deposits early in the 20th century. Australian forces occupied Nauru in World War I; it subsequently became a League of Nations mandate. After the Second World War–and a brutal occupation by Japan–Nauru became a UN trust territory. It achieved independence in 1968 and joined the UN in 1999 as the world's smallest independent republic.

GEOGRAPHY

Location: Oceania, island in the South Pacific Ocean, south of the Marshall Islands
Geographic coordinates: 0 32 S, 166 55 E
Map references: Oceania
Area: *total:* 21 sq km
country comparison to the world: 238
land: 21 sq km
water: 0 sq km
Area—comparative: about 0.1 times the size of Washington, DC
Land boundaries: 0 km
Coastline: 30 km
Maritime claims: *territorial sea:* 12 nm
contiguous zone: 24 nm
exclusive economic zone: 200 nm
Climate: tropical with a monsoonal pattern; rainy season (November to February)
Terrain: sandy beach rises to fertile ring around raised coral reefs with phosphate plateau in center
Elevation extremes: *lowest point:* Pacific Ocean 0 m
highest point: unnamed elevation along plateau rim 61 m
Natural resources: phosphates, fish
Land use: *arable land:* 0%
permanent crops: 0%
other: 100% (2005)
Irrigated land: NA
Natural hazards: periodic droughts
Environment—current issues: limited natural freshwater resources, roof storage tanks collect rainwater but mostly dependent on a single, aging desalination plant; intensive phosphate mining during the past 90 years–mainly by a UK, Australia, and NZ consortium–has left the central 90% of Nauru a wasteland and threatens limited remaining land resources
Environment—international agreements: *party to:* Biodiversity, Climate Change, Climate Change-Kyoto Protocol, Desertification, Hazardous Wastes, Law of the Sea, Marine Dumping, Ozone Layer Protection, Whaling
signed, but not ratified: none of the selected agreements
Geography—note: Nauru is one of the three great phosphate rock islands in the Pacific Ocean–the others are Banaba (Ocean Island) in Kiribati and Makatea in French Polynesia; only 53 km south of Equator

PEOPLE

Population: 9,322 (July 2011 est.)
country comparison to the world: 224
Age structure: *0–14 years:* 33% (male 1,398/female 1,682)
15–64 years: 65.3% (male 2,996/female 3,093)
65 years and over: 1.6% (male 68/female 85) (2011 est.)
Median age: *total:* 24.2 years
male: 24.4 years
female: 23.9 years (2011 est.)
Population growth rate: 0.611% (2011 est.)
country comparison to the world: 143
Birth rate: 27.78 births/1,000 population (2011 est.)
country comparison to the world: 46
Death rate: 6.11 deaths/1,000 population (July 2011 est.)
country comparison to the world: 160
Net migration rate: -15.55 migrant(s)/1,000 population (2011 est.)
country comparison to the world: 217
Urbanization: *urban population:* 100% of total population (2010)
rate of urbanization: 0.6% annual rate of change (2010-15 est.)
Sex ratio: *at birth:* 0.837 male(s)/female
under 15 years: 1.04 male(s)/female
15–64 years: 0.97 male(s)/female
65 years and over: 1 male(s)/female
total population: 0.99 male(s)/female (2011 est.)
Infant mortality rate: *total:* 8.66 deaths/1,000 live births
country comparison to the world: 156
male: 11.15 deaths/1,000 live births
female: 6.58 deaths/1,000 live births (2011 est.)
Life expectancy at birth: *total population:* 65.35 years
country comparison to the world: 167
male: 61.27 years
female: 68.75 years (2011 est.)
Total fertility rate: 3.08 children born/woman (2011 est.)
country comparison to the world: 61
HIV/AIDS—adult prevalence rate: NA
HIV/AIDS—people living with HIV/AIDS: NA
HIV/AIDS—deaths: NA
Drinking water source: *Improved:*
urban: 90% of population
total: 90% of population
Unimproved:
urban: 10% of population
total: 10% of population (2008)
Sanitation facility access: *Improved:*
urban: 50% of population
total: 50% of population
Unimproved:
urban: 50% of population
total: 50% of population (2008)
Nationality: *noun:* Nauruan(s)
adjective: Nauruan
Ethnic groups: Nauruan 58%, other Pacific Islander 26%, Chinese 8%, European 8%
Religions: Nauru Congregational 35.4%, Roman Catholic 33.2%, Nauru Independent Church 10.4%, other 14.1%, none 4.5%, unspecified 2.4% (2002 census)
Languages: Nauruan (official, a distinct Pacific Island language), English (widely understood, spoken, and used for most government and commercial purposes)
Literacy: NA
School life expectancy (primary to tertiary education): *total:* 9 years
male: 9 years
female: 10 years (2008)
Education expenditures: NA

GOVERNMENT

Country name: *conventional long form:* Republic of Nauru
conventional short form: Nauru
local long form: Republic of Nauru
local short form: Nauru
former: Pleasant Island
Government type: republic
Capital: no official capital; government offices in Yaren District
time difference: UTC+12 (17 hours ahead of Washington, DC during Standard Time)
Administrative divisions: 14 districts; Aiwo, Anabar, Anetan, Anibare, Baiti, Boe, Buada, Denigomodu, Ewa, Ijuw, Meneng, Nibok, Uaboe, Yaren

Independence: 31 January 1968 (from the Australia-, NZ-, and UK-administered UN trusteeship)
National holiday: Independence Day, 31 January (1968)
Constitution: 29 January 1968; amended 17 May 1968
Legal system: mixed legal system of common law based on the English model and customary law
International law organization participation: has not submitted an ICJ jurisdiction declaration; accepts ICCt jurisdiction
Suffrage: 20 years of age; universal and compulsory
Executive branch: *chief of state:* President Marcus STEPHEN (since 19 December 2007); note–the president is both the chief of state and head of government
head of government: President Marcus STEPHEN (since 19 December 2007)
cabinet: Cabinet appointed by the president from among the members of parliament
(For more information visit the World Leaders website)
elections: president elected by parliament for a three-year term; election last held on 1 November 2010 (next to be held in 2013)
election results: Marcus STEPHEN reelected in a parliamentary vote of 11 to 6
Legislative branch: unicameral parliament (18 seats; members elected by popular vote to serve three-year terms)
elections: last held on 19 June 2010 (next to be held in 2013)
election results: percent of vote–NA; seats–independents 18
Judicial branch: Supreme Court
Political parties and leaders: Democratic Party [Kennan ADEANG]; Nauru First (Naoero Amo) Party; Nauru Party (informal); note–loose multiparty system
Political pressure groups and leaders: Woman Information and News Agency (women's issues)
International organization participation: ACP, ADB, AOSIS, C, FAO, ICAO, Interpol, IOC, ITU, OPCW, PIF, Sparteca, SPC, UN, UNCTAD, UNESCO, UPU, WHO
Diplomatic representation in the US: *chief of mission:* Ambassador Marlene I. MOSES
chancery: 800 2nd Avenue, Suite 400 D, New York, NY 10017
telephone: [1] (212) 937-0074
FAX: [1] (212) 937-0079
consulate(s): Agana (Guam)
Diplomatic representation from the US: the US does not have an embassy in Nauru; the US Ambassador to Fiji is accredited to Nauru
Flag description: blue with a narrow, horizontal, yellow stripe across the center and a large white 12-pointed star below the stripe on the hoist side; blue stands for the Pacific Ocean, the star indicates the country's location in relation to the Equator (the yellow stripe) and the 12 points symbolize the 12 original tribes of Nauru
National anthem: *name:* "Nauru Bwiema" (Song of Nauru)
lyrics/music: Margaret HENDRIE/ Laurence Henry HICKS
note: adopted 1968

ECONOMY

Economy—overview: Revenues of this tiny island traditionally have come from exports of phosphates. Few other resources exist, with most necessities being imported, mainly from Australia, its former occupier and later major source of support. In 2005 an Australian company entered into an agreement to exploit remaining supplies. Primary reserves of phosphates were exhausted and mining ceased in 2006, but mining of a deeper layer of "secondary phosphate" in the interior of the island began the following year. The secondary phosphate deposits may last another 30 years. The rehabilitation of mined land and the replacement of income from phosphates are serious long-term problems. In anticipation of the exhaustion of Nauru's phosphate deposits, substantial amounts of phosphate income were invested in trust funds to help cushion the transition and provide for Nauru's economic future. As a result of heavy spending from the trust funds, the government faced virtual bankruptcy. To cut costs the government has frozen wages and reduced overstaffed public service departments. Nauru lost further revenue in 2008 with the closure of Australia's refugee processing center, making it almost totally dependent on food imports and foreign aid. Housing, hospitals, and other capital plant are deteriorating. The cost to Australia of keeping the government and economy afloat continues to climb. Few comprehensive statistics on the Nauru economy exist with estimates of Nauru's GDP varying widely.
GDP (purchasing power parity): $60 million (2005 est.)
country comparison to the world: 221
GDP (official exchange rate): $NA
GDP—real growth rate: NA%
GDP—per capita (PPP): $5,000 (2005 est.)
country comparison to the world: 149
GDP—composition by sector: *agriculture:* NA%
industry: NA%
services: NA%
Labor force—by occupation: note: employed in mining phosphates, public administration, education, and transportation (1992)
Unemployment rate: 90% (2004 est.)
country comparison to the world: 199
Population below poverty line: NA%
Household income or consumption by percentage share: *lowest 10%:* NA%
highest 10%: NA%
Budget: *revenues:* $13.5 million
expenditures: $13.5 million (2005)
Inflation rate (consumer prices): NA%
Agriculture—products: coconuts
Industries: phosphate mining, offshore banking, coconut products
Industrial production growth rate: NA%
Electricity—production: 31 million kWh (2007 est.)
country comparison to the world: 203
Electricity—consumption: 28.83 million kWh (2007 est.)
country comparison to the world: 202
Electricity—exports: 0 kWh (2008 est.)
Electricity—imports: 0 kWh (2008 est.)
Oil—production: 0 bbl/day (2009 est.)
country comparison to the world: 120
Oil—consumption: 1,000 bbl/day (2009 est.)
country comparison to the world: 193
Oil—exports: 0 bbl/day (2007 est.)
country comparison to the world: 194
Oil—imports: 1,026 bbl/day (2007 est.)
country comparison to the world: 189
Oil—proved reserves: 0 bbl (1 January 2010 est.)
country comparison to the world: 173
Natural gas—production: 0 cu m (2008 est.)
country comparison to the world: 169
Natural gas—consumption: 0 cu m (2008 est.)
country comparison to the world: 207
Natural gas—exports: 0 cu m (2008 est.)
country comparison to the world: 151
Natural gas—imports: 0 cu m (2008 est.)
country comparison to the world: 100
Natural gas—proved reserves: 0 cu m (1 January 2010 est.)
country comparison to the world: 174
Exports: $64,000 (2005 est.)
country comparison to the world: 220
Exports—commodities: phosphates
Imports: $20 million (2004 est.)
country comparison to the world: 218
Imports—commodities: food, fuel, manufactures, building materials, machinery
Debt—external: $33.3 million (2004 est.)
country comparison to the world: 189
Exchange rates: Australian dollars (AUD) per US dollar–1.0902 (2010) 1.2822 (2009) 1.2059 (2008) 1.2137 (2007) 1.3285 (2006)

COMMUNICATIONS

Telephones—main lines in use: 1,900 (2009)
country comparison to the world: 224
Telephones—mobile cellular: 1,500 (2002)
country comparison to the world: 215
Telephone system: *general assessment:* adequate local and international radiotelephone communication provided via Australian facilities
domestic: NA
international: country code–674; satellite earth station–1 Intelsat (Pacific Ocean)
Broadcast media: 1 government-owned television station broadcasting programs from New Zealand sent via satellite or on videotape; 1 government-owned radio station, broadcasting on AM and FM, utilizes Australian and British programs (2009)
Internet country code: .nr
Internet hosts: 4,158 (2010)
country comparison to the world: 140

TRANSPORTATION

Airports: 1 (2010)
country comparison to the world: 228
Airports—with paved runways: *total:* 1
1,524 to 2,437 m: 1 (2010)
Roadways: *total:* 24 km
country comparison to the world: 219
paved: 24 km (2002)
Ports and terminals: Nauru

MILITARY

Military branches: no regular military forces; Nauru Police Force (2009)
Manpower available for military service:
males age 16–49: 2,542 (2010 est.)
Manpower fit for military service: *males age 16–49:* 1,823
females age 16–49: 2,034 (2010 est.)
Manpower reaching militarily significant age annually: *male:* 74
female: 78 (2010 est.)
Military expenditures: NA
Military—note: Nauru maintains no defense forces; under an informal agreement, defense is the responsibility of Australia

TRANSNATIONAL ISSUES

Disputes—international: none

NAVASSA ISLAND

(TERRITORY OF THE US)

INTRODUCTION

Background: This uninhabited island was claimed by the US in 1857 for its guano. Mining took place between 1865 and 1898. The lighthouse, built in 1917, was shut down in 1996 and administration of Navassa Island transferred from the Coast Guard to the Department of the Interior. A 1998 scientific expedition to the island described it as a unique preserve of Caribbean biodiversity; the following year it became a National Wildlife Refuge and annual scientific expeditions have continued.

GEOGRAPHY

Location: Caribbean, island in the Caribbean Sea, 35 miles west of Tiburon Peninsula of Haiti
Geographic coordinates: 18 25 N, 75 02 W
Map references: Central America and the Caribbean
Area: *total:* 5.4 sq km
country comparison to the world: 244
land: 5.4 sq km
water: 0 sq km
Area—comparative: about nine times the size of The Mall in Washington, DC
Land boundaries: 0 km
Coastline: 8 km
Maritime claims: *territorial sea:* 12 nm
exclusive economic zone: 200 nm
Climate: marine, tropical
Terrain: raised coral and limestone plateau, flat to undulating; ringed by vertical white cliffs (9 to 15 m high)
Elevation extremes: *lowest point:* Caribbean Sea 0 m
highest point: unnamed elevation on southwest side 77 m
Natural resources: guano
Land use: *arable land:* 0%
permanent crops: 0%
other: 100% (2005)
Natural hazards: hurricanes
Environment—current issues: NA
Geography—note: strategic location 160 km south of the US Naval Base at Guantanamo Bay, Cuba; mostly exposed rock with numerous solution holes (limestone sinkholes) but with enough grassland to support goat herds; dense stands of fig trees, scattered cactus

PEOPLE

Population: uninhabited
note: transient Haitian fishermen and others camp on the island

GOVERNMENT

Country name: *conventional long form:* none
conventional short form: Navassa Island
Dependency status: unorganized, unincorporated territory of the US; administered by the Fish and Wildlife Service, US Department of the Interior from the Caribbean Islands National Wildlife Refuge in Boqueron, Puerto Rico; in September 1996, the Coast Guard ceased operations and maintenance of Navassa Island Light, a 46-meter-tall lighthouse on the southern side of the island; there has also been a private claim advanced against the island
Legal system: the laws of the US, where applicable, apply
Diplomatic representation from the US: none (territory of the US)
Flag description: the flag of the US is used

ECONOMY

Economy—overview: Subsistence fishing and commercial trawling occur within refuge waters.

COMMUNICATIONS

Broadcast media: no television or radio broadcast stations (2009)

TRANSPORTATION

Ports and terminals: none; offshore anchorage only

MILITARY

Military—note: defense is the responsibility of the US

TRANSNATIONAL ISSUES

Disputes—international: claimed by Haiti, source of subsistence fishing

NEPAL

INTRODUCTION

Background: In 1951, the Nepalese monarch ended the century-old system of rule by hereditary premiers and instituted a cabinet system of government. Reforms in 1990 established a multiparty democracy within the framework of a constitutional monarchy. An insurgency led by Maoist extremists broke out in 1996. The ensuing ten-year civil war between insurgents and government forces witnessed the dissolution of the cabinet and parliament and assumption of absolute power by the king. Several weeks of mass protests in April 2006 were followed by several months of peace negotiations between the Maoists and government officials, and culminated in a November 2006 peace accord and the promulgation of an interim constitution. Following a nation-wide election in April 2008, the newly formed Constituent Assembly declared Nepal a federal democratic republic and abolished the monarchy at its first meeting the following month. The Constituent Assembly elected the country's first president in July. The Maoists, who received a plurality of votes in the Constituent Assembly election, formed a coalition government in August 2008, but resigned in May 2009 after the president overruled a decision to fire the chief of the army staff. The Communist Party of Nepal-United Marxist-Leninist and the Nepali Congress party then formed a new coalition government with several smaller parties. The prime minister's resignation in June 2010 ushered in seven months of political gridlock until Jhala Nath KHANAL was elected as replacement in February 2011. His pressing tasks are to conclude the drafting of a new constitution by the late May 2011 deadline and to determine the future of the former Maoist combatants.

GEOGRAPHY

Location: Southern Asia, between China and India
Geographic coordinates: 28 00 N, 84 00 E
Map references: Asia
Area: *total:* 147,181 sq km
country comparison to the world: 93

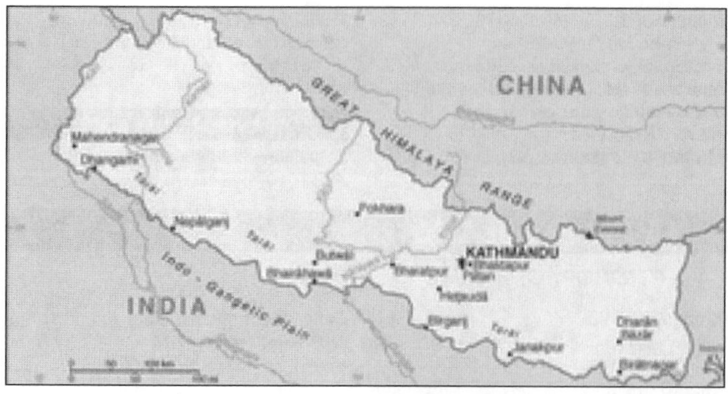

land: 143,351 sq km
water: 3,830 sq km
Area—comparative: slightly larger than Arkansas
Land boundaries: *total:* 2,926 km
border countries: China 1,236 km, India 1,690 km
Coastline: 0 km (landlocked)
Maritime claims: none (landlocked)
Climate: varies from cool summers and severe winters in north to subtropical summers and mild winters in south
Terrain: Tarai or flat river plain of the Ganges in south, central hill region, rugged Himalayas in north
Elevation extremes: *lowest point:* Kanchan Kalan 70 m
highest point: Mount Everest 8,850 m
Natural resources: quartz, water, timber, hydropower, scenic beauty, small deposits of lignite, copper, cobalt, iron ore
Land use: *arable land:* 16.07%
permanent crops: 0.85%
other: 83.08% (2005)
Irrigated land: 11,700 sq km (2003)
Total renewable water resources: 210.2 cu km (1999)
Freshwater withdrawal (domestic/industrial/agricultural): *total:* 10.18 cu km/yr (3%/1%/96%)
per capita: 375 cu m/yr (2000)
Natural hazards: severe thunderstorms; flooding; landslides; drought and famine depending on the timing, intensity, and duration of the summer monsoons
Environment—current issues: deforestation (overuse of wood for fuel and lack of alternatives); contaminated water (with human and animal wastes, agricultural runoff, and industrial effluents); wildlife conservation; vehicular emissions
Environment—international agreements: *party to:* Biodiversity, Climate Change, Climate Change-Kyoto Protocol, Desertification, Endangered Species, Hazardous Wastes, Law of the Sea, Ozone Layer Protection, Tropical Timber 83, Tropical Timber 94, Wetlands
signed, but not ratified: Marine Life Conservation
Geography—note: landlocked; strategic location between China and India; contains eight of world's 10 highest peaks, including Mount Everest and Kanchenjunga–the world's tallest and third tallest–on the borders with China and India respectively

PEOPLE

Population: 29,391,883 (July 2011 est.)
country comparison to the world: 41
Age structure: *0–14 years:* 34.6% (male 5,177,264/female 4,983,864)
15–64 years: 61.1% (male 8,607,338/female 9,344,537)
65 years and over: 4.4% (male 597,628/female 681,252) (2011 est.)
Median age: *total:* 21.6 years
male: 20.7 years
female: 22.5 years (2011 est.)
Population growth rate: 1.596% (2011 est.)
country comparison to the world: 72
Birth rate: 22.17 births/1,000 population (2011 est.)
country comparison to the world: 78
Death rate: 6.81 deaths/1,000 population (July 2011 est.)
country comparison to the world: 144
Net migration rate: 0.61 migrant(s)/1,000 population (2011 est.)
country comparison to the world: 61
Urbanization: *urban population:* 19% of total population (2010)
rate of urbanization: 4.7% annual rate of change (2010-15 est.)
Major cities—population: KATHMANDU (capital) 990,000 (2009)
Sex ratio: *at birth:* 1.04 male(s)/female
under 15 years: 1.04 male(s)/female
15–64 years: 0.92 male(s)/female
65 years and over: 0.88 male(s)/female
total population: 0.96 male(s)/female (2011 est.)
Infant mortality rate: *total:* 44.54 deaths/1,000 live births
country comparison to the world: 54
male: 44.54 deaths/1,000 live births
female: 44.55 deaths/1,000 live births (2011 est.)
Life expectancy at birth: *total population:* 66.16 years
country comparison to the world: 164
male: 64.94 years
female: 67.44 years (2011 est.)
Total fertility rate: 2.47 children born/woman (2011 est.)
country comparison to the world: 85
HIV/AIDS—adult prevalence rate: 0.4% (2009 est.)
country comparison to the world: 78
HIV/AIDS—people living with HIV/AIDS: 64,000 (2009 est.)
country comparison to the world: 52
HIV/AIDS—deaths: 4,700 (2009 est.)
country comparison to the world: 40
Major infectious diseases: *degree of risk:* high
food or waterborne diseases: bacterial diarrhea, hepatitis A, and typhoid fever
vectorborne disease: Japanese encephalitis, malaria, and dengue fever (2009)
Drinking water source: *Improved:*
urban: 93% of population
rural: 87% of population
total: 88% of population
Unimproved:
urban: 7% of population
rural: 13% of population
total: 12% of population (2008)
Sanitation facility access: *Improved:*
urban: 51% of population
rural: 27% of population
total: 31% of population
Unimproved:
urban: 49% of population
rural: 73% of population
total: 69% of population (2008)
Nationality: *noun:* Nepalese (singular and plural)
adjective: Nepalese
Ethnic groups: Chhettri 15.5%, Brahman-Hill 12.5%, Magar 7%, Tharu 6.6%, Tamang 5.5%, Newar 5.4%, Muslim 4.2%, Kami 3.9%, Yadav 3.9%, other 32.7%, unspecified 2.8% (2001 census)
Religions: Hindu 80.6%, Buddhist 10.7%, Muslim 4.2%, Kirant 3.6%, other 0.9% (2001 census)
Languages: Nepali (official) 47.8%, Maithali 12.1%, Bhojpuri 7.4%, Tharu (Dagaura/Rana) 5.8%, Tamang 5.1%, Newar 3.6%, Magar 3.3%, Awadhi 2.4%, other 10%, unspecified 2.5% (2001 census)
note: many in government and business also speak English (2001 est.)
Literacy: *definition:* age 15 and over can read and write
total population: 48.6%
male: 62.7%
female: 34.9% (2001 census)
School life expectancy (primary to tertiary education): *total:* 9 years
male: 10 years
female: 8 years (2003)
Education expenditures: 4.6% of GDP (2009)
country comparison to the world: 79

GOVERNMENT

Country name: *conventional long form:* Federal Democratic Republic of Nepal
conventional short form: Nepal
local long form: Sanghiya Loktantrik Ganatantra Nepal
local short form: Nepal
Government type: federal democratic republic
Capital: *name:* Kathmandu
geographic coordinates: 27 43 N, 85 19 E
time difference: UTC+5.75 (10.75 hours ahead of Washington, DC during Standard Time)
Administrative divisions: 14 zones (anchal, singular and plural); Bagmati, Bheri, Dhawalagiri, Gandaki, Janakpur, Karnali, Kosi, Lumbini, Mahakali, Mechi, Narayani, Rapti, Sagarmatha, Seti
Independence: 1768 (unified by Prithvi Narayan SHAH)

National holiday: Republic Day, 29 May; Democracy Day, 24 April
Constitution: 15 January 2007 (interim Constitution); note—in April 2008, a Constituent Assembly was elected to draft and promulgate a new constitution by May 2010, but the deadline has been extended to May 2011
Legal system: English common law and Hindu legal concepts
International law organization participation: has not submitted an ICJ jurisdiction declaration; non-party state to the ICCt
Suffrage: 18 years of age; universal
Executive branch: *chief of state:* President Ram Baran YADAV (since 23 July 2008); Vice President Paramananda JHA (since 23 July 2008)
head of government: Prime Minister Jhala Nath KHANAL (since 6 February 2011); Deputy Prime Ministers Bharat Mohan ADHIKHARI, Krishna MAHARA, Upendra YADAV
cabinet: cabinet was formed in May 2009 by a majority coalition made up of the Communist Party of Nepal-United Marxist-Leninist, Nepali Congress, Madhesi People's Rights Forum, Nepal-Democratic, and several smaller parties
(For more information visit the World Leaders website)
elections: president elected by Parliament; term extends until the new constitution is promulgated; election last held on 21 July 2008; date of next election NA
election results: Ram Baran YADAV elected president by the Constituent Assembly in a second round of voting on 21 July 2008; Ram Baran YADAV 308, Ram Jaja Prasad SINGH 282
Legislative branch: unicameral Constituent Assembly (601 seats; 240 members elected by direct popular vote, 335 by proportional representation, and 26 appointed by the Cabinet (Council of Ministers))
elections: last held on 10 April 2008 (next to be held NA)
election results: percent of vote by party—CPN-M 38%, NC 19%, CPN-UML 18%,Madhesi People's Right Forum 9%, other 11%; seats by party—CPN-M 229, NC 115, CPN-UML 108, Madhesi People's Rights Forum 54, Terai Madhes Democratic Party 21, other smaller parties 74; note—26 seats filled by the new Cabinet are included in the totals above
Judicial branch: Supreme Court or Sarbochha Adalat (the president appoints the chief justice on recommendation of the Constitutional Council; the chief justice appoints other judges on the recommendation of the Judicial Council)
Political parties and leaders: Chure Bhawar Rastriya Ekata Party [Keshav Prasad MAINALI]; Communist Party of Nepal-Maoist (inactive); Communist Party of Nepal-Marxist Leninist or CPN-ML [C.P. MAINALI]; Communist Party of Nepal-Unified [Raj Singh SHRIS]; Communist Party of Nepal-United [Chandra Dev JOSHI]; Communist Party of Nepal-Unified Marxist-Leninist or CPN-UML [Jhalanath KHANAL]; Dalit Janajati Party [Vishwendraman PASHWAN]; Federal Democratic National Forum; Madhesi People's Rights Forum-Democratic [Bijay Kumar GACHHADAR]; Madhesi People's Rights Forum-Nepal [Upendra YADAV]; Nepal Loktantrik Samajbadi Dal [Upendra GACHCHHADAR]; Nepal Pariwar Dal [Eknath DHAKAL]; Nepal Sadbhavana Party-Anandi Devi [Sarita GIRI]; Nepal Workers and Peasants Party or [Narayan Man BIJUKCHHE]; Nepali Congress or NC [Sushil KOIRALA]; Nepali Janata Dal [Harish Chandra SHA]; Newa Rastriya Party [Keshav Man SHAKYA]; Rastriya Janamorcha [Chitra Bahadur K.C.]; Rastriya Janamukti Party [Malwar Singh THAPA]; Rastriya Janashakti Party or RJP [Surya Bahadur THAPA]; Rastriya Prajantantra Party [Pashupati Shumsher RANA]; Rastriya Prajantantra Party Nepal [Kamal THAPA]; Sadbhavana Party [Rajendra MAHATO]; Samajbadi Prajatantrik Janata Party Nepal [Prem Bahadur SINGH]; Terai Madhes Democratic Party [Mahantha THAKUR]; Unified Communist Party of Nepal (Maoist) [Pushpa Kamal DAHAL, also known as PRACHANDA]
Political pressure groups and leaders: *other:* several small armed Madhesi groups along the southern border with India; a variety of groups advocating regional autonomy for individual ethnic groups
International organization participation: ADB, BIMSTEC, CP, FAO, G-77, IAEA, IBRD, ICAO, ICC, ICRM, IDA, IFAD, IFC, IFRCS, ILO, IMF, IMO, Interpol, IOC, IOM, IPU, ISO (correspondent), ITSO, ITU, ITUC, MIGA, MINURSO, MINUSTAH, MONUSCO, NAM, OPCW, SAARC, SACEP, UN, UNAMID, UNCTAD, UNESCO, UNIDO, UNIFIL, UNMIL, UNMIS, UNMIT, UNOCI, UNTSO, UNWTO, UPU, WCO, WFTU, WHO, WIPO, WMO, WTO
Diplomatic representation in the US: *chief of mission:* Ambassador Shankar Prasad SHARMA
chancery: 2131 Leroy Place NW, Washington, DC 20008
telephone: [1] (202) 667-4550
FAX: [1] (202) 667-5534
consulate(s) general: New York
Diplomatic representation from the US: *chief of mission:* Ambassador Scott H. DELISI
embassy: Maharajgunj, Kathmandu
mailing address: use embassy street address
telephone: [977] (1) 400-7200
FAX: [977] (1) 400-7272
Flag description: red with a blue border around the unique shape of two overlapping right triangles; the smaller, upper triangle bears a white stylized moon and the larger, lower triangle displays a white 12-pointed sun; the color red represents the rhododendron (Nepal's national flower) and is a sign of victory and bravery, the blue border signifies peace and harmony; the two right triangles are a combination of two single pennons (pennants) that originally symbolized the Himalaya Mountains while their charges represented the families of the king (upper) and the prime minister, but today they are understood to denote Hinduism and Buddhism, the country's two main religions; the moon represents the serenity of the Nepalese people and the shade and cool weather in the Himalayas, while the sun depicts the heat and higher temperatures of the lower parts of Nepal; the moon and the sun are also said to express the hope that the nation will endure as long as these heavenly bodies
note: Nepal is the only country in the world whose flag is not rectangular or square
National anthem: *name:* "Sayaun Thunga Phool Ka" (Hundreds of Flowers)
lyrics/music: Pradeep Kumar RAI/Ambar GURUNG
note: adopted 2007; after the abolition of the monarchy in 2006, a new anthem was required because of the previous anthem's praise for the king

ECONOMY

Economy—overview: Nepal is among the poorest and least developed countries in the world, with almost one-quarter of its population living below the poverty line. Agriculture is the mainstay of the economy, providing a livelihood for three-fourths of the population and accounting for about one-third of GDP. Industrial activity mainly involves the processing of agricultural products, including pulses, jute, sugarcane, tobacco, and grain. Nepal has considerable scope for exploiting its potential in hydropower, with an estimated 42,000 MW of feasible capacity, but political instability hampers foreign investment. Additional challenges to Nepal's growth include its landlocked geographic location, civil strife and labor unrest, and its susceptibility to natural disaster.
GDP (purchasing power parity): $35.81 billion (2010 est.)
country comparison to the world: 102
$34.25 billion (2009 est.)
$32.66 billion (2008 est.)
note: data are in 2010 US dollars
GDP (official exchange rate): $15.84 billion (2010 est.)
GDP—real growth rate: 4.6% (2010 est.)
country comparison to the world: 79
4.9% (2009 est.)
6.1% (2008 est.)
GDP—per capita (PPP): $1,200 (2010 est.)
country comparison to the world: 205
$1,200 (2009 est.)
$1,200 (2008 est.)
note: data are in 2010 US dollars
GDP—composition by sector: *agriculture:* 33%
industry: 15%
services: 52% (FY09 est.)
Labor force: 18 million
country comparison to the world: 32
note: severe lack of skilled labor (2009 est.)
Labor force—by occupation: *agriculture:* 75%
industry: 7%
services: 18% (2010 est.)
Unemployment rate: 46% (2008 est.)
country comparison to the world: 191
42% (2004 est.)

Population below poverty line: 24.7% (2008)
Household income or consumption by percentage share: *lowest 10%:* NA
highest 10%: 40.6% (2008)
Distribution of family income—Gini index: 47.2 (2008)
country comparison to the world: 32
36.7 (1996)
Budget: *revenues:* $3 billion
expenditures: $4.6 billion (FY10)
Inflation rate (consumer prices): 8.6% (September 2010 est.)
country comparison to the world: 190
13.2% (September 2009 est.)
Central bank discount rate: 6.5% (31 December 2010)
country comparison to the world: 64
6.5% (31 December 2009)
Commercial bank prime lending rate: 8% (31 December 2009 est.)
country comparison to the world: 115
8% (31 December 2008 est.)
Stock of narrow money: $3.03 billion (July 2010)
country comparison to the world: 110
$2.712 billion (July 2009)
Stock of broad money: $9 billion (July 2010)
country comparison to the world: 105
$7.923 billion (July 2009)
Stock of domestic credit: $6.11 billion (31 December 2009 est.)
country comparison to the world: 108
$5.556 billion (31 December 2008)
Market value of publicly traded shares: $5.2 billion (31 December 2010 est.)
country comparison to the world: 79
$5.485 billion (31 December 2009)
$4.894 billion (31 December 2008)
Agriculture—products: pulses, rice, corn, wheat, sugarcane, jute, root crops; milk, water buffalo meat
Industries: tourism, carpets, textiles; small rice, jute, sugar, and oilseed mills; cigarettes, cement and brick production
Industrial production growth rate: 1.8% (FY08)
country comparison to the world: 133
Electricity—production: 2.6 billion kWh (2009 est.)
country comparison to the world: 129
Electricity—consumption: 2.243 billion kWh (2007 est.)
country comparison to the world: 132
Electricity—exports: 0 kWh (2010 est.)
Electricity—imports: 213 million kWh (2008 est.)
Oil—production: 0 bbl/day (2010 est.)
country comparison to the world: 119
Oil—consumption: 18,000 bbl/day (2009 est.)
country comparison to the world: 130
Oil—exports: 0 bbl/day (2010 est.)
country comparison to the world: 193
Oil—imports: 16,920 bbl/day (2007 est.)
country comparison to the world: 119
Oil—proved reserves: 0 bbl (1 January 2010 est.)
country comparison to the world: 172
Natural gas—production: 0 cu m (2010 est.)
country comparison to the world: 168
Natural gas—consumption: 0 cu m (2010 est.)
country comparison to the world: 206
Natural gas—exports: 0 cu m (2010 est.)
country comparison to the world: 150
Natural gas—imports: 0 cu m (2010 est.)
country comparison to the world: 99
Natural gas—proved reserves: 0 cu m (1 January 2010 est.)
country comparison to the world: 173
Current account balance: $-449 million (2010)
country comparison to the world: 113
$537 million (2009)
Exports: $849 million (2009)
country comparison to the world: 156
$907 million (2008)
Exports—commodities: clothing, pulses, carpets, textiles, juice, pashima, jute goods
Exports—partners: India 65.6%, US 8%, Bangladesh 6.04%, Germany 5% (2009)
Imports: $5.26 billion (2009)
country comparison to the world: 113
$4.1 billion (2008)
Imports—commodities: petroleum products, machinery and equipment, gold, electrical goods, medicine
Imports—partners: India 57%, China 13% (2009)
Debt—external: $4.5 billion (2009)
country comparison to the world: 110
$3.285 billion (2008)
Stock of direct foreign investment—at home: $NA
Stock of direct foreign investment—abroad: $NA
Exchange rates: Nepalese rupees (NPR) per US dollar–72.56 (2010) 77.44 (2009) 65.21 (2008) 70.35 (2007) 72.446 (2006)

COMMUNICATIONS

Telephones—main lines in use: 820,500 (2009)
country comparison to the world: 86
Telephones—mobile cellular: 7.618 million (2009)
country comparison to the world: 77
Telephone system: *general assessment:* poor telephone and telegraph service; fair radiotelephone communication service and mobile-cellular telephone network
domestic: combined fixed-line and mobile-cellular telephone service subscribership base only about 30 per 100 persons
international: country code–977; radiotelephone communications; microwave landline to India; satellite earth station–1 Intelsat (Indian Ocean) (2008)
Broadcast media: state operates 2 television stations as well as national and regional radio stations; more than 60 independent radio stations and a small number of independent television stations (2007)
Internet country code: .np
Internet hosts: 43,928 (2010)
country comparison to the world: 91
Internet users: 577,800 (2009)
country comparison to the world: 116

TRANSPORTATION

Airports: 47 (2010)
country comparison to the world: 93
Airports—with paved runways: *total:* 11
over 3,047 m: 1
914 to 1,523 m: 9
under 914 m: 1 (2010)
Airports—with unpaved runways: *total:* 36
1,524 to 2,437 m: 1
914 to 1,523 m: 4
under 914 m: 31 (2010)
Railways: *total:* 59 km
country comparison to the world: 130
narrow gauge: 59 km 0.762-m gauge (2010)
Roadways: *total:* 17,282 km
country comparison to the world: 118
paved: 10,142 km
unpaved: 7,140 km (2007)

MILITARY

Military branches: Nepal Army (2010)
Military service age and obligation: 18 years of age for voluntary military service; 15 years of age for military training; no conscription (2011)
Manpower available for military service: *males age 16–49:* 6,941,152
females age 16–49: 7,618,397 (2010 est.)
Manpower fit for military service: *males age 16–49:* 5,260,878
females age 16–49: 5,947,512 (2010 est.)
Manpower reaching militarily significant age annually: *male:* 380,172
female: 367,103 (2010 est.)
Military expenditures: 1.6% of GDP (2006)
country comparison to the world: 94

TRANSNATIONAL ISSUES

Disputes—international: joint border commission continues to work on contested sections of boundary with India, including the 400 square kilometer dispute over the source of the Kalapani River; India has instituted a stricter border regime to restrict transit of Maoist insurgents and illegal cross-border activities; approximately 106,000 Bhutanese Lhotshampas (Hindus) have been confined in refugee camps in southeastern Nepal since 1990
Refugees and internally displaced persons: refugees (country of origin): 107,803 (Bhutan); 20,153 (Tibet/China)
IDPs: 50,000-70,000 (remaining from ten-year Maoist insurgency that officially ended in 2006; displacement spread across the country) (2007)
Illicit drugs: illicit producer of cannabis and hashish for the domestic and international drug markets; transit point for opiates from Southeast Asia to the West

NETHERLANDS

INTRODUCTION

Background: The Dutch United Provinces declared their independence from Spain in 1579; during the 17th century, they became a leading seafaring and commercial power, with settlements and colonies around the world. After a 20-year French occupation, a Kingdom of the Netherlands was formed in 1815. In 1830 Belgium seceded and formed a separate kingdom. The Netherlands remained neutral in World War I, but suffered invasion and occupation by Germany in World War II. A modern, industrialized nation, the Netherlands is also a large exporter of agricultural products. The country was a founding member of NATO and the EEC (now the EU), and participated in the introduction of the euro in 1999. In October 2010, the former Netherlands Antilles was dissolved and the three smallest islands–Bonaire, Sint Eustatius, and Saba–became special municipalities in the Netherlands administrative structure. The larger islands of Sint Maarten and Curacao joined the Netherlands and Aruba as constituent countries forming the Kingdom of the Netherlands.

GEOGRAPHY

Location: Western Europe, bordering the North Sea, between Belgium and Germany
Geographic coordinates: 52 30 N, 5 45 E
Map references: Europe
Area: *total:* 41,543 sq km
country comparison to the world: 134
land: 33,893 sq km
water: 7,650 sq km
Area—comparative: slightly less than twice the size of New Jersey
Land boundaries: *total:* 1,027 km
border countries: Belgium 450 km, Germany 577 km
Coastline: 451 km
Maritime claims: *territorial sea:* 12 nm
contiguous zone: 24 nm
exclusive fishing zone: 200 nm
Climate: temperate; marine; cool summers and mild winters
Terrain: mostly coastal lowland and reclaimed land (polders); some hills in southeast
Elevation extremes: *lowest point:* Zuidplaspolder -7 m
highest point: Mount Scenery 862 m (on the island of Saba in the Caribbean, now considered an integral part of the Netherlands following the dissolution of the Netherlands Antilles)
note: the highest point on continental Netherlands is Vaalserberg at 322 m
Natural resources: natural gas, petroleum, peat, limestone, salt, sand and gravel, arable land
Land use: *arable land:* 21.96%
permanent crops: 0.77%
other: 77.27% (2005)
Irrigated land: 5,650 sq km (2003)
Total renewable water resources: 89.7 cu km (2005)
Freshwater withdrawal (domestic/industrial/agricultural): *total:* 8.86 cu km/yr (6%/60%/34%)
per capita: 544 cu m/yr (2001)
Natural hazards: flooding
Environment—current issues: water pollution in the form of heavy metals, organic compounds, and nutrients such as nitrates and phosphates; air pollution from vehicles and refining activities; acid rain
Environment—international agreements: *party to:* Air Pollution, Air Pollution-Nitrogen Oxides, Air Pollution-Persistent Organic Pollutants, Air Pollution-Sulfur 85, Air Pollution-Sulfur 94, Air Pollution-Volatile Organic Compounds, Antarctic-Environmental Protocol, Antarctic-Marine Living Resources, Antarctic Treaty, Biodiversity, Climate Change, Climate Change-Kyoto Protocol, Desertification, Endangered Species, Environmental Modification, Hazardous Wastes, Kyoto Protocol, Law of the Sea, Marine Dumping, Marine Life Conservation, Ozone Layer Protection, Ship Pollution, Tropical Timber 83, Tropical Timber 94, Wetlands, Whaling
signed, but not ratified: none of the selected agreements
Geography—note: located at mouths of three major European rivers (Rhine, Maas or Meuse, and Schelde)

PEOPLE

Population: 16,847,007 (July 2011 est.)
country comparison to the world: 60
Age structure: *0–14 years:* 17% (male 1,466,218/female 1,398,463)
15–64 years: 67.4% (male 5,732,042/female 5,624,408)
65 years and over: 15.6% (male 1,141,507/female 1,484,369) (2011 est.)
Median age: *total:* 41.1 years
male: 40.3 years
female: 41.9 years (2011 est.)
Population growth rate: 0.371% (2011 est.)
country comparison to the world: 162
Birth rate: 10.23 births/1,000 population (2011 est.)
country comparison to the world: 188
Death rate: 8.85 deaths/1,000 population (July 2011 est.)
country comparison to the world: 75
Net migration rate: 2.33 migrant(s)/1,000 population (2011 est.)
country comparison to the world: 35
Urbanization: *urban population:* 83% of total population (2010)
rate of urbanization: 0.8% annual rate of change (2010-15 est.)
Major cities—population: AMSTERDAM (capital) 1.044 million; Rotterdam 1.008 million; The Hague (seat of government) 629,000 (2009)
Sex ratio:
at birth: 1.052 male(s)/female
under 15 years: 1.05 male(s)/female
15–64 years: 1.02 male(s)/female
65 years and over: 0.76 male(s)/female
total population: 0.98 male(s)/female (2011 est.)
Infant mortality rate: *total:* 4.59 deaths/1,000 live births
country comparison to the world: 191
male: 5.08 deaths/1,000 live births
female: 4.07 deaths/1,000 live births (2011 est.)
Life expectancy at birth: *total population:* 79.68 years
country comparison to the world: 35
male: 77.06 years
female: 82.44 years (2011 est.)
Total fertility rate: 1.66 children born/woman (2011 est.)
country comparison to the world: 174
HIV/AIDS—adult prevalence rate: 0.2% (2009 est.)
country comparison to the world: 104
HIV/AIDS—people living with HIV/AIDS: 22,000 (2009 est.)
country comparison to the world: 75
HIV/AIDS—deaths: fewer than 100 (2009 est.)
country comparison to the world: 146
Drinking water source:
Improved:
urban: 100% of population
rural: 100% of population
total: 100% of population (2008)
Sanitation facility access: *Improved:*
urban: 100% of population
rural: 100% of population
total: 100% of population (2008)
Nationality: *noun:* Dutchman(men), Dutchwoman(women)
adjective: Dutch
Ethnic groups: Dutch 80.7%, EU 5%, Indonesian 2.4%, Turkish 2.2%, Surinamese 2%, Moroccan 2%, Caribbean 0.8%, other 4.8% (2008 est.)
Religions: Roman Catholic 30%, Dutch Reformed 11%, Calvinist 6%, other Protestant 3%, Muslim 5.8%, other 2.2%, none 42% (2006)
Languages: Dutch (official), Frisian (official)
Literacy: *definition:* age 15 and over can read and write
total population: 99%
male: 99%
female: 99% (2003 est.)
School life expectancy (primary to tertiary education): *total:* 17 years
male: 17 years
female: 17 years (2008)
Education expenditures: 5.3% of GDP (2007)
country comparison to the world: 50

GOVERNMENT

Country name: *conventional long form:* Kingdom of the Netherlands
conventional short form: Netherlands
local long form: Koninkrijk der Nederlanden
local short form: Nederland
Government type: constitutional monarchy
Capital: *name:* Amsterdam
geographic coordinates: 52 23 N, 4 54 E
time difference: UTC+1 (6 hours ahead of Washington, DC during Standard Time)
daylight saving time: +1hr, begins last Sunday in March; ends last Sunday in October
note: The Hague is the seat of government; time descriptions apply to the continental Netherlands only, not to the Caribbean components
Administrative divisions: 12 provinces (provincies, singular—provincie); Drenthe, Flevoland, Fryslan (Friesland), Gelderland, Groningen, Limburg, Noord-Brabant (North Brabant), Noord-Holland (North Holland), Overijssel, Utrecht, Zeeland (Zealand), Zuid-Holland (South Holland)
Dependent areas: Aruba, Curacao, Sint Maarten
Independence: 23 January 1579 (the northern provinces of the Low Countries conclude the Union of Utrecht breaking with Spain; on 26 July 1581 they formally declared their independence with an Act of Abjuration; however, it was not until 30 January 1648 and the Peace of Westphalia that Spain recognized this independence)
National holiday: Queen's Day (Birthday of deceased Queen-Mother JULIANA and accession to the throne of her oldest daughter BEATRIX), 30 April (1909 and 1980)
Constitution: adopted 1815; amended many times, most recently in 2002
Legal system: civil law system based on the French system; constitution does not permit judicial review of acts of the States General
International law organization participation: accepts compulsory ICJ jurisdiction with reservations; accepts ICCt jurisdiction
Suffrage: 18 years of age; universal
Executive branch: *chief of state:* Queen BEATRIX (since 30 April 1980); Heir Apparent WILLEM-ALEXANDER (born 27 April 1967), son of the monarch
head of government: Prime Minister Mark RUTTE (since 14 October 2010); Deputy Prime Minister Maxime VERHAGEN (since 14 October 2010)
cabinet: Council of Ministers appointed by the monarch
(For more information visit the World Leaders website)
elections: the monarchy is hereditary; following Second Chamber elections, the leader of the majority party or leader of a majority coalition usually appointed prime minister by the monarch; deputy prime ministers appointed by the monarch
note: there is also a Council of State composed of the monarch, heir apparent, and councilors that provides consultations to the cabinet on legislative and administrative policy
Legislative branch: bicameral States General or Staten Generaal consists of the First Chamber or Eerste Kamer (75 seats; members indirectly elected by the country's 12 provincial councils to serve four-year terms) and the Second Chamber or Tweede Kamer (150 seats; members elected by popular vote to serve four-year terms)
elections: First Chamber—last held on 29 May 2007 (next to be held in May 2011); Second Chamber—last held on 9 June 2010 (next to be held by May 2015)
election results: First Chamber—percent of vote by party—NA%; seats by party—CDA 21, PvdA 14, VVD 14, Socialist Party 11, Christian Union 4, Green Left Party 4, D66 2, other 5; Second Chamber—percent of vote by party—VVD 20.5%, PvdA 19.6%, PVV, 15.4%, CDA 13.6%, SP 9.8%, D66 6.9%, GL 6.7%, CU 3.2, other 4.3%; seats by party—VVD 31, PvdA 30, PVV 24, CDA 21, SP 15, D66 10, GL 10, CU 5, other 4
Judicial branch: Supreme Court or Hoge Raad (justices are nominated for life by the monarch)
Political parties and leaders: Christian Democratic Appeal or CDA [Maxime VERHAGEN]; Christian Union or CU [Andre ROUVOET]; Democrats 66 or D66 [Alexander PECHTOLD]; Green Left or GL [Jolande SAP]; Labor Party or PvdA [Job COHEN]; Party for Freedom or PVV [Geert WILDERS]; Party for the Animals or PvdD [Marianne THIEME]; People's Party for Freedom and Democracy or VVD [Mark RUTTE] (Liberal); Reformed Political Party of SGP [Kees VAN DER STAAIJ]; Socialist Party of SP [Emile ROEMER]; plus a few minor parties
Political pressure groups and leaders: Christian Trade Union Federation or CNV [Jaap SMIT]; Confederation of Netherlands Industry and Employers or VNO-NCW [Bernard WIENTJES]; Federation for Small and Medium-sized businesses or MKB [Loek HERMANS]; Netherlands Trade Union Federation or FNV [Agnes JONGERIUS]; Social Economic Council or SER [Alexander RINNOOY KAN]; Trade Union Federation of Middle and High Personnel or MHP [Richard STEENBORG]
International organization participation: ADB (nonregional member), AfDB (nonregional member), Arctic Council (observer), Australia Group, Benelux, BIS, CBSS (observer), CE, CERN, EAPC, EBRD, EIB, EMU, ESA, EU, FAO, FATF, G-10, IADB, IAEA, IBRD, ICAO, ICC, ICRM, IDA, IEA, IFAD, IFC, IFRCS, IHO, ILO, IMF, IMO, IMSO, Interpol, IOC, IOM, IPU, ISO, ITSO, ITU, ITUC, MIGA, NATO, NEA, NSG, OAS (observer), OECD, OPCW, OSCE, Paris Club, PCA, Schengen Convention, SECI (observer), UN, UNAMID, UNCTAD, UNESCO, UNHCR, UNIDO, UNMIS, UNRWA, UNTSO, UNWTO, UPU, WCO, WHO, WIPO, WMO, WTO, ZC
Diplomatic representation in the US: *chief of mission:* Ambassador Regina "Renee" JONES-BOS
chancery: 4200 Linnean Avenue NW, Washington, DC 20008
telephone: [1] (202) 244-5300, [1] 877-388-2443
FAX: [1] (202) 362-3430
consulate(s) general: Chicago, Los Angeles, Miami, New York
consulate(s): Boston
Diplomatic representation from the US: *chief of mission:* Ambassador Fay HARTOG LEVIN
embassy: Lange Voorhout 102, 2514 EJ, The Hague
mailing address: PSC 71, Box 1000, APO AE 09715
telephone: [31] (70) 310-2209
FAX: [31] (70) 361-4688
consulate(s) general: Amsterdam
Flag description: three equal horizontal bands of red (top), white, and blue; similar to the flag of Luxembourg, which uses a lighter blue and is longer; the colors were those of WILLIAM I, Prince of Orange, who led the Dutch Revolt against Spanish sovereignty in the latter half of the 16th century; originally the upper band was orange, but because it tended to fade to red over time, the red shade was eventually made the permanent color; the banner is perhaps the oldest tricolor in continuous use
National anthem: *name:* "Het Wilhelmus" (The William)
lyrics/music: Philips VAN MARNIX van Sint Aldegonde (presumed)/unknown
note: adopted 1932, in use since the 17th century, making it the oldest national anthem in the world; also known as "Wilhelmus van Nassouwe" (William of Nassau), it is in the form of an acrostic, where the first letter of each stanza spells the name of the leader of the Dutch Revolt

ECONOMY

Economy—overview: The Netherlands economy is noted for stable industrial relations, moderate unemployment and inflation, a sizable current account surplus, and an important role as a European transportation hub. Industrial activity is predominantly in food processing, chemicals, petroleum refining, and electrical machinery. A highly mechanized agricultural sector employs only 2% of the labor force but provides large surpluses for the food-processing industry and for exports. The Netherlands, along with 11 of its EU partners, began circulating the euro currency on 1 January 2002. The country has been one of the leading European nations for attracting foreign direct investment and is one of the four largest investors in the US. After 26 years of uninterrupted economic growth, the Netherlands' economy—which is highly open and dependent on foreign trade and financial services—was hard-hit by global economic crisis. Dutch GDP contracted 3.9% in 2009, while exports declined nearly 25% due to a sharp contraction in world demand. The Dutch financial sector has also suffered, due in part to the high exposure of some Dutch banks to U.S. mortgage-backed securities. In response to turmoil in financial markets, the government nationalized two

banks and injected billions of dollars into a third, to prevent further systemic risk. The government also sought to boost the domestic economy by accelerating infrastructure programs, offering corporate tax breaks for employers to retain workers, and expanding export credit facilities. The stimulus programs and bank bailouts, however, resulted in a government budget deficit of nearly 4.6% of GDP in 2009 and 5.6% in 2010 that contrasts sharply with a surplus of 0.7% of GDP in 2008. With unemployment weighing on private-sector consumption, the government of Prime Minister Mark RUTTE is likely to come under increased pressure to keep the budget deficit in check while promoting economic recovery.
GDP (purchasing power parity): $676.9 billion (2010 est.)
country comparison to the world: 22
$665.3 billion (2009 est.)
$692.4 billion (2008 est.)
note: data are in 2010 US dollars
GDP (official exchange rate): $783.3 billion (2010 est.)
GDP—real growth rate:
1.7% (2010 est.)
country comparison to the world: 157
-3.9% (2009 est.)
1.9% (2008 est.)
GDP—per capita (PPP): $40,300 (2010 est.)
country comparison to the world: 20
$39,800 (2009 est.)
$41,600 (2008 est.)
note: data are in 2010 US dollars
GDP—composition by sector: *agriculture:* 2.6%
industry: 24.9%
services: 72.4% (2010 est.)
Labor force: 7.86 million (2010 est.)
country comparison to the world: 58
Labor force—by occupation: *agriculture:* 2%
industry: 18%
services: 80% (2005 est.)
Unemployment rate: 5.5% (2010 est.)
country comparison to the world: 54
4.8% (2009 est.)
Population below poverty line: 10.5% (2005)
Household income or consumption by percentage share: *lowest 10%:* 2.5%
highest 10%: 22.9% (1999)
Distribution of family income—Gini index: 30.9 (2007)
country comparison to the world: 108
32.6 (1994)
Investment (gross fixed): 18% of GDP (2010 est.)
country comparison to the world: 113
Budget: *revenues:* $356 billion
expenditures: $399.3 billion (2010 est.)
Public debt: 64.6% of GDP (2010 est.)
country comparison to the world: 26
60.9% of GDP (2009 est.)
Inflation rate (consumer prices): 1.1% (2010 est.)
country comparison to the world: 23
1.2% (2009 est.)
Central bank discount rate: 1.75% (31 December 2010)
country comparison to the world: 120
1.75% (31 December 2009)
note: this is the European Central Bank's rate on the marginal lending facility, which offers overnight credit to banks in the euro area
Commercial bank prime lending rate: 10.01% (31 December 2009 est.)
country comparison to the world: 91
9.66% (31 December 2008 est.)
Stock of narrow money: $368.1 billion (31 December 2010 est.)
country comparison to the world: 13
$351.6 billion (31 December 2009 est.)
note: see entry for the European Union for money supply in the euro area; the European Central Bank (ECB) controls monetary policy for the 17 members of the Economic and Monetary Union (EMU); individual members of the EMU do not control the quantity of money circulating within their own borders
Stock of broad money: $1.124 trillion (31 December 2010 est.)
country comparison to the world: 16
$1.133 trillion (31 December 2009 est.)
Stock of domestic credit: $2.083 trillion (31 December 2009 est.)
country comparison to the world: 12
$1.824 trillion (31 December 2008)
Market value of publicly traded shares: $542.5 billion (31 December 2009)
country comparison to the world: 21
$387.9 billion (31 December 2008)
$956.5 billion (31 December 2007)
Agriculture—products: grains, potatoes, sugar beets, fruits, vegetables; livestock
Industries: agroindustries, metal and engineering products, electrical machinery and equipment, chemicals, petroleum, construction, microelectronics, fishing
Industrial production growth rate: 3.2% (2010 est.)
country comparison to the world: 103
Electricity—production: 108.2 billion kWh (2008 est.)
country comparison to the world: 31
Electricity—consumption: 124.1 billion kWh (2008 est.)
country comparison to the world: 27
Electricity—exports: 10.56 billion kWh (2009 est.)
Electricity—imports: 15.45 billion kWh (2009 est.)
Oil—production: 57,190 bbl/day (2009 est.)
country comparison to the world: 61
Oil—consumption: 922,800 bbl/day (2009 est.)
country comparison to the world: 21
Oil—exports: 1.66 million bbl/day (2008 est.)
country comparison to the world: 14
Oil—imports: 2.426 million bbl/day (2008 est.)
country comparison to the world: 8
Oil—proved reserves: 100 million bbl (1 January 2010 est.)
country comparison to the world: 71
Natural gas—production: 79.58 billion cu m (2009 est.)
country comparison to the world: 10
Natural gas—consumption: 48.6 billion cu m (2009 est.)
country comparison to the world: 17
Natural gas—exports: 55.59 billion cu m (2009 est.)
country comparison to the world: 6
Natural gas—imports: 24.6 billion cu m (2009 est.)
country comparison to the world: 12
Natural gas—proved reserves: 1.416 trillion cu m (1 January 2010 est.)
country comparison to the world: 24
Current account balance: $46.69 billion (2010 est.)
country comparison to the world: 8
$39.58 billion (2009 est.)
Exports: $451.3 billion (2010 est.)
country comparison to the world: 9
$421.3 billion (2009 est.)
Exports—commodities: machinery and equipment, chemicals, fuels; foodstuffs
Exports—partners: Germany 25.54%, Belgium 12.49%, France 9.27%, UK 8.17%, Italy 5.07%, US 3.97% (2009)
Imports: $408.4 billion (2010 est.)
country comparison to the world: 11
$371.9 billion (2009 est.)
Imports—commodities: machinery and transport equipment, chemicals, fuels, foodstuffs, clothing
Imports—partners: Germany 17.16%, China 11.58%, Belgium 8.68%, US 7.77%, UK 5.72%, Russia 4.47%, France 4.4% (2009)
Reserves of foreign exchange and gold: $NA (31 December 2010 est.)
$39.61 billion (31 December 2009 est.)
Debt—external: $NA (30 June 2010)
$3.733 trillion (31 December 2009)
Stock of direct foreign investment—at home: $687.8 billion (31 December 2010 est.)
country comparison to the world: 7
$654.6 billion (31 December 2009 est.)
Stock of direct foreign investment—abroad: $950.8 billion (31 December 2010 est.)
country comparison to the world: 5
$932.2 billion (31 December 2009 est.)
Exchange rates: euros (EUR) per US dollar–0.755 (2010) 0.7198 (2009) 0.6827 (2008) 0.7345 (2007) 0.7964 (2006)

COMMUNICATIONS

Telephones—main lines in use: 7.32 million (2009)
country comparison to the world: 26
Telephones—mobile cellular: 21.182 million (2009)
country comparison to the world: 39
Telephone system: *general assessment:* highly developed and well maintained
domestic: extensive fixed-line fiber-optic network; large cellular telephone system with 5 major operators utilizing the third generation of the Global System for Mobile Communications (GSM) technology; one in five households now use Voice over the Internet Protocol (VoIP) services
international: country code–31; submarine cables provide links to the US and Europe; satellite earth stations–5 (3 Intelsat–1 Indian Ocean and 2 Atlantic Ocean, 1 Eutelsat, and 1 Inmarsat (2007)
Broadcast media: more than 90% of households are connected to cable or satellite TV systems that provide a wide range of domestic and foreign channels; public service broadcast system includes multiple

broadcasters, 3 with a national reach and the remainder operating in regional and local markets; 2 major nationwide commercial television companies, each with 3 or more stations, and a large number of commercial TV stations in regional and local markets; nearly 600 radio stations operating with a mix of public and private stations providing national or regional coverage (2008)
Internet country code: .nl
Internet hosts: 12.607 million (2010)
country comparison to the world: 10
Internet users: 14.872 million (2009)
country comparison to the world: 27

TRANSPORTATION

Airports: 27 (2010)
country comparison to the world: 124
Airports—with paved runways: *total:* 20
over 3,047 m: 2
2,438 to 3,047 m: 9
1,524 to 2,437 m: 3
914 to 1,523 m: 5
under 914 m: 1 (2010)
Airports—with unpaved runways: *total:* 7
914 to 1,523 m: 3
under 914 m: 4 (2010)
Heliports: 1 (2010)
Pipelines: gas 4,413 km; oil 365 km; refined products 716 km (2010)
Railways: *total:* 2,896 km
country comparison to the world: 57
standard gauge: 2,896 km 1.435-m gauge (2,195 km electrified) (2010)
Roadways: *total:* 136,827 km (includes 2,631 km of expressways) (2010)
country comparison to the world: 35
Waterways: 6,214 km (navigable for ships of 50 tons) (2010)
country comparison to the world: 22
Merchant marine: *total:* 706
country comparison to the world: 15
by type: bulk carrier 1, cargo 464, carrier 21, chemical tanker 57, container 73, liquefied gas 19, passenger 17, passenger/cargo 15, petroleum tanker 5, refrigerated cargo 10, roll on/roll off 21, specialized tanker 3
foreign-owned: 217 (Australia 1, Denmark 36, Finland 14, France 2, Germany 92, Ireland 7, Italy 9, Japan 1, Norway 18, Sweden 18, UAE 4, US 15)
registered in other countries: 240 (Antigua and Barbuda 18, Australia 1, Bahamas 22, Belize 1, Cambodia 1, Canada 1, Cyprus 24, Gibraltar 33, Liberia 35, Luxembourg 2, Malta 2, Marshall Islands 16, former Netherlands Antilles 52, Panama 8, Paraguay 1, Philippines 18, Portugal 1, Saint Vincent and the Grenadines 2, Singapore 1, unknown 1) (2010)
Ports and terminals: Amsterdam, IJmuiden, Moerdijk, Rotterdam, Terneuzen, Vlissingen

MILITARY

Military branches: Royal Netherlands Army, Royal Netherlands Navy (includes Naval Air Service and Marine Corps), Royal Netherlands Air Force (Koninklijke Luchtmacht, KLu), Royal Military Police (2010)
Military service age and obligation: 20 years of age for an all-volunteer force (2004)
Manpower available for military service: *males age 16–49:* 3,911,098
females age 16–49: 3,817,031 (2010 est.)
Manpower fit for military service: *males age 16–49:* 3,201,328
females age 16–49: 3,122,889 (2010 est.)
Manpower reaching militarily significant age annually: *male:* 103,462
female: 98,383 (2010 est.)
Military expenditures: 1.6% of GDP (2005 est.)
country comparison to the world: 95

TRANSNATIONAL ISSUES

Disputes—international: none
Illicit drugs: major European producer of synthetic drugs, including ecstasy, and cannabis cultivator; important gateway for cocaine, heroin, and hashish entering Europe; major source of US-bound ecstasy; large financial sector vulnerable to money laundering; significant consumer of ecstasy

NEW CALEDONIA

(SELF-GOVERNING TERRITORY OF FRANCE)

INTRODUCTION

Background: Settled by both Britain and France during the first half of the 19th century, the island was made a French possession in 1853. It served as a penal colony for four decades after 1864. Agitation for independence during the 1980s and early 1990s ended in the 1998 Noumea Accord, which over a period of 15 to 20 years will transfer an increasing amount of governing responsibility from France to New Caledonia. The agreement also commits France to conduct a referendum between 2014 and 2019 to decide whether New Caledonia should assume full sovereignty and independence.

GEOGRAPHY

Location: Oceania, islands in the South Pacific Ocean, east of Australia
Geographic coordinates: 21 30 S, 165 30 E
Map references: Oceania
Area: *total:* 18,575 sq km
country comparison to the world: 155
land: 18,275 sq km
water: 300 sq km
Area—comparative: slightly smaller than New Jersey
Land boundaries: 0 km
Coastline: 2,254 km
Maritime claims: *territorial sea:* 12 nm
exclusive economic zone: 200 nm
Climate: tropical; modified by southeast trade winds; hot, humid
Terrain: coastal plains with interior mountains
Elevation extremes: *lowest point:* Pacific Ocean 0 m
highest point: Mont Panie 1,628 m
Natural resources: nickel, chrome, iron, cobalt, manganese, silver, gold, lead, copper
Land use: *arable land:* 0.32%
permanent crops: 0.22%
other: 99.46% (2005)
Irrigated land: 100 sq km (2003)
Natural hazards: cyclones, most frequent from November to March
volcanism: Matthew and Hunter Islands are historically active
Environment—current issues: erosion caused by mining exploitation and forest fires
Geography—note: consists of the main island of New Caledonia (one of the largest in the Pacific Ocean), the archipelago of Iles Loyaute, and numerous small, sparsely populated islands and atolls

PEOPLE

Population: 256,275 (July 2011 est.)
country comparison to the world: 181
Age structure: *0–14 years:* 24.9% (male 32,653/female 31,236)
15–64 years: 67.3% (male 86,865/female 85,578)
65 years and over: 7.8% (male 8,954/female 10,989) (2011 est.)
Median age: *total:* 30.1 years
male: 29.6 years
female: 30.7 years (2011 est.)
Population growth rate: 1.524% (2011 est.)
country comparison to the world: 78
Birth rate: 16.28 births/1,000 population (2011 est.)
country comparison to the world: 126
Death rate: 5.28 deaths/1,000 population (July 2011 est.)
country comparison to the world: 179
Net migration rate: 4.25 migrant(s)/1,000 population
country comparison to the world: 22
note: there has been steady emigration from Wallis and Futuna to New Caledonia (2011 est.)
Urbanization: *urban population:* 57% of total population (2010)
rate of urbanization: 1.4% annual rate of change (2010-15 est.)

Major cities—population: NOUMEA (capital) 144,000 (2009)
Sex ratio: *at birth:* 1.05 male(s)/female
under 15 years: 1.04 male(s)/female
15–64 years: 1.01 male(s)/female
65 years and over: 0.86 male(s)/female
total population: 1 male(s)/female (2011 est.)
Infant mortality rate: *total:* 5.71 deaths/1,000 live births
country comparison to the world: 178
male: 6.75 deaths/1,000 live births
female: 4.62 deaths/1,000 live births (2011 est.)
Life expectancy at birth: *total population:* 76.75 years
country comparison to the world: 70
male: 72.67 years
female: 81.03 years (2011 est.)
Total fertility rate: 2.07 children born/woman (2011 est.)
country comparison to the world: 121
HIV/AIDS—adult prevalence rate: NA
HIV/AIDS—people living with HIV/AIDS: NA
HIV/AIDS—deaths: NA
Nationality: *noun:* New Caledonian(s)
adjective: New Caledonian
Ethnic groups: Melanesian 44.1%, European 34.1%, Wallisian & Futunian 9%, Tahitian 2.6%, Indonesian 2.5%, Vietnamese 1.4%, Ni-Vanuatu 1.1%, other 5.2% (1996 census)
Religions: Roman Catholic 60%, Protestant 30%, other 10%
Languages: French (official), 33 Melanesian-Polynesian dialects
Literacy: *definition:* age 15 and over can read and write
total population: 96.2%
male: 96.8%
female: 95.5% (1996 census)
School life expectancy (primary to tertiary education): NA
Education expenditures: NA

GOVERNMENT

Country name: *conventional long form:* Territory of New Caledonia and Dependencies
conventional short form: New Caledonia
local long form: Territoire des Nouvelle-Caledonie et Dependances
local short form: Nouvelle-Caledonie
Dependency status: territorial collectivity (or a sui generis collectivity) of France since 1998
Government type: NA
Capital: *name:* Noumea
geographic coordinates: 22 16 S, 166 27 E
time difference: UTC+11 (16 hours ahead of Washington, DC during Standard Time)
Administrative divisions: none (overseas territory of France); there are no first-order administrative divisions as defined by the US Government, but there are 3 provinces named Province des Iles, Province Nord, and Province Sud
Independence: none (overseas territory of France); note—a referendum on independence was held in 1998 but did not pass; a new referendum is scheduled to take place between 2014 and 2019
National holiday: Bastille Day, 14 July (1789); note—the local holiday is New Caledonia Day, 24 September (1853)
Constitution: 4 October 1958 (French Constitution)
Legal system: civil law system based on French law; the 1988 Matignon Accords grant substantial autonomy to the islands
Suffrage: 18 years of age; universal
Executive branch: *chief of state:* President Nicolas SARKOZY (since 16 May 2007); represented by High Commissioner Albert DUPUY (since 6 October 2010)
head of government: President of the Government Harold MARTIN (since 3 March 2011); note—since 3 March 2011, three different governments of Harold MARTIN have collapsed over the choice of a flag that will be used while it is being decolonized; President MARTIN is head of a caretaker government
cabinet: Cabinet consisting of 11 members elected from and by the Territorial Congress
(For more information visit the World Leaders website)
elections: French president elected by popular vote for a five-year term; high commissioner appointed by the French president on the advice of the French Ministry of Interior; president of the government elected by the members of the Territorial Congress for a five-year term (no term limits); note—last election held on 5 June 2009 (next to be held June 2014
Legislative branch: *elections:* unicameral Territorial Congress or Congres du territoire (54 seats; members belong to the three Provincial Assemblies or Assemblees Provinciales elected by popular vote to serve five-year terms) last held on 9 May 2009 (next to be held on 10 May 2014)
election results: percent of vote by party—NA; seats by party—UMP 13, Caledonia Together 10, UC 8, UNI 8, AE 6, FLNKS 3, Labor Party 3, other 3
note: New Caledonia holds two seats in the French Senate; elections last held on 21 September 2008 (next to be held not later than September 2014); results—percent of vote by party—NA; seats by party—UMP 2; New Caledonia also elects two seats to the French National Assembly; elections last held on 10 and 17 June 2007 (next to be held in June 2012); results—percent of vote by party—NA; seats by party—UMP 2
Judicial branch: Court of Appeal or Cour d'Appel; County Courts; Joint Commerce Tribunal Court; Children's Court
Political parties and leaders: Caledonia My Country; Caledonia Together [Philippe GOMES]; Caledonian Union or UC [Nicholas ABOUT]; Communist Republican and Left Party or CRC-SPG [Nichole BORVO COHEN-SEAT]; Democratic and European Social Rally or R.D.S.E. [Yvon COLLIN]; Front National or FN [Jean-Marie LE PEN]; Kanak Socialist Front for National Liberation or FLNKS (includes PALIKA, UNI, UC, and UPM); Labor Party [Louis Kotra UREGEI]; National Union for Independence or UNI; Parti de Liberation Kanak or PALIKA [Paul NEAOUTYINE]; Renewed Caledonian Union; Socialist Group [Jean Pierre BEL]; Socialist Kanak Liberation or LKS [Nidoish NAISSELINE]; The Future Together or AE [Didier LEROUX]; The Rally or UMP [Gerard LONGUET]; Union Nationale pour l'Independance or UNI; Union of Pro-Independence Co-operation Committees [Francois BURCK]
Political pressure groups and leaders: NA
International organization participation: ITUC, PIF (associate member), SPC, UPU, WFTU
Diplomatic representation in the US: none (overseas territory of France)
Diplomatic representation from the US: none (overseas territory of France)
Flag description: the flag of France is used
National anthem: *name:* "Soyons unis, devenons freres" (Let Us Be United, Let Us Become Brothers)
lyrics/music: Chorale Melodia (a local choir)
note: adopted 2008; the anthem contains a mixture of lyrics in both French and Nengone (an indigenous language); as a self-governing territory of France, in addition to the local anthem, "La Marseillaise" is official (see France)

ECONOMY

Economy—overview: New Caledonia has about 25% of the world's known nickel resources. Only a small amount of the land is suitable for cultivation, and food accounts for about 20% of imports. In addition to nickel, substantial financial support from France—equal to more than 15% of GDP—and tourism are keys to the health of the economy. Substantial new investment in the nickel industry, combined with the recovery of global nickel prices, brightens the economic outlook for the next several years.
GDP (purchasing power parity): $3.158 billion (2003 est.)
country comparison to the world: 175
GDP (official exchange rate): $3.3 billion (2003 est.)
GDP—real growth rate: NA%
GDP—per capita (PPP): $15,000 (2003 est.)
country comparison to the world: 74
GDP—composition by sector: *agriculture:* 15%
industry: 8.8%
services: 76.2% (2003)
Labor force: 102,600 (2007)
country comparison to the world: 183
Labor force—by occupation: *agriculture:* 20%
industry: 20%
services: 60% (2002)
Unemployment rate: 17.1% (2004)
country comparison to the world: 157
Population below poverty line: NA%
Household income or consumption by percentage share: *lowest 10%:* NA%
highest 10%: NA%
Budget: *revenues:* $996 million
expenditures: $1.072 billion (2001 est.)
Inflation rate (consumer prices): 1.4% (2000 est.)
country comparison to the world: 32

Market value of publicly traded shares: $NA
Agriculture—products: vegetables; beef, deer, other livestock products; fish
Industries: nickel mining and smelting
Electricity—production: 1.825 billion kWh (2007 est.)
country comparison to the world: 135
Electricity—consumption: 1.697 billion kWh (2007 est.)
country comparison to the world: 138
Electricity—exports: 0 kWh (2008 est.)
Electricity—imports: 0 kWh (2008 est.)
Oil—production: 0 bbl/day (2009 est.)
country comparison to the world: 115
Oil—consumption: 13,000 bbl/day (2009 est.)
country comparison to the world: 141
Oil—exports: 645.3 bbl/day (2007 est.)
country comparison to the world: 123
Oil—imports: 14,430 bbl/day (2007 est.)
country comparison to the world: 127
Oil—proved reserves: 0 bbl (1 January 2010 est.)
country comparison to the world: 168
Natural gas—production: 0 cu m (2008 est.)
country comparison to the world: 164
Natural gas—consumption: 0 cu m (2008 est.)
country comparison to the world: 202
Natural gas—exports: 0 cu m (2008 est.)
country comparison to the world: 146
Natural gas—imports: 0 cu m (2008 est.)
country comparison to the world: 93
Natural gas—proved reserves: 0 cu m (1 January 2010 est.)
country comparison to the world: 169
Exports: $1.341 billion (2006)
country comparison to the world: 145
Exports—commodities: ferronickels, nickel ore, fish
Exports—partners: France 27.52%, Japan 14.87%, Taiwan 10.51%, Spain 7.74%, China 7.31%, Australia 6.64%, Belgium 5.13%, South Korea 4.2% (2009)
Imports: $1.998 billion (2006)
country comparison to the world: 151
Imports—commodities: machinery and equipment, fuels, chemicals, foodstuffs
Imports—partners: France 39.89%, Singapore 14.12%, Australia 12.5%, NZ 5.42% (2009)
Debt—external: $79 million (1998 est.)
country comparison to the world: 183
Exchange rates: Comptoirs Francais du Pacifique francs (XPF) per US dollar–87.59 (2007) 94.97 (2006) 95.89 (2005) 96.04 (2004) 105.66 (2003)

COMMUNICATIONS

Telephones—main lines in use: 65,900 (2009)
country comparison to the world: 155
Telephones—mobile cellular: 208,000 (2009)
country comparison to the world: 172
Telephone system: *general assessment:* a submarine cable network connection between New Caledonia and Australia, completed in 2007, is expected to significantly increase network capacity and improve high-speed connectivity and access to international networks
domestic: combined fixed-line and mobile-cellular telephone subscribership exceeds 100 per 100 persons
international: country code–687; satellite earth station–1 Intelsat (Pacific Ocean) (2008)
Broadcast media: the publicly-owned French Overseas Network (RFO), which operates in France's overseas departments and territories, broadcasts over the RFO Nouvelle Caledonie television and radio stations; a small number of privately-owned radio stations also broadcast (2008)
Internet country code: .nc
Internet hosts: 22,456 (2010)
country comparison to the world: 106
Internet users: 85,000 (2009)
country comparison to the world: 164

TRANSPORTATION

Airports: 25 (2010)
country comparison to the world: 130
Airports—with paved runways: *total:* 12
over 3,047 m: 1
914 to 1,523 m: 10
under 914 m: 1 (2010)
Airports—with unpaved runways: *total:* 13
914 to 1,523 m: 5
under 914 m: 8 (2010)
Heliports: 8 (2010)
Roadways: *total:* 5,622 km (2006)
country comparison to the world: 151
Merchant marine: *registered in other countries:* 3 (France 3) (2010)
country comparison to the world: 140
Ports and terminals: Noumea

MILITARY

Military branches: no regular military forces; French military, police, and gendarmerie (2009)
Manpower available for military service: *males age 16–49:* 68,219 (2010 est.)
Manpower fit for military service: *males age 16–49:* 56,233
females age 16–49: 55,983 (2010 est.)
Manpower reaching militarily significant age annually:
male: 2,272
female: 2,167 (2010 est.)
Military expenditures: NA
Military—note: defense is the responsibility of France

TRANSNATIONAL ISSUES

Disputes—international: Matthew and Hunter Islands east of New Caledonia claimed by France and Vanuatu

NEW ZEALAND

INTRODUCTION

Background: The Polynesian Maori reached New Zealand in about A.D. 800. In 1840, their chieftains entered into a compact with Britain, the Treaty of Waitangi, in which they ceded sovereignty to Queen Victoria while retaining territorial rights. In that same year, the British began the first organized colonial settlement. A series of land wars between 1843 and 1872 ended with the defeat of the native peoples. The British colony of New Zealand became an independent dominion in 1907 and supported the UK militarily in both world wars. New Zealand's full participation in a number of defense alliances lapsed by the 1980s. In recent years, the government has sought to address longstanding Maori grievances.

GEOGRAPHY

Location: Oceania, islands in the South Pacific Ocean, southeast of Australia
Geographic coordinates: 41 00 S, 174 00 E
Map references: Oceania
Area: *total:* 267,710 sq km
country comparison to the world: 75
land: 267,710 sq km
water: NA
note: includes Antipodes Islands, Auckland Islands, Bounty Islands, Campbell Island, Chatham Islands, and Kermadec Islands
Area—comparative: about the size of Colorado
Land boundaries: 0 km
Coastline: 15,134 km
Maritime claims: *territorial sea:* 12 nm
contiguous zone: 24 nm
exclusive economic zone: 200 nm
continental shelf: 200 nm or to the edge of the continental margin
Climate: temperate with sharp regional contrasts
Terrain: predominately mountainous with some large coastal plains
Elevation extremes: *lowest point:* Pacific Ocean 0 m
highest point: Aoraki-Mount Cook 3,754 m
Natural resources: natural gas, iron ore, sand, coal, timber, hydropower, gold, limestone
Land use: *arable land:* 5.54%
permanent crops: 6.92%
other: 87.54% (2005)
Irrigated land: 2,850 sq km (2003)
Total renewable water resources: 397 cu km (1995)
Freshwater withdrawal (domestic/industrial/agricultural): *total:* 2.11 cu km/yr (48%/9%/42%)
per capita: 524 cu m/yr (2000)
Natural hazards: earthquakes are common, though usually not severe; volcanic activity
volcanism: New Zealand experiences significant volcanism on North Island; Ruapehu (elev. 2,797 m), which last erupted in 2007, has a history of large eruptions in the past century; Taranaki has the potential to produce dangerous avalanches and lahars; other historically active volcanoes include Okataina, Raoul Island, Tongariro, and White Island
Environment—current issues: deforestation; soil erosion; native flora and fauna hard-hit by invasive species
Environment—international agreements: *party to:* Antarctic-Environmental Protocol, Antarctic-Marine Living Resources, Antarctic Treaty, Biodiversity, Climate Change, Climate Change-Kyoto Protocol,

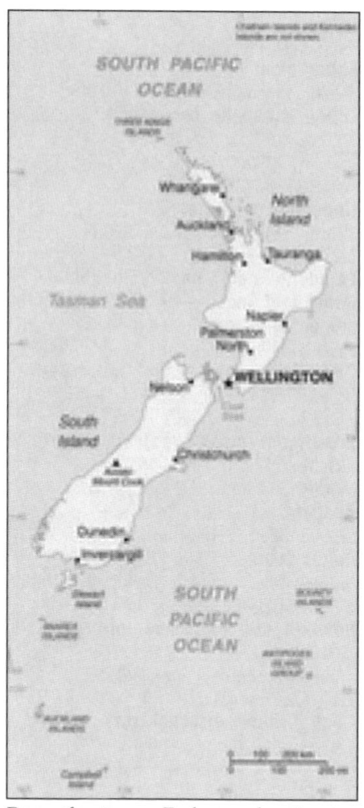

Desertification, Endangered Species, Environmental Modification, Hazardous Wastes, Law of the Sea, Marine Dumping, Ozone Layer Protection, Ship Pollution, Tropical Timber 83, Tropical Timber 94, Wetlands, Whaling
signed, but not ratified: Antarctic Seals, Marine Life Conservation
Geography—note: almost 90% of the population lives in cities; Wellington is the southernmost national capital in the world

PEOPLE

Population: 4,290,347 (July 2011 est.)
country comparison to the world: 125
Age structure: *0–14 years:* 20.4% (male 448,106/female 426,348)
15–64 years: 66.4% (male 1,426,595/female 1,420,643)
65 years and over: 13.3% (male 260,454/female 308,201) (2011 est.)
Median age: *total:* 37 years
male: 36.2 years
female: 37.8 years (2011 est.)
Population growth rate: 0.882% (2011 est.)
country comparison to the world: 126
Birth rate: 13.68 births/1,000 population (2011 est.)
country comparison to the world: 150
Death rate: 7.15 deaths/1,000 population (July 2011 est.)
country comparison to the world: 128
Net migration rate: 2.28 migrant(s)/1,000 population (2011 est.)
country comparison to the world: 38
Urbanization: *urban population:* 86% of total population (2010)
rate of urbanization: 0.9% annual rate of change (2010-15 est.)
Major cities—population: Auckland 1.36 million; WELLINGTON (capital) 391,000 (2009)
Sex ratio: *at birth:* 1.048 male(s)/female
under 15 years: 1.05 male(s)/female
15–64 years: 1 male(s)/female
65 years and over: 0.84 male(s)/female
total population: 0.99 male(s)/female (2011 est.)
Infant mortality rate: *total:* 4.78 deaths/1,000 live births
country comparison to the world: 185
male: 5.37 deaths/1,000 live births
female: 4.16 deaths/1,000 live births (2011 est.)
Life expectancy at birth: *total population:* 80.59 years
country comparison to the world: 23
male: 78.61 years
female: 82.67 years (2011 est.)
Total fertility rate: 2.08 children born/woman (2011 est.)
country comparison to the world: 119
HIV/AIDS—adult prevalence rate: 0.1% (2009 est.)
country comparison to the world: 152
HIV/AIDS—people living with HIV/AIDS: 2,500 (2009 est.)
country comparison to the world: 133
HIV/AIDS—deaths: fewer than 100 (2009 est.)
country comparison to the world: 148
Drinking water source: *Improved:*
urban: 100% of population
rural: 100% of population
total: 100% of population (2008)
Nationality: *noun:* New Zealander(s)
adjective: New Zealand
Ethnic groups: European 56.8%, Asian 8%, Maori 7.4%, Pacific islander 4.6%, mixed 9.7%, other 13.5% (2006 Census)
Religions: Anglican 13.8%, Roman Catholic 12.6%, Presbyterian, Congregational, and Reformed 10%, Christian (no denomination specified) 4.6%, Methodist 3%, Pentecostal 2%, Baptist 1.4%, other Christian 3.8%, Maori Christian 1.6%, Hindu 1.6%, Buddhist 1.3%, other religions 2.2%, none 32.2%, other or unidentified 9.9% (2006 Census)
Languages: English (official) 91.2%, Maori (official) 3.9%, Samoan 2.1%, French 1.3%, Hindi 1.1%, Yue 1.1%, Northern Chinese 1%, other 12.9%, New Zealand Sign Language (official)
note: shares sum to 114.6% due to multiple responses on census (2006 Census)
Literacy: *definition:* age 15 and over can read and write
total population: 99%
male: 99%
female: 99% (2003 est.)
School life expectancy (primary to tertiary education): *total:* 19 years
male: 19 years
female: 20 years (2008)
Education expenditures: 6.1% of GDP (2007)
country comparison to the world: 27

GOVERNMENT

Country name: *conventional long form:* none
conventional short form: New Zealand
abbreviation: NZ
Government type: parliamentary democracy and a Commonwealth realm
Capital: *name:* Wellington
geographic coordinates: 41 28 S, 174 51 E
time difference: UTC+12 (17 hours ahead of Washington, DC during Standard Time)
daylight saving time: +1hr, begins last Sunday in September; ends first Sunday in April
note: New Zealand is divided into two time zones–New Zealand standard time (12 hours in advance of UTC), and Chatham Islands time (45 minutes in advance of New Zealand standard time)
Administrative divisions: 16 regions and 1 territory*; Auckland, Bay of Plenty, Canterbury, Chatham Islands*, Gisborne, Hawke's Bay, Manawatu-Wanganui, Marlborough, Nelson, Northland, Otago, Southland, Taranaki, Tasman, Waikato, Wellington, West Coast
Dependent areas: Cook Islands, Niue, Tokelau
Independence: 26 September 1907 (from the UK)
National holiday: Waitangi Day (Treaty of Waitangi established British sovereignty over New Zealand), 6 February (1840); ANZAC Day (commemorated as the anniversary of the landing of troops of the Australian and New Zealand Army Corps during World War I at Gallipoli, Turkey), 25 April (1915)
Constitution: consists of a series of legal documents, including certain acts of the UK and New Zealand parliaments, as well as The Constitution Act 1986, which is the principal formal charter; adopted 1 January 1987, effective 1 January 1987
Legal system: common law system, based on English model, with special legislation and land courts for the Maori
International law organization participation: accepts compulsory ICJ jurisdiction with reservations; accepts ICCt jurisdiction
Suffrage: 18 years of age; universal
Executive branch: *chief of state:* Queen ELIZABETH II (since 6 February 1952); represented by Governor General Anand SATYANAND (since 23 August 2006)
head of government: Prime Minister John KEY (since 19 November 2008); Deputy Prime Minister Bill ENGLISH (since 19 November 2008)
cabinet: Executive Council appointed by the governor general on the recommendation of the prime minister
(For more information visit the World Leaders website)
elections: the monarchy is hereditary; governor general appointed by the monarch; following legislative elections, the leader of the majority party or the leader of a majority coalition usually appointed prime minister by the governor general; deputy prime minister appointed by the governor general
Legislative branch: unicameral House of Representatives–commonly called Parliament (usually 120 seats; 70 members elected by popular vote in single-member constituencies including 7 Maori constituencies, 50 proportional seats chosen from party lists; serve three-year terms)

elections: last held on 8 November 2008 (next to be held not later than 27 November 2011)
election results: percent of vote by party—National Party 44.9%, Labor Party 34%, Green Party 6.7%, NZ First 4%, ACT Party 3.7%, Maori 2.4%, Progressive 0.9%, United Front 0.9%, other 6.6%; seats by party—National Party 58, Labor Party 43, Green Party 9, ACT Party 5, Maori 5, Progressive 1, United Front 1
note: results of 2008 election saw the total number of seats increase to 122
Judicial branch: Supreme Court; Court of Appeal; High Court; note—judges appointed by the governor general
Political parties and leaders: ACT New Zealand [Rodney HIDE]; Green Party [Russel NORMAN and Metiria TUREI]; Maori Party [Tariana TURIA and Dr. Pita SHARPLES]; New Zealand National Party [John KEY]; New Zealand First Party or NZ First [Winston PETERS]; New Zealand Labor Party [Phil GOFF]; Jim Anderton's Progressive Party [James (Jim) ANDERTON]; United Future New Zealand [Peter DUNNE]
Political pressure groups and leaders: Women's Electoral Lobby or WEL
other: apartheid groups; civil rights groups; farmers groups; Maori; nuclear weapons groups; women's rights groups
International organization participation: ADB, ANZUS (US suspended security obligations to NZ on 11 August 1986), APEC, ARF, ASEAN (dialogue partner), Australia Group, BIS, C, CP, EAS, EBRD, FAO, FATF, IAEA, IBRD, ICAO, ICC, ICRM, IDA, IEA, IFAD, IFC, IFRCS, IHO, ILO, IMF, IMO, IMSO, Interpol, IOC, IOM, IPU, ISO, ITSO, ITU, ITUC, MIGA, NSG, OECD, OPCW, Paris Club (associate), PCA, PIF, Sparteca, SPC, UN, UNCTAD, UNESCO, UNHCR, UNIDO, UNMIS, UNMIT, UNTSO, UPU, WCO, WFTU, WHO, WIPO, WMO, WTO
Diplomatic representation in the US:
chief of mission: Ambassador Michael K. MOORE
chancery: 37 Observatory Circle NW, Washington, DC 20008
telephone: [1] (202) 328-4800
FAX: [1] (202) 667-5227
consulate(s) general: New York, Santa Monica
Diplomatic representation from the US: *chief of mission:* Ambassador David HUEBNER
embassy: 29 Fitzherbert Terrace, Thorndon, Wellington
mailing address: P. O. Box 1190, Wellington; PSC 467, Box 1, APO AP 96531-1034
telephone: [64] (4) 462-6000
FAX: [64] (4) 499-0490
consulate(s) general: Auckland
Flag description: blue with the flag of the UK in the upper hoist-side quadrant with four red five-pointed stars edged in white centered in the outer half of the flag; the stars represent the Southern Cross constellation
National anthem: *name:* "God Defend New Zealand"
lyrics/music: Thomas BRACKEN [English], Thomas Henry SMITH [Maori]/John Joseph WOODS
note: adopted 1940 as national song, adopted 1977 as co-national anthem; New Zealand has two national anthems with equal status; as a commonwealth realm, in addition to "God Defend New Zealand," "God Save the Queen" serves as a national anthem (see United Kingdom); "God Save the Queen" normally is played only when a member of the royal family or the governor-general is present; in all other cases, "God Defend New Zealand" is played
Government—note: while not an official symbol, the Kiwi, a small native flightless bird, represents New Zealand

ECONOMY

Economy—overview: Over the past 20 years the government has transformed New Zealand from an agrarian economy dependent on concessionary British market access to a more industrialized, free market economy that can compete globally. This dynamic growth has boosted real incomes—but left behind some at the bottom of the ladder—and broadened and deepened the technological capabilities of the industrial sector. Per capita income rose for ten consecutive years until 2007 in purchasing power parity terms, but fell in 2008-09. Debt-driven consumer spending drove robust growth in the first half of the decade, helping fuel a large balance of payments deficit that posed a challenge for economic managers. Inflationary pressures caused the central bank to raise its key rate steadily from January 2004 until it was among the highest in the OECD in 2007-08; international capital inflows attracted to the high rates further strengthened the currency and housing market, however, aggravating the current account deficit. The economy fell into recession before the start of the global financial crisis and contracted for five consecutive quarters in 2008-09. In line with global peers, the central bank cut interest rates aggressively and the government developed fiscal stimulus measures. The economy posted a 1.7% decline in 2009, but pulled out of recession late in the year, and achieved 2.1% growth in 2010. Nevertheless, key trade sectors remain vulnerable to weak external demand. The government plans to raise productivity growth and develop infrastructure, while reining in government spending.
GDP (purchasing power parity): $117.8 billion (2010 est.)
country comparison to the world: 63
$116 billion (2009 est.)
$118.5 billion (2008 est.)
note: data are in 2010 US dollars
GDP (official exchange rate): $140.4 billion (2010 est.)
GDP—real growth rate: 1.5% (2010 est.)
country comparison to the world: 162
-2.1% (2009 est.)
-0.2% (2008 est.)
GDP—per capita (PPP): $27,700 (2010 est.)
country comparison to the world: 51
$27,500 (2009 est.)
$28,400 (2008 est.)
note: data are in 2010 US dollars
GDP—composition by sector: *agriculture:* 4.6%
industry: 24%
services: 71.4% (2010 est.)
Labor force: 2.32 million (2010 est.)
country comparison to the world: 114
Labor force—by occupation: *agriculture:* 7%
industry: 19%
services: 74% (2006 est.)
Unemployment rate: 6.5% (2010 est.)
country comparison to the world: 66
6.2% (2009 est.)
Population below poverty line: NA%
Household income or consumption by percentage share: *lowest 10%:* %NA
highest 10%: %NA
Distribution of family income—Gini index: 36.2 (1997)
country comparison to the world: 83
Investment (gross fixed): 19.6% of GDP (2010 est.)
country comparison to the world: 94
Budget: *revenues:* $56.24 billion
expenditures: $62.18 billion (2010 est.)
Public debt: 25.5% of GDP (2010 est.)
country comparison to the world: 99
22.2% of GDP (2009 est.)
Inflation rate (consumer prices): 2.6% (2010 est.)
country comparison to the world: 70
2.1% (2009 est.)
Central bank discount rate: 2.5% (31 December 2009)
country comparison to the world: 83
5% (31 December 2008)
Commercial bank prime lending rate: 10.39% (31 December 2009 est.)
country comparison to the world: 67
12.21% (31 December 2008 est.)
Stock of narrow money: $24.15 billion (31 December 2010 est.)
country comparison to the world: 61
$21.81 billion (31 December 2009 est.)
Stock of broad money: $118.1 billion (31 December 2010 est.)
country comparison to the world: 48
$108.9 billion (31 December 2009 est.)
Stock of domestic credit: $206.2 billion (31 December 2010 est.)
country comparison to the world: 38
$180.5 billion (31 December 2009 est.)
Market value of publicly traded shares: $67.06 billion (31 December 2009)
country comparison to the world: 58
$24.17 billion (31 December 2008)
$47.45 billion (31 December 2007)
Agriculture—products: dairy products, lamb and mutton; wheat, barley, potatoes, pulses, fruits, vegetables; wool, beef; fish
Industries: food processing, wood and paper products, textiles, machinery, transportation equipment, banking and insurance, tourism, mining
Industrial production growth rate: 2% (2010 est.)
country comparison to the world: 131
Electricity—production: 42.4 billion kWh (2007 est.)
country comparison to the world: 53
Electricity—consumption: 39.24 billion kWh (2007 est.)
country comparison to the world: 53
Electricity—exports: 0 kWh (2008 est.)
Electricity—imports: 0 kWh (2008 est.)
Oil—production: 61,150 bbl/day (2009 est.)
country comparison to the world: 58

Oil—consumption: 154,100 bbl/day (2009 est.)
country comparison to the world: 65
Oil—exports: 54,560 bbl/day (2008 est.)
country comparison to the world: 75
Oil—imports: 143,900 bbl/day (2008 est.)
country comparison to the world: 57
Oil—proved reserves: 60 million bbl (1 January 2010 est.)
country comparison to the world: 78
Natural gas—production: 4.305 billion cu m (2009 est.)
country comparison to the world: 50
Natural gas—consumption: 4.32 billion cu m (2009 est.)
country comparison to the world: 62
Natural gas—exports: 0 cu m (2008 est.)
country comparison to the world: 154
Natural gas—imports: NA
Natural gas—proved reserves: 33.98 billion cu m (1 January 2010 est.)
country comparison to the world: 68
Current account balance: $-4.504 billion (2010 est.)
country comparison to the world: 169
$-3.693 billion (2009 est.)
Exports: $33.24 billion (2010 est.)
country comparison to the world: 62
$25.35 billion (2009 est.)
Exports—commodities: dairy products, meat, wood and wood products, fish, machinery
Exports—partners: Australia 23.36%, US 9.64%, China 9.21%, Japan 7.1%, UK 4.21% (2009)
Imports: $30.24 billion (2010 est.)
country comparison to the world: 58
$23.95 billion (2009 est.)
Imports—commodities: machinery and equipment, vehicles and aircraft, petroleum, electronics, textiles, plastics
Imports—partners: Australia 18.4%, China 15.09%, US 10.45%, Japan 7.24%, Germany 4.16%, Singapore 4.12% (2009)
Reserves of foreign exchange and gold: $17.85 billion (31 December 2010 est.)
country comparison to the world: 46
$15.59 billion (31 December 2009 est.)
Debt—external: $64.33 billion (31 December 2010 est.)
country comparison to the world: 47
$62.47 billion (31 December 2009 est.)
Stock of direct foreign investment—at home: $67.18 billion (31 December 2010 est.)
country comparison to the world: 50
$66.63 billion (31 December 2009 est.)
Stock of direct foreign investment—abroad: $NA (31 December 2009)
$59.08 billion (31 December 2008)
Exchange rates: New Zealand dollars (NZD) per US dollar–1.3874 (2010) 1.6002 (2009) 1.4151 (2008) 1.3811 (2007) 1.5408 (2006)

COMMUNICATIONS

Telephones—main lines in use: 1.87 million (2009)
country comparison to the world: 59
Telephones—mobile cellular: 4.7 million (2009)
country comparison to the world: 99
Telephone system: *general assessment:* excellent domestic and international systems
domestic: combined fixed-line and mobile-cellular telephone subscribership exceeds 150 per 100 persons
international: country code–64; the Southern Cross submarine cable system provides links to Australia, Fiji, and the US; satellite earth stations–8 (1 Inmarsat–Pacific Ocean, 7 other)
Broadcast media: state-owned Television New Zealand operates multiple television networks while state-owned Radio New Zealand operates 3 radio networks and an external shortwave radio service to the South Pacific region; a small number of national commercial television and radio stations and a large number of regional commercial television and radio stations are available; cable and satellite TV systems are accessible (2008)
Internet country code: .nz
Internet hosts: 2.47 million (2010)
country comparison to the world: 33
Internet users: 3.4 million (2009)
country comparison to the world: 62

TRANSPORTATION

Airports: 122 (2010)
country comparison to the world: 49
Airports—with paved runways: *total:* 40
over 3,047 m: 2
2,438 to 3,047 m: 1
1,524 to 2,437 m: 12
914 to 1,523 m: 24
under 914 m: 1 (2010)
Airports—with unpaved runways: *total:* 82
1,524 to 2,437 m: 3
914 to 1,523 m: 32
under 914 m: 47 (2010)
Pipelines: condensate 331 km; gas 1,838 km; liquid petroleum gas 172 km; oil 288 km; refined products 198 km (2010)
Railways: *total:* 4,128 km
country comparison to the world: 40
narrow gauge: 4,128 km 1.067-m gauge (506 km electrified) (2010)
Roadways: *total:* 93,911 km
country comparison to the world: 49
paved: 61,879 km (includes 172 km of expressways)
unpaved: 32,032 km (2009)
Merchant marine: *total:* 14
country comparison to the world: 103
by type: bulk carrier 3, cargo 3, chemical tanker 1, container 1, passenger/cargo 4, petroleum tanker 2
foreign-owned: 7 (Australia 1, Germany 2, Hong Kong 1, South Africa 1, Switzerland 2)
registered in other countries: 6 (Antigua and Barbuda 2, Cook Islands 1, France 1, Samoa 1, UK 1) (2010)
Ports and terminals: Auckland, Lyttelton, Manukau Harbor, Marsden Point, Tauranga, Wellington

MILITARY

Military branches: New Zealand Defense Force (NZDF): New Zealand Army, Royal New Zealand Navy, Royal New Zealand Air Force (Te Hokowhitu o Kahurangi, RNZAF) (2010)
Military service age and obligation: 17 years of age for voluntary military service; soldiers cannot be deployed until the age of 18; no conscription (2010)
Manpower available for military service: *males age 16–49:* 1,019,798
females age 16–49: 1,003,429 (2010 est.)
Manpower fit for military service: *males age 16–49:* 843,526
females age 16–49: 828,779 (2010 est.)
Manpower reaching militarily significant age annually: *male:* 30,846
female: 28,825 (2010 est.)
Military expenditures: 1% of GDP (2005 est.)
country comparison to the world: 129

TRANSNATIONAL ISSUES

Disputes—international: asserts a territorial claim in Antarctica (Ross Dependency)
Illicit drugs: significant consumer of amphetamines

NICARAGUA

INTRODUCTION

Background: The Pacific coast of Nicaragua was settled as a Spanish colony from Panama in the early 16th century. Independence from Spain was declared in 1821 and the country became an independent republic in 1838. Britain occupied the Caribbean Coast in the first half of the 19th century, but gradually ceded control of the region in subsequent decades. Violent opposition to governmental manipulation and corruption spread to all classes by 1978 and resulted in a short-lived civil war that brought the Marxist Sandinista guerrillas to power in 1979. Nicaraguan aid to leftist rebels in El Salvador caused the US to sponsor anti-Sandinista contra guerrillas through much of the 1980s. After losing free and fair elections in 1990, 1996, and 2001, former Sandinista President Daniel ORTEGA Saavedra was elected president in 2006. The 2008 municipal elections were marred by widespread irregularities. Nicaragua's infrastructure and economy—hard hit by the earlier civil war and by Hurricane Mitch in 1998—are slowly being rebuilt, but democratic institutions have been weakened under the ORTEGA administration.

GEOGRAPHY

Location: Central America, bordering both the Caribbean Sea and the North Pacific Ocean, between Costa Rica and Honduras
Geographic coordinates: 13 00 N, 85 00 W
Map references: Central America and the Caribbean
Area: *total:* 130,370 sq km
country comparison to the world: 97
land: 119,990 sq km
water: 10,380 sq km
Area—comparative: slightly smaller than New York state
Land boundaries: *total:* 1,231 km
border countries: Costa Rica 309 km, Honduras 922 km
Coastline: 910 km
Maritime claims: *territorial sea:* 12 nm
contiguous zone: 24 nm
continental shelf: natural prolongation
Climate: tropical in lowlands, cooler in highlands
Terrain: extensive Atlantic coastal plains rising to central interior mountains; narrow Pacific coastal plain interrupted by volcanoes
Elevation extremes: *lowest point:* Pacific Ocean 0 m
highest point: Mogoton 2,438 m
Natural resources: gold, silver, copper, tungsten, lead, zinc, timber, fish
Land use: *arable land:* 14.81%
permanent crops: 1.82%
other: 83.37% (2005)
Irrigated land: 610 sq km (2003)
Total renewable water resources: 196.7 cu km (2000)
Freshwater withdrawal (domestic/industrial/agricultural): *total:* 1.3 cu km/yr (15%/2%/83%)
per capita: 237 cu m/yr (2000)
Natural hazards: destructive earthquakes; volcanoes; landslides; extremely susceptible to hurricanes
volcanism: Nicaragua experiences significant volcanic activity; Cerro Negro (elev. 728 m), which last erupted in 1999, is one of Nicaragua's most active volcanoes; its lava flows and ash have been known to cause significant damage to farmland and buildings; other historically active volcanoes include Concepcion, Cosiguina, Las Pilas, Masaya, Momotombo, San Cristobal, and Telica
Environment—current issues: deforestation; soil erosion; water pollution
Environment—international agreements: *party to:* Biodiversity, Climate Change, Climate Change-Kyoto Protocol, Desertification, Endangered Species, Environmental Modification, Hazardous Wastes, Law of the Sea, Ozone Layer Protection, Ship Pollution, Wetlands, Whaling
signed, but not ratified: none of the selected agreements
Geography—note: largest country in Central America; contains the largest freshwater body in Central America, Lago de Nicaragua

PEOPLE

Population: 5,666,301 (July 2011 est.)
country comparison to the world: 108
Age structure: *0–14 years:* 31.7% (male 913,905/female 879,818)
15–64 years: 63.8% (male 1,743,591/female 1,874,025)
65 years and over: 4.5% (male 116,153/female 138,809) (2011 est.)
Median age: *total:* 22.9 years
male: 22.1 years
female: 23.7 years (2011 est.)
Population growth rate: 1.088% (2011 est.)
country comparison to the world: 107
Birth rate: 19.46 births/1,000 population (2011 est.)
country comparison to the world: 93
Death rate: 5.03 deaths/1,000 population (July 2011 est.)
country comparison to the world: 183
Net migration rate: -3.54 migrant(s)/1,000 population (2011 est.)
country comparison to the world: 182
Urbanization: *urban population:* 57% of total population (2010)
rate of urbanization: 2% annual rate of change (2010-15 est.)
Major cities—population: MANAGUA (capital) 934,000 (2009)
Sex ratio: *at birth:* 1.05 male(s)/female
under 15 years: 1.04 male(s)/female
15–64 years: 1 male(s)/female
65 years and over: 0.78 male(s)/female
total population: 1 male(s)/female (2011 est.)
Infant mortality rate: *total:* 22.64 deaths/1,000 live births
country comparison to the world: 88
male: 25.94 deaths/1,000 live births
female: 19.19 deaths/1,000 live births (2011 est.)
Life expectancy at birth: *total population:* 71.9 years
country comparison to the world: 130
male: 69.82 years
female: 74.09 years (2011 est.)
Total fertility rate: 2.12 children born/woman (2011 est.)
country comparison to the world: 116
HIV/AIDS—adult prevalence rate: 0.2% (2009 est.)
country comparison to the world: 105
HIV/AIDS—people living with HIV/AIDS: 6,900 (2009 est.)
country comparison to the world: 114
HIV/AIDS—deaths: fewer than 500 (2009 est.)
country comparison to the world: 84
Major infectious diseases: *degree of risk:* high
food or waterborne diseases: bacterial diarrhea, hepatitis A, and typhoid fever
vectorborne disease: dengue fever and malaria
water contact disease: leptospirosis (2009)
Drinking water source: *Improved:*
urban: 98% of population
rural: 68% of population
total: 85% of population
Unimproved:
urban: 2% of population
rural: 32% of population
total: 15% of population (2008)
Sanitation facility access: *Improved:*
urban: 63% of population
rural: 37% of population
total: 52% of population
Unimproved:
urban: 37% of population
rural: 63% of population
total: 48% of population (2008)
Nationality: *noun:* Nicaraguan(s)
adjective: Nicaraguan
Ethnic groups: mestizo (mixed Amerindian and white) 69%, white 17%, black 9%, Amerindian 5%
Religions: Roman Catholic 58.5%, Evangelical 21.6%, Moravian 1.6%, Jehovah's Witness 0.9%, other 1.7%, none 15.7% (2005 census)
Languages: Spanish (official) 97.5%, Miskito 1.7%, other 0.8% (1995 census)
note: English and indigenous languages found on the Atlantic coast
Literacy: *definition:* age 15 and over can read and write
total population: 67.5%
male: 67.2%
female: 67.8% (2003 est.)
School life expectancy (primary to tertiary education): *total:* 11 years
male: 11 years
female: 11 years (2003)
Education expenditures: 3.1% of GDP (2003)
country comparison to the world: 131

GOVERNMENT

Country name: *conventional long form:* Republic of Nicaragua
conventional short form: Nicaragua
local long form: Republica de Nicaragua
local short form: Nicaragua
Government type: republic
Capital: *name:* Managua
geographic coordinates: 12 09 N, 86 17 W
time difference: UTC-6 (1 hour behind Washington, DC during Standard Time)
Administrative divisions: 15 departments (departamentos, singular—departamento) and 2 autonomous regions* (regiones autonomistas, singular—region autonoma); Atlantico Norte*, Atlantico Sur*, Boaco, Carazo, Chinandega, Chontales, Esteli, Granada, Jinotega, Leon, Madriz, Managua, Masaya, Matagalpa, Nueva Segovia, Rio San Juan, Rivas
Independence: 15 September 1821 (from Spain)
National holiday: Independence Day, 15 September (1821)
Constitution: 9 January 1987; revised in 1995, 2000, and 2005
Legal system: civil law system; Supreme Court may review administrative acts
International law organization participation: accepts compulsory ICJ jurisdiction with reservations; non-party state to the ICCt
Suffrage: 16 years of age; universal
Executive branch: *chief of state:* President Daniel ORTEGA Saavedra (since 10 January 2007); Vice President Jaime MORALES Carazo (since 10 January 2007); note—the president is both chief of state and head of government
head of government: President Daniel ORTEGA Saavedra (since 10 January 2007); Vice President Jaime MORALES Carazo (since 10 January 2007)
cabinet: Council of Ministers appointed by the president

(For more information visit the World Leaders website)
elections: president and vice president elected on the same ticket by popular vote for a five-year term (eligible for a second term so long as it is not consecutive); election last held on 5 November 2006 (next to be held by November 2011)
election results: Daniel ORTEGA Saavedra elected president–38.1%, Eduardo MONTEALEGRE 29%, Jose RIZO 26.2%, Edmundo JARQUIN 6.4%, other 0.3%
Legislative branch: unicameral National Assembly or Asamblea Nacional (92 seats; 90 members elected by proportional representation and party lists to serve five-year terms; 1 seat for the previous president, 1 seat for the runner-up in previous presidential election)
elections: last held on 5 November 2006 (next to be held by November 2011)
election results: percent of vote by party–NA; seats by party–FSLN 38, PLC 25, ALN 24, MRS 5; note–political parties have been reorganized to reflect the following seat distribution: as of 1 March 2011–seats by party–FSLN 37, PLC 20, BDN 13, ALN 7, MRS 4, BUN 5, Independent 6
Judicial branch: Supreme Court or Corte Suprema de Justicia (16 judges elected for five-year terms by the National Assembly); note–in 2010, President Ortega directly replaced seven justices on the Supreme Court
Political parties and leaders: Conservative Party or PC [Alejandro BOLANOS Davis]; Independent Liberal Party or PLI [Indalecio RODRIGUEZ]; Liberal Constitutionalist Party or PLC [Jorge CASTILLO Quant]; Nicaraguan Liberal Alliance or ALN [Alejandro MEJIA Ferreti]; Sandinista National Liberation Front or FSLN [Daniel ORTEGA Saavedra]; Sandinista Renovation Movement or MRS [Enrique SAENZ-NAVARRETE]
Political pressure groups and leaders: National Workers Front or FNT (a Sandinista umbrella group of eight labor unions including: Farm Workers Association or ATC, Health Workers Federation or FETASALUD, Heroes and Martyrs Confederation of Professional Associations or CONAPRO, National Association of Educators of Nicaragua or ANDEN, National Union of Employees or UNE, National Union of Farmers and Ranchers or UNAG, Sandinista Workers Central or CST, and Union of Journalists of Nicaragua or UPN); Permanent Congress of Workers or CPT (an umbrella group of four non-Sandinista labor unions including: Autonomous Nicaraguan Workers Central or CTN-A, Confederation of Labor Unification or CUS, Independent General Confederation of Labor or CGT-I, and Labor Action and Unity Central or CAUS); Nicaraguan Workers' Central or CTN (an independent labor union); Superior Council of Private Enterprise or COSEP (a confederation of business groups)
International organization participation: BCIE, CACM, FAO, G-77, IADB, IAEA, IBRD, ICAO, ICRM, IDA, IFAD, IFC, IFRCS, ILO, IMF, IMO, Interpol, IOC, IOM, IPU, ITSO, ITU, ITUC, LAES, LAIA (observer), MIGA, NAM, OAS, OPANAL, OPCW, PCA, PetroCaribe, RG, SICA, UN, UNCTAD, UNESCO, UNHCR, UNIDO, Union Latina, UNWTO, UPU, WCO, WHO, WIPO, WMO, WTO
Diplomatic representation in the US: *chief of mission:* Ambassador Francisco Obadiah CAMPBELL Hooker
chancery: 1627 New Hampshire Avenue NW, Washington, DC 20009
telephone: [1] (202) 939-6570, 6573
FAX: [1] (202) 939-6545
consulate(s) general: Houston, Los Angeles, Miami, New York, San Francisco
Diplomatic representation from the US: *chief of mission:* Ambassador Robert J. CALLAHAN
embassy: Kilometer 5.5 Carretera Sur, Managua
mailing address: American Embassy Managua, APO AA 34021
telephone: [505] 252-7100, 252-7888; 252-7634 (after hours)
FAX: [505] 252-7304
Flag description: three equal horizontal bands of blue (top), white, and blue with the national coat of arms centered in the white band; the coat of arms features a triangle encircled by the words REPUBLICA DE NICARAGUA on the top and AMERICA CENTRAL on the bottom; the banner is based on the former blue-white-blue flag of the Federal Republic of Central America; the blue bands symbolize the Pacific Ocean and the Caribbean Sea, while the white band represents the land between the two bodies of water
note: similar to the flag of El Salvador, which features a round emblem encircled by the words REPUBLICA DE EL SALVADOR EN LA AMERICA CENTRAL centered in the white band; also similar to the flag of Honduras, which has five blue stars arranged in an X pattern centered in the white band
National anthem: *name:* "Salve a ti, Nicaragua" (Hail to Thee, Nicaragua)
lyrics/music: Salomon Ibarra MAYORGA/ traditional, arranged by Luis Abraham DELGADILLO
note: although only officially adopted in 1971, the music was approved in 1918 and the lyrics in 1939; the tune, originally from Spain, was used as an anthem for Nicaragua from the 1830"s until 1876

ECONOMY

Economy—overview: Nicaragua, the poorest country in Central America and the second poorest in the Hemisphere, has widespread underemployment and poverty. The US-Central America Free Trade Agreement (CAFTA) has been in effect since April 2006 and has expanded export opportunities for many agricultural and manufactured goods. Textiles and apparel account for nearly 60% of Nicaragua's exports, but increases in the minimum wage during the ORTEGA administration will likely erode its comparative advantage in this industry. ORTEGA's promotion of mixed business initiatives, owned by the Nicaraguan and Venezuelan state oil firms, together with the weak rule of law, could undermine the investment climate for domestic and international private firms in the near-term. Nicaragua relies on international economic assistance to meet internal- and external-debt financing obligations. Foreign donors have curtailed this funding, however, in response to November 2008 electoral fraud. Managua has an IMF extended Credit Facility program, which could help keep the government's fiscal deficit on target during the 2011 election year and encourage transparency in the use of Venezuelan off-budget loans and assistance. In early 2004, Nicaragua secured some $4.5 billion in foreign debt reduction under the Heavily Indebted Poor Countries (HIPC) initiative, however, Managua still struggles with a high public debt burden. Nicaragua is gradually recovering from the global economic crisis as increased exports drove positive growth in 2010. The economy is expected to grow at a rate of about 3% in 2011.
GDP (purchasing power parity): $17.71 billion (2010 est.)
country comparison to the world: 130
$16.95 billion (2009 est.)
$17.2 billion (2008 est.)
note: data are in 2010 US dollars
GDP (official exchange rate): $6.551 billion (2010 est.)
GDP—real growth rate: 4.5% (2010 est.)
country comparison to the world: 85
-1.5% (2009 est.)
2.8% (2008 est.)
GDP—per capita (PPP): $3,000 (2010 est.)
country comparison to the world: 169
$2,900 (2009 est.)
$3,000 (2008 est.)
note: data are in 2010 US dollars
GDP—composition by sector: *agriculture:* 17.6%
industry: 26.5%
services: 56% (2010 est.)
Labor force: 2.343 million (2010 est.)
country comparison to the world: 113
Labor force—by occupation: *agriculture:* 28%
industry: 19%
services: 53% (2010 est.)
Unemployment rate: 8% (2010 est.)
country comparison to the world: 91
8.2% (2009 est.)
note: underemployment was 46.5% in 2008
Population below poverty line: 48% (2005)
Household income or consumption by percentage share: *lowest 10%:* 1.4%
highest 10%: 41.8% (2005)
Distribution of family income—Gini index: 43.1 (2001)
country comparison to the world: 46
60.3 (1998)
Investment (gross fixed): 22.8% of GDP (2010 est.)
country comparison to the world: 60
Budget: *revenues:* $1.421 billion
expenditures: $1.511 billion (2010 est.)
Public debt: 78% of GDP (2010 est.)
country comparison to the world: 20
63% of GDP (2009 est.)

Inflation rate (consumer prices): 4.7% (2010 est.)
country comparison to the world: 133
3.7% (2009 est.)
Central bank discount rate: NA% (31 December 2009)
NA% (31 December 2008)
Commercial bank prime lending rate: 14.04% (31 December 2009 est.)
country comparison to the world: 58
13.17% (31 December 2008 est.)
Stock of narrow money: $1.273 billion (31 December 2010 est.)
country comparison to the world: 132
$989.5 million (31 December 2009 est.)
Stock of broad money: $2.9 billion (31 December 2010 est.)
country comparison to the world: 132
$2.586 billion (31 December 2009 est.)
Stock of domestic credit: $4.083 billion (31 December 2010 est.)
country comparison to the world: 114
$4.161 billion (31 December 2009 est.)
Market value of publicly traded shares: $NA
Agriculture—products: coffee, bananas, sugarcane, cotton, rice, corn, tobacco, sesame, soya, beans; beef, veal, pork, poultry, dairy products; shrimp, lobsters
Industries: food processing, chemicals, machinery and metal products, knit and woven apparel, petroleum refining and distribution, beverages, footwear, wood
Industrial production growth rate: 1.5% (2010 est.)
country comparison to the world: 141
Electricity—production: 3.286 billion kWh (2007 est.)
country comparison to the world: 124
Electricity—consumption: 2.569 billion kWh (2007 est.)
country comparison to the world: 131
Electricity—exports: 0 kWh (2008 est.)
Electricity—imports: 63.95 million kWh (2007 est.)
Oil—production: 0 bbl/day (2008 est.)
country comparison to the world: 121
Oil—consumption: 29,000 bbl/day (2009 est.)
country comparison to the world: 113
Oil—exports: 212.5 bbl/day (2007 est.)
country comparison to the world: 131
Oil—imports: 29,570 bbl/day (2007 est.)
country comparison to the world: 99
Oil—proved reserves: 0 bbl (1 January 2010 est.)
country comparison to the world: 174
Natural gas—production: 0 cu m (2008 est.)
country comparison to the world: 171
Natural gas—consumption: 0 cu m (2008 est.)
country comparison to the world: 113
Natural gas—exports: 0 cu m (2008 est.)
country comparison to the world: 153
Natural gas—imports: 0 cu m (2008 est.)
country comparison to the world: 102
Natural gas—proved reserves: 0 cu m (1 January 2010 est.)
country comparison to the world: 176
Current account balance: $-819 million (2010 est.)
country comparison to the world: 127
$-841.1 million (2009 est.)
Exports: $3.182 billion (2010 est.)
country comparison to the world: 119
$2.593 billion (2009 est.)
Exports—commodities: coffee, beef, shrimp and lobster, tobacco, sugar, gold, peanuts; textiles and apparel
Exports—partners: US 61.98%, El Salvador 7.74%, Costa Rica 3.67% (2009)
Imports: $4.7 billion (2010 est.)
country comparison to the world: 121
$3.481 billion (2009 est.)
Imports—commodities: consumer goods, machinery and equipment, raw materials, petroleum products
Imports—partners: US 22.63%, Venezuela 12.27%, Mexico 9.05%, Costa Rica 8.66%, China 7.16%, Guatemala 6.59%, El Salvador 5.63% (2009)
Reserves of foreign exchange and gold: $1.58 billion (31 December 2010 est.)
country comparison to the world: 112
$1.573 billion (31 December 2009 est.)
Debt—external: $4.03 billion (31 December 2010 est.)
country comparison to the world: 116
$3.633 billion (31 December 2009 est.)
Exchange rates: cordobas (NIO) per US dollar–21.35 (2010) 20.34 (2009) 19.374 (2008) 18.457 (2007) 17.582 (2006)

COMMUNICATIONS

Telephones—main lines in use: 255,000 (2009)
country comparison to the world: 122
Telephones—mobile cellular: 3.204 million (2009)
country comparison to the world: 112
Telephone system: *general assessment:* system being upgraded by foreign investment; nearly all installed telecommunications capacity now uses digital technology, owing to investments since privatization of the formerly state-owned telecommunications company
domestic: since privatization, access to fixed-line and mobile-cellular services has improved but teledensity still lags behind other Central American countries; fixed-line teledensity roughly 5 per 100 persons; mobile-cellular telephone subscribership is increasing and reached 55 per 100 persons in 2009; connected to Central American Microwave System
international: country code–505; the Americas Region Caribbean Ring System (ARCOS-1) fiber optic submarine cable provides connectivity to South and Central America, parts of the Caribbean, and the US; satellite earth stations–1 Intersputnik (Atlantic Ocean region) and 1 Intelsat (Atlantic Ocean) (2009)
Broadcast media: multiple privately-owned terrestrial television networks, supplemented by cable TV in most urban areas; of more than 100 radio broadcast stations, nearly all are privately owned; Radio Nicaragua is government-owned and Radio Sandino is controlled by the Sandinista National Liberation Front (FSLN) (2007)
Internet country code: .ni
Internet hosts: 157,162 (2010)
country comparison to the world: 70
Internet users: 199,800 (2009)
country comparison to the world: 141

TRANSPORTATION

Airports: 143 (2010)
country comparison to the world: 40
Airports—with paved runways: *total:* 11
2,438 to 3,047 m: 3
1,524 to 2,437 m: 2
914 to 1,523 m: 3
under 914 m: 3 (2010)
Airports—with unpaved runways: *total:* 132
1,524 to 2,437 m: 1
914 to 1,523 m: 16
under 914 m: 115 (2010)
Pipelines: oil 54 km (2010)
Roadways: *total:* 19,137 km
country comparison to the world: 111
paved: 2,033 km
unpaved: 17,104 km (2009)
Waterways: 2,220 km (navigable waterways as well as the use of the large Lake Managua and Lake Nicaragua; rivers serve only the sparsely populated eastern part of the country) (2010)
country comparison to the world: 40
Ports and terminals: Bluefields, Corinto

MILITARY

Military branches: National Army of Nicaragua (Ejercito Nacional de Nicaragua, ENN; includes Navy, Air Force) (2010)
Military service age and obligation: 17 years of age for voluntary military service; tour of duty 18-36 months (2008)
Manpower available for military service: *males age 16–49:* 1,452,107
females age 16–49: 1,552,698 (2010 est.)
Manpower fit for military service: *males age 16–49:* 1,227,757
females age 16–49: 1,335,653 (2010 est.)
Manpower reaching militarily significant age annually: *male:* 69,093
female: 67,522 (2010 est.)
Military expenditures: 0.6% of GDP (2006)
country comparison to the world: 157

TRANSNATIONAL ISSUES

Disputes—international: memorials and countermemorials were filed by the parties in Nicaragua's 1999 and 2001 proceedings against Honduras and Colombia at the ICJ over the maritime boundary and territorial claims in the western Caribbean Sea, final public hearings are scheduled for 2007; the 1992 ICJ ruling for El Salvador and Honduras advised a tripartite resolution to establish a maritime boundary in the Gulf of Fonseca, which considers Honduran access to the Pacific; legal dispute over navigational rights of San Juan River on border with Costa Rica
Illicit drugs: transshipment point for cocaine destined for the US and transshipment point for arms-for-drugs dealing

NIGER

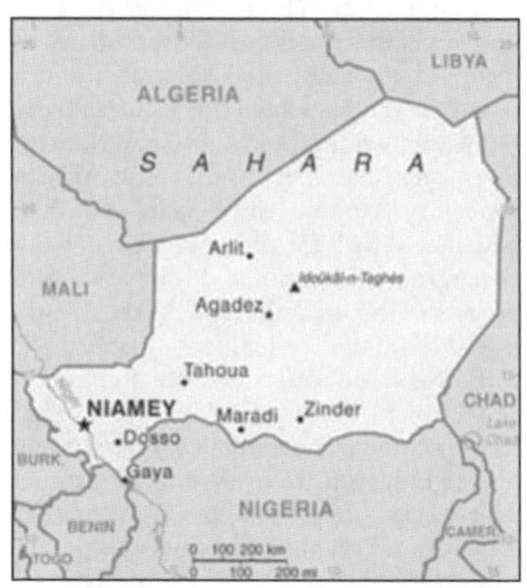

INTRODUCTION

Background: Niger became independent from France in 1960 and experienced single-party and military rule until 1991, when Gen. Ali SAIBOU was forced by public pressure to allow multiparty elections, which resulted in a democratic government in 1993. Political infighting brought the government to a standstill and in 1996 led to a coup by Col. Ibrahim BARE. In 1999, BARE was killed in a counter coup by military officers who restored democratic rule and held elections that brought Mamadou TANDJA to power in December of that year. TANDJA was reelected in 2004 and in 2009 spearheaded a constitutional amendment that would allow him to extend his term as president. In February 2010, a military coup deposed TANDJA, immediately suspended the constitution and dissolved the Cabinet, and promised that elections would be held following a transitional period of unspecified duration. Niger is one of the poorest countries in the world with minimal government services and insufficient funds to develop its resource base. The largely agrarian and subsistence-based economy is frequently disrupted by extended droughts common to the Sahel region of Africa. A predominately Tuareg ethnic group emerged in February 2007, the Nigerien Movement for Justice, and attacked several military targets in Niger's northern region throughout 2007 and 2008. Successful government offensives in 2009 limited the rebels' operational capabilities.

GEOGRAPHY

Location: Western Africa, southeast of Algeria

Geographic coordinates: 16 00 N, 8 00 E

Map references: Africa

Area: *total:* 1.267 million sq km
country comparison to the world: 22
land: 1,266,700 sq km
water: 300 sq km

Area—comparative: slightly less than twice the size of Texas

Land boundaries: *total:* 5,697 km
border countries: Algeria 956 km, Benin 266 km, Burkina Faso 628 km, Chad 1,175 km, Libya 354 km, Mali 821 km, Nigeria 1,497 km

Coastline: 0 km (landlocked)

Maritime claims: none (landlocked)

Climate: desert; mostly hot, dry, dusty; tropical in extreme south

Terrain: predominately desert plains and sand dunes; flat to rolling plains in south; hills in north

Elevation extremes: *lowest point:* Niger River 200 m
highest point: Idoukal-n-Taghes 2,022 m

Natural resources: uranium, coal, iron ore, tin, phosphates, gold, molybdenum, gypsum, salt, petroleum

Land use: *arable land:* 11.43%
permanent crops: 0.01%
other: 88.56% (2005)

Irrigated land: 730 sq km (2003)

Total renewable water resources: 33.7 cu km (2003)

Freshwater withdrawal (domestic/industrial/agricultural): *total:* 2.18 cu km/yr (4%/0%/95%)
per capita: 156 cu m/yr (2000)

Natural hazards: recurring droughts

Environment—current issues: overgrazing; soil erosion; deforestation; desertification; wildlife populations (such as elephant, hippopotamus, giraffe, and lion) threatened because of poaching and habitat destruction

Environment—international agreements: *party to:* Biodiversity, Climate Change, Climate Change-Kyoto Protocol, Desertification, Endangered Species, Environmental Modification, Hazardous Wastes, Ozone Layer Protection, Wetlands
signed, but not ratified: Law of the Sea

Geography—note: landlocked; one of the hottest countries in the world; northern four-fifths is desert, southern one-fifth is savanna, suitable for livestock and limited agriculture

PEOPLE

Population: 16,468,886 (July 2011 est.)
country comparison to the world: 62

Age structure: *0–14 years:* 49.6% (male 4,129,164/female 4,045,412)
15–64 years: 48% (male 3,944,586/female 3,964,249)
65 years and over: 2.3% (male 170,741/female 214,734) (2011 est.)

Median age: *total:* 15.2 years
male: 15 years
female: 15.4 years (2011 est.)

Population growth rate: 3.643% (2011 est.)
country comparison to the world: 2

Birth rate: 50.54 births/1,000 population (2011 est.)
country comparison to the world: 1

Death rate: 14.11 deaths/1,000 population (July 2011 est.)
country comparison to the world: 15

Net migration rate: 0 migrant(s)/1,000 population (2011 est.)
country comparison to the world: 100

Urbanization: *urban population:* 17% of total population (2010)
rate of urbanization: 4.7% annual rate of change (2010-15 est.)

Major cities—population: NIAMEY (capital) 1.004 million (2009)

Sex ratio: *at birth:* 1.03 male(s)/female
under 15 years: 1.02 male(s)/female
15–64 years: 0.99 male(s)/female
65 years and over: 0.8 male(s)/female
total population: 1 male(s)/female (2011 est.)

Infant mortality rate: *total:* 112.22 deaths/1,000 live births
country comparison to the world: 3
male: 117.19 deaths/1,000 live births
female: 107.1 deaths/1,000 live births (2011 est.)

Life expectancy at birth: *total population:* 53.4 years
country comparison to the world: 203
male: 52.13 years
female: 54.7 years (2011 est.)

Total fertility rate: 7.6 children born/woman (2011 est.)
country comparison to the world: 1

HIV/AIDS—adult prevalence rate: 0.8% (2009 est.)
country comparison to the world: 58

HIV/AIDS—people living with HIV/AIDS: 61,000 (2009 est.)
country comparison to the world: 55

HIV/AIDS—deaths: 4,300 (2009 est.)
country comparison to the world: 42

Major infectious diseases: *degree of risk:* very high
food or waterborne diseases: bacterial and protozoal diarrhea, hepatitis A, and typhoid fever
vectorborne disease: malaria
water contact disease: schistosomiasis
animal contact disease: rabies
respiratory disease: meningococcal meningitis
note: highly pathogenic H5N1 avian influenza has been identified in this country; it poses a negligible risk with extremely rare cases possible among US citizens who have close contact with birds (2009)

Drinking water source: *Improved:*
urban: 96% of population
rural: 39% of population
total: 48% of population
Unimproved:
urban: 4% of population
rural: 61% of population
total: 52% of population (2008)

Sanitation facility access: *Improved:*
urban: 34% of population
rural: 4% of population
total: 9% of population
Unimproved:
urban: 66% of population
rural: 96% of population
total: 91% of population (2008)

Nationality: *noun:* Nigerien(s)
adjective: Nigerien

Ethnic groups: Haoussa 55.4%, Djerma Sonrai 21%, Tuareg 9.3%, Peuhl 8.5%, Kanouri Manga 4.7%, other 1.2% (2001 census)

Religions: Muslim 80%, other (includes indigenous beliefs and Christian) 20%

Languages: French (official), Hausa, Djerma

Literacy: *definition:* age 15 and over can read and write

total population: 28.7%
male: 42.9%
female: 15.1% (2005 est.)
School life expectancy (primary to tertiary education): *total:* 5 years
male: 6 years
female: 5 years (2010)
Education expenditures: 4.5% of GDP (2009)
country comparison to the world: 84

GOVERNMENT

Country name: *conventional long form:* Republic of Niger
conventional short form: Niger
local long form: Republique du Niger
local short form: Niger
Government type: republic
Capital: *name:* Niamey
geographic coordinates: 13 31 N, 2 07 E
time difference: UTC+1 (6 hours ahead of Washington, DC during Standard Time)
Administrative divisions: 8 regions (regions, singular–region) includes 1 capital district* (communite urbaine); Agadez, Diffa, Dosso, Maradi, Niamey*, Tahoua, Tillaberi, Zinder
Independence: 3 August 1960 (from France)
National holiday: Republic Day, 18 December (1958); note–commemorates the founding of the Republic of Niger which predated independence from France in 1960
Constitution: adopted 31 October 2010
Legal system: mixed legal system of civil law (based on French civil law), Islamic law, and customary law
International law organization participation: has not submitted an ICJ jurisdiction declaration; accepts ICCt jurisdiction
Suffrage: 18 years of age; universal
Executive branch: *chief of state:* President Mahamadou ISSOUFOU (since 7 April 2011)
head of government: Prime Minister Brigi RAFINI (since 7 April 2011); appointed by the president and shares some executive responsibilities with the president
cabinet: 26-member Cabinet appointed by the president
(For more information visit the World Leaders website)
elections: president elected by popular vote for a five-year term (eligible for a second term); candidate must receive a majority of the votes to be elected president; a presidential election to restore civilian rule was held 31 January 2011 with a runoff election between Mahamadou ISSOUFOU and Seini OUMAROU held on 12 March 2011
election results: Mahamadou ISSOUFOU elected president in a runoff election; percent of vote–Mahamadou ISSOUFOU 58%, Seini OUMAROU 42%
Legislative branch: unicameral National Assembly (113 seats; members elected by popular vote to serve five-year terms)
elections: last held on 31 January 2011
election results: percent of vote by party–NA; seats by party–PNDS-Tarrayya 39, MNSD-Nassara 26, MODEN/FA-Lumana 24, ANDP-Zaman Lahiya 8, RDP-Jama'a 7, UDR-Tabbat 6, CDS-Rahama 2, UNI 1
Judicial branch: State Court or Cour d'Etat; Court of Appeals or Cour d'Appel
Political parties and leaders: Democratic and Social Convention-Rahama or CDS-Rahama [Mahamane OUSMANE]; National Movement for a Developing Society-Nassara or MNSD-Nassara; Niger Social Democratic Party or PSDN; Nigerien Alliance for Democracy and Social Progress-Zaman Lahiya or ANDP-Zaman Lahiya [Moumouni DJERMAKOYE]; Nigerien Democratic Movement for an African Federation or MODEN/FA Lumana; Nigerien Party for Democracy and Socialism or PNDS-Tarrayya [Mahamadou ISSOUFOU]; Rally for Democracy and Progress-Jama'a or RDP-jama'a [Hamid ALGABID]; Social and Democratic Rally or RSD-Gaskiyya [Cheiffou AMADOU]; Union for Democracy and the Republic-Tabbat or UDR-Tabbat; Union of Independent Nigeriens or UNI
Political pressure groups and leaders: The Nigerien Movement for Justice or MNJ, a predominantly Tuareg rebel group
International organization participation: ACP, AfDB, AU, ECOWAS (suspended), Entente, FAO, FZ, G-77, IAEA, IBRD, ICAO, ICRM, IDA, IDB, IFAD, IFC, IFRCS, ILO, IMF, Interpol, IOC, IOM, ITSO, ITU, ITUC, MONUSCO, NAM, OIC, OIF, OPCW, UN, UNCTAD, UNESCO, UNIDO, UNMIL, UNMIS, UNOCI, UNWTO, UPU, WADB (regional), WAEMU, WCO, WFTU, WHO, WIPO, WMO, WTO
Diplomatic representation in the US: *chief of mission:* Ambassador Aminata Djibrilla Maiga TOURE
chancery: 2204 R Street NW, Washington, DC 20008
telephone: [1] (202) 483-4224 through 4227
FAX: [1] (202)483-3169
Diplomatic representation from the US: *chief of mission:* Ambassador Bisa WILLIAMS
embassy: Rue Des Ambassades, Niamey
mailing address: B. P. 11201, Niamey
telephone: [227] 20-72-26-61 thru 64
FAX: [227] 20-73-31-67
Flag description: three equal horizontal bands of orange (top), white, and green with a small orange disk centered in the white band; the orange band denotes the drier northern regions of the Sahara; white stands for purity and innocence; green symbolizes hope and the fertile and productive southern and western areas, as well as the Niger River; the orange disc represents the sun and the sacrifices made by the people
note: similar to the flag of India, which has a blue spoked wheel centered in the white band
National anthem: *name:* "La Nigerienne" (The Nigerian)
lyrics/music: Maurice Albert THIRIET/ Robert JACQUET and Nicolas Abel Francois FRIONNET
note: adopted 1961

ECONOMY

Economy—overview: Niger is a landlocked, Sub-Saharan nation, whose economy centers on subsistence crops, livestock, and some of the world's largest uranium deposits. Drought, desertification, and strong population growth have undercut the economy. Niger shares a common currency, the CFA franc, and a common central bank, the Central Bank of West African States (BCEAO), with seven other members of the West African Monetary Union. In December 2000, Niger qualified for enhanced debt relief under the International Monetary Fund program for Highly Indebted Poor Countries (HIPC) and concluded an agreement with the Fund on a Poverty Reduction and Growth Facility (PRGF). Debt relief provided under the enhanced HIPC initiative significantly reduces Niger's annual debt service obligations, freeing funds for expenditures on basic health care, primary education, HIV/AIDS prevention, rural infrastructure, and other programs geared at poverty reduction. In December 2005, Niger received 100% multilateral debt relief from the IMF, which translates into the forgiveness of approximately US $86 million in debts to the IMF, excluding the remaining assistance under HIPC. In 2010, the Niger economy was recovering from the effects of a 2009 drought that reduced grain and cowpea production and decimated livestock herds. The economy was also hurt when the international community cut off non-humanitarian aid in response to TANDJA's moves to extend his term as president. Nearly half of the government's budget is derived from foreign donor resources. Future growth may be sustained by exploitation of oil, gold, coal, and other mineral resources.
GDP (purchasing power parity): $11.05 billion (2010 est.)
country comparison to the world: 147
$10.28 billion (2009 est.)
$10.37 billion (2008 est.)
note: data are in 2010 US dollars
GDP (official exchange rate): $5.577 billion (2010 est.)
GDP—real growth rate: 7.5% (2010 est.)
country comparison to the world: 27
-0.9% (2009 est.)
9.3% (2008 est.)
GDP—per capita (PPP): $700 (2010 est.)
country comparison to the world: 222
$700 (2009 est.)
$700 (2008 est.)
note: data are in 2010 US dollars
GDP—composition by sector:
agriculture: 39%
industry: 17%
services: 44% (2001)
Labor force: 4.688 million (2007)
country comparison to the world: 78
Labor force—by occupation:
agriculture: 90%
industry: 6%
services: 4% (1995)
Unemployment rate: NA%
Population below poverty line:
63% (1993 est.)
Household income or consumption by percentage share:
lowest 10%: 2.3%
highest 10%: 35.7% (2005)
Distribution of family income—Gini index: 50.5 (1995)

country comparison to the world: 20
Budget: *revenues:* $320 million (includes $134 million from foreign sources)
expenditures: $320 million (2002 est.)
Inflation rate (consumer prices): 0.1% (2007 est.)
country comparison to the world: 8
Central bank discount rate: 4.25% (31 December 2009)
country comparison to the world: 88
4.75% (31 December 2008)
Commercial bank prime lending rate: NA%
Stock of narrow money: $782.6 million (31 December 2009)
country comparison to the world: 146
$617.9 million (31 December 2008)
Stock of broad money: $1.038 billion (31 December 2009 est.)
country comparison to the world: 165
$844.6 million (31 December 2008 est.)
Stock of domestic credit: $683.6 million (31 December 2009)
country comparison to the world: 159
$313.5 million (31 December 2008)
Market value of publicly traded shares: $NA
Agriculture—products: cowpeas, cotton, peanuts, millet, sorghum, cassava (tapioca), rice; cattle, sheep, goats, camels, donkeys, horses, poultry
Industries: uranium mining, cement, brick, soap, textiles, food processing, chemicals, slaughterhouses
Industrial production growth rate: 5.1% (2003 est.)
country comparison to the world: 67
Electricity—production: 150 million kWh (2007 est.)
country comparison to the world: 181
Electricity—consumption: 589.5 million kWh (2007 est.)
country comparison to the world: 158
Electricity—exports: 0 kWh (2008 est.)
Electricity—imports: 450 million kWh (2007 est.)
Oil—production: 0 bbl/day (2009 est.)
country comparison to the world: 117
Oil—consumption: 6,000 bbl/day (2009 est.)
country comparison to the world: 160
Oil—exports: 0 bbl/day (2007 est.)
country comparison to the world: 191
Oil—imports: 5,367 bbl/day (2007 est.)
country comparison to the world: 156
Oil—proved reserves: 0 bbl (1 January 2010 est.)
country comparison to the world: 170
Natural gas—production: 0 cu m (2008 est.)
country comparison to the world: 166
Natural gas—consumption: 0 cu m (2008 est.)
country comparison to the world: 204
Natural gas—exports: 0 cu m (2008 est.)
country comparison to the world: 148
Natural gas—imports: 0 cu m (2008 est.)
country comparison to the world: 95
Natural gas—proved reserves: 0 cu m (1 January 2010 est.)
country comparison to the world: 171
Current account balance: $-321 million (2007 est.)
country comparison to the world: 100
Exports: $428 million (2006)
country comparison to the world: 171
Exports—commodities: uranium ore, livestock, cowpeas, onions
Exports—partners: France 52.63%, Nigeria 22.43%, US 18.24% (2009)
Imports: $800 million (2006)
country comparison to the world: 179
Imports—commodities: foodstuffs, machinery, vehicles and parts, petroleum, cereals
Imports—partners: China 16.32%, France 15.95%, Netherlands 7.66%, Algeria 7.15%, French Polynesia 6.11%, Nigeria 5.48%, Cote d'Ivoire 4.15%, US 4.05% (2009)
Debt—external: $2.1 billion (2003 est.)
country comparison to the world: 139
Exchange rates: Communaute Financiere Africaine francs (XOF) per US dollar– 495.28 (2010) 472.19 (2009) 493.51 (2007) 522.59 (2006)

COMMUNICATIONS

Telephones—main lines in use: 65,000 (2009)
country comparison to the world: 156
Telephones—mobile cellular: 2.599 million (2009)
country comparison to the world: 121
Telephone system: *general assessment:* inadequate; small system of wire, radio telephone communications, and microwave radio relay links concentrated in the southwestern area of Niger
domestic: combined fixed-line and mobile-cellular teledensity remains less than 20 per 100 persons despite a rapidly increasing cellular subscribership base; domestic satellite system with 3 earth stations and 1 planned
international: country code–227; satellite earth stations–2 Intelsat (1 Atlantic Ocean and 1 Indian Ocean) (2009)
Broadcast media: state-run TV station; 3 private TV stations provide a mix of local and foreign programming; only national radio station with national reach is state-run; about 30 private radio stations operate locally; as many as 100 community radio stations broadcast; transmissions of multiple international broadcasters are available (2007)
Internet country code: .ne
Internet hosts: 172 (2010)
country comparison to the world: 198
Internet users: 115,900 (2009)
country comparison to the world: 155

TRANSPORTATION

Airports: 27 (2010)
country comparison to the world: 125
Airports—with paved runways:
total: 10
2,438 to 3,047 m: 3
1,524 to 2,437 m: 6
914 to 1,523 m: 1 (2010)
Airports—with unpaved runways:
total: 17
1,524 to 2,437 m: 2
914 to 1,523 m: 14
under 914 m: 1 (2010)
Roadways: *total:* 18,949 km
country comparison to the world: 112
paved: 3,912 km
unpaved: 15,037 km (2008)
Waterways: 300 km (the Niger, the only major river, is navigable to Gaya between September and March) (2010)
country comparison to the world: 93

MILITARY

Military branches: Nigerien Armed Forces (Forces Armees Nigeriennes, FAN): Army, Nigerien Air Force (Force Aerienne du Niger) (2010)
Military service age and obligation: 17-21 years of age for selective compulsory or voluntary military service; enlistees must be Nigerien citizens and unmarried; 2-year service term; women may serve in health care (2009)
Manpower available for military service: *males age 16–49:* 3,329,184
females age 16–49: 3,267,669 (2010 est.)
Manpower fit for military service: *males age 16–49:* 2,194,570
females age 16–49: 2,219,416 (2010 est.)
Manpower reaching militarily significant age annually: *male:* 186,348
female: 180,779 (2010 est.)
Military expenditures: 1.3% of GDP (2006)
country comparison to the world: 118

TRANSNATIONAL ISSUES

Disputes—international: Libya claims about 25,000 sq km in a currently dormant dispute in the Tommo region; location of Benin-Niger-Nigeria tripoint is unresolved; only Nigeria and Cameroon have heeded the Lake Chad Commission's admonition to ratify the delimitation treaty that also includes the Chad-Niger and Niger-Nigeria boundaries; the dispute with Burkina Faso was referred to the ICJ in 2010
Trafficking in persons: current situation: Niger is a source, transit, and destination country for children and women trafficked for forced labor and sexual exploitation; caste-based slavery practices, rooted in ancestral master-slave relationships, continue in isolated areas of the country–an estimated 8,800 to 43,000 Nigeriens live under conditions of traditional slavery; children are trafficked within Niger for forced begging, forced labor in gold mines, domestic servitude, sexual exploitation, and possibly for forced labor in agriculture and stone quarries; women and children from neighboring states are trafficked to and through Niger for domestic servitude, sexual exploitation, forced labor in mines and on farms, and as mechanics and welders
tier rating: Tier 2 Watch List–the Government of Niger does not fully comply with the minimum standards for the elimination of trafficking and is not making any significant efforts to do so; the government demonstrated marginal efforts to combat human trafficking, including traditional slavery, during the last year (2009)

NIGERIA

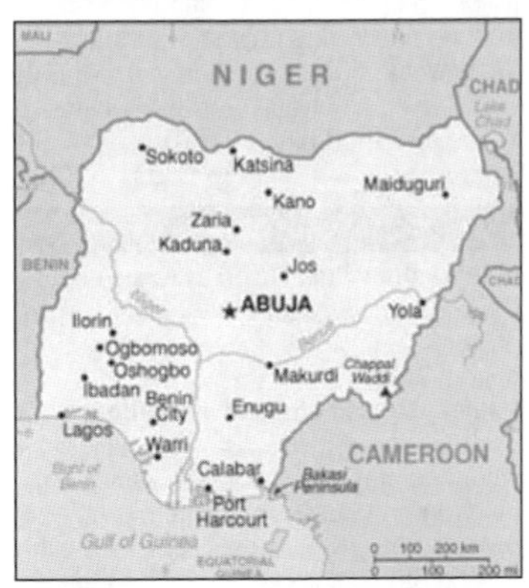

INTRODUCTION

Background: British influence and control over what would become Nigeria and Africa's most populous country grew through the 19th century. A series of constitutions after World War II granted Nigeria greater autonomy; independence came in 1960. Following nearly 16 years of military rule, a new constitution was adopted in 1999, and a peaceful transition to civilian government was completed. The government continues to face the daunting task of reforming a petroleum-based economy, whose revenues have been squandered through corruption and mismanagement, and institutionalizing democracy. In addition, Nigeria continues to experience longstanding ethnic and religious tensions. Although both the 2003 and 2007 presidential elections were marred by significant irregularities and violence, Nigeria is currently experiencing its longest period of civilian rule since independence. The general elections of April 2007 marked the first civilian-to-civilian transfer of power in the country's history. In January 2010, Nigeria assumed a nonpermanent seat on the UN Security Council for the 2010-11 term.

GEOGRAPHY

Location: Western Africa, bordering the Gulf of Guinea, between Benin and Cameroon
Geographic coordinates: 10 00 N, 8 00 E
Map references: Africa
Area: *total:* 923,768 sq km
country comparison to the world: 32
land: 910,768 sq km
water: 13,000 sq km
Area—comparative: slightly more than twice the size of California
Land boundaries: *total:* 4,047 km
border countries: Benin 773 km, Cameroon 1,690 km, Chad 87 km, Niger 1,497 km
Coastline: 853 km
Maritime claims: *territorial sea:* 12 nm
exclusive economic zone: 200 nm
continental shelf: 200 m depth or to the depth of exploitation
Climate: varies; equatorial in south, tropical in center, arid in north
Terrain: southern lowlands merge into central hills and plateaus; mountains in southeast, plains in north
Elevation extremes: *lowest point:* Atlantic Ocean 0 m
highest point: Chappal Waddi 2,419 m
Natural resources: natural gas, petroleum, tin, iron ore, coal, limestone, niobium, lead, zinc, arable land
Land use: *arable land:* 33.02%
permanent crops: 3.14%
other: 63.84% (2005)
Irrigated land: 2,820 sq km (2003)
Total renewable water resources: 286.2 cu km (2003)
Freshwater withdrawal (domestic/industrial/agricultural): *total:* 8.01 cu km/yr (21%/10%/69%)
per capita: 61 cu m/yr (2000)
Natural hazards: periodic droughts; flooding
Environment—current issues: soil degradation; rapid deforestation; urban air and water pollution; desertification; oil pollution–water, air, and soil; has suffered serious damage from oil spills; loss of arable land; rapid urbanization
Environment—international agreements: *party to:* Biodiversity, Climate Change, Climate Change-Kyoto Protocol, Desertification, Endangered Species, Hazardous Wastes, Law of the Sea, Marine Dumping, Marine Life Conservation, Ozone Layer Protection, Ship Pollution, Wetlands
signed, but not ratified: none of the selected agreements
Geography—note: the Niger enters the country in the northwest and flows southward through tropical rain forests and swamps to its delta in the Gulf of Guinea

PEOPLE

Population: 155,215,573 (July 2011 est.)
country comparison to the world: 8
note: estimates for this country explicitly take into account the effects of excess mortality due to AIDS; this can result in lower life expectancy, higher infant mortality, higher death rates, lower population growth rates, and changes in the distribution of population by age and sex than would otherwise be expected
Age structure: *0–14 years:* 40.9% (male 32,476,681/female 31,064,539)
15–64 years: 55.9% (male 44,296,228/female 42,534,542)
65 years and over: 3.1% (male 2,341,228/female 2,502,355) (2011 est.)
Median age: *total:* 19.2 years
male: 19.2 years
female: 19.3 years (2011 est.)
Population growth rate: 1.935% (2011 est.)
country comparison to the world: 59
Birth rate: 35.51 births/1,000 population (2011 est.)
country comparison to the world: 27
Death rate: 16.06 deaths/1,000 population (July 2011 est.)
country comparison to the world: 4
Net migration rate: -0.1 migrant(s)/1,000 population (2011 est.)
country comparison to the world: 122
Urbanization: *urban population:* 50% of total population (2010)
rate of urbanization: 3.5% annual rate of change (2010-15 est.)
Major cities—population: Lagos 10.203 million; Kano 3.304 million; Ibadan 2.762 million; ABUJA (capital) 1.857 million; Kaduna 1.519 million (2009)
Sex ratio: *at birth:* 1.06 male(s)/female
under 15 years: 1.05 male(s)/female
15–64 years: 1.04 male(s)/female
65 years and over: 0.94 male(s)/female
total population: 1.04 male(s)/female (2011 est.)
Infant mortality rate:
total: 91.54 deaths/1,000 live births
country comparison to the world: 9
male: 97.42 deaths/1,000 live births
female: 85.31 deaths/1,000 live births (2011 est.)
Life expectancy at birth:
total population: 47.56 years
country comparison to the world: 220
male: 46.76 years
female: 48.41 years (2011 est.)
Total fertility rate: 4.73 children born/woman (2011 est.)
country comparison to the world: 27
HIV/AIDS—adult prevalence rate: 3.6% (2009 est.)
country comparison to the world: 17
HIV/AIDS—people living with HIV/AIDS: 3.3 million (2009 est.)
country comparison to the world: 3
HIV/AIDS—deaths: 220,000 (2009 est.)
country comparison to the world: 2
Major infectious diseases: *degree of risk:* very high
food or waterborne diseases: bacterial and protozoal diarrhea, hepatitis A and E, and typhoid fever
vectorborne disease: malaria and yellow fever
respiratory disease: meningococcal meningitis
aerosolized dust or soil contact disease: one of the most highly endemic areas for Lassa fever
water contact disease: leptospirosis and schistosomiasis
animal contact disease: rabies
note: highly pathogenic H5N1 avian influenza has been identified in this country; it poses a negligible risk with extremely rare cases possible among US citizens who have close contact with birds (2009)
Drinking water source: *Improved:*
urban: 75% of population
rural: 42% of population
total: 58% of population
Unimproved:
urban: 25% of population
rural: 58% of population
total: 42% of population (2008)
Sanitation facility access: *Improved:*
urban: 36% of population
rural: 28% of population
total: 32% of population
Unimproved:
urban: 67% of population
rural: 72% of population
total: 68% of population (2008)
Nationality: *noun:* Nigerian(s)
adjective: Nigerian
Ethnic groups: Nigeria, Africa's most populous country, is composed of more

than 250 ethnic groups; the following are the most populous and politically influential: Hausa and Fulani 29%, Yoruba 21%, Igbo (Ibo) 18%, Ijaw 10%, Kanuri 4%, Ibibio 3.5%, Tiv 2.5%
Religions: Muslim 50%, Christian 40%, indigenous beliefs 10%
Languages: English (official), Hausa, Yoruba, Igbo (Ibo), Fulani, over 500 additional indigenous languages
Literacy: *definition:* age 15 and over can read and write
total population: 68%
male: 75.7%
female: 60.6% (2003 est.)
School life expectancy (primary to tertiary education): *total:* 9 years
male: 10 years
female: 8 years (2005)
Education expenditures: NA

GOVERNMENT

Country name: *conventional long form:* Federal Republic of Nigeria
conventional short form: Nigeria
Government type: federal republic
Capital: *name:* Abuja
geographic coordinates: 9 05 N, 7 32 E
time difference: UTC+1 (6 hours ahead of Washington, DC during Standard Time)
Administrative divisions: 36 states and 1 territory*; Abia, Adamawa, Akwa Ibom, Anambra, Bauchi, Bayelsa, Benue, Borno, Cross River, Delta, Ebonyi, Edo, Ekiti, Enugu, Federal Capital Territory*, Gombe, Imo, Jigawa, Kaduna, Kano, Katsina, Kebbi, Kogi, Kwara, Lagos, Nassarawa, Niger, Ogun, Ondo, Osun, Oyo, Plateau, Rivers, Sokoto, Taraba, Yobe, Zamfara
Independence:
1 October 1960 (from the UK)
National holiday: Independence Day (National Day), 1 October (1960)
Constitution: adopted 5 May 1999; effective 29 May 1999
Legal system: mixed legal system of English common law, Islamic law (in 12 northern states), and traditional law
International law organization participation: accepts compulsory ICJ jurisdiction with reservations; accepts ICCt jurisdiction
Suffrage: 18 years of age; universal
Executive branch: *chief of state:* President Goodluck JONATHAN (since 5 May 2010, acting since 9 February 2010); Vice President Mohammed Namadi SAMBO (since 19 May 2010); note–the president is both the chief of state and head of government; JONATHAN assumed the presidency on 5 May 2010 following the death of President YAR'ADUA; JONATHAN was declared Acting President on 9 February 2010 by the National Assembly during the extended illness of the former president
head of government: President Goodluck JONATHAN (since 5 May 2010, acting since 9 February 2010); Vice President Mohammed Namadi SAMBO (since 19 May 2010)
cabinet: Federal Executive Council
(For more information visit the World Leaders website)
elections: president elected by popular vote for a four-year term (eligible for a second term); election last held on 16 April 2011 (next to be held in April 2015)
election results: Goodluck JONATHAN elected president; percent of vote–Goodluck JONATHAN 58.9%, Muhammadu BUHARI 32.0%, Nuhu RIBADU 5.4%, Ibrahim SHEKARAU 2.4%, other 1.3%
Legislative branch: bicameral National Assembly consists of the Senate (109 seats, 3 from each state plus 1 from Abuja; members elected by popular vote to serve four-year terms) and House of Representatives (360 seats; members elected by popular vote to serve four-year terms)
elections: Senate–last held on 9 April 2011 (next to be held in 2015); House of Representatives–last held on 9 April 2011 (next to be held in 2015)
election results: Senate–percent of vote by party–NA; seats by party–PDP 45, ACN 13, ANPP 7, CPC 5, other 4; House of Representatives–percent of vote by party–NA; seats by party–PDP 123, ACN 47, CPC 30, ANPP 25, other 9; note–due to logistical problems elections in a number of constituencies were postponed
Judicial branch: Supreme Court (judges recommended by the National Judicial Council and appointed by the president); Federal Court of Appeal (judges are appointed by the federal government from a pool of judges recommended by the National Judicial Council)
Political parties and leaders: Accord Party [Augustine MAZIE, acting]; Action Congress of Nigeria or ACN [Bisi AKANDE]; All Nigeria Peoples Party or ANPP [Ogbonnaya ONU]; All Progressives Grand Alliance or APGA [Victor C. UMEH]; Alliance for Democracy or AD [Mojisoluwa AKINFENWA]; Conference of Nigerian Political Parities or CNPP [Abdulkadir Balarabe MUSA]; Congress for Progressive Change or CPC; Democratic Peoples Party or DPP [Jeremiah USENI]; Fresh Democratic Party [Chris OKOTIE]; Labor Party [Dan NWANYANWU]; National Democratic Party or NDP [Aliyu Habu FARI]; Peoples Democratic Party or PDP [Dr. Okwesilieze NWODO]; Peoples Progressive Alliance [Larry ESIN]
Political pressure groups and leaders: Academic Staff Union for Universities or ASUU; Campaign for Democracy or CD; Civil Liberties Organization or CLO; Committee for the Defense of Human Rights or CDHR; Constitutional Right Project or CRP; Human Right Africa; National Association of Democratic Lawyers or NADL; National Association of Nigerian Students or NANS; Nigerian Bar Association or NBA; Nigerian Labor Congress or NLC; Nigerian Medical Association or NMA; the press; Universal Defenders of Democracy or UDD
International organization participation: ACP, AfDB, AU, C, D-8, ECOWAS, FAO, G-15, G-24, G-77, IAEA, IBRD, ICAO, ICC, ICRM, IDA, IDB, IFAD, IFC, IFRCS, IHO, ILO, IMF, IMO, IMSO, Interpol, IOC, IOM, IPU, ISO, ITSO, ITU, ITUC, MIGA, MINURSO, MONUSCO, NAM, OAS (observer), OIC, OPCW, OPEC, PCA, UN, UN Security Council (temporary), UNAMID, UNCTAD, UNESCO, UNHCR, UNIDO, UNIFIL, UNITAR, UNMIL, UNOCI, UNWTO, UPU, WCO, WFTU, WHO, WIPO, WMO, WTO
Diplomatic representation in the US: *chief of mission:* Ambassador Adebowale Ibidapo ADEFUYE
chancery: 3519 International Court NW, Washington, DC 20008
telephone: [1] (202) 986-8400
FAX: [1] (202) 775-1385
consulate(s) general: Atlanta, New York
Diplomatic representation from the US: *chief of mission:* Ambassador Terence P. MCCULLEY
embassy: 1075 Diplomatic Drive, Central District Area, Abuja
mailing address: P. O. Box 5760, Garki, Abuja
telephone: [234] (9) 461-4000
FAX: [234] (9) 461-4036
Flag description: three equal vertical bands of green (hoist side), white, and green; the color green represents the forests and abundant natural wealth of the country, white stands for peace and unity
National anthem: *name:* "Arise Oh Compatriots, Nigeria's Call Obey"
lyrics/music: John A. ILECHUKWU, Eme Etim AKPAN, B. A. OGUNNAIKE, Sotu OMOIGUI and P. O. ADERIBIGBE/ Benedict Elide ODIASE
note: adopted 1978; the lyrics are a mixture of five of the top entries in a national contest

ECONOMY

Economy—overview: Oil-rich Nigeria has been hobbled by political instability, corruption, inadequate infrastructure, and poor macroeconomic management but in 2008 began pursuing economic reforms. Nigeria's former military rulers failed to diversify the economy away from its overdependence on the capital-intensive oil sector, which provides 95% of foreign exchange earnings and about 80% of budgetary revenues. Following the signing of an IMF stand-by agreement in August 2000, Nigeria received a debt-restructuring deal from the Paris Club and a $1 billion credit from the IMF, both contingent on economic reforms. Nigeria pulled out of its IMF program in April 2002, after failing to meet spending and exchange rate targets, making it ineligible for additional debt forgiveness from the Paris Club. In November 2005, Abuja won Paris Club approval for a debt-relief deal that eliminated $18 billion of debt in exchange for $12 billion in payments–a total package worth $30 billion of Nigeria's total $37 billion external debt. Since 2008 the government has begun to show the political will to implement the market-oriented reforms urged by the IMF, such as modernizing the banking system, curbing inflation by blocking excessive wage demands, and resolving regional disputes over the distribution of earnings from the oil industry. GDP rose strongly in 2007-10 because of increased oil exports and high global crude prices in 2010. President JONATHAN has pledged to continue the economic reforms of his predecessor with emphasis

on infrastructure improvements. Infrastructure is the main impediment to growth and in August 2010 JONATHAN unveiled a power sector blueprint that includes privatization of the state-run electricity generation and distribution facilities. The government also is working toward developing stronger public-private partnerships for roads. Nigeria's financial sector was hurt by the global financial and economic crises and the Central Bank governor has taken measures to strengthen that sector.
GDP (purchasing power parity): $377.9 billion (2010 est.)
country comparison to the world: 32
$348.7 billion (2009 est.)
$326 billion (2008 est.)
note: data are in 2010 US dollars
GDP (official exchange rate): $216.8 billion (2010 est.)
GDP—real growth rate: 8.4% (2010 est.)
country comparison to the world: 15
7% (2009 est.)
6% (2008 est.)
GDP—per capita (PPP):
$2,500 (2010 est.)
country comparison to the world: 178
$2,300 (2009 est.)
$2,200 (2008 est.)
note: data are in 2010 US dollars
GDP—composition by sector:
agriculture: 31.9%
industry: 32.9%
services: 35.2% (2010 est.)
Labor force: 48.33 million (2010 est.)
country comparison to the world: 11
Labor force—by occupation:
agriculture: 70%
industry: 10%
services: 20% (1999 est.)
Unemployment rate: 4.9% (2007 est.)
country comparison to the world: 45
Population below poverty line:
70% (2007 est.)
Household income or consumption by percentage share: *lowest 10%:* 2%
highest 10%: 32.4% (2004)
Distribution of family income—Gini index: 43.7 (2003)
country comparison to the world: 44
50.6 (1997)
Investment (gross fixed):
11.6% of GDP (2010 est.)
country comparison to the world: 147
Budget: *revenues:* $18.16 billion
expenditures: $29.55 billion (2010 est.)
Public debt: 13.4% of GDP (2010 est.)
country comparison to the world: 119
11.8% of GDP (2009 est.)
Inflation rate (consumer prices):
13.9% (2010 est.)
country comparison to the world: 217
11.5% (2009 est.)
Central bank discount rate:
6% (31 December 2009)
country comparison to the world: 46
9.75% (31 December 2008)
Commercial bank prime lending rate:
18.36% (31 December 2009 est.)
country comparison to the world: 38
15.48% (31 December 2008 est.)
Stock of narrow money: $40.41 billion (31 December 2010 est.)
country comparison to the world: 46
$33.61 billion (31 December 2009 est.)
Stock of broad money: $91.97 billion (31 December 2010 est.)
country comparison to the world: 55
$72.31 billion (31 December 2009 est.)
Stock of domestic credit: $77.43 billion (31 December 2010 est.)
country comparison to the world: 55
$62.18 billion (31 December 2009 est.)
Market value of publicly traded shares:
$33.32 billion (31 December 2009)
country comparison to the world: 51
$49.8 billion (31 December 2008)
$86.35 billion (31 December 2007)
Agriculture—products: cocoa, peanuts, cotton, palm oil, corn, rice, sorghum, millet, cassava (tapioca), yams, rubber; cattle, sheep, goats, pigs; timber; fish
Industries: crude oil, coal, tin, columbite; rubber products, wood; hides and skins, textiles, cement and other construction materials, food products, footwear, chemicals, fertilizer, printing, ceramics, steel
Industrial production growth rate:
4% (2010 est.)
country comparison to the world: 87
Electricity—production:
21.92 billion kWh (2007 est.)
country comparison to the world: 69
Electricity—consumption: 19.21 billion kWh (2007 est.)
country comparison to the world: 69
Electricity—exports: 0 kWh (2008 est.)
Electricity—imports: 0 kWh (2008 est.)
Oil—production: 2.211 million bbl/day (2009 est.)
country comparison to the world: 15
Oil—consumption:
280,000 bbl/day (2009 est.)
country comparison to the world: 46
Oil—exports: 2.327 million bbl/day (2007 est.)
country comparison to the world: 6
Oil—imports: 170,000 bbl/day (2007 est.)
country comparison to the world: 52
Oil—proved reserves: 37.5 billion bbl (1 January 2010 est.)
country comparison to the world: 10
Natural gas—production: 32.82 billion cu m (2008 est.)
country comparison to the world: 26
Natural gas—consumption: 12.28 billion cu m (2008 est.)
country comparison to the world: 44
Natural gas—exports: 20.55 billion cu m (2008 est.)
country comparison to the world: 11
Natural gas—imports: 0 cu m (2008 est.)
country comparison to the world: 97
Natural gas—proved reserves: 5.246 trillion cu m (1 January 2010 est.)
country comparison to the world: 8
Current account balance: $27.77 billion (2010 est.)
country comparison to the world: 14
$22.89 billion (2009 est.)
Exports: $76.33 billion (2010 est.)
country comparison to the world: 39
$59.32 billion (2009 est.)
Exports—commodities: petroleum and petroleum products 95%, cocoa, rubber
Exports—partners: US 35.08%, India 10.43%, Brazil 9.32%, Spain 7.19%, France 4.65% (2009)
Imports: $34.18 billion (2010 est.)
country comparison to the world: 55
$29.05 billion (2009 est.)
Imports—commodities: machinery, chemicals, transport equipment, manufactured goods, food and live animals
Imports—partners: China 14.89%, US 8.88%, Netherlands 8.18%, South Korea 5.46%, UK 4.63%, France 4.19% (2009)
Reserves of foreign exchange and gold:
$43.36 billion (31 December 2010 est.)
country comparison to the world: 29
$44.76 billion (31 December 2009 est.)
Debt—external: $11.02 billion (31 December 2010 est.)
country comparison to the world: 88
$10.11 billion (31 December 2009 est.)
Stock of direct foreign investment—at home:
$67.23 billion (31 December 2010 est.)
country comparison to the world: 49
$61.23 billion (31 December 2009 est.)
Stock of direct foreign investment—abroad:
$6.071 billion (31 December 2010 est.)
country comparison to the world: 57
$5.821 billion (31 December 2009 est.)
Exchange rates: nairas (NGN) per US dollar–150.88 (2010), 148.9 (2009), 117.8 (2008), 127.46 (2007), 127.38 (2006)

COMMUNICATIONS

Telephones—main lines in use: 1.419 million (2009)
country comparison to the world: 67
Telephones—mobile cellular: 73.099 million (2009)
country comparison to the world: 16
Telephone system: *general assessment:* further expansion and modernization of the fixed-line telephone network is needed; network quality remains a problem
domestic: the addition of a second fixed-line provider in 2002 resulted in faster growth but subscribership remains only about 1 per 100 persons; mobile-cellular services growing rapidly, in part responding to the shortcomings of the fixed-line network; multiple cellular providers operate nationally with subscribership reaching 50 per 100 persons in 2009
international: country code–234; landing point for the SAT-3/WASC fiber-optic submarine cable that provides connectivity to Europe and Asia; satellite earth stations–3 Intelsat (2 Atlantic Ocean and 1 Indian Ocean) (2009)
Broadcast media: nearly 70 federal-government-controlled national and regional TV stations; all 36 states operate TV stations; several private TV stations operational; cable and satellite TV subscription services are available; network of federal-government-controlled national, regional, and state radio stations; roughly 40 state-government-owned radio stations typically carry their own programs except for news broadcasts; about 20 private radio stations also operate; transmissions of international broadcasters are available (2007)
Internet country code: .ng
Internet hosts: 1,378 (2010)
country comparison to the world: 163
Internet users: 43.989 million (2009)
country comparison to the world: 9

TRANSPORTATION

Airports: 54 (2010)
country comparison to the world: 87
Airports—with paved runways:
total: 38
over 3,047 m: 9
2,438 to 3,047 m: 11
1,524 to 2,437 m: 10
914 to 1,523 m: 5
under 914 m: 3 (2010)
Airports—with unpaved runways: *total:* 16
over 3,047 m: 1
1,524 to 2,437 m: 2
914 to 1,523 m: 11
under 914 m: 2 (2010)
Heliports: 4 (2010)
Pipelines: condensate 26 km; gas 2,756 km; liquid petroleum gas 97 km; oil 3,441 km; refined products 4,090 km (2010)
Railways: *total:* 3,505 km
country comparison to the world: 49
narrow gauge: 3,505 km 1.067-m gauge (2010)
Roadways: *total:* 193,200 km
country comparison to the world: 26
paved: 28,980 km
unpaved: 164,220 km (2004)
Waterways: 8,600 km (Niger and Benue rivers and smaller rivers and creeks) (2009)
country comparison to the world: 15
Merchant marine: *total:* 98
country comparison to the world: 51
by type: cargo 4, chemical tanker 30, liquefied gas 2, passenger/cargo 1, petroleum tanker 60, specialized tanker 1
foreign-owned: 4 (India 1, Spain 1, UK 2)
registered in other countries: 37 (Bahamas 2, Belize 2, Bermuda 11, Comoros 1, Italy 1, Liberia 4, Malaysia 1, Malta 1, North Korea 1, Panama 7, Saint Vincent and the Grenadines 1, Seychelles 1, unknown 4) (2010)
Ports and terminals: Bonny Inshore Terminal, Calabar, Lagos
Transportation—note: the International Maritime Bureau reports the territorial and offshore waters in the Niger Delta and Gulf of Guinea as high risk for piracy and armed robbery against ships; in 2010, 19 commercial vessels were boarded or attacked with most occurring in the vicinity of the port of Lagos; crews were robbed and stores or cargoes stolen

MILITARY

Military branches: Nigerian Armed Forces: Army, Navy, Air Force (2008)
Military service age and obligation: 18 years of age for voluntary military service (2007)
Manpower available for military service:
males age 16–49: 37,087,711
females age 16–49: 35,232,127 (2010 est.)
Manpower fit for military service: *males age 16–49:* 20,839,976
females age 16–49: 19,867,683 (2010 est.)
Manpower reaching militarily significant age annually: *male:* 1,767,428
female: 1,687,719 (2010 est.)
Military expenditures:
1.5% of GDP (2006)
country comparison to the world: 98

TRANSNATIONAL ISSUES

Disputes—international: Joint Border Commission with Cameroon reviewed 2002 ICJ ruling on the entire boundary and bilaterally resolved differences, including June 2006 Greentree Agreement that immediately cedes sovereignty of the Bakassi Peninsula to Cameroon with a phase-out of Nigerian control within two years while resolving patriation issues; the ICJ ruled on an equidistance settlement of Cameroon-Equatorial Guinea-Nigeria maritime boundary in the Gulf of Guinea, but imprecisely defined coordinates in the ICJ decision and a sovereignty dispute between Equatorial Guinea and Cameroon over an island at the mouth of the Ntem River all contribute to the delay in implementation; only Nigeria and Cameroon have heeded the Lake Chad Commission's admonition to ratify the delimitation treaty which also includes the Chad-Niger and Niger-Nigeria boundaries; location of Benin-Niger-Nigeria tripoint is unresolved
Refugees and internally displaced persons: refugees (country of origin): 5,778 (Liberia)
IDPs: undetermined (communal violence between Christians and Muslims since President OBASANJO's election in 1999; displacement is mostly short-term) (2007)
Illicit drugs: a transit point for heroin and cocaine intended for European, East Asian, and North American markets; consumer of amphetamines; safe haven for Nigerian narcotraffickers operating worldwide; major money-laundering center; massive corruption and criminal activity; Nigeria has improved some anti-money-laundering controls, resulting in its removal from the Financial Action Task Force's (FATF's) Noncooperative Countries and Territories List in June 2006; Nigeria's anti-money-laundering regime continues to be monitored by FATF

NIUE

(SELF-GOVERNING IN FREE ASSOCIATION WITH NEW ZEALAND)

INTRODUCTION

Background: Niue's remoteness, as well as cultural and linguistic differences between its Polynesian inhabitants and those of the rest of the Cook Islands, have caused it to be separately administered. The population of the island continues to drop (from a peak of 5,200 in 1966 to an estimated 1,311 in 2011) with substantial emigration to New Zealand 2,400 km to the southwest.

GEOGRAPHY

Location: Oceania, island in the South Pacific Ocean, east of Tonga
Geographic coordinates: 19 02 S, 169 52 W
Map references: Oceania
Area: *total:* 260 sq km
country comparison to the world: 211
land: 260 sq km
water: 0 sq km
Area—comparative: 1.5 times the size of Washington, DC
Land boundaries: 0 km
Coastline: 64 km
Maritime claims: *territorial sea:* 12 nm
exclusive economic zone: 200 nm
Climate: tropical; modified by southeast trade winds
Terrain: steep limestone cliffs along coast, central plateau
Elevation extremes: *lowest point:* Pacific Ocean 0 m
highest point: unnamed elevation near Mutalau settlement 68 m
Natural resources: fish, arable land
Land use: *arable land:* 11.54%
permanent crops: 15.38%
other: 73.08% (2005)
Irrigated land: NA
Natural hazards: typhoons
Environment—current issues: increasing attention to conservationist practices to counter loss of soil fertility from traditional slash and burn agriculture
Environment—international agreements:
party to: Biodiversity, Climate Change, Climate Change-Kyoto Protocol, Desertification, Law of the Sea, Ozone Layer Protection
Geography—note: one of world's largest coral islands

PEOPLE

Population: 1,311 (July 2011 est.)
country comparison to the world: 234
Age structure: *0–14 years:* NA
15–64 years: NA
65 years and over: NA (2009 est.)
Population growth rate:
-0.032% (2011 est.)
country comparison to the world: 200
Birth rate: NA
Death rate: NA
Net migration rate: NA
Urbanization: *urban population:* 38% of total population (2010)
rate of urbanization: -1.3% annual rate of change (2010-15 est.)
Sex ratio: NA

Infant mortality rate:
total: NA
male: NA
female: NA
Life expectancy at birth:
total population: NA
male: NA
female: NA
Total fertility rate: NA
HIV/AIDS—adult prevalence rate: NA
HIV/AIDS—people living with HIV/AIDS: NA
HIV/AIDS—deaths: NA
Drinking water source: *Improved:*
urban: 100% of population
rural: 100% of population
total: 100% of population (2008)
Sanitation facility access: *Improved:*
urban: 100% of population
rural: 100% of population
total: 100% of population (2008)
Nationality: *noun:* Niuean(s)
adjective: Niuean
Ethnic groups: Niuen 78.2%, Pacific islander 10.2%, European 4.5%, mixed 3.9%, Asian 0.2%, unspecified 3% (2001 census)
Religions: Ekalesia Niue (Niuean Church–a Protestant church closely related to the London Missionary Society) 61.1%, Latter-Day Saints 8.8%, Roman Catholic 7.2%, Jehovah's Witnesses 2.4%, Seventh-Day Adventist 1.4%, other 8.4%, unspecified 8.7%, none 1.9% (2001 census)
Languages: English (official), Niuean (a Polynesian language closely related to Tongan and Samoan)
Literacy: *definition:* NA
total population: 95%
male: NA
female: NA
School life expectancy (primary to tertiary education): *total:* 13 years
male: 12 years
female: 16 years (2005)
Education expenditures: NA

GOVERNMENT

Country name: *conventional long form:* none
conventional short form: Niue
note: pronunciation falls between nyu-way and new-way, but not like new-wee
former: Savage Island
Dependency status: self-governing in free association with New Zealand since 1974; Niue fully responsible for internal affairs; New Zealand retains responsibility for external affairs and defense; however, these responsibilities confer no rights of control and are only exercised at the request of the Government of Niue
Government type: self-governing parliamentary democracy
Capital: *name:* Alofi
geographic coordinates: 19 01 S, 169 55 W
time difference: UTC-11 (6 hours behind Washington, DC during Standard Time)
Administrative divisions: none; note–there are no first-order administrative divisions as defined by the US Government, but there are 14 villages at the second order
Independence: 19 October 1974 (Niue became a self-governing parliamentary government in free association with New Zealand)
National holiday: Waitangi Day (Treaty of Waitangi established British sovereignty over New Zealand), 6 February (1840)
Constitution: 19 October 1974 (Niue Constitution Act)
Legal system: English common law
Suffrage: 18 years of age; universal
Executive branch: *chief of state:* Queen ELIZABETH II (since 6 February 1952); represented by Governor General of New Zealand Anand SATYANAND (since 23 August 2006); the UK and New Zealand are represented by New Zealand High Commissioner Mark BLUMSKY (since September 2011)
head of government: Premier Toke TALAGI (since 18 June 2008)
cabinet: Cabinet consists of the premier and 3 ministers
(For more information visit the World Leaders website)
elections: the monarchy is hereditary; premier elected by the Legislative Assembly for a three-year term; election last held on 16 May 2011 (next to be held in 2014)
election results: Toke TALAGI reelected premier in Legislative Assembly vote; Toke TALAGI–11, Togia SIONEHOLO–8
Legislative branch: unicameral Legislative Assembly (20 seats; members elected by popular vote to serve three-year terms; six elected from a common roll and 14 are village representatives)
elections: last held on 7 May 2011 (next to be held in 2014)
election results: percent of vote by party–NA; seats by party–20 independents
Judicial branch: Supreme Court of New Zealand; High Court of Niue
Political parties and leaders: Alliance of Independents or AI; Niue People's Action Party or NPP [Young VIVIAN]
Political pressure groups and leaders: NA
International organization participation: ACP, AOSIS, FAO, IFAD, OPCW, PIF, Sparteca, SPC, UNESCO, UPU, WHO, WMO
Diplomatic representation in the US: none (self-governing territory in free association with New Zealand)
Diplomatic representation from the US: none (self-governing territory in free association with New Zealand)
Flag description: yellow with the flag of the UK in the upper hoist-side quadrant; the flag of the UK bears five yellow five-pointed stars–a large star on a blue disk in the center and a smaller star on each arm of the bold red cross; the larger star stands for Niue, the smaller stars recall the Southern Cross constellation on the New Zealand flag and symbolize links with that country; yellow represents the bright sunshine of Niue and the warmth and friendship between Niue and New Zealand
National anthem: *name:* "Ko e Iki he Lagi" (The Lord in Heaven)
lyrics/music: unknown/unknown, prepared by Sioeli FUSIKATA
note: adopted 1974

ECONOMY

Economy—overview: The economy suffers from the typical Pacific island problems of geographic isolation, few resources, and a small population. Government expenditures regularly exceed revenues, and the shortfall is made up by critically needed grants from New Zealand that are used to pay wages to public employees. Niue has cut government expenditures by reducing the public service by almost half. The agricultural sector consists mainly of subsistence gardening, although some cash crops are grown for export. Industry consists primarily of small factories to process passion fruit, lime oil, honey, and coconut cream. The sale of postage stamps to foreign collectors is an important source of revenue. The island in recent years has suffered a serious loss of population because of emigration to New Zealand. Efforts to increase GDP include the promotion of tourism and financial services, although the International Banking Repeal Act of 2002 resulted in the termination of all offshore banking licenses. Economic aid from New Zealand in FY08/09 was US$5.7 million. Niue suffered a devastating typhoon in January 2004, which decimated nascent economic programs. While in the process of rebuilding, Niue has been dependent on foreign aid.
GDP (purchasing power parity):
$10.01 million (2003 est.)
country comparison to the world: 226
GDP (official exchange rate):
$10.01 million (2003)
GDP—real growth rate: 6.2% (2003 est.)
country comparison to the world: 47
GDP—per capita (PPP):
$5,800 (2003 est.)
country comparison to the world: 140
GDP—composition by sector:
agriculture: 23.5%
industry: 26.9%
services: 49.5% (2003)
Labor force: 663 (2001)
country comparison to the world: 227
Labor force—by occupation: note: most work on family plantations; paid work exists only in government service, small industry, and the Niue Development Board
Unemployment rate: 12% (2001)
country comparison to the world: 129
Population below poverty line: NA%
Household income or consumption by percentage share: *lowest 10%:* NA%
highest 10%: NA%
Budget: *revenues:* $15.07 million
expenditures: $16.33 million (FY0405)
Inflation rate (consumer prices):
4% (2005)
country comparison to the world: 115
Agriculture—products: coconuts, passion fruit, honey, limes, taro, yams, cassava (tapioca), sweet potatoes; pigs, poultry, beef cattle
Industries: handicrafts, food processing
Industrial production growth rate: NA%
Electricity—production: 3 million kWh (2007 est.)
country comparison to the world: 212
Electricity—consumption: 2.79 million kWh (2007 est.)
country comparison to the world: 212

Electricity—exports: 0 kWh (2008 est.)
Electricity—imports: 0 kWh (2008 est.)
Oil—production: 0 bbl/day (2009 est.)
country comparison to the world: 116
Oil—consumption: 0 bbl/day
country comparison to the world: 206
Oil—exports: 0 bbl/day (2007 est.)
country comparison to the world: 190
Oil—imports: 30.66 bbl/day (2007 est.)
country comparison to the world: 203
Oil—proved reserves: 0 bbl (1 January 2010 est.)
country comparison to the world: 169
Natural gas—production:
0 cu m (2008 est.)
country comparison to the world: 165
Natural gas—consumption:
0 cu m (2008 est.)
country comparison to the world: 203
Natural gas—exports: 0 cu m (2008 est.)
country comparison to the world: 147
Natural gas—imports: 0 cu m (2008 est.)
country comparison to the world: 94
Natural gas—proved reserves:
0 cu m (1 January 2010 est.)
country comparison to the world: 170
Exports: $201,400 (2004)
country comparison to the world: 219
Exports—commodities: canned coconut cream, copra, honey, vanilla, passion fruit products, pawpaws, root crops, limes, footballs, stamps, handicrafts
Imports: $9.038 million (2004)
country comparison to the world: 221
Imports—commodities: food, live animals, manufactured goods, machinery, fuels, lubricants, chemicals, drugs
Debt—external: $418,000 (2002 est.)
country comparison to the world: 194
Exchange rates: New Zealand dollars (NZD) per US dollar–1.3874 (2010), 1.6002 (2009), 1.4151 (2008), 1.3811 (2007), 1.5408 (2006)

COMMUNICATIONS

Telephones—main lines in use:
1,100 (2009)
country comparison to the world: 226
Telephones—mobile cellular: 600 (2004)
country comparison to the world: 218
Telephone system: *domestic:* single-line telephone system connects all villages on island
international: country code–683 (2001)
Broadcast media: 1 government-owned television station with many of the programs supplied by Television New Zealand; 1 government-owned radio station broadcasting in AM and FM (2009)
Internet country code: .nu
Internet hosts: 397,270 (2010)
country comparison to the world: 52
Internet users: 1,100 (2009)
country comparison to the world: 213

TRANSPORTATION

Airports: 1 (2010)
country comparison to the world: 226
Airports—with paved runways: *total:* 1
1,524 to 2,437 m: 1 (2010)
Roadways: *total:* 120 km
country comparison to the world: 212
paved: 120 km (2008)
Ports and terminals: Alofi

MILITARY

Military branches: no regular indigenous military forces; Police Force
Military—note: defense is the responsibility of New Zealand

TRANSNATIONAL ISSUES

Disputes—international: none

NORFOLK ISLAND

(TERRITORY OF AUSTRALIA)

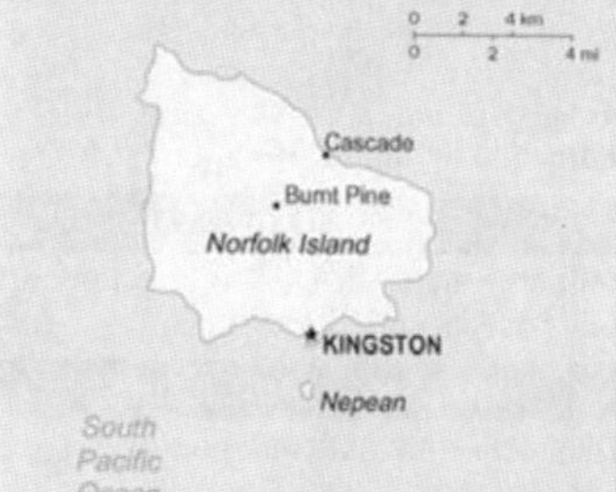

INTRODUCTION

Background: Two British attempts at establishing the island as a penal colony (1788-1814 and 1825-55) were ultimately abandoned. In 1856, the island was resettled by Pitcairn Islanders, descendants of the Bounty mutineers and their Tahitian companions.

GEOGRAPHY

Location: Oceania, island in the South Pacific Ocean, east of Australia
Geographic coordinates: 29 02 S, 167 57 E
Map references: Oceania
Area: *total:* 36 sq km
country comparison to the world: 233
land: 36 sq km
water: 0 sq km
Area—comparative: about 0.2 times the size of Washington, DC
Land boundaries: 0 km
Coastline: 32 km
Maritime claims: *territorial sea:* 12 nm
exclusive fishing zone: 200 nm
Climate: subtropical; mild, little seasonal temperature variation
Terrain: volcanic formation with mostly rolling plains
Elevation extremes: *lowest point:* Pacific Ocean 0 m
highest point: Mount Bates 319 m
Natural resources: fish
Land use: *arable land:* 0%
permanent crops: 0%
other: 100% (2005)
Irrigated land: NA
Natural hazards: typhoons (especially May to July)
Environment—current issues: NA
Geography—note: most of the 32 km coastline consists of almost inaccessible cliffs, but the land slopes down to the sea in one small southern area on Sydney Bay, where the capital of Kingston is situated

PEOPLE

Population: 2,169 (July 2011 est.)
country comparison to the world: 230
Age structure: *0–14 years:* 20.2%
15–64 years: 63.9%
65 years and over: 15.9% (2009 est.)
Population growth rate: 0.006% (2011 est.)
country comparison to the world: 193
Birth rate: NA
Death rate: NA
Net migration rate: NA
Sex ratio: NA
Infant mortality rate: *total:* NA
male: NA
female: NA
Life expectancy at birth: *total population:* NA
male: NA
female: NA
Total fertility rate: NA
HIV/AIDS—adult prevalence rate: NA
HIV/AIDS—people living with HIV/AIDS: NA
HIV/AIDS—deaths: NA
Nationality: *noun:* Norfolk Islander(s)
adjective: Norfolk Islander(s)
Ethnic groups: descendants of the Bounty mutineers, Australian, New Zealander, Polynesian
Religions: Anglican 31.8%, Roman Catholic 11.5%, Uniting Church in Australia 10.6%, Seventh-Day Adventist 3.2%, other Christian 5.6%, none 19.9%, unspecified 16.6% (2006 census)
Languages: English (official), Norfolk (a mixture of 18th century English and ancient Tahitian)
Literacy: NA
School life expectancy (primary to tertiary education): NA
Education expenditures: NA

GOVERNMENT

Country name: *conventional long form:* Territory of Norfolk Island
conventional short form: Norfolk Island
Dependency status: self-governing territory of Australia; administered from Canberra by the Department of Regional Australia, Regional Development and Local Government
Government type: NA
Capital: *name:* Kingston
geographic coordinates: 29 03 S, 167 58 E
time difference: UTC+11.5 (16.5 hours ahead of Washington, DC during Standard Time)
Administrative divisions: none (territory of Australia)
Independence: none (territory of Australia)
National holiday: Bounty Day (commemorates the arrival of Pitcairn Islanders), 8 June (1856)

Constitution: Norfolk Island Act of 1979 as amended in 2005
Legal system: English common law and the laws of Australia
Suffrage: 18 years of age; universal
Executive branch: *chief of state:* Queen ELIZABETH II (since 6 February 1952); represented by the Australian governor general
head of government: Administrator Owen WALSH (since October 2007)
cabinet: Executive Council made up of 4 of the 9 members of the Legislative Assembly; the council devises government policy and acts as an advisor to the administrator
(For more information visit the World Leaders website)
elections: the monarchy is hereditary; governor general appointed by the monarch; administrator appointed by the governor general of Australia and represents the monarch and Australia
Legislative branch: unicameral Legislative Assembly (9 seats; members elected by electors who have nine equal votes each but only four votes can be given to any one candidate; members to serve three-year terms)
elections: last held on 17 March 2010 (next to be held in 2013)
election results: seats–independents 9 (note–no political parties)
Judicial branch: Supreme Court; Court of Petty Sessions
Political parties and leaders: none
Political pressure groups and leaders: none
International organization participation: UPU
Diplomatic representation in the US: none (territory of Australia)
Diplomatic representation from the US: none (territory of Australia)
Flag description: three vertical bands of green (hoist side), white, and green with a large green Norfolk Island pine tree centered in the slightly wider white band; green stands for the rich vegetation on the island, and the pine tree–endemic to the island–is a symbol of Norfolk Island
note: somewhat reminiscent of the flag of Canada with its use of only two colors and depiction of a prominent local floral symbol in the central white band
National anthem:
name: "Come Ye Blessed"
lyrics/music: New Testament/John Prindle SCOTT
note: the local anthem, whose lyrics consist of the words from Matthew 25:34-36, 40, is also known as "The Pitcairn Anthem;" as a territory of Australia, "God Save the Queen" is official (see Australia), however, the island does not recognize "Advance Australia Fair"

ECONOMY

Economy—overview: Tourism, the primary economic activity, has steadily increased over the years and has brought a level of prosperity unusual among inhabitants of the Pacific islands. The agricultural sector has become self sufficient in the production of beef, poultry, and eggs.
GDP (purchasing power parity): $NA
Labor force: 978 (2006)
country comparison to the world: 226
Labor force—by occupation:
agriculture: 10%
industry and *services:* 90% (2000 est.)
Budget: *revenues:* $4.6 million
expenditures: $4.8 million (FY99/00)
Agriculture—products: Norfolk Island pine seed, Kentia palm seed, cereals, vegetables, fruit; cattle, poultry
Industries: tourism, light industry, ready mixed concrete
Electricity—production: NA kWh
Electricity—consumption: NA kWh
Exports: $NA (FY91/92)
Exports—commodities: postage stamps, seeds of the Norfolk Island pine and Kentia palm, small quantities of avocados
Imports: $NA
Imports—commodities: NA
Debt—external: $NA
Exchange rates: Australian dollars (AUD) per US dollar–1.0902 (2010), 1.2822 (2009), 1.2059 (2008), 1.2137 (2007), 1.3285 (2006)

COMMUNICATIONS

Telephones—main lines in use: 2,532; note–a mix of analog (2,500) and digital (32) circuits (2004)
country comparison to the world: 220
Telephones—mobile cellular: 0 (2002)
country comparison to the world: 220
Telephone system: *general assessment:* adequate
domestic: free local calls
international: country code–672; undersea coaxial cable links with Australia and New Zealand; satellite earth station–1
Broadcast media: 1 local radio station; broadcasts of several Australian radio and television stations are received via satellite (2009)
Internet country code: .nf
Internet hosts: 93 (2010)
country comparison to the world: 202

TRANSPORTATION

Airports: 1 (2010)
country comparison to the world: 227
Airports—with paved runways: *total:* 1
1,524 to 2,437 m: 1 (2010)
Roadways: *total:* 80 km
country comparison to the world: 215
paved: 53 km
unpaved: 27 km (2008)
Ports and terminals: Kingston

MILITARY

Military—note: defense is the responsibility of Australia

TRANSNATIONAL ISSUES

Disputes—international: none

NORTHERN MARIANA ISLANDS
(COMMONWEALTH IN POLITICAL UNION WITH THE US)

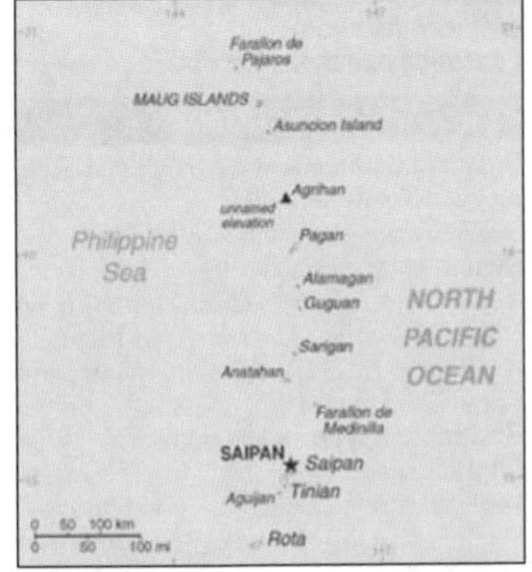

INTRODUCTION

Background: Under US administration as part of the UN Trust Territory of the Pacific, the people of the Northern Mariana Islands decided in the 1970s not to seek independence but instead to forge closer links with the US. Negotiations for territorial status began in 1972. A covenant to establish a commonwealth in political union with the US was approved in 1975, and came into force on 24 March 1976. A new government and constitution went into effect in 1978.

GEOGRAPHY

Location: Oceania, islands in the North Pacific Ocean, about three-quarters of the way from Hawaii to the Philippines
Geographic coordinates: 15 12 N, 145 45 E
Map references: Oceania
Area: *total:* 464 sq km
country comparison to the world: 195
land: 464 sq km
water: 0 sq km
note: consists of 14 islands including Saipan, Rota, and Tinian
Area—comparative: 2.5 times the size of Washington, DC
Land boundaries: 0 km
Coastline: 1,482 km
Maritime claims: *territorial sea:* 12 nm
exclusive economic zone: 200 nm
Climate: tropical marine; moderated by northeast trade winds, little seasonal temperature variation; dry season December to June, rainy season July to October
Terrain: southern islands are limestone with level terraces and fringing coral reefs; northern islands are volcanic
Elevation extremes: *lowest point:* Pacific Ocean 0 m
highest point: unnamed elevation on Agrihan 965 m
Natural resources: arable land, fish
Land use: *arable land:* 13.04%
permanent crops: 4.35%
other: 82.61% (2005)
Irrigated land: NA
Natural hazards: active volcanoes on Pagan and Agrihan; typhoons (especially August to November)

Environment—current issues: contamination of groundwater on Saipan may contribute to disease; clean-up of landfill; protection of endangered species conflicts with development
Geography—note: strategic location in the North Pacific Ocean

PEOPLE

Population: 46,050 (July 2011 est.)
country comparison to the world: 208
Age structure: *0–14 years:* 26% (male 6,349/female 5,625)
15–64 years: 70.4% (male 15,014/female 17,397)
65 years and over: 3.6% (male 790/female 875) (2011 est.)
Median age: *total:* 30 years
male: 29.8 years
female: 30.1 years (2011 est.)
Population growth rate:
-4.004% (2011 est.)
country comparison to the world: 231
Birth rate: 20.69 births/1,000 population (2011 est.)
country comparison to the world: 87
Death rate: 3.28 deaths/1,000 population (July 2011 est.)
country comparison to the world: 217
Net migration rate: -57.46 migrant(s)/1,000 population (2011 est.)
country comparison to the world: 220
Urbanization: *urban population:* 91% of total population (2010)
rate of urbanization: 1.7% annual rate of change (2010-15 est.)
Major cities—population: SAIPAN (capital) NA (2009)
Sex ratio: *at birth:* 1.061 male(s)/female
under 15 years: 1.15 male(s)/female
15–64 years: 0.85 male(s)/female
65 years and over: 0.92 male(s)/female
total population: 0.93 male(s)/female (2011 est.)
Infant mortality rate:
total: 5.79 deaths/1,000 live births
country comparison to the world: 177
male: 6.19 deaths/1,000 live births
female: 5.37 deaths/1,000 live births (2011 est.)
Life expectancy at birth: *total population:* 77.08 years
country comparison to the world: 67
male: 74.45 years
female: 79.87 years (2011 est.)
Total fertility rate: 2.13 children born/woman (2011 est.)
country comparison to the world: 114
HIV/AIDS—adult prevalence rate: NA
HIV/AIDS—people living with HIV/AIDS: NA
HIV/AIDS—deaths: NA
Drinking water source: *Improved:*
urban: 98% of population
rural: 97% of population
total: 98% of population
Unimproved:
urban: 2% of population
rural: 3% of population
total: 2% of population (2008)
Sanitation facility access: *Improved:*
urban: 92% of population
rural: 93% of population
total: 92% of population
Unimproved:
urban: 8% of population
rural: 7% of population
total: 8% of population (2000)
Nationality: *noun:* NA (US citizens)
adjective: NA
Ethnic groups: Asian 56.3%, Pacific islander 36.3%, Caucasian 1.8%, other 0.8%, mixed 4.8% (2000 census)
Religions: Christian (Roman Catholic majority, although traditional beliefs and taboos may still be found)
Languages: Philippine languages 24.4%, Chinese 23.4%, Chamorro 22.4%, English 10.8%, other Pacific island languages 9.5%, other 9.6% (2000 census)
Literacy: *definition:* age 15 and over can read and write
total population: 97%
male: 97%
female: 96% (1980 est.)
School life expectancy (primary to tertiary education): NA
Education expenditures: NA

GOVERNMENT

Country name: *conventional long form:* Commonwealth of the Northern Mariana Islands
conventional short form: Northern Mariana Islands
abbreviation: CNMI
former: Trust Territory of the Pacific Islands, Mariana Islands District
Dependency status: commonwealth in political union with the US; federal funds to the Commonwealth administered by the US Department of the Interior, Office of Insular Affairs
Government type: commonwealth; self-governing with locally elected governor, lieutenant governor, and legislature
Capital: *name:* Saipan
geographic coordinates: 15 12 N, 145 45 E
time difference: UTC+10 (15 hours ahead of Washington, DC during Standard Time)
Administrative divisions: none (commonwealth in political union with the US); there are no first-order administrative divisions as defined by the US Government, but there are four municipalities at the second order: Northern Islands, Rota, Saipan, Tinian
Independence: none (commonwealth in political union with the US)
National holiday: Commonwealth Day, 8 January (1978)
Constitution: Constitution of the Commonwealth of the Northern Mariana Islands effective 1 January 1978; Covenant Agreement fully effective 4 November 1986
Legal system: US system applies, except for customs, wages, immigration laws, and taxation
Suffrage: 18 years of age; universal; indigenous inhabitants are US citizens but do not vote in US presidential elections
Executive branch: *chief of state:* President Barack H. OBAMA (since 20 January 2009); Vice President Joseph R. BIDEN (since 20 January 2009)
head of government: Governor Benigno R. FITIAL (since 9 January 2006); Lieutenant Governor Eloy S. INOS (since 1 May 2009)
cabinet: the cabinet consists of the heads of the 10 principal departments under the executive branch who are appointed by the governor with the advice and consent of the Senate; other members include special assistants to the governor and office heads appointed by and reporting directly to the governor
(For more information visit the World Leaders website)
elections: under the US Constitution, residents of unincorporated territories, such as the Commonwealth of the Northern Mariana Islands, do not vote in elections for US president and vice president; however, they may vote in the Democratic and Republican party presidential primary elections; governor and lieutenant governor elected on the same ticket by popular vote for four-year terms (eligible for a second term); election last held on 7 November 2009 with a run-off election held on 23 November 2009 (next to be held in 2013)
election results: Benigno R. FITIAL reelected governor in a run-off election held 23 November 2009; percent of vote—Benigno R. FITIAL 51.4%, Heinz HOFSCHNEIDER 48.6%
Legislative branch: bicameral legislature consists of the Senate (9 seats; members elected by popular vote to serve four-year staggered terms) and the House of Representatives (20 seats; members elected by popular vote to serve two-year terms)
elections: Senate—last held on 7 November 2009 (next to be held in November 2011); House of Representatives—last held on 7 November 2009 (next to be held in November 2011)
election results: Senate—percent of vote by party—NA; seats by party—Covenant Party 3, Republican Party 3, Democratic Party 1, independents 2; House of Representatives—percent of vote by party—NA; seats by party—Republican Party 12, Covenant Party 4, Democratic Party 1, independents 3
note: the Northern Mariana Islands elects one nonvoting delegate to the US House of Representatives; election last held on 2 November 2010 (next to be held in November 2012); seats by party—independent 1
Judicial branch: Commonwealth Supreme Court; Superior Court; Federal District Court
Political parties and leaders: Covenant Party [Benigno R. FITIAL]; Democratic Party [Dr. Carlos S. CAMACHO]; Republican Party [Juan S. REYES]
Political pressure groups and leaders: NA
International organization participation: SPC, UPU
Flag description: blue, with a white, five-pointed star superimposed on a gray latte stone (the traditional foundation stone used in building) in the center, surrounded by a wreath; blue symbolizes the Pacific Ocean, the star represents the Commonwealth; the latte stone and the floral head wreath display elements of the native Chamorro culture
National anthem: *name:* "Gi Talo Gi Halom Tasi" (In the Middle of the Sea)

lyrics/music: Jose S. PANGELINAN [Chamoru], David PETER [Carolinian]/ Wilhelm GANZHORN
note: adopted 1996; the Carolinian version of the song is known as "Satil Matawal Pacifico;" as a commonwealth of the United States, in addition to the local anthem, "The Star-Spangled Banner" is official (see United States)

ECONOMY

Economy—overview: The economy benefits substantially from financial assistance from the US. The rate of funding has declined as locally generated government revenues have grown. The key tourist industry employs about 50% of the work force and accounts for roughly one-fourth of GDP. Japanese tourists predominate. Annual tourist entries have exceeded one-half million in recent years, but financial difficulties in Japan have caused a temporary slowdown. The agricultural sector is made up of cattle ranches and small farms producing coconuts, breadfruit, tomatoes, and melons. Garment production is by far the most important industry with the employment of 17,500 mostly Chinese workers and sizable shipments to the US under duty and quota exemptions.
GDP (purchasing power parity): $900 million (2000 est.)
country comparison to the world: 203
note: GDP estimate includes US subsidy
GDP (official exchange rate): $633.4 million (2000)
GDP—per capita (PPP): $12,500 (2000 est.)
country comparison to the world: 93
GDP—composition by sector:
agriculture: NA%
industry: NA%
services: NA%
Labor force: 38,450
country comparison to the world: 198
note: nearly 29,000 of these were foreign workers (2005 est.)
Labor force—by occupation:
agriculture: NA%
industry: NA%
services: NA%
Unemployment rate: 8% (2005 est.)
country comparison to the world: 88
3.9% (2001)
Population below poverty line: NA%
Household income or consumption by percentage share: *lowest 10%:* NA%
highest 10%: NA%
Budget: *revenues:* $193 million
expenditures: $223 million (FY01/02 est.)
Inflation rate (consumer prices): -0.8% (2000)
country comparison to the world: 6
Agriculture—products: vegetables and melons, fruits and nuts; ornamental plants; livestock, poultry and eggs; fish and aquaculture products
Industries: banking, construction, fishing, garment, tourism, handicrafts
Industrial production growth rate: NA%
Electricity—production: 60,600 kWh (January 2009)
country comparison to the world: 214
Electricity—consumption: 48,300 kWh (January 2009)
country comparison to the world: 214
Electricity—exports: 0 kWh (January 2009 est.)
Electricity—imports: 0 kWh (January 2009 est.)
Exports: $98.2 million (2008)
country comparison to the world: 195
Exports—commodities: garments
Imports: $214.4 million (2001)
country comparison to the world: 199
Imports—commodities: food, construction equipment and materials, petroleum products
Debt—external: $NA
Exchange rates: the US dollar is used

COMMUNICATIONS

Telephones—main lines in use: 25,100 (2009)
country comparison to the world: 185
Telephones—mobile cellular: 20,500 (2004)
country comparison to the world: 208
Telephone system: *general assessment:* NA
domestic: NA
international: country code–1-670; satellite earth stations–2 Intelsat (Pacific Ocean)
Broadcast media: 1 TV broadcast station on Saipan; multi-channel cable TV services are available on Saipan; 9 licensed radio broadcast stations (2009)
Internet country code: .mp
Internet hosts: 9 (2010)
country comparison to the world: 221

TRANSPORTATION

Airports: 5 (2010)
country comparison to the world: 182
Airports—with paved runways: *total:* 3
2,438 to 3,047 m: 2
1,524 to 2,437 m: 1 (2010)
Airports—with unpaved runways: *total:* 2
2,438 to 3,047 m: 1
under 914 m: 1 (2010)
Heliports: 1 (2010)
Roadways: *total:* 536 km (2008)
country comparison to the world: 192
Ports and terminals: Saipan, Tinian, Rota

MILITARY

Manpower fit for military service: *males age 16–49:* 8,793
females age 16–49: 11,569 (2010 est.)
Manpower reaching militarily significant age annually: *male:* 410
female: 306 (2010 est.)
Military—note: defense is the responsibility of the US

TRANSNATIONAL ISSUES

Disputes—international: none

NORWAY

INTRODUCTION

Background: Two centuries of Viking raids into Europe tapered off following the adoption of Christianity by King Olav TRYGGVASON in 994. Conversion of the Norwegian kingdom occurred over the next several decades. In 1397, Norway was absorbed into a union with Denmark that lasted more than four centuries. In 1814, Norwegians resisted the cession of their country to Sweden and adopted a new constitution. Sweden then invaded Norway but agreed to let Norway keep its constitution in return for accepting the union under a Swedish king. Rising nationalism throughout the 19th century led to a 1905 referendum granting Norway independence. Although Norway remained neutral in World War I, it suffered heavy losses to its shipping. Norway proclaimed its neutrality at the outset of World War II, but was nonetheless occupied for five years by Nazi Germany (1940-45). In 1949, neutrality was abandoned and Norway became a member of NATO. Discovery of oil and gas in adjacent waters in the late 1960s boosted Norway's economic fortunes. In referenda held in 1972 and 1994, Norway rejected joining the EU. Key domestic issues include immigration and integration of ethnic minorities, maintaining the country's extensive social safety net with an aging population, and preserving economic competitiveness.

GEOGRAPHY

Location: Northern Europe, bordering the North Sea and the North Atlantic Ocean, west of Sweden
Geographic coordinates: 62 00 N, 10 00 E
Map references: Europe
Area: *total:* 323,802 sq km
country comparison to the world: 67
land: 304,282 sq km
water: 19,520 sq km
Area—comparative: slightly larger than New Mexico
Land boundaries: *total:* 2,542 km
border countries: Finland 727 km, Sweden 1,619 km, Russia 196 km
Coastline: 25,148 km (includes mainland 2,650 km, as well as long fjords, numerous small islands, and minor indentations 22,498 km; length of island coastlines 58,133 km)
Maritime claims: *territorial sea:* 12 nm
contiguous zone: 10 nm
exclusive economic zone: 200 nm
continental shelf: 200 nm
Climate: temperate along coast, modified by North Atlantic Current; colder interior with increased precipitation and colder summers; rainy year-round on west coast
Terrain: glaciated; mostly high plateaus and rugged mountains broken by fertile valleys; small, scattered plains; coastline deeply indented by fjords; arctic tundra in north
Elevation extremes: *lowest point:* Norwegian Sea 0 m
highest point: Galdhopiggen 2,469 m
Natural resources: petroleum, natural gas, iron ore, copper, lead, zinc, titanium, pyrites, nickel, fish, timber, hydropower
Land use: *arable land:* 2.7%
permanent crops: 0%
other: 97.3% (2005)
Irrigated land: 1,270 sq km (2003)
Total renewable water resources: 381.4 cu km (2005)

Freshwater withdrawal (domestic/industrial/agricultural): *total:* 2.4 cu km/yr (23%/67%/10%)
per capita: 519 cu m/yr (1996)
Natural hazards: rockslides, avalanches
volcanism: Beerenberg (elev. 2,227 m) on Jan Mayen Island in the Norwegian Sea is the country's only active volcano
Environment—current issues: water pollution; acid rain damaging forests and adversely affecting lakes, threatening fish stocks; air pollution from vehicle emissions
Environment—international agreements: *party to:* Air Pollution, Air Pollution-Nitrogen Oxides, Air Pollution-Persistent Organic Pollutants, Air Pollution-Sulfur 85, Air Pollution-Sulfur 94, Air Pollution-Volatile Organic Compounds, Antarctic-Environmental Protocol, Antarctic-Marine Living Resources, Antarctic Seals, Antarctic Treaty, Biodiversity, Climate Change, Climate Change-Kyoto Protocol, Desertification, Endangered Species, Environmental Modification, Hazardous Wastes, Law of the Sea, Marine Dumping, Ozone Layer Protection, Ship Pollution, Tropical Timber 83, Tropical Timber 94, Wetlands, Whaling
signed, but not ratified: none of the selected agreements
Geography—note: about two-thirds mountains; some 50,000 islands off its much-indented coastline; strategic location adjacent to sea lanes and air routes in North Atlantic; one of the most rugged and longest coastlines in the world

PEOPLE

Population: 4,691,849 (July 2011 est.)
country comparison to the world: 118
Age structure: *0–14 years:* 18% (male 431,111/female 412,864)
15–64 years: 66% (male 1,568,729/female 1,529,799)
65 years and over: 16% (male 326,711/female 422,635) (2011 est.)
Median age: *total:* 40 years
male: 39.1 years
female: 40.8 years (2011 est.)
Population growth rate: 0.329% (2011 est.)
country comparison to the world: 168
Birth rate: 10.84 births/1,000 population (2011 est.)
country comparison to the world: 177
Death rate: 9.24 deaths/1,000 population (July 2011 est.)
country comparison to the world: 66
Net migration rate: 1.7 migrant(s)/1,000 population (2011 est.)
country comparison to the world: 43
Urbanization: *urban population:* 79% of total population (2010)
rate of urbanization: 1.2% annual rate of change (2010-15 est.)
Major cities—population: OSLO (capital) 875,000 (2009)
Sex ratio: *at birth:* 1.054 male(s)/female
under 15 years: 1.04 male(s)/female
15–64 years: 1.03 male(s)/female
65 years and over: 0.76 male(s)/female
total population: 0.98 male(s)/female (2011 est.)
Infant mortality rate:
total: 3.52 deaths/1,000 live births
country comparison to the world: 209
male: 3.85 deaths/1,000 live births
female: 3.17 deaths/1,000 live births (2011 est.)
Life expectancy at birth:
total population: 80.2 years
country comparison to the world: 25
male: 77.53 years
female: 83.02 years (2011 est.)
Total fertility rate: 1.77 children born/woman (2011 est.)
country comparison to the world: 160
HIV/AIDS—adult prevalence rate:
0.1% (2009 est.)
country comparison to the world: 151
HIV/AIDS—people living with HIV/AIDS: 4,000 (2009 est.)
country comparison to the world: 122
HIV/AIDS—deaths:
fewer than 100 (2009 est.)
country comparison to the world: 147
Drinking water source: *Improved:*
urban: 100% of population
rural: 100% of population
total: 100% of population (2008)
Sanitation facility access: *Improved:*
urban: 100% of population
rural: 100% of population
total: 100% of population (2008)
Nationality: *noun:* Norwegian(s)
adjective: Norwegian
Ethnic groups: Norwegian 94.4% (includes Sami, about 60,000), other European 3.6%, other 2% (2007 estimate)
Religions: Church of Norway 85.7%, Pentecostal 1%, Roman Catholic 1%, other Christian 2.4%, Muslim 1.8%, other 8.1% (2004)
Languages: Bokmal Norwegian (official), Nynorsk Norwegian (official), small Sami- and Finnish-speaking minorities
note: Sami is official in six municipalities
Literacy: *definition:* age 15 and over can read and write
total population: 100%
male: 100%
female: 100%
School life expectancy (primary to tertiary education): *total:* 17 years
male: 17 years
female: 18 years (2008)
Education expenditures:
6.8% of GDP (2007)
country comparison to the world: 18

GOVERNMENT

Country name: *conventional long form:* Kingdom of Norway
conventional short form: Norway
local long form: Kongeriket Norge
local short form: Norge
Government type:
constitutional monarchy
Capital: *name:* Oslo
geographic coordinates: 59 55 N, 10 45 E
time difference: UTC+1 (6 hours ahead of Washington, DC during Standard Time)
daylight saving time: +1hr, begins last Sunday in March; ends last Sunday in October
Administrative divisions: 19 counties (fylker, singular—fylke); Akershus, Aust-Agder, Buskerud, Finnmark, Hedmark, Hordaland, More og Romsdal, Nordland, Nord-Trondelag, Oppland, Oslo, Ostfold, Rogaland, Sogn og Fjordane, Sor-Trondelag, Telemark, Troms, Vest-Agder, Vestfold
Dependent areas: Bouvet Island, Jan Mayen, Svalbard
Independence: 7 June 1905 (Norway declared the union with Sweden dissolved); 26 October 1905 (Sweden agreed to the repeal of the union)
National holiday: Constitution Day, 17 May (1814)
Constitution: 17 May 1814; amended many times
Legal system: mixed legal system of civil, common, and customary law; Supreme Court can advise on legislative acts
International law organization participation: accepts compulsory ICJ jurisdiction with reservations; accepts ICCt jurisdiction
Suffrage: 18 years of age; universal
Executive branch: *chief of state:* King HARALD V (since 17 January 1991); Heir Apparent Crown Prince HAAKON MAGNUS, son of the monarch (born 20 July 1973)
head of government: Prime Minister Jens STOLTENBERG (since 17 October 2005)
cabinet: State Council appointed by the monarch with the approval of parliament (For more information visit the World Leaders website)
elections: the monarchy is hereditary; following parliamentary elections, the leader of the majority party or the leader of the majority coalition usually appointed prime minister by the monarch with the approval of the parliament
Legislative branch: modified unicameral Parliament or Storting (169 seats; members elected by popular vote by proportional representation to serve four-year terms)

elections: last held on 14 September 2009 (next to be held in September 2013)
election results: percent of vote by party–DNA 35.4%, FrP 22.9%, H 17.2%, SV 6.2%, Sp 6.2%, KrF 5.5%, V 3.9%, other 2.7%; seats by party–DNA 64, FrP 41, H 30, SV 11, Sp 11, KrF 10, V 2
Judicial branch: Supreme Court or Hoyesterett (justices appointed by the monarch)
Political parties and leaders: Center Party (Senterpartiet or Sp) [Liv Signe NAVARSETE]; Christian People's Party (Kristelig Folkeparti or KrF) [Dagfinn HOYBRATEN]; Conservative Party (Hoyre or H) [Erna SOLBERG]; Labor Party (Det norske Arbeiderpartiet or DNA) [Jens STOLTENBERG]; Liberal Party (Venstre or V) [Trine SKEI-GRANDE]; Progress Party (Framstegspartiet or FrP) [Siv JENSEN]; Socialist Left Party (Sosialistisk Venstreparti or SV) [Kristin HALVORSEN]
Political pressure groups and leaders: Norwegian Aid Committee or NORWAC; Norwegian Association of the Disabled; Pure Salmon Campaign; The Consumer Council (consumer advocacy group)
other: environmental groups; media; reform movements
International organization participation: ADB (nonregional member), AfDB (nonregional member), Arctic Council, Australia Group, BIS, CBSS, CE, CERN, EAPC, EBRD, EFTA, ESA, FAO, FATF, IADB, IAEA, IBRD, ICAO, ICC, ICRM, IDA, IEA, IFAD, IFC, IFRCS, IHO, ILO, IMF, IMO, IMSO, Interpol, IOC, IOM, IPU, ISO, ITSO, ITU, ITUC, MIGA, MONUSCO, NATO, NC, NEA, NIB, NSG, OAS (observer), OECD, OPCW, OSCE, Paris Club, PCA, Schengen Convention, UN, UNCTAD, UNESCO, UNHCR, UNIDO, UNITAR, UNMIS, UNRWA, UNTSO, UNWTO, UPU, WCO, WHO, WIPO, WMO, WTO, ZC
Diplomatic representation in the US:
chief of mission: Ambassador Wegger C. STROMMEN
chancery: 2720 34th Street NW, Washington, DC 20008
telephone: [1] (202) 333-6000
FAX: [1] (202) 337-0870
consulate(s) general: Houston, New York, San Francisco
Diplomatic representation from the US: *chief of mission:* Ambassador Barry B. WHITE
embassy: Henrik Ibsens gate 48, 0244 Oslo; note–the embassy will move to Huseby in the near future
mailing address: PSC 69, Box 1000, APO AE 09707
telephone: [47] 22 44 85 50
FAX: [47] 22 44 33 63, 22 56 27 51
Flag description: red with a blue cross outlined in white that extends to the edges of the flag; the vertical part of the cross is shifted to the hoist side in the style of the Dannebrog (Danish flag); the colors recall Norway's past political unions with Denmark (red and white) and Sweden (blue)
National anthem: *name:* "Ja, vi elsker dette landet" (Yes, We Love This Country)
lyrics/music: Bjornstjerne BJORNSON/ Rikard NORDRAAK
note: adopted 1864; in addition to the national anthem, "Kongesangen" (Song of the King), which uses the tune of "God Save the Queen," serves as the royal anthem

ECONOMY

Economy—overview: The Norwegian economy is a prosperous bastion of welfare capitalism, featuring a combination of free market activity and government intervention. The government controls key areas, such as the vital petroleum sector, through large-scale state-majority-owned enterprises. The country is richly endowed with natural resources–petroleum, hydropower, fish, forests, and minerals–and is highly dependent on the petroleum sector, which accounts for nearly half of exports and over 30% of state revenue. Norway is the world's second-largest gas exporter; its position as an oil exporter has slipped to ninth-largest as production has begun to decline. Norway opted to stay out of the EU during a referendum in November 1994; nonetheless, as a member of the European Economic Area, it contributes sizably to the EU budget. In anticipation of eventual declines in oil and gas production, Norway saves state revenue from the petroleum sector in the world's second largest sovereign wealth fund, valued at over $500 billion in 2010. After solid GDP growth in 2004-07, the economy slowed in 2008, and contracted in 2009, before returning to positive growth in 2010.
GDP (purchasing power parity):
$255.3 billion (2010 est.)
country comparison to the world: 47
$254.2 billion (2009 est.)
$257.9 billion (2008 est.)
note: data are in 2010 US dollars
GDP (official exchange rate):
$414.5 billion (2010 est.)
GDP—real growth rate: 0.4% (2010 est.)
country comparison to the world: 183
-1.4% (2009 est.)
0.8% (2008 est.)
GDP—per capita (PPP):
$54,600 (2010 est.)
country comparison to the world: 7
$54,500 (2009 est.)
$55,500 (2008 est.)
note: data are in 2010 US dollars
GDP—composition by sector:
agriculture: 2.1%
industry: 40.1%
services: 57.8% (2010 est.)
Labor force: 2.59 million (2010 est.)
country comparison to the world: 111
Labor force—by occupation:
agriculture: 2.9%
industry: 21.1%
services: 76% (2008)
Unemployment rate: 3.6% (2010 est.)
country comparison to the world: 30
3.2% (2009 est.)
Population below poverty line: NA%
Household income or consumption by percentage share: *lowest 10%:* 3.9%
highest 10%: 21% (2008)
Distribution of family income—Gini index: 25 (2008)
country comparison to the world: 134
25.8 (1995)
Investment (gross fixed):
18.6% of GDP (2010 est.)
country comparison to the world: 104
Budget: *revenues:* $226.8 billion
expenditures: $187 billion (2010 est.)
Public debt: 47.7% of GDP (2010 est.)
country comparison to the world: 55
49.8% of GDP (2009 est.)
Inflation rate (consumer prices):
2.4% (2010 est.)
country comparison to the world: 60
2.1% (2009 est.)
Central bank discount rate:
4% (31 December 2008)
country comparison to the world: 69
6.25% (31 December 2007)
Commercial bank prime lending rate:
4.28% (31 December 2009 est.)
country comparison to the world: 121
7.28% (31 December 2008 est.)
Stock of narrow money: $122.2 billion (31 December 2010 est.)
country comparison to the world: 28
$118.3 billion (31 December 2009 est.)
Stock of broad money: $256.3 billion (31 December 2010 est.)
country comparison to the world: 32
$243.3 billion (31 December 2009 est.)
Stock of domestic credit: $414.5 billion (31 December 2010 est.)
country comparison to the world: 26
$379.6 billion (31 December 2009 est.)
Market value of publicly traded shares:
$227.2 billion (31 December 2009)
country comparison to the world: 34
$125.9 billion (31 December 2008)
$357.4 billion (31 December 2007)
Agriculture—products: barley, wheat, potatoes; pork, beef, veal, milk; fish
Industries: petroleum and gas, food processing, shipbuilding, pulp and paper products, metals, chemicals, timber, mining, textiles, fishing
Industrial production growth rate:
0.3% (2010 est.)
country comparison to the world: 152
Electricity—production:
142.7 billion kWh (2008 est.)
country comparison to the world: 26
Electricity—consumption: 128.8 billion kWh (2008 est.)
country comparison to the world: 26
Electricity—exports: 17.29 billion kWh (2008 est.)
Electricity—imports: 3.414 billion kWh (2008 est.)
Oil—production: 2.35 million bbl/day (2009 est.)
country comparison to the world: 14
Oil—consumption: 204,100 bbl/day (2009 est.)
country comparison to the world: 56
Oil—exports: 2.15 million bbl/day (2009 est.)
country comparison to the world: 9
Oil—imports: 107,500 bbl/day (2008 est.)
country comparison to the world: 62
Oil—proved reserves: 6.68 billion bbl (1 January 2010 est.)
country comparison to the world: 21
Natural gas—production: 103.5 billion cu m (2009 est.)
country comparison to the world: 6

Natural gas—consumption: 4.62 billion cu m (2009 est.)
country comparison to the world: 60
Natural gas—exports: 98.85 billion cu m (2009 est.)
country comparison to the world: 2
Natural gas—imports:
0 cu m (2008 est.)
country comparison to the world: 98
Natural gas—proved reserves: 2.313 trillion cu m (1 January 2010 est.)
country comparison to the world: 17
Current account balance: $60.23 billion (2010 est.)
country comparison to the world: 5
$53.53 billion (2009 est.)
Exports: $137 billion (2010 est.)
country comparison to the world: 31
$122 billion (2009 est.)
Exports—commodities: petroleum and petroleum products, machinery and equipment, metals, chemicals, ships, fish
Exports—partners: UK 24.28%, Germany 13.4%, Netherlands 10.87%, France 8.55%, Sweden 5.76%, US 4.82% (2009)
Imports: $74.02 billion (2010 est.)
country comparison to the world: 37
$66.68 billion (2009 est.)
Imports—commodities: machinery and equipment, chemicals, metals, foodstuffs
Imports—partners: Sweden 13.86%, Germany 12.89%, China 7.8%, Denmark 6.78%, US 6.16%, UK 6.01% (2009)
Reserves of foreign exchange and gold: $NA (31 December 2010 est.)
$48.86 billion (31 December 2009 est.)
Debt—external:
$2.232 trillion (30 June 2010)
country comparison to the world: 8
note: Norway is a net external creditor
Stock of direct foreign investment—at home:
$132.8 billion (31 December 2010 est.)
country comparison to the world: 27
$128.4 billion (31 December 2009 est.)
Stock of direct foreign investment—abroad:
$226.6 billion (31 December 2010 est.)
country comparison to the world: 18
$206 billion (31 December 2009 est.)
Exchange rates: Norwegian kroner (NOK) per US dollar–6.044 (2010), 6.288 (2009), 5.6361 (2008), 5.86 (2007), 6.418 (2006)

COMMUNICATIONS

Telephones—main lines in use:
1.9 million (2009)
country comparison to the world: 58
Telephones—mobile cellular:
5.336 million (2009)
country comparison to the world: 93
Telephone system: *general assessment:* modern in all respects; one of the most advanced telecommunications networks in Europe
domestic: Norway has a domestic satellite system; the prevalence of rural areas encourages the wide use of mobile-cellular systems
international: country code–47; 2 buried coaxial cable systems; submarine cables provide links to other Nordic countries and Europe; satellite earth stations–NA Eutelsat, NA Intelsat (Atlantic Ocean), and 1 Inmarsat (Atlantic and Indian Ocean regions); note–Norway shares the Inmarsat earth station with the other Nordic countries (Denmark, Finland, Iceland, and Sweden) (1999)
Broadcast media: state-owned public radio-TV broadcaster operates 3 nationwide television stations, 3 nationwide radio stations, and 16 regional radio stations; roughly a dozen privately-owned television stations broadcast nationally and roughly another 25 local TV stations are available; nearly 75% of households have access to multi-channel cable or satellite TV systems; 2 privately-owned radio stations broadcast nationwide and another 240 stations operate locally (2008)
Internet country code: .no
Internet hosts: 3.352 million (2010)
country comparison to the world: 27
Internet users: 4.431 million (2009)
country comparison to the world: 53

TRANSPORTATION

Airports: 98 (2010)
country comparison to the world: 61
Airports—with paved runways:
total: 67
over 3,047 m: 1
2,438 to 3,047 m: 12
1,524 to 2,437 m: 11
914 to 1,523 m: 18
under 914 m: 25 (2010)
Airports—with unpaved runways:
total: 31
914 to 1,523 m: 6
under 914 m: 25 (2010)
Heliports: 1 (2010)
Pipelines: condensate 31 km; gas 64 km (2010)
Railways: *total:* 4,114 km
country comparison to the world: 41
standard gauge: 4,114 km 1.435-m gauge (2,552 km electrified) (2009)
Roadways: *total:* 92,946 km
country comparison to the world: 51
paved: 72,033 km (includes 664 km of expressways)
unpaved: 20,913 km (2007)
Waterways: 1,577 km (2010)
country comparison to the world: 52
Merchant marine: *total:* 632
country comparison to the world: 19
by type: bulk carrier 43, cargo 133, carrier 5, chemical tanker 139, combination ore/oil 12, container 1, liquefied gas 53, passenger 3, passenger/cargo 116, petroleum tanker 58, refrigerated cargo 14, roll on/roll off 9, vehicle carrier 46
foreign-owned: 104 (Bermuda 5, Canada 1, China 25, Cyprus 1, Denmark 11, Estonia 1, Finland 1, France 4, Iceland 3, Italy 3, Lithuania 1, Monaco 1, Poland 2, Saudi Arabia 3, Sweden 33, US 9)
registered in other countries: 940 (Antigua and Barbuda 9, Australia 1, Bahamas 198, Barbados 41, Belize 3, Bermuda 5, Brazil 3, Canada 4, Chile 1, Comoros 2, Cook Islands 6, Croatia 2, Cyprus 12, Denmark 2, Dominica 1, Equatorial Guinea 1, Estonia 2, Faroe Islands 6, Finland 2, France 1, Gibraltar 42, Hong Kong 49, Indonesia 4, Ireland 3, Isle of Man 26, Italy 6, Japan 1, Liberia 42, Libya 1, Malta 84, Marshall Islands 57, Netherlands 18, former Netherlands Antilles 2, Panama 89, Portugal 1, Saint Kitts and Nevis 1, Saint Vincent and the Grenadines 12, Singapore 132, Spain 10, Sweden 3, UK 39, US 10, Vanuatu 1, Venezuela 1, unknown 4) (2010)
Ports and terminals: Bergen, Haugesund, Maaloy, Mongstad, Narvik, Sture

MILITARY

Military branches: Norwegian Army (Haeren), Royal Norwegian Navy (Kongelige Norske Sjoeforsvaret, RNoN; includes Coastal Rangers and Coast Guard (Kystvakt)), Royal Norwegian Air Force (Kongelige Norske Luftforsvaret, RNoAF), Home Guard (Heimevernet, HV) (2010)
Military service age and obligation: 18-44 years of age for male compulsory military service; 16 years of age in wartime; 17 years of age for male volunteers; 18 years of age for women; 12-month service obligation, in practice shortened to 8 to 9 months; although all males between ages of 18 and 44 are liable for service, in practice they are seldom called to duty after age 30; reserve obligation to age 35-60; 16 years of age for volunteers to the Home Guard, who serve 6-month duty tours (2009)
Manpower available for military service:
males age 16–49: 1,079,043
females age 16–49: 1,051,210 (2010 est.)
Manpower fit for military service: *males age 16–49:* 888,761
females age 16–49: 865,697 (2010 est.)
Manpower reaching militarily significant age annually: *male:* 32,290
female: 30,777 (2010 est.)
Military expenditures:
1.9% of GDP (2005 est.)
country comparison to the world: 78

TRANSNATIONAL ISSUES

Disputes—international: Norway asserts a territorial claim in Antarctica (Queen Maud Land and its continental shelf); Denmark (Greenland) and Norway have made submissions to the Commission on the Limits of the Continental shelf (CLCS) and Russia is collecting additional data to augment its 2001 CLCS submission; Norway and Russia signed a comprehensive maritime boundary agreement in 2010

OMAN

INTRODUCTION

Background: The inhabitants of the area of Oman have long prospered on Indian Ocean trade. In the late 18th century, a newly established sultanate in Muscat signed the first in a series of friendship treaties with Britain. Over time, Oman's dependence on British political and military advisors increased, but it never became a British colony. In 1970, QABOOS bin Said Al-Said overthrew the restrictive rule of his father; he has ruled as sultan ever since. His extensive modernization program has opened the country to the outside world while preserving the longstanding close ties with the UK. Oman's moderate, independent foreign policy has sought to maintain good relations with all Middle Eastern countries. Inspired by the popular uprisings that swept the Middle East and North Africa in 2010-11, Omanis began staging marches and demonstrations–a small number of which turned violent in clashes with government security forces–to demand economic benefits, an end to corruption, and greater political rights. In February and March 2011, in response to protester demands, QABOOS reshuffled his cabinet, pledged to create more government jobs, and promised to implement economic and political reforms, such as granting legislative and regulatory powers to the Council of Oman.

GEOGRAPHY

Location: Middle East, bordering the Arabian Sea, Gulf of Oman, and Persian Gulf, between Yemen and UAE
Geographic coordinates: 21 00 N, 57 00 E
Map references: Middle East
Area: *total:* 309,500 sq km
country comparison to the world: 70
land: 309,500 sq km
water: 0 sq km
Area—comparative: slightly smaller than Kansas
Land boundaries: *total:* 1,374 km
border countries: Saudi Arabia 676 km, UAE 410 km, Yemen 288 km
Coastline: 2,092 km
Maritime claims: territorial sea: 12 nm
contiguous zone: 24 nm
exclusive economic zone: 200 nm
Climate: dry desert; hot, humid along coast; hot, dry interior; strong southwest summer monsoon (May to September) in far south
Terrain: central desert plain, rugged mountains in north and south
Elevation extremes: *lowest point:* Arabian Sea 0 m
highest point: Jabal Shams 2,980 m
Natural resources: petroleum, copper, asbestos, some marble, limestone, chromium, gypsum, natural gas
Land use: *arable land:* 0.12%
permanent crops: 0.14%
other: 99.74% (2005)
Irrigated land: 720 sq km (2003)
Total renewable water resources: 1 cu km (1997)
Freshwater withdrawal (domestic/industrial/agricultural): *total:* 1.36 cu km/yr (7%/2%/90%)
per capita: 529 cu m/yr (2000)
Natural hazards: summer winds often raise large sandstorms and dust storms in interior; periodic droughts
Environment—current issues: rising soil salinity; beach pollution from oil spills; limited natural freshwater resources
Environment—international agreements: *party to:* Biodiversity, Climate Change, Climate Change-Kyoto Protocol, Desertification, Hazardous Wastes, Law of the Sea, Marine Dumping, Ozone Layer Protection, Ship Pollution, Whaling
signed, but not ratified: none of the selected agreements
Geography—note: strategic location on Musandam Peninsula adjacent to Strait of Hormuz, a vital transit point for world crude oil

PEOPLE

Population: 3,027,959 (July 2011 est.)
country comparison to the world: 135
note: includes 577,293 non-nationals
Age structure: *0–14 years:* 31.2% (male 484,292/female 460,066)
15–64 years: 65.7% (male 1,133,329/female 856,701)
65 years and over: 3.1% (male 47,786/female 45,785) (2011 est.)
Median age: *total:* 24.1 years
male: 25.5 years
female: 22.4 years (2011 est.)
Population growth rate: 2.023% (2011 est.)
country comparison to the world: 48
Birth rate: 24.15 births/1,000 population (2011 est.)
country comparison to the world: 67
Death rate: 3.45 deaths/1,000 population (July 2011 est.)
country comparison to the world: 212
Net migration rate: -0.48 migrant(s)/1,000 population (2011 est.)
country comparison to the world: 137
Urbanization: *urban population:* 73% of total population (2010)
rate of urbanization: 2.3% annual rate of change (2010-15 est.)
Major cities—population: MUSCAT (capital) 634,000 (2009)
Sex ratio: *at birth:* 1.05 male(s)/female
under 15 years: 1.05 male(s)/female
15–64 years: 1.34 male(s)/female
65 years and over: 1.06 male(s)/female
total population: 1.23 male(s)/female (2011 est.)
Infant mortality rate: *total:* 15.47 deaths/1,000 live births
country comparison to the world: 116
male: 15.78 deaths/1,000 live births
female: 15.15 deaths/1,000 live births (2011 est.)
Life expectancy at birth: *total population:* 74.22 years
country comparison to the world: 104
male: 72.38 years
female: 76.16 years (2011 est.)
Total fertility rate: 2.87 children born/woman (2011 est.)
country comparison to the world: 70
HIV/AIDS—adult prevalence rate: 0.1% (2009 est.)
country comparison to the world: 149
HIV/AIDS—people living with HIV/AIDS: 1,100 (2009 est.)
country comparison to the world: 139
HIV/AIDS—deaths: fewer than 100 (2009 est.)
country comparison to the world: 144
Drinking water source: *Improved:*
urban: 92% of population
rural: 77% of population
total: 88% of population
Unimproved: urban: 8% of population
rural: 23% of population
total: 12% of population (2008)
Sanitation facility access: *Improved:*
urban: 97% of population
rural: 61% of population
total: 87% of population
Unimproved: urban: 3% of population
rural: 39% of population
total: 13% of population (2000)
Nationality: *noun:* Omani(s)
adjective: Omani
Ethnic groups: Arab, Baluchi, South Asian (Indian, Pakistani, Sri Lankan, Bangladeshi), African
Religions: Ibadhi Muslim 75%, other (includes Sunni Muslim, Shia Muslim, Hindu) 25%
Languages: Arabic (official), English, Baluchi, Urdu, Indian dialects
Literacy: *definition:* age 15 and over can read and write
total population: 81.4%
male: 86.8%
female: 73.5% (2003 census)
School life expectancy (primary to tertiary education): *total:* 12 years
male: 12 years
female: 11 years (2009)
Education expenditures: 3.9% of GDP (2006)
country comparison to the world: 109

GOVERNMENT

Country name: *conventional long form:* Sultanate of Oman
conventional short form: Oman
local long form: Saltanat Uman
local short form: Uman
former: Muscat and Oman

Government type: monarchy
Capital: *name:* Muscat
geographic coordinates: 23 37 N, 58 35 E
time difference: UTC+4 (9 hours ahead of Washington, DC during Standard Time)
Administrative divisions: 5 regions (manatiq, singular–mintaqat) and 4 governorates* (muhafazat, singular–muhafazat) Ad Dakhiliyah, Al Batinah, Al Buraymi*, Al Wusta, Ash Sharqiyah, Az Zahirah, Masqat (Muscat)*, Musandam*, Zufar (Dhofar)*
Independence: 1650 (expulsion of the Portuguese)
National holiday: Birthday of Sultan QABOOS, 18 November (1940)
Constitution: none; note–on 6 November 1996, Sultan QABOOS issued a royal decree promulgating a basic law considered by the government to be a constitution which, among other things, clarifies the royal succession, provides for a prime minister, bars ministers from holding interests in companies doing business with the government, establishes a bicameral legislature, and guarantees basic civil liberties for Omani citizens
Legal system: mixed legal system of Anglo-Saxon law and Islamic law
International law organization participation: has not submitted an ICJ jurisdiction declaration; non-party state to the ICCt
Suffrage: 21 years of age; universal; note–members of the military and security forces are not allowed to vote
Executive branch: *chief of state:* Sultan and Prime Minister QABOOS bin Said Al-Said (sultan since 23 July 1970 and prime minister since 23 July 1972); note–the monarch is both the chief of state and head of government
head of government: Sultan and Prime Minister QABOOS bin Said Al-Said (sultan since 23 July 1970 and prime minister since 23 July 1972)
cabinet: Cabinet appointed by the monarch (For more information visit the World Leaders website)
elections: the monarchy is hereditary
Legislative branch: bicameral Majlis Oman consists of Majlis al-Dawla or upper chamber (71 seats; members appointed by the monarch; has only advisory powers and Majlis al-Shura or lower chamber (84 seats; members elected by popular vote to serve four-year terms; body has only advisory powers)
elections: last held on 27 October 2007 (next to be held in 2011)
election results: new candidates won 46 seats and 38 members of the outgoing Majlis kept their positions; none of the 20 female candidates was elected
Judicial branch: Supreme Court
note: the nascent civil court system, administered by region, has judges who practice secular and sharia law
Political parties and leaders: none
Political pressure groups and leaders: none
International organization participation: ABEDA, AFESD, AMF, FAO, G-77, GCC, IAEA, IBRD, ICAO, IDA, IDB, IFAD, IFC, IHO, ILO, IMF, IMO, IMSO, Interpol, IOC, IPU, ISO, ITSO, ITU, LAS, MIGA, NAM, OIC, OPCW, UN, UNCTAD, UNESCO, UNIDO, UNWTO, UPU, WCO, WFTU, WHO, WIPO, WMO, WTO
Diplomatic representation in the US: *chief of mission:* Ambassador Hunaina bint Sultan bin Ahmad al-MUGHAIRI
chancery: 2535 Belmont Road, NW, Washington, DC 20008
telephone: [1] (202) 387-1980
FAX: [1] (202) 745-4933
Diplomatic representation from the US: *chief of mission:* Ambassador Richard J. SCHMIERER
embassy: Jameat A'Duwal Al Arabiya Street, Al Khuwair area, Muscat
mailing address: P. O. Box 202, P.C. 115, Madinat Sultan Qaboos, Muscat
telephone: [968] 24-643-400
FAX: [968] 24-699771
Flag description: three horizontal bands of white, red, and green of equal width with a broad, vertical, red band on the hoist side; the national emblem (a khanjar dagger in its sheath superimposed on two crossed swords in scabbards) in white is centered near the top of the vertical band; white represents peace and prosperity, red recalls battles against foreign invaders, and green symbolizes the Jebel Akhdar (Green Mountains) and fertility
National anthem: *name:* "Nashid as-Salaam as-Sultani" (The Sultan's Anthem)
lyrics/music: Rashid bin Uzayyiz al KHUSAIDI/James Frederick MILLS, arranged by Bernard EBBINGHAUS
note: adopted 1932; new words were written after QABOOS bin Said al Said gained power in 1970; the anthem was first performed by the band of a British ship as a salute to the Sultan during a 1932 visit to Muscat; the bandmaster of the HMS Hawkins was asked to write a salutation to the Sultan on the occasion of his visiting the ship

ECONOMY

Economy—overview: Oman is a middle-income economy that is heavily dependent on dwindling oil resources. Because of declining reserves, Muscat has actively pursued a development plan that focuses on diversification, industrialization, and privatization, with the objective of reducing the oil sector's contribution to GDP to 9% by 2020. Tourism and gas-based industries are key components of the government's diversification strategy. By using enhanced oil recovery techniques, Oman succeeded in increasing oil production, giving the country more time to diversify, and the increase in global oil prices throughout 2010 provides the government greater financial resources to invest in non-oil sectors.
GDP (purchasing power parity): $75.84 billion (2010 est.)
country comparison to the world: 81
$72.77 billion (2009 est.)
$71.98 billion (2008 est.)
note: data are in 2010 US dollars
GDP (official exchange rate): $55.62 billion (2010 est.)
GDP—real growth rate: 4.2% (2010 est.)
country comparison to the world: 91
1.1% (2009 est.)
12.9% (2008 est.)
GDP—per capita (PPP): $25,600 (2010 est.)
country comparison to the world: 53
$25,000 (2009 est.)
$25,200 (2008 est.)
note: data are in 2010 US dollars
GDP—composition by sector:
agriculture: 1.4%
industry: 48.2%
services: 50.3% (2010 est.)
Labor force: 968,800
country comparison to the world: 142
note: about 60% of the labor force is non-national (2007)
Labor force—by occupation:
agriculture: NA%
industry: NA%
services: NA%
Unemployment rate: 15% (2004 est.)
country comparison to the world: 151
Population below poverty line: NA%
Household income or consumption by percentage share: *lowest 10%:* NA%
highest 10%: NA%
Investment (gross fixed): 26.3% of GDP (2010 est.)
country comparison to the world: 35
Budget: *revenues:* $20.5 billion
expenditures: $20.1 billion (2010 est.)
Public debt: 4.4% of GDP (2010 est.)
country comparison to the world: 130
5.5% of GDP (2009 est.)
Inflation rate (consumer prices): 4% (2010 est.)
country comparison to the world: 114
3.5% (2009 est.)
Central bank discount rate: 0.05% (31 December 2009)
country comparison to the world: 135
0.91% (31 December 2008)
Commercial bank prime lending rate: 7.44% (31 December 2009 est.)
country comparison to the world: 125
7.1% (31 December 2008 est.)
Stock of narrow money: $7.257 billion (31 December 2010 est.)
country comparison to the world: 78
$6.15 billion (31 December 2009 est.)
Stock of broad money: $22.35 billion (31 December 2010 est.)
country comparison to the world: 80
$20.52 billion (31 December 2009 est.)
Stock of domestic credit: $22.05 billion (31 December 2010 est.)
country comparison to the world: 75
$19.34 billion (31 December 2009 est.)
Market value of publicly traded shares: $17.3 billion (31 December 2009)
country comparison to the world: 64
$14.91 billion (31 December 2008)
$23.06 billion (31 December 2007)
Agriculture—products: dates, limes, bananas, alfalfa, vegetables; camels, cattle; fish
Industries: crude oil production and refining, natural and liquefied natural gas (LNG) production; construction, cement, copper, steel, chemicals, optic fiber
Industrial production growth rate: 4.5% (2010 est.)
country comparison to the world: 76
Electricity—production: 13.58 billion kWh (2007 est.)
country comparison to the world: 82
Electricity—consumption: 11.36 billion kWh (2007 est.)

country comparison to the world: 82
Electricity—exports: 0 kWh (2008 est.)
Electricity—imports: 0 kWh (2008 est.)
Oil—production: 816,000 bbl/day (2009 est.)
country comparison to the world: 25
Oil—consumption: 84,000 bbl/day (2009 est.)
country comparison to the world: 83
Oil—exports: 593,700 bbl/day (2008 est.)
country comparison to the world: 25
Oil—imports: 17,290 bbl/day (2007 est.)
country comparison to the world: 117
Oil—proved reserves: 5.5 billion bbl (1 January 2010 est.)
country comparison to the world: 24
Natural gas—production: 24 billion cu m (2008 est.)
country comparison to the world: 28
Natural gas—consumption: 13.46 billion cu m (2008 est.)
country comparison to the world: 41
Natural gas—exports: 10.89 billion cu m (2008 est.)
country comparison to the world: 19
Natural gas—imports: 350 million cu m (2008 est.)
country comparison to the world: 62
Natural gas—proved reserves: 849.5 billion cu m (1 January 2010 est.)
country comparison to the world: 27
Current account balance: $2.724 billion (2010 est.)
country comparison to the world: 40
$-2.143 billion (2009 est.)
Exports: $36.12 billion (2010 est.)
country comparison to the world: 60
$27.65 billion (2009 est.)
Exports—commodities: petroleum, reexports, fish, metals, textiles
Exports—partners: China 26.98%, South Korea 17.19%, Japan 12.12%, UAE 11.23%, Thailand 7.64% (2009)
Imports:
$19.3 billion (2010 est.)
country comparison to the world: 73
$16.13 billion (2009 est.)
Imports—commodities: machinery and transport equipment, manufactured goods, food, livestock, lubricants
Imports—partners: UAE 22.9%, Japan 13.99%, US 6.46%, China 5.64%, India 5.27%, France 5.19%, South Korea 4.65% (2009)
Reserves of foreign exchange and gold: $14 billion (31 December 2010 est.)
country comparison to the world: 51
$12.2 billion (31 December 2009 est.)
Debt—external: $8.829 billion (31 December 2010 est.)
country comparison to the world: 89
$7.061 billion (31 December 2009 est.)
Stock of direct foreign investment—at home: $NA
Stock of direct foreign investment—abroad: $NA
Exchange rates: Omani rials (OMR) per US dollar–0.3845 (2010), 0.3845 (2009), 0.3845 (2008), 0.3845 (2007), 0.3845 (2006)

COMMUNICATIONS

Telephones—main lines in use: 300,100 (2009)
country comparison to the world: 115
Telephones—mobile cellular: 3.971 million (2009)
country comparison to the world: 105
Telephone system: *general assessment:* modern system consisting of open-wire, microwave, and radiotelephone communication stations; limited coaxial cable; domestic satellite system with 8 earth stations
domestic: fixed-line and mobile-cellular subscribership both increasing with fixed-line phone service gradually being introduced to remote villages using wireless local loop systems
international: country code–968; the Fiber-Optic Link Around the Globe (FLAG) and the SEA-ME-WE-3 submarine cable provide connectivity to Asia, the Middle East, and Europe; satellite earth stations–2 Intelsat (Indian Ocean), 1 Arabsat (2008)
Broadcast media: 1 state-run TV broadcaster; TV stations transmitting from Saudi Arabia, the UAE, and Yemen are accessible via satellite TV; state-run radio operates multiple stations; first private radio station began operation in 2007 and 2 additional stations now operating (2007)
Internet country code: .om
Internet hosts: 9,114 (2010)
country comparison to the world: 126
Internet users: 1.465 million (2009)
country comparison to the world: 83

TRANSPORTATION

Airports: 130 (2010)
country comparison to the world: 45
Airports—with paved runways: *total:* 11
over 3,047 m: 6
2,438 to 3,047 m: 4
914 to 1,523 m: 1 (2010)
Airports—with unpaved runways: *total:* 119
over 3,047 m: 2
2,438 to 3,047 m: 7
1,524 to 2,437 m: 51
914 to 1,523 m: 33
under 914 m: 26 (2010)
Heliports: 3 (2010)
Pipelines: condensate 107 km; gas 4,209 km; oil 3,558 km; refined products 263 km (2010)
Roadways: *total:* 53,430 km
country comparison to the world: 78
paved: 23,223 km (includes 1,384 km of expressways)
unpaved: 30,207 km (2008)
Merchant marine: *total:* 4
country comparison to the world: 134
by type: chemical tanker 1, passenger 1, passenger/cargo 2
registered in other countries: 9 (Panama 8, Saint Vincent and the Grenadines 1) (2010)
Ports and terminals: Mina' Qabus, Salalah, Suhar

MILITARY

Military branches: Sultan's Armed Forces (SAF): Royal Army of Oman, Royal Navy of Oman, Royal Air Force of Oman (al-Quwwat al-Jawwiya al-Sultanat) (2010)
Military service age and obligation: 18-30 years of age for voluntary military service; no conscription (2010)
Manpower available for military service: *males age 16–49:* 985,957
females age 16–49: 737,812 (2010 est.)
Manpower fit for military service: *males age 16–49:* 837,886
females age 16–49: 642,427 (2010 est.)
Manpower reaching militarily significant age annually: *male:* 31,959
female: 30,264 (2010 est.)
Military expenditures: 11.4% of GDP (2005 est.)
country comparison to the world: 1

TRANSNATIONAL ISSUES

Disputes—international: boundary agreement reportedly signed and ratified with UAE in 2003 for entire border, including Oman's Musandam Peninsula and Al Madhah exclave, but details of the alignment have not been made public

PACIFIC OCEAN

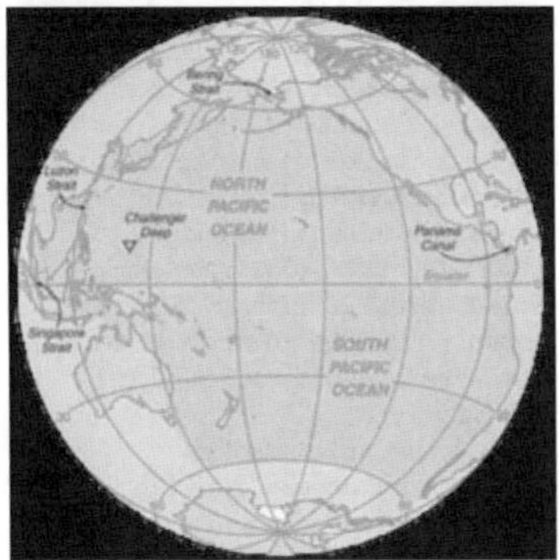

INTRODUCTION

Background: The Pacific Ocean is the largest of the world's five oceans (followed by the Atlantic Ocean, Indian Ocean, Southern Ocean, and Arctic Ocean). Strategically important access waterways include the La Perouse, Tsugaru, Tsushima, Taiwan, Singapore, and Torres Straits. The decision by the International Hydrographic Organization in the spring of 2000 to delimit a fifth ocean, the Southern Ocean, removed the portion of the Pacific Ocean south of 60 degrees south.

GEOGRAPHY

Location: body of water between the Southern Ocean, Asia, Australia, and the Western Hemisphere

Geographic coordinates: 0 00 N, 160 00 W

Map references: Political Map of the World

Area: *total:* 155.557 million sq km

note: includes Bali Sea, Bering Sea, Bering Strait, Coral Sea, East China Sea, Gulf of Alaska, Gulf of Tonkin, Philippine Sea, Sea of Japan, Sea of Okhotsk, South China Sea, Tasman Sea, and other tributary water bodies

Area—comparative: about 15 times the size of the US; covers about 28% of the global surface; almost equal to the total land area of the world

Coastline: 135,663 km

Climate: planetary air pressure systems and resultant wind patterns exhibit remarkable uniformity in the south and east; trade winds and westerly winds are well-developed patterns, modified by seasonal fluctuations; tropical cyclones (hurricanes) may form south of Mexico from June to October and affect Mexico and Central America; continental influences cause climatic uniformity to be much less pronounced in the eastern and western regions at the same latitude in the North Pacific Ocean; the western Pacific is monsoonal—a rainy season occurs during the summer months, when moisture-laden winds blow from the ocean over the land, and a dry season during the winter months, when dry winds blow from the Asian landmass back to the ocean; tropical cyclones (typhoons) may strike southeast and east Asia from May to December

Terrain: surface currents in the northern Pacific are dominated by a clockwise, warm-water gyre (broad circular system of currents) and in the southern Pacific by a counterclockwise, cool-water gyre; in the northern Pacific, sea ice forms in the Bering Sea and Sea of Okhotsk in winter; in the southern Pacific, sea ice from Antarctica reaches its northernmost extent in October; the ocean floor in the eastern Pacific is dominated by the East Pacific Rise, while the western Pacific is dissected by deep trenches, including the Mariana Trench, which is the world's deepest

Elevation extremes: *lowest point:* Challenger Deep in the Mariana Trench -10,924 m

highest point: sea level 0 m

Natural resources: oil and gas fields, polymetallic nodules, sand and gravel aggregates, placer deposits, fish

Natural hazards: surrounded by a zone of violent volcanic and earthquake activity sometimes referred to as the "Pacific Ring of Fire"; subject to tropical cyclones (typhoons) in southeast and east Asia from May to December (most frequent from July to October); tropical cyclones (hurricanes) may form south of Mexico and strike Central America and Mexico from June to October (most common in August and September); cyclical El Nino/La Nina phenomenon occurs in the equatorial Pacific, influencing weather in the Western Hemisphere and the western Pacific; ships subject to superstructure icing in extreme north from October to May; persistent fog in the northern Pacific can be a maritime hazard from June to December

Environment—current issues: endangered marine species include the dugong, sea lion, sea otter, seals, turtles, and whales; oil pollution in Philippine Sea and South China Sea

Geography—note: the major chokepoints are the Bering Strait, Panama Canal, Luzon Strait, and the Singapore Strait; the Equator divides the Pacific Ocean into the North Pacific Ocean and the South Pacific Ocean; dotted with low coral islands and rugged volcanic islands in the southwestern Pacific Ocean

ECONOMY

Economy—overview: The Pacific Ocean is a major contributor to the world economy and particularly to those nations its waters directly touch. It provides low-cost sea transportation between East and West, extensive fishing grounds, offshore oil and gas fields, minerals, and sand and gravel for the construction industry. In 1996, over 60% of the world's fish catch came from the Pacific Ocean. Exploitation of offshore oil and gas reserves is playing an ever-increasing role in the energy supplies of the US, Australia, NZ, China, and Peru. The high cost of recovering offshore oil and gas, combined with the wide swings in world prices for oil since 1985, has led to fluctuations in new drillings.

TRANSPORTATION

Ports and terminals: Bangkok (Thailand), Hong Kong (China), Kao-hsiung (Taiwan), Los Angeles (US), Manila (Philippines), Pusan (South Korea), San Francisco (US), Seattle (US), Shanghai (China), Singapore, Sydney (Australia), Vladivostok (Russia), Wellington (NZ), Yokohama (Japan)

Transportation—note: the Inside Passage offers protected waters from southeast Alaska to Puget Sound (Washington state); the International Maritime Bureau reports the territorial waters of littoral states and offshore waters in the South China Sea as high risk for piracy and armed robbery against ships; numerous commercial vessels have been attacked and hijacked both at anchor and while underway; hijacked vessels are often disguised and cargoes stolen; crew and passengers are often held for ransom, murdered, or cast adrift

TRANSNATIONAL ISSUES

Disputes—international: some maritime disputes (see littoral states)

PAKISTAN

INTRODUCTION

Background: The Indus Valley civilization, one of the oldest in the world and dating back at least 5,000 years, spread over much of what is presently Pakistan. During the second millennium B.C., remnants of this culture fused with the migrating Indo-Aryan peoples. The area underwent successive invasions in subsequent centuries from the Persians, Greeks, Scythians, Arabs (who brought Islam), Afghans, and Turks. The Mughal Empire flourished in the 16th and 17th centuries; the British came to dominate the region in the 18th century. The separation in 1947 of British India into the Muslim state of Pakistan (with West and East sections) and largely Hindu India was never satisfactorily resolved, and India and Pakistan fought two wars—in 1947-48 and 1965—over the disputed Kashmir territory. A third war between these countries in 1971—in which India capitalized on Islamabad's marginalization of Bengalis in Pakistani politics—resulted in East Pakistan becoming the separate nation of Bangladesh. In response to Indian nuclear weapons testing, Pakistan conducted its own tests in 1998. India-Pakistan relations have been rocky since the November 2008 Mumbai attacks, but both countries are

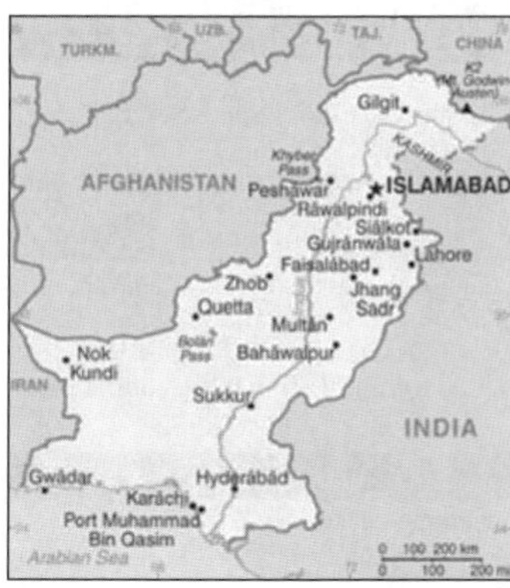

taking small steps to put relations back on track. In February 2008, Pakistan held parliamentary elections and in September 2008, after the resignation of former President MUSHARRAF, elected Asif Ali ZARDARI to the presidency. Pakistani government and military leaders are struggling to control domestic insurgents, many of whom are located in the tribal areas adjacent to the border with Afghanistan.

GEOGRAPHY

Location: Southern Asia, bordering the Arabian Sea, between India on the east and Iran and Afghanistan on the west and China in the north
Geographic coordinates: 30 00 N, 70 00 E
Map references: Asia
Area: *total:* 796,095 sq km
country comparison to the world: 36
land: 770,875 sq km
water: 25,220 sq km
Area—comparative: slightly less than twice the size of California
Land boundaries: *total:* 6,774 km
border countries: Afghanistan 2,430 km, China 523 km, India 2,912 km, Iran 909 km
Coastline: 1,046 km
Maritime claims: territorial sea: 12 nm
contiguous zone: 24 nm
exclusive economic zone: 200 nm
continental shelf: 200 nm or to the edge of the continental margin
Climate: mostly hot, dry desert; temperate in northwest; arctic in north
Terrain: flat Indus plain in east; mountains in north and northwest; Balochistan plateau in west
Elevation extremes: *lowest point:* Indian Ocean 0 m
highest point: K2 (Mt. Godwin-Austen) 8,611 m
Natural resources: land, extensive natural gas reserves, limited petroleum, poor quality coal, iron ore, copper, salt, limestone
Land use: *arable land:* 24.44%
permanent crops: 0.84%
other: 74.72% (2005)
Irrigated land: 182,300 sq km (2003)
Total renewable water resources: 233.8 cu km (2003)
Freshwater withdrawal (domestic/industrial/agricultural): *total:* 169.39 cu km/yr (2%/2%/96%)
per capita: 1,072 cu m/yr (2000)
Natural hazards: frequent earthquakes, occasionally severe especially in north and west; flooding along the Indus after heavy rains (July and August)
Environment—current issues: water pollution from raw sewage, industrial wastes, and agricultural runoff; limited natural freshwater resources; most of the population does not have access to potable water; deforestation; soil erosion; desertification
Environment—international agreements: *party to:* Biodiversity, Climate Change, Climate Change-Kyoto Protocol, Desertification, Endangered Species, Environmental Modification, Hazardous Wastes, Law of the Sea, Marine Dumping, Ozone Layer Protection, Ship Pollution, Wetlands
signed, but not ratified: Marine Life Conservation
Geography—note: controls Khyber Pass and Bolan Pass, traditional invasion routes between Central Asia and the Indian Subcontinent

PEOPLE

Population: 187,342,721 (July 2011 est.)
country comparison to the world: 6
Age structure: *0–14 years:* 35.4% (male 34,093,853/female 32,278,462)
15–64 years: 60.4% (male 58,401,016/female 54,671,873)
65 years and over: 4.2% (male 3,739,647/female 4,157,870) (2011 est.)
Median age:
total: 21.6 years
male: 21.5 years
female: 21.6 years (2011 est.)
Population growth rate: 1.573% (2011 est.)
country comparison to the world: 75
Birth rate: 24.81 births/1,000 population (2011 est.)
country comparison to the world: 62
Death rate: 6.92 deaths/1,000 population (July 2011 est.)
country comparison to the world: 138
Net migration rate: -2.17 migrant(s)/1,000 population (2011 est.)
country comparison to the world: 167
Urbanization: *urban population:* 36% of total population (2010)
rate of urbanization: 3.1% annual rate of change (2010-15 est.)
Major cities—population: Karachi 13.125 million; Lahore 7.132 million; Faisalabad 2.849 million; Rawalpindi 2.026 million; ISLAMABAD (capital) 832,000 (2009)
Sex ratio: *at birth:* 1.05 male(s)/female
under 15 years: 1.06 male(s)/female
15–64 years: 1.09 male(s)/female
65 years and over: 0.92 male(s)/female
total population: 1.07 male(s)/female (2011 est.)
Infant mortality rate:
total: 63.26 deaths/1,000 live births
country comparison to the world: 25
male: 66.52 deaths/1,000 live births
female: 59.85 deaths/1,000 live births (2011 est.)
Life expectancy at birth: *total population:* 65.99 years
country comparison to the world: 166
male: 64.18 years
female: 67.9 years (2011 est.)
Total fertility rate: 3.17 children born/woman (2011 est.)
country comparison to the world: 54
HIV/AIDS—adult prevalence rate: 0.1% (2009 est.)
country comparison to the world: 153
HIV/AIDS—people living with HIV/AIDS: 98,000 (2009 est.)
country comparison to the world: 43
HIV/AIDS—deaths: 5,800 (2009 est.)
country comparison to the world: 35
Major infectious diseases: *degree of risk:* high
food or waterborne diseases: bacterial diarrhea, hepatitis A and E, and typhoid fever
vectorborne diseases: dengue fever and malaria
animal contact disease: rabies
note: highly pathogenic H5N1 avian influenza has been identified in this country; it poses a negligible risk with extremely rare cases possible among US citizens who have close contact with birds (2009)
Drinking water source: *Improved:*
urban: 95% of population
rural: 87% of population
total: 90% of population
Unimproved: urban: 5% of population
rural: 13% of population
total: 10% of population (2008)
Sanitation facility access: *Improved:*
urban: 72% of population
rural: 29% of population
total: 45% of population
Unimproved: urban: 28% of population
rural: 71% of population
total: 55% of population (2008)
Nationality: *noun:* Pakistani(s)
adjective: Pakistani
Ethnic groups: Punjabi 44.68%, Pashtun (Pathan) 15.42%, Sindhi 14.1%, Sariaki 8.38%, Muhajirs 7.57%, Balochi 3.57%, other 6.28%
Religions: Muslim 95% (Sunni 75%, Shia 20%), other (includes Christian and Hindu) 5%
Languages: Punjabi 48%, Sindhi 12%, Siraiki (a Punjabi variant) 10%, Pashtu 8%, Urdu (official) 8%, Balochi 3%, Hindko 2%, Brahui 1%, English (official; lingua franca of Pakistani elite and most government ministries), Burushaski, and other 8%
Literacy: *definition:* age 15 and over can read and write
total population: 49.9%
male: 63%
female: 36% (2005 est.)
School life expectancy (primary to tertiary education):
total: 7 years
male: 8 years
female: 6 years (2009)
Education expenditures: 2.7% of GDP (2009)
country comparison to the world: 143

GOVERNMENT

Country name: *conventional long form:* Islamic Republic of Pakistan
conventional short form: Pakistan
local long form: Jamhuryat Islami Pakistan
local short form: Pakistan
former: West Pakistan
Government type: federal republic
Capital: *name:* Islamabad
geographic coordinates: 33 42 N, 73 10 E

time difference: UTC+5 (10 hours ahead of Washington, DC during Standard Time)

Administrative divisions: 4 provinces, 1 territory*, and 1 capital territory**; Balochistan, Federally Administered Tribal Areas*, Islamabad Capital Territory**, Khyber Pakhtunkhwa (formerly North-West Frontier Province), Punjab, Sindh

note: the Pakistani-administered portion of the disputed Jammu and Kashmir region consists of two administrative entities: Azad Kashmir and Gilgit-Baltistan

Independence: 14 August 1947 (from British India)

National holiday: Republic Day, 23 March (1956)

Constitution: 12 April 1973; suspended 5 July 1977, restored 30 December 1985; suspended 15 October 1999, restored in stages in 2002; amended 31 December 2003; suspended 3 November 2007; restored 15 December 2007; amended 19 April 2010

Legal system: common law system with Islamic law influence

International law organization participation: accepts compulsory ICJ jurisdiction with reservations; non-party state to the ICCt

Suffrage: 18 years of age; universal; joint electorates and reserved parliamentary seats for women and non-Muslims

Executive branch: *chief of state:* President Asif Ali ZARDARI (since 9 September 2008)

head of government: Prime Minister Syed Yousuf Raza GILANI (since 25 March 2008)

cabinet: Cabinet appointed by the president upon the advice of the prime minister

(For more information visit the World Leaders website)

elections: the president elected by secret ballot through an Electoral College comprising the members of the Senate, National Assembly, and the provincial assemblies for a five-year term; election last held on 6 September 2008 (next to be held not later than 2013); note—any person who is a Muslim and not less than 45 years of age and is qualified to be elected as a member of the National Assembly can contest the presidential election; the prime minister selected by the National Assembly

election results: Asif Ali ZARDARI elected president; ZARDARI 481 votes, SIDDIQUE 153 votes, SYED 44 votes; Syed Yousuf Raza GILANI elected prime minister; GILANI 264 votes, Pervaiz ELAHI 42 votes; several abstentions

Legislative branch: bicameral parliament or Majlis-e-Shoora consists of the Senate (100 seats; members indirectly elected by provincial assemblies and the territories' representatives in the National Assembly to serve six-year terms; one half are elected every three years) and the National Assembly (342 seats; 272 members elected by popular vote; 60 seats reserved for women; 10 seats reserved for non-Muslims; members serve five-year terms)

elections: Senate—last held on 3 March 2009 (next to be held in March 2012); National Assembly—last held on 18 February 2008 with by-elections on 26 June 2008 (next to be held in 2013)

election results: Senate—percent of vote by party—NA; seats by party—PPPP 27, PML 21, MMA 9, PML-N 7, ANP 6, MQM 6, JUI-F 4, BNP-A 2, JWP 1, NPP 1, PKMAP 1, PML-F 1, PPP 1, independents 13; National Assembly—percent of votes by party—NA; seats by party as of October 2010—PPPP 127, PML-N 90, PML 51, MQM 25, ANP 13, JUI-F 8, PML-F 5, BNP-A 1, NPP 1, PPP-S 1, independents 18, unfilled seats—2

Judicial branch: Supreme Court (justices appointed by the president); Federal Islamic or Sharia Court

Political parties and leaders: Awami National Party or ANP [Asfandyar Wali KHAN]; Balochistan National Party-Awami or BNP-A [Moheem Khan BALOCH]; Balochistan National Party-Hayee Group or BNP-H [Dr. Hayee BALOCH]; Balochistan National Party-Mengal or BNP-M [Sardar Ataullah MENGAL]; Jamaat-i Islami or JI [Syed Munawar HASAN]; Jamhoori Watan Party or JWP; Jamiat Ahle Hadith or JAH [Sajid MIR]; Jamiat Ulema-i Islam Fazl-ur Rehman or JUI-F [Fazl-ur REHMAN]; Jamiat Ulema-i Islam Sami-ul HAQ or JUI-S [Sami ul-HAQ]; Jamiat Ulema-i Pakistan or JUP [Shah Faridul HAQ]; Muttahida Majlis-e Amal or MMA [Qazi Hussain AHMED]; Muttahida Qaumi Movement or MQM [Altaf HUSSAIN]; National Alliance or NA [Ghulam Mustapha JATOI] (merged with PML); National Peoples Party or NPP; Pakhtun Khwa Milli Awami Party or PKMAP [Mahmood Khan ACHAKZAI]; Pakistan Awami Tehrik or PAT [Tahir ul QADRI]; Pakistan Muslim League or PML [Chaudhry Shujaat HUSSAIN]; Pakistan Muslim League-Functional or PML-F [Pir PAGARO]; Pakistan Muslim League-Nawaz or PML-N [Nawaz SHARIF]; Pakistan Peoples Party Parliamentarians or PPPP [Bilawal Bhutto ZARDARI, chairman; Asif Ali ZARDARI, co-chairman]; Pakistan Peoples Party-SHERPAO or PPP-S [Aftab Ahmed Khan SHERPAO]; Pakistan Tehrik-e Insaaf or PTI [Imran KHAN]; Tehrik-i Islami [Allama Sajid NAQVI]

note: political alliances in Pakistan can shift frequently

Political pressure groups and leaders: *other:* military (most important political force); ulema (clergy); landowners; industrialists; small merchants

International organization participation: ADB, ARF, ASEAN (dialogue partner), C, CICA, CP, D-8, ECO, FAO, G-11, G-24, G-77, IAEA, IBRD, ICAO, ICC, ICRM, IDA, IDB, IFAD, IFC, IFRCS, IHO, ILO, IMF, IMO, IMSO, Interpol, IOC, IOM, IPU, ISO, ITSO, ITU, ITUC, MIGA, MINURSO, MONUSCO, NAM, OAS (observer), OIC, OPCW, PCA, SAARC, SACEP, SCO (observer), UN, UNAMID, UNCTAD, UNESCO, UNHCR, UNIDO, UNMIL, UNMIS, UNMIT, UNOCI, UNWTO, UPU, WCO, WFTU, WHO, WIPO, WMO, WTO

Diplomatic representation in the US:
chief of mission: Ambassador Husain HAQQANI

chancery: 3517 International Court, Washington, DC 20008

telephone: [1] (202) 243-6500

FAX: [1] (202) 686-1544

consulate(s) general: Boston (Honorary Consulate General), Chicago, Houston, Los Angeles, New York

consulate(s): Chicago, Houston

Diplomatic representation from the US:
chief of mission: Ambassador Cameron MUNTER

embassy: Diplomatic Enclave, Ramna 5, Islamabad

mailing address: P. O. Box 1048, Unit 62200, APO AE 09812-2200

telephone: [92] (51) 208-0000

FAX: [92] (51) 2276427

consulate(s) general: Karachi

consulate(s): Lahore, Peshawar

Flag description: green with a vertical white band (symbolizing the role of religious minorities) on the hoist side; a large white crescent and star are centered in the green field; the crescent, star, and color green are traditional symbols of Islam

National anthem: *name:* "Qaumi Tarana" (National Anthem)

lyrics/music: Abu-Al-Asar Hafeez JULLANDHURI/Ahmed Ghulamali CHAGLA

note: adopted 1954; the anthem is also known as "Pak sarzamin shad bad" (Blessed Be the Sacred Land)

ECONOMY

Economy—overview: Pakistan, an impoverished and underdeveloped country, has suffered from decades of internal political disputes and low levels of foreign investment. Between 2001-07, however, poverty levels decreased by 10%, as Islamabad steadily raised development spending. During 2004-07, GDP growth in the 5-8% range was spurred by gains in the industrial and service sectors—despite severe electricity shortfalls—but growth slowed in 2008-09 and unemployment rose. Inflation remains the top concern among the public, climbing from 7.7% in 2007 to more than 13% in 2010. In addition, the Pakistani rupee has depreciated since 2007 as a result of political and economic instability. The government agreed to an International Monetary Fund Standby Arrangement in November 2008 in response to a balance of payments crisis, but during 2009-10 its current account strengthened and foreign exchange reserves stabilized—largely because of lower oil prices and record remittances from workers abroad. Record floods in July-August 2010 lowered agricultural output and contributed to a jump in inflation, and reconstruction costs will strain the limited resources of the government. Textiles account for most of Pakistan's export earnings, but Pakistan's failure to expand a viable export base for other manufactures has left the country vulnerable to shifts in world demand. Other long term challenges include expanding investment in education, healthcare, and electricity production, and reducing dependence on foreign donors.

GDP (purchasing power parity): $464.9 billion (2010 est.)
country comparison to the world: 28
$443.6 billion (2009 est.)
$429.2 billion (2008 est.)
note: data are in 2010 US dollars
GDP (official exchange rate): $174.9 billion (2010 est.)
GDP—real growth rate: 4.8% (2010 est.)
country comparison to the world: 77
3.4% (2009 est.)
1.6% (2008 est.)
GDP—per capita (PPP): $2,500 (2010 est.)
country comparison to the world: 179
$2,400 (2009 est.)
$2,400 (2008 est.)
note: data are in 2010 US dollars
GDP—composition by sector:
agriculture: 21.8%
industry: 23.6%
services: 54.6% (2010 est.)
Labor force: 55.77 million
country comparison to the world: 10
note: extensive export of labor, mostly to the Middle East, and use of child labor (2010 est.)
Labor force—by occupation: *agriculture:* 43%
industry: 20.3%
services: 36.6% (2005 est.)
Unemployment rate: 15% (2010 est.)
country comparison to the world: 152
14% (2009 est.)
note: substantial underemployment exists
Population below poverty line: 24% (FY05/06 est.)
Household income or consumption by percentage share: *lowest 10%:* 3.9%
highest 10%: 26.5% (2005)
Distribution of family income—Gini index: 30.6 (FY07/08)
country comparison to the world: 109
41 (FY98/99)
Investment (gross fixed): 15% of GDP (2010 est.)
country comparison to the world: 132
Budget:
revenues: $25.33 billion
expenditures: $36.24 billion (2010 est.)
Public debt: 49.9% of GDP (2010 est.)
country comparison to the world: 52
49.3% of GDP (2009 est.)
Inflation rate (consumer prices): 13.4% (2010 est.)
country comparison to the world: 215
13.6% (2009 est.)
Central bank discount rate: 12.5% (31 December 2009)
country comparison to the world: 21
15% (31 December 2008)
Commercial bank prime lending rate: NA%
Stock of narrow money: $59.75 billion (31 December 2010 est.)
country comparison to the world: 43
$47.23 billion (31 December 2009 est.)
Stock of broad money:
$85.22 billion (31 December 2010 est.)
country comparison to the world: 57
$65.13 billion (31 December 2009 est.)
Stock of domestic credit:
$71.45 billion (31 December 2010 est.)
country comparison to the world: 56
$63.1 billion (31 December 2009 est.)
Market value of publicly traded shares: $33.24 billion (31 December 2009)
country comparison to the world: 59
$23.49 billion (31 December 2008)
$70.26 billion (31 December 2007)
Agriculture—products: cotton, wheat, rice, sugarcane, fruits, vegetables; milk, beef, mutton, eggs
Industries: textiles and apparel, food processing, pharmaceuticals, construction materials, paper products, fertilizer, shrimp
Industrial production growth rate: 4.9% (2010 est.)
country comparison to the world: 71
Electricity—production: 90.8 billion kWh (2007 est.)
country comparison to the world: 34
Electricity—consumption: 72.2 billion kWh (2007 est.)
country comparison to the world: 37
Electricity—exports: 0 kWh (2008 est.)
Electricity—imports: 0 kWh (2008 est.)
Oil—production: 59,140 bbl/day (2009 est.)
country comparison to the world: 59
Oil—consumption: 373,000 bbl/day (2009 est.)
country comparison to the world: 35
Oil—exports: 30,090 bbl/day (2007 est.)
country comparison to the world: 84
Oil—imports: 319,500 bbl/day (2007 est.)
country comparison to the world: 33
Oil—proved reserves: 436.2 million bbl (1 January 2010 est.)
country comparison to the world: 50
Natural gas—production: 37.5 billion cu m (2008 est.)
country comparison to the world: 23
Natural gas—consumption: 37.5 billion cu m (2008 est.)
country comparison to the world: 22
Natural gas—exports: 0 cu m (2008 est.)
country comparison to the world: 156
Natural gas—imports: 0 cu m (2008 est.)
country comparison to the world: 105
Natural gas—proved reserves: 840.2 billion cu m (1 January 2010 est.)
country comparison to the world: 29
Current account balance: $-2.641 billion (2010 est.)
country comparison to the world: 165
$-3.583 billion (2009 est.)
Exports: $20.29 billion (2010 est.)
country comparison to the world: 68
$18.33 billion (2009 est.)
Exports—commodities: textiles (garments, bed linen, cotton cloth, yarn), rice, leather goods, sports goods, chemicals, manufactures, carpets and rugs
Exports—partners: US 15.87%, UAE 12.35%, Afghanistan 8.48%, UK 4.7%, China 4.44% (2009)
Imports:
$32.71 billion (2010 est.)
country comparison to the world: 56
$28.53 billion (2009 est.)
Imports—commodities: petroleum, petroleum products, machinery, plastics, transportation equipment, edible oils, paper and paperboard, iron and steel, tea
Imports—partners: China 15.35%, Saudi Arabia 10.54%, UAE 9.8%, US 4.81%, Kuwait 4.73%, Malaysia 4.43%, India 4.02% (2009)
Reserves of foreign exchange and gold: $16.1 billion (31 December 2010 est.)
country comparison to the world: 49
$13.77 billion (31 December 2009 est.)
Debt—external: $57.21 billion (31 December 2010 est.)
country comparison to the world: 52
$53.62 billion (31 December 2009 est.)
Stock of direct foreign investment—at home: $30.09 billion (31 December 2010 est.)
country comparison to the world: 61
$28.09 billion (31 December 2009 est.)
Stock of direct foreign investment—abroad: $1.047 billion (31 December 2010 est.)
country comparison to the world: 73
$1.017 billion (31 December 2009 est.)
Exchange rates: Pakistani rupees (PKR) per US dollar–85.27 (2010), 81.71 (2009) 70.64 (2008), 60.6295 (2007), 60.35 (2006)

COMMUNICATIONS

Telephones—main lines in use: 4.058 million (2009)
country comparison to the world: 39
Telephones—mobile cellular: 103 million (2009)
country comparison to the world: 9
Telephone system: *general assessment:* the telecommunications infrastructure is improving dramatically with foreign and domestic investments in fixed-line and mobile-cellular networks; system consists of microwave radio relay, coaxial cable, fiber-optic cable, cellular, and satellite networks;
domestic: mobile-cellular subscribership has skyrocketed, exceeding 100 million in 2009, up from only about 300,000 in 2000; approximately 90 percent of Pakistanis live within areas that have cell phone coverage and more than half of all Pakistanis have access to a cell phone; fiber systems are being constructed throughout the country to aid in network growth; fixed line availability has risen only marginally over the same period and there are still difficulties getting fixed-line service to rural areas
international: country code–92; landing point for the SEA-ME-WE-3 and SEA-ME-WE-4 submarine cable systems that provide links to Asia, the Middle East, and Europe; satellite earth stations–3 Intelsat (1 Atlantic Ocean and 2 Indian Ocean); 3 operational international gateway exchanges (1 at Karachi and 2 at Islamabad); microwave radio relay to neighboring countries (2009)
Broadcast media: media is government regulated; 1 dominant state-owned TV broadcaster, Pakistan Television Corporation (PTV), operates a network consisting of 6 channels; private TV broadcasters are permitted and some foreign satellite channels are carried by cable TV operators; the state-owned radio network operates more than 40 stations; privately-owned radio stations mostly limit programming to music and talk shows (2007)
Internet country code: .pk
Internet hosts: 330,466 (2010)
country comparison to the world: 57
Internet users: 20.431 million (2009)

country comparison to the world: 20

TRANSPORTATION

Airports: 148 (2010)
country comparison to the world: 37
Airports—with paved runways: *total:* 101
over 3,047 m: 15
2,438 to 3,047 m: 20
1,524 to 2,437 m: 39
914 to 1,523 m: 18
under 914 m: 9 (2010)
Airports—with unpaved runways: *total:* 47
1,524 to 2,437 m: 11
914 to 1,523 m: 11
under 914 m: 25 (2010)
Heliports: 20 (2010)
Pipelines: gas 10,514 km; oil 2,013 km; refined products 787 km (2010)
Railways: *total:* 7,791 km
country comparison to the world: 27
broad gauge: 7,479 km 1.676-m gauge (293 km electrified)
narrow gauge: 312 km 1.000-m gauge (2007)
Roadways: *total:* 259,197 km
country comparison to the world: 20
paved: 172,827 km (includes 711 km of expressways)
unpaved: 86,370 km (2007)
Merchant marine: *total:* 10
country comparison to the world: 116
by type: bulk carrier 1, cargo 4, petroleum tanker 5
registered in other countries: 14 (Comoros 3, Georgia 1, Marshall Islands 1, Panama 5, Saint Kitts and Nevis 3, Saint Vincent and the Grenadines 1) (2010)
Ports and terminals: Karachi, Port Muhammad Bin Qasim

MILITARY

Military branches: Army (includes National Guard), Navy (includes Marines and Maritime Security Agency), Pakistan Air Force (Pakistan Fiza'ya) (2010)
Military service age and obligation: 17-23 years of age for voluntary military service; soldiers cannot be deployed for combat until age 18; the Pakistani Air Force and Pakistani Navy have inducted their first female pilots and sailors (2009)
Manpower available for military service: *males age 16–49:* 48,453,305
females age 16–49: 44,898,096 (2010 est.)
Manpower fit for military service: *males age 16–49:* 37,945,440
females age 16–49: 37,381,549 (2010 est.)
Manpower reaching militarily significant age annually: *male:* 2,237,723
female: 2,104,906 (2010 est.)
Military expenditures: 3% of GDP (2007 est.)
country comparison to the world: 42

TRANSNATIONAL ISSUES

Disputes—international: various talks and confidence-building measures cautiously have begun to defuse tensions over Kashmir, particularly since the October 2005 earthquake in the region; Kashmir nevertheless remains the site of the world's largest and most militarized territorial dispute with portions under the de facto administration of China (Aksai Chin), India (Jammu and Kashmir), and Pakistan (Azad Kashmir and Northern Areas); UN Military Observer Group in India and Pakistan (UNMOGIP) has maintained a small group of peacekeepers since 1949; India does not recognize Pakistan's ceding historic Kashmir lands to China in 1964; India and Pakistan have maintained their 2004 cease fire in Kashmir and initiated discussions on defusing the armed stand-off in the Siachen glacier region; Pakistan protests India's fencing the highly militarized Line of Control and construction of the Baglihar Dam on the Chenab River in Jammu and Kashmir, which is part of the larger dispute on water sharing of the Indus River and its tributaries; to defuse tensions and prepare for discussions on a maritime boundary, India and Pakistan seek technical resolution of the disputed boundary in Sir Creek estuary at the mouth of the Rann of Kutch in the Arabian Sea; Pakistani maps continue to show the Junagadh claim in India's Gujarat State; by 2005, Pakistan, with UN assistance, repatriated 2.3 million Afghan refugees leaving slightly more than a million, many of whom remain at their own choosing; Pakistan has sent troops across and built fences along some remote tribal areas of its treaty-defined Durand Line border with Afghanistan, which serve as bases for foreign terrorists and other illegal activities; Afghan, Coalition, and Pakistan military meet periodically to clarify the alignment of the boundary on the ground and on maps
Refugees and internally displaced persons: refugees (country of origin): 1,043,984 (Afghanistan)
IDPs: undetermined (government strikes on Islamic militants in South Waziristan); 34,000 (October 2005 earthquake; most of those displaced returned to their home villages in the spring of 2006) (2007)
Illicit drugs: significant transit area for Afghan drugs, including heroin, opium, morphine, and hashish, bound for Iran, Western markets, the Gulf States, Africa, and Asia; financial crimes related to drug trafficking, terrorism, corruption, and smuggling remain problems; opium poppy cultivation estimated to be 2,300 hectares in 2007 with 600 of those hectares eradicated; federal and provincial authorities continue to conduct anti-poppy campaigns that utilizes forced eradication, fines, and arrests

PALAU

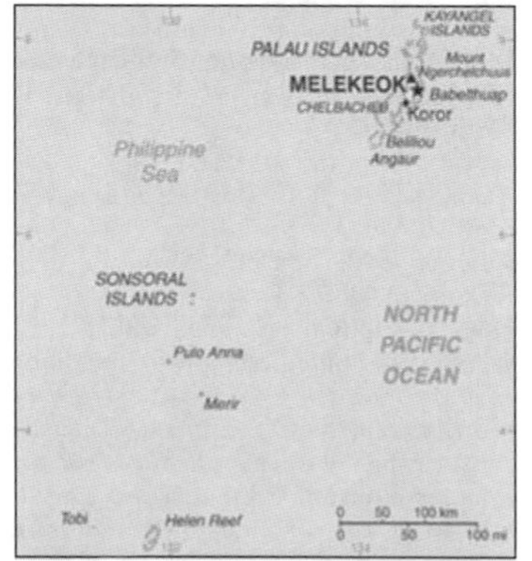

INTRODUCTION

Background: After three decades as part of the UN Trust Territory of the Pacific under US administration, this westernmost cluster of the Caroline Islands opted for independence in 1978 rather than join the Federated States of Micronesia. A Compact of Free Association with the US was approved in 1986 but not ratified until 1993. It entered into force the following year when the islands gained independence.

GEOGRAPHY

Location: Oceania, group of islands in the North Pacific Ocean, southeast of the Philippines
Geographic coordinates: 7 30 N, 134 30 E
Map references: Oceania
Area: *total:* 459 sq km
country comparison to the world: 196
land: 459 sq km
water: 0 sq km
Area—comparative: slightly more than 2.5 times the size of Washington, DC
Land boundaries: 0 km
Coastline: 1,519 km
Maritime claims:
territorial sea: 3 nm
exclusive fishing zone: 200 nm
Climate: tropical; hot and humid; wet season May to November
Terrain: varying geologically from the high, mountainous main island of Babelthuap to low, coral islands usually fringed by large barrier reefs
Elevation extremes: *lowest point:* Pacific Ocean 0 m
highest point: Mount Ngerchelchuus 242 m
Natural resources: forests, minerals (especially gold), marine products, deep-seabed minerals
Land use: *arable land:* 8.7%
permanent crops: 4.35%
other: 86.95% (2005)
Irrigated land: NA
Natural hazards: typhoons (June to December)
Environment—current issues: inadequate facilities for disposal of solid waste; threats to the marine ecosystem from sand and coral dredging, illegal fishing practices, and overfishing
Environment—international agreements: *party to:* Biodiversity, Climate Change, Climate Change-Kyoto Protocol,

Desertification, Law of the Sea, Ozone Layer Protection, Wetlands, Whaling
signed, but not ratified: none of the selected agreements
Geography—note: westernmost archipelago in the Caroline chain, consists of six island groups totaling more than 300 islands; includes World War II battleground of Beliliou (Peleliu) and world-famous rock islands

PEOPLE

Population: 20,956 (July 2011 est.)
country comparison to the world: 217
Age structure: *0–14 years:* 21.5% (male 2,329/female 2,187)
15–64 years: 72% (male 8,355/female 6,724)
65 years and over: 6.5% (male 402/female 959) (2011 est.)
Median age: *total:* 32.6 years
male: 32.3 years
female: 33.3 years (2011 est.)
Population growth rate: 0.363% (2011 est.)
country comparison to the world: 164
Birth rate: 10.74 births/1,000 population (2011 est.)
country comparison to the world: 178
Death rate: 7.87 deaths/1,000 population (July 2011 est.)
country comparison to the world: 110
Net migration rate: 0.76 migrant(s)/1,000 population (2011 est.)
country comparison to the world: 58
Urbanization: *urban population:* 83% of total population (2010)
rate of urbanization: 1.4% annual rate of change (2010-15 est.)
Sex ratio: *at birth:* 1.064 male(s)/female
under 15 years: 1.06 male(s)/female
15–64 years: 1.25 male(s)/female
65 years and over: 0.43 male(s)/female
total population: 1.13 male(s)/female (2011 est.)
Infant mortality rate: *total:* 12.43 deaths/1,000 live births
country comparison to the world: 132
male: 14.06 deaths/1,000 live births
female: 10.7 deaths/1,000 live births (2011 est.)
Life expectancy at birth: *total population:* 71.78 years
country comparison to the world: 131
male: 68.63 years
female: 75.12 years (2011 est.)
Total fertility rate: 1.73 children born/woman (2011 est.)
country comparison to the world: 164
HIV/AIDS—adult prevalence rate: NA
HIV/AIDS—people living with HIV/AIDS: NA
HIV/AIDS—deaths: NA
Drinking water source: *Improved:*
urban: 78% of population
rural: 95% of population
total: 83% of population
Unimproved: urban: 22% of population
rural: 5% of population
total: 17% of population (2000)
Sanitation facility access: *Improved:*
urban: 92% of population
rural: 52% of population
total: 80% of population
Unimproved: urban: 8% of population
rural: 48% of population
total: 20% of population (2000)
Nationality: *noun:* Palauan(s)
adjective: Palauan
Ethnic groups: Palauan (Micronesian with Malayan and Melanesian admixtures) 69.9%, Filipino 15.3%, Chinese 4.9%, other Asian 2.4%, white 1.9%, Carolinian 1.4%, other Micronesian 1.1%, other or unspecified 3.2% (2000 census)
Religions: Roman Catholic 41.6%, Protestant 23.3%, Modekngei 8.8% (indigenous to Palau), Seventh-Day Adventist 5.3%, Jehovah's Witness 0.9%, Latter-Day Saints 0.6%, other 3.1%, unspecified or none 16.4% (2000 census)
Languages: Palauan (official on most islands) 64.7%, Filipino 13.5%, English 9.4%, Chinese 5.7%, Carolinian 1.5%, Japanese 1.5%, other Asian 2.3%, other languages 1.5% (2000 census)
note: Sonsoral (Sonsoralese and English are official), Tobi (Tobi and English are official), and Angaur (Angaur, Japanese, and English are official)
Literacy: *definition:* age 15 and over can read and write
total population: 92%
male: 93%
female: 90% (1980 est.)
School life expectancy (primary to tertiary education): *total:* 15 years
male: 14 years
female: 15 years (2001)
Education expenditures: NA

GOVERNMENT

Country name: *conventional long form:* Republic of Palau
conventional short form: Palau
local long form: Beluu er a Belau
local short form: Belau
former: Trust Territory of the Pacific Islands, Palau District
Government type: constitutional government in free association with the US; the Compact of Free Association entered into force on 1 October 1994
Capital: *name:* Melekeok
geographic coordinates: 7 29 N, 134 38 E
time difference: UTC+9 (14 hours ahead of Washington, DC during Standard Time)
Administrative divisions: 16 states; Aimeliik, Airai, Angaur, Hatohobei, Kayangel, Koror, Melekeok, Ngaraard, Ngarchelong, Ngardmau, Ngatpang, Ngchesar, Ngeremlengui, Ngiwal, Peleliu, Sonsorol
Independence: 1 October 1994 (from the US-administered UN trusteeship)
National holiday: Constitution Day, 9 July (1979)
Constitution: 1 January 1981
Legal system: mixed legal system of civil, common, and customary law
International law organization participation: has not submitted an ICJ jurisdiction declaration; non-party state to the ICCt
Suffrage: 18 years of age; universal
Executive branch: *chief of state:* President Johnson TORIBIONG (since 15 January 2009); Vice President Kerai MARIUR (since 15 January 2009); note—the president is both the chief of state and head of government
head of government: President Johnson TORIBIONG (since 15 January 2009); Vice President Kerai MARIUR (since 15 January 2009)
cabinet: NA
(For more information visit the World Leaders website)
elections: president and vice president elected on separate tickets by popular vote for four-year terms (eligible for a second term); election last held on 4 November 2008 (next to be held in November 2012)
election results: Johnson TORIBIONG elected president; percent of vote—Johnson TORIBIONG 51%, Elias Camsek CHIN 49%
Legislative branch: bicameral National Congress or Olbiil Era Kelulau (OEK) consists of the Senate (9 seats; members elected by popular vote on a population basis to serve four-year terms) and the House of Delegates (16 seats; members elected by popular vote to serve four-year terms)
elections: Senate—last held on 4 November 2008 (next to be held in November 2012); House of Delegates—last held on 4 November 2008 (next to be held in November 2012)
election results: Senate—percent of vote—NA; seats—independents 9; House of Delegates—percent of vote—NA; seats—independents 16
Judicial branch: Supreme Court; Court of Common Pleas; Land Court
Political parties and leaders: none
Political pressure groups and leaders: NA
International organization participation: ACP, ADB, AOSIS, FAO, IAEA, IBRD, ICAO, ICRM, IDA, IFC, IFRCS, IMF, IOC, IPU, MIGA, OPCW, PIF, Sparteca, SPC, UN, UNCTAD, UNESCO, WHO
Diplomatic representation in the US: *chief of mission:* Ambassador Hersey KYOTA
chancery: 1701 Pennsylvania Avenue NW, Suite 300, Washington, DC 20006
telephone: [1] (202) 452-6814
FAX: [1] (202) 452-6281
consulate(s): Tamuning (Guam)
Diplomatic representation from the US: *chief of mission:* Ambassador Helen P. REED-ROWE
embassy: Koror (no street address)
mailing address: P. O. Box 6028, Republic of Palau 96940
telephone: [680] 488-2920, 2990
FAX: [680] 488-2911
Flag description: light blue with a large yellow disk shifted slightly to the hoist side; the blue color represents the ocean, the disk represents the moon; Palauans consider the full moon to be the optimum time for human activity; it is also considered a symbol of peace, love, and tranquility
National anthem: *name:* "Belau rekid" (Our Palau)
lyrics/music: multiple/Ymesei O. EZEKIEL
note: adopted 1980

ECONOMY

Economy—overview: The economy consists primarily of tourism, subsistence

agriculture, and fishing. The government is the major employer of the work force relying heavily on financial assistance from the US. The Compact of Free Association with the US, entered into after the end of the UN trusteeship on 1 October 1994, provided Palau with up to $700 million in US aid for the following 15 years in return for furnishing military facilities. Business and tourist arrivals numbered 85,000 in 2007. The population enjoys a per capita income roughly 50% higher than that of the Philippines and much of Micronesia. Long-run prospects for the key tourist sector have been greatly bolstered by the expansion of air travel in the Pacific, the rising prosperity of leading East Asian countries, and the willingness of foreigners to finance infrastructure development.
GDP (purchasing power parity): $164 million (2008 est.)
country comparison to the world: 217
$124.5 million (2004 est.)
note: data are in 2010 US dollars
GDP estimate includes US subsidy
GDP (official exchange rate): $164 million (2008)
GDP—real growth rate: NA% (2009)
5.5% (2005 est.)
GDP—per capita (PPP): $8,100 (2008 est.)
country comparison to the world: 122
$7,600 (2005 est.)
GDP—composition by sector: *agriculture:* 6.2%
industry: 12%
services: 81.8% (2003)
Labor force: 9,777 (2005)
country comparison to the world: 215
Labor force—by occupation:
agriculture: 20%
industry: NA%
services: NA% (1990)
Unemployment rate: 4.2% (2005 est.)
country comparison to the world: 38
Population below poverty line: NA%
Household income or consumption by percentage share: *lowest 10%:* NA%
highest 10%: NA%
Budget: *revenues:* $114.8 million
expenditures: $99.5 million (2008 est.)
Inflation rate (consumer prices): 2.7% (2005 est.)
country comparison to the world: 74
Market value of publicly traded shares: $NA
Agriculture—products: coconuts, copra, cassava (tapioca), sweet potatoes; fish
Industries: tourism, craft items (from shell, wood, pearls), construction, garment making
Industrial production growth rate: NA%
Current account balance: $15.09 million (FY03/04)
country comparison to the world: 60
Exports: $5.882 million (2004 est.)
country comparison to the world: 214
Exports—commodities: shellfish, tuna, copra, garments
Imports: $107.3 million (2004 est.)
country comparison to the world: 209
Imports—commodities: machinery and equipment, fuels, metals; foodstuffs
Debt—external: $0 (FY99/00)
country comparison to the world: 197
Exchange rates: the US dollar is used

COMMUNICATIONS

Telephones—main lines in use: 7,100 (2009)
country comparison to the world: 207
Telephones—mobile cellular: 13,200 (2009)
country comparison to the world: 209
Telephone system: *general assessment:* NA
domestic: fixed-line and mobile-cellular services available with a combined subscribership of roughly 100 per 100 persons
international: country code–680; satellite earth station–1 Intelsat (Pacific Ocean) (2008)
Broadcast media: no television broadcast stations; a cable television network covers the major islands and provides access to rebroadcasts, on a delayed basis, of a number of US stations as well as access to a number of real-time satellite TV channels; about a half dozen radio stations with 1 government-owned (2009)
Internet country code: .pw
Internet hosts: 3 (2010)
country comparison to the world: 227

TRANSPORTATION

Airports: 3 (2010)
country comparison to the world: 192
Airports—with paved runways:
total: 1
1,524 to 2,437 m: 1 (2010)
Airports—with unpaved runways: *total:* 2
1,524 to 2,437 m: 2 (2010)
Ports and terminals: Koror

MILITARY

Military branches: no regular military forces; Palau National Police (2009)
Manpower available for military service: *males age 16–49:* 6,987 (2010 est.)
Manpower fit for military service: *males age 16–49:* 5,272
females age 16–49: 3,969 (2010 est.)
Manpower reaching militarily significant age annually: *male:* 216
female: 222 (2010 est.)
Military expenditures: NA
Military—note: defense is the responsibility of the US; under a Compact of Free Association between Palau and the US, the US military is granted access to the islands for 50 years, but it has not stationed any military forces there (2008)

TRANSNATIONAL ISSUES

Disputes—international: maritime delineation negotiations continue with Philippines, Indonesia

PANAMA

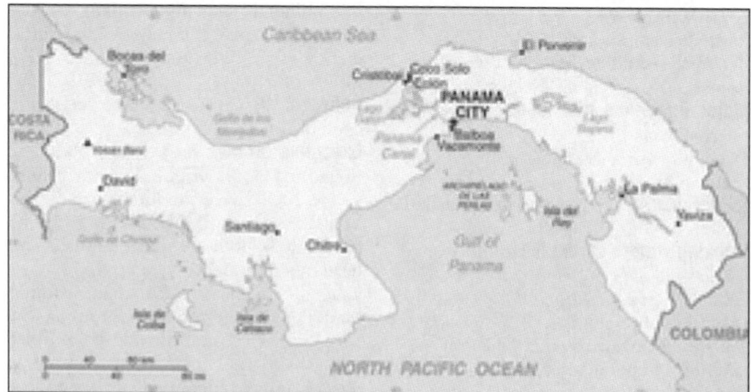

INTRODUCTION

Background: Explored and settled by the Spanish in the 16th century, Panama broke with Spain in 1821 and joined a union of Colombia, Ecuador, and Venezuela–named the Republic of Gran Colombia. When the latter dissolved in 1830, Panama remained part of Colombia. With US backing, Panama seceded from Colombia in 1903 and promptly signed a treaty with the US allowing for the construction of a canal and US sovereignty over a strip of land on either side of the structure (the Panama Canal Zone). The Panama Canal was built by the US Army Corps of Engineers between 1904 and 1914. In 1977, an agreement was signed for the complete transfer of the Canal from the US to Panama by the end of the century. Certain portions of the Zone and increasing responsibility over the Canal were turned over in the subsequent decades. With US help, dictator Manuel NORIEGA was deposed in 1989. The entire Panama Canal, the area supporting the Canal, and remaining US military bases were transferred to Panama by the end of 1999. In October 2006, Panamanians approved an ambitious plan (estimated to cost $5.3 billion) to expand the Canal. The project, which began in 2007 and could double the Canal's capacity, is expected to be completed in 2014-15.

GEOGRAPHY

Location: Central America, bordering both the Caribbean Sea and the North Pacific Ocean, between Colombia and Costa Rica
Geographic coordinates: 9 00 N, 80 00 W
Map references: Central America and the Caribbean
Area: *total:* 75,420 sq km
country comparison to the world: 117
land: 74,340 sq km
water: 1,080 sq km
Area—comparative: slightly smaller than South Carolina
Land boundaries: *total:* 555 km
border countries: Colombia 225 km, Costa Rica 330 km
Coastline: 2,490 km
Maritime claims: territorial sea: 12 nm
contiguous zone: 24 nm
exclusive economic zone: 200 nm or edge of continental margin
Climate: tropical maritime; hot, humid, cloudy; prolonged rainy season (May to January), short dry season (January to May)
Terrain: interior mostly steep, rugged mountains and dissected, upland plains; coastal areas largely plains and rolling hills
Elevation extremes: *lowest point:* Pacific Ocean 0 m
highest point: Volcan Baru 3,475 m
Natural resources: copper, mahogany forests, shrimp, hydropower
Land use:
arable land: 7.26%
permanent crops: 1.95%
other: 90.79% (2005)
Irrigated land: 430 sq km (2003)
Total renewable water resources: 148 cu km (2000)
Freshwater withdrawal (domestic/industrial/agricultural): *total:* 0.82 cu km/yr (67%/5%/28%)
per capita: 254 cu m/yr (2000)
Natural hazards: occasional severe storms and forest fires in the Darien area
Environment—current issues: water pollution from agricultural runoff threatens fishery resources; deforestation of tropical rain forest; land degradation and soil erosion threatens siltation of Panama Canal; air pollution in urban areas; mining threatens natural resources
Environment—international agreements: *party to:* Biodiversity, Climate Change, Climate Change-Kyoto Protocol, Desertification, Endangered Species, Environmental Modification, Hazardous Wastes, Law of the Sea, Marine Dumping, Ozone Layer Protection, Ship Pollution, Tropical Timber 83, Tropical Timber 94, Wetlands, Whaling
signed, but not ratified: Marine Life Conservation
Geography—note: strategic location on eastern end of isthmus forming land bridge connecting North and South America; controls Panama Canal that links North Atlantic Ocean via Caribbean Sea with North Pacific Ocean

PEOPLE

Population: 3,460,462 (July 2011 est.)
country comparison to the world: 131
Age structure: *0–14 years:* 28.6% (male 504,726/female 484,291)
15–64 years: 64.2% (male 1,123,777/female 1,098,661)
65 years and over: 7.2% (male 115,425/female 133,582) (2011 est.)
Median age: *total:* 27.5 years
male: 27.1 years
female: 27.9 years (2011 est.)
Population growth rate: 1.435% (2011 est.)
country comparison to the world: 84
Birth rate: 19.43 births/1,000 population (2011 est.)
country comparison to the world: 94
Death rate: 4.65 deaths/1,000 population (July 2011 est.)
country comparison to the world: 198
Net migration rate: -0.42 migrant(s)/1,000 population (2011 est.)
country comparison to the world: 134
Urbanization: *urban population:* 75% of total population (2010)
rate of urbanization: 2.3% annual rate of change (2010-15 est.)
Major cities—population: PANAMA CITY (capital) 1.346 million (2009)
Sex ratio: *at birth:* 1.045 male(s)/female
under 15 years: 1.04 male(s)/female
15–64 years: 1.02 male(s)/female
65 years and over: 0.87 male(s)/female
total population: 1.02 male(s)/female (2011 est.)
Infant mortality rate:
total: 11.64 deaths/1,000 live births
country comparison to the world: 138
male: 12.41 deaths/1,000 live births
female: 10.83 deaths/1,000 live births (2011 est.)
Life expectancy at birth:
total population: 77.79 years
country comparison to the world: 54
male: 75.02 years
female: 80.68 years (2011 est.)
Total fertility rate: 2.45 children born/woman (2011 est.)
country comparison to the world: 87
HIV/AIDS—adult prevalence rate: 0.9% (2009 est.)
country comparison to the world: 50
HIV/AIDS—people living with HIV/AIDS: 20,000 (2009 est.)
country comparison to the world: 79
HIV/AIDS—deaths: 1,500 (2009 est.)
country comparison to the world: 60
Major infectious diseases: *degree of risk:* intermediate
food or waterborne diseases: bacterial diarrhea
vectorborne disease: dengue fever and malaria (2009)
Drinking water source: *Improved:*
urban: 97% of population
rural: 83% of population
total: 93% of population
Unimproved: urban: 3% of population
rural: 17% of population
total: 7% of population (2008)
Sanitation facility access: *Improved:*
urban: 75% of population
rural: 51% of population
total: 69% of population
Unimproved: urban: 25% of population
rural: 49% of population
total: 31% of population (2008)
Nationality: *noun:* Panamanian(s)
adjective: Panamanian
Ethnic groups: mestizo (mixed Amerindian and white) 70%, Amerindian and mixed (West Indian) 14%, white 10%, Amerindian 6%
Religions: Roman Catholic 85%, Protestant 15%
Languages: Spanish (official), English 14%
note: many Panamanians are bilingual
Literacy: *definition:* age 15 and over can read and write
total population: 91.9%
male: 92.5%
female: 91.2% (2000 census)
School life expectancy (primary to tertiary education): *total:* 13 years
male: 13 years
female: 14 years (2008)
Education expenditures: 3.8% of GDP (2008)
country comparison to the world: 111

GOVERNMENT

Country name: *conventional long form:* Republic of Panama
conventional short form: Panama
local long form: Republica de Panama
local short form: Panama
Government type: constitutional democracy
Capital: *name:* Panama City
geographic coordinates: 8 58 N, 79 32 W
time difference: UTC-5 (same time as Washington, DC during Standard Time)
Administrative divisions: 9 provinces (provincias, singular–provincia) and 3 indigenous territories* (comarcas); Bocas del Toro, Chiriqui, Cocle, Colon, Darien, Embera-Wounaan*, Herrera, Kuna Yala*, Los Santos, Ngobe-Bugle*, Panama, Veraguas
Independence: 3 November 1903 (from Colombia; became independent from Spain on 28 November 1821)
National holiday: Independence Day, 3 November (1903)
Constitution: 11 October 1972; revised several times
Legal system: civil law system; judicial review of legislative acts in the Supreme Court of Justice
International law organization participation: accepts compulsory ICJ jurisdiction with reservations; accepts ICCt jurisdiction
Suffrage: 18 years of age; universal and compulsory
Executive branch: *chief of state:* President Ricardo MARTINELLI Berrocal (since 1 July 2009); Vice President Juan Carlos VARELA (since 1 July 2009); note–the president is both the chief of state and head of government
head of government: President Ricardo MARTINELLI Berrocal (since 1 July 2009); Vice President Juan Carlos VARELA (since 1 July 2009)
cabinet: Cabinet appointed by the president (For more information visit the World Leaders website)
elections: president and vice president elected on the same ticket by popular vote for five-year terms (not eligible for immediate reelection; president and vice president must sit out two additional terms (10

years) before becoming eligible for reelection); election last held on 3 May 2009 (next to be held in 2014)
election results: Ricardo MARTINELLI Berrocal elected president; percent of vote–Ricardo MARTINELLI Berrocal 60%, Balbina HERRERA 38%, Guillermo ENDARA Galimany 2%
note: government coalition–CD (Democratic Change), Panamenista, MOLIRENA (Nationalist Republican Liberal Movement), and UP (Patriotic Union Party)
Legislative branch: unicameral National Assembly or Asamblea Nacional (71 seats; members elected by popular vote to serve five-year terms)
elections: last held on 3 May 2009 (next to be held in May 2014)
election results: percent of vote by party–NA; seats by party–PRD 26, Panamenista 22, CD 14, UP 4, Independent 2, MOLIRENA 2, PP 1; note–changes in political affiliation now reflect the following seat distribution: as of 1 March 2011–seats by party–PRD 23, Panamenista 20, CD 23, UP 2, MOLIRENA 2, PP 1
note: legislators from outlying rural districts chosen on a plurality basis while districts located in more populous towns and cities elect multiple legislators by means of a proportion-based formula
Judicial branch: Supreme Court of Justice or Corte Suprema de Justicia (nine judges appointed for staggered 10-year terms); five superior courts; three courts of appeal
Political parties and leaders: Democratic Change or CD [Ricardo MARTINELLI]; Democratic Revolutionary Party or PRD [Francisco SANCHEZ Cardenas]; Nationalist Republican Liberal Movement or MOLIRENA [Sergio GONZALEZ-Ruiz]; Panamenista Party [Juan Carlos VARELA Rodriguez] (formerly the Arnulfista Party); Patriotic Union Party or UP (combination of the Liberal National Party or PLN and the Solidarity Party or PS)[Anibal GALINDO]; Popular Party or PP [Milton HENRIQUEZ] (formerly Christian Democratic Party or PDC)
Political pressure groups and leaders: Chamber of Commerce; National Civic Crusade; National Council of Organized Workers or CONATO; National Council of Private Enterprise or CONEP; National Union of Construction and Similar Workers (SUNTRACS); Panamanian Association of Business Executives or APEDE; Panamanian Industrialists Society or SIP; Workers Confederation of the Republic of Panama or CTRP
International organization participation: BCIE, CAN (observer), CSN (observer), FAO, G-77, IADB, IAEA, IBRD, ICAO, ICC, ICRM, IDA, IFAD, IFC, IFRCS, ILO, IMF, IMO, IMSO, Interpol, IOC, IOM, IPU, ISO, ITSO, ITU, ITUC, LAES, LAIA (observer), MIGA, NAM, OAS, OPANAL, OPCW, PCA, RG, SICA, UN, UNASUR (observer), UNCTAD, UNESCO, UNIDO, Union Latina, UNWTO, UPU, WCO, WFTU, WHO, WIPO, WMO, WTO
Diplomatic representation in the US:
chief of mission: Ambassador Mario Ernesto JARAMILLO Castillo
chancery: 2862 McGill Terrace NW, Washington, DC 20008
telephone: [1] (202) 483-1407
FAX: [1] (202) 483-8416
consulate(s) general: Atlanta, Houston, Miami, New Orleans, New York, Philadelphia, San Diego, San Francisco, Tampa
Diplomatic representation from the US:
chief of mission: Ambassador Phyllis M. POWERS
embassy: Edificio 783, Avenida Demetrio Basilio Lakas Panama, Apartado Postal 0816-02561, Zona 5, Panama City
mailing address: American Embassy Panama, Unit 0945, APO AA 34002
telephone: [507] 207-7000
FAX: [507] 317-5568
Flag description: divided into four, equal rectangles; the top quadrants are white (hoist side) with a blue five-pointed star in the center and plain red; the bottom quadrants are plain blue (hoist side) and white with a red five-pointed star in the center; the blue and red colors are those of the main political parties (Conservatives and Liberals respectively) and the white denotes peace between them; the blue star stands for the civic virtues of purity and honesty, the red star signifies authority and law
National anthem: *name:* "Himno Istemno" (Isthmus Hymn)
lyrics/music: Jeronimo DE LA OSSA/ Santos A. JORGE
note: adopted 1925

ECONOMY

Economy—overview: Panama's dollar-based economy rests primarily on a well-developed services sector that accounts for three-quarters of GDP. Services include operating the Panama Canal, logistics, banking, the Colon Free Zone, insurance, container ports, flagship registry, and tourism. Economic growth will be bolstered by the Panama Canal expansion project that began in 2007 and is scheduled to be completed by 2014 at a cost of $5.3 billion–about 25% of current GDP. The expansion project will more than double the Canal's capacity, enabling it to accommodate ships that are too large to traverse the existing canal. The United States and China are the top users of the Canal. Panama also plans to construct a metro system in Panama City, valued at $1.2 billion and scheduled to be completed by 2014. Panama's booming transportation and logistics services sectors, along with aggressive infrastructure development projects, will likely lead the economy to continued growth in 2011. Strong economic performance has not translated into broadly shared prosperity, as Panama has the second worst income distribution in Latin America. About 30% of the population lives in poverty; however, from 2006 to 2010 poverty was reduced by 10 percentage points, while unemployment dropped from 12% to 6% of the labor force. Panama and the United States signed a Trade Promotion Agreement in June 2007, which, when implemented, will help promote the country's economic growth. Seeking removal from the Organization of Economic Development's gray-list of tax havens, Panama has also recently signed various double taxation treaties with other nations.
GDP (purchasing power parity): $44.36 billion (2010 est.)
country comparison to the world: 94
$41.26 billion (2009 est.)
$39.99 billion (2008 est.)
note: data are in 2010 US dollars
GDP (official exchange rate): $26.78 billion (2010 est.)
GDP—real growth rate: 7.5% (2010 est.)
country comparison to the world: 28
3.2% (2009 est.)
10.1% (2008 est.)
GDP—per capita (PPP): $13,000 (2010 est.)
country comparison to the world: 90
$12,300 (2009 est.)
$12,100 (2008 est.)
note: data are in 2010 US dollars
GDP—composition by sector:
agriculture: 5.8%
industry: 16.6%
services: 77.6% (2010 est.)
Labor force: 1.557 million
country comparison to the world: 129
note: shortage of skilled labor, but an oversupply of unskilled labor (2010 est.)
Labor force—by occupation:
agriculture: 17.6%
industry: 8.8%
services: 73.6% (2009 est.)
Unemployment rate: 6.5% (2010 est.)
country comparison to the world: 63
6.6% (2009 est.)
Population below poverty line: 25.6% (2010 est.)
Household income or consumption by percentage share: *lowest 10%:* 0.8%
highest 10%: 41.4% (2006)
Distribution of family income—Gini index: 51 (2010 est.)
country comparison to the world: 17
56.1 (2003)
Investment (gross fixed): 26.8% of GDP (2010 est.)
country comparison to the world: 32
Budget: *revenues:* $6.944 billion
expenditures: $7.051 billion (2010 est.)
Public debt: 40% of GDP (September 2010 est.)
country comparison to the world: 71
44.6% of GDP (2009 est.)
Inflation rate (consumer prices): 3.5% (2010 est.)
country comparison to the world: 98
2.4% (2009 est.)
Commercial bank prime lending rate: 8.302% (31 November 2010 est.)
country comparison to the world: 112
8.25% (31 December 2009 est.)
Stock of narrow money: $5.04 billion (31 December 2010 est.)
country comparison to the world: 89
$4.404 billion (31 December 2009 est.)
Stock of broad money: $24.17 billion (31 December 2010 est.)
country comparison to the world: 78
$21.78 billion (31 December 2009 est.)
Stock of domestic credit: $23.2 billion (31 December 2010 est.)
country comparison to the world: 74
$20.17 billion (31 December 2009 est.)
Market value of publicly traded shares: $8.048 billion (31 December 2009)

country comparison to the world: 76
$6.568 billion (31 December 2008)
$6.219 billion (31 December 2007)
Agriculture—products: bananas, rice, corn, coffee, sugarcane, vegetables; livestock; shrimp
Industries: construction, brewing, cement and other construction materials, sugar milling
Industrial production growth rate: -1% (September 2010, year-over-year)
country comparison to the world: 158
Electricity—production: 6.546 billion kWh (2010 est.)
country comparison to the world: 104
Electricity—consumption: 5.805 billion kWh (2010 est.)
country comparison to the world: 108
Electricity—exports: 124.9 million kWh (2007 est.)
Electricity—imports: 8.74 million kWh (2007 est.)
Oil—production: 0 bbl/day (2009 est.)
country comparison to the world: 122
Oil—consumption: 93,000 bbl/day (2009 est.)
country comparison to the world: 76
Oil—exports: 4,803 bbl/day (2007 est.)
country comparison to the world: 105
Oil—imports: 87,100 bbl/day (2007 est.)
country comparison to the world: 68
Oil—proved reserves: 0 bbl (1 January 2010 est.)
country comparison to the world: 176
Natural gas—production: 0 cu m (2008 est.)
country comparison to the world: 173
Natural gas—consumption: 0 cu m (2008 est.)
country comparison to the world: 115
Natural gas—exports: 0 cu m (2008 est.)
country comparison to the world: 157
Natural gas—imports: 0 cu m (2008 est.)
country comparison to the world: 106
Natural gas—proved reserves: 0 cu m (1 January 2010 est.)
country comparison to the world: 178
Current account balance: $-2.523 billion (2010 est.)
country comparison to the world: 162
$-2.33 billion (2009 est.)
Exports: $12.52 billion (2010 est.)
country comparison to the world: 79
$10.9 billion (2009 est.)
note: includes the Colon Free Zone
Exports—commodities: bananas, shrimp, sugar, coffee, clothing
Exports—partners: Greece 21.03%, US 17.63%, Japan 9.87%, Germany 4.28%, Italy 4.27% (2009)
Imports:
$16.05 billion (2010 est.)
country comparison to the world: 77
$12.93 billion (2009 est.)
note: includes the Colon Free Zone
Imports—commodities: capital goods, foodstuffs, consumer goods, chemicals
Imports—partners: Japan 36.21%, Singapore 16.86%, US 12.3%, China 7.84% (2009)
Reserves of foreign exchange and gold: $3.525 billion (31 December 2010 est.)
country comparison to the world: 89
$3.028 billion (31 December 2009 est.)
Debt—external: $13.85 billion (31 December 2010 est.)
country comparison to the world: 79
$13.7 billion (31 December 2009 est.)
Stock of direct foreign investment—at home: $NA
Stock of direct foreign investment—abroad: $NA
Exchange rates: balboas (PAB) per US dollar–1 (2010), 1 (2009), 1 (2008), 1 (2007), 1 (2006)
note: the balboa exists alongside the dollar and may be used interchangably

COMMUNICATIONS

Telephones—main lines in use: 537,100 (2009)
country comparison to the world: 96
Telephones—mobile cellular:
5.677 million (2009)
country comparison to the world: 88
Telephone system: *general assessment:* domestic and international facilities well developed
domestic: mobile-cellular telephone subscribership has increased rapidly
international: country code–507; landing point for the Americas Region Caribbean Ring System (ARCOS-1), the MAYA-1, and PAN-AM submarine cable systems that together provide links to the US and parts of the Caribbean, Central America, and South America; satellite earth stations–2 Intelsat (Atlantic Ocean); connected to the Central American Microwave System (2008)
Broadcast media: multiple privately-owned television networks and a government-owned educational TV station; multi-channel cable and satellite TV subscription services are available; more than 100 commercial radio stations (2007)
Internet country code: .pa
Internet hosts: 9,585 (2010)
country comparison to the world: 123
Internet users: 959,800 (2009)
country comparison to the world: 104

TRANSPORTATION

Airports: 118 (2010)
country comparison to the world: 52
Airports—with paved runways: *total:* 54
over 3,047 m: 1
2,438 to 3,047 m: 1
1,524 to 2,437 m: 5
914 to 1,523 m: 17
under 914 m: 30 (2010)
Airports—with unpaved runways:
total: 64
1,524 to 2,437 m: 1
914 to 1,523 m: 11
under 914 m: 52 (2010)
Heliports: 3 (2010)
Pipelines: oil 128 km
Railways: *total:* 76 km
country comparison to the world: 127
standard gauge: 76 km 1.435-m gauge (2008)
Roadways: *total:* 11,978 km
country comparison to the world: 130
paved: 4,300 km
unpaved: 7,678 km (2002)
Waterways: 800 km (includes the 82-km Panama Canal that is being widened) (2010)
country comparison to the world: 73
Merchant marine: *total:* 6,379
country comparison to the world: 1
by type: barge carrier 1, bulk carrier 2,383, cargo 1,129, carrier 18, chemical tanker 626, combination ore/oil 3, container 751, liquefied gas 192, passenger 42, passenger/cargo 61, petroleum tanker 576, refrigerated cargo 212, roll on/roll off 100, specialized tanker 3, vehicle carrier 282
foreign-owned: 5,244 (Albania 3, Argentina 7, Australia 5, Azerbaijan 1, Bahamas 7, Bangladesh 3, Belgium 2, Bermuda 15, Brazil 3, Bulgaria 6, Burma 3, Canada 5, Chile 17, China 574, Colombia 2, Croatia 2, Cuba 4, Cyprus 8, Denmark 46, Ecuador 6, Egypt 11, Finland 2, France 13, Gabon 1, Germany 27, Gibraltar 1, Greece 402, Hong Kong 125, India 17, Indonesia 14, Iran 5, Ireland 1, Isle of Man 11, Israel 1, Italy 23, Japan 2347, Jordan 13, Kuwait 12, Latvia 4, Lebanon 2, Lithuania 4, Luxembourg 1, Malaysia 12, Maldives 3, Malta 2, Mexico 6, Monaco 14, Morocco 1, Netherlands 8, Nigeria 7, Norway 89, Oman 8, Pakistan 5, Peru 12, Philippines 6, Poland 3, Portugal 9, Qatar 1, Romania 2, Russia 39, Saudi Arabia 8, Singapore 79, South Korea 366, Spain 40, Sweden 1, Switzerland 22, Syria 42, Taiwan 337, Tanzania 2, Thailand 6, Tunisia 1, Turkey 79, UAE 83, UK 33, Ukraine 11, US 102, Venezuela 8, Vietnam 37, Yemen 4)
note: this country allows large numbers of ships owned by foreign entities to be registered in its national shipping registry and to fly its flag; these ships operate under the laws of the flag state
registered in other countries: 1 (Honduras 1) (2010)
Ports and terminals: Balboa, Colon, Cristobal

MILITARY

Military branches: no regular military forces; Panamanian public forces include: Panamanian National Police (PNP), National Air-Naval Service (SENAN), National Border Service (SENAFRONT) (2010)
Manpower available for military service:
males age 16–49: 890,006 (2010 est.)
Manpower fit for military service: *males age 16–49:* 731,254
females age 16–49: 728,329 (2010 est.)
Manpower reaching militarily significant age annually: *male:* 32,142
female: 30,879 (2010 est.)
Military expenditures: 1% of GDP (2006)
country comparison to the world: 134
Military—note: on 10 February 1990, the government of then President ENDARA abolished Panama's military and reformed the security apparatus by creating the Panamanian Public Forces; in October 1994, Panama's Legislative Assembly approved a constitutional amendment prohibiting the creation of a standing military force but allowing the temporary establishment of special police units to counter acts of "external aggression"

TRANSNATIONAL ISSUES

Disputes—international: organized illegal narcotics operations in Colombia operate within the remote border region with Panama

Illicit drugs: major cocaine transshipment point and primary money-laundering center for narcotics revenue; money-laundering activity is especially heavy in the Colon Free Zone; offshore financial center; negligible signs of coca cultivation; monitoring of financial transactions is improving; official corruption remains a major problem

PAPUA NEW GUINEA

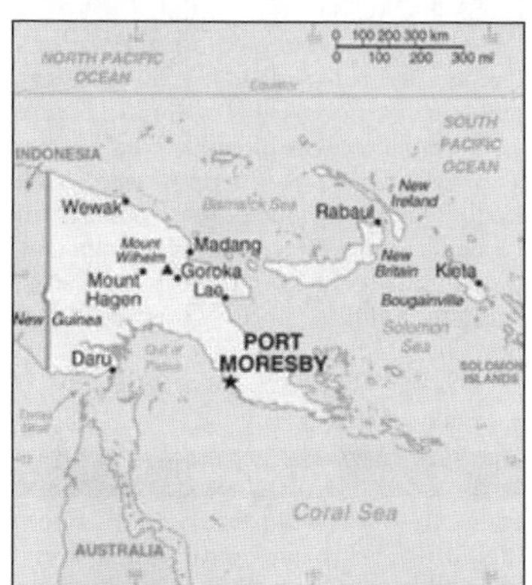

INTRODUCTION

Background: The eastern half of the island of New Guinea–second largest in the world–was divided between Germany (north) and the UK (south) in 1885. The latter area was transferred to Australia in 1902, which occupied the northern portion during World War I and continued to administer the combined areas until independence in 1975. A nine-year secessionist revolt on the island of Bougainville ended in 1997 after claiming some 20,000 lives.

GEOGRAPHY

Location: Oceania, group of islands including the eastern half of the island of New Guinea between the Coral Sea and the South Pacific Ocean, east of Indonesia
Geographic coordinates: 6 00 S, 147 00 E
Map references: Oceania
Area: *total:* 462,840 sq km
country comparison to the world: 54
land: 452,860 sq km
water: 9,980 sq km
Area—comparative: slightly larger than California
Land boundaries: *total:* 820 km
border countries: Indonesia 820 km
Coastline: 5,152 km
Maritime claims: measured from claimed archipelagic baselines
territorial sea: 12 nm
continental shelf: 200 m depth or to the depth of exploitation
exclusive fishing zone: 200 nm
Climate: tropical; northwest monsoon (December to March), southeast monsoon (May to October); slight seasonal temperature variation
Terrain: mostly mountains with coastal lowlands and rolling foothills
Elevation extremes: *lowest point:* Pacific Ocean 0 m
highest point: Mount Wilhelm 4,509 m
Natural resources: gold, copper, silver, natural gas, timber, oil, fisheries
Land use: *arable land:* 0.49%
permanent crops: 1.4%
other: 98.11% (2005)
Irrigated land: NA
Total renewable water resources: 801 cu km (1987)
Freshwater withdrawal (domestic/industrial/agricultural): *total:* 0.1 cu km/yr (56%/43%/1%)
per capita: 17 cu m/yr (1987)
Natural hazards: active volcanism; situated along the Pacific "Ring of Fire"; the country is subject to frequent and sometimes severe earthquakes; mud slides; tsunamis
volcanism: Papua New Guinea experiences severe volcanic activity; Ulawun (elev. 2,334 m), one of Papua New Guinea's potentially most dangerous volcanoes, has been deemed a "Decade Volcano" by the International Association of Volcanology and Chemistry of the Earth's Interior, worthy of study due to its explosive history and close proximity to human populations; Rabaul (elev. 688 m) destroyed the city of Rabaul in 1937 and 1994; Lamington erupted in 1951 killing 3,000 people; Manam's 2004 eruption forced the island's abandonment; other historically active volcanoes include Bam, Bagana, Garbuna, Karkar, Langila, Lolobau, Long Island, Pago, St. Andrew Strait, Victory, and Waiowa
Environment—current issues: rain forest subject to deforestation as a result of growing commercial demand for tropical timber; pollution from mining projects; severe drought
Environment—international agreements: *party to:* Antarctic Treaty, Biodiversity, Climate Change, Climate Change-Kyoto Protocol, Desertification, Endangered Species, Environmental Modification, Hazardous Wastes, Law of the Sea, Marine Dumping, Ozone Layer Protection, Ship Pollution, Tropical Timber 83, Tropical Timber 94, Wetlands
signed, but not ratified: none of the selected agreements
Geography—note: shares island of New Guinea with Indonesia; one of world's largest swamps along southwest coast

PEOPLE

Population: 6,187,591 (July 2011 est.)
country comparison to the world: 105
Age structure: *0–14 years:* 36.4% (male 1,145,946/female 1,106,705)
15–64 years: 60% (male 1,907,787/female 1,802,144)
65 years and over: 3.6% (male 121,207/female 103,802) (2011 est.)
Median age: *total:* 21.8 years
male: 22.1 years
female: 21.5 years (2011 est.)
Population growth rate: 1.985% (2011 est.)
country comparison to the world: 55
Birth rate: 26.44 births/1,000 population (2011 est.)
country comparison to the world: 52
Death rate: 6.58 deaths/1,000 population (July 2011 est.)
country comparison to the world: 148
Net migration rate: 0 migrant(s)/1,000 population (2011 est.)
country comparison to the world: 103
Urbanization: *urban population:* 13% of total population (2010)
rate of urbanization: 2.9% annual rate of change (2010-15 est.)
Major cities—population: PORT MORESBY (capital) 314,000 (2009)
Sex ratio: *at birth:* 1.05 male(s)/female
under 15 years: 1.03 male(s)/female
15–64 years: 1.06 male(s)/female
65 years and over: 1.2 male(s)/female
total population: 1.06 male(s)/female (2011 est.)
Infant mortality rate: *total:* 43.29 deaths/1,000 live births
country comparison to the world: 56
male: 47.12 deaths/1,000 live births
female: 39.28 deaths/1,000 live births (2011 est.)
Life expectancy at birth: *total population:* 66.24 years
country comparison to the world: 163
male: 64.02 years
female: 68.56 years (2011 est.)
Total fertility rate: 3.46 children born/woman (2011 est.)
country comparison to the world: 47
HIV/AIDS—adult prevalence rate: 0.9% (2009 est.)
country comparison to the world: 51
HIV/AIDS—people living with HIV/AIDS: 34,000 (2009 est.)
country comparison to the world: 67
HIV/AIDS—deaths: 1,300 (2009 est.)
country comparison to the world: 62
Major infectious diseases: *degree of risk:* very high
food or waterborne diseases: bacterial diarrhea, hepatitis A, and typhoid fever
vectorborne diseases: dengue fever and malaria (2009)
Drinking water source: *Improved:*
urban: 87% of population
rural: 33% of population
total: 40% of population
Unimproved: urban: 13% of population
rural: 67% of population
total: 60% of population (2008)
Sanitation facility access: *Improved:*
urban: 71% of population
rural: 41% of population
total: 45% of population
Unimproved: urban: 29% of population
rural: 59% of population
total: 55% of population (2008)
Nationality: *noun:* Papua New Guinean(s)
adjective: Papua New Guinean
Ethnic groups: Melanesian, Papuan, Negrito, Micronesian, Polynesian
Religions: Roman Catholic 27%, Evangelical Lutheran 19.5%, United Church 11.5%, Seventh-Day Adventist 10%, Pentecostal 8.6%, Evangelical Alliance 5.2%, Anglican 3.2%, Baptist 2.5%, other

Protestant 8.9%, Bahai 0.3%, indigenous beliefs and other 3.3% (2000 census)
Languages: Tok Pisin (official), English (official), Hiri Motu (official), some 860 indigenous languages spoken (over one-tenth of the world's total)
note: Tok Pisin, a creole language, is widely used and understood; English is spoken by 1%-2%; Hiri Motu is spoken by less than 2%
Literacy: *definition:* age 15 and over can read and write
total population: 57.3%
male: 63.4%
female: 50.9% (2000 census)
School life expectancy (primary to tertiary education): NA
Education expenditures: NA
People—note: the indigenous population of Papua New Guinea is one of the most heterogeneous in the world; PNG has several thousand separate communities, most with only a few hundred people; divided by language, customs, and tradition, some of these communities have engaged in low-scale tribal conflict with their neighbors for millennia; the advent of modern weapons and modern migrants into urban areas has greatly magnified the impact of this lawlessness

GOVERNMENT

Country name: *conventional long form:* Independent State of Papua New Guinea
conventional short form: Papua New Guinea
local short form: Papuaniugini
former: Territory of Papua and New Guinea
abbreviation: PNG
Government type: constitutional parliamentary democracy and a Commonwealth realm
Capital: *name:* Port Moresby
geographic coordinates: 9 30 S, 147 10 E
time difference: UTC+10 (15 hours ahead of Washington, DC during Standard Time)
Administrative divisions: 18 provinces, 1 autonomous region*, and 1 district**; Bougainville*, Central, Chimbu, Eastern Highlands, East New Britain, East Sepik, Enga, Gulf, Madang, Manus, Milne Bay, Morobe, National Capital**, New Ireland, Northern, Sandaun, Southern Highlands, Western, Western Highlands, West New Britain
Independence: 16 September 1975 (from the Australian-administered UN trusteeship)
National holiday: Independence Day, 16 September (1975)
Constitution: 16 September 1975
Legal system: mixed legal system of English common law and customary law
International law organization participation: has not submitted an ICJ jurisdiction declaration; non-party state to the ICCt
Suffrage: 18 years of age; universal
Executive branch: *chief of state:* Queen ELIZABETH II (since 6 February 1952); represented by Governor Michael OGIO (since 25 February 2011)
head of government: Prime Minister Sir Michael SOMARE (since 2 August 2002); Sam ABAL acting
cabinet: National Executive Council appointed by the governor general on the recommendation of the prime minister
(For more information visit the World Leaders website)
elections: the monarchy is hereditary; the governor general nominated by parliament and appointed by the chief of state; following legislative elections, the leader of the majority party or leader of the majority coalition usually appointed prime minister by the governor general acting in accordance with a decision of the parliament
Legislative branch: unicameral National Parliament (109 seats, 89 filled from open electorates and 20 from provinces and national capital district; members elected by popular vote to serve five-year terms); constitution allows up to 126 seats
elections: last held from 30 June to 10 July 2007; next to be held in June 2012
election results: percent of vote by party—NA; seats by party—NA 27, PNGP 8, PAP 6, URP 6, PANGU PATI 5, PDM 5, independents 19, others 33; note—election to 1 seat was nullified
note: 15 other parties won 4 or fewer seats; association with political parties is fluid
Judicial branch: Supreme Court (the chief justice is appointed by the governor general on the proposal of the National Executive Council after consultation with the minister responsible for justice; other judges are appointed by the Judicial and Legal Services Commission)
Political parties and leaders: National Alliance Party or NA [Michael SOMARE]; Papua and Niugini Union Party or PANGU PATI [Andrew KUMBAKOR]; Papua New Guinea Party or PNGP [Beldan NEMAH]; People's Action Party or PAP [Gabriel KAPRIS]; People's Democratic Movement or PDM; United Resources Party or URP [William DUMA]
Political pressure groups and leaders: Ahora [Andrew MAMOKO] (represents local tribes); Centre for Environment Law and Community Rights or Celcor [Damien ASE]; Community Coalition Against Corruption
International organization participation: ACP, ADB, AOSIS, APEC, ARF, ASEAN (observer), C, CP, FAO, G-77, IBRD, ICAO, ICRM, IDA, IFAD, IFC, IFRCS, IHO, ILO, IMF, IMO, Interpol, IOC, IOM (observer), IPU, ISO (correspondent), ITSO, ITU, MIGA, NAM, OPCW, PIF, Sparteca, SPC, UN, UNCTAD, UNESCO, UNIDO, UNWTO, UPU, WCO, WFTU, WHO, WIPO, WMO, WTO
Diplomatic representation in the US: *chief of mission:* Ambassador Evan Jeremy PAKI
chancery: 1779 Massachusetts Avenue NW, Suite 805, Washington, DC 20036
telephone: [1] (202) 745-3680
FAX: [1] (202) 745-3679
Diplomatic representation from the US: *chief of mission:* Ambassador Teddy B. TAYLOR
embassy: Douglas Street, Port Moresby, N.C.D.
mailing address: 4240 Port Moresby PI, US Department of State, Washington DC 20521-4240
telephone: [675] 321-1455
FAX: [675] 321-3423
Flag description: divided diagonally from upper hoist-side corner; the upper triangle is red with a soaring yellow bird of paradise centered; the lower triangle is black with five, white, five-pointed stars of the Southern Cross constellation centered; red, black, and yellow are traditional colors of Papua New Guinea; the bird of paradise—endemic to the island of New Guinea—is an emblem of regional tribal culture and represents the emergence of Papua New Guinea as a nation; the Southern Cross, visible in the night sky, symbolizes Papua New Guinea's connection with Australia and several other countries in the South Pacific
National anthem: *name:* "O Arise All You Sons"
lyrics/music: Thomas SHACKLADY
note: adopted 1975

ECONOMY

Economy—overview: Papua New Guinea (PNG) is richly endowed with natural resources, but exploitation has been hampered by rugged terrain, land tenure issues, and the high cost of developing infrastructure. The economy has a small formal sector, focused mainly on the export of those natural resources, and an informal sector, employing the majority of the population. Agriculture provides a subsistence livelihood for 85% of the people. Mineral deposits, including copper, gold, and oil, account for nearly two-thirds of export earnings. Natural gas reserves amount to an estimated 227 billion cubic meters. A consortium led by a major American oil company is constructing a liquefied natural gas (LNG) production facility that could begin exporting in 2014. As the largest investment project in the country's history, it has the potential to double GDP in the near-term and triple Papua New Guinea's export revenue. An American-owned firm also opened PNG's first oil refinery in 2004 and is building a second LNG production facility. The government faces the challenge of ensuring transparency and accountability for revenues flowing from this and other large LNG projects. The government of Prime Minister SOMARE has expended much of its energy remaining in power. He was the first prime minister ever to serve a full five-year term. The government has brought stability to the national budget, largely through expenditure control; however, it relaxed spending constraints in 2006 and 2007 as elections approached. In recent years, the government has opened up markets in telecommunications and air transport, making both more affordable to the people. Numerous challenges still face the government, including providing physical security for foreign investors, regaining investor confidence, restoring integrity to state institutions, promoting economic efficiency by privatizing moribund state institutions, and balancing relations with Australia, its former colonial ruler. Other socio-cultural challenges could upend the economy including an HIV/AIDS epidemic, with the second highest infection rate in all of East Asia and the Pacific, and chronic law and order and land tenure

issues. The global financial crisis had little impact because of continued foreign demand for PNG's commodities.
GDP (purchasing power parity): $14.95 billion (2010 est.)
country comparison to the world: 136
$13.97 billion (2009 est.)
$13.24 billion (2008 est.)
note: data are in 2010 US dollars
GDP (official exchange rate): $9.668 billion (2010 est.)
GDP—real growth rate: 7% (2010 est.)
country comparison to the world: 37
5.5% (2009 est.)
6.6% (2008 est.)
GDP—per capita (PPP): $2,500 (2010 est.)
country comparison to the world: 180
$2,400 (2009 est.)
$2,300 (2008 est.)
note: data are in 2010 US dollars
GDP—composition by sector:
agriculture: 32.2%
industry: 35.7%
services: 32.1% (2010 est.)
Labor force: 3.809 million (2010 est.)
country comparison to the world: 92
Labor force—by occupation:
agriculture: 85%
industry: NA%
services: NA% (2005 est.)
Unemployment rate: 1.8% (2004)
country comparison to the world: 10
Population below poverty line: 37% (2002 est.)
Household income or consumption by percentage share: *lowest 10%:* 1.7%
highest 10%: 40.5% (1996)
Distribution of family income—Gini index: 50.9 (1996)
country comparison to the world: 18
Investment (gross fixed): 17.3% of GDP (2010 est.)
country comparison to the world: 118
Budget:
revenues: $2.917 billion
expenditures: $2.765 billion (2010 est.)
Public debt: 27.8% of GDP (2010 est.)
country comparison to the world: 92
29.7% of GDP (2009 est.)
Inflation rate (consumer prices): 6.8% (2010 est.)
country comparison to the world: 169
6.9% (2009 est.)
Central bank discount rate: 6.92% (31 December 2009)
country comparison to the world: 53
7% (31 December 2008)
Commercial bank prime lending rate: 10.09% (31 December 2009 est.)
country comparison to the world: 97
9.2% (31 December 2008 est.)
Stock of narrow money: $2.551 billion (31 December 2010 est.)
country comparison to the world: 116
$2.263 billion (31 December 2009 est.)
Stock of broad money: $4.726 billion (31 December 2010 est.)
country comparison to the world: 122
$4.14 billion (31 December 2009 est.)
Stock of domestic credit: $2.796 billion (31 December 2010 est.)
country comparison to the world: 122
$2.424 billion (31 December 2009 est.)
Market value of publicly traded shares: $NA (31 December 2010)
country comparison to the world: 75
$6.632 billion
Agriculture—products: coffee, cocoa, copra, palm kernels, tea, sugar, rubber, sweet potatoes, fruit, vegetables, vanilla; shell fish; poultry, pork
Industries: copra crushing, palm oil processing, plywood production, wood chip production; mining of gold, silver, and copper; crude oil production, petroleum refining; construction, tourism
Industrial production growth rate: 10% (2010 est.)
country comparison to the world: 19
Electricity—production: 2.885 billion kWh (2007 est.)
country comparison to the world: 126
Electricity—consumption: 2.683 billion kWh (2007 est.)
country comparison to the world: 129
Electricity—exports: 0 kWh (2008 est.)
Electricity—imports: 0 kWh (2008 est.)
Oil—production: 35,090 bbl/day (2009 est.)
country comparison to the world: 68
Oil—consumption: 36,000 bbl/day (2009 est.)
country comparison to the world: 106
Oil—exports: 32,490 bbl/day (2007 est.)
country comparison to the world: 82
Oil—imports: 14,380 bbl/day (2007 est.)
country comparison to the world: 128
Oil—proved reserves: 170 million bbl (1 January 2010 est.)
country comparison to the world: 63
Natural gas—production: 100 million cu m (2008 est.)
country comparison to the world: 78
Natural gas—consumption: 100 million cu m (2008 est.)
country comparison to the world: 104
Natural gas—exports: 0 cu m (2008 est.)
country comparison to the world: 159
Natural gas—imports: 0 cu m (2008 est.)
country comparison to the world: 107
Natural gas—proved reserves: 226.5 billion cu m (1 January 2010 est.)
country comparison to the world: 44
Current account balance: $-99 million (2010 est.)
country comparison to the world: 79
$-446.4 million (2009 est.)
Exports:
$5.976 billion (2010 est.)
country comparison to the world: 104
$4.392 billion (2009 est.)
Exports—commodities: oil, gold, copper ore, logs, palm oil, coffee, cocoa, crayfish, prawns
Exports—partners: Australia 30.05%, Japan 7.48% (2009)
Imports: $3.547 billion (2010 est.)
country comparison to the world: 134
$2.871 billion (2009 est.)
Imports—commodities: machinery and transport equipment, manufactured goods, food, fuels, chemicals
Imports—partners: Australia 43.27%, China 13.29%, Singapore 9.59%, US 6.4%, Japan 4.62% (2009)
Reserves of foreign exchange and gold: $3.017 billion (31 December 2010 est.)
country comparison to the world: 90
$2.607 billion (31 December 2009 est.)
Debt—external: $1.548 billion (31 December 2010 est.)
country comparison to the world: 144
$1.436 billion (31 December 2009 est.)
Stock of direct foreign investment—at home: $NA
Stock of direct foreign investment—abroad: $NA
Exchange rates: kina (PGK) per US dollar–2.7517 (2010), 2.7551 (2009), 2.6956 (2008), 3.03 (2007), 3.0643 (2006)

COMMUNICATIONS

Telephones—main lines in use: 60,000 (2009)
country comparison to the world: 157
Telephones—mobile cellular: 900,000 (2009)
country comparison to the world: 147
Telephone system: *general assessment:* services are minimal; facilities provide radiotelephone and telegraph, coastal radio, aeronautical radio, and international radio communication services
domestic: access to telephone services is not widely available; combined fixed-line and mobile-cellular teledensity is about 15 per 100 persons
international: country code–675; submarine cables to Australia and Guam; satellite earth station–1 Intelsat (Pacific Ocean); international radio communication service (2009)
Broadcast media: 2 television stations, 1 commercial station operating since the late 1980s and 1 state-run station launched in 2008; satellite and cable TV services are available; state-run National Broadcasting Corporation operates 3 radio networks with multiple repeaters and about 20 provincial stations; several commercial radio stations with multiple transmission points as well as several community stations; transmissions of several international broadcasters are accessible (2009)
Internet country code: .pg
Internet hosts: 4,285 (2010)
country comparison to the world: 139
Internet users: 125,000 (2009)
country comparison to the world: 152

TRANSPORTATION

Airports: 562 (2010)
country comparison to the world: 12
Airports—with paved runways: *total:* 21
2,438 to 3,047 m: 2
1,524 to 2,437 m: 14
914 to 1,523 m: 4
under 914 m: 1 (2010)
Airports—with unpaved runways: *total:* 541
1,524 to 2,437 m: 9
914 to 1,523 m: 63
under 914 m: 469 (2010)
Heliports: 2 (2010)
Pipelines: oil 195 km (2010)
Roadways: *total:* 9,349 km
country comparison to the world: 136
paved: 3,000 km
unpaved: 6,349 km (2011)
Waterways: 11,000 km (2011)
country comparison to the world: 12
Merchant marine: *total:* 28
country comparison to the world: 87
by type: bulk carrier 2, cargo 24, petroleum tanker 2
foreign-owned: 7 (Malaysia 1, UAE 6) (2010)
Ports and terminals: Kimbe, Lae, Madang, Rabaul, Wewak

MILITARY

Military branches: Papua New Guinea Defense Force (PNGDF; includes Maritime Operations Element, Air Operations Element) (2009)
Military service age and obligation: 16 years of age for voluntary military service (with parental consent); no conscription (2010)
Manpower available for military service: *males age 16–49:* 1,568,210
females age 16–49: 1,478,965 (2010 est.)
Manpower fit for military service: *males age 16–49:* 1,130,951
females age 16–49: 1,137,753 (2010 est.)
Manpower reaching militarily significant age annually: *male:* 67,781
female: 65,820 (2010 est.)
Military expenditures: 1.4% of GDP (2005 est.)
country comparison to the world: 109

TRANSNATIONAL ISSUES

Disputes—international: relies on assistance from Australia to keep out illegal cross-border activities from primarily Indonesia, including goods smuggling, illegal narcotics trafficking, and squatters and secessionists
Refugees and internally displaced persons: refugees (country of origin): 10,177 (Indonesia) (2007)
Trafficking in persons: current situation: Papua New Guinea is a country of destination for women and children from Malaysia, the Philippines, Thailand, and China trafficked for the purpose of commercial sexual exploitation; internal trafficking of women and children for the purposes of sexual exploitation and involuntary domestic servitude occurs as well
tier rating: Tier 3—Papua New Guinea does not fully comply with the minimum standards for the elimination of trafficking and is not making significant efforts to do so; the current legal framework does not contain elements of crimes that characterize trafficking; the government lacks victim protection services or a systematic procedure to identify victims of trafficking; the government did not prosecute anyone in 2007 for trafficking; Papua New Guinea has not ratified the 2000 UN TIP Protocol (2008)
Illicit drugs: major consumer of cannabis

PARACEL ISLANDS

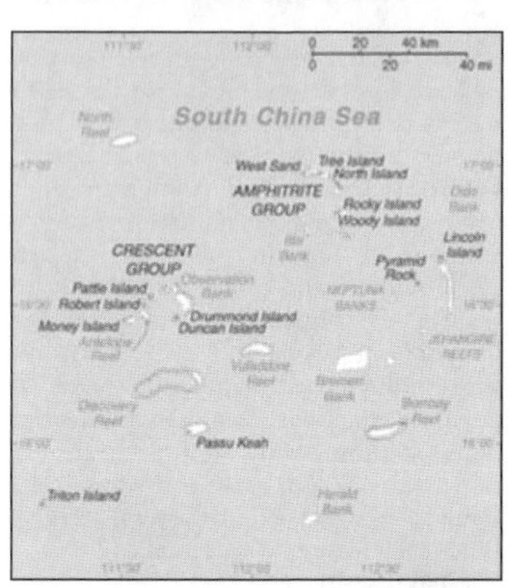

INTRODUCTION

Background: The Paracel Islands are surrounded by productive fishing grounds and by potential oil and gas reserves. In 1932, French Indochina annexed the islands and set up a weather station on Pattle Island; maintenance was continued by its successor, Vietnam. China has occupied the Paracel Islands since 1974, when its troops seized a South Vietnamese garrison occupying the western islands. China built a military installation on Woody Island with an airfield and artificial harbor. The islands also are claimed by Taiwan and Vietnam.

GEOGRAPHY

Location: Southeastern Asia, group of small islands and reefs in the South China Sea, about one-third of the way from central Vietnam to the northern Philippines
Geographic coordinates: 16 30 N, 112 00 E
Map references: Southeast Asia
Area: *total:* ca. 7.75 sq km
land: ca. 7.75 sq km
water: 0 sq km
Area—comparative: NA
Land boundaries: 0 km
Coastline: 518 km
Maritime claims: NA
Climate: tropical
Terrain: mostly low and flat
Elevation extremes: *lowest point:* South China Sea 0 m
highest point: unnamed location on Rocky Island 14 m
Natural resources: none
Land use: *arable land:* 0%
permanent crops: 0%
other: 100% (2005)
Irrigated land: 0 sq km
Natural hazards: typhoons
Environment—current issues: NA
Geography—note: composed of 130 small coral islands and reefs divided into the northeast Amphitrite Group and the western Crescent Group

PEOPLE

Population: no indigenous inhabitants
note: there are scattered Chinese garrisons
School life expectancy (primary to tertiary education): NA

GOVERNMENT

Country name: *conventional long form:* none
conventional short form: Paracel Islands

ECONOMY

Economy—overview: The islands have the potential for oil and gas development. Waters around the islands support commercial fishing, but the islands themselves are not populated on a permanent basis.

TRANSPORTATION

Airports: 1 (2010)
country comparison to the world: 229
Airports—with paved runways: *total:* 1
1,524 to 2,437 m: 1 (2010)
Ports and terminals: small Chinese port facilities on Woody Island and Duncan Island

MILITARY

Military—note: occupied by China

TRANSNATIONAL ISSUES

Disputes—international: occupied by China, also claimed by Taiwan and Vietnam

PARAGUAY

INTRODUCTION

Background: Paraguay achieved its independence from Spain in 1811. In the disastrous War of the Triple Alliance (1865-70)—between Paraguay and Argentina, Brazil, and Uruguay—Paraguay lost two-thirds of all adult males and much of its territory. The country stagnated economically for the next half century. Following the Chaco War of 1932-35 with Bolivia, Paraguay gained a large part of the Chaco lowland region. The 35-year military dictatorship of Alfredo STROESSNER ended in 1989, and, despite a marked increase in political infighting in recent years, Paraguay has held relatively free and regular presidential elections since then.

GEOGRAPHY

Location: Central South America, northeast of Argentina, southwest of Brazil
Geographic coordinates: 23 00 S, 58 00 W
Map references: South America
Area: *total:* 406,752 sq km
country comparison to the world: 59
land: 397,302 sq km
water: 9,450 sq km
Area—comparative: slightly smaller than California
Land boundaries: *total:* 3,995 km
border countries: Argentina 1,880 km, Bolivia 750 km, Brazil 1,365 km
Coastline: 0 km (landlocked)
Maritime claims: none (landlocked)
Climate: subtropical to temperate; substantial rainfall in the eastern portions, becoming semiarid in the far west
Terrain: grassy plains and wooded hills east of Rio Paraguay; Gran Chaco region west

of Rio Paraguay mostly low, marshy plain near the river, and dry forest and thorny scrub elsewhere

Elevation extremes: *lowest point:* junction of Rio Paraguay and Rio Parana 46 m
highest point: Cerro Pero 842 m

Natural resources: hydropower, timber, iron ore, manganese, limestone

Land use: *arable land:* 7.47%
permanent crops: 0.24%
other: 92.29% (2005)

Irrigated land: 670 sq km (2003)

Total renewable water resources: 336 cu km (2000)

Freshwater withdrawal (domestic/industrial/agricultural): *total:* 0.49 cu km/yr (20%/8%/71%)
per capita: 80 cu m/yr (2000)

Natural hazards: local flooding in southeast (early September to June); poorly drained plains may become boggy (early October to June)

Environment—current issues: deforestation; water pollution; inadequate means for waste disposal pose health risks for many urban residents; loss of wetlands

Environment—international agreements: *party to:* Biodiversity, Climate Change, Climate Change-Kyoto Protocol, Desertification, Endangered Species, Hazardous Wastes, Law of the Sea, Ozone Layer Protection, Wetlands
signed, but not ratified: none of the selected agreements

Geography—note: landlocked; lies between Argentina, Bolivia, and Brazil; population concentrated in southern part of country

PEOPLE

Population: 6,459,058 (July 2011 est.)
country comparison to the world: 104

Age structure: *0–14 years:* 28.5% (male 936,298/female 905,285)
15–64 years: 65.4% (male 2,121,632/female 2,100,740)
65 years and over: 6.1% (male 183,440/female 211,663) (2011 est.)

Median age: *total:* 25.4 years
male: 25.1 years
female: 25.6 years (2011 est.)

Population growth rate: 1.284% (2011 est.)
country comparison to the world: 92

Birth rate: 17.48 births/1,000 population (2011 est.)
country comparison to the world: 114

Death rate: 4.57 deaths/1,000 population (July 2011 est.)
country comparison to the world: 199

Net migration rate: -0.08 migrant(s)/1,000 population (2011 est.)
country comparison to the world: 119

Urbanization: *urban population:* 61% of total population (2010)
rate of urbanization: 2.5% annual rate of change (2010-15 est.)

Major cities—population: ASUNCION (capital) 1.977 million (2009)

Sex ratio: *at birth:* 1.05 male(s)/female
under 15 years: 1.03 male(s)/female
15–64 years: 1.01 male(s)/female
65 years and over: 0.86 male(s)/female
total population: 1.01 male(s)/female (2011 est.)

Infant mortality rate: *total:* 23.02 deaths/1,000 live births
country comparison to the world: 86
male: 26.94 deaths/1,000 live births
female: 18.91 deaths/1,000 live births (2011 est.)

Life expectancy at birth: *total population:* 76.19 years
country comparison to the world: 74
male: 73.59 years
female: 78.93 years (2011 est.)

Total fertility rate: 2.11 children born/woman (2011 est.)
country comparison to the world: 117

HIV/AIDS—adult prevalence rate: 0.3% (2009 est.)
country comparison to the world: 79

HIV/AIDS—people living with HIV/AIDS: 13,000 (2009 est.)
country comparison to the world: 91

HIV/AIDS—deaths: fewer than 500 (2009 est.)
country comparison to the world: 83

Major infectious diseases: *degree of risk:* intermediate
food or waterborne diseases: bacterial diarrhea, hepatitis A, and typhoid fever
vectorborne disease: dengue fever and malaria (2009)

Drinking water source: *Improved:*
urban: 99% of population
rural: 66% of population
total: 86% of population
Unimproved: urban: 1% of population
rural: 34% of population
total: 14% of population (2008)

Sanitation facility access: *Improved:*
urban: 90% of population
rural: 40% of population
total: 70% of population
Unimproved: urban: 10% of population
rural: 60% of population
total: 30% of population (2008)

Nationality: *noun:* Paraguayan(s)
adjective: Paraguayan

Ethnic groups: mestizo (mixed Spanish and Amerindian) 95%, other 5%

Religions: Roman Catholic 89.6%, Protestant 6.2%, other Christian 1.1%, other or unspecified 1.9%, none 1.1% (2002 census)

Languages: Spanish (official), Guarani (official)

Literacy: *definition:* age 15 and over can read and write
total population: 94%
male: 94.9%
female: 93% (2003 est.)

School life expectancy (primary to tertiary education): *total:* 12 years
male: 12 years
female: 12 years (2007)

Education expenditures: 4% of GDP (2008)
country comparison to the world: 106

GOVERNMENT

Country name: *conventional long form:* Republic of Paraguay
conventional short form: Paraguay
local long form: Republica del Paraguay
local short form: Paraguay

Government type: constitutional republic

Capital: *name:* Asuncion
geographic coordinates: 25 16 S, 57 40 W
time difference: UTC-4 (1 hour ahead of Washington, DC during Standard Time)

daylight saving time: +1hr, begins first Sunday in October; ends second Sunday in April

Administrative divisions: 17 departments (departamentos, singular—departamento) and 1 capital city*; Alto Paraguay, Alto Parana, Amambay, Asuncion*, Boqueron, Caaguazu, Caazapa, Canindeyu, Central, Concepcion, Cordillera, Guaira, Itapua, Misiones, Neembucu, Paraguari, Presidente Hayes, San Pedro

Independence: 14 May 1811 (from Spain)

National holiday: Independence Day, 14 May 1811 (observed 15 May)

Constitution: promulgated 20 June 1992

Legal system: civil law system with influences from Argentine, Spanish, Roman, and French civil law models; judicial review of legislative acts in Supreme Court of Justice

International law organization participation: accepts compulsory ICJ jurisdiction; accepts ICCt jurisdiction

Suffrage: 18 years of age; universal and compulsory up to age 75

Executive branch: *chief of state:* President Fernando Armindo LUGO Mendez (since 15 August 2008); Vice President Luis Federico FRANCO Gomez (since 15 August 2008); note—the president is both the chief of state and head of government
head of government: President Fernando Armindo LUGO Mendez (since 15 August 2008); Vice President Luis Federico FRANCO Gomez (since 15 August 2008)
cabinet: Council of Ministers appointed by the president
(For more information visit the World Leaders website)
elections: president and vice president elected on the same ticket by popular vote for a single five-year term; election last held on 20 April 2008 (next to be held in April 2013)
election results: Fernando Armindo LUGO Mendez elected president; percent of vote—Fernando Armindo LUGO Mendez 40.8%, Blanca OVELAR 30.6%, Lino OVIEDO 21.9%, Pedro FADUL 2.4%, other 4.3%

Legislative branch: bicameral National Congress or Congreso Nacional consists of the Chamber of Senators or Camara de Senadores (45 seats; members elected by popular vote to serve five-year terms) and

the Chamber of Deputies or Camara de Diputados (80 seats; members elected by popular vote to serve five-year terms)
elections: Chamber of Senators—last held on 20 April 2008 (next to be held in April 2013); Chamber of Deputies—last held on 20 April 2008 (next to be held in April 2013)
election results: Chamber of Senators—percent of vote by party—NA; seats by party—ANR 15, PLRA 14, UNACE 9, PPQ 4, other 3; Chamber of Deputies—percent of vote by party—NA; seats by party—ANR 30, PLRA 27, UNACE 15, PPQ 3, APC 2, other 3; note—as of 1 January 2010, the composition of the Chamber of Deputies is ANR 30, PLRA 29, UNACE 15, PPQ 4, other 2
Judicial branch: Supreme Court of Justice or Corte Suprema de Justicia (nine judges proposed by the Council of Magistrates or Consejo de la Magistratura, and approved by the Senate and president)
Political parties and leaders: Alianza Patriotica por el Cambio (Patriotic Alliance for Change) or APC [Fernando LUGO]; Asociacion Nacional Republicana—Colorado Party or ANR [Lilian SAMANIEGO]; Movimiento Popolar Tekojoja or Tekojoja [Sixto PEREIRA]; Movimiento Union Nacional de Ciudadanos Eticos or UNACE [Lino Cesar OVIEDO Silva]; Patria Querida (Beloved Fatherland Party) or PPQ [Pedro Nicolas Maraa FADUL Niella]; Partido del Movimiento al Socialismo or P-MAS [Camilo Ernesto SOARES Machado]; Partido Democratica Progresista or PDP [Rafael Augusto FILIZZOLA Serra]; Partido Encuentro Nacional or PEN [Fernando CAMACHO Paredes]; Partido Liberal Radical Autentico or PLRA [Amanda NUNEZ]; Partido Pais Solidario or PPS [Carlos Alberto FILIZZOLA Pallares]
Political pressure groups and leaders: Ahorristas Estafados or AE; National Coordinating Board of Campesino Organizations or MCNOC [Luis AGUAYO]; National Federation of Campesinos or FNC [Odilon ESPINOLA]; National Workers Central or CNT [Secretary General Juan TORRALES]; Paraguayan Workers Confederation or CPT; Roman Catholic Church; Unitary Workers Central or CUT [Jorge Guzman ALVARENGA Malgarejo]
International organization participation: CAN (associate), FAO, G-11, G-77, IADB, IAEA, IBRD, ICAO, ICRM, IDA, IFAD, IFC, IFRCS, ILO, IMF, IMO, Interpol, IOC, IOM, IPU, ISO (correspondent), ITSO, ITU, ITUC, LAES, LAIA, Mercosur, MIGA, MINURSO, MINUSTAH, MONUSCO, NAM (observer), OAS, OPANAL, OPCW, PCA, RG, UN, UNASUR, UNCTAD, UNESCO, UNFICYP, UNIDO, Union Latina, UNMIL, UNMIS, UNOCI, UNWTO, UPU, WCO, WHO, WIPO, WMO, WTO
Diplomatic representation in the US: *chief of mission:* Ambassador Rigoberto GAUTO Vielman
chancery: 2400 Massachusetts Avenue NW, Washington, DC 20008
telephone: [1] (202) 483-6960 through 6962
FAX: [1] (202) 234-4508
consulate(s) general: Kansas City (Kansas), Los Angeles, Miami, New York
Diplomatic representation from the US: Ambassador Liliana AYALDE
embassy: 1776 Avenida Mariscal Lopez, Casilla Postal 402, Asuncion
mailing address: Unit 4711, APO AA 34036-0001
telephone: [595] (21) 213-715
FAX: [595] (21) 228-603
Flag description: three equal, horizontal bands of red (top), white, and blue with an emblem centered in the white band; unusual flag in that the emblem is different on each side; the obverse (hoist side at the left) bears the national coat of arms (a yellow five-pointed star within a green wreath capped by the words REPUBLICA DEL PARAGUAY, all within two circles); the reverse (hoist side at the right) bears a circular seal of the treasury (a yellow lion below a red Cap of Liberty and the words PAZ Y JUSTICIA (Peace and Justice)); red symbolizes bravery and patriotism, white represents integrity and peace, and blue denotes liberty and generosity
note: the three color bands resemble those on the flag of the Netherlands; one of only three national flags that differ on their obverse and reverse sides—the others are Moldova and Saudi Arabia
National anthem: *name:* "Paraguayos, Republica o muerte!" (Paraguayans, The Republic or Death!)
lyrics/music: Francisco Esteban ACUNA de Figueroa/disputed
note: adopted 1934, in use since 1846; the anthem was officially adopted following its re-arrangement in 1934

ECONOMY

Economy—overview: Landlocked Paraguay has a market economy distinguished by a large informal sector, featuring re-export of imported consumer goods to neighboring countries, as well as the activities of thousands of microenterprises and urban street vendors. A large percentage of the population, especially in rural areas, derives its living from agricultural activity, often on a subsistence basis. Because of the importance of the informal sector, accurate economic measures are difficult to obtain. On a per capita basis, real income has stagnated at 1980 levels. The economy grew rapidly between 2003 and 2008 as growing world demand for commodities combined with high prices and favorable weather to support Paraguay's commodity-based export expansion. Paraguay is the sixth largest soy producer in the world. Drought hit in 2008, reducing agricultural exports and slowing the economy even before the onset of the global recession. The economy fell 3.8% in 2009, as lower world demand and commodity prices caused exports to contract. The government reacted by introducing fiscal and monetary stimulus packages. Growth resumed at a 14.5% level in 2010, the highest in South America. Political uncertainty, corruption, limited progress on structural reform, and deficient infrastructure are the main obstacles to growth.
GDP (purchasing power parity): $33.31 billion (2010 est.)
country comparison to the world: 105
$28.89 billion (2009 est.)
$30.05 billion (2008 est.)
note: data are in 2010 US dollars
GDP (official exchange rate): $18.48 billion (2010 est.)
GDP—real growth rate: 15.3% (2010 est.)
country comparison to the world: 2
-3.8% (2009 est.)
5.8% (2008 est.)
GDP—per capita (PPP): $5,200 (2010 est.)
country comparison to the world: 145
$4,600 (2009 est.)
$4,800 (2008 est.)
note: data are in 2010 US dollars
GDP—composition by sector: *agriculture:* 21.8%
industry: 18.2%
services: 60.1% (2010 est.)
Labor force: 3.038 million (2010 est.)
country comparison to the world: 103
Labor force—by occupation: *agriculture:* 26.5%
industry: 18.5%
services: 55% (2008)
Unemployment rate: 5.7% (2010 est.)
country comparison to the world: 58
6.4% (2009 est.)
Population below poverty line: 18.8% (2009 est.)
Household income or consumption by percentage share: *lowest 10%:* 1.1%
highest 10%: 42.3% (2007)
Distribution of family income—Gini index: 53.2 (2009)
country comparison to the world: 14
57.7 (1998)
Investment (gross fixed): 17.8% of GDP (2010 est.)
country comparison to the world: 114
Budget: *revenues:* $3.238 billion
expenditures: $3.402 billion (2010 est.)
Public debt: 22.8% of GDP (2010 est.)
country comparison to the world: 106
24% of GDP (2009 est.)
Inflation rate (consumer prices): 7.2% (2010 est.)
country comparison to the world: 177
1.9% (2009 est.)
Central bank discount rate: 20% (31 December 2008)
country comparison to the world: 11
20% (31 December 2007)
Commercial bank prime lending rate: 28.26% (31 December 2009 est.)
country comparison to the world: 7
25.81% (31 December 2008 est.)
Stock of narrow money: $2.6 billion (31 December 2010 est.)
country comparison to the world: 114
$2.107 billion (31 December 2009 est.)
Stock of broad money: $5.03 billion (31 December 2010 est.)
country comparison to the world: 118
$4.057 billion (31 December 2009 est.)
Stock of domestic credit: $4.395 billion (31 December 2010 est.)
country comparison to the world: 113
$3.607 billion (31 December 2009 est.)

Market value of publicly traded shares: $NA (31 December 2010)
country comparison to the world: 109
$409.1 million
Agriculture—products: cotton, sugarcane, soybeans, corn, wheat, tobacco, cassava (tapioca), fruits, vegetables; beef, pork, eggs, milk; timber
Industries: sugar, cement, textiles, beverages, wood products, steel, metallurgic, electric power
Industrial production growth rate: 6.5% (2010 est.)
country comparison to the world: 52
Electricity—production: 53.19 billion kWh (2007 est.)
country comparison to the world: 47
Electricity—consumption: 8.5 billion kWh (2009 est.)
country comparison to the world: 92
Electricity—exports: 45.14 billion kWh (2007 est.)
Electricity—imports: 0 kWh (2008 est.)
Oil—production: 30.96 bbl/day (2009 est.)
country comparison to the world: 112
Oil—consumption: 27,000 bbl/day (2009 est.)
country comparison to the world: 114
Oil—exports: 0 bbl/day (2009 est.)
country comparison to the world: 195
Oil—imports: 25,100 bbl/day (2009 est.)
country comparison to the world: 105
Oil—proved reserves: 0 bbl (1 January 2010 est.)
country comparison to the world: 175
Natural gas—production: 0 cu m (2009 est.)
country comparison to the world: 172
Natural gas—consumption: 0 cu m (2009 est.)
country comparison to the world: 114
Natural gas—exports: 0 cu m (2009 est.)
country comparison to the world: 155
Natural gas—imports: 0 cu m (2009 est.)
country comparison to the world: 103
Natural gas—proved reserves: 0 cu m (1 January 2010 est.)
country comparison to the world: 177
Current account balance: $-391 million (2010 est.)
country comparison to the world: 110
$-149.2 million (2009 est.)
Exports: $7.972 billion (2010 est.)
country comparison to the world: 92
$5.735 billion (2009 est.)
Exports—commodities: soybeans, feed, cotton, meat, edible oils, electricity, wood, leather
Exports—partners: Brazil 21%, Uruguay 17%, Chile 12%, Argentina 11%, Russia 4% (2009 est.)
Imports: $9.567 billion (2010 est.)
country comparison to the world: 90
$6.802 billion (2009 est.)
Imports—commodities: road vehicles, consumer goods, tobacco, petroleum products, electrical machinery, tractors, chemicals, vehicle parts
Imports—partners: China 30%, Brazil 23%, Argentina 16%, US 5% (2009 est.)
Reserves of foreign exchange and gold: $4.13 billion (31 December 2010 est.)
country comparison to the world: 76
$3.862 billion (31 December 2009 est.)
Debt—external: $2.445 billion (31 December 2010 est.)
country comparison to the world: 135
$2.388 billion (31 December 2009 est.)
Stock of direct foreign investment—at home: $3.393 million (31 December 2010)
country comparison to the world: 90
$3.053 million (31 December 2009)
Stock of direct foreign investment—abroad: $NA
Exchange rates: guarani (PYG) per US dollar–4,767.6 (2010), 4,965.4 (2009), 4,337.7 (2008), 5,031 (2007), 5,672.8 (2006)

COMMUNICATIONS

Telephones—main lines in use: 387,300 (2009)
country comparison to the world: 103
Telephones—mobile cellular: 5.619 million (2009)
country comparison to the world: 89
Telephone system: *general assessment:* the fixed-line market is a state monopoly and fixed-line telephone service is meager; principal switching center is in Asuncion
domestic: deficiencies in provision of fixed-line service have resulted in a rapid expansion of mobile-cellular services fostered by competition among multiple providers
international: country code–595; satellite earth station–1 Intelsat (Atlantic Ocean) (2008)
Broadcast media: 6 privately-owned TV stations; about 75 commercial and community radio stations broadcasting; 1 state-owned radio network (2010)
Internet country code: .py
Internet hosts: 167,281 (2010)
country comparison to the world: 69
Internet users: 1.105 million (2009)
country comparison to the world: 94

TRANSPORTATION

Airports: 800 (2010)
country comparison to the world: 9
Airports—with paved runways: *total:* 15
over 3,047 m: 3
1,524 to 2,437 m: 7
914 to 1,523 m: 5 (2010)
Airports—with unpaved runways: *total:* 785
1,524 to 2,437 m: 25
914 to 1,523 m: 290
under 914 m: 470 (2010)
Railways: *total:* 36 km
country comparison to the world: 132
standard gauge: 36 km 1.435-m gauge (2008)
Roadways: *total:* 29,500 km
country comparison to the world: 97
paved: 14,986 km
unpaved: 14,514 km (2000)
Waterways: 3,100 km (primarily on the Paraguay and Paraná river systems) (2010)
country comparison to the world: 32
Merchant marine: *total:* 23
country comparison to the world: 96
by type: cargo 15, carrier 1, container 2, passenger 1, petroleum tanker 3, roll on/roll off 1
foreign-owned: 6 (Argentina 5, Netherlands 1) (2010)
Ports and terminals: Asuncion, Villeta, San Antonio, Encarnacion

MILITARY

Military branches: Army, National Navy (Armada Nacional, includes Marine Corps, Naval Aviation), Air Force (Fuerza Aerea Paraguay, FAP) (2010)
Military service age and obligation: 18 years of age for compulsory and voluntary military service; conscript service obligation–12 months for Army, 24 months for Navy; volunteers for the Air Force must be younger than 22 years of age with a secondary school diploma (2010)
Manpower available for military service: *males age 16–49:* 1,678,335
females age 16–49: 1,675,352 (2010 est.)
Manpower fit for military service: *males age 16–49:* 1,409,859
females age 16–49: 1,433,037 (2010 est.)
Manpower reaching militarily significant age annually: *male:* 73,367
female: 71,801 (2010 est.)
Military expenditures: 1% of GDP (2006 est.)
country comparison to the world: 128

TRANSNATIONAL ISSUES

Disputes—international: unruly region at convergence of Argentina-Brazil-Paraguay borders is locus of money laundering, smuggling, arms and illegal narcotics trafficking, and fundraising for extremist organizations
Illicit drugs: major illicit producer of cannabis, most or all of which is consumed in Brazil, Argentina, and Chile; transshipment country for Andean cocaine headed for Brazil, other Southern Cone markets, and Europe; weak border controls, extensive corruption and money-laundering activity, especially in the Tri-Border Area; weak anti-money-laundering laws and enforcement

PERU

INTRODUCTION

Background: Ancient Peru was the seat of several prominent Andean civilizations, most notably that of the Incas whose empire was captured by the Spanish conquistadors in 1533. Peruvian independence was declared in 1821, and remaining Spanish forces were defeated in 1824. After a dozen years of military rule, Peru returned to democratic leadership in 1980, but experienced economic problems and the growth of a violent insurgency. President Alberto FUJIMORI's election in 1990 ushered in a decade that saw a dramatic turnaround in the economy and significant progress in curtailing guerrilla activity. Nevertheless, the president's increasing reliance on authoritarian measures and an economic

slump in the late 1990s generated mounting dissatisfaction with his regime, which led to his ouster in 2000. A caretaker government oversaw new elections in the spring of 2001, which ushered in Alejandro TOLEDO Manrique as the new head of government–Peru's first democratically elected president of Native American ethnicity. The presidential election of 2006 saw the return of Alan GARCIA Perez who, after a disappointing presidential term from 1985 to 1990, has overseen a robust macroeconomic performance.

GEOGRAPHY

Location: Western South America, bordering the South Pacific Ocean, between Chile and Ecuador
Geographic coordinates: 10 00 S, 76 00 W
Map references: South America
Area: *total:* 1,285,216 sq km
country comparison to the world: 20
land: 1,279,996 sq km
water: 5,220 sq km
Area—comparative: slightly smaller than Alaska
Land boundaries: *total:* 7,461 km
border countries: Bolivia 1,075 km, Brazil 2,995 km, Chile 171 km, Colombia 1,800 km, Ecuador 1,420 km
Coastline: 2,414 km
Maritime claims: territorial sea: 200 nm
continental shelf: 200 nm
Climate: varies from tropical in east to dry desert in west; temperate to frigid in Andes
Terrain: western coastal plain (costa), high and rugged Andes in center (sierra), eastern lowland jungle of Amazon Basin (selva)
Elevation extremes: *lowest point:* Pacific Ocean 0 m
highest point: Nevado Huascaran 6,768 m
Natural resources: copper, silver, gold, petroleum, timber, fish, iron ore, coal, phosphate, potash, hydropower, natural gas
Land use: *arable land:* 2.88%
permanent crops: 0.47%
other: 96.65% (2005)
Irrigated land: 12,000 sq km (2003)
Total renewable water resources: 1,913 cu km (2000)
Freshwater withdrawal (domestic/industrial/agricultural): *total:* 20.13 cu km/yr (8%/10%/82%)
per capita: 720 cu m/yr (2000)
Natural hazards: earthquakes, tsunamis, flooding, landslides, mild volcanic activity
volcanism: Peru experiences volcanic activity in the Andes Mountains; Ubinas (elev. 5,672 m), which last erupted in 2009, is the country's most active volcano; other historically active volcanoes include El Misti, Huaynaputina, Sabancaya, and Yucamane
Environment—current issues: deforestation (some the result of illegal logging); overgrazing of the slopes of the costa and sierra leading to soil erosion; desertification; air pollution in Lima; pollution of rivers and coastal waters from municipal and mining wastes
Environment—international agreements: *party to:* Antarctic-Environmental Protocol, Antarctic-Marine Living Resources, Antarctic Treaty, Biodiversity, Climate Change, Climate Change-Kyoto Protocol, Desertification, Endangered Species, Hazardous Wastes, Marine Dumping, Ozone Layer Protection, Ship Pollution, Tropical Timber 83, Tropical Timber 94, Wetlands, Whaling
signed, but not ratified: none of the selected agreements
Geography—note: shares control of Lago Titicaca, world's highest navigable lake, with Bolivia; a remote slope of Nevado Mismi, a 5,316 m peak, is the ultimate source of the Amazon River

PEOPLE

Population: 29,248,943 (July 2011 est.)
country comparison to the world: 42
Age structure: *0–14 years:* 28.5% (male 4,245,023/female 4,101,220)
15–64 years: 65.1% (male 9,316,128/female 9,722,258)
65 years and over: 6.4% (male 885,703/female 978,611) (2011 est.)
Median age: *total:* 26.2 years
male: 25.5 years
female: 26.8 years (2011 est.)
Population growth rate: 1.029% (2011 est.)
country comparison to the world: 115
Birth rate: 19.41 births/1,000 population (2011 est.)
country comparison to the world: 95
Death rate: 5.93 deaths/1,000 population (July 2011 est.)
country comparison to the world: 166
Net migration rate: -3.2 migrant(s)/1,000 population (2011 est.)
country comparison to the world: 176
Urbanization: *urban population:* 77% of total population (2010)
rate of urbanization: 1.6% annual rate of change (2010-15 est.)
Major cities—population: LIMA (capital) 8.769 million; Arequipa 778,000 (2009)
Sex ratio: *at birth:* 1.046 male(s)/female
under 15 years: 1.04 male(s)/female
15–64 years: 1.01 male(s)/female
65 years and over: 0.89 male(s)/female
total population: 1.01 male(s)/female (2011 est.)
Infant mortality rate: *total:* 22.18 deaths/1,000 live births
country comparison to the world: 90
male: 24.49 deaths/1,000 live births
female: 19.77 deaths/1,000 live births (2011 est.)
Life expectancy at birth: *total population:* 72.47 years
country comparison to the world: 127
male: 70.55 years
female: 74.48 years (2011 est.)
Total fertility rate: 2.32 children born/woman (2011 est.)
country comparison to the world: 97
HIV/AIDS—adult prevalence rate: 0.4% (2009 est.)
country comparison to the world: 77
HIV/AIDS—people living with HIV/AIDS: 75,000 (2009 est.)
country comparison to the world: 49
HIV/AIDS—deaths: 5,000 (2009 est.)
country comparison to the world: 38
Major infectious diseases:
degree of risk: very high
food or waterborne diseases: bacterial, hepatitis A, and typhoid fever
vectorborne disease: dengue fever, malaria, and yellow fever
water contact disease: leptospirosis (2009)
Drinking water source: *Improved:*
urban: 90% of population
rural: 61% of population
total: 82% of population
Unimproved: urban: 10% of population
rural: 39% of population
total: 18% of population (2008)
Sanitation facility access: *Improved:*
urban: 81% of population
rural: 36% of population
total: 68% of population
Unimproved: urban: 19% of population
rural: 64% of population
total: 32% of population (2008)
Nationality: *noun:* Peruvian(s)
adjective: Peruvian
Ethnic groups: Amerindian 45%, mestizo (mixed Amerindian and white) 37%, white 15%, black, Japanese, Chinese, and other 3%
Religions: Roman Catholic 81.3%, Evangelical 12.5%, other 3.3%, unspecified or none 2.9% (2007 Census)
Languages: Spanish (official) 84.1%, Quechua (official) 13%, Aymara 1.7%, Ashaninka 0.3%, other native languages (includes a large number of minor Amazonian languages) 0.7%, other 0.2% (2007 Census)
Literacy: *definition:* age 15 and over can read and write
total population: 92.9%
male: 96.4%
female: 89.4% (2007 Census)
School life expectancy (primary to tertiary education): *total:* 14 years
male: 13 years
female: 13 years (2006)
Education expenditures: 2.7% of GDP (2008)
country comparison to the world: 144

GOVERNMENT

Country name: *conventional long form:* Republic of Peru
conventional short form: Peru
local long form: Republica del Peru
local short form: Peru
Government type: constitutional republic
Capital: *name:* Lima
geographic coordinates: 12 03 S, 77 03 W
time difference: UTC-5 (same time as Washington, DC during Standard Time)

Administrative divisions: 25 regions (regiones, singular–region) and 1 province* (provincia); Amazonas, Ancash, Apurimac, Arequipa, Ayacucho, Cajamarca, Callao, Cusco, Huancavelica, Huanuco, Ica, Junin, La Libertad, Lambayeque, Lima, Lima*, Loreto, Madre de Dios, Moquegua, Pasco, Piura, Puno, San Martin, Tacna, Tumbes, Ucayali
Independence: 28 July 1821 (from Spain)
National holiday: Independence Day, 28 July (1821)
Constitution: 29 December 1993
Legal system: civil law system
International law organization participation: accepts compulsory ICJ jurisdiction with reservations; accepts ICCt jurisdiction
Suffrage: 18 years of age; universal and compulsory until the age of 70
Executive branch: *chief of state:* President Alan GARCIA Perez (since 28 July 2006); First Vice President Luis GIAMPIETRI Rojas (since 28 July 2006); Second Vice President Lourdes MENDOZA del Solar (since 28 July 2006); note–the president is both the chief of state and head of government
head of government: President Alan GARCIA Perez (since 28 July 2006); First Vice President Luis GIAMPIETRI Rojas (since 28 July 2006); Second Vice President Lourdes MENDOZA del Solar (since 28 July 2006)
note: Prime Minister Rosario FERNANDEZ Figueroa (since 18 March 2011) does not exercise executive power; this power rests with the president
cabinet: Council of Ministers appointed by the president
(For more information visit the World Leaders website)
elections: president elected by popular vote for a five-year term (eligible for nonconsecutive reelection); presidential and congressional elections last held on 9 April 2006 with runoff election held on 4 June 2006; next to be held in 10 April 2011
election results: Alan GARCIA Perez elected president in runoff election; percent of vote–Alan GARCIA Perez 52.5%, Ollanta HUMALA Tasso 47.5%
Legislative branch: unicameral Congress of the Republic of Peru or Congreso de la Republica del Peru (120 seats; members are elected by popular vote to serve five-year terms)
elections: last held on 9 April 2006 (next to be held in April 2011)
election results: percent of vote by party–UPP 21.2%, PAP 20.6%, UN 15.3%, AF 13.1%, FC 7.1%, PP 4.1%, RN 4.0%, other 14.6%; seats by party–UPP 45, PAP 36, UN 17, AF 13, FC 5, PP 2, RN 2
Judicial branch: Supreme Court of Justice or Corte Suprema de Justicia (judges are appointed by the National Council of the Judiciary)
Political parties and leaders: Alliance For Progress (Alianza Para El Progreso) [Cesar ACUNA Peralta]; Alliance For The Future (Alianza Por El Futuro) or AF (a coalition of pro-FUJIMORI parties including Cambio 90, Nueva Mayoria, and Si Cumple); Central Front (Frente Del Centro) or FC (a coalition of Accion Popular, Somos Peru, and Coordinadora Nacional de Independientes) [Victor Andres GARCIA Belaunde]; National Renovation Party (Partido Renovacion Nacional) [Rafael REY]; National Restoration Party (Restauracion Nacional) or RN [Humberto LAY Sun]; National Solidarity Party (Partido Solidaridad Nacional) or SN [Luis CASTANEDA Lossio]; Peru Possible (Peru Posible) or PP [Alejandro TOLEDO Manrique]; Peruvian Aprista Party (Partido Aprista Peruano) or PAP [Alan GARCIA Perez] (also referred to by its original name Alianza Popular Revolucionaria Americana or APRA); Peruvian Nationalist Party (Partido Nacionalista Peruano) or PNP [Ollanta HUMALA Tasso]; Popular Christian Party (Partido Popular Cristiano) or PPC [Lourdes FLORES Nano]; Union for Peru (Union por el Peru) or UPP [Aldo ESTRADA Choque]
Political pressure groups and leaders: General Workers Confederation of Peru (Confederacion General de Trabajadores del Peru) or CGTP [Mario HUAMAN]; Shining Path (Sendero Luminoso) or SL [Abimael GUZMAN Reynoso (imprisoned), Victor QUISPE Palomino (top leader at-large)] (leftist guerrilla group)
International organization participation: APEC, CAN, FAO, G-15, G-24, G-77, IADB, IAEA, IBRD, ICAO, ICRM, IDA, IFAD, IFC, IFRCS, IHO, ILO, IMF, IMO, IMSO, Interpol, IOC, IOM, IPU, ISO, ITSO, ITU, ITUC, LAES, LAIA, Mercosur (associate), MIGA, MINUSTAH, MONUSCO, NAM, OAS, OPANAL, OPCW, PCA, RG, UN, UNASUR, UNCTAD, UNESCO, UNFICYP, UNIDO, Union Latina, UNMIL, UNMIS, UNOCI, UNWTO, UPU, WCO, WFTU, WHO, WIPO, WMO, WTO
Diplomatic representation in the US: *chief of mission:* Ambassador Luis VALDIVIESO Montano
chancery: 1700 Massachusetts Avenue NW, Washington, DC 20036
telephone: [1] (202) 833-9860 through 9869
FAX: [1] (202) 659-8124
consulate(s) general: Atlanta, Boston, Chicago, Dallas, Denver, Hartford, Houston, Los Angeles, Miami, New York, Paterson (New Jersey), San Francisco
Diplomatic representation from the US: *chief of mission:* Ambassador Rose M. LIKINS
embassy: Avenida La Encalada, Cuadra 17 s/n, Surco, Lima 33
mailing address: P. O. Box 1995, Lima 1; American Embassy (Lima), APO AA 34031-5000
telephone: [51] (1) 618-2000
FAX: [51] (1) 618-2397
Flag description: three equal, vertical bands of red (hoist side), white, and red with the coat of arms centered in the white band; the coat of arms features a shield bearing a vicuna (representing fauna), a cinchona tree (the source of quinine, signifying flora), and a yellow cornucopia spilling out coins (denoting mineral wealth); red recalls blood shed for independence, white symbolizes peace
National anthem: *name:* "Himno Nacional del Peru" (National Anthem of Peru)
lyrics/music: Jose DE LA TORRE Ugarte/ Jose Bernardo ALZEDO
note: adopted 1822; the song won a national contest for an anthem

ECONOMY

Economy—overview: Peru's economy reflects its varied geography–an arid coastal region, the Andes further inland, and tropical lands bordering Colombia and Brazil. Abundant mineral resources are found in the mountainous areas, and Peru's coastal waters provide excellent fishing grounds. The Peruvian economy grew by almost 6% per year during the period 2002-06, with a stable exchange rate and low inflation. Growth jumped to nearly 9% per year in 2007 and 10% in 2008, driven by private investment and government spending, but then fell to less than 1% in 2009 in the face of the world recession, a sharp fall of private investment, and a substantial increase in counter-cyclical government spending. Growth resumed in 2010 at above 8%, due partly to a leap in private investment and continued high government spending. Peru's rapid expansion coupled with the government's conditional cash transfers and other programs have helped to reduce the national poverty rate by over 19 percentage points since 2002, though underemployment remains high. Inflation in 2010 was within the Central Bank's 1%-3% target range. Despite Peru's strong macroeconomic performance, dependence on minerals and metals exports and imported foodstuffs subjects the economy to fluctuations in world prices. Poor infrastructure hinders the spread of growth to Peru's non-coastal areas. A growing number of Peruvians are sharing in the benefits of growth but despite President GARCIA's pursuit of sound trade and macroeconomic policies, inequality persists. Nevertheless, he remains committed to Peru's free-trade path. Since 2006, Peru has signed trade deals with the United States, Canada, Singapore, China, Korea, and Japan, concluded negotiations with the European Free Trade Association (EFTA) and Chile, and begun trade talks with Central American countries and others. The US-Peru Trade Promotion Agreement (PTPA) entered into force 1 February 2009, opening the way to greater trade and investment between the two economies. Rising world prices of foodstuffs and fuel, coupled with strong domestic demand, are immediate concerns for 2011. Peru has continued to attract foreign investment. However, political disputes may impede development of some projects related to natural resource extraction.
GDP (purchasing power parity): $275.7 billion (2010 est.)
country comparison to the world: 43
$253.4 billion (2009 est.)
$251.3 billion (2008 est.)
note: data are in 2010 US dollars
GDP (official exchange rate): $152.8 billion (2010 est.)
GDP—real growth rate: 8.8% (2010 est.)
country comparison to the world: 11
0.9% (2009 est.)
9.8% (2008 est.)

GDP—per capita (PPP): $9,200 (2010 est.)
country comparison to the world: 115
$8,600 (2009 est.)
$8,600 (2008 est.)
note: data are in 2010 US dollars
GDP—composition by sector:
agriculture: 10%
industry: 35%
services: 55% (2010 est.)
Labor force: 15.68 million (mid-year 2010 est.)
country comparison to the world: 38
Labor force—by occupation:
agriculture: 0.7%
industry: 23.8%
services: 75.5% (2005)
Unemployment rate: 7.9% (2010 est.)
country comparison to the world: 85
8.4% (2009 est.)
note: data are for metropolitan Lima; widespread underemployment
Population below poverty line: 34.8% (2009)
Household income or consumption by percentage share: *lowest 10%:* 1.5%
highest 10%: 37.9% (2006)
Distribution of family income—Gini index: 49.6 (2009)
country comparison to the world: 25
46.2 (1996)
Investment (gross fixed): 25.1% of GDP (2010 est.)
country comparison to the world: 41
Budget: *revenues:* $30.53 billion
expenditures: $29.72 billion (2010 est.)
Public debt: 23.9% of GDP (2010 est.)
country comparison to the world: 102
27.2% of GDP (2009)
Inflation rate (consumer prices): 1.5% (2010)
country comparison to the world: 39
2.9% (2009 est.)
note: data are for metropolitan Lima
Central bank discount rate: 3.8% (31 December 2010)
country comparison to the world: 112
2.05% (31 December 2009)
Commercial bank prime lending rate: 3.63% (31 December 2010 est.)
country comparison to the world: 155
1.74% (31 December 2009 est.)
note: domestic currency lending rate
Stock of narrow money: $15.26 billion (31 December 2010 est.)
country comparison to the world: 65
$11.44 billion (31 December 2009 est.)
Stock of broad money: $55.2 billion (31 December 2010 est.)
country comparison to the world: 65
$43.57 billion (31 December 2009 est.)
Stock of domestic credit: $44.2 billion (31 December 2010 est.)
country comparison to the world: 67
$36.97 billion (31 December 2009 est.)
Market value of publicly traded shares: $160.9 billion (31 December 2010)
country comparison to the world: 36
$107.3 billion (31 December 2009)
$57.2 billion (31 December 2008)
Agriculture—products: asparagus, coffee, cocoa, cotton, sugarcane, rice, potatoes, corn, plantains, grapes, oranges, pineapples, guavas, bananas, apples, lemons, pears, coca, tomatoes, mango, barley, medicinal plants, palm oil, marigold, onion, wheat, dry beans; poultry, beef, dairy products; fish; guinea pigs
Industries: mining and refining of minerals; steel, metal fabrication; petroleum extraction and refining, natural gas and natural gas liquefaction; fishing and fish processing, cement, textiles, clothing, food processing
Industrial production growth rate: 8.5% (2010 est.)
country comparison to the world: 28
Electricity—production: 35.79 billion kWh (2010 est.)
country comparison to the world: 60
Electricity—consumption: 31.74 billion kWh (2009 est.)
country comparison to the world: 58
Electricity—exports: 111.9 million kWh (2010 est.)
Electricity—imports: 0 kWh (2010 est.)
Oil—production: 157,200 bbl/day (2010 est.)
country comparison to the world: 44
Oil—consumption: 150,700 bbl/day (2010 est.)
country comparison to the world: 66
Oil—exports: 68,640 bbl/day (2007 est.)
country comparison to the world: 73
Oil—imports: 88,080 bbl/day (2010 est.)
country comparison to the world: 65
Oil—proved reserves: 1.163 billion bbl (1 January 2010 est.)
country comparison to the world: 40
Natural gas—production: 7.24 billion cu m (2010)
country comparison to the world: 46
Natural gas—consumption: 3.65 billion cu m (2010)
country comparison to the world: 68
Natural gas—exports: 3.59 billion cu m
country comparison to the world: 31
note: in 2010 Peru became a net exporter of LNG (2010 est.)
Natural gas—imports: 0 cu m (2010)
country comparison to the world: 104
Natural gas—proved reserves: 453 billion cu m (1 January 2011 est.)
country comparison to the world: 33
Current account balance: $-2.315 billion (2010 est.)
country comparison to the world: 159
$211 million (2009 est.)
Exports:
$35.56 billion (2010 est.)
country comparison to the world: 61
$26.96 billion (2009 est.)
Exports—commodities: copper, gold, zinc, tin, iron ore, molybdenum; crude petroleum and petroleum products, natural gas; coffee, potatoes, asparagus and other vegetables, fruit, apparel and textiles, fishmeal
Exports—partners: US 16.98%, China 15.3%, Switzerland 14.85%, Canada 8.67%, Japan 5.14% (2010 est.)
Imports: $25.74 billion (2010 est.)
country comparison to the world: 63
$21.01 billion (2009 est.)
Imports—commodities: petroleum and petroleum products, chemicals, plastics, machinery, vehicles, color TV sets, power shovels, front-end loaders, telephones and telecommunication equipment, iron and steel, wheat, corn, soybean products, paper, cotton, vaccines and medicines
Imports—partners: US 19.72%, China 15%, Brazil 7.77%, Ecuador 4.9%, Chile 4.6%, Colombia 4.4%, Japan 4.1% (2010 est.)
Reserves of foreign exchange and gold: $44.11 billion (31 December 2010)
country comparison to the world: 28
$33.14 billion (31 December 2009)
Debt—external: $33.29 billion (31 December 2010 est.)
country comparison to the world: 62
$35.63 billion (31 December 2009)
note: public debt component of *total:* $20.6 billion (31 December 2009)
Stock of direct foreign investment—at home: $43.47 billion (31 December 2010 est.)
country comparison to the world: 56
$36.91 billion (31 December 2009 est.)
Stock of direct foreign investment—abroad: $2.12 billion (31 December 2010 est.)
country comparison to the world: 66
$1.88 billion (31 December 2009 est.)
Exchange rates: nuevo sol (PEN) per US dollar–2.8178 (2010), 3.0115 (2009), 2.91 (2008), 3.1731 (2007), 3.2742 (2006)

COMMUNICATIONS

Telephones—main lines in use: 2.965 million (2009)
country comparison to the world: 51
Telephones—mobile cellular: 24.7 million (2009)
country comparison to the world: 36
Telephone system: *general assessment:* adequate for most requirements; nationwide microwave radio relay system and a domestic satellite system with 12 earth stations
domestic: fixed-line teledensity is only about 10 per 100 persons; mobile-cellular teledensity, spurred by competition among multiple providers, has increased to roughly 85 telephones per 100 persons
international: country code–51; the South America-1 (SAM-1) and Pan American (PAN-AM) submarine cable systems provide links to parts of Central and South America, the Caribbean, and US; satellite earth stations–2 Intelsat (Atlantic Ocean) (2009)
Broadcast media: 10 major television networks of which only one, Television Nacional de Peru, is state-owned; multi-channel cable TV services are available; in excess of 2,000 radio stations including a substantial number of indigenous language stations (2010)
Internet country code: .pe
Internet hosts: 268,225 (2010)
country comparison to the world: 62
Internet users: 9.158 million (2009)
country comparison to the world: 31

TRANSPORTATION

Airports: 211 (2010)
country comparison to the world: 29
Airports—with paved runways: *total:* 58
over 3,047 m: 6
2,438 to 3,047 m: 20
1,524 to 2,437 m: 15
914 to 1,523 m: 13
under 914 m: 4 (2010)
Airports—with unpaved runways:
total: 153
2,438 to 3,047 m: 2
1,524 to 2,437 m: 24
914 to 1,523 m: 40

under 914 m: 87 (2010)
Heliports: 1 (2010)
Pipelines: extra heavy crude 533 km; gas 1,526 km; liquid petroleum gas 679 km; oil 1,033 km; refined products 15 km (2010)
Railways: *total:* 2,020 km
country comparison to the world: 73
standard gauge: 1,886 km 1.435-m gauge
narrow gauge: 134 km 0.914-m gauge (2010)
Roadways: *total:* 102,887 km
country comparison to the world: 41
note: includes 23,838 km of national roads, 19,049 km of departmental roads, and 60,000 km of local roads (2007)
Waterways: 8,808 km (there are 8,600 km of navigable tributaries on the Amazon system and 208 km on Lago Titicaca) (2010)
country comparison to the world: 14
Merchant marine: *total:* 13
country comparison to the world: 106
by type: cargo 2, chemical tanker 2, liquefied gas 2, petroleum tanker 7
foreign-owned: 1 (Bahamas 1)
registered in other countries: 13 (Belize 1, Panama 12) (2010)
Ports and terminals: Callao, Iquitos, Matarani, Paita, Pucallpa, Yurimaguas; note–Iquitos, Pucallpa, and Yurimaguas are on the upper reaches of the Amazon and its tributaries

MILITARY

Military branches: Army of Peru (Ejercito Peruano), Navy of Peru (Marina de Guerra del Peru, MGP (includes naval air, naval infantry, and Coast Guard)), Air Force of Peru (Fuerza Aerea del Peru, FAP) (2010)
Military service age and obligation: 18-30 years of age for voluntary male and female military service; no conscription (2008)
Manpower available for military service:
males age 16–49: 7,385,588
females age 16–49: 7,727,623 (2010 est.)
Manpower fit for military service: *males age 16–49:* 5,788,629
females age 16–49: 6,565,097 (2010 est.)
Manpower reaching militarily significant age annually: *male:* 304,094
female: 298,447 (2010 est.)
Military expenditures: 1.5% of GDP (2006)
country comparison to the world: 101

TRANSNATIONAL ISSUES

Disputes—international: Chile and Ecuador rejected Peru's November 2005 unilateral legislation to shift the axis of their joint treaty-defined maritime boundaries along the parallels of latitude to equidistance lines which favor Peru; organized illegal narcotics operations in Colombia have penetrated Peru's shared border; Peru rejects Bolivia's claim to restore maritime access through a sovereign corridor through Chile along the Peruvian border
Refugees and internally displaced persons: *IDPs:* 60,000-150,000 (civil war from 1980-2000; most IDPs are indigenous peasants in Andean and Amazonian regions) (2007)
Illicit drugs: until 1996 the world's largest coca leaf producer, Peru is now the world's second largest producer of coca leaf, though it lags far behind Colombia; cultivation of coca in Peru declined to 36,000 hectares in 2007; second largest producer of cocaine, estimated at 210 metric tons of potential pure cocaine in 2007; finished cocaine is shipped out from Pacific ports to the international drug market; increasing amounts of base and finished cocaine, however, are being moved to Brazil, Chile, Argentina, and Bolivia for use in the Southern Cone or transshipment to Europe and Africa; increasing domestic drug consumption

PHILIPPINES

INTRODUCTION

Background: The Philippine Islands became a Spanish colony during the 16th century; they were ceded to the US in 1898 following the Spanish-American War. In 1935 the Philippines became a self-governing commonwealth. Manuel QUEZON was elected president and was tasked with preparing the country for independence after a 10-year transition. In 1942 the islands fell under Japanese occupation during World War II, and US forces and Filipinos fought together during 1944-45 to regain control. On 4 July 1946 the Republic of the Philippines attained its independence. A 20-year rule by Ferdinand MARCOS ended in 1986, when a "people power" movement in Manila ("EDSA 1") forced him into exile and installed Corazon AQUINO as president. Her presidency was hampered by several coup attempts that prevented a return to full political stability and economic development. Fidel RAMOS was elected president in 1992. His administration was marked by increased stability and by progress on economic reforms. In 1992, the US closed its last military bases on the islands. Joseph ESTRADA was elected president in 1998. He was succeeded by his vice-president, Gloria MACAPAGAL-ARROYO, in January 2001 after ESTRADA's stormy impeachment trial on corruption charges broke down and another "people power" movement ("EDSA 2") demanded his resignation. MACAPAGAL-ARROYO was elected to a six-year term as president in May 2004. Her presidency was marred by several corruption allegations but the Philippine economy was one of the few to avoid contraction following the 2008 global financial crisis, expanding each year of her administration. Benigno AQUINO III was elected to a six-year term as president in May 2010. The Philippine Government faces threats from several groups on the US Government's Foreign Terrorist Organization list. Manila has waged a decades-long struggle against ethnic Moro insurgencies in the southern Philippines, which has led to a peace accord with the Moro National Liberation Front and on-again/off-again peace talks with the Moro Islamic Liberation Front. The decades-long Maoist-inspired New People's Army insurgency also operates through much of the country.

GEOGRAPHY

Location: Southeastern Asia, archipelago between the Philippine Sea and the South China Sea, east of Vietnam
Geographic coordinates: 13 00 N, 122 00 E
Map references: Southeast Asia
Area: *total:* 300,000 sq km
country comparison to the world: 72
land: 298,170 sq km
water: 1,830 sq km
Area—comparative: slightly larger than Arizona
Land boundaries: 0 km
Coastline: 36,289 km
Maritime claims: territorial sea: irregular polygon extending up to 100 nm from coastline as defined by 1898 treaty; since late 1970s has also claimed polygonal-shaped area in South China Sea up to 285 nm in breadth
exclusive economic zone: 200 nm
continental shelf: to depth of exploitation
Climate: tropical marine; northeast monsoon (November to April); southwest monsoon (May to October)
Terrain: mostly mountains with narrow to extensive coastal lowlands
Elevation extremes:
lowest point: Philippine Sea 0 m
highest point: Mount Apo 2,954 m

Natural resources: timber, petroleum, nickel, cobalt, silver, gold, salt, copper
Land use: *arable land:* 19%
permanent crops: 16.67%
other: 64.33% (2005)
Irrigated land: 15,500 sq km (2003)
Total renewable water resources: 479 cu km (1999)
Freshwater withdrawal (domestic/industrial/agricultural):
total: 28.52 cu km/yr (17%/9%/74%)
per capita: 343 cu m/yr (2000)
Natural hazards: astride typhoon belt, usually affected by 15 and struck by five to six cyclonic storms per year; landslides; active volcanoes; destructive earthquakes; tsunamis
volcanism: the Philippines experience significant volcanic activity; Taal (elev. 311 m), which has shown recent unrest and may erupt in the near future, has been deemed a "Decade Volcano" by the International Association of Volcanology and Chemistry of the Earth's Interior, worthy of study due to its explosive history and close proximity to human populations; Mayon (elev. 2,462 m), the country's most active volcano, erupted in 2009 forcing over 33,000 to be evacuated; other historically active volcanoes include Biliran, Babuyan Claro, Bulusan, Camiguin, Camiguin de Babuyanes, Didicas, Iraya, Jolo, Kanlaon, Makaturing, Musuan, Parker, Pinatubo and Ragang
Environment—current issues: uncontrolled deforestation especially in watershed areas; soil erosion; air and water pollution in major urban centers; coral reef degradation; increasing pollution of coastal mangrove swamps that are important fish breeding grounds
Environment—international agreements: *party to:* Biodiversity, Climate Change, Climate Change-Kyoto Protocol, Desertification, Endangered Species, Hazardous Wastes, Law of the Sea, Marine Dumping, Ozone Layer Protection, Ship Pollution, Tropical Timber 83, Tropical Timber 94, Wetlands, Whaling
signed, but not ratified: Air Pollution-Persistent Organic Pollutants
Geography—note: the Philippine archipelago is made up of 7,107 islands; favorably located in relation to many of Southeast Asia's main water bodies: the South China Sea, Philippine Sea, Sulu Sea, Celebes Sea, and Luzon Strait

PEOPLE

Population: 101,833,938 (July 2011 est.)
country comparison to the world: 12
Age structure: *0–14 years:* 34.6% (male 17,999,279/female 17,285,040)
15–64 years: 61.1% (male 31,103,967/female 31,097,203)
65 years and over: 4.3% (male 1,876,805/female 2,471,644) (2011 est.)
Median age: *total:* 22.9 years
male: 22.4 years
female: 23.4 years (2011 est.)
Population growth rate: 1.903% (2011 est.)
country comparison to the world: 60
Birth rate: 25.34 births/1,000 population (2011 est.)
country comparison to the world: 58
Death rate: 5.02 deaths/1,000 population (July 2011 est.)
country comparison to the world: 185
Net migration rate: -1.29 migrant(s)/1,000 population (2011 est.)
country comparison to the world: 157
Urbanization: *urban population:* 49% of total population (2010)
rate of urbanization: 2.3% annual rate of change (2010-15 est.)
Major cities—population: MANILA (capital) 11.449 million; Davao 1.48 million; Cebu City 845,000; Zamboanga 827,000 (2009)
Sex ratio: *at birth:* 1.05 male(s)/female
under 15 years: 1.04 male(s)/female
15–64 years: 1 male(s)/female
65 years and over: 0.76 male(s)/female
total population: 1 male(s)/female (2011 est.)
Infant mortality rate: *total:* 19.34 deaths/1,000 live births
country comparison to the world: 100
male: 21.84 deaths/1,000 live births
female: 16.71 deaths/1,000 live births (2011 est.)
Life expectancy at birth: *total population:* 71.66 years
country comparison to the world: 133
male: 68.72 years
female: 74.74 years (2011 est.)
Total fertility rate: 3.19 children born/woman (2011 est.)
country comparison to the world: 53
HIV/AIDS—adult prevalence rate: less than 0.1% (2009 est.)
country comparison to the world: 159
HIV/AIDS—people living with HIV/AIDS: 8,700 (2009 est.)
country comparison to the world: 105
HIV/AIDS—deaths: fewer than 200 (2009 est.)
country comparison to the world: 111
Major infectious diseases: *degree of risk:* high
food or waterborne diseases: bacterial diarrhea, hepatitis A, and typhoid fever
vectorborne diseases: dengue fever, malaria, and Japanese encephalitis
water contact disease: leptospirosis (2009)
Drinking water source: *Improved:*
urban: 93% of population
rural: 87% of population
total: 91% of population
Unimproved: urban: 7% of population
rural: 13% of population
total: 9% of population (2008)
Sanitation facility access: *Improved:*
urban: 80% of population
rural: 69% of population
total: 76% of population
Unimproved: urban: 20% of population
rural: 31% of population
total: 24% of population (2008)
Nationality: *noun:* Filipino(s)
adjective: Philippine
Ethnic groups: Tagalog 28.1%, Cebuano 13.1%, Ilocano 9%, Bisaya/Binisaya 7.6%, Hiligaynon Ilonggo 7.5%, Bikol 6%, Waray 3.4%, other 25.3% (2000 census)
Religions: Roman Catholic 80.9%, Muslim 5%, Evangelical 2.8%, Iglesia ni Kristo 2.3%, Aglipayan 2%, other Christian 4.5%, other 1.8%, unspecified 0.6%, none 0.1% (2000 census)
Languages: Filipino (official; based on Tagalog) and English (official); eight major dialects—Tagalog, Cebuano, Ilocano, Hiligaynon or Ilonggo, Bicol, Waray, Pampango, and Pangasinan
Literacy: *definition:* age 15 and over can read and write
total population: 92.6%
male: 92.5%
female: 92.7% (2000 census)
School life expectancy (primary to tertiary education):
total: 12 years
male: 12 years
female: 12 years (2008)
Education expenditures: 2.8% of GDP (2008)
country comparison to the world: 139

GOVERNMENT

Country name: *conventional long form:* Republic of the Philippines
conventional short form: Philippines
local long form: Republika ng Pilipinas
local short form: Pilipinas
Government type: republic
Capital: *name:* Manila
geographic coordinates: 14 35 N, 121 00 E
time difference: UTC+8 (13 hours ahead of Washington, DC during Standard Time)
Administrative divisions: 80 provinces and 120 chartered cities
provinces: Abra, Agusan del Norte, Agusan del Sur, Aklan, Albay, Antique, Apayao, Aurora, Basilan, Bataan, Batanes, Batangas, Biliran, Benguet, Bohol, Bukidnon, Bulacan, Cagayan, Camarines Norte, Camarines Sur, Camiguin, Capiz, Catanduanes, Cavite, Cebu, Compostela, Davao del Norte, Davao del Sur, Davao Oriental, Dinagat Islands, Eastern Samar, Guimaras, Ifugao, Ilocos Norte, Ilocos Sur, Iloilo, Isabela, Kalinga, Laguna, Lanao del Norte, Lanao del Sur, La Union, Leyte, Maguindanao, Marinduque, Masbate, Mindoro Occidental, Mindoro Oriental, Misamis Occidental, Misamis Oriental, Mountain Province, Negros Occidental, Negros Oriental, North Cotabato, Northern Samar, Nueva Ecija, Nueva Vizcaya, Palawan, Pampanga, Pangasinan, Quezon, Quirino, Rizal, Romblon, Samar, Sarangani, Siquijor, Sorsogon, South Cotabato, Southern Leyte, Sultan Kudarat, Sulu, Surigao del Norte, Surigao del Sur, Tarlac, Tawi-Tawi, Zambales, Zamboanga del Norte, Zamboanga del Sur, Zamboanga Sibugay
chartered cities: Alaminos, Angeles, Antipolo, Bacolod, Bago, Baguio, Bais, Balanga, Batac, Batangas, Bayawan, Bislig, Butuan, Cabadbaran, Cabanatuan, Cadiz, Cagayan de Oro, Calamba, Calapan, Calbayog, Candon, Canlaon, Cauayan, Cavite, Cebu, Cotabato, Dagupan, Danao, Dapitan, Davao, Digos, Dipolog, Dumaguete, Escalante, Gapan, General Santos, Gingoog, Himamaylan, Iligan, Iloilo, Isabela, Iriga, Kabankalan, Kalookan, Kidapawan, Koronadal, La Carlota, Laoag, Lapu-Lapu, Las Pinas, Legazpi, Ligao, Lipa, Lucena, Maasin, Makati, Malabon, Malaybalay, Malolos, Mandaluyong, Mandaue, Manila, Marawi, Marikina, Masbate, Mati,

Meycauayan, Muntinlupa, Munoz, Naga, Navotas, Olongapo, Ormoc, Oroquieta, Ozamis, Pagadian, Palayan, Panabo, Paranaque, Pasay, Pasig, Passi, Puerto Princesa, Quezon, Roxas, Sagay, Samal, San Carlos (in Negros Occidental), San Carlos (in Pangasinan), San Fernando (in La Union), San Fernando (in Pampanga), San Jose, San Jose del Monte, San Juan, San Pablo, Santa Rosa, Santiago, Silay, Sipalay, Sorsogon, Surigao, Tabaco, Tacloban, Tacurong, Tagaytay, Tagbilaran, Taguig, Tagum, Talisay (in Cebu), Talisay (in Negros Occidental), Tanauan, Tangub, Tanjay, Tarlac, Toledo, Tuguegarao, Trece Martires, Urdaneta, Valencia, Valenzuela, Victorias, Vigan, Zamboanga (2009)

Independence: 12 June 1898 (independence proclaimed from Spain); 4 July 1946 (from the US)

National holiday: Independence Day, 12 June (1898); note—12 June 1898 was date of declaration of independence from Spain; 4 July 1946 was date of independence from US

Constitution: 2 February 1987, effective 11 February 1987

Legal system: mixed legal system of civil, common, Islamic, and customary law

International law organization participation: accepts compulsory ICJ jurisdiction with reservations; non-party state to the ICCt

Suffrage: 18 years of age; universal

Executive branch: *chief of state:* President Benigno AQUINO (since 30 June 2010); Vice President Jejomar BINAY (since 30 June 2010); note—president is both chief of state and head of government
head of government: President Benigno AQUINO (since 30 June 2010)
cabinet: Cabinet appointed by the president with consent of Commission of Appointments
(For more information visit the World Leaders website)
elections: president and vice president elected on separate tickets by popular vote for a single six-year term; election held on 10 May 2010; Benigno AQUINO declared winner and took office on 30 June 2010; next election to be held in May 2016
election results: Benigno AQUINO elected president; percent of vote—Benigno AQUINO 42.1%, Joseph ESTRADA 26.3%, seven others 31.6%; Jejomar BINAY elected vice president; percent of vote Jejomar BINAY 41.6%, Manuel ROXAS 39.6%, six others 18.8%

Legislative branch: bicameral Congress or Kongreso consists of the Senate or Senado (24 seats—one-half elected every three years; members elected at large by popular vote to serve six-year terms) and the House of Representatives or Kapulungan Ng Nga Kinatawan; the House has 287 seats including 230 members in one tier representing districts and 57 sectoral party-list members in a second tier representing special minorities elected on the basis of one seat for every 2% of the total vote but with each party limited to three seats; a party represented in one tier may not hold seats in the other tier; all House members are elected by popular vote to serve three-year terms
note: the constitution limits the House of Representatives to 250 members; the number of members allowed was increased, however, through legislation when in April 2009 the Philippine Supreme Court ruled that additional party members could sit in the House of Representatives if they received the required number of votes
elections: Senate—elections last held on 10 May 2010 (next to be held in May 2013); House of Representatives—elections last held on 10 May 2010 (next to be held in May 2013)
election results: Senate—percent of vote by party—NA; seats by party—Lakas-Kampi CMD 4, LP 4, NP 4, NPC 2, PMP 2, LDP 1, PRP 1, independents 5; note—there are 23 rather than 24 sitting senators because one senator was elected mayor of Manila; House of Representatives—percent of vote by party—NA; seats by party—LP 119, Lakas-Kampi CMD 46, NPC 30, NP 22, others 10, independents 1, party-list 55; vacant seats—1 district and 2 party-list

Judicial branch: Supreme Court (15 justices are appointed by the president on the recommendation of the Judicial and Bar Council and serve until 70 years of age); Court of Appeals; Sandigan-bayan (special court for hearing corruption cases of government officials)

Political parties and leaders: Laban ng Demokratikong Pilipino (Struggle of Filipino Democrats) or LDP [Edgardo ANGARA]; Lakas ng EDSA-Christian Muslim Democrats or Lakas-CMD [Gloria MACAPAGAL-ARROYO]; Liberal Party or LP [Manuel ROXAS]; Nacionalista Party or NP [Manuel VILLAR]; Nationalist People's Coalition or NPC [Frisco SAN JUAN]; PDP-Laban [Aquilino PIMENTEL]; People's Reform Party [Miriam Defensor SANTIAGO]; Puwersa ng Masang Pilipino (Force of the Philippine Masses) or PMP [Joseph ESTRADA]

Political pressure groups and leaders: ABONO [Robert ESTRELLA]; AKBAYAN [Walden BELLO]; An Waray [Florencio NOEL];AnakMindanao[MujivHATAMIN]; ANAKPAWIS [Rafael MARIANO]; ARC [Narciso SANTIAGO III]; Association of Philippine Electric Cooperatives (APEC) [Ponciano PAYUYO]; A TEACHER [Mariano PIAMONTE]; BAGON HENERASYON [Bernadette HERRERA-DY]; Bayan Muna [Teodoro CASINO, Jr.]; Black and White Movement [Vicente ROMANO]; BUHAY [Rene VELARDE]; BUTIL [Leonila CHAVEZ]; CIBAC [Cinchoa CRUZ-GONZALES]; COOP-NATCO [Jose PING-AY]; GABRIELA [Luzviminda ILAGAN]; KABATAAN [Raymon PALATINO]; Kilosbayan [Jovito SALONGA]; YACAP [Carol LOPEZ]

International organization participation: ADB, APEC, APT, ARF, ASEAN, BIS, CD, CICA (observer), CP, EAS, FAO, G-24, G-77, IAEA, IBRD, ICAO, ICC, ICRM, IDA, IFAD, IFC, IFRCS, IHO, ILO, IMF, IMO, IMSO, Interpol, IOC, IOM, IPU, ISO, ITSO, ITU, ITUC, MIGA, MINUSTAH, NAM, OAS (observer), OPCW, PCA, PIF (partner), UN, UNCTAD, UNDOF, UNESCO, UNHCR, UNIDO, Union Latina, UNMIL, UNMIS, UNMIT, UNMOGIP, UNOCI, UNWTO, UPU, WCO, WFTU, WHO, WIPO, WMO, WTO

Diplomatic representation in the US: *chief of mission:* Ambassador Jose L. CUISIA Jr.
chancery: 1600 Massachusetts Avenue NW, Washington, DC 20036
telephone: [1] (202) 467-9300
FAX: [1] (202) 467-9417
consulate(s) general: Chicago, Honolulu, Los Angeles, New York, Saipan (Northern Mariana Islands), San Francisco, Tamuning (Guam)

Diplomatic representation from the US: *chief of mission:* Ambassador Harry K. THOMAS Jr.
embassy: 1201 Roxas Boulevard, Ermita 1000, Manila
mailing address: PSC 500, FPO AP 96515-1000
telephone: [63] (2) 301-2000
FAX: [63] (2) 301-2399

Flag description: two equal horizontal bands of blue (top) and red; a white equilateral triangle is based on the hoist side; the center of the triangle displays a yellow sun with eight primary rays; each corner of the triangle contains a small, yellow, five-pointed star; blue stands for peace and justice, red symbolizes courage, the white equal-sided triangle represents equality; the rays recall the first eight provinces that sought independence from Spain, while the stars represent the three major geographical divisions of the country: Luzon, Visayas, and Mindanao; the design of the flag dates to 1897
note: in wartime the flag is flown upside down with the red band at the top

National anthem: *name:* "Lupang Hinirang" (Chosen Land)
lyrics/music: Jose PALMA (revised by Felipe PADILLA de Leon)/Julian FELIPE
note: music adopted 1898, original Spanish lyrics adopted 1899, Filipino (Tagalog) lyrics adopted 1956; although the original lyrics were written in Spanish, later English and Filipino versions were created; today, only the Filipino version is used

ECONOMY

Economy—overview: Philippine GDP grew 7.3% in 2010, spurred by consumer demand, a rebound in exports and investments, and election-related spending. The economy weathered the 2008-09 global recession better than its regional peers due to minimal exposure to troubled international securities, lower dependence on exports, relatively resilient domestic consumption, large remittances from four- to five-million overseas Filipino workers, and a growing business process outsourcing industry. Economic growth in the Philippines averaged 4.5% during the MACAPAGAL-ARROYO administration. Despite this growth, poverty worsened, because of a high population growth rate and inequitable distribution of income. The AQUINO administration is working to reduce the government deficit from

3.9% of GDP, when it took office, to 2% of GDP by 2013. The government has had little difficulty issuing debt both locally and internationally to finance the deficits. AQUINO's first budget emphasizes education, health, conditional cash transfers for the poor, and other social spending programs, relying on the private sector to finance important infrastructure projects. Weak tax collection, exacerbated by new tax breaks and incentives, has limited the government's ability to address major challenges. The AQUINO administration has vowed to focus on improving tax collection efficiency—rather than imposing new taxes—as a part of its good governance platform.

GDP (purchasing power parity): $351.4 billion (2010 est.)
country comparison to the world: 34
$327.4 billion (2009 est.)
$323.9 billion (2008 est.)
note: data are in 2010 US dollars

GDP (official exchange rate): $188.7 billion (2010 est.)

GDP—real growth rate: 7.3% (2010 est.)
country comparison to the world: 31
1.1% (2009 est.)
3.7% (2008 est.)

GDP—per capita (PPP): $3,500 (2010 est.)
country comparison to the world: 164
$3,300 (2009 est.)
$3,400 (2008 est.)
note: data are in 2010 US dollars

GDP—composition by sector:
agriculture: 13.9%
industry: 31.3%
services: 54.8% (2010 est.)

Labor force: 38.9 million (2010 est.)
country comparison to the world: 15

Labor force—by occupation:
agriculture: 33%
industry: 15%
services: 52% (2010 est.)

Unemployment rate: 7.3% (2010 est.)
country comparison to the world: 76
7.5% (2009 est.)

Population below poverty line: 32.9% (2006 est.)

Household income or consumption by percentage share: *lowest 10%:* 2.4%
highest 10%: 31.2% (2006)

Distribution of family income—Gini index: 45.8 (2006)
country comparison to the world: 36
46.6 (2003)

Investment (gross fixed): 16% of GDP (2010 est.)
country comparison to the world: 126

Budget: *revenues:* $26.84 billion
expenditures: $33.82 billion (2010 est.)

Public debt: 56.5% of GDP (2010 est.)
country comparison to the world: 43
57.3% of GDP (2009 est.)

Inflation rate (consumer prices): 3.8% (2010 est.)
country comparison to the world: 109
3.2% (2009 est.)

Central bank discount rate: 4% (31 December 2010)
country comparison to the world: 105
3.5% (31 December 2009)

Commercial bank prime lending rate: 5.39% (31 December 2010 est.)
country comparison to the world: 128
6.89% (31 December 2009 est.)

Stock of narrow money: $30.65 billion (31 December 2010 est.)
country comparison to the world: 56
$26.35 billion (31 December 2009 est.)

Stock of broad money: $97.35 billion (31 December 2010 est.)
country comparison to the world: 53
$83.3 billion (31 December 2009 est.)

Stock of domestic credit: $97.63 billion (31 December 2010 est.)
country comparison to the world: 51
$83.3 billion (31 December 2009 est.)

Market value of publicly traded shares: $202.3 billion (31 December 2010)
country comparison to the world: 33
$130.5 billion (31 December 2009)
$85.63 billion (31 December 2008)

Agriculture—products: sugarcane, coconuts, rice, corn, bananas, cassavas, pineapples, mangoes; pork, eggs, beef; fish

Industries: electronics assembly, garments, footwear, pharmaceuticals, chemicals, wood products, food processing, petroleum refining, fishing

Industrial production growth rate: 12.1% (2010 est.)
country comparison to the world: 13

Electricity—production: 61.93 billion kWh (2009 est.)
country comparison to the world: 42

Electricity—consumption: 54.4 billion kWh (2009 est.)
country comparison to the world: 43

Electricity—exports: 0 kWh (2009 est.)

Electricity—imports: 0 kWh (2009 est.)

Oil—production: 9,671 bbl/day (July 2010 est.)
country comparison to the world: 86

Oil—consumption: 307,200 bbl/day (September 2010 est.)
country comparison to the world: 42

Oil—exports: 28,900 bbl/day (September 2010 est.)
country comparison to the world: 86

Oil—imports: 338,400 bbl/day (September 2010 est.)
country comparison to the world: 30

Oil—proved reserves: 168 million bbl (1 January 2010 est.)
country comparison to the world: 64

Natural gas—production: 2.94 billion cu m (2008 est.)
country comparison to the world: 55

Natural gas—consumption: 2.94 billion cu m (2008 est.)
country comparison to the world: 75

Natural gas—exports: 0 cu m (2008 est.)
country comparison to the world: 163

Natural gas—imports: 0 cu m (2008 est.)
country comparison to the world: 110

Natural gas—proved reserves: 108.7 billion cu m (1 January 2011 est.)
country comparison to the world: 52

Current account balance: $9.51 billion (2010 est.)
country comparison to the world: 25
$8.788 billion (2009 est.)

Exports: $50.72 billion (2010 est.)
country comparison to the world: 53
$37.6 billion (2009 est.)

Exports—commodities: semiconductors and electronic products, transport equipment, garments, copper products, petroleum products, coconut oil, fruits

Exports—partners: US 17.6%, Japan 16.2%, Netherlands 9.8%, Hong Kong 8.6%, China 7.7%, Germany 6.5%, Singapore 6.2%, South Korea 4.8% (2009 est.)

Imports:
$59.9 billion (2010 est.)
country comparison to the world: 42
$46.39 billion (2009 est.)

Imports—commodities: electronic products, mineral fuels, machinery and transport equipment, iron and steel, textile fabrics, grains, chemicals, plastic

Imports—partners: Japan 12.5%, US 12%, China 8.8%, Singapore 8.7%, South Korea 7.9%, Taiwan 7.1%, Thailand 5.7% (2009 est.)

Reserves of foreign exchange and gold: $62.37 billion (31 December 2010 est.)
country comparison to the world: 22
$44.24 billion (31 December 2009 est.)

Debt—external: $59.77 billion (30 September 2010 est.)
country comparison to the world: 48
$62.97 billion (31 December 2009 est.)

Stock of direct foreign investment—at home: $24.5 billion (31 December 2010 est.)
country comparison to the world: 64
$23.18 billion (31 December 2009 est.)

Stock of direct foreign investment—abroad: $6.495 billion (31 December 2010 est.)
country comparison to the world: 55
$6.095 billion (31 December 2009 est.)

Exchange rates: Philippine pesos (PHP) per US dollar—45.11 (2010), 47.68 (2009), 44.439 (2008), 46.148 (2007), 51.246 (2006)

COMMUNICATIONS

Telephones—main lines in use: 6.783 million (2010)
country comparison to the world: 29

Telephones—mobile cellular: 92.227 million (2010)
country comparison to the world: 11

Telephone system: *general assessment:* good international radiotelephone and submarine cable services; domestic and interisland service adequate
domestic: telecommunications infrastructure includes the following platforms: fixed line, mobile cellular, cable TV, over-the-air TV, radio and Very Small Aperture Terminal (VSAT), fiber optic cable, and satellite; mobile-cellular communications now dominate the industry
international: country code—63; a series of submarine cables together provide connectivity to Asia, US, the Middle East, and Europe; multiple international gateways (2010)

Broadcast media: multiple national private TV and radio networks; multi-channel satellite and cable TV systems available; five national or major TV networks; three government-owned networks; five major cable TV networks and a government-operated national TV and radio network; about 300 analog television stations; more than 1,000 radio stations (2010)

Internet country code: .ph

Internet hosts: 394,990 (2010)

country comparison to the world: 53
Internet users: 8.278 million (2009)
country comparison to the world: 34

TRANSPORTATION

Airports: 254 (2010)
country comparison to the world: 25
Airports—with paved runways: *total:* 85
over 3,047 m: 4
2,438 to 3,047 m: 8
1,524 to 2,437 m: 29
914 to 1,523 m: 34
under 914 m: 10 (2010)
Airports—with unpaved runways: *total:* 169
1,524 to 2,437 m: 4
914 to 1,523 m: 66
under 914 m: 99 (2010)
Heliports: 2 (2010)
Pipelines: gas 7 km; oil 107 km; refined products 181 km (2010)
Railways: *total:* 995 km
country comparison to the world: 87
narrow gauge: 995 km 1.067-m gauge (484 km are in operation) (2010)
Roadways: *total:* 213,151 km
country comparison to the world: 24
paved: 54,481 km
unpaved: 158,670 km (2009)
Waterways: 3,219 km (limited to vessels with draft less than 1.5 m) (2011)
country comparison to the world: 30
Merchant marine: *total:* 428
country comparison to the world: 24
by type: bulk carrier 75, cargo 135, carrier 16, chemical tanker 26, container 13, liquefied gas 5, passenger 7, passenger/cargo 68, petroleum tanker 45, refrigerated cargo 17, roll on/roll off 12, vehicle carrier 9
foreign-owned: 156 (Bermuda 43, China 4, Greece 4, Japan 82, Malaysia 1, Netherlands 18, Singapore 1, South Korea 1, Taiwan 1, UAE 1)
registered in other countries: 7 (Cyprus 1, Panama 6) (2010)
Ports and terminals: Batangas, Cagayan de Oro, Cebu, Davao, Liman, Manila
Transportation—note: the International Maritime Bureau reports the territorial and offshore waters in the South China Sea as high risk for piracy and armed robbery against ships; numerous commercial vessels have been attacked and hijacked both at anchor and while underway; hijacked vessels are often disguised and cargo diverted to ports in East Asia; crews have been murdered or cast adrift

MILITARY

Military branches: Armed Forces of the Philippines (AFP): Army, Navy (includes Marine Corps and Coast Guard), Air Force (2011)
Military service age and obligation: 18-25 years of age (officers 21-29) for compulsory and voluntary military service; applicants must be single male or female Philippine citizens (2010)
Manpower available for military service:
males age 16–49: 25,614,135
females age 16–49: 25,035,061 (2010 est.)
Manpower fit for military service: *males age 16–49:* 20,142,940
females age 16–49: 21,427,792 (2010 est.)
Manpower reaching militarily significant age annually: *male:* 1,060,319
female: 1,021,069 (2010 est.)
Military expenditures: 0.9% of GDP (2005 est.)
country comparison to the world: 144

TRANSNATIONAL ISSUES

Disputes—international: Philippines claims sovereignty over Scarborough Reef (also claimed by China together with Taiwan) and over certain of the Spratly Islands, known locally as the Kalayaan (Freedom) Islands, also claimed by China, Malaysia, Taiwan, and Vietnam; the 2002 "Declaration on the Conduct of Parties in the South China Sea," has eased tensions in the Spratly Islands but falls short of a legally binding "code of conduct" desired by several of the disputants; in March 2005, the national oil companies of China, the Philippines, and Vietnam signed a joint accord to conduct marine seismic activities in the Spratly Islands; Philippines retains a dormant claim to Malaysia's Sabah State in northern Borneo based on the Sultanate of Sulu's granting the Philippines Government power of attorney to pursue a sovereignty claim on his behalf; maritime delimitation negotiations continue with Palau
Refugees and internally displaced persons: *IDPs:* 300,000 (fighting between government troops and MILF and Abu Sayyaf groups) (2007)
Illicit drugs: domestic methamphetamine production has been a growing problem in recent years despite government crackdowns; major consumer of amphetamines; longstanding marijuana producer mainly in rural areas where Manila's control is limited

PITCAIRN ISLANDS

(OVERSEAS TERRITORY OF THE UK)

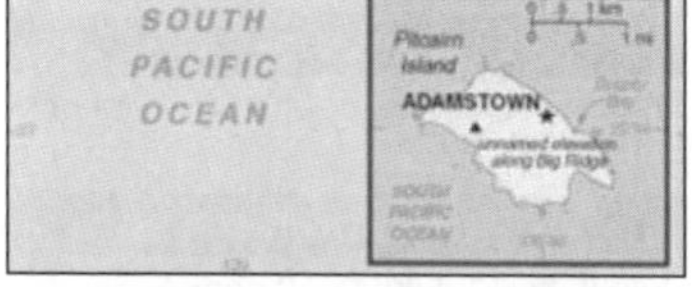

INTRODUCTION

Background: Pitcairn Island was discovered in 1767 by the British and settled in 1790 by the Bounty mutineers and their Tahitian companions. Pitcairn was the first Pacific island to become a British colony (in 1838) and today remains the last vestige of that empire in the South Pacific. Outmigration, primarily to New Zealand, has thinned the population from a peak of 233 in 1937 to less than 50 today.

GEOGRAPHY

Location: Oceania, islands in the South Pacific Ocean, about midway between Peru and New Zealand
Geographic coordinates: 25 04 S, 130 06 W
Map references: Oceania
Area: *total:* 47 sq km
country comparison to the world: 232
land: 47 sq km
water: 0 sq km
Area—comparative: about three tenths the size of Washington, DC
Land boundaries: 0 km
Coastline: 51 km
Maritime claims: territorial sea: 3 nm
exclusive economic zone: 200 nm
Climate: tropical; hot and humid; modified by southeast trade winds; rainy season (November to March)
Terrain: rugged volcanic formation; rocky coastline with cliffs
Elevation extremes: *lowest point:* Pacific Ocean 0 m
highest point: Big Ridge 347 m
Natural resources: miro trees (used for handicrafts), fish
note: manganese, iron, copper, gold, silver, and zinc have been discovered offshore
Land use: *arable land:* NA
permanent crops: NA
other: NA
Irrigated land: NA
Natural hazards: typhoons (especially November to March)
Environment—current issues: deforestation (only a small portion of the original forest remains because of burning and clearing for settlement)
Geography—note: Britain's most isolated dependency; only the larger island of Pitcairn is inhabited but it has no port or natural harbor; supplies must be transported by rowed longboat from larger ships stationed offshore

PEOPLE

Population: 48 (July 2011 est.)
country comparison to the world: 237
Age structure: *0–14 years:* NA
15–64 years: NA
65 years and over: NA (2009 est.)
Population growth rate: 0% (2011 est.)
country comparison to the world: 196
Birth rate: NA
Death rate: NA
Net migration rate: NA
Urbanization: *urban population:* 0% of total population (2010)

rate of urbanization: 0% annual rate of change (2010-15 est.)
Sex ratio: NA
Infant mortality rate: *total:* NA
male: NA
female: NA
Life expectancy at birth: *total population:* NA
male: NA
female: NA
Total fertility rate: NA
HIV/AIDS—adult prevalence rate: NA
HIV/AIDS—people living with HIV/AIDS: NA
HIV/AIDS—deaths: NA
Nationality: *noun:* Pitcairn Islander(s)
adjective: Pitcairn Islander
Ethnic groups: descendants of the Bounty mutineers and their Tahitian wives
Religions: Seventh-Day Adventist 100%
Languages: English (official), Pitkern (mixture of an 18th century English dialect and a Tahitian dialect)
Literacy: NA
School life expectancy (primary to tertiary education): NA
Education expenditures: NA

GOVERNMENT

Country name: *conventional long form:* Pitcairn, Henderson, Ducie, and Oeno Islands
conventional short form: Pitcairn Islands
Dependency status: overseas territory of the UK
Government type: NA
Capital: *name:* Adamstown
geographic coordinates: 25 04 S, 130 05 W
time difference: UTC-9 (4 hours behind Washington, DC during Standard Time)
Administrative divisions: none (overseas territory of the UK)
Independence: none (overseas territory of the UK)
National holiday: Birthday of Queen ELIZABETH II, second Saturday in June (1926)
Constitution: The Pitcairn Constitution Order 2010, effective 4 March 2010
Legal system: local island by-laws
Suffrage: 18 years of age; universal with three years residency
Executive branch: *chief of state:* Queen ELIZABETH II (since 6 February 1952); represented by UK High Commissioner to New Zealand and Governor (nonresident) of the Pitcairn Islands Victoria M. TREADELL (since May 2010); Commissioner (nonresident) Leslie JAQUES (since September 2003) serves as liaison between the governor and the Island Council
head of government: Mayor and Chairman of the Island Council Mike WARREN (since 1 January 2008)
cabinet: NA
(For more information visit the World Leaders website)
elections: the monarchy is hereditary; governor and commissioner appointed by the monarch; island mayor elected by popular vote for a three-year term; election last held in December 2010 (next to be held in December 2013)
election results: Mike WARREN reelected mayor and chairman of the Island Council
Legislative branch: unicameral Island Council (11 seats; mayor, deputy mayor, 4 members elected by popular vote, 1 member appointed by the governor, 3 ex officio members including governor, deputy governor, and commissioner; deputy mayor and elected members serve two-year terms)
elections: last held on 24 December 2009 (next to be held on 24 December 2011)
election results: percent of vote–NA; seats–4 independents
Judicial branch: Magistrate's Court; Supreme Court; Court of Appeal; judicial officers are appointed by the governor
Political parties and leaders: none
Political pressure groups and leaders: none
International organization participation: SPC, UPU
Diplomatic representation in the US: none (overseas territory of the UK)
Diplomatic representation from the US: none (overseas territory of the UK)
Flag description: blue with the flag of the UK in the upper hoist-side quadrant and the Pitcairn Islander coat of arms centered on the outer half of the flag; the green, yellow, and blue of the shield represents the island rising from the ocean; the green field features a yellow anchor surmounted by a bible (both the anchor and the bible were items found on the HMS Bounty); sitting on the crest is a Pitcairn Island wheelbarrow from which springs a slip of miro (a local plant)
National anthem: *name:* "We From Pitcairn Island"
lyrics/music: unknown/Frederick M. LEHMAN
note: serves as a local anthem; as a territory of the United Kingdom, "God Save the Queen" is official (see United Kingdom)

ECONOMY

Economy—overview: The inhabitants of this tiny isolated economy exist on fishing, subsistence farming, handicrafts, and postage stamps. The fertile soil of the valleys produces a wide variety of fruits and vegetables, including citrus, sugarcane, watermelons, bananas, yams, and beans. Bartering is an important part of the economy. The major sources of revenue are the sale of postage stamps to collectors and the sale of handicrafts to passing ships. In October 2004, more than one-quarter of Pitcairn's small labor force was arrested, putting the economy in a bind, since their services were required as lighter crew to load or unload passing ships.
GDP (purchasing power parity): $NA
Labor force: 15 able-bodied men (2004)
country comparison to the world: 229
Labor force—by occupation: note: no business community in the usual sense; some public works; subsistence farming and fishing
Budget: *revenues:* $746,000
expenditures: $1.028 million (FY04/05)
Agriculture—products: honey; wide variety of fruits and vegetables; goats, chickens, fish
Industries: postage stamps, handicrafts, beekeeping, honey
Electricity—production: NA kWh; note–electric power is provided by a small diesel-powered generator
Exports: $NA
Exports—commodities: fruits, vegetables, curios, stamps
Imports: $NA
Imports—commodities: fuel oil, machinery, building materials, flour, sugar, other foodstuffs
Exchange rates: New Zealand dollars (NZD) per US dollar–1.3874 (2010), 1.6002 (2009), 1.4151 (2008), 1.3811 (2007), 1.5408 (2006)

COMMUNICATIONS

Telephones—main lines in use: 1 (there are 17 telephones on one party line) (2004)
country comparison to the world: 229
Telephone system: *general assessment:* satellite phone services
domestic: domestic communication via radio (CB)
international: country code–872; satellite earth station–1 (Inmarsat)
Broadcast media: no local broadcast television or radio stations (2009)
Internet country code: .pn
Internet hosts: 20 (2010)
country comparison to the world: 218

TRANSPORTATION

Ports and terminals: Adamstown (on Bounty Bay)

MILITARY

Military—note: defense is the responsibility of the UK

TRANSNATIONAL ISSUES

Disputes—international: none

POLAND

INTRODUCTION

Background: Poland is an ancient nation that was conceived near the middle of the 10th century. Its golden age occurred in the 16th century. During the following century, the strengthening of the gentry and internal disorders weakened the nation. In a series of agreements between 1772 and 1795, Russia, Prussia, and Austria partitioned Poland among themselves. Poland regained its i ndependence

in 1918 only to be overrun by Germany and the Soviet Union in World War II. It became a Soviet satellite state following the war, but its government was comparatively tolerant and progressive. Labor turmoil in 1980 led to the formation of the independent trade union "Solidarity" that over time became a political force and by 1990 had swept parliamentary elections and the presidency. A "shock therapy" program during the early 1990s enabled the country to transform its economy into one of the most robust in Central Europe, but Poland still faces the lingering challenges of high unemployment, underdeveloped and dilapidated infrastructure, and a poor rural underclass. Poland joined NATO in 1999 and the European Union in 2004. With its transformation to a democratic, market-oriented country largely completed, Poland is an increasingly active member of Euro-Atlantic organizations.

GEOGRAPHY

Location: Central Europe, east of Germany
Geographic coordinates: 52 00 N, 20 00 E
Map references: Europe
Area: *total:* 312,685 sq km
country comparison to the world: 69
land: 304,255 sq km
water: 8,430 sq km
Area—comparative: slightly smaller than New Mexico
Land boundaries: *total:* 3,047 km
border countries: Belarus 605 km, Czech Republic 615 km, Germany 456 km, Lithuania 91 km, Russia (Kaliningrad Oblast) 432 km, Slovakia 420 km, Ukraine 428 km
Coastline: 440 km
Maritime claims: territorial sea: 12 nm
exclusive economic zone: defined by international treaties
Climate: temperate with cold, cloudy, moderately severe winters with frequent precipitation; mild summers with frequent showers and thundershowers
Terrain: mostly flat plain; mountains along southern border
Elevation extremes: *lowest point:* near Raczki Elblaskie -2 m
highest point: Rysy 2,499 m
Natural resources: coal, sulfur, copper, natural gas, silver, lead, salt, amber, arable land
Land use: *arable land:* 40.25%
permanent crops: 1%
other: 58.75% (2005)
Irrigated land: 1,000 sq km (2003)
Total renewable water resources: 63.1 cu km (2005)
Freshwater withdrawal (domestic/industrial/agricultural): *total:* 11.73 cu km/yr (13%/79%/8%)
per capita: 304 cu m/yr (2002)
Natural hazards: flooding
Environment—current issues: situation has improved since 1989 due to decline in heavy industry and increased environmental concern by post-Communist governments; air pollution nonetheless remains serious because of sulfur dioxide emissions from coal-fired power plants, and the resulting acid rain has caused forest damage; water pollution from industrial and municipal sources is also a problem, as is disposal of hazardous wastes; pollution levels should continue to decrease as industrial establishments bring their facilities up to EU code, but at substantial cost to business and the government
Environment—international agreements: *party to:* Air Pollution, Antarctic-Environmental Protocol, Antarctic-Marine Living Resources, Antarctic Seals, Antarctic Treaty, Biodiversity, Climate Change, Climate Change-Kyoto Protocol, Desertification, Endangered Species, Environmental Modification, Hazardous Wastes, Kyoto Protocol, Law of the Sea, Marine Dumping, Ozone Layer Protection, Ship Pollution, Wetlands
signed, but not ratified: Air Pollution-Nitrogen Oxides, Air Pollution-Persistent Organic Pollutants, Air Pollution-Sulfur 94
Geography—note: historically, an area of conflict because of flat terrain and the lack of natural barriers on the North European Plain

PEOPLE

Population: 38,441,588 (July 2011 est.)
country comparison to the world: 34
Age structure: *0–14 years:* 14.7% (male 2,910,324/female 2,748,546)
15–64 years: 71.6% (male 13,698,363/female 13,834,779)
65 years and over: 13.7% (male 2,004,550/female 3,245,026) (2011 est.)
Median age: *total:* 38.5 years
male: 36.8 years
female: 40.3 years (2011 est.)
Population growth rate: -0.062% (2011 est.)
country comparison to the world: 201
Birth rate: 10.01 births/1,000 population (2011 est.)
country comparison to the world: 192
Death rate: 10.17 deaths/1,000 population (July 2011 est.)
country comparison to the world: 52
Net migration rate: -0.47 migrant(s)/1,000 population (2011 est.)
country comparison to the world: 135
Urbanization: *urban population:* 61% of total population (2010)
rate of urbanization: -0.1% annual rate of change (2010-15 est.)
Major cities—population: WARSAW (capital) 1.71 million; Krakow 756,000 (2009)
Sex ratio: *at birth:* 1.061 male(s)/female
under 15 years: 1.06 male(s)/female
15–64 years: 0.99 male(s)/female
65 years and over: 0.62 male(s)/female
total population: 0.94 male(s)/female (2011 est.)
Infant mortality rate: *total:* 6.54 deaths/1,000 live births
country comparison to the world: 170
male: 7.25 deaths/1,000 live births
female: 5.79 deaths/1,000 live births (2011 est.)
Life expectancy at birth: *total population:* 76.05 years
country comparison to the world: 76
male: 72.1 years
female: 80.25 years (2011 est.)
Total fertility rate: 1.3 children born/woman (2011 est.)
country comparison to the world: 209
HIV/AIDS—adult prevalence rate: 0.1%; 0.1% note–no country specific models provided (2009 est.)
country comparison to the world: 154
HIV/AIDS—people living with HIV/AIDS: 27,000 (2009 est.)
country comparison to the world: 71
HIV/AIDS—deaths: fewer than 200 (2009 est.)
country comparison to the world: 109
Major infectious diseases: *degree of risk:* intermediate
food or waterborne diseases: bacterial diarrhea
vectorborne disease: tickborne encephalitis
note: highly pathogenic H5N1 avian influenza has been identified in this country; it poses a negligible risk with extremely rare cases possible among US citizens who have close contact with birds (2009)
Drinking water source: *Improved:*
urban: 100% of population
rural: 100% of population
total: 100% of population (2008)
Sanitation facility access: *Improved:*
urban: 96% of population
rural: 80% of population
total: 90% of population
Unimproved:
urban: 4% of population
rural: 20% of population
total: 10% of population (2008)
Nationality: *noun:* Pole(s)
adjective: Polish
Ethnic groups: Polish 96.7%, German 0.4%, Belarusian 0.1%, Ukrainian 0.1%, other and unspecified 2.7% (2002 census)
Religions: Roman Catholic 89.8% (about 75% practicing), Eastern Orthodox 1.3%, Protestant 0.3%, other 0.3%, unspecified 8.3% (2002)
Languages: Polish (official) 97.8%, other and unspecified 2.2% (2002 census)
Literacy: *definition:* age 15 and over can read and write
total population: 99.8%
male: 99.8%
female: 99.7% (2003 est.)
School life expectancy (primary to tertiary education): *total:* 15 years
male: 15 years
female: 16 years (2008)
Education expenditures: 4.9% of GDP (2007)
country comparison to the world: 62

GOVERNMENT

Country name: *conventional long form:* Republic of Poland
conventional short form: Poland
local long form: Rzeczpospolita Polska
local short form: Polska
Government type: republic
Capital: *name:* Warsaw
geographic coordinates: 52 15 N, 21 00 E
time difference: UTC+1 (6 hours ahead of Washington, DC during Standard Time)
daylight saving time: +1hr, begins last Sunday in March; ends last Sunday in October
Administrative divisions: 16 provinces (wojewodztwa, singular–wojewodztwo); Dolnoslaskie (Lower Silesia), Kujawsko-Pomorskie (Kuyavia-Pomerania), Lodzkie, Lubelskie (Lublin), Lubuskie (Lubusz), Malopolskie (Lesser Poland), Mazowieckie (Masovia), Opolskie, Podkarpackie (Subcarpathia), Podlaskie, Pomorskie (Pomerania), Slaskie (Silesia), Swietokrzyskie, Warminsko-Mazurskie (Warmia-Masuria), Wielkopolskie (Greater Poland), Zachodniopomorskie (West Pomerania)
Independence: 11 November 1918 (republic proclaimed); notable earlier dates: A.D. 966 (adoption of Christianity, traditional founding date), 1 July 1569 (Polish-Lithuanian Commonwealth created)
National holiday: Constitution Day, 3 May (1791)
Constitution: adopted by the National Assembly 2 April 1997; passed by national referendum 25 May 1997; effective 17 October 1997
Legal system: civil law system; changes gradually being introduced as part of broader democratization process; limited judicial review of legislative acts, but rulings of the Constitutional Tribunal are final
International law organization participation: accepts compulsory ICJ jurisdiction with reservations; accepts ICCt jurisdiction
Suffrage: 18 years of age; universal
Executive branch: *chief of state:* President Bronislaw KOMOROWSKI (since 6 August 2010)
head of government: Prime Minister Donald TUSK (since 16 November 2007); Deputy Prime Minister Waldemar PAWLAK (since 16 November 2007)
cabinet: Council of Ministers responsible to the prime minister and the Sejm; the prime minister proposes, the president appoints, and the Sejm approves the Council of Ministers
(For more information visit the World Leaders website)
elections: president elected by popular vote for a five-year term (eligible for a second term); election last held on 20 June and 4 July 2010 (next to be held in 2015); prime minister and deputy prime ministers appointed by the president and confirmed by the Sejm
election results: Bronislaw KOMOROWSKI elected president; percent of popular vote–Bronislaw KOMOROWSKI 53%, Jaroslaw KACZYNSKI 47%
Legislative branch: bicameral legislature consists of an upper house, the Senate or Senat (100 seats; members elected by a majority vote on a provincial basis to serve four-year terms), and a lower house, the Sejm (460 seats; members elected under a complex system of proportional representation to serve four-year terms); the designation of National Assembly or Zgromadzenie Narodowe is only used on those rare occasions when the two houses meet jointly
elections: Senate–last held on 21 October 2007 (next to be held by October 2011); Sejm–last held on 21 October 2007 (next to be held by October 2011)
election results: Senate–percent of vote by party–NA; seats by party–PO 60, PiS 39, independents 1; Sejm–percent of vote by party–PO 41.5%, PiS 32.1%, LiD 13.2%, PSL 8.9%, other 4.3%; seats by party–PO 209, PiS 166, LiD 53, PSL 31, German minorities 1; note–seats by party as of December 2010–PO 203, PiS 147, SLD 44, PSL 31, PJN 17, SDPL 4, DKP-SD 3, German minorities 1, independents 9, vacant 1
note: one seat is assigned to ethnic minority parties in the Sejm only
Judicial branch: Supreme Court (judges are appointed by the president on the recommendation of the National Council of the Judiciary for an indefinite period); Constitutional Tribunal (judges are chosen by the Sejm for nine-year terms)
Political parties and leaders: Civic Platform or PO [Donald TUSK, chairman; Tomasz TOMCZYKIEWICZ, parliamentary caucus leader]; Democratic Caucus of the Democratic Party (SD) or DKP SD [Bogdan LIS, parliamentary caucus leader]; Democratic Left Alliance or SLD [Grzegorz NAPIERALSKI, chairman, parliamentary caucus leader]; Democratic Party or PD [Brygida KUZNIAK, chairwoman]; Democratic Party or SD [Pawel PISKORSKI, chairman]; German Minority of Lower Silesia or MNSO [Ryszard GALLA, representative]; Law and Justice or PiS [Jaroslaw KACZYNSKI, chairman; Mariusz BLASZCZAK, parliamentary caucus leader]; League of Polish Families or LPR [Witold BALAZAK, chairman]; Poland Comes First or PJN [Joanna KLUZIK-ROSTKOWSKA, chairwoman, parliamentary caucus leader]; Polish People's Party or PSL [Waldemar PAWLAK, chairman; Stanislaw ZELICHOWSKI, parliamentary caucus leader]; Samoobrona or SO [Andrzej LEPPER, chairman]; Social Democratic Party of Poland or SDPL [Wojciech FILEMONOWICZ, chairman; Marek BOROWSKI, parliamentary caucus leader]; Union of Labor or UP [Waldemar WITKOWSKI, chairman]
Political pressure groups and leaders: All Poland Trade Union Alliance or OPZZ (trade union) [Jan GUZ]; Roman Catholic Church [Cardinal Stanislaw DZIWISZ, Archbishop Jozef MICHALIK]; Solidarity Trade Union [Piotr DUDA]
International organization participation: Arctic Council (observer), Australia Group, BIS, BSEC (observer), CBSS, CD, CE, CEI, CERN, EAPC, EBRD, EIB, ESA (cooperating state), EU, FAO, IAEA, IBRD, ICAO, ICC, ICRM, IDA, IEA, IFC, IFRCS, IHO, ILO, IMF, IMO, IMSO, Interpol, IOC, IOM, IPU, ISO, ITSO, ITU, ITUC, MIGA, MINURSO, MONUSCO, NATO, NEA, NSG, OAS (observer), OECD, OIF (observer), OPCW, OSCE, PCA, Schengen Convention, UN, UNCTAD, UNESCO, UNHCR, UNIDO, UNMIL, UNMIS, UNOCI, UNWTO, UPU, WCO, WFTU, WHO, WIPO, WMO, WTO, ZC
Diplomatic representation in the US: *chief of mission:* Ambassador Robert KUPIECKI
chancery: 2640 16th Street NW, Washington, DC 20009
telephone: [1] (202) 234-3800 through 3802
FAX: [1] (202) 328-6271
consulate(s) general: Chicago, Los Angeles, New York
Diplomatic representation from the US: *chief of mission:* Ambassador Lee FEINSTEIN
embassy: Aleje Ujazdowskie 29/31 00-540 Warsaw
mailing address: American Embassy Warsaw, US Department of State, Washington, DC 20521-5010 (pouch)
telephone: [48] (22) 504-2000
FAX: [48] (22) 504-2688
consulate(s) general: Krakow
Flag description: two equal horizontal bands of white (top) and red; colors derive from the Polish emblem–a white eagle on a red field
note: similar to the flags of Indonesia and Monaco which are red (top) and white
National anthem: *name:* "Mazurek Dabrowskiego" (Dabrowski's Mazurka)
lyrics/music: Jozef WYBICKI/traditional
note: adopted 1927; the anthem, commonly known as "Jeszcze Polska nie zginela" (Poland Has Not Yet Perished), was written in 1797; the lyrics resonate strongly with Poles because they reflect the numerous occasions in which the nation's lands have been occupied

ECONOMY

Economy—overview: Poland has pursued a policy of economic liberalization since 1990 and today stands out as a success story among transition economies. It is the only country in the European Union to maintain positive GDP growth through the 2008-2009 economic downturn. GDP per capita is still much below the EU average, but is similar to that of the three Baltic states. Since 2004, EU membership and access to EU structural funds have provided a major boost to the economy. Unemployment fell rapidly to 6.4% in October 2008, but climbed back to 11.8% for the year 2010, exceeding the EU average by more than 2%. Inflation reached a low of about 2.6% in 2010 due to the global economic slowdown but has since climbed and is expected to remain around 3%, and close to the upper limit of the National Bank of Poland's target rate. Poland's economic performance could improve over the longer term if the country addresses some of the remaining deficiencies in its road and rail infrastructure and its business environment. An inefficient commercial court system, a rigid labor code, bureaucratic red tape, burdensome tax system, and persistent

low-level corruption keep the private sector from performing up to its full potential. Rising demands to fund health care, education, and the state pension system caused the public sector budget deficit to rise to 7.9% of GDP in 2010. The PO/PSL coalition government, which came to power in November 2007, has planned to reduce the budget deficit in 2011 and has also announced its intention to enact business-friendly reforms, increase workforce participation, reduce public sector spending growth, lower taxes, and accelerate privatization. The government has moved slowly on most major reforms, but has sped up privatization.
GDP (purchasing power parity): $721.3 billion (2010 est.)
country comparison to the world: 21
$694.8 billion (2009 est.)
$683.5 billion (2008 est.)
note: data are in 2010 US dollars
GDP (official exchange rate): $468.5 billion (2010 est.)
GDP—real growth rate: 3.8% (2010 est.)
country comparison to the world: 103
1.7% (2009 est.)
5.1% (2008 est.)
GDP—per capita (PPP): $18,800 (2010 est.)
country comparison to the world: 65
$18,100 (2009 est.)
$17,800 (2008 est.)
note: data are in 2010 US dollars
GDP—composition by sector: *agriculture:* 4%
industry: 32%
services: 64% (2010 est.)
Labor force: 17 million (2010 est.)
country comparison to the world: 35
Labor force—by occupation: *agriculture:* 17.4%
industry: 29.2%
services: 53.4% (2005)
Unemployment rate: 11.8% (2010 est.)
country comparison to the world: 126
11% (2009 est.)
Population below poverty line: 17% (2003 est.)
Household income or consumption by percentage share: *lowest 10%:* 3%
highest 10%: 27.2% (2005)
Distribution of family income—Gini index: 34.9 (2005)
country comparison to the world: 87
31.6 (1998)
Investment (gross fixed): 19.5% of GDP (2010 est.)
country comparison to the world: 95
Budget: *revenues:* $91.23 billion
expenditures: $128.4 billion (2010 est.)
Public debt: 53.6% of GDP (2010 est.)
country comparison to the world: 48
49.8% of GDP (2009 est.)
Inflation rate (consumer prices): 2.6% (2010 est.)
country comparison to the world: 72
3.5% (2009 est.)
Central bank discount rate: 3.5% (31 December 2010)
country comparison to the world: 114
1.75% (31 December 2009)
Commercial bank prime lending rate: 5.99% (31 December 2008 est.)
country comparison to the world: 136
5.72% (31 December 2007 est.)
Stock of narrow money: $138.7 billion (31 December 2010 est.)
country comparison to the world: 24
$124.6 billion (31 December 2009 est.)
Stock of broad money: $251.9 billion (31 December 2010 est.)
country comparison to the world: 36
$229.2 billion (31 December 2009 est.)
Stock of domestic credit: $288.7 billion (31 December 2010 est.)
country comparison to the world: 35
$264.1 billion (31 December 2009 est.)
Market value of publicly traded shares: $135.3 billion (31 December 2009)
country comparison to the world: 42
$90.23 billion (31 December 2008)
$207.3 billion (31 December 2007)
Agriculture—products: potatoes, fruits, vegetables, wheat; poultry, eggs, pork, dairy
Industries: machine building, iron and steel, coal mining, chemicals, shipbuilding, food processing, glass, beverages, textiles
Industrial production growth rate: 6.5% (2010 est.)
country comparison to the world: 51
Electricity—production: 149.1 billion kWh (2007 est.)
country comparison to the world: 23
Electricity—consumption: 129.3 billion kWh (2007 est.)
country comparison to the world: 25
Electricity—exports: 9.703 billion kWh (2008)
Electricity—imports: 8.48 billion kWh (2008 est.)
Oil—production: 34,140 bbl/day (2009 est.)
country comparison to the world: 69
Oil—consumption: 545,400 bbl/day (2009 est.)
country comparison to the world: 30
Oil—exports: 65,280 bbl/day (2008 est.)
country comparison to the world: 74
Oil—imports: 553,900 bbl/day (2008 est.)
country comparison to the world: 22
Oil—proved reserves: 96.38 million bbl (1 January 2010 est.)
country comparison to the world: 72
Natural gas—production: 5.842 billion cu m (2009 est.)
country comparison to the world: 49
Natural gas—consumption: 16.33 billion cu m (2009 est.)
country comparison to the world: 40
Natural gas—exports: 40 million cu m (2009 est.)
country comparison to the world: 44
Natural gas—imports: 9.954 billion cu m (2009 est.)
country comparison to the world: 20
Natural gas—proved reserves: 164.8 billion cu m (1 January 2010 est.)
country comparison to the world: 48
Current account balance: $-12.33 billion (2010 est.)
country comparison to the world: 178
$-9.598 billion (2009 est.)
Exports: $160.8 billion (2010 est.)
country comparison to the world: 28
$142.1 billion (2009 est.)
Exports—commodities: machinery and transport equipment 37.8%, intermediate manufactured goods 23.7%, miscellaneous manufactured goods 17.1%, food and live animals 7.6%
Exports—partners: Germany 26.06%, Italy 6.84%, France 6.78%, UK 6.38%, Czech Republic 5.85%, Netherlands 4.14% (2009)
Imports:
$167.4 billion (2010 est.)
country comparison to the world: 24
$146.4 billion (2009 est.)
Imports—commodities: machinery and transport equipment 38%, intermediate manufactured goods 21%, chemicals 15%, minerals, fuels, lubricants, and related materials 9%
Imports—partners: Germany 28.08%, Russia 8.65%, Italy 6.5%, Netherlands 5.59%, China 5.27%, France 4.6%, Czech Republic 4.05% (2009)
Reserves of foreign exchange and gold: $99.76 billion (31 December 2010 est.)
country comparison to the world: 17
$79.58 billion (31 December 2009 est.)
Debt—external: $252.9 billion (31 December 2010 est.)
country comparison to the world: 28
$239.6 billion (31 December 2009 est.)
Stock of direct foreign investment—at home: $198.8 billion (31 December 2010 est.)
country comparison to the world: 22
$182.8 billion (31 December 2009 est.)
Stock of direct foreign investment—abroad: $30.71 billion (31 December 2010 est.)
country comparison to the world: 37
$26.21 billion (31 December 2009 est.)
Exchange rates: zlotych (PLN) per US dollar–3.0718 (2010), 3.1214 (2009), 2.3 (2008), 2.81 (2007), 3.1032 (2006)

COMMUNICATIONS

Telephones—main lines in use: 9.556 million (2009)
country comparison to the world: 23
Telephones—mobile cellular: 44.553 million (2009)
country comparison to the world: 28
Telephone system: *general assessment:* modernization of the telecommunications network has accelerated with market-based competition; fixed-line service, dominated by the former state-owned company, is dwarfed by the growth in mobile-cellular services
domestic: mobile-cellular service available since 1993 and provided by three nationwide networks with a fourth provider beginning operations in late 2006; coverage is generally good with some gaps in the east; fixed-line service lags in rural areas
international: country code–48; international direct dialing with automated exchanges; satellite earth station–1 with access to Intelsat, Eutelsat, Inmarsat, and Intersputnik (2009)
Broadcast media: state-run public television operates 2 national channels supplemented by 16 regional channels and several niche channels; privately-owned entities operate several national TV broadcast networks and a number of special interest channels; large number of privately-owned channels broadcasting locally; roughly half of all households are linked to either satellite or cable TV systems providing access to foreign television networks; state-run public radio operates 5 national networks and 17

regional radio stations; 2 privately-owned national radio networks, several commercial stations broadcasting to multiple cities, and a large number of privately-owned local radio stations (2007)
Internet country code: .pl
Internet hosts: 10.51 million (2010)
country comparison to the world: 11
Internet users: 22.452 million (2009)
country comparison to the world: 19

TRANSPORTATION

Airports: 129 (2010)
country comparison to the world: 46
Airports—with paved runways: *total:* 86
over 3,047 m: 4
2,438 to 3,047 m: 30
1,524 to 2,437 m: 39
914 to 1,523 m: 7
under 914 m: 6 (2010)
Airports—with unpaved runways: *total:* 43
2,438 to 3,047 m: 1
1,524 to 2,437 m: 5
914 to 1,523 m: 16
under 914 m: 21 (2010)
Heliports: 7 (2010)
Pipelines: gas 13,860 km; oil 1,384 km; refined products 777 km; unknown 35 km (2010)
Railways: *total:* 22,314 km
country comparison to the world: 12
broad gauge: 633 km 1.524-m gauge
standard gauge: 21,681 km 1.435-m gauge (11,769 km electrified) (2007)
Roadways: *total:* 423,997 km
country comparison to the world: 15
paved: 295,356 km (includes 765 km of expressways)
unpaved: 128,641 km (2008)
Waterways: 3,997 km (navigable rivers and canals) (2009)
country comparison to the world: 28
Merchant marine: *total:* 10
country comparison to the world: 115
by type: cargo 6, chemical tanker 3, passenger/cargo 1
registered in other countries: 104 (Antigua and Barbuda 2, Bahamas 32, Cyprus 20, Liberia 13, Malta 22, Norway 2, Panama 3, Saint Vincent and the Grenadines 1, Slovakia 2, Vanuatu 7) (2010)
Ports and terminals: Gdansk, Gdynia, Swinoujscie, Szczecin

MILITARY

Military branches: Polish Armed Forces: Land Forces, Navy, Air and Air Defense Aviation Forces, Special Forces (2010)
Military service age and obligation: 18-28 years of age for male voluntary or compulsory military service; service obligation shortened from 12 to 9 months in 2005; conscription is to end in 2012; only soldiers who have completed their conscript service are allowed to volunteer for professional service; as of April 2004, women are only allowed to serve as officers and noncommissioned officers; reserve obligation to age 50 (2009)
Manpower available for military service: *males age 16–49:* 9,531,855
females age 16–49: 9,298,593 (2010 est.)
Manpower fit for military service: *males age 16–49:* 7,817,556
females age 16–49: 7,766,361 (2010 est.)
Manpower reaching militarily significant age annually: *male:* 221,889
female: 211,172 (2010 est.)
Military expenditures: 1.71% of GDP (2005 est.)
country comparison to the world: 86

TRANSNATIONAL ISSUES

Disputes—international: as a member state that forms part of the EU's external border, Poland has implemented the strict Schengen border rules to restrict illegal immigration and trade along its eastern borders with Belarus and Ukraine
Illicit drugs: despite diligent counternarcotics measures and international information sharing on cross-border crimes, a major illicit producer of synthetic drugs for the international market; minor transshipment point for Southwest Asian heroin and Latin American cocaine to Western Europe

PORTUGAL

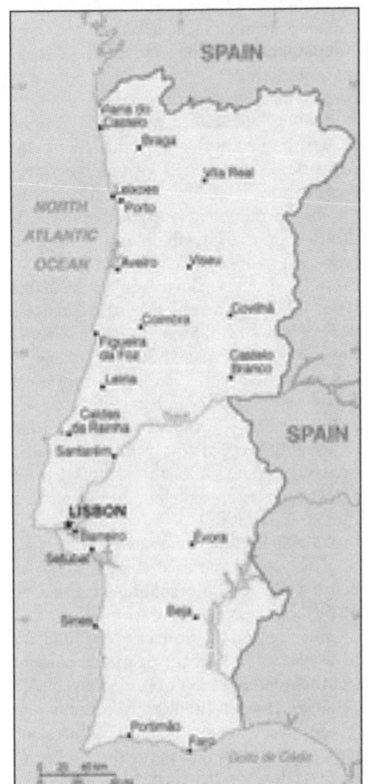

INTRODUCTION

Background: Following its heyday as a global maritime power during the 15th and 16th centuries, Portugal lost much of its wealth and status with the destruction of Lisbon in a 1755 earthquake, occupation during the Napoleonic Wars, and the independence of its wealthiest colony of Brazil in 1822. A 1910 revolution deposed the monarchy; for most of the next six decades, repressive governments ran the country. In 1974, a left-wing military coup installed broad democratic reforms. The following year, Portugal granted independence to all of its African colonies. Portugal is a founding member of NATO and entered the EC (now the EU) in 1986. In January 2011, Portugal assumed a nonpermanent seat on the UN Security Council for the 2011-12 term.

GEOGRAPHY

Location: Southwestern Europe, bordering the North Atlantic Ocean, west of Spain
Geographic coordinates: 39 30 N, 8 00 W
Map references: Europe
Area: *total:* 92,090 sq km
country comparison to the world: 110
land: 91,470 sq km
water: 620 sq km
note: includes Azores and Madeira Islands
Area—comparative: slightly smaller than Indiana
Land boundaries: *total:* 1,214 km
border countries: Spain 1,214 km
Coastline: 1,793 km
Maritime claims: territorial sea: 12 nm
contiguous zone: 24 nm
exclusive economic zone: 200 nm
continental shelf: 200 m depth or to the depth of exploitation
Climate: maritime temperate; cool and rainy in north, warmer and drier in south
Terrain: mountainous north of the Tagus River, rolling plains in south
Elevation extremes: *lowest point:* Atlantic Ocean 0 m
highest point: Ponta do Pico (Pico or Pico Alto) on Ilha do Pico in the Azores 2,351 m
Natural resources: fish, forests (cork), iron ore, copper, zinc, tin, tungsten, silver, gold, uranium, marble, clay, gypsum, salt, arable land, hydropower
Land use: *arable land:* 17.29%
permanent crops: 7.84%
other: 74.87% (2005)
Irrigated land: 6,500 sq km (2003)
Total renewable water resources: 73.6 cu km (2005)
Freshwater withdrawal (domestic/industrial/agricultural): *total:* 11.09 cu km/yr (10%/12%/78%)
per capita: 1,056 cu m/yr (1998)
Natural hazards: Azores subject to severe earthquakes
volcanism: Portugal experiences limited volcanic activity in the Azores Islands;

Fayal or Faial (elev. 1,043 m) last erupted in 1958; most volcanoes have not erupted in centuries; historically active volcanoes include Agua de Pau, Furnas, Pico, Picos Volcanic System, San Jorge, Sete Cidades, and Terceira

Environment—current issues: soil erosion; air pollution caused by industrial and vehicle emissions; water pollution, especially in coastal areas

Environment—international agreements: *party to:* Air Pollution, Biodiversity, Climate Change, Climate Change-Kyoto Protocol, Desertification, Endangered Species, Hazardous Wastes, Law of the Sea, Marine Dumping, Marine Life Conservation, Ozone Layer Protection, Ship Pollution, Tropical Timber 83, Tropical Timber 94, Wetlands, Whaling

signed, but not ratified: Air Pollution-Persistent Organic Pollutants, Air Pollution-Volatile Organic Compounds, Environmental Modification

Geography—note: Azores and Madeira Islands occupy strategic locations along western sea approaches to Strait of Gibraltar

PEOPLE

Population: 10,760,305 (July 2011 est.)
country comparison to the world: 75

Age structure: *0–14 years:* 16.2% (male 910,012/female 835,025)
15–64 years: 65.8% (male 3,539,457/female 3,541,989)
65 years and over: 18% (male 791,950/female 1,141,872) (2011 est.)

Median age: *total:* 40 years
male: 38 years
female: 42.3 years (2011 est.)

Population growth rate: 0.212% (2011 est.)
country comparison to the world: 180

Birth rate: 9.94 births/1,000 population (2011 est.)
country comparison to the world: 195

Death rate: 10.8 deaths/1,000 population (July 2011 est.)
country comparison to the world: 42

Net migration rate: 2.98 migrant(s)/1,000 population (2011 est.)
country comparison to the world: 29

Urbanization: *urban population:* 61% of total population (2010)
rate of urbanization: 1% annual rate of change (2010-15 est.)

Major cities—population: LISBON (capital) 2.808 million; Porto 1.344 million (2009)

Sex ratio: *at birth:* 1.067 male(s)/female
under 15 years: 1.09 male(s)/female
15–64 years: 1 male(s)/female
65 years and over: 0.7 male(s)/female
total population: 0.95 male(s)/female (2011 est.)

Infant mortality rate:
total: 4.66 deaths/1,000 live births
country comparison to the world: 188
male: 5.11 deaths/1,000 live births
female: 4.18 deaths/1,000 live births (2011 est.)

Life expectancy at birth: *total population:* 78.54 years
country comparison to the world: 49
male: 75.28 years
female: 82.01 years (2011 est.)

Total fertility rate: 1.5 children born/woman (2011 est.)
country comparison to the world: 185

HIV/AIDS—adult prevalence rate: 0.6% (2009 est.)
country comparison to the world: 64

HIV/AIDS—people living with HIV/AIDS: 42,000 (2009 est.)
country comparison to the world: 61

HIV/AIDS—deaths: fewer than 500 (2009 est.)
country comparison to the world: 82

Drinking water source: *Improved:*
urban: 99% of population
rural: 100% of population
total: 99% of population
Unimproved: urban: 1% of population
rural: 0% of population
total: 1% of population (2008)

Sanitation facility access: *Improved:*
urban: 100% of population
rural: 100% of population
total: 100% of population (2008)

Nationality: *noun:* Portuguese (singular and plural)
adjective: Portuguese

Ethnic groups: homogeneous Mediterranean stock; citizens of black African descent who immigrated to mainland during decolonization number less than 100,000; since 1990 East Europeans have entered Portugal

Religions: Roman Catholic 84.5%, other Christian 2.2%, other 0.3%, unknown 9%, none 3.9% (2001 census)

Languages: Portuguese (official), Mirandese (official, but locally used)

Literacy: *definition:* age 15 and over can read and write
total population: 93.3%
male: 95.5%
female: 91.3% (2003 est.)

School life expectancy (primary to tertiary education): *total:* 16 years
male: 16 years
female: 16 years (2008)

Education expenditures: 4.4% of GDP (2008)
country comparison to the world: 89

GOVERNMENT

Country name: *conventional long form:* Portuguese Republic
conventional short form: Portugal
local long form: Republica Portuguesa
local short form: Portugal

Government type: republic; parliamentary democracy

Capital: *name:* Lisbon
geographic coordinates: 38 43 N, 9 08 W
time difference: UTC 0 (5 hours ahead of Washington, DC during Standard Time)
daylight saving time: +1hr, begins last Sunday in March; ends last Sunday in October

Administrative divisions: 18 districts (distritos, singular—distrito) and 2 autonomous regions* (regioes autonomas, singular—regiao autonoma); Aveiro, Acores (Azores)*, Beja, Braga, Braganca, Castelo Branco, Coimbra, Evora, Faro, Guarda, Leiria, Lisboa (Lisbon), Madeira*, Portalegre, Porto, Santarem, Setubal, Viana do Castelo, Vila Real, Viseu

Independence: 1143 (Kingdom of Portugal recognized); 5 October 1910 (republic proclaimed)

National holiday: Portugal Day (Dia de Portugal), 10 June (1580); note—also called Camoes Day, the day that revered national poet Luis de Camoes (1524-80) died

Constitution: adopted 2 April 1976; subsequently revised
note: the revisions placed the military under strict civilian control, trimmed the powers of the president, and laid the groundwork for a stable, pluralistic liberal democracy; they allowed for the privatization of nationalized firms and government-owned communications media

Legal system: civil law system; Constitutional Tribunal review of legislative acts

International law organization participation: accepts compulsory ICJ jurisdiction with reservations; accepts ICCt jurisdiction

Suffrage: 18 years of age; universal

Executive branch: *chief of state:* President Anibal CAVACO SILVA (since 9 March 2006)
head of government: Prime Minister Jose SOCRATES Carvalho Pinto de Sousa (since 12 March 2005); note—Prime Minister SOCRATES resigned 23 March 2011; new legislative elections are called for on 5 June
cabinet: Council of Ministers appointed by the president on the recommendation of the prime minister
(For more information visit the World Leaders website)
note: there is also a Council of State that acts as a consultative body to the president
elections: president elected by popular vote for a five-year term (eligible for a second term); election last held on 23 January 2011 (next to be held in January 2016); following legislative elections, the leader of the majority party or leader of a majority coalition usually appointed prime minister by the president
election results: Anibal CAVACO SILVA reelected president; percent of vote—Anibal CAVACO SILVA 53%, Manuel ALEGRE 19.8%, Fernando NOBRE 14.1%, Francisco LOPES 7.1%, Manuel COELHO 4.5%, Defensor MOURA 1.6%

Legislative branch: unicameral Assembly of the Republic or Assembleia da Republica (230 seats; members elected by popular vote to serve four-year terms)
elections: last held on 27 September 2009 (next to be held in fall 2013)
election results: percent of vote by party—PS 42%, PSD 35%, CDS/PP 9%, BE 7%, CDU 7%; seats by party—PS 97, PSD 81, CDS/PP 21, BE 16, CDU 15

Judicial branch: Supreme Court (Supremo Tribunal de Justica); judges appointed for life by the Conselho Superior da Magistratura

Political parties and leaders: Democratic and Social Center/Popular Party or CDS/PP [Paulo PORTAS]; Social Democratic Party or PSD [Pedro Manuel PASSOS COELHO]; Socialist Party or PS [Jose SOCRATES Carvalho Pinto de Sousa]; The Left Bloc or BE [Francisco Anacleto LOUCA]; Unitarian Democratic Coalition or CDU [Jeronimo DE SOUSA] (includes Portuguese Communist Party or PCP and Ecologist Party ("The Greens") or PEV)

Political pressure groups and leaders: the media; labor unions

International organization participation: ADB (nonregional member), AfDB (nonregional member), Australia Group, BIS, CD, CE, CERN, CPLP, EAPC, EBRD, EIB, EMU, ESA, EU, FAO, FATF, IADB, IAEA, IBRD, ICAO, ICC, ICRM, IDA, IEA, IFAD, IFC, IFRCS, IHO, ILO, IMF, IMO, IMSO, Interpol, IOC, IOM, IPU, ISO, ITSO, ITU, ITUC, LAIA (observer), MIGA, NATO, NEA, NSG, OAS (observer), OECD, OPCW, OSCE, Paris Club (associate), PCA, Schengen Convention, SECI (observer), UN, UN Security Council (temporary), UNCTAD, UNESCO, UNHCR, UNIDO, UNIFIL, Union Latina, UNMIT, UNWTO, UPU, WCO, WFTU, WHO, WIPO, WMO, WTO, ZC

Diplomatic representation in the US: *chief of mission:* Ambassador Nuno Filipe Alves Salvador e BRITO
chancery: 2012 Massachusetts Avenue NW, Washington, DC 20036
telephone: [1] (202) 328-8610
FAX: [1] (202) 462-3726
consulate(s) general: Boston, New York, Newark (New Jersey), San Francisco
consulate(s): New Bedford (Massachusetts), Providence (Rhode Island)

Diplomatic representation from the US: *chief of mission:* Ambassador Allan J. KATZ
embassy: Avenida das Forcas Armadas, 1600-081 Lisbon
mailing address: Apartado 43033, 1601-301 Lisboa; PSC 83, APO AE 09726
telephone: [351] (21) 727-3300
FAX: [351] (21) 726-9109
consulate(s): Ponta Delgada (Azores)

Flag description: two vertical bands of green (hoist side, two-fifths) and red (three-fifths) with the national coat of arms (armillary sphere and Portuguese shield) centered on the dividing line; explanations for the color meanings are ambiguous, but a popular interpretation has green symbolizing hope and red the blood of those defending the nation

National anthem: *name:* "A Portugesa" (The Song of the Portuguese)
lyrics/music: Henrique LOPES DE MENDOCA/Alfredo KEIL
note: adopted 1910; "A Portuguesa" was originally written to protest the Portuguese monarchy's acquiescence to the 1890 British ultimatum forcing Portugal to give up areas of Africa; the lyrics refer to the "insult" that resulted from the event

ECONOMY

Economy—overview: Portugal has become a diversified and increasingly service-based economy since joining the European Community–the EU's predecessor–in 1986. Over the past two decades, successive governments have privatized many state-controlled firms and liberalized key areas of the economy, including the financial and telecommunications sectors. The country qualified for the Economic and Monetary Union (EMU) in 1998 and began circulating the euro on 1 January 2002 along with 11 other EU members. The economy had grown by more than the EU average for much of the 1990s, but fell back in 2001-08, and contracted 2.6% in 2009, before growing 1% in 2010. GDP per capita stands at roughly two-thirds of the EU-27 average. A poor educational system and a rigid labor market have been obstacles to greater productivity and growth. Portugal also has been increasingly overshadowed by lower-cost producers in Central Europe and Asia as a destination for foreign direct investment. Portugal's low competitiveness, low growth prospects, and high levels of public debt have made it vulnerable to bond market turbulence. The government is implementing austerity measures, including a 5% public salary cut which went into effect on January 1, 2011 and a 2% increase in the value-added tax, to reduce the budget deficit from 9.3% of GDP in 2009 to 4.6% in 2011, but some investors have expressed concern about the government's ability to achieve these targets and cover its sovereign debt. Without the option for stimulus measures, the government is focusing instead on boosting exports and implementing labor market reforms to try to raise GDP growth and increase Portugal's competitiveness–which, over time, may help mitigate investor concerns.

GDP (purchasing power parity): $247 billion (2010 est.)
country comparison to the world: 50
$243.6 billion (2009 est.)
$249.8 billion (2008 est.)
note: data are in 2010 US dollars

GDP (official exchange rate): $229.3 billion (2010 est.)

GDP—real growth rate: 1.4% (2010 est.)
country comparison to the world: 163
-2.5% (2009 est.)
0% (2008 est.)

GDP—per capita (PPP): $23,000 (2010 est.)
country comparison to the world: 57
$22,800 (2009 est.)
$23,400 (2008 est.)
note: data are in 2010 US dollars

GDP—composition by sector:
agriculture: 2.6%
industry: 23%
services: 74.5% (2010 est.)

Labor force: 5.57 million (2010 est.)
country comparison to the world: 67

Labor force—by occupation: *agriculture:* 11.7%
industry: 28.5%
services: 59.8% (2009 est.)

Unemployment rate: 10.7% (2010 est.)
country comparison to the world: 117
9.5% (2009 est.)

Population below poverty line: 18% (2006)

Household income or consumption by percentage share: *lowest 10%:* 3.1%
highest 10%: 28.4% (1995 est.)

Distribution of family income—Gini index: 38.5 (2007)
country comparison to the world: 70
35.6 (1995)

Investment (gross fixed): 19% of GDP (2010 est.)
country comparison to the world: 100

Budget: *revenues:* $93.61 billion
expenditures: $110.2 billion (2010 est.)

Public debt: 83.2% of GDP (2010 est.)
country comparison to the world: 15
76.8% of GDP (2009 est.)

Inflation rate (consumer prices): 1.1% (2010 est.)
country comparison to the world: 22
-0.8% (2009 est.)

Central bank discount rate: 1.75% (31 December 2010)
country comparison to the world: 115
1.75% (31 December 2009)
note: this is the European Central Bank's rate on the marginal lending facility, which offers overnight credit to banks in the euro area

Commercial bank prime lending rate: 6.12% (31 December 2009 est.)
country comparison to the world: 111
8.35% (31 December 2008 est.)

Stock of narrow money: $98.23 billion (31 December 2010 est.)
country comparison to the world: 32
$100.9 billion (31 December 2009 est.)
note: see entry for the European Union for money supply in the euro area; the European Central Bank (ECB) controls monetary policy for the 17 members of the Economic and Monetary Union (EMU); individual members of the EMU do not control the quantity of money circulating within their own borders

Stock of broad money: $282 billion (31 December 2010 est.)
country comparison to the world: 30
$302.3 billion (31 December 2009 est.)

Stock of domestic credit: $556.3 billion (31 December 2009 est.)
country comparison to the world: 23
$490.8 billion (31 December 2008 est.)

Market value of publicly traded shares: $98.65 billion (31 December 2009)
country comparison to the world: 46
$68.71 billion (31 December 2008)
$132.3 billion (31 December 2007)

Agriculture—products: grain, potatoes, tomatoes, olives, grapes; sheep, cattle, goats, pigs, poultry, dairy products; fish

Industries: textiles, clothing, footwear, wood and cork, paper, chemicals, auto-parts manufacturing, base metals, dairy products, wine and other foods, porcelain and ceramics, glassware, technology, telecommunications; ship construction and refurbishment; tourism

Industrial production growth rate: 0.9% (2010 est.)
country comparison to the world: 148

Electricity—production: 44.47 billion kWh (2007 est.)
country comparison to the world: 52

Electricity—consumption: 48.78 billion kWh (2007 est.)
country comparison to the world: 47

Electricity—exports: 1.313 billion kWh (2008 est.)

Electricity—imports: 10.74 billion kWh (2008 est.)

Oil—production: 4,721 bbl/day (2009 est.)
country comparison to the world: 96

Oil—consumption: 272,200 bbl/day (2009 est.)
country comparison to the world: 48

Oil—exports: 53,660 bbl/day (2008 est.)
country comparison to the world: 76
Oil—imports: 323,000 bbl/day (2008 est.)
country comparison to the world: 32
Oil—proved reserves: 0 bbl (1 January 2010 est.)
country comparison to the world: 177
Natural gas—production: NA
Natural gas—consumption: 4.846 billion cu m (2009 est.)
country comparison to the world: 59
Natural gas—exports: 0 cu m (2008 est.)
country comparison to the world: 158
Natural gas—imports: 4.895 billion cu m (2009 est.)
country comparison to the world: 32
Natural gas—proved reserves: 0 cu m (1 January 2010 est.)
country comparison to the world: 179
Current account balance: $-19.03 billion (2010 est.)
country comparison to the world: 181
$-23.95 billion (2009 est.)
Exports: $46.27 billion (2010 est.)
country comparison to the world: 57
$44.49 billion (2009 est.)
Exports—commodities: agricultural products, food products, wine, oil products, chemical products, plastics and rubber, hides, leather, wood and cork, wood pulp and paper, textile materials, clothing, footwear, machinery and tools
Exports—partners: Spain 26.25%, Germany 12.99%, France 12.04%, Angola 7.21%, UK 5.54% (2009)
Imports: $68.22 billion (2010 est.)
country comparison to the world: 40
$68.9 billion (2009 est.)
Imports—commodities: agricultural products, chemical products, vehicles and other transport material, and optical and precision instruments, computer accessories and parts, semi-conductors and related devices
Imports—partners: Spain 31.58%, Germany 12.41%, France 8.58%, Italy 5.55%, Netherlands 5.31% (2009)
Reserves of foreign exchange and gold: $NA (31 December 2010 est.)
$16.03 billion (31 December 2009 est.)
Debt—external: $497.8 billion (30 June 2010)
country comparison to the world: 21
$507 billion (30 June 2009)
Stock of direct foreign investment—at home: $105.7 billion (31 December 2010 est.)
country comparison to the world: 32
$102.6 billion (31 December 2009 est.)
Stock of direct foreign investment—abroad: $63.64 billion (31 December 2010 est.)
country comparison to the world: 28
$63.64 billion (31 December 2009 est.)
Exchange rates: euros (EUR) per US dollar–0.755 (2010), 0.7198 (2009), 0.6827 (2008), 0.7345 (2007), 0.7964 (2006)

COMMUNICATIONS

Telephones—main lines in use: 4.049 million (2009)
country comparison to the world: 40
Telephones—mobile cellular: 15.178 million (2009)
country comparison to the world: 48
Telephone system: *general assessment:* Portugal's telephone system has a state-of-the-art network with broadband, high-speed capabilities
domestic: integrated network of coaxial cables, open-wire, microwave radio relay, and domestic satellite earth stations
international: country code–351; a combination of submarine cables provide connectivity to Europe, North and East Africa, South Africa, the Middle East, Asia, and the US; satellite earth stations–3 Intelsat (2 Atlantic Ocean and 1 Indian Ocean), NA Eutelsat; tropospheric scatter to Azores (2008)
Broadcast media: the publicly-owned TV broadcaster operates 2 domestic channels and external service channels to Africa; overall, roughly 40 domestic TV stations; viewers have widespread access to international broadcasters with more than half of all households connected to multi-channel cable or satellite TV systems; publicly-owned radio operates 3 national networks and provides regional and external services; several privately-owned national radio stations and some 300 regional and local commercial radio stations (2008)
Internet country code: .pt
Internet hosts: 3.267 million (2010)
country comparison to the world: 28
Internet users: 5.168 million (2009)
country comparison to the world: 45

TRANSPORTATION

Airports: 65 (2010)
country comparison to the world: 75
Airports—with paved runways: *total:* 43
over 3,047 m: 5
2,438 to 3,047 m: 8
1,524 to 2,437 m: 7
914 to 1,523 m: 13
under 914 m: 10 (2010)
Airports—with unpaved runways: *total:* 22
914 to 1,523 m: 1
under 914 m: 21 (2010)
Pipelines: gas 1,307 km; oil 11 km; refined products 188 km (2010)
Railways: *total:* 2,786 km
country comparison to the world: 58
broad gauge: 2,603 km 1.668-m gauge (1,351 km electrified)
narrow gauge: 183 km 1.000-m gauge (2008)
Roadways: *total:* 82,900 km
country comparison to the world: 56
paved: 71,294 km (includes 2,613 km of expressways)
unpaved: 11,606 km (2008)
Waterways: 210 km (on Douro River from Porto) (2010)
country comparison to the world: 96
Merchant marine: *total:* 111
country comparison to the world: 47
by type: bulk carrier 8, cargo 33, carrier 1, chemical tanker 17, container 8, liquefied gas 9, passenger 13, passenger/cargo 5, petroleum tanker 7, roll on/roll off 1, vehicle carrier 9
foreign-owned: 80 (Belgium 8, Denmark 4, Germany 13, Greece 5, Italy 10, Japan 9, Mexico 1, Netherlands 1, Norway 1, Spain 15, Sweden 6, Switzerland 3, US 4)
registered in other countries: 14 (Cyprus 2, Malta 3, Panama 9) (2010)
Ports and terminals: Leixoes, Lisbon, Setubal, Sines

MILITARY

Military branches: Portuguese Army (Exercito Portuguesa), Portuguese Navy (Marinha Portuguesa; includes Marine Corps), Portuguese Air Force (Forca Aerea Portuguesa, FAP) (2010)
Military service age and obligation: 18 years of age for voluntary military service; no compulsory military service; women serve in the armed forces, on naval ships since 1993, but are prohibited from serving in some combatant specialties; reserve obligation to age 35 (2010)
Manpower available for military service: *males age 16–49:* 2,566,264
females age 16–49: 2,458,297 (2010 est.)
Manpower fit for military service: *males age 16–49:* 2,103,080
females age 16–49: 2,018,004 (2010 est.)
Manpower reaching militarily significant age annually: *male:* 62,208
female: 54,786 (2010 est.)
Military expenditures: 2.3% of GDP (2005 est.)
country comparison to the world: 66

TRANSNATIONAL ISSUES

Disputes—international: Portugal does not recognize Spanish sovereignty over the territory of Olivenza based on a difference of interpretation of the 1815 Congress of Vienna and the 1801 Treaty of Badajoz
Illicit drugs: seizing record amounts of Latin American cocaine destined for Europe; a European gateway for Southwest Asian heroin; transshipment point for hashish from North Africa to Europe; consumer of Southwest Asian heroin

PUERTO RICO

(TERRITORY OF THE US WITH COMMONWEALTH STATUS)

INTRODUCTION

Background: Populated for centuries by aboriginal peoples, the island was claimed by the Spanish Crown in 1493 following COLUMBUS' second voyage to the Americas. In 1898, after 400 years of colonial rule that saw the indigenous population nearly exterminated and African slave labor introduced, Puerto Rico was ceded to the US as a result of the Spanish-American War. Puerto Ricans were granted US citizenship in 1917. Popularly-elected

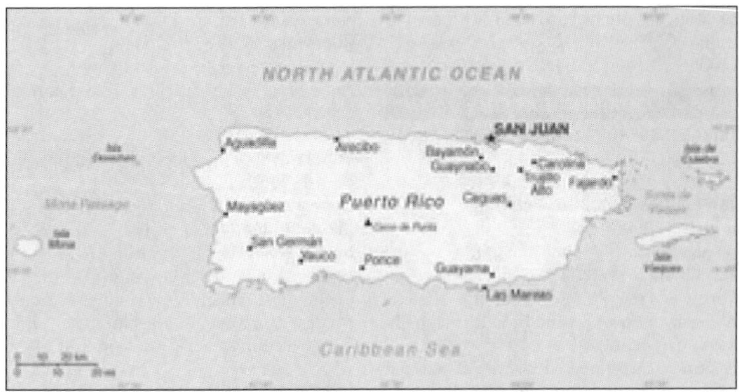

governors have served since 1948. In 1952, a constitution was enacted providing for internal self government. In plebiscites held in 1967, 1993, and 1998, voters chose not to alter the existing political status.

GEOGRAPHY

Location: Caribbean, island between the Caribbean Sea and the North Atlantic Ocean, east of the Dominican Republic
Geographic coordinates: 18 15 N, 66 30 W
Map references: Central America and the Caribbean
Area: *total:* 13,790 sq km
country comparison to the world: 162
land: 8,870 sq km
water: 4,921 sq km
Area—comparative: slightly less than three times the size of Rhode Island
Land boundaries: 0 km
Coastline: 501 km
Maritime claims: territorial sea: 12 nm
exclusive economic zone: 200 nm
Climate: tropical marine, mild; little seasonal temperature variation
Terrain: mostly mountains with coastal plain belt in north; mountains precipitous to sea on west coast; sandy beaches along most coastal areas
Elevation extremes: *lowest point:* Caribbean Sea 0 m
highest point: Cerro de Punta 1,338 m
Natural resources: some copper and nickel; potential for onshore and offshore oil
Land use: *arable land:* 3.69%
permanent crops: 5.59%
other: 90.72% (2005)
Irrigated land: 400 sq km (2003)
Natural hazards: periodic droughts; hurricanes
Environment—current issues: erosion; occasional drought causing water shortages
Geography—note: important location along the Mona Passage–a key shipping lane to the Panama Canal; San Juan is one of the biggest and best natural harbors in the Caribbean; many small rivers and high central mountains ensure land is well watered; south coast relatively dry; fertile coastal plain belt in north

PEOPLE

Population: 3,989,133 (July 2011 est.)
country comparison to the world: 128
Age structure: *0–14 years:* 18.8% (male 383,748/female 367,484)
15–64 years: 66.1% (male 1,270,557/female 1,366,417)
65 years and over: 15.1% (male 258,570/female 342,357) (2011 est.)
Median age:
total: 37.3 years
male: 35.5 years
female: 39 years (2011 est.)
Population growth rate: 0.254% (2011 est.)
country comparison to the world: 172
Birth rate: 11.35 births/1,000 population (2011 est.)
country comparison to the world: 172
Death rate: 7.95 deaths/1,000 population (July 2011 est.)
country comparison to the world: 107
Net migration rate: -0.86 migrant(s)/1,000 population (2011 est.)
country comparison to the world: 148
Urbanization: *urban population:* 99% of total population (2010)
rate of urbanization: 0.5% annual rate of change (2010-15 est.)
Major cities—population: SAN JUAN (capital) 2.73 million (2009)
Sex ratio: *at birth:* 1.05 male(s)/female
under 15 years: 1.04 male(s)/female
15–64 years: 0.93 male(s)/female
65 years and over: 0.75 male(s)/female
total population: 0.92 male(s)/female (2011 est.)
Infant mortality rate: *total:* 8.07 deaths/1,000 live births
country comparison to the world: 161
male: 8.83 deaths/1,000 live births
female: 7.26 deaths/1,000 live births (2011 est.)
Life expectancy at birth: *total population:* 78.92 years
country comparison to the world: 43
male: 75.31 years
female: 82.71 years (2011 est.)
Total fertility rate: 1.62 children born/woman (2011 est.)
country comparison to the world: 176
HIV/AIDS—adult prevalence rate: NA
HIV/AIDS—people living with HIV/AIDS: 7,397 (1997)
country comparison to the world: 111
HIV/AIDS—deaths: NA
Nationality: *noun:* Puerto Rican(s) (US citizens)
adjective: Puerto Rican
Ethnic groups: white (mostly Spanish origin) 76.2%, black 6.9%, Asian 0.3%, Amerindian 0.2%, mixed 4.4%, other 12% (2007)
Religions: Roman Catholic 85%, Protestant and other 15%
Languages: Spanish, English
Literacy: *definition:* age 15 and over can read and write
total population: 94.1%
male: 93.9%
female: 94.4% (2002 est.)
School life expectancy (primary to tertiary education): NA
Education expenditures: NA

GOVERNMENT

Country name: *conventional long form:* Commonwealth of Puerto Rico
conventional short form: Puerto Rico
Dependency status: unincorporated, organized territory of the US with commonwealth status; policy relations between Puerto Rico and the US conducted under the jurisdiction of the Office of the President
Government type: commonwealth
Capital:
name: San Juan
geographic coordinates: 18 28 N, 66 07 W
time difference: UTC-4 (1 hour ahead of Washington, DC during Standard Time)
Administrative divisions: none (territory of the US with commonwealth status); there are no first-order administrative divisions as defined by the US Government, but there are 78 municipalities (municipios, singular–municipio) at the second order; Adjuntas, Aguada, Aguadilla, Aguas Buenas, Aibonito, Anasco, Arecibo, Arroyo, Barceloneta, Barranquitas, Bayamon, Cabo Rojo, Caguas, Camuy, Canovanas, Carolina, Catano, Cayey, Ceiba, Ciales, Cidra, Coamo, Comerio, Corozal, Culebra, Dorado, Fajardo, Florida, Guanica, Guayama, Guayanilla, Guaynabo, Gurabo, Hatillo, Hormigueros, Humacao, Isabela, Jayuya, Juana Diaz, Juncos, Lajas, Lares, Las Marias, Las Piedras, Loiza, Luquillo, Manati, Maricao, Maunabo, Mayaguez, Moca, Morovis, Naguabo, Naranjito, Orocovis, Patillas, Penuelas, Ponce, Quebradillas, Rincon, Rio Grande, Sabana Grande, Salinas, San German, San Juan, San Lorenzo, San Sebastian, Santa Isabel, Toa Alta, Toa Baja, Trujillo Alto, Utuado, Vega Alta, Vega Baja, Vieques, Villalba, Yabucoa, Yauco
Independence: none (territory of the US with commonwealth status)
National holiday: US Independence Day, 4 July (1776); Puerto Rico Constitution Day, 25 July (1952)
Constitution: ratified 3 March 1952; approved by US Congress 3 July 1952; effective 25 July 1952
Legal system: civil law system based on the Spanish civil code and within the framework of the US federal system
Suffrage: 18 years of age; universal; island residents are US citizens but do not vote in US presidential elections
Executive branch: *chief of state:* President Barack H. OBAMA (since 20 January 2009); Vice President Joseph R. BIDEN (since 20 January 2009)

head of government: Governor Luis FORTUNO (since 2 January 2009)
cabinet: Cabinet appointed by the governor with the consent of the legislature
(For more information visit the World Leaders website)
elections: under the US Constitution, residents of unincorporated territories, such as Puerto Rico, do not vote in elections for US president and vice president; however, they may vote in Democratic and Republican party presidential primary elections; governor elected by popular vote for a four-year term (no term limits); election last held on 4 November 2008 (next to be held in November 2012)
election results: Luis FORTUNO elected governor with 52.8% of the vote
Legislative branch: bicameral Legislative Assembly consists of the Senate (at least 27 seats; members directly elected by popular vote to serve four-year terms) and the House of Representatives (51 seats; members elected by popular vote to serve four-year terms)
elections: Senate–last held on 4 November 2008 (next to be held in November 2012); House of Representatives–last held on 4 November 2008 (next to be held in November 2012)
election results: Senate–percent of vote by party–PNP 81.5%, PPD 18.5%; seats by party–PNP 22, PPD 5; House of Representatives–percent of vote by party–PNP 72.5%, PPD 27.5%; seats by party–PNP 37, PPD 14
note: Puerto Rico elects, by popular vote, a resident commissioner to serve a four-year term as a nonvoting representative in the US House of Representatives; aside from not voting on the House floor, he enjoys all the rights of a member of Congress; elections last held 4 November 2008 (next to be held in November 2012); results–percent of vote by party–NA; seats by party–PNP 1
Judicial branch: Supreme Court; Appellate Court; Court of First Instance composed of two sections: a Superior Court and a Municipal Court (justices for all these courts appointed by the governor with the consent of the Senate)
Political parties and leaders: National Democratic Party [Roberto PRATS]; National Republican Party of Puerto Rico [Dr. Tiody FERRE]; New Progressive Party or PNP [Pedro ROSSELLO] (pro-US statehood); Popular Democratic Party or PPD [Anibal ACEVEDO-VILA] (pro-commonwealth); Puerto Rican Independence Party or PIP [Ruben BERRIOS Martinez] (pro-independence)
Political pressure groups and leaders: Boricua Popular Army or EPB (a revolutionary group also known as Los Macheteros); note–the following radical groups are considered dormant by Federal law enforcement: Armed Forces for National Liberation or FALN, Armed Forces of Popular Resistance, Volunteers of the Puerto Rican Revolution
International organization participation: Caricom (observer), Interpol (subbureau), IOC, ITUC, UNWTO (associate), UPU
Diplomatic representation in the US: none (territory of the US with commonwealth status)
Diplomatic representation from the US: none (territory of the US with commonwealth status)
Flag description: five equal horizontal bands of red (top and bottom) alternating with white; a blue isosceles triangle based on the hoist side bears a large, white, five-pointed star in the center; the white star symbolizes Puerto Rico; the three sides of the triangle signify the executive, legislative and judicial parts of the government; blue stands for the sky and the coastal waters; red symbolizes the blood shed by warriors, while white represents liberty, victory, and peace
note: design initially influenced by the US flag, but similar to the Cuban flag, with the colors of the bands and triangle reversed
National anthem: *name:* "La Borinquena" (The Puerto Rican)
lyrics/music: Manuel Fernandez JUNCOS/ Felix Astol ARTES
note: music adopted 1952, lyrics adopted 1977; the local anthem's name is a reference to the indigenous name of the island, Borinquen; the music was originally composed as a dance in 1867 and gained popularity in the early 20th century; there is some evidence that the music was written by Francisco RAMIREZ; as a commonwealth of the United States, "The Star-Spangled Banner" is official (see United States)

ECONOMY

Economy—overview: Puerto Rico has one of the most dynamic economies in the Caribbean region. A diverse industrial sector has far surpassed agriculture as the primary locus of economic activity and income. Encouraged by duty-free access to the US and by tax incentives, US firms have invested heavily in Puerto Rico since the 1950s. US minimum wage laws apply. Sugar production has lost out to dairy production and other livestock products as the main source of income in the agricultural sector. Tourism has traditionally been an important source of income with estimated arrivals of more than 3.6 million tourists in 2008.
GDP (purchasing power parity): $64.84 billion (2010 est.)
country comparison to the world: 84
$68.84 billion (2009 est.)
$71.51 billion (2008 est.)
note: data are in 2010 US dollars
GDP (official exchange rate): $93.52 billion (2010 est.)
GDP—real growth rate: -5.8% (2010 est.)
country comparison to the world: 214
-3.7% (2009 est.)
-2.8% (2008 est.)
GDP—per capita (PPP): $16,300 (2010 est.)
country comparison to the world: 69
$17,400 (2009 est.)
$18,100 (2008 est.)
note: data are in 2010 US dollars
GDP—composition by sector:
agriculture: 1%
industry: 45%
services: 54% (2005 est.)
Labor force: 1.479 million (2007)
country comparison to the world: 133
Labor force—by occupation:
agriculture: 2.1%
industry: 19%
services: 79% (2005)
Unemployment rate: 12% (2002)
country comparison to the world: 127
Population below poverty line: NA%
Household income or consumption by percentage share: *lowest 10%:* NA%
highest 10%: NA%
Budget: *revenues:* $6.7 billion
expenditures: $9.6 billion (FY99/00)
Inflation rate (consumer prices): 6.5% (2003 est.)
country comparison to the world: 167
Market value of publicly traded shares: $NA
Agriculture—products: sugarcane, coffee, pineapples, plantains, bananas; livestock products, chickens
Industries: pharmaceuticals, electronics, apparel, food products, tourism
Industrial production growth rate: NA%
Electricity—production: 23.72 billion kWh (2007 est.)
country comparison to the world: 67
Electricity—consumption: 22.06 billion kWh (2007 est.)
country comparison to the world: 67
Electricity—exports: 0 kWh (2008 est.)
Electricity—imports: 0 kWh (2008 est.)
Oil—production: 1,783 bbl/day (2009 est.)
country comparison to the world: 104
Oil—consumption: 164,100 bbl/day (2009 est.)
country comparison to the world: 62
Oil—exports: 16,520 bbl/day (2007 est.)
country comparison to the world: 91
Oil—imports: 225,000 bbl/day (2007 est.)
country comparison to the world: 41
Oil—proved reserves: 0 bbl (1 January 2010 est.)
country comparison to the world: 179
Natural gas—production: 0 cu m (2008 est.)
country comparison to the world: 175
Natural gas—consumption: 806.6 million cu m (2008 est.)
country comparison to the world: 90
Natural gas—exports: 0 cu m (2008 est.)
country comparison to the world: 164
Natural gas—imports: 806.6 million cu m (2008 est.)
country comparison to the world: 56
Natural gas—proved reserves: 0 cu m (1 January 2010 est.)
country comparison to the world: 181
Exports: $46.9 billion (2001)
country comparison to the world: 56
Exports—commodities: chemicals, electronics, apparel, canned tuna, rum, beverage concentrates, medical equipment
Imports: $29.1 billion (2001)
country comparison to the world: 61
Imports—commodities: chemicals, machinery and equipment, clothing, food, fish, petroleum products
Debt—external: $NA
Exchange rates: the US dollar is used

COMMUNICATIONS

Telephones—main lines in use: 870,100 (2009)
country comparison to the world: 84
Telephones—mobile cellular: 2.716 million (2009)
country comparison to the world: 118
Telephone system: *general assessment:* modern system integrated with that of the US by high-capacity submarine cable and Intelsat with high-speed data capability
domestic: digital telephone system; mobile-cellular services
international: country code–1-787, 939; submarine cables provide connectivity to the US, Caribbean, Central and South America; satellite earth station–1 Intelsat
Broadcast media: more than 30 television stations operating; cable TV subscription services are available; roughly 125 radio stations operating (2007)
Internet country code: .pr
Internet hosts: 482 (2010)
country comparison to the world: 182
Internet users: 1 million (2009)
country comparison to the world: 98

TRANSPORTATION

Airports: 29 (2010)
country comparison to the world: 116
Airports—with paved runways: *total:* 17
over 3,047 m: 3
1,524 to 2,437 m: 2
914 to 1,523 m: 7
under 914 m: 5 (2010)
Airports—with unpaved runways:
total: 12
1,524 to 2,437 m: 1
914 to 1,523 m: 1
under 914 m: 10 (2010)
Roadways: *total:* 26,670 km
country comparison to the world: 101
paved: 25,337 km (includes 427 km of expressways)
unpaved: 1,333 km (2008)
Merchant marine: *total:* 3
country comparison to the world: 139
by type: roll on/roll off 3
foreign-owned: 3 (US 3)
registered in other countries: 1 (Saint Vincent and the Grenadines 1) (2008)
Ports and terminals: Ensenada Honda, Mayaguez, Playa de Guayanilla, Playa de Ponce, San Juan

MILITARY

Military branches: no regular indigenous military forces; paramilitary National Guard, Police Force
Manpower fit for military service: *males age 16–49:* 700,443
females age 16–49: 786,035 (2010 est.)
Manpower reaching militarily significant age annually: *male:* 30,517
female: 29,010 (2010 est.)
Military—note: defense is the responsibility of the US

TRANSNATIONAL ISSUES

Disputes—international: increasing numbers of illegal migrants from the Dominican Republic cross the Mona Passage to Puerto Rico each year looking for work

QATAR

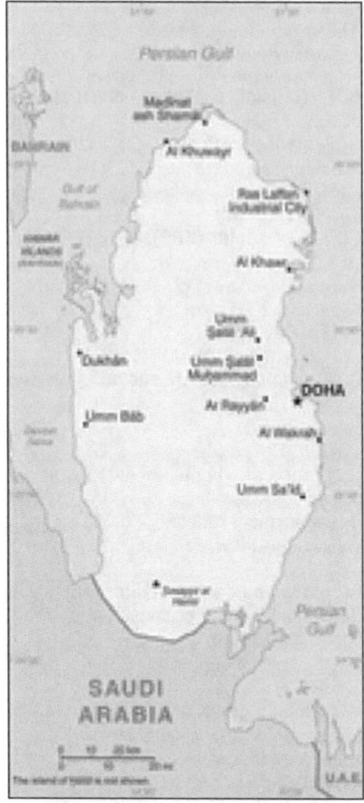

INTRODUCTION

Background: Ruled by the Al Thani family since the mid-1800s, Qatar transformed itself from a poor British protectorate noted mainly for pearling into an independent state with significant oil and natural gas revenues. During the late 1980s and early 1990s, the Qatari economy was crippled by a continuous siphoning off of petroleum revenues by the Amir, who had ruled the country since 1972. His son, the current Amir HAMAD bin Khalifa Al Thani, overthrew him in a bloodless coup in 1995. In 2001, Qatar resolved its longstanding border disputes with both Bahrain and Saudi Arabia. As of 2007, oil and natural gas revenues had enabled Qatar to attain the second-highest per capita income in the world.

GEOGRAPHY

Location: Middle East, peninsula bordering the Persian Gulf and Saudi Arabia
Geographic coordinates: 25 30 N, 51 15 E
Map references: Middle East
Area: *total:* 11,586 sq km
country comparison to the world: 165
land: 11,586 sq km
water: 0 sq km
Area—comparative: slightly smaller than Connecticut
Land boundaries: *total:* 60 km
border countries: Saudi Arabia 60 km
Coastline: 563 km
Maritime claims: *territorial sea:* 12 nm
contiguous zone: 24 nm
exclusive economic zone: as determined by bilateral agreements or the median line
Climate: arid; mild, pleasant winters; very hot, humid summers
Terrain: mostly flat and barren desert covered with loose sand and gravel
Elevation extremes: *lowest point:* Persian Gulf 0 m
highest point: Tuwayyir al Hamir 103 m
Natural resources: petroleum, natural gas, fish
Land use: *arable land:* 1.64%
permanent crops: 0.27%
other: 98.09% (2005)
Irrigated land: 130 sq km (2002)
Total renewable water resources: 0.1 cu km (1997)
Freshwater withdrawal (domestic/industrial/agricultural): *total:* 0.29 cu km/yr (24%/3%/72%)
per capita: 358 cu m/yr (2000)
Natural hazards: haze, dust storms, sandstorms common
Environment—current issues: limited natural freshwater resources are increasing dependence on large-scale desalination facilities
Environment—international agreements: *party to:* Biodiversity, Climate Change, Climate Change-Kyoto Protocol, Desertification, Endangered Species, Hazardous Wastes, Law of the Sea, Ozone Layer Protection, Ship Pollution
signed, but not ratified: none of the selected agreements
Geography—note: strategic location in central Persian Gulf near major petroleum deposits

PEOPLE

Population: 848,016 (July 2011 est.)
country comparison to the world: 160
Age structure: *0–14 years:* 21.8% (male 95,240/female 89,446)
15–64 years: 76.7% (male 460,673/female 189,914)
65 years and over: 1.5% (male 7,311/female 5,432) (2011 est.)
Median age: *total:* 30.8 years
male: 32.9 years
female: 25.5 years (2011 est.)
Population growth rate: 0.81% (2011 est.)
country comparison to the world: 134
Birth rate: 15.48 births/1,000 population (2011 est.)
country comparison to the world: 132
Death rate: 2.43 deaths/1,000 population (July 2011 est.)
country comparison to the world: 222
Net migration rate:
-4.94 migrant(s)/1,000 population (2011 est.)
country comparison to the world: 191
Urbanization: *urban population:* 96% of total population (2010)
rate of urbanization: 1.6% annual rate of change (2010-15 est.)
Major cities—population: DOHA (capital) 427,000 (2009)
Sex ratio: *at birth:* 1.056 male(s)/female
under 15 years: 1.06 male(s)/female
15–64 years: 2.44 male(s)/female
65 years and over: 1.36 male(s)/female
total population: 1.99 male(s)/female (2011 est.)
Infant mortality rate: *total:* 12.05 deaths/1,000 live births
country comparison to the world: 133
male: 12.83 deaths/1,000 live births
female: 11.22 deaths/1,000 live births (2011 est.)
Life expectancy at birth: *total population:* 75.7 years
country comparison to the world: 84
male: 73.96 years
female: 77.53 years (2011 est.)
Total fertility rate: 2.43 children born/woman (2011 est.)
country comparison to the world: 89
HIV/AIDS—adult prevalence rate: less than 0.1% (2009 est.)
country comparison to the world: 156
HIV/AIDS—people living with HIV/AIDS: fewer than 200 (2009 est.)
country comparison to the world: 160
HIV/AIDS—deaths: fewer than 100 (2009 est.)
country comparison to the world: 149
Drinking water source: *Improved:*
urban: 100% of population
rural: 100% of population
total: 100% of population (2008)
Sanitation facility access: *Improved:*
urban: 100% of population
rural: 100% of population
total: 100% of population (2008)
Nationality: *noun:* Qatari(s)
adjective: Qatari
Ethnic groups: Arab 40%, Indian 18%, Pakistani 18%, Iranian 10%, other 14%
Religions: Muslim 77.5%, Christian 8.5%, other 14% (2004 census)
Languages: Arabic (official), English commonly used as a second language
Literacy: *definition:* age 15 and over can read and write
total population: 89%
male: 89.1%
female: 88.6% (2004 census)
School life expectancy (primary to tertiary education): *total:* 12 years
male: 11 years
female: 14 years (2009)
Education expenditures: 3.3% of GDP (2005)
country comparison to the world: 121

GOVERNMENT

Country name: *conventional long form:* State of Qatar
conventional short form: Qatar
local long form: Dawlat Qatar
local short form: Qatar
note: closest approximation of the native pronunciation falls between cutter and gutter, but not like guitar
Government type: emirate
Capital: *name:* Doha
geographic coordinates: 25 17 N, 51 32 E
time difference: UTC+3 (8 hours ahead of Washington, DC during Standard Time)
Administrative divisions: 7 municipalities (baladiyat, singular–baladiyah); Ad Dawhah, Al Khawr wa adh Dhakhirah, Al Wakrah, Ar Rayyan, Ash Shamal, Az Za'ayin, Umm Salal

Independence: 3 September 1971 (from the UK)
National holiday: Independence Day, 3 September (1971); also observed is National Day, 18 December (anniversary of Al Thani family accession to the throne)
Constitution: ratified by public referendum 29 April 2003; endorsed by the Amir 8 June 2004, effective 9 June 2005
Legal system: mixed legal system of civil law and Islamic law (in family and personal matters)
International law organization participation: has not submitted an ICJ jurisdiction declaration; non-party state to the ICCt
Suffrage: 18 years of age; universal
Executive branch: *chief of state:* Amir HAMAD bin Khalifa Al Thani (since 27 June 1995 when, as heir apparent, he ousted his father, Amir KHALIFA bin Hamad Al Thani, in a bloodless coup); Heir Apparent TAMIM bin Hamad bin Khalifa Al Thani, fourth son of the amir (selected Heir Apparent by the amir on 5 August 2003); note–Amir HAMAD also holds the positions of Minister of Defense and Commander-in-Chief of the Armed Forces
head of government: Prime Minister HAMAD bin Jasim bin Jabir Al Thani (since 3 April 2007); Deputy Prime Minister Abdallah bin Hamad al-ATIYAH (since 3 April 2007)
cabinet: Council of Ministers appointed by the amir
(For more information visit the World Leaders website)
elections: the amir is hereditary
note: in April 2007, Qatar held nationwide elections for a 29-member Central Municipal Council (CMC), which has limited consultative powers aimed at improving the provision of municipal services; the first election for the CMC was held in March 1999
Legislative branch: unicameral Advisory Council or Majlis al-Shura (35 seats; members appointed)
note: no legislative elections have been held since 1970 when there were partial elections to the body; Council members have had their terms extended every year since the new constitution came into force on 9 June 2005; the constitution provides for a new 45-member Advisory Council or Majlis al-Shura; the public would elect 30 members and the Amir would appoint 15; elections to the Majlis al-Shura are tentatively scheduled for June 2010
Judicial branch: Courts of First Instance, Appeal, and Cassation; an Administrative Court and a Constitutional Court were established in 2007; note–all judges are appointed by Amiri Decree based on the recommendation of the Supreme Judiciary Council for renewable three-year terms
Political parties and leaders: none
Political pressure groups and leaders: none
International organization participation: ABEDA, AFESD, AMF, CICA (observer), FAO, G-77, GCC, IAEA, IBRD, ICAO, ICC, ICRM, IDA, IDB, IFAD, IFC, IFRCS, IHO, ILO, IMF, IMO, IMSO, Interpol, IOC, IOM (observer), IPU, ISO, ITSO, ITU, LAS, MIGA, NAM, OAPEC, OAS (observer), OIC, OPCW, OPEC, PCA, UN, UNCTAD, UNESCO, UNIDO, UNIFIL, UNMIS, UNWTO, UPU, WCO, WHO, WIPO, WMO, WTO
Diplomatic representation in the US: *chief of mission:* Ambassador Ali Fahad al-Shahwany al-HAJRI
chancery: 2555 M Street NW, Washington, DC 20037
telephone: [1] (202) 274-1600 and 274-1603
FAX: [1] (202) 237-0061
consulate(s) general: Houston
Diplomatic representation from the US: *chief of mission:* Ambassador Joseph E. LEBARON
embassy: Al-Luqta District, 22 February Road, Doha
mailing address: P. O. Box 2399, Doha
telephone: [974] 488 4161
FAX: [974] 488 4150
Flag description: maroon with a broad white serrated band (nine white points) on the hoist side; maroon represents the blood shed in Qatari wars, white stands for peace; the nine-pointed serrated edge signifies Qatar as the ninth member of the "reconciled emirates" in the wake of the Qatari-British treaty of 1916
note: the other eight emirates are the seven that compose the UAE and Bahrain; according to some sources, the dominant color was formerly red, but this darkened to maroon upon exposure to the sun and the new shade was eventually adopted
National anthem: *name:* "Al-Salam Al-Amiri" (The Peace for the Anthem)
lyrics/music: Sheikh MUBARAK bin Saif al-Thani/Abdul Aziz Nasser OBAIDAN
note: adopted 1996; the anthem was first performed that year at a meeting of the Gulf Cooperative Council hosted by Qatar

ECONOMY

Economy—overview: Despite the global financial crisis, Qatar has prospered in the last several years–in 2010 Qatar had the world's highest growth rate. Qatari authorities throughout the crisis sought to protect the local banking sector with direct investments into domestic banks. GDP rebounded in 2010 largely due to the increase in oil prices. Economic policy is focused on developing Qatar's nonassociated natural gas reserves and increasing private and foreign investment in non-energy sectors, but oil and gas still account for more than 50% of GDP, roughly 85% of export earnings, and 70% of government revenues. Oil and gas likely have made Qatar the highest per-capita income country–ahead of Liechtenstein–and the country with the lowest unemployment. Proved oil reserves of 25 billion barrels should enable continued output at current levels for 57 years. Qatar's proved reserves of natural gas exceed 25 trillion cubic meters, about 14% of the world total and third largest in the world. Qatar's successful 2022 world cup bid will likely accelerate large-scale infrastructure projects such as Qatar's metro system and the Qatar-Bahrain causeway.
GDP (purchasing power parity): $150.6 billion (2010 est.)
country comparison to the world: 59
$129.5 billion (2009 est.)
$119.2 billion (2008 est.)
note: data are in 2010 US dollars
GDP (official exchange rate): $129.5 billion (2010 est.)
GDP—real growth rate: 16.3% (2010 est.)
country comparison to the world: 1
8.6% (2009 est.)
25.4% (2008 est.)
GDP—per capita (PPP): $179,000 (2010 est.)
country comparison to the world: 1
$155,400 (2009 est.)
$144,500 (2008 est.)
note: data are in 2010 US dollars
GDP—composition by sector: *agriculture:* 0.1%
industry: 78.8%
services: 21.1% (2010 est.)
Labor force: 1.254 million (2010 est.)
country comparison to the world: 137
Unemployment rate: 0.5% (2010 est.)
country comparison to the world: 2
0.5% (2009 est.)
Population below poverty line: NA%
Household income or consumption by percentage share:
lowest 10%: NA%
highest 10%: NA%
Investment (gross fixed): 33% of GDP (2010 est.)
country comparison to the world: 13
Budget: *revenues:* $44.62 billion
expenditures: $29.69 billion (2010 est.)
Public debt: 10.3% of GDP (2010 est.)
country comparison to the world: 121
14% of GDP (2009 est.)
Inflation rate (consumer prices): 1.1% (2010 est.)
country comparison to the world: 21
-4.9% (2009 est.)
Central bank discount rate: 5.5% (31 December 2009)
country comparison to the world: 75
5.5% (31 December 2008)
Commercial bank prime lending rate: 7.04% (31 December 2009 est.)
country comparison to the world: 129
6.84% (31 December 2008 est.)
Stock of narrow money: $15.98 billion (31 December 2010 est.)
country comparison to the world: 64
$14.59 billion (31 December 2009 est.)
Stock of broad money: $65.95 billion (31 December 2010 est.)
country comparison to the world: 61
$59.09 billion (31 December 2009 est.)
Stock of domestic credit: $70.9 billion (31 December 2010 est.)
country comparison to the world: 57
$69.21 billion (31 December 2009)
Market value of publicly traded shares: $87.86 billion (31 December 2009)
country comparison to the world: 44
$76.31 billion (31 December 2008)
$95.49 billion (31 December 2007)
Agriculture—products: fruits, vegetables; poultry, dairy products, beef; fish
Industries: liquefied natural gas, crude oil production and refining, ammonia, fertilizers, petrochemicals, steel reinforcing bars, cement, commercial ship repair

Industrial production growth rate: 27.1% (2010 est.)
country comparison to the world: 1
Electricity—production: 15.11 billion kWh (2007 est.)
country comparison to the world: 79
Electricity—consumption: 13.73 billion kWh (2007 est.)
country comparison to the world: 78
Electricity—exports: 0 kWh (2008 est.)
Electricity—imports: 0 kWh (2008 est.)
Oil—production: 1.213 million bbl/day (2009 est.)
country comparison to the world: 21
Oil—consumption: 142,000 bbl/day (2009 est.)
country comparison to the world: 68
Oil—exports: 753,000 bbl/day (2008 est.)
country comparison to the world: 22
Oil—imports: 0 bbl/day (2008 est.)
country comparison to the world: 207
Oil—proved reserves: 25.41 billion bbl (1 January 2010 est.)
country comparison to the world: 12
Natural gas—production: 76.98 billion cu m (2008 est.)
country comparison to the world: 12
Natural gas—consumption: 20.2 billion cu m (2008 est.)
country comparison to the world: 33
Natural gas—exports: 56.78 billion cu m (2008 est.)
country comparison to the world: 5
Natural gas—imports: 0 cu m (2008 est.)
country comparison to the world: 109
Natural gas—proved reserves: 25.47 trillion cu m (1 January 2010 est.)
country comparison to the world: 3
Current account balance: $20.11 billion (2010 est.)
country comparison to the world: 17
$809 million (2009 est.)
Exports: $57.82 billion (2010 est.)
country comparison to the world: 48
$33.28 billion (2009 est.)
Exports—commodities: liquefied natural gas (LNG), petroleum products, fertilizers, steel
Exports—partners: Japan 34.68%, South Korea 22.44%, Singapore 10.03%, India 4.86% (2009)
Imports: $23.38 billion (2010 est.)
country comparison to the world: 66
$20.89 billion (2009 est.)
Imports—commodities: machinery and transport equipment, food, chemicals
Imports—partners: US 13.43%, Italy 8.34%, South Korea 8.33%, Japan 8.04%, Germany 7.31%, France 6.26%, UK 5.59%, China 5%, UAE 4.67%, Saudi Arabia 3.96% (2009)
Reserves of foreign exchange and gold: $22.41 billion (31 December 2010 est.)
country comparison to the world: 43
$18.81 billion (31 December 2009 est.)
Debt—external: $71.38 billion (31 December 2010 est.)
country comparison to the world: 46
$70.37 billion (31 December 2009 est.)
Stock of direct foreign investment—at home: $26.38 billion (31 December 2010 est.)
country comparison to the world: 63
$20.75 billion (31 December 2009 est.)
Stock of direct foreign investment—abroad: $19.49 billion (31 December 2010 est.)
country comparison to the world: 42
$14.27 billion (31 December 2009 est.)
Exchange rates: Qatari rials (QAR) per US dollar–3.64 (2010) 3.64 (2009) 3.64 (2008) 3.64 (2007) 3.64 (2006)

COMMUNICATIONS

Telephones—main lines in use: 285,300 (2009)
country comparison to the world: 117
Telephones—mobile cellular: 2.472 million (2009)
country comparison to the world: 122
Telephone system: *general assessment:* modern system centered in Doha
domestic: combined fixed and mobile-cellular telephone subscribership exceeds 300 telephones per 100 persons
international: country code–974; landing point for the Fiber-Optic Link Around the Globe (FLAG) submarine cable network that provides links to Asia, Middle East, Europe, and the US; tropospheric scatter to Bahrain; microwave radio relay to Saudi Arabia and the UAE; satellite earth stations–2 Intelsat (1 Atlantic Ocean and 1 Indian Ocean) and 1 Arabsat (2009)
Broadcast media: television and radio broadcast media are state controlled; home of the satellite TV channel Al-Jazeera, which was originally owned and financed by the Qatari Government; Al-Jazeera claims editorial independence in broadcasting; transmissions of several international broadcasters are accessible on FM in Doha (2007)
Internet country code: .qa
Internet hosts: 822 (2010)
country comparison to the world: 170
Internet users: 563,800 (2009)
country comparison to the world: 117

TRANSPORTATION

Airports: 6 (2010)
country comparison to the world: 171
Airports—with paved runways:
total: 4
over 3,047 m: 3
1,524 to 2,437 m: 1 (2010)
Airports—with unpaved runways: *total:* 2
914 to 1,523 m: 1
under 914 m: 1 (2010)
Heliports: 1 (2010)
Pipelines: condensate 145 km; condensate/gas 132 km; gas 980 km; liquid petroleum gas 90 km; oil 382 km (2010)
Roadways: *total:* 7,790 km (2006)
country comparison to the world: 144
Merchant marine: *total:* 29
country comparison to the world: 85
by type: bulk carrier 3, chemical tanker 2, container 14, liquefied gas 6, petroleum tanker 4
foreign-owned: 7 (Kuwait 7)
registered in other countries: 30 (Liberia 5, Marshall Islands 24, Panama 1) (2010)
Ports and terminals: Doha, Mesaieed (Umaieed), Ra's Laffan

MILITARY

Military branches: Qatari Amiri Land Force (QALF), Qatari Amiri Navy (QAN), Qatari Amiri Air Force (QAAF) (2009)
Military service age and obligation: 18 years of age for voluntary military service; no conscription (2010)
Manpower available for military service: *males age 16–49:* 389,487
females age 16–49: 165,572 (2010 est.)
Manpower fit for military service: *males age 16–49:* 321,974
females age 16–49: 140,176 (2010 est.)
Manpower reaching militarily significant age annually: *male:* 6,429
female: 5,162 (2010 est.)
Military expenditures: 10% of GDP (2005 est.)
country comparison to the world: 3

TRANSNATIONAL ISSUES

Disputes—international: none
Trafficking in persons: current situation: Qatar is a destination country for men and women from South and Southeast Asia who migrate willingly, but are subsequently trafficked into involuntary servitude as domestic workers and laborers, and, to a lesser extent, commercial sexual exploitation; the most common offense was forcing workers to accept worse contract terms than those under which they were recruited; other conditions include bonded labor, withholding of pay, restrictions on movement, arbitrary detention, and physical, mental, and sexual abuse
tier rating: Tier 2 Watch List–the Government of Qatar does not fully comply with the minimum standards for the elimination of trafficking; however, it is making significant efforts to do so; in February 2009, Qatar enacted a new migrant worker sponsorship law that criminalizes some practices commonly used by trafficking offenders, and it announced plans to use that law effectively to prevent human trafficking; punishment for offenses related to trafficking in persons remains lower than that for crimes such as rape and kidnapping, and the Qatari government has yet to take significant action to investigate, prosecute, and punish trafficking offenses; the government continues to lack formal victim identification procedures and, as a result, victims of trafficking are likely punished for acts committed as a direct result of being trafficked (2009)

ROMANIA

INTRODUCTION

Background: The principalities of Wallachia and Moldavia–for centuries under the suzerainty of the Turkish Ottoman Empire–secured their autonomy in 1856; they were de facto linked in 1859 and formally united in 1862 under the new name of Romania. The country gained recognition of its independence in 1878. It joined the Allied Powers in World War I and acquired new territories–most notably Transylvania–following the conflict. In 1940, Romania allied with the Axis powers and participated in the 1941 German invasion of the USSR. Three years later, overrun by the Soviets, Romania signed an armistice. The post-war Soviet occupation led to the formation of a Communist "people's republic" in 1947 and the abdication of the king. The decades-long rule of dictator Nicolae CEAUSESCU, who took power in 1965, and his Securitate police state became increasingly oppressive and draconian through the 1980s. CEAUSESCU was overthrown and executed in late 1989. Former Communists dominated the government until 1996 when they were swept from power. Romania joined NATO in 2004 and the EU in 2007.

GEOGRAPHY

Location: Southeastern Europe, bordering the Black Sea, between Bulgaria and Ukraine
Geographic coordinates: 46 00 N, 25 00 E
Map references: Europe
Area: *total:* 238,391 sq km
country comparison to the world: 82
land: 229,891 sq km
water: 8,500 sq km
Area—comparative: slightly smaller than Oregon
Land boundaries: *total:* 2,508 km
border countries: Bulgaria 608 km, Hungary 443 km, Moldova 450 km, Serbia 476 km, Ukraine (north) 362 km, Ukraine (east) 169 km
Coastline: 225 km
Maritime claims: *territorial sea:* 12 nm
contiguous zone: 24 nm
exclusive economic zone: 200 nm
continental shelf: 200 m depth or to the depth of exploitation
Climate: temperate; cold, cloudy winters with frequent snow and fog; sunny summers with frequent showers and thunderstorms
Terrain: central Transylvanian Basin is separated from the Moldavian Plateau on the east by the Eastern Carpathian Mountains and separated from the Walachian Plain on the south by the Transylvanian Alps
Elevation extremes: *lowest point:* Black Sea 0 m
highest point: Moldoveanu 2,544 m
Natural resources: petroleum (reserves declining), timber, natural gas, coal, iron ore, salt, arable land, hydropower
Land use: *arable land:* 39.49%
permanent crops: 1.92%
other: 58.59% (2005)
Irrigated land: 30,770 sq km (2003)
Total renewable water resources: 42.3 cu km (2003)
Freshwater withdrawal (domestic/industrial/agricultural): *total:* 6.5 cu km/yr (9%/34%/57%)
per capita: 299 cu m/yr (2003)
Natural hazards: earthquakes, most severe in south and southwest; geologic structure and climate promote landslides
Environment—current issues: soil erosion and degradation; water pollution; air pollution in south from industrial effluents; contamination of Danube delta wetlands
Environment—international agreements: *party to:* Air Pollution, Air Pollution-Persistent Organic Pollutants, Antarctic-Environmental Protocol, Antarctic Treaty, Biodiversity, Climate Change, Climate Change-Kyoto Protocol, Desertification, Endangered Species, Environmental Modification, Hazardous Wastes, Law of the Sea, Ozone Layer Protection, Ship Pollution, Wetlands
signed, but not ratified: none of the selected agreements
Geography—note: controls most easily traversable land route between the Balkans, Moldova, and Ukraine

PEOPLE

Population: 21,904,551 (July 2011 est.)
country comparison to the world: 54
Age structure: *0–14 years:* 14.8% (male 1,667,894/female 1,579,458)
15–64 years: 70.4% (male 7,684,514/female 7,725,957)
65 years and over: 14.8% (male 1,314,132/female 1,932,596) (2011 est.)
Median age: *total:* 38.7 years
male: 37.3 years
female: 40.2 years (2011 est.)
Population growth rate: -0.252% (2011 est.)
country comparison to the world: 213
Birth rate: 9.55 births/1,000 population (2011 est.)
country comparison to the world: 202
Death rate: 11.81 deaths/1,000 population (July 2011 est.)
country comparison to the world: 31
Net migration rate: -0.26 migrant(s)/1,000 population (2011 est.)
country comparison to the world: 125
Urbanization: *urban population:* 57% of total population (2010)
rate of urbanization: 0.6% annual rate of change (2010-15 est.)
Major cities—population: BUCHAREST (capital) 1.933 million (2009)
Sex ratio: *at birth:* 1.06 male(s)/female
under 15 years: 1.05 male(s)/female
15–64 years: 0.99 male(s)/female
65 years and over: 0.69 male(s)/female
total population: 0.95 male(s)/female (2011 est.)
Infant mortality rate: *total:* 11.02 deaths/1,000 live births
country comparison to the world: 143
male: 12.44 deaths/1,000 live births
female: 9.52 deaths/1,000 live births (2011 est.)
Life expectancy at birth: *total population:* 73.98 years
country comparison to the world: 109
male: 70.5 years
female: 77.66 years (2011 est.)
Total fertility rate: 1.29 children born/woman (2011 est.)
country comparison to the world: 211
HIV/AIDS—adult prevalence rate: 0.1% (2009 est.)
country comparison to the world: 158
HIV/AIDS—people living with HIV/AIDS: 16,000 (2009 est.)
country comparison to the world: 84
HIV/AIDS—deaths: fewer than 1,000 (2009 est.)
country comparison to the world: 78
Sanitation facility access:
Improved:
urban: 88% of population
rural: 54% of population
total: 72% of population
Unimproved:
urban: 12% of population
rural: 46% of population
total: 28% of population (2008)
Nationality: *noun:* Romanian(s)
adjective: Romanian
Ethnic groups: Romanian 89.5%, Hungarian 6.6%, Roma 2.5%, Ukrainian 0.3%, German 0.3%, Russian 0.2%, Turkish 0.2%, other 0.4% (2002 census)
Religions: Eastern Orthodox (including all sub-denominations) 86.8%, Protestant (various denominations including Reformate and Pentecostal) 7.5%, Roman Catholic 4.7%, other (mostly Muslim) and unspecified 0.9%, none 0.1% (2002 census)
Languages: Romanian (official) 91%, Hungarian 6.7%, Romany (Gypsy) 1.1%, other 1.2%
Literacy: *definition:* age 15 and over can read and write
total population: 97.3%
male: 98.4%
female: 96.3% (2002 census)
School life expectancy (primary to tertiary education): *total:* 15 years
male: 14 years
female: 15 years (2008)
Education expenditures: 4.3% of GDP (2007)

country comparison to the world: 93

GOVERNMENT

Country name: *conventional long form:* none
conventional short form: Romania
local long form: none
local short form: Romania
Government type: republic
Capital: *name:* Bucharest
geographic coordinates: 44 26 N, 26 06 E
time difference: UTC+2 (7 hours ahead of Washington, DC during Standard Time)
daylight saving time: +1hr, begins last Sunday in March; ends last Sunday in October
Administrative divisions: 41 counties (judete, singular–judet) and 1 municipality* (municipiu); Alba, Arad, Arges, Bacau, Bihor, Bistrita-Nasaud, Botosani, Braila, Brasov, Bucuresti (Bucharest)*, Buzau, Calarasi, Caras-Severin, Cluj, Constanta, Covasna, Dimbovita, Dolj, Galati, Gorj, Giurgiu, Harghita, Hunedoara, Ialomita, Iasi, Ilfov, Maramures, Mehedinti, Mures, Neamt, Olt, Prahova, Salaj, Satu Mare, Sibiu, Suceava, Teleorman, Timis, Tulcea, Vaslui, Vilcea, Vrancea
Independence: 9 May 1877 (independence proclaimed from the Ottoman Empire; independence recognized on 13 July 1878 by the Treaty of Berlin); 26 March 1881 (kingdom proclaimed); 30 December 1947 (republic proclaimed)
National holiday: Unification Day (of Romania and Transylvania), 1 December (1918)
Constitution: 8 December 1991; revised 29 October 2003
Legal system: civil law system
International law organization participation: has not submitted an ICJ jurisdiction declaration; accepts ICCt jurisdiction
Suffrage: 18 years of age; universal
Executive branch: *chief of state:* President Traian BASESCU (since 20 December 2004)
head of government: Prime Minister Emil BOC (since 22 December 2008); Deputy Prime Minister Marko BELA (since 23 December 2009)
cabinet: Council of Ministers appointed by the prime minister
(For more information visit the World Leaders website)
elections: president elected by popular vote for a five-year term (eligible for a second term); election last held on 22 November 2009 with runoff on 6 December 2009 (next to be held in November-December 2014); prime minister appointed by the president with the consent of the Parliament
election results: Traian BASESCU reelected president; percent of vote–Traian BASESCU 50.3%, Mircea GEOANA 49.7%
Legislative branch: bicameral Parliament or Parlament consists of the Senate or Senat (137 seats; members elected by popular vote in a mixed electoral system to serve four-year terms) and the Chamber of Deputies or Camera Deputatilor (334 seats; members elected by popular vote in a mixed electoral system to serve four-year terms)
elections: Senate–last held on 30 November 2008 (next expected to be held in November 2012); Chamber of Deputies–last held on 30 November 2008 (next expected to be held in November 2012)
election results: Senate–percent of vote by alliance/party–PSD-PC 34.2%, PDL 33.6%, PNL 18.7%, UDMR 6.4%, other 7.1%; seats by alliance/party–PSD-PC 49, PDL 51, PNL 28, UDMR 9; Chamber of Deputies–percent of vote by alliance/party–PSD-PC 33.1%, PDL 32.4%, PNL 18.6%, UDMR 6.2%, ethnic minorities 3.6%, other 6.1%; seats by alliance/party–PDL 115, PSD-PC 114, PNL 65, UDMR 22, ethnic minorities 18
Judicial branch: Supreme Court of Justice (comprised of 11 judges appointed for three-year terms by the president in consultation with the Superior Council of Magistrates, which is comprised of the minister of justice, the prosecutor general, two civil society representatives appointed by the Senate, and 14 judges and prosecutors elected by their peers); a separate body, the Constitutional Court, validates elections and makes decisions regarding the constitutionality of laws, treaties, ordinances, and internal rules of the Parliament; it is comprised of nine members serving nine-year terms, with three members each appointed by the president, the Senate, and the Chamber of Deputies
Political parties and leaders: Conservative Party or PC [Daniel CONSTANTIN] (formerly Humanist Party or PUR); Democratic Liberal Party or PDL [Emil BOC]; Democratic Union of Hungarians in Romania or UDMR [Bela MARKO]; National Liberal Party or PNL [Crin ANTONESCU]; National Union for Romania's Progress or UNPR [Gabriel OPREA]; Social Democratic Party or PSD [Victor PONTA] (formerly Party of Social Democracy in Romania or PDSR)
Political pressure groups and leaders: *other:* various human rights and professional associations
International organization participation: Australia Group, BIS, BSEC, CBSS (observer), CE, CEI, EAPC, EBRD, EIB, ESA (cooperating state), EU, FAO, G-9, IAEA, IBRD, ICAO, ICC, ICRM, IDA, IFAD, IFC, IFRCS, IHO, ILO, IMF, IMO, IMSO, Interpol, IOC, IOM, IPU, ISO, ITSO, ITU, ITUC, LAIA (observer), MIGA, MONUSCO, NATO, NSG, OAS (observer), OIF, OPCW, OSCE, PCA, SECI, UN, UNCTAD, UNESCO, UNHCR, UNIDO, Union Latina, UNMIL, UNMIS, UNOCI, UNWTO, UPU, WCO, WFTU, WHO, WIPO, WMO, WTO, ZC
Diplomatic representation in the US: *chief of mission:* Ambassador Adrian Cosmin VIERITA
chancery: 1607 23rd Street NW, Washington, DC 20008
telephone: [1] (202) 332-4846, 4848, 4851, 4852
FAX: [1] (202) 232-4748
consulate(s) general: Chicago, Los Angeles, New York
Diplomatic representation from the US: *chief of mission:* Ambassador Mark GITENSTEIN
embassy: Strada Tudor Arghezi 7-9, Bucharest
mailing address: pouch: American Embassy Bucharest, US Department of State, 5260 Bucharest Place, Washington, DC 20521-5260 (pouch)
telephone: [40] (21) 200-3300
FAX: [40] (21) 200-3442
Flag description: three equal vertical bands of blue (hoist side), yellow, and red; modeled after the flag of France, the colors are those of the principalities of Walachia (red and yellow) and Moldavia (red and blue), which united in 1862 to form Romania; the national coat of arms that used to be centered in the yellow band has been removed
note: now similar to the flag of Chad, whose blue band is darker; also resembles the flags of Andorra and Moldova
National anthem: *name:* "Desteapta-te romane!" (Wake up, Romanian!)
lyrics/music: Andrei MURESIANU/Anton PANN
note: adopted 1990; the anthem was written during the 1848 Revolution

ECONOMY

Economy—overview: Romania, which joined the European Union on 1 January 2007, began the transition from Communism in 1989 with a largely obsolete industrial base and a pattern of output unsuited to the country's needs. The country emerged in 2000 from a punishing three-year recession thanks to strong demand in EU export markets. Domestic consumption and investment have fueled strong GDP growth in recent years, but have led to large current account imbalances. Romania's macroeconomic gains have only recently started to spur creation of a middle class and address Romania's widespread poverty. Corruption and red tape continue to handicap its business environment. Inflation rose in 2007-08, driven in part by strong consumer demand and high wage growth, rising energy costs, a nationwide drought affecting food prices, and a relaxation of fiscal discipline. Romania's GDP contracted markedly in the last quarter of 2008 as the country began to feel the effects of a global downturn in financial markets and trade, and GDP fell more than 7% in 2009, prompting Bucharest to seek a $26 billion emergency assistance package from the IMF, the EU, and other international lenders. Drastic austerity measures, as part of Romania's IMF-led agreement led to a further 1.9% GDP contraction in 2010. The economy is expected to return to positive growth in 2011.
GDP (purchasing power parity): $254.2 billion (2010 est.)
country comparison to the world: 48
$257.4 billion (2009 est.)
$277 billion (2008 est.)
note: data are in 2010 US dollars
GDP (official exchange rate): $161.6 billion (2010 est.)
GDP—real growth rate: -1.3% (2010 est.)
country comparison to the world: 201
-7.1% (2009 est.)
7.3% (2008 est.)

GDP—per capita (PPP): $11,600 (2010 est.)
country comparison to the world: 96
$11,700 (2009 est.)
$12,600 (2008 est.)
note: data are in 2010 US dollars
GDP—composition by sector: *agriculture:* 12.8%
industry: 36%
services: 51.2% (2010 est.)
Labor force: 9.35 million (2010 est.)
country comparison to the world: 52
Labor force—by occupation: *agriculture:* 29.7%
industry: 23.2%
services: 47.1% (2006)
Unemployment rate: 8.2% (2010 est.)
country comparison to the world: 92
7.8% (2009 est.)
Population below poverty line: 25% (2005 est.)
Household income or consumption by percentage share: *lowest 10%:* 1.2%
highest 10%: 20.8% (2006)
Distribution of family income—Gini index: 32 (2008)
country comparison to the world: 104
28.8 (2003)
Investment (gross fixed): 21.1% of GDP (2010 est.)
country comparison to the world: 76
Budget: *revenues:* $50.89 billion
expenditures: $62 billion (2010 est.)
Public debt: 34.8% of GDP (2010 est.)
country comparison to the world: 83
24% of GDP (2009 est.)
Inflation rate (consumer prices): 6% (2010 est.)
country comparison to the world: 160
5.6% (2009 est.)
Central bank discount rate: NA%
Commercial bank prime lending rate: 17.28% (31 December 2009 est.)
country comparison to the world: 42
14.99% (31 December 2008 est.)
Stock of narrow money: $24.39 billion (31 December 2010 est.)
country comparison to the world: 60
$26.03 billion (31 December 2009 est.)
Stock of broad money: $63.67 billion (31 December 2010 est.)
country comparison to the world: 62
$61.66 billion (31 December 2009 est.)
Stock of domestic credit: $77.46 billion (31 December 2010 est.)
country comparison to the world: 54
$72.45 billion (31 December 2009 est.)
Market value of publicly traded shares: $30.32 billion (31 December 2009)
country comparison to the world: 63
$19.92 billion (31 December 2008)
$44.93 billion (31 December 2007)
Agriculture—products: wheat, corn, barley, sugar beets, sunflower seed, potatoes, grapes; eggs, sheep
Industries: electric machinery and equipment, textiles and footwear, light machinery and auto assembly, mining, timber, construction materials, metallurgy, chemicals, food processing, petroleum refining
Industrial production growth rate: 1.5% (2010 est.)
country comparison to the world: 140
Electricity—production: 58.28 billion kWh (2007 est.)
country comparison to the world: 45
Electricity—consumption: 49.44 billion kWh (2007 est.)
country comparison to the world: 46
Electricity—exports: 5.169 billion kWh (2008 est.)
Electricity—imports: 921 million kWh (2008 est.)
Oil—production: 117,000 bbl/day (2009 est.)
country comparison to the world: 50
Oil—consumption: 214,000 bbl/day (2009 est.)
country comparison to the world: 53
Oil—exports: 115,600 bbl/day (2007 est.)
country comparison to the world: 65
Oil—imports: 217,000 bbl/day (2007 est.)
country comparison to the world: 43
Oil—proved reserves: 600 million bbl (1 January 2010 est.)
country comparison to the world: 46
Natural gas—production: 11.42 billion cu m (2008 est.)
country comparison to the world: 40
Natural gas—consumption: 16.92 billion cu m (2008 est.)
country comparison to the world: 38
Natural gas—exports: 0 cu m (2008 est.)
country comparison to the world: 162
Natural gas—imports: 5.5 billion cu m (2008 est.)
country comparison to the world: 30
Natural gas—proved reserves: 63 billion cu m (1 January 2010 est.)
country comparison to the world: 60
Current account balance: $-7.934 billion (2010 est.)
country comparison to the world: 176
$-7.139 billion (2009 est.)
Exports: $51.91 billion (2010 est.)
country comparison to the world: 51
$40.6 billion (2009 est.)
Exports—commodities: machinery and equipment, textiles and footwear, metals and metal products, machinery and equipment, minerals and fuels, chemicals, agricultural products
Exports—partners: Germany 18.76%, Italy 15.42%, France 8.2%, Turkey 4.99%, Hungary 4.33% (2009)
Imports: $59.84 billion (2010 est.)
country comparison to the world: 43
$50.03 billion (2009 est.)
Imports—commodities: machinery and equipment, fuels and minerals, chemicals, textile and products, metals, agricultural products
Imports—partners: Germany 17.3%, Italy 11.78%, Hungary 8.36%, France 6.14%, China 4.91%, Austria 4.75% (2009)
Reserves of foreign exchange and gold: $50.51 billion (31 December 2010 est.)
country comparison to the world: 24
$44.11 billion (31 December 2009 est.)
Debt—external: $108.9 billion (31 December 2010 est.)
country comparison to the world: 35
$110 billion (31 December 2009 est.)
Stock of direct foreign investment—at home: $80.16 billion (31 December 2010 est.)
country comparison to the world: 44
$73.96 billion (31 December 2009 est.)
Stock of direct foreign investment—abroad: $1.831 billion (31 December 2010 est.)
country comparison to the world: 69
$1.731 billion (31 December 2009 est.)
Exchange rates: lei (RON) per US dollar–3.2 (2010) 3.0493 (2009) 2.5 (2008) 2.43 (2007) 2.809 (2006)

COMMUNICATIONS

Telephones—main lines in use: 5.313 million (2009)
country comparison to the world: 31
Telephones—mobile cellular: 25.377 million (2009)
country comparison to the world: 34
Telephone system: *general assessment:* the telecommunications sector is being expanded and modernized; domestic and international service improving rapidly, especially mobile-cellular services
domestic: more than 90 percent of telephone network is automatic; fixed-line teledensity exceeds 20 telephones per 100 persons; mobile-cellular teledensity, expanding rapidly, roughly 115 telephones per 100 persons
international: country code–40; the Black Sea Fiber Optic System provides connectivity to Bulgaria and Turkey; satellite earth stations–10; digital, international, direct-dial exchanges operate in Bucharest (2009)
Broadcast media: a mixture of public and private TV stations; the public broadcaster operates multiple stations; roughly 100 private national, regional, and local stations operating; more than 75% of households are connected to multi-channel cable or satellite TV systems that provide access to Romanian, European, and international stations; state-owned public radio broadcaster operates 4 national networks and regional and local stations; more than 100 private radio stations broadcasting (2008)
Internet country code: .ro
Internet hosts: 2.464 million (2010)
country comparison to the world: 34
Internet users: 7.787 million (2009)
country comparison to the world: 37

TRANSPORTATION

Airports: 54 (2010)
country comparison to the world: 88
Airports—with paved runways: *total:* 26
over 3,047 m: 4
2,438 to 3,047 m: 10
1,524 to 2,437 m: 11
under 914 m: 1 (2010)
Airports—with unpaved runways:
total: 28
914 to 1,523 m: 7
under 914 m: 21 (2010)
Heliports: 3 (2010)
Pipelines: gas 3,652 km; oil 2,424 km (2010)
Railways: *total:* 10,784 km
country comparison to the world: 21
broad gauge: 57 km 1.524-m gauge
standard gauge: 10,645 km 1.435-m gauge (4,002 km electrified)
narrow gauge: 5 km 1.000-m gauge (2010)
Roadways: *total:* 81,713 km (does not include urban roads)
country comparison to the world: 57
paved: 66,632 km (includes 321 km of expressways)
unpaved: 15,081 km (2009)

Waterways: 1,731 km (includes 1,075 km on the Danube River, 524 km on secondary branches, and 132 km on canals) (2006)
country comparison to the world: 47

Merchant marine: *total:* 15
country comparison to the world: 102
by type: cargo 10, passenger/cargo 2, petroleum tanker 2, roll on/roll off 1
registered in other countries: 35 (Cambodia 1, Georgia 7, Liberia 3, Malta 8, Marshall Islands 2, Moldova 2, North Korea 1, Panama 2, Saint Vincent and the Grenadines 1, Sierra Leone 4, Syria 1, Togo 1, unknown 1) (2010)

Ports and terminals: Braila, Constanta, Galati (Galatz), Mancanului (Giurgiu), Midia, Tulcea

MILITARY

Military branches: Land Forces, Naval Forces, Romanian Air Force (Fortele Aeriene Romane, FAR), Special Operations (2010)

Military service age and obligation: 18-35 years of age for male and female voluntary military service; conscription officially ended October 2006; all military inductees (including women) contract for an initial 5-year term of service, with subsequent successive contracts for 3-year terms until age 36 (2009)

Manpower available for military service: *males age 16–49:* 5,601,234
females age 16–49: 5,428,939 (2010 est.)

Manpower fit for military service: *males age 16–49:* 4,550,409
females age 16–49: 4,507,880 (2010 est.)

Manpower reaching militarily significant age annually: *male:* 117,798
female: 111,607 (2010 est.)

Military expenditures: 1.9% of GDP (2007 est.)
country comparison to the world: 79

TRANSNATIONAL ISSUES

Disputes—international: the ICJ gave Ukraine until December 2006 to reply, and Romania until June 2007 to issue a rejoinder, in their dispute submitted in 2004 over Ukrainian-administered Zmiyinyy/Serpilor (Snake) Island and Black Sea maritime boundary delimitation; Romania also opposes Ukraine's reopening of a navigation canal from the Danube border through Ukraine to the Black Sea

Illicit drugs: major transshipment point for Southwest Asian heroin transiting the Balkan route and small amounts of Latin American cocaine bound for Western Europe; although not a significant financial center, role as a narcotics conduit leaves it vulnerable to laundering, which occurs via the banking system, currency exchange houses, and casinos

RUSSIA

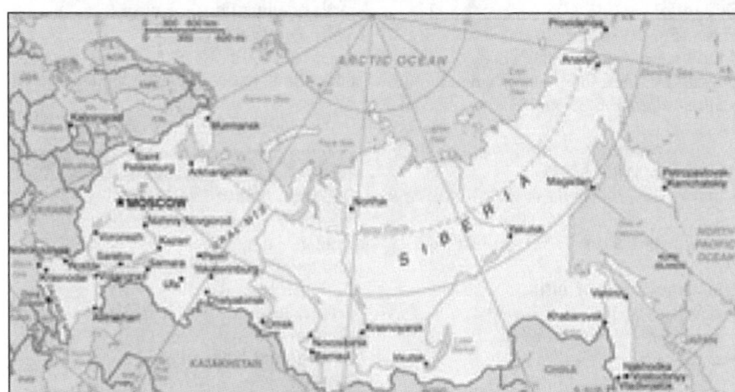

INTRODUCTION

Background: Founded in the 12th century, the Principality of Muscovy, was able to emerge from over 200 years of Mongol domination (13th-15th centuries) and to gradually conquer and absorb surrounding principalities. In the early 17th century, a new Romanov Dynasty continued this policy of expansion across Siberia to the Pacific. Under PETER I (ruled 1682-1725), hegemony was extended to the Baltic Sea and the country was renamed the Russian Empire. During the 19th century, more territorial acquisitions were made in Europe and Asia. Defeat in the Russo-Japanese War of 1904-05 contributed to the Revolution of 1905, which resulted in the formation of a parliament and other reforms. Repeated devastating defeats of the Russian army in World War I led to widespread rioting in the major cities of the Russian Empire and to the overthrow in 1917 of the imperial household. The Communists under Vladimir LENIN seized power soon after and formed the USSR. The brutal rule of Iosif STALIN (1928-53) strengthened Communist rule and Russian dominance of the Soviet Union at a cost of tens of millions of lives. The Soviet economy and society stagnated in the following decades until General Secretary Mikhail GORBACHEV (1985-91) introduced glasnost (openness) and perestroika (restructuring) in an attempt to modernize Communism, but his initiatives inadvertently released forces that by December 1991 splintered the USSR into Russia and 14 other independent republics. Since then, Russia has shifted its post-Soviet democratic ambitions in favor of a centralized semi-authoritarian state whose legitimacy is buttressed, in part, by carefully managed national elections, former President PUTIN's genuine popularity, and the prudent management of Russia's windfall energy wealth. Russia has severely disabled a Chechen rebel movement, although violence still occurs throughout the North Caucasus.

GEOGRAPHY

Location: Northern Asia (the area west of the Urals is considered part of Europe), bordering the Arctic Ocean, between Europe and the North Pacific Ocean

Geographic coordinates: 60 00 N, 100 00 E

Map references: Asia

Area: *total:* 17,098,242 sq km
country comparison to the world: 1
land: 16,377,742 sq km
water: 720,500 sq km

Area—comparative: approximately 1.8 times the size of the US

Land boundaries: *total:* 20,241.5 km
border countries: Azerbaijan 284 km, Belarus 959 km, China (southeast) 3,605 km, China (south) 40 km, Estonia 290 km, Finland 1,313 km, Georgia 723 km, Kazakhstan 6,846 km, North Korea 17.5 km, Latvia 292 km, Lithuania (Kaliningrad Oblast) 227 km, Mongolia 3,441 km, Norway 196 km, Poland (Kaliningrad Oblast) 432 km, Ukraine 1,576 km

Coastline: 37,653 km

Maritime claims: *territorial sea:* 12 nm
contiguous zone: 24 nm
exclusive economic zone: 200 nm
continental shelf: 200 m depth or to the depth of exploitation

Climate: ranges from steppes in the south through humid continental in much of European Russia; subarctic in Siberia to tundra climate in the polar north; winters vary from cool along Black Sea coast to frigid in Siberia; summers vary from warm in the steppes to cool along Arctic coast

Terrain: broad plain with low hills west of Urals; vast coniferous forest and tundra in Siberia; uplands and mountains along southern border regions

Elevation extremes: *lowest point:* Caspian Sea -28 m
highest point: Gora El'brus 5,633 m

Natural resources: wide natural resource base including major deposits of oil, natural gas, coal, and many strategic minerals, reserves of rare earth elements, timber
note: formidable obstacles of climate, terrain, and distance hinder exploitation of natural resources

Land use: *arable land:* 7.17%
permanent crops: 0.11%
other: 92.72% (2005)

Irrigated land: 46,000 sq km (2003)
Total renewable water resources: 4,498 cu km (1997)
Freshwater withdrawal (domestic/industrial/agricultural): *total:* 76.68 cu km/yr (19%/63%/18%)
per capita: 535 cu m/yr (2000)
Natural hazards: permafrost over much of Siberia is a major impediment to development; volcanic activity in the Kuril Islands; volcanoes and earthquakes on the Kamchatka Peninsula; spring floods and summer/autumn forest fires throughout Siberia and parts of European Russia
volcanism: Russia experiences significant volcanic activity on the Kamchatka Peninsula and Kuril Islands; the peninsula alone is home to some 29 historically active volcanoes, with dozens more in the Kuril Islands; Kliuchevskoi (elev. 4,835 m), which erupted in 2007 and 2010, is Kamchatka's most active volcano; Avachinsky and Koryaksky volcanoes, which pose a threat to the city of Petropavlovsk-Kamchatskiy, have been deemed "Decade Volcanoes" by the International Association of Volcanology and Chemistry of the Earth's Interior, worthy of study due to their explosive history and close proximity to human populations; other notable historically active volcanoes include Bezymianny, Chikurachki, Ebeko, Gorely, Grozny, Karymsky, Ketoi, Kronotsky, Ksudach, Medvezhia, Mutnovsky, Sarychev Peak, Shiveluch, Tiatia, Tolbachik, and Zheltovsky
Environment—current issues: air pollution from heavy industry, emissions of coal-fired electric plants, and transportation in major cities; industrial, municipal, and agricultural pollution of inland waterways and seacoasts; deforestation; soil erosion; soil contamination from improper application of agricultural chemicals; scattered areas of sometimes intense radioactive contamination; groundwater contamination from toxic waste; urban solid waste management; abandoned stocks of obsolete pesticides
Environment—international agreements: *party to:* Air Pollution, Air Pollution-Nitrogen Oxides, Air Pollution-Sulfur 85, Antarctic-Environmental Protocol, Antarctic-Marine Living Resources, Antarctic Seals, Antarctic Treaty, Biodiversity, Climate Change, Climate Change-Kyoto Protocol, Desertification, Endangered Species, Environmental Modification, Hazardous Wastes, Law of the Sea, Marine Dumping, Ozone Layer Protection, Ship Pollution, Tropical Timber 83, Wetlands, Whaling
signed, but not ratified: Air Pollution-Sulfur 94
Geography—note: largest country in the world in terms of area but unfavorably located in relation to major sea lanes of the world; despite its size, much of the country lacks proper soils and climates (either too cold or too dry) for agriculture; Mount El'brus is Europe's tallest peak

PEOPLE

Population: 138,739,892 (July 2011 est.)
country comparison to the world: 9
Age structure: *0–14 years:* 15.2% (male 10,818,203/female 10,256,611)
15–64 years: 71.8% (male 47,480,851/female 52,113,279)
65 years and over: 13% (male 5,456,639/female 12,614,309) (2011 est.)
Median age: *total:* 38.7 years
male: 35.5 years
female: 41.9 years (2011 est.)
Population growth rate: -0.47% (2011 est.)
country comparison to the world: 223
Birth rate: 11.05 births/1,000 population (2011 est.)
country comparison to the world: 174
Death rate: 16.04 deaths/1,000 population (July 2011 est.)
country comparison to the world: 5
Net migration rate: 0.29 migrant(s)/1,000 population (2011 est.)
country comparison to the world: 68
Urbanization: *urban population:* 73% of total population (2010)
rate of urbanization: -0.2% annual rate of change (2010-15 est.)
Major cities—population: MOSCOW (capital) 10.523 million; Saint Petersburg 4.575 million; Novosibirsk 1.397 million; Yekaterinburg 1.344 million; Nizhniy Novgorod 1.267 million (2009)
Sex ratio: *at birth:* 1.06 male(s)/female
under 15 years: 1.06 male(s)/female
15–64 years: 0.92 male(s)/female
65 years and over: 0.44 male(s)/female
total population: 0.85 male(s)/female (2011 est.)
Infant mortality rate: *total:* 10.08 deaths/1,000 live births
country comparison to the world: 147
male: 11.58 deaths/1,000 live births
female: 8.49 deaths/1,000 live births (2011 est.)
Life expectancy at birth: *total population:* 66.29 years
country comparison to the world: 162
male: 59.8 years
female: 73.17 years (2011 est.)
Total fertility rate: 1.42 children born/woman (2011 est.)
country comparison to the world: 197
HIV/AIDS—adult prevalence rate: 1% (2009 est.)
country comparison to the world: 48
HIV/AIDS—people living with HIV/AIDS: 980,000 (2009 est.)
country comparison to the world: 11
HIV/AIDS—deaths: NA
Major infectious diseases: *degree of risk:* intermediate
food or waterborne diseases: bacterial diarrhea
vectorborne disease: tickborne encephalitis
note: highly pathogenic H5N1 avian influenza has been identified in this country; it poses a negligible risk with extremely rare cases possible among US citizens who have close contact with birds (2009)
Drinking water source: *Improved:*
urban: 98% of population
rural: 89% of population
total: 96% of population
Unimproved:
urban: 2% of population
rural: 11% of population
total: 4% of population (2008)
Sanitation facility access: *Improved:*
urban: 93% of population
rural: 70% of population
total: 87% of population
Unimproved:
urban: 7% of population
rural: 30% of population
total: 13% of population (2008)
Nationality: *noun:* Russian(s)
adjective: Russian
Ethnic groups: Russian 79.8%, Tatar 3.8%, Ukrainian 2%, Bashkir 1.2%, Chuvash 1.1%, other or unspecified 12.1% (2002 census)
Religions: Russian Orthodox 15-20%, Muslim 10-15%, other Christian 2% (2006 est.)
note: estimates are of practicing worshipers; Russia has large populations of non-practicing believers and non-believers, a legacy of over seven decades of Soviet rule
Languages: Russian (official), many minority languages
Literacy: *definition:* age 15 and over can read and write
total population: 99.4%
male: 99.7%
female: 99.2% (2002 census)
School life expectancy (primary to tertiary education): *total:* 14 years
male: 14 years
female: 15 years (2008)
Education expenditures: 3.9% of GDP (2006)
country comparison to the world: 108

GOVERNMENT

Country name: *conventional long form:* Russian Federation
conventional short form: Russia
local long form: Rossiyskaya Federatsiya
local short form: Rossiya
former: Russian Empire, Russian Soviet Federative Socialist Republic
Government type: federation
Capital: *name:* Moscow
geographic coordinates: 55 45 N, 37 35 E
time difference: UTC+3 (8 hours ahead of Washington, DC during Standard Time)
daylight saving time: +1hr; note—Russia has announced that it will remain on daylight saving time permanently, which began on 27 March 2011
note: Russia is divided into 9 time zones
Administrative divisions: 46 provinces (oblastey, singular—oblast), 21 republics (respublik, singular—respublika), 4 autonomous okrugs (avtonomnykh okrugov, singular—avtonomnyy okrug), 9 krays (krayev, singular—kray), 2 federal cities (goroda, singular—gorod), and 1 autonomous oblast (avtonomnaya oblast')
oblasts: Amur (Blagoveshchensk), Arkhangel'sk, Astrakhan', Belgorod, Bryansk, Chelyabinsk, Irkutsk, Ivanovo, Kaliningrad, Kaluga, Kemerovo, Kirov, Kostroma, Kurgan, Kursk, Leningrad, Lipetsk, Magadan, Moscow, Murmansk, Nizhniy Novgorod, Novgorod, Novosibirsk, Omsk, Orenburg, Orel, Penza, Pskov, Rostov, Ryazan', Sakhalin (Yuzhno-Sakhalinsk), Samara, Saratov, Smolensk, Sverdlovsk (Yekaterinburg), Tambov, Tomsk, Tula, Tver', Tyumen', Ul'yanovsk, Vladimir, Volgograd, Vologda, Voronezh, Yaroslavl'
republics: Adygeya (Maykop), Altay (Gorno-Altaysk), Bashkortostan (Ufa),

Buryatiya (Ulan-Ude), Chechnya (Groznyy), Chuvashiya (Cheboksary), Dagestan (Makhachkala), Ingushetiya (Magas), Kabardino-Balkariya (Nal'chik), Kalmykiya (Elista), Karachayevo-Cherkesiya (Cherkessk), Kareliya (Petrozavodsk), Khakasiya (Abakan), Komi (Syktyvkar), Mariy-El (Yoshkar-Ola), Mordoviya (Saransk), North Ossetia (Vladikavkaz), Sakha [Yakutiya] (Yakutsk), Tatarstan (Kazan'), Tyva (Kyzyl), Udmurtiya (Izhevsk)

autonomous okrugs: Chukotka (Anadyr'), Khanty-Mansi (Khanty-Mansiysk), Nenets (Nar'yan-Mar), Yamalo-Nenets (Salekhard)

krays: Altay (Barnaul), Kamchatka (Petropavlovsk-Kamchatskiy), Khabarovsk, Krasnodar, Krasnoyarsk, Perm', Primorskiy [Maritime] (Vladivostok), Stavropol', Zabaykal'sk (Chita)

federal cities: Moscow [Moskva], Saint Petersburg [Sankt-Peterburg]

autonomous oblast: Yevrey [Jewish] (Birobidzhan)

note: administrative divisions have the same names as their administrative centers (exceptions have the administrative center name following in parentheses)

Independence: 24 August 1991 (from the Soviet Union); notable earlier dates: 1157 (Principality of Vladimir-Suzdal created); 16 January 1547 (Tsardom of Muscovy established); 22 October 1721 (Russian Empire proclaimed); 30 December 1922 (Soviet Union established)

National holiday: Russia Day, 12 June (1990)

Constitution: adopted 12 December 1993

Legal system: civil law system; judicial review of legislative acts

International law organization participation: has not submitted an ICJ jurisdiction declaration; non-party state to the ICCt

Suffrage: 18 years of age; universal

Executive branch: *chief of state:* President Dmitriy Anatolyevich MEDVEDEV (since 7 May 2008)

head of government: Premier Vladimir Vladimirovich PUTIN (since 8 May 2008); First Deputy Premiers Igor Ivanovich SHUVALOV and Viktor Alekseyevich ZUBKOV (since 12 May 2008); Deputy Premiers Sergey Borisovich IVANOV (since 12 May 2008), Aleksandr Gennadiyevich KHLOPONIN (since 19 January 2010), Dmitriy Nikolayevich KOZAK (since 14 October 2008), Aleksey Leonidovich KUDRIN (since 24 September 2007), Igor Ivanovich SECHIN (since 12 May 2008), Vyacheslav Viktorovich VOLODIN (since 21 October 2010), Aleksandr Dmitriyevich ZHUKOV (since 9 March 2004)

cabinet: the "Government" is composed of the premier, his deputies, and ministers; all are appointed by the president, and the premier is also confirmed by the Duma

(For more information visit the World Leaders website)

note: there is also a Presidential Administration (PA) that provides staff and policy support to the president, drafts presidential decrees, and coordinates policy among government agencies; a Security Council also reports directly to the president

elections: president elected by popular vote for a four-year term (eligible for a second term); election last held 2 March 2008 (next to be held in March 2012); note—the term length was extended to six years in late 2008, to go into effect following the 2012 presidential election; there is no vice president; if the president dies in office, cannot exercise his powers because of ill health, is impeached, or resigns, the premier serves as acting president until a new presidential election is held, which must be within three months; premier appointed by the president with the approval of the Duma

election results: Dmitriy MEDVEDEV elected president; percent of vote—Dmitriy MEDVEDEV 70.2%, Gennady ZYUGANOV 17.7%, Vladimir ZHIRINOVSKY 9.4%, Andrey BOGDANOV 1.3%, other 1.4%

Legislative branch: bicameral Federal Assembly or Federalnoye Sobraniye consists of an upper house, the Federation Council or Sovet Federatsii (166 seats; members appointed by the top executive and legislative officials in each of the 83 federal administrative units—oblasts, krays, republics, autonomous okrugs and oblasts, and the federal cities of Moscow and Saint Petersburg; members to serve four-year terms) and a lower house, the State Duma or Gosudarstvennaya Duma (450 seats; as of 2007, all members elected by proportional representation from party lists winning at least 7% of the vote; members elected by popular vote to serve four-year terms)

elections: State Duma—last held on 2 December 2007 (next to be held in December 2011)

election results: State Duma—United Russia 64.3%, CPRF 11.5%, LDPR 8.1%, Just Russia 7.7%, other 8.4%; total seats by party—United Russia 315, CPRF 57, LDPR 40, Just Russia 38

Judicial branch: Constitutional Court; Supreme Court; Supreme Arbitration Court; judges for all courts are appointed for life by the Federation Council on the recommendation of the president

Political parties and leaders: A Just Russia [Sergey MIRONOV]; Communist Party of the Russian Federation or CPRF [Gennadiy Andreyevich ZYUGANOV]; Liberal Democratic Party of Russia or LDPR [Vladimir Volfovich ZHIRINOVSKIY]; Patriots of Russia [Gennadiy SEMIGIN]; Right Cause [Leonid Yakovlevich GOZMAN, Boris Yuriyevich TITOV, and Georgiy Georgiyevich BOVT] (formed from merger of Civic Force, Democratic Party of Russia, and Union of Right Forces); United Russia [Vladimir Vladimirovich PUTIN]; Yabloko Party [Sergey Sergeyevich MITROKHIN]

Political pressure groups and leaders: Association of Citizens with Initiative of Russia (TIGR); Confederation of Labor of Russia (KTR); Federation of Independent Labor Unions of Russia; Freedom of Choice Interregional Organization of Automobilists; Glasnost Defense Foundation; Golos Association in Defense of Voters' Rights; Greenpeace Russia; Human Rights Watch (Russian chapter); Institute for Collective Action; Memorial (human rights group); Movement Against Illegal Migration; Pamjat (preservation of historical monuments and recording of history); Russian Orthodox Church; Russian Federation of Car Owners; Russian-Chechen Friendship Society; SOVA Analytical-Information Center; Union of the Committees of Soldiers' Mothers; World Wildlife Fund (Russian chapter)

International organization participation: APEC, Arctic Council, ARF, ASEAN (dialogue partner), BIS, BSEC, CBSS, CE, CERN (observer), CICA, CIS, CSTO, EAEC, EAPC, EBRD, FAO, FATF, G-20, G-8, GCTU, IAEA, IBRD, ICAO, ICC, ICRM, IDA, IFC, IFRCS, IHO, ILO, IMF, IMO, IMSO, Interpol, IOC, IOM (observer), IPU, ISO, ITSO, ITU, ITUC, LAIA (observer), MIGA, MINURSO, MONUSCO, NSG, OAS (observer), OECD (accession state), OIC (observer), OPCW, OSCE, Paris Club, PCA, PFP, SCO, UN, UN Security Council, UNCTAD, UNESCO, UNHCR, UNIDO, UNITAR, UNMIL, UNMIS, UNOCI, UNTSO, UNWTO, UPU, WCO, WFTU, WHO, WIPO, WMO, WTO (observer), ZC

Diplomatic representation in the US: *chief of mission:* Ambassador Sergey Ivanovich KISLYAK

chancery: 2650 Wisconsin Avenue NW, Washington, DC 20007

telephone: [1] (202) 298-5700, 5701, 5704, 5708

FAX: [1] (202) 298-5735

consulate(s) general: Houston, New York, San Francisco, Seattle

Diplomatic representation from the US: *chief of mission:* Ambassador John R. BEYRLE

embassy: Bolshoy Deviatinskiy Pereulok No. 8, 121099 Moscow

mailing address: PSC-77, APO AE 09721

telephone: [7] (495) 728-5000

FAX: [7] (495) 728-5090

consulate(s) general: Saint Petersburg, Vladivostok, Yekaterinburg

Flag description: three equal horizontal bands of white (top), blue, and red

note: the colors may have been based on those of the Dutch flag; despite many popular interpretations, there is no official meaning assigned to the colors of the Russian flag; this flag inspired other Slav countries to adopt horizontal tricolors of the same colors but in different arrangements, and so red, blue, and white became the Pan-Slav colors

National anthem: *name:* "Gimn Rossiyskoy Federatsii" (National Anthem of the Russian Federation)

lyrics/music: Sergei Vladimirovich MIKHALKOV/Alexandr Vasilievich ALEXANDROV

note: in 2000, Russia adopted the tune of the anthem of the former Soviet Union (composed in 1939); the lyrics, also adopted in 2000, were written by the same person who authored the Soviet lyrics in 1943

ECONOMY

Economy—overview: Russia has undergone significant changes since the collapse of the Soviet Union, moving from a globally-isolated, centrally-planned economy to a more market-based and globally-integrated economy. Economic reforms in the 1990s privatized most industry, with notable exceptions in the energy and defense-related sectors. The protection of property rights is still weak and the private sector remains subject to heavy state interference. Russian industry is primarily split between globally-competitive commodity producers–in 2009 Russia was the world's largest exporter of natural gas, the second largest exporter of oil, and the third largest exporter of steel and primary aluminum–and other less competitive heavy industries that remain dependent on the Russian domestic market. This reliance on commodity exports makes Russia vulnerable to boom and bust cycles that follow the highly volatile swings in global commodity prices. The government since 2007 has embarked on an ambitious program to reduce this dependency and build up the country's high technology sectors, but with few results so far. The economy had averaged 7% growth since the 1998 Russian financial crisis, resulting in a doubling of real disposable incomes and the emergence of a middle class. The Russian economy, however, was one of the hardest hit by the 2008-09 global economic crisis as oil prices plummeted and the foreign credits that Russian banks and firms relied on dried up. The Central Bank of Russia spent one-third of its $600 billion international reserves, the world's third largest, in late 2008 to slow the devaluation of the ruble. The government also devoted $200 billion in a rescue plan to increase liquidity in the banking sector and aid Russian firms unable to roll over large foreign debts coming due. The economic decline bottomed out in mid-2009 and the economy began to grow in the first quarter of 2010. However, a severe drought and fires in central Russia reduced agricultural output, prompting a ban on grain exports for part of the year, and slowed growth in other sectors such as manufacturing and retail trade. High oil prices buoyed Russian growth in the first quarter of 2011 and could help Russia reduce the budget deficit inherited from the lean years of 2008-09, but inflation and increased government expenditures may limit the positive impact of these revenues. Russia's long-term challenges include a shrinking workforce, a high level of corruption, difficulty in accessing capital for smaller, non-energy companies, and poor infrastructure in need of large investments.

GDP (purchasing power parity): $2.223 trillion (2010 est.)
country comparison to the world: 7
$2.138 trillion (2009 est.)
$2.319 trillion (2008 est.)
note: data are in 2010 US dollars

GDP (official exchange rate): $1.465 trillion (2010 est.)

GDP—real growth rate: 4% (2010 est.)
country comparison to the world: 100
-7.8% (2009 est.)
5.2% (2008 est.)

GDP—per capita (PPP): $15,900 (2010 est.)
country comparison to the world: 71
$15,300 (2009 est.)
$16,500 (2008 est.)
note: data are in 2010 US dollars

GDP—composition by sector:
agriculture: 4.2%
industry: 33.8%
services: 62% (2010 est.)

Labor force: 75.55 million (2010 est.)
country comparison to the world: 7

Labor force—by occupation:
agriculture: 10%
industry: 31.9%
services: 58.1% (2008)

Unemployment rate: 7.6% (2010 est.)
country comparison to the world: 82
8.4% (2009)

Population below poverty line: 13.1% (2009)

Household income or consumption by percentage share:
lowest 10%: 1.9%
highest 10%: 30.4% (September 2007)

Distribution of family income—Gini index: 42.2 (2009)
country comparison to the world: 51
39.9 (2001)

Investment (gross fixed): 18.9% of GDP (2010 est.)
country comparison to the world: 101

Budget:
revenues: $262 billion
expenditures: $341.1 billion (2010 est.)

Public debt: 9.5% of GDP (2010 est.)
country comparison to the world: 123
8.3% of GDP (2009 est.)

Inflation rate (consumer prices): 6.7% (2010 est.)
country comparison to the world: 168
11.7% (2009)

Central bank discount rate: 8.75% (31 December 2009)
country comparison to the world: 25
13% (31 December 2008)

Commercial bank prime lending rate: 15.31% (31 December 2009 est.)
country comparison to the world: 66
12.23% (31 December 2008 est.)

Stock of narrow money: $269.1 billion (31 December 2010 est.)
country comparison to the world: 16
$203.7 billion (31 December 2009 est.)

Stock of broad money: $650.7 billion (31 December 2010 est.)
country comparison to the world: 20
$645.5 billion (31 December 2009)

Stock of domestic credit: $549.9 billion (31 December 2010 est.)
country comparison to the world: 24
$420.4 billion (31 December 2009 est.)

Market value of publicly traded shares: $861.4 billion (31 December 2009 est.)
country comparison to the world: 8
$1.322 trillion (31 December 2008)
$1.503 trillion (31 December 2007 est.)

Agriculture—products: grain, sugar beets, sunflower seed, vegetables, fruits; beef, milk

Industries: complete range of mining and extractive industries producing coal, oil, gas, chemicals, and metals; all forms of machine building from rolling mills to high-performance aircraft and space vehicles; defense industries including radar, missile production, and advanced electronic components, shipbuilding; road and rail transportation equipment; communications equipment; agricultural machinery, tractors, and construction equipment; electric power generating and transmitting equipment; medical and scientific instruments; consumer durables, textiles, foodstuffs, handicrafts

Industrial production growth rate: 8.3% (2010 est.)
country comparison to the world: 29

Electricity—production: 925.9 billion kWh (2009)
country comparison to the world: 5

Electricity—consumption: 857.6 billion kWh (2009)
country comparison to the world: 5

Electricity—exports: 17.7 billion kWh (2009 est.)

Electricity—imports: 3.066 billion kWh (2009)

Oil—production: 10.12 million bbl/day (2010 est.)
country comparison to the world: 1

Oil—consumption: 2.74 million bbl/day (2010 est.)
country comparison to the world: 6

Oil—exports: 5.43 million bbl/day (2009)
country comparison to the world: 2

Oil—imports: 42,000 bbl/day (2009 est.)
country comparison to the world: 94

Oil—proved reserves: 74.2 billion bbl (1 January 2009 est.)
country comparison to the world: 8

Natural gas—production: 583.6 billion cu m (2009)
country comparison to the world: 2

Natural gas—consumption: 439.6 billion cu m (2009)
country comparison to the world: 3

Natural gas—exports: 179.1 billion cu m (2009)
country comparison to the world: 1

Natural gas—imports: 35.1 billion cu m (2009)
country comparison to the world: 8

Natural gas—proved reserves: 47.57 trillion cu m (1 January 2010 est.)
country comparison to the world: 1

Current account balance: $68.85 billion (2010 est.)
country comparison to the world: 4
$48.97 billion (2009 est.)

Exports: $376.7 billion (2010 est.)
country comparison to the world: 13
$303.4 billion (2009 est.)

Exports—commodities: petroleum and petroleum products, natural gas, metals, wood and wood products, chemicals, and a wide variety of civilian and military manufactures

Exports—partners: Netherlands 10.62%, Italy 6.46%, Germany 6.24%, China 5.69%, Turkey 4.3%, Ukraine 4.01% (2009)

Imports: $237.3 billion (2010 est.)
country comparison to the world: 19
$191.8 billion (2009 est.)

Imports—commodities: machinery, vehicles, pharmaceutical products, plastic, semi-finished metal products, meat, fruits

and nuts, optical and medical instruments, iron, steel
Imports—partners: Germany 14.39%, China 13.98%, Ukraine 5.48%, Italy 4.84%, US 4.46% (2009)
Reserves of foreign exchange and gold: $483.1 billion (30 November 2010)
country comparison to the world: 3
$439.4 billion (31 December 2009)
Debt—external: $480.2 billion (30 November 2010 est.)
country comparison to the world: 22
$467.2 billion (31 December 2009)
Stock of direct foreign investment—at home: $306.8 billion (31 December 2010 est.)
country comparison to the world: 17
$256.8 billion (31 December 2009 est.)
Stock of direct foreign investment—abroad: $260.5 billion (31 December 2010 est.)
country comparison to the world: 16
$224.5 billion (31 December 2009 est.)
Exchange rates: Russian rubles (RUB) per US dollar–30 (2010) 31.74 (2009) 24.853 (2008) 25.581 (2007) 27.191 (2006)

COMMUNICATIONS

Telephones—main lines in use: 44.802 million (2009)
country comparison to the world: 4
Telephones—mobile cellular: 230.5 million (2009)
country comparison to the world: 4
Telephone system: *general assessment:* the telephone system is experiencing significant changes; there are more than 1,000 companies licensed to offer communication services; access to digital lines has improved, particularly in urban centers; Internet and e-mail services are improving; Russia has made progress toward building the telecommunications infrastructure necessary for a market economy; the estimated number of mobile subscribers jumped from fewer than 1 million in 1998 to some 230 million in 2009; a large demand for fixed line service remains unsatisfied
domestic: cross-country digital trunk lines run from Saint Petersburg to Khabarovsk, and from Moscow to Novorossiysk; the telephone systems in 60 regional capitals have modern digital infrastructures; cellular services, both analog and digital, are available in many areas; in rural areas, the telephone services are still outdated, inadequate, and low density
international: country code–7; Russia is connected internationally by undersea fiber optic cables; digital switches in several cities provide more than 50,000 lines for international calls; satellite earth stations provide access to Intelsat, Intersputnik, Eutelsat, Inmarsat, and Orbita systems (2008)
Broadcast media: 6 national TV stations with the federal government owning 1 and holding a controlling interest in a second; state-owned Gazprom maintains a controlling interest in a third national channel; government-affiliated Bank Rossiya owns controlling interest in a fourth and fifth, while the sixth national channel is owned by the Moscow city administration; roughly 3,300 national, regional, and local TV stations operating with over two-thirds completely or partially controlled by the federal or local governments; satellite TV services are available; 2 state-run national radio networks with a third majority-owned by Gazprom; roughly 2,400 public and commercial radio stations (2007)
Internet country code: .ru; note–Russia also has responsibility for a legacy domain ".su" that was allocated to the Soviet Union and is being phased out
Internet hosts: 10.382 million (2010)
country comparison to the world: 12
Internet users: 40.853 million (2009)
country comparison to the world: 10

TRANSPORTATION

Airports: 1,213 (2010)
country comparison to the world: 5
Airports—with paved runways: *total:* 593
over 3,047 m: 51
2,438 to 3,047 m: 201
1,524 to 2,437 m: 126
914 to 1,523 m: 98
under 914 m: 117 (2010)
Airports—with unpaved runways: *total:* 620
over 3,047 m: 3
2,438 to 3,047 m: 13
1,524 to 2,437 m: 68
914 to 1,523 m: 84
under 914 m: 452 (2010)
Heliports: 50 (2010)
Pipelines: condensate 122 km; gas 160,952 km; liquid petroleum gas 127 km; oil 77,630 km; oil/gas/water 38 km; refined products 13,658 km (2010)
Railways:
total: 87,157 km
country comparison to the world: 2
broad gauge: 86,200 km 1.520-m gauge (40,300 km electrified)
narrow gauge: 957 km 1.067-m gauge (on Sakhalin Island)
note: an additional 30,000 km of non-common carrier lines serve industries (2006)
Roadways:
total: 982,000 km
country comparison to the world: 8
paved: 776,000 km (includes 30,000 km of expressways)
unpaved: 206,000 km
note: includes public, local, and departmental roads (2009)
Waterways: 102,000 km (including 48,000 km with guaranteed depth; the 72,000 km system in European Russia links Baltic Sea, White Sea, Caspian Sea, Sea of Azov, and Black Sea) (2009)
country comparison to the world: 2
Merchant marine: *total:* 1,097
country comparison to the world: 11
by type: bulk carrier 22, cargo 634, carrier 2, chemical tanker 38, combination ore/oil 39, container 13, passenger 15, passenger/cargo 6, petroleum tanker 236, refrigerated cargo 77, roll on/roll off 11, specialized tanker 4
foreign-owned: 145 (Belgium 4, Cyprus 11, Italy 9, South Korea 1, Switzerland 4, Turkey 104, Ukraine 12)
registered in other countries: 443 (Antigua and Barbuda 3, Belize 32, Bulgaria 2, Cambodia 60, Comoros 21, Cook Islands 1, Cyprus 47, Dominica 6, Georgia 7, Hong Kong 1, Liberia 108, Malaysia 2, Malta 47, Marshall Islands 6, Moldova 5, Mongolia 4, Panama 39, Saint Kitts and Nevis 11, Saint Vincent and the Grenadines 15, Sierra Leone 6, Vanuatu 1, unknown 19) (2010)
Ports and terminals: Kaliningrad, Kavkaz, Nakhodka, Novorossiysk, Primorsk, Saint Petersburg, Vostochnyy

MILITARY

Military branches: Ground Forces (Sukhoputnyye Voyskia, SV), Navy (Voyenno-Morskoy Flot, VMF), Air Forces (Voyenno-Vozdushniye Sily, VVS); Airborne Troops (VDV), Strategic Rocket Forces (Raketnyye Voyska Strategicheskogo Naznacheniya, RVSN), and Space Troops (Kosmicheskiye Voyska, KV) are independent "combat arms," not subordinate to any of the three branches; Russian Ground Forces include the following combat arms: motorized-rifle troops, tank troops, missile and artillery troops, air defense of the ground troops (2010)
Military service age and obligation: 18-27 years of age for compulsory or voluntary military service; males are registered for the draft at 17 years of age; service obligation–1 year (conscripts can only be sent to combat zones after 6 months training); reserve obligation to age 50
note: over 60% of draft-age Russian males receive some type of deferment–generally health related–each draft cycle (2009)
Manpower available for military service:
males age 16–49: 34,132,156
females age 16–49: 34,985,115 (2010 est.)
Manpower fit for military service: *males age 16–49:* 20,431,035
females age 16–49: 26,381,518 (2010 est.)
Manpower reaching militarily significant age annually:
male: 693,843
female: 660,359 (2010 est.)
Military expenditures: 3.9% of GDP (2005)
country comparison to the world: 26

TRANSNATIONAL ISSUES

Disputes—international: Russia remains concerned about the smuggling of poppy derivatives from Afghanistan through Central Asian countries; China and Russia have demarcated the once disputed islands at the Amur and Ussuri confluence and in the Argun River in accordance with the 2004 Agreement, ending their centuries-long border disputes; the sovereignty dispute over the islands of Etorofu, Kunashiri, Shikotan, and the Habomai group, known in Japan as the "Northern Territories" and in Russia as the "Southern Kurils," occupied by the Soviet Union in 1945, now administered by Russia, and claimed by Japan, remains the primary sticking point to signing a peace treaty formally ending World War II hostilities; Russia's military support and subsequent recognition of Abkhazia and South Ossetia independence in 2008 continue to sour relations with Georgia; Azerbaijan, Kazakhstan, and Russia ratified Caspian seabed delimitation treaties based on equidistance, while Iran continues to insist on a one-fifth slice of the lake; Norway and Russia signed a comprehensive maritime boundary agreement in 2010; various groups in Finland advocate restoration of Karelia (Kareliya) and other areas ceded to the Soviet Union following the Second World War but the Finnish Government asserts no territorial demands; in May 2005, Russia recalled its signatures to the 1996 border agreements with Estonia (1996) and Latvia (1997), when the two Baltic states announced issuance of unilateral declarations referencing Soviet occupation and ensuing territorial losses; Russia demands better treatment of ethnic Russians in Estonia and Latvia; Estonian citizen groups continue to press for realignment of the boundary based on the 1920 Tartu Peace Treaty that would bring the now divided ethnic Setu people and parts of the Narva region within Estonia; Lithuania and Russia committed to demarcating their boundary in 2006 in accordance with the land and maritime treaty ratified by Russia in May 2003 and by Lithuania in 1999; Lithuania operates a simplified transit regime for Russian nationals traveling from the Kaliningrad coastal exclave into Russia, while still conforming, as an EU member state with an EU external border, where strict Schengen border rules apply; preparations for the demarcation delimitation of land boundary with Ukraine have commenced; the dispute over the boundary between Russia and Ukraine through the Kerch Strait and Sea of Azov remains unresolved despite a December 2003 framework agreement and on-going expert-level discussions; Kazakhstan and Russia boundary delimitation was ratified on November 2005 and field demarcation should commence in 2007; Russian Duma has not yet ratified 1990 Bering Sea Maritime Boundary Agreement with the US; Denmark (Greenland) and Norway have made submissions to the Commission on the Limits of the Continental shelf (CLCS) and Russia is collecting additional data to augment its 2001 CLCS submission

Refugees and internally displaced persons: *IDPs:* 18,000-160,000 (displacement from Chechnya and North Ossetia) (2007)

Trafficking in persons: current situation: Russia is a source, transit, and destination country for men, women, and children trafficked for various purposes; it remains a significant source of women trafficked to over 50 countries for commercial sexual exploitation; Russia is also a transit and destination country for men and women trafficked from Central Asia, Eastern Europe, and North Korea to Central and Western Europe and the Middle East for purposes of forced labor and sexual exploitation; internal trafficking remains a problem in Russia with women trafficked from rural areas to urban centers for commercial sexual exploitation, and men trafficked internally and from Central Asia for forced labor in the construction and agricultural industries; debt bondage is common among trafficking victims, and child sex tourism remains a concern

tier rating: Tier 2 Watch List—Russia is on the Tier 2 Watch List for a fifth consecutive year for its failure to show evidence of increasing efforts to combat trafficking over the previous year, particularly in providing assistance to victims of trafficking; comprehensive trafficking victim assistance legislation, which would address key deficiencies, has been pending before the Duma since 2003 and was neither passed nor enacted in 2007 (2008)

Illicit drugs: limited cultivation of illicit cannabis and opium poppy and producer of methamphetamine, mostly for domestic consumption; government has active illicit crop eradication program; used as transshipment point for Asian opiates, cannabis, and Latin American cocaine bound for growing domestic markets, to a lesser extent Western and Central Europe, and occasionally to the US; major source of heroin precursor chemicals; corruption and organized crime are key concerns; major consumer of opiates

RWANDA

INTRODUCTION

Background: In 1959, three years before independence from Belgium, the majority ethnic group, the Hutus, overthrew the ruling Tutsi king. Over the next several years, thousands of Tutsis were killed, and some 150,000 driven into exile in neighboring countries. The children of these exiles later formed a rebel group, the Rwandan Patriotic Front (RPF), and began a civil war in 1990. The war, along with several political and economic upheavals, exacerbated ethnic tensions, culminating in April 1994 in a state-orchestrated genocide, in which Rwandans killed up to a million of their fellow citizens, including approximately three-quarters of the Tutsi population. The genocide ended later that same year when the predominantly Tutsi RPF, operating out of Uganda and northern Rwanda, defeated the national army and Hutu militias, and established an RPF-led government of national unity. Approximately 2 million Hutu refugees—many fearing Tutsi retribution—fled to neighboring Burundi, Tanzania, Uganda, and Zaire. Since then, most of the refugees have returned to Rwanda, but several thousand remained in the neighboring Democratic Republic of the Congo (DRC; the former Zaire) and formed an extremist insurgency bent on retaking Rwanda, much as the RPF tried in 1990. Rwanda held its first local elections in 1999 and its first post-genocide presidential and legislative elections in 2003. Rwanda in 2009 staged a joint military operation with the Congolese Army in DRC to rout out the Hutu extremist insurgency there and Kigali and Kinshasa restored diplomatic relations. Rwanda also joined the Commonwealth in late 2009.

GEOGRAPHY

Location: Central Africa, east of Democratic Republic of the Congo

Geographic coordinates: 2 00 S, 30 00 E

Map references: Africa

Area: *total:* 26,338 sq km

country comparison to the world: 148

land: 24,668 sq km

water: 1,670 sq km

Area—comparative: slightly smaller than Maryland

Land boundaries: *total:* 893 km

border countries: Burundi 290 km, Democratic Republic of the Congo 217 km, Tanzania 217 km, Uganda 169 km

Coastline: 0 km (landlocked)

Maritime claims: none (landlocked)

Climate: temperate; two rainy seasons (February to April, November to January); mild in mountains with frost and snow possible

Terrain: mostly grassy uplands and hills; relief is mountainous with altitude declining from west to east

Elevation extremes: *lowest point:* Rusizi River 950 m

highest point: Volcan Karisimbi 4,519 m

Natural resources: gold, cassiterite (tin ore), wolframite (tungsten ore), methane, hydropower, arable land

Land use:

arable land: 45.56%

permanent crops: 10.25%

other: 44.19% (2005)

Irrigated land: 90 sq km (2003)

Total renewable water resources: 5.2 cu km (2003)

Freshwater withdrawal (domestic/industrial/agricultural): *total:* 0.15 cu km/yr (24%/8%/68%)

per capita: 17 cu m/yr (2000)

Natural hazards: periodic droughts; the volcanic Virunga mountains are in the northwest along the border with Democratic Republic of the Congo
volcanism: Visoke (elev. 3,711 m), located on the border with the Democratic Republic of the Congo, is the country's only historically active volcano
Environment—current issues: deforestation results from uncontrolled cutting of trees for fuel; overgrazing; soil exhaustion; soil erosion; widespread poaching
Environment—international agreements: *party to:* Biodiversity, Climate Change, Climate Change-Kyoto Protocol, Desertification, Endangered Species, Hazardous Wastes, Ozone Layer Protection, Wetlands
signed, but not ratified: Law of the Sea
Geography—note: landlocked; most of the country is savanna grassland with the population predominantly rural

PEOPLE

Population: 11,370,425 (July 2011 est.)
country comparison to the world: 73
note: estimates for this country explicitly take into account the effects of excess mortality due to AIDS; this can result in lower life expectancy, higher infant mortality, higher death rates, lower population growth rates, and changes in the distribution of population by age and sex than would otherwise be expected
Age structure: *0–14 years:* 42.9% (male 2,454,924/female 2,418,504)
15–64 years: 54.7% (male 3,097,956/female 3,123,910)
65 years and over: 2.4% (male 110,218/female 164,913) (2011 est.)
Median age: *total:* 18.7 years
male: 18.5 years
female: 19 years (2011 est.)
Population growth rate: 2.792% (2011 est.)
country comparison to the world: 16
Birth rate: 36.74 births/1,000 population (2011 est.)
country comparison to the world: 22
Death rate: 9.88 deaths/1,000 population (July 2011 est.)
country comparison to the world: 58
Net migration rate: 1.06 migrant(s)/1,000 population (2011 est.)
country comparison to the world: 52
Urbanization: *urban population:* 19% of total population (2010)
rate of urbanization: 4.4% annual rate of change (2010-15 est.)
Major cities—population: KIGALI (capital) 909,000 (2009)
Sex ratio: *at birth:* 1.03 male(s)/female
under 15 years: 1.01 male(s)/female
15–64 years: 0.99 male(s)/female
65 years and over: 0.67 male(s)/female
total population: 0.99 male(s)/female (2011 est.)
Infant mortality rate: *total:* 64.04 deaths/1,000 live births
country comparison to the world: 24
male: 67.64 deaths/1,000 live births
female: 60.32 deaths/1,000 live births (2011 est.)
Life expectancy at birth: *total population:* 58.02 years
country comparison to the world: 193
male: 56.57 years
female: 59.52 years (2011 est.)
Total fertility rate: 4.9 children born/woman (2011 est.)
country comparison to the world: 23
HIV/AIDS—adult prevalence rate: 2.9% (2009 est.)
country comparison to the world: 24
HIV/AIDS—people living with HIV/AIDS: 170,000 (2009 est.)
country comparison to the world: 32
HIV/AIDS—deaths: 4,100 (2009 est.)
country comparison to the world: 43
Major infectious diseases: *degree of risk:* very high
food or waterborne diseases: bacterial diarrhea, hepatitis A, and typhoid fever
vectorborne disease: malaria
animal contact disease: rabies (2009)
Drinking water source:
Improved:
urban: 77% of population
rural: 62% of population
total: 65% of population
Unimproved:
urban: 23% of population
rural: 38% of population
total: 35% of population (2008)
Sanitation facility access: *Improved:*
urban: 50% of population
rural: 55% of population
total: 54% of population
Unimproved:
urban: 50% of population
rural: 45% of population
total: 46% of population (2008)
Nationality:
noun: Rwandan(s)
adjective: Rwandan
Ethnic groups: Hutu (Bantu) 84%, Tutsi (Hamitic) 15%, Twa (Pygmy) 1%
Religions: Roman Catholic 56.5%, Protestant 26%, Adventist 11.1%, Muslim 4.6%, indigenous beliefs 0.1%, none 1.7% (2001)
Languages: Kinyarwanda (official, universal Bantu vernacular), French (official), English (official), Kiswahili (Swahili, used in commercial centers)
Literacy: *definition:* age 15 and over can read and write
total population: 70.4%
male: 76.3%
female: 64.7% (2003 est.)
School life expectancy (primary to tertiary education):
total: 11 years
male: 11 years
female: 11 years (2009)
Education expenditures: 4.1% of GDP (2008)
country comparison to the world: 102
People—note: Rwanda is the most densely populated country in Africa

GOVERNMENT

Country name: *conventional long form:* Republic of Rwanda
conventional short form: Rwanda
local long form: Republika y'u Rwanda
local short form: Rwanda
former: Ruanda, German East Africa
Government type: republic; presidential, multiparty system
Capital: *name:* Kigali
geographic coordinates: 1 57 S, 30 04 E
time difference: UTC+2 (7 hours ahead of Washington, DC during Standard Time)
Administrative divisions: 4 provinces (in French–provinces, singular–province; in Kinyarwanda–intara for singular and plural) and 1 city* (in French–ville; in Kinyarwanda–umujyi); Est (Eastern), Kigali*, Nord (Northern), Ouest (Western), Sud (Southern)
Independence: 1 July 1962 (from Belgium-administered UN trusteeship)
National holiday: Independence Day, 1 July (1962)
Constitution: new constitution passed by referendum 26 May 2003
Legal system: mixed legal system of civil law, based on German and Belgian models, and customary law; judicial review of legislative acts in the Supreme Court
International law organization participation: has not submitted an ICJ jurisdiction declaration; non-party state to the ICCt
Suffrage: 18 years of age; universal
Executive branch: *chief of state:* President Paul KAGAME (since 22 April 2000)
head of government: Prime Minister Bernard MAKUZA (since 8 March 2000)
cabinet: Council of Ministers appointed by the president
(For more information visit the World Leaders website)
elections: President elected by popular vote for a seven-year term (eligible for a second term); elections last held on 9 August 2010 (next to be held in 2017)
election results: Paul KAGAME elected to a second term as president; Paul KAGAME 93.1%, Jean NTAWUKURIRYAYO 5.1%, Prosper HIGIRO 1.4%, Alvera MUKABAR 0.4%
Legislative branch: bicameral Parliament consists of Senate (26 seats; 12 members elected by local councils, 8 appointed by the president, 4 appointed by the Political Organizations Forum, 2 represent institutions of higher learning; members to serve eight-year terms) and Chamber of Deputies (80 seats; 53 members elected by popular vote, 24 women elected by local bodies, 3 selected by youth and disability organizations; members to serve five-year terms)
elections: Senate–NA; Chamber of Deputies–last held on 15 September 2008 (next to be held in September 2013)
election results: percent of vote by party–RPF 78.8%, PSD 13.1%, PL 7.5%; seats by party–RPF 42, PSD 7, PL 4, additional 27 members indirectly elected
Judicial branch: Supreme Court; High Courts of the Republic; Provincial Courts; District Courts; mediation committees
Political parties and leaders: Centrist Democratic Party or PDC [Agnes MUKABARANGA]; Democratic Popular Union of Rwanda or UDPR [Gonzague RWIGEMA]; Democratic Republican Movement or MDR [Celestin KABANDA] (officially banned); Islamic Democratic Party or PDI [Musa Fazil HARERIMANA]; Liberal Party or PL [Protais MITALI]; Party for Democratic Renewal (officially banned); Party for Progress and

Concord or PPC [Alvera MUKABARAMBA]; Rwandan Patriotic Front or RPF [Paul KAGAME]; Rwandan Socialist Party or PSR [Jean Baptist RUCIBIGANGO]; Social Democratic Party or PSD [Vincent BIRUTA]; Socialist Party-Imberakuri or PS-Imberakuri [Christine MUKABUNANI]; Solidarity and Prosperity Party or PSP [Pheobe KANYANGE]

Political pressure groups and leaders: IBUKA (association of genocide survivors)

International organization participation: ACP, AfDB, AU, C, CEPGL, COMESA, EAC, EADB, FAO, G-77, IBRD, ICAO, ICRM, IDA, IFAD, IFC, IFRCS, ILO, IMF, Interpol, IOC, IOM, IPU, ISO (correspondent), ITSO, ITU, ITUC, MIGA, NAM, OIF, OPCW, UN, UNAMID, UNCTAD, UNESCO, UNIDO, UNMIS, UNWTO, UPU, WCO, WHO, WIPO, WMO, WTO

Diplomatic representation in the US: *chief of mission:* Ambassador James KIMONYO
chancery: 1714 New Hampshire Avenue NW, Washington, DC 20009
telephone: [1] (202) 232-2882
FAX: [1] (202) 232-4544

Diplomatic representation from the US: *chief of mission:* Ambassador W. Stuart SYMINGTON
embassy: 2657 Avenue de la Gendarmerie, Kigali
mailing address: B. P. 28, Kigali
telephone: [250] 596-400
FAX: [250] 596-591

Flag description: three horizontal bands of sky blue (top, double width), yellow, and green, with a golden sun with 24 rays near the fly end of the blue band; blue represents happiness and peace, yellow economic development and mineral wealth, green hope of prosperity and natural resources; the sun symbolizes unity, as well as enlightenment and transparency from ignorance

National anthem:
name: "Rwanda nziza" (Rwanda, Our Beautiful Country)
lyrics/music: Faustin MURIGO/Jean-Bosco HASHAKAIMANA
note: adopted 2001

ECONOMY

Economy—overview: Rwanda is a poor rural country with about 90% of the population engaged in (mainly subsistence) agriculture and some mineral and agro-processing. Tourism is now Rwanda's primary foreign exchange earner and in 2008, minerals overtook coffee and tea as Rwanda's primary export. Minerals exports declined 40% in 2009-10 due to the global economic downturn. The 1994 genocide decimated Rwanda's fragile economic base, severely impoverished the population, particularly women, and temporarily stalled the country's ability to attract private and external investment. However, Rwanda has made substantial progress in stabilizing and rehabilitating its economy to pre-1994 levels. GDP has rebounded with an average annual growth of 7-8% since 2003 and inflation has been reduced to single digits. Nonetheless, a significant percent of the population still live below the official poverty line. Despite Rwanda's fertile ecosystem, food production often does not keep pace with demand, requiring food imports. Agricultural production has increased significantly over the last three years and last year Rwanda was self sufficient in food production. Rwanda continues to receive substantial aid money and obtained IMF-World Bank Heavily Indebted Poor Country (HIPC) initiative debt relief in 2005-06. In recognition of Rwanda's successful management of its macro economy, in 2010, the IMF graduated Rwanda to a Policy Support Instrument (PSI). Rwanda also received a Millennium Challenge Threshold Program in 2008. Africa's most densely populated country is trying to overcome the limitations of its small, landlocked economy by leveraging regional trade. Rwanda joined the East African Community and is aligning its budget, trade, and immigration policies with its regional partners. The government has embraced an expansionary fiscal policy to reduce poverty by improving education, infrastructure, and foreign and domestic investment and pursuing market-oriented reforms. Energy shortages, instability in neighboring states, and lack of adequate transportation linkages to other countries continue to handicap private sector growth. The Rwandan government is seeking to become regional leader in information and communication technologies. In 2010, Rwanda neared completion of the first modern Special Economic Zone (SEZ) in Kigali. The SEZ seeks to attract investment in all sectors, but specifically in agribusiness, information and communications technologies, trade and logistics, mining, and construction. The global downturn hurt export demand and tourism, but economic growth is recovering, driven in large part by the services sector, and inflation has been contained. On the back of this growth, government is gradually ending its fiscal stimulus policy while protecting aid to the poor.

GDP (purchasing power parity): $12.16 billion (2010 est.)
country comparison to the world: 142
$11.42 billion (2009 est.)
$10.97 billion (2008 est.)
note: data are in 2010 US dollars

GDP (official exchange rate): $5.622 billion (2010 est.)

GDP—real growth rate: 6.5% (2010 est.)
country comparison to the world: 45
4.1% (2009 est.)
11.2% (2008 est.)

GDP—per capita (PPP): $1,100 (2010 est.)
country comparison to the world: 209
$1,100 (2009 est.)
$1,100 (2008 est.)
note: data are in 2010 US dollars

GDP—composition by sector:
agriculture: 42.1%
industry: 14.3%
services: 43.6% (2010 est.)

Labor force: 4.446 million (2007)
country comparison to the world: 82

Labor force—by occupation:
agriculture: 90%
industry and *services:* 10% (2000)

Unemployment rate: NA%

Population below poverty line: 60% (2001 est.)

Household income or consumption by percentage share: *lowest 10%:* 2.1%
highest 10%: 38.2% (2000)

Distribution of family income—Gini index: 46.8 (2000)
country comparison to the world: 35
28.9 (1985)

Investment (gross fixed): 20% of GDP (2010 est.)
country comparison to the world: 88

Budget: *revenues:* $1.169 billion
expenditures: $1.366 billion (2010 est.)

Inflation rate (consumer prices): 6.4% (2010 est.)
country comparison to the world: 166
10.4% (2009 est.)

Central bank discount rate: 11.25% (31 December 2008)
country comparison to the world: 27
12.5% (31 December 2007)

Commercial bank prime lending rate: NA% (31 December 2010)
country comparison to the world: 34
16.51% (31 December 2008 est.)

Stock of narrow money: $602.3 million (31 December 2010 est.)
country comparison to the world: 155
$537.6 million (31 December 2009 est.)

Stock of broad money: $1.243 billion (31 December 2010 est.)
country comparison to the world: 156
$1.068 billion (31 December 2009 est.)

Stock of domestic credit: $600.4 million (31 December 2010 est.)
country comparison to the world: 162
$515.5 million (31 December 2009 est.)

Market value of publicly traded shares: $NA

Agriculture—products: coffee, tea, pyrethrum (insecticide made from chrysanthemums), bananas, beans, sorghum, potatoes; livestock

Industries: cement, agricultural products, small-scale beverages, soap, furniture, shoes, plastic goods, textiles, cigarettes

Industrial production growth rate: 7.5% (2010 est.)
country comparison to the world: 39

Electricity—production: 120 million kWh (2007 est.)
country comparison to the world: 186

Electricity—consumption: 231.6 million kWh (2007 est.)
country comparison to the world: 175

Electricity—exports: 10 million kWh (2007)

Electricity—imports: 130 million kWh (2007 est.)

Oil—production: 0 bbl/day (2009 est.)
country comparison to the world: 124

Oil—consumption: 6,000 bbl/day (2009 est.)
country comparison to the world: 159

Oil—exports: 0 bbl/day (2007 est.)
country comparison to the world: 197

Oil—imports: 5,623 bbl/day (2007 est.)
country comparison to the world: 154

Oil—proved reserves: 0 bbl (1 January 2010 est.)

country comparison to the world: 180
Natural gas—production: 0 cu m (2008 est.)
country comparison to the world: 176
Natural gas—consumption: 0 cu m (2008 est.)
country comparison to the world: 117
Natural gas—exports: 0 cu m (2008 est.)
country comparison to the world: 165
Natural gas—imports: 0 cu m (2008 est.)
country comparison to the world: 111
Natural gas—proved reserves: 56.63 billion cu m (1 January 2010 est.)
country comparison to the world: 63
Current account balance: $-489 million (2010 est.)
country comparison to the world: 116
$-379 million (2009 est.)
Exports: $226 million (2010 est.)
country comparison to the world: 179
$193 million (2009 est.)
Exports—commodities: coffee, tea, hides, tin ore
Exports—partners: Kenya 33.88%, Democratic Republic of the Congo 13.56%, Thailand 6.22%, China 5.49%, US 5.47%, Swaziland 5.43%, Belgium 5.19% (2009)
Imports: $1.047 billion (2010 est.)
country comparison to the world: 171
$961 million (2009 est.)
Imports—commodities: foodstuffs, machinery and equipment, steel, petroleum products, cement and construction material
Imports—partners: Kenya 16.53%, Uganda 14.92%, China 7.92%, UAE 6.89%, Belgium 5.54%, Germany 5.19%, Tanzania 4.81%, Sweden 4% (2009)
Reserves of foreign exchange and gold: $816 million (31 December 2010 est.)
country comparison to the world: 119
$742.7 million (31 December 2009 est.)
Debt—external: $NA
Exchange rates: Rwandan francs (RWF) per US dollar–586.25 (2010), 568.18 (2009), 550 (2008), 585 (2007), 560 (2006)

COMMUNICATIONS

Telephones—main lines in use: 33,500 (2009)
country comparison to the world: 175
Telephones—mobile cellular: 2.429 million (2009)
country comparison to the world: 123
Telephone system: *general assessment:* small, inadequate telephone system primarily serves business, education, and government
domestic: the capital, Kigali, is connected to the centers of the provinces by microwave radio relay and, recently, by cellular telephone service; much of the network depends on wire and HF radiotelephone; combined fixed-line and mobile-cellular telephone density has increased to about 25 telephones per 100 persons
international: country code–250; international connections employ microwave radio relay to neighboring countries and satellite communications to more distant countries; satellite earth stations–1 Intelsat (Indian Ocean) in Kigali (includes telex and telefax service)
Broadcast media: government owns and operates the only TV station; government-owned and operated Radio Rwanda has a national reach; 9 private radio stations; transmissions of multiple international broadcasters are available (2007)
Internet country code: .rw
Internet hosts: 815 (2010)
country comparison to the world: 171
Internet users: 450,000 (2009)
country comparison to the world: 118

TRANSPORTATION

Airports: 9 (2010)
country comparison to the world: 160
Airports—with paved runways: *total:* 4
over 3,047 m: 1
914 to 1,523 m: 2
under 914 m: 1 (2010)
Airports—with unpaved runways:
total: 5
914 to 1,523 m: 2
under 914 m: 3 (2010)
Roadways:
total: 14,008 km
country comparison to the world: 124
paved: 2,662 km
unpaved: 11,346 km (2004)
Waterways: (Lac Kivu navigable by shallow-draft barges and native craft) (2009)
Ports and terminals: Cyangugu, Gisenyi, Kibuye

MILITARY

Military branches: Rwandan Defense Force (RDF): Rwandan Army (Rwandan Land Force), Rwandan Air Force (2011)
Military service age and obligation: 18 years of age for voluntary military service; no conscription; Rwandan citizenship required (2011)
Manpower available for military service:
males age 16–49: 2,625,917
females age 16–49: 2,608,110 (2010 est.)
Manpower fit for military service: *males age 16–49:* 1,685,066
females age 16–49: 1,749,580 (2010 est.)
Manpower reaching militarily significant age annually: *male:* 110,736
female: 110,328 (2010 est.)
Military expenditures: 2.9% of GDP (2006 est.)
country comparison to the world: 48

TRANSNATIONAL ISSUES

Disputes—international: Burundi and Rwanda dispute two sq km (0.8 sq mi) of Sabanerwa, a farmed area in the Rukurazi Valley where the Akanyaru/Kanyaru River shifted its course southward after heavy rains in 1965; fighting among ethnic groups–loosely associated political rebels, armed gangs, and various government forces in Great Lakes region transcending the boundaries of Burundi, Democratic Republic of the Congo, Rwanda, and Uganda–abated substantially from a decade ago due largely to UN peacekeeping, international mediation, and efforts by local governments to create civil societies; nonetheless, 57,000 Rwandan refugees still reside in 21 African states, including Zambia, Gabon, and 20,000 who fled to Burundi in 2005 and 2006 to escape drought and recriminations from traditional courts investigating the 1994 massacres; the 2005 DROC and Rwanda border verification mechanism to stem rebel actions on both sides of the border remains in place
Refugees and internally displaced persons: refugees (country of origin): 46,272 (Democratic Republic of the Congo); 4,400 (Burundi) (2007)

SAINT BARTHELEMY

(OVERSEAS COLLECTIVITY OF FRANCE)

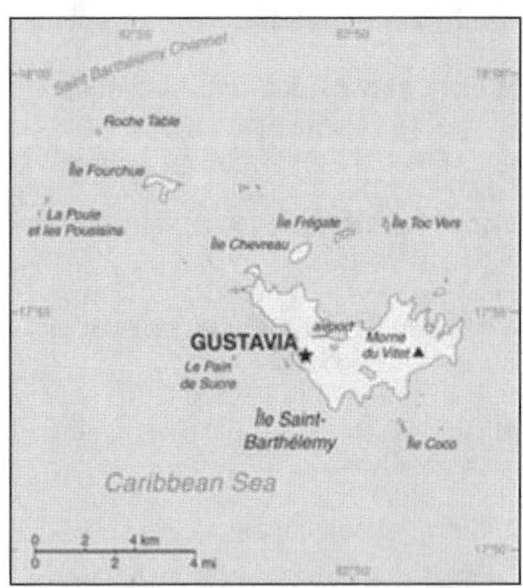

INTRODUCTION

Background: Discovered in 1493 by Christopher COLUMBUS who named it for his brother Bartolomeo, Saint Barthelemy was first settled by the French in 1648. In 1784, the French sold the island to Sweden, who renamed the largest town Gustavia, after the Swedish King GUSTAV III, and made it a free port; the island prospered as a trade and supply center during the colonial wars of the 18th century. France repurchased the island in 1878 and placed it under the administration of Guadeloupe. Saint Barthelemy retained its free port status along with various Swedish appellations such as Swedish street and town names, and the three-crown symbol on the coat of arms. In 2003, the populace of the island voted to secede from Guadeloupe and in 2007, the island became a French overseas collectivity.

GEOGRAPHY

Location: Caribbean, island between the Caribbean Sea and the North Atlantic Ocean; located in the Leeward Islands (northern) group; Saint Barthelemy lies east of the US Virgin Islands
Geographic coordinates: 17 90 N, 62 85 W
Map references: Central America and the Caribbean
Area: 21 sq km
Area—comparative: less than an eighth of the size of Washington, DC
Land boundaries: 0 km
Climate: tropical, with practically no variation in temperature; has two seasons (dry and humid)
Terrain: hilly, almost completely surrounded by shallow-water reefs, with plentiful beaches
Elevation extremes: *lowest point:* Caribbean Ocean 0 m
highest point: Morne du Vitet 286 m
Natural resources: has few natural resources, its beaches being the most important
Environment—current issues: with no natural rivers or streams, fresh water is in short supply, especially in summer, and provided by desalinization of sea water, collection of rain water, or imported via water tanker

PEOPLE

Population: 7,367 (July 2011 est.)
country comparison to the world: 226
Age structure: *0–14 years:* 18.8% (male 712/female 675)
15–64 years: 69.5% (male 2,779/female 2,342)
65 years and over: 11.7% (male 428/female 431) (2011 est.)
Median age: *total:* 40.4 years
male: 40.6 years
female: 40.3 years (2011 est.)
Sex ratio: *at birth:* 1.027 male(s)/female
under 15 years: 1.06 male(s)/female
15–64 years: 1.19 male(s)/female
65 years and over: 0.98 male(s)/female
total population: 1.14 male(s)/female (2011 est.)
Ethnic groups: white, Creole (mulatto), black, Guadeloupe Mestizo (French-East Asia)
Religions: Roman Catholic, Protestant, Jehovah's Witness
Languages: French (primary), English
School life expectancy (primary to tertiary education): NA
Education expenditures: NA

GOVERNMENT

Country name: *conventional long form:* Overseas Collectivity of Saint Barthelemy
conventional short form: Saint Barthelemy
local long form: Collectivite d'outre mer de Saint-Barthelemy
local short form: Saint-Barthelemy
Dependency status: overseas collectivity of France
Capital: *name:* Gustavia
geographic coordinates: 17 53 N, 62 51 W
time difference: UTC-4 (1 hour ahead of Washington, DC, during Standard Time)
Independence: none (overseas collectivity of France)
National holiday: Bastille Day, 14 July (1789); note—local holiday is St. Barthelemy Day, 24 August
Constitution: 4 October 1958 (French Constitution)
Legal system: French civil law
Suffrage: 18 years of age, universal
Executive branch: *chief of state:* President Nicolas SARKOZY (since 16 May 2007), represented by Prefect Dominique LACROIX (since 21 March 2007)
head of government: President of the Territorial Council Bruno MAGRAS (since 16 July 2007)
cabinet: Executive Council; note—there is also an advisory, economic, social, and cultural council
(For more information visit the World Leaders website)
elections: French president elected by popular vote for a five-year term; prefect appointed by the French president on the advice of the French Ministry of Interior; president of the Territorial Council elected by the members of the Council for a five-year term
election results: Bruno MAGRAS unanimously elected president by the Territorial Council on 16 July 2007
Legislative branch: unicameral Territorial Council (19 seats; members elected by popular vote to serve five-year terms)
elections: last held on 1 and 8 July 2007 (next to be held in July 2012)
election results: percent of vote by party—SBA 72.2%, Action-Equilibre-Transparence 9.9%, Ensemble pour Saint-Barthelemy 7.9%, Tous Unis pour Saint-Barthelemy 9.9%; seats by party—SBA 16, Action-Equilibre-Transparence 1, Ensemble pour Saint-Barthelemy 1, Tous Unis pour Saint-Barthelemy 1
note: Saint Barthelemy elects one seat to the French Senate; election last held on 21 September 2008 (next to be held in September 2014); results—percent of vote by party—NA; seats by party—UMP 1
Political parties and leaders: Action-Equilibre-Transparence [Maxime DESOUCHES]; Ensemble pour Saint-Barthelemy [Benoit CHAUVIN]; Saint-Barth d'Abord! or SBA [Bruno MAGRAS]; Tous Unis pour Saint-Barthelemy [Karine MIOT-RICHARD]
Political pressure groups and leaders: The Marine Reserve (protection of fish); Rotary Club
International organization participation: UPU
Diplomatic representation in the US: none (overseas collectivity of France)
Diplomatic representation from the US: none (overseas collectivity of France)
Flag description: the flag of France is used
National anthem: *name:* "L'Hymne a St. Barthelemy" (Hymn to St. Barthelemy)
lyrics/music: Isabelle Massart DERAVIN/ Michael VALENTI
note: local anthem in use since 1999; as a collectivity of France, "La Marseillaise" is official (see France)

ECONOMY

Economy—overview: The economy of Saint Barthelemy is based upon high-end tourism and duty-free luxury commerce, serving visitors primarily from North America. The luxury hotels and villas host 70,000 visitors each year with another 130,000 arriving by boat. The relative isolation and high cost of living inhibits mass tourism. The construction and public sectors also enjoy significant investment in support of tourism. With limited fresh water resources, all food must be imported, as must all energy resources and most manufactured goods. Employment is strong and attracts labor from Brazil and Portugal.
Exchange rates: euros (EUR) per US dollar—0.7715 (2010), 0.7338 (2009), 0.6827 (2008), 0.7345 (2007), 0.7964 (2006)

COMMUNICATIONS

Telephone system: *general assessment:* fully integrated access
domestic: direct dial capability with both fixed and wireless systems
international: country code—590; undersea fiber-optic cable provides voice and data

connectivity to Puerto Rico and Guadeloupe
Broadcast media: no local TV broadcasters; 3 FM radio channels (2 via repeater)
Internet country code: .bl; note–.gp, the Internet country code for Guadeloupe, and .fr, the Internet country code for France, might also be encountered

TRANSPORTATION

Airports: 1 (2010)
country comparison to the world: 233
Airports—with paved runways: *total:* 1
under 914 m: 1 (2010)
Ports and terminals: Gustavia
Transportation—note: nearest airport for international flights is Princess Juliana International Airport (SXM) located in Sint Maarten

MILITARY

Manpower fit for military service: *males age 16–49:* 1,495
females age 16–49: 1,263 (2010 est.)
Manpower reaching militarily significant age annually: *male:* 23
female: 21 (2010 est.)
Military—note: defense is the responsibility of France

SAINT HELENA, ASCENSION, AND TRISTAN DA CUNHA

(OVERSEAS TERRITORY OF THE UK)

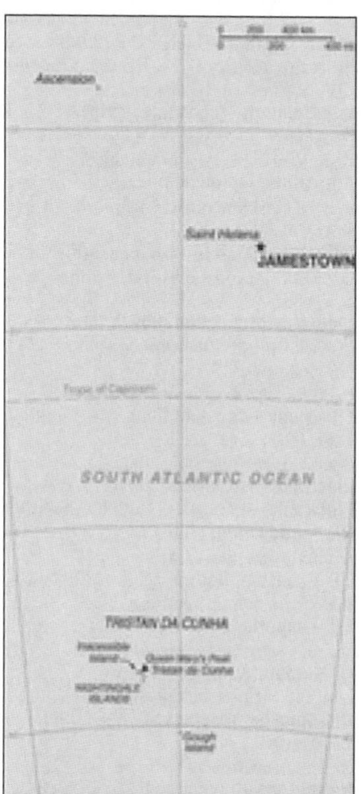

INTRODUCTION

Background: Saint Helena is a British Overseas Territory consisting of Saint Helena and Ascension Islands, and the island group of Tristan da Cunha.
Saint Helena: Uninhabited when first discovered by the Portuguese in 1502, Saint Helena was garrisoned by the British during the 17th century. It acquired fame as the place of Napoleon BONAPARTE's exile from 1815 until his death in 1821, but its importance as a port of call declined after the opening of the Suez Canal in 1869. During the Anglo-Boer War in South Africa, several thousand Boer prisoners were confined on the island between 1900 and 1903.
Ascension Island: This barren and uninhabited island was discovered and named by the Portuguese in 1503. The British garrisoned the island in 1815 to prevent a rescue of Napoleon from Saint Helena. It served as a provisioning station for the Royal Navy's West Africa Squadron on anti-slavery patrol. The island remained under Admiralty control until 1922, when it became a dependency of Saint Helena. During World War II, the UK permitted the US to construct an airfield on Ascension in support of trans-Atlantic flights to Africa and anti-submarine operations in the South Atlantic. In the 1960s the island became an important space tracking station for the US. In 1982, Ascension was an essential staging area for British forces during the Falklands War. It remains a critical refueling point in the air-bridge from the UK to the South Atlantic.
Tristan da Cunha: The island group consists of the islands of Tristan da Cunha, Nightingale, Inaccessible, and Gough. Tristan da Cunha is named after its Portuguese discoverer (1506); it was garrisoned by the British in 1816 to prevent any attempt to rescue Napoleon from Saint Helena. Gough and Inaccessible Islands have been designated World Heritage Sites. South Africa leases a site for a meteorological station on Gough Island.

GEOGRAPHY

Location: islands in the South Atlantic Ocean, about midway between South America and Africa; Ascension Island lies 700 nm northwest of Saint Helena; Tristan da Cunha lies 2,300 nm southwest of Saint Helena
Geographic coordinates: Saint Helena: 15 57 S, 5 42 W
Ascension Island: 7 57 S, 14 22 W
Tristan da Cunha island group: 37 15 S, 12 30 W
Map references: Africa
Area: *total:* 308 sq km
country comparison to the world: 207
land: Saint Helena Island 122 sq km; Ascension Island 88 sq km; Tristan da Cunha island group 98 sq km
water: 0 sq km
Area—comparative: slightly more than twice the size of Washington, DC
Land boundaries: 0 km
Coastline: Saint Helena: 60 km
Ascension Island: NA
Tristan da Cunha: 40 km
Maritime claims: territorial sea: 12 nm
exclusive fishing zone: 200 nm
Climate: Saint Helena: tropical marine; mild, tempered by trade winds
Ascension Island: tropical marine; mild, semi-arid
Tristan da Cunha: temperate marine; mild, tempered by trade winds (tends to be cooler than Saint Helena)
Terrain: the islands of this group result from volcanic activity associated with the Atlantic Mid-Ocean Ridge
Saint Helena: rugged, volcanic; small scattered plateaus and plains
Ascension: surface covered by lava flows and cinder cones of 44 dormant volcanoes; ground rises to the east
Tristan da Cunha: sheer cliffs line the coastline of the nearly circular island; the flanks of the central volcanic peak are deeply dissected; narrow coastal plain lies between The Peak and the coastal cliffs
Elevation extremes: *lowest point:* Atlantic Ocean 0 m
highest point: Queen Mary's Peak on Tristan da Cunha 2,060 m; Green Mountain on Ascension Island 859 m; Mount Actaeon on Saint Helena Island 818 m
Natural resources: fish, lobster
Land use:
arable land: 12.9%
permanent crops: 0%
other: 87.1% (2005)
Irrigated land: NA
Natural hazards: active volcanism on Tristan da Cunha
volcanism: the island volcanoes of Tristan da Cunha (elev. 2,060 m) and Nightingale Island (elev. 365 m) experience volcanic activity; Tristan da Cunha erupted in 1962 and Nightingale in 2004
Environment—current issues: NA
Geography—note: Saint Helena harbors at least 40 species of plants unknown elsewhere in the world; Ascension is a breeding ground for sea turtles and sooty terns; Queen Mary's Peak on Tristan da Cunha is the highest island mountain in the South Atlantic and a prominent landmark on the sea lanes around southern Africa

PEOPLE

Population: 7,700 (July 2011 est.)
country comparison to the world: 225
note: only Saint Helena, Ascension, and Tristan da Cunha islands are inhabited
Age structure: *0–14 years:* 17.8% (male 697/female 671)
15–64 years: 70% (male 2,731/female 2,656)
65 years and over: 12.3% (male 468/female 477) (2011 est.)
Median age: *total:* 38.8 years
male: 38.8 years
female: 38.7 years (2011 est.)
Population growth rate: 0.377% (2011 est.)
country comparison to the world: 160
Birth rate: 10.65 births/1,000 population (2011 est.)

country comparison to the world: 181
Death rate: 6.88 deaths/1,000 population (July 2011 est.)
country comparison to the world: 140
Net migration rate: 0 migrant(s)/1,000 population (2011 est.)
country comparison to the world: 106
Urbanization: *urban population:* 40% of total population (2010)
rate of urbanization: -0.3% annual rate of change (2010-15 est.)
Major cities—population: JAMESTOWN (capital) 1,000 (2009)
Sex ratio: *at birth:* 1.05 male(s)/female
under 15 years: 1.04 male(s)/female
15–64 years: 1.04 male(s)/female
65 years and over: 0.94 male(s)/female
total population: 1.02 male(s)/female (2011 est.)
Infant mortality rate: *total:* 16.38 deaths/1,000 live births
country comparison to the world: 110
male: 19.28 deaths/1,000 live births
female: 13.34 deaths/1,000 live births (2011 est.)
Life expectancy at birth: *total population:* 78.76 years
country comparison to the world: 46
male: 75.83 years
female: 81.83 years (2011 est.)
Total fertility rate: 1.57 children born/woman (2011 est.)
country comparison to the world: 179
HIV/AIDS—adult prevalence rate: NA
HIV/AIDS—people living with HIV/AIDS: NA
HIV/AIDS—deaths: NA
Nationality: *noun:* Saint Helenian(s)
adjective: Saint Helenian
note: referred to locally as "Saints"
Ethnic groups: African descent 50%, white 25%, Chinese 25%
Religions: Anglican (majority), Baptist, Seventh-Day Adventist, Roman Catholic
Languages: English
Literacy: *definition:* age 20 and over can read and write
total population: 97%
male: 97%
female: 98% (1987 est.)
School life expectancy (primary to tertiary education): NA
Education expenditures: NA

GOVERNMENT

Country name: *conventional long form:* Saint Helena, Ascension, and Tristan da Cunha
conventional short form: none
Dependency status: overseas territory of the UK
Government type: NA
Capital:
name: Jamestown
geographic coordinates: 15 56 S, 5 44 W
time difference: UTC 0 (5 hours ahead of Washington, DC during Standard Time)
Administrative divisions: 3 administrative areas; Ascension, Saint Helena, Tristan da Cunha
Independence: none (overseas territory of the UK)
National holiday: Birthday of Queen ELIZABETH II, second Saturday in June (1926)
Constitution: The Saint Helena, Ascension and Tristan da Cunha Constitution Order 2009, effective 1 September 2009
Legal system: English common law and local statutes
Suffrage: NA
Executive branch: *chief of state:* Queen ELIZABETH II (since 6 February 1952)
head of government: Governor Andrew GURR (since 11 November 2007)
cabinet: Executive Council consists of the governor, 3 ex-officio officers, and 5 elected members of the Legislative Council
(For more information visit the World Leaders website)
elections: none; the monarchy is hereditary; governor appointed by the monarch
Legislative branch: unicameral Legislative Council (17 seats, including a speaker and deputy speaker, 12 elected, and three ex officio members; members elected by popular vote to serve four-year terms)
note: the Constitution Order provides for separate Island Councils for both Ascension and Tristan da Cunha
elections: last held on 4 November 2009 (next to be held in 2013)
election results: percent of vote—NA; seats—independents 12
Judicial branch: Supreme Court; Court of Appeal
Political parties and leaders: none
Political pressure groups and leaders: *other:* private sector; unions
International organization participation: UPU
Diplomatic representation in the US: none (overseas territory of the UK)
Diplomatic representation from the US: none (overseas territory of the UK)
Flag description: blue with the flag of the UK in the upper hoist-side quadrant and the Saint Helenian shield centered on the outer half of the flag; the upper third of the shield depicts a white plover (wire bird) on a yellow field; the remainder of the shield depicts a rocky coastline on the left, offshore is a three-masted sailing ship with sails furled but flying an English flag
National anthem: note: as a territory of the United Kingdom, "God Save the Queen" is official (see United Kingdom)

ECONOMY

Economy—overview: The economy depends largely on financial assistance from the UK, which amounted to about $27 million in FY06/07 or more than twice the level of annual budgetary revenues. The local population earns income from fishing, raising livestock, and sales of handicrafts. Because there are few jobs, 25% of the work force has left to seek employment on Ascension Island, on the Falklands, and in the UK.
GDP (purchasing power parity): $18 million (1998 est.)
country comparison to the world: 225
GDP (official exchange rate): $NA
GDP—real growth rate: NA%
GDP—per capita (PPP): $2,500 (1998 est.)
country comparison to the world: 182
GDP—composition by sector:
agriculture: NA%
industry: NA%
services: NA%
Labor force: 2,486
country comparison to the world: 223
note: 1,200 work offshore (1998 est.)
Labor force—by occupation: *agriculture:* 6%
industry: 48%
services: 46% (1987 est.)
Unemployment rate: 14% (1998 est.)
country comparison to the world: 141
Population below poverty line: NA%
Household income or consumption by percentage share: *lowest 10%:* NA%
highest 10%: NA%
Budget: *revenues:* $10.23 million
expenditures: $25.14 million
note: revenue data reflect locally raised revenues only; the budget deficit is resolved by grant aid from the United Kingdom (FY06/07 est.)
Inflation rate (consumer prices): 3.2% (1997 est.)
country comparison to the world: 89
Agriculture—products: coffee, corn, potatoes, vegetables; timber; fish, lobster; livestock
Industries: construction, crafts (furniture, lacework, fancy woodwork), fishing, philatelic sales
Industrial production growth rate: NA%
Electricity—production: 8 million kWh (2007 est.)
country comparison to the world: 211
Electricity—consumption: 7.44 million kWh (2007 est.)
country comparison to the world: 211
Electricity—exports: 0 kWh (2008 est.)
Electricity—imports: 0 kWh (2008 est.)
Oil—production: 0 bbl/day (2009 est.)
country comparison to the world: 129
Oil—consumption: 0 bbl/day (2009 est.)
country comparison to the world: 207
Oil—exports: 0 bbl/day (2007 est.)
country comparison to the world: 201
Oil—imports: 79.73 bbl/day (2007 est.)
country comparison to the world: 202
Oil—proved reserves: 0 bbl (1 January 2010 est.)
country comparison to the world: 185
Natural gas—production: 0 cu m (2008 est.)
country comparison to the world: 180
Natural gas—consumption: 0 cu m (2008 est.)
country comparison to the world: 121
Natural gas—exports: 0 cu m (2008 est.)
country comparison to the world: 172
Natural gas—imports: 0 cu m (2008 est.)
country comparison to the world: 117
Natural gas—proved reserves: 0 cu m (1 January 2010 est.)
country comparison to the world: 185
Exports: $19 million (2004 est.)
country comparison to the world: 208
Exports—commodities: fish (frozen, canned, and salt-dried skipjack, tuna), coffee, handicrafts
Imports: $45 million (2004 est.)
country comparison to the world: 217
Imports—commodities: food, beverages, tobacco, fuel oils, animal feed, building materials, motor vehicles and parts, machinery and parts
Debt—external: $NA
Exchange rates: Saint Helenian pounds (SHP) per US dollar–0.6388 (2010), 0.6175 (2009), 0.4993 (2007), 0.5418 (2006)

COMMUNICATIONS

Telephones—main lines in use: 2,900 (2009)
country comparison to the world: 218
Telephone system: *general assessment:* can communicate worldwide
domestic: automatic digital network
international: country code (Saint Helena)–290, (Ascension Island)–247; international direct dialing; satellite voice and data communications; satellite earth stations–5 (Ascension Island–4, Saint Helena–1)
Broadcast media: St. Helena has no local TV station; 2 local radio stations, one of which is relayed to Ascension Island; satellite TV stations rebroadcast terrestrially; Ascension Island has no local TV station, but has 1 local radio station and receives relays of broadcasts from 1 St. Helena radio station; broadcasts from the British Forces Broadcasting Service (BFBS) are available, as well as TV services for the US military; Tristan da Cunha has 1 local radio station and receives BFBS TV and radio broadcasts (2007)
Internet country code: .sh; note–Ascension Island assigned .ac
Internet hosts: 6,873 (2010)
country comparison to the world: 136
Internet users: 900 (2009)
country comparison to the world: 214
Communications—note: South Africa maintains a meteorological station on Gough Island

TRANSPORTATION

Airports: 1 (2010)
country comparison to the world: 231
Airports—with paved runways: *total:* 1
over 3,047 m: 1 (2010)
Roadways: *total:* 198 km (Saint Helena 138 km, Ascension 40 km, Tristan da Cunha 20 km)
country comparison to the world: 208
paved: 168 km (Saint Helena 118 km, Ascension 40 km, Tristan da Cunha 10 km)
unpaved: 30 km (Saint Helena 20 km, Ascension 0 km, Tristan da Cunha 10 km) (2002)
Ports and terminals: Saint Helena: Jamestown
Ascension Island: Georgetown
Tristan da Cunha: Calshot Harbor (Edinburgh)
Transportation—note: there is no air connection to Saint Helena or Tristan da Cunha; following a pause in 2009, an international airport for Saint Helena continues in development with a goal of beginning construction in late 2011 or 2012

MILITARY

Manpower fit for military service: *males age 16–49:* 1,565
females age 16–49: 1,579 (2010 est.)
Manpower reaching militarily significant age annually: *male:* 49
female: 48 (2010 est.)
Military—note: defense is the responsibility of the UK

TRANSNATIONAL ISSUES

Disputes—international: none

SAINT KITTS AND NEVIS

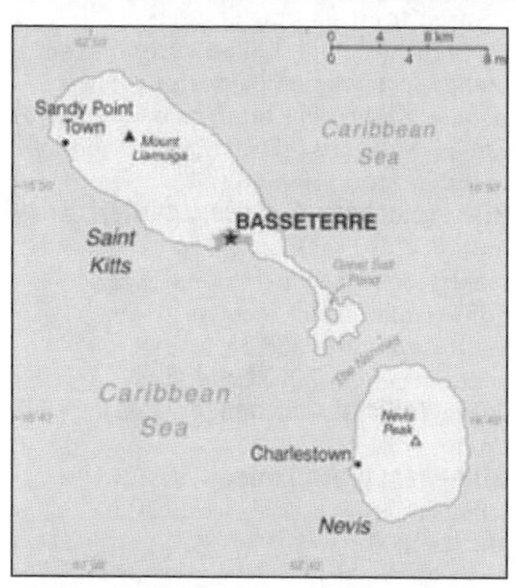

INTRODUCTION

Background: Carib Indians occupied the islands for hundreds of years before the British began settlement in 1623. The islands became an associated state of the UK with full internal autonomy in 1967. The island of Anguilla rebelled and was allowed to secede in 1971. Saint Kitts and Nevis achieved independence in 1983. In 1998, a vote in Nevis on a referendum to separate from Saint Kitts fell short of the two-thirds majority needed. Nevis continues in its efforts to separate from Saint Kitts.

GEOGRAPHY

Location: Caribbean, islands in the Caribbean Sea, about one-third of the way from Puerto Rico to Trinidad and Tobago
Geographic coordinates: 17 20 N, 62 45 W
Map references: Central America and the Caribbean
Area: *total:* 261 sq km (Saint Kitts 168 sq km; Nevis 93 sq km)
country comparison to the world: 210
land: 261 sq km
water: 0 sq km
Area—comparative: one and a half times the size of Washington, DC
Land boundaries: 0 km
Coastline: 135 km
Maritime claims: territorial sea: 12 nm
contiguous zone: 24 nm
exclusive economic zone: 200 nm
continental shelf: 200 nm or to the edge of the continental margin
Climate: tropical, tempered by constant sea breezes; little seasonal temperature variation; rainy season (May to November)
Terrain: volcanic with mountainous interiors
Elevation extremes: *lowest point:* Caribbean Sea 0 m
highest point: Mount Liamuiga 1,156 m
Natural resources: arable land
Land use: *arable land:* 19.44%
permanent crops: 2.78%
other: 77.78% (2005)
Irrigated land: NA
Total renewable water resources: 0.02 cu km (2000)
Natural hazards: hurricanes (July to October)
Environment—current issues: NA
Environment—international agreements: *party to:* Biodiversity, Climate Change, Climate Change-Kyoto Protocol, Desertification, Endangered Species, Hazardous Wastes, Law of the Sea, Marine Dumping, Ozone Layer Protection, Ship Pollution, Whaling
signed, but not ratified: none of the selected agreements
Geography—note: with coastlines in the shape of a baseball bat and ball, the two volcanic islands are separated by a 3-km-wide channel called The Narrows; on the southern tip of long, baseball bat-shaped Saint Kitts lies the Great Salt Pond; Nevis Peak sits in the center of its almost circular namesake island and its ball shape complements that of its sister island

PEOPLE

Population: 50,314 (July 2011 est.)
country comparison to the world: 206
Age structure: *0–14 years:* 22.8% (male 5,701/female 5,755)
15–64 years: 69.6% (male 17,740/female 17,297)
65 years and over: 7.6% (male 1,681/female 2,140) (2011 est.)
Median age: *total:* 32 years
male: 32.1 years
female: 32 years (2011 est.)
Population growth rate: 0.823% (2011 est.)
country comparison to the world: 132
Birth rate: 14.07 births/1,000 population (2011 est.)
country comparison to the world: 146
Death rate: 7.1 deaths/1,000 population (July 2011 est.)
country comparison to the world: 130
Net migration rate: 1.25 migrant(s)/1,000 population (2011 est.)
country comparison to the world: 50
Urbanization: *urban population:* 32% of total population (2010)
rate of urbanization: 1.8% annual rate of change (2010-15 est.)
Major cities—population: BASSETERRE (capital) 13,000 (2009)
Sex ratio: *at birth:* 1.02 male(s)/female
under 15 years: 0.99 male(s)/female
15–64 years: 1.03 male(s)/female
65 years and over: 0.78 male(s)/female
total population: 1 male(s)/female (2011 est.)
Infant mortality rate: *total:* 9.66 deaths/1,000 live births
country comparison to the world: 152
male: 6.36 deaths/1,000 live births
female: 13.02 deaths/1,000 live births (2011 est.)
Life expectancy at birth: *total population:* 74.6 years
country comparison to the world: 96
male: 72.25 years
female: 77.01 years (2011 est.)

Total fertility rate: 1.79 children born/woman (2011 est.)
country comparison to the world: 156
HIV/AIDS—adult prevalence rate: NA
HIV/AIDS—people living with HIV/AIDS: NA
HIV/AIDS—deaths: NA
Drinking water source: *Improved:*
urban: 99% of population
rural: 99% of population
total: 99% of population
Unimproved:
urban: 1% of population
rural: 1% of population
total: 1% of population (2008)
Sanitation facility access: *Improved:*
urban: 96% of population
rural: 96% of population
total: 96% of population
Unimproved:
urban: 4% of population
rural: 4% of population
total: 4% of population (2008)
Nationality: *noun:* Kittitian(s), Nevisian(s)
adjective: Kittitian, Nevisian
Ethnic groups: predominantly black; some British, Portuguese, and Lebanese
Religions: Anglican, other Protestant, Roman Catholic
Languages: English (official)
Literacy: *definition:* age 15 and over has ever attended school
total population: 97.8%
male: NA
female: NA (2003 est.)
School life expectancy (primary to tertiary education): *total:* 13 years
male: 12 years
female: 13 years (2008)
Education expenditures: 9.6% of GDP (2005)
country comparison to the world: 6

GOVERNMENT

Country name: *conventional long form:* Federation of Saint Kitts and Nevis
conventional short form: Saint Kitts and Nevis
former: Federation of Saint Christopher and Nevis
Government type: parliamentary democracy and a Commonwealth realm
Capital: *name:* Basseterre
geographic coordinates: 17 18 N, 62 43 W
time difference: UTC-4 (1 hour ahead of Washington, DC during Standard Time)
Administrative divisions: 14 parishes; Christ Church Nichola Town, Saint Anne Sandy Point, Saint George Basseterre, Saint George Gingerland, Saint James Windward, Saint John Capesterre, Saint John Figtree, Saint Mary Cayon, Saint Paul Capesterre, Saint Paul Charlestown, Saint Peter Basseterre, Saint Thomas Lowland, Saint Thomas Middle Island, Trinity Palmetto Point
Independence: 19 September 1983 (from the UK)
National holiday: Independence Day, 19 September (1983)
Constitution: 19 September 1983
Legal system: English common law
International law organization participation: has not submitted an ICJ jurisdiction declaration; accepts ICCt jurisdiction
Suffrage: 18 years of age; universal
Executive branch: *chief of state:* Queen ELIZABETH II (since 6 February 1952); represented by Governor General Cuthbert Montraville SEBASTIAN (since 1 January 1996)
head of government: Prime Minister Dr. Denzil DOUGLAS (since 6 July 1995); Deputy Prime Minister Sam CONDOR (since 6 July 1995)
cabinet: Cabinet appointed by the governor general in consultation with the prime minister
(For more information visit the World Leaders website)
elections: the monarchy is hereditary; the governor general appointed by the monarch; following legislative elections, the leader of the majority party or leader of a majority coalition usually appointed prime minister by the governor general; deputy prime minister appointed by the governor general
Legislative branch: unicameral National Assembly (14 seats, 3 appointed and 11 popularly elected from single-member constituencies; members serve five-year terms)
elections: last held on 25 January 2010 (next to be held by 2015)
election results: percent of vote by party—NA; seats by party—SKNLP 6, CCM 2, PAM 2, NRP 1
Judicial branch: Eastern Caribbean Supreme Court (consisting of a Court of Appeal and a High Court; based on Saint Lucia; two judges of the Supreme Court reside in Saint Kitts and Nevis); member of the Caribbean Court of Justice (CCJ)
Political parties and leaders: Concerned Citizens Movement or CCM [Vance AMORY]; Nevis Reformation Party or NRP [Joseph PARRY]; People's Action Movement or PAM [Lindsay GRANT]; Saint Kitts and Nevis Labor Party or SKNLP [Dr. Denzil DOUGLAS]
Political pressure groups and leaders: NA
International organization participation: ACP, AOSIS, C, Caricom, CDB, FAO, G-77, IBRD, ICAO, ICRM, IDA, IFAD, IFC, IFRCS, ILO, IMF, IMO, Interpol, IOC, ITU, MIGA, NAM, OAS, OECS, OPANAL, OPCW, PetroCaribe, UN, UNCTAD, UNESCO, UNIDO, UPU, WHO, WIPO, WTO
Diplomatic representation in the US: *chief of mission:* Ambassador Jacinth HENRY-MARTIN
chancery: 3216 New Mexico Avenue NW, Washington, DC 20016
telephone: [1] (202) 686-2636
FAX: [1] (202) 686-5740
consulate(s) general: New York
Diplomatic representation from the US: the US does not have an embassy in Saint Kitts and Nevis; the US Ambassador to Barbados is accredited to Saint Kitts and Nevis
Flag description: divided diagonally from the lower hoist side by a broad black band bearing two white, five-pointed stars; the black band is edged in yellow; the upper triangle is green, the lower triangle is red; green signifies the island's fertility, red symbolizes the struggles of the people from slavery, yellow denotes year-round sunshine, and black represents the African heritage of the people; the white stars stand for the islands of Saint Kitts and Nevis, but can also express hope and liberty, or independence and optimism
National anthem: *name:* "Oh Land of Beauty!"
lyrics/music: Kenrick Anderson GEORGES
note: adopted 1983

ECONOMY

Economy—overview: The economy of Saint Kitts and Nevis is heavily dependent upon tourism revenues, which has replaced sugar, the traditional mainstay of the economy until the 1970s. Following the 2005 harvest, the government closed the sugar industry after decades of losses of 3-4% of GDP annually. To compensate for employment losses, the government has embarked on a program to diversify the agricultural sector and to stimulate other sectors of the economy, such as tourism, export-oriented manufacturing, and offshore banking. More than 200,000 tourists visited the islands in 2009. Like other tourist destinations in the Caribbean, St. Kitts and Nevis is vulnerable to damage from natural disasters and shifts in tourism demand. The current government is constrained by one of the world's highest public debt burdens equivalent to roughly 185% of GDP, largely attributable to public enterprise losses.
GDP (purchasing power parity): $684 million (2010 est.)
country comparison to the world: 209
$694.6 million (2009 est.)
$768.2 million (2008 est.)
note: data are in 2010 US dollars
GDP (official exchange rate): $531 million (2010 est.)
GDP—real growth rate: -1.5% (2010 est.)
country comparison to the world: 205
-9.6% (2009 est.)
4.6% (2008 est.)
GDP—per capita (PPP): $13,700 (2010 est.)
country comparison to the world: 87
$14,000 (2009 est.)
$15,700 (2008 est.)
note: data are in 2010 US dollars
GDP—composition by sector:
agriculture: 3.5%
industry: 25.8%
services: 70.7% (2001)
Labor force: 18,170 (June 1995)
country comparison to the world: 209
Unemployment rate: 4.5% (1997)
country comparison to the world: 42
Population below poverty line: NA%
Household income or consumption by percentage share: *lowest 10%:* NA%
highest 10%: NA%
Budget: *revenues:* $212.4 million
expenditures: $232.1 million (2008 est.)
Public debt: 185% of GDP (2009 est.)
country comparison to the world: 2
Inflation rate (consumer prices): 4.5% (2007 est.)
country comparison to the world: 127
Central bank discount rate: 6.5% (31 December 2009)
country comparison to the world: 63
6.5% (31 December 2008)

Commercial bank prime lending rate: 8.75% (31 December 2009 est.)
country comparison to the world: 102
8.75% (31 December 2008 est.)
Stock of narrow money: $94.45 million (31 December 2009)
country comparison to the world: 181
$93.23 million (31 December 2008)
Stock of broad money: $823.8 million (31 December 2009)
country comparison to the world: 168
$787.8 million (31 December 2008)
Stock of domestic credit: $790.8 million (31 December 2008 est.)
country comparison to the world: 156
$782.4 million (31 December 2007 est.)
Market value of publicly traded shares: $648 million (31 December 2009)
country comparison to the world: 105
$595.2 million (31 December 2008)
$439.7 million (31 December 2007)
Agriculture—products: sugarcane, rice, yams, vegetables, bananas; fish
Industries: tourism, cotton, salt, copra, clothing, footwear, beverages
Industrial production growth rate: NA%
Electricity—production: 130 million kWh (2007 est.)
country comparison to the world: 184
Electricity—consumption: 120.9 million kWh (2007 est.)
country comparison to the world: 187
Electricity—exports: 0 kWh (2008 est.)
Electricity—imports: 0 kWh (2008 est.)
Oil—production: 0 bbl/day (2009 est.)
country comparison to the world: 126
Oil—consumption: 1,000 bbl/day (2009 est.)
country comparison to the world: 191
Oil—exports: 0 bbl/day (2007 est.)
country comparison to the world: 199
Oil—imports: 1,225 bbl/day (2007 est.)
country comparison to the world: 185
Oil—proved reserves: 0 bbl (1 January 2010 est.)
country comparison to the world: 182
Natural gas—production: 0 cu m (2008 est.)
country comparison to the world: 178
Natural gas—consumption: 0 cu m (2008 est.)
country comparison to the world: 119
Natural gas—exports: 0 cu m (2008 est.)
country comparison to the world: 168
Natural gas—imports: 0 cu m (2008 est.)
country comparison to the world: 114
Natural gas—proved reserves: 0 cu m (1 January 2010 est.)
country comparison to the world: 183
Current account balance: $-163 million (2007 est.)
country comparison to the world: 86
Exports: $84 million (2006)
country comparison to the world: 198
Exports—commodities: machinery, food, electronics, beverages, tobacco
Exports—partners: US 62.3%, Canada 7.93%, Azerbaijan 6.72% (2009)
Imports: $383 million (2006)
country comparison to the world: 190
Imports—commodities: machinery, manufactures, food, fuels
Imports—partners: US 43.37%, Trinidad and Tobago 15.26%, Italy 11.83% (2009)
Debt—external: $314 million (2004)
country comparison to the world: 171
Exchange rates: East Caribbean dollars (XCD) per US dollar–2.7 (2010), 2.7 (2009), 2.7 (2005), 2.7 (2004), 2.7 (2003)

COMMUNICATIONS

Telephones—main lines in use: 20,500 (2009)
country comparison to the world: 194
Telephones—mobile cellular: 83,000 (2009)
country comparison to the world: 189
Telephone system: *general assessment:* good interisland and international connections
domestic: interisland links via Eastern Caribbean Fiber Optic cable; construction of enhanced wireless infrastructure launched in November 2004; fixed-line teledensity about 40 per 100 persons; mobile-cellular teledensity is roughly 160 per 100 persons
international: country code–1-869; connected internationally by the East Caribbean Fiber Optic System (ECFS) and Southern Caribbean fiber optic system (SCF) submarine cables
Broadcast media: the government operates a national television network that broadcasts on 2 channels; cable subscription services provide access to local and international channels; the government operates a national radio network; a mix of government-owned and privately-owned broadcasters operate roughly 15 radio stations (2007)
Internet country code: .kn
Internet hosts: 51 (2010)
country comparison to the world: 210
Internet users: 17,000 (2009)
country comparison to the world: 195

TRANSPORTATION

Airports: 2 (2010)
country comparison to the world: 204
Airports—with paved runways: *total:* 2
1,524 to 2,437 m: 1
914 to 1,523 m: 1 (2010)
Railways: *total:* 50 km
country comparison to the world: 131
narrow gauge: 50 km 0.762-m gauge on Saint Kitts for tourists (2008)
Roadways: *total:* 383 km
country comparison to the world: 199
paved: 163 km
unpaved: 220 km (2002)
Merchant marine: *total:* 160
country comparison to the world: 41
by type: bulk carrier 20, cargo 92, chemical tanker 4, combination ore/oil 1, container 3, liquefied gas 4, passenger/cargo 5, petroleum tanker 24, refrigerated cargo 4, roll on/roll off 3
foreign-owned: 94 (Bahrain 1, Belgium 1, China 1, Estonia 3, Italy 1, Japan 3, Kuwait 3, Latvia 2, Malaysia 1, Norway 1, Pakistan 3, Russia 11, Singapore 5, Syria 5, Turkey 22, UAE 17, UK 2, Ukraine 10, US 1, Yemen 1) (2010)
Ports and terminals: Basseterre, Charlestown

MILITARY

Military branches: Royal Saint Kitts and Nevis Defense Force (includes Coast Guard), Royal Saint Kitts and Nevis Police Force; for national security, Saint Kitts and Nevis relies on the Regional Security System, headquartered in Barbados (2010)
Military service age and obligation: 18 years of age for voluntary military service; no conscription (2010)
Manpower available for military service: *males age 16–49:* 13,506
females age 16–49: 13,089 (2010 est.)
Manpower fit for military service: *males age 16–49:* 10,742
females age 16–49: 10,923 (2010 est.)
Manpower reaching militarily significant age annually: *male:* 380
female: 422 (2010 est.)
Military expenditures: NA

TRANSNATIONAL ISSUES

Disputes—international: joins other Caribbean states to counter Venezuela's claim that Aves Island sustains human habitation, a criterion under UNCLOS, which permits Venezuela to extend its EEZ/continental shelf over a large portion of the eastern Caribbean Sea
Illicit drugs: transshipment point for South American drugs destined for the US and Europe; some money-laundering activity

SAINT LUCIA

INTRODUCTION

Background: The island, with its fine natural harbor at Castries, was contested between England and France throughout the 17th and early 18th centuries (changing possession 14 times); it was finally ceded to the UK in 1814. Even after the abolition of slavery on its plantations in 1834, Saint Lucia remained an agricultural island, dedicated to producing tropical commodity crops. Self-government was granted in 1967 and independence in 1979.

GEOGRAPHY

Location: Caribbean, island between the Caribbean Sea and North Atlantic Ocean, north of Trinidad and Tobago
Geographic coordinates: 13 53 N, 60 58 W
Map references: Central America and the Caribbean
Area: *total:* 616 sq km
country comparison to the world: 192
land: 606 sq km
water: 10 sq km
Area—comparative: three and a half times the size of Washington, DC
Land boundaries: 0 km
Coastline: 158 km
Maritime claims: territorial sea: 12 nm
contiguous zone: 24 nm
exclusive economic zone: 200 nm
continental shelf: 200 nm or to the edge of the continental margin
Climate: tropical, moderated by northeast trade winds; dry season January to April, rainy season May to August

Terrain: volcanic and mountainous with some broad, fertile valleys
Elevation extremes: *lowest point:* Caribbean Sea 0 m
highest point: Mount Gimie 950 m
Natural resources: forests, sandy beaches, minerals (pumice), mineral springs, geothermal potential
Land use: *arable land:* 6.45%
permanent crops: 22.58%
other: 70.97% (2005)
Irrigated land: 30 sq km (2003)
Freshwater withdrawal (domestic/industrial/agricultural): *total:* 0.01
per capita: 81 cu m/yr (1997)
Natural hazards: hurricanes; volcanic activity
Environment—current issues: deforestation; soil erosion, particularly in the northern region
Environment—international agreements: *party to:* Biodiversity, Climate Change, Climate Change-Kyoto Protocol, Desertification, Endangered Species, Environmental Modification, Hazardous Wastes, Law of the Sea, Marine Dumping, Ozone Layer Protection, Ship Pollution, Wetlands, Whaling
signed, but not ratified: none of the selected agreements
Geography—note: the twin Pitons (Gros Piton and Petit Piton), striking cone-shaped peaks south of Soufriere, are one of the scenic natural highlights of the Caribbean

PEOPLE

Population: 161,557 (July 2011 est.)
country comparison to the world: 185
Age structure: *0–14 years:* 22.8% (male 18,925/female 17,945)
15–64 years: 67.5% (male 52,859/female 56,173)
65 years and over: 9.7% (male 7,074/female 8,581) (2011 est.)
Median age: *total:* 31 years
male: 29.9 years
female: 32.1 years (2011 est.)
Population growth rate: 0.389% (2011 est.)
country comparison to the world: 159
Birth rate: 14.63 births/1,000 population (2011 est.)
country comparison to the world: 138
Death rate: 7 deaths/1,000 population (July 2011 est.)
country comparison to the world: 135
Net migration rate: -3.73 migrant(s)/1,000 population (2011 est.)
country comparison to the world: 185
Urbanization: *urban population:* 28% of total population (2010)
rate of urbanization: 1.6% annual rate of change (2010-15 est.)
Major cities—population: CASTRIES (capital) 15,000 (2009)
Sex ratio: *at birth:* 1.055 male(s)/female
under 15 years: 1.05 male(s)/female
15–64 years: 0.94 male(s)/female
65 years and over: 0.82 male(s)/female
total population: 0.95 male(s)/female (2011 est.)
Infant mortality rate: *total:* 12.72 deaths/1,000 live births
country comparison to the world: 130
male: 11.92 deaths/1,000 live births
female: 13.58 deaths/1,000 live births (2011 est.)
Life expectancy at birth: *total population:* 76.84 years
country comparison to the world: 69
male: 74.15 years
female: 79.68 years (2011 est.)
Total fertility rate: 1.81 children born/woman (2011 est.)
country comparison to the world: 153
HIV/AIDS—adult prevalence rate: NA
HIV/AIDS—people living with HIV/AIDS: NA
HIV/AIDS—deaths: NA
Drinking water source: *Improved:*
urban: 98% of population
rural: 98% of population
total: 98% of population
Unimproved:
urban: 2% of population
rural: 2% of population
total: 2% of population (2008)
Sanitation facility access: *Improved:*
urban: 89% of population
rural: 89% of population
total: 89% of population
Unimproved:
urban: 11% of population
rural: 11% of population
total: 11% of population (2000)
Nationality:
noun: Saint Lucian(s)
adjective: Saint Lucian
Ethnic groups: black 82.5%, mixed 11.9%, East Indian 2.4%, other or unspecified 3.1% (2001 census)
Religions: Roman Catholic 67.5%, Seventh Day Adventist 8.5%, Pentecostal 5.7%, Rastafarian 2.1%, Anglican 2%, Evangelical 2%, other Christian 5.1%, other 1.1%, unspecified 1.5%, none 4.5% (2001 census)
Languages: English (official), French patois
Literacy: *definition:* age 15 and over has ever attended school
total population: 90.1%
male: 89.5%
female: 90.6% (2001 est.)
School life expectancy (primary to tertiary education): *total:* 13 years
male: 13 years
female: 14 years (2009)
Education expenditures: 4.5% of GDP (2009)
country comparison to the world: 85

GOVERNMENT

Country name: *conventional long form:* none
conventional short form: Saint Lucia
Government type: parliamentary democracy and a Commonwealth realm
Capital: *name:* Castries
geographic coordinates: 14 01 N, 61 00 W
time difference: UTC-4 (1 hour ahead of Washington, DC during Standard Time)
Administrative divisions: 11 quarters; Anse-la-Raye, Castries, Choiseul, Dauphin, Dennery, Gros-Islet, Laborie, Micoud, Praslin, Soufriere, Vieux-Fort
Independence: 22 February 1979 (from the UK)
National holiday: Independence Day, 22 February (1979)
Constitution: 22 February 1979
Legal system: English common law
International law organization participation: has not submitted an ICJ jurisdiction declaration; accepts ICCt jurisdiction
Suffrage: 18 years of age; universal
Executive branch: *chief of state:* Queen ELIZABETH II (since 6 February 1952); represented by Governor General Dame Pearlette LOUISY (since September 1997)
head of government: Prime Minister Stephenson KING (since 9 September 2007); note–Sir John COMPTON died in office 7 September 2007
cabinet: Cabinet appointed by the governor general on the advice of the prime minister (For more information visit the World Leaders website)
elections: the monarchy is hereditary; the governor general appointed by the monarch; following legislative elections, the leader of the majority party or the leader of a majority coalition usually appointed prime minister by the governor general; deputy prime minister appointed by the governor general
Legislative branch: bicameral Parliament consists of the Senate (11 seats; six members appointed on the advice of the prime minister, three on the advice of the leader of the opposition, and two after consultation with religious, economic, and social groups) and the House of Assembly (17 seats; members elected by popular vote to serve five-year terms)
elections: House of Assembly–last held on 11 December 2006 (next to be held in December 2011)
election results: House of Assembly–percent of vote by party–UWP 50%, SLP 46.9%, other 3.1%; seats by party–UWP 11, SLP 6
Judicial branch: Eastern Caribbean Supreme Court (consists of a High Court and a Court of Appeals; based on Saint Lucia; three judges of the Supreme Court reside in Saint Lucia); member of the Caribbean Court of Justice (CCJ)
Political parties and leaders: National Alliance or NA [George ODLUM]; Saint Lucia Freedom Party or SFP [Martinus FRANCOIS]; Saint Lucia Labor Party or SLP [Kenneth ANTHONY]; Sou Tout Apwe Fete Fini or STAFF [Christopher HUNTE]; United Workers Party or UWP [Stephenson KING]
Political pressure groups and leaders: NA
International organization participation: ACP, AOSIS, C, Caricom, CDB, FAO, G-77, IBRD, ICAO, ICRM, IDA, IFAD, IFC, IFRCS, ILO, IMF, IMO, Interpol, IOC, ISO, ITU, ITUC, MIGA, NAM, OAS, OECS, OIF, OPANAL, OPCW, PetroCaribe, UN, UNCTAD, UNESCO, UNIDO, UPU, WCO, WFTU, WHO, WIPO, WMO, WTO
Diplomatic representation in the US: *chief of mission:* Ambassador Michael LOUIS
chancery: 3216 New Mexico Avenue NW, Washington, DC 20016
telephone: [1] (202) 364-6792 through 6795
FAX: [1] (202) 364-6723
consulate(s) general: Miami, New York
Diplomatic representation from the US: the US does not have an embassy in Saint

Lucia; the US Ambassador to Barbados is accredited to Saint Lucia
Flag description: blue, with a gold isosceles triangle below a black arrowhead; the upper edges of the arrowhead have a white border; the blue color represents the sky and sea, gold stands for sunshine and prosperity, and white and black the racial composition of the island (with the latter being dominant); the two major triangles invoke the twin Pitons (Gros Piton and Petit Piton), cone-shaped volcanic plugs that are a symbol of the island
National anthem: *name:* "Sons and Daughters of St. Lucia"
lyrics/music: Charles JESSE/Leton Felix THOMAS
note: adopted 1967

ECONOMY

Economy—overview: The island nation has been able to attract foreign business and investment, especially in its offshore banking and tourism industries, with a surge in foreign direct investment in 2006, attributed to the construction of several tourism projects. Although crops such as bananas, mangos, and avocados continue to be grown for export, tourism provides Saint Lucia's main source of income and the industry is the island's biggest employer. Tourism is the main source of foreign exchange, although tourism sector revenues declined with the global economic downturn as US and European travel dropped in 2009. The manufacturing sector is the most diverse in the Eastern Caribbean area, and the government is trying to revitalize the banana industry, although recent hurricanes have caused exports to contract. Saint Lucia is vulnerable to a variety of external shocks including volatile tourism receipts, natural disasters, and dependence on foreign oil. The public debt-to-GDP ratio is about 77% and high debt servicing obligations constrain the KING administration's ability to respond to adverse external shocks. Economic fundamentals remain solid, even though unemployment needs to be reduced.
GDP (purchasing power parity): $1.798 billion (2010 est.)
country comparison to the world: 188
$1.783 billion (2009 est.)
$1.849 billion (2008 est.)
note: data are in 2010 US dollars
GDP (official exchange rate): $985 million (2010 est.)
GDP—real growth rate: 0.8% (2010 est.)
country comparison to the world: 178
-3.6% (2009 est.)
0.7% (2008 est.)
GDP—per capita (PPP): $11,200 (2010 est.)
country comparison to the world: 100
$11,100 (2009 est.)
$11,600 (2008 est.)
note: data are in 2010 US dollars
GDP—composition by sector:
agriculture: 5%
industry: 15%
services: 80% (2005 est.)
Labor force: 79,700 (2007)
country comparison to the world: 184
Labor force—by occupation:
agriculture: 21.7%
industry: 24.7%
services: 53.6% (2002 est.)
Unemployment rate: 20% (2003 est.)
country comparison to the world: 167
Population below poverty line: NA%
Household income or consumption by percentage share: *lowest 10%:* NA%
highest 10%: NA%
Budget: *revenues:* $320.9 million
expenditures: $393.3 million (2010 est.)
Inflation rate (consumer prices): 1.9% (2007 est.)
country comparison to the world: 49
Central bank discount rate: 6.5% (31 December 2009)
country comparison to the world: 62
6.5% (31 December 2008)
Commercial bank prime lending rate: 10.58% (31 December 2009 est.)
country comparison to the world: 84
10.08% (31 December 2008 est.)
Stock of narrow money: $244.3 million (31 December 2009)
country comparison to the world: 169
$245 million (31 December 2008)
Stock of broad money: $1.094 billion (31 December 2009)
country comparison to the world: 162
$1.061 billion (31 December 2008)
Stock of domestic credit: $1.378 billion (31 December 2008 est.)
country comparison to the world: 137
$1.217 billion (31 December 2007 est.)
Agriculture—products: bananas, coconuts, vegetables, citrus, root crops, cocoa
Industries: clothing, assembly of electronic components, beverages, corrugated cardboard boxes, tourism; lime processing, coconut processing
Industrial production growth rate: NA%
Electricity—production: 325 million kWh (2007 est.)
country comparison to the world: 165
Electricity—consumption: 302.2 million kWh (2007 est.)
country comparison to the world: 168
Electricity—exports: 0 kWh (2008 est.)
Electricity—imports: 0 kWh (2008 est.)
Oil—production: 0 bbl/day (2009 est.)
country comparison to the world: 130
Oil—consumption: 3,000 bbl/day (2009 est.)
country comparison to the world: 175
Oil—exports: 0 bbl/day (2007 est.)
country comparison to the world: 203
Oil—imports: 2,747 bbl/day (2007 est.)
country comparison to the world: 170
Oil—proved reserves: 0 bbl (1 January 2010 est.)
country comparison to the world: 190
Natural gas—production: 0 cu m (2008 est.)
country comparison to the world: 185
Natural gas—consumption: 0 cu m (2008 est.)
country comparison to the world: 124
Natural gas—exports: 0 cu m (2008 est.)
country comparison to the world: 177
Natural gas—imports: 0 cu m (2008 est.)
country comparison to the world: 120
Natural gas—proved reserves: 0 cu m (1 January 2010 est.)
country comparison to the world: 189
Current account balance: $-199 million (2007 est.)
country comparison to the world: 90
Exports: $288 million (2006)
country comparison to the world: 176
Exports—commodities: bananas 41%, clothing, cocoa, vegetables, fruits, coconut oil
Exports—partners: Spain 29.41%, UK 15.28%, South Korea 10.54%, US 9.75%, India 9.52% (2009)
Imports: $791 million (2006)
country comparison to the world: 181
Imports—commodities: food 23%, manufactured goods 21%, machinery and transportation equipment 19%, chemicals, fuels
Imports—partners: Brazil 83.44%, US 4.67%, Trinidad and Tobago 4.56% (2009)
Debt—external: $257 million (2004)
country comparison to the world: 172
Exchange rates: East Caribbean dollars (XCD) per US dollar—2.7 (2010), 2.7 (2009), 2.7 (2005), 2.7 (2004), 2.7 (2003)

COMMUNICATIONS

Telephones—main lines in use: 41,000 (2009)
country comparison to the world: 167
Telephones—mobile cellular: 176,000 (2009)
country comparison to the world: 173
Telephone system: *general assessment:* an adequate system that is automatically switched
domestic: fixed-line teledensity is 25 per 100 persons and mobile-cellular teledensity is roughly 110 per 100 persons
international: country code—1-758; the East Caribbean Fiber Optic System (ECFS) and Southern Caribbean fiber optic system (SCF) submarine cables, along with Intelsat from Martinique, carry calls internationally; direct microwave radio relay link with Martinique and Saint Vincent and the Grenadines; tropospheric scatter to Barbados
Broadcast media: 3 privately-owned television stations; 1 public television station operating on a cable network; multi-channel cable TV service is obtainable; a mix of state-owned and privately-owned broadcasters operate nearly 25 radio stations including repeater transmission stations (2007)
Internet country code: .lc
Internet hosts: 106 (2010)
country comparison to the world: 201
Internet users: 142,900 (2009)
country comparison to the world: 149

TRANSPORTATION

Airports: 2 (2010)
country comparison to the world: 205
Airports—with paved runways: *total:* 2
2,438 to 3,047 m: 1
1,524 to 2,437 m: 1 (2010)
Roadways:
total: 1,210 km (2002)
country comparison to the world: 180
Ports and terminals: Castries, Cul-de-Sac, Vieux-Fort

MILITARY

Military branches: no regular military forces; Royal Saint Lucia Police Force (includes Special Service Unit and Coast Guard) (2010)

Manpower available for military service: *males age 16–49:* 41,414 (2010 est.)
Manpower fit for military service: *males age 16–49:* 32,688
females age 16–49: 36,289 (2010 est.)
Manpower reaching militarily significant age annually: *male:* 1,574
female: 1,502 (2010 est.)
Military expenditures: NA

TRANSNATIONAL ISSUES

Disputes—international: joins other Caribbean states to counter Venezuela's claim that Aves Island sustains human habitation, a criterion under UNCLOS, which permits Venezuela to extend its EEZ/continental shelf over a large portion of the eastern Caribbean Sea
Illicit drugs: transit point for South American drugs destined for the US and Europe

SAINT MARTIN
(OVERSEAS COLLECTIVITY OF FRANCE)

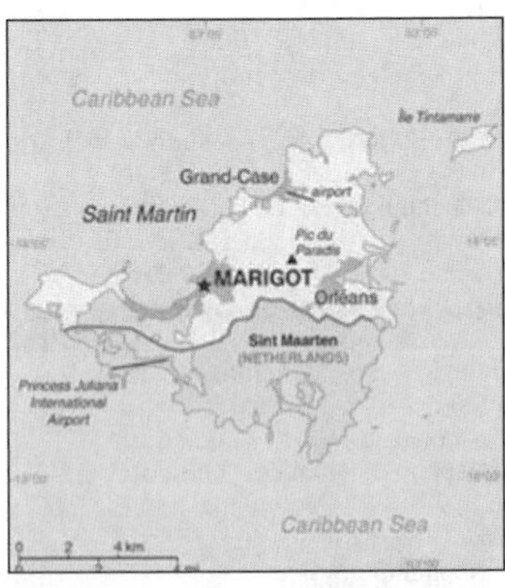

INTRODUCTION

Background: Although sighted by Christopher COLUMBUS in 1493 and claimed for Spain, it was the Dutch who occupied the island in 1631 and set about exploiting its salt deposits. The Spanish retook the island in 1633, but continued to be harassed by the Dutch. The Spanish finally relinquished Saint Martin to the French and Dutch, who divided it between themselves in 1648. Friction between the two sides caused the border to frequently fluctuate over the next two centuries, with the French eventually holding the greater portion of the island (about 57%). The cultivation of sugar cane introduced slavery to the island in the late 18th century; the practice was not abolished until 1848. The island became a free port in 1939; the tourism industry was dramatically expanded during the 1970s and 1980s. In 2003, the populace of Saint Martin voted to secede from Guadeloupe and in 2007, the northern portion of the island became a French overseas collectivity. In 2010, the Dutch portion of the island became an independent nation within the Kingdom of the Netherlands.

GEOGRAPHY

Location: Caribbean, located in the Leeward Islands (northern) group; French part of the island of Saint Martin in the Caribbean Sea; Saint Martin lies east of the US Virgin Islands
Geographic coordinates: 18 05 N, 63 57 W
Map references: Central America and the Caribbean
Area: *total:* 54.4 sq km
country comparison to the world: 229
land: 54.4 sq km
water: NEGL
Area—comparative: more than one-third the size of Washington, DC
Land boundaries: *total:* 15 km
border countries: Sint Maarten 15 km
Coastline: 58.9 km (for entire island)
Climate: temperature averages 80-85 degrees all year long; low humidity, gentle trade winds, brief, intense rain showers; July-November is the hurricane season
Elevation extremes: *lowest point:* Caribbean Ocean 0 m
highest point: Pic du Paradis 424 m
Natural resources: salt
Natural hazards: subject to hurricanes from July to November
Environment—current issues: freshwater supply is dependent on desalinization of sea water
Geography—note: the island of Saint Martin is the smallest landmass in the world shared by two independent states, the French territory of Saint Martin and the Dutch territory of Sint Maarten

PEOPLE

Population: 30,615 (July 2011 est.)
country comparison to the world: 213
Age structure: *0–14 years:* 26.9% (male 4,086/female 4,139)
15–64 years: 67.1% (male 9,807/female 10,737)
65 years and over: 6% (male 829/female 1,017) (2011 est.)
Median age: *total:* 31 years
male: 29.9 years
female: 31.8 years (2011 est.)
Sex ratio: *at birth:* 1.04 male(s)/female
under 15 years: 0.99 male(s)/female
15–64 years: 0.91 male(s)/female
65 years and over: 0.82 male(s)/female
total population: 0.93 male(s)/female (2011 est.)
Ethnic groups: creole (mulatto), black, Guadeloupe Mestizo (French-East Asia), white, East Indian
Religions: Roman Catholic, Jehovah's Witness, Protestant, Hindu
Languages: French (official), English, Dutch, French Patois, Spanish, Papiamento (dialect of Netherlands Antilles)
School life expectancy (primary to tertiary education): NA
Education expenditures: NA

GOVERNMENT

Country name: *conventional long form:* Overseas Collectivity of Saint Martin
conventional short form: Saint Martin
local long form: Collectivity d'outre mer de Saint-Martin
local short form: Saint-Martin
Dependency status: overseas collectivity of France
Capital: *name:* Marigot
geographical coordinates: 18 04 N, 63 05 W
time difference: UTC-4 (1 hour ahead of Washington, DC during Standard Time)
Independence: none (overseas collectivity of France)
National holiday: Bastille Day, 14 July (1789); note—local holiday is Schoalcher Day (Slavery Abolition Day) 12 July (1848)
Constitution: 4 October 1958 (French Constitution)
Legal system: French civil law
Suffrage: 18 years of age, universal
Executive branch: *chief of state:* President Nicolas SARKOZY (since 16 May 2007), represented by Prefect Dominique LACROIX (since 21 March 2007)
head of government: President of the Territorial Council Frantz GUMBS (since 5 May 2009)
cabinet: Executive Council; note—there is also an advisory economic, social, and cultural council
(For more information visit the World Leaders website)
election: French president elected by popular vote to a five-year term; prefect appointed by the French president on the advice of the French Ministry of Interior; president of the Territorial Council elected by the members of the Council for a five-year term
election results: Frantz GUMBS elected president by the Territorial Council on 7 August 2008 but election was declared invalid on 10 April 2009
Legislative branch: unicameral Territorial Council (23 seats; members are elected by popular vote to serve five-year terms)
elections: last held on 1 and 8 July 2007 (next to be held in July 2012)
election results: percent of seats by party—UPP 49%, RRR 42.2%, Reussir Saint-Martin 8.9%; seats by party—UPP 16, RRR 6, Reussir Saint-Martin 1
note: Saint Martin elects one member to the French Senate; election last held on 21 September 2008 (next to be held in September 2014); results—percent of vote by party—NA; seats by party—UMP 1
Political parties and leaders: Union Pour le Progres or UPP [Louis-Constant FLEMING]; Rassemblement Responsabilite Reussite or RRR [Alain RICHARDSON]; Reussir Saint-Martin [Jean-Luc HAMLET]
Political pressure groups and leaders: NA
International organization participation: UPU
Diplomatic representation in the US: none (overseas collectivity of France)
Diplomatic representation from the US: none (overseas collectivity of France)

Flag description: the flag of France is used
National anthem: *name:* "O Sweet Saint Martin's Land"
lyrics/music: Gerard KEMPS
note: the song, written in 1958, is used as an unofficial anthem for the entire island (both French and Dutch sides); as a collectivity of France, in addition to the local anthem, "La Marseillaise" remains official on the French side (see France); as a constituent part of the Kingdom of the Netherlands, in addition to the local anthem, "Het Wilhelmus" remains official on the Dutch side (see Netherlands)

ECONOMY

Economy—overview: The economy of Saint Martin centers around tourism with 85% of the labor force engaged in this sector. Over one million visitors come to the island each year with most arriving through the Princess Juliana International Airport in Sint Maarten. No significant agriculture and limited local fishing means that almost all food must be imported. Energy resources and manufactured goods are also imported, primarily from Mexico and the United States. Saint Martin is reported to have the highest per capita income in the Caribbean.
GDP—composition by sector:
agriculture: 1%
industry: 15%
services: 84% (2000)
Labor force—by occupation: 85% directly or indirectly employed in tourist industry
Industries: tourism, light industry and manufacturing, heavy industry
Imports—commodities: crude petroleum, food, manufactured items
Exchange rates: euros (EUR) per US dollar–0.7715 (2010), 0.7338 (2009), 0.6827 (2008), 0.7345 (2007), 0.7964 (2006)

COMMUNICATIONS

Telephone system: *general assessment:* fully integrated access
domestic: direct dial capability with both fixed and wireless systems
international: country code–590; undersea fiber-optic cable provides voice and data connectivity to Puerto Rico and Guadeloupe
Broadcast media: 1 local TV station; receives television broadcasts from the Netherlands Antilles; access to about 20 radio stations, including RFO Guadeloupe radio broadcasts via repeater (2008)
Internet country code: .mf; note–.gp, the Internet country code for Guadeloupe, and .fr, the Internet country code for France, might also be encountered

TRANSPORTATION

Airports: 1 (2010)
country comparison to the world: 230
Airports—with paved runways: *total:* 1
914 to 1,523 m: 1 (2010)
Transportation—note: nearest airport for international flights is Princess Juliana International Airport (SXM) located in Sint Maarten

MILITARY

Manpower fit for military service: *males age 16–49:* 6,435
females age 16–49: 6,967 (2010 est.)
Manpower reaching militarily significant age annually: *male:* 168
female: 168 (2010 est.)
Military—note: defense is the responsibility of France

SAINT PIERRE AND MIQUELON

(TERRITORIAL OVERSEAS COLLECTIVITY OF FRANCE)

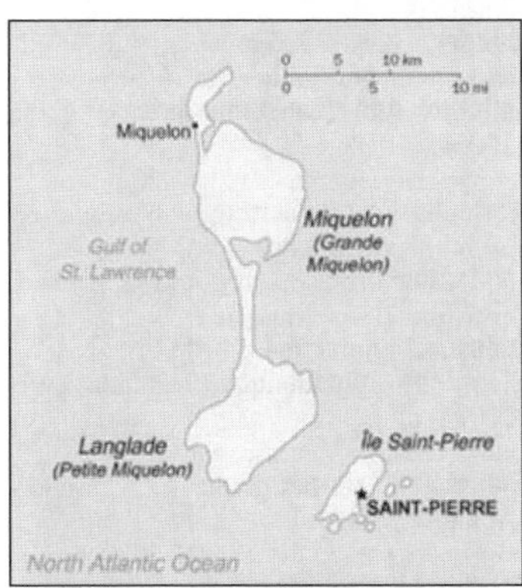

INTRODUCTION

Background: First settled by the French in the early 17th century, the islands represent the sole remaining vestige of France's once vast North American possessions.

GEOGRAPHY

Location: Northern North America, islands in the North Atlantic Ocean, south of Newfoundland (Canada)
Geographic coordinates: 46 50 N, 56 20 W
Map references: North America
Area: *total:* 242 sq km
country comparison to the world: 212
land: 242 sq km
water: 0 sq km
note: includes eight small islands in the Saint Pierre and the Miquelon groups
Area—comparative: one and half times the size of Washington, DC
Land boundaries: 0 km
Coastline: 120 km
Maritime claims: territorial sea: 12 nm
exclusive economic zone: 200 nm
Climate: cold and wet, with considerable mist and fog; spring and autumn are often windy
Terrain: mostly barren rock
Elevation extremes: *lowest point:* Atlantic Ocean 0 m
highest point: Morne de la Grande Montagne 240 m
Natural resources: fish, deepwater ports
Land use: *arable land:* 12.5%
permanent crops: 0%
other: 87.5% (2005)
Irrigated land: NA
Natural hazards: persistent fog throughout the year can be a maritime hazard
Environment—current issues: recent test drilling for oil in waters around Saint Pierre and Miquelon may bring future development that would impact the environment
Geography—note: vegetation scanty

PEOPLE

Population: 5,888 (July 2011 est.)
country comparison to the world: 227
Age structure: *0–14 years:* 17.1% (male 518/female 487)
15–64 years: 67.1% (male 2,004/female 1,949)
65 years and over: 15.8% (male 379/female 551) (2011 est.)
Median age: *total:* 42.6 years
male: 42.2 years
female: 43 years (2011 est.)
Population growth rate: -0.968% (2011 est.)
country comparison to the world: 229
Birth rate: 8.32 births/1,000 population (2011 est.)
country comparison to the world: 218
Death rate: 8.83 deaths/1,000 population (July 2011 est.)
country comparison to the world: 77
Net migration rate: -9.17 migrant(s)/1,000 population (2011 est.)
country comparison to the world: 208
Urbanization: *urban population:* 91% of total population (2010)
rate of urbanization: 0.1% annual rate of change (2010-15 est.)
Major cities—population: SAINT-PIERRE (capital) 5,000 (2009)
Sex ratio: *at birth:* 1.042 male(s)/female
under 15 years: 1.06 male(s)/female
15–64 years: 1.03 male(s)/female
65 years and over: 0.68 male(s)/female
total population: 0.98 male(s)/female (2011 est.)
Infant mortality rate: *total:* 7.47 deaths/1,000 live births
country comparison to the world: 162
male: 8.71 deaths/1,000 live births
female: 6.15 deaths/1,000 live births (2011 est.)
Life expectancy at birth: *total population:* 79.87 years
country comparison to the world: 31
male: 77.61 years
female: 82.26 years (2011 est.)
Total fertility rate: 1.55 children born/woman (2011 est.)
country comparison to the world: 180
HIV/AIDS—adult prevalence rate: NA
HIV/AIDS—people living with HIV/AIDS: NA
HIV/AIDS—deaths: NA
Nationality: *noun:* Frenchman(men), Frenchwoman(women)
adjective: French
Ethnic groups: Basques and Bretons (French fishermen)

Religions: Roman Catholic 99%, other 1%
Languages: French (official)
Literacy: *definition:* age 15 and over can read and write
total population: 99%
male: 99%
female: 99% (1982 est.)
School life expectancy (primary to tertiary education): NA
Education expenditures: NA

GOVERNMENT

Country name: *conventional long form:* Territorial Collectivity of Saint Pierre and Miquelon
conventional short form: Saint Pierre and Miquelon
local long form: Departement de Saint-Pierre et Miquelon
local short form: Saint-Pierre et Miquelon
Dependency status: self-governing territorial overseas collectivity of France
Government type: NA
Capital: *name:* Saint-Pierre
geographic coordinates: 46 46 N, 56 11 W
time difference: UTC-3 (2 hours ahead of Washington, DC during Standard Time)
daylight saving time: +1hr, begins second Sunday in March; ends first Sunday in November
Administrative divisions: none (territorial overseas collectivity of France); note–there are no first-order administrative divisions as defined by the US Government, but there are two communes–Saint Pierre, Miquelon at the second order
Independence: none (territorial collectivity of France; has been under French control since 1763)
National holiday: Bastille Day, 14 July (1789)
Constitution: 4 October 1958 (French Constitution)
Legal system: French civil law
Suffrage: 18 years of age; universal
Executive branch: *chief of state:* President Nicolas SARKOZY (since 16 May 2007); represented by Prefect Jean-Regis BORIUS (since 29 October 2009)
head of government: President of the Territorial Council Stephane ARTANO (since 21 February 2007)
cabinet: NA
(For more information visit the World Leaders website)
elections: French president elected by popular vote for a five-year term; election last held on 6 May 2007 (next to be held in 2012); prefect appointed by the French president on the advice of the French Ministry of Interior; president of the Territorial Council elected by the members of the council
Legislative branch: unicameral Territorial Council or Conseil Territorial (19 seats, 15 from Saint Pierre and four from Miquelon; members elected by popular vote to serve six-year terms)
elections: elections last held on 19 and 26 March 2006 (next to be held in March 2012)
election results: percent of vote by party–NA; seats by party–AD 16, Cap sur l'Avenir 2, SPM 2000/AM 1
note: Saint Pierre and Miquelon elect one member to the French Senate; elections last held on 21 September 2008 (next to be held in September 2014); results–percent of vote by party–NA; seats by party–UMP 1; Saint Pierre and Miquelon also elects one member to the French National Assembly; elections last held on, first round–10 June 2007, second round–17 June 2007 (next to be held in 2012); results–percent of vote by party–NA; seats by party–PRG 1
Judicial branch: Superior Tribunal of Appeals or Tribunal Superieur d'Appel
Political parties and leaders: Archipelago Tomorrow or AD (affiliated with UDF/RPR list); Cap sur l'Avenir (affiliated with PRG); Left Radical Party or PRG; Rassemblement pour la Republique or RPR (now UMP); Saint Pierre and Miquelon 2000/Avenir Miquelon or SPM 2000/AM; Socialist Party or PS; Union pour la Democratie Francaise or UDF
Political pressure groups and leaders: NA
International organization participation: UPU, WFTU
Diplomatic representation in the US: none (territorial overseas collectivity of France)
Diplomatic representation from the US: none (territorial overseas collectivity of France)
Flag description: a yellow three-masted sailing ship facing the hoist side rides on a blue background with scattered, white, wavy lines under the ship; a continuous black-over-white wavy line divides the ship from the white wavy lines; on the hoist side, a vertical band is divided into three parts: the top part (called ikkurina) is red with a green diagonal cross extending to the corners overlaid by a white cross dividing the rectangle into four sections; the middle part has a white background with an ermine pattern; the third part has a red background with two stylized yellow lions outlined in black, one above the other; these three heraldic arms represent settlement by colonists from the Basque Country (top), Brittany, and Normandy; the blue on the main portion of the flag symbolizes the Atlantic Ocean and the stylized ship represents the Grande Hermine in which Jacques Cartier "discovered" the islands in 1536
note: the flag of France used for official occasions
National anthem: note: as a collectivity of France, "La Marseillaise" is official (see France)

ECONOMY

Economy—overview: The inhabitants have traditionally earned their livelihood by fishing and by servicing fishing fleets operating off the coast of Newfoundland. The economy has been declining, however, because of disputes with Canada over fishing quotas and a steady decline in the number of ships stopping at Saint Pierre. In 1992, an arbitration panel awarded the islands an exclusive economic zone of 12,348 sq km to settle a longstanding territorial dispute with Canada, although it represents only 25% of what France had sought. France heavily subsidizes the islands to the great betterment of living standards. The government hopes an expansion of tourism will boost economic prospects. Fish farming, crab fishing, and agriculture are being developed to diversify the local economy. Recent test drilling for oil may pave the way for development of the energy sector.
GDP (purchasing power parity): $48.3 million (2003 est.)
country comparison to the world: 222
note: supplemented by annual payments from France of about $60 million
GDP (official exchange rate): $NA
GDP—real growth rate: NA%
GDP—per capita (PPP): $7,000 (2001 est.)
country comparison to the world: 131
GDP—composition by sector:
agriculture: NA%
industry: NA%
services: NA%
Labor force: 3,450 (2005)
country comparison to the world: 221
Labor force—by occupation:
agriculture: 18%
industry: 41%
services: 41% (1996 est.)
Unemployment rate: 10.3% (1999)
country comparison to the world: 111
Population below poverty line: NA%
Household income or consumption by percentage share: *lowest 10%:* NA%
highest 10%: NA%
Budget: *revenues:* $70 million
expenditures: $60 million (1996 est.)
Inflation rate (consumer prices): 8.1% (2005)
country comparison to the world: 187
Agriculture—products: vegetables; poultry, cattle, sheep, pigs; fish
Industries: fish processing and supply base for fishing fleets; tourism
Industrial production growth rate: NA%
Electricity—production: 53 million kWh (2007 est.)
country comparison to the world: 198
Electricity—consumption: 49.29 million kWh (2007 est.)
country comparison to the world: 197
Electricity—exports: 0 kWh (2008 est.)
Electricity—imports: 0 kWh (2008 est.)
Oil—production: 0 bbl/day (2009 est.)
country comparison to the world: 125
Oil—consumption: 1,000 bbl/day (2009 est.)
country comparison to the world: 192
Oil—exports: 0 bbl/day (2007 est.)
country comparison to the world: 198
Oil—imports: 563.6 bbl/day (2007 est.)
country comparison to the world: 196
Oil—proved reserves: 0 bbl (1 January 2010 est.)
country comparison to the world: 181
Natural gas—production: 0 cu m (2008 est.)
country comparison to the world: 177
Natural gas—consumption: 0 cu m (2008 est.)
country comparison to the world: 118
Natural gas—exports: 0 cu m (2008 est.)
country comparison to the world: 167
Natural gas—imports: 0 cu m (2008 est.)
country comparison to the world: 113
Natural gas—proved reserves: 0 cu m (1 January 2010 est.)

country comparison to the world: 182
Exports: $5.5 million (2005 est.)
country comparison to the world: 215
Exports—commodities: fish and fish products, soybeans, animal feed, mollusks and crustaceans, fox and mink pelts
Imports: $68.2 million (2005 est.)
country comparison to the world: 214
Imports—commodities: meat, clothing, fuel, electrical equipment, machinery, building materials
Debt—external: $NA
Exchange rates: euros (EUR) per US dollar–0.755 (2010), 0.7198 (2009), 0.6734 (2008), 0.7345 (2007), 0.7964 (2006)

COMMUNICATIONS

Telephones—main lines in use: 4,800 (2009)
country comparison to the world: 212
Telephone system: *general assessment:* adequate
domestic: NA
international: country code–508; radiotelephone communication with most countries in the world; satellite earth station–1 in French domestic satellite system
Broadcast media: 2 television stations with a third repeater station, all part of the French Overseas Network; has radio stations on St. Pierre and on Miquelon that are part of the French Overseas Network (2007)
Internet country code: .pm
Internet hosts: 0 (2010)
country comparison to the world: 231

TRANSPORTATION

Airports: 2 (2010)
country comparison to the world: 203
Airports—with paved runways: *total:* 2
1,524 to 2,437 m: 1
914 to 1,523 m: 1 (2010)
Roadways: *total:* 117 km
country comparison to the world: 213
paved: 80 km
unpaved: 37 km (2000)
Ports and terminals: Saint-Pierre

MILITARY

Manpower fit for military service: *males age 16–49:* 1,064
females age 16–49: 1,069 (2010 est.)
Manpower reaching militarily significant age annually: *male:* 34
female: 32 (2010 est.)
Military—note: defense is the responsibility of France

TRANSNATIONAL ISSUES

Disputes—international: none

SAINT VINCENT AND THE GRENADINES

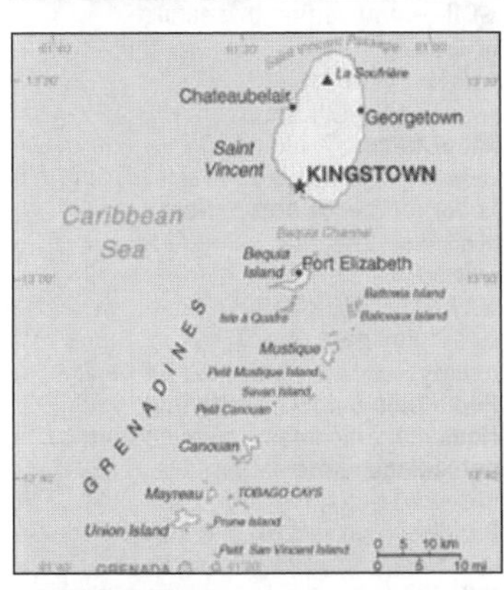

INTRODUCTION

Background: Resistance by native Caribs prevented colonization on Saint Vincent until 1719. Disputed between France and the United Kingdom for most of the 18th century, the island was ceded to the latter in 1783. Between 1960 and 1962, Saint Vincent and the Grenadines was a separate administrative unit of the Federation of the West Indies. Autonomy was granted in 1969 and independence in 1979.

GEOGRAPHY

Location: Caribbean, islands between the Caribbean Sea and North Atlantic Ocean, north of Trinidad and Tobago
Geographic coordinates: 13 15 N, 61 12 W
Map references: Central America and the Caribbean
Area: *total:* 389 sq km (Saint Vincent 344 sq km)
country comparison to the world: 202
land: 389 sq km
water: 0 sq km
Area—comparative: twice the size of Washington, DC
Land boundaries: 0 km
Coastline: 84 km
Maritime claims: territorial sea: 12 nm
contiguous zone: 24 nm
exclusive economic zone: 200 nm
continental shelf: 200 nm
Climate: tropical; little seasonal temperature variation; rainy season (May to November)
Terrain: volcanic, mountainous
Elevation extremes: *lowest point:* Caribbean Sea 0 m
highest point: La Soufriere 1,234 m
Natural resources: hydropower, cropland
Land use: *arable land:* 17.95%
permanent crops: 17.95%
other: 64.1% (2005)
Irrigated land: 10 sq km (2003)
Freshwater withdrawal (domestic/industrial/agricultural): *total:* 0.01
per capita: 83 cu m/yr (1995)
Natural hazards: hurricanes; Soufriere volcano on the island of Saint Vincent is a constant threat
Environment—current issues: pollution of coastal waters and shorelines from discharges by pleasure yachts and other effluents; in some areas, pollution is severe enough to make swimming prohibitive
Environment—international agreements: *party to:* Biodiversity, Climate Change, Climate Change-Kyoto Protocol, Desertification, Endangered Species, Environmental Modification, Hazardous Wastes, Law of the Sea, Marine Dumping, Ozone Layer Protection, Ship Pollution, Whaling
signed, but not ratified: none of the selected agreements
Geography—note: the administration of the islands of the Grenadines group is divided between Saint Vincent and the Grenadines and Grenada; Saint Vincent and the Grenadines is comprised of 32 islands and cays

PEOPLE

Population: 103,869 (July 2011 est.)
country comparison to the world: 192
Age structure: *0–14 years:* 24.5% (male 12,842/female 12,638)
15–64 years: 67.4% (male 36,042/female 33,985)
65 years and over: 8.1% (male 3,807/female 4,555) (2011 est.)
Median age: *total:* 30.1 years
male: 30.2 years
female: 30 years (2011 est.)
Population growth rate: -0.327% (2011 est.)
country comparison to the world: 218
Birth rate: 14.62 births/1,000 population (2011 est.)
country comparison to the world: 139
Death rate: 6.98 deaths/1,000 population (July 2011 est.)
country comparison to the world: 136
Net migration rate: -10.92 migrant(s)/1,000 population (2011 est.)
country comparison to the world: 210
Urbanization: *urban population:* 49% of total population (2010)
rate of urbanization: 1% annual rate of change (2010-15 est.)
Major cities—population: KINGSTOWN (capital) 28,000 (2009)
Sex ratio: *at birth:* 1.031 male(s)/female
under 15 years: 1.02 male(s)/female
15–64 years: 1.06 male(s)/female
65 years and over: 0.83 male(s)/female
total population: 1.03 male(s)/female (2011 est.)
Infant mortality rate: *total:* 14.27 deaths/1,000 live births
country comparison to the world: 124
male: 15.54 deaths/1,000 live births
female: 12.97 deaths/1,000 live births (2011 est.)
Life expectancy at birth: *total population:* 74.15 years
country comparison to the world: 107
male: 72.26 years
female: 76.09 years (2011 est.)
Total fertility rate: 1.92 children born/woman (2011 est.)
country comparison to the world: 136
HIV/AIDS—adult prevalence rate: NA
HIV/AIDS—people living with HIV/AIDS: NA
HIV/AIDS—deaths: NA
Nationality: *noun:* Saint Vincentian(s) or Vincentian(s)
adjective: Saint Vincentian or Vincentian

Ethnic groups: black 66%, mixed 19%, East Indian 6%, European 4%, Carib Amerindian 2%, other 3%
Religions: Anglican 47%, Methodist 28%, Roman Catholic 13%, other (includes Hindu, Seventh-Day Adventist, other Protestant) 12%
Languages: English, French patois
Literacy: *definition:* age 15 and over has ever attended school
total population: 96%
male: 96%
female: 96% (1970 est.)
School life expectancy (primary to tertiary education):
total: 14 years
male: 13 years
female: 13 years (2005)
Education expenditures: 6.6% of GDP (2009)
country comparison to the world: 20

GOVERNMENT

Country name: *conventional long form:* none
conventional short form: Saint Vincent and the Grenadines
Government type: parliamentary democracy and a Commonwealth realm
Capital: *name:* Kingstown
geographic coordinates: 13 09 N, 61 14 W
time difference: UTC-4 (1 hour ahead of Washington, DC during Standard Time)
Administrative divisions: 6 parishes; Charlotte, Grenadines, Saint Andrew, Saint David, Saint George, Saint Patrick
Independence: 27 October 1979 (from the UK)
National holiday: Independence Day, 27 October (1979)
Constitution: 27 October 1979
Legal system: English common law
International law organization participation: has not submitted an ICJ jurisdiction declaration; accepts ICCt jurisdiction
Suffrage: 18 years of age; universal
Executive branch: *chief of state:* Queen ELIZABETH II (since 6 February 1952); represented by Governor General Sir Fredrick Nathaniel BALLANTYNE (since 2 September 2002)
head of government: Prime Minister Ralph E. GONSALVES (since 29 March 2001)
cabinet: Cabinet appointed by the governor general on the advice of the prime minister (For more information visit the World Leaders website)
elections: the monarchy is hereditary; the governor general appointed by the monarch; following legislative elections, the leader of the majority party usually appointed prime minister by the governor general; deputy prime minister appointed by the governor general on the advice of the prime minister
Legislative branch: unicameral House of Assembly (21 seats, 15 elected representatives and 6 appointed senators; representatives elected by popular vote to serve five-year terms)
elections: last held on 13 December 2010 (next to be held in 2015)
election results: percent of vote by party–ULP 51.6%, NDP 47.8%; seats by party–ULP 8, NDP 7
Judicial branch: Eastern Caribbean Supreme Court (consisting of a High Court and Court of Appeals; based on Saint Lucia; two judges of the Supreme Court reside in Saint Vincent and the Grenadines)
Political parties and leaders: New Democratic Party or NDP [Arnhim EUSTACE]; Unity Labor Party or ULP [Ralph GONSALVES] (formed by the coalition of Saint Vincent Labor Party or SVLP and the Movement for National Unity or MNU)
Political pressure groups and leaders: NA
International organization participation: ACP, AOSIS, C, Caricom, CDB, FAO, G-77, IBRD, ICAO, ICRM, IDA, IFAD, IFRCS, ILO, IMF, IMO, Interpol, IOC, ISO (subscriber), ITU, ITUC, MIGA, NAM, OAS, OECS, OPANAL, OPCW, PetroCaribe, UN, UNCTAD, UNESCO, UNIDO, UPU, WFTU, WHO, WIPO, WTO
Diplomatic representation in the US: *chief of mission:* Ambassador La Celia A. PRINCE
chancery: 3216 New Mexico Avenue NW, Washington, DC 20016
telephone: [1] (202) 364-6730
FAX: [1] (202) 364-6736
consulate(s) general: New York
Diplomatic representation from the US: the US does not have an embassy in Saint Vincent and the Grenadines; the US Ambassador to Barbados is accredited to Saint Vincent and the Grenadines
Flag description: three vertical bands of blue (hoist side), gold (double width), and green; the gold band bears three green diamonds arranged in a V pattern, which stands for Vincent; the diamonds recall the islands as the "Gems of the Antilles"; blue conveys the colors of a tropical sky and crystal waters, yellow signifies the golden Grenadine sands, and green represents lush vegetation
National anthem: *name:* "St. Vincent! Land So Beautiful!"
lyrics/music: Phyllis Joyce MCCLEAN PUNNETT/Joel Bertram MIGUEL
note: adopted 1967

ECONOMY

Economy—overview: Success of the economy hinges upon seasonal variations in agriculture, tourism, and construction activity as well as remittance inflows. Much of the workforce is employed in banana production and tourism, but persistent high unemployment has prompted many to leave the islands. This lower-middle-income country is vulnerable to natural disasters–tropical storms wiped out substantial portions of crops in 1994, 1995, and 2002. In 2008, the islands had more than 200,000 tourist arrivals, mostly to the Grenadines, a drop of nearly 20% from 2007. Saint Vincent is home to a small offshore banking sector and has moved to adopt international regulatory standards. The government's ability to invest in social programs and respond to external shocks is constrained by its high public debt burden, which was over 90% of GDP at the end of 2010. Following the global downturn, St. Vincent and the Grenadines saw an economic decline in 2009, after slowing since 2006, when GDP growth reached a 10-year high of nearly 7%. The GONSALVES administration is directing government resources to infrastructure projects, including a new international airport that is expected to be completed in 2011.
GDP (purchasing power parity): $1.069 billion (2010 est.)
country comparison to the world: 199
$1.093 billion (2009 est.)
$1.105 billion (2008 est.)
note: data are in 2010 US dollars
GDP (official exchange rate): $561 million (2010 est.)
GDP—real growth rate: -2.3% (2010 est.)
country comparison to the world: 209
-1.1% (2009 est.)
-0.6% (2008 est.)
GDP—per capita (PPP): $10,300 (2010 est.)
country comparison to the world: 107
$10,500 (2009 est.)
$10,500 (2008 est.)
note: data are in 2010 US dollars
GDP—composition by sector:
agriculture: 10%
industry: 26%
services: 64% (2001 est.)
Labor force: 57,520 (2007 est.)
country comparison to the world: 187
Labor force—by occupation:
agriculture: 26%
industry: 17%
services: 57% (1980 est.)
Unemployment rate: 15% (2001 est.)
country comparison to the world: 149
Population below poverty line: NA%
Household income or consumption by percentage share: *lowest 10%:* NA%
highest 10%: NA%
Budget: *revenues:* $192.2 million
expenditures: $218.1 million (2010 est.)
Inflation rate (consumer prices): 0.8% (2009 est.)
country comparison to the world: 16
5.3% (2008 est.)
Central bank discount rate: 6.5% (31 December 2009)
country comparison to the world: 61
6.5% (31 December 2008)
Commercial bank prime lending rate: 9.19% (31 December 2009 est.)
country comparison to the world: 93
9.52% (31 December 2008 est.)
Stock of narrow money: $133 million (31 December 2009)
country comparison to the world: 177
$138.7 million (31 December 2008)
Stock of broad money: $444.4 million (31 December 2009)
country comparison to the world: 174
$453.5 million (31 December 2008)
Stock of domestic credit: $417.4 million (31 December 2008 est.)
country comparison to the world: 166
$387.8 million (31 December 2007 est.)
Agriculture—products: bananas, coconuts, sweet potatoes, spices; small numbers of cattle, sheep, pigs, goats; fish
Industries: food processing, cement, furniture, clothing, starch
Electricity—production: 133.8 million kWh (2007 est.)
country comparison to the world: 183

Electricity—consumption: 124.4 million kWh (2007 est.)
country comparison to the world: 186
Electricity—exports: 0 kWh (2008 est.)
Electricity—imports: 0 kWh (2008 est.)
Oil—production: 0 bbl/day (2009 est.)
country comparison to the world: 138
Oil—consumption: 2,000 bbl/day (2009 est.)
country comparison to the world: 182
Oil—exports: 0 bbl/day (2007 est.)
country comparison to the world: 138
Oil—imports: 1,451 bbl/day (2007 est.)
country comparison to the world: 182
Oil—proved reserves: 0 bbl (1 January 2010 est.)
country comparison to the world: 200
Natural gas—production: 0 cu m (2008 est.)
country comparison to the world: 197
Natural gas—consumption: 0 cu m (2008 est.)
country comparison to the world: 133
Natural gas—exports: 0 cu m (2008 est.)
country comparison to the world: 195
Natural gas—imports: 0 cu m (2008 est.)
country comparison to the world: 133
Natural gas—proved reserves: 0 cu m (1 January 2010 est.)
country comparison to the world: 199
Current account balance: $-149 million (2007 est.)
country comparison to the world: 84
Exports: $193 million (2006)
country comparison to the world: 182
Exports—commodities: bananas, eddoes and dasheen (taro), arrowroot starch; tennis racquets
Exports—partners: Greece 40.04%, Poland 11.78%, France 9.05%, China 8.53%, India 4.71% (2009)
Imports: $578 million (2006)
country comparison to the world: 186
Imports—commodities: foodstuffs, machinery and equipment, chemicals and fertilizers, minerals and fuels
Imports—partners: Singapore 16.16%, Trinidad and Tobago 13.71%, US 13.41%, China 10.9%, Italy 8.89%, Turkey 6.6%, France 5.64%, Romania 4.44% (2009)
Debt—external: $479 million (2010 est.)
country comparison to the world: 162
$223 million (2004)
Exchange rates: East Caribbean dollars (XCD) per US dollar–2.7 (2010), 2.7 (2009), 2.7 (2005), 2.7 (2004), 2.7 (2003)

COMMUNICATIONS

Telephones—main lines in use: 23,000 (2009)
country comparison to the world: 188
Telephones—mobile cellular: 121,100 (2009)
country comparison to the world: 182
Telephone system: *general assessment:* adequate islandwide, fully automatic telephone system
domestic: fixed-line teledensity exceeds 20 per 100 persons and mobile-cellular teledensity exceeds 100 per 100 persons
international: country code–1-784; the East Caribbean Fiber Optic System (ECFS) and Southern Caribbean fiber optic system (SCF) submarine cables carry international calls; connectivity also provided by VHF/UHF radiotelephone from Saint Vincent to Barbados; SHF radiotelephone to Grenada and Saint Lucia; access to Intelsat earth station in Martinique through Saint Lucia
Broadcast media: St. Vincent and the Grenadines Broadcasting Corporation operates 1 television station and 5 repeater stations that give near total coverage to the multi-island state; multi-channel cable TV service is obtainable; a partially government-funded national radio service broadcasts on 1 station and has 2 repeater stations; about a dozen privately-owned radio stations and repeater stations operate (2007)
Internet country code: .vc
Internet hosts: 211 (2010)
country comparison to the world: 192
Internet users: 76,000 (2009)
country comparison to the world: 168

TRANSPORTATION

Airports: 6 (2010)
country comparison to the world: 175
Airports—with paved runways:
total: 5
1,524 to 2,437 m: 1
914 to 1,523 m: 3
under 914 m: 1 (2010)
Airports—with unpaved runways: *total:* 1
under 914 m: 1 (2010)
Roadways: *total:* 829 km
country comparison to the world: 185
paved: 580 km
unpaved: 249 km (2003)
Merchant marine: *total:* 444
country comparison to the world: 23
by type: bulk carrier 76, cargo 274, carrier 16, chemical tanker 4, container 21, liquefied gas 3, passenger 2, passenger/cargo 9, petroleum tanker 10, refrigerated cargo 12, roll on/roll off 15, specialized tanker 2
foreign-owned: 382 (Austria 2, Bangladesh 1, Belgium 6, Bermuda 1, Bulgaria 10, China 82, Croatia 8, Cyprus 2, Czech Republic 1, Denmark 19, Dominica 1, Egypt 4, Estonia 10, France 2, Germany 2, Greece 63, Guyana 2, Hong Kong 4, Israel 3, Italy 5, Japan 3, Kenya 2, Latvia 15, Lebanon 4, Lithuania 10, Monaco 3, Netherlands 2, Nigeria 1, Norway 12, Oman 1, Pakistan 1, Poland 1, Romania 1, Russia 15, Slovenia 2, Sweden 2, Switzerland 5, Syria 13, Turkey 18, UAE 4, UK 7, Ukraine 12, US 19, Venezuela 1)
note: this country allows large numbers of ships owned by foreign entities to be registered in its national shipping registry and to fly its flag; these ships operate under the laws of the flag state (2010)
Ports and terminals: Kingstown

MILITARY

Military branches: no regular military forces; Royal Saint Vincent and the Grenadines Police Force, Coast Guard; for national defense, Saint Vincent relies on the Regional Security System, headquartered in Barbados (2010)
Manpower available for military service: *males age 16–49:* 27,809 (2010 est.)
Manpower fit for military service: *males age 16–49:* 22,875
females age 16–49: 22,015 (2010 est.)
Manpower reaching militarily significant age annually: *male:* 964
female: 953 (2010 est.)
Military expenditures: NA

TRANSNATIONAL ISSUES

Disputes—international: joins other Caribbean states to counter Venezuela's claim that Aves Island sustains human habitation, a criterion under UNCLOS, which permits Venezuela to extend its EEZ/continental shelf over a large portion of the eastern Caribbean Sea
Illicit drugs: transshipment point for South American drugs destined for the US and Europe; small-scale cannabis cultivation

SAMOA

INTRODUCTION

Background: New Zealand occupied the German protectorate of Western Samoa at the outbreak of World War I in 1914. It continued to administer the islands as a mandate and then as a trust territory until 1962, when the islands became the first Polynesian nation to reestablish independence in the 20th century. The country dropped the "Western" from its name in 1997.

GEOGRAPHY

Location: Oceania, group of islands in the South Pacific Ocean, about half way between Hawaii and New Zealand
Geographic coordinates: 13 35 S, 172 20 W
Map references: Oceania
Area: *total:* 2,831 sq km
country comparison to the world: 177
land: 2,821 sq km
water: 10 sq km
Area—comparative: slightly smaller than Rhode Island
Land boundaries: 0 km
Coastline: 403 km
Maritime claims: territorial sea: 12 nm
contiguous zone: 24 nm
exclusive economic zone: 200 nm
Climate: tropical; rainy season (November to April), dry season (May to October)
Terrain: two main islands (Savaii, Upolu) and several smaller islands and uninhabited islets; narrow coastal plain with volcanic, rocky, rugged mountains in interior
Elevation extremes: *lowest point:* Pacific Ocean 0 m
highest point: Mount Silisili 1,857 m
Natural resources: hardwood forests, fish, hydropower
Land use: *arable land:* 21.13%
permanent crops: 24.3%
other: 54.57% (2005)
Irrigated land: NA
Natural hazards: occasional typhoons; active volcanism

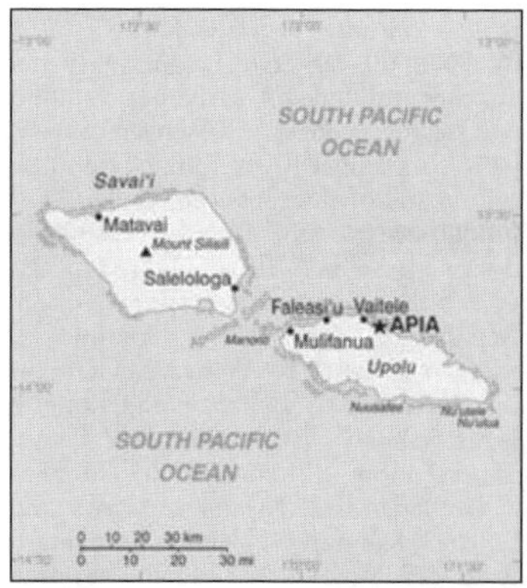

volcanism: Savai'I Island (elev. 1,858 m), which last erupted in 1911, is historically active

Environment—current issues: soil erosion, deforestation, invasive species, overfishing

Environment—international agreements: *party to:* Biodiversity, Climate Change, Climate Change-Kyoto Protocol, Desertification, Hazardous Wastes, Law of the Sea, Ozone Layer Protection, Ship Pollution, Wetlands
signed, but not ratified: none of the selected agreements

Geography—note: occupies an almost central position within Polynesia

PEOPLE

Population: 193,161 (July 2011 est.)
country comparison to the world: 183
note: prior estimates used official net migration data by sex, but a highly unusual pattern for 1993 lead to a significant imbalance in the sex ratios (more men and fewer women) and a seeming reduction in the female population; the revised total was calculated using a 1993 number that was an average of the 1992 and 1994 migration figures

Age structure: *0–14 years:* 35.4% (male 35,233/female 33,060)
15–64 years: 59.4% (male 59,366/female 55,376)
65 years and over: 5.2% (male 4,472/female 5,654) (2011 est.)

Median age:
total: 22.1 years
male: 22 years
female: 22.3 years (2011 est.)

Population growth rate: 0.6% (2011 est.)
country comparison to the world: 144

Birth rate: 22.5 births/1,000 population (2011 est.)
country comparison to the world: 75

Death rate: 5.34 deaths/1,000 population (July 2011 est.)
country comparison to the world: 177

Net migration rate: -11.16 migrant(s)/1,000 population (2011 est.)
country comparison to the world: 211

Urbanization: *urban population:* 20% of total population (2010)
rate of urbanization: 0% annual rate of change (2010-15 est.)

Major cities—population: APIA (capital) 36,000 (2009)

Sex ratio: *at birth:* 1.051 male(s)/female
under 15 years: 1.07 male(s)/female
15–64 years: 1.08 male(s)/female
65 years and over: 0.8 male(s)/female
total population: 1.06 male(s)/female (2011 est.)

Infant mortality rate: *total:* 22.74 deaths/1,000 live births
country comparison to the world: 87
male: 26.88 deaths/1,000 live births
female: 18.39 deaths/1,000 live births (2011 est.)

Life expectancy at birth: *total population:* 72.4 years
country comparison to the world: 128
male: 69.55 years
female: 75.39 years (2011 est.)

Total fertility rate: 3.22 children born/woman (2011 est.)
country comparison to the world: 51

HIV/AIDS—adult prevalence rate: NA

HIV/AIDS—people living with HIV/AIDS: NA

HIV/AIDS—deaths: NA

Drinking water source: *Improved:*
urban: 92% of population
rural: 88% of population
total: 89% of population
Unimproved:
urban: 8% of population
rural: 12% of population
total: 11% of population (2000)

Sanitation facility access: *Improved:*
urban: 100% of population
rural: 100% of population
total: 100% of population (2008)

Nationality: *noun:* Samoan(s)
adjective: Samoan

Ethnic groups: Samoan 92.6%, Euronesians (persons of European and Polynesian blood) 7%, Europeans 0.4% (2001 census)

Religions: Congregationalist 34.8%, Roman Catholic 19.6%, Methodist 15%, Latter-Day Saints 12.7%, Assembly of God 6.6%, Seventh-Day Adventist 3.5%, Worship Centre 1.3%, other Christian 4.5%, other 1.9%, unspecified 0.1% (2001 census)

Languages: Samoan (Polynesian) (official), English

Literacy: *definition:* age 15 and over can read and write
total population: 99.7%
male: 99.6%
female: 99.7% (2003 est.)

School life expectancy (primary to tertiary education): *total:* 12 years
male: 12 years
female: 13 years (2005)

Education expenditures: 5.7% of GDP (2008)
country comparison to the world: 37

GOVERNMENT

Country name: *conventional long form:* Independent State of Samoa
conventional short form: Samoa
local long form: Malo Sa'oloto Tuto'atasi o Samoa
local short form: Samoa
former: Western Samoa

Government type: parliamentary democracy

Capital: *name:* Apia
geographic coordinates: 13 50 S, 171 44 W
time difference: UTC-11 (6 hours behind Washington, DC during Standard Time) +1hr, begins last Sunday in September; ends first Sunday in April

Administrative divisions: 11 districts; A'ana, Aiga-i-le-Tai, Atua, Fa'asaleleaga, Gaga'emauga, Gagaifomauga, Palauli, Satupa'itea, Tuamasaga, Va'a-o-Fonoti, Vaisigano

Independence: 1 January 1962 (from New Zealand-administered UN trusteeship)

National holiday: Independence Day Celebration, 1 June (1962); note—1 January 1962 is the date of independence from the New Zealand-administered UN trusteeship; it is observed in June

Constitution: 1 January 1962

Legal system: mixed legal system of English common law and customary law; judicial review of legislative acts with respect to fundamental rights of the citizen

International law organization participation: has not submitted an ICJ jurisdiction declaration; accepts ICCt jurisdiction

Suffrage: 21 years of age; universal

Executive branch: *chief of state:* TUIATUA Tupua Tamasese Efi (since 20 June 2007)
head of government: Prime Minister Sailele Malielegaoi TUILA'EPA (since 1998); Deputy Prime Minister MISA Telefoni (since 2001)
cabinet: Cabinet consists of 12 members appointed by the chief of state on the prime minister's advice
(For more information visit the World Leaders website)
elections: chief of state elected by the Legislative Assembly to serve a five-year term (no term limits); election last held on 15 June 2007 (next to be held in 2012); following legislative elections, the leader of the majority party usually appointed prime minister by the chief of state with the approval of the Legislative Assembly
election results: TUIATUA Tupua Tamasese Efi unanimously elected by the Legislative Assembly

Legislative branch: unicameral Legislative Assembly or Fono (49 seats, 47 members elected by voters affiliated with traditional village-based electoral districts, 2 elected by independent, mostly non-Samoan or part-Samoan, voters who cannot, (or choose not to) establish a village affiliation; only chiefs (matai) may stand for election to the Fono from the 47 village-based electorates; members serve five-year terms)
elections: election last held on 4 March 2011 (next election to be held not later than March 2016)
election results: percent of vote by party—NA; seats by party—HRPP 29, Tautua Samoa 13, independents 7

Judicial branch: Court of Appeal; Supreme Court; District Court; Land and Titles Court

Political parties and leaders: Human Rights Protection Party or HRPP [Sailele Malielegaoi TUILA'EPA]; Samoa Christian Party or TCP [Tuala Tiresa MALIETOA]; Samoa Democratic United Party or SDUP [LE MAMEA Ropati]; Samoa Progressive Political Party or SPPP [Toeolesulusulu SIUEVA]; Tautua Samoa [Vaelua Eti ALESANA]

Political pressure groups and leaders: NA

International organization participation: ACP, ADB, AOSIS, C, FAO, G-77, IBRD, ICAO, ICRM, IDA, IFAD, IFC, IFRCS, ILO, IMF, IMO, Interpol, IOC, IPU, ITU, ITUC, MIGA, OPCW, PIF, Sparteca, SPC, UN, UNCTAD, UNESCO, UNIDO, UPU, WCO, WHO, WIPO, WMO, WTO (observer)

Diplomatic representation in the US: *chief of mission:* Ambassador Aliioaiga Feturi ELISAIA
chancery: 800 Second Avenue, Suite 400D, New York, NY 10017
telephone: [1] (212) 599-6196 through 6197
FAX: [1] (212) 599-0797
consulate(s) general: Pago Pago (American Samoa)

Diplomatic representation from the US: *chief of mission:* US Ambassador to New Zealand is accredited to Samoa
embassy: Accident Corporation Building, 5th Floor, Matafele, Apia
mailing address: P. O. Box 3430, Matafele, Apia
telephone: [685] 21436/21631/21452/22696
FAX: [685] 22030

Flag description: red with a blue rectangle in the upper hoist-side quadrant bearing five white five-pointed stars representing the Southern Cross constellation; red stands for courage, blue represents freedom, and white signifies purity

National anthem: *name:* "O le Fu"a o le Sa"olotoga o Samoa" (The Banner of Freedom)
lyrics/music: Sauni Iiga KURESA
note: adopted 1962; the anthem is also known as "Samoa Tula'i" (Samoa Arise)

ECONOMY

Economy—overview: The economy of Samoa has traditionally been dependent on development aid, family remittances from overseas, agriculture, and fishing. The country is vulnerable to devastating storms. Agriculture employs two-thirds of the labor force and furnishes 90% of exports, featuring coconut cream, coconut oil, and copra. The manufacturing sector mainly processes agricultural products. One factory in the Foreign Trade Zone employs 3,000 people to make automobile electrical harnesses for an assembly plant in Australia. Tourism is an expanding sector accounting for 25% of GDP; 122,000 tourists visited the islands in 2007. In late September 2009, an earthquake and the resulting tsunami severely damaged Samoa, and nearby American Samoa, disrupting transportation and power generation, and resulting in about 200 deaths. The Samoan Government has called for deregulation of the financial sector, encouragement of investment, and continued fiscal discipline, while at the same time protecting the environment. Observers point to the flexibility of the labor market as a basic strength for future economic advances. Foreign reserves are in a relatively healthy state, the external debt is stable, and inflation is low.

GDP (purchasing power parity): $1.055 billion (2010 est.)
country comparison to the world: 200
$1.055 billion (2009 est.)
$1.112 billion (2008 est.)
note: data are in 2010 US dollars

GDP (official exchange rate): $556 million (2010 est.)

GDP—real growth rate: 0% (2010 est.)
country comparison to the world: 189
-5.1% (2009 est.)
5.1% (2008 est.)

GDP—per capita (PPP): $5,500 (2010 est.)
country comparison to the world: 143
$5,500 (2009 est.)
$5,900 (2008 est.)
note: data are in 2010 US dollars

GDP—composition by sector:
agriculture: 11.6%
industry: 13.1%
services: 75.3% (2004 est.)

Labor force: 66,270 (2007 est.)
country comparison to the world: 185

Labor force—by occupation:
agriculture: NA%
industry: NA%
services: NA%

Unemployment rate: NA%

Population below poverty line: NA%

Household income or consumption by percentage share: *lowest 10%:* NA%
highest 10%: NA%

Budget: *revenues:* $171.3 million
expenditures: $78.1 million (FY04/05 est.)

Inflation rate (consumer prices): 6% (2007 est.)
country comparison to the world: 161

Commercial bank prime lending rate: 12.08% (31 December 2009 est.)
country comparison to the world: 61
12.66% (31 December 2008 est.)

Stock of narrow money: $80.56 million (31 December 2009)
country comparison to the world: 182
$60.13 million (31 December 2008)

Stock of broad money: $283.2 million (31 December 2009)
country comparison to the world: 179
$222.9 million (31 December 2008)

Stock of domestic credit: $243 million (31 December 2009)
country comparison to the world: 173
$208.9 million (31 December 2008)

Market value of publicly traded shares: $NA

Agriculture—products: coconuts, bananas, taro, yams, coffee, cocoa

Industries: food processing, building materials, auto parts

Industrial production growth rate: 2.8% (2000)
country comparison to the world: 116

Electricity—production: 109 million kWh (2007 est.)
country comparison to the world: 189

Electricity—consumption: 101.4 million kWh (2007 est.)
country comparison to the world: 190

Electricity—exports: 0 kWh (2008 est.)

Electricity—imports: 0 kWh (2008 est.)

Oil—production: 0 bbl/day (2009 est.)
country comparison to the world: 143

Oil—consumption: 1,000 bbl/day (2009 est.)
country comparison to the world: 202

Oil—exports: 0 bbl/day (2007 est.)
country comparison to the world: 142

Oil—imports: 1,105 bbl/day (2007 est.)
country comparison to the world: 188

Oil—proved reserves: 0 bbl (1 January 2010 est.)
country comparison to the world: 206

Natural gas—production: 0 cu m (2008 est.)
country comparison to the world: 203

Natural gas—consumption: 0 cu m (2008 est.)
country comparison to the world: 139

Natural gas—exports: 0 cu m (2008 est.)
country comparison to the world: 203

Natural gas—imports: 0 cu m (2008 est.)
country comparison to the world: 139

Natural gas—proved reserves: 0 cu m (1 January 2010 est.)
country comparison to the world: 204

Current account balance: $-24 million (2007 est.)
country comparison to the world: 67

Exports: $131 million (2006)
country comparison to the world: 188

Exports—commodities: fish, coconut oil and cream, copra, taro, automotive parts, garments, beer

Exports—partners: American Samoa 41.12%, Australia 24.74%, Taiwan 6.24%, China 5.61%, US 4.07% (2009)

Imports: $324 million (2006)
country comparison to the world: 193

Imports—commodities: machinery and equipment, industrial supplies, foodstuffs

Imports—partners: NZ 24.13%, Fiji 17.34%, Singapore 12.54%, China 10.02%, Australia 9.85%, US 5.95% (2009)

Reserves of foreign exchange and gold: $70.15 million (FY03/04)
country comparison to the world: 136

Debt—external: $177 million (2004)
country comparison to the world: 175

Exchange rates: tala (SAT) per US dollar–NA (2007), 2.7594 (2006), 2.7103 (2005), 2.7807 (2004), 2.9732 (2003)

COMMUNICATIONS

Telephones—main lines in use: 31,900 (2009)
country comparison to the world: 177

Telephones—mobile cellular: 151,000 (2009)
country comparison to the world: 176

Telephone system: *general assessment:* adequate
domestic: combined fixed-line and mobile-cellular teledensity roughly 85 telephones per 100 persons; coverage extended to roughly 95 percent of the country
international: country code–685; satellite earth station–1 Intelsat (Pacific Ocean)

Broadcast media: state-owned television station privatized in 2008; 4 privately-owned television broadcast stations; about a half dozen privately-owned radio stations and one state-owned radio station; television and radio broadcasts of several stations from American Samoa are available (2009)

Internet country code: .ws

Internet hosts: 17,044 (2010)
country comparison to the world: 114

Internet users: 9,000 (2009)
country comparison to the world: 201

TRANSPORTATION

Airports: 4 (2010)
country comparison to the world: 186

Airports—with paved runways: *total:* 1
2,438 to 3,047 m: 1 (2010)
Airports—with unpaved runways: *total:* 3
under 914 m: 3 (2010)
Roadways:
total: 2,337 km
country comparison to the world: 170
paved: 332 km
unpaved: 2,005 km (2001)
Merchant marine: *total:* 2
country comparison to the world: 144
by type: passenger/cargo 1, cargo 1
foreign-owned: 1 (NZ 1) (2010)
Ports and terminals: Apia

MILITARY

Military branches: no regular military forces; Samoa Police Force (2008)
Manpower available for military service: *males age 16–49:* 47,906 (2010 est.)
Manpower fit for military service: *males age 16–49:* 38,260
females age 16–49: 38,032 (2010 est.)
Manpower reaching militarily significant age annually: *male:* 2,221
female: 2,062 (2010 est.)
Military expenditures: NA
Military—note: Samoa has no formal defense structure or regular armed forces; informal defense ties exist with NZ, which is required to consider any Samoan request for assistance under the 1962 Treaty of Friendship

TRANSNATIONAL ISSUES

Disputes—international: none

SAN MARINO

INTRODUCTION

Background: The third smallest state in Europe (after the Holy See and Monaco), San Marino also claims to be the world's oldest republic. According to tradition, it was founded by a Christian stonemason named Marinus in A.D. 301. San Marino's foreign policy is aligned with that of the European Union, although it is not a member; social and political trends in the republic track closely with those of its larger neighbor, Italy.

GEOGRAPHY

Location: Southern Europe, an enclave in central Italy
Geographic coordinates: 43 46 N, 12 25 E
Map references: Europe
Area: *total:* 61 sq km
country comparison to the world: 227
land: 61 sq km
water: 0 sq km
Area—comparative: about one third times the size of Washington, DC
Land boundaries: *total:* 39 km
border countries: Italy 39 km
Coastline: 0 km (landlocked)
Maritime claims: none (landlocked)
Climate: Mediterranean; mild to cool winters; warm, sunny summers
Terrain: rugged mountains
Elevation extremes: *lowest point:* Torrente Ausa 55 m
highest point: Monte Titano 755 m
Natural resources: building stone
Land use: *arable land:* 16.67%
permanent crops: 0%
other: 83.33% (2005)
Irrigated land: NA
Natural hazards: NA
Environment—current issues: air pollution; urbanization decreasing rural farmlands
Environment—international agreements: *party to:* Biodiversity, Climate Change, Desertification, Whaling
signed, but not ratified: Air Pollution
Geography—note: landlocked; smallest independent state in Europe after the Holy See and Monaco; dominated by the Apennines

PEOPLE

Population: 31,817 (July 2011 est.)
country comparison to the world: 212
Age structure: *0–14 years:* 16.6% (male 2,821/female 2,474)
15–64 years: 65.4% (male 10,076/female 10,734)
65 years and over: 18% (male 2,537/female 3,175) (2011 est.)
Median age: *total:* 42.5 years
male: 41.7 years
female: 43.2 years (2011 est.)
Population growth rate: 1.043% (2011 est.)
country comparison to the world: 114
Birth rate: 9.02 births/1,000 population (2011 est.)
country comparison to the world: 210
Death rate: 7.89 deaths/1,000 population (July 2011 est.)
country comparison to the world: 109
Net migration rate: 9.3 migrant(s)/1,000 population (2011 est.)
country comparison to the world: 10
Urbanization: *urban population:* 94% of total population (2010)
rate of urbanization: 0.6% annual rate of change (2010-15 est.)
Sex ratio: *at birth:* 1.095 male(s)/female
under 15 years: 1.14 male(s)/female
15–64 years: 0.94 male(s)/female
65 years and over: 0.8 male(s)/female
total population: 0.94 male(s)/female (2011 est.)
Infant mortality rate: *total:* 4.72 deaths/1,000 live births
country comparison to the world: 186
male: 4.9 deaths/1,000 live births
female: 4.52 deaths/1,000 live births (2011 est.)
Life expectancy at birth: *total population:* 83.01 years
country comparison to the world: 3
male: 80.5 years
female: 85.74 years (2011 est.)
Total fertility rate: 1.47 children born/woman (2011 est.)
country comparison to the world: 189
HIV/AIDS—adult prevalence rate: NA
HIV/AIDS—people living with HIV/AIDS: NA
HIV/AIDS—deaths: NA
Nationality: *noun:* Sammarinese (singular and plural)
adjective: Sammarinese
Ethnic groups: Sammarinese, Italian
Religions: Roman Catholic
Languages: Italian
Literacy: *definition:* age 10 and over can read and write
total population: 96%
male: 97%
female: 95%
School life expectancy (primary to tertiary education): NA
Education expenditures: NA

GOVERNMENT

Country name: *conventional long form:* Republic of San Marino
conventional short form: San Marino
local long form: Repubblica di San Marino
local short form: San Marino
Government type: republic
Capital: *name:* San Marino
geographic coordinates: 43 56 N, 12 25 E
time difference: UTC+1 (6 hours ahead of Washington, DC during Standard Time)
daylight saving time: +1hr, begins last Sunday in March; ends last Sunday in October
Administrative divisions: 9 municipalities (castelli, singular–castello); Acquaviva, Borgo Maggiore, Chiesanuova, Domagnano, Faetano, Fiorentino, Montegiardino, San Marino Citta, Serravalle
Independence: 3 September 301
National holiday: Founding of the Republic, 3 September (A.D. 301)
Constitution: 8 October 1600; electoral law of 1926 serves some of the functions of a constitution
Legal system: civil law system with Italian civil law influences
International law organization participation: has not submitted an ICJ jurisdiction declaration; accepts ICCt jurisdiction
Suffrage: 18 years of age; universal
Executive branch: *chief of state:* Co-chiefs of State Captain Regent Maria Luisa BERTI and Captain Regent Filippo TAMAGNINI (for the period 1 April 2011-1 October 2011)

head of government: Secretary of State for Foreign and Political Affairs Antonella MULARONI (since 3 December 2008)
cabinet: Congress of State elected by the Great and General Council for a five-year term
(For more information visit the World Leaders website)
elections: co-chiefs of state (captains regent) elected by the Great and General Council for a six-month term; election last held in September 2010 (next to be held in March 2011); secretary of state for foreign and political affairs elected by the Great and General Council for a five-year term; election last held on 9 November 2008 (next to be held by 2013)
election results: Giovanni Francesco UGOLINI and Andrea ZAFFERANI elected captains regent; percent of legislative vote—NA; Antonella MULARONI elected secretary of state for foreign and political affairs; percent of legislative vote—NA
note: the popularly elected parliament (Grand and General Council) selects two of its members to serve as the Captains Regent (co-chiefs of state) for a six-month period; they preside over meetings of the Grand and General Council and its cabinet (Congress of State), which has 10 other members, all are selected by the Grand and General Council; assisting the captains regent are 10 secretaries of state; the secretary of state for Foreign Affairs has assumed some prime ministerial roles
Legislative branch: unicameral Grand and General Council or Consiglio Grande e Generale (60 seats; members elected by popular vote to serve five-year terms)
elections: last held on 9 November 2008 (next to be held by June 2013)
election results: percent of vote by party—Pact for San Marino coalition 54.2% (PDCS 31.9%, AP 11.5%, Freedom List 6.3%, San Marino Union of Moderates 4.2%), Reforms and Freedom coalition 45.8% (Party of Socialists and Democrats 32%, United Left 8.6%, Democrats of the Center 4.9%); seats by party—Pact for San Marino coalition 35 (PDCS 22, AP 7, the Freedom List 4, San Marino Union of Moderates 2), Reforms and Freedom coalition 25 (Party of Socialists and Democrats 18, United Left 5, Democrats of the Center 2)
Judicial branch: Council of Twelve or Consiglio dei XII
Political parties and leaders: Christian Democrats or PDCS [Marco GATTI]; Communist Refoundation or RC [Ivan FOSHI]; Democrats of the Center or DdC [Giovanni LONFERNINI]; Freedom List (including NPS and We Sammarinesi) or NS [Gabriele GATTEI]; New Socialist Party or NPS [Augusto CASALI]; Party of Socialists and Democrats or PDS [Paride ANDREOLI]; Popular Alliance or AP [Carlo FRANCIOSI]; Union of Moderates (including National Alliance or ANS [Glauco SANSOVINI] and San Marino Populars or POP [Romeo MORRI and Angela VENTURINI]; United Left or SU [Alessandro ROSSI]
Political pressure groups and leaders: NA
International organization participation: CE, FAO, IBRD, ICAO, ICRM, IFRCS, ILO, IMF, IMO, Interpol, IOC, IOM (observer), IPU, ITU, ITUC, LAIA (observer), OPCW, OSCE, Schengen Convention (de facto member), UN, UNCTAD, UNESCO, Union Latina, UNWTO, UPU, WHO, WIPO
Diplomatic representation in the US: *chief of mission:* Ambassador Paolo RONDELLI
chancery: 888 27th Street NW, Suite 900, Washington, DC 20006
telephone: 202-337-2260
Diplomatic representation from the US: the US does not have an embassy in San Marino; the ambassador to Italy is accredited to San Marino
Flag description: two equal horizontal bands of white (top) and light blue with the national coat of arms superimposed in the center; the main colors derive from the shield of the coat of arms, which features three white towers on three peaks on a blue field; the towers represent three castles built on San Marino's highest feature Mount Titano: Guaita, Cesta, and Montale; the coat of arms is flanked by a wreath, below a crown and above a scroll bearing the word LIBERTAS (Liberty); the white and blue colors are also said to stand for peace and liberty respectively
National anthem: *name:* "Inno Nazionale della Repubblica" (National Anthem of the Republic)
lyrics/music: none/Federico CONSOLO
note: adopted 1894; the music for the lyricless anthem is based on a 10th century chorale piece

ECONOMY

Economy—overview: San Marino's economy relies heavily on its tourism and banking industries, as well as on the manufacture and export of ceramics, clothing, fabrics, furniture, paints, spirits, tiles, and wine. The per capita level of output and standard of living are comparable to those of the most prosperous regions of Italy, which supplies much of its food. San Marino boasts the world's longest life expectancy for men with 80 years. The economy benefits from foreign investment due to its relatively low corporate taxes and low taxes on interest earnings. San Marino has recently faced increased international pressure to improve cooperation with foreign tax authorities and transparency within its own banking sector, which generates about one-fifth of the country's tax revenues. Italy's implementation in October 2009 of a tax amnesty to repatriate untaxed funds held abroad has resulted in financial outflows from San Marino to Italy worth more than $4.5 billion. Such outflows, combined with a money-laundering scandal at San Marino's largest financial institution and the recent global economic downturn, have contributed to a deep recession and growing budget deficit. Industrial production declined sharply in 2010, especially in the textile sector. However, San Marino has little national debt, and an unemployment rate less than half the size of Italy's. The San Marino government has adopted measures to counter the downturn, including subsidized credit to businesses. San Marino also continues to work towards harmonizing its fiscal laws with EU members and international standards. In September 2009, the OECD removed San Marino from its list of tax havens that have yet to fully implement global tax standards, and in 2010 San Marino signed Tax Information Exchange Agreements with most major countries. The future of the country's economy will be heavily influenced by the signing of a financial information exchange agreement with Italy, which many Italian investors see as fundamental for their business operations with San Marino.
GDP (purchasing power parity): $1.137 billion (2009)
country comparison to the world: 197
$850 million (2004 est.)
GDP (official exchange rate): $1.535 billion (2009)
GDP—real growth rate: -13% (2009 est.)
country comparison to the world: 216
4.3% (2007 est.)
GDP—per capita (PPP): $36,200 (2009)
country comparison to the world: 31
$41,900 (2007)
GDP—composition by sector:
agriculture: 0.1%
industry: 39.2%
services: 60.7% (2009)
Labor force: 22,950 (June 2010)
country comparison to the world: 208
Labor force—by occupation:
agriculture: 0.2%
industry: 36.3%
services: 63.5% (June 2010 est.)
Unemployment rate: 3.8% (November 2010)
country comparison to the world: 33
3.1% (2008)
Population below poverty line: NA%
Household income or consumption by percentage share: *lowest 10%:* NA%
highest 10%: NA%
Budget: *revenues:* $882.1 million
expenditures: $940.4 million (2009)
Inflation rate (consumer prices): 2.8% (June 2010)
country comparison to the world: 78
-3.5% (2008)
Commercial bank prime lending rate: 5.39% (September 2010)
country comparison to the world: 141
5.35% (31 December 2009 est.)
Stock of narrow money: $NA (31 December 2008)
$1.326 billion (31 December 2007)
Stock of broad money: $NA (31 December 2008)
$4.584 billion (31 December 2007)
Stock of domestic credit: $8.822 billion (30 September 2010)
country comparison to the world: 97
$8.008 billion (31 December 2009)
Market value of publicly traded shares: $NA
Agriculture—products: wheat, grapes, corn, olives; cattle, pigs, horses, beef, cheese, hides
Industries: tourism, banking, textiles, electronics, ceramics, cement, wine
Industrial production growth rate: -4.9% (2009)

country comparison to the world: 164
Exports: $2.436 billion (2009)
country comparison to the world: 128
$4.628 billion (2007)
Exports—commodities: building stone, lime, wood, chestnuts, wheat, wine, baked goods, hides, ceramics
Imports: $2.165 billion (2009)
country comparison to the world: 150
$3.744 billion (2007)
Imports—commodities: wide variety of consumer manufactures, food
Debt—external: $NA
Exchange rates: euros (EUR) per US dollar–0.755 (2010), 0.7179 (2009), 0.6734 (2008), 0.7345 (2007), 0.7964 (2006)

COMMUNICATIONS

Telephones—main lines in use: 21,500 (2009)
country comparison to the world: 192
Telephones—mobile cellular: 24,000 (2009)
country comparison to the world: 205
Telephone system: *general assessment:* automatic telephone system completely integrated into Italian system
domestic: combined fixed-line and mobile-cellular teledensity 150 telephones per 100 persons
international: country code–378; connected to Italian international network
Broadcast media: state-owned public broadcaster operates 1 TV station and 2 radio stations; receives radio and TV broadcasts from Italy (2008)
Internet country code: .sm
Internet hosts: 8,895 (2010)
country comparison to the world: 130
Internet users: 17,000 (2009)
country comparison to the world: 194

TRANSPORTATION

Roadways: *total:* 292 km
country comparison to the world: 204
paved: 292 km (2006)

MILITARY

Military branches: no regular military forces; voluntary Military Force (Corpi Militari) performs ceremonial duties and limited police support functions (2010)
Military service age and obligation: 16-55 for voluntary service in Voluntary Military Force (2006)
Manpower available for military service: *males age 16–49:* 6,892 (2010 est.)
Manpower fit for military service: *males age 16–49:* 5,565
females age 16–49: 6,067 (2010 est.)
Manpower reaching militarily significant age annually: *male:* 186
female: 166 (2010 est.)
Military expenditures: NA
Military—note: defense is the responsibility of Italy

TRANSNATIONAL ISSUES

Disputes—international: none

SAO TOME AND PRINCIPE

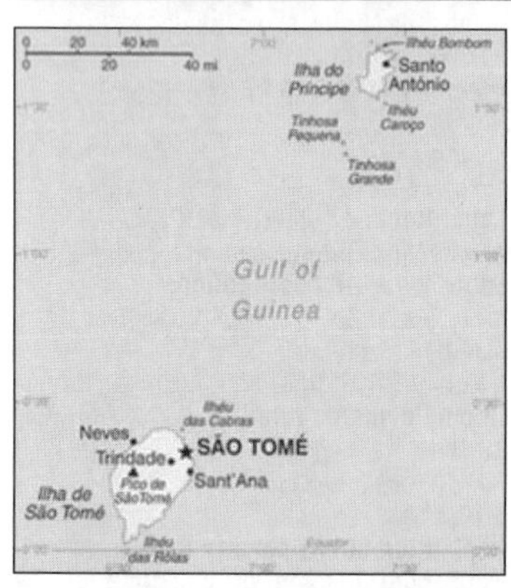

INTRODUCTION

Background: Discovered and claimed by Portugal in the late 15th century, the islands' sugar-based economy gave way to coffee and cocoa in the 19th century–all grown with plantation slave labor, a form of which lingered into the 20th century. While independence was achieved in 1975, democratic reforms were not instituted until the late 1980s. The country held its first free elections in 1991, but frequent internal wrangling between the various political parties precipitated repeated changes in leadership and two failed coup attempts in 1995 and 2003. The recent discovery of oil in the Gulf of Guinea promises to attract increased attention to the small island nation.

GEOGRAPHY

Location: Western Africa, islands in the Gulf of Guinea, straddling the Equator, west of Gabon
Geographic coordinates: 1 00 N, 7 00 E
Map references: Africa
Area: *total:* 964 sq km
country comparison to the world: 184
land: 964 sq km
water: 0 sq km
Area—comparative: more than five times the size of Washington, DC
Land boundaries: 0 km
Coastline: 209 km
Maritime claims: measured from claimed archipelagic baselines
territorial sea: 12 nm
exclusive economic zone: 200 nm
Climate: tropical; hot, humid; one rainy season (October to May)
Terrain: volcanic, mountainous
Elevation extremes: *lowest point:* Atlantic Ocean 0 m
highest point: Pico de Sao Tome 2,024 m
Natural resources: fish, hydropower
Land use: *arable land:* 8.33%
permanent crops: 48.96%
other: 42.71% (2005)
Irrigated land: 100 sq km (2003)
Natural hazards: NA
Environment—current issues: deforestation; soil erosion and exhaustion
Environment—international agreements: *party to:* Biodiversity, Climate Change, Climate Change-Kyoto Protocol, Desertification, Endangered Species, Environmental Modification, Law of the Sea, Ozone Layer Protection, Ship Pollution, Wetlands
signed, but not ratified: none of the selected agreements
Geography—note: the smallest country in Africa; the two main islands form part of a chain of extinct volcanoes and both are mountainous

PEOPLE

Population: 179,506 (July 2011 est.)
country comparison to the world: 184
Age structure: *0–14 years:* 44.7% (male 40,777/female 39,386)
15–64 years: 52.2% (male 46,114/female 47,509)
65 years and over: 3.2% (male 2,634/female 3,086) (2011 est.)
Median age: *total:* 17.5 years
male: 17.1 years
female: 18 years (2011 est.)
Population growth rate: 2.052% (2011 est.)
country comparison to the world: 46
Birth rate: 38.03 births/1,000 population (2011 est.)
country comparison to the world: 16
Death rate: 8.18 deaths/1,000 population (July 2011 est.)
country comparison to the world: 95
Net migration rate: -9.33 migrant(s)/1,000 population (2011 est.)
country comparison to the world: 209
Urbanization: *urban population:* 62% of total population (2010)
rate of urbanization: 2.8% annual rate of change (2010-15 est.)
Major cities—population: SAO TOME (capital) 60,000 (2009)
Sex ratio: *at birth:* 1.03 male(s)/female
under 15 years: 1.03 male(s)/female
15–64 years: 0.97 male(s)/female
65 years and over: 0.86 male(s)/female
total population: 1 male(s)/female (2011 est.)
Infant mortality rate: *total:* 53.21 deaths/1,000 live births
country comparison to the world: 42
male: 55.3 deaths/1,000 live births
female: 51.07 deaths/1,000 live births (2011 est.)
Life expectancy at birth: *total population:* 63.11 years
country comparison to the world: 176
male: 61.93 years
female: 64.33 years (2011 est.)
Total fertility rate: 5.08 children born/woman (2011 est.)
country comparison to the world: 18
HIV/AIDS—adult prevalence rate: NA
HIV/AIDS—people living with HIV/AIDS: NA
HIV/AIDS—deaths: NA
Major infectious diseases:
degree of risk: high
food or waterborne diseases: bacterial diarrhea, hepatitis A, and typhoid fever
vectorborne disease: malaria
animal contact disease: rabies (2009)

Drinking water source: *Improved:*
urban: 89% of population
rural: 88% of population
total: 89% of population
Unimproved:
urban: 11% of population
rural: 12% of population
total: 11% of population (2008)
Sanitation facility access: *Improved:*
urban: 30% of population
rural: 19% of population
total: 26% of population
Unimproved:
urban: 70% of population
rural: 81% of population
total: 74% of population (2008)
Nationality: *noun:* Sao Tomean(s)
adjective: Sao Tomean
Ethnic groups: mestico, angolares (descendants of Angolan slaves), forros (descendants of freed slaves), servicais (contract laborers from Angola, Mozambique, and Cape Verde), tongas (children of servicais born on the islands), Europeans (primarily Portuguese)
Religions: Catholic 70.3%, Evangelical 3.4%, New Apostolic 2%, Adventist 1.8%, other 3.1%, none 19.4% (2001 census)
Languages: Portuguese (official)
Literacy: *definition:* age 15 and over can read and write
total population: 84.9%
male: 92.2%
female: 77.9% (2001 census)
School life expectancy (primary to tertiary education): *total:* 11 years
male: 11 years
female: 11 years (2010)
Education expenditures: NA

GOVERNMENT

Country name: *conventional long form:* Democratic Republic of Sao Tome and Principe
conventional short form: Sao Tome and Principe
local long form: Republica Democratica de Sao Tome e Principe
local short form: Sao Tome e Principe
Government type: republic
Capital: *name:* Sao Tome
geographic coordinates: 0 12 N, 6 39 E
time difference: UTC 0 (5 hours ahead of Washington, DC during Standard Time)
Administrative divisions: 2 provinces; Principe, Sao Tome
note: Principe has had self government since 29 April 1995
Independence: 12 July 1975 (from Portugal)
National holiday: Independence Day, 12 July (1975)
Constitution: approved March 1990, effective 10 September 1990
Legal system: mixed legal system of civil law base on the Portuguese model and customary law
International law organization participation: has not submitted an ICJ jurisdiction declaration; non-party state to the ICCt
Suffrage: 18 years of age; universal
Executive branch: *chief of state:* President Fradique Bandiera Melo DE MENEZES (since 3 September 2001)
head of government: Prime Minister Joaquim Rafael BRANCO (since 22 June 2008)
cabinet: Council of Ministers appointed by the president on the proposal of the prime minister
(For more information visit the World Leaders website)
elections: president elected by popular vote for a five-year term (eligible for a second term); election last held on 30 July 2006 (next to be held on 17 July 2011); prime minister chosen by the National Assembly and approved by the president
election results: Fradique DE MENEZES elected president; percent of vote—Fradique DE MENEZES 60%, Patrice TROVOADA 38.5%, other 1.5%
Legislative branch: unicameral National Assembly or Assembleia Nacional (55 seats; members elected by popular vote to serve four-year terms)
elections: last held on 1 August 2010 (next to be held in 2014)
election results: percent of vote by party—NA; seats by party—ADI 26, MLSTP-PSD 21, PCD 7, MDFM 1
Judicial branch: Supreme Court (judges are appointed by the National Assembly)
Political parties and leaders: Force for Change Democratic Movement or MDFM [Tome Soares da VERA CRUZ]; Independent Democratic Action or ADI [Patrice TROVOADA]; Movement for the Liberation of Sao Tome and Principe-Social Democratic Party or MLSTP-PSD [Rafael BRANCO]; New Way Movement or NR; Party for Democratic Convergence or PCD [Delfim NEVES]; Ue-Kedadji coalition; other small parties
Political pressure groups and leaders: Association of Sao Tome and Principe NGOs or FONG
other: the media
International organization participation: ACP, AfDB, AOSIS, AU, CPLP, FAO, G-77, IBRD, ICAO, ICRM, IDA, IFAD, IFC, IFRCS, ILO, IMF, IMO, Interpol, IOC, IOM (observer), IPU, ITU, ITUC, NAM, OIF, OPCW, UN, UNCTAD, UNESCO, UNIDO, Union Latina, UNWTO, UPU, WCO, WHO, WIPO, WMO, WTO (observer)
Diplomatic representation in the US: *chief of mission:* Ambassador Ovidio Manuel Barbosa PEQUENO
chancery: 1211 Connecticut Avenue NW, Suite 300, Washington, DC 20036
telephone: [1] (202) 775-2075, 2076
FAX: [1] (202) 775-2077
Diplomatic representation from the US: the US does not have an embassy in Sao Tome and Principe; the Ambassador to Gabon is accredited to Sao Tome and Principe on a nonresident basis and makes periodic visits to the islands
Flag description: three horizontal bands of green (top), yellow (double width), and green with two black five-pointed stars placed side by side in the center of the yellow band and a red isosceles triangle based on the hoist side; green stands for the country's rich vegetation, red recalls the struggle for independence, and yellow represents cocoa, one of the country's main agricultural products; the two stars symbolize the two main islands
note: uses the popular Pan-African colors of Ethiopia
National anthem: *name:* "Independencia total" (Total Independence)
lyrics/music: Alda Neves DA GRACA do Espirito Santo/Manuel dos Santos Barreto de Sousa e ALMEIDA
note: adopted 1975

ECONOMY

Economy—overview: This small, poor island economy has become increasingly dependent on cocoa since independence in 1975. Cocoa production has substantially declined in recent years because of drought and mismanagement. Sao Tome and Principe has to import all fuels, most manufactured goods, consumer goods, and a substantial amount of food. Over the years, it has had difficulty servicing its external debt and has relied heavily on concessional aid and debt rescheduling. Sao Tome and Principe benefited from $200 million in debt relief in December 2000 under the Highly Indebted Poor Countries (HIPC) program, which helped bring down the country's $300 million debt burden. In August 2005, the government signed on to a new 3-year IMF Poverty Reduction and Growth Facility (PRGF) program worth $4.3 million. Considerable potential exists for development of a tourist industry, and the government has taken steps to expand facilities in recent years. The government also has attempted to reduce price controls and subsidies. Potential exists for the development of petroleum resources in Sao Tome and Principe's territorial waters in the oil-rich Gulf of Guinea, which are being jointly developed in a 60-40 split with Nigeria, but any actual production is at least several years off. The first production licenses were sold in 2004, though a dispute over licensing with Nigeria delayed the country's receipt of more than $20 million in signing bonuses for almost a year.
GDP (purchasing power parity): $311 million (2010 est.)
country comparison to the world: 212
$297.8 million (2009 est.)
$286.3 million (2008 est.)
note: data are in 2010 US dollars
GDP (official exchange rate): $196 million (2010 est.)
GDP—real growth rate: 4.5% (2010 est.)
country comparison to the world: 84
4% (2009 est.)
5.8% (2008 est.)
GDP—per capita (PPP): $1,800 (2010 est.)
country comparison to the world: 194
$1,700 (2009 est.)
$1,700 (2008 est.)
note: data are in 2010 US dollars
GDP—composition by sector: *agriculture:* 14.7%
industry: 22.9%
services: 62.4% (2010 est.)
Labor force: 52,490 (2007)
country comparison to the world: 189
Labor force—by occupation: note: population mainly engaged in subsistence agriculture and fishing; shortages of skilled workers
Unemployment rate: NA%

Population below poverty line: 54% (2004 est.)
Household income or consumption by percentage share: *lowest 10%:* NA%
highest 10%: NA%
Investment (gross fixed): 41% of GDP (2010 est.)
country comparison to the world: 3
Budget: *revenues:* $35.56 million
expenditures: $38.64 million (2010 est.)
Inflation rate (consumer prices): 13% (2010 est.)
country comparison to the world: 212
16.7% (2009 est.)
Central bank discount rate: 16% (31 December 2009)
country comparison to the world: 4
28% (31 December 2008)
Commercial bank prime lending rate: 32.4% (31 December 2009 est.)
country comparison to the world: 5
32.4% (31 December 2008 est.)
Stock of narrow money: $17.18 million (31 December 2010 est.)
country comparison to the world: 187
$19.1 million (31 December 2009 est.)
Stock of broad money: $82.2 million (31 December 2009)
country comparison to the world: 186
$64.79 million (31 December 2008)
Stock of domestic credit: $17.14 million (31 December 2010 est.)
country comparison to the world: 184
$16.57 million (31 December 2009 est.)
Market value of publicly traded shares: $NA
Agriculture—products: cocoa, coconuts, palm kernels, copra, cinnamon, pepper, coffee, bananas, papayas, beans; poultry; fish
Industries: light construction, textiles, soap, beer, fish processing, timber
Industrial production growth rate: 7% (2010 est.)
country comparison to the world: 45
Electricity—production: 19 million kWh (2007 est.)
country comparison to the world: 207
Electricity—consumption: 17.67 million kWh (2007 est.)
country comparison to the world: 207
Electricity—exports: 0 kWh (2008)
Electricity—imports: 0 kWh (2008 est.)
Oil—production: 0 bbl/day (2009 est.)
country comparison to the world: 134
Oil—consumption: 1,000 bbl/day (2009 est.)
country comparison to the world: 189
Oil—exports: 0 bbl/day (2007 est.)
country comparison to the world: 205
Oil—imports: 725.5 bbl/day (2007 est.)
country comparison to the world: 192
Oil—proved reserves: 0 bbl (1 January 2010 est.)
country comparison to the world: 196
Natural gas—production: 0 cu m (2008 est.)
country comparison to the world: 192
Natural gas—consumption: 0 cu m (2008 est.)
country comparison to the world: 129
Natural gas—exports: 0 cu m (2008 est.)
country comparison to the world: 187
Natural gas—imports: 0 cu m (2008 est.)
country comparison to the world: 126
Natural gas—proved reserves: 0 cu m (1 January 2010 est.)
country comparison to the world: 195
Current account balance: $-73 million (2010 est.)
country comparison to the world: 74
$-49 million (2009 est.)
Exports: $13 million (2010 est.)
country comparison to the world: 212
$10 million (2009 est.)
Exports—commodities: cocoa 80%, copra, coffee, palm oil
Exports—partners: UK 32.99%, Netherlands 26.93%, Belgium 21.04%, Portugal 4.31% (2009)
Imports: $99 million (2010 est.)
country comparison to the world: 210
$80 million (2009 est.)
Imports—commodities: machinery and electrical equipment, food products, petroleum products
Imports—partners: Portugal 58.9%, Brazil 6.68%, US 4.71%, Japan 4.49% (2009)
Reserves of foreign exchange and gold: $46 million (31 December 2010 est.)
country comparison to the world: 137
$39 million (31 December 2009 est.)
Debt—external: $318 million (2002)
country comparison to the world: 170
Exchange rates: dobras (STD) per US dollar–19,641 (2010), 16,209 (2009), 14,900 (2008). 13,700 (2007), 12,050 (2006)

COMMUNICATIONS

Telephones—main lines in use: 7,800 (2009)
country comparison to the world: 205
Telephones—mobile cellular: 64,000 (2009)
country comparison to the world: 193
Telephone system: *general assessment:* local telephone network of adequate quality with most lines connected to digital switches
domestic: combined fixed-line and mobile-cellular teledensity roughly 35 telephones per 100 persons
international: country code–239; satellite earth station–1 Intelsat (Atlantic Ocean) (2008)
Broadcast media: 1 government-owned TV station; 1 government-owned radio station; 3 independent local radio stations authorized in 2005 with 2 operating at the end of 2006; transmissions of multiple international broadcasters are available (2007)
Internet country code: .st
Internet hosts: 1,514 (2010)
country comparison to the world: 158
Internet users: 26,700 (2009)
country comparison to the world: 183

TRANSPORTATION

Airports: 2 (2010)
country comparison to the world: 206
Airports—with paved runways:
total: 2
1,524 to 2,437 m: 1
914 to 1,523 m: 1 (2010)
Roadways: *total:* 320 km
country comparison to the world: 203
paved: 218 km
unpaved: 102 km (2000)
Merchant marine: *total:* 3
country comparison to the world: 138
by type: bulk carrier 1, cargo 2
foreign-owned: 1 (Greece 1) (2010)
Ports and terminals: Sao Tome

MILITARY

Military branches: Armed Forces of Sao Tome and Principe (Forcas Armadas de Sao Tome e Principe, FASTP): Army, Coast Guard of Sao Tome e Principe (Guarda Costeira de Sao Tome e Principe, GCSTP), Presidential Guard (2010)
Military service age and obligation: 18 years of age (est.) (2004)
Manpower available for military service: *males age 16–49:* 39,182
females age 16–49: 39,845 (2010 est.)
Manpower fit for military service: *males age 16–49:* 27,310
females age 16–49: 29,279 (2010 est.)
Manpower reaching militarily significant age annually: *male:* 2,076
female: 2,003 (2010 est.)
Military expenditures: 0.8% of GDP (2006)
country comparison to the world: 148
Military—note: Sao Tome and Principe's army is a tiny force with almost no resources at its disposal and would be wholly ineffective operating unilaterally; infantry equipment is considered simple to operate and maintain but may require refurbishment or replacement after 25 years in tropical climates; poor pay, working conditions, and alleged nepotism in the promotion of officers have been problems in the past, as reflected in the 1995 and 2003 coups; these issues are being addressed with foreign assistance aimed at improving the army and its focus on realistic security concerns; command is exercised from the president, through the Minister of Defense, to the Chief of the Armed Forces staff (2005)

TRANSNATIONAL ISSUES

Disputes—international: none

SAUDI ARABIA

INTRODUCTION

Background: Saudi Arabia is the birthplace of Islam and home to Islam's two holiest shrines in Mecca and Medina. The king's official title is the Custodian of the Two Holy Mosques. The modern Saudi state was founded in 1932 by ABD AL-AZIZ bin Abd al-Rahman Al SAUD (Ibn Saud) after a 30-year campaign to unify most of the Arabian Peninsula. A male descendent of Ibn Saud, his son ABDALLAH bin Abd al-Aziz, rules the country today as required by the country's 1992 Basic Law. Following Iraq's invasion of Kuwait in 1990, Saudi

Arabia accepted the Kuwaiti royal family and 400,000 refugees while allowing Western and Arab troops to deploy on its soil for the liberation of Kuwait the following year. The continuing presence of foreign troops on Saudi soil after the liberation of Kuwait became a source of tension between the royal family and the public until all operational US troops left the country in 2003. Major terrorist attacks in May and November 2003 spurred a strong on-going campaign against domestic terrorism and extremism. King ABDALLAH has continued the cautious reform program begun when he was crown prince. To promote increased political participation, the government held elections nationwide from February through April 2005 for half the members of 179 municipal councils. In December 2005, King ABDALLAH completed the process by appointing the remaining members of the advisory municipal councils. The king instituted an Inter-Faith Dialogue initiative in 2008 to encourage religious tolerance on a global level; in February 2009, he reshuffled the cabinet, which led to more moderates holding ministerial and judicial positions, and appointed the first female to the cabinet. The country remains a leading producer of oil and natural gas and holds more than 20% of the world's proven oil reserves. The government continues to pursue economic reform and diversification, particularly since Saudi Arabia's accession to the WTO in December 2005, and promotes foreign investment in the kingdom. A burgeoning population, aquifer depletion, and an economy largely dependent on petroleum output and prices are all ongoing governmental concerns. The 2010-11 uprising across Middle Eastern and North African countries sparked modest incidents in Saudi cities, predominantly by Shia demonstrators calling for the release of detainees and the withdrawal from Bahrain of the Gulf Cooperation Council's Peninsula Shield Force. Other relatively minor, non-Shia demonstrations focused on labor, prisoner, and infrastructure complaints. Protests in general were met by a strong police presence, with some arrests, but not the bloodshed seen in protests elsewhere in the region. King ABDALLAH in February and March 2011 announced a series of benefits to Saudi citizens including funds to build affordable housing, salary increases for government workers, and unemployment benefits. The King also announced that Riyadh would begin preparations for a second round of municipal elections in September 2011.

GEOGRAPHY

Location: Middle East, bordering the Persian Gulf and the Red Sea, north of Yemen
Geographic coordinates: 25 00 N, 45 00 E
Map references: Middle East
Area: *total:* 2,149,690 sq km
country comparison to the world: 14
land: 2,149,690 sq km
water: 0 sq km
Area—comparative: slightly more than one-fifth the size of the US
Land boundaries: *total:* 4,431 km
border countries: Iraq 814 km, Jordan 744 km, Kuwait 222 km, Oman 676 km, Qatar 60 km, UAE 457 km, Yemen 1,458 km
Coastline: 2,640 km
Maritime claims: territorial sea: 12 nm
contiguous zone: 18 nm
continental shelf: not specified
Climate: harsh, dry desert with great temperature extremes
Terrain: mostly uninhabited, sandy desert
Elevation extremes: *lowest point:* Persian Gulf 0 m
highest point: Jabal Sawda' 3,133 m
Natural resources: petroleum, natural gas, iron ore, gold, copper
Land use: *arable land:* 1.67%
permanent crops: 0.09%
other: 98.24% (2005)
Irrigated land: 16,200 sq km (2003)
Total renewable water resources: 2.4 cu km (1997)
Freshwater withdrawal (domestic/industrial/agricultural): *total:* 17.32 cu km/yr (10%/1%/89%)
per capita: 705 cu m/yr (2000)
Natural hazards: frequent sand and dust storms
volcanism: Despite Saudi Arabia's many volcanic formations, there has been little activity in the past few centuries; volcanoes include Harrat Rahat, Harrat Khaybar, Harrat Lunayyir, and Jabal Yar
Environment—current issues: desertification; depletion of underground water resources; the lack of perennial rivers or permanent water bodies has prompted the development of extensive seawater desalination facilities; coastal pollution from oil spills
Environment—international agreements: *party to:* Biodiversity, Climate Change, Climate Change-Kyoto Protocol, Desertification, Endangered Species, Hazardous Wastes, Law of the Sea, Marine Dumping, Ozone Layer Protection, Ship Pollution
signed, but not ratified: none of the selected agreements
Geography—note: extensive coastlines on Persian Gulf and Red Sea provide great leverage on shipping (especially crude oil) through Persian Gulf and Suez Canal

PEOPLE

Population: 26,131,703 (July 2011 est.)
country comparison to the world: 46
note: includes 5,576,076 non-nationals
Age structure: *0–14 years:* 29.4% (male 3,939,377/female 3,754,020)
15–64 years: 67.6% (male 9,980,253/female 7,685,328)
65 years and over: 3% (male 404,269/female 368,456) (2011 est.)
Median age: *total:* 25.3 years
male: 26.4 years
female: 23.9 years (2011 est.)
Population growth rate: 1.536% (2011 est.)
country comparison to the world: 77
Birth rate: 19.34 births/1,000 population (2011 est.)
country comparison to the world: 96
Death rate: 3.33 deaths/1,000 population (July 2011 est.)
country comparison to the world: 215
Net migration rate: -0.64 migrant(s)/1,000 population (2011 est.)
country comparison to the world: 142
Urbanization: *urban population:* 82% of total population (2010)
rate of urbanization: 2.2% annual rate of change (2010-15 est.)
Major cities—population: RIYADH (capital) 4.725 million; Jeddah 3.234 million; Mecca 1.484 million; Medina 1.104 million; Ad Dammam 902,000 (2009)
Sex ratio: *at birth:* 1.05 male(s)/female
under 15 years: 1.04 male(s)/female
15–64 years: 1.27 male(s)/female
65 years and over: 1.03 male(s)/female
total population: 1.17 male(s)/female (2011 est.)
Infant mortality rate: *total:* 16.16 deaths/1,000 live births
country comparison to the world: 111
male: 18.54 deaths/1,000 live births
female: 13.65 deaths/1,000 live births (2011 est.)
Life expectancy at birth: *total population:* 74.11 years
country comparison to the world: 108
male: 72.15 years
female: 76.16 years (2011 est.)
Total fertility rate: 2.31 children born/woman (2011 est.)
country comparison to the world: 99
HIV/AIDS—adult prevalence rate: 0.01% (2001 est.)
country comparison to the world: 167
HIV/AIDS—people living with HIV/AIDS: NA
HIV/AIDS—deaths: NA
Drinking water source: *Improved:*
urban: 97% of population
rural: 63% of population
total: 89% of population
Unimproved:
urban: 3% of population
rural: 37% of population
total: 11% of population (1990)
Nationality: *noun:* Saudi(s)
adjective: Saudi or Saudi Arabian
Ethnic groups: Arab 90%, Afro-Asian 10%
Religions: Muslim 100%
Languages: Arabic (official)
Literacy: *definition:* age 15 and over can read and write
total population: 78.8%
male: 84.7%
female: 70.8% (2003 est.)
School life expectancy (primary to tertiary education): *total:* 14 years
male: 14 years
female: 13 years (2009)
Education expenditures: 5.6% of GDP (2008)
country comparison to the world: 40

GOVERNMENT

Country name: *conventional long form:* Kingdom of Saudi Arabia
conventional short form: Saudi Arabia
local long form: Al Mamlakah al Arabiyah as Suudiyah
local short form: Al Arabiyah as Suudiyah
Government type: monarchy
Capital: *name:* Riyadh
geographic coordinates: 24 38 N, 46 43 E

time difference: UTC+3 (8 hours ahead of Washington, DC during Standard Time)
Administrative divisions: 13 provinces (mintaqat, singular–mintaqah); Al Bahah, Al Hudud ash Shamaliyah (Northern Border), Al Jawf, Al Madinah (Medina), Al Qasim, Ar Riyad (Riyadh), Ash Sharqiyah (Eastern), 'Asir, Ha'il, Jizan, Makkah (Mecca), Najran, Tabuk
Independence: 23 September 1932 (unification of the kingdom)
National holiday: Unification of the Kingdom, 23 September (1932)
Constitution: governed according to Islamic law; the Basic Law that articulates the government's rights and responsibilities was promulgated by royal decree in 1992
Legal system: Islamic (sharia) legal system with some elements of Egyptian, French, and customary law; note–several secular codes have been introduced; commercial disputes handled by special committees
International law organization participation: has not submitted an ICJ jurisdiction declaration; non-party state to the ICCt
Suffrage: 21 years of age; male
Executive branch: *chief of state:* King and Prime Minister ABDALLAH bin Abd al-Aziz Al Saud (since 1 August 2005); Heir Apparent Crown Prince SULTAN bin Abd al- Aziz Al Saud (half brother of the monarch); note–the monarch is both the chief of state and head of government
head of government: King and Prime Minister ABDALLAH bin Abd al-Aziz Al Saud (since 1 August 2005); Deputy Prime Minister SULTAN bin Abd al-Aziz Al Saud; Second Deputy Prime Minister NAYIF bin Abd Al-Aziz Al Saud
cabinet: Council of Ministers appointed by the monarch every four years and includes many royal family members
(For more information visit the World Leaders website)
elections: none; the monarchy is hereditary; note–an Allegiance Commission created by royal decree in October 2006 established a committee of Saudi princes that will play a role in selecting future Saudi kings, but the system will not take effect until after Crown Prince SULTAN becomes king
Legislative branch: Consultative Council or Majlis al-Shura (150 members and a chairman appointed by the monarch to serve four-year terms); note–though the Council of Ministers announced in October 2003 its intent to introduce elections for a third of the Majlis al-Shura incrementally over a period of four to five years, to date no such elections have been held or announced
Judicial branch: Supreme Council of Justice
Political parties and leaders: none
Political pressure groups and leaders: Ansar Al Marah (supports women's rights)
other: gas companies; religious groups
International organization participation: ABEDA, AfDB (nonregional member), AFESD, AMF, BIS, FAO, G-20, G-77, GCC, IAEA, IBRD, ICAO, ICC, ICRM, IDA, IDB, IFAD, IFC, IFRCS, IHO, ILO, IMF, IMO, IMSO, Interpol, IOC, IOM (observer), IPU, ISO, ITSO, ITU, LAS, MIGA, NAM, OAPEC, OAS (observer), OIC, OPCW, OPEC, PCA, UN, UNCTAD, UNESCO, UNIDO, UNRWA, UNWTO, UPU, WCO, WFTU, WHO, WIPO, WMO, WTO
Diplomatic representation in the US:
chief of mission: Ambassador Adil al-Ahmad al-JUBAYR
chancery: 601 New Hampshire Avenue NW, Washington, DC 20037
telephone: [1] (202) 342-3800
FAX: [1] (202) 944-3113
consulate(s) general: Houston, Los Angeles, New York
Diplomatic representation from the US:
chief of mission: Ambassador James B. SMITH
embassy: Collector Road M, Diplomatic Quarter, Riyadh
mailing address: American Embassy, Unit 61307, APO AE 09803-1307; International Mail: P. O. Box 94309, Riyadh 11693
telephone: [966] (1) 488-3800
FAX: [966] (1) 488-7360
consulate(s) general: Dhahran, Jiddah (Jeddah)
Flag description: green, a traditional color in Islamic flags, with the Shahada or Muslim creed in large white Arabic script (translated as "There is no god but God; Muhammad is the Messenger of God") above a white horizontal saber (the tip points to the hoist side); design dates to the early twentieth century and is closely associated with the Al Saud family which established the kingdom in 1932; the flag is manufactured with differing obverse and reverse sides so that the Shahada reads–and the sword points–correctly from right to left on both sides
note: one of only three national flags that differ on their obverse and reverse sides–the others are Moldova and Paraguay
National anthem: *name:* "Aash Al Maleek" (Long Live Our Beloved King)
lyrics/music: Ibrahim KHAFAJI/Abdul Rahman al-KHATEEB
note: music adopted 1947, lyrics adopted 1984

ECONOMY

Economy—overview: Saudi Arabia has an oil-based economy with strong government controls over major economic activities. It possesses about 20% of the world's proven petroleum reserves, ranks as the largest exporter of petroleum, and plays a leading role in OPEC. The petroleum sector accounts for roughly 80% of budget revenues, 45% of GDP, and 90% of export earnings. Saudi Arabia is encouraging the growth of the private sector in order to diversify its economy and to employ more Saudi nationals. Diversification efforts are focusing on power generation, telecommunications, natural gas exploration, and petrochemical sectors. Almost 6 million foreign workers play an important role in the Saudi economy, particularly in the oil and service sectors, while Riyadh is struggling to reduce unemployment among its own nationals. Saudi officials are particularly focused on employing its large youth population, which generally lacks the education and technical skills the private sector needs. Riyadh has substantially boosted spending on job training and education, most recently with the opening of the King Abdallah University of Science and Technology–Saudi Arabia's first co-educational university. As part of its effort to attract foreign investment, Saudi Arabia acceded to the WTO in December 2005 after many years of negotiations. The government has begun establishing six "economic cities" in different regions of the country to promote foreign investment and plans to spend $373 billion between 2010 and 2014 on social development and infrastructure projects to advance Saudi Arabia's economic development.
GDP (purchasing power parity): $622 billion (2010 est.)
country comparison to the world: 23
$599.5 billion (2009 est.)
$596 billion (2008 est.)
note: data are in 2010 US dollars
GDP (official exchange rate): $443.7 billion (2010 est.)
GDP—real growth rate: 3.7% (2010 est.)
country comparison to the world: 106
0.6% (2009 est.)
4.2% (2008 est.)
GDP—per capita (PPP): $24,200 (2010 est.)
country comparison to the world: 55
$23,700 (2009 est.)
$23,900 (2008 est.)
note: data are in 2010 US dollars
GDP—composition by sector:
agriculture: 2.7%
industry: 61.9%
services: 35.4% (2010 est.)
Labor force: 7.337 million
country comparison to the world: 62
note: about 80% of the labor force is non-national (2010 est.)
Labor force—by occupation:
agriculture: 6.7%
industry: 21.4%
services: 71.9% (2005 est.)
Unemployment rate: 10.8% (2010 est.)
country comparison to the world: 119
10.5% (2009 est.)
note: data are for Saudi males only (local bank estimates; some estimates range as high as 25%)
Population below poverty line: NA%
Household income or consumption by percentage share: *lowest 10%:* NA%
highest 10%: NA%
Investment (gross fixed): 24.5% of GDP (2010 est.)
country comparison to the world: 47
Budget: *revenues:* $185.1 billion
expenditures: $173.1 billion (2010 est.)
Public debt: 16.7% of GDP (2010 est.)
country comparison to the world: 114
22.6% of GDP (2009 est.)
Inflation rate (consumer prices): 5.7% (2010 est.)
country comparison to the world: 152
5.1% (2009 est.)
Central bank discount rate: 2.5% (31 December 2008)
Commercial bank prime lending rate: NA%
Stock of narrow money: $166.9 billion (31 December 2010 est.)

country comparison to the world: 20
$139.1 billion (31 December 2009 est.)
Stock of broad money: $286.9 billion (31 December 2010 est.)
country comparison to the world: 29
$274.4 billion (31 December 2009 est.)
Stock of domestic credit: $11.24 billion (31 December 2010 est.)
country comparison to the world: 91
$2.248 billion (31 December 2009 est.)
Market value of publicly traded shares: $318.8 billion (31 December 2009)
country comparison to the world: 23
$246.3 billion (31 December 2008)
$515.1 billion (31 December 2007)
Agriculture—products: wheat, barley, tomatoes, melons, dates, citrus; mutton, chickens, eggs, milk
Industries: crude oil production, petroleum refining, basic petrochemicals, ammonia, industrial gases, sodium hydroxide (caustic soda), cement, fertilizer, plastics, metals, commercial ship repair, commercial aircraft repair, construction
Industrial production growth rate: 3.1% (2010 est.)
country comparison to the world: 105
Electricity—production: 179.1 billion kWh (2007 est.)
country comparison to the world: 21
Electricity—consumption: 165.1 billion kWh (2007 est.)
country comparison to the world: 21
Electricity—exports: 0 kWh (2008 est.)
Electricity—imports: 0 kWh (2008 est.)
Oil—production: 9.764 million bbl/day (2009 est.)
country comparison to the world: 2
Oil—consumption: 2.43 million bbl/day (2009 est.)
country comparison to the world: 9
Oil—exports: 8.728 million bbl/day (2007 est.)
country comparison to the world: 1
Oil—imports: 79,250 bbl/day (2007 est.)
country comparison to the world: 73
Oil—proved reserves: 264.6 billion bbl (1 January 2010 est.)
country comparison to the world: 1
Natural gas—production: 77.1 billion cu m (2009 est.)
country comparison to the world: 11
Natural gas—consumption: 77.1 billion cu m (2009 est.)
country comparison to the world: 11
Natural gas—exports: 0 cu m (2008 est.)
country comparison to the world: 166
Natural gas—imports: 0 cu m (2008 est.)
country comparison to the world: 112
Natural gas—proved reserves: 7.461 trillion cu m (1 January 2010 est.)
country comparison to the world: 5
Current account balance: $52.03 billion (2010 est.)
country comparison to the world: 6
$22.77 billion (2009 est.)
Exports: $235.3 billion (2010 est.)
country comparison to the world: 19
$192.3 billion (2009 est.)
Exports—commodities: petroleum and petroleum products 90%
Exports—partners: Japan 15.33%, South Korea 12.71%, US 12.2%, China 10.38%, India 7.12%, Taiwan 4.54%, Singapore 4.25% (2009)
Imports: $99.17 billion (2010 est.)
country comparison to the world: 32
$87.1 billion (2009 est.)
Imports—commodities: machinery and equipment, foodstuffs, chemicals, motor vehicles, textiles
Imports—partners: US 12.32%, China 12.06%, Germany 7.67%, Japan 6.15%, South Korea 5.32%, India 4.99%, UK 4.72%, France 4.05% (2009)
Reserves of foreign exchange and gold: $456.2 billion (31 December 2010 est.)
country comparison to the world: 4
$410.1 billion (31 December 2009 est.)
Debt—external: $82.92 billion (31 December 2010 est.)
country comparison to the world: 42
$72.77 billion (31 December 2009 est.)
Stock of direct foreign investment—at home: $204.3 billion (31 December 2010 est.)
country comparison to the world: 20
$167 billion (31 December 2009 est.)
Stock of direct foreign investment—abroad: $18 billion (31 December 2010 est.)
country comparison to the world: 45
$11.41 billion (31 December 2009 est.)
Exchange rates: Saudi riyals (SAR) per US dollar–3.75 (2010), 3.75 (2009), 3.75 (2008), 3.745 (2007), 3.745 (2006)

COMMUNICATIONS

Telephones—main lines in use: 4.171 million (2009)
country comparison to the world: 38
Telephones—mobile cellular: 44.864 million (2009)
country comparison to the world: 27
Telephone system: *general assessment:* modern system including a combination of extensive microwave radio relays, coaxial cables, and fiber-optic cables
domestic: mobile-cellular subscribership has been increasing rapidly
international: country code–966; landing point for the international submarine cable Fiber-Optic Link Around the Globe (FLAG) and for both the SEA-ME-WE-3 and SEA-ME-WE-4 submarine cable networks providing connectivity to Asia, Middle East, Europe, and US; microwave radio relay to Bahrain, Jordan, Kuwait, Qatar, UAE, Yemen, and Sudan; coaxial cable to Kuwait and Jordan; satellite earth stations–5 Intelsat (3 Atlantic Ocean and 2 Indian Ocean), 1 Arabsat, and 1 Inmarsat (Indian Ocean region) (2008)
Broadcast media: broadcast media are state-controlled; state-run TV operates 4 networks; Saudi Arabia is a major market for pan-Arab satellite TV broadcasters; state-run radio operates several networks; multiple international broadcasters are available (2007)
Internet country code: .sa
Internet hosts: 488,598 (2010)
country comparison to the world: 51
Internet users: 9.774 million (2009)
country comparison to the world: 30

TRANSPORTATION

Airports: 217 (2010)
country comparison to the world: 27
Airports—with paved runways: *total:* 81
over 3,047 m: 33
2,438 to 3,047 m: 15
1,524 to 2,437 m: 27
914 to 1,523 m: 2
under 914 m: 4 (2010)
Airports—with unpaved runways: *total:* 136
2,438 to 3,047 m: 8
1,524 to 2,437 m: 71
914 to 1,523 m: 41
under 914 m: 16 (2010)
Heliports: 9 (2010)
Pipelines: condensate 212 km; gas 2,846 km; liquid petroleum gas 1,183 km; oil 4,232 km; refined products 1,151 km (2010)
Railways: *total:* 1,392 km
country comparison to the world: 81
standard gauge: 1,392 km 1.435-m gauge (with branch lines and sidings) (2008)
Roadways: *total:* 221,372 km
country comparison to the world: 23
paved: 47,529 km (includes 3,891 km of expressways)
unpaved: 173,843 km (2006)
Merchant marine: *total:* 74
country comparison to the world: 58
by type: cargo 2, chemical tanker 22, container 4, liquefied gas 2, passenger/cargo 11, petroleum tanker 22, refrigerated cargo 3, roll on/roll off 8
foreign-owned: 15 (Egypt 1, Greece 4, Kuwait 4, UAE 6)
registered in other countries: 55 (Bahamas 16, Dominica 3, Liberia 24, Norway 3, Panama 8) (2010)
Ports and terminals: Ad Dammam, Al Jubayl, Jeddah, Yanbu al Bahr

MILITARY

Military branches: Ministry of Defense and Aviation Forces: Royal Saudi Land Forces, Royal Saudi Naval Forces (includes Marine Forces and Special Forces), Royal Saudi Air Force (Al-Quwwat al-Jawwiya al-Malakiya as-Sa'udiya), Royal Saudi Air Defense Forces, Royal Saudi Strategic Rocket Forces, Saudi Arabian National Guard (SANG)
Military service age and obligation: 18 years of age (est.); no conscription (2004)
Manpower available for military service: *males age 16–49:* 8,644,522
females age 16–49: 6,601,985 (2010 est.)
Manpower fit for military service: *males age 16–49:* 7,365,624
females age 16–49: 5,677,819 (2010 est.)
Manpower reaching militarily significant age annually: *male:* 261,105
female: 244,763 (2010 est.)
Military expenditures: 10% of GDP (2005 est.)
country comparison to the world: 2

TRANSNATIONAL ISSUES

Disputes—international: Saudi Arabia has reinforced its concrete-filled security barrier along sections of the now fully demarcated border with Yemen to stem illegal cross-border activities; Kuwait and Saudi Arabia continue discussions on a maritime boundary with Iran; Saudi Arabia claims Egyptian-administered islands of Tiran and Sanafir
Refugees and internally displaced persons: refugees (country of origin): 240,015 (Palestinian Territories) (2007)

Trafficking in persons: current situation: Saudi Arabia is a destination country for workers from South and Southeast Asia who are subjected to conditions that constitute involuntary servitude including being subjected to physical and sexual abuse, non-payment of wages, confinement, and withholding of passports as a restriction on their movement; domestic workers are particularly vulnerable because some are confined to the house in which they work unable to seek help; Saudi Arabia is also a destination country for Nigerian, Yemeni, Pakistani, Afghan, Somali, Malian, and Sudanese children trafficked for forced begging and involuntary servitude as street vendors; some Nigerian women were reportedly trafficked into Saudi Arabia for commercial sexual exploitation
tier rating: Tier 3–Saudi Arabia does not fully comply with the minimum standards for the elimination of trafficking and is not making significant efforts to do so; the government continues to lack adequate anti-trafficking laws and, despite evidence of widespread trafficking abuses, did not report any criminal prosecutions, convictions, or prison sentences for trafficking crimes committed against foreign domestic workers (2008)
Illicit drugs: death penalty for traffickers; improving anti-money-laundering legislation and enforcement

SENEGAL

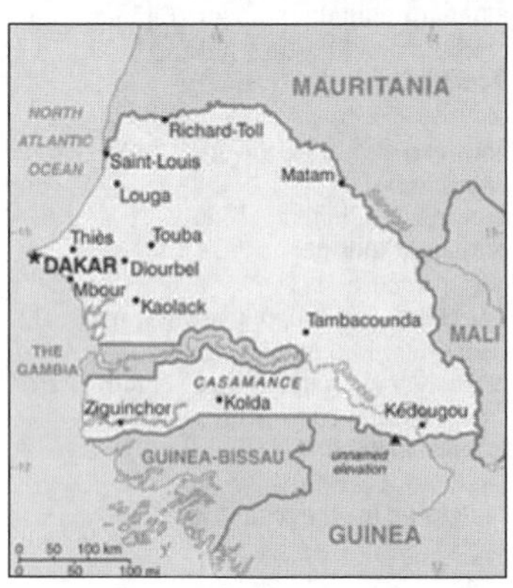

INTRODUCTION

Background: The French colonies of Senegal and the French Sudan were merged in 1959 and granted their independence as the Mali Federation in 1960. The union broke up after only a few months. Senegal joined with The Gambia to form the nominal confederation of Senegambia in 1982. The envisaged integration of the two countries was never carried out, and the union was dissolved in 1989. The Movement of Democratic Forces in the Casamance (MFDC) has led a low-level separatist insurgency in southern Senegal since the 1980s, and several peace deals have failed to resolve the conflict. Nevertheless, Senegal remains one of the most stable democracies in Africa. Senegal was ruled by a Socialist Party for 40 years until current President Abdoulaye WADE was elected in 2000. He was reelected in February 2007 and has amended Senegal's constitution over a dozen times to increase executive power and to weaken the opposition, part of the president's increasingly autocratic governing style. Senegal has a long history of participating in international peacekeeping and regional mediation.

GEOGRAPHY

Location: Western Africa, bordering the North Atlantic Ocean, between Guinea-Bissau and Mauritania
Geographic coordinates: 14 00 N, 14 00 W
Map references: Africa
Area: *total:* 196,722 sq km
country comparison to the world: 87
land: 192,530 sq km
water: 4,192 sq km
Area—comparative: slightly smaller than South Dakota
Land boundaries: *total:* 2,640 km
border countries: The Gambia 740 km, Guinea 330 km, Guinea-Bissau 338 km, Mali 419 km, Mauritania 813 km
Coastline: 531 km
Maritime claims: territorial sea: 12 nm
contiguous zone: 24 nm
exclusive economic zone: 200 nm
continental shelf: 200 nm or to the edge of the continental margin
Climate: tropical; hot, humid; rainy season (May to November) has strong southeast winds; dry season (December to April) dominated by hot, dry, harmattan wind
Terrain: generally low, rolling, plains rising to foothills in southeast
Elevation extremes: *lowest point:* Atlantic Ocean 0 m
highest point: unnamed elevation southwest of Kedougou 581 m
Natural resources: fish, phosphates, iron ore
Land use: *arable land:* 12.51%
permanent crops: 0.24%
other: 87.25% (2005)
Irrigated land: 1,200 sq km (2003)
Total renewable water resources: 39.4 cu km (1987)
Freshwater withdrawal (domestic/industrial/agricultural): *total:* 2.22 cu km/yr (4%/3%/93%)
per capita: 190 cu m/yr (2002)
Natural hazards: lowlands seasonally flooded; periodic droughts
Environment—current issues: wildlife populations threatened by poaching; deforestation; overgrazing; soil erosion; desertification; overfishing
Environment—international agreements: *party to:* Biodiversity, Climate Change, Climate Change-Kyoto Protocol, Desertification, Endangered Species, Hazardous Wastes, Law of the Sea, Marine Life Conservation, Ozone Layer Protection, Ship Pollution, Wetlands, Whaling
signed, but not ratified: none of the selected agreements
Geography—note: westernmost country on the African continent; The Gambia is almost an enclave within Senegal

PEOPLE

Population: 12,643,799 (July 2011 est.)
country comparison to the world: 71
Age structure: *0–14 years:* 43.3% (male 2,748,457/female 2,722,633)
15–64 years: 53.9% (male 3,200,056/female 3,611,173)
65 years and over: 2.9% (male 166,577/female 194,903) (2011 est.)
Median age: *total:* 18 years
male: 17.2 years
female: 18.9 years (2011 est.)
Population growth rate: 2.557% (2011 est.)
country comparison to the world: 26
Birth rate: 36.73 births/1,000 population (2011 est.)
country comparison to the world: 23
Death rate: 9.26 deaths/1,000 population (July 2011 est.)
country comparison to the world: 65
Net migration rate: -1.9 migrant(s)/1,000 population (2011 est.)
country comparison to the world: 161
Urbanization: *urban population:* 42% of total population (2010)
rate of urbanization: 3.3% annual rate of change (2010-15 est.)
Major cities—population: DAKAR (capital) 2.777 million (2009)
Sex ratio: *at birth:* 1.03 male(s)/female
under 15 years: 1.01 male(s)/female
15–64 years: 0.98 male(s)/female
65 years and over: 0.87 male(s)/female
total population: 0.99 male(s)/female (2011 est.)
Infant mortality rate: *total:* 56.42 deaths/1,000 live births
country comparison to the world: 36
male: 62.94 deaths/1,000 live births
female: 49.7 deaths/1,000 live births (2011 est.)
Life expectancy at birth: *total population:* 59.78 years
country comparison to the world: 188
male: 57.85 years
female: 61.77 years (2011 est.)
Total fertility rate: 4.78 children born/woman (2011 est.)
country comparison to the world: 25
HIV/AIDS—adult prevalence rate: 0.9% (2009 est.)
country comparison to the world: 52
HIV/AIDS—people living with HIV/AIDS: 59,000 (2009 est.)
country comparison to the world: 57
HIV/AIDS—deaths: 2,600 (2009 est.)
country comparison to the world: 50
Major infectious diseases: *degree of risk:* very high
food or waterborne diseases: bacterial and protozoal diarrhea, hepatitis A, and typhoid fever
vectorborne diseases: Crimean-Congo hemorrhagic fever, dengue fever, malaria, Rift Valley fever, and yellow fever
water contact disease: schistosomiasis

respiratory disease: meningococcal meningitis
animal contact disease: rabies (2009)
Drinking water source: *Improved:*
urban: 92% of population
rural: 52% of population
total: 69% of population
Unimproved:
urban: 8% of population
rural: 48% of population
total: 31% of population (2008)
Sanitation facility access: *Improved:*
urban: 69% of population
rural: 38% of population
total: 51% of population
Unimproved:
urban: 31% of population
rural: 62% of population
total: 49% of population (2008)
Nationality: *noun:* Senegalese (singular and plural)
adjective: Senegalese
Ethnic groups: Wolof 43.3%, Pular 23.8%, Serer 14.7%, Jola 3.7%, Mandinka 3%, Soninke 1.1%, European and Lebanese 1%, other 9.4%
Religions: Muslim 94%, Christian 5% (mostly Roman Catholic), indigenous beliefs 1%
Languages: French (official), Wolof, Pulaar, Jola, Mandinka
Literacy: *definition:* age 15 and over can read and write
total population: 39.3%
male: 51.1%
female: 29.2% (2002 est.)
School life expectancy (primary to tertiary education): *total:* 8 years
male: 8 years
female: 7 years (2008)
Education expenditures: 5.8% of GDP (2009)
country comparison to the world: 34

GOVERNMENT

Country name: *conventional long form:* Republic of Senegal
conventional short form: Senegal
local long form: Republique du Senegal
local short form: Senegal
former: Senegambia (along with The Gambia), Mali Federation
Government type: republic
Capital: *name:* Dakar
geographic coordinates: 14 40 N, 17 26 W
time difference: UTC 0 (5 hours ahead of Washington, DC during Standard Time)
Administrative divisions: 14 regions (regions, singular—region); Dakar, Diourbel, Fatick, Kaffrine, Kaolack, Kedougou, Kolda, Louga, Matam, Saint-Louis, Sedhiou, Tambacounda, Thies, Ziguinchor
Independence: 4 April 1960 (from France); note—complete independence achieved upon dissolution of federation with Mali on 20 August 1960
National holiday: Independence Day, 4 April (1960)
Constitution: adopted 7 January 2001; amended many times
Legal system: civil law system based on French law; judicial review of legislative acts in Constitutional Court
International law organization participation: accepts compulsory ICJ jurisdiction with reservations; accepts ICCt jurisdiction
Suffrage: 18 years of age; universal
Executive branch: *chief of state:* President Abdoulaye WADE (since 1 April 2000)
head of government: Prime Minister Soulayemane Ndene NDIAYE (since 1 May 2009)
cabinet: Council of Ministers appointed by the prime minister in consultation with the president
(For more information visit the World Leaders website)
elections: president elected by popular vote for a five-year term (eligible for a second term) under new constitution; election last held on 25 February 2007 (next to be held in 2012); prime minister appointed by the president
election results: Abdoulaye WADE reelected president; percent of vote—Abdoulaye WADE 55.9%, Idrissa SECK 14.9%, Ousmane Tanor DIENG 13.6%, Moustapha NIASSE 5.9%, other 9.7%
Legislative branch: bicameral Parliament consisting of the Senate, reinstituted in 2007, (100 seats; 35 members indirectly elected and 65 members appointed by the president) and the National Assembly or Assemblee Nationale (150 seats; 90 members elected by direct popular vote and 60 elected by proportional representation from party lists to serve five-year terms)
elections: Senate—last held on 19 August 2007 (next to be held—NA); National Assembly—last held on 3 June 2007 (next to be held in 2012); note—the National Assembly in December 2005 voted to postpone legislative elections originally scheduled for 2006; legislative elections were first rescheduled to coincide with the 25 February 2007 presidential elections and later for 3 June 2007; the election was boycotted by 12 opposition parties, including the former ruling Socialist Party, which resulted in a record-low 35% voter turnout
election results: Senate results—percent of vote by party—NA; seats by party—PDS 34, AJ/PADS 1, 65 appointed by the president; National Assembly results—percent of vote by party—NA; seats by party—SOPI Coalition 131, other 19
Judicial branch: Constitutional Court; Council of State; Court of Final Appeals or Cour de Cassation; Court of Appeals
Political parties and leaders: African Party of Independence [Majhemout DIOP]; Alliance for the Republic-Yakaar [Macky SALL]; Alliance of Forces of Progress or AFP [Moustapha NIASSE]; And-Jef/African Party for Democracy and Socialism or AJ/PADS [Landing SAVANE]; Democratic League-Labor Party Movement or LD-MPT [Dr. Abdoulaye BATHILY]; Front for Socialism and Democracy/Benno Jubel or FSD/BJ [Cheikh Abdoulaye Bamba DIEYE]; Gainde Centrist Bloc or BGC [Jean-Paul DIAS]; Independence and Labor Party or PIT [Amath DANSOKHO]; Jef-Jel [Talla SYLLA]; National Democratic Rally or RND [Madior DIOUF]; People's Labor Party or PTP [El Hadji DIOUF]; Reform Party or PR [Abdourahim AGNE]; Rewmi Party [Idrissa Seck]; Senegalese Democratic Party or PDS [Abdoulaye WADE]; Socialist Party or PS [Ousmane Tanor DIENG]; SOPI Coalition [Abdoulaye WADE] (a coalition led by the PDS); Union for Democratic Renewal or URD [Djibo Leyti KA]
Political pressure groups and leaders: *other:* labor; students; Sufi brotherhoods, including the Mourides and Tidjanes; teachers
International organization participation: ACP, AfDB, AU, CPLP (associate), ECOWAS, FAO, FZ, G-15, G-77, IAEA, IBRD, ICAO, ICC, ICRM, IDA, IDB, IFAD, IFC, IFRCS, ILO, IMF, IMO, IMSO, Interpol, IOC, IOM, IPU, ISO, ITSO, ITU, ITUC, MIGA, MONUSCO, NAM, OIC, OIF, OPCW, PCA, UN, UNAMID, UNCTAD, UNESCO, UNIDO, Union Latina, UNMIL, UNOCI, UNWTO, UPU, WADB (regional), WAEMU, WCO, WFTU, WHO, WIPO, WMO, WTO
Diplomatic representation in the US: *chief of mission:* Ambassador Fatou Danielle DIAGNE
chancery: 2112 Wyoming Avenue NW, Washington, DC 20008
telephone: [1] (202) 234-0540
FAX: [1] (202) 332-6315
consulate(s) general: Houston, New York
Diplomatic representation from the US: *chief of mission:* Ambassador Marcia S. BERNICAT
embassy: Avenue Jean XXIII at the corner of Rue Kleber, Dakar
mailing address: B. P. 49, Dakar
telephone: [221] 33-829-2100
FAX: [221] 33-822-2991
Flag description: three equal vertical bands of green (hoist side), yellow, and red with a small green five-pointed star centered in the yellow band; green represents Islam, progress, and hope; yellow signifies natural wealth and progress; red symbolizes sacrifice and determination; the star denotes unity and hope
note: uses the popular Pan-African colors of Ethiopia; the colors from left to right are the same as those of neighboring Mali and the reverse of those on the flag of neighboring Guinea
National anthem: *name:* "Pincez Tous vos Koras, Frappez les Balafons" (Pluck Your Koras, Strike the Balafons)
lyrics/music: Leopold Sedar SENGHOR/Herbert PEPPER
note: adopted 1960; the lyrics were written by Leopold Sedar SENGHOR, Senegal's first president; the anthem is sometimes played incorporating the Koras (harp-like stringed instruments) and Balafons (types of xylophones) mentioned in the title

ECONOMY

Economy—overview: Senegal relies heavily on donor assistance. The country's key export industries are phosphate mining, fertilizer production, and commercial fishing. The country is also working on iron ore and oil exploration projects. In January 1994, Senegal undertook a bold and ambitious economic reform program with the support of the international donor community. Government price controls and subsidies have been steadily dismantled. After seeing its economy contract by 2.1% in 1993, Senegal made an important turnaround, thanks to the reform program, with real growth in GDP averaging over 5% annually during 1995-2007. Annual inflation had been pushed down to the single digits. The country was adversely affected by the global economic downturn in 2009 and GDP growth fell below 2%. As a member of the West African Economic and Monetary Union (WAEMU), Senegal is working toward greater regional integration with a unified external tariff and a more stable monetary policy. High unemployment, however, continues to prompt illegal migrants to flee Senegal in search of better job opportunities in Europe. Under the IMF's Highly Indebted Poor Countries (HIPC) debt relief program, Senegal benefited from eradication of two-thirds of its bilateral, multilateral, and private-sector debt. In 2007, Senegal and the IMF agreed to a new, non-disbursing, Policy Support Initiative program which was completed in 2010. Senegal received its first disbursement from the $540 million Millennium Challenge Account compact it signed in September 2009 for infrastructure and agriculture development. In 2010, the Senegalese people protested against frequent power cuts. The government pledged to expand capacity by 2012 and to promote renewable energy but until Senegal has more capacity, more protests are likely and economic activity will be hindered. During the year, bakers protested government price controls on bread. Foreign investment in Senegal is constrained by Senegal's business environment, which has slipped in recent years, and by perceptions of corruption.

GDP (purchasing power parity): $23.88 billion (2010 est.)
country comparison to the world: 114
$22.91 billion (2009 est.)
$22.42 billion (2008 est.)
note: data are in 2010 US dollars

GDP (official exchange rate): $12.88 billion (2010 est.)

GDP—real growth rate: 4.2% (2010 est.)
country comparison to the world: 93
2.2% (2009 est.)
3.2% (2008 est.)

GDP—per capita (PPP): $1,900 (2010 est.)
country comparison to the world: 192
$1,900 (2009 est.)
$1,900 (2008 est.)
note: data are in 2010 US dollars

GDP—composition by sector:
agriculture: 14.9%
industry: 21.4%
services: 63.6% (2010 est.)

Labor force: 5.53 million (2010 est.)
country comparison to the world: 68

Labor force—by occupation:
agriculture: 77.5%
industry and *services:* 22.5% (2007 est.)

Unemployment rate: 48% (2007 est.)
country comparison to the world: 192

Population below poverty line: 54% (2001 est.)

Household income or consumption by percentage share: *lowest 10%:* 2.5%
highest 10%: 30.1% (2005)

Distribution of family income—Gini index: 41.3 (2001)
country comparison to the world: 54
41.3 (1995)

Investment (gross fixed): 25.9% of GDP (2010 est.)
country comparison to the world: 38

Budget: *revenues:* $2.726 billion
expenditures: $3.315 billion (2010 est.)

Public debt: 32.1% of GDP (2010 est.)
country comparison to the world: 89
29.6% of GDP (2009 est.)

Inflation rate (consumer prices): 1.2% (2010 est.)
country comparison to the world: 27
-1% (2009 est.)

Central bank discount rate: 4.25% (31 December 2009)
country comparison to the world: 92
4.75% (31 December 2008)

Commercial bank prime lending rate: NA%

Stock of narrow money: $2.8 billion (31 December 2010 est.)
country comparison to the world: 112
$2.903 billion (31 December 2009 est.)

Stock of broad money: $4.603 billion (31 December 2010 est.)
country comparison to the world: 123
$4.745 billion (31 December 2009 est.)

Stock of domestic credit: $3.516 billion (31 December 2010 est.)
country comparison to the world: 116
$3.412 billion (31 December 2009 est.)

Market value of publicly traded shares: $NA

Agriculture—products: peanuts, millet, corn, sorghum, rice, cotton, tomatoes, green vegetables; cattle, poultry, pigs; fish

Industries: agricultural and fish processing, phosphate mining, fertilizer production, petroleum refining; iron ore, zircon, and gold mining, construction materials, ship construction and repair

Industrial production growth rate: 3.8% (2010 est.)
country comparison to the world: 92

Electricity—production: 1.88 billion kWh (2007 est.)
country comparison to the world: 134

Electricity—consumption: 1.384 billion kWh (2007 est.)
country comparison to the world: 142

Electricity—exports: 0 kWh (2008 est.)

Electricity—imports: 0 kWh (2008 est.)

Oil—production: 0 bbl/day (2008 est.)
country comparison to the world: 128

Oil—consumption: 39,000 bbl/day (2009 est.)
country comparison to the world: 102

Oil—exports: 5,653 bbl/day (2007 est.)
country comparison to the world: 101

Oil—imports: 42,850 bbl/day (2007 est.)
country comparison to the world: 93

Oil—proved reserves: 0 bbl (1 January 2010 est.)
country comparison to the world: 184

Natural gas—production: 50 million cu m (2008 est.)
country comparison to the world: 82

Natural gas—consumption: 50 million cu m (2008 est.)
country comparison to the world: 107

Natural gas—exports: 0 cu m (2008 est.)
country comparison to the world: 171

Natural gas—imports: 0 cu m (2008 est.)
country comparison to the world: 116

Natural gas—proved reserves: NA cu m

Current account balance: $-1.046 billion (2010 est.)
country comparison to the world: 135
$-1.356 billion (2009 est.)

Exports: $2.112 billion (2010 est.)
country comparison to the world: 131
$1.902 billion (2009 est.)

Exports—commodities: fish, groundnuts (peanuts), petroleum products, phosphates, cotton

Exports—partners: Mali 20.12%, India 9.84%, The Gambia 5.58%, France 5.02%, Italy 4.23% (2009)

Imports: $4.474 billion (2010 est.)
country comparison to the world: 126
$4.549 billion (2009 est.)

Imports—commodities: food and beverages, capital goods, fuels

Imports—partners: France 19.58%, UK 9.64%, China 8.08%, Netherlands 5.64%, Thailand 4.75%, US 3.97% (2009)

Reserves of foreign exchange and gold: $2.2 billion (31 December 2010 est.)
country comparison to the world: 100
$2.123 billion (31 December 2009 est.)

Debt—external: $3.885 billion (31 December 2010 est.)
country comparison to the world: 117
$3.462 billion (31 December 2009 est.)

Exchange rates: Communaute Financiere Africaine francs (XOF) per US dollar– 495.28 (2010), 472.19 (2009), 447.81 (2008), 481.83 (2007), 522.89 (2006)

COMMUNICATIONS

Telephones—main lines in use: 278,800 (2009)
country comparison to the world: 119

Telephones—mobile cellular: 6.902 million (2009)
country comparison to the world: 82

Telephone system: *general assessment:* good system with microwave radio relay, coaxial cable and fiber-optic cable in trunk system
domestic: above-average urban system with a fiber-optic network; nearly two-thirds of all fixed-line connections are in Dakar where a call-center industry is emerging; expansion of fixed-line services in rural areas needed; mobile-cellular service is expanding rapidly
international: country code–221; the SAT-3/WASC fiber-optic cable provides connectivity to Europe and Asia while Atlantis-2 provides connectivity to South America; satellite earth station–1 Intelsat (Atlantic Ocean) (2007)

Broadcast media: state-run Radiodiffusion Television Senegalaise (RTS) operates 2 TV stations; a few private TV subscription channels rebroadcast foreign channels without providing any local news or programs; RTS operates a national radio network and a number of regional FM stations; a large number of community and private-broadcast radio stations are available; transmissions of at least 2 international broadcasters are accessible on FM in Dakar (2007)

Internet country code: .sn

Internet hosts: 241 (2010)
country comparison to the world: 190

Internet users: 1.818 million (2009)
country comparison to the world: 76

TRANSPORTATION

Airports: 20 (2010)
country comparison to the world: 134
Airports—with paved runways: *total:* 10
over 3,047 m: 2
1,524 to 2,437 m: 7
914 to 1,523 m: 1 (2010)
Airports—with unpaved runways: *total:* 10
1,524 to 2,437 m: 6
914 to 1,523 m: 3
under 914 m: 1 (2010)
Pipelines: gas 43 km; refined products 8 km (2010)
Railways: *total:* 906 km
country comparison to the world: 92
narrow gauge: 906 km 1.000-m gauge (2008)
Roadways: *total:* 13,576 km
country comparison to the world: 127
paved: 3,972 km (includes 7 km of expressways)
unpaved: 9,604 km (2003)
Waterways: 1,000 km (primarily on the Senegal, Saloum, and Casamance rivers) (2010)
country comparison to the world: 64
Merchant marine: *total:* 1
country comparison to the world: 158
by type: passenger/cargo 1 (2010)
Ports and terminals: Dakar

MILITARY

Military branches: Senegalese Armed Forces: Army, Senegalese Navy (Marine Senegalaise), Senegalese Air Force (Armee de l'Air du Senegal) (2009)
Military service age and obligation: 18 years of age for compulsory and voluntary military service; conscript service obligation–2 years (2004)
Manpower available for military service:
males age 16–49: 2,699,196
females age 16–49: 3,018,565 (2010 est.)
Manpower fit for military service: *males age 16–49:* 1,788,493
females age 16–49: 2,133,370 (2010 est.)
Manpower reaching militarily significant age annually: *male:* 145,509
female: 145,064 (2010 est.)
Military expenditures: 1.4% of GDP (2005 est.)
country comparison to the world: 106

TRANSNATIONAL ISSUES

Disputes—international: The Gambia and Guinea-Bissau attempt to stem separatist violence, cross border raids, and arms smuggling into their countries from Senegal's Casamance region, and in 2006, respectively accepted 6,000 and 10,000 Casamance residents fleeing the conflict; 2,500 Guinea-Bissau residents fled into Senegal in 2006 to escape armed confrontations along the border
Refugees and internally displaced persons: refugees (country of origin): 19,630 (Mauritania)
IDPs: 22,400 (approximately 65% of the IDP population returned in 2005, but new displacement is occurring due to clashes between government troops and separatists in Casamance region) (2007)
Illicit drugs: transshipment point for Southwest and Southeast Asian heroin and South American cocaine moving to Europe and North America; illicit cultivator of cannabis

SERBIA

INTRODUCTION

Background: The Kingdom of Serbs, Croats, and Slovenes was formed in 1918; its name was changed to Yugoslavia in 1929. Various paramilitary bands resisted Nazi Germany's occupation and division of Yugoslavia from 1941 to 1945, but fought each other and ethnic opponents as much as the invaders. The military and political movement headed by Josip "TITO" Broz (Partisans) took full control of Yugoslavia when German and Croatian separatist forces were defeated in 1945. Although Communist, TITO's new government and his successors (he died in 1980) managed to steer their own path between the Warsaw Pact nations and the West for the next four and a half decades. In 1989, Slobodan MILOSEVIC became president of the Republic of Serbia and his ultranationalist calls for Serbian domination led to the violent breakup of Yugoslavia along ethnic lines. In 1991, Croatia, Slovenia, and Macedonia declared independence, followed by Bosnia in 1992. The remaining republics of Serbia and Montenegro declared a new Federal Republic of Yugoslavia (FRY) in April 1992 and under MILOSEVIC's leadership, Serbia led various military campaigns to unite ethnic Serbs in neighboring republics into a "Greater Serbia." These actions were ultimately unsuccessful and led to the signing of the Dayton Peace Accords in 1995. MILOSEVIC retained control over Serbia and eventually became president of the FRY in 1997. In 1998, an ethnic Albanian insurgency in the formerly autonomous Serbian province of Kosovo provoked a Serbian counterinsurgency campaign that resulted in massacres and massive expulsions of ethnic Albanians living in Kosovo. The MILOSEVIC government's rejection of a proposed international settlement led to NATO's bombing of Serbia in the spring of 1999, to the withdrawal of Serbian military and police forces from Kosovo in June 1999, and to the stationing of a NATO-led force in Kosovo to provide a safe and secure environment for the region's ethnic communities. FRY elections in late 2000 led to the ouster of MILOSEVIC and the installation of democratic government. In 2003, the FRY became Serbia and Montenegro, a loose federation of the two republics. Widespread violence predominantly targeting ethnic Serbs in Kosovo in March 2004 caused the international community to open negotiations on the future status of Kosovo in January 2006. In June 2006, Montenegro seceded from the federation and declared itself an independent nation. Serbia subsequently gave notice that it was the successor state to the union of Serbia and Montenegro. In February 2008, after nearly two years of inconclusive negotiations, the UN-administered province of Kosovo declared itself independent of Serbia–an action Serbia refuses to recognize. At Serbia's request, the UN General Assembly (UNGA) in October 2008 sought an advisory opinion from the International Court of Justice (ICJ) on whether Kosovo's unilateral declaration of independence was in accordance with international law. In a ruling considered unfavorable to Serbia, the ICJ issued an advisory opinion in July 2010 stating that international law did not prohibit declarations of independence. In late 2010, Serbia agreed to an EU-drafted UNGA Resolution acknowledging the ICJ's decision and calling for a new round of talks between Serbia and Kosovo.

GEOGRAPHY

Location: Southeastern Europe, between Macedonia and Hungary
Geographic coordinates: 44 00 N, 21 00 E
Map references: Europe
Area: *total:* 77,474 sq km
country comparison to the world: 116
land: 77,474 sq km
water: 0 sq km
Area—comparative: slightly smaller than South Carolina
Land boundaries: *total:* 2,026 km
border countries: Bosnia and Herzegovina 302 km, Bulgaria 318 km, Croatia 241 km, Hungary 151 km, Kosovo 352 km, Macedonia 62 km, Montenegro 124 km, Romania 476 km
Coastline: 0 km (landlocked)
Maritime claims: none (landlocked)
Climate: in the north, continental climate (cold winters and hot, humid summers with well distributed rainfall); in other parts, continental and Mediterranean climate (relatively cold winters with heavy snowfall and hot, dry summers and autumns)
Terrain: extremely varied; to the north, rich fertile plains; to the east, limestone ranges and basins; to the southeast, ancient mountains and hills

Elevation extremes: *lowest point:* Danube and Timok Rivers 35 m
highest point: Midzor 2,169 m
Natural resources: oil, gas, coal, iron ore, copper, zinc, antimony, chromite, gold, silver, magnesium, pyrite, limestone, marble, salt, arable land
Land use: *arable land:* NA
permanent crops: NA
other: NA
Irrigated land: NA
Total renewable water resources: 208.5 cu km (note–includes Kosovo) (2003)
Natural hazards: destructive earthquakes
Environment—current issues: air pollution around Belgrade and other industrial cities; water pollution from industrial wastes dumped into the Sava which flows into the Danube
Environment—international agreements: *party to:* Air Pollution, Biodiversity, Climate Change, Climate Change-Kyoto Protocol, Desertification, Endangered Species, Hazardous Wastes, Law of the Sea, Marine Dumping, Marine Life Conservation, Ozone Layer Protection, Ship Pollution, Wetlands
signed, but not ratified: none of the selected agreements
Geography—note: controls one of the major land routes from Western Europe to Turkey and the Near East

PEOPLE

Population: 7,310,555 (July 2011 est.)
country comparison to the world: 97
note: does not include the population of Kosovo
Age structure: *0–14 years:* 15.1% (male 567,757/female 532,604)
15–64 years: 68.5% (male 2,503,490/female 2,500,949)
65 years and over: 16.5% (male 493,436/female 712,319) (2011 est.)
Median age: *total:* 41.3 years
male: 39.6 years
female: 43.1 years (2011 est.)
Population growth rate: -0.467% (2011 est.)
country comparison to the world: 222
Birth rate: 9.19 births/1,000 population (2011 est.)
country comparison to the world: 207
Death rate: 13.85 deaths/1,000 population (July 2011 est.)
country comparison to the world: 16
Net migration rate: 0 migrant(s)/1,000 population (2011 est.)
country comparison to the world: 105
Urbanization: *urban population:* 56% of total population (2010)
rate of urbanization: 0.6% annual rate of change (2010-15 est.)
Major cities—population: BELGRADE (capital) 1.115 million (2009)
Sex ratio: *at birth:* 1.065 male(s)/female
under 15 years: 1.07 male(s)/female
15–64 years: 1 male(s)/female
65 years and above: 0.7 male(s)/female
total population: 0.95 male(s)/female (2011 est.)
Infant mortality rate: *total:* 6.52 deaths/1,000 live births
country comparison to the world: 171
male: 7.53 deaths/1,000 live births
female: 5.45 deaths/1,000 live births (2011 est.)
Life expectancy at birth: *total population:* 74.32 years
country comparison to the world: 102
male: 71.49 years
female: 77.34 years (2011 est.)
Total fertility rate: 1.4 children born/woman (2011 est.)
country comparison to the world: 201
HIV/AIDS—adult prevalence rate: 0.1% (2009 est.)
country comparison to the world: 157
HIV/AIDS—people living with HIV/AIDS: 6,400 (2009 est.)
country comparison to the world: 116
HIV/AIDS—deaths: fewer than 100 (2009 est.)
country comparison to the world: 150
Major infectious diseases: *degree of risk:* intermediate
food or waterborne diseases: bacterial diarrhea
vectorborne disease: Crimean Congo hemorrhagic fever
note: highly pathogenic H5N1 avian influenza has been identified in this country; it poses a negligible risk with extremely rare cases possible among US citizens who have close contact with birds (2009)
Drinking water source: *Improved:*
urban: 99% of population
rural: 98% of population
total: 99% of population
Unimproved:
urban: 1% of population
rural: 2% of population
total: 1% of population (2008)
Sanitation facility access: *Improved:*
urban: 96% of population
rural: 88% of population
total: 92% of population
Unimproved:
urban: 4% of population
rural: 12% of population
total: 8% of population (2008)
Nationality: *noun:* Serb(s)
adjective: Serbian
Ethnic groups: Serb 82.9%, Hungarian 3.9%, Romany (Gypsy) 1.4%, Yugoslavs 1.1%, Bosniaks 1.8%, Montenegrin 0.9%, other 8% (2002 census)
Religions: Serbian Orthodox 85%, Catholic 5.5%, Protestant 1.1%, Muslim 3.2%, unspecified 2.6%, other, unknown, or atheist 2.6% (2002 census)
Languages: Serbian (official) 88.3%, Hungarian 3.8%, Bosniak 1.8%, Romany (Gypsy) 1.1%, other 4.1%, unknown 0.9% (2002 census)
note: Romanian, Hungarian, Slovak, Ukrainian, and Croatian all official in Vojvodina
Literacy: *definition:* age 15 and over can read and write
total population: 96.4%
male: 98.9%
female: 94.1% (2003 census)
note: includes Montenegro
School life expectancy (primary to tertiary education): *total:* 14 years
male: 13 years
female: 14 years (2009)
Education expenditures: 4.7% of GDP (2008)
country comparison to the world: 73

GOVERNMENT

Country name: *conventional long form:* Republic of Serbia
conventional short form: Serbia
local long form: Republika Srbija
local short form: Srbija
former: People's Republic of Serbia, Socialist Republic of Serbia
Government type: republic
Capital: *name:* Belgrade (Beograd)
geographic coordinates: 44 50 N, 20 30 E
time difference: UTC+1 (6 hours ahead of Washington, DC during Standard Time)
daylight saving time: +1hr, begins last Sunday in March; ends last Sunday in October
Administrative divisions: 167 municipalities (opcstine, singular–opcstina)
Serbia Proper: Belgrade City (Beograd): Barajevo, Cukarica, Grocka, Lazarevac, Mladenovac, Novi Beograd, Obrenovac, Palilula, Rakovica, Savski Venac, Sopot, Stari Grad, Surcin, Vozdovac, Vracar, Zemun, Zvezdara; Bor: Bor, Kladovo, Majdanpek, Negotin; Branicevo: Golubac, Kucevo, Malo Crnice, Petrovac, Pozarevac, Veliko Gradiste, Zabari, Zagubica; Grad Nis: Crveni Krst, Mediana, Niska Banja, Palilula, Pantelej; Jablanica: Bojnik, Crna Trava, Lebane, Leskovac, Medveda, Vlasotince; Kolubara: Lajkovac, Ljig, Mionica, Osecina, Ub, Valjevo; Macva: Bogatic, Koceljeva, Krupanj, Ljubovija, Loznica, Mali Zvornik, Sabac, Vladimirci; Moravica: Cacak, Gornkji Milanovac, Ivanjica, Lucani; Nisava: Aleksinac, Doljevac, Gadzin Han, Merosina, Nis, Razanj, Svrljig; Pcinja: Bosilegrad, Bujanovac, Presevo, Surdulica, Trgoviste, Vladicin Han, Vranje; Pirot: Babusnica, Bela Palanka, Dimitrovgrad, Pirot; Podunavlje: Smederevo, Smederevskia Palanka, Velika Plana; Pomoravlje: Cuprija, Despotovac, Jagodina, Paracin, Rekovac, Svilajnac; Rasina: Aleksandrovac, Brus, Cicevac, Krusevac, Trstenik, Varvarin; Raska: Kraljevo, Novi Pazar, Raska, Tutin, Vrnjacka Banja; Sumadija: Arandelovac, Batocina, Knic, Kragujevac, Lapovo, Raca, Topola; Toplica: Blace, Kursumlija, Prokuplje, Zitorada; Zajecar: Boljevac, Knjazevac, Sokobanja, Zajecar; Zlatibor: Arilje, Bajina Basta, Cajetina, Kosjeric, Nova Varos, Pozega, Priboj, Prijepolje, Sjenica, Uzice
Vojvodina Autonomous Province: South Backa: Bac, Backa Palanka, Backi Petrovac, Becej, Beocin, Novi Sad, Sremski Karlovci, Srobobran, Temerin, Titel, Vrbas, Zabalj; South Banat: Alibunar, Bela Crkva, Kovacica, Kovin, Opovo, Pancevo, Plandiste, Vrsac; North Backa: Backa Topola, Mali Idjos, Subotica; North Banat: Ada, Coka, Kanjiza, Kikinda, Novi Knezevac, Senta; Central Banat: Nova Crnja, Novi Becej, Secanj, Zitiste, Zrenjanin; Srem: Indija, Irig, Pecinci, Ruma, Sid, Sremska Mitrovica, Stara Pazova; West Backa: Apatin, Kula, Odzaci, Sombor
Independence: 5 June 2006 (from Serbia and Montenegro)
National holiday: National Day, 15 February
Constitution: adopted 8 November 2006; effective 10 November 2006

Legal system: civil law system
International law organization participation: has not submitted an ICJ jurisdiction declaration; accepts ICCt jurisdiction
Suffrage: 18 years of age; universal
Executive branch: *chief of state:* President Boris TADIC (since 11 July 2004)
head of government: Prime Minister Mirko CVETKOVIC (since 7 July 2008)
cabinet: Republican Ministries act as cabinet (For more information visit the World Leaders website)
elections: president elected by direct vote for a five-year term (eligible for a second term); election last held on 3 February 2008 (next to be held in 2013); prime minister elected by the National Assembly
election results: Boris TADIC elected president in the second round of voting; Boris TADIC received 51.2% of the vote and Tomislav NIKOLIC 48.8%
Legislative branch: unicameral National Assembly (250 seats; deputies elected according to party lists to serve four-year terms)
elections: last held on 11 May 2008 (next to be held in May 2012)
election results: percent of vote by party—For a European Serbia coalition 38.4%, SRS 29.5%, DSS-NS 11.6%, SPS-led coalition 7.6%, LPD 5.2%, other 7.7%; seats by party—For a European Serbia coalition 102, SRS 57, DSS-NS 30, SNS 21, SPS-led coalition 20, LDP 13, other 7
Judicial branch: courts of general jurisdiction (municipal courts, district courts, Appellate Courts, the Supreme Court of Cassation); courts of special jurisdiction (commercial courts, the High Commercial Court, the High Magistrates Court, the Administrative Court)
Political parties and leaders: Coalition for Sandzak or KZS [Sulejman UGLJANIN]; Democratic Party or DS [Boris TADIC]; Democratic Party of Albanians or PDSh [Ragmi MUSTAFA]; Democratic Party of Serbia or DSS [Vojislav KOSTUNICA]; Democratic Union of the Valley or BDL [Skender DESTANI]; Force of Serbia Movement or PSS [Bogoljub KARIC]; G17 Plus [Mladjan DINKIC]; League of Social Democrats of Vojvodina or LSV [Nenad CANAK]; League of Vojvodina Hungarians or SVM [Istvan PASTOR]; Liberal Democratic Party or LDP [Cedomir JOVANOVIC]; Movement for Democratic Progress or LPD [Jonuz MUSLIU]; New Serbia or NS [Velimir ILIC]; Party of Democratic Action or PVD [Riza HALIMI]; Party of United Pensioners of Serbia or PUPS [Jovan KRKOBABIC]; People's Party or NS [Maja GOJKOVIC]; Roma Party or RP [Srdjan SAJN]; Sandzak Democratic Party or SDP [Resad HODZIC]; Serbian Progressive Party or SNS [Tomislav NIKOLIC]; Serbian Radical Party or SRS [Vojislav SESELJ (currently on trial at The Hague), with Dragan TODOROVIC as acting leader]; Serbian Renewal Movement or SPO [Vuk DRASKOVIC]; Social Democratic Party of Serbia or SDPS [Rasim LJAJIC]; Socialist Party of Serbia or SPS [Ivica DACIC]; Union of Roma of Serbia or URS [Rajko DJURIC]; United Serbia or JS [Dragan "Palma" MARKOVIC]
Political pressure groups and leaders: Obraz (Orthodox clero-fascist organization); 1389 (Serbian nationalist movement)
International organization participation: BIS, BSEC, CE, CEI, EAPC, EBRD, FAO, G-9, IAEA, IBRD, ICAO, ICC, ICRM, IDA, IFC, IFRCS, IHO, ILO, IMF, IMO, IMSO, Interpol, IOC, IOM, IPU, ISO, ITSO, ITU, ITUC, MIGA, MONUSCO, NAM (observer), OAS (observer), OIF (observer), OPCW, OSCE, PCA, PFP, SECI, UN, UNCTAD, UNESCO, UNFICYP, UNHCR, UNIDO, UNMIL, UNOCI, UNWTO, UPU, WCO, WHO, WIPO, WMO, WTO (observer)
Diplomatic representation in the US: *chief of mission:* Ambassador Vladimir PETROVIC
chancery: 2134 Kalorama Road NW, Washington, DC 20008
telephone: [1] (202) 332-0333
FAX: [1] (202) 332-3933
consulate(s) general: Chicago, New York
Diplomatic representation from the US: *chief of mission:* Ambassador Mary WARLICK
embassy: Kneza Milosa 50, 11000 Belgrade
mailing address: 5070 Belgrade Place, Washington, DC 20521-5070
telephone: [381] (11) 361-9344
FAX: [381] (11) 361-8230
Flag description: three equal horizontal stripes of red (top), blue, and white—the Pan-Slav colors representing freedom and revolutionary ideals; charged with the coat of arms of Serbia shifted slightly to the hoist side; the principal field of the coat of arms represents the Serbian state and displays a white two-headed eagle on a red shield; a smaller red shield on the eagle represents the Serbian nation, and is divided into four quarters by a white cross; a white Cyrillic "C" in each quarter stands for the phrase "Only Unity Saves the Serbs"; a royal crown surmounts the coat of arms
note: the Pan-Slav colors were inspired by the 19th-century flag of Russia
National anthem: *name:* "Boze pravde" (God of Justice)
lyrics/music: Jovan DORDEVIC/Davorin JENKO
note: adopted 1904; the song was originally written as part of a play in 1872 and has been used as an anthem by the Serbian people throughout the 20th and 21st centuries

ECONOMY

Economy—overview: MILOSEVIC-era mismanagement of the economy, an extended period of international economic sanctions, and the damage to Yugoslavia's infrastructure and industry during the NATO airstrikes in 1999 left the economy only half the size it was in 1990. After the ousting of former Federal Yugoslav President MILOSEVIC in September 2000, the Democratic Opposition of Serbia (DOS) coalition government implemented stabilization measures and embarked on a market reform program. After renewing its membership in the IMF in December 2000, Yugoslavia continued to reintegrate into the international community by rejoining the World Bank (IBRD) and the European Bank for Reconstruction and Development (EBRD). Belgrade has made progress in trade liberalization and enterprise restructuring and privatization, including telecommunications and small- and medium-size firms. It has made some progress towards EU membership, signing a Stabilization and Association Agreement with Brussels in May 2008, and with full implementation of the Interim Trade Agreement with the EU in February 2010. Serbia is also pursuing membership in the World Trade Organization. Structural economic reforms needed to ensure the country's long-term viability have largely stalled since the onset of the global financial crisis. Serbia, however, is slowly recovering from the crisis. Economic growth in 2010 was a modest 1.7%, following a 3.1% contraction in 2009, but exports rose by over 16% and manufacturing output increased 3.2%. High unemployment and stagnant household incomes are ongoing political and economic problems. Serbia signed an augmented $4 billion Stand By Arrangement with the IMF in May 2009 that expires in April 2011. IMF conditions on Serbia constrain the use of stimulus efforts to revive the economy, while Serbia's concerns about inflation and exchange rate stability preclude the use of expansionary monetary policy. Serbia adopted a new long-term economic growth plan in 2010 that calls for a quadrupling of exports over ten years and heavy investments in basic infrastructure. Serbia is still a transitional economy with unfinished privatization and incomplete structural reforms. Major challenges ahead include: high government expenditures for salaries, pensions and unemployment; a growing need for new government borrowing; rising public and private foreign debt; and stagnant levels of foreign direct investment. Privatization revenues have fallen precipitously in recent years, while a high percentage of economic activity remains in the hands of the state. Other serious challenges include an inefficient judicial system, high levels of corruption, and an aging population. Factors favorable to Serbia's economic growth include a strategic location, a relatively inexpensive and skilled labor force, and a generous package of incentives for foreign investments.
GDP (purchasing power parity): $80.1 billion (2010 est.)
country comparison to the world: 78
$78.72 billion (2009 est.)
$81.26 billion (2008 est.)
note: data are in 2010 US dollars
GDP (official exchange rate): $38.71 billion (2010 est.)
GDP—real growth rate: 1.8% (2010 est.)
country comparison to the world: 156
-3.1% (2009 est.)
5.5% (2008 est.)
GDP—per capita (PPP): $10,900 (2010 est.)
country comparison to the world: 102
$10,700 (2009 est.)
$11,000 (2008 est.)

note: data are in 2010 US dollars
GDP—composition by sector: *agriculture:* 12.6%
industry: 21.9%
services: 65.5% (2010 est.)
Labor force:
2.95 million (2010 est.)
country comparison to the world: 104
Labor force—by occupation: *agriculture:* 23.9%
industry: 20.5%
services: 55.6% (October 2009)
Unemployment rate: 19.2% (2010 est.)
country comparison to the world: 163
16.6% (2009 est.)
Population below poverty line: 8.8% (2010 est.)
Distribution of family income—Gini index: 26 (2008)
country comparison to the world: 133
30 (2003)
Investment (gross fixed): 25.9% of GDP (2010 est.)
country comparison to the world: 39
Budget: *revenues:* $16.47 billion
expenditures: $18.48 billion (2010 est.)
Public debt: 41.5% of GDP (2010 est.)
country comparison to the world: 65
32.9% of GDP (2009 est.)
Inflation rate (consumer prices): 10.3% (2010 est.)
country comparison to the world: 197
8.4% (2009 est.)
Central bank discount rate: 12% (17 January 2011)
country comparison to the world: 45
9.92% (31 December 2009)
Commercial bank prime lending rate: 13.28% (30 November 2010)
country comparison to the world: 72
11.78% (31 December 2009 est.)
Stock of narrow money: $3.06 billion (31 December 2010 est.)
country comparison to the world: 108
$3.821 billion (31 December 2009 est.)
Stock of broad money: $17.68 billion (31 December 2010 est.)
country comparison to the world: 85
$17.82 billion (31 December 2009 est.)
Stock of domestic credit: $20.03 billion (30 November 2010 est.)
country comparison to the world: 78
$19.25 billion (31 December 2009 est.)
Market value of publicly traded shares: $12.37 billion (24 January 2011)
country comparison to the world: 66
$11.52 billion (31 December 2009)
$12.17 billion (31 December 2008)
Agriculture—products: wheat, maize, sugar beets, sunflower, raspberries; beef, pork, milk
Industries: base metals, furniture, food processing, machinery, chemicals, sugar, tires, clothes, pharmaceuticals
Industrial production growth rate: 3.2% (2010 est.)
country comparison to the world: 104
Electricity—production: 35.9 billion kWh (2010)
country comparison to the world: 59
Electricity—consumption: 34.1 billion kWh (2010)
country comparison to the world: 57
Electricity—exports: 1.3 billion kWh (2010)
Electricity—imports: 770 million kWh (2010)
Oil—production: 11,400 bbl/day (2010 est.)
country comparison to the world: 81
Oil—consumption: 90,000 bbl/day (2009 est.)
country comparison to the world: 79
Oil—exports: 5,045 bbl/day (2008)
country comparison to the world: 103
Oil—imports: 78,600 bbl/day (2010 est.)
country comparison to the world: 74
Oil—proved reserves: 77.5 million bbl (1 January 2010 est.)
country comparison to the world: 76
Natural gas—production: 356 million cu m (2010 est.)
country comparison to the world: 70
Natural gas—consumption: 2.35 billion cu m (2010 est.)
country comparison to the world: 78
Natural gas—exports: 0 cu m (2010 est.)
country comparison to the world: 161
Natural gas—imports: 2 billion cu m (2010 est.)
country comparison to the world: 45
Natural gas—proved reserves: 48.14 billion cu m (1 January 2010 est.)
country comparison to the world: 66
Current account balance: $-1.046 billion (2010 est.)
country comparison to the world: 134
$-1.356 billion (2009 est.)
Exports: $9.7 billion (2010 est.)
country comparison to the world: 88
$8.368 billion (2009 est.)
Exports—commodities: iron and steel, rubber, clothes, wheat, fruit and vegetables, nonferrous metals, electric appliances, metal products, weapons and ammunition
Exports—partners: Italy 11.5%, Bosnia and Herzegovina 11.2%, Germany 10.5%, Montenegro 8.4%, Romania 6.3%, Russia 5.4%, Macedonia 4.9%, Slovenia 4.4% (2010 est.)
Imports: $15.78 billion (2010 est.)
country comparison to the world: 78
$15.03 billion (2009 est.)
Imports—partners: Russia 12.8%, Germany 10.6%, Italy 8.5%, China 7.2%, Hungary 4.9% (2010 est.)
Reserves of foreign exchange and gold: $15.1 billion (30 November 2010 est.)
country comparison to the world: 50
$15.22 billion (31 December 2009 est.)
Debt—external: $30.9 billion (30 November 2010 est.)
country comparison to the world: 63
$32.01 billion (31 December 2009 est.)
Stock of direct foreign investment—at home: $23.52 billion (31 December 2009 est.)
country comparison to the world: 65
$11.95 billion (2006 est.)
Stock of direct foreign investment—abroad: $NA
Exchange rates: Serbian dinars (RSD) per US dollar–79.979 (2010), 62.9 (2008), 54.5 (2007), 59.98 (2006)

COMMUNICATIONS

Telephones—main lines in use: 3.106 million (2009)
country comparison to the world: 49
Telephones—mobile cellular: 9.912 million (2009)
country comparison to the world: 64
Telephone system: *general assessment:* replacements of, and upgrades to, telecommunications equipment damaged during the 1999 war has resulted in a modern telecommunications system more than 95% digitalized in 2009
domestic: wireless service, available through multiple providers with national coverage, is growing very rapidly; best telecommunications services are centered in urban centers; 3G mobile network launched in 2007
international: country code–381 (2009)
Internet country code: .rs
Internet hosts: 528,253 (2010)
country comparison to the world: 49
Internet users: 4.107 million (2009)
country comparison to the world: 57

TRANSPORTATION

Airports: 29 (2010)
country comparison to the world: 117
Airports—with paved runways: *total:* 11
over 3,047 m: 2
2,438 to 3,047 m: 3
1,524 to 2,437 m: 3
914 to 1,523 m: 3 (2010)
Airports—with unpaved runways: *total:* 18
1,524 to 2,437 m: 1
914 to 1,523 m: 9
under 914 m: 8 (2010)
Heliports: 2 (2010)
Railways: *total:* 3,379 km
country comparison to the world: 51
standard gauge: 3,379 km 1.435-m gauge (1,254 km electrified) (2006)
Roadways: *total:* 36,884 km
country comparison to the world: 91
paved: 31,938 km
unpaved: 4,946 km (2007)
Waterways: 587 km (primarily on the Danube and Sava rivers) (2009)
country comparison to the world: 81

MILITARY

Military branches: Serbian Armed Forces (Vojska Srbije, VS): Land Forces Command (includes Riverine Component, consisting of a river flotilla on the Danube), Air and Air Defense Forces Command (2010)
Military service age and obligation: 17 years of age for male compulsory military service; 18 years of age for voluntary service; conscription to be abolished effective 2011; 6-month service obligation, with a reserve obligation to age 60 for men and 50 for women (2010)
Manpower fit for military service: *males age 16–49:* 1,395,426
females age 16–49: 1,356,415 (2010 est.)
Manpower reaching militarily significant age annually: *male:* 43,945
female: 41,080 (2010 est.)

TRANSNATIONAL ISSUES

Disputes—international: Serbia with several other states protest the U.S. and other states' recognition of Kosovo's declaring itself as a sovereign and independent state in February 2008; ethnic Serbian municipalities along Kosovo's northern border challenge final status of Kosovo-Serbia boundary; several thousand NATO-led KFOR peacekeepers under UNMIK authority continue to keep the peace within Kosovo between the ethnic Albanian majority and the Serb minority in Kosovo; Serbia delimited about half of the boundary with Bosnia and Herzegovina, but sections along the Drina River remain in dispute

Refugees and internally displaced persons: refugees (country of origin): 71,111 (Croatia); 27,414 (Bosnia and Herzegovina); 206,000 (Kosovo), note–mostly ethnic Serbs and Roma who fled Kosovo in 1999 (2007)

Illicit drugs: transshipment point for Southwest Asian heroin moving to Western Europe on the Balkan route; economy vulnerable to money laundering

SEYCHELLES

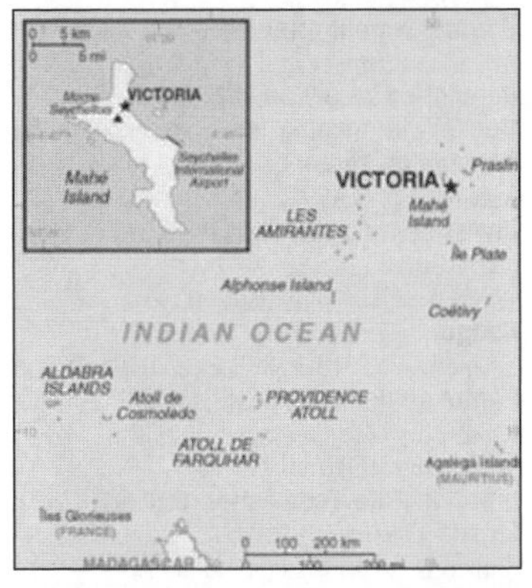

INTRODUCTION

Background: A lengthy struggle between France and Great Britain for the islands ended in 1814, when they were ceded to the latter. Independence came in 1976. Socialist rule was brought to a close with a new constitution and free elections in 1993. President France-Albert RENE, who had served since 1977, was re-elected in 2001, but stepped down in 2004. Vice President James MICHEL took over the presidency and in July 2006 was elected to a new five-year term.

GEOGRAPHY

Location: archipelago in the Indian Ocean, northeast of Madagascar
Geographic coordinates: 4 35 S, 55 40 E
Map references: Africa
Area: *total:* 455 sq km
country comparison to the world: 197
land: 455 sq km
water: 0 sq km
Area—comparative: 2.5 times the size of Washington, DC
Land boundaries: 0 km
Coastline: 491 km
Maritime claims: territorial sea: 12 nm
contiguous zone: 24 nm
exclusive economic zone: 200 nm
continental shelf: 200 nm or to the edge of the continental margin
Climate: tropical marine; humid; cooler season during southeast monsoon (late May to September); warmer season during northwest monsoon (March to May)
Terrain: Mahe Group is granitic, narrow coastal strip, rocky, hilly; others are coral, flat, elevated reefs
Elevation extremes: *lowest point:* Indian Ocean 0 m
highest point: Morne Seychellois 905 m
Natural resources: fish, copra, cinnamon trees
Land use: *arable land:* 2.17%
permanent crops: 13.04%
other: 84.79% (2005)
Irrigated land: NA
Natural hazards: lies outside the cyclone belt, so severe storms are rare; short droughts possible
Environment—current issues: water supply depends on catchments to collect rainwater
Environment—international agreements: *party to:* Biodiversity, Climate Change, Climate Change-Kyoto Protocol, Desertification, Endangered Species, Hazardous Wastes, Law of the Sea, Marine Dumping, Ozone Layer Protection, Ship Pollution, Wetlands
signed, but not ratified: none of the selected agreements
Geography—note: 41 granitic and about 75 coralline islands

PEOPLE

Population: 89,188 (July 2011 est.)
country comparison to the world: 195
Age structure: *0–14 years:* 21.9% (male 9,987/female 9,501)
15–64 years: 71% (male 33,044/female 30,277)
65 years and over: 7.2% (male 2,399/female 3,980) (2011 est.)
Median age: *total:* 32.5 years
male: 32 years
female: 33 years (2011 est.)
Population growth rate: 0.945% (2011 est.)
country comparison to the world: 120
Birth rate: 15.33 births/1,000 population (2011 est.)
country comparison to the world: 134
Death rate: 6.91 deaths/1,000 population (July 2011 est.)
country comparison to the world: 139
Net migration rate: 1.03 migrant(s)/1,000 population (2011 est.)
country comparison to the world: 53
Urbanization: *urban population:* 55% of total population (2010)
rate of urbanization: 1.3% annual rate of change (2010-15 est.)
Major cities—population: VICTORIA (capital) 26,000 (2009)
Sex ratio: *at birth:* 1.031 male(s)/female
under 15 years: 1.05 male(s)/female
15–64 years: 1.09 male(s)/female
65 years and over: 0.6 male(s)/female
total population: 1.03 male(s)/female (2011 est.)
Infant mortality rate: *total:* 11.66 deaths/1,000 live births
country comparison to the world: 137
male: 14.62 deaths/1,000 live births
female: 8.6 deaths/1,000 live births (2011 est.)
Life expectancy at birth: *total population:* 73.52 years
country comparison to the world: 115
male: 68.87 years
female: 78.32 years (2011 est.)
Total fertility rate: 1.91 children born/woman (2011 est.)
country comparison to the world: 138
HIV/AIDS—adult prevalence rate: NA
HIV/AIDS—people living with HIV/AIDS: NA
HIV/AIDS—deaths: NA
Nationality: *noun:* Seychellois (singular and plural)
adjective: Seychellois
Ethnic groups: mixed French, African, Indian, Chinese, and Arab
Religions: Roman Catholic 82.3%, Anglican 6.4%, Seventh Day Adventist 1.1%, other Christian 3.4%, Hindu 2.1%, Muslim 1.1%, other non-Christian 1.5%, unspecified 1.5%, none 0.6% (2002 census)
Languages: Creole 91.8%, English (official) 4.9%, other 3.1%, unspecified 0.2% (2002 census)
Literacy: *definition:* age 15 and over can read and write
total population: 91.8%
male: 91.4%
female: 92.3% (2002 census)
School life expectancy (primary to tertiary education):
total: 15 years
male: 13 years
female: 14 years (2008)
Education expenditures: 5% of GDP (2006)
country comparison to the world: 56

GOVERNMENT

Country name: *conventional long form:* Republic of Seychelles
conventional short form: Seychelles
local long form: Republic of Seychelles
local short form: Seychelles
Government type: republic
Capital: *name:* Victoria
geographic coordinates: 4 38 S, 55 27 E
time difference: UTC+4 (9 hours ahead of Washington, DC during Standard Time)
Administrative divisions: 23 administrative districts; Anse aux Pins, Anse Boileau, Anse Etoile, Anse Louis, Anse Royale, Baie Lazare, Baie Sainte Anne, Beau Vallon, Bel Air, Bel Ombre, Cascade, Glacis, Grand' Anse (on Mahe), Grand' Anse (on Praslin), La Digue, La Riviere Anglaise, Mont Buxton, Mont Fleuri, Plaisance, Pointe La Rue, Port Glaud, Saint Louis, Takamaka
Independence: 29 June 1976 (from the UK)
National holiday: Constitution Day (National Day), 18 June (1993)
Constitution: 18 June 1993
Legal system: mixed legal system of English common law, French civil law, and customary law

International law organization participation: has not submitted an ICJ jurisdiction declaration; accepts ICCt jurisdiction
Suffrage: 17 years of age; universal
Executive branch: *chief of state:* President James Alix MICHEL (since 14 April 2004); note—the president is both the chief of state and head of government
head of government: President James Alix MICHEL (since 14 April 2004)
cabinet: Council of Ministers appointed by the president
(For more information visit the World Leaders website)
elections: president elected by popular vote for a five-year term (eligible for two more terms); election last held on 19-21 May 2011 (next to be held in 2016)
election results: President James MICHEL elected president; percent of vote—James MICHEL 55.5%, Wavel RAMKALAWAN 41.4%, Philippe BOULLE 1.7%, Ralph VOLCERE 1.5%; note—this was the second election in which President James MICHEL participated; he was originally sworn in as president after former president France Albert RENE stepped down in April 2004
Legislative branch: unicameral National Assembly or Assemblee Nationale (34 seats; 25 members elected by popular vote, 9 allocated on a proportional basis to parties winning at least 10% of the vote; members to serve five-year terms)
elections: last held on 10-12 May 2007 (next to be held in 2012)
election results: percent of vote by party—SPPF 56.2%, SNP 43.8%; seats by party—SPPF 23, SNP 11
Judicial branch: Court of Appeal; Supreme Court; judges for both courts are appointed by the president
Political parties and leaders: Democratic Party or DP [James MANCHAM, Paul CHOW]; People's Party (Parti Lepep) or PL [France Albert RENE]; Seychelles People's Progressive Front or SPPF [James MICHEL] (the governing party); Seychelles National Party or SNP [Wavel RAMKALAWAN] (formerly the United Opposition or UO)
Political pressure groups and leaders: Roman Catholic Church
other: trade unions
International organization participation: ACP, AfDB, AOSIS, AU, C, COMESA, FAO, G-77, IAEA, IBRD, ICAO, ICRM, IFAD, IFC, IFRCS, ILO, IMF, IMO, InOC, Interpol, IOC, IPU, ISO (correspondent), ITU, ITUC, MIGA, NAM, OIF, OPCW, SADC, UN, UNCTAD, UNESCO, UNIDO, UNWTO, UPU, WCO, WHO, WIPO, WMO, WTO (observer)
Diplomatic representation in the US: *chief of mission:* Ambassador Jean Ronald JUMEAU
chancery: 800 Second Avenue, Suite 400C, New York, NY 10017
telephone: [1] (212) 972-1785
FAX: [1] (212) 972-1786
Diplomatic representation from the US: the US does not have an embassy in Seychelles; the ambassador to Mauritius is accredited to Seychelles
Flag description: five oblique bands of blue (hoist side), yellow, red, white, and green (bottom) radiating from the bottom of the hoist side; the oblique bands are meant to symbolize a dynamic new country moving into the future; blue represents sky and sea, yellow the sun giving light and life, red the peoples' determination to work for the future in unity and love, white social justice and harmony, green the land and natural environment
National anthem: *name:* "Koste Seselwa" (Seychellois Unite)
lyrics/music: David Francois Marc ANDRE and George Charles Robert PAYET
note: adopted 1996

ECONOMY

Economy—overview: Since independence in 1976, per capita output in this Indian Ocean archipelago has expanded to roughly seven times the pre-independence, near-subsistence level, moving the island into the upper-middle income group of countries. Growth has been led by the tourist sector, which employs about 30% of the labor force and provides more than 70% of hard currency earnings, and by tuna fishing. In recent years, the government has encouraged foreign investment to upgrade hotels and other services. At the same time, the government has moved to reduce the dependence on tourism by promoting the development of farming, fishing, and small-scale manufacturing. GDP grew about 7-8% per year in 2006-07, driven by tourism and a boom in tourism-related construction. The Seychelles rupee was allowed to depreciate in 2006 after being overvalued for years and fell by 10% in the first 9 months of 2007. Despite these actions, the Seychelles economy has struggled to maintain its gains and in 2008 suffered from food and oil price shocks, a foreign exchange shortage, high inflation, large financing gaps, and the global recession. In July 2008 the government defaulted on a Euro amortizing note worth roughly US$80 million, leading to a downgrading of Seychelles credit rating, but in October 2010 the EU approved a $2.9 million grant as part of a larger four-year program for Seychelles. In response to Seychelles successful implementation of tighter monetary and fiscal policies, the IMF upgraded Seychelles to a three-year extended fund facility (EFF) of $31 million in December 2009. In 2008, GDP fell more than 1% due to declining tourism, but the economy recovered in 2009-10 with a notable increase in tourist numbers for 2010.
GDP (purchasing power parity): $2.053 billion (2010 est.)
country comparison to the world: 185
$1.932 billion (2009 est.)
$1.919 billion (2008 est.)
note: data are in 2010 US dollars
GDP (official exchange rate): $936 million (2010 est.)
GDP—real growth rate: 6.2% (2010 est.)
country comparison to the world: 48
0.7% (2009 est.)
-1.3% (2008 est.)
GDP—per capita (PPP): $23,200 (2010 est.)
country comparison to the world: 56
$22,100 (2009 est.)
$22,200 (2008 est.)
note: data are in 2010 US dollars
GDP—composition by sector:
agriculture: 2.9%
industry: 30.8%
services: 66.2% (2009 est.)
Labor force: 39,560 (2006)
country comparison to the world: 196
Labor force—by occupation: *agriculture:* 3%
industry: 23%
services: 74% (2006)
Unemployment rate: 2% (2006 est.)
country comparison to the world: 14
Population below poverty line: NA%
Household income or consumption by percentage share: *lowest 10%:* NA%
highest 10%: NA%
Investment (gross fixed): 36.2% of GDP (2010 est.)
country comparison to the world: 6
Budget: *revenues:* $316.5 million
expenditures: $310.3 million (2010 est.)
Public debt: 58.8% of GDP (2010 est.)
country comparison to the world: 38
58.7% of GDP (2009 est.)
Inflation rate (consumer prices): -2.2% (2010 est.)
country comparison to the world: 1
31.8% (2009 est.)
Central bank discount rate: NA% (31 December 2009)
country comparison to the world: 80
5.13% (31 December 2007)
Commercial bank prime lending rate: 15.35% (31 December 2009 est.)
country comparison to the world: 71
11.81% (31 December 2008 est.)
Stock of narrow money: $274.2 million (31 December 2010 est.)
country comparison to the world: 168
$240.5 million (31 December 2009 est.)
Stock of broad money: $415 million (31 December 2010 est.)
country comparison to the world: 176
$352 million (31 December 2009 est.)
Stock of domestic credit: $678.5 million (31 December 2010 est.)
country comparison to the world: 160
$582.5 million (31 December 2009 est.)
Market value of publicly traded shares: $NA
Agriculture—products: coconuts, cinnamon, vanilla, sweet potatoes, cassava (tapioca), copra, bananas; poultry; tuna
Industries: fishing, tourism, processing of coconuts and vanilla, coir (coconut fiber) rope, boat building, printing, furniture; beverages
Industrial production growth rate: 2% (2010 est.)
country comparison to the world: 128
Electricity—production: 250 million kWh (2007 est.)
country comparison to the world: 173
Electricity—consumption: 232.5 million kWh (2007 est.)
country comparison to the world: 174
Electricity—exports: 0 kWh (2008 est.)
Electricity—imports: 0 kWh (2008 est.)
Oil—production: 0 bbl/day (2009 est.)
country comparison to the world: 127
Oil—consumption: 7,000 bbl/day (2009 est.)
country comparison to the world: 155
Oil—exports: 0 bbl/day (2007 est.)

country comparison to the world: 200
Oil—imports: 7,653 bbl/day (2007 est.)
country comparison to the world: 146
Oil—proved reserves: 0 bbl (1 January 2010 est.)
country comparison to the world: 183
Natural gas—production: 0 cu m (2008 est.)
country comparison to the world: 179
Natural gas—consumption: 0 cu m (2008 est.)
country comparison to the world: 120
Natural gas—exports: 0 cu m (2008 est.)
country comparison to the world: 169
Natural gas—imports: 0 cu m (2008 est.)
country comparison to the world: 115
Natural gas—proved reserves: 0 cu m (1 January 2010 est.)
country comparison to the world: 184
Current account balance: $-351 million (2010 est.)
country comparison to the world: 104
$-284.2 million (2009 est.)
Exports: $464 million (2010 est.)
country comparison to the world: 169
$432.5 million (2009 est.)
Exports—commodities: canned tuna, frozen fish, cinnamon bark, copra, petroleum products (reexports)
Exports—partners: UK 24.84%, France 18.53%, Italy 9.45%, Mauritius 9.03%, Japan 6.98%, Spain 4.92% (2009)
Imports: $831 million (2010 est.)
country comparison to the world: 178
$759.1 million (2009 est.)
Imports—commodities: machinery and equipment, foodstuffs, petroleum products, chemicals, other manufactured goods
Imports—partners: Saudi Arabia 16.44%, India 8.33%, Spain 7.49%, South Africa 6.72%, France 6.39%, Brazil 6.07%, Singapore 5.07% (2009)
Reserves of foreign exchange and gold: $193 million (31 December 2010 est.)
country comparison to the world: 134
$190.6 million (31 December 2009 est.)
Debt—external: $1.374 billion (31 December 2010 est.)
country comparison to the world: 145
$1.321 billion (31 December 2009 est.)
Exchange rates: Seychelles rupees (SCR) per US dollar–12.221 (2010), 13.61 (2009), 8 (2008), 6.5 (2007), 5.5 (2006)

COMMUNICATIONS

Telephones—main lines in use: 22,100 (2009)
country comparison to the world: 189
Telephones—mobile cellular: 92,300 (2009)
country comparison to the world: 186
Telephone system: *general assessment:* effective system
domestic: combined fixed-line and mobile-cellular teledensity is 130 telephones per 100 persons; radiotelephone communications between islands in the archipelago
international: country code–248; direct radiotelephone communications with adjacent island countries and African coastal countries; satellite earth station–1 Intelsat (Indian Ocean)
Broadcast media: the government operates the only terrestrial TV station, which provides local programming and airs broadcasts from international services; multichannel cable and satellite TV are available via subscription; the government operates 1 AM and 1 FM radio station; transmissions of 2 international broadcasters are accessible in Victoria (2007)
Internet country code: .sc
Internet hosts: 256 (2010)
country comparison to the world: 187
Internet users: 32,000 (2008)
country comparison to the world: 179

TRANSPORTATION

Airports: 14 (2010)
country comparison to the world: 149
Airports—with paved runways: *total:* 8
2,438 to 3,047 m: 1
914 to 1,523 m: 6
under 914 m: 1 (2010)
Airports—with unpaved runways:
total: 6
914 to 1,523 m: 1
under 914 m: 5 (2010)
Roadways: *total:* 458 km
country comparison to the world: 195
paved: 440 km
unpaved: 18 km (2003)
Merchant marine: *total:* 9
country comparison to the world: 119
by type: cargo 1, carrier 1, chemical tanker 6, petroleum tanker 1
foreign-owned: 3 (Hong Kong 1, Nigeria 1, South Africa 1) (2010)
Ports and terminals: Victoria

MILITARY

Military branches: Seychelles Defense Force: Army, Coast Guard (includes Naval Wing, Air Wing), National Guard (2005)
Military service age and obligation: 18 years of age for voluntary military service (younger with parental consent); no conscription (2010)
Manpower available for military service:
males age 16–49: 26,257
females age 16–49: 23,996 (2010 est.)
Manpower fit for military service: *males age 16–49:* 20,231
females age 16–49: 19,891 (2010 est.)
Manpower reaching militarily significant age annually: *male:* 686
female: 650 (2010 est.)
Military expenditures: 2% of GDP (2006 est.)
country comparison to the world: 71

TRANSNATIONAL ISSUES

Disputes—international: Mauritius and Seychelles claim the Chagos Islands (UK-administered British Indian Ocean Territory)

SIERRA LEONE

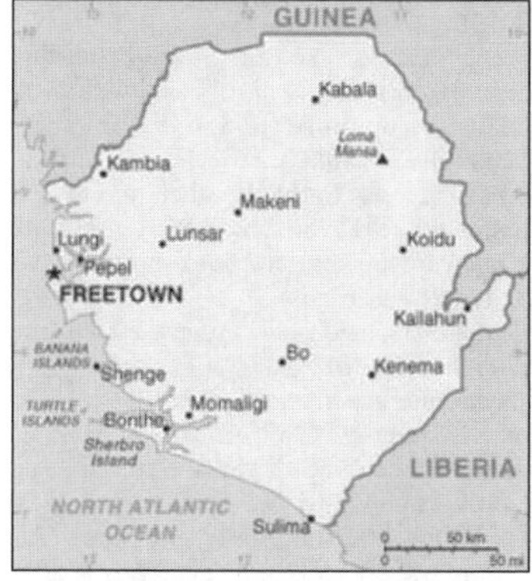

INTRODUCTION

Background: Democracy is slowly being reestablished after the civil war from 1991 to 2002 that resulted in tens of thousands of deaths and the displacement of more than 2 million people (about a third of the population). The military, which took over full responsibility for security following the departure of UN peacekeepers at the end of 2005, is increasingly developing as a guarantor of the country's stability. The armed forces remained on the sideline during the 2007 presidential election, but still look to the UN Integrated Office in Sierra Leone (UNIOSIL)–a civilian UN mission–to support efforts to consolidate peace. The new government's priorities include furthering development, creating jobs, and stamping out endemic corruption.

GEOGRAPHY

Location: Western Africa, bordering the North Atlantic Ocean, between Guinea and Liberia
Geographic coordinates: 8 30 N, 11 30 W
Map references: Africa
Area: *total:* 71,740 sq km
country comparison to the world: 118
land: 71,620 sq km
water: 120 sq km
Area—comparative: slightly smaller than South Carolina
Land boundaries:
total: 958 km
border countries: Guinea 652 km, Liberia 306 km
Coastline: 402 km
Maritime claims: territorial sea: 12 nm
contiguous zone: 24 nm
exclusive economic zone: 200 nm
continental shelf: 200 nm
Climate: tropical; hot, humid; summer rainy season (May to December); winter dry season (December to April)
Terrain: coastal belt of mangrove swamps, wooded hill country, upland plateau, mountains in east
Elevation extremes: *lowest point:* Atlantic Ocean 0 m
highest point: Loma Mansa (Bintimani) 1,948 m
Natural resources: diamonds, titanium ore, bauxite, iron ore, gold, chromite
Land use: *arable land:* 7.95%
permanent crops: 1.05%
other: 91% (2005)
Irrigated land: 300 sq km (2003)

Total renewable water resources: 160 cu km (1987)
Freshwater withdrawal (domestic/industrial/agricultural): *total:* 0.38 cu km/yr (5%/3%/92%)
per capita: 69 cu m/yr (2000)
Natural hazards: dry, sand-laden harmattan winds blow from the Sahara (December to February); sandstorms, dust storms
Environment—current issues: rapid population growth pressuring the environment; overharvesting of timber, expansion of cattle grazing, and slash-and-burn agriculture have resulted in deforestation and soil exhaustion; civil war depleted natural resources; overfishing
Environment—international agreements: *party to:* Biodiversity, Climate Change, Climate Change-Kyoto Protocol, Desertification, Endangered Species, Law of the Sea, Marine Life Conservation, Ozone Layer Protection, Ship Pollution, Wetlands
signed, but not ratified: Environmental Modification
Geography—note: rainfall along the coast can reach 495 cm (195 inches) a year, making it one of the wettest places along coastal, western Africa

PEOPLE

Population: 5,363,669 (July 2011 est.)
country comparison to the world: 112
Age structure: *0–14 years:* 41.8% (male 1,113,528/female 1,130,112)
15–64 years: 54.5% (male 1,401,907/female 1,522,335)
65 years and over: 3.7% (male 86,614/female 109,173) (2011 est.)
Median age:
total: 19.1 years
male: 18.6 years
female: 19.5 years (2011 est.)
Population growth rate: 2.249% (2011 est.)
country comparison to the world: 36
Birth rate: 38.46 births/1,000 population (2011 est.)
country comparison to the world: 14
Death rate: 11.73 deaths/1,000 population (July 2011 est.)
country comparison to the world: 32
Net migration rate: -4.25 migrant(s)/1,000 population
country comparison to the world: 190
note: refugees currently in surrounding countries are slowly returning (2011 est.)
Urbanization: *urban population:* 38% of total population (2010)
rate of urbanization: 3.3% annual rate of change (2010-15 est.)
Major cities—population: FREETOWN (capital) 875,000 (2009)
Sex ratio:
at birth: 1.03 male(s)/female
under 15 years: 0.98 male(s)/female
15–64 years: 0.92 male(s)/female
65 years and over: 0.81 male(s)/female
total population: 0.94 male(s)/female (2011 est.)
Infant mortality rate: *total:* 78.38 deaths/1,000 live births
country comparison to the world: 14
male: 87.2 deaths/1,000 live births
female: 69.3 deaths/1,000 live births (2011 est.)
Life expectancy at birth: *total population:* 56.13 years
country comparison to the world: 197
male: 53.69 years
female: 58.65 years (2011 est.)
Total fertility rate: 4.94 children born/woman (2011 est.)
country comparison to the world: 21
HIV/AIDS—adult prevalence rate: 1.6% (2009 est.)
country comparison to the world: 33
HIV/AIDS—people living with HIV/AIDS: 49,000 (2009 est.)
country comparison to the world: 59
HIV/AIDS—deaths: 2,800 (2009 est.)
country comparison to the world: 47
Major infectious diseases: *degree of risk:* very high
food or waterborne diseases: bacterial and protozoal diarrhea, hepatitis A, and typhoid fever
vectorborne diseases: malaria and yellow fever
water contact disease: schistosomiasis
aerosolized dust or soil contact disease: Lassa fever (2009)
Drinking water source: *Improved:*
urban: 86% of population
rural: 26% of population
total: 49% of population
Unimproved:
urban: 14% of population
rural: 74% of population
total: 51% of population (2008)
Sanitation facility access: *Improved:*
urban: 24% of population
rural: 6% of population
total: 13% of population
Unimproved:
urban: 76% of population
rural: 94% of population
total: 87% of population (2008)
Nationality: *noun:* Sierra Leonean(s)
adjective: Sierra Leonean
Ethnic groups: Temne 35%, Mende 31%, Limba 8%, Kono 5%, Kriole 2% (descendants of freed Jamaican slaves who were settled in the Freetown area in the late-18th century; also known as Krio), Mandingo 2%, Loko 2%, other 15% (includes refugees from Liberia's recent civil war, and small numbers of Europeans, Lebanese, Pakistanis, and Indians) (2008 census)
Religions: Muslim 60%, Christian 10%, indigenous beliefs 30%
Languages: English (official, regular use limited to literate minority), Mende (principal vernacular in the south), Temne (principal vernacular in the north), Krio (English-based Creole, spoken by the descendants of freed Jamaican slaves who were settled in the Freetown area, a lingua franca and a first language for 10% of the population but understood by 95%)
Literacy: *definition:* age 15 and over can read and write English, Mende, Temne, or Arabic
total population: 35.1%
male: 46.9%
female: 24.4% (2004 est.)
School life expectancy (primary to tertiary education):
total: 12 years
male: 13 years
female: 11 years (2007)
Education expenditures: 4.3% of GDP (2009)
country comparison to the world: 94

GOVERNMENT

Country name: *conventional long form:* Republic of Sierra Leone
conventional short form: Sierra Leone
local long form: Republic of Sierra Leone
local short form: Sierra Leone
Government type: constitutional democracy
Capital:
name: Freetown
geographic coordinates: 8 30 N, 13 15 W
time difference: UTC 0 (5 hours ahead of Washington, DC during Standard Time)
Administrative divisions: 3 provinces and 1 area*; Eastern, Northern, Southern, Western*
Independence: 27 April 1961 (from the UK)
National holiday: Independence Day, 27 April (1961)
Constitution: 1 October 1991; amended several times
Legal system: mixed legal system of English common law and customary law
International law organization participation: has not submitted an ICJ jurisdiction declaration; accepts ICCt jurisdiction
Suffrage: 18 years of age; universal
Executive branch: *chief of state:* President Ernest Bai KOROMA (since 17 September 2007); note–the president is both the chief of state and head of government
head of government: President Ernest Bai KOROMA (since 17 September 2007)
cabinet: Ministers of State appointed by the president with the approval of the House of Representatives; the cabinet is responsible to the president
(For more information visit the World Leaders website)
elections: president elected by popular vote for a five-year term (eligible for a second term); election last held on 11 August 2007 and 8 September 2007 (next to be held in 2012)
election results: second round results; percent of vote–Ernest Bai KOROMA 54.6%, Solomon BEREWA 45.4%
Legislative branch: unicameral Parliament (124 seats; 112 members elected by popular vote, 12 filled by paramount chiefs elected in separate elections; members to serve five-year terms)
elections: last held on 11 August 2007 (next to be held in 2012)
election results: percent of vote by party–NA; seats by party–APC 59, SLPP 43, PMDC 10
Judicial branch: Supreme Court; Appeals Court; High Court
Political parties and leaders: All People's Congress or APC [Ernest Bai KOROMA]; Peace and Liberation Party or PLP [Darlington MORRISON]; People's Movement for Democratic Change or PMDC [Charles MARGAI]; Sierra Leone People's Party or SLPP [John BENJAMIN]; numerous others
Political pressure groups and leaders: *other:* student unions; trade unions

International organization participation: ACP, AfDB, AU, C, ECOWAS, FAO, G-77, IAEA, IBRD, ICAO, ICRM, IDA, IDB, IFAD, IFC, IFRCS, ILO, IMF, IMO, Interpol, IOC, IOM, IPU, ISO (correspondent), ITU, ITUC, MIGA, NAM, OIC, OPCW, UN, UNAMID, UNCTAD, UNESCO, UNIDO, UNIFIL, UNMIS, UNMIT, UNWTO, UPU, WCO, WFTU, WHO, WIPO, WMO, WTO
Diplomatic representation in the US: *chief of mission:* Ambassador Bockari Kortu STEVENS
chancery: 1701 19th Street NW, Washington, DC 20009
telephone: [1] (202) 939-9261 through 9263
FAX: [1] (202) 483-1793
Diplomatic representation from the US: *chief of mission:* Ambassador Michael S. OWEN
embassy: Southridge-Hill Station, Freetown
mailing address: use embassy street address
telephone: [232] (22) 515 000 or (76) 515 000
FAX: [232] (22) 515 355
Flag description: three equal horizontal bands of green (top), white, and blue; green symbolizes agriculture, mountains, and natural resources, white represents unity and justice, and blue the sea and the natural harbor in Freetown
National anthem: *name:* "High We Exalt Thee, Realm of the Free"
lyrics/music: Clifford Nelson FYLE/John Joseph AKA
note: adopted 1961

ECONOMY

Economy—overview: Sierra Leone is an extremely poor nation with tremendous inequality in income distribution. While it possesses substantial mineral, agricultural, and fishery resources, its physical and social infrastructure has yet to recover from the civil war, and serious social disorders continue to hamper economic development. Nearly half of the working-age population engages in subsistence agriculture. Manufacturing consists mainly of the processing of raw materials and of light manufacturing for the domestic market. Alluvial diamond mining remains the major source of hard currency earnings, accounting for nearly half of Sierra Leone's exports. The fate of the economy depends upon the maintenance of domestic peace and the continued receipt of substantial aid from abroad, which is essential to offset the severe trade imbalance and supplement government revenues. The IMF has completed a Poverty Reduction and Growth Facility program that helped stabilize economic growth and reduce inflation and in 2010 approved a new program worth $45 million over three years. Political stability has led to a revival of economic activity such as the rehabilitation of bauxite and rutile mining, which are set to benefit from planned tax incentives. A number of offshore oil discoveries were announced in 2009 and 2010. The development on these reserves, which could be significant, is still several years away.
GDP (purchasing power parity): $4.72 billion (2010 est.)
country comparison to the world: 163
$4.498 billion (2009 est.)
$4.358 billion (2008 est.)
note: data are in 2010 US dollars
GDP (official exchange rate): $1.905 billion (2010 est.)
GDP—real growth rate: 5% (2010 est.)
country comparison to the world: 74
3.2% (2009 est.)
5.5% (2008 est.)
GDP—per capita (PPP): $900 (2010 est.)
country comparison to the world: 219
$900 (2009 est.)
$900 (2008 est.)
note: data are in 2010 US dollars
GDP—composition by sector: *agriculture:* 49%
industry: 31%
services: 21% (2005 est.)
Labor force: 2.207 million (2007 est.)
country comparison to the world: 116
Labor force—by occupation:
agriculture: NA%
industry: NA%
services: NA%
Unemployment rate: NA%
Population below poverty line: 70.2% (2004)
Household income or consumption by percentage share: *lowest 10%:* 2.6%
highest 10%: 33.6% (2003)
Distribution of family income—Gini index: 62.9 (1989)
country comparison to the world: 5
Budget: *revenues:* $96 million
expenditures: $351 million (2000 est.)
Inflation rate (consumer prices): 11.7% (2007 est.)
country comparison to the world: 202
Central bank discount rate: NA%
Commercial bank prime lending rate: NA% (31 December 2010)
country comparison to the world: 9
24.5% (31 December 2008 est.)
Stock of narrow money: $209.4 million (31 December 2009)
country comparison to the world: 173
$219.1 million (31 December 2008)
Stock of broad money: $437 million (31 December 2009)
country comparison to the world: 175
$434.3 million (31 December 2008)
Stock of domestic credit: $178.4 million (31 December 2009)
country comparison to the world: 176
$140.9 million (31 December 2008)
Market value of publicly traded shares: $NA
Agriculture—products: rice, coffee, cocoa, palm kernels, palm oil, peanuts; poultry, cattle, sheep, pigs; fish
Industries: diamond mining; small-scale manufacturing (beverages, textiles, cigarettes, footwear); petroleum refining, small commercial ship repair
Industrial production growth rate: NA%
Electricity—production: 80 million kWh (2007 est.)
country comparison to the world: 195
Electricity—consumption: 74.4 million kWh (2007 est.)
country comparison to the world: 194
Electricity—exports: 0 kWh (2008 est.)
Electricity—imports: 0 kWh (2008 est.)
Oil—production: 28.98 bbl/day (2009 est.)
country comparison to the world: 113
Oil—consumption: 9,000 bbl/day (2009 est.)
country comparison to the world: 152
Oil—exports: 502.4 bbl/day (2007 est.)
country comparison to the world: 125
Oil—imports: 8,316 bbl/day (2007 est.)
country comparison to the world: 144
Oil—proved reserves: 0 bbl (1 January 2010 est.)
country comparison to the world: 187
Natural gas—production: 0 cu m (2008 est.)
country comparison to the world: 182
Natural gas—consumption: 0 cu m (2008 est.)
country comparison to the world: 122
Natural gas—exports: 0 cu m (2008 est.)
country comparison to the world: 174
Natural gas—imports: 0 cu m (2008 est.)
country comparison to the world: 118
Natural gas—proved reserves: 0 cu m (1 January 2010 est.)
country comparison to the world: 187
Current account balance: $-63 million (2007 est.)
country comparison to the world: 72
Exports: $216 million (2006)
country comparison to the world: 180
Exports—commodities: diamonds, rutile, cocoa, coffee, fish
Exports—partners: Belgium 26.56%, US 11.87%, Netherlands 7.91%, UK 7.4%, India 6.67%, Cote d'Ivoire 6.13%, Greece 4.05% (2009)
Imports: $560 million (2006)
country comparison to the world: 187
Imports—commodities: foodstuffs, machinery and equipment, fuels and lubricants, chemicals
Imports—partners: South Africa 14.61%, China 7.58%, US 5.87%, Cote d'Ivoire 5.65%, India 5.19%, Malaysia 5.19%, France 5.08%, UK 4.48%, Netherlands 4.06% (2009)
Debt—external: $1.61 billion (2003 est.)
country comparison to the world: 143
Exchange rates: leones (SLL) per US dollar–NA (2007), 2,961.7 (2006), 2,889.6 (2005), 2,701.3 (2004), 2,347.9 (2003)

COMMUNICATIONS

Telephones—main lines in use: 32,800 (2009)
country comparison to the world: 176
Telephones—mobile cellular: 1.16 million (2009)
country comparison to the world: 143
Telephone system: *general assessment:* marginal telephone service with poor infrastructure
domestic: the national microwave radio relay trunk system connects Freetown to Bo and Kenema; while mobile-cellular service is growing rapidly from a small base, service area coverage remains limited
international: country code–232; satellite earth station–1 Intelsat (Atlantic Ocean) (2008)
Broadcast media: 1 government-owned TV station; 1 private TV station began operating in 2005; a pay-per-view TV service began operations in late 2007; 1 government-owned national radio broadcast station; about two dozen private radio stations primarily clustered in major cities; transmissions of several international broadcasters are available (2007)
Internet country code: .sl

Internet hosts: 281 (2010)
country comparison to the world: 184
Internet users: 14,900 (2009)
country comparison to the world: 198

TRANSPORTATION

Airports: 9 (2010)
country comparison to the world: 159
Airports—with paved runways: *total:* 1
over 3,047 m: 1 (2010)
Airports—with unpaved runways: *total:* 8
914 to 1,523 m: 7
under 914 m: 1 (2010)
Heliports: 2 (2010)
Roadways: *total:* 11,300 km
country comparison to the world: 132
paved: 904 km
unpaved: 10,396 km (2002)
Waterways: 800 km (600 km navigable year round) (2009)
country comparison to the world: 72
Merchant marine:
total: 189
country comparison to the world: 35
by type: bulk carrier 7, cargo 131, carrier 1, chemical tanker 12, container 3, liquefied gas 3, passenger 1, passenger/cargo 6, petroleum tanker 20, refrigerated cargo 1, roll on/roll off 3, vehicle carrier 1
foreign-owned: 91 (Bangladesh 1, China 12, Cyprus 1, Egypt 2, Estonia 1, Hong Kong 4, Japan 3, Malaysia 1, North Korea 1, Romania 4, Russia 6, Singapore 5, Syria 20, Taiwan 1, Turkey 14, UAE 6, UK 1, Ukraine 5, US 1, Yemen 2) (2010)
Ports and terminals: Freetown, Pepel, Sherbro Islands

MILITARY

Military branches: Republic of Sierra Leone Armed Forces (RSLAF): Army (includes Maritime Wing and Air Wing) (2010)
Military service age and obligation: 17 years 6 months of age for male and female voluntary military service (younger with parental consent); no conscription; candidates must be HIV negative (2009)
Manpower available for military service: *males age 16–49:* 1,183,093 (2010 est.)
Manpower fit for military service: *males age 16–49:* 731,898
females age 16–49: 838,032 (2010 est.)
Manpower reaching militarily significant age annually: *male:* 54,212
female: 57,154 (2010 est.)
Military expenditures: 2.3% of GDP (2006)
country comparison to the world: 65

TRANSNATIONAL ISSUES

Disputes—international: as domestic fighting among disparate ethnic groups, rebel groups, warlords, and youth gangs in Cote d'Ivoire, Guinea, Liberia, and Sierra Leone gradually abates, the number of refugees in border areas has begun to slowly dwindle; Sierra Leone considers excessive Guinea's definition of the flood plain limits to define the left bank boundary of the Makona and Moa rivers and protests Guinea's continued occupation of these lands including the hamlet of Yenga occupied since 1998
Refugees and internally displaced persons: refugees (country of origin): 27,311 (Liberia) (2007)

SINGAPORE

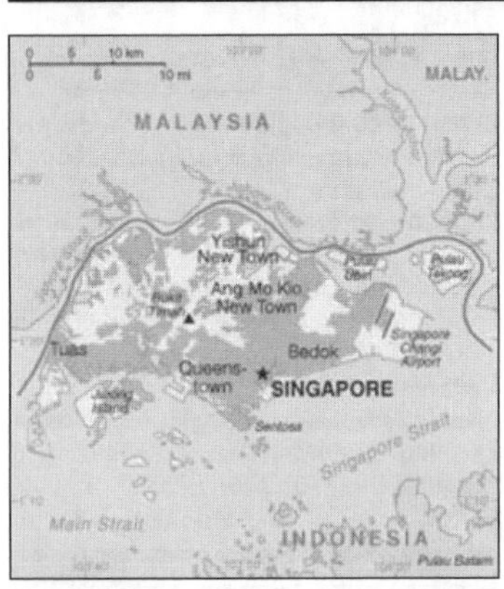

INTRODUCTION

Background: Singapore was founded as a British trading colony in 1819. It joined the Malaysian Federation in 1963 but separated two years later and became independent. Singapore subsequently became one of the world's most prosperous countries with strong international trading links (its port is one of the world's busiest in terms of tonnage handled) and with per capita GDP equal to that of the leading nations of Western Europe.

GEOGRAPHY

Location: Southeastern Asia, islands between Malaysia and Indonesia
Geographic coordinates: 1 22 N, 103 48 E
Map references: Southeast Asia
Area: *total:* 697 sq km
country comparison to the world: 191
land: 687 sq km
water: 10 sq km
Area—comparative: slightly more than 3.5 times the size of Washington, DC
Land boundaries: 0 km
Coastline: 193 km
Maritime claims: territorial sea: 3 nm
exclusive fishing zone: within and beyond territorial sea, as defined in treaties and practice
Climate: tropical; hot, humid, rainy; two distinct monsoon seasons—Northeastern monsoon (December to March) and Southwestern monsoon (June to September); inter-monsoon—frequent afternoon and early evening thunderstorms
Terrain: lowland; gently undulating central plateau contains water catchment area and nature preserve
Elevation extremes: *lowest point:* Singapore Strait 0 m
highest point: Bukit Timah 166 m
Natural resources: fish, deepwater ports
Land use: *arable land:* 1.47%
permanent crops: 1.47%
other: 97.06% (2005)
Irrigated land: NA
Total renewable water resources: 0.6 cu km (1975)
Freshwater withdrawal (domestic/industrial/agricultural): *total:* 0.19 cu km/yr (45%/51%/4%)
per capita: 44 cu m/yr (1975)
Natural hazards: NA
Environment—current issues: industrial pollution; limited natural freshwater resources; limited land availability presents waste disposal problems; seasonal smoke/haze resulting from forest fires in Indonesia
Environment—international agreements: *party to:* Biodiversity, Climate Change, Climate Change-Kyoto Protocol, Desertification, Endangered Species, Hazardous Wastes, Law of the Sea, Ozone Layer Protection, Ship Pollution
signed, but not ratified: none of the selected agreements
Geography—note: focal point for Southeast Asian sea routes

PEOPLE

Population: 4,740,737 (July 2011 est.)
country comparison to the world: 117
Age structure:
0–14 years: 13.8% (male 338,419/female 314,704)
15–64 years: 77% (male 1,774,444/female 1,874,985)
65 years and over: 9.2% (male 196,101/female 242,084) (2011 est.)
Median age:
total: 40.1 years
male: 39.6 years
female: 40.6 years (2011 est.)
Population growth rate: 0.817% (2011 est.)
country comparison to the world: 133
Birth rate: 8.5 births/1,000 population (2011 est.)
country comparison to the world: 217
Death rate: 4.95 deaths/1,000 population (July 2011 est.)
country comparison to the world: 188
Net migration rate: 4.63 migrant(s)/1,000 population (2011 est.)
country comparison to the world: 20
Urbanization: *urban population:* 100% of total population (2010)
rate of urbanization: 0.9% annual rate of change (2010-15 est.)
Sex ratio:
at birth: 1.077 male(s)/female
under 15 years: 1.08 male(s)/female
15–64 years: 0.95 male(s)/female
65 years and over: 0.81 male(s)/female
total population: 0.95 male(s)/female (2011 est.)
Infant mortality rate: *total:* 2.32 deaths/1,000 live births
country comparison to the world: 221
male: 2.52 deaths/1,000 live births
female: 2.11 deaths/1,000 live births (2011 est.)
Life expectancy at birth: *total population:* 82.14 years

country comparison to the world: 7
male: 79.53 years
female: 84.96 years (2011 est.)
Total fertility rate: 1.11 children born/woman (2011 est.)
country comparison to the world: 221
HIV/AIDS—adult prevalence rate: 0.1% (2009 est.)
country comparison to the world: 161
HIV/AIDS—people living with HIV/AIDS: 3,400 (2009 est.)
country comparison to the world: 128
HIV/AIDS—deaths: fewer than 100 (2009 est.)
country comparison to the world: 152
Drinking water source: *Improved:*
urban: 100% of population
total: 100% of population (2008)
Sanitation facility access: *Improved:*
urban: 100% of population
total: 100% of population
Unimproved:
urban: 0% of population
total: 0% of population (2008)
Nationality: *noun:* Singaporean(s)
adjective: Singapore
Ethnic groups: Chinese 76.8%, Malay 13.9%, Indian 7.9%, other 1.4% (2000 census)
Religions: Buddhist 42.5%, Muslim 14.9%, Taoist 8.5%, Hindu 4%, Catholic 4.8%, other Christian 9.8%, other 0.7%, none 14.8% (2000 census)
Languages: Mandarin (official) 35%, English (official) 23%, Malay (official) 14.1%, Hokkien 11.4%, Cantonese 5.7%, Teochew 4.9%, Tamil (official) 3.2%, other Chinese dialects 1.8%, other 0.9% (2000 census)
Literacy: *definition:* age 15 and over can read and write
total population: 92.5%
male: 96.6%
female: 88.6% (2000 census)
School life expectancy (primary to tertiary education): NA
Education expenditures: 3% of GDP (2009)
country comparison to the world: 132

GOVERNMENT

Country name: *conventional long form:* Republic of Singapore
conventional short form: Singapore
local long form: Republic of Singapore
local short form: Singapore
Government type: parliamentary republic
Capital: *name:* Singapore
geographic coordinates: 1 17 N, 103 51 E
time difference: UTC+8 (13 hours ahead of Washington, DC during Standard Time)
Administrative divisions: none
Independence: 9 August 1965 (from Malaysian Federation)
National holiday: National Day, 9 August (1965)
Constitution: 3 June 1959; amended 1965 (based on pre-independence State of Singapore Constitution)
Legal system: English common law
International law organization participation: has not submitted an ICJ jurisdiction declaration; non-party state to the ICCt
Suffrage: 21 years of age; universal and compulsory
Executive branch: *chief of state:* President S R NATHAN (since 1 September 1999)
note: uses S R NATHAN but his full name and the one used in formal communications is Sellapan Ramanathan
head of government: Prime Minister LEE Hsien Loong (since 12 August 2004); Deputy Prime Minister TEO Chee Hean (since 1 April 2009) and Deputy Prime Minister THARMAN Shanmugaratnam (since 21 May 2011); Senior Minister HENG Chee How (since 21 May 2011)
cabinet: appointed by president, responsible to parliament
(For more information visit the World Leaders website)
elections: president elected by popular vote for six-year term; appointed on 17 August 2005 (next election to be held by August 2011); following legislative elections, leader of majority party or leader of majority coalition usually appointed prime minister by president; deputy prime ministers appointed by president
election results: Sellapan Rama (S R) NATHAN was appointed president in August 2005 after Presidential Elections Committee disqualified three other would-be candidates; scheduled election not held
Legislative branch: unicameral Parliament (87 seats; members elected by popular vote to serve five-year terms); note–in addition, there are up to nine nominated members; up to three losing opposition candidates who came closest to winning seats may be appointed as "nonconstituency" members
elections: last held on 7 May 2011 (next to be held in May 2016)
election results: percent of vote by party–PAP 60.1%, WP 12.8%, NSP 12.1%, others 15%; seats by party–PAP 81, WP 6
Judicial branch: Supreme Court (chief justice is appointed by the president with the advice of the prime minister, other judges are appointed by the president with the advice of the chief justice); Court of Appeals
Political parties and leaders: National Solidarity Party or NSP [GOH Meng Seng]; People's Action Party or PAP [LEE Hsien Loong]; Reform Party [NG Teck Siong]; Singapore Democratic Alliance or SDA [CHIAM See Tong]; Singapore Democratic Party or SDP [CHEE Soon Juan]; Workers' Party or WP [Sylvia LIM Swee Lian]
note: SDA includes Singapore Justice Party or SJP, Singapore National Malay Organization or PKMS, Singapore People's Party or SPP
Political pressure groups and leaders: none
International organization participation: ADB, AOSIS, APEC, ARF, ASEAN, BIS, C, CP, EAS, FATF, G-77, IAEA, IBRD, ICAO, ICC, ICRM, IDA, IFC, IFRCS, IHO, ILO, IMF, IMO, IMSO, Interpol, IOC, IPU, ISO, ITSO, ITU, ITUC, MIGA, NAM, OPCW, PCA, UN, UNCTAD, UNESCO, UNMIT, UPU, WCO, WHO, WIPO, WMO, WTO
Diplomatic representation in the US: *chief of mission:* Ambassador CHAN Heng Chee
chancery: 3501 International Place NW, Washington, DC 20008
telephone: [1] (202) 537-3100
FAX: [1] (202) 537-0876
consulate(s) general: San Francisco
consulate(s): New York
Diplomatic representation from the US: *chief of mission:* Ambassador David I. ADELMAN
embassy: 27 Napier Road, Singapore 258508
mailing address: FPO AP 96507-0001
telephone: [65] 6476-9100
FAX: [65] 6476-9340
Flag description: two equal horizontal bands of red (top) and white; near the hoist side of the red band, there is a vertical, white crescent (closed portion is toward the hoist side) partially enclosing five white five-pointed stars arranged in a circle; red denotes brotherhood and equality; white signifies purity and virtue; the waxing crescent moon symbolizes a young nation on the ascendancy; the five stars represent the nation's ideals of democracy, peace, progress, justice, and equality
National anthem: *name:* "Majulah Singapura" (Onward Singapore)
lyrics/music: ZUBIR Said
note: adopted 1965; the anthem, which was first performed in 1958 at the Victoria Theatre, is sung only in Malay

ECONOMY

Economy—overview: Singapore has a highly developed and successful free-market economy. It enjoys a remarkably open and corruption-free environment, stable prices, and a per capita GDP higher than that of most developed countries. The economy depends heavily on exports, particularly in consumer electronics, information technology products, pharmaceuticals, and on a growing financial services sector. Real GDP growth averaged 7.1% between 2004 and 2007. The economy contracted 1.3% in 2009 as a result of the global financial crisis, but rebounded nearly 14.7% in 2010, on the strength of renewed exports. Over the longer term, the government hopes to establish a new growth path that focuses on raising productivity, which has sunk to 1% growth per year in the last decade. Singapore has attracted major investments in pharmaceuticals and medical technology production and will continue efforts to establish Singapore as Southeast Asia's financial and high-tech hub.
GDP (purchasing power parity): $291.9 billion (2010 est.)
country comparison to the world: 41
$255.1 billion (2009 est.)
$257 billion (2008 est.)
note: data are in 2010 US dollars
GDP (official exchange rate): $222.7 billion (2010 est.)
GDP—real growth rate: 14.5% (2010 est.)
country comparison to the world: 3
-0.8% (2009 est.)
1.5% (2008 est.)
GDP—per capita (PPP): $62,100 (2010 est.)
country comparison to the world: 5
$54,800 (2009 est.)
$55,800 (2008 est.)
note: data are in 2010 US dollars
GDP—composition by sector:
agriculture: 0%

industry: 27.2%
services: 72.8% (2010 est.)
Labor force: 3.075 million (2010 est.)
country comparison to the world: 102
Labor force—by occupation:
agriculture: 0.1%
industry: 30.2%
services: 69.7% (2010)
Unemployment rate: 2.1% (2010 est.)
country comparison to the world: 16
3% (2009 est.)
Population below poverty line: NA%
Household income or consumption by percentage share: *lowest 10%:* 4.4%
highest 10%: 23.2% (2008)
Distribution of family income—Gini index: 47.8 (2009)
country comparison to the world: 29
48.1 (2008)
Investment (gross fixed): 27.2% of GDP (2010 est.)
country comparison to the world: 30
Budget: *revenues:* $29.87 billion
expenditures: $34.01 billion
note: expenditures include both operational and development expenditures (2010 est.)
Public debt: 102.4% of GDP (2010 est.)
country comparison to the world: 9
110% of GDP (2009 est.)
note: for Singapore, public debt consists largely of Singapore Government Securities (SGS) issued to assist the Central Provident Fund (CPF), which administers Singapore's defined contribution pension fund; special issues of SGS are held by the CPF, and are non-tradeable; the government has not borrowed to finance deficit expenditures since the 1980s
Inflation rate (consumer prices): 2.8% (2010 est.)
country comparison to the world: 76
0.6% (2009 est.)
Commercial bank prime lending rate: 5.38% (31 December 2010 est.)
country comparison to the world: 140
5.38% (31 December 2009 est.)
Stock of narrow money: $82.13 billion (31 December 2010 est.)
country comparison to the world: 35
$64.26 billion (31 December 2009 est.)
Stock of broad money: $294.4 billion (31 December 2010 est.)
country comparison to the world: 27
$255.2 billion (31 December 2009 est.)
Stock of domestic credit: $199.8 billion (31 December 2010 est.)
country comparison to the world: 39
$249 billion (31 December 2009 est.)
Market value of publicly traded shares: $620.5 billion (31 December 2010)
country comparison to the world: 20
$474.3 billion (31 December 2009)
$268.6 billion (31 December 2008)
Agriculture—products:
orchids, vegetables; poultry, eggs; fish, ornamental fish
Industries: electronics, chemicals, financial services, oil drilling equipment, petroleum refining, rubber processing and rubber products, processed food and beverages, ship repair, offshore platform construction, life sciences, entrepot trade
Industrial production growth rate: 25% (2010 est.)
country comparison to the world: 3
Electricity—production: 39.21 billion kWh (2008 est.)
country comparison to the world: 54
Electricity—consumption: 37.11 billion kWh (2008 est.)
country comparison to the world: 55
Electricity—exports: 0 kWh (2009 est.)
Electricity—imports: 0 kWh (2009 est.)
Oil—production: 10,910 bbl/day (2009 est.)
country comparison to the world: 84
Oil—consumption: 927,000 bbl/day (2009 est.)
country comparison to the world: 20
Oil—exports: 1.374 million bbl/day (2007 est.)
country comparison to the world: 18
Oil—imports: 1.195 million bbl/day (2007 est.)
country comparison to the world: 14
Oil—proved reserves: 0 bbl (1 January 2010 est.)
country comparison to the world: 188
Natural gas—production: 0 cu m (2009 est.)
country comparison to the world: 183
Natural gas—consumption: 8.341 billion cu m (2009 est.)
country comparison to the world: 50
Natural gas—exports: 0 cu m (2009 est.)
country comparison to the world: 175
Natural gas—imports: 8.341 billion cu m (2009 est.)
country comparison to the world: 26
Natural gas—proved reserves: 0 cu m (1 January 2010 est.)
country comparison to the world: 188
Current account balance: $44.08 billion (2010 est.)
country comparison to the world: 9
$23.94 billion (2009 est.)
Exports: $351.2 billion (2010 est.)
country comparison to the world: 14
$268.9 billion (2009 est.)
Exports—commodities: machinery and equipment (including electronics), consumer goods, pharmaceuticals and other chemicals, mineral fuels
Exports—partners: Hong Kong 11.6%, Malaysia 11.5%, US 11.2%, Indonesia 9.7%, China 9.7%, Japan 4.6%, Hong Kong 11.6% (2009 est.)
Imports: $310.4 billion (2010 est.)
country comparison to the world: 15
$245 billion (2009 est.)
Imports—commodities: machinery and equipment, mineral fuels, chemicals, foodstuffs, consumer goods
Imports—partners: US 14.7%, Malaysia 11.6%, China 10.5%, Japan 7.6%, Indonesia 5.8%, South Korea 5.7% (2009 est.)
Reserves of foreign exchange and gold: $225.8 billion (31 December 2010 est.)
country comparison to the world: 11
$187.8 billion (31 December 2009 est.)
Debt—external: $21.66 billion (31 December 2010 est.)
country comparison to the world: 70
$20.3 billion (31 December 2009 est.)
Stock of direct foreign investment—at home: $274.6 billion (31 December 2010 est.)
country comparison to the world: 18
$260.5 billion (31 December 2009 est.)
Stock of direct foreign investment—abroad: $172.1 billion (31 December 2010 est.)
country comparison to the world: 20
$167.4 billion (31 December 2009 est.)
Exchange rates: Singapore dollars (SGD) per US dollar–1.3702 (2010), 1.4545 (2009), 1.415 (2008), 1.507 (2007), 1.5889 (2006)

COMMUNICATIONS

Telephones—main lines in use: 1.852 million (2009)
country comparison to the world: 62
Telephones—mobile cellular: 6.652 million (2009)
country comparison to the world: 83
Telephone system: *general assessment:* excellent service
domestic: excellent domestic facilities; launched 3G wireless service in February 2005; combined fixed-line and mobile-cellular teledensity is more than 180 telephones per 100 persons
international: country code–65; numerous submarine cables provide links throughout Asia, Australia, the Middle East, Europe, and US; satellite earth stations–4; supplemented by VSAT coverage (2008)
Broadcast media: state controls broadcast media; 8 domestic TV stations operated by MediaCorp, wholly owned by a state investment company; broadcasts from Malaysian and Indonesian stations available; satellite dishes banned; multi-channel cable TV service is accessible; a total of 18 domestic radio stations broadcasting with MediaCorp operating more than a dozen and another 4 stations are closely linked to the ruling party or controlled by the Singapore Armed Forces Reservists Association; large number of Malaysian and Indonesian radio stations are available (2008)
Internet country code: .sg
Internet hosts: 992,786 (2010)
country comparison to the world: 44
Internet users: 3.235 million (2009)
country comparison to the world: 65

TRANSPORTATION

Airports: 8 (2010)
country comparison to the world: 165
Airports—with paved runways: *total:* 8
over 3,047 m: 2
2,438 to 3,047 m: 1
1,524 to 2,437 m: 4
914 to 1,523 m: 1 (2010)
Pipelines: gas 111 km (2010)
Roadways: *total:* 3,356 km
country comparison to the world: 163
paved: 3,356 km (includes 161 km of expressways) (2009)
Merchant marine: *total:* 1,422
country comparison to the world: 6
by type: bulk carrier 183, cargo 88, carrier 6, chemical tanker 233, container 321, liquefied gas 117, petroleum tanker 404, refrigerated cargo 5, roll on/roll off 13, vehicle carrier 52
foreign-owned: 850 (Australia 11, Bangladesh 2, Bermuda 21, Chile 7, China 26, Cyprus 3, Denmark 125, France 3, Germany 30, Greece 19, Hong Kong 38, India 19, Indonesia 53, Italy 3, Japan 146, Malaysia 27, Netherlands 1, Norway 132, Slovenia 1, South Africa 3, South Korea 9, Sweden 9, Switzerland 4, Taiwan 79, Thailand 30, UAE 10, UK 6, US 33)
note: this country allows large numbers of ships owned by foreign entities to be regis-

tered in its national shipping registry and to fly its flag; these ships operate under the laws of the flag state
registered in other countries: 327 (Australia 2, Bahamas 7, Bangladesh 3, Belize 7, Cambodia 4, Cyprus 1, Dominica 1, France 3, Gibraltar 1, Honduras 12, Hong Kong 13, Indonesia 42, Isle of Man 1, Kiribati 11, Liberia 27, Malaysia 19, Malta 3, Marshall Islands 28, Mongolia 1, North Korea 2, Panama 79, Philippines 1, Saint Kitts and Nevis 5, Sierra Leone 5, Thailand 1, Tuvalu 25, US 17, unknown 6) (2010)
Ports and terminals: Singapore
Transportation—note: the International Maritime Bureau reports the territorial and offshore waters in the South China Sea as high risk for piracy and armed robbery against ships; numerous commercial vessels have been attacked and hijacked both at anchor and while underway; hijacked vessels are often disguised and cargo diverted to ports in East Asia; crews have been murdered or cast adrift

MILITARY

Military branches: Singapore Armed Forces: Army, Navy, Air Force (includes Air Defense) (2010)
Military service age and obligation: 18-21 years of age for male compulsory military service; 16 years of age for volunteers; 2-year conscript service obligation, with a reserve obligation to age 40 (enlisted) or age 50 (officers) (2008)
Manpower available for military service: *males age 16–49:* 1,255,902 (2010 est.)
Manpower fit for military service: *males age 16–49:* 1,018,839
females age 16–49: 1,087,134 (2010 est.)
Manpower reaching militarily significant age annually: *male:* 27,098
female: 25,368 (2010 est.)
Military expenditures: 4.9% of GDP (2005 est.)
country comparison to the world: 17

TRANSNATIONAL ISSUES

Disputes—international: disputes persist with Malaysia over deliveries of fresh water to Singapore, Singapore's extensive land reclamation works, bridge construction, and maritime boundaries in the Johor and Singapore Straits; in 2008, ICJ awards sovereignty of Pedra Branca (Pulau Batu Puteh/Horsburgh Island) to Singapore, and Middle Rocks to Malaysia, but does not rule on maritime regimes, boundaries, or disposition of South Ledge; Indonesia and Singapore continue to work on finalization of their 1973 maritime boundary agreement by defining unresolved areas north of Indonesia's Batam Island; piracy remains a problem in the Malacca Strait
Illicit drugs: drug abuse limited because of aggressive law enforcement efforts; as a transportation and financial services hub, Singapore is vulnerable, despite strict laws and enforcement, as a venue for money laundering

SINT MAARTEN

(PART OF THE KINGDOM OF THE NETHERLANDS)

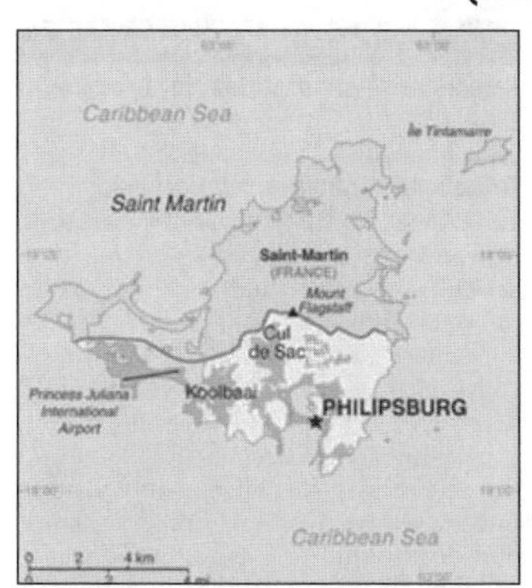

INTRODUCTION

Background: Although sighted by Christopher COLUMBUS in 1493 and claimed for Spain, it was the Dutch who occupied the island in 1631 and set about exploiting its salt deposits. The Spanish retook the island in 1633, but continued to be harassed by the Dutch. The Spanish finally relinquished the island of Saint Martin to the French and Dutch, who divided it amongst themselves in 1648. The establishment of cotton, tobacco, and sugar plantations dramatically expanded slavery on the island in the 18th and 19th centuries; the practice was not abolished in the Dutch half until 1863. The island's economy declined until 1939 when it became a free port; the tourism industry was dramatically expanded beginning in the 1950s. In 1954, Sint Maarten and several other Dutch Caribbean possessions became part of the Kingdom of the Netherlands as the Netherlands Antilles. In a 2000 referendum, the citizens of Sint Maarten voted to become a self-governing country within the Kingdom of the Netherlands. The change in status became effective in October of 2010 with the dissolution of the Netherlands Antilles.

GEOGRAPHY

Location: Caribbean, located in the Leeward Islands (northern) group; Dutch part of the island of Saint Martin in the Caribbean Sea; Sint Maarten lies east of the US Virgin Islands
Geographic coordinates: 18 4 N, 63 4 W
Map references: Central America and the Caribbean
Area: *total:* 34 sq km
country comparison to the world: 234
land: 34 sq km
water: 0 sq km
note: Dutch part of the island of Saint Martin
Area—comparative: one-fifth the size of Washington, DC
Land boundaries: *total:* 15 km
border countries: Saint Martin (France) 15 km
Coastline: 364 km
Maritime claims: territorial sea: 12 nm
exclusive fishing zone: 12 nm
Climate: tropical marine climate, ameliorated by northeast trade winds, results in moderate temperatures; average rainfall of 1500 mm/year; July-November is the hurricane season
Terrain: low, hilly terrain, volcanic origin
Elevation extremes: *lowest point:* Caribbean Sea 0 m
highest point: Mount Flagstaff 386 m
Natural resources: fish, salt
Land use:
arable land: 10%
permanent crops: 0%
other: 90%
Irrigated land: NA
Natural hazards: subject to hurricanes from July to November
Environment—current issues: NA
Geography—note: the northern border is shared with the French overseas collectivity of Saint Martin; together, these two entities make up the smallest landmass in the world shared by two self-governing states

PEOPLE

Population: 37,429 (January 2010 est.)
country comparison to the world: 210
Age structure: *0–14 years:* 23.4% (male 4,299/female 4,455)
15–64 years: 73% (male 13,053/female 14,259)
65 years and over: 3.6% (male 608/female 755) (2010 est.)
Population growth rate: NA
Birth rate: NA
Death rate: 3 deaths/1,000 population (2009)
country comparison to the world: 218
Net migration rate: 14.24 migrant(s)/1,000 population (2008)
country comparison to the world: 6
Sex ratio: *at birth:* 0.98 male(s)/female
under 15 years: 0.96 male(s)/female
15–64 years: 0.92 male(s)/female
65 years and over: 0.81 male(s)/female
total population: 0.92 male(s)/female (2010)
Life expectancy at birth: *total population:* NA
male: 73.1 years
female: 78.2 years (2009)
Total fertility rate: 1.7 children born/woman (2009)
country comparison to the world: 168
HIV/AIDS—adult prevalence rate: NA
HIV/AIDS—people living with HIV/AIDS: NA
HIV/AIDS—deaths: NA
Religions: Roman Catholic 39%, Protestant 27%, Pentecostal 11.6%, none 6.7%, Seventh Day Adventist 6.2%, other 5.4%, Jewish 3.4%, not reported 0.7% (2001 census)
Languages: English (official) 67.5%, Spanish 12.9%, Creole 8.2%, Dutch (official) 4.2%, Papiamento (a Spanish-Portu-

guese-Dutch-English dialect) 2.2%, French 1.5%, other 3.5% (2001 census)
School life expectancy (primary to tertiary education): NA
Education expenditures: NA

GOVERNMENT

Country name: Dutch long form: Land Sint Maarten
Dutch short form: Sint Maarten
English long form: Country of Sint Maarten
English short form: Sint Maarten
former: Netherlands Antilles; Curacao and Dependencies
Dependency status: constituent country within the Kingdom of the Netherlands; full autonomy in internal affairs granted in 2010; Dutch Government responsible for defense and foreign affairs
Government type: parliamentary
Capital: *name:* Philipsburg
geographic coordinates: 18 1 N, 63 2 W
time difference: UTC-4 (1 hour ahead of Washington, DC during Standard Time)
Administrative divisions: none (part of the Kingdom of the Netherlands)
Independence: none (part of the Kingdom of the Netherlands)
National holiday: Queen's Day (Birthday of Queen-Mother JULIANA and accession to the throne of her oldest daughter BEATRIX), 30 April (1909 and 1980)
Constitution: Staatsregeling, 10 October 2010; revised Kingdom Charter pending
Legal system: based on Dutch civil law system with some English common law influence
Suffrage: 18 years of age; universal
Executive branch: *chief of state:* Queen BEATRIX of the Netherlands (since 30 April 1980); represented by Governor General Eugene HOLIDAY (since 10 October 2010)
head of government: Sarah WESCOTT-WILLIAMS (since 10 October 2010)
cabinet: Cabinet
(For more information visit the World Leaders website)
elections: the monarch is hereditary; governor general appointed by the monarch for a six-year term; following legislative elections, the leader of the majority party is usually elected prime minister by the legislature
Legislative branch: unicameral parliament or Staten (15 seats; members elected by popular vote for four year term)
elections: last held 17 September 2010 (next to be held in 2014)
election results: percent of vote by party–National Alliance 45.9%, UPP 36.1%, Democratic Party 17.1%; seats by party–National Alliance 7, UPP 6, Democratic Party 2
Judicial branch: Common Court of Justice, Joint High Court of Justice (judges appointed by the monarch)
Political parties and leaders: Democratic Party or DP [Sarah WESCOTT-WILLIAMS]; National Alliance or NA [William MARLIN]; United People's Party or UPP [Theodore HEYLIGER]; Concordia Political Alliance or CPA [Jeffery RICHARDSON]
Diplomatic representation in the US: none (represented by the Kingdom of the Netherlands)
Diplomatic representation from the US: the US does not have an embassy in Sint Maarten; the Consul General to Curacao is accredited to Sint Maarten
Flag description: two equal horizontal bands of red (top) and blue with a white isosceles triangle based on the hoist side; the center of the triangle displays the Sint Maarten coat of arms; the arms consist of an orange-bordered blue shield prominently displaying the white court house in Philipsburg, as well as a bouquet of yellow sage (the national flower) in the upper left, and the silhouette of a Dutch-French friendship monument in the upper right; the shield is surmounted by a yellow rising sun in front of which is a Brown Pelican in flight; a yellow scroll below the shield bears the motto: SEMPER PROGREDIENS (Always Progressing); the three main colors are identical to those on the Dutch flag
note: the flag somewhat resembles that of the Philippines, but with the main red and blue bands reversed; the banner more closely evokes the wartime Philippine flag
National anthem: *name:* "O Sweet Saint Martin's Land"
lyrics/music: Gerard KEMPS
note: the song, written in 1958, is used as an unofficial anthem for the entire island (both French and Dutch sides); as a collectivity of France, in addition to the local anthem, "La Marseillaise" is official on the French side (see France); as a constituent part of the Kingdom of the Netherlands, in addition to the local anthem, "Het Wilhelmus" is official on the Dutch side (see Netherlands)

ECONOMY

Economy—overview: The economy of Sint Maarten centers around tourism with nearly four-fifths of the labor force engaged in this sector. Over one million visitors come to the island each year–1.3 million in 2008–with most arriving through the Princess Juliana International Airport. Cruise ships and yachts also call on Sint Maarten's numerous ports and harbors. No significant agriculture and limited local fishing means that almost all food must be imported. Energy resources and manufactured goods are also imported. Sint Maarten had the highest per capita income among the five islands that formerly comprised the Netherlands Antilles.
GDP (purchasing power parity): $794.7 million (2008 est.)
country comparison to the world: 206
$748.9 million (2007 est.)
$703.2 million (2006 est.)
note: data are in 2008 US dollars
GDP (official exchange rate): $794.7 million (2008)
GDP—real growth rate: 1.6% (2008 est.)
country comparison to the world: 158
4.5% (2007 est.)
GDP—per capita (PPP): $15,400 (2008 est.)
country comparison to the world: 73
GDP—composition by sector:
agriculture: 0.4%
industry: 18.3%
services: 81.3% (2008 est.)
Labor force: 23,200 (2008 est.)
country comparison to the world: 207
Labor force—by occupation:
agriculture: 1.1%
industry: 15.2%
services: 83.7% (2008 est.)
Unemployment rate: 10.6% (2008 est.)
country comparison to the world: 114
Inflation rate (consumer prices): 0.7% (2009 est.)
country comparison to the world: 11
4.6% (2008 est.)
Agriculture—products: sugar
Industries: tourism, light industry, and manufacturing
Electricity—production: 304.3 million kWh (2008 est.)
country comparison to the world: 167
Exports—commodities: sugar
Exports—partners: China 23.49%, US 10.91%, Japan 5.92% (2009)
Imports—partners: China 17.35%, Japan 14.79%, US 8.96%, Saudi Arabia 6.89% (2009)
Exchange rates: Netherlands Antillean guilders (ANG) per US dollar–1.79 (2010), 1.79 (2009), 1.79 (2008), 1.79 (2007), 1.79 (2006)

COMMUNICATIONS

Telephones—main lines in use: 5,153 (2001)
country comparison to the world: 210
Telephones—mobile cellular: NA
Telephone system: *general assessment:* generally adequate facilities
domestic: extensive interisland microwave radio relay links
international: country code–599 (country code changes to 1-721 effective 30 September 2011); the Americas Region Caribbean Ring System (ARCOS-1) and the Americas-2 submarine cable systems provide connectivity to Central America, parts of South America and the Caribbean, and the US; satellite earth stations–2 Intelsat (Atlantic Ocean)
Internet country code: .sx
Internet hosts: NA
Internet users: NA

TRANSPORTATION

Airports: 1
country comparison to the world: 232
Airports—with paved runways: *total:* 1
1,524 to 2,437 m: 1 (2010)
Roadways: *total:* 53 km
country comparison to the world: 216
Ports and terminals: Philipsburg

MILITARY

Military branches: the Royal Netherlands Navy maintains a permanent and active presence in the region from its main operating base on Curacao and through a detachment on Sint Maarten; other local security forces include a coast guard, paramilitary National Guard (Vrijwilligers Korps Sint Maarten), and Police Force (KPSM) (2010)
Military service age and obligation: no conscription (2010)
Military—note: defense is the responsibility of the Kingdom of the Netherlands

SLOVAKIA

INTRODUCTION

Background: Slovakia's roots can be traced to the 9th century state of Great Moravia. Subsequently, the Slovaks became part of the Hungarian Kingdom, where they remained for the next 1,000 years. Following the formation of the dual Austro-Hungarian monarchy in 1867, language and education policies favoring the use of Hungarian (Magyarization) resulted in a strengthening of Slovak nationalism and a cultivation of cultural ties with the closely related Czechs, who were themselves ruled by the Austrians. After the dissolution of the Austro-Hungarian Empire at the close of World War I, the Slovaks joined the Czechs to form Czechoslovakia. Following the chaos of World War II, Czechoslovakia became a Communist nation within Soviet-dominated Eastern Europe. Soviet influence collapsed in 1989 and Czechoslovakia once more became free. The Slovaks and the Czechs agreed to separate peacefully on 1 January 1993. Slovakia joined both NATO and the EU in the spring of 2004 and the euro area on 1 January 2009.

GEOGRAPHY

Location: Central Europe, south of Poland
Geographic coordinates: 48 40 N, 19 30 E
Map references: Europe
Area: *total:* 49,035 sq km
country comparison to the world: 130
land: 48,105 sq km
water: 930 sq km
Area—comparative: about twice the size of New Hampshire
Land boundaries: *total:* 1,474 km
border countries: Austria 91 km, Czech Republic 197 km, Hungary 676 km, Poland 420 km, Ukraine 90 km
Coastline: 0 km (landlocked)
Maritime claims: none (landlocked)
Climate: temperate; cool summers; cold, cloudy, humid winters
Terrain: rugged mountains in the central and northern part and lowlands in the south
Elevation extremes: *lowest point:* Bodrok River 94 m
highest point: Gerlachovsky Stit 2,655 m
Natural resources: brown coal and lignite; small amounts of iron ore, copper and manganese ore; salt; arable land
Land use: *arable land:* 29.23%
permanent crops: 2.67%
other: 68.1% (2005)
Irrigated land: 1,830 sq km (2003)
Total renewable water resources: 50.1 cu km (2003)
Freshwater withdrawal (domestic/industrial/agricultural): *total:* 1.04
per capita: 193 cu m/yr (2003)
Natural hazards: NA
Environment—current issues: air pollution from metallurgical plants presents human health risks; acid rain damaging forests
Environment—international agreements: *party to:* Air Pollution, Air Pollution-Nitrogen Oxides, Air Pollution-Persistent Organic Pollutants, Air Pollution-Sulfur 85, Air Pollution-Sulfur 94, Air Pollution-Volatile Organic Compounds, Antarctic Treaty, Biodiversity, Climate Change, Climate Change-Kyoto Protocol, Desertification, Endangered Species, Environmental Modification, Hazardous Wastes, Law of the Sea, Ozone Layer Protection, Ship Pollution, Wetlands, Whaling
signed, but not ratified: none of the selected agreements
Geography—note: landlocked; most of the country is rugged and mountainous; the Tatra Mountains in the north are interspersed with many scenic lakes and valleys

PEOPLE

Population: 5,477,038 (July 2011 est.)
country comparison to the world: 111
Age structure: *0–14 years:* 15.6% (male 437,755/female 417,797)
15–64 years: 71.6% (male 1,955,031/female 1,965,554)
65 years and over: 12.8% (male 262,363/female 438,538) (2011 est.)
Median age: *total:* 37.6 years
male: 36.1 years
female: 39.2 years (2011 est.)
Population growth rate: 0.117% (2011 est.)
country comparison to the world: 184
Birth rate: 10.48 births/1,000 population (2011 est.)
country comparison to the world: 182
Death rate: 9.6 deaths/1,000 population (July 2011 est.)
country comparison to the world: 61
Net migration rate: 0.29 migrant(s)/1,000 population (2011 est.)
country comparison to the world: 67
Urbanization: *urban population:* 55% of total population (2010)
rate of urbanization: 0.1% annual rate of change (2010-15 est.)
Major cities—population: BRATISLAVA (capital) 428,000 (2009)
Sex ratio: *at birth:* 1.051 male(s)/female
under 15 years: 1.05 male(s)/female
15–64 years: 0.99 male(s)/female
65 years and over: 0.6 male(s)/female
total population: 0.94 male(s)/female (2011 est.)
Infant mortality rate: *total:* 6.59 deaths/1,000 live births
country comparison to the world: 169
male: 7.69 deaths/1,000 live births
female: 5.43 deaths/1,000 live births (2011 est.)
Life expectancy at birth: *total population:* 75.83 years
country comparison to the world: 79
male: 71.92 years
female: 79.93 years (2011 est.)
Total fertility rate: 1.37 children born/woman (2011 est.)
country comparison to the world: 204
HIV/AIDS—adult prevalence rate: less than 0.1% (2009 est.)
country comparison to the world: 144
HIV/AIDS—people living with HIV/AIDS: fewer than 500 (2009 est.)
country comparison to the world: 151
HIV/AIDS—deaths: fewer than 100 (2009 est.)
country comparison to the world: 139
Drinking water source: *Improved:*
urban: 100% of population
rural: 100% of population
total: 100% of population (2008)
Sanitation facility access: *Improved:*
urban: 100% of population
rural: 99% of population
total: 100% of population
Unimproved:
urban: 0% of population
rural: 1% of population
total: 0% of population (2008)
Nationality:
noun: Slovak(s)
adjective: Slovak
Ethnic groups: Slovak 85.8%, Hungarian 9.7%, Roma 1.7%, Ruthenian/Ukrainian 1%, other and unspecified 1.8% (2001 census)
Religions: Roman Catholic 68.9%, Protestant 10.8%, Greek Catholic 4.1%, other or unspecified 3.2%, none 13% (2001 census)
Languages: Slovak (official) 83.9%, Hungarian 10.7%, Roma 1.8%, Ukrainian 1%, other or unspecified 2.6% (2001 census)
Literacy: *definition:* age 15 and over can read and write
total population: 99.6%
male: 99.7%
female: 99.6% (2004)
School life expectancy (primary to tertiary education): *total:* 15 years
male: 14 years
female: 16 years (2008)
Education expenditures: 3.6% of GDP (2007)

country comparison to the world: 115

GOVERNMENT

Country name: *conventional long form:* Slovak Republic
conventional short form: Slovakia
local long form: Slovenska Republika
local short form: Slovensko
Government type: parliamentary democracy
Capital:
name: Bratislava
geographic coordinates: 48 09 N, 17 07 E
time difference: UTC+1 (6 hours ahead of Washington, DC during Standard Time)
daylight saving time: +1hr, begins last Sunday in March; ends last Sunday in October
Administrative divisions: 8 regions (kraje, singular–kraj); Banskobystricky, Bratislavsky, Kosicky, Nitriansky, Presovsky, Trenciansky, Trnavsky, Zilinsky
Independence: 1 January 1993 (Czechoslovakia split into the Czech Republic and Slovakia)
National holiday: Constitution Day, 1 September (1992)
Constitution: ratified 1 September 1992, effective 1 January 1993; changed September 1998; amended February 2001
note: the change in September 1998 allowed direct election of the president; the amendment of February 2001 allowed Slovakia to apply for NATO and EU membership
Legal system: civil law system based on Austro-Hungarian codes; note–legal code modified to comply with the obligations of Organization on Security and Cooperation in Europe and to expunge Marxist-Leninist legal system
International law organization participation: accepts compulsory ICJ jurisdiction with reservations; accepts ICCt jurisdiction
Suffrage: 18 years of age; universal
Executive branch: *chief of state:* President Ivan GASPAROVIC (since 15 June 2004)
head of government: Prime Minister Iveta RADICOVA (since 8 July 2010); Deputy Prime Ministers Jan FIGEL, Ivan MIKLOS, Jozef MIHAL, Rudolf CHMEL (since 9 July 2010)
cabinet: Cabinet appointed by the president on the recommendation of the prime minister
(For more information visit the World Leaders website)
elections: president elected by popular vote for a five-year term (eligible for a second term); election last held on 21 March and 4 April 2009 (next to be held no later than April 2014); following National Council elections, the leader of the majority party or the leader of a majority coalition usually appointed prime minister by the president
election results: Ivan GASPAROVIC reelected president in runoff; percent of vote–Ivan GASPAROVIC 55.5%, Iveta RADICOVA 44.5%
Legislative branch: unicameral National Council of the Slovak Republic or Narodna Rada Slovenskej Republiky (150 seats; members elected on the basis of proportional representation to serve four-year terms)
elections: last held on 12 June 2010 (next to be held in June 2014)
election results: percent of vote by party–Smer 34.8%, SDKU-DS 15%, SaS 12.1%, KDH 8.5%, Most-Hid 8.1%, SNS 5.1%, other 16.2%; seats by party–Smer 62, SDKU-DS 28, SaS 22, KDH 15, Most-Hid 14, SNS 9
Judicial branch: Supreme Court (judges are elected by the National Council); Constitutional Court (judges appointed by president from group of nominees approved by the National Council); Special Court (judges elected by a council of judges and appointed by president)
Political parties and leaders: parties in the Parliament:: Bridge or Most-Hid [Bela BUGAR]; Christian Democratic Movement or KDH [Jan FIGEL]; Direction-Social Democracy or Smer-SD [Robert FICO]; Freedom and Solidarity or SaS [Richard SULIK]; Slovak Democratic and Christian Union-Democratic Party or SDKU-DS [Mikulas DZURINDA]; Slovak National Party or SNS [Jan SLOTA]
selected parties outside the Parliament:: Alliance for a Europe of Nations or AZEN [Milan URBANI]; Association of Slovak Workers or ZRS [Jan LUPTAK]; Civic Conservative Party or OKS [Peter ZAJAC]; Green Party or SZ [Peter PILINSKY]; Party of the Democratic Left or SDL [Marek BLAHA]; Party of the Hungarian Coalition or SMK [Jozsef BERENYI]; People's Party–Movement for a Democratic Slovakia or LS-HZDS [Vladimir MECIAR]; People's Party–Our Slovakia or LSNS [Marian KOTLEBA]; Slovak Communist Party or KSS [Jozef HRDLICKA]; Union–Party for Slovakia or Unia [Milan CELIK]
Political pressure groups and leaders: Association of Towns and Villages or ZMOS; Confederation of Trade Unions or KOZ; Entrepreneurs Association of Slovakia or ZPS; Federation of Employers' Associations of the Slovak Republic; National Union of Employers or RUZ; Slovak Chamber of Commerce and Industry or SOPK; Slovenska Pospolitost; The Business Alliance of Slovakia or PAS
International organization participation: Australia Group, BIS, BSEC (observer), CBSS (observer), CE, CEI, CERN, EAPC, EBRD, EIB, EMU, EU, FAO, IAEA, IBRD, ICAO, ICC, ICRM, IDA, IEA, IFC, IFRCS, ILO, IMF, IMO, IMSO, Interpol, IOC, IOM, IPU, ISO, ITU, ITUC, MIGA, NATO, NEA, NSG, OAS (observer), OECD, OIF (observer), OPCW, OSCE, PCA, Schengen Convention, SECI (observer), UN, UNCTAD, UNESCO, UNFICYP, UNIDO, UNTSO, UNWTO, UPU, WCO, WFTU, WHO, WIPO, WMO, WTO, ZC
Diplomatic representation in the US: *chief of mission:* Ambassador Peter BURIAN
chancery: 3523 International Court NW, Washington, DC 20008
telephone: [1] (202) 237-1054
FAX: [1] (202) 237-6438
consulate(s) general: Los Angeles, New York
Diplomatic representation from the US: *chief of mission:* Ambassador Theodore SEDGWICK
embassy: Hviezdoslavovo Namestie 4, 81102 Bratislava
mailing address: P.O. Box 309, 814 99 Bratislava
telephone: [421] (2) 5443-3338
FAX: [421] (2) 5441-8861
Flag description: three equal horizontal bands of white (top), blue, and red derive from the Pan-Slav colors; the Slovakian coat of arms (consisting of a red shield bordered in white and bearing a white Cross of Lorraine surmounting three blue hills) is centered over the bands but offset slightly to the hoist side
note: the Pan-Slav colors were inspired by the 19th-century flag of Russia
National anthem: *name:* "Nad Tatrou sa blyska" (Storm Over the Tatras)
lyrics/music: Janko MATUSKA/traditional
note: adopted 1993, in use since 1844; the anthem's music is based on the Slovak folk song "Kopala studienku"

ECONOMY

Economy—overview: Slovakia has made significant economic reforms since its separation from the Czech Republic in 1993. Reforms to the taxation, healthcare, pension, and social welfare systems helped Slovakia consolidate its budget and get on track to join the EU in 2004 and to adopt the euro in January 2009. Major privatizations are nearly complete, the banking sector is almost entirely in foreign hands, and the government has helped facilitate a foreign investment boom with business friendly policies. Slovakia's economic growth exceeded expectations in 2001-08 despite a general European slowdown. Unemployment, at an unacceptable 18% in 2003-04, dropped to 7.7% in 2008 but remains the economy's Achilles heel. Foreign direct investment (FDI) accounted for much of the growth until 2008. Cheap and skilled labor, low taxes, a 19% flat tax for corporations and individuals, no dividend taxes, a relatively liberal labor code and a favorable geographical location are Slovakia's main advantages for foreign investors. Foreign investment in the automotive and electronic sectors has been especially strong. To maintain a stable operating environment for investors, the European Bank for Reconstruction and Development advised the Slovak government to refrain from intervening in important sectors of the economy. However, Bratislava's approach to mitigating the economic slowdown has included substantial government intervention and the option to nationalize strategic companies. RADICOVA's government, in power since July 2010, has allowed the budget deficit to rise slightly, to 7.4% of GDP in 2010. GDP fell nearly 5% in 2009 before gaining back 4% in 2010, and unemployment rose above 12% in 2010, as the global recession impacted many segments of the economy.
GDP (purchasing power parity): $120.2 billion (2010 est.)
country comparison to the world: 62
$115.5 billion (2009 est.)
$121.3 billion (2008 est.)
note: data are in 2010 US dollars

GDP (official exchange rate): $87.45 billion (2010 est.)
GDP—real growth rate: 4% (2010 est.)
country comparison to the world: 99
-4.8% (2009 est.)
5.8% (2008 est.)
GDP—per capita (PPP): $22,000 (2010 est.)
country comparison to the world: 58
$21,100 (2009 est.)
$22,200 (2008 est.)
note: data are in 2010 US dollars
GDP—composition by sector:
agriculture: 2.7%
industry: 35.6%
services: 61.8% (2010 est.)
Labor force: 2.673 million (2010 est.)
country comparison to the world: 108
Labor force—by occupation:
agriculture: 3.5%
industry: 27%
services: 69.4% (December 2009)
Unemployment rate: 13.5% (2010 est.)
country comparison to the world: 138
11.4% (2009 est.)
Population below poverty line: 21% (2002)
Household income or consumption by percentage share: *lowest 10%:* 3.1%
highest 10%: 20.9% (1996)
Distribution of family income—Gini index: 26 (2005)
country comparison to the world: 128
26.3 (1996)
Investment (gross fixed): 22.2% of GDP (2010 est.)
country comparison to the world: 66
Budget: *revenues:* $28.45 billion
expenditures: $35.01 billion (2010 est.)
Public debt: 41% of GDP (2010 est.)
country comparison to the world: 67
35.7% of GDP (2009 est.)
Inflation rate (consumer prices): 1% (2010 est.)
country comparison to the world: 19
1.6% (2009 est.)
Central bank discount rate: 1% (31 December 2010)
country comparison to the world: 108
3% (31 December 2008)
note: this is the European Central Bank's rate on the marginal lending facility, which offers overnight credit to banks from the euro area; as of 1 January 2009 Slovakia became a member of the Economic and Monetary Union (EMU)
Commercial bank prime lending rate: NA%
Stock of narrow money: $34.37 billion (31 December 2010 est.)
country comparison to the world: 52
$34.1 billion (31 December 2009 est.)
note: this figure represents the US dollar value of Slovak koruny in circulation prior to Slovakia joining the Economic and Monetary Union (EMU); see entry for the European Union for money supply in the euro area; the European Central Bank (ECB) controls monetary policy for the 17 members of the Economic and Monetary Union (EMU); individual members of the EMU do not control the quantity of money circulating within their own borders
Stock of broad money: $52.63 billion (31 December 2010 est.)
country comparison to the world: 66
$52.68 billion (31 December 2009 est.)
Stock of domestic credit: $65.09 billion (31 December 2010 est.)
country comparison to the world: 58
$64.25 billion (31 December 2009 est.)
Market value of publicly traded shares: $4.672 billion (31 December 2009)
country comparison to the world: 81
$5.079 billion (31 December 2008)
$6.971 billion (31 December 2007)
Agriculture—products: grains, potatoes, sugar beets, hops, fruit; pigs, cattle, poultry; forest products
Industries: metal and metal products; food and beverages; electricity, gas, coke, oil, nuclear fuel; chemicals and manmade fibers; machinery; paper and printing; earthenware and ceramics; transport vehicles; textiles; electrical and optical apparatus; rubber products
Industrial production growth rate: 7.5% (2010 est.)
country comparison to the world: 40
Electricity—production: 25.9 billion kWh (2009 est.)
country comparison to the world: 64
Electricity—consumption: 28.75 billion kWh (2009 est.)
country comparison to the world: 60
Electricity—exports: 8.891 billion kWh (2008 est.)
Electricity—imports: 9.412 billion kWh (2008 est.)
Oil—production: 4,114 bbl/day (2009 est.)
country comparison to the world: 97
Oil—consumption: 79,930 bbl/day (2009 est.)
country comparison to the world: 85
Oil—exports: 75,110 bbl/day (2008 est.)
country comparison to the world: 71
Oil—imports: 144,000 bbl/day (2008 est.)
country comparison to the world: 56
Oil—proved reserves: 9 million bbl (1 January 2010 est.)
country comparison to the world: 92
Natural gas—production: 103 million cu m (2009 est.)
country comparison to the world: 77
Natural gas—consumption: 6.493 billion cu m (2009 est.)
country comparison to the world: 54
Natural gas—exports: 15 million cu m (2009 est.)
country comparison to the world: 45
Natural gas—imports: 6.974 billion cu m (2009 est.)
country comparison to the world: 28
Natural gas—proved reserves: 14.16 billion cu m (1 January 2010 est.)
country comparison to the world: 78
Current account balance: $-1.93 billion (2010 est.)
country comparison to the world: 154
$-2.819 billion (2009 est.)
Exports: $64.18 billion (2010 est.)
country comparison to the world: 46
$55.32 billion (2009 est.)
Exports—commodities: machinery and electrical equipment 35.9%, vehicles 21%, base metals 11.3%, chemicals and minerals 8.1%, plastics 4.9% (2009 est.)
Exports—partners: Germany 20.1%, Czech Republic 12.9%, France 7.8%, Poland 7.2%, Hungary 6.3%, Italy 6.1%, Austria 5.8%, UK 4.8% (2009 est.)
Imports: $62.43 billion (2010 est.)
country comparison to the world: 41
$53.67 billion (2009 est.)
Imports—commodities: machinery and transport equipment 31%, mineral products 13%, vehicles 12%, base metals 9%, chemicals 8%, plastics 6% (2009 est.)
Imports—partners: Germany 16.8%, Czech Republic 12.3%, Russia 9%, South Korea 6.8%, China 5.8%, Hungary 5.3%, Poland 4% (2009 est.)
Reserves of foreign exchange and gold: $1.16 billion (31 January 2010 est.)
country comparison to the world: 114
$1.16 billion (31 January 2010 est.)
Debt—external: $59.33 billion (30 June 2010 est.)
country comparison to the world: 50
$52.53 billion (31 December 2008)
Stock of direct foreign investment—at home: $52.2 billion (31 December 2010 est.)
country comparison to the world: 53
$50.26 billion (31 December 2009 est.)
Stock of direct foreign investment—abroad: $2.643 billion (31 December 2010 est.)
country comparison to the world: 63
$2.743 billion (31 December 2009 est.)
Exchange rates: Slovak koruny (SKK) per US dollar–0.755 (2010), 0.7198 (2009), 21.05 (2008), 24.919 (2007), 29.611 (2006)

COMMUNICATIONS

Telephones—main lines in use: 1.022 million (2009)
country comparison to the world: 78
Telephones—mobile cellular: 5.498 million (2009)
country comparison to the world: 92
Telephone system: *general assessment:* Slovakia has a modern telecommunications system that has expanded dramatically in recent years with the growth in cellular services
domestic: analog system is now receiving digital equipment and is being enlarged with fiber-optic cable, especially in the larger cities; 3 companies provide nationwide cellular services
international: country code—421; 3 international exchanges (1 in Bratislava and 2 in Banska Bystrica) are available; Slovakia is participating in several international telecommunications projects that will increase the availability of external services
Broadcast media: state-owned public broadcaster, Slovak Television (STV), operates 3 national TV stations; roughly 35 privately-owned television broadcast stations operating nationally, regionally, and locally; about 40% of households are connected to multi-channel cable or satellite TV systems; channels from the Czech Republic and Hungary are widely viewed; state-owned public radio operates multiple national and regional networks; more than 20 privately-owned radio stations (2008)
Internet country code: .sk
Internet hosts: 1.133 million (2010)
country comparison to the world: 41
Internet users: 4.063 million (2009)
country comparison to the world: 58

TRANSPORTATION

Airports: 36 (2010)

country comparison to the world: 107
Airports—with paved runways: *total:* 20
over 3,047 m: 1
2,438 to 3,047 m: 3
1,524 to 2,437 m: 3
914 to 1,523 m: 3
under 914 m: 10 (2010)
Airports—with unpaved runways: *total:* 16
914 to 1,523 m: 9
under 914 m: 7 (2010)
Heliports: 1 (2010)
Pipelines: gas 6,769 km; oil 416 km (2010)
Railways: *total:* 3,622 km
country comparison to the world: 48
broad gauge: 99 km 1.520-m gauge
standard gauge: 3,473 km 1.435-m gauge (1,577 km electrified)
narrow gauge: 50 km 1.000-m or 0.750-m gauge (2008)
Roadways: *total:* 43,761 km
country comparison to the world: 84
paved: 38,085 km (includes 384 km of expressways)
unpaved: 5,676 km (2008)
Waterways: 172 km (on Danube River) (2009)
country comparison to the world: 100
Merchant marine: *total:* 23
country comparison to the world: 97
by type: bulk carrier 1, cargo 19, refrigerated cargo 3
foreign-owned: 21 (Germany 4, Greece 1, Ireland 1, Italy 2, Montenegro 1, Poland 2, Slovenia 1, Turkey 2, Ukraine 7) (2010)
Ports and terminals: Bratislava, Komarno

MILITARY

Military branches: Armed Forces of the Slovak Republic (Ozbrojene Sily Slovenskej Republiky): Land Forces (Pozemne Sily), Air Forces (Vzdusne Sily) (2010)
Military service age and obligation: 18-30 years of age for voluntary military service; conscription abolished in 2006; women are eligible to serve (2010)
Manpower available for military service: *males age 16–49:* 1,405,310
females age 16–49: 1,369,897 (2010 est.)
Manpower fit for military service: *males age 16–49:* 1,156,113
females age 16–49: 1,139,380 (2010 est.)
Manpower reaching militarily significant age annually: *male:* 31,646
female: 30,219 (2010 est.)
Military expenditures: 1.87% of GDP (2005 est.)
country comparison to the world: 80

TRANSNATIONAL ISSUES

Disputes—international: bilateral government, legal, technical and economic working group negotiations continued in 2006 between Slovakia and Hungary over Hungary's completion of its portion of the Gabcikovo-Nagymaros hydroelectric dam project along the Danube; as a member state that forms part of the EU's external border, Slovakia has implemented the strict Schengen border rules
Illicit drugs: transshipment point for Southwest Asian heroin bound for Western Europe; producer of synthetic drugs for regional market; consumer of ecstasy

SLOVENIA

INTRODUCTION

Background: The Slovene lands were part of the Austro-Hungarian Empire until the latter's dissolution at the end of World War I. In 1918, the Slovenes joined the Serbs and Croats in forming a new multinational state, which was named Yugoslavia in 1929. After World War II, Slovenia became a republic of the renewed Yugoslavia, which though Communist, distanced itself from Moscow's rule. Dissatisfied with the exercise of power by the majority Serbs, the Slovenes succeeded in establishing their independence in 1991 after a short 10-day war. Historical ties to Western Europe, a strong economy, and a stable democracy have assisted in Slovenia's transformation to a modern state. Slovenia acceded to both NATO and the EU in the spring of 2004; it joined the eurozone in 2007.

GEOGRAPHY

Location: south Central Europe, Julian Alps between Austria and Croatia
Geographic coordinates: 46 07 N, 14 49 E
Map references: Europe
Area: *total:* 20,273 sq km
country comparison to the world: 154
land: 20,151 sq km
water: 122 sq km
Area—comparative: slightly smaller than New Jersey
Land boundaries: *total:* 1,086 km
border countries: Austria 330 km, Croatia 455 km, Hungary 102 km, Italy 199 km
Coastline: 46.6 km
Maritime claims: territorial sea: 12 nm
Climate: Mediterranean climate on the coast, continental climate with mild to hot summers and cold winters in the plateaus and valleys to the east
Terrain: a short coastal strip on the Adriatic, an alpine mountain region adjacent to Italy and Austria, mixed mountains and valleys with numerous rivers to the east
Elevation extremes: *lowest point:* Adriatic Sea 0 m
highest point: Triglav 2,864 m
Natural resources: lignite coal, lead, zinc, building stone, hydropower, forests
Land use: *arable land:* 8.53%
permanent crops: 1.43%
other: 90.04% (2005)
Irrigated land: 30 sq km (2003)
Total renewable water resources: 32.1 cu km (2005)
Freshwater withdrawal (domestic/industrial/agricultural): *total:* 0.9
per capita: 457 cu m/yr (2002)
Natural hazards: flooding; earthquakes
Environment—current issues: Sava River polluted with domestic and industrial waste; pollution of coastal waters with heavy metals and toxic chemicals; forest damage near Koper from air pollution (originating at metallurgical and chemical plants) and resulting acid rain
Environment—international agreements: *party to:* Air Pollution, Air Pollution-Nitrogen Oxides, Air Pollution-Persistent Organic Pollutants, Air Pollution-Sulfur 94, Biodiversity, Climate Change, Climate Change-Kyoto Protocol, Desertification, Endangered Species, Environmental Modification, Hazardous Wastes, Law of the Sea, Marine Dumping, Ozone Layer Protection, Ship Pollution, Wetlands, Whaling
signed, but not ratified: none of the selected agreements
Geography—note: despite its small size, this eastern Alpine country controls some of Europe's major transit routes

PEOPLE

Population: 2,000,092 (July 2011 est.)
country comparison to the world: 145
Age structure: *0–14 years:* 13.4% (male 138,604/female 130,337)
15–64 years: 69.8% (male 703,374/female 692,640)
65 years and over: 16.8% (male 132,069/female 203,068) (2011 est.)
Median age: *total:* 42.4 years
male: 40.7 years
female: 44.1 years (2011 est.)
Population growth rate: -0.163% (2011 est.)
country comparison to the world: 210
Birth rate: 8.85 births/1,000 population (2011 est.)
country comparison to the world: 213
Death rate: 10.87 deaths/1,000 population (July 2011 est.)
country comparison to the world: 40
Net migration rate: 0.39 migrant(s)/1,000 population (2011 est.)
country comparison to the world: 65
Urbanization: *urban population:* 50% of total population (2010)
rate of urbanization: 0.2% annual rate of change (2010-15 est.)
Major cities—population: LJUBLJANA (capital) 260,000 (2009)
Sex ratio: *at birth:* 1.066 male(s)/female
under 15 years: 1.06 male(s)/female
15–64 years: 1.02 male(s)/female
65 years and over: 0.65 male(s)/female
total population: 0.95 male(s)/female (2011 est.)

Infant mortality rate: *total:* 4.17 deaths/1,000 live births
country comparison to the world: 197
male: 4.71 deaths/1,000 live births
female: 3.58 deaths/1,000 live births (2011 est.)
Life expectancy at birth: *total population:* 77.3 years
country comparison to the world: 62
male: 73.64 years
female: 81.2 years (2011 est.)
Total fertility rate: 1.3 children born/woman (2011 est.)
country comparison to the world: 208
HIV/AIDS—adult prevalence rate: less than 0.1% (2009 est.)
country comparison to the world: 160
HIV/AIDS—people living with HIV/AIDS: fewer than 1,000 (2009 est.)
country comparison to the world: 141
HIV/AIDS—deaths: fewer than 100 (2009 est.)
country comparison to the world: 151
Drinking water source: *Improved:*
urban: 100% of population
rural: 99% of population
total: 99% of population
Unimproved:
urban: 0% of population
rural: 1% of population
total: 1% of population (2008)
Sanitation facility access: *Improved:*
urban: 100% of population
rural: 100% of population
total: 100% of population (2008)
Nationality: *noun:* Slovene(s)
adjective: Slovenian
Ethnic groups: Slovene 83.1%, Serb 2%, Croat 1.8%, Bosniak 1.1%, other or unspecified 12% (2002 census)
Religions: Catholic 57.8%, Muslim 2.4%, Orthodox 2.3%, other Christian 0.9%, unaffiliated 3.5%, other or unspecified 23%, none 10.1% (2002 census)
Languages: Slovenian (official) 91.1%, Serbo-Croatian 4.5%, other or unspecified 4.4%, Italian (official, only in municipalities where Italian national communities reside, Hungarian (official, only in municipalities where Hungarian national communities reside (2002 census)
Literacy: *definition:* NA
total population: 99.7%
male: 99.7%
female: 99.6% (2000 est.)
School life expectancy (primary to tertiary education): *total:* 17 years
male: 16 years
female: 18 years (2008)
Education expenditures: 5.2% of GDP (2007)
country comparison to the world: 51

GOVERNMENT

Country name: *conventional long form:* Republic of Slovenia
conventional short form: Slovenia
local long form: Republika Slovenija
local short form: Slovenija
former: People's Republic of Slovenia, Socialist Republic of Slovenia
Government type: parliamentary republic
Capital: *name:* Ljubljana
geographic coordinates: 46 03 N, 14 31 E
time difference: UTC+1 (6 hours ahead of Washington, DC during Standard Time)
daylight saving time: +1hr, begins last Sunday in March; ends last Sunday in October
Administrative divisions: 199 municipalities (obcine, singular–obcina) and 11 urban municipalities (mestne obcine, singular–mestna obcina)
municipalities: Ajdovscina, Apace, Beltinci, Benedikt, Bistrica ob Sotli, Bled, Bloke, Bohinj, Borovnica, Bovec, Braslovce, Brda, Brezice, Brezovica, Cankova, Cerklje na Gorenjskem, Cerknica, Cerkno, Cerkvenjak, Cirkulane, Crensovci, Crna na Koroskem, Crnomelj, Destrnik, Divaca, Dobje, Dobrepolje, Dobrna, Dobrova-Polhov Gradec, Dobrovnik/Dobronak, Dolenjske Toplice, Dol pri Ljubljani, Domzale, Dornava, Dravograd, Duplek, Gorenja Vas-Poljane, Gorisnica, Gorje, Gornja Radgona, Gornji Grad, Gornji Petrovci, Grad, Grosuplje, Hajdina, Hoce-Slivnica, Hodos, Horjul, Hrastnik, Hrpelje-Kozina, Idrija, Ig, Ilirska Bistrica, Ivancna Gorica, Izola/Isola, Jesenice, Jezersko, Jursinci, Kamnik, Kanal, Kidricevo, Kobarid, Kobilje, Kocevje, Komen, Komenda, Kosanjevica na Krki, Kostel, Kozje, Kranjska Gora, Krizevci, Krsko, Kungota, Kuzma, Lasko, Lenart, Lendava/Lendva, Litija, Ljubno, Ljutomer, Log-Dragomer, Logatec, Loska Dolina, Loski Potok, Lovrenc na Pohorju, Luce, Lukovica, Majsperk, Makole, Markovci, Medvode, Menges, Metlika, Mezica, Miklavz na Dravskem Polju, Miren-Kostanjevica, Mirna Pec, Mislinja, Mokronog-Trebelno, Moravce, Moravske Toplice, Mozirje, Muta, Naklo, Nazarje, Odranci, Oplotnica, Ormoz, Osilnica, Pesnica, Piran/Pirano, Pivka, Podcetrtek, Podlehnik, Podvelka, Poljcane, Polzela, Postojna, Prebold, Preddvor, Prevalje, Puconci, Race-Fram, Radece, Radenci, Radlje ob Dravi, Radovljica, Ravne na Koroskem, Razkrizje, Recica ob Savinji, Rence-Vogrsko, Ribnica, Ribnica na Pohorju, Rogaska Slatina, Rogasovci, Rogatec, Ruse, Selnica ob Dravi, Semic, Sevnica, Sezana, Slovenska Bistrica, Slovenske Konjice, Sodrazica, Solcava, Sredisce ob Dravi, Starse, Straza, Sveta Ana, Sveta Trojica v Slovenskih Goricah, Sveti Andraz v Slovenskih Goricah, Sveti Jurij, Sveti Jurij v Slovenskih Goricah, Sveti Tomaz, Salovci, Sempeter-Vrtojba, Sencur, Sentilj, Sentjernej, Sentjur, Sentrupert, Skocjan, Skofja Loka, Skofljica, Smarje pri Jelsah, Smarjeske Toplice, Smartno ob Paki, Smartno pri Litiji, Sostanj, Store, Tabor, Tisina, Tolmin, Trbovlje, Trebnje, Trnovska Vas, Trzic, Trzin, Turnisce, Velika Polana, Velike Lasce, Verzej, Videm, Vipava, Vitanje, Vodice, Vojnik, Vransko, Vrhnika, Vuzenica, Zagorje ob Savi, Zalec, Zavrc, Zelezniki, Zetale, Ziri, Zirovnica, Zrece, Zuzemberk
urban municipalities: Celje, Koper-Capodistria, Kranj, Ljubljana, Maribor, Murska Sobota, Nova Gorica, Novo Mesto, Ptuj, Slovenj Gradec, Velenje
Independence: 25 June 1991 (from Yugoslavia)
National holiday: Independence Day/Statehood Day, 25 June (1991)
Constitution: adopted 23 December 1991, amended 14 July 1997 and 25 July 2000
Legal system: civil law system
International law organization participation: has not submitted an ICJ jurisdiction declaration; accepts ICCt jurisdiction
Suffrage: 18 years of age, 16 if employed; universal
Executive branch: *chief of state:* President Danilo TURK (since 22 December 2007)
head of government: Prime Minister Borut PAHOR (since 7 November 2008)
cabinet: Council of Ministers nominated by the prime minister and elected by the National Assembly
(For more information visit the World Leaders website)
elections: president elected by popular vote for a five-year term (eligible for a second term); election last held on 21 October and 11 November 2007 (next to be held on 8 October 2012); following National Assembly elections, the leader of the majority party or the leader of a majority coalition usually nominated to become prime minister by the president and elected by the National Assembly
election results: Danilo TURK elected president; percent of vote–Danilo TURK 68.2%, Alojze PETERLE 31.8%; Borut PAHOR elected prime minister by National Assembly vote
Legislative branch: bicameral Parliament consists of a National Council or Drzavni Svet (40 seats; members indirectly elected by an electoral college to serve five-year terms; note–this is primarily an advisory body with limited legislative powers; it may propose laws, ask to review any National Assembly decision, and call national referenda) and the National Assembly or Drzavni Zbor (90 seats; 40 members directly elected and 50 are elected on a proportional basis; note–the number of directly elected and proportionally elected seats varies with each election; the constitution mandates 1 seat each for Slovenia's Hungarian and Italian minorities; members elected by popular vote to serve four-year terms)
elections: National Assembly–last held on 21 September 2008 (next to be held on 8 October 2012)
election results: percent of vote by party–SD 30.5%, SDS 29.3%, ZARES 9.4%, DeSUS 7.5%, SNS 5.5%, SLS+SMS 5.2%, LDS 5.2%, other 7.4%; seats by party–SD 29, SDS 28, ZARES 9, DeSUS 7, SNS 5, SLS+SMS 5, LDS 5, Hungarian minority 1, Italian minority 1
Judicial branch: Supreme Court (judges are elected by the National Assembly on the recommendation of the Judicial Council); Constitutional Court (judges elected for nine-year terms by the National Assembly and nominated by the president)
Political parties and leaders: Democratic Party of Pensioners of Slovenia or DeSUS [Karl ERJAVEC]; Liberal Democracy of Slovenia or LDS [Katarina KRESAL]; New Slovenia or NSi [Ljudmila NOVAK (acting)]; Slovene National Party or SNS [Zmago JELINCIC]; Slovene People's Party or SLS [Radovan ZERJAV]; Slovene Youth Party or SMS [Darko KRANJC]; Slovenian Democratic Party or SDS [Janez JANSA]; Social Democrats or SD [Borut PAHOR] (formerly ZLSD); ZARES [Gregor GOLOBIC]

Political pressure groups and leaders: Slovenian Roma Association [Jozek Horvat MUC]
other: Catholic Church
International organization participation: Australia Group, BIS, CE, CEI, EAPC, EBRD, EIB, EMU, ESA (cooperating state), EU, FAO, IADB, IAEA, IBRD, ICAO, ICC, ICRM, IDA, IFC, IFRCS, IHO, ILO, IMF, IMO, Interpol, IOC, IOM, IPU, ISO, ITU, MIGA, NATO, NSG, OAS (observer), OECD, OIF (observer), OPCW, OSCE, PCA, Schengen Convention, SECI, UN, UNCTAD, UNESCO, UNHCR, UNIDO, UNIFIL, UNTSO, UNWTO, UPU, WCO, WHO, WIPO, WMO, WTO, ZC
Diplomatic representation in the US: *chief of mission:* Ambassador Roman KIRN
chancery: 2410 California Street N.W., Washington, DC 20008
telephone: [1] (202) 386-6601
FAX: [1] (202) 386-6633
consulate(s) general: Cleveland, New York
Diplomatic representation from the US: *chief of mission:* Ambassador Joseph A. MUSSOMELI
embassy: Presernova 31, 1000 Ljubljana
mailing address: American Embassy Ljubljana, US Department of State, 7140 Ljubljana Place, Washington, DC 20521-7140
telephone: [386] (1) 200-5500
FAX: [386] (1) 200-5555
Flag description: three equal horizontal bands of white (top), blue, and red, derive from the medieval coat of arms of the Duchy of Carniola; the Slovenian seal (a shield with the image of Triglav, Slovenia's highest peak, in white against a blue background at the center; beneath it are two wavy blue lines depicting seas and rivers, and above it are three six-pointed stars arranged in an inverted triangle, which are taken from the coat of arms of the Counts of Celje, the great Slovene dynastic house of the late 14th and early 15th centuries) appears in the upper hoist side of the flag centered on the white and blue bands
National anthem: *name:* "Zdravljica" (A Toast)
lyrics/music: France PRESEREN/Stanko PREMRL
note: adopted 1989; the anthem was originally written in 1848; the full poem, whose seventh verse is used as the anthem, speaks of pan-Slavic nationalism

ECONOMY

Economy—overview: Slovenia became the first 2004 European Union entrant to adopt the euro (on 1 January 2007) and has become a model of economic success and stability for the region. With the highest per capita GDP in Central Europe, Slovenia has excellent infrastructure, a well-educated work force, and a strategic location between the Balkans and Western Europe. Privatization has lagged since 2002, and the economy has one of highest levels of state control in the EU. Structural reforms to improve the business environment have allowed for somewhat greater foreign participation in Slovenia's economy and have helped to lower unemployment. In March 2004, Slovenia became the first transition country to graduate from borrower status to donor partner at the World Bank. In December 2007, Slovenia was invited to begin the accession process for joining the OECD. Despite its economic success, foreign direct investment (FDI) in Slovenia has lagged behind the region average, and taxes remain relatively high. Furthermore, the labor market is often seen as inflexible, and legacy industries are losing sales to more competitive firms in China, India, and elsewhere. In 2009, the world recession caused the economy to contract–through falling exports and industrial production–by more than 8%, and unemployment to rise above 9%. Although growth resumed in 2010, the unemployment rate continued to rise, topping 10%.
GDP (purchasing power parity): $56.58 billion (2010 est.)
country comparison to the world: 90
$55.91 billion (2009 est.)
$60.85 billion (2008 est.)
note: data are in 2010 US dollars
GDP (official exchange rate): $47.85 billion (2010 est.)
GDP—real growth rate: 1.2% (2010 est.)
country comparison to the world: 167
-8.1% (2009 est.)
3.7% (2008 est.)
GDP—per capita (PPP): $28,200 (2010 est.)
country comparison to the world: 50
$27,900 (2009 est.)
$30,300 (2008 est.)
note: data are in 2010 US dollars
GDP—composition by sector:
agriculture: 2.4%
industry: 31%
services: 66.6% (2010 est.)
Labor force: 930,000 (2010 est.)
country comparison to the world: 144
Labor force—by occupation:
agriculture: 2.2%
industry: 35%
services: 62.8% (2009)
Unemployment rate: 10.6% (2010 est.)
country comparison to the world: 113
9.2% (2009 est.)
Population below poverty line: 12.3% (2008)
Household income or consumption by percentage share:
lowest 10%: 3.4%
highest 10%: 24.6% (2004)
Distribution of family income—Gini index: 28.4 (2008)
country comparison to the world: 120
23.8 (2004)
Investment (gross fixed): 18.7% of GDP (2010 est.)
country comparison to the world: 103
Budget: *revenues:* $22.56 billion
expenditures: $25.53 billion (2010 est.)
Public debt: 35.5% of GDP (2010 est.)
country comparison to the world: 82
31.3% of GDP (2009 est.)
Inflation rate (consumer prices): 2.1% (2010 est.)
country comparison to the world: 53
0.9% (2009 est.)
Central bank discount rate: 1.75% (31 December 2010)
country comparison to the world: 118
1.75% (31 December 2009)
note: this is the European Central Bank's rate on the marginal lending facility, which offers overnight credit to banks in the euro area
Commercial bank prime lending rate: 5.47% (31 December 2009 est.)
country comparison to the world: 120
7.41% (31 December 2008 est.)
Stock of narrow money: $10.47 billion (31 December 2010 est.)
country comparison to the world: 71
$10.33 billion (31 December 2009 est.)
note: see entry for the European Union for money supply in the euro area; the European Central Bank (ECB) controls monetary policy for the 17 members of the Economic and Monetary Union (EMU); individual members of the EMU do not control the quantity of money circulating within their own borders
Stock of broad money: $24.03 billion (31 December 2010 est.)
country comparison to the world: 79
$25.65 billion (31 December 2009 est.)
Stock of domestic credit: $52.67 billion (31 December 2009 est.)
country comparison to the world: 63
$50.46 billion (31 December 2008 est.)
Market value of publicly traded shares: $11.77 billion (31 December 2009)
country comparison to the world: 60
$22.1 billion (31 December 2008)
$28.96 billion (31 December 2007)
Agriculture—products: potatoes, hops, wheat, sugar beets, corn, grapes; cattle, sheep, poultry
Industries: ferrous metallurgy and aluminum products, lead and zinc smelting; electronics (including military electronics), trucks, automobiles, electric power equipment, wood products, textiles, chemicals, machine tools
Industrial production growth rate: 1% (2010 est.)
country comparison to the world: 146
Electricity—production: 13 billion kWh (2009 est.)
country comparison to the world: 83
Electricity—consumption: 14.7 billion kWh (2009 est.)
country comparison to the world: 76
Electricity—exports: 7.82 billion kWh (2008 est.)
Electricity—imports: 6.218 billion kWh (2008 est.)
Oil—production: 5 bbl/day (2009 est.)
country comparison to the world: 114
Oil—consumption: 60,000 bbl/day (2009 est.)
country comparison to the world: 91
Oil—exports: 0 bbl/day (2009 est.)
country comparison to the world: 202
Oil—imports: 57,000 bbl/day (2009 est.)
country comparison to the world: 82
Oil—proved reserves: 0 bbl (1 January 2010 est.)
country comparison to the world: 186
Natural gas—production: 0 cu m (2009 est.)
country comparison to the world: 181
Natural gas—consumption: 1.05 billion cu m (2009 est.)
country comparison to the world: 88
Natural gas—exports: 0 cu m (2009 est.)
country comparison to the world: 173
Natural gas—imports: 1.05 billion cu m (2009 est.)

country comparison to the world: 54
Natural gas—proved reserves: 0 cu m (1 January 2010 est.)
country comparison to the world: 186
Current account balance: $-598 million (2010 est.)
country comparison to the world: 122
$-732.4 million (2009 est.)
Exports: $24.97 billion (2010 est.)
country comparison to the world: 65
$22.53 billion (2009 est.)
Exports—commodities: manufactured goods, machinery and transport equipment, chemicals, food
Exports—partners: Germany 19.36%, Italy 11.31%, Croatia 7.75%, Austria 7.42%, France 7.35% (2009)
Imports: $25.96 billion (2010 est.)
country comparison to the world: 62
$23.5 billion (2009 est.)
Imports—commodities: machinery and transport equipment, manufactured goods, chemicals, fuels and lubricants, food
Imports—partners: Germany 16.46%, Italy 15.89%, Austria 11.81%, France 4.98%, Croatia 4.32% (2009)
Reserves of foreign exchange and gold: $NA (31 December 2010 est.)
$1.08 billion (31 December 2009 est.)
Debt—external: $51.57 billion (30 June 2010)
country comparison to the world: 56
$54.61 billion (31 December 2008)
Stock of direct foreign investment—at home: $15.73 billion (31 December 2010 est.)
country comparison to the world: 75
$15.13 billion (31 December 2009 est.)
Stock of direct foreign investment—abroad: $9.001 billion (31 December 2010 est.)
country comparison to the world: 50
$7.901 billion (31 December 2009 est.)
Exchange rates: euros (EUR) per US dollar–0.755 (2010), 0.7198 (2009), 0.6827 (2008), 0.7345 (2007)

COMMUNICATIONS

Telephones—main lines in use: 1.034 million (2009)
country comparison to the world: 77
Telephones—mobile cellular: 2.1 million (2009)
country comparison to the world: 132
Telephone system: *general assessment:* well-developed telecommunications infrastructure
domestic: combined fixed-line and mobile-cellular teledensity roughly 150 telephones per 100 persons
international: country code–386
Broadcast media: public television broadcaster, Radiotelevizija Slovenija (RTV), operates a system of national and regional TV stations; 35 domestic commercial television stations operating nationally, regionally, and locally; about 60% of households are connected to multi-channel cable TV systems; public radio broadcaster operates 3 national and 4 regional stations; more than 75 regional and local commercial and non-commercial radio stations (2007)
Internet country code: .si
Internet hosts: 137,494 (2010)
country comparison to the world: 72
Internet users: 1.298 million (2009)
country comparison to the world: 92

TRANSPORTATION

Airports: 16 (2010)
country comparison to the world: 142
Airports—with paved runways: *total:* 7
over 3,047 m: 1
2,438 to 3,047 m: 1
1,524 to 2,437 m: 1
914 to 1,523 m: 3
under 914 m: 1 (2010)
Airports—with unpaved runways: *total:* 9
1,524 to 2,437 m: 1
914 to 1,523 m: 3
under 914 m: 5 (2010)
Pipelines: gas 840 km; oil 5 km (2010)
Railways: *total:* 1,228 km
country comparison to the world: 83
standard gauge: 1,228 km 1.435-m gauge (503 km electrified) (2007)
Roadways: *total:* 38,873 km
country comparison to the world: 89
paved: 38,873 km (includes 696 km of expressways) (2008)
Waterways: (there is some transport on the Drava River) (2010)
Merchant marine: registered in other countries: 25 (Antigua and Barbuda 1, Bahamas 1, Cyprus 4, Liberia 5, Malta 4, Marshall Islands 6, Saint Vincent and the Grenadines 2, Singapore 1, Slovakia 1) (2010)
country comparison to the world: 92
Ports and terminals: Koper

MILITARY

Military branches: Slovenian Army (includes air and naval forces)
Military service age and obligation: 18-25 years of age for voluntary military service; conscription abolished in 2003 (2010)
Manpower available for military service: *males age 16–49:* 477,592
females age 16–49: 464,301 (2010 est.)
Manpower fit for military service: *males age 16–49:* 392,075
females age 16–49: 380,077 (2010 est.)
Manpower reaching militarily significant age annually: *male:* 9,818
female: 9,395 (2010 est.)
Military expenditures: 1.7% of GDP (2005 est.)
country comparison to the world: 87

TRANSNATIONAL ISSUES

Disputes—international: the Croatia-Slovenia land and maritime boundary agreement, which would have ceded most of Piran Bay and maritime access to Slovenia and several villages to Croatia, remains unratified and in dispute; Slovenia also protests Croatia's 2003 claim to an exclusive economic zone in the Adriatic; as a member state that forms part of the EU's external border, Slovenia has implemented the strict Schengen border rules to curb illegal migration and commerce through southeastern Europe while encouraging close cross-border ties with Croatia
Illicit drugs: minor transit point for cocaine and Southwest Asian heroin bound for Western Europe, and for precursor chemicals

SOMALIA

INTRODUCTION

Background: Britain withdrew from British Somaliland in 1960 to allow its protectorate to join with Italian Somaliland and form the new nation of Somalia. In 1969, a coup headed by Mohamed SIAD Barre ushered in an authoritarian socialist rule characterized by the persecution, jailing and torture of political opponents and dissidents. After the regime's collapse early in 1991, Somalia descended into turmoil, factional fighting, and anarchy. In May 1991, northern clans declared an independent Republic of Somaliland that now includes the administrative regions of Awdal, Woqooyi Galbeed, Togdheer, Sanaag, and Sool. Although not recognized by any government, this entity has maintained a stable existence and continues efforts to establish a constitutional democracy, including holding municipal, parliamentary, and presidential elections. The regions of Bari, Nugaal, and northern Mudug comprise a neighboring semi-autonomous state of Puntland, which has been self-governing since 1998 but does not aim at independence; it has also made strides toward reconstructing a legitimate, representative government but has suffered some civil strife. Puntland disputes its border with Somaliland as it also claims portions of eastern Sool and Sanaag. Beginning in 1993, a two-year UN humanitarian effort (primarily in the south) was able to alleviate famine conditions, but when the UN withdrew in 1995, having suffered significant casualties, order still had not been restored. In 2000, the Somalia National Peace Conference (SNPC) held in Djibouti resulted in the formation of an interim government, known as the Transitional National Government (TNG). When the TNG failed to establish adequate security or governing institutions, the Government of Kenya, under the auspices of the Intergovernmental Authority on Development (IGAD), led a subsequent peace process that concluded in October 2004 with the election of Abdullahi YUSUF Ahmed as President of a second interim government, known as the Transitional Federal Government (TFG) of the Somali Republic. The TFG included a 275-member parliamentary body, known as the Transitional Federal Parliament (TFP). President YUSUF resigned late in 2008 while United Nations-sponsored talks between the TFG and the opposition Alliance for the Re-Liberation of Somalia (ARS) were underway in Djibouti. In January 2009, following the creation of a TFG-ARS unity government, Ethiopian military forces, which had entered Somalia in December

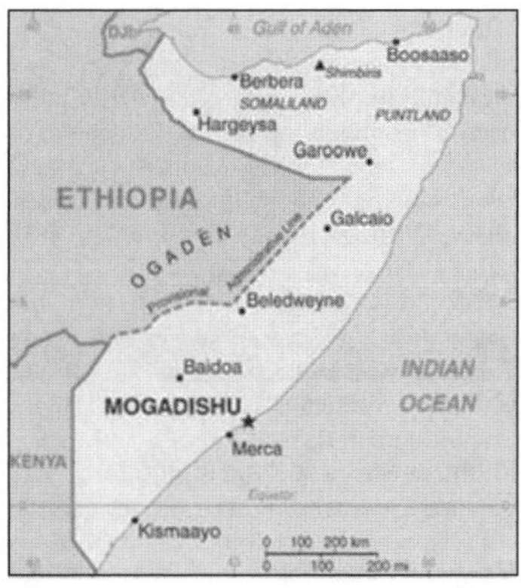

2006 to support the TFG in the face of advances by the opposition Islamic Courts Union (ICU), withdrew from the country. The TFP was increased to 550 seats with the addition of 200 ARS and 75 civil society members of parliament. The expanded parliament elected Sheikh SHARIF Sheikh Ahmed, the former CIC and ARS chairman as president on 31 January 2009, in Djibouti. Subsequently, President SHARIF appointed Omar Abdirashid ali SHARMARKE, son of a former president of Somalia, as prime minister on 13 February 2009. SHARMARKE resigned in September 2010 and was replaced by Mohamed Abdullahi MOHAMED, aka Farmajo, a dual US-Somali citizen who lived in the United States from 1985 until his return to Somalia in October 2010. The creation of the TFG was based on the Transitional Federal Charter (TFC), which outlines a five-year mandate leading to the establishment of a new Somali constitution and a transition to a representative government following national elections. However, in January 2009 the TFP amended the TFC to extend TFG's mandate until 2011.

GEOGRAPHY

Location: Eastern Africa, bordering the Gulf of Aden and the Indian Ocean, east of Ethiopia
Geographic coordinates: 10 00 N, 49 00 E
Map references: Africa
Area: *total:* 637,657 sq km
country comparison to the world: 43
land: 627,337 sq km
water: 10,320 sq km
Area—comparative: slightly smaller than Texas
Land boundaries:
total: 2,340 km
border countries: Djibouti 58 km, Ethiopia 1,600 km, Kenya 682 km
Coastline: 3,025 km
Maritime claims: territorial sea: 200 nm
Climate: principally desert; northeast monsoon (December to February), moderate temperatures in north and hot in south; southwest monsoon (May to October), torrid in the north and hot in the south, irregular rainfall, hot and humid periods (tangambili) between monsoons
Terrain: mostly flat to undulating plateau rising to hills in north
Elevation extremes: *lowest point:* Indian Ocean 0 m
highest point: Shimbiris 2,416 m
Natural resources: uranium and largely unexploited reserves of iron ore, tin, gypsum, bauxite, copper, salt, natural gas, likely oil reserves
Land use: *arable land:* 1.64%
permanent crops: 0.04%
other: 98.32% (2005)
Irrigated land: 2,000 sq km (2003)
Total renewable water resources: 15.7 cu km (1997)
Freshwater withdrawal (domestic/industrial/agricultural): *total:* 3.29 cu km/yr (0%/0%/100%)
per capita: 400 cu m/yr (2000)
Natural hazards: recurring droughts; frequent dust storms over eastern plains in summer; floods during rainy season
Environment—current issues: famine; use of contaminated water contributes to human health problems; deforestation; overgrazing; soil erosion; desertification
Environment—international agreements: *party to:* Biodiversity, Desertification, Endangered Species, Law of the Sea, Ozone Layer Protection
signed, but not ratified: none of the selected agreements
Geography—note: strategic location on Horn of Africa along southern approaches to Bab el Mandeb and route through Red Sea and Suez Canal

PEOPLE

Population: 9,925,640 (July 2011 est.)
country comparison to the world: 86
note: this estimate was derived from an official census taken in 1975 by the Somali Government; population counting in Somalia is complicated by the large number of nomads and by refugee movements in response to famine and clan warfare
Age structure: *0–14 years:* 44.7% (male 2,217,890/female 2,217,063)
15–64 years: 52.9% (male 2,663,729/female 2,588,716)
65 years and over: 2.4% (male 95,859/female 142,383) (2011 est.)
Median age: *total:* 17.8 years
male: 17.8 years
female: 17.7 years (2011 est.)
Population growth rate: 1.603% (2011 est.)
country comparison to the world: 71
Birth rate: 42.71 births/1,000 population (2011 est.)
country comparison to the world: 8
Death rate: 14.87 deaths/1,000 population (July 2011 est.)
country comparison to the world: 11
Net migration rate: -11.81 migrant(s)/1,000 population (2011 est.)
country comparison to the world: 213
Urbanization: *urban population:* 37% of total population (2010)
rate of urbanization: 4.1% annual rate of change (2010-15 est.)
Major cities—population: MOGADISHU (capital) 1.353 million (2009)
Sex ratio: *at birth:* 1.03 male(s)/female
under 15 years: 1 male(s)/female
15–64 years: 1 male(s)/female
65 years and over: 0.7 male(s)/female
total population: 1 male(s)/female (2011 est.)
Infant mortality rate: *total:* 105.56 deaths/1,000 live births
country comparison to the world: 5
male: 114.53 deaths/1,000 live births
female: 96.31 deaths/1,000 live births (2011 est.)
Life expectancy at birth: *total population:* 50.4 years
country comparison to the world: 213
male: 48.49 years
female: 52.37 years (2011 est.)
Total fertility rate: 6.35 children born/woman (2011 est.)
country comparison to the world: 4
HIV/AIDS—adult prevalence rate: 0.7% (2009 est.)
country comparison to the world: 59
HIV/AIDS—people living with HIV/AIDS: 34,000 (2009 est.)
country comparison to the world: 66
HIV/AIDS—deaths: 1,600 (2009 est.)
country comparison to the world: 58
Major infectious diseases: *degree of risk:* high
food or waterborne diseases: bacterial and protozoal diarrhea, hepatitis A and E, and typhoid fever
vectorborne diseases: dengue fever, malaria, and Rift Valley fever
water contact disease: schistosomiasis
animal contact disease: rabies (2009)
Drinking water source: *Improved:*
urban: 67% of population
rural: 9% of population
total: 30% of population
Unimproved:
urban: 33% of population
rural: 91% of population
total: 70% of population (2008)
Sanitation facility access: *Improved:*
urban: 52% of population
rural: 6% of population
total: 23% of population
Unimproved:
urban: 48% of population
rural: 94% of population
total: 77% of population (2008)
Nationality:
noun: Somali(s)
adjective: Somali
Ethnic groups: Somali 85%, Bantu and other non-Somali 15% (including 30,000 Arabs)
Religions: Sunni Muslim
Languages: Somali (official), Arabic, Italian, English
Literacy: *definition:* age 15 and over can read and write
total population: 37.8%
male: 49.7%
female: 25.8% (2001 est.)
School life expectancy (primary to tertiary education): *total:* 3 years
male: 3 years
female: 2 years (2007)
Education expenditures: NA

GOVERNMENT

Country name: *conventional long form:* none
conventional short form: Somalia
local long form: Jamhuuriyada Demuqraadiga Soomaaliyeed
local short form: Soomaaliya
former: Somali Republic, Somali Democratic Republic
Government type: no permanent national government; transitional, parliamentary federal government

Capital: *name:* Mogadishu
geographic coordinates: 2 04 N, 45 22 E
time difference: UTC+3 (8 hours ahead of Washington, DC during Standard Time)
Administrative divisions: 18 regions (plural–NA, singular–gobolka); Awdal, Bakool, Banaadir, Bari, Bay, Galguduud, Gedo, Hiiraan, Jubbada Dhexe (Middle Jubba), Jubbada Hoose (Lower Jubba), Mudug, Nugaal, Sanaag, Shabeellaha Dhexe (Middle Shabeelle), Shabeellaha Hoose (Lower Shabeelle), Sool, Togdheer, Woqooyi Galbeed
Independence: 1 July 1960 (from a merger of British Somaliland that became independent from the UK on 26 June 1960 and Italian Somaliland that became independent from the Italian-administered UN trusteeship on 1 July 1960 to form the Somali Republic)
National holiday: Foundation of the Somali Republic, 1 July (1960); note–26 June (1960) in Somaliland
Constitution: 25 August 1979, presidential approval 23 September 1979
note: the formation of transitional governing institutions, known as the Transitional Federal Government, is currently ongoing
Legal system: mixed legal system of civil law, Islamic law, and customary law (referred to as Xeer)
International law organization participation: accepts compulsory ICJ jurisdiction with reservations; non-party state to the ICCt
Suffrage: 18 years of age; universal
Executive branch: *chief of state:* Transitional Federal President Sheikh SHARIF Sheikh Ahmed (since 31 January 2009); note–a transitional governing entity with a five-year mandate, known as the Transitional Federal Institutions (TFIs), was established in October 2004; the TFIs relocated to Somalia in June 2004; in 2009, the TFIs were given a two-year extension to October 2011
head of government: Prime Minister Mohamed Abdullahi Mohamed FARMAJO (since 1 November 2010)
cabinet: Cabinet appointed by the prime minister and approved by the Transitional Federal Assembly
(For more information visit the World Leaders website)
election results: Sheikh SHARIF Sheikh Ahmed elected president by the expanded Transitional Federal Assembly in Djibouti
Legislative branch: unicameral National Assembly
note: unicameral Transitional Federal Assembly (TFA) (550 seats; 475 members appointed according to the 4.5 clan formula, with the remaining 75 seats reserved for civil society and business persons)
Judicial branch: following the breakdown of the central government, most regions have reverted to local forms of conflict resolution, either secular, traditional Somali customary law, or sharia (Islamic) law with a provision for appeal of all sentences
Political parties and leaders: none
Political pressure groups and leaders: *other:* numerous clan and sub-clan factions exist both in support and in opposition to the transitional government
International organization participation: ACP, AfDB, AFESD, AMF, AU, CAEU, FAO, G-77, IBRD, ICAO, ICRM, IDA, IDB, IFAD, IFC, IFRCS, IGAD, ILO, IMF, IMO, Interpol, IOC, IOM, ITSO, ITU, LAS, NAM, OIC, UN, UNCTAD, UNESCO, UNHCR, UNIDO, UPU, WFTU, WHO, WIPO, WMO
Diplomatic representation in the US: Somalia does not have an embassy in the US (ceased operations on 8 May 1991); note–the Transitional Federal Government is represented in the United States through its Permanent Mission to the United Nations
Diplomatic representation from the US: the US does not have an embassy in Somalia; US interests are represented by the US Embassy in Nairobi, Kenya at United Nations Avenue, Nairobi; *mailing address:* Unit 64100, Nairobi; APO AE 09831; *telephone:* [254] (20) 363-6000; FAX [254] (20) 363-6157
Flag description: light blue with a large white five-pointed star in the center; the blue field was originally influenced by the flag of the UN, but today is said to denote the sky and the neighboring Indian Ocean; the five points of the star represent the five regions in the horn of Africa that are inhabited by Somali people: the former British Somaliland and Italian Somaliland (which together make up Somalia), Djibouti, Ogaden (Ethiopia), and the Northern Frontier District (Kenya)
National anthem: *name:* "Soomaaliyeey toosoo" (Somalia Wake Up)
lyrics/music: Ali Mire AWALE and Yuusuf Xaaji Aadan Cilmi QABILLE
note: adopted 2000; written in 1947, the lyrics speak of creating unity and an end to fighting
Government—note: although an interim government was created in 2004, other regional and local governing bodies continue to exist and control various regions of the country, including the self-declared Republic of Somaliland in northwestern Somalia and the semi-autonomous State of Puntland in northeastern Somalia

ECONOMY

Economy—overview: Despite the lack of effective national governance, Somalia has maintained a healthy informal economy, largely based on livestock, remittance/money transfer companies, and telecommunications. Agriculture is the most important sector with livestock normally accounting for about 40% of GDP and more than 50% of export earnings. Nomads and semi-pastoralists, who are dependent upon livestock for their livelihood, make up a large portion of the population. Livestock, hides, fish, charcoal, and bananas are Somalia's principal exports, while sugar, sorghum, corn, qat, and machined goods are the principal imports. Somalia's small industrial sector, based on the processing of agricultural products, has largely been looted and the machinery sold as scrap metal. Somalia's service sector also has grown. Telecommunication firms provide wireless services in most major cities and offer the lowest international call rates on the continent. In the absence of a formal banking sector, money transfer/remittance services have sprouted throughout the country, handling up to $1.6 billion in remittances annually. Mogadishu's main market offers a variety of goods from food to the newest electronic gadgets. Hotels continue to operate and are supported with private-security militias. Due to armed attacks on and threats to humanitarian aid workers, the World Food Programme partially suspended its operations in southern Somalia in early January 2010 pending improvement in the security situation. Somalia's arrears to the IMF have continued to grow.
GDP (purchasing power parity): $5.896 billion (2010 est.)
country comparison to the world: 158
$5.75 billion (2009 est.)
$5.607 billion (2008 est.)
note: data are in 2010 US dollars
GDP (official exchange rate): $2.372 billion (2010 est.)
GDP—real growth rate: 2.6% (2010 est.)
country comparison to the world: 135
2.6% (2009 est.)
2.6% (2008 est.)
GDP—per capita (PPP): $600 (2010 est.)
country comparison to the world: 224
$600 (2009 est.)
$600 (2008 est.)
note: data are in 2010 US dollars
GDP—composition by sector:
agriculture: 65%
industry: 10%
services: 25% (2005 est.)
Labor force: 3.447 million (few skilled laborers) (2007)
country comparison to the world: 97
Labor force—by occupation:
agriculture: 71%
industry and *services:* 29% (1975)
Unemployment rate: NA%
Population below poverty line: NA%
Household income or consumption by percentage share:
lowest 10%: NA%
highest 10%: NA%
Budget: *revenues:* $NA
expenditures: $NA
Inflation rate (consumer prices): NA%
note: businesses print their own money, so inflation rates cannot be easily determined
Central bank discount rate: NA%
Commercial bank prime lending rate: NA%
Agriculture—products: bananas, sorghum, corn, coconuts, rice, sugarcane, mangoes, sesame seeds, beans; cattle, sheep, goats; fish
Industries: a few light industries, including sugar refining, textiles, wireless communication
Industrial production growth rate: NA%
Electricity—production: 280 million kWh (2007 est.)
country comparison to the world: 169
Electricity—consumption: 260.4 million kWh (2007 est.)
country comparison to the world: 172
Electricity—exports: 0 kWh (2008 est.)

Electricity—imports: 0 kWh (2008 est.)
Oil—production: 108.1 bbl/day (2009 est.)
country comparison to the world: 111
Oil—consumption: 5,000 bbl/day (2009 est.)
country comparison to the world: 163
Oil—exports: 1,475 bbl/day (2007 est.)
country comparison to the world: 119
Oil—imports: 6,387 bbl/day (2007 est.)
country comparison to the world: 150
Oil—proved reserves: 0 bbl (1 January 2010 est.)
country comparison to the world: 189
Natural gas—production: 0 cu m (2008 est.)
country comparison to the world: 184
Natural gas—consumption: 0 cu m (2008 est.)
country comparison to the world: 123
Natural gas—exports: 0 cu m (2008 est.)
country comparison to the world: 176
Natural gas—imports: 0 cu m (2008 est.)
country comparison to the world: 119
Natural gas—proved reserves: 5.663 billion cu m (1 January 2009 est.)
country comparison to the world: 89
Exports: $300 million (2006)
country comparison to the world: 174
Exports—commodities: livestock, bananas, hides, fish, charcoal, scrap metal
Exports—partners: UAE 58.27%, Yemen 20.32%, Saudi Arabia 3.78% (2009)
Imports: $798 million (2006)
country comparison to the world: 180
Imports—commodities: manufactures, petroleum products, foodstuffs, construction materials, qat
Imports—partners: Djibouti 30.84%, Kenya 8.06%, India 7.86%, China 6.97%, Brazil 6.59%, Yemen 4.97%, Oman 4.72%, UAE 4.6% (2009)
Debt—external: $3 billion (2001 est.)
country comparison to the world: 127
Exchange rates: Somali shillings (SOS) per US dollar–NA (2007-10)
1,438.3 (2006) official rate; the unofficial black market rate was about 23,000 shillings per dollar as of February 2007, the Republic of Somaliland, a self-declared independent country not recognized by any foreign government, issues its own currency, the Somaliland shilling

COMMUNICATIONS

Telephones—main lines in use: 100,000 (2009)
country comparison to the world: 145
Telephones—mobile cellular: 641,000 (2009)
country comparison to the world: 155
Telephone system: *general assessment:* the public telecommunications system was almost completely destroyed or dismantled during the civil war; private companies offer limited local fixed-line service and private wireless companies offer service in most major cities while charging the lowest international rates on the continent
domestic: local cellular telephone systems have been established in Mogadishu and in several other population centers
international: country code–252; international connections are available from Mogadishu by satellite
Broadcast media: 2 private TV stations rebroadcast Al-Jazeera and CNN; Somaliland has 1 government-operated TV station and Puntland has 1 private TV station; Radio Mogadishu operated by the transitional government; 1 SW and roughly 10 private FM radio stations broadcast in Mogadishu; several radio stations operate in central and southern regions; Somaliland has 1 government-operated radio station; Puntland has roughly a half dozen private radio stations; transmissions of at least 2 international broadcasters are available (2007)
Internet country code: .so
Internet hosts: 3 (2010)
country comparison to the world: 229
Internet users: 106,000 (2009)
country comparison to the world: 159

TRANSPORTATION

Airports: 59 (2010)
country comparison to the world: 80
Airports—with paved runways: *total:* 7
over 3,047 m: 4
2,438 to 3,047 m: 2
1,524 to 2,437 m: 1 (2010)
Airports—with unpaved runways: *total:* 52
2,438 to 3,047 m: 4
1,524 to 2,437 m: 19
914 to 1,523 m: 23
under 914 m: 6 (2010)
Roadways: *total:* 22,100 km
country comparison to the world: 106
paved: 2,608 km
unpaved: 19,492 km (2000)
Merchant marine: *total:* 1
country comparison to the world: 159
by type: cargo 1
foreign-owned: 1 (UAE 1) (2008)
Ports and terminals: Berbera, Kismaayo
Transportation—note: the International Maritime Bureau reports the territorial and offshore waters in the Gulf of Aden and Indian Ocean remain the region of greatest risk for piracy and armed robbery against ships accounting for 50% of all attacks in 2010; 217 vessels, including commercial shipping and pleasure craft, were attacked or hijacked both at anchor and while underway; hijackings off the coast of Somalia accounted for 92% of all ship seizures in 2010; as of May 2011, 26 vessels and 522 hostages were being held for ransom by Somali pirates; the presence of several naval task forces in the Gulf of Aden and additional anti-piracy measures on the part of ship operators have reduced piracy incidents in that body of water; in response Somali-based pirates, using hijacked fishing trawlers as "mother ships" to extend their range, shifted operations as far south as the Mozambique Channel and eastward to the vicinity of the Maldives

MILITARY

Military branches: National Security Force (NSF): Somali Army (2011)
Military service age and obligation: note: since 2005, the UN has listed the Transitional Federal Government and its allied militias as persistent violators in recruiting children (2010)
Manpower available for military service:
males age 16–49: 2,260,175
females age 16–49: 2,159,293 (2010 est.)
Manpower fit for military service: *males age 16–49:* 1,331,894
females age 16–49: 1,357,051 (2010 est.)
Manpower reaching militarily significant age annually: *male:* 101,634
female: 101,072 (2010 est.)
Military expenditures: 0.9% of GDP (2005 est.)
country comparison to the world: 138

TRANSNATIONAL ISSUES

Disputes—international: Ethiopian forces invaded southern Somalia and routed Islamist Courts from Mogadishu in January 2007; "Somaliland" secessionists provide port facilities in Berbera to landlocked Ethiopia and have established commercial ties with other regional states; "Puntland" and "Somaliland" "governments" seek international support in their secessionist aspirations and overlapping border claims; the undemarcated former British administrative line has little meaning as a political separation to rival clans within Ethiopia's Ogaden and southern Somalia's Oromo region; Kenya works hard to prevent the clan and militia fighting in Somalia from spreading south across the border, which has long been open to nomadic pastoralists
Refugees and internally displaced persons: *IDPs:* 1.1 million (civil war since 1988, clan-based competition for resources) (2007)

SOUTH AFRICA

INTRODUCTION

Background: Dutch traders landed at the southern tip of modern day South Africa in 1652 and established a stopover point on the spice route between the Netherlands and the Far East, founding the city of Cape Town. After the British seized the Cape of Good Hope area in 1806, many of the Dutch settlers (the Boers) trekked north to found their own republics. The discovery of diamonds (1867) and gold (1886) spurred wealth and immigration and intensified the subjugation of the native inhabitants. The Boers resisted British encroachments but were defeated in the Boer War (1899-1902); however, the British and the Afrikaners, as the Boers became known, ruled together beginning in 1910 under the Union of South Africa, which became a republic in 1961 after a whites-only referendum. In 1948, the National Party was voted into power and instituted a policy of apartheid–the separate development of the races–which favored the white minority at the expense of the black majority. The African National Congress (ANC) led the opposition to

apartheid and many top ANC leaders, such as Nelson MANDELA, spent decades in South Africa's prisons. Internal protests and insurgency, as well as boycotts by some Western nations and institutions, led to the regime's eventual willingness to negotiate a peaceful transition to majority rule. The first multi-racial elections in 1994 brought an end to apartheid and ushered in majority rule under an ANC-led government. South Africa since then has struggled to address apartheid-era imbalances in decent housing, education, and health care. ANC infighting, which has grown in recent years, came to a head in September 2008 when President Thabo MBEKI resigned, and Kgalema MOTLANTHE, the party's General-Secretary, succeeded him as interim president. Jacob ZUMA became president after the ANC won general elections in April 2009. In January 2011, South Africa assumed a nonpermanent seat on the UN Security Council for the 2011-12 term.

GEOGRAPHY

Location: Southern Africa, at the southern tip of the continent of Africa
Geographic coordinates: 29 00 S, 24 00 E
Map references: Africa
Area: *total:* 1,219,090 sq km
country comparison to the world: 25
land: 1,214,470 sq km
water: 4,620 sq km
note: includes Prince Edward Islands (Marion Island and Prince Edward Island)
Area—comparative: slightly less than twice the size of Texas
Land boundaries: *total:* 4,862 km
border countries: Botswana 1,840 km, Lesotho 909 km, Mozambique 491 km, Namibia 967 km, Swaziland 430 km, Zimbabwe 225 km
Coastline: 2,798 km
Maritime claims: territorial sea: 12 nm
contiguous zone: 24 nm
exclusive economic zone: 200 nm
continental shelf: 200 nm or to edge of the continental margin
Climate: mostly semiarid; subtropical along east coast; sunny days, cool nights
Terrain: vast interior plateau rimmed by rugged hills and narrow coastal plain
Elevation extremes: *lowest point:* Atlantic Ocean 0 m
highest point: Njesuthi 3,408 m
Natural resources: gold, chromium, antimony, coal, iron ore, manganese, nickel, phosphates, tin, rare earth elements, uranium, gem diamonds, platinum, copper, vanadium, salt, natural gas
Land use: *arable land:* 12.1%
permanent crops: 0.79%
other: 87.11% (2005)
Irrigated land: 14,980 sq km (2003)
Total renewable water resources: 50 cu km (1990)
Freshwater withdrawal (domestic/industrial/agricultural): *total:* 12.5 cu km/yr (31%/6%/63%)
per capita: 264 cu m/yr (2000)
Natural hazards: prolonged droughts
volcanism: the volcano forming Marion Island in the Prince Edward Islands, which last erupted in 2004, is South Africa's only active volcano
Environment—current issues: lack of important arterial rivers or lakes requires extensive water conservation and control measures; growth in water usage outpacing supply; pollution of rivers from agricultural runoff and urban discharge; air pollution resulting in acid rain; soil erosion; desertification
Environment—international agreements: *party to:* Antarctic-Environmental Protocol, Antarctic-Marine Living Resources, Antarctic Seals, Antarctic Treaty, Biodiversity, Climate Change, Climate Change-Kyoto Protocol, Desertification, Endangered Species, Hazardous Wastes, Law of the Sea, Marine Dumping, Marine Life Conservation, Ozone Layer Protection, Ship Pollution, Wetlands, Whaling
signed, but not ratified: none of the selected agreements
Geography—note: South Africa completely surrounds Lesotho and almost completely surrounds Swaziland

PEOPLE

Population: 49,004,031 (July 2011 est.)
country comparison to the world: 25
note: estimates for this country explicitly take into account the effects of excess mortality due to AIDS; this can result in lower life expectancy, higher infant mortality, higher death rates, lower population growth rates, and changes in the distribution of population by age and sex than would otherwise be expected
Age structure: *0–14 years:* 28.5% (male 6,998,726/female 6,959,542)
15–64 years: 65.8% (male 16,287,314/female 15,972,046)
65 years and over: 5.7% (male 1,125,709/female 1,660,694) (2011 est.)
Median age: *total:* 25 years
male: 24.7 years
female: 25.3 years (2011 est.)
Population growth rate: -0.38% (2011 est.)
country comparison to the world: 220
Birth rate: 19.48 births/1,000 population (2011 est.)
country comparison to the world: 92
Death rate: 17.09 deaths/1,000 population (July 2011 est.)
country comparison to the world: 3
Net migration rate: -6.19 migrant(s)/1,000 population
country comparison to the world: 200
note: there is an increasing flow of Zimbabweans into South Africa and Botswana in search of better economic opportunities (2011 est.)
Urbanization: *urban population:* 62% of total population (2010)
rate of urbanization: 1.2% annual rate of change (2010-15 est.)
Major cities—population: Johannesburg 3.607 million; Cape Town 3.353 million; Ekurhuleni (East Rand) 3.144 million; Durban 2.837 million; PRETORIA (capital) 1.404 million (2009)
Sex ratio: *at birth:* 1.02 male(s)/female
under 15 years: 1 male(s)/female
15–64 years: 1.02 male(s)/female
65 years and over: 0.68 male(s)/female
total population: 0.99 male(s)/female (2011 est.)
Infant mortality rate: *total:* 43.2 deaths/1,000 live births
country comparison to the world: 57
male: 47.19 deaths/1,000 live births
female: 39.14 deaths/1,000 live births (2011 est.)
Life expectancy at birth: *total population:* 49.33 years
country comparison to the world: 216
male: 50.24 years
female: 48.39 years (2011 est.)
Total fertility rate: 2.3 children born/woman (2011 est.)
country comparison to the world: 100
HIV/AIDS—adult prevalence rate: 17.8% (2009 est.)
country comparison to the world: 4
HIV/AIDS—people living with HIV/AIDS: 5.6 million (2009 est.)
country comparison to the world: 2
HIV/AIDS—deaths: 310,000 (2009 est.)
country comparison to the world: 1
Major infectious diseases: *degree of risk:* intermediate
food or waterborne diseases: bacterial diarrhea, hepatitis A, and typhoid fever
water contact disease: schistosomiasis (2009)
Drinking water source: *Improved:*
urban: 99% of population
rural: 78% of population
total: 91% of population
Unimproved:
urban: 1% of population
rural: 22% of population
total: 9% of population (2008)
Sanitation facility access: *Improved:*
urban: 84% of population
rural: 65% of population
total: 77% of population
Unimproved:
urban: 16% of population
rural: 35% of population
total: 23% of population (2008)
Nationality: *noun:* South African(s)
adjective: South African
Ethnic groups: black African 79%, white 9.6%, colored 8.9%, Indian/Asian 2.5% (2001 census)
Religions: Zion Christian 11.1%, Pentecostal/Charismatic 8.2%, Catholic 7.1%, Methodist 6.8%, Dutch Reformed 6.7%, Anglican 3.8%, Muslim 1.5%, other Christian 36%, other 2.3%, unspecified 1.4%, none 15.1% (2001 census)
Languages: IsiZulu (official) 23.8%, IsiXhosa (official) 17.6%, Afrikaans (official) 13.3%, Sepedi (offcial) 9.4%, English

(official) 8.2%, Setswana (official) 8.2%, Sesotho (official) 7.9%, Xitsonga (official) 4.4%, other 7.2%, isiNdebele (official), Tshivenda (official), siSwati (official) (2001 census)

Literacy: *definition:* age 15 and over can read and write
total population: 86.4%
male: 87%
female: 85.7% (2003 est.)

School life expectancy (primary to tertiary education): *total:* 13 years
male: 13 years
female: 13 years (2004)

Education expenditures: 5.4% of GDP (2009)
country comparison to the world: 45

GOVERNMENT

Country name: *conventional long form:* Republic of South Africa
conventional short form: South Africa
former: Union of South Africa
abbreviation: RSA

Government type: republic

Capital: *name:* Pretoria (administrative capital)
geographic coordinates: 25 42 S, 28 13 E
time difference: UTC+2 (7 hours ahead of Washington, DC during Standard Time)
note: Cape Town (legislative capital); Bloemfontein (judicial capital)

Administrative divisions: 9 provinces; Eastern Cape, Free State, Gauteng, KwaZulu-Natal, Limpopo, Mpumalanga, Northern Cape, North-West, Western Cape

Independence: 31 May 1910 (Union of South Africa formed from four British colonies: Cape Colony, Natal, Transvaal, and Orange Free State); 31 May 1961 (republic declared); 27 April 1994 (majority rule)

National holiday: Freedom Day, 27 April (1994)

Constitution: 10 December 1996; note—certified by the Constitutional Court 4 December 1996; was signed by then President MANDELA 10 December 1996; and entered into effect 4 February 1997

Legal system: mixed legal system of Roman-Dutch civil law, English common law, and customary law

International law organization participation: has not submitted an ICJ jurisdiction declaration; accepts ICCt jurisdiction

Suffrage: 18 years of age; universal

Executive branch: *chief of state:* President Jacob ZUMA (since 9 May 2009); Deputy President Kgalema MOTLANTHE (since 11 May 2009); note—the president is both the chief of state and head of government
head of government: President Jacob ZUMA (since 9 May 2009); Deputy President Kgalema MOTLANTHE (since 11 May 2009)
cabinet: Cabinet appointed by the president
(For more information visit the World Leaders website)
elections: president elected by the National Assembly for a five-year term (eligible for a second term); election last held on 6 May 2009 (next to be held in 2014)
election results: Jacob ZUMA elected president; National Assembly vote—Jacob ZUMA 277, Mvume DANDALA 47, other 76

Legislative branch: bicameral Parliament consisting of the National Council of Provinces (90 seats; 10 members elected by each of the nine provincial legislatures for five-year terms; has special powers to protect regional interests, including the safeguarding of cultural and linguistic traditions among ethnic minorities) and the National Assembly (400 seats; members elected by popular vote under a system of proportional representation to serve five-year terms)
elections: National Assembly and National Council of Provinces—last held on 22 April 2009 (next to be held in April 2014)
election results: National Council of Provinces—percent of vote by party—NA; seats by party—NA; National Assembly—percent of vote by party—ANC 65.9%, DA 16.7%, COPE 7.4%, IFP 4.6%, other 5.4%; seats by party—ANC 264, DA 67, COPE 30, IFP 18, other 21

Judicial branch: Constitutional Court; Supreme Court of Appeals; High Courts; Magistrate Courts

Political parties and leaders: African Christian Democratic Party or ACDP [Kenneth MESHOE]; African National Congress or ANC [Jacob ZUMA]; Congress of the People or COPE [Mosiuoa LEKOTA]; Democratic Alliance or DA [Helen ZILLE]; Freedom Front Plus or FF+ [Pieter MULDER]; Independent Democrats or ID [Patricia DE LILLE]; Inkatha Freedom Party or IFP [Mangosuthu BUTHELEZI]; Pan-Africanist Congress or PAC [Motsoko PHEKO]; United Christian Democratic Party or UCDP [Lucas MANGOPE]; United Democratic Movement or UDM [Bantu HOLOMISA]

Political pressure groups and leaders: Congress of South African Trade Unions or COSATU [Zwelinzima VAVI, general secretary]; South African Communist Party or SACP [Blade NZIMANDE, general secretary]; South African National Civics Organization or SANCO [Mlungisi HLONGWANE, national president]
note: note—COSATU and SACP are in a formal alliance with the ANC

International organization participation: ACP, AfDB, AU, BIS, C, CD, FAO, FATF, G-20, G-24, G-77, IAEA, IBRD, ICAO, ICC, ICRM, IDA, IFAD, IFC, IFRCS, IHO, ILO, IMF, IMO, IMSO, Interpol, IOC, IOM, IPU, ISO, ITSO, ITU, ITUC, MIGA, MONUSCO, NAM, NSG, OPCW, Paris Club (associate), PCA, SACU, SADC, UN, UN Security Council (temporary), UNAMID, UNCTAD, UNESCO, UNHCR, UNIDO, UNITAR, UNWTO, UPU, WCO, WFTU, WHO, WIPO, WMO, WTO, ZC

Diplomatic representation in the US: *chief of mission:* Ambassador Ebrahim RASOOL
chancery: 3051 Massachusetts Avenue NW, Washington, DC 20008
telephone: [1] (202) 232-4400
FAX: [1] (202) 265-1607
consulate(s) general: Chicago, Los Angeles, New York

Diplomatic representation from the US: *chief of mission:* Ambassador Donald H. GIPS
embassy: 877 Pretorius Street, Pretoria
mailing address: P. O. Box 9536, Pretoria 0001
telephone: [27] (12) 431-4000
FAX: [27] (12) 342-2299
consulate(s) general: Cape Town, Durban, Johannesburg

Flag description: two equal width horizontal bands of red (top) and blue separated by a central green band that splits into a horizontal Y, the arms of which end at the corners of the hoist side; the Y embraces a black isosceles triangle from which the arms are separated by narrow yellow bands; the red and blue bands are separated from the green band and its arms by narrow white stripes; the flag colors do not have any official symbolism, but the Y stands for the "convergence of diverse elements within South African society, taking the road ahead in unity"; black, yellow, and green are found on the flag of the African National Congress, while red, white, and blue are the colors in the flags of the Netherlands and the UK, whose settlers ruled South Africa during the colonial era
note: the South African flag is the only national flag to display six colors as part of its primary design

National anthem: *name:* "National Anthem of South Africa"
lyrics/music: Enoch SONTONGA and Cornelius Jacob LANGENHOVEN/ Enoch SONTONGA and Marthinus LOURENS de Villiers
note: adopted 1994; the anthem is a combination of "N'kosi Sikelel' iAfrica" (God Bless Africa) and "Die Stem van Suid Afrika" (The Call of South Africa), which were respectively the anthems of the non-white and white communities under apartheid; the official lyrics contain a mixture of Xhosa, Zulu, Sesotho, Afrikaans, and English; the music incorporates the melody used in the Tanzanian and Zambian anthems

ECONOMY

Economy—overview: South Africa is a middle-income, emerging market with an abundant supply of natural resources; well-developed financial, legal, communications, energy, and transport sectors; a stock exchange that is the 18th largest in the world; and modern infrastructure supporting a relatively efficient distribution of goods to major urban centers throughout the region. At the end of 2007, South Africa began to experience an electricity crisis. State power supplier Eskom encountered problems with aged plants, necessitating "load-shedding" cuts to residents and businesses in the major cities. Growth was robust from 2004 to 2007 as South Africa reaped the benefits of macroeconomic stability and a global commodities boom, but began to slow in the second half of 2007 due to the electricity crisis and the subsequent global financial crisis' impact on commodity prices and demand. GDP fell nearly 2% in 2009. Unemployment remains high and outdated infrastructure has constrained growth. Daunting economic problems remain from the apartheid era—especially poverty, lack of economic empowerment among the disadvantaged groups, and a shortage of public transportation. South

Africa's former economic policy was fiscally conservative, focusing on controlling inflation, and attaining a budget surplus. The current government largely follows the same prudent policies, but must contend with the impact of the global crisis and is facing growing pressure from special interest groups to use state-owned enterprises to deliver basic services to low-income areas and to increase job growth. More than a quarter of South Africa's population currently receives social grants.
GDP (purchasing power parity): $524 billion (2010 est.)
country comparison to the world: 26
$509.8 billion (2009 est.)
$518.5 billion (2008 est.)
note: data are in 2010 US dollars
GDP (official exchange rate): $357.3 billion (2010 est.)
GDP—real growth rate: 2.8% (2010 est.)
country comparison to the world: 128
-1.7% (2009 est.)
3.6% (2008 est.)
GDP—per capita (PPP): $10,700 (2010 est.)
country comparison to the world: 104
$10,400 (2009 est.)
$10,600 (2008 est.)
note: data are in 2010 US dollars
GDP—composition by sector:
agriculture: 3%
industry: 31.2%
services: 65.8% (2010 est.)
Labor force: 17.32 million economically active (2010 est.)
country comparison to the world: 34
Labor force—by occupation: *agriculture:* 9%
industry: 26%
services: 65% (2007 est.)
Unemployment rate: 23.3% (2010 est.)
country comparison to the world: 174
24% (2009 est.)
Population below poverty line: 50% (2000 est.)
Household income or consumption by percentage share: *lowest 10%:* 1.3%
highest 10%: 44.7% (2000)
Distribution of family income—Gini index: 65 (2005)
country comparison to the world: 2
59.3 (1994)
Investment (gross fixed): 19.9% of GDP (2010 est.)
country comparison to the world: 91
Budget: *revenues:* $103.1 billion
expenditures: $126.2 billion (2010 est.)
Public debt: 33.2% of GDP (2010 est.)
country comparison to the world: 88
29.7% of GDP (2009 est.)
Inflation rate (consumer prices): 4.5% (2010 est.)
country comparison to the world: 130
7.2% (2009 est.)
Central bank discount rate: 7% (31 December 2009)
country comparison to the world: 35
11.5% (31 December 2008)
Commercial bank prime lending rate: 11.71% (31 December 2009 est.)
country comparison to the world: 40
15.13% (31 December 2008 est.)
Stock of narrow money: $65.87 billion (31 December 2010 est.)
country comparison to the world: 41
$52.04 billion (31 December 2009 est.)
Stock of broad money: $256.2 billion (31 December 2010 est.)
country comparison to the world: 33
$199.8 billion (31 December 2009 est.)
Stock of domestic credit: $328.3 billion (31 December 2010 est.)
country comparison to the world: 32
$255.2 billion (31 December 2009 est.)
Market value of publicly traded shares: $704.8 billion (31 December 2009)
country comparison to the world: 19
$491.3 billion (31 December 2008)
$833.5 billion (31 December 2007)
Agriculture—products: corn, wheat, sugarcane, fruits, vegetables; beef, poultry, mutton, wool, dairy products
Industries: mining (world's largest producer of platinum, gold, chromium), automobile assembly, metalworking, machinery, textiles, iron and steel, chemicals, fertilizer, foodstuffs, commercial ship repair
Industrial production growth rate: 3% (2010 est.)
country comparison to the world: 107
Electricity—production: 240.3 billion kWh (2007 est.)
country comparison to the world: 16
Electricity—consumption: 215.1 billion kWh (2007 est.)
country comparison to the world: 17
Electricity—exports: 14.16 billion kWh (2008 est.)
Electricity—imports: 10.57 billion kWh (2008 est.)
Oil—production: 191,000 bbl/day (2009 est.)
country comparison to the world: 43
Oil—consumption: 579,000 bbl/day (2009 est.)
country comparison to the world: 29
Oil—exports: 128,500 bbl/day (2007 est.)
country comparison to the world: 63
Oil—imports: 490,500 bbl/day (2007 est.)
country comparison to the world: 25
Oil—proved reserves: 15 million bbl (1 January 2010 est.)
country comparison to the world: 87
Natural gas—production: 3.25 billion cu m (2008 est.)
country comparison to the world: 53
Natural gas—consumption: 6.45 billion cu m (2008 est.)
country comparison to the world: 55
Natural gas—exports: 0 cu m (2008 est.)
country comparison to the world: 170
Natural gas—imports: 3.2 billion cu m (2008 est.)
country comparison to the world: 39
Natural gas—proved reserves: 27.16 million cu m (1 January 2006 est.)
country comparison to the world: 102
Current account balance: $-16.51 billion (2010 est.)
country comparison to the world: 179
$-11.3 billion (2009 est.)
Exports: $76.86 billion (2010 est.)
country comparison to the world: 38
$66.54 billion (2009 est.)
Exports—commodities: gold, diamonds, platinum, other metals and minerals, machinery and equipment
Exports—partners: China 10.34%, US 9.19%, Japan 7.59%, Germany 7.01%, UK 5.54%, Switzerland 4.72% (2009)
Imports: $77.04 billion (2010 est.)
country comparison to the world: 36
$66.01 billion (2009 est.)
Imports—commodities: machinery and equipment, chemicals, petroleum products, scientific instruments, foodstuffs
Imports—partners: China 17.21%, Germany 11.24%, US 7.38%, Saudi Arabia 4.87%, Japan 4.67%, Iran 3.95% (2009)
Reserves of foreign exchange and gold: $45.52 billion (31 December 2010 est.)
country comparison to the world: 26
$39.68 billion (31 December 2009 est.)
Debt—external: $80.52 billion (30 June 2010 est.)
country comparison to the world: 44
$73.84 billion (30 June 2009)
Stock of direct foreign investment—at home: $83.08 billion (31 December 2010 est.)
country comparison to the world: 41
$73.61 billion (31 December 2009 est.)
Stock of direct foreign investment—abroad: $53.38 billion (31 December 2010 est.)
country comparison to the world: 32
$51.58 billion (31 December 2009 est.)
Exchange rates: rand (ZAR) per US dollar–7.38 (2010), 8.42 (2009), 7.9576 (2008), 7.05 (2007), 6.7649 (2006)

COMMUNICATIONS

Telephones—main lines in use: 4.32 million (2009)
country comparison to the world: 34
Telephones—mobile cellular: 46.436 million (2009)
country comparison to the world: 26
Telephone system: *general assessment:* the system is the best developed and most modern in Africa
domestic: combined fixed-line and mobile-cellular teledensity roughly 105 telephones per 100 persons; consists of carrier-equipped open-wire lines, coaxial cables, microwave radio relay links, fiber-optic cable, radiotelephone communication stations, and wireless local loops; key centers are Bloemfontein, Cape Town, Durban, Johannesburg, Port Elizabeth, and Pretoria
international: country code–27; the SAT-3/WASC and SAFE fiber optic cable systems connect South Africa to Europe and Asia; satellite earth stations–3 Intelsat (1 Indian Ocean and 2 Atlantic Ocean)
Broadcast media: the South African Broadcasting Corporation (SABC) operates 4 TV stations, 3 are free-to-air and 1 is pay TV; e.tv, a private station, is accessible to more than half the population; multiple subscription TV services provide a mix of local and international channels; well developed mix of public and private radio stations at the national, regional, and local levels; the SABC radio network, state-owned and controlled but nominally independent, operates 18 stations, one for each of the 11 official languages, 4 community stations, and 3 commercial stations; more than 100 community-based stations extend coverage to rural areas (2007)
Internet country code: .za
Internet hosts: 3.751 million (2010)
country comparison to the world: 24

Internet users:
4.42 million (2009)
country comparison to the world: 54

TRANSPORTATION

Airports: 578 (2010)
country comparison to the world: 11
Airports—with paved runways: *total:* 147
over 3,047 m: 11
2,438 to 3,047 m: 6
1,524 to 2,437 m: 53
914 to 1,523 m: 67
under 914 m: 10 (2010)
Airports—with unpaved runways:
total: 431
2,438 to 3,047 m: 1
1,524 to 2,437 m: 32
914 to 1,523 m: 261
under 914 m: 137 (2010)
Heliports: 1 (2010)
Pipelines: condensate 11 km; gas 908 km; oil 980 km; refined products 1,382 km (2010)
Railways: *total:* 20,872 km
country comparison to the world: 14
narrow gauge: 20,436 km 1.065-m gauge (8,271 km electrified); 436 km 0.610-m gauge (2008)
Roadways: *total:* 362,099 km
country comparison to the world: 18
paved: 73,506 km (includes 239 km of expressways)
unpaved: 288,593 km (2002)
Merchant marine: *total:* 4
country comparison to the world: 133
by type: container 1, petroleum tanker 3
foreign-owned: 1 (Denmark 1)
registered in other countries: 11 (Mexico 1, NZ 1, Seychelles 1, Singapore 3, UK 5) (2010)
Ports and terminals: Cape Town, Durban, Port Elizabeth, Richards Bay, Saldanha Bay

MILITARY

Military branches: South African National Defense Force (SANDF): South African Army, South African Navy (SAN), South African Air Force (SAAF), Joint Operations Command, Military Intelligence, South African Military Health Services (2009)
Military service age and obligation: 18 years of age for voluntary military service; women are eligible to serve in noncombat roles; 2-year service obligation (2007)
Manpower available for military service:
males age 16–49: 13,439,781
females age 16–49: 12,473,641 (2010 est.)
Manpower fit for military service: *males age 16–49:* 7,617,063
females age 16–49: 6,476,264 (2010 est.)
Manpower reaching militarily significant age annually: *male:* 482,122
female: 485,017 (2010 est.)
Military expenditures: 1.7% of GDP (2006)
country comparison to the world: 88
Military—note: with the end of apartheid and the establishment of majority rule, former military, black homelands forces, and ex-opposition forces were integrated into the South African National Defense Force (SANDF); as of 2003 the integration process was considered complete

TRANSNATIONAL ISSUES

Disputes—international: South Africa has placed military along the border to apprehend the thousands of Zimbabweans fleeing economic dysfunction and political persecution; as of January 2007, South Africa also supports large numbers of refugees and asylum seekers from the Democratic Republic of the Congo (33,000), Somalia (20,000), Burundi (6,500), and other states in Africa (26,000); managed dispute with Namibia over the location of the boundary in the Orange River; in 2006, Swazi king advocates resort to ICJ to claim parts of Mpumalanga and KwaZulu-Natal from South Africa
Refugees and internally displaced persons: refugees (country of origin): 10,772 (Democratic Republic of Congo); 7,818 (Somalia); 5,759 (Angola) (2007)
Illicit drugs: transshipment center for heroin, hashish, and cocaine, as well as a major cultivator of marijuana in its own right; cocaine and heroin consumption on the rise; world's largest market for illicit methaqualone, usually imported illegally from India through various east African countries, but increasingly producing its own synthetic drugs for domestic consumption; attractive venue for money launderers given the increasing level of organized criminal and narcotics activity in the region and the size of the South African economy

SOUTH GEORGIA AND SOUTH SANDWICH ISLANDS

(OVERSEAS TERRITORY OF THE UK, ALSO CLAIMED BY ARGENTINA)

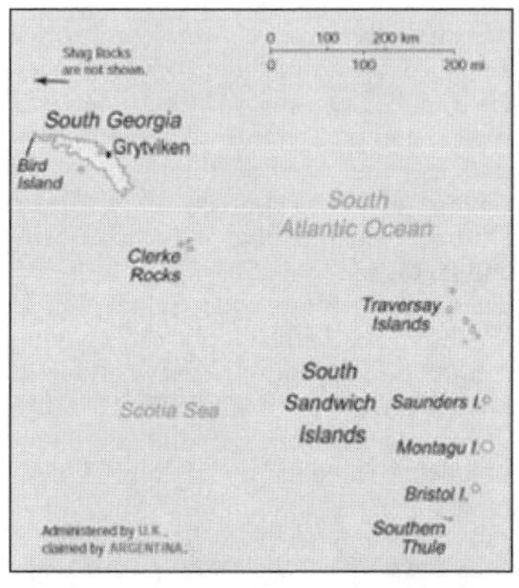

INTRODUCTION

Background: The islands, which have large bird and seal populations, lie approximately 1,000 km east of the Falkland Islands and have been under British administration since 1908—except for a brief period in 1982 when Argentina occupied them. Grytviken, on South Georgia, was a 19th and early 20th century whaling station. Famed explorer Ernest SHACKLETON stopped there in 1914 en route to his ill-fated attempt to cross Antarctica on foot. He returned some 20 months later with a few companions in a small boat and arranged a successful rescue for the rest of his crew, stranded off the Antarctic Peninsula. He died in 1922 on a subsequent expedition and is buried in Grytviken. Today, the station houses scientists from the British Antarctic Survey. Recognizing the importance of preserving the marine stocks in adjacent waters, the UK, in 1993, extended the exclusive fishing zone from 12 nm to 200 nm around each island.

GEOGRAPHY

Location: Southern South America, islands in the South Atlantic Ocean, east of the tip of South America
Geographic coordinates:
54 30 S, 37 00 W
Map references: South America
Area: *total:* 3,903 sq km
country comparison to the world: 176
land: 3,903 sq km
water: 0 sq km
note: includes Shag Rocks, Black Rock, Clerke Rocks, South Georgia Island, Bird Island, and the South Sandwich Islands, which consist of 11 islands
Area—comparative: slightly larger than Rhode Island
Land boundaries: 0 km
Coastline: NA
Maritime claims: territorial sea: 12 nm
exclusive fishing zone: 200 nm
Climate: variable, with mostly westerly winds throughout the year interspersed with periods of calm; nearly all precipitation falls as snow
Terrain: most of the islands, rising steeply from the sea, are rugged and mountainous; South Georgia is largely barren and has steep, glacier-covered mountains; the South Sandwich Islands are of volcanic origin with some active volcanoes
Elevation extremes: *lowest point:* Atlantic Ocean 0 m
highest point: Mount Paget (South Georgia) 2,934 m
Natural resources: fish
Land use: *arable land:* 0%
permanent crops: 0%
other: 100% (largely covered by permanent ice and snow with some sparse vegetation consisting of grass, moss, and lichen) (2005)
Irrigated land: 0 sq km
Natural hazards: the South Sandwich Islands have prevailing weather conditions that generally make them difficult to approach by ship; they are also subject to active volcanism
Environment—current issues: NA
Geography—note: the north coast of South Georgia has several large bays, which provide good anchorage; reindeer, introduced early in the 20th century, live on South Georgia

PEOPLE

Population: no indigenous inhabitants
note: the small military garrison on South Georgia withdrew in March 2001 replaced by a permanent group of scientists of the British Antarctic Survey, which also has a biological station on Bird Island; the South Sandwich Islands are uninhabited

GOVERNMENT

Country name: *conventional long form:* South Georgia and the South Sandwich Islands
conventional short form: South Georgia and South Sandwich Islands
abbreviation: SGSSI
Dependency status: overseas territory of the UK, also claimed by Argentina; administered from the Falkland Islands by a commissioner, who is concurrently governor of the Falkland Islands, representing Queen ELIZABETH II
Legal system: the laws of the UK where applicable apply; the senior magistrate from the Falkland Islands presides over the Magistrates Court
Diplomatic representation in the US: none (overseas territory of the UK, also claimed by Argentina)
Diplomatic representation from the US: none (overseas territory of the UK, also claimed by Argentina)
Flag description: blue, with the flag of the UK in the upper hoist-side quadrant and the South Georgia and South Sandwich Islands coat of arms centered on the outer half of the flag; the coat of arms features a shield with a golden lion rampant, holding a torch; the shield is supported by a fur seal on the left and a Macaroni penguin on the right; a reindeer appears above the crest, and below the shield on a scroll is the motto LEO TERRAM PROPRIAM PROTEGAT (Let the Lion Protect its Own Land)); the lion with the torch represents the UK and discovery; the background of the shield, blue and white estoiles, are found in the coat of arms of James Cook, discoverer of the islands; all the outer supporting animals represented are native to the islands

ECONOMY

Economy—overview: Some fishing takes place in adjacent waters. There is a potential source of income from harvesting finfish and krill. The islands receive income from postage stamps produced in the UK, sale of fishing licenses, and harbor and landing fees from tourist vessels. Tourism from specialized cruise ships is increasing rapidly.

COMMUNICATIONS

Telephone system: *general assessment:* NA
domestic: NA
international: coastal radiotelephone station at Grytviken
Internet country code: .gs
Internet hosts: 320 (2010)
country comparison to the world: 183

TRANSPORTATION

Ports and terminals: Grytviken

MILITARY

Military—note: defense is the responsibility of the UK

TRANSNATIONAL ISSUES

Disputes—international: Argentina, which claims the islands in its constitution and briefly occupied them by force in 1982, agreed in 1995 to no longer seek settlement by force

SOUTHERN OCEAN

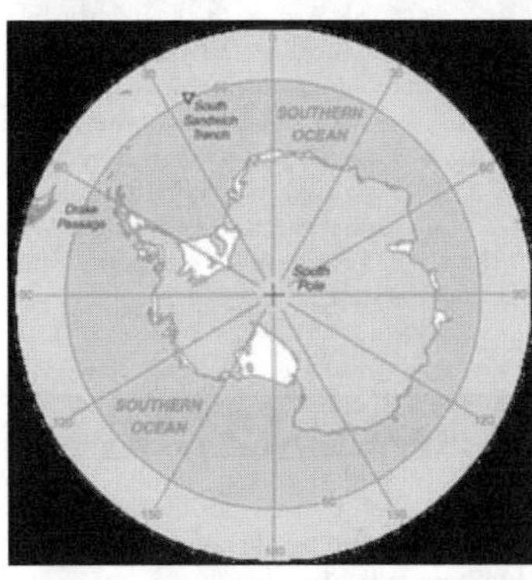

INTRODUCTION

Background: A large body of recent oceanographic research has shown that the Antarctic Circumpolar Current (ACC), an ocean current that flows from west to east around Antarctica, plays a crucial role in global ocean circulation. The region where the cold waters of the ACC meet and mingle with the warmer waters of the north defines a distinct border–the Antarctic Convergence–which fluctuates with the seasons, but which encompasses a discrete body of water and a unique ecologic region. The Convergence concentrates nutrients, which promotes marine plant life, and which in turn allows for a greater abundance of animal life. In the spring of 2000, the International Hydrographic Organization decided to delimit the waters within the Convergence as a fifth world ocean–the Southern Ocean–by combining the southern portions of the Atlantic Ocean, Indian Ocean, and Pacific Ocean. The Southern Ocean extends from the coast of Antarctica north to 60 degrees south latitude, which coincides with the Antarctic Treaty Limit and which approximates the extent of the Antarctic Convergence. As such, the Southern Ocean is now the fourth largest of the world's five oceans (after the Pacific Ocean, Atlantic Ocean, and Indian Ocean, but larger than the Arctic Ocean). It should be noted that inclusion of the Southern Ocean does not imply recognition of this feature as one of the world's primary oceans by the US Government.

GEOGRAPHY

Location: body of water between 60 degrees south latitude and Antarctica
Geographic coordinates: 60 00 S, 90 00 E (nominally), but the Southern Ocean has the unique distinction of being a large circumpolar body of water totally encircling the continent of Antarctica; this ring of water lies between 60 degrees south latitude and the coast of Antarctica and encompasses 360 degrees of longitude
Map references: Antarctic Region
Area: *total:* 20.327 million sq km
note: includes Amundsen Sea, Bellingshausen Sea, part of the Drake Passage, Ross Sea, a small part of the Scotia Sea, Weddell Sea, and other tributary water bodies
Area—comparative: slightly more than twice the size of the US
Coastline: 17,968 km
Climate: sea temperatures vary from about 10 degrees Celsius to -2 degrees Celsius; cyclonic storms travel eastward around the continent and frequently are intense because of the temperature contrast between ice and open ocean; the ocean area from about latitude 40 south to the Antarctic Circle has the strongest average winds found anywhere on Earth; in winter the ocean freezes outward to 65 degrees south latitude in the Pacific sector and 55 degrees south latitude in the Atlantic sector, lowering surface temperatures well below 0 degrees Celsius; at some coastal points intense persistent drainage winds from the interior keep the shoreline ice-free throughout the winter
Terrain: the Southern Ocean is deep, 4,000 to 5,000 m over most of its extent with only limited areas of shallow water; the Antarctic continental shelf is generally narrow and unusually deep, its edge lying at depths of 400 to 800 m (the global mean is 133 m); the Antarctic icepack grows from an average minimum of 2.6 million sq km in March to about 18.8 million sq km in September, better than a sixfold increase in area; the Antarctic Circumpolar Current (21,000 km in length) moves perpetually eastward; it is the world's largest ocean current, transporting 130 million cubic meters of water per second–100 times the flow of all the world's rivers
Elevation extremes: *lowest point:* -7,235 m at the southern end of the South Sandwich Trench
highest point: sea level 0 m
Natural resources: probable large and possible giant oil and gas fields on the continental margin; manganese nodules, possible placer deposits, sand and gravel, fresh water as icebergs; squid, whales, and seals–none exploited; krill, fish
Natural hazards: huge icebergs with drafts up to several hundred meters; smaller bergs and iceberg fragments; sea ice (generally 0.5 to 1 m thick) with sometimes dynamic short-term variations and with large annual and interannual variations; deep continental shelf floored by glacial deposits

varying widely over short distances; high winds and large waves much of the year; ship icing, especially May-October; most of region is remote from sources of search and rescue
Environment—current issues: increased solar ultraviolet radiation resulting from the Antarctic ozone hole in recent years, reducing marine primary productivity (phytoplankton) by as much as 15% and damaging the DNA of some fish; illegal, unreported, and unregulated fishing in recent years, especially the landing of an estimated five to six times more Patagonian toothfish than the regulated fishery, which is likely to affect the sustainability of the stock; large amount of incidental mortality of seabirds resulting from long-line fishing for toothfish
note: the now-protected fur seal population is making a strong comeback after severe overexploitation in the 18th and 19th centuries
Environment—international agreements: the Southern Ocean is subject to all international agreements regarding the world's oceans; in addition, it is subject to these agreements specific to the Antarctic region: International Whaling Commission (prohibits commercial whaling south of 40 degrees south [south of 60 degrees south between 50 degrees and 130 degrees west]); Convention on the Conservation of Antarctic Seals (limits sealing); Convention on the Conservation of Antarctic Marine Living Resources (regulates fishing)
note: many nations (including the US) prohibit mineral resource exploration and exploitation south of the fluctuating Polar Front (Antarctic Convergence), which is in the middle of the Antarctic Circumpolar Current and serves as the dividing line between the cold polar surface waters to the south and the warmer waters to the north
Geography—note: the major chokepoint is the Drake Passage between South America and Antarctica; the Polar Front (Antarctic Convergence) is the best natural definition of the northern extent of the Southern Ocean; it is a distinct region at the middle of the Antarctic Circumpolar Current that separates the cold polar surface waters to the south from the warmer waters to the north; the Front and the Current extend entirely around Antarctica, reaching south of 60 degrees south near New Zealand and near 48 degrees south in the far South Atlantic coinciding with the path of the maximum westerly winds

ECONOMY

Economy—overview: Fisheries in 2006-07 landed 126,976 metric tons, of which 82% (104,586 tons) was krill (Euphausia superba) and 9.5% (12,027 tons) Patagonian toothfish (Dissostichus eleginoides—also known as Chilean sea bass), compared to 127,910 tons in 2005-06 of which 83% (106,591 tons) was krill and 9.7% (12,396 tons) Patagonian toothfish (estimated fishing from the area covered by the Convention of the Conservation of Antarctic Marine Living Resources (CCAMLR), which extends slightly beyond the Southern Ocean area). International agreements were adopted in late 1999 to reduce illegal, unreported, and unregulated fishing, which in the 2000-01 season landed, by one estimate, 8,376 metric tons of Patagonian and Antarctic toothfish. In the 2007-08 Antarctic summer, 45,213 tourists visited the Southern Ocean, compared to 35,552 in 2006-07, and 29,799 in 2005-06 (estimates provided to the Antarctic Treaty by the International Association of Antarctica Tour Operators (IAATO), and does not include passengers on overflights and those flying directly in and out of Antarctica).

TRANSPORTATION

Ports and terminals: McMurdo, Palmer, and offshore anchorages in Antarctica
note: few ports or harbors exist on the southern side of the Southern Ocean; ice conditions limit use of most to short periods in midsummer; even then some cannot be entered without icebreaker escort; most Antarctic ports are operated by government research stations and, except in an emergency, are not open to commercial or private vessels
Transportation—note: Drake Passage offers alternative to transit through the Panama Canal

TRANSNATIONAL ISSUES

Disputes—international: Antarctic Treaty defers claims (see Antarctica entry), but Argentina, Australia, Chile, France, NZ, Norway, and UK assert claims (some overlapping), including the continental shelf in the Southern Ocean; several states have expressed an interest in extending those continental shelf claims under the United Nations Convention on the Law of the Sea (UNCLOS) to include undersea ridges; the US and most other states do not recognize the land or maritime claims of other states and have made no claims themselves (the US and Russia have reserved the right to do so); no formal claims exist in the waters in the sector between 90 degrees west and 150 degrees west

SPAIN

INTRODUCTION

Background: Spain's powerful world empire of the 16th and 17th centuries ultimately yielded command of the seas to England. Subsequent failure to embrace the mercantile and industrial revolutions caused the country to fall behind Britain, France, and Germany in economic and political power. Spain remained neutral in World Wars I and II but suffered through a devastating civil war (1936-39). A peaceful transition to democracy following the death of dictator Francisco FRANCO in 1975, and rapid economic modernization (Spain joined the EU in 1986) gave Spain a dynamic and rapidly growing economy and made it a global champion of freedom and human rights. The government continues to battle the Basque Fatherland and Liberty (ETA) terrorist organization, but its major focus for the immediate future will be on measures to reverse the severe economic recession that started in mid-2008.

GEOGRAPHY

Location: Southwestern Europe, bordering the Mediterranean Sea, North Atlantic Ocean, Bay of Biscay, and Pyrenees Mountains; southwest of France
Geographic coordinates: 40 00 N, 4 00 W
Map references: Europe
Area: *total:* 505,370 sq km
country comparison to the world: 51
land: 498,980 sq km
water: 6,390 sq km
note: there are two autonomous cities—Ceuta and Melilla—and 17 autonomous communities including Balearic Islands and Canary Islands, and three small Spanish possessions off the coast of Morocco—Islas Chafarinas, Penon de Alhucemas, and Penon de Velez de la Gomera
Area—comparative: slightly more than twice the size of Oregon
Land boundaries: *total:* 1,917.8 km
border countries: Andorra 63.7 km, France 623 km, Gibraltar 1.2 km, Portugal 1,214 km, Morocco (Ceuta) 6.3 km, Morocco (Melilla) 9.6 km
Coastline: 4,964 km
Maritime claims: territorial sea: 12 nm
contiguous zone: 24 nm
exclusive economic zone: 200 nm (applies only to the Atlantic Ocean)
Climate: temperate; clear, hot summers in interior, more moderate and cloudy along coast; cloudy, cold winters in interior, partly cloudy and cool along coast
Terrain: large, flat to dissected plateau surrounded by rugged hills; Pyrenees Mountains in north
Elevation extremes: *lowest point:* Atlantic Ocean 0 m
highest point: Pico de Teide (Tenerife) on Canary Islands 3,718 m
Natural resources: coal, lignite, iron ore, copper, lead, zinc, uranium, tungsten, mercury, pyrites, magnesite, fluorspar,

gypsum, sepiolite, kaolin, potash, hydropower, arable land
Land use: *arable land:* 27.18%
permanent crops: 9.85%
other: 62.97% (2005)
Irrigated land: 37,800 sq km (2003)
Total renewable water resources: 111.1 cu km (2005)
Freshwater withdrawal (domestic/industrial/agricultural): *total:* 37.22 cu km/yr (13%/19%/68%)
per capita: 864 cu m/yr (2002)
Natural hazards: periodic droughts, occasional flooding
volcanism: Spain experiences volcanic activity in the Canary Islands, located off Africa's northwest coast; Teide (elev. 3,715 m) has been deemed a "Decade Volcano" by the International Association of Volcanology and Chemistry of the Earth's Interior, worthy of study due to its explosive history and close proximity to human populations; La Palma (elev. 2,426 m), which last erupted in 1971, is the most active of the Canary Islands volcanoes; Lanzarote is the only other historically active volcano
Environment—current issues: pollution of the Mediterranean Sea from raw sewage and effluents from the offshore production of oil and gas; water quality and quantity nationwide; air pollution; deforestation; desertification
Environment—international agreements: *party to:* Air Pollution, Air Pollution-Nitrogen Oxides, Air Pollution-Sulfur 94, Air Pollution-Volatile Organic Compounds, Antarctic-Environmental Protocol, Antarctic-Marine Living Resources, Antarctic Treaty, Biodiversity, Climate Change, Climate Change-Kyoto Protocol, Desertification, Endangered Species, Environmental Modification, Hazardous Wastes, Law of the Sea, Marine Dumping, Marine Life Conservation, Ozone Layer Protection, Ship Pollution, Tropical Timber 83, Tropical Timber 94, Wetlands, Whaling
signed, but not ratified: Air Pollution-Persistent Organic Pollutants
Geography—note: strategic location along approaches to Strait of Gibraltar; Spain controls a number of territories in northern Morocco including the enclaves of Ceuta and Melilla, and the islands of Penon de Velez de la Gomera, Penon de Alhucemas, and Islas Chafarinas

PEOPLE

Population: 46,754,784 (July 2011 est.)
country comparison to the world: 27
Age structure: *0–14 years:* 15.1% (male 3,646,614/female 3,435,311)
15–64 years: 67.7% (male 16,036,556/female 15,637,090)
65 years and over: 17.1% (male 3,389,681/female 4,609,532) (2011 est.)
Median age: *total:* 40.5 years
male: 39.3 years
female: 41.9 years (2011 est.)
Population growth rate: 0.574% (2011 est.)
country comparison to the world: 146
Birth rate: 10.66 births/1,000 population (2011 est.)
country comparison to the world: 180
Death rate: 8.8 deaths/1,000 population (July 2011 est.)
country comparison to the world: 78
Net migration rate: 3.89 migrant(s)/1,000 population (2011 est.)
country comparison to the world: 25
Urbanization: *urban population:* 77% of total population (2010)
rate of urbanization: 1% annual rate of change (2010-15 est.)
Major cities—population: MADRID (capital) 5.762 million; Barcelona 5.029 million; Valencia 812,000 (2009)
Sex ratio: *at birth:* 1.065 male(s)/female
under 15 years: 1.06 male(s)/female
15–64 years: 1.01 male(s)/female
65 years and over: 0.72 male(s)/female
total population: 0.96 male(s)/female (2011 est.)
Infant mortality rate: *total:* 3.39 deaths/1,000 live births
country comparison to the world: 212
male: 3.74 deaths/1,000 live births
female: 3.03 deaths/1,000 live births (2011 est.)
Life expectancy at birth: *total population:* 81.17 years
country comparison to the world: 14
male: 78.16 years
female: 84.37 years (2011 est.)
Total fertility rate: 1.47 children born/woman (2011 est.)
country comparison to the world: 188
HIV/AIDS—adult prevalence rate: 0.4% (2009 est.)
country comparison to the world: 76
HIV/AIDS—people living with HIV/AIDS: 130,000 (2009 est.)
country comparison to the world: 36
HIV/AIDS—deaths: 1,600 (2009 est.)
country comparison to the world: 59
Drinking water source: *Improved:*
urban: 100% of population
rural: 100% of population
total: 100% of population (2008)
Sanitation facility access: *Improved:*
urban: 100% of population
rural: 100% of population
total: 100% of population (2008)
Nationality: *noun:* Spaniard(s)
adjective: Spanish
Ethnic groups: composite of Mediterranean and Nordic types
Religions: Roman Catholic 94%, other 6%
Languages: Castilian Spanish (official) 74%, Catalan 17%, Galician 7%, and Basque 2% (official regionally)
Literacy: *definition:* age 15 and over can read and write
total population: 97.9%
male: 98.7%
female: 97.2% (2003 est.)
School life expectancy (primary to tertiary education): *total:* 16 years
male: 16 years
female: 17 years (2008)
Education expenditures: 4.3% of GDP (2007)
country comparison to the world: 95

GOVERNMENT

Country name: *conventional long form:* Kingdom of Spain
conventional short form: Spain
local long form: Reino de Espana
local short form: Espana
Government type: parliamentary monarchy
Capital: *name:* Madrid
geographic coordinates: 40 24 N, 3 41 W
time difference: UTC+1 (6 hours ahead of Washington, DC during Standard Time)
daylight saving time: +1hr, begins last Sunday in March; ends last Sunday in October
note: Spain is divided into two time zones including the Canary Islands
Administrative divisions: 17 autonomous communities (comunidades autonomas, singular—comunidad autonoma) and 2 autonomous cities* (ciudades autonomas, singular—ciudad autonoma); Andalucia, Aragon, Asturias, Baleares (Balearic Islands), Ceuta*, Canarias (Canary Islands), Cantabria, Castilla-La Mancha, Castilla y Leon, Cataluna (Catalonia), Comunidad Valenciana (Valencian Community), Extremadura, Galicia, La Rioja, Madrid, Melilla*, Murcia, Navarra, Pais Vasco (Basque Country)
note: the autonomous cities of Ceuta and Melilla plus three small islands of Islas Chafarinas, Penon de Alhucemas, and Penon de Velez de la Gomera, administered directly by the Spanish central government, are all along the coast of Morocco and are collectively referred to as Places of Sovereignty (Plazas de Soberania)
Independence: 1492; the Iberian peninsula was characterized by a variety of independent kingdoms prior to the Muslim occupation that began in the early 8th century A.D. and lasted nearly seven centuries; the small Christian redoubts of the north began the reconquest almost immediately, culminating in the seizure of Granada in 1492; this event completed the unification of several kingdoms and is traditionally considered the forging of present-day Spain
National holiday: National Day, 12 October (1492); year when Columbus first set foot in the Americas
Constitution: approved by legislature 31 October 1978; passed by referendum 6 December 1978; signed by the king 27 December 1978
Legal system: civil law system with regional variations
International law organization participation: accepts compulsory ICJ jurisdiction with reservations; accepts ICCt jurisdiction
Suffrage: 18 years of age; universal
Executive branch: *chief of state:* King JUAN CARLOS I (since 22 November 1975); Heir Apparent Prince FELIPE, son of the monarch, born 30 January 1968
head of government: President of the Government (Prime Minister equivalent) Jose Luis Rodriguez ZAPATERO (since 17 April 2004); First Vice President (and Minister of the Interior) Alfredo Perez RUBALCABA (since 20 October 2010), Second Vice President (and Minister of Economy and Finance) Elena SALGADO Mendez (since 8 April 2009), and Third Vice President (and Minister of Regional Affairs) Manuel CHAVES Gonzalez (since 8 April 2009)
cabinet: Council of Ministers designated by the president

(For more information visit the World Leaders website)
note: there is also a Council of State that is the supreme consultative organ of the government, but its recommendations are non-binding
elections: the monarchy is hereditary; following legislative elections, the leader of the majority party or the leader of the majority coalition usually proposed president by the monarch and elected by the National Assembly; election last held on 9 and 11 April 2008 (next to be held in March 2012); vice presidents appointed by the monarch on the proposal of the president
election results: Jose Luis Rodriguez ZAPATERO reelected President of the Government; percent of National Assembly vote—46.9%
Legislative branch: bicameral; General Courts or Las Cortes Generales (National Assembly) consists of the Senate or Senado (264 seats as of 2008; 208 members directly elected by popular vote and the other 56—as of 2008—appointed by the regional legislatures; members to serve four-year terms) and the Congress of Deputies or Congreso de los Diputados (350 seats; each of the 50 electoral provinces fills a minimum of two seats and the North African enclaves of Ceuta and Melilla fill one seat each with members serving a four-year term; the other 248 members are determined by proportional representation based on popular vote on block lists who serve four-year terms)
elections: Senate—last held on 9 March 2008 (next to be held by March 2012); Congress of Deputies—last held on 9 March 2008 (next to be held by March 2012)
election results: Senate—percent of vote by party—NA; seats by party—PP 101, PSOE 88, Entesa Catalona de Progress 12, CiU 4, PNV 2, CC 1, members appointed by regional legislatures 56; Congress of Deputies—percent of vote by party—PSOE 43.6%, PP 40.1%, CiU 3.1%, PNV 1.2%, ERC 1.2%, other 10.8%; seats by party—PSOE 169, PP 154, CiU 10, PNV 6, ERC 3, other 8; note—seats by party in the Congress of Deputies as of 15 December 2009—PSOE 169, PP 153, CiU 10, PNV 6, ERC 3, other 9
Judicial branch: Supreme Court or Tribunal Supremo
Political parties and leaders: Basque Nationalist Party or PNV or EAJ [Inigo URKULLU Renteria]; Canarian Coalition or CC [Claudina MORALES Rodriquez] (a coalition of five parties); Convergence and Union or CiU [Artur MAS i Gavarro] (a coalition of the Democratic Convergence of Catalonia or CDC [Artur MAS i Gavarro] and the Democratic Union of Catalonia or UDC [Josep Antoni DURAN i LLEIDA]); Entesa Catalonia de Progress (a Senate coalition grouping four Catalan parties—PSC, ERC, ICV, EUA); Galician Nationalist Bloc or BNG [Guillerme VAZQUEZ Vazquez]; Initiative for Catalonia Greens or ICV [Joan SAURA i Laporta]; Navarra Yes or NaBai [collective leadership] (a coalition of four Navarran parties); Popular Party or PP [Mariano RAJOY Brey]; Republican Left of Catalonia or ERC [Joan PUIGCERCOS i Boixassa]; Spanish Socialist Workers Party or PSOE [Jose Luis Rodriguez ZAPATERO]; Union of People of Navarra or UPN [Yolanda BARCINA Angulo]; Union, Progress and Democracy or UPyD [Rosa DIEZ Gonzalez]; United Left or IU [Cayo LARA Moya] (a coalition of parties including the Communist Party of Spain or PCE and other small parties)
Political pressure groups and leaders: Association for Victims of Terrorism or AVT (grassroots organization devoted primarily to opposing ETA terrorist attacks and supporting its victims); Basta Ya (Spanish for "Enough is Enough"); grassroots organization devoted primarily to opposing ETA terrorist attacks and supporting its victims); Nunca Mais (Galician for "Never Again"; formed in response to the oil Tanker Prestige oil spill); Socialist General Union of Workers or UGT and the smaller independent Workers Syndical Union or USO; Trade Union Confederation of Workers' Commissions or CC.OO.
other: business and landowning interests; Catholic Church; free labor unions (authorized in April 1977); university students
International organization participation: ADB (nonregional member), AfDB (nonregional member), Arctic Council (observer), Australia Group, BCIE, BIS, CBSS (observer), CE, CERN, EAPC, EBRD, EIB, EMU, ESA, EU, FAO, FATF, IADB, IAEA, IBRD, ICAO, ICC, ICRM, IDA, IEA, IFAD, IFC, IFRCS, IHO, ILO, IMF, IMO, IMSO, Interpol, IOC, IOM, IPU, ISO, ITSO, ITU, ITUC, LAIA (observer), MIGA, MONUSCO, NATO, NEA, NSG, OAS (observer), OECD, OPCW, OSCE, Paris Club, PCA, Schengen Convention, SECI (observer), SICA (observer), UN, UNCTAD, UNESCO, UNHCR, UNIDO, UNIFIL, Union Latina, UNMIS, UNRWA, UNWTO, UPU, WCO, WHO, WIPO, WMO, WTO, ZC
Diplomatic representation in the US: *chief of mission:* Ambassador Jorge DEZCALLAR de Mazarredo
chancery: 2375 Pennsylvania Avenue NW, Washington, DC 20037
telephone: [1] (202) 452-0100, 728-2340
FAX: [1] (202) 833-5670
consulate(s) general: Boston, Chicago, Houston, Los Angeles, Miami, New Orleans, New York, San Francisco, San Juan (Puerto Rico)
Diplomatic representation from the US: *chief of mission:* Ambassador Alan D. SOLOMONT
embassy: Serrano 75, 28006 Madrid
mailing address: PSC 61, APO AE 09642
telephone: [34] (91) 587-2200
FAX: [34] (91) 587-2303
consulate(s) general: Barcelona
Flag description: three horizontal bands of red (top), yellow (double width), and red with the national coat of arms on the hoist side of the yellow band; the coat of arms is quartered to display the emblems of the traditional kingdoms of Spain (clockwise from upper left, Castile, Leon, Navarre, and Aragon) while Granada is represented by the stylized pomegranate at the bottom of the shield; the arms are framed by two columns representing the Pillars of Hercules, which are the two promontories (Gibraltar and Ceuta) on either side of the eastern end of the Strait of Gibraltar; the red scroll across the two columns bears the imperial motto of "Plus Ultra" (further beyond) referring to Spanish lands beyond Europe; the triband arrangement with the center stripe twice the width of the outer dates to the 18th century
note: the red and yellow colors are related to those of the oldest Spanish kingdoms: Aragon, Castile, Leon, and Navarre
National anthem: *name:* "Himno Nacional Espanol" (National Anthem of Spain)
lyrics/music: none/unknown
note: officially in use between 1770 and 1931, restored in 1939; the Spanish anthem has no lyrics; in the years prior to 1931 it became known as "Marcha Real" (The Royal March); it first appeared in a 1761 military bugle call book and was replaced by "Himno de Riego" in the years between 1931 and 1939; the long version of the anthem is used for the king, while the short version is used for the prince, prime minister, and occasions such as sporting events

ECONOMY

Economy—overview: Spain's mixed capitalist economy is the 12th largest in the world, and its per capita income roughly matches that of Germany and France. However, after almost 15 years of above average GDP growth, the Spanish economy began to slow in late 2007 and entered into a recession in the second quarter of 2008. GDP contracted by 3.7% in 2009, ending a 16-year growth trend, and by another 0.2% in 2010, making Spain the last major economy to emerge from the global recession. The reversal in Spain's economic growth reflected a significant decline in construction amid an oversupply of housing and falling consumer spending, while exports actually have begun to grow. Government efforts to boost the economy through stimulus spending, extended unemployment benefits, and loan guarantees did not prevent a sharp rise in the unemployment rate, which rose from a low of about 8% in 2007 to 20% in 2010. The government budget deficit worsened from 3.8% of GDP in 2008 to about 9.7% of GDP in 2010, more than three times the euro-zone limit. Spain's large budget deficit and poor economic growth prospects have made it vulnerable to financial contagion from other highly-indebted euro zone members despite the government's efforts to cut spending, privatize industries, and boost competitiveness through labor market reforms. Spanish banks' high exposure to the collapsed domestic construction and real estate market also poses a continued risk for the sector. The government oversaw a restructuring of the savings bank sector in 2010, and provided some $15 billion in capital to various institutions. Investors remain concerned that Madrid may need to bail out more trou-

bled banks. The Bank of Spain, however, is seeking to boost confidence in the financial sector by pressuring banks to come clean about their losses and consolidate into stronger groups.
GDP (purchasing power parity): $1.369 trillion (2010 est.)
country comparison to the world: 14
$1.371 trillion (2009 est.)
$1.424 trillion (2008 est.)
note: data are in 2010 US dollars
GDP (official exchange rate): $1.41 trillion (2010 est.)
GDP—real growth rate: -0.1% (2010 est.)
country comparison to the world: 191
-3.7% (2009 est.)
0.9% (2008 est.)
GDP—per capita (PPP): $29,400 (2010 est.)
country comparison to the world: 48
$29,600 (2009 est.)
$31,000 (2008 est.)
note: data are in 2010 US dollars
GDP—composition by sector:
agriculture: 2.9%
industry: 25.5%
services: 71.6% (2010 est.)
Labor force: 22.96 million (2010 est.)
country comparison to the world: 27
Labor force—by occupation:
agriculture: 4.2%
industry: 24%
services: 71.7% (2009 est.)
Unemployment rate: 20% (2010 est.)
country comparison to the world: 164
18.1% (2009 est.)
Population below poverty line: 19.8% (2005)
Household income or consumption by percentage share: *lowest 10%:* 2.6%
highest 10%: 26.6% (2000)
Distribution of family income—Gini index: 32 (2005)
country comparison to the world: 103
32.5 (1990)
Investment (gross fixed): 22.9% of GDP (2010 est.)
country comparison to the world: 59
Budget: *revenues:* $515.8 billion
expenditures: $648.6 billion (2010 est.)
Public debt: 63.4% of GDP (2010 est.)
country comparison to the world: 27
53.2% of GDP (2009 est.)
Inflation rate (consumer prices): 1.3% (2010 est.)
country comparison to the world: 28
-0.3% (2009 est.)
Central bank discount rate: 1.75% (31 December 2010)
country comparison to the world: 119
1.75% (31 December 2009)
note: this is the European Central Bank's rate on the marginal lending facility, which offers overnight credit to banks in the euro area
Commercial bank prime lending rate: 10.72% (31 December 2009 est.)
country comparison to the world: 78
11.02% (31 December 2008 est.)
Stock of narrow money: $849.2 billion (31 December 2010 est.)
country comparison to the world: 9
$856.5 billion (31 December 2009 est.)
note: see entry for the European Union for money supply in the euro area; the European Central Bank (ECB) controls monetary policy for the 17 members of the Economic and Monetary Union (EMU); individual members of the EMU do not control the quantity of money circulating within their own borders
Stock of broad money: $2.264 trillion (31 December 2010 est.)
country comparison to the world: 9
$2.451 trillion (31 December 2009 est.)
Stock of domestic credit: $3.683 trillion (31 December 2009 est.)
country comparison to the world: 8
$3.451 trillion (31 December 2008 est.)
Market value of publicly traded shares: $1.297 trillion (31 December 2009)
country comparison to the world: 11
$946.1 billion (31 December 2008)
$1.8 trillion (31 December 2007)
Agriculture—products: grain, vegetables, olives, wine grapes, sugar beets, citrus; beef, pork, poultry, dairy products; fish
Industries: textiles and apparel (including footwear), food and beverages, metals and metal manufactures, chemicals, shipbuilding, automobiles, machine tools, tourism, clay and refractory products, footwear, pharmaceuticals, medical equipment
Industrial production growth rate: -2% (2010 est.)
country comparison to the world: 161
Electricity—production: 300.5 billion kWh (2008 est.)
country comparison to the world: 13
Electricity—consumption: 276.1 billion kWh (2008 est.)
country comparison to the world: 14
Electricity—exports: 16.92 billion kWh (2008 est.)
Electricity—imports: 5.88 billion kWh (2008 est.)
Oil—production: 27,230 bbl/day (2009 est.)
country comparison to the world: 71
Oil—consumption: 1.482 million bbl/day (2009 est.)
country comparison to the world: 17
Oil—exports: 218,600 bbl/day (2008 est.)
country comparison to the world: 53
Oil—imports: 1.716 million bbl/day (2008 est.)
country comparison to the world: 11
Oil—proved reserves: 150 million bbl (1 January 2010 est.)
country comparison to the world: 66
Natural gas—production: 13 million cu m (2009 est.)
country comparison to the world: 88
Natural gas—consumption: 33.88 billion cu m (2009 est.)
country comparison to the world: 26
Natural gas—exports: 975 million cu m (2009 est.)
country comparison to the world: 36
Natural gas—imports: 34.67 billion cu m (2009 est.)
country comparison to the world: 9
Natural gas—proved reserves: 2.548 billion cu m (1 January 2010 est.)
country comparison to the world: 95
Current account balance: $-66.74 billion (2010 est.)
country comparison to the world: 190
$-80.38 billion (2009 est.)
Exports: $268.3 billion (2010 est.)
country comparison to the world: 18
$224 billion (2009 est.)
Exports—commodities: machinery, motor vehicles; foodstuffs, pharmaceuticals, medicines, other consumer goods
Exports—partners: France 19.27%, Germany 11.11%, Portugal 9.21%, Italy 8.24%, UK 6.18% (2009)
Imports: $324.6 billion (2010 est.)
country comparison to the world: 14
$286.8 billion (2009 est.)
Imports—commodities: machinery and equipment, fuels, chemicals, semifinished goods, foodstuffs, consumer goods, measuring and medical control instruments
Imports—partners: Germany 15.02%, France 12.82%, Italy 7.17%, China 5.8%, Netherlands 5.22%, UK 4.7% (2009)
Reserves of foreign exchange and gold: $NA (31 December 2010 est.)
$28.2 billion (31 December 2009 est.)
Debt—external: $2.166 trillion (30 June 2010)
country comparison to the world: 10
$2.317 trillion (31 December 2008)
Stock of direct foreign investment—at home: $668.5 billion (31 December 2010 est.)
country comparison to the world: 8
$664 billion (31 December 2009 est.)
Stock of direct foreign investment—abroad: $641 billion (31 December 2010 est.)
country comparison to the world: 9
$634.4 billion (31 December 2009 est.)
Exchange rates: euros (EUR) per US dollar–0.755 (2010), 0.7198 (2009), 0.6827 (2008), 0.7345 (2007), 0.7964 (2006)

COMMUNICATIONS

Telephones—main lines in use: 20.057 million (2009)
country comparison to the world: 13
Telephones—mobile cellular: 50.991 million (2009)
country comparison to the world: 23
Telephone system: *general assessment:* well developed, modern facilities; fixed-line teledensity is roughly 50 per 100 persons
domestic: combined fixed-line and mobile-cellular teledensity is nearly 175 telephones per 100 persons
international: country code–34; submarine cables provide connectivity to Europe, Middle East, Asia, and US; satellite earth stations–2 Intelsat (1 Atlantic Ocean and 1 Indian Ocean), NA Eutelsat; tropospheric scatter to adjacent countries
Broadcast media: a mixture of both publicly-operated and privately-owned TV and radio stations broadcasting; overall, hundreds of TV channels are available including national, regional, local, public, and international channels; satellite and cable TV systems are accessible; multiple national radio networks, a large number of regional radio networks, and a larger number of local radio stations broadcasting; overall, hundreds of radio stations operating (2008)
Internet country code: .es
Internet hosts: 3.822 million (2010)
country comparison to the world: 23
Internet users: 28.119 million (2009)
country comparison to the world: 14

TRANSPORTATION

Airports: 154 (2010)
country comparison to the world: 35
Airports—with paved runways: *total:* 97
over 3,047 m: 18
2,438 to 3,047 m: 13
1,524 to 2,437 m: 18
914 to 1,523 m: 24
under 914 m: 24 (2010)
Airports—with unpaved runways: *total:* 57
1,524 to 2,437 m: 3
914 to 1,523 m: 16
under 914 m: 38 (2010)
Heliports: 9 (2010)
Pipelines: gas 9,359 km; oil 560 km; refined products 3,441 km (2010)
Railways: *total:* 15,288 km
country comparison to the world: 18
broad gauge: 11,919 km 1.668-m gauge (6,950 km electrified)
standard gauge: 1,392 km 1.435-m gauge (1,054 km electrified)
narrow gauge: 1,949 km 1.000-m gauge (815 km electrified); 28 km 0.914-m gauge (28 km electrified) (2008)
Roadways: *total:* 681,298 km
country comparison to the world: 10
paved: 681,298 km (includes 15,152 km of expressways) (2008)
Waterways: 1,000 km (2009)
country comparison to the world: 65
Merchant marine: *total:* 138
country comparison to the world: 44
by type: bulk carrier 7, cargo 17, chemical tanker 12, container 8, liquefied gas 13, passenger 1, passenger/cargo 40, petroleum tanker 17, refrigerated cargo 5, roll on/roll off 13, vehicle carrier 5
foreign-owned: 26 (Canada 5, Denmark 2, Germany 5, Italy 1, Mexico 2, Norway 10, Switzerland 1)
registered in other countries: 107 (Angola 1, Argentina 3, Bahamas 9, Belize 1, Brazil 12, Cape Verde 1, Cyprus 7, France 1, Malta 10, Nigeria 1, Panama 40, Portugal 15, Uruguay 5, Venezuela 1) (2010)
Ports and terminals: Algeciras, Barcelona, Bilbao, Cartagena, Huelva, Tarragona, Valencia (Spain); Las Palmas, Santa Cruz de Tenerife (Canary Islands)

MILITARY

Military branches: Spanish Armed Forces: Army (Ejercito de Tierra), Spanish Navy (Armada Espanola, AE; includes Marine Corps), Spanish Air Force (Ejercito del Aire Espanola, EdA) (2010)
Military service age and obligation: 20 years of age (2004)
Manpower available for military service: *males age 16–49:* 11,759,557
females age 16–49: 11,204,688 (2010 est.)
Manpower fit for military service: *males age 16–49:* 9,603,939
females age 16–49: 9,116,928 (2010 est.)
Manpower reaching militarily significant age annually: *male:* 217,244
female: 205,278 (2010 est.)
Military expenditures: 1.2% of GDP (2005 est.)
country comparison to the world: 124

TRANSNATIONAL ISSUES

Disputes—international: in 2002, Gibraltar residents voted overwhelmingly by referendum to reject any "shared sovereignty" arrangement; the government of Gibraltar insists on equal participation in talks between the UK and Spain; Spain disapproves of UK plans to grant Gibraltar greater autonomy; Morocco protests Spain's control over the coastal enclaves of Ceuta, Melilla, and the islands of Penon de Velez de la Gomera, Penon de Alhucemas, and Islas Chafarinas, and surrounding waters; both countries claim Isla Perejil (Leila Island); Morocco serves as the primary launching site of illegal migration into Spain from North Africa; Portugal does not recognize Spanish sovereignty over the territory of Olivenza based on a difference of interpretation of the 1815 Congress of Vienna and the 1801 Treaty of Badajoz
Illicit drugs: despite rigorous law enforcement efforts, North African, Latin American, Galician, and other European traffickers take advantage of Spain's long coastline to land large shipments of cocaine and hashish for distribution to the European market; consumer for Latin American cocaine and North African hashish; destination and minor transshipment point for Southwest Asian heroin; money-laundering site for Colombian narcotics trafficking organizations and organized crime

SPRATLY ISLANDS

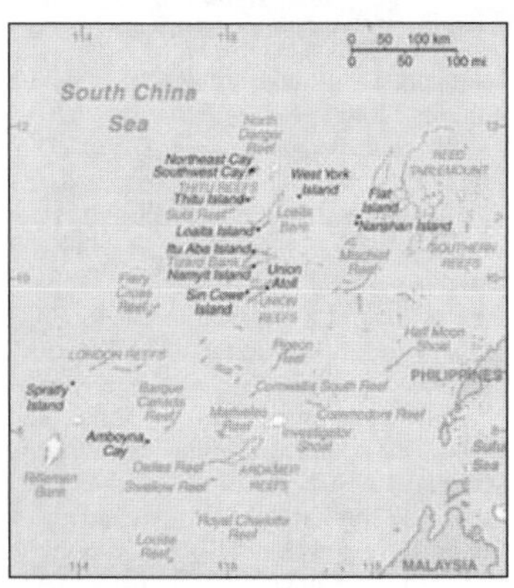

INTRODUCTION

Background: The Spratly Islands consist of more than 100 small islands or reefs. They are surrounded by rich fishing grounds and potentially by gas and oil deposits. They are claimed in their entirety by China, Taiwan, and Vietnam, while portions are claimed by Malaysia and the Philippines. About 45 islands are occupied by relatively small numbers of military forces from China, Malaysia, the Philippines, Taiwan, and Vietnam. Brunei has established a fishing zone that overlaps a southern reef but has not made any formal claim.

GEOGRAPHY

Location: Southeastern Asia, group of reefs and islands in the South China Sea, about two-thirds of the way from southern Vietnam to the southern Philippines
Geographic coordinates: 8 38 N, 111 55 E
Map references: Southeast Asia
Area: *total:* less than 5 sq km
country comparison to the world: 245
land: less than 5 sq km
water: 0 sq km
note: includes 100 or so islets, coral reefs, and sea mounts scattered over an area of nearly 410,000 sq km of the central South China Sea
Area—comparative: NA
Land boundaries: 0 km
Coastline: 926 km
Maritime claims: NA
Climate: tropical
Terrain: flat
Elevation extremes: *lowest point:* South China Sea 0 m
highest point: unnamed location on Southwest Cay 4 m
Natural resources: fish, guano, undetermined oil and natural gas potential
Land use: *arable land:* 0%
permanent crops: 0%
other: 100% (2005)
Irrigated land: 0 sq km
Natural hazards: typhoons; numerous reefs and shoals pose a serious maritime hazard
Environment—current issues: NA
Geography—note: strategically located near several primary shipping lanes in the central South China Sea; includes numerous small islands, atolls, shoals, and coral reefs

PEOPLE

Population: no indigenous inhabitants
note: there are scattered garrisons occupied by military personnel of several claimant states

GOVERNMENT

Country name: *conventional long form:* none
conventional short form: Spratly Islands

ECONOMY

Economy—overview: Economic activity is limited to commercial fishing. The proximity to nearby oil- and gas-producing sedimentary basins suggests the potential for oil and gas deposits, but the region is largely unexplored. There are no reliable estimates of potential reserves. Commercial exploitation has yet to be developed.

TRANSPORTATION

Airports: 4 (2010)

country comparison to the world: 189
Airports—with paved runways: *total:* 3
914 to 1,523 m: 2
under 914 m: 1 (2010)
Airports—with unpaved runways: *total:* 1
914 to 1,523 m: 1 (2010)
Heliports: 3 (2010)
Ports and terminals: none; offshore anchorage only

MILITARY

Military—note: Spratly Islands consist of more than 100 small islands or reefs of which about 45 are claimed and occupied by China, Malaysia, the Philippines, Taiwan, and Vietnam

TRANSNATIONAL ISSUES

Disputes—international: all of the Spratly Islands are claimed by China (including Taiwan) and Vietnam; parts of them are claimed by Brunei, Malaysia and the Philippines; despite no public territorial claim to Louisa Reef, Brunei implicitly lays claim by including it within the natural prolongation of its continental shelf and basis for a seabed median with Vietnam; claimants in November 2002 signed the "Declaration on the Conduct of Parties in the South China Sea," which has eased tensions but falls short of a legally binding "code of conduct"; in March 2005, the national oil companies of China, the Philippines, and Vietnam signed a joint accord to conduct marine seismic activities in the Spratly Islands

SRI LANKA

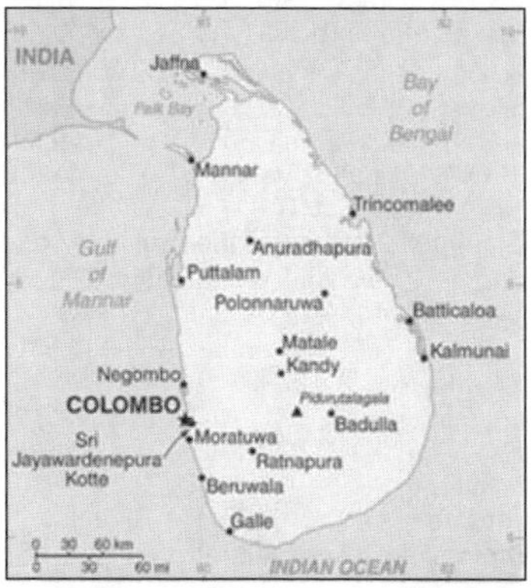

INTRODUCTION

Background: The first Sinhalese arrived in Sri Lanka late in the 6th century B.C., probably from northern India. Buddhism was introduced in about the mid-third century B.C., and a great civilization developed at the cities of Anuradhapura (kingdom from circa 200 B.C. to circa A.D. 1000) and Polonnaruwa (from about 1070 to 1200). In the 14th century, a south Indian dynasty established a Tamil kingdom in northern Sri Lanka. The coastal areas of the island were controlled by the Portuguese in the 16th century and by the Dutch in the 17th century. The island was ceded to the British in 1796, became a crown colony in 1802, and was formally united under British rule by 1815. As Ceylon, it became independent in 1948; its name was changed to Sri Lanka in 1972. Tensions between the Sinhalese majority and Tamil separatists erupted into war in 1983. After two decades of fighting, the government and Liberation Tigers of Tamil Eelam (LTTE) formalized a cease-fire in February 2002 with Norway brokering peace negotiations. Violence between the LTTE and government forces intensified in 2006, but the government regained control of the Eastern Province in 2007. By May 2009, the government announced that its military had defeated the remnants of the LTTE. Since the end of the conflict, the government has resettled tens of thousands of internally displaced persons and has undertaken a number of massive infrastructure projects to reconstruct its economy.

GEOGRAPHY

Location: Southern Asia, island in the Indian Ocean, south of India
Geographic coordinates: 7 00 N, 81 00 E
Map references: Asia
Area: *total:* 65,610 sq km
country comparison to the world: 121
land: 64,630 sq km
water: 980 sq km
Area—comparative: slightly larger than West Virginia
Land boundaries: 0 km
Coastline: 1,340 km
Maritime claims: territorial sea: 12 nm
contiguous zone: 24 nm
exclusive economic zone: 200 nm
continental shelf: 200 nm or to the edge of the continental margin
Climate: tropical monsoon; northeast monsoon (December to March); southwest monsoon (June to October)
Terrain: mostly low, flat to rolling plain; mountains in south-central interior
Elevation extremes: *lowest point:* Indian Ocean 0 m
highest point: Pidurutalagala 2,524 m
Natural resources: limestone, graphite, mineral sands, gems, phosphates, clay, hydropower
Land use: *arable land:* 13.96%
permanent crops: 15.24%
other: 70.8% (2005)
Irrigated land: 7,430 sq km (2003)
Total renewable water resources: 50 cu km (1999)
Freshwater withdrawal (domestic/industrial/agricultural): *total:* 12.61 cu km/yr (2%/2%/95%)
per capita: 608 cu m/yr (2000)
Natural hazards: occasional cyclones and tornadoes
Environment—current issues: deforestation; soil erosion; wildlife populations threatened by poaching and urbanization; coastal degradation from mining activities and increased pollution; freshwater resources being polluted by industrial wastes and sewage runoff; waste disposal; air pollution in Colombo
Environment—international agreements: *party to:* Biodiversity, Climate Change, Climate Change-Kyoto Protocol, Desertification, Endangered Species, Environmental Modification, Hazardous Wastes, Law of the Sea, Ozone Layer Protection, Ship Pollution, Wetlands
signed, but not ratified: Marine Life Conservation
Geography—note: strategic location near major Indian Ocean sea lanes

PEOPLE

Population: 21,283,913 (July 2011 est.)
country comparison to the world: 57
Age structure: *0–14 years:* 24.9% (male 2,705,953/female 2,599,717)
15–64 years: 67.2% (male 6,993,668/female 7,313,440)
65 years and over: 7.9% (male 720,219/female 950,916) (2011 est.)
Median age: *total:* 30.8 years
male: 29.7 years
female: 31.8 years (2011 est.)
Population growth rate: 0.934% (2011 est.)
country comparison to the world: 122
Birth rate: 17.42 births/1,000 population (2011 est.)
country comparison to the world: 116
Death rate: 5.92 deaths/1,000 population (July 2011 est.)
country comparison to the world: 167
Net migration rate: -2.16 migrant(s)/1,000 population (2011 est.)
country comparison to the world: 166
Urbanization: *urban population:* 14% of total population (2010)
rate of urbanization: 1.1% annual rate of change (2010-15 est.)
Major cities—population: COLOMBO (capital) 681,000 (2009)
Sex ratio: *at birth:* 1.04 male(s)/female
under 15 years: 1.04 male(s)/female
15–64 years: 0.96 male(s)/female
65 years and over: 0.86 male(s)/female
total population: 0.97 male(s)/female (2011 est.)
Infant mortality rate: *total:* 9.7 deaths/1,000 live births
country comparison to the world: 149
male: 10.68 deaths/1,000 live births
female: 8.68 deaths/1,000 live births (2011 est.)
Life expectancy at birth: *total population:* 75.73 years
country comparison to the world: 81
male: 72.21 years
female: 79.38 years (2011 est.)
Total fertility rate: 2.2 children born/woman (2011 est.)
country comparison to the world: 106
HIV/AIDS—adult prevalence rate: less than 0.1% (2009 est.)

country comparison to the world: 121
HIV/AIDS—people living with HIV/AIDS: 2,800 (2009 est.)
country comparison to the world: 130
HIV/AIDS—deaths: fewer than 200 (2009 est.)
country comparison to the world: 101
Major infectious diseases: *degree of risk:* high
food or waterborne diseases: bacterial diarrhea and hepatitis A
vectorborne disease: dengue fever and chikungunya
water contact disease: leptospirosis
animal contact disease: rabies (2009)
Drinking water source: *Improved:*
urban: 98% of population
rural: 88% of population
total: 90% of population
Unimproved:
urban: 2% of population
rural: 12% of population
total: 10% of population (2008)
Sanitation facility access: *Improved:*
urban: 88% of population
rural: 92% of population
total: 91% of population
Unimproved:
urban: 12% of population
rural: 8% of population
total: 9% of population (2008)
Nationality: *noun:* Sri Lankan(s)
adjective: Sri Lankan
Ethnic groups: Sinhalese 73.8%, Sri Lankan Moors 7.2%, Indian Tamil 4.6%, Sri Lankan Tamil 3.9%, other 0.5%, unspecified 10% (2001 census provisional data)
Religions: Buddhist 69.1%, Muslim 7.6%, Hindu 7.1%, Christian 6.2%, unspecified 10% (2001 census provisional data)
Languages: Sinhala (official and national language) 74%, Tamil (national language) 18%, other 8%
note: English is commonly used in government and is spoken competently by about 10% of the population
Literacy: *definition:* age 15 and over can read and write
total population: 90.7%
male: 92.3%
female: 89.1% (2001 census)
School life expectancy (primary to tertiary education): *total:* 13 years
male: 12 years
female: 13 years (2004)
Education expenditures: NA

GOVERNMENT

Country name: *conventional long form:* Democratic Socialist Republic of Sri Lanka
conventional short form: Sri Lanka
local long form: Shri Lamka Prajatantrika Samajaya di Janarajaya/Ilankai Jananayaka Choshalichak Kutiyarachu
local short form: Shri Lamka/Ilankai
former: Serendib, Ceylon
Government type: republic
Capital: *name:* Colombo
geographic coordinates: 6 56 N, 79 51 E
time difference: UTC+5.5 (10.5 hours ahead of Washington, DC during Standard Time)
note: Sri Jayewardenepura Kotte (legislative capital)
Administrative divisions: 9 provinces; Central, Eastern, North Central, Northern, North Western, Sabaragamuwa, Southern, Uva, Western
Independence: 4 February 1948 (from the UK)
National holiday: Independence Day, 4 February (1948)
Constitution: adopted 16 August 1978, certified 31 August 1978; amended 20 December 2001
Legal system: mixed legal system of Roman-Dutch civil law, English common law, and Jaffna Tamil customary law
International law organization participation: has not submitted an ICJ jurisdiction declaration; non-party state to the ICCt
Suffrage: 18 years of age; universal
Executive branch: *chief of state:* President Mahinda Percy RAJAPAKSA (since 19 November 2005); note–the president is both the chief of state and head of government; Dissanayake Mudiyanselage JAYARATNE holds the largely ceremonial title of prime minister (since 21 April 2010)
head of government: President Mahinda Percy RAJAPAKSA (since 19 November 2005)
cabinet: Cabinet appointed by the president in consultation with the prime minister (For more information visit the World Leaders website)
elections: president elected by popular vote for a six-year term (two-term limit); election last held on 26 January 2010 (next to be held in 2016)
election results: Mahinda RAJAPAKSA reelected president for second term; percent of vote–Mahinda RAJAPAKSA 57.88%, Sarath FONSEKA 40.15%, other 1.97%
Legislative branch: unicameral Parliament (225 seats; members elected by popular vote on the basis of an open-list, proportional representation system by electoral district to serve six-year terms)
elections: last held on 8 April 2010 with a repoll in two electorates held on 20 April 2010 (next to be held by April 2016)
election results: percent of vote by alliance or party–United People's Freedom Alliance 60.93%, United National Party 29.34%, Democratic National Alliance 5.49%, Tamil National Alliance 2.9%, other 1.94%; seats by alliance or party–United People's Freedom Alliance 144, United National Party 60, Tamil National Alliance 14, Democratic National Alliance 7
Judicial branch: Supreme Court; Court of Appeals; judges for both courts are appointed by the president
Political parties and leaders: Coalitions and leaders: Democratic National Alliance led by Janatha Vimukthi Peramuna or JVP [Somawansa AMARASINGHE]; Tamil National Alliance led by Illandai Tamil Arasu Kachchi [R. SAMPANTHAN]; United National Front led by United National Party [Ranil WICKREMESINGHE]; United People's Freedom Alliance led by Sri Lanka Freedom Party [Mahinda RAJAPAKSA]
Political pressure groups and leaders: Liberation Tigers of Tamil Eelam or LTTE [P. SIVAPARAN, Chief of International Secretariat; V. RUDRAKUMARAN, legal advisor]; note–this insurgent group suffered military defeat in May 2009; some cadres remain scattered throughout country;
other: Buddhist clergy; labor unions; radical chauvinist Sinhalese groups such as the National Movement Against Terrorism; Sinhalese Buddhist lay groups
International organization participation: ADB, ARF, BIMSTEC, C, CP, FAO, G-11, G-15, G-24, G-77, IAEA, IBRD, ICAO, ICC, ICRM, IDA, IFAD, IFC, IFRCS, IHO, ILO, IMF, IMO, IMSO, Interpol, IOC, IOM, IPU, ISO, ITSO, ITU, ITUC, MIGA, MINURSO, MINUSTAH, MONUSCO, NAM, OAS (observer), OPCW, PCA, SAARC, SACEP, SCO (dialogue member), UN, UNCTAD, UNESCO, UNIDO, UNWTO, UPU, WCO, WFTU, WHO, WIPO, WMO, WTO
Diplomatic representation in the US: *chief of mission:* Ambassador Jaliya Chitran WICKRAMASURIYA
chancery: 2148 Wyoming Avenue NW, Washington, DC 20008
telephone: [1] (202) 483-4025 through 4028
FAX: [1] (202) 232-7181
consulate(s) general: Los Angeles
consulate(s): New York
Diplomatic representation from the US: *chief of mission:* Ambassador Patricia A. BUTENIS
embassy: 210 Galle Road, Colombo 3
mailing address: P. O. Box 106, Colombo
telephone: [94] (11) 249-8500
FAX: [94] (11) 243-7345
Flag description: yellow with two panels; the smaller hoist-side panel has two equal vertical bands of green (hoist side) and orange; the other larger panel depicts a yellow lion holding a sword on a dark red rectangular field that also displays a yellow bo leaf in each corner; the yellow field appears as a border around the entire flag and extends between the two panels; the lion represents Sinhalese ethnicity, the strength of the nation, and bravery; the sword demonstrates the sovereignty of the nation; the four bo leaves–symbolizing Buddhism and its influence on the country–stand for the four virtues of kindness, friendliness, happiness, and equanimity; orange signifies Sri Lankan Tamils, green the Sri Lankan Moors; dark red represents the European Burghers, but also refers to the rich colonial background of the country; yellow denotes other ethnic groups; also referred to as the Lion Flag
National anthem: *name:* "Sri Lanka Matha" (Mother Sri Lanka)
lyrics/music: Ananda SAMARKONE
note: adopted 1951

ECONOMY

Economy—overview: Sri Lanka is engaging in large-scale reconstruction and development projects following the end of the 26-year conflict with the LTTE, including increasing electricity access and rebuilding its road and rail network. Addi-

tionally, Sri Lanka seeks to reduce poverty by using a combination of state directed policies and private investment promotion to spur growth in disadvantaged areas, develop small and medium enterprises, and promote increased agriculture. High levels of government funding may be difficult, as the government already is faced with high debt interest payments, a bloated civil service, and historically high budget deficits. The 2008-09 global financial crisis and recession exposed Sri Lanka's economic vulnerabilities and nearly caused a balance of payments crisis, which was alleviated by a $2.6 billion IMF standby agreement in July 2009. The end of the civil war and the IMF loan, however, have largely restored investors' confidence, reflected in part by the Sri Lankan stock market's recognition as one of the best performing markets in the world. Sri Lankan growth rates averaged nearly 5% in during the war, but increased government spending on development and fighting the LTTE in the final years spurred GDP growth to around 6-7% per year in 2006-08. After experiencing 3.5% growth in 2009, Sri Lanka's economy is poised to achieve high growth rates in the postwar period.
GDP (purchasing power parity): $106.5 billion (2010 est.)
country comparison to the world: 69
$97.6 billion (2009 est.)
$94.02 billion (2008 est.)
note: data are in 2010 US dollars
GDP (official exchange rate): $49.68 billion (2010 est.)
GDP—real growth rate: 9.1% (2010 est.)
country comparison to the world: 8
3.8% (2009 est.)
6% (2008 est.)
GDP—per capita (PPP): $5,000 (2010 est.)
country comparison to the world: 148
$4,600 (2009 est.)
$4,400 (2008 est.)
note: data are in 2010 US dollars
GDP—composition by sector:
agriculture: 12.6%
industry: 29.8%
services: 57.6% (2010 est.)
Labor force: 8.1 million (2010 est.)
country comparison to the world: 56
Labor force—by occupation:
agriculture: 32.7%
industry: 26.3%
services: 41% (December 2008 est.)
Unemployment rate: 5.4% (2010 est.)
country comparison to the world: 51
5.9% (2009 est.)
Population below poverty line: 23% (2008 est.)
Household income or consumption by percentage share: *lowest 10%:* 1.1%
highest 10%: 39.7% (2004)
Distribution of family income—Gini index: 49 (2007)
country comparison to the world: 26
46 (1995)
Investment (gross fixed): 23.6% of GDP (2010 est.)
country comparison to the world: 52
Budget: *revenues:* $7.415 billion
expenditures: $11.18 billion (2010 est.)
Public debt: 86.7% of GDP (2010 est.)
country comparison to the world: 13
85.8% of GDP (2009 est.)
Inflation rate (consumer prices): 5.6% (2010 est.)
country comparison to the world: 150
3% (2009 est.)
Central bank discount rate: 7.5% (31 December 2009)
country comparison to the world: 20
15% (31 December 2008)
Commercial bank prime lending rate: 10.91% (31 December 2009)
country comparison to the world: 23
18.89% (31 December 2008 est.)
Stock of narrow money: $4.4 billion (31 December 2010 est.)
country comparison to the world: 97
$3.628 billion (31 December 2009 est.)
Stock of broad money: $19.72 billion (31 December 2010 est.)
country comparison to the world: 84
$16.41 billion (31 December 2009 est.)
Stock of domestic credit: $18.34 billion (31 December 2010 est.)
country comparison to the world: 83
$16.64 billion (31 December 2009 est.)
Market value of publicly traded shares: $8.133 billion (31 December 2009)
country comparison to the world: 85
$4.326 billion (31 December 2008)
$7.553 billion (31 December 2007)
Agriculture—products: rice, sugarcane, grains, pulses, oilseed, spices, vegetables, fruit, tea, rubber, coconuts; milk, eggs, hides, beef; fish
Industries: processing of rubber, tea, coconuts, tobacco and other agricultural commodities; telecommunications, insurance, banking; tourism, shipping; clothing, textiles; cement, petroleum refining, information technology services, construction
Industrial production growth rate: 6.9% (2010 est.)
country comparison to the world: 49
Electricity—production: 9.901 billion kWh (2008 est.)
country comparison to the world: 90
Electricity—consumption: 8.417 billion kWh (2008 est.)
country comparison to the world: 93
Electricity—exports: 0 kWh (2008 est.)
Electricity—imports: 0 kWh (2008 est.)
Oil—production: 0 bbl/day (2008 est.)
country comparison to the world: 158
Oil—consumption: 90,000 bbl/day (2009 est.)
country comparison to the world: 77
Oil—exports: 968.4 bbl/day (2007 est.)
country comparison to the world: 121
Oil—imports: 87,690 bbl/day (2007 est.)
country comparison to the world: 66
Oil—proved reserves: 0 bbl (1 January 2010 est.)
country comparison to the world: 114
Natural gas—production: 0 cu m (2008 est.)
country comparison to the world: 109
Natural gas—consumption: 0 cu m (2008 est.)
country comparison to the world: 156
Natural gas—exports: 0 cu m (2008 est.)
country comparison to the world: 71
Natural gas—proved reserves: 0 cu m (1 January 2010 est.)
country comparison to the world: 119
Current account balance: $-1.784 billion (2010 est.)
country comparison to the world: 152
$-291 million (2009 est.)
Exports: $7.908 billion (2010 est.)
country comparison to the world: 93
$7.085 billion (2009 est.)
Exports—commodities: textiles and apparel, tea and spices; rubber manufactures; precious stones; coconut products, fish
Exports—partners: US 20.59%, UK 12.87%, Italy 5.51%, Germany 5.29%, India 4.54%, Belgium 4.43% (2009)
Imports: $11.6 billion (2010 est.)
country comparison to the world: 86
$9.186 billion (2009 est.)
Imports—commodities: petroleum, textiles, machinery and transportation equipment, building materials, mineral products, foodstuffs
Imports—partners: India 20.73%, China 13.45%, Singapore 7.26%, Iran 6.7%, South Korea 5.23% (2009)
Reserves of foreign exchange and gold: $5.63 billion (31 December 2010 est.)
country comparison to the world: 70
$5.358 billion (31 December 2009 est.)
Debt—external: $17.97 billion (31 December 2010 est.)
country comparison to the world: 75
$17.44 billion (31 December 2009 est.)
Stock of direct foreign investment—at home: $NA
Stock of direct foreign investment—abroad: $NA
Exchange rates: Sri Lankan rupees (LKR) per US dollar–113.36 (2010), 114.95 (2009), 108.33 (2008), 110.78 (2007)n 103.99 (2006)

COMMUNICATIONS

Telephones—main lines in use: 3.523 million (2010)
country comparison to the world: 45
Telephones—mobile cellular: 15.868 million (2010)
country comparison to the world: 46
Telephone system: *general assessment:* telephone services have improved significantly and are available in most parts of the country
domestic: national trunk network consists mostly of digital microwave radio relay; fiber-optic links now in use in Colombo area and fixed wireless local loops have been installed; competition is strong in mobile cellular systems and mobile cellular subscribership is increasing
international: country code–94; the SEA-ME-WE-3 and SEA-ME-WE-4 submarine cables provide connectivity to Asia, Australia, Middle East, Europe, US; satellite earth stations–2 Intelsat (Indian Ocean)
Broadcast media: government operates 2 television channels and a radio network; multi-channel satellite and cable TV subscription services are obtainable; 8 private TV stations and about a dozen private radio stations in operation (2008)
Internet country code: .lk
Internet hosts: 8,865 (2010)
country comparison to the world: 131

Internet users: 1.777 million (2009)
country comparison to the world: 77

TRANSPORTATION

Airports: 18 (2010)
country comparison to the world: 139
Airports—with paved runways: *total:* 14
over 3,047 m: 1
1,524 to 2,437 m: 6
914 to 1,523 m: 7 (2010)
Airports—with unpaved runways: *total:* 4
914 to 1,523 m: 1
under 914 m: 3 (2010)
Railways: *total:* 1,449 km
country comparison to the world: 80
broad gauge: 1,449 km 1.676-m gauge (2007)
Roadways: *total:* 91,907 km (2008)
country comparison to the world: 53
Waterways: 160 km (primarily on rivers in southwest) (2008)
country comparison to the world: 101
Merchant marine: *total:* 22
country comparison to the world: 98
by type: bulk carrier 4, cargo 14, chemical tanker 1, petroleum tanker 3
foreign-owned: 5 (Germany 5) (2010)
Ports and terminals: Colombo

MILITARY

Military branches: Sri Lanka Army, Sri Lanka Navy, Sri Lanka Air Force (2010)
Military service age and obligation: 18 years of age for voluntary military service; 5-year service obligation (2010)
Manpower available for military service: *males age 16–49:* 5,342,147
females age 16–49: 5,466,409 (2010 est.)
Manpower fit for military service: *males age 16–49:* 4,177,432
females age 16–49: 4,574,833 (2010 est.)
Manpower reaching militarily significant age annually: *male:* 167,026
female: 162,587 (2010 est.)
Military expenditures: 2.6% of GDP (2006)
country comparison to the world: 54

TRANSNATIONAL ISSUES

Disputes—international: none
Refugees and internally displaced persons: *IDPs:* 460,000 (both Tamils and non-Tamils displaced due to long-term civil war between the government and the separatist Liberation Tigers of Tamil Eelam (LTTE)) (2007)
Trafficking in persons: current situation: Sri Lanka is a source and destination country for men and women trafficked for the purposes of involuntary servitude and commercial sexual exploitation; Sri Lankan men and women migrate willingly to the Persian Gulf, Middle East, and East Asia to work as construction workers, domestic servants, or garment factory workers, where some find themselves in situations of involuntary servitude when faced with restrictions on movement, withholding of passports, threats, physical or sexual abuse, and debt bondage; children are trafficked internally for commercial sexual exploitation and, less frequently, for forced labor
tier rating: Tier 2 Watch List–for a fourth consecutive year, Sri Lanka is on the Tier 2 Watch List for failing to provide evidence of increasing efforts to combat severe forms of human trafficking, particularly in the area of law enforcement; the government failed to arrest, prosecute, or convict any person for trafficking offenses and continued to punish some victims of trafficking for crimes committed as a result of being trafficked; Sri Lanka has not ratified the 2000 UN TIP Protocol (2008)

SUDAN

INTRODUCTION

Background: Military regimes favoring Islamic-oriented governments have dominated national politics since independence from the UK in 1956. Sudan was embroiled in two prolonged civil wars during most of the remainder of the 20th century. These conflicts were rooted in northern economic, political, and social domination of largely non-Muslim, non-Arab southern Sudanese. The first civil war ended in 1972 but broke out again in 1983. The second war and famine-related effects resulted in more than four million people displaced and, according to rebel estimates, more than two million deaths over a period of two decades. Peace talks gained momentum in 2002-04 with the signing of several accords. The final North/South Comprehensive Peace Agreement (CPA), signed in January 2005, granted the southern rebels autonomy for six years followed by a referendum on independence for Southern Sudan. The referendum was held in January 2011 and indicated overwhelming support for independence. A separate conflict, which broke out in the western region of Darfur in 2003, has displaced nearly two million people and caused an estimated 200,000 to 400,000 deaths. The UN took command of the Darfur peacekeeping operation from the African Union in December 2007. Peacekeeping troops have struggled to stabilize the situation, which has become increasingly regional in scope and has brought instability to eastern Chad. Sudan also has faced large refugee influxes from neighboring countries primarily Ethiopia and Chad. Armed conflict, poor transport infrastructure, and lack of government support have chronically obstructed the provision of humanitarian assistance to affected populations.

GEOGRAPHY

Location: Northern Africa, bordering the Red Sea, between Egypt and Eritrea
Geographic coordinates: 15 00 N, 30 00 E
Map references: Africa
Area: *total:* 2,505,813 sq km
country comparison to the world: 10
land: 2.376 million sq km
water: 129,813 sq km
Area—comparative: slightly more than one-quarter the size of the US
Land boundaries: *total:* 7,687 km
border countries: Central African Republic 1,165 km, Chad 1,360 km, Democratic Republic of the Congo 628 km, Egypt 1,273 km, Eritrea 605 km, Ethiopia 1,606 km, Kenya 232 km, Libya 383 km, Uganda 435 km
Coastline: 853 km
Maritime claims: territorial sea: 12 nm
contiguous zone: 18 nm
continental shelf: 200 m depth or to the depth of exploitation
Climate: tropical in south; arid desert in north; rainy season varies by region (April to November)
Terrain: generally flat, featureless plain; mountains in far south, northeast, and west; desert dominates the north
Elevation extremes: *lowest point:* Red Sea 0 m
highest point: Kinyeti 3,187 m
Natural resources: petroleum; small reserves of iron ore, copper, chromium ore, zinc, tungsten, mica, silver, gold; hydropower
Land use: *arable land:* 6.78%
permanent crops: 0.17%
other: 93.05% (2005)
Irrigated land: 18,630 sq km (2003)
Total renewable water resources: 154 cu km (1997)
Freshwater withdrawal (domestic/industrial/agricultural): *total:* 37.32 cu km/yr (3%/1%/97%)
per capita: 1,030 cu m/yr (2000)
Natural hazards: dust storms and periodic persistent droughts
Environment—current issues: inadequate supplies of potable water; wildlife populations threatened by excessive hunting; soil erosion; desertification; periodic drought
Environment—international agreements: *party to:* Biodiversity, Climate Change, Climate Change-Kyoto Protocol, Desertification, Endangered Species, Hazardous Wastes, Law of the Sea, Ozone Layer Protection, Wetlands
signed, but not ratified: none of the selected agreements
Geography—note: largest country in Africa; dominated by the Nile and its tributaries

PEOPLE

Population: 45,047,502 (July 2011 est.)
country comparison to the world: 29
Age structure: *0–14 years:* 42.1% (male 9,696,726/female 9,286,894)
15–64 years: 55.2% (male 12,282,082/female 12,571,424)
65 years and over: 2.7% (male 613,817/female 596,559) (2011 est.)
Median age: *total:* 18.5 years
male: 18.1 years
female: 18.9 years (2011 est.)
Population growth rate: 2.484% (2011 est.)
country comparison to the world: 27
Birth rate: 36.12 births/1,000 population (2011 est.)
country comparison to the world: 25
Death rate: 11 deaths/1,000 population (July 2011 est.)
country comparison to the world: 38
Net migration rate: -0.29 migrant(s)/1,000 population (2011 est.)
country comparison to the world: 127
Urbanization: *urban population:* 40% of total population (2010)
rate of urbanization: 3.7% annual rate of change (2010-15 est.)
Major cities—population: KHARTOUM (capital) 5.021 million (2009)
Sex ratio: *at birth:* 1.05 male(s)/female
under 15 years: 1.04 male(s)/female
15–64 years: 1.01 male(s)/female
65 years and over: 1.05 male(s)/female
total population: 1.03 male(s)/female (2011 est.)
Infant mortality rate: *total:* 68.07 deaths/1,000 live births
country comparison to the world: 20
male: 68.77 deaths/1,000 live births
female: 67.34 deaths/1,000 live births (2011 est.)
Life expectancy at birth: *total population:* 55.42 years
country comparison to the world: 198
male: 54.18 years
female: 56.71 years (2011 est.)
Total fertility rate: 4.84 children born/woman (2011 est.)
country comparison to the world: 24
HIV/AIDS—adult prevalence rate: 1.1% (2009 est.)
country comparison to the world: 43
HIV/AIDS—people living with HIV/AIDS: 260,000 (2009 est.)
country comparison to the world: 23
HIV/AIDS—deaths: 12,000 (2009 est.)
country comparison to the world: 23
Major infectious diseases: *degree of risk:* very high
food or waterborne diseases: bacterial and protozoal diarrhea, hepatitis A and E, and typhoid fever
vectorborne diseases: malaria, dengue fever, African trypanosomiasis (sleeping sickness)
water contact disease: schistosomiasis
respiratory disease: meningococcal meningitis
animal contact disease: rabies
note: highly pathogenic H5N1 avian influenza has been identified in this country; it poses a negligible risk with extremely rare cases possible among US citizens who have close contact with birds (2009)
Drinking water source: *Improved:*
urban: 64% of population
rural: 52% of population
total: 57% of population
Unimproved:
urban: 36% of population
rural: 48% of population
total: 43% of population (2008)
Sanitation facility access: *Improved:*
urban: 55% of population
rural: 18% of population
total: 34% of population
Unimproved:
urban: 45% of population
rural: 82% of population
total: 66% of population (2008)
Nationality: *noun:* Sudanese (singular and plural)
adjective: Sudanese
Ethnic groups: black 52%, Arab 39%, Beja 6%, foreigners 2%, other 1%
Religions: Sunni Muslim 70% (in north), Christian 5% (mostly in south and Khartoum), indigenous beliefs 25%
Languages: Arabic (official), English (official), Nubian, Ta Bedawie, diverse dialects of Nilotic, Nilo-Hamitic, Sudanic languages
note: program of "Arabization" in process
Literacy: *definition:* age 15 and over can read and write
total population: 61.1%
male: 71.8%
female: 50.5% (2003 est.)
School life expectancy (primary to tertiary education): *total:* 4 years (2000)
Education expenditures: NA

GOVERNMENT

Country name: *conventional long form:* Republic of the Sudan
conventional short form: Sudan
local long form: Jumhuriyat as-Sudan
local short form: As-Sudan
former: Anglo-Egyptian Sudan
Government type: Government of National Unity (GNU)–the National Congress Party (NCP) and Sudan People's Liberation Movement (SPLM) formed a power-sharing government under the 2005 Comprehensive Peace Agreement (CPA); the NCP, which came to power by military coup in 1989, is the majority partner; the agreement stipulated national elections in 2009, but these were subsequently rescheduled; elections took place in April 2010 and the NCP was elected as the majority party; due to the CPA stipulations, there is also an autonomous government in Southern Sudan where SPLM holds the majority of positions.
Capital: *name:* Khartoum
geographic coordinates: 15 36 N, 32 32 E
time difference: UTC+3 (8 hours ahead of Washington, DC during Standard Time)
Administrative divisions: 25 states (wilayat, singular–wilayah); A'ali an Nil (Upper Nile), Al Bahr al Ahmar (Red Sea), Al Buhayrat (Lakes), Al Jazira (Gezira), Al Khartoum (Khartoum), Al Qadarif (Gedaref), Al Wahda (Unity), An Nil al Abyad (White Nile), An Nil al Azraq (Blue Nile), Ash Shimaliyya (Northern), Bahr al Jabal (Central Equatoria), Gharb al Istiwa'iyya (Western Equatoria), Gharb Bahr al Ghazal (Western Bahr el Ghazal), Gharb Darfur (Western Darfur), Janub Darfur (Southern Darfur), Janub Kurdufan (Southern Kordofan), Junqoley (Jonglei), Kassala (Kassala), Nahr an Nil (River Nile), Shimal Bahr al Ghazal (Northern Bahr el Ghazal), Shimal Darfur (Northern Darfur), Shimal Kurdufan (Northern Kordofan), Sharq al Istiwa'iyya (Eastern Equatoria), Sinnar (Sinnar), Warab (Warrap)
Independence: 1 January 1956 (from Egypt and the UK)
National holiday: Independence Day, 1 January (1956)
Constitution: Interim National Constitution ratified 5 July 2005
note: under the Comprehensive Peace Agreement, the Interim National Constitution was ratified 5 July 2005; Constitution of Southern Sudan was signed December 2005
Legal system: mixed legal system of Islamic law and English common law in the north, and primarily customary law in the south
International law organization participation: accepts compulsory ICJ jurisdiction with reservations; withdrew acceptance of ICCt jurisdiction in 2008
Suffrage: 17 years of age; universal
Executive branch: *chief of state:* President Umar Hassan Ahmad al-BASHIR (since 16 October 1993); note–the president is both the chief of state and head of government
head of government: President Umar Hassan Ahmad al-BASHIR (since 16 October 1993)
cabinet: Council of Ministers appointed by the president; note–the National Congress Party or NCP (formerly the National Islamic Front or NIF) dominates al-BASHIR's cabinet
(For more information visit the World Leaders website)
elections: election on 11-15 April 2010; next to be held in 2015
election results: Umar Hassan Ahmad al-BASHIR reelected president; percent of vote–Umar Hassan Ahmad al-BASHIR 68.2%, Yasir ARMAN 21.7%, Abdullah Deng NHIAL 3.9%, others 6.2%
note: al-BASHIR assumed power as chairman of Sudan's Revolutionary Command Council for National Salvation (RCC) in June 1989 and served concurrently as chief of state, chairman of the RCC, prime minister, and minister of defense until mid-October 1993 when he was appointed president by the RCC; he was elected president by popular vote for the first time in March 1996
Legislative branch: bicameral National Legislature consists of a Council of States (50 seats; members indirectly elected by state legislatures to serve six-year terms) and a National Assembly (450 seats; 60% from geographic constituencies, 25% from a women's list, and 15% from party lists; members to serve six-year terms)
elections: last held on 11-15 April 2010 (next to be held in 2016)
election results: National Assembly–percent of vote by party–NA; seats by party–NCP 323, SPLM 99, PCP 4, DUP 4, UFP 3, URRP 2, DUPO 2, SPLM-DC 2, other 7, vacant 4

Judicial branch: Constitutional Court of nine justices; National Supreme Court; National Courts of Appeal; other national courts; National Judicial Service Commission will undertake overall management of the National Judiciary
Political parties and leaders: Democratic Unionist Party or DUP [Hatim al-SIR]; Democratic Unionist Party-Original or DUPO; National Congress Party or NCP [Umar Hassan al-BASHIR]; Popular Congress Party or PCP [Hassan al-TURABI]; Sudan People's Liberation Movement or SPLM [Salva KIIR]; Sudan People's Liberation Movement-Democratic Change or SPLM-DC; Umma Federal Party or UFP; Umma Renewal and Reform Party or URRP
Political pressure groups and leaders: Umma Party [SADIQ Siddiq al-Mahdi]; Popular Congress Party or PCP [Hassan al-TURABI]; Darfur rebel groups including the Justice and Equality Movement or JEM [Khalil IBRAHIM] and the Sudan Liberation Movement or SLM [various factional leaders]
International organization participation: ABEDA, ACP, AfDB, AFESD, AMF, AU, CAEU, COMESA, FAO, G-77, IAEA, IBRD, ICAO, ICRM, IDA, IDB, IFAD, IFC, IFRCS, IGAD, ILO, IMF, IMO, Interpol, IOC, IOM, IPU, ISO, ITSO, ITU, LAS, MIGA, NAM, OIC, OPCW, PCA, UN, UNCTAD, UNESCO, UNHCR, UNIDO, UNWTO, UPU, WCO, WFTU, WHO, WIPO, WMO, WTO (observer)
Diplomatic representation in the US: *chief of mission:* Ambassador (vacant); Charge d'Affaires FATAHELRAHMAN Ali Mohamed
chancery: 2210 Massachusetts Avenue NW, Washington, DC 20008
telephone: [1] (202) 338-8565
FAX: [1] (202) 667-2406
Diplomatic representation from the US: *chief of mission:* Ambassador (vacant); Charge d'Affaires Robert E. WHITEHEAD
embassy: Sharia Ali Abdul Latif Street, Khartoum
mailing address: P.O. Box 699, Khartoum; APO AE 09829
telephone: [249] (183) 774700 through 704
FAX: [249] (183) 774137
Flag description: three equal horizontal bands of red (top), white, and black with a green isosceles triangle based on the hoist side; colors and design based on the Arab Revolt flag of World War I, but the meanings of the colors are expressed as follows: red signifies the struggle for freedom, white is the color of peace, light, and love, black represents Sudan itself (in Arabic 'Sudan' means black), green is the color of Islam, agriculture, and prosperity
National anthem: *name:* "Nahnu Djundulla Djundulwatan" (We Are the Army of God and of Our Land)
lyrics/music: Sayed Ahmad Muhammad SALIH/Ahmad MURJAN
note: adopted 1956; the song originally served as the anthem of the Sudanese military

ECONOMY

Economy—overview: Since 1997, Sudan has been working with the IMF to implement macroeconomic reforms including a managed float of the exchange rate and a large reserve of foreign exchange. A new currency, the Sudanese Pound, was introduced in January 2007 at an initial exchange rate of $1.00 equals 2 Sudanese Pounds. Sudan began exporting crude oil in the last quarter of 1999 and the economy boomed on the back of increases in oil production, high oil prices, and significant inflows of foreign direct investment until the second half of 2008. The Darfur conflict, the aftermath of two decades of civil war in the south, the lack of basic infrastructure in large areas, and a reliance by much of the population on subsistence agriculture ensure much of the population will remain at or below the poverty line for years to come despite rapid rises in average per capita income. Sudan's real GDP expanded by 5.2% during 2010, an improvement over 2009's 4.2% growth but significantly below the more than 10% per year growth experienced prior to the global financial crisis in 2006 and 2007. While the oil sector continues to drive growth, services and utilities play an increasingly important role in the economy with agriculture production remaining important as it employs 80% of the work force and contributes a third of GDP. In the lead up to the referendum on southern secession, which took place in January 2011, Sudan saw its currency depreciate considerably on the black market with the Central Bank's official rate also losing value as the Sudanese people started to hoard foreign currency. The Central Bank of Sudan intervened heavily in the currency market to defend the value of the pound and the Sudanese government introduced a number of measures to restrain excess local demand for hard currency, but uncertainty about the secession has meant that foreign exchange remains in heavy demand.
GDP (purchasing power parity): $100 billion (2010 est.)
country comparison to the world: 71
$95.18 billion (2009 est.)
$89.81 billion (2008 est.)
note: data are in 2010 US dollars
GDP (official exchange rate): $68.44 billion (2010 est.)
GDP—real growth rate: 5.1% (2010 est.)
country comparison to the world: 71
6% (2009 est.)
6.8% (2008 est.)
GDP—per capita (PPP): $2,300 (2010 est.)
country comparison to the world: 185
$2,200 (2009 est.)
$2,200 (2008 est.)
note: data are in 2010 US dollars
GDP—composition by sector:
agriculture: 32.1%
industry: 29%
services: 38.9% (2010 est.)
Labor force: 11.92 million (2007 est.)
country comparison to the world: 44
Labor force—by occupation:
agriculture: 80%
industry: 7%
services: 13% (1998 est.)
Unemployment rate: 18.7% (2002 est.)
country comparison to the world: 162
Population below poverty line: 40% (2004 est.)
Household income or consumption by percentage share: *lowest 10%:* NA%
highest 10%: NA%
Investment (gross fixed): 20.2% of GDP (2010 est.)
country comparison to the world: 85
Budget: *revenues:* $11.06 billion
expenditures: $13.15 billion (2010 est.)
Public debt: 94.2% of GDP (2010 est.)
country comparison to the world: 12
105.1% of GDP (2009 est.)
Inflation rate (consumer prices): 11.8% (2010 est.)
country comparison to the world: 203
11.2% (2009 est.)
Stock of narrow money: $7.713 billion (31 December 2010 est.)
country comparison to the world: 76
$7.003 billion (31 December 2009 est.)
Stock of broad money: $13.5 billion (31 December 2010 est.)
country comparison to the world: 92
$12.31 billion (31 December 2009 est.)
Stock of domestic credit: $10.15 billion (31 December 2010 est.)
country comparison to the world: 93
$9.307 billion (31 December 2009 est.)
Market value of publicly traded shares: $NA
Agriculture—products: cotton, groundnuts (peanuts), sorghum, millet, wheat, gum arabic, sugarcane, cassava (tapioca), mangos, papaya, bananas, sweet potatoes, sesame; sheep and other livestock
Industries: oil, cotton ginning, textiles, cement, edible oils, sugar, soap distilling, shoes, petroleum refining, pharmaceuticals, armaments, automobile/light truck assembly
Industrial production growth rate: 3.5% (2010 est.)
country comparison to the world: 96
Electricity—production: 4.341 billion kWh (2007 est.)
country comparison to the world: 118
Electricity—consumption: 3.438 billion kWh (2007 est.)
country comparison to the world: 120
Electricity—exports: 0 kWh (2008 est.)
Electricity—imports: 0 kWh (2008 est.)
Oil—production: 486,700 bbl/day (2009 est.)
country comparison to the world: 31
Oil—consumption: 84,000 bbl/day (2009 est.)
country comparison to the world: 82
Oil—exports: 303,800 bbl/day (2007 est.)
country comparison to the world: 41
Oil—imports: 11,400 bbl/day (2007 est.)
country comparison to the world: 137
Oil—proved reserves: 6.8 billion bbl (1 January 2010 est.)
country comparison to the world: 20
Natural gas—production: 0 cu m (2008 est.)
country comparison to the world: 186
Natural gas—consumption: 0 cu m (2008 est.)
country comparison to the world: 125
Natural gas—exports: 0 cu m (2008 est.)
country comparison to the world: 178
Natural gas—imports: 0 cu m (2008 est.)

country comparison to the world: 121
Natural gas—proved reserves: 84.95 billion cu m (1 January 2010 est.)
country comparison to the world: 56
Current account balance: $-2.595 billion (2010 est.)
country comparison to the world: 163
$-2.817 billion (2009 est.)
Exports: $9.777 billion (2010 est.)
country comparison to the world: 87
$7.56 billion (2009 est.)
Exports—commodities: oil and petroleum products; cotton, sesame, livestock, groundnuts, gum arabic, sugar
Exports—partners: China 58.29%, Japan 14.7%, Indonesia 8.83%, India 4.86% (2009)
Imports: $8.483 billion (2010 est.)
country comparison to the world: 95
$8.253 billion (2009 est.)
Imports—commodities: foodstuffs, manufactured goods, refinery and transport equipment, medicines and chemicals, textiles, wheat
Imports—partners: China 21.87%, Saudi Arabia 7.22%, Egypt 6.1%, India 5.53%, UAE 5.3% (2009)
Reserves of foreign exchange and gold: $2.063 billion (31 December 2010 est.)
country comparison to the world: 102
$897 million (31 December 2009 est.)
Debt—external: $37.98 billion (31 December 2010 est.)
country comparison to the world: 58
$35.71 billion (31 December 2009 est.)
Exchange rates: Sudanese pounds (SDG) per US dollar–2.36 (2010), 2.3 (2009), 2.1 (2008), 2.06 (2007), 2.172 (2006)

COMMUNICATIONS

Telephones—main lines in use: 370,400 (2009)
country comparison to the world: 106
Telephones—mobile cellular: 15.34 million (2009)
country comparison to the world: 47
Telephone system: *general assessment:* well-equipped system by regional standards and being upgraded; cellular communications started in 1996 and have expanded substantially with wide coverage of most major cities
domestic: consists of microwave radio relay, cable, fiber optic, radiotelephone communications, tropospheric scatter, and a domestic satellite system with 14 earth stations
international: country code–249; linked to international submarine cable Fiber-Optic Link Around the Globe (FLAG); satellite earth stations–1 Intelsat (Atlantic Ocean), 1 Arabsat (2000)
Broadcast media: in the north, the Sudanese Government directly controls TV and radio, requiring that both media reflect government policies; TV has a permanent military censor; a private radio station is in operation; in southern Sudan, TV is controlled by the regional government; several private FM stations are operational in southern Sudan; some foreign radio broadcasts are available (2007)
Internet country code: .sd
Internet hosts: 70 (2010)
country comparison to the world: 207
Internet users: 4.2 million (2008)
country comparison to the world: 56

TRANSPORTATION

Airports: 140 (2010)
country comparison to the world: 41
Airports—with paved runways:
total: 19
over 3,047 m: 3
2,438 to 3,047 m: 10
1,524 to 2,437 m: 5
under 914 m: 1 (2010)
Airports—with unpaved runways:
total: 121
1,524 to 2,437 m: 21
914 to 1,523 m: 62
under 914 m: 38 (2010)
Heliports: 5 (2010)
Pipelines: gas 156 km; oil 4,070 km; refined products 1,613 km (2010)
Railways: *total:* 5,978 km
country comparison to the world: 30
narrow gauge: 4,578 km 1.067-m gauge; 1,400 km 0.600-m gauge for cotton plantations (2008)
Roadways: *total:* 11,900 km
country comparison to the world: 131
paved: 4,320 km
unpaved: 7,580 km (2000)
Waterways: 4,068 km (1,723 km open year round on White and Blue Nile rivers) (2008)
country comparison to the world: 25
Merchant marine: *total:* 2
country comparison to the world: 147
by type: cargo 2 (2010)
Ports and terminals: Port Sudan

MILITARY

Military branches: Sudanese Armed Forces (SAF): Land Forces, Navy (includes Marines), Sudanese Air Force (Sikakh al-Jawwiya as-Sudaniya), Popular Defense Forces; Sudan People's Liberation Army (SPLA): Popular Army, Air Force (2011)
Military service age and obligation: 18-33 years of age for male and female compulsory and voluntary military service; 12-24 month service obligation; a requirement that completion of national sevice was mandatory before entering public or private sector employment has been cancelled (2011)
Manpower available for military service: *males age 16–49:* 10,433,973
females age 16–49: 10,411,443 (2010 est.)
Manpower fit for military service: *males age 16–49:* 6,475,530
females age 16–49: 6,840,885 (2010 est.)
Manpower reaching militarily significant age annually: *male:* 532,030
female: 512,476 (2010 est.)
Military expenditures: 3% of GDP (2005 est.)
country comparison to the world: 43

TRANSNATIONAL ISSUES

Disputes—international: the effects of Sudan's almost constant ethnic and rebel militia fighting since the mid-20th century have penetrated all of the neighboring states; as of 2006, Chad, Ethiopia, Kenya, Central African Republic, Democratic Republic of the Congo, and Uganda provided shelter for over half a million Sudanese refugees, which includes 240,000 Darfur residents driven from their homes by Janjawid armed militia and the Sudanese military forces; Sudan, in turn, hosted about 116,000 Eritreans, 20,000 Chadians, and smaller numbers of Ethiopians, Ugandans, Central Africans, and Congolese as refugees; in February 2006, Sudan and DROC signed an agreement to repatriate 13,300 Sudanese and 6,800 Congolese; Sudan accuses Eritrea of supporting Sudanese rebel groups; efforts to demarcate the porous boundary with Ethiopia proceed slowly due to civil and ethnic fighting in eastern Sudan; the boundary that separates Kenya and Sudan's sovereignty is unclear in the "Ilemi Triangle," which Kenya has administered since colonial times; Sudan claims but Egypt de facto administers security and economic development of Halaib region north of the 22nd parallel boundary; periodic violent skirmishes with Sudanese residents over water and grazing rights persist among related pastoral populations along the border with the Central African Republic
Refugees and internally displaced persons: refugees (country of origin): 157,220 (Eritrea); 25,023 (Chad); 11,009 (Ethiopia); 7,895 (Uganda); 5,023 (Central African Republic)
IDPs: 5.3–6.2 million (civil war 1983-2005; ongoing conflict in Darfur region) (2007)
Trafficking in persons: current situation: Sudan is a source country for men, women, and children trafficked internally for the purposes of forced labor and sexual exploitation; Sudan is also a transit and destination country for Ethiopian women trafficked abroad for domestic servitude; Sudanese women and girls are trafficked within the country as well as possibly to Middle Eastern countries for domestic servitude; the terrorist rebel organization, Lord's Resistance Army, continues to harbor small numbers of Sudanese and Ugandan children in the southern part of the country for use as cooks, porters, and combatants; some of these children are also trafficked across borders into Uganda or the Democratic Republic of the Congo; militia groups in Darfur, some of which are linked to the government, abduct women for short periods of forced labor and to perpetrate sexual violence; during the two decades-long north-south civil war, thousands of Dinka women and children were abducted and subsequently enslaved by members of the Missiriya and Rezeigat tribes; while there have been no known new abductions of Dinka by members of Baggara tribes in the last few years, intertribal abductions continue in southern Sudan
tier rating: Tier 3–Sudan does not fully comply with the minimum standards for the elimination of trafficking and is not making significant efforts to do so; combating human trafficking through law enforcement or prevention measures was not a priority for the government in 2007 (2008)

SURINAME

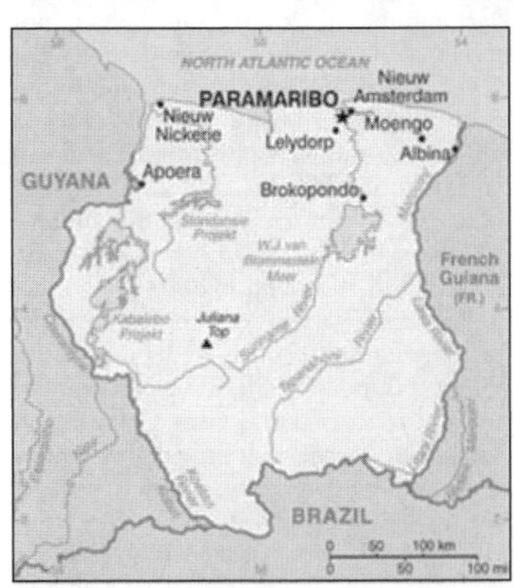

INTRODUCTION

Background: First explored by the Spaniards in the 16th century and then settled by the English in the mid-17th century, Suriname became a Dutch colony in 1667. With the abolition of slavery in 1863, workers were brought in from India and Java. Independence from the Netherlands was granted in 1975. Five years later the civilian government was replaced by a military regime that soon declared a socialist republic. It continued to exert control through a succession of nominally civilian administrations until 1987, when international pressure finally forced a democratic election. In 1990, the military overthrew the civilian leadership, but a democratically elected government–a four-party coalition–returned to power in 1991. The coalition expanded to eight parties in 2005 and ruled until August 2010, when voters returned former military leader Desire BOUTERSE and his opposition coalition to power.

GEOGRAPHY

Location: Northern South America, bordering the North Atlantic Ocean, between French Guiana and Guyana
Geographic coordinates: 4 00 N, 56 00 W
Map references: South America
Area: *total:* 163,820 sq km
country comparison to the world: 91
land: 156,000 sq km
water: 7,820 sq km
Area—comparative: slightly larger than Georgia
Land boundaries: *total:* 1,703 km
border countries: Brazil 593 km, French Guiana 510 km, Guyana 600 km
Coastline: 386 km
Maritime claims: territorial sea: 12 nm
exclusive economic zone: 200 nm
Climate: tropical; moderated by trade winds
Terrain: mostly rolling hills; narrow coastal plain with swamps
Elevation extremes: *lowest point:* unnamed location in the coastal plain -2 m
highest point: Juliana Top 1,230 m
Natural resources: timber, hydropower, fish, kaolin, shrimp, bauxite, gold, and small amounts of nickel, copper, platinum, iron ore
Land use: *arable land:* 0.36%
permanent crops: 0.06%
other: 99.58% (2005)
Irrigated land: 510 sq km (2003)
Total renewable water resources: 122 cu km (2003)
Freshwater withdrawal (domestic/industrial/agricultural): *total:* 0.67 cu km/yr (4%/3%/93%)
per capita: 1,489 cu m/yr (2000)
Natural hazards: NA
Environment—current issues: deforestation as timber is cut for export; pollution of inland waterways by small-scale mining activities
Environment—international agreements: *party to:* Biodiversity, Climate Change, Climate Change-Kyoto Protocol, Desertification, Endangered Species, Law of the Sea, Marine Dumping, Ozone Layer Protection, Ship Pollution, Tropical Timber 94, Wetlands, Whaling
signed, but not ratified: none of the selected agreements
Geography—note: smallest independent country on South American continent; mostly tropical rain forest; great diversity of flora and fauna that, for the most part, is increasingly threatened by new development; relatively small population, mostly along the coast

PEOPLE

Population: 491,989 (July 2011 est.)
country comparison to the world: 172
Age structure: *0–14 years:* 26.4% (male 66,440/female 63,469)
15–64 years: 67.3% (male 164,739/female 166,139)
65 years and over: 6.3% (male 13,300/female 17,902) (2011 est.)
Median age: *total:* 28.7 years
male: 28.3 years
female: 29.1 years (2011 est.)
Population growth rate: 1.087% (2011 est.)
country comparison to the world: 108
Birth rate: 16.42 births/1,000 population (2011 est.)
country comparison to the world: 124
Death rate: 5.54 deaths/1,000 population (July 2011 est.)
country comparison to the world: 175
Net migration rate: 0 migrant(s)/1,000 population (2011 est.)
country comparison to the world: 102
Urbanization: *urban population:* 69% of total population (2010)
rate of urbanization: 1.5% annual rate of change (2010-15 est.)
Major cities—population: PARAMARIBO (capital) 259,000 (2009)
Sex ratio: *at birth:* 1.068 male(s)/female
under 15 years: 1.04 male(s)/female
15–64 years: 0.99 male(s)/female
65 years and over: 0.75 male(s)/female
total population: 0.99 male(s)/female (2011 est.)
Infant mortality rate: *total:* 17.61 deaths/1,000 live births
country comparison to the world: 103
male: 20.79 deaths/1,000 live births
female: 14.2 deaths/1,000 live births (2011 est.)
Life expectancy at birth: *total population:* 74.22 years
country comparison to the world: 103
male: 71.47 years
female: 77.16 years (2011 est.)
Total fertility rate: 1.95 children born/woman (2011 est.)
country comparison to the world: 134
HIV/AIDS—adult prevalence rate: 1% (2009 est.)
country comparison to the world: 46
HIV/AIDS—people living with HIV/AIDS: 3,700 (2009 est.)
country comparison to the world: 124
HIV/AIDS—deaths: fewer than 200 (2009 est.)
country comparison to the world: 104
Major infectious diseases: *degree of risk:* high
food or waterborne diseases: bacterial and protozoal diarrhea, hepatitis A, and typhoid fever
vectorborne disease: dengue fever, Mayaro virus, and malaria
water contact disease: leptospirosis (2009)
Drinking water source: *Improved:*
urban: 97% of population
rural: 81% of population
total: 93% of population
Unimproved:
urban: 3% of population
rural: 19% of population
total: 7% of population (2008)
Sanitation facility access: *Improved:*
urban: 90% of population
rural: 66% of population
total: 84% of population
Unimproved:
urban: 10% of population
rural: 34% of population
total: 16% of population (2008)
Nationality: *noun:* Surinamer(s)
adjective: Surinamese
Ethnic groups: Hindustani (also known locally as "East Indians"; their ancestors emigrated from northern India in the latter part of the 19th century) 37%, Creole (mixed white and black) 31%, Javanese 15%, "Maroons" (their African ancestors were brought to the country in the 17th and 18th centuries as slaves and escaped to the interior) 10%, Amerindian 2%, Chinese 2%, white 1%, other 2%
Religions: Hindu 27.4%, Protestant 25.2% (predominantly Moravian), Roman Catholic 22.8%, Muslim 19.6%, indigenous beliefs 5%
Languages: Dutch (official), English (widely spoken), Sranang Tongo (Surinamese, sometimes called Taki-Taki, is native language of Creoles and much of the younger population and is lingua franca among others), Caribbean Hindustani (a dialect of Hindi), Javanese
Literacy: *definition:* age 15 and over can read and write
total population: 89.6%
male: 92%
female: 87.2% (2004 census)
School life expectancy (primary to tertiary education): *total:* 13 years (2006)
Education expenditures: NA

GOVERNMENT

Country name: *conventional long form:* Republic of Suriname
conventional short form: Suriname
local long form: Republiek Suriname
local short form: Suriname
former: Netherlands Guiana, Dutch Guiana
Government type: constitutional democracy
Capital:
name: Paramaribo
geographic coordinates: 5 50 N, 55 10 W
time difference: UTC-3 (2 hours ahead of Washington, DC during Standard Time)
Administrative divisions: 10 districts (distrikten, singular–distrikt); Brokopondo, Commewijne, Coronie, Marowijne, Nickerie, Para, Paramaribo, Saramacca, Sipaliwini, Wanica
Independence: 25 November 1975 (from the Netherlands)
National holiday: Independence Day, 25 November (1975)
Constitution: ratified 30 September 1987; effective 30 October 1987
Legal system: civil law system influenced by Dutch civil law; note–the Commissie Nieuw Surinaamse Burgerlijk Wetboek completed drafting a new civil code in February 2009
International law organization participation: accepts compulsory ICJ jurisdiction with reservations; accepts ICCt jurisdiction
Suffrage: 18 years of age; universal
Executive branch: *chief of state:* President Desire Delano BOUTERSE (since 12 August 2010); Vice President Robert AMEERALI (since 12 August 2010); note–the president is both the chief of state and head of government
head of government: President Desire Delano BOUTERSE (since 12 August 2010); Vice President Robert AMEERALI (since 12 August 2010)
cabinet: Cabinet of Ministers appointed by the president
(For more information visit the World Leaders website)
elections: president and vice president elected by the National Assembly or, if no presidential or vice presidential candidate receives a two-thirds constitutional majority in the National Assembly after two votes, by a simple majority in the larger United People's Assembly (893 representatives from the national, local, and regional councils), for five-year terms (no term limits); election last held on 19 July 2010 (next to be held in 2015)
election results: Desire Delano BOUTERSE elected president; percent of vote–Desire Delano BOUTERSE 70.6%, Chandrikapersad SATOKHI 25.5%, other 3.9%
Legislative branch: unicameral National Assembly or Nationale Assemblee (51 seats; members elected by popular vote to serve five-year terms)
elections: last held on 25 May 2010 (next to be held in May 2015)
election results: percent of vote by party–Mega Combination 45.1%, New Front 27.5%, A-Com 13.7%, People's Alliance 11.8%, DOE 1.9%; seats by party–Mega Combination 23, New Front 14, A-Com 7, People's Alliance 6, DOE 1
Judicial branch: Cantonal Courts and a Court of Justice as an appellate court (justices are nominated for life); member of the Caribbean Court of Justice (CCJ)
Political parties and leaders: A-Combination (a coalition that includes the General Liberation and Development Party ABOP [Ronnie BRUNSWIJK], SEEKA [Paul ABENA], Union of Brotherhood and Unity in Politics BEP [Caprino ALENDY]; Basic Party for Renewal and Democracy or BVD [Dilip SARDJOE]; Basic Party for Renewal and Democracy or PVF [Soedeschand JAIRAM]; Democratic Union Suriname or DUS [Japhet DIEKO]; Mega-Combination-Ruling Coalition (a coalition that joined with A-Combination and the PL to form a majority in Parliament in 2010–includes the National Democratic Party or NDP [Desire BOUTERSE] (largest party in the coalition), Progressive Worker and Farmer's Union or PALU [Jim HOK], Party for National Unity and Solidarity of the Highest Order or KTPI [Willy SOEMITA], DNP-2000 [Jules WIJDENBOSCH], and New Suriname or NS [Nanan PANDAY]); National Union or NU [P. VAN LEEUWAARDE]; New Front for Democracy and Development or NF (a coalition made up of the National Party of Suriname or NPS [Runaldo VENETIAAN], United Reform Party or VHP [Ramdien SARDJOE], Democratic Alternative 1991 or DA-91–an independent, business-oriented party [Winston JESSURUN], Surinamese Labor Party or SPA [Siegfried GILDS]); Party for Democracy and Development in Unity or DOE [Carl BREEVELD]; Party for the Permanent Prosperity Republic Suriname or PVRS [NA]; People's Alliance, Pertjaja Luhur's or PL [Paul SOMOHARDJO] (includes D-21 [Soewarta MOESTADJA] and Pendawa Lima [Raymond SAPEON], which merged with PL in 2010)
note: BVD and PVF participated in the elections as a coalition (BVD/PVF) in the most recent elections, but separated after the election
Political pressure groups and leaders: Association of Indigenous Village Chiefs [Ricardo PANE]; Association of Saramaccan Authorities or Maroon [Head Captain WASE]; Women's Parliament Forum or PVF [Iris GILLIAD]
International organization participation: ACP, AOSIS, Caricom, FAO, G-77, IADB, IBRD, ICAO, ICRM, IDB, IFAD, IFRCS, IHO, ILO, IMF, IMO, Interpol, IOC, IPU, ISO (correspondent), ITU, ITUC, LAES, MIGA, NAM, OAS, OIC, OPANAL, OPCW, PCA, PetroCaribe, RG, UN, UNASUR, UNCTAD, UNESCO, UNIDO, UPU, WHO, WIPO, WMO, WTO
Diplomatic representation in the US: *chief of mission:* Ambassador Subhas-Chandra MUNGRA
chancery: Suite 460, 4301 Connecticut Avenue NW, Washington, DC 20008
telephone: [1] (202) 244-7488
FAX: [1] (202) 244-5878
consulate(s) general: Miami
Diplomatic representation from the US: *chief of mission:* Ambassador John R. NAY
embassy: Dr. Sophie Redmondstraat 129, Paramaribo
mailing address: US Department of State, PO Box 1821, Paramaribo
telephone: [597] 472-900
FAX: [597] 410-025
Flag description: five horizontal bands of green (top, double width), white, red (quadruple width), white, and green (double width); a large, yellow, five-pointed star is centered in the red band; red stands for progress and love; green symbolizes hope and fertility; white signifies peace, justice, and freedom; the star represents the unity of all ethnic groups; from its yellow light the nation draws strength to bear sacrifices patiently while working toward a golden future
National anthem: *name:* "God zij met ons Suriname!" (God Be With Our Suriname)
lyrics/music: Cornelis Atses HOEKSTRA and Henry DE ZIEL/Johannes Corstianus DE PUY
note: adopted 1959; the anthem, originally adapted from a Sunday school song written in 1893, contains lyrics in both Dutch and Sranan Tongo

ECONOMY

Economy—overview: The economy is dominated by the mining industry, with exports of alumina, gold, and oil accounting for about 85% of exports and 25% of government revenues, making the economy highly vulnerable to mineral price volatility. In 2000, the government of Ronald VENETIAAN, returned to office and inherited an economy with inflation of over 100% and a growing fiscal deficit. He quickly implemented an austerity program, raised taxes, attempted to control spending, and tamed inflation. Economic growth reached about 7% in 2008, owing to sizeable foreign investment in mining and oil. Suriname has received aid for projects in the bauxite and gold mining sectors from Netherlands, Belgium, and the European Development Fund. The economy slowed in 2009, however, as investment waned and the country earned less from its commodity exports when global prices for most commodities fell. Trade picked up, boosting Suriname's economic growth in 2010, but the government's budget remained strained, with increased social spending during the election. In January 2011, the government devalued the currency by 20% and raised taxes to reduce the budget deficit. Suriname's economic prospects for the medium term will depend on continued commitment to responsible monetary and fiscal policies and to the introduction of structural reforms to liberalize markets and promote competition.
GDP (purchasing power parity): $4.711 billion (2010 est.)
country comparison to the world: 165
$4.512 billion (2009 est.)
$4.378 billion (2008 est.)
note: data are in 2010 US dollars
GDP (official exchange rate): $3.682 billion (2010 est.)
GDP—real growth rate: 4.4% (2010 est.)
country comparison to the world: 87
3.1% (2009 est.)
4.7% (2008 est.)
GDP—per capita (PPP): $9,700 (2010 est.)

country comparison to the world: 113
$9,400 (2009 est.)
$9,200 (2008 est.)
note: data are in 2010 US dollars
GDP—composition by sector:
agriculture: 10.8%
industry: 24.4%
services: 64.8% (2005 est.)
Labor force: 165,600 (2007)
country comparison to the world: 176
Labor force—by occupation: *agriculture:* 8%
industry: 14%
services: 78% (2004)
Unemployment rate: 9.5% (2004)
country comparison to the world: 103
Population below poverty line: 70% (2002 est.)
Household income or consumption by percentage share: *lowest 10%:* NA%
highest 10%: NA%
Budget: *revenues:* $392.6 million
expenditures: $425.9 million (2004)
Inflation rate (consumer prices): 6.4% (2007 est.)
country comparison to the world: 165
Commercial bank prime lending rate: 11.65% (31 December 2009 est.)
country comparison to the world: 68
12.2% (31 December 2008 est.)
Stock of narrow money: $608 million (31 December 2009)
country comparison to the world: 154
$495.6 million (31 December 2008)
Stock of broad money: $1.809 billion (31 December 2009)
country comparison to the world: 145
$1.573 billion (31 December 2008)
Stock of domestic credit: $793.1 million (31 December 2008 est.)
country comparison to the world: 155
$651 million (31 December 2007 est.)
Market value of publicly traded shares: $NA
Agriculture—products: paddy rice, bananas, palm kernels, coconuts, plantains, peanuts; beef, chickens; shrimp; forest products
Industries: bauxite and gold mining, alumina production; oil, lumbering, food processing, fishing
Industrial production growth rate: 6.5% (1994 est.)
country comparison to the world: 50
Electricity—production: 1.605 billion kWh (2007 est.)
country comparison to the world: 138
Electricity—consumption: 1.467 billion kWh (2007 est.)
country comparison to the world: 140
Electricity—exports: 0 kWh (2008 est.)
Electricity—imports: 0 kWh (2008 est.)
Oil—production: 15,190 bbl/day (2009 est.)
country comparison to the world: 79
Oil—consumption: 14,000 bbl/day (2009 est.)
country comparison to the world: 137
Oil—exports: 4,308 bbl/day (2007 est.)
country comparison to the world: 107
Oil—imports: 6,296 bbl/day (2007 est.)
country comparison to the world: 151
Oil—proved reserves: 79.6 million bbl (1 January 2010 est.)
country comparison to the world: 75
Natural gas—production: 0 cu m (2008 est.)
country comparison to the world: 170
Natural gas—consumption: 0 cu m (2008 est.)
country comparison to the world: 112
Natural gas—exports: 0 cu m (2008 est.)
country comparison to the world: 152
Natural gas—imports: 0 cu m (2008 est.)
country comparison to the world: 101
Natural gas—proved reserves: 0 cu m (1 January 2010 est.)
country comparison to the world: 175
Current account balance: $24 million (2007 est.)
country comparison to the world: 59
Exports: $1.391 billion (2006 est.)
country comparison to the world: 144
Exports—commodities: alumina, gold, crude oil, lumber, shrimp and fish, rice, bananas
Exports—partners: Canada 35.47%, Belgium 14.92%, US 10.15%, UAE 9.87%, Norway 4.92%, Netherlands 4.7%, France 4.47% (2009)
Imports: $1.297 billion (2006 est.)
country comparison to the world: 168
Imports—commodities: capital equipment, petroleum, foodstuffs, cotton, consumer goods
Imports—partners: US 30.79%, Netherlands 19.17%, Trinidad and Tobago 13.04%, China 6.8%, Japan 5.85% (2009)
Reserves of foreign exchange and gold: $263.3 million (2006)
country comparison to the world: 131
Debt—external: $504.3 million (2005 est.)
country comparison to the world: 160
Exchange rates: Surinamese dollars (SRD) per US dollar–2.745 (2010), 2.745 (2009), 2.745 (2008), 2.745 (2007), 2.7438 (2006)

COMMUNICATIONS

Telephones—main lines in use: 83,700 (2009)
country comparison to the world: 147
Telephones—mobile cellular: 763,900 (2009)
country comparison to the world: 150
Telephone system: *general assessment:* international facilities are good
domestic: combined fixed-line and mobile-cellular teledensity roughly 175 telephones per 100 persons; microwave radio relay network
international: country code–597; satellite earth stations–2 Intelsat (Atlantic Ocean)
Broadcast media: 2 state-owned TV stations; 1 state-owned radio station; multiple private radio and TV stations (2007)
Internet country code: .sr
Internet hosts: 171 (2010)
country comparison to the world: 199
Internet users: 163,000 (2009)
country comparison to the world: 146

TRANSPORTATION

Airports: 51 (2010)
country comparison to the world: 91
Airports—with paved runways: *total:* 5
over 3,047 m: 1
under 914 m: 4 (2010)
Airports—with unpaved runways: *total:* 46
914 to 1,523 m: 5
under 914 m: 41 (2010)
Pipelines: oil 50 km (2010)
Roadways: *total:* 4,304 km
country comparison to the world: 154
paved: 1,130 km
unpaved: 3,174 km (2003)
Waterways: 1,200 km (most navigable by ships with drafts up to 7 m) (2010)
country comparison to the world: 60
Merchant marine: *total:* 1
country comparison to the world: 157
by type: cargo 1 (2008)
Ports and terminals: Paramaribo, Wageningen

MILITARY

Military branches: National Army (Nationaal Leger, NL; includes Marine Section and Air Wing) (2010)
Military service age and obligation: 18 years of age (est.); recruitment is voluntary, with personnel drawn almost exclusively from the Creole community (2007)
Manpower available for military service: *males age 16–49:* 134,218
females age 16–49: 134,439 (2010 est.)
Manpower fit for military service: *males age 16–49:* 109,445
females age 16–49: 112,538 (2010 est.)
Manpower reaching militarily significant age annually: *male:* 4,119
female: 4,106 (2010 est.)
Military expenditures: 0.6% of GDP (2006 est.)
country comparison to the world: 159

TRANSNATIONAL ISSUES

Disputes—international: area claimed by French Guiana between Riviere Litani and Riviere Marouini (both headwaters of the Lawa); Suriname claims a triangle of land between the New and Kutari/Koetari rivers in a historic dispute over the headwaters of the Courantyne; Guyana seeks United Nations Convention on the Law of the Sea (UNCLOS) arbitration to resolve the long-standing dispute with Suriname over the axis of the territorial sea boundary in potentially oil-rich waters
Illicit drugs: growing transshipment point for South American drugs destined for Europe via the Netherlands and Brazil; transshipment point for arms-for-drugs dealing

SVALBARD

(TERRITORY OF NORWAY)

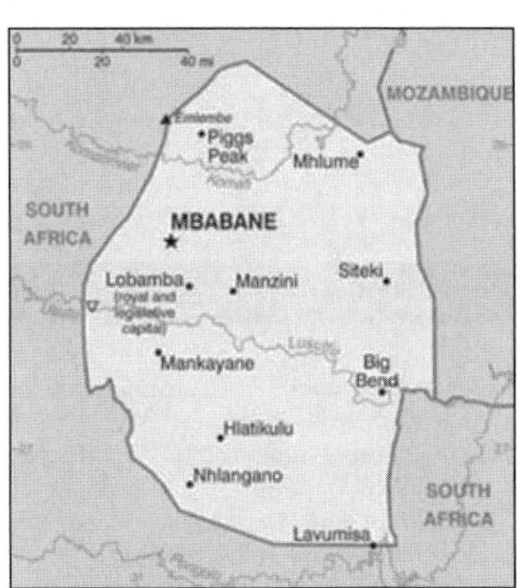

INTRODUCTION

Background: First discovered by the Norwegians in the 12th century, the islands served as an international whaling base during the 17th and 18th centuries. Norway's sovereignty was recognized in 1920; five years later it officially took over the territory.

GEOGRAPHY

Location: Northern Europe, islands between the Arctic Ocean, Barents Sea, Greenland Sea, and Norwegian Sea, north of Norway
Geographic coordinates: 78 00 N, 20 00 E
Map references: Europe
Area: *total:* 62,045 sq km
country comparison to the world: 124
land: 62,045 sq km
water: 0 sq km
note: includes Spitsbergen and Bjornoya (Bear Island)
Area—comparative: slightly smaller than West Virginia
Land boundaries: 0 km
Coastline: 3,587 km
Maritime claims: territorial sea: 4 nm
exclusive fishing zone: 200 nm unilaterally claimed by Norway but not recognized by Russia
Climate: arctic, tempered by warm North Atlantic Current; cool summers, cold winters; North Atlantic Current flows along west and north coasts of Spitsbergen, keeping water open and navigable most of the year
Terrain: wild, rugged mountains; much of high land ice covered; west coast clear of ice about one-half of the year; fjords along west and north coasts
Elevation extremes: *lowest point:* Arctic Ocean 0 m
highest point: Newtontoppen 1,717 m
Natural resources: coal, iron ore, copper, zinc, phosphate, wildlife, fish
Land use:
arable land: 0%
permanent crops: 0%
other: 100% (no trees; the only bushes are crowberry and cloudberry) (2005)
Irrigated land: NA
Natural hazards: ice floes often block the entrance to Bellsund (a transit point for coal export) on the west coast and occasionally make parts of the northeastern coast inaccessible to maritime traffic
Environment—current issues: NA
Geography—note: northernmost part of the Kingdom of Norway; consists of nine main islands; glaciers and snowfields cover 60% of the total area; Spitsbergen Island is the site of the Svalbard Global Seed Vault, a seed repository established by the Global Crop Diversity Trust and the Norwegian Government

PEOPLE

Population: 2,019 (July 2011 est.)
country comparison to the world: 231
Age structure: *0–14 years:* NA
15–64 years: NA
65 years and over: NA (2009 est.)
Population growth rate: -0.024% (2011 est.)
country comparison to the world: 199
Birth rate: NA
Death rate: NA
Net migration rate: NA
Sex ratio: NA
Infant mortality rate: *total:* NA
male: NA
female: NA
Life expectancy at birth: *total population:* NA
male: NA
female: NA
Total fertility rate: NA
HIV/AIDS—adult prevalence rate: 0% (2001)
country comparison to the world: 169
HIV/AIDS—people living with HIV/AIDS: 0 (2001)
country comparison to the world: 164
HIV/AIDS—deaths: 0 (2001)
country comparison to the world: 153
Ethnic groups: Norwegian 55.4%, Russian and Ukrainian 44.3%, other 0.3% (1998)
Languages: Norwegian, Russian
Literacy: NA
Education expenditures: NA

GOVERNMENT

Country name: *conventional long form:* none
conventional short form: Svalbard (sometimes referred to as Spitzbergen)
Dependency status: territory of Norway; administered by the Polar Department of the Ministry of Justice, through a governor (sysselmann) residing in Longyearbyen, Spitsbergen; by treaty (9 February 1920) sovereignty was awarded to Norway
Government type: NA
Capital: *name:* Longyearbyen
geographic coordinates: 78 13 N, 15 33 E
time difference: UTC+1 (6 hours ahead of Washington, DC during Standard Time)
daylight saving time: +1hr, begins last Sunday in March; ends last Sunday in October
Independence: none (territory of Norway)
Legal system: the laws of Norway where applicable apply
Executive branch: *chief of state:* King HARALD V of Norway (since 17 January 1991)
head of government: Governor Odd Olsen INGERO (since September 2009); Assistant Governor Lars FAUSE (since September 2008)
elections: none; the monarchy is hereditary; governor and assistant governor responsible to the Polar Department of the Ministry of Justice
Political pressure groups and leaders: NA
International organization participation: none
Flag description: the flag of Norway is used
National anthem: note: as a territory of Norway, "Ja, vi elsker dette landet" is official (see Norway)

ECONOMY

Economy—overview: Coal mining, tourism, and international research are the major revenue sources on Svalbard. Coal mining is the dominant economic activity and a treaty of 9 February 1920 gave the 41 signatories equal rights to exploit mineral deposits, subject to Norwegian regulation. Although US, UK, Dutch, and Swedish coal companies have mined in the past, the only companies still engaging in this are Norwegian and Russian. The settlements on Svalbard are essentially company towns. The Norwegian state-owned coal company employs nearly 60% of the Norwegian population on the island, runs many of the local services, and provides most of the local infrastructure. There is also some hunting of seal, reindeer, and fox.
GDP (purchasing power parity): $NA
GDP—real growth rate: NA%
Labor force: 1,234 in Norwegian settlements (2003)
country comparison to the world: 225
Budget:
revenues: $25.07 million
expenditures: $NA (2004 est.)
Exports: $197.6 million (2000)
Imports: $NA
Exchange rates: Norwegian kroner (NOK) per US dollar–6.044 (2010), 6.288 (2009), 5.86 (2007), 6.418 (2006)

COMMUNICATIONS

Telephones—main lines in use: NA
Telephone system: *general assessment:* probably adequate
domestic: local telephone service
international: country code–47-790; satellite earth station–1 of unknown type (for communication with Norwegian mainland only)
Broadcast media: the Norwegian Broadcasting Corporation (NRK) began direct television transmission to Svalbard via satellite in 1984; Longyearbyen households have access to 3 NRK radio and 2 television stations (2008)
Internet country code: .sj

TRANSPORTATION

Airports: 4 (2010)
country comparison to the world: 187
Airports—with paved runways:
total: 1
2,438 to 3,047 m: 1 (2010)

Airports—with unpaved runways:
total: 3
under 914 m: 3 (2010)
Heliports: 1 (2010)
Ports and terminals: Barentsburg, Longyearbyen, Ny-Alesund, Pyramiden

MILITARY

Military branches: no regular military forces
Military—note: Svalbard is a territory of Norway, demilitarized by treaty on 9 February 1920; Norwegian military activity is limited to fisheries surveillance by the Norwegian Coast Guard

TRANSNATIONAL ISSUES

Disputes—international: despite recent discussions, Russia and Norway dispute their maritime limits in the Barents Sea and Russia's fishing rights beyond Svalbard's territorial limits within the Svalbard Treaty zone

SWAZILAND

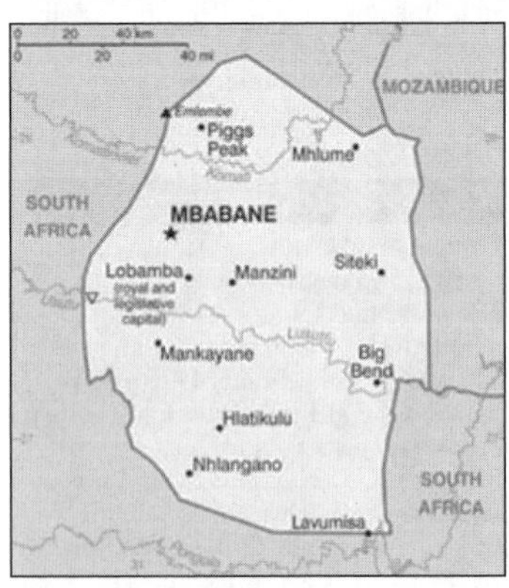

INTRODUCTION

Background: Autonomy for the Swazis of southern Africa was guaranteed by the British in the late 19th century; independence was granted in 1968. Student and labor unrest during the 1990s pressured King MSWATI III, the world's last absolute monarch, to grudgingly allow political reform and greater democracy, although he has backslid on these promises in recent years. A constitution came into effect in 2006, but the legal status of political parties remains unclear. The African United Democratic Party tried unsuccessfully to register as an official political party in mid 2006. Talks over the constitution broke down between the government and progressive groups in 2007. Swaziland recently surpassed Botswana as the country with the world's highest known HIV/AIDS prevalence rate.

GEOGRAPHY

Location: Southern Africa, between Mozambique and South Africa
Geographic coordinates: 26 30 S, 31 30 E
Map references: Africa
Area: *total:* 17,364 sq km
country comparison to the world: 158
land: 17,204 sq km
water: 160 sq km
Area—comparative: slightly smaller than New Jersey
Land boundaries: *total:* 535 km
border countries: Mozambique 105 km, South Africa 430 km
Coastline: 0 km (landlocked)
Maritime claims: none (landlocked)
Climate: varies from tropical to near temperate
Terrain: mostly mountains and hills; some moderately sloping plains
Elevation extremes: *lowest point:* Great Usutu River 21 m
highest point: Emlembe 1,862 m
Natural resources: asbestos, coal, clay, cassiterite, hydropower, forests, small gold and diamond deposits, quarry stone, and talc
Land use: *arable land:* 10.25%
permanent crops: 0.81%
other: 88.94% (2005)
Irrigated land: 500 sq km (2003)
Total renewable water resources: 4.5 cu km (1987)
Freshwater withdrawal (domestic/industrial/agricultural): *total:* 1.04 cu km/yr (2%/1%/97%)
per capita: 1,010 cu m/yr (2000)
Natural hazards: drought
Environment—current issues: limited supplies of potable water; wildlife populations being depleted because of excessive hunting; overgrazing; soil degradation; soil erosion
Environment—international agreements: *party to:* Biodiversity, Climate Change, Climate Change-Kyoto Protocol, Desertification, Endangered Species, Hazardous Wastes, Ozone Layer Protection
signed, but not ratified: Law of the Sea
Geography—note: landlocked; almost completely surrounded by South Africa

PEOPLE

Population: 1,370,424 (July 2011 est.)
country comparison to the world: 152
note: estimates for this country explicitly take into account the effects of excess mortality due to AIDS; this can result in lower life expectancy, higher infant mortality, higher death rates, lower population growth rates, and changes in the distribution of population by age and sex than would otherwise be expected
Age structure: *0–14 years:* 37.8% (male 261,762/female 255,828)
15–64 years: 58.6% (male 399,746/female 403,681)
65 years and over: 3.6% (male 20,472/female 28,935) (2011 est.)
Median age: *total:* 20.3 years
male: 19.9 years
female: 20.7 years (2011 est.)
Population growth rate: 1.204% (2011 est.)
country comparison to the world: 97
Birth rate: 26.63 births/1,000 population (2011 est.)
country comparison to the world: 51
Death rate: 14.6 deaths/1,000 population (July 2011 est.)
country comparison to the world: 12
Net migration rate: 0 migrant(s)/1,000 population (2011 est.)
country comparison to the world: 113
Urbanization: *urban population:* 21% of total population (2010)
rate of urbanization: 1.5% annual rate of change (2010-15 est.)
Major cities—population: MBABANE (capital) 74,000 (2009)
Sex ratio: *at birth:* 1.03 male(s)/female
under 15 years: 1.02 male(s)/female
15–64 years: 0.98 male(s)/female
65 years and over: 0.72 male(s)/female
total population: 0.99 male(s)/female (2011 est.)
Infant mortality rate: *total:* 63.09 deaths/1,000 live births
country comparison to the world: 26
male: 67.14 deaths/1,000 live births
female: 58.93 deaths/1,000 live births (2011 est.)
Life expectancy at birth: *total population:* 48.66 years
country comparison to the world: 218
male: 48.93 years
female: 48.39 years (2011 est.)
Total fertility rate: 3.11 children born/woman (2011 est.)
country comparison to the world: 59
HIV/AIDS—adult prevalence rate: 25.9% (2009 est.)
country comparison to the world: 1
HIV/AIDS—people living with HIV/AIDS: 180,000 (2009 est.)
country comparison to the world: 30
HIV/AIDS—deaths: 7,000 (2009 est.)
country comparison to the world: 31
Major infectious diseases: *degree of risk:* high
food or waterborne diseases: bacterial diarrhea, hepatitis A, and typhoid fever
vectorborne disease: malaria
water contact disease: schistosomiasis (2009)
Drinking water source: *Improved:*
urban: 92% of population
rural: 61% of population
total: 69% of population
Unimproved:
urban: 8% of population
rural: 39% of population
total: 31% of population (2008)
Sanitation facility access: *Improved:*
urban: 61% of population
rural: 53% of population
total: 55% of population
Unimproved:
urban: 39% of population
rural: 47% of population
total: 45% of population (2008)
Nationality: *noun:* Swazi(s)
adjective: Swazi
Ethnic groups: African 97%, European 3%
Religions: Zionist 40% (a blend of Christianity and indigenous ancestral worship), Roman Catholic 20%, Muslim 10%, other

(includes Anglican, Bahai, Methodist, Mormon, Jewish) 30%
Languages: English (official, used for government business), siSwati (official)
Literacy: *definition:* age 15 and over can read and write
total population: 81.6%
male: 82.6%
female: 80.8% (2003 est.)
School life expectancy (primary to tertiary education): *total:* 11 years
male: 11 years
female: 10 years (2007)
Education expenditures: 7.8% of GDP (2008)
country comparison to the world: 12

GOVERNMENT

Country name: *conventional long form:* Kingdom of Swaziland
conventional short form: Swaziland
local long form: Umbuso weSwatini
local short form: eSwatini
Government type: monarchy
Capital: *name:* Mbabane
geographic coordinates: 26 18 S, 31 06 E
time difference: UTC+2 (7 hours ahead of Washington, DC during Standard Time)
note: Lobamba (royal and legislative capital)
Administrative divisions: 4 districts; Hhohho, Lubombo, Manzini, Shiselweni
Independence: 6 September 1968 (from the UK)
National holiday: Independence Day, 6 September (1968)
Constitution: signed by the King July 2005; went into effect 8 February 2006
Legal system: mixed legal system of civil, common, and customary law
International law organization participation: accepts compulsory ICJ jurisdiction with reservations; non-party state to the ICCt
Suffrage: 18 years of age
Executive branch: *chief of state:* King MSWATI III (since 25 April 1986)
head of government: Prime Minister Barnabas Sibusiso DLAMINI (since 16 October 2008)
cabinet: Cabinet recommended by the prime minister and confirmed by the monarch
(For more information visit the World Leaders website)
elections: none; the monarchy is hereditary; prime minister appointed by the monarch from among the elected members of the House of Assembly
Legislative branch: bicameral Parliament or Libandla consists of the Senate (30 seats; 10 members appointed by the House of Assembly and 20 appointed by the monarch; members to serve five-year terms) and the House of Assembly (65 seats; 10 members appointed by the monarch and 55 elected by popular vote; members to serve five-year terms)
elections: House of Assembly—last held on 19 September 2008 (next to be held in 2013)
election results: House of Assembly—balloting is done on a nonparty basis; candidates for election nominated by the local council of each constituency and for each constituency the three candidates with the most votes in the first round of voting are narrowed to a single winner by a second round
Judicial branch: High Court; Supreme Court; judges for both courts are appointed by the monarch
Political parties and leaders: the status of political parties, previously banned, is unclear under the 2006 Constitution and currently being debated; the following are considered political associations; African United Democratic Party or AUDP [Stanley MAUNDZISA, president]; Imbokodvo National Movement or INM; Ngwane National Liberatory Congress or NNLC [Obed DLAMINI, president]; People's United Democratic Movement or PUDEMO [Mario MASUKU, president]
Political pressure groups and leaders: Swaziland Democracy Campaign; Swaziland Federation of Trade Unions; Swaziland Solidarity Network or SSN
International organization participation: ACP, AfDB, AU, C, COMESA, FAO, G-77, IBRD, ICAO, ICRM, IDA, IFAD, IFC, IFRCS, ILO, IMF, IMO, Interpol, IOC, IOM, ISO (correspondent), ITSO, ITU, ITUC, MIGA, NAM, OPCW, PCA, SACU, SADC, UN, UNCTAD, UNESCO, UNIDO, UNWTO, UPU, WCO, WHO, WIPO, WMO, WTO
Diplomatic representation in the US: *chief of mission:* Ambassador Abednigo Mandla NTSHANGASE
chancery: 1712 New Hampshire Avenue, NW, Washington, DC 20009
telephone: [1] (202) 234-5002
FAX: [1] (202) 234-8254
Diplomatic representation from the US: *chief of mission:* Ambassador Earl M. IRVING
embassy: 2350 Mbabane Place, Mbabane
mailing address: P. O. Box 199, Mbabane
telephone: [268] 404-2445
FAX: [268] 404-2059
Flag description: three horizontal bands of blue (top), red (triple width), and blue; the red band is edged in yellow; centered in the red band is a large black and white shield covering two spears and a staff decorated with feather tassels, all placed horizontally; blue stands for peace and stability, red represents past struggles, and yellow the mineral resources of the country; the shield, spears, and staff symbolize protection from the country's enemies, while the black and white of the shield are meant to portray black and white people living in peaceful coexistence
National anthem: *name:* "Nkulunkulu Mnikati wetibusiso temaSwati" (Oh God, Bestower of the Blessings of the Swazi)
lyrics/music: Andrease Enoke Fanyana SIMELANE/David Kenneth RYCROFT
note: adopted 1968; the anthem uses elements of both ethnic Swazi and Western music styles

ECONOMY

Economy—overview: In this small, landlocked economy, subsistence agriculture occupies approximately 70% of the population. The manufacturing sector has diversified since the mid-1980s. Sugar and wood pulp were major foreign exchange earners; however, the wood pulp producer closed in January 2010, and sugar is now the main export earner. In 2007, the sugar industry increased efficiency and diversification efforts, in response to a 17% decline in EU sugar prices. Mining has declined in importance in recent years with only coal and quarry stone mines remaining active. Surrounded by South Africa, except for a short border with Mozambique, Swaziland is heavily dependent on South Africa from which it receives more than nine-tenths of its imports and to which it sends 60% of its exports. Swaziland's currency is pegged to the South African rand, subsuming Swaziland's monetary policy to South Africa. The government is heavily dependent on customs duties from the Southern African Customs Union (SACU), and worker remittances from South Africa substantially supplement domestically earned income. The government has also legislated that 30% of local pension funds need to be invested in Swaziland, boosting demand for government bonds. Customs revenues plummeted due to the global economic crisis and a drop in South African imports. The resulting decline in revenue has pushed the country into a fiscal crisis. The government has requested assistance from the IMF and from the African Development Bank. With an estimated 40% unemployment rate, Swaziland's need to increase the number and size of small and medium enterprises and attract foreign direct investment is acute. Overgrazing, soil depletion, drought, and floods persist as problems for the future. More than one-fourth of the population needed emergency food aid in 2006-07 because of drought, and more than one-quarter of the adult population has been infected by HIV/AIDS.
GDP (purchasing power parity): $6.067 billion (2010 est.)
country comparison to the world: 156
$5.949 billion (2009 est.)
$5.881 billion (2008 est.)
note: data are in 2010 US dollars
GDP (official exchange rate): $3.553 billion (2010 est.)
GDP—real growth rate: 2% (2010 est.)
country comparison to the world: 151
1.2% (2009 est.)
3.1% (2008 est.)
GDP—per capita (PPP): $4,500 (2010 est.)
country comparison to the world: 154
$4,400 (2009 est.)
$4,500 (2008 est.)
note: data are in 2010 US dollars
GDP—composition by sector:
agriculture: 8.6%
industry: 42%
services: 49.4% (2010 est.)
Labor force: 457,900 (2007)
country comparison to the world: 156
Labor force—by occupation: *agriculture:* 70%
industry: NA%
services: NA%
Unemployment rate: 40% (2006 est.)
country comparison to the world: 186
Population below poverty line: 69% (2006)
Household income or consumption by percentage share: *lowest 10%:* 1.6%

highest 10%: 40.7% (2001)
Distribution of family income—Gini index: 50.4 (2001)
country comparison to the world: 21
Investment (gross fixed): 12.6% of GDP (2010 est.)
country comparison to the world: 143
Budget: *revenues:* $961.7 million
expenditures: $1.379 billion (2010 est.)
Inflation rate (consumer prices): 5% (2010 est.)
country comparison to the world: 141
7.3% (2009 est.)
Central bank discount rate: 6.5% (31 December 2009)
country comparison to the world: 36
11% (31 December 2008)
Commercial bank prime lending rate: 11.38% (31 December 2009 est.)
country comparison to the world: 45
14.83% (31 December 2008 est.)
Stock of narrow money: $335.7 million (31 December 2010 est.)
country comparison to the world: 163
$273.9 million (31 December 2009 est.)
Stock of broad money: $1.266 billion (31 December 2010 est.)
country comparison to the world: 154
$920.7 million (31 December 2009 est.)
Stock of domestic credit: $258.5 million (31 December 2010 est.)
country comparison to the world: 172
$274.5 million (31 December 2009 est.)
Market value of publicly traded shares: $NA (31 December 2009)
country comparison to the world: 113
$203.1 million (31 December 2007)
$199.9 million (31 December 2006)
Agriculture—products: sugarcane, cotton, corn, tobacco, rice, citrus, pineapples, sorghum, peanuts; cattle, goats, sheep
Industries: coal, wood pulp, sugar, soft drink concentrates, textiles and apparel
Industrial production growth rate: 1% (2010 est.)
country comparison to the world: 144
Electricity—production: 441 million kWh (2007 est.)
country comparison to the world: 161
Electricity—consumption: 1.266 billion kWh (2007 est.)
country comparison to the world: 144
Electricity—exports: 0 kWh (2008)
Electricity—imports: 770 million kWh; note—electricity supplied by South Africa (2008 est.)
Oil—production: 0 bbl/day (2009 est.)
country comparison to the world: 144
Oil—consumption: 4,000 bbl/day (2009 est.)
country comparison to the world: 174
Oil—exports: 0 bbl/day (2007 est.)
country comparison to the world: 143
Oil—imports: 4,100 bbl/day (2007 est.)
country comparison to the world: 167
Oil—proved reserves: 0 bbl (1 January 2010 est.)
country comparison to the world: 207
Natural gas—production: 0 cu m (2008 est.)
country comparison to the world: 204
Natural gas—consumption: 0 cu m (2008 est.)
country comparison to the world: 140
Natural gas—exports: 0 cu m (2008 est.)
country comparison to the world: 204
Natural gas—imports: 0 cu m (2008 est.)
country comparison to the world: 72
Natural gas—proved reserves: 0 cu m (1 January 2010 est.)
country comparison to the world: 205
Current account balance: $-374 million (2010 est.)
country comparison to the world: 107
$-213 million (2009 est.)
Exports: $1.417 billion (2010 est.)
country comparison to the world: 141
$1.338 billion (2009 est.)
Exports—commodities: soft drink concentrates, sugar, wood pulp, cotton yarn, refrigerators, citrus and canned fruit
Imports: $1.643 billion (2010 est.)
country comparison to the world: 158
$1.585 billion (2009 est.)
Imports—commodities: motor vehicles, machinery, transport equipment, foodstuffs, petroleum products, chemicals
Reserves of foreign exchange and gold: $708 million (31 December 2010 est.)
country comparison to the world: 121
$959 million (31 December 2009 est.)
Debt—external: $497 million (31 December 2010 est.)
country comparison to the world: 161
$411 million (31 December 2009 est.)
Stock of direct foreign investment—at home: $NA
Stock of direct foreign investment—abroad: $NA
Exchange rates: emalangeni per US dollar—7.57 (2010), 8.42 (2009), 7.75 (2008), 7.4 (2007)
6.85 (2006)

COMMUNICATIONS

Telephones—main lines in use: 44,000 (2009)
country comparison to the world: 166
Telephones—mobile cellular: 656,000 (2009)
country comparison to the world: 154
Telephone system: *general assessment:* a somewhat modern but not an advanced system
domestic: single source for mobile-cellular service with a geographic coverage of about 90% and a rising subscribership base; combined fixed-line and mobile cellular teledensity exceeded 60 telephones per 100 persons in 2009; telephone system consists of carrier-equipped, open-wire lines and low-capacity, microwave radio relay
international: country code—268; satellite earth station—1 Intelsat (Atlantic Ocean) (2009)
Broadcast media: state-owned TV station; satellite dishes are able to access South African providers; state-owned radio network with 3 channels; 1 private radio station (2007)
Internet country code: .sz
Internet hosts: 2,335 (2010)
country comparison to the world: 152
Internet users: 90,100 (2009)
country comparison to the world: 162

TRANSPORTATION

Airports: 15 (2010)
country comparison to the world: 145
Airports—with paved runways: *total:* 2
over 3,047 m: 1
2,438 to 3,047 m: 1 (2010)
Airports—with unpaved runways: *total:* 13
914 to 1,523 m: 6
under 914 m: 7 (2010)
Railways: *total:* 301 km
country comparison to the world: 120
narrow gauge: 301 km 1.067-m gauge (2008)
Roadways: *total:* 3,594 km
country comparison to the world: 160
paved: 1,078 km
unpaved: 2,516 km (2002)

MILITARY

Military branches: Umbutfo Swaziland Defense Force (USDF): Ground Force (includes Air Wing) (2010)
Military service age and obligation: 18-30 years of age for male and female voluntary military service; no conscription; compulsory HIV testing required, only HIV-negative applicants accepted (2010)
Manpower available for military service: *males age 16–49:* 344,038 (2010 est.)
Manpower fit for military service: *males age 16–49:* 201,853
females age 16–49: 175,477 (2010 est.)
Manpower reaching militarily significant age annually: *male:* 16,168
female: 15,763 (2010 est.)
Military expenditures: 4.7% of GDP (2006)
country comparison to the world: 18

TRANSNATIONAL ISSUES

Disputes—international: in 2006, Swazi king advocates resort to ICJ to claim parts of Mpumalanga and KwaZulu-Natal from South Africa
Trafficking in persons: current situation: Swaziland is a source, destination, and transit country for women and children trafficked internally and transnationally for the purposes of commercial sexual exploitation, domestic servitude, and forced labor in agriculture; Swazi girls, particularly orphans, are trafficked internally for commercial sexual exploitation and domestic servitude, as well as to South Africa and Mozambique; Swazi boys are trafficked for forced labor in commercial agriculture and market vending; some Swazi women are forced into prostitution in South Africa and Mozambique after voluntarily migrating to these countries in search of work
tier rating: Tier 2 Watch List—the government of Swaziland does not comply with the minimum standards for the elimination of trafficking and is not making significant efforts to do so; the government believes that trafficking probably does occur, but does not know the extent of the problem; the government does not judge trafficking to be an "important" problem and chooses to direct its limited resources towards other issues, a judgment which significantly limited the government's current efforts to eliminate human trafficking, or to plan anti-trafficking activities or initiatives for the future (2010)

SWEDEN

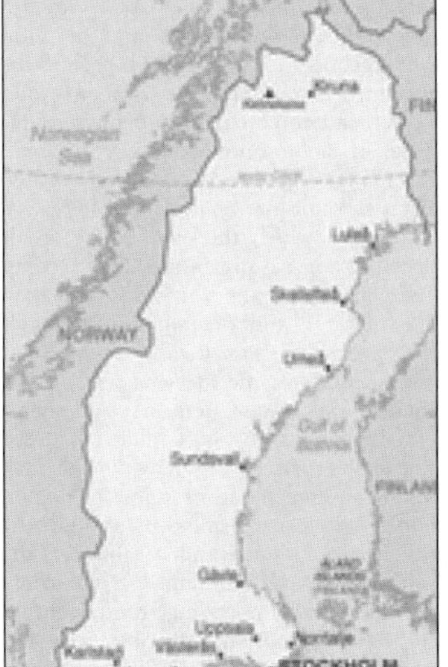

INTRODUCTION

Background: A military power during the 17th century, Sweden has not participated in any war for almost two centuries. An armed neutrality was preserved in both world wars. Sweden's long-successful economic formula of a capitalist system interlarded with substantial welfare elements was challenged in the 1990s by high unemployment and in 2000-02 and 2009 by the global economic downturns, but fiscal discipline over the past several years has allowed the country to weather economic vagaries. Sweden joined the EU in 1995, but the public rejected the introduction of the euro in a 2003 referendum.

GEOGRAPHY

Location: Northern Europe, bordering the Baltic Sea, Gulf of Bothnia, Kattegat, and Skagerrak, between Finland and Norway
Geographic coordinates: 62 00 N, 15 00 E
Map references: Europe
Area: *total:* 450,295 sq km
country comparison to the world: 55
land: 410,335 sq km
water: 39,960 sq km
Area—comparative: slightly larger than California
Land boundaries: *total:* 2,233 km
border countries: Finland 614 km, Norway 1,619 km
Coastline: 3,218 km
Maritime claims: territorial sea: 12 nm (adjustments made to return a portion of straits to high seas)
exclusive economic zone: agreed boundaries or midlines
continental shelf: 200 m depth or to the depth of exploitation
Climate: temperate in south with cold, cloudy winters and cool, partly cloudy summers; subarctic in north
Terrain: mostly flat or gently rolling lowlands; mountains in west
Elevation extremes: *lowest point:* reclaimed bay of Lake Hammarsjon, near Kristianstad -2.4 m
highest point: Kebnekaise 2,111 m
Natural resources: iron ore, copper, lead, zinc, gold, silver, tungsten, uranium, arsenic, feldspar, timber, hydropower
Land use: *arable land:* 5.93%
permanent crops: 0.01%
other: 94.06% (2005)
Irrigated land: 1,150 sq km (2003)
Total renewable water resources: 179 cu km (2005)
Freshwater withdrawal (domestic/industrial/agricultural): *total:* 2.68 cu km/yr (37%/54%/9%)
per capita: 296 cu m/yr (2002)
Natural hazards: ice floes in the surrounding waters, especially in the Gulf of Bothnia, can interfere with maritime traffic
Environment—current issues: acid rain damage to soils and lakes; pollution of the North Sea and the Baltic Sea
Environment—international agreements: *party to:* Air Pollution, Air Pollution-Nitrogen Oxides, Air Pollution-Persistent Organic Pollutants, Air Pollution-Sulfur 85, Air Pollution-Sulfur 94, Air Pollution-Volatile Organic Compounds, Antarctic-Environmental Protocol, Antarctic-Marine Living Resources, Antarctic Treaty, Biodiversity, Climate Change, Climate Change-Kyoto Protocol, Desertification, Endangered Species, Environmental Modification, Hazardous Wastes, Law of the Sea, Marine Dumping, Ozone Layer Protection, Ship Pollution, Tropical Timber 83, Tropical Timber 94, Wetlands, Whaling
signed, but not ratified: none of the selected agreements
Geography—note: strategic location along Danish Straits linking Baltic and North Seas

PEOPLE

Population: 9,088,728 (July 2011 est.)
country comparison to the world: 90
Age structure: *0–14 years:* 15.4% (male 722,558/female 680,933)
15–64 years: 64.8% (male 2,982,268/female 2,910,135)
65 years and over: 19.7% (male 800,169/female 992,665) (2011 est.)
Median age:
total: 42 years
male: 40.8 years
female: 43.1 years (2011 est.)
Population growth rate: 0.163% (2011 est.)
country comparison to the world: 183
Birth rate: 10.18 births/1,000 population (2011 est.)
country comparison to the world: 189
Death rate: 10.2 deaths/1,000 population (July 2011 est.)
country comparison to the world: 50
Net migration rate: 1.65 migrant(s)/1,000 population (2011 est.)
country comparison to the world: 44
Urbanization: *urban population:* 85% of total population (2010)
rate of urbanization: 0.6% annual rate of change (2010-15 est.)
Major cities—population: STOCKHOLM (capital) 1.279 million (2009)
Sex ratio: *at birth:* 1.061 male(s)/female
under 15 years: 1.06 male(s)/female
15–64 years: 1.02 male(s)/female
65 years and over: 0.8 male(s)/female
total population: 0.98 male(s)/female (2011 est.)
Infant mortality rate: *total:* 2.74 deaths/1,000 live births
country comparison to the world: 219
male: 2.9 deaths/1,000 live births
female: 2.57 deaths/1,000 live births (2011 est.)
Life expectancy at birth: *total population:* 81.07 years
country comparison to the world: 16
male: 78.78 years
female: 83.51 years (2011 est.)
Total fertility rate: 1.67 children born/woman (2011 est.)
country comparison to the world: 171
HIV/AIDS—adult prevalence rate: 0.1% (2009 est.)
country comparison to the world: 162
HIV/AIDS—people living with HIV/AIDS: 8,100 (2009 est.)
country comparison to the world: 108
HIV/AIDS—deaths: fewer than 100 (2009 est.)
country comparison to the world: 115
Drinking water source: *Improved:*
urban: 100% of population
rural: 100% of population
total: 100% of population (2008)
Sanitation facility access: *Improved:*
urban: 100% of population
rural: 100% of population
total: 100% of population (2008)
Nationality: *noun:* Swede(s)
adjective: Swedish
Ethnic groups: indigenous population: Swedes with Finnish and Sami minorities; foreign-born or first-generation immigrants: Finns, Yugoslavs, Danes, Norwegians, Greeks, Turks
Religions: Lutheran 87%, other (includes Roman Catholic, Orthodox, Baptist, Muslim, Jewish, and Buddhist) 13%
Languages: Swedish (official), small Sami- and Finnish-speaking minorities
Literacy: *definition:* age 15 and over can read and write
total population: 99%
male: 99%
female: 99% (2003 est.)
School life expectancy (primary to tertiary education): *total:* 16 years
male: 15 years
female: 16 years (2008)
Education expenditures: 6.6% of GDP (2007)
country comparison to the world: 21

GOVERNMENT

Country name: *conventional long form:* Kingdom of Sweden
conventional short form: Sweden
local long form: Konungariket Sverige
local short form: Sverige
Government type: constitutional monarchy
Capital: *name:* Stockholm
geographic coordinates: 59 20 N, 18 03 E
time difference: UTC+1 (6 hours ahead of Washington, DC during Standard Time)
daylight saving time: +1hr, begins last Sunday in March; ends last Sunday in October
Administrative divisions: 21 counties (lan, singular and plural); Blekinge, Dalarna, Gavleborg, Gotland, Halland, Jamtland, Jonkoping, Kalmar, Kronoberg, Norrbotten, Orebro, Ostergotland, Skane, Sodermanland, Stockholm, Uppsala, Varmland, Vasterbotten, Vasternorrland, Vastmanland, Vastra Gotaland
Independence: 6 June 1523 (Gustav VASA elected king)
National holiday: Swedish Flag Day, 6 June (1916); National Day, 6 June (1983)
Constitution: 1 January 1975
Legal system: civil law system influenced by Roman-Germanic law and customary law
International law organization participation: accepts compulsory ICJ jurisdiction with reservations; accepts ICCt jurisdiction
Suffrage: 18 years of age; universal
Executive branch: *chief of state:* King CARL XVI GUSTAF (since 19 September 1973); Heir Apparent Princess VICTORIA Ingrid Alice Desiree, daughter of the monarch (born 14 July 1977)
head of government: Prime Minister Fredrik REINFELDT (since 5 October 2006); Deputy Prime Minister Jan BJORKLUND (since 5 October 2010)
cabinet: Cabinet appointed by the prime minister
(For more information visit the World Leaders website)
elections: the monarchy is hereditary; following legislative elections, the leader of the majority party or the leader of the majority coalition usually becomes the prime minister
Legislative branch: unicameral Parliament or Riksdag (349 seats; members are elected by popular vote on a proportional representation basis to serve four-year terms)
elections: last held on 19 September 2010 (next to be held in September 2014)
election results: percent of vote by party—Social Democrats 30.7%, Moderates 30.1%, Greens 7.3%, Liberal People's Party 7.1%, Center Party 6.6%, Sweden Democrats 5.7%, Christian Democrats 5.6%, Left Party 5.6%, others 1.3%; seats by party—Social Democrats 112, Moderates 107, Greens 25, Liberal People's Party 24, Center Party 23, Sweden Democrats 20, Christian Democrats 19, Left Party 19
Judicial branch: Supreme Court or Hogsta Domstolen (judges are appointed by the prime minister and the cabinet)
Political parties and leaders: Center Party [Maud OLOFSSON]; Christian Democratic Party [Goran HAGGLUND]; Environment Party the Greens [no formal leader but party spokespersons are Maria WETTERSTRAND and Peter ERIKSSON]; Left Party or V (formerly Communist) [Lars OHLY]; Liberal People's Party [Jan BJORKLUND]; Moderate Party [Fredrik REINFELDT]; Social Democratic Party [Mona SAHLIN]; Sweden Democrats [Jimmie AKESSON]
Political pressure groups and leaders: Children's Rights in Society; Swedish Confederation of Professional Employees or TCO; Swedish Federation of Trade Unions or LO
other: media
International organization participation: ADB (nonregional member), AfDB (nonregional member), Arctic Council, Australia Group, BIS, CBSS, CE, CERN, EAPC, EBRD, EIB, ESA, EU, FAO, FATF, G-9, G-10, IADB, IAEA, IBRD, ICAO, ICC, ICRM, IDA, IEA, IFAD, IFC, IFRCS, IHO, ILO, IMF, IMO, IMSO, Interpol, IOC, IOM, IPU, ISO, ITSO, ITU, ITUC, MIGA, MONUSCO, NC, NEA, NIB, NSG, OAS (observer), OECD, OPCW, OSCE, Paris Club, PCA, PFP, Schengen Convention, UN, UNCTAD, UNESCO, UNHCR, UNIDO, UNMIS, UNMOGIP, UNRWA, UNTSO, UPU, WCO, WFTU, WHO, WIPO, WMO, WTO, ZC
Diplomatic representation in the US: *chief of mission:* Ambassador Jonas HAFSTROM
chancery: The House of Sweden, 2900 K Street NW, Washington, DC 20007
telephone: [1] (202) 467-2600
FAX: [1] (202) 467-2699
consulate(s) general: New York
Diplomatic representation from the US: *chief of mission:* Ambassador Matthew W. BARZUN
embassy: Dag Hammarskjolds Vag 31, SE-11589 Stockholm
mailing address: American Embassy Stockholm, US Department of State, 5750 Stockholm Place, Washington, DC 20521-5750
telephone: [46] (08) 783 53 00
FAX: [46] (08) 661 19 64
Flag description: blue with a golden yellow cross extending to the edges of the flag; the vertical part of the cross is shifted to the hoist side in the style of the Dannebrog (Danish flag); the colors reflect those of the Swedish coat of arms–three gold crowns on a blue field
National anthem: *name:* "Du Gamla, Du Fria" (Thou Ancient, Thou Free)
lyrics/music: Richard DYBECK/traditional
note: in use since 1844; the anthem, also known as "Sang till Norden" (Song of the North), is based on a Swedish folk tune; it has never been officially adopted by the government; "Kungssangen" (The King's Song) serves as the royal anthem and is played in the presence of the royal family and during certain state ceremonies

ECONOMY

Economy—overview: Aided by peace and neutrality for the whole of the 20th century, Sweden has achieved an enviable standard of living under a mixed system of high-tech capitalism and extensive welfare benefits. It has a modern distribution system, excellent internal and external communications, and a skilled labor force. In September 2003, Swedish voters turned down entry into the euro system concerned about the impact on the economy and sovereignty. Timber, hydropower, and iron ore constitute the resource base of an economy heavily oriented toward foreign trade. Privately owned firms account for about 90% of industrial output, of which the engineering sector accounts for 50% of output and exports. Agriculture accounts for little more than 1% of GDP and of employment. Until 2008, Sweden was in the midst of a sustained economic upswing, boosted by increased domestic demand and strong exports. This and robust finances offered the center-right government considerable scope to implement its reform program aimed at increasing employment, reducing welfare dependence, and streamlining the state's role in the economy. Despite strong finances and underlying fundamentals, the Swedish economy slid into recession in the third quarter of 2008 and growth continued downward in 2009 as deteriorating global conditions reduced export demand and consumption. Strong exports of commodities and a return to profitability by Sweden's banking sector drove the strong rebound in 2010.
GDP (purchasing power parity): $354.7 billion (2010 est.)
country comparison to the world: 33
$336.1 billion (2009 est.)
$355.1 billion (2008 est.)
note: data are in 2010 US dollars
GDP (official exchange rate): $455.8 billion (2010 est.)
GDP—real growth rate: 5.5% (2010 est.)
country comparison to the world: 62
-5.3% (2009 est.)
-0.6% (2008 est.)
GDP—per capita (PPP): $39,100 (2010 est.)
country comparison to the world: 23
$37,100 (2009 est.)
$39,300 (2008 est.)
note: data are in 2010 US dollars
GDP—composition by sector:
agriculture: 1.7%
industry: 26.1%
services: 72.2% (2010 est.)
Labor force: 4.93 million (2010 est.)
country comparison to the world: 76
Labor force—by occupation:
agriculture: 1.1%
industry: 28.2%
services: 70.7% (2008 est.)
Unemployment rate: 8.3% (2010 est.)
country comparison to the world: 95
8.3% (2009 est.)
Population below poverty line: NA%
Household income or consumption by percentage share: *lowest 10%:* 3.6%
highest 10%: 22.2% (2000)
Distribution of family income—Gini index: 23 (2005)
country comparison to the world: 136
25 (1992)
Investment (gross fixed): 18.1% of GDP (2010 est.)
country comparison to the world: 110
Budget: *revenues:* $230.1 billion

expenditures: $236.6 billion (2010 est.)
Public debt: 40.8% of GDP (2010 est.)
country comparison to the world: 69
41.6% of GDP (2009 est.)
Inflation rate (consumer prices): 1.4% (2010 est.)
country comparison to the world: 31
-0.3% (2009 est.)
Central bank discount rate: 2% (31 December 2008)
country comparison to the world: 106
3.5% (31 December 2007)
Commercial bank prime lending rate: NA%
NA%
Stock of narrow money: $225 billion (31 December 2010 est.)
country comparison to the world: 17
$205.2 billion (31 December 2009 est.)
Stock of broad money: $293.2 billion (31 December 2010 est.)
country comparison to the world: 28
$260.3 billion (31 December 2009 est.)
Stock of domestic credit: $640.2 billion (31 December 2010 est.)
country comparison to the world: 21
$583.8 billion (31 December 2009 est.)
Market value of publicly traded shares: $432.3 billion (31 December 2009)
country comparison to the world: 22
$252.5 billion (31 December 2008)
$612.5 billion (31 December 2007)
Agriculture—products: barley, wheat, sugar beets; meat, milk
Industries: iron and steel, precision equipment (bearings, radio and telephone parts, armaments), wood pulp and paper products, processed foods, motor vehicles
Industrial production growth rate: 8% (2010 est.)
country comparison to the world: 31
Electricity—production: 144 billion kWh (2007 est.)
country comparison to the world: 25
Electricity—consumption: 134.5 billion kWh (2007 est.)
country comparison to the world: 23
Electricity—exports: 14.71 billion kWh (2008 est.)
Electricity—imports: 12.75 billion kWh (2008 est.)
Oil—production: 4,833 bbl/day (2009 est.)
country comparison to the world: 95
Oil—consumption: 328,100 bbl/day (2009 est.)
country comparison to the world: 38
Oil—exports: 248,500 bbl/day (2008 est.)
country comparison to the world: 47
Oil—imports: 589,900 bbl/day (2008 est.)
country comparison to the world: 21
Oil—proved reserves: 0 bbl (1 January 2010 est.)
country comparison to the world: 191
Natural gas—production: 0 cu m (2008 est.)
country comparison to the world: 187
Natural gas—consumption: 1.229 billion cu m (2009 est.)
country comparison to the world: 86
Natural gas—exports: 0 cu m (2008 est.)
country comparison to the world: 179
Natural gas—imports: 1.229 billion cu m (2009 est.)
country comparison to the world: 52
Natural gas—proved reserves: 0 cu m (1 January 2010 est.)
country comparison to the world: 190
Current account balance: $21.68 billion (2010 est.)
country comparison to the world: 16
$30.23 billion (2009 est.)
Exports: $162.6 billion (2010 est.)
country comparison to the world: 27
$133.3 billion (2009 est.)
Exports—commodities: machinery 35%, motor vehicles, paper products, pulp and wood, iron and steel products, chemicals
Exports—partners: Norway 10.61%, Germany 10.2%, UK 7.45%, Denmark 7.35%, Finland 6.44%, US 6.36%, France 5.05%, Netherlands 4.67% (2009)
Imports: $158.6 billion (2010 est.)
country comparison to the world: 27
$120.5 billion (2009 est.)
Imports—commodities: machinery, petroleum and petroleum products, chemicals, motor vehicles, iron and steel; foodstuffs, clothing
Imports—partners: Germany 17.9%, Denmark 8.9%, Norway 8.7%, Netherlands 6.17%, UK 5.56%, Finland 5.14%, France 5.06%, China 4.79% (2009)
Reserves of foreign exchange and gold: $NA (31 December 2010 est.)
$47.29 billion (31 December 2009 est.)
Debt—external: $853.3 billion (30 June 2010)
country comparison to the world: 16
$617.3 billion (31 December 2008)
Stock of direct foreign investment—at home: $321.4 billion (31 December 2010 est.)
country comparison to the world: 16
$304.5 billion (31 December 2009 est.)
Stock of direct foreign investment—abroad: $383.9 billion (31 December 2010 est.)
country comparison to the world: 13
$367.4 billion (31 December 2009 est.)
Exchange rates: Swedish kronor (SEK) per US dollar–7.5077 (2010), 7.6529 (2009), 6.4074 (2008), 6.7629 (2007), s7.3731 (2006)

COMMUNICATIONS

Telephones—main lines in use: 5.146 million (2009)
country comparison to the world: 32
Telephones—mobile cellular: 11.426 million (2009)
country comparison to the world: 60
Telephone system: *general assessment:* highly developed telecommunications infrastructure; ranked among leading countries for fixed-line, mobile-cellular, Internet and broadband penetration
domestic: coaxial and multiconductor cables carry most of the voice traffic; parallel microwave radio relay systems carry some additional telephone channels
international: country code–46; submarine cables provide links to other Nordic countries and Europe; satellite earth stations–1 Intelsat (Atlantic Ocean), 1 Eutelsat, and 1 Inmarsat (Atlantic and Indian Ocean regions); note–Sweden shares the Inmarsat earth station with the other Nordic countries (Denmark, Finland, Iceland, and Norway)
Broadcast media: publicly-owned television broadcaster operates 2 terrestrial networks plus regional stations; multiple privately-owned television broadcasters operating nationally, regionally, and locally; about 50 local TV stations; widespread access to pan-Nordic and international broadcasters through multi-channel cable and satellite TV systems; publicly-owned radio broadcaster operates 3 national stations and a network of 25 regional channels; nearly a hundred privately-owned local radio stations with some consolidating into near national networks; an estimated 900 community and neighborhood radio stations broadcast intermittently (2008)
Internet country code: .se
Internet hosts: 4.396 million (2010)
country comparison to the world: 20
Internet users: 8.398 million (2009)
country comparison to the world: 33

TRANSPORTATION

Airports: 249 (2010)
country comparison to the world: 26
Airports—with paved runways: *total:* 152
over 3,047 m: 3
2,438 to 3,047 m: 12
1,524 to 2,437 m: 76
914 to 1,523 m: 25
under 914 m: 36 (2010)
Airports—with unpaved runways:
total: 97
914 to 1,523 m: 5
under 914 m: 92 (2010)
Heliports: 2 (2010)
Pipelines: gas 786 km (2010)
Railways:
total: 11,633 km
country comparison to the world: 20
standard gauge: 11,568 km 1.435-m gauge (7,531 km electrified)
narrow gauge: 65 km 1.000-m gauge (65 km electrified) (2008)
Roadways: *total:* 572,900 km (includes 1,855 km of expressways)
country comparison to the world: 12
note: (includes 98,400 km of state roads, 433,500 km of private roads, and 41,000 km of municipal roads; 215,700 km of these are open to public traffic) (2009)
Waterways: 2,052 km (2010)
country comparison to the world: 42
Merchant marine: *total:* 163
country comparison to the world: 39
by type: bulk carrier 4, cargo 20, carrier 1, chemical tanker 31, passenger 5, passenger/cargo 37, petroleum tanker 12, roll on/roll off 32, vehicle carrier 21
foreign-owned: 46 (Denmark 15, Estonia 3, Finland 16, Germany 3, Ireland 1, Italy 5, Norway 3)
registered in other countries: 194 (Antigua and Barbuda 1, Bahamas 6, Barbados 6, Bermuda 17, Cook Islands 3, Cyprus 5, Denmark 16, Faroe Islands 5, France 6, Germany 1, Gibraltar 12, Isle of Man 1, Italy 1, Liberia 10, Malta 3, Netherlands 18, former Netherlands Antilles 1, Norway 33, Panama 1, Portugal 6, Saint Vincent and the Grenadines 2, Singapore 9, UK 25, US 5, unknown 1) (2010)
Ports and terminals: Brofjorden, Goteborg, Helsingborg, Karlshamn, Lulea, Malmo, Stockholm, Trelleborg, Visby

MILITARY

Military branches: Swedish Armed Forces (Forsvarsmakten): Army (Armen), Royal Swedish Navy (Marinen), Swedish Air Force (Svenska Flygvapnet) (2010)
Military service age and obligation: 18-47 years of age for male and female voluntary military service; service obligation: 7.5 months (Army), 7-15 months (Navy), 8-12 months (Air Force); the Swedish Parliament has abolished compulsory military service, with exclusively voluntary recruitment as of July 2010; conscription remains an option in emergencies; after completing initial service, soldiers have a reserve commitment until age 47 (2010)
Manpower available for military service: *males age 16–49:* 2,065,691
females age 16–49: 1,996,764 (2010 est.)
Manpower fit for military service: *males age 16–49:* 1,709,055
females age 16–49: 1,650,432 (2010 est.)
Manpower reaching militarily significant age annually: *male:* 54,960
female: 52,275 (2010 est.)
Military expenditures: 1.5% of GDP (2005 est.)
country comparison to the world: 100

TRANSNATIONAL ISSUES

Disputes—international: none

SWITZERLAND

INTRODUCTION

Background: The Swiss Confederation was founded in 1291 as a defensive alliance among three cantons. In succeeding years, other localities joined the original three. The Swiss Confederation secured its independence from the Holy Roman Empire in 1499. A constitution of 1848, subsequently modified in 1874, replaced the confederation with a centralized federal government. Switzerland's sovereignty and neutrality have long been honored by the major European powers, and the country was not involved in either of the two world wars. The political and economic integration of Europe over the past half century, as well as Switzerland's role in many UN and international organizations, has strengthened Switzerland's ties with its neighbors. However, the country did not officially become a UN member until 2002. Switzerland remains active in many UN and international organizations but retains a strong commitment to neutrality.

GEOGRAPHY

Location: Central Europe, east of France, north of Italy
Geographic coordinates: 47 00 N, 8 00 E
Map references: Europe
Area: *total:* 41,277 sq km
country comparison to the world: 135
land: 39,997 sq km
water: 1,280 sq km
Area—comparative: slightly less than twice the size of New Jersey
Land boundaries: *total:* 1,852 km
border countries: Austria 164 km, France 573 km, Italy 740 km, Liechtenstein 41 km, Germany 334 km
Coastline: 0 km (landlocked)
Maritime claims: none (landlocked)
Climate: temperate, but varies with altitude; cold, cloudy, rainy/snowy winters; cool to warm, cloudy, humid summers with occasional showers
Terrain: mostly mountains (Alps in south, Jura in northwest) with a central plateau of rolling hills, plains, and large lakes
Elevation extremes: *lowest point:* Lake Maggiore 195 m
highest point: Dufourspitze 4,634 m
Natural resources: hydropower potential, timber, salt
Land use: *arable land:* 9.91%
permanent crops: 0.58%
other: 89.51% (2005)
Irrigated land: 250 sq km (2003)
Total renewable water resources: 53.3 cu km (2005)
Freshwater withdrawal (domestic/industrial/agricultural): *total:* 2.52 cu km/yr (24%/74%/2%)
per capita: 348 cu m/yr (2002)
Natural hazards: avalanches, landslides; flash floods
Environment—current issues: air pollution from vehicle emissions and open-air burning; acid rain; water pollution from increased use of agricultural fertilizers; loss of biodiversity
Environment—international agreements: *party to:* Air Pollution, Air Pollution-Nitrogen Oxides, Air Pollution-Persistent Organic Pollutants, Air Pollution-Sulfur 85, Air Pollution-Sulfur 94, Air Pollution-Volatile Organic Compounds, Antarctic Treaty, Biodiversity, Climate Change, Climate Change-Kyoto Protocol, Desertification, Endangered Species, Environmental Modification, Hazardous Wastes, Marine Dumping, Marine Life Conservation, Ozone Layer Protection, Ship Pollution, Tropical Timber 83, Tropical Timber 94, Wetlands, Whaling
signed, but not ratified: Law of the Sea
Geography—note: landlocked; crossroads of northern and southern Europe; along with southeastern France, northern Italy, and southwestern Austria, has the highest elevations in the Alps

PEOPLE

Population: 7,639,961 (July 2011 est.)
country comparison to the world: 94
Age structure: *0–14 years:* 15.2% (male 602,894/female 560,175)
15–64 years: 67.8% (male 2,612,557/female 2,569,318)
65 years and over: 17% (male 543,074/female 751,943) (2011 est.)
Median age: *total:* 41.7 years
male: 40.6 years
female: 42.8 years (2011 est.)
Population growth rate: 0.21% (2011 est.)
country comparison to the world: 181
Birth rate: 9.53 births/1,000 population (2011 est.)
country comparison to the world: 203
Death rate: 8.72 deaths/1,000 population (July 2011 est.)
country comparison to the world: 82
Net migration rate: 1.29 migrant(s)/1,000 population (2011 est.)
country comparison to the world: 48
Urbanization: *urban population:* 74% of total population (2010)
rate of urbanization: 0.5% annual rate of change (2010-15 est.)
Major cities—population: Zurich 1.143 million; BERN (capital) 346,000 (2009)
Sex ratio: *at birth:* 1.054 male(s)/female
under 15 years: 1.08 male(s)/female
15–64 years: 1.02 male(s)/female
65 years and over: 0.72 male(s)/female
total population: 0.97 male(s)/female (2011 est.)
Infant mortality rate: *total:* 4.08 deaths/1,000 live births
country comparison to the world: 201
male: 4.53 deaths/1,000 live births
female: 3.6 deaths/1,000 live births (2011 est.)
Life expectancy at birth: *total population:* 81.07 years
country comparison to the world: 15
male: 78.24 years
female: 84.05 years (2011 est.)
Total fertility rate: 1.46 children born/woman (2011 est.)
country comparison to the world: 190

HIV/AIDS—adult prevalence rate: 0.4% (2009 est.)
country comparison to the world: 75
HIV/AIDS—people living with HIV/AIDS: 18,000 (2009 est.)
country comparison to the world: 82
HIV/AIDS—deaths: fewer than 100 (2009 est.)
country comparison to the world: 116
Drinking water source: *Improved:*
urban: 100% of population
rural: 100% of population
total: 100% of population (2008)
Sanitation facility access: *Improved:*
urban: 100% of population
rural: 100% of population
total: 100% of population (2008)
Nationality: *noun:* Swiss (singular and plural)
adjective: Swiss
Ethnic groups: German 65%, French 18%, Italian 10%, Romansch 1%, other 6%
Religions: Roman Catholic 41.8%, Protestant 35.3%, Muslim 4.3%, Orthodox 1.8%, other Christian 0.4%, other 1%, unspecified 4.3%, none 11.1% (2000 census)
Languages: German (official) 63.7%, French (official) 20.4%, Italian (official) 6.5%, Serbo-Croatian 1.5%, Albanian 1.3%, Portuguese 1.2%, Spanish 1.1%, English 1%, Romansch (official) 0.5%, other 2.8% (2000 census)
note: German, French, Italian, and Romansch are all national and official languages
Literacy: *definition:* age 15 and over can read and write
total population: 99%
male: 99%
female: 99% (2003 est.)
School life expectancy (primary to tertiary education):
total: 16 years
male: 16 years
female: 15 years (2008)
Education expenditures: 5.2% of GDP (2007)
country comparison to the world: 53

GOVERNMENT

Country name: *conventional long form:* Swiss Confederation
conventional short form: Switzerland
local long form: Schweizerische Eidgenossenschaft (German); Confederation Suisse (French); Confederazione Svizzera (Italian); Confederaziun Svizra (Romansh)
local short form: Schweiz (German); Suisse (French); Svizzera (Italian); Svizra (Romansh)
Government type: formally a confederation but similar in structure to a federal republic
Capital: *name:* Bern
geographic coordinates: 46 57 N, 7 26 E
time difference: UTC+1 (6 hours ahead of Washington, DC during Standard Time)
daylight saving time: +1hr, begins last Sunday in March; ends last Sunday in October
Administrative divisions: 26 cantons (cantons, singular—canton in French; cantoni, singular—cantone in Italian; Kantone, singular—Kanton in German); Aargau, Appenzell Ausser-Rhoden, Appenzell Inner-Rhoden, Basel-Landschaft, Basel-Stadt, Bern, Fribourg, Geneve, Glarus, Graubunden, Jura, Luzern, Neuchatel, Nidwalden, Obwalden, Sankt Gallen, Schaffhausen, Schwyz, Solothurn, Thurgau, Ticino, Uri, Valais, Vaud, Zug, Zurich
note: 6 of the cantons—Appenzell Ausser-Rhoden, Appenzell-Inner-Rhoden, Basel-Landschaft, Basel-Stadt, Nidwalden, Obwalden—are refered to as half cantons because they elect only one member to the Council of States and, in popular referendums where a majority of popular votes and a majority of cantonal votes are required, these six cantons only have a half vote
Independence: 1 August 1291 (founding of the Swiss Confederation)
National holiday: Founding of the Swiss Confederation, 1 August (1291)
Constitution: revision of Constitution of 1874 approved by the Federal Parliament 18 December 1998, adopted by referendum 18 April 1999, officially entered into force 1 January 2000
Legal system: civil law system; judicial review of legislative acts, except for federal decrees of a general obligatory character
International law organization participation: accepts compulsory ICJ jurisdiction with reservations; accepts ICCt jurisdiction
Suffrage: 18 years of age; universal
Executive branch: *chief of state:* President of the Swiss Confederation Micheline CALMY-REY (since 1 January 2011); Vice President Eveline WIDMER-SCHLUMPF (since 1 January 2011); note—the president is both the chief of state and head of government representing the Federal Council; the Federal Council is the formal chief of state and head of government whose council members, rotating in one-year terms as federal president, represent the Council
head of government: President of the Swiss Confederation Micheline CALMY-REY (since 1 January 2011); Vice President Eveline WIDMER-SCHLUMPF (since 1 January 2011)
cabinet: Federal Council or Bundesrat (in German), Conseil Federal (in French), Consiglio Federale (in Italian) is elected by the Federal Assembly usually from among its members for a four-year term
(For more information visit the World Leaders website)
elections: president and vice president elected by the Federal Assembly from among the members of the Federal Council for a one-year term (they may not serve consecutive terms); election last held on 8 December 2010 (next to be held in early December 2011)
election results: Micheline CALMY-REY elected president; number of Federal Assembly votes—106 of 189; Eveline WIDMER-SCHLUMPF elected vice president; current Vice President Eveline WIDMER-SCHLUMPF is slated to become president on 1 January 2012
Legislative branch: bicameral Federal Assembly or Bundesversammlung (in German), Assemblee Federale (in French), Assemblea Federale (in Italian) consists of the Council of States or Staenderat (in German), Conseil des Etats (in French), Consiglio degli Stati (in Italian) (46 seats; membership consists of 2 representatives from each canton and 1 from each half canton; members serve four-year terms) and the National Council or Nationalrat (in German), Conseil National (in French), Consiglio Nazionale (in Italian) (200 seats; members elected by popular vote on the basis of proportional representation serve four-year terms)
elections: Council of States—last held in most cantons in October 2007 (each canton determines when the next election will be held); National Council—last held on 21 October 2007 (next to be held in October 2011)
election results: Council of States—percent of vote by party—NA; seats by party—CVP 15, FDP 12, SVP 7, SPS 9, other 3; National Council—percent of vote by party—SVP 29%, SPS 19.5%, FDP 15.6%, CVP 14.6%, Greens 9.6%, other 11.7%; seats by party—SVP 62, SPS 43, FDP 31, CVP 31, Green Party 20, other small parties 13
Judicial branch: Federal Supreme Court (judges elected for six-year terms by the Federal Assembly)
Political parties and leaders: Green Party (Gruene Partei der Schweiz or Gruene, Parti Ecologiste Suisse or Les Verts, Partito Ecologista Svizzero or I Verdi, Partida Ecologica Svizra or La Verda) [Ueli LEUENBERGER]; Christian Democratic People's Party (Christlichdemokratische Volkspartei der Schweiz or CVP, Parti Democrate-Chretien Suisse or PDC, Partito Popolare Democratico Svizzero or PPD, Partida Cristiandemocratica dalla Svizra or PCD) [Christophe DARBELLAY]; Conservative Democratic Party (Buergerlich–Demokratische Partei der Schweiz or BDP, Parti Bourgeois Democratique Suisse or PBD, Partito Borghese Democratico Svizzero or PBD) [Hans GRUNDER]; Free Democratic Party or FDP.The Liberals (FDP.Die Liberalen, PLR.Les Liberaux-Radicaux, PLR.I Liberali) [Fulvio PELLI]; Social Democratic Party (Sozialdemokratische Partei der Schweiz or SPS, Parti Socialiste Suisse or PSS, Partito Socialista Svizzero or PSS, Partida Socialdemocratica de la Svizra or PSS) [Christian LEVRAT]; Swiss People's Party (Schweizerische Volkspartei or SVP, Union Democratique du Centre or UDC, Unione Democratica di Centro or UDC, Uniun Democratica dal Center or UDC) [Toni BRUNNER]; and other minor parties
Political pressure groups and leaders: NA
International organization participation: ADB (nonregional member), AfDB (nonregional member), Australia Group, BIS, CE, CERN, EAPC, EBRD, EFTA, ESA, FAO, FATF, G-10, IADB, IAEA, IBRD, ICAO, ICC, ICRM, IDA, IEA, IFAD, IFC, IFRCS, ILO, IMF, IMO, IMSO, Interpol, IOC, IOM, IPU, ISO, ITSO, ITU, ITUC, LAIA (observer), MIGA, MONUSCO, NEA, NSG, OAS (observer), OECD, OIF, OPCW, OSCE, Paris Club, PCA, PFP, Schengen Convention, UN, UNCTAD, UNESCO, UNHCR, UNIDO, UNITAR, UNMIS, UNRWA, UNTSO, UNWTO, UPU, WCO, WHO, WIPO, WMO, WTO, ZC

Diplomatic representation in the US: *chief of mission:* Ambassador Manuel SAGER
chancery: 2900 Cathedral Avenue NW, Washington, DC 20008
telephone: [1] (202) 745-7900
FAX: [1] (202) 387-2564
consulate(s) general: Atlanta, Chicago, Los Angeles, New York, San Francisco
consulate(s): Boston
Diplomatic representation from the US: *chief of mission:* Ambassador Donald S. BEYER, Jr.
embassy: Sulgeneckstrasse 19, CH-3007 Bern
mailing address: use embassy street address
telephone: [41] (031) 357 70 11
FAX: [41] (031) 357 73 44
Flag description: red square with a bold, equilateral white cross in the center that does not extend to the edges of the flag; various medieval legends purport to describe the origin of the flag; a white cross used as identification for troops of the Swiss Confederation is first attested at the Battle of Laupen (1339)
National anthem: *name:* "Schweizerpsalm" [German] "Cantique Suisse" [French] "Salmo svizzero," [Italian] "Psalm svizzer" [Romansch] (Swiss Psalm)
lyrics/music: Leonhard WIDMER [German], Charles CHATELANAT [French], Camillo VALSANGIACOMO [Italian], and Flurin CAMATHIAS [Romansch]/Alberik ZWYSSIG
note: unofficially adopted 1961, official adoption 1981; the anthem has been popular in a number of Swiss cantons since its composition (in German) in 1841; translated into the other three official languages of the country (French, Italian, and Romansch), it is official in each of those languages

ECONOMY

Economy—overview: Switzerland is a peaceful, prosperous, and modern market economy with low unemployment, a highly skilled labor force, and a per capita GDP among the highest in the world. Switzerland's economy benefits from a highly developed service sector, led by financial services, and a manufacturing industry that specializes in high-technology, knowledge-based production. The Swiss have brought their economic practices largely into conformity with the EU's, in order to enhance their international competitiveness, but some trade protectionism remains, particularly for its small agricultural sector. The global financial crisis and resulting economic downturn put Switzerland in a recession in 2009 as global export demand stalled. The Swiss National Bank during this period effectively implemented a zero-interest rate policy in a bid to boost the economy and prevent appreciation of the franc. Switzerland's economy grew by 2.7% in 2010, when Bern implemented a third fiscal stimulus program, but its prized banking sector has recently faced significant challenges. The country's largest banks suffered sizable losses in 2008-09, leading its largest bank to accept a government rescue deal in late 2008. Switzerland has also come under increasing pressure from individual neighboring countries, the EU, the US, and international institutions to reform its banking secrecy laws. Consequently, the government agreed to conform to OECD regulations on administrative assistance in tax matters, including tax evasion. The government has renegotiated its double taxation agreements with numerous countries, including the US, to incorporate the OECD standard, and it is working with Germany and the UK to resolve outstanding issues, particularly the possibility of imposing taxes on bank deposits held by foreigners. Parliament passed the first five double-taxation agreements, including that with the US, in March 2010. The agreement with the US awaits US Senate approval. In 2009, Swiss financial regulators ordered the country's largest bank to reveal at Washington's behest the names of US account-holders suspected of using the bank to commit tax fraud. These steps will have a lasting impact on Switzerland's long history of bank secrecy.
GDP (purchasing power parity): $324.5 billion (2010 est.)
country comparison to the world: 38
$316.4 billion (2009 est.)
$322.6 billion (2008 est.)
note: data are in 2010 US dollars
GDP (official exchange rate): $523.8 billion (2010 est.)
GDP—real growth rate: 2.6% (2010 est.)
country comparison to the world: 134
-1.9% (2009 est.)
1.9% (2008 est.)
GDP—per capita (PPP): $42,600 (2010 est.)
country comparison to the world: 17
$41,600 (2009 est.)
$42,600 (2008 est.)
note: data are in 2010 US dollars
GDP—composition by sector:
agriculture: 1.3%
industry: 27.5%
services: 71.2% (2010 est.)
Labor force: 4.62 million (2010)
country comparison to the world: 79
Labor force—by occupation:
agriculture: 3.4%
industry: 23.4%
services: 73.2% (2010)
Unemployment rate: 3.9% (2010 est.)
country comparison to the world: 35
3.7% (2009 est.)
Population below poverty line: 6.9% (2010)
Household income or consumption by percentage share: *lowest 10%:* 7.5%
highest 10%: 19% (2007)
Distribution of family income—Gini index: 33.7 (2008)
country comparison to the world: 93
33.1 (1992)
Investment (gross fixed): 19.9% of GDP (2010 est.)
country comparison to the world: 90
Budget: *revenues:* $188.1 billion
expenditures: $192.7 billion
note: includes federal, cantonal, and municipal accounts (2011 est.)
Public debt: 38.2% of GDP (2010 est.)
country comparison to the world: 79
40.5% of GDP (2009 est.)
Inflation rate (consumer prices): 0.7% (2010 est.)
country comparison to the world: 13
-0.5% (2009 est.)
Central bank discount rate: 0.04% (31 December 2010)
country comparison to the world: 141
0.05% (31 December 2009)
Commercial bank prime lending rate: 0.54% (31 December 2010 est.)
country comparison to the world: 151
2.75% (31 December 2009 est.)
Stock of narrow money: $384.2 billion (31 December 2010 est.)
country comparison to the world: 11
$328.7 billion (31 December 2009 est.)
Stock of broad money: $834.6 billion (31 December 2010 est.)
country comparison to the world: 19
$777.8 billion (31 December 2009 est.)
Stock of domestic credit: $879.7 billion (30 November 2010 est.)
country comparison to the world: 16
$992.6 billion (31 December 2009 est.)
Market value of publicly traded shares: $1.071 trillion (31 December 2009)
country comparison to the world: 12
$862.7 billion (31 December 2008)
$1.275 trillion (31 December 2007)
Agriculture—products: grains, fruits, vegetables; meat, eggs
Industries: machinery, chemicals, watches, textiles, precision instruments, tourism, banking, and insurance
Industrial production growth rate: 2.4% (2010 est.)
country comparison to the world: 124
Electricity—production: 66.5 billion kWh (2009)
country comparison to the world: 41
Electricity—consumption: 57.5 billion kWh (2009)
country comparison to the world: 41
Electricity—exports: 54.2 billion kWh (2009 est.)
Electricity—imports: 52 billion kWh (2009)
Oil—production: 3,488 bbl/day (2009 est.)
country comparison to the world: 101
Oil—consumption: 280,000 bbl/day (2009 est.)
country comparison to the world: 44
Oil—exports: 10,680 bbl/day (2009 est.)
country comparison to the world: 92
Oil—imports: 263,600 bbl/day (2009 est.)
country comparison to the world: 38
Oil—proved reserves: 0 bbl (1 January 2010 est.)
country comparison to the world: 192
Natural gas—production: 0 cu m (2009 est.)
country comparison to the world: 188
Natural gas—consumption: 3.042 billion cu m (2009)
country comparison to the world: 73
Natural gas—exports: 0 cu m (2009 est.)
country comparison to the world: 181
Natural gas—imports: 3.042 billion cu m (2009)
country comparison to the world: 41
Natural gas—proved reserves: 0 cu m (1 January 2009 est.)
country comparison to the world: 191
Current account balance: $49.35 billion (2010 est.)
country comparison to the world: 7
$54.01 billion (2009 est.)
Exports: $232.6 billion (2010 est.)

country comparison to the world: 20
$208.5 billion (2009 est.)
Exports—commodities: machinery, chemicals, metals, watches, agricultural products
Exports—partners: Germany 20.98%, US 9.09%, France 8.62%, Italy 8.08%, Austria 5.38% (2009)
Imports: $226.3 billion (2010 est.)
country comparison to the world: 20
$192.8 billion (2009 est.)
Imports—commodities: machinery, chemicals, vehicles, metals; agricultural products, textiles
Imports—partners: Germany 27.19%, Italy 10.42%, US 9.61%, France 7.69%, Netherlands 4.35% (2009)
Reserves of foreign exchange and gold: $236.6 billion (31 December 2010)
country comparison to the world: 10
$135.3 billion (31 December 2009 est.)
Debt—external: $1.2 trillion (30 September 2010)
country comparison to the world: 13
$1.305 trillion (31 December 2008)
Stock of direct foreign investment—at home: $514 billion (31 December 2010 est.)
country comparison to the world: 11
$471.9 billion (31 December 2009 est.)
Stock of direct foreign investment—abroad: $814.6 billion (31 December 2010 est.)
country comparison to the world: 7
$796.2 billion (31 December 2009 est.)
Exchange rates: Swiss francs (CHF) per US dollar–1.0429 (2010), 1.0881 (2009), 1.0774 (2008), 1.1973 (2007), 1.2539 (2006)

COMMUNICATIONS

Telephones—main lines in use: 4.65 million (2009)
country comparison to the world: 33
Telephones—mobile cellular: 9.255 million (2009)
country comparison to the world: 69
Telephone system: *general assessment:* highly developed telecommunications infrastructure with excellent domestic and international services
domestic: ranked among leading countries for fixed-line teledensity and infrastructure; mobile-cellular subscribership roughly 120 per 100 persons; extensive cable and microwave radio relay networks
international: country code–41; satellite earth stations–2 Intelsat (Atlantic Ocean and Indian Ocean)
Broadcast media: the publicly-owned radio and television broadcaster, Swiss Broadcasting Corporation (SRG/SSR), operates 7 national television networks, 3 broadcasting in German, 2 in Italian, and 2 in French; private commercial television stations broadcast regionally and locally; television broadcasts from stations in Germany, Italy, and France are widely accessed using multi-channel cable and satellite TV services; SRG/SSR operates 18 radio stations that, along with private broadcasters, provide national to local coverage (2009)
Internet country code: .ch
Internet hosts: 4.816 million (2010)
country comparison to the world: 17
Internet users: 6.152 million (2009)
country comparison to the world: 42

TRANSPORTATION

Airports: 65 (2010)
country comparison to the world: 76
Airports—with paved runways: *total:* 42
over 3,047 m: 3
2,438 to 3,047 m: 3
1,524 to 2,437 m: 14
914 to 1,523 m: 5
under 914 m: 17 (2010)
Airports—with unpaved runways: *total:* 23
under 914 m: 23 (2010)
Heliports: 1 (2010)
Pipelines: gas 1,681 km; oil 94 km; refined products 7 km (2010)
Railways: *total:* 4,888 km
country comparison to the world: 37
standard gauge: 3,397 km 1.435-m gauge (3,142 km electrified)
narrow gauge: 1,481 km 1.000-m gauge (1,378 km electrified); 10 km 0.800-m gauge (10 km electrified) (2008)
Roadways: *total:* 71,454 km
country comparison to the world: 66
paved: 71,454 km (includes 1,790 of expressways) (2010)
Waterways: 1,299 km (there are 1,227 km of waterways on lakes and rivers for public transport and another 65 km on the Rhine River between Basel-Rheinfelden and Schaffhausen-Bodensee used for the transport of commercial goods) (2010)
country comparison to the world: 57
Merchant marine:
total: 35
country comparison to the world: 80
by type: bulk carrier 15, cargo 9, chemical tanker 6, container 4, petroleum tanker 1
registered in other countries: 109 (Antigua and Barbuda 7, Bahamas 2, Cayman Islands 1, France 5, Germany 1, Italy 6, Liberia 17, Luxembourg 1, Malta 14, Marshall Islands 12, NZ 2, Panama 22, Portugal 3, Russia 4, Saint Vincent and the Grenadines 5, Singapore 4, Spain 1, Tonga 1, Tuvalu 1) (2010)
Ports and terminals: Basel

MILITARY

Military branches: Swiss Armed Forces: Land Forces, Swiss Air Force (Schweizer Luftwaffe) (2010)
Military service age and obligation: 19-26 years of age for male compulsory military service; 18 years of age for voluntary male and female military service; every Swiss male has to serve at least 260 days in the armed forces; conscripts receive 18 weeks of mandatory training, followed by seven 3-week intermittent recalls for training during the next 10 years (2010)
Manpower available for military service: *males age 16–49:* 1,828,043
females age 16–49: 1,786,552 (2010 est.)
Manpower fit for military service: *males age 16–49:* 1,493,509
females age 16–49: 1,459,450 (2010 est.)
Manpower reaching militarily significant age annually: *male:* 46,562
female: 42,585 (2010 est.)
Military expenditures: 1% of GDP (2005 est.)
country comparison to the world: 133

TRANSNATIONAL ISSUES

Disputes—international: none
Illicit drugs: a major international financial center vulnerable to the layering and integration stages of money laundering; despite significant legislation and reporting requirements, secrecy rules persist and nonresidents are permitted to conduct business through offshore entities and various intermediaries; transit country for and consumer of South American cocaine, Southwest Asian heroin, and Western European synthetics; domestic cannabis cultivation and limited ecstasy production

SYRIA

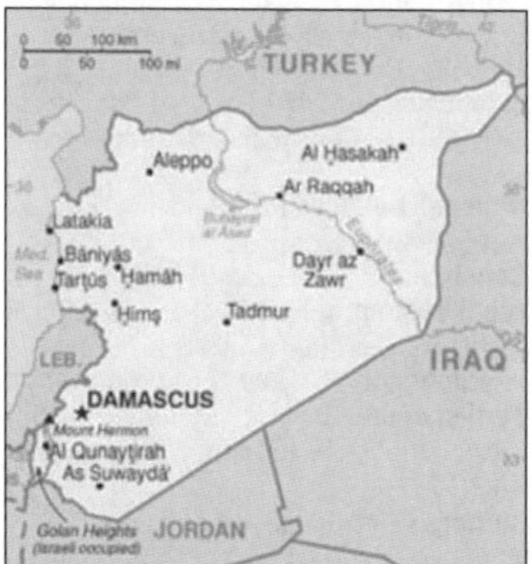

INTRODUCTION

Background: Following World War I, France acquired a mandate over the northern portion of the former Ottoman Empire province of Syria. The French administered the area as Syria until granting it independence in 1946. The new country lacked political stability, however, and experienced a series of military coups during its first decades. Syria united with Egypt in February 1958 to form the United Arab Republic. In September 1961, the two entities separated, and the Syrian Arab Republic was reestablished. In November 1970, Hafiz al-ASAD, a member of the Socialist Ba'th Party and the minority Alawite sect, seized power in a bloodless coup and brought political stability to the country. In the 1967 Arab-Israeli War, Syria lost the Golan Heights to Israel. During the 1990s, Syria and Israel held occasional peace talks over its return. Following the death of President al-ASAD, his son, Bashar al-ASAD, was approved as president by popular referendum in July 2000. Syrian troops–stationed in Lebanon since 1976 in an ostensible peacekeeping role–were withdrawn in April 2005. During the July-August 2006 conflict between Israel and Hizballah, Syria placed its military forces on alert but did not intervene directly on behalf of its ally Hizballah. In May 2007 Bashar al-ASAD was elected to his second term as president. Influenced by major uprisings that began elsewhere in the region, antigovernment protests broke out in the southern province of Da'ra in March 2011 and spread to other Syrian cities. Protesters called for the repeal of the restrictive Emergency Law allowing arrests without charge, the legalization of political parties, and the removal of corrupt local officials. The government responded with a mix of force and concessions, including the repeal of the Emergency Law, but as of mid-April 2011 had not succeeded in quelling protests.

GEOGRAPHY

Location: Middle East, bordering the Mediterranean Sea, between Lebanon and Turkey
Geographic coordinates: 35 00 N, 38 00 E
Map references: Middle East
Area: *total:* 185,180 sq km
country comparison to the world: 88
land: 183,630 sq km
water: 1,550 sq km
note: includes 1,295 sq km of Israeli-occupied territory
Area—comparative: slightly larger than North Dakota
Land boundaries: *total:* 2,253 km
border countries: Iraq 605 km, Israel 76 km, Jordan 375 km, Lebanon 375 km, Turkey 822 km
Coastline: 193 km
Maritime claims: territorial sea: 12 nm
contiguous zone: 24 nm
Climate: mostly desert; hot, dry, sunny summers (June to August) and mild, rainy winters (December to February) along coast; cold weather with snow or sleet periodically in Damascus
Terrain: primarily semiarid and desert plateau; narrow coastal plain; mountains in west
Elevation extremes: *lowest point:* unnamed location near Lake Tiberias -200 m
highest point: Mount Hermon 2,814 m
Natural resources: petroleum, phosphates, chrome and manganese ores, asphalt, iron ore, rock salt, marble, gypsum, hydropower
Land use: *arable land:* 24.8%
permanent crops: 4.47%
other: 70.73% (2005)
Irrigated land: 13,330 sq km (2003)
Total renewable water resources: 46.1 cu km (1997)
Freshwater withdrawal (domestic/industrial/agricultural): *total:* 19.95 cu km/yr (3%/2%/95%)
per capita: 1,048 cu m/yr (2000)
Natural hazards: dust storms, sandstorms
volcanism: Syria's two historically active volcanoes, Es Safa and an unnamed volcano near the Turkish border have not erupted in centuries
Environment—current issues: deforestation; overgrazing; soil erosion; desertification; water pollution from raw sewage and petroleum refining wastes; inadequate potable water
Environment—international agreements: *party to:* Biodiversity, Climate Change, Climate Change-Kyoto Protocol, Desertification, Endangered Species, Hazardous Wastes, Ozone Layer Protection, Ship Pollution, Wetlands
signed, but not ratified: Environmental Modification
Geography—note: there are 41 Israeli settlements and civilian land use sites in the Israeli-occupied Golan Heights (2010 est.)

PEOPLE

Population: 22,517,750 (July 2010 est.)
country comparison to the world: 52
note: approximately 19,100 Israeli settlers live in the Golan Heights (2008 est.)
Age structure: *0–14 years:* 35.2% (male 4,066,109/female 3,865,817)
15–64 years: 61% (male 6,985,067/female 6,753,619)
65 years and over: 3.8% (male 390,802/female 456,336) (2011 est.)
Median age: *total:* 21.9 years
male: 21.7 years
female: 22.1 years (2011 est.)
Population growth rate: 0.913% (2011 est.)
country comparison to the world: 125
Birth rate: 23.99 births/1,000 population (2011 est.)
country comparison to the world: 69
Death rate: 3.68 deaths/1,000 population (July 2011 est.)
country comparison to the world: 210
Net migration rate: -11.18 migrant(s)/1,000 population (2011 est.)
country comparison to the world: 212
Urbanization: *urban population:* 56% of total population (2010)
rate of urbanization: 2.5% annual rate of change (2010-15 est.)
Major cities—population: Aleppo 2.985 million; DAMASCUS (capital) 2.527 million; Hims 1.276 million; Hamah 854,000 (2009)
Sex ratio: *at birth:* 1.06 male(s)/female
under 15 years: 1.05 male(s)/female
15–64 years: 1.03 male(s)/female
65 years and over: 0.86 male(s)/female
total population: 1.03 male(s)/female (2011 est.)
Infant mortality rate: *total:* 15.62 deaths/1,000 live births
country comparison to the world: 115
male: 17.96 deaths/1,000 live births
female: 13.14 deaths/1,000 live births (2011 est.)
Life expectancy at birth: *total population:* 74.69 years
country comparison to the world: 94
male: 72.31 years
female: 77.21 years (2011 est.)
Total fertility rate: 2.94 children born/woman (2011 est.)
country comparison to the world: 67
HIV/AIDS—adult prevalence rate: less than 0.1% (2001 est.)
country comparison to the world: 163
HIV/AIDS—people living with HIV/AIDS: fewer than 500 (2003 est.)
country comparison to the world: 154
HIV/AIDS—deaths: fewer than 200 (2003 est.)
country comparison to the world: 105
Drinking water source: *Improved:*
urban: 94% of population
rural: 84% of population
total: 89% of population
Unimproved:
urban: 6% of population
rural: 16% of population
total: 11% of population (2008)
Sanitation facility access: *Improved:*
urban: 96% of population
rural: 95% of population
total: 96% of population
Unimproved:
urban: 4% of population
rural: 5% of population
total: 4% of population (2008)
Nationality: *noun:* Syrian(s)
adjective: Syrian
Ethnic groups: Arab 90.3%, Kurds, Armenians, and other 9.7%
Religions: Sunni Muslim 74%, other Muslim (includes Alawite, Druze) 16%, Christian (various denominations) 10%, Jewish (tiny communities in Damascus, Al Qamishli, and Aleppo)
Languages: Arabic (official), Kurdish, Armenian, Aramaic, Circassian (widely understood); French, English (somewhat understood)
Literacy: *definition:* age 15 and over can read and write
total population: 79.6%
male: 86%
female: 73.6% (2004 census)
School life expectancy (primary to tertiary education): *total:* 11 years
male: 12 years
female: 11 years (2007)
Education expenditures: 4.9% of GDP (2007)
country comparison to the world: 60

GOVERNMENT

Country name: *conventional long form:* Syrian Arab Republic
conventional short form: Syria
local long form: Al Jumhuriyah al Arabiyah as Suriyah
local short form: Suriyah
former: United Arab Republic (with Egypt)
Government type: republic under an authoritarian regime
Capital: *name:* Damascus
geographic coordinates: 33 30 N, 36 18 E
time difference: UTC+2 (7 hours ahead of Washington, DC during Standard Time)
daylight saving time: +1hr, begins first Friday in April; ends last Friday in October
Administrative divisions: 14 provinces (muhafazat, singular–muhafazah); Al Hasakah, Al Ladhiqiyah (Latakia), Al Qunaytirah, Ar Raqqah, As Suwayda', Dar'a, Dayr az Zawr, Dimashq (Damascus), Halab, Hamah, Hims, Idlib, Rif Dimashq, Tartus
Independence: 17 April 1946 (from League of Nations mandate under French administration)
National holiday: Independence Day, 17 April (1946)
Constitution: 13 March 1973
Legal system: mixed legal system of civil and Islamic law (for family courts)
International law organization participation: has not submitted an ICJ jurisdiction declaration; non-party state to the ICCt
Suffrage: 18 years of age; universal

Executive branch: *chief of state:* President Bashar al-ASAD (since 17 July 2000) Vice President Farouk al-SHARA (since 11 February 2006) oversees foreign policy; Vice President Najah al-ATTAR (since 23 March 2006) oversees cultural policy
head of government: Prime Minister Adil SAFR (since 14 April 2011)
cabinet: Council of Ministers appointed by the president; note—new Council appointed on 14 April 2011
(For more information visit the World Leaders website)
elections: president approved by popular referendum for a second seven-year term (no term limits); referendum last held on 27 May 2007 (next to be held in May 2014); the president appoints the vice presidents, prime minister, and deputy prime ministers
election results: Bashar al-ASAD approved as president; percent of vote—Bashar al-ASAD 97.6%, other 2.4%
Legislative branch: unicameral People's Council or Majlis al-Shaab (250 seats; members elected by popular vote to serve four-year terms)
elections: last held on 22-23 April 2007 (next to be held in 2011)
election results: percent of vote by party—NA; seats by party—NPF 172, independents 78
Judicial branch: Supreme Judicial Council (appoints and dismisses judges; headed by the president); national level—Supreme Constitutional Court (adjudicates electoral disputes and rules on constitutionality of laws and decrees; justices appointed for four-year terms by the president); Court of Cassation; Appeals Courts (Appeals Courts represent an intermediate level between the Court of Cassation and local level courts); local level—Magistrate Courts; Courts of First Instance; Juvenile Courts; Customs Courts; specialized courts—Economic Security Courts (hear cases related to economic crimes); Supreme State Security Court (hear cases related to national security); Personal Status Courts (religious; hear cases related to marriage and divorce)
Political parties and leaders: legal parties: National Progressive Front or NPF [President Bashar al-ASAD, Dr. Suleiman QADDAH] (includes Arab Socialist Renaissance (Ba'th) Party [President Bashar al-ASAD]; Socialist Unionist Democratic Party [Fadlallah Nasr Al-DIN]; Syrian Arab Socialist Union or ASU [Safwan al-QUDSI]; Syrian Communist Party (two branches) [Wissal Farha BAKDASH, Yusuf Rashid FAYSAL]; Syrian Social Nationalist Party [As'ad HARDAN]; Unionist Socialist Party [Fayez ISMAIL])
opposition parties not legally recognized: Communist Action Party [Fateh al-JAMOUS]; National Democratic Rally [Hasan ABDUL-AZIM, spokesman] (includes five parties—Arab Democratic Socialist Union Party [Hasan ABDUL-AZIM], Arab Socialist Movement, Democratic Ba'th Party [Ibrahim MAKHOS], Democratic People's Party [Riad al TURK], Revolutionary Workers' Party [Abdul Hafez al HAFEZ])
Kurdish parties (considered illegal): Azadi Party [Kheirudin MURAD]; Future Party [Masha'l TAMMO]; Kurdish Democratic Alliance (includes four parties); Kurdish Democratic Front (includes three parties); Yekiti Party [Fu'ad ALEYKO]
other parties: Syrian Democratic Party [Mustafa QALAAJI]
Political pressure groups and leaders: Arab Human Rights Organization in Syria or AHRO; Damascus Declaration Group (a broad alliance of secular, religious, and Kurdish opposition groups); National Salvation Front (alliance between former Vice President Abd al-Halim KHADDAM and other small opposition groups in exile; formerly included the Syrian Muslim Brotherhood); Syrian Center for Media and Freedom of Expression [Mazin DARWISH]; Syrian Human Rights Organization [Muhanad al-HASANI]; Syrian Human Rights Society or HRAS [Fayez FAWAZ]; Syrian Muslim Brotherhood or SMB [Muhammad Riyad al-SHAQFAH] (operates in exile in London)
International organization participation: ABEDA, AFESD, AMF, CAEU, FAO, G-24, G-77, IAEA, IBRD, ICAO, ICC, ICRM, IDA, IDB, IFAD, IFC, IFRCS, IHO, ILO, IMF, IMO, Interpol, IOC, IPU, ISO, ITSO, ITU, LAS, MIGA, NAM, OAPEC, OIC, UN, UNCTAD, UNESCO, UNIDO, UNRWA, UNWTO, UPU, WCO, WFTU, WHO, WIPO, WMO, WTO (observer)
Diplomatic representation in the US: *chief of mission:* Ambassador Imad MOUSTAPHA
chancery: 2215 Wyoming Avenue NW, Washington, DC 20008
telephone: [1] (202) 232-6313
FAX: [1] (202) 265-4585
Diplomatic representation from the US: *chief of mission:* Ambassador Robert S. FORD
embassy: Abou Roumaneh, Al-Mansour Street, No. 2, Damascus
mailing address: P. O. Box 29, Damascus
telephone: [963] (11) 3391-4444
FAX: [963] (11) 3391-3999
Flag description: three equal horizontal bands of red (top), white, and black; two small, green, five-pointed stars in a horizontal line centered in the white band; the band colors derive from the Arab Liberation flag and represent oppression (black), overcome through bloody struggle (red), to be replaced by a bright future (white); identical to the former flag of the United Arab Republic (1958-1961) where the two stars represented the constituent states of Syria and Egypt; the current design dates to 1980
note: similar to the flag of Yemen, which has a plain white band, Iraq, which has an Arabic inscription centered in the white band, and that of Egypt, which has a gold Eagle of Saladin centered in the white band
National anthem: *name:* "Humat ad-Diyar" (Guardians of the Homeland)
lyrics/music: Khalil Mardam BEY/ Mohammad Salim FLAYFEL and Ahmad Salim FLAYFEL
note: adopted 1936, restored 1961; between 1958 and 1961, while Syria was a member of the United Arab Republic with Egypt, the country had a different anthem

ECONOMY

Economy—overview: Syrian economic growth remained in the 4-5% range in 2008-10 even though the global economic crisis affected oil prices and the economies of Syria's key export partners and sources of investment. Damascus has implemented modest economic reforms in the past few years, including cutting lending interest rates, opening private banks, consolidating all of the multiple exchange rates, raising prices on some subsidized items, most notably gasoline and cement, and establishing the Damascus Stock Exchange—which began operations in 2009. In addition, President ASAD signed legislative decrees to encourage corporate ownership reform, and to allow the Central Bank to issue Treasury bills and bonds for government debt. Nevertheless, the economy remains highly controlled by the government. Long-run economic constraints include declining oil production, high unemployment, rising budget deficits, and increasing pressure on water supplies caused by heavy use in agriculture, rapid population growth, industrial expansion, and water pollution.
GDP (purchasing power parity): $107.4 billion (2010 est.)
country comparison to the world: 67
$104 billion (2009 est.)
$98.13 billion (2008 est.)
note: data are in 2010 US dollars
GDP (official exchange rate): $59.33 billion (2010 est.)
GDP—real growth rate: 3.2% (2010 est.)
country comparison to the world: 118
6% (2009 est.)
4.5% (2008 est.)
GDP—per capita (PPP): $4,800 (2010 est.)
country comparison to the world: 153
$4,800 (2009 est.)
$4,600 (2008 est.)
note: data are in 2010 US dollars
GDP—composition by sector:
agriculture: 17.6%
industry: 26.8%
services: 55.6% (2010 est.)
Labor force: 5.527 million (2010 est.)
country comparison to the world: 69
Labor force—by occupation:
agriculture: 17%
industry: 16%
services: 67% (2008 est.)
Unemployment rate: 8.3% (2010 est.)
country comparison to the world: 94
8.5% (2009 est.)
Population below poverty line: 11.9% (2006 est.)
Household income or consumption by percentage share: *lowest 10%:* NA%
highest 10%: NA%
Investment (gross fixed): 16.6% of GDP (2010 est.)
country comparison to the world: 122
Budget: *revenues:* $12.53 billion
expenditures: $15.3 billion (2010 est.)
Public debt: 29.8% of GDP (2010 est.)
country comparison to the world: 90
28.1% of GDP (2009 est.)
Inflation rate (consumer prices): 5.9% (2010 est.)
country comparison to the world: 155

2.6% (2009 est.)
Central bank discount rate: 5% (31 December 2009)
country comparison to the world: 82
5% (31 December 2008)
Commercial bank prime lending rate: 10.04% (31 December 2009 est.)
country comparison to the world: 83
10.19% (31 December 2008 est.)
Stock of narrow money: $21.6 billion (31 December 2010 est.)
country comparison to the world: 62
$19.53 billion (31 December 2009 est.)
Stock of broad money: $161 billion (31 December 2009)
country comparison to the world: 43
$147.5 billion (31 December 2008)
Stock of domestic credit: $27.14 billion (31 December 2010 est.)
country comparison to the world: 71
$23.58 billion (31 December 2009 est.)
Market value of publicly traded shares: $NA
Agriculture—products: wheat, barley, cotton, lentils, chickpeas, olives, sugar beets; beef, mutton, eggs, poultry, milk
Industries: petroleum, textiles, food processing, beverages, tobacco, phosphate rock mining, cement, oil seeds crushing, car assembly
Industrial production growth rate: 6% (2010 est.)
country comparison to the world: 55
Electricity—production: 36.5 billion kWh (2007 est.)
country comparison to the world: 57
Electricity—consumption: 27.35 billion kWh (2007 est.)
country comparison to the world: 63
Electricity—exports: 0 kWh (2008)
Electricity—imports: 1.4 billion kWh (2007)
Oil—production: 400,400 bbl/day (2009 est.)
country comparison to the world: 33
Oil—consumption: 252,000 bbl/day (2009 est.)
country comparison to the world: 49
Oil—exports: 155,000 bbl/day (2008 est.)
country comparison to the world: 56
Oil—imports: 58,710 bbl/day (2007 est.)
country comparison to the world: 81
Oil—proved reserves: 2.5 billion bbl (1 January 2010 est.)
country comparison to the world: 33
Natural gas—production: 6.04 billion cu m (2008 est.)
country comparison to the world: 48
Natural gas—consumption: 6.18 billion cu m (2008 est.)
country comparison to the world: 56
Natural gas—exports: 0 cu m (2008 est.)
country comparison to the world: 180
Natural gas—imports: 140 million cu m (2008 est.)
country comparison to the world: 66
Natural gas—proved reserves: 240.7 billion cu m (1 January 2010 est.)
country comparison to the world: 43
Current account balance: $649 million (2010 est.)
country comparison to the world: 49
$394 million (2009 est.)
Exports: $12.84 billion (2010 est.)
country comparison to the world: 78
$11.76 billion (2009 est.)
Exports—commodities: crude oil, minerals, petroleum products, fruits and vegetables, cotton fiber, textiles, clothing, meat and live animals, wheat
Exports—partners: Iraq 30.22%, Lebanon 12.21%, Germany 8.89%, Egypt 6.8%, Saudi Arabia 5.04%, Italy 4.55% (2009)
Imports: $13.57 billion (2010 est.)
country comparison to the world: 80
$12.62 billion (2009 est.)
Imports—commodities: machinery and transport equipment, electric power machinery, food and livestock, metal and metal products, chemicals and chemical products, plastics, yarn, paper
Imports—partners: Saudi Arabia 10.1%, China 9.95%, Turkey 6.97%, Egypt 6.44%, UAE 4.97%, Italy 4.93%, Russia 4.92%, Germany 4.38%, Lebanon 4.12% (2009)
Reserves of foreign exchange and gold: $17.96 billion (31 December 2010 est.)
country comparison to the world: 45
$17.44 billion (31 December 2009 est.)
Debt—external: $7.682 billion (31 December 2010 est.)
country comparison to the world: 93
$7.359 billion (31 December 2009 est.)
Exchange rates: Syrian pounds (SYP) per US dollar–46.456 (2010), 46.708 (2009) 46.5281 (2008), 50.0085 (2007), 51.689 (2006)

COMMUNICATIONS

Telephones—main lines in use: 3.871 million (2009)
country comparison to the world: 42
Telephones—mobile cellular: 9.697 million (2009)
country comparison to the world: 66
Telephone system: *general assessment:* fair system currently undergoing significant improvement and digital upgrades, including fiber-optic technology and expansion of the network to rural areas
domestic: the number of fixed-line connections has increased markedly since 2000; mobile-cellular service growing with telephone subscribership reaching nearly 50 per 100 persons in 2009
international: country code–963; submarine cable connection to Egypt, Lebanon, and Cyprus; satellite earth stations–1 Intelsat (Indian Ocean) and 1 Intersputnik (Atlantic Ocean region); coaxial cable and microwave radio relay to Iraq, Jordan, Lebanon, and Turkey; participant in Medarabtel
Broadcast media: state-run television and radio broadcast networks; state operates 2 TV networks and a satellite channel; roughly two-thirds of Syrian homes have a satellite dish providing access to foreign TV broadcasts; 3 state-run radio channels; first private radio station launched in 2005; private radio broadcasters prohibited from transmitting news or political content (2007)
Internet country code: .sy
Internet hosts: 8,114 (2010)
country comparison to the world: 133
Internet users: 4.469 million (2009)
country comparison to the world: 52

TRANSPORTATION

Airports: 104 (2010)
country comparison to the world: 56
Airports—with paved runways: *total:* 29
over 3,047 m: 5
2,438 to 3,047 m: 15
1,524 to 2,437 m: 1
914 to 1,523 m: 3
under 914 m: 5 (2010)
Airports—with unpaved runways: *total:* 75
1,524 to 2,437 m: 1
914 to 1,523 m: 15
under 914 m: 59 (2010)
Heliports: 7 (2010)
Pipelines: gas 3,161 km; oil 1,997 km (2010)
Railways: *total:* 2,052 km
country comparison to the world: 72
standard gauge: 1,801 km 1.435-m gauge
narrow gauge: 251 km 1.050-m gauge (2008)
Roadways: *total:* 97,401 km
country comparison to the world: 44
paved: 19,490 km (includes 1,103 km of expressways)
unpaved: 77,911 km (2006)
Waterways: 900 km (navigable but not economically significant) (2010)
country comparison to the world: 69
Merchant marine: *total:* 41
country comparison to the world: 76
by type: bulk carrier 7, cargo 30, carrier 3, container 1
foreign-owned: 5 (Jordan 2, Lebanon 2, Romania 1)
registered in other countries: 199 (Barbados 1, Belize 2, Bolivia 4, Cambodia 22, Comoros 6, Cyprus 1, Dominica 2, Georgia 35, Lebanon 3, Liberia 1, Libya 2, Malta 5, Moldova 3, North Korea 6, Panama 42, Saint Kitts and Nevis 5, Saint Vincent and the Grenadines 13, Sierra Leone 20, Togo 5, unknown 8) (2010)
Ports and terminals: Baniyas, Latakia, Tartus

MILITARY

Military branches: Syrian Armed Forces: Syrian Arab Army, Syrian Arab Navy, Syrian Arab Air and Air Defense Forces (includes Air Defense Command) (2008)
Military service age and obligation: 18 years of age for compulsory military service; conscript service obligation–21 months (18 months in the Syrian Arab Navy); women are not conscripted but may volunteer to serve (2010)
Manpower available for military service: *males age 16–49:* 5,889,837
females age 16–49: 5,660,751 (2010 est.)
Manpower fit for military service: *males age 16–49:* 5,055,510
females age 16–49: 4,884,151 (2010 est.)
Manpower reaching militarily significant age annually: *male:* 256,698
female: 244,712 (2010 est.)
Military expenditures: 5.9% of GDP (2005 est.)
country comparison to the world: 10

TRANSNATIONAL ISSUES

Disputes—international: Golan Heights is Israeli-occupied with the almost 1,000-strong UN Disengagement Observer Force

(UNDOF) patrolling a buffer zone since 1964; lacking a treaty or other documentation describing the boundary, portions of the Lebanon-Syria boundary are unclear with several sections in dispute; since 2000, Lebanon has claimed Shab'a Farms in the Golan Heights; 2004 Agreement and pending demarcation settles border dispute with Jordan; approximately two million Iraqis have fled the conflict in Iraq with the majority taking refuge in Syria and Jordan

Refugees and internally displaced persons: refugees (country of origin): 1-1.4 million (Iraq); 522,100 (Palestinian Refugees (UNRWA))

IDPs: 305,000 (most displaced from Golan Heights during 1967 Arab-Israeli War) (2007)

Trafficking in persons: current situation: Syria is a destination and transit country for women and children trafficked for commercial sexual exploitation and forced labor; a significant number of women and children in the large and expanding Iraqi refugee community in Syria are reportedly forced into commercial sexual exploitation by Iraqi gangs or, in some cases, their families; women from Indonesia, Sri Lanka, the Philippines, Ethiopia, and Sierra Leone are recruited for work in Syria as domestic servants, but some face conditions of involuntary servitude, including long hours, non-payment of wages, withholding of passports, restrictions on movement, threats, and physical or sexual abuse

tier rating: Tier 2 Watch List—Syria again failed to report any law enforcement efforts to punish trafficking offenses in 2007; in addition, the government did not offer protection services to victims of trafficking and may have arrested, prosecuted, or deported some victims for prostitution or immigration violations; Syria has not ratified the 2000 UN TIP Protocol (2008)

Illicit drugs: a transit point for opiates, hashish, and cocaine bound for regional and Western markets; weak anti-money-laundering controls and bank privatization may leave it vulnerable to money laundering

TAIWAN

INTRODUCTION

Background: In 1895, military defeat forced China to cede Taiwan to Japan. Taiwan reverted to Chinese control after World War II. Following the Communist victory on the mainland in 1949, 2 million Nationalists fled to Taiwan and established a government using the 1947 constitution drawn up for all of China. Over the next five decades, the ruling authorities gradually democratized and incorporated the local population within the governing structure. In 2000, Taiwan underwent its first peaceful transfer of power from the Nationalist to the Democratic Progressive Party. Throughout this period, the island prospered and became one of East Asia's economic "Tigers." The dominant political issues continue to be the relationship between Taiwan and China—specifically the question of Taiwan's eventual status—as well as domestic political and economic reform.

GEOGRAPHY

Location: Eastern Asia, islands bordering the East China Sea, Philippine Sea, South China Sea, and Taiwan Strait, north of the Philippines, off the southeastern coast of China
Geographic coordinates:
23 30 N, 121 00 E
Map references: Southeast Asia
Area: *total:* 35,980 sq km
country comparison to the world: 138
land: 32,260 sq km
water: 3,720 sq km
note: includes the Pescadores, Matsu, and Quemoy islands
Area—comparative: slightly smaller than Maryland and Delaware combined
Land boundaries: 0 km
Coastline: 1,566.3 km
Maritime claims: territorial sea: 12 nm
exclusive economic zone: 200 nm
Climate: tropical; marine; rainy season during southwest monsoon (June to August); cloudiness is persistent and extensive all year
Terrain: eastern two-thirds mostly rugged mountains; flat to gently rolling plains in west
Elevation extremes: *lowest point:* South China Sea 0 m
highest point: Yu Shan 3,952 m
Natural resources: small deposits of coal, natural gas, limestone, marble, and asbestos
Land use: *arable land:* 24%
permanent crops: 1%
other: 75% (2001)
Irrigated land: NA
Total renewable water resources: 67 cu km (2000)
Natural hazards: earthquakes; typhoons
volcanism: Kueishantao Island (elev. 401 m), east of Taiwan, is its only historically active volcano, although it has not erupted in centuries
Environment—current issues: air pollution; water pollution from industrial emissions, raw sewage; contamination of drinking water supplies; trade in endangered species; low-level radioactive waste disposal
Environment—international agreements: *party to:* none of the selected agreements because of Taiwan's international status
Geography—note: strategic location adjacent to both the Taiwan Strait and the Luzon Strait

PEOPLE

Population: 23,071,779 (July 2011 est.)
country comparison to the world: 50
Age structure: *0–14 years:* 15.6% (male 1,875,359/female 1,732,007)
15–64 years: 73.4% (male 8,538,881/female 8,406,716)
65 years and over: 10.9%
(male 1,198,591/female 1,320,225) (2011 est.)
Median age: *total:* 37.6 years
male: 36.9 years
female: 38.3 years (2011 est.)
Population growth rate:
0.193% (2011 est.)
country comparison to the world: 182
Birth rate: 8.9 births/1,000 population (2011 est.)
country comparison to the world: 211
Death rate: 7 deaths/1,000 population (July 2011 est.)
country comparison to the world: 134
Net migration rate: 0.03 migrant(s)/1,000 population (2011 est.)
country comparison to the world: 71
Sex ratio: *at birth:* 1.084 male(s)/female
under 15 years: 1.08 male(s)/female
15–64 years: 1.02 male(s)/female
65 years and over: 0.92 male(s)/female
total population: 1.02 male(s)/female (2011 est.)
Infant mortality rate: *total:* 5.18 deaths/1,000 live births
country comparison to the world: 181
male: 5.46 deaths/1,000 live births
female: 4.88 deaths/1,000 live births (2011 est.)
Life expectancy at birth: *total population:* 78.32 years
country comparison to the world: 51
male: 75.5 years
female: 81.36 years (2011 est.)
Total fertility rate: 1.15 children born/woman (2011 est.)
country comparison to the world: 220
HIV/AIDS—adult prevalence rate: NA
HIV/AIDS—people living with HIV/AIDS: NA
HIV/AIDS—deaths: NA
Nationality: *noun:* Taiwan (singular and plural)
note: example—he or she is from Taiwan; they are from Taiwan
adjective: Taiwan
Ethnic groups: Taiwanese (including Hakka) 84%, mainland Chinese 14%, indigenous 2%
Religions: mixture of Buddhist and Taoist 93%, Christian 4.5%, other 2.5%
Languages: Mandarin Chinese (official), Taiwanese (Min), Hakka dialects
Literacy: *definition:* age 15 and over can read and write
total population: 96.1%
male: NA
female: NA (2003)
School life expectancy (primary to tertiary education): NA
Education expenditures: NA

GOVERNMENT

Country name: *conventional long form:* none
conventional short form: Taiwan
local long form: none
local short form: Taiwan
former: Formosa
Government type: multiparty democracy
Capital: *name:* Taipei
geographic coordinates: 25 03 N, 121 30 E
time difference: UTC+8 (13 hours ahead of Washington, DC during Standard Time)
Administrative divisions: includes main island of Taiwan plus smaller islands nearby and off coast of China's Fujian Province; Taiwan is divided into 14 counties (hsien, singular and plural), 3 municipalities (shih, singular and plural), and 5 special municipalities (chih-hsia-shih, singular and plural)
note: Taiwan uses a variety of romanization systems; while a modified Wade-Giles system still dominates, the city of Taipei has adopted a Pinyin romanization for street and place names within its boundaries; other local authorities use different romanization systems; names for administrative divisions that follow are taken from the Taiwan Yearbook 2007 published by the Government Information Office in Taipei.
counties: Changhua, Chiayi (county), Hsinchu (county), Hualien, Kinmen, Lienchiang, Miaoli, Nantou, Penghu, Pingtung, Taitung, Taoyuan, Yilan, Yunlin
municipalities: Chiayi (city), Hsinchu (city), Keelung
special municipalities: Kaohsiung, New Taipei, Taichung, Tainan, Taipei
National holiday: Republic Day (Anniversary of the Chinese Revolution), 10 October (1911)
Constitution: adopted 25 December 1946; promulgated 1 January 1947; effective 25 December 1947; amended many times
Legal system: civil law system
International law organization participation: has not submitted an ICJ jurisdiction declaration; non-party state to the ICCt

Suffrage: 20 years of age; universal
Executive branch: *chief of state:* President MA Ying-jeou (since 20 May 2008); Vice President Vincent SIEW (since 20 May 2008)
head of government: Premier (President of the Executive Yuan) WU Den-yih (since 10 September 2009); Vice Premier (Vice President of Executive Yuan) Sean CHEN (since 17 May 2010)
cabinet: Executive Yuan—ministers appointed by president on recommendation of premier
(For more information visit the World Leaders website)
elections: president and vice president elected on the same ticket by popular vote for four-year terms (eligible for a second term); election last held on 22 March 2008 (next to be held on 14 January 2012); premier appointed by the president; vice premiers appointed by the president on the recommendation of the premier
election results: MA Ying-jeou elected president; percent of vote—MA Ying-jeou 58.45%, Frank HSIEH 41.55%
Legislative branch: unicameral Legislative Yuan (113 seats—73 district members elected by popular vote, 34 at-large members elected on basis of proportion of islandwide votes received by participating political parties, 6 elected by popular vote among aboriginal populations; members to serve four-year terms); parties must receive 5% of vote to qualify for at-large seats
elections: Legislative Yuan—last held on 12 January 2008 (next to be held on 14 January 2012)
election results: Legislative Yuan—percent of vote by party—KMT 53.5%, DPP 38.2%, NPSU 2.4%, PFP 0.3%, others 1.6%, independents 4%; seats by party—KMT 81, DPP 27, NPSU 3, PFP 1, independent 1; note—following the 2008 elections, several rounds of byelections were held to fill seats vacated as a result of corruption changes; seats by party as of January 2011—KMT 74, DPP 31, NPSU 3, independent 2, vacant 3
Judicial branch: Judicial Yuan (justices appointed by the president with consent of the Legislative Yuan)
Political parties and leaders: Democratic Progressive Party or DPP [TSAI Ing-wen]; Kuomintang or KMT (Nationalist Party) [MA Ying-jeou]; Non-Partisan Solidarity Union or NPSU [LIN Pin-kuan]; People First Party or PFP [James SOONG]
Political pressure groups and leaders: environmental groups; independence movement; various business groups
note: debate on Taiwan independence has become acceptable within the mainstream of domestic politics on Taiwan; public opinion polls consistently show a substantial majority of Taiwan people supports maintaining Taiwan's status quo for the foreseeable future; advocates of Taiwan independence oppose the stand that the island will eventually unify with mainland China; advocates of eventual unification predicate their goal on the democratic transformation of the mainland
International organization participation: ADB, APEC, BCIE, ICC, IOC, ITUC, WTO
Diplomatic representation in the US: none; commercial and cultural relations with the people in the United States are maintained through an unofficial instrumentality, the Taipei Economic and Cultural Representative Office in the United States (TECRO), a private nonprofit corporation that performs citizen and consular services similar to those at diplomatic posts
representative: Jason C. YUAN
office: 4201 Wisconsin Avenue NW, Washington, DC 20016
telephone: [1] 202 895-1800
Taipei Economic and Cultural Offices (branch offices): Atlanta, Boston, Chicago, Guam, Houston, Honolulu, Kansas City, Los Angeles, Miami, New York, San Francisco, Seattle
Diplomatic representation from the US: none; commercial and cultural relations with the people on Taiwan are maintained through an unofficial instrumentality, the American Institute in Taiwan (AIT), a private nonprofit corporation that performs citizen and consular services similar to those at diplomatic posts
director: William A. STANTON
office: #7 Lane 134, Hsin Yi Road, Section 3, Taipei, Taiwan
telephone: [1] [886] (02) 2162-2000
FAX: [1] [886] (07) 238-7744
other offices: Kaohsiung
Flag description: red field with a dark blue rectangle in the upper hoist-side corner bearing a white sun with 12 triangular rays; the blue and white design of the canton (symbolizing the sun of progress) dates to 1895; it was later adopted as the flag of the Kuomintang Party; blue signifies liberty, justice, and democracy; red stands for fraternity, sacrifice, and nationalism, white represents equality, frankness, and the people's livelihood; the 12 rays of the sun are those of the months and the twelve traditional Chinese hours (each ray equals two hours)
National anthem: *name:* "Zhonghua Minguo guoge" (National Anthem of the Republic of China)
lyrics/music: HU Han-min, TAI Chi-t'ao, and LIAO Chung-k'ai/CHENG Mao-Yun
note: adopted 1930; the anthem is also the song of the Kuomintang Party; it is informally known as "San Min Chu I" or "San Min Zhu Yi" (Three Principles of the People); because of political pressure from China, "Guo Qi Ge" (National Banner Song) is used at international events rather than the official anthem of Taiwan; the "National Banner Song" has gained popularity in Taiwan and is commonly used during flag raisings

ECONOMY

Economy—overview: Taiwan has a dynamic capitalist economy with gradually decreasing government guidance of investment and foreign trade. In keeping with this trend, some large, state-owned banks and industrial firms have been privatized. Exports, led by electronics and machinery, generate about 70% of Taiwan's GDP growth, and have provided the primary impetus for economic development. This heavy dependence on exports exposes the economy to upturns and downturns in world demand. In 2009, Taiwan's GDP contracted 1.9%, due primarily to a 20% year-on-year decline in exports. In 2010 GDP grew 10.5%, as exports returned to the level of previous years. Taiwan's diplomatic isolation, low birth rate, and rapidly aging population are major long-term challenges. Free trade agreements have proliferated in East Asia over the past several years, but so far Taiwan has been excluded from this greater economic integration, largely because of its diplomatic status. Taiwan's Total Fertility rate of just over one child per woman is among the lowest in the world, raising the prospect of future labor shortages, falling domestic demand, and declining tax revenues. Taiwan's population is aging quickly, with the number of people over 65 accounting for 10.9% of the island's total population as 2011. The island runs a large trade surplus, and its foreign reserves are the world's fourth largest, behind China, Japan, and Russia. Since President MA Ying-jeou took office in May 2008, cross-Strait economic ties have increased significantly. Since 2005 China has overtaken the US to become Taiwan's second-largest source of imports after Japan. China is also the island's number one destination for foreign direct investment. Taiwan has focused much of its efforts on improving the cross-Strait economic relationship. Three financial memorandums of understanding, covering banking, securities, and insurance, took effect in mid-January 2010, opening the island to greater investments from the mainland's financial firms and institutional investors, and providing new opportunities for Taiwan financial firms to operate in China. Taiwan and the mainland in June 2010 signed the landmark Economic Cooperation Framework Agreement (ECFA), an agreement that the Taiwan authorities hope will eventually lead to a free-trade arrangement that will increase cross-Strait economic ties by lowering tariffs on a number of goods and by reducing market access barriers for services. The Taiwan authorities have said that the ECFA will serve as a stepping stone toward trade pacts with other regional partners and they announced that formal negotiations towards an economic cooperation agreement with Singapore would begin in 2011. Closer economic links with the mainland brings greater opportunities for the Taiwan economy, but also poses new challenges. For example, FDI in China has resulted in Chinese import substitution away from Taiwan's exports and a restriction of potential job creation in Taiwan.
GDP (purchasing power parity): $821.8 billion (2010 est.)
country comparison to the world: 19
$741.5 billion (2009 est.)
$756.1 billion (2008 est.)
note: data are in 2010 US dollars
GDP (official exchange rate): $430.6 billion (2010 est.)
GDP—real growth rate:
10.8% (2010 est.)
country comparison to the world: 4

-1.9% (2009 est.)
0.7% (2008 est.)
GDP—per capita (PPP):
$35,700 (2010 est.)
country comparison to the world: 32
$32,300 (2009 est.)
$33,000 (2008 est.)
note: data are in 2010 US dollars
GDP—composition by sector:
agriculture: 1.4%
industry: 31.1%
services: 67.5% (2010 est.)
Labor force: 11.07 million (2010 est.)
country comparison to the world: 47
Labor force—by occupation:
agriculture: 5.2%
industry: 35.9%
services: 58.8% (2010 est.)
Unemployment rate: 5.2% (2010 est.)
country comparison to the world: 50
5.9% (2009 est.)
Population below poverty line: 1.16% (2010 est.)
Household income or consumption by percentage share: *lowest 10%:* 6.4%
highest 10%: 40.3% (2010)
Distribution of family income—Gini index: 32.6 (2000)
country comparison to the world: 100
Investment (gross fixed): 21.8% of GDP (2010 est.)
country comparison to the world: 70
Budget: *revenues:* $72.24 billion
expenditures: $79.65 billion (2010 est.)
Public debt: 33.9% of GDP (2010 est.)
country comparison to the world: 87
33.1% of GDP (2009 est.)
Inflation rate (consumer prices): 1% (2010 est.)
country comparison to the world: 20
-0.9% (2009)
Central bank discount rate: 1.625% (31 December 2010)
country comparison to the world: 132
1.25% (February 2009)
Commercial bank prime lending rate: 2.68% (31 December 2010)
country comparison to the world: 152
2.56% (31 December 2009 est.)
Stock of narrow money: $377.3 billion (31 December 2010 est.)
country comparison to the world: 12
$328.2 billion (31 December 2009 est.)
Stock of broad money: $1.022 trillion (31 December 2010 est.)
country comparison to the world: 17
$911.9 billion (31 December 2009 est.)
Stock of domestic credit: $751.5 billion (31 December 2010 est.)
country comparison to the world: 18
$671 billion (31 December 2009 est.)
Market value of publicly traded shares: $784.1 billion (31 December 2010)
country comparison to the world: 14
$657.3 billion (31 December 2009)
$354.7 billion (31 December 2008)
Agriculture—products:
rice, vegetables, fruit, tea, flowers; pigs, poultry; fish
Industries: electronics, communications and information technology products, petroleum refining, armaments, chemicals, textiles, iron and steel, machinery, cement, food processing, vehicles, consumer products, pharmaceuticals
Industrial production growth rate: 26.4% (2010 est.)
country comparison to the world: 2
Electricity—production: 229.1 billion kWh (2009)
country comparison to the world: 18
Electricity—consumption: 220.8 billion kWh (2009)
country comparison to the world: 16
Electricity—exports: 0 kWh (2009 est.)
Electricity—imports:
0 kWh (2009 est.)
Oil—production: 276,800 bbl/day (2009 est.)
country comparison to the world: 38
Oil—consumption: 834,000 bbl/day (2010 est.)
country comparison to the world: 22
Oil—exports: 303,000 bbl/day (2010 est.)
country comparison to the world: 42
Oil—imports: 876,300 bbl/day (2010 est.)
country comparison to the world: 17
Oil—proved reserves: 2.8 million bbl (1 January 2010 est.)
country comparison to the world: 95
Natural gas—production: 350.7 million cu m (2009 est.)
country comparison to the world: 71
Natural gas—consumption: 11.63 billion cu m (2009 est.)
country comparison to the world: 45
Natural gas—exports: 0 cu m (2008 est.)
country comparison to the world: 190
Natural gas—imports: 11.59 billion cu m (2008 est.)
country comparison to the world: 19
Natural gas—proved reserves: 6.229 billion cu m (1 January 2010 est.)
country comparison to the world: 85
Current account balance: $39 billion (2010 est.)
country comparison to the world: 10
$42.92 billion (2009 est.)
Exports: $274.6 billion (2010 est.)
country comparison to the world: 17
$203.4 billion (2009 est.)
Exports—commodities: electronics, flat panels, machinery; metals; textiles, plastics, chemicals; optical, photographic, measuring, and medical instruments
Exports—partners: China 28.1%, Hong Kong 13.8%, US 11.5%, Japan 6.6%, Singapore 4.4% (2010 est.)
Imports: $251.4 billion (2010 est.)
country comparison to the world: 18
$174.4 billion (2009 est.)
Imports—commodities: electronics, machinery, crude petroleum, precision instruments, organic chemicals, metals
Imports—partners: Japan 20.7%, China 14.2%, US 10%, South Korea 6.4%, Saudi Arabia 4.7% (2010 est.)
Reserves of foreign exchange and gold: $387.2 billion (31 December 2010 est.)
country comparison to the world: 5
$353 billion (31 December 2009 est.)
Debt—external: $91.41 billion (31 December 2010 est.)
country comparison to the world: 38
$82.02 billion (31 December 2009 est.)
Stock of direct foreign investment—at home: $111.1 billion (31 December 2010 est.)
country comparison to the world: 31
$107.2 billion (31 December 2009 est.)
Stock of direct foreign investment—abroad: $162.9 billion (31 December 2010 est.)
country comparison to the world: 21
$145.3 billion (31 December 2009 est.)
Exchange rates: New Taiwan dollars (TWD) per US dollar–31.642 (2010), 33.061 (2009), 31.53 (2008), 32.84 (2007), 32.534 (2006)

COMMUNICATIONS

Telephones—main lines in use: 14.596 million (2009)
country comparison to the world: 19
Telephones—mobile cellular: 26.959 million (2009)
country comparison to the world: 33
Telephone system: *general assessment:* provides telecommunications service for every business and private need
domestic: thoroughly modern; completely digitalized
international: country code–886; roughly 15 submarine fiber cables provide links throughout Asia, Australia, the Middle East, Europe, and the US; satellite earth stations–2
Broadcast media: 5 free-to-air nationwide television networks operating roughly 75 TV stations; about 85% of households utilize multi-channel cable TV; national and regional radio networks with about 170 radio stations broadcasting (2008)
Internet country code: .tw
Internet hosts: 6.336 million (2010)
country comparison to the world: 15
Internet users: 16.147 million (2009)
country comparison to the world: 24

TRANSPORTATION

Airports: 41 (2010)
country comparison to the world: 102
Airports—with paved runways:
total: 38
over 3,047 m: 8
2,438 to 3,047 m: 8
1,524 to 2,437 m: 11
914 to 1,523 m: 7
under 914 m: 4 (2010)
Airports—with unpaved runways: *total:* 3
1,524 to 2,437 m: 2
under 914 m: 1 (2010)
Heliports: 4 (2010)
Pipelines: gas 412 km (2010)
Railways: *total:* 1,482 km
country comparison to the world: 79
standard gauge:
340 km 1.435-m gauge
narrow gauge: 1,085 km 1.067-m gauge; 57 km 0.762-m gauge
note: the 0.762 gauge track belong to three entities, Taiwan Cement, TaiPower, and the Forestry Bureau (2009)
Roadways: *total:* 41,475 km
country comparison to the world: 87
paved: 41,033 km (includes 720 km of expressways)
unpaved: 442 km (2009)
Merchant marine: *total:* 101
country comparison to the world: 50
by type: bulk carrier 28, cargo 19, chemical tanker 2, container 27, passenger/cargo 4, petroleum tanker 12, refrigerated cargo 7, roll on/roll off 2
foreign-owned:2 (France 1, Vietnam 1)

registered in other countries: 574 (Cambodia 1, Honduras 2, Hong Kong 26, Indonesia 1, Italy 11, Kiribati 5, Liberia 88, Marshall Islands 2, Panama 337, Philippines 1, Sierra Leone 1, Singapore 79, Thailand 1, UK 11, unknown 8) (2010)
Ports and terminals: Chilung (Keelung), Kaohsiung, Hualian, Taichung

MILITARY

Military branches: Army, Navy (includes Marine Corps), Air Force, Coast Guard Administration, Armed Forces Reserve Command, Combined Service Forces Command, Armed Forces Police Command
Military service age and obligation: 19-35 years of age for male compulsory military service; service obligation—2 years; women may enlist; women in Air Force service are restricted to noncombat roles; reserve obligation to age 30 (Army); the Ministry of Defense is in the process of implementing a voluntary enlistment system over the period 2010-2015, although nonvolunteers will still be required to perform alternative service or go through 4 months of military training (2010)
Manpower available for military service: *males age 16–49:* 6,183,567
females age 16–49: 6,006,676 (2010 est.)
Manpower fit for military service: *males age 16–49:* 5,074,173
females age 16–49: 4,951,088 (2010 est.)
Manpower reaching militarily significant age annually: *male:* 166,190
female: 155,306 (2010 est.)
Military expenditures: 2.2% of GDP; note—in 2009, the Taiwanese president pledged to maintain defense spending at 3.0% or higher; projected 2.73% for 2011 (2009)

TRANSNATIONAL ISSUES

Disputes—international: involved in complex dispute with Brunei, China, Malaysia, the Philippines, and Vietnam over the Spratly Islands, and with China and the Philippines over Scarborough Reef; the 2002 "Declaration on the Conduct of Parties in the South China Sea" has eased tensions but falls short of a legally binding "code of conduct" desired by several of the disputants; Paracel Islands are occupied by China, but claimed by Taiwan and Vietnam; in 2003, China and Taiwan became more vocal in rejecting both Japan's claims to the uninhabited islands of the Senkaku-shoto (Diaoyu Tai) and Japan's unilaterally declared exclusive economic zone in the East China Sea where all parties engage in hydrocarbon prospecting
Illicit drugs: regional transit point for heroin, methamphetamine, and precursor chemicals; transshipment point for drugs to Japan; major problem with domestic consumption of methamphetamine and heroin; rising problems with use of ketamine and club drugs

TAJIKISTAN

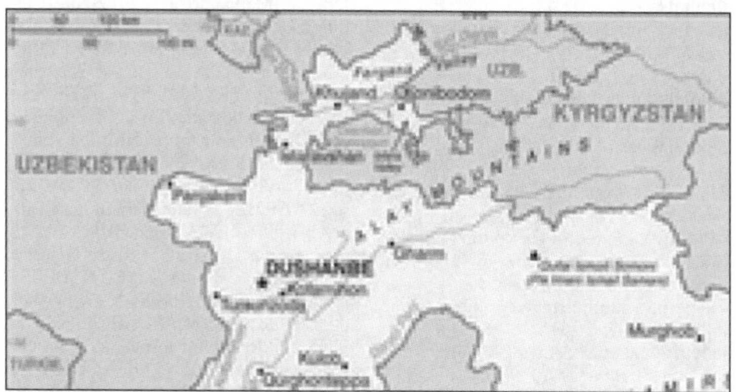

INTRODUCTION

Background: The Tajik people came under Russian rule in the 1860s and 1870s, but Russia's hold on Central Asia weakened following the Revolution of 1917. Bolshevik control of the area was fiercely contested and not fully reestablished until 1925. Much of present-day Sughd province was transferred from the Uzbek SSR to the newly formed Tajik SSR in 1929. Ethnic Uzbeks form a substantial minority in Tajikistan. Tajikistan became independent in 1991 following the breakup of the Soviet Union, and experienced a civil war between regional factions from 1992-97. Tajikistan experienced several security incidents in 2010, including a mass prison-break from a Dushanbe detention facility, the country's first suicide car bombing in Khujand, and armed conflict between government forces and opposition militants in the Rasht Valley. The country remains the poorest in the former Soviet sphere. Attention by the international community since the beginning of the NATO intervention in Afghanistan has brought increased economic development and security assistance, which could create jobs and strengthen stability in the long term. Tajikistan is seeking WTO membership and has joined NATO's Partnership for Peace.

GEOGRAPHY

Location: Central Asia, west of China, south of Kyrgyzstan
Geographic coordinates:
39 00 N, 71 00 E
Map references: Asia
Area: *total:* 143,100 sq km
country comparison to the world: 95
land: 141,510 sq km
water: 2,590 sq km
Area—comparative: slightly smaller than Wisconsin
Land boundaries: *total:* 3,651 km
border countries: Afghanistan 1,206 km, China 414 km, Kyrgyzstan 870 km, Uzbekistan 1,161 km
Coastline: 0 km (landlocked)
Maritime claims: none (landlocked)
Climate: midlatitude continental, hot summers, mild winters; semiarid to polar in Pamir Mountains
Terrain: Pamir and Alay Mountains dominate landscape; western Fergana Valley in north, Kofarnihon and Vakhsh Valleys in southwest
Elevation extremes: *lowest point:* Syr Darya (Sirdaryo) 300 m
highest point: Qullai Ismoili Somoni (Pik Imeni Ismail Samani) 7,495 m
Natural resources: hydropower, some petroleum, uranium, mercury, brown coal, lead, zinc, antimony, tungsten, silver, gold
Land use: *arable land:* 6.52%
permanent crops: 0.89%
other: 92.59% (2005)
Irrigated land:
7,220 sq km (2003)
Total renewable water resources: 99.7 cu km (1997)
Freshwater withdrawal (domestic/industrial/agricultural): *total:* 11.96 cu km/yr (4%/5%/92%)
per capita: 1,837 cu m/yr (2000)
Natural hazards: earthquakes; floods
Environment—current issues: inadequate sanitation facilities; increasing levels of soil salinity; industrial pollution; excessive pesticides
Environment—international agreements: *party to:* Biodiversity, Climate Change, Climate Change-Kyoto Protocol, Desertification, Environmental Modification, Ozone Layer Protection, Wetlands
signed, but not ratified: none of the selected agreements
Geography—note: landlocked; mountainous region dominated by the Trans-Alay Range in the north and the Pamirs in the southeast; highest point, Qullai Ismoili Somoni (formerly Communism Peak), was the tallest mountain in the former USSR

PEOPLE

Population: 7,627,200 (July 2011 est.)
country comparison to the world: 95

Age structure: *0–14 years:* 33.9% (male 1,316,623/female 1,270,899)
15–64 years: 62.7% (male 2,368,554/female 2,413,982)
65 years and over: 3.4% (male 108,896/female 148,246) (2011 est.)
Median age: *total:* 22.6 years
male: 22.1 years
female: 23.1 years (2011 est.)
Population growth rate: 1.846% (2011 est.)
country comparison to the world: 62
Birth rate: 26.29 births/1,000 population (2011 est.)
country comparison to the world: 54
Death rate: 6.6 deaths/1,000 population (July 2011 est.)
country comparison to the world: 147
Net migration rate: -1.24 migrant(s)/1,000 population (2011 est.)
country comparison to the world: 155
Urbanization: *urban population:* 26% of total population (2010)
rate of urbanization: 2.2% annual rate of change (2010-15 est.)
Major cities—population: DUSHANBE (capital) 704,000 (2009)
Sex ratio: *at birth:* 1.05 male(s)/female
under 15 years: 1.04 male(s)/female
15–64 years: 0.98 male(s)/female
65 years and over: 0.74 male(s)/female
total population: 0.99 male(s)/female (2011 est.)
Infant mortality rate: *total:* 38.54 deaths/1,000 live births
country comparison to the world: 65
male: 43.21 deaths/1,000 live births
female: 33.63 deaths/1,000 live births (2011 est.)
Life expectancy at birth: *total population:* 66.03 years
country comparison to the world: 165
male: 62.97 years
female: 69.25 years (2011 est.)
Total fertility rate: 2.89 children born/woman (2011 est.)
country comparison to the world: 69
HIV/AIDS—adult prevalence rate: 0.2% (2009 est.)
country comparison to the world: 106
HIV/AIDS—people living with HIV/AIDS: 9,100 (2009 est.)
country comparison to the world: 102
HIV/AIDS—deaths: fewer than 500 (2009 est.)
country comparison to the world: 81
Major infectious diseases: *degree of risk:* high
food or waterborne diseases: bacterial diarrhea, hepatitis A, and typhoid fever
vectorborne disease: malaria (2009)
Drinking water source: *Improved:*
urban: 94% of population
rural: 61% of population
total: 70% of population
Unimproved: urban: 6% of population
rural: 39% of population
total: 30% of population (2008)
Sanitation facility access: *Improved:*
urban: 95% of population
rural: 94% of population
total: 94% of population
Unimproved: urban: 5% of population
rural: 6% of population
total: 6% of population (2008)
Nationality: *noun:* Tajikistani(s)
adjective: Tajikistani
Ethnic groups: Tajik 79.9%, Uzbek 15.3%, Russian 1.1%, Kyrgyz 1.1%, other 2.6% (2000 census)
Religions: Sunni Muslim 85%, Shia Muslim 5%, other 10% (2003 est.)
Languages: Tajik (official), Russian widely used in government and business
Literacy: *definition:* age 15 and over can read and write
total population: 99.5%
male: 99.7%
female: 99.2% (2000 census)
School life expectancy (primary to tertiary education): *total:* 11 years
male: 12 years
female: 10 years (2008)
Education expenditures: 3.5% of GDP (2008)
country comparison to the world: 119

GOVERNMENT

Country name: *conventional long form:* Republic of Tajikistan
conventional short form: Tajikistan
local long form: Jumhurii Tojikiston
local short form: Tojikiston
former: Tajik Soviet Socialist Republic
Government type: republic
Capital: *name:* Dushanbe
geographic coordinates: 38 35 N, 68 48 E
time difference: UTC+5 (10 hours ahead of Washington, DC during Standard Time)
Administrative divisions: 2 provinces (viloyatho, singular–viloyat) and 1 autonomous province* (viloyati mukhtor); Viloyati Khatlon (Qurghonteppa), Viloyati Mukhtori Kuhistoni Badakhshon [Gorno-Badakhshan]* (Khorugh), Viloyati Sughd (Khujand); the rest of the country consists of "districts under republican subordination," ruled directly from Dushanbe
note: the administrative center name follows in parentheses
Independence: 9 September 1991 (from the Soviet Union)
National holiday: Independence Day (or National Day), 9 September (1991)
Constitution: 6 November 1994
Legal system: civil law system
International law organization participation: has not submitted an ICJ jurisdiction declaration; accepts ICCt jurisdiction
Suffrage: 18 years of age; universal
Executive branch: *chief of state:* President Emomali RAHMON (since 6 November 1994; head of state and Supreme Assembly chairman since 19 November 1992)
head of government: Prime Minister Oqil OQILOV (since 20 January 1999)
cabinet: Council of Ministers appointed by the president, approved by the Supreme Assembly
(For more information visit the World Leaders website)
elections: president elected by popular vote for a seven-year term (eligible for a second term); election last held on 6 November 2006 (next to be held in November 2013); prime minister appointed by the president
election results: Emomali RAHMON reelected president; percent of vote—Emomali RAHMON 79.3%, Olimjon BOBOEV 6.2%, other 14.5%
Legislative branch: bicameral Supreme Assembly or Majlisi Oli consists of the National Assembly (upper chamber) or Majlisi Milli (34 seats; 25 members selected by local deputies, 8 appointed by the president; 1 seat reserved for the former president; members serve five-year terms) and the Assembly of Representatives (lower chamber) or Majlisi Namoyandagon (63 seats; members elected by popular vote to serve five-year terms)
elections: National Assembly—last held on 28 February 2010 (next to be held in February 2015); Assembly of Representatives—last held on 28 February 2010 (next to be held in February 2015)
election results: National Assembly—percent of vote by party—NA; seats by party—NA; Assembly of Representatives—percent of vote by party—PDPT 71%, Islamic Revival Party 8.2%, CPT 7%, APT 5.1%, PER 5.1%, other 3.6%; seats by party—PDPT 55, Islamic Revival Party 2, CPT 2, APT 2, PER 2
Judicial branch: Supreme Court (judges are appointed by the president)
Political parties and leaders: Agrarian Party of Tajikistan or APT [Amir QARAQULOV]; Democratic Party or DPT [Mahmadruzi ISKANDAROV (imprisoned October 2005); Rahmatullo VALIYEV, deputy]; Islamic Revival Party [Muhiddin KABIRI]; Party of Economic Reform or PER [Olimjon BOBOEV]; People's Democratic Party of Tajikistan or PDPT [Emomali RAHMON]; Social Democratic Party or SDPT [Rahmatullo ZOYIROV]; Socialist Party or SPT [Mirhuseyn NARZIEV]; Tajik Communist Party or CPT [Shodi SHABDOLOV]
note: for the DPT, the Ministry of Justice named a new chairman, Masud SOBIROV, in 2006; Mr. ISKANDAROV's supporters do not recognize Mr. SOBIROV; for the SPT, the Ministry of Justice named a new chairman, Abduhalim GHAFAROV, in 2004; Mr. NARZIEV's supporters do not recognize Mr. GHAFAROV
Political pressure groups and leaders: splinter parties recognized by the government but not by the base of the party: Democratic Party or DPT [Masud SOBIROV] (splintered from ISKANDAROV's DPT); Socialist Party or SPT [Abduhalim GHAFFOROV] (splintered from NARZIEV's SPT)
unregistered political parties: Progressive Party [Sulton QUVVATOV]; Unity Party [Hikmatullo SAIDOV]
International organization participation: ADB, CICA, CIS, CSTO, EAEC, EAPC, EBRD, ECO, FAO, G-77, GCTU, IAEA, IBRD, ICAO, ICRM, IDA, IDB, IFAD, IFC, IFRCS, ILO, IMF, Interpol, IOC, IOM, IPU, ISO (correspondent), ITSO, ITU, MIGA, NAM (observer), OIC, OPCW, OSCE, PFP, SCO, UN, UNCTAD, UNESCO, UNIDO, UNWTO, UPU, WCO, WFTU, WHO, WIPO, WMO, WTO (observer)
Diplomatic representation in the US: *chief of mission:* Ambassador Abdujabbor SHIRINOV
chancery: 1005 New Hampshire Avenue NW, Washington, DC 20037

telephone: [1] (202) 223-6090
FAX: [1] (202) 223-6091
Diplomatic representation from the US:
chief of mission: Ambassador Kenneth GROSS
embassy: 109-A Ismoili Somoni Avenue, Dushanbe 734019
mailing address: 7090 Dushanbe Place, Dulles, VA 20189
telephone: [992] (37) 229-20-00
FAX: [992] (37) 229-20-50
Flag description: three horizontal stripes of red (top), a wider stripe of white, and green; a gold crown surmounted by seven gold, five-pointed stars is located in the center of the white stripe; red represents the sun, victory, and the unity of the nation, white stands for purity, cotton, and mountain snows, while green is the color of Islam and the bounty of nature; the crown symbolizes the Tajik people; the seven stars signify the Tajik magic number "seven"—a symbol of perfection and the embodiment of happiness
National anthem: *name:* "Surudi milli" (National Anthem)
lyrics/music: Gulnazar KELDI/Suleiman YUDAKOV
note: adopted 1991; after the fall of the Soviet Union, Tajikistan kept the music of the anthem from its time as a Soviet republic but adopted new lyrics

ECONOMY

Economy—overview: Tajikistan has one of the lowest per capita GDPs among the 15 former Soviet republics. Because of a lack of employment opportunities in Tajikistan, as many as a million Tajik citizens work abroad, almost all of them in Russia, supporting families in Tajikistan through remittances. Less than 7% of the land area is arable. Cotton is the most important crop, and its production is closely monitored, and in many cases controlled, by the government. In the wake of the National Bank of Tajikistan's admission in December 2007 that it had improperly lent money to investors in the cotton sector, the IMF canceled its program in Tajikistan. A reform agenda is underway, according to which over half a billion dollars in farmer debt is being forgiven, and IMF assistance has been reinstated. Mineral resources include silver, gold, uranium, and tungsten. Industry consists only of a large aluminum plant, hydropower facilities, and small obsolete factories mostly in light industry and food processing. The civil war (1992-97) severely damaged the already weak economic infrastructure and caused a sharp decline in industrial and agricultural production. Tajikistan's economic situation remains fragile due to uneven implementation of structural reforms, corruption, weak governance, seasonal power shortages, and the external debt burden. A debt restructuring agreement was reached with Russia in December 2002, including a $250 million write-off of Tajikistan's $300 million debt. Electricity output expanded with the completion of the Sangtuda I hydropower dam—finished in 2009 with Russian investment. The smaller Sangtuda-2, built with Iranian investment, is scheduled for completion in 2012. The government of Tajikistan is pinning major hopes on the massive Roghun dam which, if finished according to Tajik plans, will be the tallest dam in the world. The World Bank has agreed to fund technical, economic, social, and environmental feasibility studies for the dam. Favorable reports from these studies could create investor interest in the project, which is currently moving forward with domestic funding. In January 2010, the government began selling shares in the Roghun enterprise to its population, ultimately raising over $180 million. According to numerous reports, many Tajik individuals and businesses were forced to buy shares. The coerced share sales finally ended in mid-2010 under intense criticism from donors, particularly the IMF. Tajikistan has received substantial infrastructure development loans from the Chinese government to improve roads and electricity transmission. To help increase north-south trade, the US funded a $36 million bridge which opened in August 2007 linking Tajikistan with Afghanistan. While Tajikistan has experienced steady economic growth since 1997, more than half of the population continues to live in poverty. Economic growth reached 10.6% in 2004, but dropped below 8% in 2005-08, as the effects of higher oil prices and then the international financial crisis began to register—mainly in the form of lower prices for key export commodities and lower remittances from Tajiks working abroad.
GDP (purchasing power parity): $14.74 billion (2010 est.)
country comparison to the world: 137
$13.84 billion (2009 est.)
$13.32 billion (2008 est.)
note: data are in 2010 US dollars
GDP (official exchange rate): $5.642 billion (2010 est.)
GDP—real growth rate:
6.5% (2010 est.)
country comparison to the world: 44
3.9% (2009 est.)
7.9% (2008 est.)
GDP—per capita (PPP): $2,000 (2010 est.)
country comparison to the world: 190
$1,900 (2009 est.)
$1,800 (2008 est.)
note: data are in 2010 US dollars
GDP—composition by sector:
agriculture: 19.2%
industry: 22.6%
services: 58.1% (2010 est.)
Labor force: 2.1 million (2009)
country comparison to the world: 120
Labor force—by occupation:
agriculture: 49.8%
industry: 12.8%
services: 37.4% (2009 est.)
Unemployment rate: 2.2% (2009 est.)
country comparison to the world: 18
2.3% (2008 est.)
note: official rates; actual unemployment is much higher
Population below poverty line: 53% (2009 est.)
Household income or consumption by percentage share: *lowest 10%:* 3.3%
highest 10%: 25.6% (2007 est.)
Distribution of family income—Gini index: 32.6 (2006)
country comparison to the world: 99 34.7 (1998)
Investment (gross fixed): 20.9% of GDP (2010 est.)
country comparison to the world: 79
Budget: *revenues:* $1.482 billion
expenditures: $1.538 billion (2010 est.)
Inflation rate (consumer prices):
5.8% (2010 est.)
country comparison to the world: 153 6.4% (2009 est.)
Central bank discount rate: 8% (31 December 2009)
country comparison to the world: 24
13.5% (31 December 2008)
Commercial bank prime lending rate:
22.91% (31 December 2009 est.)
country comparison to the world: 10
23.7% (31 December 2008 est.)
Stock of narrow money: $863 million (31 December 2010 est.)
country comparison to the world: 143
$712.3 million (31 December 2009 est.)
Stock of broad money: $1.095 billion (31 December 2010 est.)
country comparison to the world: 161
$851.4 million (31 December 2009 est.)
Stock of domestic credit: $1.209 billion (31 December 2010 est.)
country comparison to the world: 144
$939.7 million (31 December 2009 est.)
Market value of publicly traded shares: $NA
Agriculture—products: cotton, grain, fruits, grapes, vegetables; cattle, sheep, goats
Industries: aluminum, cement, vegetable oil
Industrial production growth rate: 7.5% (2010 est.)
country comparison to the world: 38
Electricity—production: 16.1 billion kWh (2009 est.)
country comparison to the world: 75
Electricity—consumption: 16.7 billion kWh (2009)
country comparison to the world: 73
Electricity—exports: NA kWh (2008 est.)
Electricity—imports: 338.5 million kWh (2010 est.)
Oil—production: 221 bbl/day (2009 est.)
country comparison to the world: 108
Oil—consumption: 38,000 bbl/day (2009 est.)
country comparison to the world: 104
Oil—exports: 348.9 bbl/day (2007 est.)
country comparison to the world: 127
Oil—imports: 10,100 bbl/day (2008)
country comparison to the world: 141
Oil—proved reserves: 12 million bbl (1 January 2010 est.)
country comparison to the world: 89
Natural gas—production: 16.1 million cu m (2009 est.)
country comparison to the world: 87
Natural gas—consumption: 266.1 million cu m (2009 est.)
country comparison to the world: 99
Natural gas—exports: 0 cu m (2009 est.)
country comparison to the world: 183
Natural gas—imports: 250 million cu m (2009 est.)
country comparison to the world: 64

Natural gas—proved reserves: 5.663 billion cu m (1 January 2010 est.)
country comparison to the world: 88
Current account balance: $-330 million (2010 est.)
country comparison to the world: 102
$-179.9 million (2009 est.)
Exports: $1.318 billion (2010 est.)
country comparison to the world: 146
$1.039 billion (2009 est.)
Exports—commodities: aluminum, electricity, cotton, fruits, vegetable oil, textiles
Exports—partners: Russia 19.16%, China 18.38%, Turkey 12.09%, Iran 11.11%, Uzbekistan 7.92%, Norway 6.17%, Greece 4.32% (2009)
Imports: $3.301 billion (2010 est.)
country comparison to the world: 136
$2.77 billion (2009 est.)
Imports—commodities: petroleum products, aluminum oxide, machinery and equipment, foodstuffs
Imports—partners: Russia 23.92%, China 23.74%, Kazakhstan 8.92%, Turkey 4.96%, Uzbekistan 4.73% (2009)
Reserves of foreign exchange and gold: $303 million (31 December 2010 est.)
country comparison to the world: 128
$227 million (31 December 2009 est.)
Debt—external: $1.997 billion (31 December 2010 est.)
country comparison to the world: 141
$1.771 billion (31 December 2009 est.)
Stock of direct foreign investment—at home: $100.3 billion (31 December 2009 est.)
country comparison to the world: 34
$93.05 billion (31 December 2008 est.)
Stock of direct foreign investment—abroad: $18.5 billion (31 December 2010 est.)
country comparison to the world: 44
$16.3 billion (31 December 2009 est.)
Exchange rates: Tajikistani somoni (TJS) per US dollar–4.3788 (2010), 4.1428 (2009), 3.4563 (2008), 3.4418 (2007), 3.3 (2006)

COMMUNICATIONS

Telephones—main lines in use: 290,000 (2009)
country comparison to the world: 116
Telephones—mobile cellular: 4.9 million (2009)
country comparison to the world: 97
Telephone system: *general assessment:* foreign investment in the telephone system has resulted in major improvements; conversion of the existing fixed network from analogue to digital more than 90% complete by 2009
domestic: fixed line availability has not changed significantly since 1998 while mobile cellular subscribership, aided by competition among multiple operators, has expanded rapidly; coverage now extends to all major cities and towns
international: country code–992; linked by cable and microwave radio relay to other CIS republics and by leased connections to the Moscow international gateway switch; Dushanbe linked by Intelsat to international gateway switch in Ankara (Turkey); satellite earth stations–3 (2 Intelsat and 1 Orbita) (2009)
Broadcast media: state-run television broadcaster transmits nationally on 4 stations and regionally on 4 stations; 11 independent TV stations broadcast locally and regionally; some households are able to receive Russian and other foreign stations via cable and satellite; state-run radio broadcaster operates Radio Tajikistan, Voice of Dushanbe, and several regional stations; a small number of independent radio stations also broadcast (2010)
Internet country code: .tj
Internet hosts: 1,504 (2010)
country comparison to the world: 160
Internet users: 700,000 (2009)
country comparison to the world: 110

TRANSPORTATION

Airports: 26 (2010)
country comparison to the world: 127
Airports—with paved runways: *total:* 17
over 3,047 m: 2
2,438 to 3,047 m: 4
1,524 to 2,437 m: 5
914 to 1,523 m: 3
under 914 m: 3 (2010)
Airports—with unpaved runways: *total:* 9
1,524 to 2,437: 1
914 to 1,523 m: 1
under 914 m: 7 (2010)
Pipelines: gas 549 km; oil 38 km (2010)
Railways: *total:* 680 km
country comparison to the world: 103
broad gauge: 680 km 1.520-m gauge (2008)
Roadways: *total:* 27,767 km (2000)
country comparison to the world: 99
Waterways: 200 km (along Vakhsh River) (2010)
country comparison to the world: 99

MILITARY

Military branches: Ground Forces, Air and Air Defense Forces, Mobile Forces (2010)
Military service age and obligation: 18 years of age for compulsory military service; 2-year conscript service obligation (2009)
Manpower available for military service: *males age 16–49:* 2,012,790
females age 16–49: 2,020,618 (2010 est.)
Manpower fit for military service: *males age 16–49:* 1,490,267
females age 16–49: 1,675,083 (2010 est.)
Manpower reaching militarily significant age annually: *male:* 76,430
female: 74,038 (2010 est.)
Military expenditures: 1.5% of GDP (2010)
country comparison to the world: 96

TRANSNATIONAL ISSUES

Disputes—international: in 2006, China and Tajikistan pledged to commence demarcation of the revised boundary agreed to in the delimitation of 2002; talks continue with Uzbekistan to delimit border and remove minefields; disputes in Isfara Valley delay delimitation with Kyrgyzstan
Trafficking in persons: current situation: Tajikistan is a source country for women trafficked through Kyrgyzstan and Russia to the UAE, Turkey, and Russia for the purpose of commercial sexual exploitation; men are trafficked to Russia and Kazakhstan for the purpose of forced labor, primarily in the construction and agricultural industries; boys and girls are trafficked internally for various purposes, including forced labor and forced begging
tier rating: Tier 2 Watch List–Tajikistan is on the Tier 2 Watch List for its failure to provide evidence of increasing efforts to combat human trafficking, especially efforts to investigate, prosecute, convict, and sentence traffickers; despite evidence of low- and mid-level officials' complicity in trafficking, the government did not punish any public officials for trafficking complicity during 2007; lack of capacity and poor coordination between government institutions remained key obstacles to effective anti-trafficking efforts (2008)
Illicit drugs: major transit country for Afghan narcotics bound for Russian and, to a lesser extent, Western European markets; limited illicit cultivation of opium poppy for domestic consumption; Tajikistan seizes roughly 80% of all drugs captured in Central Asia and stands third worldwide in seizures of opiates (heroin and raw opium); significant consumer of opiates

TANZANIA

INTRODUCTION

Background: Shortly after achieving independence from Britain in the early 1960s, Tanganyika and Zanzibar merged to form the nation of Tanzania in 1964. One-party rule ended in 1995 with the first democratic elections held in the country since the 1970s. Zanzibar's semi-autonomous status and popular opposition have led to two contentious elections since 1995, which the ruling party won despite international observers' claims of voting irregularities. The formation of a government of national unity between Zanzibar's two leading parties succeeded in minimizing electoral tension in 2010

GEOGRAPHY

Location: Eastern Africa, bordering the Indian Ocean, between Kenya and Mozambique
Geographic coordinates: 6 00 S, 35 00 E
Map references: Africa
Area: *total:* 947,300 sq km
country comparison to the world: 31

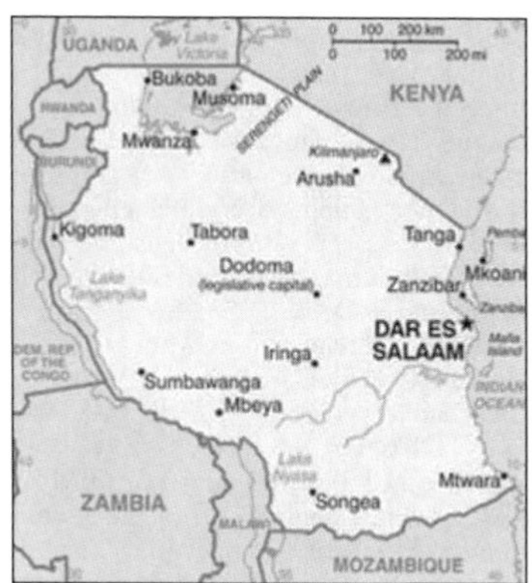

land: 885,800 sq km
water: 61,500 sq km
note: includes the islands of Mafia, Pemba, and Zanzibar
Area—comparative: slightly larger than twice the size of California
Land boundaries: *total:* 3,861 km
border countries: Burundi 451 km, Democratic Republic of the Congo 459 km, Kenya 769 km, Malawi 475 km, Mozambique 756 km, Rwanda 217 km, Uganda 396 km, Zambia 338 km
Coastline: 1,424 km
Maritime claims: territorial sea: 12 nm
exclusive economic zone: 200 nm
Climate: varies from tropical along coast to temperate in highlands
Terrain: plains along coast; central plateau; highlands in north, south
Elevation extremes: *lowest point:* Indian Ocean 0 m
highest point: Kilimanjaro 5,895 m
Natural resources: hydropower, tin, phosphates, iron ore, coal, diamonds, gemstones, gold, natural gas, nickel
Land use: *arable land:* 4.23%
permanent crops: 1.16%
other: 94.61% (2005)
Irrigated land: 1,840 sq km (2003)
Total renewable water resources: 91 cu km (2001)
Freshwater withdrawal (domestic/industrial/agricultural): *total:* 5.18 cu km/yr (10%/0%/89%)
per capita: 135 cu m/yr (2000)
Natural hazards: flooding on the central plateau during the rainy season; drought
volcanism: Tanzania experiences limited volcanic activity; Ol Doinyo Lengai (elev. 2,962 m) has emitted lava in recent years; other historically active volcanoes include Kieyo and Meru
Environment—current issues: soil degradation; deforestation; desertification; destruction of coral reefs threatens marine habitats; recent droughts affected marginal agriculture; wildlife threatened by illegal hunting and trade, especially for ivory
Environment—international agreements: *party to:* Biodiversity, Climate Change, Climate Change-Kyoto Protocol, Desertification, Endangered Species, Hazardous Wastes, Law of the Sea, Ozone Layer Protection, Wetlands
signed, but not ratified: none of the selected agreements
Geography—note: Kilimanjaro is highest point in Africa; bordered by three of the largest lakes on the continent: Lake Victoria (the world's second-largest freshwater lake) in the north, Lake Tanganyika (the world's second deepest) in the west, and Lake Nyasa (Lake Malawi) in the southwest

PEOPLE

Population: 42,746,620 (July 2011 est.)
country comparison to the world: 31
note: estimates for this country explicitly take into account the effects of excess mortality due to AIDS; this can result in lower life expectancy, higher infant mortality, higher death rates, lower population growth rates, and changes in the distribution of population by age and sex than would otherwise be expected
Age structure: *0–14 years:* 42% (male 9,003,152/female 8,949,061)
15–64 years: 55.1% (male 11,633,721/female 11,913,951)
65 years and over: 2.9% (male 538,290/female 708,445) (2011 est.)
Median age: *total:* 18.5 years
male: 18.2 years
female: 18.7 years (2011 est.)
Population growth rate: 2.002% (2011 est.)
country comparison to the world: 50
Birth rate: 32.64 births/1,000 population (2011 est.)
country comparison to the world: 39
Death rate: 12.09 deaths/1,000 population (July 2011 est.)
country comparison to the world: 28
Net migration rate: -0.53 migrant(s)/1,000 population (2011 est.)
country comparison to the world: 139
Urbanization: *urban population:* 26% of total population (2010)
rate of urbanization: 4.7% annual rate of change (2010-15 est.)
Major cities—population: DAR ES SALAAM (capital) 3.207 million (2009)
Sex ratio: *at birth:* 1.03 male(s)/female
under 15 years: 1.01 male(s)/female
15–64 years: 0.98 male(s)/female
65 years and over: 0.77 male(s)/female
total population: 0.98 male(s)/female (2011 est.)
Infant mortality rate: *total:* 66.93 deaths/1,000 live births
country comparison to the world: 21
male: 73.7 deaths/1,000 live births
female: 59.95 deaths/1,000 live births (2011 est.)
Life expectancy at birth: *total population:* 52.85 years
country comparison to the world: 205
male: 51.34 years
female: 54.42 years (2011 est.)
Total fertility rate: 4.16 children born/woman (2011 est.)
country comparison to the world: 40
HIV/AIDS—adult prevalence rate: 5.6% (2009 est.)
country comparison to the world: 12
HIV/AIDS—people living with HIV/AIDS: 1.4 million (2009 est.)
country comparison to the world: 7
HIV/AIDS—deaths: 86,000 (2009 est.)
country comparison to the world: 4
Major infectious diseases: *degree of risk:* very high
food or waterborne diseases: bacterial diarrhea, hepatitis A, and typhoid fever
vectorborne diseases: malaria and plague
water contact disease: schistosomiasis
animal contact disease: rabies (2009)
Drinking water source: *Improved:*
urban: 80% of population
rural: 45% of population
total: 54% of population
Unimproved:
urban: 20% of population
rural: 55% of population
total: 46% of population (2008)
Sanitation facility access: *Improved:*
urban: 32% of population
rural: 21% of population
total: 24% of population
Unimproved:
urban: 68% of population
rural: 79% of population
total: 76% of population (2008)
Nationality: *noun:* Tanzanian(s)
adjective: Tanzanian
Ethnic groups: mainland–African 99% (of which 95% are Bantu consisting of more than 130 tribes), other 1% (consisting of Asian, European, and Arab); Zanzibar–Arab, African, mixed Arab and African
Religions: mainland–Christian 30%, Muslim 35%, indigenous beliefs 35%; Zanzibar–more than 99% Muslim
Languages: Kiswahili or Swahili (official), Kiunguja (name for Swahili in Zanzibar), English (official, primary language of commerce, administration, and higher education), Arabic (widely spoken in Zanzibar), many local languages
note: Kiswahili (Swahili) is the mother tongue of the Bantu people living in Zanzibar and nearby coastal Tanzania; although Kiswahili is Bantu in structure and origin, its vocabulary draws on a variety of sources including Arabic and English; it has become the lingua franca of central and eastern Africa; the first language of most people is one of the local languages
Literacy: *definition:* age 15 and over can read and write Kiswahili (Swahili), English, or Arabic
total population: 69.4%
male: 77.5%
female: 62.2% (2002 census)
School life expectancy (primary to tertiary education):
total: 9 years
male: 9 years
female: 9 years (2007)
Education expenditures: 6.8% of GDP (2008)
country comparison to the world: 17

GOVERNMENT

Country name: *conventional long form:* United Republic of Tanzania
conventional short form: Tanzania
local long form: Jamhuri ya Muungano wa Tanzania
local short form: Tanzania
former: United Republic of Tanganyika and Zanzibar
Government type: republic
Capital: *name:* Dar es Salaam
geographic coordinates: 6 48 S, 39 17 E
time difference: UTC+3 (8 hours ahead of Washington, DC during Standard Time)

note: legislative offices have been transferred to Dodoma, which is planned as the new national capital, and the National Assembly now meets there on a regular basis; the Executive Branch with all ministries and diplomatic representation remains located in Dar es Salaam

Administrative divisions: 26 regions; Arusha, Dar es Salaam, Dodoma, Iringa, Kagera, Kigoma, Kilimanjaro, Lindi, Manyara, Mara, Mbeya, Morogoro, Mtwara, Mwanza, Pemba North, Pemba South, Pwani, Rukwa, Ruvuma, Shinyanga, Singida, Tabora, Tanga, Zanzibar Central/South, Zanzibar North, Zanzibar Urban/West

Independence: 26 April 1964; Tanganyika became independent on 9 December 1961 (from UK-administered UN trusteeship); Zanzibar became independent on 19 December 1963 (from UK); Tanganyika united with Zanzibar on 26 April 1964 to form the United Republic of Tanganyika and Zanzibar; renamed United Republic of Tanzania on 29 October 1964

National holiday: Union Day (Tanganyika and Zanzibar), 26 April (1964)

Constitution: 25 April 1977; major revisions October 1984

Legal system: English common law; judicial review of legislative acts limited to matters of interpretation

International law organization participation: has not submitted an ICJ jurisdiction declaration; accepts ICCt jurisdiction

Suffrage: 18 years of age; universal

Executive branch: *chief of state:* President Jakaya KIKWETE (since 21 December 2005); Vice President Mohammed Gharib BILAL (since 6 November 2010); note–the president is both chief of state and head of government
head of government: President Jakaya KIKWETE (since 21 December 2005); Vice President Mohammed Gharib BILAL (since 6 November 2010)
note: Zanzibar elects a president who is head of government for matters internal to Zanzibar; Ali Mohamed SHEIN elected to that office on 31 October 2010, sworn in 3 November 2010
cabinet: Cabinet appointed by the president from among the members of the National Assembly
(For more information visit the World Leaders website)
elections: president and vice president elected on the same ballot by popular vote for five-year terms (eligible for a second term); election last held on 31 October 2010 (next to be held in 2015); prime minister appointed by the president
election results: Jakaya KIKWETE elected president; percent of vote–Jakaya KIKWETE 61.2%, Willibrod SLAA 26.3%, Ibrahim LIPUMBA 8.1%, other 4.4%

Legislative branch: unicameral National Assembly or Bunge (357 seats; 239 members elected by popular vote, 102 allocated to women nominated by the president, 5 to members of the Zanzibar House of Representatives; members serve five-year terms, up to 10 additional members appointed by the president, 1 seat reserved for the Attorney General); note–in addition to enacting laws that apply to the entire United Republic of Tanzania, the Assembly enacts laws that apply only to the mainland; Zanzibar has its own House of Representatives with jurisdiction exclusive to Zanzibar (the Zanzibar House of Representatives has 50 seats; members elected by universal suffrage to serve five-year terms)
elections: last held on 31 October 2010 (next to be held in 2015)
election results: National Assembly–percent of vote by party–NA; seats by party–CCM 259, CHADEMA 48, CUF 34, NCCR-M 4, other 7, Zanzibar representatives 5; Zanzibar House of Representatives–percent of vote by party–NA; seats by party–CCM 28, CUF 22

Judicial branch: Permanent Commission of Enquiry (official ombudsman); Court of Appeal (consists of a chief justice and four judges); High Court (consists of a Jaji Kiongozi and 29 judges appointed by the president; holds regular sessions in all regions); District Courts; Primary Courts (limited jurisdiction and appeals can be made to the higher courts)

Political parties and leaders: Chama Cha Demokrasia na Maendeleo (Party of Democracy and Development) or CHADEMA [Willibrod SLAA]; Chama Cha Mapinduzi or CCM (Revolutionary Party) [Jakaya Mrisho KIKWETE]; Civic United Front or CUF [Ibrahim LIPUMBA]; Democratic Party [Christopher MTIKLA] (unregistered); National Convention for Construction and Reform–Mageuzi [Hashim RUNGWE]; Tanzania Labor Party or TLP [Mutamwega MUGAHWYA]; United Democratic Party or UDP [Fahma DOVUTWA]

Political pressure groups and leaders: Economic and Social Research Foundation or ESRF; Free Zanzibar; Tanzania Media Women's Association or TAMWA

International organization participation: ACP, AfDB, AU, C, EAC, EADB, FAO, IAEA, IBRD, ICAO, ICRM, IDA, IFAD, IFC, IFRCS, ILO, IMF, IMO, IMSO, Interpol, IOC, IOM, IPU, ISO, ITSO, ITU, ITUC, MIGA, MONUSCO, NAM, OPCW, SADC, UN, UNAMID, UNCTAD, UNESCO, UNHCR, UNIDO, UNIFIL, UNMIS, UNOCI, UNWTO, UPU, WCO, WFTU, WHO, WIPO, WMO, WTO

Diplomatic representation in the US: *chief of mission:* Ambassador Mwandaidi Sinare MAAJAR
chancery: 2139 R Street NW, Washington, DC 20008
telephone: [1] (202) 939-6125
FAX: [1] (202) 797-7408

Diplomatic representation from the US: *chief of mission:* Ambassador Alfonso E. LENHARDT
embassy: 686 Old Bagamoyo Road, Msasani, Dar es Salaam
mailing address: P. O. Box 9123, Dar es Salaam
telephone: [255] (22) 266-8001
FAX: [255] (22) 266-8238, 266-8373

Flag description: divided diagonally by a yellow-edged black band from the lower hoist-side corner; the upper triangle (hoist side) is green and the lower triangle is blue; the banner combines colors found on the flags of Tanganyika and Zanzibar; green represents the natural vegetation of the country, gold its rich mineral deposits, black the native Swahili people, and blue the country's many lakes and rivers, as well as the Indian Ocean

National anthem: *name:* "Mungu ibariki Afrika" (God Bless Africa)
lyrics/music: collective/Enoch Mankayi SONTONGA
note: adopted 1961; the anthem, which is also a popular song in Africa, shares the same melody with that of Zambia, but has different lyrics; the melody is also incorporated into South Africa's anthem

ECONOMY

Economy—overview: Tanzania is one of the world's poorest economies in terms of per capita income, however, Tanzania average 7% GDP growth per year between 2000 and 2008 on strong gold production and tourism. The economy depends heavily on agriculture, which accounts for more than 40% of GDP, provides 85% of exports, and employs about 80% of the work force. The World Bank, the IMF, and bilateral donors have provided funds to rehabilitate Tanzania's aging economic infrastructure, including rail and port infrastructure that are important trade links for inland countries. Recent banking reforms have helped increase private-sector growth and investment, and the government has increased spending on agriculture to 7% of its budget. Continued donor assistance and solid macroeconomic policies supported a positive growth rate, despite the world recession. In 2008, Tanzania received the world's largest Millennium Challenge Compact grant, worth $698 million. Dar es Salaam used fiscal stimulus and loosened monetary policy to ease the impact of the global recession. GDP growth in 2009-10 was a respectable 6% per year due to high gold prices and increased production.

GDP (purchasing power parity): $58.44 billion (2010 est.)
country comparison to the world: 88
$54.88 billion (2009 est.)
$51.43 billion (2008 est.)
note: data are in 2010 US dollars

GDP (official exchange rate): $22.67 billion (2010 est.)

GDP—real growth rate: 6.5% (2010 est.)
country comparison to the world: 43
6.7% (2009 est.)
7.3% (2008 est.)

GDP—per capita (PPP): $1,400 (2010 est.)
country comparison to the world: 202
$1,300 (2009 est.)
$1,300 (2008 est.)
note: data are in 2010 US dollars

GDP—composition by sector:
agriculture: 42%
industry: 18%
services: 40% (2010 est.)

Labor force: 21.86 million (2010 est.)
country comparison to the world: 29

Labor force—by occupation: *agriculture:* 80%
industry and *services:* 20% (2002 est.)

Unemployment rate: NA%
Population below poverty line: 36% (2002 est.)
Household income or consumption by percentage share: *lowest 10%:* 2.9%
highest 10%: 26.9% (2000)
Distribution of family income—Gini index: 34.6 (2000)
country comparison to the world: 88
38.2 (1993)
Investment (gross fixed): 17.4% of GDP (2010 est.)
country comparison to the world: 116
Budget: *revenues:* $4.263 billion
expenditures: $5.644 billion (2010 est.)
Public debt: 23.3% of GDP (2010 est.)
country comparison to the world: 104
21.4% of GDP (2009 est.)
Inflation rate (consumer prices): 7.2% (2010 est.)
country comparison to the world: 179
12.1% (2009 est.)
Central bank discount rate: 3.7% (31 December 2009)
country comparison to the world: 16
15.99% (31 December 2008)
Commercial bank prime lending rate: 15.03% (31 December 2009 est.)
country comparison to the world: 43
14.98% (31 December 2008 est.)
Stock of narrow money: $3.394 billion (31 December 2010 est.)
country comparison to the world: 106
$2.972 billion (31 December 2009 est.)
Stock of broad money: $7.44 billion (31 December 2010 est.)
country comparison to the world: 111
$6.65 billion (31 December 2009 est.)
Stock of domestic credit: $4.163 million (31 December 2010 est.)
country comparison to the world: 186
$3.878 million (31 December 2009 est.)
Market value of publicly traded shares: $NA (31 December 2009)
country comparison to the world: 102
$1.293 billion (31 December 2008)
$541.1 million (31 December 2006)
Agriculture—products: coffee, sisal, tea, cotton, pyrethrum (insecticide made from chrysanthemums), cashew nuts, tobacco, cloves, corn, wheat, cassava (tapioca), bananas, fruits, vegetables; cattle, sheep, goats
Industries: agricultural processing (sugar, beer, cigarettes, sisal twine); diamond, gold, and iron mining, salt, soda ash; cement, oil refining, shoes, apparel, wood products, fertilizer
Industrial production growth rate: 7% (2010 est.)
country comparison to the world: 44
Electricity—production: 3.786 billion kWh (2007 est.)
country comparison to the world: 120
Electricity—consumption: 3.182 billion kWh (2007 est.)
country comparison to the world: 123
Electricity—exports: 0 kWh (2008)
Electricity—imports: 200 million kWh (2007 est.)
Oil—production: 0 bbl/day (2009 est.)
country comparison to the world: 135
Oil—consumption: 34,000 bbl/day (2009 est.)
country comparison to the world: 110
Oil—exports: 0 bbl/day (2007 est.)
country comparison to the world: 206
Oil—imports: 28,070 bbl/day (2007 est.)
country comparison to the world: 102
Oil—proved reserves: 0 bbl (1 January 2010 est.)
country comparison to the world: 197
Natural gas—production: 560.7 million cu m (2008 est.)
country comparison to the world: 67
Natural gas—consumption: 560.7 million cu m (2008 est.)
country comparison to the world: 93
Natural gas—exports: 0 cu m (2008 est.)
country comparison to the world: 191
Natural gas—imports: 0 cu m (2008 est.)
country comparison to the world: 129
Natural gas—proved reserves: 6.513 billion cu m (1 January 2010 est.)
country comparison to the world: 84
Current account balance: $-1.523 billion (2010 est.)
country comparison to the world: 151
$-1.746 billion (2009 est.)
Exports: $3.809 billion (2010 est.)
country comparison to the world: 116
$3.365 billion (2009 est.)
Exports—commodities: gold, coffee, cashew nuts, manufactures, cotton
Exports—partners: India 8.51%, China 7.55%, Japan 7.12%, Netherlands 6.21%, UAE 5.71%, Germany 5.17% (2009)
Imports:
$6.334 billion (2010 est.)
country comparison to the world: 105
$5.834 billion (2009 est.)
Imports—commodities: consumer goods, machinery and transportation equipment, industrial raw materials, crude oil
Imports—partners: India 13.97%, China 13.71%, South Africa 7.8%, Kenya 6.89%, UAE 4.65%, Japan 4.34% (2009)
Reserves of foreign exchange and gold: $3.687 billion (31 December 2010 est.)
country comparison to the world: 86
$3.206 billion (31 December 2009 est.)
note: excludes gold
Debt—external: $7.576 billion (31 December 2010 est.)
country comparison to the world: 95
$6.879 billion (31 December 2009 est.)
Stock of direct foreign investment—at home: $NA
Stock of direct foreign investment—abroad: $NA
Exchange rates: Tanzanian shillings (TZS) per US dollar–1,423.3 (2010), 1,320.3 (2009), 1,178.1 (2008), 1,255 (2007), 1,251.9 (2006)

COMMUNICATIONS

Telephones—main lines in use: 173,552 (2010)
country comparison to the world: 131
Telephones—mobile cellular: 17.677 million (2010)
country comparison to the world: 42
Telephone system: *general assessment:* telecommunications services are marginal; system operating below capacity and being modernized for better service; small aperture terminal (VSAT) system under construction
domestic: fixed-line telephone network inadequate with less than 1 connection per 100 persons; mobile-cellular service, aided by multiple providers, is increasing rapidly; trunk service provided by open-wire, microwave radio relay, tropospheric scatter, and fiber-optic cable; some links being made digital
international: country code–255; satellite earth stations–2 Intelsat (1 Indian Ocean, 1 Atlantic Ocean)
Broadcast media: a state-owned TV station and multiple privately-owned TV stations; state-owned national radio station supplemented by more than 40 privately-owned radio stations; transmissions of several international broadcasters are available (2007)
Internet country code: .tz
Internet hosts: 24,182 (2010)
country comparison to the world: 103
Internet users: 678,000 (2009)
country comparison to the world: 111

TRANSPORTATION

Airports: 124 (2010)
country comparison to the world: 48
Airports—with paved runways: *total:* 9
over 3,047 m: 2
2,438 to 3,047 m: 2
1,524 to 2,437 m: 4
914 to 1,523 m: 1 (2010)
Airports—with unpaved runways: *total:* 115
1,524 to 2,437 m: 19
914 to 1,523 m: 63
under 914 m: 33 (2010)
Pipelines: gas 254 km; oil 888 km; refined products 8 km (2010)
Railways: *total:* 3,689 km
country comparison to the world: 45
narrow gauge: 969 km 1.067-m gauge; 2,720 km 1.000-m gauge (2008)
Roadways: *total:* 78,892 km
country comparison to the world: 61
paved: 4,741 km
unpaved: 74,151 km (2007)
Waterways: (Lake Tanganyika, Lake Victoria, and Lake Nyasa (Lake Malawi) are the principal avenues of commerce with neighboring countries; the rivers are not navigable) (2009)
Merchant marine: *total:* 72
country comparison to the world: 59
by type: bulk carrier 4, cargo 43, carrier 4, chemical tanker 2, container 1, passenger 1, passenger/cargo 2, petroleum tanker 15
foreign-owned: 25 (Greece 1, Romania 1, Saudi Arabia 1, Syria 13, Turkey 7, UAE 1, United States 1)
registered in other countries: 3 (Honduras 1, Panama 2) (2010)
Ports and terminals: Dar es Salaam, Zanzibar
Transportation—note: the International Maritime Bureau reports that shipping in territorial and offshore waters in the Indian Ocean remain at risk for piracy and armed robbery against ships, especially as Somali-based pirates extend their activities south; numerous commercial vessels have been attacked and hijacked both at anchor and while underway; crews have been robbed and stores or cargoes stolen

MILITARY

Military branches: Tanzanian People's Defense Force (Jeshi la Wananchi la

Tanzania, JWTZ): Army, Naval Wing (includes Coast Guard), Air Defense Command (includes Air Wing), National Service (2007)
Military service age and obligation: 18 years of age for voluntary military service (2007)
Manpower available for military service: *males age 16–49:* 9,985,445 (2010 est.)
Manpower fit for military service: *males age 16–49:* 5,860,339
females age 16–49: 5,882,279 (2010 est.)
Manpower reaching militarily significant age annually: *male:* 512,294
female: 514,164 (2010 est.)
Military expenditures: 0.2% of GDP (2005 est.)
country comparison to the world: 170

TRANSNATIONAL ISSUES

Disputes—international: Tanzania still hosts more than a half-million refugees, more than any other African country, mainly from Burundi and the Democratic Republic of the Congo, despite the international community's efforts at repatriation; disputes with Malawi over the boundary in Lake Nyasa (Lake Malawi) and the meandering Songwe River remain dormant
Refugees and internally displaced persons: refugees (country of origin): 352,640 (Burundi); 127,973 (Democratic Republic of the Congo) (2007)
Illicit drugs: targeted by traffickers moving hashish, Afghan heroin, and South American cocaine transported down the East African coastline, through airports, or overland through Central Africa; Zanzibar likely used by traffickers for drug smuggling; traffickers in the past have recruited Tanzanian couriers to move drugs through Iran into East Asia.

THAILAND

INTRODUCTION

Background: A unified Thai kingdom was established in the mid-14th century. Known as Siam until 1939, Thailand is the only Southeast Asian country never to have been taken over by a European power. A bloodless revolution in 1932 led to a constitutional monarchy. In alliance with Japan during World War II, Thailand became a US treaty ally in 1954 after sending troops to Korea and fighting alongside the US in Vietnam. A military coup in September 2006 ousted then Prime Minister THAKSIN Chinnawat. December 2007 elections saw the pro-THAKSIN People's Power Party (PPP) emerge at the head of a coalition government that took office in February 2008. The anti-THAKSIN People's Alliance for Democracy (PAD, aka yellow-shirts) in May 2008 began street demonstrations against the new government, eventually occupying the prime minister's office in August and Bangkok's two international airports in November. After an early December 2008 court ruling that dissolved the ruling PPP and two other coalition parties for election violations, the Democrat Party formed a new coalition government and ABHISIT Wetchachiwa became prime minister. In October 2008 THAKSIN fled abroad in advance of an abuse of power conviction and has agitated his followers from abroad since then. THAKSIN supporters under the banner of the United Front for Democracy Against Dictatorship (UDD, aka red-shirts) rioted in April 2009, shutting down an ASEAN meeting in Pattaya. Following a February 2010 court verdict confiscating half of THAKSIN's frozen assets, the UDD staged large protests between March and May 2010, and occupied several blocks of downtown Bangkok. Clashes between security forces and protesters, elements of which were armed, resulted in at least 92 deaths and an estimated $1.5 billion in arson-related property losses. These protests exposed major cleavages in the Thai body politic that continue to hamper the current government. The ABHISIT administration has announced a plan for a general election some time in 2011 ahead of its full term by the year-end. Since January 2004, thousands have been killed as separatists in Thailand's southern ethnic Malay-Muslim provinces increased the violence associated with their cause.

GEOGRAPHY

Location: Southeastern Asia, bordering the Andaman Sea and the Gulf of Thailand, southeast of Burma
Geographic coordinates: 15 00 N, 100 00 E
Map references: Southeast Asia
Area: *total:* 513,120 sq km
country comparison to the world: 50
land: 510,890 sq km
water: 2,230 sq km
Area—comparative: slightly more than twice the size of Wyoming
Land boundaries: *total:* 4,863 km
border countries: Burma 1,800 km, Cambodia 803 km, Laos 1,754 km, Malaysia 506 km
Coastline: 3,219 km
Maritime claims: territorial sea: 12 nm
exclusive economic zone: 200 nm
continental shelf: 200 m depth or to the depth of exploitation
Climate: tropical; rainy, warm, cloudy southwest monsoon (mid-May to September); dry, cool northeast monsoon (November to mid-March); southern isthmus always hot and humid
Terrain: central plain; Khorat Plateau in the east; mountains elsewhere
Elevation extremes: *lowest point:* Gulf of Thailand 0 m
highest point: Doi Inthanon 2,576 m
Natural resources: tin, rubber, natural gas, tungsten, tantalum, timber, lead, fish, gypsum, lignite, fluorite, arable land
Land use: *arable land:* 27.54%
permanent crops: 6.93%
other: 65.53% (2005)
Irrigated land: 49,860 sq km (2003)
Total renewable water resources: 409.9 cu km (1999)
Freshwater withdrawal (domestic/industrial/agricultural): *total:* 82.75 cu km/yr (2%/2%/95%)
per capita: 1,288 cu m/yr (2000)
Natural hazards: land subsidence in Bangkok area resulting from the depletion of the water table; droughts
Environment—current issues: air pollution from vehicle emissions; water pollution from organic and factory wastes; deforestation; soil erosion; wildlife populations threatened by illegal hunting
Environment—international agreements: *party to:* Biodiversity, Climate Change, Climate Change-Kyoto Protocol, Desertification, Endangered Species, Hazardous Wastes, Marine Life Conservation, Ozone Layer Protection, Tropical Timber 83, Tropical Timber 94, Wetlands
signed, but not ratified: Law of the Sea
Geography—note: controls only land route from Asia to Malaysia and Singapore

PEOPLE

Population: 66,720,153 (July 2011 est.)
country comparison to the world: 20
note: estimates for this country explicitly take into account the effects of excess mortality due to AIDS; this can result in lower life expectancy, higher infant mortality, higher death rates, lower population growth rates, and changes in the

distribution of population by age and sex than would otherwise be expected
Age structure: *0–14 years:* 19.9% (male 6,779,723/female 6,466,625)
15–64 years: 70.9% (male 23,410,091/ female 23,913,499)
65 years and over: 9.2% (male 2,778,012/ female 3,372,203) (2011 est.)
Median age: *total:* 34.2 years
male: 33.3 years
female: 35.2 years (2011 est.)
Population growth rate: 0.566% (2011 est.)
country comparison to the world: 147
Birth rate: 12.95 births/1,000 population (2011 est.)
country comparison to the world: 153
Death rate: 7.29 deaths/1,000 population (July 2011 est.)
country comparison to the world: 123
Net migration rate: 0 migrant(s)/1,000 population (2011 est.)
country comparison to the world: 107
Urbanization: *urban population:* 34% of total population (2010)
rate of urbanization: 1.8% annual rate of change (2010-15 est.)
Major cities—population: BANGKOK (capital) 6.902 million (2009)
Sex ratio: *at birth:* 1.054 male(s)/female
under 15 years: 1.05 male(s)/female
15–64 years: 0.98 male(s)/female
65 years and over: 0.82 male(s)/female
total population: 0.98 male(s)/female (2011 est.)
Infant mortality rate: *total:* 16.39 deaths/1,000 live births
country comparison to the world: 108
male: 17.38 deaths/1,000 live births
female: 15.35 deaths/1,000 live births (2011 est.)
Life expectancy at birth: *total population:* 73.6 years
country comparison to the world: 113
male: 71.24 years
female: 76.08 years (2011 est.)
Total fertility rate: 1.66 children born/ woman (2011 est.)
country comparison to the world: 172
HIV/AIDS—adult prevalence rate: 1.3% (2009 est.)
country comparison to the world: 38
HIV/AIDS—people living with HIV/AIDS: 530,000 (2009 est.)
country comparison to the world: 16
HIV/AIDS—deaths: 28,000 (2009 est.)
country comparison to the world: 13
Major infectious diseases: *degree of risk:* high
food or waterborne diseases: bacterial diarrhea
vectorborne diseases: dengue fever, Japanese encephalitis, and malaria
animal contact disease: rabies
water contact disease: leptospirosis
note: highly pathogenic H5N1 avian influenza has been identified in this country; it poses a negligible risk with extremely rare cases possible among US citizens who have close contact with birds (2009)
Drinking water source: *Improved:*
urban: 99% of population
rural: 98% of population
total: 98% of population
Unimproved:
urban: 1% of population
rural: 2% of population
total: 2% of population (2008)
Sanitation facility access:
Improved:
urban: 95% of population
rural: 96% of population
total: 96% of population
Unimproved:
urban: 5% of population
rural: 4% of population
total: 4% of population (2008)
Nationality: *noun:* Thai (singular and plural)
adjective: Thai
Ethnic groups: Thai 75%, Chinese 14%, other 11%
Religions: Buddhist 94.6%, Muslim 4.6%, Christian 0.7%, other 0.1% (2000 census)
Languages: Thai, English (secondary language of the elite), ethnic and regional dialects
Literacy: *definition:* age 15 and over can read and write
total population: 92.6%
male: 94.9%
female: 90.5% (2000 census)
School life expectancy (primary to tertiary education): *total:* 12 years
male: 12 years
female: 13 years (2010)
Education expenditures: 4.1% of GDP (2009)
country comparison to the world: 103

GOVERNMENT

Country name: *conventional long form:* Kingdom of Thailand
conventional short form: Thailand
local long form: Ratcha Anachak Thai
local short form: Prathet Thai
former: Siam
Government type: constitutional monarchy
Capital: *name:* Bangkok
geographic coordinates: 13 45 N, 100 31 E
time difference: UTC+7 (12 hours ahead of Washington, DC during Standard Time)
Administrative divisions: 76 provinces (changwat, singular and plural); Amnat Charoen, Ang Thong, Buriram, Chachoengsao, Chai Nat, Chaiyaphum, Chanthaburi, Chiang Mai, Chiang Rai, Chon Buri, Chumphon, Kalasin, Kamphaeng Phet, Kanchanaburi, Khon Kaen, Krabi, Krung Thep Mahanakhon (Bangkok), Lampang, Lamphun, Loei, Lop Buri, Mae Hong Son, Maha Sarakham, Mukdahan, Nakhon Nayok, Nakhon Pathom, Nakhon Phanom, Nakhon Ratchasima, Nakhon Sawan, Nakhon Si Thammarat, Nan, Narathiwat, Nong Bua Lamphu, Nong Khai, Nonthaburi, Pathum Thani, Pattani, Phangnga, Phatthalung, Phayao, Phetchabun, Phetchaburi, Phichit, Phitsanulok, Phra Nakhon Si Ayutthaya, Phrae, Phuket, Prachin Buri, Prachuap Khiri Khan, Ranong, Ratchaburi, Rayong, Roi Et, Sa Kaeo, Sakon Nakhon, Samut Prakan, Samut Sakhon, Samut Songkhram, Sara Buri, Satun, Sing Buri, Sisaket, Songkhla, Sukhothai, Suphan Buri, Surat Thani, Surin, Tak, Trang, Trat, Ubon Ratchathani, Udon Thani, Uthai Thani, Uttaradit, Yala, Yasothon
Independence: 1238 (traditional founding date; never colonized)
National holiday: Birthday of King PHUMIPHON (BHUMIBOL), 5 December (1927)
Constitution: 24 August 2007
Legal system: civil law system with common law influences
International law organization participation: has not submitted an ICJ jurisdiction declaration; non-party state to the ICCt
Suffrage: 18 years of age; universal and compulsory
Executive branch: *chief of state:* King PHUMIPHON Adunyadet, also spelled BHUMIBOL Adulyadej (since 9 June 1946)
head of government: Prime Minister ABHISIT Wechachiwa, also spelled ABHISIT Vejjajiva (since 17 December 2008); Deputy Prime Minister SANAN Kachornprasat, also spelled SANAN Kachornparsart (since 7 February 2008); Deputy Prime Minister SUTHEP Thueaksuban, also spelled SUTHEP Thaugsuban (since 22 December 2008); Deputy Prime Minister TRAIRONG Suwannakhiri (since 18 January 2010)
cabinet: Council of Ministers
(For more information visit the World Leaders website)
note: there is also a Privy Council advising the king
elections: the monarchy is hereditary; according to 2007 constitution, the prime minister elected from among members of House of Representatives; following national elections for House of Representatives, the leader of the party positioned to organize a majority coalition usually becomes prime minister by appointment by the king; the prime minister limited to two four-year terms
Legislative branch: bicameral National Assembly or Rathasapha consisted of the Senate or Wuthisapha (150 seats; 76 members elected by popular vote representing 76 provinces, 74 appointed by judges and independent government bodies; members serve six-year terms) and the House of Representatives or Sapha Phuthaen Ratsadon (480 seats; 400 members elected from 157 multi-seat constituencies and 80 elected on proportional party-list basis of 10 per eight zones or groupings of provinces; members serve four-year terms)
elections: Senate—last held on 2 March 2008 (next to be held in March 2014); House of Representatives—last election held on 23 December 2007 (next to be held on 3 July 2011)
election results: Senate—percent of vote by party—NA; seats by party—NA; House of Representatives—percent of vote by party—NA; seats by party—PPP 233, DP 164, TNP 34, Motherland 24, Middle Way 11, Unity 9, Royalist People's 5; following the PPP's dissolution in December 2008, most of the party's seats were assumed by its successor, the Phuea Thai Party
note: 74 senators were appointed on 19 February 2008 by a seven-member committee headed by the chief of the Constitutional Court; 76 senators were

elected on 2 March 2008; elections to the Senate are non-partisan; registered political party members are disqualified from being senators

Judicial branch: Constitutional Court, Supreme Court of Justice, and Supreme Administrative Court; all judges are appointed by the king; the king's appointments to the Constitutional Courtare made upon the advice of the Senate; the nine Constitutional Court judges are drawn from the Supreme Court of Justice and Supreme Administrative Court as well as from among substantive experts in law and social sciences outside the judiciary

Political parties and leaders: Chat Thai Phattana Party or CP (Thai Nation Development Party) [CHUMPON Silpa-archa]; Democrat Party or DP (Prachathipat Party) [ABHISIT Wetchachiwa, also spelled ABHISIT Vejjajiva]; Motherland Party (Phuea Phaendin Party) [CHANCHAI Chairungrueng]; Phuea Thai Party (For Thais Party) or PTP [YONGYUTH Wichaidit]; Phumjai (Bhumjai) Thai Party or PJT (Thai Pride) [CHAWARAT Chanvirakun]; Royalist People's Party (Pracharaj) [SANOH Thienthong]; Ruam Jai Thai Party (Thai Unity Party) [WANNARAT Channukun]

Political pressure groups and leaders: People's Alliance for Democracy or PAD; United Front for Democracy Against Dictatorship or UDD

International organization participation: ADB, APEC, ARF, ASEAN, BIMSTEC, BIS, CICA, CP, EAS, FAO, G-77, IAEA, IBRD, ICAO, ICC, ICRM, IDA, IFAD, IFC, IFRCS, IHO, ILO, IMF, IMO, IMSO, Interpol, IOC, IOM, IPU, ISO, ITSO, ITU, ITUC, MIGA, NAM, OAS (observer), OIC (observer), OIF (observer), OPCW, OSCE (partner), PCA, PIF (partner), UN, UNAMID, UNCTAD, UNESCO, UNHCR, UNIDO, UNMIS, UNWTO, UPU, WCO, WFTU, WHO, WIPO, WMO, WTO

Diplomatic representation in the US:
chief of mission: Ambassador Kittiphong Na RANONG
chancery: 1024 Wisconsin Avenue NW, Suite 401, Washington, DC 20007
telephone: [1] (202) 944-3600
FAX: [1] (202) 944-3611
consulate(s) general: Chicago, Los Angeles, New York

Diplomatic representation from the US:
chief of mission: Ambassador Kristie A. KENNEY
embassy: 120-122 Wireless Road, Bangkok 10330
mailing address: APO AP 96546
telephone: [66] (2) 205-4000
FAX: [66] (2) 254-2990, 205-4131
consulate(s) general: Chiang Mai

Flag description: five horizontal bands of red (top), white, blue (double width), white, and red; the red color symbolizes the nation and the blood of life; white represents religion and the purity of Buddhism; blue stands for the monarchy
note: similar to the flag of Costa Rica but with the blue and red colors reversed

National anthem: *name:* "Phleng Chat Thai" (National Anthem of Thailand)
lyrics/music: LUANG Saranuprapan/ PHRA Jenduriyang
note: music adopted 1932, lyrics adopted 1939; by law, people are required to stand for the national anthem at 0800 and 1800 every day; the anthem is played in schools, offices, theaters, and on television and radio during this time; "Phleng Sansasoen Phra Barami" (A Salute to the Monarch) serves as the royal anthem and is played in the presence of the royal family and during certain state ceremonies

ECONOMY

Economy—overview: With a well-developed infrastructure, a free-enterprise economy, generally pro-investment policies, and strong export industries, Thailand enjoyed solid growth from 2000 to 2007—averaging more than 4% per year—as it recovered from the Asian financial crisis of 1997-98. Thai exports—mostly machinery and electronic components, agricultural commodities, and jewelry—continue to drive the economy, accounting for more than half of GDP. The global financial crisis of 2008-09 severely cut Thailand's exports, with most sectors experiencing double-digit drops. In 2009, the economy contracted 2.2%. In 2010, Thailand's economy expanded 7.6%, its fastest pace since 1995, as exports rebounded from their depressed 2009 level. Antigovernment protests during March-May and the country's polarized political situation had—at most—a temporary impact on business and consumer confidence. Although tourism was hit hard during the protests, its quick recovery helped boost consumer confidence to new highs. Moreover, business and investor sentiment remained buoyant as Thailand's stock market grew almost 5% during the three-month period. The economy probably will continue to experience high grow well into 2011.

GDP (purchasing power parity): $586.9 billion (2010 est.)
country comparison to the world: 25
$544.4 billion (2009 est.)
$557.4 billion (2008 est.)
note: data are in 2010 US dollars

GDP (official exchange rate): $318.9 billion (2010 est.)

GDP—real growth rate: 7.8% (2010 est.)
country comparison to the world: 21
-2.3% (2009 est.)
2.5% (2008 est.)

GDP—per capita (PPP): $8,700 (2010 est.)
country comparison to the world: 118
$8,200 (2009 est.)
$8,400 (2008 est.)
note: data are in 2010 US dollars

GDP—composition by sector:
agriculture: 10.4%
industry: 45.6%
services: 44% (2010 est.)

Labor force: 38.7 million (2010 est.)
country comparison to the world: 16

Labor force—by occupation:
agriculture: 42.4%
industry: 19.7%
services: 37.9% (2008 est.)

Unemployment rate: 1.2% (2010 est.)
country comparison to the world: 7
1.5% (2009)

Population below poverty line: 9.6% (2006 est.)

Household income or consumption by percentage share: *lowest 10%:* 1.6%
highest 10%: 33.7% (2006)

Distribution of family income—Gini index: 43 (2006)
country comparison to the world: 47 42 (2002)

Investment (gross fixed): 24.9% of GDP (2010 est.)
country comparison to the world: 45

Budget: *revenues:* $56.33 billion
expenditures: $56.87 billion (2010 est.)

Public debt: 42.3% of GDP (2010 est.)
country comparison to the world: 63
44.9% of GDP (2009)

Inflation rate (consumer prices): 3.3% (2010 est.)
country comparison to the world: 93
-0.9% (2009 est.)

Central bank discount rate: 1.75% (31 December 2010)
country comparison to the world: 131
1.25% (31 December 2009)

Commercial bank prime lending rate: 6.1% (31 December 2010)
country comparison to the world: 134
5.96% (31 December 2009)

Stock of narrow money: $38 billion (31 December 2010 est.)
country comparison to the world: 48
$34.26 billion (31 December 2009 est.)

Stock of broad money: $354.5 billion (31 December 2010 est.)
country comparison to the world: 25
$309.7 billion (31 December 2009 est.)

Stock of domestic credit: $336 billion (31 December 2010 est.)
country comparison to the world: 31
$292.4 billion (31 December 2009 est.)

Market value of publicly traded shares: $138.2 billion (31 December 2009)
country comparison to the world: 39
$102.6 billion (31 December 2008)
$196 billion (31 December 2007)

Agriculture—products: rice, cassava (tapioca), rubber, corn, sugarcane, coconuts, soybeans

Industries: tourism, textiles and garments, agricultural processing, beverages, tobacco, cement, light manufacturing such as jewelry and electric appliances, computers and parts, integrated circuits, furniture, plastics, automobiles and automotive parts; world's second-largest tungsten producer and third-largest tin producer

Industrial production growth rate: 14.5% (2010 est.)
country comparison to the world: 9

Electricity—production: 148.2 billion kWh (2008 est.)
country comparison to the world: 24

Electricity—consumption: 134.4 billion kWh (2008 est.)
country comparison to the world: 24

Electricity—exports: 846 million kWh (2009 est.)

Electricity—imports: 2.313 billion kWh (2009 est.)

Oil—production: 380,000 bbl/day (2010 est.)

country comparison to the world: 34
Oil—consumption: 356,000 bbl/day (2009 est.)
country comparison to the world: 36
Oil—exports: 269,100 bbl/day (2009 est.)
country comparison to the world: 45
Oil—imports: 1.695 million bbl/day (2009 est.)
country comparison to the world: 12
Oil—proved reserves: 430 million bbl (1 January 2010 est.)
country comparison to the world: 51
Natural gas—production: 28.76 billion cu m (2008 est.)
country comparison to the world: 27
Natural gas—consumption: 37.31 billion cu m (2008 est.)
country comparison to the world: 23
Natural gas—exports: 0 cu m (2008 est.)
country comparison to the world: 182
Natural gas—imports: 8.55 billion cu m (2008 est.)
country comparison to the world: 24
Natural gas—proved reserves: 342 billion cu m (1 January 2010 est.)
country comparison to the world: 39
Current account balance: $12.29 billion (2010 est.)
country comparison to the world: 22
$21.86 billion (2009)
Exports: $191.3 billion (2010 est.)
country comparison to the world: 26
$151.9 billion (2009 est.)
Exports—commodities: textiles and footwear, fishery products, rice, rubber, jewelry, automobiles, computers and electrical appliances
Exports—partners: US 10.9%, China 10.6%, Japan 10.3%, Hong Kong 6.2%, Australia 5.6%, Malaysia 5% (2009 est.)
Imports: $156.9 billion (2010 est.)
country comparison to the world: 28
$118 billion (2009 est.)
Imports—commodities: capital goods, intermediate goods and raw materials, consumer goods, fuels
Imports—partners: Japan 18.7%, China 12.7%, Malaysia 6.4%, US 6.3%, UAE 5%, Singapore 4.3%, South Korea 4.1% (2009 est.)
Reserves of foreign exchange and gold: $176.1 billion (31 December 2010 est.)
country comparison to the world: 12
$138.4 billion (31 December 2009)
Debt—external: $82.5 billion (31 December 2010 est.)
country comparison to the world: 43
$70.3 billion (31 December 2009 est.)
Stock of direct foreign investment—at home: $117.9 billion (31 December 2010 est.)
country comparison to the world: 29
$109.6 billion (31 December 2009)
Stock of direct foreign investment—abroad: $20.3 billion (31 December 2010 est.)
country comparison to the world: 40
$18.2 billion (31 December 2009 est.)
Exchange rates: baht per US dollar–31.663 (2010), 34.286 (2009), 33.37 (2008), 34.52 (2007), 37.882 (2006)

COMMUNICATIONS

Telephones—main lines in use: 7.024 million (2009)
country comparison to the world: 27
Telephones—mobile cellular: 83.057 million (2009)
country comparison to the world: 14
Telephone system: *general assessment:* high quality system, especially in urban areas like Bangkok
domestic: fixed line system provided by both a government owned and commercial provider; wireless service expanding rapidly
international: country code–66; connected to major submarine cable systems providing links throughout Asia, Australia, Middle East, Europe, and US; satellite earth stations–2 Intelsat (1 Indian Ocean, 1 Pacific Ocean)
Broadcast media: 6 terrestrial TV stations in Bangkok broadcast nationally via relay stations–2 of the networks are owned by the military, the other 4 are government-owned or controlled, leased to private enterprise, and are all required to broadcast government-produced news programs twice a day; multi-channel satellite and cable TV subscription services are available; radio frequencies have been allotted for more than 500 government and commercial radio stations; many small community radio stations operate with low-power transmitters (2008)
Internet country code: .th
Internet hosts: 1.335 million (2010)
country comparison to the world: 37
Internet users: 17.483 million (2009)
country comparison to the world: 23

TRANSPORTATION

Airports: 105 (2010)
country comparison to the world: 55
Airports—with paved runways: *total:* 64
over 3,047 m: 8
2,438 to 3,047 m: 11
1,524 to 2,437 m: 24
914 to 1,523 m: 15
under 914 m: 6 (2010)
Airports—with unpaved runways:
total: 41
1,524 to 2,437 m: 1
914 to 1,523 m: 13
under 914 m: 27 (2010)
Heliports: 4 (2010)
Pipelines: gas 1,889 km; liquid petroleum gas 85 km; refined products 1,099 km (2010)
Railways: *total:* 4,071 km
country comparison to the world: 42
standard gauge: 29 km 1.435-m gauge
narrow gauge: 4,042 km 1.000-m gauge (2008)
Roadways: *total:* 180,053 km (includes 450 km of expressways) (2006)
country comparison to the world: 27
Waterways: 4,000 km (3,701 km navigable by boats with drafts up to 0.9 m) (2010)
country comparison to the world: 26
Merchant marine: *total:* 382
country comparison to the world: 27
by type: bulk carrier 30, cargo 116, chemical tanker 23, container 19, liquefied gas 36, passenger 1, passenger/cargo 10, petroleum tanker 120, refrigerated cargo 27
foreign-owned: 15 (China 1, Hong Kong 1, Japan 2, Malaysia 3, Singapore 1, Taiwan 1, UK 6)
registered in other countries: 41 (Bahamas 4, Panama 6, Singapore 30, Tuvalu 1) (2010)
Ports and terminals: Bangkok, Laem Chabang, Map Ta Phut, Prachuap Port, Si Racha

MILITARY

Military branches: Royal Thai Army (Kongthap Bok Thai, RTA), Royal Thai Navy (Kongthap Ruea Thai, RTN, includes Royal Thai Marine Corps), Royal Thai Air Force (Kongthap Agard Thai, RTAF) (2010)
Military service age and obligation: 21 years of age for compulsory military service; 18 years of age for voluntary military service; males register at 18 years of age; 2-year conscript service obligation (2009)
Manpower available for military service:
males age 16–49: 17,689,921
females age 16–49: 17,754,795 (2010 est.)
Manpower fit for military service: *males age 16–49:* 13,308,372
females age 16–49: 14,182,567 (2010 est.)
Manpower reaching militarily significant age annually: *male:* 533,424
female: 509,780 (2010 est.)
Military expenditures: 1.8% of GDP (2005 est.)
country comparison to the world: 82

TRANSNATIONAL ISSUES

Disputes—international: separatist violence in Thailand's predominantly Muslim southern provinces prompt border closures and controls with Malaysia to stem terrorist activities; Southeast Asian states have enhanced border surveillance to check the spread of avian flu; talks continue on completion of demarcation with Laos but disputes remain over several islands in the Mekong River; despite continuing border committee talks, Thailand must deal with Karen and other ethnic rebels, refugees, and illegal cross-border activities, and as of 2006, over 116,000 Karen, Hmong, and other refugees and asylum seekers from Burma; Cambodia and Thailand dispute sections of boundary; in 2011 Thailand and Cambodia resorted to arms in the dispute over the location of the boundary on the precipice surmounted by Preah Vihear temple ruins, awarded to Cambodia by ICJ decision in 1962 and part of a planned UN World Heritage site; Thailand is studying the feasibility of jointly constructing the Hatgyi Dam on the Salween river near the border with Burma; in 2004, international environmentalist pressure prompted China to halt construction of 13 dams on the Salween River that flows through China, Burma, and Thailand; 140,000 mostly Karen refugees fleeing civil strife, political upheaval and economic stagnation in Burma live in remote camps in Thailand near the border
Refugees and internally displaced persons: refugees (country of origin): 132,241 (Burma) (2007)
Illicit drugs: a minor producer of opium, heroin, and marijuana; transit point for illicit heroin en route to the international drug market from Burma and Laos; eradi-

cation efforts have reduced the area of cannabis cultivation and shifted some production to neighboring countries; opium poppy cultivation has been reduced by eradication efforts; also a drug money-laundering center; minor role in methamphetamine production for regional consumption; major consumer of methamphetamine since the 1990s despite a series of government crackdowns

TIMOR-LESTE

INTRODUCTION

Background: The Portuguese began to trade with the island of Timor in the early 16th century and colonized it in mid-century. Skirmishing with the Dutch in the region eventually resulted in an 1859 treaty in which Portugal ceded the western portion of the island. Imperial Japan occupied Portuguese Timor from 1942 to 1945, but Portugal resumed colonial authority after the Japanese defeat in World War II. East Timor declared itself independent from Portugal on 28 November 1975 and was invaded and occupied by Indonesian forces nine days later. It was incorporated into Indonesia in July 1976 as the province of Timor Timur (East Timor). An unsuccessful campaign of pacification followed over the next two decades, during which an estimated 100,000 to 250,000 individuals lost their lives. On 30 August 1999, in a UN-supervised popular referendum, an overwhelming majority of the people of Timor-Leste voted for independence from Indonesia. Between the referendum and the arrival of a multinational peacekeeping force in late September 1999, anti-independence Timorese militias–organized and supported by the Indonesian military–commenced a large-scale, scorched-earth campaign of retribution. The militias killed approximately 1,400 Timorese and forcibly pushed 300,000 people into western Timor as refugees. Most of the country's infrastructure, including homes, irrigation systems, water supply systems, and schools, and nearly 100% of the country's electrical grid were destroyed. On 20 September 1999, the Australian-led peacekeeping troops of the International Force for East Timor (INTERFET) deployed to the country and brought the violence to an end. On 20 May 2002, Timor-Leste was internationally recognized as an independent state. In late April 2006, internal tensions threatened the new nation's security when a military strike led to violence and a near breakdown of law and order. At Dili's request, an Australian-led International Stabilization Force (ISF) deployed to Timor-Leste in late May. In August, the UN Security Council established the UN Integrated Mission in Timor-Leste (UNMIT), which included an authorized police presence of over 1,600 personnel. The ISF and UNMIT restored stability, allowing for presidential and parliamentary elections in April and June 2007 in a largely peaceful atmosphere. In February 2008, a rebel group staged an unsuccessful attack against the president and prime minister. The ringleader was killed in the attack and most of the rebels surrendered in April 2008. Since the unsuccessful attacks the government has enjoyed one of its longest periods of post-independence stability.

GEOGRAPHY

Location: Southeastern Asia, northwest of Australia in the Lesser Sunda Islands at the eastern end of the Indonesian archipelago; note–Timor-Leste includes the eastern half of the island of Timor, the Oecussi (Ambeno) region on the northwest portion of the island of Timor, and the islands of Pulau Atauro and Pulau Jaco
Geographic coordinates: 8 50 S, 125 55 E
Map references: Southeast Asia
Area: *total:* 14,874 sq km
country comparison to the world: 159
land: 14,874 sq km
water: 0 sq km
Area—comparative: slightly larger than Connecticut
Land boundaries: *total:* 228 km
border countries: Indonesia 228 km
Coastline: 706 km
Maritime claims: territorial sea: 12 nm
contiguous zone: 24 nm
exclusive fishing zone: 200 nm
Climate: tropical; hot, humid; distinct rainy and dry seasons
Terrain: mountainous
Elevation extremes: *lowest point:* Timor Sea, Savu Sea, and Banda Sea 0 m
highest point: Foho Tatamailau 2,963 m
Natural resources: gold, petroleum, natural gas, manganese, marble
Land use: *arable land:* 8.2%
permanent crops: 4.57%
other: 87.23% (2005)
Irrigated land: 1,065 sq km (2003)
Natural hazards: floods and landslides are common; earthquakes; tsunamis; tropical cyclones
Environment—current issues: widespread use of slash and burn agriculture has led to deforestation and soil erosion
Environment—international agreements: *party to:* Biodiversity, Climate Change, Climate Change-Kyoto Protocol, Desertification
signed, but not ratified: none of the selected agreements
Geography—note: Timor comes from the Malay word for "East"; the island of Timor is part of the Malay Archipelago and is the largest and easternmost of the Lesser Sunda Islands

PEOPLE

Population: 1,177,834
country comparison to the world: 157
note: other estimates range as low as 800,000 (July 2011 est.)
Age structure: *0–14 years:* 33.8% (male 202,431/female 195,895)
15–64 years: 62.5% (male 374,659/female 361,983)
65 years and over: 3.6% (male 20,160/ female 22,706) (2011 est.)
Median age: *total:* 22.5 years
male: 22.5 years
female: 22.5 years (2011 est.)
Population growth rate: 1.981% (2011 est.)
country comparison to the world: 56
Birth rate: 25.7 births/1,000 population (2011 est.)
country comparison to the world: 56
Death rate: 5.89 deaths/1,000 population (July 2011 est.)
country comparison to the world: 169
Net migration rate: 0 migrant(s)/1,000 population (2011 est.)
country comparison to the world: 109
Urbanization: *urban population:* 28% of total population (2010)
rate of urbanization: 5% annual rate of change (2010-15 est.)
Major cities—population: DILI (capital) 166,000 (2009)
Sex ratio: *at birth:* 1.05 male(s)/female
under 15 years: 1.03 male(s)/female
15–64 years: 1.04 male(s)/female
65 years and over: 0.9 male(s)/female
total population: 1.03 male(s)/female (2011 est.)
Infant mortality rate: *total:* 38.01 deaths/1,000 live births
country comparison to the world: 66
male: 43.79 deaths/1,000 live births
female: 31.95 deaths/1,000 live births (2011 est.)
Life expectancy at birth: *total population:* 67.95 years
country comparison to the world: 155
male: 65.54 years
female: 70.47 years (2011 est.)
Total fertility rate: 3.13 children born/ woman (2011 est.)
country comparison to the world: 57
HIV/AIDS—adult prevalence rate: NA
HIV/AIDS—people living with HIV/AIDS: NA
HIV/AIDS—deaths: NA
Major infectious diseases: *degree of risk:* very high

food or waterborne diseases: bacterial and protozoal diarrhea, hepatitis A, and typhoid fever
vectorborne diseases: chikungunya, dengue fever and malaria (2009)
Drinking water source: *Improved:*
urban: 86% of population
rural: 63% of population
total: 69% of population
Unimproved: urban: 14% of population
rural: 37% of population
total: 31% of population (2008)
Sanitation facility access: *Improved:*
urban: 76% of population
rural: 40% of population
total: 50% of population
Unimproved: urban: 24% of population
rural: 60% of population
total: 50% of population (2008)
Nationality: *noun:* Timorese
adjective: Timorese
Ethnic groups: Austronesian (Malayo-Polynesian), Papuan, small Chinese minority
Religions: Roman Catholic 98%, Muslim 1%, Protestant 1% (2005)
Languages: Tetum (official), Portuguese (official), Indonesian, English
note: there are about 16 indigenous languages; Tetum, Galole, Mambae, and Kemak are spoken by a significant portion of the population
Literacy: *definition:* age 15 and over can read and write
total population: 58.6%
male: NA
female: NA (2002)
School life expectancy (primary to tertiary education): *total:* 11 years (2004)
Education expenditures: 16.8% of GDP (2009)
country comparison to the world: 1

GOVERNMENT

Country name: *conventional long form:* Democratic Republic of Timor-Leste (pronounced TEE-mor LESS-tay)
conventional short form: Timor-Leste
local long form: Republika Demokratika Timor Lorosa'e [Tetum]; Republica Democratica de Timor-Leste [Portuguese]
local short form: Timor Lorosa'e [Tetum]; Timor-Leste [Portuguese]
former: East Timor, Portuguese Timor
Government type: republic
Capital: *name:* Dili
geographic coordinates: 8 35 S, 125 36 E
time difference: UTC+9 (14 hours ahead of Washington, DC during Standard Time)
Administrative divisions: 13 administrative districts; Aileu, Ainaro, Baucau, Bobonaro (Maliana), Cova-Lima (Suai), Dili, Ermera (Gleno), Lautem (Los Palos), Liquica, Manatuto, Manufahi (Same), Oecussi (Ambeno), Viqueque
note: administrative divisions have the same names as their administrative centers (exceptions have the administrative center name following in parentheses)
Independence: 28 November 1975 (independence proclaimed from Portugal); note—20 May 2002 is the official date of international recognition of Timor-Leste's independence from Indonesia
National holiday: Independence Day, 28 November (1975)
Constitution: 20 May 2002 (effective date)
Legal system: civil law system based on the Indonesian model; note—new penal code based on the Portuguese model was passed by Parliament and promulgated in 2009; new civil code expected to be promulgated in 2011
International law organization participation: has not submitted an ICJ jurisdiction declaration; accepts ICCt jurisdiction
Suffrage: 17 years of age; universal
Executive branch: *chief of state:* President Jose RAMOS-HORTA (since 20 May 2007); note—the president plays a largely symbolic role but is able to veto legislation, dissolve parliament, and call national elections
head of government: Prime Minister Kay Rala Xanana GUSMAO (since 8 August 2007), note—he formerly used the name Jose Alexandre GUSMAO; Vice Prime Minister Jose Luis GUTERRES (since 8 August 2007)
cabinet: Council of Ministers
(For more information visit the World Leaders website)
elections: president elected by popular vote for a five-year term (eligible for a second term); election last held on 9 April 2007 with run-off on 8 May 2007 (next to be held in May 2012); following elections, president appoints leader of majority party or majority coalition as prime minister
election results: Jose RAMOS-HORTA elected president; percent of vote—Jose RAMOS-HORTA 69.2%, Francisco GUTTERES 30.8%
Legislative branch: unicameral National Parliament (number of seats can vary from 52 to 65; members elected by popular vote to serve five-year terms)
elections: last held on 30 June 2007 (next elections due by June 2012)
election results: percent of vote by party—FRETILIN 29%, CNRT 24.1%, ASDT-PSD 15.8%, PD 11.3%, PUN 4.5%, KOTA-PPT (Democratic Alliance) 3.2%, UNDERTIM 3.2%, others 8.9%; seats by party—FRETILIN 21, CNRT 18, ASDT-PSD 11, PD 8, PUN 3, KOTA-PPT 2, UNDERTIM 2
Judicial branch: Supreme Court of Justice—constitution calls for one judge to be appointed by National Parliament and rest appointed by Superior Council for Judiciary; note—until Supreme Court is established, Court of Appeals is highest court
Political parties and leaders: Democratic Party or PD [Fernando de ARAUJO]; National Congress for Timorese Reconstruction or CNRT [Xanana GUSMAO]; National Democratic Union of Timorese Resistance or UNDERTIM [Cornelio DA Conceicao GAMA]; National Unity Party or PUN [Fernanda BORGES]; People's Party of Timor or PPT [Jacob XAVIER]; Revolutionary Front of Independent Timor-Leste or FRETILIN [Mari ALKATIRI]; Social Democratic Association of Timor or ASDT [Francisco Xavier do AMARAL]; Social Democratic Party or PSD [Zacarias Albano da COSTA]; Sons of the Mountain Warriors or KOTA [Manuel TILMAN] (also known as Association of Timorese Heroes)
Political pressure groups and leaders: NA
International organization participation: ACP, ADB, AOSIS, ARF, ASEAN (observer), CPLP, FAO, G-77, IBRD, ICAO, ICRM, IDA, IFAD, IFC, IFRCS, ILO, IMF, IMO, Interpol, IOC, IOM, IPU, ITU, MIGA, NAM, OPCW, PIF (observer), UN, UNCTAD, UNESCO, UNIDO, Union Latina, UNWTO, UPU, WCO, WHO, WMO
Diplomatic representation in the US: *chief of mission:* Ambassador Constancio da Conceicao PINTO
chancery: 4201 Connecticut Avenue NW, Suite 504, Washington, DC 20008
telephone: [1] (202) 966-3202
FAX: [1] (202) 966-3205
Diplomatic representation from the US: *chief of mission:* Ambassador Judith FERGIN
embassy: Avenida de Portugal, Praia dos Conqueiros, Dili
mailing address: US Department of State, 8250 Dili Place, Washington, DC 20521-8250
telephone: (670) 332-4684
FAX: (670) 331-3206
Flag description: red, with a black isosceles triangle (based on the hoist side) superimposed on a slightly longer yellow arrowhead that extends to the center of the flag; a white star—pointing to the upper hoist-side corner of the flag—is in the center of the black triangle; yellow denotes the colonialism in Timor-Leste's past; black represents the obscurantism that needs to be overcome; red stands for the national liberation struggle; the white star symbolizes peace and serves as a guiding light
National anthem: *name:* "Patria" (Fatherland)
lyrics/music: Fransisco Borja DA COSTA/Afonso DE ARAUJO
note: adopted 2002; the song was first used as an anthem when Timor-Leste declared its independence from Portugal in 1975; the lyricist, Fransisco Borja DA COSTA, was killed in an Indonesian invasion just days after independence was declared

ECONOMY

Economy—overview: In late 1999, about 70% of the economic infrastructure of Timor-Leste was laid waste by Indonesian troops and anti-independence militias. Three hundred thousand people fled westward. Over the next three years a massive international program, manned by 5,000 peacekeepers (8,000 at peak) and 1,300 police officers, led to substantial reconstruction in both urban and rural areas. By the end of 2005, refugees had returned or had settled in Indonesia. The country continues to face great challenges in rebuilding its infrastructure, strengthening the civil administration, and generating jobs for young people entering the work force. The development of oil and gas resources in offshore waters has greatly supplemented government revenues. This technology-intensive industry, however, has done little to create jobs for the unemployed because there are no production facilities in Timor-Leste. Gas is piped to Australia. In June 2005, the National Parliament unanimously approved the creation of a Petroleum Fund to serve as a repository for all petroleum revenues and to preserve the

value of Timor-Leste's petroleum wealth for future generations. The Fund held assets of US$6.6 billion as of October 2010. The economy continues to recover strongly from the mid-2006 outbreak of violence and civil unrest, which disrupted both private and public sector economic activity. The government in 2008 resettled tens of thousands of an estimated 100,000 internally displaced persons (IDPs); most IDPs returned home by early 2009. Government spending increased markedly in 2009 and 2010, primarily on basic infrastructure, including electricity and roads. Limited experience in procurement and infrastructure building has hampered these projects. The underlying economic policy challenge the country faces remains how best to use oil-and-gas wealth to lift the non-oil economy onto a higher growth path and to reduce poverty.
GDP (purchasing power parity): $3.051 billion (2010 est.)
country comparison to the world: 176
$2.877 billion (2009 est.)
$2.547 billion (2008 est.)
note: data are in 2010 US dollars
GDP (official exchange rate): $628 million (2010 est.)
GDP—real growth rate: 6.1% (2010 est.)
country comparison to the world: 51
12.9% (2009 est.)
11% (2008 est.)
GDP—per capita (PPP): $2,600 (2010 est.)
country comparison to the world: 174
$2,500 (2009 est.)
$2,300 (2008 est.)
note: data are in 2010 US dollars
GDP—composition by sector:
agriculture: 32.2%
industry: 12.8%
services: 55% (2005)
Labor force: 414,200 (2007)
country comparison to the world: 157
Labor force—by occupation:
agriculture: 90%
industry: NA%
services: NA% (2006 est.)
Unemployment rate:
| 20% (2006 est.)
country comparison to the world: 166
note: data are for rural areas, unemployment rises to more than 40% among urban youth
Population below poverty line: 42% (2003 est.)
Household income or consumption by percentage share: *lowest 10%:* 2.9%
highest 10%: 31.3% (2001)
Distribution of family income—Gini index: 38 (2002 est.)
country comparison to the world: 72
Budget: *revenues:* $1.481 billion
expenditures: $838 million (2010 est.)
Inflation rate (consumer prices): 7.8% (2007 est.)
country comparison to the world: 183
Commercial bank prime lending rate: 11.17% (31 December 2009 est.)
country comparison to the world: 59
13.11% (31 December 2008 est.)
Stock of narrow money: $102.8 million (31 December 2008)
country comparison to the world: 180
$74.94 million (31 December 2007)
Stock of broad money: $268.4 million (31 December 2009)
country comparison to the world: 180
$192.7 million (31 December 2008)
Stock of domestic credit: $127.1 million (31 December 2008 est.)
country comparison to the world: 180
$118.1 million (31 December 2007 est.)
Market value of publicly traded shares: $NA
Agriculture—products: coffee, rice, corn, cassava, sweet potatoes, soybeans, cabbage, mangoes, bananas, vanilla
Industries: printing, soap manufacturing, handicrafts, woven cloth
Industrial production growth rate: 8.5% (2004 est.)
country comparison to the world: 27
Electricity—production: NA kWh (2009 est.)
Electricity—consumption: NA kWh
Electricity—exports: 0 kWh (2009 est.)
Electricity—imports: 0 kWh (2009 est.)
Oil—production: 96,270 bbl/day (2009 est.)
country comparison to the world: 53
Oil—consumption: 2,500 bbl/day (2009 est.)
country comparison to the world: 180
Oil—exports: 100,900 bbl/day (2007 est.)
country comparison to the world: 67
Oil—proved reserves: 553.8 million bbl (1 January 2008)
country comparison to the world: 48
Natural gas—production: 0 cu m (2008 est.)
country comparison to the world: 193
Natural gas—consumption: 0 cu m (2008 est.)
country comparison to the world: 130
Natural gas—exports: 0 cu m (2008 est.)
country comparison to the world: 189
Natural gas—imports: 0 cu m (2008 est.)
country comparison to the world: 127
Natural gas—proved reserves: 200 billion cu m (1 January 2006 est.)
country comparison to the world: 45
Current account balance: $1.161 billion (2007 est.)
country comparison to the world: 45
Exports: $10 million (2005 est.); note–excludes oil
country comparison to the world: 213
Exports—commodities: coffee, sandalwood, marble; note–potential for oil and vanilla exports
Imports: $202 million (2004 est.)
country comparison to the world: 200
Imports—commodities: food, gasoline, kerosene, machinery
Exchange rates: the US dollar is used

COMMUNICATIONS

Telephones—main lines in use: 2,400 (2009)
country comparison to the world: 221
Telephones—mobile cellular: 116,000 (2009)
country comparison to the world: 183
Telephone system: *general assessment:* rudimentaryserviceinurbanandsomeruralareas
domestic: system suffered significant damage during the violence associated with independence; limited fixed-line services; mobile-cellular services and coverage available in urban and some rural areas
international: country code–670; international service is available in major urban centers
Broadcast media: 1 public TV broadcast station broadcasting nationally and 1 public radio broadcaster with stations in each of the 13 administrative districts; one commercial TV broadcast station broadcasting in parts of Dili only, a few commercial radio stations, and roughly a dozen community radio stations (2009)
Internet country code: .tl
Internet hosts: 206 (2010)
country comparison to the world: 193
Internet users: 2,100 (2009)
country comparison to the world: 210

TRANSPORTATION

Airports: 6 (2010)
country comparison to the world: 174
Airports—with paved runways: *total:* 2
2,438 to 3,047 m: 1
1,524 to 2,437 m: 1 (2010)
Airports—with unpaved runways:
total: 4
914 to 1,523 m: 2
under 914 m: 2 (2010)
Heliports: 8 (2010)
Roadways: *total:* 6,040 km
country comparison to the world: 149
paved: 2,600 km
unpaved: 3,440 km (2005)
Merchant marine: *total:* 1
country comparison to the world: 161
by type: passenger/cargo 1 (2010)
Ports and terminals: Dili

MILITARY

Military branches: Timor-Leste Defense Force (Forcas de Defesa de Timor-L'este, Falintil (F-FDTL)): Army, Navy (Armada) (2010)
Military service age and obligation: 18 years of age for voluntary military service; no conscription; 18-month service obligation (2011)
Manpower available for military service:
males age 16–49: 305,643
females age 16–49: 293,052 (2010 est.)
Manpower fit for military service: *males age 16–49:* 243,120
females age 16–49: 251,061 (2010 est.)
Manpower reaching militarily significant age annually: *male:* 12,737
female: 12,389 (2010 est.)
Military expenditures: NA

TRANSNATIONAL ISSUES

Disputes—international: Timor-Leste-Indonesia Boundary Committee has resolved all but some sections of border along Timor-Leste's Oecussi exclave; maritime boundaries with Indonesia remain unresolved; many refugees who left Timor-Leste in 2003 still reside in Indonesia and refuse repatriation; in 2007, Australia and Timor-Leste signed a 50-year development zone and revenue sharing agreement in lieu of a maritime boundary
Refugees and internally displaced persons: *IDPs:* 100,000 (2007)
Illicit drugs: NA

TOGO

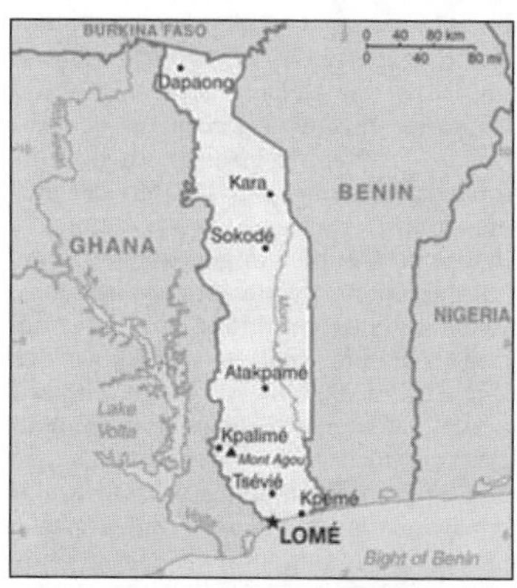

INTRODUCTION

Background: French Togoland became Togo in 1960. Gen. Gnassingbe EYADEMA, installed as military ruler in 1967, ruled Togo with a heavy hand for almost four decades. Despite the facade of multiparty elections instituted in the early 1990s, the government was largely dominated by President EYADEMA, whose Rally of the Togolese People (RPT) party has maintained power almost continually since 1967 and maintains a majority of seats in today's legislature. Upon EYADEMA's death in February 2005, the military installed the president's son, Faure GNASSINGBE, and then engineered his formal election two months later. Democratic gains since then allowed Togo to hold its first relatively free and fair legislative elections in October 2007. After years of political unrest and condemnation from international organizations for human rights abuses, Togo is finally being re-welcomed into the international community.

GEOGRAPHY

Location: Western Africa, bordering the Bight of Benin, between Benin and Ghana
Geographic coordinates: 8 00 N, 1 10 E
Map references: Africa
Area: *total:* 56,785 sq km
country comparison to the world: 125
land: 54,385 sq km
water: 2,400 sq km
Area—comparative: slightly smaller than West Virginia
Land boundaries: *total:* 1,647 km
border countries: Benin 644 km, Burkina Faso 126 km, Ghana 877 km
Coastline: 56 km
Maritime claims: territorial sea: 30 nm
exclusive economic zone: 200 nm
Climate: tropical; hot, humid in south; semiarid in north
Terrain: gently rolling savanna in north; central hills; southern plateau; low coastal plain with extensive lagoons and marshes
Elevation extremes: *lowest point:* Atlantic Ocean 0 m
highest point: Mont Agou 986 m
Natural resources: phosphates, limestone, marble, arable land
Land use: *arable land:* 44.2%
permanent crops: 2.11%
other: 53.69% (2005)
Irrigated land: 70 sq km (2003)
Total renewable water resources: 14.7 cu km (2001)
Freshwater withdrawal (domestic/industrial/agricultural): *total:* 0.17 cu km/yr (53%/2%/45%)
per capita: 28 cu m/yr (2000)
Natural hazards: hot, dry harmattan wind can reduce visibility in north during winter; periodic droughts
Environment—current issues: deforestation attributable to slash-and-burn agriculture and the use of wood for fuel; water pollution presents health hazards and hinders the fishing industry; air pollution increasing in urban areas
Environment—international agreements: *party to:* Biodiversity, Climate Change, Climate Change-Kyoto Protocol, Desertification, Endangered Species, Law of the Sea, Ozone Layer Protection, Ship Pollution, Tropical Timber 83, Tropical Timber 94, Wetlands, Whaling
signed, but not ratified: none of the selected agreements
Geography—note: the country's length allows it to stretch through six distinct geographic regions; climate varies from tropical to savanna

PEOPLE

Population: 6,771,993 (July 2011 est.)
country comparison to the world: 100
note: estimates for this country explicitly take into account the effects of excess mortality due to AIDS; this can result in lower life expectancy, higher infant mortality, higher death rates, lower population growth rates, and changes in the distribution of population by age and sex than would otherwise be expected
Age structure: *0–14 years:* 40.9% (male 1,387,537/female 1,381,040)
15–64 years: 56% (male 1,878,114/female 1,912,132)
65 years and over: 3.1% (male 92,689/female 120,481) (2011 est.)
Median age: *total:* 19.3 years
male: 19 years
female: 19.5 years (2011 est.)
Population growth rate: 2.762% (2011 est.)
country comparison to the world: 18
Birth rate: 35.58 births/1,000 population (2011 est.)
country comparison to the world: 26
Death rate: 7.96 deaths/1,000 population (July 2011 est.)
country comparison to the world: 106
Net migration rate: 0 migrant(s)/1,000 population (2011 est.)
country comparison to the world: 108
Urbanization: *urban population:* 43% of total population (2010)
rate of urbanization: 3.9% annual rate of change (2010-15 est.)
Major cities—population: LOME (capital) 1.593 million (2009)
Sex ratio: *at birth:* 1.03 male(s)/female
under 15 years: 1 male(s)/female
15–64 years: 0.96 male(s)/female
65 years and over: 0.63 male(s)/female
total population: 0.97 male(s)/female (2011 est.)
Infant mortality rate: *total:* 51.48 deaths/1,000 live births
country comparison to the world: 44
male: 58.43 deaths/1,000 live births
female: 44.32 deaths/1,000 live births (2011 est.)
Life expectancy at birth: *total population:* 62.71 years
country comparison to the world: 177
male: 60.19 years
female: 65.3 years (2011 est.)
Total fertility rate: 4.69 children born/woman (2011 est.)
country comparison to the world: 29
HIV/AIDS—adult prevalence rate: 3.2% (2009 est.)
country comparison to the world: 22
HIV/AIDS—people living with HIV/AIDS: 120,000 (2009 est.)
country comparison to the world: 39
HIV/AIDS—deaths: 7,700 (2009 est.)
country comparison to the world: 28
Major infectious diseases: *degree of risk:* very high
food or waterborne diseases: bacterial and protozoal diarrhea, hepatitis A, and typhoid fever
vectorborne diseases: malaria and yellow fever
water contact disease: schistosomiasis
respiratory disease: meningococcal meningitis
animal contact disease: rabies
note: highly pathogenic H5N1 avian influenza has been identified in this country; it poses a negligible risk with extremely rare cases possible among US citizens who have close contact with birds (2009)
Drinking water source: *Improved:*
urban: 87% of population
rural: 41% of population
total: 60% of population
Unimproved: urban: 13% of population
rural: 59% of population
total: 40% of population (2008)
Sanitation facility access: *Improved:*
urban: 24% of population
rural: 3% of population
total: 12% of population
Unimproved: urban: 76% of population
rural: 97% of population
total: 88% of population (2008)
Nationality: *noun:* Togolese (singular and plural)
adjective: Togolese
Ethnic groups: African (37 tribes; largest and most important are Ewe, Mina, and Kabre) 99%, European and Syrian-Lebanese less than 1%
Religions: Christian 29%, Muslim 20%, indigenous beliefs 51%
Languages: French (official, the language of commerce), Ewe and Mina (the two major African languages in the south), Kabye (sometimes spelled Kabiye) and Dagomba (the two major African languages in the north)

Literacy: *definition:* age 15 and over can read and write
total population: 60.9%
male: 75.4%
female: 46.9% (2003 est.)
School life expectancy (primary to tertiary education): *total:* 10 years
male: 11 years
female: 8 years (2007)
Education expenditures: 4.6% of GDP (2009)
country comparison to the world: 78

GOVERNMENT

Country name: *conventional long form:* Togolese Republic
conventional short form: Togo
local long form: Republique togolaise
local short form: none
former: French Togoland
Government type: republic under transition to multiparty democratic rule
Capital: *name:* Lome
geographic coordinates: 6 08 N, 1 13 E
time difference: UTC 0 (5 hours ahead of Washington, DC during Standard Time)
Administrative divisions: 5 regions (regions, singular—region); Centrale, Kara, Maritime, Plateaux, Savanes
Independence: 27 April 1960 (from French-administered UN trusteeship)
National holiday: Independence Day, 27 April (1960)
Constitution: adopted by public referendum 27 September 1992
Legal system: customary law system
International law organization participation: accepts compulsory ICJ jurisdiction with reservations; non-party state to the ICC
Suffrage: 18 years of age; universal
Executive branch: *chief of state:* President Faure GNASSINGBE (since 4 May 2005); *head of government:* Prime Minister Gilbert HOUNGBO (since 7 September 2008)
cabinet: Council of Ministers appointed by the president and the prime minister
(For more information visit the World Leaders website)
elections: president elected by popular vote for a five-year term (no term limits); election last held on 4 March 2010 (next to be held in 2015); prime minister appointed by the president
election results: Faure GNASSINGBE reelected president; percent of vote—Faure GNASSINGBE 60.9%, Jean-Pierre FABRE 33.9%, Yawovi AGBOYIBO 3%, other 2.2%
Legislative branch: unicameral National Assembly (81 seats; members elected by popular vote to serve five-year terms)
elections: last held on 14 October 2007 (next to be held in 2012)
election results: percent of vote by party—RPT 39.4%, UFC 37.0%, CAR 8.2%, independents 2.5%, other 12.9%; seats by party—RPT 50, UFC 27, CAR 4
Judicial branch: Court of Appeal or Cour d'Appel; Supreme Court or Cour Supreme
Political parties and leaders: Action Committee for Renewal or CAR [Yawovi AGBOYIBO]; Democratic Convention of African Peoples or CDPA; Democratic Party for Renewal or PDR; Juvento [Monsilia DJATO]; Movement of the Believers of Peace and Equality or MOCEP; National Alliance for Change or ANC [Jean-Pierre FABRE]; Pan-African Patriotic Convergence or CPP; Rally for the Support for Development and Democracy or RSDD [Harry OLYMPIO]; Rally of the Togolese People or RPT [Faure GNASSINGBE]; Socialist Pact for Renewal or PSR; Union for Democracy and Social Progress or UDPS [Gagou KOKOU]; Union of Forces for Change or UFC [Gilchrist OLYMPIO]
Political pressure groups and leaders: NA
International organization participation: ACP, AfDB, AU, ECOWAS, Entente, FAO, FZ, G-77, IAEA, IBRD, ICAO, ICC, ICRM, IDA, IDB, IFAD, IFC, IFRCS, ILO, IMF, IMO, Interpol, IOC, IOM, IPU, ISO (correspondent), ITSO, ITU, ITUC, MIGA, NAM, OIC, OIF, OPCW, PCA, UN, UNAMID, UNCTAD, UNESCO, UNIDO, UNMIL, UNOCI, UNWTO, UPU, WADB (regional), WAEMU, WCO, WFTU, WHO, WIPO, WMO, WTO
Diplomatic representation in the US: *chief of mission:* Ambassador Kadangha Limbiya BARIKI
chancery: 2208 Massachusetts Avenue NW, Washington, DC 20008
telephone: [1] (202) 234-4212
FAX: [1] (202) 232-3190
Diplomatic representation from the US: *chief of mission:* Ambassador Patricia McMahon HAWKINS
embassy: 4332 Blvd. Gnassingbe Eyadema, Cite OUA, Lome
mailing address: B. P. 852, Lome; 2300 Lome Place, Washington, DC 20512-2300
telephone: [228] 261-5470
FAX: [228] 261-5501
Flag description: five equal horizontal bands of green (top and bottom) alternating with yellow; a white five-pointed star on a red square is in the upper hoist-side corner; the five horizontal stripes stand for the five different regions of the country; the red square is meant to express the loyalty and patriotism of the people; green symbolizes hope, fertility, and agriculture; yellow represents mineral wealth and faith that hard work and strength will bring prosperity; the star symbolizes life, purity, peace, dignity, and Togo's independence
note: uses the popular Pan-African colors of Ethiopia
National anthem: *name:* "Salut a toi, pays de nos aieux" (Hail to Thee, Land of Our Forefathers)
lyrics/music: Alex CASIMIR-DOSSEH
note: adopted 1960, restored 1992; this anthem was replaced by another during one-party rule between 1979 and 1992

ECONOMY

Economy—overview: This small, sub-Saharan economy suffers from anemic economic growth and depends heavily on both commercial and subsistence agriculture, which provides employment for 65% of the labor force. Some basic foodstuffs must still be imported. Cocoa, coffee, and cotton generate about 40% of export earnings with cotton being the most important cash crop. Togo is the world's fourth-largest producer of phosphate. The government's decade-long effort, supported by the World Bank and the IMF, to implement economic reform measures, encourage foreign investment, and bring revenues in line with expenditures has moved slowly. Progress depends on follow through on privatization, increased openness in government financial operations, progress toward legislative elections, and continued support from foreign donors. Togo is on track with its IMF Extended Credit Facility and reached a HIPC debt relief completion point in 2010 at which 95% of the country's debt was forgiven. Economic growth prospects remain marginal due to declining cotton production and underinvestment in phosphate mining.
GDP (purchasing power parity): $5.974 billion (2010 est.)
country comparison to the world: 157
$5.778 billion (2009 est.)
$5.596 billion (2008 est.)
note: data are in 2010 US dollars
GDP (official exchange rate): $3.194 billion (2010 est.)
GDP—real growth rate: 3.4% (2010 est.)
country comparison to the world: 113
3.2% (2009 est.)
2.4% (2008 est.)
GDP—per capita (PPP): $900 (2010 est.)
country comparison to the world: 216
$900 (2009 est.)
$900 (2008 est.)
note: data are in 2010 US dollars
GDP—composition by sector:
agriculture: 47.4%
industry: 25.4%
services: 27.2% (2009 est.)
Labor force: 2.595 million (2007)
country comparison to the world: 110
Labor force—by occupation:
agriculture: 65%
industry: 5%
services: 30% (1998 est.)
Unemployment rate: NA%
Population below poverty line:
32% (1989 est.)
Household income or consumption by percentage share: *lowest 10%:* 3.3%
highest 10%: 27.1% (2006)
Investment (gross fixed): 18.2% of GDP (2010 est.)
country comparison to the world: 109
Budget: *revenues:* $602.3 million
expenditures: $692.1 million (2010 est.)
Inflation rate (consumer prices):
2.6% (2010 est.)
country comparison to the world: 71
2% (2009 est.)
Central bank discount rate:
4.25% (31 December 2009)
country comparison to the world: 96
4.75% (31 December 2008)
Commercial bank prime lending rate:
NA%
Stock of narrow money: $754.5 million (31 December 2010 est.)
country comparison to the world: 147
$789.7 million (31 December 2009 est.)
Stock of broad money: $1.238 billion (31 December 2010 est.)

country comparison to the world: 158
$1.306 billion (31 December 2009 est.)
Stock of domestic credit: $817.7 million (31 December 2010 est.)
country comparison to the world: 154
$862.4 million (31 December 2009 est.)
Market value of publicly traded shares: $NA
Agriculture—products: coffee, cocoa, cotton, yams, cassava (tapioca), corn, beans, rice, millet, sorghum; livestock; fish
Industries: phosphate mining, agricultural processing, cement, handicrafts, textiles, beverages
Industrial production growth rate: 2.5% (2010 est.)
country comparison to the world: 120
Electricity—production: 230 million kWh (2007 est.)
country comparison to the world: 175
Electricity—consumption: 640 million kWh (2007 est.)
country comparison to the world: 152
Electricity—exports: 0 kWh (2008 est.)
Electricity—imports: 514 million kWh; note—electricity supplied by Ghana (2007 est.)
Oil—production: 0 bbl/day (2009 est.)
country comparison to the world: 133
Oil—consumption: 21,000 bbl/day (2009 est.)
country comparison to the world: 123
Oil—exports: 1,547 bbl/day (2005)
country comparison to the world: 117
Oil—imports: 15,270 bbl/day (2007 est.)
country comparison to the world: 124
Oil—proved reserves: 0 bbl (1 January 2010 est.)
country comparison to the world: 195
Natural gas—production: 0 cu m (2008 est.)
country comparison to the world: 191
Natural gas—consumption: 0 cu m (2008 est.)
country comparison to the world: 128
Natural gas—exports: 0 cu m (2008 est.)
country comparison to the world: 186
Natural gas—imports: 0 cu m (2008 est.)
country comparison to the world: 125
Natural gas—proved reserves: 0 cu m (1 January 2010 est.)
country comparison to the world: 194
Current account balance: $-339 million (2010 est.)
country comparison to the world: 103
$-236 million (2009 est.)
Exports: $859 million (2010 est.)
country comparison to the world: 155
$818 million (2009 est.)
Exports—commodities: reexports, cotton, phosphates, coffee, cocoa
Exports—partners: Germany 17.57%, Ghana 12.74%, Burkina Faso 11.02%, India 10.22%, Belgium 7.1%, Benin 6.92%, Netherlands 5.94%, Mali 4.41% (2009)
Imports: $1.337 billion (2010 est.)
country comparison to the world: 167
$1.261 billion (2009 est.)
Imports—commodities: machinery and equipment, foodstuffs, petroleum products
Imports—partners: China 36.58%, France 8.64%, Netherlands 6.76%, India 5.06%, US 4.4% (2009)
Reserves of foreign exchange and gold: $686 million (31 December 2010 est.)
country comparison to the world: 122
$703.2 million (31 December 2009 est.)
Debt—external: $NA (31 December 2010)
$1.573 billion (31 December 2008 est.)
Exchange rates: Communaute Financiere Africaine francs (XOF) per US dollar–495.28 (2010), 472.19 (2009), 447.81 (2008), 482.71 (2007), 522.59 (2006)

COMMUNICATIONS

Telephones—main lines in use: 178,700 (2009)
country comparison to the world: 129
Telephones—mobile cellular: 2.187 million (2009)
country comparison to the world: 129
Telephone system: *general assessment:* fair system based on a network of microwave radio relay routes supplemented by open-wire lines and a mobile-cellular system
domestic: microwave radio relay and open-wire lines for conventional system; combined fixed-line and mobile-cellular teledensity roughly 40 telephones per 100 persons with mobile-cellular use predominating
international: country code–228; satellite earth stations–1 Intelsat (Atlantic Ocean), 1 Symphonie
Broadcast media: 2 state-owned TV stations with multiple transmission sites; 5 private TV stations broadcast locally; cable TV service is available; state-owned radio network with multiple stations; several dozen private radio stations and a few community radio stations; transmissions of multiple international broadcasters are obtainable (2007)
Internet country code: .tg
Internet hosts: 860 (2010)
country comparison to the world: 168
Internet users: 356,300 (2009)
country comparison to the world: 123

TRANSPORTATION

Airports: 8 (2010)
country comparison to the world: 163
Airports—with paved runways: *total:* 2
2,438 to 3,047 m: 2 (2010)
Airports—with unpaved runways: *total:* 6
914 to 1,523 m: 4
under 914 m: 2 (2010)
Railways: *total:* 532 km
country comparison to the world: 110
narrow gauge: 532 km 1.000-m gauge (2008)
Roadways:
total: 7,520 km
country comparison to the world: 146
paved: 2,376 km
unpaved: 5,144 km (2000)
Waterways: 50 km (seasonally navigable by small craft on the Mono River depending on rainfall) (2009)
country comparison to the world: 103
Merchant marine: *total:* 53
country comparison to the world: 69
by type: bulk carrier 5, cargo 40, chemical tanker 2, container 2, petroleum tanker 2, refrigerated cargo 1, roll on/roll off 1
foreign-owned: 23 (China 2, Greece 1, Lebanon 6, Romania 1, Syria 5, Turkey 4, UAE 1, UK 3) (2010)
Ports and terminals: Kpeme, Lome

MILITARY

Military branches: Togolese Armed Forces (Forces Armees Togolaise, FAT): Ground Forces, Togolese Navy (Marine du Togo), Togolese Air Force (Force Aerienne Togolaise, TAF), National Gendarmerie (2011)
Military service age and obligation: 18 years of age for selective compulsory and voluntary military service; 2-year service obligation (2006)
Manpower available for military service:
males age 16–49: 1,577,572
females age 16–49: 1,589,715 (2010 est.)
Manpower fit for military service:
males age 16–49: 1,104,536
females age 16–49: 1,158,061 (2010 est.)
Manpower reaching militarily significant age annually: *male:* 74,036
female: 73,515 (2010 est.)
Military expenditures: 1.6% of GDP (2005 est.)
country comparison to the world: 93

TRANSNATIONAL ISSUES

Disputes—international: in 2001, Benin claimed Togo moved boundary monuments–joint commission continues to resurvey the boundary; in 2006, 14,000 Togolese refugees remain in Benin and Ghana out of the 40,000 who fled there in 2005; talks continue between Benin and Togo on funding the Adjrala hydroelectric dam on the Mona River
Refugees and internally displaced persons: refugees (country of origin): 5,000 (Ghana)
IDPs: 1,500 (2007)
Illicit drugs: transit hub for Nigerian heroin and cocaine traffickers; money laundering not a significant problem

TOKELAU

(TERRITORY OF NEW ZEALAND)

INTRODUCTION

Background: Originally settled by Polynesian emigrants from surrounding island groups, the Tokelau Islands were made a British protectorate in 1889. They were transferred to New Zealand administration in 1925. Referenda held in 2006 and 2007 to change the status of the islands from that of a New Zealand territory to one of free association with New Zealand did not meet the needed threshold for approval.

GEOGRAPHY

Location: Oceania, group of three atolls in the South Pacific Ocean, about one-half of the way from Hawaii to New Zealand
Geographic coordinates: 9 00 S, 172 00 W

Map references: Oceania
Area: *total:* 12 sq km
country comparison to the world: 240
land: 12 sq km
water: 0 sq km
Area—comparative: about 17 times the size of The Mall in Washington, DC
Land boundaries: 0 km
Coastline: 101 km
Maritime claims: territorial sea: 12 nm
exclusive economic zone: 200 nm
Climate: tropical; moderated by trade winds (April to November)
Terrain: low-lying coral atolls enclosing large lagoons
Elevation extremes: *lowest point:* Pacific Ocean 0 m
highest point: unnamed location 5 m
Natural resources: NEGL
Land use: *arable land:* 0% (soil is thin and infertile)
permanent crops: 0%
other: 100% (2005)
Irrigated land: NA
Natural hazards: lies in Pacific typhoon belt
Environment—current issues: limited natural resources and overcrowding are contributing to emigration to New Zealand
Geography—note: consists of three atolls (Atafu, Fakaofo, Nukunonu), each with a lagoon surrounded by a number of reef-bound islets of varying length and rising to over 3 m above sea level

PEOPLE

Population: 1,384 (July 2011 est.)
country comparison to the world: 233
Age structure: *0–14 years:* 42%
15–64 years: 53%
65 years and over: 5% (2009 est.)
Population growth rate: -0.011% (2011 est.)
country comparison to the world: 198
Birth rate: NA
Death rate: NA
Net migration rate: NA
Urbanization: *urban population:* 0% of total population (2010)
rate of urbanization: 0% annual rate of change (2010-15 est.)
Sex ratio: NA
Infant mortality rate: *total:* NA
male: NA
female: NA
Life expectancy at birth: *total population:* NA
male: NA
female: NA
Total fertility rate: NA
HIV/AIDS—adult prevalence rate: NA
HIV/AIDS—people living with HIV/AIDS: NA
HIV/AIDS—deaths: NA
Drinking water source: *Improved:*
rural: 97% of population
total: 97% of population
Unimproved:
rural: 3% of population
total: 3% of population (2008)
Sanitation facility access: *Improved:*
rural: 93% of population
total: 93% of population
Unimproved:
rural: 7% of population
total: 7% of population (2008)
Nationality: *noun:* Tokelauan(s)
adjective: Tokelauan
Ethnic groups: Polynesian
Religions: Congregational Christian Church 70%, Roman Catholic 28%, other 2%
note: on Atafu, all Congregational Christian Church of Samoa; on Nukunonu, all Roman Catholic; on Fakaofo, both denominations, with the Congregational Christian Church predominant
Languages: Tokelauan (a Polynesian language), English
Literacy: NA
School life expectancy (primary to tertiary education): *total:* 12 years
male: 12 years
female: 13 years (2003)
Education expenditures: NA

GOVERNMENT

Country name: *conventional long form:* none
conventional short form: Tokelau
Dependency status: self-administering territory of New Zealand; note—Tokelau and New Zealand have agreed to a draft constitution as Tokelau moves toward free association with New Zealand; a UN-sponsored referendum on self governance in October 2007 did not produce the two-thirds majority vote necessary for changing the political status
Government type: NA
Capital: none; each atoll has its own administrative center
time difference: UTC-11 (6 hours behind Washington, DC during Standard Time)
Administrative divisions: none (territory of New Zealand)
Independence: none (territory of New Zealand)
National holiday: Waitangi Day (Treaty of Waitangi established British sovereignty over New Zealand), 6 February (1840)
Constitution: administered under the Tokelau Islands Act of 1948; amended in 1970
Legal system: common law system of New Zealand
Suffrage:
21 years of age; universal
Executive branch: *chief of state:* Queen ELIZABETH II (since 6 February 1952); represented by Governor General of New Zealand Anand SATYANAND (since 23 August 2006); New Zealand is represented by Administrator David PAYTON (since 17 October 2006)
head of government: Foua TOLOA (since 21 February 2009); note—position rotates annually among the 3 Faipule (village leaders)
cabinet: the Council for the Ongoing Government of Tokelau, consisting of 3 Faipule (village leaders) and 3 Pulenuku (village mayors), functions as a cabinet (For more information visit the World Leaders website)
elections: the monarchy is hereditary; governor general appointed by the monarch; administrator appointed by the Minister of Foreign Affairs and Trade in New Zealand; the head of government chosen from the Council of Faipule and serves a one-year term
Legislative branch: unicameral General Fono (20 seats; members elected by popular vote to serve three-year terms based upon proportional representation from the three islands—Atafu has 7 seats, Fakaofo has 7 seats, Nukunonu has 6 seats); note—the Tokelau Amendment Act of 1996 confers limited legislative power to the General Fono
elections: last held on 17-19 January 2008 (next to be held in 2011)
election results: independents 20
Judicial branch: Supreme Court in New Zealand exercises civil and criminal jurisdiction in Tokelau
Political parties and leaders:
none
Political pressure groups and leaders:
none
International organization participation: PIF (observer), SPC, UNESCO (associate), UPU
Diplomatic representation in the US: none (territory of New Zealand)
Diplomatic representation from the US: none (territory of New Zealand)
Flag description: a yellow stylized Tokelauan canoe on a dark blue field sails toward the manu—the Southern Cross constellation of four, white, five-pointed stars at the hoist side; the Southern Cross represents the role of Christianity in Tokelauan culture and symbolizes the country's navigating into the future, the color yellow indicates happiness and peace, and the blue field represents the ocean on which the community relies
National anthem: *name:* "Te Atua" (For the Almighty)
lyrics/music: unknown/Falani KALOLO
note: adopted 2008; in preparation for eventual self governance, Tokelau held a national contest to choose an anthem; as a territory of New Zealand, "God Defend New Zealand" and "God Save the Queen" are official (see New Zealand)

ECONOMY

Economy—overview: Tokelau's small size (three villages), isolation, and lack of resources greatly restrain economic development and confine agriculture to the subsistence level. The people rely heavily on aid from New Zealand—about $10 million annually in 2008 and 2009—to maintain public services. New Zealand's support amounts to 80% of Tokelau's recurrent government budget. An international trust fund, currently worth nearly US$32 million, was established in 2004 to provide Tokelau an independent source of revenue. The principal sources of revenue come from sales of copra, postage stamps, souvenir coins, and handicrafts. Money is also remitted to families from relatives in New Zealand.
GDP (purchasing power parity): $1.5 million (1993 est.)
country comparison to the world: 227
GDP (official exchange rate):$NA
GDP—real growth rate: NA%

GDP—per capita (PPP): $1,000 (1993 est.)
country comparison to the world: 215
GDP—composition by sector: *agriculture:* NA%
industry: NA%
services: NA%
Labor force: 440 (2001)
country comparison to the world: 228
Unemployment rate: NA%
Population below poverty line: NA%
Budget:
revenues: $430,800
expenditures: $2.8 million (1987 est.)
Inflation rate (consumer prices): NA%
Agriculture—products: coconuts, copra, breadfruit, papayas, bananas; pigs, poultry, goats; fish
Industries: small-scale enterprises for copra production, woodworking, plaited craft goods; stamps, coins; fishing
Electricity—production: NA kWh
Electricity—consumption: NA kWh
Exports: $0 (2002)
country comparison to the world: 222
Exports—commodities: stamps, copra, handicrafts
Imports: $969,200 (2002)
country comparison to the world: 222
Imports—commodities: foodstuffs, building materials, fuel
Exchange rates: New Zealand dollars (NZD) per US dollar– 1.3874 (2010), 1.6002 (2009), 1.4151 (2008), 1.3811 (2007), 1.5408 (2006)

COMMUNICATIONS

Telephones—main lines in use: 300 (2009)
country comparison to the world: 227
Telephone system: *general assessment:* modern satellite-based communications system
domestic: radiotelephone service between islands
international: country code–690; radiotelephone service to Samoa; government-regulated telephone service (TeleTok); satellite earth stations–3
Broadcast media: no broadcast television stations; each atoll operates a radio service that provides shipping news and weather reports (2009)
Internet country code: .tk
Internet hosts: 526 (2010)
country comparison to the world: 178
Internet users: 800 (2008)
country comparison to the world: 215

TRANSPORTATION

Ports and terminals: none; offshore anchorage only

MILITARY

Military—note: defense is the responsibility of New Zealand

TRANSNATIONAL ISSUES

Disputes—international: Tokelau included American Samoa's Swains Island (Olohega) in its 2006 draft independence constitution

TONGA

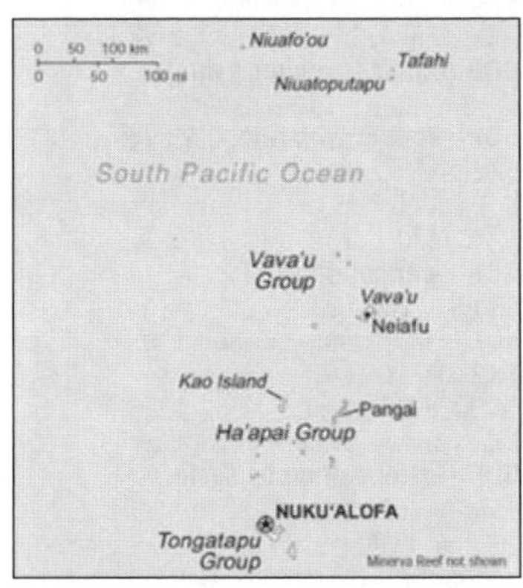

INTRODUCTION

Background: Tonga–unique among Pacific nations–never completely lost its indigenous governance. The archipelagos of "The Friendly Islands" were united into a Polynesian kingdom in 1845. Tonga became a constitutional monarchy in 1875 and a British protectorate in 1900; it withdrew from the protectorate and joined the Commonwealth of Nations in 1970. Tonga remains the only monarchy in the Pacific.

GEOGRAPHY

Location: Oceania, archipelago in the South Pacific Ocean, about two-thirds of the way from Hawaii to New Zealand
Geographic coordinates: 20 00 S, 175 00 W
Map references: Oceania
Area: *total:* 747 sq km
country comparison to the world: 189
land: 717 sq km
water: 30 sq km
Area—comparative: four times the size of Washington, DC
Land boundaries: 0 km
Coastline: 419 km
Maritime claims: territorial sea: 12 nm
exclusive economic zone: 200 nm
continental shelf: 200 m depth or to the depth of exploitation
Climate: tropical; modified by trade winds; warm season (December to May), cool season (May to December)
Terrain: most islands have limestone base formed from uplifted coral formation; others have limestone overlying volcanic base
Elevation extremes: *lowest point:* Pacific Ocean 0 m
highest point: unnamed elevation on Kao Island 1,033 m
Natural resources: fish, fertile soil
Land use: *arable land:* 20%
permanent crops: 14.67%
other: 65.33% (2005)
Irrigated land: NA
Natural hazards: cyclones (October to April); earthquakes and volcanic activity on Fonuafo'ou
volcanism: the Tonga Islands experience volcanic activity; Fonualei (elev. 180 m) has shown frequent activity in recent years, while Niuafo'ou (elev. 260 m), which last erupted in 1985, has forced evacuations; other historically active volcanoes include Late and Tofua
Environment—current issues: deforestation results as more and more land is being cleared for agriculture and settlement; some damage to coral reefs from starfish and indiscriminate coral and shell collectors; overhunting threatens native sea turtle populations
Environment—international agreements: *party to:* Biodiversity, Climate Change, Climate Change-Kyoto Protocol, Desertification, Law of the Sea, Marine Dumping, Marine Life Conservation, Ozone Layer Protection, Ship Pollution
signed, but not ratified: none of the selected agreements
Geography—note: archipelago of 169 islands (36 inhabited)

PEOPLE

Population: 105,916 (July 2011 est.)
country comparison to the world: 191
Age structure: *0–14 years:* 37.2% (male 20,023/female 19,393)
15–64 years: 56.7% (male 30,125/female 29,959)
65 years and over: 6.1% (male 2,986/female 3,430) (2011 est.)
Median age: *total:* 21.4 years
male: 21 years
female: 21.8 years (2011 est.)
Population growth rate: 0.243% (2011 est.)
country comparison to the world: 176
Birth rate: 25.27 births/1,000 population (2011 est.)
country comparison to the world: 60
Death rate: 4.9 deaths/1,000 population (July 2011 est.)
country comparison to the world: 190
Net migration rate: -17.94 migrant(s)/1,000 population (2011 est.)
country comparison to the world: 218
Urbanization: *urban population:* 23% of total population (2010)
rate of urbanization: 0.8% annual rate of change (2010-15 est.)
Sex ratio: *at birth:* 1.03 male(s)/female
under 15 years: 1.04 male(s)/female
15–64 years: 0.99 male(s)/female
65 years and over: 0.7 male(s)/female
total population: 0.99 male(s)/female (2011 est.)
Infant mortality rate: *total:* 13.65 deaths/1,000 live births
country comparison to the world: 125
male: 14.07 deaths/1,000 live births
female: 13.21 deaths/1,000 live births (2011 est.)
Life expectancy at birth:
total population: 75.16 years
country comparison to the world: 87
male: 73.79 years
female: 76.58 years (2011 est.)
Total fertility rate: 3.65 children born/woman (2011 est.)

country comparison to the world: 43
HIV/AIDS—adult prevalence rate: NA
HIV/AIDS—people living with HIV/AIDS: NA
HIV/AIDS—deaths: NA
Drinking water source: *Improved:*
urban: 100% of population
rural: 100% of population
total: 100% of population (2008)
Sanitation facility access: *Improved:*
urban: 98% of population
rural: 96% of population
total: 96% of population
Unimproved: urban: 2% of population
rural: 4% of population
total: 4% of population (2008)
Nationality: *noun:* Tongan(s)
adjective: Tongan
Ethnic groups: Polynesian, Europeans
Religions: Christian (Free Wesleyan Church claims over 30,000 adherents)
Languages: Tongan (official), English (official)
Literacy: *definition:* can read and write Tongan and/or English
total population: 98.9%
male: 98.8%
female: 99% (1999 est.)
School life expectancy (primary to tertiary education):
total: 14 years
male: 14 years
female: 14 years (2007)
Education expenditures: 3.9% of GDP (2004)
country comparison to the world: 107

GOVERNMENT

Country name: *conventional long form:* Kingdom of Tonga
conventional short form: Tonga
local long form: Pule'anga Tonga
local short form: Tonga
former: Friendly Islands
Government type: constitutional monarchy
Capital: *name:* Nuku'alofa
geographic coordinates: 21 08 S, 175 12 W
time difference: UTC+13 (18 hours ahead of Washington, DC during Standard Time)
Administrative divisions: 3 island groups; Ha'apai, Tongatapu, Vava'u
Independence: 4 June 1970 (from UK protectorate)
National holiday: National Day, 4 November (1875)
Constitution: 4 November 1875; revised 1 January 1967
Legal system: English common law
International law organization participation: has not submitted an ICJ jurisdiction declaration; non-party state to the ICCt
Suffrage: 21 years of age; universal
Executive branch: *chief of state:* King George TUPOU V (since 11 September 2006)
head of government: Prime Minister Lord Siale'ataonga TU'IVAKANO (since 22 December 2010)
cabinet: Cabinet is nominated by the prime minister and appointed by the monarch
(For more information visit the World Leaders website)
note: there is also a Privy Council that advises the monarch
elections: the monarchy is hereditary; prime minister and deputy prime minister elected by and from the members of parliament and appointed by the monarch
election results: Lord Siale'ataonga TU'IVAKANO elected by parliament on 21 December 2010 with 14 of 26 votes
Legislative branch: unicameral Legislative Assembly or Fale Alea (26 seats—9 for nobles elected from among the country's 29 nobles, 17 members elected by popular vote to serve four-year terms)
elections: last held on 25 November 2010 (next to be held in 2014)
election results: Peoples Representatives: percent of vote—independents 67.3%, Democratic Party 28.5%; seats—Democratic Party 12, independents 5
Judicial branch: Supreme Court (judges are appointed by the monarch); Court of Appeal (Chief Justice and high court justices from overseas chosen and approved by Privy Council)
Political parties and leaders: Democratic Party of the Friendly Islands [Samuela 'Akilisi POHIVA]; People's Democratic Party or PDP [Tesina FUKO]; Sustainable Nation-Building Party [Sione FONUA]; Tonga Democratic Labor Party [NA]; Tonga Human Rights and Democracy Movement or THRDM [Uliti UATA]
Political pressure groups and leaders: Human Rights and Democracy Movement Tonga or HRDMT [Rev. Simote VEA, chairman]; Public Servant's Association [Finau TUTONE]
International organization participation: ACP, ADB, AOSIS, C, FAO, G-77, IBRD, ICAO, ICRM, IDA, IFAD, IFC, IFRCS, IHO, IMF, IMO, IMSO, Interpol, IOC, ITU, ITUC, OPCW, PIF, Sparteca, SPC, UN, UNCTAD, UNESCO, UNIDO, UPU, WCO, WHO, WIPO, WMO, WTO
Diplomatic representation in the US: *chief of mission:* Ambassador Sonatane Tu'akinamolahi TAUMOEPEAU-TUPOU
chancery: 250 East 51st Street, New York, NY 10022
telephone: [1] (917) 369-1025
FAX: [1] (917) 369-1024
consulate(s) general: San Francisco
Diplomatic representation from the US: the US does not have an embassy in Tonga; the US ambassador to Fiji is accredited to Tonga
Flag description: red with a bold red cross on a white rectangle in the upper hoist-side corner; the cross reflects the deep-rooted Christianity in Tonga; red represents the blood of Christ and his sacrifice; white signifies purity
National anthem: *name:* "Ko e fasi `o e tu"i `o e `Otu Tonga" (Song of the King of the Tonga Islands)
lyrics/music: Uelingatoni Ngu TUPOUMALOHI/Karl Gustavus SCHMITT
note: in use since 1875; the anthem is more commonly known as "Fasi Fakafonua" (National Song)

ECONOMY

Economy—overview: Tonga has a small, open, South Pacific island economy. It has a narrow export base in agricultural goods. Squash, vanilla beans, and yams are the main crops. Agricultural exports, including fish, make up two-thirds of total exports. The country must import a high proportion of its food, mainly from New Zealand. The country remains dependent on external aid and remittances from Tongan communities overseas to offset its trade deficit. Tourism is the second-largest source of hard currency earnings following remittances. Tonga had 39,000 visitors in 2006. The government is emphasizing the development of the private sector, especially the encouragement of investment, and is committing increased funds for health and education. Tonga has a reasonably sound basic infrastructure and well developed social services. High unemployment among the young, a continuing upturn in inflation, pressures for democratic reform, and rising civil service expenditures are major issues facing the government.
GDP (purchasing power parity): $751 million (2010 est.)
country comparison to the world: 208
$749.1 million (2009 est.)
$750.9 million (2008 est.)
note: data are in 2010 US dollars
GDP (official exchange rate): $363 million (2010 est.)
GDP—real growth rate: 0.3% (2010 est.)
country comparison to the world: 184
-0.3% (2009 est.)
1.3% (2008 est.)
GDP—per capita (PPP): $6,100 (2010 est.)
country comparison to the world: 139
$6,200 (2009 est.)
$6,300 (2008 est.)
note: data are in 2010 US dollars
GDP—composition by sector:
agriculture: 25%
industry: 17%
services: 57% (FY05/06 est.)
Labor force: 39,960 (2007)
country comparison to the world: 194
Labor force—by occupation:
agriculture: 31.8%
industry: 30.6%
services: 2,003% (2003 est.)
Unemployment rate: 13% (FY03/04 est.)
country comparison to the world: 134
Population below poverty line: 24% (FY03/04)
Household income or consumption by percentage share: *lowest 10%:* NA%
highest 10%: NA%
Budget:
revenues: $80.48 million
expenditures: $109.8 million (FY07/08 est.)
Inflation rate (consumer prices): 5.9% (2007 est.)
country comparison to the world: 156
Commercial bank prime lending rate: 12.47% (31 December 2009 est.)
country comparison to the world: 63
12.46% (31 December 2008 est.)
Stock of narrow money: $44.64 million (31 December 2009)
country comparison to the world: 184
$36.16 million (31 December 2008)
Stock of broad money: $153.8 million (31 December 2009)

country comparison to the world: 184
$136.9 million (31 December 2008)
Stock of domestic credit: $149.2 million (31 December 2008 est.)
country comparison to the world: 179
$163.1 million (31 December 2007 est.)
Market value of publicly traded shares: $NA
Agriculture—products: squash, coconuts, copra, bananas, vanilla beans, cocoa, coffee, ginger, black pepper; fish
Industries: tourism, construction, fishing
Industrial production growth rate: 1% (2003 est.)
country comparison to the world: 145
Electricity—production: 43 million kWh (2007 est.)
country comparison to the world: 200
Electricity—consumption: 39.99 million kWh (2007 est.)
country comparison to the world: 199
Electricity—exports: 0 kWh (2008)
Electricity—imports: 0 kWh (2008 est.)
Oil—production: 0 bbl/day (2009 est.)
country comparison to the world: 132
Oil—consumption: 1,000 bbl/day (2009 est.)
country comparison to the world: 190
Oil—exports: 0 bbl/day (2007 est.)
country comparison to the world: 204
Oil—imports: 1,173 bbl/day (2007 est.)
country comparison to the world: 186
Oil—proved reserves: 0 bbl (1 January 2010 est.)
country comparison to the world: 194
Natural gas—production: 0 cu m (2008 est.)
country comparison to the world: 190
Natural gas—consumption: 0 cu m (2008 est.)
country comparison to the world: 127
Natural gas—exports: 0 cu m (2008 est.)
country comparison to the world: 185
Natural gas—imports: 0 cu m (2008 est.)
country comparison to the world: 124
Natural gas—proved reserves: 0 cu m (1 January 2010 est.)
country comparison to the world: 193
Current account balance: $-23 million (2007 est.)
country comparison to the world: 66
Exports: $22 million (2006)
country comparison to the world: 206
Exports—commodities: squash, fish, vanilla beans, root crops
Exports—partners: Hong Kong 25.42%, US 22.65%, Japan 12.21%, NZ 7.31%, Fiji 7.2%, Samoa 6.06%, South Korea 4.48% (2009)
Imports: $139 million (2006)
country comparison to the world: 207
Imports—commodities: foodstuffs, machinery and transport equipment, fuels, chemicals
Imports—partners: Fiji 34.37%, NZ 25.03%, US 9.43%, Australia 7.53%, China 5.64% (2009)
Reserves of foreign exchange and gold: $40.83 million (FY04/05)
country comparison to the world: 138
Debt—external: $80.7 million (2004)
country comparison to the world: 182
Exchange rates: pa'anga (TOP) per US dollar–NA (2007), 2.0277 (2006), 1.96 (2005), 1.9716 (2004), 2.142 (2003)

COMMUNICATIONS

Telephones—main lines in use: 31,000 (2009)
country comparison to the world: 180
Telephones—mobile cellular: 53,000 (2009)
country comparison to the world: 196
Telephone system: *general assessment:* competition between Tonga Telecommunications Corporation (TCC) and Shoreline Communications Tonga (SCT) is accelerating expansion of telecommunications; SCT granted approval to introduce high-speed digital service for telephone, Internet, and television while TCC has exclusive rights to operate the mobile-phone network; international telecom services are provided by government-owned Tonga Telecommunications International (TTI)
domestic: combined fixed-line and mobile-cellular teledensity about 70 telephones per 100 persons; fully automatic switched network
international: country code–676; satellite earth station–1 Intelsat (Pacific Ocean) (2009)
Broadcast media: 2 state-owned television stations and 2 privately-owned stations; satellite and cable TV services are available; 2 state-owned and 3 privately-owned radio stations; Radio Australia broadcasts obtainable via a satellite feed (2009)
Internet country code: .to
Internet hosts: 20,847 (2010)
country comparison to the world: 110
Internet users: 8,400 (2009)
country comparison to the world: 202

TRANSPORTATION

Airports: 6 (2010)
country comparison to the world: 176
Airports—with paved runways: *total:* 1
2,438 to 3,047 m: 1 (2010)
Airports—with unpaved runways: *total:* 5
1,524 to 2,437 m: 1
914 to 1,523 m: 3
under 914 m: 1 (2010)
Roadways: *total:* 680 km
country comparison to the world: 188
paved: 184 km
unpaved: 496 km (2000)
Merchant marine: *total:* 10
country comparison to the world: 114
by type: bulk carrier 1, cargo 6, carrier 1, liquefied gas 1, passenger/cargo 1
foreign-owned: 3 (Australia 1, Switzerland 1, UK 1) (2010)
Ports and terminals: Nuku'alofa, Neiafu, Pangai

MILITARY

Military branches: Tonga Defense Services (TDS): Land Force (Royal Guard), Maritime Force (includes Royal Marines, Air Wing) (2010)
Military service age and obligation: 16 years of age for voluntary enlistment (with parental consent); no conscription (2010)
Manpower available for military service: *males age 16–49:* 24,460
females age 16–49: 24,041 (2010 est.)
Manpower fit for military service: *males age 16–49:* 20,956
females age 16–49: 20,577 (2010 est.)
Manpower reaching militarily significant age annually: *male:* 1,196
female: 1,134 (2010 est.)
Military expenditures: 0.9% of GDP (2006 est.)
country comparison to the world: 137

TRANSNATIONAL ISSUES

Disputes—international: none

TRINIDAD AND TOBAGO

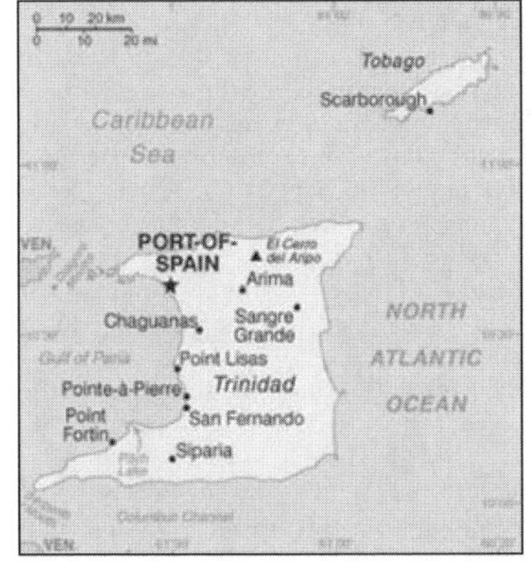

INTRODUCTION

Background: First colonized by the Spanish, the islands came under British control in the early 19th century. The islands' sugar industry was hurt by the emancipation of the slaves in 1834. Manpower was replaced with the importation of contract laborers from India between 1845 and 1917, which boosted sugar production as well as the cocoa industry. The discovery of oil on Trinidad in 1910 added another important export. Independence was attained in 1962. The country is one of the most prosperous in the Caribbean thanks largely to petroleum and natural gas production and processing. Tourism, mostly in Tobago, is targeted for expansion and is growing. The government is coping with a rise in violent crime.

GEOGRAPHY

Location: Caribbean, islands between the Caribbean Sea and the North Atlantic Ocean, northeast of Venezuela
Geographic coordinates: 11 00 N, 61 00 W
Map references: Central America and the Caribbean
Area: *total:* 5,128 sq km
country comparison to the world: 173
land: 5,128 sq km
water: 0 sq km
Area—comparative: slightly smaller than Delaware

Land boundaries: 0 km
Coastline: 362 km
Maritime claims: measured from claimed archipelagic baselines
territorial sea: 12 nm
contiguous zone: 24 nm
exclusive economic zone: 200 nm
continental shelf: 200 nm or to the outer edge of the continental margin
Climate: tropical; rainy season (June to December)
Terrain: mostly plains with some hills and low mountains
Elevation extremes: *lowest point:* Caribbean Sea 0 m
highest point: El Cerro del Aripo 940 m
Natural resources: petroleum, natural gas, asphalt
Land use: *arable land:* 14.62%
permanent crops: 9.16%
other: 76.22% (2005)
Irrigated land: 40 sq km (2003)
Total renewable water resources: 3.8 cu km (2000)
Freshwater withdrawal (domestic/industrial/agricultural): *total:* 0.31 cu km/yr (68%/26%/6%)
per capita: 237 cu m/yr (2000)
Natural hazards: outside usual path of hurricanes and other tropical storms
Environment—current issues: water pollution from agricultural chemicals, industrial wastes, and raw sewage; oil pollution of beaches; deforestation; soil erosion
Environment—international agreements: *party to:* Biodiversity, Climate Change, Climate Change-Kyoto Protocol, Desertification, Endangered Species, Hazardous Wastes, Law of the Sea, Marine Dumping, Marine Life Conservation, Ozone Layer Protection, Ship Pollution, Tropical Timber 83, Tropical Timber 94, Wetlands
signed, but not ratified: none of the selected agreements
Geography—note: Pitch Lake, on Trinidad's southwestern coast, is the world's largest natural reservoir of asphalt

PEOPLE

Population: 1,227,505 (July 2011 est.)
country comparison to the world: 155
Age structure: *0–14 years:* 19.5% (male 122,044/female 116,859)
15–64 years: 72.1% (male 455,148/female 429,990)
65 years and over: 8.4% (male 44,439/female 59,025) (2011 est.)
Median age: *total:* 33.1 years
male: 32.6 years
female: 33.6 years (2011 est.)
Population growth rate: -0.087% (2011 est.)
country comparison to the world: 205
Birth rate: 14.35 births/1,000 population (2011 est.)
country comparison to the world: 143
Death rate: 8.29 deaths/1,000 population (July 2011 est.)
country comparison to the world: 90
Net migration rate: -6.93 migrant(s)/1,000 population (2011 est.)
country comparison to the world: 202
Urbanization: *urban population:* 14% of total population (2010)
rate of urbanization: 3% annual rate of change (2010-15 est.)
Major cities—population: PORT-OF-SPAIN (capital) 57,000 (2009)
Sex ratio: *at birth:* 1.028 male(s)/female
under 15 years: 1.05 male(s)/female
15–64 years: 1.06 male(s)/female
65 years and over: 0.75 male(s)/female
total population: 1.02 male(s)/female (2011 est.)
Infant mortality rate: *total:* 27.69 deaths/1,000 live births
country comparison to the world: 73
male: 28.93 deaths/1,000 live births
female: 26.41 deaths/1,000 live births (2011 est.)
Life expectancy at birth: *total population:* 71.37 years
country comparison to the world: 135
male: 68.51 years
female: 74.3 years (2011 est.)
Total fertility rate: 1.72 children born/woman (2011 est.)
country comparison to the world: 166
HIV/AIDS—adult prevalence rate: 1.5% (2009 est.)
country comparison to the world: 35
HIV/AIDS—people living with HIV/AIDS: 15,000 (2009 est.)
country comparison to the world: 86
HIV/AIDS—deaths: fewer than 1,000 (2009 est.)
country comparison to the world: 79
Drinking water source:
Improved:
urban: 98% of population
rural: 93% of population
total: 34% of population
Unimproved: urban: 2% of population
rural: 7% of population
total: 6% of population (2008)
Sanitation facility access: *Improved:*
urban: 92% of population
rural: 92% of population
total: 92% of population
Unimproved:
urban: 8% of population
rural: 8% of population
total: 8% of population (2008)
Nationality: *noun:* Trinidadian(s), Tobagonian(s)
adjective: Trinidadian, Tobagonian
Ethnic groups: Indian (South Asian) 40%, African 37.5%, mixed 20.5%, other 1.2%, unspecified 0.8% (2000 census)
Religions: Roman Catholic 26%, Hindu 22.5%, Anglican 7.8%, Baptist 7.2%, Pentecostal 6.8%, Muslim 5.8%, Seventh Day Adventist 4%, other Christian 5.8%, other 10.8%, unspecified 1.4%, none 1.9% (2000 census)
Languages: English (official), Caribbean Hindustani (a dialect of Hindi), French, Spanish, Chinese
Literacy: *definition:* age 15 and over can read and write
total population: 98.6%
male: 99.1%
female: 98% (2003 est.)
School life expectancy (primary to tertiary education): *total:* 12 years
male: 11 years
female: 12 years (2007)
Education expenditures: NA
People—note: in 2007, the government of Trinidad and Tobago estimated the population to be 1.3 million

GOVERNMENT

Country name: *conventional long form:* Republic of Trinidad and Tobago
conventional short form: Trinidad and Tobago
Government type: parliamentary democracy
Capital: *name:* Port-of-Spain
geographic coordinates: 10 39 N, 61 31 W
time difference: UTC-4 (1 hour ahead of Washington, DC during Standard Time)
Administrative divisions: 9 regional corporations, 2 city corporations, 3 borough corporations, 1 ward
regional corporations: Couva/Tabaquite/Talparo, Diego Martin, Mayaro/Rio Claro, Penal/Debe, Princes Town, Sangre Grande, San Juan/Laventille, Siparia, Tunapuna/Piarco
city corporations: Port-of-Spain, San Fernando
borough corporations: Arima, Chaguanas, Point Fortin
ward: Tobago
Independence: 31 August 1962 (from the UK)
National holiday: Independence Day, 31 August (1962)
Constitution: 1 August 1976
Legal system: English common law; judicial review of legislative acts in the Supreme Court
International law organization participation: has not submitted an ICJ jurisdiction declaration; accepts ICCt jurisdiction
Suffrage: 18 years of age; universal
Executive branch: *chief of state:* President George Maxwell RICHARDS (since 17 March 2003)
head of government: Prime Minister Kamla PERSAD-BISSESSAR (since 26 May 2010)
cabinet: Cabinet appointed from among the members of Parliament
(For more information visit the World Leaders website)
elections: president elected by an electoral college, which consists of the members of the Senate and House of Representatives, for a five-year term (eligible for a second term); election last held on 11 February 2008 (next to be held by February 2013); the president usually appoints as prime minister the leader of the majority party in the House of Representatives
election results: George Maxwell RICHARDS reelected president; percent of electoral college vote—NA
Legislative branch: bicameral Parliament consists of the Senate (31 seats; 16 members appointed by the ruling party, 9 by the President, 6 by the opposition party to serve a maximum term of five years) and the House of Representatives (41 seats; members are elected by popular vote to serve five-year terms)
elections: House of Representatives—last held on 24 May 2010 (next to be held in 2015)
election results: House of Representatives—percent of vote—NA; seats by party—UNC 21, PNM 12, COP 6, TOP 2
note: Tobago has a unicameral House of Assembly with 12 members serving four-

year terms; last election held in January 2005; seats by party–PNM 11, DAC 1
Judicial branch: Supreme Court of Judicature (comprised of the High Court of Justice and the Court of Appeals; the chief justice is appointed by the president after consultation with the prime minister and the leader of the opposition; other justices are appointed by the president on the advice of the Judicial and Legal Service Commission); the highest court of appeal is the Privy Council in London; member of the Caribbean Court of Justice (CCJ)
Political parties and leaders: Congress of the People or COP [Winston DOOKERAN]; Democratic Action Congress or DAC [Hochoy CHARLES] (only active in Tobago); Democratic National Alliance or DNA [Gerald YETMING] (coalition of NAR, DDPT, MND); Movement for National Development or MND [Garvin NICHOLAS]; National Alliance for Reconstruction or NAR [Dr. Carson CHARLES]; People's National Movement or PNM [Patrick MANNING]; Tobago Organization of the People or TOP [Ashworth JACK]; United National Congress or UNC [Kamla PERSAD-BISSESSAR]
Political pressure groups and leaders: Jamaat-al Muslimeen [Yasin ABU BAKR]
International organization participation: ACP, AOSIS, C, Caricom, CDB, FAO, G-24, G-77, IADB, IBRD, ICAO, ICRM, IDA, IFAD, IFC, IFRCS, IHO, ILO, IMF, IMO, Interpol, IOC, IOM, ISO, ITSO, ITU, ITUC, LAES, MIGA, NAM, OAS, OPANAL, OPCW, Paris Club (associate), UN, UNCTAD, UNESCO, UNIDO, UPU, WCO, WFTU, WHO, WIPO, WMO, WTO
Diplomatic representation in the US: *chief of mission:* Ambassador Neil PARSAN
chancery: 1708 Massachusetts Avenue NW, Washington, DC 20036
telephone: [1] (202) 467-6490
FAX: [1] (202) 785-3130
consulate(s) general: Miami, New York
Diplomatic representation from the US: *chief of mission:* Ambassador Beatrice W. WELTERS
embassy: 15 Queen's Park West, Port-of-Spain
mailing address: P. O. Box 752, Port-of-Spain
telephone: [1] (868) 622-6371 through 6376
FAX: [1] (868) 822-5905
Flag description: red with a white-edged black diagonal band from the upper hoist side to the lower fly side; the colors represent the elements of earth, water, and fire; black stands for the wealth of the land and the dedication of the people; white symbolizes the sea surrounding the islands, the purity of the country's aspirations, and equality; red symbolizes the warmth and energy of the sun, the vitality of the land, and the courage and friendliness of its people
National anthem: *name:* "Forged From the Love of Liberty"
lyrics/music: Patrick Stanislaus CASTAGNE
note: adopted 1962; the song was originally created to serve as an anthem for the West Indies Federation; it was adopted by Trinidad and Tobago following the Federation's dissolution in 1962

ECONOMY

Economy—overview: Trinidad and Tobago has earned a reputation as an excellent investment site for international businesses and has one of the highest growth rates and per capita incomes in Latin America. Economic growth between 2000 and 2007 averaged slightly over 8%, significantly above the regional average of about 3.7% for that same period; however, GDP has slowed down since then and contracted about 3.5% in 2009, before rising more than 2% in 2010. Growth has been fueled by investments in liquefied natural gas (LNG), petrochemicals, and steel. Additional petrochemical, aluminum, and plastics projects are in various stages of planning. Trinidad and Tobago is the leading Caribbean producer of oil and gas, and its economy is heavily dependent upon these resources but it also supplies manufactured goods, notably food products and beverages, as well as cement to the Caribbean region. Oil and gas account for about 40% of GDP and 80% of exports, but only 5% of employment. The country is also a regional financial center, and tourism is a growing sector, although it is not as important domestically as it is to many other Caribbean islands. The economy benefits from a growing trade surplus. The previous MANNING administration benefited from fiscal surpluses fueled by the dynamic export sector; however, declines in oil and gas prices have reduced government revenues which will challenge the new government's commitment to maintaining high levels of public investment.
GDP (purchasing power parity): $26.1 billion (2010 est.)
country comparison to the world: 112
$26.09 billion (2009 est.)
$27.05 billion (2008 est.)
note: data are in 2010 US dollars
GDP (official exchange rate): $20.59 billion (2010 est.)
GDP—real growth rate: 0% (2010 est.)
country comparison to the world: 190
-3.5% (2009 est.)
2.4% (2008 est.)
GDP—per capita (PPP): $21,200 (2010 est.)
country comparison to the world: 61
$21,200 (2009 est.)
$22,000 (2008 est.)
note: data are in 2010 US dollars
GDP—composition by sector:
agriculture: 0.5%
industry: 59.4%
services: 40.1% (2010 est.)
Labor force: 631,000 (2010 est.)
country comparison to the world: 153
Labor force—by occupation:
agriculture: 3.8%
manufacturing, mining, and quarrying: 12.8%
construction and utilities: 20.4%
services: 62.9% (2007 est.)
Unemployment rate: 6.4% (2010 est.)
country comparison to the world: 62
5.8% (2009 est.)
Population below poverty line: 17% (2007 est.)
Household income or consumption by percentage share:
lowest 10%: NA%
highest 10%: NA%
Investment (gross fixed): 11.6% of GDP (2010 est.)
country comparison to the world: 146
Budget: *revenues:* $6.614 billion
expenditures: $7.24 billion (2010 est.)
Public debt: 26.4% of GDP (2010 est.)
country comparison to the world: 94
29% of GDP (2009 est.)
Inflation rate (consumer prices): 11.3% (2010 est.)
country comparison to the world: 200
7% (2009 est.)
Central bank discount rate: 7.25% (31 December 2009)
country comparison to the world: 39
10.75% (31 December 2008)
Commercial bank prime lending rate: 11.94% (31 December 2009 est.)
country comparison to the world: 64
12.44% (31 December 2008 est.)
Stock of narrow money: $3.734 billion (31 December 2010 est.)
country comparison to the world: 103
$3.407 billion (31 December 2009 est.)
Stock of broad money: $12.47 billion (31 December 2010 est.)
country comparison to the world: 94
$11.35 billion (31 December 2009 est.)
Stock of domestic credit: $2.924 billion (31 December 2010 est.)
country comparison to the world: 120
$2.823 billion (31 December 2009 est.)
Market value of publicly traded shares: $11.15 billion (31 December 2009)
country comparison to the world: 65
$12.16 billion (31 December 2008)
$15.61 billion (31 December 2007)
Agriculture—products: cocoa, rice, citrus, coffee, vegetables; poultry
Industries: petroleum and petroleum products, liquefied natural gas (LNG), methanol, ammonia, urea, steel products, beverages, food processing, cement, cotton textiles
Industrial production growth rate: 2.5% (2010 est.)
country comparison to the world: 121
Electricity—production: 7.202 billion kWh (2007 est.)
country comparison to the world: 100
Electricity—consumption: 7.034 billion kWh (2007 est.)
country comparison to the world: 99
Electricity—exports: 0 kWh (2008 est.)
Electricity—imports: 0 kWh (2008 est.)
Oil—production: 151,600 bbl/day (2009 est.)
country comparison to the world: 46
Oil—consumption: 43,000 bbl/day (2009 est.)
country comparison to the world: 99
Oil—exports: 248,300 bbl/day (2007 est.)
country comparison to the world: 48
Oil—imports: 92,480 bbl/day (2007 est.)
country comparison to the world: 64
Oil—proved reserves: 728.3 million bbl (1 January 2010 est.)
country comparison to the world: 44

Natural gas—production: 39.3 billion cu m (2008 est.)
country comparison to the world: 21
Natural gas—consumption: 21.94 billion cu m (2008 est.)
country comparison to the world: 32
Natural gas—exports: 17.36 billion cu m (2008 est.)
country comparison to the world: 13
Natural gas—imports: 0 cu m (2008 est.)
country comparison to the world: 122
Natural gas—proved reserves: 436.1 billion cu m (1 January 2010 est.)
country comparison to the world: 34
Current account balance: $3.363 billion (2010 est.)
country comparison to the world: 38
$1.702 billion (2009 est.)
Exports: $12.06 billion (2010 est.)
country comparison to the world: 80
$9.312 billion (2009 est.)
Exports—commodities: petroleum and petroleum products, liquefied natural gas (LNG), methanol, ammonia, urea, steel products, beverages, cereal and cereal products, sugar, cocoa, coffee, citrus fruit, vegetables, flowers
Exports—partners: US 38.53%, Jamaica 8.86%, Spain 6.88%, Mexico 6.23% (2009)
Imports: $8.234 billion (2010 est.)
country comparison to the world: 98
$7.161 billion (2009 est.)
Imports—commodities: mineral fuels, lubricants, machinery, transportation equipment, manufactured goods, food, chemicals, live animals
Imports—partners: US 30.87%, Colombia 7.1%, Venezuela 7.01%, Russia 6.64%, Brazil 5.53%, China 4.19% (2009)
Reserves of foreign exchange and gold: $9.659 billion (31 December 2010 est.)
country comparison to the world: 60
$9.246 billion (31 December 2009 est.)
Debt—external: $4.303 billion (31 December 2010 est.)
country comparison to the world: 112
$3.895 billion (31 December 2009 est.)
Stock of direct foreign investment—at home: $102 billion (31 December 2008 est.)
country comparison to the world: 33
$12.44 billion (2007)
Stock of direct foreign investment—abroad: $3.829 billion (2007)
country comparison to the world: 62
Exchange rates: Trinidad and Tobago dollars (TTD) per US dollar–6.3337 (2010), 6.3099 (2009), 6.2896 (2008), 6.3275 (2007), 6.3107 (2006)

COMMUNICATIONS

Telephones—main lines in use: 314,800 (2009)
country comparison to the world: 112
Telephones—mobile cellular: 1.97 million (2009)
country comparison to the world: 133
Telephone system: *general assessment:* excellent international service; good local service
domestic: mobile-cellular teledensity roughly 185 telephones per 100 persons
international: country code–1-868; submarine cable systems provide connectivity to US and parts of the Caribbean and South America; satellite earth station–1 Intelsat (Atlantic Ocean); tropospheric scatter to Barbados and Guyana
Broadcast media: 5 TV networks each broadcasting on multiple stations; one of the networks is state-owned; multiple cable TV subscription service providers; multiple radio networks, one state-owned, broadcast over about 35 stations (2007)
Internet country code: .tt
Internet hosts: 168,876 (2010)
country comparison to the world: 68
Internet users: 593,000 (2009)
country comparison to the world: 115

TRANSPORTATION

Airports: 6 (2010)
country comparison to the world: 172
Airports—with paved runways: *total:* 3
over 3,047 m: 1
2,438 to 3,047 m: 1
1,524 to 2,437 m: 1 (2010)
Airports—with unpaved runways: *total:* 3
914 to 1,523 m: 1
under 914 m: 2 (2010)
Pipelines: gas 671 km; oil 334 km (2010)
Roadways: *total:* 8,320 km
country comparison to the world: 140
paved: 4,252 km
unpaved: 4,068 km (2000)
Merchant marine: *total:* 6
country comparison to the world: 128
by type: passenger 1, passenger/cargo 4, petroleum tanker 1
registered in other countries: 2 (Bahamas 1, unknown 1) (2010)
Ports and terminals: Point Fortin, Point Lisas, Port-of-Spain, Scarborough

MILITARY

Military branches: Trinidad and Tobago Defense Force (TTDF): Trinidad and Tobago Army, Coast Guard, Air Guard, Trinidad and Tobago Police Service (2010)
Military service age and obligation: 18 years of age for voluntary military service (16 years of age with parental consent); no conscription (2010)
Manpower available for military service: *males age 16–49:* 341,764
females age 16–49: 317,899 (2010 est.)
Manpower fit for military service: *males age 16–49:* 269,824
females age 16–49: 261,735 (2010 est.)
Manpower reaching militarily significant age annually: *male:* 8,164
female: 7,503 (2010 est.)
Military expenditures:
0.3% of GDP (2006)
country comparison to the world: 168

TRANSNATIONAL ISSUES

Disputes—international: Barbados and Trinidad and Tobago abide by the April 2006 Permanent Court of Arbitration decision delimiting a maritime boundary and limiting catches of flying fish in Trinidad and Tobago's exclusive economic zone; in 2005, Barbados and Trinidad and Tobago agreed to compulsory international arbitration under UNCLOS challenging whether the northern limit of Trinidad and Tobago's and Venezuela's maritime boundary extends into Barbadian waters; Guyana has also expressed its intention to include itself in the arbitration as the Trinidad and Tobago-Venezuela maritime boundary may extend into its waters as well
Illicit drugs: transshipment point for South American drugs destined for the US and Europe; producer of cannabis

TUNISIA

INTRODUCTION

Background: Rivalry between French and Italian interests in Tunisia culminated in a French invasion in 1881 and the creation of a protectorate. Agitation for independence in the decades following World War I was finally successful in getting the French to recognize Tunisia as an independent state in 1956. The country's first president, Habib BOURGUIBA, established a strict one-party state. He dominated the country for 31 years, repressing Islamic fundamentalism and establishing rights for women unmatched by any other Arab nation. In November 1987, BOURGUIBA was removed from office and replaced by Zine el Abidine BEN ALI in a bloodless coup. Street protests that began in Tunis in December 2010 over high unemployment, corruption, widespread poverty, and high food prices escalated in January 2011, culminating in rioting that led to hundreds of deaths. On 14 January 2011, the same day BEN ALI dismissed the government, he fled the country, and by late January 2011, Prime Minister Mohamed GHANNOUCHI announced the formation of a "national unity government" with the head of the Chamber of Deputies, Fouad M'BAZAA, as the interim president.

GEOGRAPHY

Location: Northern Africa, bordering the Mediterranean Sea, between Algeria and Libya
Geographic coordinates: 34 00 N, 9 00 E
Map references: Africa
Area: *total:* 163,610 sq km
country comparison to the world: 92
land: 155,360 sq km
water: 8,250 sq km
Area—comparative: slightly larger than Georgia
Land boundaries: *total:* 1,424 km
border countries: Algeria 965 km, Libya 459 km
Coastline: 1,148 km

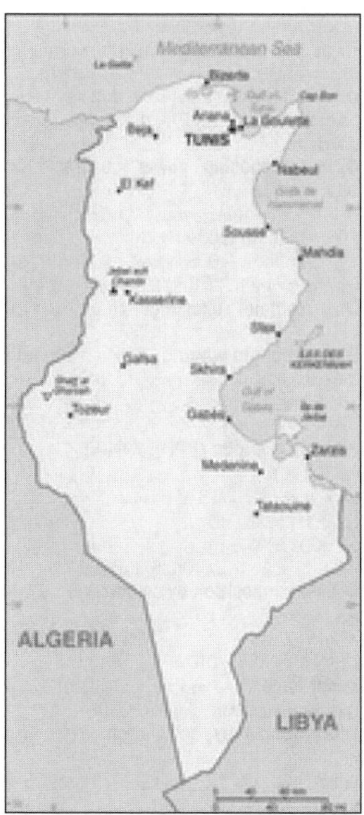

Maritime claims: territorial sea: 12 nm
contiguous zone: 24 nm
exclusive economic zone: 12 nm
Climate: temperate in north with mild, rainy winters and hot, dry summers; desert in south
Terrain: mountains in north; hot, dry central plain; semiarid south merges into the Sahara
Elevation extremes: *lowest point:* Shatt al Gharsah -17 m
highest point: Jebel ech Chambi 1,544 m
Natural resources: petroleum, phosphates, iron ore, lead, zinc, salt
Land use: *arable land:* 17.05%
permanent crops: 13.08%
other: 69.87% (2005)
Irrigated land: 3,940 sq km (2003)
Total renewable water resources: 4.6 cu km (2003)
Freshwater withdrawal (domestic/industrial/agricultural): *total:* 2.64 cu km/yr (14%/4%/82%)
per capita: 261 cu m/yr (2000)
Natural hazards: NA
Environment—current issues: toxic and hazardous waste disposal is ineffective and poses health risks; water pollution from raw sewage; limited natural freshwater resources; deforestation; overgrazing; soil erosion; desertification
Environment—international agreements: *party to:* Biodiversity, Climate Change, Climate Change-Kyoto Protocol, Desertification, Endangered Species, Environmental Modification, Hazardous Wastes, Law of the Sea, Marine Dumping, Ozone Layer Protection, Ship Pollution, Wetlands
signed, but not ratified: Marine Life Conservation
Geography—note: strategic location in central Mediterranean; Malta and Tunisia are discussing the commercial exploitation of the continental shelf between their countries, particularly for oil exploration

PEOPLE

Population: 10,629,186 (July 2011 est.)
country comparison to the world: 78
Age structure: *0–14 years:* 23.2% (male 1,274,348/female 1,193,131)
15–64 years: 69.3% (male 3,638,014/female 3,728,294)
65 years and over: 7.5% (male 390,055/female 405,344) (2011 est.)
Median age: *total:* 30 years
male: 29.6 years
female: 30.4 years (2011 est.)
Population growth rate: 0.978% (2011 est.)
country comparison to the world: 118
Birth rate: 17.4 births/1,000 population (2011 est.)
country comparison to the world: 117
Death rate: 5.83 deaths/1,000 population (July 2011 est.)
country comparison to the world: 171
Net migration rate: -1.79 migrant(s)/1,000 population (2011 est.)
country comparison to the world: 160
Urbanization: *urban population:* 67% of total population (2010)
rate of urbanization: 1.5% annual rate of change (2010-15 est.)
Major cities—population: TUNIS (capital) 759,000 (2009)
Sex ratio: *at birth:* 1.07 male(s)/female
under 15 years: 1.07 male(s)/female
15–64 years: 1.01 male(s)/female
65 years and over: 0.86 male(s)/female
total population: 1.01 male(s)/female (2011 est.)
Infant mortality rate: *total:* 25.92 deaths/1,000 live births
country comparison to the world: 79
male: 29.6 deaths/1,000 live births
female: 21.97 deaths/1,000 live births (2011 est.)
Life expectancy at birth: *total population:* 75.01 years
country comparison to the world: 91
male: 73 years
female: 77.17 years (2011 est.)
Total fertility rate: 2.03 children born/woman (2011 est.)
country comparison to the world: 125
HIV/AIDS—adult prevalence rate: less than 0.1% (2009 est.)
country comparison to the world: 164
HIV/AIDS—people living with HIV/AIDS: 2,400 (2009 est.)
country comparison to the world: 134
HIV/AIDS—deaths: fewer than 100 (2009 est.)
country comparison to the world: 117
Drinking water source: *Improved:*
urban: 99% of population
rural: 84% of population
total: 94% of population
Unimproved:
urban: 1% of population
rural: 16% of population
total: 6% of population (2008)
Sanitation facility access: *Improved:*
urban: 96% of population
rural: 64% of population
total: 85% of population
Unimproved:
urban: 4% of population
rural: 36% of population
total: 15% of population (2008)
Nationality: *noun:* Tunisian(s)
adjective: Tunisian
Ethnic groups: Arab 98%, European 1%, Jewish and other 1%
Religions: Muslim 98%, Christian 1%, Jewish and other 1%
Languages: Arabic (official, one of the languages of commerce), French (commerce)
Literacy: *definition:* age 15 and over can read and write
total population: 74.3%
male: 83.4%
female: 65.3% (2004 census)
School life expectancy (primary to tertiary education): *total:* 15 years
male: 14 years
female: 15 years (2008)
Education expenditures: 7.1% of GDP (2007)
country comparison to the world: 15

GOVERNMENT

Country name: *conventional long form:* Tunisian Republic
conventional short form: Tunisia
local long form: Al Jumhuriyah at Tunisiyah
local short form: Tunis
Government type: republic
Capital: *name:* Tunis
geographic coordinates: 36 48 N, 10 11 E
time difference: UTC+1 (6 hours ahead of Washington, DC during Standard Time)
Administrative divisions: 24 governorates; Ariana (Aryanah), Beja (Bajah), Ben Arous (Bin 'Arus), Bizerte (Banzart), Gabes (Qabis), Gafsa (Qafsah), Jendouba (Jundubah), Kairouan (Al Qayrawan), Kasserine (Al Qasrayn), Kebili (Qibili), Kef (Al Kaf), Mahdia (Al Mahdiyah), Manouba (Manubah), Medenine (Madanin), Monastir (Al Munastir), Nabeul (Nabul), Sfax (Safaqis), Sidi Bou Zid (Sidi Bu Zayd), Siliana (Silyanah), Sousse (Susah), Tataouine (Tatawin), Tozeur (Tawzar), Tunis, Zaghouan (Zaghwan)
Independence: 20 March 1956 (from France)
National holiday: Independence Day, 20 March (1956); also the anniversary of BEN ALI's assumption of the presidency, 7 November (1987)
Constitution: 1 June 1959; amended 1988, 2002
Legal system: mixed legal system of civil law, based on the French civil code, and Islamic law; some judicial review of legislative acts in the Supreme Court in joint session
International law organization participation: has not submitted an ICJ jurisdiction declaration; non-party state to the ICCt
Suffrage: 18 years of age; universal except for active government security forces (including the police and the military), people with mental disabilities, people who have served more than three months

in prison (criminal cases only), and people given a suspended sentence of more than six months

Executive branch: *chief of state:* Interim President Fouad M'BAZAA (since 15 January 2011); note–an interim government took office on 17 January 2011 to replace the government of former President Zine el Abidine BEN ALI

head of government: Prime Minister Beji Caid ESSEBSI (since 27 February 2011)

cabinet: Council of Ministers appointed by the president; note–the formation of a new cabinet was announced on 17 January 2011

(For more information visit the World Leaders website)

elections: president elected by popular vote for a five-year term (no term limits); election last held on 25 October 2009 (next to be held in October 2014); prime minister appointed by the president

election results: President Zine El Abidine BEN ALI reelected for a fifth term; percent of vote–Zine El Abidine BEN ALI 89.6%, Mohamed BOUCHIHA 5%, Ahmed INOUBLI 3.8%, Ahmed BRAHIM 1.6%; voter turnout 89.4%

Legislative branch: bicameral system consists of the Chamber of Advisors (126 seats; 85 members elected by municipal counselors, deputies, mayors, and professional associations and trade unions; 41 members are presidential appointees; members serve six-year terms); and the Chamber of Deputies or Majlis al-Nuwaab (214 seats; members elected by popular vote to serve five-year terms)

elections: Chamber of Advisors–last held on 3 July 2005 (next to be held in July 2011); Chamber of Deputies–last held on 25 October 2009 (next to be held in October 2014);

election results: Chamber of Deputies–percent of vote by party–RCD 84.6%, MDS 4.6%, PUP 3.4%, UDU 2.6%, PSL 2.2%, PVP 1.7%, Al-Tajdid 0.5%; seats by party–RCD 161, MDS 16, PUP 12, UDU 9, PSL 8, PVP 6, Al-Tajdid 2

Judicial branch: Court of Cassation or Cour de Cassation

Political parties and leaders: Et-Tajdid Movement [Ahmed IBRAHIM]; Constitutional Democratic Rally (Rassemblement Constitutionnel Democratique) or RCD, the ruling party of ex-president Zine al-Abidene BEN ALI dissolved on 9 March 2011 by Tunisian court; Democratic Forum for Labor and Liberties or FDTL [Mustapha Ben JAFAAR]; Green Party for Progress or PVP [Mongi KHAMASSI]; Liberal Social Party or PSL [Mondher THABET]; Movement of Socialist Democrats or MDS [Ismail BOULAHYA]; Popular Unity Party or PUP [Mohamed BOUCHIHA]; Progressive Democratic Party [Maya JERIBI]; Unionist Democratic Union or UDU [Ahmed INOUBLI]; al-Nahda (Renaissance) [Rachid GHANNOUCHI]

Political pressure groups and leaders: 18 October Group [collective leadership]; Tunisian League for Human Rights or LTDH [Mokhtar TRIFI]

International organization participation: ABEDA, AfDB, AFESD, AMF, AMU, AU, BSEC (observer), FAO, G-11, G-77, IAEA, IBRD, ICAO, ICC, ICRM, IDA, IDB, IFAD, IFC, IFRCS, IHO, ILO, IMF, IMO, IMSO, Interpol, IOC, IOM, IPU, ISO, ITSO, ITU, ITUC, LAS, MIGA, MONUSCO, NAM, OAPEC, OAS (observer), OIC, OIF, OPCW, OSCE (partner), UN, UNCTAD, UNESCO, UNHCR, UNIDO, UNOCI, UNWTO, UPU, WCO, WFTU, WHO, WIPO, WMO, WTO

Diplomatic representation in the US: *chief of mission:* Ambassador Mohamed Salah TEKAYA

chancery: 1515 Massachusetts Avenue NW, Washington, DC 20005

telephone: [1] (202) 862-1850

FAX: [1] (202) 862-1858

Diplomatic representation from the US: *chief of mission:* Ambassador Gordon GRAY

embassy: Zone Nord-Est des Berges du Lac Nord de Tunis 1053

mailing address: use embassy street address

telephone: [216] 71 107-000

FAX: [216] 71 963-263

Flag description: red with a white disk in the center bearing a red crescent nearly encircling a red five-pointed star; resembles the Ottoman flag (red banner with white crescent and star) and recalls Tunisia's history as part of the Ottoman Empire; red represents the blood shed by martyrs in the struggle against oppression, white stands for peace; the crescent and star are traditional symbols of Islam

note: the flag is based on that of Turkey, itself a successor state to the Ottoman Empire

National anthem: *name:* "Humat Al Hima" (Defenders of the Homeland)

lyrics/music: Mustafa Sadik AL-RAFII and Aboul-Qacem ECHEBBI/Mohamad Abdel WAHAB

note: adopted 1957, replaced 1958, restored 1987; Mohamad Abdel WAHAB also composed the music for the anthem of the United Arab Emirates

ECONOMY

Economy—overview: Tunisia has a diverse economy, with important agricultural, mining, tourism, and manufacturing sectors. Governmental control of economic affairs while still heavy has gradually lessened over the past decade with increasing privatization, simplification of the tax structure, and a prudent approach to debt. Progressive social policies also have helped raise living conditions in Tunisia relative to the region. Real growth, which averaged almost 5% over the past decade, declined to 4.6% in 2008 and to 3-4% in 2009-10 because of economic contraction and slowing of import demand in Europe–Tunisia's largest export market. However, development of non-textile manufacturing, a recovery in agricultural production, and strong growth in the services sector somewhat mitigated the economic effect of slowing exports. Tunisia will need to reach even higher growth levels to create sufficient employment opportunities for an already large number of unemployed as well as the growing population of university graduates. The challenges ahead include: privatizing industry, liberalizing the investment code to increase foreign investment, improving government efficiency, reducing the trade deficit, and reducing socioeconomic disparities in the impoverished south and west.

GDP (purchasing power parity): $100 billion (2010 est.)

country comparison to the world: 70

$96.43 billion (2009 est.)

$93.54 billion (2008 est.)

note: data are in 2010 US dollars

GDP (official exchange rate): $44.29 billion (2010 est.)

GDP—real growth rate: 3.7% (2010 est.)

country comparison to the world: 107

3.1% (2009 est.)

4.5% (2008 est.)

GDP—per capita (PPP): $9,400 (2010 est.)

country comparison to the world: 114

$9,200 (2009 est.)

$9,000 (2008 est.)

note: data are in 2010 US dollars

GDP—composition by sector:

agriculture: 10.6%

industry: 34.6%

services: 54.8% (2010 est.)

Labor force: 3.83 million (2010 est.)

country comparison to the world: 91

Labor force—by occupation: *agriculture:* 18.3%

industry: 31.9%

services: 49.8% (2009 est.)

Unemployment rate: 14% (2010 est.)

country comparison to the world: 142

13.3% (2009 est.)

Population below poverty line: 3.8% (2005 est.)

Household income or consumption by percentage share: *lowest 10%:* 2.3%

highest 10%: 31.5% (2000)

Distribution of family income—Gini index: 40 (2005 est.)

country comparison to the world: 61

41.7 (1995 est.)

Investment (gross fixed): 26.1% of GDP (2010 est.)

country comparison to the world: 37

Budget: *revenues:* $9.806 billion

expenditures: $11.76 billion (2010 est.)

Public debt: 49.5% of GDP (2010 est.)

country comparison to the world: 53

47.1% of GDP (2009 est.)

Inflation rate (consumer prices): 4.5% (2010 est.)

country comparison to the world: 129

3.5% (2009 est.)

Central bank discount rate: NA%

Commercial bank prime lending rate: NA%

Stock of narrow money: $11.49 billion (31 December 2010 est.)

country comparison to the world: 70

$11.02 billion (31 December 2009 est.)

Stock of broad money: $29.39 billion (31 December 2010 est.)

country comparison to the world: 75

$26.88 billion (31 December 2009 est.)

Stock of domestic credit: $31.1 billion (31 December 2010 est.)

country comparison to the world: 69

$28.45 billion (31 December 2009 est.)

Market value of publicly traded shares: $9.12 billion (31 December 2009)

country comparison to the world: 77
$6.374 billion (31 December 2008)
$5.355 billion (31 December 2007)
Agriculture—products: olives, olive oil, grain, tomatoes, citrus fruit, sugar beets, dates, almonds; beef, dairy products
Industries: petroleum, mining (particularly phosphate and iron ore), tourism, textiles, footwear, agribusiness, beverages
Industrial production growth rate: 1.6% (2010 est.)
country comparison to the world: 136
Electricity—production: 11.08 billion kWh (2008 est.)
country comparison to the world: 87
Electricity—consumption: 11.8 billion kWh (2008 est.)
country comparison to the world: 81
Electricity—exports: 130 million kWh (2007 est.)
Electricity—imports: 145 million kWh (2007 est.)
Oil—production: 91,380 bbl/day (2009 est.)
country comparison to the world: 54
Oil—consumption: 89,000 bbl/day (2009 est.)
country comparison to the world: 80
Oil—exports: 77,130 bbl/day (2007 est.)
country comparison to the world: 69
Oil—imports: 87,300 bbl/day (2007 est.)
country comparison to the world: 67
Oil—proved reserves: 425 million bbl (1 January 2010 est.)
country comparison to the world: 52
Natural gas—production:
2.97 billion cu m (2008 est.)
country comparison to the world: 54
Natural gas—consumption: 4.22 billion cu m (2008 est.)
country comparison to the world: 64
Natural gas—exports: 0 cu m (2008 est.)
country comparison to the world: 188
Natural gas—imports: 1.25 billion cu m (2008 est.)
country comparison to the world: 51
Natural gas—proved reserves: 65.13 billion cu m (1 January 2010 est.)
country comparison to the world: 59
Current account balance: $-1.389 billion (2010 est.)
country comparison to the world: 145
$-1.234 billion (2009 est.)
Exports: $16.11 billion (2010 est.)
country comparison to the world: 74
$14.42 billion (2009 est.)
Exports—commodities: clothing, semi-finished goods and textiles, agricultural products, mechanical goods, phosphates and chemicals, hydrocarbons, electrical equipment
Exports—partners: France 29.6%, Italy 21%, Germany 8.8%, Libya 5.8%, Spain 5%, UK 4.8% (2009 est.)
Imports: $20.02 billion (2010 est.)
country comparison to the world: 72
$18.12 billion (2009 est.)
Imports—commodities: textiles, machinery and equipment, hydrocarbons, chemicals, foodstuffs
Imports—partners: France 20.1%, Italy 16.4%, Germany 8.8%, China 5%, Spain 4.5%, US 4% (2009 est.)
Reserves of foreign exchange and gold: $11.23 billion (31 December 2010 est.)
country comparison to the world: 55
$11.06 billion (31 December 2009 est.)
Debt—external: $18.76 billion (31 December 2010 est.)
country comparison to the world: 72
$19.6 billion (31 December 2009 est.)
Stock of direct foreign investment—at home: $33.56 billion (31 December 2010 est.)
country comparison to the world: 60
$31.86 billion (31 December 2009 est.)
Stock of direct foreign investment—abroad: $251 million (31 December 2010 est.)
country comparison to the world: 79
$233 million (31 December 2009 est.)
Exchange rates: Tunisian dinars (TND) per US dollar–1.4367 (2010), 1.3503 (2009), 1.211 (2008), 1.2776 (2007), 1.331 (2006)

COMMUNICATIONS

Telephones—main lines in use: 1.279 million (2009)
country comparison to the world: 70
Telephones—mobile cellular: 9.754 million (2009)
country comparison to the world: 65
Telephone system: *general assessment:* above the African average and continuing to be upgraded; key centers are Sfax, Sousse, Bizerte, and Tunis; telephone network is completely digitized; Internet access available throughout the country
domestic: in an effort to jumpstart expansion of the fixed-line network, the government has awarded a concession to build and operate a VSAT network with international connectivity; rural areas are served by wireless local loops; competition between the two mobile-cellular service providers has resulted in lower activation and usage charges and a strong surge in subscribership; a third mobile, fixed, and ISP operator was licensed in 2009 and will begin offering services in 2010; expansion of mobile-cellular services to include multimedia messaging and e-mail and Internet to mobile phone services also leading to a surge in subscribership; overall fixed-line and mobile-cellular teledensity is about 100 telephones per 100 persons
international: country code–216; a landing point for the SEA-ME-WE-4 submarine cable system that provides links to Europe, Middle East, and Asia; satellite earth stations–1 Intelsat (Atlantic Ocean) and 1 Arabsat; coaxial cable and microwave radio relay to Algeria and Libya; participant in Medarabtel; 2 international gateway digital switches
Broadcast media: broadcast media is mainly government-controlled; the state-run Tunisian Radio and Television Establishment (ERTT) operates 2 national television networks, several national radio networks, and a number of regional radio stations; 1 TV and 3 radio stations are privately-owned and report domestic news stories directly from the official Tunisian news agency; the state retains control of broadcast facilities and transmitters through L'Office National de la Telediffusion; Tunisians also have access to Egyptian, pan-Arab, and European satellite TV channels (2007)
Internet country code: .tn
Internet hosts: 490 (2010)
country comparison to the world: 181
Internet users: 3.5 million (2009)
country comparison to the world: 60

TRANSPORTATION

Airports: 32 (2010)
country comparison to the world: 112
Airports—with paved runways: *total:* 16
over 3,047 m: 4
2,438 to 3,047 m: 6
1,524 to 2,437 m: 2
914 to 1,523 m: 4 (2010)
Airports—with unpaved runways: *total:* 16
1,524 to 2,437 m: 2
914 to 1,523 m: 7
under 914 m: 7 (2010)
Pipelines: gas 2,386 km; oil 1,323 km; refined products 453 km (2010)
Railways: *total:* 2,167 km
country comparison to the world: 68
standard gauge: 471 km 1.435-m gauge
narrow gauge: 1,688 km 1.000-m gauge (65 km electrified)
dual gauge: 8 km (2008)
Roadways: *total:* 19,232 km
country comparison to the world: 110
paved: 12,655 km (includes 262 km of expressways)
unpaved: 6,577 km (2006)
Merchant marine: *total:* 11
country comparison to the world: 110
by type: bulk carrier 1, cargo 2, chemical tanker 2, passenger/cargo 4, roll on/roll off 2
registered in other countries: 1 (Panama 1) (2010)
Ports and terminals: Bizerte, Gabes, Rades, Sfax, Skhira

MILITARY

Military branches: Tunisian Armed Forces (Forces Armees Tunisiens, FAT): Army, Navy, Republic of Tunisia Air Force (Al-Quwwat al-Jawwiya al-Jamahiriyah At'Tunisia) (2011)
Military service age and obligation: 20 years of age for compulsory military service, 18 years of age for voluntary military service; 1-year conscript service obligation (2007)
Manpower available for military service: *males age 16–49:* 2,846,572
females age 16–49: 2,952,180 (2010 est.)
Manpower fit for military service: *males age 16–49:* 2,397,716
females age 16–49: 2,484,097 (2010 est.)
Manpower reaching militarily significant age annually: *male:* 90,436
female: 87,346 (2010 est.)
Military expenditures: 1.4% of GDP (2006)
country comparison to the world: 108

TRANSNATIONAL ISSUES

Disputes—international: none

TURKEY

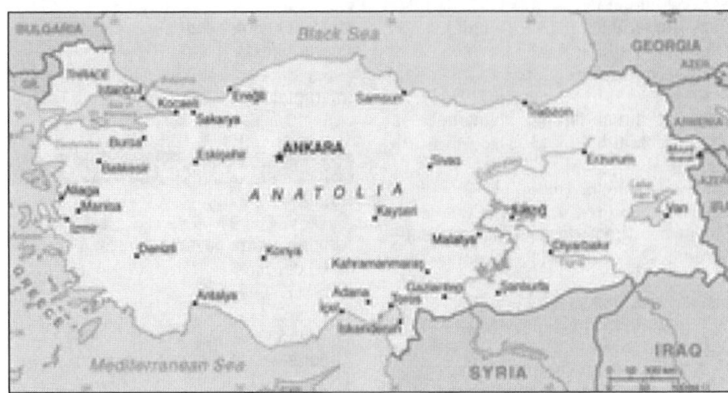

INTRODUCTION

Background: Modern Turkey was founded in 1923 from the Anatolian remnants of the defeated Ottoman Empire by national hero Mustafa KEMAL, who was later honored with the title Ataturk or "Father of the Turks." Under his authoritarian leadership, the country adopted wide-ranging social, legal, and political reforms. After a period of one-party rule, an experiment with multi-party politics led to the 1950 election victory of the opposition Democratic Party and the peaceful transfer of power. Since then, Turkish political parties have multiplied, but democracy has been fractured by periods of instability and intermittent military coups (1960, 1971, 1980), which in each case eventually resulted in a return of political power to civilians. In 1997, the military again helped engineer the ouster–popularly dubbed a "post-modern coup"–of the then Islamic-oriented government. Turkey intervened militarily on Cyprus in 1974 to prevent a Greek takeover of the island and has since acted as patron state to the "Turkish Republic of Northern Cyprus," which only Turkey recognizes. A separatist insurgency begun in 1984 by the Kurdistan Workers' Party (PKK)–now known as the People's Congress of Kurdistan or Kongra-Gel (KGK)–has dominated the Turkish military's attention and claimed more than 30,000 lives. After the capture of the group's leader in 1999, the insurgents largely withdrew from Turkey mainly to northern Iraq. In 2004, KGK announced an end to its ceasefire and attacks attributed to the KGK increased. Turkey joined the UN in 1945 and in 1952 it became a member of NATO. In 1964, Turkey became an associate member of the European Community. Over the past decade, it has undertaken many reforms to strengthen its democracy and economy; it began accession membership talks with the European Union in 2005.

GEOGRAPHY

Location: Southeastern Europe and Southwestern Asia (that portion of Turkey west of the Bosporus is geographically part of Europe), bordering the Black Sea, between Bulgaria and Georgia, and bordering the Aegean Sea and the Mediterranean Sea, between Greece and Syria
Geographic coordinates: 39 00 N, 35 00 E
Map references: Middle East
Area: *total:* 783,562 sq km
country comparison to the world: 37
land: 769,632 sq km
water: 13,930 sq km
Area—comparative: slightly larger than Texas
Land boundaries: *total:* 2,648 km
border countries: Armenia 268 km, Azerbaijan 9 km, Bulgaria 240 km, Georgia 252 km, Greece 206 km, Iran 499 km, Iraq 352 km, Syria 822 km
Coastline: 7,200 km
Maritime claims: territorial sea: 6 nm in the Aegean Sea; 12 nm in Black Sea and in Mediterranean Sea
exclusive economic zone: in Black Sea only: to the maritime boundary agreed upon with the former USSR
Climate: temperate; hot, dry summers with mild, wet winters; harsher in interior
Terrain: high central plateau (Anatolia); narrow coastal plain; several mountain ranges
Elevation extremes: *lowest point:* Mediterranean Sea 0 m
highest point: Mount Ararat 5,166 m
Natural resources: coal, iron ore, copper, chromium, antimony, mercury, gold, barite, borate, celestite (strontium), emery, feldspar, limestone, magnesite, marble, perlite, pumice, pyrites (sulfur), clay, arable land, hydropower
Land use: *arable land:* 29.81%
permanent crops: 3.39%
other: 66.8% (2005)
Irrigated land: 52,150 sq km (2003)
Total renewable water resources: 234 cu km (2003)
Freshwater withdrawal (domestic/industrial/agricultural): *total:* 39.78 cu km/yr (15%/11%/74%)
per capita: 544 cu m/yr (2001)
Natural hazards: severe earthquakes, especially in northern Turkey, along an arc extending from the Sea of Marmara to Lake Van
volcanism: Turkey experiences little volcanic activity; its three historically active volcanoes; Ararat, Nemrut Dagi, and Tendurek Dagi have not erupted since the 19th century or earlier
Environment—current issues: water pollution from dumping of chemicals and detergents; air pollution, particularly in urban areas; deforestation; concern for oil spills from increasing Bosporus ship traffic
Environment—international agreements: *party to:* Air Pollution, Antarctic Treaty, Biodiversity, Climate Change, Desertification, Endangered Species, Hazardous Wastes, Ozone Layer Protection, Ship Pollution, Wetlands
signed, but not ratified: Environmental Modification
Geography—note: strategic location controlling the Turkish Straits (Bosporus, Sea of Marmara, Dardanelles) that link Black and Aegean Seas; Mount Ararat, the legendary landing place of Noah's ark, is in the far eastern portion of the country

PEOPLE

Population: 78,785,548 (July 2011 est.)
country comparison to the world: 17
Age structure: *0–14 years:* 26.6% (male 10,707,793/female 10,226,999)
15–64 years: 67.1% (male 26,741,332/female 26,162,757)
65 years and over: 6.3% (male 2,259,422/female 2,687,245) (2011 est.)
Median age: *total:* 28.5 years
male: 28.1 years
female: 28.8 years (2011 est.)
Population growth rate: 1.235% (2011 est.)
country comparison to the world: 95
Birth rate: 17.93 births/1,000 population (2011 est.)
country comparison to the world: 106
Death rate: 6.1 deaths/1,000 population (July 2011 est.)
country comparison to the world: 161
Net migration rate: 0.51 migrant(s)/1,000 population (2011 est.)
country comparison to the world: 64
Urbanization: *urban population:* 70% of total population (2010)
rate of urbanization: 1.7% annual rate of change (2010-15 est.)
Major cities—population: Istanbul 10.378 million; ANKARA (capital) 3.846 million; Izmir 2.679 million; Bursa 1.559 million; Adana 1.339 million (2009)
Sex ratio: *at birth:* 1.05 male(s)/female
under 15 years: 1.05 male(s)/female
15–64 years: 1.02 male(s)/female
65 years and over: 0.84 male(s)/female
total population: 1.02 male(s)/female (2011 est.)
Infant mortality rate: *total:* 23.94 deaths/1,000 live births
country comparison to the world: 84
male: 25 deaths/1,000 live births
female: 22.82 deaths/1,000 live births (2011 est.)

Life expectancy at birth: *total population:* 72.5 years
country comparison to the world: 126
male: 70.61 years
female: 74.49 years (2011 est.)
Total fertility rate: 2.15 children born/woman (2011 est.)
country comparison to the world: 113
HIV/AIDS—adult prevalence rate: less than 0.1%; less than 0.1% note—no country specific models provided (2009 est.)
country comparison to the world: 111
HIV/AIDS—people living with HIV/AIDS: 4,600 (2009 est.)
country comparison to the world: 121
HIV/AIDS—deaths: fewer than 200 (2009 est.)
country comparison to the world: 110
Drinking water source: *Improved:*
urban: 100% of population
rural: 96% of population
total: 99% of population
Unimproved: urban: 0% of population
rural: 4% of population
total: 1% of population (2008)
Sanitation facility access: *Improved:*
urban: 97% of population
rural: 75% of population
total: 90% of population
Unimproved:
urban: 3% of population
rural: 25% of population
total: 10% of population (2008)
Nationality: *noun:* Turk(s)
adjective: Turkish
Ethnic groups: Turkish 70-75%, Kurdish 18%, other minorities 7-12% (2008 est.)
Religions: Muslim 99.8% (mostly Sunni), other 0.2% (mostly Christians and Jews)
Languages: Turkish (official), Kurdish, other minority languages
Literacy: *definition:* age 15 and over can read and write
total population: 87.4%
male: 95.3%
female: 79.6% (2004 est.)
School life expectancy (primary to tertiary education): *total:* 12 years
male: 12 years
female: 11 years (2008)
Education expenditures: 2.9% of GDP (2006)
country comparison to the world: 137

GOVERNMENT

Country name: *conventional long form:* Republic of Turkey
conventional short form: Turkey
local long form: Turkiye Cumhuriyeti
local short form: Turkiye
Government type: republican parliamentary democracy
Capital: *name:* Ankara
geographic coordinates: 39 56 N, 32 52 E
time difference: UTC+2 (7 hours ahead of Washington, DC during Standard Time)
daylight saving time: +1hr, begins last Monday in March; ends last Sunday in October
Administrative divisions: 81 provinces (iller, singular—ili); Adana, Adiyaman, Afyonkarahisar, Agri, Aksaray, Amasya, Ankara, Antalya, Ardahan, Artvin, Aydin, Balikesir, Bartin, Batman, Bayburt, Bilecik, Bingol, Bitlis, Bolu, Burdur, Bursa, Canakkale, Cankiri, Corum, Denizli, Diyarbakir, Duzce, Edirne, Elazig, Erzincan, Erzurum, Eskisehir, Gaziantep, Giresun, Gumushane, Hakkari, Hatay, Igdir, Isparta, Istanbul, Izmir (Smyrna), Kahramanmaras, Karabuk, Karaman, Kars, Kastamonu, Kayseri, Kilis, Kirikkale, Kirklareli, Kirsehir, Kocaeli, Konya, Kutahya, Malatya, Manisa, Mardin, Mersin, Mugla, Mus, Nevsehir, Nigde, Ordu, Osmaniye, Rize, Sakarya, Samsun, Sanliurfa, Siirt, Sinop, Sirnak, Sivas, Tekirdag, Tokat, Trabzon (Trebizond), Tunceli, Usak, Van, Yalova, Yozgat, Zonguldak
Independence: 29 October 1923 (successor state to the Ottoman Empire)
National holiday: Republic Day, 29 October (1923)
Constitution: 7 November 1982; amended several times; note—amendment passed by referendum 21 October 2007 concerning presidential elections
Legal system: civil law system based on various European legal systems notably the Swiss civil code; note—member of the European Court of Human Rights (ECHR), although Turkey claims limited derogations on the ratified European Convention on Human Rights
International law organization participation: has not submitted an ICJ jurisdiction declaration; non-party state to the ICCt
Suffrage: 18 years of age; universal
Executive branch: *chief of state:* President Abdullah GUL (since 28 August 2007)
head of government: Prime Minister Recep Tayyip ERDOGAN (since 14 March 2003)
cabinet: Council of Ministers appointed by the president on the nomination of the prime minister
(For more information visit the World Leaders website)
elections: president elected directly for a five-year term (eligible for a second term); prime minister appointed by the president from among members of parliament
election results: on 28 August 2007 the National Assembly elected Abdullah GUL president on the third ballot; National Assembly vote—339
note: in October 2007 Turkish voters approved a referendum package of constitutional amendments including a provision for direct presidential elections
Legislative branch: unicameral Grand National Assembly of Turkey or Turkiye Buyuk Millet Meclisi (550 seats; members elected by popular vote to serve four-year terms)
elections: last held on 22 July 2007 (next to be held by 12 June 2011)
election results: percent of vote by party—AKP 46.7%, CHP 20.8%, MHP 14.3%, independents 5.2%, other 13.0%; seats by party—AKP 341, CHP 112, MHP 71, independents 26; note—seats by party as of 15 November 2010—AKP 335, CHP 101, MHP 70, BDP 20, DSP 6, DP 1, TP 1, independents 7, vacant 9 (BDP entered parliament as independents; DSP entered parliament on CHP's party list; DP and TP switched to their respective parties after having been elected to parliament as an independent or on the list of another party); only parties surpassing the 10% threshold are entitled to parliamentary seats
Judicial branch: Constitutional Court; High Court of Appeals (Yargitay); Council of State (Danistay); Court of Accounts (Sayistay); Military High Court of Appeals; Military High Administrative Court
Political parties and leaders: Democratic Left Party or DSP [Masum TURKER]; Democratic Party or DP [Namik Kemal ZEYBEK]; Equality and Democracy Party or EDP [Ziva HALIS]; Felicity Party or SP [Necmettin ERBAKAN] (sometimes translated as Contentment Party); Freedom and Solidarity Party or ODP [Alper TAS]; Grand Unity Party or BBP [Yalcin TOPCU]; Justice and Development Party or AKP [Recep Tayyip ERDOGAN]; Nationalist Movement Party or MHP [Devlet BAHCELI]; Peace and Democracy Party or BDP [Selahattin DEMIRTAS]; People's Voice Party or HSP [Numan KURTULMUS]; Republican People's Party or CHP [Kemal KILICDAROGLU]; Turkey Party or TP [Abdullatif SENER]
note: the parties listed above are some of the more significant of the 61 parties that Turkey had according to the Ministry of Interior statistics current as of May 2009
Political pressure groups and leaders: Confederation of Businessmen and Industrialists of Turkey or TUSKON [Rizanur MERAL]; Confederation of Public Sector Unions or KESK [Sami EVREN]; Confederation of Revolutionary Workers Unions or DISK [Suleyman CELEBI]; Independent Industrialists' and Businessmen's Association or MUSIAD [Omer Cihad VARDAN]; Moral Rights Workers Union or Hak-Is [Salim USLU]; Turkish Confederation of Employers' Unions or TISK [Tugrul KUDATGOBILIK]; Turkish Confederation of Labor or Turk-Is [Mustafa KUMLU]; Turkish Confederation of Tradesmen and Craftsmen or TESK [Bendevi PALANDOKEN]; Turkish Industrialists' and Businessmen's Association or TUSIAD [Umit BOYNER]; Turkish Union of Chambers of Commerce and Commodity Exchanges or TOBB [M. Rifat HISARCIKLIOGLU]
International organization participation: ADB (nonregional member), Australia Group, BIS, BSEC, CE, CERN (observer), CICA, D-8, EAPC, EBRD, ECO, EU (candidate country), FAO, FATF, G-20, IAEA, IBRD, ICAO, ICC, ICRM, IDA, IDB, IEA, IFAD, IFC, IFRCS, IHO, ILO, IMF, IMO, IMSO, Interpol, IOC, IOM, IPU, ISO, ITSO, ITU, ITUC, MIGA, NATO, NEA, NSG, OAS (observer), OECD, OIC, OPCW, OSCE, Paris Club (associate), PCA, SECI, UN, UNCTAD, UNESCO, UNHCR, UNIDO, UNIFIL, UNMIS, UNRWA, UNWTO, UPU, WCO, WFTU, WHO, WIPO, WMO, WTO, ZC
Diplomatic representation in the US: *chief of mission:* Ambassador Namik TAN
chancery: 2525 Massachusetts Avenue NW, Washington, DC 20008
telephone: [1] (202) 612-6700
FAX: [1] (202) 612-6744
consulate(s) general: Chicago, Houston, Los Angeles, New York

Diplomatic representation from the US: *chief of mission:* Ambassador (vacant); Charge d'Affaires Douglas A. SILLIMAN
embassy: 110 Ataturk Boulevard, Kavaklidere, 06100 Ankara
mailing address: PSC 93, Box 5000, APO AE 09823
telephone: [90] (312) 455-5555
FAX: [90] (312) 467-0019
consulate(s) general: Istanbul
consulate(s): Adana; note–there is a Consular Agent in Izmir
Flag description: red with a vertical white crescent moon (the closed portion is toward the hoist side) and white five-pointed star centered just outside the crescent opening; the flag colors and designs closely resemble those on the banner of Ottoman Empire, which preceded modern-day Turkey; the crescent moon and star serve as insignia for the Turks, as well as being traditional symbols of Islam; according to legend, the flag represents the reflection of the moon and a star in a pool of blood of Turkish warriors
National anthem: *name:* "Istiklal Marsi" (Independence March)
lyrics/music: Mehmet Akif ERSOY/Zeki UNGOR
note: lyrics adopted 1921, music adopted 1932; the anthem's original music was adopted in 1924; a new composition was agreed upon in 1932

ECONOMY

Economy—overview: Turkey's economy is increasingly driven by its industry and service sectors, although its traditional agriculture sector still accounts for about 30% of employment. An aggressive privatization program has reduced state involvement in basic industry, banking, transport, and communication, and an emerging cadre of middle-class entrepreneurs is adding dynamism to the economy. Turkey's traditional textiles and clothing sectors still account for one-third of industrial employment, despite stiff competition in international markets that resulted from the end of the global quota system. Other sectors, notably the automotive, construction, and electronics industries, are rising in importance and have surpassed textiles within Turkey's export mix. Oil began to flow through the Baku-Tbilisi-Ceyhan pipeline in May 2006, marking a major milestone that will bring up to 1 million barrels per day from the Caspian to market. Several gas pipelines also are being planned to help move Central Asian gas to Europe via Turkey, which will help address Turkey's dependence on energy imports over the long term. After Turkey experienced a severe financial crisis in 2001, Ankara adopted financial and fiscal reforms as part of an IMF program. The reforms strengthened the country's economic fundamentals and ushered in an era of strong growth–averaging more than 6% annually until 2008, when global economic conditions and tighter fiscal policy caused GDP to contract in 2009, reduced inflation to 6.3%–a 34-year low–and cut the public sector debt-to-GPD ratio below 50%. Turkey's well-regulated financial markets and banking system weathered the global financial crisis and GDP rebounded strongly to 7.3% in 2010, as exports returned to normal levels following the recession. The economy, however, continues to be burdened by a high current account deficit and remains dependent on often volatile, short-term investment to finance its trade deficit. The stock value of FDI stood at $174 billion at year-end 2010, but inflows have slowed considerably in light of continuing economic turmoil in Europe, the source of much of Turkey's FDI. Further economic and judicial reforms and prospective EU membership are expected to boost Turkey's attractiveness to foreign investors. However, Turkey's relatively high current account deficit, uncertainty related to policy-making, and fiscal imbalances leave the economy vulnerable to destabilizing shifts in investor confidence.
GDP (purchasing power parity): $960.5 billion (2010 est.)
country comparison to the world: 17
$887.7 billion (2009 est.)
$931.4 billion (2008 est.)
note: data are in 2010 US dollars
GDP (official exchange rate): $741.9 billion (2010 est.)
GDP—real growth rate: 8.2% (2010 est.)
country comparison to the world: 16
-4.7% (2009 est.)
0.7% (2008 est.)
GDP—per capita (PPP): $12,300 (2010 est.)
country comparison to the world: 94
$11,600 (2009 est.)
$12,300 (2008 est.)
note: data are in 2010 US dollars
GDP—composition by sector:
agriculture: 8.8%
industry: 25.7%
services: 65.5% (2010 est.)
Labor force: 24.73 million
country comparison to the world: 24
note: about 1.2 million Turks work abroad (2010 est.)
Labor force—by occupation:
agriculture: 29.5%
industry: 24.7%
services: 45.8% (2005)
Unemployment rate: 12.4% (2010 est.)
country comparison to the world: 131
14.1% (2009 est.)
note: underemployment amounted to 4% in 2008
Population below poverty line: 17.11% (2008)
Household income or consumption by percentage share: *lowest 10%:* 1.9%
highest 10%: 33.2% (2005)
Distribution of family income—Gini index: 41 (2007)
country comparison to the world: 55
43.6 (2003)
Investment (gross fixed): 18% of GDP (2010 est.)
country comparison to the world: 111
Budget:
revenues: $159.4 billion
expenditures: $189.6 billion (2010 est.)
Public debt: 48.1% of GDP (2010 est.)
country comparison to the world: 54
46.3% of GDP (2009 est.)
Inflation rate (consumer prices): 8.7% (2010 est.)
country comparison to the world: 191
6.3% (2009 est.)
Central bank discount rate: 15% (22 December 2009)
country comparison to the world: 6
25% (31 December 2008)
Commercial bank prime lending rate: NA%
Stock of narrow money: $57.02 billion (31 December 2010 est.)
country comparison to the world: 44
$44.94 billion (31 December 2009 est.)
Stock of broad money: $255.5 billion (31 December 2010 est.)
country comparison to the world: 34
$202.2 billion (31 December 2009 est.)
Stock of domestic credit: $401.8 billion (31 December 2010 est.)
country comparison to the world: 27
$373.1 billion (31 December 2009 est.)
Market value of publicly traded shares: $225.7 billion (31 December 2009)
country comparison to the world: 35
$117.9 billion (31 December 2008)
$286.6 billion (31 December 2007)
Agriculture—products: tobacco, cotton, grain, olives, sugar beets, hazelnuts, pulse, citrus; livestock
Industries: textiles, food processing, autos, electronics, mining (coal, chromate, copper, boron), steel, petroleum, construction, lumber, paper
Industrial production growth rate: 6% (2010 est.)
country comparison to the world: 56
Electricity—production: 198.4 billion kWh (2008 est.)
country comparison to the world: 20
Electricity—consumption: 198.1 billion kWh (2008 est.)
country comparison to the world: 19
Electricity—exports: 1.12 billion kWh (2008 est.)
Electricity—imports: 790 million kWh (2008 est.)
Oil—production: 52,980 bbl/day (2009 est.)
country comparison to the world: 62
Oil—consumption: 579,500 bbl/day (2009 est.)
country comparison to the world: 28
Oil—exports: 133,100 bbl/day (2008 est.)
country comparison to the world: 61
Oil—imports: 734,600 bbl/day (2008 est.)
country comparison to the world: 18
Oil—proved reserves: 262.2 million bbl (1 January 2010 est.)
country comparison to the world: 56
Natural gas—production: 1.014 billion cu m (2009 est.)
country comparison to the world: 63
Natural gas—consumption: 35.07 billion cu m (2009 est.)
country comparison to the world: 24
Natural gas—exports: 708 million cu m (2009 est.)
country comparison to the world: 39
Natural gas—imports: 35.77 billion cu m (2009 est.)
country comparison to the world: 7
Natural gas—proved reserves: 6.088 billion cu m (1 January 2010 est.)
country comparison to the world: 86

Current account balance: $-38.82 billion (2010 est.)
country comparison to the world: 184
$-13.94 billion (2009 est.)
Exports: $117.4 billion (2010 est.)
country comparison to the world: 32
$109.6 billion (2009 est.)
Exports—commodities: apparel, foodstuffs, textiles, metal manufactures, transport equipment
Exports—partners: Germany 9.6%, France 6.1%, UK 5.8%, Italy 5.8%, Iraq 5% (2009 est.)
Imports: $166.3 billion (2010 est.)
country comparison to the world: 25
$134.5 billion (2009 est.)
Imports—commodities: machinery, chemicals, semi-finished goods, fuels, transport equipment
Imports—partners: Russia 14%, Germany 10%, China 9%, US 6.1%, Italy 5.4%, France 5% (2009 est.)
Reserves of foreign exchange and gold: $78 billion (31 December 2010 est.)
country comparison to the world: 19
$75 billion (31 December 2009 est.)
Debt—external: $270.7 billion (31 December 2010 est.)
country comparison to the world: 27
$268.3 billion (31 December 2009 est.)
Stock of direct foreign investment—at home: $84.45 billion (31 December 2010 est.)
country comparison to the world: 39
$174 billion (31 December 2008 est.)
Stock of direct foreign investment—abroad: $16.42 billion (31 December 2010 est.)
country comparison to the world: 47
$15.42 billion (31 December 2009 est.)
Exchange rates: Turkish liras (TRY) per US dollar–1.5181 (2010), 1.55 (2009), 1.3179 (2008), 1.319 (2007), 1.4286 (2006)

COMMUNICATIONS

Telephones—main lines in use: 16.534 million (2009)
country comparison to the world: 18
Telephones—mobile cellular: 62.78 million (2009)
country comparison to the world: 17
Telephone system: *general assessment:* comprehensive telecommunications network undergoing rapid modernization and expansion especially in mobile-cellular services
domestic: additional digital exchanges are permitting a rapid increase in subscribers; the construction of a network of technologically advanced intercity trunk lines, using both fiber-optic cable and digital microwave radio relay, is facilitating communication between urban centers; remote areas are reached by a domestic satellite system; the number of subscribers to mobile-cellular telephone service is growing rapidly
international: country code–90; international service is provided by the SEA-ME-WE-3 submarine cable and by submarine fiber-optic cables in the Mediterranean and Black Seas that link Turkey with Italy, Greece, Israel, Bulgaria, Romania, and Russia; satellite earth stations–12 Intelsat; mobile satellite terminals–328 in the Inmarsat and Eutelsat systems (2002)
Broadcast media: national public broadcaster Turkish Radio and Television Corporation (TRT) operates multiple TV and radio networks and stations; multiple privately-owned national television stations and up to 300 private regional and local television stations; multi-channel cable TV subscriptions are obtainable; more than 1,000 private radio broadcast stations (2009)
Internet country code: .tr
Internet hosts: 3.433 million (2010)
country comparison to the world: 26
Internet users: 27.233 million (2009)
country comparison to the world: 15

TRANSPORTATION

Airports: 99 (2010)
country comparison to the world: 60
Airports—with paved runways: *total:* 88
over 3,047 m: 16
2,438 to 3,047 m: 33
1,524 to 2,437 m: 19
914 to 1,523 m: 16
under 914 m: 4 (2010)
Airports—with unpaved runways:
total: 11
1,524 to 2,437 m: 1
914 to 1,523 m: 6
under 914 m: 4 (2010)
Heliports: 20 (2010)
Pipelines: gas 10,706 km; oil 3,636 km (2010)
Railways: *total:* 8,697 km
country comparison to the world: 24
standard gauge: 8,697 km 1.435-m gauge (1,920 km electrified) (2008)
Roadways: *total:* 352,046 km
country comparison to the world: 19
paved: 313,151 km (includes 2,010 km of expressways)
unpaved: 38,895 km (2008)
Waterways: 1,200 km (2008)
country comparison to the world: 59
Merchant marine: *total:* 645
country comparison to the world: 18
by type: bulk carrier 95, cargo 290, chemical tanker 85, combination ore/oil 1, container 40, liquefied gas 6, passenger 1, passenger/cargo 59, petroleum tanker 31, refrigerated cargo 1, roll on/roll off 34, specialized tanker 2
foreign-owned: 3 (Germany 1, Italy 2)
registered in other countries: 686 (Albania 1, Antigua and Barbuda 7, Azerbaijan 1, Bahamas 3, Barbados 1, Belize 18, Cambodia 26, Comoros 16, Cook Islands 4, Dominica 1, Georgia 22, Italy 3, Kiribati 3, Liberia 15, Malta 211, Marshall Islands 72, Moldova 18, Mongolia 1, former Netherlands Antilles 8, Panama 79, Russia 104, Saint Kitts and Nevis 22, Saint Vincent and the Grenadines 18, Sierra Leone 14, Slovakia 2, Tanzania 7, Togo 4, Turkmenistan 1, Tuvalu 1, UK 1, unknown 2) (2010)
Ports and terminals: Aliaga, Diliskelesi, Eregli, Izmir, Izmit (Kocaeli), Mercin (Icel), Limani, Yarimca

MILITARY

Military branches: Turkish Armed Forces (TSK): Turkish Land Forces (Turk Kara Kuvvetleri), Turkish Naval Forces (Turk Deniz Kuvvetleri; includes naval air and naval infantry), Turkish Air Force (Turk Hava Kuvvetleri) (2010)
Military service age and obligation: 19-41 years of age for male compulsory military service; 18 years of age for voluntary service; 15 months conscript obligation for non-university graduates, 6-12 months for university graduates; women serve in the Turkish Armed Forces only as officers; reserve obligation to age 41 (2010)
Manpower available for military service:
males age 16–49: 21,079,077
females age 16–49: 20,558,696 (2010 est.)
Manpower fit for military service: *males age 16–49:* 17,664,510
females age 16–49: 17,340,816 (2010 est.)
Manpower reaching militarily significant age annually: *male:* 700,079
female: 670,328 (2010 est.)
Military expenditures: 5.3% of GDP (2005 est.)
country comparison to the world: 15
Military—note: a "National Security Policy Document" adopted in October 2005 increases the Turkish Armed Forces (TSK) role in internal security, augmenting the General Directorate of Security and Gendarmerie General Command (Jandarma); the TSK leadership continues to play a key role in politics and considers itself guardian of Turkey's secular state; in April 2007, it warned the ruling party about any pro-Islamic appointments; despite on-going negotiations on EU accession since October 2005, progress has been limited in establishing required civilian supremacy over the military; primary domestic threats are listed as fundamentalism (with the definition in some dispute with the civilian government), separatism (the Kurdish problem), and the extreme left wing; Ankara strongly opposed establishment of an autonomous Kurdish region; an overhaul of the Turkish Land Forces Command (TLFC) taking place under the "Force 2014" program is to produce 20-30% smaller, more highly trained forces characterized by greater mobility and firepower and capable of joint and combined operations; the TLFC has taken on increasing international peacekeeping responsibilities, and took charge of a NATO International Security Assistance Force (ISAF) command in Afghanistan in April 2007; the Turkish Navy is a regional naval power that wants to develop the capability to project power beyond Turkey's coastal waters; the Navy is heavily involved in NATO, multinational, and UN operations; its roles include control of territorial waters and security for sea lines of communications; the Turkish Air Force adopted an "Aerospace and Missile Defense Concept" in 2002 and has initiated project work on an integrated missile defense system; Air Force priorities include attaining a modern deployable, survivable, and sustainable force structure, and establishing a sustainable command and control system (2008)

TRANSNATIONAL ISSUES

Disputes—international: complex maritime, air, and territorial disputes with Greece in the Aegean Sea; status of north

Cyprus question remains; Syria and Iraq protest Turkish hydrological projects to control upper Euphrates waters; Turkey has expressed concern over the status of Kurds in Iraq; in 2009, Swiss mediators facilitated an accord reestablishing diplomatic ties between Armenia and Turkey, but neither side has ratified the agreement and the rapprochement effort has faltered; Turkish authorities have complained that blasting from quarries in Armenia might be damaging the medieval ruins of Ani, on the other side of the Arpacay valley;
Refugees and internally displaced persons: *IDPs:* 1-1.2 million (fighting 1984-99 between Kurdish PKK and Turkish military; most IDPs in southeastern provinces) (2007)
Illicit drugs: key transit route for Southwest Asian heroin to Western Europe and, to a lesser extent, the US–via air, land, and sea routes; major Turkish and other international trafficking organizations operate out of Istanbul; laboratories to convert imported morphine base into heroin exist in remote regions of Turkey and near Istanbul; government maintains strict controls over areas of legal opium poppy cultivation and over output of poppy straw concentrate; lax enforcement of money-laundering controls

TURKMENISTAN

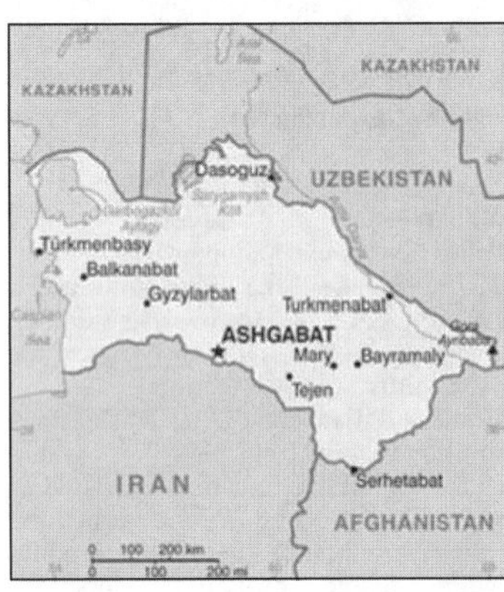

INTRODUCTION

Background: Eastern Turkmenistan for centuries formed part of the Persian province of Khurasan; in medieval times Merv (today known as Mary) was one of the great cities of the Islamic world and an important stop on the Silk Road. Annexed by Russia between 1865 and 1885, Turkmenistan became a Soviet republic in 1924. It achieved independence upon the dissolution of the USSR in 1991. Extensive hydrocarbon/natural gas reserves could prove a boon to this underdeveloped country once extraction and delivery projects are expanded. The Turkmen Government is actively working to diversify its gas export routes beyond the still dominant Russian pipeline network. In 2010, new gas export pipelines that carry Turkmen gas to China and to northern Iran began operating, effectively ending the Russian monopoly on Turkmen gas exports. President for Life Saparmurat NYYAZOW died in December 2006, and Turkmenistan held its first multi-candidate presidential election in February 2007. Gurbanguly BERDIMUHAMEDOW, a deputy cabinet chairman under NYYAZOW, emerged as the country's new president.

GEOGRAPHY

Location: Central Asia, bordering the Caspian Sea, between Iran and Kazakhstan
Geographic coordinates: 40 00 N, 60 00 E
Map references: Asia
Area: *total:* 488,100 sq km
country comparison to the world: 52
land: 469,930 sq km
water: 18,170 sq km
Area—comparative: slightly larger than California
Land boundaries: *total:* 3,736 km
border countries: Afghanistan 744 km, Iran 992 km, Kazakhstan 379 km, Uzbekistan 1,621 km
Coastline: 0 km; note–Turkmenistan borders the Caspian Sea (1,768 km)
Maritime claims: none (landlocked)
Climate: subtropical desert
Terrain: flat-to-rolling sandy desert with dunes rising to mountains in the south; low mountains along border with Iran; borders Caspian Sea in west
Elevation extremes: *lowest point:* Vpadina Akchanaya -81 m
note: Sarygamysh Koli is a lake in northern Turkmenistan with a water level that fluctuates above and below the elevation of Vpadina Akchanaya (the lake has dropped as low as -110 m)
highest point: Gora Ayribaba 3,139 m
Natural resources: petroleum, natural gas, sulfur, salt
Land use: *arable land:* 4.51%
permanent crops: 0.14%
other: 95.35% (2005)
Irrigated land: 18,000 sq km (2003)
Total renewable water resources: 60.9 cu km (1997)
Freshwater withdrawal (domestic/industrial/agricultural): *total:* 24.65 cu km/yr (2%/1%/98%)
per capita: 5,104 cu m/yr (2000)
Natural hazards: NA
Environment—current issues: contamination of soil and groundwater with agricultural chemicals, pesticides; salination, water logging of soil due to poor irrigation methods; Caspian Sea pollution; diversion of a large share of the flow of the Amu Darya into irrigation contributes to that river's inability to replenish the Aral Sea; desertification
Environment—international agreements: *party to:* Biodiversity, Climate Change, Climate Change-Kyoto Protocol, Desertification, Hazardous Wastes, Ozone Layer Protection
signed, but not ratified: none of the selected agreements
Geography—note: landlocked; the western and central low-lying desolate portions of the country make up the great Garagum (Kara-Kum) desert, which occupies over 80% of the country; eastern part is plateau

PEOPLE

Population: 4,997,503 (July 2011 est.)
country comparison to the world: 115
Age structure: *0–14 years:* 27.5% (male 696,749/female 679,936)
15–64 years: 68.4% (male 1,692,885/female 1,724,019)
65 years and over: 4.1% (male 88,590/female 115,324) (2011 est.)
Median age:
total: 25.3 years
male: 24.9 years
female: 25.8 years (2011 est.)
Population growth rate: 1.138% (2011 est.)
country comparison to the world: 103
Birth rate: 19.54 births/1,000 population (2011 est.)
country comparison to the world: 91
Death rate: 6.24 deaths/1,000 population (July 2011 est.)
country comparison to the world: 158
Net migration rate: -1.92 migrant(s)/1,000 population (2011 est.)
country comparison to the world: 163
Urbanization: *urban population:* 50% of total population (2010)
rate of urbanization: 2.2% annual rate of change (2010-15 est.)
Major cities—population: ASHGABAT (capital) 637,000 (2009)
Sex ratio: *at birth:* 1.05 male(s)/female
under 15 years: 1.02 male(s)/female
15–64 years: 0.98 male(s)/female
65 years and over: 0.77 male(s)/female
total population: 0.98 male(s)/female (2011 est.)
Infant mortality rate: *total:* 42.34 deaths/1,000 live births
country comparison to the world: 58
male: 50.42 deaths/1,000 live births
female: 33.85 deaths/1,000 live births (2011 est.)
Life expectancy at birth: *total population:* 68.52 years
country comparison to the world: 151
male: 65.57 years
female: 71.63 years (2011 est.)
Total fertility rate: 2.16 children born/woman (2011 est.)
country comparison to the world: 111
HIV/AIDS—adult prevalence rate: less than 0.1% (2007 est.)
country comparison to the world: 110
HIV/AIDS—people living with HIV/AIDS: fewer than 200 (2007 est.)
country comparison to the world: 157
HIV/AIDS—deaths: fewer than 100 (2004 est.)

country comparison to the world: 118
Drinking water source: *Improved:*
urban: 97% of population
rural: 72% of population
total: 83% of population
Unimproved:
urban: 3% of population
rural: 28% of population
total: 17% of population (2000)
Sanitation facility access: *Improved:*
urban: 99% of population
rural: 97% of population
total: 98% of population
Unimproved: urban: 1% of population
rural: 3% of population
total: 2% of population (2008)
Nationality: *noun:* Turkmen(s)
adjective: Turkmen
Ethnic groups: Turkmen 85%, Uzbek 5%, Russian 4%, other 6% (2003)
Religions: Muslim 89%, Eastern Orthodox 9%, unknown 2%
Languages: Turkmen (official) 72%, Russian 12%, Uzbek 9%, other 7%
Literacy: *definition:* age 15 and over can read and write
total population: 98.8%
male: 99.3%
female: 98.3% (1999 est.)
School life expectancy (primary to tertiary education): NA
Education expenditures: NA

GOVERNMENT

Country name: *conventional long form:* none
conventional short form: Turkmenistan
local long form: none
local short form: Turkmenistan
former: Turkmen Soviet Socialist Republic
Government type: defines itself as a secular democracy and a presidential republic; in actuality displays authoritarian presidential rule, with power concentrated within the presidential administration
Capital: *name:* Ashgabat (Ashkhabad)
geographic coordinates: 37 57 N, 58 23 E
time difference: UTC+5 (10 hours ahead of Washington, DC during Standard Time)
Administrative divisions: 5 provinces (welayatlar, singular–welayat) and 1 independent city*: Ahal Welayaty (Anew), Ashgabat*, Balkan Welayaty (Balkanabat), Dashoguz Welayaty, Lebap Welayaty (Turkmenabat), Mary Welayaty
note: administrative divisions have the same names as their administrative centers (exceptions have the administrative center name following in parentheses)
Independence: 27 October 1991 (from the Soviet Union)
National holiday: Independence Day, 27 October (1991)
Constitution: adopted 26 September 2008
Legal system: civil law system with Islamic law influences
International law organization participation: has not submitted an ICJ jurisdiction declaration; non-party state to the ICCt
Suffrage: 18 years of age; universal
Executive branch: *chief of state:* President Gurbanguly BERDIMUHAMEDOW (since 14 February 2007); note–the president is both the chief of state and head of government
head of government: President Gurbanguly BERDIMUHAMEDOW (since 14 February 2007)
cabinet: Cabinet of Ministers appointed by the president
(For more information visit the World Leaders website)
elections: president elected by popular vote for a five-year term; election last held on 11 February 2007 (next to be held in February 2012)
election results: Gurbanguly BERDIMUHAMEDOW elected president; percent of vote–Gurbanguly BERDIMUHAMEDOW 89.2%, Amanyaz ATAJYKOW 3.2%, other candidates 7.6%
Legislative branch: unicameral parliament known as the National Assembly (Mejlis) (125 seats; members elected by popular vote to serve five-year terms)
elections: last held on 14 December 2008 (next to be held in December 2013)
election results: 100% of elected officials are members of either the Democratic Party of Turkmenistan or its pseudo-civil society parent organization, the Revival Movement, and are preapproved by the president
note: in 26 September 2008, a new constitution of Turkmenistan abolished a second, 2,507-member legislative body known as the People's Council and expanded the number of deputies in the National Assembly from 65 to 125; the powers formerly held by the People's Council were divided up between the president and the National Assembly
Judicial branch: Supreme Court (judges are appointed by the president)
Political parties and leaders: Democratic Party of Turkmenistan or DPT [Gurbanguly BERDIMUHAMEDOW is chairman; Kasymguly BABAYEW is DPT Political Council First Secretary]
note: formal opposition parties are outlawed; unofficial, small opposition movements exist abroad; the three most prominent opposition groups-in-exile are the National Democratic Movement of Turkmenistan (NDMT), the Republican Party of Turkmenistan, and the Watan (Fatherland) Party; the NDMT was led by former Foreign Minister Boris SHIKHMURADOV until his arrest and imprisonment in the wake of the 25 November 2002 attack on President NYYAZOW's motorcade
Political pressure groups and leaders: none
International organization participation: ADB, CIS (associate member, has not ratified the 1993 CIS charter although it participates in meetings), EAPC, EBRD, ECO, FAO, G-77, IBRD, ICAO, ICRM, IDA, IDB, IFC, IFRCS, ILO, IMF, IMO, Interpol, IOC, IOM (observer), ISO (correspondent), ITU, MIGA, NAM, OIC, OPCW, OSCE, PFP, UN, UNCTAD, UNESCO, UNIDO, UNWTO, UPU, WCO, WFTU, WHO, WIPO, WMO
Diplomatic representation in the US: *chief of mission:* Ambassador Meret Bairamovich ORAZOW
chancery: 2207 Massachusetts Avenue NW, Washington, DC 20008
telephone: [1] (202) 588-1500
FAX: [1] (202) 588-0697
Diplomatic representation from the US: *chief of mission:* Ambassador Eileen A. MALLOY
embassy: No. 9 1984 Street (formerly Pushkin Street), Ashgabat, Turkmenistan 744000
mailing address: 7070 Ashgabat Place, Washington, DC 20521-7070
telephone: [993] (12) 35-00-45
FAX: [993] (12) 39-26-14
Flag description: green field with a vertical red stripe near the hoist side, containing five tribal guls (designs used in producing carpets) stacked above two crossed olive branches; five white stars and a white crescent moon appear in the upper corner of the field just to the fly side of the red stripe; the green color and crescent moon represent Islam; the five stars symbolize the regions or welayats of Turkmenistan; the guls reflect the national identity of Turkmenistan where carpet-making has long been a part of traditional nomadic life
note: the flag of Turkmenistan is the most intricate of all national flags
National anthem: *name:* "Garassyz, Bitarap Turkmenistanyn" (Independent, Neutral, Turkmenistan State Anthem)
lyrics/music: collective/Veli MUKHATOV
note: adopted 1997, lyrics revised 2008; following the death of the President Saparmurat NYYAZOW, the lyrics were altered to eliminate references to the former president

ECONOMY

Economy—overview: Turkmenistan is largely a desert country with intensive agriculture in irrigated oases and sizeable gas and oil resources. The two largest crops are cotton, most of which is produced for export, and wheat, which is domestically consumed. Although agriculture accounts for roughly 10% of GDP, it continues to employ nearly half of the country's workforce. With an authoritarian ex-Communist regime in power and a tribally based social structure, Turkmenistan has taken a cautious approach to economic reform, hoping to use gas and cotton export revenues to sustain its inefficient economy. Privatization goals remain limited. From 1998-2005, Turkmenistan suffered from the continued lack of adequate export routes for natural gas and from obligations on extensive short-term external debt. At the same time, however, total exports rose by an average of roughly 15% per year from 2003-08, largely because of higher international oil and gas prices. New pipelines to China and Iran, that began operation in early 2010, have given Turkmenistan additional export routes for its gas, although these new routes have not offset the sharp drop in export revenue since early 2009 from decreased gas exports to Russia. Overall prospects in the near future are discouraging because of widespread internal poverty, endemic corruption, a poor educational system, government misuse of oil and gas revenues, and Ashgabat's reluctance to adopt market-oriented reforms. In the past,

Turkmenistan's economic statistics were state secrets. The new government has established a State Agency for Statistics, but GDP numbers and other figures are subject to wide margins of error. In particular, the rate of GDP growth is uncertain. Since his election, President BERDIMUHAMEDOW unified the country's dual currency exchange rate, ordered the redenomination of the manat, reduced state subsidies for gasoline, and initiated development of a special tourism zone on the Caspian Sea. Although foreign investment is encouraged, numerous bureaucratic obstacles impede international business activity.
GDP (purchasing power parity): $36.9 billion (2010 est.)
country comparison to the world: 101
$33.79 billion (2009 est.)
$31.85 billion (2008 est.)
note: data are in 2010 US dollars
GDP (official exchange rate): $27.96 billion (2010 est.)
GDP—real growth rate: 9.2% (2010 est.)
country comparison to the world: 7
6.1% (2009 est.)
14.7% (2008 est.)
GDP—per capita (PPP): $7,500 (2010 est.)
country comparison to the world: 127
$6,900 (2009 est.)
$6,600 (2008 est.)
note: data are in 2010 US dollars
GDP—composition by sector:
agriculture: 10.2%
industry: 30%
services: 59.8% (2010 est.)
Labor force: 2.3 million (2008 est.)
country comparison to the world: 115
Labor force—by occupation:
agriculture: 48.2%
industry: 14%
services: 37.8% (2004 est.)
Unemployment rate: 60% (2004 est.)
country comparison to the world: 196
Population below poverty line: 30% (2004 est.)
Household income or consumption by percentage share: *lowest 10%:* 2.6%
highest 10%: 31.7% (1998)
Distribution of family income—Gini index: 40.8 (1998)
country comparison to the world: 58
Investment (gross fixed): 12.4% of GDP (2010 est.)
country comparison to the world: 145
Budget: *revenues:* $1.97 billion
expenditures: $1.878 billion (2010 est.)
Inflation rate (consumer prices): 12% (2010 est.)
country comparison to the world: 206
10% (2009 est.)
Stock of narrow money: $573 million (31 December 2010 est.)
country comparison to the world: 159
$469.5 million (31 December 2009 est.)
Stock of broad money: $1.053 billion (31 December 2010 est.)
country comparison to the world: 164
$912.3 million (31 December 2009 est.)
Stock of domestic credit: $2.089 billion (31 December 2010 est.)
country comparison to the world: 126
$1.811 billion (31 December 2009 est.)
Market value of publicly traded shares: $NA
Agriculture—products: cotton, grain; livestock
Industries: natural gas, oil, petroleum products, textiles, food processing
Industrial production growth rate: 7.3% (2010 est.)
country comparison to the world: 41
Electricity—production: 15.5 billion kWh (2009 est.)
country comparison to the world: 78
Electricity—consumption: 13 billion kWh (2009 est.)
country comparison to the world: 79
Electricity—exports: 2.5 billion kWh (2009 est.)
Electricity—imports: 0 kWh (2009 est.)
Oil—production: 197,700 bbl/day (2009 est.)
country comparison to the world: 42
Oil—consumption: 120,000 bbl/day (2009 est.)
country comparison to the world: 72
Oil—exports: 38,360 bbl/day (2009 est.)
country comparison to the world: 81
Oil—imports: 0 bbl/day (2009 est.)
country comparison to the world: 206
Oil—proved reserves: 600 million bbl (1 January 2010 est.)
country comparison to the world: 45
Natural gas—production: 34 billion cu m (2009 est.)
country comparison to the world: 25
Natural gas—consumption: 20 billion cu m (2009 est.)
country comparison to the world: 35
Natural gas—exports: 14 billion cu m (2009 est.)
country comparison to the world: 15
Natural gas—imports: 0 cu m (2008 est.)
country comparison to the world: 128
Natural gas—proved reserves: 7.504 trillion cu m (1 January 2010 est.)
country comparison to the world: 4
Current account balance: $3.081 billion (2010 est.)
country comparison to the world: 39
$1.065 billion (2009 est.)
Exports: $9.672 billion (2010 est.)
country comparison to the world: 89
$6.737 billion (2009 est.)
Exports—commodities: gas, crude oil, petrochemicals, textiles, cotton fiber
Exports—partners: Ukraine 22.3%, Turkey 10.27%, Hungary 6.75%, UAE 6.25%, Poland 6.16%, Afghanistan 5.79%, Iran 5.17% (2009)
Imports: $4.888 billion (2010 est.)
country comparison to the world: 118
$4.109 billion (2009 est.)
Imports—commodities: machinery and equipment, chemicals, foodstuffs
Imports—partners: China 18.03%, Turkey 16.49%, Russia 16.45%, Germany 5.91%, UAE 5.81%, Ukraine 5.67%, US 5.41%, France 4.32% (2009)
Reserves of foreign exchange and gold: $10.81 billion (31 December 2010 est.)
country comparison to the world: 56
$9.551 billion (31 December 2009 est.)
Debt—external: $5 billion (2009 est.)
country comparison to the world: 107
$1.4 billion (2004 est.)
Exchange rates: Turkmen manat (TMM) per US dollar–2.85 (2010), 2.85 (2009), 14,250 (2008)

COMMUNICATIONS

Telephones—main lines in use: 478,000 (2009)
country comparison to the world: 100
Telephones—mobile cellular: 1.5 million (2009)
country comparison to the world: 139
Telephone system: *general assessment:* telecommunications network remains underdeveloped and progress toward improvement is slow; strict government control and censorship inhibits liberalization and modernization
domestic: Turkmentelekom, in cooperation with foreign partners, has installed high speed fiber-optic lines and has upgraded most of the country's telephone exchanges and switching centers with new digital technology; mobile telephone usage is expanding with Russia's Mobile Telesystems (MTS) the primary service provider; combined fixed-line and mobile teledensity is about 40 per 100 persons
international: country code–993; linked by fiber-optic cable and microwave radio relay to other CIS republics and to other countries by leased connections to the Moscow international gateway switch; an exchange in Ashgabat switches international traffic through Turkey via Intelsat; satellite earth stations–1 Orbita and 1 Intelsat (2008)
Broadcast media: broadcast media is government controlled and censored; 4 state-owned TV and 4 state-owned radio networks; satellite dishes and programming provide an alternative to the state-run media; officials sometimes limit access to satellite TV by seizing satellite dishes (2007)
Internet country code: .tm
Internet hosts: 794 (2010)
country comparison to the world: 172
Internet users: 80,400 (2009)
country comparison to the world: 165

TRANSPORTATION

Airports: 27 (2010)
country comparison to the world: 123
Airports—with paved runways: *total:* 22
over 3,047 m: 1
2,438 to 3,047 m: 10
1,524 to 2,437 m: 9
914 to 1,523 m: 2 (2010)
Airports—with unpaved runways: *total:* 5
1,524 to 2,437 m: 1
under 914 m: 4 (2010)
Heliports: 1 (2010)
Pipelines: gas 7,352 km; oil 1,457 km (2010)
Railways: *total:* 2,980 km
country comparison to the world: 55
broad gauge: 2,980 km 1.520-m gauge (2008)
Roadways: *total:* 58,592 km
country comparison to the world: 75
paved: 47,577 km
unpaved: 11,015 km (2002)
Waterways: 1,300 km (Amu Darya and Kara Kum canal are important inland waterways) (2008)
country comparison to the world: 56

Merchant marine: *total:* 9
country comparison to the world: 118
by type: cargo 4, petroleum tanker 4, refrigerated cargo 1
foreign-owned: 1 (Turkey 1) (2010)
Ports and terminals: Turkmenbasy

MILITARY

Military branches: Army, Navy, Air and Air Defense Forces (2010)
Military service age and obligation: 18-30 years of age for compulsory military service; 2-year conscript service obligation (2009)
Manpower available for military service: *males age 16–49:* 1,380,794
females age 16–49: 1,387,211 (2010 est.)
Manpower fit for military service: *males age 16–49:* 1,066,649
females age 16–49: 1,185,538 (2010 est.)
Manpower reaching militarily significant age annually: *male:* 53,829
female: 52,988 (2010 est.)
Military expenditures: 3.4% of GDP (2005 est.)
country comparison to the world: 36

TRANSNATIONAL ISSUES

Disputes—international: cotton monoculture in Uzbekistan and Turkmenistan creates water-sharing difficulties for Amu Darya river states; field demarcation of the boundaries with Kazakhstan commenced in 2005, but Caspian seabed delimitation remains stalled with Azerbaijan, Iran, and Kazakhstan due to Turkmenistan's indecision over how to allocate the sea's waters and seabed; bilateral talks continue with Azerbaijan on dividing the seabed and contested oilfields in the middle of the Caspian
Refugees and internally displaced persons: refugees (country of origin): 11,173 (Tajikistan); less than 1,000 (Afghanistan) (2007)
Illicit drugs: transit country for Afghan narcotics bound for Russian and Western European markets; transit point for heroin precursor chemicals bound for Afghanistan

TURKS AND CAICOS ISLANDS

(OVERSEAS TERRITORY OF THE UK)

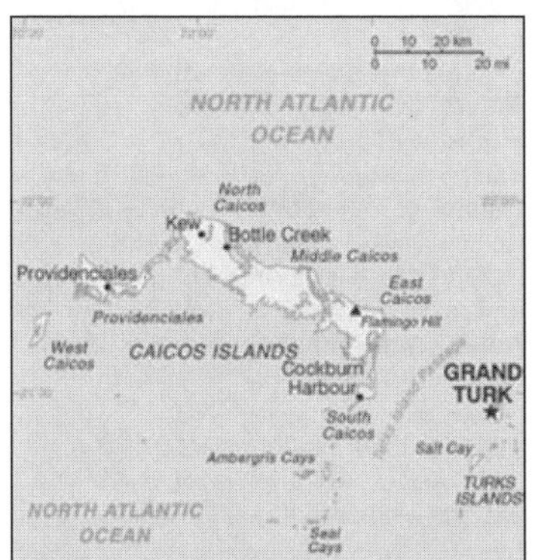

INTRODUCTION

Background: The islands were part of the UK's Jamaican colony until 1962, when they assumed the status of a separate crown colony upon Jamaica's independence. The governor of The Bahamas oversaw affairs from 1965 to 1973. With Bahamian independence, the islands received a separate governor in 1973. Although independence was agreed upon for 1982, the policy was reversed and the islands remain a British overseas territory.

GEOGRAPHY

Location: Caribbean, two island groups in the North Atlantic Ocean, southeast of The Bahamas, north of Haiti
Geographic coordinates: 21 45 N, 71 35 W
Map references: Central America and the Caribbean
Area: *total:* 948 sq km
country comparison to the world: 185
land: 948 sq km
water: 0 sq km
Area—comparative: 2.5 times the size of Washington, DC
Land boundaries: 0 km
Coastline: 389 km
Maritime claims: territorial sea: 12 nm
exclusive fishing zone: 200 nm
Climate: tropical; marine; moderated by trade winds; sunny and relatively dry
Terrain: low, flat limestone; extensive marshes and mangrove swamps
Elevation extremes: *lowest point:* Caribbean Sea 0 m
highest point: Flamingo Hill 48 m
Natural resources: spiny lobster, conch
Land use: *arable land:* 2.33%
permanent crops: 0%
other: 97.67% (2005)
Irrigated land: NA
Natural hazards: frequent hurricanes
Environment—current issues: limited natural freshwater resources, private cisterns collect rainwater
Geography—note: about 40 islands (eight inhabited)

PEOPLE

Population: 44,819 (July 2011 est.)
country comparison to the world: 209
Age structure: *0–14 years:* 22.7% (male 5,188/female 5,008)
15–64 years: 73.2% (male 16,653/female 16,156)
65 years and over: 4% (male 813/female 1,001) (2011 est.)
Median age: *total:* 30.4 years
male: 30.8 years
female: 30.1 years (2011 est.)
Population growth rate: 3.485% (2011 est.)
country comparison to the world: 4
Birth rate: 17.76 births/1,000 population (2011 est.)
country comparison to the world: 110
Death rate: 2.99 deaths/1,000 population (July 2011 est.)
country comparison to the world: 219
Net migration rate: 20.08 migrant(s)/1,000 population (2011 est.)
country comparison to the world: 2
Urbanization: *urban population:* 93% of total population (2010)
rate of urbanization: 1.6% annual rate of change (2010-15 est.)
Major cities—population: GRAND TURK (capital) 6,000 (2009)
Sex ratio: *at birth:* 1.052 male(s)/female
under 15 years: 1.04 male(s)/female
15–64 years: 1.1 male(s)/female
65 years and over: 0.98 male(s)/female
total population: 1.07 male(s)/female (2011 est.)
Infant mortality rate: *total:* 11.97 deaths/1,000 live births
country comparison to the world: 134
male: 15.02 deaths/1,000 live births
female: 8.77 deaths/1,000 live births (2011 est.)
Life expectancy at birth: *total population:* 79.11 years
country comparison to the world: 40
male: 76.39 years
female: 81.97 years (2011 est.)
Total fertility rate: 1.7 children born/woman (2011 est.)
country comparison to the world: 169
HIV/AIDS—adult prevalence rate: NA
HIV/AIDS—people living with HIV/AIDS: NA
HIV/AIDS—deaths: NA
Drinking water source: *Improved:*
urban: 100% of population
rural: 100% of population
total: 100% of population (2008)
Sanitation facility access: *Improved:*
urban: 98% of population
rural: 94% of population
total: 96% of population
Unimproved:
urban: 2% of population
rural: 6% of population
total: 4% of population (2000)
Nationality: *noun:* none
adjective: none
Ethnic groups: black 90%, mixed, European, or North American 10%
Religions: Baptist 40%, Anglican 18%, Methodist 16%, Church of God 12%, other 14% (1990)
Languages: English (official)
Literacy: *definition:* age 15 and over has ever attended school
total population: 98%
male: 99%
female: 98% (1970 est.)
School life expectancy (primary to tertiary education): *total:* 11 years
male: 11 years
female: 12 years (2005)
Education expenditures: NA

People—note: destination and transit point for illegal Haitian immigrants bound for the Turks and Caicos Islands, The Bahamas, and the US

GOVERNMENT

Country name: *conventional long form:* none
conventional short form: Turks and Caicos Islands
abbreviation: TCI
Dependency status: overseas territory of the UK
Government type: NA
Capital: *name:* Grand Turk (Cockburn Town)
geographic coordinates: 21 28 N, 71 08 W
time difference: UTC-5 (same time as Washington, DC during Standard Time)
daylight saving time: +1hr, begins second Sunday in March; ends first Sunday in November
Administrative divisions: none (overseas territory of the UK)
Independence: none (overseas territory of the UK)
National holiday: Constitution Day, 30 August (1976)
Constitution: Turks and Caicos Islands Constitution (Interim Amendment) Order 2009, S.I. 2009/701–effective 14 August 2009–suspended Ministerial government, the House of Assembly, and the constitutional right to trial by jury, and imposed direct British rule
Legal system: mixed legal system of English common law and civil law
Suffrage:
18 years of age; universal
Executive branch: *chief of state:* Queen ELIZABETH II (since 6 February 1952); represented by Governor Gordon WETHERELL (since 5 August 2008)
head of government: Governor Gordon WETHERELL (since 14 August 2009); note–the office of premier is suspended by the Order in Council, effective 14 August 2009
cabinet: under provisions of the Order in Council, the cabinet is suspended effective 14 August 2009 and replaced by an Advisory Council appointed by the governor
(For more information visit the World Leaders website)
elections: the monarchy is hereditary; governor appointed by the monarch
note: following an investigation into allegations of widespread corruption and misconduct within the Turks and Caicos Government, the UK foreign minister directed the governor to bring into effect on 14 August 2009 an Order in Council suspending Ministerial government and the House of Assembly, and imposing direct rule for a period of up to two years
Legislative branch: under provisions of the Order in Council, the unicameral House of Assembly is dissolved and all seats vacated for a period of up to two years; in the interim, a Consultative Forum, appointed by the governor, will be established
elections: last held on 9 February 2007 (next to be held by July 2011)
election results: under provisions of the Order in Council, all seats in the House of Assembly are vacated
Judicial branch: Supreme Court; Court of Appeal
Political parties and leaders: People's Democratic Movement or PDM [Floyd SEYMOUR]; Progressive National Party or PNP [Michael Eugene MISICK]
Political pressure groups and leaders: NA
International organization participation: Caricom (associate), CDB, Interpol (subbureau), UPU
Diplomatic representation in the US: none (overseas territory of the UK)
Diplomatic representation from the US: none (overseas territory of the UK)
Flag description: blue, with the flag of the UK in the upper hoist-side quadrant and the colonial shield centered on the outer half of the flag; the shield is yellow and displays a conch shell, a spiny lobster, and Turks Head cactus–three common elements of the islands' biota
National anthem: *name:* "This Land of Ours"
lyrics/music: Conrad HOWELL
note: serves as a local anthem; as a territory of the United Kingdom, "God Save the Queen" is the official anthem (see United Kingdom)

ECONOMY

Economy—overview: The Turks and Caicos economy is based on tourism, offshore financial services, and fishing. Most capital goods and food for domestic consumption are imported. The US is the leading source of tourists, accounting for more than three-quarters of the 175,000 visitors that arrived in 2004. Major sources of government revenue also include fees from offshore financial activities and customs receipts.
GDP (purchasing power parity): $216 million (2002 est.)
country comparison to the world: 214
GDP (official exchange rate): $NA
GDP—real growth rate: 4.9% (2000 est.)
country comparison to the world: 75
GDP—per capita (PPP): $11,500 (2002 est.)
country comparison to the world: 97
GDP—composition by sector:
agriculture: NA%
industry: NA%
services: NA%
Labor force: 4,848 (1990 est.)
country comparison to the world: 219
Labor force—by occupation: note: about 33% in government and 20% in agriculture and fishing; significant numbers in tourism, financial, and other services
Unemployment rate: 10% (1997 est.)
country comparison to the world: 110
Population below poverty line: NA%
Household income or consumption by percentage share: *lowest 10%:* NA%
highest 10%: NA%
Budget:
revenues: $47 million
expenditures: $33.6 million (1997-98 est.)
Inflation rate (consumer prices): 4% (1995)
country comparison to the world: 116
Agriculture—products: corn, beans, cassava (tapioca), citrus fruits; fish
Industries: tourism, offshore financial services
Industrial production growth rate: NA%
Electricity—production: 12 million kWh (2007 est.)
country comparison to the world: 210
Electricity—consumption: 11.16 million kWh (2007 est.)
country comparison to the world: 210
Electricity—exports: 0 kWh (2008 est.)
Electricity—imports: 0 kWh (2008 est.)
Oil—production: 0 bbl/day (2009 est.)
country comparison to the world: 131
Oil—consumption: NA bbl/day (2009 est.)
Oil—exports: NA (2007 est.)
Oil—imports: NA bbl/day (2007 est.)
Oil—proved reserves: 0 bbl (1 January 2010 est.)
country comparison to the world: 193
Natural gas—production: 0 cu m (2008 est.)
country comparison to the world: 189
Natural gas—consumption: 0 cu m (2008 est.)
country comparison to the world: 126
Natural gas—exports: 0 cu m (2008 est.)
country comparison to the world: 184
Natural gas—imports: 0 cu m (2008 est.)
country comparison to the world: 123
Natural gas—proved reserves: 0 cu m (1 January 2010 est.)
country comparison to the world: 192
Exports: $169.2 million (2000)
country comparison to the world: 184
Exports—commodities: lobster, dried and fresh conch, conch shells
Imports: $175.6 million (2000)
country comparison to the world: 203
Imports—commodities: food and beverages, tobacco, clothing, manufactures, construction materials
Debt—external: $NA
Exchange rates: the US dollar is used

COMMUNICATIONS

Telephones—main lines in use: 3,700 (2009)
country comparison to the world: 216
Telephones—mobile cellular: 25,100 (2004)
country comparison to the world: 204
Telephone system: *general assessment:* fully digital system with international direct dialing
domestic: full range of services available; GSM wireless service available
international: country code–1-649; the Americas Region Caribbean Ring System (ARCOS-1) fiber optic telecommunications submarine cable provides connectivity to South and Central America, parts of the Caribbean, and the US; satellite earth station–1 Intelsat (Atlantic Ocean)
Broadcast media: while there are no local terrestrial TV stations, broadcasts from the Bahamas can be received; multi-channel cable and satellite TV services are available; government-run radio network operates alongside private broadcasters with a total of about 15 stations broadcasting (2007)

Internet country code: .tc
Internet hosts: 8,969 (2010)
country comparison to the world: 127

TRANSPORTATION

Airports: 8 (2010)
country comparison to the world: 164
Airports—with paved runways: *total:* 7
1,524 to 2,437 m: 4
914 to 1,523 m: 1
under 914 m: 2 (2010)
Airports—with unpaved runways: *total:* 1
under 914 m: 1 (2010)
Roadways: *total:* 121 km
country comparison to the world: 211
paved: 24 km
unpaved: 97 km (2003)
Merchant marine: registered in other countries: 1 (Panama 1) (2008)
country comparison to the world: 160
Ports and terminals: Cockburn Harbour, Grand Turk, Providenciales

MILITARY

Manpower fit for military service: *males age 16–49:* 11,842
females age 16–49: 11,755 (2010 est.)
Manpower reaching militarily significant age annually: *male:* 338
female: 342 (2010 est.)
Military—note: defense is the responsibility of the UK

TRANSNATIONAL ISSUES

Disputes—international: have received Haitians fleeing economic and civil disorder
Illicit drugs: transshipment point for South American narcotics destined for the US and Europe

TUVALU

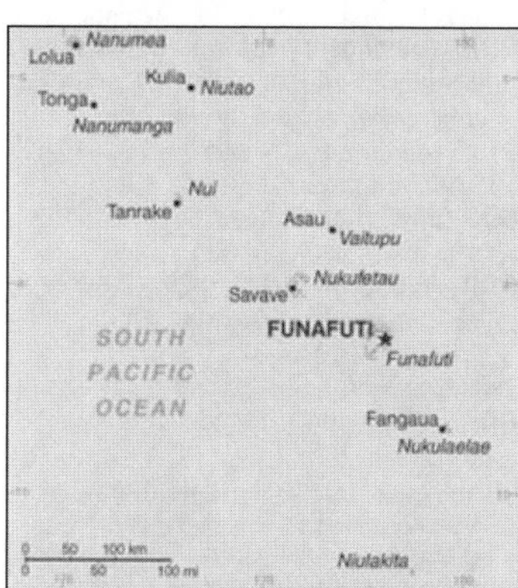

INTRODUCTION

Background: In 1974, ethnic differences within the British colony of the Gilbert and Ellice Islands caused the Polynesians of the Ellice Islands to vote for separation from the Micronesians of the Gilbert Islands. The following year, the Ellice Islands became the separate British colony of Tuvalu. Independence was granted in 1978. In 2000, Tuvalu negotiated a contract leasing its Internet domain name ".tv" for $50 million in royalties over a 12-year period.

GEOGRAPHY

Location: Oceania, island group consisting of nine coral atolls in the South Pacific Ocean, about one-half of the way from Hawaii to Australia
Geographic coordinates: 8 00 S, 178 00 E
Map references: Oceania
Area: *total:* 26 sq km
country comparison to the world: 236
land: 26 sq km
water: 0 sq km
Area—comparative: 0.1 times the size of Washington, DC
Land boundaries: 0 km
Coastline: 24 km
Maritime claims: territorial sea: 12 nm
contiguous zone: 24 nm
exclusive economic zone: 200 nm
Climate: tropical; moderated by easterly trade winds (March to November); westerly gales and heavy rain (November to March)
Terrain: low-lying and narrow coral atolls
Elevation extremes: *lowest point:* Pacific Ocean 0 m
highest point: unnamed location 5 m
Natural resources: fish
Land use: *arable land:* 0%
permanent crops: 66.67%
other: 33.33% (2005)
Irrigated land: NA
Natural hazards: severe tropical storms are usually rare, but in 1997 there were three cyclones; low level of islands make them sensitive to changes in sea level
Environment—current issues: since there are no streams or rivers and groundwater is not potable, most water needs must be met by catchment systems with storage facilities (the Japanese Government has built one desalination plant and plans to build one other); beachhead erosion because of the use of sand for building materials; excessive clearance of forest undergrowth for use as fuel; damage to coral reefs from the spread of the Crown of Thorns starfish; Tuvalu is concerned about global increases in greenhouse gas emissions and their effect on rising sea levels, which threaten the country's underground water table; in 2000, the government appealed to Australia and New Zealand to take in Tuvaluans if rising sea levels should make evacuation necessary
Environment—international agreements: *party to:* Biodiversity, Climate Change, Climate Change-Kyoto Protocol, Desertification, Law of the Sea, Ozone Layer Protection, Ship Pollution, Whaling
signed, but not ratified: none of the selected agreements
Geography—note: one of the smallest and most remote countries on Earth; six of the nine coral atolls—Nanumea, Nui, Vaitupu, Nukufetau, Funafuti, and Nukulaelae—have lagoons open to the ocean; Nanumaya and Niutao have landlocked lagoons; Niulakita does not have a lagoon

PEOPLE

Population: 10,544 (July 2011 est.)
country comparison to the world: 223
Age structure: *0–14 years:* 30.6% (male 1,656/female 1,569)
15–64 years: 64% (male 3,294/female 3,459)
65 years and over: 5.4% (male 238/female 328) (2011 est.)
Median age: *total:* 24.1 years
male: 22.7 years
female: 26.1 years (2011 est.)
Population growth rate: 0.702% (2011 est.)
country comparison to the world: 140
Birth rate: 23.24 births/1,000 population (2011 est.)
country comparison to the world: 71
Death rate: 9.2 deaths/1,000 population (July 2011 est.)
country comparison to the world: 67
Net migration rate: -7.02 migrant(s)/1,000 population (2011 est.)
country comparison to the world: 203
Urbanization: *urban population:* 50% of total population (2010)
rate of urbanization: 1.4% annual rate of change (2010-15 est.)
Sex ratio: *at birth:* 1.042 male(s)/female
under 15 years: 1.06 male(s)/female
15–64 years: 0.95 male(s)/female
65 years and over: 0.73 male(s)/female
total population: 0.97 male(s)/female (2011 est.)
Infant mortality rate: *total:* 34.52 deaths/1,000 live births
country comparison to the world: 69
male: 37.56 deaths/1,000 live births
female: 31.33 deaths/1,000 live births (2011 est.)
Life expectancy at birth: *total population:* 64.75 years
country comparison to the world: 169
male: 62.7 years
female: 66.9 years (2011 est.)
Total fertility rate: 3.11 children born/woman (2011 est.)
country comparison to the world: 58
HIV/AIDS—adult prevalence rate: NA
HIV/AIDS—people living with HIV/AIDS: NA
HIV/AIDS—deaths: NA
Drinking water source: *Improved:*
urban: 98% of population
rural: 97% of population
total: 97% of population
Unimproved: urban: 2% of population
rural: 3% of population
total: 3% of population (2008)
Sanitation facility access: *Improved:*
urban: 88% of population
rural: 81% of population
total: 84% of population
Unimproved:
urban: 12% of population
rural: 19% of population
total: 16% of population (2008)
Nationality: *noun:* Tuvaluan(s)

adjective: Tuvaluan
Ethnic groups: Polynesian 96%, Micronesian 4%
Religions: Church of Tuvalu (Congregationalist) 97%, Seventh-Day Adventist 1.4%, Baha'i 1%, other 0.6%
Languages: Tuvaluan (official), English (official), Samoan, Kiribati (on the island of Nui)
Literacy: NA
School life expectancy (primary to tertiary education): *total:* 11 years
male: 10 years
female: 11 years (2001)
Education expenditures: NA

GOVERNMENT

Country name: *conventional long form:* none
conventional short form: Tuvalu
local long form: none
local short form: Tuvalu
former: Ellice Islands
note: "Tuvalu" means "group of eight" referring to the country's eight traditionally inhabited islands
Government type: parliamentary democracy and a Commonwealth realm
Capital: *name:* Funafuti
geographic coordinates: 8 30 S, 179 12 E
time difference: UTC+12 (17 hours ahead of Washington, DC during Standard Time)
note: administrative offices are in Vaiaku Village on Fongafale Islet
Administrative divisions: none
Independence: 1 October 1978 (from the UK)
National holiday: Independence Day, 1 October (1978)
Constitution: 1 October 1978
Legal system: mixed legal system of English common law and local customary law
International law organization participation: has not submitted an ICJ jurisdiction declaration; non-party state to the ICCt
Suffrage:
18 years of age; universal
Executive branch: *chief of state:* Queen ELIZABETH II (since 6 February 1952); represented by Governor General Iakoba TAEIA Italeli (since May 2010)
head of government: Prime Minister Willie TELAVI (since 24 December 2010)
cabinet: Cabinet appointed by the governor general on the recommendation of the prime minister
(For more information visit the World Leaders website)
elections: the monarchy is hereditary; governor general appointed by the monarch on the recommendation of the prime minister; prime minister and deputy prime minister elected by and from the members of parliament following parliamentary elections
election results: Willie TELAVI elected prime minister in a parliamentary election on 24 December 2010 following a no-confidence vote on 21 December 2010 that ousted Maatia TOAFA
Legislative branch: unicameral Parliament or Fale I Fono, also called House of Assembly (15 seats; members elected by popular vote to serve four-year terms)
elections: last held on 16 September 2010 (next to be held in 2014)
election results: percent of vote—NA; seats—independents 15; 10 members reelected
Judicial branch: High Court (a chief justice visits twice a year to preside over its sessions; its rulings can be appealed to the Court of Appeal in Fiji); eight Island Courts (with limited jurisdiction)
Political parties and leaders: there are no political parties but members of parliament usually align themselves in informal groupings
Political pressure groups and leaders: none
International organization participation: ACP, ADB, AOSIS, C, FAO, IBRD, IDA, IFRCS (observer), ILO, IMF, IMO, IOC, ITU, OPCW, PIF, Sparteca, SPC, UN, UNCTAD, UNESCO, UPU, WHO
Diplomatic representation in the US: Tuvalu does not have an embassy in the US—the country's only diplomatic post is in Fiji—Tuvalu does, however, have a UN office located at 800 2nd Avenue, Suite 400D, New York, NY 10017, *telephone:* [1] (212) 490-0534, *fax:* [1] (212) 937-0692
Diplomatic representation from the US: the US does not have an embassy in Tuvalu; the US ambassador to Fiji is accredited to Tuvalu
Flag description: light blue with the flag of the UK in the upper hoist-side quadrant; the outer half of the flag represents a map of the country with nine yellow, five-pointed stars on a blue field symbolizing the nine atolls in the ocean
National anthem: *name:* "Tuvalu mo te Atua" (Tuvalu for the Almighty)
lyrics/music: Afaese MANOA
note: adopted 1978; the anthem's name is also the nation's motto

ECONOMY

Economy—overview: Tuvalu consists of a densely populated, scattered group of nine coral atolls with poor soil. The country has no known mineral resources and few exports and is almost entirely dependent upon imported food and fuel. Subsistence farming and fishing are the primary economic activities. Fewer than 1,000 tourists, on average, visit Tuvalu annually. Job opportunities are scarce and public sector workers make up most of those employed. About 15% of the adult male population work as seamen on merchant ships abroad, and remittances are a vital source of income contributing around $2 million in 2007. Substantial income is received annually from the Tuvalu Trust Fund (TTF) an international trust fund established in 1987 by Australia, NZ, and the UK and supported also by Japan and South Korea. Thanks to wise investments and conservative withdrawals, this fund grew from an initial $17 million to an estimated value of $77 million in 2006. The TTF contributed nearly $9 million towards the government budget in 2006 and is an important cushion for meeting shortfalls in the government's budget. The US Government is also a major revenue source for Tuvalu because of payments from a 1988 treaty on fisheries. In an effort to ensure financial stability and sustainability, the government is pursuing public sector reforms, including privatization of some government functions and personnel cuts. Tuvalu also derives royalties from the lease of its ".tv" Internet domain name with revenue of more than $2 million in 2006. A minor source of government revenue comes from the sale of stamps and coins. With merchandise exports only a fraction of merchandise imports, continued reliance must be placed on fishing and telecommunications license fees, remittances from overseas workers, official transfers, and income from overseas investments. Growing income disparities and the vulnerability of the country to climatic change are among leading concerns for the nation.
GDP (purchasing power parity): $36 million (2010 est.)
country comparison to the world: 223
$36.34 million (2009 est.)
$36.68 million (2008 est.)
GDP (official exchange rate): $32 million (2010 est.)
GDP—real growth rate: 0.2% (2010 est.)
country comparison to the world: 186
-1.7% (2009 est.)
7% (2008 est.)
GDP—per capita (PPP): $3,400 (2010 est.)
country comparison to the world: 166
$2,900 (2009 est.)
$3,000 (2008 est.)
GDP—composition by sector:
agriculture: 16.6%
industry: 27.2%
services: 56.2% (2002)
Labor force: 3,615 (2004 est.)
country comparison to the world: 220
Labor force—by occupation: note: people make a living mainly through exploitation of the sea, reefs, and atolls and from wages sent home by those abroad (mostly workers in the phosphate industry and sailors)
Unemployment rate: NA%
Population below poverty line: NA%
Household income or consumption by percentage share: *lowest 10%:* NA%
highest 10%: NA%
Budget: *revenues:* $21.54 million
expenditures: $23.05 million (2006)
Inflation rate (consumer prices): 3.8% (2006 est.)
country comparison to the world: 110
Agriculture—products:
coconuts; fish
Industries: fishing, tourism, copra
Industrial production growth rate: NA%
Current account balance: $-11.68 million (2003)
country comparison to the world: 63
Exports: $1 million (2004 est.)
country comparison to the world: 217
Exports—commodities: copra, fish
Imports: $12.91 million (2005)
country comparison to the world: 220
Imports—commodities: food, animals, mineral fuels, machinery, manufactured goods
Debt—external: $NA
Exchange rates: Tuvaluan dollars or Australian dollars (AUD) per US dollar—

1.0902 (2010), 1.2822 (2009), 1.2137 (2007), 1.3285 (2006)

COMMUNICATIONS

Telephones—main lines in use: 1,700 (2009)
country comparison to the world: 225
Telephones—mobile cellular: 2,000 (2009)
country comparison to the world: 214
Telephone system: *general assessment:* serves particular needs for internal communications
domestic: radiotelephone communications between islands
international: country code–688; international calls can be made by satellite
Broadcast media: no television broadcast stations; many households use satellite dishes to watch foreign TV stations; 1 government-owned radio station, Radio Tuvalu, includes relays of programming from international broadcasters (2009)
Internet country code: .tv
Internet hosts: 109,478 (2010)
country comparison to the world: 77
Internet users: 4,200 (2008)
country comparison to the world: 205

TRANSPORTATION

Airports: 1 (2010)
country comparison to the world: 234
Airports—with unpaved runways: *total:* 1
1,524 to 2,437 m: 1 (2010)
Roadways: *total:* 8 km
country comparison to the world: 221
paved: 8 km (2002)
Merchant marine: *total:* 66
country comparison to the world: 64
by type: bulk carrier 7, cargo 20, chemical tanker 16, container 3, passenger 2, passenger/cargo 1, petroleum tanker 15, refrigerated cargo 1, vehicle carrier 1
foreign-owned: 49 (Thailand 1, Vietnam 6, Turkey 1, Switzerland 1, South Korea 1, Singapore 25, Maldives 1, Malaysia 1, Kenya 1, Hong Kong 1, China 9, Ukraine 1) (2010)
Ports and terminals: Funafuti

MILITARY

Military branches: no regular military forces; Tuvalu Police Force (2009)
Manpower fit for military service: *males age 16–49:* 2,021
females age 16–49: 2,026 (2010 est.)
Manpower reaching militarily significant age annually: *male:* 119
female: 111 (2010 est.)
Military expenditures: NA

TRANSNATIONAL ISSUES

Disputes—international: none

UGANDA

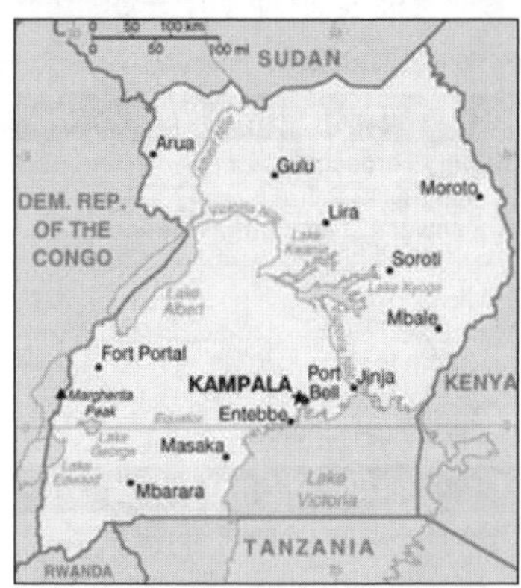

INTRODUCTION

Background: The colonial boundaries created by Britain to delimit Uganda grouped together a wide range of ethnic groups with different political systems and cultures. These differences prevented the establishment of a working political community after independence was achieved in 1962. The dictatorial regime of Idi AMIN (1971-79) was responsible for the deaths of some 300,000 opponents; guerrilla war and human rights abuses under Milton OBOTE (1980-85) claimed at least another 100,000 lives. The rule of Yoweri MUSEVENI since 1986 has brought relative stability and economic growth to Uganda. During the 1990s, the government promulgated non-party presidential and legislative elections.

GEOGRAPHY

Location: Eastern Africa, west of Kenya, east of the Democratic Republic of the Congo
Geographic coordinates: 1 00 N, 32 00 E
Map references: Africa
Area: *total:* 241,038 sq km
country comparison to the world: 80
land: 197,100 sq km
water: 43,938 sq km
Area—comparative: slightly smaller than Oregon
Land boundaries: *total:* 2,698 km
border countries: Democratic Republic of the Congo 765 km, Kenya 933 km, Rwanda 169 km, Sudan 435 km, Tanzania 396 km
Coastline: 0 km (landlocked)
Maritime claims: none (landlocked)
Climate: tropical; generally rainy with two dry seasons (December to February, June to August); semiarid in northeast
Terrain: mostly plateau with rim of mountains
Elevation extremes: *lowest point:* Lake Albert 621 m
highest point: Margherita Peak on Mount Stanley 5,110 m
Natural resources: copper, cobalt, hydropower, limestone, salt, arable land, gold
Land use: *arable land:* 21.57%
permanent crops: 8.92%
other: 69.51% (2005)
Irrigated land: 90 sq km (2003)
Total renewable water resources: 66 cu km (1970)
Freshwater withdrawal (domestic/industrial/agricultural): *total:* 0.3 cu km/yr (43%/17%/40%)
per capita: 10 cu m/yr (2002)
Natural hazards: NA
Environment—current issues: draining of wetlands for agricultural use; deforestation; overgrazing; soil erosion; water hyacinth infestation in Lake Victoria; widespread poaching
Environment—international agreements: *party to:* Biodiversity, Climate Change, Climate Change-Kyoto Protocol, Desertification, Endangered Species, Hazardous Wastes, Law of the Sea, Marine Life Conservation, Ozone Layer Protection, Wetlands
signed, but not ratified: Environmental Modification
Geography—note: landlocked; fertile, well-watered country with many lakes and rivers

PEOPLE

Population: 34,612,250 (July 2011 est.)
country comparison to the world: 36
note: estimates for this country explicitly take into account the effects of excess mortality due to AIDS; this can result in lower life expectancy, higher infant mortality, higher death rates, lower population growth rates, and changes in the distribution of population by age and sex than would otherwise be expected
Age structure: *0–14 years:* 49.9% (male 8,692,239/female 8,564,571)
15–64 years: 48.1% (male 8,383,548/female 8,255,473)
65 years and over: 2.1% (male 291,602/female 424,817) (2011 est.)
Median age: *total:* 15.1 years
male: 15 years
female: 15.1 years (2011 est.)
Population growth rate: 3.576% (2011 est.)
country comparison to the world: 3
Birth rate: 47.49 births/1,000 population (2011 est.)
country comparison to the world: 2
Death rate: 11.71 deaths/1,000 population (July 2011 est.)
country comparison to the world: 33
Net migration rate: -0.02 migrant(s)/1,000 population (2011 est.)
country comparison to the world: 116
Urbanization: *urban population:* 13% of total population (2010)
rate of urbanization: 4.8% annual rate of change (2010-15 est.)
Major cities—population: KAMPALA (capital) 1.535 million (2009)
Sex ratio: *at birth:* 1.03 male(s)/female
under 15 years: 1.01 male(s)/female
15–64 years: 1.01 male(s)/female
65 years and over: 0.7 male(s)/female
total population: 1.01 male(s)/female (2011 est.)
Infant mortality rate: *total:* 62.47 deaths/1,000 live births
country comparison to the world: 28
male: 66.05 deaths/1,000 live births
female: 58.77 deaths/1,000 live births (2011 est.)
Life expectancy at birth: *total population:* 53.24 years
country comparison to the world: 204
male: 52.17 years
female: 54.33 years (2011 est.)
Total fertility rate: 6.69 children born/woman (2011 est.)
country comparison to the world: 2
HIV/AIDS—adult prevalence rate: 6.5% (2009 est.)
country comparison to the world: 10
HIV/AIDS—people living with HIV/AIDS: 1.2 million (2009 est.)
country comparison to the world: 9
HIV/AIDS—deaths: 64,000 (2009 est.)
country comparison to the world: 8
Major infectious diseases: *degree of risk:* very high
food or waterborne diseases: bacterial diarrhea, hepatitis A, and typhoid fever
vectorborne diseases: malaria, plague, and African trypanosomiasis (sleeping sickness)
water contact disease: schistosomiasis
animal contact disease: rabies (2009)
Drinking water source: *Improved:*
urban: 91% of population
rural: 64% of population
total: 67% of population
Unimproved:
urban: 9% of population
rural: 36% of population
total: 33% of population (2008)
Sanitation facility access: *Improved:*
urban: 38% of population
rural: 49% of population
total: 48% of population
Unimproved urban: 62% of population
rural: 51% of population
total: 52% of population (2008)
Nationality: *noun:* Ugandan(s)
adjective: Ugandan
Ethnic groups: Baganda 16.9%, Banyakole 9.5%, Basoga 8.4%, Bakiga 6.9%, Iteso 6.4%, Langi 6.1%, Acholi 4.7%, Bagisu 4.6%, Lugbara 4.2%, Bunyoro 2.7%, other 29.6% (2002 census)
Religions: Roman Catholic 41.9%, Protestant 42% (Anglican 35.9%, Pentecostal 4.6%, Seventh Day Adventist 1.5%), Muslim 12.1%, other 3.1%, none 0.9% (2002 census)
Languages: English (official national language, taught in grade schools, used in courts of law and by most newspapers and some radio broadcasts), Ganda or Luganda (most widely used of the Niger-Congo languages, preferred for native language publications in the capital and may be taught in school), other Niger-Congo languages, Nilo-Saharan languages, Swahili, Arabic
Literacy: *definition:* age 15 and over can read and write
total population: 66.8%
male: 76.8%
female: 57.7% (2002 census)
School life expectancy (primary to tertiary education): *total:* 11 years

male: 11 years
female: 11 years (2009)
Education expenditures: 3.2% of GDP (2009)
country comparison to the world: 124

GOVERNMENT

Country name: *conventional long form:* Republic of Uganda
conventional short form: Uganda
Government type: republic
Capital: *name:* Kampala
geographic coordinates: 0 19 N, 32 25 E
time difference: UTC+3 (8 hours ahead of Washington, DC during Standard Time)
Administrative divisions: 80 districts; Abim, Adjumani, Amolatar, Amuria, Amuru, Apac, Arua, Budaka, Bududa, Bugiri, Bukedea, Bukwa, Bulisa, Bundibugyo, Bushenyi, Busia, Butaleja, Dokolo, Gulu, Hoima, Ibanda, Iganga, Isingiro, Jinja, Kaabong, Kabale, Kabarole, Kaberamaido, Kalangala, Kaliro, Kampala, Kamuli, Kamwenge, Kanungu, Kapchorwa, Kasese, Katakwi, Kayunga, Kibale, Kiboga, Kiruhara, Kisoro, Kitgum, Koboko, Kotido, Kumi, Kyenjojo, Lira, Luwero, Lyantonde, Manafwa, Maracha, Masaka, Masindi, Mayuge, Mbale, Mbarara, Mityana, Moroto, Moyo, Mpigi, Mubende, Mukono, Nakapiripirit, Nakaseke, Nakasongola, Namutumba, Nebbi, Ntungamo, Oyam, Pader, Pallisa, Rakai, Rukungiri, Sembabule, Sironko, Soroti, Tororo, Wakiso, Yumbe
Independence: 9 October 1962 (from the UK)
National holiday: Independence Day, 9 October (1962)
Constitution: 8 October 1995; amended 2005
note: the amendments in 2005 removed presidential term limits and legalized a multiparty political system
Legal system: mixed legal system of English common law and customary law
International law organization participation: accepts compulsory ICJ jurisdiction with reservations; accepts ICCt jurisdiction
Suffrage: 18 years of age; universal
Executive branch: *chief of state:* President Lt. Gen. Yoweri Kaguta MUSEVENI (since seizing power on 26 January 1986); note—the president is both chief of state and head of government
head of government: President Lt. Gen. Yoweri Kaguta MUSEVENI (since seizing power on 26 January 1986); Prime Minister Amama MBABAZI (since 24 May 2011); note—the prime minister assists the president in the supervision of the cabinet
cabinet: Cabinet appointed by the president from among elected legislators
(For more information visit the World Leaders website)
elections: president reelected by popular vote for a five-year term; election last held on 18 February 2011 (next to be held in 2016)
election results: Lt. Gen. Yoweri Kaguta MUSEVENI elected president; percent of vote—Lt. Gen. Yoweri Kaguta MUSEVENI 68.4%, Kizza BESIGYE 26.0%, other 5.6%
Legislative branch: unicameral National Assembly (372 seats; 215 members elected by popular vote, 104 nominated by legally established special interest groups [women 79, army 10, disabled 5, youth 5, labor 5], 13 ex-officio members; members to serve five-year terms); note—the composition of the National Assembly has changed but the the details are not yet available
elections: last held on 18 February 2011 (next to be held in 2016)
election results: percent of vote by party—NA; seats by party—NRM 279, FDC 34, DP 11, UPC 9, CP 1, JEEMA 1, independents 37
Judicial branch: Court of Appeal (judges are appointed by the president and approved by the legislature); High Court (judges are appointed by the president)
Political parties and leaders: Conservative Party or CP [Ken LUKYAMUZI]; Democratic Party or DP [Kizito SSEBAANA]; Forum for Democratic Change or FDC [Kizza BESIGYE]; Inter-Party Co-operation or IPC (a coalition of opposition groups); Justice Forum or JEEMA [Muhammad Kibirige MAYANJA]; National Resistance Movement or NRM [Yoweri MUSEVENI]; Peoples Progressive Party or PPP [Bidandi SSALI]; Ugandan People's Congress or UPC [Miria OBOTE]
note: a national referendum in July 2005 opened the way for Uganda's transition to a multi-party political system
Political pressure groups and leaders: Lord's Resistance Army or LRA [Joseph KONY]; Young Parliamentary Association [Henry BANYENZAKI]; Parliamentary Advocacy Forum or PAFO; National Association of Women Organizations in Uganda or NAWOU [Florence NEKYON]; The Ugandan Coalition for Political Accountability to Women or COPAW
International organization participation: ACP, AfDB, AU, C, COMESA, EAC, EADB, FAO, G-77, IAEA, IBRD, ICAO, ICRM, IDA, IDB, IFAD, IFC, IFRCS, IGAD, ILO, IMF, Interpol, IOC, IOM, IPU, ISO (correspondent), ITSO, ITU, ITUC, MIGA, NAM, OIC, OPCW, PCA, UN, UNAMID, UNCTAD, UNESCO, UNHCR, UNIDO, UNMIS, UNOCI, UNWTO, UPU, WCO, WFTU, WHO, WIPO, WMO, WTO
Diplomatic representation in the US: *chief of mission:* Ambassador Perezi Karukubiro KAMUNANWIRE
chancery: 5911 16th Street NW, Washington, DC 20011
telephone: [1] (202) 726-7100 through 7102, 0416
FAX: [1] (202) 726-1727
Diplomatic representation from the US: *chief of mission:* Ambassador Jerry P. LANIER
embassy: 1577 Ggaba Road, Kampala
mailing address: P. O. Box 7007, Kampala
telephone: [256] (414) 259 791 through 93, 95
FAX: [256] (414) 258-794
Flag description: six equal horizontal bands of black (top), yellow, red, black, yellow, and red; a white disk is superimposed at the center and depicts a red-crested crane (the national symbol) facing the hoist side; black symbolizes the African people, yellow sunshine and vitality, red African brotherhood; the crane was the military badge of Ugandan soldiers under the UK
National anthem: *name:* "Oh Uganda, Land of Beauty!"
lyrics/music: George Wilberforce KAKOMOA
note: adopted 1962

ECONOMY

Economy—overview: Uganda has substantial natural resources, including fertile soils, regular rainfall, small deposits of copper, gold, and other minerals, and recently discovered oil. Uganda has never conducted a national minerals survey. Agriculture is the most important sector of the economy, employing over 80% of the work force. Coffee accounts for the bulk of export revenues. Since 1986, the government—with the support of foreign countries and international agencies—has acted to rehabilitate and stabilize the economy by undertaking currency reform, raising producer prices on export crops, increasing prices of petroleum products, and improving civil service wages. The policy changes are especially aimed at dampening inflation and boosting production and export earnings. Since 1990 economic reforms ushered in an era of solid economic growth based on continued investment in infrastructure, improved incentives for production and exports, lower inflation, better domestic security, and the return of exiled Indian-Ugandan entrepreneurs. Uganda has received about $2 billion in multilateral and bilateral debt relief. In 2007 Uganda received $10 million for a Millennium Challenge Account Threshold Program. The global economic downturn has hurt Uganda's exports; however, Uganda's GDP growth is still relatively strong due to past reforms and sound management of the downturn. Oil revenues and taxes will become a larger source of government funding as oil comes on line in the next few years. Instability in southern Sudan is the biggest risk for the Ugandan economy in 2011 because Uganda's main export partner is Sudan, and Uganda is a key destination for Sudanese refugees.
GDP (purchasing power parity): $42.15 billion (2010 est.)
country comparison to the world: 97
$40.08 billion (2009 est.)
$37.37 billion (2008 est.)
note: data are in 2010 US dollars
GDP (official exchange rate): $17.01 billion (2010 est.)
GDP—real growth rate: 5.2% (2010 est.)
country comparison to the world: 66
7.2% (2009 est.)
8.7% (2008 est.)
GDP—per capita (PPP): $1,300 (2010 est.)
country comparison to the world: 204
$1,200 (2009 est.)
$1,200 (2008 est.)
note: data are in 2010 US dollars
GDP—composition by sector:
agriculture: 23.6%
industry: 24.5%
services: 51.9% (2010 est.)

Labor force: 15.51 million (2010 est.)
country comparison to the world: 39
Labor force—by occupation:
agriculture: 82%
industry: 5%
services: 13% (1999 est.)
Unemployment rate: NA%
Population below poverty line: 35% (2001 est.)
Household income or consumption by percentage share: *lowest 10%:* 2.6%
highest 10%: 34.1% (2005)
Distribution of family income—Gini index: 45.7 (2002)
country comparison to the world: 37
37.4 (1996)
Investment (gross fixed): 20.9% of GDP (2010 est.)
country comparison to the world: 80
Budget: *revenues:* $2.457 billion
expenditures: $2.938 billion (2010 est.)
Public debt: 20.4% of GDP (2010 est.)
country comparison to the world: 109
20.2% of GDP (2009 est.)
Inflation rate (consumer prices): 9.4% (2010 est.)
country comparison to the world: 193
14.2% (2009 est.)
Central bank discount rate: 9.65% (31 December 2009)
country comparison to the world: 13
19.42% (31 December 2008)
Commercial bank prime lending rate: 20.96% (31 December 2009 est.)
country comparison to the world: 16
20.45% (31 December 2008 est.)
Stock of narrow money: $1.997 billion (31 December 2010 est.)
country comparison to the world: 119
$1.603 billion (31 December 2009 est.)
Stock of broad money: $3.905 billion (31 December 2010 est.)
country comparison to the world: 127
$3.322 billion (31 December 2009 est.)
Stock of domestic credit: $1.882 billion (31 December 2010 est.)
country comparison to the world: 127
$1.716 billion (31 December 2009 est.)
Market value of publicly traded shares: $NA (31 December 2010)
country comparison to the world: 116
$116.3 million
Agriculture—products: coffee, tea, cotton, tobacco, cassava (tapioca), potatoes, corn, millet, pulses, cut flowers; beef, goat meat, milk, poultry
Industries: sugar, brewing, tobacco, cotton textiles; cement, steel production
Industrial production growth rate: 6% (2010 est.)
country comparison to the world: 57
Electricity—production: 2.256 billion kWh (2007 est.)
country comparison to the world: 131
Electricity—consumption: 2.068 billion kWh (2007 est.)
country comparison to the world: 135
Electricity—exports: 30 million kWh (2007)
Electricity—imports: 0 kWh (2008 est.)
Oil—production: 0 bbl/day (2009 est.)
country comparison to the world: 136
Oil—consumption: 13,000 bbl/day (2009 est.)
country comparison to the world: 140
Oil—exports: 0 bbl/day (2007 est.)
country comparison to the world: 207
Oil—imports: 13,090 bbl/day (2007 est.)
country comparison to the world: 134
Oil—proved reserves: 1.5 billion bbl (1 January 2010 est.)
country comparison to the world: 38
Natural gas—production: 0 cu m (2008 est.)
country comparison to the world: 194
Natural gas—consumption: 0 cu m (2008 est.)
country comparison to the world: 131
Natural gas—exports: 0 cu m (2008 est.)
country comparison to the world: 192
Natural gas—imports: 0 cu m (2008 est.)
country comparison to the world: 130
Natural gas—proved reserves: 0 cu m (1 January 2010 est.)
country comparison to the world: 196
Current account balance: $-784 million (2010 est.)
country comparison to the world: 126
$-451 million (2009 est.)
Exports: $2.941 billion (2010 est.)
country comparison to the world: 123
$2.7 billion (2009 est.)
Exports—commodities: coffee, fish and fish products, tea, cotton, flowers, horticultural products; gold
Exports—partners: Sudan 13.47%, Kenya 8.98%, UAE 7.52%, Rwanda 7.5%, Switzerland 7.42%, Democratic Republic of the Congo 6.85%, Netherlands 5.67%, Belgium 5.66%, Germany 5.18%, Italy 4.33% (2009)
Imports: $4.474 billion (2010 est.)
country comparison to the world: 127
$3.844 billion (2009 est.)
Imports—commodities: capital equipment, vehicles, petroleum, medical supplies; cereals
Imports—partners: Kenya 13.9%, India 12.79%, UAE 11.16%, China 8.91%, South Africa 5.08%, France 4.6%, Japan 4.37%, US 4.07% (2009)
Reserves of foreign exchange and gold: $3.743 billion (31 December 2010 est.)
country comparison to the world: 85
$2.995 billion (31 December 2009 est.)
note: excludes gold
Debt—external: $2.888 billion (31 December 2010 est.)
country comparison to the world: 130
$2.554 billion (31 December 2009 est.)
Stock of direct foreign investment—at home: $NA
Stock of direct foreign investment—abroad: $NA
Exchange rates: Ugandan shillings (UGX) per US dollar–2,166 (2010), 2,030 (2009), 1,658.1 (2008), 1,685.8 (2007), 1,834.9 (2006)

COMMUNICATIONS

Telephones—main lines in use: 233,500 (2009)
country comparison to the world: 125
Telephones—mobile cellular: 9.384 million (2009)
country comparison to the world: 68
Telephone system: *general assessment:* mobile cellular service is increasing rapidly, but the number of main lines is still deficient; work underway on a national backbone information and communications technology infrastructure; international phone networks and Internet connectivity provided through satellite and VSAT applications
domestic: intercity traffic by wire, microwave radio relay, and radiotelephone communication stations, fixed and mobile-cellular systems for short-range traffic; mobile-cellular teledensity about 30 per 100 persons in 2009
international: country code–256; satellite earth stations–1 Intelsat (Atlantic Ocean) and 1 Inmarsat; analog links to Kenya and Tanzania
Broadcast media: public broadcaster, Uganda Broadcasting Corporation (UBC), operates radio and television networks; Uganda first began licensing privately-owned stations in the 1990s; by 2007 there were nearly 150 radio and 35 TV stations, mostly based in and around Kampala; transmissions of multiple international broadcasters are available in Kampala (2007)
Internet country code: .ug
Internet hosts: 19,927 (2010)
country comparison to the world: 111
Internet users: 3.2 million (2009)
country comparison to the world: 66

TRANSPORTATION

Airports: 46 (2010)
country comparison to the world: 94
Airports—with paved runways: *total:* 5
over 3,047 m: 3
1,524 to 2,437 m: 1
914 to 1,523 m: 1 (2010)
Airports—with unpaved runways:
total: 41
over 3,047 m: 1
1,524 to 2,437 m: 7
914 to 1,523 m: 25
under 914 m: 8 (2010)
Railways: *total:* 1,244 km
country comparison to the world: 82
narrow gauge: 1,244 km 1.000-m gauge (2008)
Roadways:
total: 70,746 km
country comparison to the world: 68
paved: 16,272 km
unpaved: 54,474 km (2003)
Waterways: (there are no long navigable stretches of river in Uganda; parts of the Albert Nile that flow out of Lake Albert in the northwestern part of the country are navigable; several lakes including Lake Victoria and Lake Kyoga have substantial traffic; Lake Albert is navigable along a 200-km stretch from its northern tip to its southern shores) (2009)
Ports and terminals: Entebbe, Jinja, Port Bell

MILITARY

Military branches: Uganda Peoples Defense Force (UPDF): Army (includes Marine Unit), Uganda Air Force (2010)
Military service age and obligation: 18-26 years of age for voluntary military duty; 18-30 years of age for professionals; no conscription; 9-year service obligation; the government has stated that recruitment under 18 years of age could occur

with proper consent and that "no person under the apparent age of 13 years shall be enrolled in the armed forces"; Ugandan citizenship and secondary education required (2010)
Manpower available for military service: *males age 16–49:* 7,249,271
females age 16–49: 7,025,439 (2010 est.)
Manpower fit for military service: *males age 16–49:* 4,313,068
females age 16–49: 4,200,901 (2010 est.)
Manpower reaching militarily significant age annually: *male:* 423,923
female: 420,236 (2010 est.)
Military expenditures: 2.2% of GDP (2006)
country comparison to the world: 67

TRANSNATIONAL ISSUES

Disputes—international: Uganda is subject to armed fighting among hostile ethnic groups, rebels, armed gangs, militias, and various government forces that extend across its borders; Uganda hosts 209,860 Sudanese, 27,560 Congolese, and 19,710 Rwandan refugees, while Ugandan refugees as well as members of the Lord's Resistance Army (LRA) seek shelter in southern Sudan and the Democratic Republic of the Congo's Garamba National Park; LRA forces have also attacked Kenyan villages across the border
Refugees and internally displaced persons: refugees (country of origin): 215,700 (Sudan); 28,880 (Democratic Republic of Congo); 24,900 (Rwanda)
IDPs: 1.27 million (350,000 IDPs returned in 2006 following ongoing peace talks between the Lord's Resistance Army (LRA) and the Government of Uganda) (2007)

UKRAINE

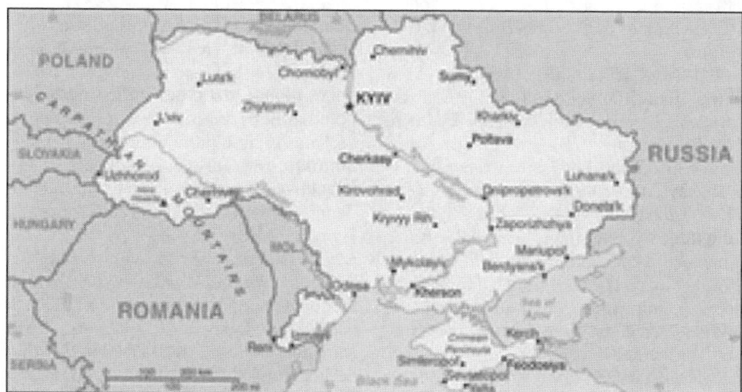

INTRODUCTION

Background: Ukraine was the center of the first eastern Slavic state, Kyivan Rus, which during the 10th and 11th centuries was the largest and most powerful state in Europe. Weakened by internecine quarrels and Mongol invasions, Kyivan Rus was incorporated into the Grand Duchy of Lithuania and eventually into the Polish-Lithuanian Commonwealth. The cultural and religious legacy of Kyivan Rus laid the foundation for Ukrainian nationalism through subsequent centuries. A new Ukrainian state, the Cossack Hetmanate, was established during the mid-17th century after an uprising against the Poles. Despite continuous Muscovite pressure, the Hetmanate managed to remain autonomous for well over 100 years. During the latter part of the 18th century, most Ukrainian ethnographic territory was absorbed by the Russian Empire. Following the collapse of czarist Russia in 1917, Ukraine was able to achieve a short-lived period of independence (1917-20), but was reconquered and forced to endure a brutal Soviet rule that engineered two forced famines (1921-22 and 1932-33) in which over 8 million died. In World War II, German and Soviet armies were responsible for some 7 to 8 million more deaths. Although final independence for Ukraine was achieved in 1991 with the dissolution of the USSR, democracy and prosperity remained elusive as the legacy of state control and endemic corruption stalled efforts at economic reform, privatization, and civil liberties. A peaceful mass protest "Orange Revolution" in the closing months of 2004 forced the authorities to overturn a rigged presidential election and to allow a new internationally monitored vote that swept into power a reformist slate under Viktor YUSHCHENKO. Subsequent internal squabbles in the YUSHCHENKO camp allowed his rival Viktor YANUKOVYCH to stage a comeback in parliamentary elections and become prime minister in August of 2006. An early legislative election, brought on by a political crisis in the spring of 2007, saw Yuliya TYMOSHENKO, as head of an "Orange" coalition, installed as a new prime minister in December 2007. Viktor YANUKOVUYCH was elected president in a February 2010 run-off election that observers assessed as meeting most international standards. The following month, the Rada approved a vote of no-confidence prompting Yuliya TYMOSHENKO to resign from her post as prime minister.

GEOGRAPHY

Location: Eastern Europe, bordering the Black Sea, between Poland, Romania, and Moldova in the west and Russia in the east
Geographic coordinates:
49 00 N, 32 00 E
Map references: Europe
Area: *total:* 603,550 sq km
country comparison to the world: 45
land: 579,330 sq km
water: 24,220 sq km
Area—comparative: slightly smaller than Texas
Land boundaries: *total:* 4,566 km
border countries: Belarus 891 km, Hungary 103 km, Moldova 940 km, Poland 428 km, Romania (south) 176 km, Romania (southwest) 362 km, Russia 1,576 km, Slovakia 90 km
Coastline: 2,782 km
Maritime claims: territorial sea: 12 nm
exclusive economic zone: 200 nm
continental shelf: 200 m or to the depth of exploitation
Climate: temperate continental; Mediterranean only on the southern Crimean coast; precipitation disproportionately distributed, highest in west and north, lesser in east and southeast; winters vary from cool along the Black Sea to cold farther inland; summers are warm across the greater part of the country, hot in the south
Terrain: most of Ukraine consists of fertile plains (steppes) and plateaus, mountains being found only in the west (the Carpathians), and in the Crimean Peninsula in the extreme south
Elevation extremes:
lowest point: Black Sea 0 m
highest point: Hora Hoverla 2,061 m
Natural resources: iron ore, coal, manganese, natural gas, oil, salt, sulfur, graphite, titanium, magnesium, kaolin, nickel, mercury, timber, arable land
Land use: *arable land:* 53.8%
permanent crops: 1.5%
other: 44.7% (2005)
Irrigated land:
22,080 sq km (2003)
Total renewable water resources:
139.5 cu km (1997)
Freshwater withdrawal (domestic/industrial/agricultural): *total:* 37.53 cu km/yr (12%/35%/52%)
per capita: 807 cu m/yr (2000)
Natural hazards: NA
Environment—current issues: inadequate supplies of potable water; air and water pollution; deforestation; radiation contamination in the northeast from 1986 accident at Chornobyl' Nuclear Power Plant
Environment—international agreements: *party to:* Air Pollution, Air Pollution-Nitrogen Oxides, Air Pollution-Sulfur 85, Antarctic-Environmental Protocol, Antarctic-Marine Living Resources, Antarctic Treaty, Biodiversity, Climate Change, Climate Change-Kyoto Protocol,

Desertification, Endangered Species, Environmental Modification, Hazardous Wastes, Law of the Sea, Marine Dumping, Ozone Layer Protection, Ship Pollution, Wetlands
signed, but not ratified: Air Pollution-Persistent Organic Pollutants, Air Pollution-Sulfur 94, Air Pollution-Volatile Organic Compounds
Geography—note: strategic position at the crossroads between Europe and Asia; second-largest country in Europe

PEOPLE

Population: 45,134,707 (July 2011 est.)
country comparison to the world: 28
Age structure: *0–14 years:* 13.7% (male 3,186,606/female 3,014,069)
15–64 years: 70.8% (male 15,282,749/female 16,673,641)
65 years and over: 15.5% (male 2,294,777/female 4,682,865) (2011 est.)
Median age: *total:* 39.9 years
male: 36.7 years
female: 43.1 years (2011 est.)
Population growth rate: -0.622% (2011 est.)
country comparison to the world: 225
Birth rate: 9.62 births/1,000 population (2011 est.)
country comparison to the world: 199
Death rate: 15.74 deaths/1,000 population (July 2011 est.)
country comparison to the world: 6
Net migration rate: -0.09 migrant(s)/1,000 population (2011 est.)
country comparison to the world: 120
Urbanization: *urban population:* 69% of total population (2010)
rate of urbanization: -0.1% annual rate of change (2010-15 est.)
Major cities—population: KYIV (capital) 2.779 million; Kharkiv 1.455 million; Dnipropetrovsk 1.013 million; Odesa 1.009 million; Donetsk 971,000 (2009)
Sex ratio: *at birth:* 1.065 male(s)/female
under 15 years: 1.06 male(s)/female
15–64 years: 0.92 male(s)/female
65 years and over: 0.49 male(s)/female
total population: 0.85 male(s)/female (2011 est.)
Infant mortality rate: *total:* 8.54 deaths/1,000 live births
country comparison to the world: 158
male: 10.71 deaths/1,000 live births
female: 6.23 deaths/1,000 live births (2011 est.)
Life expectancy at birth: *total population:* 68.58 years
country comparison to the world: 150
male: 62.79 years
female: 74.75 years (2011 est.)
Total fertility rate: 1.28 children born/woman (2011 est.)
country comparison to the world: 212
HIV/AIDS—adult prevalence rate: 1.1% (2009 est.)
country comparison to the world: 44
HIV/AIDS—people living with HIV/AIDS: 350,000 (2009 est.)
country comparison to the world: 18
HIV/AIDS—deaths: 24,000 (2009 est.)
country comparison to the world: 15
Drinking water source: *Improved:*
urban: 98% of population
rural: 97% of population
total: 98% of population
Unimproved: urban: 2% of population
rural: 3% of population
total: 2% of population (2008)
Sanitation facility access: *Improved:*
urban: 97% of population
rural: 90% of population
total: 95% of population
Unimproved: urban: 3% of population
rural: 10% of population
total: 5% of population (2008)
Nationality: *noun:* Ukrainian(s)
adjective: Ukrainian
Ethnic groups: Ukrainian 77.8%, Russian 17.3%, Belarusian 0.6%, Moldovan 0.5%, Crimean Tatar 0.5%, Bulgarian 0.4%, Hungarian 0.3%, Romanian 0.3%, Polish 0.3%, Jewish 0.2%, other 1.8% (2001 census)
Religions: Ukrainian Orthodox–Kyiv Patriarchate 50.4%, Ukrainian Orthodox–Moscow Patriarchate 26.1%, Ukrainian Greek Catholic 8%, Ukrainian Autocephalous Orthodox 7.2%, Roman Catholic 2.2%, Protestant 2.2%, Jewish 0.6%, other 3.2% (2006 est.)
Languages: Ukrainian (official) 67%, Russian 24%, other (includes small Romanian-, Polish-, and Hungarian-speaking minorities) 9%
Literacy: *definition:* age 15 and over can read and write
total population: 99.4%
male: 99.7%
female: 99.2% (2001 census)
School life expectancy (primary to tertiary education): *total:* 15 years
male: 14 years
female: 15 years (2008)
Education expenditures: 5.3% of GDP (2007)
country comparison to the world: 48

GOVERNMENT

Country name: *conventional long form:* none
conventional short form: Ukraine
local long form: none
local short form: Ukrayina
former: Ukrainian National Republic, Ukrainian State, Ukrainian Soviet Socialist Republic
Government type: republic
Capital: *name:* Kyiv (Kiev)
note: pronounced KAY-yiv
geographic coordinates: 50 26 N, 30 31 E
time difference: UTC+2 (7 hours ahead of Washington, DC during Standard Time)
daylight saving time: +1hr, begins last Sunday in March; ends last Sunday in October
Administrative divisions: 24 provinces (oblasti, singular–oblast'), 1 autonomous republic* (avtonomna respublika), and 2 municipalities (mista, singular–misto) with oblast status**; Cherkasy, Chernihiv, Chernivtsi, Crimea or Avtonomna Respublika Krym* (Simferopol'), Dnipropetrovs'k, Donets'k, Ivano-Frankivs'k, Kharkiv, Kherson, Khmel'nyts'kyy, Kirovohrad, Kyiv**, Kyiv, Luhans'k, L'viv, Mykolayiv, Odesa, Poltava, Rivne, Sevastopol'**, Sumy, Ternopil', Vinnytsya, Volyn' (Luts'k), Zakarpattya (Uzhhorod), Zaporizhzhya, Zhytomyr
note: administrative divisions have the same names as their administrative centers (exceptions have the administrative center name following in parentheses)
Independence: 24 August 1991 (from the Soviet Union); notable earlier dates: ca. A.D. 982 (VOLODYMYR I consolidates Kyivan Rus), 1648 (establishment of Cossack Hetmanate)
National holiday: Independence Day, 24 August (1991); note—22 January 1918, the day Ukraine first declared its independence (from Soviet Russia) and the day the short-lived Western and Greater (Eastern) Ukrainian republics united (1919), is now celebrated as Unity Day
Constitution: adopted 28 June 1996
Legal system: civil law system; judicial review of legislative acts
International law organization participation: has not submitted an ICJ jurisdiction declaration; non-party state to the ICCt
Suffrage: 18 years of age; universal
Executive branch: *chief of state:* President Viktor YANUKOVYCH (since 25 February 2010)
head of government: Prime Minister Mykola AZAROV (since 11 March 2010); First Deputy Prime Minister Andriy KLYUYEV (since 11 March 2010); Deputy Prime Ministers Borys KOLESNIKOV and Serhiy TIHIPKO (both since 11 March 2010)
cabinet: Cabinet of Ministers nominated by the president and approved by the Rada (For more information visit the World Leaders website)
note: there is also a National Security and Defense Council or NSDC originally created in 1992 as the National Security Council; the NSDC staff is tasked with developing national security policy on domestic and international matters and advising the president; a Presidential Administration helps draft presidential edicts and provides policy support to the president
elections: president elected by popular vote for a five-year term (eligible for a second term); election last held on 17 January 2010 with runoff on 7 February 2010 (next to be held in 2015)
election results: Viktor YANUKOVYCH elected president; percent of vote—Viktor YANUKOVYCH 48.95%, Yuliya TYMOSHENKO 45.5%, other 5.6%
Legislative branch: unicameral Supreme Council or Verkhovna Rada (450 seats; members allocated on a proportional basis to those parties that gain 3% or more of the national electoral vote; members to serve five-year terms)
elections: last held on 30 September 2007 (next must be held in 2012 or sooner if a ruling coalition cannot be formed in the Rada)
election results: percent of vote by party/bloc—Party of Regions 34.4%, Block of Yuliya Tymoshenko 30.7%, Our Ukraine-People's Self Defense Bloc 14.2%, CPU 5.4%, Lytvyn Bloc 4%, other parties 11.3%;

seats by party/bloc—Party of Regions 175, Block of Yuliya Tymoshenko 156, Our Ukraine-People's Self Defense 72, CPU 27, Lytvyn Bloc 20

Judicial branch: Supreme Court; Constitutional Court

Political parties and leaders: Block of Yuliya Tymoshenko-Batkivshchyna (BYuT-Batkivshchyna) [Yuliya TYMOSHENKO]; Communist Party of Ukraine or CPU [Petro SYMONENKO]; European Party of Ukraine [Mykola KATERYNCHUK]; Forward Ukraine! [Viktor MUSIYAKA]; Front of Change [Arseniy YATSENYUK]; Lytvyn Bloc (composed of People's Party and Labor Party of Ukraine) [Volodymyr LYTVYN]; Our Ukraine [Viktor YUSHCHENKO]; Party of Industrialists and Entrepreneurs [Anatoliy KINAKH]; Party of Regions [Viktor YANUKOVYCH]; Party of the Defenders of the Fatherland [Yuriy KARMAZIN]; People's Movement of Ukraine (Rukh) [Borys TARASYUK]; People's Party [Volodymyr LYTVYN]; Peoples' Self-Defense [Yuriy LUTSENKO]; PORA! (It's Time!) party [Vladyslav KASKIV]; Progressive Socialist Party [Natalya VITRENKO]; Reforms and Order Party [Viktor PYNZENYK]; Sobor [Anatoliy MATVIYENKO]; Social Democratic Party [Yevhen KORNICHUK]; Social Democratic Party (United) or SDPU(o) [Yuriy ZAHORODNIY]; Socialist Party of Ukraine or SPU [Oleksandr MOROZ]; Strong Ukraine [SERHIY TIHIPKO]; Ukrainian People's Party [Yuriy KOSTENKO]; United Center [Viktor BALOHA]; Viche [Inna BOHOSLOVSKA]

Political pressure groups and leaders: Committee of Voters of Ukraine [Aleksandr CHERNENKO]; OPORA [Olha AIVAZOVSKA]

International organization participation: Australia Group, BSEC, CBSS (observer), CE, CEI, CICA (observer), CIS (participating member, has not signed the 1993 CIS charter although it participates in meetings), EAEC (observer), EAPC, EBRD, FAO, GCTU, GUAM, IAEA, IBRD, ICAO, ICC, ICRM, IDA, IFC, IFRCS, IHO, ILO, IMF, IMO, IMSO, Interpol, IOC, IOM, IPU, ISO, ITU, ITUC, LAIA (observer), MIGA, MONUSCO, NAM (observer), NSG, OAS (observer), OIF (observer), OPCW, OSCE, PCA, PFP, SECI (observer), UN, UNCTAD, UNESCO, UNIDO, UNMIL, UNMIS, UNWTO, UPU, WCO, WFTU, WHO, WIPO, WMO, WTO, ZC

Diplomatic representation in the US: *chief of mission:* Ambassador Oleksandr MOTSYK
chancery: 3350 M Street NW, Washington, DC 20007
telephone: [1] (202) 333-0606
FAX: [1] (202) 333-0817
consulate(s) general: Chicago, New York, San Francisco

Diplomatic representation from the US: *chief of mission:* Ambassador John F. TEFFT
embassy: 10 Yurii Kotsiubynsky Street, 01901 Kyiv
mailing address: 5850 Kyiv Place, Washington, DC 20521-5850
telephone: [380] (44) 490-4000
FAX: [380] (44) 490-4085

Flag description: two equal horizontal bands of azure (top) and golden yellow represent grain fields under a blue sky

National anthem: *name:* "Sche ne vmerla Ukraina" (Ukraine Has Not Yet Perished)
lyrics/music: Paul CHUBYNSKYI/Mikhail VERBYTSKYI
note: music adopted 1991, lyrics adopted 2003; the song was first performed in 1864 at the Ukraine Theatre in Lviv; the lyrics, originally written in 1862, were revised in 2003

ECONOMY

Economy—overview: After Russia, the Ukrainian republic was far and away the most important economic component of the former Soviet Union, producing about four times the output of the next-ranking republic. Its fertile black soil generated more than one-fourth of Soviet agricultural output, and its farms provided substantial quantities of meat, milk, grain, and vegetables to other republics. Likewise, its diversified heavy industry supplied the unique equipment (for example, large diameter pipes) and raw materials to industrial and mining sites (vertical drilling apparatus) in other regions of the former USSR. Shortly after independence in August 1991, the Ukrainian Government liberalized most prices and erected a legal framework for privatization, but widespread resistance to reform within the government and the legislature soon stalled reform efforts and led to some backtracking. Output by 1999 had fallen to less than 40% of the 1991 level. Ukraine's dependence on Russia for energy supplies and the lack of significant structural reform have made the Ukrainian economy vulnerable to external shocks. Ukraine depends on imports to meet about three-fourths of its annual oil and natural gas requirements and 100% of its nuclear fuel needs. After a two-week dispute that saw gas supplies cutoff to Europe, Ukraine agreed to 10-year gas supply and transit contracts with Russia in January 2009 that brought gas prices to "world" levels. The strict terms of the contracts have further hobbled Ukraine's cash-strapped state gas company, Naftohaz. Outside institutions—particularly the IMF—have encouraged Ukraine to quicken the pace and scope of reforms. Ukrainian Government officials eliminated most tax and customs privileges in a March 2005 budget law, bringing more economic activity out of Ukraine's large shadow economy, but more improvements are needed, including fighting corruption, developing capital markets, and improving the legislative framework. Ukraine's economy was buoyant despite political turmoil between the prime minister and president until mid-2008. Real GDP growth exceeded 7% in 2006-07, fueled by high global prices for steel—Ukraine's top export—and by strong domestic consumption, spurred by rising pensions and wages. Ukraine reached an agreement with the IMF for a $16.4 billion Stand-By Arrangement in November 2008 to deal with the economic crisis, but the Ukrainian Government's lack of progress in implementing reforms has twice delayed the release of IMF assistance funds. The drop in steel prices and Ukraine's exposure to the global financial crisis due to aggressive foreign borrowing lowered growth in 2008 and the economy contracted more than 15% in 2009, among the worst economic performances in the world; growth resumed in 2010, buoyed by exports. External conditions are likely to hamper efforts for economic recovery in 2011.

GDP (purchasing power parity): $305.2 billion (2010 est.)
country comparison to the world: 40
$292.9 billion (2009 est.)
$343.8 billion (2008 est.)
note: data are in 2010 US dollars

GDP (official exchange rate): $136.4 billion (2010 est.)

GDP—real growth rate:
4.2% (2010 est.)
country comparison to the world: 89
-14.8% (2009 est.)
1.9% (2008 est.)

GDP—per capita (PPP): $6,700 (2010 est.)
country comparison to the world: 134
$6,400 (2009 est.)
$7,500 (2008 est.)
note: data are in 2010 US dollars

GDP—composition by sector:
agriculture: 9.8%
industry: 32.3%
services: 57.9% (2010 est.)

Labor force: 22.06 million (2010 est.)
country comparison to the world: 28

Labor force—by occupation:
agriculture: 15.8%
industry: 18.5%
services: 65.7% (2008)

Unemployment rate: 8.4% (2010 est.)
country comparison to the world: 98
8.8% (2009 est.)
note: officially registered; large number of unregistered or underemployed workers

Population below poverty line: 35% (2009)

Household income or consumption by percentage share: *lowest 10%:* 3.4%
highest 10%: 25.7% (2006)

Distribution of family income—Gini index: 31 (2006)
country comparison to the world: 107
29 (1999)

Investment (gross fixed): 16.1% of GDP (2010 est.)
country comparison to the world: 125

Budget: *revenues:* $41.18 billion
expenditures: $49.79 billion
note: this is the planned, consolidated budget (2010 est.)

Public debt: 38.4% of GDP (2010 est.)
country comparison to the world: 78
30% of GDP (2009 est.)

Inflation rate (consumer prices): 9.8% (2010 est.)
country comparison to the world: 195

15.9% (2009 est.)
Central bank discount rate: 10.25% (31 December 2009)
country comparison to the world: 32
12% (31 December 2008)
Commercial bank prime lending rate: 20.86% (31 December 2009 est.)
country comparison to the world: 27
17.49% (31 December 2008 est.)
Stock of narrow money: $34.97 billion (31 December 2010 est.)
country comparison to the world: 51
$30 billion (31 December 2009 est.)
Stock of broad money: $73.91 billion (31 December 2010 est.)
country comparison to the world: 59
$62.22 billion (31 December 2009 est.)
Stock of domestic credit: $110.8 billion (31 December 2010 est.)
country comparison to the world: 48
$103.9 billion (31 December 2009 est.)
Market value of publicly traded shares: $16.79 billion (31 December 2009)
country comparison to the world: 57
$24.36 billion (31 December 2008)
$111.8 billion (31 December 2007)
Agriculture—products: grain, sugar beets, sunflower seeds, vegetables; beef, milk
Industries: coal, electric power, ferrous and nonferrous metals, machinery and transport equipment, chemicals, food processing
Industrial production growth rate: 8% (2010 est.)
country comparison to the world: 30
Electricity—production: 172.9 billion kWh (2009 est.)
country comparison to the world: 22
Electricity—consumption: 134.6 billion kWh (2009 est.)
country comparison to the world: 22
Electricity—exports: 4 billion kWh (2009 est.)
Electricity—imports: 0 kWh (2009 est.)
Oil—production: 99,930 bbl/day (2009 est.)
country comparison to the world: 52
Oil—consumption: 348,000 bbl/day (2009 est.)
country comparison to the world: 37
Oil—exports: 154,400 bbl/day (2009 est.)
country comparison to the world: 57
Oil—imports: 147,600 bbl/day (2009 est.)
country comparison to the world: 55
Oil—proved reserves: 395 million bbl (1 January 2010 est.)
country comparison to the world: 54
Natural gas—production:
21.2 billion cu m (2009 est.)
country comparison to the world: 31
Natural gas—consumption:
52 billion cu m (2009 est.)
country comparison to the world: 15
Natural gas—exports: 5 billion cu m (2009 est.)
country comparison to the world: 28
Natural gas—imports: 26.83 billion cu m (2009 est.)
country comparison to the world: 11
Natural gas—proved reserves: 1.104 trillion cu m (1 January 2010 est.)
country comparison to the world: 25
Current account balance: $603 million (2010 est.)
country comparison to the world: 50
$-1.732 billion (2009 est.)
Exports: $49.71 billion (2010 est.)
country comparison to the world: 54
$40.39 billion (2009 est.)
Exports—commodities: ferrous and nonferrous metals, fuel and petroleum products, chemicals, machinery and transport equipment, food products
Exports—partners: Russia 21.1%, Turkey 5.3%, China 3.8% (2009 est.)
Imports:
$53.54 billion (2010 est.)
country comparison to the world: 48
$45.05 billion (2009 est.)
Imports—commodities: energy, machinery and equipment, chemicals
Imports—partners: Russia 28%, Germany 8.6%, China 6.1%, Kazakhstan 4.9%, Poland 4.9% (2009 est.)
Reserves of foreign exchange and gold: $32.91 billion (31 December 2010 est.)
country comparison to the world: 35
$26.51 billion (31 December 2009 est.)
Debt—external: $97.5 billion (31 December 2010 est.)
country comparison to the world: 36
$94.3 billion (31 December 2009 est.)
Stock of direct foreign investment—at home: $52.31 billion (31 December 2010 est.)
country comparison to the world: 52
$46.81 billion (31 December 2009 est.)
Stock of direct foreign investment—abroad: $2.327 billion (31 December 2010 est.)
country comparison to the world: 65
$2.067 billion (31 December 2009 est.)
Exchange rates: hryvnia (UAH) per US dollar–7.9111 (2010), 7.7912 (2009), 4.9523 (2008), 5.05 (2007), 5.05 (2006)

COMMUNICATIONS

Telephones—main lines in use: 13.026 million (2009)
country comparison to the world: 20
Telephones—mobile cellular: 55.333 million (2009)
country comparison to the world: 20
Telephone system: *general assessment:* Ukraine's telecommunication development plan emphasizes improving domestic trunk lines, international connections, and the mobile-cellular system
domestic: at independence in December 1991, Ukraine inherited a telephone system that was antiquated, inefficient, and in disrepair; more than 3.5 million applications for telephones could not be satisfied; telephone density is rising and the domestic trunk system is being improved; about one-third of Ukraine's networks are digital and a majority of regional centers now have digital switching stations; improvements in local networks and local exchanges continue to lag; the mobile-cellular telephone system's expansion has slowed, largely due to saturation of the market which has reached 120 mobile phones per 100 people
international: country code–380; 2 new domestic trunk lines are a part of the fiber-optic Trans-Asia-Europe (TAE) system and 3 Ukrainian links have been installed in the fiber-optic Trans-European Lines (TEL) project that connects 18 countries; additional international service is provided by the Italy-Turkey-Ukraine-Russia (ITUR) fiber-optic submarine cable and by an unknown number of earth stations in the Intelsat, Inmarsat, and Intersputnik satellite systems
Broadcast media: TV coverage is provided by Ukraine's state-controlled nationwide broadcast channel (UT1) and a number of privately-owned television broadcast networks; Russian television broadcasts have a small audience nationwide, but larger audiences in the eastern and southern regions; multi-channel cable and satellite TV services are available; Ukraine's radio broadcast market, a mix of independent and state-owned networks, is comprised of some 300 stations (2007)
Internet country code: .ua
Internet hosts: 1.098 million (2010)
country comparison to the world: 42
Internet users: 7.77 million (2009)
country comparison to the world: 38

TRANSPORTATION

Airports: 425 (2010)
country comparison to the world: 19
Airports—with paved runways: *total:* 189
over 3,047 m: 12
2,438 to 3,047 m: 51
1,524 to 2,437 m: 24
914 to 1,523 m: 5
under 914 m: 97 (2010)
Airports—with unpaved runways: *total:* 236
2,438 to 3,047 m: 3
1,524 to 2,437 m: 7
914 to 1,523 m: 12
under 914 m: 214 (2010)
Heliports: 7 (2010)
Pipelines: gas 36,493 km; oil 4,514 km; refined products 4,211 km (2010)
Railways: *total:* 21,658 km
country comparison to the world: 13
broad gauge: 21,658 km 1.524-m gauge (9,729 km electrified) (2009)
Roadways: *total:* 169,495 km
country comparison to the world: 30
paved: 165,820 km (includes 15 km of expressways)
unpaved: 3,675 km (2009)
Waterways: 2,150 km (most on Dnieper River) (2009)
country comparison to the world: 41
Merchant marine: *total:* 160
country comparison to the world: 40
by type: bulk carrier 4, cargo 123, chemical tanker 1, passenger 5, passenger/cargo 5, petroleum tanker 9, refrigerated cargo 11, specialized tanker 2
foreign-owned:
1 (Iran 1)
registered in other countries: 197 (Belize 6, Cambodia 37, Comoros 10, Cyprus 2, Dominica 2, Georgia 15, Liberia 16, Malta 30, Marshall Islands 1, Moldova 12, Mongolia 1, Panama 11, Russia 12, Saint Kitts and Nevis 10, Saint Vincent and the Grenadines 12, Sierra Leone 5, Slovakia 7, Tuvalu 1, Vanuatu 3, unknown 4) (2010)
Ports and terminals: Feodosiya (Theodosia), Illichivsk, Mariupol', Mykolayiv, Odesa, Yuzhnyy

MILITARY

Military branches: Ground Forces, Naval Forces, Air and Air Defense Forces (Viyskovo-Povitryani Syly, VPS) (2010)
Military service age and obligation: 18-25 years of age for compulsory and voluntary military service; conscript service obligation–12 months for Army and Air Force, 18 months for Navy (2010)
Manpower available for military service: *males age 16–49:* 10,984,394
females age 16–49: 11.26 million (2010 est.)
Manpower fit for military service: *males age 16–49:* 6,893,551
females age 16–49: 8,792,504 (2010 est.)
Manpower reaching militarily significant age annually: *male:* 246,397
female: 234,916 (2010 est.)
Military expenditures: 1.4% of GDP (2005 est.)
country comparison to the world: 107

TRANSNATIONAL ISSUES

Disputes—international: 1997 boundary delimitation treaty with Belarus remains un-ratified due to unresolved financial claims, stalling demarcation and reducing border security; delimitation of land boundary with Russia is complete with preparations for demarcation underway; the dispute over the boundary between Russia and Ukraine through the Kerch Strait and Sea of Azov remains unresolved despite a December 2003 framework agreement and ongoing expert-level discussions; Moldova and Ukraine operate joint customs posts to monitor transit of people and commodities through Moldova's break-away Transnistria Region, which remains under OSCE supervision; the ICJ gave Ukraine until December 2006 to reply, and Romania until June 2007 to rejoin, in their dispute submitted in 2004 over Ukrainian-administered Zmiyinyy/Serpilor (Snake) Island and Black Sea maritime boundary; Romania opposes Ukraine's reopening of a navigation canal from the Danube border through Ukraine to the Black Sea
Illicit drugs: limited cultivation of cannabis and opium poppy, mostly for CIS consumption; some synthetic drug production for export to the West; limited government eradication program; used as transshipment point for opiates and other illicit drugs from Africa, Latin America, and Turkey to Europe and Russia; Ukraine has improved anti-money-laundering controls, resulting in its removal from the Financial Action Task Force's (FATF's) Noncooperative Countries and Territories List in February 2004; Ukraine's anti-money-laundering regime continues to be monitored by FATF

UNITED ARAB EMIRATES

INTRODUCTION

Background: The Trucial States of the Persian Gulf coast granted the UK control of their defense and foreign affairs in 19th century treaties. In 1971, six of these states–Abu Zaby, 'Ajman, Al Fujayrah, Ash Shariqah, Dubayy, and Umm al Qaywayn–merged to form the United Arab Emirates (UAE). They were joined in 1972 by Ra's al Khaymah. The UAE's per capita GDP is on par with those of leading West European nations. Its generosity with oil revenues and its moderate foreign policy stance have allowed the UAE to play a vital role in the affairs of the region. For more than three decades, oil and global finance drove the UAE's economy. However, in 2008-09, the confluence of falling oil prices, collapsing real estate prices, and the international banking crisis hit the UAE especially hard. In March 2011, about 100 Emirati activists and intellectuals posted on the Internet and sent to the government a petition calling for greater political reform, including the establishment of a parliament with full legislative powers and the further expansion of the electorate and the rights of the Federal National Council, the UAE's quasi-legislature. In April 2011, the Emirati Government arrested four activists–all of whom signed the petition–for their alleged criticisms of the UAE system of government.

GEOGRAPHY

Location: Middle East, bordering the Gulf of Oman and the Persian Gulf, between Oman and Saudi Arabia
Geographic coordinates: 24 00 N, 54 00 E
Map references: Middle East
Area: *total:* 83,600 sq km
country comparison to the world: 114
land: 83,600 sq km
water: 0 sq km
Area—comparative: slightly smaller than Maine
Land boundaries: *total:* 867 km
border countries: Oman 410 km, Saudi Arabia 457 km
Coastline: 1,318 km
Maritime claims: territorial sea: 12 nm
contiguous zone: 24 nm
exclusive economic zone: 200 nm
continental shelf: 200 nm or to the edge of the continental margin
Climate: desert; cooler in eastern mountains
Terrain: flat, barren coastal plain merging into rolling sand dunes of vast desert wasteland; mountains in east
Elevation extremes: *lowest point:* Persian Gulf 0 m
highest point: Jabal Yibir 1,527 m
Natural resources: petroleum, natural gas
Land use: *arable land:* 0.77%
permanent crops: 2.27%
other: 96.96% (2005)
Irrigated land: 760 sq km (2003)
Total renewable water resources: 0.2 cu km (1997)
Freshwater withdrawal (domestic/industrial/agricultural): *total:* 2.3 cu km/yr (23%/9%/68%)
per capita: 511 cu m/yr (2000)
Natural hazards: frequent sand and dust storms
Environment—current issues: lack of natural freshwater resources compensated by desalination plants; desertification; beach pollution from oil spills
Environment—international agreements: *party to:* Biodiversity, Climate Change, Climate Change-Kyoto Protocol, Desertification, Endangered Species, Hazardous Wastes, Marine Dumping, Ozone Layer Protection
signed, but not ratified: Law of the Sea
Geography—note: strategic location along southern approaches to Strait of Hormuz, a vital transit point for world crude oil

PEOPLE

Population: 5,148,664 (July 2011 est.)
country comparison to the world: 114
note: estimate is based on the results of the 2005 census that included a significantly higher estimate of net immigration of non-citizens than previous estimates
Age structure: *0–14 years:* 20.4% (male 537,925/female 513,572)
15–64 years: 78.7% (male 2,968,958/female 1,080,717)
65 years and over: 0.9% (male 30,446/female 17,046)
note: 73.9% of the population in the 15-64 age group is non-national (2011 est.)
Median age: *total:* 30.2 years
male: 32.1 years

female: 24.9 years (2011 est.)
Population growth rate: 3.282% (2011 est.)
country comparison to the world: 6
Birth rate: 15.87 births/1,000 population (2011 est.)
country comparison to the world: 129
Death rate: 2.06 deaths/1,000 population (July 2011 est.)
country comparison to the world: 224
Net migration rate: 19 migrant(s)/1,000 population (2011 est.)
country comparison to the world: 3
Urbanization: *urban population:* 84% of total population (2010)
rate of urbanization: 2.3% annual rate of change (2010-15 est.)
Major cities—population: ABU DHABI (capital) 666,000 (2009)
Sex ratio: *at birth:* 1.05 male(s)/female
under 15 years: 1.05 male(s)/female
15–64 years: 2.75 male(s)/female
65 years and over: 1.8 male(s)/female
total population: 2.2 male(s)/female (2011 est.)
Infant mortality rate: *total:* 11.94 deaths/1,000 live births
country comparison to the world: 135
male: 13.96 deaths/1,000 live births
female: 9.82 deaths/1,000 live births (2011 est.)
Life expectancy at birth:
total population: 76.51 years
country comparison to the world: 71
male: 73.94 years
female: 79.22 years (2011 est.)
Total fertility rate: 2.4 children born/woman (2011 est.)
country comparison to the world: 93
HIV/AIDS—adult prevalence rate: 0.2% (2001 est.)
country comparison to the world: 92
HIV/AIDS—people living with HIV/AIDS: NA
HIV/AIDS—deaths: NA
Drinking water source: *Improved:*
urban: 100% of population
rural: 100% of population
total: 100% of population (2008)
Sanitation facility access: *Improved:*
urban: 98% of population
rural: 95% of population
total: 97% of population
Unimproved: urban: 2% of population
rural: 5% of population
total: 3% of population (2008)
Nationality: *noun:* Emirati(s)
adjective: Emirati
Ethnic groups: Emirati 19%, other Arab and Iranian 23%, South Asian 50%, other expatriates (includes Westerners and East Asians) 8% (1982)
note: less than 20% are UAE citizens (1982)
Religions: Muslim 96% (Shia 16%), other (includes Christian, Hindu) 4%
Languages: Arabic (official), Persian, English, Hindi, Urdu
Literacy: *definition:* age 15 and over can read and write
total population: 77.9%
male: 76.1%
female: 81.7% (2003 est.)
School life expectancy (primary to tertiary education): *total:* 13 years
male: 13 years
female: 14 years (2009)
Education expenditures:
1.2% of GDP (2009)
country comparison to the world: 162

GOVERNMENT

Country name: *conventional long form:* United Arab Emirates
conventional short form: none
local long form: Al Imarat al Arabiyah al Muttahidah
local short form: none
former: Trucial Oman, Trucial States
abbreviation: UAE
Government type: federation with specified powers delegated to the UAE federal government and other powers reserved to member emirates
Capital: name: Abu Dhabi
geographic coordinates: 24 28 N, 54 22 E
time difference: UTC+4 (9 hours ahead of Washington, DC during Standard Time)
Administrative divisions: 7 emirates (imarat, singular–imarah); Abu Zaby (Abu Dhabi), 'Ajman, Al Fujayrah, Ash Shariqah (Sharjah), Dubayy (Dubai), Ra's al Khaymah, Umm al Qaywayn (Quwain)
Independence: 2 December 1971 (from the UK)
National holiday: Independence Day, 2 December (1971)
Constitution: 2 December 1971; made permanent in 1996
Legal system: mixed legal system of Islamic law and civil law
International law organization participation: has not submitted an ICJ jurisdiction declaration; non-party state to the ICCt
Suffrage: none
Executive branch: *chief of state:* President KHALIFA bin Zayid Al-Nuhayyan (since 3 November 2004), ruler of Abu Zaby (Abu Dhabi) (since 4 November 2004); Vice President and Prime Minister MUHAMMAD BIN RASHID Al-Maktum (since 5 January 2006)
head of government: Prime Minister and Vice President MUHAMMAD bin Rashid Al-Maktum (since 5 January 2006); Deputy Prime Ministers SAIF bin Zayid Al-Nuhayyan (since 11 May 2009) and MANSUR bin Zayid Al-Nuhayyan (since 11 May 2009)
cabinet: Council of Ministers appointed by the president
(For more information visit the World Leaders website)
note: there is also a Federal Supreme Council (FSC) composed of the seven emirate rulers; the FSC is the highest constitutional authority in the UAE; establishes general policies and sanctions federal legislation; meets four times a year; Abu Zaby (Abu Dhabi) and Dubayy (Dubai) rulers have effective veto power
elections: president and vice president elected by the FSC for five-year terms (no term limits) from among the seven FSC members; election last held 3 November 2009 upon the death of the UAE's Founding Father and first President ZAYID bin Sultan Al Nuhayyan (next election NA); prime minister and deputy prime minister appointed by the president
election results: KHALIFA bin Zayid Al-Nuhayyan elected president by a unanimous vote of the FSC; MUHAMMAD BIN RASHID Al-Maktum unanimously affirmed vice president after the 2006 death of his brother Sheikh MAKTUM bin Rashid Al-Maktum
Legislative branch: unicameral Federal National Council (FNC) or Majlis al-Ittihad al-Watani (40 seats; 20 members appointed by the rulers of the constituent states, 20 members elected to serve four-year terms)
elections: elections for one half of the FNC (the other half remains appointed) held on 18-20 December 2006; the new electoral college–a body of 6,689 Emiratis (including 1,189 women) appointed by the rulers of the seven emirates–were the only eligible voters and candidates; 456 candidates including 65 women ran for 20 contested FNC seats; one female from the Emirate of Abu Dhabi won a seat and 8 women were among the 20 appointed members
note: the FNC reviews legislation but cannot change or veto
Judicial branch: Union Supreme Court (judges are appointed by the president)
Political parties and leaders: none; political parties are not allowed
Political pressure groups and leaders: NA
International organization participation: ABEDA, AFESD, AMF, CAEU, CICA, FAO, G-77, GCC, IAEA, IBRD, ICAO, ICC, ICRM, IDA, IDB, IFAD, IFC, IFRCS, IHO, ILO, IMF, IMO, IMSO, Interpol, IOC, IPU, ISO, ITSO, ITU, LAS, MIGA, NAM, OAPEC, OIC, OPCW, OPEC, PCA, UN, UNCTAD, UNESCO, UNIDO, UPU, WCO, WHO, WIPO, WMO, WTO
Diplomatic representation in the US: *chief of mission:* Ambassador Yusif bin Mani bin Said al-UTAYBA
chancery: 3522 International Court NW, Suite 400, Washington, DC 20008
telephone: [1] (202) 243-2400
FAX: [1] (202) 243-2432
Diplomatic representation from the US: *chief of mission:* Ambassador Richard G. OLSON, Jr.
embassy: Embassies District, Plot 38 Sector W59-02, Street No. 4, Abu Dhabi
mailing address: P. O. Box 4009, Abu Dhabi
telephone: [971] (2) 414-2200
FAX: [971] (2) 414-2603
consulate(s) general: Dubai
Flag description: three equal horizontal bands of green (top), white, and black with a wider vertical red band on the hoist side; the flag incorporates all four Pan-Arab colors, which in this case represent fertility (green), neutrality (white), petroleum resources (black), and unity (red); red was the traditional color incorporated into all flags of the emirates before their unification
National anthem: name: "Nashid al-watani al-imarati" (National Anthem of the UAE)
lyrics/music: AREF Al Sheikh Abdullah Al Hassan/Mohamad Abdel WAHAB

note: music adopted 1971, lyrics adopted 1996; Mohamad Abdel WAHAB also composed the music for the anthem of Tunisia

ECONOMY

Economy—overview: The UAE has an open economy with a high per capita income and a sizable annual trade surplus. Successful efforts at economic diversification have reduced the portion of GDP based on oil and gas output to 25%. Since the discovery of oil in the UAE more than 30 years ago, the UAE has undergone a profound transformation from an impoverished region of small desert principalities to a modern state with a high standard of living. The government has increased spending on job creation and infrastructure expansion and is opening up utilities to greater private sector involvement. In April 2004, the UAE signed a Trade and Investment Framework Agreement with Washington and in November 2004 agreed to undertake negotiations toward a Free Trade Agreement with the US, however, those talks have not moved forward. The country's Free Trade Zones–offering 100% foreign ownership and zero taxes–are helping to attract foreign investors. The global financial crisis, tight international credit, and deflated asset prices constricted the economy in 2009 and 2010. UAE authorities tried to blunt the crisis by increasing spending and boosting liquidity in the banking sector. The crisis hit Dubai hardest, as it was heavily exposed to depressed real estate prices. Dubai lacked sufficient cash to meet its debt obligations, prompting global concern about its solvency. The UAE Central Bank and Abu Dhabi-based banks bought the largest shares. In December 2009 Dubai received an additional $10 billion loan from the emirate of Abu Dhabi. The economy is expected to continue a slow rebound. Dependence on oil, a large expatriate workforce, and growing inflation pressures are significant long-term challenges. The UAE's strategic plan for the next few years focuses on diversification and creating more opportunities for nationals through improved education and increased private sector employment.

GDP (purchasing power parity):
$246.8 billion (2010 est.)
country comparison to the world: 51
$239.1 billion (2009 est.)
$246.9 billion (2008 est.)
note: data are in 2010 US dollars

GDP (official exchange rate):
$301.9 billion (2010 est.)

GDP—real growth rate: 3.2% (2010 est.)
country comparison to the world: 117
-3.2% (2009 est.)
5.3% (2008 est.)

GDP—per capita (PPP):
$49,600 (2010 est.)
country comparison to the world: 9
$49,800 (2009 est.)
$53,400 (2008 est.)
note: data are in 2010 US dollars

GDP—composition by sector:
agriculture: 0.9%
industry: 51.5%
services: 47.6% (2010 est.)

Labor force: 3.908 million
country comparison to the world: 89
note: expatriates account for about 85% of the work force (2010 est.)

Labor force—by occupation:
agriculture: 7%
industry: 15%
services: 78% (2000 est.)

Unemployment rate: 2.4% (2001)
country comparison to the world: 20

Population below poverty line:
19.5% (2003)

Household income or consumption by percentage share: *lowest 10%:* NA%
highest 10%: NA%

Investment (gross fixed):
26.8% of GDP (2010 est.)
country comparison to the world: 31

Budget: *revenues:* $65.02 billion
expenditures: $60.02 billion (2010 est.)

Public debt: 44.6% of GDP (2010 est.)
country comparison to the world: 61
48.9% of GDP (2009 est.)

Inflation rate (consumer prices):
2.2% (2010 est.)
country comparison to the world: 55
1.6% (2009 est.)

Central bank discount rate: NA%

Stock of narrow money: $68.76 billion (31 December 2010 est.)
country comparison to the world: 39
$60.85 billion (31 December 2009 est.)

Stock of broad money: $228.5 billion (31 December 2010 est.)
country comparison to the world: 38
$201.6 billion (31 December 2009 est.)

Stock of domestic credit: $290 billion (31 December 2010 est.)
country comparison to the world: 34
$263.6 billion (31 December 2009 est.)

Market value of publicly traded shares:
$109.6 billion (31 December 2009)
country comparison to the world: 40
$97.85 billion (31 December 2008)
$224.7 billion (31 December 2007)

Agriculture—products: dates, vegetables, watermelons; poultry, eggs, dairy products; fish

Industries: petroleum and petrochemicals; fishing, aluminum, cement, fertilizers, commercial ship repair, construction materials, some boat building, handicrafts, textiles

Industrial production growth rate:
3.2% (2010 est.)
country comparison to the world: 100

Electricity—production:
71.54 billion kWh (2007 est.)
country comparison to the world: 39

Electricity—consumption:
65.98 billion kWh (2007 est.)
country comparison to the world: 38

Electricity—exports: 0 kWh (2008 est.)

Electricity—imports: 0 kWh (2008 est.)

Oil—production: 2.798 million bbl/day (2009 est.)
country comparison to the world: 8

Oil—consumption: 435,000 bbl/day (2009 est.)
country comparison to the world: 32

Oil—exports: 2.7 million bbl/day (2007 est.)
country comparison to the world: 3

Oil—imports: 192,900 bbl/day (2007 est.)
country comparison to the world: 46

Oil—proved reserves: 97.8 billion bbl (1 January 2010 est.)
country comparison to the world: 6

Natural gas—production: 50.24 billion cu m (2008 est.)
country comparison to the world: 18

Natural gas—consumption: 59.42 billion cu m (2008 est.)
country comparison to the world: 13

Natural gas—exports: 7.567 billion cu m (2008 est.)
country comparison to the world: 24

Natural gas—imports: 16.75 billion cu m (2008 est.)
country comparison to the world: 15

Natural gas—proved reserves: 6.071 trillion cu m (1 January 2010 est.)
country comparison to the world: 7

Current account balance:
$3.409 billion (2010 est.)
country comparison to the world: 36
$7.871 billion (2009 est.)

Exports: $195.8 billion (2010 est.)
country comparison to the world: 25
$192.2 billion (2009 est.)

Exports—commodities: crude oil 45%, natural gas, reexports, dried fish, dates

Exports—partners: Japan 17.27%, South Korea 10.49%, India 9.96%, Iran 6.82%, Thailand 5.11% (2009)

Imports: $159 billion (2010 est.)
country comparison to the world: 26
$150 billion (2009 est.)

Imports—commodities: machinery and transport equipment, chemicals, food

Imports—partners: China 15.03%, India 14.27%, US 8.44%, Germany 5.81%, Japan 4.52% (2009)

Reserves of foreign exchange and gold:
$39.1 billion (31 December 2010 est.)
country comparison to the world: 32
$36.1 billion (31 December 2009 est.)

Debt—external:
$122.7 billion (31 December 2010 est.)
country comparison to the world: 34
$122.5 billion (31 December 2009 est.)

Stock of direct foreign investment—at home:
$76.38 billion (31 December 2010 est.)
country comparison to the world: 47
$70.18 billion (31 December 2009 est.)

Stock of direct foreign investment—abroad:
$54.91 billion (31 December 2010 est.)
country comparison to the world: 31
$51.41 billion (31 December 2009 est.)

Exchange rates: Emirati dirhams (AED) per US dollar–3.673 (2010), 3.673 (2009), 3.6725 (2008), 3.6725 (2007), 3.6725 (2006)

COMMUNICATIONS

Telephones—main lines in use:
1.561 million (2009)
country comparison to the world: 63

Telephones—mobile cellular:
10.672 million (2009)
country comparison to the world: 61

Telephone system: *general assessment:* modern fiber-optic integrated services; digital network with rapidly growing use of mobile-cellular telephones; key centers are Abu Dhabi and Dubai
domestic: microwave radio relay, fiber optic and coaxial cable

international: country code–971; linked to the international submarine cable FLAG (Fiber-Optic Link Around the Globe); landing point for both the SEA-ME-WE-3 and SEA-ME-WE-4 submarine cable networks; satellite earth stations–3 Intelsat (1 Atlantic Ocean and 2 Indian Ocean) and 1 Arabsat; tropospheric scatter to Bahrain; microwave radio relay to Saudi Arabia
Broadcast media: except for the many organizations now operating in Dubai's Media Free Zone, most television and radio stations remain government-owned; widespread use of satellite dishes provides access to pan-Arab and other international broadcasts (2007)
Internet country code: .ae
Internet hosts: 379,309 (2010)
country comparison to the world: 54
Internet users: 3.449 million (2009)
country comparison to the world: 61

TRANSPORTATION

Airports: 41 (2010)
country comparison to the world: 104
Airports—with paved runways:
total: 25
over 3,047 m: 12
2,438 to 3,047 m: 3
1,524 to 2,437 m: 4
914 to 1,523 m: 4
under 914 m: 2 (2010)
Airports—with unpaved runways:
total: 16
over 3,047 m: 1
2,438 to 3,047 m: 1
1,524 to 2,437 m: 4
914 to 1,523 m: 5
under 914 m: 5 (2010)
Heliports: 5 (2010)
Pipelines: condensate 458 km; refined products 212 km; gas 2,352 km; liquid petroleum gas 220 km; oil 1,437 km (2010)
Roadways: *total:* 4,080 km
country comparison to the world: 156
paved: 4,080 km (includes 253 km of expressways) (2008)
Merchant marine: *total:* 57
country comparison to the world: 68
by type: bulk carrier 4, cargo 9, chemical tanker 7, container 7, liquefied gas 1, passenger/cargo 1, petroleum tanker 24, roll on/roll off 4
foreign-owned: 13 (Greece 3, Kuwait 10)
registered in other countries: 278 (Bahamas 27, Belize 5, Cambodia 2, Comoros 11, Cyprus 5, Georgia 1, Gibraltar 5, Hong Kong 2, India 4, Iran 1, Jordan 7, Liberia 27, Malta 1, Marshall Islands 17, Mexico 1, Netherlands 4, North Korea 6, Panama 83, Papua New Guinea 6, Philippines 1, Saint Kitts and Nevis 17, Saint Vincent and the Grenadines 4, Saudi Arabia 6, Sierra Leone 6, Singapore 10, Tanzania 1, Togo 1, UK 9, Vanuatu 1, unknown 7) (2010)
Ports and terminals: Al Fujayrah, Mina' Jabal 'Ali (Dubai), Khawr Fakkan (Khor Fakkan), Mubarraz Island, Mina' Rashid (Dubai), Mina' Saqr (Ra's al Khaymah)

MILITARY

Military branches: United Arab Emirates Armed Forces: Army, Navy (includes Marines), Air Force and Air Defense, Border and Coast Guard Directorate (BCGD) (2009)
Military service age and obligation: 18 years of age (est.) for voluntary military service; 18 years of age for officers and women; no conscription (2009)
Manpower available for military service:
males age 16–49: 2,676,928 (includes non-nationals)
females age 16–49: 981,649 (2010 est.)
Manpower fit for military service: *males age 16–49:* 2,229,366
females age 16–49: 842,759 (2010 est.)
Manpower reaching militarily significant age annually: *male:* 27,439
female: 24,419 (2010 est.)
Military expenditures:
3.1% of GDP (2005 est.)
country comparison to the world: 39

TRANSNATIONAL ISSUES

Disputes—international: boundary agreement was signed and ratified with Oman in 2003 for entire border, including Oman's Musandam Peninsula and Al Madhah enclaves, but contents of the agreement and detailed maps showing the alignment have not been published; Iran and UAE dispute Tunb Islands and Abu Musa Island, which Iran occupies
Illicit drugs: the UAE is a drug transshipment point for traffickers given its proximity to Southwest Asian drug-producing countries; the UAE's position as a major financial center makes it vulnerable to money laundering; anti-money-laundering controls improving, but informal banking remains unregulated

UNITED KINGDOM

INTRODUCTION

Background: The United Kingdom has historically played a leading role in developing parliamentary democracy and in advancing literature and science. At its zenith in the 19th century, the British Empire stretched over one-fourth of the earth's surface. The first half of the 20th century saw the UK's strength seriously depleted in two world wars and the Irish republic withdraw from the union. The second half witnessed the dismantling of the Empire and the UK rebuilding itself into a modern and prosperous European nation. As one of five permanent members of the UN Security Council, a founding member of NATO, and of the Commonwealth, the UK pursues a global approach to foreign policy. The UK is also an active member of the EU, although it chose to remain outside the Economic and Monetary Union. The Scottish Parliament, the National Assembly for Wales, and the Northern Ireland Assembly were established in 1999. The latter was suspended until May 2007 due to wrangling over the peace process, but devolution was fully completed in March 2010.

GEOGRAPHY

Location: Western Europe, islands–including the northern one-sixth of the island of Ireland–between the North Atlantic Ocean and the North Sea; northwest of France
Geographic coordinates: 54 00 N, 2 00 W
Map references: Europe
Area: *total:* 243,610 sq km
country comparison to the world: 79
land: 241,930 sq km
water: 1,680 sq km
note: includes Rockall and Shetland Islands
Area—comparative: slightly smaller than Oregon
Land boundaries: *total:* 360 km
border countries: Ireland 360 km
Coastline: 12,429 km
Maritime claims: territorial sea: 12 nm
exclusive fishing zone: 200 nm
continental shelf: as defined in continental shelf orders or in accordance with agreed upon boundaries
Climate: temperate; moderated by prevailing southwest winds over the North Atlantic Current; more than one-half of the days are overcast
Terrain: mostly rugged hills and low mountains; level to rolling plains in east and southeast
Elevation extremes:
lowest point: The Fens -4 m
highest point: Ben Nevis 1,343 m
Natural resources: coal, petroleum, natural gas, iron ore, lead, zinc, gold, tin, limestone, salt, clay, chalk, gypsum, potash, silica sand, slate, arable land
Land use: *arable land:* 23.23%
permanent crops: 0.2%
other: 76.57% (2005)
Irrigated land: 1,700 sq km (2003)
Total renewable water resources: 160.6 cu km (2005)
Freshwater withdrawal (domestic/industrial/agricultural): *total:* 11.75 cu km/yr (22%/75%/3%)
per capita: 197 cu m/yr (1994)
Natural hazards: winter windstorms; floods
Environment—current issues: continues to reduce greenhouse gas emissions (has met Kyoto Protocol target of a 12.5% reduction from 1990 levels and intends to meet the legally binding target and move toward a domestic goal of a 20% cut in

emissions by 2010); by 2005 the government reduced the amount of industrial and commercial waste disposed of in landfill sites to 85% of 1998 levels and recycled or composted at least 25% of household waste, increasing to 33% by 2015

Environment—international agreements: *party to:* Air Pollution, Air Pollution-Nitrogen Oxides, Air Pollution-Persistent Organic Pollutants, Air Pollution-Sulfur 94, Air Pollution-Volatile Organic Compounds, Antarctic-Environmental Protocol, Antarctic-Marine Living Resources, Antarctic Seals, Antarctic Treaty, Biodiversity, Climate Change, Climate Change-Kyoto Protocol, Desertification, Endangered Species, Environmental Modification, Hazardous Wastes, Law of the Sea, Marine Dumping, Marine Life Conservation, Ozone Layer Protection, Ship Pollution, Tropical Timber 83, Tropical Timber 94, Wetlands, Whaling
signed, but not ratified: none of the selected agreements

Geography—note: lies near vital North Atlantic sea lanes; only 35 km from France and linked by tunnel under the English Channel; because of heavily indented coastline, no location is more than 125 km from tidal waters

PEOPLE

Population: 62,698,362 (July 2011 est.)
country comparison to the world: 22

Age structure: *0–14 years:* 17.3% (male 5,575,119/female 5,301,301)
15–64 years: 66.2% (male 20,979,401/female 20,500,913)
65 years and over: 16.5% (male 4,564,375/female 5,777,253) (2011 est.)

Median age: *total:* 40 years
male: 38.8 years
female: 41.1 years (2011 est.)

Population growth rate: 0.557% (2011 est.)
country comparison to the world: 148

Birth rate: 12.29 births/1,000 population (2011 est.)
country comparison to the world: 162

Death rate: 9.33 deaths/1,000 population (July 2011 est.)
country comparison to the world: 64

Net migration rate: 2.6 migrant(s)/1,000 population (2011 est.)
country comparison to the world: 32

Urbanization: *urban population:* 80% of total population (2010)
rate of urbanization: 0.7% annual rate of change (2010-15 est.)

Major cities—population: LONDON (capital) 8.615 million; Birmingham 2.296 million; Manchester 2.247 million; West Yorkshire 1.541 million; Glasgow 1.166 million (2009)

Sex ratio: *at birth:* 1.052 male(s)/female
under 15 years: 1.05 male(s)/female
15–64 years: 1.03 male(s)/female
65 years and over: 0.76 male(s)/female
total population: 0.98 male(s)/female (2011 est.)

Infant mortality rate:
total: 4.62 deaths/1,000 live births
country comparison to the world: 189
male: 5.07 deaths/1,000 live births
female: 4.15 deaths/1,000 live births (2011 est.)

Life expectancy at birth: *total population:* 80.05 years
country comparison to the world: 28
male: 77.95 years
female: 82.25 years (2011 est.)

Total fertility rate: 1.91 children born/woman (2011 est.)
country comparison to the world: 139

HIV/AIDS—adult prevalence rate: 0.2% (2009 est.)
country comparison to the world: 107

HIV/AIDS—people living with HIV/AIDS: 85,000 (2009 est.)
country comparison to the world: 45

HIV/AIDS—deaths: fewer than 1,000 (2009 est.)
country comparison to the world: 80

Drinking water source: *Improved:*
urban: 100% of population
rural: 100% of population
total: 100% of population (2008)

Sanitation facility access: *Improved:*
urban: 100% of population
rural: 100% of population
total: 100% of population (2008)

Nationality: *noun:* Briton(s), British (collective plural)
adjective: British

Ethnic groups: white (of which English 83.6%, Scottish 8.6%, Welsh 4.9%, Northern Irish 2.9%) 92.1%, black 2%, Indian 1.8%, Pakistani 1.3%, mixed 1.2%, other 1.6% (2001 census)

Religions: Christian (Anglican, Roman Catholic, Presbyterian, Methodist) 71.6%, Muslim 2.7%, Hindu 1%, other 1.6%, unspecified or none 23.1% (2001 census)

Languages: English
note: the following are recognized regional languages: Scots (about 30% of the population of Scotland), Scottish Gaelic (about 60,000 in Scotland), Welsh (about 20% of the population of Wales), Irish (about 10% of the population of Northern Ireland), Cornish (some 2,000 to 3,000 in Cornwall)

Literacy: *definition:* age 15 and over has completed five or more years of schooling
total population: 99%
male: 99%
female: 99% (2003 est.)

School life expectancy (primary to tertiary education): *total:* 16 years
male: 16 years
female: 17 years (2008)

Education expenditures:
5.5% of GDP (2007)
country comparison to the world: 44

GOVERNMENT

Country name: *conventional long form:* United Kingdom of Great Britain and Northern Ireland; note—Great Britain includes England, Scotland, and Wales
conventional short form: United Kingdom

abbreviation: UK

Government type: constitutional monarchy and Commonwealth realm
Capital: name: London
geographic coordinates: 51 30 N, 0 10 W
time difference: UTC 0 (5 hours ahead of Washington, DC during Standard Time)

daylight saving time: +1hr, begins last Sunday in March; ends last Sunday in October
note: applies to the United Kingdom proper, not to its overseas dependencies or territories

Administrative divisions: England: 27 two-tier counties, 32 London boroughs and 1 City of London or Greater London, 36 metropolitan districts, 56 unitary authorities (including 4 single-tier counties*)

two-tier counties: Buckinghamshire, Cambridgeshire, Cumbria, Derbyshire, Devon, Dorset, East Sussex, Essex, Gloucestershire, Hampshire, Hertfordshire, Kent, Lancashire, Leicestershire, Lincolnshire, Norfolk, North Yorkshire, Northamptonshire, Nottinghamshire, Oxfordshire, Somerset, Staffordshire, Suffolk, Surrey, Warwickshire, West Sussex, Worcestershire

London boroughs and City of London or Greater London: Barking and Dagenham, Barnet, Bexley, Brent, Bromley, Camden, Croydon, Ealing, Enfield, Greenwich, Hackney, Hammersmith and Fulham, Haringey, Harrow, Havering, Hillingdon, Hounslow, Islington, Kensington and Chelsea, Kingston upon Thames, Lambeth, Lewisham, City of London, Merton, Newham, Redbridge, Richmond upon Thames, Southwark, Sutton, Tower Hamlets, Waltham Forest, Wandsworth, Westminster

metropolitan districts: Barnsley, Birmingham, Bolton, Bradford, Bury, Calderdale, Coventry, Doncaster, Dudley, Gateshead, Kirklees, Knowsley, Leeds, Liverpool, Manchester, Newcastle upon Tyne, North Tyneside, Oldham, Rochdale, Rotherham, Salford, Sandwell, Sefton, Sheffield, Solihull, South Tyneside, St. Helens, Stockport, Sunderland, Tame-

side, Trafford, Wakefield, Walsall, Wigan, Wirral, Wolverhampton
unitary authorities: Bath and North East Somerset, Blackburn with Darwen, Bedford, Blackpool, Bournemouth, Bracknell Forest, Brighton and Hove, City of Bristol, Central Bedfordshire, Cheshire East, Cheshire West and Chester, Cornwall, Darlington, Derby, Durham County*, East Riding of Yorkshire, Halton, Hartlepool, Herefordshire*, Isle of Wight*, Isles of Scilly*, City of Kingston upon Hull, Leicester, Luton, Medway, Middlesbrough, Milton Keynes, North East Lincolnshire, North Lincolnshire, North Somerset, Northumberland*, Nottingham, Peterborough, Plymouth, Poole, Portsmouth, Reading, Redcar and Cleveland, Rutland, Shropshire, Slough, South Gloucestershire, Southampton, Southend-on-Sea, Stockton-on-Tees, Stoke-on-Trent, Swindon, Telford and Wrekin, Thurrock, Torbay, Warrington, West Berkshire, Wiltshire, Windsor and Maidenhead, Wokingham, York
Northern Ireland: 26 district council areas
district council areas: Antrim, Ards, Armagh, Ballymena, Ballymoney, Banbridge, Belfast, Carrickfergus, Castlereagh, Coleraine, Cookstown, Craigavon, Derry, Down, Dungannon, Fermanagh, Larne, Limavady, Lisburn, Magherafelt, Moyle, Newry and Mourne, Newtownabbey, North Down, Omagh, Strabane
Scotland: 32 council areas
council areas: Aberdeen City, Aberdeenshire, Angus, Argyll and Bute, Clackmannanshire, Dumfries and Galloway, Dundee City, East Ayrshire, East Dunbartonshire, East Lothian, East Renfrewshire, City of Edinburgh, Eilean Siar (Western Isles), Falkirk, Fife, Glasgow City, Highland, Inverclyde, Midlothian, Moray, North Ayrshire, North Lanarkshire, Orkney Islands, Perth and Kinross, Renfrewshire, Shetland Islands, South Ayrshire, South Lanarkshire, Stirling, The Scottish Borders, West Dunbartonshire, West Lothian
Wales: 22 unitary authorities
unitary authorities: Blaenau Gwent; Bridgend; Caerphilly; Cardiff; Carmarthenshire; Ceredigion; Conwy; Denbighshire; Flintshire; Gwynedd; Isle of Anglesey; Merthyr Tydfil; Monmouthshire; Neath Port Talbot; Newport; Pembrokeshire; Powys; Rhondda Cynon Taff; Swansea; The Vale of Glamorgan; Torfaen; Wrexham
Dependent areas: Anguilla, Bermuda, British Indian Ocean Territory, British Virgin Islands, Cayman Islands, Falkland Islands, Gibraltar, Montserrat, Pitcairn Islands, Saint Helena, Ascension, and Tristan da Cunha, South Georgia and the South Sandwich Islands, Turks and Caicos Islands
Independence: 12 April 1927 (Royal and Parliamentary Titles Act establishes current name of the United Kingdom of Great Britain and Northern Ireland); notable earlier dates: 927 (minor English kingdoms united); 3 March 1284 (enactment of the Statute of Rhuddlan uniting England and Wales); 1536 (Act of Union formally incorporates England and Wales); 1 May 1707 (Acts of Union formally unite England and Scotland as Great Britain); 1 January 1801 (Acts of Union formally unite Great Britain and Ireland as the United Kingdom of Great Britain and Ireland); 6 December 1921 (Anglo-Irish Treaty formalizes partition of Ireland; six counties remain part of the United Kingdom as Northern Ireland)
National holiday: the UK does not celebrate one particular national holiday
Constitution: unwritten; partly statutes, partly common law and practice
Legal system: common law system; has nonbinding judicial review of Acts of Parliament under the Human Rights Act of 1998
International law organization participation: accepts compulsory ICJ jurisdiction with reservations; accepts ICCt jurisdiction
Suffrage: 18 years of age; universal
Executive branch: *chief of state:* Queen ELIZABETH II (since 6 February 1952); Heir Apparent Prince CHARLES (son of the queen, born 14 November 1948)
head of government: Prime Minister David CAMERON (since 11 May 2010)
cabinet: Cabinet of Ministers appointed by the prime minister
(For more information visit the World Leaders website)
elections: the monarchy is hereditary; following legislative elections, the leader of the majority party or the leader of the majority coalition usually the prime minister
Legislative branch: bicameral Parliament consists of House of Lords (741 seats; consisting of approximately 625 life peers, 91 hereditary peers, and 25 clergy—as of 15 December 2010) and House of Commons (650 seats since 2010 elections; members elected by popular vote to serve five-year terms unless the House is dissolved earlier)
elections: House of Lords—no elections (note—in 1999, as provided by the House of Lords Act, elections were held in the House of Lords to determine the 92 hereditary peers who would remain there; elections are held only as vacancies in the hereditary peerage arise); House of Commons—last held on 6 May 2010 (next to be held by June 2015)
election results: House of Commons—percent of vote by party—Conservative 36.1%, Labor 29%, Liberal Democrats 23%, other 11.9%; seats by party—Conservative 305, Labor 258, Liberal Democrat 57, other 30
note: in 1998 elections were held for a Northern Ireland Assembly (because of unresolved disputes among existing parties, the transfer of power from London to Northern Ireland came only at the end of 1999 and has been suspended four times, the latest occurring in October 2002 and lasting until 8 May 2007); in 1999, the UK held the first elections for a Scottish Parliament and a Welsh Assembly
Judicial branch: Supreme Court of the UK (established in October 2009 taking over appellate jurisdiction formerly vested in the House of Lords); Senior Courts of England and Wales (comprising the Court of Appeal, the High Court of Justice, and the Crown Courts); Court of Judicature (Northern Ireland); Scotland's Court of Session and High Court of the Justiciary
Political parties and leaders: Conservative [David CAMERON]; Democratic Unionist Party or DUP (Northern Ireland) [Peter ROBINSON]; Labor Party [Ed MILIBAND]; Liberal Democrats (Lib Dems) [Nick CLEGG]; Party of Wales (Plaid Cymru) [Ieuan Wyn JONES]; Scottish National Party or SNP [Alex SALMOND]; Sinn Fein (Northern Ireland) [Gerry ADAMS]; Social Democratic and Labor Party or SDLP (Northern Ireland) [Margaret RICHIE]; Ulster Unionist Party (Northern Ireland) [Tom ELLIOTT]
Political pressure groups and leaders: Campaign for Nuclear Disarmament; Confederation of British Industry; National Farmers' Union; Trades Union Congress
International organization participation: ADB (nonregional member), AfDB (nonregional member), Arctic Council (observer), Australia Group, BIS, C, CBSS (observer), CDB, CE, CERN, EAPC, EBRD, EIB, ESA, EU, FAO, FATF, G-20, G-5, G-7, G-8, G-10, IADB, IAEA, IBRD, ICAO, ICC, ICRM, IDA, IEA, IFAD, IFC, IFRCS, IHO, ILO, IMF, IMO, IMSO, Interpol, IOC, IOM, IPU, ISO, ITSO, ITU, ITUC, MIGA, MONUSCO, NATO, NEA, NSG, OAS (observer), OECD, OPCW, OSCE, Paris Club, PCA, PIF (partner), SECI (observer), UN, UN Security Council, UNCTAD, UNESCO, UNFICYP, UNHCR, UNIDO, UNMIS, UNRWA, UPU, WCO, WHO, WIPO, WMO, WTO, ZC
Diplomatic representation in the US: *chief of mission:* Ambassador Sir Nigel E. SHEINWALD
chancery: 3100 Massachusetts Avenue NW, Washington, DC 20008
telephone: [1] (202) 588-6500
FAX: [1] (202) 588-7870
consulate(s) general: Atlanta, Boston, Chicago, Houston, Los Angeles, Miami, New York, San Francisco
consulate(s): Dallas, Denver, Orlando
Diplomatic representation from the US: *chief of mission:* Ambassador Louis B. SUSMAN
embassy: 24 Grosvenor Square, London, W1A 1AE note—a new embassy is scheduled to open by the end of 2017 in the Nine Elms area of Wandsworth (architect Kiernan TIMBERLAKE)
mailing address: PSC 801, Box 40, FPO AE 09498-4040
telephone: [44] (0) 20 7499-9000
FAX: [44] (0) 20 7629-9124
consulate(s) general: Belfast, Edinburgh
Flag description: blue field with the red cross of Saint George (patron saint of England) edged in white superimposed on the diagonal red cross of Saint Patrick (patron saint of Ireland), which is superimposed on the diagonal white cross of Saint Andrew (patron saint of Scotland); properly known as the Union Flag, but commonly called the Union Jack; the design and colors (especially the Blue Ensign) have been the basis for a number of other flags including other Commonwealth countries and their constituent

states or provinces, and British overseas territories
National anthem: name: "God Save the Queen"
lyrics/music: unknown
note: in use since 1745; by tradition, the song serves as both the national and royal anthem of the United Kingdom; it is known as either "God Save the Queen" or "God Save the King," depending on the gender of the reigning monarch; it also serves as the royal anthem of many Commonwealth nations

ECONOMY

Economy—overview: The UK, a leading trading power and financial center, is the third largest economy in Europe after Germany and France. Over the past two decades, the government has greatly reduced public ownership and contained the growth of social welfare programs. Agriculture is intensive, highly mechanized, and efficient by European standards, producing about 60% of food needs with less than 2% of the labor force. The UK has large coal, natural gas, and oil resources, but its oil and natural gas reserves are declining and the UK became a net importer of energy in 2005. Services, particularly banking, insurance, and business services, account by far for the largest proportion of GDP while industry continues to decline in importance. After emerging from recession in 1992, Britain's economy enjoyed the longest period of expansion on record during which time growth outpaced most of Western Europe. In 2008, however, the global financial crisis hit the economy particularly hard, due to the importance of its financial sector. Sharply declining home prices, high consumer debt, and the global economic slowdown compounded Britain's economic problems, pushing the economy into recession in the latter half of 2008 and prompting the then BROWN government to implement a number of measures to stimulate the economy and stabilize the financial markets; these include nationalizing parts of the banking system, cutting taxes, suspending public sector borrowing rules, and moving forward public spending on capital projects. Facing burgeoning public deficits and debt levels, the CAMERON government in 2010 initiated a five-year austerity program, which aims to lower London's budget deficit from over 11% of GDP in 2010 to nearly 1% by 2015. The Bank of England periodically coordinates interest rate moves with the European Central Bank, but Britain remains outside the European Economic and Monetary Union (EMU).
GDP (purchasing power parity):
$2.173 trillion (2010 est.)
country comparison to the world: 8
$2.146 trillion (2009 est.)
$2.256 trillion (2008 est.)
note: data are in 2010 US dollars
GDP (official exchange rate): $2.247 trillion (2010 est.)
GDP—real growth rate: 1.3% (2010 est.)
country comparison to the world: 164
-4.9% (2009 est.)
-0.1% (2008 est.)
GDP—per capita (PPP):
$34,800 (2010 est.)
country comparison to the world: 37
$34,600 (2009 est.)
$36,600 (2008 est.)
note: data are in 2010 US dollars
GDP—composition by sector:
agriculture: 0.9%
industry: 22.1%
services: 77.1% (2010 est.)
Labor force: 31.45 million (2010 est.)
country comparison to the world: 19
Labor force—by occupation:
agriculture: 1.4%
industry: 18.2%
services: 80.4% (2006 est.)
Unemployment rate: 7.9% (2010 est.)
country comparison to the world: 86
7.6% (2009 est.)
Population below poverty line:
14% (2006 est.)
Household income or consumption by percentage share: *lowest 10%:* 2.1%
highest 10%: 28.5% (1999)
Distribution of family income—Gini index: 34 (2005)
country comparison to the world: 92
36.8 (1999)
Investment (gross fixed):
14.4% of GDP (2010 est.)
country comparison to the world: 137
Budget: *revenues:* $926.7 billion
expenditures: $1.154 trillion (2010 est.)
Public debt: 76.5% of GDP (2010 est.)
country comparison to the world: 23
68.2% of GDP (2009 est.)
Inflation rate (consumer prices):
3.3% (2010 est.)
country comparison to the world: 92
2.2% (2009 est.)
Central bank discount rate:
NA% (31 December 2009)
country comparison to the world: 137
0.86% (31 December 2008)
Commercial bank prime lending rate:
0.63% (31 December 2009 est.)
country comparison to the world: 148
4.63% (31 December 2008 est.)
Stock of narrow money: $88.62 billion (31 December 2010 est.)
country comparison to the world: 34
$84.92 billion (31 December 2009 est.)
Stock of broad money: $3.344 trillion (31 December 2010 est.)
country comparison to the world: 7
$3.199 trillion (31 December 2009 est.)
Stock of domestic credit: $5.151 trillion (31 December 2009)
country comparison to the world: 6
$4.436 trillion (31 December 2008)
Market value of publicly traded shares:
$2.796 trillion (31 December 2009)
country comparison to the world: 6
$1.852 trillion (31 December 2008)
$3.859 trillion (31 December 2007)
Agriculture—products: cereals, oilseed, potatoes, vegetables; cattle, sheep, poultry; fish
Industries: machine tools, electric power equipment, automation equipment, railroad equipment, shipbuilding, aircraft, motor vehicles and parts, electronics and communications equipment, metals, chemicals, coal, petroleum, paper and paper products, food processing, textiles, clothing, other consumer goods
Industrial production growth rate:
1.9% (2010 est.)
country comparison to the world: 132
Electricity—production: 368.6 billion kWh (2007 est.)
country comparison to the world: 12
Electricity—consumption: 345.8 billion kWh (2007 est.)
country comparison to the world: 12
Electricity—exports: 1.272 billion kWh (2008 est.)
Electricity—imports: 12.29 billion kWh (2008 est.)
Oil—production: 1.502 million bbl/day (2009 est.)
country comparison to the world: 20
Oil—consumption: 1.669 million bbl/day (2009 est.)
country comparison to the world: 15
Oil—exports: 1.393 million bbl/day (2008 est.)
country comparison to the world: 17
Oil—imports: 1.491 million bbl/day (2008 est.)
country comparison to the world: 13
Oil—proved reserves: 3.084 billion bbl (1 January 2010 est.)
country comparison to the world: 31
Natural gas—production: 58.56 billion cu m (2009 est.)
country comparison to the world: 16
Natural gas—consumption: 87.45 billion cu m (2009 est.)
country comparison to the world: 8
Natural gas—exports: 12.17 billion cu m (2009 est.)
country comparison to the world: 17
Natural gas—imports: 41.06 billion cu m (2009 est.)
country comparison to the world: 6
Natural gas—proved reserves: 292 billion cu m (1 January 2010 est.)
country comparison to the world: 40
Current account balance: $-40.34 billion (2010 est.)
country comparison to the world: 186
$-23.65 billion (2009 est.)
Exports: $405.6 billion (2010 est.)
country comparison to the world: 11
$356.2 billion (2009 est.)
Exports—commodities: manufactured goods, fuels, chemicals; food, beverages, tobacco
Exports—partners: US 14.71%, Germany 11.06%, France 8%, Netherlands 7.79%, Ireland 6.89%, Belgium 4.65%, Spain 4% (2009)
Imports: $546.5 billion (2010 est.)
country comparison to the world: 7
$483.9 billion (2009 est.)
Imports—commodities: manufactured goods, machinery, fuels; foodstuffs
Imports—partners: Germany 12.87%, US 9.74%, China 8.88%, Netherlands 6.94%, France 6.64%, Belgium 4.86%, Norway 4.84%, Ireland 4.01%, Italy 3.99% (2009)
Reserves of foreign exchange and gold:
$NA (31 December 2010 est.)
$66.72 billion (31 December 2009 est.)
Debt—external: $8.981 trillion (30 June 2010)
country comparison to the world: 3

$9.041 trillion (31 December 2008)
Stock of direct foreign investment—at home:
$1.169 trillion (31 December 2010 est.)
country comparison to the world: 3
$1.125 trillion (31 December 2009 est.)
Stock of direct foreign investment—abroad: $1.705 trillion (31 December 2010 est.)
country comparison to the world: 3
$1.652 trillion (31 December 2009 est.)
Exchange rates: British pounds (GBP) per US dollar–0.6388 (2010), 0.6175 (2009), 0.5302 (2008), 0.4993 (2007), 0.5418 (2006)

COMMUNICATIONS

Telephones—main lines in use: 32.117 million (2009)
country comparison to the world: 10
Telephones—mobile cellular: 80.375 million (2009)
country comparison to the world: 15
Telephone system: *general assessment:* technologically advanced domestic and international system
domestic: equal mix of buried cables, microwave radio relay, and fiber-optic systems
international: country code—44; numerous submarine cables provide links throughout Europe, Asia, Australia, the Middle East, and US; satellite earth stations—10 Intelsat (7 Atlantic Ocean and 3 Indian Ocean), 1 Inmarsat (Atlantic Ocean region), and 1 Eutelsat; at least 8 large international switching centers
Broadcast media: public service broadcaster BBC is the largest broadcasting corporation in the world; BBC operates multiple TV networks with regional and local TV service; a mixed system of public and commercial TV broadcasters along with satellite and cable systems provide access to hundreds of TV stations throughout the world; BBC operates multiple national, regional, and local radio networks with multiple transmission sites; a large number of commercial radio stations as well as satellite radio services are available (2008)
Internet country code: .uk
Internet hosts: 7.03 million (2010)
country comparison to the world: 14
Internet users: 51.444 million (2009)
country comparison to the world: 7

TRANSPORTATION

Airports: 505 (2010)
country comparison to the world: 14
Airports—with paved runways: *total:* 306
over 3,047 m: 9
2,438 to 3,047 m: 32
1,524 to 2,437 m: 124
914 to 1,523 m: 77
under 914 m: 64 (2010)
Airports—with unpaved runways: *total:* 199
over 3,047 m: 1
1,524 to 2,437 m: 3
914 to 1,523 m: 22
under 914 m: 173 (2010)
Heliports: 11 (2010)
Pipelines: condensate 8 km; gas 14,071 km; liquid petroleum gas 59 km; oil 595 km; refined products 4,907 km (2010)
Railways: *total:* 16,454 km
country comparison to the world: 17
broad gauge: 303 km 1.600-m gauge (in Northern Ireland)
standard gauge: 16,151 km 1.435-m gauge (5,248 km electrified) (2008)
Roadways: *total:* 394,428 km
country comparison to the world: 16
paved: 394,428 km (includes 3,519 km of expressways) (2009)
Waterways: 3,200 km (620 km used for commerce) (2009)
country comparison to the world: 31
Merchant marine: *total:* 527
country comparison to the world: 22
by type: bulk carrier 30, cargo 70, carrier 3, chemical tanker 71, container 190, liquefied gas 10, passenger 7, passenger/cargo 67, petroleum tanker 20, refrigerated cargo 6, roll on/roll off 29, vehicle carrier 24
foreign-owned: 271 (Australia 1, Bermuda 9, China 15, Denmark 40, France 32, Germany 78, Greece 1, Hong Kong, Italy 4, Japan 4, Netherlands 1, Norway 32, NZ 1, South Africa 5, Spain 7, Sweden 21, Taiwan 1, Turkey 1, UAE 7, United States 11)
note: this country allows large numbers of ships owned by foreign entities to be registered in its national shipping registry and to fly its flag; these ships operate under the laws of the flag state
registered in other countries: 275 (Algeria 12, Antigua and Barbuda 2, Argentina 2, Australia 5, Bahamas 24, Barbados 7, Belgium 2, Belize 4, Bermuda 11, Cambodia 3, Cape Verde 2, Cayman Islands 2, Comoros 1, Cook Islands 2, Cyprus 7, Georgia 4, Gibraltar 4, Greece 27, Honduras 1, Hong Kong 27, Italy 2, Liberia 44, Libya 1, Luxembourg 5, Malta 16, Marshall Islands 9, Moldova 6, Nigeria 2, Panama 44, Saint Kitts and Nevis 2, Saint Vincent and the Grenadines 7, Sierra Leone 1, Singapore 6, Thailand 6, Togo 3, Tonga 1, US 4, unknown 1) (2010)
Ports and terminals: Dover, Felixstowe, Immingham, Liverpool, London, Southampton, Teesport (England); Forth Ports, Hound Point (Scotland); Milford Haven (Wales)

MILITARY

Military branches: Army, Royal Navy (includes Royal Marines), Royal Air Force (2010)
Military service age and obligation: 16-33 years of age (officers 17-28) for voluntary military service (with parental consent under 18); women serve in military services, but are excluded from ground combat positions and some naval postings; as of October 2009, women comprised 12.1% of officers and 9% of enlisted personnel in the regular forces; must be citizen of the UK, Commonwealth, or Republic of Ireland; reservists serve a minimum of 3 years, to age 45 or 55; 16 years of age for voluntary military service by Nepalese citizens in the Brigade of Gurkhas; 16-34 years of age for voluntary military service by Papua New Guinean citizens (2009)
Manpower available for military service: *males age 16–49:* 14,856,917
females age 16–49: 14,307,316 (2010 est.)
Manpower fit for military service: *males age 16–49:* 12,255,452
females age 16–49: 11,779,679 (2010 est.)
Manpower reaching militarily significant age annually: *male:* 383,989
female: 365,491 (2010 est.)
Military expenditures: 2.4% of GDP (2005 est.)
country comparison to the world: 63

TRANSNATIONAL ISSUES

Disputes—international: in 2002, Gibraltar residents voted overwhelmingly by referendum to reject any "shared sovereignty" arrangement between the UK and Spain; the Government of Gibraltar insists on equal participation in talks between the two countries; Spain disapproves of UK plans to grant Gibraltar greater autonomy; Mauritius and Seychelles claim the Chagos Archipelago (British Indian Ocean Territory); in 2001, the former inhabitants of the archipelago, evicted 1967–1973, were granted U.K. citizenship and the right of return, followed by Orders in Council in 2004 that banned rehabitation, a High Court ruling reversing the ban, a Court of Appeal refusal to hear the case, and a Law Lords' decision in 2008 denying the right of return; in addition, the United Kingdom created the world's largest marine protection area around the Chagos islands prohibiting the extraction of any natural resources therein; UK rejects sovereignty talks requested by Argentina, which still claims the Falkland Islands (Islas Malvinas) and South Georgia and the South Sandwich Islands; territorial claim in Antarctica (British Antarctic Territory) overlaps Argentine claim and partially overlaps Chilean claim; Iceland, the UK, and Ireland dispute Denmark's claim that the Faroe Islands' continental shelf extends beyond 200 nm
Illicit drugs: producer of limited amounts of synthetic drugs and synthetic precursor chemicals; major consumer of Southwest Asian heroin, Latin American cocaine, and synthetic drugs; money-laundering center

UNITED STATES

INTRODUCTION

Background: Britain's American colonies broke with the mother country in 1776 and were recognized as the new nation of the United States of America following the Treaty of Paris in 1783. During the 19th and 20th centuries, 37 new states were added to the original 13 as the nation expanded across the North American continent and acquired a number of overseas possessions. The two most traumatic experiences in the nation's history were the Civil War (1861-65), in which a northern Union of states defeated a secessionist Confederacy of 11 southern slave states, and the Great Depression of the 1930s, an economic downturn during which about a quarter of the labor force lost its jobs. Buoyed by victories in World Wars I and II and the end of the Cold War in 1991, the US remains the world's most powerful nation state. Over a span of more than five decades, the economy has achieved steady growth, low unemployment and inflation, and rapid advances in technology.

GEOGRAPHY

Location: North America, bordering both the North Atlantic Ocean and the North Pacific Ocean, between Canada and Mexico
Geographic coordinates: 38 00 N, 97 00 W
Map references: North America
Area: *total:* 9,826,675 sq km
country comparison to the world: 3
land: 9,161,966 sq km
water: 664,709 sq km
note: includes only the 50 states and District of Columbia
Area—comparative: about half the size of Russia; about three-tenths the size of Africa; about half the size of South America (or slightly larger than Brazil); slightly larger than China; more than twice the size of the European Union
Land boundaries: *total:* 12,034 km
border countries: Canada 8,893 km (including 2,477 km with Alaska), Mexico 3,141 km
note: US Naval Base at Guantanamo Bay, Cuba is leased by the US and is part of Cuba; the base boundary is 28 km
Coastline: 19,924 km
Maritime claims: territorial sea: 12 nm
contiguous zone: 24 nm
exclusive economic zone: 200 nm
continental shelf: not specified
Climate: mostly temperate, but tropical in Hawaii and Florida, arctic in Alaska, semiarid in the great plains west of the Mississippi River, and arid in the Great Basin of the southwest; low winter temperatures in the northwest are ameliorated occasionally in January and February by warm chinook winds from the eastern slopes of the Rocky Mountains
Terrain: vast central plain, mountains in west, hills and low mountains in east; rugged mountains and broad river valleys in Alaska; rugged, volcanic topography in Hawaii
Elevation extremes: *lowest point:* Death Valley -86 m
highest point: Mount McKinley 6,194 m
note: the peak of Mauna Kea (4,207 m above sea level) on the island of Hawaii rises about 10,200 m above the Pacific Ocean floor; by this measurement, it is the world's tallest mountain–higher than Mount Everest, which is recognized as the tallest mountain above sea level
Natural resources: coal, copper, lead, molybdenum, phosphates, rare earth elements, uranium, bauxite, gold, iron, mercury, nickel, potash, silver, tungsten, zinc, petroleum, natural gas, timber
note: the US has the world's largest coal reserves with 491 billion short tons accounting for 27% of the world's total
Land use: *arable land:* 18.01%
permanent crops: 0.21%
other: 81.78% (2005)
Irrigated land: 223,850 sq km (2003)
Total renewable water resources: 3,069 cu km (1985)
Freshwater withdrawal (domestic/industrial/agricultural): *total:* 477 cu km/yr (13%/46%/41%)
per capita: 1,600 cu m/yr (2000)
Natural hazards: tsunamis; volcanoes; earthquake activity around Pacific Basin; hurricanes along the Atlantic and Gulf of Mexico coasts; tornadoes in the Midwest and Southeast; mud slides in California; forest fires in the west; flooding; permafrost in northern Alaska, a major impediment to development
volcanism: the United States experiences volcanic activity in the Hawaiian Islands, Western Alaska, the Pacific Northwest, and in the Northern Mariana Islands; both Mauna Loa (elev. 4,170 m) in Hawaii and Mount Rainier (elev. 4,392 m) in Washington have been deemed "Decade Volcanoes" by the International Association of Volcanology and Chemistry of the Earth's Interior, worthy of study due to their explosive history and close proximity to human populations; Pavlof (elev. 2,519 m) is the most active volcano in Alaska's Aleutian Arc and poses a significant threat to air travel since the area constitutes a major flight path between North America and East Asia; St. Helens (elev. 2,549 m, famous for the devastating 1980 eruption, remains active today; numerous other historically active volcanoes exist, mostly concentrated in the Aleutian arc and Hawaii; they include: in Alaska: Aniakchak, Augustine, Chiginagak, Fourpeaked, Iliamna, Katmai, Kupreanof, Martin, Novarupta, Redoubt, Spurr, Wrangell; in Hawaii: Trident, Ugashik-Peulik, Ukinrek Maars, Veniaminof; in the Northern Mariana Islands: Anatahan; and in the Pacific Northwest: Mount Baker, Mount Hood
Environment—current issues: air pollution resulting in acid rain in both the US and Canada; the US is the largest single emitter of carbon dioxide from the burning of fossil fuels; water pollution from runoff of pesticides and fertilizers; limited natural freshwater resources in much of the western part of the country require careful management; desertification
Environment—international agreements: *party to:* Air Pollution, Air Pollution-Nitrogen Oxides, Antarctic-Environmental Protocol, Antarctic-Marine Living Resources, Antarctic Seals, Antarctic Treaty, Climate Change, Desertification, Endangered Species, Environmental Modification, Marine Dumping, Marine Life Conservation, Ozone Layer Protection, Ship Pollution, Tropical Timber 83, Tropical Timber 94, Wetlands, Whaling
signed, but not ratified: Air Pollution-Persistent Organic Pollutants, Air Pollution-Volatile Organic Compounds, Biodiversity, Climate Change-Kyoto Protocol, Hazardous Wastes
Geography—note: world's third-largest country by size (after Russia and Canada) and by population (after China and India); Mt. McKinley is highest point in North America and Death Valley the lowest point on the continent

PEOPLE

Population: 313,232,044 (July 2011 est.)
country comparison to the world: 3
Age structure: *0–14 years:* 20.1% (male 32,107,900/female 30,781,823)
15–64 years: 66.8% (male 104,411,352/female 104,808,064)
65 years and over: 13.1% (male 17,745,363/female 23,377,542) (2011 est.)
Median age: *total:* 36.9 years
male: 35.6 years
female: 38.2 years (2011 est.)

Population growth rate: 0.963% (2011 est.)
country comparison to the world: 119
Birth rate: 13.83 births/1,000 population (2011 est.)
country comparison to the world: 149
Death rate: 8.38 deaths/1,000 population (July 2011 est.)
country comparison to the world: 89
Net migration rate: 4.18 migrant(s)/1,000 population (2011 est.)
country comparison to the world: 23
Urbanization: *urban population:* 82% of total population (2010)
rate of urbanization: 1.2% annual rate of change (2010-15 est.)
Major cities—population: New York-Newark 19.3 million; Los Angeles-Long Beach-Santa Ana 12.675 million; Chicago 9.134 million; Miami 5.699 million; WASHINGTON, D.C. (capital) 4.421 million (2009)
Sex ratio: *at birth:* 1.047 male(s)/female
under 15 years: 1.04 male(s)/female
15–64 years: 1 male(s)/female
65 years and over: 0.75 male(s)/female
total population: 0.97 male(s)/female (2011 est.)
Infant mortality rate:
total: 6.06 deaths/1,000 live births
country comparison to the world: 176
male: 6.72 deaths/1,000 live births
female: 5.37 deaths/1,000 live births (2011 est.)
Life expectancy at birth:
total population: 78.37 years
country comparison to the world: 50
male: 75.92 years
female: 80.93 years (2011 est.)
Total fertility rate: 2.06 children born/woman (2011 est.)
country comparison to the world: 123
HIV/AIDS—adult prevalence rate:
0.6% (2009 est.)
country comparison to the world: 63
HIV/AIDS—people living with HIV/AIDS:
1.2 million (2009 est.)
country comparison to the world: 10
HIV/AIDS—deaths: 17,000 (2009 est.)
country comparison to the world: 18
Drinking water source: *Improved:*
urban: 100% of population
rural: 94% of population
total: 99% of population
Unimproved:
urban: 0% of population
rural: 6% of population
total: 1% of population (2008)
Sanitation facility access: *Improved:*
urban: 100% of population
rural: 99% of population
total: 100% of population
Unimproved:
urban: 0% of population
rural: 1% of population
total: 0% of population (2008)
Nationality: *noun:* American(s)
adjective: American
Ethnic groups: white 79.96%, black 12.85%, Asian 4.43%, Amerindian and Alaska native 0.97%, native Hawaiian and other Pacific islander 0.18%, two or more races 1.61% (July 2007 estimate)
note: a separate listing for Hispanic is not included because the US Census Bureau considers Hispanic to mean persons of Spanish/Hispanic/Latino origin including those of Mexican, Cuban, Puerto Rican, Dominican Republic, Spanish, and Central or South American origin living in the US who may be of any race or ethnic group (white, black, Asian, etc.); about 15.1% of the total US population is Hispanic
Religions: Protestant 51.3%, Roman Catholic 23.9%, Mormon 1.7%, other Christian 1.6%, Jewish 1.7%, Buddhist 0.7%, Muslim 0.6%, other or unspecified 2.5%, unaffiliated 12.1%, none 4% (2007 est.)
Languages: English 82.1%, Spanish 10.7%, other Indo-European 3.8%, Asian and Pacific island 2.7%, other 0.7% (2000 census)
note: Hawaiian is an official language in the state of Hawaii
Literacy: *definition:* age 15 and over can read and write
total population: 99%
male: 99%
female: 99% (2003 est.)
School life expectancy (primary to tertiary education): *total:* 16 years
male: 15 years
female: 17 years (2008)
Education expenditures:
5.5% of GDP (2007)
country comparison to the world: 43

GOVERNMENT

Country name: *conventional long form:* United States of America
conventional short form: United States
abbreviation: US or USA
Government type: Constitution-based federal republic; strong democratic tradition
Capital: name: Washington, DC
geographic coordinates: 38 53 N, 77 02 W
time difference: UTC-5 (during Standard Time)
daylight saving time: +1hr, begins second Sunday in March; ends first Sunday in November
note: the 50 United States cover six time zones
Administrative divisions: 50 states and 1 district*; Alabama, Alaska, Arizona, Arkansas, California, Colorado, Connecticut, Delaware, District of Columbia*, Florida, Georgia, Hawaii, Idaho, Illinois, Indiana, Iowa, Kansas, Kentucky, Louisiana, Maine, Maryland, Massachusetts, Michigan, Minnesota, Mississippi, Missouri, Montana, Nebraska, Nevada, New Hampshire, New Jersey, New Mexico, New York, North Carolina, North Dakota, Ohio, Oklahoma, Oregon, Pennsylvania, Rhode Island, South Carolina, South Dakota, Tennessee, Texas, Utah, Vermont, Virginia, Washington, West Virginia, Wisconsin, Wyoming
Dependent areas: American Samoa, Baker Island, Guam, Howland Island, Jarvis Island, Johnston Atoll, Kingman Reef, Midway Islands, Navassa Island, Northern Mariana Islands, Palmyra Atoll, Puerto Rico, Virgin Islands, Wake Island
note: from 18 July 1947 until 1 October 1994, the US administered the Trust Territory of the Pacific Islands; it entered into a political relationship with all four political entities: the Northern Mariana Islands is a commonwealth in political union with the US (effective 3 November 1986); the Republic of the Marshall Islands signed a Compact of Free Association with the US (effective 21 October 1986); the Federated States of Micronesia signed a Compact of Free Association with the US (effective 3 November 1986); Palau concluded a Compact of Free Association with the US (effective 1 October 1994)
Independence: 4 July 1776 (declared); 3 September 1783 (recognized by Great Britain)
National holiday: Independence Day, 4 July (1776)
Constitution: 17 September 1787, effective 4 March 1789
Legal system: common law system based on English common law at the federal level; state legal systems based on common law except Louisiana, which is based on Napoleonic civil code; judicial review of legislative acts
International law organization participation: withdrew acceptance of compulsory ICJ jurisdiction in 2005; withdrew acceptance of ICCt jurisdiction in 2002
Suffrage: 18 years of age; universal
Executive branch: *chief of state:* President Barack H. OBAMA (since 20 January 2009); Vice President Joseph R. BIDEN (since 20 January 2009); note—the president is both the chief of state and head of government
head of government: President Barack H. OBAMA (since 20 January 2009); Vice President Joseph R. BIDEN (since 20 January 2009)
cabinet: Cabinet appointed by the president with Senate approval
(For more information visit the World Leaders website)
elections: president and vice president elected on the same ticket by a college of representatives who are elected directly from each state; president and vice president serve four-year terms (eligible for a second term); election last held 4 November 2008 (next to be held on 6 November 2012)
election results: Barack H. OBAMA elected president; percent of popular vote—Barack H. OBAMA 52.4%, John MCCAIN 46.3%, other 1.3%;
Legislative branch: bicameral Congress consists of the Senate (100 seats, 2 members elected from each state by popular vote to serve six-year terms; one-third elected every two years) and the House of Representatives (435 seats; members directly elected by popular vote to serve two-year terms)
elections: Senate—last held on 2 November 2010 (next to be held in November 2012); House of Representatives—last held on 2 November 2010 (next to be held in November 2012)
election results: Senate—percent of vote by party—NA; seats by party—Democratic Party 51, Republican Party 47, independent 2; House of Representatives—percent of vote by party—NA; seats by party—Democratic Party 192, Republican Party 243
Judicial branch: Supreme Court (nine justices; nominated by the president and confirmed with the advice and consent

of the Senate; appointed to serve for life); United States Courts of Appeal; United States District Courts; State and County Courts

Political parties and leaders: Democratic Party [Timothy KAINE]; Green Party; Libertarian Party [William (Bill) REDPATH]; Republican Party [Reince PRIEBUS]

Political pressure groups and leaders: environmentalists; business groups; labor unions; churches; ethnic groups; political action committees or PAC; health groups; education groups; civic groups; youth groups; transportation groups; agricultural groups; veterans groups; women's groups; reform lobbies

International organization participation: ADB (nonregional member), AfDB (nonregional member), ANZUS, APEC, Arctic Council, ARF, ASEAN (dialogue partner), Australia Group, BIS, BSEC (observer), CBSS (observer), CD, CERN (observer), CICA (observer), CP, EAPC, EBRD, FAO, FATF, G-20, G-5, G-7, G-8, G-10, IADB, IAEA, IBRD, ICAO, ICC, ICRM, IDA, IEA, IFAD, IFC, IFRCS, IHO, ILO, IMF, IMO, IMSO, Interpol, IOC, IOM, ISO, ITSO, ITU, ITUC, MIGA, MINUSTAH, MONUSCO, NAFTA, NATO, NEA, NSG, OAS, OECD, OPCW, OSCE, Paris Club, PCA, PIF (partner), SAARC (observer), SECI (observer), SPC, UN, UN Security Council, UNCTAD, UNESCO, UNHCR, UNITAR, UNMIL, UNRWA, UNTSO, UPU, WCO, WHO, WIPO, WMO, WTO, ZC

Flag description: 13 equal horizontal stripes of red (top and bottom) alternating with white; there is a blue rectangle in the upper hoist-side corner bearing 50 small, white, five-pointed stars arranged in nine offset horizontal rows of six stars (top and bottom) alternating with rows of five stars; the 50 stars represent the 50 states, the 13 stripes represent the 13 original colonies; the blue stands for loyalty, devotion, truth, justice, and friendship; red symbolizes courage, zeal, and fervency, while white denotes purity and rectitude of conduct; commonly referred to by its nickname of Old Glory

note: the design and colors have been the basis for a number of other flags, including Chile, Liberia, Malaysia, and Puerto Rico

National anthem: name: "The Star-Spangled Banner"

lyrics/music: Francis Scott KEY/John Stafford SMITH

note: adopted 1931; during the War of 1812, after witnessing the successful American defense of Fort McHenry in Baltimore following British naval bombardment, Francis Scott KEY wrote the lyrics to what would become the national anthem; the lyrics were set to the tune of "The Anacreontic Song;" only the first verse is sung

ECONOMY

Economy—overview: The US has the largest and most technologically powerful economy in the world, with a per capita GDP of $47,200. In this market-oriented economy, private individuals and business firms make most of the decisions, and the federal and state governments buy needed goods and services predominantly in the private marketplace. US business firms enjoy greater flexibility than their counterparts in Western Europe and Japan in decisions to expand capital plant, to lay off surplus workers, and to develop new products. At the same time, they face higher barriers to enter their rivals' home markets than foreign firms face entering US markets. US firms are at or near the forefront in technological advances, especially in computers and in medical, aerospace, and military equipment; their advantage has narrowed since the end of World War II. The onrush of technology largely explains the gradual development of a "two-tier labor market" in which those at the bottom lack the education and the professional/technical skills of those at the top and, more and more, fail to get comparable pay raises, health insurance coverage, and other benefits. Since 1975, practically all the gains in household income have gone to the top 20% of households. The war in March-April 2003 between a US-led coalition and Iraq, and the subsequent occupation of Iraq, required major shifts in national resources to the military. Soaring oil prices between 2005 and the first half of 2008 threatened inflation and unemployment, as higher gasoline prices ate into consumers' budgets. Imported oil accounts for about 60% of US consumption. Long-term problems include inadequate investment in economic infrastructure, rapidly rising medical and pension costs of an aging population, sizable trade and budget deficits, and stagnation of family income in the lower economic groups. The merchandise trade deficit reached a record $840 billion in 2008 before shrinking to $506 billion in 2009, and ramping back up to $630 billion in 2010. The global economic downturn, the sub-prime mortgage crisis, investment bank failures, falling home prices, and tight credit pushed the United States into a recession by mid-2008. GDP contracted until the third quarter of 2009, making this the deepest and longest downturn since the Great Depression. To help stabilize financial markets, the US Congress established a $700 billion Troubled Asset Relief Program (TARP) in October 2008. The government used some of these funds to purchase equity in US banks and other industrial corporations, much of which had been returned to the government by early 2011. In January 2009 the US Congress passed and President Barack OBAMA signed a bill providing an additional $787 billion fiscal stimulus to be used over 10 years—two-thirds on additional spending and one-third on tax cuts—to create jobs and to help the economy recover. Approximately two-thirds of these funds were injected into the economy by the end of 2010. In March 2010, President OBAMA signed a health insurance reform bill into law that will extend coverage to an additional 32 million American citizens by 2016, through private health insurance for the general population and Medicaid for the impoverished. In July 2010, the president signed the DODD-FRANK Wall Street Reform and Consumer Protection Act, a bill designed to promote financial stability by protecting consumers from financial abuses, ending taxpayer bailouts of financial firms, dealing with troubled banks that are "too big to fail," and improving accountability and transparency in the financial system—in particular, by requiring certain financial derivatives to be traded in markets that are subject to government regulation and oversight. In November 2010, in an attempt to keep interest rates from rising and snuffing out the nascent recovery, the US Federal Reserve Bank (The Fed) announced that it would purchase $600 billion worth of US Government bonds by June 2011.

GDP (purchasing power parity):
$14.66 trillion (2010 est.)
country comparison to the world: 2
$14.25 trillion (2009 est.)
$14.64 trillion (2008 est.)
note: data are in 2010 US dollars

GDP (official exchange rate):
$14.66 trillion (2010 est.)

GDP—real growth rate: 2.8% (2010 est.)
country comparison to the world: 129
-2.6% (2009 est.)
0% (2008 est.)

GDP—per capita (PPP):
$47,200 (2010 est.)
country comparison to the world: 11
$46,400 (2009 est.)
$48,100 (2008 est.)
note: data are in 2010 US dollars

GDP—composition by sector:
agriculture: 1.2%
industry: 22.2%
services: 76.7% (2010 est.)

Labor force: 154.9 million
country comparison to the world: 4
note: includes unemployed (2010 est.)

Labor force—by occupation: farming, forestry, and fishing: 0.7%
manufacturing, extraction, transportation, and crafts: 20.3%
managerial, professional, and technical: 37.3%
sales and office: 24.2%
other *services:* 17.6%
note: figures exclude the unemployed (2009)

Unemployment rate: 9.7% (2010 est.)
country comparison to the world: 106
9.3% (2009 est.)

Population below poverty line:
12% (2004 est.)

Household income or consumption by percentage share: *lowest 10%:* 2%
highest 10%: 30% (2007 est.)

Distribution of family income—Gini index: 45 (2007)
country comparison to the world: 39
40.8 (1997)

Investment (gross fixed): 12.8% of GDP (2010 est.)
country comparison to the world: 142

Budget: *revenues:* $2.092 trillion
expenditures: $3.397 trillion (2010 est.)

Public debt: 58.9% of GDP (2010 est.)
country comparison to the world: 37
53.5% of GDP (2009 est.)
note: data cover only what the United States Treasury denotes as "Debt Held by the Public," which includes all debt instruments issued by the Treasury that are owned by non-US Government enti-

ties; the data include Treasury debt held by foreign entities; the data exclude debt issued by individual US states, as well as intra-governmental debt; intra-governmental debt consists of Treasury borrowings from surpluses in the trusts for Federal Social Security, Federal Employees, Hospital Insurance (Medicare and Medicaid), Disability and Unemployment, and several other smaller trusts; if data for intra-government debt were added, "Gross Debt" would increase by about 30% of GDP
Inflation rate (consumer prices):
1.4% (2010 est.)
country comparison to the world: 35
-0.3% (2009 est.)
Central bank discount rate:
0.5% (31 December 2009)
country comparison to the world: 136
0.86% (31 December 2008)
Commercial bank prime lending rate:
3.25% (31 December 2009 est.)
country comparison to the world: 145
5.09% (31 December 2008 est.)
Stock of narrow money: $1.74 trillion (31 December 2010 est.)
country comparison to the world: 5
$1.722 trillion (31 December 2009 est.)
Stock of broad money: $12.39 trillion (31 December 2009)
country comparison to the world: 3
$12.46 trillion (31 December 2008)
Stock of domestic credit: $32.61 trillion (31 December 2009 est.)
country comparison to the world: 1
$31.53 trillion (31 December 2008 est.)
Market value of publicly traded shares:
$15.08 trillion (31 December 2009)
country comparison to the world: 1
$11.74 trillion (31 December 2008)
$19.95 trillion (31 December 2007)
Agriculture—products: wheat, corn, other grains, fruits, vegetables, cotton; beef, pork, poultry, dairy products; fish; forest products
Industries: highly diversified, world leading, high-technology innovator, second largest industrial output in world; petroleum, steel, motor vehicles, aerospace, telecommunications, chemicals, electronics, food processing, consumer goods, lumber, mining
Industrial production growth rate: 3.3% (2010 est.)
country comparison to the world: 98
Electricity—production: 4.11 trillion kWh (2008 est.)
country comparison to the world: 1
Electricity—consumption: 3.873 trillion kWh (2008 est.)
country comparison to the world: 1
Electricity—exports: 24.08 billion kWh (2008 est.)
Electricity—imports: 57.02 billion kWh (2008 est.)
Oil—production: 9.056 million bbl/day (2009 est.)
country comparison to the world: 3
Oil—consumption: 18.69 million bbl/day (2009 est.)
country comparison to the world: 1
Oil—exports: 1.704 million bbl/day (2008 est.)
country comparison to the world: 13
Oil—imports: 11.31 million bbl/day (2008 est.)
country comparison to the world: 1
Oil—proved reserves: 19.12 billion bbl (1 January 2010 est.)
country comparison to the world: 14
Natural gas—production: 593.4 billion cu m (2009 est.)
country comparison to the world: 1
Natural gas—consumption: 646.6 billion cu m (2009 est.)
country comparison to the world: 1
Natural gas—exports: 30.35 billion cu m (2009 est.)
country comparison to the world: 9
Natural gas—imports: 106.1 billion cu m (2009 est.)
country comparison to the world: 1
Natural gas—proved reserves: 6.928 trillion cu m (1 January 2010 est.)
country comparison to the world: 6
Current account balance: $-561 billion (2010 est.)
country comparison to the world: 191
$-378.4 billion (2009 est.)
Exports: $1.27 trillion (2010 est.)
country comparison to the world: 4
$1.069 trillion (2009 est.)
Exports—commodities: agricultural products (soybeans, fruit, corn) 9.2%, industrial supplies (organic chemicals) 26.8%, capital goods (transistors, aircraft, motor vehicle parts, computers, telecommunications equipment) 49.0%, consumer goods (automobiles, medicines) 15.0%
Exports—partners: Canada 19.37%, Mexico 12.21%, China 6.58%, Japan 4.84%, UK 4.33%, Germany 4.1% (2009)
Imports: $1.903 trillion (2010 est.)
country comparison to the world: 1
$1.575 trillion (2009 est.)
Imports—commodities: agricultural products 4.9%, industrial supplies 32.9% (crude oil 8.2%), capital goods 30.4% (computers, telecommunications equipment, motor vehicle parts, office machines, electric power machinery), consumer goods 31.8% (automobiles, clothing, medicines, furniture, toys)
Imports—partners: China 19.3%, Canada 14.24%, Mexico 11.12%, Japan 6.14%, Germany 4.53% (2009)
Reserves of foreign exchange and gold:
$NA (31 December 2010 est.)
$130.8 billion (31 December 2009 est.)
Debt—external: $13.98 trillion (30 June 2010)
country comparison to the world: 1
$13.75 trillion (31 December 2008)
note: approximately 4/5ths of US external debt is denominated in US dollars; foreign lenders have been willing to hold US dollar denominated debt instruments because they view the dollar as the world's reserve currency
Stock of direct foreign investment—at home: $2.581 trillion (31 December 2010 est.)
country comparison to the world: 1
$2.41 trillion (31 December 2009 est.)
Stock of direct foreign investment—abroad: $3.597 trillion (31 December 2010 est.)
country comparison to the world: 1
$3.367 trillion (31 December 2009 est.)
Exchange rates: British pounds per US dollar: 0.6388 (2010), 0.6494 (2009), 0.5302 (2008), 0.4993 (2007), 0.5418 (2006)
Canadian dollars per US dollar: 1.0346 (2010), 1.1431 (2009), 1.0364 (2008), 1.0724 (2007), 1.1334 (2006)
Chinese yuan per US dollar: 6.7852 (2010), 6.8314 (2009), 6.9385 (2008), 7.61 (2007), 7.97 (2006)
euros per US dollar: 0.755 (2010), 0.7198 (2009), 0.6827 (2008), 0.7345 (2007), 0.7964 (2006)
Japanese yen per US dollar: 1.0346 (2010), 1.1548 (2009), 1.0364 (2008), 1.0724 (2007), 1.1334 (2006), 87.78 (2010), 93.57 (2009), 103.58 (2008), 117.99 (2007), 116.18 (2006)

COMMUNICATIONS

Telephones—main lines in use:
141 million (2009)
country comparison to the world: 2
Telephones—mobile cellular:
286 million (2009)
country comparison to the world: 3
Telephone system: *general assessment:* a large, technologically advanced, multipurpose communications system
domestic: a large system of fiber-optic cable, microwave radio relay, coaxial cable, and domestic satellites carries every form of telephone traffic; a rapidly growing cellular system carries mobile telephone traffic throughout the country
international: country code–1; multiple ocean cable systems provide international connectivity; satellite earth stations–61 Intelsat (45 Atlantic Ocean and 16 Pacific Ocean), 5 Intersputnik (Atlantic Ocean region), and 4 Inmarsat (Pacific and Atlantic Ocean regions) (2000)
Broadcast media: 4 major terrestrial television networks with affiliate stations throughout the country, plus cable and satellite networks, independent stations, and a limited public broadcasting sector that is largely supported by private grants; overall, thousands of TV stations broadcasting; multiple national radio networks with large numbers of affiliate stations; while most stations are commercial, National Public Radio (NPR) has a network of some 600 member stations; satellite radio available; overall, nearly 15,000 radio stations operating (2008)
Internet country code: .us
Internet hosts: 439 million (2010); note–the US Internet total host count includes the following top level domain host addresses: .us, .com, .edu, .gov, .mil, .net, and .org
country comparison to the world: 1
Internet users: 245 million (2009)
country comparison to the world: 2

TRANSPORTATION

Airports: 15,079 (2010)
country comparison to the world: 1
Airports—with paved runways: *total:* 5,194
over 3,047 m: 189
2,438 to 3,047 m: 235
1,524 to 2,437 m: 1,479
914 to 1,523 m: 2,316
under 914 m: 975 (2010)

Airports—with unpaved runways: *total:* 9,885
2,438 to 3,047 m: 7
1,524 to 2,437 m: 155
914 to 1,523 m: 1,752
under 914 m: 7,971 (2010)
Heliports: 126 (2010)
Pipelines: petroleum products 244,620 km; natural gas 548,665 km (2010)
Railways: *total:* 226,427 km
country comparison to the world: 1
standard gauge: 226,427 km 1.435-m gauge (2007)
Roadways: *total:* 6,506,204 km
country comparison to the world: 1
paved: 4,374,784 km (includes 75,238 km of expressways)
unpaved: 2,131,420 km (2008)
Waterways: 41,009 km (19,312 km used for commerce; Saint Lawrence Seaway of 3,769 km, including the Saint Lawrence River of 3,058 km, is shared with Canada) (2008)
country comparison to the world: 4
Merchant marine: *total:* 418
country comparison to the world: 26
by type: barge carrier 6, bulk carrier 58, cargo 58, carrier 3, chemical tanker 30, container 87, passenger 18, passenger/cargo 56, petroleum tanker 45, refrigerated cargo 3, roll on/roll off 27, vehicle carrier 27
foreign-owned: 86 (Australia 1, Bermuda 5, Canada 1, Denmark 34, France 4, Germany 3, Malaysia 2, Norway 10, Singapore 17, Sweden 5, UK 4)
registered in other countries: 734 (Antigua and Barbuda 6, Australia 2, Bahamas 100, Belgium 2, Bermuda 25, Cambodia 4, Canada 9, Cayman Islands 54, Comoros 2, Cyprus 7, Georgia 1, Greece 7, Hong Kong 31, Indonesia 2, Ireland 2, Isle of Man 2, Italy 21, Liberia 39, Luxembourg 3, Malta 35, Marshall Islands 168, Netherlands 15, Norway 9, Panama 102, Portugal 4, Saint Kitts and Nevis 1, Saint Vincent and the Grenadines 19, Sierra Leone 1, Singapore 33, South Korea 8, UK 11, unknown 8) (2010)
Ports and terminals: cargo ports (tonnage): Baton Rouge, Corpus Christi, Houston, Long Beach, Los Angeles, New Orleans, New York, Plaquemines, Tampa, Texas City
container ports (TEUs): Los Angeles (7,849,985), Long Beach (6,350,125), New York/New Jersey (5,265,058), Savannah (2,616,126), Oakland (2,236,244), Hampton Roads (2,083,278) (2008)
cruise departure ports (passengers): Miami (2,032,000), Port Everglades (1,277,000), Port Canaveral (1,189,000), Seattle (430,000), Long Beach (415,000) (2009)

MILITARY

Military branches: United States Armed Forces: US Army, US Navy (includes Marine Corps), US Air Force, US Coast Guard; note–Coast Guard administered in peacetime by the Department of Homeland Security, but in wartime reports to the Department of the Navy (2009)
Military service age and obligation: 18 years of age (17 years of age with parental consent) for male and female voluntary service; maximum enlistment age 42 (Army), 27 (Air Force), 34 (Navy), 28 (Marines); service obligation 8 years, including 2-5 years active duty (Army), 2 years active (Navy), 4 years active (Air Force, Marines) (2010)
Manpower available for military service: *males age 16–49:* 73,270,043
females age 16–49: 71,941,969 (2010 est.)
Manpower fit for military service: *males age 16–49:* 60,620,143
females age 16–49: 59,401,941 (2010 est.)
Manpower reaching militarily significant age annually: *male:* 2,161,727
female: 2,055,685 (2010 est.)
Military expenditures: 4.06% of GDP (2005 est.)
country comparison to the world: 24

TRANSNATIONAL ISSUES

Disputes—international: the US has intensified domestic security measures and is collaborating closely with its neighbors, Canada and Mexico, to monitor and control legal and illegal personnel, transport, and commodities across the international borders; abundant rainfall in recent years along much of the Mexico-US border region has ameliorated periodically strained water-sharing arrangements; 1990 Maritime Boundary Agreement in the Bering Sea still awaits Russian Duma ratification; Canada and the United States dispute how to divide the Beaufort Sea and the status of the Northwest Passage but continue to work cooperatively to survey the Arctic continental shelf; The Bahamas and US have not been able to agree on a maritime boundary; US Naval Base at Guantanamo Bay is leased from Cuba and only mutual agreement or US abandonment of the area can terminate the lease; Haiti claims US-administered Navassa Island; US has made no territorial claim in Antarctica (but has reserved the right to do so) and does not recognize the claims of any other states; Marshall Islands claims Wake Island; Tokelau included American Samoa's Swains Island among the islands listed in its 2006 draft constitution
Refugees and internally displaced persons: refugees (country of origin): the US admitted 62,643 refugees during FY04/05 including; 10,586 (Somalia); 8,549 (Laos); 6,666 (Russia); 6,479 (Cuba); 3,100 (Haiti); 2,136 (Iran) (2006)
Illicit drugs: world's largest consumer of cocaine (shipped from Colombia through Mexico and the Caribbean), Colombian heroin, and Mexican heroin and marijuana; major consumer of ecstasy and Mexican methamphetamine; minor consumer of high-quality Southeast Asian heroin; illicit producer of cannabis, marijuana, depressants, stimulants, hallucinogens, and methamphetamine; money-laundering center

UNITED STATES PACIFIC ISLAND WILDLIFE REFUGES

(TERRITORIES OF THE US)

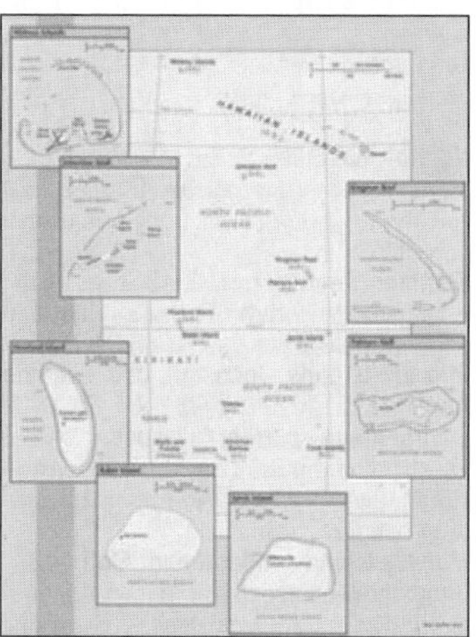

INTRODUCTION

Background: All of the following US Pacific island territories except Midway Atoll constitute the Pacific Remote Islands National Wildlife Refuge (NWR) Complex and as such are managed by the Fish and Wildlife Service of the US Department of the Interior. Midway Atoll NWR has been included in a Refuge Complex with the Hawaiian Islands NWR and also designated as part of Papahanaumokuakea Marine National Monument. These remote refuges are the most widespread collection of marine- and terrestrial-life protected areas on the planet under a single country's jurisdiction. They sustain many endemic species including corals, fish, shellfish, marine mammals, seabirds, water birds, land birds, insects, and vegetation not found elsewhere.
Baker Island: The US took possession of the island in 1857. Its guano deposits were mined by US and British companies during the second half of the 19th century. In 1935, a short-lived attempt at colonization began on this island but was disrupted by World War II and thereafter abandoned. The island was established as a NWR in 1974.
Howland Island: Discovered by the US early in the 19th century, the uninhabited atoll was officially claimed by the US in 1857. Both US and British companies mined for guano deposits until about 1890. In 1935, a short-lived attempt at colonization began on this island, similar to the effort on nearby Baker Island, but was

disrupted by World War II and thereafter abandoned. The famed American aviatrix Amelia EARHART disappeared while seeking out Howland Island as a refueling stop during her 1937 round-the-world flight; Earhart Light, a day beacon near the middle of the west coast, was named in her memory. The island was established as a NWR in 1974.
Jarvis Island: First discovered by the British in 1821, the uninhabited island was annexed by the US in 1858 but abandoned in 1879 after tons of guano had been removed. The UK annexed the island in 1889 but never carried out plans for further exploitation. The US occupied and reclaimed the island in 1935. It was abandoned in 1942 during World War II. The island was established as a NWR in 1974.
Johnston Atoll: Both the US and the Kingdom of Hawaii annexed Johnston Atoll in 1858, but it was the US that mined the guano deposits until the late 1880s. Johnston and Sand Islands were designated wildlife refuges in 1926. The US Navy took over the atoll in 1934. Subsequently, the US Air Force assumed control in 1948. The site was used for high-altitude nuclear tests in the 1950s and 1960s. Until late in 2000 the atoll was maintained as a storage and disposal site for chemical weapons. Munitions destruction, cleanup, and closure of the facility were completed by May 2005. The Fish and Wildlife Service and the US Air Force are currently discussing future management options; in the interim, Johnston Atoll and the three-mile Naval Defensive Sea around it remain under the jurisdiction and administrative control of the US Air Force.
Kingman Reef: The US annexed the reef in 1922. Its sheltered lagoon served as a way station for flying boats on Hawaii-to-American Samoa flights during the late 1930s. There are no terrestrial plants on the reef, which is frequently awash, but it does support abundant and diverse marine fauna and flora. In 2001, the waters surrounding the reef out to 12 nm were designated a US NWR.
Midway Islands: The US took formal possession of the islands in 1867. The laying of the trans-Pacific cable, which passed through the islands, brought the first residents in 1903. Between 1935 and 1947, Midway was used as a refueling stop for trans-Pacific flights. The US naval victory over a Japanese fleet off Midway in 1942 was one of the turning points of World War II. The islands continued to serve as a naval station until closed in 1993. Today the islands are a NWR and are the site of the world's largest Laysan albatross colony.
Palmyra Atoll: The Kingdom of Hawaii claimed the atoll in 1862, and the US included it among the Hawaiian Islands when it annexed the archipelago in 1898. The Hawaii Statehood Act of 1959 did not include Palmyra Atoll, which is now partly privately owned by the Nature Conservancy with the rest owned by the Federal government and managed by the US Fish and Wildlife Service. These organizations are managing the atoll as a wildlife refuge. The lagoons and surrounding waters within the 12 nm US territorial seas were transferred to the US Fish and Wildlife Service and designated as a NWR in January 2001.

GEOGRAPHY

Location: Oceania
Baker Island: atoll in the North Pacific Ocean 1,830 nm (3,389 km) southwest of Honolulu, about half way between Hawaii and Australia
Howland Island: island in the North Pacific Ocean 1,815 nm (3,361 km) southwest of Honolulu, about half way between Hawaii and Australia
Jarvis Island: island in the South Pacific Ocean 1,305 nm (2,417 km) south of Honolulu, about half way between Hawaii and Cook Islands
Johnston Atoll: atoll in the North Pacific Ocean 717 nm (1,328 km) southwest of Honolulu, about one-third of the way from Hawaii to the Marshall Islands
Kingman Reef: reef in the North Pacific Ocean 930 nm (1,722 km) south of Honolulu, about half way between Hawaii and American Samoa
Midway Islands: atoll in the North Pacific Ocean 1,260 nm (2,334 km) northwest of Honolulu near the end of the Hawaiian Archipelago, about one-third of the way from Honolulu to Tokyo
Palmyra Atoll: atoll in the North Pacific Ocean 960 nm (1,778 km) south of Honolulu, about half way between Hawaii and American Samoa
Geographic coordinates: Baker Island: 0 13 N, 176 28 W
Howland Island: 0 48 N, 176 38 W
Jarvis Island: 0 23 S, 160 01 W
Johnston Atoll: 16 45 N, 169 31 W
Kingman Reef: 6 23 N, 162 25 W
Midway Islands: 28 12 N, 177 22 W
Palmyra Atoll: 5 53 N, 162 05 W
Map references: Oceania
Area: total–6,959.41 sq km; emergent land–22.41 sq km; submerged–6,937 sq km
country comparison to the world: 237
Baker Island: total–129.1 sq km; emergent land–2.1 sq km; submerged–127 sq km
Howland Island: total–138.6 sq km; emergent land–2.6 sq km; submerged–136 sq km
Jarvis Island: total–152 sq km; emergent land–5 sq km; submerged–147 sq km
Johnston Atoll: total–276.6 sq km; emergent land–2.6 sq km; submerged–274 sq km
Kingman Reef: total–1,958.01 sq km; emergent land–0.01 sq km; submerged–1,958 sq km
Midway Islands: total–2,355.2 sq km; emergent land–6.2 sq km; submerged–2,349 sq km
Palmyra Atoll: total–1,949.9 sq km; emergent land–3.9 sq km; submerged–1,946 sq km
Area—comparative: Baker Island: about two and a half times the size of The Mall in Washington, DC
Howland Island: about three times the size of The Mall in Washington, DC
Jarvis Island: about eight times the size of The Mall in Washington, DC
Johnston Atoll: about four and a half times the size of The Mall in Washington, DC
Kingman Reef: a little more than one and a half times the size of The Mall in Washington, DC
Midway Islands: about nine times the size of The Mall in Washington, DC
Palmyra Atoll: about 20 times the size of The Mall in Washington, DC
Land boundaries: none
Coastline: Baker Island: 4.8 km
Howland Island: 6.4 km
Jarvis Island: 8 km
Johnston Atoll: 34 km
Kingman Reef: 3 km
Midway Islands: 15 km
Palmyra Atoll: 14.5 km
Maritime claims: territorial sea: 12 nm
exclusive economic zone: 200 nm
Climate: Baker, Howland, and Jarvis Islands: equatorial; scant rainfall, constant wind, burning sun
Johnston Atoll and Kingman Reef: tropical, but generally dry; consistent northeast trade winds with little seasonal temperature variation
Midway Islands: subtropical with cool, moist winters (December to February) and warm, dry summers (May to October); moderated by prevailing easterly winds; most of the 1,067 mm (42 in) of annual rainfall occurs during the winter
Palmyra Atoll: equatorial, hot; located within the low pressure area of the Intertropical Convergence Zone (ITCZ) where the northeast and southeast trade winds meet, it is extremely wet with between 4,000-5,000 mm (160-200 in) of rainfall each year
Terrain: low and nearly level sandy coral islands with narrow fringing reefs that have developed at the top of submerged volcanic mountains, which in most cases rise steeply from the ocean floor
Elevation extremes: *lowest point:* Pacific Ocean 0 m
highest point: Baker Island, unnamed location–8 m; Howland Island, unnamed location–3 m; Jarvis Island, unnamed location–7 m; Johnston Atoll, Sand Island–10 m; Kingman Reef, unnamed location–less than 2 m; Midway Islands, unnamed location–13 m; Palmyra Atoll, unnamed location–3 m
Natural resources: terrestrial and aquatic wildlife
Land use: *arable land:* 0%
permanent crops: 0%
other: 100% (2008)
Natural hazards: Baker, Howland, and Jarvis Islands: the narrow fringing reef surrounding the island can be a maritime hazard
Kingman Reef: wet or awash most of the time, maximum elevation of less than 2 m makes Kingman Reef a maritime hazard
Midway Islands, Johnston, and Palmyra Atolls: NA
Environment—current issues: Baker, Howland, and Jarvis Islands, and Johnston Atoll: no natural freshwater resources
Kingman Reef: none
Midway Islands and Palmyra Atoll: NA
Geography—note: Baker, Howland, and Jarvis Islands: scattered vegetation consisting

of grasses, prostrate vines, and low growing shrubs; primarily a nesting, roosting, and foraging habitat for seabirds, shorebirds, and marine wildlife; closed to the public
Johnston Atoll: Johnston Island and Sand Island are natural islands, which have been expanded by coral dredging; North Island (Akau) and East Island (Hikina) are manmade islands formed from coral dredging; the egg-shaped reef is 34 km in circumference; closed to the public
Kingman Reef: barren coral atoll with deep interior lagoon; closed to the public
Midway Islands: a coral atoll managed as a NWR and open to the public for wildlife-related recreation in the form of wildlife observation and photography
Palmyra Atoll: the high rainfall and resulting lush vegetation make the environment of this atoll unique among the US Pacific Island territories; supports a large undisturbed stand of Pisonia beach forest

PEOPLE

Population: no indigenous inhabitants
note: public entry is by special-use permit from US Fish and Wildlife Service only and generally restricted to scientists and educators; visited annually by US Fish and Wildlife Service
Johnston Atoll: in previous years, an average of 1,100 US military and civilian contractor personnel were present; as of May 2005, all US government personnel had left the island
Midway Islands: approximately 40 people make up the staff of US Fish and Wildlife Service and their services contractor living at the atoll
Palmyra Atoll: four to 20 Nature Conservancy, US Fish and Wildlife staff, and researchers

GOVERNMENT

Country name: *conventional long form:* none
conventional short form: Baker Island; Howland Island; Jarvis Island; Johnston Atoll; Kingman Reef; Midway Islands; Palmyra Atoll
Dependency status: unincorporated territories of the US; administered from Washington, DC by the Fish and Wildlife Service of the US Department of the Interior as part of the National Wildlife Refuge system
note on Palmyra Atoll: incorporated Territory of the US; partly privately owned and partly federally owned; administered from Washington, DC by the Fish and Wildlife Service of the US Department of the Interior; the Office of Insular Affairs of the US Department of the Interior continues to administer nine excluded areas comprising certain tidal and submerged lands within the 12 nm territorial sea or within the lagoon
Legal system: the laws of the US where applicable apply
Diplomatic representation from the US: none (territories of the US)
Flag description: the flag of the US is used

ECONOMY

Economy—overview: no economic activity

TRANSPORTATION

Airports: Baker Island: one abandoned World War II runway of 1,665 m covered with vegetation and unusable
Howland Island: airstrip constructed in 1937 for scheduled refueling stop on the round-the-world flight of Amelia EARHART and Fred NOONAN; the aviators left Lae, New Guinea, for Howland Island but were never seen again; the airstrip is no longer serviceable
Johnston Atoll: one closed and not maintained
Kingman Reef: lagoon was used as a halfway station between Hawaii and American Samoa by Pan American Airways for flying boats in 1937 and 1938
Midway Islands: 3—one operational (2,409 m paved); no fuel for sale except emergencies
Palmyra Atoll: 1–1,846 m unpaved runway; privately owned (2010)
Ports and terminals: Baker, Howland, and Jarvis Islands, and Kingman Reef: none; offshore anchorage only
Johnston Atoll: Johnston Island
Midway Islands: Sand Island
Palmyra Atoll: West Lagoon

MILITARY

Military—note: defense is the responsibility of the US

TRANSNATIONAL ISSUES

Disputes—international: none

URUGUAY

INTRODUCTION

Background: Montevideo, founded by the Spanish in 1726 as a military stronghold, soon took advantage of its natural harbor to become an important commercial center. Claimed by Argentina but annexed by Brazil in 1821, Uruguay declared its independence four years later and secured its freedom in 1828 after a three-year struggle. The administrations of President Jose BATLLE in the early 20th century established widespread political, social, and economic reforms that established a statist tradition. A violent Marxist urban guerrilla movement named the Tupamaros, launched in the late 1960s, led Uruguay's president to cede control of the government to the military in 1973. By yearend, the rebels had been crushed, but the military continued to expand its hold over the government. Civilian rule was not restored until 1985. In 2004, the left-of-center Frente Amplio Coalition won national elections that effectively ended 170 years of political control previously held by the Colorado and Blanco parties. Uruguay's political and labor conditions are among the freest on the continent.

GEOGRAPHY

Location: Southern South America, bordering the South Atlantic Ocean, between Argentina and Brazil
Geographic coordinates: 33 00 S, 56 00 W
Map references: South America
Area: *total:* 176,215 sq km
country comparison to the world: 90
land: 175,015 sq km
water: 1,200 sq km
Area—comparative: slightly smaller than the state of Washington
Land boundaries: *total:* 1,648 km
border countries: Argentina 580 km, Brazil 1,068 km
Coastline: 660 km
Maritime claims: territorial sea: 12 nm
contiguous zone: 24 nm
exclusive economic zone: 200 nm
continental shelf: 200 nm or edge of continental margin
Climate: warm temperate; freezing temperatures almost unknown
Terrain: mostly rolling plains and low hills; fertile coastal lowland
Elevation extremes: *lowest point:* Atlantic Ocean 0 m
highest point: Cerro Catedral 514 m
Natural resources: arable land, hydropower, minor minerals, fish
Land use: *arable land:* 7.77%
permanent crops: 0.24%
other: 91.99% (2005)
Irrigated land: 2,100 sq km (2003)
Total renewable water resources: 139 cu km (2000)
Freshwater withdrawal (domestic/industrial/agricultural): *total:* 3.15 cu km/yr (2%/1%/96%)
per capita: 910 cu m/yr (2000)
Natural hazards: seasonally high winds (the pampero is a chilly and occasional violent wind that blows north from the Argentine pampas), droughts, floods; because of the absence of mountains, which act as weather barriers, all locations are particularly vulnerable to rapid changes from weather fronts

Environment—current issues: water pollution from meat packing/tannery industry; inadequate solid/hazardous waste disposal
Environment—international agreements: *party to:* Antarctic-Environmental Protocol, Antarctic-Marine Living Resources, Antarctic Treaty, Biodiversity, Climate Change, Climate Change-Kyoto Protocol, Desertification, Endangered Species, Environmental Modification, Hazardous Wastes, Law of the Sea, Ozone Layer Protection, Ship Pollution, Wetlands
signed, but not ratified: Marine Dumping, Marine Life Conservation
Geography—note: second-smallest South American country (after Suriname); most of the low-lying landscape (three-quarters of the country) is grassland, ideal for cattle and sheep raising

PEOPLE

Population: 3,308,535 (July 2011 est.)
country comparison to the world: 132
Age structure: *0–14 years:* 22.2% (male 373,613/female 361,160)
15–64 years: 64.1% (male 1,042,163/female 1,078,357)
65 years and over: 13.7% (male 180,729/female 272,513) (2011 est.)
Median age: *total:* 33.6 years
male: 32 years
female: 35.2 years (2011 est.)
Population growth rate: 0.231% (2011 est.)
country comparison to the world: 177
Birth rate: 13.52 births/1,000 population (2011 est.)
country comparison to the world: 151
Death rate: 9.58 deaths/1,000 population (July 2011 est.)
country comparison to the world: 62
Net migration rate: -1.63 migrant(s)/1,000 population (2011 est.)
country comparison to the world: 159
Urbanization: *urban population:* 92% of total population (2010)
rate of urbanization: 0.4% annual rate of change (2010-15 est.)
Major cities—population: MONTEVIDEO (capital) 1.633 million (2009)
Sex ratio: *at birth:* 1.037 male(s)/female
under 15 years: 1.03 male(s)/female
15–64 years: 0.99 male(s)/female
65 years and over: 0.67 male(s)/female
total population: 0.95 male(s)/female (2011 est.)
Infant mortality rate:
total: 9.69 deaths/1,000 live births
country comparison to the world: 150
male: 10.73 deaths/1,000 live births
female: 8.61 deaths/1,000 live births (2011 est.)
Life expectancy at birth:
total population: 76.21 years
country comparison to the world: 73
male: 73.07 years
female: 79.46 years (2011 est.)
Total fertility rate: 1.89 children born/woman (2011 est.)
country comparison to the world: 140
HIV/AIDS—adult prevalence rate:
0.5% (2009 est.)
country comparison to the world: 66
HIV/AIDS—people living with HIV/AIDS:
9,900 (2009 est.)
country comparison to the world: 97
HIV/AIDS—deaths: NA
Drinking water source: *Improved:*
urban: 100% of population
rural: 100% of population
total: 100% of population (2008)
Sanitation facility access: *Improved:*
urban: 100% of population
rural: 99% of population
total: 100% of population
Unimproved:
urban: 0% of population
rural: 1% of population
total: 0% of population (2008)
Nationality: *noun:* Uruguayan(s)
adjective: Uruguayan
Ethnic groups: white 88%, mestizo 8%, black 4%, Amerindian (practically nonexistent)
Religions: Roman Catholic 47.1%, non-Catholic Christians 11.1%, nondenominational 23.2%, Jewish 0.3%, atheist or agnostic 17.2%, other 1.1% (2006)
Languages: Spanish (official), Portunol, Brazilero (Portuguese-Spanish mix on the Brazilian frontier)
Literacy: *definition:* age 15 and over can read and write
total population: 98%
male: 97.6%
female: 98.4% (2003 est.)
School life expectancy (primary to tertiary education): *total:* 16 years
male: 14 years
female: 17 years (2008)
Education expenditures:
2.9% of GDP (2006)
country comparison to the world: 135

GOVERNMENT

Country name: *conventional long form:* Oriental Republic of Uruguay
conventional short form: Uruguay
local long form: Republica Oriental del Uruguay
local short form: Uruguay
former: Banda Oriental, Cisplatine Province
Government type: constitutional republic
Capital: name: Montevideo
geographic coordinates: 34 53 S, 56 11 W
time difference: UTC-3 (2 hours ahead of Washington, DC during Standard Time)
daylight saving time: +1hr, begins first Sunday in October; ends second Sunday in March
Administrative divisions: 19 departments (departamentos, singular—departamento); Artigas, Canelones, Cerro Largo, Colonia, Durazno, Flores, Florida, Lavalleja, Maldonado, Montevideo, Paysandu, Rio Negro, Rivera, Rocha, Salto, San Jose, Soriano, Tacuarembo, Treinta y Tres
Independence: 25 August 1825 (from Brazil)
National holiday: Independence Day, 25 August (1825)
Constitution: 27 November 1966; effective 15 February 1967; suspended 27 June 1973; revised 26 November 1989 and 7 January 1997
Legal system: civil law system based on the Spanish civil code
International law organization participation: accepts compulsory ICJ jurisdiction; accepts ICCt jurisdiction
Suffrage: 18 years of age; universal and compulsory
Executive branch: *chief of state:* President Jose "Pepe" MUJICA Cordano (since 1 March 2010); Vice President Danilo ASTORI Saragoza (since 1 March 2010); note—the president is both the chief of state and head of government
head of government: President Jose "Pepe" MUJICA Cordano (since 1 March 2010); Vice President Danilo ASTORI Saragoza (since 1 March 2010)
cabinet: Council of Ministers appointed by the president with parliamentary approval (For more information visit the World Leaders website)
elections: president and vice president elected on the same ticket by popular vote for five-year terms (may not serve consecutive terms); election last held on 29 November 2009 (next to be held in October 2014)
election results: Jose "Pepe" MUJICA elected president; percent of vote—Jose "Pepe" MUJICA 54.8%, Luis Alberto LACALLE 45.2%
Legislative branch: bicameral General Assembly or Asamblea General consists of Chamber of Senators or Camara de Senadores (30 seats; members elected by popular vote to serve five-year terms; vice president has one vote in the Senate) and Chamber of Representatives or Camara de Representantes (99 seats; members elected by popular vote to serve five-year terms)
elections: Chamber of Senators—last held on 25 October 2009 (next to be held in October 2014); Chamber of Representatives—last held on 25 October 2009 (next to be held in October 2014)
election results: Chamber of Senators—percent of vote by party—NA; seats by party—Frente Amplio 16, Blanco 9, Colorado Party 5; Chamber of Representatives—percent of vote by party—NA; seats by party—Frente Amplio 50, Blanco 30, Colorado Party 17, Independent Party 2
Judicial branch: Supreme Court (judges are nominated by the president and elected for 10-year terms by the General Assembly)
Political parties and leaders: Broad Front (Frente Amplio)—formerly known as the Progressive Encounter/Broad Front Coalition or EP-FA [Jorge BROVETTO] (a broad governing coalition that includes Movement of the Popular Participation or MPP, New Space Party (Nuevo Espacio) [Rafael MICHELINI], Progressive Alliance (Alianza Progresista) [Rodolfo NIN NOVOA], Socialist Party [Eduardo FERNANDEZ and Reinaldo GARGANO], Communist Party [Eduardo LORIER], Uruguayan Assembly (Asamblea Uruguay) [Danilo ASTORI], and Vertiente Artiguista [Mariano ARANA]); Colorado Party (Foro Batllista) [Pedro BORDABERRY and Julio Maria SANGUINETTI]; National Party or Blanco [Luis Alberto LACALLE and Jorge LARRANAGA]

Political pressure groups and leaders: Architect's Society of Uruguay (professional organization); Chamber of Uruguayan Industries (manufacturer's association); Chemist and Pharmaceutical Association (professional organization); PIT/CNT (powerful federation of Uruguayan Unions–umbrella labor organization); Rural Association of Uruguay (rancher's association); Uruguayan Construction League; Uruguayan Network of Political Women
other: Catholic Church; students
International organization participation: CAN (associate), FAO, G-77, IADB, IAEA, IBRD, ICAO, ICC, ICRM, IDA, IFAD, IFC, IFRCS, IHO, ILO, IMF, IMO, Interpol, IOC, IOM, IPU, ISO, ITSO, ITU, LAES, LAIA, Mercosur, MIGA, MINURSO, MINUSTAH, MONUSCO, NAM (observer), OAS, OPANAL, OPCW, PCA, RG, UN, UNASUR, UNCTAD, UNESCO, UNIDO, Union Latina, UNMOGIP, UNOCI, UNWTO, UPU, WCO, WHO, WIPO, WMO, WTO
Diplomatic representation in the US:
chief of mission: Ambassador Carlos Alberto GIANELLI Derois
chancery: 1913 I Street NW, Washington, DC 20006
telephone: [1] (202) 331-1313 through 1316
FAX: [1] (202) 331-8142
consulate(s) general: Chicago, Los Angeles, Miami, New York
consulate(s): San Juan (Puerto Rico)
Diplomatic representation from the US: *chief of mission:* Ambassador David NELSON
embassy: Lauro Muller 1776, Montevideo 11200
mailing address: APO AA 34035
telephone: [598] (2) 418-7777
FAX: [598] (2) 418-8611
Flag description: nine equal horizontal stripes of white (top and bottom) alternating with blue; a white square in the upper hoist-side corner with a yellow sun bearing a human face known as the Sun of May with 16 rays that alternate between triangular and wavy; the stripes represent the nine original departments of Uruguay; the sun symbol evokes the legend of the sun breaking through the clouds on 25 May 1810 as independence was first declared from Spain (Uruguay subsequently won its independence from Brazil)
note: the banner was inspired by the national colors of Argentina and by the design of the US flag
National anthem: name: "Himno Nacional" (National Anthem of Uruguay)
lyrics/music: Francisco Esteban ACUNA de Figueroa/Francisco Jose DEBALI
note: adopted 1848; the anthem is also known as "Orientales, la Patria o la tumba!" ("Uruguayans, the Fatherland or Death!"); it is the world's longest national anthem in terms of music (105 bars; almost five minutes); generally only the first verse and chorus are sung

ECONOMY

Economy—overview: Uruguay's economy is characterized by an export-oriented agricultural sector, a well-educated work force, and high levels of social spending. Following financial difficulties in the late 1990s and early 2000s, economic growth for Uruguay averaged 8% annually during the period 2004-08. The 2008-09 global financial crisis put a brake on Uruguay's vigorous growth, which decelerated to 2.9% in 2009. Nevertheless, the country managed to avoid a recession and keep positive growth rates, mainly through higher public expenditure and investment, and GDP growth exceeded 8% in 2010.
GDP (purchasing power parity): $47.99 billion (2010 est.)
country comparison to the world: 92
$44.24 billion (2009 est.)
$43.13 billion (2008 est.)
note: data are in 2010 US dollars
GDP (official exchange rate): $40.27 billion (2010 est.)
GDP—real growth rate:
8.5% (2010 est.)
country comparison to the world: 14
2.6% (2009 est.)
8.6% (2008 est.)
GDP—per capita (PPP): $13,700 (2010 est.)
country comparison to the world: 86
$12,700 (2009 est.)
$12,400 (2008 est.)
note: data are in 2010 US dollars
GDP—composition by sector:
agriculture: 9.3%
industry: 22.8%
services: 67.9% (2010 est.)
Labor force: 1.637 million (2010 est.)
country comparison to the world: 127
Labor force—by occupation:
agriculture: 13%
industry: 14%
services: 73% (2010 est.)
Unemployment rate: 6.8% (2010 est.)
country comparison to the world: 67
7.3% (2009 est.)
Population below poverty line: 20.9% of households (2009)
Household income or consumption by percentage share: *lowest 10%:* 1.6%
highest 10%: 35.5% (2007)
Distribution of family income—Gini index: 47.1 (2007)
country comparison to the world: 33
44.8 (1999)
Investment (gross fixed):
19.4% of GDP (2010 est.)
country comparison to the world: 98
Budget: *revenues:* $17.5 billion
expenditures: $17.9 billion (2010)
Public debt: 56% of GDP (September 2010 est.)
country comparison to the world: 44
60% of GDP (2009 est.)
Inflation rate (consumer prices):
6.9% (2010)
country comparison to the world: 171
5.9% (2009)
Central bank discount rate:
20% (31 December 2010)
country comparison to the world: 10
20% (31 December 2009)
Commercial bank prime lending rate:
10.6% (31 December 2010 est.)
country comparison to the world: 39
15.28% (31 December 2009 est.)
Stock of narrow money: $3.98 billion (31 December 2010 est.)
country comparison to the world: 101
$3.14 billion (31 December 2009 est.)
Stock of broad money: $14.22 billion (31 December 2009)
country comparison to the world: 90
$11.78 billion (31 December 2008)
Stock of domestic credit: $9.55 billion (31 December 2010)
country comparison to the world: 96
$8.43 billion (31 December 2009)
Market value of publicly traded shares:
$238 million (31 December 2010)
country comparison to the world: 112
$219 million (31 December 2009)
$NA (31 December 2008)
Agriculture—products: beef, soybeans, cellulose, rice, wheat, lumber, dairy products; fish
Industries: food processing, electrical machinery, transportation equipment, petroleum products, textiles, chemicals, beverages
Industrial production growth rate:
16.5% (2010 est.)
country comparison to the world: 6
Electricity—production: 9.89 billion kWh (2010 est.)
country comparison to the world: 91
Electricity—consumption:
7.57 billion kWh (2010 est.)
country comparison to the world: 96
Electricity—exports: 711.4 million kWh (2010 est.)
Electricity—imports: 386.8 million kWh (2010 est.)
Oil—production:
40,110 bbl/day (November 2010 est.)
country comparison to the world: 67
Oil—consumption:
34,670 bbl/day (November 2010 est.)
country comparison to the world: 109
Oil—exports: 4,353 bbl/day (November 2010 est.)
country comparison to the world: 106
Oil—imports: 53,110 bbl/day (November 2010 est.)
country comparison to the world: 83
Oil—proved reserves: 0 bbl (1 January 2010 est.)
country comparison to the world: 199
Natural gas—production: 0 cu m (2008 est.)
country comparison to the world: 196
Natural gas—consumption: 70 million cu m (2008 est.)
country comparison to the world: 106
Natural gas—exports: 0 cu m (2008 est.)
country comparison to the world: 194
Natural gas—imports: 76.6 million cu m (2010 est.)
country comparison to the world: 68
Natural gas—proved reserves: 0 cu m (1 January 2010 est.)
country comparison to the world: 198
Current account balance: $-377 million (2010 est.)
country comparison to the world: 108
$178 million (2009 est.)
Exports: $6.7 billion (2010 est.)
country comparison to the world: 101
$5.5 billion (2009 est.)
Exports—commodities: beef, soybeans, cellulose, rice, wheat, wood, dairy products; wool

Exports—partners: Brazil 21%, Nueva Palmira Free Zone 10.2%, Argentina 7.5%, Chile 5.5%, Russia 5.3% (2010 est.)
Imports: $8.3 billion (2010 est.)
country comparison to the world: 97
$6.7 billion (2009 est.)
Imports—commodities: crude oil (13.4%), refined oil (4.9%), passenger vehicles (3.5%), transportation vehicles (2.7%), vehicles parts (2.2%), cellular phones (2.1%), insecticides (1.7%)
Imports—partners: Brazil 18.6%, Argentina 16.7%, China 13.5%, Venezuela 9.1%, US 8.3%, Russia 4.2% (2010 est.)
Reserves of foreign exchange and gold: $7.7 billion (31 December 2010 est.)
country comparison to the world: 63
$8.038 billion (31 December 2009 est.)
Debt—external: $13.39 billion (31 December 2010 est.)
country comparison to the world: 81
$14.1 billion (31 December 2009)
Stock of direct foreign investment—at home: $9.1 billion (30 September 2010)
country comparison to the world: 81
$7.7 billion (31 December 2009)
Stock of direct foreign investment—abroad: $300 million (2009 est.)
country comparison to the world: 77
$156 million (2007 est.)
Exchange rates: Uruguayan pesos (UYU) per US dollar–20.051 (2010), 22.568 (2009), 20.936 (2008), 23.947 (2007), 24.048 (2006)

COMMUNICATIONS

Telephones—main lines in use: 953,400 (2009)
country comparison to the world: 82
Telephones—mobile cellular: 4.112 million (2009)
country comparison to the world: 103
Telephone system: *general assessment:* fully digitalized
domestic: most modern facilities concentrated in Montevideo; new nationwide microwave radio relay network; overall fixed-line and mobile-cellular teledensity approaching 150 telephones per 100 persons
international: country code–598; the UNISOR submarine cable system provides direct connectivity to Brazil and Argentina; satellite earth stations–2 Intelsat (Atlantic Ocean) (2009)
Broadcast media: mixture of privately-owned and state-run broadcast media; more than 100 commercial radio stations and about 20 television channels broadcasting; cable TV is available; large number of community radio and TV stations; adopted the hybrid Japanese/Brazilian HDTV standard (ISDB-T) in December 2010 (2010)
Internet country code: .uy
Internet hosts: 765,525 (2010)
country comparison to the world: 47
Internet users: 1.405 million (2009)
country comparison to the world: 86

TRANSPORTATION

Airports: 58 (2010)
country comparison to the world: 81
Airports—with paved runways: *total:* 9
over 3,047 m: 1
1,524 to 2,437 m: 4
914 to 1,523 m: 2
under 914 m: 2 (2010)
Airports—with unpaved runways: *total:* 49
1,524 to 2,437 m: 3
914 to 1,523 m: 20
under 914 m: 26 (2010)
Pipelines: gas 226 km; oil 155 km (2010)
Railways: *total:* 1,641 km
country comparison to the world: 77
standard gauge: 1,641 km 1.435-m gauge (1,200 km operational) (2010)
Roadways: *total:* 77,732 km
country comparison to the world: 63
paved: 7,743 km
unpaved: 69,989 km (2010)
Waterways: 1,600 km (2010)
country comparison to the world: 50
Merchant marine: *total:* 18
country comparison to the world: 101
by type: bulk carrier 1, cargo 2, chemical tanker 4, passenger/cargo 7, petroleum tanker 3, roll on/roll off 1
foreign-owned: 9 (Argentina 2, Denmark 1, Greece 1, Spain 5)
registered in other countries: 1 (Liberia 1) (2010)
Ports and terminals: Montevideo

MILITARY

Military branches: Uruguayan Armed Forces: Uruguayan National Army (Ejercito Nacional Uruguaya, ENU), Uruguayan National Navy (Armada Nacional del Uruguay; includes naval air arm, Naval Rifle Corps (Cuerpo de Fusileros Navales, Fusna), Maritime Prefecture in wartime), Uruguayan Air Force (Fuerza Aerea Uruguaya, FAU) (2010)
Military service age and obligation: 18-30 years of age for voluntary military service; up to 40 years of age for specialists; enlistment is voluntary in peacetime, but the government has the authority to conscript in emergencies; minimum 6-year education (2009)
Manpower available for military service: *males age 16–49:* 771,159
females age 16–49: 780,932 (2010 est.)
Manpower fit for military service: *males age 16–49:* 649,025
females age 16–49: 654,903 (2010 est.)
Manpower reaching militarily significant age annually: *male:* 27,564
female: 26,811 (2010 est.)
Military expenditures: 1.6% of GDP (2006)
country comparison to the world: 92

TRANSNATIONAL ISSUES

Disputes—international: in 2010, the ICJ ruled in favor of Uruguay's operation of two paper mills on the Uruguay River, which forms the border with Argentina; the two countries formed a joint pollution monitoring regime; uncontested boundary dispute between Brazil and Uruguay over Braziliera/Brasiliera Island in the Quarai/Cuareim River leaves the tripoint with Argentina in question; smuggling of firearms and narcotics continues to be an issue along the Uruguay-Brazil border
Illicit drugs: small-scale transit country for drugs mainly bound for Europe, often through sea-borne containers; law enforcement corruption; money laundering because of strict banking secrecy laws; weak border control along Brazilian frontier; increasing consumption of cocaine base and synthetic drugs

UZBEKISTAN

INTRODUCTION

Background: Russia conquered the territory of present-day Uzbekistan in the late 19th century. Stiff resistance to the Red Army after the Bolshevik Revolution was eventually suppressed and a socialist republic established in 1924. During the Soviet era, intensive production of "white gold" (cotton) and grain led to overuse of agrochemicals and the depletion of water supplies, which have left the land poisoned and the Aral Sea and certain rivers half dry. Independent since 1991, the country seeks to gradually lessen its dependence on agriculture while developing its mineral and petroleum reserves. Current concerns include terrorism by Islamic militants, economic stagnation, and the curtailment of human rights and democratization.

GEOGRAPHY

Location: Central Asia, north of Turkmenistan, south of Kazakhstan
Geographic coordinates: 41 00 N, 64 00 E
Map references: Asia
Area: *total:* 447,400 sq km
country comparison to the world: 56
land: 425,400 sq km
water: 22,000 sq km
Area—comparative: slightly larger than California
Land boundaries: *total:* 6,221 km
border countries: Afghanistan 137 km, Kazakhstan 2,203 km, Kyrgyzstan 1,099 km, Tajikistan 1,161 km, Turkmenistan 1,621 km
Coastline: 0 km (doubly landlocked); note—Uzbekistan includes the southern portion of the Aral Sea with a 420 km shoreline
Maritime claims: none (doubly landlocked)
Climate: mostly midlatitude desert, long, hot summers, mild winters; semiarid grassland in east
Terrain: mostly flat-to-rolling sandy desert with dunes; broad, flat intensely irrigated river valleys along course of Amu Darya, Syr Darya (Sirdaryo), and Zarafshon; Fergana Valley in east surrounded by mountainous Tajikistan and Kyrgyzstan; shrinking Aral Sea in west
Elevation extremes:
lowest point: Sariqamish Kuli -12 m
highest point: Adelunga Toghi 4,301 m

Natural resources: natural gas, petroleum, coal, gold, uranium, silver, copper, lead and zinc, tungsten, molybdenum
Land use: *arable land:* 10.51%
permanent crops: 0.76%
other: 88.73% (2005)
Irrigated land: 42,810 sq km (2003)
Total renewable water resources: 72.2 cu km (2003)
Freshwater withdrawal (domestic/industrial/agricultural): *total:* 58.34 cu km/yr (5%/2%/93%)
per capita: 2,194 cu m/yr (2000)
Natural hazards: NA
Environment—current issues: shrinkage of the Aral Sea is resulting in growing concentrations of chemical pesticides and natural salts; these substances are then blown from the increasingly exposed lake bed and contribute to desertification; water pollution from industrial wastes and the heavy use of fertilizers and pesticides is the cause of many human health disorders; increasing soil salination; soil contamination from buried nuclear processing and agricultural chemicals, including DDT
Environment—international agreements: *party to:* Biodiversity, Climate Change, Climate Change-Kyoto Protocol, Desertification, Endangered Species, Environmental Modification, Hazardous Wastes, Ozone Layer Protection, Wetlands
signed, but not ratified: none of the selected agreements
Geography—note: along with Liechtenstein, one of the only two doubly landlocked countries in the world

PEOPLE

Population: 28,128,600 (July 2011 est.)
country comparison to the world: 44
Age structure: *0–14 years:* 26.5% (male 3,817,755/female 3,635,142)
15–64 years: 68.8% (male 9,620,356/female 9,742,818)
65 years and over: 4.7% (male 560,574/female 751,955) (2011 est.)
Median age: *total:* 25.7 years
male: 25.2 years
female: 26.3 years (2011 est.)
Population growth rate: 0.94% (2011 est.)
country comparison to the world: 121
Birth rate: 17.43 births/1,000 population (2011 est.)
country comparison to the world: 115
Death rate: 5.29 deaths/1,000 population (July 2011 est.)
country comparison to the world: 178
Net migration rate: -2.74 migrant(s)/1,000 population (2011 est.)
country comparison to the world: 173
Urbanization: *urban population:* 36% of total population (2010)
rate of urbanization: 1.4% annual rate of change (2010-15 est.)
Major cities—population: TASHKENT (capital) 2.201 million (2009)
Sex ratio: *at birth:* 1.06 male(s)/female
under 15 years: 1.05 male(s)/female
15–64 years: 0.99 male(s)/female
65 years and over: 0.75 male(s)/female
total population: 0.99 male(s)/female (2011 est.)
Infant mortality rate: *total:* 21.92 deaths/1,000 live births
country comparison to the world: 92
male: 25.95 deaths/1,000 live births
female: 17.64 deaths/1,000 live births (2011 est.)
Life expectancy at birth:
total population: 72.51 years
country comparison to the world: 125
male: 69.48 years
female: 75.71 years (2011 est.)
Total fertility rate: 1.89 children born/woman (2011 est.)
country comparison to the world: 141
HIV/AIDS—adult prevalence rate: 0.1% (2009 est.)
country comparison to the world: 109
HIV/AIDS—people living with HIV/AIDS: 28,000 (2009 est.)
country comparison to the world: 70
HIV/AIDS—deaths:
fewer than 500 (2009 est.)
country comparison to the world: 97
Drinking water source: *Improved:*
urban: 98% of population
rural: 81% of population
total: 87% of population
Unimproved:
urban: 2% of population
rural: 19% of population
total: 13% of population (2008)
Sanitation facility access: *Improved:*
urban: 100% of population
rural: 100% of population
total: 100% of population (2008)
Nationality:
noun: Uzbekistani
adjective: Uzbekistani
Ethnic groups: Uzbek 80%, Russian 5.5%, Tajik 5%, Kazakh 3%, Karakalpak 2.5%, Tatar 1.5%, other 2.5% (1996 est.)
Religions: Muslim 88% (mostly Sunnis), Eastern Orthodox 9%, other 3%
Languages: Uzbek (official) 74.3%, Russian 14.2%, Tajik 4.4%, other 7.1%
Literacy: *definition:* age 15 and over can read and write
total population: 99.3%
male: 99.6%
female: 99% (2003 est.)
School life expectancy (primary to tertiary education): *total:* 11 years
male: 12 years
female: 11 years (2009)
Education expenditures: NA

GOVERNMENT

Country name: *conventional long form:* Republic of Uzbekistan
conventional short form: Uzbekistan
local long form: Ozbekiston Respublikasi
local short form: Ozbekiston
former: Uzbek Soviet Socialist Republic
Government type: republic; authoritarian presidential rule, with little power outside the executive branch
Capital: name: Tashkent (Toshkent)
geographic coordinates: 41 20 N, 69 18 E
time difference: UTC+5 (10 hours ahead of Washington, DC during Standard Time)
Administrative divisions: 12 provinces (viloyatlar, singular–viloyat), 1 autonomous republic* (respublika), and 1 city** (shahar); Andijon Viloyati, Buxoro Viloyati, Farg'ona Viloyati, Jizzax Viloyati, Namangan Viloyati, Navoiy Viloyati, Qashqadaryo Viloyati (Qarshi), Qoraqalpog'iston Respublikasi [Karakalpakstan Republic]* (Nukus), Samarqand Viloyati, Sirdaryo Viloyati (Guliston), Surxondaryo Viloyati (Termiz), Toshkent Shahri [Tashkent City]**, Toshkent Viloyati [Tashkent province], Xorazm Viloyati (Urganch)
note: administrative divisions have the same names as their administrative centers (exceptions have the administrative center name following in parentheses)
Independence: 1 September 1991 (from the Soviet Union)
National holiday: Independence Day, 1 September (1991)
Constitution: adopted 8 December 1992
Legal system: civil law system
International law organization participation: has not submitted an ICJ jurisdiction declaration; non-party state to the ICCt
Suffrage: 18 years of age; universal
Executive branch: *chief of state:* President Islom KARIMOV (since 24 March 1990, when he was elected president by the then Supreme Soviet; elected president of independent Uzbekistan in 1991)
head of government: Prime Minister Shavkat MIRZIYOYEV (since 11 December 2003); First Deputy Prime Minister Rustam AZIMOV (since 2 January 2008)
cabinet: Cabinet of Ministers appointed by the president with approval of the Supreme Assembly
(For more information visit the World Leaders website)
elections: president elected by popular vote for a seven-year term (eligible for a second term; previously was a five-year term, extended by constitutional amendment in 2002); election last held

on 23 December 2007 (next to be held in 2014); prime minister, ministers, and deputy ministers appointed by the president
election results: Islom KARIMOV reelected president; percent of vote—Islom KARIMOV 88.1%, Asliddin RUSTAMOV 3.2%, Dilorom TOSHMUHAMEDOVA 2.9%, Akmal SAIDOV 2.6%, other 3.2%
Legislative branch: bicameral Supreme Assembly or Oliy Majlis consists of an upper house or Senate (100 seats; 84 members elected by regional governing councils and 16 appointed by the president; members to serve five-year terms) and a lower house or Legislative Chamber (150 seats; 135 members elected by popular vote to serve five-year terms, while 15 spots reserved for the new Ecological Movement of Uzbekistan)
elections: last held on 27 December 2009 and 10 January 2010 (next to be held in December 2014)
election results: Senate—percent of vote by party—NA; seats by party—NA; Legislative Chamber—percent of vote by party—NA; seats by party—LDPU 53, NDP 32, National Rebirth Party 31, Adolat 19
note: all parties in the Supreme Assembly support President KARIMOV
Judicial branch: Supreme Court (judges are nominated by the president and confirmed by the Supreme Assembly)
Political parties and leaders: Adolat (Justice) Social Democratic Party [Ismoil SAIFNAZAROV]; Ecological Movement of Uzbekistan [Boriy ALIXONOV, chairman]; Liberal Democratic Party of Uzbekistan or LDPU [Muhammadjon AHMADJONOV]; National Rebirth Party (Milliy Tiklanish) [Ahtam TURSUNOV]; People's Democratic Party or NDP (formerly Communist Party) [Latif GULOMOV]
Political pressure groups and leaders: there are no significant opposition political parties or pressure groups operating in Uzbekistan
International organization participation: ADB, CICA, CIS, EAPC, EBRD, ECO, FAO, IAEA, IBRD, ICAO, ICRM, IDA, IDB, IFC, IFRCS, ILO, IMF, Interpol, IOC, ISO, ITSO, ITU, MIGA, NAM, OIC, OPCW, OSCE, PFP, SCO, UN, UNCTAD, UNESCO, UNIDO, UNWTO, UPU, WCO, WFTU, WHO, WIPO, WMO, WTO (observer)
Diplomatic representation in the US:
chief of mission: Ambassador Ilxamdjan NEMATOV
chancery: 1746 Massachusetts Avenue NW, Washington, DC 20036
telephone: [1] (202) 887-5300
FAX: [1] (202) 293-6804
consulate(s) general: New York
Diplomatic representation from the US:
chief of mission: Ambassador Richard B. NORLAND
embassy: 3 Moyqo'rq'on, 5th Block, Yunusobod District, Tashkent 100093
mailing address: use embassy street address
telephone: [998] (71) 120-5450
FAX: [998] (71) 120-6335
Flag description: three equal horizontal bands of blue (top), white, and green separated by red fimbriations with a white crescent moon (closed side to the hoist) and 12 white stars shifted to the hoist on the top band; blue is the color of the Turkic peoples and of the sky, white signifies peace and the striving for purity in thoughts and deeds, while green represents nature and is the color of Islam; the red stripes are the vital force of all living organisms that links good and pure ideas with the eternal sky and with deeds on earth; the crescent represents Islam and the 12 stars the months and constellations of the Uzbek calendar
National anthem: name: "O'zbekiston Respublikasining Davlat Madhiyasi" (National Anthem of the Republic of Uzbekistan)
lyrics/music: Abdulla ARIPOV/Mutal BURHANOV
note: adopted 1992; after the fall of the Soviet Union, Uzbekistan kept the music of the anthem from its time as a Soviet Republic but adopted new lyrics

ECONOMY

Economy—overview: Uzbekistan is a dry, landlocked country; 11% of the land is intensely cultivated, in irrigated river valleys. More than 60% of the population lives in densely populated rural communities. Export of hydrocarbons, including natural gas and petroleum, provided about 40% of foreign exchange earnings in 2009. Other major export earners include gold and cotton. Uzbekistan is now the world's second-largest cotton exporter and fifth largest producer; it has come under increasing international criticism for the use of child labor in its annual cotton harvest. Nevertheless, Uzbekistan enjoyed a bumper cotton crop in 2010 amidst record high prices. Following independence in September 1991, the government sought to prop up its Soviet-style command economy with subsidies and tight controls on production and prices. While aware of the need to improve the investment climate, the government still sponsors measures that often increase, not decrease, its control over business decisions. A sharp increase in the inequality of income distribution has hurt the lower ranks of society since independence. In 2003, the government accepted Article VIII obligations under the IMF, providing for full currency convertibility. However, strict currency controls and tightening of borders have lessened the effects of convertibility and have also led to some shortages that have further stifled economic activity. The Central Bank often delays or restricts convertibility, especially for consumer goods. Potential investment by Russia and China in Uzbekistan's gas and oil industry, as well as increased cooperation with South Korea in the realm of civil aviation, may boost growth prospects. However, decreased demand for natural gas in Europe and Russia in the wake of the global financial crisis could reduce energy-related revenues in the near term. In November 2005, Russian President Vladimir PUTIN and Uzbekistan President KARIMOV signed an "alliance," which included provisions for economic and business cooperation. Russian businesses have shown increased interest in Uzbekistan, especially in mining, telecom, and oil and gas. In 2006, Uzbekistan took steps to rejoin the Collective Security Treaty Organization (CSTO) and the Eurasian Economic Community (EurASEC), which it subsequently left in 2008, both organizations dominated by Russia. In the past Uzbek authorities had accused US and other foreign companies operating in Uzbekistan of violating Uzbek tax laws and have frozen their assets, but no new expropriations occurred in 2008-09. Instead, the Uzbek Government has actively courted several major U.S. and international corporations, offering attractive financing and tax advantages, and has landed a significant US investment in the automotive industry. Although growth slowed in 2009-10, Uzbekistan has seen few other effects from the global economic downturn, primarily due to its relative isolation from the global financial markets.
GDP (purchasing power parity):
$85.85 billion (2010 est.)
country comparison to the world: 77
$79.12 billion (2009 est.)
$73.19 billion (2008 est.)
note: data are in 2010 US dollars
GDP (official exchange rate):
$38.99 billion (2010 est.)
GDP—real growth rate: 8.5% (2010 est.)
country comparison to the world: 13
8.1% (2009 est.)
9% (2008 est.)
GDP—per capita (PPP): $3,100 (2010 est.)
country comparison to the world: 168
$2,900 (2009 est.)
$2,700 (2008 est.)
note: data are in 2010 US dollars
GDP—composition by sector:
agriculture: 21.2%
industry: 32.3%
services: 46.4% (2010 est.)
Labor force: 16 million (2010 est.)
country comparison to the world: 37
Labor force—by occupation:
agriculture: 44%
industry: 20%
services: 36% (1995)
Unemployment rate: 1.1% (2010 est.)
country comparison to the world: 6
1.1% (2009 est.)
note: officially measured by the Ministry of Labor, plus another 20% underemployed
Population below poverty line: 26% (2008 est.)
Household income or consumption by percentage share: *lowest 10%:* 2.8%
highest 10%: 29.6% (2003)
Distribution of family income—Gini index: 36.8 (2003)
country comparison to the world: 80
44.7 (1998)
Budget: *revenues:* $12.25 billion
expenditures: $12.3 billion (2010 est.)
Public debt: 9% of GDP (2010 est.)
country comparison to the world: 124
9.6% of GDP (2009 est.)
Inflation rate (consumer prices):
15% (2010 est.)

country comparison to the world: 219
14.1% (2009 est.)
note: official data; based on independent analysis of consumer prices, inflation reached 38% in 2008
Stock of narrow money: $4.895 billion (31 December 2010 est.)
country comparison to the world: 91
$3.829 billion (31 December 2009 est.)
Stock of broad money: $7.197 billion (31 December 2010 est.)
country comparison to the world: 112
$5.648 billion (31 December 2009 est.)
Stock of domestic credit: $6.482 billion (31 December 2010 est.)
country comparison to the world: 107
$5.484 billion (31 December 2009 est.)
Market value of publicly traded shares: $NA (31 December 2010)
country comparison to the world: 103
$715.3 million
Agriculture—products: cotton, vegetables, fruits, grain; livestock
Industries: textiles, food processing, machine building, metallurgy, gold, petroleum, natural gas, chemicals
Industrial production growth rate:
8% (2010 est.)
country comparison to the world: 32
Electricity—production: 44.8 billion kWh (2009 est.)
country comparison to the world: 51
Electricity—consumption:
40.1 billion kWh (2009 est.)
country comparison to the world: 52
Electricity—exports: 11.52 billion kWh (2009 est.)
Electricity—imports: 11.44 billion kWh (2009 est.)
Oil—production: 70,910 bbl/day (2009 est.)
country comparison to the world: 56
Oil—consumption:
145,000 bbl/day (2009 est.)
country comparison to the world: 67
Oil—exports: 6,104 bbl/day (2007 est.)
country comparison to the world: 98
Oil—imports: 35,810 bbl/day (2007 est.)
country comparison to the world: 95
Oil—proved reserves: 594 million bbl (1 January 2010 est.)
country comparison to the world: 47
Natural gas—production: 67.6 billion cu m (2008 est.)
country comparison to the world: 13
Natural gas—consumption: 52.6 billion cu m (2008 est.)
country comparison to the world: 14
Natural gas—exports: 15 billion cu m (2008 est.)
country comparison to the world: 14
Natural gas—imports: 0 cu m (2008 est.)
country comparison to the world: 132
Natural gas—proved reserves: 1.841 trillion cu m (1 January 2010 est.)
country comparison to the world: 19
Current account balance: $5.588 billion (2010 est.)
country comparison to the world: 32
$3.595 billion (2009 est.)
Exports: $13.13 billion (2010 est.)
country comparison to the world: 77
$10.74 billion (2009 est.)
Exports—commodities: energy products, cotton, gold, mineral fertilizers, ferrous and nonferrous metals, textiles, food products, machinery, automobiles
Exports—partners: Ukraine 29.91%, Russia 13.94%, Turkey 7.53%, Kazakhstan 7.26%, Bangladesh 6.83%, China 5.69%, South Korea 4.19% (2009)
Imports: $9.44 billion (2010 est.)
country comparison to the world: 91
$9.023 billion (2009 est.)
Imports—commodities: machinery and equipment, foodstuffs, chemicals, ferrous and nonferrous metals
Imports—partners: Russia 23.72%, China 20.36%, South Korea 13.03%, Germany 6.09%, Ukraine 5.39%, Kazakhstan 4.68% (2009)
Reserves of foreign exchange and gold: $10.5 billion (31 December 2010 est.)
country comparison to the world: 58
$9 billion (31 December 2009 est.)
Debt—external: $4.236 billion (31 December 2010 est.)
country comparison to the world: 114
$4.053 billion (31 December 2009 est.)
Stock of direct foreign investment—at home: $NA
Stock of direct foreign investment—abroad: $NA
Exchange rates: Uzbekistani soum (UZS) per US dollar–1,588.1 (2010), 1,466.7 (2009), 1,317 (2008), 1,263.8 (2007), 1,219.8 (2006)

COMMUNICATIONS

Telephones—main lines in use: 1.857 million (2009)
country comparison to the world: 61
Telephones—mobile cellular: 16.418 million (2009)
country comparison to the world: 45
Telephone system: *general assessment:* digital exchanges in large cities but still antiquated and inadequate in rural areas
domestic: the state-owned telecommunications company, Uzbektelecom, owner of the fixed line telecommunications system, has used loans from the Japanese government and the China Development Bank to upgrade fixed-line services including conversion to digital exchanges; mobile-cellular services are growing rapidly, with the subscriber base exceeding 16 million in 2009
international: country code–998; linked by fiber-optic cable or microwave radio relay with CIS member states and to other countries by leased connection via the Moscow international gateway switch; after the completion of the Uzbek link to the Trans-Asia-Europe (TAE) fiber-optic cable, Uzbekistan plans to establish a fiber-optic connection to Afghanistan (2009)
Broadcast media: government controls media; 8 state-owned broadcasters–4 TV and 4 radio–provide service to virtually the entire country; about 20 privately-owned TV stations, overseen by local officials, broadcast to local markets; privately-owned TV stations are required to lease transmitters from the government-owned Republic TV and Radio Industry Corporation and are prohibited from broadcasting live; about 15 privately-owned radio broadcasters; programming content includes news updates, music, call-in talk shows, and other entertainment in a half-Russian, half-Uzbek format mandated for private radio (2007)
Internet country code: .uz
Internet hosts: 47,718 (2010)
country comparison to the world: 89
Internet users: 4.689 million (2009)
country comparison to the world: 50

TRANSPORTATION

Airports: 54 (2010)
country comparison to the world: 86
Airports—with paved runways: *total:* 33
over 3,047 m: 6
2,438 to 3,047 m: 13
1,524 to 2,437 m: 6
914 to 1,523 m: 4
under 914 m: 4 (2010)
Airports—with unpaved runways:
total: 21
2,438 to 3,047 m: 2
under 914 m: 19 (2010)
Pipelines: gas 10,253 km; oil 868 km (2010)
Railways: *total:* 3,645 km
country comparison to the world: 47
broad gauge: 3,645 km 1.520-m gauge (620 km electrified) (2008)
Roadways: *total:* 86,496 km
country comparison to the world: 55
paved: 75,511 km
unpaved: 10,985 km (2000)
Waterways: 1,100 km (2009)
country comparison to the world: 63
Ports and terminals: Termiz (Amu Darya)

MILITARY

Military branches: Army, Air and Air Defense Forces, National Guard
Military service age and obligation: 18 years of age for compulsory military service; 1-year conscript service obligation; moving toward a professional military, but conscription will continue; the military cannot accommodate everyone who wishes to enlist, and competition for entrance into the military is similar to the competition for admission to universities (2009)
Manpower available for military service:
males age 16–49: 7,887,292
females age 16–49: 7,886,459 (2010 est.)
Manpower fit for military service: *males age 16–49:* 6,566,118
females age 16–49: 6,745,818 (2010 est.)
Manpower reaching militarily significant age annually: *male:* 306,404
female: 295,456 (2010 est.)
Military expenditures:
3.5% of GDP (2010)
country comparison to the world: 33

TRANSNATIONAL ISSUES

Disputes—international: prolonged drought and cotton monoculture in Uzbekistan and Turkmenistan creates water-sharing difficulties for Amu Darya river states; field demarcation of the boundaries with Kazakhstan commenced in 2004; border delimitation of 130 km of border with Kyrgyzstan is hampered by serious disputes around enclaves and other areas

Refugees and internally displaced persons: refugees (country of origin): 39,202 (Tajikistan); 1,060 (Afghanistan)
IDPs: 3,400 (forced population transfers by government from villages near Tajikistan border) (2007)
Trafficking in persons: current situation: Uzbekistan is a source country for women and girls trafficked to Kazakhstan, Russia, Middle East, and Asia for the purpose of commercial sexual exploitation; men are trafficked to Kazakhstan and Russia for purposes of forced labor in the construction, cotton, and tobacco industries; men and women are also trafficked internally for the purposes of domestic servitude, forced labor in the agricultural and construction industries, and for commercial sexual exploitation
tier rating: Tier 2 Watch List–Uzbekistan is on the Tier 2 Watch List for its failure to provide evidence of increasing efforts to combat severe forms of trafficking in 2007; the government did not amend its criminal code to increase penalties for convicted traffickers; in March 2008, Uzbekistan adopted ILO Conventions on minimum age of employment and on the elimination of the worst forms of child labor and is working with the ILO on implementation; the government also demonstrated its increasing commitment to combat trafficking in March 2008 by adopting a comprehensive anti-trafficking law; Uzbekistan has not ratified the 2000 UN TIP Protocol (2008)
Illicit drugs: transit country for Afghan narcotics bound for Russian and, to a lesser extent, Western European markets; limited illicit cultivation of cannabis and small amounts of opium poppy for domestic consumption; poppy cultivation almost wiped out by government crop eradication program; transit point for heroin precursor chemicals bound for Afghanistan

VANUATU

INTRODUCTION

Background: Multiple waves of colonizers, each speaking a distinct language, migrated to the New Hebrides in the millennia preceding European exploration in the 18th century. This settlement pattern accounts for the complex linguistic diversity found on the archipelago to this day. The British and French, who settled the New Hebrides in the 19th century, agreed in 1906 to an Anglo-French Condominium, which administered the islands until independence in 1980, when the new name of Vanuatu was adopted.

GEOGRAPHY

Location: Oceania, group of islands in the South Pacific Ocean, about three-quarters of the way from Hawaii to Australia
Geographic coordinates: 16 00 S, 167 00 E
Map references: Oceania
Area: *total:* 12,189 sq km
country comparison to the world: 163
land: 12,189 sq km
water: 0 sq km
note: includes more than 80 islands, about 65 of which are inhabited
Area—comparative: slightly larger than Connecticut
Land boundaries: 0 km
Coastline: 2,528 km
Maritime claims: measured from claimed archipelagic baselines
territorial sea: 12 nm
contiguous zone: 24 nm
exclusive economic zone: 200 nm
continental shelf: 200 nm or to the edge of the continental margin
Climate: tropical; moderated by southeast trade winds from May to October; moderate rainfall from November to April; may be affected by cyclones from December to April
Terrain: mostly mountainous islands of volcanic origin; narrow coastal plains
Elevation extremes: *lowest point:* Pacific Ocean 0 m
highest point: Tabwemasana 1,877 m
Natural resources: manganese, hardwood forests, fish
Land use: *arable land:* 1.64%
permanent crops: 6.97%
other: 91.39% (2005)
Irrigated land: NA
Natural hazards: tropical cyclones or typhoons (January to April); volcanic eruption on Aoba (Ambae) island began on 27 November 2005, volcanism also causes minor earthquakes; tsunamis
volcanism: Vanuatu experiences significant volcanic activity, with many volcanoes erupting in recent years; Yasur (elev. 361 m), one of the world's most active volcanoes, has experienced continuous activity in recent centuries; other historically active volcanoes include, Aoba, Ambrym, Epi, Gaua, Kuwae, Lopevi, Suretamatai, and Traitor's Head
Environment—current issues: most of the population does not have access to a reliable supply of potable water; deforestation
Environment—international agreements: *party to:* Antarctic-Marine Living Resources, Biodiversity, Climate Change, Climate Change-Kyoto Protocol, Desertification, Endangered Species, Law of the Sea, Marine Dumping, Ozone Layer Protection, Ship Pollution, Tropical Timber 94
signed, but not ratified: none of the selected agreements
Geography—note: a Y-shaped chain of four main islands and 80 smaller islands; several of the islands have active volcanoes

PEOPLE

Population: 224,564 (July 2011 est.)
country comparison to the world: 182
Age structure: *0–14 years:* 29.6% (male 33,968/female 32,550)
15–64 years: 66.1% (male 75,559/female 72,919)
65 years and over: 4.3% (male 4,862/female 4,706) (2011 est.)
Median age: *total:* 24.9 years
male: 24.9 years
female: 25 years (2011 est.)
Population growth rate: 1.343% (2011 est.)
country comparison to the world: 87
Birth rate: 20.86 births/1,000 population (2011 est.)
country comparison to the world: 86
Death rate: 7.43 deaths/1,000 population (July 2011 est.)
country comparison to the world: 119
Net migration rate: 0 migrant(s)/1,000 population (2011 est.)
country comparison to the world: 101
Urbanization: *urban population:* 26% of total population (2010)
rate of urbanization: 4.2% annual rate of change (2010-15 est.)
Sex ratio: *at birth:* 1.05 male(s)/female
under 15 years: 1.04 male(s)/female
15–64 years: 1.04 male(s)/female
65 years and over: 1.05 male(s)/female
total population: 1.04 male(s)/female (2011 est.)
Infant mortality rate: *total:* 46.85 deaths/1,000 live births
country comparison to the world: 52
male: 49.3 deaths/1,000 live births
female: 44.28 deaths/1,000 live births (2011 est.)
Life expectancy at birth: *total population:* 64.7 years
country comparison to the world: 170
male: 63.04 years
female: 66.44 years (2011 est.)
Total fertility rate: 2.39 children born/woman (2011 est.)
country comparison to the world: 94
HIV/AIDS—adult prevalence rate: NA
HIV/AIDS—people living with HIV/AIDS: NA
HIV/AIDS—deaths: NA
Drinking water source: *Improved:*
urban: 96% of population
rural: 79% of population
total: 83% of population
Unimproved:
urban: 4% of population
rural: 21% of population
total: 17% of population (2008)
Sanitation facility access: *Improved:*
urban: 66% of population
rural: 48% of population
total: 52% of population
Unimproved:
urban: 34% of population
rural: 52% of population
total: 48% of population (2008)
Nationality: *noun:* Ni-Vanuatu (singular and plural)
adjective: Ni-Vanuatu
Ethnic groups: Ni-Vanuatu 98.5%, other 1.5% (1999 Census)
Religions: Presbyterian 31.4%, Anglican 13.4%, Roman Catholic 13.1%, Seventh-Day Adventist 10.8%, other Christian 13.8%, indigenous beliefs 5.6% (including Jon Frum cargo cult), other 9.6%, none 1%, unspecified 1.3% (1999 Census)
Languages: local languages (more than 100) 72.6%, pidgin (known as Bislama or Bichelama) 23.1%, English (official) 1.9%, French (official) 1.4%, other 0.3%, unspecified 0.7% (1999 Census)
Literacy: *definition:* age 15 and over can read and write
total population: 74%
male: NA
female: NA (1999 census)
School life expectancy (primary to tertiary education): *total:* 10 years
male: 11 years
female: 10 years (2004)
Education expenditures: 4.8% of GDP (2009)
country comparison to the world: 67

GOVERNMENT

Country name: *conventional long form:* Republic of Vanuatu
conventional short form: Vanuatu
local long form: Ripablik blong Vanuatu
local short form: Vanuatu
former: New Hebrides
Government type: parliamentary republic
Capital: *name:* Port-Vila (on Efate)
geographic coordinates: 17 44 S, 168 19 E
time difference: UTC+11 (16 hours ahead of Washington, DC during Standard Time)
Administrative divisions: 6 provinces; Malampa, Penama, Sanma, Shefa, Tafea, Torba
Independence: 30 July 1980 (from France and the UK)
National holiday: Independence Day, 30 July (1980)
Constitution: 30 July 1980
Legal system: mixed legal system of English common law, French law, and customary law
International law organization participation: has not submitted an ICJ jurisdiction declaration; non-party state to the ICCt
Suffrage: 18 years of age; universal
Executive branch: *chief of state:* President Iolu Johnson ABBIL (since 3 September 2009)
head of government: Prime Minister Sato KILMAN (since 13 May 2011); note–appeals court ruling on 13 May 2011 overturned the election of Serge VOHOR following a no confidence vote on 24 April 2011 against Sato KILMAN
cabinet: Council of Ministers appointed by the prime minister, responsible to parliament
(For more information visit the World Leaders website)
elections: president elected for a five-year term by an electoral college consisting of parliament and the presidents of the regional councils; election for president last held on 2 September 2009 (next to be held in 2014); following legislative elections, the leader of the majority party or majority coalition usually elected prime minister by parliament from among its members; election for prime minister last held on 24 April 2011 following the no-confidence vote against KILMAN (next to be held following general elections in 2012)
election results: Iolu Johnson ABBIL elected president, with 41 votes out of 58, on the third ballot on 2 September 2009
Legislative branch: unicameral Parliament (52 seats; members elected by popular vote to serve four-year terms)
elections: last held on 2 September 2008 (next to be held in 2012)
election results: percent of vote by party–NA; seats by party–VP 11, NUP 8, UMP 7, VRP 7, PPP 4, GC 2, MPP 1, NA 1, NAG 1, PAP 1, Shepherds Alliance 1, VFFP 1, VLP 1, VNP 1, VPRFP 1, and independent 4; note–political party associations are fluid
note: the National Council of Chiefs advises on matters of culture and language
Judicial branch: Supreme Court (chief justice is appointed by the president after consultation with the prime minister and the leader of the opposition, three other justices are appointed by the president on the advice of the Judicial Service Commission)
Political parties and leaders: Greens Confederation or GC [Moana CARCASSES]; Jon Frum Movement or JF [Song KEASPAI]; Melanesian Progressive Party or MPP [Barak SOPE]; Nagriamel movement or NAG [Havo MOLI]; Namangi Aute or NA [Paul TELUKLUK]; National United Party or NUP [Ham LINI]; People's Action Party or PAP [Peter VUTA]; People's Progressive Party or PPP [Sato KILMAN]; Shepherds Alliance Party [leader NA]; Union of Moderate Parties or UMP [Serge VOHOR]; Vanua'aku Pati (Our Land Party) or VP [Edward NATAPEI]; Vanuatu Family First Party or VFFP [Eta RORI]; Vanuatu Labor Party or VLP [Joshua KALSAKAU]; Vanuatu National Party or VNP [Issac HAMARILIU]; Vanuatu Republican Farmers Party or VPRFP [Jean RAVOU]; Vanuatu Republican Party or VRP [Maxime Carlot KORMAN]
Political pressure groups and leaders: NA
International organization participation: ACP, ADB, AOSIS, C, FAO, G-77, IBRD, ICAO, ICRM, IDA, IFC, IFRCS, ILO, IMF, IMO, IMSO, IOC, ITU, ITUC, MIGA, NAM, OAS (observer), OIF, OPCW, PIF, Sparteca, SPC, UN, UNCTAD, UNESCO, UNIDO, UNWTO, UPU, WCO, WFTU, WHO, WMO, WTO (observer)
Diplomatic representation in the US: Vanuatu does not have an embassy in the US; it does, however, have a Permanent Mission to the UN
Diplomatic representation from the US: the US does not have an embassy in Vanuatu; the US ambassador to Papua New Guinea is accredited to Vanuatu
Flag description: two equal horizontal bands of red (top) and green with a black isosceles triangle (based on the hoist side) all separated by a black-edged yellow stripe in the shape of a horizontal Y (the two points of the Y face the hoist side and enclose the triangle); centered in the triangle is a boar's tusk encircling two crossed namele fern fronds, all in yellow; red represents the blood of boars and men, green the richness of the islands, and black the ni-Vanuatu people; the yellow Y-shape–which reflects the pattern of the islands in the Pacific Ocean–symbolizes the light of the Gospel spreading through the islands; the boar's tusk is a symbol of prosperity frequently worn as a pendant on the islands; the fern fronds represent peace
National anthem: *name:* "Yumi, Yumi, Yumi" (We, We, We)
lyrics/music: Francois Vincent AYSSAV
note: adopted 1980, the anthem is written in Bislama, a Creole language that mixes Pidgin English and French

ECONOMY

Economy—overview: This South Pacific island economy is based primarily on small-scale agriculture, which provides a living for about two-thirds of the population. Fishing, offshore financial services, and tourism, with nearly 197,000 visitors in 2008, are other mainstays of the economy. Mineral deposits are negligible; the country has no known petroleum deposits. A small light industry sector caters to the local market. Tax revenues come mainly from import duties. Economic development is hindered by dependence on relatively few commodity exports, vulnerability to natural disasters, and long distances from main markets and between constituent islands. In response to foreign concerns, the government has promised to tighten regulation of its offshore financial center. In mid-2002, the government stepped up efforts to boost tourism through improved air connections, resort development, and cruise ship facilities. Agriculture, especially livestock farming, is a second target for growth. Australia and New Zealand are the main suppliers of tourists and foreign aid.
GDP (purchasing power parity): $1.137 billion (2010 est.)
country comparison to the world: 196
$1.113 billion (2009 est.)
$1.075 billion (2008 est.)
note: data are in 2010 US dollars
GDP (official exchange rate): $693 million (2010 est.)
GDP—real growth rate: 2.2% (2010 est.)
country comparison to the world: 142
3.5% (2009 est.)
6.2% (2008 est.)
GDP—per capita (PPP): $5,100 (2010 est.)
country comparison to the world: 147
$5,100 (2009 est.)
$5,000 (2008 est.)
note: data are in 2010 US dollars
GDP—composition by sector:
agriculture: 26%
industry: 12%
services: 62% (2000 est.)
Labor force: 115,900 (2007)
country comparison to the world: 181
Labor force—by occupation:
agriculture: 65%
industry: 5%
services: 30% (2000 est.)
Unemployment rate: 1.7% (1999)
country comparison to the world: 9
Population below poverty line: NA%
Household income or consumption by percentage share: *lowest 10%:* NA%
highest 10%: NA%
Budget: *revenues:* $78.7 million
expenditures: $72.23 million (2005 est.)
Inflation rate (consumer prices): 3.9% (2007 est.)
country comparison to the world: 111
Central bank discount rate:
6% (31 December 2009)
country comparison to the world: 71
6% (31 December 2008)
Commercial bank prime lending rate: 5.5% (31 December 2009 est.)
country comparison to the world: 143
5.29% (31 December 2008 est.)
Stock of narrow money: $229.2 million (31 December 2009)
country comparison to the world: 171
$177.7 million (31 December 2008)
Stock of broad money: $614.2 million (31 December 2009)
country comparison to the world: 170
$531.6 million (31 December 2008)

Stock of domestic credit: $274 million (31 December 2008 est.)
country comparison to the world: 171
$229.5 million (31 December 2007 est.)
Market value of publicly traded shares: $NA
Agriculture—products: copra, coconuts, cocoa, coffee, taro, yams, fruits, vegetables; beef; fish
Industries: food and fish freezing, wood processing, meat canning
Electricity—production: 42 million kWh (2007 est.)
country comparison to the world: 201
Electricity—consumption: 39.06 million kWh (2007 est.)
country comparison to the world: 200
Electricity—exports: 0 kWh (2008 est.)
Electricity—imports: 0 kWh (2008 est.)
Oil—production: 0 bbl/day (2009 est.)
country comparison to the world: 118
Oil—consumption: 1,000 bbl/day (2009 est.)
country comparison to the world: 194
Oil—exports: 0 bbl/day (2007 est.)
country comparison to the world: 192
Oil—imports: 653.6 bbl/day (2007 est.)
country comparison to the world: 194
Oil—proved reserves: 0 bbl (1 January 2010 est.)
country comparison to the world: 171
Natural gas—production: 0 cu m (2008 est.)
country comparison to the world: 167
Natural gas—consumption: 0 cu m (2008 est.)
country comparison to the world: 205
Natural gas—exports: 0 cu m (2008 est.)
country comparison to the world: 149
Natural gas—imports: 0 cu m (2008 est.)
country comparison to the world: 96
Natural gas—proved reserves: 0 cu m (1 January 2010 est.)
country comparison to the world: 172
Current account balance: $-60 million (2007 est.)
country comparison to the world: 71
Exports: $40 million (2006)
country comparison to the world: 201
Exports—commodities: copra, beef, cocoa, timber, kava, coffee
Exports—partners: Thailand 53.15%, Japan 12.22%, Poland 11.78% (2009)
Imports: $156 million (2006)
country comparison to the world: 204
Imports—commodities: machinery and equipment, foodstuffs, fuels
Imports—partners: Japan 17.3%, Australia 13.46%, China 12.26%, Singapore 12%, NZ 6.88%, Poland 6.61%, France 5.86%, Fiji 5.52% (2009)
Reserves of foreign exchange and gold: $40.54 million (2003)
country comparison to the world: 139
Debt—external: $81.2 million (2004)
country comparison to the world: 181
Exchange rates: vatu (VUV) per US dollar–97.93 (2009), NA (2007), 111.93 (2006), NA (2005), 111.79 (2004)

COMMUNICATIONS

Telephones—main lines in use: 7,200 (2009)
country comparison to the world: 206
Telephones—mobile cellular: 126,500 (2009)
country comparison to the world: 181
Telephone system: *general assessment:* NA
domestic: NA
international: country code–678; satellite earth station–1 Intelsat (Pacific Ocean)
Broadcast media: 1 state-owned television station; multi-channel pay TV is available; state-owned Radio Vanuatu operates 2 radio stations; 2 privately-owned radio broadcasters; programming from multiple international broadcasters is accessible (2008)
Internet country code: .vu
Internet hosts: 1,347 (2010)
country comparison to the world: 164
Internet users: 17,000 (2009)
country comparison to the world: 196

TRANSPORTATION

Airports: 31 (2010)
country comparison to the world: 114
Airports—with paved runways: *total:* 3
2,438 to 3,047 m: 1
1,524 to 2,437 m: 1
914 to 1,523 m: 1 (2010)
Airports—with unpaved runways: *total:* 28
914 to 1,523 m: 5
under 914 m: 23 (2010)
Roadways: *total:* 1,070 km
country comparison to the world: 183
paved: 256 km
unpaved: 814 km (1999)
Merchant marine: *total:* 72
country comparison to the world: 60
by type: bulk carrier 35, cargo 5, container 1, liquefied gas 2, passenger 2, refrigerated cargo 26, vehicle carrier 1
foreign-owned: 70 (Australia 2, Belgium 1, Canada 5, Greece 4, Japan 44, Monaco 1, Norway 1, Poland 7, Russia 1, UAE 1, Ukraine 3) (2010)
Ports and terminals: Forari Bay, Luganville (Santo, Espiritu Santo), Port-Vila

MILITARY

Military branches: no regular military forces; Vanuatu Police Force (VPF), Vanuatu Mobile Force (VMF; includes Police Maritime Wing (PMW)) (2009)
Manpower available for military service: *males age 16–49:* 62,216 (2010 est.)
Manpower fit for military service: *males age 16–49:* 43,331
females age 16–49: 44,927 (2010 est.)
Manpower reaching militarily significant age annually: *male:* 2,323
female: 2,230 (2010 est.)
Military expenditures: NA

TRANSNATIONAL ISSUES

Disputes—international: Matthew and Hunter Islands east of New Caledonia claimed by Vanuatu and France

VENEZUELA

INTRODUCTION

Background: Venezuela was one of three countries that emerged from the collapse of Gran Colombia in 1830 (the others being Ecuador and New Granada, which became Colombia). For most of the first half of the 20th century, Venezuela was ruled by generally benevolent military strongmen, who promoted the oil industry and allowed for some social reforms. Democratically elected governments have held sway since 1959. Hugo CHAVEZ, president since 1999, seeks to implement his "21st Century Socialism," which purports to alleviate social ills while at the same time attacking capitalist globalization and existing democratic institutions. Current concerns include: a weakening of democratic institutions, political polarization, a politicized military, drug-related violence along the Colombian border, overdependence on the petroleum industry with its price fluctuations, and irresponsible mining operations that are endangering the rain forest and indigenous peoples.

GEOGRAPHY

Location: Northern South America, bordering the Caribbean Sea and the North Atlantic Ocean, between Colombia and Guyana
Geographic coordinates: 8 00 N, 66 00 W
Map references: South America
Area:
total: 912,050 sq km
country comparison to the world: 33
land: 882,050 sq km
water: 30,000 sq km
Area—comparative: slightly more than twice the size of California
Land boundaries:
total: 4,993 km
border countries: Brazil 2,200 km, Colombia 2,050 km, Guyana 743 km
Coastline: 2,800 km
Maritime claims: territorial sea: 12 nm
contiguous zone: 15 nm
exclusive economic zone: 200 nm
continental shelf: 200 m depth or to the depth of exploitation
Climate: tropical; hot, humid; more moderate in highlands

Terrain: Andes Mountains and Maracaibo Lowlands in northwest; central plains (llanos); Guiana Highlands in southeast
Elevation extremes: *lowest point:* Caribbean Sea 0 m
highest point: Pico Bolivar 5,007 m
Natural resources: petroleum, natural gas, iron ore, gold, bauxite, other minerals, hydropower, diamonds
Land use: *arable land:* 2.85%
permanent crops: 0.88%
other: 96.27% (2005)
Irrigated land: 5,750 sq km (2003)
Total renewable water resources: 1,233.2 cu km (2000)
Freshwater withdrawal (domestic/industrial/agricultural): *total:* 8.37 cu km/yr (6%/7%/47%)
per capita: 313 cu m/yr (2000)
Natural hazards: subject to floods, rockslides, mudslides; periodic droughts
Environment—current issues: sewage pollution of Lago de Valencia; oil and urban pollution of Lago de Maracaibo; deforestation; soil degradation; urban and industrial pollution, especially along the Caribbean coast; threat to the rainforest ecosystem from irresponsible mining operations
Environment—international agreements: *party to:* Antarctic Treaty, Biodiversity, Climate Change, Climate Change-Kyoto Protocol, Desertification, Endangered Species, Hazardous Wastes, Marine Life Conservation, Ozone Layer Protection, Ship Pollution, Tropical Timber 83, Tropical Timber 94, Wetlands
signed but not ratified:: none of the selected agreements
Geography—note: on major sea and air routes linking North and South America; Angel Falls in the Guiana Highlands is the world's highest waterfall

PEOPLE

Population: 27,635,743 (July 2011 est.)
country comparison to the world: 45
Age structure: *0–14 years:* 29.5% (male 4,149,781/female 4,002,931)
15–64 years: 65.1% (male 8,846,945/female 9,130,561)
65 years and over: 5.4% (male 665,436/female 840,089) (2011 est.)
Median age: *total:* 26.1 years
male: 25.4 years
female: 26.8 years (2011 est.)
Population growth rate: 1.493% (2011 est.)
country comparison to the world: 79
Birth rate: 20.1 births/1,000 population (2011 est.)
country comparison to the world: 88
Death rate: 5.17 deaths/1,000 population (July 2011 est.)
country comparison to the world: 181
Net migration rate: 0 migrant(s)/1,000 population (2011 est.)
country comparison to the world: 111
Urbanization: *urban population:* 93% of total population (2010)
rate of urbanization: 1.7% annual rate of change (2010-15 est.)
Major cities—population: CARACAS (capital) 3.051 million; Maracaibo 2.153 million; Valencia 1.738 million; Barquisimeto 1.159 million; Maracay 1.04 million (2009)
Sex ratio: *at birth:* 1.05 male(s)/female
under 15 years: 1.03 male(s)/female
15–64 years: 0.97 male(s)/female
65 years and over: 0.79 male(s)/female
total population: 0.98 male(s)/female (2011 est.)
Infant mortality rate: *total:* 20.62 deaths/1,000 live births
country comparison to the world: 95
male: 24.12 deaths/1,000 live births
female: 16.95 deaths/1,000 live births (2011 est.)
Life expectancy at birth: *total population:* 73.93 years
country comparison to the world: 110
male: 70.84 years
female: 77.17 years (2011 est.)
Total fertility rate: 2.42 children born/woman (2011 est.)
country comparison to the world: 92
HIV/AIDS—adult prevalence rate: NA; NA note—no country specific models provided
HIV/AIDS—people living with HIV/AIDS: NA
HIV/AIDS—deaths: NA
Major infectious diseases: *degree of risk:* high
food or waterborne diseases: bacterial diarrhea
vectorborne disease: dengue fever and malaria (2009)
Drinking water source: *Improved:*
urban: 94% of population
rural: 74% of population
total: 92% of population
Unimproved:
urban: 6% of population
rural: 26% of population
total: 8% of population (2000)
Sanitation facility access: *Improved:*
urban: 93% of population
rural: 54% of population
total: 89% of population
Unimproved:
urban: 7% of population
rural: 46% of population
total: 11% of population (2000)
Nationality: *noun:* Venezuelan(s)
adjective: Venezuelan
Ethnic groups: Spanish, Italian, Portuguese, Arab, German, African, indigenous people
Religions: nominally Roman Catholic 96%, Protestant 2%, other 2%
Languages: Spanish (official), numerous indigenous dialects
Literacy: *definition:* age 15 and over can read and write
total population: 93%
male: 93.3%
female: 92.7% (2001 census)
School life expectancy (primary to tertiary education): *total:* 14 years
male: 13 years
female: 15 years (2008)
Education expenditures: 3.7% of GDP (2007)
country comparison to the world: 113

GOVERNMENT

Country name: *conventional long form:* Bolivarian Republic of Venezuela
conventional short form: Venezuela
local long form: Republica Bolivariana de Venezuela
local short form: Venezuela
Government type: federal republic
Capital: *name:* Caracas
geographic coordinates: 10 30 N, 66 56 W
time difference: UTC-4.5 (half an hour ahead of Washington, DC during Standard Time)
Administrative divisions: 23 states (estados, singular—estado), 1 capital district* (distrito capital), and 1 federal dependency** (dependencia federal); Amazonas, Anzoategui, Apure, Aragua, Barinas, Bolivar, Carabobo, Cojedes, Delta Amacuro, Dependencias Federales (Federal Dependencies)**, Distrito Capital (Capital District)*, Falcon, Guarico, Lara, Merida, Miranda, Monagas, Nueva Esparta, Portuguesa, Sucre, Tachira, Trujillo, Vargas, Yaracuy, Zulia
note: the federal dependency consists of 11 federally controlled island groups with a total of 72 individual islands
Independence: 5 July 1811 (from Spain)
National holiday: Independence Day, 5 July (1811)
Constitution: 30 December 1999
Legal system: civil law system based on the Spanish civil code
International law organization participation: has not submitted an ICJ jurisdiction declaration; accepts ICCt jurisdiction
Suffrage: 18 years of age; universal
Executive branch: *chief of state:* President Hugo CHAVEZ Frias (since 2 February 1999); Executive Vice President Elias JAUA Milano (since 26 January 2010); note—the president is both the chief of state and head of government
head of government: President Hugo CHAVEZ Frias (since 2 February 1999); Executive Vice President Elias JAUA Milano (since 26 January 2010)
cabinet: Council of Ministers appointed by the president
(For more information visit the World Leaders website)
elections: president elected by popular vote for a six-year term (eligible for unlimited reelection); election last held on 3 December 2006 (next to be held in December 2012)
note: in 1999, a National Constituent Assembly drafted a new constitution that increased the presidential term to six years; an election was subsequently held on 30 July 2000 under the terms of this constitution; in 2009, a national referendum approved the elimination of term limits on all elected officials, including the presidency
election results: Hugo CHAVEZ Frias reelected president; percent of vote—Hugo CHAVEZ Frias 62.9%, Manuel ROSALES 36.9%, other 0.2%
Legislative branch: unicameral National Assembly or Asamblea Nacional (165 seats; members elected by popular vote to serve five-year terms; three seats reserved for the indigenous peoples of Venezuela)
elections: last held on 26 September 2010 (next to be held in 2015)
election results: percent of vote by party—pro-government 48.9%, opposition coalition 47.9%, other 3.2%; seats by party—pro-government 98, opposition 65, other 2

Judicial branch: Supreme Tribunal of Justice or Tribunal Supremo de Justicia (32 magistrates are elected by the National Assembly for a single 12-year term)

Political parties and leaders: A New Time or UNT [Omar BARBOZA]; Brave People's Alliance or ABP [Oscar PEREZ, currently in exile]; Christian Democrats or COPEI [Roberto ENRIQUEZ]; Communist Party of Venezuela or PCV [Oscar FIGUERA]; Democratic Action or AD [Henry RAMOS Allup]; Fatherland for All or PPT [Jose ALBORNOZ]; For Social Democracy or PODEMOS [Ismael GARCIA]; Justice First [Julio BORGES]; Movement Toward Socialism or MAS [Nicolas SOSA]; The Radical Cause [Daniel SANTOLO]; United Socialist Party of Venezuela or PSUV [Hugo CHAVEZ]; Venezuela Project or PV [Henrique SALAS Romer]

Political pressure groups and leaders: FEDECAMARAS, a conservative business group; VECINOS groups; Venezuelan Confederation of Workers or CTV (labor organization dominated by the Democratic Action); various civil society groups and human rights organizations

International organization participation: Caricom (observer), CDB, FAO, G-15, G-24, G-77, IADB, IAEA, IBRD, ICAO, ICRM, IDA, IFAD, IFC, IFRCS, IHO, ILO, IMF, IMO, IMSO, Interpol, IOC, IOM, IPU, ITSO, ITU, ITUC, LAES, LAIA, LAS (observer), Mercosur (associate), MIGA, NAM, OAS, OPANAL, OPCW, OPEC, PCA, PetroCaribe, RG, UN, UNASUR, UNCTAD, UNESCO, UNHCR, UNIDO, Union Latina, UNWTO, UPU, WCO, WFTU, WHO, WIPO, WMO, WTO

Diplomatic representation in the US: *chief of mission:* Ambassador (vacant); Charge d'Affaires Angelo RIVERO Santos
chancery: 1099 30th Street NW, Washington, DC 20007
telephone: [1] (202) 342-2214
FAX: [1] (202) 342-6820
consulate(s) general: Boston, Chicago, Houston, Miami, New Orleans, New York, San Francisco, San Juan (Puerto Rico)

Diplomatic representation from the US: *chief of mission:* Ambassador (vacant); Charge d'Affaires John CAULFIELD
embassy: Calle F con Calle Suapure, Urbanizacion Colinas de Valle Arriba, Caracas 1080
mailing address: P. O. Box 62291, Caracas 1060-A; APO AA 34037
telephone: [58] (212) 975-6411, 907-8400 (after hours)
FAX: [58] (212) 907-8199

Flag description: three equal horizontal bands of yellow (top), blue, and red with the coat of arms on the hoist side of the yellow band and an arc of eight white five-pointed stars centered in the blue band; the flag retains the three equal horizontal bands and three main colors of the banner of Gran Colombia, the South American republic that broke up in 1830; yellow is interpreted as standing for the riches of the land, blue for the courage of its people, and red for the blood shed in attaining independence; the seven stars on the original flag represented the seven provinces in Venezuela that united in the war of independence; in 2006, President Hugo CHAVEZ ordered an eighth star added to the star arc–a decision that sparked much controversy–to conform with the flag proclaimed by Simon Bolivar in 1827 and to represent the province of Guayana

National anthem: *name:* "Gloria al bravo pueblo" (Glory to the Brave Nation)
lyrics/music: Vicente SALIAS/Juan Jose LANDAETA
note: adopted 1881; the lyrics were written in 1810, the music some years later; both SALIAS and LANDAETA were executed in 1814 during Venezuela's struggle for independence

ECONOMY

Economy—overview: Venezuela remains highly dependent on oil revenues, which account for roughly 95% of export earnings, about 55% of the federal budget revenues, and around 30% of GDP. A nationwide strike between December 2002 and February 2003 had far-reaching economic consequences–real GDP declined by around 9% in 2002 and 8% in 2003–but economic output recovered strongly through 2008. Fueled by high oil prices, record government spending helped to boost GDP by about 10% in 2006, 8% in 2007, and nearly 5% in 2008, before a sharp drop in oil prices caused a contraction in 2009-10. This spending, combined with recent minimum wage hikes and improved access to domestic credit, created a consumption boom which came at the cost of higher inflation–roughly 32% in 2008, and slowing only slightly to 30% in 2010, despite the lengthy downturn. Imports also jumped significantly before the recession of 2009. President Hugo CHAVEZ's continued efforts to increase the government's control of the economy by nationalizing firms in the agribusiness, financial, construction, oil, and steel sectors have hurt the private investment environment, reduced productive capacity, and slowed non-petroleum exports. In the first half of 2010 Venezuela faced the prospect of lengthy nationwide blackouts when its main hydroelectric power plant–which provides more than 35% of the country's electricity–nearly shut down. In May, 2010, CHAVEZ closed the unofficial foreign exchange market–the "parallel" market–in an effort to stem inflation and slow the currency's depreciation. In June 2010, the government created the "Transaction System for Foreign Currency Denominated Securities" (SITME) to replace the "parallel" market. In December 2010, CHAVEZ eliminated the dual exchange rate system and unified the exchange rate at 4.3 bolivars per dollar. In January 2011, CHAVEZ announced the second devaluation of the bolivar within twelve months. In December 2010, the National Assembly passed a package of five organic laws designed to complete the transformation of the Venezuelan economy in line with CHAVEZ's vision of 21st century socialism. These laws likely will be implemented in 2011. Venezuela began 2011 wrestling with macroeconomic imbalances resulting from the government's unorthodox economic policies, a housing crisis, and a continuing electricity crisis.

GDP (purchasing power parity): $345.2 billion (2010 est.)
country comparison to the world: 35
$351.9 billion (2009 est.)
$363.9 billion (2008 est.)
note: data are in 2010 US dollars

GDP (official exchange rate): $290.7 billion (2010 est.)

GDP—real growth rate: -1.9% (2010 est.)
country comparison to the world: 206
-3.3% (2009 est.)
4.8% (2008 est.)

GDP—per capita (PPP):
$12,700 (2010 est.)
country comparison to the world: 91
$13,100 (2009 est.)
$13,800 (2008 est.)
note: data are in 2010 US dollars

GDP—composition by sector:
agriculture: 4.1%
industry: 34.9%
services: 61.1% (2010 est.)

Labor force: 13.3 million (2010 est.)
country comparison to the world: 41

Labor force—by occupation:
agriculture: 13%
industry: 23%
services: 64% (1997 est.)

Unemployment rate: 12.1% (2010 est.)
country comparison to the world: 130
7.9% (2009 est.)

Population below poverty line: 37.9% (2005 est.)

Household income or consumption by percentage share: *lowest 10%:* 1.7%
highest 10%: 32.7% (2006)

Distribution of family income—Gini index: 41 (2009)
country comparison to the world: 56
49.5 (1998)

Investment (gross fixed): 16.4% of GDP (2010 est.)
country comparison to the world: 124

Budget: *revenues:* $50.12 billion
expenditures: $56.53 billion (2010 est.)

Public debt: 25.5% of GDP (2010 est.)
country comparison to the world: 98
18% of GDP (2009 est.)

Inflation rate (consumer prices): 29.8% (2010 est.)
country comparison to the world: 223
27.1% (2009 est.)

Central bank discount rate: 29.5% (31 December 2009)
country comparison to the world: 3
33.5% (31 December 2008)

Commercial bank prime lending rate: 19.89% (31 December 2009 est.)
country comparison to the world: 11
22.37% (31 December 2008 est.)

Stock of narrow money: $69.36 billion (31 December 2010 est.)
country comparison to the world: 38
$93.19 billion (31 December 2009 est.)

Stock of broad money: $78.11 billion (31 December 2010 est.)
country comparison to the world: 58
$107 billion (31 December 2009 est.)

Stock of domestic credit: $54.22 billion (31 December 2010 est.)
country comparison to the world: 62
$75.87 billion (31 December 2009 est.)
Market value of publicly traded shares: $NA (31 December 2010)
country comparison to the world: 69
$8.251 billion
Agriculture—products: corn, sorghum, sugarcane, rice, bananas, vegetables, coffee; beef, pork, milk, eggs; fish
Industries: petroleum, construction materials, food processing, textiles; iron ore mining, steel, aluminum; motor vehicle assembly
Industrial production growth rate: -8% (2010 est.)
country comparison to the world: 166
Electricity—production: 113.3 billion kWh (2007 est.)
country comparison to the world: 29
Electricity—consumption: 83.02 billion kWh (2007 est.)
country comparison to the world: 35
Electricity—exports: 540 million kWh (2007 est.)
Electricity—imports: 1.651 billion kWh (2007 est.)
Oil—production: 2.472 million bbl/day (2009 est.)
country comparison to the world: 11
Oil—consumption:
740,000 bbl/day (2009 est.)
country comparison to the world: 23
Oil—exports: 2.182 million bbl/day (2007 est.)
country comparison to the world: 8
Oil—imports: 0 bbl/day (2007 est.)
country comparison to the world: 205
Oil—proved reserves: 97.77 billion bbl (1 January 2010 est.)
country comparison to the world: 7
Natural gas—production: 23.06 billion cu m (2009 est.)
country comparison to the world: 29
Natural gas—consumption: 24.86 billion cu m (2009 est.)
country comparison to the world: 31
Natural gas—exports: 0 cu m (2008 est.)
country comparison to the world: 196
Natural gas—imports: 1.8 billion cu m (2009 est.)
country comparison to the world: 47
Natural gas—proved reserves: 4.983 trillion cu m (1 January 2010 est.)
country comparison to the world: 9
Current account balance: $22.07 billion (2010 est.)
country comparison to the world: 15
$8.561 billion (2009 est.)
Exports: $64.87 billion (2010 est.)
country comparison to the world: 44
$57.6 billion (2009 est.)
Exports—commodities: petroleum, bauxite and aluminum, minerals, chemicals, agricultural products, basic manufactures
Exports—partners: US 27.3%, Columbia 11.4%, China 12.8%, Brazil 8.8% (2009 est.)
Imports: $31.37 billion (2010 est.)
country comparison to the world: 57
$38.44 billion (2009 est.)
Imports—commodities: agricultural products, raw materials, machinery and equipment, transport equipment, construction materials
Imports—partners: US 20.6%, Columbia 17.8%, China 12.8%, Mexico 8.7%, Brazil 4.7% (2009 est.)
Reserves of foreign exchange and gold: $29.49 billion (31 December 2010 est.)
country comparison to the world: 37
$35 billion (31 December 2009 est.)
Debt—external: $55.61 billion (31 December 2010 est.)
country comparison to the world: 54
$53.58 billion (31 December 2009 est.)
Stock of direct foreign investment—at home:
$37.71 billion (31 December 2010 est.)
country comparison to the world: 58
$41.21 billion (31 December 2009 est.)
Stock of direct foreign investment—abroad:
$20.97 billion (31 December 2010 est.)
country comparison to the world: 39
$17.67 billion (31 December 2009 est.)
Exchange rates: bolivars (VEB) per US dollar–4.3039 (2010), 2.147 (2009), 2.147 (2008), 2,147 (2007), 2,147 (2006)

COMMUNICATIONS

Telephones—main lines in use: 6.867 million (2009)
country comparison to the world: 28
Telephones—mobile cellular: 28.124 million (2009)
country comparison to the world: 32
Telephone system: *general assessment:* modern and expanding
domestic: domestic satellite system with 3 earth stations; recent substantial improvement in telephone service in rural areas; substantial increase in digitalization of exchanges and trunk lines; installation of a national interurban fiber-optic network capable of digital multimedia services; combined fixed and mobile-cellular telephone subscribership 130 per 100 persons
international: country code–58; submarine cable systems provide connectivity to the Caribbean, Central and South America, and US; satellite earth stations–1 Intelsat (Atlantic Ocean) and 1 PanAmSat; participating with Colombia, Ecuador, Peru, and Bolivia in the construction of an international fiber-optic network; constructing submarine cable to provide connectivity to Cuba with an estimated date of completion in late 2011 (2010)
Broadcast media: government supervises a mixture of state-run and private broadcast media; 1 state-run television network, 4 privately-owned TV networks, a privately-owned news channel with limited national coverage, and a government-backed pan-American channel; state-run radio network includes 65 news stations and roughly another 30 stations targeted at specific audiences; state-sponsored community broadcasters include 244 radio stations and 36 television stations; the number of private broadcast radio stations has been declining, but a large number remain in operation (2010)
Internet country code: .ve
Internet hosts: 238,665 (2010)
country comparison to the world: 64
Internet users: 8.918 million (2009)
country comparison to the world: 32

TRANSPORTATION

Airports: 409 (2010)
country comparison to the world: 20
Airports—with paved runways: *total:* 129
over 3,047 m: 5
2,438 to 3,047 m: 10
1,524 to 2,437 m: 34
914 to 1,523 m: 63
under 914 m: 17 (2010)
Airports—with unpaved runways:
total: 280
2,438 to 3,047 m: 1
1,524 to 2,437 m: 16
914 to 1,523 m: 91
under 914 m: 172 (2010)
Heliports: 4 (2010)
Pipelines: extra heavy crude 980 km; gas 5,347 km; oil 6,694 km; refined products 1,620 km (2010)
Railways: *total:* 806 km
country comparison to the world: 98
standard gauge: 806 km 1.435-m gauge (2008)
Roadways: *total:* 96,155 km
country comparison to the world: 46
paved: 32,308 km
unpaved: 63,847 km (2002)
Waterways: 7,100 km (the Orinoco River (400 km) and Lake de Maracaibo are navigable by oceangoing vessels) (2010)
country comparison to the world: 21
Merchant marine: *total:* 59
country comparison to the world: 66
by type: bulk carrier 5, cargo 15, chemical tanker 5, liquefied gas 5, passenger 1, passenger/cargo 12, petroleum tanker 16
foreign-owned: 10 (Denmark 1, Estonia 1, Germany 1, Greece 4, Mexico 1, Norway 1, Spain 1)
registered in other countries: 9 (Panama 8, Saint Vincent and the Grenadines 1) (2010)
Ports and terminals: La Guaira, Maracaibo, Puerto Cabello, Punta Cardon
Transportation—note: the International Maritime Bureau reports the territorial and offshore waters in the Caribbean Sea as a significant risk for piracy and armed robbery against ships; numerous vessels, including commercial shipping and pleasure craft, have been attacked and hijacked both at anchor and while underway; crews have been robbed and stores or cargoes stolen

MILITARY

Military branches: Bolivarian National Armed Forces (Fuerza Armada Nacional Bolivariana, FANB): Bolivarian Army (Ejercito Bolivariano, EB), Bolivarian Navy (Armada Bolivariana, AB; includes Naval Infantry, Coast Guard, Naval Aviation), Bolivarian National Military Aviation (Aviacion Militar Nacional Bolivariana, AMNB), Bolivarian National Guard (Guardia Nacional Bolivaria, GNB), Bolivarian Militia (Milicia Bolivariana, MB) (2011)
Military service age and obligation: 18-30 years of age for compulsory and voluntary military service; 30-month conscript service obligation; Navy requires 6th-grade education for enlisted personnel; all citizens of military service age (18-60 years

old) are obligated to register for military service (2011)
Manpower available for military service: *males age 16–49:* 7,013,854
females age 16–49: 7,165,661 (2010 est.)
Manpower fit for military service: *males age 16–49:* 5,614,743
females age 16–49: 6,074,834 (2010 est.)
Manpower reaching militarily significant age annually: *male:* 277,210
female: 273,353 (2010 est.)
Military expenditures: 1.2% of GDP (2005 est.)
country comparison to the world: 122

TRANSNATIONAL ISSUES

Disputes—international: claims all of the area west of the Essequibo River in Guyana, preventing any discussion of a maritime boundary; Guyana has expressed its intention to join Barbados in asserting claims before the United Nations Convention on the Law of the Sea (UNCLOS) that Trinidad and Tobago's maritime boundary with Venezuela extends into their waters; dispute with Colombia over maritime boundary and Venezuelan-administered Los Monjes islands near the Gulf of Venezuela; Colombian-organized illegal narcotics and paramilitary activities penetrate Venezuela's shared border region; in 2006, an estimated 139,000 Colombians sought protection in 150 communities along the border in Venezuela; US, France, and the Netherlands recognize Venezuela's granting full effect to Aves Island, thereby claiming a Venezuelan EEZ/continental shelf extending over a large portion of the eastern Caribbean Sea; Dominica, Saint Kitts and Nevis, Saint Lucia, and Saint Vincent and the Grenadines protest Venezuela's full effect claim
Trafficking in persons: current situation: Venezuela is a source, transit, and destination country for men, women, and children trafficked for the purposes of commercial sexual exploitation and forced labor; Venezuelan women and girls are trafficked within the country for sexual exploitation, lured from the nation's interior to urban and tourist areas; child prostitution in urban areas and child sex tourism in resort destinations appear to be growing; Venezuelan women and girls are trafficked for commercial sexual exploitation to Western Europe, Mexico, and Caribbean destinations
tier rating: Tier 2 Watch List—Venezuela is placed on the Tier 2 Watch List, up from Tier 3, as it showed greater resolve to address trafficking through law enforcement measures and prevention efforts in 2007, although stringent punishment of offenders and victim assistance remain lacking (2008)
Illicit drugs: small-scale illicit producer of opium and coca for the processing of opiates and coca derivatives; however, large quantities of cocaine, heroin, and marijuana transit the country from Colombia bound for US and Europe; significant narcotics-related money-laundering activity, especially along the border with Colombia and on Margarita Island; active eradication program primarily targeting opium; increasing signs of drug-related activities by Colombian insurgents on border

VIETNAM

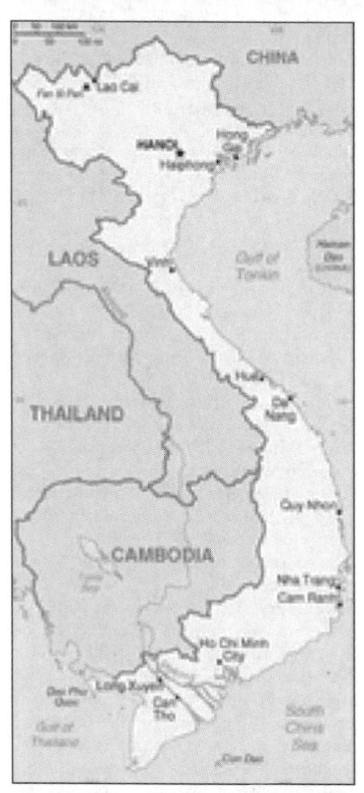

INTRODUCTION

Background: The conquest of Vietnam by France began in 1858 and was completed by 1884. It became part of French Indochina in 1887. Vietnam declared independence after World War II, but France continued to rule until its 1954 defeat by Communist forces under Ho Chi MINH. Under the Geneva Accords of 1954, Vietnam was divided into the Communist North and anti-Communist South. US economic and military aid to South Vietnam grew through the 1960s in an attempt to bolster the government, but US armed forces were withdrawn following a cease-fire agreement in 1973. Two years later, North Vietnamese forces overran the South reuniting the country under Communist rule. Despite the return of peace, for over a decade the country experienced little economic growth because of conservative leadership policies, the persecution and mass exodus of individuals—many of them successful South Vietnamese merchants—and growing international isolation. However, since the enactment of Vietnam's "doi moi" (renovation) policy in 1986, Vietnamese authorities have committed to increased economic liberalization and enacted structural reforms needed to modernize the economy and to produce more competitive, export-driven industries. The Communist leaders, however, maintain control on political expression and have resisted outside calls to improve human rights. The country continues to experience small-scale protests from various groups, the vast majority connected to land-use issues, calls for increased political space and the lack of equitable mechanisms for resolving disputes. Various ethnic minorities, such as the Montagnards of the Central Highlands and the Khmer Krom in the southern delta region, have also held protests.

GEOGRAPHY

Location: Southeastern Asia, bordering the Gulf of Thailand, Gulf of Tonkin, and South China Sea, as well as China, Laos, and Cambodia
Geographic coordinates: 16 10 N, 107 50 E
Map references: Southeast Asia
Area: *total:* 331,210 sq km
country comparison to the world: 65
land: 310,070 sq km
water: 21,140 sq km
Area—comparative: slightly larger than New Mexico
Land boundaries: *total:* 4,639 km
border countries: Cambodia 1,228 km, China 1,281 km, Laos 2,130 km
Coastline: 3,444 km (excludes islands)
Maritime claims: territorial sea: 12 nm
contiguous zone: 24 nm
exclusive economic zone: 200 nm
continental shelf: 200 nm or to the edge of the continental margin
Climate: tropical in south; monsoonal in north with hot, rainy season (May to September) and warm, dry season (October to March)
Terrain: low, flat delta in south and north; central highlands; hilly, mountainous in far north and northwest
Elevation extremes: *lowest point:* South China Sea 0 m
highest point: Fan Si Pan 3,144 m
Natural resources: phosphates, coal, manganese, rare earth elements, bauxite, chromate, offshore oil and gas deposits, timber, hydropower
Land use: *arable land:* 20.14%
permanent crops: 6.93%
other: 72.93% (2005)
Irrigated land: 30,000 sq km (2003)
Total renewable water resources: 891.2 cu km (1999)
Freshwater withdrawal (domestic/industrial/agricultural): *total:* 71.39 cu km/yr (8%/24%/68%)
per capita: 847 cu m/yr (2000)
Natural hazards: occasional typhoons (May to January) with extensive flooding, especially in the Mekong River delta

Environment—current issues: logging and slash-and-burn agricultural practices contribute to deforestation and soil degradation; water pollution and overfishing threaten marine life populations; groundwater contamination limits potable water supply; growing urban industrialization and population migration are rapidly degrading environment in Hanoi and Ho Chi Minh City
Environment—international agreements: *party to:* Biodiversity, Climate Change, Climate Change-Kyoto Protocol, Desertification, Endangered Species, Environmental Modification, Hazardous Wastes, Law of the Sea, Ozone Layer Protection, Ship Pollution, Wetlands
signed, but not ratified: none of the selected agreements
Geography—note: extending 1,650 km north to south, the country is only 50 km across at its narrowest point

PEOPLE

Population: 90,549,390 (July 2011 est.)
country comparison to the world: 14
Age structure: *0–14 years:* 25.2% (male 11,945,354/female 10,868,610)
15–64 years: 69.3% (male 31,301,879/female 31,419,306)
65 years and over: 5.5% (male 1,921,652/female 3,092,589) (2011 est.)
Median age: *total:* 27.8 years
male: 26.8 years
female: 28.9 years (2011 est.)
Population growth rate: 1.077% (2011 est.)
country comparison to the world: 110
Birth rate: 17.07 births/1,000 population (2011 est.)
country comparison to the world: 119
Death rate: 5.96 deaths/1,000 population (July 2011 est.)
country comparison to the world: 164
Net migration rate: -0.35 migrant(s)/1,000 population (2011 est.)
country comparison to the world: 132
Urbanization: *urban population:* 30% of total population (2010)
rate of urbanization: 3% annual rate of change (2010-15 est.)
Major cities—population: Ho Chi Minh City 5.976 million; HANOI (capital) 2.668 million; Haiphong 1.941 million; Da Nang 807,000 (2009)
Sex ratio: *at birth:* 1.117 male(s)/female
under 15 years: 1.1 male(s)/female
15–64 years: 0.99 male(s)/female
65 years and over: 0.62 male(s)/female
total population: 0.99 male(s)/female (2011 est.)
Infant mortality rate: *total:* 20.9 deaths/1,000 live births
country comparison to the world: 94
male: 21.27 deaths/1,000 live births
female: 20.48 deaths/1,000 live births (2011 est.)
Life expectancy at birth: *total population:* 72.18 years
country comparison to the world: 129
male: 69.72 years
female: 74.92 years (2011 est.)
Total fertility rate: 1.91 children born/woman (2011 est.)
country comparison to the world: 137
HIV/AIDS—adult prevalence rate: 0.4% (2009 est.)
country comparison to the world: 74
HIV/AIDS—people living with HIV/AIDS: 280,000 (2009 est.)
country comparison to the world: 22
HIV/AIDS—deaths: 14,000 (2009 est.)
country comparison to the world: 22
Major infectious diseases: *degree of risk:* high
food or waterborne diseases: bacterial diarrhea, hepatitis A, and typhoid fever
vectorborne diseases: dengue fever, malaria, Japanese encephalitis, and plague
water contact disease: leptospirosis
note: highly pathogenic H5N1 avian influenza has been identified in this country; it poses a negligible risk with extremely rare cases possible among US citizens who have close contact with birds (2009)
Drinking water source: *Improved:*
urban: 99% of population
rural: 92% of population
total: 94% of population
Unimproved:
urban: 1% of population
rural: 8% of population
total: 6% of population (2008)
Sanitation facility access: *Improved:*
urban: 94% of population
rural: 67% of population
total: 75% of population
Unimproved:
urban: 6% of population
rural: 33% of population
total: 25% of population (2008)
Nationality: *noun:* Vietnamese (singular and plural)
adjective: Vietnamese
Ethnic groups: Kinh (Viet) 86.2%, Tay 1.9%, Thai 1.7%, Muong 1.5%, Khome 1.4%, Hoa 1.1%, Nun 1.1%, Hmong 1%, others 4.1% (1999 census)
Religions: Buddhist 9.3%, Catholic 6.7%, Hoa Hao 1.5%, Cao Dai 1.1%, Protestant 0.5%, Muslim 0.1%, none 80.8% (1999 census)
Languages: Vietnamese (official), English (increasingly favored as a second language), some French, Chinese, and Khmer, mountain area languages (Mon-Khmer and Malayo-Polynesian)
Literacy: *definition:* age 15 and over can read and write
total population: 90.3%
male: 93.9%
female: 86.9% (2002 est.)
School life expectancy (primary to tertiary education): *total:* 10 years
male: 11 years
female: 10 years (2001)
Education expenditures: 5.3% of GDP (2008)
country comparison to the world: 49

GOVERNMENT

Country name: *conventional long form:* Socialist Republic of Vietnam
conventional short form: Vietnam
local long form: Cong Hoa Xa Hoi Chu Nghia Viet Nam
local short form: Viet Nam
abbreviation: SRV
Government type: Communist state
Capital: *name:* Hanoi (Ha Noi)
geographic coordinates: 21 02 N, 105 51 E
time difference: UTC+7 (12 hours ahead of Washington, DC during Standard Time)
Administrative divisions: 58 provinces (tinh, singular and plural) and 5 municipalities (thanh pho, singular and plural)
provinces: An Giang, Bac Giang, Bac Kan, Bac Lieu, Bac Ninh, Ba Ria-Vung Tau, Ben Tre, Binh Dinh, Binh Duong, Binh Phuoc, Binh Thuan, Ca Mau, Cao Bang, Dac Lak, Dac Nong, Dien Bien, Dong Nai, Dong Thap, Gia Lai, Ha Giang, Ha Nam, Ha Tinh, Hai Duong, Hau Giang, Hoa Binh, Hung Yen, Khanh Hoa, Kien Giang, Kon Tum, Lai Chau, Lam Dong, Lang Son, Lao Cai, Long An, Nam Dinh, Nghe An, Ninh Binh, Ninh Thuan, Phu Tho, Phu Yen, Quang Binh, Quang Nam, Quang Ngai, Quang Ninh, Quang Tri, Soc Trang, Son La, Tay Ninh, Thai Binh, Thai Nguyen, Thanh Hoa, Thua Thien-Hue, Tien Giang, Tra Vinh, Tuyen Quang, Vinh Long, Vinh Phuc, Yen Bai
municipalities: Can Tho, Da Nang, Ha Noi, Hai Phong, Ho Chi Minh City
Independence: 2 September 1945 (from France)
National holiday: Independence Day, 2 September (1945)
Constitution: 15 April 1992
Legal system: civil law system; note—the civil code of 2005 reflects a European style civil law
International law organization participation: has not submitted an ICJ jurisdiction declaration; non-party state to the ICCt
Suffrage: 18 years of age; universal
Executive branch: *chief of state:* President Nguyen Minh TRIET (since 27 June 2006); Vice President Nguyen Thi DOAN (since 25 July 2007)
head of government: Prime Minister Nguyen Tan DUNG (since 27 June 2006); Permanent Deputy Prime Minister Nguyen Sinh HUNG (since 28 June 2006), Deputy Prime Minister Hoang Trung HAI (since 2 August 2007), Deputy Prime Minister Nguyen Thien NHAN (since 2 August 2007), Deputy Prime Minister Pham Gia KHIEM (since 28 June 2006), and Deputy Prime Minister Truong Vinh TRONG (since 28 June 2006)
cabinet: Cabinet appointed by president based on proposal of prime minister and confirmed by National Assembly
(For more information visit the World Leaders website)
elections: president elected by the National Assembly from among its members for five-year term; last election held 27 June 2006 (next to be held in July 2011); prime minister appointed by the president from among the members of the National Assembly; deputy prime ministers appointed by the prime minister; appointment of prime minister and deputy prime ministers confirmed by National Assembly
election results: Nguyen Minh TRIET elected president; percent of National Assembly vote—94%; Nguyen Tan DUNG elected prime minister; percent of National Assembly vote—92%
Legislative branch: unicameral National Assembly or Quoc Hoi (500 seats;

members elected by popular vote to serve five-year terms)
elections: last held on 22 May 2011 (next to be held in May 2016)
election results: percent of vote by party–NA; seats by party–CPV 458, non-party CPV-approved 38, self-nominated 4; note–500 candidates were elected; CPV and non-party CPV-approved delegates were members of the Vietnamese Fatherland Front
Judicial branch: Supreme People's Court (chief justice is elected by the National Assembly on the recommendation of the president for a five-year term)
Political parties and leaders: Communist Party of Vietnam or CPV [Nguyen Phu TRONG]; other parties proscribed
Political pressure groups and leaders: 8406 Bloc; Democratic Party of Vietnam or DPV; People's Democratic Party Vietnam or PDP-VN; Alliance for Democracy
note: these groups advocate democracy but are not recognized by the government
International organization participation: ADB, APEC, ARF, ASEAN, CICA, CP, EAS, FAO, G-77, IAEA, IBRD, ICAO, ICRM, IDA, IFAD, IFC, IFRCS, ILO, IMF, IMO, IMSO, Interpol, IOC, IOM, IPU, ISO, ITSO, ITU, MIGA, NAM, OIF, OPCW, UN, UNCTAD, UNESCO, UNIDO, UNWTO, UPU, WCO, WFTU, WHO, WIPO, WMO, WTO
Diplomatic representation in the US: *chief of mission:* Ambassador Nguyen Quoc CUONG
chancery: 1233 20th Street NW, Suite 400, Washington, DC 20036
telephone: [1] (202) 861-0737
FAX: [1] (202) 861-0917
consulate(s) general: Houston, New York, San Francisco
Diplomatic representation from the US: *chief of mission:* Ambassador (vacant); Charge d'Affaires Virginia E. PALMER
embassy: 7 Lang Ha Street, Ba Dinh District, Hanoi
mailing address: PSC 461, Box 400, FPO AP 96521-0002
telephone: [84] (4) 3850-5000
FAX: [84] (4) 3850-5010
consulate(s) general: Ho Chi Minh City
Flag description: red field with a large yellow five-pointed star in the center; red symbolizes revolution and blood, the five-pointed star represents the five elements of the populace–peasants, workers, intellectuals, traders, and soldiers–that unite to build socialism
National anthem: *name:* "Tien quan ca" (The Song of the Marching Troops)
lyrics/music: Nguyen Van CAO
note: adopted as the national anthem of the Democratic Republic of Vietnam in 1945; it became the national anthem of the unified Socialist Republic of Vietnam in 1976; although it consists of two verses, only the first is used as the official anthem

ECONOMY

Economy—overview: Vietnam is a densely-populated developing country that in the last 30 years has had to recover from the ravages of war, the loss of financial support from the old Soviet Bloc, and the rigidities of a centrally-planned economy. While Vietnam's economy remains dominated by state-owned enterprises (SOEs), which still produce about 40% of GDP, Vietnamese authorities have reaffirmed their commitment to economic liberalization and international integration. They have moved to implement the structural reforms needed to modernize the economy and to produce more competitive export-driven industries. Vietnam joined the WTO in January 2007 following more than a decade-long negotiation process. Vietnam became an official negotiating partner in the developing Trans-Pacific Partnership trade agreement in 2010. Agriculture's share of economic output has continued to shrink from about 25% in 2000 to about 20% in 2010, while industry's share increased from 36% to 41% in the same period. Deep poverty has declined significantly and Vietnam is working to create jobs to meet the challenge of a labor force that is growing by more than one million people every year. The global recession has hurt Vietnam's export-oriented economy, with GDP in 2009-10 growing less than the 7% per annum average achieved during the last decade. In 2010, exports increased by more than 25%, year-on-year, but the trade deficit remained high, prompting the government to consider administrative measures to limit the trade deficit. Vietnam's managed currency, the dong, continues to face downward pressure due to a persistent trade imbalance, and, since 2008, the government devalued it by 20% through a series of small devaluations. Foreign donors pledged nearly $8 billion in new development assistance for 2011. However, the government's strong growth-oriented economic policies have caused it to struggle to control one of the region's highest inflation rates, which reached 11.8% in 2010. Vietnam's economy also faces challenges from falling foreign exchange reserves, an undercapitalized banking sector, and high borrowing costs. The near-bankruptcy and subsequent default of the SOE Vinashin, a leading shipbuilder, led to a ratings downgrade of Vietnam's sovereign debt, exacerbating Vietnam's borrowing difficulties.
GDP (purchasing power parity): $276.6 billion (2010 est.)
country comparison to the world: 42
$259 billion (2009 est.)
$245.9 billion (2008 est.)
note: data are in 2010 US dollars
GDP (official exchange rate): $103.6 billion (2010 est.)
GDP—real growth rate: 6.8% (2010 est.)
country comparison to the world: 39
5.3% (2009 est.)
6.3% (2008 est.)
GDP—per capita (PPP): $3,100 (2010 est.)
country comparison to the world: 167
$2,900 (2009 est.)
$2,800 (2008 est.)
note: data are in 2010 US dollars
GDP—composition by sector:
agriculture: 20.6%
industry: 41.1%
services: 38.3% (2010 est.)
Labor force: 46.21 million (2010 est.)
country comparison to the world: 13
Labor force—by occupation:
agriculture: 53.9%
industry: 20.3%
services: 25.8% (2009)
Unemployment rate: 2.9% (2010 est.)
country comparison to the world: 24
6.5% (2009 est.)
Population below poverty line: 10.6% (2010 est.)
Household income or consumption by percentage share: *lowest 10%:* 3.1%
highest 10%: 29.8% (2006)
Distribution of family income—Gini index: 37 (2004)
country comparison to the world: 77
36.1 (1998)
Investment (gross fixed): 35.1% of GDP (2010 est.)
country comparison to the world: 8
Budget: *revenues:* $27.08 billion
expenditures: $29.65 billion (2010 est.)
Public debt: 56.7% of GDP (2010 est.)
country comparison to the world: 42
52.6% of GDP (2009 est.)
Inflation rate (consumer prices): 11.8% (2010 est.)
country comparison to the world: 205
6.5% (2009 est.)
Central bank discount rate: 7% (31 December 2010)
country comparison to the world: 72
6% (31 December 2009)
Commercial bank prime lending rate: 15.78% (31 December 2008)
country comparison to the world: 77
11.18% (31 December 2007)
Stock of narrow money: $33.76 billion (31 December 2010 est.)
country comparison to the world: 53
$31.75 billion (31 December 2009 est.)
Stock of broad money: $118.8 billion (31 December 2010 est.)
country comparison to the world: 47
$107.3 billion (31 December 2009 est.)
Stock of domestic credit: $132.1 billion (31 December 2010 est.)
country comparison to the world: 44
$114.6 billion (31 December 2009 est.)
Market value of publicly traded shares: $37 billion (31 December 2010 est.)
country comparison to the world: 61
$21.2 billion (31 December 2009)
$9.589 billion (31 December 2008)
Agriculture—products: paddy rice, coffee, rubber, cotton, tea, pepper, soybeans, cashews, sugar cane, peanuts, bananas; poultry; fish, seafood
Industries: food processing, garments, shoes, machine-building; mining, coal, steel; cement, chemical fertilizer, glass, tires, oil, paper
Industrial production growth rate: 14% (2010 est.)
country comparison to the world: 10
Electricity—production: 97.3 billion kWh (2010 est.)
country comparison to the world: 33
Electricity—consumption: 85.6 billion kWh (2010 est.)
country comparison to the world: 33
Electricity—exports: 535 million kWh (2009 est.)
Electricity—imports: 3.85 billion kWh (2009 est.)

Oil—production:
300,600 bbl/day (2010 est.)
country comparison to the world: 36
Oil—consumption:
311,400 bbl/day (2010 est.)
country comparison to the world: 41
Oil—exports: 171,500 bbl/day (2010 est.)
country comparison to the world: 54
Oil—imports: 182,300 bbl/day (2010 est.)
country comparison to the world: 48
Oil—proved reserves: 4.7 billion bbl (1 January 2010 est.)
country comparison to the world: 26
Natural gas—production: 9.4 billion cu m (2010 est.)
country comparison to the world: 42
Natural gas—consumption: 10.3 billion cu m (2010 est.)
country comparison to the world: 47
Natural gas—exports: 0 cu m (2010 est.)
country comparison to the world: 198
Natural gas—imports: 905,800 cu m (2010 est.)
country comparison to the world: 69
Natural gas—proved reserves: 680 billion cu m (1 January 2010 est.)
country comparison to the world: 31
Current account balance: $-12.22 billion (2010 est.)
country comparison to the world: 177
$-7.44 billion (2009 est.)
Exports: $72.03 billion (2010 est.)
country comparison to the world: 41
$57.1 billion (2009 est.)
Exports—commodities: clothes, shoes, marine products, crude oil, electronics, wooden products, rice, machinery
Exports—partners: US 20%, Japan 10.7%, China 9.8%, South Korea 4.3% (2010 est.)
Imports: $84.3 billion (2010 est.)
country comparison to the world: 35
$65.4 billion (2009 est.)
Imports—commodities: machinery and equipment, petroleum products, steel products, raw materials for the clothing and shoe industries, electronics, plastics, automobiles
Imports—partners: China 23.8%, South Korea 11.6%, Japan 10.8%, Taiwan 8.4%, Thailand 6.7%, Singapore 4.9% (2010 est.)
Reserves of foreign exchange and gold: $13 billion (31 December 2010 est.)
country comparison to the world: 53
$16.8 billion (31 December 2009 est.)
Debt—external: $33.45 billion (31 December 2010 est.)
country comparison to the world: 61
$27.84 billion (31 December 2009 est.)
Stock of direct foreign investment—at home:
$77.95 billion (31 December 2010 est.)
country comparison to the world: 45
$66.95 billion (31 December 2009 est.)
Stock of direct foreign investment—abroad: $7.7 billion (31 December 2009 est.)
country comparison to the world: 52
$5.3 billion (31 December 2008)
Exchange rates: dong (VND) per US dollar–19,148.9 (2010), 17,799.6 (2009), 16,548.3 (2008), 16,119 (2007), 15,983 (2006)

COMMUNICATIONS

Telephones—main lines in use:
17.427 million (2009)
country comparison to the world: 17
Telephones—mobile cellular: 98.224 million (2009)
country comparison to the world: 10
Telephone system: *general assessment:* Vietnam is putting considerable effort into modernization and expansion of its telecommunication system
domestic: all provincial exchanges are digitalized and connected to Hanoi, Da Nang, and Ho Chi Minh City by fiber-optic cable or microwave radio relay networks; main lines have been increased, and the use of mobile telephones is growing rapidly
international: country code–84; a landing point for the SEA-ME-WE-3, the C2C, and Thailand-Vietnam-Hong Kong submarine cable systems; the Asia-America Gateway submarine cable system, scheduled for completion by the end of 2008, will provide new access links to Asia and the US; satellite earth stations–2 Intersputnik (Indian Ocean region)
Broadcast media: government controls all broadcast media exercising oversight through the Ministry of Information and Communication (MIC); government-controlled national television provider, Vietnam Television (VTV), operates a network of 9 channels with several regional broadcasting centers; programming is relayed nationwide via a network of provincial and municipal TV stations; law limits access to satellite TV but many households are able to access foreign programming via home satellite equipment; government-controlled Voice of Vietnam, the national radio broadcaster, broadcasts on 6 channels and is repeated on AM, FM, and shortwave stations throughout Vietnam (2008)
Internet country code: .vn
Internet hosts: 129,318 (2010)
country comparison to the world: 73
Internet users: 23.382 million (2009)
country comparison to the world: 17

TRANSPORTATION

Airports: 44 (2010)
country comparison to the world: 98
Airports—with paved runways: *total:* 37
over 3,047 m: 9
2,438 to 3,047 m: 5
1,524 to 2,437 m: 14
914 to 1,523 m: 9 (2010)
Airports—with unpaved runways: *total:* 7
1,524 to 2,437 m: 1
914 to 1,523 m: 3
under 914 m: 3 (2010)
Heliports: 1 (2010)
Pipelines: condensate 28 km; condensate/gas 10 km; gas 216 km; refined products 206 km (2010)
Railways: *total:* 2,347 km
country comparison to the world: 65
standard gauge: 178 km 1.435-m gauge
narrow gauge: 2,169 km 1.000-m gauge (2008)
Roadways: *total:* 171,392 km
country comparison to the world: 29
paved: 125,789 km
unpaved: 45,603 km (2008)
Waterways: 17,702 km (5,000 km are navigable by vessels up to 1.8 m draft) (2011)
country comparison to the world: 7
Merchant marine: *total:* 537
country comparison to the world: 21
by type: barge carrier 1, bulk carrier 103, cargo 330, chemical tanker 24, container 20, liquefied gas 7, passenger 1, passenger/cargo 1, petroleum tanker 46, refrigerated cargo 2, roll on/roll off 1, specialized tanker 1
registered in other countries: 84 (Cambodia 1, Honduras 1, Liberia 3, Mongolia 34, Panama 37, Taiwan 1, Tuvalu 6, unknown 1) (2010)
Ports and terminals: Cam Pha Port, Da Nang, Hai Phong, Ho Chi Minh City, Phu My, Quy Nhon
Transportation—note: the International Maritime Bureau reports the territorial and offshore waters in the South China Sea as high risk for piracy and armed robbery against ships; numerous commercial vessels have been attacked and hijacked both at anchor and while underway; hijacked vessels are often disguised and cargo diverted to ports in East Asia; crews have been murdered or cast adrift

MILITARY

Military branches: People's Armed Forces: People's Army of Vietnam (PAVN; includes People's Navy Command (with Naval Infantry, Coast Guard), Air and Air Defense Force (Khong Quan Nhan Dan), Border Defense Command), People's Public Security Forces, Militia Force, Self-Defense Forces (2010)
Military service age and obligation: 18 years of age for male compulsory military service; females may volunteer for active duty military service; conscript service obligation–2 years (3 to 4 years in the navy); 18-45 years of age (male) or 18-40 years of age (female) for Militia Force or Self Defense Forces (2006)
Manpower available for military service:
males age 16–49: 25,649,738
females age 16–49: 24,995,692 (2010 est.)
Manpower fit for military service: *males age 16–49:* 20,405,847
females age 16–49: 21,098,102 (2010 est.)
Manpower reaching militarily significant age annually: *male:* 847,743
female: 787,341 (2010 est.)
Military expenditures:
2.5% of GDP (2005 est.)
country comparison to the world: 60

TRANSNATIONAL ISSUES

Disputes—international: southeast Asian states have enhanced border surveillance to check the spread of avian flu; Cambodia and Laos protest Vietnamese squatters and armed encroachments along border; Cambodia accuses Vietnam of a wide variety of illicit cross-border activities; Progress on a joint development area with Cambodia is hampered by an unresolved dispute over sovereignty of offshore islands; an estimated 300,000 Vietnamese refugees reside in China; establishment of a maritime boundary with Cambodia is hampered by unresolved dispute over the sovereignty of offshore islands; the decade-long demarcation of the China-Vietnam land boundary was completed in 2009; China occupies the Paracel Islands also claimed by Vietnam and Taiwan; Brunei claims a maritime boundary extending beyond as far as a median with Vietnam, thus asserting an implicit claim to Lousia Reef; the 2002 "Declaration on the Conduct of Parties in the South China Sea" has eased tensions but falls short of a legally binding "code of conduct" desired by several of the disputants; Vietnam continues to expand construction of facilities in the Spratly Islands; in March 2005, the national oil companies of China, the Philippines, and Vietnam signed a joint accord to conduct marine seismic activities in the Spratly Islands

Illicit drugs: minor producer of opium poppy; probable minor transit point for Southeast Asian heroin; government continues to face domestic opium/heroin/methamphetamine addiction problems despite longstanding crackdowns

VIRGIN ISLANDS

(TERRITORY OF THE US)

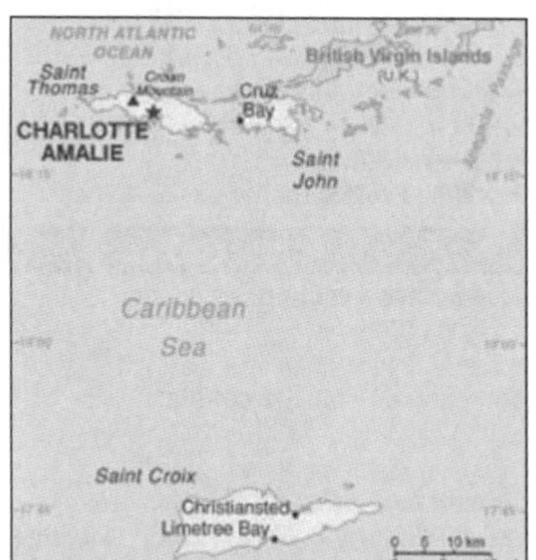

INTRODUCTION

Background: During the 17th century, the archipelago was divided into two territorial units, one English and the other Danish. Sugarcane, produced by slave labor, drove the islands' economy during the 18th and early 19th centuries. In 1917, the US purchased the Danish portion, which had been in economic decline since the abolition of slavery in 1848.

GEOGRAPHY

Location: Caribbean, islands between the Caribbean Sea and the North Atlantic Ocean, east of Puerto Rico
Geographic coordinates: 18 20 N, 64 50 W
Map references: Central America and the Caribbean
Area: *total:* 1,910 sq km
country comparison to the world: 181
land: 346 sq km
water: 1,564 sq km
Area—comparative: twice the size of Washington, DC
Land boundaries: 0 km
Coastline: 188 km
Maritime claims: territorial sea: 12 nm
exclusive economic zone: 200 nm
Climate: subtropical, tempered by easterly trade winds, relatively low humidity, little seasonal temperature variation; rainy season September to November
Terrain: mostly hilly to rugged and mountainous with little level land
Elevation extremes:
lowest point: Caribbean Sea 0 m
highest point: Crown Mountain 474 m
Natural resources: sun, sand, sea, surf
Land use: *arable land:* 5.71%
permanent crops: 2.86%
other: 91.43% (2005)
Irrigated land: NA
Natural hazards: several hurricanes in recent years; frequent and severe droughts and floods; occasional earthquakes
Environment—current issues: lack of natural freshwater resources
Geography—note: important location along the Anegada Passage–a key shipping lane for the Panama Canal; Saint Thomas has one of the best natural deepwater harbors in the Caribbean

PEOPLE

Population: 109,666 (July 2011 est.)
country comparison to the world: 187
Age structure: *0–14 years:* 19.5% (male 10,886/female 10,470)
15–64 years: 65.2% (male 33,280/female 38,210)
65 years and over: 15.3% (male 7,626/female 9,194) (2011 est.)
Median age: *total:* 40.4 years
male: 39.9 years
female: 40.8 years (2011 est.)
Population growth rate: -0.08% (2011 est.)
country comparison to the world: 204
Birth rate: 11.41 births/1,000 population (2011 est.)
country comparison to the world: 171
Death rate: 7.17 deaths/1,000 population (July 2011 est.)
country comparison to the world: 127
Net migration rate: -5.04 migrant(s)/1,000 population (2011 est.)
country comparison to the world: 192
Urbanization: *urban population:* 95% of total population (2010)
rate of urbanization: -0.1% annual rate of change (2010-15 est.)
Major cities—population: CHARLOTTE AMALIE (capital) 54,000 (2009)
Sex ratio: *at birth:* 1.058 male(s)/female
under 15 years: 1.04 male(s)/female
15–64 years: 0.87 male(s)/female
65 years and over: 0.83 male(s)/female
total population: 0.9 male(s)/female (2011 est.)
Infant mortality rate:
total: 7.24 deaths/1,000 live births
country comparison to the world: 165
male: 7.97 deaths/1,000 live births
female: 6.47 deaths/1,000 live births (2011 est.)
Life expectancy at birth: *total population:* 79.33 years
country comparison to the world: 38
male: 76.29 years
female: 82.55 years (2011 est.)
Total fertility rate: 1.8 children born/woman (2011 est.)
country comparison to the world: 154
HIV/AIDS—adult prevalence rate: NA
HIV/AIDS—people living with HIV/AIDS: NA
HIV/AIDS—deaths: NA
Nationality: *noun:* Virgin Islander(s) (US citizens)
adjective: Virgin Islander
Ethnic groups: black 76.2%, white 13.1%, Asian 1.1%, other 6.1%, mixed 3.5% (2000 census)
Religions: Baptist 42%, Roman Catholic 34%, Episcopalian 17%, other 7%
Languages: English 74.7%, Spanish or Spanish Creole 16.8%, French or French Creole 6.6%, other 1.9% (2000 census)
Literacy: *definition:* age 15 and over can read and write
total population: 90-95% est.
male: NA
female: NA (2005 est.)
School life expectancy (primary to tertiary education): NA
Education expenditures: NA

GOVERNMENT

Country name: *conventional long form:* United States Virgin Islands
conventional short form: Virgin Islands
former: Danish West Indies
abbreviation: USVI
Dependency status: organized, unincorporated territory of the US with policy relations between the Virgin Islands and the US under the jurisdiction of the Office of Insular Affairs, US Department of the Interior
Government type: NA
Capital: *name:* Charlotte Amalie
geographic coordinates: 18 21 N, 64 56 W
time difference: UTC-4 (1 hour ahead of Washington, DC during Standard Time)
Administrative divisions: none (territory of the US); there are no first-order administrative divisions as defined by the US Government, but there are three islands at the second order; Saint Croix, Saint John, Saint Thomas
Independence: none (territory of the US)
National holiday: Transfer Day (from Denmark to the US), 31 March (1917)
Constitution: Revised Organic Act of 22 July 1954
Legal system: US common law
Suffrage: 18 years of age; universal; island residents are US citizens but do not vote in US presidential elections
Executive branch: *chief of state:* President Barack H. OBAMA (since 20 January 2009); Vice President Joseph R. BIDEN (since 20 January 2009)
head of government: Governor John DeJONGH (since 1 January 2007)

cabinet: NA
(For more information visit the World Leaders website)
elections: under the US Constitution, residents of unincorporated territories, such as the Virgin Islands, do not vote in elections for US president and vice president; however, they may vote in the Democratic and Republican presidential primary elections; governor and lieutenant governor elected on the same ticket by popular vote for four-year terms (eligible for a second term); election last held on 2 November 2010 (next to be held in November 2014)
election results: John DeJONGH reelected governor; percent of vote–John DeJONGH 56.3%, Kenneth MAPP 43.6%
Legislative branch: unicameral Senate (15 seats; members are elected by popular vote to serve two-year terms)
elections: last held on 2 November 2010 (next to be held in November 2012)
election results: percent of vote by party–NA; seats by party–Democratic Party 10, ICM 2, independent 3
note: the Virgin Islands elects one non-voting representative to the US House of Representatives; election last held 2 November 2010 (next to be held on November 2012)
Judicial branch: US District Court of the Virgin Islands (under Third Circuit jurisdiction); Superior Court of the Virgin Islands (judges appointed by the governor for 10-year terms)
Political parties and leaders: Democratic Party [Arturo WATLINGTON]; Independent Citizens' Movement or ICM [Usie RICHARDS]; Republican Party [Gary SPRAUVE]
Political pressure groups and leaders: NA
International organization participation: IOC, UPU, WFTU
Diplomatic representation in the US: none (territory of the US)
Diplomatic representation from the US: none (territory of the US)
Flag description: white field with a modified US coat of arms in the center between the large blue initials V and I; the coat of arms shows a yellow eagle holding an olive branch in its right talon and three arrows in the left with a superimposed shield of seven red and six white vertical stripes below a blue panel; white is a symbol of purity, the letters stand for the Virgin Islands
National anthem: *name:* "Virgin Islands March"
lyrics/music: multiple/Alton Augustus ADAMS, Sr.
note: adopted 1963; serves as a local anthem; as a territory of the United States, "The Star-Spangled Banner" is official (see United States)

ECONOMY

Economy—overview: Tourism is the primary economic activity, accounting for 80% of GDP and employment. The islands hosted 2.4 million visitors in 2008. The manufacturing sector consists of petroleum refining, rum distilling, textiles, electronics, pharmaceuticals, and watch assembly. One of the world's largest petroleum refineries is at Saint Croix. The agricultural sector is small, with most food being imported. International business and financial services are small but growing components of the economy. The islands are vulnerable to substantial damage from storms. The government is working to improve fiscal discipline, to support construction projects in the private sector, to expand tourist facilities, to reduce crime, and to protect the environment.
GDP (purchasing power parity): $1.577 billion (2004 est.)
country comparison to the world: 193
GDP (official exchange rate): $NA
GDP—real growth rate: 2% (2002 est.)
country comparison to the world: 147
GDP—per capita (PPP): $14,500 (2004 est.)
country comparison to the world: 79
GDP—composition by sector:
agriculture: 1%
industry: 19%
services: 80% (2003 est.)
Labor force: 49,820 (2007 est.)
country comparison to the world: 190
Labor force—by occupation: *agriculture:* 1%
industry: 19%
services: 80% (2003 est.)
Unemployment rate: 6.2% (2004)
country comparison to the world: 60
Population below poverty line: 28.9% (2002)
Household income or consumption by percentage share: *lowest 10%:* NA%
highest 10%: NA%
Budget: *revenues:* $837 million
expenditures: $837 million (FY08/09)
Inflation rate (consumer prices): 2.2% (2003)
country comparison to the world: 56
Agriculture—products: fruit, vegetables, sorghum; Senepol cattle
Industries: tourism, petroleum refining, watch assembly, rum distilling, construction, pharmaceuticals, textiles, electronics
Industrial production growth rate: NA%
Electricity—production: 776.4 million kWh (2007 est.)
country comparison to the world: 151
Electricity—consumption: 722 million kWh (2007 est.)
country comparison to the world: 150
Electricity—exports: 0 kWh (2008 est.)
Electricity—imports: 0 kWh (2008 est.)
Oil—production: 16,870 bbl/day (2009 est.)
country comparison to the world: 76
Oil—consumption: 88,820 bbl/day (2009 est.)
country comparison to the world: 81
Oil—exports: 388,000 bbl/day (2007 est.)
country comparison to the world: 32
Oil—imports: 480,600 bbl/day (2007 est.)
country comparison to the world: 26
Oil—proved reserves: 0 bbl (1 January 2010 est.)
country comparison to the world: 202
Natural gas—production: 0 cu m (2008 est.)
country comparison to the world: 199
Natural gas—consumption: 0 cu m (2008 est.)
country comparison to the world: 135
Natural gas—exports: 0 cu m (2008 est.)
country comparison to the world: 199
Natural gas—imports: 0 cu m (2008 est.)
country comparison to the world: 135
Natural gas—proved reserves: 0 cu m (1 January 2010 est.)
country comparison to the world: 201
Exports: $4.234 billion (2001)
country comparison to the world: 115
Exports—commodities: refined petroleum products
Imports: $4.609 billion (2001)
country comparison to the world: 122
Imports—commodities: crude oil, foodstuffs, consumer goods, building materials
Debt—external: $NA
Exchange rates: the US dollar is used

COMMUNICATIONS

Telephones—main lines in use: 75,000 (2009)
country comparison to the world: 151
Telephones—mobile cellular: 80,300 (2005)
country comparison to the world: 190
Telephone system: *general assessment:* modern system with total digital switching, uses fiber-optic cable and microwave radio relay
domestic: full range of services available
international: country code–1-340; submarine cable connections to US, the Caribbean, Central and South America; satellite earth stations–NA
Broadcast media: about a dozen television broadcast stations including 1 public TV station; multi-channel cable and satellite TV services are available; 24 radio stations broadcasting (2009)
Internet country code: .vi
Internet hosts: 8,933 (2010)
country comparison to the world: 129
Internet users: 30,000 (2009)
country comparison to the world: 180

TRANSPORTATION

Airports: 2 (2010)
country comparison to the world: 207
Airports—with paved runways: *total:* 2
over 3,047 m: 1
1,524 to 2,437 m: 1 (2010)
Roadways: *total:* 1,260 km (2008)
country comparison to the world: 179
Ports and terminals: Charlotte Amalie, Christiansted, Cruz Bay, Frederiksted, Limetree Bay

MILITARY

Manpower fit for military service: *males age 16–49:* 17,542
females age 16–49: 20,946 (2010 est.)
Manpower reaching militarily significant age annually: *male:* 744
female: 788 (2010 est.)
Military—note: defense is the responsibility of the US

TRANSNATIONAL ISSUES

Disputes—international: none

WAKE ISLAND

(TERRITORY OF THE US)

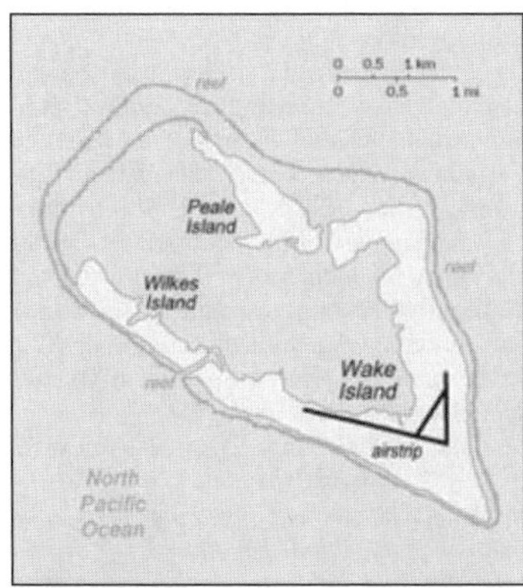

INTRODUCTION

Background: The US annexed Wake Island in 1899 for a cable station. An important air and naval base was constructed in 1940-41. In December 1941, the island was captured by the Japanese and held until the end of World War II. In subsequent years, Wake was developed as a stopover and refueling site for military and commercial aircraft transiting the Pacific. Since 1974, the island's airstrip has been used by the US military, as well as for emergency landings. Operations on the island were suspended and all personnel evacuated in 2006 with the approach of super typhoon IOKE (category 5), but resultant damage was comparatively minor. A US Air Force repair team restored full capability to the airfield and facilities, which remains a vital strategic link in the Pacific region.

GEOGRAPHY

Location: Oceania, atoll in the North Pacific Ocean, about two-thirds of the way from Hawaii to the Northern Mariana Islands
Geographic coordinates: 19 17 N, 166 39 E
Map references: Oceania
Area: *total:* 6.5 sq km
country comparison to the world: 241
land: 6.5 sq km
water: 0 sq km
Area—comparative: about 11 times the size of The Mall in Washington, DC
Land boundaries: 0 km
Coastline: 19.3 km
Maritime claims: territorial sea: 12 nm
exclusive economic zone: 200 nm
Climate: tropical
Terrain: atoll of three low coral islands, Peale, Wake, and Wilkes, built up on an underwater volcano; central lagoon is former crater, islands are part of the rim
Elevation extremes: *lowest point:* Pacific Ocean 0 m
highest point: unnamed location 6 m
Natural resources: none
Land use: *arable land:* 0%
permanent crops: 0%
other: 100% (2005)
Irrigated land: 0 sq km
Natural hazards: occasional typhoons
Environment—current issues: NA
Geography—note: strategic location in the North Pacific Ocean; emergency landing location for transpacific flights

PEOPLE

Population: no indigenous inhabitants
note: approximately 150 military personnel and civilian contractors maintain and operate the airfield and communications facilities (2009)
School life expectancy (primary to tertiary education): NA
Education expenditures: NA

GOVERNMENT

Country name:
conventional long form: none
conventional short form: Wake Island
Dependency status: unorganized, unincorporated territory of the US; administered from Washington, DC by the Department of the Interior; activities in the atoll are currently conducted by the US Air Force
Legal system: US common law
Flag description: the flag of the US is used

ECONOMY

Economy—overview: Economic activity is limited to providing services to military personnel and contractors located on the island. All food and manufactured goods must be imported.
Electricity—production: NA kWh

COMMUNICATIONS

Telephone system: *general assessment:* satellite communications; 2 DSN circuits off the Overseas Telephone System (OTS); located in the Hawaii area code–808
domestic: NA
international: NA
Broadcast media: American Armed Forces Radio and Television Service (AFRTS) provides satellite radio/television broadcasts (2009)

TRANSPORTATION

Airports: 1 (2010)
country comparison to the world: 211
Airports—with paved runways: *total:* 1
2,438 to 3,047 m: 1 (2010)
Ports and terminals: none; two offshore anchorages for large ships
Transportation—note: there are no commercial or civilian flights to and from Wake Island, except in direct support of island missions; emergency landing is available

MILITARY

Military—note: defense is the responsibility of the US; the US Air Force is responsible for overall administration and operation of the island facilities; the launch support facility is administered by the US Missile Defense Agency (MDA)

TRANSNATIONAL ISSUES

Disputes—international: claimed by Marshall Islands

WALLIS AND FUTUNA

(OVERSEAS COLLECTIVITY OF FRANCE)

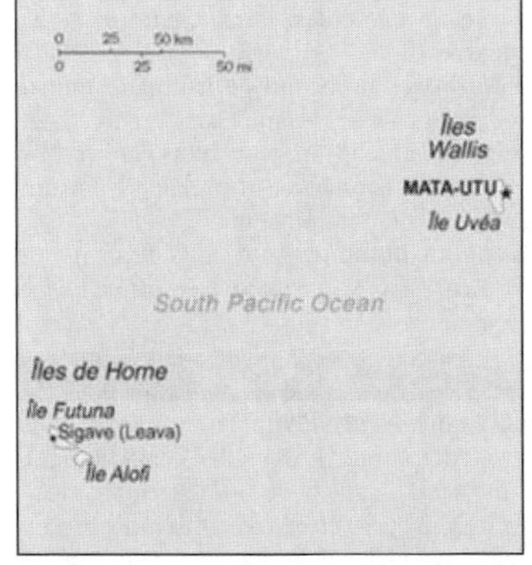

INTRODUCTION

Background: The Futuna island group was discovered by the Dutch in 1616 and Wallis by the British in 1767, but it was the French who declared a protectorate over the islands in 1842, and took official control of them between 1886 and 1888. Notably, Wallis and Futuna was the only French colony to side with the Vichy regime during World War II, a phase that ended in May of 1942 with the arrival of 2,000 American troops. In 1959, the inhabitants of the islands voted to become a French overseas territory and officially assumed this status in July 1961.

GEOGRAPHY

Location: Oceania, islands in the South Pacific Ocean, about two-thirds of the way from Hawaii to New Zealand
Geographic coordinates: 13 18 S, 176 12 W
Map references: Oceania
Area: *total:* 142 sq km
country comparison to the world: 219
land: 142 sq km
water: 0 sq km
note: includes Ile Uvea (Wallis Island), Ile Futuna (Futuna Island), Ile Alofi, and 20 islets
Area—comparative: 1.5 times the size of Washington, DC
Land boundaries: 0 km

Coastline: 129 km
Maritime claims: territorial sea: 12 nm
exclusive economic zone: 200 nm
Climate: tropical; hot, rainy season (November to April); cool, dry season (May to October); rains 2,500-3,000 mm per year (80% humidity); average temperature 26.6 degrees C
Terrain: volcanic origin; low hills
Elevation extremes: *lowest point:* Pacific Ocean 0 m
highest point: Mont Singavi (on Futuna) 765 m
Natural resources: NEGL
Land use: *arable land:* 7.14%
permanent crops: 35.71%
other: 57.15% (2005)
Irrigated land: NA
Natural hazards: NA
Environment—current issues: deforestation (only small portions of the original forests remain) largely as a result of the continued use of wood as the main fuel source; as a consequence of cutting down the forests, the mountainous terrain of Futuna is particularly prone to erosion; there are no permanent settlements on Alofi because of the lack of natural freshwater resources
Geography—note: both island groups have fringing reefs

PEOPLE

Population: 15,398 (July 2011 est.)
country comparison to the world: 220
Age structure: *0–14 years:* 25.1% (male 2,018/female 1,846)
15–64 years: 67.1% (male 5,191/female 5,137)
65 years and over: 7.8% (male 563/female 643) (2011 est.)
Median age: *total:* 28.5 years
male: 27.5 years
female: 29.7 years (2011 est.)
Population growth rate: 0.357% (2011 est.)
country comparison to the world: 165
Birth rate: 13.96 births/1,000 population NA (2011 est.)
country comparison to the world: 148
Death rate: 4.68 deaths/1,000 population NA (July 2011 est.)
country comparison to the world: 197
Net migration rate: -5.72 migrant(s)/1,000 population
country comparison to the world: 197
note: there has been steady emigration from Wallis and Futuna to New Caledonia (2011 est.)
Urbanization: *urban population:* 0% of total population (2008)
rate of urbanization: 0% annual rate of change (2005-10 est.)
Sex ratio: *at birth:* 1.048 male(s)/female
under 15 years: 1.1 male(s)/female
15–64 years: 1.01 male(s)/female
65 years and over: 0.85 male(s)/female
total population: 1.02 male(s)/female (2011 est.)
Infant mortality rate:
total: 4.67 deaths/1,000 live births
country comparison to the world: 187
male: 4.91 deaths/1,000 live births
female: 4.42 deaths/1,000 live births (2011 est.)
Life expectancy at birth: *total population:* 78.98 years
country comparison to the world: 42
male: 76 years
female: 82.11 years (2011 est.)
Total fertility rate: 1.81 children born/woman (2011 est.)
country comparison to the world: 151
HIV/AIDS—adult prevalence rate: NA
HIV/AIDS—people living with HIV/AIDS: NA
HIV/AIDS—deaths: NA
Drinking water source: *Improved:*
rural: 100% of population
total: 100% of population (2008)
Sanitation facility access: *Improved:*
rural: 96% of population
total: 96% of population
Unimproved:
rural: 4% of population
total: 4% of population (2008)
Nationality: *noun:* Wallisian(s), Futunan(s), or Wallis and Futuna Islanders
adjective: Wallisian, Futunan, or Wallis and Futuna Islander
Ethnic groups: Polynesian
Religions: Roman Catholic 99%, other 1%
Languages: Wallisian (indigenous Polynesian language) 58.9%, Futunian 30.1%, French (official) 10.8%, other 0.2% (2003 census)
Literacy: *definition:* age 15 and over can read and write
total population: 50%
male: 50%
female: 50% (1969 est.)
School life expectancy (primary to tertiary education): NA
Education expenditures: NA

GOVERNMENT

Country name: *conventional long form:* Territory of the Wallis and Futuna Islands
conventional short form: Wallis and Futuna
local long form: Territoire des Iles Wallis et Futuna
local short form: Wallis et Futuna
Dependency status: overseas territory of France
Government type: NA
Capital: *name:* Mata-Utu (on Ile Uvea)
geographic coordinates: 13 57 S, 171 56 W
time difference: UTC+12 (17 hours ahead of Washington, DC during Standard Time)
Administrative divisions: none (overseas territory of France); there are no first-order administrative divisions as defined by the US Government, but there are three kingdoms at the second order named Alo, Sigave, Wallis
Independence: none (overseas territory of France)
National holiday:
Bastille Day, 14 July (1789)
Constitution: 4 October 1958 (French Constitution)
Legal system: French civil law
Suffrage: 18 years of age; universal
Executive branch: *chief of state:* President Nicolas SARKOZY (since 16 May 2007); represented by High Administrator Michel JEANJEAN (since 10 June 2010)
head of government: President of the Territorial Assembly Victor BRIAL (since 11 December 2007)
cabinet: Council of the Territory consists of 3 kings and 3 members appointed by the high administrator on the advice of the Territorial Assembly
(For more information visit the World Leaders website)
note: there are 3 traditional kings with limited powers
elections: French president elected by popular vote for a five-year term; high administrator appointed by the French president on the advice of the French Ministry of the Interior; the presidents of the Territorial Government and the Territorial Assembly elected by the members of the assembly
Legislative branch: unicameral Territorial Assembly or Assemblee Territoriale (20 seats; members elected by popular vote to serve five-year terms)
elections: last held on 1 April 2007 (next to be held in April 2012)
election results: percent of vote by party–NA; seats by party–UMP 13, other 7
note: Wallis and Futuna elects one senator to the French Senate and one deputy to the French National Assembly; French Senate–elections last held on 21 September 2008 (next to be held by September 2014); results–percent of vote by party–NA; seats–UMP 1; French National Assembly–elections last held on 17 June 2007 (next to be held by 2012); results–percent of vote by party–NA; seats–PS 1
Judicial branch: justice generally administered under French law by the high administrator, but the three traditional kings administer customary law and there is a magistrate in Mata-Utu; a court of appeal is located in Noumea, New Caledonia
Political parties and leaders: Lua Kae Tahi (Giscardians); Mouvement des Radicaux de Gauche or MRG; Rally for the Republic or RPR (UMP); Socialist Party or PS; Taumu'a Lelei; Union Populaire Locale or UPL; Union Pour la Democratie Francaise or UDF
Political pressure groups and leaders: NA
International organization participation: PIF (observer), SPC, UPU
Diplomatic representation in the US: none (overseas territory of France)
Diplomatic representation from the US: none (overseas territory of France)
Flag description: unofficial, local flag has a red field with four white isosceles triangles in the middle, representing the three native kings of the islands and the French administrator; the apexes of the triangles are oriented inward and at right angles to each other; the flag of France, outlined in white on two sides, is in the upper hoist quadrant
note: the design is derived from an original red banner with a white cross pattee that was introduced in the 19th century by French missionaries; the flag of France used for official occasions
National anthem: note: as a territory of France, "La Marseillaise" is official (see France)

ECONOMY

Economy—overview: The economy is limited to traditional subsistence agriculture, with about 80% of labor force earnings from agriculture (coconuts and vegeta-

bles), livestock (mostly pigs), and fishing. About 4% of the population is employed in government. Revenues come from French Government subsidies, licensing of fishing rights to Japan and South Korea, import taxes, and remittances from expatriate workers in New Caledonia.
GDP (purchasing power parity): $60 million (2004 est.)
country comparison to the world: 220
GDP (official exchange rate): $NA
GDP—real growth rate: NA%
GDP—per capita (PPP): $3,800 (2004 est.)
country comparison to the world: 160
GDP—composition by sector:
agriculture: NA%
industry: NA%
services: NA%
Labor force: 3,104 (2003)
country comparison to the world: 222
Labor force—by occupation:
agriculture: 80%
industry: 4%
services: 16% (2001 est.)
Unemployment rate: 15.2% (2003)
country comparison to the world: 153
Population below poverty line: NA%
Household income or consumption by percentage share: *lowest 10%:* NA%
highest 10%: NA%
Budget: *revenues:* $29,730
expenditures: $31,330 (2004)
Public debt: 5.6% of GDP (2004 est.)
country comparison to the world: 128
Inflation rate (consumer prices): 2.8% (2005)
country comparison to the world: 80
Agriculture—products: coconuts, breadfruit, yams, taro, bananas; pigs, goats; fish
Industries: copra, handicrafts, fishing, lumber
Industrial production growth rate: NA%
Electricity—production: NA kWh
Electricity—consumption: NA kWh
Electricity—exports: 0 kWh (2002)
Electricity—imports: 0 kWh (2002)
Exports: $47,450 (2004)
country comparison to the world: 221
Exports—commodities: copra, chemicals, construction materials
Imports: $61.17 million (2004)
country comparison to the world: 216
Imports—commodities: chemicals, machinery, consumer goods
Debt—external: $3.67 million (2004)
country comparison to the world: 193
Exchange rates: Comptoirs Francais du Pacifique francs (XPF) per US dollar—87.59 (2007), 94.97 (2006), 95.89 (2005), 96.04 (2004), 105.66 (2003)

COMMUNICATIONS

Telephones—main lines in use: 3,000 (2009)
country comparison to the world: 217
Telephones—mobile cellular: NA
Telephone system: *general assessment:* NA
domestic: NA
international: country code—681
Broadcast media: the publicly-owned French Overseas Network (RFO), which broadcasts to France's overseas departments and territories, is carried on the RFO Wallis and Fortuna television and radio stations (2008)
Internet country code: .wf
Internet hosts: 1,734 (2010)
country comparison to the world: 156
Internet users: 1,300 (2009)
country comparison to the world: 211

TRANSPORTATION

Airports: 2 (2010)
country comparison to the world: 202
Airports—with paved runways: *total:* 2
1,524 to 2,437 m: 1
914 to 1,523 m: 1 (2010)
Merchant marine: *total:* 8
country comparison to the world: 122
by type: chemical tanker 2, passenger 6
foreign-owned: 8 (France 6, French Polynesia 2) (2008)
Ports and terminals: Leava, Mata-Utu

MILITARY

Manpower fit for military service: *males age 16–49:* 3,376
females age 16–49: 3,314 (2010 est.)
Manpower reaching militarily significant age annually: *male:* 168
female: 139 (2010 est.)
Military—note: defense is the responsibility of France

TRANSNATIONAL ISSUES

Disputes—international: none

WEST BANK

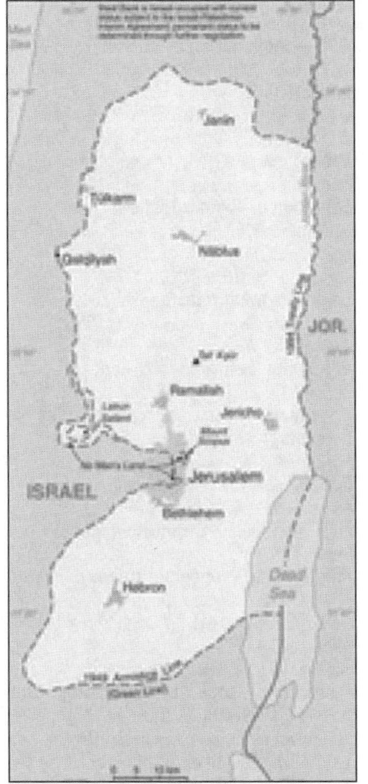

INTRODUCTION

Background: The September 1993 Israel-PLO Declaration of Principles on Interim Self-Government Arrangements provided for a transitional period of Palestinian self-rule in the West Bank and the Gaza Strip. Under a series of agreements signed between May 1994 and September 1999, Israel transferred to the Palestinian Authority (PA) security and civilian responsibility for many Palestinian-populated areas of the West Bank and Gaza Strip. Negotiations to determine the permanent status of the West Bank and Gaza Strip stalled following the outbreak of an intifada in September 2000. In April 2003, the Quartet (US, EU, UN, and Russia) presented a roadmap to a final settlement of the conflict by 2005 based on reciprocal steps by the two parties leading to two states, Israel and a democratic Palestine. Following Palestinian leader Yasir ARAFAT's death in late 2004, Mahmud ABBAS was elected PA president in January 2005. A month later, Israel and the PA agreed to the Sharm el-Sheikh Commitments in an effort to move the peace process forward. In September 2005, Israel unilaterally withdrew all its settlers and soldiers and dismantled its military facilities in the Gaza Strip and withdrew settlers and redeployed soldiers from four small northern West Bank settlements. Nonetheless, Israel still controls maritime, airspace, and most access to the Gaza Strip. In January 2006, the Islamic Resistance Movement, HAMAS, won control of the Palestinian Legislative Council (PLC). HAMAS took control of the PA government in March 2006, but President ABBAS had little success negotiating with HAMAS to present a political platform acceptable to the international community so as to lift economic sanctions on Palestinians. Violent clashes between Fatah and HAMAS supporters in the Gaza Strip in 2006 and early 2007 resulted in numerous Palestinian deaths and injuries. In February 2007, ABBAS and HAMAS Political Bureau Chief MISHAL signed the Mecca Agreement in Saudi Arabia that resulted in the formation of a Palestinian National Unity Government (NUG) headed by HAMAS member Ismail HANIYA. However, fighting continued in the Gaza Strip, and in June 2007, HAMAS militants succeeded in a violent takeover of all military and governmental institutions in the Gaza Strip. ABBAS dismissed the NUG and through a series of presidential decrees formed a PA government in the West Bank led by independent Salam FAYYAD. HAMAS rejected the NUG's dismissal, and despite multiple rounds of Egyptian-brokered reconciliation negotiations, the two groups have failed to bridge their differences. The status quo remains with HAMAS in control of the Gaza Strip and ABBAS and the Fatah-dominated PA governing the West Bank. FAYYAD and his PA government continue to implement a series of security and economic reforms to improve conditions in the West Bank. ABBAS has said he will not resume negotiations with current Prime Minister NETANYAHU until Israel halts all settlement activity in the West Bank and Jerusalem.

GEOGRAPHY

Location: Middle East, west of Jordan, east of Israel
Geographic coordinates: 32 00 N, 35 15 E
Map references: Middle East
Area: *total:* 5,860 sq km
country comparison to the world: 171
land: 5,640 sq km
water: 220 sq km
note: includes West Bank, Latrun Salient, and the northwest quarter of the Dead Sea, but excludes Mt. Scopus; East Jerusalem and Jerusalem No Man's Land are also included only as a means of depicting the entire area occupied by Israel in 1967
Area—comparative: slightly smaller than Delaware
Land boundaries: *total:* 404 km
border countries: Israel 307 km, Jordan 97 km
Coastline: 0 km (landlocked)
Maritime claims: none (landlocked)
Climate: temperate; temperature and precipitation vary with altitude, warm to hot summers, cool to mild winters
Terrain: mostly rugged dissected upland, some vegetation in west, but barren in east
Elevation extremes:
lowest point: Dead Sea -408 m
highest point: Tall Asur 1,022 m
Natural resources: arable land
Land use: *arable land:* 16.9%
permanent crops: 18.97%
other: 64.13% (2001)
Irrigated land: 150 sq km; note—includes Gaza Strip (2003)
Natural hazards: droughts
Environment—current issues: adequacy of freshwater supply; sewage treatment
Geography—note: landlocked; highlands are main recharge area for Israel's coastal aquifers; there are about 355 Israeli civilian sites including about 145 small outpost communities in the West Bank and 32 sites in East Jerusalem (2010 est.)

PEOPLE

Population: 2,568,555 (July 2010 est.)
country comparison to the world: 140
note: approximately 296,700 Israeli settlers live in the West Bank (2009 est.); approximately 192,800 Israeli settlers live in East Jerusalem (2008 est.) (July 2011 est.)
Age structure: *0–14 years:* 35.8% (male 471,908/female 447,816)
15–64 years: 60.5% (male 796,421/female 757,227)
65 years and over: 3.7% (male 39,546/female 55,637) (2011 est.)
Median age: *total:* 21.3 years
male: 21.1 years
female: 21.5 years (2011 est.)
Population growth rate: 2.097% (2011 est.)
country comparison to the world: 42
Birth rate: 24.56 births/1,000 population (2011 est.)
country comparison to the world: 65
Death rate: 3.58 deaths/1,000 population (July 2011 est.)
country comparison to the world: 211
Net migration rate: 0 migrant(s)/1,000 population (2011 est.)
country comparison to the world: 112
Urbanization: *urban population:* 72% of total population (2008)
rate of urbanization: 3.3% annual rate of change (2005-10 est.)
Sex ratio: *at birth:* 1.06 male(s)/female
under 15 years: 1.05 male(s)/female
15–64 years: 1.05 male(s)/female
65 years and over: 0.71 male(s)/female
total population: 1.04 male(s)/female (2011 est.)
Infant mortality rate: *total:* 14.92 deaths/1,000 live births
country comparison to the world: 120
male: 16.73 deaths/1,000 live births
female: 13.01 deaths/1,000 live births (2011 est.)
Life expectancy at birth: *total population:* 75.01 years
country comparison to the world: 89
male: 72.97 years
female: 77.17 years (2011 est.)
Total fertility rate: 3.05 children born/woman (2011 est.)
country comparison to the world: 63
HIV/AIDS—adult prevalence rate: NA
HIV/AIDS—people living with HIV/AIDS: NA
HIV/AIDS—deaths: NA
Nationality: *noun:* NA
adjective: NA
Ethnic groups: Palestinian Arab and other 83%, Jewish 17%
Religions: Muslim 75% (predominantly Sunni), Jewish 17%, Christian and other 8%
Languages: Arabic, Hebrew (spoken by Israeli settlers and many Palestinians), English (widely understood)
Literacy: *definition:* age 15 and over can read and write
total population: 92.4%
male: 96.7%
female: 88% (2004 est.)
School life expectancy (primary to tertiary education): *total:* 14 years
male: 13 years
female: 14 years (2006)
Education expenditures: NA

GOVERNMENT

Country name: *conventional long form:* none
conventional short form: West Bank

ECONOMY

Economy—overview: The West Bank—the larger of the two areas comprising the Palestinian territories—experienced a high single-digit economic growth rate in 2010 as a result of inflows of donor aid, the Palestinian Authority's (PA) implementation of economic and security reforms, and the easing of some movement and access restrictions by the Israeli Government. Nevertheless, overall standard-of-living measures remain near levels seen prior to the start of the second intifada in 2000. The almost decade-long downturn largely has been a result of Israeli closure policies—a steady increase in movement and access restrictions across the West Bank in response to Israeli security concerns which have disrupted labor and trade flows, industrial capacity, and basic commerce, both external and internal. Since 2008, the PA under President Mahmoud ABBAS and Prime Minister Salam FAYYAD has implemented a largely successful campaign of institutional reforms that has contributed to increased security and economic performance, supported by more than $3 billion in direct foreign donor assistance to the PA's budget since 2007. An easing of some Israeli restrictions on West Bank movement and access since 2008 also has contributed to an uptick in retail activity in larger cities. The biggest impediments to economic improvements in the West Bank remain Palestinians' lack of access to land and resources in Israeli-controlled areas, import and export restrictions, and a high-cost capital structure. Absent robust private sector growth, the PA will continue to rely heavily on donor aid for its budgetary needs.
GDP (purchasing power parity): $12.79 billion (2009 est.)
country comparison to the world: 141
$11.95 billion (2008)
note: data are in 2010 US dollars
GDP (official exchange rate): $6.641 billion (2008 est.)
GDP—real growth rate: 7% (2009 est.)
country comparison to the world: 36
2.3% (2008 est.)
GDP—per capita (PPP): $2,900 (2008 est.)
country comparison to the world: 171
note: data are in 2008 US dollars
GDP—composition by sector:
agriculture: 5%
industry: 14%
services: 81% (includes Gaza Strip) (2008 est.)
Labor force: 694,000 (2009)
country comparison to the world: 149
Labor force—by occupation:
agriculture: 12%
industry: 23%
services: 65% (June 2008)
Unemployment rate: 16.5% (2010 est.)
country comparison to the world: 156
19% (2009 est.)
Population below poverty line:
46% (2007 est.)
Household income or consumption by percentage share: *lowest 10%:* NA%
highest 10%: NA%
Budget: *revenues:* $1.87 billion
expenditures: $3.1 billion
note: Palestinian Authority, includes Gaza Strip (2010)
Inflation rate (consumer prices):
3.5% (2010 est.)
country comparison to the world: 99
9.9% (2009 est.)
note: includes Gaza Strip
Commercial bank prime lending rate: 5.78% (31 December 2009 est.)
country comparison to the world: 123
7.19% (31 December 2008 est.)

Stock of narrow money:
$NA (31 December 2008)
$1.574 billion (31 December 2007)
Stock of broad money: $5.567 billion (31 December 2009)
country comparison to the world: 116
$5.251 billion (31 December 2008)
Stock of domestic credit: $1.8 billion (31 December 2010 est.)
country comparison to the world: 129
$1.367 billion (31 December 2008 est.)
Market value of publicly traded shares:
$NA (31 December 2009)
country comparison to the world: 96
$2.123 billion (31 December 2008)
$2.475 billion (31 December 2007)
Agriculture—products: olives, citrus fruit, vegetables; beef, dairy products
Industries: small-scale manufacturing, quarrying, textiles, soap, olive-wood carvings, and mother-of-pearl souvenirs
Industrial production growth rate: NA%
Electricity—production: 500 million kWh
country comparison to the world: 160
note: most imported electricity is from Israel; Jerusalem District Electric Company buys and distributes electricity to Palestinians in east Jerusalem and its concession in the West Bank; the Israel Electric Company directly supplies electricity to most Jewish residents and military facilities (2007 est.)
Electricity—consumption: 3.265 billion kWh (2007 est.)
country comparison to the world: 121
Electricity—exports: 0 kWh (2008)
Electricity—imports: 2.8 billion kWh (2007 est.)
Oil—production: 0 bbl/day (2009 est.)
country comparison to the world: 141
Oil—consumption:
24,000 bbl/day (2009 est.)
country comparison to the world: 117
Oil—exports: 511 bbl/day (2007 est.)
country comparison to the world: 124
Oil—imports: 22,150 bbl/day (2007 est.)
country comparison to the world: 108
Oil—proved reserves:
0 bbl (1 January 2009 est.)
country comparison to the world: 204
Natural gas—production: 0 cu m (2008 est.)
country comparison to the world: 201
Natural gas—consumption:
0 cu m (2008 est.)
country comparison to the world: 137
Natural gas—exports: 0 cu m (2008 est.)
country comparison to the world: 201
Natural gas—imports: 0 cu m (2008 est.)
country comparison to the world: 137
Natural gas—proved reserves:
0 cu m (1 January 2009 est.)
country comparison to the world: 202
Exports: $529 million (2008)
country comparison to the world: 165
$339 million (2006)
note: includes Gaza Strip
Exports—commodities: stone, olives, fruit, vegetables, limestone
Imports: $3.772 billion (2008)
country comparison to the world: 130
$2.84 billion (2006)
Imports—commodities: food, consumer goods, construction materials, petroleum, chemicals
Debt—external: $1.04 billion (2010 est.)
country comparison to the world: 149
$1.3 billion (2007 est.)
Exchange rates: new Israeli shekels (ILS) per US dollar–3.739 (2010), 3.932 (2009), 3.56 (2008), 4.14 (2007), 4.4565 (2006)

COMMUNICATIONS

Telephones—main lines in use: 360,400 (includes Gaza Strip) (2010)
country comparison to the world: 110
Telephones—mobile cellular: 2.405 million (includes Gaza Strip) (2010)
country comparison to the world: 125
Telephone system: *general assessment:* continuing political and economic instability has impeded significant liberalization of the telecommunications industry
domestic: Israeli company BEZEK and the Palestinian company PALTEL are responsible for fixed line services; PALTEL plans to establish a fiber-optic connection to Jordan to route domestic mobile calls; the Palestinian JAWWAL company and WATANIYA PALESTINE provide cellular services
international: country code–970; 1 international switch in Ramallah (2009) (2009)
Broadcast media: the Palestinian Authority operates 1 television and 1 radio station; about 30 independent TV and 25 radio stations operating; Jordanian TV is available; satellite TV is accessible (2008)
Internet country code: .ps; note–same as Gaza Strip
Internet users: 1.379 million (includes Gaza Strip) (2009)
country comparison to the world: 88

TRANSPORTATION

Airports: 2 (2010)
country comparison to the world: 208
Airports—with paved runways: *total:* 2
1,524 to 2,437 m: 1
under 914 m: 1 (2010)
Roadways: *total:* 5,147 km
country comparison to the world: 153
paved: 5,147 km
note: includes Gaza Strip (2006)

MILITARY

Manpower fit for military service: *males age 16–49:* 579,248
females age 16–49: 547,782 (2010 est.)
Manpower reaching militarily significant age annually: *male:* 30,925
female: 29,440 (2010 est.)
Military expenditures: NA

TRANSNATIONAL ISSUES

Disputes—international: West Bank and Gaza Strip are Israeli-occupied with current status subject to the Israeli-Palestinian Interim Agreement–permanent status to be determined through further negotiation; Israel continues construction of a "seam line" separation barrier along parts of the Green Line and within the West Bank; Israel withdrew from four settlements in the northern West Bank in August 2005; since 1948, about 350 peacekeepers from the UN Truce Supervision Organization (UNTSO), headquartered in Jerusalem, monitor ceasefires, supervise armistice agreements, prevent isolated incidents from escalating, and assist other UN personnel in the region
Refugees and internally displaced persons: refugees (country of origin): 722,000 (Palestinian Refugees (UNRWA)) (2007)

WESTERN SAHARA

INTRODUCTION

Background: Morocco annexed the northern two-thirds of Western Sahara (formerly Spanish Sahara) in 1976 and claimed the rest of the territory in 1979, following Mauritania's withdrawal. A guerrilla war with the Polisario Front contesting Morocco's sovereignty ended in a 1991 UN-brokered cease-fire; a UN-organized referendum on the territory's final status has been repeatedly postponed. The UN since 2007 has sponsored intermittent talks between representatives of the Government of Morocco and the Polisario Front to negotiate the status of Western Sahara. Morocco has put forward an autonomy proposal for the territory, which would allow for some local administration while maintaining Moroccan sovereignty. The Polisario, with Algeria's support, demands a popular referendum that includes the option of independence.

GEOGRAPHY

Location: Northern Africa, bordering the North Atlantic Ocean, between Mauritania and Morocco
Geographic coordinates: 24 30 N, 13 00 W
Map references: Africa
Area: *total:* 266,000 sq km
country comparison to the world: 77
land: 266,000 sq km
water: 0 sq km

Area—comparative: about the size of Colorado
Land boundaries: *total:* 2,046 km
border countries: Algeria 42 km, Mauritania 1,561 km, Morocco 443 km
Coastline: 1,110 km
Maritime claims: contingent upon resolution of sovereignty issue
Climate: hot, dry desert; rain is rare; cold offshore air currents produce fog and heavy dew
Terrain: mostly low, flat desert with large areas of rocky or sandy surfaces rising to small mountains in south and northeast
Elevation extremes: *lowest point:* Sebjet Tah -55 m
highest point: unnamed elevation 805 m
Natural resources: phosphates, iron ore
Land use: *arable land:* 0.02%
permanent crops: 0%
other: 99.98% (2005)
Irrigated land: NA
Natural hazards: hot, dry, dust/sand-laden sirocco wind can occur during winter and spring; widespread harmattan haze exists 60% of time, often severely restricting visibility
Environment—current issues: sparse water and lack of arable land
Environment—international agreements: *party to:* none of the selected agreements
Geography—note: the waters off the coast are particularly rich fishing areas

PEOPLE

Population: 507,160 (July 2011 est.)
country comparison to the world: 170
note: estimate is based on projections by age, sex, fertility, mortality, and migration; fertility and mortality are based on data from neighboring countries
Age structure: *0–14 years:* 38.9% (male 99,797/female 97,700)
15–64 years: 57.5% (male 143,808/female 147,823)
65 years and over: 3.6% (male 7,918/female 10,114) (2011 est.)
Median age: *total:* 20.3 years
male: 19.8 years
female: 20.8 years (2011 est.)
Population growth rate: 3.097% (2011 est.)
country comparison to the world: 9
Birth rate: 32.1 births/1,000 population (2011 est.)
country comparison to the world: 40
Death rate: 8.96 deaths/1,000 population (July 2011 est.)
country comparison to the world: 71
Urbanization: *urban population:* 82% of total population (2010)
rate of urbanization: 3.5% annual rate of change (2010-15 est.)
Major cities—population: EL AAIUN (capital) 213,000 (2009)
Sex ratio: *at birth:* 1.04 male(s)/female
under 15 years: 1.02 male(s)/female
15–64 years: 0.97 male(s)/female
65 years and over: 0.78 male(s)/female
total population: 0.98 male(s)/female (2011 est.)
Infant mortality rate:
total: 60.44 deaths/1,000 live births
country comparison to the world: 33
male: 65.55 deaths/1,000 live births
female: 55.13 deaths/1,000 live births (2011 est.)
Life expectancy at birth: *total population:* 61.13 years
country comparison to the world: 185
male: 58.94 years
female: 63.41 years (2011 est.)
Total fertility rate: 4.3 children born/woman (2011 est.)
country comparison to the world: 36
HIV/AIDS—adult prevalence rate: NA
HIV/AIDS—people living with HIV/AIDS: NA
HIV/AIDS—deaths: NA
Nationality: *noun:* Sahrawi(s), Sahraoui(s)
adjective: Sahrawi, Sahrawian, Sahraouian
Ethnic groups: Arab, Berber
Religions: Muslim
Languages: Hassaniya Arabic, Moroccan Arabic
Literacy: NA
School life expectancy (primary to tertiary education): NA
Education expenditures: NA

GOVERNMENT

Country name:
conventional long form: none
conventional short form: Western Sahara
former: Rio de Oro, Saguia el Hamra, Spanish Sahara
Government type: legal status of territory and issue of sovereignty unresolved; territory contested by Morocco and Polisario Front (Popular Front for the Liberation of the Saguia el Hamra and Rio de Oro), which in February 1976 formally proclaimed a government-in-exile, the Sahrawi Arab Democratic Republic (SADR), near Tindouf, Algeria, led by President Mohamed ABDELAZIZ; territory partitioned between Morocco and Mauritania in April 1976 when Spain withdrew, with Morocco acquiring northern two-thirds; Mauritania, under pressure from Polisario guerrillas, abandoned all claims to its portion in August 1979; Morocco moved to occupy that sector shortly thereafter and has since asserted administrative control; the Polisario's government-in-exile was seated as an Organization of African Unity (OAU) member in 1984; Morocco between 1980 and 1987 built a fortified sand berm delineating the roughly 80 percent of Western Sahara west of the barrier that currently is controlled by Morocco; guerrilla activities continued sporadically until a UN-monitored cease-fire was implemented on 6 September 1991 (Security Council Resolution 690) by the United Nations Mission for the Referendum in Western Sahara (MINURSO)
Capital: none
time difference: UTC 0 (5 hours ahead of Washington, DC during Standard Time)
daylight saving time: +1hr, begins first Sunday in April; ends last Sunday in July
Administrative divisions: none (territory west of the berm under de facto Moroccan control)
Suffrage: none; (residents of Moroccan-controlled Western Sahara participate in Moroccan elections)
Executive branch: none
Political pressure groups and leaders: none
International organization participation: AU, WFTU
Diplomatic representation in the US: none
Diplomatic representation from the US: none

ECONOMY

Economy—overview: Western Sahara has a small market-based economy whose main industries are fishing, phosphate mining, and pastoral nomadism. The territory's arid desert climate makes sedentary agriculture difficult, and Western Sahara imports much of its food. The Moroccan Government administers Western Sahara's economy and is a source of employment, infrastructure development, and social spending in the territory. Western Sahara's unresolved legal status makes the exploitation of its natural resources a contentious issue between Morocco and the Polisario. Morocco and the EU in July 2006 signed a four-year agreement allowing European vessels to fish off the coast of Morocco, including the disputed waters off the coast of Western Sahara. Oil has never been found in Western Sahara in commercially significant quantities, but Morocco and the Polisario have quarreled over who has the right to authorize and benefit from oil exploration in the territory. Western Sahara's main long-term economic challenge is the development of a more diverse set of industries capable of providing greater employment and income to the territory.
GDP (purchasing power parity): $900 million (2007 est.)
country comparison to the world: 202
GDP (official exchange rate): $NA
GDP—real growth rate: NA%
GDP—per capita (PPP): $2,500 (2007 est.)
country comparison to the world: 183
GDP—composition by sector:
agriculture: NA%
industry: NA%
services: 40% (2007 est.)
Labor force: 144,000 (2010 est.)
country comparison to the world: 178
Labor force—by occupation:
agriculture: 50%
industry and *services:* 50% (2005 est.)
Unemployment rate: NA%
Population below poverty line: NA%
Household income or consumption by percentage share:
lowest 10%: NA%
highest 10%: NA%
Budget: *revenues:* $NA
expenditures: $NA
Inflation rate (consumer prices): NA%
Agriculture—products: fruits and vegetables (grown in the few oases); camels, sheep, goats (kept by nomads); fish
Industries: phosphate mining, handicrafts
Industrial production growth rate: NA%
Electricity—production: 90 million kWh (2007 est.)
country comparison to the world: 193
Electricity—consumption: 83.7 million kWh (2007 est.)
country comparison to the world: 192

Electricity—exports: 0 kWh (2008 est.)
Electricity—imports: 0 kWh (2008 est.)
Oil—production: 0 bbl/day (2009 est.)
country comparison to the world: 142
Oil—consumption:
2,000 bbl/day (2009 est.)
country comparison to the world: 185
Oil—exports: 0 bbl/day (2007 est.)
country comparison to the world: 141
Oil—imports: 1,702 bbl/day (2007 est.)
country comparison to the world: 178
Oil—proved reserves: 0 bbl (1 January 2010 est.)
country comparison to the world: 205
Natural gas—production:
0 cu m (2008 est.)
country comparison to the world: 202
Natural gas—consumption:
0 cu m (2008 est.)
country comparison to the world: 138
Natural gas—exports: 0 cu m (2008 est.)
country comparison to the world: 202
Natural gas—imports: 0 cu m (2008 est.)
country comparison to the world: 138
Natural gas—proved reserves: 0 cu m (1 January 2010 est.)
country comparison to the world: 203
Exports: $NA
Exports—commodities: phosphates 62%
Imports: $NA
Imports—commodities: fuel for fishing fleet, foodstuffs
Debt—external: $NA
Exchange rates: Moroccan dirhams (MAD) per US dollar–8.3619 (2010), 8.0571 (2009), 8.3563 (2007), 8.7722 (2006)

COMMUNICATIONS

Telephones—main lines in use: about 2,000 (1999 est.)
Telephones—mobile cellular: 0 (1999)
country comparison to the world: 219
Telephone system: *general assessment:* sparse and limited system
domestic: NA
international: country code–212; tied into Morocco's system by microwave radio relay, tropospheric scatter, and satellite; satellite earth stations–2 Intelsat (Atlantic Ocean) linked to Rabat, Morocco
Broadcast media: Morocco's state-owned broadcaster, Radio-Television Marocaine (RTM), operates a radio service from Laayoune and relays TV service; a Polisario-backed radio station also broadcasts (2008)
Internet country code: .eh

TRANSPORTATION

Airports: 6 (2010)
country comparison to the world: 173
Airports—with paved runways: *total:* 3
2,438 to 3,047 m: 3 (2010)
Airports—with unpaved runways: *total:* 3
1,524 to 2,437 m: 1
914 to 1,523 m: 1
under 914 m: 1 (2010)
Ports and terminals: Ad Dakhla, Laayoune (El Aaiun)

MILITARY

Manpower fit for military service: *males age 16–49:* 79,489
females age 16–49: 87,362 (2010 est.)
Manpower reaching militarily significant age annually: *male:* 5,523
female: 5,429 (2010 est.)

TRANSNATIONAL ISSUES

Disputes—international: Many neighboring states reject Moroccan administration of Western Sahara; several states have extended diplomatic relations to the "Sahrawi Arab Democratic Republic" represented by the Polisario Front in exile in Algeria, while others recognize Moroccan sovereignty over Western Sahara; most of the approximately 102,000 Sahrawi refugees are sheltered in camps in Tindouf, Algeria

YEMEN

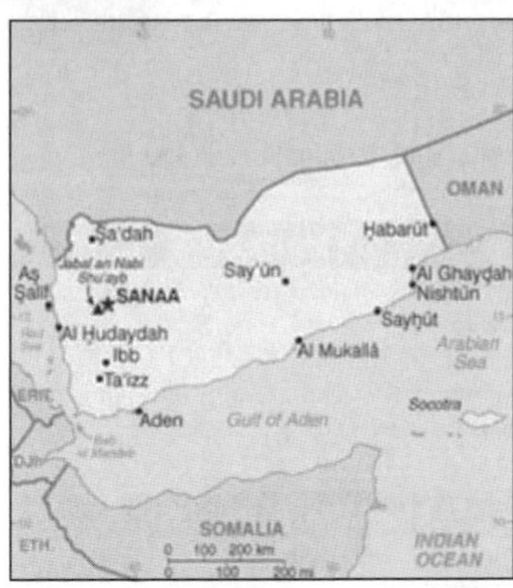

INTRODUCTION

Background: North Yemen became independent of the Ottoman Empire in 1918. The British, who had set up a protectorate area around the southern port of Aden in the 19th century, withdrew in 1967 from what became South Yemen. Three years later, the southern government adopted a Marxist orientation. The massive exodus of hundreds of thousands of Yemenis from the south to the north contributed to two decades of hostility between the states. The two countries were formally unified as the Republic of Yemen in 1990. A southern secessionist movement in 1994 was quickly subdued. In 2000, Saudi Arabia and Yemen agreed to a delimitation of their border. Fighting in the northwest between the government and Huthi rebels, a group seeking a return to traditional Zaydi Islam, began in 2004 and has since resulted in seven rounds of fighting–the last ended in early 2010 with a tentative ceasefire. The southern secessionist movement was revitalized in 2008 when a popular socioeconomic protest movement initiated the prior year took on political goals including secession. Public rallies in Sana'a against President SALIH–inspired by similar demonstrations in Tunisia and Egypt–slowly built momentum starting in late January 2011 fueled by complaints over high unemployment, poor economic conditions, and corruption. By the following month, some protests had resulted in violence, and the demonstrations had spread to other major cities. By March the opposition had hardened its demands and was unifying behind calls for SALIH's immediate ouster. Media reports indicated that as many as 100 protesters had been killed and many more injured amid the protests. Domestic and international efforts to mediate a resolution to the political crisis had not yielded a deal as of mid April.

GEOGRAPHY

Location: Middle East, bordering the Arabian Sea, Gulf of Aden, and Red Sea, between Oman and Saudi Arabia
Geographic coordinates: 15 00 N, 48 00 E
Map references: Middle East
Area: *total:* 527,968 sq km
country comparison to the world: 49
land: 527,968 sq km
water: 0 sq km
note: includes Perim, Socotra, the former Yemen Arab Republic (YAR or North Yemen), and the former People's Democratic Republic of Yemen (PDRY or South Yemen)
Area—comparative: slightly larger than twice the size of Wyoming
Land boundaries: *total:* 1,746 km
border countries: Oman 288 km, Saudi Arabia 1,458 km
Coastline: 1,906 km
Maritime claims: territorial sea: 12 nm
contiguous zone: 24 nm
exclusive economic zone: 200 nm
continental shelf: 200 nm or to the edge of the continental margin
Climate: mostly desert; hot and humid along west coast; temperate in western mountains affected by seasonal monsoon; extraordinarily hot, dry, harsh desert in east
Terrain: narrow coastal plain backed by flat-topped hills and rugged mountains; dissected upland desert plains in center slope into the desert interior of the Arabian Peninsula
Elevation extremes: *lowest point:* Arabian Sea 0 m
highest point: Jabal an Nabi Shu'ayb 3,760 m
Natural resources: petroleum, fish, rock salt, marble; small deposits of coal, gold, lead, nickel, and copper; fertile soil in west
Land use: *arable land:* 2.91%
permanent crops: 0.25%
other: 96.84% (2005)
Irrigated land: 5,500 sq km (2003)
Total renewable water resources:
4.1 cu km (1997)
Freshwater withdrawal (domestic/industrial/agricultural): *total:* 6.63 cu km/yr (4%/1%/95%)
per capita: 316 cu m/yr (2000)
Natural hazards: sandstorms and dust storms in summer
volcanism: Yemen experiences limited volcanic activity; Jebel at Tair (Jabal al-Tair, Jebel Teir, Jabal al-Tayr, Jazirat at-Tair) (elev. 244 m), which forms an island in the Red Sea, erupted in 2007 after awakening from dormancy; other historically active volcanoes include Harra of Arhab, Harras of Dhamar, Harra es-Sawad, and Jebel Zubair, although many of these have not erupted in over a century
Environment—current issues: limited natural freshwater resources; inadequate supplies of potable water; overgrazing; soil erosion; desertification
Environment—international agreements: *party to:* Biodiversity, Climate Change, Climate Change-Kyoto Protocol, Desertification, Endangered Species, Environmental Modification, Hazardous Wastes, Law of the Sea, Ozone Layer Protection
signed, but not ratified: none of the selected agreements
Geography—note: strategic location on Bab el Mandeb, the strait linking the Red Sea and the Gulf of Aden, one of world's most active shipping lanes

PEOPLE

Population: 24,133,492 (July 2011 est.)
country comparison to the world: 49
Age structure: *0–14 years:* 43% (male 5,285,218/female 5,094,736)
15–64 years: 54.4% (male 6,666,600/female 6,459,414)
65 years and over: 2.6% (male 298,175/female 329,349) (2011 est.)
Median age: *total:* 18.1 years
male: 18 years
female: 18.2 years (2011 est.)
Population growth rate: 2.647% (2011 est.)
country comparison to the world: 21
Birth rate: 33.49 births/1,000 population (2011 est.)
country comparison to the world: 35
Death rate: 7.02 deaths/1,000 population (July 2011 est.)
country comparison to the world: 133
Net migration rate: 0 migrant(s)/1,000 population (2011 est.)
country comparison to the world: 73
Urbanization: *urban population:* 32% of total population (2010)
rate of urbanization: 4.6% annual rate of change (2010-15 est.)
Major cities—population: SANAA (capital) 2.229 million (2009)
Sex ratio:
at birth: 1.05 male(s)/female
under 15 years: 1.04 male(s)/female
15–64 years: 1.03 male(s)/female
65 years and over: 0.91 male(s)/female
total population: 1.03 male(s)/female (2011 est.)
Infant mortality rate:
total: 55.11 deaths/1,000 live births
country comparison to the world: 38
male: 59.7 deaths/1,000 live births
female: 50.29 deaths/1,000 live births (2011 est.)
Life expectancy at birth:
total population: 63.74 years
country comparison to the world: 173
male: 61.7 years
female: 65.87 years (2011 est.)
Total fertility rate: 4.63 children born/woman (2011 est.)
country comparison to the world: 30
HIV/AIDS—adult prevalence rate:
0.1% (2001 est.)
country comparison to the world: 112
HIV/AIDS—people living with HIV/AIDS:
12,000 (2001 est.)
country comparison to the world: 92
HIV/AIDS—deaths: NA
Major infectious diseases:
degree of risk: high
food or waterborne diseases: bacterial diarrhea, hepatitis A, and typhoid fever
vectorborne diseases: dengue fever and malaria
water contact disease: schistosomiasis (2009)
Drinking water source: *Improved:*
urban: 72% of population
rural: 57% of population
total: 62% of population

Unimproved:
urban: 28% of population
rural: 43% of population
total: 38% of population (2008)
Sanitation facility access:
Improved:
urban: 94% of population
rural: 33% of population
total: 52% of population
Unimproved:
urban: 6% of population
rural: 67% of population
total: 48% of population (2008)
Nationality: *noun:* Yemeni(s)
adjective: Yemeni
Ethnic groups: predominantly Arab; but also Afro-Arab, South Asians, Europeans
Religions: Muslim including Shaf'i (Sunni) and Zaydi (Shia), small numbers of Jewish, Christian, and Hindu
Languages: Arabic (official)
Literacy: *definition:* age 15 and over can read and write
total population: 50.2%
male: 70.5%
female: 30% (2003 est.)
School life expectancy (primary to tertiary education): *total:* 9 years
male: 11 years
female: 7 years (2005)
Education expenditures:
5.2% of GDP (2008)
country comparison to the world: 54

GOVERNMENT

Country name: *conventional long form:* Republic of Yemen
conventional short form: Yemen
local long form: Al Jumhuriyah al Yamaniyah
local short form: Al Yaman
former: Yemen Arab Republic [Yemen (Sanaa) or North Yemen] and People's Democratic Republic of Yemen [Yemen (Aden) or South Yemen]
Government type: republic
Capital: *name:* Sanaa
geographic coordinates: 15 21 N, 44 12 E
time difference: UTC+3 (8 hours ahead of Washington, DC during Standard Time)
Administrative divisions: 21 governorates (muhafazat, singular—muhafazah); Abyan, 'Adan (Aden), Ad Dali', Al Bayda', Al Hudaydah, Al Jawf, Al Mahrah, Al Mahwit, Amanat al 'Asimah, 'Amran, Dhamar, Hadramawt, Hajjah, Ibb, Lahij, Ma'rib, Raymah, Sa'dah, San'a' (Sanaa), Shabwah, Ta'izz
Independence: 22 May 1990 (Republic of Yemen was established with the merger of the Yemen Arab Republic [Yemen (Sanaa) or North Yemen] and the Marxist-dominated People's Democratic Republic of Yemen [Yemen (Aden) or South Yemen]); note—previously North Yemen became independent in November 1918 (from the Ottoman Empire) and became a republic with the overthrow of the theocratic Imamate in 1962; South Yemen became independent on 30 November 1967 (from the UK)
National holiday: Unification Day, 22 May (1990)
Constitution: 16 May 1991; amended 29 September 1994 and February 2001
Legal system: mixed legal system of Islamic law, English common law, and customary law
International law organization participation: has not submitted an ICJ jurisdiction declaration; non-party state to the ICCt
Suffrage: 18 years of age; universal
Executive branch: *chief of state:* President Ali Abdallah SALIH (since 22 May 1990, the former president of North Yemen, assumed office upon the merger of North and South Yemen); Vice President Maj. Gen. Abd al-Rab Mansur HADI (since 3 October 1994)
head of government: Prime Minister Ali Muhammad MUJAWWAR (since 31 March 2007)
cabinet: Council of Ministers appointed by the president on the advice of the prime minister; note—President Ali Abdallah SALIH dimissed the Council of Ministers on 20 March 2011
(For more information visit the World Leaders website)
elections: president elected by popular vote for a seven-year term; election last held on 20 September 2006 (next to be held in September 2013); vice president appointed by the president; prime minister and deputy prime ministers appointed by the president
election results: Ali Abdallah SALIH elected president; percent of vote—Ali Abdallah SALIH 77.2%, Faysal BIN SHAMLAN 21.8%, other 1%
Legislative branch: bicameral legislature consisting of a Shura Council (111 seats; members appointed by the president) and House of Representatives (301 seats; members elected by popular vote to serve eight-year terms)
elections: last held on 27 April 2003 (scheduled April 2009 election postponed for two years)
election results: percent of vote by party—NA; seats by party—GPC 238, Islah 47, YSP 6, Nasserite Unionist Party 3, National Arab Socialist Ba'th Party 2, independents 5
Judicial branch: Supreme Court
Political parties and leaders: General People's Congress or GPC [Abdul-Kader BAJAMMAL]; Islamic Reform Grouping or Islah [Muhammed Abdallah AL-YADUMI]; Nasserite Unionist Party [Abd al-Malik al-MAKHLAFI]; National Arab Socialist Ba'th Party [Dr. Qasim SALAM]; Yemeni Socialist Party or YSP [Yasin Said NU'MAN]; note—there are at least seven more active political parties
Political pressure groups and leaders: Muslim Brotherhood; Women National Committee
other: conservative tribal groups; Huthis, southern secessionist groups; al-Qa'ida in the Arabian Peninsula (AQAP)
International organization participation: AFESD, AMF, CAEU, FAO, G-77, IAEA, IBRD, ICAO, ICRM, IDA, IDB, IFAD, IFC, IFRCS, ILO, IMF, IMO, Interpol, IOC, IOM, IPU, ISO, ITSO, ITU, ITUC, LAS, MIGA, MINURSO, MONUSCO, NAM, OAS (observer), OIC, OPCW, UN, UNAMID, UNCTAD, UNESCO, UNHCR, UNIDO, UNMIL, UNMIS, UNOCI, UNWTO, UPU, WCO, WFTU, WHO, WIPO, WMO, WTO (observer)
Diplomatic representation in the US: *chief of mission:* Ambassador Abd al-Wahab Abdallah al-HAJRI
chancery: 2319 Wyoming Avenue NW, Washington, DC 20008
telephone: [1] (202) 965-4760
FAX: [1] (202) 337-2017
Diplomatic representation from the US: *chief of mission:* Ambassador Gerald M. FEIERSTEIN
embassy: Sa'awan Street, Sanaa
mailing address: P. O. Box 22347, Sanaa
telephone: [967] (1) 755-2000 ext. 2153 or 2266
FAX: [967] (1) 303-182
Flag description: three equal horizontal bands of red (top), white, and black; the band colors derive from the Arab Liberation flag and represent oppression (black), overcome through bloody struggle (red), to be replaced by a bright future (white)
note: similar to the flag of Syria, which has two green stars in the white band, and of Iraq, which has an Arabic inscription centered in the white band; also similar to the flag of Egypt, which has a heraldic eagle centered in the white band
National anthem: *name:* "al-qumhuriyatu l-muttahida" (United Republic)
lyrics/music: Abdullah Abdulwahab NOA'MAN/Ayyoab Tarish ABSI
note: adopted 1990; the music first served as the anthem for South Yemen before unification with North Yemen in 1990

ECONOMY

Economy—overview: Yemen is a low income country that is highly dependent on declining oil resources for revenue. Petroleum accounts for roughly 25% of GDP and 70% of government revenue. Yemen has tried to counter the effects of its declining oil resources by diversifying its economy through an economic reform program initiated in 2006 that is designed to bolster non-oil sectors of the economy and foreign investment. In October 2009, Yemen exported its first liquefied natural gas as part of this diversification effort. In January 2010, the international community established the Friends of Yemen group that aims to support Yemen's efforts towards economic and political reform, and in August 2010 the IMF approved a three-year $370 million program to further this effort. Despite these ambitious endeavors, Yemen continues to face difficult long term challenges, including declining water resources and a high population growth rate.
GDP (purchasing power parity): $63.4 billion (2010 est.)
country comparison to the world: 85
$58.69 billion (2009 est.)
$56.51 billion (2008 est.)
note: data are in 2010 US dollars
GDP (official exchange rate):
$31.27 billion (2010 est.)
GDP—real growth rate: 8% (2010 est.)
country comparison to the world: 20
3.9% (2009 est.)
3.6% (2008 est.)
GDP—per capita (PPP): $2,700 (2010 est.)
country comparison to the world: 173

$2,600 (2009 est.)
$2,500 (2008 est.)
note: data are in 2010 US dollars
GDP—composition by sector:
agriculture: 8.2%
industry: 38.8%
services: 53% (2010 est.)
Labor force: 6.832 million (2010 est.)
country comparison to the world: 63
Labor force—by occupation: note: most people are employed in agriculture and herding; services, construction, industry, and commerce account for less than one-fourth of the labor force
Unemployment rate: 35% (2003 est.)
country comparison to the world: 181
Population below poverty line:
45.2% (2003)
Household income or consumption by percentage share: *lowest 10%:* 2.9%
highest 10%: 30.8% (2005)
Distribution of family income—Gini index: 37.7 (2005)
country comparison to the world: 73
33.4 (1998)
Investment (gross fixed): 19.4% of GDP (2010 est.)
country comparison to the world: 96
Budget: *revenues:* $7.581 billion
expenditures: $9.345 billion (2010 est.)
Public debt: 39.1% of GDP (2010 est.)
country comparison to the world: 76
36.7% of GDP (2009 est.)
Inflation rate (consumer prices):
12.2% (2010 est.)
country comparison to the world: 207
5.4% (2009 est.)
Central bank discount rate: NA%
Commercial bank prime lending rate: 18% (31 December 2009 est.)
country comparison to the world: 25
18% (31 December 2008 est.)
Stock of narrow money: $3.551 billion (31 December 2010 est.)
country comparison to the world: 105
$3.74 billion (31 December 2009 est.)
Stock of broad money: $9.739 billion (31 December 2010 est.)
country comparison to the world: 99
$9.552 billion (31 December 2009 est.)
Stock of domestic credit: $5.297 billion (31 December 2010 est.)
country comparison to the world: 110
$5.098 billion (31 December 2009 est.)
Market value of publicly traded shares: $NA
Agriculture—products: grain, fruits, vegetables, pulses, qat, coffee, cotton; dairy products, livestock (sheep, goats, cattle, camels), poultry; fish
Industries: crude oil production and petroleum refining; small-scale production of cotton textiles and leather goods; food processing; handicrafts; small aluminum products factory; cement; commercial ship repair; natural gas production
Industrial production growth rate:
9% (2010 est.)
country comparison to the world: 23
Electricity—production:
5.665 billion kWh (2007 est.)
country comparison to the world: 107
Electricity—consumption:
4.133 billion kWh (2007 est.)
country comparison to the world: 118
Electricity—exports: 0 kWh (2008)
Electricity—imports: 0 kWh (2008 est.)
Oil—production:
288,400 bbl/day (2009 est.)
country comparison to the world: 37
Oil—consumption:
155,000 bbl/day (2009 est.)
country comparison to the world: 64
Oil—exports: 274,400 bbl/day (2007 est.)
country comparison to the world: 44
Oil—imports: 65,860 bbl/day (2007 est.)
country comparison to the world: 78
Oil—proved reserves: 3.16 billion bbl (1 January 2010 est.)
country comparison to the world: 30
Natural gas—production:
454,700 cu m (2009 est.)
country comparison to the world: 91
Natural gas—consumption:
0 cu m (2008 est.)
country comparison to the world: 141
Natural gas—exports:
454,700 cu m (2009 est.)
country comparison to the world: 46
Natural gas—imports: 0 cu m (2008 est.)
country comparison to the world: 73
Natural gas—proved reserves: 478.5 billion cu m (1 January 2010 est.)
country comparison to the world: 32
Current account balance: $-2.181 billion (2010 est.)
country comparison to the world: 156
$-2.328 billion (2009 est.)
Exports: $7.462 billion (2010 est.)
country comparison to the world: 96
$5.812 billion (2009 est.)
Exports—commodities: crude oil, coffee, dried and salted fish, liquefied natural gas
Exports—partners: China 36%, Thailand 17.63%, India 13.54%, South Africa 6.16%, Japan 5.49%, UAE 4.99% (2009)
Imports: $8.35 billion (2010 est.)
country comparison to the world: 96
$7.518 billion (2009 est.)
Imports—commodities: food and live animals, machinery and equipment, chemicals
Imports—partners: China 13.98%, UAE 12.3%, India 8.63%, Saudi Arabia 5.8%, US 4.52%, Brazil 4.51%, Turkey 4.51%, Kuwait 4.33%, France 4.24% (2009)
Reserves of foreign exchange and gold: $5.744 billion (31 December 2010 est.)
country comparison to the world: 68
$6.993 billion (31 December 2009 est.)
Debt—external: $7.147 billion (31 December 2010 est.)
country comparison to the world: 96
$6.552 billion (31 December 2009 est.)
Stock of direct foreign investment—at home: $NA
Exchange rates: Yemeni rials (YER) per US dollar–220.05 (2010), 202.85 (2009), 199.76 (2008), 199.14 (2007), 197.18 (2006)

COMMUNICATIONS

Telephones—main lines in use: 997,000 (2009)
country comparison to the world: 80
Telephones—mobile cellular: 8.313 million (2009)
country comparison to the world: 72
Telephone system: *general assessment:* since unification in 1990, efforts have been made to create a national telecommunications network
domestic: the national network consists of microwave radio relay, cable, tropospheric scatter, GSM and CDMA mobile-cellular telephone systems; fixed-line and mobile-cellular teledensity remains low by regional standards
international: country code–967; landing point for the international submarine cable Fiber-Optic Link Around the Globe (FLAG); satellite earth stations–3 Intelsat (2 Indian Ocean and 1 Atlantic Ocean), 1 Intersputnik (Atlantic Ocean region), and 2 Arabsat; microwave radio relay to Saudi Arabia and Djibouti
Broadcast media: state-run TV with 2 stations; state-run radio with 2 national radio stations and 5 local stations; stations from Oman and Saudi Arabia can be accessed (2007)
Internet country code: .ye
Internet hosts: 255 (2010)
country comparison to the world: 188
Internet users: 2.349 million (2009)
country comparison to the world: 71

TRANSPORTATION

Airports: 55 (2010)
country comparison to the world: 84
Airports—with paved runways:
total: 17
over 3,047 m: 4
2,438 to 3,047 m: 9
1,524 to 2,437 m: 3
914 to 1,523 m: 1 (2010)
Airports—with unpaved runways:
total: 38
over 3,047 m: 3
2,438 to 3,047 m: 5
1,524 to 2,437 m: 6
914 to 1,523 m: 14
under 914 m: 10 (2010)
Pipelines: gas 423 km; liquid petroleum gas 22 km; oil 1,367 km (2010)
Roadways: *total:* 71,300 km
country comparison to the world: 67
paved: 6,200 km
unpaved: 65,100 km (2005)
Merchant marine: *total:* 6
country comparison to the world: 127
by type: cargo 1, chemical tanker 2, petroleum tanker 2, roll on/roll off 1
registered in other countries: 14 (Moldova 1, Panama 4, Saint Kitts and Nevis 1, Sierra Leone 2, unknown 6) (2010)
Ports and terminals: Aden, Al Hudaydah, Al Mukalla
Transportation—note: the International Maritime Bureau reports offshore waters in the Gulf of Aden are high risk for piracy; numerous vessels, including commercial shipping and pleasure craft, have been attacked and hijacked both at anchor and while underway; crew, passengers, and cargo are held for ransom; the presence of several naval task forces in the Gulf of Aden and additional anti-piracy measures on the part of ship operators reduced the incidence of piracy in that body of water by more than half in 2010

MILITARY

Military branches: Army, Navy (includes Marines), Yemen Air Force (Al Quwwat

al Jawwiya al Jamahiriya al Yemeniya; includes Air Defense Force), Republican Guard (2010)
Military service age and obligation: voluntary military service program authorized in 2001; 2-year service obligation (2006)
Manpower available for military service:
males age 16–49: 5,652,256
females age 16–49: 5,387,160 (2010 est.)
Manpower fit for military service:
males age 16–49: 4,056,944
females age 16–49: 4,116,895 (2010 est.)
Manpower reaching militarily significant age annually: *male:* 287,141
female: 277,612 (2010 est.)
Military expenditures: 6.6% of GDP (2006)
country comparison to the world: 7
Military—note: a Coast Guard was established in 2002

TRANSNATIONAL ISSUES

Disputes—international: Saudi Arabia has reinforced its concrete-filled security barrier along sections of the fully demarcated border with Yemen to stem illegal cross-border activities
Refugees and internally displaced persons: refugees (country of origin): 91,587 (Somalia) (2007)

ZAMBIA

INTRODUCTION

Background: The territory of Northern Rhodesia was administered by the [British] South Africa Company from 1891 until it was taken over by the UK in 1923. During the 1920s and 1930s, advances in mining spurred development and immigration. The name was changed to Zambia upon independence in 1964. In the 1980s and 1990s, declining copper prices, economic mismanagement and a prolonged drought hurt the economy. Elections in 1991 brought an end to one-party rule, but the subsequent vote in 1996 saw blatant harassment of opposition parties. The election in 2001 was marked by administrative problems with three parties filing a legal petition challenging the election of ruling party candidate Levy MWANAWASA. The new president launched an anticorruption investigation in 2002 to probe high-level corruption during the previous administration. In 2006-07, this task force successfully prosecuted four cases, including a landmark civil case in the UK in which former President CHILUBA and numerous others were found liable for more than USD 41 million. MWANAWASA was reelected in 2006 in an election that was deemed free and fair. Upon his abrupt death in August 2008, he was succeeded by his Vice President Rupiah BANDA, who subsequently won a special presidential by-election in October 2008. Under President BANDA, the Task Force on Corruption was abolished, President CHILUBA and his wife were acquitted in their criminal cases, and the government declined to register the UK civil verdict.

GEOGRAPHY

Location: Southern Africa, east of Angola, south of the Democratic Republic of the Congo
Geographic coordinates: 15 00 S, 30 00 E
Map references: Africa
Area: *total:* 752,618 sq km
country comparison to the world: 39
land: 743,398 sq km
water: 9,220 sq km
Area—comparative: slightly larger than Texas
Land boundaries:
total: 5,664 km
border countries: Angola 1,110 km, Democratic Republic of the Congo 1,930 km, Malawi 837 km, Mozambique 419 km, Namibia 233 km, Tanzania 338 km, Zimbabwe 797 km
Coastline: 0 km (landlocked)
Maritime claims: none (landlocked)
Climate: tropical; modified by altitude; rainy season (October to April)
Terrain: mostly high plateau with some hills and mountains
Elevation extremes: *lowest point:* Zambezi river 329 m
highest point: unnamed elevation in Mafinga Hills 2,301 m
Natural resources: copper, cobalt, zinc, lead, coal, emeralds, gold, silver, uranium, hydropower
Land use: *arable land:* 6.99%
permanent crops: 0.04%
other: 92.97% (2005)
Irrigated land: 1,560 sq km (2003)
Total renewable water resources: 105.2 cu km (2001)
Freshwater withdrawal (domestic/industrial/agricultural): *total:* 1.74 cu km/yr (17%/7%/76%)
per capita: 149 cu m/yr (2000)
Natural hazards: periodic drought; tropical storms (November to April)
Environment—current issues: air pollution and resulting acid rain in the mineral extraction and refining region; chemical runoff into watersheds; poaching seriously threatens rhinoceros, elephant, antelope, and large cat populations; deforestation; soil erosion; desertification; lack of adequate water treatment presents human health risks
Environment—international agreements: *party to:* Biodiversity, Climate Change, Climate Change-Kyoto Protocol, Desertification, Endangered Species, Hazardous Wastes, Law of the Sea, Ozone Layer Protection, Wetlands
signed, but not ratified: none of the selected agreements
Geography—note: landlocked; the Zambezi forms a natural riverine boundary with Zimbabwe

PEOPLE

Population: 13,881,336 (July 2011 est.)
country comparison to the world: 68
note: estimates for this country explicitly take into account the effects of excess mortality due to AIDS; this can result in lower life expectancy, higher infant mortality, higher death rates, lower population growth rates, and changes in the distribution of population by age and sex than would otherwise be expected
Age structure: *0–14 years:* 46.7% (male 3,253,125/female 3,228,844)
15–64 years: 50.8% (male 3,544,640/female 3,508,344)
65 years and over: 2.5% (male 148,531/female 197,852) (2011 est.)
Median age: *total:* 16.5 years
male: 16.5 years
female: 16.6 years (2011 est.)
Population growth rate: 3.062% (2011 est.)
country comparison to the world: 11
Birth rate: 44.08 births/1,000 population (2011 est.)
country comparison to the world: 4
Death rate: 12.61 deaths/1,000 population (July 2011 est.)
country comparison to the world: 27
Net migration rate: -0.84 migrant(s)/1,000 population (2011 est.)
country comparison to the world: 147
Urbanization: *urban population:* 36% of total population (2010)
rate of urbanization: 3.2% annual rate of change (2010-15 est.)
Major cities—population: LUSAKA (capital) 1.413 million (2009)
Sex ratio: *at birth:* 1.03 male(s)/female
under 15 years: 1.01 male(s)/female
15–64 years: 1.01 male(s)/female
65 years and over: 0.68 male(s)/female
total population: 1 male(s)/female (2011 est.)
Infant mortality rate: *total:* 66.6 deaths/1,000 live births
country comparison to the world: 22
male: 71.27 deaths/1,000 live births
female: 61.78 deaths/1,000 live births (2011 est.)
Life expectancy at birth: *total population:* 52.36 years
country comparison to the world: 208
male: 51.13 years
female: 53.63 years (2011 est.)
Total fertility rate: 5.98 children born/woman (2011 est.)
country comparison to the world: 8
HIV/AIDS—adult prevalence rate: 13.5% (2009 est.)
country comparison to the world: 6
HIV/AIDS—people living with HIV/AIDS: 980,000 (2009 est.)
country comparison to the world: 12
HIV/AIDS—deaths: 45,000 (2009 est.)
country comparison to the world: 10
Major infectious diseases: *degree of risk:* very high
food or waterborne diseases: bacterial and protozoal diarrhea, hepatitis A, and typhoid fever
vectorborne diseases: malaria and plague are high risks in some locations
water contact disease: schistosomiasis
animal contact disease: rabies (2009)
Drinking water source:
Improved:
urban: 87% of population
rural: 46% of population
total: 60% of population
Unimproved:
urban: 13% of population
rural: 54% of population
total: 40% of population (2008)
Sanitation facility access:
Improved:
urban: 59% of population
rural: 43% of population
total: 49% of population
Unimproved:
urban: 41% of population
rural: 47% of population
total: 51% of population (2008)
Nationality: *noun:* Zambian(s)
adjective: Zambian

Ethnic groups: African 99.5% (includes Bemba, Tonga, Chewa, Lozi, Nsenga, Tumbuka, Ngoni, Lala, Kaonde, Lunda, and other African groups), other 0.5% (includes Europeans, Asians, and Americans) (2000 Census)
Religions: Christian 50%-75%, Muslim and Hindu 24%-49%, indigenous beliefs 1%
Languages: Bemba (official) 30.1%, Nyanja (official) 10.7%, Tonga (official) 10.6%, Lozi (official) 5.7%, Chewa 4.9%, Nsenga 3.4%, Tumbuka 2.5%, Lunda (official) 2.2%, Kaonde (official) 2%, Lala 2%, Luvale (official) 1.7%, English (official) 1.7%, other 22.5% (2000 Census)
Literacy: *definition:* age 15 and over can read and write English
total population: 80.6%
male: 86.8%
female: 74.8% (2003 est.)
School life expectancy (primary to tertiary education): *total:* 7 years
male: 8 years
female: 7 years (2000)
Education expenditures:
1.3% of GDP (2008)
country comparison to the world: 160

GOVERNMENT

Country name: *conventional long form:* Republic of Zambia
conventional short form: Zambia
former: Northern Rhodesia
Government type: republic
Capital: *name:* Lusaka
geographic coordinates: 15 25 S, 28 17 E
time difference: UTC+2 (7 hours ahead of Washington, DC during Standard Time)
Administrative divisions: 9 provinces; Central, Copperbelt, Eastern, Luapula, Lusaka, Northern, North-Western, Southern, Western
Independence: 24 October 1964 (from the UK)
National holiday: Independence Day, 24 October (1964)
Constitution: 24 August 1991; amended in 1996 to establish presidential term limits
Legal system: mixed legal system of English common law and customary law; judicial review of legislative acts in an ad hoc constitutional council
International law organization participation: has not submitted an ICJ jurisdiction declaration; accepts ICCt jurisdiction
Suffrage: 18 years of age; universal
Executive branch: *chief of state:* President Rupiah BANDA (since 19 August 2008); Vice President George KUNDA (since 14 November 2008); the president is both the chief of state and head of government
head of government: President Rupiah BANDA (since 19 August 2008); Vice President George KUNDA (since 14 November 2008)
cabinet: Cabinet appointed by the president from among the members of the National Assembly
(For more information visit the World Leaders website)
elections: president elected by popular vote for a five-year term (eligible for a second term); election last held on 30 October 2008 (next to be held by October 2011); vice president appointed by the president; note—due to the death of former President Levy MWANAWASA, early elections were held to identify a replacement to serve out the remainder of his term
election results: Rupiah BANDA elected president; percent of vote—Rupiah BANDA 40.1%, Michael SATA 38.1%, Hakainde HICHILEMA 19.7%, Godfrey MIYANDA 0.8%, other 1.3%
Legislative branch: unicameral National Assembly (158 seats; 150 members are elected by popular vote, 8 members appointed by the president, to serve five-year terms)
elections: last held on 28 September 2006 (next to be held by October 2011)
election results: percent of vote by party—NA; seats by party—MMD 72, PF 44, UDA 27, ULP 2, NDF 1, independents 2; seats not determined 2
Judicial branch: Supreme Court (the final court of appeal; justices are appointed by the president); High Court (has unlimited jurisdiction to hear civil and criminal cases)
Political parties and leaders: Alliance for Democracy and Development or ADD [Charles MILUPI]; Forum for Democracy and Development or FDD [Edith NAWAKWI]; Heritage Party or HP [Godfrey MIYANDA]; Movement for Multiparty Democracy or MMD [Rupiah BANDA]; National Restoration Party or NAREP [Elias CHIPIMO]; Party of Unity for Democracy and Development or PUDD [Dan PULE]; Patriotic Front or PF [Michael SATA]; Reform Party [Nevers MUMBA]; United Democratic Alliance or UDA (a coalition of RP, ZADECO, PUDD, and ZRP); United Liberal Party or ULP [Sakwiba SIKOTA]; United National Independence Party or UNIP [Tilyenji KAUNDA]; United Party for National Development or UPND [Hakainde HICHILEMA]; Zambia Democratic Congress or ZADECO [Langton SICHONE]; Zambian Republican Party or ZRP [Benjamin MWILA]
Political pressure groups and leaders: NA
International organization participation: ACP, AfDB, AU, C, COMESA, FAO, G-77, IAEA, IBRD, ICAO, ICRM, IDA, IFAD, IFC, IFRCS, ILO, IMF, Interpol, IOC, IOM, IPU, ISO (correspondent), ITSO, ITU, ITUC, MIGA, MONUSCO, NAM, OPCW, PCA, SADC, UN, UNAMID, UNCTAD, UNESCO, UNHCR, UNIDO, UNMIL, UNMIS, UNOCI, UNWTO, UPU, WCO, WHO, WIPO, WMO, WTO
Diplomatic representation in the US: *chief of mission:* Ambassador Sheila Z. SIWELA
chancery: 2419 Massachusetts Avenue NW, Washington, DC 20008
telephone: [1] (202) 265-9717 through 9719
FAX: [1] (202) 332-0826
Diplomatic representation from the US: *chief of mission:* Ambassador Mark C. STORELLA
embassy: corner of Independence and United Nations Avenues, Lusaka
mailing address: P. O. Box 31617, Lusaka
telephone: [260] (211) 357-000
FAX: [260] (211) 357-224
Flag description: green field with a panel of three vertical bands of red (hoist side), black, and orange below a soaring orange eagle, on the outer edge of the flag; green stands for the country's natural resources and vegetation, red symbolizes the struggle for freedom, black the people of Zambia, and orange the country's mineral wealth; the eagle represents the people's ability to rise above the nation's problems
National anthem: *name:* "Lumbanyeni Zambia" (Stand and Sing of Zambia, Proud and Free)
lyrics/music: multiple/Enoch Mankayi SONTONGA
note: adopted 1964; the melody, from the popular song "God Bless Africa," is the same as that of Tanzania but with different lyrics; the melody is also incorporated into South Africa's anthem

ECONOMY

Economy—overview: Zambia's economy has experienced strong growth in recent years, with real GDP growth in 2005-10 about 6% per year. Privatization of government-owned copper mines in the 1990s relieved the government from covering mammoth losses generated by the industry and greatly increased copper mining output and profitability to spur economic growth. Copper output has increased steadily since 2004, due to higher copper prices and foreign investment. In 2005, Zambia qualified for debt relief under the Highly Indebted Poor Country Initiative, consisting of approximately USD 6 billion in debt relief. Poverty remains a significant problem in Zambia, despite a stronger economy. Zambia's dependency on copper makes it vulnerable to depressed commodity prices, but record high copper prices and a bumper maize crop in 2010 helped Zambia rebound quickly from the world economic slowdown that began in 2008. A high birth rate, relatively high HIV/AIDS burden, and market distorting agricultural policies have meant that Zambia's economic growth has not dramatically decreased the stubbornly high poverty rates.
GDP (purchasing power parity):
$20.04 billion (2010 est.)
country comparison to the world: 124
$18.63 billion (2009 est.)
$17.5 billion (2008 est.)
note: data are in 2010 US dollars
GDP (official exchange rate):
$16.19 billion (2010 est.)
GDP—real growth rate: 7.6% (2010 est.)
country comparison to the world: 24
6.4% (2009 est.)
5.7% (2008 est.)
GDP—per capita (PPP): $1,500 (2010 est.)
country comparison to the world: 201
$1,400 (2009 est.)
$1,400 (2008 est.)
note: data are in 2010 US dollars
GDP—composition by sector:
agriculture: 19.7%
industry: 33.7%
services: 46.6% (2010 est.)
Labor force: 5.524 million (2010 est.)
country comparison to the world: 70
Labor force—by occupation:
agriculture: 85%

industry: 6%
services: 9% (2004)
Unemployment rate: 14% (2006 est.)
country comparison to the world: 143
50% (2000 est.)
Population below poverty line:
64% (2006)
Household income or consumption by percentage share: *lowest 10%:* 1.2%
highest 10%: 38.8% (2004)
Distribution of family income—Gini index: 50.8 (2004)
country comparison to the world: 19
52.6 (1998)
Investment (gross fixed):
20.5% of GDP (2010 est.)
country comparison to the world: 83
Budget: *revenues:* $3.2 billion
expenditures: $3.743 billion (2010 est.)
Public debt: 24.1% of GDP (2010 est.)
country comparison to the world: 101
25.9% of GDP (2009 est.)
Inflation rate (consumer prices):
8.5% (2010 est.)
country comparison to the world: 189
13.4% (2009 est.)
Central bank discount rate:
8.39% (31 December 2009)
country comparison to the world: 22
14.49% (31 December 2008)
Commercial bank prime lending rate:
22.06% (31 December 2009 est.)
country comparison to the world: 22
19.06% (31 December 2008 est.)
Stock of narrow money: $1.234 billion (31 December 2010 est.)
country comparison to the world: 133
$983.3 million (31 December 2009 est.)
Stock of broad money: $3.573 billion (31 December 2010 est.)
country comparison to the world: 129
$2.744 billion (31 December 2009 est.)
Stock of domestic credit: $2.992 billion (31 December 2010 est.)
country comparison to the world: 119
$2.373 billion (31 December 2009 est.)
Market value of publicly traded shares:
$NA (31 December 2010)
country comparison to the world: 94
$2.346 billion (31 December 2007)
$1.186 billion (31 December 2006)
Agriculture—products: corn, sorghum, rice, peanuts, sunflower seed, vegetables, flowers, tobacco, cotton, sugarcane, cassava (tapioca), coffee; cattle, goats, pigs, poultry, milk, eggs, hides
Industries: copper mining and processing, construction, foodstuffs, beverages, chemicals, textiles, fertilizer, horticulture
Industrial production growth rate:
12.1% (2010 est.)
country comparison to the world: 12
Electricity—production:
9.752 billion kWh (2007 est.)
country comparison to the world: 92
Electricity—consumption:
8.838 billion kWh (2007 est.)
country comparison to the world: 91
Electricity—exports:
268 million kWh (2007)
Electricity—imports:
222 million kWh (2008 est.)
Oil—production: 159.5 bbl/day (2009 est.)
country comparison to the world: 109
Oil—consumption: 16,000 bbl/day (2009 est.)
country comparison to the world: 132
Oil—exports: 275.3 bbl/day (2007 est.)
country comparison to the world: 129
Oil—imports: 14,730 bbl/day (2007 est.)
country comparison to the world: 126
Oil—proved reserves: 0 bbl (1 January 2010 est.)
country comparison to the world: 208
Natural gas—production: 0 cu m (2008 est.)
country comparison to the world: 205
Natural gas—consumption: 0 cu m (2008 est.)
country comparison to the world: 142
Natural gas—exports: 0 cu m (2008 est.)
country comparison to the world: 205
Natural gas—imports: 0 cu m (2008 est.)
country comparison to the world: 74
Natural gas—proved reserves: 0 cu m (1 January 2010 est.)
country comparison to the world: 206
Current account balance: $-99 million (2010 est.)
country comparison to the world: 78
$-174 million (2009 est.)
Exports: $6.463 billion (2010 est.)
country comparison to the world: 102
$4.203 billion (2009 est.)
Exports—commodities: copper/cobalt 64%, cobalt, electricity; tobacco, flowers, cotton
Exports—partners: China 21.37%, Saudi Arabia 8.93%, Democratic Republic of the Congo 8.55%, South Korea 8.32%, Egypt 8.08%, South Africa 6.96%, India 5% (2009)
Imports: $4.949 billion (2010 est.)
country comparison to the world: 117
$3.735 billion (2009 est.)
Imports—commodities: machinery, transportation equipment, petroleum products, electricity, fertilizer, foodstuffs, clothing
Imports—partners: South Africa 51.78%, UAE 7.7%, China 5.85%, Democratic Republic of the Congo 4.22% (2009)
Reserves of foreign exchange and gold:
$2.287 billion (31 December 2010 est.)
country comparison to the world: 97
$1.892 billion (31 December 2009 est.)
Debt—external: $3.495 billion (31 December 2010 est.)
country comparison to the world: 120
$3.091 billion (31 December 2009 est.)
Stock of direct foreign investment—at home: $NA
Stock of direct foreign investment—abroad: $NA
Exchange rates: Zambian kwacha (ZMK) per US dollar–4,823.6 (2010), 5,046.1 (2009), 3,512.9 (2008), 3,990.2 (2007), 3,601.5 (2006)

COMMUNICATIONS

Telephones—main lines in use: 90,300 (2009)
country comparison to the world: 146
Telephones—mobile cellular: 4.407 million (2009)
country comparison to the world: 101
Telephone system: *general assessment:* among the best in Sub-Saharan Africa
domestic: high-capacity microwave radio relay connects most larger towns and cities; several cellular telephone services in operation and network coverage is improving; domestic satellite system being installed to improve telephone service in rural areas; Internet service is widely available; very small aperture terminal (VSAT) networks are operated by private firms
international: country code–260; satellite earth stations–2 Intelsat (1 Indian Ocean and 1 Atlantic Ocean), 3 owned by Zamtel
Broadcast media: state-owned Zambia National Broadcasting Corporation (ZNBC) operates 1 TV station and is the principal local-content provider; several private TV stations are available; multi-channel subscription TV services are obtainable; ZNBC operates 3 radio networks; about 2 dozen private radio stations also broadcasting; relays of at least 2 international broadcasters are accessible in Lusaka and Kitwe (2007)
Internet country code: .zm
Internet hosts: 14,771 (2010)
country comparison to the world: 118
Internet users: 816,200 (2009)
country comparison to the world: 105

TRANSPORTATION

Airports: 94 (2010)
country comparison to the world: 64
Airports—with paved runways: *total:* 8
over 3,047 m: 1
2,438 to 3,047 m: 3
1,524 to 2,437 m: 3
914 to 1,523 m: 1 (2010)
Airports—with unpaved runways:
total: 86
2,438 to 3,047 m: 1
1,524 to 2,437 m: 4
914 to 1,523 m: 63
under 914 m: 18 (2010)
Pipelines: oil 771 km (2010)
Railways: *total:* 2,157 km
country comparison to the world: 69
narrow gauge: 2,157 km 1.067-m gauge
note: includes 891 km of the Tanzania-Zambia Railway Authority (TAZARA) (2008)
Roadways: *total:* 91,440 km
country comparison to the world: 54
paved: 20,117 km
unpaved: 71,323 km (2001)
Waterways: 2,250 km (includes Lake Tanganyika and the Zambezi and Luapula rivers) (2010)
country comparison to the world: 39
Ports and terminals: Mpulungu

MILITARY

Military branches: Zambian National Defense Force (ZNDF): Zambian Army, Zambian Air Force, National Service (2009)
Military service age and obligation: 18-27 years of age for male and female voluntary military service (16 years of age with parental consent); no conscription; Zambian citizenship required; mandatory HIV testing upon enlistment; mandatory retirement for officers at age 65 (Army, Air Force) (2010)
Manpower available for military service:
males age 16–49: 3,041,069
females age 16–49: 2,948,291 (2010 est.)

Manpower fit for military service: *males age 16–49:* 1,745,656
females age 16–49: 1,688,670 (2010 est.)
Manpower reaching militarily significant age annually: *male:* 158,592
female: 158,805 (2010 est.)
Military expenditures:
1.8% of GDP (2005 est.)
country comparison to the world: 84

TRANSNATIONAL ISSUES

Disputes—international: in 2004, Zimbabwe dropped objections to plans between Botswana and Zambia to build a bridge over the Zambezi River, thereby de facto recognizing a short, but not clearly delimited, Botswana-Zambia boundary in the river
Refugees and internally displaced persons: refugees (country of origin): 42,565 (Angola); 60,874 (Democratic Republic of the Congo); 4,100 (Rwanda) (2007)
Illicit drugs: transshipment point for moderate amounts of methaqualone, small amounts of heroin, and cocaine bound for southern Africa and possibly Europe; a poorly developed financial infrastructure coupled with a government commitment to combating money laundering make it an unattractive venue for money launderers; major consumer of cannabis

ZIMBABWE

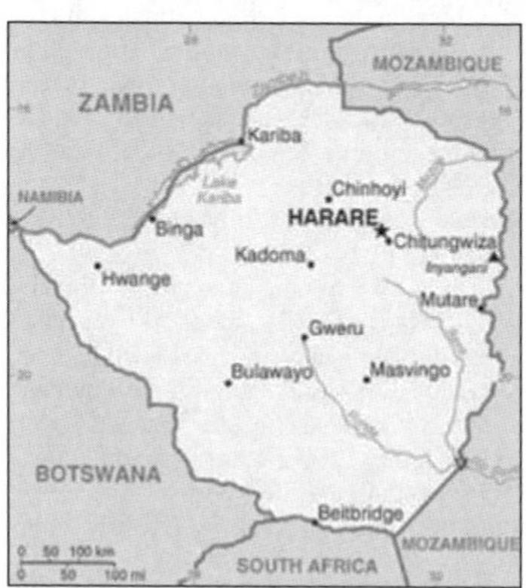

INTRODUCTION

Background: The UK annexed Southern Rhodesia from the [British] South Africa Company in 1923. A 1961 constitution was formulated that favored whites in power. In 1965 the government unilaterally declared its independence, but the UK did not recognize the act and demanded more complete voting rights for the black African majority in the country (then called Rhodesia). UN sanctions and a guerrilla uprising finally led to free elections in 1979 and independence (as Zimbabwe) in 1980. Robert MUGABE, the nation's first prime minister, has been the country's only ruler (as president since 1987) and has dominated the country's political system since independence. His chaotic land redistribution campaign, which began in 2000, caused an exodus of white farmers, crippled the economy, and ushered in widespread shortages of basic commodities. Ignoring international condemnation, MUGABE rigged the 2002 presidential election to ensure his reelection. The ruling ZANU-PF party used fraud and intimidation to win a two-thirds majority in the March 2005 parliamentary election, allowing it to amend the constitution at will and recreate the Senate, which had been abolished in the late 1980s. In April 2005, Harare embarked on Operation Restore Order, ostensibly an urban rationalization program, which resulted in the destruction of the homes or businesses of 700,000 mostly poor supporters of the opposition. President MUGABE in June 2007 instituted price controls on all basic commodities causing panic buying and leaving store shelves empty for months. General elections held in March 2008 contained irregularities but still amounted to a censure of the ZANU-PF-led government with the opposition winning a majority of seats in parliament. MDC opposition leader Morgan TSVANGIRAI won the most votes in the presidential polls, but not enough to win outright. In the lead up to a run-off election in late June 2008, considerable violence enacted against opposition party members led to the withdrawal of TSVANGIRAI from the ballot. Extensive evidence of violence and intimidation resulted in international condemnation of the process. Difficult negotiations over a power-sharing government, in which MUGABE remained president and TSVANGIRAI became prime minister, were finally settled in February 2009, although the leaders have yet failed to agree upon many key outstanding governmental issues. Mugabe in October publicly called for early elections in 2011-two years before his term ends-but no election date has been set.

GEOGRAPHY

Location: Southern Africa, between South Africa and Zambia
Geographic coordinates: 20 00 S, 30 00 E
Map references: Africa
Area: *total:* 390,757 sq km
country comparison to the world: 60
land: 386,847 sq km
water: 3,910 sq km
Area—comparative: slightly larger than Montana
Land boundaries: *total:* 3,066 km
border countries: Botswana 813 km, Mozambique 1,231 km, South Africa 225 km, Zambia 797 km
Coastline: 0 km (landlocked)
Maritime claims: none (landlocked)
Climate: tropical; moderated by altitude; rainy season (November to March)
Terrain: mostly high plateau with higher central plateau (high veld); mountains in east
Elevation extremes: *lowest point:* junction of the Runde and Save Rivers 162 m
highest point: Inyangani 2,592 m
Natural resources: coal, chromium ore, asbestos, gold, nickel, copper, iron ore, vanadium, lithium, tin, platinum group metals
Land use: *arable land:* 8.24%
permanent crops: 0.33%
other: 91.43% (2005)
Irrigated land: 1,740 sq km (2003)
Total renewable water resources:
20 cu km (1987)
Freshwater withdrawal (domestic/industrial/agricultural): *total:* 4.21 cu km/yr (14%/7%/79%)
per capita: 324 cu m/yr (2002)
Natural hazards: recurring droughts; floods and severe storms are rare
Environment—current issues: deforestation; soil erosion; land degradation; air and water pollution; the black rhinoceros herd—once the largest concentration of the species in the world—has been significantly reduced by poaching; poor mining practices have led to toxic waste and heavy metal pollution
Environment—international agreements: *party to:* Biodiversity, Climate Change, Desertification, Endangered Species, Law of the Sea, Ozone Layer Protection
signed, but not ratified: none of the selected agreements
Geography—note: landlocked; the Zambezi forms a natural riverine boundary with Zambia; in full flood (February-April) the massive Victoria Falls on the river forms the world's largest curtain of falling water

PEOPLE

Population: 12,084,304 (July 2011 est.)
country comparison to the world: 72
note: estimates for this country explicitly take into account the effects of excess mortality due to AIDS; this can result in lower life expectancy, higher infant mortality, higher death rates, lower population growth rates, and changes in the distribution of population by age and sex than would otherwise be expected
Age structure: *0–14 years:* 41.9% (male 2,555,916/female 2,504,947)
15–64 years: 54.3% (male 3,063,580/female 3,500,366)
65 years and over: 3.8% (male 193,380/female 266,115) (2011 est.)
Median age: *total:* 18.3 years
male: 17.4 years
female: 19.2 years (2011 est.)
Population growth rate: 4.31% (2011 est.)
country comparison to the world: 1

Birth rate: 31.86 births/1,000 population (2011 est.)
country comparison to the world: 41
Death rate: 13.58 deaths/1,000 population (July 2011 est.)
country comparison to the world: 19
Net migration rate: 24.83 migrant(s)/1,000 population
country comparison to the world: 1
note: there is an increasing flow of Zimbabweans into South Africa and Botswana in search of better economic opportunities (2011 est.)
Urbanization: *urban population:* 38% of total population (2010)
rate of urbanization: 3.4% annual rate of change (2010-15 est.)
Major cities—population: HARARE (capital) 1.606 million (2009)
Sex ratio: *at birth:* 1.03 male(s)/female
under 15 years: 1.02 male(s)/female
15–64 years: 0.83 male(s)/female
65 years and over: 0.75 male(s)/female
total population: 0.91 male(s)/female (2011 est.)
Infant mortality rate: *total:* 29.5 deaths/1,000 live births
country comparison to the world: 70
male: 31.98 deaths/1,000 live births
female: 26.94 deaths/1,000 live births (2011 est.)
Life expectancy at birth: *total population:* 49.64 years
country comparison to the world: 215
male: 49.93 years
female: 49.34 years (2011 est.)
Total fertility rate: 3.63 children born/woman (2011 est.)
country comparison to the world: 44
HIV/AIDS—adult prevalence rate: 14.3% (2009 est.)
country comparison to the world: 5
HIV/AIDS—people living with HIV/AIDS: 1.2 million (2009 est.)
country comparison to the world: 8
HIV/AIDS—deaths:
83,000 (2009 est.)
country comparison to the world: 5
Major infectious diseases:
degree of risk: high
food or waterborne diseases: bacterial and protozoal diarrhea, hepatitis A, and typhoid fever
vectorborne disease: malaria
water contact disease: schistosomiasis
animal contact disease: rabies (2009)
Drinking water source: *Improved:*
urban: 99% of population
rural: 72% of population
total: 82% of population
Unimproved:
urban: 1% of population
rural: 28% of population
total: 18% of population (2008)
Sanitation facility access: *Improved:*
urban: 56% of population
rural: 37% of population
total: 44% of population
Unimproved:
urban: 44% of population
rural: 63% of population
total: 56% of population (2008)
Nationality: *noun:* Zimbabwean(s)
adjective: Zimbabwean
Ethnic groups: African 98% (Shona 82%, Ndebele 14%, other 2%), mixed and Asian 1%, white less than 1%
Religions: syncretic (part Christian, part indigenous beliefs) 50%, Christian 25%, indigenous beliefs 24%, Muslim and other 1%
Languages: English (official), Shona, Sindebele (the language of the Ndebele, sometimes called Ndebele), numerous but minor tribal dialects
Literacy: *definition:* age 15 and over can read and write English
total population: 90.7%
male: 94.2%
female: 87.2% (2003 est.)
School life expectancy (primary to tertiary education): *total:* 9 years
male: 10 years
female: 9 years (2003)
Education expenditures: NA

GOVERNMENT

Country name: *conventional long form:* Republic of Zimbabwe
conventional short form: Zimbabwe
former: Southern Rhodesia, Rhodesia
Government type: parliamentary democracy
Capital: *name:* Harare
geographic coordinates: 17 50 S, 31 03 E
time difference: UTC+2 (7 hours ahead of Washington, DC during Standard Time)
Administrative divisions: 8 provinces and 2 cities* with provincial status; Bulawayo*, Harare*, Manicaland, Mashonaland Central, Mashonaland East, Mashonaland West, Masvingo, Matabeleland North, Matabeleland South, Midlands
Independence: 18 April 1980 (from the UK)
National holiday: Independence Day, 18 April (1980)
Constitution: 21 December 1979
Legal system: mixed legal system of English common law, Roman-Dutch civil law, and customary law
International law organization participation: has not submitted an ICJ jurisdiction declaration; non-party state to the ICCt
Suffrage: 18 years of age; universal
Executive branch: *chief of state:* Executive President Robert Gabriel MUGABE (since 31 December 1987); Vice President John NKOMO (since December 2009) and Vice President Joyce MUJURU (since 6 December 2004)
head of government: Prime Minister Morgan TSVANGIRAI (since 11 February 2009); Deputy Prime Minister Arthur MUTAMBARA
cabinet: Cabinet appointed by the president and prime minister; responsible to the House of Assembly
(For more information visit the World Leaders website)
elections: presidential candidates nominated with a nomination paper signed by at least 10 registered voters (at least one from each province) and elected by popular vote for a five-year term (no term limits); elections last held on 28 March 2008 followed by a run-off on 27 June 2008 (next to be held in 2013); co-vice presidents drawn from party leadership
election results: Robert Gabriel MUGABE reelected president; percent of vote—Robert Gabriel MUGABE 85.5%, Morgan TSVANGIRAI 9.3%, other 5.2%; note—first round voting results—Morgan TSVANGIRAI 47.9%, Robert Gabriel MUGABE 43.2%, Simba MAKONI 8.3%, other 0.6%; first-round round polls were deemed to be flawed suppressing TSVANGIRAI's results; the 27 June 2008 run-off between MUGABE and TSVANGIRAI was severely flawed and internationally condemned
Legislative branch: bicameral Parliament consists of a Senate (93 seats—60 members elected by popular vote for a five-year term, 10 provincial governors nominated by the president and the prime minister, 16 traditional chiefs elected by the Council of Chiefs, 2 seats held by the president and deputy president of the Council of Chiefs, and 5 members appointed by the president) and a House of Assembly (210 seats—members elected by popular vote for five-year terms)
elections: last held on 28 March 2008 (next to be held in 2013)
election results: Senate—percent of vote by party—MDC 51.6%, ZANU-PF 45.8%, other 2.6%; seats by party—MDC 30, ZANU-PF 30; House of Assembly—percent of vote by party—MDC 51.3%, ZANU-PF 45.8%, other 2.9%; seats by party—MDC 109, ZANU-PF 97, other 4
Judicial branch: Supreme Court; High Court
Political parties and leaders: African National Party or ANP [Egypt DZINEMUNHENZVA]; Movement for Democratic Change or MDC [Morgan TSVANGIRAI]; Movement for Democratic Change—Mutambara or MDC-M [Arthur MUTAMBARA] (splinter faction of the MDC); Peace Action is Freedom for All or PAFA; United Parties [Abel MUZOREWA]; United People's Party or UPP [Daniel SHUMBA]; Zimbabwe African National Union-Ndonga or ZANU-Ndonga [Wilson KUMBULA]; Zimbabwe African National Union-Patriotic Front or ZANU-PF [Robert Gabriel MUGABE]; Zimbabwe African Peoples Union or ZAPU [Agrippa MADLELA]; Zimbabwe Youth in Alliance or ZIYA
Political pressure groups and leaders: Crisis in Zimbabwe Coalition; National Constitutional Assembly or NCA [Lovemore MADHUKU]; Women of Zimbabwe Arise or WOZA [Jenny WILLIAMS]; Zimbabwe Congress of Trade Unions or ZCTU [Wellington CHIBEBE]
International organization participation: ACP, AfDB, AU, COMESA, FAO, G-15, G-77, IAEA, IBRD, ICAO, ICRM, IDA, IFAD, IFC, IFRCS, ILO, IMF, IMO, Interpol, IOC, IOM, IPU, ISO (correspondent), ITSO, ITU, ITUC, MIGA, NAM, OPCW, PCA, SADC, UN, UNAMID, UNCTAD, UNESCO, UNIDO, UNMIL, UNMIS, UNOCI, UNWTO, UPU, WCO, WFTU, WHO, WIPO, WMO, WTO
Diplomatic representation in the US: *chief of mission:* Ambassador Machivenyika MAPURANGA

chancery: 1608 New Hampshire Avenue NW, Washington, DC 20009
telephone: [1] (202) 332-7100
FAX: [1] (202) 483-9326

Diplomatic representation from the US:
chief of mission: Ambassador Charles A. RAY
embassy: 172 Herbert Chitepo Avenue, Harare
mailing address: P. O. Box 3340, Harare
telephone: [263] (4) 250-593 through 250-594
FAX: [263] (4) 796-488, or 722-618

Flag description: seven equal horizontal bands of green, yellow, red, black, red, yellow, and green with a white isosceles triangle edged in black with its base on the hoist side; a yellow Zimbabwe bird representing the long history of the country is superimposed on a red five-pointed star in the center of the triangle, which symbolizes peace; green represents agriculture, yellow mineral wealth, red the blood shed to achieve independence, and black stands for the native people

National anthem: *name:* "Kalibusiswe Ilizwe leZimbabwe" [Northern Ndebele language] "Simudzai Mureza WeZimbabwe" [Shona] (Blessed Be the Land of Zimbabwe)
lyrics/music: Solomon MUTSWAIRO/ Fred Lecture CHANGUNDEGA
note: adopted 1994

ECONOMY

Economy—overview: Zimbabwe's economy is growing at a brisk pace despite continuing political uncertainty. Following a decade of contraction, Zimbabwe's economy recorded real growth of 5.9% in 2010. But the government of Zimbabwe still faces a number of difficult economic problems, including a large external debt burden and insufficient formal employment. Zimbabwe's 1998-2002 involvement in the war in the Democratic Republic of the Congo drained hundreds of millions of dollars from the economy. The government's land reform program, characterized by chaos and violence, has badly damaged the commercial farming sector, the traditional source of exports and foreign exchange and the provider of 400,000 jobs, turning Zimbabwe into a net importer of food products. The EU and the US provide food aid on humanitarian grounds, though on a smaller scale than before. Until early 2009, the Reserve Bank of Zimbabwe routinely printed money to fund the budget deficit, causing hyperinflation. The power-sharing government formed in February 2009 has led to some economic improvements, including the cessation of hyperinflation by eliminating the use of the Zimbabwe dollar and removing price controls. The economy is registering its first growth in a decade, but will be reliant on further political improvement for greater growth.

GDP (purchasing power parity): $5.457 billion (2010 est.)
country comparison to the world: 159
$5.006 billion (2009 est.)
$4.723 billion (2008 est.)
note: data are in 2010 US dollars

GDP (official exchange rate):
$7.474 billion
note: in 2009, the Zimbabwean dollar was taken out of circulation, making Zimbabwe's GDP at the official exchange rate a highly inaccurate statistic (2010 est.)

GDP—real growth rate: 9% (2010 est.)
country comparison to the world: 10
6% (2009 est.)
-17.7% (2008 est.)

GDP—per capita (PPP): $500 (2010 est.)
country comparison to the world: 225
$400 (2009 est.)
$400 (2008 est.)
note: data are in 2010 US dollars

GDP—composition by sector:
agriculture: 19.5%
industry: 24%
services: 56.5% (2010 est.)

Labor force: 3.848 million (2010 est.)
country comparison to the world: 90

Labor force—by occupation:
agriculture: 66%
industry: 10%
services: 24% (1996)

Unemployment rate: 95% (2009 est.)
country comparison to the world: 200
80% (2005 est.)
note: figures reflect underemployment; true unemployment is unknown and, under current economic conditions, unknowable

Population below poverty line: 68% (2004)

Household income or consumption by percentage share: *lowest 10%:* 2%
highest 10%: 40.4% (1995)

Distribution of family income—Gini index: 50.1 (2006)
country comparison to the world: 23
50.1 (1995)

Investment (gross fixed): 21% of GDP (2010 est.)
country comparison to the world: 77

Budget: *revenues:* $2.25 billion
expenditures: $2.25 billion (2010 est.)

Public debt: 149% of GDP (2010 est.)
country comparison to the world: 4
162.5% of GDP (2009 est.)

Inflation rate (consumer prices):
5.03% (2010 est.)
country comparison to the world: 142
5.1% (2009 est.)

Central bank discount rate:
NA% (31 December 2010)
country comparison to the world: 1
975% (31 December 2007)

Commercial bank prime lending rate: 10% (January 2011 est.)
country comparison to the world: 1
578.96% (31 December 2007)

Stock of narrow money: $NA (31 December 2010 est.)
$2.151 million (31 December 2008 est.)
note: Zimbabwe's central bank no longer publishes data on monetary aggregates, except for bank deposits, which amounted to $2.1 billion in November 2010; the Zimbabwe dollar stopped circulating in early 2009; since then, the US dollar and South African rand have been the most frequently used currencies; there are no reliable estimates of the amount of foreign currency circulating in Zimbabwe

Stock of broad money: $3.057 million (31 December 2010 est.)
country comparison to the world: 188
$NA (31 December 2009 est.)

Stock of domestic credit: $NA

Market value of publicly traded shares: $4.47 billion (10 January 2011)
country comparison to the world: 80
$5.333 billion (31 December 2007)
$26.56 billion (31 December 2006)

Agriculture—products: corn, cotton, tobacco, wheat, coffee, sugarcane, peanuts; sheep, goats, pigs

Industries: mining (coal, gold, platinum, copper, nickel, tin, diamonds, clay, numerous metallic and nonmetallic ores), steel; wood products, cement, chemicals, fertilizer, clothing and footwear, foodstuffs, beverages

Industrial production growth rate: 4% (2010 est.)
country comparison to the world: 90

Electricity—production: 8.89 billion kWh (2007 est.)
country comparison to the world: 94

Electricity—consumption: 10.89 billion kWh (2007 est.)
country comparison to the world: 83

Electricity—exports: 32 million kWh (2007 est.)

Electricity—imports: 2.691 billion kWh (2007 est.)

Oil—production: 0 bbl/day (2009 est.)
country comparison to the world: 145

Oil—consumption: 11,000 bbl/day (2009 est.)
country comparison to the world: 145

Oil—exports: 0 bbl/day (2007 est.)
country comparison to the world: 144

Oil—imports: 13,830 bbl/day (2007 est.)
country comparison to the world: 132

Oil—proved reserves: 0 bbl (1 January 2010 est.)
country comparison to the world: 209

Natural gas—production:
0 cu m (2008 est.)
country comparison to the world: 206

Natural gas—consumption:
0 cu m (2008 est.)
country comparison to the world: 143

Natural gas—exports: 0 cu m (2008 est.)
country comparison to the world: 206

Natural gas—imports: 0 cu m (2008 est.)
country comparison to the world: 75

Natural gas—proved reserves: 0 cu m (1 January 2010 est.)
country comparison to the world: 207

Current account balance:
$-1.503 billion (2010 est.)
country comparison to the world: 150
$-807.5 million (2009 est.)

Exports: $2.54 billion (2010 est.)
country comparison to the world: 126
$1.213 billion (2009 est.)

Exports—commodities: platinum, cotton, tobacco, gold, ferroalloys, textiles/clothing

Exports—partners: Democratic Republic of the Congo 14.82%, South Africa 13.39%, Botswana 13.23%, China 7.82%, Zambia 7.3%, Netherlands 5.39%, UK 4.93% (2009)

Imports: $4.043 billion (2010 est.)
country comparison to the world: 128
$2.413 billion (2009 est.)

Imports—commodities: machinery and transport equipment, other manufactures, chemicals, fuels, food products
Imports—partners: South Africa 62.24%, China 4.2% (2009)
Reserves of foreign exchange and gold: $376 million (31 December 2010 est.)
country comparison to the world: 126
$351 million (31 December 2009 est.)
Debt—external: $7.662 billion (31 December 2010 est.)
country comparison to the world: 94
$5.667 billion (31 December 2009 est.)
Stock of direct foreign investment—at home: $NA
Stock of direct foreign investment—abroad: $NA
Exchange rates: Zimbabwean dollars (ZWD) per US dollar–234.25 (2010), 234 (2009), 30,000 (2007), 162 (2006), 78 (2005)

COMMUNICATIONS

Telephones—main lines in use: 385,100 (2009)
country comparison to the world: 104
Telephones—mobile cellular: 2.991 million (2009)
country comparison to the world: 113
Telephone system: *general assessment:* system was once one of the best in Africa, but now suffers from poor maintenance
domestic: consists of microwave radio relay links, open-wire lines, radiotelephone communication stations, fixed wireless local loop installations, and a substantial mobile-cellular network; Internet connection is available in Harare and planned for all major towns and for some of the smaller ones
international: country code–263; satellite earth stations–2 Intelsat; 2 international digital gateway exchanges (in Harare and Gweru)
Broadcast media: government owns all local radio and television stations; foreign shortwave broadcasts and satellite television are available to those who can afford antennas and receivers; in rural areas, access to television broadcasts is extremely limited (2007)
Internet country code: .zw
Internet hosts: 29,866 (2010)
country comparison to the world: 98
Internet users: 1.423 million (2009)
country comparison to the world: 84

TRANSPORTATION

Airports: 216 (2010)
country comparison to the world: 28
Airports—with paved runways:
total: 19
over 3,047 m: 3
2,438 to 3,047 m: 2
1,524 to 2,437 m: 5
914 to 1,523 m: 9 (2010)
Airports—with unpaved runways:
total: 197
1,524 to 2,437 m: 3
914 to 1,523 m: 120
under 914 m: 74 (2010)
Pipelines:
refined products 270 km (2010)
Railways: *total:* 3,077 km
country comparison to the world: 54
narrow gauge: 3,077 km 1.067-m gauge (313 km electrified) (2008)
Roadways: *total:* 97,267 km
country comparison to the world: 45
paved: 18,481 km
unpaved: 78,786 km (2002)
Waterways: (some navigation possible on Lake Kariba) (2009)
Ports and terminals: Binga, Kariba

MILITARY

Military branches: Zimbabwe Defense Forces (ZDF): Zimbabwe National Army (ZNA), Air Force of Zimbabwe (AFZ), Zimbabwe Republic Police (ZRP) (2009)
Military service age and obligation: 18-24 years of age for compulsory military service; women are eligible to serve (2010)
Manpower available for military service:
males age 16–49: 2,616,051
females age 16–49: 2,868,376 (2010 est.)
Manpower fit for military service: *males age 16–49:* 1,528,166
females age 16–49: 1,646,041 (2010 est.)
Manpower reaching militarily significant age annually: *male:* 154,870
female: 152,550 (2010 est.)
Military expenditures:
3.8% of GDP (2006)
country comparison to the world: 27

TRANSNATIONAL ISSUES

Disputes—international: Botswana built electric fences and South Africa has placed military along the border to stem the flow of thousands of Zimbabweans fleeing to find work and escape political persecution; Namibia has supported, and in 2004 Zimbabwe dropped objections to, plans between Botswana and Zambia to build a bridge over the Zambezi River, thereby de facto recognizing a short, but not clearly delimited, Botswana-Zambia boundary in the river
Refugees and internally displaced persons: refugees (country of origin): 2,500 (Democratic Republic of Congo)
IDPs: 569,685 (MUGABE-led political violence, human rights violations, land reform, and economic collapse) (2007)
Trafficking in persons: current situation: Zimbabwe is a source, transit, and destination country for men, women, and children trafficked for the purposes of forced labor and sexual exploitation; large scale migration of Zimbabweans to surrounding countries–as they flee a progressively more desperate situation at home–has increased; rural Zimbabwean men, women, and children are trafficked internally to farms for agricultural labor and domestic servitude and to cities for domestic labor and commercial sexual exploitation; NGOs believe internal trafficking increased during the year, largely due to the closure of schools, worsening political violence, and a faltering economy; young men and boys are trafficked to South Africa for farm work, often laboring for months in South Africa without pay before "employers" have them arrested and deported as illegal immigrants; young women and girls are lured abroad with false employment offers that result in involuntary domestic servitude or commercial sexual exploitation; men, women, and children from neighboring states are trafficked through Zimbabwe en route to South Africa
tier rating: Tier 3–the Government of Zimbabwe does not fully comply with the minimum standards for the elimination of trafficking and is not making significant efforts to do so; the government made minimal progress in combating trafficking in 2008, and members of its military and the former ruling party's youth militias perpetrated acts of trafficking on local populations; anti-trafficking efforts were further weakened as it failed to address Zimbabwe's economic and social problems during the reporting period, thus increasing the population's vulnerability to trafficking within and outside of the country (2009)
Illicit drugs: transit point for cannabis and South Asian heroin, mandrax, and methamphetamines en route to South Africa

APPENDIX A

ABBREVIATIONS

ABEDA	Arab Bank for Economic Development in Africa
ACP Group	African, Caribbean, and Pacific Group of States
ADB	Asian Development Bank
AfDB	African Development Bank
AFESD	Arab Fund for Economic and Social Development
AG	Australia Group
Air Pollution	Convention on Long-Range Transboundary Air Pollution
Air Pollution-Nitrogen Oxides	Protocol to the 1979 Convention on Long-Range Transboundary Air Pollution Concerning the Control of Emissions of Nitrogen Oxides or Their Transboundary Fluxes
Air Pollution-Persistent Organic Pollutants	Protocol to the 1979 Convention on Long-Range Transboundary Air Pollution on Persistent Organic Pollutants
Air Pollution-Sulphur 85	Protocol to the 1979 Convention on Long-Range Transboundary Air Pollution on the Reduction of Sulphur Emissions or Their Transboundary Fluxes by at Least 30%
Air Pollution-Sulphur 94	Protocol to the 1979 Convention on Long-Range Transboundary Air Pollution on Further Reduction of Sulphur Emissions
Air Pollution-Volatile Organic Compounds	Protocol to the 1979 Convention on Long-Range Transboundary Air Pollution Concerning the Control of Emissions of Volatile Organic Compounds or Their Transboundary Fluxes
AMF	Arab Monetary Fund
AMU	Arab Maghreb Union
Antarctic Marine Living Resources	Convention on the Conservation of Antarctic Marine Living Resources
Antarctic Seals	Convention for the Conservation of Antarctic Seals
Antarctic-Environmental Protocol	Protocol on Environmental Protection to the Antarctic Treaty
ANZUS	Australia-New Zealand-United States Security Treaty
OSIS	Alliance of Small Island States
APEC	Asia-Pacific Economic Cooperation
Arabsat	Arab Satellite Communications Organization
ARF	ASEAN Regional Forum
ASEAN	Association of Southeast Asian Nations
AU	African Union
Autodin	Automatic Digital Network
BA	Baltic Assembly
bbl/day	barrels per day
BCIE	Central American Bank for Economic Integration
BDEAC	Central African States Development Bank
Benelux	Benelux Union
BGN	United States Board on Geographic Names
BIMSTEC	Bay of Bengal Initiative for Multisectoral Technical and Economic Cooperation
Biodiversity	Convention on Biological Diversity
BIS	Bank for International Settlements
BSEC	Black Sea Economic Cooperation Zone
C	Commonwealth
c.i.f.	cost, insurance, and freight
CACM	Central American Common Market
CAEU	Council of Arab Economic Unity
CAN	Andean Community of Nations
Caricom	Caribbean Community and Common Market
CB	citizen's band mobile radio communications
CBSS	Council of the Baltic Sea States
CCC	Customs Cooperation Council
CD	Community of Democracies
CDB	Caribbean Development Bank
CE	Council of Europe
CEI	Central European Initiative
CEMA	Council for Mutual Economic Assistance
CEMAC	Economic and Monetary Community of Central Africa
CEPGL	Economic Community of the Great Lakes Countries
CEPT	Conference Europeanne des Poste et Telecommunications
CERN	European Organization for Nuclear Research
CIA	Central Intelligence Agency
CICA	Conference of Interaction and Confidence-Building Measures in Asia
CIS	Commonwealth of Independent States
CITES	see Endangered Species
Climate Change	United Nations Framework Convention on Climate Change
Climate Change-Kyoto Protocol	Kyoto Protocol to the United Nations Framework Convention on Climate Change
COCOM	Coordinating Committee on Export Controls
COMESA	Common Market for Eastern and Southern Africa
Comsat	Communications Satellite Corporation
CP	Colombo Plan
CPLP	Comunidade dos Paises de Lingua Portuguesa

CSN	South American Community of Nations became UNASUL - Union of South American Nations
CSN	Union of South American Nations
CSTO	Collective Security Treaty Organization
CTBTO	Preparation commission for the Nuclear-Ban-Treaty Operation
CY	calendar year
D-8	Developing Eight
DC	developed country
DDT	dichloro-diphenyl-trichloro-ethane
Desertification	United Nations Convention to Combat Desertification in Those Countries Experiencing Serious Drought and/or Desertification, Particularly in Africa
DIA	United States Defense Intelligence Agency
DSN	Defense Switched Network
DST	daylight savings time
DWT	deadweight ton
EAC	East African Community
EADB	East African Development Bank
EAEC	Eurasian Economic Community
EAPC	Euro-Atlantic Partnership Council
EAS	East Asia Summit
EBRD	European Bank for Reconstruction and Development
EC	European Community or European Commission
ECA	Economic Commission for Africa
ECE	Economic Commission for Europe
ECLAC	Economic Commission for Latin America and the Caribbean
ECO	Economic Cooperation Organization
ECOSOC	Economic and Social Council
ECOWAS	Economic Community of West African States
ECSC	European Coal and Steel Community
EE	Eastern Europe
EEC	European Economic Community
EEZ	exclusive economic zone
EFTA	European Free Trade Association
EIB	European Investment Bank
EMU	European Monetary Union
Endangered Species	Convention on the International Trade in Endangered Species of Wild Flora and Fauna (CITES)
Entente	Council of the Entente
Environmental Modification	Convention on the Prohibition of Military or Any Other Hostile Use of Environmental Modification Techniques
ESA	European Space Agency
ESCAP	Economic and Social Commission for Asia and the Pacific
ESCWA	Economic and Social Commission for Western Asia
est.	estimate
EU	European Union
Euratom	European Atomic Energy Community
Eutelsat	European Telecommunications Satellite Organization
Ex-Im	Export-Import Bank of the United States
f.o.b.	free on board
FAO	Food and Agriculture Organization
FATF	Financial Action Task Force
FAX	facsimile
FLS	Front Line States
FOC	flags of convenience
FSU	former Soviet Union
FY	fiscal year
FZ	Franc Zone
G-2	Group of 2
G-3	Group of 3
G-5	Group of 5
G-6	Group of 6
G-7	Group of 7
G-8	Group of 8
G-9	Group of 9
G-10	Group of 10
G-15	Group of 15
G-11	Group of 11
G-24	Group of 24
G-77	Group of 77
GATT	General Agreement on Tariffs and Trade; now WTO
GCC	Gulf Cooperation Council
GCTU	General Confederation of Trade Unions
GDP	gross domestic product
GMT	Greenwich Mean Time
GNP	gross national product

GRT	gross register ton
GSM	global system for mobile cellular communications
GUAM	Organization for Democracy and Economic Development; acronym for member states - Georgia, Ukraine, Azerbaijan, Moldova
GWP	gross world product
Hazardous Wastes	Basel Convention on the Control of Transboundary Movements of Hazardous Wastes and Their Disposal
HF	high-frequency
HIV/AIDS	human immunodeficiency virus/acquired immune deficiency syndrome
IADB	Inter-American Development Bank
IAEA	International Atomic Energy Agency
IANA	Internet Assigned Numbers Authority
IBRD	International Bank for Reconstruction and Development (World Bank)
ICAO	International Civil Aviation Organization
ICC	International Chamber of Commerce
ICCt	International Criminal Court
ICJ	International Court of Justice (World Court)
ICRC	International Committee of the Red Cross
ICRM	International Red Cross and Red Crescent Movement
ICSID	International Center for Secretariat of Investment Disputes
ICTR	International Criminal Tribunal for Rwanda
ICTY	International Criminal Tribunal for the former Yugoslavia
IDA	International Development Association
IDB	Islamic Development Bank
IDP	Internally Displaced Person
IEA	International Energy Agency
IFAD	International Fund for Agricultural Development
IFC	International Finance Corporation
IFRCS	International Federation of Red Cross and Red Crescent Societies
IGAD	Inter-Governmental Authority on Development
IHO	International Hydrographic Organization
ILO	International Labor Organization
IMF	International Monetary Fund
IMO	International Maritime Organization
IMSO	International Mobile Satellite Organization
Inmarsat	International Maritime Satellite Organization
InOC	Indian Ocean Commission
INSTRAW	International Research and Training Institute for the Advancement of Women
Intelsat	International Telecommunications Satellite Organization
Interpol	International Criminal Police Organization
Intersputnik	International Organization of Space Communications
IOC	International Olympic Committee
IOM	International Organization for Migration
IPU	Inter-parliamentary Union
ISO	International Organization for Standardization
ISP	Internet Service Provider
ITSO	International Telecommunications Satellites Organization
ITU	International Telecommunication Union
ITUC	International Trade Union Confederation, the successor to ICFTU (International Confederation of Free Trade Unions) and the WCL (World Confederation of Labor)
kHz	kilohertz
km	kilometer
kW	kilowatt
kWh	kilowatt-hour
LAES	Latin American Economic System
LAIA	Latin American Integration Association
LAS	League of Arab States
Law of the Sea	United Nations Convention on the Law of the Sea (LOS)
LDC	less developed country
LLDC	least developed country
London Convention	see Marine Dumping
LOS	see Law of the Sea
m	meter
Marecs	Maritime European Communications Satellite
Marine Dumping	Convention on the Prevention of Marine Pollution by Dumping Wastes and Other Matter
Marine Life Conservation	Convention on Fishing and Conservation of Living Resources of the High Seas
MARPOL	see Ship Pollution
Medarabtel	Middle East Telecommunications Project of the International Telecommunications Union
Mercosur	Southern Cone Common Market
MHz	megahertz
MICAH	International Civilian Support Mission in Haiti
MIGA	Multilateral Investment Geographic Agency
MINURSO	United Nations Mission for the Referendum in Western Sahara
MINUSTAH	United Nations Stabilization Mission in Haiti
MONUSCO	United Nations Organization Stabilization Mission in the Democratic Republic of the Congo

NA	not available
NAFTA	North American Free Trade Agreement
NAM	Nonaligned Movement
NATO	North Atlantic Treaty Organization
NC	Nordic Council
NEA	Nuclear Energy Agency
NEGL	negligible
NGA	National Geospatial-Intelligence Agency
NGO	nongovernmental organization
NIB	Nordic Investment Bank
NIC	newly industrializing country
NIE	newly industrializing economy
NIS	new independent states
nm	nautical mile
NMT	Nordic Mobile Telephone
NSG	Nuclear Suppliers Group
Nuclear Test Ban	Treaty Banning Nuclear Weapons Tests in the Atmosphere, in Outer Space, and Under Water
NZ	New Zealand
OAPEC	Organization of Arab Petroleum Exporting Countries
OAS	Organization of American States
OAU	Organization of African Unity; see African Union
ODA	official development assistance
OECD	Organization for Economic Cooperation and Development
OECS	Organization of Eastern Caribbean States
OHCHR	Office of the United Nations High Commissioner for Human Rights
OIC	Organization of the Islamic Conference
OIF	International Organization of the French-speaking World
OOF	other official flows
OPANAL	Agency for the Prohibition of Nuclear Weapons in Latin America and the Caribbean
OPCW	Organization for the Prohibition of Chemical Weapons
OPEC	Organization of Petroleum Exporting Countries
OSCE	Organization for Security and Cooperation in Europe
Ozone Layer Protection	Montreal Protocol on Substances That Deplete the Ozone Layer
PCA	Permanent Court of Arbitration
PFP	Partnership for Peace
PIF	Pacific Islands Forum
PPP	purchasing power parity
Ramsar	see Wetlands
RG	Rio Group
SAARC	South Asian Association for Regional Cooperation
SACEP	South Asia Co-operative Environment Programme
SACU	Southern African Customs Union
SADC	Southern African Development Community
SAFE	South African Far East Cable
SCO	Shanghai Cooperation Organization
SECI	Southeast European Cooperative Initiative
SHF	super-high-frequency
Ship Pollution	Protocol of 1978 Relating to the International Convention for the Prevention of Pollution From Ships, 1973 (MARPOL)
SICA	Central American Integration System
Sparteca	South Pacific Regional Trade and Economic Cooperation Agreement
SPC	Secretariat of the Pacific Communities
SPF	South Pacific Forum
sq km	square kilometer
sq mi	square mile
TAT	Trans-Atlantic Telephone
TEU	Twenty-Foot Equivalent Unit, a unit of measure for containerized cargo capacity
Tropical Timber 83	International Tropical Timber Agreement, 1983
Tropical Timber 94	International Tropical Timber Agreement, 1994
UAE	United Arab Emirates
UDEAC	Central African Customs and Economic Union
UHF	ultra-high-frequency
UK	United Kingdom
UN	United Nations
UN-AIDS	Joint United Nations Program on HIV/AIDS
UNAMID	African Union/United Nations Hybrid Operation in Darfur
UNASUR	Union of South American Nations
UNCLOS	United Nations Convention on the Law of the Sea, also know as LOS
UNCTAD	United Nations Conference on Trade and Development
UNDCP	United Nations Drug Control Program
UNDEF	United Nations Democracy Fund
UNDOF	United Nations Disengagement Observer Force
UNDP	United Nations Development Program
UNEP	United Nations Environment Program
UNESCO	United Nations Educational, Scientific, and Cultural Organization

UNFICYP	United Nations Peace-keeping Force in Cyprus
UNFIP	United Nations Fund for International Partnerships
UNFPA	United Nations Population Fund
UN-Habitat	United Nations Center for Human Settlements
UNHCR	United Nations High Commissioner for Refugees
UNICEF	United Nations Children's Fund
UNICRI	United Nations Interregional Crime and Justice Research Institute
UNIDIR	United Nations Institute for Disarmament Research
UNIDO	United Nations Industrial Development Organization
UNIFIL	United Nations Interim Force in Lebanon
UN-INSTRAW	International Research and Training Institute for the Advancement of Women
UNITAR	United Nations Institute for Training and Research
UNMIK	United Nations Interim Administration Mission in Kosovo
UNMIL	United Nations Mission in Liberia
UNMIS	United Nations Mission in the Sudan
UNMIT	United Nations Integrated Mission in Timor-Leste
UNMOGIP	United Nations Military Observer Group in India and Pakistan
UNOCI	United Nations Operation in Cote d'Ivoire
UNOPS	United Nations Office of Project Services
UNRISD	United Nations Research Institute for Social Development
UNRWA	United Nations Relief and Works Agency for Palestine Refugees in the Near East
UNSC	United Nations Security Council
UNSSC	Untied Nations System Staff College
UNTSO	United Nations Truce Supervision Organization
UNU	United Nations University
UNWTO	World Tourism Organization
UPU	Universal Postal Union
US	United States
USSR	Union of Soviet Socialist Republics (Soviet Union); used for information dated before 25 December 1991
UTC	Coordinated Universal Time
UV	ultra violet
VHF	very-high-frequency
VSAT	very small aperture terminal
WADB	West African Development Bank
WAEMU	West African Economic and Monetary Union
WCL	World Confederation of Labor
WCO	World Customs Organization
Wetlands	Convention on Wetlands of International Importance Especially As Waterfowl Habitat
WEU	Western European Union
WFP	World Food Program
WFTU	World Federation of Trade Unions
Whaling	International Convention for the Regulation of Whaling
WHO	World Health Organization
WIPO	World Intellectual Property Organization
WMO	World Meteorological Organization
WP	Warsaw Pact
WTO	World Trade Organization
Z-	
ZC	Zangger Committee

APPENDIX B

INTERNATIONAL ORGANIZATIONS AND GROUPS

advanced developing countries
another term for those less developed countries (LDCs) with particularly rapid industrial development; see newly industrializing economies (NIEs)

advanced economies
a term used by the International Monetary FUND (IMF) for the top group in its hierarchy of advanced economies, countries in transition, and developing countries; it includes the following 33 advanced economies: Australia, Austria, Belgium, Canada, Cyprus, Czech Republic, Denmark, Finland, France, Germany, Greece, Hong Kong, Iceland, Ireland, Israel, Italy, Japan, South Korea, Luxembourg, Malta, Netherlands, NZ, Norway, Portugal, Singapore, Slovak Republic, Slovenia, Spain, Sweden, Switzerland, Taiwan, UK, US; note—this group would presumably also cover the following nine smaller countries of Andorra, Bermuda, Faroe Islands, Guernsey, Holy See, Jersey, Liechtenstein, Monaco, and San Marino that are included in the more comprehensive group of "developed countries"

African Development Bank Group (AfDB)
note—regional multilateral development finance institution temporarily located in Tunis, Tunisia; the Bank Group consists of the African Development Bank, the African Development Fund, and the Nigerian Trust Fund
established—10 September 1964
aim—to promote economic development and social progress
regional members—(53) Algeria, Angola, Benin, Botswana, Burkina Faso, Burundi, Cameroon, Cape Verde, Central African Republic, Chad, Comoros, Democratic Republic of the Congo, Republic of the Congo, Cote d'Ivoire, Djibouti, Egypt, Equatorial Guinea, Eritrea, Ethiopia, Gabon, The Gambia, Ghana, Guinea, Guinea-Bissau, Kenya, Lesotho, Liberia, Libya, Madagascar, Malawi, Mali, Mauritania, Mauritius, Morocco, Mozambique, Namibia, Niger, Nigeria, Rwanda, Sao Tome and Principe, Senegal, Seychelles, Sierra Leone, Somalia, South Africa, Sudan, Swaziland, Tanzania, Togo, Tunisia, Uganda, Zambia, Zimbabwe
nonregional members—(24) Argentina, Austria, Belgium, Brazil, Canada, China, Denmark, Finland, France, Germany, India, Italy, Japan, South Korea, Kuwait, Netherlands, Norway, Portugal, Saudi Arabia, Spain, Sweden, Switzerland, UK, US

African Union (AU)
note—replaces Organization of African Unity (OAU)
established—8 July 2001
aim—to achieve greater unity among African States; to defend states' integrity and independence; to accelerate political, social, and economic integration; to encourage international cooperation; to promote democratic principles and institutions
members—(53) Algeria, Angola, Benin, Botswana, Burkina Faso, Burundi, Cameroon, Cape Verde, Central African Republic, Chad, Comoros, Democratic Republic of the Congo, Republic of the Congo, Cote d'Ivoire, Djibouti, Egypt, Equatorial Guinea, Eritrea, Ethiopia, Gabon, The Gambia, Ghana, Guinea (suspended), Guinea-Bissau, Kenya, Lesotho, Liberia, Libya, Madagascar, Malawi, Mali, Mauritania, Mauritius, Mozambique, Namibia, Niger, Nigeria, Rwanda, Sahrawi Arab Democratic Republic (Western Sahara), Sao Tome and Principe, Senegal, Seychelles, Sierra Leone, Somalia, South Africa, Sudan, Swaziland, Tanzania, Togo, Tunisia, Uganda, Zambia, Zimbabwe

African Union/United Nations Hybrid Operation in Darfur (UNAMID)
established—31 July 2007
aim—to contribute to the restoration of security conditions which will allow safe humanitarian assistance throughout Darfur, to contribute to the protection of civilian populations under imminent threat of physical attack, to monitor, observe compliance with, and verify the implementation of various ceasefire agreements
members—(35) Bangladesh, Burkina Faso, Burundi, Cameroon, Canada, China, Egypt, Ethiopia, The Gambia, Germany, Ghana, Indonesia, Italy, Jordan, Kenya, South Korea, Malawi, Malaysia, Mali, Namibia, Nepal, Netherlands, Nigeria, Pakistan, Rwanda, Senegal, Sierra Leone, South Africa, Tanzania, Thailand, Togo, Uganda, Yemen, Zambia, Zimbabwe

African, Caribbean, and Pacific Group of States (ACP Group)
established—6 June 1975
aim—to manage their preferential economic and aid relationship with the EU
members—(79) Angola, Antigua and Barbuda, The Bahamas, Barbados, Belize, Benin, Botswana, Burkina Faso, Burundi, Cameroon, Cape Verde, Central African Republic, Chad, Comoros, Democratic Republic of the Congo, Republic of the Congo, Cook Islands, Cote d'Ivoire, Cuba, Djibouti, Dominica, Dominican Republic, Equatorial Guinea, Eritrea, Ethiopia, Fiji, Gabon, The Gambia, Ghana, Grenada, Guinea, Guinea-Bissau, Guyana, Haiti, Jamaica, Kenya, Kiribati, Lesotho, Liberia, Madagascar, Malawi, Mali, Marshall Islands, Mauritania, Mauritius, Federated States of Micronesia, Mozambique, Namibia, Nauru, Niger, Nigeria, Niue, Palau, Papua New Guinea, Rwanda, Saint Kitts and Nevis, Saint Lucia, Saint Vincent and the Grenadines, Samoa, Sao Tome and Principe, Senegal, Seychelles, Sierra Leone, Solomon Islands, Somalia, South Africa, Sudan, Suriname, Swaziland, Tanzania, Timor-Leste, Togo, Tonga, Trinidad and Tobago, Tuvalu, Uganda, Vanuatu, Zambia, Zimbabwe

Agency for the Prohibition of Nuclear Weapons in Latin America and the Caribbean (OPANAL)
note—acronym from Organismo para la Proscripcion de las Armas Nucleares en la America Latina y el Caribe (OPANAL)
established—14 February 1967 under the Treaty of Tlatelolco; effective—25 April 1969 on the 11th ratification
aim—to encourage the peaceful uses of atomic energy and prohibit nuclear weapons
members—(33) Antigua and Barbuda, Argentina, The Bahamas, Barbados, Belize, Bolivia, Brazil, Chile, Colombia, Costa Rica, Cuba, Dominica, Dominican Republic, Ecuador, El Salvador, Grenada, Guatemala, Guyana, Haiti, Honduras, Jamaica, Mexico, Nicaragua, Panama, Paraguay, Peru, Saint Kitts and Nevis, Saint Lucia, Saint Vincent and the Grenadines, Suriname, Trinidad and Tobago, Uruguay, Venezuela

Alliance of Small Island States (AOSIS)
established—November 1990
aim—to call attention to threats of sea-level rise and coral bleaching to small islands and lowlying coastal developing states from global warming; to emphasize the importance of information and information technology in the process of achieving sustainable development
members—(39) Antigua and Barbuda, The Bahamas, Barbados, Belize, Cape Verde, Comoros, Cook Islands, Cuba, Dominica, Dominican Republic, Fiji, Grenada, Guinea-Bissau, Guyana, Haiti, Jamaica, Kiribati, Maldives, Marshall Islands, Mauritius, Federated

States of Micronesia, Nauru, Niue, Palau, Papua New Guinea, St. Kitts and Nevis, St. Lucia, St. Vincent and the Grenadines, Samoa, Sao Tome and Principe, Seychelles, Singapore, Solomon Islands, Suriname, Timor-Leste, Tonga, Trinidad and Tobago, Tuvalu, Vanuatu
observers–(3) American Samoa, Guam, U.S. Virgin Islands

Andean Community of Nations (CAN)
note–formerly known as the Andean Group (AG) and the Andean Common Market (Ancom)
established–26 May 1969; present name established 1 October 1992; effective–16 October 1969
aim–to promote harmonious development through economic integration
members–(4) Bolivia, Colombia, Ecuador, Peru
associate members–(5) Argentina, Brazil, Chile, Paraguay, Uruguay
observers–(2) Mexico, Panama

Arab Bank for Economic Development in Africa (ABEDA)
note–also known as Banque Arabe de Developpement Economique en Afrique (BADEA)
established–18 February 1974; effective–16 September 1974
aim–to promote economic development
members–(17 plus the Palestine Liberation Organization) Algeria, Bahrain, Egypt, Iraq, Jordan, Kuwait, Lebanon, Libya, Mauritania, Morocco, Oman, Qatar, Saudi Arabia, Sudan, Syria, Tunisia, UAE, Palestine Liberation Organization; note–these are all the members of the Arab League excluding Comoros, Djibouti, Somalia, Yemen

Arab Fund for Economic and Social Development (AFESD)
established–16 May 1968
aim–to promote economic and social development
members–(20 plus the Palestine Liberation Organization) Algeria, Bahrain, Djibouti, Egypt, Iraq, Jordan, Kuwait, Lebanon, Libya, Mauritania, Morocco, Oman, Qatar, Saudi Arabia, Somalia (suspended 1993), Sudan, Syria, Tunisia, UAE, Yemen, Palestine Liberation Organization

Arab Maghreb Union (AMU)
established–17 February 1989
aim–to promote cooperation and integration among the Arab states of northern Africa
members–(5) Algeria, Libya, Mauritania, Morocco, Tunisia

Arab Monetary Fund (AMF)
established–27 April 1976; effective–2 February 1977
aim–to promote Arab cooperation, development, and integration in monetary and economic affairs
members–(21 plus the Palestine Liberation Organization) Algeria, Bahrain, Comoros, Djibouti, Egypt, Iraq, Jordan, Kuwait, Lebanon, Libya, Mauritania, Morocco, Oman, Qatar, Saudi Arabia, Somalia, Sudan, Syria, Tunisia, UAE, Yemen, Palestine Liberation Organization

Arctic Council
established–18 September 1996
aim–to address the common concerns and challenges faced by Arctic governments and the people of the Arctic; to protect the Arctic environment
members–(8) Canada, Denmark (Greenland, Faroe Islands), Finland, Iceland, Norway, Russia, Sweden, US
permanent participants–(6) Aleut International Association, Arctic Athabaskan Council, Gwich'in Council International, Inuit Circumpolar Conference, Russian Association of Indigenous People of the North, Saami Council
observers–(6) **France, Germany, Netherlands, Poland, Spain, UK**

ASEAN Regional Forum (ARF)
established–25 July 1994
aim–to foster constructive dialogue and consultation on political and security issues of common interest and concern
members–(27) Australia, Bangladesh, Brunei, Burma, Cambodia, Canada, China, EU, India, Indonesia, Japan, North Korea, South Korea, Laos, Malaysia, Mongolia, NZ, Pakistan, Papua New Guinea, Philippines, Russia, Singapore, Sri Lanka, Thailand, Timor-Leste, US, Vietnam

Asia-Pacific Economic Cooperation (APEC)
established–7 November 1989
aim–to promote trade and investment in the Pacific basin
members–(21) Australia, Brunei, Canada, Chile, China, Hong Kong, Indonesia, Japan, South Korea, Malaysia, Mexico, NZ, Papua New Guinea, Peru, Philippines, Russia, Singapore, Taiwan, Thailand, US, Vietnam
observers–(3) Association of Southeast Asian Nations, Pacific Economic Cooperation Council, Pacific Islands Forum Secretariat

Asian Development Bank (ADB)
established–19 December 1966
aim–to promote regional economic cooperation
members–(48) Afghanistan, Armenia, Australia, Azerbaijan, Bangladesh, Bhutan, Brunei, Burma, Cambodia, China, Cook Islands, Fiji, Georgia, Hong Kong, India, Indonesia, Japan, Kazakhstan, Kiribati, South Korea, Kyrgyzstan, Laos, Malaysia, Maldives, Marshall Islands, Federated States of Micronesia, Mongolia, Nauru, Nepal, NZ, Pakistan, Palau, Papua New Guinea, Philippines, Samoa, Singapore, Solomon Islands, Sri Lanka, Taiwan, Tajikistan, Thailand, Timor-Leste, Tonga, Turkmenistan, Tuvalu, Uzbekistan, Vanuatu, Vietnam
nonregional members–(19) Austria, Belgium, Canada, Denmark, Finland, France, Germany, Ireland, Italy, Luxembourg, Netherlands, Norway, Portugal, Spain, Sweden, Switzerland, Turkey, UK, US

Association of Southeast Asian Nations (ASEAN)
established–8 August 1967
aim–to encourage regional economic, social, and cultural cooperation among the non-Communist countries of Southeast Asia
members–(10) Brunei, Burma, Cambodia, Indonesia, Laos, Malaysia, Philippines, Singapore, Thailand, Vietnam
dialogue partners–(11) Australia, Canada, China, EU, India, Japan, South Korea, NZ, Pakistan, Russia, US
observers–(1) Papua New Guinea

Australia Group (AG)

established–June 1985
aim–to consult on and coordinate export controls related to chemical and biological weapons
members–(41) Argentina, Australia, Austria, Belgium, Bulgaria, Canada, Croatia, Cyprus, Czech Republic, Denmark, Estonia, European Commission, Finland, France, Germany, Greece, Hungary, Iceland, Ireland, Italy, Japan, South Korea, Latvia, Lithuania, Luxembourg, Malta, Netherlands, NZ, Norway, Poland, Portugal, Romania, Slovakia, Slovenia, Spain, Sweden, Switzerland, Turkey, Ukraine, UK, US

Australia-New Zealand-United States Security Treaty (ANZUS)

established–1 September 1951; effective–29 April 1952
aim–to implement a trilateral mutual security agreement, although the US suspended security obligations to NZ on 11 August 1986; Australia and the US continue to hold annual meetings
members–(3) Australia, NZ, US

Baltic Assembly (BA)

established–12 May 1990
aim–to thoroughly discuss various cooperation issues between Baltic states
members–(3) Estonia, Latvia, Lithuania

Bank for International Settlements (BIS)

established–20 January 1930; effective–17 March 1930
aim–to promote cooperation among central banks in international financial settlements
members–(56) Algeria, Argentina, Australia, Austria, Belgium, Bosnia and Herzegovina, Brazil, Bulgaria, Canada, Chile, China, Croatia, Czech Republic, Denmark, European Central Bank, Estonia, Finland, France, Germany, Greece, Hong Kong, Hungary, Iceland, India, Indonesia, Ireland, Israel, Italy, Japan, South Korea, Latvia, Lithuania, Macedonia, Malaysia, Mexico, Netherlands, NZ, Norway, Philippines, Poland, Portugal, Romania, Russia, Saudi Arabia, Serbia, Singapore, Slovakia, Slovenia, South Africa, Spain, Sweden, Switzerland, Thailand, Turkey, UK, US; note–Montenegro has a separate central bank; its links with BIS are currently under review

Bay of Bengal Initiative for Multi-Sectoral Technical and Economic Cooperation (BIMSTEC)

established–June 1997
aim–to foster socio-economic cooperation among members
members–(7) Bangladesh, Bhutan, Burma, India, Nepal, Sri Lanka, Thailand

Benelux Union (Benelux)

note–acronym from Belgium, Netherlands, and Luxembourg; was formerly known as Benelux Economic Union
established–3 February 1958; effective–1 November 1960; changed names 17 June 2008
aim–to develop closer economic and legal cooperation and integration
members–(3) Belgium, Luxembourg, Netherlands

Big Seven

note–membership is the same as the Group of 7
established–1975
aim–to discuss and coordinate major economic policies
members–(7) Big Six (Canada, France, Germany, Italy, Japan, UK) plus the US

Black Sea Economic Cooperation Zone (BSEC)

established–25 June 1992
aim–to enhance regional stability through economic cooperation
members–(12) Albania, Armenia, Azerbaijan, Bulgaria, Georgia, Greece, Moldova, Romania, Russia, Serbia, Turkey, Ukraine; note–Macedonia is in the process of joining
observers–(17) Austria, Belarus, Black Sea Commission, Commission of the EC, Croatia, Czech Republic, Egypt, Energy Charter Secretariat, France, Germany, International Black Sea Club, Israel, Italy, Poland, Slovakia, Tunisia, US; note–Bosnia and Herzegovina and Slovenia have applied for observer status

Caribbean Community and Common Market (Caricom)

established–4 July 1973; effective–1 August 1973
aim–to promote economic integration and development, especially among the less developed countries
members–(15) Antigua and Barbuda, The Bahamas, Barbados, Belize, Dominica, Grenada, Guyana, Haiti, Jamaica, Montserrat, Saint Kitts and Nevis, Saint Lucia, Saint Vincent and the Grenadines, Suriname, Trinidad and Tobago
associate members–(5) Anguilla, Bermuda, British Virgin Islands, Cayman Islands, Turks and Caicos Islands
observers–(8) Aruba, Colombia, Curacao, Dominican Republic, Mexico, Puerto Rico, Sint Maarten, Venezuela

Caribbean Development Bank (CDB)

established–18 October 1969; effective–26 January 1970
aim–to promote economic development and cooperation
regional members–(21) Anguilla, Antigua and Barbuda, The Bahamas, Barbados, Belize, British Virgin Islands, Cayman Islands, Colombia, Dominica, Grenada, Guyana, Haiti, Jamaica, Mexico, Montserrat, Saint Kitts and Nevis, Saint Lucia, Saint Vincent and the Grenadines, Trinidad and Tobago, Turks and Caicos Islands, Venezuela
nonregional members–(5) Canada, China, Germany, Italy, UK

Central African Customs and Economic Union (UDEAC)

see Economic and Monetary Community of Central Africa (CEMAC)

Central African States Development Bank (BDEAC)

note–acronym from Banque de Developpement des Etats de l'Afrique Centrale
established–3 December 1975
aim–to provide loans for economic development
members–(10) African Development Bank (AfDB), Cameroon, Central African States Bank (BEAC), Central African Republic, Chad, Republic of the Congo, Equatorial Guinea, France, Gabon, Kuwait

Central American Bank for Economic Integration (BCIE)
note–acronym from Banco Centroamericano de Integracion Economico
established–13 December 1960 signature of Articles of Agreement; 31 May 1961 began operations
aim–to promote economic integration and development
members–(5) Costa Rica, El Salvador, Guatemala, Honduras, Nicaragua
nonregional members–(7) Argentina, Colombia, Dominican Republic, Mexico, Panama, Spain, Taiwan

Central American Common Market (CACM)
established–13 December 1960, collapsed in 1969, reinstated in 1991
aim–to promote establishment of a Central American Common Market
members–(5) Costa Rica, El Salvador, Guatemala, Honduras, Nicaragua

Central American Integration System (SICA)
established–13 December 1991; operational 1 February 1993
aim–to strengthen democracy; to set up a new model of regional security; to promote freedom; to achieve a regional system of welfare and economic and social justice; to attain economic unity and strengthen the area as an economic bloc; to act as a bloc in international matters
members–(7) Belize, Costa Rica, El Salvador, Guatemala, Honduras, Nicaragua, Panama
associated member–(1) Dominican Republic
observers–(8) Argentina, Brazil, Chile, China, Germany, Italy, Mexico, Spain

Central European Initiative (CEI)
note–evolved from the Quadrilateral Initiative and the Hexagonal Initiative
established–11 November 1989 as the Quadrilateral Initiative, 27 July 1991 became the Hexagonal Initiative, July 1992 its present name was adopted
aim–to form an economic and political cooperation group for the region between the Adriatic and the Baltic Seas
members–(18) Albania, Austria, Belarus, Bosnia and Herzegovina, Bulgaria, Croatia, Czech Republic, Hungary, Italy, Macedonia, Moldova, Montenegro, Poland, Romania, Serbia, Slovakia, Slovenia, Ukraine

centrally planned economies
a term applied mainly to the traditionally Communist states that looked to the former USSR for leadership; most are now evolving toward more democratic and market-oriented systems; also known formerly as the Second World or as the Communist countries; through the 1980s, this group included Albania, Bulgaria, Cambodia, China, Cuba, Czechoslovakia, German Democratic Republic, Hungary, North Korea, Laos, Mongolia, Poland, Romania, USSR, Vietnam, Yugoslavia, but now is limited to Cuba and North Korea, and less so to China

Collective Security Treaty Organization (CSTO)
established–7 October 2002
aim–to coordinate military and political cooperation, to develop multilateral structures and mechanisms of cooperation for ensuring national security of the member states
members–(6) Armenia, Belarus, Kazakhstan, Kyrgyzstan, Russia, Tajikistan

Colombo Plan (CP)
established–May 1950 proposal was adopted; 1 July 1951 commenced full operations
aim–to promote economic and social development in Asia and the Pacific
members–(26) Afghanistan, Australia, Bangladesh, Bhutan, Brunei, Burma, Fiji, India, Indonesia, Iran, Japan, South Korea, Laos, Malaysia, Maldives, Mongolia, Nepal, NZ, Pakistan, Papua New Guinea, Philippines, Singapore, Sri Lanka, Thailand, US, Vietnam

Common Market for Eastern and Southern Africa (COMESA)
note–formerly known as Preferential Trade Area for Eastern and Southern Africa (PTA)
established–treaty signed 5 November 1993; treaty ratified 8 December 1994
aim–recognizing, promoting and protecting fundamental human rights, commitment to the principles of liberty and rule of law, maintaining peace and stability through the promotion and strengthening of good neighborliness, commitment to peaceful settlement of disputes among member states
members–(19) Burundi, Comoros, Democratic Republic of the Congo, Djibouti, Egypt, Eritrea, Ethiopia, Kenya, Libya, Madagascar, Malawi, Mauritius, Rwanda, Seychelles, Sudan, Swaziland, Uganda, Zambia, Zimbabwe

Commonwealth (C)
note–also known as Commonwealth of Nations
established–31 December 1931
aim–to foster multinational cooperation and assistance, as a voluntary association that evolved from the British Empire
members–(54) Antigua and Barbuda, Australia, The Bahamas, Bangladesh, Barbados, Belize, Botswana, Brunei, Cameroon, Canada, Cyprus, Dominica, Fiji (suspended), The Gambia, Ghana, Grenada, Guyana, India, Jamaica, Kenya, Kiribati, Lesotho, Malawi, Malaysia, Maldives, Malta, Mauritius, Mozambique, Namibia, Nauru, NZ, Nigeria, Pakistan (reinstated 2004), Papua New Guinea, Rwanda, Saint Kitts and Nevis, Saint Lucia, Saint Vincent and the Grenadines, Samoa, Seychelles, Sierra Leone, Singapore, Solomon Islands, South Africa, Sri Lanka, Swaziland, Tanzania, Tonga, Trinidad and Tobago, Tuvalu, Uganda, UK, Vanuatu, Zambia; note–on 7 December 2003 Zimbabwe withdrew its membership from the Commonwealth

Commonwealth of Independent States (CIS)
established–8 December 1991; effective–21 December 1991
aim–to coordinate intercommonwealth relations and to provide a mechanism for the orderly dissolution of the USSR
members–(9) Armenia, Azerbaijan, Belarus, Kazakhstan, Kyrgyzstan, Moldova, Russia, Tajikistan, Uzbekistan; note–neither Ukraine as a participating member nor Turkmenistan as an associate member have signed the 1993 CIS charter, although both participate in meetings; Georgia left the organization in August 2009

Communist countries
traditionally the Marxist-Leninist states with authoritarian governments and command economies based on the Soviet model; most of the original and the successor states are no longer Communist; see centrally planned economies

Community of Democracies (CD)
established–27 June 2000
aim–"to respect and uphold core democratic principals and practices" including free and fair elections, freedom of speech and expression, equal access to education, rule of law, and freedom of peaceful assembly
members–(17) Cape Verde, Chile, Czech Republic, El Salvador, India, Italy, South Korea, Lithuania, Mali, Mexico, Mongolia, Morocco, Philippines, Poland, Portugal, South Africa, US

Comuinidade dos Paises de Lingua Portuguesa (CPLP)
established–1996
aim–to establish a forum for friendship among Portuguese-speaking nations where Portuguese is an official language
members–(8) Angola, Brazil, Cape Verde, Guinea-Bissau, Mozambique, Portugal, Sao Tome and Principe, Timor-Leste
associate observers–(3) Equatorial Guinea, Mauritius, Senegal

Conference of Interaction and Confidence-Building Measures in Asia (CICA)
established–proposed 5 October 1992; established 14 September 1999
aim–promoting a multi-national forum for enhancing cooperation towards promoting peace, security, and stability in Asia
members–(22 and the Palestine Liberation Organization) Afghanistan, Azerbaijan, Bahrain, China, Egypt, India, Iraq, Iran, Israel, Jordan, Kazakhstan, Kyrgyzstan, Mongolia, Pakistan, South Korea, Russia, Tajikistan, Thailand, Turkey, UAE, Uzbekistan, Vietnam, and the Palestine Liberation Organization
observers–(12) Bangladesh, Cambodia, Indonesia, Japan, League of Arab States, Malaysia, OSCE, Philippines, Qatar, Ukraine, UN, US

Coordinating Committee on Export Controls (COCOM)
established in 1949 to control the export of strategic products and technical data from member countries to proscribed destinations; members were: Australia, Belgium, Canada, Denmark, France, Germany, Greece, Italy, Japan, Luxembourg, Netherlands, Norway, Portugal, Spain, Turkey, UK, US; abolished 31 March 1994; COCOM members established a new organization, the Wassenaar Arrangement, with expanded membership on 12 July 1996 that focuses on nonproliferation export controls as opposed to East-West control of advanced technology

Council for Mutual Economic Assistance (CEMA)
note–also known as CMEA or Comecon
established 25 January 1949 to promote the development of socialist economies and abolished 1 January 1991; members included Afghanistan (observer), Albania (had not participated since 1961 break with USSR), Angola (observer), Bulgaria, Cuba, Czechoslovakia, Ethiopia (observer), GDR, Hungary, Laos (observer), Mongolia, Mozambique (observer), Nicaragua (observer), Poland, Romania, USSR, Vietnam, Yemen (observer), Yugoslavia (associate)

Council of Arab Economic Unity (CAEU)
established–3 June 1957; effective–30 May 1964
aim–to promote economic integration among Arab nations
members–(11 plus the Palestine Liberation Organization) Egypt, Iraq, Jordan, Kuwait, Libya, Mauritania, Somalia, Sudan, Syria, UAE, Yemen, Palestine Liberation Organization

Council of Europe (CE)
established–5 May 1949; effective–3 August 1949
aim–to promote increased unity and quality of life in Europe
members–(47) Albania, Andorra, Armenia, Austria, Azerbaijan, Belgium, Bosnia and Herzegovina, Bulgaria, Croatia, Cyprus, Czech Republic, Denmark, Estonia, Finland, France, Georgia, Germany, Greece, Hungary, Iceland, Ireland, Italy, Latvia, Liechtenstein, Lithuania, Luxembourg, Macedonia, Malta, Moldova, Monaco, Montenegro, Netherlands, Norway, Poland, Portugal, Romania, Russia, San Marino, Serbia, Slovakia, Slovenia, Spain, Sweden, Switzerland, Turkey, Ukraine, UK

Council of the Baltic Sea States (CBSS)
established–6 March 1992
aim–to promote cooperation among the Baltic Sea states in the areas of aid to new democratic institutions, economic development, humanitarian aid, energy and the environment, cultural programs and education, and transportation and communication
members–(12) Denmark, Estonia, EC, Finland, Germany, Iceland, Latvia, Lithuania, Norway, Poland, Russia, Sweden
observers–(10) Belarus, France, Italy, Netherlands, Romania, Spain, Slovakia, Ukraine, UK, US

Council of the Entente (Entente)
established–29 May 1959
aim–to promote economic, social, and political coordination
members–(5) Benin, Burkina Faso, Cote d'Ivoire, Niger, Togo

countries in transition
a term used by the International Monetary Fund (IMF) for the middle group in its hierarchy of formerly centrally planned economies; IMF statistics include the following 28 countries in transition: Albania, Armenia, Azerbaijan, Belarus, Bosnia and Herzegovina, Bulgaria, Croatia, Czech Republic, Estonia, Georgia, Hungary, Kazakhstan, Kyrgyzstan, Latvia, Lithuania, Macedonia, Moldova, Mongolia, Montenegro, Poland, Romania, Russia, Serbia, Slovakia, Slovenia, Tajikistan, Turkmenistan, Ukraine, Uzbekistan; note–this group is identical to the group traditionally referred to as the "former USSR/Eastern Europe" except for the addition of Mongolia

Customs Cooperation Council (CCC)
note–see World Customs Organization (WCO)

developed countries (DCs)
the top group in the hierarchy of developed countries (DCs), former USSR/Eastern Europe (former USSR/EE), and less developed countries (LDCs); includes the market-oriented economies of the mainly democratic nations in the Organization for Economic Cooperation and Development (OECD), Bermuda, Israel, South Africa, and the European ministates; also known as the First World, high-income countries, the North, industrial countries; generally have a per capita GDP in excess of $15,000 although four OECD countries and South Africa have figures well under $15,000 and eight of the excluded OPEC countries have figures of more than $20,000; the DCs include: Andorra, Australia, Austria, Belgium, Bermuda, Canada, Denmark, Faroe Islands, Finland, France, Germany,

Greece, Holy See, Iceland, Ireland, Israel, Italy, Japan, Liechtenstein, Luxembourg, Malta, Monaco, Netherlands, NZ, Norway, Portugal, San Marino, South Africa, Spain, Sweden, Switzerland, Turkey, UK, US; note–similar to the new International Monetary Fund (IMF) term "advanced economies" that adds Hong Kong, South Korea, Singapore, and Taiwan but drops Malta, Mexico, South Africa, and Turkey

developing countries

a term used by the International Monetary Fund (IMF) for the bottom group in its hierarchy of advanced economies, countries in transition, and developing countries; IMF statistics include the following 126 developing countries: Afghanistan, Algeria, Angola, Antigua and Barbuda, Argentina, Aruba, The Bahamas, Bahrain, Bangladesh, Barbados, Belize, Benin, Bhutan, Bolivia, Botswana, Brazil, Burkina Faso, Burma, Burundi, Cambodia, Cameroon, Cape Verde, Central African Republic, Chad, Chile, China, Colombia, Comoros, Democratic Republic of the Congo, Republic of the Congo, Costa Rica, Cote d'Ivoire, Cyprus, Djibouti, Dominica, Dominican Republic, Ecuador, Egypt, El Salvador, Equatorial Guinea, Ethiopia, Fiji, Gabon, The Gambia, Ghana, Grenada, Guatemala, Guinea, Guinea-Bissau, Guyana, Haiti, Honduras, India, Indonesia, Iran, Iraq, Jamaica, Jordan, Kenya, Kiribati, Kuwait, Laos, Lebanon, Lesotho, Liberia, Libya, Madagascar, Malawi, Malaysia, Maldives, Mali, Malta, Marshall Islands, Mauritania, Mauritius, Mexico, Federated States of Micronesia, Morocco, Mozambique, Namibia, Nepal, Netherlands Antilles, Nicaragua, Niger, Nigeria, Oman, Pakistan, Panama, Papua New Guinea, Paraguay, Peru, Philippines, Qatar, Rwanda, Saint Kitts and Nevis, Saint Lucia, Saint Vincent and the Grenadines, Samoa, Sao Tome and Principe, Saudi Arabia, Senegal, Seychelles, Sierra Leone, Solomon Islands, Somalia, South Africa, Sri Lanka, Sudan, Suriname, Swaziland, Syria, Tanzania, Thailand, Togo, Trinidad and Tobago, Tunisia, Turkey, UAE, Uganda, Uruguay, Vanuatu, Venezuela, Vietnam, Yemen, Zambia, Zimbabwe; note–this category would presumably also cover the following 46 other countries that are traditionally included in the more comprehensive group of "less developed countries": American Samoa, Anguilla, British Virgin Islands, Brunei, Cayman Islands, Christmas Island, Cocos Islands, Cook Islands, Cuba, Eritrea, Falkland Islands, French Guiana, French Polynesia, Gaza Strip, Gibraltar, Greenland, Grenada, Guadeloupe, Guam, Guernsey, Isle of Man, Jersey, North Korea, Macau, Martinique, Mayotte, Montserrat, Nauru, New Caledonia, Niue, Norfolk Island, Northern Mariana Islands, Palau, Pitcairn Islands, Puerto Rico, Reunion, Saint Helena, Ascension, and Tristan da Cunha, Saint Pierre and Miquelon, Tokelau, Tonga, Turks and Caicos Islands, Tuvalu, Virgin Islands, Wallis and Futuna, West Bank, Western Sahara

Developing Eight (D-8)

established–15 June 1997
aim–to improve developing countries' positions in the world economy, diversify and create new opportunities in trade relations, enhance participation in decision-making at the international level, provide better standards of living
member–(8) Bangladesh, Egypt, Indonesia, Iran, Malaysia, Nigeria, Pakistan, Turkey

East African Community (EAC)

note–originally established in 1967, it was disbanded in 1977
established–January 2001
aim–to establish a political and economic union among the countries
members–(5) Burundi, Kenya, Rwanda, Tanzania, Uganda

East African Development Bank (EADB)

established–6 June 1967; effective–1 December 1967
aim–to promote economic development
members–(4) Kenya, Rwanda, Tanzania, Uganda

East Asia Summit (EAS)

established–14 December 2005
aim–to promote cooperation in political and security issues; to promote development, financial stability, energy security, economic integration and growth; to eradicate poverty and narrow the development gap in East Asia, and to promote deeper cultural understanding
members–(16) Australia, Brunei, Burma, Cambodia, China, India, Indonesia, Japan, South Korea, Laos, Malaysia, NZ, Philippines, Singapore, Thailand, Vietnam

Economic and Monetary Community of Central Africa (CEMAC)

note–was formerly the Central African Customs and Economic Union (UDEAC)
established–8 December 1964; effective–1 January 1966
aim–to promote the establishment of a Central African Common Market
members–(6) Cameroon, Central African Republic, Chad, Republic of the Congo, Equatorial Guinea, Gabon

Economic and Monetary Union (EMU)

note–an integral part of the European Union; also known as the European Economic and Monetary Union
established–1-2 December 1969 (proposed at summit conference of heads of government; 7 February 1992 (Maastricht Treaty signed)
aim–to promote a single market by creating a single currency, the euro; timetable–2 May 1998: European exchange rates fixed for 1 January 1999; 1 January 1999: all banks and stock exchanges begin using euros; 1 January 2002: the euro goes into circulation; 1 July 2002 local currencies no longer accepted
members–(17) Austria, Belgium, Cyprus, Estonia, Finland, France, Germany, Greece, Ireland, Italy, Luxembourg, Malta, Netherlands, Portugal, Slovakia, Slovenia, Spain

Economic and Social Council (ECOSOC)

established–26 June 1945; effective–24 October 1945
aim–to coordinate the economic and social work of the UN; includes five regional commissions (Economic Commission for Africa, Economic Commission for Europe, Economic Commission for Latin America and the Caribbean, Economic and Social Commission for Asia and the Pacific, Economic and Social Commission for Western Asia) and nine functional commissions (Commission for Social Development, Commission on Human Rights, Commission on Narcotic Drugs, Commission on the Status of Women, Commission on Population and Development, Statistical Commission, Commission on Science and Technology for Development, Commission on Sustainable Development, and Commission on Crime Prevention and Criminal Justice)
members–(54) selected on a rotating basis from all regions

Economic Community of the Great Lakes Countries (CEPGL)

note–acronym from Communaute Economique des Pays des Grands Lacs

established–20 September 1976
aim–to promote regional economic cooperation and integration
members–(3) Burundi, Democratic Republic of the Congo, Rwanda; note–organization collapsed because of fighting in 1998; reactivated in 2006

Economic Community of West African States (ECOWAS)

established–28 May 1975
aim–to promote regional economic cooperation
members–(15) Benin, Burkina Faso, Cape Verde, Cote d'Ivoire (suspended), The Gambia, Ghana, Guinea (suspended), Guinea-Bissau, Liberia, Mali, Niger (suspended), Nigeria, Senegal, Sierra Leone, Togo

Economic Cooperation Organization (ECO)

established–27-29 January 1985
aim–to promote regional cooperation in trade, transportation, communications, tourism, cultural affairs, and economic development
members–(10) Afghanistan, Azerbaijan, Iran, Kazakhstan, Kyrgyzstan, Pakistan, Tajikistan, Turkey, Turkmenistan, Uzbekistan

Eurasian Economic Community (EAEC or EurasEC)

note–merged with Central Asian Cooperation Organization (CACO) in 2005
established–May 2001
aim–to create a common economic and energy policy
members–(5) Belarus, Kazakhstan, Kyrgyzstan, Russia, Tajikistan
observers–(3) Armenia, Moldova, Ukraine

Euro-Atlantic Partnership Council (EAPC)

note–began as the North Atlantic Cooperation Council (NACC); an extension of NATO
established–8 November 1991; effective–20 December 1991
aim–to discuss cooperation on mutual political and security issues
members–(50) Albania, Armenia, Austria, Azerbaijan, Belarus, Belgium, Bosnia and Herzegovina, Bulgaria, Canada, Croatia, Czech Republic, Denmark, Estonia, Finland, France, Georgia, Germany, Greece, Hungary, Iceland, Ireland, Italy, Kazakhstan, Kyrgyzstan, Latvia, Lithuania, Luxembourg, Macedonia, Malta, Moldova, Montenegro, Netherlands, Norway, Poland, Portugal, Romania, Russia, Serbia, Slovakia, Slovenia, Spain, Sweden, Switzerland, Tajikistan, Turkey, Turkmenistan, Ukraine, UK, US, Uzbekistan

European Bank for Reconstruction and Development (EBRD)

established–8-9 January 1990 (proposals made); 15 April 1991 (bank inaugurated)
aim–to facilitate the transition of seven centrally planned economies in Europe (Bulgaria, former Czechoslovakia, Hungary, Poland, Romania, former USSR, and former Yugoslavia) to market economies by committing 60% of its loans to privatization
members–(63) Albania, Armenia, Australia, Austria, Azerbaijan, Belarus, Belgium, Bosnia and Herzegovina, Bulgaria, Canada, Croatia, Cyprus, Czech Republic, Denmark, Egypt, EC, European Investment Bank (EIB), Estonia, Finland, France, Georgia, Germany, Greece, Hungary, Iceland, Ireland, Israel, Italy, Japan, Kazakhstan, South Korea, Kyrgyzstan, Latvia, Liechtenstein, Lithuania, Luxembourg, Macedonia, Malta, Mexico, Moldova, Mongolia, Montenegro, Morocco, Netherlands, NZ, Norway, Poland, Portugal, Romania, Russia, Serbia, Slovakia, Slovenia, Spain, Sweden, Switzerland, Tajikistan, Turkey, Turkmenistan, Ukraine, UK, US, Uzbekistan

European Community (or European Communities, EC)

established 8 April 1965 to integrate the European Atomic Energy Community (Euratom), the European Coal and Steel Community (ECSC), the European Economic Community (EEC or Common Market), and to establish a completely integrated common market and an eventual federation of Europe; merged into the European Union (EU) on 7 February 1992; member states at the time of merger were Belgium, Denmark, France, Germany, Greece, Ireland, Italy, Luxembourg, Netherlands, Portugal, Spain, UK

European Free Trade Association (EFTA)

established–4 January 1960; effective–3 May 1960
aim–to promote expansion of free trade
members–(4) Iceland, Liechtenstein, Norway, Switzerland

European Investment Bank (EIB)

established–25 March 1957; effective–1 January 1958
aim–to promote economic development of the EU and its predecessors, the EEC and the EC
members–(27) Austria, Belgium, Bulgaria, Cyprus, Czech Republic, Denmark, Estonia, Finland, France, Germany, Greece, Hungary, Ireland, Italy, Latvia, Lithuania, Luxembourg, Malta, Netherlands, Poland, Portugal, Romania, Slovakia, Slovenia, Spain, Sweden, UK

European Organization for Nuclear Research (CERN)

note–acronym retained from the predecessor organization Conseil Europeenne pour la Recherche Nucleaire
established–1 July 1953; effective–29 September 1954
aim–to foster nuclear research for peaceful purposes only
members–(20) Austria, Belgium, Bulgaria, Czech Republic, Denmark, Finland, France, Germany, Greece, Hungary, Italy, Netherlands, Norway, Poland, Portugal, Slovakia, Spain, Sweden, Switzerland, UK
observers–(8) EC, India, Israel, Japan, Russia, Turkey, United Nations Educational, Scientific, and Cultural Organization (UNESCO), US

European Space Agency (ESA)

established–31 May 1975
aim–to promote peaceful cooperation in space research and technology
members–(18) Austria, Belgium, Czech Republic, Denmark, Finland, France, Germany, Greece, Ireland, Italy, Luxembourg, Netherlands, Norway, Portugal, Spain, Sweden, Switzerland, UK
cooperating states–(5) Estonia, Hungary, Poland, Romania, Slovenia

European Union (EU)

note–see European Union entry at the end of the "country" listings

Financial Action Task Force (FATF)

established–by G-7 Summit in Paris in 1989

aim–to develop and promote policies to combat money laundering and terrorist financing
members–(36) Argentina, Australia, Austria, Belgium, Brazil, Canada, China, Denmark, EC, Finland, France, Germany, Greece, Gulf Cooperation Council, Hong Kong, Iceland, India, Ireland, Italy, Japan, South Korea, Luxembourg, Mexico, Netherlands (Netherland Antilles and Aruba), NZ, Norway, Portugal, Russia, Singapore, South Africa, Spain, Sweden, Switzerland, Turkey, UK, US

First World

another term for countries with advanced, industrialized economies; this term is fading from use; see developed countries (DCs)

Food and Agriculture Organization (FAO)

established–16 October 1945
aim–to raise living standards and increase availability of agricultural products; a UN specialized agency
members–(192) includes all UN member countries except Brunei, Liechtenstein, and Singapore (189 total); plus Cook Islands, EU, and Niue

former Soviet Union (FSU)

former term often used to identify as a group the successor nations to the Soviet Union or USSR; this group of 15 countries consists of: Armenia, Azerbaijan, Belarus, Estonia, Georgia, Kazakhstan, Kyrgyzstan, Latvia, Lithuania, Moldova, Russia, Tajikistan, Turkmenistan, Ukraine, Uzbekistan

former USSR/Eastern Europe (former USSR/EE)

the middle group in the hierarchy of developed countries (DCs), former USSR/Eastern Europe (former USSR/EE), and less developed countries (LDCs); these countries are in political and economic transition and may well be grouped differently in the near future; this group of 27 countries consists of: Albania, Armenia, Azerbaijan, Belarus, Bosnia and Herzegovina, Bulgaria, Croatia, Czech Republic, Estonia, Georgia, Hungary, Kazakhstan, Kyrgyzstan, Latvia, Lithuania, Macedonia, Moldova, Poland, Romania, Russia, Slovakia, Slovenia, Tajikistan, Turkmenistan, Ukraine, Uzbekistan, Yugoslavia; this group is identical to the IMF group "countries in transition" except for the IMF's inclusion of Mongolia

Four Dragons

the four small Asian less developed countries (LDCs) that have experienced unusually rapid economic growth; also known as the Four Tigers; this group consists of Hong Kong, South Korea, Singapore, Taiwan; these countries are included in the IMF's "advanced economies" group

Franc Zone (FZ)

note–also known as Conference des Ministres des Finances des Pays de la Zone Franc
established–1964
aim–to form a monetary union among countries whose currencies were linked to the French franc
members–(16) Benin, Burkina Faso, Cameroon, Central African Republic, Chad, Comoros, Republic of the Congo, Cote d'Ivoire, Equatorial Guinea, France, Gabon, Guinea-Bissau, Mali, Niger, Senegal, Togo

Front Line States (FLS)

established to achieve black majority rule in South Africa; has since gone out of existence; members included Angola, Botswana, Mozambique, Namibia, Tanzania, Zambia, Zimbabwe

General Agreement on Tariffs and Trade (GATT)

see the World Trade Organization (WTO)

General Confederation of Trade Unions (GCTU)

established–16 April 1992
aim–to consolidate trade union actions to protect citizens' social and labor rights and interests, to help secure trade unions' rights and guarantees, and to strengthen international trade union solidarity
members–(11) Armenia, Azerbaijan, Belarus, Georgia, Kazakhstan, Kyrgyzstan, Moldova, Russia, Tajikistan, Ukraine, Uzbekistan

Group of 10 (G-10)

note–also known as the Paris Club; includes the wealthiest members of the IMF who provide most of the money to be loaned and act as the informal steering committee; name persists despite increased membership
established–October 1962
aim–to coordinate credit policy
members–(11) Belgium, Canada, France, Germany, Italy, Japan, Netherlands, Sweden, Switzerland, UK, US
observers–(4) BIS, EC, IMF, OECD

Group of 11 (G-11)

established–2006
aim–to narrow the income gap with the world's richest nations
members–(12) Croatia, Ecuador, El Salvador, Georgia, Honduras, Indonesia, Jordan, Morocco, Pakistan, Paraguay, Sri Lanka, Tunisia

Group of 15 (G-15)

note–byproduct of the Nonaligned Movement; name persists despite increased membership
established–September 1989
aim–to promote economic cooperation among developing nations; to act as the main political organ for the Nonaligned Movement
members–(18) Algeria, Argentina, Brazil, Chile, Egypt, India, Indonesia, Iran, Jamaica, Kenya, Malaysia, Mexico, Nigeria, Peru, Senegal, Sri Lanka, Venezuela, Zimbabwe

Group of 20 (G-20)

established–created 1999; inaugurated 15-16 December 1999
aim–to promote open and constructive discussion between industrial and emerging-market countries on any issues related to global economic stability; helps to support growth and development across the globe
members–(20) Argentina, Australia, Brazil, Canada, China, EU, France, Germany, India, Indonesia, Italy, Japan, South Korea, Mexico, Russia, Saudi Arabia, South Africa, Turkey, UK, US

Group of 24 (G-24)
established–1 August 1989
aim–to promote the interests of developing countries in Africa, Asia, and Latin America within the IMF
members–(24) Algeria, Argentina, Brazil, Colombia, Democratic Republic of the Congo, Cote d'Ivoire, Egypt, Ethiopia, Gabon, Ghana, Guatemala, India, Iran, Lebanon, Mexico, Nigeria, Pakistan, Peru, Philippines, South Africa, Sri Lanka, Syria, Trinidad and Tobago, Venezuela
observers–(1) China

Group of 3 (G-3)
established–September 1990
aim–mechanism for policy coordination
members–(2) Colombia, Mexico; note–Panama shows interest in joining

Group of 5 (G-5)
established–22 September 1985
aim–to coordinate the economic policies of five major noncommunist economic powers
members–(5) France, Germany, Japan, UK, US

Group of 6 (G-6)
also known as Groupe des Six Sur le Desarmement (not to be confused with the Big Six) was established in 22 May 1984 with the aim of achieving nuclear disarmament; its members were Argentina, Greece, India, Mexico, Sweden, Tanzania

Group of 7 (G-7)
note–membership is the same as the Big Seven
established–22 September 1985
aim–to facilitate economic cooperation among the seven major noncommunist economic powers
members–(7) Group of 5 (France, Germany, Japan, UK, US) plus Canada and Italy

Group of 77 (G-77)
established–15 June1964; October 1967 first ministerial meeting
aim–to promote economic cooperation among developing countries; name persists in spite of increased membership
members–(129 plus the Palestine Liberation Organization) Afghanistan, Algeria, Angola, Antigua and Barbuda, Argentina, The Bahamas, Bahrain, Bangladesh, Barbados, Belize, Benin, Bhutan, Bolivia, Bosnia and Herzegovina, Botswana, Brazil, Brunei, Burkina Faso, Burma, Burundi, Cambodia, Cameroon, Cape Verde, Central African Republic, Chad, Chile, China, Colombia, Comoros, Democratic Republic of the Congo, Republic of the Congo, Costa Rica, Cote d'Ivoire, Cuba, Djibouti, Dominica, Dominican Republic, Ecuador, Egypt, El Salvador, Equatorial Guinea, Eritrea, Ethiopia, Fiji, Gabon, The Gambia, Ghana, Grenada, Guatemala, Guinea, Guinea-Bissau, Guyana, Haiti, Honduras, India, Indonesia, Iran, Iraq, Jamaica, Jordan, Kenya, North Korea, Kuwait, Laos, Lebanon, Lesotho, Liberia, Libya, Madagascar, Malawi, Malaysia, Maldives, Mali, Marshall Islands, Mauritania, Mauritius, Federated States of Micronesia, Mongolia, Morocco, Mozambique, Namibia, Nepal, Nicaragua, Niger, Nigeria, Oman, Pakistan, Panama, Papua New Guinea, Paraguay, Peru, Philippines, Qatar, Rwanda, Saint Kitts and Nevis, Saint Lucia, Saint Vincent and the Grenadines, Samoa, Sao Tome and Principe, Saudi Arabia, Senegal, Seychelles, Sierra Leone, Singapore, Solomon Islands, Somalia, South Africa, Sri Lanka, Sudan, Suriname, Swaziland, Syria, Tajikistan, Thailand, Timor-Leste, Togo, Tonga, Trinidad and Tobago, Tunisia, Turkmenistan, Uganda, UAE, Uruguay, Vanuatu, Venezuela, Vietnam, Yemen, Zambia, Zimbabwe, Palestine Liberation Organization

Group of 8 (G-8)
established–October 1975
aim–to facilitate economic cooperation among the developed countries (DCs) that participated in the Conference on International Economic Cooperation (CIEC), held in several sessions between December 1975 and 3 June 1977
members–(9) Canada, EC, France, Germany, Italy, Japan, Russia, UK, US

Group of 9 (G-9)
established–NA
aim–to discuss matters of mutual interest on an informal basis
members–(9) Austria, Belgium, Bulgaria, Denmark, Finland, Hungary, Romania, Serbia, Sweden

Gulf Cooperation Council (GCC)
note–also known as the Cooperation Council for the Arab States of the Gulf
established–25 May 1981
aim–to promote regional cooperation in economic, social, political, and military affairs
members–(6) Bahrain, Kuwait, Oman, Qatar, Saudi Arabia, UAE

high income countries
another term for the industrialized countries with high per capita GDPs; see developed countries (DCs)

Indian Ocean Commission (InOC)
established–21 December 1982
aim–to organize and promote regional cooperation in all sectors, especially economic
members–(5) Comoros, France (for Reunion and Mayotte), Madagascar, Mauritius, Seychelles

industrial countries
another term for the developed countries; see developed countries (DCs)

Inter-American Development Bank (IADB)
note–also known as Banco Interamericano de Desarrollo (BID)
established–8 April 1959; effective–30 December 1959
aim–to promote economic and social development in Latin America
members–(48) Argentina, Austria, The Bahamas, Barbados, Belgium, Belize, Bolivia, Brazil, Canada, Chile, China, Colombia, Costa Rica, Croatia, Denmark, Dominican Republic, Ecuador, El Salvador, Finland, France, Germany, Guatemala, Guyana, Haiti, Honduras,

Israel, Italy, Jamaica, Japan, South Korea, Mexico, Netherlands, Nicaragua, Norway, Panama, Paraguay, Peru, Portugal, Slovenia, Spain, Suriname, Sweden, Switzerland, Trinidad and Tobago, UK, US, Uruguay, Venezuela

Inter-Governmental Authority on Development (IGAD)

note–formerly known as Inter-Governmental Authority on Drought and Development (IGADD)
established–15-16 January 1986 as the Inter-Governmental Authority on Drought and Development; revitalized–21 March 1996 as the Inter-Governmental Authority on Development
aim–to promote a social, economic, and scientific community among its members
members–(6) Djibouti, Ethiopia, Kenya, Somalia, Sudan, Uganda; note–Eritrea declared its suspension in 2007

Inter-Parliamentary Union (IPU)

established–1889
aim–fosters contacts among parliamentarians, considers and expresses views of international interest and concern with the purpose of bringing about action by parliaments and parliamentarians, contributes to the defense and promotion of human rights, contributes to better knowledge of representative institutions
members–(154 and the Palestine Liberation Organization) Afghanistan, Albania, Algeria, Andorra, Angola, Argentina, Armenia, Australia, Austria, Azerbaijan, Bahrain, Bangladesh, Belarus, Belgium, Benin, Bolivia, Bosnia and Herzegovina, Botswana, Brazil, Bulgaria, Burkina Faso, Burundi, Cambodia, Cameroon, Canada, Cape Verde, Chile, China, Colombia, Comoros, Democratic Republic of the Congo, Republic of the Congo, Costa Rica, Cote d'Ivoire, Croatia, Cuba, Cyprus, Czech Republic, Denmark, Djibouti, Dominican Republic, Ecuador, Egypt, El Salvador, Estonia, Ethiopia, Finland, France, Gabon, The Gambia, Georgia, Germany, Ghana, Greece, Guatemala, Guinea-Bissau, Hungary, Iceland, India, Indonesia, Iran, Iraq, Ireland, Israel, Italy, Japan, Jordan, Kazakhstan, Kenya, North Korea, South Korea, Kuwait, Kyrgyzstan, Laos, Latvia, Lebanon, Lesotho, Liberia, Libya, Liechtenstein, Lithuania, Luxembourg, Macedonia, Malawi, Malaysia, Maldives, Mali, Malta, Mauritania, Mauritius, Mexico, Moldova, Monaco, Mongolia, Montenegro, Morocco, Mozambique, Namibia, Nepal, Netherlands, NZ, Nicaragua, Nigeria, Norway, Oman, Pakistan, Palau, Panama, Papua New Guinea, Paraguay, Peru, Philippines, Poland, Portugal, Qatar, Romania, Russia, Rwanda, Samoa, San Marino, Sao Tome and Principe, Saudi Arabia, Senegal, Serbia, Seychelles, Sierra Leone, Singapore, Slovakia, Slovenia, South Africa, Spain, Sri Lanka, Sudan, Suriname, Sweden, Switzerland, Syria, Tanzania, Tajikistan, Thailand, Timor-Leste, Togo, Tunisia, Turkey, Uganda, Ukraine, UAE, UK, Uruguay, Venezuela, Vietnam, Yemen, Zambia, Zimbabwe, Palestine Liberation Organization
associate members–(9) Andean Parliament, Central American Parliament, East African Legislative Assembly, European Parliament, Inter-Parliamentary Committee of the West African Economic and Monetary Union, Latin American Parliament, Parliament of the Economic Community of West African States, Parliamentary Assembly of the Council of Europe, Transitional Arab Parliament

International Atomic Energy Agency (IAEA)

established–26 October 1956; effective–29 July 1957
aim–to promote peaceful uses of atomic energy
members–(150) Afghanistan, Albania, Algeria, Angola, Argentina, Armenia, Australia, Austria, Azerbaijan, Bahrain, Bangladesh, Belarus, Belgium, Belize, Benin, Bolivia, Bosnia and Herzegovina, Botswana, Brazil, Bulgaria, Burkina Faso, Burma, Burundi, Cameroon, Canada, Central African Republic, Chad, Chile, China, Colombia, Democratic Republic of the Congo, Republic of the Congo, Costa Rica, Cote d'Ivoire, Croatia, Cuba, Cyprus, Czech Republic, Denmark, Dominican Republic, Ecuador, Egypt, El Salvador, Eritrea, Estonia, Ethiopia, Finland, France, Gabon, Georgia, Germany, Ghana, Greece, Guatemala, Haiti, Holy See, Honduras, Hungary, Iceland, India, Indonesia, Iran, Iraq, Ireland, Israel, Italy, Jamaica, Japan, Jordan, Kazakhstan, Kenya, South Korea, Kuwait, Kyrgyzstan, Latvia, Lebanon, Lesotho, Liberia, Libya, Liechtenstein, Lithuania, Luxembourg, Macedonia, Madagascar, Malawi, Malaysia, Mali, Malta, Marshall Islands, Mauritania, Mauritius, Mexico, Moldova, Monaco, Mongolia, Montenegro, Morocco, Mozambique, Namibia, Nepal, Netherlands, NZ, Nicaragua, Niger, Nigeria, Norway, Oman, Pakistan, Palau, Panama, Paraguay, Peru, Philippines, Poland, Portugal, Qatar, Romania, Russia, Saudi Arabia, Senegal, Serbia, Seychelles, Sierra Leone, Singapore, Slovakia, Slovenia, South Africa, Spain, Sri Lanka, Sudan, Sweden, Switzerland, Syria, Tajikistan, Tanzania, Thailand, Tunisia, Turkey, Uganda, Ukraine, UAE, UK, US, Uruguay, Uzbekistan, Venezuela, Vietnam, Yemen, Zambia, Zimbabwe; note–membership pending for Cambodia, Cape Verde, Papua New Guinea, Rwanda, and Togo until the necessary legal instruments are deposited with the IAEA

International Bank for Reconstruction and Development (IBRD)

note–also known as the World Bank
established–22 July 1944; effective–27 December 1945
aim–to provide economic development loans; a UN specialized agency
members–(187) includes all UN member countries except Andorra, Cuba, North Korea, Liechtenstein, Monaco, Nauru; plus Kosovo

International Chamber of Commerce (ICC)

established–1919
aim–to promote free trade and private enterprise and to represent business interests at national and international levels
members–(89 national committees) Algeria, Argentina, Australia, Austria, Bahrain, Bangladesh, Belgium, Bolivia, Brazil, Bulgaria, Cameroon, Canada, Caribbean, Chile, China, Colombia, Costa Rica, Croatia, Cuba, Cyprus, Czech Republic, Denmark, Dominican Republic, Ecuador, Egypt, El Salvador, Finland, France, Georgia, Germany, Ghana, Greece, Guatemala, Hong Kong, Hungary, Iceland, India, Indonesia, Iran, Ireland, Israel, Italy, Japan, Jordan, South Korea, Kuwait, Lebanon, Lithuania, Luxembourg, Madagascar, Malaysia, Mexico, Monaco, Morocco, Nepal, Netherlands, NZ, Nigeria, Norway, Pakistan, Panama, Philippines, Poland, Portugal, Qatar, Romania, Russia, Saudi Arabia, Senegal, Serbia, Singapore, Slovakia, Slovenia, South Africa, Spain, Sri Lanka, Sweden, Switzerland, Syria, Taiwan, Thailand, Togo, Tunisia, Turkey, Ukraine, UAE, UK, US, Uruguay; note–Peru is restructuring

International Civil Aviation Organization (ICAO)

established–7 December 1944; effective–4 April 1947
aim–to promote international cooperation in civil aviation; a UN specialized agency
members–(190) includes all UN member countries except Dominica, Liechtenstein, and Tuvalu (189 total); plus Cook Islands

International Civilian Support Mission in Haiti (MICAH)

established 17 December 1999 to promote respect for human rights; members included Argentina, Benin, Canada, France, India, Mali, Niger, Senegal, Togo, Tunisia, US; closed 2001

International Committee of the Red Cross (ICRC)

established–17 February 1863

aim–to provide humanitarian aid in wartime
members–(15-25 individuals) all Swiss nationals

International Court of Justice (ICJ)

note–also known as the World Court; primary judicial organ of the UN
established–26 June 1945 with the signing of the UN Charter (inaugural sitting of the Court was on 18 April 1946); superseded Permanent Court of International Justice (attached to the League of Nations)
aim–to settle, in accordance with international law, legal disputes submitted to it by States and to give advisory opinions on legal questions referred to it by authorized UN organs and specialized agencies
members–(15 judges) elected by the UN General Assembly and Security Council to represent all principal legal systems; judges elected to nine-year terms (eligible for two additional terms); elections held every three years for one-third of the judges
jurisdiction–based on the principle of consent in contentious issues; consent to compulsory jurisdiction is outlined in Statute 36 of the ICJ; states provide declarations of consent to compulsory jurisdiction of the ICJ either with or without reservations (date in parens after each state is when the declaration was deposited with the UN Secretary-General); Haiti, Luxembourg, Nicaragua, and Uruguay deposited declarations with the Permanent Court of International Justice prior to 1945 and these were later transferred to the ICJ)
states accepting compulsory jurisdiction with reservations–(55) Australia (22 March 2002), Barbados (1 August 1980), Belgium (17 June 1958), Botswana (16 March 1970), Bulgaria (21 June 1992), Cambodia 19 September 1957), Canada (10 May 1994), Democratic Republic of the Congo (8 February 1989), Cote d'Ivoire (29 September 2001), Cyprus (3 September 2002), Denmark (10 December 1956), Djibouti (2 September 2005), Egypt (22 July 1957), Estonia (31 October 1991), Finland (25 June 1958), The Gambia (22 June 1966), Germany (30 April 2008), Greece (10 January 1994), Guinea (4 December 1998), Honduras (6 June 1986), Hungary (22 October 1992), India (18 September 1974), Japan (9 July 2007), Kenya (19 April 1965), Lesotho (6 September 2000), Liberia (20 March 1952), Liechtenstein (29 March 1950), Madagascar (2 July 1992), Malawi (12 December 1966), Malta (2 September 1983), Mauritius (23 September 1968), Mexico (28 October 1947), Netherlands (1 August 1956), New Zealand (23 September 1977), Nicaragua (24 September 1929), Nigeria (30 April 1998), Norway (25 June 1996), Pakistan (13 September 1960), Panama (25 October 1921), Peru (7 July 2003), Philippines (18 January 1972), Poland (25 March 1996), Portugal (25 February 2005), Senegal (2 December 1985), Slovakia (28 May 2004), Somalia (11 April 1963), Spain (20 October 1990), Sudan (2 January 1958), Suriname (31 August 1987), Swaziland (26 May 1969), Sweden (6 April 1957), Switzerland (28 July 1948), Togo (25 October 1979), Uganda (3 October 1963), United Kingdom (5 July 2004)
states accepting compulsory jurisdiction without reservations–(11) Austria (19 May 1971), Cameroon (3 March 1994), Costa Rica (20 February 1973), Dominica (31 March 2006), Dominican Republic (30 September 1924), Georgia (20 June 1995), Guinea-Bissau (7 August 1989), Haiti (4 October 1921), Luxembourg (15 September 1930), Paraguay (25 September 1996), Uruguay (28 January 1921)

International Criminal Court (ICCt)

established–1 July 2002
aim–to hold all individuals and countries accountable to international laws of conduct; to specify international standards of conduct; to provide an important mechanism for implementing these standards; to ensure that perpetrators are brought to justice
members–21 judges (three judges form the Presidency) and six judges each in the Pre-trial, Trial, and Appeals Divisions; judges elected by secret ballot by the Assembly of States Parties to the Rome Statute for nine-year terms (not eligible for reelection)
jurisdiction–governed by the Statute of the International Criminal Court treaty (or Rome Statute), adopted 17 July 1998 at the UN Conference of Plenipotentiaries in Rome and entered into force 1 July 2002
states accepting jurisdiction–(114) Afghanistan, Albania, Andorra, Antigua and Barbuda, Argentina, Australia, Austria, Bangladesh, Barbados, Belgium, Belize, Benin, Bolivia, Bosnia and Herzegovina, Botswana, Brazil, Bulgaria, Burkina Faso, Burundi, Cambodia, Canada, Central African Republic, Chad, Chile, Colombia, Comoros, Cook Islands, Democratic Republic of the Congo, Republic of the Congo, Costa Rica, Croatia, Cyprus, Czech Republic, Denmark, Djibouti, Dominica, Dominican Republic, Ecuador, Estonia, Fiji, Finland, France, Gabon, The Gambia, Georgia, Germany, Ghana, Greece, Guinea, Guyana, Honduras, Hungary, Iceland, Ireland, Italy, Japan, Jordan, Kenya, South Korea, Latvia, Lesotho, Liberia, Liechtenstein, Lithuania, Luxembourg, Macedonia, Madagascar, Malawi, Mali, Malta, Marshall Islands, Mauritius, Mexico, Moldova, Mongolia, Montenegro, Namibia, Nauru, Netherlands, NZ, Niger, Nigeria, Norway, Panama, Paraguay, Peru, Poland, Portugal, Romania, Saint Kitts and Nevis, Saint Lucia, Saint Vincent and the Grenadines, Samoa, San Marino, Senegal, Seychelles, Serbia, Sierra Leone, Slovakia, Slovenia, South Africa, Spain, Suriname, Sweden, Switzerland, Tajikistan, Tanzania, Timor-Leste, Trinidad and Tobago, Uganda, UK, Uruguay, Venezuela, Zambia

International Criminal Police Organization (Interpol)

established–September 1923 set up as the International Criminal Police Commission; 13 June 1956 constitution modified and present name adopted
aim–to promote international cooperation among police authorities in fighting crime
members–(188) Afghanistan, Albania, Algeria, Andorra, Angola, Antigua and Barbuda, Argentina, Armenia, Aruba, Australia, Austria, Azerbaijan, The Bahamas, Bahrain, Bangladesh, Barbados, Belarus, Belgium, Belize, Benin, Bhutan, Bolivia, Bosnia and Herzegovina, Botswana, Brazil, Brunei, Bulgaria, Burkina Faso, Burma, Burundi, Cambodia, Cameroon, Canada, Cape Verde, Central African Republic, Chad, Chile, China, Colombia, Comoros, Democratic Republic of the Congo, Republic of the Congo, Costa Rica, Cote d'Ivoire, Croatia, Cuba, Cyprus, Czech Republic, Denmark, Djibouti, Dominica, Dominican Republic, Ecuador, Egypt, El Salvador, Equatorial Guinea, Eritrea, Estonia, Ethiopia, Fiji, Finland, France, Gabon, The Gambia, Georgia, Germany, Ghana, Greece, Grenada, Guatemala, Guinea, Guinea-Bissau, Guyana, Haiti, Holy See, Honduras, Hungary, Iceland, India, Indonesia, Iran, Iraq, Ireland, Israel, Italy, Jamaica, Japan, Jordan, Kazakhstan, Kenya, South Korea, Kuwait, Kyrgyzstan, Laos, Latvia, Lebanon, Lesotho, Liberia, Libya, Liechtenstein, Lithuania, Luxembourg, Macedonia, Madagascar, Malawi, Malaysia, Maldives, Mali, Malta, Marshall Islands, Mauritania, Mauritius, Mexico, Moldova, Monaco, Mongolia, Montenegro, Morocco, Mozambique, Namibia, Nauru, Nepal, Netherlands, NZ, Nicaragua, Niger, Nigeria, Norway, Oman, Pakistan, Panama, Papua New Guinea, Paraguay, Peru, Philippines, Poland, Portugal, Qatar, Romania, Russia, Rwanda, Saint Kitts and Nevis, Saint Lucia, Saint Vincent and the Grenadines, Samoa, San Marino, Sao Tome and Principe, Saudi Arabia, Senegal, Serbia, Seychelles, Sierra Leone, Singapore, Slovakia, Slovenia, Somalia, South Africa, Spain, Sri Lanka, Sudan, Suriname, Swaziland, Sweden, Switzerland, Syria, Tajikistan, Tanzania, Thailand, Timor-Leste, Togo, Tonga, Trinidad and Tobago, Tunisia, Turkey, Turkmenistan, Uganda, Ukraine, UAE, UK, US, Uruguay, Uzbekistan, Venezuela, Vietnam, Yemen, Zambia, Zimbabwe
subbureaus–(11) American Samoa, Anguilla, Bermuda, British Virgin Islands, Cayman Islands, Gibraltar, Hong Kong, Macau, Montserrat, Puerto Rico, Turks and Caicos Islands

International Development Association (IDA)

established–26 January 1960; effective–24 September 1960

aim–to provide economic loans for low-income countries; UN specialized agency and IBRD affiliate
members–(171) Afghanistan, Albania, Algeria, Angola, Argentina, Armenia, Australia, Austria, Azerbaijan, The Bahamas, Bangladesh, Barbados, Belgium, Belize, Benin, Bhutan, Bolivia, Bosnia and Herzegovina, Botswana, Brazil, Burkina Faso, Burma, Burundi, Cambodia, Cameroon, Canada, Cape Verde, Central African Republic, Chad, Chile, China, Colombia, Comoros, Democratic Republic of the Congo, Republic of the Congo, Costa Rica, Cote d'Ivoire, Croatia, Cyprus, Czech Republic, Denmark, Djibouti, Dominica, Dominican Republic, Ecuador, Egypt, El Salvador, Equatorial Guinea, Eritrea, Estonia, Ethiopia, EU, Fiji, Finland, France, Gabon, The Gambia, Georgia, Germany, Ghana, Greece, Grenada, Guatemala, Guinea, Guinea-Bissau, Guyana, Haiti, Honduras, Hungary, Iceland, India, Indonesia, Iran, Iraq, Ireland, Israel, Italy, Japan, Jordan, Kazakhstan, Kenya, Kiribati, South Korea, Kosovo, Kuwait, Kyrgyzstan, Laos, Latvia, Lebanon, Lesotho, Liberia, Libya, Luxembourg, Macedonia, Madagascar, Malawi, Malaysia, Maldives, Mali, Marshall Islands, Mauritania, Mauritius, Mexico, Federated States of Micronesia, Moldova, Mongolia, Montenegro, Morocco, Mozambique, Nepal, Netherlands, NZ, Nicaragua, Niger, Nigeria, Norway, Oman, Pakistan, Palau, Panama, Papua New Guinea, Paraguay, Peru, Philippines, Poland, Portugal, Russia, Rwanda, Saint Kitts and Nevis, Saint Lucia, Saint Vincent and the Grenadines, Samoa, Sao Tome and Principe, Saudi Arabia, Senegal, Serbia, Sierra Leone, Singapore, Slovakia, Slovenia, Solomon Islands, Somalia, South Africa, Spain, Sri Lanka, Sudan, Swaziland, Sweden, Switzerland, Syria, Tajikistan, Tanzania, Thailand, Timor-Leste, Togo, Tonga, Trinidad and Tobago, Tunisia, Turkey, Tuvalu, Uganda, Ukraine, UAE, UK, US, Uzbekistan, Vanuatu, Vietnam, Yemen, Zambia, Zimbabwe

International Energy Agency (IEA)

established–15 November 1974
aim–to promote cooperation on energy matters, especially emergency oil sharing and relations between oil consumers and oil producers; established by the OECD
members–(29) Australia, Austria, Belgium, Canada, Czech Republic, Denmark, EC, Finland, France, Germany, Greece, Hungary, Ireland, Italy, Japan, South Korea, Luxembourg, Netherlands, NZ, Norway, Poland, Portugal, Slovakia, Spain, Sweden, Switzerland, Turkey, UK, US

International Federation of Red Cross and Red Crescent Societies (IFRCS)

note–formerly known as League of Red Cross and Red Crescent Societies (LORCS)
established–5 May 1919
aim–to organize, coordinate, and direct international relief actions; to promote humanitarian activities; to represent and encourage the development of National Societies; to bring help to victims of armed conflicts, refugees, and displaced people; to reduce the vulnerability of people through development programs
members–(185 plus the Palestine Liberation Organization) Afghanistan, Albania, Algeria, Andorra, Angola, Antigua and Barbuda, Argentina, Armenia, Australia, Austria, Azerbaijan, The Bahamas, Bahrain, Bangladesh, Barbados, Belarus, Belgium, Belize, Benin, Bolivia, Bosnia and Herzegovina, Botswana, Brazil, Brunei, Bulgaria, Burkina Faso, Burma, Burundi, Cambodia, Cameroon, Canada, Cape Verde, Central African Republic, Chad, Chile, China, Colombia, Comoros, Democratic Republic of the Congo, Republic of the Congo, Cook Islands, Costa Rica, Cote d'Ivoire, Croatia, Cuba, Czech Republic, Denmark, Djibouti, Dominica, Dominican Republic, Ecuador, Egypt, El Salvador, Equatorial Guinea, Estonia, Ethiopia, Fiji, Finland, France, Gabon, The Gambia, Georgia, Germany, Ghana, Greece, Grenada, Guatemala, Guinea, Guinea-Bissau, Guyana, Haiti, Honduras, Hungary, Iceland, India, Indonesia, Iran, Iraq, Ireland, Israel, Italy, Jamaica, Japan, Jordan, Kazakhstan, Kenya, Kiribati, North Korea, South Korea, Kuwait, Kyrgyzstan, Laos, Latvia, Lebanon, Lesotho, Liberia, Libya, Liechtenstein, Lithuania, Luxembourg, Macedonia, Madagascar, Malawi, Malaysia, Mali, Malta, Mauritania, Mauritius, Mexico, Federated States of Micronesia, Moldova, Monaco, Mongolia, Montenegro, Morocco, Mozambique, Namibia, Nepal, Netherlands, NZ, Nicaragua, Niger, Nigeria, Norway, Pakistan, Palau, Panama, Papua New Guinea, Paraguay, Peru, Philippines, Poland, Portugal, Qatar, Romania, Russia, Rwanda, Saint Kitts and Nevis, Saint Lucia, Saint Vincent and the Grenadines, Samoa, San Marino, Sao Tome and Principe, Saudi Arabia, Senegal, Serbia, Seychelles, Sierra Leone, Singapore, Slovakia, Slovenia, Solomon Islands, Somalia, South Africa, Spain, Sri Lanka, Sudan, Suriname, Swaziland, Sweden, Switzerland, Syria, Tajikistan, Tanzania, Thailand, Timor-Leste, Togo, Tonga, Trinidad and Tobago, Tunisia, Turkey, Turkmenistan, Uganda, Ukraine, UAE, UK, US, Uruguay, Uzbekistan, Vanuatu, Venezuela, Vietnam, Yemen, Zambia, Zimbabwe, Palestine Liberation Organization
observers–(3) Eritrea, Maldives, and Tuvalu

International Finance Corporation (IFC)

established–25 May 1955; effective–24 July 1956
aim–to support private enterprise in international economic development; a UN specialized agency and IBRD affiliate
members–(182) includes all UN member countries except Andorra, Brunei, Cuba, North Korea, Liechtenstein, Monaco, Nauru, Saint Vincent and the Grenadines, San Marino, Suriname, Tuvalu; plus Kosovo

International Fund for Agricultural Development (IFAD)

established–November 1974
aim–to promote agricultural development; a UN specialized agency
members–(165)
List A–(23 industrialized aid contributors) Austria, Belgium, Canada, Denmark, Finland, France, Germany, Greece, Iceland, Ireland, Italy, Japan, Luxembourg, Netherlands, NZ, Norway, Portugal, Spain, Sweden, Switzerland, UK, US
List B–(12 petroleum-exporting aid contributors) Algeria, Gabon, Indonesia, Iran, Iraq, Kuwait, Libya, Nigeria, Qatar, Saudi Arabia, UAE, Venezuela
List C–(130 aid recipients) Afghanistan, Albania, Angola, Antigua and Barbuda, Argentina, Armenia, Azerbaijan, The Bahamas, Bangladesh, Barbados, Belize, Benin, Bhutan, Bolivia, Bosnia and Herzegovina, Botswana, Brazil, Burkina Faso, Burma, Burundi, Cambodia, Cameroon, Cape Verde, Central African Republic, Chad, Chile, China, Colombia, Comoros, Democratic Republic of the Congo, Republic of the Congo, Cook Islands, Costa Rica, Cote d'Ivoire, Croatia, Cuba, Cyprus, Djibouti, Dominica, Dominican Republic, Ecuador, Egypt, El Salvador, Equatorial Guinea, Eritrea, Ethiopia, Fiji, The Gambia, Georgia, Ghana, Grenada, Guatemala, Guinea, Guinea-Bissau, Guyana, Haiti, Honduras, India, Israel, Jamaica, Jordan, Kazakhstan, Kenya, Kiribati, North Korea, South Korea, Kyrgyzstan, Laos, Lebanon, Lesotho, Liberia, Macedonia, Madagascar, Malawi, Malaysia, Maldives, Mali, Malta, Marshall Islands, Mauritania, Mauritius, Mexico, Moldova, Mongolia, Morocco, Mozambique, Namibia, Nepal, Nicaragua, Niger, Niue, Oman, Pakistan, Panama, Papua New Guinea, Paraguay, Peru, Philippines, Romania, Rwanda, Saint Kitts and Nevis, Saint Lucia, Saint Vincent and the Grenadines, Samoa, Sao Tome and Principe, Senegal, Seychelles, Sierra Leone, Solomon Islands, Somalia, South Africa, Sri Lanka, Sudan, Suriname, Swaziland, Syria, Tajikistan, Tanzania, Thailand, Timor-Leste, Togo, Tonga, Trinidad and Tobago, Tunisia, Turkey, Uganda, Uruguay, Vietnam, Yemen, Zambia, Zimbabwe

International Hydrographic Organization (IHO)

note–name changed from International Hydrographic Bureau on 22 September 1970

established–June 1919; effective–June 1921
aim–to train hydrographic surveyors and nautical cartographers to achieve standardization in nautical charts and electronic chart displays; to provide advice on nautical cartography and hydrography; to develop the sciences in the field of hydrography and techniques used for descriptive oceanography
members–(80) Algeria, Argentina, Australia, Bahrain, Bangladesh, Belgium, Brazil, Burma, Canada, Chile, China (including Hong Kong and Macau), Colombia, Democratic Republic of the Congo, Croatia, Cuba, Cyprus, Denmark, Dominican Republic, Ecuador, Egypt, Estonia, Fiji, Finland, France, Germany, Greece, Guatemala, Iceland, India, Indonesia, Iran, Ireland, Italy, Jamaica, Japan, North Korea, South Korea, Kuwait, Latvia, Malaysia, Mauritius, Mexico, Monaco, Morocco, Mozambique, Netherlands, NZ, Nigeria, Norway, Oman, Pakistan, Papua New Guinea, Peru, Philippines, Poland, Portugal, Qatar, Romania, Russia, Saudi Arabia, Serbia, Singapore, Slovenia, South Africa, Spain, Sri Lanka, Suriname, Sweden, Syria, Thailand, Tonga, Trinidad and Tobago, Tunisia, Turkey, Ukraine, UAE, UK, US, Uruguay, Venezuela

International Labor Organization (ILO)

established–28 June 1919 set up as part of Treaty of Versailles; 11 April 1919 became operative; 14 December 1946 affiliated with the UN
aim–to deal with world labor issues; a UN specialized agency
members–(183) includes all UN member countries except Andorra, Bhutan, North Korea, Liechtenstein, Federated States of Micronesia, Monaco, Nauru, Palau, Tonga; note–includes the following dependencies: Netherlands (Aruba)

International Maritime Organization (IMO)

note–name changed from Intergovernmental Maritime Consultative Organization (IMCO) on 22 May 1982
established–6 March 1948 set up as the Inter-Governmental Maritime Consultative Organization; effective–17 March 1958
aim–to deal with international maritime affairs; a UN specialized agency
members–(168) includes all UN member countries except Afghanistan, Andorra, Armenia, Belarus, Bhutan, Botswana, Burkina Faso, Burundi, Central African Republic, Chad, Kyrgyzstan, Laos, Lesotho, Liechtenstein, Mali, Federated States of Micronesia, Nauru, Niger, Palau, Rwanda, Swaziland, Tajikistan, Uzbekistan, Zambia; and Cook Islands
associate members–(3) Faroe Islands, Hong Kong, Macau

International Mobile Satellite Organization (IMSO)

established–15 April 1999
aim–acts as watchdog over Inmarsat (International Maritime Satellite Organization), a private company, to make sure it follows ICAO standards and recommended practices; plays an active role in the development of international telecommunications policies
members–(94) Algeria, Antigua and Barbuda, Argentina, Australia, The Bahamas, Bahrain, Bangladesh, Belarus, Belgium, Bosnia and Herzegovina, Brazil, Brunei, Bulgaria, Cameroon, Canada, Chile, China, Colombia, Comoros, Cook Islands, Costa Rica, Croatia, Cuba, Cyprus, Czech Republic, Denmark, Egypt, Finland, France, Gabon, Germany, Ghana, Greece, Hungary, Iceland, India, Indonesia, Iran, Iraq, Israel, Italy, Japan, Kenya, South Korea, Kuwait, Latvia, Lebanon, Liberia, Libya, Malaysia, Malta, Marshall Islands, Mauritius, Mexico, Monaco, Montenegro, Morocco, Mozambique, Netherlands, NZ, Nigeria, Norway, Oman, Pakistan, Panama, Peru, Philippines, Poland, Portugal, Qatar, Romania, Russia, Saudi Arabia, Senegal, Serbia, Singapore, Slovakia, South Africa, Spain, Sri Lanka, Sweden, Switzerland, Tanzania, Thailand, Tonga, Tunisia, Turkey, Ukraine, UAE, UK, US, Vanuatu, Venezuela, Vietnam

International Monetary Fund (IMF)

established–22 July 1944; effective–27 December 1945
aim–to promote world monetary stability and economic development; a UN specialized agency
members–(187) includes all UN member countries except Andorra, Cuba, North Korea, Liechtenstein, Monaco, Nauru; plus Kosovo; note–includes the following dependencies or areas of special interest: China (Hong Kong and Macau), Netherlands (Aruba)

International Olympic Committee (IOC)

established–23 June 1894
aim–to promote the Olympic ideals and administer the Olympic games: 2012 Summer Olympics in London, UK; 2014 Winter Olympics in Sochi, Russia
National Olympic Committees–(204 and the Palestine Liberation Organization) Afghanistan, Albania, Algeria, American Samoa, Andorra, Angola, Antigua and Barbuda, Argentina, Armenia, Aruba, Australia, Austria, Azerbaijan, The Bahamas, Bahrain, Bangladesh, Barbados, Belarus, Belgium, Belize, Benin, Bermuda, Bhutan, Bolivia, Bosnia and Herzegovina, Botswana, Brazil, British Virgin Islands, Brunei, Bulgaria, Burkina Faso, Burma, Burundi, Cambodia, Cameroon, Canada, Cape Verde, Cayman Islands, Central African Republic, Chad, Chile, China, Colombia, Comoros, Democratic Republic of the Congo, Republic of the Congo, Cook Islands, Costa Rica, Cote d'Ivoire, Croatia, Cuba, Cyprus, Czech Republic, Denmark, Djibouti, Dominica, Dominican Republic, Ecuador, Egypt, El Salvador, Equatorial Guinea, Eritrea, Estonia, Ethiopia, Fiji, Finland, France, Gabon, The Gambia, Georgia, Germany, Ghana, Greece, Grenada, Guam, Guatemala, Guinea, Guinea-Bissau, Guyana, Haiti, Honduras, Hong Kong, Hungary, Iceland, India, Indonesia, Iran, Iraq, Ireland, Israel, Italy, Jamaica, Japan, Jordan, Kazakhstan, Kenya, Kiribati, North Korea, South Korea, Kuwait, Kyrgyzstan, Laos, Latvia, Lebanon, Lesotho, Liberia, Libya, Liechtenstein, Lithuania, Luxembourg, Macedonia, Madagascar, Malawi, Malaysia, Maldives, Mali, Malta, Marshall Islands, Mauritania, Mauritius, Mexico, Federated States of Micronesia, Moldova, Monaco, Mongolia, Montenegro, Morocco, Mozambique, Namibia, Nauru, Nepal, Netherlands, NZ, Nicaragua, Niger, Nigeria, Norway, Oman, Pakistan, Palau, Panama, Papua New Guinea, Paraguay, Peru, Philippines, Poland, Portugal, Puerto Rico, Qatar, Romania, Russia, Rwanda, Saint Kitts and Nevis, Saint Lucia, Saint Vincent and the Grenadines, Samoa, San Marino, Sao Tome and Principe, Saudi Arabia, Senegal, Serbia, Seychelles, Sierra Leone, Singapore, Slovakia, Slovenia, Solomon Islands, Somalia, South Africa, Spain, Sri Lanka, Sudan, Suriname, Swaziland, Sweden, Switzerland, Syria, Taiwan, Tajikistan, Tanzania, Thailand, Timor-Leste, Togo, Tonga, Trinidad and Tobago, Tunisia, Turkey, Turkmenistan, Tuvalu, Uganda, Ukraine, UAE, UK, US, Uruguay, Uzbekistan, Vanuatu, Venezuela, Vietnam, Virgin Islands, Yemen, Zambia, Zimbabwe, Palestine Liberation Organization

International Organization for Migration (IOM)

note–established as Provisional Intergovernmental Committee for the Movement of Migrants from Europe; renamed Intergovernmental Committee for European Migration (ICEM) on 15 November 1952; renamed Intergovernmental Committee for Migration (ICM) in November 1980; current name adopted 14 November 1989
established–5 December 1951
aim–to facilitate orderly international emigration and immigration

members–(132) Afghanistan, Albania, Algeria, Angola, Argentina, Armenia, Australia, Austria, Azerbaijan, The Bahamas, Bangladesh, Belarus, Belgium, Belize, Benin, Bolivia, Bosnia and Herzegovina, Botswana, Brazil, Bulgaria, Burkina Faso, Burundi, Cambodia, Cameroon, Canada, Cape Verde, Central African Republic, Chile, Colombia, Democratic Republic of the Congo, Republic of the Congo, Costa Rica, Cote d'Ivoire, Croatia, Cyprus, Czech Republic, Denmark, Dominican Republic, Ecuador, Egypt, El Salvador, Estonia, Finland, France, Gabon, The Gambia, Georgia, Germany, Ghana, Greece, Guatemala, Guinea, Guinea-Bissau, Haiti, Honduras, Hungary, India, Iran, Ireland, Israel, Italy, Jamaica, Japan, Jordan, Kazakhstan, Kenya, South Korea, Kyrgyzstan, Latvia, Lesotho, Liberia, Libya, Lithuania, Luxembourg, Madagascar, Mali, Malta, Mauritania, Mauritius, Mexico, Moldova, Mongolia, Montenegro, Morocco, Namibia, Nepal, Netherlands, NZ, Nicaragua, Niger, Nigeria, Norway, Pakistan, Panama, Paraguay, Peru, Philippines, Poland, Portugal, Romania, Rwanda, Senegal, Serbia, Sierra Leone, Slovakia, Slovenia, Somalia, South Africa, Spain, Sri Lanka, Sudan, Swaziland, Sweden, Switzerland, Tajikistan, Tanzania, Thailand, Timor-Leste, Togo, Trinidad and Tobago, Tunisia, Turkey, Uganda, Ukraine, UK, US, Uruguay, Venezuela, Vietnam, Yemen, Zambia, Zimbabwe
observers–(17) Bahrain, Bhutan, China, Cuba, Ethiopia, Guyana, Holy See, Indonesia, Macedonia, Mozambique, Papua New Guinea, Qatar, Russia, San Marino, Sao Tome and Principe, Saudi Arabia, Turkmenistan

International Organization for Standardization (ISO)

established–February 1947
aim–to promote the development of international standards with a view to facilitating international exchange of goods and services and to developing cooperation in the sphere of intellectual, scientific, technological and economic activity
members–(109 national standards organizations) Algeria, Argentina, Armenia, Australia, Austria, Azerbaijan, Bahrain, Bangladesh, Barbados, Belarus, Belgium, Bosnia and Herzegovina, Botswana, Brazil, Bulgaria, Cameroon, Canada, Chile, China, Colombia, Democratic Republic of the Congo, Costa Rica, Cote d'Ivoire, Croatia, Cuba, Cyprus, Czech Republic, Denmark, Ecuador, Egypt, Estonia, Ethiopia, Fiji, Finland, France, Germany, Ghana, Greece, Hungary, Iceland, India, Indonesia, Iran, Iraq, Ireland, Israel, Italy, Jamaica, Japan, Jordan, Kazakhstan, Kenya, North Korea, South Korea, Kuwait, Lebanon, Libya, Lithuania, Luxembourg, Macedonia, Malaysia, Mali, Malta, Mauritius, Mexico, Mongolia, Morocco, Namibia, Netherlands, NZ, Nigeria, Norway, Oman, Pakistan, Panama, Peru, Philippines, Poland, Portugal, Qatar, Romania, Russia, Saint Lucia, Saudi Arabia, Senegal, Serbia, Singapore, Slovakia, Slovenia, South Africa, Spain, Sri Lanka, Sudan, Sweden, Switzerland, Syria, Tanzania, Thailand, Trinidad and Tobago, Tunisia, Turkey, Ukraine, UAE, UK, US, Uruguay, Uzbekistan, Vietnam, Yemen
correspondent members–(43 plus the Palestine Liberation Organization) Afghanistan, Albania, Angola, Benin, Bhutan, Bolivia, Brunei, Burkina Faso, Burma, Republic of the Congo, Dominican Republic, El Salvador, Gabon, The Gambia, Georgia, Guatemala, Guinea, Hong Kong, Kyrgyzstan, Latvia, Lesotho, Liberia, Macau, Madagascar, Malawi, Mauritania, Moldova, Montenegro, Mozambique, Nepal, Papua New Guinea, Paraguay, Rwanda, Seychelles, Sierra Leone, Suriname, Swaziland, Tajikistan, Togo, Turkmenistan, Uganda, Zambia, Zimbabwe, Palestine Liberation Organization
subscriber members–(10) Antigua and Barbuda, Burundi, Cambodia, Central African Republic, Dominica, Eritrea, Guyana, Honduras, Laos, Saint Vincent and the Grenadines

International Organization of the French-speaking World (OIF)

note–name changed from Agency of Cultural and Technical Cooperation (ACCT) in 1997; also known as Organisation Internationale de la Francophonie
established–20 March 1970
aim–founded around a common language to promote and spread the cultures of its members and to reinforce cultural and technical cooperation between them
members–(53) Albania, Andorra, Belgium, Benin, Bulgaria, Burkina Faso, Burundi, Cambodia, Cameroon, Canada, Canada–New Brunswick, Canada–Quebec, Cape Verde, Central African Republic, Chad, Comoros, Democratic Republic of Congo, Republic of Congo, Cote d'Ivoire, Djibouti, Dominica, Egypt, Equatorial Guinea, France, French Community of Belgium, Gabon, Greece, Guinea, Guinea-Bissau, Haiti, Laos, Lebanon, Luxembourg, Macedonia, Madagascar, Mali, Mauritania, Mauritius, Moldova, Monaco, Morocco, Niger, Romania, Rwanda, Saint Lucia, Sao Tome and Principe, Senegal, Seychelles, Switzerland, Togo, Tunisia, Vanuatu, Vietnam
associates–(3) Armenia, Cyprus, Ghana
observers–(14) Austria, Croatia, Czech Republic, Georgia, Hungary, Latvia, Lithuania, Mozambique, Poland, Serbia, Slovakia, Slovenia, Thailand, Ukraine

International Red Cross and Red Crescent Movement (ICRM)

established–1928
aim–to promote worldwide humanitarian aid through the International Committee of the Red Cross (ICRC) in wartime, and International Federation of Red Cross and Red Crescent Societies (IFRCS; formerly League of Red Cross and Red Crescent Societies or LORCS) in peacetime
National Societies–(185 countries and the Palestine Liberation Organization); note–same as membership for International Federation of Red Cross and Red Crescent Societies (IFRCS)

International Telecommunication Union (ITU)

established–17 May 1865 set up as the International Telegraph Union; 9 December 1932 adopted present name; effective–1 January 1934; affiliated with the UN–15 November 1947
aim–to deal with world telecommunications issues; a UN specialized agency
members–(192) includes all UN member countries except Palau (191 total); plus Holy See

International Telecommunications Satellites Organization (ITSO)

established–August 1964
aim–to act as a watchdog over Intelsat, Ltd., a private company, to make sure it provides on a global and non-discriminatory basis public telecommunication services
members–(150) Afghanistan, Algeria, Angola, Argentina, Armenia, Australia, Austria, Azerbaijan, The Bahamas, Bahrain, Bangladesh, Barbados, Belgium, Benin, Bhutan, Bolivia, Bosnia and Herzegovina, Botswana, Brazil, Brunei, Bulgaria, Burkina Faso, Cameroon, Canada, Cape Verde, Central African Republic, Chad, Chile, China, Colombia, Comoros, Democratic Republic of the Congo, Republic of the Congo, Costa Rica, Cote d'Ivoire, Croatia, Cuba, Cyprus, Czech Republic, Denmark, Dominican Republic, Ecuador, Egypt, El Salvador, Equatorial Guinea, Estonia, Ethiopia, Fiji, Finland, France, Gabon, The Gambia, Georgia, Germany, Ghana, Greece, Guatemala, Guinea, Guinea-Bissau, Haiti, Holy See, Honduras, Hungary, Iceland, India, Indonesia, Iran, Iraq, Ireland, Israel, Italy, Jamaica, Japan, Jordan, Kazakhstan, Kenya, North Korea, South Korea, Kuwait, Kyrgyzstan, Lebanon, Libya, Liechtenstein, Luxembourg, Madagascar, Malawi, Malaysia, Mali, Malta, Mauritania, Mauritius, Mexico, the Federated States of Micronesia, Monaco, Mongolia, Montenegro, Morocco, Mozambique, Namibia, Nepal, Netherlands, NZ, Nicaragua, Niger, Nigeria, Norway, Oman, Pakistan, Panama,

Papua New Guinea, Paraguay, Peru, Philippines, Poland, Portugal, Qatar, Romania, Russia, Rwanda, Saudi Arabia, Senegal, Serbia, Singapore, Somalia, South Africa, Spain, Sri Lanka, Sudan, Swaziland, Sweden, Switzerland, Syria, Tajikistan, Tanzania, Thailand, Togo, Trinidad and Tobago, Tunisia, Turkey, Uganda, UAE, UK, US, Uruguay, Uzbekistan, Venezuela, Vietnam, Yemen, Zambia, Zimbabwe

International Trade Union Confederation (ITUC)

note–its predecessors were the International Confederation of Free Trade Unions (ICFTU) and the World Confederation of Labor (WCL)
established–3 November 2006
aim–to promote the trade union movement
members–(312 affiliated organizations in the following 156 countries and the Palestine Liberation Organization as of December 2007) Albania, Algeria, Angola, Antigua and Barbuda, Aruba, Argentina, Australia, Austria, Azerbaijan, Bahrain, Bangladesh, Barbados, Belarus, Belgium, Belize, Benin, Bermuda, Bonaire, Bosnia and Herzegovina, Botswana, Brazil, Bulgaria, Burkina Faso, Burundi, Cameroon, Canada, Cape Verde, Central African Republic, Comoros, Chad, Chile, Colombia, Democratic Republic of the Congo, Republic of the Congo, Cook Islands, Costa Rica, Cote d'Ivoire, Croatia, Curacao, Cyprus, Czech Republic, Denmark, Djibouti, Dominica, Dominican Republic, Ecuador, El Salvador, Eritrea, Estonia, Ethiopia, Fiji, Finland, France, French Polynesia, Gabon, The Gambia, Georgia, Germany, Ghana, Greece, Grenada, Guatemala, Guinea, Guinea-Bissau, Guyana, Haiti, Holy See, Honduras, Hong Kong, Hungary, Iceland, India, Indonesia, Ireland, Israel, Italy, Japan, Jordan, Kenya, Kiribati, South Korea, Kosovo, Kuwait, Latvia, Liberia, Liechtenstein, Lithuania, Luxembourg, Macedonia, Madagascar, Malawi, Malaysia, Mali, Malta, Mauritania, Mauritius, Mexico, Mongolia, Montenegro, Morocco, Mozambique, Namibia, Nepal, Netherlands, New Caledonia, NZ, Nicaragua, Niger, Nigeria, Norway, Pakistan, Panama, Paraguay, Peru, Philippines, Poland, Portugal, Puerto Rico, Romania, Russia, Rwanda, Saint Lucia, Saint Vincent and the Grenadines, Samoa, San Marino, Sao Tome and Principe, Senegal, Serbia, Seychelles, Sierra Leone, Singapore, Slovakia, South Africa, Spain, Sri Lanka, Suriname, Swaziland, Sweden, Switzerland, Taiwan, Tanzania, Thailand, Togo, Tonga, Trinidad and Tobago, Tunisia, Turkey, Uganda, Ukraine, UK, US, Vanuatu, Venezuela, Yemen, Zambia, Zimbabwe, and the Palestine Liberation Organization

Islamic Development Bank (IDB)

established–15 December 1973 by declaration of intent; effective–12 August 1974
aim–to promote Islamic economic aid and social development
members–(56 plus the Palestine Liberation Organization) Afghanistan, Albania, Algeria, Azerbaijan, Bahrain, Bangladesh, Benin, Bosnia and Herzegovina, Brunei, Burkina Faso, Cameroon, Chad, Comoros, Cote d'Ivoire, Djibouti, Egypt, Gabon, The Gambia, Guinea, Guinea-Bissau, Indonesia, Iran, Iraq, Jordan, Kazakhstan, Kuwait, Kyrgyzstan, Lebanon, Libya, Malaysia, Maldives, Mali, Mauritania, Morocco, Mozambique, Niger, Nigeria, Oman, Pakistan, Qatar, Saudi Arabia, Senegal, Sierra Leone, Somalia, Sudan, Suriname, Syria, Tajikistan, Togo, Tunisia, Turkey, Turkmenistan, Uganda, UAE, Uzbekistan, Yemen, Palestine Liberation Organization

Latin American Economic System (LAES)

note–also known as Sistema Economico Latinoamericana (SELA)
established–17 October 1975
aim–to promote economic and social development through regional cooperation
members–(28) Argentina, the Bahamas, Barbados, Belize, Bolivia, Brazil, Chile, Colombia, Costa Rica, Cuba, Dominican Republic, Ecuador, El Salvador, Grenada, Guatemala, Guyana, Haiti, Honduras, Jamaica, Mexico, Nicaragua, Panama, Paraguay, Peru, Secretaria General Iberoamericana, Suriname, Trinidad and Tobago, Uruguay, Venezuela

Latin American Integration Association (LAIA)

note–also known as Asociacion Latinoamericana de Integracion (ALADI)
established–12 August 1980; effective–18 March 1981
aim–to promote freer regional trade
members–(12) Argentina, Bolivia, Brazil, Chile, Colombia, Cuba, Ecuador, Mexico, Paraguay, Peru, Uruguay, Venezuela
observers–(30) China, Corporacion Andina de Fomento, Costa Rica, Dominican Republic, EC, El Salvador, Guatemala, Honduras, Inter-American Development Bank, Inter-American Institute for Cooperation on Agriculture, Italy, Japan, South Korea, Latin America Economic System, Nicaragua, Organizacion Panamericana de la Salud, Organizacion Mundial de la Salud, Organization of American States, Panama, Pan-American Health Organization, Portugal, Romania, Russia, San Marino, Secretaria General Iberoamericana, Spain, Switzerland, Ukraine, United Nations Development Program, United Nations Economic Commission for Latin America and the Caribbean

League of Arab States (LAS)

note–also known as Arab League (AL)
established–22 March 1945
aim–aim–to promote economic, social, political, and military cooperation
members–(21 plus the Palestine Liberation Organization) Algeria, Bahrain, Comoros, Djibouti, Egypt, Iraq, Jordan, Kuwait, Lebanon, Libya, Mauritania, Morocco, Oman, Qatar, Saudi Arabia, Somalia, Sudan, Syria, Tunisia, UAE, Yemen, Palestine Liberation Organization
observers–(4) Brazil, Eritrea, India, Venezuela

least developed countries (LLDCs)

that subgroup of the less developed countries (LDCs) initially identified by the UN General Assembly in 1971 as having no significant economic growth, per capita GDPs normally less than $1,000, and low literacy rates; also known as the undeveloped countries; the 44 LLDCs are: Afghanistan, Bangladesh, Benin, Bhutan, Burkina Faso, Burma, Burundi, Cambodia, Cameroon, Central African Republic, Chad, Comoros, Democratic Republic of the Congo, Cote d'Ivoire, Equatorial Guinea, Eritrea, Ethiopia, The Gambia, Ghana, Guinea, Guinea-Bissau, Haiti, Kenya, Lesotho, Liberia, Malawi, Mali, Moldova, Mozambique, Nepal, Niger, Rwanda, Sao Tome and Principe, Senegal, Sierra Leone, Somalia, Sudan, Tajikistan, Tanzania, Togo, Tokelau, Tuvalu, Uganda, Zambia

less developed countries (LDCs)

the bottom group in the hierarchy of developed countries (DCs), former USSR/Eastern Europe (former USSR/EE), and less developed countries (LDCs); mainly countries and dependent areas with low levels of output, living standards, and technology; per capita GDPs are generally below $5,000 and often less than $1,500; however, the group also includes a number of countries with high per capita incomes, areas of advanced technology, and rapid rates of growth; includes the advanced developing countries, developing countries, Four Dragons (Four Tigers), least developed countries (LLDCs), low-income countries, middle-income countries, newly industrializing economies (NIEs), the South, Third World, underdeveloped countries, undeveloped countries; the 172 LDCs are: Afghanistan, Algeria, American Samoa, Angola, Anguilla, Antigua and Barbuda, Argentina, Aruba, The Bahamas, Bahrain, Bangladesh, Barbados, Belize, Benin,

Bhutan, Bolivia, Botswana, Brazil, British Virgin Islands, Brunei, Burkina Faso, Burma, Burundi, Cambodia, Cameroon, Cape Verde, Cayman Islands, Central African Republic, Chad, Chile, China, Christmas Island, Cocos Islands, Colombia, Comoros, Democratic Republic of the Congo, Republic of the Congo, Cook Islands, Costa Rica, Cote d'Ivoire, Cuba, Cyprus, Djibouti, Dominica, Dominican Republic, Ecuador, Egypt, El Salvador, Equatorial Guinea, Eritrea, Ethiopia, Falkland Islands, Fiji, French Guiana, French Polynesia, Gabon, The Gambia, Gaza Strip, Ghana, Gibraltar, Greenland, Grenada, Guadeloupe, Guam, Guatemala, Guernsey, Guinea, Guinea-Bissau, Guyana, Haiti, Honduras, Hong Kong, India, Indonesia, Iran, Iraq, Isle of Man, Jamaica, Jersey, Jordan, Kenya, Kiribati, North Korea, South Korea, Kuwait, Laos, Lebanon, Lesotho, Liberia, Libya, Macau, Madagascar, Malawi, Malaysia, Maldives, Mali, Marshall Islands, Martinique, Mauritania, Mauritius, Mayotte, Federated States of Micronesia, Mongolia, Montserrat, Morocco, Mozambique, Namibia, Nauru, Nepal, Netherlands Antilles, New Caledonia, Nicaragua, Niger, Nigeria, Niue, Norfolk Island, Northern Mariana Islands, Oman, Palau, Pakistan, Panama, Papua New Guinea, Paraguay, Peru, Philippines, Pitcairn Islands, Puerto Rico, Qatar, Reunion, Rwanda, Saint Helena, Ascension, and Tristan da Cunha, Saint Kitts and Nevis, Saint Lucia, Saint Pierre and Miquelon, Saint Vincent and the Grenadines, Samoa, Sao Tome and Principe, Saudi Arabia, Senegal, Seychelles, Sierra Leone, Singapore, Solomon Islands, Somalia, Sri Lanka, Sudan, Suriname, Swaziland, Syria, Taiwan, Tanzania, Thailand, Togo, Tokelau, Tonga, Trinidad and Tobago, Tunisia, Turks and Caicos Islands, Tuvalu, UAE, Uganda, Uruguay, Vanuatu, Venezuela, Vietnam, Virgin Islands, Wallis and Futuna, West Bank, Western Sahara, Yemen, Zambia, Zimbabwe; note–similar to the new International Monetary Fund (IMF) term "developing countries" which adds Malta, Mexico, South Africa, and Turkey but omits in its recently published statistics American Samoa, Anguilla, British Virgin Islands, Brunei, Cayman Islands, Christmas Island, Cocos Islands, Cook Islands, Cuba, Eritrea, Falkland Islands, French Guiana, French Polynesia, Gaza Strip, Gibraltar, Greenland, Grenada, Guadeloupe, Guam, Guernsey, Isle of Man, Jersey, North Korea, Macau, Martinique, Mayotte, Montserrat, Nauru, New Caledonia, Niue, Norfolk Island, Northern Mariana Islands, Palau, Pitcairn Islands, Puerto Rico, Reunion, Saint Helena, Ascension, and Tristan da Cunha, Saint Pierre and Miquelon, Tokelau, Tonga, Turks and Caicos Islands, Tuvalu, Virgin Islands, Wallis and Futuna, West Bank, Western Sahara

low-income countries
another term for those less developed countries with below-average per capita GDPs; see less developed countries (LDCs)

middle-income countries
another term for those less developed countries with above-average per capita GDPs; see less developed countries (LDCs)

Multilateral Investment Guarantee Agency (MIGA)
established–12 April 1988
aim–encourages flow of foreign direct investment among member countries by offering investment insurance, consultation, and negotiation on conditions for foreign investment and technical assistance; a UN specialized agency
members–(175) includes all UN member countries except Andorra, Bhutan, Brunei, Burma, Comoros, Cuba, Kiribati, North Korea, Liechtenstein, Marshall Islands, Monaco, Nauru, Niger, San Marino, Sao Tome and Principe, Somalia, Tonga, Tuvalu; plus Kosovo

Near Abroad
Russian term for the 14 non-Russian successor states of the USSR, in which 25 million ethnic Russians live and in which Moscow has expressed a strong national security interest; the 14 countries are Armenia, Azerbaijan, Belarus, Estonia, Georgia, Kazakhstan, Kyrgyzstan, Latvia, Lithuania, Moldova, Tajikistan, Turkmenistan, Ukraine, Uzbekistan

new independent states (NIS)
a term referring to all the countries of the FSU except the Baltic countries (Estonia, Latvia, Lithuania)

newly industrializing countries (NICs)
former term for the newly industrializing economies; see newly industrializing economies (NIEs)

newly industrializing economies (NIEs)
that subgroup of the less developed countries (LDCs) that has experienced particularly rapid industrialization of their economies; formerly known as the newly industrializing countries (NICs); also known as advanced developing countries; usually includes the Four Dragons (Hong Kong, South Korea, Singapore, Taiwan), and Brazil

Nonaligned Movement (NAM)
established–1-6 September 1961
aim–to establish political and military cooperation apart from the traditional East or West blocs
members–(117 plus the Palestine Liberation Organization) Afghanistan, Algeria, Angola, Antigua and Barbuda, The Bahamas, Bahrain, Bangladesh, Barbados, Belarus, Belize, Benin, Bhutan, Bolivia, Botswana, Brunei, Burkina Faso, Burma, Burundi, Cambodia, Cameroon, Cape Verde, Central African Republic, Chad, Chile, Colombia, Comoros, Democratic Republic of the Congo, Republic of the Congo, Cote d'Ivoire, Cuba, Djibouti, Dominica, Dominican Republic, Ecuador, Egypt, Equatorial Guinea, Eritrea, Ethiopia, Gabon, The Gambia, Ghana, Grenada, Guatemala, Guinea, Guinea-Bissau, Guyana, Haiti, Honduras, India, Indonesia, Iran, Iraq, Jamaica, Jordan, Kenya, North Korea, Kuwait, Laos, Lebanon, Lesotho, Liberia, Libya, Madagascar, Malawi, Malaysia, Maldives, Mali, Mauritania, Mauritius, Mongolia, Morocco, Mozambique, Namibia, Nepal, Nicaragua, Niger, Nigeria, Oman, Pakistan, Panama, Papua New Guinea, Peru, Philippines, Qatar, Rwanda, Saint Kitts and Nevis, Saint Lucia, Saint Vincent and the Grenadines, Sao Tome and Principe, Saudi Arabia, Senegal, Seychelles, Sierra Leone, Singapore, Somalia, South Africa, Sri Lanka, Sudan, Suriname, Swaziland, Syria, Tanzania, Thailand, Timor-Leste, Togo, Trinidad and Tobago, Tunisia, Turkmenistan, Uganda, UAE, Uzbekistan, Vanuatu, Venezuela, Vietnam, Yemen, Zambia, Zimbabwe, Palestine Liberation Organization
observers–(18) Argentina, Armenia, Azerbaijan, Bosnia and Herzegovina, Brazil, China, Costa Rica, Croatia, El Salvador, Kazakhstan, Kyrgyzstan, Mexico, Montenegro, Paraguay, Serbia, Tajikistan, Ukraine, Uruguay

Nordic Council (NC)
established–16 March 1952; effective–12 February 1953
aim–to promote regional economic, cultural, and environmental cooperation
members–(5) Denmark (including Faroe Islands and Greenland), Finland (including Aland Islands), Iceland, Norway, Sweden
observers–(3) the Sami (Lapp) local parliaments of Finland, Norway, and Sweden

Nordic Investment Bank (NIB)
established–4 December 1975; effective–1 June 1976
aim–to promote economic cooperation and development
members–(8) Denmark (including Faroe Islands and Greenland), Estonia, Finland (including Aland Islands), Iceland, Latvia, Lithuania, Norway, Sweden

North

a popular term for the rich industrialized countries generally located in the northern portion of the Northern Hemisphere; the counterpart of the South; see developed countries (DCs)

North American Free Trade Agreement (NAFTA)

established–17 December 1992
aim–to eliminate trade barriers, promote fair competition, increase investment opportunities, provide protection of intellectual property rights, and create procedures to settle disputes
members–(3) Canada, Mexico, US

North Atlantic Treaty Organization (NATO)

established–4 April 1949
aim–to promote mutual defense and cooperation
members–(28) Albania, Belgium, Bulgaria, Canada, Croatia, Czech Republic, Denmark, Estonia, France, Germany, Greece, Hungary, Iceland, Italy, Latvia, Lithuania, Luxembourg, Netherlands, Norway, Poland, Portugal, Romania, Slovakia, Slovenia, Spain, Turkey, UK, US

Nuclear Energy Agency (NEA)

note–also known as OECD Nuclear Energy Agency
established–1 February 1958
aim–to promote the peaceful uses of nuclear energy; associated with OECD
members–(29) Australia, Austria, Belgium, Canada, Czech Republic, Denmark, Finland, France, Germany, Greece, Hungary, Iceland, Ireland, Italy, Japan, South Korea, Luxembourg, Mexico, Netherlands, Norway, Poland, Portugal, Slovakia, Spain, Sweden, Switzerland, Turkey, UK, US

Nuclear Suppliers Group (NSG)

note–also known as the London Suppliers Group or the London Group
established–1974; effective–1975
aim–to establish guidelines for exports of nuclear materials, processing equipment for uranium enrichment, and technical information to countries of proliferation concern and regions of conflict and instability
members–(46) Argentina, Australia, Austria, Belarus, Belgium, Brazil, Bulgaria, Canada, China, Croatia, Cyprus, Czech Republic, Denmark, Estonia, Finland, France, Germany, Greece, Hungary, Iceland, Ireland, Italy, Japan, Kazakhstan, South Korea, Latvia, Lithuania, Luxembourg, Malta, Netherlands, NZ, Norway, Poland, Portugal, Romania, Russia, Slovakia, Slovenia, South Africa, Spain, Sweden, Switzerland, Turkey, Ukraine, UK, US
observer–(1) European Commission (a policy-planning body for the EU)

Organization for Democracy and Economic Development (GUAM)

note-acronym standing for the member countries, Georgia, Ukraine, Azerbaijan, Moldova; formerly known as GUUAM before Uzbekistan withdrew in 5 May 2005
established–7 June 2001
aim–commits the countries to cooperation and assistance in social and economic development, the strengthening and broadening of trade and economic relations, and the development and effective use of transport and communications, highways, and related infrastructure crossing the boundaries of the member states
members–(4) Azerbaijan, Georgia, Moldova, Ukraine

Organization for Economic Cooperation and Development (OECD)

established–14 December 1960; effective–30 September 1961
aim–to promote economic cooperation and development
members–(34) Australia, Austria, Belgium, Canada, Chile, Czech Republic, Denmark, Estonia, Finland, France, Germany, Greece, Hungary, Iceland, Ireland, Israel, Italy, Japan, South Korea, Luxembourg, Mexico, Netherlands, NZ, Norway, Poland, Portugal, Slovakia, Slovenia, Spain, Sweden, Switzerland, Turkey, UK, US
special member–(1) EC
accession states–(1) Russia

Organization for Security and Cooperation in Europe (OSCE)

note–formerly the Conference on Security and Cooperation in Europe (CSCE) established 3 July 1975
established–1 January 1995
aim–to foster the implementation of human rights, fundamental freedoms, democracy, and the rule of law; to act as an instrument of early warning, conflict prevention, and crisis management; and to serve as a framework for conventional arms control and confidence building measures
members–(56) Albania, Andorra, Armenia, Austria, Azerbaijan, Belarus, Belgium, Bosnia and Herzegovina, Bulgaria, Canada, Croatia, Cyprus, Czech Republic, Denmark, Estonia, Finland, France, Georgia, Germany, Greece, Holy See, Hungary, Iceland, Ireland, Italy, Kazakhstan, Kyrgyzstan, Latvia, Liechtenstein, Lithuania, Luxembourg, Macedonia, Malta, Moldova, Monaco, Montenegro, Netherlands, Norway, Poland, Portugal, Romania, Russia, San Marino, Serbia, Slovakia, Slovenia, Spain, Sweden, Switzerland, Tajikistan, Turkey, Turkmenistan, Ukraine, UK, US, Uzbekistan
partners for cooperation–(12) Afghanistan, Algeria, Australia, Egypt, Israel, Japan, Jordan, South Korea, Mongolia, Morocco, Thailand, Tunisia

Organization for the Prohibition of Chemical Weapons (OPCW)

established–29 April 1997
aim–to enforce the Convention on the Prohibition of the Development, Production, Stockpiling, and Use of Chemical Weapons and on Their Destruction; to provide a forum for consultation and cooperation among the signatories of the Convention
members (countries that have ratified the Convention)–(188) Afghanistan, Albania, Algeria, Andorra, Antigua and Barbuda, Argentina, Armenia, Australia, Austria, Azerbaijan, The Bahamas, Bahrain, Bangladesh, Barbados, Belarus, Belgium, Belize, Benin, Bhutan, Bolivia, Bosnia and Herzegovina, Botswana, Brazil, Brunei, Bulgaria, Burkina Faso, Burundi, Cambodia, Cameroon, Canada, Cape Verde, Central African Republic, Chad, Chile, China, Colombia, Comoros, Democratic Republic of the Congo, Republic of the Congo, Cook Islands, Costa Rica, Cote d'Ivoire, Croatia, Cuba, Cyprus, Czech Republic, Denmark, Dominica, Dominican Republic, Djibouti, Ecuador, El Salvador, Equatorial Guinea, Eritrea, Estonia, Ethiopia, Fiji, Finland, France, Gabon, The Gambia, Georgia,

Germany, Ghana, Greece, Grenada, Guatemala, Guinea, Guinea-Bissau, Guyana, Haiti, Holy See, Honduras, Hungary, Iceland, India, Indonesia, Iran, Iraq, Ireland, Italy, Jamaica, Japan, Jordan, Kazakhstan, Kenya, Kiribati, South Korea, Kuwait, Kyrgyzstan, Laos, Latvia, Lebanon, Lesotho, Liberia, Libya, Liechtenstein, Lithuania, Luxembourg, Macedonia, Madagascar, Malawi, Malaysia, Maldives, Mali, Malta, Marshall Islands, Mauritania, Mauritius, Mexico, Federated States of Micronesia, Moldova, Monaco, Mongolia, Montenegro, Morocco, Mozambique, Namibia, Nauru, Nepal, Netherlands, NZ, Nicaragua, Niger, Nigeria, Niue, Norway, Oman, Pakistan, Palau, Panama, Papua New Guinea, Paraguay, Peru, Philippines, Poland, Portugal, Qatar, Romania, Russia, Rwanda, Saint Kitts and Nevis, Saint Lucia, Saint Vincent and the Grenadines, Samoa, San Marino, Sao Tome and Principe, Saudi Arabia, Senegal, Serbia, Seychelles, Sierra Leone, Singapore, Slovakia, Slovenia, Solomon Islands, South Africa, Spain, Sri Lanka, Sudan, Suriname, Swaziland, Sweden, Switzerland, Tajikistan, Tanzania, Thailand, Timor-Leste, Togo, Tonga, Trinidad and Tobago, Tunisia, Turkey, Turkmenistan, Tuvalu, Uganda, Ukraine, UAE, UK, US, Uruguay, Uzbekistan, Vanuatu, Venezuela, Vietnam, Yemen, Zambia, Zimbabwe
signatory states (countries that have signed, but not ratified, the Convention)–(2) Burma, Israel

Organization of African Unity (OAU)
see African Union

Organization of American States (OAS)
established–14 April 1890 as the International Union of American Republics; 30 April 1948 adopted present charter; effective–13 December 1951
aim–to promote regional peace and security as well as economic and social development
members–(35) Antigua and Barbuda, Argentina, The Bahamas, Barbados, Belize, Bolivia, Brazil, Canada, Chile, Colombia, Costa Rica, Cuba (excluded from formal participation since 1962), Dominica, Dominican Republic, Ecuador, El Salvador, Grenada, Guatemala, Guyana, Haiti, Honduras, Jamaica, Mexico, Nicaragua, Panama, Paraguay, Peru, Saint Kitts and Nevis, Saint Lucia, Saint Vincent and the Grenadines, Suriname, Trinidad and Tobago, US, Uruguay, Venezuela
observers–(65) Algeria, Angola, Armenia, Austria, Azerbaijan, Belgium, Benin, Bosnia and Herzegovina, Bulgaria, China, Croatia, Cyprus, Czech Republic, Denmark, Egypt, Equatorial Guinea, Estonia, EU, Finland, France, Georgia, Germany, Ghana, Greece, Holy See, Hungary, Iceland, India, Ireland, Israel, Italy, Japan, Kazakhstan, South Korea, Latvia, Lebanon, Lithuania, Luxembourg, Monaco, Morocco, Netherlands, Nigeria, Norway, Pakistan, Philippines, Poland, Portugal, Qatar, Romania, Russia, Saudi Arabia, Serbia, Slovakia, Slovenia, Spain, Sri Lanka, Sweden, Switzerland, Thailand, Tunisia, Turkey, Ukraine, UK, Vanuatu, Yemen

Organization of Arab Petroleum Exporting Countries (OAPEC)
established–9 January 1968
aim–to promote cooperation in the petroleum industry
members–(11) Algeria, Bahrain, Egypt, Iraq, Kuwait, Libya, Qatar, Saudi Arabia, Syria, Tunisia (suspended), UAE

Organization of Eastern Caribbean States (OECS)
established–18 June 1981; effective–4 July 1981
aim–to promote political, economic, and defense cooperation
members–(9) Anguilla, Antigua and Barbuda, British Virgin Islands, Dominica, Grenada, Montserrat, Saint Kitts and Nevis, Saint Lucia, Saint Vincent and the Grenadines

Organization of Petroleum Exporting Countries (OPEC)
established–14 September 1960
aim–to coordinate petroleum policies
members–(12) Algeria, Angola, Ecuador, Iran, Iraq, Kuwait, Libya, Nigeria, Qatar, Saudi Arabia, UAE, Venezuela; note–Indonesia left OPEC in 2008

Organization of the Islamic Conference (OIC)
established–22-25 September 1969
aim–to promote Islamic solidarity in economic, social, cultural, and political affairs
members–(56 plus the Palestine Liberation Organization) Afghanistan, Albania, Algeria, Azerbaijan, Bahrain, Bangladesh, Benin, Brunei, Burkina Faso, Cameroon, Chad, Comoros, Cote d'Ivoire, Djibouti, Egypt, Gabon, The Gambia, Guinea, Guinea-Bissau, Guyana, Indonesia, Iran, Iraq, Jordan, Kazakhstan, Kuwait, Kyrgyzstan, Lebanon, Libya, Malaysia, Maldives, Mali, Mauritania, Morocco, Mozambique, Niger, Nigeria, Oman, Pakistan, Qatar, Saudi Arabia, Senegal, Sierra Leone, Somalia, Sudan, Suriname, Syria, Tajikistan, Togo, Tunisia, Turkey, Turkmenistan, Uganda, UAE, Uzbekistan, Yemen, Palestine Liberation Organization
observers–(12) AU, Bosnia and Herzegovina, Central African Republic, ECO, LAS, Moro National Liberation Front, NAM, Parliamentary Union of the OIC Member States, Russia, Thailand, Turkish Muslim Community of Kibris, UN

Pacific Community (SPC)
local name of the Secretariat of the Pacific Community

Pacific Islands Forum (PIF)
note–formerly known as South Pacific Forum (SPF)
established–5 August 1971
aim–to promote regional cooperation in political matters
members–(16) Australia, Cook Islands, Fiji, Kiribati, Marshall Islands, Federated States of Micronesia, Nauru, NZ, Niue, Palau, Papua New Guinea, Samoa, Solomon Islands, Tonga, Tuvalu, Vanuatu
associate members–(2) French Polynesia, New Caledonia
partners–(14) Canada, China, EU, France, India, Indonesia, Italy, Japan, South Korea, Malaysia, Philippines, Thailand, UK, US
observers–(6) Asia Development Bank, The Commonwealth, Timor-Leste, Tokelau, Wallis and Futuna, the World Bank

Paris Club
established–1956
aim–to provide a forum for debtor countries to negotiate rescheduling of debt service payments or loans extended by governments or official agencies of participating countries; to help restore normal trade and project finance to debtor countries
members–(19) Australia, Austria, Belgium, Canada, Denmark, Finland, France, Germany, Ireland, Italy, Japan, Netherlands, Norway, Russia, Spain, Sweden, Switzerland, UK, US
associate members–(13) Abu Dhabi, Argentina, Brazil, Israel, South Korea, Kuwait, Mexico, Morocco, NZ, Portugal, South Africa, Trinidad and Tobago, Turkey

Partnership for Peace (PFP)

established–10-11 January 1994
aim–to expand and intensify political and military cooperation throughout Europe, increase stability, diminish threats to peace, and build relationships by promoting the spirit of practical cooperation and commitment to democratic principles that underpin NATO; program under the auspices of NATO
members–(22) Armenia, Austria, Azerbaijan, Belarus, Bosnia and Herzegovina, Finland, Georgia, Ireland, Kazakhstan, Kyrgyzstan, Macedonia, Malta, Moldova, Montenegro, Russia, Serbia, Sweden, Switzerland, Tajikistan, Turkmenistan, Ukraine, Uzbekistan; note–a nation that becomes a member of NATO is no longer a member of PFP

Permanent Court of Arbitration (PCA)

established–29 July 1899
aim–to facilitate the settlement of international disputes
members–(111) Argentina, Australia, Austria, Bahrain, Belarus, Belgium, Belize, Benin, Bolivia, Brazil, Bulgaria, Burkina Faso, Cambodia, Cameroon, Canada, Chile, China, Colombia, Democratic Republic of the Congo, Costa Rica, Croatia, Cuba, Cyprus, Czech Republic, Denmark, Dominican Republic, Ecuador, Egypt, El Salvador, Eritrea, Estonia, Ethiopia, Fiji, Finland, France, Germany, Greece, Guatemala, Guyana, Haiti, Honduras, Hungary, Iceland, India, Iran, Iraq, Ireland, Israel, Italy, Japan, Jordan, Kenya, South Korea, Kuwait, Kyrgyzstan, Laos, Latvia, Lebanon, Libya, Liechtenstein, Lithuania, Luxembourg, Macedonia, Madagascar, Malaysia, Malta, Mauritius, Mexico, Montenegro, Morocco, Netherlands, NZ, Nicaragua, Nigeria, Norway, Pakistan, Panama, Paraguay, Peru, Philippines, Poland, Portugal, Qatar, Romania, Russia, Saudi Arabia, Senegal, Serbia, Singapore, Slovakia, Slovenia, South Africa, Spain, Sri Lanka, Sudan, Suriname, Swaziland, Sweden, Switzerland, Thailand, Togo, Turkey, Uganda, Ukraine, UAE, UK, US, Uruguay, Venezuela, Zambia, Zimbabwe

PetroCaribe

established–29 June 2005
aim–to eliminate existing social inequities, to foster high standards of living, to promote effective people's participation in shaping their own destiny
members–(18) Antigua and Barbuda, The Bahamas, Belize, Cuba, Dominica, Dominican Republic, Grenada, Guatemala, Guyana, Haiti, Honduras, Jamaica, Nicaragua, St. Kitts and Nevis, St. Lucia, St. Vincent and the Grenadines, Suriname, Venezuela

Rio Group (RG)

note–formerly known as Grupo de los Ocho, established NA December 1986; composed of the Contadora Group and the Lima Group
established–1988
aim–to consult on regional Latin American issues
members–(24) Argentina, Belize, Bolivia, Brazil, Chile, Colombia, Costa Rica, Cuba, Dominican Republic, Ecuador, El Salvador, Guatemala, Guyana, Haiti, Honduras, Jamaica (representing CARICOM), Mexico, Nicaragua, Panama, Paraguay, Peru, Suriname, Uruguay, Venezuela

Schengen Convention

established–signed June 1990; effective March 1995
aim–to allow free movement within an area without internal border controls
members–(25) Austria, Belgium, Czech Republic, Denmark, Estonia, Finland, France, Germany, Greece, Hungary, Iceland, Italy, Latvia, Lithuania, Luxembourg, Malta, Monaco, Netherlands, Norway, Poland, Portugal, Slovakia, Slovenia, Spain, Sweden, Switzerland; note–UK and Ireland have not joined; Liechtenstein and Cyprus will probably join in the near future; Bulgaria and Romania are still not fully implemented
De Facto members–(3) Holy See, Monaco, San Marino

Second World

another term for the traditionally Marxist-Leninist states of the USSR and Eastern Europe, with authoritarian governments and command economies based on the Soviet model; the term is fading from use; see centrally planned economies

Secretariat of the Pacific Community (SPC)

established–6 February 1947; effective 29 July 1948
aim–to serve island development in 22 Pacific countries; to develop technical assistance and professional, scientific, and research support; to build planning and management capability
members–(26) America Samoa, Australia, Cook Islands, Fiji, France, French Polynesia, Guam, Kiribati, Marshall Islands, Federated States of Micronesia, Nauru, New Caledonia, Niue, Northern Mariana Islands, NZ, Palau, Papua New Guinea, Pitcairn Islands, Samoa, Solomon Islands, Tokelau, Tonga, Tuvalu, Vanuatu, US, Wallis and Futuna

Shanghai Cooperation Organization (SCO)

established–15 June 2001
aim–to combat terrorism, extremism, and separatism; to safeguard regional security through mutual trust, disarmament, and cooperative security; and to increase cooperation in political, trade, economic, scientific and technological, cultural, and educational fields
members–(6) China, Kazakhstan, Kyrgyzstan, Russia, Tajikistan, Uzbekistan
dialogue members–(2) Belarus, Sri Lanka
observers–(4) India, Iran, Mongolia, Pakistan

socialist countries

in general, countries in which the government owns and plans the use of the major factors of production; note–the term is sometimes used incorrectly as a synonym for Communist countries

South

a popular term for the poorer, less industrialized countries generally located south of the developed countries; the counterpart of the North; see less developed countries (LDCs)

South American Community of Nations (CSN)

established on 9 December 2004; its aim was to coordinate common policies regarding multilateral organizations, to integrate physical infrastructure, and to consolidate the merger of CAN and Mercosur; the members were Argentina, Bolivia, Brazil, Chile, Colombia, Ecuador, Guyana, Paraguay, Peru, Surinam, Uruguay, Venezuela; in 2008 it became Union of South American Nations (UNASUR)

South Asia Co-operative Environment Program (SACEP)

established–January 1983
aim–to promote regional cooperation in South Asia in the field of environment, both natural and human, and on issues of economic and social development; to support conservation and management of natural resources of the region
members–(8) Afghanistan, Bangladesh, Bhutan, India, Maldives, Nepal, Pakistan, Sri Lanka

South Asian Association for Regional Cooperation (SAARC)

established–8 December 1985
aim–to promote economic, social, and cultural cooperation
members–(8) Afghanistan. Bangladesh, Bhutan, India, Maldives, Nepal, Pakistan, Sri Lanka
observers–(9) Australia, Burma, China, EU, Iran, Japan, South Korea, Mauritius, US

South Pacific Forum (SPF)

note–see Pacific Island Forum

South Pacific Regional Trade and Economic Cooperation Agreement (Sparteca)

established–1981
aim–to redress unequal trade relationships of Australia and New Zealand with small island economies in the Pacific region
members–(16) Australia, Cook Islands, Fiji, Kiribati, Marshall Islands, Federated States of Micronesia, Nauru, NZ, Niue, Palau, Papua New Guinea, Samoa, Solomon Islands, Tonga, Tuvalu, Vanuatu

Southeast European Cooperative Initiative (SECI)

established–6 December 1996
aim–to encourage cooperation among participating states and to facilitate their integration into European structures
members–(13) Albania, Bosnia and Herzegovina, Bulgaria, Croatia, Greece, Hungary, Macedonia, Moldova, Montenegro, Romania, Serbia, Slovenia, Turkey
observers–(18) Austria, Azerbaijan, Belgium, Canada, Czech Republic, France, Georgia, Germany, Israel, Italy, Japan, Netherlands, Portugal, Slovakia, Spain, Ukraine, UK, US

Southern African Customs Union (SACU)

established–11 December 1969
aim–to promote free trade and cooperation in customs matters
members–(5) Botswana, Lesotho, Namibia, South Africa, Swaziland

Southern African Development Community (SADC)

note–evolved from the Southern African Development Coordination Conference (SADCC)
established–17 August 1992
aim–to promote regional economic development and integration
members–(14) Angola, Botswana, Democratic Republic of the Congo, Lesotho, Madagascar, Malawi, Mauritius, Mozambique, Namibia, Seychelles, South Africa, Swaziland, Tanzania, Zambia, Zimbabwe

Southern Cone Common Market (Mercosur) or Southern Common Market

note–also known as Mercado Comun del Cono Sur (Mercosur)
established–26 March 1991
aim–to increase regional economic cooperation
members–(4) Argentina, Brazil, Paraguay, Uruguay
associate members–(5) Bolivia, Chile, Colombia, Ecuador, Peru, Venezuela

Third World

another term for the less developed countries; the term is obsolescent; see less developed countries (LDCs)

underdeveloped countries

refers to those less developed countries with the potential for above-average economic growth; see less developed countries (LDCs)

undeveloped countries

refers to those extremely poor less developed countries (LDCs) with little prospect for economic growth; see least developed countries (LLDCs)

Union Latina

established–15 May 1954; became functional 1983
aim–to project, protect, and promote the common heritage and unifying identities of the Latin, and Latin-influenced, world
members–(36) Andorra, Angola, Bolivia, Brazil, Cape Verde, Chile, Colombia, Cote d'Ivoire, Costa Rica, Cuba, Dominican Republic, Ecuador, El Salvador, France, Guatemala, Guinea-Bissau, Haiti, Honduras, Italy, Moldova, Monaco, Mozambique, Nicaragua, Panama, Paraguay, Peru, Philippines, Portugal, Romania, San Marino, Sao Tome and Principe, Senegal, Spain, Timor-Leste, Uruguay, Venezuela
observers–(3) Argentina, Holy See, Order of Malta

Union of South American Nations (UNASUR—Spanish; UNASUL—Portuguese)

formerly South American Community of Nations (CSN) which terminated on 16 April 2007
established–23 May 2008
aim–to model a community after the European Union which will include a common currency, parliament, passport, and defense policy
members–(12) Argentina, Bolivia, Brazil, Chile, Colombia, Ecuador, Guyana, Paraguay, Peru, Suriname, Uruguay, Venezuela
observers–(2) Mexico, Panama

United Nations (UN)

established–26 June 1945; effective–24 October 1945
aim–to maintain international peace and security and to promote cooperation involving economic, social, cultural, and humanitarian problems
constituent organizations–the UN is composed of six principal organs and numerous subordinate agencies and bodies as follows:
1) Secretariat

2) **General Assembly:** Joint United Nations Program on HIV/AIDS (UN-AIDS), International Research and Training Institute for the Advancement of Women (UN-INSTRAW), Office of the United Nations High Commissioner for Human Rights (OHCHR), Organization for the Prohibition of Chemical Weapons (OPCW), Preparation Committee for the Nuclear-Test-Ban Treaty Organization (CTBTO), United Nations Center for Human Settlements (UN-Habitat), United Nations Children's Fund (UNICEF), United Nations Conference on Trade and Development (UNCTAD), United Nations Democracy Fund (UNDEF), United Nations Development Program (UNDP), United Nations Drug Control Program (UNDCP), United Nations Environment Program (UNEP),United Nations Fund for International Partnerships (UNFIP), United Nations High Commissioner for Refugees (UNHCR), United Nations Institute for Disarmament Research (UNIDIR), United Nations Institute for Training and Research (UNITAR), United Nations Interregional Crime and Justice Research Institute (UNICRI), United Nations Population Fund (UNFPA), United Nations Office of Project Services (UNOPS), United Nations Relief and Works Agency for Palestine Refugees in the Near East (UNRWA), United Nations Research Institute for Social Development (UNRISD), United Nations System Staff College (UNSSC), United Nations University (UNU), World Food Program (WFP)
3) **Security Council:** International Criminal Tribunal for the Former Yugoslavia (ICTY), International Criminal Tribunal for Rwanda (ICTR), United Nations Compensation Commission, United Nations Disengagement Observer Force (UNDOF), African Union/ United Nations Hybrid Operation in Darfur (UNAMID), United Nations Assistance Mission in Afghanistan (UNAMA), United Nations Integrated Office in Burundi (BINUB), United Nations Integrated Mission in Timor-Leste (UNMIT), United Nations Interim Administration Mission in Kosovo (UNMIK), United Nations Interim Force in Lebanon (UNIFIL), United Nations Mission in Liberia (UNMIL), United Nations Military Observer Group in India and Pakistan (UNMOGIP), United Nations Operation in Cote d'Ivoire (UNOCI), United Nations Mission for the Referendum in Western Sahara (MINURSO), United Nations Mission in the Sudan (UNMIS), United Nations Organization Stabilization Mission in the Democratic Republic of the Congo (MONUSCO), United Nations Peace-Keeping Force in Cyprus (UNFICYP), United Nations Stabilization Mission in Haiti (MINUSTAH), and United Nations Truce Supervision Organization (UNTSO)
4) **Economic and Social Council (ECOSOC):** Commission for Social Development, Commission on Crime Prevention and Criminal Justice, Commission on Narcotics Drugs, Commission on Population and Development, Commission on Science and Technology for Development, Commission on Sustainable Development, Commission on the Status of Women, Economic and Social Commission for Asia and the Pacific (ESCAP), Economic and Social Commission for Western Asia (ESCWA), Economic Commission for Africa (ECA), Economic Commission for Europe (ECE), Economic Commission for Latin America and the Caribbean (ECLAC), Food and Agriculture Organization of the United Nations (FAO), International Atomic Energy Agency (IAEA), International Bank for Reconstruction and Development (IBRD), International Center for Secretariat of Investment Disputes (ICSID), International Civil Aviation Organization (ICAO), International Development Association (IDA), International Finance Corporation (IFC), International Fund for Agricultural Development (IFAD), International Labor Organization (ILO), International Maritime Organization (IMO), International Monetary Fund (IMF), International Telecommunication Union (ITU), Multilateral Investment Geographic Agency (MIGA), Statistical Commission, United Nations Educational, Scientific, and Cultural Organization (UNESCO), United Nations Forum on Forests, United Nations Industrial Development Organization (UNIDO), Universal Postal Union (UPU), World Health Organization (WHO), World Intellectual Property Organization (WIPO), World Meteorological Organization (WMO), World Tourism Organization (UNWTO), and World Trade Organization (WTO)5) **Trusteeship Council** (inactive; no trusteeships at this time) 6) **International Court of Justice (ICJ)**

United Nations Children's Fund (UNICEF)
note–acronym retained from the predecessor organization, UN International Children's Emergency Fund
established–11 December 1946
aim–to help establish child health and welfare services
executive board members–(36) selected on a rotating basis from all regions

United Nations Conference on Trade and Development (UNCTAD)
established–30 December 1964
aim–to promote international trade
members–(193) all UN members plus Holy See

United Nations Development Program (UNDP)
established–22 November 1965
aim–to provide technical assistance to stimulate economic and social development
members (executive board)–(36) selected on a rotating basis from all regions

United Nations Disengagement Observer Force (UNDOF)
established–31 May 1974
aim–to observe the 1973 Arab-Israeli cease-fire; established by the UN Security Council
members–(6) Austria, Canada, Croatia, India, Japan, Philippines

United Nations Educational, Scientific, and Cultural Organization (UNESCO)
established–16 November 1945; effective–4 November 1946
aim–to promote cooperation in education, science, and culture
members–(193) includes all UN member countries except Liechtenstein (191 total); plus Cook Islands and Niue
associate members–(7) Aruba, British Virgin Islands, Cayman Islands, Faroe Islands, Macau, Tokelau

United Nations Environment Program (UNEP)
established–15 December 1972
aim–to promote international cooperation on all environmental matters
members–(58) selected on a rotating basis from all regions

United Nations General Assembly
established–26 June 1945; effective–24 October 1945
aim–to function as the primary deliberative organ of the UN
members–(192) all UN members are represented in the General Assembly

United Nations High Commissioner for Refugees (UNHCR)
established–3 December 1949; effective–1 January 1951
aim–to ensure the humanitarian treatment of refugees and find permanent solutions to refugee problems

members (executive committee)–(79) Algeria, Argentina, Australia, Austria, Bangladesh, Belgium, Benin, Brazil, Canada, Chile, China, Colombia, Democratic Republic of the Congo, Costa Rica, Cote d'Ivoire, Cyprus, Denmark, Djibouti, Ecuador, Egypt, Estonia, Ethiopia, Finland, France, Germany, Ghana, Greece, Guinea, Holy See, Hungary, India, Iran, Ireland, Israel, Italy, Japan, Jordan, Kenya, South Korea, Lebanon, Lesotho, Luxembourg, Macedonia, Madagascar, Mexico, Moldova, Montenegro, Morocco, Mozambique, Namibia, Netherlands, NZ, Nicaragua, Nigeria, Norway, Pakistan, Philippines, Poland, Portugal, Romania, Russia, Serbia, Slovenia, Somalia, South Africa, Spain, Sudan, Sweden, Switzerland, Tanzania, Thailand, Tunisia, Turkey, Uganda, UK, US, Venezuela, Yemen, Zambia

United Nations Industrial Development Organization (UNIDO)
established–17 November 1966; effective–1 January 1967
aim–UN specialized agency that promotes industrial development especially among the members
members–(173) includes all UN member countries except Andorra, Antigua and Barbuda, Australia, Brunei, Canada, Estonia, Iceland, Kiribati, Latvia, Liechtenstein, Marshall Islands, Federated States of Micronesia, Nauru, Palau, San Marino, Singapore, Solomon Islands, Tuvalu, US

United Nations Institute for Training and Research (UNITAR)
established–11 December 1963 adoption of the resolution establishing the Institute; effective–24 March 1965
aim–to help the UN become more effective through training and research
members (Board of Trustees)–(16) Algeria, Brazil, Burkina Faso, China, Republic of the Congo, France, Guatemala, India, Iran, Jamaica, Nigeria, Norway, Russia, South Africa, Switzerland, US; note–the UN Secretary General can appoint up to 30 members

United Nations Integrated Mission in Timor-Leste (UNMIT)
established–25 August 2006
aim–to support the Government, to support the electoral process, to ensure the restoration and maintenance of public security
members–(14) Australia, Bangladesh, Brazil, China, Fiji, India, Malaysia, Nepal, NZ, Pakistan, Philippines, Portugal, Sierra Leone, Singapore

United Nations Interim Administration Mission in Kosovo (UNMIK)
established–10 June 1999
aim–to promote the establishment of substantial autonomy and self-government in Kosovo; to perform basic civilian administrative functions; to support the reconstruction of key infrastructure and humanitarian and disaster relief
note–gives civilian support only; works closely with NATO Kosovo Force (KFOR)

United Nations Interim Force in Lebanon (UNIFIL)
established–19 March 1978
aim–to confirm the withdrawal of Israeli forces, and assist in reestablishing Lebanese authority in southern Lebanon; established by the UN Security Council
members–(31) Bangladesh, Belgium, Brunei, Cambodia, China, Croatia, Cyprus, Denmark, El Salvador, France, Germany, Ghana, Greece, Guatemala, Hungary, India, Indonesia, Ireland, Italy, South Korea, Macedonia, Malaysia, Nepal, Nigeria, Portugal, Qatar, Sierra Leone, Slovenia, Spain, Tanzania, Turkey

United Nations Military Observer Group in India and Pakistan (UNMOGIP)
established–24 January 1949
aim–to observe the 1949 India-Pakistan cease-fire; established by the UN Security Council
members–(8) Chile, Croatia, Finland, Italy, South Korea, Philippines, Sweden, Uruguay

United Nations Mission for the Referendum in Western Sahara (MINURSO)
established–29 April 1991
aim–to supervise the cease-fire and conduct a referendum in Western Sahara; established by the UN Security Council
members–(30) Argentina, Austria, Bangladesh, Brazil, China, Croatia, Djibouti, Egypt, El Salvador, France, Ghana, Greece, Guinea, Honduras, Hungary, Ireland, Italy, Jordan, South Korea, Malaysia, Mongolia, Nepal, Nigeria, Pakistan, Paraguay, Poland, Russia, Sri Lanka, Uruguay, Yemen

United Nations Mission in Liberia (UNMIL)
established–19 September 2003
aim–to support the cease-fire agreement and peace process, protect UN facilities and people, support humanitarian activities, and assist in national security reform
members–(44) Bangladesh, Benin, Bolivia, Brazil, Bulgaria, China, Croatia, Denmark, Ecuador, Egypt, El Salvador, Ethiopia, Finland, France, The Gambia, Ghana, Indonesia, Jordan, South Korea, Kyrgyzstan, Malaysia, Mali, Moldova, Mongolia, Montenegro, Namibia, Nepal, Niger, Nigeria, Pakistan, Paraguay, Peru, Philippines, Poland, Romania, Russia, Senegal, Serbia, Togo, Ukraine, US, Yemen, Zambia, Zimbabwe

United Nations Mission in the Central African Republic and Chad (MINURCAT)
established on 25 September 2007; to create the security and conditions which will to contribute to the protection of refugees, displaced persons, and citizens in danger, to facilitate the provision of humanitarian assistance in eastern Chad and the northeastern Central African Republic, to create favorable conditions for the reconstruction and economic and social development of these areas; members were Bangladesh, Benin, Burkina Faso, Democratic Republic of the Congo, Egypt, Ethiopia, Ghana, Ireland, Kenya, Mali, Mongolia, Namibia, Nepal, Nigeria, Norway, Pakistan, Poland, Russia, Rwanda, Senegal, Serbia, Sri Lanka, Togo, Tunisia, US; MINURCAT was dissolved in December 2010

United Nations Mission in the Sudan (UNMIS)
established–March 2005
aim–to support implementation of the comprehensive Peace Agreement by Monitoring and verifying the implementation of the Cease Fire Agreement, by observing and monitoring movements of armed groups, and by helping disarm, demobilizing and reintegrating armed bands
members–(60) Australia, Bangladesh, Belgium, Benin, Bolivia, Brazil, Burkina Faso, Cambodia, Canada, China, Croatia, Denmark, Ecuador, Egypt, El Salvador, Fiji, Finland, Germany, Greece, Guatemala, Guinea, India, Indonesia, Iran, Japan, Jordan, Kenya, Kyrgyzstan, Malaysia, Moldova, Mongolia, Morocco, Namibia, Nepal, Netherland, NZ, Niger, Norway, Pakistan, Paraguay, Peru, Philippines, Poland, Qatar, Romania, Russia, Rwanda, Sierra Leone, Spain, Sweden, Switzerland, Tanzania, Thailand, Turkey, Uganda, Ukraine, UK, Yemen, Zambia, Zimbabwe

United Nations Operation in Cote d'Ivoire (UNOCI)

established–27 February 2004
aim–to facilitate the implementation by the Ivorian parties of the peace agreement signed by them in January 2003
members–(42) Bangladesh, Benin, Bolivia, Brazil, Chad, China, Ecuador, Egypt, El Salvador, Ethiopia, France, The Gambia, Ghana, Guatemala, Guinea, India, Ireland, Jordan, South Korea, Moldova, Morocco, Namibia, Nepal, Niger, Nigeria, Pakistan, Paraguay, Peru, Philippines, Poland, Romania, Russia, Senegal, Serbia, Tanzania, Togo, Tunisia, Uganda, Uruguay, Yemen, Zambia, Zimbabwe

United Nations Organization Stabilization Mission in the Democratic Republic of the Congo (MONUSCO)

established–28 May 2010
aim–to protect the civilians; to assist the government in the areas of stabilization and peace consolidation
members–(52) Algeria, Bangladesh, Belgium, Benin, Bolivia, Bosnia and Herzegovine, Burkina Faso, Cameroon, Canada, China, Czech Republic, Denmark, Egypt, France, Ghana, Guatemala, India, Indonesia, Ireland, Jordan, Kenya, Malawi, Malaysia, Mali, Mongolia, Morocco, Mozambique, Nepal, Niger, Nigeria, Norway, Pakistan, Paraguay, Peru, Poland, Romania, Russia, Senegal, Serbia, South Africa, Spain, Sri Lanka, Sweden, Switzerland, Tanzania, Tunisia, Ukraine, UK, US, Uruguay, Yemen, Zambia

United Nations Peacekeeping Force in Cyprus (UNFICYP)

established–4 March 1964
aim–to serve as a peacekeeping force between Greek Cypriots and Turkish Cypriots in Cyprus; established by the UN Security Council
members–(11) Argentina, Austria, Canada, Chile, Croatia, Hungary, Paraguay, Peru, Serbia, Slovakia, UK

United Nations Population Fund (UNFPA)

note–acronym retained from predecessor organization UN Fund for Population Activities
established–July 1967
aim–to assist both developed and developing countries to deal with their population problems
members (executive board)–(36) selected on a rotating basis from all regions

United Nations Relief and Works Agency for Palestine Refugees in the Near East (UNRWA)

established–8 December 1949
aim–to provide assistance to Palestinian refugees
members (advisory commission)–(23) Australia, Belgium, Canada, Denmark, Egypt, Finland, France, Germany, Ireland, Italy, Japan, Jordan, Lebanon, Netherlands, Norway, Saudi Arabia, Spain, Sweden, Switzerland, Syria, Turkey, UK, US
observers–(3) EC, LAS, Palestine Liberation Organization

United Nations Research Institute for Social Development (UNRISD)

established–1963
aim–to conduct research into the problems of economic development during different phases of economic growth
members–no country members, but a Board of Directors consisting of a chairman appointed by the UN secretary general and 12 individual members

United Nations Secretariat

established–26 June 1945; effective–24 October 1945
aim–to serve as the primary administrative organ of the UN; a Secretary General is appointed for a five-year term by the General Assembly on the recommendation of the Security Council
members–the UN Secretary General and staff

United Nations Security Council (UNSC)

established–26 June 1945; effective–24 October 1945
aim–to maintain international peace and security
permanent members–(5) China, France, Russia, UK, US
nonpermanent members–(10) elected for two-year terms by the UN General Assembly; Bosnia and Herzegovina (2010-11), Brazil (2010-11), Colombia (2011-12), Gabon (2010-11), Germany (2011-12), India (2011-12), Lebanon (2010-11), Nigeria (2010-11), Portugal (2011-12), South Africa (2011-12)

United Nations Truce Supervision Organization (UNTSO)

established–June 1948
aim–to supervise the 1948 Arab-Israeli cease-fire; currently supports timely deployment of reinforcements to other peacekeeping operations in the region as needed; initially established by the UN Security Council
members–(23) Argentina, Australia, Austria, Belgium, Canada, Chile, China, Denmark, Estonia, Finland, France, Ireland, Italy, Nepal, Netherlands, NZ, Norway, Russia, Slovakia, Slovenia, Sweden, Switzerland, US

United Nations Trusteeship Council

established on 26 June 1945, effective on 24 October 1945, to supervise the administration of the 11 UN trust territories; members were China, France, Russia, UK, US; it formally suspended operations 1 November 1995 after the Trust Territory of the Pacific Islands (Palau) became the Republic of Palau, a constitutional government in free association with the US; the Trusteeship Council was not dissolved

United Nations University (UNU)

established–3 December 1973
aim–to conduct research in development, welfare, and human survival and to train scholars
members–(24 members of UNU Council and the Rector are appointed by the Secretary General of the United Nations and the Director General of UNESCO)

Universal Postal Union (UPU)

established–9 October 1874, affiliated with the UN 15 November 1947; effective–1 July 1948
aim–to promote international postal cooperation; a UN specialized agency
members–(191) includes all UN member countries except Andorra, Marshall Islands, Federated States of Micronesia, and Palau (189 total); plus Holy See and Overseas Territories of the UK; note–includes the following dependencies or areas of special interest: Australia (Norfolk Island), China (Hong Kong, Macau), Denmark (Faroe Islands, Greenland), France (French Polynesia including Clipperton Island, French Southern and Antarctic Lands, Mayotte, New Caledonia, Saint Barthelemy, Saint Martin, Saint Pierre and Miquelon, Scattered Islands [Bassas da India, Europe, Juan de Nova, Glorioso Islands, Tromelin], Wallis and Futuna), Netherlands (Aruba), NZ

(Cook Island, Niue, Tokelau), UK (Guernsey, Isle of Man, Jersey; Anguilla, Bermuda, British Indian Ocean Territory, British Virgin Islands, Cayman Islands, Falkland Islands, Gibraltar, Montserrat, Pitcairn Islands, Saint Helena, Ascension, and Tristan da Cunha, South Georgia and South Sandwich Islands, Turks and Caicos), US (American Samoa, Guam, Northern Mariana Islands, Puerto Rico, Virgin Islands)

Warsaw Pact (WP)
established 14 May 1955 to promote mutual defense; members met 1 July 1991 to dissolve the alliance; member states at the time of dissolution were: Bulgaria, Czechoslovakia, Hungary, Poland, Romania, and the USSR; earlier members included German Democratic Republic (GDR) and Albania

West African Development Bank (WADB)
note–also known as Banque Ouest-Africaine de Developpement (BOAD); is a financial institution of WAEMU
established–14 November 1973
aim–to promote regional economic development and integration
regional members–(8) Benin, Burkina Faso, Cote d'Ivoire, Guinea-Bissau, Mali, Niger, Senegal, Togo

West African Economic and Monetary Union (WAEMU)
note–also known as Union Economique et Monetaire Ouest Africaine (UEMOA)
established–1 August 1994
aim–to increase competitiveness of members' economic markets; to create a common market
members–(8) Benin, Burkina Faso, Cote d'Ivoire, Guinea-Bissau, Mali, Niger, Senegal, Togo

Western European Union (WEU)
established 23 October 1954; effective–6 May 1955; aim to provide mutual defense and to move toward political unification; 10 members: Belgium, France, Germany, Greece, Italy, Luxembourg, Netherlands, Portugal, Spain, UK; 6 associate members: Czech Republic, Hungary, Iceland, Norway, Poland, Turkey; 7 associate partners: Bulgaria, Estonia, Latvia, Lithuania, Romania, Slovakia, Slovenia; 5 observers: Austria, Denmark, Finland, Ireland, Sweden; note–to cease existence completely by June 2011

World Bank Group
includes International Bank for Reconstruction and Development (IBRD), International Development Association (IDA), International Finance Corporation (IFC), and Multilateral Investment Guarantee Agency (MIGA)

World Confederation of Labor (WCL)
established 19 June 1920 as the International Federation of Christian Trade Unions (IFCTU), renamed 4 October 1968; aim was to promote the trade union movement; on 31 October 2006 it merged with the International Confederation of Free Trade Unions (ICFTU) to form the International Trade Union Confederation (ITUC); members were (105 national organizations) Antigua and Barbuda, Argentina, Aruba, Austria, Bangladesh, Belgium, Belize, Benin, Bolivia, Brazil, Bulgaria, Burkina Faso, Cameroon, Canada, Central African Republic, Chad, Chile, Colombia, Democratic Republic of the Congo, Republic of the Congo, Costa Rica, Cote d'Ivoire, Cuba, Cyprus, Czech Republic, Denmark, Dominica, Dominican Republic, Ecuador, El Salvador, France, French Guiana, Gabon, The Gambia, Ghana, Guadeloupe, Guatemala, Guinea, Guyana, Haiti, Honduras, Hong Kong, Hungary, India, Indonesia, Iran, Italy, Japan, Kazakhstan, South Korea, Liberia, Libya, Liechtenstein, Lithuania, Luxembourg, Macedonia, Madagascar, Malawi, Malaysia, Malta, Martinique, Mauritania, Mauritius, Mexico, Morocco, Namibia, Nepal, Netherlands, Nicaragua, Niger, Pakistan, Panama, Paraguay, Peru, Philippines, Poland, Portugal, Puerto Rico, Romania, Rwanda, Saint Lucia, Saint Vincent and the Grenadines, Sao Tome and Principe, Senegal, Serbia, Sierra Leone, Singapore, Slovakia, South Africa, Spain, Sri Lanka, Suriname, Switzerland, Taiwan, Thailand, Togo, Trinidad and Tobago, Ukraine, US, Uruguay, Venezuela, Vietnam, Zambia, Zimbabwe

World Customs Organization (WCO)
note–began as the Customs Cooperation Council (CCC)
established–15 December 1950
aim–to promote international cooperation in customs matters
members–(178) Afghanistan, Albania, Algeria, Andorra, Angola, Argentina, Armenia, Australia, Austria, Azerbaijan, The Bahamas, Bahrain, Bangladesh, Barbados, Belarus, Belgium, Belize, Benin, Bermuda, Bhutan, Bolivia, Bosnia and Herzegovina, Botswana, Brazil, Brunei, Bulgaria, Burkina Faso, Burma, Burundi, Cambodia, Cameroon, Canada, Cape Verde, Central African Republic, Chad, Chile, China, Colombia, Comoros, Democratic Republic of the Congo, Republic of the Congo, Costa Rica, Cote d'Ivoire, Croatia, Cuba, Cyprus, Czech Republic, Denmark, Djibouti, Dominican Republic, EC, Ecuador, Egypt, El Salvador, Eritrea, Estonia, Ethiopia, Fiji, Finland, France, Gabon, The Gambia, Georgia, Germany, Ghana, Greece, Guatemala, Guinea, Guinea-Bissau, Guyana, Haiti, Honduras, Hong Kong, Hungary, Iceland, India, Indonesia, Iran, Iraq, Ireland, Israel, Italy, Jamaica, Japan, Jordan, Kazakhstan, Kenya, South Korea, Kuwait, Kyrgyzstan, Laos, Latvia, Lebanon, Lesotho, Liberia, Libya, Lithuania, Luxembourg, Macau, Macedonia, Madagascar, Malawi, Malaysia, Maldives, Mali, Malta, Mauritania, Mauritius, Mexico, Moldova, Mongolia, Montenegro, Morocco, Mozambique, Namibia, Nepal, Netherlands, NZ, Nicaragua, Niger, Nigeria, Norway, Oman, Pakistan, Panama, Papua New Guinea, Paraguay, Peru, Philippines, Poland, Portugal, Qatar, Romania, Russia, Rwanda, Saint Lucia, Samoa, Sao Tome and Principe, Saudi Arabia, Senegal, Serbia, Seychelles, Sierra Leone, Singapore, Slovakia, Slovenia, South Africa, Spain, Sri Lanka, Sudan, Swaziland, Sweden, Switzerland, Syria, Tajikistan, Tanzania, Thailand, Timor-Leste, Togo, Tonga, Trinidad and Tobago, Tunisia, Turkey, Turkmenistan, Uganda, Ukraine, UAE, UK, US, Uruguay, Uzbekistan, Vanuatu, Venezuela, Vietnam, Yemen, Zambia, Zimbabwe

World Federation of Trade Unions (WFTU)
established–3 October 1945
aim–to promote the trade union movement
members–(in 2010 there were 109 participating nations and the Palestine Liberation Organization; the nations were not listed); (in 2009 there were 125 nations and the Palestine Liberation Organization) Afghanistan, Albania, Angola, Antigua and Barbuda, Argentina, Armenia, Australia, Austria, Azerbaijan, Bahrain, Bangladesh, Barbados, Belarus, Benin, Bolivia, Botswana, Brazil, Bulgaria, Burkina Faso, Cambodia, Cameroon, Canada, Chile, Colombia, Democratic Republic of the Congo, Republic of the Congo, Costa Rica, Cote d'Ivoire, Cuba, Cyprus, Czech Republic, Djibouti, Dominican Republic, Ecuador, Egypt, El Salvador, Eritrea, Ethiopia, Fiji, Finland, France, French Guiana, The Gambia, Ghana, Greece, Guadeloupe, Guatemala, Guinea, Guinea-Bissau, Guyana, Haiti, Honduras, Hungary, India, Indonesia, Iran, Iraq, Jamaica, Japan, Jordan, Kazakhstan, North Korea, Kuwait, Kyrgyzstan, Laos, Lebanon, Lesotho, Liberia, Libya, Madagascar, Malawi, Malaysia, Mali, Martinique, Mauritius, Mexico, Mozambique, Nepal, New Caledonia, NZ, Niger, Nigeria, Oman, Pakistan, Panama, Papua New Guinea, Peru, Philippines, Poland, Portugal, Puerto Rico, Reunion, Romania, Russia,

Saint Lucia, Saint Pierre and Miquelon, Saint Vincent and the Grenadines, Saudi Arabia, Senegal, Sierra Leone, Slovakia, Solomon Islands, Somalia, South Africa, Sri Lanka, Sudan, Sweden, Syria, Tajikistan, Tanzania, Thailand, Togo, Trinidad and Tobago, Tunisia, Turkey, Turkmenistan, Uganda, Ukraine, Uruguay, Uzbekistan, Vanuatu, Venezuela, Vietnam, Yemen, Zimbabwe, Palestine Liberation Organization

World Food Program (WFP)

established–24 November 1961
aim–to provide food aid in support of economic development or disaster relief; an ECOSOC organization
members–(36) selected on a rotating basis from all regions

World Health Organization (WHO)

established–22 July 1946; effective–7 April 1948
aim–to deal with health matters worldwide; a UN specialized agency
members–(193) includes all UN member countries except Liechtenstein (191 total); plus Cook Islands and Niue

World Intellectual Property Organization (WIPO)

established–14 July 1967; effective–26 April 1970
aim–to furnish protection for literary, artistic, and scientific works; a UN specialized agency
members–(184) includes all UN member countries except Kiribati, Marshall Islands, Federated States of Micronesia, Nauru, Palau, Solomon Islands, Timor-Leste, Tuvalu, Vanuatu (183 total); plus Holy See

World Meteorological Organization (WMO)

established–11 October 1947; effective–4 April 1951
aim–to sponsor meteorological cooperation; a UN specialized agency
members–(183) includes all UN member countries except Andorra, Equatorial Guinea, Grenada, Liechtenstein, Marshall Islands, Nauru, Palau, Saint Kitts and Nevis, Saint Vincent and the Grenadines, San Marino, Tuvalu (181 total); plus Cook Islands and Niue

World Tourism Organization (UNWTO)

established–2 January 1975
aim–to promote tourism as a means of contributing to economic development, international understanding, and peace
members–(154) Afghanistan, Albania, Algeria, Andorra, Angola, Argentina, Armenia, Australia, Austria, Azerbaijan, The Bahamas, Bahrain, Bangladesh, Belarus, Benin, Bhutan, Bolivia, Bosnia and Herzegovina, Botswana, Brazil, Brunei, Bulgaria, Burkina Faso, Burundi, Cambodia, Cameroon, Canada, Cape Verde, Central African Republic, Chad, Chile, China, Colombia, Democratic Republic of the Congo, Republic of the Congo, Costa Rica, Cote d'Ivoire, Croatia, Cuba, Cyprus, Czech Republic, Djibouti, Dominican Republic, Ecuador, Egypt, El Salvador, Equatorial Guinea, Eritrea, Ethiopia, Fiji, France, Gabon, The Gambia, Georgia, Germany, Ghana, Greece, Guatemala, Guinea, Guinea-Bissau, Haiti, Honduras, Hungary, India, Indonesia, Iran, Iraq, Israel, Italy, Jamaica, Japan, Jordan, Kazakhstan, Kenya, North Korea, South Korea, Kuwait, Kyrgyzstan, Laos, Latvia, Lebanon, Lesotho, Libya, Lithuania, Macedonia, Madagascar, Malawi, Malaysia, Maldives, Mali, Malta, Mauritania, Mauritius, Mexico, Moldova, Monaco, Mongolia, Montenegro, Morocco, Mozambique, Namibia, Nepal, Netherlands, Nicaragua, Niger, Nigeria, Norway, Oman, Pakistan, Panama, Papua New Guinea, Paraguay, Peru, Philippines, Poland, Portugal, Qatar, Romania, Russia, Rwanda, San Marino, Sao Tome and Principe, Saudi Arabia, Senegal, Serbia, Seychelles, Sierra Leone, Slovakia, Slovenia, South Africa, Spain, Sri Lanka, Sudan, Swaziland, Switzerland, Syria, Tajikistan, Tanzania, Thailand, Timor-Leste, Togo, Tunisia, Turkey, Turkmenistan, Uganda, Ukraine, Uruguay, Uzbekistan, Vanuatu, Venezuela, Vietnam, Yemen, Zambia, Zimbabwe
associate members–(6) Aruba, Flanders, Hong Kong, Macau, Madeira Islands, Puerto Rico
observers–(1 plus Palestine Liberation Organization) Holy See, Palestine Liberation Organization

World Trade Organization (WTO)

note–succeeded General Agreement on Tariff and Trade (GATT)
established–15 April 1994; effective–1 January 1995
aim–to provide a forum to resolve trade conflicts between members and to carry on negotiations with the goal of further lowering and/or eliminating tariffs and other trade barriers
members–(153) Albania, Angola, Antigua and Barbuda, Argentina, Armenia, Australia, Austria, Bahrain, Bangladesh, Barbados, Belgium, Belize, Benin, Bolivia, Botswana, Brazil, Brunei, Bulgaria, Burkina Faso, Burma, Burundi, Cambodia, Cameroon, Canada, Cape Verde, Central African Republic, Chad, Chile, China, Colombia, Democratic Republic of the Congo, Republic of the Congo, Costa Rica, Cote d'Ivoire, Croatia, Cuba, Cyprus, Czech Republic, Denmark, Djibouti, Dominica, Dominican Republic, Ecuador, Egypt, El Salvador, Estonia, EU, Fiji, Finland, France, Gabon, The Gambia, Georgia, Germany, Ghana, Greece, Grenada, Guatemala, Guinea, Guinea-Bissau, Guyana, Haiti, Honduras, Hong Kong, Hungary, Iceland, India, Indonesia, Ireland, Israel, Italy, Jamaica, Japan, Jordan, Kenya, South Korea, Kuwait, Kyrgyzstan, Latvia, Lesotho, Liechtenstein, Lithuania, Luxembourg, Macau, Macedonia, Madagascar, Malawi, Malaysia, Maldives, Mali, Malta, Mauritania, Mauritius, Mexico, Moldova, Mongolia, Morocco, Mozambique, Namibia, Nepal, Netherlands, NZ, Nicaragua, Niger, Nigeria, Norway, Oman, Pakistan, Panama, Papua New Guinea, Paraguay, Peru, Philippines, Poland, Portugal, Qatar, Romania, Rwanda, Saint Kitts and Nevis, Saint Lucia, Saint Vincent and the Grenadines, Saudi Arabia, Senegal, Sierra Leone, Singapore, Slovakia, Slovenia, Solomon Islands, South Africa, Spain, Sri Lanka, Suriname, Swaziland, Sweden, Switzerland, Taiwan, Tanzania, Thailand, Togo, Tonga, Trinidad and Tobago, Tunisia, Turkey, Uganda, Ukraine, UAE, UK, US, Uruguay, Venezuela, Vietnam, Zambia, Zimbabwe
observers–(31) Afghanistan, Algeria, Andorra, Azerbaijan, The Bahamas, Belarus, Bhutan, Bosnia and Herzegovina, Comoros, Equatorial Guinea, Ethiopia, Holy See, Iran, Iraq, Kazakhstan, Laos, Lebanon, Liberia, Libya, Montenegro, Russia, Samoa, Sao Tome and Principe, Serbia, Seychelles, Sudan, Syria, Tajikistan, Uzbekistan, Vanuatu, Yemen; note–with the exception of the Holy See, an observer must start accession negotiations within five years of becoming observers

Zangger Committee (ZC)

established–early 1970s
aim–to establish guidelines for the export control provisions of the Nonproliferation of Nuclear Weapons Treaty (NPT)
members–(37) Argentina, Australia, Austria, Belgium, Bulgaria, Canada, China, Croatia, Czech Republic, Denmark, Finland, France, Germany, Greece, Hungary, Ireland, Italy, Japan, Kazakhstan, South Korea, Luxembourg, Netherlands, Norway, Poland, Portugal, Romania, Russia, Slovakia, Slovenia, South Africa, Spain, Sweden, Switzerland, Turkey, Ukraine, UK, US
observers–(1) European Commission

APPENDIX C

SELECTED INTERNATIONAL ENVIRONMENTAL AGREEMENTS

Air Pollution
see Convention on Long-Range Transboundary Air Pollution

Air Pollution-Nitrogen Oxides
see Protocol to the 1979 Convention on Long-Range Transboundary Air Pollution Concerning the Control of Emissions of Nitrogen Oxides or Their Transboundary Fluxes

Air Pollution-Persistent Organic Pollutants
see Protocol to the 1979 Convention on Long-Range Transboundary Air Pollution on Persistent Organic Pollutants

Air Pollution-Sulphur 85
see Protocol to the 1979 Convention on Long-Range Transboundary Air Pollution on the Reduction of Sulphur Emissions or Their Transboundary Fluxes by at least 30%

Air Pollution-Sulphur 94
see Protocol to the 1979 Convention on Long-Range Transboundary Air Pollution on Further Reduction of Sulphur Emissions

Air Pollution-Volatile Organic Compounds
see Protocol to the 1979 Convention on Long-Range Transboundary Air Pollution Concerning the Control of Emissions of Volatile Organic Compounds or Their Transboundary Fluxes

Antarctic—Environmental Protocol
see Protocol on Environmental Protection to the Antarctic Treaty

Antarctic Treaty
opened for signature–1 December 1959
entered into force–23 June 1961
objective–to ensure that Antarctica is used for peaceful purposes only (such as international cooperation in scientific research); to defer the question of territorial claims asserted by some nations and not recognized by others; to provide an international forum for management of the region; applies to land and ice shelves south of 60 degrees south latitude
parties–(46) Argentina, Australia, Austria, Belarus, Belgium, Brazil, Bulgaria, Canada, Chile, China, Colombia, Cuba, Czech Republic, Denmark, Ecuador, Estonia, Finland, France, Germany, Greece, Guatemala, Hungary, India, Italy, Japan, North Korea, South Korea, Netherlands, NZ, Norway, Papua New Guinea, Peru, Poland, Romania, Russia, Slovakia, South Africa, Spain, Sweden, Switzerland, Turkey, Ukraine, UK, US, Uruguay, Venezuela

Basel Convention on the Control of Transboundary Movements of Hazardous Wastes and Their Disposal
note–abbreviated as Hazardous Wastes
opened for signature–22 March 1989
entered into force–5 May 1992
objective–to reduce transboundary movements of wastes subject to the Convention to a minimum consistent with the environmentally sound and efficient management of such wastes; to minimize the amount and toxicity of wastes generated and ensure their environmentally sound management as closely as possible to the source of generation; and to assist LDCs in environmentally sound management of the hazardous and other wastes they generate
parties–(172) Albania, Algeria, Andorra, Antigua and Barbuda, Argentina, Armenia, Australia, Austria, Azerbaijan, The Bahamas, Bahrain, Bangladesh, Barbados, Belarus, Belgium, Belize, Benin, Bhutan, Bolivia, Bosnia and Herzegovina, Botswana, Brazil, Brunei, Bulgaria, Burkina Faso, Burundi, Cambodia, Cameroon, Canada, Cape Verde, Central African Republic, Chad, Chile, China, Colombia, Comoros, Democratic Republic of the Congo, Republic of the Congo, Cook Islands, Costa Rica, Cote d'Ivoire, Croatia, Cuba, Cyprus, Czech Republic, Denmark, Djibouti, Dominica, Dominican Republic, Ecuador, Egypt, El Salvador, Equatorial Guinea, Eritrea, Estonia, Ethiopia, EU, Finland, France, Gabon, The Gambia, Georgia, Germany, Ghana, Greece, Guatemala, Guinea, Guinea-Bissau, Guyana, Honduras, Hungary, Iceland, India, Indonesia, Iran, Ireland, Israel, Italy, Jamaica, Japan, Jordan, Kazakhstan, Kenya, Kiribati, North Korea, South Korea, Kuwait, Kyrgyzstan, Latvia, Lebanon, Lesotho, Liberia, Libya, Liechtenstein, Lithuania, Luxembourg, Macedonia, Madagascar, Malawi, Malaysia, Maldives, Mali, Malta, Marshall Islands, Mauritania, Mauritius, Mexico, Federated States of Micronesia, Moldova, Monaco, Mongolia, Montenegro, Morocco, Mozambique, Namibia, Nauru, Nepal, Netherlands, NZ, Nicaragua, Niger, Nigeria, Norway, Oman, Pakistan, Panama, Papua New Guinea, Paraguay, Peru, Philippines, Poland, Portugal, Qatar, Romania, Russia, Rwanda, Saint Kitts and Nevis, Saint Lucia, Saint Vincent and the Grenadines, Samoa, Saudi Arabia, Senegal, Serbia, Seychelles, Singapore, Slovakia, Slovenia, South Africa, Spain, Sri Lanka, Sudan, Swaziland, Sweden, Switzerland, Syria, Tanzania, Thailand, Trinidad and Tobago, Tunisia, Turkey, Turkmenistan, Uganda, Ukraine, UAE, UK, Uruguay, Uzbekistan, Venezuela, Vietnam, Yemen, Zambia
countries that have signed, but not yet ratified–(3) Afghanistan, Haiti, US

Biodiversity
see Convention on Biological Diversity

Climate Change
see United Nations Framework Convention on Climate Change

Climate Change-Kyoto Protocol
see Kyoto Protocol to the United Nations Framework Convention on Climate Change

Convention for the Conservation of Antarctic Seals
note–abbreviated as Antarctic Seals
opened for signature–1 June 1972
entered into force–11 March 1978
objective–to promote and achieve the protection, scientific study, and rational use of Antarctic seals, and to maintain a satisfactory balance within the ecological system of Antarctica
parties–(16) Argentina, Australia, Belgium, Brazil, Canada, Chile, France, Germany, Italy, Japan, Norway, Poland, Russia, South Africa, UK, US

countries that have signed, but not yet ratified–(1) NZ

Convention on Biological Diversity

note–abbreviated as Biodiversity
opened for signature–5 June 1992
entered into force–29 December 1993
objective–to develop national strategies for the conservation and sustainable use of biological diversity
parties–(191) Afghanistan, Albania, Algeria, Andorra, Angola, Antigua and Barbuda, Argentina, Armenia, Australia, Austria, Azerbaijan, The Bahamas, Bahrain, Bangladesh, Barbados, Belarus, Belgium, Belize, Benin, Bhutan, Bolivia, Bosnia and Herzegovina, Botswana, Brazil, Brunei, Bulgaria, Burkina Faso, Burma, Burundi, Cambodia, Cameroon, Canada, Cape Verde, Central African Republic, Chad, Chile, China, Colombia, Comoros, Democratic Republic of the Congo, Republic of the Congo, Cook Islands, Costa Rica, Cote d'Ivoire, Croatia, Cuba, Cyprus, Czech Republic, Denmark, Djibouti, Dominica, Dominican Republic, Ecuador, Egypt, El Salvador, Equatorial Guinea, Eritrea, Estonia, Ethiopia, EU, Fiji, Finland, France, Gabon, The Gambia, Georgia, Germany, Ghana, Greece, Grenada, Guatemala, Guinea, Guinea-Bissau, Guyana, Haiti, Honduras, Hungary, Iceland, India, Indonesia, Iran, Ireland, Israel, Italy, Jamaica, Japan, Jordan, Kazakhstan, Kenya, Kiribati, North Korea, South Korea, Kuwait, Kyrgyzstan, Laos, Latvia, Lebanon, Lesotho, Liberia, Libya, Liechtenstein, Lithuania, Luxembourg, Macedonia, Madagascar, Malawi, Malaysia, Maldives, Mali, Malta, Marshall Islands, Mauritania, Mauritius, Mexico, Federated States of Micronesia, Moldova, Monaco, Mongolia, Montenegro, Morocco, Mozambique, Namibia, Nauru, Nepal, Netherlands, NZ, Nicaragua, Niger, Nigeria, Niue, Norway, Oman, Pakistan, Palau, Panama, Papua New Guinea, Paraguay, Peru, Philippines, Poland, Portugal, Qatar, Romania, Russia, Rwanda, Saint Kitts and Nevis, Saint Lucia, Saint Vincent and the Grenadines, Samoa, San Marino, Sao Tome and Principe, Saudi Arabia, Senegal, Serbia, Seychelles, Sierra Leone, Singapore, Slovakia, Slovenia, Solomon Islands, South Africa, Spain, Sri Lanka, Sudan, Suriname, Swaziland, Sweden, Switzerland, Syria, Tajikistan, Tanzania, Thailand, Timor-Leste, Togo, Tonga, Trinidad and Tobago, Tunisia, Turkey, Turkmenistan, Tuvalu, Uganda, Ukraine, UAE, UK, Uruguay, Uzbekistan, Vanuatu, Venezuela, Vietnam, Yemen, Zambia, Zimbabwe
countries that have signed, but not yet ratified–(1) US

Convention on Fishing and Conservation of Living Resources of the High Seas

note–abbreviated as Marine Life Conservation
opened for signature–29 April 1958
entered into force–20 March 1966
objective–to solve through international cooperation the problems involved in the conservation of living resources of the high seas, considering that because of the development of modern technology some of these resources are in danger of being overexploited
parties–(38) Australia, Belgium, Bosnia and Herzegovina, Burkina Faso, Cambodia, Colombia, Denmark, Dominican Republic, Fiji, Finland, France, Haiti, Jamaica, Kenya, Lesotho, Madagascar, Malawi, Malaysia, Mauritius, Mexico, Montenegro, Netherlands, Nigeria, Portugal, Senegal, Serbia, Sierra Leone, Solomon Islands, South Africa, Spain, Switzerland, Thailand, Tonga, Trinidad and Tobago, Uganda, UK, US, Venezuela
countries that have signed, but not yet ratified–(21) Afghanistan, Argentina, Bolivia, Canada, Costa Rica, Cuba, Ghana, Iceland, Indonesia, Iran, Ireland, Israel, Lebanon, Liberia, Nepal, NZ, Pakistan, Panama, Sri Lanka, Tunisia, Uruguay

Convention on Long-Range Transboundary Air Pollution

note–abbreviated as Air Pollution
opened for signature–13 November 1979
entered into force–16 March 1983
objective–to protect the human environment against air pollution and to gradually reduce and prevent air pollution, including long-range transboundary air pollution
parties–(51) Albania, Armenia, Austria, Azerbaijan, Belarus, Belgium, Bosnia and Herzegovina, Bulgaria, Canada, Croatia, Cyprus, Czech Republic, Denmark, Estonia, EU, Finland, France, Georgia, Germany, Greece, Hungary, Iceland, Ireland, Italy, Kazakhstan, Kyrgyzstan, Latvia, Liechtenstein, Lithuania, Luxembourg, Macedonia, Malta, Moldova, Monaco, Montenegro, Netherlands, Norway, Poland, Portugal, Romania, Russia, Serbia, Slovakia, Slovenia, Spain, Sweden, Switzerland, Turkey, Ukraine, UK, US
countries that have signed, but not yet ratified–(2) Holy See, San Marino

Convention on Wetlands of International Importance Especially as Waterfowl Habitat (Ramsar)

note–abbreviated as Wetlands
opened for signature–2 February 1971
entered into force–21 December 1975
objective–to stem the progressive encroachment on and loss of wetlands now and in the future, recognizing the fundamental ecological functions of wetlands and their economic, cultural, scientific, and recreational value
parties–(154) Albania, Algeria, Antigua and Barbuda, Argentina, Armenia, Australia, Austria, Azerbaijan, The Bahamas, Bahrain, Bangladesh, Barbados, Belarus, Belgium, Belize, Benin, Bolivia, Bosnia and Herzegovina, Botswana, Brazil, Bulgaria, Burkina Faso, Burma, Burundi, Cambodia, Cameroon, Canada, Cape Verde, Central African Republic, Chad, Chile, China, Colombia, Comoros, Democratic Republic of the Congo, Republic of the Congo, Costa Rica, Cote d'Ivoire, Croatia, Cuba, Cyprus, Czech Republic, Denmark, Djibouti, Dominican Republic, Ecuador, Egypt, El Salvador, Equatorial Guinea, Estonia, Fiji, Finland, France, Gabon, The Gambia, Georgia, Germany, Ghana, Greece, Guatemala, Guinea, Guinea-Bissau, Honduras, Hungary, Iceland, India, Indonesia, Iran, Ireland, Israel, Italy, Jamaica, Japan, Jordan, Kazakhstan, Kenya, South Korea, Kyrgyzstan, Latvia, Lebanon, Lesotho, Liberia, Libya, Liechtenstein, Lithuania, Luxembourg, Macedonia, Madagascar, Malawi, Malaysia, Mali, Malta, Marshall Islands, Mauritania, Mauritius, Mexico, Moldova, Monaco, Mongolia, Morocco, Mozambique, Namibia, Nepal, Netherlands, NZ, Nicaragua, Niger, Nigeria, Norway, Pakistan, Palau, Panama, Papua New Guinea, Paraguay, Peru, Philippines, Poland, Portugal, Romania, Russia, Rwanda, Saint Lucia, Samoa, Sao Tome and Principe, Senegal, Serbia, Seychelles, Sierra Leone, Slovakia, Slovenia, South Africa, Spain, Sri Lanka, Sudan, Suriname, Sweden, Switzerland, Syria, Tanzania, Tajikistan, Thailand, Togo, Trinidad and Tobago, Tunisia, Turkey, Uganda, Ukraine, UK, US, Uruguay, Uzbekistan, Venezuela, Vietnam, Zambia

Convention on the Conservation of Antarctic Marine Living Resources

note–abbreviated as Antarctic-Marine Living Resources
opened for signature–5 May 1980
entered into force–7 April 1982

objective–to safeguard the environment and protect the integrity of the ecosystem of the seas surrounding Antarctica, and to conserve Antarctic marine living resources
parties–(31) Argentina, Australia, Belgium, Brazil, Bulgaria, Canada, Chile, EU, Finland, France, Germany, Greece, India, Italy, Japan, South Korea, Mauritius, Namibia, Netherlands, NZ, Norway, Peru, Poland, Russia, South Africa, Spain, Sweden, Ukraine, UK, US, Uruguay, Vanuatu

Convention on the International Trade in Endangered Species of Wild Flora and Fauna (CITES)
note–abbreviated as Endangered Species
opened for signature–3 March 1973
entered into force–1 July 1975
objective–to protect certain endangered species from overexploitation by means of a system of import/export permits
parties–(170) Afghanistan, Albania, Algeria, Antigua and Barbuda, Argentina, Australia, Austria, Azerbaijan, The Bahamas, Bangladesh, Barbados, Belarus, Belgium, Belize, Benin, Bhutan, Bolivia, Botswana, Brazil, Brunei, Bulgaria, Burkina Faso, Burma, Burundi, Cambodia, Cameroon, Canada, Cape Verde, Central African Republic, Chad, Chile, China, Colombia, Comoros, Democratic Republic of the Congo, Republic of the Congo, Costa Rica, Cote d'Ivoire, Croatia, Cuba, Cyprus, Czech Republic, Denmark, Djibouti, Dominica, Dominican Republic, Ecuador, Egypt, El Salvador, Equatorial Guinea, Eritrea, Estonia, Ethiopia, Fiji, Finland, France, Gabon, The Gambia, Georgia, Germany, Ghana, Greece, Grenada, Guatemala, Guinea, Guinea-Bissau, Guyana, Honduras, Hungary, Iceland, India, Indonesia, Iran, Ireland, Israel, Italy, Jamaica, Japan, Jordan, Kazakhstan, Kenya, South Korea, Kuwait, Laos, Latvia, Lesotho, Liberia, Libya, Liechtenstein, Lithuania, Luxembourg, Macedonia, Madagascar, Malawi, Malaysia, Mali, Malta, Mauritania, Mauritius, Mexico, Moldova, Monaco, Mongolia, Montenegro, Morocco, Mozambique, Namibia, Nepal, Netherlands, NZ, Nicaragua, Niger, Nigeria, Norway, Palau, Pakistan, Panama, Papua New Guinea, Paraguay, Peru, Philippines, Poland, Portugal, Qatar, Romania, Russia, Rwanda, Saint Kitts and Nevis, Saint Lucia, Saint Vincent and the Grenadines, Samoa, San Marino, Sao Tome and Principe, Saudi Arabia, Senegal, Serbia, Seychelles, Sierra Leone, Singapore, Slovakia, Slovenia, Somalia, South Africa, Spain, Sri Lanka, Sudan, Suriname, Swaziland, Sweden, Switzerland, Syria, Tanzania, Thailand, Togo, Trinidad and Tobago, Tunisia, Turkey, Uganda, Ukraine, UAE, UK, US, Uruguay, Uzbekistan, Vanuatu, Venezuela, Vietnam, Yemen, Zambia, Zimbabwe

Convention on the Prevention of Marine Pollution by Dumping Wastes and Other Matter (London Convention)
note–abbreviated as Marine Dumping
opened for signature–29 December 1972
entered into force–30 August 1975
objective–to control pollution of the sea by dumping and to encourage regional agreements supplementary to the Convention; the London Convention came into force in 1996
parties–(88) Afghanistan, Angola, Antigua and Barbuda, Argentina, Australia, Azerbaijan, Barbados, Belarus, Belgium, Bolivia, Brazil, Bulgaria, Canada, Cape Verde, Chile, China, Democratic Republic of the Congo, Costa Rica, Cote d'Ivoire, Croatia, Cuba, Cyprus, Denmark, Dominican Republic, Egypt, Equatorial Guinea, Finland, France, Gabon, Germany, Greece, Guatemala, Haiti, Honduras, Hong Kong (associate member), Hungary, Iceland, Iran, Ireland, Italy, Jamaica, Japan, Jordan, Kenya, Kiribati, South Korea, Libya, Luxembourg, Malta, Mexico, Monaco, Montenegro, Morocco, Nauru, Netherlands, NZ, Nigeria, Norway, Oman, Pakistan, Panama, Papua New Guinea, Peru, Philippines, Poland, Portugal, Russia, Saint Kitts and Nevis, Saint Lucia, Saint Vincent and the Grenadines, Saudi Arabia, Serbia, Seychelles, Slovenia, Solomon Islands, South Africa, Spain, Suriname, Sweden, Switzerland, Tonga, Trinidad and Tobago, Tunisia, Ukraine, UAE, UK, US, Vanuatu
associate members to the London Convention–(2) Faroe Islands, Macau *countries that have signed, but not yet ratified*–(3) Chad, Kuwait, Uruguay

Convention on the Prohibition of Military or Any Other Hostile Use of Environmental Modification Techniques
note–abbreviated as Environmental Modification
opened for signature–10 December 1976
entered into force–5 October 1978
objective–to prohibit the military or other hostile use of environmental modification techniques in order to further world peace and trust among nations
parties–(73) Afghanistan, Algeria, Antigua and Barbuda, Argentina, Armenia, Australia, Austria, Bangladesh, Belarus, Belgium, Benin, Brazil, Bulgaria, Canada, Cape Verde, Chile, China, Costa Rica, Cuba, Cyprus, Czech Republic, Denmark, Dominica, Egypt, Finland, Germany, Ghana, Greece, Guatemala, Hungary, India, Ireland, Italy, Japan, Kazakhstan, North Korea, South Korea, Kuwait, Laos, Lithuania, Malawi, Mauritius, Mongolia, Netherlands, NZ, Nicaragua, Niger, Norway, Pakistan, Panama, Papua New Guinea, Poland, Romania, Russia, Saint Lucia, Saint Vincent and the Grenadines, Sao Tome and Principe, Slovakia, Slovenia, Solomon Islands, Spain, Sri Lanka, Sweden, Switzerland, Tajikistan, Tunisia, Ukraine, UK, US, Uruguay, Uzbekistan, Vietnam, Yemen
countries that have signed, but not yet ratified–(16) Bolivia, Democratic Republic of the Congo, Ethiopia, Holy See, Iceland, Iran, Iraq, Lebanon, Liberia, Luxembourg, Morocco, Portugal, Sierra Leone, Syria, Turkey, Uganda

Desertification
see United Nations Convention to Combat Desertification in those Countries Experiencing Serious Drought and/or Desertification, Particularly in Africa

Endangered Species
see Convention on the International Trade in Endangered Species of Wild Flora and Fauna (CITES)

Environmental Modification
see Convention on the Prohibition of Military or Any Other Hostile Use of Environmental Modification Techniques

Hazardous Wastes
see Basel Convention on the Control of Transboundary Movements of Hazardous Wastes and Their Disposal

International Convention for the Regulation of Whaling
note–abbreviated as Whaling
opened for signature–2 December 1946
entered into force–10 November 1948

objective–to protect all species of whales from overhunting; to establish a system of international regulation for the whale fisheries to ensure proper conservation and development of whale stocks; and to safeguard for future generations the great natural resources represented by whale stocks
parties–(84) Antigua and Barbuda, Argentina, Australia, Austria, Belgium, Belize, Benin, Brazil, Cambodia, Cameroon, Chile, China, Republic of the Congo, Costa Rica, Cote D'Ivoire, Croatia, Cyprus, Czech Republic, Denmark, Dominica, Ecuador, Eritrea, Estonia, Finland, France, Gabon, The Gambia, Germany, Greece, Grenada, Guatemala, Guinea, Guinea-Bissau, Hungary, Iceland, India, Ireland, Israel, Italy, Japan, Kenya, Kiribati, South Korea, Laos, Lithuania, Luxembourg, Mali, Marshall Islands, Mauritania, Mexico, Monaco, Mongolia, Morocco, Nauru, Netherlands, NZ, Nicaragua, Norway, Oman, Palau, Panama, Peru, Portugal, Romania, Russia, Saint Kitts and Nevis, Saint Lucia, Saint Vincent and the Grenadines, San Marino, Senegal, Slovakia, Slovenia, Solomon Islands, South Africa, Spain, Suriname, Sweden, Switzerland, Tanzania, Togo, Tuvalu, UK, US, Uruguay

International Tropical Timber Agreement, 1983

note–abbreviated as Tropical Timber 83
opened for signature–18 November 1983
entered into force–1 April 1985; this agreement expired when the International Tropical Timber Agreement, 1994, went into force
objective–to provide an effective framework for cooperation between tropical timber producers and consumers and to encourage the development of national policies aimed at sustainable utilization and conservation of tropical forests and their genetic resources
parties–(59) Australia, Austria, Belgium, Bolivia, Brazil, Burma, Cambodia, Cameroon, Canada, Central African Republic, China, Colombia, Democratic Republic of the Congo, Republic of the Congo, Cote d'Ivoire, Denmark, Ecuador, Egypt, EU, Fiji, Finland, France, Gabon, Germany, Ghana, Greece, Guatemala, Guyana, Honduras, India, Indonesia, Ireland, Italy, Japan, South Korea, Liberia, Luxembourg, Malaysia, Mexico, Nepal, Netherlands, NZ, Nigeria, Norway, Panama, Papua New Guinea, Peru, Philippines, Portugal, Russia, Spain, Suriname, Sweden, Switzerland, Thailand, Togo, Trinidad and Tobago, UK, US, Vanuatu, Venezuela

International Tropical Timber Agreement, 1994

note–abbreviated as Tropical Timber 94
opened for signature–26 January 1994
entered into force–1 January 1997
objective–to ensure that by the year 2000 exports of tropical timber originate from sustainably managed sources; to establish a fund to assist tropical timber producers in obtaining the resources necessary to reach this objective
parties–(61) Australia, Austria, Belgium, Bolivia, Brazil, Burma, Cambodia, Cameroon, Canada, Central African Republic, China, Colombia, Democratic Republic of the Congo, Republic of the Congo, Cote d'Ivoire, Denmark, Ecuador, Egypt, EU, Fiji, Finland, France, Gabon, Germany, Ghana, Greece, Guatemala, Guyana, Honduras, India, Indonesia, Ireland, Italy, Japan, South Korea, Liberia, Luxembourg, Malaysia, Mexico, Nepal, Netherlands, NZ, Nigeria, Norway, Panama, Papua New Guinea, Peru, Philippines, Poland, Portugal, Spain, Suriname, Sweden, Switzerland, Thailand, Togo, Trinidad and Tobago, UK, US, Vanuatu, Venezuela

Kyoto Protocol to the United Nations Framework Convention on Climate Change

note–abbreviated as Climate Change-Kyoto Protocol
opened for signature–16 March 1998
entered into force–23 February 2005
objective–to further reduce greenhouse gas emissions by enhancing the national programs of developed countries aimed at this goal and by establishing percentage reduction targets for the developed countries
parties–(184) Albania, Algeria, Angola, Antigua and Barbuda, Argentina, Armenia, Australia, Austria, Azerbaijan, The Bahamas, Bahrain, Bangladesh, Barbados, Belarus, Belgium, Belize, Benin, Bhutan, Bolivia, Bosnia and Herzegovina, Botswana, Brazil, Bulgaria, Burkina Faso, Burma, Burundi, Cambodia, Cameroon, Canada, Cape Verde, Central African Republic, Chile, China, Colombia, Comoros, Democratic Republic of the Congo, Republic of the Congo, Cook Island, Costa Rica, Cote d'Ivoire, Croatia, Cuba, Cyprus, Czech Republic, Denmark, Djibouti, Dominica, Dominican Republic, Ecuador, Egypt, El Salvador, Equatorial Guinea, Eritrea, Estonia, Ethiopia, EU, Fiji, Finland, France, Gabon, The Gambia, Georgia, Germany, Ghana, Greece, Grenada, Guatemala, Guinea, Guinea-Bissau, Guyana, Haiti, Honduras, Hungary, Iceland, India, Indonesia, Iran, Ireland, Israel, Italy, Jamaica, Japan, Jordan, Kenya, Kiribati, North Korea, South Korea, Kuwait, Kyrgyzstan, Laos, Latvia, Lebanon, Lesotho, Liberia, Libya, Liechtenstein, Lithuania, Luxembourg, Macedonia, Madagascar, Malawi, Malaysia, Maldives, Mali, Malta, Marshall Islands, Mauritania, Mauritius, Mexico, Federated States of Micronesia, Moldova, Monaco, Mongolia, Montenegro, Morocco, Mozambique, Namibia, Nauru, Nepal, Netherlands, NZ, Nicaragua, Niger, Nigeria, Niue, Norway, Oman, Pakistan, Palau, Panama, Papua New Guinea, Paraguay, Peru, Philippines, Poland, Portugal, Qatar, Romania, Russia, Rwanda, Saint Kitts and Nevis, Saint Lucia, Saint Vincent and the Grenadines, Samoa, Sao Tome and Principe, Saudi Arabia, Senegal, Serbia, Seychelles, Sierra Leone, Singapore, Slovakia, Slovenia, Solomon Islands, South Africa, Spain, Sri Lanka, Sudan, Suriname, Swaziland, Sweden, Switzerland, Syria, Tajikistan, Tanzania, Thailand, Timor-Leste, Togo, Tonga, Trinidad and Tobago, Tunisia, Turkmenistan, Tuvalu, Uganda, Ukraine, UAE, UK, Uruguay, Uzbekistan, Vanuatu, Venezuela, Vietnam, Yemen, Zambia
countries that have signed, but not yet ratified–(2) Kazakhstan, US

Law of the Sea

see United Nations Convention on the Law of the Sea (LOS)

Marine Dumping

see Convention on the Prevention of Marine Pollution by Dumping Wastes and Other Matter (London Convention)

Marine Life Conservation

see Convention on Fishing and Conservation of Living Resources of the High Seas

Montreal Protocol on Substances That Deplete the Ozone Layer

note–abbreviated as Ozone Layer Protection
opened for signature–16 September 1987
entered into force–1 January 1989
objective–to protect the ozone layer by controlling emissions of substances that deplete it
parties–(194) Afghanistan, Albania, Algeria, Andorra, Angola, Antigua and Barbuda, Argentina, Armenia, Australia, Austria, Azerbaijan, The Bahamas, Bahrain, Bangladesh, Barbados, Belarus, Belgium, Belize, Benin, Bhutan, Bolivia, Bosnia and Herzegovina, Botswana, Brazil, Brunei, Bulgaria, Burkina Faso, Burma, Burundi, Cambodia, Cameroon, Canada, Cape Verde, Central African Republic, Chad, Chile, China, Colombia, Comoros, Democratic Republic of the Congo, Republic of the Congo, Cook Islands, Costa Rica, Cote d'Ivoire, Croatia, Cuba, Cyprus, Czech Republic, Denmark, Djibouti, Dominica, Dominican Republic, Ecuador, Egypt, El Salvador,

Equatorial Guinea, Eritrea, Estonia, Ethiopia, EU, Fiji, Finland, France, Gabon, The Gambia, Georgia, Germany, Ghana, Greece, Grenada, Guatemala, Guinea, Guinea-Bissau, Guyana, Haiti, Holy See, Honduras, Hungary, Iceland, India, Indonesia, Iran, Iraq, Ireland, Israel, Italy, Jamaica, Japan, Jordan, Kazakhstan, Kenya, Kiribati, North Korea, South Korea, Kuwait, Kyrgyzstan, Laos, Latvia, Lebanon, Lesotho, Liberia, Libya, Liechtenstein, Lithuania, Luxembourg, Macedonia, Madagascar, Malawi, Malaysia, Maldives, Mali, Malta, Marshall Islands, Mauritania, Mauritius, Mexico, Federated States of Micronesia, Moldova, Monaco, Mongolia, Montenegro, Morocco, Mozambique, Namibia, Nauru, Nepal, Netherlands, NZ, Nicaragua, Niger, Nigeria, Niue, Norway, Oman, Pakistan, Palau, Panama, Papua New Guinea, Paraguay, Peru, Philippines, Poland, Portugal, Qatar, Romania, Russia, Rwanda, Saint Kitts and Nevis, Saint Lucia, Saint Vincent and the Grenadines, Samoa, Sao Tome and Principe, Saudi Arabia, Senegal, Serbia, Seychelles, Sierra Leone, Singapore, Slovakia, Slovenia, Solomon Islands, Somalia, South Africa, Spain, Sri Lanka, Sudan, Suriname, Swaziland, Sweden, Switzerland, Syria, Tajikistan, Tanzania, Thailand, Togo, Tonga, Trinidad and Tobago, Tunisia, Turkey, Turkmenistan, Tuvalu, Uganda, Ukraine, UAE, UK, US, Uruguay, Uzbekistan, Vanuatu, Venezuela, Vietnam, Yemen, Zambia, Zimbabwe

Nuclear Test Ban
see Treaty Banning Nuclear Weapons Tests in the Atmosphere, in Outer Space, and Under Water

Ozone Layer Protection
see Montreal Protocol on Substances That Deplete the Ozone Layer

Protocol of 1978 Relating to the International Convention for the Prevention of Pollution From Ships, 1973 (MARPOL)
note–abbreviated as Ship Pollution
opened for signature–17 February 1978
entered into force–2 October 1983
objective–to preserve the marine environment through the complete elimination of pollution by oil and other harmful substances and the minimization of accidental discharge of such substances
parties–(139) Algeria, Angola, Antigua and Barbuda, Argentina, Australia, Austria, Azerbaijan, The Bahamas, Bangladesh, Barbados, Belarus, Belgium, Belize, Benin, Bolivia, Brazil, Brunei, Bulgaria, Burma, Cambodia, Canada, Cape Verde, Chile, China, Colombia, Comoros, Republic of Congo, Cote d'Ivoire, Croatia, Cuba, Cyprus, Czech Republic, Denmark, Djibouti, Dominica, Dominican Republic, Ecuador, Egypt, Equatorial Guinea, Estonia, Faroe Islands, Finland, France, Gabon, The Gambia, Georgia, Germany, Ghana, Greece, Guatemala, Guinea, Guyana, Honduras, Hong Kong, Hungary, Iceland, India, Indonesia, Iran, Ireland, Israel, Italy, Jamaica, Japan, Kazakhstan, Kenya, North Korea, South Korea, Latvia, Lebanon, Liberia, Lithuania, Luxembourg, Libya, Macau, Madagascar, Malawi, Malaysia, Maldives, Malta, Marshall Islands, Mauritania, Mauritius, Mexico, Moldova, Monaco, Mongolia, Montenegro, Morocco, Mozambique, Netherlands, NZ, Nicaragua, Nigeria, Norway, Oman, Pakistan, Panama, Papua New Guinea, Peru, Philippines, Poland, Portugal, Qatar Romania, Russia, Saint Kitts and Nevis, Saint Lucia, Saint Vincent and the Grenadines, Samoa, Sao Tome and Principe, Saudi Arabia, Senegal, Serbia, Seychelles, Sierra Leone, Singapore, Slovakia, Slovenia, South Africa, Spain, Sri Lanka, Suriname, Sweden, Switzerland, Syria, Togo, Tonga, Trinidad and Tobago, Tunisia, Turkey, Tuvalu, Ukraine, UK, US, Uruguay, Vanuatu, Venezuela, Vietnam

Protocol on Environmental Protection to the Antarctic Treaty
note–abbreviated as Antarctic-Environmental Protocol
opened for signature–4 October 1991
entered into force–14 January 1998
objective–to provide for comprehensive protection of the Antarctic environment and dependent and associated ecosystems; applies to the area covered by the Antarctic Treaty
consultative parties–(31) Argentina, Australia, Belgium, Brazil, Bulgaria, Canada, Chile, China, Czech Republic, Ecuador, Finland, France, Germany, India, Italy, Japan, South Korea, Netherlands, NZ, Norway, Peru, Poland, Romania, Russia, South Africa, Spain, Sweden, Ukraine, UK, US, Uruguay
non consultative parties–(12) Austria, Colombia, Cuba, Denmark, Greece, Guatemala, Hungary, North Korea, Papua New Guinea, Slovakia, Switzerland, Turkey

Protocol to the 1979 Convention on Long-Range Transboundary Air Pollution Concerning the Control of Emissions of Nitrogen Oxides or Their Transboundary Fluxes
note–abbreviated as Air Pollution-Nitrogen Oxides
opened for signature–31 October 1988
entered into force–14 February 1991
objective–to provide for the control or reduction of nitrogen oxides and their transboundary fluxes
parties–(32) Austria, Belarus, Belgium, Bulgaria, Canada, Croatia, Cyprus, Czech Republic, Denmark, Estonia, EU, Finland, France, Germany, Greece, Hungary, Ireland, Italy, Liechtenstein, Lithuania, Luxembourg, Netherlands, Norway, Russia, Slovakia, Slovenia, Spain, Sweden, Switzerland, Ukraine, UK, US
countries that have signed, but not yet ratified–(1) Poland

Protocol to the 1979 Convention on Long-Range Transboundary Air Pollution Concerning the Control of Emissions of Volatile Organic Compounds or Their Transboundary Fluxes
note–abbreviated as Air Pollution-Volatile Organic Compounds
opened for signature–18 November 1991
entered into force–29 September 1997
objective–to provide for the control and reduction of emissions of volatile organic compounds in order to reduce their transboundary fluxes so as to protect human health and the environment from adverse effects
parties–(23) Austria, Belgium, Bulgaria, Croatia, Czech Republic, Denmark, Estonia, Finland, France, Germany, Hungary, Italy, Liechtenstein, Lithuania, Luxembourg, Monaco, Netherlands, Norway, Slovakia, Spain, Sweden, Switzerland, UK
countries that have signed, but not yet ratified–(6) Canada, EU, Greece, Portugal, Ukraine, US

Protocol to the 1979 Convention on Long-Range Transboundary Air Pollution on Further Reduction of Sulphur Emissions
note–abbreviated as Air Pollution-Sulphur 94
opened for signature–14 June 1994
entered into force–5 August 1998

objective–to provide for a further reduction in sulfur emissions or transboundary fluxes
parties–(28) Austria, Belgium, Bulgaria, Canada, Croatia, Cyprus, Czech Republic, Denmark, EU, Finland, France, Germany, Greece, Hungary, Ireland, Italy, Liechtenstein, Lithuania, Luxembourg, Monaco, Netherlands, Norway, Slovakia, Slovenia, Spain, Sweden, Switzerland, UK
countries that have signed, but not yet ratified–(3) Poland, Russia, Ukraine

Protocol to the 1979 Convention on Long-Range Transboundary Air Pollution on Persistent Organic Pollutants

note–abbreviated as Air Pollution-Persistent Organic Pollutants
opened for signature–24 June 1998
entered into force–23 October 2003
objective–to provide for the control and reduction of emissions of persistent organic pollutants in order to reduce their transboundary fluxes so as to protect human health and the environment from adverse effects
parties–(29) Austria, Belgium, Bulgaria, Canada, Croatia, Cyprus, Czech Republic, Denmark, Estonia, EU, Finland, France, Germany, Hungary, Iceland, Italy, Latvia, Liechtenstein, Lithuania, Luxembourg, Moldova, Netherlands, Norway, Romania, Slovakia, Slovenia, Sweden, Switzerland, UK
countries that have signed, but not yet ratified–(8) Armenia, Greece, Ireland, Poland, Portugal, Spain, Ukraine, US

Protocol to the 1979 Convention on Long-Range Transboundary Air Pollution on the Reduction of Sulphur Emissions or Their Transboundary Fluxes by at Least 30%

note–abbreviated as Air Pollution-Sulphur 85
opened for signature–8 July 1985
entered into force–2 September 1987
objective–to provide for a 30% reduction in sulfur emissions or transboundary fluxes by 1993
parties–(23) Austria, Belarus, Belgium, Bulgaria, Canada, Czech Republic, Denmark, Estonia, Finland, France, Germany, Hungary, Italy, Liechtenstein, Lithuania, Luxembourg, Netherlands, Norway, Russia, Slovakia, Sweden, Switzerland, Ukraine

Ship Pollution

see Protocol of 1978 Relating to the International Convention for the Prevention of Pollution From Ships, 1973 (MARPOL)

Treaty Banning Nuclear Weapon Tests in the Atmosphere, in Outer Space, and Under Water

note–abbreviated as Nuclear Test Ban
opened for signature–5 August 1963
entered into force–10 October 1963
objective–to obtain an agreement on general and complete disarmament under strict international control in accordance with the objectives of the United Nations; to put an end to the armaments race and eliminate incentives for the production and testing of all kinds of weapons, including nuclear weapons
parties–(113) Afghanistan, Antigua and Barbuda, Argentina, Armenia, Australia, Austria, The Bahamas, Bangladesh, Belgium, Benin, Bhutan, Bolivia, Bosnia and Herzegovina, Botswana, Brazil, Bulgaria, Burma, Canada, Central African Republic, Chad, China, Colombia, Democratic Republic of the Congo, Costa Rica, Cote d'Ivoire, Croatia, Cyprus, Czech Republic, Denmark, Dominican Republic, Ecuador, Egypt, El Salvador, Fiji, Finland, Gabon, The Gambia, Germany, Ghana, Greece, Guatemala, Honduras, Hungary, Iceland, India, Indonesia, Iran, Iraq, Ireland, Israel, Italy, Jamaica, Japan, Jordan, Kenya, South Korea, Kuwait, Laos, Lebanon, Liberia, Luxembourg, Madagascar, Malawi, Malaysia, Malta, Mauritania, Mauritius, Mexico, Morocco, Nepal, Netherlands, NZ, Nicaragua, Niger, Nigeria, Norway, Panama, Papua New Guinea, Peru, Philippines, Poland, Romania, Russia, Rwanda, Samoa, San Marino, Senegal, Serbia, Seychelles, Sierra Leone, Singapore, Slovakia, Slovenia, South Africa, Spain, Sri Lanka, Sudan, Suriname, Swaziland, Sweden, Switzerland, Syria, Thailand, Togo, Tonga, Trinidad and Tobago, Tunisia, Turkey, Uganda, UK, US, Venezuela, Zambia
countries that have signed, but not yet ratified–(17) Algeria, Burkina Faso, Burundi, Cameroon, Chile, Ethiopia, Haiti, Libya, Mali, Pakistan, Paraguay, Portugal, Somalia, Tanzania, Uruguay, Vietnam, Yemen

Tropical Timber 83

see International Tropical Timber Agreement, 1983

Tropical Timber 94

see International Tropical Timber Agreement, 1994

United Nations Convention on the Law of the Sea (LOS)

note–abbreviated as Law of the Sea
opened for signature–10 December 1982
entered into force–16 November 1994
objective–to set up a comprehensive new legal regime for the sea and oceans; to include rules concerning environmental standards as well as enforcement provisions dealing with pollution of the marine environment
parties–(157) Albania, Algeria, Angola, Antigua and Barbuda, Argentina, Armenia, Australia, Austria, The Bahamas, Bahrain, Bangladesh, Barbados, Belarus, Belgium, Belize, Benin, Bolivia, Bosnia and Herzegovina, Botswana, Brazil, Brunei, Bulgaria, Burkina Faso, Burma, Cameroon, Canada, Cape Verde, Chile, China, Comoros, Democratic Republic of the Congo, Republic of the Congo, Cook Islands, Costa Rica, Cote d'Ivoire, Croatia, Cuba, Cyprus, Czech Republic, Denmark, Djibouti, Dominica, Egypt, Equatorial Guinea, Estonia, EU, Fiji, Finland, France, Gabon, The Gambia, Georgia, Germany, Ghana, Greece, Grenada, Guatemala, Guinea, Guinea-Bissau, Guyana, Haiti, Honduras, Hungary, Iceland, India, Indonesia, Iraq, Ireland, Italy, Jamaica, Japan, Jordan, Kenya, Kiribati, South Korea, Kuwait, Laos, Latvia, Lebanon, Lesotho, Liberia, Lithuania, Luxembourg, Macedonia, Madagascar, Malaysia, Maldives, Mali, Malta, Marshall Islands, Mauritania, Mauritius, Mexico, Federated States of Micronesia, Moldova, Monaco, Mongolia, Montenegro, Morocco, Mozambique, Namibia, Nauru, Nepal, Netherlands, NZ, Nicaragua, Nigeria, Niue, Norway, Oman, Pakistan, Palau, Panama, Papua New Guinea, Paraguay, Philippines, Poland, Portugal, Qatar, Romania, Russia, Saint Kitts and Nevis, Saint Lucia, Saint Vincent and the Grenadines, Samoa, Sao Tome and Principe, Saudi Arabia, Senegal, Serbia, Seychelles, Sierra Leone, Singapore, Slovakia, Slovenia, Solomon Islands, Somalia, South Africa, Spain, Sri Lanka, Sudan, Suriname, Sweden, Tanzania, Togo, Tonga, Trinidad and Tobago, Tunisia, Tuvalu, Uganda, Ukraine, UK, Uruguay, Vanuatu, Vietnam, Yemen, Zambia, Zimbabwe

countries that have signed, but not yet ratified–(21) Afghanistan, Bhutan, Burundi, Cambodia, Central African Republic, Chad, Colombia, Dominican Republic, El Salvador, Ethiopia, Iran, North Korea, Libya, Liechtenstein, Malawi, Niger, Rwanda, Swaziland, Switzerland, Thailand, UAE

United Nations Convention to Combat Desertification in Those Countries Experiencing Serious Drought and/or Desertification, Particularly in Africa

note–abbreviated as Desertification
opened for signature–14 October 1994
entered into force–26 December 1996
objective–to combat desertification and mitigate the effects of drought through national action programs that incorporate long-term strategies supported by international cooperation and partnership arrangements
parties–(193) Afghanistan, Albania, Algeria, Andorra, Angola, Antigua and Barbuda, Argentina, Armenia, Australia, Austria, Azerbaijan, The Bahamas, Bahrain, Bangladesh, Barbados, Belarus, Belgium, Belize, Benin, Bhutan, Bolivia, Bosnia and Herzegovina, Botswana, Brazil, Brunei, Bulgaria, Burkina Faso, Burma, Burundi, Cambodia, Cameroon, Canada, Cape Verde, Central African Republic, Chad, Chile, China, Colombia, Comoros, Democratic Republic of the Congo, Republic of the Congo, Cook Islands, Costa Rica, Cote d'Ivoire, Croatia, Cuba, Cyprus, Czech Republic, Denmark, Djibouti, Dominica, Dominican Republic, Ecuador, Egypt, El Salvador, Equatorial Guinea, Eritrea, Ethiopia, EU, Fiji, Finland, France, Gabon, The Gambia, Georgia, Germany, Ghana, Greece, Grenada, Guatemala, Guinea, Guinea-Bissau, Guyana, Haiti, Honduras, Hungary, Iceland, India, Indonesia, Iran, Ireland, Israel, Italy, Jamaica, Japan, Jordan, Kazakhstan, Kenya, Kiribati, North Korea, South Korea, Kuwait, Kyrgyzstan, Laos, Latvia, Lebanon, Lesotho, Liberia, Libya, Liechtenstein, Lithuania, Luxembourg, Macedonia, Madagascar, Malawi, Malaysia, Maldives, Mali, Malta, Marshall Islands, Mauritania, Mauritius, Mexico, Federated States of Micronesia, Moldova, Monaco, Mongolia, Montenegro, Morocco, Mozambique, Namibia, Nauru, Nepal, Netherlands, NZ, Nicaragua, Niger, Nigeria, Niue, Norway, Oman, Pakistan, Palau, Panama, Papua New Guinea, Paraguay, Peru, Philippines, Poland, Portugal, Qatar, Romania, Russia, Rwanda, Saint Kitts and Nevis, Saint Lucia, Saint Vincent and the Grenadines, Samoa, San Marino, Sao Tome and Principe, Saudi Arabia, Senegal, Serbia, Seychelles, Sierra Leone, Singapore, Slovakia, Slovenia, Solomon Islands, Somalia, South Africa, Spain, Sri Lanka, Sudan, Suriname, Swaziland, Sweden, Switzerland, Syria, Tajikistan, Thailand, Tanzania, Timor-Leste, Togo, Tonga, Trinidad and Tobago, Tunisia, Turkey, Turkmenistan, Tuvalu, Uganda, Ukraine, UAE, UK, US, Uruguay, Uzbekistan, Vanuatu, Venezuela, Vietnam, Yemen, Zambia, Zimbabwe

United Nations Framework Convention on Climate Change

note–abbreviated as Climate Change
opened for signature–9 May 1992
entered into force–21 March 1994
objective–to achieve stabilization of greenhouse gas concentrations in the atmosphere at a low enough level to prevent dangerous anthropogenic interference with the climate system
parties–(192) Afghanistan, Albania, Algeria, Angola, Antigua and Barbuda, Argentina, Armenia, Australia, Austria, Azerbaijan, The Bahamas, Bahrain, Bangladesh, Barbados, Belarus, Belgium, Belize, Benin, Bhutan, Bolivia, Bosnia and Herzegovina, Botswana, Brazil, Brunei, Bulgaria, Burkina Faso, Burma, Burundi, Cambodia, Cameroon, Canada, Cape Verde, Central African Republic, Chad, Chile, China, Colombia, Comoros, Democratic Republic of the Congo, Republic of the Congo, Cook Islands, Costa Rica, Cote d'Ivoire, Croatia, Cuba, Cyprus, Czech Republic, Denmark, Djibouti, Dominica, Dominican Republic, Ecuador, Egypt, El Salvador, Equatorial Guinea, Eritrea, Estonia, Ethiopia, EU, Fiji, Finland, France, Gabon, The Gambia, Georgia, Germany, Ghana, Greece, Grenada, Guatemala, Guinea, Guinea-Bissau, Guyana, Haiti, Honduras, Hungary, Iceland, India, Indonesia, Iran, Ireland, Israel, Italy, Jamaica, Japan, Jordan, Kazakhstan, Kenya, Kiribati, North Korea, South Korea, Kuwait, Kyrgyzstan, Laos, Latvia, Lebanon, Lesotho, Liberia, Libya, Liechtenstein, Lithuania, Luxembourg, Macedonia, Madagascar, Malawi, Malaysia, Maldives, Mali, Malta, Marshall Islands, Mauritania, Mauritius, Mexico, Federated States of Micronesia, Moldova, Monaco, Mongolia, Montenegro, Morocco, Mozambique, Namibia, Nauru, Nepal, Netherlands, NZ, Nicaragua, Niger, Nigeria, Niue, Norway, Oman, Pakistan, Palau, Panama, Papua New Guinea, Paraguay, Peru, Philippines, Poland, Portugal, Qatar, Romania, Russia, Rwanda, Saint Kitts and Nevis, Saint Lucia, Saint Vincent and the Grenadines, Samoa, San Marino, Sao Tome and Principe, Saudi Arabia, Senegal, Serbia, Seychelles, Sierra Leone, Singapore, Slovakia, Slovenia, Solomon Islands, South Africa, Spain, Sri Lanka, Sudan, Suriname, Swaziland, Sweden, Switzerland, Syria, Tajikistan, Tanzania, Thailand, Timor-Leste, Togo, Tonga, Trinidad and Tobago, Tunisia, Turkey, Turkmenistan, Tuvalu, Uganda, Ukraine, UAE, UK, US, Uruguay, Uzbekistan, Vanuatu, Venezuela, Vietnam, Yemen, Zambia, Zimbabwe

Wetlands

see Convention on Wetlands of International Importance Especially As Waterfowl Habitat (Ramsar)

Whaling

see International Convention for the Regulation of Whaling

APPENDIX D

CROSS-REFERENCE LIST OF COUNTRY DATA CODES

FIPS 10: Countries, Dependencies, Areas of Special Sovereignty, and Their Principal Administrative Divisions (FIPS 10) is maintained by the Office of Targeting and Transnational Issues, National Geospatial-Intelligence Agency, and published by the National Institute of Standards and Technology (Department of Commerce). FIPS 10 codes are intended for general use throughout the US Government, especially in activities associated with the mission of the Department of State and national defense programs.

ISO 3166: Codes for the Representation of Names of Countries (ISO 3166) is prepared by the International Organization for Standardization. ISO 3166 includes two- and three-character alphabetic codes and three-digit numeric codes that may be needed for activities involving exchange of data with international organizations that have adopted that standard. Except for the numeric codes, ISO 3166 codes have been adopted in the US as FIPS 104-1: American National Standard Codes for the Representation of Names of Countries, Dependencies, and Areas of Special Sovereignty for Information Interchange.

STANAG 1059: Letter Codes for Geographical Entities (8th edition, 2004) is a Standardization Agreement (STANAG) established and maintained by the North Atlantic Treaty Organization (NATO/OTAN) for the purpose of providing a common set of geo-spatial identifiers for countries, territories, and possessions. The 8th edition established trigraph codes for each country based upon the ISO 3166-1 alpha-3 character sets. These codes are used throughout NATO.

Internet: The Internet country code is the two-letter digraph maintained by the International Organization for Standardization (ISO) in the ISO 3166 Alpha-2 list and used by the Internet Assigned Numbers Authority (IANA) to establish country-coded top-level domains (ccTLDs).

ENTITY	FIPS 10	ISO 3166			STANAG	INTERNET	COMMENT
Afghanistan	AF	AF	AFG	004	AFG	.af	
Albania	AL	AL	ALB	008	ALB	.al	
Algeria	AG	DZ	DZA	012	DZA	.dz	
American Samoa	AQ	AS	ASM	016	ASM	.as	
Andorra	AN	AD	AND	020	AND	.ad	
Angola	AO	AO	AGO	024	AGO	.ao	
Anguilla	AV	AI	AIA	660	AIA	.ai	
Antarctica	AY	AQ	ATA	010	ATA	.aq	ISO defines as the territory south of 60 degrees south latitude
Antigua and Barbuda	AC	AG	ATG	028	ATG	.ag	
Argentina	AR	AR	ARG	032	ARG	.ar	
Armenia	AM	AM	ARM	051	ARM	.am	
Aruba	AA	AW	ABW	533	ABW	.aw	
Ashmore and Cartier Islands	AT	-	-	-	AUS	-	ISO includes with Australia
Australia	AS	AU	AUS	036	AUS	.au	ISO includes Ashmore and Cartier Islands, Coral Sea Islands
Austria	AU	AT	AUT	040	AUT	.at	
Azerbaijan	AJ	AZ	AZE	031	AZE	.az	
Bahamas, The	BF	BS	BHS	044	BHS	.bs	
Bahrain	BA	BH	BHR	048	BHR	.bh	
Baker Island	FQ	-	-	-	UMI	-	ISO includes with the US Minor Outlying Islands
Bangladesh	BG	BD	BGD	050	BGD	.bd	
Barbados	BB	BB	BRB	052	BRB	.bb	
Bassas da India	BS	-	-	-	-	-	administered as part of French Southern and Antarctic Lands; no ISO codes assigned
Belarus	BO	BY	BLR	112	BLR	.by	
Belgium	BE	BE	BEL	056	BEL	.be	
Belize	BH	BZ	BLZ	084	BLZ	.bz	
Benin	BN	BJ	BEN	204	BEN	.bj	
Bermuda	BD	BM	BMU	060	BMU	.bm	
Bhutan	BT	BT	BTN	064	BTN	.bt	
Bolivia	BL	BO	BOL	068	BOL	.bo	
Bosnia and Herzegovina	BK	BA	BIH	070	BIH	.ba	
Botswana	BC	BW	BWA	072	BWA	.bw	
Bouvet Island	BV	BV	BVT	074	BVT	.bv	
Brazil	BR	BR	BRA	076	BRA	.br	
British Indian Ocean Territory	IO	IO	IOT	086	IOT	.io	
British Virgin Islands	VI	VG	VGB	092	VGB	.vg	

ENTITY	FIPS 10	ISO 3166			STANAG	INTERNET	COMMENT
Brunei	BX	BN	BRN	096	BRN	.bn	
Bulgaria	BU	BG	BGR	100	BGR	.bg	
Burkina Faso	UV	BF	BFA	854	BFA	.bf	
Burma	BM	MM	MMR	104	MMR	.mm	ISO uses the name Myanmar
Burundi	BY	BI	BDI	108	BDI	.bi	
Cambodia	CB	KH	KHM	116	KHM	.kh	
Cameroon	CM	CM	CMR	120	CMR	.cm	
Canada	CA	CA	CAN	124	CAN	.ca	
Cape Verde	CV	CV	CPV	132	CPV	.cv	
Cayman Islands	CJ	KY	CYM	136	CYM	.ky	
Central African Republic	CT	CF	CAF	140	CAF	.cf	
Chad	CD	TD	TCD	148	TCD	.td	
Chile	CI	CL	CHL	152	CHL	.cl	
China	CH	CN	CHN	156	CHN	.cn	see also Taiwan
Christmas Island	KT	CX	CXR	162	CXR	.cx	
Clipperton Island	IP	-	-	-	FYP	-	ISO includes with French Polynesia
Cocos (Keeling) Islands	CK	CC	CCK	166	AUS	.cc	
Colombia	CO	CO	COL	170	COL	.co	
Comoros	CN	KM	COM	174	COM	.km	
Congo, Democratic Republic of the	CG	CD	COD	180	COD	.cd	formerly Zaire
Congo, Republic of the	CF	CG	COG	178	COG	.cg	
Cook Islands	CW	CK	COK	184	COK	.ck	
Coral Sea Islands	CR	-	-	-	AUS	-	ISO includes with Australia
Costa Rica	CS	CR	CRI	188	CRI	.cr	
Cote d'Ivoire	IV	CI	CIV	384	CIV	.ci	
Croatia	HR	HR	HRV	191	HRV	.hr	
Cuba	CU	CU	CUB	192	CUB	.cu	
Curacao	UC	CW	CUW	531	-	.cw	
Cyprus	CY	CY	CYP	196	CYP	.cy	
Czech Republic	EZ	CZ	CZE	203	CZE	.cz	
Denmark	DA	DK	DNK	208	DNK	.dk	
Djibouti	DJ	DJ	DJI	262	DJI	.dj	
Dominica	DO	DM	DMA	212	DMA	.dm	
Dominican Republic	DR	DO	DOM	214	DOM	.do	
Ecuador	EC	EC	ECU	218	ECU	.ec	
Egypt	EG	EG	EGY	818	EGY	.eg	
El Salvador	ES	SV	SLV	222	SLV	.sv	
Equatorial Guinea	EK	GQ	GNQ	226	GNQ	.gq	
Eritrea	ER	ER	ERI	232	ERI	.er	
Estonia	EN	EE	EST	233	EST	.ee	
Ethiopia	ET	ET	ETH	231	ETH	.et	
Europa Island	EU	-	-	-	-	-	administered as part of French Southern and Antarctic Lands; no ISO codes assigned
Falkland Islands (Islas Malvinas)	FK	FK	FLK	238	FLK	.fk	
Faroe Islands	FO	FO	FRO	234	FRO	.fo	
Fiji	FJ	FJ	FJI	242	FJI	.fj	
Finland	FI	FI	FIN	246	FIN	.fi	
France	FR	FR	FRA	250	FRA	.fr	ISO includes metropolitan France along with the dependencies of Clipperton Island, French Guiana, French Polynesia, French Southern and Antarctic Lands, Guadeloupe, Martinique, Mayotte, New Caledonia, Reunion, Saint Pierre and Miquelon, Wallis and Futuna
France, Metropolitan	-	FX	FXX	249	-	.fx	ISO limits to the European part of France
French Guiana	FG	GF	GUF	254	GUF	.gf	

ENTITY	FIPS 10	ISO 3166			STANAG	INTERNET	COMMENT
French Polynesia	FP	PF	PYF	258	PYF	.pf	
French Southern and Antarctic Lands	FS	TF	ATF	260	ATF	.tf	FIPS 10-4 does not include the French-claimed portion of Antarctica (Terre Adelie)
Gabon	GB	GA	GAB	266	GAB	.ga	
Gambia, The	GA	GM	GMB	270	GMB	.gm	
Gaza Strip	GZ	PS	PSE	275	PSE	.ps	ISO identifies as Occupied Palestinian Territory
Georgia	GG	GE	GEO	268	GEO	.ge	
Germany	GM	DE	DEU	276	DEU	.de	
Ghana	GH	GH	GHA	288	GHA	.gh	
Gibraltar	GI	GI	GIB	292	GIB	.gi	
Glorioso Islands	GO	-	-	-	-	-	administered as part of French Southern and Antarctic Lands; no ISO codes assigned
Greece	GR	GR	GRC	300	GRC	.gr	For its internal communications, the European Union recommends the use of the code EL in lieu of the ISO 3166-2 code of GR
Greenland	GL	GL	GRL	304	GRL	.gl	
Grenada	GJ	GD	GRD	308	GRD	.gd	
Guadeloupe	GP	GP	GLP	312	GLP	.gp	
Guam	GQ	GU	GUM	316	GUM	.gu	
Guatemala	GT	GT	GTM	320	GTM	.gt	
Guernsey	GK	GG	GGY	831	UK	.gg	
Guinea	GV	GN	GIN	324	GIN	.gn	
Guinea-Bissau	PU	GW	GNB	624	GNB	.gw	
Guyana	GY	GY	GUY	328	GUY	.gy	
Haiti	HA	HT	HTI	332	HTI	.ht	
Heard Island and McDonald Islands	HM	HM	HMD	334	HMD	.hm	
Holy See (Vatican City)	VT	VA	VAT	336	VAT	.va	
Honduras	HO	HN	HND	340	HND	.hn	
Hong Kong	HK	HK	HKG	344	HKG	.hk	
Howland Island	HQ	-	-	-	UMI	-	ISO includes with the US Minor Outlying Islands
Hungary	HU	HU	HUN	348	HUN	.hu	
Iceland	IC	IS	ISL	352	ISL	.is	
India	IN	IN	IND	356	IND	.in	
Indonesia	ID	ID	IDN	360	IDN	.id	
Iran	IR	IR	IRN	364	IRN	.ir	
Iraq	IZ	IQ	IRQ	368	IRQ	.iq	
Ireland	EI	IE	IRL	372	IRL	.ie	
Isle of Man	IM	IM	IMN	833	UK	.im	
Israel	IS	IL	ISR	376	ISR	.il	
Italy	IT	IT	ITA	380	ITA	.it	
Jamaica	JM	JM	JAM	388	JAM	.jm	
Jan Mayen	JN	-	-	-	SJM	-	ISO includes with Svalbard
Japan	JA	JP	JPN	392	JPN	.jp	
Jarvis Island	DQ	-	-	-	UMI	-	ISO includes with the US Minor Outlying Islands
Jersey	JE	JE	JEY	832	UK	.je	
Johnston Atoll	JQ	-	-	-	UMI	-	ISO includes with the US Minor Outlying Islands
Jordan	JO	JO	JOR	400	JOR	.jo	
Juan de Nova Island	JU	-	-	-	-	-	administered as part of French Southern and Antarctic Lands; no ISO codes assigned
Kazakhstan	KZ	KZ	KAZ	398	KAZ	.kz	
Kenya	KE	KE	KEN	404	KEN	.ke	

ENTITY	FIPS 10	ISO 3166			STANAG	INTERNET	COMMENT
Kingman Reef	KQ	-	-	-	UMI	-	ISO includes with the US Minor Outlying Islands
Kiribati	KR	KI	KIR	296	KIR	.ki	
Korea, North	KN	KP	PRK	408	PRK	.kp	
Korea, South	KS	KR	KOR	410	KOR	.kr	
Kosovo	KV	-	-	-	-	-	ISO codes have not been designated
Kuwait	KU	KW	KWT	414	KWT	.kw	
Kyrgyzstan	KG	KG	KGZ	417	KGZ	.kg	
Laos	LA	LA	LAO	418	LAO	.la	
Latvia	LG	LV	LVA	428	LVA	.lv	
Lebanon	LE	LB	LBN	422	LBN	.lb	
Lesotho	LT	LS	LSO	426	LSO	.ls	
Liberia	LI	LR	LBR	430	LBR	.lr	
Libya	LY	LY	LBY	434	LBY	.ly	
Liechtenstein	LS	LI	LIE	438	LIE	.li	
Lithuania	LH	LT	LTU	440	LTU	.lt	
Luxembourg	LU	LU	LUX	442	LUX	.lu	
Macau	MC	MO	MAC	446	MAC	.mo	
Macedonia	MK	MK	MKD	807	FYR	.mk	
Madagascar	MA	MG	MDG	450	MDG	.mg	
Malawi	MI	MW	MWI	454	MWI	.mw	
Malaysia	MY	MY	MYS	458	MYS	.my	
Maldives	MV	MV	MDV	462	MDV	.mv	
Mali	ML	ML	MLI	466	MLI	.ml	
Malta	MT	MT	MLT	470	MLT	.mt	
Marshall Islands	RM	MH	MHL	584	MHL	.mh	
Martinique	MB	MQ	MTQ	474	MTQ	.mq	
Mauritania	MR	MR	MRT	478	MRT	.mr	
Mauritius	MP	MU	MUS	480	MUS	.mu	
Mayotte	MF	YT	MYT	175	FRA	.yt	
Mexico	MX	MX	MEX	484	MEX	.mx	
Micronesia, Federated States of	FM	FM	FSM	583	FSM	.fm	
Midway Islands	MQ	-	-	-	UMI	-	ISO includes with the US Minor Outlying Islands
Moldova	MD	MD	MDA	498	MDA	.md	
Monaco	MN	MC	MCO	492	MCO	.mc	
Mongolia	MG	MN	MNG	496	MNG	.mn	
Montenegro	MJ	ME	MNE	499	MNE	.me	
Montserrat	MH	MS	MSR	500	MSR	.ms	
Morocco	MO	MA	MAR	504	MAR	.ma	
Mozambique	MZ	MZ	MOZ	508	MOZ	.mz	
Myanmar	-	-	-	-	-	-	see Burma
Namibia	WA	NA	NAM	516	NAM	.na	
Nauru	NR	NR	NRU	520	NRU	.nr	
Navassa Island	BQ	-	-	-	US	-	ISO includes with the US Minor Outlying Islands
Nepal	NP	NP	NPL	524	NPL	.np	
Netherlands	NL	NL	NLD	528	NLD	.nl	
Netherlands Antilles	NT	AN	ANT	530	ANT	.an	
New Caledonia	NC	NC	NCL	540	NCL	.nc	
New Zealand	NZ	NZ	NZL	554	NZL	.nz	
Nicaragua	NU	NI	NIC	558	NIC	.ni	
Niger	NG	NE	NER	562	NER	.ne	
Nigeria	NI	NG	NGA	566	NGA	.ng	
Niue	NE	NU	NIU	570	NIU	.nu	
Norfolk Island	NF	NF	NFK	574	NFK	.nf	
Northern Mariana Islands	CQ	MP	MNP	580	MNP	.mp	
Norway	NO	NO	NOR	578	NOR	.no	

ENTITY	FIPS 10	ISO 3166			STANAG	INTERNET	COMMENT
Oman	MU	OM	OMN	512	OMN	.om	
Pakistan	PK	PK	PAK	586	PAK	.pk	
Palau	PS	PW	PLW	585	PLW	.pw	
Palmyra Atoll	LQ	-	-	-	UMI	-	ISO includes with the US Minor Outlying Islands
Panama	PM	PA	PAN	591	PAN	.pa	
Papua New Guinea	PP	PG	PNG	598	PNG	.pg	
Paracel Islands	PF	-	-	-	-	-	
Paraguay	PA	PY	PRY	600	PRY	.py	
Peru	PE	PE	PER	604	PER	.pe	
Philippines	RP	PH	PHL	608	PHL	.ph	
Pitcairn Islands	PC	PN	PCN	612	PCN	.pn	
Poland	PL	PL	POL	616	POL	.pl	
Portugal	PO	PT	PRT	620	PRT	.pt	
Puerto Rico	RQ	PR	PRI	630	PRI	.pr	
Qatar	QA	QA	QAT	634	QAT	.qa	
Reunion	RE	RE	REU	638	REU	.re	
Romania	RO	RO	ROU	642	ROU	.ro	
Russia	RS	RU	RUS	643	RUS	.ru	
Rwanda	RW	RW	RWA	646	RWA	.rw	
Saint Barthelemy	TB	BL	BLM	652	-	.bl	ccTLD .fr and .gp may also be used
Saint Helena	SH	SH	SHN	654	SHN	.sh	
Saint Kitts and Nevis	SC	KN	KNA	659	KNA	.kn	
Saint Lucia	ST	LC	LCA	662	LCA	.lc	
Saint Martin	RN	MF	MAF	663	-	.mf	ccTLD .fr and .gp may also be used
Saint Pierre and Miquelon	SB	PM	SPM	666	SPM	.pm	
Saint Vincent and the Grenadines	VC	VC	VCT	670	VCT	.vc	
Samoa	WS	WS	WSM	882	WSM	.ws	
San Marino	SM	SM	SMR	674	SMR	.sm	
Sao Tome and Principe	TP	ST	STP	678	STP	.st	
Saudi Arabia	SA	SA	SAU	682	SAU	.sa	
Senegal	SG	SN	SEN	686	SEN	.sn	
Serbia	RI	RS	SRB	688	-	.rs	
Seychelles	SE	SC	SYC	690	SYC	.sc	
Sierra Leone	SL	SL	SLE	694	SLE	.sl	
Singapore	SN	SG	SGP	702	SGP	.sg	
Sint Maarten	NN	SX	SXM	534	-	.sx	
Slovakia	LO	SK	SVK	703	SVK	.sk	
Slovenia	SI	SI	SVN	705	SVN	.si	
Solomon Islands	BP	SB	SLB	090	SLB	.sb	
Somalia	SO	SO	SOM	706	SOM	.so	
South Africa	SF	ZA	ZAF	710	ZAF	.za	
South Georgia and the Islands	SX	GS	SGS	239	SGS	.gs	
Spain	SP	ES	ESP	724	ESP	.es	
Spratly Islands	PG	-	-	-	-	-	
Sri Lanka	CE	LK	LKA	144	LKA	.lk	
Sudan	SU	SD	SDN	736	SDN	.sd	
Suriname	NS	SR	SUR	740	SUR	.sr	
Svalbard	SV	SJ	SJM	744	SJM	.sj	ISO includes Jan Mayen
Swaziland	WZ	SZ	SWZ	748	SWZ	.sz	
Sweden	SW	SE	SWE	752	SWE	.se	
Switzerland	SZ	CH	CHE	756	CHE	.ch	
Syria	SY	SY	SYR	760	SYR	.sy	
Taiwan	TW	TW	TWN	158	TWN	.tw	
Tajikistan	TI	TJ	TJK	762	TJK	.tj	
Tanzania	TZ	TZ	TZA	834	TZA	.tz	

ENTITY	FIPS 10	ISO 3166			STANAG	INTERNET	COMMENT
Thailand	TH	TH	THA	764	THA	.th	
Timor-Leste	TT	TL	TLS	626	TLS	.tl	
Togo	TO	TG	TGO	768	TGO	.tg	
Tokelau	TL	TK	TKL	772	TKL	.tk	
Tonga	TN	TO	TON	776	TON	.to	
Trinidad and Tobago	TD	TT	TTO	780	TTO	.tt	
Tromelin Island	TE	-	-	-	-	-	administered as part of French Southern and Antarctic Lands; no ISO codes assigned
Tunisia	TS	TN	TUN	788	TUN	.tn	
Turkey	TU	TR	TUR	792	TUR	.tr	
Turkmenistan	TX	TM	TKM	795	TKM	.tm	
Turks and Caicos Islands	TK	TC	TCA	796	TCA	.tc	
Tuvalu	TV	TV	TUV	798	TUV	.tv	
Uganda	UG	UG	UGA	800	UGA	.ug	
Ukraine	UP	UA	UKR	804	UKR	.ua	
United Arab Emirates	AE	AE	ARE	784	ARE	.ae	
United Kingdom	UK	GB	GBR	826	GBR	.uk	For its internal communications, the European Union recommends the use of the code UK in lieu of the ISO 3166-2 code of GB
United States	US	US	USA	840	USA	.us	
United States Minor Outlying Islands	-	UM	UMI	581	-	.um	ISO includes Baker Island, Howland Island, Jarvis Island, Johnston Atoll, Kingman Reef, Midway Islands, Navassa Island, Palmyra Atoll, Wake Island
Uruguay	UY	UY	URY	858	URY	.uy	
Uzbekistan	UZ	UZ	UZB	860	UZB	.uz	
Vanuatu	NH	VU	VUT	548	VUT	.vu	
Venezuela	VE	VE	VEN	862	VEN	.ve	
Vietnam	VM	VN	VNM	704	VNM	.vn	
Virgin Islands	VQ	VI	VIR	850	VIR	.vi	
Virgin Islands (UK)	-	-	-	-	-	.vg	see British Virgin Islands
Virgin Islands (US)	-	-	-	-	-	.vi	see Virgin Islands
Wake Island	WQ	-	-	-	UMI	-	ISO includes with the US Minor Outlying Islands
Wallis and Futuna	WF	WF	WLF	876	WLF	.wf	
West Bank	WE	PS	PSE	275	PSE	.ps	ISO identifies as Occupied Palestinian Territory
Western Sahara	WI	EH	ESH	732	ESH	.eh	
Western Samoa	-	-	-	-	-	.ws	see Samoa
World	-	-	-	-	-	-	the Factbook uses the W data code from DIAM 65-18 Geopolitical Data Elements and Related Features, Data Standard No. 3, December 1994, published by the Defense Intelligence Agency
Yemen	YM	YE	YEM	887	YEM	.ye	
Zaire	-	-	-	-	-	-	see Democratic Republic of the Congo
Zambia	ZA	ZM	ZMB	894	ZMB	.zm	
Zimbabwe	ZI	ZW	ZWE	716	ZWE	.zw	

APPENDIX E

CROSS-REFERENCE LIST OF HYDROGRAPHIC DATA CODES

IHO 23-4th: *Limits of Oceans and Seas*, Special Publication 23, Draft 4th Edition 1986, published by the International Hydrographic Bureau of the International Hydrographic Organization.

IHO 23-3rd: *Limits of Oceans and Seas*, Special Publication 23, 3rd Edition 1953, published by the International Hydrographic Organization.

ACIC M 49-1: *Chart of Limits of Seas and Oceans*, revised January 1958, published by the Aeronautical Chart and Information Center (ACIC), United States Air Force.

DIAM 65-18: *Geopolitical Data Elements and Related Features*, Data Standard No. 4, Defense Intelligence Agency Manual 65-18, December 1994, published by the Defense Intelligence Agency. The US Government has not yet adopted a standard for hydrographic codes similar to the Federal Information Processing Standards (FIPS) 10-4 country codes. The names and limits of the following oceans and seas are not always directly comparable because of differences in the customers, needs, and requirements of the individual organizations. Even the number of principal water bodies varies from organization to organization. *Factbook* users, for example, find the Atlantic Ocean and Pacific Ocean entries useful, but none of the following standards include those oceans in their entirety. Nor is there any provision for combining codes or overcodes to aggregate water bodies. The recently delimited Southern Ocean is not included.

Principal Oceans and Seas of the World With Hydrographic Codes by Institution

	IHO 23-4th	IHO 23-3rd*	ACIC M 49-1	DIAM 65-18
Arctic Ocean	9	17	A	5A
Atlantic Ocean	-	-	-	-
Baltic Sea	2	1	B26	7B
Eastern Mediterranean	3.1.2	28 B	-	8E
Indian Ocean	5	45	F	6A
Mediterranean Sea	3.1	28	B11	-
North Atlantic Ocean	1	23	B	1A
North Pacific Ocean	7	57	D	3A
Pacific Ocean	-	-	-	-
South Atlantic Ocean	4	32	C	2A
South China and Eastern Archipelagic Seas	6	49, 48	D18 plus others	3U plus others
South Pacific Ocean	8	61	E	4A
Western Mediterranean	3.1.1	28 A	-	8W

*The letters after the numbers are subdivisions, not footnotes.

APPENDIX F

CROSS-REFERENCE LIST OF GEOGRAPHIC NAMES

NAME	ENTRY INTHE WORLD FACTBOOK	LATITUDE(DEG MIN)	LONGITUDE(DEG MIN)
Abidjan (capital)	Cote d'Ivoire	5 19 N	4 02 W
Abkhazia (region)	Georgia	43 00 N	41 00 E
Abu Dhabi (capital)	United Arab Emirates	24 28 N	54 22 E
Abu Musa (island)	Iran	25 52 N	55 03 E
Abuja (capital)	Nigeria	9 12 N	7 11 E
Abyssinia (former name for Ethiopia)	Ethiopia	8 00 N	38 00 E
Acapulco (city)	Mexico	16 51 N	99 55 W
Accra (capital)	Ghana	5 33 N	0 13 W
Adamstown (capital)	Pitcairn Islands	25 04 S	130 05 W
Addis Ababa (capital)	Ethiopia	9 02 N	38 42 E
Adelie Land (claimed by France; also Terre Adelie)	Antarctica	66 30 S	139 00 E
Aden (city)	Yemen	12 46 N	45 01 E
Aden, Gulf of	Indian Ocean	12 30 N	48 00 E
Admiralty Island	United States (Alaska)	57 44 N	134 20 W
Admiralty Islands	Papua New Guinea	2 10 S	147 00 E
Adriatic Sea	Atlantic Ocean	42 30 N	16 00 E
Adygey (region)	Russia	44 30 N	40 10 E
Aegean Islands	Greece	38 00 N	25 00 E
Aegean Sea	Atlantic Ocean	38 30 N	25 00 E
Afars and Issas, French Territory of the (or FTAI; former name for Djibouti)	Djibouti	11 30 N	43 00 E
Afghanestan (local name for Afghanistan)	Afghanistan	33 00 N	65 00 E
Agalega Islands	Mauritius	10 25 S	56 40 E
Agana (city; former name for Hagatna)	Guam	13 28 N	144 45 E
Ajaccio (city)	France (Corsica)	41 55 N	8 44 E
Ajaria (region)	Georgia	41 45 N	42 10 E
Akmola (city; former name for Astana)	Kazakhstan	51 10 N	71 30 E
Aksai Chin (region)	China (de facto), India (claimed)	35 00 N	79 00 E
Al Arabiyah as Suudiyah (local name for Saudi Arabia)	Saudi Arabia	25 00 N	45 00 E
Al Bahrayn (local name for Bahrain)	Bahrain	26 00 N	50 33 E
Al Imarat al Arabiyah al Muttahidah (local name for the United Arab Emirates)	United Arab Emirates	24 00 N	54 00 E
Al Iraq (local name for Iraq)	Iraq	33 00 N	44 00 E
Al Jaza'ir (local name for Algeria)	Algeria	28 00 N	3 00 E
Al Kuwayt (local name for Kuwait)	Kuwait	29 30 N	45 45 E
Al Maghrib (local name for Morocco)	Morocco	32 00 N	5 00 W
Al Urdun (local name for Jordan)	Jordan	31 00 N	36 00 E
Al Yaman (local name for Yemen)	Yemen	15 00 N	48 00 E
Aland Islands	Finland	60 15 N	20 00 E
Alaska (state)	United States	65 00 N	153 00 W
Alaska, Gulf of	Pacific Ocean	58 00 N	145 00 W
Alboran Sea	Atlantic Ocean	36 00 N	2 30 W
Aldabra Islands (Groupe d'Aldabra)	Seychelles	9 25 S	46 22 E
Alderney (island)	Guernsey	49 43 N	2 12 W
Aleutian Islands	United States (Alaska)	52 00 N	176 00 W
Alexander Archipelago (island group)	United States (Alaska)	57 00 N	134 00 W
Alexander Island	Antarctica	71 00 S	70 00 W
Alexandretta (region; former name for Iskenderun)	Turkey	36 34 N	36 08 E
Alexandria (city)	Egypt	31 12 N	29 54 E
Algiers (capital)	Algeria	36 47 N	2 03 E
Alhucemas, Penon de (island group)	Spain	35 13 N	3 53 W

NAME	ENTRY INTHE WORLD FACTBOOK	LATITUDE(DEG MIN)	LONGITUDE(DEG MIN)
Alma-Ata (city; former name for Almaty)	Kazakhstan	43 15 N	76 57 E
Almaty (former capital)	Kazakhstan	43 15 N	76 57 E
Alofi (capital)	Niue	19 01 S	169 55 W
Alphonse Island	Seychelles	7 01 S	52 45 E
Alsace (region)	France	48 30 N	7 20 E
Amami Strait	Pacific Ocean	28 40 N	129 30 E
Amindivi Islands (former name for Laccadive Islands)	India	11 30 N	72 30 E
Amirante Isles (island group; also Les Amirantes)	Seychelles	6 00 S	53 10 E
Amman (capital)	Jordan	31 57 N	35 56 E
Amsterdam (capital)	Netherlands	52 23 N	4 54 E
Amsterdam Island (Ile Amsterdam)	French Southern and Antarctic Lands	37 52 S	77 32 E
Amundsen Sea	Southern Ocean	72 30 S	112 00 W
Amur River	China, Russia	52 56 N	141 10 E
Amurskiy Liman (strait)	Pacific Ocean	53 00 N	141 30 E
Anadyrskiy Zaliv (gulf)	Pacific Ocean	64 00 N	177 00 E
Anatolia (region)	Turkey	39 00 N	35 00 E
Andaman Islands	India	12 00 N	92 45 E
Andaman Sea	Indian Ocean	10 00 N	95 00 E
Andorra la Vella (capital)	Andorra	42 30 N	1 30 E
Andros (island)	Greece	37 45 N	24 42 E
Andros Island	The Bahamas	24 26 N	77 57 W
Anegada Passage	Atlantic Ocean	18 30 N	63 40 W
Angkor Wat (ruins)	Cambodia	13 26 N	103 50 E
Anglo-Egyptian Sudan (former name for Sudan)	Sudan	15 00 N	30 00 E
Anjouan (island)	Comoros	12 15 S	44 25 E
Ankara (capital)	Turkey	39 56 N	32 52 E
Annobon (island)	Equatorial Guinea	1 25 S	5 36 E
Antananarivo (capital)	Madagascar	18 52 S	47 30 E
Antigua (island)	Antigua and Barbuda	14 34 N	90 44 W
Antipodes Islands	New Zealand	49 41 S	178 43 E
Antwerp (city)	Belgium	51 13 N	4 25 E
Aomen (local Chinese short-form name for Macau)	Macau	22 10 N	113 33 E
Aozou Strip (region)	Chad	22 00 N	18 00 E
Apia (capital)	Samoa	13 50 S	171 44 W
Aqaba, Gulf of	Indian Ocean	29 00 N	34 30 E
Arab, Shatt al (river)	Iran, Iraq	29 57 N	48 34 E
Arabian Sea	Indian Ocean	15 00 N	65 00 E
Arafura Sea	Pacific Ocean	9 00 S	133 00 E
Aral Sea	Kazakhstan, Uzbekistan	45 00 N	60 00 E
Argun River	China, Russia	53 20 N	121 28 E
Aru Sea	Pacific Ocean	6 15 S	135 00 E
As-Sudan (local name for Sudan)	Sudan	15 00 N	30 00 E
Ascension Island	Saint Helena, Ascension, and Tristan da Cunha	7 57 S	14 22 W
Ashgabat, Ashkhabad (capital)	Turkmenistan	37 57 N	58 23 E
Asmara, Asmera (capital)	Eritrea	15 20 N	38 53 E
Assumption Island	Seychelles	9 46 S	46 34 E
Astana (capital; formerly Akmola)	Kazakhstan	51 10 N	71 30 E
Asuncion (capital)	Paraguay	25 16 S	57 40 W
Asuncion Island	Northern Mariana Islands	19 40 N	145 24 E
Atacama (desert)	Chile	23 00 S	70 10 W
Atacama (region)	Chile	24 30 S	69 15 W
Athens (capital)	Greece	37 59 N	23 44 E

NAME	ENTRY INTHE WORLD FACTBOOK	LATITUDE(DEG MIN)	LONGITUDE(DEG MIN)
Attu Island	United States	52 55 N	172 57 E
Auckland (city)	New Zealand	36 52 S	174 46 E
Auckland Islands	New Zealand	51 00 S	166 30 E
Australes, Iles (island group; also Iles Tubuai)	French Polynesia	23 20 S	151 00 W
Avarua (capital)	Cook Islands	21 12 S	159 46 W
Axel Heiberg Island	Canada	79 30 N	90 00 W
Azad Kashmir (region)	Pakistan	34 30 N	74 00 E
Azarbaycan, Azerbaidzhan (local name for Azerbaijan)	Azerbaijan	40 30 N	47 30 E
Azores (islands)	Portugal	38 30 N	28 00 W
Azov, Sea of	Atlantic Ocean	49 00 N	36 00 E
Bab el Mandeb (strait)	Indian Ocean	12 40 N	43 20 E
Babuyan Channel	Pacific Ocean	18 44 N	121 40 E
Babuyan Islands	Philippines	19 10 N	121 40 E
Baffin Bay	Arctic Ocean	73 00 N	66 00 W
Baffin Island	Canada	68 00 N	70 00 W
Baghdad (capital)	Iraq	33 21 N	44 25 E
Baku (capital; also Baki, Baky)	Azerbaijan	40 23 N	49 51 E
Balabac Strait	Pacific Ocean	7 35 N	117 00 E
Balearic Islands	Spain	39 30 N	3 00 E
Balearic Sea (Iberian Sea)	Atlantic Ocean (Mediterranean Sea)	40 30 N	2 00 E
Bali (island)	Indonesia	8 20 S	115 00 E
Bali Sea	Indian Ocean	7 45 S	115 30 E
Balintang Channel	Pacific Ocean	19 49 N	121 40 E
Balintang Islands	Philippines	19 55 N	122 10 E
Balkan Peninsula	Albania, Bosnia and Herzegovina, Bulgaria, Croatia, Greece, Kosovo, Macedonia, Montenegro, Romania, Serbia, Slovenia, Turkey (European part)	42 00 N	23 00 E
Balleny Islands	Antarctica	67 00 S	163 00 E
Balochistan (region)	Pakistan	28 00 N	63 00 E
Baltic Sea	Atlantic Ocean	57 00 N	19 00 E
Bamako (capital)	Mali	12 39 N	8 00 W
Banaba (Ocean Island)	Kiribati	0 52 S	169 35 E
Banat (region)	Hungary, Romania, Serbia	45 30 N	21 00 E
Banda Sea	Pacific Ocean	5 00 S	128 00 E
Bandar Seri Begawan (capital)	Brunei	4 53 N	114 56 E
Bangka (island)	Indonesia	2 30 S	106 00 E
Bangkok (capital)	Thailand	13 45 N	100 31 E
Bangui (capital)	Central African Republic	4 22 N	18 35 E
Banjul (capital)	The Gambia	13 28 N	16 39 W
Banks Island	Canada	75 15 N	121 30 W
Banks Island	Australia	10 12 S	142 16 E
Banks Islands (Iles Banks)	Vanuatu	14 00 S	167 30 E
Barbuda (island)	Antigua and Barbuda	17 38 N	61 48 W
Barcelona (city)	Spain	41 25 N	2 13 E
Barents Sea	Arctic Ocean	74 00 N	36 00 E
Barranquilla (city)	Colombia	10 59 N	74 48 W
Bashi Channel	Pacific Ocean	22 00 N	121 00 E
Basilan Strait	Pacific Ocean	6 49 N	122 05 E
Basque Provinces	Spain	43 00 N	2 30 W
Bass Strait	Pacific Ocean	39 20 S	145 30 E
Bassas da India	Indian Ocean	21 30 S	39 50 E
Basse-Terre (capital)	France (Guadeloupe)	16 00 N	61 44 W
Basseterre (capital)	Saint Kitts and Nevis	17 18 N	62 43 W
Bastia (city)	France (Corsica)	42 42 N	9 27 E

NAME	ENTRY INTHE WORLD FACTBOOK	LATITUDE(DEG MIN)	LONGITUDE(DEG MIN)
Basutoland (former name for Lesotho)	Lesotho	29 30 S	28 30 E
Batan Islands	Philippines	20 30 N	121 50 E
Bavaria (region; also Bayern)	Germany	48 30 N	11 30 E
Beagle Channel	Atlantic Ocean	54 53 S	68 10 W
Bear Island (see Bjornoya)	Svalbard	74 26 N	19 05 E
Beaufort Sea	Arctic Ocean	73 00 N	140 00 W
Bechuanaland (former name for Botswana)	Botswana	22 00 S	24 00 E
Beijing (capital)	China	39 56 N	116 24 E
Beirut (capital)	Lebanon	33 53 N	35 30 E
Bekaa Valley	Lebanon	34 00 N	36 05 E
Belau (Palau Islands)	Palau	7 30 N	134 30 E
Belep Islands (Iles Belep)	New Caledonia	19 45 S	163 40 E
Belfast (city)	United Kingdom	54 36 N	5 55 W
Belgian Congo (former name for Democratic Republic of the Congo)	Democratic Republic of the Congo	0 00 N	25 00 E
Belgie, Belgique (local name for Belgium)	Belgium	50 50 N	4 00 E
Belgrade (capital)	Serbia	44 50 N	20 30 E
Belize City	Belize	17 30 N	88 12 W
Belle Isle, Strait of	Atlantic Ocean	51 35 N	56 30 W
Bellingshausen Sea	Southern Ocean	71 00 S	85 00 W
Belmopan (capital)	Belize	17 15 N	88 46 W
Belorussia (former name for Belarus)	Belarus	53 00 N	28 00 E
Benadir (region; former name of Italian Somaliland)	Somalia	4 00 N	46 00 E
Bengal (region)	Bangladesh, India	24 30 N	88 15 E
Bengal, Bay of	Indian Ocean	15 00 N	90 00 E
Berau, Gulf of	Pacific Ocean	2 30 S	132 30 E
Bering Island	Russia	55 00 N	166 30 E
Bering Sea	Pacific Ocean	60 00 N	175 00 W
Bering Strait	Pacific Ocean	65 30 N	169 00 W
Berkner Island	Antarctica	79 30 S	49 30 W
Berlin (capital)	Germany	52 31 N	13 24 E
Berlin, East (former name for eastern sector of Berlin)	Germany	52 30 N	13 33 E
Berlin, West (former name for western sector of Berlin)	Germany	52 30 N	13 20 E
Bern (capital)	Switzerland	46 57 N	7 26 E
Bessarabia (region)	Moldova, Romania, Ukraine	47 00 N	28 30 E
Bharat (local name for India)	India	20 00 N	77 00 E
Bhopal (city)	India	23 16 N	77 24 E
Biafra (region)	Nigeria	5 30 N	7 30 E
Big Diomede Island	Russia	65 46 N	169 06 W
Bijagos, Arquipelago dos (island group)	Guinea-Bissau	11 25 N	16 20 W
Bikini Atoll	Marshall Islands	11 35 N	165 23 E
Bilbao (city)	Spain	43 15 N	2 58 W
Bioko (island)	Equatorial Guinea	3 30 N	8 42 E
Biscay, Bay of	Atlantic Ocean	44 00 N	4 00 W
Bishkek (capital)	Kyrgyzstan	42 54 N	74 36 E
Bishop Rock	United Kingdom	49 52 N	6 27 W
Bismarck Archipelago (island group)	Papua New Guinea	5 00 S	150 00 E
Bismarck Sea	Pacific Ocean	4 00 S	148 00 E
Bissau (capital)	Guinea-Bissau	11 51 N	15 35 W
Bjornoya (Bear Island)	Svalbard	74 26 N	19 05 E
Black Forest (region)	Germany	48 00 N	8 15 E
Black Rock (island)	South Georgia and the South Sandwich Islands	53 39 S	41 48 W
Black Sea	Atlantic Ocean	43 00 N	35 00 E

NAME	ENTRY INTHE WORLD FACTBOOK	LATITUDE(DEG MIN)	LONGITUDE(DEG MIN)
Bloemfontein (judicial capital)	South Africa	29 12 S	26 07 E
Bo Hai (gulf)	Pacific Ocean	38 00 N	120 00 E
Boa Vista (island)	Cape Verde	16 05 N	22 50 W
Bogota (capital)	Colombia	4 36 N	74 05 W
Bohemia (region)	Czech Republic	50 00 N	14 30 E
Bombay (city; see Mumbai)	India	18 58 N	72 50 E
Bonaire (island)	Netherlands	12 10 N	68 15 W
Bonifacio, Strait of	Atlantic Ocean (Mediterranean Sea)	41 01 N	14 00 E
Bonin Islands	Japan	27 00 N	142 10 E
Bonn (former capital)	Germany	50 44 N	7 05 E
Bophuthatswana (region; enclave)	South Africa	26 30 S	25 30 E
Bora-Bora (island)	French Polynesia	16 30 S	151 45 W
Bordeaux (city)	France	44 50 N	0 34 W
Borneo (island)	Brunei, Indonesia, Malaysia	0 30 N	114 00 E
Bornholm (island)	Denmark	55 10 N	15 00 E
Bosna i Hercegovina (local name for Bosnia and Herzegovina)	Bosnia and Herzegovina	44 00 N	18 00 E
Bosnia (political region)	Bosnia and Herzegovina	44 00 N	18 00 E
Bosporus (strait)	Atlantic Ocean	41 00 N	29 00 E
Bothnia, Gulf of	Atlantic Ocean	63 00 N	20 00 E
Bougainville (island)	Papua New Guinea	6 00 S	155 00 E
Bougainville Strait	Pacific Ocean	6 40 S	156 10 E
Bounty Islands	New Zealand	47 43 S	174 00 E
Bourbon Island (former name of Reunion)	Reunion	21 06 S	55 36 E
Brasilia (capital)	Brazil	15 47 S	47 55 W
Bratislava (capital)	Slovakia	48 09 N	17 07 E
Brazzaville (capital)	Republic of the Congo	4 16 S	15 17 E
Bridgetown (capital)	Barbados	13 06 N	59 37 W
Brisbane (city)	Australia	27 28 S	153 02 E
Bristol Bay	Pacific Ocean	57 00 N	160 00 W
Bristol Channel	Atlantic Ocean	51 18 N	3 30 W
Britain (see Great Britain)	United Kingdom	54 00 N	2 00 W
British Bechuanaland (region; former name for northwest South Africa)	South Africa	27 30 S	23 30 E
British Central African Protectorate (former name of Nyasaland)	Malawi	13 30 S	34 00 E
British East Africa (former name for British possessions in eastern Africa)	Kenya, Tanzania, Uganda	1 00 N	38 00 E
British Guiana (former name for Guyana)	Guyana	5 00 N	59 00 W
British Honduras (former name for Belize)	Belize	17 15 N	88 45 W
British Solomon Islands (former name for Solomon Islands)	Solomon Islands	8 00 S	159 00 E
British Somaliland (former name for northern Somalia)	Somalia	10 00 N	49 00 E
Brussels (capital)	Belgium	50 50 N	4 20 E
Bubiyan (island)	Kuwait	29 47 N	48 10 E
Bucharest (capital)	Romania	44 26 N	26 06 E
Budapest (capital)	Hungary	47 30 N	19 05 E
Buenos Aires (capital)	Argentina	34 36 S	58 27 W
Bujumbura (capital)	Burundi	3 23 S	29 22 E
Bukovina (region)	Romania, Ukraine	48 00 N	26 00 E
Byelarus (local name for Belarus)	Belarus	53 00 N	28 00 E
Byelorussia (former name for Belarus)	Belarus	53 00 N	28 00 E
Cabinda (province)	Angola	5 33 S	12 12 E
Cabo Verde (local name for Cape Verde)	Cape Verde	16 00 N	24 00 W
Cabot Strait	Atlantic Ocean	47 20 N	59 30 W
Caicos Islands	Turks and Caicos Islands	21 56 N	71 58 W
Cairo (capital)	Egypt	30 03 N	31 15 E

NAME	ENTRY INTHE WORLD FACTBOOK	LATITUDE(DEG MIN)	LONGITUDE(DEG MIN)
Calcutta (city)	India	22 32 N	88 21 E
Calgary (city)	Canada	51 02 N	114 04 W
California, Gulf of	Pacific Ocean	28 00 N	112 00 W
Cameroun (local name for Cameroon)	Cameroon	6 00 N	12 00 E
Campbell Island	New Zealand	52 33 S	169 09 E
Campeche, Bay of	Atlantic Ocean (Gulf of Mexico)	20 00 N	94 00 W
Canal Zone (former name for US possessions in Panama)	Panama	9 00 N	79 45 W
Canarias Sea	Atlantic Ocean	28 00 N	16 00 W
Canary Islands	Spain	28 00 N	15 30 W
Canberra (capital)	Australia	35 17 S	149 08 E
Cancun (city)	Mexico	21 10 N	86 50 W
Canton (city; now Guangzhou)	China	23 06 N	113 16 E
Canton Island (Kanton Island)	Kiribati	2 49 S	171 40 W
Cape Juby (region; former name for Southern Morocco)	Morocco	27 53 N	12 58 W
Cape Province (region; former name for Northern, Western, and Eastern Cape Provinces of South Africa)	South Africa	31 30 S	22 30 E
Cape Town (legislative capital)	South Africa	33 57 S	18 25 E
Cape of Good Hope (cape; also alternate name for Cape Province of South Africa)	South Africa	34 15 S	18 20 E
Caracas (capital)	Venezuela	10 30 N	66 56 W
Cargados Carajos Shoals	Mauritius	16 25 S	59 38 E
Caribbean Sea	Atlantic Ocean	15 00 N	73 00 W
Caroiine isiands	Federated States of Micronesia, Palau	7 30 N	148 00 E
Carpatho-Ukraine (region; former name for Zakarpats'ka oblast')	Ukraine	48 22 N	23 32 E
Carpentaria, Gulf of	Pacific Ocean	14 00 S	139 00 E
Casablanca (city)	Morocco	33 35 N	7 34 W
Castries (capital)	Saint Lucia	14 01 N	61 00 W
Catalonia (region)	Spain	42 00 N	2 00 E
Cato Island	Australia	23 15 S	155 32 E
Caucasus (region)	Russia	42 00 N	45 00 E
Cayenne (capital)	French Guiana	4 56 N	52 20 W
Celebes (island)	Indonesia	2 00 S	121 00 E
Celebes Sea	Pacific Ocean	3 00 N	122 00 E
Celtic Sea	Atlantic Ocean	51 00 N	6 30 W
Central African Empire (former name for Central African Republic)	Central African Republic	7 00 N	21 00 E
Ceram (Seram) Sea	Pacific Ocean	2 30 S	129 30 E
Ceska Republika (local name for Czech Republic)	Czech Republic	49 45 N	15 30 E
Ceskoslovensko (former local name for Czechoslovakia)	Czech Republic, Slovakia	49 00 N	17 30 E
Cetinje (capital city)	Montenegro	42 24 N	18 55 E
Ceuta (city)	Spain	35 53 N	5 19 W
Ceylon (former name for Sri Lanka)	Sri Lanka	7 00 N	81 00 E
Chafarinas, Islas (island)	Spain	35 12 N	2 26 W
Chagos Archipelago (Oil Islands)	British Indian Ocean Territory	6 00 S	71 30 E
Challenger Deep (Mariana Trench)	Pacific Ocean	11 22 N	142 36 E
Channel Islands	Guernsey, Jersey	49 20 N	2 20 W
Charlotte Amalie (capital)	Virgin Islands	18 21 N	64 56 W
Chatham Islands	New Zealand	44 00 S	176 30 W
Chechnya (region; also Chechnia)	Russia	43 15 N	45 40 E
Cheju Strait	Pacific Ocean	34 00 N	126 30 E
Cheju-do (island)	Korea, South	33 20 N	126 30 E
Chengdu (city)	China	30 43 N	104 04 E

NAME	ENTRY INTHE WORLD FACTBOOK	LATITUDE(DEG MIN)	LONGITUDE(DEG MIN)
Chennai (city; also Madras)	India	13 04 N	80 16 E
Chesterfield Islands (Iles Chesterfield)	New Caledonia	19 52 S	158 15 E
Chihli, Gulf of (see Bo Hai)	Pacific Ocean	38 30 N	120 00 E
Chiloe (island)	Chile	42 50 S	74 00 W
China, People's Republic of	China	35 00 N	105 00 E
China, Republic of	Taiwan	23 30 N	121 00 E
Chisinau (capital; also Kishinev)	Moldova	47 00 N	28 50 E
Choiseul (island)	Solomon Islands	7 05 S	121 00 E
Choson (local name for North Korea)	North Korea	40 00 N	127 00 E
Christmas Island (Indian Ocean)	Australia	10 25 S	105 39 E
Christmas Island (Pacific Ocean; also Kiritimati)	Kiribati	1 52 N	157 20 W
Chukchi Sea	Arctic Ocean	69 00 N	171 00 W
Chuuk Islands (Truk Islands)	Federated States of Micronesia	7 25 N	151 47 W
Cilicia (region)	Turkey	36 50 N	34 30 E
Ciskei (enclave)	South Africa	33 00 S	27 00 E
Citta del Vaticano (local name for Vatican City)	Holy See	41 54 N	12 27 E
Cochin China (region)	Vietnam	11 00 N	107 00 E
Coco, Isla del (island)	Costa Rica	5 32 N	87 04 W
Cocos Islands	Cocos (Keeling) Islands	12 30 S	96 50 E
Colombo (capital)	Sri Lanka	6 56 N	79 51 E
Colon, Archipielago de (Galapagos Islands)	Ecuador	0 00 N	90 30 W
Commander Islands (Komandorskiye Ostrova)	Russia	55 00 N	167 00 E
Comores (local name for Comoros)	Comoros	12 10 S	44 15 E
Con Son (islands)	Vietnam	8 43 N	106 36 E
Conakry (capital)	Guinea	9 31 N	13 43 W
Confederatio Helvetica (local name for Switzerland)	Switzerland	47 00 N	8 00 E
Congo (Brazzaville) (former name for Republic of the Congo)	Republic of the Congo	1 00 S	15 00 E
Congo (Leopoldville) (former name for the Democratic Republic of the Congo)	Democratic Republic of the Congo	0 00 N	25 00 E
Constantinople (city; former name for Istanbul)	Turkey	41 01 N	28 58 E
Cook Strait	Pacific Ocean	41 15 S	174 30 E
Copenhagen (capital)	Denmark	55 40 N	12 35 E
Coral Sea	Pacific Ocean	15 00 S	150 00 E
Corfu (island)	Greece	39 40 N	19 45 E
Corinth (region)	Greece	37 56 N	22 56 E
Corisco (island)	Equatorial Guinea	0 55 N	9 19 E
Corn Islands (Islas del Maiz)	Nicaragua	12 15 N	83 00 W
Corocoro Island	Guyana, Venezuela	3 38 N	66 50 W
Corsica (island; also Corse)	France	42 00 N	9 00 E
Cosmoledo Group (island group; also Atoll de Cosmoledo)	Seychelles	9 43 S	47 35 E
Cotonou (former capital)	Benin	6 21 N	2 26 E
Cotopaxi (volcano)	Ecuador	0 39 S	78 26 W
Courantyne River	Guyana, Suriname	5 57 N	57 06 W
Cozumel (island)	Mexico	20 30 N	86 55 W
Crete (island)	Greece	35 15 N	24 45 E
Crimea (region)	Ukraine	45 00 N	34 00 E
Crimean Peninsula	Ukraine	45 00 N	34 00 E
Crooked Island Passage	Atlantic Ocean	22 55 N	74 35 W
Crozet Islands (Iles Crozet)	French Southern and Antarctic Lands	46 30 S	51 00 E
Cyclades (island group)	Greece	37 00 N	25 10 E

NAME	ENTRY INTHE WORLD FACTBOOK	LATITUDE(DEG MIN)	LONGITUDE(DEG MIN)
Cyrenaica (region)	Libya	31 00 N	22 00 E
Czechoslovakia (former name for the entity that subsequently split into the Czech Republic and Slovakia)	Czech Republic, Slovakia	49 00 N	18 00 E
D'Entrecasteaux Islands	Papua New Guinea	9 30 S	150 40 E
Dagestan (region)	Russia	43 00 N	47 00 E
Dahomey (former name for Benin)	Benin	9 30 N	2 15 E
Daito Islands	Japan	43 00 N	17 00 E
Dakar (capital)	Senegal	14 40 N	17 26 W
Dalmatia (region)	Croatia	43 00 N	17 00 E
Daman (city; also Damao)	India	20 10 N	73 00 E
Damascus (capital)	Syria	33 30 N	36 18 E
Danger Islands (see Pukapuka Atoll)	Cook Islands	10 53 S	165 49 W
Danish Straits	Atlantic Ocean	58 00 N	11 00 E
Danish West Indies (former name for the Virgin Islands)	Virgin Islands	18 20 N	64 50 W
Danmark (local name)	Denmark	56 00 N	10 00 E
Danzig (city; former name for Gdansk)	Poland	54 23 N	18 40 E
Dao Bach Long Vi (island)	Vietnam	20 08 N	107 44 E
Dar es Salaam (capital)	Tanzania	6 48 S	39 17 E
Dardanelles (strait)	Atlantic Ocean	40 15 N	26 25 E
Davis Strait	Atlantic Ocean	67 00 N	57 00 W
Dead Sea	Israel, Jordan, West Bank	32 30 N	35 30 E
Deception Island	Antarctica	62 56 S	60 34 W
Denmark Strait	Atlantic Ocean	67 00 N	24 00 W
Desolation Islands (Isles Kerguelen)	French Southern and Antarctic Lands	49 30 S	69 30 E
Deutschland (local name for Germany)	Germany	51 00 N	9 00 E
Devils Island (Ile du Diable)	French Guiana	5 17 N	52 35 W
Devon Island	Canada	76 00 N	87 00 W
Dhaka (capital)	Bangladesh	23 43 N	90 25 E
Dhivehi Raajje (local name for Maldives)	Maldives	3 15 N	73 00 E
Dhofar (region)	Oman	17 00 N	54 10 E
Diego Garcia (island)	British Indian Ocean Territory	7 20 S	72 25 E
Diego Ramirez (islands)	Chile	56 30 S	68 43 W
Dili (capital)	Timor-Leste	8 35 S	125 36 E
Dilmun (former name for Bahrain)	Bahrain	7 00 N	81 00 E
Diomede Islands	Russia (Big Diomede), United States (Little Diomede)	65 47 N	169 00 W
Diu (region)	India	20 42 N	70 59 E
Djibouti (capital)	Djibouti	11 30 N	43 15 E
Dnieper (river)	Belarus, Russia, Ukraine (Dnyapro, Dnepr, Dnipro)	46 30 N	32 18 E
Dniester (river)	Moldova, Ukraine (Nistru, Dnister)	46 18 N	30 17 E
Dobruja (region)	Bulgaria, Romania	43 30 N	28 00 E
Dodecanese (island group)	Greece	36 00 N	27 05 E
Dodoma (city)	Tanzania	6 11 S	35 45 E
Doha (capital)	Qatar	25 17 N	51 32 E
Donets Basin	Russia, Ukraine	48 15 N	38 30 E
Douala (city)	Cameroon	4 03 N	9 42 E
Douglas (capital)	Man, Isle of	54 09 N	4 28 W
Dover, Strait of	Atlantic Ocean	51 00 N	1 30 E
Drake Passage	Atlantic Ocean, Southern Ocean	60 00 S	60 00 W
Druk Yul (local name for Bhutan)	Bhutan	27 30 N	90 30 E
Dubai, Dubayy (city)	United Arab Emirates	25 18 N	55 18 E
Dublin (capital)	Ireland	53 20 N	6 15 W
Duesseldorf (city)	Germany	51 13 N	6 47 E
Durban (city)	South Africa	29 51 S	31 02 E

NAME	ENTRY INTHE WORLD FACTBOOK	LATITUDE(DEG MIN)	LONGITUDE(DEG MIN)
Dushanbe (capital)	Tajikistan	38 35 N	68 48 E
Dutch Antilles (former name for the Netherlands Antilles)	Aruba, Curacao, Sint Maarten	12 10 N	68 30 W
Dutch East Indies (former name for Indonesia)	Indonesia	5 00 S	120 00 E
Dutch Guiana (former name for Suriname)	Suriname	4 00 N	56 00 W
Dutch West Indies (former name for the Netherlands Antilles)	Aruba, Curacao, Sint Maarten	12 10 N	68 30 W
Dzungarian Gate (valley)	China, Kazakhstan	45 25 N	82 25 E
East China Sea	Pacific Ocean	30 00 N	126 00 E
East Frisian Islands	Germany	53 44 N	7 25 E
East Germany (German Democratic Republic; former name for eastern portion of Germany)	Germany	52 00 N	13 00 E
East Korea Strait (Eastern Channel or Tsushima Strait)	Pacific Ocean	34 00 N	129 00 E
East Pakistan (former name for Bangladesh)	Bangladesh	24 00 N	90 00 E
East Siberian Sea	Arctic Ocean	74 00 N	166 00 E
Easter Island (Isla de Pascua)	Chile	27 07 S	109 22 W
Eastern Channel (East Korea Strait or Tsushima Strait)	Pacific Ocean	34 00 N	129 00 E
Eastern Samoa (former name for American Samoa)	American Samoa	14 20 S	170 00 W
Edinburgh (city)	United Kingdom	55 57 N	3 11 W
Eesti (local name for Estonia)	Estonia	59 00 N	26 00 E
Eire (local name for Ireland)	Ireland	53 00 N	8 00 W
Elba (island)	Italy	42 46 N	10 17 E
Elemi Triangle (region)	Ethiopia (claimed), Kenya (de facto), Sudan (claimed)	5 00 N	35 30 E
Ellada, Ellas (local name for Greece)	Greece	39 00 N	22 00 E
Ellef Ringnes Island	Canada	78 00 N	103 00 W
Ellesmere Island	Canada	81 00 N	80 00 W
Ellice Islands	Tuvalu	8 00 S	178 00 E
Ellsworth Land (region)	Antarctica	75 00 S	92 00 W
Elobey, Islas de (island group)	Equatorial Guinea	0 59 N	9 33 E
Enderbury Island	Kiribati	3 08 S	171 05 W
Enewetak Atoll (Eniwetok Atoll)	Marshall Islands	11 30 N	162 15 E
England (region)	United Kingdom	52 30 N	1 30 W
English Channel	Atlantic Ocean	50 20 N	1 00 W
Eniwetok Atoll (see Enewetak Atoll)	Marshall Islands	11 30 N	162 15 E
Eolie, Isole (island group)	Italy	38 30 N	15 00 E
Epirus, Northern (region)	Albania, Greece	40 00 N	20 30 E
Episkopi Cantonment (capital)	Akrotiri, Dhekelia	34 40 N	32 51 E
Ertra (local name for Eritrea)	Eritrea	15 00 N	39 00 E
Espana	Spain	40 00 N	4 00 W
Essequibo (region; claimed by Venezuela)	Guyana	6 59 N	58 23 W
Etorofu (island; also Iturup)	Russia (de facto)	44 55 N	147 40 E
Europa Island	Indian Ocean	22 20 S	40 22 E
Farquhar Group (island group; also Atoll de Farquhar)	Seychelles	10 10 S	51 10 E
Fergana Valley	Kyrgyzstan, Tajikistan, Uzbekistan	41 00 N	72 00 E
Fernando Po (island; see Bioko)	Equatorial Guinea	3 30 N	8 42 E
Fernando de Noronha (island group)	Brazil	3 51 S	32 25 W
Filipinas (local name for the Philippines; also Pilipinas)	Philippines	13 00 N	122 00 E
Finland, Gulf of	Atlantic Ocean (Baltic Sea)	60 00 N	27 00 E
Fiume (city; former name for Rijeka)	Croatia	45 19 N	14 25 E
Florence (city)	Italy	43 46 N	11 16 E

NAME	ENTRY INTHE WORLD FACTBOOK	LATITUDE(DEG MIN)	LONGITUDE(DEG MIN)
Flores (island)	Indonesia	8 45 S	121 00 E
Flores Sea	Pacific Ocean	7 40 S	119 45 E
Florida, Straits of	Atlantic Ocean	25 00 N	79 45 W
Fongafale (largest island of Funafuti)	Tuvalu	8 30 S	179 12 E
Former Soviet Union (FSU)	Armenia, Azerbaijan, Belarus, Estonia, Georgia, Kazakhstan, Kyrgyzstan, Latvia, Lithuania, Moldova, Russia, Tajikistan, Turkmenistan, Ukraine, Uzbekistan		
Formosa (island)	Taiwan	23 30 N	121 00 E
Formosa Strait (see Taiwan Strait)	Pacific Ocean	24 00 N	119 00 E
Foroyar (local name for Faroe Islands)	Faroe Islands	62 00 N	7 00 W
Fort-de-France (capital)	Martinique	14 36 N	61 05 W
Frankfurt am Main (city)	Germany	50 07 N	8 41 E
Franz Josef Land (island group)	Russia	81 00 N	55 00 E
Freetown (capital)	Sierra Leone	8 30 N	13 15 W
French Cameroon (former name for Cameroon)	Cameroon	6 00 N	12 00 E
French Guinea (former name for Guinea)	Guinea	11 00 N	10 00 W
French Indochina (former name for French possessions in southeast Asia)	Cambodia, Laos, Vietnam	15 00 N	107 00 E
French Morocco (former name for Morocco)	Morocco	32 00 N	5 00 W
French Somaliland (former name for Djibouti)	Djibouti	11 30 N	43 00 E
French Sudan (former name for Mali)	Mali	17 00 N	4 00 W
French Territory of the Afars and Issas (or FTAI; former name for Djibouti)	Djibouti	11 30 N	43 00 E
French Togoland (former name for Togo)	Togo	8 00 N	1 10 E
French West Indies (former name for French possessions in the West Indies)	Guadeloupe, Martinique	16 30 N	62 00 W
Friendly Islands	Tonga	20 00 S	175 00 W
Frisian Islands	Denmark, Germany, Netherlands	53 35 N	6 40 E
Frunze (city; former name for Bishkek)	Kyrgyzstan	42 54 N	74 36 E
Funafuti (capital, atoll)	Tuvalu	8 30 S	179 12 E
Fundy, Bay of	Atlantic Ocean	45 00 N	66 00 W
Futuna Islands (Hoorn Islands/Iles de Horne)	Wallis and Futuna	14 19 S	178 05 W
Fyn (island)	Denmark	55 20 N	10 25 E
Gaborone (capital)	Botswana	24 45 S	25 55 E
Galapagos Islands (Archipielago de Colon)	Ecuador	0 00 N	90 30 W
Galicia (region)	Spain	42 45 N	8 10 E
Galicia (region)	Poland, Ukraine	49 30 N	23 00 E
Galilee (region)	Israel	32 54 N	35 20 E
Galleons Passage	Atlantic Ocean	11 00 N	60 55 W
Gambier Islands (Iles Gambier)	French Polynesia	23 09 S	134 58 W
Gaspar Strait	Pacific Ocean	3 00 S	107 00 E
Gdansk (city; formerly Danzig)	Poland	54 23 N	18 40 E
Geneva (city)	Switzerland	46 12 N	6 10 E
Genoa (city)	Italy	44 25 N	8 57 E
George Town (capital)	Cayman Islands	19 20 N	81 23 W
George Town (city)	Malaysia	5 26 N	100 16 E
George Town (city)	The Bahamas	23 30 N	75 46 W
Georgetown (capital)	Guyana	6 48 N	58 10 W
Georgetown (city)	The Gambia	13 30 N	14 47 W
German Democratic Republic (East Germany; former name for eastern portion of Germany)	Germany	52 00 N	13 00 E

NAME	ENTRY INTHE WORLD FACTBOOK	LATITUDE(DEG MIN)	LONGITUDE(DEG MIN)
German Southwest Africa (former name for Namibia)	Namibia	22 00 S	17 00 E
Germany, Federal Republic of	Germany	51 00 N	9 00 E
Gibraltar (city, peninsula)	Gibraltar	36 11 N	5 22 W
Gibraltar, Strait of	Atlantic Ocean	35 57 N	5 36 W
Gidi Pass	Egypt	30 13 N	33 09 E
Gilbert Islands	Kiribati	1 25 N	173 00 E
Glorioso Islands	Indian Ocean	11 30 S	47 20 E
Goa (state)	India	15 20 N	74 00 E
Gobi (desert)	China, Mongolia	42 30 N	107 00 E
Godthab (capital; also Nuuk)	Greenland	64 11 N	51 44 W
Golan Heights (region)	Syria	33 00 N	35 45 E
Gold Coast (former name for Ghana)	Ghana	8 00 N	2 00 W
Golfo San Jorge (gulf)	Atlantic Ocean	46 00 S	66 00 W
Golfo San Matias (gulf)	Atlantic Ocean	41 30 S	64 00 W
Good Hope, Cape of	South Africa	34 24 S	18 30 E
Goteborg (city)	Sweden	57 43 N	11 58 E
Gotland (island)	Sweden	57 30 N	18 33 E
Gough Island	Saint Helena, Ascension, and Tristan da Cunha	40 20 S	9 55 W
Graham Land (region)	Antarctica	65 00 S	64 00 W
Gran Chaco (region)	Argentina, Paraguay	24 00 S	60 00 W
Grand Bahama (island)	The Bahamas	26 40 N	78 35 W
Grand Banks (fishing ground)	Atlantic Ocean	47 06 N	55 48 W
Grand Cayman (island)	Cayman Islands	19 20 N	81 20 W
Grand Turk (capital; also Cockburn Town)	Turks and Caicos Islands	21 28 N	71 08 W
Great Australian Bight	Indian Ocean	35 00 S	130 00 E
Great Belt (strait; also Store Baelt)	Atlantic Ocean	55 30 N	11 00 E
Great Bitter Lake	Egypt	30 20 N	32 23 E
Great Britain (island)	United Kingdom	54 00 N	2 00 W
Great Channel	Indian Ocean	6 25 N	94 20 E
Great Inagua (island)	The Bahamas	21 00 N	73 20 W
Great Rift Valley	Ethiopia, Kenya	0 30 N	36 00 E
Greater Sunda Islands	Brunei, Indonesia, Malaysia	2 00 S	110 00 E
Green Islands	Papua New Guinea	4 30 S	154 10 E
Greenland Sea	Arctic Ocean	79 00 N	5 00 W
Grenadines, Northern (island group)	Saint Vincent and the Grenadines	13 15 N	61 12 W
Grenadines, Southern (island group)	Grenada	12 07 N	61 40 W
Grytviken (town; on South Georgia)	South Georgia and the South Sandwich Islands	54 15 S	36 45 W
Guadalahara (city)	Mexico	20 40 N	103 24 W
Guadalcanal (island)	Solomon Islands	9 32 S	160 12 E
Guadalupe, Isla de (island)	Mexico	29 11 N	118 17 W
Guangzhou (city; also Canton)	China	23 09 N	113 21 E
Guantanamo Bay (US Naval Base)	Cuba	20 00 N	75 08 W
Guatemala (capital)	Guatemala	14 38 N	90 31 W
Guine-Bissau (local name for Guinea-Bissau)	Guinea-Bissau	12 00 N	15 00 W
Guinea Ecuatorial (local name for Equatorial Guinea)	Equatorial Guinea	2 00 N	10 00 E
Guinea, Gulf of	Atlantic Ocean	3 00 N	2 30 E
Guinee (local name for Guinea)	Guinea	11 00 N	10 00 W
Gustavia (capital)	Saint Barthelemy	17 53 N	62 51 W
Guyane Francaise (local name for French Guiana)	French Guiana	4 00 N	53 00 W
Ha'apai Group (island group)	Tonga	19 42 S	174 29 W
Habomai Islands	Russia (de facto)	43 30 N	146 10 E
Hadhramaut (region)	Yemen	15 00 N	50 00 E

NAME	ENTRY INTHE WORLD FACTBOOK	LATITUDE(DEG MIN)	LONGITUDE(DEG MIN)
Hagatna (capital; formerly Agana)	Guam	13 28 N	144 45 E
Hague, The (seat of government)	Netherlands	52 05 N	4 18 E
Haifa (city)	Israel	32 50 N	35 00 E
Hainan Dao (island)	China	19 00 N	109 30 E
Haiphong (city)	Vietnam	20 52 N	106 41 E
Hala'ib Triangle (region)	Egypt (claimed), Sudan (de facto)	22 30 N	35 00 E
Halifax (city)	Canada	44 39 N	63 36 W
Halmahera (island)	Indonesia	1 00 N	128 00 E
Halmahera Sea	Pacific Ocean	0 30 S	129 00 E
Hamburg (city)	Germany	53 34 N	9 59 E
Hamilton (capital)	Bermuda	32 17 N	64 46 W
Han-guk (local name for South Korea	South Korea	37 00 N	127 30 E
Hanoi (capital)	Vietnam	21 02 N	105 51 E
Harare (capital)	Zimbabwe	17 50 S	31 03 E
Harvey Islands (former name for Cook Islands)	Cook Islands	21 14 S	159 46 W
Hatay (province)	Turkey	36 30 N	36 15 E
Havana (capital)	Cuba	23 08 N	82 22 W
Hawaii (island)	United States	19 45 N	155 45 W
Hawaiian Islands	United States	21 00 N	157 45 W
Hawar (island)	Bahrain	25 40 N	50 47 E
Hayastan (local name for Armenia)	Armenia	40 00 N	45 00 E
Heard Island	Heard Island and McDonald Islands	53 06 S	73 30 E
Hejaz (region)	Saudi Arabia	24 30 N	38 30 E
Helsinki (capital)	Finland	60 10 N	24 58 E
Herzegovina (political region)	Bosnia and Herzegovina	44 00 N	18 00 E
Hiiumaa (island)	Estonia	58 50 N	22 30 E
Hispaniola (island)	Dominican Republic, Haiti	18 45 N	71 00 W
Ho Chi Minh City (formerly Saigon)	Vietnam	10 45 N	106 40 E
Hokkaido (island)	Japan	44 00 N	143 00 E
Holland (region)	Netherlands	52 30 N	5 45 E
Hong Kong (special administrative region)	Hong Kong	22 15 N	114 10 E
Honiara (capital)	Solomon Islands	9 26 S	159 57 E
Honshu (island)	Japan	36 00 N	138 00 E
Hormuz, Strait of	Indian Ocean	26 34 N	56 15 E
Horn of Africa (region)	Djibouti, Eritrea, Ethiopia, Somalia	8 00 N	48 00 E
Horn, Cape (Cabo de Hornos)	Chile	55 59 S	67 16 W
Horne, Iles de (island group)	Wallis and Futuna	14 19 S	178 05 W
Hrvatska (local name for Croatia)	Croatia	45 10 N	15 30 E
Hudson Bay	Arctic Ocean	60 00 N	86 00 W
Hudson Strait	Arctic Ocean	62 00 N	71 00 W
Hunter Island	New Caledonia, Vanuatu	22 24 S	172 06 E
Iberian Peninsula	Portugal, Spain	40 00 N	5 00 W
Iceland Sea	Arctic Ocean	68 00 N	20 00 W
Ifni (region; former name of part of Spanish West Africa)	Morocco	29 22 N	10 09 W
Inaccessible Island	Saint Helena, Ascension, and Tristan da Cunha	37 17 S	12 40 W
Indochina (region)	Cambodia, Laos, Vietnam	15 00 N	107 00 E
Ingushetia (region)	Russia	43 15 N	45 00 E
Inhambane (region)	Mozambique	22 30 S	34 30 E
Inini (former name for French Guiana)	French Guiana	4 00 N	53 00 W
Inland Sea	Japan	34 20 N	133 30 E
Inner Hebrides (islands)	United Kingdom	56 30 N	6 20 W
Inner Mongolia (region; also Nei Mongol)	China	42 00 N	113 00 E
Ionian Islands	Greece	38 30 N	20 30 E

NAME	ENTRY INTHE WORLD FACTBOOK	LATITUDE(DEG MIN)	LONGITUDE(DEG MIN)
Ionian Sea	Atlantic Ocean	38 30 N	18 00 E
Irian Jaya (province)	Indonesia	5 00 S	138 00 E
Irish Sea	Atlantic Ocean	53 30 N	5 20 W
Iron Gate (river gorge)	Romania, Serbia	44 41 N	22 31 E
Iskenderun (region; formerly Alexandretta)	Turkey	36 34 N	36 08 E
Islamabad (capital)	Pakistan	33 42 N	73 10 E
Island (local name for Iceland)	Iceland	65 00 N	18 00 W
Islas Malvinas (island group)	Falkland Islands (Islas Malvinas)	51 45 S	59 00 W
Istanbul (city)	Turkey	41 01 N	28 58 E
Istrian Peninsula	Croatia, Slovenia	45 00 N	14 00 E
Italia (local name for Italy)	Italy	42 50 N	12 50 E
Italian East Africa (former name for Italian possessions in eastern Africa)	Eritrea, Ethiopia, Somalia	8 00 N	38 00 E
Italian Somaliland (former name for southern Somalia)	Somalia	10 00 N	49 00 E
Ittihad al-Imarat al-Arabiyah (local name for the United Arab Emirates)	United Arab Emirates	24 00 N	54 00 E
Iturup (island; see Etorofu)	Russia (de facto)	44 55 N	147 40 E
Ityop'iya (local name for Ethiopia)	Ethiopia	8 00 N	38 00 E
Ivory Coast (former name for Cote d'Ivoire)	Cote d'Ivoire	8 00 N	5 00 W
Iwo Jima (island)	Japan	24 47 N	141 20 E
Izmir (region)	Turkey	38 25 N	27 10 E
Jakarta (capital)	Indonesia	6 10 S	106 48 E
James Bay	Arctic Ocean	54 00 N	80 00 W
Jamestown (capital)	Saint Helena, Ascension, and Tristan da Cunha	15 56 S	5 44 W
Jammu (city)	India	32 42 N	74 52 E
Jammu and Kashmir (region)	India, Pakistan	34 00 N	76 00 E
Japan, Sea of	Pacific Ocean	40 00 N	135 00 E
Jars, Plain of	Laos	19 27 N	103 10 E
Java (island)	Indonesia	7 30 S	110 00 E
Java Sea	Pacific Ocean	5 00 S	110 00 E
Jerusalem (capital, proclaimed)	Israel, West Bank	31 47 N	35 14 E
Jiddah, Jeddah (city)	Saudi Arabia	21 30 N	39 12 E
Johannesburg (city)	South Africa	26 15 S	28 00 E
Joseph Bonaparte Gulf	Pacific Ocean	14 00 S	128 45 E
Juan Fernandez, Islas de (island group)	Chile	33 00 S	80 00 W
Juan de Fuca, Strait of	Pacific Ocean	48 18 N	124 00 W
Juan de Nova Island	Indian Ocean	17 03 S	42 45 E
Jubal, Strait of	Indian Ocean	27 40 N	33 55 E
Judaea (region)	Israel, West Bank	31 35 N	35 00 E
Jugoslavia, Jugoslavija (local names for Yugoslavia, a former Balkan federation)	Bosnia and Herzegovina, Croatia, Macedonia, Montenegro, Serbia, Slovenia	43 00 N	21 00 E
Jutland (region)	Denmark	56 00 N	9 15 E
Juventud, Isla de la (Isle of Youth)	Cuba	21 40 N	82 50 W
Kabardino-Balkaria (region)	Russia	43 30 N	43 30 E
Kabul (capital)	Afghanistan	34 31 N	69 12 E
Kaduna (city)	Nigeria	10 33 N	7 27 E
Kailas Range	China, India	30 00 N	82 00 E
Kalaallit Nunaat (local name for Greenland)	Greenland	72 00 N	40 00 W
Kalahari (desert)	Botswana, Namibia	24 30 S	21 00 E
Kalimantan (region)	Indonesia	0 00 N	115 00 E
Kaliningrad (region; formerly part of East Prussia)	Russia	54 30 N	21 00 E
Kamaran (island)	Yemen	15 21 N	42 34 E
Kamchatka Peninsula (Poluostrov Kamchatka)	Russia	56 00 N	160 00 E
Kampala (capital)	Uganda	0 19 N	32 25 E

NAME	ENTRY INTHE WORLD FACTBOOK	LATITUDE(DEG MIN)	LONGITUDE(DEG MIN)
Kampuchea (former name for Cambodia)	Cambodia	13 00 N	105 00 E
Kane Basin (portion of channel)	Arctic Ocean	79 30 N	68 00 W
Kanton Island	Kiribati	2 49 S	171 40 W
Kara Sea	Arctic Ocean	76 00 N	80 00 E
Karachevo-Cherkessia (region)	Russia	43 40 N	41 50 E
Karachi (city)	Pakistan	24 51 N	67 03 E
Karafuto (island; former name for southern Sakhalin Island)	Russia	50 00 N	143 00 E
Karakoram Pass	China, India	35 30 N	77 50 E
Karelia, Kareliya (region)	Finland, Russia	63 15 N	30 48 E
Karelian Isthmus	Russia	60 25 N	30 00 E
Karimata Strait	Pacific Ocean	2 05 S	108 40 E
Kashmir (region)	India, Pakistan	34 00 N	76 00 E
Katanga (region)	Democratic Republic of the Congo	10 00 S	26 00 E
Kathmandu (capital)	Nepal	27 43 N	85 19 E
Kattegat (strait)	Atlantic Ocean	57 00 N	11 00 E
Kauai Channel	Pacific Ocean	21 45 N	158 50 W
Kazakstan (former name for Kazakhstan)	Kazakhstan	48 00 N	68 00 E
Keeling Islands	Cocos (Keeling) Islands	12 30 S	96 50 E
Kerguelen, Iles (island group)	French Southern and Antarctic Lands	49 30 S	69 30 E
Kermadec Islands	New Zealand	29 50 S	178 15 W
Kerulen River	China, Mongolia	48 48 N	117 00 E
Khabarovsk (city)	Russia	48 27 N	135 06 E
Khanka, Lake	China, Russia	45 00 N	132 24 E
Khartoum (capital)	Sudan	15 36 N	32 32 E
Khios (island)	Greece	38 22 N	26 04 E
Khmer Republic (former name for Cambodia)	Cambodia	13 00 N	105 00 E
Khuriya Muriya Islands (Kuria Muria Islands)	Oman	17 30 N	56 00 E
Khyber Pass	Afghanistan, Pakistan	34 05 N	71 10 E
Kibris (Turkish local name for Cyprus)	Cyprus	35 00 N	33 00 E
Kiel Canal (Nord-Ostsee Kanal)	Atlantic Ocean	53 53 N	9 08 E
Kiev (city; former name for Kyiv)	Ukraine	50 26 N	30 31 E
Kigali (capital)	Rwanda	1 57 S	30 04 E
Kingston (capital)	Jamaica	18 00 N	76 48 W
Kingston (capital)	Norfolk Island	29 03 S	167 58 E
Kingstown (capital)	Saint Vincent and the Grenadines	13 09 N	61 14 W
Kinshasa (capital)	Democratic Republic of the Congo	4 18 S	15 18 E
Kipros (Greek local name for Cyprus)	Cyprus	35 00 N	33 00 E
Kirghiziya, Kirgizia (former name for Kyrgyzstan)	Kyrgyzstan	41 00 N	75 00 E
Kirguizstan (local name for Kyrgyzstan)	Kyrgyzstan	41 00 N	75 00 E
Kiritimati (Christmas Island)	Kiribati	1 52 N	157 20 W
Kishinev (see Chisinau)	Moldova	47 00 N	28 50 E
Kithira Strait	Atlantic Ocean	36 00 N	23 00 E
Kobe (city)	Japan	34 41 N	135 10 E
Kodiak Island	United States	57 49 N	152 23 W
Kola Peninsula (Kol'skiy Poluostrov)	Russia	67 20 N	37 00 E
Kolonia (town; former capital; changed to Palikir)	Federated States of Micronesia	6 58 N	158 13 E
Korea Bay	Pacific Ocean	39 00 N	124 00 E
Korea Strait	Pacific Ocean	34 00 N	129 00 E
Korea, Democratic People's Republic of	North Korea	40 00 N	127 00 E
Korea, Republic of	South Korea	37 00 N	127 30 E
Koror (capital)	Palau	7 20 N	134 29 E
Kosovo (region)	Kosovo	42 30 N	21 00 E

NAME	ENTRY INTHE WORLD FACTBOOK	LATITUDE(DEG MIN)	LONGITUDE(DEG MIN)
Kosrae (island)	Federated States of Micronesia	5 20 N	163 00 E
Kowloon (city)	Hong Kong	22 18 N	114 10 E
Kra, Isthmus of	Burma, Thailand	10 20 N	99 00 E
Krakatoa (volcano)	Indonesia	6 07 S	105 24 E
Krakow (city)	Poland	50 03 N	19 56 E
Kuala Lumpur (capital)	Malaysia	3 10 N	101 42 E
Kunashiri (island; also Kunashir)	Russia (de facto)	44 20 N	146 00 E
Kunlun Mountains	China	36 00 N	84 00 E
Kuril Islands	Russia (de facto)	46 10 N	152 00 E
Kuwait (capital)	Kuwait	29 20 N	47 59 E
Kuznetsk Basin	Russia	54 00 N	86 00 E
Kwajalein Atoll	Marshall Islands	9 05 N	167 20 E
Kyiv (capital)	Ukraine	50 26 N	30 31 E
Kyushu (island)	Japan	33 00 N	131 00 E
La Paz (administrative capital)	Bolivia	16 30 S	68 09 W
La Perouse Strait	Pacific Ocean	45 45 N	142 00 E
Labrador (peninsula, region)	Canada	54 00 N	62 00 W
Labrador Sea	Atlantic Ocean	60 00 N	55 00 W
Laccadive Islands	India	10 00 N	73 00 E
Laccadive Sea	Indian Ocean	7 00 N	76 00 E
Lagos (former capital)	Nigeria	6 27 N	3 24 E
Lahore (city)	Pakistan	31 33 N	74 23 E
Lake Erie	Atlantic Ocean	42 30 N	81 00 W
Lake Huron	Atlantic Ocean	45 00 N	83 00 W
Lake Michigan	Atlantic Ocean	43 30 N	87 30 W
Lake Ontario	Atlantic Ocean	43 30 N	78 00 W
Lake Superior	Atlantic Ocean	48 00 N	88 00 W
Lakshadweep (Laccadive Islands)	India	10 00 N	73 00 E
Lantau Island	Hong Kong	22 15 N	113 55 E
Lao (local name for Laos)	Laos	18 00 N	105 00 E
Laptev Sea	Arctic Ocean	76 00 N	126 00 E
Las Palmas (city)	Spain (Canary Islands)	28 06 N	15 24 W
Latakia (region)	Syria	36 00 N	35 50 E
Latvija (local name for Latvia)	Latvia	57 00 N	25 00 E
Lau Group (island group)	Fiji	18 20 S	178 30 E
Lefkosa (see Nicosia)	Cyprus	35 10 N	33 22 E
Leipzig (city)	Germany	51 21 N	12 23 E
Lemnos (island)	Greece	39 54 N	25 21 E
Leningrad (city; former name for Saint Petersburg)	Russia	59 55 N	30 15 E
Lesser Sunda Islands	Indonesia	9 00 S	120 00 E
Lesvos (island)	Greece	39 15 N	26 15 E
Leyte (island)	Philippines	10 50 N	124 50 E
Liancourt Rocks (claimed by Japan)	South Korea	37 15 N	131 50 E
Liaodong Wan (gulf)	Pacific Ocean	40 30 N	121 20 E
Liban (local name for Lebanon)	Lebanon	33 50 N	36 50 E
Libreville (capital)	Gabon	0 23 N	9 27 E
Lietuva (local name for Lithuania)	Lithuania	56 00 N	24 00 E
Ligurian Sea	Atlantic Ocean	43 30 N	9 00 E
Lilongwe (capital)	Malawi	13 59 S	33 44 E
Lima (capital)	Peru	12 03 S	77 03 W
Lincoln Sea	Arctic Ocean	83 00 N	56 00 W
Line Islands	Jarvis Island, Kingman Reef, Kiribati, Palmyra Atoll	0 05 N	157 00 W
Lion, Gulf of	Atlantic Ocean	43 20 N	4 00 E
Lisbon (capital)	Portugal	38 43 N	9 08 W
Little Belt (strait; also Lille Baelt)	Atlantic Ocean	55 05 N	9 55 E

NAME	ENTRY INTHE WORLD FACTBOOK	LATITUDE(DEG MIN)	LONGITUDE(DEG MIN)
Ljubljana (capital)	Slovenia	46 03 N	14 31 E
Llanos (region)	Venezuela	8 00 N	68 00 W
Lobamba (city)	Swaziland	26 27 S	31 12 E
Lombok (island)	Indonesia	8 28 S	116 40 E
Lombok Strait	Indian Ocean	8 30 S	115 50 E
Lome (capital)	Togo	6 08 N	1 13 E
London (capital)	United Kingdom	51 30 N	0 10 W
Longyearbyen (capital)	Svalbard	78 13 N	15 33 E
Lord Howe Island	Australia	31 30 S	159 00 E
Lorraine (region)	France	48 42 N	6 11 E
Louisiade Archipelago	Papua New Guinea	11 00 S	153 00 E
Lourenco Marques (city; former name for Maputo)	Mozambique	25 56 S	32 34 E
Loyalty Islands (Iles Loyaute)	New Caledonia	21 00 S	167 00 E
Luanda (capital)	Angola	8 48 S	13 14 E
Lubnan (local name for Lebanon)	Lebanon	33 50 N	36 50 E
Lubumbashi (city)	Democratic Republic of the Congo	11 40 S	27 28 E
Lusaka (capital)	Zambia	15 25 S	28 17 E
Luxembourg (capital)	Luxembourg	49 45 N	6 10 E
Luzon (island)	Philippines	16 00 N	121 00 E
Luzon Strait	Pacific Ocean	20 30 N	121 00 E
Lyakhov Islands	Russia	73 45 N	138 00 E
Macao	Macau	22 10 N	113 33 E
Macau (special administrative region)	China	22 10 N	113 33 E
Macquarie Island	Australia	54 36 S	158 54 E
Madagasikara (local name for Madagascar)	Madagascar	20 00 S	47 00 E
Maddalena, Isola	Italy	41 13 N	09 24 E
Madeira Islands	Portugal	32 40 N	16 45 W
Madras (city; see Chennai)	India	13 04 N	80 16 E
Madrid (capital)	Spain	40 24 N	3 41 W
Magellan, Strait of	Atlantic Ocean	54 00 S	71 00 W
Maghreb (region)	Algeria, Libya, Mauritania, Morocco, Tunisia	34 00 N	3 00 E
Magreb (local name for Morocco)	Morocco	32 00 N	5 00 W
Magyarorszag (local name for Hungary)	Hungary	47 00 N	20 00 E
Mahe Island	Seychelles	4 41 S	55 30 E
Maiz, Islas del (Corn Islands)	Nicaragua	12 15 N	83 00 W
Majorca Island (Isla de Mallorca)	Spain	39 30 N	3 00 E
Majuro (capital)	Marshall Islands	7 05 N	171 08 E
Makassar Strait	Pacific Ocean	2 00 S	117 30 E
Makedonija (local name for Macedonia)	Macedonia	41 50 N	22 00 E
Malabo (capital)	Equatorial Guinea	3 45 N	8 47 E
Malacca, Strait of	Indian Ocean	2 30 N	101 20 E
Malagasy Republic	Madagascar	20 00 S	47 00 E
Malay Archipelago	Brunei, Indonesia, Malaysia, Papua New Guinea, Philippines	2 30 N	120 00 E
Malay Peninsula	Malaysia, Thailand	7 10 N	100 35 E
Male (capital)	Maldives	4 10 N	73 31 E
Mallorca, Isla de (island; also Majorca)	Spain	39 30 N	3 00 E
Malmady (region)	Belgium	50 26 N	6 02 E
Malpelo, Isla de (island)	Colombia	4 00 N	90 30 W
Malta Channel	Atlantic Ocean	56 44 N	26 53 E
Malvinas, Islas (island group)	Falkland Islands (Islas Malvinas)	51 45 S	59 00 W
Mamoutzou (capital)	Mayotte	12 47 S	45 14 E
Managua (capital)	Nicaragua	12 09 N	86 17 W
Manama (capital)	Bahrain	26 13 N	50 35 E

NAME	ENTRY IN THE WORLD FACTBOOK	LATITUDE(DEG MIN)	LONGITUDE(DEG MIN)
Manchukuo (former state)	China	44 00 N	124 00 E
Manchuria (region)	China	44 00 N	124 00 E
Manila (capital)	Philippines	14 35 N	121 00 E
Manipa Strait	Pacific Ocean	3 20 S	127 23 E
Mannar, Gulf of	Indian Ocean	8 30 N	79 00 E
Manua Islands	American Samoa	14 13 S	169 35 W
Maputo (capital)	Mozambique	25 58 S	32 35 E
Marcus Island (Minami-tori-shima)	Japan	24 16 N	154 00 E
Margarita, Isla (island)	Venezuela	10 00 N	64 00 W
Mariana Islands	Guam, Northern Mariana Islands	16 00 N	145 30 E
Marie Byrd Land (region)	Antarctica	77 00 S	130 00 W
Marigot (capital)	Saint Martin	18 04 N	63 05 W
Marion Island	South Africa	46 51 S	37 52 E
Marmara, Sea of	Atlantic Ocean	40 40 N	28 15 E
Marquesas Islands (Iles Marquises)	French Polynesia	9 00 S	139 30 W
Marseille (city)	France	43 18 N	5 23 E
Martin Vaz, Ilhas (island group)	Brazil	20 30 S	28 51 W
Mas a Tierra (Robinson Crusoe Island)	Chile	33 38 S	78 52 W
Mascarene Islands	Mauritius, Reunion	21 00 S	57 00 E
Maseru (capital)	Lesotho	29 28 S	27 30 E
Mata-Utu (capital)	Wallis and Futuna	13 57 S	171 56 W
Matsu (island)	Taiwan	26 13 N	119 56 E
Matthew Island	New Caledonia, Vanuatu	22 20 S	171 20 E
Mauritanie (local name for Mauritania)	Mauritania	20 00 N	12 00 W
Mazatlan (city)	Mexico	23 13 N	106 25 W
Mbabane (capital)	Swaziland	26 18 S	31 06 E
McDonald Islands	Heard Island and McDonald Islands	53 06 S	73 30 E
Mecca (city)	Saudi Arabia	21 27 N	39 49 E
Mediterranean Sea	Atlantic Ocean	36 00 N	15 00 E
Melbourne (city)	Australia	37 49 S	144 58 E
Melilla (exclave)	Spain	35 19 N	2 58 W
Memel (region)	Lithuania	55 43 N	21 30 E
Mesopotamia (region)	Iraq	33 00 N	44 00 E
Messina, Strait of	Atlantic Ocean	38 15 N	15 35 E
Mexico City (capital)	Mexico	19 24 N	99 09 W
Mexico, Gulf of	Atlantic Ocean	25 00 N	90 00 W
Middle Congo (former name for Republic of the Congo)	Republic of the Congo	1 00 S	15 00 E
Milan (city)	Italy	45 28 N	9 11 E
Milwaukee Deep (Puerto Rico Trench)	Atlantic Ocean	19 55 N	65 27 W
Minami-tori-shima (Marcus Island)	Japan	24 16 N	154 00 E
Mindanao (island)	Philippines	8 00 N	125 00 E
Mindanao Sea	Pacific Ocean	9 15 N	124 30 E
Mindoro (island)	Philippines	12 50 N	121 05 E
Mindoro Strait	Pacific Ocean	12 20 N	120 40 E
Mingrelia (region)	Georgia	42 30 N	41 52 E
Minicoy Island	India	8 17 N	73 02 E
Minorca Island (Isla de Menorca)	Spain	40 00 N	4 00 E
Minsk (capital)	Belarus	53 54 N	27 34 E
Misr (local name for Egypt)	Egypt	27 00 N	30 00 E
Mitla Pass	Egypt	30 02 N	32 54 E
Mocambique (local name for Mozambique)	Mozambique	18 15 S	35 00 E
Mogadishu (capital)	Somalia	2 04 N	45 22 E
Moldavia (region)	Moldova, Romania	47 00 N	29 00 E
Molucca Sea	Pacific Ocean	2 00 N	127 00 E
Moluccas (Spice Islands)	Indonesia	2 00 S	128 00 E

NAME	ENTRY INTHE WORLD FACTBOOK	LATITUDE(DEG MIN)	LONGITUDE(DEG MIN)
Mombasa (city)	Kenya	4 03 S	39 40 E
Mona Passage	Atlantic Ocean	18 30 N	67 45 W
Monaco (capital)	Monaco	43 44 N	7 25 E
Mongol Uls (local name for Mongolia)	Mongolia	46 00 N	105 00 E
Monrovia (capital)	Liberia	6 18 N	10 47 W
Monterrey (city)	Mexico	25 40 N	100 19 W
Montevideo (capital)	Uruguay	34 53 S	56 11 W
Montreal (city)	Canada	45 31 N	73 34 W
Moravia (region)	Czech Republic	49 30 N	17 00 E
Moravian Gate (pass)	Czech Republic	49 35 N	17 50 E
Moroni (capital)	Comoros	11 41 S	43 16 E
Mortlock Islands (Nomoi Islands)	Federated States of Micronesia	5 30 N	153 40 E
Moscow (capital)	Russia	55 45 N	37 35 E
Mount Pinatubo (volcano)	Philippines	15 08 N	120 21 E
Mozambique Channel	Indian Ocean	19 00 S	41 00 E
Mumbai (city; also Bombay)	India	18 58 N	72 50 E
Munich, Muenchen (city)	Germany	48 08 N	11 35 E
Muritaniyah (local name for Mauritania)	Mauritania	20 00 N	12 00 W
Musandam Peninsula	Oman, United Arab Emirates	26 18 N	56 24 E
Muscat (capital)	Oman	23 37 N	58 35 E
Muscat and Oman (former name for Oman)	Oman	21 00 N	57 00 E
Myanma, Myanmar	Burma	22 00 N	98 00 E
N'Djamena (capital)	Chad	12 07 N	15 03 E
Nagorno-Karabakh (region)	Azerbaijan	40 00 N	46 40 E
Nairobi (capital)	Kenya	1 17 S	36 49 E
Namib (desert)	Namibia	24 00 S	15 00 E
Nampo-shoto (island group)	Japan	30 00 N	140 00 E
Nan Madol (ruins)	Federated States of Micronesia	6 85 N	158 35 E
Naples (city)	Italy	40 51 N	14 15 E
Nassau (capital)	The Bahamas	25 05 N	77 21 W
Natal (region)	South Africa	29 00 S	30 25 E
Natuna Besar Islands	Indonesia	3 30 N	102 30 E
Natuna Sea	Pacific Ocean	3 30 N	108 00 E
Naxcivan (region)	Azerbaijan	39 20 N	45 20 E
Naxos (island)	Greece	37 05 N	25 30 E
Nederland (local name for the Netherlands)	Netherlands	52 30 N	5 45 E
Nederlandse Antillen (local name for the former Netherlands Antilles)	Curacao, Sint Maarten	12 15 N	68 45 W
Negev (region)	Israel	30 30 N	34 55 E
Negros (island)	Philippines	10 00 N	123 00 E
Nejd (region)	Saudi Arabia	24 05 N	45 15 E
Netherlands Antilles (former name of Dutch Caribbean dependencies)	Curacao, Sint Maarten	12 15 N	68 45 W
Netherlands East Indies (former name for Indonesia)	Indonesia	5 00 S	120 00 E
Netherlands Guiana (former name for Suriname)	Suriname	4 00 N	56 00 W
Nevis (island)	Saint Kitts and Nevis	17 09 N	62 35 W
New Britain (island)	Papua New Guinea	6 00 S	150 00 E
New Delhi (capital)	India	28 36 N	77 12 E
New Guinea (island)	Indonesia, Papua New Guinea	5 00 S	140 00 E
New Hebrides (island group)	Vanuatu	16 00 S	167 00 E
New Ireland (island)	Papua New Guinea	3 20 N	152 00 E
New Siberian Islands	Russia	75 00 N	142 00 E
New Territories (mainland region)	Hong Kong	22 24 N	114 10 E
Newfoundland (island, with mainland area, and a province)	Canada	52 00 N	56 00 W
Niamey (capital)	Niger	13 31 N	2 07 E

NAME	ENTRY INTHE WORLD FACTBOOK	LATITUDE(DEG MIN)	LONGITUDE(DEG MIN)
Nicobar Islands	India	8 00 N	93 30 E
Nicosia (capital; also Lefkosia)	Cyprus	35 10 N	33 22 E
Nightingale Island	Saint Helena, Ascension, and Tristan da Cunha	37 25 S	12 30 W
Nihon, Nippon (local name for Japan)	Japan	36 00 N	138 00 E
Nomoi Islands (Mortlock Islands)	Federated States of Micronesia	5 30 N	153 40 E
Norge (local name for Norway)	Norway	62 00 N	10 00 E
Norman Isles (Channel Islands)	Guernsey, Jersey	49 20 N	2 20 W
North Atlantic Ocean	Atlantic Ocean	30 00 N	45 00 W
North Channel	Atlantic Ocean	55 10 N	5 40 W
North Frisian Islands	Denmark, Germany	54 50 N	8 12 E
North Greenland Sea	Arctic Ocean	78 00 N	5 00 W
North Island	New Zealand	39 00 S	176 00 E
North Ossetia (region)	Russia	43 00 N	44 10 E
North Pacific Ocean	Pacific Ocean	30 00 N	165 00 W
North Sea	Atlantic Ocean	56 00 N	4 00 E
North Vietnam (former name for northern portion of Vietnam)	Vietnam	23 00 N	106 00 E
North Yemen (Yemen Arab Republic; now part of Yemen)	Yemen	15 00 N	44 00 E
Northeast Providence Channel	Atlantic Ocean	25 40 N	77 09 W
Northern Areas	Pakistan	36 0 N	75 0 E
Northern Cyprus (region)	Cyprus	35 15 N	33 44 E
Northern Epirus (region)	Albania, Greece	40 00 N	20 30 E
Northern Grenadines (political region)	Saint Vincent and the Grenadines	12 45 N	61 15 W
Northern Ireland	United Kingdom	54 40 N	6 45 W
Northern Rhodesia (former name for Zambia)	Zambia	15 00 S	30 00 E
Northwest Passages	Arctic Ocean	74 40 N	100 00 W
Norwegian Sea	Atlantic Ocean	66 00 N	6 00 E
Nouakchott (capital)	Mauritania	18 06 N	15 57 W
Noumea (capital)	New Caledonia	22 16 S	166 27 E
Nouvelle-Caledonie (local name for New Caledonia)	New Caledonia	21 30 S	165 30 E
Nouvelles Hebrides (former name for Vanuatu)	Vanuatu	16 00 S	167 00 E
Novaya Zemlya (islands)	Russia	74 00 N	57 00 E
Nubia (region)	Egypt, Sudan	20 30 N	33 00 E
Nuku'alofa (capital)	Tonga	21 08 S	175 12 W
Nunavut (region)	Canada	72 00 N	90 00 W
Nuuk (capital; also Godthab)	Greenland	64 11 N	51 44 W
Nyasaland (former name for Malawi)	Malawi	13 30 S	34 00 E
Nyassa (region)	Mozambique	13 30 S	37 00 E
Oahu (island)	United States (Hawaii)	21 30 N	158 00 W
Ocean Island (Banaba)	Kiribati	0 52 S	169 35 E
Ocean Island (Kure Island)	United States	28 25 N	178 20 W
Oesterreich (local name for Austria)	Austria	47 20 N	13 20 E
Ogaden (region)	Ethiopia, Somalia	7 00 N	46 00 E
Oil Islands (Chagos Archipelago)	British Indian Ocean Territory	6 00 S	71 30 E
Okhotsk, Sea of	Pacific Ocean	53 00 N	150 00 E
Okinawa (island group)	Japan	26 30 N	128 00 E
Oland (island)	Sweden	56 45 N	16 40 E
Oman, Gulf of	Indian Ocean	24 30 N	58 30 E
Ombai Strait	Pacific Ocean	8 30 S	125 00 E
Oran (city)	Algeria	35 43 N	0 43 W
Orange River Colony (region; former name of Free State Province of South Africa)	South Africa	28 20 S	26 40 E
Oranjestad (capital)	Aruba	12 33 N	70 06 W
Oresund (The Sound) (strait)	Atlantic Ocean	55 50 N	12 40 E

NAME	ENTRY INTHE WORLD FACTBOOK	LATITUDE(DEG MIN)	LONGITUDE(DEG MIN)
Orkney Islands	United Kingdom	59 00 N	3 00 W
Osaka (city)	Japan	34 42 N	135 30 E
Oslo (capital)	Norway	59 55 N	10 45 E
Osumi Strait (Van Diemen Strait)	Pacific Ocean	31 00 N	131 00 E
Otranto, Strait of	Atlantic Ocean	40 00 N	19 00 E
Ottawa (capital)	Canada	45 25 N	75 40 W
Ouagadougou (capital)	Burkina Faso	12 22 N	1 31 W
Outer Hebrides (islands)	United Kingdom	57 45 N	7 00 W
Outer Mongolia (region)	Mongolia	46 00 N	105 00 E
P'yongyang (capital)	North Korea	39 01 N	125 45 E
Pacific Islands, Trust Territory of the (former name of a large area of the western North Pacific Ocean)	Marshall Islands, Federated States of Micronesia, Northern Mariana Islands, Palau	10 00 N	155 00 E
Pagan (island)	Northern Mariana Islands	18 08 N	145 47 E
Pago Pago (capital)	American Samoa	14 16 S	170 42 W
Palawan (island)	Philippines	9 30 N	118 30 E
Palermo (city)	Italy	38 07 N	13 21 E
Palestine (region)	Israel, West Bank	32 00 N	35 15 E
Palikir (capital)	Federated States of Micronesia	6 55 N	158 08 E
Palk Strait	Indian Ocean	10 00 N	79 45 E
Pamirs (mountains)	China, Tajikistan	38 00 N	73 00 E
Pampas (region)	Argentina	35 00 S	63 00 W
Panama (capital)	Panama	8 58 N	79 32 W
Panama Canal	Panama	9 00 N	79 45 W
Panama, Gulf of	Pacific Ocean	8 00 N	79 30 W
Panay (island)	Philippines	11 15 N	122 30 E
Pantelleria, Isola di (island)	Italy	36 47 N	12 00 E
Papeete (capital)	French Polynesia	17 32 S	149 34 W
Paramaribo (capital)	Suriname	5 50 N	55 10 W
Parece Vela (island)	Japan	20 20 N	136 00 E
Paris (capital)	France	48 52 N	2 20 E
Pascua, Isla de (Easter Island)	Chile	27 07 S	109 22 W
Pashtunistan (region)	Afghanistan, Pakistan	32 00 N	69 00 E
Passion, Ile de la (island)	Clipperton Island	10 17 N	109 13 W
Patagonia (region)	Argentina	48 00 S	61 00 W
Peking (see Beijing)	China	39 56 N	116 24 E
Pelagian Islands (Isole Pelagie)	Italy	35 40 N	12 40 E
Peleliu (Beliliou) (island)	Palau	7 01 N	134 15 E
Peloponnese (peninsula)	Greece	37 30 N	22 25 E
Pemba Island	Tanzania	5 20 S	39 45 E
Penang Island	Malaysia	5 23 N	100 15 E
Pentland Firth (channel)	Atlantic Ocean	58 44 N	3 13 W
Perim (island)	Yemen	12 39 N	43 25 E
Perouse Strait, La	Pacific Ocean	44 45 N	142 00 E
Persia (former name for Iran)	Iran	32 00 N	53 00 E
Persian Gulf	Indian Ocean	27 00 N	51 00 E
Perth (city)	Australia	31 56 S	115 50 E
Pescadores (islands)	Taiwan	23 30 N	119 30 E
Peshawar (city)	Pakistan	34 01 N	71 40 E
Peter I Island	Antarctica	68 48 S	90 35 W
Petrograd (city; former name for Saint Petersburg)	Russia	59 55 N	30 15 E
Philip Island	Norfolk Island	29 08 S	167 57 E
Philippine Sea	Pacific Ocean	20 00 N	134 00 E
Philipsburg (capital)	Sint Maarten	18 1 N	63 2 W
Phnom Penh (capital)	Cambodia	11 33 N	104 55 E
Phoenix Islands	Kiribati	3 30 S	172 00 W
Pinatubo, Mount (volcano)	Philippines	15 08 N	120 21 E

NAME	ENTRY IN THE WORLD FACTBOOK	LATITUDE(DEG MIN)	LONGITUDE(DEG MIN)
Pines, Isle of (island; former name for Isla de la Juventud)	Cuba	21 40 N	82 50 W
Pleasant Island	Nauru	0 32 S	166 55 E
Plymouth (capital)	Montserrat	16 44 N	62 14 W
Podgorica (administrative capital)	Montenegro	42 26 N	19 16 E
Polska (local name)	Poland	52 00 N	20 00 E
Polynesie Francaise (local name for French Polynesia)	French Polynesia	15 00 S	140 00 W
Pomerania (region)	Germany, Poland	53 40 N	15 35 E
Ponape (Pohnpei) (island)	Federated States of Micronesia	6 55 N	158 15 E
Port Louis (capital)	Mauritius	20 10 S	57 30 E
Port Moresby (capital)	Papua New Guinea	9 30 S	147 10 E
Port-Vila (capital)	Vanuatu	17 44 S	168 19 E
Port-au-Prince (capital)	Haiti	18 32 N	72 20 W
Port-of-Spain (capital)	Trinidad and Tobago	10 39 N	61 31 W
Porto-Novo (capital)	Benin	6 29 N	2 37 E
Portuguese East Africa (former name for Mozambique)	Mozambique	18 15 S	35 00 E
Portuguese Guinea (former name for Guinea-Bissau)	Guinea-Bissau	12 00 N	15 00 W
Portuguese Timor (former name for Timor-Leste)	Timor-Leste	9 00 S	126 00 E
Poznan (city)	Poland	52 25 N	16 55 E
Prague (capital)	Czech Republic	50 05 N	14 28 E
Praia (capital)	Cape Verde	14 55 N	23 31 W
Prathet Thai (local name for Thailand)	Thailand	15 00 N	100 00 E
Pretoria (administrative capital)	South Africa	25 42 S	28 13 E
Prevlaka peninsula	Croatia	42 24 N	18 31 E
Pribilof Islands	United States	57 00 N	170 00 W
Prince Edward Island	Canada	46 20 N	63 20 W
Prince Edward Islands	South Africa	46 35 S	38 00 E
Prince Patrick Island	Canada	76 30 N	119 00 W
Principe (island)	Sao Tome and Principe	1 38 N	7 25 E
Pristina, Prishtina, Prishtine (capital)	Kosovo	42 40 N	21 10 E
Prussia (region)	Germany, Poland, Russia	53 00 N	14 00 E
Pukapuka Atoll	Cook Islands	10 53 S	165 49 W
Punjab (region)	India, Pakistan	30 50 N	73 30 E
Puntland (region)	Somalia	8 21 N	49 08 E
Qazaqstan (local name for Kazakhstan)	Kazakhstan	48 00 N	68 00 E
Qita Ghazzah (local name Gaza Strip)	Gaza Strip	31 25 N	34 20 E
Quebec (city)	Canada	46 48 N	71 15 W
Queen Charlotte Islands	Canada	53 00 N	132 00 W
Queen Elizabeth Islands	Canada	78 00 N	95 00 W
Queen Maud Land (claimed by Norway)	Antarctica	73 30 S	12 00 E
Quemoy (island)	Taiwan	24 27 N	118 23 E
Quito (capital)	Ecuador	0 13 S	78 30 W
Rabat (capital)	Morocco	34 02 N	6 51 W
Ralik Chain (island group)	Marshall Islands	8 00 N	167 00 E
Rangoon (capital; also Yangon)	Burma	16 47 N	96 10 E
Rapa Nui (Easter Island)	Chile	27 07 S	109 22 W
Ratak Chain (island group)	Marshall Islands	9 00 N	171 00 E
Red Sea	Indian Ocean	20 00 N	38 00 E
Redonda (island)	Antigua and Barbuda	16 55 N	62 19 W
Republica Dominicana (local name for Dominican Republic)	Dominican Republic	19 00 N	70 40 W
Republique Centrafricain (local name for Central African Republic)	Central African Republic	7 00 N	21 00 E
Republique Francaise (local name for France)	France	46 00 N	2 00 E
Republique Gabonaise (local name for Gabon)	Gabon	1 00 S	11 45 E

NAME	ENTRY INTHE WORLD FACTBOOK	LATITUDE(DEG MIN)	LONGITUDE(DEG MIN)
Republique Rwandaise (local name for Rwanda)	Rwanda	2 00 S	30 00 E
Republique Togolaise (local name for Togo)	Togo	8 00 N	1 10 E
Revillagigedo Island	United States (Alaska)	55 35 N	131 06 W
Revillagigedo Islands	Mexico	19 00 N	112 45 W
Reykjavik (capital)	Iceland	64 09 N	21 57 W
Rhodes (island)	Greece	36 10 N	28 00 E
Rhodesia, Northern (former name for Zambia)	Zambia	15 00 S	30 00 E
Rhodesia, Southern (former name for Zimbabwe)	Zimbabwe	20 00 S	30 00 E
Riga (capital)	Latvia	56 57 N	24 06 E
Riga, Gulf of	Atlantic Ocean	57 30 N	23 30 E
Rio Muni (mainland region)	Equatorial Guinea	1 30 N	10 00 E
Rio de Janiero (city)	Brazil	22 55 S	43 17 W
Rio de Oro (region)	Western Sahara	23 45 N	15 45 W
Rio de la Plata (gulf)	Atlantic Ocean	35 00 S	59 00 W
Riyadh (capital)	Saudi Arabia	24 38 N	46 43 E
Road Town (capital)	British Virgin Islands	18 27 N	64 37 W
Robinson Crusoe Island (Mas a Tierra)	Chile	33 38 S	78 52 W
Rocas, Atol das (island)	Brazil	3 51 S	33 49 W
Rockall (island)	United Kingdom	57 35 N	13 48 W
Rodrigues (island)	Mauritius	19 42 S	63 25 E
Rome (capital)	Italy	41 54 N	12 29 E
Roncador Cay (island)	Colombia	13 32 N	80 03 W
Roosevelt Island	Antarctica	79 30 S	162 00 W
Roseau (capital)	Dominica	15 18 N	61 24 W
Ross Dependency (claimed by New Zealand)	Antarctica	80 00 S	180 00 E
Ross Island	Antarctica	81 30 S	175 00 W
Ross Sea	Antarctica, Southern Ocean	76 00 S	175 00 W
Rossiya (local name for Russia)	Russia	60 00 N	100 00 E
Rota (island)	Northern Mariana Islands	14 10 N	145 12 E
Rotuma (island)	Fiji	12 30 S	177 05 E
Ruanda (former name for Rwanda)	Rwanda	2 00 S	30 00 E
Rub al Khali (desert)	Saudi Arabia	19 30 N	49 00 E
Rumelia (region)	Albania, Bulgaria, Macedonia	42 00 N	22 30 E
Ruthenia (region; former name for Carpatho-Ukraine)	Ukraine	48 22 N	23 32 E
Ryukyu Islands	Japan	26 30 N	128 00 E
Saar (region)	Germany	49 25 N	7 00 E
Saaremaa (island)	Estonia	58 25 N	22 30 E
Saba (island)	Netherlands	17 38 N	63 10 W
Sabah (state)	Malaysia	5 20 N	117 10 E
Sable Island	Canada	43 55 N	59 50 W
Safety Islands (Iles du Salut)	French Guiana	5 20 N	52 37 W
Sahara Occidental (former name for Western Sahara)	Western Sahara	24 30 N	13 00 W
Sahel (region)	Burkina Faso, Chad, The Gambia, Guinea- Bissau, Mali, Mauritania, Niger, Senegal	15 00 N	8 00 W
Saigon (city; former name for Ho Chi Minh City)	Vietnam	10 45 N	106 40 E
Saint Brandon (Cargados Carajos Shoals)	Mauritius	16 25 S	59 38 E
Saint Christopher (island)	Saint Kitts and Nevis	17 20 N	62 45 W
Saint Christopher and Nevis	Saint Kitts and Nevis	17 20 N	62 45 W
Saint Eustatius (island)	Netherlands	17 30 N	63 00 W
Saint George's (capital)	Grenada	12 03 N	61 45 W
Saint George's Channel	Atlantic Ocean	52 00 N	6 00 W
Saint Helena Island	Saint Helena, Ascension, and Tristan da Cunha	15 57 S	5 42 W

NAME	ENTRY INTHE WORLD FACTBOOK	LATITUDE(DEG MIN)	LONGITUDE(DEG MIN)
Saint Helens, Mount (volcano)	United States	46 15 N	122 12 W
Saint Helier (capital)	Jersey	49 12 N	2 07 W
Saint John (city)	Canada (New Brunswick)	45 16 N	66 04 W
Saint John's (capital)	Antigua and Barbuda	17 06 N	61 51 W
Saint Lawrence Island	United States	49 30 N	67 00 W
Saint Lawrence Seaway	Atlantic Ocean	49 15 N	67 00 W
Saint Lawrence, Gulf of	Atlantic Ocean	48 00 N	62 00 W
Saint Paul Island	Canada	47 12 N	60 09 W
Saint Paul Island	United States	57 11 N	170 16 W
Saint Paul Island (Ile Saint-Paul)	French Southern and Antarctic Lands	38 43 S	77 29 E
Saint Peter Port (capital)	Guernsey	49 27 N	2 32 W
Saint Peter and Saint Paul Rocks (Penedos de Sao Pedro e Sao Paulo)	Brazil	0 23 N	29 23 W
Saint Petersburg (city; former capital)	Russia	59 55 N	30 15 E
Saint Thomas (island)	Virgin Islands	18 21 N	64 55 W
Saint Vincent Passage	Atlantic Ocean	13 30 N	61 00 W
Saint-Denis (capital)	Reunion	20 52 S	55 28 E
Saint-Pierre (capital)	Saint Pierre and Miquelon	46 46 N	56 11 W
Saipan (island)	Northern Mariana Islands	15 12 N	145 45 E
Sak'art'velo (local name for Georgia)	Georgia	42 00 N	43 30 E
Sakhalin Island (Ostrov Sakhalin)	Russia	51 00 N	143 00 E
Sakishima Islands	Japan	24 30 N	124 00 E
Sala y Gomez, Isla (island)	Chile	26 28 S	105 00 W
Salisbury (city; former name for Harare)	Zimbabwe	17 50 S	105 00 W
Salzburg (city)	Austria	47 48 N	13 02 E
Samar (island)	Philippines	12 00 N	125 00 E
Samaria (region)	West Bank	32 15 N	35 10 E
Samoa Islands	American Samoa, Samoa	14 00 S	171 00 W
Samos (island)	Greece	37 48 N	26 44 E
San Ambrosio, Isla (island)	Chile	26 21 S	79 52 W
San Andres y Providencia, Archipielago (island group)	Colombia	13 00 N	81 30 W
San Bernardino Strait	Pacific Ocean	12 32 N	124 10 E
San Felix, Isla (island)	Chile	26 17 S	80 05 W
San Jose (capital)	Costa Rica	9 56 N	84 05 W
San Juan (capital)	Puerto Rico	18 28 N	66 07 W
San Marino (capital)	San Marino	43 56 N	12 25 E
San Salvador (capital)	El Salvador	13 42 N	89 12 W
Sanaa (capital)	Yemen	15 21 N	44 12 E
Sandzak (region)	Montenegro, Serbia	43 05 N	19 45 E
Santa Cruz (city)	Bolivia	17 48 S	63 10 W
Santa Cruz Islands	Solomon Islands	11 00 S	166 15 E
Santa Sede (local name for the Holy See)	Holy See	41 54 N	12 27 E
Santiago (capital)	Chile	33 27 S	70 40 W
Santo Antao (island)	Cape Verde	17 05 N	25 10 W
Santo Domingo (capital)	Dominican Republic	18 28 N	69 54 W
Sao Paulo (city)	Brazil	23 35 S	46 43 W
Sao Pedro e Sao Paulo, Penedos de (rocks)	Brazil	0 23 N	29 23 W
Sao Tiago (island)	Cape Verde	15 05 N	23 40 W
Sao Tome (island)	Sao Tome and Principe	0 12 N	6 39 E
Sapporo (city)	Japan	43 04 N	141 20 E
Sapudi Strait	Pacific Ocean	7 05 S	114 10 E
Sarajevo (capital)	Bosnia and Herzegovina	43 52 N	18 25 E
Sarawak (state)	Malaysia	2 30 N	113 30 E
Sardinia (island)	Italy	40 00 N	9 00 E
Sargasso Sea (region)	Atlantic Ocean	30 00 N	55 00 W
Sark (island)	Guernsey	49 26 N	2 21 W

NAME	ENTRY INTHE WORLD FACTBOOK	LATITUDE(DEG MIN)	LONGITUDE(DEG MIN)
Savage Island (former name for Niue)	Niue	19 02 S	169 52 W
Savu Sea	Pacific Ocean	9 30 S	122 00 E
Saxony (region)	Germany	51 00 N	13 00 E
Schleswig-Holstein (region)	Germany	54 31 N	9 33 E
Schweiz (local German name for Switzer-land)	Switzerland	47 00 N	8 00 E
Scopus, Mount	Israel, West Bank	31 48 N	35 14 E
Scotia Sea	Atlantic Ocean, Southern Ocean	56 00 S	40 00 W
Scotland (region)	United Kingdom	57 00 N	4 00 W
Scott Island	Antarctica	67 24 S	179 55 W
Senegambia (region; former name of confederation of Senegal and The Gambia)	The Gambia, Senegal	13 50 N	15 25 W
Senyavin Islands	Federated States of Micronesia	6 55 N	158 00 E
Seoul (capital)	South Korea	37 34 N	127 00 E
Serendib (former name for Sri Lanka)	Sri Lanka	7 00 N	81 00 E
Serrana Bank (shoal)	Colombia	14 25 N	80 16 W
Serranilla Bank (shoal)	Colombia	15 51 N	79 46 W
Settlement, The (capital)	Christmas Island	10 25 S	105 43 E
Severnaya Zemlya (island group; also Northland)	Russia	79 30 N	98 00 E
Shaba (region)	Democratic Republic of the Congo	8 00 S	27 00 E
Shag Island	Heard Island and McDonald Islands	53 00 S	72 30 E
Shag Rocks	South Georgia and the South Sandwich Islands	53 33 S	42 02 W
Shanghai (city)	China	31 14 N	121 30 E
Shenyang (city; also Mukden)	China	41 46 N	123 24 E
Shetland Islands	United Kingdom	60 30 N	1 30 W
Shikoku (island)	Japan	33 45 N	133 30 E
Shikotan (island)	Russia (de facto)	43 47 N	146 45 E
Shqiperia (local name for Albania)	Albania	41 00 N	20 00 E
Siam (former name for Thailand)	Thailand	15 00 N	100 00 E
Siberia (region)	Russia	60 00 N	100 00 E
Sibutu Passage	Pacific Ocean	4 50 N	119 35 E
Sicily (island)	Italy	37 30 N	14 00 E
Sicily, Strait of	Atlantic Ocean	37 20 N	11 20 E
Sidra, Gulf of	Atlantic Ocean	31 30 N	18 00 E
Sikkim (state)	India	27 50 N	88 30 E
Silesia (region)	Czech Republic, Germany, Poland	51 00 N	17 00 E
Sinai Peninsula	Egypt	29 30 N	34 00 E
Singapore (capital)	Singapore	1 17 N	103 51 E
Singapore Strait	Pacific Ocean	1 15 N	104 00 E
Sinkiang (autonomous region; also Xinjiang)	China	42 00 N	86 00 E
Sint Eustatius (island)	Netherlands	17 29 N	62 58 W
Sint Maarten (island; also Saint-Martin)	Sint Maarten, Saint Martin	18 04 N	63 04 W
Sjaelland (island)	Denmark	55 30 N	12 00 E
Skagerrak (strait)	Atlantic Ocean	57 45 N	9 00 E
Skopje (capital)	Macedonia	41 59 N	21 26 E
Slavonia (region)	Croatia	45 27 N	18 00 E
Slovenija (local name for Slovenia)	Slovenia	46 00 N	15 00 E
Slovensko (local name for Slovakia)	Slovakia	48 40 N	19 30 E
Smyrna (region; former name for Izmir)	Turkey	38 25 N	27 10 E
Society Islands (Iles de la Societe)	French Polynesia	17 00 S	150 00 W
Socotra (island)	Yemen	12 30 N	54 00 E
Sofia (capital)	Bulgaria	42 41 N	23 19 E
Solomon Islands, northern	Papua New Guinea	6 00 S	155 00 E
Solomon Islands, southern	Solomon Islands	8 00 S	159 00 E

NAME	ENTRY INTHE WORLD FACTBOOK	LATITUDE(DEG MIN)	LONGITUDE(DEG MIN)
Solomon Sea	Pacific Ocean	8 00 S	153 00 E
Somaliland (region)	Somalia	9 30 N	46 00 E
Somers Islands (former name for Bermuda)	Bermuda	32 20 N	64 45 W
Songkhla (city)	Thailand	7 12 N	100 36 E
Sound, The (strait; also Oresund)	Atlantic Ocean	55 50 N	12 40 E
South Atlantic Ocean	Atlantic Ocean	30 00 S	15 00 W
South China Sea	Pacific Ocean	10 00 N	113 00 E
South Georgia (island)	South Georgia and the South Sandwich Islands	54 15 S	36 45 W
South Island	New Zealand	43 00 S	171 00 E
South Korea	South Korea	37 00 N	127 30 E
South Orkney Islands	Antarctica	61 00 S	45 00 W
South Ossetia (region)	Georgia	42 20 N	44 00 E
South Pacific Ocean	Pacific Ocean	30 00 S	130 00 W
South Sandwich Islands	South Georgia and the South Sandwich Islands	57 45 S	26 30 W
South Shetland Islands	Antarctica	62 00 S	59 00 W
South Tyrol (region)	Italy	46 30 N	10 30 E
South Vietnam (former name for the southern portion of Vietnam)	Vietnam	12 00 N	108 00 E
South Yemen (People's Democratic Republic of Yemen; now part of Yemen)	Yemen	14 00 N	48 00 E
South-West Africa (former name for Namibia)	Namibia	22 00 S	17 00 E
Southern Grenadines (island group)	Grenada	12 20 N	61 30 W
Southern Rhodesia (former name for Zimbabwe)	Zimbabwe	20 00 S	30 00 E
Soviet Union (former name of a large Eurasian empire, roughly coequal with the former Russian Empire)	Armenia, Azerbaijan, Belarus, Estonia, Georgia, Kazakhstan, Kyrgyzstan, Latvia, Lithuania, Moldova, Russia, Tajikistan, Turkmenistan, Ukraine, Uzbekistan		
Spanish Guinea (former name for Equatorial Guinea)	Equatorial Guinea	2 00 N	10 00 E
Spanish Morocco (former name for northern Morocco)	Morocco	32 00 N	7 00 W
Spanish North Africa (exclaves)	Spain (Ceuta, Islas Chafarinas, Melilla, Penon de Alhucemas, Penon de Velez de la Gomera)	35 15 N	4 00 W
Spanish Sahara (former name)	Western Sahara	24 30 N	13 00 W
Spanish West Africa (former name for Ifni and Spanish Sahara)	Morocco, Western Sahara	25 00 N	13 00 W
Spice Islands (Moluccas)	Indonesia	2 00 S	28 00 E
Spitsbergen (island)	Svalbard	78 00 N	20 00 E
Srbija (local name for Serbia)	Serbia	44 00 N	21 00 E
St. John's (city)	Canada (Newfoundland)	47 34 N	52 43 W
Stanley (capital)	Falkland Islands (Islas Malvinas)	51 42 S	57 41 W
Stockholm (capital)	Sweden	59 20 N	18 03 E
Strasbourg (city)	France	48 35 N	7 44 E
Stuttgart (city)	Germany	48 46 N	9 11 E
Sucre (constitutional capital)	Bolivia	19 02 S	65 17 W
Suez Canal	Egypt	29 55 N	32 33 E
Suez, Gulf of	Indian Ocean	28 10 N	33 27 E
Suisse (local French name for Switzerland)	Switzerland	47 00 N	8 00 E
Sulawesi (island; Celebes)	Indonesia	2 00 S	121 00 E
Sulawesi Sea	Pacific Ocean	3 00 N	122 00 E
Sulu Archipelago (island group)	Philippines	6 00 N	121 00 E
Sulu Sea	Pacific Ocean	8 00 N	120 00 E

NAME	ENTRY INTHE WORLD FACTBOOK	LATITUDE(DEG MIN)	LONGITUDE(DEG MIN)
Sumatra (island)	Indonesia	0 00 N	102 00 E
Sumba (island)	Indonesia	10 00 S	120 00 E
Sumba Strait	Pacific Ocean	9 10 S	120 00 E
Sumbawa (island)	Indonesia	8 30 S	118 00 E
Sunda Islands (Soenda Isles)	Indonesia, Malaysia	2 00 S	110 00 E
Sunda Strait	Indian Ocean	6 00 S	105 45 E
Suomi (local name for Finland)	Finland	64 00 N	26 00 E
Surabaya (city)	Indonesia	7 13 S	112 45 E
Surigao Strait	Pacific Ocean	10 15 N	125 23 E
Surinam (former name for Suriname)	Suriname	4 00 N	56 00 W
Suriyah (local name for Syria)	Syria	35 00 N	38 00 E
Surtsey (volcanic island)	Iceland	63 17 N	20 40 W
Suva (capital)	Fiji	18 08 S	178 25 E
Sverdlovsk (city; also Yekaterinburg)	Russia	56 50 N	60 39 E
Sverige (local name for Sweden)	Sweden	62 00 N	15 00 E
Svizzera (local Italian name for Switzerland)	Switzerland	47 00 N	8 00 E
Swains Island	American Samoa	11 03 S	171 15 W
Swan Islands	Honduras	17 25 S	83 56 W
Sydney (city)	Australia	33 53 S	151 13 E
T'bilisi (capital)	Georgia	41 43 N	44 49 E
Tadzhikistan (former name for Tajikistan)	Tajikistan	39 00 N	71 00 E
Tahiti (island)	French Polynesia	17 37 S	149 27 W
Taipei (capital)	Taiwan	25 03 N	121 30 E
Taiwan Strait	Pacific Ocean	24 00 N	119 00 E
Tallinn (capital)	Estonia	59 25 N	24 45 E
Tanganyika (former name for the mainland portion of Tanzania)	Tanzania	6 00 S	35 00 E
Tangier (city)	Morocco	35 48 N	5 45 W
Tannu-Tuva (region)	Russia	51 25 N	94 45 E
Tarawa (island)	Kiribati	1 25 N	173 00 E
Tartary, Gulf of	Pacific Ocean	50 00 N	141 00 E
Tashkent (capital)	Uzbekistan	41 20 N	69 18 E
Tasman Sea	Pacific Ocean	4 30 S	168 00 E
Tasmania (island)	Australia	43 00 S	147 00 E
Tatar Strait	Pacific Ocean	50 00 N	141 00 E
Taymyr Peninsula (Poluostrov Taymyr)	Russia	76 00 N	104 00 E
Tchad (local name for Chad)	Chad	15 00 N	19 00 E
Tegucigalpa (capital)	Honduras	14 06 N	87 13 W
Tehran (capital)	Iran	35 40 N	51 26 E
Tel Aviv (capital, de facto)	Israel	32 05 N	34 48 E
Teluk Bone (gulf)	Pacific Ocean	4 00 S	120 45 E
Teluk Tomini (gulf)	Pacific Ocean	0 30 S	121 00 E
Terre Adelie (claimed by France; also Adelie Land)	Antarctica	66 30 S	139 00 E
Terres Australes et Antarctiques Francaises (local name for the French Southern and Antarctic Lands)	French Southern and Antarctic Lands	43 00 S	67 00 E
Thailand, Gulf of	Pacific Ocean	10 00 N	101 00 E
The Former Yugoslav Republic of Macedonia	Macedonia	41 50 N	22 00 E
Thessaloniki (city; also Salonika)	Greece	40 38 N	22 57 E
Thimphu (capital)	Bhutan	27 28 N	89 39 E
Thuringia (region)	Germany	51 00 N	11 00 E
Thurston Island	Antarctica	72 20 S	99 00 W
Tiberias, Lake	Israel	32 48 N	35 35 E
Tibet (autonomous region; also Xizang)	China	32 00 N	90 00 E
Tibilisi (see T'bilisi)	Georgia	41 43 N	44 49 E
Tien Shan (mountains)	China, Kyrgyzstan	42 00 N	80 00 E
Tierra del Fuego (island, island group)	Argentina, Chile	54 00 S	69 00 W

NAME	ENTRY INTHE WORLD FACTBOOK	LATITUDE(DEG MIN)	LONGITUDE(DEG MIN)
Timor (island)	Timor-Leste, Indonesia	9 00 S	125 00 E
Timor Lorosa'e (local name for Timor-Leste)	Timor-Leste	9 00 N	126 00 E
Timor Sea	Pacific Ocean	11 00 S	128 00 E
Tinian (island)	Northern Mariana Islands	15 00 N	145 38 E
Tiran, Strait of	Indian Ocean	28 00 N	34 27 E
Tirana, Tirane (capital)	Albania	41 20 N	19 50 E
Tirol, Tyrol (region)	Austria, Italy	47 00 N	11 00 E
Tobago (island)	Trinidad and Tobago	11 15 N	60 40 W
Tokyo (capital)	Japan	35 42 N	139 46 E
Tonkin, Gulf of	Pacific Ocean	20 00 N	108 00 E
Toronto (city)	Canada	43 40 N	79 23 W
Torres Strait	Pacific Ocean	10 25 S	142 10 E
Torshavn (capital)	Faroe Islands	62 01 N	6 46 W
Toshkent (see Tashkent)	Uzbekistan	41 20 N	69 18 E
Transcarpathia (region; alternate name for Carpatho-Ukraine)	Ukraine	48 22 N	23 32 E
Transjordan (former name for Jordan)	Jordan	31 00 N	36 00 E
Transkei (enclave)	South Africa	32 15 S	28 15 E
Transvaal (region; former name for north-eastern South Africa)	South Africa	25 10 S	29 25 E
Transylvania (region)	Romania	46 30 N	24 00 E
Trindade, Ilha de (island)	Brazil	20 31 S	29 20 W
Trinidad (island)	Trinidad and Tobago	10 22 N	61 15 W
Tripoli (capital)	Libya	32 54 N	13 11 E
Tripoli (city)	Lebanon	34 26 N	35 51 E
Tripolitania (region)	Libya	31 00 N	14 00 E
Tristan da Cunha Group (island group)	Saint Helena, Ascension, and Tristan da Cunha	37 15 S	12 30 W
Trobriand Islands	Papua New Guinea	8 38 S	151 04 E
Tromelin Island	Indian Ocean	15 52 S	54 25 E
Trucial Coast (former name for the United Arab Emirates)	United Arab Emirates	24 00 N	54 00 E
Trucial Oman (former name for the United Arab Emirates)	United Arab Emirates	24 00 N	54 00 E
Trucial States (former name for the United Arab Emirates)	United Arab Emirates	24 00 N	54 00 E
Truk Islands (former name for the Chuuk Islands)	Federated States of Micronesia	7 25 N	151 47 E
Tsugaru Strait	Pacific Ocean	41 35 N	141 00 E
Tuamotu Islands (Iles Tuamotu)	French Polynesia	19 00 S	142 00 W
Tubuai Islands (Iles Tubuai)	French Polynesia	23 00 S	150 00 W
Tunb al Kubra (island)	Iran	26 14 N	55 19 E
Tunb as Sughra (island)	Iran	26 14 N	55 09 E
Tunis (capital)	Tunisia	36 48 N	10 11 E
Turin (city)	Italy	45 04 N	7 40 E
Turkish Straits (see Bosporus and Dardenelles)	Atlantic Ocean	40 40 N	28 00 E
Turkiye (local name for Turkey)	Turkey	39 00 N	35 00 E
Turkmenia, Turkmeniya (former name for Turkmenistan)	Turkmenistan	40 00 N	60 00 E
Turks Island Passage	Atlantic Ocean	21 40 N	71 00 W
Tuscany (region)	Italy	43 25 N	11 00 E
Tutuila (island)	American Samoa	14 18 S	170 42 W
Tyrrhenian Sea	Atlantic Ocean	40 00 N	12 00 E
Ubangi-Shari (former name for the Central African Republic	Central African Republic	6 38 N	20 33 E
Ukrayina (local name for Ukraine)	Ukraine	49 00 N	32 00 E
Ulaanbaatar (capital)	Mongolia	47 55 N	106 53 E
Ullung-do (island)	South Korea	37 29 N	130 52 E

NAME	ENTRY INTHE WORLD FACTBOOK	LATITUDE(DEG MIN)	LONGITUDE(DEG MIN)
Ulster (region)	Ireland, United Kingdom	54 35 N	7 00 W
Uman (local name for Oman)	Oman	21 00 N	57 00 E
Unimak Pass (strait)	Pacific Ocean	54 20 N	164 50 W
Union of Soviet Socialist Republics or USSR (former name of a large Eurasian empire, roughly coequal with the former Russian Empire)	Armenia, Azerbaijan, Belarus, Estonia, Georgia, Kazakhstan, Kyrgyzstan, Latvia, Lithuania, Moldova, Russia, Tajikistan, Turkmenistan, Ukraine, Uzbekistan		
United Arab Republic or UAR (former name for a federation between Egypt and Syria)	Egypt, Syria		
Upper Volta (former name for Burkina Faso)	Burkina Faso	13 00 N	2 00 W
Ural Mountains	Kazakhstan, Russia	60 00 N	60 00 E
Urdunn (local name for Jordan)	Jordan	31 00 N	36 00 E
Urundi (former name for Burundi)	Burundi	3 30 S	30 00 E
Ussuri River	China, Russia	48 28 N	135 02 E
Vaduz (capital)	Liechtenstein	47 09 N	9 31 E
Vakhan (Wakhan Corridor)	Afghanistan	37 00 N	73 00 E
Valletta (capital)	Malta	35 54 N	14 31 E
Valley, The (capital)	Anguilla	18 13 N	63 04 W
Van Diemen Strait (Osumi Strait)	Pacific Ocean	31 00 N	131 00 E
Vancouver (city)	Canada	49 16 N	123 08 W
Vancouver Island	Canada	49 45 N	126 00 W
Vatican City (capital)	Holy See	41 54 N	12 27 E
Velez de la Gomera, Penon de (island)	Spain	35 11 N	4 18 W
Venda (enclave)	South Africa	23 00 S	31 00 E
Verde Island Passage	Pacific Ocean	13 34 N	120 51 E
Victoria (capital)	Seychelles	4 38 S	55 27 E
Victoria (island)	Canada	71 00 N	110 00 W
Victoria Land (region)	Antarctica	72 00 S	155 00 E
Vienna (capital)	Austria	48 12 N	16 22 E
Vientiane (capital)	Laos	17 58 N	102 36 E
Vilnius (capital)	Lithuania	54 41 N	25 19 E
Viti Levu (island)	Fiji	18 00 S	178 00 E
Vladivostok (city)	Russia	43 10 N	131 56 E
Vojvodina (region)	Serbia	45 35 N	20 00 E
Volcano Islands	Japan	25 00 N	141 00 E
Vostok Island	Kiribati	10 06 S	152 23 W
Wake Atoll	Wake Island	19 17 N	166 39 E
Wakhan Corridor (see Vakhan)	Afghanistan	37 00 N	73 00 E
Walachia (region)	Romania	44 45 N	26 05 E
Wales (region)	United Kingdom	52 30 N	3 30 W
Wallis Islands	Wallis and Futuna	13 17 S	176 10 W
Walvis Bay (city; former exclave)	Namibia	22 59 S	14 31 E
Warsaw (capital)	Poland	52 15 N	21 00 E
Washington, DC (capital)	United States	38 53 N	77 02 W
Weddell Sea	Southern Ocean	72 00 S	45 00 W
Wellington (capital)	New Zealand	41 28 S	174 51 E
West Frisian Islands	Netherlands	53 26 N	5 30 E
West Germany (Federal Republic of Germany; former name for western portion of Germany)	Germany	53 22 N	5 20 E
West Island (capital)	Cocos (Keeling) Islands	12 10 S	96 55 E
West Korea Strait (Western Channel)	Pacific Ocean	34 40 N	129 00 E
West Pakistan (former name for present-day Pakistan)	Pakistan	30 00 N	70 00 E
West Siberian Plain	Russia	60 00 N	75 00 E
Western Channel (West Korea Strait)	Pacific Ocean	34 40 N	129 00 E

NAME	ENTRY INTHE WORLD FACTBOOK	LATITUDE(DEG MIN)	LONGITUDE(DEG MIN)
Western Samoa (former name for Samoa)	Samoa	13 35 S	172 20 W
Wetar Strait	Pacific Ocean	8 20 S	126 30 E
White Sea	Arctic Ocean	65 30 N	38 00 E
Wilkes Land (region)	Antarctica	71 00 S	120 00 E
Willemstad (capital)	Curacao	12 06 N	68 56 W
Windhoek (capital)	Namibia	22 34 S	17 06 E
Windward Passage	Atlantic Ocean	20 00 N	73 50 W
Winnipeg (city)	Canada	49 53 N	97 10 W
Wrangel Island (Ostrov Vrangelya)	Russia	71 14 N	179 36 W
Xianggang (local name for Hong Kong)	Hong Kong	22 15 N	114 10 E
Y'israel (local name for Israel)	Israel	31 30 N	34 45 E
Yaitopya (local name for Ethiopia)	Ethiopia	8 00 N	38 00 E
Yalu River	China, North Korea	39 55 N	124 20 E
Yamoussoukro (capital)	Cote d'Ivoire	6 49 N	5 17 W
Yangon (see Rangoon)	Burma	16 47 N	96 10 E
Yaounde (capital)	Cameroon	3 52 N	11 31 E
Yap Islands	Federated States of Micronesia	9 30 N	138 00 E
Yaren (governmental center)	Nauru	0 32 S	166 55 E
Yekaterinburg (city; formerly Sverdlovsk)	Russia	56 50 N	60 39 E
Yellow Sea	Pacific Ocean	36 00 N	123 00 E
Yemen Arab Republic (also Yemen (Sanaa); former name for northern portion of Yemen)	Yemen	15 00 N	44 00 E
Yemen, People's Democratic Republic of (also Yemen (Aden); former name for southern portion of Yemen)	Yemen	14 00 N	46 00 E
Yerevan (capital)	Armenia	40 11 N	44 30 E
Yokohama (city)	Japan	35 26 N	139 37 E
Youth, Isle of (Isla de la Juventud)	Cuba	21 40 N	82 50 W
Yucatan Channel	Atlantic Ocean	21 45 N	85 45 W
Yucatan Peninsula	Mexico	19 30 N	89 00 W
Yugoslavia (former name for a federation of Serbia and Montenegro)	Montenegro, Serbia	43 00 N	21 00 E
Yugoslavia, Kingdom of (former name for a Balkan federation)	Bosnia and Herzegovina, Croatia, Macedonia, Montenegro, Serbia, Slovenia	43 00 N	19 00 E
Yugoslavia, Socialist Federal Republic of (former name for a Balkan federation)	Bosnia and Herzegovina, Croatia, Macedonia, Montenegro, Serbia, Slovenia	43 00 N	19 00 E
Zagreb (capital)	Croatia	45 48 N	15 58 E
Zaire (former name for the Democratic Republic of the Congo)	Democratic Republic of the Congo	15 00 S	30 00 E
Zakhalinskiy Zaliv (bay)	Pacific Ocean	54 00 N	142 00 E
Zaliv Shelikhova (bay)	Pacific Ocean	60 00 N	157 30 E
Zambezia (region)	Mozambique	16 00 S	37 00 E
Zanzibar (island)	Tanzania	6 10 S	39 11 E
Zhong Guo, Zhonghua (local name for China)	China	35 00 N	105 00 E
Zion, Mount (locale in Jerusalem)	Israel, West Bank	31 46 N	35 14 E
Zurich (city)	Switzerland	47 23 N	8 32 E

APPENDIX G

WEIGHTS AND MEASURES

Note: At this time, only three countries - Burma, Liberia, and the US - have not adopted the International System of Units (SI, or metric system) as their official system of weights and measures. Although use of the metric system has been sanctioned by law in the US since 1866, it has been slow in displacing the American adaptation of the British Imperial System known as the US Customary System. The US is the only industrialized nation that does not mainly use the metric system in its commercial and standards activities, but there is increasing acceptance in science, medicine, government, and many sectors of industry.

MATHEMATICAL NOTATION MATHEMATICAL POWER	NAME
10^{18} or 1,000,000,000,000,000,000	one quintillion
10^{15} or 1,000,000,000,000,000	one quadrillion
10^{12} or 1,000,000,000,000	one trillion
10^{9} or 1,000,000,000	one billion
10^{6} or 1,000,000	one million
10^{3} or 1,000	one thousand
10^{2} or 100	one hundred
10^{1} or 10	ten
10^{0} or 1	one
10^{-1} or 0.1	one-tenth
10^{-2} or 0.01	one-hundredth
10^{-3} or 0.001	one-thousandth
10^{-6} or 0.000 001	one-millionth
10^{-9} or 0.000 000 001	one-billionth
10^{-12} or 0.000 000 000 001	one-trillionth
10^{-15} or 0.000 000 000 000 001	one-quadrillionth
10^{-18} or 0.000 000 000 000 000 001	one-quintillionth

Metric Interrelationships

PREFIX	SYMBOL	LENGTH, WEIGHT, OR CAPACITY
yotta	Y	10^{24}
zetta	Z	10^{21}
exa	E	10^{18}
peta	P	10^{15}
tera	T	10^{12}
giga	G	10^{9}
mega	M	10^{6}
kilo	k	10^{3}
hecto	h	10^{2}
deka	da	10^{1}
basic unit	-	1 meter, 1 gram, 1 liter
deci	d	10^{-1}
centi	c	10^{-2}
milli	m	10^{-3}
micro	u	10^{-6}
nano	n	10^{-9}
pico	p	10^{-12}
femto	f	10^{-15}
atto	a	10^{-18}
zepto	z	10^{-21}
yocto	y	10^{-24}

CONVERSION FACTORS TO CONVERT FROM	TO	MULTIPLY BY
acres	ares	40.468 564 224
acres	hectares	0.404 685 642 24
acres	square feet	43,560
acres	square kilometers	0.004 046 856 422 4
acres	square meters	4,046.856 422 4

CONVERSION FACTORS TO CONVERT FROM	TO	MULTIPLY BY
acres	square miles (statute)	0.001 562 50
acres	square yards	4,840
ares	square meters	100
ares	square yards	119.599
barrels, US beer	gallons	31
barrels, US beer	liters	117.347 77
barrels, US petroleum	gallons (British)	34.97
barrels, US petroleum	gallons (US)	42
barrels, US petroleum	liters	158.987 29
barrels, US proof spirits	gallons	40
barrels, US proof spirits	liters	151.416 47
bushels (US)	bushels (British)	0.968 9
bushels (US)	cubic feet	1.244 456
bushels (US)	cubic inches	2,150.42
bushels (US)	cubic meters	0.035 239 07
bushels (US)	cubic yards	0.046 090 96
bushels (US)	dekaliters	3.523 907
bushels (US)	dry pints	64
bushels (US)	dry quarts	32
bushels (US)	liters	35.239 070 17
bushels (US)	pecks	4
cables	fathoms	120
cables	meters	219.456
cables	yards	240
carat	milligrams	200
centimeters	feet	0.032 808 40
centimeters	inches	0.393 700 8
centimeters	meters	0.01
centimeters	yards	0.010 936 13
centimeters, cubic	cubic inches	0.061 023 744
centimeters, square	square feet	0.001 076 39
centimeters, square	square inches	0.155 000 31
centimeters, square	square meters	0.000 1
centimeters, square	square yards	0.000 119 599
chains, square surveyor's	ares	4.046 86
chains, square surveyor's	square feet	4,356
chains, surveyor's	feet	66
chains, surveyor's	meters	20.116 8
chains, surveyor's	rods	4
cords of wood	cubic feet	128
cords of wood	cubic meters	3.624 556
cords of wood	cubic yards	4.740 7
cups	liquid ounces (US)	8
cups	liters	0.236 588 2
degrees Celsius	degrees Fahrenheit	multiply by 1.8 and add 32
degrees Fahrenheit	degrees Celsius	subtract 32 and divide by 1.8
dekaliters	bushels	0.283 775 9
dekaliters	cubic feet	0.353 146 7
dekaliters	cubic inches	610.237 4
dekaliters	dry pints	18.161 66
dekaliters	dry quarts	9.080 829 8
dekaliters	liters	10
dekaliters	pecks	1.135 104
drams, avoirdupois	avoirdupois ounces	0.062 55
drams, avoirdupois	grains	27.344
drams, avoirdupois	grams	1.771 845 2
drams, troy	grains	60
drams, troy	grams	3.887 934 6
drams, troy	scruples	3
drams, troy	troy ounces	0.125

CONVERSION FACTORS TO CONVERT FROM	TO	MULTIPLY BY
drams, liquid (US)	cubic inches	0.226
drams, liquid (US)	liquid drams (British)	1.041
drams, liquid (US)	liquid ounces	0.125
drams, liquid (US)	milliliters	3.696 69
drams, liquid (US)	minims	60
fathoms	feet	6
fathoms	meters	1.828 8
feet	centimeters	30.48
feet	inches	12
feet	kilometers	0.000 304 8
feet	meters	0.304 8
feet	statute miles	0.000 189 39
feet	yards	0.333 333 3
feet, cubic	bushels	0.803 563 95
feet, cubic	cubic decimeters	28.316 847
feet, cubic	cubic inches	1,728
feet, cubic	cubic meters	0.028 316 846 592
feet, cubic	cubic yards	0.037 037 04
feet, cubic	dry pints	51.428 09
feet, cubic	dry quarts	25.714 05
feet, cubic	gallons	7.480 519
feet, cubic	gills	239.376 6
feet, cubic	liquid ounces	957.506 5
feet, cubic	liquid pints	59.844 16
feet, cubic	liquid quarts	29.922 08
feet, cubic	liters	28.316 846 592
feet, cubic	pecks	3.214 256
feet, square	acres	0.000 022 956 8
feet, square	square centimeters	929.030 4
feet, square	square decimeters	9.290 304
feet, square	square inches	144
feet, square	square meters	0.092 903 04
feet, square	square yards	0.111 111 1
furlongs	feet	660
furlongs	inches	7,920
furlongs	meters	201.168
furlongs	statute miles	0.125
furlongs	yards	220
gallons, liquid (US)	cubic feet	0.133 680 6
gallons, liquid (US)	cubic inches	231
gallons, liquid (US)	cubic meters	0.003 785 411 784
gallons, liquid (US)	cubic yards	0.004 951 13
gallons, liquid (US)	gills (US)	32
gallons, liquid (US)	liquid gallons (British)	0.832 67
gallons, liquid (US)	liquid ounces	128
gallons, liquid (US)	liquid pints	8
gallons, liquid (US)	liquid quarts	4
gallons, liquid (US)	liters	3.785 411 784
gallons, liquid (US)	milliliters	3,785.411 784
gallons, liquid (US)	minims	61,440
gills (US)	centiliters	11.829 4
gills (US)	cubic feet	0.004 177 517
gills (US)	cubic inches	7.218 75
gills (US)	gallons	0.031 25
gills (US)	gills (British)	0.832 67
gills (US)	liquid ounces	4
gills (US)	liquid pints	0.25
gills (US)	liquid quarts	0.125
gills (US)	liters	0.118 294 118 25
gills (US)	milliliters	118.294 118 25

CONVERSION FACTORS TO CONVERT FROM	TO	MULTIPLY BY
gills (US)	minims	1,920
grains	avoirdupois drams	0.036 571 43
grains	avoirdupois ounces	0.002 285 71
grains	avoirdupois pounds	0.000 142 86
grains	grams	0.064 798 91
grains	kilograms	0.000 064 798 91
grains	milligrams	64.798 910
grains	pennyweights	0.042
grains	scruples	0.05
grains	troy drams	0.016 6
grains	troy ounces	0.002 083 33
grains	troy pounds	0.000 173 61
grams	avoirdupois drams	0.564 383 39
grams	avoirdupois ounces	0.035 273 961
grams	avoirdupois pounds	0.002 204 622 6
grams	grains	15.432 361
grams	kilograms	0.001
grams	milligrams	1,000
grams	troy ounces	0.032 150 746 6
grams	troy pounds	0.002 679 23
hands (height of horse)	centimeters	10.16
hands (height of horse)	inches	4
hectares	acres	2.471 053 8
hectares	square feet	107,639.1
hectares	square kilometers	0.01
hectares	square meters	10,000
hectares	square miles	0.003 861 02
hectares	square yards	11,959.90
hundredweights, long	avoirdupois pounds	112
hundredweights, long	kilograms	50.802 345
hundredweights, long	long tons	0.05
hundredweights, long	metric tons	0.050 802 345
hundredweights, long	short tons	0.056
hundredweights, short	avoirdupois pounds	100
hundredweights, short	kilograms	45.359 237
hundredweights, short	long tons	0.044 642 86
hundredweights, short	metric tons	0.045 359 237
hundredweights, short	short tons	0.05
inches	centimeters	2.54
inches	feet	0.083 333 33
inches	meters	0.025 4
inches	millimeters	25.4
inches	yards	0.027 777 78
inches, cubic	bushels	0.000 465 025
inches, cubic	cubic centimeters	16.387 064
inches, cubic	cubic feet	0.000 578 703 7
inches, cubic	cubic meters	0.000 016 387 064
inches, cubic	cubic yards	0.000 021 433 47
inches, cubic	dry pints	0.029 761 6
inches, cubic	dry quarts	0.014 880 8
inches, cubic	gallons	0.004 329 0
inches, cubic	gills	0.138 528 1
inches, cubic	liquid ounces	0.554 112 6
inches, cubic	liquid pints	0.034 632 03
inches, cubic	liquid quarts	0.017 316 02
inches, cubic	liters	0.016 387 064
inches, cubic	milliliters	16.387 064
inches, cubic	minims (US)	265.974 0
inches, cubic	pecks	0.001 860 10
inches, square	square centimeters	6.451 600

CONVERSION FACTORS TO CONVERT FROM	TO	MULTIPLY BY
inches, square	square feet	0.006 944 44
inches, square	square meters	0.000 645 16
inches, square	square yards	0.000 771 605
kilograms	avoirdupois drams	564.383 4
kilograms	avoirdupois ounces	35.273 962
kilograms	avoirdupois pounds	2.204 622 622
kilograms	grains	15,432.36
kilograms	grams	1,000
kilograms	long tons	0.000 984 2
kilograms	metric tons	0.001
kilograms	short hundredweights	0.022 046 23
kilograms	short tons	0.001 102 31
kilograms	troy ounces	32.150 75
kilograms	troy pounds	2.679 229
kilometers	meters	1,000
kilometers	statute miles	0.621 371 192
kilometers, square	acres	247.105 38
kilometers, square	hectares	100
kilometers, square	square meters	1,000,000
kilometers, square	statute miles	0.386 102 16
knots (nautical mi/hr)	kilometers/hour	1.852
knots (nautical mi/hr)	statute miles/hour	1.151
leagues, nautical	kilometers	5.556
leagues, nautical	nautical miles	3
leagues, statute	kilometers	4.828 032
leagues, statute	statute miles	3
links, square surveyor's	square centimeters	404.686
links, square surveyor's	square inches	62.726 4
links, surveyor's	centimeters	20.116 8
links, surveyor's	chains	0.01
links, surveyor's	inches	7.92
liters	bushels	0.028 377 59
liters	cubic feet	0.035 314 67
liters	cubic inches	61.023 74
liters	cubic meters	0.001
liters	cubic yards	0.001 307 95
liters	dekaliters	0.1
liters	dry pints	1.816 166
liters	dry quarts	0.908 082 98
liters	gallons	0.264 172 052
liters	gills (US)	8.453 506
liters	liquid ounces	33.814 02
liters	liquid pints	2.113 376
liters	liquid quarts	1.056 688 2
liters	milliliters	1,000
liters	pecks	0.113 510 4
meters	centimeters	100
meters	feet	3.280 839 895
meters	inches	39.370 079
meters	kilometers	0.001
meters	millimeters	1,000
meters	statute miles	0.000 621 371
meters	yards	1.093 613 298
meters, cubic	bushels	28.377 59
meters, cubic	cubic feet	35.314 666 7
meters, cubic	cubic inches	61,023.744
meters, cubic	cubic yards	1.307 950 619
meters, cubic	gallons	264.172 05
meters, cubic	liters	1,000
meters, cubic	pecks	113.510 4

CONVERSION FACTORS TO CONVERT FROM	TO	MULTIPLY BY
meters, square	acres	0.000 247 105 38
meters, square	hectares	0.000 1
meters, square	square centimeters	10,000
meters, square	square feet	10.763 910 4
meters, square	square inches	1,550.003 1
meters, square	square yards	1.195 990 046
microns	meters	0.000 001
microns	inches	0.000 039 4
mils	inches	0.001
mils	millimeters	0.025 4
miles, nautical	kilometers	1.852 0
miles, nautical	statute miles	1.150 779 4
miles, statute	centimeters	160,934.4
miles, statute	feet	5,280
miles, statute	furlongs	8
miles, statute	inches	63,360
miles, statute	kilometers	1.609 344
miles, statute	meters	1,609.344
miles, statute	rods	320
miles, statute	yards	1,760
miles, square nautical	square kilometers	3.429 904
miles, square nautical	square statute miles	1.325
miles, square statute	acres	640
miles, square statute	hectares	258.998 811 033 6
miles, square statute	sections	1
miles, square statute	square kilometers	2.589 988 110 336
miles, square statute	square nautical miles	0.755 miles
miles, square statute	square rods	102,400
milligrams	grains	0.015 432 358 35
milliliters	cubic inches	0.061 023 744
milliliters	gallons	0.000 264 17
milliliters	gills (US)	0.008 453 5
milliliters	liquid ounces	0.033 814 02
milliliters	liquid pints	0.002 113 4
milliliters	liquid quarts	0.001 056 7
milliliters	liters	0.001
milliliters	minims	16.230 73
millimeters	inches	0.039 370 078 7
minims (US)	cubic inches	0.003 759 77
minims (US)	gills (US)	0.000 520 83
minims (US)	liquid ounces	0.002 083 33
minims (US)	milliliters	0.061 611 52
minims (US)	minims (British)	1.041
ounces, avoirdupois	avoirdupois drams	16
ounces, avoirdupois	avoirdupois pounds	0.062 5
ounces, avoirdupois	grains	437.5
ounces, avoirdupois	grams	28.349 523 125
ounces, avoirdupois	kilograms	0.028 349 523 125
ounces, avoirdupois	troy ounces	0.911 458 3
ounces, avoirdupois	troy pounds	0.075 954 86
ounces, liquid (US)	cubic feet	0.001 044 38
ounces, liquid (US)	centiliters	2.957 35
ounces, liquid (US)	cubic inches	1.804 687 5
ounces, liquid (US)	gallons	0.007 812 5
ounces, liquid (US)	gills (US)	0.25
ounces, liquid (US)	liquid drams	8
ounces, liquid (US)	liquid ounces (British)	1.041
ounces, liquid (US)	liquid pints	0.062 5
ounces, liquid (US)	liquid quarts	0.031 25
ounces, liquid (US)	liters	0.029 573 53

CONVERSION FACTORS TO CONVERT FROM	TO	MULTIPLY BY
ounces, liquid (US)	milliliters	29.573 529 6
ounces, liquid (US)	minims	480
ounces, troy	avoirdupois drams	17.554 29
ounces, troy	avoirdupois ounces	1.097 143
ounces, troy	avoirdupois pounds	0.068 571 43
ounces, troy	grains	480
ounces, troy	grams	31.103 476 8
ounces, troy	pennyweights	20
ounces, troy	troy drams	8
ounces, troy	troy pounds	0.083 333 3
paces (US)	centimeters	76.2
paces (US)	inches	30
pecks (US)	bushels	0.25
pecks (US)	cubic feet	0.311 114
pecks (US)	cubic inches	537.605
pecks (US)	cubic meters	0.008 809 77
pecks (US)	cubic yards	0.011 522 74
pecks (US)	dekaliters	0.880 976 75
pecks (US)	dry pints	16
pecks (US)	dry quarts	8
pecks (US)	liters	8.809 767 5
pecks (US)	pecks (British)	0.968 9
pennyweights	grains	24
pennyweights	grams	1.555 173 84
pennyweights	troy ounces	0.05
pints, dry (US)	bushels	0.015 625
pints, dry (US)	cubic feet	0.019 444 63
pints, dry (US)	cubic inches	33.600 312 5
pints, dry (US)	dekaliters	0.055 061 05
pints, dry (US)	dry pints (British)	0.968 9
pints, dry (US)	dry quarts	0.5
pints, dry (US)	liters	0.550 610 47
pints, liquid (US)	cubic feet	0.016 710 07
pints, liquid (US)	cubic inches	28.875
pints, liquid (US)	deciliters	4.731 76
pints, liquid (US)	gallons	0.125
pints, liquid (US)	gills (US)	4
pints, liquid (US)	liquid ounces	16
pints, liquid (US)	liquid pints (British)	0.832 67
pints, liquid (US)	liquid quarts	0.5
pints, liquid (US)	liters	0.473 176 473
pints, liquid (US)	milliliters	473.176 473
pints, liquid (US)	minims	7,680
points (typographical)	inches	0.013 837
points (typographical)	millimeters	0.351 459 8
pounds, avoirdupois	avoirdupois drams	256
pounds, avoirdupois	avoirdupois ounces	16
pounds, avoirdupois	grains	7,000
pounds, avoirdupois	grams	453.592 37
pounds, avoirdupois	kilograms	0.453 592 37
pounds, avoirdupois	long tons	0.000 446 428 6
pounds, avoirdupois	metric tons	0.000 453 592 37
pounds, avoirdupois	quintals	0.004 535 92
pounds, avoirdupois	short tons	0.000 5
pounds, avoirdupois	troy ounces	14.583 33
pounds, avoirdupois	troy pounds	1.215 278
pounds, troy	avoirdupois drams	210.651 4
pounds, troy	avoirdupois ounces	13.165 71
pounds, troy	avoirdupois pounds	0.822 857 1
pounds, troy	grains	5,760

CONVERSION FACTORS TO CONVERT FROM	TO	MULTIPLY BY
pounds, troy	grams	373.241 721 6
pounds, troy	kilograms	0.373 241 721 6
pounds, troy	pennyweights	240
pounds, troy	troy ounces	12
quarts, dry (US)	bushels	0.031 25
quarts, dry (US)	cubic feet	0.038 889 25
quarts, dry (US)	cubic inches	67.200 625
quarts, dry (US)	dekaliters	0.110 122 1
quarts, dry (US)	dry pints	2
quarts, dry (US)	dry quarts (British)	0.968 9
quarts, dry (US)	liters	1.101 221
quarts, dry (US)	pecks	0.125
quarts, dry (US)	pints, dry (US)	2
quarts, liquid (US)	cubic feet	0.033 420 14
quarts, liquid (US)	cubic inches	57.75
quarts, liquid (US)	deciliters	9.463 53
quarts, liquid (US)	gallons	0.25
quarts, liquid (US)	gills (US)	8
quarts, liquid (US)	liquid ounces	32
quarts, liquid (US)	liquid pints (US)	2
quarts, liquid (US)	liquid quarts (British)	0.832 67
quarts, liquid (US)	liters	0.946 352 946
quarts, liquid (US)	milliliters	946.352 946
quarts, liquid (US)	minims	15,360
quintals	avoirdupois pounds	220.462 26
quintals	kilograms	100
quintals	metric tons	0.1
rods	feet	16.5
rods	meters	5.029 2
rods	yards	5.5
rods, square	acres	0.006 25
rods, square	square meters	25.292 85
rods, square	square yards	30.25
scruples	grains	20
scruples	grams	1.295 978 2
scruples	troy drams	0.333
sections (US)	square kilometers	2.589 988 1
sections (US)	square statute miles	1
spans	centimeters	22.86
spans	inches	9
steres	cubic meters	1
steres	cubic yards	1.307 95
tablespoons	milliliters	14.786 76
tablespoons	teaspoons	3
teaspoons	milliliters	4.928 922
teaspoons	tablespoons	0.333 333
ton-miles, long	metric ton-kilometers	1.635 169
ton-miles, short	metric ton-kilometers	1.459 972
tons, gross register	cubic feet of permanently enclosed space	100
tons, gross register	cubic meters of permanently enclosed space	2.831 684 7
tons, long (deadweight)	avoirdupois ounces	35,840
tons, long (deadweight)	avoirdupois pounds	2,240
tons, long (deadweight)	kilograms	1,016.046 909 8
tons, long (deadweight)	long hundredweights	20
tons, long (deadweight)	metric tons	1.016 046 908 8
tons, long (deadweight)	short hundredweights	22.4
tons, long (deadweight)	short tons	1.12
tons, metric	avoirdupois pounds	2,204.623
tons, metric	kilograms	1,000
tons, metric	long hundredweights	19.684 130 3

CONVERSION FACTORS TO CONVERT FROM	TO	MULTIPLY BY
tons, metric	long tons	0.984 206 5
tons, metric	quintals	10
tons, metric	short hundredweights	22.046 23
tons, metric	short tons	1.102 311 3
tons, metric	troy ounces	32,150.75
tons, net register	cubic feet of permanently enclosed space for cargo and passengers	100
tons, net register	cubic meters of permanently enclosed space for cargo and passengers	2.831 684 7
tons, shipping	cubic feet of permanently enclosed cargo space	42
tons, shipping	cubic meters of permanently enclosed cargo space	1.189 307 574
tons, short	avoirdupois pounds	2,000
tons, short	kilograms	907.184 74
tons, short	long hundredweights	17.857 14
tons, short	long tons	0.892 857 1
tons, short	metric tons	0.907 184 74
tons, short	short hundredweights	20
townships (US)	sections	36
townships (US)	square kilometers	93.239 572
townships (US)	square statute miles	36
miles, square statute	acres	640
miles, square statute	hectares	258.998 811 033 6
miles, square statute	square feet	27,878,400
miles, square statute	square meters	2,589,988.110 336
miles, square statute	square yards	3,097,600
yards	centimeters	91.44
yards	feet	3
yards	inches	36
yards	meters	0.914 4
yards	miles	0.000 568 18
yards, cubic	bushels	21.696 227
yards, cubic	cubic feet	27
yards, cubic	cubic inches	46,656
yards, cubic	cubic meters	0.764 554 857 984
yards, cubic	gallons	201.974 0
yards, cubic	liters	764.554 857 984
yards, cubic	pecks	86.784 91
yards, square	acres	0.000 206 611 6
yards, square	hectares	0.000 083 612 736
yards, square	square centimeters	8,361.273 6
yards, square	square feet	9
yards, square	square inches	1,296
yards, square	square meters	0.836 127 36
yards, square	square miles	0.000 000 322 830 6

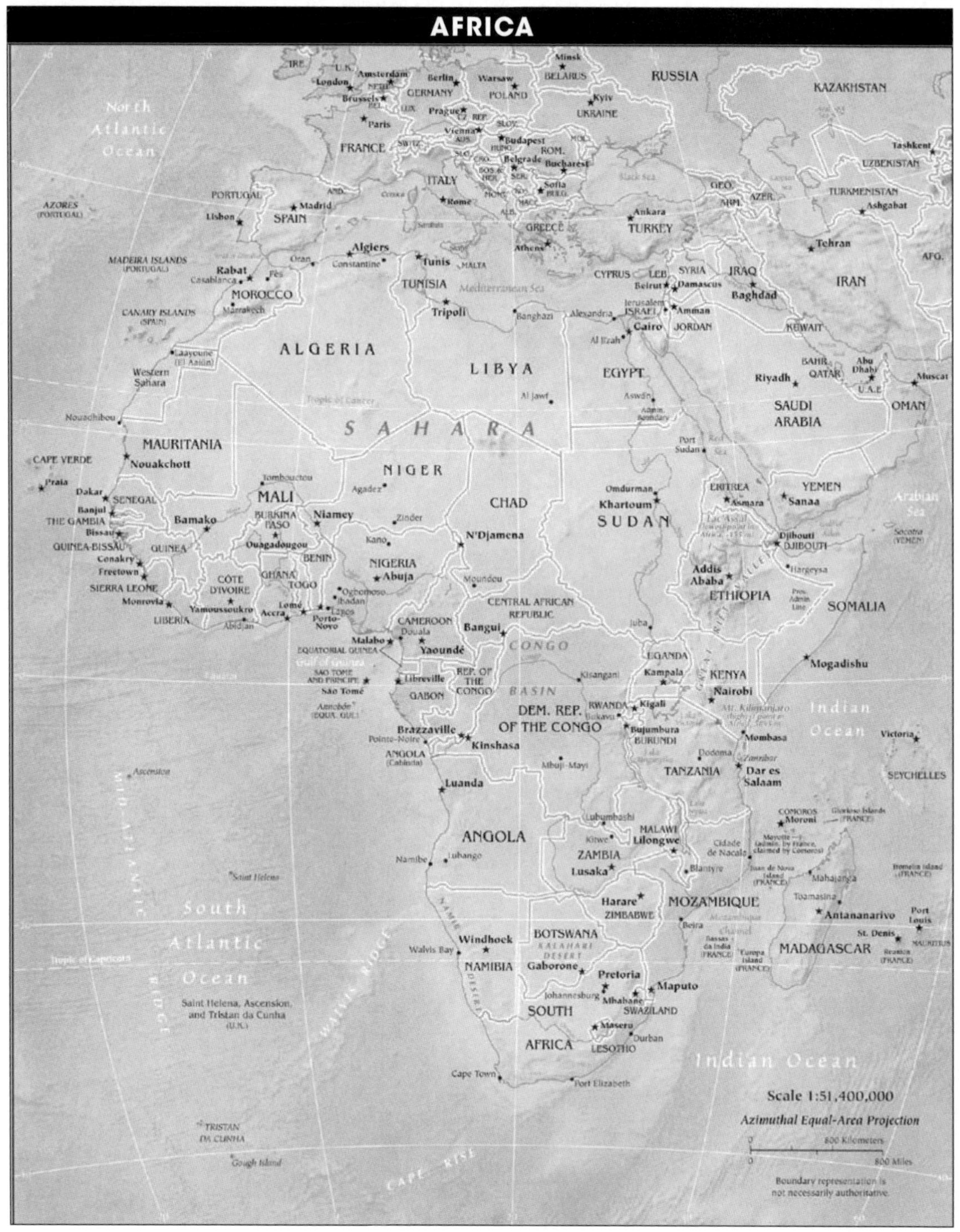
AFRICA
RUSSIA
KAZAKHSTAN
UZBEKISTAN
TURKMENISTAN
Tashkent
Ashgabat
Minsk
BELARUS
Kyiv
UKRAINE
Warsaw
POLAND
Berlin
GERMANY
Amsterdam
London
Brussels
Paris
FRANCE
Prague
Vienna
Budapest
ROM.
Belgrade
Bucharest
Sofia
ITALY
Rome
Madrid
SPAIN
Lisbon
PORTUGAL
AZORES
(PORTUGAL)
North Atlantic Ocean
Ankara
TURKEY
GREECE
Athens
Tehran
IRAN
IRAQ
Baghdad
SYRIA
Damascus
Beirut
CYPRUS
Jerusalem
ISRAEL
Amman
JORDAN
KUWAIT
BAHR.
QATAR
Abu Dhabi
U.A.E.
Muscat
OMAN
Riyadh
SAUDI ARABIA
YEMEN
Sanaa
MADEIRA ISLANDS
(PORTUGAL)
Algiers
Tunis
TUNISIA
Mediterranean Sea
Rabat
Casablanca
Fès
MOROCCO
Marrakech
CANARY ISLANDS
(SPAIN)
Tripoli
Banghazi
Alexandria
Cairo
EGYPT
ALGERIA
LIBYA
Laayoune
(El Aaiún)
Western Sahara
Tropic of Cancer
SAHARA
Nouadhibou
MAURITANIA
Nouakchott
CAPE VERDE
Praia
Dakar
SENEGAL
Banjul
THE GAMBIA
Bissau
GUINEA-BISSAU
GUINEA
Conakry
Freetown
SIERRA LEONE
Monrovia
LIBERIA
MALI
Tombouctou
Bamako
BURKINA FASO
Ouagadougou
NIGER
Niamey
Agadez
Zinder
Kano
NIGERIA
Abuja
CHAD
N'Djamena
SUDAN
Khartoum
Omdurman
Port Sudan
ERITREA
Asmara
Djibouti
DJIBOUTI
Addis Ababa
ETHIOPIA
SOMALIA
Mogadishu
CÔTE D'IVOIRE
Yamoussoukro
Abidjan
GHANA
Accra
TOGO
Lomé
BENIN
Porto-Novo
Lagos
Ibadan
CAMEROON
Douala
Yaoundé
Malabo
EQUATORIAL GUINEA
CENTRAL AFRICAN REPUBLIC
Bangui
Juba
SAO TOME AND PRINCIPE
São Tomé
Libreville
GABON
REP. OF THE CONGO
CONGO BASIN
DEM. REP. OF THE CONGO
Kisangani
UGANDA
Kampala
KENYA
Nairobi
RWANDA
Kigali
Bujumbura
BURUNDI
Brazzaville
Kinshasa
Pointe-Noire
ANGOLA
(Cabinda)
Luanda
Mbuji-Mayi
TANZANIA
Dodoma
Dar es Salaam
Mombasa
Indian Ocean
Victoria
SEYCHELLES
COMOROS
Moroni
Lubumbashi
MALAWI
Lilongwe
ZAMBIA
Lusaka
Kitwe
Blantyre
Namibe
Lubango
Harare
ZIMBABWE
MOZAMBIQUE
Beira
Antananarivo
MADAGASCAR
Port Louis
St. Denis
Windhoek
Walvis Bay
NAMIBIA
BOTSWANA
Gaborone
Pretoria
Johannesburg
Mbabane
Maputo
SWAZILAND
SOUTH AFRICA
Maseru
LESOTHO
Durban
Cape Town
Port Elizabeth
Ascension
Saint Helena
South Atlantic Ocean
Tropic of Capricorn
Saint Helena, Ascension, and Tristan da Cunha (U.K.)
TRISTAN DA CUNHA
Gough Island
CAPE RISE
WALVIS RIDGE
MID-ATLANTIC RIDGE
Scale 1:51,400,000
Azimuthal Equal-Area Projection
800 Kilometers
800 Miles
Boundary representation is not necessarily authoritative.

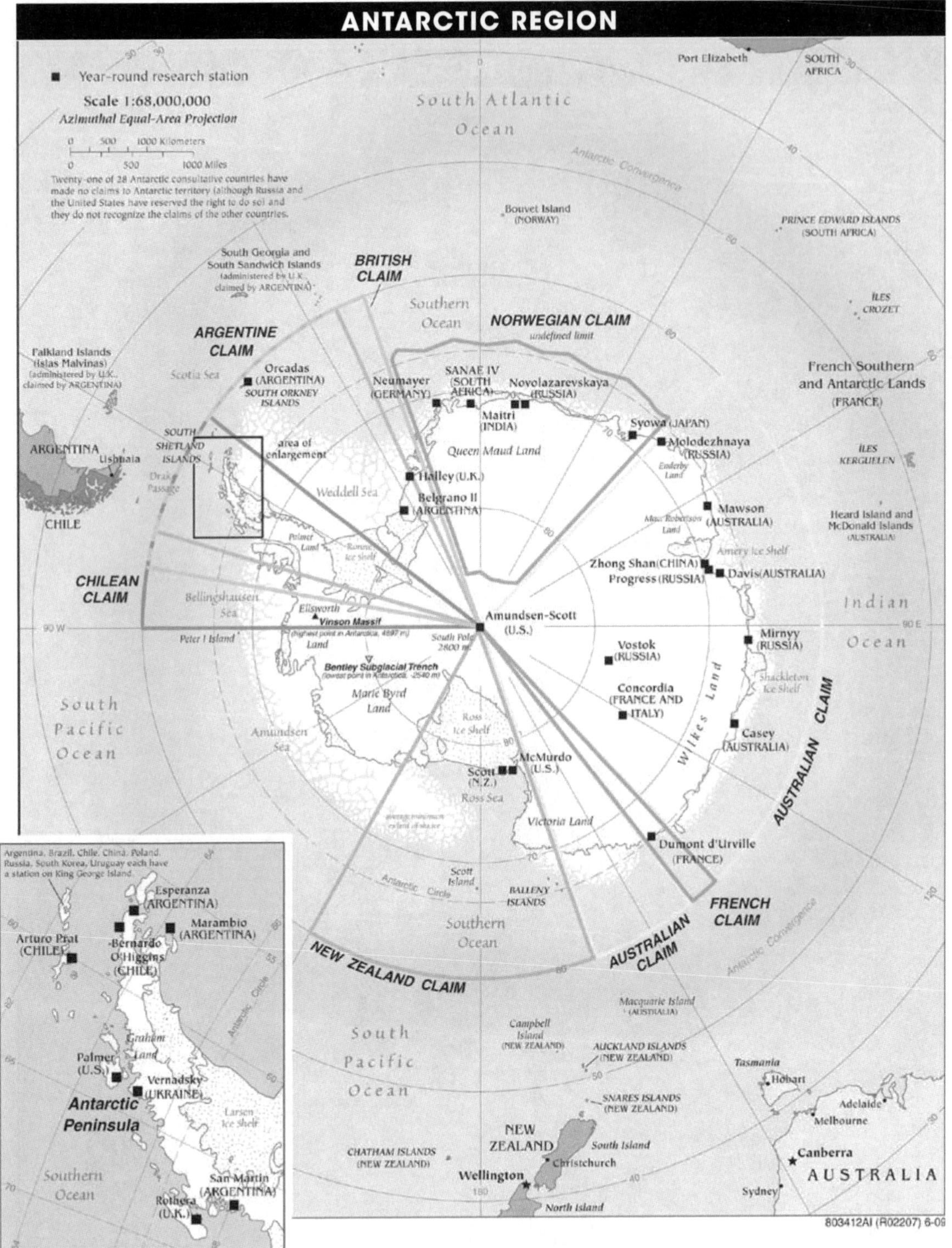
ANTARCTIC REGION
Year-round research station
Scale 1:68,000,000
Azimuthal Equal-Area Projection
Twenty-one of 28 Antarctic consultative countries have made no claims to Antarctic territory (although Russia and the United States have reserved the right to do so) and they do not recognize the claims of the other countries.
South Atlantic Ocean
Southern Ocean
Indian Ocean
South Pacific Ocean
BRITISH CLAIM
ARGENTINE CLAIM
NORWEGIAN CLAIM
undefined limit
CHILEAN CLAIM
AUSTRALIAN CLAIM
FRENCH CLAIM
NEW ZEALAND CLAIM
Amundsen-Scott (U.S.)
Vostok (RUSSIA)
Concordia (FRANCE AND ITALY)
McMurdo (U.S.)
Scott (N.Z.)
Dumont d'Urville (FRANCE)
Casey (AUSTRALIA)
Mirnyy (RUSSIA)
Davis (AUSTRALIA)
Zhong Shan (CHINA)
Progress (RUSSIA)
Mawson (AUSTRALIA)
Molodezhnaya (RUSSIA)
Syowa (JAPAN)
Novolazarevskaya (RUSSIA)
Maitri (INDIA)
SANAE IV (SOUTH AFRICA)
Neumayer (GERMANY)
Halley (U.K.)
Belgrano II (ARGENTINA)
Orcadas (ARGENTINA)
Queen Maud Land
Wilkes Land
Marie Byrd Land
Victoria Land
Ellsworth Land
Vinson Massif
Bentley Subglacial Trench
Weddell Sea
Ross Sea
Amundsen Sea
Bellingshausen Sea
Scotia Sea
Ross Ice Shelf
Amery Ice Shelf
Shackleton Ice Shelf
Antarctic Convergence
Antarctic Circle
Bouvet Island (NORWAY)
South Georgia and South Sandwich Islands (administered by U.K., claimed by ARGENTINA)
Falkland Islands (Islas Malvinas) (administered by U.K., claimed by ARGENTINA)
ARGENTINA
CHILE
Ushuaia
Port Elizabeth
SOUTH AFRICA
PRINCE EDWARD ISLANDS (SOUTH AFRICA)
ÎLES CROZET
French Southern and Antarctic Lands (FRANCE)
ÎLES KERGUELEN
Heard Island and McDonald Islands (AUSTRALIA)
Peter I Island
Scott Island
BALLENY ISLANDS
Macquarie Island (AUSTRALIA)
Campbell Island (NEW ZEALAND)
AUCKLAND ISLANDS (NEW ZEALAND)
SNARES ISLANDS (NEW ZEALAND)
CHATHAM ISLANDS (NEW ZEALAND)
NEW ZEALAND
Wellington
Christchurch
South Island
North Island
Tasmania
Hobart
Melbourne
Adelaide
Canberra
Sydney
AUSTRALIA
area of enlargement
Argentina, Brazil, Chile, China, Poland, Russia, South Korea, Uruguay each have a station on King George Island.
Esperanza (ARGENTINA)
Marambio (ARGENTINA)
Arturo Prat (CHILE)
Bernardo O'Higgins (CHILE)
Palmer (U.S.)
Vernadsky (UKRAINE)
Antarctic Peninsula
Graham Land
Larsen Ice Shelf
San Martín (ARGENTINA)
Rothera (U.K.)
803412AI (R02207) 6-09

ARCTIC REGION

ASIA
Arctic Ocean
RUSSIA
SIBERIA
URAL MOUNTAINS
NORWAY
SWEDEN
FINLAND
Oslo
Stockholm
Copenhagen
Helsinki
Tallinn
Riga
Vilnius
Warsaw
Minsk
BELARUS
Moscow
Kyiv
UKRAINE
Saint Petersburg
Murmansk
NOVAYA ZEMLYA
Nizhniy Novgorod
Kazan'
Perm'
Yekaterinburg
Chelyabinsk
Samara
Saratov
Volgograd
Voronezh
Omsk
Novosibirsk
Krasnoyarsk
Noril'sk
Irkutsk
Chita
Yakutsk
Magadan
Khabarovsk
Vladivostok
Harbin
Changchun
Anadyr'
Providenya
Wrangel Island
Bering Sea
Sea of Okhotsk
Sakhalin
Petropavlovsk-Kamchatskiy
Astana
KAZAKHSTAN
UZBEKISTAN
TURKMENISTAN
Ashgabat
Tashkent
Bishkek
KYRGYZSTAN
Dushanbe
TAJIKISTAN
Almaty
Ürümqi
Tbilisi
Yerevan
Baku
Tehran
IRAN
Esfahan
Shiraz
Mashhad
AFGHANISTAN
Kabul
Kandahar
Islamabad
Faisalabad
Lahore
Quetta
PAKISTAN
Karachi
Doha
QATAR
Abu Dhabi
U.A.E.
Muscat
OMAN
New Delhi
Jaipur
Lucknow
Kanpur
Patna
Kathmandu
Thimphu
Ahmadabad
Indore
Surat
INDIA
Mumbai
Pune
DECCAN
Hyderabad
Nagpur
Kolkata
Dhaka
Chittagong
Vishakhapatnam
Bengaluru
Chennai
Cochin
Jaffna
SRI LANKA
Colombo
MALDIVES
Male
Arabian Sea
Bay of Bengal
Indian Ocean
TAKLA MAKAN DESERT
Kashi
MONGOLIA
Ulaanbaatar
GOBI DESERT
CHINA
Beijing
Tianjin
Taiyuan
Baotou
Qingdao
Jinan
Lanzhou
Xi'an
Zhengzhou
Nanjing
Shanghai
Hangzhou
Wuhan
Nanchang
Chengdu
Chongqing
Changsha
Guiyang
Kunming
Guangzhou
Nanning
Hong Kong S.A.R.
Macau S.A.R.
Shenyang
Dalian
NORTH KOREA
Pyongyang
Seoul
SOUTH KOREA
Busan
JAPAN
Tokyo
Osaka
Nagoya
Fukuoka
Sapporo
East China Sea
Okinawa
Taipei
Taiwan
Philippine Sea
Luzon
Manila
PHILIPPINES
Mindanao
Davao
BURMA
Mandalay
Rangoon
LAOS
Vientiane
Hanoi
Haiphong
Da Nang
THAILAND
Bangkok
VIETNAM
CAMBODIA
Phnom Penh
Ho Chi Minh City
South China Sea
MALAYSIA
Kuala Lumpur
Medan
SINGAPORE
Singapore
Bandar Seri Begawan
BRUNEI
Borneo
Pontianak
Sumatra
Palembang
INDONESIA
Jakarta
Semarang
Surabaya
Bandung
Jawa
Dili
TIMOR-LESTE
Timor
Makassar
Celebes Sea
AUSTRALIA
Christmas Island (AUSTL.)
Cocos (Keeling) Islands (AUSTL.)
Scale 1:48,000,000
Azimuthal Equal-Area Projection
Boundary representation is not necessarily authoritative.

CENTRAL AMERICA AND THE CARIBBEAN

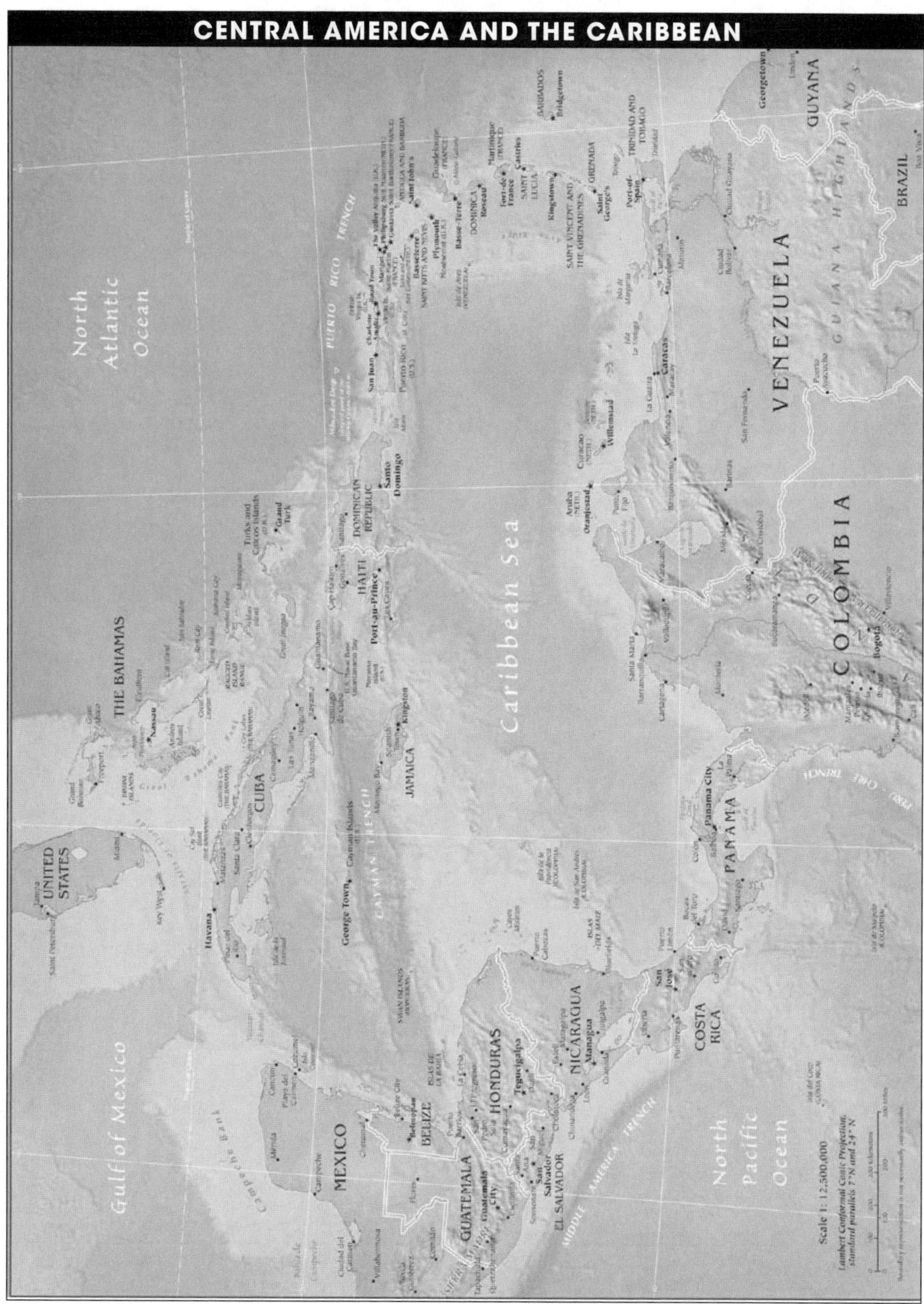

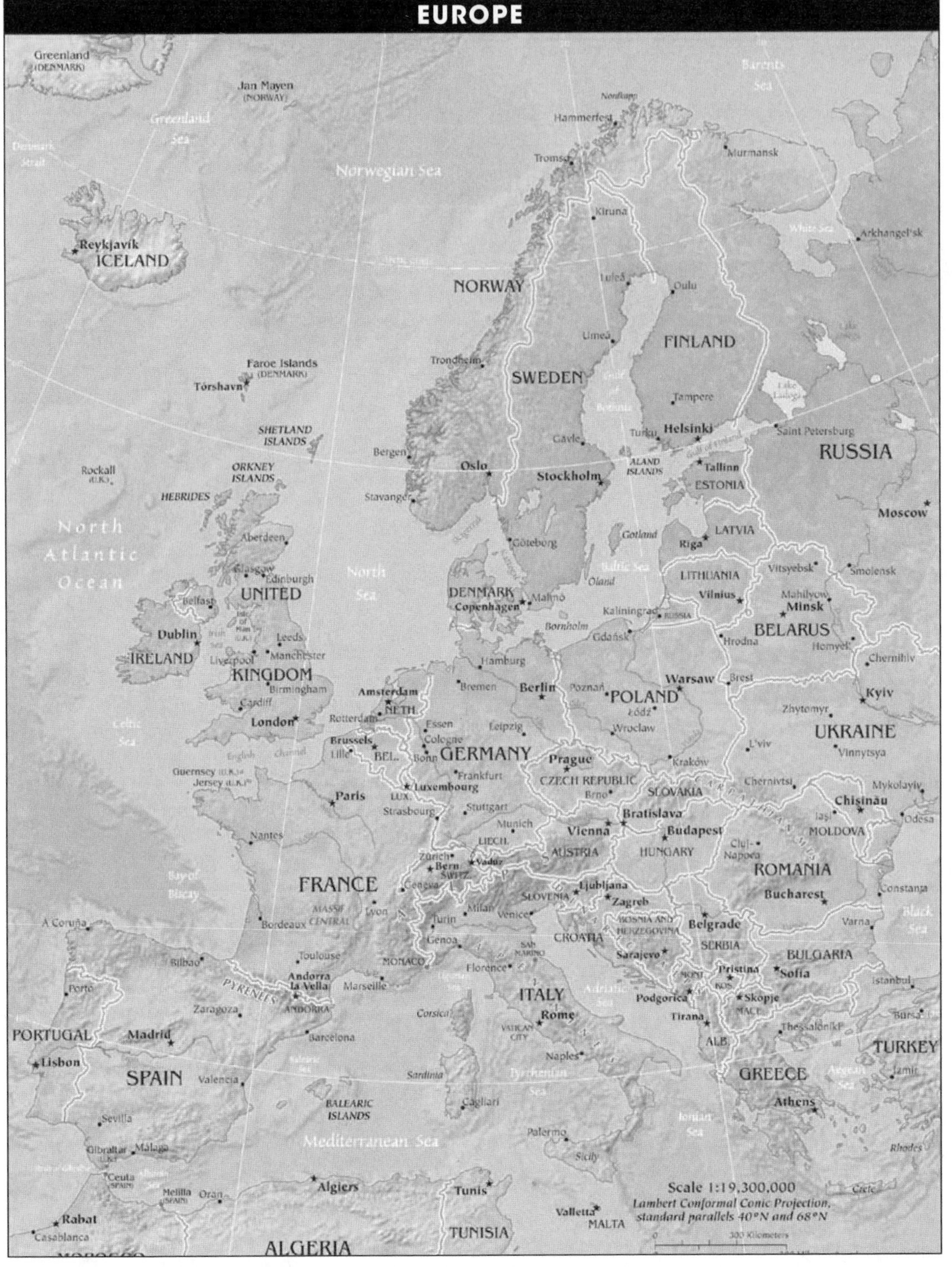
EUROPE
Greenland (DENMARK)
Jan Mayen (NORWAY)
Greenland Sea
Norwegian Sea
Hammerfest
Tromsø
Murmansk
Kiruna
Arkhangel'sk
Reykjavík
ICELAND
NORWAY
Luleå
Oulu
FINLAND
Umeå
Trondheim
SWEDEN
Faroe Islands (DENMARK)
Tórshavn
Tampere
Helsinki
Turku
Saint Petersburg
SHETLAND ISLANDS
Bergen
Gävle
RUSSIA
ORKNEY ISLANDS
Oslo
Stockholm
ALAND ISLANDS
Tallinn
Rockall (U.K.)
HEBRIDES
Stavanger
ESTONIA
Moscow
North Atlantic Ocean
Aberdeen
Göteborg
Gotland
Riga
LATVIA
Glasgow
Edinburgh
North Sea
Smolensk
UNITED
LITHUANIA
Vitsyebsk
Belfast
DENMARK
Copenhagen
Malmö
Vilnius
Minsk
Öland
Kaliningrad
RUSSIA
Dublin
Leeds
Bornholm
BELARUS
IRELAND
Liverpool
Manchester
Gdańsk
Hrodna
Homyel'
KINGDOM
Hamburg
Chernihiv
Birmingham
Amsterdam
Bremen
Berlin
Poznań
Warsaw
Brest
Cardiff
NETH.
POLAND
Kyiv
London
Rotterdam
Łódź
Zhytomyr
Essen
Leipzig
Wrocław
UKRAINE
Brussels
Cologne
L'viv
Lille
BEL.
Bonn
GERMANY
Prague
Vinnytsya
Guernsey (U.K.)
Jersey (U.K.)
Frankfurt
Kraków
Luxembourg
CZECH REPUBLIC
Chernivtsi
Mykolayiv
Paris
LUX.
Brno
SLOVAKIA
Chişinău
Strasbourg
Stuttgart
Bratislava
Iaşi
Odesa
Munich
Vienna
Budapest
MOLDOVA
Nantes
LIECH.
Cluj-Napoca
Zurich
Vaduz
AUSTRIA
HUNGARY
Bern
ROMANIA
SWITZ.
FRANCE
Geneva
Ljubljana
Constanţa
SLOVENIA
Zagreb
Bucharest
MASSIF CENTRAL
Lyon
Milan
Venice
A Coruña
Bordeaux
Turin
BOSNIA AND HERZEGOVINA
Belgrade
Varna
CROATIA
Genoa
SERBIA
SAN MARINO
Toulouse
Sarajevo
BULGARIA
Bilbao
MONACO
Florence
Pristina
Sofia
Andorra la Vella
Marseille
KOS.
Istanbul
Porto
PYRENEES
ITALY
Podgorica
Skopje
Zaragoza
ANDORRA
Corsica
MACE.
Bursa
Rome
Tirana
PORTUGAL
Madrid
VATICAN CITY
Thessaloniki
Barcelona
ALB.
TURKEY
Naples
Lisbon
GREECE
İzmir
SPAIN
Valencia
Sardinia
Tyrrhenian Sea
BALEARIC ISLANDS
Cagliari
Athens
Sevilla
Ionian Sea
Palermo
Mediterranean Sea
Sicily
Rhodes
Gibraltar
Málaga
Ceuta (SPAIN)
Melilla (SPAIN)
Oran
Algiers
Tunis
Scale 1:19,300,000
Lambert Conformal Conic Projection, standard parallels 40°N and 68°N
Crete
Rabat
Valletta
MALTA
Casablanca
TUNISIA
ALGERIA

MIDDLE EAST
RUSSIA
KAZAKHSTAN
UZBEKISTAN
TURKMENISTAN
GEORGIA
ARMENIA
AZERBAIJAN
BULGARIA
GREECE
TURKEY
CYPRUS
SYRIA
LEBANON
ISRAEL
JORDAN
IRAQ
IRAN
KUWAIT
BAHRAIN
QATAR
UNITED ARAB EMIRATES
OMAN
SAUDI ARABIA
YEMEN
EGYPT
SUDAN
ERITREA
DJIBOUTI
ETHIOPIA
SOMALIA
AFG.
PAK.
Black Sea
Caspian Sea
Mediterranean Sea
Red Sea
Persian Gulf
Gulf of Aden
Arabian Sea
Dead Sea
Bucharest
Sofia
Athens
Ankara
Nicosia
Beirut
Damascus
Amman
Baghdad
Tehran
Tbilisi
Yerevan
Baku
Ashgabat
Kuwait
Manama
Doha
Abu Dhabi
Muscat
Riyadh
Cairo
Khartoum
Asmara
Sanaa
Djibouti
Addis Ababa
Istanbul
İzmir
Konya
Kayseri
Adana
Aleppo
Mosul
Kirkuk
Basrah
Tabriz
Mashhad
Shiraz
Eşfahan
Jiddah
Mecca
Medina
Aden
Alexandria
Port Said
Suez
Aswan
Luxor
Port Sudan
Omdurman
Golan Heights
West Bank
Gaza Strip
Tel Aviv-Yafo
Jerusalem
RUB AL KHALI
Scale 1:21,000,000
Lambert Conformal Conic Projection, standard parallels 12°N and 38°N
0 300 Kilometers
0 300 Miles
Boundary representation is not necessarily authoritative
Golan Heights is Israeli-occupied Syria.
West Bank and Gaza Strip are Israeli-occupied with current status subject to the Israeli-Palestinian Interim Agreement — permanent status to be determined through further negotiation.
Israel proclaimed Jerusalem as its capital in 1950, but the US, like nearly all other countries, maintains its Embassy in Tel Aviv.

NORTH AMERICA

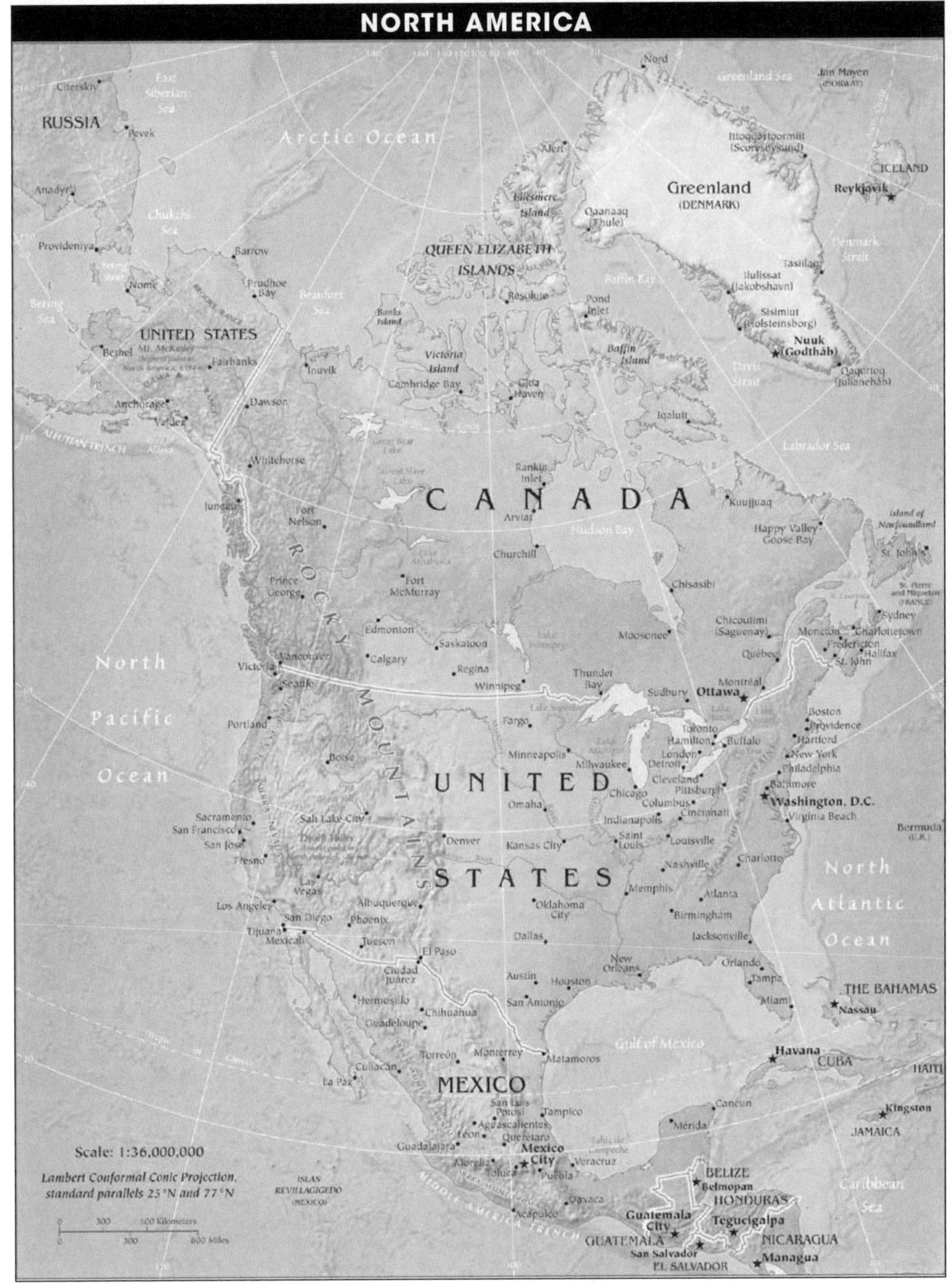

OCEANIA

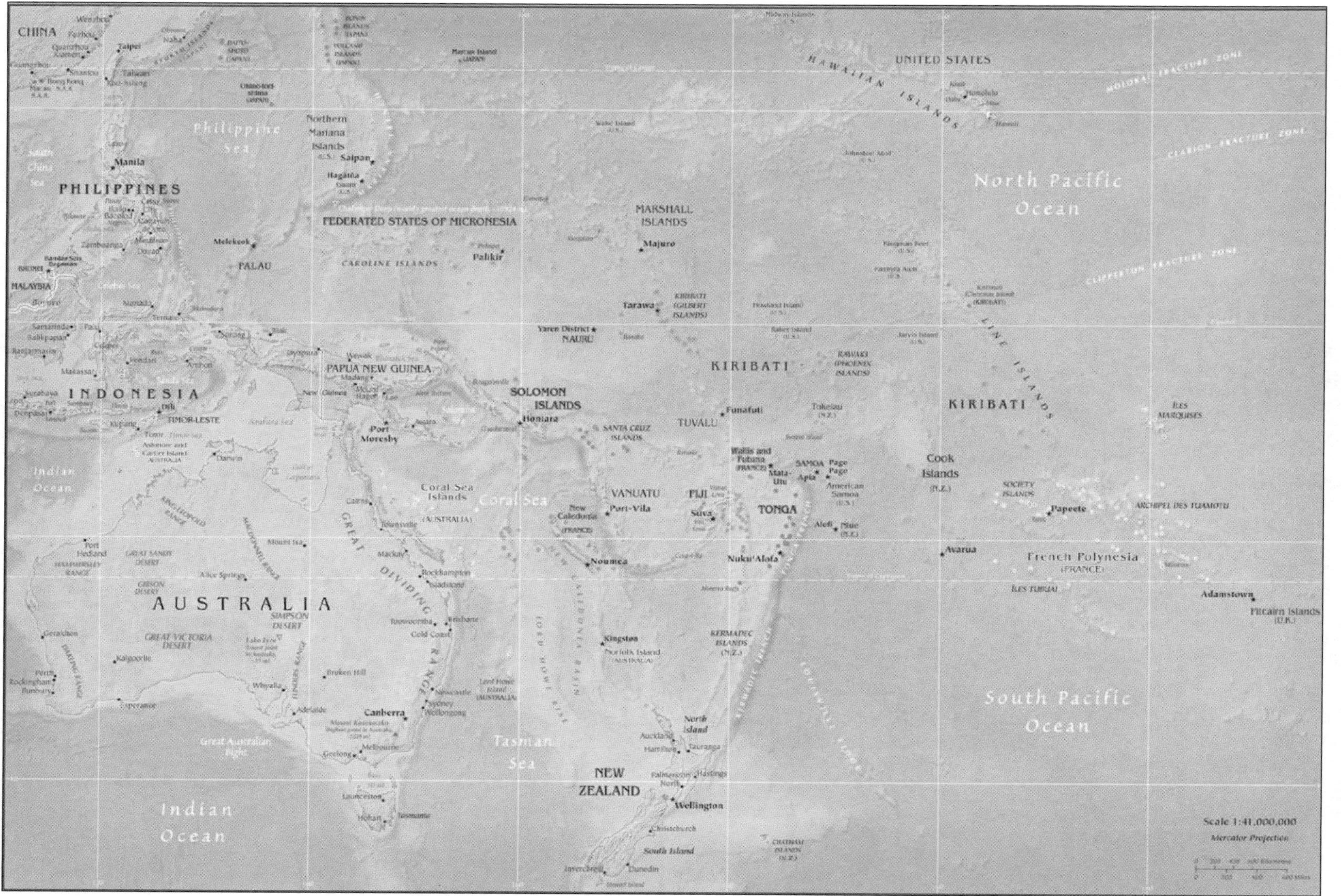
CHINA
Taipei
Taiwan
PHILIPPINES
Manila
Philippine Sea
PALAU
Melekeok
Northern Mariana Islands (U.S.)
Saipan
Hagåtña
FEDERATED STATES OF MICRONESIA
Palikir
CAROLINE ISLANDS
MARSHALL ISLANDS
Majuro
Yaren District
NAURU
Tarawa
KIRIBATI
UNITED STATES
HAWAIIAN ISLANDS
LINE ISLANDS
North Pacific Ocean
INDONESIA
TIMOR-LESTE
Dili
PAPUA NEW GUINEA
Port Moresby
SOLOMON ISLANDS
Honiara
TUVALU
Funafuti
VANUATU
Port-Vila
FIJI
Suva
Wallis and Futuna (FRANCE)
SAMOA
Apia
TONGA
Nuku'Alofa
Cook Islands (N.Z.)
Avarua
Papeete
French Polynesia (FRANCE)
Adamstown
Pitcairn Islands (U.K.)
South Pacific Ocean
New Caledonia (FRANCE)
Noumea
Coral Sea Islands (AUSTRALIA)
Coral Sea
Norfolk Island (AUSTRALIA)
Kingston
AUSTRALIA
GREAT DIVIDING RANGE
Canberra
Tasman Sea
NEW ZEALAND
North Island
South Island
Wellington
Indian Ocean
Scale 1:41,000,000
Mercator Projection

PHYSICAL MAP OF THE WORLD

POLITICAL MAP OF THE WORLD

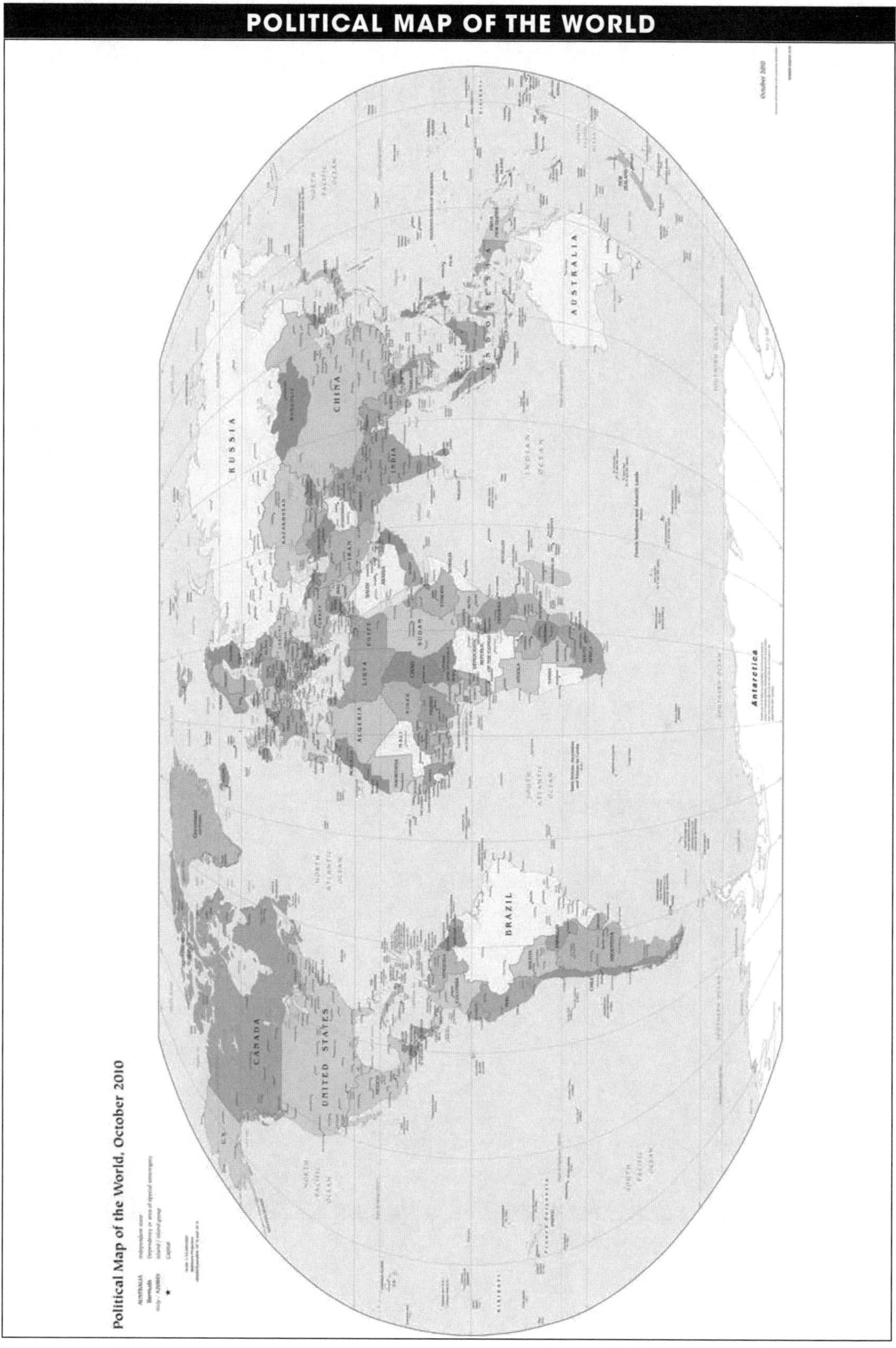

SOUTH AMERICA
Caribbean Sea
HONDURAS
Tegucigalpa
NICARAGUA
Managua
San José
COSTA RICA
Panama City
PANAMA
Barranquilla
Cartagena
Maracaibo
Barquisimeto
Valencia
Caracas
Barcelona
Port-of-Spain
TRINIDAD AND TOBAGO
BARBADOS
North Atlantic Ocean
Cúcuta
San Cristóbal
VENEZUELA
Ciudad Guayana
Georgetown
GUYANA
Paramaribo
SURINAME
Cayenne
French Guiana (FRANCE)
Medellín
Bucaramanga
Pereira
Ibagué
Bogotá
Cali
COLOMBIA
Boa Vista
Macapá
Quito
ECUADOR
Guayaquil
Cuenca
AMAZON
BASIN
Iquitos
Manaus
Santarém
Belém
São Luís
Fortaleza
Teresina
Natal
João Pessoa
Recife
Maceió
Aracaju
Salvador
Piura
Chiclayo
Trujillo
Pucallpa
Huánuco
Rio Branco
Porto Velho
PERU
BRAZIL
BRAZILIAN
HIGHLANDS
MATO GROSSO PLATEAU
Huancayo
Lima
Cusco
Ica
Arequipa
South Pacific Ocean
PERU-CHILE TRENCH
Trinidad
La Paz
BOLIVIA
Cochabamba
Santa Cruz
Sucre
Potosí
Cuiabá
Brasília
Goiânia
Contagem
Belo Horizonte
Uberlândia
Arica
Iquique
Campo Grande
Vitória
Campos
PARAGUAY
Londrina
Campinas
São Paulo
Santos
Rio de Janeiro
Antofagasta
ATACAMA DESERT
Asunción
Ciudad del Este
Curitiba
Joinville
Salta
San Miguel de Tucumán
Resistencia
Florianópolis
Isla San Félix (CHILE)
Isla San Ambrosio (CHILE)
CHILE
Porto Alegre
Córdoba
Santa Fe
Salto
URUGUAY
Mendoza
Rosario
ARCHIPIÉLAGO JUAN FERNÁNDEZ (CHILE)
Valparaíso
Santiago
Buenos Aires
La Plata
Montevideo
South Atlantic Ocean
RIO GRANDE RISE
Concepción
ARGENTINA
Bahía Blanca
Temuco
San Carlos de Bariloche
Puerto Montt
Comodoro Rivadavia
PATAGONIA
Laguna del Carbón
Scale 1:35,000,000
Azimuthal Equal-Area Projection
SOUTH GEORGIA RIDGE
Río Gallegos
Stanley
Falkland Islands (Islas Malvinas)
(administered by U.K., claimed by ARGENTINA)
Punta Arenas
Ushuaia
Cape Horn
South Georgia and South Sandwich Islands
(administered by U.K., claimed by ARGENTINA)
500 Kilometers
500 Miles
Boundary representation is not necessarily authoritative.

SOUTHEAST ASIA
CHINA
INDIA
BURMA
LAOS
THAILAND
VIETNAM
CAMBODIA
PHILIPPINES
MALAYSIA
BRUNEI
SINGAPORE
INDONESIA
TIMOR-LESTE
PALAU
AUSTRALIA
JAPAN
Chengdu
Chongqing
Wuhan
Hefei
Shanghai
Xingang
Hangzhou
Ningbo
Changsha
Fuzhou
Kunming
Taipei
Xiamen
Guangzhou
Shantou
Nanning
Hong Kong S.A.R.
Macau S.A.R.
Taiwan
Hanoi
Haiphong
Zhanjiang
Haikou
Myitkyina
Chittagong
Mandalay
Akyab
Nay Pyi Taw
Rangoon
Pathein
Mawlamyine
Chiang Mai
Vientiane
Udon Thani
Nakhon Ratchasima
Bangkok
Dawei
Hue
Da Nang
Phnom Penh
Nha Trang
Ho Chi Minh City
Can Tho
Phuket
Hat Yai
Kota Baharu
George Town
Kuala Terengganu
Ipoh
Kuantan
Medan
Kuala Lumpur
Johor Bahru
Singapore
Pekanbaru
Padang
Sumatra
Palembang
Tanjungkarang-Telukbetung
Jakarta
Bandung
Semarang
Java
Surbaya
Malang
Denpasar
Baguio
Luzon
Quezon City
Manila
Iloilo
Bacolod
Cebu City
Cagayan de Oro
Mindanao
Zamboanga
Davao
Kota Kinabalu
Sandakan
Bandar Seri Begawan
Tawau
Sibu
Kuching
Borneo
Pontianak
Samarinda
Palu
Sulawesi (Celebes)
Banjarmasin
Kendari
Makassar
Ambon
Manado
Sorong
Jayapura
Merauke
Dili
Kupang
Darwin
Melekeok
Philippine Sea
South China Sea
North Pacific Ocean
Indian Ocean
Java Sea
Timor Sea
Arafura Sea
PARACEL ISLANDS
SPRATLY ISLANDS
KEPULAUAN NATUNA
KEPULAUAN MENTAWAI
KEPULAUAN ARU
ANDAMAN ISLANDS (INDIA)
NICOBAR ISLANDS (INDIA)
BABUYAN ISLANDS
BONIN ISLANDS
VOLCANO ISLANDS
FED. STATES OF MICRONESIA
PHILIPPINE TRENCH
JAVA TRENCH
Christmas Island (AUSTRALIA)
Cocos (Keeling) Islands (AUSTRALIA)
Ashmore and Cartier Islands (AUSTRALIA)
Port Hedland
Karratha
GREAT SANDY DESERT
Mount Isa
Scale 1:32,000,000
Mercator Projection
Boundary representation is not necessarily authoritative.
Names in Vietnam are shown without diacritical marks.

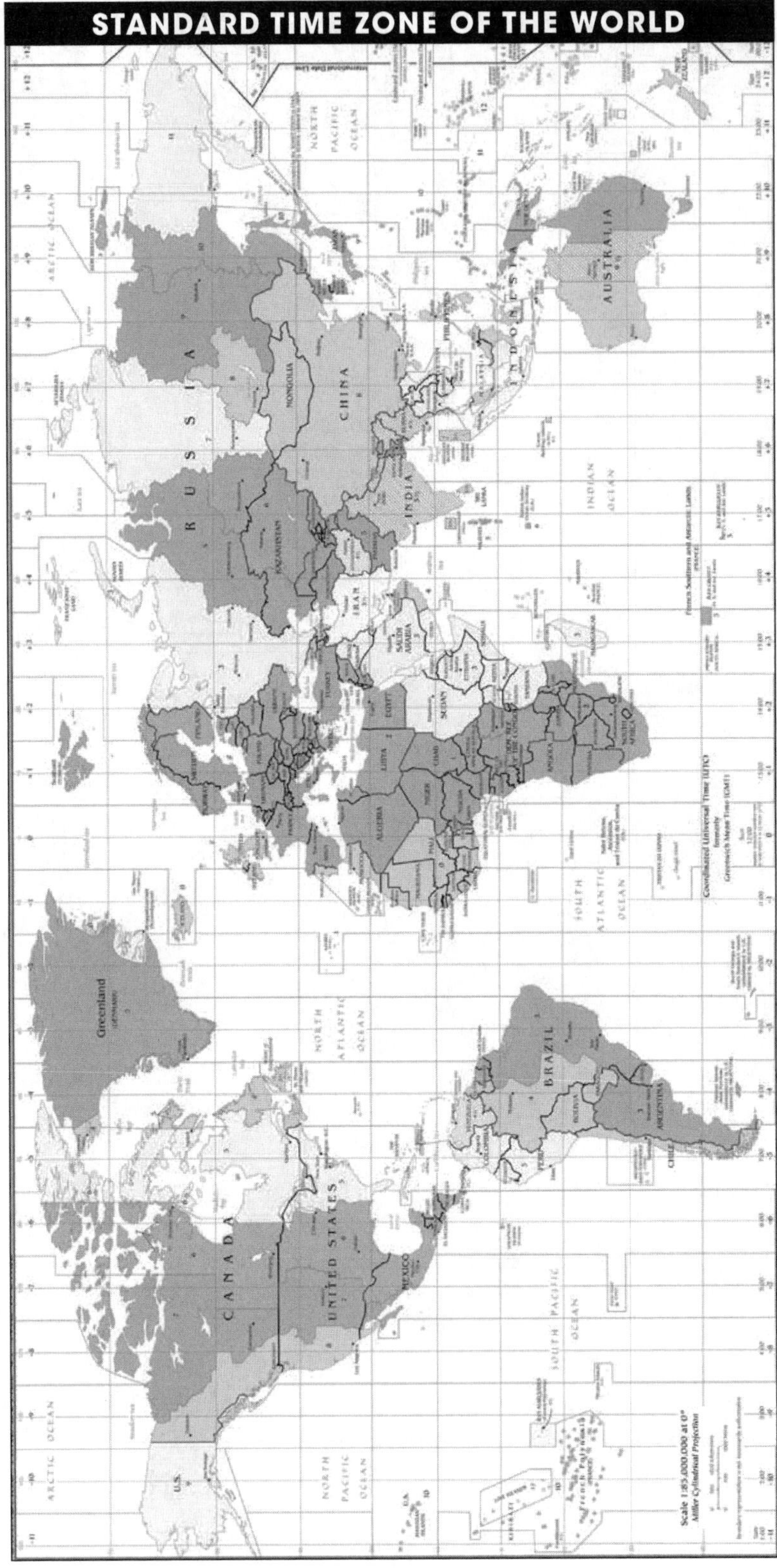
STANDARD TIME ZONE OF THE WORLD
RUSSIA
CHINA
MONGOLIA
KAZAKHSTAN
INDIA
IRAN
SAUDI ARABIA
TURKEY
EGYPT
LIBYA
SUDAN
CHAD
NIGER
ALGERIA
INDONESIA
AUSTRALIA
Greenland
CANADA
UNITED STATES
MEXICO
BRAZIL
BOLIVIA
ARGENTINA
CHILE
PERU
NORTH PACIFIC OCEAN
SOUTH PACIFIC OCEAN
NORTH ATLANTIC OCEAN
SOUTH ATLANTIC OCEAN
INDIAN OCEAN
ARCTIC OCEAN

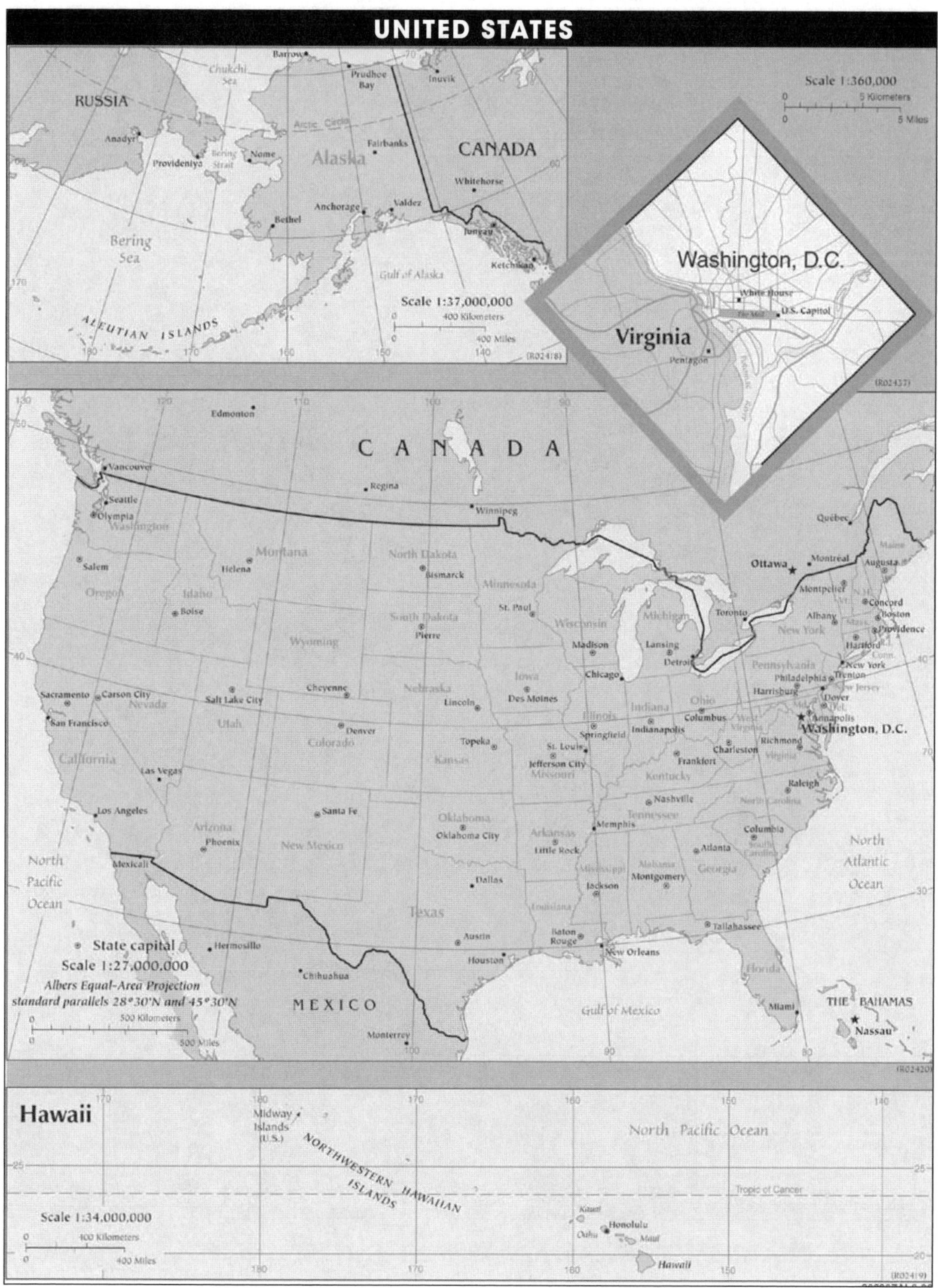
UNITED STATES
RUSSIA
Chukchi Sea
Barrow
Prudhoe Bay
Inuvik
Anadyr
Provideniya
Bering Strait
Nome
Arctic Circle
Fairbanks
Alaska
CANADA
Whitehorse
Anchorage
Valdez
Bethel
Bering Sea
Juneau
Ketchikan
Gulf of Alaska
ALEUTIAN ISLANDS
Scale 1:37,000,000
400 Kilometers
400 Miles
Scale 1:360,000
5 Kilometers
5 Miles
Washington, D.C.
White House
U.S. Capitol
Virginia
Pentagon
Edmonton
C A N A D A
Vancouver
Regina
Winnipeg
Seattle
Olympia
Washington
Montana
North Dakota
Bismarck
Minnesota
Salem
Helena
Oregon
Idaho
Boise
South Dakota
Pierre
St. Paul
Wisconsin
Michigan
Wyoming
Madison
Lansing
Detroit
Chicago
Iowa
Des Moines
Cheyenne
Nebraska
Lincoln
Salt Lake City
Sacramento
Carson City
Nevada
San Francisco
Utah
Denver
Colorado
Topeka
Kansas
St. Louis
Jefferson City
Missouri
Springfield
Illinois
Indiana
Indianapolis
Ohio
Columbus
California
Las Vegas
Los Angeles
Santa Fe
Arizona
Phoenix
New Mexico
Oklahoma
Oklahoma City
Arkansas
Little Rock
Memphis
Tennessee
Nashville
Kentucky
Frankfort
Charleston
West Virginia
Richmond
Virginia
Raleigh
North Carolina
Columbia
South Carolina
Atlanta
Georgia
Alabama
Montgomery
Mississippi
Jackson
Louisiana
Baton Rouge
New Orleans
Tallahassee
Florida
Miami
Dallas
Texas
Austin
Houston
Mexicali
Hermosillo
Chihuahua
MEXICO
Monterrey
Gulf of Mexico
North Pacific Ocean
North Atlantic Ocean
THE BAHAMAS
Nassau
Québec
Ottawa
Montréal
Toronto
Maine
Augusta
Montpelier
Concord
Boston
Providence
Albany
New York
Hartford
Pennsylvania
Harrisburg
Philadelphia
Trenton
New Jersey
Dover
Annapolis
Washington, D.C.
State capital
Scale 1:27,000,000
Albers Equal-Area Projection
standard parallels 28°30'N and 45°30'N
500 Kilometers
500 Miles
Hawaii
Midway Islands (U.S.)
NORTHWESTERN HAWAIIAN ISLANDS
North Pacific Ocean
Tropic of Cancer
Kauai
Oahu
Honolulu
Maui
Hawaii
Scale 1:34,000,000
400 Kilometers
400 Miles

CENTRAL BALKAN REGION